GROUND WARFARE

An International Encyclopedia

INTERNATIONAL WARFARE ENCYCLOPEDIAS FROM ABC-CLIO

Spencer C. Tucker, General Editor

Air Warfare: An International Encyclopedia, Walter J. Boyne, Editor

Naval Warfare: An International Encyclopedia, Spencer C. Tucker, Editor

Ground Warfare: An International Encyclopedia, Stanley Sandler, Editor

GROUND WARFARE

An International Encyclopedia

VOLUME ONE, A–G

EDITED BY

Stanley Sandler

ASSOCIATE EDITORS

Michael Ashkenazi

Paul D. Buell

FOREWORD BY

General Henry H. Shelton, USA (R)

ABC CLIO

Santa Barbara, California   Denver, Colorado   Oxford, England

Library of Congress Cataloging-in-Publication Data
Sandler, Stanley, 1937–
Ground warfare : an international encyclopedia / Stanley Sandler ; foreword by Henry H. Shelton.
p. cm
Includes bibliographical references and index.
ISBN 1-57607-344-0 (Hardcover: alk. paper) — Ebook ISBN 1-57607-733-0
1. Battles—Encyclopedias. 2. Military history—Encyclopedias. 3. Military biography—Encyclopedias. I. Title.
D25.A2 S26 2002
355'.003—dc21

2002004568

07 06 05 04 03 02 10 9 8 7 6 5 4 3 2 1

This book is also available on the World Wide Web as an e-book. Visit abc-clio.com for details.

ABC-CLIO, Inc.
130 Cremona Drive, P.O. Box 1911
Santa Barbara, California 93116-1911

This book is printed on acid-free paper.
Manufactured in the United States of America

CONTENTS

A-TO-Z LIST OF ENTRIES

Nuremberg Principle
Nurhaci (1559–1626)

October War (1973)
Oda, Nobunaga (1534–1582)
Offa's Wars (771–796)
Office of Strategic Services
Ögödei (c. 1186–1241)
Okinawa (1 April–21 June 1945)
Omani Conquest of East Africa (1622–1730)
Omdurman (1898)
Onin War (1467–1477)
Orleans, Siege of (12 October 1428–8 May 1429)
Osaka Castle, Siege of (1614–1615)
Osan, Battle of (5 July 1950)
Ostende, Siege of (1601–1604)
Ostrogoths
Otto I, the "Great" (912–973)
Ottoman Empire (1300s–1922)
Oudenaarde, Battle of (11 July 1708)
Oudinot, Nicholas-Charles, Duc de Reggio (1767–1847)

Pachacutec Yupanqui (r. 1438–1471)
Pacific, War of the (1879–1884)
Pacifism/War Resistance
Paekche (attributed 18 B.C.E.–660 C.E.)
Pagan Kingdom (1044–c. 1300)
Palo Alto (8 May 1846)
Panama Incursion (1989–1990)
Panipat, Battles of (21 April 1526, 5 November 1556, 14 January 1761)
Paramilitary Organizations
Paris, Siege of (1870–1871)
Parma and Piacenza, Alessandro Farnese, Duke of (1545–1592)
Parthian Empire (247 B.C.E.–226 C.E.)
Patton, George Smith, Jr. (1885–1945)
Pavia, Battle of (24 February 1525)
Pearl Harbor Attack (1941)
Peleliu (15 September–27 November 1944)
Peloponnesian Wars (460–456, 431–404 B.C.E.)
Peng Dehuai (1898–1974)
Pequot War (1636–1637)
Pericles (495–429 B.C.E.)
Pershing, John J. (1860–1948)
Persian Civil Wars (1725–1794)
Persian Empire (550 B.C.E.–642 C.E.)
Persian Wars of Expansion (559–509 B.C.E.)
Persian-Afghan Wars (1726–1857)
Peru-Bolivia Confederation, War of the (1836–1839)
Peru-Ecuador Conflict (1941–1999)
Peruvian Guerrilla War (1980–2000)
Pétain, Henri-Philippe (1856–1951)
Peter I, Romanov, Czar of Russia ("The Great") (1672–1725)
Petersburg, Siege of (June 1864–April 1865)
Pharsalus, Battle of (48 B.C.E.)
Philip, King (Metacomet)(1639–1676)
Philip II Augustus (1165–1223)
Philip II of Macedon (382?–336 B.C.E.)
Philippi, Battle of (42 B.C.E.)
Philippine Insurrection (1899–1902)
Philippines, U.S. Loss of (7 December 1941–9 June 1942)
Philippines, U.S. Retaking of (20 October 1944–2 September 1945)
Pickett, George Edward (1825–1875)
Pilsudski, Józef Klemens (1867–1935)
Pinkie (10 September 1547)
Pitt, William, the Elder (1708–1778)
Pizarro, Francisco (c. 1478–1541)
Plains of Abraham (13 September 1759)
Plassey, Battle of (23 June 1757)
Plataea, Battle of (479 B.C.E.)
Plattsburgh Movement (1915–1918)
Pleven/Plevna, Siege of (20 July–10 December 1877)
Poitiers, Battle of (18 September 1356)
Polish Campaign of 1939
Polish Wars of Expansion (1386–1498)
Poltava (8 July 1709)
Pompey the Great (Gnaeus Pompeius Magnus) (106–48 B.C.E.)
Pontiac's Rebellion (1763–1766)
Pope, John (1822–1892)
Porkchop Hill (16–18 April 1953)
Port Arthur, Siege of (May 1904–January 1905)
Portuguese-Castilian War (1369–1385)
Potemkin, Prince Grigory Aleksandrovich (1739–1791)
Powell, Colin L. (1937–)
Powhatan War (1622, 1644)
Prague, Siege of (1420)
Preston (17 August 1648)
Princeton, Battle of (3 January 1777)
Prisoners of War
Propellants
Psychological Operations
Ptolemy I Soter (c. 367–283 B.C.E.)
Pugachev's Revolt (1773–1774)
Pulaski, Count Kazimierz (1747–1779)
Punic Wars (264–146 B.C.E.)
Pusan Perimeter (August–September 1950)
Pyramids (21 July 1798)
Pyrrhus (319–272 B.C.E.)

FOREWORD

The ABC-CLIO *Ground Warfare: An International Encyclopedia* should be required reading for those trying to understand the *nature* and *conduct* of war. The study of war is relevant to all, not just to the practitioner of war, like me for the past 38 years, but to those who have never heard a shot fired in anger and who just want to become better educated on a subject that has been with us since our creation. Don't be fooled or misled by those who say there will be no more wars. There will always be wars and rumors of war for as long as humankind shall live.

Having had the honor of serving this great country as a soldier for 38 years, I trained for war every day of those 38 years while simultaneously praying that I would never have to put into practice what I was trained for. That was not to be the case. My career was bookended with conflict, from Vietnam as a young infantry lieutenant in 1964 to the recent 11 September 2001 crisis as the 14th Chairman of the Joint Chiefs of Staff. However, it is inevitable that soldiers, sailors, airmen, and marines will have to fight when called upon. The conduct of war is not a question of when, but merely where and for how long, because war will occur, and war and those who fight it will forever continue to be a part of this world. An old Japanese saying captures this theme: Warriors and gold may be idle, but they never rust.

From walking patrols in Vietnam as a young officer to employing our nation's armed forces to achieve our national strategic goals as the chairman of the Joint Chiefs of Staff, I have found that the study of the nature and conduct of war is always relevant to both the soldier and the layperson. Ardant du Picq stated, "Only study of the past can give us a sense of reality, and show us how the soldier will fight in the future." *Ground Warfare* provides the tools for one to study war and understand how the soldier will fight. The encyclopedia's organization and structure capture the essence of what is required of the student or novice of war to fully comprehend war's nature and conduct.

The nature of war is constant. It is a violent clash of wills to achieve a purpose, using violent means to accomplish that purpose. As Carl von Clausewitz states in *On War,* "War is thus an act of force to compel our enemy to do our will." As I advised the president of the United States regarding the employment of military force as the chairman, there are other tools available to employ besides military means to accomplish the country's objectives. Economic, diplomatic, and political tools, or a combination thereof, are applicable in many situations without having to employ force. My experiences have shown, however, that force will compel an enemy to meet your will when applied appropriately. Many of our adversaries respect only force and power. *Ground Warfare* captures a history of employing that force on the will of others to achieve a purpose, thus providing a thorough understanding of the nature of war. However, the volume goes beyond just understanding the nature of war; it is organized in such a manner to help the reader understand its conduct as well.

Ground Warfare shows that the nature of war is constant, that war is a brutal clash of peoples spanning the world and time from the ancient wars to the most current conflicts. There are literally hundreds of examples depicting the true nature of war. The narratives of each event capture this clash of wills and the violence invoked on its participants and the peoples involved.

The encyclopedia goes beyond just a description of the conflicts themselves and provides a comprehensive understanding of war's true nature. It also captures the impact of thousands of influential individuals on the nature of war, who changed its conduct through their personalities and influence and who often had an impact not just on warfare, but on the history of the world. The influences of Attila the Hun and William the Conqueror, or General Douglas MacArthur or General Colin Powell, are outlined, providing the reader an insight into the conduct of war as shaped by different personalities and their influences. More importantly, the work reminds readers that *people* will continue to influence the conduct of war for many more centuries to come.

The work also outlines the creation of state structures through military means and the influence of different and numerous ethnic cultures and civilizations on the conduct of war. Not only is it important to understand the leadership of Attila the Hun, but the culture of the Hun warriors he led during this time frame is needed fully to understand how a culture or a civilization could affect the conduct of war. The same is true in understanding, for example, the differences in the American soldier and America's culture and civilization during General MacArthur's era of World War II and that same soldier and American culture of today. There have been changes that influence the conduct of the American way of war. Just in my 38 years of service, I have seen this change and its impact on the conduct of war. Such changes in our culture and civilization, let alone the cultures and civilizations of the rest of the world, have changed the manner in which we conduct war. Civilizations varied over time and throughout the world, shaping the conduct of war, having an impact on each other. It is evident this will continue to be a reality.

Finally, the encyclopedia covers numerous topics about warfare, from Airborne Operations to Women in the World's Militaries, all from a historical perspective. There is no doubt that certain techniques and tactics and technologies will continue to influence the conduct of war.

But, as we have seen in the past, these topics might merely transform the manner in which we conduct war, but they will never change the true nature of war.

I am always reminded of T. R. Fehrenbach's book entitled *This Kind of War,* when it comes to the discussion of land warfare. He states: "You may fly over a land forever; you may bomb it, atomize it, pulverize it and wipe it clean of life—but if you desire to defend it, protect it, and keep it for civilization, you must do this on the ground, the way the Roman legions did, by putting your young men into the mud." This was certainly true for the Korean War, about which Fehrenbach writes so passionately, as well as for the legions of Rome, or the combat in Afghanistan and our fight against terrorism. Land warfare will always eventually come down to a young soldier, sailor, airman, or marine standing on the ground you want to control or influence. *Ground Warfare* provides you that historical perspective so desperately needed to ensure the successful lessons of the past are remembered and the mistakes of the past are not forgotten and made again, paid for in the needless loss of resources, lives, and blood. The totality of *Ground Warfare* makes it ideal for the individual who wants to study warfare. Not only is it ideal for the practitioner of war, but for the layperson as well. For the historian, it is a needed reference to ensure warfare is never forgotten as a significant element making its impact on the course of history. It is an ideal reference that captures the significant events, people, and topical areas key to fully understanding war. All-encompassing, *Ground Warfare* captures the full essence of warfare, its *nature* and *conduct.*

Henry H. Shelton
General, U.S. Army (Retired)
14th Chairman of the Joint Chiefs of Staff

PREFACE

"What are these essays but grotesque bodies pieced together of different members, without any definite shape, without any order, coherence or proportion, except they be accidental." So wrote the sixteenth-century French essayist Michel Eyquem de Montaigne of his own collection of essays. While our authors would undoubtedly object to any characterization of their works herein as "grotesque bodies," Montaigne did hit upon the task of the editor of such works. That is to impose "order, coherence or proportion" upon vastly disparate articles so that in the end the work does cohere. No historical work requires more intense collaborative effort than an encyclopedia. Literally hundreds of contributors must be solicited, assigned their articles, the laggards chivvied along, and their articles edited when they finally arrive. Putative authors sometimes drop out for a variety of reasons, and replacements must be located, often at the last minute. (It is also not unknown for authors to die during some more protracted encyclopedia projects.)

Encyclopedias are exercises in brevity. Editors must thus make a paragraph do for a page, a sentence for a paragraph, a phrase for a sentence. Even good authors, fascinated by their topics, often run over their assigned word lengths to some degree or other and must be reined in. Encyclopedia editors have their work cut out for them. These volumes are written and edited with a broad public in mind: high school or undergraduate students who need reliable information for research papers; even university graduate scholars looking for factual information; and the general public in search of a quick, authoritative source of military history information.

The reader will find within these volumes a broad range of military history from earliest times to the present; from the Egyptians at Kadesh (1274 B.C.E.) to the Alliance in the Gulf War (1991), the great captains, their battles, campaigns, strategies, tactics, defeats, and triumphs.

But this is not simply a battle-and-king military history that focuses almost exclusively on overt military clashes. Although the editor eschews the unoriginal term *New History,* this work does offer many topics previously slighted or ignored in past decades, such as women in war, logistics, military spending, psychological warfare, minorities, weapons development, civilians, etc.

I am indebted to the editor-in-chief of these volumes, Professor Spencer Tucker, who first broached the idea; and to editors at ABC-CLIO, especially Allison Miller, who persevered with me through the interminable process. A particular thanks is owed our mapmaker, Professor Don Frazier. Also, my thanks to my two associate editors, Michael Ashkenazi and Paul Buell, who contributed their extensive knowledge of the ancient and medieval worlds, respectively. And a special note of thanks is richly merited by our production editor, Michelle Trader, and copy editors Beth Partin and Anais Scott. My greatest debt of gratitude, however, must go, of course, to the several hundred contributors who performed the near-miracle of distilling down the contents of their topics to fit the limitations of any encyclopedia article—somehow pouring a quart into a pint. It is upon their contributions that this work will be judged. Perhaps even the shade of the aloof Montaigne might dip into these volumes to some profit or pleasure.

Stanley L. Sandler

INTRODUCTION

Military history has come into its own in the last few decades. Courses have multiplied on university and college campuses; books on the subject proliferate, for buffs and scholars alike; and war films like *Hope and Glory, Braveheart, Saving Private Ryan, Enemy at the Gates, Pearl Harbor,* and *Black Hawk Down* make money for their producers and on occasion even enjoy favorable critical reviews. Such popularity may well be simply the result of a long peace, at least as far as the industrialized nations are concerned. Much of even the recent military past has receded into myth: little Round Top, the Red Baron (immortalized by an American pizza concern), dawn at Pearl Harbor, Spitfires over the white cliffs of Dover, D day, raising the flag at Iwo Jima, Pork Chop Hill, and, less positively, the Somme, the Bataan Death March, Auschwitz, and the Hanoi Hilton. Any night's viewing of the History Channel should confirm war's popularity in the United States, and reenactors in the industrial nations try to re-create the militaries of times past, from the classical Romans, through the Crusaders, to World War II. (There do not seem to be any Vietnam reenactor troops yet, but there is a group of British imaginary soldiers who play at being Yanks in Britain during World War II. They drive about the southern English countryside in U.S. Army Jeeps and "deuce-and-a-half"-ton trucks, with their authentic U.S. Army uniform pockets stuffed with Wrigley's Spearmint Gum and Life Savers.)

More seriously, the world's militaries—land, sea, and air—have established professional history organizations staffed with historians who spend most of their time studying and explicating their nations' past wars, and whose qualifications, research, teaching, and publishing would not disgrace a first-class academic history department. Their work, and that of military historians in academe (whose class enrollments have held up over the recent decades), have sparked a renewed interest in military history that is seen in such phenomena as the History Channel, the historic preservation and battlefield reenactor movements, and the fact that each federal government agency of any size currently has its own history office. Entire publishing firms, journals, and book clubs devote themselves to military history, as witness an outpouring of books and articles in a wide variety of military themes, from the coffee table and buff variety to the profoundly scholarly. And, as is often the case in the history profession in general, military history publications can be both vivid and documented.

Indeed, as again in the history profession generally, military history has moved away from what has become known uncharitably as the drum-and-trumpet chronicles (the civil equivalent would be battle-and-king history) that had dominated the field since the nineteenth century, and emerged by the 1960s into a broadening of work in logistics, home fronts, financing, technologies, minorities, women, and so on, in war and in the military. In fact, some of these works have wandered so far from war itself that one feels compelled to remind the authors of such (almost literally) bloodless studies that the purpose of a military is not so much to improve the nation's economy and technology, streamline management, or integrate its minorities and women as it is to wage and win wars, and that war is humankind's most nasty and dirty business. We still need to bear in mind George Orwell's reminder that soldiers since Marathon have had "lice crawling over their testicles." (Orwell, a frontline soldier who survived being shot in the throat during the Spanish Civil War, would know.)

Indeed, an increasing number of military historians, again both within and without the military establishments, have in the last decade or so taken a particular interest in the face of battle itself, the experience of the individual soldier facing the ultimate test. Historians do try to determine how it was that masses of young men in, say, Pickett's Charge could persevere in the face of such a hail of large-caliber bullets tearing away at comrades to the right and left of them that a fine mist of blood and tissue (according to some accounts) preceded the advanced line.

And yet, paradoxically, by the latter half of the twentieth century, war itself had become a major taboo. It may well be that photography had more to do with this state of affairs than all of the disarmament conferences and peace literature combined. Americans were stunned by the Brady and Sullivan glass plate depictions of the human detritus in the wake of the battles of Gettysburg and Antietam. No artist, with the possible exception of Goya, had been able remotely to capture the horrors of war as did the camera. The general antiwar feeling increased as photography became more widely dispersed, intensified with images of the Somme or Verdun in World War I, and peaked with the living room war of Vietnam, as military horrors in the field were presented by avuncular television newscasters only a day or so removed from the event. By the twenty-first century, the image of a hut in flames and of dead civilians would be transmitted around the globe in real time—and protests could develop almost as swiftly.

Of course, wars did erupt in this period, but with decreasing bloodshed and destruction—and this in the nuclear era, which threatened mass destruction. But no one actually declared war and rarely even admitted that they were engaged in anything as horrid as an actual war. Belligerents were engaged in what they would term perhaps *liberation struggles* or in *repelling aggression.* The term *war,* with the exception of the war on terrorism and the Cold War (which, by definition, was no war), seemed confined to such domestic concerns as the war on drugs, the war on poverty, or the war on inflation. In fact, historians have puzzled over the question of the last declaration of war. (Was it the Soviet Union's on Japan in the waning days of World War II?)

The reason for such shying away from any declaration of war was to some extent tied to the fear of nuclear Armageddon, much prophesied after 1949. But it was at least equally due to the fact that wars from the second half of the twentieth century on tended to be between ethnic and racial groups, and mostly fought guerrilla-style or through terrorism. But even the few conventional post–World War II conflicts were fought in the absence of formal declarations of war as the belligerents studiously avoided making any gesture of recognition to their opponents. The Democratic People's Republic of Korea was not about to dignify the "Lick-Spittle Rhee Puppet Regime" (otherwise known as the Republic of Korea) with the amenity of passing a formal note that a state of war now existed between the two nations. Nor would the latter dignify the "Tyrannical Communist Regime" (People's Republic of Korea) with the same. Even when the anti–Saddam Hussein coalition threw the Iraqi dictator's forces bodily out of Kuwait in 1991, there was no declaration of war, although tens of thousands of unfortunate Iraqi troops perished. In fact, most belligerent parties go to some lengths in their official statements and in their propaganda carefully to insist that they mean the people on the other side no harm; only their evil leaders are to blame for the current sad state of affairs. Of course, declarations of war or not, mothers still find themselves mourning their lost children, cities are eviscerated, and young men are mutilated and die ghastly deaths.

War at this writing seems increasingly confined to the periphery; to the dispossessed, to the radical student dreamers (the very word *Taliban* means "students"), the fringes, the losers. They are those who wish to split off a state (e.g., Québécois, Basque, Croatian, Chechen, Kurdish, or Palestinian separatists), to purify a state (Algerian, Egyptian, Afghan, Iranian Islamists), to unite separated brethren (e.g., Serbs, Irish, Kashmiri nationalists), to impose their ideology on their own people (Chinese Civil War, Korean War, Vietnamese War, Wars of National Liberation of the 1960s and 1970s)—or varying combinations of some of the above. (The war on terrorism is an effort to combat the methods used by so many of these groups.) They generally succeed in bringing down far more misery on the inhabitants of the lands that they are trying to liberate than what the unfortunates suffered at the hands of the existing governments. (In fact, that is precisely the cold-blooded aim of numerous such revolutionary groups: to cause through their own deliberate actions and the counteractions of the government such dislocation and terror that the people will lose confidence in the legitimacy of that government.) But, by the standards of the two world wars, or even Korea or Vietnam, these struggles are still of the small war variety.

No one can foresee how long this relatively tranquil state of affairs will endure; a nuclear conflict between India and Pakistan is a horrific possibility, Greece and Turkey (midsize military powers) occasionally threaten military action, the two Koreas (one possibly with some nuclear capability) are divided by the most dangerous piece of real estate on earth, and it is increasingly difficult to isolate even the most obscure of conflicts in this day of global mass media. No era is forever, and we would do well to learn of past wars, if for no other reason than to appreciate better the current long peace—while we can.

MAPS

ACRONYMS

ABM Antiballistic Missile Treaty
ACP automatic cartridge pistol
AEW Airborne Early Warning
AFVs armored fighting vehicles
AIF Australian Imperial Force
ALB Air Land Battle Doctrine
ANZAC Australian and New Zealand Army Corps
ARVN Army of the Republic of Vietnam
ASDIC Anti-Submarine Detection Investigation Committee
ASW antisubmarine warfare
ATC American Tobacco Company
AWACS airborne warning and control system
BEF British Expeditionary Force
C2 Command and Control system
C3I command, control, communications, and intelligence
Cheka Chrezvychainaya Komissariat po bor'be s kontrarevoliutsiei i sabotazhem (All-Russian Commission for Struggle against Counterrevolution and Sabotage)
CI counterintelligence
CIA Central Intelligence Agency
CIGS chief of the Imperial General Staff
CORDS American Office of Civil Operations of Rural Development Support
CPB Communist Party of Burma
CTBT Comprehensive Test Ban Treaty
DIA Defense Intelligence Agency
DK Democratic Kampuchea
DPRK Democratic People's Republic of Korea
DSM Distinguished Service Medal
EAM-ELAS National Liberation Front–National Popular Liberation Army
EDES Greek Democratic National Army
ELN Ejercito de Liberación Nacional (Army of National Liberation)
EOKA National Organization of Greek Fighters
EPL Ejercito Popular de Liberaci (Army of Popular Liberation)
EVA Ever Victorious Army
FARC Fuerzas Armadas Revolucionarias de Colombia (Colombian Revolutionary Armed Forces)
FLEC Front for the Liberation of Cabinda
FLN National Liberation Front
FNLA National Front for the Liberation of Angola
GCNG Greek Cypriot National Guard
GDP gross domestic product
GDR German Democratic Republic
GPU Gosudarstvennoe Politicheskoe Upravlenie (State Political Administration)
GULAG Glavnoe Upravlenie Ispravitel'no-trudovykh Lagerei (Chief Administration of Corrective Labor Camps)
HUMINT human intelligence
I&W Indicators and Warnings
IDF Israeli Defense Forces
IFF Identification Friend-or-Foe transponders
IMINT (aerial and satellite photo and radar) imagery intelligence
INA Indian National Army (Azad Hind Fauj)
INF intermediate-range nuclear forces
IR infrared (radar)
IRA Irish Republican Army
IRB Irish Republican Brotherhood
JDA Japanese Defense Academy
JTF Joint Task Force
KGB Komitet Gosudarstvennoi Bezopastnosti (Committee of State Security)
KNU Karen National Union

KPA Korean Peoples Army
KUFNS Khmer United Front for National Salvation
M-19 Movimento 19 de Abril (19th of April Movement)
MACV U.S. Military Assistance Command
MAUD Military Application of Uranium Detonation
MED Manhattan Engineer District
MFA armed forces movement
MID Military Intelligence Division of the Army General Staff
MLRS Multiple Launch Rocket System
MPLA Popular Movement for the Liberation of Angola
MVD Ministerstvo Vnutrennikh Del (Ministry of Internal Affairs)
NATO North Atlantic Treaty Organization
NF Intermediate Nuclear Forces Treaty
NKGB Narodnaya Komissariat Gosudarstvennoi Bezopastnosti (People's Commissariat for State Security)
NKVD Narodnaya Komissariat Vnutrennikh Del
NSA National Security Agency
NSC National Security Council
NVA North Vietnamese Army
OAS Secret Army Organization
OECS Organization of Eastern Caribbean States
OFS Orange Free State
OGPU Unified State Political Administration
OKW Wehrmacht High Command
OPEC Organization of Petroleum Exporting Countries
ORBAT Order of Battle information
OSRD Office of Scientific Research and Development
OSS Office of Strategic Services
PAIGC African Party for the Independence of Guinea and Cape Verde
PG Provisional Government
PLO Palestinian Liberation Organization
PNI Partai Nasional Indonesia (Indonesian Nationalist Party)
PPA Planters Protective Association
PRA People's Revolutionary Army
PRD Dominican Revolutionary Party
PRK People's Republic of Kampuchea
RADAR Radio Detecting and Ranging
RAR Royal Australian Regiment
RDF Radio Direction Finding
RKKA Workers and Peasants Red Army
ROK Republic of Korea
RPF Assembly of the French People Party
SALT Strategic Arms Limitation I Interim Agreement
SAMS surface-to-air missile defenses
SAS Special Air Service
SBS Special Boat Service
SEAL sea, air, and land team
SIGINT signals intelligence
SLORC State Law and Order Restoration Council
SONAR Sound Navigation and Ranging
START Strategic Arms Reduction Talks I and II
TECHINT technical intelligence
TOW target on wire
UN United Nations
UNC United Nations Command
UNEF United Nations Emergency Force
UNITA National Union for the Total Independence of Angola
VOC Vereenigde Oost-Indische Compagnie (Dutch East India Company)

GLOSSARY

(Recurring topics. Each listing in this glossary has its own article in the text.)

Airborne Operations The insertion of troops and equipment on the battlefield by means of parachute, helicopter, or glider.

Armor Body protection for soldiers.

Armored Fighting Vehicles Protected vehicles used on the battlefield.

Arms Control Limits, usually set by treaty, on the number and types of weapons; usually refers to nuclear weapons.

Artillery Basically, a heavy metal tube from which a missile is discharged violently by explosive force.

Awards and Honors Recognition granted by authorities for meritorious service by soldiers.

Ballistics The science of projectiles, divided into interior and exterior ballistics. Its aim is to improve the design of shells/projectiles so that increased accuracy and predictability are the result. It deals also with rockets and ballistic missiles.

Bayonet Metal blade or spike that, when fixed to a musket or rifle, facilitates its use in hand-to-hand combat. In recent decades, more likely to be used for opening ration tins.

Bazooka Shaped-charge, smoothbore, man-portable, antitank, and pillbox weapon.

Buffalo Soldiers African-American troops of the late-nineteenth-century U.S. regular army.

Catapult Engine for throwing a heavy weight, using an arm released from tension; the artillery of the ancient world.

Cavalry The noble, mobile arm of battle; the traditional horse-borne arm of mobility; can now refer to a motor vehicle–mounted unit.

Chaplains Military officers who tend to the spiritual, moral, and physical needs of troops in the field and in camp; pastors in uniform.

Chemical and Biological Warfare The deliberate use of chemical or biological agents against an enemy.

Civil Affairs/Military Government Those activities of a commander that embrace the relationship between the military forces and civil authorities and people in a friendly country or area (Civil Affairs) or occupied country or area (Military Government).

Coastal Defense The defense of a nation's coast from an enemy sea invasion or blockade, accomplished with heavy artillery, mines, small warships, and nets.

Cold War (1946–1989) Period of tension between the Soviet Union and the People's Republic of China and their allies, representing communism (the East); and the United States and its allies, representing capitalism and democratic socialism (the West); punctuated with several "hot" wars, the most significant being the Korean War (1950–1953) and the Vietnam War (c. 1955–1975). The Cold War ended unexpectedly with the fall of the Berlin Wall and the bloodless victory of the West.

Communications, Military The application of technology to the transmission of military orders and intelligence.

Conscription The selection of persons for involuntary military service.

Death Squads Clandestine and usually irregular organizations, often paramilitary in nature, that carry out extrajudicial executions and other violent acts against clearly defined individuals or groups of people.

Disarmament The removal or drastic reduction by nation-states of major weapons.

Economic Warfare Compelling an enemy to submit either by direct action against its economic basis or indirectly through blockade or boycott.

Electronic Warfare The use of the electromagnetic

spectrum to gain knowledge of the presence and movement of an opposing force and also to deny any opposing force the use of that spectrum.

Engineering, Military The application of science and technology for military purposes, primarily through the use of civil engineering.

Ethics of War Rules, principles, or virtues applied to warfare.

Firearm A tube, closed at one end, that has in it an explosive with a projectile above it, nearer to the open end of the tube. An ignition system fires the explosive charge, which forces the projectile along the tube by means of the gases from the explosion; can include both artillery and small arms.

History, Military History dealing with the use of organized armed force, either on behalf of some form of recognized state authority or against it.

Infantry Lightly armed ground troops; the backbone of any army; not only the most numerous of the fighting arms but the only one that can actually take and hold ground.

Intelligence, Military Military specialty that provides a commander and staff with the knowledge of the enemy and of weather and terrain required for the planning and conduct of operations. (There is no truth to the assertion that "military intelligence" is an oxymoron.)

Laws of War International laws, enforced sometimes by nations after war and sometimes by commanders in battle, governing both the decision to engage in war and the manner of its conduct, particularly the forms of violence used, the definition of combatants, the treatment of prisoners, and the treatment of neutrals and noncombatants.

Logistics Largely an American usage, encompassing military supply, transportation, medical service, and construction-maintenance.

Machine Gun Rapid-firing small arm that can maintain a high rate of fire without the requirement of reloading after each round, which today means a fully automatic weapon; either man-portable ("machine pistol") or heavier.

Maps and Cartography The result of the utilization of cartography and topographical reproduction for military strategy and operations.

Medals and Decorations Tangible recognition of faithful military service or success awarded to individual soldiers.

Medicine, Military The medical and surgical specialty concerned with the ailments of soldiers and sailors.

Mercenaries Hired professional soldiers who fight for a state or entity without regard to political interests or issues.

Military-Industrial Complex The institutions and people that plan, procure, and fight a war and that supposedly shape the economy, the political realm, and the wider society, even in peacetime. Term first used by outgoing U.S. President Dwight D. Eisenhower in 1961.

Mortar Muzzle-loaded weapon firing its bomb at a high angle to attack protected positions and trenches.

Pacifism/War Resistance The organized opposition to war, killing, or violence; usually divided into two segments: religious (e.g., Quakers, Mennonites) and secular.

Paramilitary Organizations Unofficial groups organized along military lines yet lacking the traditional role or legitimization of conventional or "genuine" military organizations.

Propellants Compounds used to move a projectile from the firing device to the target.

Psychological Operations The use of psychology and propaganda by military units to persuade target audiences to adopt at least some of their views and possibly to modify their behavior.

Rank, Military Official indication of a soldier's length and quality of service in organized militaries.

Rifles Firearms designed with barrel grooves to impart a spin and thus far greater accuracy; either small arms or artillery.

Tactics The theory and practice of using military forces in combat.

Terrorism Acts of violence intended for a wide audience in order to create an environment of fear for political reasons.

Theory, Military That body of knowledge usually published in books and journals that examines the nature of wars and the art of war on an abstract level.

Unarmored Fighting Vehicles Unprotected military vehicles, either specially built for the military or adapted from commercial models and used in combat-support roles (e.g., Jeeps, trucks, ambulances).

Uniforms Military clothing worn by organized bodies of troops to distinguish them from the uniformed personnel of other armed forces and to strengthen morale.

War Crimes Actions in wartime that violate the laws or usages of war.

A

Abbas the Great (1571–1629)

One of Persia's greatest shahs, presiding over a period of national strengthening and growing Iranian military power. When Abbas became shah in 1587, his Safavid Dynasty was in serious trouble. The Qizilbash tribal confederation, the regime's foundation, had become fractious and undisciplined. Many provinces were virtually independent. In 1590, to gain a free hand for domestic consolidation, Abbas signed a humiliating peace with the Ottoman Empire, recognizing the occupation of much of northern and western Iran by Persia's great enemy. Equally menacing, the hostile Uzbeks under the Shaybanid state controlled the central Asian and Afghan frontiers.

To balance off the Qizilbash, Abbas organized a separate corps of 10,000 cavalry and 12,000 infantry, composed of Armenian and Georgian slaves, prisoners-of-war, and converts to Shia Islam—the Ghulamans. Many Qizilbash leaders lost their fortunes and freedom to Abbas's need for revenue. The Ghulamans also proved their mettle in repressing domestic revolt and coercing disloyal provinces into submission. By 1596, Abbas was pushing back the northern frontier, cowing the Shaybanids.

In 1598, an English trade delegation arrived, seeking commercial and military exchanges. The shah persuaded them to help him modernize the army. With English coaching, the Safavids expanded their army to include 12,000 gunners, 12,000 mounted musketeers, and 500 bronze and brass cannons. At the same time, the shah enlarged his Ghulaman forces through regular expeditions into the Caucasus, taking thousands of prisoner-recruits.

By 1602, despite losing many cannons and troops, Abbas drove the Shaybanids out of Herat, Mashhad, and Khurasan province. The next year, he moved westward and launched an offensive against the Ottomans. In 1605, at the Battle of Sufiyan, Abbas crushed the Turks and regained Tabriz. Despite frequent Ottoman counterthrusts, the shah's forces routinely pushed them out. The 1618 Treaty of Sarab compelled Istanbul to evacuate Kurdistan and Azerbaijan. Persian forces also crossed the Straits of Hormuz and occupied Oman. In 1623, as domestic turmoil struck the Ottoman Empire, Abbas invaded Iraq and Anatolia. Diyarbakir, Baghdad, and Mosul fell in rapid succession, bringing the Safavid state to its greatest territorial extent. When Abbas died in 1629, Iran was stable, powerful, and respected.

Weston F. Cook, Jr.

See also: Ottoman Empire

References and further reading:

Monshi, Iskandar Beg. *History of Shah Abbas the Great.* 2 vols. Trans. R. M. Savory. Boulder, CO: Westview Press, 1978.

Savory, R. M. *Iran under the Safavids.* Cambridge, UK: Cambridge University Press, 1980.

Abbasid Revolution (747–751)

A violent revolution in medieval Islam that replaced the Arab-dominated Umayyad Caliphate with a more open, multicultural system that symbolized the unity of Islam for five centuries.

The Umayyad family had seized control of the Islamic empire in 661. Over the next 70 years, Umayyad caliphs presided over the conquest of an empire that spread from the Indus Valley to the Pyrenees Mountains. They created a unified state, a vigorous commerce, and a resolute military based upon Arab tribes. Their leaders ruled as patrons of the Islamic religion, and of Arabic culture, but by 740 Umayyad policies had severely polarized their subjects and generated widespread hostility to their authority. Arab armies on the frontiers resented their low pay, constant campaigning, and the privileges of more favored tribes. Thousands of non-Arab converts to Islam seethed at discriminatory taxes, mistreatment, and Umayyad hypocrisy. Shia Muslims, committed to the rule of the descendants of the Prophet Muhammad's son-in-law Ali, saw the Umayyads as oppressive usurpers.

The success of the Abbasids derived from their ability to mobilize these animosities and create a revolutionary movement. Unlike the Umayyads, the Abbasids could claim a family tie to the Prophet through his uncle Abbas. They also asserted that a member of Ali's family willed his right to the caliphate to the Abbasids. Their anti-Umayyad propaganda proved especially appealing in Khurasan, the northeastern frontier of Iran, where both Arab and non-Arab Muslims harbored intense grudges against the state. Conspiratorial cells also made recruits in Iraq, Palestine, and Syria. Additionally, the Umayyads had just passed through a dynastic war, and the new caliph, Marwan II, had alienated traditional friends of the regime.

In 747, the Abbasids launched their revolt in Khurasan. To rally support, they unfurled black flags, symbols of the Mahdi, a messianic figure in popular Islam. They spoke of their movement in terms of a millennial, divinely ordained upheaval, to sweep away the wicked Marwan.

Seizing the frontier town of Marv in February 748, the Abbasid commanders Abu Salama and Abu Muslim began their westward advance. Picking up momentum and support as they advanced, the Abbasids reached central Iran in August. The next year, most of Iraq fell and the head of the movement, Abu al-'Abbas, declared himself caliph at Kufa. In January 750, Abbasid forces met Marwan's army along the banks of the Zab River in northern Iraq, and shattered the dispirited Umayyads. The defeated caliph, hounded from Syria to Egypt, finally fell into Abbasid hands, and was killed seven months later.

The establishment of the Abbasid Caliphate was a true revolution, not just a change of administrations. Caliph Abu al-'Abbas and his successor, Caliph al-Mansur, ended ethnic and economic discriminations against non-Arab Muslims, establishing the fundamental principle that all Muslims were equal before the state as well as before God. Freed of Umayyad elitism, Islam experienced a dramatic surge in conversions. They also massacred Umayyads, and former supporters of questionable loyalty, with revolutionary zeal. Their Shiite partners were honored, but denied power. To dramatize the newness and purity of their government, the Abbasids abandoned the Umayyad capital of Damascus, and built themselves a new center in Iraq, Baghdad. Here the Abbasid court welcomed Muslims of all ethnic backgrounds, laying the foundations of an intellectual, philosophical, and scientific renaissance.

Weston F. Cook, Jr.

See also: Harun al-Raschid

References and further reading:

Kennedy, Hugh. *The Prophet and the Age of the Caliphs.* London: Longman Press, 1986.

Shaban, M. A. *The Abbasid Revolution.* Cambridge, UK: Cambridge University Press, 1970.

Sharon, Moshe. *Black Banners from the East.* Jerusalem: Magnes Press, 1983.

Abd-el Krim, Mohammed (1882–1963)

Riff chieftain who helped spark the Spanish Civil War. In 1921 Abd-el Krim, the university-educated son of a Riff nobleman, revolted against Spanish rule in Morocco. After capturing the frontier post at Abaran, the former journalist and his Beni-Ouriaghel tribesmen attacked a column of 20,000 Spanish troops at Anual on 21 July 1921, killing some 16,000. The uprising brought Spain under virtual military dictatorship in 1923 and, a year later, squeezed the Spanish in Morocco into two fortified strongholds at Tetuan and Mililla on the Mediterranean.

Krim then moved south on 12 April 1925 against a threat from French Morocco. Armed with captured Spanish artillery, machine guns, and rifles, he eliminated 43 of 66 French blockhouses over a 50-mile span until checked outside Fez.

Joining France in a counteroffensive, Spain landed 50,000 troops in the Bay of Alhucemas on 8 September 1925. Three columns—including a Spanish Foreign Legion contingent under Francisco Franco—pushed south into Riff territory as World War I hero Marshal Henri-Philippe Pétain led six mutually supporting French forces on a wide front north to Ajdir, Krim's headquarters. The city fell on 2 October 1925.

Weakened by heavy casualties, poor harvests, disease, and diverse tribal interests, Abd-el Krim surrendered on 26 May 1926, and was exiled to a French island for 20 years. Although finally suppressed in 1934, the Riff uprising contributed to the Spanish Civil War two years later.

When released in 1947 Abd-el Krim accepted Egypt's protection and later became leader of a North African nationalist movement opposed to European rule.

Gary J. Komar

See also: Franco, Francisco; Pétain, Henri-Philippe

References and further reading:

Asprey, Robert B. *War in the Shadows.* Garden City: Doubleday and Co., 1975.

McLeave, Hugh. *The Damned Die Hard.* New York: Saturday Review Press, 1973.

Abdelkader (1808–1883)

Algerian leader, fought against French colonialism. Born 6 September 1808 near Mascara, Algeria, he became emir of Mascara when his father, Mahi-el-Din, died in 1832. He furthered jihad against the French and shrewdly increased his power. Although he won a battle at La Macta on 28 June 1835, he failed to prevent French forces from sacking Mascara in December of that year. Abdelkader met defeat at Sikkah on 6 July 1836 and was on the verge of losing support of the tribes. He began peace talks with the French, eventually agreeing to the peace of the Tafna on 30 May 1837.

Abdelkader ran an efficient government over most of Algeria until French attempts to expand their holdings prompted him to revive the jihad in November 1839. His forces struggled, losing Tlemcen in 1842, and suffering a major defeat at Smala on 10 May 1843 when his 40,000-man army lost to a French army of 2,000 led by the Duke of Aumale. He then fled to Morocco and raised another army. But he was defeated at the hands of a smaller French army at the Battle of the Isly River on 14 August 1844, a defeat that resulted in the loss of Moroccan support. He then moved to suppress a rebellion in the Dahra region of Algeria, winning several battles. Later he was forced by the French to return to Morocco, where he again attempted to raise an army. Compelled to fight against Morocco, he eventually was defeated, left for Algeria, and surrendered to the Duke of Aumale on 23 December 1847 in exchange for safe passage to the Levant. He was held in a French prison from 1848 to 1852, freed by Napoleon III, then moved to Damascus where he died on 26 May 1883.

Harold Wise

See also: French Colonial Wars

References and further reading:

Danziger, Raphael. *Abd-al Qadir and the Algerians: Resistance to the French Internal Consolidation.* New York: Holmes & Meier, 1977.

Abercromby, Sir Ralph (1734–1801)

British general whose leadership restored prestige to the army in the 1790s, setting the stage for the defeat of Napoleon in Egypt. Born in Clackmannanshire, Scotland, on 7 October 1734, Abercromby, unusually, studied at Rugby, Edinburgh, and Leipzig, then entered the army in 1756. Service in the Seven Years' War introduced him to the doctrine of Frederick the Great. Although he rose in rank, his criticism of government policies kept him away from active service; he retired from the army in 1783 and became a member of Parliament for Clackmannan.

He returned to service in 1793, commanding a brigade against revolutionary France. During the retreat from Flanders in the winter of 1794–1795, Abercromby commanded the rear column, securing the army's retreat from a disastrous campaign. In recognition of his skills he was knighted and given command of an expedition to the West Indies, successfully taking Grenada, St. Lucia, St. Vincent, and Trinidad from the French.

Returning to Britain victorious in 1797, Abercromby was appointed commander in chief in Ireland. He successfully reorganized the forces there, but his zealous efforts to establish the supremacy of civil rather than military power met with resistance from the government and led him to resign his post.

In 1800 he was ordered to oust French forces in Egypt. His direction of the army to a successful landing at Aboukir Bay in early March 1801 in the face of stiff opposition is remembered as one of the British army's most daring exploits. Abercromby faced a counterattack by French forces on 21 March; it was repelled with heavy losses, but he was mortally wounded in the fighting. He died at sea on board the HMS *Foudroyant* and was buried at Malta, a future Wellington cutoff.

Don N. Hagist

See also: Alexandria; French Revolutionary Wars; Irish Uprising

Aboukir (25 July 1799)

A battle between the French army in Egypt, about 10,000 strong under Napoleon Bonaparte, and the Ottoman army of Rhodes, 15,000 strong, under Mustafa Pasha. In October 1798, the Ottoman sultan declared war on France, and began a two-pronged advance against the French army of Egypt. The army of Damascus was to march through Syria and Palestine and into Egypt by way of the Sinai Desert. The army of Rhodes was to be transported to Alexandria by a combined Turkish and British fleet.

Bonaparte sought to deal with the first of these thrusts by invading Syria. Although he was unable to obtain any permanent advantage, and suffered heavy losses besieging Acre, he did scatter the forces of the pasha of Damascus at the Battle of Mount Tabor on 16 April 1799. Bonaparte then withdrew to Cairo, arriving on 14 June 1799. On 11 July, the infantry and artillery of the army of Rhodes arrived off Aboukir, near Alexandria, and quickly seized the town, but not the castle, which resisted for several days before submitting. This delay allowed Bonaparte, by dint of vigorous movement, to bring 10,000 men, including 1,000 cavalry, to Aboukir by 24 July.

Bonaparte found the Turkish army deployed in two lines, in front of Aboukir castle, which was strongly garrisoned with Turkish forces. Though the French infantry badly mauled the Turkish forces, the issue was decided by the French cavalry commanded by Joachim Murat. A cavalry charge by Murat forced its way through the Turkish lines, and into the Ottoman headquarters. Overwhelmed by the French cavalry, the Turkish army fled to the beaches, or toward the castle. Turkish losses were in excess of 2,000, further increased by the subsequent surrender of Aboukir Castle. French losses were 220 killed and 750 wounded.

As a result of the battle, Bonaparte secured the position of the French army in Egypt. However, both Bonaparte and the Directory independently decided to abandon the Egyptian campaign. Bonaparte, without actually receiving orders to do so, left the Army of Egypt on 22 August 1799, with only a few staff officers. He arrived in France on 9 October, where he began preparations to overthrow the Directory, which culminated in the coup of the 18th Brumaire.

Joseph M. Isenberg

See also: Alexandria; French Revolutionary Wars; Murat, Joachim, Grand Duke of Cleves-Berg, King of Naples; Napoleon I; Pyramids

References and further reading:

Chandler, David G. *Campaigns of Napoleon.* New York: Scribner, 1966.
Connelley, Owen. *Blundering to Glory.* Wilmington, DE: Scholarly Resources, 1999.

Abrams, Creighton William, Jr. (1914–1974)

U.S. Army commander during World War II, Korea, and Vietnam. Creighton Abrams was born 15 September 1914 in Springfield, Massachusetts. He graduated from the U.S. Military Academy at West Point in 1936 and was commissioned a cavalry officer. During World War II, he commanded the 37th Tank Battalion and demonstrated aggressive leadership as the vanguard of the drive across Europe. During the Korean War, Abrams gained valuable staff experience serving as the chief of staff for the I, IX, and X Corps.

During the early 1960s, Abrams found himself involved in two crises—as commander of the 3d Armored Division in Europe during the Berlin Crisis and as commander of federal troops deployed to maintain order after the integration of the University of Mississippi.

In 1967, Abrams was assigned as deputy commander of the U.S. Military Assistance Command (MACV), Vietnam. During the Tet Offensive, he supervised operations in northern Vietnam, including the recapture of Hue. Abrams assumed command of MACV in the aftermath of Tet. He changed American tactics to emphasis the defense of populated areas and replaced search-and-destroy missions with small-unit patrols.

Abrams left Vietnam in 1972 after the North Vietnamese Easter Offensive and returned to the United States to serve as army chief of staff. During this assignment he worked to rebuild the army, instituting the all-volunteer army and working to integrate reserve components. Abrams died in 1974 following a battle with cancer. He was respected as both a combat soldier and an advisor to civilian leaders. The outstanding performance of the U.S. Army during the triumphant Gulf War can be considered his memorial.

William Hartley

See also: U.S. Army; Vietnam Conflict

References and further reading:

Millet, Alan. *A Short History of the Vietnam War.* Bloomington: Indiana University Press, 1978.
Sorley, Lewis. *Thunderbolt: General Creighton Abrams and the Army of His Times.* New York: Simon and Schuster, 1992.

Abu al-'Abbas (722–754)

Overthrew Umayyads and established the Abbasid Caliphate. The Abbasids—a branch of the family of the Prophet Muhammad descended from his uncle 'Abbas—had been seeking to overthrow the reigning Umayyad Caliphate since 718. They eventually recruited a large cadre of supporters in Khurasan (northeastern Iran), raising the banners of revolt in 747. Abbasid forces swept through Iran and reached Kufa, Iraq, in September 749, but that same year their leader, Ibrahim ibn Muhammad, was caught and executed. His obscure brother, Abu al-'Abbas, then assumed leadership of the movement.

Abu al-'Abbas declared himself caliph at Kufa in November 749. In his inauguration, he warned his listeners he would spill whatever blood was required to eliminate any opponents. Two months later, his forces crushed the main Umayyad army along the Zab River. Marwan II, the Umayyad caliph, fled westward first to Syria, then Palestine, and finally Egypt. At the same time, the Abbasids began a campaign of massacres against other Umayyad family members and their supporters. When Yazid ibn Hubayra, Umayyad governor of Wasit, surrendered after a year of siege, the Abbasids promised him safe conduct, but shortly after his capitulation Yazid's captors broke their pledge and killed him. Even the tombs of Umayyad caliphs were desecrated. Marwan II was finally slain in August 750.

When several Shia Muslim groups attempted to revolt, Abu al-'Abbas crushed them pitilessly. In the end, just before his death, the first Abbasid caliph was killing some of the men who had brought his family to power.

Weston F. Cook Jr.

See also: Abbasid Revolution

References and further reading:

Kennedy, Hugh. *The Early Abbasid Caliphate: A Political History.* London: Oxford University Press, 1981.

Shaban, M. A. *The Abbasid Revolution.* Cambridge, UK: Cambridge University Press, 1970.

Abu Klea (19 January 1885)

Battle during the Gordon Relief Expedition, 1885. Sir Garnet Wolseley, sent to rescue General Charles Gordon from his entrapment at Khartoum, dispatched two columns of about 2,000 men each to the city. One was the river column under General William Earle, and the other was the desert column under General Herbert Stewart. As Earle's column worked its way up the Nile, Stewart's men set off across the desert. The desert march became a desperate race between oases both against heat and thirst and against the Mahdi's dervishes. On 19 January, the Mahdi (Muhammad Ahmad) placed about 10,000 of his men on the track between Stewart and the oasis at Abu Klea. Stewart formed a square but left outriders well ahead of the formation and allowed Colonel Fred Burnaby to weaken the rear of the square by moving sailors and their naval guns out of position. The Mahdi's forces launched a massive, determined, and well-paced attack against the square. To avoid shooting their comrades outside, the men on the front of the square held fire until the outriders could get in, by which time the dervishes were dangerously close. While sustained fire kept the front secure, the Mahdists worked their way around and smashed into the back corner Burnaby had dismantled, and broke the square itself. The square was restored, and terrible losses were inflicted on the dervishes, who were forced to retire. This represents, however, the only time a British square was broken. Burnaby and 65 British soldiers and 800 of the Mahdi's men were killed. Two days later, Stewart himself was mortally wounded, Wolseley had to send his chief of staff Colonel Redvers Buller across the desert to extract the column, and Gordon was not rescued.

James B. Thomas

See also: Buller, Sir Redvers Henry; Gordon, Charles George; Khartoum, Siege of; Muhammad Ahmad; Wolseley, Garnet Joseph, Viscount

References and further reading:

Holt, P. M., and M. W. Daly. *A History of the Sudan from the Coming of Islam to the Present Day.* London: Longman, 2000.

Pakenham, Thomas. *The Scramble for Africa.* New York: Random House, 1991.

Academies, Military

In the ancient world, the training of cadres for military leadership was the province of a small military elite that was responsible for the preservation and transmission of military doctrine. In the West, the collapse of Roman political institutions severely inhibited this transmittal, and in the Middle Ages, warfare had to be all but reinvented with the aid of Vegetius's treatise, *De re militari,* dating from the fourth century.

By the end of the Renaissance, powerful kingdoms were rapidly solving the problems associated with making military power a dependable school of statecraft. The need for trained military leaders combined with the application of reason to social problems produced the first military academy, the Schola Militaris, at Seigen, the Netherlands, founded in 1617 to teach the military doctrine of the Nassau family. Though it lasted only six years, its utility was evident, and other schools followed, most notably the military academy Louis XIV founded at Metz in 1668 and the Royal Military Academy founded in 1741 at Woolwich to train engineering and artillery officers. The majority of officers still bought their commissions and learned their duties by exercise or were promoted from the ranks of noncommissioned officers, who already knew the basics of military science. The experience of the Napoleonic Wars, including dramatic changes in the size of armies, tactical doctrine, and military technology, as well as a heady dose of national fervor, promoted the proliferation of military academies for the training of competent and patriotic junior officers, as well as the development of higher schools for advanced training in specialized areas and for staff officers.

The U.S. Military Academy, founded in 1802 in West Point, New York, was the first (and for decades the only) school of engineering in the United States. Sylvanus Thayer, superintendent from 1817 to 1833, modeled the institution on France's École Polytechnique and introduced modes of organization and discipline that would become normative over the institution's history. Though its graduates proved their engineering skills in the country's rapid expansion and their military skills in the Mexican-American and Civil Wars, the academy languished somewhat in the late nineteenth century.

After World War I, the academy's curriculum was reformed and updated, and in the 20 years following World War II, the teaching of humanities and social sciences was significantly augmented to comprise nearly 50 percent of the curriculum, which leads to a bachelor's degree and commission as a second lieutenant in the army. The Corps of Cadets avenges a little over 4,000. Entrants are selected on the basis of prior academic performance from nominees presented by the president and members of Congress (a process unique to the United States) and from army personnel.

Britain supplemented its foundation at Woolwich by establishing the Royal Military Academy in 1802; it has been located at Sandhurst since 1812. In 1947, Woolwich was merged into Sandhurst. The curriculum, divided into six terms of 13 weeks, comprises military subjects, mathematics, science, and modern languages, with some students prepared for examination in mechanics at Cambridge University or degree studies at the Royal Military College of Science at Shrivenham. The student body averages 1,000. Upon graduation, they receive commissions on the basis of vacancies existing in the various army branches.

Napoleon founded the École Speciale Militaire in 1802 and housed it at St. Cyr, near Versailles, in 1808. Though it moved to Brittany in 1947 and absorbed other schools for cavalry, engineers' artillery, and infantry, and was renamed the École Speciale Militaire Interarmes, it is still popularly known as St. Cyr. Cadets are organized into three battalions, two of persons holding the baccalaureat and one chosen from noncommissioned officers. The former are commissioned as second lieutenants after two years of study and go on to a third year of specialized study in a specific branch of the army. The latter receive their commissions at the end of a single year.

The German army system of cadet training began with the Prussian efforts to reorganize and modernize during the Napoleonic Wars and constituted one of the most highly effective officer training systems ever developed. After a hiatus at the end of World War II, West Germany developed officer training schools at Hanover, Munich, and Koblenz based on the historic system. Candidates who were secondary school graduates and had completed army basic training (or noncommissioned officers who passed a special examination) put in a year at school, followed by regimental service and schooling in a special arm, such as artillery or infantry, and were commissioned when their commanding officer was satisfied with their progress. After reunification, the training methods and institutions of West Germany prevailed over those of East Germany.

Czarist Russia opened a modern military academy in 1855 (and an air academy as early as 1910!). The Soviets closed these as antidemocratic and relied on a system of three-year officer candidate schools operated by the various branches of the army and navy. During World War II, they founded 30 junior military schools (Suvarov schools) throughout the nation and two naval schools (Nakhimov schools) at Leningrad and Baku. Open only to children of armed service veterans, these gave children a seven-year secondary school preparation for officer candidate school. After the collapse of the Soviet Union, Russia maintained eight military academies and one military university, none of which functioned at full enrollment.

Despite drastic reductions in most of the world's military forces, military academies enjoy unprecedented prestige and can select among the most highly qualified of applicants.

Joseph M. McCarthy

References and further reading:

Ambrose, Stephen. *Duty, Honor, Country: A History of West Point.* Baltimore: Johns Hopkins University Press, 1999.

Cope, John N. *International Military Education and Training: An Assessment.* Washington, DC: U.S. Government Printing Office, 1995.

Crackell, Theodore J. *The Illustrated History of West Point.* New York: H. N. Abrams, 1991.

Preston, Richard A. *Perspectives in the History of Military Education and Professionalism.* Colorado Springs: U.S. Air Force Academy, 1980.

Ruggero, Ed. *Duty First: West Point and the Making of American Leaders.* New York: HarperCollins, 2001.

Adowa (1896)

The most decisive defeat inflicted on a colonial expeditionary force by native African troops. In 1895, newly enthroned Ethiopian king Menelik II became infuriated when he discovered that Italy had assumed a virtual protectorate over his country in the altered international version of the recently signed Italo-Abyssian treaty. Menelik mounted a hastily organized raid on an Italian advance post at Coatit, which the Italians repulsed. This incident gave the Italians the impression that the Ethiopians could not stand up to European firepower.

Rome, in turn, was infuriated by Menelik's stance and ordered Eritrea's Italian governor Oreste Baratieri to punish the Africans' insolence. The general captured the towns of Makalle, Adrigat, and Adowa, and planned to fortify these outposts as a natural line of defense. Once again, vastly underestimating the military capabilities of infuriated African warriors, Baratieri's mere 25,000 men were outnumbered by Menelik's forces, which had swelled to an accretion of 196,000 men gathered at their capital, Addis Ababa, more than half of whom were half armed with modern rifles and supported with mobile field pieces.

Apprehensive, Baratieri withdrew to Adigat where Menelik besieged his force for 45 days. When Menelik offered the surrounded Italian garrison safe passage, the gesture only further stung Rome, whereupon the Francesco Crispi régime dispatched reinforcements and money, urging Baratieri to press the war home.

Apprehensive, Baratieri expected that Menelik would attack his dug-in troops, but the Ethiopian commander instead concentrated his forces at Adowa and waited for Baratieri to advance. This waiting game continued through February 1896, as supplies on both sides dwindled. Menelik's system of depots was running low and he began to consider falling back. Baratieri's food, even on half rations, would last only another week. Also, he was stung by a telegram from Crispi, accusing him of cowardice, and the exasperation of his subordinates.

Hastily, Baratieri organized his 15,000 men for a march in three independent brigade columns, which were to converge on peaks overlooking Adowa to the rear of Menelik's concentration after a one-day hike (assuming that Menelik would feel compelled to assault their defensive positions on the heights). Unexpectedly forbidding terrain, combined with outdated maps supplied from headquarters, resulted in the isolated Italian columns becoming confused and disorganized. They engaged Menelik's advance guards piecemeal, opening great gaps in their disjointed line. At this juncture, Menelik's 85,000-plus multitude was reinforced by the 30,000 troops of Ras Makkonen. Swarms of African troops threw themselves upon the divided Italian brigades. One of the Italian units, in the absence of orders from Italian headquarters, nevertheless mounted a valiant, methodical fighting withdrawal. Baratieri tardily ordered the other beleaguered brigade to retreat.

Ultimately, General Baratieri's botched venture lost 6,600 Italian soldiers and 268 officers, along with an unknown number of allied Eritrean troops. Another 1,700 were taken prisoner. Abyssinian losses are variously estimated at 2,000 to 7,000 dead and 10,000 wounded, with 900 captured. On 25 October 1896 the ensuing Treaty of Addis Ababa recognized the independence of Abyssinia. But forty years later, under Benito Mussolini, the Italians would be back.

Jim Bloom

See also: Italian Colonial Wars

References and further reading:

Berkeley, F. H. *The Campaign of Adowa and the Rise of Menelik.* 2d ed. London: Constable, 1935.

Brown, Pamela S., and Fassil Yirgu. *One House: The Battle of Adowa 1896—100 Years.* 4th ed. Chicago: Nyala Publishing, 1996.

Marcus, Harold G. *The Life and Times of Menelik II: Ethiopia 1844–1913.* Lawrenceville, NJ: Red Sea Press, 1995.

Ofeansky, Thomas P., and LaVerle Berry, eds. *Ethiopia: A Country Study.* 4th ed. Washington, DC: Federal Research Division of the U.S. Library of Congress, 1993.

Adrianople, Battle of (Thrace, 9 August 378)

Disastrous Roman defeat by Visigoths. The Eastern emperor Valens (r. 364–378) and many high-ranking officers died in the battle. Roman historian Ammianus Marcellinus wrote of the event, "Scarcely one-third of the whole army escaped."

In oppressive heat and dust, Valens marched 11 miles from Adrianople to attack the Goths. His 20,000 cavalry and 40,000 infantry outnumbered the 50,000 Gothic infantry, but the latter held a good defensive position in a wagon laager atop a low hill. Such Gothic wagon circles were no hastily arranged affairs. Ammianus says that the laager at Adrianople was formed as perfectly as if "turned on a lathe." The poet Claudian tells of another with a double moat and stakes placed along the top "two deep." The wagons themselves were covered in ox-hide and seemed "like a wall."

To gain time and allow the return of his cavalry away foraging, Visigothic king Fritigern fired the grass on the plain below and dispatched emissaries to Valens, who refused to be delayed and attacked. At first the battle went his way. Valens used his infantry to hold the center while his cavalry closed on the laager and engaged the Goths on the right. Suddenly, some 50,000 Gothic cavalry appeared on the right and attacked the Romans "like a thunderbolt." Valens's army was checked and his cavalry routed, the infantry trapped between the Gothic horse and foot. The latter now sallied from their wagons and overwhelmed the Roman cavalry, redeploying on the Roman left wing. The infantry broke. Valens's efforts to halt the rout with two units of the *auxilia palatina* were to no avail.

The early Christian scholar St. Ambrose called the battle "the end of all humanity, the end of the world." Although a major defeat and a significant blow to Roman prestige, the consequences of Adrianople should not be exaggerated. The defeat was the result of Valens's impetuosity and not of the army's inefficiency.

Nic Fields

See also: Goths

References and further reading:

Burns, T. S. "The Battle of Adrianople." *Historia* 22 (1973), 336–345.

Æthelbald's Wars (733–750)

Mercia establishes hegemony, paving the way for English unification. The early history of the kingdom of Mercia is obscure, but this changed with the reign of Penda, son of Pybba (r. 632–654). He extended his writ over Wessex (645) and East Anglia (650), gaining control of all England south of the Humber. Although Mercia declined after his death, Mercian expansion was resumed during the next century under Æthelbald, son of Alweo (r. 716–757), a grandson of Eowa, brother of Penda.

By 731, Æthelbald controlled all England south of the Humber. In a 736 charter he is styled "king not only of all the Mercians but also of all the provinces, which are called by the general name South English." Since Æthelbald issued coins bearing his image in a crowned war-helm (*cynehelm*), this hegemony was undoubtedly acquired through conquest.

In 737, Æthelbald raided north of the Humber, possibly with the idea of conquering Northumbria. In 740, he sacked York, while King Eadberht was absent campaigning against the Picts. Other campaigns included one against the Welsh (743).

Æthelbald's only reversal came in 752 at the hands of Cuthred, king of the West Saxons. He encountered Æthelbald at Beorhford (Burford, Oxfordshire?) and, in the words of the *Anglo-Saxon Chronicle,* "put him to flight." In the end, this victory did little more than postpone the inevitable. In 757, Cuthred's successor, Cynewulf (r. 757–786), acknowledged his subordination by coming to the Mercian court.

Æthelbald's supremacy was neither easily won nor stable; he also made many enemies. St. Boniface, archbishop of Canterbury, for example, reproached him for not taking a wife, and instead committing fornication with nuns, and for seizing the lands of the Mercian church. This was a common theme with the churchmen of the time.

As an old man Æthelbald was still violent and prone to lascivious behavior. His bodyguard murdered him at Seckington near Tamworth, in a terrible (though not unique) breach of the Anglo-Saxon code of fidelity. Who gave the orders is unclear. Æthelbald's heir, Beornred, succeeded him, but civil war followed and his cousin Offa (r. 757–796) seized the throne.

No other king maintained so general an ascendancy for so long, and it is significant that a contemporary chronicler describes Æthelbald as a "royal tyrant." As a consequence, Æthelbald and his successor, Offa, thus paved the way for the future unification of the English. On his death Æthelbald was Rex Britanniae, Bretwalda ("Britain-ruler"), a king to whom other kings were subject. They attended his court, paid him tribute, and fought under his leadership.

Nic Fields

See also: Offa's Wars

References and further reading:

Walker, I. W. *Mercia and the Making of England.* Stroud, UK: Sutton, 2001.

Aëtius, Flavius (c. 395–454)

Victor over the Huns. Born son of a *magister equitum* (master of cavalry) at Durostorum, Moesia Inferior, Aëtius was appointed *magister utriusque militiae praesentalis* (master of both services with the emperor) in 430 after successful battles against Visigoths and Franks in Gaul. On the death of rival Bonifatius, Aëtius secured profound influence over Valentinian III (r. 425–455). He was consul three times (432, 437, 446), an unprecedented number for one not of the imperial house; it was said that envoys were no longer sent to the emperor, but to Aëtius. Appointed patrician (*patricius:* a title now denoting generalissimo) in 433, he became the effective ruler of the western empire. He held absolute power for over 20 years until treacherously assassinated by Valentinian himself (21 September 454). One of Valentinian's advisers later remarked: "You have cut off your right hand with your left."

Aëtius's power was based not so much on military ability, undeniably great, as on his close relations with the Huns. He had lived as a hostage among them as a youth. He had learned much about their customs and had established ties of friendship with the Hun royal family. In 424, Aëtius raised a large force of Huns on behalf of the usurper Ioannes. Again, in 433, he recovered his hold over the imperial court in Ravenna with the aid of Hunnic mercenaries. For the next five or six years he employed them regularly. For instance, he commanded Huns during the recovery of Gaul, when his main achievement was the merciless destruction of the Burgundian kingdom. Hunnic mercenaries are not mentioned again after 439. In 451, Aëtius combined his armies with the Visigoths to defeat Attila at Châlons.

Nic Fields

See also: Châlons, Battle of

References and further reading:

Moss, J. R. "The Effects of the Policies of Aetius on the History of Western Europe." *Historia* 22 (1973), 711–731.

Agathocles (361–289 B.C.E.)

Ancient Sicilian warlord, notable for his ruthlessness. Agathocles was born in 361 B.C.E. in Thermae Himerenses (present Términi Imerese) on the east bank of the mouth of the Himera (present San Leonardo) River on the north coast of Sicily. About 343 he moved to Syracuse (present Siracusa), the major seaport on the east coast of Sicily, joined the army, and rose through the ranks. Frustrated by local politics, greedy for power, and seeking to subdue and unify the rival Sicilian towns, he tried three times after 323 to overthrow the government, was banished the first two times, but finally led a successful coup d'état in 317. He quickly consolidated his authority, reigned as Tyrant of Syracuse from 316 to 304, and as king thereafter until his death, probably from cancer, in 289.

Under Agathocles, Syracuse conquered most of northern and western Sicily. Worried about this growth of Syracusan military domination, Carthage, the main power in North Africa, launched a preemptive invasion of Sicily in 311. Carthaginian general Hamilcar Grisgo defeated the Sicilians at Himera, just across the river from Agathocles' birthplace, and besieged Syracuse. Escaping from the siege with 13,000 men, Agathocles went on the counteroffensive in 310. He invaded Africa, disrupted Hamilcar's logistics, and defeated a superior Carthaginian force near Carthage. Usually successful against the Carthaginians on both fronts, he concluded a peace treaty in 306.

Agathocles invaded southern Italy about 305 and stayed until 302. For the rest of his life, he enlarged both the army and the navy, built fortifications and public works, and continued pursuing his territorial ambitions throughout the Sicilian countryside. The Romans later honored his memory as they fought Carthage in the Punic Wars.

Eric v. d. Luft

See also: Ancient Warfare; Punic Wars

References and further reading:

Guido, Margaret. *Syracuse: A Handbook to Its History and Principal Monuments.* London: Parrish, 1961.

Randall-MacIver, David. *Greek Cities in Italy and Sicily.* Westport, CT: Greenwood Press, 1970.

Tillyard, Henry Julius Wetenhall. *Agathocles.* Cambridge, UK: Cambridge University Press, 1908.

Agincourt, Battle of (25 October 1415)

Great English victory during Hundred Years' War. Following England's declaration of war, an army under Henry V (r. 1387–1422) landed in Normandy in August and laid siege to Harfleur, which surrendered in September. On 12 October, the 6,000-strong English army began a march to Calais. Hampered by torrential rains, wrecked bridges, and heavily defended fords, it did not cross the upper Somme until 19 October. By that time French forces, numbering some 25,000 to 30,000 men, had crossed the lower Somme. They moved to stand astride the English line of advance at Agincourt on 24 October.

The following morning found the armies at opposite ends of a narrow defile between heavily wooded areas. After occupying the narrowest part of the gorge, the English were

King Henry V of England taking part in the Battle of Agincourt. (Hulton/Archive)

forced to fight defensively in prepared positions. The French assembled in three formations, each behind the other. This deployment negated their numerical advantage. They also failed to employ cavalry against English archers either on the flanks or in the rear, forcing them to undertake successive frontal assaults. The French attacks were broken one after the other. Only the first threatened to break into English defenses. With the second- and third-wave attacks faltering amid thick mud and suffering heavy losses, an English charge dispersed what remained of the French army.

The battle cost the French perhaps 5,000 to 6,000 dead, with hundreds taken prisoner. The English losses, allegedly, were a little more than a hundred. The battle had no immediate military result: The war continued the treaty of Troyes in 1420. By that time the English-French struggle formed part of a wider context of civil war within France. The battle's enduring fame undoubtedly owes much to Shakespeare's play, *King Henry the Fifth.*

H. P. Willmott

See also: Crécy, Battle of

References and further reading:

Curry, Anne. *The Battle of Agincourt: Sources and Interpretations.* Rochester, NY: Boydell Press, 2000.

Hibbert, Christopher. *Agincourt.* London: Batsford, 1964.

Agricola, Gnaeus Julius (40–93)

Model governor of Britain, father-in-law of historian Tacitus. Agricola was born in the colony of Forum Iulii in Gallia Narbonensis. He had gone over to Vespasian even before the would-be emperor had publicly declared his hand (69). Agricola was later a loyal supporter of the Flavians and in 77, aged 37, he was appointed governor (*legatus praetorius*) of Britain. He came to his post with considerable local knowledge and experience, something unusual for a Roman governor of the time. Agricola had twice served in the province previously, as a legionary tribune (*tribunus laticlavius*) during the Boudican Revolt (60–61), and as legate (*legatus legionis*) of legio XX Valeria Victrix (70–73).

Agricola arrived in his province late in 77. His first action was suppression of the Ordovices of central and north Wales. The following summer (78), his first full campaigning season, seems to have found him in the territory of the Brigantes where, according to Tacitus, he built forts. He may also have spent some time north of the Solway, in what is now southern Scotland. He also operated there during his third season (79), ravaging tribes as far north as the estuary of the Tay. Again, he built forts. The next two years (80–81), Agricola consolidated the Forth-Clyde line. In a sixth season (82), spent north of the Forth, victory narrowly eluded him against the Caledonians. A seventh (83) culminated in the Battle of Mons Graupius. Recalled in spring 84, he was, according to Tacitus, denied further appointments because of the jealousy of Emperor Domitian (r. 81–96).

Nic Fields

See also: Mons Graupius, Battle of

References and further reading:

Hanson, W. S. *Agricola and the Conquest of the North.* London: Batsford, 1991.

Levick, Barbara. *Vespasian.* London and New York: Routledge, 1999.

Aguinaldo, Emilio (1869–1964)

Filipino nationalist. Born on 23 March 1869 near Cavite on Luzon and educated at the University of Santo Tomas in Manila, Aguinaldo became the mayor of Cavite Viejo in 1896. At the same time, he led a local faction of the Katipunan, a revolutionary group, in the Filipino insurrection against Spanish rule. In 1897, Aguinaldo agreed to go into exile in return for a pension and the promise of reform within the Philippines. During the Spanish-American War, he aided the American forces in the Philippine Islands, but then broke with the Americans on the issue of Philippine independence. Proclaimed president of the Republic of the Philippines by a provisional congress in 1899, Aguinaldo fought an epochal and costly guerrilla war against American forces until his capture by a ruse on 23 March 1901. Persuaded to take an oath of allegiance to the United States, he was granted a pension by the American government and retired to private life.

During World War II, the elderly Aguinaldo was forced to participate in anti-American propaganda by the Japanese government. Following the liberation of the Philippines, he was taken into custody by American troops and held on suspicion of collaboration with the enemy, but he was subsequently exonerated. Aguinaldo died in Manila on 6 February 1964.

Alexander M. Bielakowski

See also: Philippine Insurrection

References and further reading:

Gates, John M. *Schoolbooks and Krags: The United States Army in the Philippines, 1899–1902.* Westport, CT: Greenwood Press, 1973.

Linn, Brian McAllister. *Philippine War, 1899–1902.* Lawrence: University Press of Kansas, 2000.

Airborne Operations

The use of parachute troops began during World War II. The first operational use was in May 1940 when German glider-borne troops attacked the strategic fort of Eben Emael. The British and American armies soon developed parachute and glider-borne troops as well. The German paratroops' last operation was the assault on Crete in 1941, after which German Fallschirmjäger saw action in the infantry role only.

Allied paratroops were in action from early in the war, but the greatest operations were Operation OVERLORD (the invasion of Europe in June 1944), Operation MARKET GARDEN (Montgomery's ill-conceived assault on Holland in September 1944), and the Rhine crossing in 1945. Since the war there have been other parachute operations, most notably the Anglo-French Suez Canal operation in November 1956, the Mitla Pass operation by Israeli airborne forces in the same year, and operations by American air cavalry in the Vietnam conflict.

Since their inception airborne troops have possessed an aura of superiority in comparison with regular infantry. They have always been seen as fit, intelligent, and aggressive troops, whose surprise and shock effect far outweigh their actual strength on the ground. Due to their method of arriving quickly in a battle area, they have great strategic mobility, although this is severely limited once they are on the ground due to their diminished scales of equipment. A paratrooper starts and finishes the battle with what he can take with him, or what can be air-dropped to him.

The greatest single advantage of airborne operations is surprise—in the time it takes the aircraft to fly over the drop zone, troops are landing on the ground, forming up, and fighting. In rear areas this can have a devastating effect, as long as planning and reconnaissance have placed the drop zone in a relatively undefended area, or the drop zone has been secured beforehand by an advance party.

Air-mobile troops—carried into battle by rotary-wing aircraft—have the added advantage that once used they can be retrieved, returned to base, and reused as necessary. The air cavalry concept is a good one, but such troops should not be committed against defended landing zones or for tasks that the ground forces are capable of performing equally well. Airborne and air-mobile troops are too expensive to train and maintain to allow them to be squandered casually.

World War II also created the glider-borne soldier. Although paratroops can be widely scattered over a drop zone (for instance the U.S. 272d and 101st Airborne Divisions' drops west of the Utah beach lodgement area on 6 June 1944), glider-borne troops are able to be landed in tactical groups exactly where they are needed (Eben Emael 1940, Caen Canal and River Orne operation, 6 June 1944). They can also bring in more and heavier equipment. But gliders soon earned an evil reputation for their landing characteristics—grimly derided as "controlled crashes." The end of World War II also saw the end of glider operations.

Perhaps the most complicated airborne operation was Operation MARKET GARDEN in September 1944. The commander of the British Twenty-first Army Group, General Bernard Montgomery, wanted a narrow thrust operation aimed at Berlin, a plan only reluctantly approved by General Eisenhower.

The operation, opening on 17 September 1944, involved three simultaneous airborne operations and was conducted by the U.S. 101st and 272d Airborne Divisions, and the British 1st Airborne. This air assault showed the characteristics of airborne troops in their best light—adaptable, intelligent, and above all of an indomitable spirit and a high esprit de corps. However, most particularly at Arnhem, the operation also demonstrated the limitations of such troops when faced with serious opposition. The drop zones were too far from the Arnhem bridges for the paratroops to have any reasonable chance of getting to them easily. Further, they were dropped into the middle of two German SS Panzer divisions; paratroops have a limited ability to fight tanks because they cannot carry heavy antitank weapons with them.

In all cases where airborne, glider-borne, or air-mobile troops are to be used, it has been vital to the operation's success that the airborne troops land in areas where there is no strong enemy presence, yet which are within easy striking distance of the objective. Once the landed troops have taken their objective, they must be relieved by ground forces as soon as possible, for vertical envelopment is only of short duration.

David Westwood

See also: Armored Fighting Vehicles; Eisenhower, Dwight David; Infantry; Montgomery, Bernard Law; Student, Kurt

References and further reading:

Bando, M. A. *101st Airborne at Normandy.* Orseolo, WI: Motor Books, 1998.

Harclerode, Peter. *Para!* London: Brockhampton, 1999.

Stimpel, H-M. *Die deutschen Fallschimtruppe 1942–45,* Vols. 1 and 2. Mannheim: Motor Books, 1999.

Akbar the Great (1542–1605)

The most powerful and cultured emperor of India's Mogul Dynasty, a true renaissance ruler and a brilliant warrior. Inheriting Mogul India from his opium-addled father, Humayun, Akbar came to the throne as an adolescent. The Sur Dynasty of the Punjab greeted his accession with an invasion of Hindu, Afghan, and Muslim warriors under the Hindu general Hemu. Although his army was much smaller than Hemu's, Akbar and his advisers scattered the Surid forces at the Battle of Panipat in November 1556. Within two years, the Mogul counteroffensive captured Lahore, Multan, and other Surid strongholds. From 1561 to 1564, Akbar conquered the Hindu rajas of Mmalwa and Gondwana, gaining control of northern central India. He also began the massive defensive complex, the "Red Fort," at Agra.

After a four-year pause, the emperor took the field against the rajas of Chitor, Ranthambor, and Kalanjar and, in 1573, intervened in the sultanate of Gujarat. Intervention turned to annexation. Gujarat gave Akbar direct access to the Arabian Sea. He next turned east and subjugated Bihar and Bengal, then pivoted west in 1581 to subdue Afghanistan. One observer at this time remarked that his army consisted of 50,000 cavalry, 500 war elephants, 28 cannons, and uncountable infantry—Turks, Persians, Uzbeks, Afghans, and all kinds of Hindu warriors.

In 1585, the desire to control access to central Asian trade routes led Akbar to occupy Kashmir and Swat in the Indus Valley north. To secure Bengal, he then turned eastward in 1592 to take Orissa and its coastline. New disturbances in the east brought Akbar back to the Indus and, to anchor his power there, he imposed his power over Sindh, Multan, and Baluchistan. With these areas pacified by 1595, he spent the next several years trying to expand southward in the Deccan, the heartland of central India. After 1601, however, Akbar's final military efforts had to focus on his rebellious heir, Salim, who became master of the empire on his father's death in 1605.

Weston F. Cook, Jr.

See also: Panipat, Battles of

References and further reading:

Richards, John F. *The Mughal Empire.* Cambridge, UK: Cambridge University Press, 1993.

Streusand, Douglas E. *The Formation of the Mughal Empire.* Delhi: Oxford University Press, 1989.

Alamo (23 February–6 March 1836)

A minor siege operation of the Texan war for independence that became an effective rallying cry for Texas forces and a permanent national talisman. On 23 February 1836, General Antonio Lopez de Santa Anna arrived near San Antonio and the Mission San Antonio de Valero and began a siege of the Texas volunteers who had refused to retreat from this exposed position. Earlier, in December 1835, Texas volunteers had driven out a detachment of the Mexican army and occupied San Antonio, and now they faced Santa Anna and his army, which grew ever larger as more and more troops arrived in San Antonio. Eventually the army probably numbered somewhere between 2,000 and 6,000 men; the Texans, even with a few reinforcements who answered the "call" from the Alamo defenders and made their way through the Mexican lines, never numbered more than 190 men.

The siege lasted for 13 days from 23 February through 6 March 1836. The Mexicans subjected the defenders to bombardment from across the San Antonio River, and as the days passed they continually moved their artillery closer, coming within 800 yards of the Alamo. By 4 March, the Mexicans had moved some batteries within 200 yards of the Alamo's north wall, and the next day Santa Anna made plans to take the Alamo by storm. Two previous attempts had failed, and Santa Anna was determined that this one would succeed; the delay at the Alamo was giving valuable time to Texas insurgents to prepare defenses further north.

On 6 March 1836, Mexican troops moved into positions shortly after midnight and at 5:00 A.M. they received the signal to attack. Santa Anna had indicated there will be no quarter given, no mercy shown. The Mexicans attacked—one column at the northwest corner, one column in the center of the north wall, a third column toward the northeast, and a final column along the southern defenses. Twice the outnumbered defenders repelled the invaders, who took heavy casualties.

Then the attackers managed to climb the north walls and the defenders retreated toward the Mission. Santa Anna's men broke through on the south, and then generally penetrated the outer walls. The defenders retreated to the Long Barracks, which they had fortified somewhat, and the Mexicans turned the defenders' own cannons around to batter the walls. By 6:30 A.M. the battle was over and the Mexican army had secured the Alamo.

All the defenders were killed, including Jim Bowie, William Travis, who assumed command when Bowie became too ill to continue, and Davy Crockett from Tennessee. However Santa Anna spared the wife of one soldier, Susan Dickerson, her baby, their Mexican nurse, and an African-American youth. Santa Anna later had the bodies of the dead defenders cremated. However, the Alamo became a rallying cry for the Texas insurgents (Remember the Alamo!) who used the delay at the Alamo to declare independence, approve a constitution, form a government, and prepare for continued fighting.

Charles M. Dobbs

See also: Santa Anna, Antonio López de; Texas War of Independence

References and further reading:

Davis, William L. *Three Roads to the Alamo.* New York: HarperCollins Publishers, 1998.

Hardin, Stephen L. *Texan Iliad: A Military History of the Texas Revolution, 1835–1836.* Austin: University of Texas Press, 1996.

Hutton, Paul Andrew. "The Alamo as Icon." In *The Texas Military Experience,* ed. Joseph G. Dawson III. College Station: Texas A&M University Press, 1995.

Mexican Army soldiers under General Antonio López de Santa Anna besiege Texans barricaded inside the Alamo during the Texas Revolution. (Library of Congress)

Alanbrooke, First Viscount (Alan Francis Brooke) (1883–1963)

British field marshal, chief of the Imperial General Staff (CIGS) during World War II. Brooke was born on 23 July 1883 in Bagneres-de-Bigorre, France, to an Irish family. He attended the Royal Military Academy, Woolwich, and in 1902 was commissioned into the Royal Field Artillery. During World War I, Brooke served in France and was promoted to lieutenant colonel. After assignments as an instructor at both the Staff College and Imperial Defence College, Brooke assumed command of the Mobile Division in 1937. The following year he became the commander of the Anti-Aircraft Corps.

In 1939 Brooke was appointed to command II Corps of the British Expeditionary Force. During the campaign in France he skillfully withdrew his formation to Dunkirk where it was evacuated to England. In 1940 Brooke became commander in chief, Home Forces. In this appointment he prepared the army to meet the threatened German invasion of Britain.

In December 1941 Brooke succeeded John Dill as chief of the Imperial General Staff. As the head of the army, he worked closely with Winston Churchill to shape British military strategy. On many occasions Brooke had great difficulty in discouraging the prime minister's more unsound military schemes. Brooke also engaged in often heated negotiations with his American counterparts to formulate overall Allied strategy.

After 1940 Brooke had no further opportunities for field command. In August 1942 he declined an offer to be commander in chief, Middle East, as he believed he was of more use as CIGS than he would be in Egypt.

Brooke was promoted to field marshal in 1944, and was created Viscount Alanbrooke in 1946. He died in Hampshire on 17 June 1963.

Bradley P. Tolppanen

See also: Churchill, Sir Winston

References and further reading:

Bryant, Arthur. *Triumph in the West.* London: Collins, 1959.
———. *The Turn of the Tide.* Garden City, NY: Doubleday, 1957.
Fraser, David. *Alanbrooke.* New York: Atheneum, 1982.

Alaric (c. 370–410)

King of the Visigoths (395–410) who invaded Italy and sacked Rome (410). Alaric began his career as the leader of auxiliary mercenary troops under Emperor Theodosius I (r. 379–395). Following Theodosius's death, the Visigoths elected Alaric king. He led the Visigoth army first into Greece. There it sacked Corinth, Megara, Argos, and Sparta as well as the Athenian port of Piraeus, sparing Athens itself only after the city offered and paid a large ransom. In 400 Alaric invaded the Po Valley, but the Roman general Flavius Stilicho defeated his troops at Pollentia in 402, and again at Verona in 403. Alaric then retired from military campaigning after being appointed prefect of the Roman province of Illyricum. Two years later (404–405) he joined forces with Honorius (r. 395–423), emperor of the western empire, against his brother Arcadius (r. 383–408) who ruled the eastern empire. When Arcadius unexpectedly died, Honorius canceled the planned invasion. Alaric demanded payment of a heavy indemnity that Stilicho agreed to pay. Honorius subsequently ordered the execution of Stilicho and Roman troops massacred the families of Stilicho's barbarian auxiliaries (408). These resulted in the defection of large numbers of Stilicho's former soldiers to Alaric, who again invaded Italy. Since Honorius refused to pay a 4,000-pound gold ransom, Alaric responded by surrounding Rome on three separate occasions before allies within the city opened the city's gates to him on 24 August 410. For three days the Visigoths pillaged the city, for the first time since antiquity. Next Alaric marched south to Messina in preparation for an invasion of Sicily and North Africa, but his ships never left port due to a violent storm. He died shortly thereafter and was succeeded by his brother Ataulf.

Cynthia Clark Northrup

See also: Roman Army; Stilicho, Flavius; Visigoths

References and further reading:

Ferrill, Arthur. *The Fall of the Roman Empire: The Military Explanation.* New York: Thames & Hudson, 1986.
Heather, Peter. *The Goths.* Oxford, UK, and Cambridge, MA: Blackwell Publishers, 1996.

Alba, Fernando Alvarez de Toledo, Duque de (1507–1582)

Accomplished Spanish commander, known for his severity, and political figure during much of the nineteenth century. Alvarez de Toledo, known by his noble title Duque de Alba, was born on 29 October 1507 at Piedrahita, Castille. His father died in battle in 1510, so he was raised by his grandfather, Rodrique de Toledo. At age 16, he joined the army of King Charles I of Spain against France. His defense of the Catalan coast and Perpignan during the Franco-Spanish War earned him his reputation as a great commander. By 1555, he was named generalissimo of the army of Italy. In 1567, he was named captain general of the Low Countries. He undertook a march with 10,000 troops from Italy to Brussels to put down a rebellion. Alba was able to capture many of those involved with the rebellion. He established the Council of the Disturbances, which condemned to death without appeal hundreds of citizens, including various noblemen.

In 1573 he was recalled to the court by the king, and experienced some ostracism from the king. However, in 1580 Phillip II of Spain wanted Alba to reconquer Portugal, which he claimed by inheritance. Although the duke accomplished his mission swiftly and completely, he died on 11 December 1582 in Lisbon. By this time he had become viceroy of Portugal.

Peter Carr

See also: Dutch War of Independence

References and further reading:

Maltby, William S. *Alba: A Biography of Fernando Alvarez de Toledo, Third Duke of Alba, 1507–1582.* Berkeley: University of California Press, 1983.

Alcibiades (c. 450–404 B.C.E.)

Athenian commander and traitor. A ward of Pericles and student of Socrates, Alcibiades ranked among the most popular, handsome young men of Athens and the most notorious turncoat in Greek history. Known for winning chariot races at Olympia, Alcibiades persuaded the assembly to let him assume control of the Peloponnesian War after the leader Nicias experienced several military setbacks. Alcibiades reopened the war and initiated a campaign against Syracuse in 416 B.C.E.

Just after the Athenian fleet sailed, all the Hermeses (stone pillars bearing the head and genitals of Hermes, god of travel) across Athens were defaced and castrated and Alcibiades was implicated. The assembly issued orders recalling him to Athens. He managed to escape and fled to Sparta where he betrayed the Athenian plans for attacking Syracuse. He also informed the Spartans that they could fortify an area outside of Athens called Decelea, denying the Athenians access to their agricultural fields at Euboea and their silver-mining operations. During this campaign over 20,000 Athenian slaves deserted and joined the Spartans. In 412 B.C.E. Alcibiades fled to Persia after the Spartan king learned that he had been sleeping with his wife. Once in Persia he offered to negotiate for the Athenians but part of the arrangements involved Athenian acceptance of a modified form of government.

Alcibiades returned to Athens in 409 B.C.E. and accepted his former position as commander of the Athenian fleet. Three years later the navy experienced a major defeat as a result of Alcibiades' placing an inexperienced acquaintance in charge. For neglecting his duties the assembly exiled him. He retired to a private villa in the Hellespont where he remained until 404 B.C.E., when he was assassinated by his enemies, who could have been Spartan, Athenian, or Persian.

Cynthia Clark Northrup

See also: Peloponnesian Wars; Pericles

References and further reading:

Ellis, Walter M. *Alcibiades.* New York: Routledge, 1989.

Forde, Steven. *The Ambition to Rule: Alcibiades and the Politics of Imperialism in Thucydides.* Ithaca, NY: Cornell University Press, 1989.

Alesia, Siege of (52 B.C.E.)

Decisive battle of the Gaulic rebellion against Rome. Led by the charismatic Vercingetorix, most tribes in Gaul rebelled against Roman rule in 52 B.C.E. Provincial Roman governor Julius Caesar responded by concentrating his legions, then forcing Vercingetorix and 80,000 men into the fortress city of Alesia after three pitched battles.

Commanding fewer than 50,000 legionnaires, Caesar nevertheless began a siege. Vercingetorix then dispatched his cavalry to rally reinforcements from across Gaul, and in turn the Romans constructed a double wall of fortifications around Alesia facing toward *and* away from the city. The walls stretched for 25 miles, connecting with more than 50 miles of trenches, 23 forts, breastworks, palisades, turrets, and extensive obstacles to slow the approach of Gaulic warriors.

When the Gaulic relief force arrived, the Romans faced 80,000 men in Alesia plus an estimated 250,000 foot soldiers and 8,000 cavalry attacking from outside the city. Caesar skillfully used interior lines, his fortifications, and the greater discipline of his men to offset the Gaulic advantage, but after two days of heavy fighting found his army pushed almost to the breaking point. On the third day the Gauls captured the Roman camp and Mount Rea, which formed a crucial point in the Roman defense. In desperation, Caesar personally led the last of his reserves in a climactic counterattack, and when his German cavalry outflanked the Gauls and attacked them from behind, the battle decisively turned to his advantage. With all hope of victory gone, Vercingetorix surrendered the next day, and Roman power in Gaul quickly recovered.

Lance Janda

See also: Caesar, Julius; Roman Army

References and further reading:

Meier, Christian. *Caesar: A Biography.* Trans. David McLintock. New York: Basic Books, 1982.

Alexander, Field Marshal Earl the Hon Harold Rupert Leofric George (1891–1969)

British field marshal, commander of Allied forces during World War II. Born 10 December 1891 in London, Alexander was commissioned into the army in 1911.

He served in France during World War I, commanded a Baltic unit in Latvia, and led a brigade on the Indian frontier. In 1937 Alexander became the youngest major general in the British army.

Alexander served with the British Expeditionary Force in France in 1939, and in 1940 commanded the rear guard during the retreat to Dunkirk. He directed the evacuation of the troops and was the last British soldier to leave the beaches. In 1942 Alexander was sent to Burma during the Japanese invasion. Unable to retrieve an already desperate situation, he conducted a retreat into India.

In August 1942 Alexander was appointed commander in chief, Middle East. Under his direction, Bernard Montgomery and the Eighth British Army won a decisive victory at El Alamein and advanced to the Tunisian frontier. In 1943 Alexander was selected to command the Eighteenth Army Group under Dwight Eisenhower. He used his immense charm to establish an excellent relationship with his American superior. As Army Group commander, Alexander coordinated the capture of Tunis, the conquest of Sicily, the invasion of the Italian mainland, and the liberation of Rome.

In 1944, Alexander was promoted to field marshal and appointed supreme allied commander in the Mediterranean. He accepted the surrender of all German forces in Italy on 29 April 1945.

After the war, Alexander served as governor-general of Canada (1946–1952) and as Britain's minister of Defense (1952–1954). He was made a viscount in 1946, and ennobled as Earl Alexander of Tunis after his Canadian duty. He died at Slough on 16 June 1969.

Bradley P. Tolppanen

See also: British Military, Twentieth-Century Organization and Structure; Eisenhower, Dwight David; Montgomery, Bernard Law

References and further reading:

Jackson, W. G. F. *Alexander of Tunis as Military Commander.* London: Batsford, 1971.

Nicolson, Nigel. *Alex: The Life of Field Marshal Earl Alexander of Tunis.* London: Weidenfeld & Nicolson, 1973.

Alexander the Great (July 365–June 323 B.C.E.)

Recognized as perhaps the foremost military commander of history. Alexander was born to Philip II of Macedon and Olympias, and spent three years under the tutelage of Aristotle at Mieza.

At age 16, Alexander was appointed regent while Philip fought in Byzantium. Alexander was appointed an army commander during the Macedonian expedition against Athenian-Theban troops at Chaeronea in 338 B.C.E. He succeeded in bringing the battle to a successful conclusion by defeating the Theban guard known as the Sacred Band.

A succession struggle between Alexander's supporters and those of other contenders followed after Philip's assassination in 336 B.C.E. Alexander eliminated the opposition by purge and execution, and attempted to consolidate his rule through a crusade against the Persian Empire.

In 335 B.C.E., Alexander marched against the Triballians, Getae, and Illyrians, who had rebelled against Philip. Restoring Macedonian control in that region, Alexander concluded a 240-mile march in 13 days to crush the revolt of Thebes.

In 334 B.C.E. an army under Alexander, including 2,000 of the 3,000 famed Companion Cavalry, joined a Macedonian force under General Parmenio that was stationed near Troy. The combined Macedonian force of 40,000 infantry and 5,000 cavalry encountered a Persian force of 40,000 under General Memnon at the Granicus River in May. Alexander attacked along a steep river embankment, which the Persians had thought was unassailable. Greek mercenary infantry in Persian service fled, and the Persian cavalry was defeated.

Alexander continued through Asia Minor, reducing the Persian naval ports of Miletus and Halicarnassus after siege. To speed the march and provide better security against marauding Persian forces, the Macedonian forces were divided: Parmenio advanced inland while Alexander followed the coast. A linkup was effected in April 333 at Gordium.

Emperor Darius III Condomannus and a Persian force of 140,000 met the Macedonians at the Pinarus River along the Gulf of Issus in November. Alexander displayed a fluid command by attacking off the march and shuffling his formations according to battlefield needs and developments. This, combined with a concentration of brute force against a weak point in the Persian defenses, won the battle.

After rejecting an armistice proposed by Darius, Alexander spent 332–331 eliminating the remnants of Persian coastal naval power. This expedition culminated in November 332, at Tyre, after a seven-month siege. Alexander then proceeded to capture Egypt, founding Alexandria.

Realizing that his manpower and supplies were running out, he called in June 331 for reinforcements from Macedonia. Logistic difficulties were minimized because Macedonians were trained to march with their own pack requirements, eliminating an extravagant baggage train.

In October, Persian and Macedonian forces met at Gaugamela. The Persians again held numerical superiority and attempted to envelop both flanks of the Macedonian line. However, Alexander executed the oblique order of attack, which suddenly concentrated his forces from its line formation, to attack and overwhelm the Persian left flank.

Alexander pursued the remnants of the Persian force until July 330 B.C.E., when Darius was assassinated by one of his own satraps, effectively terminating Persian rule of the empire.

Alexander continued his conquests with six sieges in 329, as he advanced through northeastern Persia, concluding in 328 with the battles at Sogdian Rock and Rock of Chorienes. Thereafter he attempted to legitimize his presence by marriage to Roxane, daughter of a Sogdian prince. Alexander also adopted Achaemenian courtly dress and customs.

He continued through the Indus Valley to capture Ora, Rock of Aornos, and Multan in 327, defeating King Porus at Hyphasis in 326.

Alexander attempted to justify his leadership and consolidate his gains by advocating a joint Macedonian-Persian elite to administer the empire. During 330, 328, and 327, Alexander defeated plots against his leadership.

But it was a mutiny in 326 that forced him to call off his Indian campaign and return to Persepolis. He died of fever in June 323 while preparing a campaign against Carthage.

A drawing depicting Alexander's address to his officers before the Battle of Issus. (Library of Congress)

Alexander's accomplishments did not survive his death. The quarrels of Alexander's generals, the Diadochi 323–280, split the empire into the monarchies of Macedonia and Asia Minor, Egypt, and Persia. Internal disharmony in Persia, ruled by the Seleucids, created an environment for resurgent Persian power during the Parthian era, 247 B.C.E.–228 C.E. But Alexander the Great did achieve a semidivine status in the ancient world and is still remembered for his great military and administrative skills.

Neville G. Panthaki

See also: Alexander's Wars of Conquest; Gaugamela, Battle of; Issus, Battle of; Persian Empire

References and further reading:

Green, Peter. *Alexander the Great.* London: Weidenfeld & Nicolson, 1970.

Keegan, John. *The Mask of Command.* London: Penguin Books, 1988.

Lane Fox, Robin. *Alexander the Great.* London: Folio Society, 1997.

Snyder, John. *Alexander the Great.* New York: Twayne Publishers, 1966.

Alexander's Wars of Conquest (334–323 B.C.E.)

One of history's greatest military leaders, Alexander of Macedon conquered an empire that stretched from Greece to India. The son and heir of Philip of Macedon, Alexander inherited the kingdom at age 20 in the year 336 B.C.E. Within two years he consolidated his rule over the Greek city-states and turned his attention to Asia, more specifically to the rich and powerful empire of Persia. He commanded perhaps the most formidable fighting force in the world. The Macedonian army was a marvel for its time, with revolutionary formations of infantry (the phalanx) and the innovative use of cavalry; the Macedonians fought with great precision and discipline and were augmented by Greek allies and mercenaries. They were not citizen-soldiers but a professional force that constantly trained and drilled. Alexander himself was a daring cavalry warrior, often leading the charge and inspiring his men with his personal bravery, suffering several serious wounds in the process. Among his qualities was a wide strategic vision along with a talent for the tiny details of warfare. With his soldiers he brought scientists and men of the liberal arts; one of his goals was to spread Greek thought across the known world. The first step of his war of conquest was the invasion of Asia Minor, once home to the already legendary city of Troy and now controlled by the Persian Empire.

In an act heavy with symbolism, Alexander crossed the Hellespont into Asia Minor in the spring of 334 B.C.E. with slightly fewer than 40,000 men. His immediate goal was to free the Greek cities along the Ionian coast, presenting himself to the people as a liberator, and to swell his ranks in the process, as well as seizing ports to neutralize the Persian naval advantage. The armies of the Persian provincial satraps who held local command in the region's armies met Alexander's forces in battle at the Granicus River (modern-day western Turkey) in May 334 B.C.E. The Greeks won a dramatic victory, with Alexander himself leading a daring cavalry charge across the river. This victory set a precedent in that it established Alexander as a bold commander whose risks paid off and it inspired fanatical devotion in his troops. After that victory, Alexander moved southward into Asia Minor, laying siege to Halicarnassus in September and November 334. During the winter of 334–333 he conquered Phrygia (central Turkey). In April 333 Alexander captured Gordium and severed the legendary "Gordian Knot" with his sword. Moving farther west, Alexander then conquered Cappodocia and began moving toward Syria, surprising the defenders at the Cilician Gates (near Balkar Daghari, Turkey) and forcing them to surrender without fighting.

The Persian emperor Darius III, who had ignored previous advice from his officers on how to best handle the invasion, led an imperial army into southern Turkey in an attempt to stop Alexander from reaching the sea. Darius managed to get behind Alexander, whose troops were greatly outnumbered when the armies met in battle at the Pinarus River near the town of Issus in early November 333 B.C.E. De-

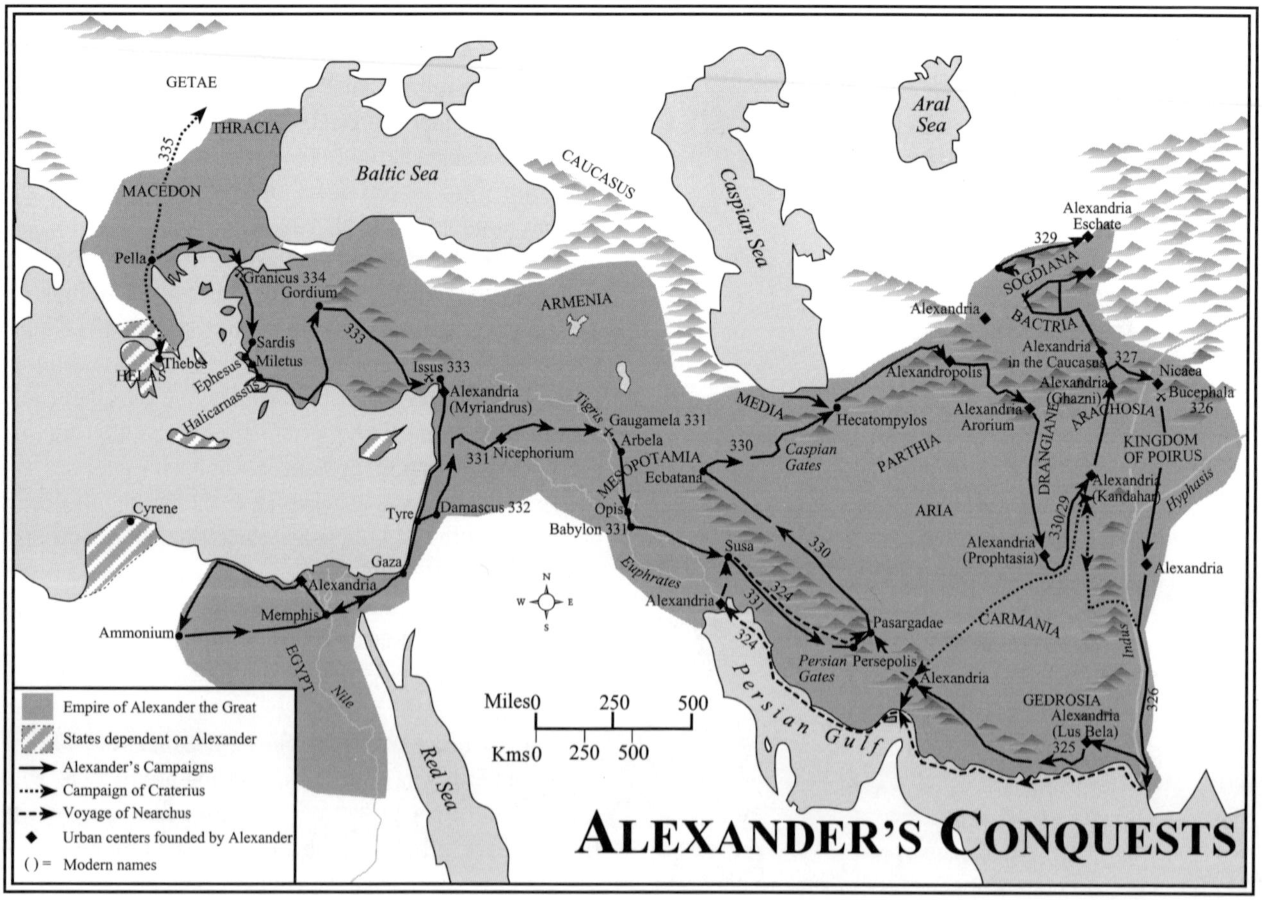

spite his numerical superiority, Darius could not bring his full force to bear on Alexander because of the mountainous terrain. Still, the battle was in doubt until Alexander charged directly toward Darius, who fled the field. Upon seeing their leader retreating in his golden chariot, the Persian morale was broken and the Macedonians won the day. After this battle, Alexander made the strategic decision not to pursue the fleeing Darius but to capture more territory and build up his strength and wealth for an invasion of Persia itself.

From Issus, Alexander moved south along the Phoenician coast. The city of Tyre fell in July 332 B.C.E. after a seven-month siege and Gaza fell in October after a shorter siege. Following a precedent set with Thebes, any city that resisted Alexander was sacked. At Tyre, Alexander received a peace offer from Darius that presented all the territory west of the Euphrates River as well as marriage to a Persian princess in exchange for an end to the invasion. Alexander refused. In December 332, Alexander arrived in Egypt to be welcomed as a liberator and proclaimed as pharaoh. He founded the city of Alexandria, one of many new cities left in his wake, and was hailed as the son of Re, an Egyptian god. While Alexander may have already considered himself divine, the battle-weary veterans that had followed him from Macedonia grumbled at this development. Alexander remained in Egypt for some time, ensuring his supply lines and gaining recruits until he felt confident of his ability to face Darius once again. In the summer of 331, he set out for Mesopotamia. The army crossed the Tigris River on 19 September 331 and met the Persian forces on 1 October 331.

Determined not to repeat the failure of Issus, Darius wanted to fight on open plains to take advantage of his greatly superior numbers. The armies clashed at Gaugamela in northern Iraq near the ancient city of Nineveh. Once again, however, Alexander charged directly toward Darius, who again panicked and fled. This time, the Persian army was thoroughly defeated but Darius himself escaped. Alexander entered Babylon in triumph as the new king of Asia. In November 331 B.C.E., Alexander occupied Susa in modern-day Iran, and in late December he captured a strongly defended mountain pass called the Persian Gates with a brilliant night surprise maneuver. That victory cleared the way to the capital of Persepolis. Alexander occupied the city in January 330 and seized the enormous Persian treasury. The immense wealth put into circulation had economic ramifications that lasted for centuries. After some discussion, Alexander decided to burn the palace of Darius, partly in retribution for the burning of the Acropolis by Xerxes in 480 B.C.E. As in other conquered lands, Alexander allowed local officials who pledged loyalty to remain in their positions, a move designed to gain Alexander favor with the public and ensure some continuity of government under his rule. Darius remained alive in Media, south of the Caspian Sea, trying to raise another army. In the summer of 330 B.C.E., Alexander pursued Darius, who failed to find a protector among the remaining Persian nobility. In July Persians fleeing from Alexander's army murdered Darius. Alexander was disappointed because he wanted to catch Darius alive and have him continue as the king of Persia under Alexander's overall rule. Upon Darius's death, however, Alexander himself assumed the trappings of the Persian throne, a move that further separated him from his officers.

With Persia defeated and only scattered Persian forces still offering resistance, Alexander looked further to the east. His geographers assured him that the great ocean and the end of the world lay just beyond India, which they assumed to be merely a peninsula. The Greek and Thessalian troops were sent home, leaving the Macedonians to continue this unparalleled campaign. After subduing the tribes on the south shore of the Caspian Sea in September 330 B.C.E., Alexander occupied Parthia and Aria later that fall. In October, there was an alleged conspiracy among the officers to overthrow Alexander, which resulted in the execution of Philotas and his father Parmenio, who had been one of Alexander's most valuable generals. In early 329, Alexander invaded Arachosia in southern Afghanistan and then moved north to Bactria and Sogdiana. He spent two years campaigning in this rough mountainous terrain, advancing as far north as the Jaxartes River and founding cities along the way. These expeditions contributed greatly to Europe's knowledge of the geography of the east. It was around this time that Alexander ordered that everyone should abase himself, Persian style, in his presence. This practice drove a deeper wedge between Alexander and his troops and he was forced to deal with further plots and conspiracies against him.

In early 327 B.C.E., he laid siege to the heavily fortified stronghold at the Chiorenes Rocks in Sogdiana. After the victory that cleared the way to India, he married the daughter of the Sogdianian leader and, as in other lands, incorporated locals into his army. Despite the misgivings of the troops at taking on another campaign so far from home, Alexander entered India through the Khyber Pass in mid-327 and in May 326 faced the army of King Porus in battle near the Hydaspes River. For the first time, Europeans met elephants on the battlefield. Alexander won the battle and for the first time, the troops refused to go any farther and threatened open mutiny. They wanted to go home and no amount of persuasion from the charismatic "King of Asia" could convince them to continue. Reluctantly, Alexander gave the order to return but he did not simply backtrack along his original route. He decided to build a large fleet of ships, numbering 800 to 1,000, to help transport his army down the Indus River toward the Arabian Sea. When the journey began, half of the troops rode in the ships while the other half marched along both riverbanks. Alexander incorporated war elephants into his own army and along the way there was hard fighting with local tribes, including the Malli. When they reached the Indus Delta, Alexander chose to march some of his men back overland toward the cities of Persia, while others sailed in some of the ships to the Persian Gulf. In the fall of 325, Alexander's army underwent a harsh passage across the desert region of Gedrosia (Baluchistan), beset by rough weather and poor terrain. Many of them died during this journey.

Alexander reached Susa in the spring of 325 B.C.E. and retook the task of administering his newly won empire. During this period, he deposed more than a third of the provincial satraps he had previously appointed and furthered a policy that aimed to unite the Macedonian and Persian people through intermarriage. This, along with his practice of admitting eastern recruits into the army, was not popular among the rank and file, who finally erupted into open rebellion at Opis in the summer of 324. Alexander defused the situation by threatening to replace his entire army with Persians but Alexander's views on his own divinity remained a point of contention. Most accounts say he thought himself a god and requested divine honors be given to him. Historians have raised questions regarding his mental state and at the very least most conclude that Alexander was very unstable at this time. In late 324, he sent an army into Luristan for a punitive strike against the Cossaeans but he conquered no new lands after this point. Some speculate that he planned eventually to spread his empire toward the west and invade Italy and the rest of Europe. His immediate strategy was to travel south into Arabia and he was preparing for that campaign when he fell ill.

Alexander died in Babylon on 10 June 323 B.C.E. at the young age of 32. The legacy of his conquests shaped the grand pattern of world history for centuries. He was the first to truly bridge the gap between East and West, creating a two-way cultural flow that influenced both continents. Upon the framework of his Hellenistic empire were built the Roman and Byzantine Empires. Later, while much knowledge was lost in the West during the Dark Ages, the Greek thought that Alexander carried to the East survived to be reintroduced to Europe, contributing to the renaissance that spread Western civilization around the world.

Harold Wise

See also: Alexander the Great; Gaugamela, Battle of; Granicus, Battle of the; Hydaspes, Battle of the; Issus, Battle of; Persian Empire; Philip II of Macedon; Tyre, Siege of

References and further reading:

Brunt, P. A. *Arrian with an English translation,* Vol. 1. Cambridge, MA: Harvard University Press, 1976; Vol. 2, 1983.

Green, Peter. *Alexander of Macedon 356–323 B.C.: A Historical Biography.* Berkeley: University of California Press, 1992.

Hammond, N. G. L. *The Genius of Alexander the Great.* Chapel Hill: University of North Carolina Press, 1997.

Tarn, William Woodthorpe. *Alexander the Great.* 2 vols. Cambridge, UK: Cambridge University Press, 1948.

Alexandria (20–21 March 1801)

Battle for Egypt between British forces, about 14,000 strong, under Sir Ralph Abercromby, and French troops numbering about 10,000, under Adolphe Menou. Once Bonaparte landed in Egypt in 1798, the defeat and removal of the French army became a major concern for the British government. Because of difficulties in Ireland and elsewhere, no large force could be spared until December 1800, when the British commander at Gibraltar, Sir Ralph Abercromby, was ordered to prepare and lead an expedition of 15,000 men to Egypt. The expedition arrived off Aboukir on 1 March 1801, but the British were prevented by bad weather from landing for several days. This gave Menou the opportunity to collect the widely scattered French garrisons.

On 8 March, Abercromby finally landed, and quickly drove away the small French force at Aboukir. He was then prevented from immediately exploiting his success by his need to land horses and supplies. The landing of stores was impeded by more bad weather, and Abercromby halted 11 miles from Alexandria.

On 13 March, Abercromby defeated a sizable French force under General Friant at the battle of the Roman Camp, about two-thirds of a mile from Alexandria. The British advanced on the same day to the main French position, the Heights of Nicopolis. An attempt to storm the heights late on 13 March failed, and Abercromby set about fortifying his own position, and on landing heavy guns from his ships.

On 19 March, Menou arrived from Cairo, bringing reinforcements. Because the French expected the arrival of both a British-allied Turkish army and a second British expedition from India, under Sir David Baird, Menou decided to strike first.

The French plan called for a night assault on 20–21 March. The French began a rather noisy demonstration on the left of the British line. At the same time, an attack on the right flank, to be followed by an attack on the center, was supposed to drive the British from their position. Both attacks were defeated. Menou then launched an attack on the British left, which failed, and a cavalry charge, which reached the British lines. In the confusion, Abercromby was captured by the French cavalry and rescued by the British. He was also wounded in the leg, which subsequently proved fatal on 28 March. By 9 A.M. the French retreated in good order, with losses of 1,600 killed and wounded and 200 captured. British losses were around 1,500.

As a result of this action, the British were able to besiege Alexandria, and in the company of Turkish forces to march on Cairo. Cairo surrendered to the combined British and Turkish force on 27 June, and with this all French forces except those in Alexandria capitulated. Alexandria surrendered on 2 September. The French effort to conquer Egypt and threaten India thus ended.

Joseph M. Isenberg

See also: Aboukir; French Revolutionary Wars; Kléber, Jean-Baptiste; Napoleon I; Pyramids

References and further reading:

Fortescue, J. W. *A History of the British Army.* Vol. 4. London: Macmillan, 1915.

Alexius I Comnenus (1048–1118)

Byzantine general and emperor, founder of the dynasty of the Comnenoi. As the nephew of the emperor Issac (r. 1057–1059) and brother of the Grand Domestic of the East, also named Issac, Alexius Comnenus received important military commands at an early age. In 1075 he was assigned the task of fighting a rebellious Norman mercenary, Russell of Ballieul, who had defected to the Turks. Alexius paid a group of Turks to kidnap Russell, for delivery to the Byzantine government.

In 1081, he was appointed Grand Domestic of the West by Emperor Nicephorus III and directed to defeat two pretenders to the throne, Nicephorus Bryennius and Nicephorus Basilacius. Alexius accomplished this, and also turned back a Petcheneg raid across the Danube. Later in 1081, after being threatened by Nicephorus III, Alexius entered into a conspiracy with his brother and in-laws to seize power, which was easily accomplished. Alexius was crowned emperor.

Almost immediately Alexius I was forced to face an invasion by the Norman duke Robert Guiscard, allegedly upholding the rights of another deposed Byzantine emperor, Michael VII. Alexius sent diplomatic missions to the Holy Roman Empire and Venice, which were successful. He also ordered the fortification of strong points, and marched out with the Byzantine army to oppose the Normans. In this he was less successful, and was repeatedly defeated in the campaigns of 1081 and 1082. The war with the Normans continued with uneven results until the death of Robert in 1085. Robert's sons declined to continue the war, preferring to argue over the succession to their own lands.

As a consequence of this war, Alexius cashiered a unit of Manichean, or Bogomil, troops. These rebelled, along with their coreligionists. In 1086, the rebels invited a group of Petcheneg nomads across the Danube. Alexius marched on the rebels in 1087, but was defeated at Dristra. Petcheneg raids continued until 1091, when a Cuman force, invited by Alexius, defeated the Petchenegs at Lebunium. The Bogomil Revolt then collapsed.

In 1093 and 1094 Alexius campaigned, with some success, against the Serbian principality of Raska. In 1095, he sent an embassy to Pope Urban III requesting mercenaries; Urban took the opportunity to preach a crusade, the First Crusade, against the Turks. Alexius spent most of the next two years coordinating the passage of the crusading armies across Byzantine territory, and attempting to persuade the leaders of those armies to swear fealty to him. As the crusading armies reached the Muslim-held territories of Syria, Lebanon, and Palestine, Alexius used the opportunity to recover parts of Anatolia in 1098 and 1099. From 1100 on, Alexius sought the submission of the crusader states to Byzantine governance, finally securing that of Antioch in 1111. He died in 1118 and was succeeded by his son John II.

Alexius also reformed the currency, and thus left the empire stronger both financially and militarily. He also sought to strengthen his government by assigning offices and revenues to members of his family. As a result, he made many enemies, both inside and outside of the empire. His work, while ensuring the stable succession of members of his family for many years, may thus not have resulted in the long-term advantage to the empire.

Joseph M. Isenberg

See also: Antioch, Battle of; Bogomils' Revolt; Crusades; Dorylaeum, Battle of; John II Comnenus; Norman-Byzantine Wars

References and further reading:

Anna Comnena. *The Alexiad of Anna Comnena.* Trans. E. R. A. Sewter. New York: Penguin Books, 1969.

Ostrogorsky, George. *History of the Byzantine State.* Trans. Joan Hussey. Rev. ed. New Brunwick, NJ: Rutgers University Press, 1969.

Alfonso VIII (1155–1214)

Christian king of Castile, whose victory at Las Navas de Tolosa began the demise of Muslim rule in Spain. Alfonso succeeded his father, Sancho III, as king of Castile in 1158. His regency was a period of political unrest, but in 1169 he took firm control of the government and made an important alliance the following year by marrying Eleanor, the daughter of King Henry II of England.

In 1188 Alfonso combined forces with his tributary, King Alfonso IX of Léon, to attack the Muslims in southern Spain, but was decisively defeated at Alarcos on 19 July 1195 by Abu Yusuf Ya'qub al-Mansur. Taking advantage of Alfonso's momentary weakness after Alarcos, the Muslims from the south, Léon from the northwest, and Navarre from the northeast all threatened to invade Castile. Alfonso placated his Christian rivals with diplomacy, and renewed his plans to drive the Almohad Muslims out of Spain.

In 1211 Alfonso asked Pope Innocent III to authorize the archbishop of Toledo, Ximénes de Rada, to preach a Spanish Crusade against the Almohad regime. The response was enormous, with 70,000 volunteers from Italy, France, and Germany joining 60,000 Spaniards under Alfonso and King Pedro II of Aragon. This army, huge by medieval standards, easily took Malagón on 24 June 1212 and Calatrava on 1 July. Many troops died from disease or deserted, but Alfonso was reinforced by Sancho VII of Navarre by 13 July. Encountering the main Muslim force under Muhammad al-Nasir, emir of Morocco, at Las Navas de Tolosa on 16 July 1212, Pedro commanded the left, Alfonso the center, and Sancho the right. They caught the Muslims in a classic pincer formation and routed them, killing thousands. By 1260 the Muslims in Spain were confined to a small area of Andalusia around Granada. Alfonso died on 6 October 1214.

Eric v. d. Luft

See also: Almohad Conquest of Muslim Spain; Almoravid Empire; Reconquest of Spain; Religion and War

References and further reading:

Fletcher, Richard. *Moorish Spain.* Berkeley: University of California Press, 1993.

The Latin Chronicle of the Kings of Castile Trans. with an introduction and notes by Joseph F. O'Callaghan. Tempe: Arizona Center for Medieval and Renaissance Studies, 2001.

O'Callaghan, Joseph F. *A History of Medieval Spain.* Ithaca, NY: Cornell University Press, 1975.

Reilly, Bernard F. *The Medieval Spains.* New York: Cambridge University Press, 1993.

Alfred the Great (849–899)

King of Wessex, laid the foundation for a united Christian Anglo-Saxon kingdom. Alfred was born in Wantage, Berkshire, and grew up in Wessex, the major Saxon kingdom in southwestern England. After 865, he was involved in a series of wars with pagan Danish invaders who had conquered most of northern and eastern England. In 871, with his brother King Ethelred, he defeated the Danes at Ashdown. Alfred succeeded Ethelred, but was defeated by the Danes at Meretum later in the year. The Danes took Mercia, the region to the north, and invaded Wessex in 876. Alfred took prominent hostages and forced peace with the Danes, only to be invaded again in 878.

Alfred successfully defended his fortress at Athelney and in 886 the Danish king, Guthrum, converted to Christianity and withdrew from Wessex for good. For the remainder of his reign, Alfred consolidated his power in Anglo-Saxon England, keeping the Danes at bay. He reorganized the army of Freemen, the *fyrd,* placing half permanently on duty and half at home and in the fields. He also established more than 30 new fortifications called *burhs* throughout Wessex. These *burhs* featured large walls; the more hides, or large acreages of land, a person had in the countryside, the more wall that person was responsible for protecting. As a result of this reorganization, the Danish army broke up in 896. Alfred died on 26 October 899, succeeded by his son Edward.

Christopher P. Goedert

See also: Æthelbald's Wars; Viking Raids

References and further reading:

Abels, Richard. *Alfred the Great: War, Kingship and Culture in Anglo-Saxon England.* London and New York: Longman, 1998.
Keynes, Simon, and Michael Lapidge, eds. *Alfred the Great: Asser's Life of King Alfred and Other Contemporary Sources.* Harmondsworth, UK: Penguin Books, 1983.
Peddie, John. *Alfred the Good Soldier: His Life and Campaigns.* Bath, UK: Millstream, 1989.

Algiers, Battle of (7 January–24 September 1957)

An example of how military success can fail to bring political and diplomatic victory. After more than a century of French occupation, organized Algerian insurrection began on 1 November 1954, All Saints' Day, with a series of explosions and assassinations. The underlying cause was the exclusion of the Muslim populations from political and economic affairs and a series of unstable French governments' inability to effect any meaningful reforms. The National Liberation Front (FLN) organized by Ahmed Ben-Bella and eight others spearheaded the uprising.

After an inconclusive, three-year campaign in the Algerian countryside, both the French and FLN concentrated their efforts on Algiers, one of the largest cities of France (Algeria was legally a part of metropolitan France itself).

Following a series of bombings at public sites and the call for a general strike on the part of the native population, the 10th Paratroop Division, with units of the French Foreign Legion, occupied Algiers in January 1957. The military quickly divided the city into zones that were swept by the soldiers, arresting and interrogating detainees without judicial authority. French operations were successful and the FLN was driven out of the city by September.

The French military sealed the Algerian borders with Tunis and Morocco with electrified fortifications, and intensified the use of helicopters to transport assault forces rapidly to rebel opposition points. By 1959 French forces were on the verge of military victory; however, for political reasons, President Charles de Gaulle offered the insurrectionists self-determination. In 1962, Algeria was granted independence. The cost of the war was appalling. Casualties on both sides totaled almost 3 million or nearly one-third the population of Algeria.

William E. Wingfield

See also: Ben-Bella, Ahmed; French Foreign Legion; de Gaulle, General Charles

References and further reading:

Horne, Alistair. *A Savage War of Peace: Algeria, 1954–1962.* New York: Viking Press, 1978. Rev. ed., New York: Penguin Books, 1987.
Porch, Douglas. *The French Foreign Legion: A Complete History of the Legendary Fighting Force.* New York: HarperCollins, 1991.

Allen, Ethan (1738–1789)

American guerrilla leader in the French and Indian War, the conflict between Vermont and New York, and the American Revolution. Allen was born in the frontier community of Litchfield, Connecticut, on 21 January 1738. A voracious reader and freethinker, later a Deist, he expected to enroll at Yale College, but his father's death in 1755 made that hope financially impossible. Based at Fort William Henry in the French and Indian War, he became familiar with the Lake Champlain Valley and what is now Vermont.

In the 1760s, Allen and four of his brothers settled in Vermont, then called "The Grants," disputed territory between New Hampshire and New York. Taking the side of New Hampshire, Allen raised a guerrilla militia of about 400 farmers, the "Green Mountain Boys," and waged an effective defensive campaign against intrusions from New York, fighting to wound and to humiliate, not to kill. New York governor William Tryon offered £100 for his capture, but in vain.

On 10 May 1775, accompanied by Benedict Arnold, Allen led the Boys in the bloodless capture of Forts Ticonderoga and Crown Point. His bombastic personality suffered when the Boys joined the Continental army and voted Seth Warner their colonel instead of him. He volunteered for Arnold's and Richard Montgomery's invasion of Lower Canada. Outnumbered, outmaneuvered, and captured in Montreal, he was a prisoner of war from 25 September 1775 until exchanged for a British officer on 6 May 1778. Upon his release, he reported to Valley Forge, where George Washington brevetted him colonel in the Continental army. Returning to Vermont that summer, he heard about Warner's success at Bennington.

Disappointed by the refusal of the Continental Congress to consider Vermont the fourteenth state, he negotiated fruitlessly in 1779 and 1780 with the British in Quebec for a separate peace. He died, embittered, in Burlington on 11 February 1789.

Eric v. d. Luft

See also: American Revolution; Arnold, Benedict; Bennington; Fort Ticonderoga; French and Indian War; Quebec, Battle of; Seven Years' War; Washington, George

References and further reading:

Bellesiles, Michael A. *Revolutionary Outlaws: Ethan Allen and the Struggle for Independence on the Early American Frontier.* Charlottesville: University Press of Virginia, 1993.

Hoyt, Edwin Palmer. *The Damndest Yankees: Ethan Allen and His Clan.* Brattleboro, VT: Greene, 1976.

Jellison, Charles Albert. *Ethan Allen: Frontier Rebel.* Taftsville, VT: Countryman, 1974.

Pell, John. *Ethan Allen.* Freeport, NY: Books for Libraries, 1972.

Allenby, Edmund Henry Hynman, Viscount (1861–1936)

British cavalryman and field commander in the Middle Eastern theater of World War I. Born in Brackenhurst, Nottinghamshire, England, on 23 April 1861, Allenby graduated from the Royal Military Academy at Sandhurst; joined the Inniskilling Dragoons in 1882; fought in South Africa in 1884, 1885, and 1888; and served as both staff officer and field officer under Horatio Herbert Kitchener in the Second Boer War. He commanded the 5th Lancers in England from 1902 to 1905 and was inspector general of Cavalry from 1910 to 1914.

Allenby commanded the cavalry of the British Expeditionary Force in France in 1914, was soon promoted to command the Fifth Corps, and by October 1915 he was commanding the Third Army. Despite his several successes, notably at Arras on 9–15 April 1917, his cavalry background and especially his preference for the tactics of mobility and swift assault made him unsuitable for trench warfare. His attitude irritated his superiors, and in June they got rid of him by reassigning him to replace General Sir Archibald James Murray as head of the British Expeditionary Force in Egypt.

Allenby's decisive leadership and inspirational personality immediately restored British morale in the Near East. At last free to take advantage of cavalry tactics, he quickly mobilized and pushed north toward Damascus, exploiting Lawrence of Arabia's capture of Aqaba on 6 July and coordinating his strategy with both Lawrence's guerrilla harassment of the Turks in the east and Frederick S. Maude's successes in Mesopotamia. He won the third battle of Gaza by a brilliant outflanking maneuver at Beersheba on 31 October, routed the Turks throughout November, and entered Jerusalem in triumph on 10 December after two days of heavy fighting. In early 1918 his offensive campaign stalled only because his superiors sent many of his troops to France. He received enough reinforcements to resume full operations in July. Relentless in pursuit and matchless as a tactician, he beat the Turks decisively at Megiddo on 19–21 September, Damascus on 1 October, Homs on 16 October, and Aleppo on 25 October. His conquest of Palestine, Syria, and nearby lands forced Turkey out of the war on 30 October. Allenby was rewarded by promotion to field marshal and elevation to viscount. He served as high commissioner for Egypt from 1919 to 1925 and died in London on 14 May 1936.

Eric v. d. Luft

See also: Baghdad; Beersheba; Boer Wars; Kitchener, Horatio Herbert; Lawrence, Thomas Edward; Megiddo; World War I

References and further reading:

Bullock, David L. *Allenby's War: The Palestine-Arabian Campaigns, 1916–1918.* London: Blandford, 1988.

Hughes, Matthew. *Allenby and British Strategy in the Middle East, 1917–1919.* London: Cass, 1999.

James, Lawrence. *Imperial Warrior: The Life and Times of Field-Marshal Viscount Allenby, 1861–1936.* London: Weidenfeld & Nicolson, 1993.

Savage, Raymond. *Allenby of Armageddon: A Record of the Career and Campaigns of Field-Marshal Viscount Allenby.* Indianapolis: Bobbs-Merrill, 1926.

Alma (20 September 1854)

First major engagement in the Sevastopol campaign of the Crimean War, decisive allied victory over Russia. Marching south toward Sevastopol from their landing at Eupatoria, 60,000 allied troops found 35,000 Russians under Prince Alexandr Sergeevich Menshikov in a good defensive position on high ground south of the Alma River, about 20 miles north of Sevastopol, blocking their line of march. The allies deployed on the north bank, in sight of the Russians, but out of range. Omer Pasha held the far right with 7,000 Turks on the shore of the Black Sea. General Armand Jacques Leroy de Saint-Arnaud commanded 37,000 French on the right. General Fitzroy James Henry Somerset, Baron Raglan, led 26,000 British in the center and on the left. The Light Brigade of Brigadier General James Thomas Brudenell, seventh earl of Cardigan, secured the allied far left, about eight miles inland. Among the troops on the British left were three Highland regiments under Major General Colin Campbell, the 79th, 93d, and 42d.

The allied plan was to charge uphill along the whole front after the French had turned the Russian left flank, but Raglan, in his first battle since Waterloo, became confused and advanced into the center too early. Campbell, perceiving the danger, saved the day by attacking with a moving firing line 2,000 yards long and only two ranks deep. Firing muskets while on quick march was difficult, but Campbell's soldiers were the best trained and best disciplined in the British army. They routed the Suzdal Regiment, captured a 12-gun redoubt, and opened the Russian right. After that, despite Raglan's tactical error, superior firepower prevailed. Soon the entire Russian force was in full retreat. Raglan wanted to pursue, but Saint-Arnaud, citing logistical concerns, refused.

The British lost about 2,000, the French and Turks about 1,000, and the Russians about 6,000.

Eric v. d. Luft

See also: Balaklava; Campbell, Colin; Crimean War; Inkerman, Battle of the; Light Brigade, Charge of the; Sevastopol, Siege of

References and further reading:

Gibbs, Peter. *The Battle of the Alma.* Philadelphia, PA: Lippincott, 1963.

Harris, Henry Edward David. *The Alma, 1854.* London: Knight, 1971.

Lambert, Andrew D., and Stephen Badsey. *The Crimean War.* Dover, NH: Sutton, 1994.

Palmer, Alan Warwick. *The Banner of Battle: The Story of the Crimean War.* New York: St. Martin's Press, 1987.

Almohad Conquest of Muslim Spain (1146–1172)

Rescued Muslim Spain from the forces of the Christian Spanish Reconquest for at least another century. When the Almohad movement overthrew the Almoravids in Morocco in 1147, Muslim Spain was thrown into turmoil. The Almoravid governors in Granada, the Banu Ghaniya, managed to hold on there and in the Balearic islands, but most towns declared their independence under governments led by Muslim judges, military emirs, or sectarians. Northern Christians quickly took advantage of these conditions. Alfonso VII of Léon-Castile made Cordova his vassal in 1146, and went on to occupy Calatrava, Andujar, and Almeria. Portuguese forces entered Lisbon in 1147, aided by English and Lowland crusaders.

Abd al-Mu'min, the Almohad caliph, watched the Christian advance with grave concern. Islamic messianic revivalists, the Almohads had declared jihad against the Almoravids for letting Muslim security in Spain deteriorate. Therefore, Abd al-Mu'min knew he had to intervene as soon as his forces had pacified Morocco and completed the conquest of Algeria and Tunisia. His general, Abu Ishaq Bazzaz, crossed over to the Algarve in 1148, received allegiance from several cities, and drove the Almoravids out of Seville, but Almohad harshness inflamed revolts in Morocco that spread to Spain. Nonetheless, Castilian threats to Cordova soon forced the Andalusis to return to Almohad again in 1149. After reinforcing Cordova, the caliph eased his treatment of Spanish Muslims.

In 1157, with Maghrib affairs settled, the Almohads returned to Spain. They wrested Almeria back from the Castilians, and fortified Gibraltar. Muhammad ibn Mardanish of Valencia and Ibn Hamushk of Jaen, Muslim freebooters and sometime allies of Aragon and Castile, bedeviled the caliph's efforts, and in 1162 Ibn Hamushk took Granada by ruse. The Almohads proved just as wily, and slipped into the city fortress through their own deceptions and chased the rebels out, inflicting heavy losses.

Abd al-Mu'min died in 1163 while preparing a massive expedition to invade Portugal, Castile, and Léon simultaneously (one observer placed his forces at 100,000 horsemen and 100,000 infantry). The new caliph, Yusuf I, sent many of these forces into Spain in 1165, where they humiliated the army of Ibn Mardanish near Murcia. The Almohads also managed to relieve a Portuguese siege of Badajoz and cow Ibn Hamushk into submission. Caliph Yusuf I arrived in Seville in 1171, bringing in Berber and Arab troops from as far as Tunisia. Shortly thereafter, Ibn Mardanish died, and his family submitted to Yusuf's authority. The caliph passed the year campaigning against Heute, Cuenca, and faced down a Castilian expedition. By August, the army had returned to Murcia. Islamic Spain was now in Almohad hands, but the struggle between Christian and Muslim over Spain would very soon resume.

Weston F. Cook Jr.

See also: Almoravid Empire

References and further reading:

Abun-Nasr, Jamil M. *A History of the Maghrib in the Islamic Period.* 3d ed. Cambridge, UK: Cambridge University Press, 1993.

Kennedy, Hugh. *Muslim Spain and Portugal.* London and New York: Longman Press, 1996.

Almoravid Empire (1050–1148)

The Almoravid movement conquered an empire that extended from central Iberia south to the Senegal River, spreading Islam deeper into west Africa and rescuing Muslim Spain from the Spanish Crusades.

The Almoravid Empire began in the 1030s, with the efforts of one Abdullah ibn Yasin to reform the practice of Islam among the Sanhaja Berbers of Morocco. Appalled at Sanhaja ignorance and moral laxity, he imposed heavy fines, severe punishments, and outlawed many traditional customs. Expelled by one Sanhaja tribe, Abdullah turned to another, the Lamtuna. Here he found a more sympathetic reception, especially from Abu Bakr ibn Umar, head of the Targut clan. Training at a fortress (*rabat*) in Mauritania in both warfare and Abdullah's puritanical vision of Islam, these Lamtuna warriors came to be known as the Almoravids ("Those of the rabat"). By 1042, the Lamtuna had persuaded or coerced the rest of their Sanhaja kin to enlist in their ranks.

Proclaiming the slogan "Spread righteousness, correct injustice, and end evil [non-Quranic] taxation," the Almoravids launched their jihad around 1050. In 1054, Abu Bakr captured the Moroccan trade center at Sijilmassa, then pivoted south to drive the pagan Soninke out of Awghadust and capture the gold routes from the Niger basin. To control the High Atlas region, he set up a fortified camp on the site of what would become Marrakech under his cousin, Yusuf ibn Tashfin. These two men would divide the empire between them, Abu Bakr campaigning south in Mali-Mauritania, and Ibn Tashfin subjugating the cities and tribes of northern Morocco. Abdullah ibn Yasin had died in 1060 and, when Abu Bakr died in 1087, Yusuf ibn Tashfin became ruler of the burgeoning Almoravid Empire.

In 1085, the king of Castile captured Muslim Toledo, foraged at will throughout the little emirates of Islamic Spain, and laid siege to Zaragoza. Desperate, the emirs appealed to Yusuf to rescue them from the Christian crusaders. The Almoravids crossed into Spain in 1086 and, at the Battle of Badajoz, inflicted a crushing defeat on the Castilians. Continued threats from Christian warriors forced ibn Tashfin to return to Andalus several times. Finally, encouraged by Andalusi clergy and popular support, the Almoravids annexed most of Muslim Spain to their empire in 1099.

Almoravid forces do not seem to have numbered more than 30,000. They rode to battle on horse and camel, but the bulk of their fighters were infantry. They advanced on foot in ranks, the lead elements using long spears and the rear forces carrying javelins. Distinctive for their blue garb and facial coverings, their cohesive forces proved extremely formidable, although a lack of technical skills did put them at a severe disadvantage against towns and fortifications. In such circumstances, they usually turned to Spanish Muslims or other experts. Thus, while the Almoravids could defeat Spanish Christian forces in the field, they repeatedly fared poorly in siege craft.

Yusuf died in 1106. His sons, Ali and Tamim, took over, with Ali remaining in Africa and Tamim handling Spain. In Spain, Almoravid forces took Coimbra, Ucles, Lisbon, and Santarem, and ranged north to the Pyrenees, but Toledo remained unattainable. Then, in 1118, the balance of forces shifted as Alfonso I of Aragon captured Zaragoza, and mauled Muslim relief forces. Shortly thereafter, in 1120, Muhammad ibn Tumart of Morocco, founder of the Almohad movement, denounced the Almoravids, and launched his rebellion against them at Tinmal, in the hills above Marrakech. Confronting foreign invasions and internal revolt simultaneously, defeats in Spain alienated the Andalusian population, and seemed to validate Almohad accusations that God had turned against the regime. Frantic, Ali fortified Marrakech and hired Christian mercenaries. Ali died in 1143; his son, Tashfin, was slain fighting in the Tlemcen-Oran region only two years later. In 1147, Almohad troops entered Marrakech, and a series of revolts in Spain completed the Almoravid collapse.

Weston F. Cook Jr.

See also: Almohad Conquest of Muslim Spain; El Cid, Rodrigo Diaz de Vivar

References and further reading:

Abun-Nasr, Jamil M. *A History of the Maghrib in the Islamic Period.* 3d ed. Cambridge, UK: Cambridge University Press, 1993.

Kennedy, Hugh. *Muslim Spain and Portugal.* London and New York: Longman Press, 1996.

Amazons

Race of warrior women described by the Greeks. The Amazons were thought to have been fierce fighters who from their birth were brought up to be warriors. The name Amazon is believed to derive from the Greek word *amazos* (breastless), referring to the legend that the Amazons had their left breast seared during childhood to facilitate the use of a bow. In addition to the bow, the Amazons, who usually fought from horseback, used swords, double axes, and crescent-shaped shields. Various Greek myths and works of literature refer to encounters between Greeks and Amazons, such as *The Iliad* and *The Labors of Hercules.*

Though their place of origin remains in dispute, the lands most associated with the Amazons are Thermiscrya in the mouth of River Thermodon (in modern-day Turkey), the Black Sea region, and Libya.

Until very recently, the Amazons were seen only as a mythological phenomenon. Archaeological work in Kazakhstan, however, has brought to light female burials accompanied by weapons, suggesting that the Greek myths may have had some basis in fact. In particular, seven female graves contained iron swords and daggers, bronze arrowheads, and whetstones for sharpening the weapons. In addition, the curved leg bones of a teenage girl attest to a life on

Amazons after a hunt. Nineteenth-century lithograph. (Library of Congress)

horseback, while an arrowhead found in the skeleton of another female suggests that she had been killed in battle. Although these women, who were members of the Sarmatian tribe, cannot have been the Amazons of Greek myth (who were said to have lived far to the west), they may have been members of similar nomadic tribes who occupied the Eurasian steppes in the Iron Age.

Ioannis Georganas

See also: Ancient Warfare

References and further reading:

Davis-Kimball, Jeannine. "Warrior Women of the Eurasian Steppes." *Archaeology* 50:1 (1997), 44–48.

Tyrell, William Blake. *Amazons: A Study in Athenian Mythmaking.* Baltimore and London: Johns Hopkins University Press, 1984.

American Civil War (1861–1865)

The bloodiest war ever fought in the Americas. Ultimately, the new lands acquired from Mexico proved the undoing of the Union. Increasingly, politicians in the slave and nonslave states were unwilling to compromise, and a new political party, the Republican Party, dedicated to halting slavery's expansion, rose to power in the North. As the two major parties prepared to nominate candidates for the 1860 presidential election, the fissures and fault lines were obvious.

There were four major candidates for the presidency in 1860. The Republicans nominated Abraham Lincoln as their standard-bearer; combining many regional issues and emphasizing their resistance to the expansion of slavery since it threatened "free" labor, the Republicans were assured a majority in the North. The Democrats fractured into three groups. Senator Stephen Douglas was the candidate of the northern Democrats; John C. Breckinridge was the candidate of the Deep South; and John Bell became the candidate of Democrats living in the uplands South, which had a stronger commitment to the idea of the Union.

Lincoln's election set the process into motion. Once the electoral college in December confirmed the popular vote of November 1860, South Carolina seceded from the Union. In the next several months before Lincoln became president (which was 4 March 1861), six other Deep South states followed South Carolina. The new Confederate States of America chose Montgomery, Alabama, as its capital.

The first key battle of the Civil War would be over the fate of the remaining eight slave states. Washington, D.C., after all, rested between the slave states of Virginia and Maryland; if all the slave states left the Union, the North would have a difficult time indeed! However, if the eight remaining slave states remained in the Union, the South might not have a defendable border with the North.

Fort Sumter helped firm the battle lines. An artificial fort in the middle of Charleston Harbor, South Carolina, it was one of the few federal facilities in the Deep South that remained under northern control after secession. Lincoln wanted to hold the fort, but not appear aggressive; the South needed to gain control over the fort. Ultimately, the South fired on Sumter, which soon surrendered, and the eight remaining slave states divided in half, with Missouri, Kentucky, Maryland, and Delaware somewhat uneasily remaining in the Union.

Both the North and South had advantages (and disadvantages) in the coming conflict. The North had some 20 million white citizens; the South about 6 million, and white immigration clearly favored the North. The North had a clear superiority in manufacturing, banking, transportation—indeed, in industrial power. It had 110,000 manufacturing establishments to the South's 18,000; the North produced more pig iron and coal—the basis of industry. The North had more than 22,000 miles of railroad track, and longer lines and more common gauge; the South had but 9,000 miles of frequently shorter lines with different gauges. And the main transportation system of the South—rivers—provided easy access for northern invasions into the South's heartland. If the war became a long, drawn-out, and costly affair, the North had the industrial might and population numbers to prevail.

However, the South did have advantages. It was on the defensive, in a conflict that to some extent did pit families, friends, and business partners against one another. It did not need to win, only to continue to exist to emerge victorious. The South had many experienced and talented military leaders and more of a martial tradition, it believed, than the North. And it had so-called King Cotton. The South believed that Great Britain and, to a lesser extent, France were so dependent on southern cotton for their growing textile industries that the two countries would soon recognize southern belligerency and come to the South's aid.

And so both sides stumbled into the Battle of First Bull Run. President Lincoln initially did not accept a proposal from the North's leading general, Winfield Scott, victor of the Mexican War, that his "Anaconda" Plan would take time and require marshalling of much resources. But it would squeeze the South through a naval blockade and then drive through the natural invasion corridors—the Mississippi River, eastern Tennessee to Atlanta to the coast—until the South ceased to exist as an organized entity.

Instead, General Irwin McDowell marched an ill-trained army to defeat as it met another ill-trained army under Generals P. G. T. Beauregard and Joseph Johnston. The July 1861 battle was significant for the conclusions both sides drew. Lincoln began to accept that the war would be costly and take time, and made plans for such a contingency. His counterpart, Jefferson Davis, and other southerners were emboldened by the results and never really organized the South—which, as a confederation, had weaker central government than the North—for an extensive conflict.

The fighting began in the West, the area between the Appalachian Mountains and the Mississippi River. The South probably made a strategic mistake in moving its capital to Richmond, Virginia, because it focused attention on the battlefields between Washington, D.C., and Richmond—only 100 miles apart. However, the war for the South was lost—and perhaps could have been won—in the broad area between the mountains and the great river. In February 1862, General Ulysses S. Grant and Flag Officer Andrew Foote gained control over Forts Henry and Donelson and thus access to the Tennessee and Cumberland Rivers; CSA General Albert Sidney Johnston then retreated into Mississippi to recoup. The ensuing battle at Shiloh was, to that date, the bloodiest day of war on the North American continent, but its inconclusive result left Grant in possession of the battlefield and helped the North continue its drive to divide the South along the Mississippi River (New Orleans would soon fall, adding to the pressure).

In the East, Robert E. Lee demonstrated his mastery over a series of inferior northern commanders. Lee was a Napoleonic general in that he wanted battlefield victory, but defensive firepower was evident in this war—rifled muskets that were accurate at hundreds of yards, as well as mortars and cannon; to gain such victory meant sustaining huge losses in manpower—which the South could hardly spare. Still, after Stonewall Jackson confounded and then eluded several Union generals in the Shenandoah Valley, Lee struck against George McClellan, who had used the Union navy to transport a huge army to the peninsula between the York and James Rivers and very slowly advance on Richmond. In the so-called Seven Days' Battles, Lee tried to envelop and crush isolated parts of the Union army, but while he pushed that army back to Harrison's Landing, he could not destroy it and he sustained huge losses. Committed to an offensive-defensive strategy, Lee moved north, badly defeated John Pope's Army of Virginia, and then cartwheeled around Pope to "invade" Maryland. Ultimately Lee had to concentrate at Sharpsburg, Maryland, where his Army of Northern Virginia sustained an uncoordinated series of attacks from McClellan's larger force. Imagine the results if either McClellan had committed the 20,000 fresh troops he held from battle or he had attacked the next day as Lee prepared to cross the Potomac River with his baggage and his wounded. The campaign season in the East ended with Burnside's unfortunate assault at Fredericksburg, after his quick march to get behind Lee failed when pontooning equipment did not arrive to allow a quick movement across the Rappahannock River.

In the West, Confederate general Braxton Bragg, with Kirby Smith guarding his right flank, moved north threatening both Louisville and Cincinnati, and then fought indecisively at Perryville and retreated south soon after Lee left Sharpsburg. The year ended and the next began with inconclusive fighting at Murfreesboro between Nashville and Chattanooga. While not broadly recognized at the time, the two failed southern offensives—Lee into Maryland and Bragg into Kentucky—represented the high tide of the Confederacy and perhaps the last opportunity for foreign recognition. Lincoln announced the Emancipation Proclamation, foreign support for the South muted, the North grimly applied its manpower and industrial advantages, and the blockade began to bite deeply.

The next year, 1863, saw the tide turn to the North. Grant had tried to take Vicksburg, the last major southern strongpoint on the Mississippi River, and failed. However, while waiting for winter rains to subside, he came upon a brilliant strategy. He moved his men to the west side of the river opposite Vicksburg; he would have the U.S. Navy transport his army across the river below Vicksburg; they would live off the land and drive inland to repel any relief for the army at Vicksburg, and then turn to the city and either besiege it or take it. The strategy was brilliant, and worked, as Grant drove John Pemberton back into Vicksburg and in May settled into a siege that ended with Vicksburg's surrender on 4 July 1863.

Meanwhile, another Union commander felt he could defeat "Bobby Lee." Joseph ("Fighting Joe") Hooker revived the Army of the Potomac, and then set into motion a broad turning movement that would force Lee to retreat or else to be trapped between the Union corps remaining at Fredericksburg or the superior forces that had marched north and west and then southeast to get behind him. Lee, however, divided his army, leaving 10,000 men under Jubal Early at Fredericksburg, and then divided again, keeping but 17,000 in front of Hooker's 70,000-plus men and sending Jackson with about 26,000 to find the hanging flank. It was a brilliant conception, but Jackson was fatally wounded by a southern sharpshooter. Most of Hooker's men did not even see battle. Lee then received permission once again to invade the North; promised his senior subordinate, James Longstreet, that he would avoid battle; and stumbled into Gettysburg, the greatest battle ever fought in the Americas, and a Union defensive victory. Once again the North could have won a huge victory if George Meade had attacked as Lee retreated back to Virginia.

Still, attention turned to the campaign around Chattanooga. William Rosecrans maneuvered well and forced Braxton Bragg to evacuate Chattanooga in September 1863. With a lull in fighting in the East, Lee sent Longstreet and his corps to help Bragg. The battle at Chickamauga ("bloody creek" in Cherokee) was one of the few times the South outnumbered the North; Rosecrans lost track of his units and moved a division out of the line he felt was in reserve—and this occurred just as Longstreet's veterans attacked that part of the front. Most of the northern troops fled to Chattanooga and Bragg slowly followed, frittering away the advantage he had won in costly fashion at Chickamauga. Lincoln appointed Grant to command, and Grant came to Chattanooga, brought in reinforcements, opened a more secure supply line, tried to roll the Confederate left and then right, and was surprised by the performance of troops assigned to demonstrate against the Confederate center. The Army of Tennessee fled the field and Grant was appointed to command all Union armies.

The Union blockade was also having its destructive effect. Most southern ports were closed—either seized by the Union navy or guarded by Union gunboats; while some blockade runners did get through, the high cost of such cargoes attested to the declining frequency of their success. And southern commerce raiders never imposed the cost to northern merchants that American privateers exacted from British merchants during the American Revolution.

The war ground on in 1864. Grant had a strategic vision. He would accompany (but not directly command) the Army of the Potomac as it confronted Lee; he would lock onto Lee and not release his grip. Meanwhile William T. Sherman would maneuver from Chattanooga to Atlanta and eventually to the coast, cutting the South in half again; Union forces at Mobile, Alabama, would cut through the Deep South to meet Sherman at Atlanta while Benjamin Butler and the Army of the James would move up the peninsula to threaten Richmond and Petersburg while Grant (and Meade) occupied Lee.

So Grant and Lee fought a series of bloody campaigns from Wilderness to Spotsylvania Court House to North Anna to Cold Harbor to the siege of Petersburg and Richmond. While Grant took many casualties, he ended Lee's ability to take the offensive, ground down under the weight of fighting a total war. Meanwhile, Sherman and Joseph Johnston maneuvered brilliantly to the outskirts of Atlanta where President Davis replaced Johnston with John Hood. Hood attacked somewhat rashly, lost, and moved north through Alabama hoping to tempt Sherman to follow. Sherman decided to give up his long supply line (the relief drive from Mobile never took place; instead Union troops moved into Arkansas), and with 62,000 troops, he set off to march through and destroy the Deep South. Hood moved north, won a costly victory at Franklin (12 generals killed), and then was destroyed by George Thomas at Nashville in mid-December. Sherman reached the Atlantic coast at Savannah, Georgia soon thereafter, and began moving north through the Carolinas.

In spring 1865, Lee recognized the desperateness of his position. Grant was about to cut the last Confederate link to the south from the siege at Richmond and Petersburg; Sherman was moving north against a small army commanded by Joseph Johnston. Lee tried to disengage and march south, but the large and well-armed Union cavalry cut off his retreat, forcing his weakening army to march west, where he decided to surrender to Grant at Appomattox Court House, Virginia. Johnston soon surrendered to Sherman at Durham Station, North Carolina, and the remaining Confederate armies followed.

Ultimately, the South tried to fight as the North fought, and became caught up in a strategy of annihilation where it simply did not have the numbers, the strength, or the organization. A confederation did not provide for sufficient central power to resist. Moreover, too many generals did not understand the tremendous advances in killing power—the greater accuracy of rifled muskets, the introduction of breechloading repeating rifles, rifled cannons, and siege mortars; too often, generals sent troops straight ahead into well-prepared defenses and entrenched troops.

The American Civil War might be termed the first "modern" war, for both sides employed the major products of the Industrial Revolution: railroads, telegraphs, steamships, ironclads, rifled small arms and artillery, photography. None were used for the first time in this conflict, but all were employed on a far larger scale than ever before.

The Civil War was also by far America's bloodiest conflict; more than 600,000 died—nearly 2 percent of the population. But, as in the aftermath of so many of its conflicts, the United States emerged far stronger economically and politically than it had been at its beginning.

Charles M. Dobbs

See also: Antietam/Sharpsburg; Bull Run, First/Manassas; Bull Run, Second/Manassas Junction; Chattanooga, Battle of; Gettysburg; Grant, Ulysses Simpson; Lee, Robert Edward; Lincoln, Abraham; Seven Days' Battles; Sherman, William Tecumseh; Shiloh; Wilderness

References and further reading:

Catton, Bruce. *The Centennial History of the Civil War.* Garden City, NY: Doubleday, 1961, 1963, 1965.

Editors of Military Affairs. *Military Analysis of the Civil War.* Millwood, NY: KTO Press, 1977.

Griffith, Paddy. *Battle Tactics of the Civil War.* New Haven, CT: Yale University Press, 1989.

Williams, T. Harry. *Lincoln and His Generals.* New York: Knopf, 1952.

American Indian Wars

The continuous military confrontation between the European invaders and the indigenous population that lasted from the early seventeenth to the late nineteenth century. Those wars represented a fundamental struggle over land and resources. Throughout those campaigns Euro-American colonists demonstrated their technological superiority and justified the military expansion through social Darwinist ideologies of manifest destiny and religious evangelism. Most of the indigenous societies defended their homelands in fierce struggles and attempted to compensate for demographic and technological disadvantages through elaborate techniques of guerilla warfare and the military experience gained in intertribal conflicts. Ultimately, the Indian wars resulted in the complete military defeat of the indigenous societies and brought them to the verge of extinction.

Fighting already accompanied the colonial period, which was marked by almost continuous warfare. The Spanish colonists encountered heavy military resistance against their missionary efforts in the Southwest. In 1680 the rebellion of the Pueblo tribes drove the Spaniards out of the Rio Grande province for more than a decade. The French colonizers were engaged in frequent military confrontations with the Iroquois Confederacy while French traders and missionaries maintained friendly relations with other cultures of the Northeast. The English settlement efforts sparked war almost from the beginning. Prominent campaigns and battles were the Pequot War (1636–1637); the uprisings of the Wampanoag and Narragansett against the New England colonies, known as King Philip's War (1675–1676, proportionally the bloodiest conflict in American history); and Pontiac's Rebellion in the Northwest Territory in 1763.

Those conflicts in the colonial period were part of a larger imperial contest between Britain and France in which the Indian tribes served as respective allies. Their expertise as scouts but also their manpower was important as both sides struggled for dominance of the North American possessions. This practice of instrumentalization did not end with the founding of the United States. Although the military power of the Indians east of the Mississippi River had already substantially declined by 1776, tribes north of the Ohio River continued with British support to protect their homeland against the further encroachment of white settlement. Their efforts were repelled in the Battles of Fallen Timbers (1794) and Tippecanoe (1811). The army's victory ended the military ability of the tribes in the Northwest Territory effectively to challenge white intrusion. British support for Indian military campaigns ended in 1813 with the Battle of Thames in southern Ontario.

Between 1812 and the 1840s the remaining eastern tribes

were relocated by the federal government to territories west of the Mississippi so far untouched by white settlement. Washington hoped that the removal would end the military confrontation between Native Americans and the U.S. Army. The resettlement of the tribes was accompanied by occasional military resistance. Most prominent campaigns were the Florida Seminole Wars (1817–1818, 1835–1842, 1856–1858) and the Black Hawk War sparked by the refusal of the Sac and Fox to leave their homeland.

With the further territorial expansion of the United States after the Mexican War, the Indian territory no longer marked the effective western boundary of the United States, but divided the country in two. The effects of this division on Indian land became obvious as the California gold rush of 1848–1849 massively increased migration through tribal territories. This migration violated the Indian Intercourse Act of 1834 and set the stage for an ecological disaster as intruders increasingly decimated the bison herds that served not only as the foundation of the economic life but also the cultural existence for many of the Plains tribes.

As military confrontations remained a constant experience, the army concentrated on keeping the travel routes to the West open through a series of military outposts. In addition to this primary military objective, hostile tribes were crushed in a number of campaigns such as the Rouge River War, Yakima War, and the campaign of 1858, which eliminated Indian military resistance in the Oregon Territory.

The Civil War temporarily diverted the military energies of the government. As regular troops departed for the battlefields in the East they were replaced with local volunteers. Those regiments often displayed little discipline and training and frequently contributed to the escalation of Indian-white confrontation. During the Civil War many tribes actively participated on both sides. Particularly the Confederacy gained the support of numerous tribes, or tribal factions, such as the Cherokees, Creeks, Choctaws, Chickasaws, and Seminoles. They were motivated by the location of their tribal lands, which were surrounded by Confederate states, and their lack of confidence in the seriousness of Washington's Indian policy. On whatever side they fought, the Indian nations suffered terrible economic and social disruption to their territories as troops from both sides devastated their land and the continued fighting resulted in food shortages and famine.

While a number of tribes participated in the war effort, others saw the diversion of the U.S. Army as an opportunity to defeat a supposedly weakened enemy. The western tribes began their campaign in 1862 when Sioux attacked New Ulm in Minnesota, killing approximately 700 settlers. The Minnesota militia retaliated and captured more than 1,500 Indians. President Lincoln prevented the execution of at least 300 prisoners, a measure favored by the militia's commander, General John Pope. Warfare in the West escalated dramatically during those years. Army records indicate that fighting peaked during the years 1864–1867. But not only the frequency of campaigns increased. The dehumanization and brutalization, inherent in any war, reached new lows. The Indians themselves, understandably, held a low opinion of whites. Fostered by racism and derogatory stereotyping of Indians as savages, Indian-fighters took a savage stand. Instructive in this respect was the Sand Creek massacre.

On 29 November 1,200 troops of the First and Third Colorado Cavalry under the command of Colonel John M. Chivington attacked a partially disarmed and surrendered camp of Cheyenne Indians and killed anywhere from 150 to 500 men, women, and children. Formal investigations into the massacre did result in the condemnation of the attack through the U.S. government, an early forerunner of the guilty conscience Americans were beginning to develop, particularly in the East and, as westerners were quick to point out, away from the Indian wars.

In the years following the Civil War, the U.S. Army underwent dramatic reorganization. The troops were reduced in numbers. Continuous conflicts between the federal government and local commanders and the lack of proper doctrine highlighted the constraints imposed by demobilization. The army's task was to clear the way for settlement of the West by forcing Indians onto reservations.

By 1868 military commanders, such as Major General Philip Sheridan, embarked on a new strategy of total warfare that carried the war into the winter camps of the western nations. The army attacked the Indians during the winter when their food supply and thus their mobility were at a low point. This approach further obscured the division between combatants and noncombatants and highlighted the totality of warfare as carried out by both sides. Overall, however, the troops and commanders were more sympathetic to the plight of the Indians, much to the disgust of the local whites. The army, after all, was not after the Indians' lands.

War continued for another two decades. Outstanding battles and campaigns included the Little Bighorn (1876), the Red River War (1874–1875), the Modoc War of 1872–1873, and the Apache Wars that ended with the defeat of Geronimo in 1886. The final military engagements took place at Wounded Knee, Dakota Territory (1890), and on a military expedition against the Ojibwa in Minnesota in 1898. Despite occasional Indian victories in the years after the Civil War, the tide had long since turned against the indigenous Americans. Despite their increased military activities as fighting peaked between 1866 and 1869, resistance became futile as the tribes were outnumbered and outgunned. Massive western migration created an atmosphere in which Indian military actions could cause temporary delays but could not halt their defeat. With the escalation of military commanders to a strategy of annihilation, warfare caused increasing numbers of casualties. The number of Indian victims is unknown. The U.S. Army suffered 932 killed and more than 1,000 wounded between 1866 and 1891.

With those last engagements, more than two centuries of military confrontation between Native Americans and Euro-American settlers had ended. The military defeat threatened the very survival of Indian nations and ended tribal control over the trans-Mississippi West. The army gained valuable experience in those decades of war, although its doctrine remained fixed on the wars of Napoleon. Many famous Indian-fighters rose to prominence in military and civilian life as the United States entered the stage of world powers. The expertise of those who militarily matured during the Indian wars provided the backbone of colonial warfare in America's emerging colonial empire.

Frank Schumacher

See also: Little Bighorn; Sand Creek; Sheridan, Philip Henry

References and further reading:

Keenan, Jerry. *Encyclopedia of American Indian Wars 1492–1890.* New York: Norton, 1999.

McDermott, John D. *A Guide to the Indian Wars of the West.* Lincoln: University of Nebraska Press, 1998.

Starkey Armstrong. *European and Native American Warfare, 1675–1815.* Norman: University of Oklahoma Press, 1998.

Utley, Robert M. *Frontier Regulars: The United States Army and the Indian, 1866–1891.* New York: Macmillan, 1973.

———. *The Indian Frontier and the American West, 1846–1891.* Albuquerque: University of New Mexico Press, 1997.

Wooster, Robert. *The Military and United States Indian Policy, 1865–1903.* New Haven: Yale University Press, 1988.

American Revolution (1775–1783)

The war for independence of the former British North American colonies.

The American Revolution in many ways was a consequence of the Seven Years' War (1756–1763)—called the French and Indian War in America. When that conflict ended, Great Britain had gained much territory from France but had also incurred vast wartime expenses. In seeking to rearrange governance of their territories, especially in North America among the 13 British colonies, formerly French Quebec, and the Native American tribes west of the Appalachian Mountains, and in seeking new sources of taxation to repay the vast debt, the British perhaps inevitably would have angered their American cousins. The king and Parliament wanted to separate the peoples of now larger British North America; they wanted to reinforce the Navigation Acts and end the long era of "salutary neglect" to strengthen the mercantilist trading empire; and they wanted the colonists who benefited from the results of the Seven Years' War to bear a fair share of the resulting financial costs of the conflict.

Between 1763 and 1774, the king and Parliament in Great Britain and revolutionary leaders in America increasingly and more bitterly disagreed on the meaning of a series of acts; the consequence would be war. At first, the colonists tried to make a case about the rights of Englishmen. This increasingly separate British society in North America interpreted the Proclamation of 1763, the Sugar Act, Stamp Act, Declaratory Act, and Townsend Duties as violating their rights, especially the right of taxation only by one's own representatives—ultimately a call for a separate parliament in America and an anticipation of the dominion theory of government.

Thereafter, the American revolutionaries viewed the so-called Intolerable Acts (really the Coercive Acts and the unrelated Quebec Act) as severe infringements on their basic rights. In 1774, the colonists engaged in actual rebellion by calling for and convening a Continental Congress, an alternate authority to the Crown. Later, in the Declaration of Independence (July 1776), they would hold "these truths as self-evident," including "life, liberty and the pursuit of happiness."

When fighting broke out in April 1775 with the British march from Boston to Lexington and Concord and the colonial militia raids on the column on its retreat, the British not unnaturally assumed the fighting would reflect the style of fighting in Europe in the late eighteenth century. Warfare featured small, highly trained armies and navies; the goal was maneuver and demonstration of the helplessness of the enemy—not his destruction, since there was no value in destroying royalty. Armies relied on muskets that were not accurate at long distances. So armies marched in columns; flankers as much sought to keep the columns under control as to seek out the enemy. Having marched in columns typically along well-traveled routes in Europe, armies deployed into lines in a relatively narrow field for battle. The two opposing forces (most often professional lifetime soldiers) would come relatively close—50 or 100 yards—fire a volley or two, fix bayonets, and then charge one another. It was a test of discipline and courage, reflecting brutal parade-ground conditions. Within the confines of this theory of warfare the British excelled, winning all of their eighteenth-century wars to date and almost all of their battles, land or sea.

Indeed, the advantages, based on this style of fighting in mid-eighteenth-century Europe, seemingly rested with the British. An island nation of more than 7.5 million, it outnumbered the 13 colonies in population by three to one.

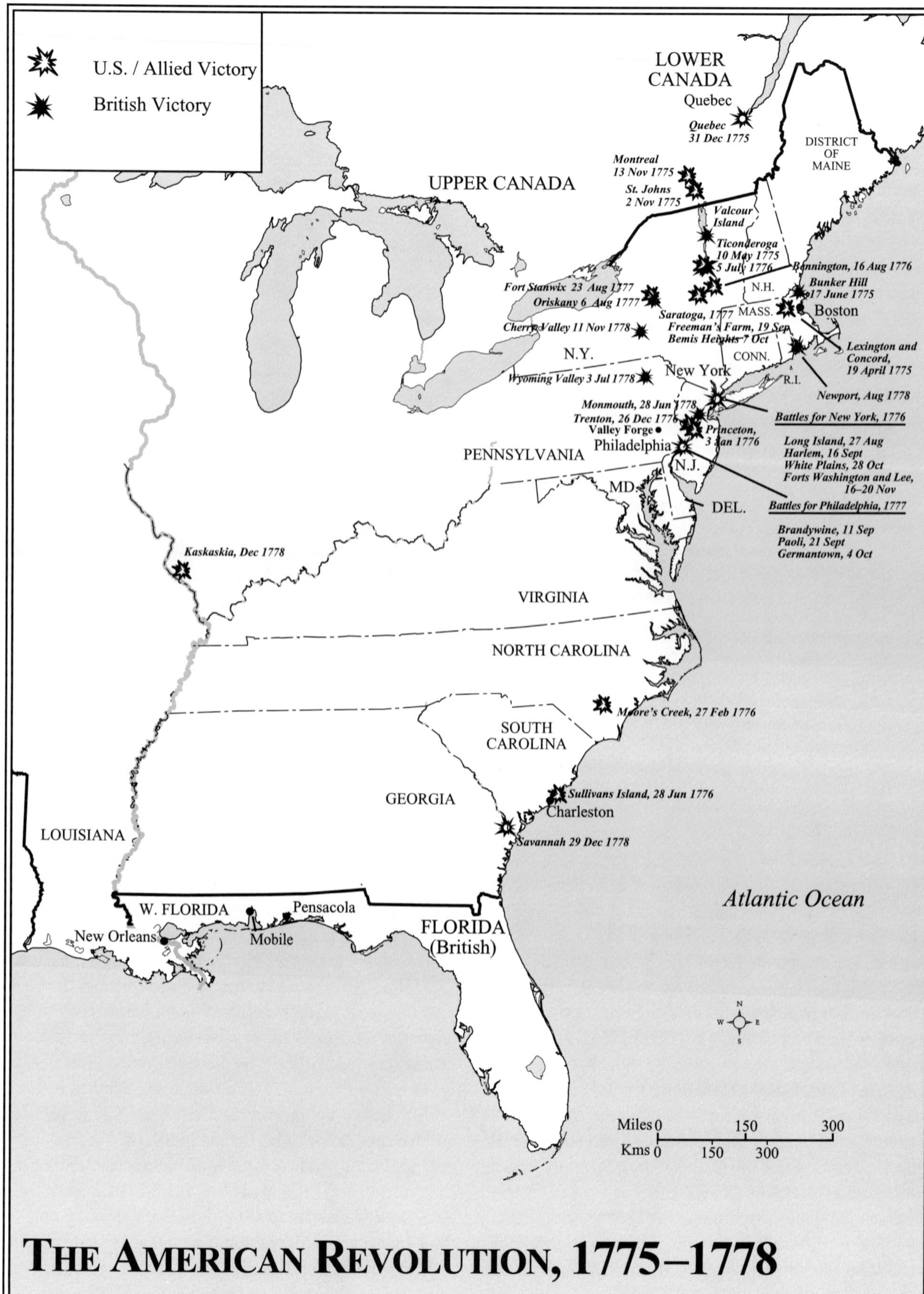

THE AMERICAN REVOLUTION, 1775–1778

Great Britain had great wealth; emerging industries capable of supplying the needs of the world's most powerful navy; a highly trained army and the financial resources and royal family connections to hire German mercenaries to augment its forces; and a trained if not evenly effective officer corps.

By the same analysis, the rebellious colonies had real difficulties. Population was thinly spread along 1,200 miles of Atlantic coastline; there were few cities, and they were more locations for trade than for the industry and finance necessary to fight a war. The American colonies had neither army nor navy nor trained officers; there was no central governing authority with real power. Moreover there were many Loyalists in the United States—the number will always be in dispute—who favored the king and continuing ties to the Mother Country. In theory and in practice, they represented manpower to augment trained troops, sources of supply, and information. In addition to the Loyalists, most numerous in New York and the South, there were many Americans who took no side in the conflict, for they wished to avoid authority in all its guises.

However, there were real advantages for the revolutionaries if they would stay the course. Several million Americans occupied a vast land; it was sparsely settled; there were few roads or waterways to connect isolated towns surrounded mostly by independent farmers. The nature of the conflict would differ from the tradition of fighting in mid-eighteenth-century Europe. Indeed, save for the continuing existence of a colonial army, there was no center of gravity whose capture or destruction would cause the American resistance to collapse. Moreover, in this war of attrition, the British did not have unlimited resources. They would have to maintain a secure seaward connection between home and the war front; they would have to support thousands of soldiers many thousands of leagues from home. Taxes were already high to pay for the costs of the Seven Years' War; there was a degree of war weariness in the British population, and there was a risk of imperial overstretch. Committing too many resources to the revolutionary war could mean drawing down defenses in the West Indies, India, Africa, and perhaps even the British Isles. The longer the war continued, the greater the pressure from various domestic interest groups to end it.

Whatever the theoretical advantages of the British, they faced very real and practical problems in winning. They had problems settling on an effective strategy that maintained the initiative. Not only did the British have difficulty finding a center of gravity of colonial resistance, they also changed objectives several times—perhaps an indication they could not understand how to win in this different kind of war. In 1775 at Bunker Hill and in 1776 during the Battles for New York City, the British believed to some extent that the mere demonstration of their military prowess in battlefield tactics would cause the Americans to concede they could not win, and to return to the fold. The brothers Admiral Richard Howe and General William Howe not only were charged with defeating the colonists in war, they also were commissioners to seek a peace—somewhat contradictory goals that may have caused them to avoid landing the true knockout blow that was within their grasp around New York City, and to ease a reconciliation that, unknown to them, was no longer possible. In 1776 and again in 1777 the British believed that the taking of apparently geographically significant sites would cripple the colonial war effort and compel surrender, as indeed it would in similar circumstances in more populated and more developed western and central Europe. By 1779–1780, the British believed the objective was to locate centers of Loyalism and so they turned south to win at Savannah, Charleston, and Camden, but to lose and ultimately face disaster at King's Mountain, Cowpens, Guilford Court House, and ultimately Yorktown.

No discussion of the difficulties for Great Britain in settling upon an appropriate strategy—at an affordable cost with a convincing explanation to the men at war and the people at home—should underestimate the considerable difficulties faced by the rebellious United States. There was no central authority; the Continental Congress lacked a true executive; and the new states were sources of contending power reflecting the fear of might threatening liberty that motivated the Revolution in the first place. The lack of an effective central government impeded the development of a national economy. Reflecting the mercantilist system the British had sought to establish, the colonies had little industry, few banking or financial resources, and thus little ability to provide for the fiscal needs of war. The colonial militia had not proven particularly effective in the Seven Years' War, and there were few experienced military leaders. Of course, the people were not unified: Somewhere between 20 and 33 percent of the population favored continued ties with Britain; an unknown percentage wanted to keep power at a distance and avoid authority—British or American.

The conflict proved costly for both sides. The British people suffered another conflict at great cost in blood and wealth. About 1 percent of the American population—approximately 25,000 people in a nation of around 2.5 million—died as a result of the war, a higher percentage of American dead than any other conflict save for the American Civil War.

There were several phases to the conflict. From 1775 to 1778, the conflict was largely in the North. In April 1775, the British, wanting to end the rebellion quickly, marched the relatively short distance from Boston past Lexington to Concord to seize colonial weapons and ammunition. The retreat to Boston displayed the advantages of a militia, armed with relatively effective rifled muskets, firing at a distance from the woods and behind stone walls at British troops marching in formation and easily located in their brightly colored uniforms. Soon thereafter, a huge gathering of New England militia and the hauling in of cannon from captured Fort Ticonderoga caused the British to quit Boston for Halifax, Nova Scotia, and to plan for an offensive the following year.

In summer 1776, a large British navy transported a vast army to New York City and by October the British had defeated the Americans at Long Island, Manhattan, Harlem Heights, and White Plains. The British had followed their strategy of fighting, outflanking one American position after another, and generally made clear the hopelessness of colonial defense. But the American army survived, and George Washington secured vital, morale-boosting victories in December 1776 and January 1777 in the Trenton and Princeton campaign. He crossed the Delaware River on Christmas night, compelled the surrender of German troops in Trenton, surprised Charles Lord Cornwallis, and won at Princeton before retiring to Morristown, New Jersey, for the winter.

In 1777, the British violated the principle of mass, and divided their forces. One army sought to march, row, and portage from Montreal to Albany, where it would join with forces coming from the west along the Mohawk River and up the Hudson River from New York City on the assumption that cutting the colonies in two would result in colonial surrender. At the same time, the British navy transported most of the army in New York City southward for an advance on Philadelphia, the largest city in North America. The result was not what the British had sought. Burgoyne surrendered his bogged-down army at Saratoga in October, having no option of retreat or resupply; the British withdrew from Philadelphia in 1778, demonstrating, perhaps, the uselessness of occupying it in 1777; and finally France, impressed with the infant republic's military successes, entered into a more formal and overt alliance, which made British victory in North America problematic.

After a lull for the better part of a year from mid-1778 to mid-1779, the British looked to the southern colonies to salvage a victory in the increasingly expensive conflict. The French-American alliance after the victory at Saratoga meant increased obligations on the Royal Navy and a dispersion of British strength from North America. Thus, the British army that went south to Savannah in December 1779 was smaller than the British army that invaded New York City in 1776 or that seized Philadelphia in 1777.

The British enjoyed early victories in the southern campaign. After taking Savannah and Augusta, the British quickly regained control over sparsely settled Georgia, protecting British interests in adjacent Florida. The advance into South Carolina in 1780 brought two major victories and several minor ones. Indecision among civilian leaders in Charleston afforded the British an opportunity to blockade an American army commanded by General Benjamin Lincoln and to compel its surrender. Soon thereafter, the British smashed an American army at Camden in central South Carolina whose commander, Horatio Gates, had foolishly assigned equal combat responsibilities to raw militia as to trained Continental troops. The inexperienced militia fled the field at the first British bayonet charge, and the outnumbered Continental troops fought valiantly but hopelessly.

The Americans rebounded with some luck and the emergence of outstanding leaders who reconciled European military tactics to the reality of the colonial scene. Militia fought militia at King's Mountain in northwest South Carolina, and that American victory threatened the British hold over the recently subdued colony. A classic victory at Cowpens by General Dan Morgan over Lieutenant Colonel Banastre Tarleton in western South Carolina suggested a way to use inexperienced militia to advantage. He only asked the militia to use the greater range of their rifled muskets to fire a few rounds at the advancing British and then retreat; in the end, the Continentals held firm, the militia reentered the fighting, and the British lost in a classic double envelopment. The battle strategy at Cowpens became a campaign strategy of retreat to the Dan River and Virginia and a subsequent battle at Guilford Court House. General Nathanael Greene engaged in a careful retreat, drawing the British under Cornwallis farther and farther from their supplies. Greene then fought on a field he had previously selected. While the British technically held the field, they soon had to abandon it and retreat toward the coast to obtain needed supplies.

Finally, the bankruptcy of the British search for a strategy became clear when Lord Cornwallis, after retreating to Hillsboro and Wilmington, North Carolina, marched into Virginia seeking to destroy a smaller American army commanded by the Marquis de Lafayette and others and then retreated down the York River peninsula to Yorktown. A nearly unique coordination of forces saw General Washington marching down from his siege of New York City, the French supplying payment in gold to American troops—who had not been paid in a year in many cases—and the French fleet winning one of its very few victories over the British navy. For the first time and only briefly, the trident of seapower passed briefly to the French. Cornwallis was besieged by land and by sea and forced to capitulate.

The American Revolution demonstrated the importance of many of the Nine Principles of War. The length of the war front from Massachusetts to South Carolina and Georgia made economy of force difficult. The lack of a true capital or center to the colonial economy made it more difficult to select a center of gravity to conquer or destroy. The distance of the conflict from the Mother Country prevented the British from employing real economy of force to achieve a continuing mass. And given the sympathies of British and Americans alike, it was hard to maintain security and secrecy.

The conflict also called into question the military tactics that proved effective in Europe. Marching in tight columns with flankers more concerned with keeping men in columns than in defending those columns exposed the British to harassing attacks. The vastness of the country and the distance from England permitted harassing guerrilla attacks on supplies by sea, privateers, and by land, foraging parties. Indeed, the main contribution of the American navy was the actions of private ship captains, so-called privateers, seizing British merchant vessels and selling the cargoes and ships in foreign ports; it was costly and weakened support at home for the British war effort. Finally, it is difficult to see how Great Britain could have held on much longer to a stretch of colonies whose main city, Philadelphia, was the second largest metropolis in the British Empire.

Charles M. Dobbs

See also: Camden, Battle of; Cornwallis, Sir Charles; Cowpens; Greene, Nathanael; Guilford Court House; King's Mountain; Lafayette, Marie Joseph Paul Yves Roch Gilbert du Motier, Marquis de; Lexington and Concord; Long Island, Battle of; Marion, Francis; Monmouth; Princeton, Battle of; Savannah, Siege and Taking of; Trenton; Washington, George; Yorktown

References and further reading:

Black, Jeremy. *War for America: The Fight for Independence, 1775–1783.* Stroud, UK: Alan Sutton, 1991.

Bowler, R. Arthur. *Logistics and the Failure of the British Army in America.* Princeton: Princeton University Press, 1975.

Mackesy, Piers. *The War for America, 1775–1783.* London: Longmans, 1964.

Palmer, Dave R. *The Way of the Fox: American Strategy in the War for America, 1775–1783.* Westport, CT: Greenwood Press, 1975.

Wood, W. J. *Battles of the Revolutionary War, 1775–1781.* New York: Da Capo Press, 1990.

Amiens (1918)

Two offensives from 8 August to 3 September 1918, fought between the Allies under Marshal Ferdinand Foch and the Germans under General Erich Ludendorff. Its aim was to disengage Amiens and the Paris-Amiens railway line and open the northern coalfields to the Allies. Field Marshal Douglas Haig's plan involved the consolidation of Allied positions along the old French front line, which extended from Mericourt to Hangest. While the British were to press the enemy in the direction of Chaulnes, the French First Army was to move toward Roye. Meanwhile, General Rawlinson's Fourth Army, consisting of three Canadian, two Australian, one American, and two British cavalry divisions, was to form the main assault group. The French and Fourth Army, preceded by tanks, caught the Germans off guard, and by nightfall the Allies were 10 miles inside enemy lines, having captured 15,000 prisoners and 400 guns. This breakthrough could be termed the turning point of the war; as German units collapsed, Ludendorff announced that "the war must be ended," and termed 8 August the "Black day" for the German army. By the time the second offensive began on 21 August the Germans had already evacuated Mondidier, thus freeing the Paris-Amiens railway line. As the Allied forces penetrated across the Somme, taking Peronne, St.-Quentin, Queant, Meautte, and the Arras-Albert railway line, the Germans, whose morale and defenses were low, retreated back to the Siegfried Line. German losses during the Battle of Amiens were 50,000 killed and wounded and 33,000 prisoners. British and Colonial losses were 22,000 and French losses 24,000.

Margaret Hardy

See also: Foch, Ferdinand; Haig, Douglas; Ludendorff, Erich Friedrich Wilhelm; World War II

References and further reading:

Edmonds, James Edward, H. R. Davies, and R. Maxwell-Hyslop. *Great Britain.* Committee of Imperial Defence. Historical Section, Military Operations, Italy, 1915–1919. Compiled by Sir James E. Edmonds and H. R. Davies. London: World Microfilms Publications in association with Her Majesty's Stationery Office, c. 1974.

Amin, Idi (1925–)

Military ruler of Uganda (r. 1971–1979). Amin was born in northern Uganda. He joined the King's African Rifles (formerly the Uganda Rifles) and proved successful as a noncommissioned officer and a boxer.

In the Africanization of the officer corps that followed Uganda's independence in 1962, Amin was promoted to captain. Promotions were rapid in the new army, and by 1964 Amin was a colonel. Civilian control of the military proved a serious problem for the Obote government, resulting in numerous attempts at army reorganization. Amin won favor by leading the government's assault on the powerful Kabaka Mutetsa of the Buganda in 1966. By 1968, he was a major general and commander of the army.

Amin began to promote a disproportionate number of northerners, straining relations with Obote. In January 1971 Amin staged a coup. Initially pro-British and pro-Israeli, Amin proved to be an unstable though canny leader.

His foreign policy drew Uganda closer to the Arab world,

offending Israel, and increasing internal repression included the expulsion of almost all of Uganda's Asian population in late 1972. He skillfully exploited ethnic tensions within the army, which he used to crush dissent and threaten Uganda's neighbors.

Amin, ill equipped to lead a nation, became progressively more erratic. In 1976, he allowed a hijacked El Al jetliner to land in Kampala, resulting in the humiliating Israeli commando raid at Entebbe Airport on 4 July. Condemned internationally for human rights violations, Amin was increasingly isolated diplomatically. In March 1978 he attempted to seize part of Tanzania. The resulting counterattack, aided by Ugandan exiles, took Kampala on 11 April 1979 and forced Amin into exile. He eventually settled in Saudi Arabia.

Amin's career is an example of the problems of civil-military relations in the developing world, and his policies had a long-term destabilizing effect on Uganda and the region.

Adam Seipp

References and further reading:

Avirgan, Tony, and Martha Honey. *War in Uganda: The Legacy of Idi Amin.* Westport, CT: Lawrence Hill & Co., 1982.

Mazrui, Ali A. *Soldiers and Kinsmen in Uganda: The Making of a Military Ethnocracy.* Beverly Hills, CA: Sage Publications, 1975.

Omara-Otunnu, Amii. *Politics and the Military in Uganda, 1890–1985.* New York: St. Martin's Press, 1987.

Amoaful, Battle of (Ghana, 31 January 1874)

Decisive victory by British over the Ashanti tribes. The Second Ashanti War was fought over trade routes to the interior and influence on the coast. In February 1873, the Ashanti invaded the British protectorate to safeguard their interests and prevent the further extension of British administrative control. When British allies, the Fante, failed to stop the advance, Gladstone's government committed itself to a military offensive and appointed Garnet Wolseley as commander of the expedition.

On 14 January 1874, Wolseley and 3,500 British, West Indian, and locally raised Hausa troops crossed the Prah River and entered Ashantiland. The Ashanti force, numbering upwards of 15,000, concentrated in and around the village of Amoaful. On 31 January, the British attacked. The artillery laid down barrages as the infantry advanced in loose square formations. By noon, due to the heavy firepower, the Ashanti were forced to abandon the village. Rather than retreat, they launched a determined counterattack. Carrying nothing but muskets and other outdated weapons, the extremely mobile attackers were able to break into many of the squares. British reinforcements, however, threw back the enemy. British casualties were light: 4 dead and about 200 wounded. Ashanti losses were very heavy; probably more than 2,000 were killed. After the victory at Amoaful, the British advance continued. On 4 February, Kumasi, the political seat of the Ashanti kingdom, fell with little resistance, and the Second Ashanti War was over. The British protectorate over the Gold Coast was further extended and strengthened.

James Thomas

See also: Buller, Sir Redvers Henry; Wolseley, Garnet Joseph, Viscount

References and further reading:

Edgerton, Robert. *Fall of the Asante Empire.* New York: Free Press, 1995.

Fynn, J. K. "Ghana-Asante." In *West African Resistance,* ed. Michael Crowder. New York: Africana Publishing, 1972.

Keegan, John. "The Ashanti Campaign, 1873–4." In *Victorian Military Campaigns,* ed. Brian Bond. New York: Praeger, 1967.

'Amr ibn al-'As (al-Aasi) (c. 585–664)

Arab general who conquered Egypt during the early Arab invasions. A contemporary of the Prophet Muhammad who converted to Islam before the fall of Mecca, 'Amr ibn al-'As rose to become one of the leading Arab generals during the initial Arab conquests. Throughout his career, 'Amr served on numerous important missions. His first major expedition took him to Oman on behalf of the Prophet to convince the local rulers to convert to Islam. 'Amr succeeded on this mission, but during his stay in Oman the Prophet died, prompting 'Amr to return to Medina.

Abu Bakr, the successor of Muhammad, gave 'Amr command of the army to invade Palestine in 633. Although reports of this invasion were conflicting, 'Amr was responsible for the conquest of the Byzantine territories west of the Jordan River. In addition, 'Amr took part in the battles of Yarmuk and during the capture of Damascus.

'Amr's major achievement was yet to come. In 640, 'Amr led another army of conquest into Egypt. There remains some debate on whether 'Amr did this on his own initiative or whether the caliph 'Umar directed him to invade Egypt. In either case, 'Umar ostensibly approved of it, as 'Amr did receive reinforcements and the conquest ended in 642 with the capture of Alexandria. Afterwards, 'Amr contributed greatly to the administration of Egypt and built the city that became Cairo. His career in Egypt, however, was short-lived as the caliph 'Uthman recalled him to Medina.

After this, 'Amr remained absent from major military events until the Battle of Siffin in 657, when Mu'awiyya and 'Ali battled for the caliphate. 'Amr sided with Mu'awiyya and led the cavalry. The battle ended more or less in a draw with the dispute settled through arbitration. Before the decision came, however, 'Amr was able to occupy Egypt and remove

'Ali's factions from power there in 658. 'Amr remained governor of Egypt until his death.

Timothy May

See also: Byzantine-Muslim Wars; Charles Martel; Heraclius; Khalid ibn al-Walid; Muslim Conquests; Sassanid Empire; Tariq ibn Ziyad; Yarmuk, Battle of

References and further reading:

Belyaev, E. A. *Arabs, Islam, and the Arab Caliphate in the Early Middle Ages.* London: Pall Mall Press, 1969.

Donner, Fred M. *The Early Islamic Conquests.* Princeton, NJ: Princeton University Press, 1981.

Shaban, M. A. *Islamic History: A New Interpretation.* London: Cambridge University Press, 1971.

Anaconda Plan (1861–1862)

The strategy adopted early in the American Civil War to strangle and cut up the Confederacy. The commanding general of Union armies, Winfield Scott, proposed the Anaconda Plan, realizing that the war would be long and costly, and that victory would reflect the Union's superiority in manpower, industry, transportation, etc.—the elements of total war.

He proposed a multipart plan that would take time to put into effect and would require considerable resources. First, the U.S. Navy would blockade southern ports. The Confederacy was deficient in manufacturing and war material, and thus required markets to sell its cotton and other commercial crops. Scott proposed cutting off such contact, depriving the South of trade and income, and squeezing it—hence "Anaconda."

He then proposed a series of offensives designed to cut the South in half and half again, until its ability to resist was destroyed. He called for a drive along the Mississippi River, to cut off Texas and Arkansas from the rest of the Confederacy; he further proposed a drive through the breadbasket of the South—into Kentucky (which was "neutral" at the time) and Tennessee, into Georgia, and then to the coast. And, if such action did not compel the South's capitulation, Scott proposed cutting again and yet again.

Initially, this idea met a cool response. Too many northerners (and too many southerners) believed that after one battle and one big victory the other side would give up and there was no need for such a long-term strategy.

Eventually, however, the North very much came to adopt Scott's idea, with a highly effective blockade reducing the South to occasional blockade "runners" and some raiders, and with a series of offensives in 1863 and 1864 that cut the South into parts and made defeat inevitable.

Charles M. Dobbs

See also: American Civil War; Scott, Winfield

References and further reading:

Barney, William L. *Flawed Victory: A New Perspective on the Civil War.* New York: Praeger, 1975.

Jones, Archer. *Civil War Command and Strategy.* New York: Free Press, 1992.

Anawrahta (d. 1077)

First king of Burma. His exact birth date is unknown. By 1044, he ruled the Kingdom of Pagan on the Irrawaddy River in modern-day central Burma. Through military force and political skill, he united the formerly separate fiefdoms of central Burma and moved to take possession of Arakan and Lower Burma between 1044 and 1056. In 1057, he conquered the kingdom of Thaton, which introduced Theravada Buddhism to Burma. He raided Thailand perhaps as far as the Chao Phraya Valley and guarded his frontier by building a series of forts on the Thai border. He died in 1077 after being gored by a wild buffalo near the gates of Pagan. Anawrahta in effect created the Burmese state and his dynasty ruled until 1287 when Burma was invaded by Chinese forces under Kublai Khan.

Harold Wise

See also: Kublai Khan

References and further reading:

Maring, J. M., and E. G. Maring, eds. *Historical and Cultural Dictionary of Burma.* Metuchen, NJ: Scarecrow Press, 1973.

Ancient Warfare

A form of warfare practiced exclusively until the development of firearms in most human societies, and more sparsely into the twentieth century. It is characterized by varied forms of organization focused on the use of weapons and mobility motivated by human and animal muscle.

Ancient warfare can be divided into a number of types. These types roughly (but not wholly) correlate with the ability of human societies to organize themselves into more organizationally complex systems, which in turn is highly dependent on the society's ability to produce material and energy surpluses.

Band

Band warfare is characterized by small-scale operations of face-to-face (all members of a band are known to one another) bands, usually led by one or more experienced or politically powerful war leaders. This form of warfare is the most ancient, limited usually by the small size and resources

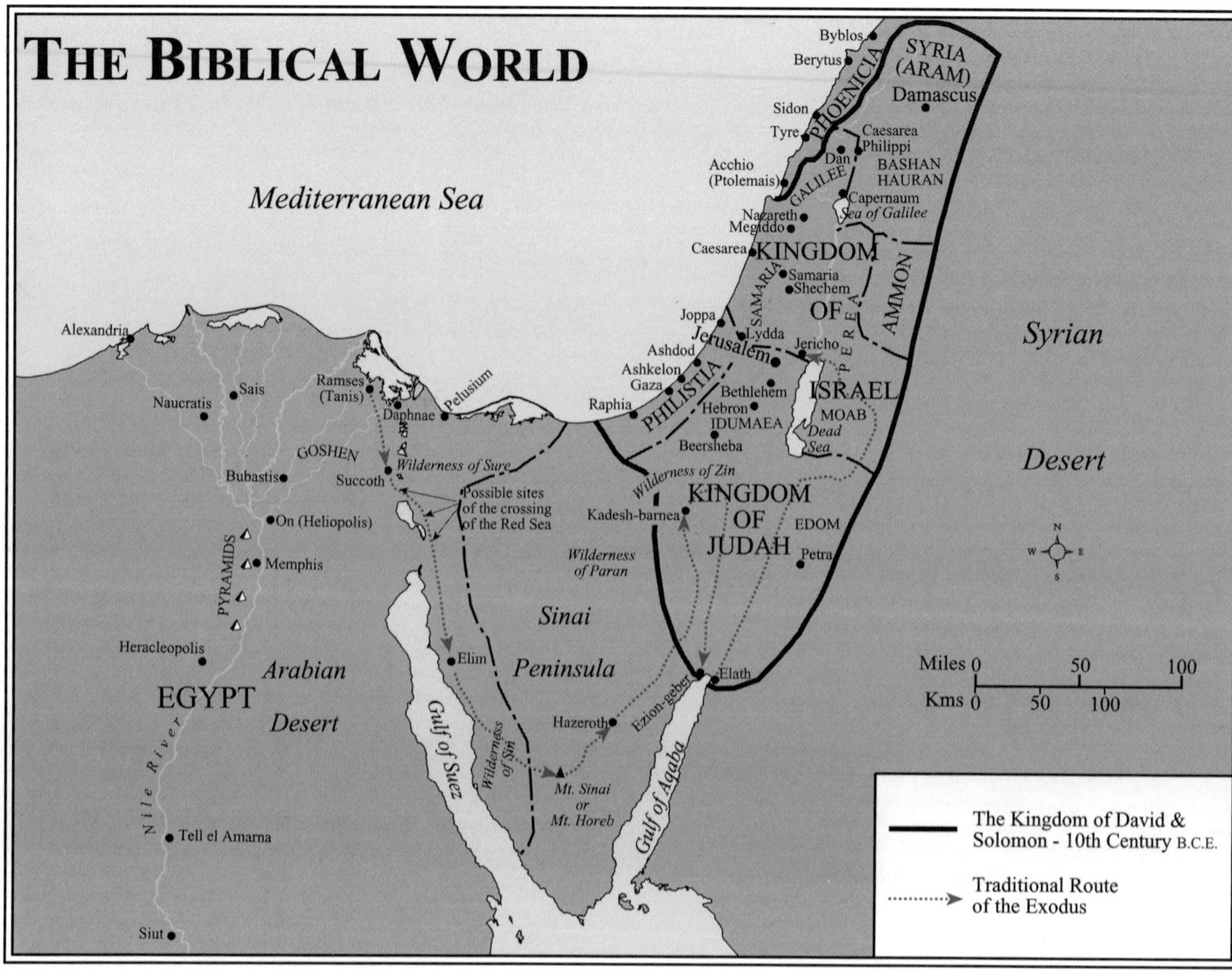

of groups who practiced it. As a form of warfare it was practiced well into the twentieth century by native people in areas such as New Guinea and Melanesia, Borneo, and Africa. Larger temporary assemblies of such bands could overwhelm much more sophisticated organizational forms, as Arminius demonstrated at Teutoburger Wald.

The motivation for such war was usually defensive, or raids for women, goods, or trophies.

Mass warfare

With the rise of societies that had sufficient surpluses for permanent leaders and large populations came mass warfare. This was usually characterized by the use of citizen or peasant levies who engaged in warfare at the command of, or in response to, the needs of the city-state. Early Roman military formations, the armies of Ur and early Egypt, and the Athenian *hoplitoi* were of this type. Such armies suffered from an inability to keep in the field during important agricultural periods, as well as a low level of training, uneven equipment distribution (often a soldier had to supply his own kit), and often weak motivation except in defense.

Champion/chivalrous warfare

Practiced in a number of ancient cultures, a heavily (and expensively!) equipped individual with superior training and usually some form of support by servant-soldiers fought his opposite number in a duel-like event. The loser's side forfeited the battle, sometimes being pursued to destruction. Such warfare could be practiced by two sides having the same cultural and religious matrix (as in the case of Homeric Greek champions) or one, the stronger, forcing another side to accept its definition of how warfare was to be conducted. Thus the battle between David and Goliath and the Aztec "Flower Wars" where one side forced a form of warfare on another.

The resources for a culture in champion warfare required pinpoint investment and training. Thus Chinese warring-states champions fought as archers mounted on expensive chariots drawn by expensive horses; Homeric champions required a full set of bronze armor; and Japanese cataphract-knights dueled cap-a-pie with long bows and swords, while the rest of the army was poorly equipped.

Champion warfare was utterly helpless against organized, objective-oriented armies (although they almost always held such armies in contempt), as the Japanese mounted samurai learned when faced by an organized Mongol invasion in the thirteenth century, and the Native American nations' warriors learned in the nineteenth.

Organized and professional warfare

Organized professional armies emerged in parallel with the other types. They came in a variety of forms, either a core of professionals supplemented by volunteers and levies (as in most Fertile Crescent states, including Egypt and Assyria) or wholly professional armies, such as the Roman army of the Imperial period, and possibly the Chinese army of Qin Shihuangdi (third century B.C.E.). Professional armies were characterized by internal organization into detachments, lines of command, clear ideas of tactics (and often formal manuals), and political-military objectives. Such armies could only be raised and maintained by states with large surpluses, which if depleted, as in the case of the later Roman, became effectively moribund.

Unit types and technology

Two main types of units dominated ancient warfare: infantry and cavalry. Later, ancient armies included engineering and fire-support elements to a limited degree. The type of unit relied on by any particular culture depended on two main factors: available technology and organizational innovations.

Infantry

Infantry was the mainstay of most ancient armies, and tactics and strategy were geared to the extensive exploitation of infantry. Two factors dominate the type of infantry and its capabilities: metallurgy and training, both individual and in formation.

The availability of metals or other hard materials determined the individual soldier's armaments and armor, and thus the possible range of tactics that could be employed. At one end of the scale, stone, wood, and obsidian, along with salt-saturated cotton armor, were the fundamental tools of Aztec and Meso-American warfare. Ancient and Middle Kingdom Egyptian armies relied heavily on stone-headed maces. At the other end were the sophisticated armor and steel weapons of Rome, India, and China.

Major weapons were blades (daggers, axes, and swords) and polearms (spears, pikes, and javelins). Slings and bows provided longer-range support fire. Although the technology did not determine the goals of warfare, it did mean that comparatively little physical damage was done to opponents. Most casualties were sustained as the losing side was slaughtered in flight.

Due to the limitations on a standing army, most ancient armies varied in the degree of training of their components. Thus the army of Xerxes, the great king of Persia, was composed of discrete national units with training ranging from none to professional, and battle tactics that depended largely on mass effects. The emergence of organized city-states and nations brought ideas of unit cohesion and organization. This allowed armies to apply force multipliers in the form of organized mass tactics. Ur, c. 2500 B.C.E., was first to utilize a phalanx of well-protected spearmen, an art brought to its apotheosis by the Greeks a millennium and a half later. Roman and Chinese organization into flexible company-sized units, who attacked after a barrage of missile fire, was even more effective.

Cavalry

Many armies utilized some form of cavalry (except in the New World) as a shock, scouting, or maneuver element. Cavalry use in ancient warfare was determined by two technological innovations. Until about 400 B.C.E. horses were used more often to pull chariots than in an actual cavalry role. Chariot cavalry evolved in parallel in the Middle East, East Asia, and South Asia, possibly spreading from some common Central Asian source. Egyptian, Assyrian, Indus, and Chinese tactics relied heavily on chariot cavalry. The breeding of horses strong enough and with sufficient wind to carry a man arrived on the scene fairly late in the Middle East and China, earlier, apparently in Central Asia, and quickly superseded unwieldy chariots.

The stirrup allowed a cavalryman a reasonably stable platform for shooting a bow, couching a lance, or delivering a sword cut so that his entire weight lay behind the point. Cavalry was used extensively by Central Asian peoples who invented the stirrup, and those people, the Parthians, Indians, and Chinese, in direct contact with them. By about the fourth century C.E. major armies in Europe were heavily cavalry oriented. This was much less the case in China, where great reserves of manpower meant greater reliance on peasant levies, and even less so in India and Southeast Asia, though even in those areas, cavalry had great impact. Elephants and camels, a form of cavalry, were utilized as well, though rarely to any great effect.

Engineering and siege warfare

Siege warfare was a haphazard affair, and not undertaken lightly, as the records of Thutmose III indicate. The first to set about siege warfare in an orderly and efficient manner were the Assyrians, and a number of records indicate the sophistication of Assyrian siege practices. These included towers, protected wheeled rams, mining, armored archers, and deceit. Less attention was paid to siege warfare in East, South, and Southeast Asia, where doctrine emphasized mobility, and where sieges were usually terminated by stratagem or storm.

Engineers could also be used in the field, and the Roman army excelled in the use of support field artillery, including single-shot and repeating catapults, and ballistae, to support cohorts in the field.

Command and control

Command and control in ancient warfare were extremely chancy due to technical limitations. Sun-tzu's *Bingfa* implies that standard battle communication by flags and trumpet calls was common in the Chinese armies of the Warring States period (fifth to third centuries B.C.E.). The Roman army used a variety of signaling devices.

The band form of ancient warfare invested command in a well-known and experienced war leader, who knew his followers and their capacities and could plan and act accordingly. With the development of mass, and later of organized ancient warfare, this process changed dramatically. Command, often of very large formations, was given to "gentlemen," that is, to representatives of whatever power elite was in control. Though many of these served successfully in what amounted to a training regime, gaining experience gradually, it was not uncommon in any ancient army to appoint a general simply on the basis of his position in the regime. Thus Nebuchadnezzar II of Chaldean Babylon sent his master chef (!), Nebuzaradan, to capture Jerusalem.

Supply

Most ancient armies lived off the country by pillaging as necessary; but the Egyptian army under Thutmose III gives some evidence of a supply train. Assyrian and Chinese armies were accompanied by a well-organized baggage train of supplies, though they too were encouraged to live off the locals for at least part of their support. The Roman army had an efficient supply service, providing troops with standard measures of grain, oil, and wine. At the other end of the scale were ancient armies such as the Mongols, who were able to move rapidly in part because they had no supply train at all: in addition to living off the country, soldiers would sustain themselves by bleeding their horses, or by consumption of dead or lame animals from the strings of horses that each soldier maintained.

Doctrine and manuals

A number of ancient warfare manuals have survived. They represent a vast sweep of ancient military thinking, indicating that in the armies of the great states at least, there were clear standards of military doctrine. Three of these deserve mention, from three different military traditions: Sun-tzu's *Bingfa* (c. fifth century B.C.E.), Kautilya's *Arthashastra* (c. third century B.C.E.) and Vegetius's *Epitoma Rei Militaria* (c. third century).

To summarize very briefly, each of these three manuals or guides reflects the doctrinal imperatives of the society that created them. For the Chinese, warfare was embodied in the imaginative and independent application of general principles, emphasizing dynamism, intelligence work, and strategic oversight. For the Indians, warfare was to be conducted with a firm eye on politics, and on the exploitation, in imaginative ways, of variations on set-piece battles or maneuvers that had been worked out over lengthy periods of time. For the Romans, success in war was a matter of meticulous planning and preparation, and close, even finicky, attention to the minutiae and details of an army's functioning.

Those who believe that military history (and history in general, for that matter) moves from the simple to the complex in a Darwinian arc of progress that flatters our own times might note that illiterate hill peasants in the late twentieth century could be trained in about six weeks on the use of the high-tech American Stinger missile. But it took a lifetime to master the medieval English longbow. The complexity and sophistication of ancient warfare were, on balance, comparable to the conduct of conflict in the modern world.

Michael Ashkenazi

See also: Animals in War; Assyria; Aztecs; Babylonian Empire; Megiddo, Battle of; Sun-tzu; Teutoburger Wald, Battle of; Vegetius Renatus, Flavius

References and further reading:

Boudet, Jacques. *The Ancient Art of Warfare.* London: Barrie & Rockliff, Cresset Press, 1969.

Connolly, Peter. *Greece and Rome at War.* London: Macdonald Hall, 1981.

Humble, Richard. *Warfare in the Ancient World.* London: Cassell, 1980.

Swan, Emma. *The Pharaoh Smites His Enemies. A Comparative Study.* Munchner Äegyptologische Studien. Munchen: Deutscher Kunstverlag, 1986.

Wales, Horace Geoffrey Quaritch. *Ancient South-East Asian Warfare.* London: Bernard Quaritch, 1952.

Yadin, Yigal. *The Art of Warfare in Biblical Lands.* London: Weidenfeld & Nicolson, 1963.

Anders, Wladyslaw (1892–1970)

Polish World War II commander. Born in Blonie, Russian-occupied Poland, on 11 August 1892, Anders served as a cavalry officer in the Russian Imperial Army during World War I. After the war he was commissioned in the newly formed Polish army and fought against the Bolsheviks during the Russo-Polish War, 1919–1921. The outbreak of World War II found Anders commanding the Novogrodek Cavalry Brigade. He was wounded twice as he led his men against the advancing German and Soviet armies. After being captured by the Soviets, he was denied medical treatment while being pressured over a period of several weeks to join the Red Army, but continually refused. Anders was then moved to Lubyanka Prison in Moscow, where he was kept under inhuman conditions for more than a year.

After the German invasion of the Soviet Union, Joseph Stalin formally recognized the Polish government-in-exile in London. After being removed from his prison cell, a stunned Anders was informed that he had been promoted to lieutenant general and appointed the commander of all Polish troops in the Soviet Union. But Anders was hampered by Soviet unwillingness both to provide proper supplies and to disclose the location of approximately 15,000 vanished Polish military officers. As a result of both Polish and British pressure, Anders was allowed to move his force into Iran in 1942 so that they could be supplied by the British. In 1943, the Germans discovered the mass graves of approximately 15,000 Polish officers in the Katyn Forrest, which confirmed Anders's worst fears—that the Soviets had murdered all of the missing officers.

The Polish II Corps, as Anders's force was designated, served with distinction during the Italian Campaign, eventually capturing Monte Cassino. At the end of the war, the Polish II Corps was disbanded and the vast majority of its men refused to return to Soviet-dominated Poland. As an ardent anti-Communist, Anders lived in the United Kingdom after the war, and died in London on 12 May 1970. At his request, he was buried alongside his men in the Polish cemetery at Monte Cassino.

Alexander M. Bielakowski

See also: Bor-Komorowski, Tadeusz; Maczek, Stanislaw; Sikorski, Wladyslaw Eugeniusz

References and further reading:

Anders, Wladyslaw. *An Army in Exile: The Story of the Second Polish Corps.* London: Macmillan & Co., 1949.

Sarner, Harvey. *General Anders and the Soldiers of the Second Polish Corps.* Cathedral City, CA: Brunswick Press, 1997.

Angles, Saxons, and Jutes

Germanic peoples from the Baltic that took part in the settlement of lowland Britain. The Angles are mentioned by the Roman historian Tacitus. During his time (first century) they formed part of the Suevi confederation. The Saxons are called raiders of the empire by later Roman historians, including Ammianus Marcellinus. They came from lands around the lower Elbe and were closely linked with the Angles, who lived immediately to the north. The Jutes have often been associated with Jutland, but archaeological evidence suggests links with the Rhineland Franks. The collective term *Anglo-Saxon,* covering all three peoples, was coined by the Normans.

By the end of the sixth century, Anglo-Saxon war bands and federations in Britain had begun to coalesce into protokingdoms: Northumbria, Lindsey, Mercia, Hwicce, Mid-Anglia, East Anglia, Essex, Wessex, Sussex, and Kent. Starting with Bede (673–735), chroniclers identified Seven Kingdoms (the "Heptarchy"), although this may overstate the degree of political distinctions among the various groups. Three of the seven, Essex, Sussex—the kingdoms of the East and South Saxons respectively—and predominately Jutish Kent, but named after the Celtic Cantiaci, are remembered in the names of the English counties. The fourth kingdom, that of the East Angles, lay in the extreme east of Britain. This region is still known as East Anglia. The two other Angle kingdoms were those of Northumbria (the land north of the Humber), and Mercia. Mercia had its origins in the upper and middle Trent Valley but expanded gradually at the expense of its neighbors, the Middle Angles and Hwicce. It eventually took in the whole of what is now the English Midlands. Among these kingdoms, it was Wessex, the kingdom of the West Saxons, that was destined to unite the Anglo-Saxons under a single crown.

In the early archaeology of the period it is often very difficult to distinguish between the graves of late Roman soldiers of Germanic descent and those of continental Saxons. Weapons, buckles, and shields are often identical. There was a Saxon saying that warfare was proper "for a nobleman," and male burials in Britain during the pagan period were often accompanied by war gear. Chieftains and more important noblemen would possess a mail-shirt (*byrne*) and a crested helmet (the *Spangenhelm* type being common), a sword, shield, and spear(s). The early Anglo-Saxon mail-shirt reached to just below the waist and had short sleeves. Noblemen of middling rank may have possessed a helm, perhaps a sword (an expensive item requiring skill in its production), and a shield and spear(s). The lowest-ranking warriors would have been equipped with just a shield and spear(s), and perhaps also a secondary weapon such as an ax or *seax.* This was the long single-edged knife from which the name *Saxon* derives. Primarily an everyday tool, it could also be used to finish off a felled opponent in battle. Although the main weapon was the spear, not only for the peasant but also for the professional soldier and even the nobility, all warriors carried the *seax.* The wearing of a knife may have actually been a symbol of freemanship. Saxon shields, round or near-round ovals in shape, were stoutly made of solid planks of linden wood and with heavy projecting iron bosses. Although bows were widely used by the continental Saxons, the Anglo-Saxons seem to have used the bow mainly for hunting, displaying certain disdain for its use in battle. Bows were mainly made of yew, elm, or ash.

Nic Fields

See also: Æthelbald's Wars; Saxon Raids

References and further reading:

Bartholomew, P. "Fourth-Century Saxons." *Britannia* 15 (1984), 169–185.

Hawkes, S. C., ed. *Weapons and Warfare in Anglo-Saxon England.* Oxford, UK: Oxford University Committee for Archaeology, 1989.

Underwood, R. *Anglo-Saxon Weapons and Warfare.* Stroud, Gloucestershire, UK: Tempus, 1999.

Anglo-French Wars (1542–1628)

Conflicts between the English and French between the reigns of Henry VIII and Louis XIII. These were not the most decisive wars in European history, but the fighting did exemplify the changes taking place in warfare at the time. The fortress design revolution, the greater reliance on gunpowder weapons in battle, and the increasing size of armies made this period of warfare between England and France an important transition in European warfare. In addition, the administrative improvements that took place during the sixteenth and early seventeenth centuries enabled the French and English to fight nearly constant warfare, spreading across four dynasties (Valois and Bourbon, Tudor and Stuart).

French preoccupation with its own religious civil wars meant that it had little energy to devote to war with its main Protestant rival. English preoccupation with the Spanish meant that France was not England's main focus. However, the two nations did differ constantly, participating on either side of no less than three major series of wars in Europe, including the German Wars of Religion, the Revolt of the Netherlands, and the Thirty Years' War.

In the Anglo-French War of 1542–1546, Henry VIII joined the Hapsburg Dynasty in a war against the Valois. The English captured the port of Boulogne. Though they had gained a French port, the war cost England 2 million pounds. Years of border skirmishes short of all-out war followed. When hostilities reached all-out war in 1549, the French king Henry II declared war with the intention of retaking Boulogne, which reverted to French control in 1550.

In 1557, England's Queen Mary drew her country into a war allied with Spain, whose king was her husband, though the war was very unpopular with the Protestant English people. During the Anglo-French War of 1557–1560, Mary managed to lose England's last continental foothold in the port of Calais on French territory. When Elizabeth succeeded Mary to the throne, religious and political differences split the fragile Anglo-Spanish alliance.

In the Anglo-French War of 1589–1593, Elizabeth I embroiled England in the great Protestant-Catholic wars on the Continent. Protestant England sided with the Protestant Dutch rebels against Spain while France supported fellow Catholic power Spain. Later, England sided with the Huguenot (Protestant) French against the Catholic Valois in the French Wars of Religion, a series of French civil wars primarily over religion. In 1589, after defeating the Spanish Armada, Elizabeth I sent troops to aid the French Protestants. In the Anglo-French War of 1627–1628, known in France as the Third Bearnese Revolt, England again came to the aid of Huguenot rebels fighting the French government.

The Anglo-French Wars between 1542 and 1628 may have centered on the religious differences between Catholic France and Protestant England, but major changes taking place during the military revolution permitted the two major European powers to battle each other constantly and to little conclusion.

David C. Arnold

See also: French Wars of Religion

References and further reading:

Dunn, Richard S. *The Age of Religious Wars, 1559–1715.* 2d ed. New York: W. W. Norton, 1979.

Holt, Mack P. *The French Wars of Religion, 1562–1629.* Cambridge, UK: Cambridge University Press, 1995.

Anglo-Scots Wars (1290–1388)

England tries to subdue Scotland. The thirteenth century witnessed many wars within England and Scotland, but no conflict between the two kingdoms until its last decade. The deaths of Alexander III, in 1286, and of his successor, Margaret, the Maid of Norway, in 1290, left Scotland exposed to the hazards of a disputed succession. With a dozen claimants to the throne, the Scots asked Edward I, king of England, to mediate. His nomination of John Baliol as king was a reasonable decision but for the fact that Edward used his position to exact recognition of English overlordship. Further demands pushed the Scots into an alliance with France. In 1296, despite the distraction of a Welsh rebellion, the English "conquered the Kingdom of Scotland and searched it through in twenty-one weeks." Baliol was forced to abdicate and Edward assumed for himself the title of king of Scotland, but, to use the Clausewitzian dictum, the English found it "easy to conquer but hard to occupy."

English behavior aroused bitter hatred and this led to the 1297 uprising. Raids into England, in turn, provoked the English invasion of 1298. Despite the overwhelming victory at Falkirk in that year, the English were unable to subjugate Scotland until 1305 when William Wallace was captured and executed. Thereafter Robert Bruce emerged as Scotland's leader and king, and between 1307 and 1314 the Scots largely cleared their country of the English: only Stirling, Dunbar, and Berwick remained under English control.

The attempt to relieve Stirling resulted in the disaster at Bannockburn (24 June 1314). A renewed English attempt to invade Scotland in 1322 again ended in emphatic defeat, this time at Byland. The Peace of Northampton, concluded in 1328, brought English recognition of Scotland's independence.

This independence survived the disastrous Scottish defeats in 1333 (at Halidon Hill) and in 1346 (at Neville's Cross), but in the 60 years after Northampton, Anglo-Scottish conflict was not so much national as cross-border raiding by local magnates as the outbreak of the Hundred Years' War forced the English to adopt a defensive posture. Thus the disastrous English defeat at Otterburn in 1388 was less a Scottish as opposed to a Douglas victory. It was part of the continuing cross-border raiding that plagued the area. Likewise, the Scottish defeat at Homildon near Woolmer in 1402 was a Percy victory.

Cross-border raiding continued intermittently over the next century with both sides seeking to profit from the other's distractions and civil wars. After Homildon there was no conflict between the English and Scottish kingdoms until 1513, and the Battle of Flodden.

H. P. Willmott

See also: Anglo-Scots Wars (1513–1560); Bannockburn, Battle of; Falkirk, Battle of; Flodden, Battle of

References and further reading:

Hooper, Nicholas, and Matthew Bennett. *The Cambridge Illustrated Atlas of Warfare: The Middle Ages, 768–1487.* Cambridge, UK: Cambridge University Press, 1996.

Prestwich, Michael. *The Three Edwards: War and State in England, 1272–1377.* New York: St. Martin's Press, 1980.

———. *Edward I.* Berkeley: University of California Press, 1988.

Anglo-Scots Wars (1513–1560)

Intermittent conflicts in Scotland that eventually ended the French presence and ensured English dominance. In 1503, James IV of Scotland, having temporarily tamed the McDonald clan and set his domestic affairs in order, married Margaret Tudor, sister of Henry VIII, as part of a truce that would allow Scotland to develop in peace with its southern neighbor. Instead, under the guise of building a fleet to fight the Turks, James IV engaged in an arms race with England, while still earning the name "Rex Pacificator" for not joining with France. In 1513, however, he renewed the "Auld Alliance" when France was attacked by Spain, England, the Pope, and Venice, and took an army over the Tweed, successfully besieging four English castles before being routed at the Battle of Flodden on 9 September 1513, at which James IV and most of the Scottish nobility were slaughtered.

Scotland was now in the hands of regents, ruling in the name of James V, a child, and policy shifted from pro-French to pro-English until 1528, when James came of age. He then made peace with England in order to restore law and order to Scotland before making a marital alliance with France by marrying first Madeline de Valois and then Marie de Guise in 1537. War flared again when several Irish chiefs offered James V the crown of Ireland, enraging Henry VIII, who sent an army to lay waste. When Scottish nobles, remembering Flodden, refused to go to war, James V led a small army to doom at Solway Moss on 24 November 1542, after which he died of exhaustion and defeat, leaving Mary Stuart, an infant, as queen under the regency of Marie de Guise.

Henry VIII wanted the infant queen as a bride for his son, Edward VI, a marriage that would unite the kingdoms, and sent an army to seize her. In the subsequent "Rough Wooing," carried out after Henry VIII's death by Protector Somerset, the English, with help from the McDonalds, marauded the borders for four years, until French troops arrived to assist the regent by reducing the Castle of St. Andrews by sea with naval guns, where English-backed rebel assassins were hiding. Peace was signed in 1549, and the child queen was smuggled to France as the wife of the dauphin, Francis II. During the reign of Mary I, Scotland and England remained at peace, but internally Protestants assailed the regent and her French garrison, led by John Knox and many of the nobles.

The accession of Elizabeth I in 1559 led to open English support for the Protestants against the regent, including an English fleet that arrived at Leith to harass the French troops. The war ended in 1560, when Marie de Guise died, and Mary, Queen of Scots, widowed in France, returned to Scotland to rule and signed the Treaty of Edinburgh, recognizing Elizabeth I's right to rule England, and stipulating the withdrawal of English and French troops from Scottish soil.

Margaret Sankey

See also: Flodden, Battle of

References and further reading:

Bingham, Caroline. *James V.* London: Collins, 1971.

Nicholls, Mark. *History of the Modern British Isles 1529–1603.* Malden, MA: Blackwell, 1999.

Philips, Gervase. *The Anglo-Scots Wars 1513–1550.* Woodbridge, Suffolk, UK: Boydell Press, 1999.

Anglo-Sikh Wars (1845–1849)

Two conflicts between the Sikh nation and the British East India Company, resulting in Britain annexing the Punjab, the Sikh homeland. In 1845 the largest and best-trained native army on the Indian subcontinent was the Khalsa, which Maharaja Ranjit Singh, the founder of the Sikh kingdom, had built on the Napoleonic model. A fierce power struggle erupted upon Ranjit Singh's death in 1839, because his son and heir, Dalip Singh, was only two years old. The Khalsa was the dominant political force, but it was essentially leaderless. The British grew fearful of its 45,000 infantry, 26,000 cavalry, and 376 cannon. Governor-General Sir Henry Hardinge mobilized south of the River Sutlej, the boundary of the Punjab.

Around 11 December 1845, five divisions of Sikhs crossed the Sutlej, forcing the British to withdraw southeast. Lal Singh and Sir Hugh Gough fought to a bloody draw at Mudki on 18 December. At Ferozeshah on 21–22 December, Gough's 22,000 routed 35,000–50,000 Sikhs under the split command of Lal Singh and Tej Singh. The two Sikh commanders unwittingly aided the British by not trusting each other, not communicating, and keeping their forces separate. At Aliwal on 28 January 1846, Sir Harry Smith's 10,000 caught an undetermined number of Sikhs under Runjoor Singh in a classic pincer attack.

After Aliwal, the Sikhs abandoned all territory south of the Sutlej except a heavily fortified three-mile stretch of shoreline at Sobraon, where Tej Singh garrisoned a nest of concentric entrenchments. On the north shore, Lal Singh commanded artillery. A bridge of boats spanned the river. Gough's 30,000 advanced against the 25,000–40,000 Sikhs just after midnight on 10 February, began heavy bombardment at dawn, and broke through in midmorning. As the Sikhs tried to flee, the bridge collapsed and thousands drowned.

The British occupied Lahore, the Sikh capital, on 20 February. By the terms of the treaty imposed on 11 March, the Sikhs had to cede much territory, pay reparations, limit the Khalsa to 20,000 men, and recognize the joint authority of Dalip Singh as raja, his mother Jindan as regent, Lal Singh as vizier, and Sir Henry Lawrence as British resident.

The Second Anglo-Sikh War began in April 1848 when Diwan Mul Raj, a minor Hindu leader, revolted. Governor-General James Andrew Broun Ramsay, Earl and Marquis of Dalhousie, sent three British columns and one Sikh column to capture Mul Raj's fortress at Multan. The Sikh column, led by Shere Singh, defected on 14 September, forcing the British to raise the siege and await reinforcements. Dalhousie sent Gough to the northwest frontier of the Punjab, but ordered him to undertake no offensive operations north of the River Chenab until after Multan fell. Shere Singh outmaneuvered and ambushed Gough at Ramnagar on 22 November, inflicting significant casualties. The British completed their siege of Multan on 4 January 1849, sacked the town, and massacred the inhabitants.

At Chillianwallah on 13 January, Gough's 14,000 faced Shere Singh's 30,000–40,000 along a five-mile front. After an hour of only artillery engagement, Gough attacked through dense jungle where hand-to-hand fighting favored the Sikhs. Three hours later Gough ordered retreat.

At Gujerat on 21 February, Gough, now reinforced and commanding 23,000 men, was more careful than he had been at Chillianwallah. He gathered sufficient intelligence about the strength and position of Shere Singh's 60,000, neutralized the Sikh batteries with artillery, broke the Sikh lines with well-coordinated infantry attacks, and pursued with cavalry. The Khalsa surrendered. Dalhousie annexed the Punjab in March, but scattered resistance continued through April.

Eric v. d. Luft

See also: British-Indian Army; British Military, Twentieth Century Organization and Structure

References and further reading:

Bruce, George. *Six Battles for India: The Anglo-Sikh Wars, 1845–6, 1848–9.* London: Arthur Barker, 1969.

Cook, Hugh C. B. *The Sikh Wars: The British Army in the Punjab, 1845–1849.* London: Cooper, 1975.

Featherstone, Donald F. *At Them with the Bayonet! The First Sikh War.* London: Jarrolds, 1968.

Gough, Charles, and Arthur D. Innes. *The Sikhs and the Sikh Wars: The Rise, Conquest, and Annexation of the Punjab State.* Punjab: State Languages Department, 1970.

Innes, Arthur D., and Charles Gough. *Annexation of Punjab.* Delhi: National Book Shop, 1984.

Anglo-Spanish War (1585–1604)

Conflict resulting from growing rivalry between Spain and England, once traditional allies, encouraged by the temporary neutering of their common enemy, France, as a result of the French Civil Wars. The issues included the English desire to enter the lucrative New World trade, which Spain was attempting to monopolize. Religious differences also separated the two nations in the wake of the English Reformation. However, the main cause of this conflict was undoubtedly the Spanish deployment of its powerful Army of

Flanders in the Low Countries. While its primary mission was to end the revolt of the Netherlands against Philip II, the presence of this large professional force, especially near the North Sea ports, was considered a security threat by the English. Following the arrival of this force in the Netherlands in 1567, Elizabeth I allowed English privateers to begin raiding the Spanish New World, and English "volunteers" (often raised and supported by the Crown) to serve against the Spanish forces in the Low Countries. This initiated a two-decade-long balancing act, during which Spain and England moved gradually toward open war.

Open warfare effectively began in August 1585, when Elizabeth I signed the Treaty of Nonsuch, guaranteeing English military support to the Dutch rebels. At the same time she released Sir Francis Drake to carry out a large expedition against the Spanish New World, and mobilized her home defense forces in case of Spanish attack. The latter was a wise move, for with the Treaty of Nonsuch Philip II finally resolved to undertake an invasion of England, and began mobilization of forces that would eventually comprise the Great Armada of 1588.

The war was waged on both land and sea. In the later 1580s English forces under the Earl of Leicester and then Lord Willoughby faced the Spanish in the Netherlands under the Prince of Parma in a series of largely indecisive actions that tied down significant numbers of Spanish troops and bought time for the Dutch to reorganize their defenses. At sea, English forces under Drake effectively raided the New World, and then the Spanish coast itself in 1587, delaying the Spanish attempt to invade England until 1588.

The defeat of Philip's Armada in 1588, due to a combination of poor planning, bad weather, and English naval activity, is often considered the decisive point of the war, but in fact England's attempt to counterattack with an invasion of Portugal, commanded by Drake and Sir John Norreys, also failed, and Spanish sea power gradually recovered. An extended stalemate followed, as both sides were drawn into the final round of the French Wars of Religion in the early 1590s. Both Spain and England deployed expeditions to Brittany in this period, enterprises that ended in 1594 when Norreys destroyed Spanish fortifications threatening the Brest harbor. This was an important victory, as it deprived the Spanish of a base for a renewed attempt to invade England. The Earl of Essex and Lord Admiral Charles Howard further forestalled the Spanish by launching a major expedition to Cadiz in 1596, successfully taking the town and inflicting substantial damage on Spanish shipping.

Nevertheless, despite these preventative actions, England faced "Armada scares" in 1596, 1597, and 1599. The Spanish fleet approached the English coast on each of these occasions, only to be dispersed yet again by bad weather. In 1600, Spanish forces actually managed to land at Kinsale in Ireland in support of Hugh O'Neill's rebellion there, but were contained and subsequently defeated by the rapid reaction of Charles Blount, Lord Mountjoy.

By the late 1590s, the war was proving to be a serious financial drain on both sides, and the source of growing internal dissent in England. The death of Philip II in 1598 removed one obstacle to a negotiated settlement, but believing they were near victory, the Spanish continued the war until the defeat of O'Neill's rebellion in Ireland convinced them otherwise. The almost simultaneous death of Elizabeth I, and the succession of the pacifistic James I in early 1603, cleared the final barrier to a negotiated peace, which was concluded in August 1604 with the Treaty of London.

John S. Nolan

See also: Dutch War of Independence; Parma and Piacenza, Alessandro Farnese, Duke of

References and further reading:

Fissel, Mark Charles. *English Warfare, 1511–1642.* London: Routledge, 2001.

Wernham, R. B. *After the Armada.* Oxford, UK: Oxford University Press, 1984.

———. *Return of the Armadas.* Oxford, UK: Oxford University Press, 1994.

Anglo-Zulu War (11 January–1 September 1879)

Battle that shattered the military power of the Zulu kingdom, leaving it vulnerable to colonial dismemberment. On 11 January 1879, British and colonial forces under Lieutenant General Lord Chelmsford invaded the Zulu kingdom with the objective of eliminating it as a military power in southern Africa. Three columns totaling 18,000 men were to converge on oNdini, King Cetshwayo's capital, rapidly forcing a battle where their superior military technology would secure a decisive conclusion to the campaign.

However, inadequate transport and supply bogged down the British advance, and Chelmsford's strategy was dislocated on 22 January when the main Zulu army of 24,000 men outmaneuvered, divided, and annihilated the British center column at Isandlwana. The Zulu failure that same night to capture the center column's depot at Rorke's Drift, and the inability at Nyezane of the Zulu coastal army of 6,000 men to prevent the right column from fighting its way through to Eshowe, did not alter the situation. Chelmsford was forced onto the defensive, and had to raise more than 8,000 Black levies to defend the borders of Natal. Meanwhile, the Zulu blockaded the right column in Fort Eshowe, and only the left column under Colonel Evelyn Wood based at

Khambula in northwestern Zululand retained the initiative through mounted raids.

Lack of commissariat arrangements and the requirements of ritual purification after battle meant the Zulu were unable to mount a sustained campaign and press their advantage. But they were ready for a second round by March. On 29 March the main Zulu army assaulted Khambula, but the British routed them in the most tenaciously fought and decisive battle of the war. At Gingindlovu on 2 April the Eshowe Relief Column broke another Zulu army and evacuated the Eshowe garrison. Zulu morale never recovered from these two defeats, for they reconfirmed the lesson of the war of 1838 against the Voortrekkers that warriors armed mainly with sharp-edged weapons were helpless against concentrated firepower from behind all-round defenses.

Chelmsford launched his second invasion in May, offering lenient terms of surrender, which increasingly detached Zulu support from Cetshwayo. While the 1st Division moved cumbersomely up the coastal plain, the Second Division advanced into the Zulu heartland from the northwest in cooperation with the left (now flying) column. Drawn up in an impenetrable infantry square, this joint force conclusively routed the Zulu army at Ulundi on 4 July. Flying columns during July and August completed the pacification of Zululand, and any lingering resistance ended with the capture of the fugitive King Cetshwayo on 28 August. The Zulu formally surrendered on 1 September and the British withdrew. Zulu military power never recovered.

John Laband

See also: Isandlwana, Khambula; Rorke's Drift; Zibhebhu kaMaphitha Zulu; Zulu Civil Wars and Rebellion

References and further reading:

Knight, Ian. *Brave Men's Blood: The Epic of the Zulu War, 1879.* London: Greenhill Books, 1990.

Laband, John. *Kingdom in Crisis: The Zulu Response to the British Invasion of 1879.* Manchester, UK, and New York: Manchester University Press and St Martin's Press, 1992.

———, ed. *Lord Chelmsford's Zululand Campaign 1878–1879.* Dover, NH: Alan Sutton for the Army Records Society, 1994.

Laband, John, and Paul Thompson. *The Illustrated Guide to the Anglo-Zulu War.* Pietermaritzburg, South Africa: University of Natal Press, 2000.

Angolan Civil War (1975–1991)

Following the overthrow of the fascist regime in Portugal in 1974 by the armed forces movement (MFA), negotiations began on a program for Angolan independence. A transitional government was established, consisting of Portugal and the three nationalist groups: the Popular Movement for the Liberation of Angola (MPLA), the National Front for the Liberation of Angola (FNLA), and the National Union for the Total Independence of Angola (UNITA). However, this arrangement broke down and the country was plunged into civil war. Angola was invaded by regular troops from Zaire in support of the FNLA and by South African regulars in support of UNITA.

The invasions failed, and the MPLA, backed by combat troops from Cuba, Guinea, and Guinea-Bissau and with equipment from Mozambique, Nigeria, and Algeria, was able to seize control of the bulk of the country by early 1976. Many years were to pass before the FNLA and FLEC (Front for the Liberation of Cabinda, a smaller rebel faction in the far north) were crushed. UNITA continued to wreak havoc in the south and southeast, though its fortunes waxed and waned according to the extent of South African and U.S. involvement at any particular time; UNITA was receiving up to U.S. $50 million a year in covert aid from the United States at one point.

Even at the height of the Cold War, though, the internal conflict never prevented the United States or other Western countries from doing business with Angola or exploiting its oil reserves. Oil companies operated in the Cabinda enclave for many years, more or less unaffected by the turmoil in the rest of the country other than a South African–inspired attempt to destroy their installations in 1985. Though thwarted by the Angolan armed forces, had it succeeded it would have crippled the Luanda government (around 95 percent of Angola's exports are oil-based). Additionally it would have been a diplomatic coup for the South Africans because they would then have been able to claim that the Luanda government was obviously a communist client state.

A high-level meeting was held in Luanda in early 1988 among American, Cuban, and Angolan government officials in an attempt to hammer out a settlement—the withdrawal of the Cuban troops in return for peace in Angola and the independence of Namibia. This proposal was certainly a step in the right direction, but in practice did not result in serious disengagement until 1990, following the independence of Namibia and the demise of communism in Eastern Europe and the USSR.

Despite having to rely heavily on assistance from its socialist allies, Angola has been reestablishing ties with the U.S. and other Western countries since the early 1990s. A cease-fire was signed in Lisbon in June 1991 between the Angolan president, Jose Eduardo dos Santos, and Jonas Savimbi, the leader of UNITA. The agreement provided for the integration of government and rebel troops into a new, reduced national army of some 50,000, the withdrawal of all foreign troops, and multiparty elections to be held under UN supervision. The agreement has been holding, more or less, into the twenty-first century.

James Corbin

Marxist MPLA fighters equipped with shoulder rocket launchers and Soviet assault rifles during the civil war in Angola, 1976. (Hulton/Archive)

See also: Angolan War of Independence; Mozambican War of Independence

References and further reading:

Brittain, Victoria. *Death of Dignity: Angola's Civil War.* Trenton, NJ: Africa World Press, 1986.
Ciment, James. *Angola and Mozambique: Postcolonial Wars in Southern Africa.* New York: Facts on File, 1997.
Henderson, Lawrence W. *Angola: Five Centuries of Conflict.* London: Cornell University Press, 1976.

Angolan War of Independence (1962–1975)

Angola has a history of occupation dating back to the Portuguese settlement established at Luanda in 1575. Popular resistance to colonial rule had its roots in the system of forced labor; after World War II spontaneous clashes between the various African communities and the colonial administration became increasingly frequent. In the 1960s this resentment flared into full-scale war.

Three indigenous Angolan political movements rose to the forefront in this time. The Popular Movement for the Liberation of Angola (MPLA), receiving substantial assistance from the USSR and its allies, emphasized the importance of transcending tribalism. The National Front for the Liberation of Angola (FNLA) appealed to tribal allegiances in the country's north and was supported by Zaire and a number of Western countries opposed to a communist takeover of Angola. Finally the National Union for the Total Independence of Angola (UNITA), coming onto the scene in the late 1960s, drew the bulk of its support from the south, forming an open alliance with right-wing Portuguese forces and later South Africa.

The war began in early 1961 with attacks on prisons in Luanda and coffee plantations in northwest Angola. For four months the area under attack increased until it covered all of northwest Angola. These actions caught the Portuguese government of Antonio Salazar off guard. Salazar realized that these were more than isolated peasant rebellions, and large contingents of regular troops soon arrived in Luanda by steamship.

The ground forces quickly occupied the areas in question, and by 1 July 1961, Portuguese troop strength in Angola had reached 17,000. The military instituted a Strategic Hamlet policy similar to that used in Vietnam. Meanwhile, the MPLA forces had to pull back over the border into newly independent Congo, while the FNLA operated out of Zaire.

International pressure, both from the United Nations and a strange coalition of the United States and newly independent African countries, was quickly tying Portugal's hands. The MPLA and FNLA were waging an increasingly successful media campaign, but by 1965 the war had reached a stalemate.

Portugal had succeeded in recapturing control of all the towns and plantations in the north. But Angolan nationalists, with added pressure from other rebel groups, were forcing Portugal to spend almost half its national budget on defense. Fifty thousand Portuguese troops were tied down in Angola alone. The end of the 1960s brought UNITA to the fore, and saw the three Angolan independence movements fighting as much with each other, or seeking a solution to their internecine warfare, as with opposing the Portuguese.

By the 1970s, with MPLA and FNLA conducting cross-border attacks from Congo and Zaire, respectively, and UNITA operating in the center of the country, the Portuguese were growing weary. In both Lisbon and Luanda, there was a growing sense that the African Empire had passed the peak of its importance. With the Portuguese coup of 1974, the new Lisbon government sought immediately to transition the country to independence.

The Portuguese tried to form a transitional power-sharing government with all three warring parties. However, when it became obvious that the parties were intent on seizing power by force, the Portuguese government quietly left on 10 November 1975, handing power to the MPLA and setting the stage for the 15-plus years of civil war to come.

James Corbin

See also: Angolan Civil War; Mozambican War of Independence

References and further reading:

Brittain, Victoria. *Death of Dignity: Angola's Civil War.* Trenton, NJ: Africa World Press, 1986.

Ciment, James. *Angola and Mozambique: Postcolonial Wars in Southern Africa.* New York: Facts on File, 1997.

Henderson, Lawrence W. *Angola: Five Centuries of Conflict.* London: Cornell University Press, 1976.

Animals in War

Man's companions in battle. If war is defined as organized violence in the service of the state, arguably animals have been participants from the beginning: horses in both the combat and logistical roles, mules and oxen and camels as beasts of burden, dogs as guards and fighters, birds as messengers, etc.

Though animals have been admired throughout military history, their limitations in war are often the most striking thing about their use. In animal-dependent logistical systems, there is the tyranny of fodder. Under nineteenth-century criteria a man needed three or four pounds of supplies on a daily basis, as compared to a mule or horse, which needed more than twenty pounds of grain and hay to function. Assuming such food is not available on the march and a given supply train has to carry its own supplies, the reasonable radius an army could operate from its source of supply was estimated to be a hundred miles.

Even with the rise of the railroad, providing strategic logistical support, armies were still dependent on horses and mules for supply at the operational and tactical levels of war into the middle of the twentieth century. Even in World War II, only the American and British armies were able to forgo the use of animal transport except in certain specialized circumstances, such as mountain and jungle warfare.

For example, the German army in its prime of 1941 had some 600,000 horses to support a field force of more than a million men during the invasion of the Soviet Union. Both sides used large numbers of horses in the eastern campaigns. The irony is that Germany, the power that symbolized mechanized warfare, came to be ever more dependent on literal horsepower as the war progressed, due to fuel shortages, disruption of industrial production due to the Allied strategic bombing campaign, and disjointed industrial planning.

With the decimation of animal populations in two world wars and ever-increasing industrial output, the post–World War II period has seen the near demise of animal transport. Still, there is always the specialized circumstance, such as in 2000 when an Ethiopian army, using pack animals as transport, was able to forge mountainous territory in its final offensive against Eritrea.

Besides being the mainstay of logistics until relatively recently, there is also the application of animals to communications, mostly in the form of dogs and pigeons.

While there are traditions of pigeons being used to carry messages going back to the Bronze Age, the real boom in the use of carrier pigeons occurred during the Franco-Prussian War. With that conflict, the birds were a mainstay, keeping communications open during the siege of Paris. This example having been set, various armies set up pigeon services that were maintained until the rise of the radio; by 1887 Germany is said to have had eight regional lofts with some 400 birds at each installation. During World War I, as many as 500,000 of the birds may have been used in the course of the war.

While there have been experiments with ocean mammals, such as dolphins and sea lions, in guard and recovery roles, the other most notable use of animals in war has been that of the dog.

While it was no doubt an obvious choice to make to take dogs to war, considering that war probably first evolved from the hunt, some traditions are better documented than others. In the Bronze Age, Assyrians, Babylonians, Greeks, and Persians all made use of war dogs, not simply as sentry animals, but also as packs of implacable attackers in the front ranks of battle. The Romans then acted as a channel for this tradition to be transferred to western Europe, where it was maintained until the gun made this role impractical, except in an environment such as the colonial wars against traditional peoples. A reprise of this role was the desperate use by the Soviets of antitank dogs: animals trained to run under German tanks in World War II while bearing explosive charges.

More typical is the use of dogs as scouts and sentries, where their alertness and sense of smell make them of value as trackers. Related to this is the use of the animals as retrievers in World War I, where Red Cross dogs were sent into no-man's-land to help wounded soldiers. Even with advances in robotic technology, the dog at least still seems to have a future in specialized military roles.

George R. Shaner

See also: Cavalry; Logistics

References and further reading:

DiNardo, R. L. *Mechanized Juggernaut or Military Anachronism? Horses and the German Army of World War II.* New York: Greenwood Press, 1991.

Huston, James A. *The Sinews of War: Army Logistics, 1775–1953.* Washington, DC: United States Army, 1966.

Lemish, Michael G. *War Dogs: Canines in Combat.* Washington, DC: Brassey, 1996.

Lubow, Robert E. *The War Animals.* Garden City, NY: Doubleday, 1977.

Antietam/Sharpsburg (17 September 1862)

The bloodiest single day in U.S. history. Confederate general Robert E. Lee believed in an offensive/defensive strategy for the Confederacy, and thus proposed to follow up the victory at Second Bull Run/Manassas Junction with an invasion of the North. Using cavalry along the passes to screen movement of infantry units, Lee had an ambitious plan to seize Harper's Ferry, take several other towns for their supplies, and eventually wheel the Army of Northern Virginia around Washington, D.C., into Pennsylvania.

In one of the strange twists of fate, one set of orders were lost, and eventually came to the attention of Union general

Battle of Antietam, 1862. (Library of Congress)

George McClellan. Moving slowly, McClellan missed an opportunity to put the larger Army of the Potomac between the smaller units of Lee's badly divided army.

Lee drew his army together along Antietam Creek in Sharpsburg, Maryland, with the Potomac River to his rear—hardly an ideal defensive position. D. H. Hill's division had slowed the Union advance at Turner's Gap while Stonewall Jackson captured Harper's Ferry and 12,500 Union troops. McClellan tossed away his advantage by having Joseph Hooker's corps attack from the Union right in early morning, his center under Edwin Sumner attack at noon, and Ambrose Burnside's corps attack from the Union left in late afternoon, permitting Lee to shift his few reserves to meet each threatened Union breakthrough. Indeed, at the crucial moment, A. P. Hill's division arrived from Harper's Ferry and drove back Burnside's men, who were about to break through the weakened Confederate lines. McClellan held more than 20,000 troops in reserve (who might have affected the outcome). But the irresolute commander, fearing a Confederate trap, did not continue the battle on the following day or pursue Lee as the Confederates crossed the Potomac River and retreated to Virginia.

Antietam was the single bloodiest day in American history, with more than 23,000 casualties on both sides. Combined with Bragg's ineffectual battle at Perryville, Kentucky, this truly was the Confederate high tide.

Charles M. Dobbs

See also: American Civil War; Lee, Robert Edward; McClellan, George Brinton

References and further reading:

Gallagher, Gary W., ed. *The Antietam Campaign.* Chapel Hill: University of North Carolina Press, 1999.

Harsh, Joseph L. *Taken at the Flood: Robert E. Lee and Confederate Strategy in the Maryland Campaign of 1862.* Kent, OH: Kent State University Press, 1999.

Sears, Stephen W. *Landscape Turned Red: The Battle of Antietam.* New Haven, CT: Ticknor & Fields, 1983.

Antioch, Battle of (Syria, 1098)

The First Crusade's victory at Antioch established a Latin state in the Middle East and divided the crusader leaders by petty jealousies. The Battle of Antioch existed in two stages. The first part consisted of the crusaders' siege of Antioch, the second of their defense of the city.

On 20 October 1098 the crusaders, led by Bohemund of Taranto and Raymond of Toulouse, entered the territory of Yaghi Siyan, ruler of Antioch. The crusaders did not immediately assault the city, as Bohemund hoped to win the city through subterfuge and then keep it for himself.

Despite sorties from the Yaghi Siyan's armies and diminishing supplies, the crusaders slowly continued to surround the city through the construction of their own fortifications. During this period they also routed two relief armies.

Bohemund eventually secured an ally within the city. In accordance with their plan, Bohemund led the crusader army away. Then on 3 June, 60 knights entered a tower, which the traitor Firuz commanded, by climbing a ladder. These knights then seized two other towers and lowered ladders for the rest of the infantry as well as opening the gates. The crusaders charged in and by nightfall the Turkic garrison was defeated.

Despite their victory, the approach of Kerbogha of Mosul loomed over their heads. On 7 June 1099, Kerbogha camped before Antioch. After failed negotiations on 28 June 1098, the crusaders marshaled their army before Antioch. The knights advanced through a hail of arrows from the Turkic horse archers. This failed to stop them and panic seized the Muslim army. Kerbogha attempted to turn a flank, but Bohemund countered it adeptly. Kerbogha's forces began to desert. In a rare episode of discipline, the crusaders did not pillage their camp but pressed home their advantage and completely routed the Muslims.

Timothy May

See also: Armor, Ancient and Medieval; Byzantine-Muslim Wars; Crusades; Jerusalem, Siege of; John II Comnenus; Norman-Byzantine Wars; Religion and War; Seljuqs

References and further reading:

Holt, P. M. *The Age of the Crusades.* London: Longman Group, 1990.

Mayer, Hans Eberhard. *The Crusades.* Oxford, UK: Oxford University Press, 1988.

Riley-Smith, Jonathan. *The Crusades: A Short History.* New Haven, CT: Yale University Press, 1987.

Runciman, Steven. *A History of the Crusades.* Vol 1. Cambridge, UK: Cambridge University Press, 1951.

Antwerp, Siege of (1585)

Central episode in the revolt of the Low Countries against Spain. The siege was conducted by Philip's general Alexander Farnese, the prince of Parma, between September 1584 and August 1585. In taking the supposedly impregnable city, Parma would establish himself as one of the great military engineers of all time.

Conditions for such an attempt on the strongest citadel of the Dutch rebels were good in the summer of 1584, as the assassination of William the Silent on 10 July 1584 had left the rebels leaderless. It was believed that the fall of Antwerp would finally break the back of the rebellion while restoring to Philip II's control the largest port in Europe.

Parma, knowing the defenses were extremely strong and surrounded by low-lying country, which the defenders flooded, eschewed a standard close siege and settled for an extended blockade, building forts on all routes into the city. The final piece of his investment was the construction of a vast fortified bridge across the Scheldt River, which effectively cut off the city from the sea. A Dutch small-boat attack against this construction failed on 22 December. Likewise, an attempt to destroy the completed ridge with two incendiary ships designed by Francisco Giambelli also narrowly failed on 5 April, although heavy casualties were inflicted on the Spanish, with Parma himself narrowly escaping death. The final hope of the city was for outside intervention: It was believed that Queen Elizabeth I of England would be reluctant to let this important port fall into Spanish hands. Ironically, the city was forced to surrender by starvation on 17 August, only one day before Queen Elizabeth finally decided to commit and ordered Sir John Norreys to take 4,450 English troops to relieve the city.

John S. Nolan

See also: Anglo-Spanish War; Dutch War of Independence

References and further reading:

Arnold, Thomas. *The Renaissance at War.* London: Cassell, 2001.
Parker, Geoffrey. *The Dutch Revolt.* London: Penguin, 1979.

Anzio, Battle of (22 January–23 May 1944)

Allied amphibious attempt on the west coast of Italy to outflank the German defenses along the Gustav Line and at Monte Cassino 60 miles to the south. Anzio was some 35 miles south of Rome. A near-disaster, Anzio (Operation SHINGLE) became one of the most controversial operations of World War II. What some consider to be one of the great missed opportunities of that war others regard as an ill-conceived gamble. Winston Churchill later wrote that he had hoped the Allies were hurling a wild cat onto the shore but that all they got was a stranded whale.

Commanded by Major General John P. Lucas, units of the VI Corps (Fifth Army) came ashore at Anzio and nearby Nettuno on 22 January 1944 against practically no opposition. With only two divisions in the first wave, Fifth Army commander General Mark Clark told Lucas not to take risks. The German commander in Italy, Field Marshal Albert Kesselring, reacted quickly and within days he had no less than six divisions of a hastily improvised Fourteenth Army, commanded by General Eberhard von Mackensen, besieging the Allied forces. The Germans made vicious but unsuccessful attempts to drive the Allies back into the sea, and Anglo-American forces reciprocated with attempts to break out of the vise. Although SHINGLE's primary purpose was not achieved, the operation was not a failure. Anzio helped to draw into Italy German troops from as far away as the Balkans, France, and Germany itself.

It was not until May that VI Corps broke out and linked up with Fifth Army, and began the victorious march on Rome. Allied casualties were approximately 7,000 killed and 36,000 wounded or missing in action; 44,000 were disabled with injuries or sickness. German losses were estimated to have been 40,000, including 5,000 killed and 4,500 captured.

Colin F. Baxter

See also: Clark, General Mark Wayne; World War II

References and further reading:

D'Este, Carlo. *Fatal Decision: Anzio and the Battle for Rome.* New York: HarperCollins, 1990.

Apache Wars (1860–1886)

Sporadic but bloody outbreaks of violence across the desert Southwest as the U.S. Army stamped out resistance to white settlement in the area. In 1848, after American rule replaced that of Mexico following the Mexican-American War, relations with most Apache tribes failed to improve. American trappers, miners, and settlers continued to penetrate the region, especially after the California Gold Rush. At the same time, the Apaches continued to raid both north and south of the new border. Conflict was almost inevitable.

Several military outposts were established in the region during the 1850s but failed to bring the Apaches under effective control. In 1860, the Apaches and their relatives, the Navajo, commenced wide-scale raids and depredations. This Apache war continued until 1865. With regular army troops sent east at the start of the American Civil War, militia troops attempted to hunt the Apaches led by Cochise. In late 1862, having made no apparent progress using unreliable irregulars, regular troops had to be recalled from the East. In 1863, the noted Indian-fighter Christopher "Kit" Carson implemented a scorched-earth campaign in which Apache males were simply shot, regardless of the circumstances, and Apache women and children were rounded up and imprisoned. By 1865, those Apaches who were not already dead or captured had been driven into the mountains.

In 1871, the Apaches banded together to cut another swath across Arizona and New Mexico. U.S. troops under General George Crook successfully duplicated Carson's feat. Cochise's band was pacified by 1872, but other warring bands under the war leaders Victorio and Geronimo continued to operate sporadically. By 1873, the Apaches had had enough. Hunger had taken its toll, and the tribe, especially the women and children, were exhausted. The Apaches were unceremoniously installed on a nearby reservation.

By 1876, the war had returned. Led again by Victorio and Geronimo, the Apaches once more spread terror across the Southwest. The U.S. Army chased them into Mexico, but the raids continued. Crook finally hunted down the Apaches for a second time, pinning Geronimo in his mountain refuge in 1883 and forcing his surrender.

Geronimo's nomadic band remained only two years on the reservation before breaking out for a last attempt at freedom. In 1886, Geronimo was finally run to ground, this time by troops under General Nelson Miles. Geronimo and his tribesmen were transported first to Florida, then Oklahoma. The Apache Wars were finally over. By the end, the names of fierce Apache warriors like Geronimo, Cochise, Victorio, Mangas Coloradas, and Juh were known nationwide.

Michael S. Casey

See also: American Indian Wars

References and further reading:

Basso, Keith H., ed. *Western Apache Raiding and Warfare.* Tucson: University of Arizona Press, 1971.

Lummis, Charles F. *General Crook and the Apache Wars.* Flagstaff, AZ: Northland Press, 1966.

Reedstrom, Ernest Lisle. *Apache Wars: An Illustrated Battle History.* New York: Sterling Publishing, 1990.

Thrapp, Dan L. *Victorio and the Mimbres Apaches.* Norman: University of Oklahoma Press, 1974.

Appomattox Court House (9 April 1865)

Site of the surrender of General Robert Lee and the Army of Northern Virginia. When Robert E. Lee evacuated Richmond and Petersburg he had intended to move south, join with Joseph Johnston's army in North Carolina, defeat William Sherman's Union army, and then turn to face the forces of U. S. Grant. But Grant and the Army of the Potomac blocked the line south, and Lee was forced to move westward to escape Grant and to secure supplies.

On 7 April 1865, the two generals exchanged letters exploring possible terms of surrender. Lee still had hopes that, as he approached Appomattox Court House, his forces could arrive at Appomattox Station, about three miles away, first and thereby obtain vital supplies and an escape route south. Many southern soldiers had deserted or fallen out from the army from hunger, exhaustion, and defeatism.

That evening Lee saw campfires to the southwest, which meant that Philip Sheridan's cavalry had arrived first, and the Army of Northern Virginia had few viable options. On Sunday morning, 9 April, southern infantry sought to break through Sheridan's cavalry and failed; the cavalry were too numerous, and the remainder of the Army of the Potomac was beginning to arrive on the scene.

Lee nobly declined a suggestion from his subordinates that the army retreat to the Appalachian Mountains and engage in partisan warfare. He met that afternoon with Grant in the McLean family home in Appomattox Court House. (Paradoxically Mr. McLean had moved to this supposedly peaceful corner of Virginia to escape the war.) Grant offered generous surrender terms, and had his supply officers arrange to feed southern troops and their animals; Lee then surrendered and the fighting in Virginia ended. Several weeks later, Joseph Johnston would surrender to William Sherman in North Carolina, and the Civil War was over.

Charles M. Dobbs

See also: American Civil War; Grant, Ulysses Simpson; Lee, Robert Edward

References and further reading:

Calkins, Chris M. *The Appomattox Campaign: March 29–April 9, 1865.* Conshohocken, PA: Combined Books, 1997.

Korn, Jerry. *Pursuit to Appomattox: The Last Battles.* Alexandria, VA: Time-Life Books, 1989.

Rodick, Burleigh Cushing. *Appomattox: The Last Campaign.* New York: Philosophical Library, 1965.

Arabi Pasha (Ahmad Urabi Pasha) (1839–1911)

An Egyptian army officer who led a revolt against foreign domination of Egypt. He became the embodiment of the growing spirit of Egyptian nationalism.

At age 17, Arabi joined an army in which Turks, Circassians, and Albanians held almost all senior positions; in 1880, there were only eight native Egyptian officers. The army was torn by strife because military administration was in the firm grip of foreigners, and promotions were almost impossible for Egyptians. Similarly, foreigners controlled civil administration. Europeans drew comfortable, regular annual salaries, while the bureaucracy labored under salary arrears.

Egypt was near bankrupt in 1876 due to khedive Ismail's ambitious development program and wasteful spending. To safeguard their financial investment in Egypt, the French and British forced Ismail's deposition in 1879, and replaced him with the ineffectual Tawfiq.

Before his deposition, Ismail had planned to increase the army to 60,000 men. The *firman* (Ottoman decree) that made Tawfiq khedive stipulated a reduction to 18,000, but the Anglo-French controllers insisted on further reduction, resulting in wholesale retrenchment of officers and men, mostly Egyptians.

The disaffected military elements organized under the leadership of Colonels Ahmad Arabi, Ali Fahmi, and Sami al-Barudi. In 1879, they drafted a manifesto to the khedive repudiating foreign influence, guaranteeing payment of debts, and demanding the return of railroad revenues to Egypt. Radicalized over time, Arabi and his colleagues demanded the right to debate the Egyptian budget in Parliament. The unwillingness of the European powers to grant this demand led to an impasse.

Riots broke out in Lower Egypt, and to restore order, preserve the European bondholders, and protect the Suez Canal (a European investment), the British bombarded Alexandria from 11 to 14 July 1882, landed an expeditionary force of 10,000 soldiers, and defeated Egyptian troops at the Battle of Tel-el-Kebir. Arabi and the leaders were court-martialed, tried, convicted, and on 26 December 1882, put on a train at the Kasr el-Nil barracks bound for Port Suez, en route to exile in Ceylon (Sri Lanka).

The British would remain a dominant military force in Egypt until the early 1950s.

Edmund Abaka

See also: Tel-el-Kebir

References and further reading:

Afigbo, A. E., et al. *The Making of Modern Africa.* Vol. 1, *The Nineteenth Century.* Harlow, Essex, UK: Longman, 1986.

July, Robert W. *A History of the African People.* Prospect Heights, IL: Waveland Press, 1998.

Ardennes, Battle of (Belgium, 16–26 December 1944)

Last major German offensive on the western front during World War II. Nearly bedridden from the effects of the 20 July 1944 assassination plot, Adolf Hitler planned to regain the initiative against the Anglo-American forces on the western front. At the least, he hoped to stall the Anglo-American advance in time to fortify the West Wall; at best, he wanted to capture Antwerp and encircle a large number of enemy troops. Hitler believed that 30 divisions on the east front against the Russians would not matter, but in the west they could prove decisive.

The German offensive began on 16 December 1944 in the Ardennes Forest when 200,000 German soldiers of the newly created Sixth Panzer Army and the Fifth Panzer Army struck an area defended by 80,000 American troops. Dressed as American military police, Otto Skorzeny's German commandos spread confusion throughout the American rear areas. The following day, troops of the 1st SS Panzer Division murdered unarmed American prisoners of war at Malmedy. Allied commander General Dwight Eisenhower grasped the immediacy of the situation, and ordered all reserves into the "Bulge." After holding a conference with his commanders, he ordered General Patton to drive his Third U.S. Army north to strike the southern pincer of the German advance.

After overrunning the outlying positions, the Germans encountered the stiff defense of the 101st U.S. Airborne Division in the crossroads town of Bastogne, Belgium. Muddy roads and insufficient supplies of gasoline hampered the German advance. On 20 December the weather cleared, allowing the Americans to utilize their superior airpower to destroy German tanks. Against the advice of his field commanders, Hitler did not permit retreat. On 26 December elements of Patton's Third Army arrived to break the German siege of Bastogne. Over the next several days, the American army drove the Germans back. A major Russian offensive launched on 12 January 1945 ended any possibility that the Germans would renew their offensive.

In all, the Germans suffered over 80,000 and the Americans 75,000 casualties; the loss of more than 300 tanks, however, proved the most devastating to the German army's ability to defend the Fatherland.

Gregory Dehler

See also: Eisenhower, Dwight David; Hitler, Adolf; Patton, George Smith, Jr.; Rundstedt, Karl Rudolph Gerd von; World War II

References and further reading:

Cole, Hugh M. *The Ardennes: The Battle of the Bulge.* Washington, DC: Government Printing Office, 1965.

Merriam, Robert E. *The Battle of the Bulge.* New York: Ballantine Books, 1947.

Argentine Dirty War (1976–1983)

Internal clash involving radical left factions and the military governments of Argentina. Argentina experienced internal strife as early as May 1969 when leftist groups clashed with government troops and police in the city of Cordoba. Some of the groups, such as the Montoneros, whose military wings included the Peronist Armed Forces and the Revolutionary Armed Forces, were loyal to former president Juan Domingo Peron. Another group was the Trotskyite People's Revolutionary Army. The radicals carried out a number of bank robberies, kidnappings, and murders as they sought to gain attention and support for their cause. The most famous of the early incidents was the kidnapping and murder of former president Pedro Aramburu. Peron returned to Argentina in 1973, was elected president, and the violence briefly subsided. His death in 1974 and the ineptness of his successor, his widow, Isabel Peron, opened the door to renewed strife.

One of the Madres de Plaza de Mayo ("Mothers of the Disappeared") wearing a photograph of her son, who disappeared during the years of political oppression in Argentina. Buenos Aires, 1981. (Owen Franken/Corbis)

When General Jorge Rafael Videla became president by overthrowing Isabel Peron in 1976, the antiguerrilla units of Argentine armed forces were unleashed. In a two-year period Videla's regime used kidnapping, torture, imprisonment, rape, and exile as weapons to wipe out not only suspected terrorists but all political and intellectual opposition. Many victims simply disappeared without a trace. International attention was focused on the Dirty War by the Mothers of the Plaza de Mayo who assembled weekly in the central plaza of Buenos Aires to protest the disappearance of their children and grandchildren.

The Argentine military was discredited by its disastrous performance (except for the naval air arm) in the Falklands/Malvinas War of 1982. Argentina returned to civilian rule in 1983 with the election of Raul Alfonson. Under Alfonson, the National Commission on Disappeared People investigated the Dirty War, and some of the military officials responsible for the atrocities were later prosecuted. It is estimated that more than 13,000 people lost their lives at the hands of military governments during the Dirty War, while the radical left killed approximately 200.

George M. Lauderbaugh

See also: Falkland Islands War

References and further reading:

Anderson, Martin Edwin. *Dossier Secreto: Argentina's Desaparecidos and the Myth of the Dirty War.* Boulder, CO: Westview Press, 1993.

Hodges, Donald. *Argentina's Dirty War.* Austin: University of Texas Press, 1991.

Armies: Organization and Tactics

The earliest wars have no written record; nor does the organization of the participants or their tactics. Fortifications at Jericho are evidence that the inhabitants needed to defend the city against attackers, but the details of the assaulting party are unknown to us. It is not until the middle of the second millennium B.C.E. that we begin to see the details of warfare.

Armies around 1500 B.C.E. consisted of masses of infantry, unprotected in terms of body armor, and armed variously with spears, axes, swords, or shields. This body of men was probably drawn from the poor element of society; in addition there were more specialist foot troops with ballistic weapons—archers and slingers. Finally, there was the elite of these armies: the charioteers, carried in horse-drawn, wheeled carts, armed again with ballistic weapons, the precursors of the present-day armored forces.

Chariots appeared in Sumer in about 2500 B.C.E., and were dominant in war until about 1200 B.C.E. Egyptian charioteers were armed with bows, Hittite charioteers with spears, and Assyrian charioteers—the real experts—used light vehicles for archers and heavier chariots for spear men. These were then superseded by cavalry, formed from the lesser nobles who had enough money to buy horses and to arm themselves. Cavalry took the measure of chariots because of one simple factor: greater mobility.

The organization of the warring groups—for they were not yet armies—was related purely to social standing, with chariot owners (the aristocracy) having the most say in deciding where and when to engage the enemy—or, by their ineptitude, when to run in the face of enemy surprise or superior force.

Tactics of the time revolved around the fact that operations of war were no more than raids upon neighbors (whether tribes or states) in which an area of land was overrun. The enemy, if defeated, was slaughtered to a man, settlements were looted and burned, and slaves taken. There was little or no political motive at this time, just the elemental emotions of fear, envy, and greed.

Battles were primitive but some tactics seemed to work, and these were naturally repeated. Armies began a battle by approaching each other, with the infantry in the center and the chariots or cavalry on the flanks. Sometimes the infantry just clashed with their opposite numbers, and the slaughter began. On other occasions the mobile troops would advance through their infantry and attack the enemy infantry, often leading to the attacked infantry breaking, and the slaughter beginning again.

But warfare was a haphazard business until the Egyptians and then the Assyrians began to impose organization and discipline. One man in particular stands out: Philip II of Macedon (father of Alexander the Great). He created the first combined arms force in history, within which his heavy missile engines coordinated with cavalry and infantry to give a battle-winning formula. He also invented the light infantryman, or *hypaspist,* who was a disciplined soldier on the lines of the heavy infantry hoplite, but who combined this discipline with the mobility and flexibility of the irregulars.

At this time artillery appeared, and catapults and ballistas had accurate ranges of perhaps 200 yards, with a maximum range of 500 yards. Such siege engines also had their more mobile counterparts, when it became apparent that artillery could be used against men as well as walls. Defensively, walls of ever-increasing strength became necessary, creating the first strategic arms races. Naturally measures were developed specifically for use in siege warfare, such as siege towers, mural hooks, and battering rams.

Slowly but surely the various components of military force began to appear, and with the arrival of the Romans, military organization gained acceptability, for it won battles. Maneuver had already shown its worth, initially by chance, but later by design; however, the problem still remained of how to train the enormous mass of the Greek-type phalanx to turn, for its very size made it difficult to maneuver. The answer came in the Roman legion, designed to allow ease of movement in rough country.

Lessons taught by Hannibal and learned by Scipio soon refined the art of maneuver for a force that held sway throughout the civilized Western world for centuries. The heart of Roman tactics came from the discipline of the legion—a force consisting of officers who knew their job, seasoned veterans who always fought well, and a leavening of new recruits who learned from those who stood at either side of them in battle.

The legions were originally made up of Roman citizens who were well trained, highly disciplined, and commanded by men who always took the offensive, if possible. They were almost always accompanied by legionary cavalry, which doubled as a reconnaissance force and as shock troops.

The Romans left a legacy of the national army. In some cases this was perpetuated, but there were still roving tribal bands, such as those of Genghis Khan, who practiced mobile warfare by means of mounted archers. The horse had enormous effect upon tactics in the first two millennia of the Christian era. Mounted bowmen evolved into mounted knights, whose armor and sheer weight on the battlefield were often a winning combination, although the Battle of Crécy (1346) saw a remarkable event.

The British earl Marshal Warwick was faced by the French on their advance not far from Abbeville. Warwick held his infantry in the center, his archers on his flanks. Heavy cavalry were held in reserve. The whole consisted of some 20,000 men. Against them was ranged nearly 60,000 French, including 12,000 heavy cavalry, 6,000 Genoese crossbowmen, and 17,000 light cavalry.

The battle started with the Genoese firing at the British, but their bolts fell short. The British archers then replied and destroyed the Genoese line with a storm of cloth yard arrows. Into the retreating Genoese moved the ponderous French heavy cavalry, who soon came within range of the longbowmen. Some 15 or 16 times, despite tremendous losses, the French cavalry continued to advance into the British fire. The result was utter defeat for the French, losing at least one-third of their total force. The British casualties amounted to about 200 dead and wounded.

From earliest times fortifications have been important, and no more so than during the period in Europe following the demise of the Romans. Wooden and then stone defenses sprang up everywhere, partly as a sign of increased social standing, but also as local strong points. The armies of the medieval period were raised by local barons who had the right to levy troops when needed, but the castles had permanent defenders raised from the immediate retinue of the local potentate.

The problems facing the attacker were simple: how to breach the walls or how to starve out the defenders. Engineering techniques developed whereby trenches were dug under castle walls to collapse the stone, and many subterfuges were practiced to either gain entrance or deprive the defenders of the will to fight. Sieges were part of the military art, and even the trench warfare of World War I can be seen as little more than a prolonged siege.

The feudal system allowed local lords to raise armies from their subjects, but one disadvantage to the state was that the lords also had to pay these armies, which they were loath to do at times of harvest. Further, winter warfare was never popular, and most troops returned home at the end of the campaigning period. National armies, however, emerged in the seventeenth century, although there were isolated examples well before this.

As society developed and prosperity increased, more money was available from taxation to fund permanent armed forces. They began to be organized, normally on a territorial basis. At the same time firearms became more and more effective, although pikemen, archers, and similarly armed troops were still present on the battlefield until the late eighteenth century.

At the same time, as cannon were seen to be very effective against fortifications, fortress designers increased the sophistication of the plans, giving defenders better fields of fire and creating many more impediments to attacking forces.

Warfare came into its modern phase with the operations carried out by Napoleon and those who eventually defeated him. Maneuver became the chosen method, rather than mere head-to-head conflict. Napoleon successfully attacked his opponents piecemeal, and his rate of movement was a continual advantage against the more ponderous armies he was pitted against.

Napoleon also benefited from earlier developments by Gustavus Adolphus of Sweden, who trained his men in flexible tactics; and artillery benefited from the work of Gribeauval, who designed field artillery carriages that were also extremely mobile. In the final act, however, Napoleon was defeated by tactics that would not have been out of place 500 years previously.

Napoleon appeared to have left a legacy that maneuver was supreme in tactics. In the Boer War the British army was consistently outmaneuvered by the Boers, but weight of numbers told in the end. World War I, however, despite many technical advances in small arms, artillery, organization, and tactics before the war started, and the subsequent development of aircraft and tanks, resembled nothing so much as a protracted siege in Europe. Only in the Middle East was there some maneuver, but misconceived plans often led to failure until the Gaza campaign.

On the western front a four-year stalemate was interrupted by frontal, near-suicidal attacks until the Germans revived mobile infantry warfare in 1918.

It is startling to realize that throughout the operations planned by Field Marshal Haig the aim was to penetrate the enemy lines and then let loose cavalry for the pursuit. When this strategy failed, Haig settled down to win a war of attrition in which the nation with the last man standing would be the victor.

The arrival of tanks and aircraft on the battlefield created a school of thought that saw that, properly combined, these two elements would be able to help mobile infantry break through enemy front lines and then exploit the rear areas, wherein lay the higher command centers, supply depots, and administrative elements of the modern field army.

The result of this thinking led the Germans to create a tactical air force and the panzer (armored) division. The combined effect of aerial bombardment in support of the ground forces, the penetrative effect of massed armor, and the backup of artillery made progress into the enemy rear areas relatively easy for the infantry, whose task was to support tanks against antitank weapons, but primarily to take and hold the ground, which none of the other elements of Blitzkrieg could do.

Germany's enemies learned from the early successes of the German army in World War II, and by 1944 the Allies were conducting a most sophisticated version of lightning war. Aircraft were on immediate call by ground forces, artillery fired tasks on request, tanks were supported by mobile infantry. The exceptional fighting machine that was General Patton's Third Army could only be stopped by fuel shortage, never by the inventors of the technique he practiced so ably.

Airborne warfare also came of age during this period, culminating in the air assault across the Rhine in 1945. In the east air resupply was fundamental to the Chindit operations, although German efforts to resupply the Sixth Army in Stalingrad failed. For such operations, command of the air is fundamental.

In the twentieth century, war had become three-dimensional, and operations by the Israelis in the Six-Day War (1967) show how important and decisive it is to eliminate the enemy air element. Similarly, in the Gulf War, the main threat to the coalition forces was the Iraqi air force, which surprisingly withdrew itself from combat. In the Falklands the most serious damage done to the British was with air-launched missiles from Argentinean aircraft.

The organization of armies has developed slowly, from the massed bands of prehistory via the legions of Rome, to the feudal forces with their time-outs for harvest and the winter. The size of national armies increased exponentially from the seventeenth century, culminating in the massive forces of World Wars I and II. Army groups, armies, and corps controlled divisions throughout the two world wars, and even in the Korean War large armies took part. The development of the air-land battle concept has, however, meant that smaller numbers can have much greater effect, especially in view of the increased power of air and land artillery (as well as supporting naval artillery when required).

In the third millennium it seems that most armies in NATO (North Atlantic Treaty Organization) will be engaged in low-intensity warfare, meaning that, with the aid of technology, fewer men will be able to do far more than their predecessors, who were always dependent upon sheer weight of numbers. This is undoubtedly as much the result of the exponentially rising costs of warfare and weapons as of strategic decisions.

David Westwood

See also: Armor, Ancient and Medieval; Armored Fighting Vehicles; Artillery

References and further reading:

Diagram Group. *Weapons.* Leicester, UK: Galley Press (WH Smith Group), 1980.

Gerber, J. ed. *Landkriegfuehrung.* Osnabruck, Germany: Biblio Verlag, 1992.

Gies, Francis, and Joseph Gies. *Cathedral, Forge and Waterwheel.* New York: HarperCollins, 1994.

Hale, J. R., et al. *War and Society.* 5 vols. Stroud, UK: Fontana Paperbacks, 1985.

Middeldorf, Eike. *Taktik im Russlandfeldzug.* Frankfurt am Main: Mittler und Sohn, 1957.

Armor, Ancient and Medieval

In the early times, protection for soldiers was often limited to a helmet and shield with little use of armor to protect the body. The Standard of Ur, circa 2900 B.C.E., depicts Sumerian soldiers wearing pointed helmets probably of leather, and kilts and cloaks, the latter reinforced with circular metal plates. At roughly the same time, Egyptian soldiers were depicted in figurines, temple carvings, monuments, and other sources wearing only kilts and protected by long shields of animal hide with flat bottoms and pointed tops. The pharaohs were depicted helmeted in the Blue Crown of Egypt, or in the Crowns of Upper and Lower Egypt. Assyrians in the eighth century B.C.E., here following the lead of the Hittites, discovered the utility of iron for weaponry and armor. Their armor consisted of helmets and chain mail, or mail of overlapping iron scales.

Early Greek plate armor from the fifteenth century B.C.E. has been found at Dendra, near Mycenae, consisting of four or five large segments of bronze covering the lower face and reaching to the knees, with articulated shoulder pieces. Some sources in the Homeric and other periods refer to gold armor, but most armor was of bronze. By the ninth century B.C.E., Greek soldiers wore brass "bell" corselets, greaves, and Illyrian-style helmets, but by the eighth century B.C.E., the *hoplon* (which gave the heavy foot soldiers of Greece their name *hoplites*) had evolved. This was a large round shield with a wooden core and bronze facing that remained in use through the classical period (fifth and early fourth centuries B.C.E.).

Hoplites wore bronze "muscled" cuirasses or, more commonly, a *linothorax,* that is, a cuirass made of layers of linen or canvas glued together, and sometimes reinforced with metal plates or scales. The linothorax was cheaper, and its lighter weight allowed more freedom of movement. Hoplites also wore bronze greaves that protected the knees and calves. They covered their heads with bronze helmets of various styles, often with a decorative horsehair crest. Spartan soldiers also wore a characteristic red cloak that was laid aside in battle. Greek soldiers are sometimes depicted going into battle naked, protected only by a shield and helmet.

Early Roman armor (prior to the third century B.C.E.) was provided by each individual soldier, and varied from chain mail to breastplates, or square plates worn across the back and chest. Greaves were sometimes worn, and soldiers also carried a *scutum,* a four-foot-long shield of planks glued together with a binding of iron and covered in leather, with a wooden, bronze, or iron boss. Helmets were of bronze, and often of Etruscan or Greek style. Poorer soldiers were defended only by wicker shields, and perhaps a helmet.

By the second century B.C.E., Marius had reformed the legion, and each legionary was issued a mailed cuirass of either closed rings or open rings that were riveted shut. The groin was afforded extra protection by a belt, the *cingulum,* complete with a dangling apron reinforced by round metal plates. Soldiers wore bronze helmets, often of the Montefortino type, with hinged cheekpieces. Greaves remained in use only by centurions.

The chain-mail cuirass continued in use into the first century of the empire and beyond. For Roman soldiers, it was replaced by the *lorica segmentata,* an iron-segmented armor covering the shoulders, chest, and torso. It remained in use until the third century. Auxiliary troops continued to use chain mail, and muscled cuirasses were often seen, especially among officers, the Pretorian Guards, and wealthy officials. Toward the end of the second century, shields became more rounded, and continued in that shape until the end of the empire. By the time of Constantine the Great (d. 337), scale armor came into use among cuirassed Roman cavalry troops, the *clibanarii,* though it is unclear to what extent armor, other than the oval shield, was used among late imperial infantry.

Early medieval armor derived from armor of the late imperial period. In the eastern empire, soldiers were clad in chain mail and protected by round shields; cuirassed cavalry continued on armed horses as the *cataphractarii.* In the West, the barbarians had adopted Roman styles of armor, mixing it with their own, with scale armor of bronze, iron, or horn. Soldiers wore *spangenhelms* of iron, bronze, or horn plates attached to iron or bronze straps radiating from an apical point, with a nasal protecting the upper face. Soldiers also carried round or oval shields.

In the tenth and eleventh centuries, oblong Viking or Norman "kite-shields" were widely used, as were conical helmets fashioned of one piece of metal (rather than segmented like the spangenhelm), with nasals to protect the nose and face. Horned or winged Viking helmets have never been found. A mail coat (*byrnie*) falling to the knees was worn, and a chain-mail *coif* protected head and neck underneath the helmet.

By the twelfth century, while chain mail continued in use, the arm and hand came to be completely covered, and the coif was attached to the body armor. This type of armor came to be called a *hauberk,* rather than a *byrnie.* The helmet came to cover the entire head, with two eye-slits and small breathing holes in front. Unrecognizable in this armor, knights began to employ heraldic symbols so as to recognize friend and foe. The shield, no longer needed to cover the exposed face, became shorter and more triangular. It too was covered with heraldic symbols.

Chain mail, or scale armor, was used almost exclusively into the fourteenth century. Plate armor began to be used first to cover the chest, elbows, and knees. It was often riveted to strong fabric or leather, or tied together by leather straps. Later the entire body, including the arms and legs, were covered. Chain mail continued in use beneath the plate armor. Fully articulated armor covering the entire body made the shield superfluous, and it fell out of use in battle, though targes were used in tournaments. Helmet styles changed considerably, and various styles were used simultaneously, with the chain-mail coif often attached to the helmet itself. A hinged visor was attached to the helmet, which could be brought down to protect the face during combat, but kept up for better vision, and ventilation, when not needed.

By the late fifteenth and early sixteenth centuries, armor had become more ornament than protection. Artillery and handguns, as well as improved projectile weapons such as the crossbow, had rendered armor ineffectual. By then the helmet and breastplate were all the armor that remained in use on the battlefield, though highly ornamental tournament armor also remained.

Michael C. Paul

See also: Firearms; Macedonian Wars; Roman Republic, Wars of the; Viking Raids

References and further reading:

Connolly, Peter. *Greece and Rome at War.* Englewood Cliffs, NJ: Prentice Hall, 1981.

Nickel, Helmut. *Warriors and Worthies: Arms and Armor through the Ages.* New York: Atheneum, 1969.

Nicolle, David. *Arms & Armour of the Crusading Era, 1050–1350.* 2 vols. London: Greenhill Books, 1999.

Oakeshott, R. Ewart. *The Archaeology of Weapons: Arms and Armour from Prehistory to the Age of Chivalry.* Woodbridge, Suffolk, UK: Boydell Press, 1994.

Snodgress, Anthony. *Early Greek Armour and Weapons.* Edinburgh, Scotland: Edinburgh University Press, 1964.

Armored Fighting Vehicles

The armored fighting vehicle includes both tanks and other vehicles, but it is with the tank that the story begins. Stalemate in World War I had led to thousands upon thousands of deaths. Infantry were unable to penetrate barbed-wire obstacles protected by rifle, machine gun, and artillery fire, and the British in particular were desperate for a new weapon to break through the wire and its supporting weapons. The aim was to penetrate the German lines, get into the rear areas, and then, astonishingly, release the cavalry to mop up.

The first tanks (so called as a code name) were sent into operation near Bapaume on 15 September 1916. Only a few were capable even of getting into battle, and the results were not remarkable. However, surprisingly, this relatively ineffective beginning did spark further interest in the British army hierarchy, and on 20 November 1917 some 200 tanks broke through the German Second Army's front, closely followed by infantry. The assault gained six miles, but as there were no armor reserves and few infantry to back up the breakthrough, the battle ended, as so many in that war, with partial withdrawal and then continued stalemate.

However, after World War I a few enterprising officers in Britain, France, and Germany saw the tank for what it really was: a highly mobile weapon that, if used in concentration, could smash through defensive positions in a way infantry alone could not. Further, it could break out into the enemy rear area and disrupt command and supply organizations. If also supported by discriminating artillery fire and well-trained infantry (to eliminate antitank threats), the tank could well be the weapon of the future.

Military establishments by their nature are conservative, and none was more so than the British army in the 1920s and 1930s. Cavalrymen saw themselves as the social superiors of mere infantry, and tank men were no better than mechanics. For this reason, tanks in the British army rarely fared well, and even to the end of World War II British tanks were undergunned (with the exception of the Firefly). For such minds, tanks were almost always seen as aids to infantry, and only in the western desert were they occasionally used properly.

Basil Liddell Hart and J. F. C. Fuller were the main proponents of the tank in Britain in the 1930s, but they were hardly heeded. However, in Germany there were many avid readers of their works, including Heinz Guderian, who was instrumental in creating and using the German panzer arm. He saw the tanks as the spearhead of the future's mobile war. With air support, plus ground artillery, infantry, and combat engineers, the Germans brought the new scenario into being. Germany had lost World War I and was thus somewhat more willing than the victorious French and British to consider new weapons and tactics.

Surprisingly, until 1941 most German tanks were much lighter than their opponents' armored fighting vehicles (AFVs). It was Germany's tactical use that provided its early successes. The tanks operated as divisions, supported *by* infantry, rather than in small numbers, operating in support *of* infantry. The tanks thus brought a new pace onto the western front and into the North African desert. The same tactics were used in Russia, and only when the Russians learned to coordinate and control their operations did they begin to win tank battles against the Germans.

In the early part of the war the Red Army lost seven tanks for every German tank lost. Parity was not achieved until as late as 1944. But in 1943 the biggest tank battle the world has ever seen took place around Kursk in Russia. The Germans sought to cut the Soviet salient off and destroy the Russians with attacks from north and south. Delayed by Hitler, the attack took place only after the Soviets had prepared what was to be a deathtrap for the Germans. In the south, three SS panzer divisions attacked, and were, as in the north, ground to pieces against the Red Army's defenses and tanks. The end of the battle was also the end of German hopes for victory in the east. U.S. Army tank development got off to a late start, and the Americans had to make do with the Sherman (M-4), an adequate vehicle with the advantage of mechanical reliability and a powered turret, but which was always undergunned.

Since the war, the tank-versus-tank tactical problem has seen many technological solutions; but for every answer in the military art, there is soon another question. Tanks are mobile, but they need to be reliable. Tanks themselves are good tank-killers, but are vulnerable to enemy antitank weapons, so they need better defenses. Weight increases demand increased engine power; increased weapons performance requires either better crews or more crewmen.

In World War II antitank aircraft appeared, and they proved very effective. The Germans had the Ju-87 armed with 3.7 cm guns, and then the Henschel Hs 130B-2, similarly armed. The Allies flew against German armor in the Falaise pocket in France with rocket-armed fighter aircraft. In the last two decades tank-killing helicopters have arrived, as well as the American A-10 antitank aircraft, armed with the formidable 30 mm Gatling-type rotary cannon.

The post–World War II years have seen tank battles on any scale only in the Middle East, with Israeli armor proving superior in nearly every instance. In the Gulf War (1991), Iraqi armor was so outclassed that the opposing coalition forces, singly and en masse, were totally victorious in each engagement.

Alongside the tank, there have been developments in armored reconnaissance vehicles, armored personnel carriers, and various other protected command, engineering, and support vehicles. Artillery is now armored, and the Multiple Launch Rocket System (MLRS) is fired from a tracked armored vehicle. The main battle tank may eventually prove an anachronism, like the battleship before it, but this will only happen when tanks are generally seen to be ineffective. Even in peacekeeping roles there is great morale effect in seeing a Challenger II or an Abrams tank supporting infantry, and for this reason alone it is doubtful if the end of the tank is yet in sight.

David Westwood

See also: Artillery; Infantry

References and further reading:

Gelbart, Marsh. *Tanks: Main Battle and Light Tanks.* London: Brassey, 1996.

Glantz, David M., and Jonathan M. House. *The Battle of Kursk.* Lawrence: University of Kansas Press, 1999.

Kens, Karlheinz, and Heinz J. Nowarra. *Die deutschen Flugzeuge 1933–1945.* Munich: Lehmann, 1968.

Macksey, K., and J. H. Batchelor. *Tank.* London: Macdonald, 1970.

Arms Control

Limiting, usually by treaty, the number and types of nuclear weapons. As a result of the destructive power of atomic bombs, arms control became an increasingly important means of preventing nuclear war, particularly after the start of the Cold War and the subsequent development of thermonuclear weapons.

Arms control is a concept distinct from disarmament in that arms control efforts seek to reduce, but not necessarily eliminate, certain weapons. Disarmament seeks the elimination of all weapons of a specified type. In addition to the quantitative goal of reducing the numbers of nuclear weapons, arms control efforts served as an effective diplomatic communications tool, especially during the Cold War. Arms control agreements, particularly those between the United States and the Soviet Union, communicated the mutual goal of survival. The willingness to enter into arms control agreements continually emphasized the goal of avoiding nuclear war.

Arms control evolved with the twentieth century. Beginning with the Hague Conferences of 1899 and 1907, when arms control became recognized by international law, and continuing through the Treaty of Versailles and founding of the League of Nations, diplomatic efforts to control the number and types of weapons became the primary means of avoiding war. Aerial bombardment and the use of poison gas were prohibited in the earlier treaties. And the reasoning of the Versailles Treaty was that a Germany without arms could not start another war. The Washington Conference of 1922 provided for the scrapping of more battleships than had been lost in all the naval engagements to date. Yet World War I saw the extensive use of poison gas, World War II brought the horrors of aerial bombardment to scores of cities in Europe and Asia, and even the Washington Treaty battleship destruction paved the way for the far more destructive aircraft carrier. There were also attempts to outlaw the submarine, primarily because Germany had used this weapon with such success.

With the advent of nuclear weapons, the primary focus of arms control efforts shifted from controlling conventional weapons to restricting weapons of mass destruction. If nuclear weapons could be controlled, then the potential to wage a nuclear war would be lessened as well.

Because all countries are concerned with their survival, national security considerations dictate arms control agreements. Such agreements, to be effective, must be negotiated directly between nations, freely derived, and protect the rights and interests of the negotiating countries. Arms control agreements can prevent war, slow down an arms race, ease taxpayers' burdens, and, most importantly, promote trust between countries when negotiated in this fashion. Trust between countries, developed through the process of confidence building, is particularly effective in making strong arms control agreements. Early attempts to control nuclear weapons, such as the Acheson-Lilienthal and the Baruch plans, failed in large part because of national security interests. The Soviet Union, for instance, would not accept these early control plans because it did not yet possess nuclear weapons. Once the Soviet Union developed nuclear weapons, it was in a better security position to think about arms control measures. However, the intensity of the Cold War militated against such measures. Hence, the period beginning with the end of World War II and lasting to 1960 was marked by great hostility and reluctance to enter into any arms limitation agreements.

The Cuban Missile Crisis changed superpower relations. The very real threat of a nuclear exchange started both the United States and the Soviet Union down a productive arms control path beginning with the installation of the hot line and a defense early warning system. These confidence-building measures led in no small part to the first major arms control agreement, the Partial Test Ban Treaty, which was signed in 1963. A decade later, the ABM (Antiballistic Missile) Treaty and the Strategic Arms Limitation (SALT) I Interim Agreement came into being. SALT I was particularly significant in that it was the first agreement limiting strategic weapons. Although never ratified by the United States Senate, the follow-on SALT II agreement was followed in practice and led to cuts in the strategic arsenals of both the superpowers. In 1987, the INF (intermediate-range nuclear forces) Treaty eliminated shorter-range missile systems, the first and only agreement to eliminate an entire category of weapons. In the late 1980s and early 1990s two new arms control agreements were negotiated: Strategic Arms Reduction Talks (START) I and II. These treaties provided for deep cuts across the entire spectrum of strategic nuclear weapons in both the United States and the former Soviet Union and set the stage for current nuclear arms control activities. The success in negotiating these treaties came about in large measure because of reduced tensions between the two superpowers. As the Cold War wound down, both the United States and the Soviet Union found it less and less necessary to maintain their mammoth stockpiles of nuclear weapons. Reduced political tensions drove arms control success in the Reagan and Bush years. When the Soviet Union broke up, this need, although not eradicated, lessened even further. Today, stockpiles of nuclear weapons remain, but in reduced numbers, with further reductions possible, one of the most promising developments since 1945.

With the end of the Cold War and the reduction of the number and types of nuclear weapons, arms control activities have turned from concentrating almost exclusively on such weapons to an increasing emphasis on chemical and biological weapons, which have the potential to be weapons of mass destruction. Such weapons are easy to construct and deliver and represent a growing threat, particularly because of the existence of rogue states and an increase in terrorism. Nuclear-weapons-production activities require an extensive infrastructure, making them easy to monitor by technical surveillance. Chemical and biological weapons do not. Such weapons and their production facilities can be easily hidden and disguised as innocent chemical facilities. Recognizing the destructive power of chemical weapons, the Chemical Weapons Convention was chartered in 1997 as an international legal body dedicated to ending all activity related to chemical weapons production. Similar work has been accomplished in the field of biological weapons, with the creation of the Biological Weapons Convention in 1972. Much more work needs to be done in this area.

The goal of arms control is the reduction, not necessarily the elimination, of weapons of mass destruction. Disarmament is the goal of the United Nations, but arms control efforts are important steps in this direction.

Roger A. Meade

See also: Disarmament

References and further reading:

Adler, Emanuel, ed. *The International Practice of Arms Control.* Baltimore: Johns Hopkins University Press, 1991.

Goldblat, Jozef. *Arms Control: A Guide to Negotiations and Agreements.* London: Sage Publications, 1994.

Arnhem (1944)

A failed Allied airborne operation. In the summer of 1944, as the Allied advance slowed to a crawl, the Allies launched the largest airborne operation of the war, to capture a series of key bridges across the Rhine River. The focal point of this operation of the war was the road bridge at Arnhem. With the capture of this bridge, the Allies could flank the Western Wall and advance in the Ruhr Valley. Operation MARKET GARDEN, under the command of Field Marshal Bernard Montgomery, called for a number of airborne drops. The British 1st Airborne Division was charged with the capture of the Arnhem bridge. Unlike most airborne attacks, the 1st Airborne Division would not drop on the target itself, but rather would drop eight miles away and advance as regular infantry. They were then supposed to hold the bridge until a linkup with American and British troops.

A somewhat rushed operation, MARKET GARDEN saw a failing of intelligence gathering. The 1st Airborne Division fell directly on the II SS Panzer Corps and faced much stiffer opposition than expected. The British forces fought valiantly, but suffered from a lack of heavy weapons and the failure of airdropped supplies to reach the troops. After 10 days, the surviving British troops were evacuated. Of the 10,000 British troops who dropped on Arnhem, only 2,000 escaped. Montgomery later claimed that if he had been given all the resources he had requested, MARKET GARDEN would have worked and perhaps shortened the war. Arnhem pointed out the shortcomings of airborne troops, as they lacked the heavy weapons and fire support that regular infantry troops enjoyed.

Drew Philip Halévy

See also: Airborne Operations; Montgomery, Bernard Law; World War II

References and further reading:

Ryan, Cornelius. *A Bridge Too Far.* New York: Simon & Schuster, 1974.

Weinberg, Gerhard L. *A World at Arms.* New York: Cambridge University Press, 1994.

Arnold, Benedict (1741–1801)

Colonial American field commander, tactical genius, and traitor. Arnold was born into a prominent family in Norwich, Connecticut, on 14 January 1741. He deserted both militia units to which he was assigned in the French and Indian War (perhaps revealing some of his future lack of character). From 1762 until 1775 he was a successful merchant and smuggler. He brought a company of Massachusetts militia into the Boston campaign on 29 April 1775 and was soon promoted to colonel. Despite the protests of Ethan Allen's Green Mountain Boys, who considered him too arrogant, Arnold accompanied Allen during the capture of Fort Ticonderoga on 10 May. As leader of the doomed 1775–1776 expedition to Quebec, he was promoted to brigadier general in January 1776. His tactics stalled the British at Valcour Island on 11 October and Split Rock on 13 October during the Lake Champlain campaign, and at Danbury, Connecticut, on 23–28 April 1777.

British lieutenant colonel Barry St. Leger, leading an expedition of regulars, Tories, and Mohawks from Oswego, New York, toward Albany that summer, threatened Peter Gansevoort at Fort Stanwix. Although American brigadier general Nicholas Herkimer stopped the Tories and Mohawks on 6 August at Oriskany, he suffered 75 percent casualties. Arnold, by now a major general, rescued Gansevoort on 23 August by a clever combination of tactics and guile.

Arnold's actions at Freeman's Farm on 19 September and Bemis Heights on 7 October won the Saratoga campaign for Horatio Gates, but Gates did not sufficiently appreciate or reward his skill and heroism.

After recovering from a serious leg wound suffered at Saratoga, Arnold became military governor of Philadelphia, where he married a Tory, Peggy Shippen, in April 1779. Court-martialed for consorting with the enemy, he was convicted and reprimanded in January 1780, but still retained Washington's trust. He plotted with his wife, Major John André, and General Sir Henry Clinton to desert to the British and deliver West Point to them, but was discovered and had to flee on 25 September. André was hanged as a spy on 2 October. As a commissioned British brigadier general, Arnold had minor successes in the 1781 Virginia campaign; committed several atrocities in his New London, Connecticut, campaign; then retired to London in December, where he died on 14 June 1801. Despite the unconvincing efforts of some "revisionist" historians, Arnold's reputation has remained so odious through the intervening centuries that the very name "Benedict" is rarely given in the United States.

Eric v. d. Luft

See also: Allen, Ethan; American Revolution; Fort Ticonderoga; French and Indian War; Gates, Horatio; Greene, Nathanael; Knox, Henry; Quebec, Battle of; Saratoga; Washington, George

The attack of Benedict Arnold's men in Quebec. (From the collections of Library of Congress)

References and further reading:

Boylan, Brian Richard. *Benedict Arnold: The Dark Eagle.* New York: Norton, 1973.

Brandt, Clare. *The Man in the Mirror: A Life of Benedict Arnold.* New York: Random House, 1994.

Martin, James Kirby. *Benedict Arnold, Revolutionary Hero: An American Warrior Reconsidered.* New York: New York University Press, 1997.

Randall, Willard Sterne. *Benedict Arnold: Patriot and Traitor.* New York: Morrow, 1990.

Aroghee, Battle of (10 April 1868)

The key event of the British expedition to Abyssinia (modern-day Ethiopia). Theodore III, the Coptic Christian emperor of Abyssinia, had become emotionally unstable and erratic after the death of his first wife. Theodore had started as a petty Abyssinian chieftain named Kassa. He claimed royal blood and direct lineage to Solomon and Alexander the Great. This was not true, but he crowned himself Emperor Theodore of Ethiopia in 1855 after overrunning the territory of rival chieftains. His rule started out well and he reformed his country's legal and administrative policies while competing with Egypt and other local powers for trade and political influence. He considered himself a crusader against the Muslims. Theodore was a man of personal charm, a ruler of talent, and a courageous warrior, but the ignorance and superstition of nineteenth-century Abyssinia surrounded him. He was a bestial, half-demented madman when drunk, which was often after the death of his first wife.

In 1862, Theodore wished to conclude a formal treaty with Queen Victoria—as one monarch to another. He sent a personal letter to Her Majesty that the Foreign Office read, filed, and failed to answer. After waiting two years for an answer, Theodore imprisoned his two British consuls along with other resident Europeans, a total of 58 hostages. A diplomatic cat-and-mouse game followed for three and a half years as Theodore alternately released and rejailed his hostages according to his mood swings, keeping them from leaving the country. Finally, after numerous exchanges of diplomatic letters, some signed by Victoria herself, the dignity of the British Empire demanded military action.

In August 1867, London gave orders for war. Lieutenant General Sir Robert Napier, commander of the Bombay Army of India, brought an Anglo-Indian army to the Red Sea port of Zula. Napier's army consisted of 44 elephants to carry the artillery, a portable railroad, telegraph lines, fresh water condensers, well-drilling equipment, hospital ships, and construction equipment for port infrastructure at Zula. Other modern equipment included new Snider-Enfield rifles, 6-pounder Hales's War Rockets, breach-loading 12-pounder cannon, steel 7-pounder mountain guns, and photography cameras. The famous explorer Henry Morton Stanley, the press, and military observers went along.

Sir Robert's force cornered Theodore deep in Abyssinia's interior on the Aroghee Plain, before the mountain fortress of Magdala. Of a total force of 37,000, a strike force of 5,000 made it to the Aroghee Plain. They faced Theodore's army of 7,000 warriors and some outmoded artillery. Theodore opened the battle with ineffective artillery and his battle force charged down from Magdala to attack the British. Napier's force quickly killed 700 and wounded 1,200 of the attackers, killing many of Theodore's battle chiefs. The survivors were driven off the Aroghee Plain with minimal British losses.

After this defeat, Theodore attempted suicide, released some hostages, and tried to bargain. Napier refused further bargaining attempts. On 13 April, the British shelled Magdala and seized it by a coup de main. Theodore shot himself—with a pistol originally given him as a gift from Queen Victoria—and his troops surrendered. All hostages were released and Stanley concluded, "Though a little war, it was a great campaign."

Thomas D. Morgan

See also: British Military, Twentieth-Century Organization and Structure

References and further reading:

Chandler, David G. "The Expedition to Abyssinia 1867–8." In *Victorian Military Campaigns,* ed. Brian Bond. New York: Frederick A. Praeger, 1967.

Holland, T. J. Major, and H. M. Hozier. *Record of the Expedition to Abyssinia.* Vols. 1 and 2. London: Her Majesty's Stationery Office, 1879.

Morehead, Alan. *The Blue Nile.* New York: Random House, 1962.

Arsuf, Battle of (Palestine, 7 September 1191)

Solidified military reputation of Richard I. After securing Acre, King Richard sought to regain Jerusalem. On 22 August 1191, his army, consisting mainly of the forces that he had brought from England, along with Hospitallers, Templars, and a French contingent, marched from Acre. It moved south along the coast road. This enabled Richard to protect one flank while being supplied by his fleet. Saladin's army traveled a parallel route and harassed the crusaders.

By 30 August 1191 skirmishes between the two armies had become more frequent. The crusaders continued their march, and Richard rode up and down the road encouraging his army. Eventually Saladin forced the issue. He chose Arsuf, a wide plain with forests on both flanks, for a battle that began on the morning of 7 September 1191.

The crusader first line was composed of archers and crossbowmen, backed by infantry. The knights made up a third line. Knights Templar and Hospitallers held the right and left flanks. Richard held the center.

Around midmorning, Muslim infantry made an initial foray, which was followed up by continuous waves of Turkish horse archers. Despite the constant barrage of arrows, in which the left flank suffered the most, Richard was able to maintain his position.

The master of the Hospitallers asked to charge. Richard refused as he waited for more Turks to commit to battle, but the Hospitallers' discipline broke and they charged the enemy. The rest of the knights followed.

The heavy cavalry routed the Muslims. The victory enhanced Richard's prestige, but Arsuf was not decisive. Saladin still had an army and blocked Richard's march on Jerusalem.

Timothy May

See also: Armor, Ancient and Medieval; Crusades; Frederick I Barbarossa; Hattin, Battle of; Ramleh, Battle of; Richard I; Saladin

References and further reading:

Holt, P. M. *The Age of the Crusades.* London: Longman Group, 1990.

Mayer, Hans Eberhard. *The Crusades.* Oxford, UK: Oxford University Press, 1988.

Riley-Smith, Jonathan. *The Crusades: A Short History.* New Haven, CT: Yale University Press, 1987.

Runciman, Steven. *A History of the Crusades.* 3 vols. Cambridge, UK: Cambridge University Press, 1951–1954.

Art in War

War art functions on two distinct but related levels. In wartime, state-sponsored images of war, such as propaganda posters, operate within a defined set of goals calculated to create emotions on the home front favorable to a war effort. These can serve to dehumanize the enemy and engender hatred, distort truth and reality, promote patriotism, and encourage recruitment. Napoleon saw art as a mechanism for manipulating the minds of the French populace and as a means of legitimizing his regime. The most famous example of this is Gros's *Napoleon at Eylau,* which sought to establish him as a great leader.

Other forms of art operate during wartime on a more mundane level, such as Norman Wilkinson's "dazzle" painting as camouflage on ships during World War I. Wars also create unique commercial opportunities for printmakers, an example being Currier and Ives, which capitalized on public fascination with the American Civil War by producing numerous lithographs of the battles and the leaders. Frequently at odds with this kind of forced imagery are pictures by frontline soldiers themselves that paint a considerably different perspective of war, one usually not nearly so positive as that in the popular realm.

However, it is on the second level, in the art produced after the events, where we see the greatest visual evocation of war. This postwar art is the product of many different ideals and motivations, some of which mirror the wartime creations. Memory and recollection, national pride, class, racism, colonialism, political legitimacy, and commercial enterprise have all influenced the creation of war art at one time or another. In ancient times, the depiction of the leader as military hero served to define his or her rule and power over the masses. The larger-than-life representations of the pharaohs Rameses II and Seti towering over their enemies on the walls of their palaces in Egypt sent a clear message to their people as well as to any perspective enemies, as did *Trajan's Column in Rome,* which recounted that emperor's victories in the Dacian War. Louis XIV employed artists to highlight his martial successes in the late seventeenth century, while modern states have used images of war to cement their position and create national memory. Examples include John Trumbull's portrayal of battles and scenes of the American Revolution around the Rotunda of the Capitol in Washington, D.C., the Galerie de Batailles at Versailles created by Louis Philippe to enhance his position and ambition as natural successor to Napoleon, and Daniel Maclise's massive murals of Trafalgar and Waterloo on the walls of the Houses of Parliament in London, images befitting the Victorian colonial imagination. Similarly, Benjamin West's *Death of Wolfe* epitomized the supremacy of Britain and the power of the ruling class. Other artists have sought to condemn war through their art, the classic examples being Jacques Callot's *Miseries of War,* Goya's images of the Spanish massacres, Picasso's *Guernica,* and Otto Dix's vivid and cathartic images of the hell of the Great War.

In the twentieth century some of the most lasting war art has come, perhaps improbably, from official war artists, commissioned by their respective governments and taken into the military. These artists realized that the traditional heroic images of thundering cavalry charges and heaps of enemy dead in tasteful arrangement would be rejected by the contemporary public, and thus they concentrated on the everyday lives of the soldiers, even behind-the-scenes depictions of military engineers or supply lines, and on individual soldiers or small groups. Even Nazi war art emphasized individual or small-unit determination, rather than the movement's vile ideology. The Japanese seemed to be the only official war artists still to use triumphalist themes, showing, for example, long lines of disheveled, surrendered Allied troops early in the war, marching off to become "guests of the emperor." Despite the dominance of still and motion-picture photography, the world's major armies still give an important place to war art. Whether the work of official or individual artists, war art continues to document and interpret humankind's closest approach to hell.

Peter Harrington

See also: Film and War; Psychological Operations

References and further reading:

Harrington, Peter. *British Artists and War: The Face of Battle in Paintings and Prints 1700–1914.* London: Greenhill, 1993.

Paret, Peter. *Imagined Battles: Reflections of War in European Art.* Chapel Hill: University of North Carolina Press, 1997.

Perlmutter, David D. *Visions of War: Picturing Warfare from the Stone Age to the Cyber Age.* New York: St. Martin's Press, 1999.

Artillery

One of the most effective and long-lasting weapons in military history. The earliest history of cannon and gunpowder is not certain, but cannon are first mentioned in a Florentine decree of 1326, and Walter de Milimete wrote in the same year of an early hand cannon. Again, in 1327 Edward III used "crackys of war" against the Scots, but these were probably more fireworks than explosive devices. Certainly cannon were used by the English against the French at the Battle of Crécy in 1346, the French having had cannon since at least 1338.

Antipersonnel use of cannon was superseded by the need for a means of defeating the extremely thick walled defenses of castles and other strong points of the time. This need led to the creation of the siege gun, an altogether cumbersome and weighty device. Tests had shown that thin barrels exploded when used, and so thick barrels were forged for these pieces.

Some early artillery pieces were actually breech-loaded, but the difficulty in obtaining a tight breech seal led to the majority of artillery designs being for muzzle-loading weapons. Because the whole barrel could then be cast as one piece, there were fewer weaknesses than in breech-loading systems. Breech-loaders, however, had a higher rate of fire, for the breech could be loaded as a separate part, and fitted to the gun when needed. So, a supply of breeches effectively became cartridges, loaded into the gun when needed, while empty breeches could be loaded away from the gun.

Early projectiles were arrows, and only later was shot (often round stone) fired. Stone was effective against fortifications, but gunners began to examine antipersonnel shells.

This consisted of a number of stones (the "langridge" shell) or of balls of iron or lead ("case" shot), which were loaded as a whole into the gun, and which burst on being fired at infantry or cavalry.

Artillery and its use were much more an art than a science in the early days of the weapon, but it was soon realized that by putting wheels on guns they were more mobile, and by 1542 Ralphe Hogg was producing cast guns in Sussex. Henry VIII then created the post of master gunner, assisted by 12 gunners, based at the Tower of London. Artillery was coming of age, but the process was slow. Only in 1716 were guns, transport, drivers, and horses assembled, rather than being made up of civilians called upon when the need arose.

Artillery was still, however, a luxury; further, the absence of good roads and the heavy guns meant that artillery often failed to arrive in time for the battle. Even if it did, it took so long to get into position and prepare to fire that the guns were almost always ineffective tactically. The gunners persisted, however, and by 1680 gun limbers, a Prussian invention, were introduced into England. Limbers ensured that the gun was properly maneuverable, having a double set of wheels rather than merely the wheels upon which the gun was mounted. This in turn meant that more than one horse could be harnessed to pull the gun.

By 1700, ballistic tables were giving the performance to be expected from artillery pieces. However, despite the improved mobility of the guns following the introduction of the limber, artillery could still not be guaranteed to turn up in time for the battle. Siege trains consisted of about 100 guns, 60 mortars, and more than 3,000 wagons and 15,000 horses, plus gunners, drivers, and the myriad other hangers-on. And the length of the column was some 15 miles.

By 1770, however, it was realized that guns needed to be lighter to be more mobile, and by 1855 artillery had come of age. Guns were light in both weight and weight of shot, but a combination of the new mobility and infantry-artillery cooperation led to great advances in the science of gunnery. Certainly skillful use of field and horse artillery (field artillery is of a heavier caliber) and the esprit de corps of the British gunners in the face of repeated French cavalry attacks helped in gaining the victory over Napoleon.

The nineteenth century saw breech-loading guns, time fuses, air-burst shells, and, above all, rifled guns. By the beginning of the twentieth century, most European armies used breech-loading rifled artillery in a range of calibers. The mortar made a rejuvenated appearance in the form of the Howitzer—a gun designed to fire at high angles of fire, allowing the shell to penetrate deeply for greater effect, or to allow indirect fire over height obstacles between gun and target.

Field artillery, artillery that fought with the infantry, also benefited from the French development of pneumatic recoil systems. Previously the whole of the gun had recoiled on firing—barrel, trunnion, wheels. Now only the barrel recoiled, and was returned to the original firing position by means of the pneumatic recoil system. This made gunnery easier, and increased the rate of fire, especially when at full recoil the breech was opened automatically and the empty shell case ejected.

By the start of World War I in Europe, all armies knew the value of artillery and had a range of guns for different purposes. Howitzers fired at high angles to overcome walls and other fortifications, field artillery shot at enemy infantry and gun positions, coastal artillery took on shipping, and the guns were even moving in the mountains. But on the western front artillery and the machine gun proved supreme for most of the war. Shells were a problem, however, for as the mud deepened in France, the shells just sank below the surface and either created a shower of mud or simply did not explode.

The stalemate of World War I led to some careful thinking among gunners, and the outbreak of World War II should have seen artillery operating with complete maneuverability, able to put down fire at a moment's notice anywhere within range. In fact the guns were once more horse-drawn, although there were some few effective artillery tractors. The Germans, however, went a stage further in their thinking, and used their Junkers Ju-87 dive-bomber aircraft as artillery, the revelation that, in concert with their imaginative use of tanks, brought about defeats for Poland, France, and Great Britain on the Continent.

In Russia the effect of blitzkrieg seemed to be leading to defeat in 1941 and 1942, but the Soviet army, despite enormous losses of men and materiel, were the artillerymen of that war. They had lost some 40,000 guns and 60,000 mortars in the second half of 1941, and had concluded that the organization and tactics they (and the Germans) were using did not let artillery realize the power it possessed, so it concentrated artillery at army level and above. This meant that if a regimental commander needed artillery support he could (theoretically) have available to him the fire of up to two artillery divisions, a possible total of 72 or more heavy guns.

The Germans, in desperation at the losses sustained in tanks, began to install artillery into tank chassis, often using captured chassis for the purpose. This self-propelled artillery served well, and all the nations in the West adopted the principle. American self-propelled artillery gained a deserved reputation for being there with the shell whenever required. In the war against the Japanese, artillery was often used for bunker-busting at point-blank range.

The second half of the twentieth century saw a number of developments that were first used during World War II. One was the continued use of aircraft as artillery—what else is, after all, a bombing attack? Aircraft are used nowadays as flying guns and antitank weapons, as are helicopters. Rocketry, which started the whole process, evolved through the Russian *katyusha* and the German Nebelwerfer into the Multiple Launch Rocket System (MLRS), which proved of extreme value in the Gulf War in 1991. Artillery proper still plays a significant role in conventional and unconventional war.

Artillery has always been regarded as one of the three fighting arms, and there is little doubt that it will continue to be so for the foreseeable future. Aircraft and helicopters are extremely valuable antitank and antipersonnel weapons, but they do not have the endurance on the battlefield of the gun. Further, they are far more vulnerable to light antiaircraft fire than the gun is to small-arms fire. The artillery duels of the two world wars may never be fought again, but the ability to use field artillery on call is something no experienced battlefield commander will willingly dispense with.

David Westwood

See also: Armored Fighting Vehicles; Infantry

References and further reading:

Blackmore, H. L. *The Armouries of the Tower of London.* Vol. 1, *Ordnance.* London: Her Majesty's Stationery Office, 1976.

Chamberlain, P., and T. Gander. *Heavy Artillery.* London: Macdonald and Jane's, 1975.

Hogg, Ian V. *Illustrated Encyclopedia of Artillery.* London: Hutchinson, 1987.

Aryan Conquest of India (c. 1500 B.C.E.)

Invasions between 1500 and 1000 B.C.E. of the collapsing Harappan civilization by waves of Indo-European-speaking Aryans, leading to the emergence of India proper. The nomadic Aryans (nobles) migrated into the Punjab of west India and then into the Ganges River plain of north India, subjugating the Harappans into the Hindu caste system. The conquests described in the Vedic epics brought the Sanskrit language, Hinduism, horses, cattle, and chariots to India.

According to the Rig Veda, Harappan (*dasa*) warfare centered on citadels like Mohenjo-Daro and Harappa, from which infantry sallied forth armed with copper and bronze swords, axes, spears, and bows. The Aryans used chariot mobility and bow fire to defeat the Harappa in open battle. Aryan tribes moved slowly from the Afghan mountain passes onto the Punjab flatlands, deploying their *grama* or wagon trains in a *samgrama* or circular war formation to protect them against marauding Harappan forces until the Aryan chariots could be put together and employed with devastating effect against the Harappans in the field. Although Harappan forces organized by village might number between 30,000 and 100,000, against as little as a few thousand Aryans organized by tribal units in grama formations, it appears Aryans rarely failed in open battle. In one undated battle, Aryan chariots completely crush and push the Harappan militia into the river, a favorite tactic employed by the Aryans.

Harappans eventually stayed inside their earth-and-timber-walled citadels (*puras*), but well-armed Aryan infantry utilized fire arrows and thunderous battering rams to break through. In some cases waterways were also diverted against Harappa strongholds. The Harappa king, Shambara, stiffly resisted Aryan expansion but eventually lost more than 100 citadels to the Puru Aryan tribe and was captured and thrown off a mountain. After the appearance of Aryan iron weapons further resistance by the Harappans was futile. In fact, initial Aryan expansion was slowed not so much by Harappan resistance as by the huge size of India, and by further waves of Aryans who then fought for control with earlier Aryan invaders.

The Battle of Ten Kings saw newly arrived Aryan tribes conquer existing Aryan tribes like the Puru. King Sumbasa led the heavily armored knights of the new Aryans against the chariots of older Aryan groups. Attempts to divert water channels against Sumbasa failed, and he crushed the existing Aryans, took seven of their citadels, and usurped their power in the Punjab.

Such infighting among the Aryans greatly delayed their eventual domination of India. It was the Hindu castes, and not warfare, that allowed for Aryan domination of west and north India by 600 B.C.E. in the form of small, Aryan, Gangetic kingdoms. This Hindu caste system was made up of incasts (priests, warriors, merchants, laborers) and outcasts (conquered Harappans). The warrior *varna,* or *kshatriya* caste, formed the hereditary military arm of Aryan society and developed the ability to incorporate invading groups into its eclectic military system during the Vedic period of 1500–600 B.C.E.

Christopher Howell

See also: Alexander's Wars of Conquest; Aryans; Chandragupta Maurya; Mauryan Empire, Conquests of

References and further reading:

Griffith, Ralph, trans. *Hinduism: The Rig Veda.* Vol. 5, Sacred Writings Series, ed. J. Pelikan. New York: Quality Paperback Books, 1992.

Kulke, H., and D. Rothermund. *A History of India.* New York: Barnes & Noble Books, 1986.

Singh, Sarva Daman. *Ancient Indian Warfare.* Delhi: Motilal Banarsidass Publishers, 1997. Reprint.

Aryans

A prehistoric people, also called Indo-Europeans, putatively ancestral to major Old World civilizations, whose existence has been extrapolated from the linguistic, literary, and cultural evidence of modern populations. Most reconstructions of Indo-European society based on linguistic evidence and on the earliest texts of Indic, Iranian, and Greco-Roman civilizations portray a patriarchal, tripartite social system of commoners, priests, and a war-oriented aristocracy. The prominence of horse- and wheeled-vehicle terms in the reconstructed common language is among the evidence for attributing to Indo-European speakers early mastery of horse-based warfare and wheeled mobility.

While language development is plausibly recoverable based on more than two centuries of scholarship, no substantive archeological remains are universally attributed nor can their original area of occupation be defined with any certainty. If the Aryans did exist as a definable entity from approximately 4500 to 2000 B.C.E., military significance would plausibly include extensive cultural transformation throughout Europe and southwestern Asia, horse-based military elites in "daughter" cultures, and perhaps a more militarized and patriarchal civilization in the West.

The apparent need to explain their success in the absence of material relics led to depictions of the Aryans as conquerors who swept from an unknown homeland to become an enduring elite in much of Eurasia. This hypothesis intersected with nineteenth-century political and social developments to produce the popular, scholarly, and political cult of *Aryanism* and the imagined superiority of their descendants, which was a core tenet of the racialist and militaristic ideology of the Third Reich.

Much modern scholarship continues to attribute the expansion of Aryan culture to traits enabling greater success in war, but a minority view instead associates Indo-European languages with the Neolithic agricultural revolution and resulting population growth. While a plausible explanation for the dramatic expansion of one of many prehistoric language families, the latter requires a time depth not generally supported by linguistic analysis and does not address evidence of post-Neolithic cultural change.

Anne L. Angstadt

See also: Ancient Warfare; Animals in War; Aryan Conquest of India; Dorian Invasion; Sea Peoples

References and further reading:

Childe, V. Gordon. *The Aryans.* New York: Dorset Press, 1987. Reprint.

Gimbutas, Marija. *The Civilization of the Goddess.* New York: HarperCollins, 1991.

Mallory, J. P. *In Search of the Indo-Europeans.* New York: Thames & Hudson, 1989.

Renfrew, Colin. *Archeology & Language: The Puzzle of Indo-European Origins.* New York: Cambridge University Press, 1988.

Ashurnasirpal II (r. 883–859 B.C.E.)

King who forged Assyria into one of the dominant powers in the Near East. Though his annals are not in a good state of preservation, it is apparent that Ashurnasirpal II (Akkadian, Ashur-nasir-apli) campaigned continuously during his reign, directing his efforts to the north against the Aramean states of Bit Zamani, Bit Adini, and Neo-Hittite states located in Anatolia. All of these states, in one form or another, became vassals to Assyria. Moreover, Ashurnasirpal established a long line of fortresses to protect Assyrian trade routes.

At some point during his reign, Ashurnasirpal II crossed the Euphrates River and had his army ceremoniously wash their weapons in the Mediterranean Sea, and collected luxury items, exotic goods, and even native troops from the Neo-Hittite state of Carchemish, as well as the Phoenician and Aramean states. Ashurnasirpal even used many deportees from these campaigns to populate the new Assyrian capital of Kalhu. Ashurnasirpal's palace at Kalhu exhibits finished relief sculptures that are influenced by Neo-Hittite and Phoenician artistic forms.

Ashurnasirpal restructured the Assyrian state and army (which had been weak for centuries), created a large bureaucracy, continued a policy of deporting conquered peoples, and claims in his annals to have used psychological warfare on his enemies by performing public displays of cruelty, mass executions, and the burning of disloyal vassal cities.

Mark W. Chavalas

See also: Assyria

References and further reading:

Grayson, Albert Kirk, *Assyrian Royal Inscriptions.* Vol. 2. Wiesbaden, Germany: Otto Harrassowitz, 1976.

———. *Assyrian Rulers of the Early First Millennium BCE I (1184–859 BCE).* Toronto, ON: University of Toronto Press, 1991.

Liverani, Mario. *Studies on the Annals of Ashurnasirpal II.* Rome: Herder, 1982.

Paley, Samuel M. *King of the World: Ashur-nasir-pal II of Assyria 883–859 BCE.* New York: Brooklyn Museum, 1976.

Assaye

The first and most hard-fought victory of Arthur Wellesley (later duke of Wellington). Eighteenth-century India witnessed the rise and decay of the Maratha Empire. The Maratha War of 1775–1782 had resulted in a humiliating defeat of the East India Company's forces. The two major opponents of the British during the second Anglo Maratha War were Scindia, maharajah of Gwalior, and Bhoonsla, rajah of Berar. Arthur Wellesley's force was part of an army assembled on the northwestern Mysore border. By early August, negotiations with Scindia and Governor General Wellesley having failed, the latter moved against the two principal Maratha forces. After storming the city of Ahmednuggur, Wellesley's tiny army of 6,000 men advanced against Scindia's army of 50,000 men. (Scindia's infantry has been trained by European officers and was commanded by a German officer.) Early on 23 September, Wellesley received intelligence that the enemy were camped behind the Kaitna River, close to its junction with the River Jooee. He decided to outflank the Indian camp and crossed the Kaitna to deploy in the V formed by the two rivers. Maratha and Mysore allied cavalry (of dubious loyalty) were left to face Scindia's cavalry on the southern bank of the Kaitna River. Maratha's reactions were swift and Wellesley's attack sustained heavy casualties in their frontal advance against Indian artillery. But because the front narrowed as the two rivers flowed toward their confluence, Maratha's troops were able to deploy only a fraction of their strength. The British initiative was checked for three hours before a devastating volley from the 74th and 78th Foot routed the regular Indian infantry.

By 6 P.M., the battle was over, but Wellesley's crippled army could not pursue the fleeing Marathas. Wellesley was later to contend that Assaye was his greatest victory and proportionally the bloodiest action he had ever witnessed. His casualties amounted to one-third of his force. The defeat of the Maratha army caused Scindia to sue for peace, but hostilities continued until 1805.

Assaye revealed Wellesley's potential as a field tactician and he was soon recalled in Europe to deal with less exotic opponents who more directly threatened Great Britain.

Gilles Boué

See also: Maratha Wars; Wellington, Arthur Wellesley, Duke of

References and further reading:

Asquith, Stuart, and Charles Grant. *Wellington in India.* Upavon: CSG, 1995.

Weller, Jim. *Wellington in India.* London: Greenhill Books, 1993.

Assyria (c. 2000–612 B.C.E.)

The world's largest state during the early first millennium B.C.E. (c. 900–612 B.C.E.), ranging from Iran to Egypt and from Babylonia to Anatolia and the Mediterranean Sea. It originated from a Semitic-speaking people of what is today northern Iraq.

The early kings of Assyria (c. 1900–1750 B.C.E.) concentrated on mercantile activity (primarily with Anatolia), but apparently did not expand militarily. Military expansion began with Shalmaneser I (r. 1273–1244 B.C.E.), who went east and conquered Khanigalbat. Assyria again had command of trade routes leading to Syria and Anatolia, adding rich agricultural land and prosperous cities to its territory, and a large population for military use. However, Shalmaneser's successor, Tukulti-Ninurta I (r. 1244–1208 B.C.E.), attacked Babylon, a continuous military effort that diverted Assyrian energies and resources from the task of securing other areas, and thus Assyria declined. Assyrian military prestige was reasserted by Tiglath-Pileser I (r. 1115–1077 B.C.E.), who made campaigns across the Euphrates River to fight against the Arameans.

Within a century, the Aramean tribes had begun to infiltrate the Assyrian state, once again forcing the Assyrians into an offensive militarism in order to survive. Ashurnasirpal II (r. 884–859 B.C.E.) directed many of his campaigns against the Aramean and Neo-Hittite states east of Assyria, reaching the Mediterranean Sea and forming a defensive ring to protect Assyrian mercantile activity. His successor, Shalmaneser III (r. 858–824 B.C.E.), was beset with internal problems because of the difficulties of keeping together Ashurnasirpal's conquests. He crossed the Euphrates River on at least two occasions to meet a large coalition of Syro-Canaanite states (including Ahab of Israel and Hadadezer of Damascus), apparently fighting to a draw, although Assyria continued to receive tribute from these regions.

Once again, Assyria collapsed from the weight of keeping such an enormous state and did not expand again until the reign of Tiglath-Pileser III (r. 745–727 B.C.E.), who defeated Urartu to the north and ended the Aramean state of Damascus. The next major successor, Sargon II (r. 721–705 B.C.E.), also fought against a Syro-Canaanite coalition, and completed the conquest of Israelite Samaria, deporting many of its citizens. He fought battles against Elam (in southwestern Iran) and Babylon, and took the throne of Babylon for himself. His successor, Sennacherib (r. 705–681 B.C.E.), suffered through a major revolt throughout the empire, culminating in his campaign in Syria and Judea in 701 B.C.E. The Babylonians (Assyrian vassals in southern Mesopotamia) rebelled against Assyria at Sennacherib's accession, culminating in Babylon's utter destruction by the Assyrians in 689 B.C.E. Sennacherib's successor, Esarhaddon (r. 681–669 B.C.E.), became the first Assyrian king to enter and conquer Egypt as far as Thebes (671 B.C.E.). Ashurbanipal (r. 668–631 B.C.E.), however, lost Egypt (655 B.C.E.) and endured a devastating civil war (652–648 B.C.E.) and economic destruction (after 631 B.C.E.). With the empire in disarray, the capital city of Nineveh was destroyed by a combined force of Medes, Chaldeans, and northern tribes in 612 B.C.E., ending the Assyrian state.

The Assyrian army was composed primarily of native agricultural workers, hillsmen, and seminomads. Foreigners were employed as royal bodyguards and in later periods in

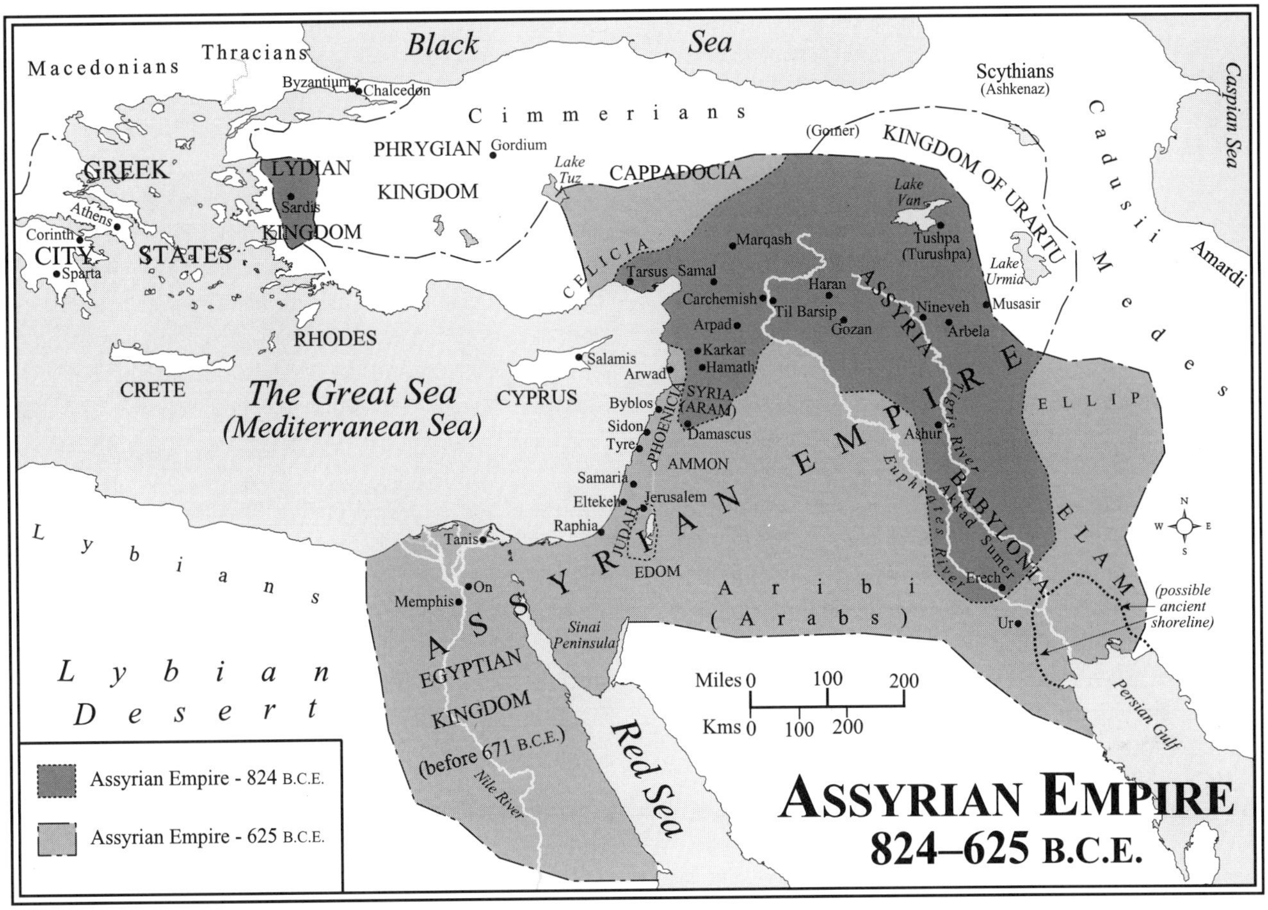

infantry units. Both Assyrian kings and military commanders are known to have led the army into the field of battle.

The military strength of the army was due in part to its chariot force. By the ninth century, the Assyrians used a wooden-framed vehicle with a rear-wheel axis, allowing for more maneuverability. It contained an archer/lancer, a driver, and a shield bearer. The chariots were used in the center of attack as shock troops. Horses were used to draw chariots and for light cavalry who rode bareback. Since horses were not in good supply in northern Iraq, the Assyrians often raided the Iranian plateau to replenish their stock. The compound bow—effective over 250–650 meters—was a major weapon. Individual armor, usually metal links sewn on leather, and simple helmets were used for defense.

The army was fed on the march with travel rations or at provincial centers that supplied food. In enemy territory the army was trained to live off the land. Communication between units was by cavalry and runners. Even mountainous territory did not deter the Assyrian army; the annals describe in detail the movement of the army on rafts across rivers, and carts were used to transport heavy equipment in mountainous terrain.

Siege techniques were often employed by the Assyrians. A ramp of piled-up earth was used to gain access to the upper walls of an enemy city or fortress. Battering rams were used to smash down gateways. Many of their sieges were very long; Israelite Samaria took nearly three years to overthrow.

Assyrian battle tactics were sophisticated, versatile, and anticipate classical Greek hoplite tactics. The annals describe a number of battles in detail. The Assyrians took advantage of topographic positions, tactical superiority, guerrilla warfare, and psychological terror tactics. The most enduring of their tactics was the policy of deportation, which caused as many as 5 million people to be uprooted during a period of three centuries (c. 900–612 B.C.E.).

Mark W. Chavalas

See also: Sargon of Akkad

References and further reading:

Oded, Bustenay. *Mass Deportations and Deportees in the Neo-Assyrian Empire.* Wiesbaden: Ludwig Reichert, 1979.

Postgate, Nicolas. *Taxation and Conscription in the Assyrian Empire.* Rome: Pontifical Biblical Institute, 1974.

Reade, Julian. "The Neo-Assyrian Court and Army: Evidence from the Sculptures." *Iraq* 34 (1972), 87–112.

Saggs, Harry F. "Assyrian Warfare in the Sargonid Period." *Iraq* 25 (1963), 145–154.

Scurlock, JoAnn. "Neo-Assyrian Battle Tactics." In *Crossing*

Boundaries and Linking Horizons, ed. Gordon Young, Mark W. Chavalas, and Richard Averbeck, 491–517. Bethesda, MD: CDL Press, 1997.

Ataturk, (Mustafa) Kemal (1881–1938)

First president of the Republic of Turkey, and hero of Gallipoli. Ataturk was born in Salonika in 1881, the son of a customs officer. Moving quickly through the Ottoman Empire's military schools (where he gained the nickname Kemal, or "perfection"), Mustafa, who had rejected religious training to become an officer, developed a keen appreciation for foreign modernization and progressive reform, especially after seeing his country lose territory to Italy in 1911 and to Greece in 1912. Although chief of staff to the Young Turk army, which had taken over Constantinople from reactionaries, his rivalry with Enver Pasha caused him to be sent to outposts like Syria before becoming military attaché to the embassy in Bulgaria, where he carefully studied European armies. At the outbreak of World War I, Mustafa advocated neutrality, but performed brilliantly, holding Gallipoli against British attack in April 1915 by personally inspiring the men with his bravery to hold their lines, even at great cost. Afterwards, he commanded a division on the Russian front, advancing to the rank of general, although he refused to take part in a German plan to attack Baghdad.

Mustafa Kemal Ataturk. (Library of Congress)

After the war, Mustafa did his best to encourage nationalist feelings and the secreting of weapons by the army, a plan that bore fruit as he helped to organize a rival legislature in Ankara and pushed this body to accept his National Pact, limiting the country to Turkish majority areas and a program of reform. In 1921, he was given supreme command of the Turkish army and successfully fought the Battle of Sakarya in August to push the Greeks out, followed by a coup de grace at Dumlupina in August 1922. In recognition of his role in the foundation of the Republic of Turkey as an independent nation, he was elected its first president in October 1923 and voted the name Ataturk, or "Father of Turkey." As a political leader, he advocated a program of literacy, women's rights, military and social Westernization, as well as secularism and state investment in infrastructure. (He even abolished the traditional fez male headgear as too redolent of the past.) He died, still in office, on 10 November 1938.

Margaret Sankey

See also: Greco-Turkish War; World War I

References and further reading:

Lewis, Bernard. *The Emergence of Modern Turkey.* Oxford, UK: Oxford University Press, 1968.
Tachau, Frank. *Kemal Ataturk.* New York: Chelsea House, 1987.
Volkan, Vamik, and Morman Itzkowitz. *The Immortal Ataturk.* Chicago: University of Chicago Press, 1984.

Atlanta, Battles Around (20–22 July 1864)

Battles that unhinged the Confederacy and paved the way for Sherman's "March to the Sea." In spring 1864, Union general William T. Sherman and Confederate general Joseph Johnston engaged in a classic and brilliant campaign of maneuver from the Tennessee-Georgia border to Atlanta. Johnston selected strong defensive positions in the mountains that ran roughly northeast to southwest, while Sherman mostly maneuvered by threatening to swing west and south and cut off Johnston's retreat. Sherman did test Johnston's defenses at Kennesaw Mountain and was repulsed with great loss of life. It took about 75 days for the Union armies to advance the approximately 100 miles between the two cities.

Sherman had divided his force into three for the final attack on Atlanta, and Johnston waited along Peachtree Creek (which was just north of the city) to strike at George Thomas's more isolated army. Before he could attack, Johnston was relieved, and John B. Hood attacked Thomas as the

Army of the Cumberland crossed Peachtree Creek on July 20. The attack was not well coordinated, Thomas's men held their good defensive positions, and Hood's men fell back.

Hood then retreated into Atlanta's inner defenses, and turned his attention to James McPherson's Army of the Tennessee on the Union left. Hood planned for an attack at daybreak on July 22, but Confederate infantry could not get to their assault positions until noon. Meanwhile, McPherson had sent reserves to strengthen his left, and the reserves held up the Confederate attack, which later threatened to break through only to be halted by massed Union artillery and the commitment of additional northern reserves.

When the battle ended, Hood realized he could no longer defend the city, and he sought to lure Sherman away by threatening to tear up the long northern supply line to Chattanooga and Nashville. Sherman refused to take the bait, and eventually decided on the famous March to the Sea to Savannah.

Charles M. Dobbs

See also: American Civil War; Hood, John Bell; Johnston, Joseph Eggleston; Sherman, William Tecumseh

References and further reading:

Carter, Samuel III. *The Siege of Atlanta, 1864.* New York: St. Martin's Press, 1973.

Castel, Albert. *Decision in the West: The Atlanta Campaign of 1864.* Lawrence: University of Kansas Press, 1992.

Kennett, Lee. *Marching through Georgia.* New York: HarperCollins, 1995.

Atomic Bomb, Development of

The most expensive and complicated weapon development through 1945. The development of the atomic bomb spans several decades and extends across continents, starting in the late nineteenth century with a series of scientific discoveries, later followed by several political developments. The initial, scientific phase took place in Europe, primarily in Great Britain and Germany. The discovery by physicist J. J. Thompson of the electron, a subatomic particle with a negative charge, began the process. Based on this discovery, James Rutherford conceived of the atom as a sphere of a positive charge, with an equal amount of electrons scattered throughout to balance it. This allowed him in 1919 to create oxygen from nitrogen by shifting the number of electrons, then to speculate about the existence of the neutron the following year. The existence of the neutron was confirmed in 1932. A year later, Hungarian-born physicist Leo Szilard realized that if an element could be forced to emit two neutrons when it swallowed one (instead of a simple exchange with another element), then fission might occur. This idea became the subject of a classified British patent in 1935. Other investigative strands included the discovery of artificial radioactivity by French and Italian research teams. In 1938, German scientists discovered that uranium could be fissioned.

The consequences of such a discovery were obvious: The huge amount of energy that could be released might lead to the devising of not only new power sources, but also weapons. In the United States, awareness of fission had been the subject of hundreds of articles, but there was only limited governmental involvement in such matters. Things changed when on 2 August 1939, Albert Einstein, informed by Leo Szilard that Nazi Germany had banned all exports of uranium, sent a letter to President Roosevelt warning him of the potential dangers a uranium-based bomb could pose. In light of the German threat, Einstein and Szilard asked the American leader to order the development of a counter-project.

The race to build the bomb thus became a contest between the United States and Nazi Germany, though neither nation was yet at war with the other. In Germany, the project was spearheaded by Werner Heisenberg, but the structure of Nazi bureaucracy and the budgetary appropriation process emphasized the need for weapons that could be developed quickly under war conditions. Thus, despite a clear interest in some circles, budgetary restrictions combined with certain theoretical mistakes slowed the German bomb effort. (Oddly, the German nuclear effort seemed to have centered around the need for a new method of electrical generation.)

The United States and the United Kingdom knew nothing of the slowed German pace. Consequently, committees were set up to study the feasibility of a bomb. The British MAUD committee (a code name later transformed into the acronym for Military Application of Uranium Detonation) concluded in 1941 that a uranium-based bomb was feasible. Meanwhile, in the United States Vannevar Bush, President Roosevelt's science adviser, led the Office of Scientific Research and Development (OSRD) until 1942 for the same purpose. He enlisted the assistance of several other high-level scientists, including Arthur Conant, future president of Harvard. The U.S. effort gained momentum when a copy of the MAUD report was obtained in 1942. That year, General Leslie R. Groves was appointed head of the Manhattan Engineer District (MED), later known as the Manhattan Project. In parallel, Arthur Compton oversaw the plutonium program, which, it was discovered, could have an even greater yield than uranium 235. In December, working in Chicago, scientist Enrico Fermi was to produce the first self-sustaining chain reaction.

The following year saw several developments at the political and military levels. President Roosevelt and British prime minister Winston Churchill agreed on atomic policy that included the transfer of some 35 British scientists to the United States to work on the atom bomb project. Leslie Groves established the Oak Ridge National Laboratory in Tennessee, where the manufacture of the fissile matter began. Throughout 1943 and 1944, several procedures were tested for proper design and implementation of a chain reaction, including the firing of a portion of uranium into another (later implemented on the "Little Boy" A-bomb dropped on Hiroshima) and the implosion concept for plutonium (used in "Fat Man," dropped on Nagasaki). The first test, however, was carried out in the New Mexico desert on 16 July 1945: The Trinity test was a success, and cleared the way for implementation of the atomic bomb operation.

President Truman, who had been informed of the Manhattan Project's existence 13 days after taking office in April 1945, had a special Interim Committee formed on 9 May 1945 to advise him on the use of this weaponry. On 25 July, the president directed General Carl Spaatz to have the specially trained 509th Composite Group, flying modified B-29 bombers, deliver its first bomb on or after August 3. On 6 and 9 August, B-29s *Enola Gay* and *Bok's Car* dropped bombs on Hiroshima and Nagasaki, respectively. The death toll exceeded 100,000 at the first target, while over 70,000 people died in the second bombing.

Guillaume de Syon

See also: World War II

References and further reading:

Alperovitz, Gar. *The Decision to Use the Atomic Bomb and the Architecture of an American Myth.* New York: Knopf, 1995.

Rhodes, Richard. *The Making of the Atomic Bomb.* New York: Simon & Schuster, 1986.

Sherwin, Martin J. *A World Destroyed: Hiroshima and the Origins of the Arms Race.* New York: Vintage Books, 1987.

Takaki, Ronald T. *Hiroshima: Why America Dropped the Atomic Bomb.* Boston: Little Brown & Co., 1995.

Attila the Hun (406?–453)

King of the Huns, invader of Europe with Germanic allies. At first ruled jointly with his elder brother Bleda, whom he murdered in 445. The brothers were quite unlike and had always detested each other. Attila, however nefarious, had the attributes of greatness, whereas Bleda's principal occupation, so says Priscus of Panium, a contemporary Greek writer, was laughing at his court buffoon, a grotesque Moorish dwarf. The brothers were members of a dynastic lineage that had united previously separated Hunnic groups around itself, together with many subject peoples (the majority Germanic) to create a substantial empire in central Europe, mostly north of the Danube. It is important to note here that Hunnic society was now much more sedentary than when described by Roman historian Ammianus Marcellinus several generations before, when the Huns avoided all permanent housing scrupulously.

Roman ambassadors, Priscus among their number, who tried to negotiate with Attila noted that even when gold was freely available the king himself still wore plain clothes, ate off wooden plates, and never touched bread. The ambassadors found Attila sullen, capricious, and arrogant, but as he was confronted with treachery on all sides, this moodiness is hardly surprising.

Under Attila's rule the Hunnic empire occupied an impressive area. In the north it extended to the Baltic, where, according to Priscus, "he ruled all the islands." It did not quite stretch to the Rhine, for the Franks and Burgundians lay in between, but Attila was said to rule "all Scythia." He delighted in war, but after he had ascended the throne, his head, rather than his arm, had achieved the conquests toward the north. During this period the Romans had successfully bought off their formidable neighbors, but now injudiciously allowed their payments of tribute to fall in arrears.

In 441 and 443, Attila invaded the Balkan provinces of the eastern empire, defeating the Roman armies with deplorable ease. In 447, favored by recent earthquakes that had devastated Asia Minor, he marched on Constantinople itself. The walls had suffered severe damage. Fortunately for the eastern empire, the fortifications were repaired and strengthened just before the arrival of the Huns. Attila turned aside and drove south into Greece, and was only checked at Thermopylai.

His next campaign was that of 451, when he turned west and invaded Gaul; he was defeated at Châlons by Roman and allied forces under Aëtius. In the spring of the following year, he invaded Italy, sacking several northern Italian cities (Aquileia at the tip of the Adriatic was utterly destroyed), and was compelled to withdraw, short of Ravenna, only by a combination of famine and pestilence. He clearly intended to invade east again in 453, but died unexpectedly during the night after his marriage to a young girl called Ildico.

Attila was a charismatic and powerful figure who demonstrated considerable ability as a general, but his successes were limited. He could lay waste with fire and sword the Thracian and Illyrian provinces, but he could not penetrate further into the empire. His campaigns were thus pursued in support of a diplomatic policy whose main aim seems to have been the extraction of vast sums of gold as blackmail. In 443, when Roman armies had failed to stem his advance, Attila's terms had had to be accepted. They

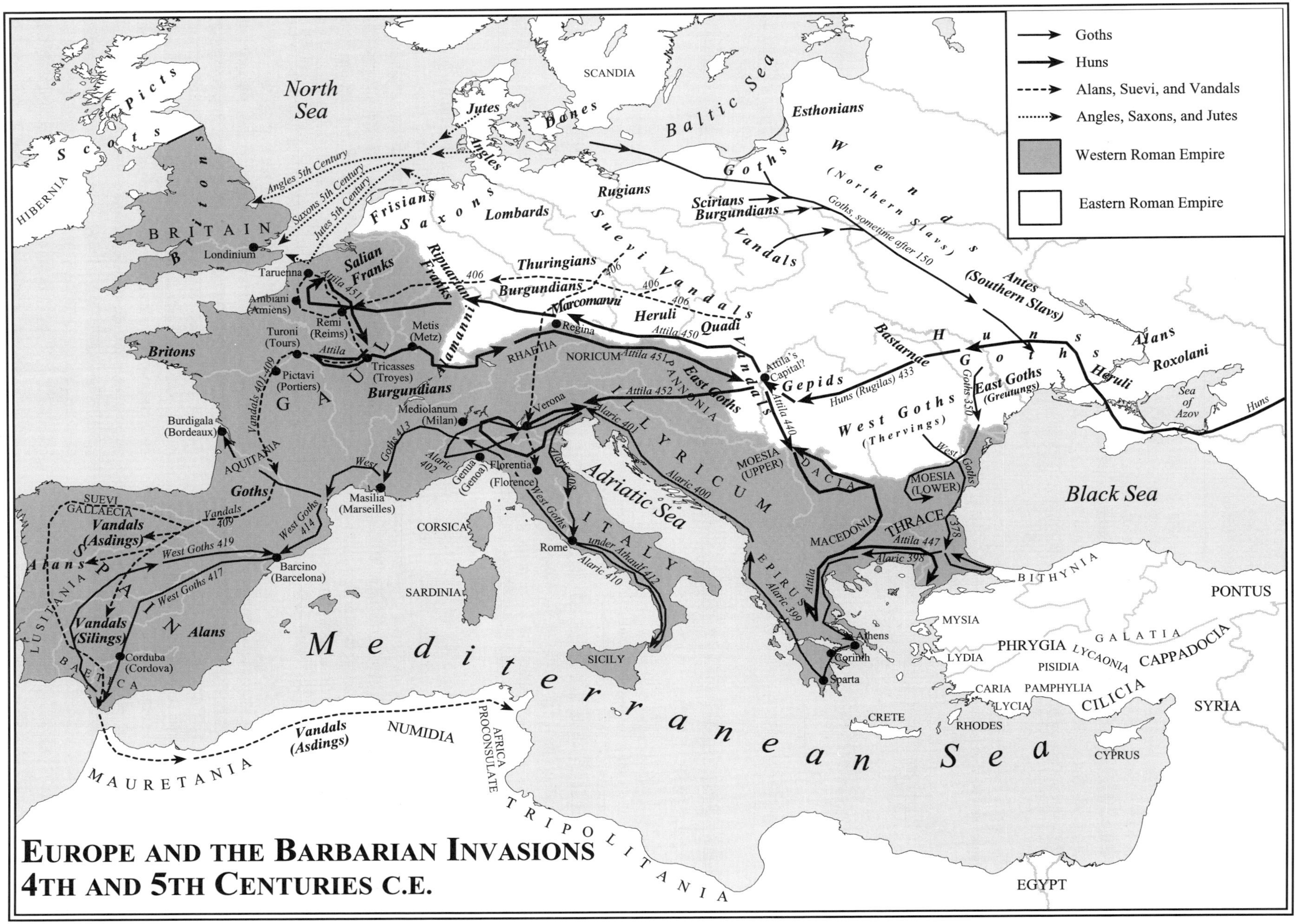

Europe and the Barbarian Invasions 4th and 5th Centuries C.E.

called for the immediate payment of 6,000 pounds of gold and future annual payments of 2,100 pounds of gold each.

On Attila's death, his realm was divided between his sons, but his Hunnic empire soon fell to pieces. The subject German peoples rebelled and defeated their overlords. The Huns broke up into small hordes and never regained the unity that had made them a serious menace to the Roman Empire, although they continued to be recognizable components of other steppe states into the sixth century and lineages continued to be traced back to the glory days of empire.

Nic Fields

See also: Aëtius, Flavius; Châlons, Battle of; Huns

References and further reading:

Gordon, C. D. *The Age of Attila: Fifth Century Byzantium and the Barbarians.* Ann Arbor: University of Michigan Press, 1960.

Maenchen-Helfen, Otto J. *The World of the Huns: Studies in Their History and Culture.* Berkeley, Los Angeles, and London: University of California Press, 1973.

Thompson, E. A. *The Huns.* Revised and with an afterword by Peter Heather. Oxford, UK, and Cambridge, MA: Blackwell Publishers, 1996.

Aurangzeb (1618–1707)

Aurangzeb, the third son of Shah Jahan and Mumtaz Mahal, was the emperor of Mogul India between 1658 and 1707. He was an experienced military commander and administrator, and prior to his ascension to the throne had served on important appointments as the governor of the Mogul Deccan for eight years, as governor of Gujarat for three years, and as the commander of the Mogul armies for the Central Asian invasions of Balkh and the first two sieges of Kandahar Fort. He rose to power through a dramatic civil war, wherein he imprisoned his father and sentenced his brothers, son, and nephew to death. His rule provided his European portrait through the publication of Bernier's *Travels* in 1670.

Following tradition, Aurangzeb retained Shahjahanabad Delhi as his capital for the first 20 years of his reign. In this period, he was also preoccupied with safeguarding the northwestern frontiers of the empire against the attacks of the Persians and Central Asian Turks. Meanwhile, deep within the Mogul territories, the Maratha chief Sivaji twice plundered the great imperial port at Surat in 1664 and 1670, with little resistance. When reconciliation and agreement with such factions failed, Aurangzeb decided to completely change his approach. In the next decade, the grand encampment or tent city became the movable capital of the empire for Aurangzeb. The emperor campaigned actively in parts of Rajasthan and the Deccan. He added the word *Alamgir* (world-seizer) to his titles, and dedicated himself to fostering a more conservative Islamic regime (versus the more liberal approach of his forefathers) and to an aggressive expansion of the empire's frontiers. The Hindus and the Marathas were no longer colleagues, but subordinates, and like several other southern Muslim kingdoms, were marked for annexation rather than containment.

Following this policy, Aurangzeb conquered the Deccan kingdoms of Bijapur and Golconda in 1686–1687. Also, he made efforts to quell the Maratha war, which included the capture and execution of Sivaji's son Sambhaji. Though the Maratha kingdom was broken up following these events, the guerrilla tactics of the Marathas had now spread throughout south India. The last years of Aurangzeb and his weary armies were therefore spent in laborious and fruitless sieges of countless forts in Maratha hill country.

Manu P. Sobti

See also: Mogul-Persian Wars

References and further reading:

Bhadra, Gautam. "Two Frontier Uprisings in Mughal India." In *Subaltern Studies II: Writings on South Asian History and Society,* ed. Ranjit Guha. Delhi: n.p., 1983.

Richards, John F. *The New Cambridge History of India: The Mughal Empire.* Cambridge, UK: Cambridge University Press, 1993.

Sarkar, Jadunath. *History of Aurangzib.* 5 vols. Calcutta: M. C. Sarkar, 1912–1924.

Aurelian, Lucius Domitus (214–275)

Coarse but successful Roman emperor who expelled barbarians, subdued rebellious provinces, and reunited the empire. Aurelian is believed to have been born at Moesia, in the northern Balkans, around 214. Attracted to a military life, he joined the Roman army and rose through the ranks by virtue of skill, bravery, and single-minded determination. He served as a high-ranking cavalry officer under the emperor Claudius II (Gothicus) and distinguished himself in several victories over the Goths. When Claudius died of the plague in 270, Aurelian was elevated to the purple by his troops. He then disposed of several minor contenders before hastily transferring forces to Italy and fending off large incursions by the Juthungi and Alamanni. Having routed the invaders, Aurelian fortified Rome by erecting the famous walls around that city. By now his ruthless nature and relentless emphasis on military discipline had given rise to the nickname Manu ad ferrum ("Hand on hilt").

Aurelian's greatest challenge came in 272, when various eastern provinces of the empire declared their support for the celebrated Queen Zenobia of Palmyra. Accordingly, that year he assembled a large army in Asia Minor, and marched south against the usurpers. He quickly vanquished Zenobia's dreaded heavy cavalry in two battles and besieged the capital of Palmyra. The Romans subsequently captured the queen, but Aurelian spared her along with the city. The following year, after successfully defeating the Goths and Carpi along the Danube, the emperor was incensed that Palmyra had revolted and had slaughtered the Roman garrison installed there. He then executed a well-conducted foray that surprised the defenders, captured Palmyra, and mercilessly razed it.

With affairs of the eastern empire secure, Aurelian turned his attention to the province of Gaul, which had been in revolt for over a decade. He easily defeated troops of the Gallic Empire at Châlons in 274, and spared the life of Tetricus, their leader. Tetricus and Zenobia were subsequently brought to Rome and featured in a magnificent triumph. Both were allowed to live out their lives in relative luxury. With the empire restored and a host of enemies vanquished, the senate and people of Rome hailed their burly peasant-emperor as Restitutor Orbis (Restorer of the World). He is regarded as the most successful of the Barracks emperors.

Determined to restabilize social conditions, Aurelian also displayed considerable skill as an administrator. He canceled debts, pardoned political crimes, openly distributed bread to the poor, and overhauled the currency. One of his last military acts was preparing a final, military showdown against the Persian Empire. En route to his destination, Aurelian was assassinated near Byzantium by his own officers in 275.

John C. Fredriksen

See also: Roman Army

References and further reading:

Southern, Pat. *Late Roman Army.* London: Routledge, 2000.
Stoneman, Richard. *Palmyra and Its Empire: Revolt against Rome.* Ann Arbor: University of Michigan Press, 1992.
Watson, Alaric. *Aurelian and the Third Century.* New York: Routledge, 1999.

Austerlitz, Battle of (Moravia, 2 December 1805)

Decisive victory by Napoleon over the Austro-Russian armies. After Napoleon had defeated Karl Mack at Ulm (20 October 1805), Czar Alexander I and Emperor Francis II decided to renew the attack. Napoleon's forces, numbering 73,000, were grouped around Brünn.

On 27 November the 86,000 allied troops, mostly Russian, moved toward Brünn, hoping to turn Napoleon's right flank, cut him off from Vienna, and drive him against the Bohemian mountains, despite allied commander Mikhail Kutusov's objections.

Almost throughout the entire battle, the initiative was wholly Napoleon's. The allied army disintegrated, suffering some 16,000–25,000 casualties. French losses numbered around 9,000. An armistice on 3 December between France and Austria led to the Peace of Pressburg (26 December), ending the Third Coalition.

Austerlitz was one of Napoleon's most extensively planned battles, reflecting the peak of his tactical skill. It was an excellent example of flexibility and meeting the unexpected, and it was perhaps his greatest tactical victory.

James K. Kieswetter

See also: Davout, Louis-Nicolas, Duke of Auerstädt, Prince of Eckmühl; Kutuzov, Prince Mikhail Illarionovich Golenishchev; Lannes, Jean, Duke of Montebello; Masséna, André, Duc de Rivoli, Prince d'Essling; Murat, Joachim, Grand Duke of Cleves-Berg, King of Naples; Napoleon I; Soult, Nicolas-Jean de Dieu

References and further reading:

Chandler, David G. *The Campaigns of Napoleon.* New York: Macmillan, 1966.
Duffy, Christopher. *Austerlitz 1805.* London: Leo Cooper, 1977.
Esposito, Vincent J., and John Robert Elting. *A Military History and Atlas of the Napoleonic Wars.* New York: Frederick A. Praeger, 1964.
Manceron, Claude. *Austerlitz: The Story of a Battle.* Trans. George Unwin. New York: Norton, 1966.

Australian Military

Australian troops at the opening of the twentieth century were heavily committed in the Boer War in South Africa; at the century's close they were heavily engaged in peace keeping in East Timor. It is a reflection of both the near-continuous overseas service and the changing emphasis of Australian army operations between the beginning of the twentieth century and its end.

When Australia federated as a nation in 1901, troops from the individual Australian colonies were already fighting in the Boer War in aid of Britain. The new federal government lost little time in raising an Australian Commonwealth Horse unit, eventually 4,000 strong, and dispatching it to South Africa in February 1902.

At federation, the army was 28,836 strong: composed of 1,500 permanent soldiers, 16,000 militia, and 11,200 volunteers. Britain had demanded that 9,000 troops be available for overseas imperial service, and Australia complied, but stipulated that Australian troops be always under Australian control, and that they only be committed overseas at Australian government discretion. These conditions were influenced by executions of Australians by the British army during the Boer War, which caused such resentment that Australians in the military have never again been subject to capital punishment or generally subjected to non-Australian military discipline.

A perceived threat to Australia from Japanese and German naval expansion saw compulsory military training introduced for all Australian men in 1907. Australia was the first English-speaking nation to do so by law. All boys between 12 and 18 years were required to drill, and then subsequently complete two years of formal military training. In 1910 the military training age was extended to 26 years. By 1913 Australia had nearly 90,000 trained soldiers, 56,000 school-age cadets, and 48,000 army-sponsored Rifle Club members under arms. In 1911 the Royal Military College was established at Duntroon.

When Britain declared war on Germany on 4 August 1914, Australia responded with a promise to assist Britain to "the last man and last shilling." However, the Defence Act did not allow the army to engage in overseas service, so a new expeditionary army, the Australian Imperial Force (AIF), had to be created from volunteers. This force eventually amounted to five divisions on the western front in 1919. Australians fought with distinction at Gallipoli in 1915, the tragic battles of the Somme in 1916, and Ypres in 1917; engaged successfully in the last great cavalry charge in history at Beersheba in Palestine in 1917; and were instrumental in the decisive final assaults against the Hindenberg Line in late 1918. Ultimately 330,000 Australians served overseas during World War I (nearly 7 percent of Australia's total population), and of these 59,000 lost their lives and 152,000 were wounded. All had been volunteers, as the Australian people had twice rejected conscription during the course of the war. Many Australian veterans felt that the lives of their comrades had been squandered by incompetent British commanders, who tended to use the Australians as shock troops, and there was considerable rancor between the Australians and the British ("pommies") well before the end of the war.

The interwar years saw the Australian army languish because of Britain's policy of Asian defense from a strong Singapore naval base. Compulsory training was discontinued in 1929, and army strength fell to fewer than 28,000 men. Until World War II began, all equipment in use was World War I surplus.

Australia declared war on Germany on 3 September 1939, formed the Second AIF, and then dispatched the 6th Infantry Division to the Middle East, where it participated in the Libyan campaign against Italian forces in early 1940. It then fought in both Greece and Crete, suffering heavy losses of men and equipment. In 1941 the newly arrived 9th Australian Division comprised the main force in the successful defense of Tobruk, while the 7th Division invaded Syria to prevent its coming under German control.

Japanese threats to Southeast Asia necessitated the newly formed 8th Infantry Division being posted to Malaya, while the still-equipping 1st Armoured Division remained in Australia. The rapid southward advance of the Japanese caused great concern and the 6th and 7th Divisions were recalled to Australia. En route, Churchill tried every pressure to have them diverted to Burma, but the Australian government refused. In 1942 the 8th Division went into captivity with the fall of Singapore, while the 9th Division remained in the Middle East and figured prominently in the El Alamein battles.

The Malaya-Singapore-Burma British-led military fiascos and the disastrous Japanese bombing of the northern town of Darwin led to a profound reorientation of Australian policy in the midst of war, from Great Britain to the United States. Australians noted that while the British fell apart before the Japanese, the Americans at least went down fighting in the Philippines. For a while, General Douglas MacArthur was about as popular in Australia as he was in the United States. Further, the vast military power of the United States military was obvious to all Australians by 1943, even though the Pacific was a secondary theater so far as Washington was concerned.

The Japanese advance in New Guinea over the Kokoda Track in 1942 was countered by brigade-level actions of militia units, until the 7th Division was dispatched in mid-1942. By 1943 five further divisions were in action in the Southwest Pacific theater, and operations in the Solomons, New Guinea, and New Britain culminated in large-scale amphibious landings on Borneo in the final months of the war. Australia pushed hard politically for an army corps to participate in the planned landings on the Japanese homeland, an event eventually made unnecessary by the atomic bomb attacks. Total Australian army casualties during World War II were 219,500, including 20,000 POWs.

At the end of World War II the Australian government set target establishments for the postwar army, but they were not met, and by 1949 strength was only 15,000 regulars and 23,000 reserves, reflecting on Australia being initially only able to contribute a single battalion to UN forces at the outbreak of the Korean War. In October 1951, under U.S. pressure, this combat commitment was increased to two battalions and subsequent Australian command of the Commonwealth Force. Commonwealth forces were rated by the UN Command as among the best troops in Korea. At the same time, conscription was introduced—but then allowed to lapse after the end of the Korean commitment.

A policy of forward defense in Southeast Asia led to commitment of Australian troops to Vietnam in 1962, first as advisers and then as combatants with the arrival of the Royal

Australian Regiment (RAR) in mid-1965. Participation peaked during 1968 and 1969 at three logistically self-supporting infantry battalions with armor support, which succeeded in repressing most Vietcong activity in Phoc Tuy province during the period. Australian combat troops had also been continuously committed alongside British forces in Malaya and Borneo since 1955 fighting Indonesian insurgency, and conscription had to be reintroduced to meet the added Vietnam commitment. Australian forces were eventually withdrawn from Vietnam in 1972 and conscription ended once again.

Since 1947 the Australian army has participated in 44 peacekeeping operations globally, and in 2000 deployed a substantial force to the former Indonesian province of East Timor during its transition to an independent nation.

Michael Hyde

See also: Beersheba; Gallipoli; Kokoda Trail; Milne Bay; Tobruk, Battle of; World War I; World War II

References and further reading:

Adam-Smith, P. *The ANZACS.* Melbourne: Thomas Nelson, 1978.

Grey, G. *A Military History of Australia.* Melbourne: Cambridge University Press, 1999.

Odgers, G. *Army Australia: An Illustrated History.* Sydney: Child and Associates Publishers, 1988.

Tanner, T. W. *Compulsory Citizen Soldiers.* Sydney: Alternative Publishing Cooperative, 1980.

Austrian Civil Wars (1934)

The authoritarian Austrian government's suppression of both a socialist and a national socialist (Nazi) uprising. One month after Hitler's seizure of power in Germany, the Christian Conservative government of Chancellor Engelbert Dollfuß dissolved on 4 March 1933. Weak in the interior, Dollfuß gave in to Mussolini's pressure to establish a pure authoritarian regime. Right-wing Socialists such as Karl Renner tried to avoid a clash with the government, while the left wing prepared for resistance. After dissolution of the socialist paramilitary organization Schutzbund (31 March 1933), semifascist formations within the government bloc tried to complete disarmament of their opponents and to provoke them. After an incident at Linz, fighting spread immediately to the industrialized areas of eastern Austria. Government forces dispatched about 100,000 soldiers against (in theory) 40,000 poorly armed Schutzbund members. The latter hardly stood a chance. Socialist appeals for a general strike as well as for the full mobilization of the Schutzbund failed totally. Police and army forces encircled their opponents in workers' districts, and artillery fire even on civilian targets broke the last resistance on 15 February. Government forces suffered 42 dead and 123 wounded. More than 100 civilians were killed. Schutzbund casualties are estimated at more than 1,000. The Socialist party and trade unions were outlawed, their leaders imprisoned or driven into exile. Nine Schutzbund fighters were sentenced to death and executed. "February 1934" was a government-provoked action to complete the abolition of democracy.

Now it was the turn of the Austrian Nazis. Backed by their German comrades, they organized a coup d'état after various methods to abolish the Dollfuß regime, the main obstacle to the intended Anschluß to Germany, had failed.

On 25 July Viennese SS squads stormed the chancellery. Dollfuß was wounded and died. His death was the only success enjoyed by the badly organized rioters; not even all Nazi units participated. The army—contrary to the Nazis' calculations—stayed loyal to the government. In the capital, army and police regained control quickly. Fighting outside Vienna, different from events in February, concentrated on rural areas, especially in Styria and Carinthia, and lasted up to 27 July. Left without support—Italian troops at the border prevented German intervention—the Nazis' struggle was hopeless. This new civil war left 269 dead; 13 rioters were executed. German and Austrian Nazis had to change their tactics completely but time and the dividing of Austrians after two civil wars worked in the Nazis' favor when Anschluß—union with Germany—came in 1938 without resistance.

Martin Moll

See also: Hitler, Adolf

References and further reading:

Fröschl, Erich-Zoitl Helge. *Februar 1934: Ursachen, Fakten, Folgen.* Vienna: Verlag der Wiener: Volksbuchhandlung, 1984.

Jagschitz, Gerhard. *Der Putsch: Die Nationalsozialisten 1934 in Österreich.* Graz, Vienna, and Cologne: Styria, 1976.

Austrian Succession, War of the (1740–1748)

A conflict that saw the emergence of Prussia as a great power. The War of the Austrian Succession was a series of smaller wars in the Habsburg-Bourbon feud and the first struggle for German hegemony between Austria and Prussia.

Charles VI, Holy Roman Emperor, was determined to avoid the same fate that had befallen the Spanish Habsburgs in 1700. His daughter and heiress, Maria Theresa, was disqualified from election as Holy Roman Emperor, but she could inherit the Habsburg lands in Austria, Bohemia, Silesia, and Hungary.

Charles drew up an agreement, the Pragmatic Sanction, and sought acceptance from the states of Europe for this solution. Most states agreed, but Bavaria and Saxony refused. When Charles died in October 1740, the German states began plotting with Austria's rival, France, on how to take advantage of the situation.

Frederick II, the new Prussian king, was an ambitious young man with a full treasury and a well-trained army. He coveted the resource-rich Austrian province of Silesia. On 16 December, Frederick invaded, completing his conquest in six weeks.

Austria also faced threats to its Italian territories from Spain and in southern Germany from France and the German states, but was saved thanks to Maria Theresa's courage and Frederick's pragmatism. The queen, her children in her arms, made a dramatic appearance before the assembly of Hungarian noblemen, winning their support. Frederick, satisfied in his ambition, agreed to peace with Austria. This left the Austrian armies free to tackle the French and Spanish threats.

By mid-1742, Great Britain, already at war with Spain over trade and colonial affairs, decided to play a more active role on the Continent. A combined British-Hanoverian-Austrian army defeated the French at Dettingen in June 1743 (the last battle in which a British monarch, George II, personally took part). The war might have ended at this point as Prussia was sated, Austria had secured the rest of its territories, and France was disillusioned by its German allies and war-weary.

The balance of power was upset in September 1743 when Sardinia joined the Anglo-Habsburg coalition. A Franco-Spanish expedition was sent to crush Sardinia, while France agreed to join Spain in its war against Britain. The French war effort concentrated on the Netherlands and northern Italy. The war in the Netherlands was successful, but the Italian campaign was marred by disagreements with the Spanish over objectives and protocol.

Supporters of Charles Stuart (Bonnie Prince Charlie) launched a rebellion in Scotland, but were defeated at Culloden (the last battle on British soil). The British also gained the French fortress of Louisbourg on Cape Breton Island, and captured New England militia.

The Treaty of Aix-la-Chapelle, which officially ended the war, settled nothing. Aside from the recognition of Prussia's conquest of Silesia, all else was returned to prewar status. The treaty was a truce, postponing the resumption of the Anglo-French and Austro-Prussian wars.

David H. Olivier

See also: Culloden, Battle of; Fontenoy; Frederick the Great, King of Prussia; Saxe, Hermann Maurice, Comte de

References and further reading:

Anderson, M. S. *The War of the Austrian Succession, 1740–1748.* London: Longman, 1995.

Browning, Reed. *The War of the Austrian Succession.* New York: St. Martin's Press, 1993.

Woodfine, Philip. *Britannia's Glories: The Walpole Ministry and the 1739 War with Spain.* Woodbridge: The Boydell Press, 1998.

Austro-Swiss Wars (1315–1499)

Wars for Swiss independence and consolidation. The Austro-Swiss conflict, lasting nearly 200 years, resulted in the formation of the Swiss Confederation and eventual independence from Habsburg feudal rule. During this period, the Swiss fought against Habsburg Austria, imperial German cities, and Burgundy.

In 1273, Rudolf I, the Habsburg Holy Roman Emperor, attempted to reassert his family's authority over the cantons of Uri and Schwyz. Both regions had received letters of charter from the previous emperor, Frederick II.

In 1291, Uri and Schwyz were joined by the Canton of Unterwalden to form the Swiss Confederation. They pledged themselves as a military alliance to defend their charters within the empire, but to resist Habsburg rule.

In November 1315, Duke Leopold of Austria led an expedition to assert Habsburg authority. The Swiss of Uri and Schwyz fielded a small force of peasants armed mostly with halberds. These men occupied a defensive position at Morgarten. They had picked their place well: The Austrian army had to file along a narrow road, which the Swiss blocked. On the Austrian left was a steep slope, on their right the freezing waters of Lake Egeri.

Duke Leopold made no attempt to scout the defile ahead, and his mounted troops rode forward confidently. Without warning, Swiss soldiers swept down from the hills, swinging their terrible pole-arms. The mounted knights, hemmed in by their own numbers, could face certain death in the lake or an ugly yet just as certain death against the halberd. Leopold's army disintegrated and the Habsburgs grudgingly let the Swiss manage their own affairs.

Swiss expansion during the fourteenth and fifteenth centuries took the form of accepting new members into the Confederation. The Canton of Bern entered into a military agreement with the forest cantons of the Confederation. Bern's expansion brought it into conflict with its Burgundian neighbors, who launched a punitive expedition in 1339. At the Battle of Laupen, the Bernese, aided by other Swiss allies, won a decisive victory over mounted knights.

In 1386, the Swiss, confident in their fighting abilities, provoked a renewal of hostilities with the Habsburgs. The Battles of Sempach (1386) and Nafels (1388) broke the Habsburg effort to bring the Swiss back under imperial dynastic control. In fact, the growing Confederation secured a series of treaties in 1389, 1394, and 1412. This last treaty established peace between the Swiss and the Habsburgs for 50 years.

During this period of peace in the east, the Swiss became more embroiled in military adventures in Italy and Burgundy. They established their reputation for cohesion, aggressiveness, and disregard for death. Battles such as those at Arbedo (1422) and St. Jacob en Birs (1444) made the Swiss masters of the field against the traditional means of chivalric combat.

The Habsburgs could endure the situation no longer, and in 1469 Sigismund, duke of Austria, mortgaged the province of Alsace to Charles the Bold of Burgundy for 50,000 guilders. Sigismund hoped to fight his war against the Swiss by proxy. Unfortunately for Sigismund, Charles took advantage of his new obsession to expand northward toward the Netherlands. Sigismund backtracked and formed a league with the Swiss and the French to fight Burgundy. Once the Swiss marched into the theater along the Rhine, both Austria and France left the war.

The Swiss faced the full power of Burgundy, one of the strongest states in Europe at that time. Charles possessed a number of cannon, and he had in his army large contingents of men-at-arms and crossbowmen. Despite these advantages, Charles (also called "The Rash") possessed little tactical skill or strategic sense. He lost three battles, Granson (1476), Murten (1476), and Nancy (1477). At Nancy, Charles fell mortally wounded into a frozen trench, a halberd gash splitting his head from temple to cheek.

After Nancy, the Habsburgs acquired Burgundy by marriage, but they never succeeded in reestablishing their authority over the cantons. In 1499 the Habsburgs made one last attempt, at the Battle of Frastenz. The Swiss prevailed because Maximilian of Austria failed to adequately guard a cliff that flanked his line. The Swiss charged up this cliffside and won their last desperate battle for independence.

The Treaty of Basel concluded nearly 200 years of conflict, and it ended in fact, as well as in theory, Habsburg hegemony over the Swiss. The Swiss had achieved their independence.

Bryan R. Gibby

See also: Grandson and Morat, Battles of; Laupen, Battle of; Sempach, Battle of

References and further reading:

Delbruck, Hans. *Medieval Warfare.* Trans. Walter J. Renfroe Jr. Lincoln: University of Nebraska Press, 1990.

Oman, Charles W. *A History of the Art of War in the Middle Ages.* London: Greenhill Books, 1991.

Austro-Turk Wars (1529–1739)

No fewer than eight wars fought between the Austrian and Ottoman Turkish Empires, 1529–1739. The Turks sought to expand into Europe proper and the Austrians stood in the way, while harboring expansionist dreams of their own.

The war of 1529–1533 was a direct result of the Ottoman defeat of Hungary in the Hungarian-Turk War of 1521–1526. The Hungarian king John Zapolya, now a subject of the Turk sultan Süleyman the Magnificent, requested help against the Austrians. The sultan took more than four months to move his huge army from Constantinople to Vienna, allowing the Archduke Ferdinand time to build his defenses. The unsuccessful siege lasted from 27 September to 14 October 1529. The sultan tried again in 1532 but a monthlong siege at the Austrian fortress of Guns failed. A truce was called because of the Turk-Persian Wars of 1526–1555 but this agreement did not stop the Turkish army from pillaging and plundering.

The sultan Süleyman, reacting to the attack by 24,000 Austrian and Bohemian troops on the Turk fort at Essek in 1537, renewed the war of 1537–1547. In 1543, issues of succession for the Hungarian throne led to a well-planned Turkish expedition that left Belgrade and captured the large forts of Stuhlweissenberg and Grau, then occupied Croatia as well as Buda and Pest, the capitals of Hungary. In 1545 Ferdinand offered a truce and an annual tribute of 30,000 ducats for Austrian Hungary. Again, the Turk-Persian Wars of 1526–1555 played a role in this 1547 truce being signed at Adrianople.

The war of 1551–1553 involved Austrian and Turk disputes over Transylvania. Ferdinand besieged its capital, Lippa, in 1551 while an Ottoman army captured three fortresses in the nearby Temsvar region, soon made into a new Turkish province. However, the Turks failed to take the fortress at Erlau (Eger), and the army was recalled for the Turk-Persian war, yet again! An armistice restarted the 1547 truce of Adrianople.

The war of 1566 saw Süleyman repulsed at Malta in 1565 by the Knights Hospitallers. The sultan, near the end of his life, sought one vindicating victory over the Austrians and their new emperor, Maximillian II. A Turkish army of several hundred thousand crushed the Croatian fortress town of Szigetvar but the sultan died of natural causes during the battle. And some 3,000 Turks were blown up when timed powder bombs exploded as they breached the last defenses. The Turkish army returned with the body of Süleyman to Constantinople, effectively ending the war.

The "Long War" of 1591–1606 began with the defeat of the Bosnian Ottomans by Croatians at Sissek in 1593. The Porte (Ottoman government) suffered its worst losses against Vienna in the longest war between the two. The Porte lost much of Hungary, Romania, Moldavia, Walachia, and Transylvania both on the field and through defection to Vienna. Attacks by Dnieper Cossacks and losses of Esztergom and Giurgiu forced the sultan Muhammed III to take the field with the Prophet's standard and rally his retreating infantry for an unlikely victory at Mezokersztes. Here 30,000 Germans and Hungarians died. Fortress and siege warfare became the norm, with the Austrians taking Raab but not Buda in 1598 and the Turks failing to take Varazdin and Pest in 1599 and 1603. The Turks regained lost territory in alliance with the Transylvanian prince Stephen Bocksay, and the fluctuating Long War ended in the Treaty of Zsitva-Torok of 1606, with the Austrians as clear winners over the Porte, now busy with yet another Turk-Persian war, 1603–1612.

The war of 1663–1664 stemmed from the success of the Turks in the Transylvanian-Turk War of 1657–1662. The Turks, led by Grand Vizier Fazil Pasha, were now seen as liberators by Wallachian and Romanian Christians against the Habsburg Austrian reformation and the Thirty Years' War in Europe. Buda was captured in 1663, as was Neuhasel, a great victory for the Turks. After winter in Belgrade, the Turkish-led forces captured forts on the road to Vienna, forcing the Austrian Holy Roman Emperor Leopold I to start peace talks. Unfortunately for the Turks, during the final battle at the Raab River on 1 August 1664, flooding allowed only half of their army to cross the river, and that half was defeated by the Austrian cavalry under Montecuccoli. The Treaty of Vasvar, a 20-year truce, was signed afterwards.

The war of 1683–1699 began with renewed hopes by the Turks that the tide had turned against the Austrians. Some 70,000 Austrian and Polish troops under King Jan III Sobieski repulsed 138,000 Turks led by Kara Mustapha Pasha. This last invasion of Austria and siege of Vienna was a disaster for the Turks. Pope Innocent XI started a crusading Holy League in 1686, composed of the Holy Roman Empire, Venice, Poland, and Moscow, to combat the Turks. Buda was taken from the Turks in 1686, as was Transylvania in 1687. The sultan Süleyman II sent a Turk army that captured Serbia and Belgrade in 1690. Turk forces invaded Transylvania in 1691, but were decisively defeated. Austria became involved with France in the War of the Grand Alliance, and a fixed border between the Turks and Austrians remained stable for five years. In 1697, a large Ottoman Turk army left Belgrade to invade Hungary and was met by the imperial army under Prince Eugene of Savoy. The Turks suffered a crushing defeat on 11 September 1697, at the Battle of Zenta. The war ended with the Treaty of Karlowits in 1699, as the Turks were now occupied with the Russo-Turk War of 1695–1700 and the Venetian-Turkish War of 1685–1699.

The war of 1716–1718 began with 60,000 troops under Eugene Savoy decisively defeating the Turks at the Battle of Peterwardein on the Danube River, on 5 August 1716. The Turks lost 6,000 men, 100 artillery pieces, and their grand vizier. Eugene then besieged Belgrade, the strongest city of the Turks in the Balkans. A large Turkish relief force was initially victorious, but was finally routed by Eugene's cavalry charge, forcing the surrender of Belgrade. The Austrian forces then marched on Constantinople, and with most of the Balkans lost, the Sublime Porte sued for peace in 1718.

Christopher Howell

See also: Mohács, Battles of; Süleyman I; Vienna, Sieges of

References and further reading:

Hourani, Albert. *A History of the Arab Peoples.* Cambridge, MA: Harvard University Press, 1991.

Rothenburg, Gunther. *The Austrian Military Border in Croatia, 1522–1747.* Urbana: University of Illinois Press, 1960.

Avars

Central Asian nomads threatening Byzantium and Western Europe. The Avars were obscure before their emissaries first appeared in Constantinople in 558. Three years later, having absorbed Hun and other fragmented steppe groups long present in eastern Europe, they moved from the Ukraine into the Byzantine Balkans, from which they freely raided west and east, ultimately settling in the strategically located Hungarian plains. Adept horsemen and fierce warriors, they probably introduced iron stirrups to the West, giving their horsemen a considerable advantage, and carried swords, bows, and long lances. Enlisting conquered or vassal Slavic peoples to enlarge their forces, the Avars created a formidable army, capable of occupying lands and taking cities.

In 582, Avar chieftain Baian crossed the Danube in force, taking the city of Sirmium and menacing Thessalonica in 586. The Byzantine emperor Maurice (r. 582–602) drove the Avars out of the Balkans in 599, but this success was but a pause. In 611, when the Persian Empire launched a massive invasion of Byzantine Asia, the opportunistic Avars poured southward, reoccupying the Balkans and again besieging Thessalonica. Byzantium was embattled on two sides.

In 625, the Byzantines' worst terror came true when the Sassanid shah made an alliance with the Avars. The following year the two allies marched on Constantinople. For months, their forces encircled the Byzantine capital, but city defenses remained impregnable. Finally, abandoned by the Persians and exhausted by horrendous casualties, Avar forces disintegrated. Their army in ruins, the Avar state in the southern Balkans evaporated as embittered Slavs revolted or embraced Byzantine patronage. After 635, references to the Avars in Byzantine sources dwindle, but they remained an important enemy of the Franks, until their very rich state in Hungary was totally crushed by Charlemagne at the end of the eighth century. Surviving Avars were apparently absolved by local Slavs, Bulgarians, and Magyars.

Weston F. Cook Jr.

See also: Heraclius; Mauricius Flavius Tiberius

References and further reading:

Treadgold, Warren. *A History of the Byzantine State and Society.* Stanford, CA: Stanford University Press, 1997.

Whitby, L. Michael. *The Emperor Maurice and His Historian.* Oxford, UK: Clarendon Press, 1988.

ʻAyn Jalut, Battle of (1260)

The battle ended Mongol expansion in the west.

Under Khan Möngke (r. 1251–1259), Mongol armies led by his younger brother Hüle'ü (d. 1265) reconquered Iran and advanced beyond, culminating in the capture of Baghdad in 1258. This brought the Mongols into close contact not only with the crusading world, but also with Mamluk Egypt, a direct competitor with the Mongols for control of Syria. The crusaders—lured by the image of Priester John, the legend of a powerful Christian king who would come out of Asia to liberate the Holy Land from the Muslims once and for all, and well aware that the coming of the Mongols entirely changed political relationships in the Near East—submitted in some cases, or at least endeavored to enter into active negotiations with the conquerors, whose ranks did include Nestorian Christians. The Mamluks, on the other hand, potential victims of any Christian-Mongol alliance, decided to resist.

At this point Khan Möngke died, forcing Hüle'ü to redeploy his main armies back to Iran to safeguard his power base and to resist incursions by forces of the antagonistic Golden Horde. Left behind in Syria, which was overrun by the Mongols in 1260, was the Nestorian general Kit-buqa with a small force. It was this force that came into contact with Mamluk forces led by Qutuz at ʻAyn Jalut in Galilee on 3 September 1260. Details of the battle are sparse and the sources are contradictory but the result was total defeat of the Mongols, their first in the west, and the death of Kit-buqa. The Mamluks or their allies reoccupied Syria, and although the war to control it went on almost to the end of Mongol rule in Iran, the Mongols were never again as strong a position in the extreme west as they were in 1260.

Paul D. Buell

See also: Mongol Empire

References and further reading:

Morgan, David. *The Mongols.* Oxford, UK, and New York: Basil Blackwell, 1986.

Smith, J. M. "ʻAyn Jalut: Mamluk Success or Mongol Failure?" *Harvard Journal of Asiatic Studies* 44: 2 (1984), 307–345.

Aztecs

Warlike tribe situated in Mexico, also known as Tenocha or Mexica. The Tenochas were Nahuatl speakers, descended from northern wandering barbarian peoples given to skinning their sacrificial victims. Entering the Valley of Mexico in c. 1168, they became a formidable military power with great talent for engineering. Eventually they developed a large island at the marshy western end of Lake Texcoco by attaching chinampas, floating mud rafts of soil, to one another. These eventually took root, and from c. 1325, they built their city of Tenochtitlan (City of the Tenochas), dominated by temples and other structures, with several causeways connecting it to the mainland. Over two centuries, their control was extended over a substantial territory in central and southern Mexico. They had small hairless dogs, but no horses, mules, sheep, goats, or chickens. Their diet, poor in proteins, consisted principally of corn and beans with some fish and a few birds. They did have the maguey plant, which provided honey, sugar, sewing needles, parchment, vinegar, and pulque, a milky and fiery liquor. Their principal leader, known as Chief Speaker, presided over a council and a congeries of nobles, warriors, priests, and ordinary citizens. Ahuizotl, a tyrant king (r. 1486–1502), had doubled the size of the Aztec domain. His successor was Moctezuma Xocoyotzin (born c. 1468, r. 1502–1520), a son of Axayacatl (r. 1469–1486), Ahuizotl's predecessor. Moctezuma ruled between 5 and 15 million (estimates vary) Aztecs and client peoples. The Aztecs were more or less continually at war with their neighbors. Their objective was to secure sacrificial victims for their gods, principally the sun god and Huitzilopochtli, their war god. These deities were thought to require frequent offerings of human blood to defeat the powers of darkness. Wars were often provoked by the arrogance of Aztec emissaries, including merchants, or by Aztec demands that their deities be added to a neighboring state's panoply of gods. Conquered tribes were never absorbed into the Aztec state, but remained subject peoples. They were expected to provide tribute, men, and supplies to help conquer other enemies of the Aztecs. Thus resentment of their Aztec overlords steadily grew. When wars were not being waged for cause, prearranged *xochiyaoyotl* (flower wars) were conducted with neighboring peoples so that each might capture sufficient sacrificial victims for their respective deities. The flayed skins of some prisoners were worn by priests or captors for ceremonial purposes. Weapons included bows and arrows, obsidian swords, spears and spear throwers, stone clubs, and lances. Iron weapons were unknown in pre-Columbian Mexico. Gourds filled with angry hornets were sometimes hurled at attackers. Armor consisted of thick cotton padding, shields of wood, hides, and sugarcane stalks. Helmets were made of various combinations of wood, animal hides, bone, reeds, feathers, paper, and cloth. Public education was geared to military necessity. Boys were taught the use of arms, tactics, and the elements of strategy. The Aztec nation was divided into four military quarters, made up of several *calpulli* (clans). Each quarter had its own captain-general. Armies, usually of about 8,000 men, were subdivided into corps, divisions, squadrons, and squads. Several armies, under a *tlacatecatcuhti,* or principal commander, would take to the field in wartime. A *calpixque* acted in a logistical capacity, supplying arms and food, and sent messengers to summon soldiers from other towns. He also counted remaining supplies after a battle. *Quachics* had charge of the rear guard in battle, and taught students how to capture prisoners. There were several elite warrior elements, notably the Eagle, Arrow, and Jaguar units. Soldiers were not paid, but received prizes for battle performance. Outstanding warriors received land grants. Mobilization was swift, and near Tenochtitlan, could be accomplished within hours. There was no standing army, because the educational system and frequent military exercises prepared each man for his role in wartime. Discipline in battle was strong. Aztec warriors produced many council members, judges, governors, and tax collectors. When 555 Spanish conquistadors under Hernan Cortez landed at Vera Cruz in 1519, Moctezuma and others thought he might be Quetzalcoatl (Plumed Serpent), a semimythical personage, making his promised return. The Spanish conquest owed much to their initiative, modern arms, technology, and the horse. Most importantly, many subject peoples viewed the Spanish as liberators, and actively supported the Aztecs' overthrow in 1521.

Keir B. Sterling

See also: Cortez, Hernando de

References and further reading:

Diaz del Castillo, Bernal. *True History of the Conquest of New Spain.* 5 vols. Ed. and trans. A. P. Maudley. London: Hakluyt Society, 1908.

Clendinnon, Inga. *The Aztecs: An Interpretation.* Cambridge, UK: Cambridge University Press, 1991.

Peterson, Frederick. *Ancient Mexico.* New York: Capricorn Books, 1962.

Thomas, Hugh. *Conquest: Montezuma, Cortes, and the Fall of Old Mexico.* New York: Simon & Schuster, 1993.

B

Babur (Bäbr), Zahir ud-Din Muhammad Babur Mirza (1483–1530)

Conqueror of northern India and founder of the Mogul Empire. Babur, or "the Tiger," was descended from Genghis Khan (c. 1162–1227) on his mother's side and from Tamerlane (1336–1405) on his father's. Babur spent his early career trying unsuccessfully to conquer the city of Samarkand in central Asia, controlled by the Uzbeks. Although the city had been the capital of Tamerlane's empire, Babur was unable to hold it, and, after his third failed siege attempt in 1512, he turned, in frustration, to India. Beginning in 1519, Babur led a series of raids into northern India from his own capital in Kabul. During this period, Babur tried to take control of the sultanate of Delhi by making dubious dynastic claims on the basis of his lineage from Tamerlane, who had established the Sayyids as the ruling family of the sultanate. However, the sultan of Delhi, Ibrahim Lodi, had overthrown the Sayyids and refused to recognize Babur's claim. Rather than continue to press the issue, Babur launched an invasion of northern India in an attempt to overthrow Lodi. On 21 April 1526, Babur led a small but highly mobile force against Lodi at Panipat, 80 kilometers north of Delhi. Despite the numerical superiority of Lodi's forces, perhaps as much as ten to one, Babur won a crushing victory, largely through skillful diplomacy and the use of artillery and firearms, which completely surprised the sultan's forces. Indeed, Babur was the first Muslim conqueror to employ such weapons. With significantly more military experience than Ibrahim Lodi, Babur deftly lured his opponent into an ambush, utterly destroying the sultan's forces.

After his, Babur quickly moved to take Agra and Delhi, and, on 15 March 1527, turned against the sultanate's primary rival in India, the confederation of Rajput states led by Rana Sangha of Mewar. Babur's defeat of Rana Sangha at Khanua, roughly 150 kilometers south of Delhi, left him in control of northern India. For the next three years, until his sudden death in 1530, Babur consolidated his control of the sultanate of Delhi and established the Mogul Empire. Although Babur was not fully a Mongol, the word *Mogul* is a Persian variation of "Mongol." Babur left the Mogul throne to his farcical son, Humayan, who almost lost the empire.

Babur the Conqueror, founder of the Mogul Empire. (Hulton/Archive)

Babur wrote a brilliantly detailed and candid narrative of his adventures in central Asia and India, the *Baburnama*.

Eric Pullin

See also: Delhi Sultanate, Wars of; Genghis Khan; Panipat, Battles of; Tamerlane

References and further reading:

Babur. *Babur-nama.* Trans. A. S. Beveridge. London: Luzac, 1922.
Eraly, A. *The Great Mughals.* Delhi: Viking, 1997.
Keay, John. *India: A History.* New York: Grove Press, 2000.
Richards, John F. *The Mughal Empire.* Cambridge, UK: Cambridge University Press, 1993.
Zaman, M. K. *Mughal Artillery.* Delhi: Idarah-I Adabiyat-I Dellii, 1983.

Babylonian Empire (c. 1900–539 B.C.E.)

The preeminent civilization of southern Mesopotamia for nearly two millennia, with two periods of major political and military expansion (during the reigns of Hammurabi, c. 1792–1750 B.C.E., and Nebuchadnezzar II, 605–562 B.C.E.).

The city of Babylon (founded c. 2300 B.C.E.) did not become preeminent in southern Mesopotamia until the migration of the Amorites from the northwest (c. 1900 B.C.E.). The Amorite chieftain Hammurabi (or Hammurapi) was successful in consolidating control of central Mesopotamia, subduing the city-states of Larsa, Uruk, Isin, Mari, and Eshnunna. At his death, all of Mesopotamia proper was under the control of Babylon. The empire was short-lived, and the remaining rulers of the Old Babylonian kingdom (c. 1750–1595 B.C.E.) were little more than rulers of the city of Babylon itself.

The next major political dynasty of Babylon was that of the Kassites, an ethnic group of unknown origin who ruled Babylon for over four centuries (c. 1570–1155 B.C.E.). For most of this period, Babylon controlled much of southern Mesopotamia. The Kassites engaged in diplomatic relations and marriage alliances with other major powers of the Near East, including Egypt, Mitanni, Assyria, and the Hittites.

Although a series of relatively minor dynasties followed the Kassites, one ruler stood out as a military strategist, Nebuchadnezzar I (r. 1124–1103 B.C.E.) of the Second Dynasty of Isin. He avenged a previous sack of Babylon by the Elamites (a people from southwestern Iran), launching a surprise attack against Susa, the Elamite capital, and recovering the statue of the patron deity of Babylon, Marduk.

Babylonian political and military power was not again apparent until the rise of the Chaldeans, a tribal group that emerged as the dominant role player in Babylonia and competed with Assyria for control of central Mesopotamia. Although Mukin-zeri (c. 731–721 B.C.E.) and Merodach-Baladan II (c. 721–703 B.C.E.) were successful in keeping Assyrian military presence out of Babylonia, Sennacherib of Assyria (r. 705–681 B.C.E.) sacked Babylon in 689 B.C.E. and destroyed the city. Assyrian annals describe many of these campaigns in some detail, but one does not get a clear picture of the nature of Chaldean military tactics, except that they often engaged in guerrilla warfare against their more powerful enemy to the north.

With the decline of Assyria (626–612 B.C.E.), the Chaldeans once again achieved political autonomy under Nabopolassar (r. 626–605 B.C.E.), who founded the Neo-Babylonian Empire. The Chaldeans and Medes were successful in ending the Assyrian state and destroying the city of Nineveh in 612 B.C.E. (in part by diverting the Tigris River into the city). However, in 605 B.C.E. Egypt marched through Judea to assist the remnants of the Assyrian army at Carchemish in northern Syria, where Nebuchadnezzar II (r. 605–562 B.C.E.), then Nabopolassar's crown prince, crushed the enemy and enlarged the Neo-Babylonian Empire from the borders of Iran in the east to the borders of Egypt in the west. The most famous military venture undertaken by Nebuchadnezzar II was the conquest of Jerusalem in 597 B.C.E. and its destruction 10 years later, an event that is described in detail both in the Old Testament and in later Jewish tradition. The Chaldean Empire lasted only a generation after the death of Nebuchadnezzar II, falling prey to the armies of Cyrus II of Persia in 539 B.C.E.

Mark W. Chavalas

See also: Assyria; Cyrus II the Great

References and further reading:

Brinkman, J. *A Political History of Post-Kassite Babylonia 1158–722 BCE.* Rome: Pontifical Biblical Institute, 1968.
———. *Materials and Studies for Kassite History.* Chicago: Oriental Institute, 1976.
Frame, G. *Babylonia 689–627 B.C.: A Political History.* Leiden: Nederlands Historisch-Archaeologisch Instituut, 1992.
Voth, S. "Analysis of Military Titles and Functions in Published Texts of the Old Babylonian Period." Ph.D. diss., Hebrew Union College, 1981.

Bacon, Nathaniel (1647–1676)

Colonial rebellion leader. Born in England in 1647, Bacon was the son of minor English nobility who immigrated to Virginia in 1674 and became a successful planter. Bacon was ambitious, and through his relationship to prominent colonial leaders he became a member of the governor's council. He supported the growing dissent of colonial farmers toward Governor William Berkeley, whom they blamed for making restrictive treaties with the Indians, thus monopolizing land for the elite. Bacon became the leader of a faction that denounced the governor's policies and conducted unauthorized raids against the Indians. Berkeley branded Bacon an outlaw, though he had just been elected to the House of Burgesses. When Bacon went to claim his seat, he was captured and brought before Berkeley in shackles.

Bacon apologized on his knees and was pardoned, but still sought a commission legally to lead colonial forces against the Indians. Berkeley refused the commission and Bacon threatened Jamestown with an armed force, causing Berkeley to flee. Bacon issued a "Declaration of the People" to justify his rebellion. The conflict escalated as Bacon and Berkeley vied for control of the government. Bacon even besieged and burned Jamestown to deny its use as a fortification.

In October 1676 Bacon suddenly fell ill with typhus and died, leaving his rebel cohorts to be hunted by Berkeley and newly arrived English troops. Though some view Bacon as an early proponent of democratic reforms, others portray him as an opportunist who sought power, prestige, and wealth at the expense of the rights of the people, and who murdered Indians without compunction.

Steven J. Rauch

See also: American Indian Wars

References and further reading:

Washburn, Wilcomb E. *The Governor and the Rebel.* Chapel Hill: University of North Carolina Press, 1957.

Webb, Stephen S. *1676: The End of American Independence.* New York: Alfred A. Knopf, 1984.

Badajoz, Siege of (16 March–6 April 1812)

A successful siege that paved the way to Madrid for the British. After securing Portugal for the British, Arthur Wellesley, the future Duke of Wellington, regarded the Spanish fortresses of Ciudad Rodrigo and Badajoz, both near the Portuguese border, as key to the liberation of Spain. He captured Ciudad Rodrigo from the French on 19 January 1812. With 32,000 men on 16 March, he laid siege to Badajoz, which Nicolas Jean de Dieu Soult had gained for the French by besieging it from 26 January to 11 March 1811. Twice previously, on 16 May 1811 and from 24 May to 19 June 1811, Wellesley and William Carr Beresford had failed to recapture it.

The garrison commander at Badajoz, Armand Phillipon, had 5,000 French and Hessians inside a reliable citadel with nine bastions, elaborate breastworks, ditches, salients, obstacles, and six strong outlying redoubts, three on each side of the Guadiana River. Phillipon was a crafty defender. On 19 March he sent a successful sortie against 3,000 British who were digging in around Fort Picurina, one of the redoubts.

By 6 April, sappers and heavy artillery had created three breaches. Believing that a French relief army was only two days' march away, Wellesley ordered an immediate assault. At about 10 P.M. the British stormed the fortress. Savage hand-to-hand fighting left 3,500 British and 1,400 French casualties. The British had already lost 1,500 before the assault itself. Phillipon tried to escape to the north bank of the Guadiana, but surrendered on 7 April.

Once inside the walled city, the victorious but exasperated British troops released their pent-up anger, looting, burning, murdering, and raping. Wellesley could not regain control of his drunk and riotous soldiers until 11 April.

Wellesley's victories at Ciudad Rodrigo and Badajoz prepared his way to crush Auguste Frederic Viesse de Marmont at Salamanca on 22 July, inflicting three times as many casualties as he suffered. He entered Madrid in triumph on 12 August.

Eric v. d. Luft

See also: Napoleonic Wars; Soult, Nicolas-Jean de Dieu; Wellington, Arthur Wellesley, Duke of

References and further reading:

Fletcher, Ian. *In Hell before Daylight: The Siege and Storming of the Fortress of Badajoz, 16 March to 6 April 1812.* Tunbridge Wells, UK: Baton, 1984.

———. *Badajoz 1812: Wellington's Bloodiest Siege.* Oxford, UK: Osprey, 1999.

Fryer, Mary Beacock. *"Our Young Soldier": Lieutenant Francis Simcoe, 6 June 1791–6 April 1812.* Toronto, ON: Dundurn, 1996.

Robinson, Thomas Gerald. *Los sitios de Badajoz y la batalla de la Albuera [Sieges of Badajoz and the Battle of Albuera: Spanish and English].* Badajoz: Universitas Editorial, 1998.

Baghdad (1916–1917)

The penultimate operation of the ill-starred Anglo-British Mesopotamian Campaign, capped in a bitter victory with the capture of the preponderance of the remaining Turkish forces in the Near Eastern Theater.

After the British defeat at Kut-al-Amara at the end of April 1916, British military leaders conducted a top-down review of their overall drab performance. They deduced that the neglected Mesopotamian front required more detailed planning and devotion of better resources than originally presumed. The troops, two-thirds of whom were Indian, were poorly armed and equipped.

London took direct charge of directing operations and appointed Lieutenant General Sir Frederick S. Maude to replace the sickly General Nixon. Maude continued the buildup already begun and further developed a communications network and transportation facilities suitable to support an offensive deep into Mesopotamia.

In consequence of an expansion of the shipping capacity at Basra, the buildup accelerated during the summer and fall of 1916. Between April and November 1916 British engineers completed a major railway linking the northern approaches of Basra with Amara. In keeping with the revised, offensively oriented view of the army's role in Mesopotamia, London expanded the river fleet. Further strengthening the combined-arms approach, new aircraft were sent to the Royal Flying Corps squadrons, which had recently gained air superiority over the Germans in the theater. By late summer Maude had at his disposal two corps containing two divisions each, another division in the rear, and two cavalry brigades. Combat elements numbered 166,000 out of the total of 340,000 troops in Iraq, opposed by Khalil Pasha's 42,000-man Turkish Sixth Army.

On 13 December 1916, Maude began his methodical, resolute advance toward Baghdad, systematically eliminating Turkish detachments along both banks of the Tigris River. Early in February 1917, his joint army, navy, and air force retook Kut after a series of well-planned combined strikes against the fortress's 12,000 defenders. Later that month they captured Turkish defensive positions at Sanaiyat. The British gained the upper hand following several days of fighting along the Diyala River south of Baghdad and forced the Turks to withdraw during the second week of March. On 11 March, advance elements of the Black Watch regiment entered the town center on the heels of Arab looters and raised the Union Jack over Baghdad, taking 9,000 prisoners.

Jim Bloom

See also: Kut-al-Amara; World War I

References and further reading:

Barker, A. J. *The Neglected War: Mesopotamia 1914–1918.* London: Faber & Faber, 1967.

Burne, Alfred H. *Mesopotamia: The Last Phase.* London: Gale & Polden, 1936.

Evans, R. *A Brief Outline of the Campaign in Mesopotamia, 1914–18.* London: Sifton, Praed, 1926.

Moberly, Sir F. J. *Official History of the Great War: The Campaign in Mesopotamia 1914–1918.* 4 vols. London: Her Majesty's Stationery Office, 1923–1930.

BAGRATION, Operation (23 June–29 August 1944)

Soviet offensive that destroyed German Army Group Center. Since Stalingrad (autumn 1942–spring 1943) and Kursk (summer 1943), the Red Army's offensive momentum had been against German Army Group South. Soviet success resulted in a salient held by Field Marshal Ernst von Busch's Army Group Center (replaced by Field Marshal Model, 29 June, also commanding Army Group North Ukraine).

The front line was 15–60 kilometers east of Polotsk-Vitebsk-Orsha-Mogilev-Bobruisk-Pripiat Marshes-Kovel. Army Group Center and flank forces numbered 63 divisions and three brigades, totaling 1.2 million men, 17,000 guns, 1,500 tanks and assault guns, and 2,100 aircraft. Wehrmacht fortifications and defensive lines were built to a depth of 250–270 kilometers.

Soviet planning to destroy this force, Operation BAGRATION, began in April 1944. Soviet forces comprised 166 divisions, 12 tank and motorized corps, and 21 infantry brigades, totaling 1.5 million men, 31,000 guns, 5,200 tanks and assault guns, and 5,000 aircraft. The Red Army built up manpower and material superiority along the breakthrough sectors.

BAGRATION began with a massive partisan sabotage campaign (led by Byelorussian Communist Party secretary P. K. Ponomarenko), 19–22 June, to disrupt German logistics. Air strikes were launched against the German rear (21 June). The 1st Baltic Front (General I. Bagramyan) and 3d Byelorussian Front (General I. Chernyakhovskii) attacked Wehrmacht forces around Vitebsk as part of a deception regarding the main direction of attack (beginning 10 June on the Finnish border), and to encircle and destroy them.

On 23 June the assault groups/forward detachments of the 3d and 2d (General G. Zakharov) Byelorussian Fronts attacked to encircle and destroy Wehrmacht forces around Orsha-Mogilev and Bobruisk. The main offensive developed from 24 June as pincer movements by the 3d and 1st (General K. Rokossovskii) Byelorussian Fronts to capture Minsk (4 July).

The 1st Ukrainian Front (Marshal Konev) drove toward Lvov (13–27 July) and Lublin (18–23 July) to bar retreat of Army Group Center. The 2d and 3d Ukrainian Fronts (Generals Malinovskii and Tolbukhin) attacked Romania, capturing Ploesti (30 August) and Bucharest (31 August).

BAGRATION propelled the Red Army 550–600 kilometers (roughly, Tartu-Riga-Warsaw-Bucharest), destroying 17 German divisions and 3 brigades, reducing another 50 divisions to half-strength. While the Red Army halted its offensive to regroup and reinforce, the Wehrmacht stiffened its defense of Warsaw, the gateway to Berlin.

Neville G. Panthaki

See also: Berlin, Soviet Drive on; Kursk, Battle of; Russian and Soviet Armies; Stalingrad

References and further reading:

Erickson, John. *Road to Berlin: Stalin's War with Germany.* New Haven, CT: Yale University Press, 1999.

Great Patriotic War of the Soviet Union, 1941–1945: A General Outline. Moscow: Progress Publishers, 1974.
Overy, Richard. *Russia's War: Blood upon the Snow.* New York: Penguin Putnam, 1997.
Werth, Alexander. *Russia at War, 1941–1945.* New York: Carroll & Graf, 2000.
Zaloga, Steve. *Operation Bagration 1944.* New York: Osprey Trade Editions, 2000.

Baker, Newton D. (1871–1937)

American secretary of war during World War I. Born on 3 December 1871 in Martinsburg, West Virginia, the bookish Baker took a law degree from Washington and Lee University in 1894. After a brief period of legal practice, he served as private secretary for William Wilson, postmaster general under Democratic president Grover Cleveland. He returned to the bar in 1897 and moved to Cleveland, Ohio. A Progressive reformer, Baker was elected mayor in 1912.

In 1916 President Woodrow Wilson summoned Baker to Washington to serve as secretary of war. Wilson chose Baker, an outspoken pacifist, in order to appeal to those who opposed military preparedness. When America joined the Allies in World War I, Baker found himself in a difficult position. Criticized from all sides in the political arena, he proved unable to establish order in the War Department or to define the relationship between the chief of staff, General Peyton March, and the American Expeditionary Force under General John Pershing. To the dismay of the general staff, Baker frequently sided with Pershing. Baker fought to keep procurement of supplies in the hands of the War Department, but lost the battle when President Woodrow Wilson formed the War Industries Board under Bernard Baruch. Baker successfully oversaw the first American conscription. He believed that the draft would provide a socially democratic army, as well as avoid unnecessary wartime hysteria. To protect the men from the forces of vice, Baker appointed a commission to oversee moral conditions in army camps throughout the country.

Baker left office in 1921 when Warren Harding replaced Wilson as president. Baker returned to Cleveland where he practiced law as a corporate attorney. Although briefly mentioned as a presidential candidate for the Democratic Party in 1932, he never again entered politics. He died in Cleveland on 25 December 1937.

Gregory Dehler

See also: March, Peyton; Pershing, John J.; World War I

References and further reading:

Beaver, Daniel R. *Newton D. Baker and the American War Effort, 1917–1919.* Lincoln: University of Nebraska Press, 1966.
Cramer, C. H. *Newton Baker: A Biography.* Cleveland: World Publishing Company, 1961.

Balaklava (24–25 October 1854)

Allied repulse of Russian attack in the Crimean War. The Allies used the town of Balaklava on the Black Sea coast, eight miles southeast of Sevastopol, as a supply base during the siege of Sevastopol. Commanded by Lieutenant General Ivan Petrovich Liprandi, 25,000 Russian troops launched a surprise attack the night of 24 October 1854. Before dawn on 25 October they had routed the Turkish garrisons from six high-ground redoubts on and east of Causeway Heights. The Russians deployed on Causeway Heights, on Fedioukine Heights, and at the eastern ends of both the north and south valleys, mainly with cavalry and artillery. The Allies (6,000 British under General Fitzroy James Henry Somerset, Baron Raglan; 7,700 French under General François Certain Canrobert; and 4,000 Turks under British General Sir Colin Campbell) were scattered around the south valley, the western end of the north valley, and Sapoune Heights.

Attempting to reach the town and isolate the Allies, four squadrons of Russian cavalry charged into the south valley from the northeast. Campbell rallied the 93d Highlanders, ordered them into a line two ranks deep, and held fire until the last possible moment. His "Thin Red Line" of infantry stopped the Russian charge completely with only three volleys.

General James Yorke Scarlett perceived a column of 3,000 Russian cavalry about three miles long, heading west atop Causeway Heights north of Balaklava. Interpreting this movement as the first step of a second attempt to take the town, and oblivious to the danger of being drawn into the center and outflanked on both sides, Scarlett successfully preempted by leading "The Charge of the Heavy Brigade" uphill into the advancing Russians.

Bad communication created the disastrous "Charge of the Light Brigade" at midday. Just after the charge, the 4th Chasseurs d'Afrique attacked into the Fedioukine Heights, enabling the remnant of the Light Brigade, about 200, to escape.

Eric v. d. Luft

See also: Alma; Campbell, Colin; Crimean War; Inkerman, Battle of the; Light Brigade, Charge of the; Sevastopol, Siege of

References and further reading:

Compton, Piers. *Cardigan of Balaclava.* London: Hale, 1972.
Mollo, John, and Boris Mollo. *Into the Valley of Death: The British Cavalry Division at Balaclava, 1854.* London: Windrow & Greene, 1991.

Turkish soldiers, Army of Salonica, 1912–1913. (Library of Congress)

Selby, John Millin. *Balaclava: Gentlemen's Battle.* New York: Athenaeum, 1970.

———. *The Thin Red Line of Balaclava.* London: Hamilton, 1970.

Balkan War, First (1912–1913)

A conflict fought by the Balkan countries against the Ottoman Empire, for independence and aggrandizement. It increased European diplomatic tensions and, together with the Second Balkan War, is cited as the precursor of World War I.

Austria-Hungary's annexation of Bosnia-Herzegovina on 6 October 1908 caused concern among the Balkan nations and Russia. A coalition of Balkan nations, the Balkan League, was formed: Serbia, Bulgaria, Greece, and Montenegro. However, the agenda of the Balkan nations was not merely mutual protection against Austria-Hungary, but further dismemberment of the Ottoman Empire in Europe.

On 13 March 1912, Serbia arranged a treaty with Bulgaria. Greece concluded a military convention with Bulgaria on 29 May. On 14 August, Bulgaria dispatched a note to the Turks demanding that Macedonia be granted autonomy. The league ignored a joint Russo-Austrian declaration of 8 October, calling for restraint and condemning any disruption in the balance of power and the Balkan status quo. That day, Montenegro declared war on the Ottoman Empire. On 18 October the league followed suit. Allied strength was 750,000 men.

The Allies won a series of decisive victories over the Turks during the next two months, forcing them to relinquish Albania, Macedonia, and practically all of their other holdings in southeast Europe. In Thrace, the Bulgarians defeated the Ottoman forces at Kirk Kilise on 22–23 October, while the Serbs won the Battle of Kumanovo on 24 October, captured Bitola, and then linked with Montenegrin forces to enter Skopje. By 8 November, the Greeks captured Salonika/Thessaloníki and advanced on Ioánnina.

By the end of November, the Turks had been pushed back to the Tchataldja line on the outskirts of Constantinople/Istanbul itself. The only Ottoman holdouts were the garrisons of Scutari, Yanina, and Adrianople/Edirne.

An armistice was concluded on 3 December 1912. However, a peace conference in London could not bring about agreement and ended in failure.

On 23 January 1913 a coup by Enver Bey's Young Turks occurred, and war resumed. In the subsequent fighting the allies captured Ioánnina, and Adrianople/Edirne on 26 March. The Turks obtained an armistice with Bulgaria, Greece, and Serbia on 19 April 1913. Montenegro accepted the armistice a few days later. Another peace conference, with the major European powers again acting as mediators, met at London on 20 May.

The political questions to be resolved were: Serbia's and Montenegro's attempt to gain ports on the Adriatic, which

Austria-Hungary wished to prevent; the territorial composition of Albania, which conflicted with Greece and Serbia; Romania's sudden demand for compensation from Bulgaria; and Serb-Bulgarian rivalry over Macedonia.

By the terms of the Treaty of London, concluded on 30 May, the Turks ceded the island of Crete/Kríti to Greece and relinquished all territories in Europe west of the Enos-Midia line. Creation of an independent Albania was agreed upon but the issue of its borders was forwarded to a European commission. Romania was compensated with Silistria.

It was the issue of Macedonia that broke apart the Balkan partnership. Serbia and Greece formed an alliance against Bulgaria on 1 June 1913, which led to the Second Balkan War.

Neville G. Panthaki

See also: Italo-Turkish War; Russo-Turkish War; World War I

References and further reading:

Anderson, M. S. *The Eastern Question.* London: Macmillan, 1970.
Menning, Bruce. *Bayonets before Bullets: The Imperial Russian Army, 1861–1914.* Bloomington: Indiana University Press, 1992.
Seaton-Watson, Hugh. *The Decline of Imperial Russia.* London: Methuen, 1966.
Taylor, A. J. P. *The Struggle for Mastery in Europe.* Oxford, UK: Oxford University Press, 1991.

Balkans Campaign (1941)

One of the quickest, most one-sided victories in modern military history. By early 1941 Hitler was almost ready to invade Russia. Italy, his ally in Europe, had proved a weak link in North Africa and when Mussolini decided to show off the prowess of his forces in Albania, and then in Greece, the resulting debacle forced the Germans into planning an operation to clean up the Italian mess. The Germans did not want to divert troops to the Balkans who could be better used in the attack on Russia. However, Mussolini's incompetence and arrogance and Allied subterfuge drew the Germans into the conflict to protect their southern flank.

The Germans had to take over where Mussolini had failed, not only for sound military reasons, but also to back up their ally. The Italian attack had forced Greece to give up its neutrality, but the Germans could count on Rumania and Bulgaria for support, and believed that Yugoslavia too was on their side. However, the Yugoslavs were subjected to a coup d'état on 27 March 1941, and so the German high command now had to plan on an attack on Yugoslavia as well as on Greece.

Operation MARITA was aimed originally at occupying the Aegean north coast and the Salonika Basin, but events in Yugoslavia meant this country would also have to be attacked. The changed stance of Yugoslavia gave rise to German fears that not only would the Italian front collapse, but the operation against the Aegean north coast would fail because of the Yugoslavian threat to the German flanks and rear. As it turned out, the Germans had really no grounds for such fears.

The essential German aim was defensive, because any offensive operational plans were rendered impossible by the enormous manpower demands of the oncoming operation against Russia.

The plan as put into operation was a hasty one, and the Second Army, part of the German force, was thrown together in just 10 days. However hasty, nevertheless the plan worked. The invasion of Yugoslavia commenced on 16 April 1941, and by 17 April all resistance was over. On 6 April Greece was also invaded and despite valiant efforts by Greek forces, the Metaxas Lien was quickly reduced and German forces were fighting in central Greece by mid-April. The Greek army surrendered on 23 April, and British forces, which had been sent to Greece to support the Royal Hellenic Army, were forced to quit Thermopylae on 24 April, pulling back into the Peloponnesus for pickup by the Royal Navy.

The rapid reduction of both Yugoslavia and Greece released many of the troops involved for the next great operation, BARBAROSSA. However, German troops had to be left as occupation forces in both countries. Soon resistance forces were established that caused increasing problems to the Germans and to their Bulgarian and Italian allies. Yugoslavian and Greek resistance forces, increasingly aided by the Allies, made life more and more uncomfortable for the occupiers, and there were many atrocities committed by both sides through the rest of the war.

David Westwood

See also: Crete; World War II

References and further reading:

Lanz, Hubert, et al. *The German Campaigns in the Balkans (Spring 1941).* MS No B-525. Washington, DC: U.S. Army Center of Military History, undated; Supplementary Report. Washington, DC: U.S. Army Center of Military History, 1989.

Ballistics

The science of projectiles. This science is divided into interior and exterior ballistics. Its aim is to improve the design of shells/projectiles so that increased accuracy and predictability are the result. It deals also with rockets and ballistic missiles.

Interior ballistics deals with the design of projectiles, their propellants, and the motion of the projectile within the gun or firing device. Propellant types, their composition, and their burning are part of internal ballistics, with the aim being to provide optimum internal gun pressure to move the projectile along the barrel to the muzzle, or, in the case of rockets, sufficient speed to overcome inertia, gravity, and other external effects.

Exterior ballistics examines the effects of such matters as gravity, air resistance, wind, and sometimes earth rotation upon the shell/projectile. The projectile flight path is examined and in combination with propellant calculations this can be lengthened to increase range. The parabola of the shell in flight is tested, as is the stability of the shell/projectile to ensure accuracy on target.

Rifling or its absence (particularly nowadays) is examined to provide the projectile with the required spin on emergence to allow a good firing trajectory. The shell/projectile design is also studied, to ensure that the rifling can grip the body of the shell (to eliminate windage and to impart rotation), and to ensure that the required effect at the target is achieved, and at the same time to eliminate waste deposits in the barrel.

Modern artillery relies upon firing tables, which are the result of ballistics; the type and weight of the propellant charge, the type of projectile, and above all its accuracy are consolidated into range tables that are used by the gunners to organize their shoots. Much of the work is now done by field artillery fire-control computers, which allows the gunners to achieve optimum effect at the target for minimum ammunition expenditure.

Ballistics now includes rockets, missiles, and ballistic missiles. The trajectory is particularly important in fire-and-forget missiles, and with the aid of targeting computers extremely high accuracy can result.

All of the work of the ballistic scientists and the gunners can, however, be rendered useless without strict quality control and testing by manufacturers, something that increasingly needs strict supervision by governmental bodies. As weapons become more complex, quality control becomes ever more vital.

David Westwood

See also: Artillery; Rifles and Rifling

References and further reading:

Deutsch, F. W. *Waffenlehre.* Berlin: Mittler & Sohn, 1939.

Handbook of Ballistics. 2 vols. London: Her Majesty's Stationery Office, 1921.

Baltimore (12–14 September 1814)

Successful American defense of an important seaport against British land and sea attacks in the War of 1812. After British troops under Major General Robert Ross burned Washington, D.C., on 24–25 August 1814, Americans under Major General Samuel Smith immediately strengthened the defenses of nearby Baltimore. When Ross landed 5,000 men at North Point, Maryland, on 11–12 September, Smith had more than 9,000 militia around Baltimore, including Major George Armistead's 1,000 troops garrisoning Fort McHenry and the Lazaretto Battery.

Brigadier General John Stricker's 3,200 Maryland militia opposed Ross's landing on 12 September. British artillery and frontal infantry assault carried the day, but at a cost of 340 British casualties to 215 American. Ross was among the British dead. Colonel Arthur Brooke assumed command and continued the march toward Baltimore, making headquarters on the Philadelphia Road on 13 September.

The British knew that they could not capture a well-fortified city of 40,000 armed and angry citizens unless they could first neutralize Fort McHenry. Admiral Alexander Forrester Inglis Cochrane positioned his fleet in the Patapsco River just out of range of McHenry's guns and lobbed about 1,500 mortar rounds, Congreve rockets, and other ordnance at the fort the night of 13–14 September. Only about 400 found their mark. Little damage to the fort occurred. The garrison lost only four killed and 24 wounded. As a last resort, Cochrane tried to land 1,200 regulars from barges southwest of Baltimore on 14 September, but artillery fire from shore batteries drove them back. When Brooke learned that the naval shelling and the amphibious attack had both failed, he retreated, took ship again, and on 14 October sailed for Jamaica.

American lawyer Francis Scott Key, temporarily detained aboard the American sloop *Minden* anchored amidst the British fleet, observed the entire naval bombardment and wrote on the back of an envelope the next morning a commemorative poem, "The Defense of Fort McHenry," later retitled "The Star Spangled Banner," which did not officially become the U.S. national anthem until 1931.

Eric v. d. Luft

See also: War of 1812; Washington, Burning of

References and further reading:

Muller, Charles G. *The Darkest Day: 1814: The Washington-Baltimore Campaign.* Philadelphia: Lippincott, 1963.

Sheads, Scott S. *The Rockets' Red Glare: The Maritime Defense of Baltimore in 1814.* Centreville, MD: Tidewater, 1986.

Whitehorne, Joseph A. *The Battle for Baltimore, 1814.* Baltimore: Nautical & Aviation Publishing Company of America, 1997.

Ban Chao (31–101)

Restorer of Chinese control in Turkistan. After Wudi (141–87 B.C.E.), Chinese control over Turkistan was gradually lost and with it control over the lucrative Silk Route trade. Although Guangwudi (r. 25–57), founder of latter Han, rebuilt the power of the dynasty after the interregnum of Wang Mang (r. 9–23), he and his successors showed little interested in restoring Han power in Central Asia. This was in spite of a considerable weakening of China's northern rival, the Xiongnu, during the period. Largely to blame was the predominance of a "reformist" element within Chinese government that condemned the expensive forward position assumed in the north and west under Wudi.

This situation changed largely due to the efforts of a single individual, Ban Chao. Part of an army sent out under Dou Gu (died 88) against the Xiongnu, Ban was dispatched in 73 as an envoy to the kingdom of Shanshan, one of many small states located along the Silk Road. He not only fulfilled his mission successfully, but was able to intercept and destroy a Xiongnu embassy and return Shanshan to Chinese control. Buoyed by this success, Ban was sent out again with a force of about 30, this time to Khoten, which he was also able to return to Chinese obedience. His force expanded by local auxiliaries, Ban then proceeded to a general conquest not only of those areas claimed by the Chinese in the past, but many beyond. This accomplished, he went on to defend his new empire for almost a quarter of a century (until his return to China in 100 after 31 years of service along the frontiers and in Central Asia), largely single-handed, with little or no support from successive pacifist courts. This was an achievement unparalleled in Chinese history.

Paul D. Buell

See also: Han Wudi

References and further reading:

Barfield, Thomas J. *The Perilous Frontier: Nomadic Empires and China.* Cambridge, MA, and Oxford, UK: Basil Blackwell, 1989.

de Crespigny, Rafe. *Northern Frontier: The Policies and Strategy of the Later Han Empire.* Faculty of Asian Studies Monographs: New Series No 4. Canberra: The Australian National University, 1984.

Banana Wars (1898–1933)

U.S. interventions in the Caribbean in the early twentieth century. After the Spanish-American War, 1898, the United States began to assert hegemony over the Caribbean and intervened militarily in nine countries on 34 separate occasions between 1898 and 1933. American military and naval contingents administered Cuba, the Dominican Republic, Haiti, Nicaragua, and Panama for extended periods. Costa Rica, Honduras, Guatemala, and Mexico experienced briefer interventions. These interventions were dubbed the Banana Wars by the marines, who were often deployed to carry them out.

The interventions were initiated by President Theodore Roosevelt, who declared his desire to make the Caribbean an "American Lake" and to build a canal somewhere in Central America. Strategic concern about protecting a transisthmian canal, while exaggerated, was the main reason for the interventions.

In 1903 Panama revolted against Colombia in a bid to become independent. Although the United States did not initiate the revolt, it ensured that the rebels were successful by sending the cruiser *Nashville* to Colon to prevent the transit of Colombian troops to Panama City. The United States soon negotiated a treaty with Panama that established the American Canal Zone, which subsequently was often used as a base to meddle in Panama's domestic affairs.

The Dominican Republic had long suffered corrupt government and larcenous public officials. As a result, the country was continually in economic and political chaos and threatened by its European creditors. In 1904 the U.S. Navy patrolled Dominican waters and briefly sent marines and bluejackets ashore. The policy of the Roosevelt and William Howard Taft administrations was to limit military intervention to providing order in customs collection. President Woodrow Wilson, on the other hand, directed longer interventions with the aim of restructuring the political systems along the lines of American democracy. In 1915 American marines occupied the Dominican Republic and did not leave until 1924. The marines were ordered to provide law and order and to train a Dominican constabulary. A similar policy was enacted in neighboring Haiti, which was also occupied in 1915. Haiti proved difficult to pacify and American forces did not withdraw until 1934.

Nicaragua was another troublesome country for the United States. In August 1912 President Taft, concerned about civil war in the Central American republic, ordered Major Smedley Butler and a battalion of 354 marines to restore order. Butler's marines completed their task in a few weeks but left a detachment of 100 marines at the American legation, a presence that would be maintained until 1933. In 1925 the marines were briefly withdrawn but soon returned as a result of civil war. Henry Stimson, President Calvin Coolidge's special envoy, brokered a settlement among the warring factions. However, one rebel, Augusto Cesar Sandino, rejected the deal and became the object of an extensive campaign by the marines. Sandino frustrated all efforts to subdue him and the marines withdrew on 1 February 1933.

The Mexican Revolution, which began in 1910, was the cause of two significant interventions, ordered by the Wilson administration. The first was the occupation of the port of Veracruz between April and November 1914. The second was the Mexican Punitive Expedition (March 1916–February 1917) led by General John Pershing in pursuit of the Mexican revolutionary and bandit Pancho Villa.

Other noteworthy interventions by the United States in the Caribbean include Cuba 1906–1909 and 1912, and Honduras 1907, 1911, and 1924.

The U.S. interventions in other nations' affairs in Latin America restored peace and, in the longer-lasting interventions, built up something of an infrastructure with railroads, roads, port facilities, telegraphs, water and sewage systems, and so on. And, of course, they provided security for the Panama Canal. But the canal meant nothing to Latin American peasants, and the American interventions did not begin to address the endemic poverty of these areas and fostered among the nationalist elites a deep resentment and, often for their own purposes, a convenient rallying cry against "Yanqui imperialismo."

George M. Lauderbaugh

See also: Mexico, U.S. Punitive Expedition in; Nicaraguan Civil War (1925–1933); Sandino, Augusto César; Spanish-American War; Veracruz, U.S. Landings at; Villa, Francisco "Pancho"

References and further reading:

Fuller, Stephen, and Graham Cosmas. *Marines in the Dominican Republic 1916–1924.* Washington: USMA History and Museums Division, 1974.

Langley, Lester D. *The Banana Wars: An Inner History of American Empire 1900–1934.* Lexington: University Press of Kentucky, 1983.

Musicant, Ivan. *The Banana Wars.* New York: Macmillan, 1990.

Perkins, Whitney. *Constraint of Empire: The United States and Caribbean Interventions.* Westport, CT: Greenwood, 1981.

Quirk, Robert. *An Affair of Honor: Woodrow Wilson and the Occupation of Veracruz.* Lexington: University Press of Kentucky, 1961.

Sandler, Stanley. *Glad to See Them Come and Sorry to See Them Go: A History of U.S. Army Tactical Civil Affairs/Military Government, 1775–1991.* Fort Bragg, NC: US Army Civil Affairs and Psychological Operations Command, 1998.

Schmidt, Hans. *The United States' Occupation of Haiti, 1915–1934.* New Brunswick, NJ: Rutgers University Press, 1971.

Bannockburn, Battle of (24 June 1314)

Decisive English defeat in Anglo-Scot wars of the fourteenth century. Following the death of Edward I in 1307, the new English king, Edward II, abandoned the war against Scotland. This respite allowed Robert the Bruce, king of the Scots, to consolidate power and slowly reduce the English garrisons north of the River Tweed. By 1314, only Berwick and Stirling Castle were still in English hands. Bruce laid siege to Stirling in the summer of 1314, and the castle's governor promised to surrender if not relieved by June 24. Stirling was an important strategic site, commanding access to the Highlands. Edward was determined to relieve the castle and unfurled the royal standard at Newcastle. He quickly raised a force of approximately 3,000 armored knights and 15,000 foot, and marched north to do battle.

Bruce had a much smaller army. It numbered fewer than 8,000 men, including perhaps 500 mounted knights, but the Scottish king did have the advantage of choosing the site of the battle. He deployed his forces on a small rise overlooking Bannock burn (brook), a tributary of the Firth of Forth. His right flank was secured by a stream and his left anchored by a thick forest. The infantry, armed primarily with long spears, was organized into *schiltrons,* hollow, circular formations. The horsemen were kept as a mounted reserve. The larger English army would have to attack on wet, marshy ground, and would be hemmed in by the streams and by woodlands. This inhibited the charge of the English mounted troops and negated their numerical superiority.

On June 24, the English army crossed the Bannock burn in preparation for an all-out assault on the Scottish position. Bruce seized upon the disorganization of the enemy and ordered four of his schiltrons to attack the English flank. A fierce hand-to-hand battle ensued with neither side giving ground. Edward ordered his archers to flank the Scots on their left, but this movement was countered by Bruce. He ordered Sir Robert Keith to launch a charge with the mounted reserve, driving the English back in disorder. The Scots pressed their advantage, and the English front gave way and a rout ensued. Many of the English became bogged down in the marshy ground and were killed or captured. Others were drowned in the burn as they tried to flee the field. Edward himself barely escaped capture, fleeing with the remnant of his army to Dunbar. English losses included two earls, 60 barons, and perhaps 10,000 other ranks. Scottish losses were about 4,000.

The defeat at Bannockburn ended English hopes for conquering Scotland by military force. Subsequently, both Stirling and Berwick fell to the Scots, removing the remaining vestiges of English military presence. Edward II never recovered the prestige lost to the crown in this defeat, ultimately losing his throne and his life. Scotland and England would continue to fight incessantly until the unification of the thrones by the Act of Union in 1707.

Barry P. Neville

See also: Anglo-Scots Wars (1290–1388); Anglo-Scots Wars (1513–1560); Falkirk, Battle of; Flodden, Battle of

References and further reading:

Linday, Robert. *Edward the Second.* New York: W. W. Norton, 1997.

Reese, Peter. *Bannockburn.* Edinburgh: Canongate Books, 2000.

Scott, Ronald McNair. *Robert the Bruce.* New York: Carroll & Graf Publishers, 1996.

Barons' War (1263–1285)

Barons led by Simon de Montfort press their call for reform in the field. The reign of Henry III was increasingly difficult for the English nobility to bear: Henry had begun his rule as a minor child under a regency, but as an adult proved to be a spendthrift and an incompetent, losing what was left of the Angevin Empire in France, engaging in a guerrilla war with Wales, and refuting the Magna Carta in 1232, before engaging in an expensive and foolish plan to make his brother Holy Roman Emperor, and his younger son king of Sicily. The leader of the resistance was the king's brother-in-law, Simon de Montfort, Earl of Leicester, a noted soldier and crusader who had been undercut in Gascony by Henry's favorites. In 1258, Parliament met at Oxford, armed and demanding a say in revenue spending and a redress of grievances, such as the violation of London's charter. In 1259, Henry III signed the Provisions of Westminster, guaranteeing court reform and representation, but by 1260 the barons were quarrelling and the king, prompted by his son, Edward, went to the French to intervene.

De Montfort led an army through England in response to the king's voiding of the Westminster agreement, and on 12 May 1264 caught the royal army at Lewes, where he routed them, after Prince Edward had left the field to chase a party of infantry. Unusually for the time, de Montfort deployed his army so that he had a "fourth" left in reserve. Now in control of the king and the country, de Montfort braced for an attack from France, and held a parliament in January 1265, which demanded that two knights from each shire be called regularly to advise the king and monitor spending.

Despite an alliance with the prince of Wales, de Montfort could not hold up against the defection of Gilbert de Clare to the king, nor the growing party of Prince Edward, who successfully kept the barons' forces away from supplies and reinforcement at Kenilworth. Moving to relieve them, de Montfort, probably exhausted, allowed himself to be trapped at Evesham, where, on 3 August 1265, he and his army were massacred by troops led by Prince Edward. After 1265, Edward, the future Edward I, ruled as acting king for his father, and repudiated the concessions made to the barons. Nonetheless, by 1267 he had approved the Statutes of Marlborough, which legalized reform along the lines of the baron's campaign, and Parliament began to meet regularly. Despite his defeat in the field, Simon de Montfort eventually became the father of the English Parliament and of representative government.

Margaret Sankey

See also: Edward I; English Civil War

References and further reading:

Beamish, Tufton. *Battle Royal.* London: Frederick Muller, 1965.

Maddicott, J. R. *Simon de Montfort.* Cambridge, UK: Cambridge University Press, 1994.

Treharne, R. F. *Simon de Montfort and Baronial Reform.* Ed. E. B. Fryde. London: Hambledon Press, 1986.

Barton, Clarissa ("Clara") (1821–1912)

"The Angel of the Battlefield," founder of the American Red Cross. Clara Barton, in full Clarissa Harlowe Barton, was born in Oxford, Massachusetts, on 25 December 1821, the youngest daughter of Captain Stephen Barton, a veteran of frontier conflicts in the 1790s under "Mad Anthony" Wayne and a comrade-in-arms of William Henry Harrison. Her family educated her at home to be a teacher. She taught for 18 years, eventually becoming principal of a free school in Bordentown, New Jersey, in 1852. When the townspeople objected to a woman holding such a position of authority, she resigned, moved to Washington, D.C., and in 1855 found a job in the U.S. Patent Office.

At the outbreak of the Civil War, she volunteered to help the U.S. Army Medical Department with logistics, arranging transport of medical supplies and soldiers' personal belongings. Although she had no training as a nurse, she attended

Clara Barton: a wartime photo by Matthew Brady, c. 1890–1910. (Library of Congress)

wounded at First Bull Run, Cedar Mountain, Second Bull Run, Chantilly, Harpers Ferry, and South Mountain. At Antietam, where U.S. Army surgeon James Dunn gave her the nickname, Angel of the Battlefield, a Confederate bullet tore through her sleeve and killed the man she was nursing. Undaunted, she continued her errand of mercy at Fredericksburg, Charleston, the Wilderness, Spotsylvania, and Petersburg. From 1865 to 1869 she led efforts to discover the fates of MIAs and POWs.

As she happened to be in Europe when the Franco-Prussian War began in 1870, she volunteered for the International Red Cross. After returning home in 1873, she lobbied for the United States to sign the Geneva agreements until President Chester A. Arthur did so in 1882. On 21 May 1881 she founded the American Association of the Red Cross, which she ruled arbitrarily as president for 23 years. She wrote several books, including *The Red Cross in Peace and War* (1899). After Congress chartered the American Red Cross in 1900, antiauthoritarian factions within the organization forced her to retire on 14 May 1904. She died in Glen Echo, Maryland, on 12 April 1912.

Eric v. d. Luft

See also: American Civil War; Ethics of Warfare; Geneva Conventions; Logistics; Medicine, Military; Military and Society; Nightingale, Florence; Prisoners of War; Red Cross; Spanish-American War; War Crimes

References and further reading:

Buckingham, Clyde E. *Clara Barton: A Broad Humanity: Philanthropic Efforts on Behalf of the Armed Forces and Disaster Victims.* Self-published: Buckingham, 1980.

Burton, David Henry. *Clara Barton in the Service of Humanity.* Westport, CT: Greenwood, 1995.

Oates, Stephen B. *A Woman of Valor: Clara Barton and the Civil War.* New York: Free Press, 1994.

Pryor, Elizabeth Brown. *Clara Barton: Professional Angel.* Philadelphia: University of Pennsylvania Press, 1987.

Basil II Bulgaroctonus (r. 10 January 976–15 December 1025)

His rule marked medieval high-water mark of Byzantine Empire. In almost 50 years, Basil II, beset by years of revolt and defeat at the start of his reign, destroyed the Bulgar state and reestablished the Byzantine Empire on the Danube and Drava for the first time in four centuries. With its position in southern Italy maintained, and Basil able to effect a series of annexations among the Caucasian kingdoms, the Byzantine Empire came into possession of a security unknown for hundreds of years.

Two episodes best illustrate Basil II and war. First, in April 995, the Aleppo Caliphate's attempt to recover Antioch by taking advantage of imperial distraction in the Haemus resulted in one of the most remarkable victories in history. Basil marched an army from Constantinople to Antioch in 16 days, and on the seventeenth fell upon a Muslim army, not so much surprised as positively amazed by the imperial appearance: The distance between the two cities is over 770 miles. The second was the slow deliberation of the effort whereby the Byzantines destroyed the Bulgar state, and the cataclysmic defeat incurred by the Bulgars at Balathista (Kleidion Pass) on 29 July 1014. Basil secured his name, Bulgaroctonus, "Slayer of Bulgars," from this victory. The battle resulted in the capture of some 14,000 Bulgarians. Basil blinded 99 in every 100 and, leaving the one man with a single eye, sent them back to Bulgaria. Csar Samuel of Bulgaria collapsed on seeing the survivors return. He died two days later. The Bulgarian war, which had lasted over two centuries, continued for another four years, but was a broken-back affair after Balathista.

Basil ruled the empire and commanded the army. Merciless in war, he was most moderate to his subjects in peacetime, but brutal in the suppression of corruption and the Anatolian aristocracy's attempts to promote its interests at the expense of the state. Unglamorous and one of the least attractive of all the emperors in terms of physical appearance, lack of cultural interests, and utter disdain for the trappings of power, he was trusted by army and subjects alike.

H. P. Willmott

See also: John I Tzimisces

References and further reading:

Haldon, John. *Warfare, State and Society in the Byzantine World, 565–1204.* London: UCL Press, 1999.

Norwich, John Julius. *Byzantium: The Apogee.* London: Penguin, 1993.

Ostrogorsky, George. *History of the Byzantine State.* Trans. Joan Hussey. Oxford, UK: Blackwell, 1993.

Treadgold, Warren. *Byzantium and Its Army, 284–1081.* Stanford, CA: Stanford University Press, 1995.

Bataan Death March (April 1942)

The 60-mile movement, under often deliberately horrific conditions, of 76,000 American and Filipino prisoners of war immediately following the Allied surrender in April 1942. In groups of between 500 and 1,000 men, the exhausted and disease-ridden prisoners were marched with little or no food and water to a prisoner of war camp under construction in central Luzon. Anyone who could not keep up was bayoneted or beaten to death. The slightest offenses, such as possession of unauthorized food or water, were met with draconian punishment, including many beheadings (an old Japanese military custom). Some who could not con-

American prisoners of war herded together by the Japanese at Bataan; about 10,000 died. (Hulton/Archive)

tinue the march were buried alive. In more studied cruelty Japanese guards amused themselves by forcing their prisoners to sit in the hot sun near sparkling artesian wells. Anyone who moved toward the wells would be shot or bayoneted.

Adding to the misery was the fact that most of the prisoners had already endured months of malnutrition and disease when they surrendered and were thus from the beginning in poor physical shape. Many had advanced cases of dysentery, with major cramping and diarrhea.

Predictably, the route of the march was littered with dead and dying, and all number of ghastly sights. As many as 10,000 men died as a result of the death march. Those who survived could look forward to continued similar treatment in hellish prison camps.

Some Japanese commanders after the war argued that the Japanese authorities had not expected nearly the number of prisoners who actually fell into their hands, and thus their arrangements for the feeding and transport of the prisoners broke down. This is a valid enough point, but the intense, studied individual cruelties inflicted upon helpless Allied prisoners cannot be so easily excused.

John C. McManus

See also: MacArthur, Douglas; Philippines, U.S. Loss of; Wainwright, Jonathan Mayhew, IV; World War II

References and further reading:

Coleman, John. *Bataan and Beyond: Memories of an American POW.* College Station: Texas A&M Press, 1978.

Falk, Stanley. *Bataan: The March of Death.* New York: W. W. Norton, 1962.

Knox, Donald. *Death March.* New York: Harcourt, Brace, Jovanovich, 1981.

Stewart, Sidney. *Give Us This Day.* New York: W. W. Norton, 1999.

Bay of Pigs Invasion (17 April 1961)

The United States sponsored the invasion of Cuba by expatriates with the hope of toppling the regime of Fidel Castro. Castro had seized power in January 1959, but the following year, the United States was increasingly opposed to Castro's policies and actions, which included show trials of supporters of his predecessor, dictator Fulgencio Batista, expropriation of U.S. property, condemnation of the Latin American policy of the United States, and friendly relations with the Soviet Union and other members of the Warsaw Pact. The United States broke diplomatic relations with Cuba in the waning days of the administration of Dwight D. Eisenhower.

When John F. Kennedy assumed the presidency, he was briefed on a Central Intelligence Agency (CIA) plan to oust Castro. Kennedy approved the plan but added the stipulation that U.S. forces not be directly involved. The American-trained invasion force of 1,300 men landed on the southeastern shore of Cuba in a swampy area but was soon overwhelmed by superior Cuban regular and militia forces personally commanded by Castro. The debacle resulted in 1,189 prisoners and 114 killed from the exile-manned Brigade 2506, while nearly 1,800 on the island lost their lives.

Several factors contributed to the invasion's failure: First, air strikes on 15 April failed to completely destroy the Cuban air force and alerted Castro to the impending attack. Second, U.S. naval and air support on the day of the invasion was withheld. Finally, U.S. intelligence failed to discern the popular support for Castro and underestimated the readiness of Cuban forces. The failure of the invasion greatly embarrassed the Kennedy administration and concomitantly strengthened Castro's hold on the island. Castro used his victory to openly declare Cuba a socialist state firmly aligned with the Soviet bloc. The Kennedy administration took the defeat as a personal offense and spent the rest of its term devising increasingly improbable, even bizarre, methods of disposing of the Cuban dictator. The Bay of Pigs fiasco also set the stage for the Cuban Missile Crisis of October 1962, the closest the world has come to nuclear war.

George M. Lauderbaugh

See also: Castro Ruz, Fidel; Castro-Cuban Revolution; Guevara de la Serna, Ernesto "Che"

References and further reading:

Higgins, Trumbull. *The Perfect Failure: Kennedy, Eisenhower, and the CIA at the Bay of Pigs.* New York: W. W. Norton, 1987.

Kornbluh, Peter, ed. *Bay of Pigs Declassified: The Secret CIA Report on the Invasion of Cuba.* New York: New York Press, 1998.

Wyden, Peter. *Bay of Pigs: The Untold Story.* New York: Simon & Schuster, 1979.

Bayinnaung (r. 1551–1581)

Third ruler of the Burmese Toungoo Dynasty, King Bayinnaung took up the sword at the beginning of his 30-year reign and never put it down. He became king after the assassination of his brother-in-law Tabinshwehti (r. 1531–1550) by an ethnic Mon prince, and built an empire that extended far beyond the borders of modern Burma. Portuguese mercenaries with superior firearms played a major role in his conquests.

Bayinnaung retook the royal capital of Pegu in southern Burma from the Mons and captured the city of Ava (located near present-day Mandalay) in 1555 from the Shans, an ethnolinguistic group related to the Tais of Siam (Thailand). He subjugated their homeland, the Shan states (now in eastern Burma) in 1556, and gained the allegiance of the Tai state of Lanna (Chiang Mai, in northern Thailand); but found himself at war with another Tai polity, Luang Prabang, whose ruler repeatedly challenged Bayinnaung's ambitions in what are now Laos, Shan State, and northern Thailand.

Bayinnaung invaded Siam and captured its rich capital, Ayuthaya, in 1564. A Mon revolt that year necessitated his return to Pegu and he had to recapture Ayuthaya in 1569 from Siamese patriots, pillaging it thoroughly. He failed to crush the resistance of the Lao states, Luang Prabang, and Vientiane, though this goal preoccupied him through the 1570s. When he died in 1581, he was planning conquest of the kingdom of Arakan (now Burma's Arakan State).

Bayinnaung's empire, won by the sword, was short-lived, but he remains a symbol of Burmese imperial glory. The hard-fisted military regime that came to power in 1988, the State Law and Order Restoration Council, erected a statue of him in the border town of Tachilek. It is no coincidence that the sword-clad statue faces southward toward Thailand—his old enemy, Siam.

Donald M. Seekins

See also: Burmese Civil Wars (c. 1300–1599)

References and further reading:

Cady, John F. *The United States and Burma.* The American Foreign Policy Library. Cambridge, MA: Harvard University Press, 1976.

Hall, D. G. E. *A History of South-East Asia.* 3d ed. London: Macmillan, 1964.

Bayonet

Metal blade or spike that, when affixed to a musket or rifle, facilitates its use in hand-to-hand combat. The adoption of the bayonet in the late seventeenth century allowed the elimination of the pike-armed infantryman who had theretofore been the only means of protecting musketeers against cavalry. The first plug bayonets were affixed by inserting the round wooden handle of a knife into the mouth of the gun barrel. Because the bayonet thus rendered the gun unable to fire, it could not always be attached, leading to several instances in which infantry was overrun by cavalry while attempting to insert their plug bayonets. These difficulties were overcome by the introduction in the beginning of the eighteenth century of the obvious solution, the socket bayonet. Consisting of a metal spike instead of the earlier blade, the socket bayonet was affixed by means of a tubular sleeve that fit around the end of the barrel, allowing the musketeer to fire the weapon with the bayonet attached. Thus armed, the musketeer could both provide firepower and defend himself, and bayonet-armed musketeers quickly became the standard infantrymen throughout Europe. The final development of the weapon came about with the introduction in the early nineteenth century of the sword bayonet, which combined the functions of the bayonet and short sword carried by most infantryman of the time. The weapon was essentially a fully hilted short sword that could thus be used as a sidearm, the hilt of which attached to the firearm by means of a lug on the side of the barrel or the stock of a rifle. By the late nineteenth century almost all European armies were equipped with sword bayonets and the forces in both world wars used a shorter variation, the knife bayonet, which is still used by most modern armies.

Christopher C. W. Bauermeister

See also: Cavalry; Firearms; Infantry

References and further reading:

Evans, R. D. C., and Frederick J. Stephens. *The Bayonet: An Evolution and History.* Milton Keynes, UK: Militaria Publications, 1985.

Bazookas

Shaped-charge antitank and pillbox weapon.

The appearance of tanks on the battlefield created the need for antitank weapons. Artillery was the first antidote, closely followed by the antitank rifle. Antitank guns were designed in the 1930s and are still in use, despite the increasing use of antitank rockets.

The shaped charge was invented in the 1930s as well, and the Americans bought the knowledge from the Swiss. The bazooka was first designed in 1942 by an American army officer, who knew that there was a stockpile of shaped-charge projectiles looking for a weapon to fire them. He had already designed a mortar to fire these projectiles, which were launched with the aid of a small rocket stage. The American antitank rifle program was well under way, but was not doing well. A demonstration of the new weapon led immediately to further development work and adoption of the bazooka, as it was called, after a comic strip device.

The first bazookas were found to be effective, mainly due to the shaped-charge effect, which concentrated all the explosive force of the charge forward in a tight stream, allowing a relatively small-caliber weapon to penetrate armor effectively.

Various versions appeared, and the weapon is still in use today with some armies. American troops found themselves at a near-fatal disadvantage during the early weeks of the Korean War, when their World War II–vintage bazooka rounds bounced harmlessly off the armor of Soviet-manufactured North Korean tanks. Postwar-model bazookas were air-rushed to the battlefield and began to score heavily against North Korean armor from their first use.

David Westwood

See also: Armored Fighting Vehicles

References and further reading:

Hogg, Ian, ed. *Jane's Infantry Weapons.* London: Jane's, 1975.
Norris, John. *Anti-tank Weapons.* London: Brassey, 1996.

Beauregard, Pierre Gustave Toutant ("P.T.") (1818–1893)

Confederate field commander who captured Fort Sumter and shared credit for the victory at First Bull Run. Beauregard was born in St. Bernard Parish, Louisiana, on 28 May 1818. Graduating second in his class at West Point in 1838, he was assigned first to the artillery, then to the engineers. On Winfield Scott's staff in the Mexican-American War, he was brevetted twice and wounded twice. From 23 to 28 January 1861, he was superintendent of West Point, but was fired for his explicit southern sympathies.

As brigadier general of Confederate forces around Charleston harbor, he started the Civil War by ordering the bombardment of Fort Sumter on 12 April 1861. Although Joseph E. Johnston was his superior officer, they operated as equals at First Bull Run and together defeated Irvin McDowell. After 21 July 1861 Beauregard held the rank of full general.

Reassigned to serve under Albert Sidney Johnston in the Mississippi Valley, Beauregard took command when Johnston was killed at Shiloh on 6 April 1862. His hesitancy to consolidate his attacks that evening allowed Ulysses S. Grant and Don Carlos Buell to gain ground the next day. Henry W. Halleck pursued him to Corinth, Mississippi, and gradually forced him to abandon that important supply center. Braxton Bragg replaced him as commander of the Army of the Mississippi on 17 June 1862.

Thereafter Jefferson Davis allowed Beauregard only minor assignments. As commander of the Atlantic coastal defenses from South Carolina to Florida, he withstood attacks on Charleston until April 1864. Commanding the Army of North Carolina and Southern Virginia from April to September 1864, he isolated and beat Benjamin Butler at Bermuda Hundred and harassed Union forces around Petersburg. At the end of the war, he was again Joseph Johnston's second-in-command.

Because he admired Napoleon's strategy and tactics, Beauregard was sometimes called "The Little Napoleon." He is supposed to have designed the familiar Confederate battle flag with 13 white stars inside a diagonal blue cross on a red field.

He died in New Orleans on 20 February 1893.

Eric v. d. Luft

See also: American Civil War; Bragg, Braxton; Bull Run, First/Manassas; Fort Sumter; Grant, Ulysses Simpson; Halleck, Henry Wager; Hood, John Bell; Johnston, Albert Sidney; Johnston, Joseph Eggleston; Lee, Robert Edward; McDowell, Irvin; Petersburg, Siege of; Scott, Winfield; Sherman, William Tecumseh; Shiloh

References and further reading:

Basso, Hamilton. *Beauregard: The Great Creole.* New York: Scribner's, 1933.

Roman, Alfred. *The Military Operations of General Beauregard in the War between the States, 1861 to 1865.* New York: Da Capo, 1994.

Williams, T. Harry. *P.G.T. Beauregard: Napoleon in Gray.* Baton Rouge: Louisiana State University Press, 1995.

Beersheba (1917)

A severe Turkish defeat in World War I. A British force of 88,000 men, including Australian and New Zealand Army Corps (ANZAC) and Indian divisions under General Sir Edmund Allenby, attacked the Turkish force of 35,000 troops, commanded by Field Marshal Erich Von Falkenhayn, on the Gaza-Beersheba line at Beersheba on 31 October 1917. Motivated by its geographical and strategic location, Allenby chose Beersheba for its abundant water supply and because its location would be of service as a rendezvous point to develop the principal attack on Hareira, Sheria, and Gaza itself. Surprise and secrecy were of the essence, as success was predicated on the ability of the British forces to capture the city's wells; failure would mean the collapse of the mounted divisions and possibly the entire offense. The Turks, already predisposed to expect that Gaza would be the objective of the main attack, were further encouraged to maintain this view as the British embarked on an eight-day decoy bombardment of Gaza. Meanwhile British troops moved into position by night. Two mounted divisions took up their positions to the northeast, the east, and the southeast, and two infantry divisions marched to the west and southwest, while a third infantry division covered the left of the main infantry attack. All day the battle raged, until dusk when a mounted ANZAC cavalry brigade charged through Turkish wire and trenches into Beersheba, capturing 1,100 Turkish prisoners and securing the coveted water supply. Those soldiers of the Turkish Seventh Army that had hastily evacuated Beersheba were pursued toward Jerusalem, while the Turkish Eighth Army evacuated Gaza and retreated up the coast. British and colonial forces sustained 1,200 casualties in the taking of Beersheba; the Turks sustained 64 casualties and 1,100 were taken prisoner.

Margaret Hardy

See also: World War I

References and further reading:

Falls, Cyril, and Archibald Frank Becke. *Military Operations, Egypt & Palestine: From June 1917 to the End of the War.* London: His Majesty's Stationery Office, 1930.

Keogh, Eustace Graham. *Suez to Aleppo.* Melbourne: Published for the Directorate of Military Training by Wilke & Co., 1955.

MacMunn, George, Cyril Falls, and Archibald F. Becke. *Military Operations, Egypt & Palestine from the Outbreak of War with Germany to June 1917.* Great Britain. Committee of Imperial Defence. Historical Section. London: His Majesty's Stationery Office, 1928–1930.

Wilson, Arnold Talbot. *Loyalties: Mesopotamia, 1914–1917: A Personal and Historical Record.* London: Oxford University Press, H. Milford, 1930.

Wilson, Robert Henry, and Helen Millgate. *Palestine 1917.* Tunbridge Wells, UK: Costello, 1987.

Belgium, Invasion of (August–October 1914)

The forgotten campaign that disrupted Germany's Schlieffen plan. Belgium was a small country whose neutrality had been guaranteed by the European powers in 1839. The deployment of the six Belgian field divisions numbering fewer than 100,000 troops was ordered as early as 31 July 1914, with two divisions garrisoning fortifications, one facing England, one Germany, and two facing France. King Albert I, the greatest asset of the poorly trained and neglected Belgian army, understood very quickly that his country was to be invaded by vastly superior German forces. On 4 August, Von Kluk's First Army attacked the fortifications around Liège. The forts defended by General Leman slowed the German offensive until 15 August. This desperate resistance gave time for the field army to concentrate in the Antwerp area. The German cavalry advance guard faced a major setback at Haelen on 12 August that further slowed the invading forces. From 20 August to the end of September, the Belgian army defending the Antwerp fortified lines was a flanking threat to the Germans now invading northern France. Two reserve corps had to be left to prevent any Belgian sally. On 25 August and 9 September, the Belgians tried to break through the besieging forces. These attacks forced the German high command to send reinforcements toward Belgium at a time when these troops were desperately needed in the first Battle of the Marne. They also had to divert their heavy artillery toward Antwerp to destroy methodically the two lines of forts.

In early October, the Germans launched 12,000 troops against the 7,000 survivors helped by 2,000 British marines. The king decided to retire his army along the coast to join the marching Allied armies that were spreading north in the "race to the sea." The Belgians left Antwerp on 9 October for the River Yser, 100 miles to the southwest. Then Albert ordered a halt to the retreat and a dug-in determination to keep at least a part of Belgium free. The Belgian legacy to the final victory is their unexpectedly stubborn defense of their country, which gave time to the French to redeploy their armies northward and brought the British into the war.

Gilles Boué

See also: Marne, Battle of the; World War I

References and further reading:

How Belgium Saved Europe. Philadelphia: J. B. Lippincott, 1915.
Keegan, John. *The First World War.* London: Pimlico, 1998.
Rapport du commandement: L'action de l'armée belge. Paris: Chapelot, 1915.

Belisarius (c. 505–c. 565)

Byzantine general and conqueror. Belisarius began his career as a bodyguard for Justinian I (r. 527–565) when the future emperor was himself commander of the emperor's guards. Belisarius is first mentioned as leading a raid against the Persians in 526 or 527. By 529 he was commanding the Byzantine army fighting the Persians, with the rank of *magister militium,* or commander of forces, when he won an important victory over the Persians at Dara. Upon the conclusion of peace with the Persians, he was recalled to Constantinople in 531.

In 532 he commanded troops loyal to Justinian I, and suppressed the Nika revolt. In 533 he was appointed to command an expedition to attack the Vandal kingdom of Africa. He defeated the Vandal king in two decisive battles, the Battle of the Tenth Milepost and the Battle of Tricamarium, thus completing the initial conquest of the province of Africa. In 534 he invaded Corsica, and, in 535, upon his return from the African expedition, he was granted a triumphal entry into Constantinople and made consul.

Late in 535, Belisarius invaded Sicily, the empire having declared war upon the Gothic kingdom of Italy. On 31 December, the Byzantines took Syracuse. In 536, he crossed to the mainland, taking Reggio Calabria, and besieged Naples. In December 536 he entered Rome, which had been temporarily abandoned by the Goths, and began making preparations for a siege. The Goths attacked the city in March 537, and continued the blockade for an entire year, until March 538. As the Goths withdrew, Belisarius defeated their rear guard in battle at the Milvian Bridge. The Goths were forced to break the siege by Byzantine raiding parties, one of which occupied the town of Rimini, just 30 miles from the Gothic capital of Ravenna.

Gothic forces besieged Rimini, and the Byzantine forces, reinforced with troops brought by the eunuch Narses, marched to the rescue. A dispute between Narses and Belisarius led to a divided command, which resulted in the failure to rescue the besieged town of Milan before its fall in early 539. Belisarius continued to attack Gothic strongholds, aided by an opportune Frankish invasion, until he finally reached Ravenna. The city held out until April 540, when the Goths, in an attempt to suborn Belisarius, offered to acknowledge him as emperor of the West. This he refused, but by pretending to accept he persuaded the Goths to open the gates to the city. Once inside, he arrested the Gothic leaders and sent them to Constantinople, and took possession of the city in the name of the emperor.

A Persian invasion in 540 caused his dispatch to the east, where he fought indecisive campaigns in 541 and 542. Later, in 542, accused of plotting against the regime by the Empress Theodora, he was removed from office. Once restored to favor, he was sent, with the lower rank of Count of the Stables, to suppress a rebellion in Italy in 544. Although he relieved the sieges of several small towns, he was unable to rescue the city of Rome, which fell to the Goths in December 546. Although Byzantine forces quickly reoccupied the city, a stalemate was reached, and Belisarius returned to Constantinople in 549.

In 551, he headed a party of ambassadors appointed by Justinian I to negotiate with Pope Vigilius, then a prisoner in Constantinople, over ongoing theological disputes. This failed.

In 559 and 560, Belisarius defeated an invading force of Kutrigur Hun in Thrace. In 562, accused again of plotting against Justinian, he was deprived of his offices. Although he again returned to favor, he retired and died around 565.

Joseph M. Isenberg

See also: Byzantine-Persian Wars; Franks; Gothic War; Goths; Huns; Narses; Persian Empire

References and further reading:

Norwich, John Julius. *Byzantium: The Early Centuries.* New York: Alfred A. Knopf, 1989.

Procopius, with an English Translation by H. B.Dewing. 7 vols. Loeb Classical Library. London: W. Heinemann, 1914–1940.

Ben-Bella, Ahmed (1916–)

Algerian nationalist and guerrilla commander. Ahmed Ben-Bella was born on 15 December 1916, near the Moroccan frontier. After being conscripted in 1937, he served with distinction in the French army during World War II, rising to the rank of sergeant. Following the war, a politically active nationalist, he helped establish the headquarters of the groups preparing an armed revolt, out of which grew the National Liberation Front (FLN). The FLN began an armed revolt in November 1954, with Ben-Bella as the most prominent leader, by coordinated attacks on public buildings, military and police posts, and communications installations. In October 1956, while flying from Morocco to Tunis, the French secret services arranged for the pilot to land at Algiers in order to arrest Ben-Bella and four other rebel leaders. Consequently, Ben-Bella was incarcerated during the most intense period of the Algerian conflict. He was released from prison in March 1962 as an aspect of the Evian agreements and cease-fire, and immediately resumed active leadership. In September 1962, the provisional arrangements were ended and Ben-Bella formed the government of independent Algeria.

Ben-Bella began suppressing all opposing factions within and without the FLN, concentrating power on himself. His rigid policies and his suppression of opposition leaders led to his overthrow and imprisonment in June 1965. When finally released in October 1980, he left the country. After the liberalization of Algeria's regime, Ben-Bella returned to Algeria. Taking an extreme militant line, he tried, with very limited success, to rebuild a party of his own and a position of influence, but failed. Currently, at age 85, he resides in Switzerland.

William E. Wingfield

See also: Algiers, Battle of

References and further reading:

Horne, Alistair. *A Savage War of Peace: Algeria, 1954–1962.* New York: Viking Press, 1978. Rev. ed., New York: Penguin Books, 1987.

Bennington (16 August 1777)

A fierce little battle in the American Revolution that frustrated British attempts to sever New England from the remainder of the revolting British North American colonies.

In spring 1777, British general John Burgoyne had led an army southward from Montreal toward Albany, where he expected to connect with a supporting force coming across Lake Ontario and down the Mohawk River and the main British army moving northward from New York City. However, by late summer it was clear that the British in New York City were moving on Philadelphia, and the small force to the west had retreated to Canada. Burgoyne's own force was running low on beef, wagons, and draft animals.

In this situation, Burgoyne sent about 800 men on 11 August on a foraging expedition about 40 miles due east toward Bennington, Vermont (then a part of Massachusetts), which was an American supply point. Several days later, the group, mostly German troops commanded by Colonel Friedrich Baum, paused near Bennington to wait for reinforcements. Meanwhile, small groups of New England–area militia commanded by General John Stark, acting as if they were loyal to the Crown, worked their way around and behind the Germans' positions.

Without warning, the American militia turned on the surrounded troops, killing many of them and pursuing others into the woods. For two hours the battle raged, and eventually the Germans ran out of ammunition.

Around this time, reinforcements—another group of about 700 commanded by Lieutenant Colonel Heinrich von Breymann—were detached from Burgoyne's force and, as they entered the battle, Vermont militia arrived, and so the Hessian reinforcements suffered the same fate as the original foraging party.

As a consequence, the loss of so many men weakened Burgoyne and denied him needed food and supplies; the Americans were emboldened by their victory at Bennington. Eventually, fearing that he was outnumbered and could not retreat to Montreal before winter, Burgoyne would fight twice at Saratoga, and surrender. Burgoyne's capitulation was a triumph that impelled the French to cast off their cautious support for the Americans and to sign a treaty of alliance, which probably provided the margin for American victory.

Charles M. Dobbs

See also: American Revolution; Burgoyne, John

References and further reading:

Hargrove, Richard J. *General John Burgoyne.* Newark: University of Delaware Press, 1983.

Pancake, John S. *1777: The Year of the Hangman.* Tuscaloosa: University of Alabama Press, 1977.

Resch, Tyler. *Bennington's Battle Monument: Massive and Lofty; with a Picture Essay by Julius Rosenwald.* Bennington: Beech Seal Press, 1993.

Berezina River, Battle of (26–29 November 1812)

The Russian pursuit of Napoleon's Grande Armée, and its destruction during this engagement. Napoleon and 110,000 troops retreated from Moscow on 19 October 1812, retracing their path to the Nieman River. They dispersed on the march, abandoning their baggage and also their weapons. Russian forces continued to harass, giving battle at Maloyaroslavets, Krasnoie, and Orsha.

Csar Alexander I directed Marshal Mikhail Kutuzov with 27,000 troops, Admiral Paul Chichagov with 24,000, and Prince Peter Wittgenstein with 34,000 to converge and prevent the French from crossing the Berezina River. On 22 November, Russian forces liberated Minsk, and destroyed French bridgeheads at Borissov on 24 November.

Napoleon conceived a ruse that kept Russian forces south of Borissov, allowing the emperor and 30,000 of his troops to cross the Berezina over two bridges his engineers had constructed at Studianka from 26 to 27 November. Fighting on both sides of the river continued through 28 November. On the morning of 29 November, Napoleon ordered the bridges burnt. A "second" French army of 30,000 stragglers and civilians was abandoned.

Within a week, the number of French effectives dropped to 13,000 as they marched the 160-mile distance from the Berezina River to Vilna. The Russians gave up the pursuit, except for advanced guard action and Cossack raids.

At Smorgoni (40 miles away from Vilna) on 5 December, Napoleon announced that he was leaving his troops to return to Paris. Marshal Murat assumed command of an army that reached Vilna on 8 December. The town had been prepared to feed 100,000 men for 40 days, but the army that arrived was in shambles. Murat could not keep the troops in formation or organized. They looted and pillaged while Russian advance guards and Cossacks threatened attack. Murat and 10,000 men evacuated Vilna on 10 December.

Meanwhile, Napoleon reached Warsaw on 10 December, Dresden on 14 December, Mainz on 16 December, and arrived in Paris before midnight on 18 December.

Approximately 5,000 of the original 600,000 Grande Armée recrossed the Neimen on 14 December (an additional 25,000 straggled across later).

Neville G. Panthaki

See also: Borodino; Kutuzov, Prince Mikhail Illarionovich Golenishchev; Moscow, Retreat from; Murat, Joachim, Grand Duke of Cleves-Berg, King of Naples; Napoleon I; Napoleonic Wars

References and further reading:

Austin, Paul Britten. *1812 Napoleon's Invasion of Russia.* London: Greenhill Books, 2000.

Cate, Curtis. *The War of the Two Emperors.* New York: Random House, 1985.

Nafziger, George F. *Napoleon's Invasion of Russia.* Novato, CA: Presidio Press, 1988.

Nicolson, Nigel. *Napoleon 1812.* New York: Harper & Row, 1985.

Berlin, Soviet Drive on (16 April–2 May 1945)

The defeat of Germany through the capture of the administrative and military heart of Nazi power, by the Red Army. By the end of 1944 the Soviet-German front was along the Narew and Vistula Rivers and Carpathian Mountains. The initial success of Hitler's Ardennes offensive, launched 16 December 1944, prompted Churchill to ask Stalin on 6 January 1945 for Soviet assistance to relieve the pressure, by way of an immediate offensive. Stalin agreed to aid the Allies by launching the Soviet offensive prematurely despite bad weather conditions and incomplete resupply efforts.

Soviet operations toward Berlin began 12 January, as forces of Marshal Ivan Konev's 1st Ukrainian Front struck from bridgeheads at Sandomierz toward Silesia. Forces of Marshal Georgii Zhukov's 1st Belorussian Front and Konstantin Rokossovskii's 2d Belorussian Front followed on 14 January, from bridgeheads near Warsaw toward Berlin and just north of Warsaw on the Narew River toward Danzig, respectively.

The ruins of Warsaw were liberated (after the German crushing of the Polish uprising in the city) on 17 January, and on 19 January Zhukov's armored spearheads drove into Lodz. On 18 January, General Cherniakhovskii's 3d Belorussian Front attacked East Prussia, while General Petrov's 4th Ukrainian Front struck through the Carpathians. By the end of the first week, the Soviet offensive had been carried 220 kilometers deep and was 900 kilometers wide.

Hitler halted the Ardennes offensive, but redeployed the Sixth Panzer Army to Budapest, and did not evacuate 30 German divisions isolated in Courland. Hence, only 12 German armored divisions and 50 understrength infantry divisions defended a 1,500-kilometer front.

By 31 January, Zhukov's forces were at Küstrin on the Oder, 82 kilometers from Berlin. Budapest capitulated on 13 February.

From Vistula-Oder, the Red Army overcame six German defense lines and encountered fortressed cities such as Königsberg, Breslau, Posnen, and Torun, which held out as the Soviet offensive moved westward. Zhukov captured Posnen on 23 February after nearly a month of siege; Königsberg surrendered 9 April to Marshal Aleksandr Vasilevskii, who replaced Cherniakhovskii as commander of the 3d Belorussian Front. Millions of Germans fled ahead of the Soviet advance, and hundreds of thousands died in the winter cold. The Soviet offensive temporarily halted to allow Red Army logistics to reinforce the troops.

Vienna was captured by Generals Malinovskii's and Tolbukhin's troops on 13 April. Zhukov and Konev advanced from the Oder-Neisse Rivers on 16 April, overcoming Berlin's seven defense lines and encircling it by 25 April, the day Soviet and American forces met on the Elbe River.

Hitler committed suicide on 30 April, the Nazi central command dissolved, and Berlin capitulated on 2 May.

Neville G. Panthaki

See also: Ardennes, Battle of, BAGRATION, Operation; Hitler, Adolf; Konev, Ivan Stepanovich; Rokossovsky, Konstantin Konstantinovich; World War II; Zhukov, Georgy Konstantinovich

References and further reading:

Erickson, John. *Road to Berlin.* Boulder, CO: Westview Press, 1983.
Le Tissier, Tony. *Race for the Reichstag.* London: Frank Cass, 1999.
Read, Anthony, and David Fisher. *Fall of Berlin.* New York: W. W. Norton, 1995.
Werth, Alexander. *Russia at War, 1941–1945.* London: Pan Books, 1964.

Bernadotte, Jean Baptiste Jules (1763–1844)

French field commander and diplomat, marshal of France, and founder of a Swedish royal dynasty. Born the son of a lawyer in Pau, France, on 26 January 1763, Bernadotte enlisted in the army in 1780. After affiliating with the Revolution in 1790, he rose rapidly through the ranks, becoming a brigadier general in 1793 and fighting with distinction at Fleurus in 1794 and Tagliamento in 1797. With Jacobin leanings, he remained neutral in the coup of 18th Brumaire in 1799. After holding a variety of civil and military offices, Napoleon made him one of the 18 original marshals of France in 1804.

Bernadotte commanded a corps at Austerlitz in 1805, but failed to engage as ordered at Jena in 1806. His victory over the Russians at Mohrungen on 25 January 1807 did not fully restore him to Napoleon's favor. After his inadequate generalship caused unacceptably high casualties at Wagram in 1809, Napoleon dismissed him on 24 September.

Turning against Napoleon and in command of an allied army in Germany in 1813, Bernadotte defeated Nicolas Charles Oudinot at Grossbeeren on 23 August and Michel Ney at Dennewitz on 6 September and commanded one of the three victorious allied armies at Leipzig. He participated in the push toward Paris in the spring of 1814, but did not oppose Napoleon in the Hundred Days.

Elected crown prince of Sweden in 1810 so that the childless Charles XIII could be peacefully succeeded, Bernadotte, as King Charles XIV John, founded the Swedish royal line that survives into the twenty-first century.

Eric v. d. Luft

See also: Austerlitz, Battle of; Blücher, Gebhard Leberecht von; French Revolutionary Wars; Jena and Auerstädt; Murat, Joachim, Grand Duke of Cleves-Berg, King of Naples; Napoleon I; Napoleonic Wars; Ney, Michel, Duc d'Elchingen, Prince de la Moskova; Oudinot, Nicolas-Charles, Duc de Reggio; Russo-Swedish Wars, Schwarzenberg, Karl Philipp zu; Wagram

References and further reading:

Barton, Dunbar Plunket. *Bernadott.* Vol. 1, *Bernadotte: The First Phase, 1763–1799;* Vol. 2, *Bernadotte and Napoleon, 1799–1810;* Vol. 3, *Bernadotte: Prince and King, 1810–1844.* London: Murray, 1914, 1921, and 1925.
Muriel, John St. Clair. *Sergeant Belle-Jambe: The Life of Marshal Bernadotte, King of Sweden.* London: Rich & Cowan, 1943.
Scott, Franklin Daniel. *Bernadotte and the Fall of Napoleon.* Cambridge, MA: Harvard University Press, 1935.
Wencker, Friedrich. *Bernadotte: A Biography.* London: Jarrolds, 1936.

Berthier, Louis-Alexandre, Prince of Neuchâtel and Valangin, Prince of Wagram (1753–1815)

Marshal of France, commander during the wars of the French Revolution and Empire. Berthier was born on 20 November 1753 at Versailles. He joined the French army as an engineer in 1766. During the American Revolution, he served on the staff of the Comte de Rochambeau. Thus by the outbreak of the French Revolution he had already served 20 years and risen to the rank of lieutenant colonel.

In the early part of the Wars of the French Revolution, Berthier served first on the staff of Rochambeau, and then as chief of staff, first to the Marquis de Lafayette, and then to Count Luckner. He was suspended from the army in 1792 at the outset of the Terror. Reinstated in 1795, he was promoted to *general de brigade* and made chief of staff of the Army of the Alps and Italy. He was promoted in June 1795 to *general de division.* When Napoleon Bonaparte assumed command of the Army of Italy, he retained Berthier as chief of staff.

Berthier would serve Bonaparte as chief of staff almost without interruption throughout the remainder of the Wars of the French Revolution and Empire, finally leaving Bonaparte at the end of the 1814 campaign.

Berthier thus made his reputation as a staff officer, rather than as a field commander. The staff of the French Army of the Consulate and Empire was a model of efficiency for the period and served as an example to be imitated by other European armies. In large part, this success was due to Berthier, who was one of the few individuals who could both reduce the rambling directives of Napoleon Bonaparte to a coherent set of orders and supervise the subsequent execution of those orders.

On the two occasions when Berthier served separately from Napoleon, the results were far less fortunate. Left in command of the forces in Germany in 1809, Berthier proved incapable of responding to the Austrian offensive. Disaster was averted only by the timely arrival of Napoleon. At the conclusion of the 1812 campaign, Berthier was left as chief of staff to Marshal Murat, in command of the army after Napoleon's return to Paris. Berthier proved incapable of steadying the king of Naples, who quickly panicked, and eventually abandoned his command to Eugene Beauharnais.

As a result of his faithful service to the empire, Berthier was created prince of Neuchâtel and Valangin in 1806, and vice-constable of the empire. In 1809, he was created prince of Wagram. In 1814, he accepted the Bourbon Restoration, and was made a peer of France and commander of the Order of St. Louis. After Napoleon's escape from Elba, he, together with other marshals loyal to the Bourbons, accompanied the flight of Louis XVIII from Paris. He then went to Bavaria, where he died on 1 June 1815. Sources disagree on whether Berthier was murdered or committed suicide.

Joseph Isenberg

See also: Aboukir; Austerlitz, Battle of; Berezina River, Battle of; Borodino; Dresden, Battle of; French Revolutionary Wars; Leipzig, Battle of; Lodi; Moscow, Retreat from; Napoleon I; Napoleonic Wars; Pyramids; Rivoli

References and further reading:

Chandler, David G. *Napoleon's Marshals.* London: Weidenfeld & Nicolson, 1987.

Young, Peter. *Napoleon's Marshals.* Reading, UK: Osprey, 1973.

Bismarck, Otto von (1 April 1815–30 July 1898)

Master of Realpolitik, who unified Germany under the leadership of Prussia. Otto von Bismarck was a Prussian Junker, born at Schönhausen, conservative and monarchical in outlook. He was appointed Prussian delegate to the Frankfurt Diet on 15 July 1851, Prussian minister to St. Petersburg in 1859, and Prussian ambassador to Paris in 1862. From 1862 to 1890, Bismarck served as foreign minister and minister president of Prussia.

Bismarck was not a German nationalist; he led the process of German unification to ensure Prussian control of the outcome. His only concerns were to make Prussia great in Germany and make Germany great in Europe.

On 30 September 1862, Bismarck delivered his notorious speech, stating that debate would not solve the issue of German unification, but "blood and iron" would. He manipulated European politics and became chancellor of a unified Germany under the leadership of King Wilhelm I after three wars: Schlesswig-Holstein (Denmark), January–July 1864; Austro-Prussian War, June–July 1866; and Franco-Prussian War, July 1870–January 1871.

Bismarck restrained General Count Helmuth von Moltke after the Austrian defeat, and disallowed a Prussian military procession through Vienna or the levying of a war indemnity.

However, Bismarck did not act with similar prudence at the conclusion of the Franco-Prussian war. There was a Prussian military parade through Paris; a war indemnity against France of 5 billion francs; the occupation of northern France until the indemnity was paid in 1873; the proclamation of Wilhelm I as emperor of Germany in the Hall of Mirrors in Versailles; and, most galling for the French, the annexation of Alsace and Lorraine to Germany. War, the destruction of Metternich's Congress System, and the rearrangement of the balance of power had made Germany a unified nation. From 1871 to 1890, Bismarck worked to ensure the survival of his creation within the new status quo.

Domestically, Bismarck consolidated Prussia's political, military, and economic dominance within Germany. He became the "honest broker of peace" in Europe, attempting to prevent war or otherwise to localize conflict. Bismarck did not wish to embroil Germany in a proxy war against Russia on Austria's behalf in the Balkans. Bismarck also strongly sought to avoid a two-front war involving Germany, against France and Russia.

Bismarck also promoted colonial expansion in Africa to ease tension in Europe by diverting attention from growing German influence and providing an outlet for the grandeur of France and England. He also staunchly believed that strong German-Russian relations would keep France isolated.

Bismarck's multiple priorities complicated his diplomacy, committing Germany in a series of alliance treaties. In these arrangements, Germany was "one of three so long as there are five great European powers."

Bismarck was a diplomatic wizard; however, his successors were unable to maintain Germany's alliance web. The contradictory nature of Germany's multiple contractual ties, and its breakdown, was one of the causes of World War I.

As Prussian and German chancellor, Bismarck constantly empowered the army with state revenue and patronage. Force had created Germany and would defend it, but Bismarck ensured that the military remained subservient to the state.

Wilhelm I allowed Bismarck to rule Germany by proxy. Bismarck resigned in March 1890, because he was not granted the same privilege by Wilhelm II.

Neville G. Panthaki

See also: Franco-Prussian War; German Wars of Unification; Königgrätz, Battle of; Revolutions of 1848; Russo-Turkish Wars

References and further reading:
Eyck, Erich. *Bismarck and the German Empire.* London: Unwin University Books, 1968.
Gall, Lothar. *Bismarck, the White Revolutionary.* Boston: Allen and Unwin, 1986.
Pflanze, Otto. *Bismarck and the Development of Germany.* Princeton, NJ: Princeton University Press, 1990.
Taylor, Alan John Percivale. *Bismarck.* London: Hamish Hamilton, 1966.

Black Patch War (1904–1909)

A conflict that broke out between tobacco farmers in the Black Patch region of Kentucky and Tennessee, and the American Tobacco Company (ATC) and other tobacco companies. The area had taken its name from the dark tobacco leaves produced for European markets. The tobacco market had increasingly come under the control of a few giant firms, led by the ATC, that had reached an understanding in order to avoid competition and reduce prices paid for tobacco. Prices had declined by 48 percent between 1874 and 1894. In order to defend themselves, farmers organized themselves into the Planters Protective Association (PPA) in 1904. The PPA, led by wealthy planter Flex Ewing and other elite landholders, sent organizers to educate and recruit other farmers in order to form cooperative pool crops, and withhold them from the market until better prices were offered. By 1907 PPA members had pooled 50,000 hogshead of tobacco and nearly caused the price the ATC paid for their crop to double.

The PPA ultimately had to resort to violence in order to ensure solidarity among tobacco producers. By 1905, some of the members of the PPA formed a secret organization called the night riders. They terrorized, whipped, and on occasion murdered farmers who sold their tobacco to the tobacco trusts and others whose interests stood in opposition to the PPA. With the silent support of law enforcement officials and private citizens, they even intimidated local courts. Their campaign culminated on 6 December 1907 in a raid by 500 masked men on the city of Hopkinsville, Kentucky, that resulted in more than $200,000 worth of property damage when several tobacco warehouses were set afire. Major Erskine Birch and the town's militia pursued the vigilantes and engaged a small group of them, killing a teenager. In reaction, Kentucky governor Augustus Wilson sought to penetrate the vigilantes with informers and to send troops where they were needed. Yet, this strategy met with little success and Wilson then established a defensive strategy of placing troops near likely targets.

By 1908, the target of night riders' violence had begun to change. A second wave of violence was directed at black people, as well as prostitutes, bootleggers, and others who offended community morals. The source of this second wave came not from the tobacco farmers but was led by local white ironworkers. Planters soon became concerned that these actions were driving off the laborers they needed. As the violence broadened and became more chaotic, Black Patchers turned against vigilantism. After threats about burning the town of Murray, Judge A. J. G. Wells asked Wilson to dispatch troops to guard Calloway County. State and local courts began to assert their authority and ended the lawlessness. Under financial strains, the PPA buckled and local militias ensured that order was maintained.

T. Jason Soderstrum

See also: Guerrilla/Partisan/Irregular Warfare

References and further reading:
Campbell, Tracy. *The Politics of Despair: Power and Resistance in the Tobacco Wars.* Lexington: University Press of Kentucky, 1993.
Marshall, Suzanne. *Violence: In the Black Patch of Kentucky and Tennessee.* Columbia: University of Missouri Press, 1994.
Waldrep, Christopher. *Night Riders: Defending Community in the Black Patch, 1890–1915.* Durham, NC: Duke University Press, 1993.

Blenheim-Höchstädt, Battle of (13 August 1704)

Marlborough's masterpiece. The year 1704 saw the rising star of both Marlborough and Prince Eugene de Savoie; Marlborough had decided to come and help the imperials in Bavaria. The strategic options were significant: the allies had to protect Vienna and a victory would give them Bavaria; a defeat meant a Franco-Bavarian invasion of Austria. On 21 May, Marlborough joined Eugene's army. Together they won a first battle at Donauwörth in early July. This victory gave them a base for further operations in Bavaria. The elector of Bavaria avoided any clash with the invading armies. He needed first to join Marechal de Tallard and his French army.

The month of July saw allied troops committing the Rape of Bavaria, as this period of burning and looting was known. Now the Franco-Bavarians tried to draw Marlborough and Eugene to battle, as their armies were separated by the Danube River. By a series of rapid marches, Marlborough came to help Eugene, which was withdrawing slowly to protect Donauwörth. Tallard and the elector thought they had achieved a strategic success in bringing Marlborough out of Bavaria. In Marlborough's mind, it was time for a decisive battle.

The Franco-Bavarian forces (60,000) were positioned east of Höchstädt. Their deployment was covered by the Danube on their right and by wooded hills on their left; the center was protected by a marshy river (the Nebel). Three villages were occupied: left to right, Lutzingen, Oberglau, and Blindheim (Bleinheim). Marlborough's opponents were unaware of his position, and orders were given to send out foragers. At 6 A.M. on 13 August, nine allied columns (52,000) appeared out of the woods, heading toward the Nebel. Until noon, the two facing armies deployed for battle; the Franco-Bavarian army separated in two groups, the elector and Marechal Marsin on the left, and Tallard on the right. Their infantry had deployed in the three villages with cavalry on their flanks making the center of the line, between Oberglau and Blindheim, a weak spot. Marsin deployed to fight by the Nebel, trying to prevent any crossing; Tallard chose to hold a defensive line more than 500 yards further with Blindheim as a strong point. Marlborough ordered a general advance at 12:30, and Blindheim was assaulted by Lord Cutt's troops, who were repulsed with heavy losses. This attack was so fierce that the French general Clerambaut decided to move more infantry into Blindheim, cramming 27 battalions in the village. The French center was left in charge of the cavalry. Marsin and Eugene faced each other in a fight lasting until 4 P.M. When the Franco-Bavarians retired behind Lutzingen, Eugene's troops were too exhausted to go further. Marsin then tried to attack the flanks of Marlborough's cavalry squadrons and infantry lines advancing across the Nebel. The furious cavalry fight lasted with various degrees of success till a general charge led by Marlborough separated the French in two parts. Marsin escaped slowly into the night. Tallard was taken prisoner by Hessian cavalry he thought to be French. At 9 P.M., to the disgrace of France, more than 10,000 soldiers surrendered in overcrowded Blindheim. The Franco-Bavarian army was destroyed, as more than 30,000 were either killed, prisoners or drowned trying to cross the Danube.

Austria was safe, the allies overran Bavaria, and Louis XIV's army might belonged to the past.

Gilles Boué

See also: Malplaquet, Battle of; Marlborough, John Churchill, First Duke of; Oudenaarde, Battle of; Ramillies, Battle of; Spanish Succession, War of the

References and further reading:

Chandler, David. *Marlborough as a Military Commander.* London: 1973.

Grant, Charles. *From Pike to Shot.* Lancing, UK: WRG Publishing, 1986.

Jones, J. R. *Marlborough.* Cambridge, UK: Cambridge University Press, 1993.

Bloch, Jean de (1836–1902)

Industrialist, philanthropist, and antiwar activist, also known as Jan Gotlib Bloch and Ivan Stanislavovich Bliokh. Born on 30 July 1836 in Radom, (Russian) Poland, Jean de Bloch amassed a very large fortune in the railroad, banking, insurance, sugar, bakery, and forestry businesses. An enlightened entrepreneur, Bloch proposed a pension plan for railway workers—over strong opposition from other business leaders, and even from established charitable and educational institutions. Bloch also wrote the first modern "systems analysis" of warfare. He was not without honor in his own country, and Czar Alexander II ennobled him and gave him the title State Councillor of the Empire.

Bloch also had a notable research and literary career, winning many gold medals at international expositions. He pioneered the use of research institutions to examine social and political questions, including a Statistical Bureau, which he founded in Warsaw, that employed prominent scholars and writers.

Bloch began to think about warfare while organizing railway supplies for the Russo-Turkish War. In 1888, he started work on *Memorial on the Defense of Warsaw.* With the encouragement of Josif Hurko, the czar's governor-general of Poland, Bloch began to write articles explaining the social and economic effects of war to officers, and the military aspects of war to civilians. Bloch's first studies on war were published in Warsaw's monthly *Biblioteka Warszawska* between March 1893 and September 1894. Translations of these chapters were published in Russian, French, and German journals.

In 1898, Bloch published his six-volume *Buduschaya voina [The Future of War from Its Technical, Economic and Political Points of View]* in which he argued that the modern technology of arms design substantially increased the lethality and power of defense. In particular, the magazine rifle, smokeless powder, flat trajectory bullet, smaller rifle bore, quick-firing artillery, and high-explosive artillery shells would create a 1,000-meter fire-swept zone, making a frontal assault against an entrenched defense suicidal. This zone could be crossed, but such a crossing would require the attacking forces to have a numerical advantage of eight to one, a number Bloch derived from his analysis of actual operations in the 1877–1878 Russo-Turkish War and of the 1870–1871 Franco-Prussian War, and of experiments of his own design conducted in Switzerland. The new weapons technologies, combined with mass conscription armies and railroad-generated strategic mobility, had transformed warfare into a long-term stalemate decided by attrition and the economic exhaustion of the continental powers. Bloch concluded that "under the military, social, and economic conditions of Europe at the present day, though it is quite possible that war may break out, it is almost impossible for it to be waged successfully."

Bloch's deductions on tactics and logistics contained far more realistic insights than those of his contemporaries, and he dealt with problems that no one else thought about. He alone questioned the effects of a long, drawn-out war on the ability of civilians to bear great privation and hardship and on the stability of the European social order. He also anticipated the concept of a "war economy," or the relationship of commerce, industry, and agriculture to the military effort.

His six-volume work, *The Future of War,* was approved by the Russian censor and was published only after Bloch had an audience with Czar Nicholas II. *The Future of War* was translated into German and French in full, and into English in an abridged version. In the last three years of his life, Bloch worked tirelessly for the cause of peace, and his work helped inspire Czar Nicholas II to call for the 1899 Hague Peace Conference. Bloch died on 6 January 1902 in Warsaw, some 14 years before World War I proved so many of his predictions correct.

Mark D. Mandeles

See also: Pacifism/War Resistance; World War I

References and further reading:

Bliokh, Ivan S. *The Future of War in its Technical, Economic, and Political Relations.* Russian ed. 6 vols. St Petersburg: Tipografiya I A. Efrona, 1898.

Bloch, I. S. *The Future of War in its Technical, Economic, and Political Relations.* Trans. by R. C. Long. Boston, MA: Ginn & Co., 1902.

Bloch, Jean de. *The Work of the Peace Societies: How to Widen Their Programme.* Chatham, UK: Observer Works, 1901.

Van den Dungen, Peter. *A Bibliography of the Pacifist Writings of Jean de Bloch.* London: Housmans, 1977.

Blood River (Ncome) (16 December 1838)

Opened the Zulu kingdom to white settlement, and is still a potent symbol for competing Afrikaner and Zulu nationalism. In February 1838 hostilities erupted between the Zulu and the Voortrekkers, or emigrant farmers (Boers) from the Cape Colony, who had invaded the Zulu kingdom in search of new lands to settle, free from British rule. The succeeding months of inconclusive fighting showed that the normal Zulu tactic of enveloping their foe preparatory to closing in hand-to-hand combat with the stabbing spear could not succeed against a laager, or the Boer fighting formation of wagons drawn up end to end in a circle, and defended by muskets and cannon shooting in ordered rotation to keep up an uninterrupted rate of impenetrable fire.

In late 1838 a commando of armed horsemen under Andries Pretorius advanced east into Zululand, on 15 December laagering its 64 wagons on a spit of land between the Ncome River to the east and a dry watercourse to the south. The defenders, who consisted of 472 Boers, three white traders from Port Natal, and 120 Port Natal Africans under Alexander Biggar, were thus allowed to concentrate along the west and north faces of the laager. Inside were some 700 oxen, 750 horses, 130 black wagon-drivers, and 200 grooms. The Zulu army of between 12,000 and 16,000 men under Ndlela kaSompisi Ntuli and Nzobo kaSobadli Ntombela advanced from the southeast before dawn on Sunday, 16 December. The left horn of 3,000 younger warriors came on precipitately in advance of the chest and right horn, crossed the Ncome south of the laager, and at about 6:30 A.M. charged it from the west and north. Repulsed by the defenders' fire, the left horn broke and fled, pursued some way by Boer horsemen. At about 8 A.M. the Zulu right horn, followed by the chest, attacked along the same route as the already defeated left horn, but were also unable to break through the Boer zone of fire. Some Zulu units began to withdraw in disarray at about 11 A.M. Pretorius and about 160 mounted men sallied out in a pursuit that lasted several hours. The Zulu army broke and fled in all directions. Their slaughter bloodied the waters of the Ncome, which the Boers renamed the Bloed, or Blood River. Three Boers were wounded, but well over 1,000 Zulu were killed.

His army scattered, King Dingane was soon forced to cede the Boers half his kingdom.

John Laband

See also: Ndlela kaSompisi Ntuli; Zibhebhu kaMaphitha Zulu

References and further reading:

Bird, J., compiler. *The Annals of Natal 1495–1845.* Vol. 1. Cape Town: Struik, facsimile reprint, 1966.

Laband, John. *The Rise and Fall of the Zulu Nation.* London: Arms & Armour Press, 1997.

Nathan, M. *The Voortrekkers of South Africa.* London: Gordon & Gotch, 1937.

Blücher, Gebhard Leberecht von (1742–1819)

Fiery hero of Prussia's war against Napoleon. Born 16 December in Rostock, Blücher grew up in Mecklenburg. Poorly educated, he volunteered in a Swedish Hussar regiment but was captured by Prussians in 1760. The regiment's colonel liked the youth and made him an adjutant. Blücher gained a reputation for gambling and drinking, often dueling to extricate himself from difficulties, but his behavior cost him a promotion in 1771. He appealed directly to Frederick the Great, who replied that "Blücher may go to the devil," and cashiered him. For several years, he tended his small landholdings. After the death of the king, Blücher was reinstated as major in his former regiment. He was lieutenant colonel in 1788 and served with the duke of Braunschweig on the Rhine, where he led dashing charges against the French and earned the rank of colonel in 1794.

In 1801, he was named lieutenant general and governor of Münster where he was outspoken against Napoleon. He led the advance guard at Auerstädt in 1806, and subsequently, the army's rear guard. He surrendered at Lübeck, and following the Treaty of Tilsit, Blücher was military governor of Pomerania. When a Prussian contingent assisted Napoleon's invasion of Russia in 1812, Blücher's rage forced the king to exile him temporarily to Silesia. He returned after Prussia's alliance with Russia in 1813, and received 38,000 Prussians and Russian troops.

In May, he battled Mortier's Corps at Lützen, and he commanded the allied right wing at Bautzen. He protested the armistice in June, but when the war resumed in August, Blücher's Army of Silesia numbered 90,000 men. Moreover, Gneisenau became his chief of staff and held the general's unconditional trust. On 26 August, the Army of Silesia crushed Marshal Macdonald on the Katzbach, liberating Silesia. Refusing to abide conservative strategy, Blücher crossed the Elbe, eluded Napoleon, and approached Leipzig. Now a field marshal, Blücher defeated Marmont at Möckern on 16 October, attacked Leipzig on 18 October, but failed to block the French retreat. Blücher's army crossed into France at Kaub on 1 January, and "Feldmarshal Vorwärts" raced toward Paris.

Napoleon defeated the Prussians at Brienne, Vauchamps, and Craonne, but Blücher checked the emperor at Laon before advancing into Paris on 31 March.

In 1815, Blücher moved from the Rhine to join Wellington in Belgium. At Ligny on 16 June, Blücher was extricated from beneath his dead horse while Gneisenau ordered a withdrawal toward Wavre. That decision allowed the Prussians to fall on Napoleon's right flank at Waterloo two days later. The aging hero was celebrated throughout Prussia, but he fell ill and died on his property of Krieblowitz on 12 September 1819.

Llewellyn Cook

See also: Gneisenau, August Neidhart von; Jena and Auerstädt; Leipzig, Battle of; Waterloo

References and further reading:

Britt, Albert Sidney. *The Wars of Napoleon.* Wayne, NJ: Avery Publishing, 1985.

Petre, F. Lorraine. *Napoleon's Last Campaigns in Germany.* London: J. Lane, 1912.

Weigley, Russell F. *The Age of Battles: The Quest for Decisive Warfare from Breitenfeld to Waterloo.* Bloomington: Indiana University Press, 1991.

Boer Wars (1880–1902)

Conflict of British imperialism and Afrikaner nationalism. After the discovery of diamonds in the 1860s, British politicians decided that the best way to administer South Africa and protect their commercial interests was through confederation. Britain annexed the Transvaal in 1877 to accomplish this goal. Because of the poor state of the republic's finances and the Zulu military threat, they did not anticipate resistance. But the Boers rejected confederation and declared war. The British were ill-prepared to fight the First Boer War (1880–1881), having underestimated the ability and determination of the Boer horse commandos. With fewer than 3,500 troops in South Africa, and only half of these in the Transvaal, British forces were dangerously outnumbered. In the war's defining moment, 180 Boers climbed Majuba Hill at night, where nearly 600 British soldiers sat atop a seemingly impenetrable position. The Boers took them completely by surprise. British casualties approached 250; the Boers lost one man.

Majuba Hill was the last battle of the war. Rather than commit to a protracted struggle, the British sued for peace. The Pretoria Convention of 1881 returned self-government to the Transvaal, although the British maintained suzerainty over the republic's foreign affairs. The peace agreement, revised by the London Convention (1884), made little reference to this vague notion and the exact relationship of Great Britain and the Transvaal remained unclear for the next 15 years.

With the discovery of gold near Johannesburg and the influx of thousands of migrant workers, there was renewed interest in South Africa. Cecil Rhodes, prime minister of the Cape Colony, mining magnate, and imperialist, wanted to see a unified British South Africa. Rhodes attempted to physically encircle the Boer Republics, thus ensuring their economic dependency on British, who controlled railroads and ports. When this plan failed, he tried to overthrow the Transvaal government in the ill-conceived Jameson Raid (1895). The disastrous adventure heightened anxiety and deepened mutual distrust.

Tension increased with the establishment of a German colony in southern Africa, growing pressure from British capitalists to break up the Transvaal's economic monopolies, and the inability to solve the issue of paramountcy. As the Transvaal signed a military alliance with the Orange Free State and purchased weapons, the British focused on the issue of the Uitlanders (immigrant workers). Alfred Milner, the high commissioner, believed that this disenfranchised population of mostly British and British colonial subjects could gain control of the republic through democratic means. Milner made their right to vote, whether or not a legitimate concern of the British government, the central issue of all negotiations with Paul Kruger, the Transvaal president. When negotiations failed in the summer of 1899, the British

A company of Boer soldiers; prisoners of war in Simons Town, South Africa, c. 1901. (Library of Congress)

prepared an ultimatum. The Boers forestalled them, declaring war on 12 October.

The (second) Boer War (1899–1902) did not open well for the British, who seemed to have learned nothing from the First Boer War. Due mostly to political constraints, the army lacked numbers, equipment, and intelligence. To make matters worse, strategy had been determined by an angry public. The Boers had invaded the Cape Colony and Natal, laying siege to Ladysmith, Kimberley, and Mafeking. The public demanded the immediate relief of these three towns. The British commander, Redvers Buller, complied and against his better judgment divided his forces. The result was December's "Black Week," in which three British armies were defeated at Stormberg, Magersfontein, and Colenso.

British forces were strengthened with the arrival of several divisions and the wide-scale recruitment of volunteers. Over the next year, Lord Roberts, the new commander, advanced through the Orange Free State and the Transvaal. Boer commandos refused to surrender and turned to guerrilla tactics, destroying railroads and telegraph lines and attacking isolated outposts. In 1901, Lord Kitchener, succeeding Roberts, employed controversial counterinsurgency techniques to break Boer resistance. To limit their movement, the British constructed blockhouses. To strip the Boers of their resources, the British scorched land, seized livestock, and relocated civilians.

Concentration camps eventually housed more than 150,000 men, women, and children. Poorly situated and maintained, they had an alarmingly high rate of mortality and 28,000 Boers and more than 14,000 black Africans died in them, leaving a lasting legacy of hatred for the British by the Boers.

Although Boer commandos managed to sustain a vigorous resistance in the guerrilla phase of the war, the plight of their civilians, the devastation of their lands, the growing fear of armed blacks, and the numerical superiority of the British forced them to the negotiating table. The Treaty of Vereeniging, 31 May 1902, ended the war on mild terms for the Boers.

Stephen M. Miller

See also: Botha, Louis; Buller, Sir Redvers Henry; Colenso, Battle of; De Wet, Christiaan Rudolph; French, John Denton Pinkstone, First Earl of Ypres; Hamilton, General Ian Standish Monteith; Kimberley, Siege of; Kitchener, Horatio Herbert; Kruger, Paul-Stephanus Johannes Paulus; Ladysmith, Siege of; Magersfontein, Battle of; Roberts, Frederick Sleigh, First Earl, Viscount St. Pierre of Kandahar; Smuts, Jan Christian

References and further reading:

Nasson, Bill. *The South African War.* London: Arnold, 1999.
Pakenham, Thomas. *The Boer War.* New York: Random House, 1979.
Warwick, Peter, ed. *The South African War.* London: Longman, 1980.

Bogomils' Revolt (1086–1091)

A rebellion against the Byzantine Empire in what is now Bulgaria. The Bogomils were a dualist sect present throughout the Mediterranean world, variously known as the Paulicians in Anatolia, Bogomils in eastern Europe, and Cathars or Albigensians in western Europe. Although the heresy was poorly regarded officially, such was the Byzantine Empire's need for manpower that Alexius I Comnenus (r. 1081–1118) raised a unit of Bulgarian dualists from the area around what is now Plovdiv.

This unit proved unsatisfactory and was dismissed by Alexius in 1085 or 1086. It promptly revolted, and was joined by coreligionists, as well as some Serb and Vlach freebooters.

The rebels seized the fortress of Beliatoba, where the Bogomils then invited Petcheneg nomads to join them. A force of Petchenegs crossed the Danube and marched toward the rebels. A small Byzantine force sent to stop them was defeated, but a second force compelled the Petchenegs to withdraw.

In 1087, a much larger Petcheneg force, numbering 80,000, crossed the Danube under Tzugalu. Though this force suffered an initial defeat, a large number joined the Bogomils and camped near the town of Paradunavum. Alexius I Comnenus decided to retake the town and marched against them. At Dristra, the Byzantine army was mauled and Alexius forced to flee.

For the next three years, the Petchenegs and rebels raided throughout the western parts of the empire. In 1090, another Byzantine army was defeated and destroyed.

In the spring of 1091, Alexius persuaded the Cumans, the traditional enemies of the Petchenegs, to attack the rebellious force. The Cumans heavily defeated the Petchenegs and recrossed the Danube. Alexius settled the surviving Petchenegs in the area around Salonika. A number were enrolled into the Byzantine army.

Alexius did not proceed further against the dualist heretics until the last two years of his reign. At that time, under pressure from the Church to establish religious conformity, many Bogomils were forcibly converted, and their leaders either imprisoned or executed.

Joseph M. Isenberg

See also: Alexius I Comnenus

References and further reading:

Fine, John V. A. *The Early Medieval Balkans.* Ann Arbor: University of Michigan Press, 1983.
Treadgold, Warren. *A History of the Byzantine State and Society.* Stanford CA: Stanford University Press, 1997.

Bohemian Civil Wars (1448–1478)

Wars between various claimants for the Bohemian throne. The peace established by the Compacts of Jihlava (1436) was strained by the successive deaths of Bohemian kings Sigismund (d. 1437) and Albrecht (d. 1440), and the control exercised by Holy Roman Emperor Frederick III over Albrecht's infant son and putative heir, Ladislas Posthumous. Conflict between Catholic and Hussite factions resurfaced in the interregnum, culminating in an attempt by Catholic nobles of the League of Strakonice to seize Prague. It was frustrated by the Hussite George Podebrady (1448). In 1450,

Podebrady defeated the League at Rokycany, and then invaded Saxony to cut off the Catholic party from the support of Duke Frederick.

Having secured peace, Podebrady moved to protect the Catholics from Hussite reprisals, forcing the radical stronghold of Tabor to submit to his rule in 1452. In alliance with Count Cilli of Austria and János Hunyadi of Hungary, Podebrady forced Frederick III to release Ladislas in 1453; Podebrady was named regent for Bohemia in Ladislas's name.

While visiting Prague for his wedding in 1457, Ladislas became ill and died. The Bohemian estates rejected the claims of Ladislas's relatives, and elected Podebrady king. In 1458 Podebrady defeated an invasion of Moravia by Archduke Albert Habsburg of Austria, and induced William of Saxony to recognize his election, forwarded by the Congress of Cheb in 1459.

Though he sought to pacify the country, Podebrady could not completely satisfy the baronial party, nor would he meet the Catholics' demands to eliminate Hussitism. Encouraged by Pope Paul II and by Frederick III, Catholic barons formed the League of Zelená Hora, and, together with the cities of Pilsen and Breslau, revolted against Podebrady's rule in 1467. Moving quickly, Podebrady destroyed the forces of Breslau, subdued Lusatia and Silesia, and secured a truce with the league.

Offered the Bohemian crown by the pope, Hungarian king Matthias Corvinus invaded, and drove Podebrady out of Moravia in 1468. Podebrady counterattacked the following year, defeating Matthias and forcing him to withdraw. Matthias raided Bohemia in 1470 without result. Podebrady's death the following year led to the election of Vladislav II Jagiellon by the Bohemian estates, though Matthias refused to recognize the election, and continued his war against the Hussite party. Unable to gain ground in Bohemia, he held on to the dependent territories of Moravia, Silesia, and Lusatia. Placing his heavy forces in the towns, he used his light cavalry to devastate the countryside, forcing Vladislav to withdraw in a rare example of a besieged army starving the besiegers into submission. By the Peace of Olomouc (1478), Vladislav retained Bohemia proper and Matthias gained the dependencies, with both men sharing the royal title.

Brian Hodson

See also: Hunyadi, János; Hussite Wars; Matthias I

References and further reading:

Heymann, George. *George of Bohemia: King of Heretics.* Princeton, NJ: Princeton University Press, 1965.

Rázsó, Gyula. "The Mercenary Army of King Matthias Corvinus." In *From Hunyadi to Rákoczi: War and Society in Medieval and Early Modern Hungary,* ed. János Bak and Béla Király, 125–140. Brooklyn, NY: Brooklyn College Press, 1982.

Full-length portrait of Simon Bolivar. (Library of Congress)

Bolivar, Simon (24 July 1783–17 December 1830)

South American liberator. Bolivar was primarily responsible for freeing most of Latin America from Spanish rule. Bolivar was born into a wealthy Creole family in Caracas, Venezuela. Orphaned as a child, he was raised by an uncle and tutors. He completed his education in Europe, where he married a Spanish noblewoman. They returned to Caracas in 1802, but she died within a year. Bolivar returned to Europe where he systematically studied Enlightenment thinking. He witnessed Napoleon proclaim himself emperor as well. Returning to Venezuela, Bolivar joined the movement for independence after the Bourbon monarchy of Spain was deposed by Napoleon. In 1810, the ruling junta of Venezuela sent Bolivar to England to seek assistance. Unsuccessful in that mission, Bolivar returned to command the important fortress of Puerto Cabello. Some of his officers betrayed the fortress to royalists, leading to the collapse of the independence movement in Venezuela.

Bolivar vowed to continue the fight for independence. He retreated to New Granada (modern Colombia), where he took command of a patriot army. He defeated royalists seeking to conquer New Granada, then invaded Venezuela in May 1813. Defeating the Spanish forces in six battles, Bolivar occupied Caracas on August 6. He was given the nickname El Libertador (Liberator), which he kept for the remainder of his career. Bolivar made himself dictator, but was soon overthrown by royalist forces. He fled again to New Granada, but was defeated there. Bolivar went into exile in 1815 to Jamaica and Haiti. During this time, he wrote his famous "Letter from Jamaica," outlining his hopes for independence for Latin America.

With weapons supplied by Haiti, Bolivar returned to Venezuela. He was defeated and fled to the Orinoco region in 1818. There, Bolivar built a new army. He united the irregular cavalry of the region, the *llaneros,* with thousands of British and Irish mercenaries, veterans of the Napoleonic Wars. On 11 June 1819, Bolivar left his base with 2,500 men and marched through the Andes via the icy Pisba pass. The Spanish troops in New Granada were caught by surprise and defeated. On 10 August 1819, Bolivar captured Bogota and became president of newly independent Colombia.

In December 1819, Bolivar sponsored the declaration of Gran Colombia, a union of Venezuela, Colombia, and Ecuador. A truce forestalled hostilities during 1820, but Bolivar invaded Venezuela again in 1821. He defeated the royalists at the Battle of Carabobo on 25 June 1821, freeing virtually all of Venezuela. He quickly moved his army into Ecuador. A two-pronged invasion with General Antonio de Sucre freed Ecuador by the end of 1822. A secret meeting with Jose de San Martin, who had freed southern Latin America, resulted in Bolivar becoming the single head of the liberation army. He took over command of San Martin's army in Peru and completed the liberation of that state. The region known as Upper Peru became a separate nation, and the grateful citizens named it Bolivia after the great liberator.

In 1826, Bolivar called a pan-American meeting, to sponsor unity between the newly independent states. It was sparsely attended and produced no lasting results. Bolivar's last four years were occupied with civil war and dissent. He decided to go into voluntary exile in 1830, but died of tuberculosis before he could leave. Bolivar has another claim to fame, as one of only two persons in modern times to have a nation—Bolivia—named in his honor. (Cecil Rhodes and Rhodesia was the other—and that only temporarily.)

Tim J. Watts

See also: South American Wars of Independence

References and further reading:

de Madariaga, Salvador. *Bolivar.* Westport, CT: Greenwood Press, 1979.

Masur, Gerhard. *Simon Bolivar.* Albuquerque: University of New Mexico Press, 1948.

Mijares, Augusto. *The Liberator.* Caracas: North American Association of Venezuela, 1983.

Bolshevik Revolution (1917–1921)

Political-military coup that imposed Marxist-Leninist communism on Russia from 1917 to 1991. In 1903, organizational, ideological disputes split the Marxist Russian Social Democratic Labor Party into two wings, Mensheviks and Bolsheviks. There were also numerous underground movements in exile; their divisions grew by 1917, but the dominant Bolshevik, Vladimir Lenin (né Ulaynoff), brooked no compromise with his version of Marxism.

In Russia itself, World War I privations exacerbated underlying socioeconomic tensions, which had first surfaced in the 1905–1907 Revolution. There was dissatisfaction in the countryside with the 1861 serf emancipation, and urban discontent with living conditions resulting from the 1890s' industrial growth.

A bloodless February 1917 revolution overthrew Csar Nicholas II, leading to establishment of a provisional government (PG), a self-appointed body of prowar liberals and industrialists to rule until national elections could be held for a constituent assembly. The Petrograd Soviet, comprising Russia's socialist movements, appeared, issuing Soviet Order No. 1 (1 March), declaring the Russian army's democratization, gaining its support, and effectively controlling the PG's actions, but it was not prepared to take power itself.

Lenin and his fellow revolutionaries returned from exile to Russia by the Germans in the famous "sealed railway car." But on his arrival Lenin issued his "April Theses," declaring no support for the provisional government and proclaiming the popular slogans "All Power to Soviets," and "Peace, Bread, Land," which separated the Bolsheviks from moderate socialists.

The provisional government postponed elections; and did not end the war, launching instead the disastrous June Kerenskii offensive. It also did not or could not improve urban conditions or pass land legislation. Nonetheless, moderate socialists joined the provisional government in July, and the Bolsheviks were driven underground after the "July Days" demonstrations.

General Kornilov's failed right-wing military coup in August swung the pendulum back left, with grassroots Bolsheviks organizing railway workers. Red Guards prevented Kornilov's forces from reaching Petrograd. Now with popular urban support, the Bolsheviks gained control of the Petrograd and Moscow soviets.

Without the peasant support necessary for a majority in

Bolsheviks on the street during the Russian Revolution, 1918. (Library of Congress)

the constituent assembly elections and no majority guaranteed at the 2d Congress of Soviets in November, Lenin (in hiding) pushed a divided Bolshevik Party toward insurrection. A Military Revolutionary Committee was formed, through which Trotsky directed the seizure of power on 25 October (Old Style).

Despite holding Petrograd, Moscow, and other towns, the Bolsheviks had tenuous control of Russia. They quickly passed decrees satisfying soldiers, peasants, workers, and national minorities. Their calls for international peace were ignored by the Allies (who believed them to be a tool of the Germans). Lenin's new regime therefore independently signed a cease-fire with Germany in December and the harsh Brest-Litovsk peace in March 1918.

The Communist Party's "carrot" policies toward the peasants and others were accompanied by the "stick" of the All-Russian Commission for Struggle Against Counter-Revolution and Sabotage (CHEKA) formed December 1917 under Feliks Dzierzynski. CHEKA gained more powers as Bolsheviks intensified their "dictatorship of proletariat," deemed necessary to attain true socialism in the face of a hostile coalition of the imperialistic-capitalistic powers. The November constituent assembly elections gave the Social Revolutionaries 40 percent of the vote, against only 25 percent for Bolsheviks. This minority vote meant nothing to the Bolsheviks; the assembly met only once before Red Guards closed it permanently. All opposition parties—monarchist, liberal, socialist—were banned by mid-1918.

The Workers and Peasants Red Army (RKKA), formed in January 1918 to face the Germans, was replaced by the demobilized Czarist Army. Under Trotsky, the new Red Army was built around Latvian Rifles, Red Guards, partisans, and worker detachments to a force of 5 million by 1920, running the Soviet Union as a "one-armed camp" under "War Communism" during the subsequent Russian Civil War. His political control secured by the victorious civil war, Lenin reverted to a "New Economic Policy" that permitted some limited private enterprise. But he also continued the "temporary" wartime governmental measures, which included terror as a weapon and the stifling of all political opposition, thus preparing the way for Stalin's rise to power. Nonetheless, the Communist Party did not achieve control over the peasantry until Stalin's 1930s collectivization, which resulted in the deaths of millions. The Soviet state that emerged was not that initially envisaged by the Bolsheviks or their 1917 supporters, but a police state based around a military-industrial complex that allowed victory over Germany in World War II and nuclear Cold War competition with America afterwards, but which collapsed due to its own internal contradictions in 1991.

Neil Harvey Croll

See also: Russian Civil War (1918–1921); Russo-Polish War; Stalin; Trotsky, Leon; Warsaw/Vistula; World War I

References and further reading:

Kowalski, R. *The Russian Revolution 1917–1921.* London: Routledge, 1997.

Read, C. *From Tsar to Soviets: The Russian People and Their Revolution.* London: UCL Press, 1996.

Swain, G. *The Origins of the Russian Civil War.* London: Longman, 1996.

White, J. D. *The Russian Revolution 1917–1921: A Short History.* London: Edward Arnold, 1994.

Bor-Komorowski, Tadeusz (1895–1966)

Polish commander. Born in 1895 near the city of Lwow in Russian-occupied Poland, Bor-Komorowski, while a student, was in Warsaw for the declaration of Polish independence in 1918. After enlisting in the cavalry to fight in the Russo-Polish War (1919–1921), he rose through the ranks to become an officer. Eventually he became an internationally known horseman, attending the Ecole de Guerre in Paris, and was the commander of the Polish Cavalry School during the Polish Campaign of 1939. Promoted to major general, he became a commander of the Krakow region in the Polish Home Army (Armia Krajowa) and took the pseudonym "Bor" (forest). In July 1943 following the arrest of Major General Stefan Rowecki, Bor-Komorowski was promoted to lieutenant general and became the commander of the entire Home Army. In July 1944, he made the decision, with the consent of Prime Minister Stanislaw Miklolajczyk in London, to initiate a long-planned general uprising against the Germans in order to liberate Warsaw before the arrival of Soviet forces. Despite the presence of the Red Army across the Vistula from Warsaw, Stalin ordered that no assistance be given to the obviously noncommunist revolt. From 1 August to 2 October 1944, the Home Army and the people of Warsaw fought a valiant, but futile, battle against the numerically and materially superior German forces. By the end of the battle, the city of Warsaw had been leveled and nearly 250,000 soldiers of the Home Army and civilians were killed. Surrendering at the end of the battle, Bor-Komorowski was imprisoned at the Innsbruck concentration camp, from which he was liberated in May 1945. Condemned as a criminal by the Soviet Union, he refused to return to communist-dominated Poland following the war. After serving briefly as the prime minister of the Polish government-in-exile, he then worked for the Polish Welfare Association in London. He died on 25 August 1966.

Alexander M. Bielakowski

See also: World War II

References and further reading:

Bor-Komorowski, Tadeusz. *Secret Army.* Nashville, TN: Battery Press, 1984.

Ciechoanowski, Jan M. *Warsaw Rising of 1944.* Cambridge, UK: Cambridge University Press, 1974.

Borno-Kanem Sultanate (9th–19th Centuries)

A late medieval Islamic regime that consolidated its power in central Africa during the early modern era through the adoption of gunpowder firearms. Sometime after 900, Kanuri-speaking herdsmen founded the Kanem kingdom on the northeast shore of Lake Chad. Although still pastoralists, their king (the *mai*) ruled as a secluded divinity, assisted by merchants, shamen, and female family members (especially the queen mother).

The ruling Sayfawa family embraced Islam sometime in the eleventh century, and established a capital at Njimi. Conversion to Islam opened the region to the caravan routes leading north to Egypt, and to the Tunis-Libyan coast. Sultan Dunama I created a permanent standing cavalry during the mid-twelfth century, to keep the roads through the Fezzan safe from nomadic banditry. From Kanem, Mediterranean Muslim merchants wanted camels, salt, ostrich feathers, ivory, and especially slaves. Given these demands, the mai led annual slaving raids against the various settled peoples to the south and west.

Expanding his cavalry to 40,000 troops, Sultan Dunama II (1210–1248) gave Kanem an empire by annexing the oasis towns of Fezzan, and southwestward into the territories known as Borno. He imposed a provincial structure on his state, promoted Islamic scholarship, and compelled many of his subjects to become settled farmers and taxpayers.

In the 1300s and 1400s, the Kanem state endured tumultuous changes. Factional fighting within the dynasty led to assassinations and coups. A branch of the royal family, the Bulala, seceded, and regularly launched raids against Kanem. Northern territories slipped away, and, to the south, peoples like the So, routinely victims of Kanem slave raids, became more organized and fought back. Aridity and overgrazing undermined the domestic economy.

About 1390, Mai Umar ibn Idris relocated the capital from beleaguered Kanem to Borno. Here the Sayfawa spent decades forcing the Chadic-speaking Borno people to accept their rule and taxes. Popular resistance to Kunari domination claimed the lives of several mais during this painful transition. Nonetheless, access to the city-states of the Hausa people opened Borno to traffic in crafts, gold, and other new commodities, as well as to new slave raid zones.

Borno-Kanem reached the height of its power in the sixteenth century. Its armored cavalry, backed by a conscript infantry force, slaved aggressively throughout the region, and imposed tribute on Kano, Kebbi, and other Hausa city-states. Mai Idris Alawma (1571–1603) opened direct diplomatic relations with the Ottoman Empire. He persuaded Istanbul to provide him with Janissaries, and to train his own people in the use of firearms. In an environment where mobility decided battle, firearms did not significantly enhance his offensive power, but gunpowder weapons strengthened Idris's hold on his own people and his neighbors. He imposed Islamic *shari'a* law on the realm, and made Borno a center of Islamic culture in the heart of Africa.

In 1667, Taureg nomad raiders devastated Borno territory and the empire shrank dramatically. Nonetheless, the Borno-Kanem sultanate survived to remain a significant regional power. During Africa's eighteenth- and nineteenth-century jihadist movements, Muhammad al-Kanami (d. 1837) rejuvenated Borno, but in 1846, a resentful Mai Ibrahim tried to put down al-Kanami's son, Shehu Umar (1837–1881). Umar defeated Ibrahim, executed him, and deposed the Sayfawa Dynasty. Despite Umar's efforts, the decline of Borno-Kanem accelerated. In 1893, Rabih ibn Fadlallah, seeking to rally the region against French imperialism, conquered Borno-Kanem and ended its millennium-long existence.

Weston F. Cook Jr.

References and further reading:

Ibn Fatuwa. *History of the First Twelve Years of the Reign of Mai Idris Alooma (1571–1583).* Trans. H. R. Palmer. Lagos: University of Nigeria, 1934.

July, Robert W. *A History of the African People.* 5th ed. Prospect Heights, IL: Waveland Press, 1998.

Smith, Robert S. *Warfare and Diplomacy in Pre-Colonial West Africa.* 2d ed. Madison: University of Wisconsin Press, 1989.

Borodino (5–8 September 1812)

Set-piece battle that Napoleon hoped would result in a total Russian defeat or capitulation. Appointed commander in chief of Russian forces on 20 August 1812, Mikhail Kutuzov had a mandate from Czar Alexander I to engage Napoleon before Moscow.

The battlefield lay 70 miles west of Moscow, between the new and old Smolensk-Moscow roads. Russian forces were spread over 5 miles. Barclay's First Army was positioned along the Kolocha River, the north end of the battlefield. Bagration's Second Army held the center (Borodino) with its fortifications: the 20-gun "Great Redoubt" held by Raevski, and three supporting batteries of 12 guns (flèches) held by Borozdin at Schevardino.

The French attacked at 4 P.M. and captured Schevardino by nightfall on 5 September, losing 2,000 men. The Russians lost 8,000.

On 6 September Napoleon repositioned his troops for a frontal assault. The infantry corps of Ney, Junot, and Davout were positioned 1.5 miles south of the Kolocha River, while Murat's cavalry and 20,000 Imperial Guards were held in reserve at Schevardino. One hundred thousand French infantry, 28,000 cavalry, and 590 guns faced 72,000 Russian infantry, 10,000 militia, 17,000 cavalry, 7,000 cossacks, and 640 guns.

Battle recommenced at 6 A.M. on 7 September. As 102 French guns fired, Eugène's Army of Italy attacked Borodino, while Davout attacked the flèches. Both sides reinforced, Napoleon concentrated 45,000 men and 400 guns. By noon, the Russians withdrew and regrouped closer to the Great/Raevski redoubt.

Kutuzov sent Urarov's I Cavalry Corps and Platov's cossacks to outflank the French at Borodino. Although unsuccessful, this force of 13,000 caused the French to reposition 16 regiments. Napoleon also decided to preserve the Guard (his reserve), using it neither as a shock force nor for pursuit, after the fourth attack upon the redoubt, which caused a Russian withdrawal behind the Kolocha at 1530. Russian losses were 44,000 men including 23 generals; French losses were 35,000 men including 43 generals.

Kutuzov ordered a retreat from the battlefield to Mozhiask, commencing at 3 A.M. on 8 September. The Russian rear guard of 10,000 infantry and 10,000 cavalry reoccupied the Great Redoubt, which the French abandoned, before retreating.

Neville G. Panthaki

See also: Berezina River, Battle of; Kutuzov, Prince Mikhail Illarionovich Golenishchev; Moscow, Retreat from; Napoleon I

References and further reading:

Austin, Paul. *1812: The Great Retreat.* London: Greenhill Books, 1996.

Nafziger, George. *Napoleon's Invasion of Russia.* Novato, CA: Presidio, 1988.

Nicolson, Nigel. *Napoleon 1812.* New York: Harper & Row, 1985.

Riehn, Richard. *1812: Napoleon's Russian Campaign.* New York: John Wiley & Sons, 1991.

de Ségur, Philippe-Paul. *Napoleon's Russian Campaign.* Boston: Houghton Mifflin, 1980.

Bosworth, Battle of (22 August 1485)

The culmination of the English dynastic struggle known as Wars of the Roses. Bosworth was the last great battle of medieval England. Pursuant to this battle, the Tudor Dynasty took the throne and introduced the Renaissance to England.

In 1483 Edward IV died, leaving the throne to his son Edward V. The new king's uncle, Richard of Gloucester, seized the throne, claiming his nephews were illegitimate. The coup strained the power structure of the kingdom and provided an opportunity for the inheritor of the Lancastrian claim, Henry Tudor, Earl of Richmond, to return from French exile. On 7 August 1485 Tudor landed in Wales accompanied by a few hundred adherents and approximately 2,000 French mercenaries.

Tudor marched inland, gathering supporters. By 21 August he had approximately 5,000 soldiers, and approached Richard's forces. Richard had been awaiting reinforcements, but still outnumbered the rebels, having nearly 8,000 men. Both sides appear to have possessed some artillery support.

On 22 August 1485, Tudor's army closed the gap between the two armies, a bog causing some delay, as they attempted to gain the high ground. Richard was the more experienced commander, with the better position, but this was eventually outweighed by the intercession of new Tudor supporters.

Much conjecture remains concerning the actual course of the battle, and the popularity of fictional accounts has done little to enlighten researchers.

The outcome is clear: Richard was killed, and with his death his faction collapsed. The War of the Roses came to an end, and the Tudors instituted a policy of ensuring that no dynastic challenges could exist. In doing so they broke the spirit of feudal England and helped to create a national state.

Daniel German

See also: Richard III; Wars of the Roses

References and further reading:

Seymour, William. *Battles in Britain and Their Political Background, 1066–1746.* New York: Hippocrene Books, 1975.

Botha, Louis (1862–1919)

Boer field commander in the Second Boer War. Born near Greytown, Natal, on 27 September 1862, Botha settled near Vryheid, Natal, in 1884, where he became field cornet in 1894. As a member of Parliament in Transvaal after 1898, he abstained from voting to declare war on Britain in October 1899, but went on active duty as soon as war broke out. Commandant General Piet Joubert soon noticed Botha's tactical skills and made him second-in-command.

Botha distinguished himself at Talana on 20 October, and contributed to the bottling up of Sir George Stuart White in Ladysmith on 30 October. He jointly directed the siege of Ladysmith with Joubert from 2 November until 23 November, when Joubert was disabled by falling from his horse and Botha assumed full command. As commandant general of the Boers, he defeated Redvers Henry Buller at Colenso on 15 December, Spion Kop on 24 January 1900, and Vaal Krantz on 5–7 February, but lost to him at Tugela River on 17–18 February, Biggarsberg on 14 May, Bergendal on 27 August, and Lydenberg on 6 September.

Tactically superior, but outgunned and usually outnumbered, Botha thereafter waged expert guerrilla warfare in Transvaal and Natal, attempted to make peace with Horatio Herbert Kitchener at Middelburg on 28 February 1901, and accelerated operations in September to coincide with Jan Smuts's invasion of the Cape Colony. He defeated Lieutenant Colonel Hubert Gough at Blood River Poort on 17 September, but lost to Major A. J. Chapman at Fort Itala and Fort Prospect on 26 September. His exhausted troops never recovered from their Pyrrhic victory over Lieutenant Colonel G. E. Benson at Bakenlaagte on 30 October. He cosigned the peace treaty at Vereenigung on 31 May 1902. Honored as a war hero, praised for both strategy and tactics, and valued for his diplomatic genius, he was elected prime minister of the Union of South Africa as a moderate in 1910 and held that office until his death in Pretoria on 27 August 1919.

Eric v. d. Luft

See also: Boer Wars; Buller, Sir Redvers Henry; Colenso, Battle of; Joubert, Petrus Jacobus; Kitchener, Horatio Herbert; Ladysmith, Siege of; Smuts, Jan Christian

References and further reading:

Malan, Jacques. *General Louis Botha, 1862–1919.* Pretoria: National Cultural History and Open-Air Museum, 1979.

Meintjes, Johannes. *General Louis Botha: A Biography.* London: Cassell, 1970.

Moore, Dermot Michael. *General Louis Botha's Second Expedition to Natal during the Anglo-Boer War, September–October 1901.* Cape Town: Historical Publication Society, 1979.

Boudicca's Rebellion (60–61)

The Iceni and Trinovantes, Celtic tribes that inhabited what are now East Anglia and Essex, joined forces in revolt against the Roman administration of Britain. Their leader was Boudicca, queen of the Iceni.

With the bulk of Roman forces on campaign in north Wales, the Britons faced minimal resistance. The provincial towns of Camulodunum (Colchester), Londinium (London), and Verulamium (St. Albans) were overrun and sacked, and part of *legio IX* ambushed and destroyed. The provincial governor, C. Suetonius Paullinus, marched south down Watling Street (modern A5 highway). At his disposal were *legio XX* and part of *legio XIV* and some auxiliaries, a total of some 10,000 men.

Elated by her earlier victories, Boudicca staked all on one battle. According to Tacitus, Suetonius drew up his forces along a defile—legionaries in the center with auxiliary infantry alongside and cavalry on the wings—with dense woodland protecting his rear. When the battle was joined, the legionaries discharged their *pila* into the oncoming Britons, then charged. Confident of victory, the Britons had brought along their womenfolk to watch the spectacle from wagons positioned behind the warbands. The Britons soon found themselves crushed against the wagons. Tacitus says 80,000 of the enemy fell, including many of the spectator women, for the loss of only 400 Romans. The whereabouts of the battlefield are unknown, although a case has been made for the village of Mancetter near Nuneaton in Warwickshire.

Soon after the battle Boudicca took her own life. Suetonius, now heavily reinforced by units from the Rhine frontier, concentrated his efforts against the Iceni and Trinovantes. Their territory was laid waste and a chain of forts constructed across East Anglia.

Nic Fields

See also: Celts

References and further reading:

Sealey, P. R. *The Boudican Revolt against Rome.* Princes Risborough, UK: Shire, 1997.

Webster, G. *Boudicca: The British Revolt against Rome AD 60.* London: Batsford, 1993.

Boulogne, Siege of (1544)

Delayed final act of the Hundred Years War. Determined to press English claims on French territory, Henry VIII joined Habsburg emperor Charles V in 1544 in a combined invasion of France. Henry fixed on the city of Boulogne as his first major objective.

Henry's forces, consisting mainly of continental mercenaries, invested the city on 19 July, and the lower town and harbor were taken by assault on 21 July, but the upper town was assailable from only one side. Its French garrison of 2,000 held out until 14 September, when prolonged bombardment by the English artillery forced their surrender. This delay led Charles V to suspect that Henry was not serious about their plan for a joint march on Paris, and the emperor signed a separate peace on 18 September, leaving Henry to defend his new conquest against an army of 30,000 French.

Some 4,000 English and mercenaries repulsed a furious French assault on 9 October, and after this failure the French settled into a siege of the city that was conducted without enthusiasm. It continued until June 1546. During this period, Henry VIII invested large sums in building modern defenses for the city, an exercise that provided an excellent training ground for English military engineers such as John Rogers.

The standoff finally ended in June 1546, when the Treaty of Ardes sanctioned English possession of Boulogne for eight years. After that it was to be sold back to the French. In fact, it was returned in 1549, when King Henry II of France agreed to "ransom" the city for a substantial sum of money. Of England's once French empire, this left only Calais, and it was conquered in 1558, bringing the era of English adventurism in France definitively to an end.

John S. Nolan

See also: Anglo-French Wars; Calais, Siege of; Guise, François de Lorraine, Second Duke of; Hundred Years War

References and further reading:

Fissel, Mark Charles. *English Warfare, 1511–1642.* London: Routledge, 2001.

Oman, Charles W. C. *The Art of War in the Sixteenth Century.* London: Methuen, 1937.

Bouquet, Henry (1719–1765)

Swiss soldier of fortune, served with the British army in North America, won the Battle of Bushy Run (1763) during Pontiac's War, said to have saved western Pennsylvania. Born in Rolle, Switzerland, in 1719, Bouquet began his military career in April 1736, in the Swiss regiment of Constants, in the service of the Netherlands. Over the next 20 years Bouquet served in Swiss regiments in the service of Sardinia and the Netherlands. In early 1756, Bouquet accepted a commission in the British army's Royal American Regiment.

Henry Bouquet enjoyed a distinguished but unfortunately short career with the British army in North America. He began as the lieutenant colonel of the Royal American Regiment, but he quickly assumed more challenging duties. In 1758 he was second-in-command of Brigadier General John Forbes's successful attack on Fort Duquesne, at the site of present-day Pittsburgh. Because Forbes was often ill, Bouquet provided much of the leadership for the expedition. Bouquet went on to lead an expedition to northwestern Pennsylvania in 1760. He is best remembered for his victory at the Battle of Bushy Run, 5–6 August 1763, against aboriginal Americans during Pontiac's War, which ended the threat to western Pennsylvania. In 1764, Bouquet commanded an expedition against the Ohio tribes that led to peace in that region. Henry Bouquet was promoted to brigadier general and sent to British Florida, but he died there of yellow fever on 2 September 1765.

Bouquet's life illustrates the pan-European military world of the mid-eighteenth century. Bouquet seems to have been competent, but aloof. He successfully adapted to the conditions of warfare in North America, and he saved western Pennsylvania at the Battle of Bushy Run.

Scott N. Hendrix

See also: Bushy Run, Battle of; Pontiac's Rebellion; Seven Years' War

References and further reading:

Anderson, Fred. *Crucible of War: The Seven Years War and the Fate of Empire in British North America, 1754–1766.* New York: Alfred A. Knopf, 2000.

Bouquet, Henry. *The Papers of Henry Bouquet.* Vols. 1–6. Donald Kent, Autumn Leonard, S. K. Stevens, Louis M. Waddell, eds. Harrisburg: Pennsylvania Historical and Museum Commission, 1951–1994.

Nester, William R. *"Haughty Conquerors": Amherst and the Great Indian Uprising of 1763.* Westport, CT: Praeger, 2000.

Boxer Rebellion (1900–1901)

An antiforeign uprising in China. The Boxer Rebellion occurred in the aftermath of an alliance of sorts between the Qing court and various anti-Qing, antiforeign, and anti-

The fall of the Peking Castle; the Boxers being driven away by British and Japanese troops. Lithograph, 1900. (Library of Congress)

Christian movements in China. It was a time of ferment and great change, and the so-called Boxers—more correctly, Yihequan or Righteous and Harmonious Fists—sought to stem this tide of change. A young emperor, guided by several brilliant Confucian scholars, tried a "Hundred Days Reform" in 1898 to help China resist the encroachments of the "foreign barbarians." Previous efforts had failed and the Chinese envied the apparent ability of the Japanese to modernize and gain Western respect. But many in China opposed such dramatic and revolutionary change, and they made an informal alliance with the Boxers to demonstrate how to resist the foreigners.

By fall 1899, the Boxers, with tacit approval of the Qing court in Beijing, began moving against foreigners and, more importantly, Chinese Christian converts. Chinese Christians at that time had to forsake their Chinese names and adopt Western names, and hence appeared to be turning their backs on their culture and values; this set them at odds with the Boxers, who were very traditional. The Boxers began in more central China and then moved north to the capital.

The Boxers laid siege to the foreign legations in Beijing and Tianjin; they cut road and rail communications as well as telegraph lines between the two cities, isolating the foreigners—diplomats, their families, and troops—in the foreign legation area. Nearly 500 foreign troops helped to defend the legations against the growing mass of Chinese protestors. A careful study of these sieges makes clear that these Chinese conservatives (who believed that their philosophy enabled them to resist Western bullets) sought to destroy those sections of the various foreign legations where Chinese Christian converts were hiding.

Foreign troops resisted and matters settled down to siege warfare. The initial relief force only numbered some 2,000 marines and seamen; they could not force their way past Chinese defenses near the coast, and on 26 June 1900 they returned to their ships. Nonetheless, Chinese defenses were badly outdated, reflecting a level of military technology reaching back several centuries to the introduction of cannon and muskets into China. A combined international relief force in which the Japanese contingent was the largest took

Tianjin on 14 July 1900 and reached Beijing, the Qing capital, on 4 August. While French troops guarded the line of communications to the Gulf of Chihli, the international expedition gained control over Beijing 10 days later. The Chinese resisted a Japanese attack at one gate; but meanwhile a combined American-Russian force broke through at another portal, and a British force made it through a third gate; they drove through Chinese defenders to reach the British legation where the Westerners and their Chinese Christian converts were holding out. Eventually, the relief expedition gained control of the capital; the empress dowager, the emperor, and much of the court escaped to Xian; and the foreign troops generally sacked the Forbidden City, the Qing court section of Beijing. Interestingly, the relief force had no overall commander, and the commanders of each national contingent acted somewhat independently of the other forces, probably limiting the effectiveness of the overall relief expedition.

The Qing court appointed a well-regarded Chinese government official, Li Hongzhang, to negotiate a settlement, which included a large indemnity, increased foreign control over China, and punishment for many Chinese officials who had encouraged the Boxers to rebel. Indeed, the United States feared that as a major consequence of the failure of the Boxer Rebellion, the foreign powers involved would carve China into territories. To forestall this mutilation of China, the U.S. secretary of state issued his famous "Open Door" notes.

Charles M. Dobbs

See also: Chinese Military

References and further reading:

Purcell, Victor. *The Boxer Uprising: A Background Study.* Hamden, CT: Archon Books, 1963.

Tan, Chester. *The Boxer Catastrophe.* New York: Octagon, 1955.

Wehrle, Edmund S. *Britain, China, and the Anti-Missionary Riots, 1891–1900.* Minneapolis: University of Minnesota Press, 1966.

Boyne (1 July 1690)

Battle that ensured Protestant ascendancy in Ireland for more than two centuries. James II, having been replaced on the British throne by his daughter, Mary, and son-in-law, William of Orange, as part of the Revolution of 1688, had fled to exile in France, but with the support of Louis XIV returned to Ireland in March 1689. Although largely to restore James II to the throne, the campaign was also key in the plans of Louis XIV to divert William and the League of Augsburg from fighting in continental Europe. Under the earl of Tyrconnell, James had a Catholic army, which he deployed first in Ulster against the city of Derry, which William relieved by sea on 1 August 1689. Unable to hold Ulster, the Jacobite forces retreated to winter quarters centered at Dublin, while the Williamite troops, augmented by contingents from Denmark, the Netherlands, and the German states, weathered the winter in Ulster.

James needed to defend Dublin, and chose the River Boyne as the only practical line of defense for the spring campaign. Taking a position at Oldbridge, Meath, with an army of 25,000, James was daring William into battle, a challenge that he accepted, arriving with 35,000 men on 30 June 1690 on the banks of the Boyne. With the advice of his general, Schomberg, William chose a frontal attack across the river at Oldbridge, supported by a flanking movement by one-third of the army upstream. After a daylong cannonade, the battle began at dawn on 1 July.

The flanking movement, arriving at 6 A.M., fought their way across the river, drawing men from Oldbridge as the Jacobites fought not to be turned and to secure a safe line of retreat. Meanwhile, William waited until 10 A.M. to begin the main attack at Oldbridge, waiting for the tidal river to ebb. The infantry fighting was harsh and both sides took heavy casualties, William himself losing the heel of his boot to a bullet. Tyrconnell and the French duke of Berwick attacked with cavalry, but were unable to break the Williamite lines. Eventually, with the Jacobite cavalry covering them with fierce fighting, the infantry began to retreat to Duleek, where the Williamite army ceased their pursuit. Rather than a great military battle, the Boyne was a political turning point, as James II once again chose to flee to France via Kinsale, leaving his Irish supporters badly disillusioned with Jacobitism and with the harsh Protestant ascendancy that followed.

Margaret Sankey

See also: English Wars in Ireland

References and further reading:

Hayes-McCoy, G. A. *Irish Battles.* London: Longmans, Green, 1969.

Kinross, John. *The Boyne and Aughrim.* Moreton-in-Marsh, Gloustershire, UK: Windrush, 1997.

Shepherd, Robert. *Ireland's Fate.* London: Aurum, 1990.

Braddock's Defeat (9 July 1755)

A major British defeat in North America. In the 1750s France began to build forts in the Ohio Territory and fortified the forks of the Ohio with Fort Duquesne (Pittsburgh). Pleas for help from Virginia brought British major general Edward Braddock with the 44th and 48th Regiments. Benjamin Franklin intervened to help supply wagons and George Washington served as a civilian volunteer.

The army marched from Fort Cumberland, Maryland, and by 9 July was closing in on Fort Duquesne. The French garrison was desperate: too small to withstand a siege, they needed the help of the Ottawa, Chippewa, Huron, and other aboriginal bands who gathered there for supplies. At the last minute the Indians agreed, and a group of about 800 led by Captain Daniel Beaujeu moved out to ambush Braddock at the crossing of the Monongahela River. Arriving too late, the French and British collided.

The British advance guard formed a line and fired several volleys, killing Beaujeu. The French wavered, yet the Canadian militia and Indians began to hit the British flanks. The advance guard fell back as support came forward, intermingling units and breaking their formations. Unable to coordinate attacks, the English were caught in a crossfire from three sides.

The French and Indians worked their way around the British flanks, surrounding the whole army and spreading panic. Braddock organized several ineffective counterattacks, yet was soon wounded. After two hours the English collapsed in confusion.

A force of 1,800 had lost 900 men, an almost unprecedented casualty ratio. Braddock died a few days later. The French and Indians lost about 100. The battle gave experience to men such as Daniel Morgan, Horatio Gates, and George Washington.

Although a stunning defeat, the battle at the Monongahela was merely the opening of the Seven Years' War (French and Indian War in North America). The conclusion of the war would see France cede over to Britain almost all of its North American empire.

Brian Dunkerly

See also: Pitt, William, the Elder; Plains of Abraham; Seven Years' War; Wolfe, James

References and further reading:

Kopperman, Paul E. *Braddock on the Monongahela.* Pittsburgh, PA: University of Pittsburgh Press, 1977.

McCardell, Lee. *Ill Starred General.* Pittsburgh, PA: University of Pittsburgh Press, 1958.

O'Meara, Walter. *Guns at the Forks.* Englewood Cliffs, NJ: Prentice Hall, 1965.

Bradley, Omar Nelson (1893–1981)

U.S. Army World War II commander. General Omar Bradley was born 12 February 1893 in Clark, Missouri. After graduating from West Point in 1915, Bradley served in various teaching assignments during the interwar period. In 1938 he was assigned to the general staff in Washington, where he quickly caught the eye of General George Marshall. In 1941, Bradley became commandant of the Infantry School, and his reputation as an excellent training officer led to his command of the 82d and 28th Divisions.

During World War II, Bradley served with II Corps, first as assistant commander, then commander, during the North African and Sicilian campaigns. During the D day invasion, Bradley led the American forces on shore as First Army commander. Bradley commanded the Twelfth Army Group, as the American troops moved across France and the Rhine, eventually linking up with the Russians in Germany. During the drive across Europe, Bradley clashed with Field Marshal Bernard Montgomery about whether to concentrate on a single quick thrust or a broad advance into the German heartland.

During his campaigns in Europe, he earned the nickname "the G.I.'s general." Following the end of World War II, Bradley became administrator of the Veterans Administration. In 1948, he was named army chief of staff and became the first chairman of the Joint Chiefs of Staff in 1949. During his tenure in this position, he faced many challenges and played a key role in establishing America's early Cold War policy. Although not as colorful as his counterparts, Bradley was at the center of many pivotal decisions and was known as a stabilizer, or, as others believed, too cautious.

William Hartley

Portrait of Omar N. Bradley, 1939–1945. (Library of Congress)

See also: Eisenhower, Dwight David; U.S. Army; World War II

References and further reading:
Bradley, Omar. *A Soldier's Story.* New York: Holt, Rinehart & Winston, 1951.
Bradley, Omar, and Clay Blair. *A General's Life.* New York: Simon & Schuster, 1983.
Whiting, Charles. *Bradley.* New York: Ballantine Books, 1971.

Bragg, Braxton (1817–1876)

Confederate field commander, excellent tactician and strategist, but a failure because of his indecision, harshness, and poor leadership. Bragg was born in Warren County, North Carolina, on 22 March 1817. Assigned to artillery after his graduation from West Point in 1837, he saw action against Indians until the Mexican War, when he distinguished himself at Monterrey and Buena Vista under Zachary Taylor. He retired as brevet lieutenant colonel to his Louisiana plantation in 1856.

As Confederate brevet major general of Louisiana militia in 1861, Bragg defended the Gulf coast between Pensacola and Mobile until promoted to regular major general in Albert S. Johnston's Army of the Mississippi in March 1862. He commanded the Confederate right flank at Shiloh, first under Albert Sidney Johnston, then under P. T. Beauregard. His performance there earned him promotion to full general on 12 April 1862. He replaced Beauregard as commander of the Army of the Mississippi on 17 June.

Coordinating with Edmund Kirby Smith, Bragg invaded Kentucky in September 1862, losing to Don Carlos Buell at Perryville on 8 October and to William S. Rosecrans at Murfreesboro. As victor over Rosecrans at Chickamauga, he failed to exploit his advantage and allowed the enemy to retreat to Chattanooga. Ulysses S. Grant and George H. Thomas routed him at Missionary Ridge, Tennessee, on 25 November. He was relieved of command on 2 December.

Bragg did not see field service again for more than a year, when, on 14 January 1865, near Wilmington, North Carolina, he lost yet another battle, this time Fort Fisher, to Admiral David Dixon Porter and Brigadier General Alfred H. Terry. He died in Galveston, Texas, on 27 September 1876.

Eric v. d. Luft

See also: American Civil War; Beauregard, Pierre Gustave Toutant; Buena Vista; Chattanooga, Battle of; Chickamauga, Battle of; Grant, Ulysses Simpson; Johnston, Albert Sidney; Monterrey; Murfreesboro; Rosecrans, William Starke; Shiloh; Thomas, George Henry

References and further reading:
McWhiney, Grady. *Braxton Bragg and Confederate Defeat.* Tuscaloosa: University of Alabama Press, 1991.
Seitz, Don Carlos. *Braxton Bragg: General of the Confederacy.* Freeport, NY: Books for Libraries, 1971.
Stout, L. H. *Reminiscences of General Braxton Bragg.* Hattiesburg, MS: Book Farm, 1942.

Brandywine (11 September 1777)

Part of General William Howe's Philadelphia campaign during the American Revolution. The Battle of the Brandywine, 11 September 1777, illustrated the weaknesses of the American army, while nevertheless proving the wisdom of General George Washington's eventual plan to outlast the British and secure a "protracted victory." Howe planned to strike against the seat of the American rebellion, Philadelphia, and sent 15,000 troops to the head of the Chesapeake Bay. Once landed, they were to progress toward the rebel capital and rally loyalists to their side. For his part, Washington struggled to keep his improvised army of regulars and militia units intact following reassignment of thousands of troops to General Horatio Gates's Saratoga campaign. Regardless, he intended to thwart Howe's progress at Brandywine Creek, 26 miles west of the city, and stationed his 12,000 remaining troops accordingly. Howe sent his German units under General William von Knyphausen against the American center at Chadd's Ford, while the greater bulk of his forces under General Charles Cornwallis circled the American left flank in an envelopment maneuver. Washington struck against the Germans while holding the left at bay under Major General John Sullivan, but Howe's rapid advance across the Brandywine against Sullivan's command forced Washington to retreat. Skillful tactics employed by General Nathanael Greene during the American withdrawal enabled Washington to preserve his army intact. In all, the British lost 576 casualties and the Americans 1,000. Howe proceeded to occupy Philadelphia, but without inflicting a major injury to the American army.

Jeffrey Webb

See also: American Revolution; Cornwallis, Sir Charles; Germantown; Greene, Nathanael; Saratoga; Washington, George

References and further reading:
Fleming, Thomas. *Liberty! The American Revolution.* New York: Viking, 1997.
Reed, John F. *Campaign to Valley Forge, July 1, 1777–December 19, 1777.* Philadelphia: University of Pennsylvania Press, 1965.
Robson, Eric. *The American Revolution in Its Political and Military Aspects, 1763–1783.* New York: Norton, 1966.

Brant, Joseph (1742–1807)

Mohawk aboriginal leader. Known by the Indian name of Thayendanegea, Joseph Brant was born in 1742, the son of a

Mohawk chief. His sister Molly married British superintendent of Indian Affairs Sir William Johnson, who had a significant influence on Brant's life. He was enrolled in Moor's Indian Charity School (later Dartmouth College) in 1761 but left in 1763 to fight for the British during Pontiac's Rebellion. He later converted to Christianity and developed excellent translation skills, which earned him the position of secretary to the superintendent. By age 32 Brant had become a prominent and respected leader of the Mohawk Nation.

Brant's strong loyalty to Great Britain was rewarded with a commission as a captain in 1775. During the American Revolution he led devastating raids against the rebels in New York and Pennsylvania. He fought at Fort Schuyler and at the Battle of Oriskany in 1777. But Brant's loyalty was tested when the Iroquois nations split their allegiance during that conflict.

The American victory shattered Brant's world, and the Mohawks were forced to relocate to Canada. In 1785 Brant traveled to England to seek land grants near the Grand River in Upper Canada. For his loyal service, Brant was provided with an estate near Burlington Bay on Lake Ontario. For the remainder of his life, Brant worked for peace to protect the Grand River settlement. He died on 24 November 1807 in Brantford, Ontario.

Brant's unique loyalty to Great Britain, combined with political and military leadership, reflects a remarkable man who sought to reconcile two different cultures.

Steven J. Rauch

See also: American Revolution

References and further reading:

Kelsay, Isabel Thompson. *Joseph Brant, 1743–1807: Man of Two Worlds.* New York: Syracuse University Press, 1990.

Van Every, Dale. *A Company of Heroes: The American Frontier: 1775–1783.* New York: Morrow, 1962.

Brazilian Revolt (1893–1895)

Naval military revolt against the nascent Brazilian republic. In 1889 the Brazilian empire came to an end and was replaced by a republic. However, some elements of Brazilian society, especially the navy, longed for the return of the monarchy. The newly created republic soon experienced economic problems, which led to political unrest. In 1893, the navy, under the leadership of Admiral Custodio de Melo, revolted and demanded that a plebiscite be held to choose between a republic and a monarchy. The revolt gained strength when the respected commandant of the Naval Academy, Admiral Luis Felipe Saldanha da Gama, lent his support to the rebel cause. Another key demand of the naval rebels was the removal of President Floriano Peixoto, who was also a marshal in the Brazilian army. In addition to the naval revolt, Floriano faced rebellion in the south from the Federalists led by Silveira Martins. Martins's Federalist army invaded the states of Santa Catarina and Parana and threatened São Paulo. At the same time the rebel navy prepared to bombard Rio de Janeiro.

The bombardment of Rio was avoided when commanders of warships from the United States, Great Britain, France, Italy, and Portugal intervened and declared their intention to protect their merchant ships as well as foreign-owned property in the capital. The Brazilian government appealed to U.S. president Grover Cleveland to provide additional protection. Cleveland dispatched more American cruisers, which positioned themselves between the rebel fleet and the harbor. The rebel admirals were unwilling to risk damage to the U.S. vessels and refrained from opening fire. Floriano then organized a loyalist fleet in Pernambuco and threatened to attack the blockade. In May 1894, the rebellious naval officers, realizing that their position was untenable, left their vessels and sought asylum on Portuguese warships. With the naval rebellion in hand, the Floriano government devoted full attention to subduing the Federalist rebels in the south. The last of the rebels surrendered in August 1895; the new Brazilian republic had weathered a significant challenge to its existence. For his steadfast determination to preserve the republic Floriano earned the sobriquet "the consolidator of the Republic."

George M. Lauderbaugh

References and further reading:

Bello, José Maria. *A History of Modern Brazil, 1889–1964.* Stanford, CA: Stanford University Press, 1966.

Burns, E. Bradford. *A History of Brazil.* New York: Columbia University Press, 1980.

Fausto, Boris. *A Concise History of Brazil.* New York: Cambridge University Press, 1999.

Skidmore, Thomas E. *Brazil: Five Centuries of Change.* New York: Oxford University Press, 1999.

Breda, Siege of (August 1624–June 1625)

One of Spain's last victories as a major military power. This siege was an episode in the struggle for Spanish political domination, but on another level a magnificent military contest between two of the seventeenth century's most outstanding generals. Genoese-born Ambrogio de Spinola (1569–1630), perhaps the most brilliant and humane general in Spanish history, laid siege with 60,000 men to the strategic United Provinces (Netherlands) city of Breda, which guarded the roads to Amsterdam and Utrecht. Spinola's adversary, Prince Maurice, Count of Nassau (1567–1625), was an equally remarkable general remembered for the modernization and professionalization of his army. Nassau garrisoned Breda with 9,000 infantry, an excellent artillery defense, and ample provisions, but avoided direct confrontation and camped outside Breda. Spinola meanwhile constructed military masterpieces, batteries on raised platforms for sufficient range. Numerous sorties led to the defeat of relief operations headed by Peter Ernst von Masfeld. To make matters worse, Nassau died on 23 April and was succeeded by his younger brother, Frederick Henry of Nassau (1584–1647). Upon the fifth try to relieve Breda a combined Netherlands and German supply convoy supported by an army of 12,000 men attacked the weakest of Spinola's points only to be defeated with appalling losses. On 1 June, Spinola established breaching batteries to the counter scarp of the moat surrounding Breda. Inadvertently becoming cognizant of the extreme privations of Breda's citizens and to avoid further bloodshed, Spinola offered unusually humane and very honorable terms and Breda capitulated on 5 June. The withdrawing army was allowed to keep their weapons and supplies, and Breda's survivors were not harmed. Four thousand men and 8,000 women and children died of hunger and privation during the 11-month siege.

Annette Richardson

See also: Dutch War of Independence; Maurice of Nassau

References and further reading:

Israel, Jonathan I. *The Dutch Republic and the Hispanic World 1606–1661.* Oxford, UK: Clarendon Press, 1986.
Wedgewood, Cicely V. *The Thirty Years' War.* London: Jonathan Cape, 1938. Reprint, New York: Methuen & Co, 1981.

Breitenfeld (17 September 1631)

Battle that marked the rising star of Gustavus Adolphus, King of Sweden, and reinvigorated the Protestant cause in the Thirty Years' War. Gustavus Adolphus began his campaign of intervention in Germany on the side of the Protestants in 1630 following a string of military disasters from Bohemia in 1620 to Denmark and the north German plains in 1626.

Gustavus landed on the Baltic coast in the fall of 1630 but made no overtly aggressive moves against the troops of the Catholic League, commanded by Jan Tzerklas, Count Tilly. The Swedish army instead concentrated to establish a secure base on the Baltic coast and down the Oder River valley.

Following the siege and sack of Magdeburg on 20 May 1631, Gustavus advanced toward the Elbe River and met the army of Tilly at Breitenfeld on 17 September. Tilly's army of nearly 40,000 men arranged themselves in the Spanish style of large squares known as *tercios,* consisting of a mix of pikemen and musketeers. The imperial cavalry formed up on either flank. Tilly emplaced his two dozen artillery pieces directly in front of the center of his array.

Gustavus had fewer than 30,000 men in his army, plus a contingent of 15,000 Saxons. Arrayed in two lines, a mixed linear formation of cavalry/infantry held the right wing, infantry/cannon occupied the middle, and another cavalry/infantry formation anchored the Swedish left. Gustavus placed the Saxons on the extreme left.

Both sides opened the battle with artillery fire. Tilly's cavalry on the left prematurely charged and threw off his coordinated plan. Tilly, chagrined at his cavalry's impetuousness, advanced his four *tercios* and remaining cavalry against the Saxons. The Saxon corps melted away under the attack, but the imperial force now had a yawning gap between the infantry and left-wing cavalry. The Swedes finally repulsed the imperial cavalry and wheeled forward to envelop Tilly's exposed right flank. The king's cavalry overran Tilly's initial position, captured all his guns, and forced his *tercios* off the field with about 7,000 dead. Swedish losses amounted to fewer than 2,000 men.

The victory invigorated the Protestant cause and allowed Gustavus to penetrate deep into German Catholic domains of Swabia and Bavaria. This reversal of fortunes ensured the continuation of the Thirty Years' War.

Bryan R. Gibby

See also: Gustavus II Adolphus; Lützen, Battle of; Nordlingen; Thirty Years' War

References and further reading:

Delbruck, Hans. *The Dawn of Modern Warfare.* Trans. Walter Renfoe. Lincoln: University of Nebraska Press, 1990.
Dodge, Theodore A. *Gustavus Adolphus.* New York: Da Capo Press, 1998.
Parker, Geoffrey, ed. *The Thirty Years War.* London: Routledge, 1997.

Brian Boru, King of Ireland (940–1014)

Successfully resisted the Vikings and united Ireland. Youngest son of Kennedy, Dalcassian king of Thomond, Brian was educated by monks at Inisfallen, after his father was killed in 951 fighting off an attack by Danish raiders based at Waterford. Brian's childhood was dominated by the struggle between the feuding Irish dynasties, headed by the O'Neill, with and against the Danish settlers at Limerick, Waterford, and Dublin. After his elder brother Marthghamhain became king of Thomond, Brian led a campaign of harassment against the Danes, operating for 18 months in 962 in the wilds of Clare, before joining the regular forces, which besieged and sacked Limerick in 968 and secured the throne of Munster for Marthghamhain. This success unfortunately stirred Irish resentment, and in 976 Marthghamhain was lured into a parlay by the king of Ui Fhidhghinte and assassinated. Taking up the crown, Brian attacked not only the perpetrator, but also their Danish ally, Ivar of Limerick, at Inis Cathaigh, and the king of Desmond at Bealach Leachta. This, again, drew the attention of rival Irish kings, especially of Leinster, which Brian attacked, in the process sacking the Danish stronghold at Waterford, after disagreement over the cattle tribute.

This consolidation of power brought Brian into conflict with the O'Neill high king, Maoil-Seachlainn, who wasted some of Munster en route to attack Leinster himself, including uprooting the Dalcassians' (literally) family tree. While Brian assembled a fleet of longboats to harass Danish settlement on the Shannon, he and the high king engaged in a long-term war of maneuver for supremacy, Brian losing twice in battle, at Meath and Aonach Teide, between 992 and 994. Only the revolt of Sitric, the Danish ruler of Dublin, brought the two men together briefly in 999, when they jointly defeated Sitric at Gleann Mama, and looted Dublin. Having added Danish cavalry to his army, Brian finally challenged the high king at Tara in 1002, demanding and getting Maoil-Seachlainn's abdication. During the following decade of peace, Brian endowed monasteries, established schools, and married his children into Danish and Irish families, in an attempt to keep his position through diplomacy and to benefit from Scandinavian trade.

In 1012, probably provoked by his fourth wife, Gormlaith of Leinster, Brian deliberately humiliated Leinster with a new cattle tax, pushing the king into alliance with the Danes of Limerick and Dublin, who also invited Orkney Vikings and pirates to fight with them. With support from his son-in-law, Malcolm II of Scotland, Brian and his army, led by his eldest son, defeated the Danes and their allies at Clontarf in 1014. Brian himself was discovered in his tent by a fleeing party of Danes, and killed in the aftermath of the victory, robbing Ireland of its unifying leader and plunging it back into civil war.

Margaret Sankey

See also: Viking Raids; Vikings

References and further reading:

Cudmore, Patrick. *The Battle of Clontarf.* New York: P. J. Kennedy, 1895.

Newman, Roger Chatterton. *Brian Boru, King of Ireland.* Dublin: Anvil, 1983.

British Dynastic Wars (1000–1066)

By the year 1000, the power of the English king Æthelred ("the Unready") had been severely eroded by Viking invasions, despite a purge of nobles in 1002 to eradicate a suspected fifth column of Viking supporters. In 1016, Cnut, the king of Denmark, defeated Æthelred's son Edmund Ironside at Ashingdon and became king of England, marrying Edmund's widow, Emma of Normandy, and murdering most of the royal family, sparing only the two young sons of Ironside. When Cnut died in 1035, his eldest son, Harthacanute, chose to stay in Scandinavia, leaving the younger, Harold "Harefoot," as king of England. To cement his power, Harold invited Emma and Edmund Ironside's two sons, Edward and Alfred, back from exile in Normandy, but upon arrival, Alfred was tricked and assassinated by Earl Godwin of Wessex. Harold "Harefoot" died in 1040 and was succeeded by Harthacanute, who died the following year.

In this dynastic crisis, Earl Godwin and the *witan,* or council of nobles, turned to Edward, the surviving son of Ironside, and crowned him king in 1042. While marrying Godwin's daughter Edith, Edward also extended friendship to Normans, to whom he gave lands in an attempt to build a secure base of support against Godwin. In 1051, Edward attempted to oust Godwin from power, briefly succeeding in forcing him and his sons into exile. Godwin, with help from the Flemish, ravaged the Isle of Wight and intimidated Edward into returning him to favor, which included earldoms for all of his sons. Godwin's son Harold waged a successful war against Gruffydd of Wales, ending decades of border warfare. In 1064, Harold traveled to Normandy, where he met Duke William, perhaps to negotiate William's acceptance of Harold's succession to the English throne or to swear a pact of friendship, although later Norman chronicles suggest that Harold was there to offer the throne to William or to swear fealty.

When Edward the Confessor died in 1066, Harold Godwinson, with the accession of the *witan,* assumed the throne. William of Normandy assembled an invasion force to contest Harold's claims, but was delayed waiting for a south wind to sail across the channel, and was instead stuck in the mouth of the Somme River. Harold had no sooner dismissed his defense force in southern England than he received word that Harold Hardrata of Norway, accompanied by Harold Godwinson's own brother Tostig, whom he had outlawed in 1065 for mismanagement of his earldom, had landed in Northumbria and had taken the city of York. Harold marched north at a brutal pace, arriving five days later and defeating Harold Hardrata at Stamford Bridge. Meanwhile, William of Normandy finally sailed, arriving in England in late September.

Harold, accompanied by his surviving brothers, marched south, again on forced march, and on 14 October 1066, met William of Normandy at Hastings, where they were defeated. William, now the "Conqueror," seized the throne of England and began a program of fortification, including the construction of the Tower of London, land confiscation, and settlement by his Norman followers. The dynastic wars had weakened Saxon England, rendering William's conquest that much easier.

Margaret Sankey

See also: William the Conqueror

References and further reading:

Abbot, Jacob. *History of William the Conqueror.* Pensacola, FL: Beka, 1999.

Butler, Denis. *1066: The Story of a Year.* London: Blonde, 1966.

Lawson, M. K. *Cnut: The Danes in England in the Early Eleventh Century.* London: Longmans, 1993.

British Military, Twentieth-Century Organization and Structure

In the twentieth century, Britain's wars have been prosecuted on land by an army, the Royal Marines of the Royal Navy, and the Royal Air Force Regiment of the Royal Air Force.

At the beginning of the twentieth century the British army was administered by cabinet through a ministry, the War Office. This ministry was engaged in a committee system to ensure close coordination in planning and procurement, but was replaced in 1964 by a single Defense Ministry. As well, in 1914–1918 and 1939–1964, procurement for the army was handled by the Ministry of Supply.

Under the Directing Ministry, the army has been administered by an Army Council that advises the minister, consisting of members for Personnel, Training, and so on (precise membership having varied over the century), presided over by a chief of staff, title again varying. The chief sits on a Chief of Staff's Committee or, since 1964, a Defense Council including civil servants and ministers, presided over by a chief of the defense staff, while the members of the Army Council sit on their own committees with their equivalents in other services, and also in other ministries. For instance, through the interwar period the triservice committee on procurement was presided over by the president of the Board of Trade (a cabinet office). Various directors (for example, the director of Tank Development) report to the members, and often sit on interservice committees in their own right.

From 1870 to 1939, the British army was a small, professional "medium service" organization in which soldiers enlisted in the regular force for six to nine years, followed by three to six years in the reserves. Conscription was in force for most of the period of the world wars, and was continued into peacetime until 1956 as the National Service. Although expected to alleviate the army's traditional difficulty in recruiting infantry, it instead entailed a massive training burden that was never satisfactorily addressed. In the periods of voluntary service, both parliamentary guidance and those same recruiting difficulties dictated that officers and men separate with a useful skill or trade, a requirement that dictated a training-centered personnel policy. For technicians and their officers in the first half of the century, this did not present a problem. They entered the army through various training schools, typically at the age of 16, some of which also served as central depots and even factories, such as Woolwich Arsenal. Infantry (and through the mid-1930s, the cavalry) presented more serious difficulties. These soldiers entered individual infantry regiments from the Royal Military Academy Sandhurst (for officers) or through regional depots and recruitment areas. Battalions of each regiment rotated in turn through an overseas tour, usually India or Germany, depending upon the period. However, the home-based force was designated for "expeditionary" duties closer to home, most notably the world wars and their troubled aftermath, the Falklands Conflict, Ireland in 1920–1922 and after 1967, and Palestine in 1936–1939. As in many armies, British infantry enlist in a specific regiment and are not normally liable for transfer to other regiments, a practice that promotes unit cohesion but which complicated manpower policy in overseas tours by comparison to the technical corps, which are formally single regiments and can practice individual rotation. In addition to the regular army, British land forces include various reserves, the volunteer "part-time soldiers of the Territorial Army, and colonial formations raised by local governments to support British garrisons. The British army also employs Nepalese mercenaries, the famous Ghurkas, in several regular battalions.

The field army has been organized into divisions of two to four brigades of two to four battalions each, supported by individual brigades (although for some of the Cold War period the division level was formally abolished in favor of the brigade group). The regiment is not used as a tactical formation, but for traditional reasons battalion-level artillery formations are known as regiments. The five "divisions" into which the United Kingdom is currently divided for recruitment purposes should not be confused with tactical divisions.

The principal corps of the British army include the Royal Artillery, Royal Engineers, Royal Ordnance and Artificers Corps, Royal Corps of Signals, Royal Armored Corps (incorporating the historic cavalry regiments and the battalions of the Royal Tank Regiment since 1938), Royal Electrical and Mechanical Engineers, and the Army Air Corps. Tables of organization often unhelpfully refer to these by their acronyms.

The Royal Marines were traditionally small contingents carried aboard navy ships for various duties, including as landed infantry in formations of up to divisional strength. However, after 1940 the marines were organized into special-purpose reinforced battalion-size amphibious units, the Royal Marine Commandos. The RAF Regiment, although active from 1918, was only formally organized in 1942 to supply antiaircraft, ground liaison support, and airbase security duties. Post-1945, the regiment has also operated antiaircraft missiles in the field.

Britain raised many Special Forces formations during World War II. These survive today as the Commandoes, the battalions of the Parachute Regiment, the Special Air Service, and the Special Boat Service of the Royal Marines.

There are many elements of continuity in the contemporary British army. The British public remains adamantly opposed to peacetime conscription (as do those in the remainder of the Commonwealth and the United States) and it remains extremely difficult to recruit soldiers, particularly when the economy is good. As the only Allied power in both world wars to "see it through" from the beginning to the end, Great Britain paid a terrible price in blood and treasure. Raising a mass conscript army, as well as a first-class air force, and maintaining a great navy proved too much of a strain on its resources, and Great Britain had slipped from the rank of a world power by 1941.

Today, the British army is a highly regarded, mobile, well-equipped professional force, inspired but not bound by its traditions, and which has demonstrated its quality in various unconventional conflicts around the globe, including its policing of Northern Ireland, as well as in the Korean War, the Falklands conflict, and the Gulf War.

Eric Lund

See also: Falkland Islands War; Liddell Hart, Sir Basil Henry; Northern Ireland, Civil War in; World War I; World War II

References and further reading:

Barnett, Correlli. *Britain and Her Army, 1509–1970.* New York: William Morrow, 1970.

Carver, Michael, Lord. *Britain's Army in the Twentieth Century.* London: Macmillan, 1998.

Ladd, James D. *By Sea, by Land: The Royal Marines 1919–1997. The Official History.* London: Jane's, 1998.

British-Indian Army

Red-jacketed, well trained, and campaigning on the northwest frontier, the British army in India remains the stereotypical image of nineteenth-century colonial armies.

The East India Company's first garrison of 35 English and 40 Indian soldiers at Madras in 1642 established the British army in India. Living in unhealthy conditions and limited in numbers (even in 1721 the total European military contingent at company towns in India was only 245 men), these troops were of varying effectiveness. They remained active in a series of small campaigns against the Dutch, Portuguese, and local Indian warlords in a shifting pattern of commercially important alliances, sometimes locally reinforced by Royal Navy ships or shore parties when necessary.

Around 1749 the French instigated a campaign along the Coromandel coast, when powerful Indian warlords bolstered by French troops began a series of actions against the outnumbered British. The French were checked by clever diplomacy and British success in battle, but the face of military organization in India changed considerably. Although at this time the Europeans usually only numbered four or five hundred among the sometimes tens of thousands of troops of the Indian warlords, their discipline and well-drilled musketry became the deciding factor in battle, a situation that continued into the twentieth century. In 1756 native troops trained by, and attached to, the British army (named sepoys) were issued red uniforms, which created a very effective esprit de corps, and European-standard training and weapons were introduced readily to these units. The British forces were clearly separated into three East India Company entities by this time: the Bombay, Madras, and Bengal Armies, each with its own command and responsibilities. Some British army units were also present in the country, but the proportion of Indian to European soldiers remained high; at the time of the Indian Mutiny in 1857 the Bengal Army consisted of approximately 225,000 Indians and 40,000 Europeans.

In 1756 the Bengal Army commenced the subjugation of the Ganges River valley, seeking permanent influence for British trade. By then the Mogul emperors had patterned their armies more closely on European models, including the deployment of field artillery, and the Bengal Army fought a series of very difficult battles against Mogul forces as large as 40,000 men. In 1764, after its success in the Battle of Buxar, the British-Indian Army became the undisputed power in India.

Then followed a series of regional conflicts at the outer edges of the Indian territories. The Marathas seized power in the northwest of India in 1775. This move led to the Maratha Wars, 1776–1805, and permanent British reestablishment throughout Hindustan. An incursion into northern India in 1814 by the Nepalese was defeated by the Bengal Army and in 1816, Nepal was added to areas of British influence. Imagined Russian influence in Afghanistan in 1838 led to a preemptive campaign and elements of the Bengal and Bombay Armies and large forces of allied Sikhs, after hard fighting, captured Kabul. Later the Russians did advance into Afghanistan, and campaigning continued until a British-Russian treaty in 1895 stabilized the Afghan frontiers. British troops were permanently withdrawn inside the Khyber Pass in 1899. In 1845, the Sikhs, who had remained independent, sensed British designs on the Punjab, and attacked Hindustan with a well-trained and organized army of 30,000, but by 1849 after two campaigns, their power was broken and the British annexed the Punjab. The British-Indian Army fought a series of campaigns in Burma lasting from 1823 until the entire country was occupied in 1886.

The Indian Mutiny of 1857–1858, when many Indian units (mostly of the Bengal Army) rebelled against the British presence in India, saw widespread campaigning with many desperate battles and eventual cruel retribution by the British army. As a result of the mutiny, the East India Company armies were entirely absorbed into the British army in India, and Indians were denied access to artillery and leadership training for many decades.

After 1900 the country remained generally peaceful and the army was gradually reorganized to reflect modern European structures. By 1921 the "Indianization" of politics saw the army declared a defense force rather than a tool of British imperial policy. The Indian army saw prominent service during World War II in North Africa, Italy, Malaya, and Burma. That all was not well with the Anglo-Indian army could be seen in the mutinies in the twentieth century, and in the modest success of the Japanese in recruiting Indian prisoners of war for service in the pro-Japanese Indian National Army after the fall of Singapore and Burma.

From 1922 onward Indian nationals began staff training in Britain and the Officer Corps became increasingly Indian before independence and partition, and separation into the Pakistani and Indian Armies in 1947, with both often-contending forces retaining many elements of their British heritage, including, in the case of India, the strict separation of the military from the civil authority.

Michael Hyde

See also: British Military, Twentieth-Century Organization and Structure; Burma, Retreat from; Chillianwallah; Gujerat; Indian Mutiny; Indian National Army

References and further reading:

Heathcote, T. A. *The Indian Army: The Garrison of British Imperial India, 1822– 1922.* London: David & Charles, 1974.

Kar, Lt. Col. H. C. *Military History of India.* Calcutta: Firma KLM Private Limited, 1980.

Lawford, J. P. *Britain's Army in India: From Its Origins to the Conquest of Bengal.* London: George Allen & Unwin, 1978.

Brunanburgh (September or October 937)

Vikings suffer grave defeat at hands of Anglo-Saxons. The site of Brunanburgh has not been satisfactorily identified. The *Anglo-Saxon Chronicle* says the battle took place "ymbe Brunanburgh [around Brown's fort]," and near the "Dinges mere [Sea of Storm]." Modern scholars suggest Bromborough in Cheshire, but popular tradition identifies Brunanburgh with Brinkburn in Northumberland.

Anlaf Guthfrithsson, King of Dublin, was seeking to exact revenge upon Æthelstan (r. 924/5–939), grandson of Alfred the Great, for ignominiously expelling his father from Northumbria 10 years earlier. He formed a coalition of the Danes of Ireland, Welsh, Strathclyde Birtons, and the Scots. Relying on Northumbrian hostility to Æthelstan's rule, especially among the Anglo-Danish aristocracy, the aim of the alliance was to crush Æthelstan with the manpower of the Celtic peoples and of the British Viking settlements.

No primary source tells us how the members of the alliance combined their armies. The only apparently reliable fact, that they landed a fleet in the Humber, has been disputed. The coalition certainly established itself in Northumbria, and raided south of the Humber so as to disaffect Danish settlers in Mercia.

Eventually, the alliance was met by Æthelstan at the head of an army drawn from both Wessex and Mercia. Celts and Danes threw themselves against West Saxon and Mercian levies in a daylong battle, with savage hand-to-hand fighting in regular battle order (the famous "shield walls"). Victory was not to be theirs. According to the *Anglo-Saxon Chronicle,* the allies lost five kings and seven earls, and a son of the king of the Scots. The northern kings retired to their own lands with difficulty. Anlaf brought the mere wreckage of an army back to Dublin.

Nic Fields

See also: Alfred the Great; Viking Raids; Vikings

References and further reading:

Stenton, F. M. *Anglo-Saxon England.* 3d ed. Oxford, UK: Clarendon, 1971.

Brunswick, Frederick William, Duke of (1771–1815)

German general, hero of Napoleonic Wars. Born on 9 October 1771, he was set to inherit the title of duke when his father, Karl Wilhelm Ferdinand, died in 1806, but was denied the title by the Treaty of Tilsit (9 July 1807), which abolished his duchy and added the territory to the Kingdom of Westphalia. In Austria and desiring revenge against Napoleon, he set about organizing a corps of Brunswick infantry and cavalry (25 February 1809), which came to be called "the Black Horde" or "the Black Brunswickers" after the color of their uniforms. The corps fought alongside Austria at Saxony, participating in the battles of Zittau (31 May 1809) and Borbitz (12 June 1809). The Austrians surrendered on 12 July of that year, but Brunswick's corps fought on, battling to the north German coast. They briefly occupied Braunschweig (Brunswick) on 31 July and skillfully escaped Westphalian forces to Elsfleth where they disembarked for England. Brunswick lived in England until 1813, when he was finally awarded his due title of Duke of Brunswick. He lived in Brunswick as duke until returning to the battlefield in 1815 as commander of Brunswick troops in the British army against Napoleon. He died from wounds inflicted during the Battle of Quatre Bras on 16 June 1815, just two days prior to the Battle of Waterloo.

Harold Wise

See also: Ferdinand, Karl Wilhelm, Duke of Brunswick; Napoleon I; Napoleonic Wars

References and further reading:

Connely, J. *Napoleon's Satellite Kingdoms.* New York: Free Press, 1965.
Pivka, Otto von. *The Black Brunswickers.* Reading, UK: Osprey, 1973.

General Aleksei Alekseevich Brusilov. (Hulton/Archive)

Brusilov, Aleksei Alekseevich (1853–1926)

Probably the best Russian commander of World War I. Born in Tiflis, Georgia, of noble origins, Brusilov graduated from Corps de Pages Imperial Guard Academy (1872). He rose to command 15th Dragoon Tver Regiment in Russo-Turkish War (1877–1878), graduated from the St. Petersburg Cavalry School (1883), and taught there, becoming head of the school in 1902.

An intensely patriotic Russian nationalist, Brusilov became increasingly disillusioned with the czarist system, from the humiliation of the lost Russo-Japanese War (1904–1905) to the World War I debacles, publicly siding with the people in February 1917. He nonetheless served his country loyally, commanding Second Guard Cavalry Division (1906–1909); XIV and XVII Army Corps; and was Warsaw District Deputy Commander (1909–1914).

Brusilov proved to be the best Russian World War I commander, leading the Eighth Army in Galicia (1914), capturing Przemysl's fortress (March–April 1915), and threatening to break through the Carpathian Mountains to Hungary. His finest achievement came as commander in chief of the Southwestern Front from March 1916, when he masterminded the best Russian offensive of the war. Launched in June 1916 along a 300-mile front, the "Brusilov Offensive" pushed the Austro-Hungarians back 60–70 miles, capturing 375,000 men. This had immense consequences—the Western Allies' situation at Verdun and the Somme was eased, with Germany withdrawing troops to stop Brusilov. The situation on the Italian front also stabilized, Rumania joined the Allies, and two European imperial dynasties collapsed. Austria-Hungary never recovered, but the czarist army's disintegration began with the 1,000,000 casualties it suffered. An outstanding military operation, the Brusilov Offensive would have been more successful had Brusilov received necessary supplies and support from flanking Russian generals.

Brusilov next served as the provisional government's military advisor (February–May 1917); Russian army supreme commander (May–July 1918), and directed the June Offensive, which stalled due to supply shortages, prompting the czarist army's disintegration.

Removed in July 1918, Brusilov sat out the Russian Civil War, wanting no involvement. He did act as Red Army chief adviser against Poland (1920), which he saw as a national conflict. He then served as cavalry general inspector until 1924, when he was relieved by S. Budienny, and retired. Relatively unmolested by the Soviet government, Brusilov wrote his memoirs and died in Moscow.

Neil Harvey Croll

See also: Russian and Soviet Armies; Russo-Japanese War; Russo-Polish War; World War I

References and further reading:

Brusilov, A. A. *Moi vospominaniia. [My Recollections].* Moscow: Gosizdat, Otdel voennogo literatury, 1929; Riga: Knigoizdat. "Mir," 1929; Paris: Hachette, 1929; Moscow: Voenizdat, 1943, 1963, and 1983.

———. *A Soldier's Notebook, 1914–1918.* London: Macmillan, 1930; Westport, CT: Greenwood Press, 1930, 1971.

———. *The Memoirs of General Brousillov, 1914–1918.* Trans. Frank J. Haronian. New York: Haronian, 1934.

Brusilov, A. A., and N. V. Brusilova. *Brusilov A. A. and Nadezhda V. Brusilova papers, ca. 1880–1940.*

Sokolov, Iu. V. *Krasnaia zvezda ili krest? Zhizn' I sud'ba generala Brusilova.* Moscow: Rossiia molodaia, 1994.

Brusilov Offensive (June 1916)

An assault on the Austrian Galician front by the Russian commander, Aleksei Brusilov, in response to the Western Allies' pleas for relief. Perhaps the most successful of the Russian generals, Brusilov stressed training and tactical flexibility as a prerequisite to breaking the stalemate of trench warfare. He directed elements of five Russian armies, totaling 600,000 men, to assault the Austrian lines along a 250-mile front.

The combined Austro-German armies numbered roughly 620,000 men, most of them Austro-Hungarian units. The Austro-Germans held well-fortified positions, trenches, bunkers, and wire entanglements. The Austrians remained ignorant of the impending attack largely due to Brusilov's skill in deploying his armies and reserves and their own lax intelligence. Brusilov used aerial photography and sapper tunnels to improve his tactical position.

On 4 June 1916 the offensive began with a series of well-concentrated artillery bombardments to soften the enemy entrenchments. The weight of the attack fell upon the Austrians at Lusk where Brusilov held a considerable numerical advantage. At the Battle of Lusk the Austrians suffered a severe defeat their Fourth Army collapsed, allowing the Russians to achieve a breakthrough. Taking advantage of the gap, the Russians sent in reserves, outflanking the Austrians. The disintegrating Fourth Army fell back, followed by the First Army in the north. Russian attacks in the south against the Austrians at Czernovitz dislodged the Austrian Seventh Army from their fortifications. In the center, after a fierce counterattack, the German Sud Armee retired with the Austrians.

Within 10 days the Galician front had crumbled, the Russians eventually forcing the Austrians back to the Carpathian Mountains. The Russian offensive stalled due to the failure of the supporting Russian commands to coordinate attacks with Brusilov. The summer victories encouraged Rumania to join the Allies on 27 August 1916.

But the casualties the Russians had suffered offset any gains. They were unable to mount any more strategic offensives for the remainder of the war and they failed to hold on to their gains, both factors playing into the eventual dissolution of the Russian army itself.

Stephen Chenault

See also: Bolshevik Revolution; Brusilov, Aleksei Alekseevich; Russian and Soviet Armies; Russian Civil War; World War I

References and further reading:

Asprey, Robert B. *The German High Command at War: Hindenburg and Ludendorff Conduct World War I.* New York: William Morrow, 1991.

Herwing, Holger H. *The First World War: Germany and Austria-Hungary, 1914–1918.* London: Arnold, 1997.

Lincoln, W. Bruce. *Passage through Armageddon: The Russians in War & Revolution 1914–1918.* New York: Simon & Schuster, 1986.

Budennyi, Semen Mikhailovich (1883–1973)

Russian Civil War and World War II Soviet commander. Of peasant origins, from the Don region, he joined the Russian army (1903), serving as a dragoon in the Russo-Japanese War. Graduating from the St. Petersburg Cavalry School (1908), Budennyi fought through World War I as platoon sergeant with the Caucasus Cavalry Division.

With the Bolshevik disbandment of the czarist army after the October Revolution he returned home, forming a Red partisan detachment, the first step toward becoming commander of the Red Army's First Cavalry Army, with whom he forged his reputation. He won victories over previously undefeated White Cavalry, halting Denikin's advance (1919) and playing a leading role in Denikin's and Wrangel's defeats (1920). However, in August 1920 he disobeyed orders to support Mikhail Tukhachevskii's western front, allowing the Polish counteroffensive that prevented a Soviet victory in the Polish-Soviet war.

A major figure in the Red Army "Tsaritsyny" clique, the military base with which Stalin backed his rise in the 1920s, and the purge of the Red Army hierarchy around Tukhachevskii in 1937, Budennyi was a judge in the trials. His lack of Marxist zeal concerned Stalin, but unswerving loyalty and a certain mental dullness allowed Budennyi to become a marshal of the Soviet Union (1935) and escape the purges.

Post–Civil War, Budennyi was initially associated with the Frunze group striving to create new revolutionary military doctrine, but became distanced by his continued belief in the supremacy of cavalry despite technological advances.

His appointment as Red Army cavalry inspector from 1924 to 1937 reflected this philosophy.

This misunderstanding of modern technological warfare was revealed in disasters in the Finno-Soviet War (1939–1940) and early in the Nazi invasion of the Soviet Union, when he was removed after poor command performances, forcing Stalin to create the post of Red Army cavalry commander for him (January 1943).

Government minister for horse breeding until 1953, Budennyi wrote memoirs and collected numerous decorations on the back of those meritoriously received during World War I and the Civil War. He died in Moscow in 1973.

Neil Harvey Croll

See also: Russian Civil War; Stalin; World War I

References and further reading:

Budennaia, M. V., P. S. Karpachev, and I. V. Stavitskii. *Stranitsy bol'shoi zhizni: Vospominaniia o Marshale Sovetskogo Soiuza S. M. Budennom. [Pages of a Great Life: Recollections about Marshal of the Soviet Union S. M. Budennyi].* Moscow: Voenizdat, 1981.

Budennyi, S. M. *Proidennyi put'.* Moscow: Voenizdat, 1958.

Kolesnikov, A. A. *Marshaly Rossii. [Marshals of Russia].* Iaroslavl': Izdat. "Niuans," 1999.

Shukman, H., ed. *Stalin's Generals.* London: Weidenfeld & Nicolson, 1993.

Buena Vista (23 February 1847)

One of the most significant American victories of the Mexican War. Honoring the armistice he had signed, General Zachary Taylor and the U.S. Army of Observation had remained near Monterrey for weeks after the September 1846 battle. Meanwhile, General Antonio López de Santa Anna rebuilt the Army of the North. Taylor eventually moved southward, but 5,000 troops were detached to General Winfield Scott for his invasion from Veracruz to Mexico City.

The Battle of Buena Vista then took place as Santa Anna sought to hurry his poorly fed and undersupplied army to attack Taylor, now weakened by the detachment of many of his regular troops. Taylor learned of Santa Anna's advance and retreated three miles from Agua Nueva to the Hacienda Buena Vista, a better defensive position.

The Americans had taken a strong position in the rugged hills through which the river passed. Santa Anna demanded surrender (something that never crossed "Old Rough and Ready" Taylor's mind) and the fighting began at dawn on 23 February. The Mexicans had placed their artillery on a higher slope against the American left, and the Mexican right, although composed of many raw recruits, pressured the American left, which yielded. Mississippi and Indiana units in reserve pounded the charging Mexicans, drove off the cavalry, resisted an infantry attack, and maintained the position. The artillery had saved the day, filling gaps in the line and subjecting the Mexicans to relentless pounding.

While the Americans expected the fighting to resume the next day, Santa Anna had retreated after suffering many casualties and desertions. He had thrown away his numerical superiority in a series of piecemeal and uncoordinated attacks.

Charles M. Dobbs

See also: Mexican-American War; Scott, Winfield; Taylor, Zachary

References and further reading:

Eisenhower, John D. *So Far from God: The U.S. War with Mexico, 1846–1848.* New York: Random House, 1989.

Henry, Robert Selph. *The Story of the Mexican War.* New York: F. Ungar, 1961.

Lavender, David S. *Climax at Buena Vista: The American Campaigns in Northeastern Mexico, 1846–1847.* Philadelphia, PA: Lippincott, 1966.

Winders, Richard Bruce. *Mr. Polk's Army: The American Military Experience in the Mexican War.* College Station: Texas A&M University Press, 1997.

Buffalo Soldiers

African-American soldiers of the U.S. late nineteenth-century regular army. On 28 July 1866, the United States Congress established two cavalry and two infantry regiments from African-American U.S. Civil War troops. The 9th and 10th U.S. Cavalry and the 24th and 25th U.S. Infantry became known as "Buffalo Soldiers" by their Cheyenne and Comanche adversaries, perhaps because of their usually thick, dark hair. Despite difficult assignments and both public and military discrimination, these units proved invaluable in engagements against Plains Indians, Mexican bandits, cattle rustlers, and other "undesirables" of the wild West between 1866 and 1944.

The 9th and 10th Cavalries were initially commanded by Colonels Edward Hatch and Benjamin Grierson, both white officers. The 9th was ordered to Texas in June 1867 to protect stagecoach routes, build forts, map the frontier, and install telegraph communications. After helping subdue Kiowa and Comanche in west Texas, the 9th and 10th moved to Arizona and New Mexico districts in the spring of 1876. There they fought against Arapaho and Apache raiders including Geronimo. The last action of the Buffalo Soldiers in the Indian Wars involved the Sioux Indians, Sitting Bull, and the Ghost Dance Campaign of 1891.

The 24th Infantry with James Morgan fought in the Battle of San Juan Hill during the 1898 Spanish-American War. The 9th and 10th Cavalries enforced neutrality laws on the Mexi-

Battle of Bull Run, Virginia, 21 July 1861. Lithograph by Currier and Ives. (Library of Congress)

can border and pursued the Mexican bandit Pancho Villa in the punitive expedition under John J. Pershing in 1916. Between World Wars I and II the 9th and 10th Cavalries became excellent horse and marksmanship service troops at the cavalry school in Fort Riley, Kansas. By World War II, cavalry had become obsolete and the units were disbanded in 1944. Buffalo Soldiers' officers were almost entirely white, and the troops were kept in the far West and away from population centers to avoid "disturbances." Although these were excellent troops by any standard, they did on occasion violently react to the unremitting racism they suffered. The worst case was in 1917 when troopers of the 24th shot up the city of Houston, Texas, killing 17 whites; 16 black soldiers were hanged before the end of the year in the largest single judicial execution in U.S. history.

The Buffalo Soldiers reflected America's racial mores of the time, and good soldiers that they were, such units have no place in the twentieth-century U.S. Army, which achieved racial integration a full decade ahead of U.S. civil society.

Christopher Howell

See also: American Indian Wars; Geronimo; Villa, Francisco "Pancho"

References and further reading:

Fowler, W. L. *Black Infantry in the West, 1869–1891.* Norman: University of Oklahoma Press, 1996.

Hill, Walter. *The Record.* National Archives and Records Administration periodical. Washington, DC: March 1998.

McNalty, Bernard, and Morriss MacGregor, eds. *Blacks in the United States Armed Forces: Basic Documents.* 13 vols. Wilmington, DE: Scholarly Resources, 1977.

Bull Run, First/Manassas (21 July 1861)

Several months after the bombardment of Fort Sumter in the Charleston, South Carolina, harbor, President Abraham Lincoln felt compelled to order the undertrained and inexperienced army commanded by General Irwin McDowell to attack Confederate positions just south of the U.S. capital, near Manassas, Virginia.

McDowell had a simple but effective plan. He wanted General Robert Patterson in the Shenandoah Valley to occupy Joseph Johnston's Confederate forces. McDowell, in turn, would feint an attack directly at General P. G. T. Beauregard's units at Blackburn's Ford and then seek to march around to the west to turn the Confederate left flank.

Beauregard soon realized McDowell's turning movement—Union fixed bayonets glinted in the morning sun—

and he readjusted his position. Attacks by green troops were met by green troops, and the battle swayed back and forth around the Henry House. At one point, Confederate general Bernard Bee rallied his troops, pointing at a colonel whose men were fighting bravely: "Look! There is Jackson standing like a stonewall! Rally behind the Virginians!"—and a legend was born on the battlefield.

In the end, Johnston easily eluded Patterson (who, admittedly, showed little energy); his men boarded trains at Piedmont, debarked at Manassas, and marched to the battlefield. The appearance of rested Conference reinforcements turned the tide, and at about 4 P.M. the Union right flank began to crumble, Beauregard ordered an attack, and troops (and many spectators who wanted to watch the supposed "triumph") fled to Washington, D.C.

The key to the battle was the conclusions the two sides drew. Lincoln recognized it would be a long and costly contest; too many Confederates thought it wise to meet Union armies on every battlefield, hence engaging in a war of annihilation they ultimately could not win.

Charles M. Dobbs

See also: American Civil War; Anaconda Plan; Beauregard, Pierre Gustave Toutant

References and further reading:

Catton, Bruce. *The American Heritage New History of the Civil War.* New York: Viking, 1996.

Davis, William C. *Battle at Bull Run.* Garden City, NY: Doubleday, 1977.

McDonald, JoAnna. *We Shall Meet Again: The First Battle of Manassas.* New York: Oxford University Press, 2000.

Bull Run, Second/Manassas Junction (28–30 August 1862)

A serious Union defeat in the American Civil War. Correctly judging that Union general George McClellan would do nothing after the so-called Seven Days' Battles, Confederate general Robert E. Lee sent half his army under Stonewall Jackson to destroy the Union army supply base at Manassas and thereby to entice Union general John Pope to attack. Lee feared that if he did not move quickly, eventually Pope and McClellan would unite, giving the Union generals overwhelming superiority in manpower and artillery.

Jackson on 28 August 1862 stationed his men along an unfinished railroad grade, and Pope, thinking Jackson vulnerable, launched a series of savage but unsupported attacks on 29 August. Late in the day, Lee and the remaining half of the Army of Northern Virginia under James Longstreet arrived and quietly took up positions on Jackson's right flank.

On 30 August 1862, after questionable actions by Union general Fitz John Porter—although he probably had some idea that Longstreet was nearing the battlefield, Porter followed what would be erroneous orders to attack Jackson, exposing his own flank and that of Pope's entire army; Porter did not like Pope and perhaps that figured into his decisions—Confederate artillery fired a mass barrage, and 28,000 men under Longstreet crushed Porter's force, and drove the Union army past the old Bull Run battlefield.

Effective rearguard defense and the arrival of the vanguard of McClellan's army from Harrison Landing held off Lee's pursuing forces, and Lee then determined to invade the North, leading several weeks later to the Battle of Antietam/ Sharpsburg.

Charles M. Dobbs

See also: American Civil War; Lee, Robert Edward; Pope, John

References and further reading:

Cozzens, Peter, and Robert I. Girardi, eds. *The Military Memoirs of General John Pope.* Chapel Hill: University of North Carolina Press, 1998.

Greene, A. Wilson. *The Second Battle of Manassas.* Conshohocken, PA: Eastern National Park and Monument Association, 1985.

Hennessy, John J. *Return to Bull Run: The Campaign and Battle of Second Manassas.* New York: Simon & Schuster, 1993.

Buller, Sir Redvers Henry (1839–1908)

British general. Buller was born in Devon 7 December 1839. He was commissioned in 1858 and fought in the Anglo-Chinese War of 1860. After several years of outpost duty in Canada, he served in the Red River campaign in 1870 and soon became a member of Sir Garnet Wolseley's "Wolseley Ring." Buller rapidly advanced in rank and experience, participating in the Ashanti War, the Kaffir War, the Zulu War (receiving the Victoria Cross), the first Anglo-Boer War, the Egyptian campaign of 1882, the Charles Gordon relief expedition, and then to the staff of the War Office. In 1886, he went to Ireland, first as a special commissioner to restore order, and then as undersecretary for Ireland. In 1887, he returned to the war office as quartermaster general and, in 1890, adjutant general. He held that position until 1897, working for army reform and creating the modern Army Service Corps. He then commanded the army training camp at Aldershot from 1897 to 1899.

In 1899, Sir Redvers Buller (he had received a knighthood in 1894) was sent to command the British forces in the second Anglo-Boer War. After severe problems, he was superseded by Lord Roberts, but he remained to command the Natal operations, where he was ultimately successful. In 1900, he returned to Aldershot, but two years later he was retired after an impolitic speech defending his military decisions in Natal. He died 2 June 1908. Buller has been severely criticized for his actions in South Africa. Recent scholarship, however, revealing political intrigue and the public need for a scapegoat, has begun to judge him in a more favorable light.

James B. Thomas

See also: Boer Wars; Wolseley, Garnet Joseph, Viscount

References and further reading:

Melville, Charles H. *The Life of General the Rt. Hon. Sir Redvers Buller, V.C., G.C.B., G.C.M.G.* 2 vols. London: Edward Arnold & Co., 1923.

Pakenham, Thomas. *The Boer War.* New York: Random House, 1979.

Bunker (Breed's) Hill

A minor American defeat of the American Revolutionary War that heartened the Americans by the large number of casualties inflicted on the British.

In the aftermath of the British march to and retreat from Concord and Lexington, American militia constructed defenses on Bunker Hill in Charlestown. Moving before the British could construct fortifications at the same point, the Americans fortified both Breed's Hill and Bunker Hill, which was above it, during the night of 16–17 June 1775. The Americans used this position and adjoining stone and rail fences to construct a defended position.

The British chose to attack head-on. Had they used their sea mobility and moved troops behind the narrow neck of land that defined then-Charlestown, they could have cut off the Americans. Instead, they wanted to demonstrate their professional superiority against irregular troops, and thought perhaps that such a frontal assault was the most humane way of dispatching them. This was an odd decision in light of the recent precipitous British retreat from Concord.

While British ships fired at the Americans, the British landed some 2,200 men on a hot and humid June day. The first two times the British attacked without firing but with fixing of bayonets. In each case, the Americans fired behind their fortifications and exacted a high toll. General William Howe realized the third attack must win the day, for he could not ask more of his exhausted men.

The third attack won the battlefield for the British. The Americans ran low on and then out of ammunition, and finally retreated after British troops forced their way into the breastworks. The British were too exhausted from the weather and the fighting to pursue.

The Americans lost only 440 men; the British suffered more than 1,000 casualties. Along with the tenacity of the American defenders, Breed's Hill meant that this would be a far more costly conflict than the British had initially envisioned. Thereafter, they would use maneuver to avoid such heavy losses of valuable professional soldiers. (Although all of the fighting took place on Breed's Hill, for some reason, perhaps euphony, this battle has gone down in American history as the Battle of Bunker Hill. In fact, during World War II a fleet aircraft carrier, USS *Bunker Hill,* fought with great distinction in the Pacific.)

Charles M. Dobbs

See also: American Revolution

References and further reading:

Chidsey, Donald Barr. *The Siege of Boston.* New York: Crown, 1966.

Elting, John R. *The Battle of Bunker's Hill.* Monmouth, NJ: Philip Freneau Press, 1975.

Burgoyne, John (1722–1792)

British general and playwright blamed by critics for losing the American Revolution. Born in Sutton, Bedfordshire, and educated at the Westminster School, Burgoyne came to prominence for his victory over the Spanish at Valencia de Alacantara, Portugal, in 1762 during the Seven Years' War. The former member of the House of Commons (elected 1761, 1768), whose demand for an investigation of the East India Trading Company led to the Reforming Act of 1772, was appointed major general the same year and sent to Boston, Massachusetts, in May 1775. On June 17, he observed British troops at the Battle of Bunker Hill and returned to England in December concerned with what he had witnessed. Returning to North America in 1776 as second-in-command to Sir Guy Carleton in Canada, he was involved in the invasion of New York from Canada. After the seizure of Crown Point, New York, on Lake Champlain, disgust over Carleton's inactivity and leadership led him to return to England. Once home, he was able to convince British officials to back an ambitious plan for putting down the American rebellion in the northern colonies that became known as the Saratoga campaign. The plan called for British troops from Canada to march across New York, meeting troops coming up from the coast, causing New England to be cut off from the rest of the colonies.

Reoccupying Crown Point with 7,000 soldiers in June 1777, poorly equipped (although he burned his supply train with many wagons of his personal goods), and unprepared for warfare on the American terrain, he advanced near Albany, New York, and lost nearly 1,000 men near Bennington, Vermont. Expecting to link with British troops under Sir William Howe that never came, he found himself surrounded by American forces under General Horatio Gates. He surrendered in Saratoga, New York, on 17 October 1777, and he and his army were held as prisoners of war until the close of hostilities. This defeat led France openly to aid American war efforts.

"Gentleman Johnny" Burgoyne ended his inglorious military career as commander of British forces in Ireland from 1782 to 1783 and died on 4 August 1792 in London.

T. Jason Soderstrum

See also: American Revolution; Arnold, Benedict; Gates, Horatio; Saratoga

References and further reading:

Glover, Michael. *General Burgoyne in Canada and America: Scapegoat for a System.* London: Gordon & Cremonesi, 1976.
Hargrove, Richard J. *General John Burgoyne.* Newark: University of Delaware Press, 1983.
Howson, Gerald. *Burgoyne of Saratoga: A Biography.* New York: Times Books, 1979.

Burgundians

Germanic invaders of Gaul. Originally from Scandinavia, the Burgundians first appeared on the Main River soon after 250, but had little contact with Rome until around 406, when they crossed the Rhine, and under King Gundichar established a kingdom in the province of Germania Prima. In 436, they tried unsuccessfully to occupy Belgica Prima, suffering an appalling defeat at the hands of Flavius Aëtius, leading an army of Huns, and only narrowly escaped destruction. This defeat later became the basis of legends retold in the twelfth-century Middle-High German epic *Nibelungenlied.*

The few Burgundian survivors fled to the territory surrounding Lake Geneva, in what is now Switzerland. Later, suffering invasion repeatedly, they moved to the valley of the Rhine and occupied eastern Gaul. Lugdunum (Lyon) became the capital of a new Burgundian kingdom. In 554, the Burgundians were attacked by their former allies, the Franks, and their kingdom was annexed.

Greatest of the Burgundian kings was Gundobad (r. 473–517). He formulated a law code (*lex Burgundinum*) for his people and later sponsored an even more significant law code (*lex Romana Burgundinum*) for his Roman subjects. The Burgundians were Arian Christians, but in 493 Clotilda, Gundobad's daughter, who had embraced orthodoxy, married Clovis, and helped convert the Frankish king to the Roman Church.

Nic Fields

See also: Aëtius, Flavius; Franks; Huns

References and further reading:

Drew, K. F. *The Burgundian Code: Book of Constitutions or Laws of Gundobad.* Philadelphia: University of Pennsylvania Press, 1949.

Burma, Retreat from (1941–1942)

The longest retreat in British army history. In late 1941 Burma was poorly defended. Land invasion was considered unlikely, and only two regular British battalions, four Burmese rifle battalions, and some supporting units constituted the military garrison. The Royal Air Force had one fighter squadron and some light bombers based in Burma, and China had pledged a fighter squadron of the American Volunteer Group to protect the Burma Road, which was China's only land link for supplies in its war against Japan.

The Japanese 15th Army (33d, 55th, and Imperial Guards Divisions and Tenth Air Brigade) was charged with the capture of Burma, and landed in Thailand on 8 December 1941, before rapidly moving northwestward to the Burmese border. In uncontested operations they captured the strategic airfields at Victoria Point, Mergui, and Tavoy along the Isthmus of Kra. They were then ordered to stand by as reserves for the Malayan Campaign.

Finally the 15th Army moved into Burma in strength in mid-January 1942. They first struck toward Moulmien, and then the Bilin River where the Indian 17th Division was ordered to hold them, in spite of the Sittang River being a much better defensive line. In the event, Japanese units quickly breached the Bilin position and reached the Sittang bridge before the 17th Division had withdrawn, and demolition was ordered with the bulk of the defenders still on the eastern side. The majority escaped by swimming, but most equipment and weapons were lost and the 17th Division was neutralized as a fighting unit. The Japanese rapidly advanced on Rangoon, capturing it on 8 March.

The Rangoon garrison joined with the newly arrived 7th Armoured Brigade, 1st Burma Division, and the depleted 17th Division to stage a fighting withdrawal up the Irrawaddy River valley into India. The Chinese 5th and 6th Armies commanded by Lieutenant General Joseph Stilwell began their withdrawal up the Sittang River valley into the northeast. Japanese reinforcements (18th and 56th Divisions) were shipped into Rangoon in mid-March to continue the momentum of attack, and the Chinese committed their 65th Army.

By 8 May the Japanese had captured Lashio, cut the Burma Road, pushed the Chinese back to Myitkyina, and nearly trapped the British and Indian forces on the upper Chindwin River. On 12 May the monsoon rains started, ending the fighting and allowing the British, Indian, and Chinese forces to retire into India and China, respectively. They left 50,000 Chinese, and 13,500 British, Indian, and Burmese casualties along the 900-mile-long battlefield.

Michael Hyde

See also: Singapore; Stilwell, Joseph Warren; Wavell, Archibald Percival, First Earl

References and further reading:

Bisheshwar, Prasad, ed. *The Retreat from Burma 1941–1942, Official History of the Indian Armed Forces in the Second World War 1939–45: Campaigns in the Eastern Theatre.* New Delhi: Combined Services Historical Section (India and Pakistan), Orient Longmans, 1954.

Japanese Monograph No. 79. *The Burma and Andaman Invasion Naval Operations (March–April 1942): The Naval Armament Program and Naval Operations* (Part II). In *War in Asia and the Pacific,* Vol. 5, ed. D. Detwiler. New York: Garland Publishing Co., 1980.

Kirby, Maj. Gen. S. Woodburn. *History of the Second World War: The War against Japan.* Vol. 2, *India's Most Dangerous Hour.* London: Her Majesty's Stationery Office, 1958.

Burmese Civil Wars (c. 1300–1599)

Mongol incursions into northern and central Burma during the late thirteenth century that contributed to the dismemberment of the Pagan kingdom. The Pagan kingdom had first unified the country in the mid-eleventh century. But the ensuing three centuries of disunity and war are often described by historians as a time of ethnic conflict between three major groups: the Burmans (Myanmars), Shans, and Mons. In fact, although differences in language, culture, and customs must have been keenly perceived, the notion of ethnic identity and ethnic politics in the modern sense was alien to the people of the time. Instead of clearly defined ethnic boundaries, borders were fluid, and members of one group rather easily assimilated into another.

Also, the Burmese lowland peoples, along with the Tais, shared a common Indo-Buddhist culture with elaborate legal-political concepts that had originated in ancient India and had been adopted in Burma by the Mons and Burmans. They included the definition of a state's legitimacy in terms of the sovereign's personal merit acquired over many lifetimes and through promotion of the Buddhist religion and other good deeds in this lifetime, and a unitary concept of power that assumed that it was accumulated by a meritorious king at the expense of his rivals. The triumph of one polity over another reflected the superior merit of the royal victor.

Thus, between the fourteenth and seventeenth centuries there was nearly continuous warfare between competing polities, including those in the neighboring kingdom of Arakan, in which victory involved the acquisition of prisoners of war (labor was perpetually in short supply), treasure, and above all prestige.

It is commonly believed that Shans entering central Burma in the twilight years of the Pagan kingdom adopted Mon/Burman culture and institutions and established Burman-style successor states at Myinsaing, Pinya, and Sagaing after the Three Shan Brothers killed the last Pagan king in the early fourteenth century. Their descendant, Thadominbya (r. 1364–1368), founded an important power center at Ava in 1364 that flourished until it was sacked by the Shan usurper Thohanbwa in 1527.

In southern (Lower) Burma, Wareru, son-in-law of the Tai ruler of Sukhotai, established a state at Martaban in 1281. His successors, rulers of the Mons, governed a prosperous kingdom centered on Pegu (Hanthawaddy) until the Burman king of Toungoo, Tabinshwehti (r. 1531–1550), subjugated it in 1539. Tabinshwehti and his brother-in-law Bayinnaung (r. 1551–1581) established the second Burmese empire, after Pagan. Its unity was short-lived, however, breaking apart under attacks from rebellious royal princes, the Arakanese, and Siamese in 1599.

Donald M. Seekins

See also: Bayinnaung; Pagan Kingdom

References and further reading:

Aung-Thwin, Michael. "The Myth of the 'Three Shan Brothers' and the Ava Period in Burmese History." *The Journal of Asian Studies* November (1996).

Hall, D. G. E. *A History of South-East Asia.* 2d ed. London: Macmillan, 1964.

Burmese Civil Wars (1948–)

Ethnic diversity; an uncomfortable geopolitical location between India, Thailand, and China; the legacy of British "divide and rule" policies; and the influx of arms and armed gangs during World War II have ensured that Burma has not enjoyed internal peace since independence in 1948. The Communist Party of Burma (CPB), once part of the united front that ruled Burma after independence, went underground in March 1948. In early 1949 the Karen National Union (KNU), which sought an independent homeland for the Karen ethnic minority, took up arms. Communist and ethnic rebels occupied most of central and southern Burma, and the administration of Prime Minister U Nu, controlling little territory outside the capital, was nicknamed the "six mile Rangoon government." Thanks to the rebels' chronic disunity and loyal contingents of the Burma army commanded by General Ne Win, the government imposed its writ over most of the Burmese heartland by the mid-1950s, though it still had to fight Chinese Kuomintang irregulars who had fled Yunnan Province into Shan State.

The military regime established by Ne Win in March 1962 used armed force freely, including the brutal treatment of civilians in insurgent areas. During the 1960s, the focus of insurgency shifted from central Burma to the Border Areas,

A civilian lies dead as a soldier keeps watch during an attack by insurgents near Rangoon, 1949. (Hulton/Archive)

and involved three overlapping categories of armed groups: noncommunist ethnic nationalists, a reinvigorated CPB generously backed by China, and apolitical local warlords in the opium/heroin trade. In the early 1980s, there were as many as 28 major communist, ethnic, and drug-dealing armed groups operating in the Border Areas. The Burma army received little or no foreign military aid between 1962 and 1988 when the armed forces received large shipments of arms from abroad, especially from China.

In early 1989 the CPB collapsed due to internal frictions, and the military regime—the State Law and Order Restoration Council (SLORC)—concluded cease-fire agreements with former CPB groups, granting them de facto autonomy. By 1996, 17 armed groups had signed cease-fires with SLORC, including Khun Sa's powerful Mong Tai Army. SLORC also succeeded in fostering divisions among the Karens, and captured their strongholds of Manerplaw and Kawmoorah along the Thai-Burma border in January–February 1995.

To date, Karen, Karenni, and Shan ethnic rebels continue their resistance. Drug-financed armies, especially the United Wa State Army, with large amounts of cash and arms, still pose a threat to the central government.

Donald M. Seekins

See also: Guerrilla/Partisan/Irregular Warfare

References and further reading:

Lintner, Bertil. *Burma in Revolt: Opium and Insurgency since 1948.* 2d ed. Chiang Mai: Silkworm Books, 1999.

Smith, Martin. *Burma: Insurgency and the Politics of Ethnicity.* 2d ed. Dhaka/Bangkok/London: Universities Press/White Lotus/Zed Books, 1999.

Burnside, Ambrose Everett (1824–1881)

Humble, genial, but usually incompetent Union field commander in the American Civil War. Burnside was born in Liberty, Indiana, on 23 May 1824. After graduating from West Point in 1847, he served as an artilleryman in the Mexican-American War and against the Apaches, then resigned in 1853 to run a firearms company in Rhode Island. He developed a breech-loading carbine in 1856, but (beginning his losing streak) lost his patent to creditors in 1857. This weapon was used during the Civil War.

As a volunteer colonel, Burnside raised a Rhode Island regiment in April and May 1861. His satisfactory command of a brigade at First Bull Run earned him his first star on 6 August and an assignment to invade the North Carolina coast. Setting sail in December with 13,000 amphibious troops in 80 ships, he captured Roanoke Island on 7 February 1862 and New Bern on 14 March. Promoted to major general on 18 March, he twice declined Abraham Lincoln's offer to replace George B. McClellan as commander of the Army of the Potomac. He commanded the Union right at South Mountain, Maryland, on 14 September and the Union left at Antietam, where his advance at the Stone Bridge was too sluggish to help the Unionists.

Even though he accurately recognized himself as unfit for high command, Burnside accepted Lincoln's third offer, probably so that it would not go to Joseph Hooker, and took over the Army of the Potomac on 9 November. After leading his troops to disaster at Fredericksburg, he was replaced by Hooker on 26 January 1863.

Reassigned to the Department of the Ohio, he had some success against John Hunt Morgan's cavalry raiders that summer but was bottled up in Knoxville, Tennessee, for most of the fall by James Longstreet. Back in Virginia in 1864 under Ulysses S. Grant, he fought poorly at the Wilderness, Spotsylvania, North Anna, Cold Harbor, and Petersburg, and failed miserably at the Battle of the Crater on 20 July. He resigned in disgrace on 15 April 1865.

Burnside served Rhode Island as governor from 1866 to 1869 and as senator from 1875 to 1881. He died in Bristol, Rhode Island, on 13 September 1881. From his bushy side-whiskers comes the word *sideburns,* perhaps Burnside's only claim to lasting fame.

Eric v. d. Luft

See also: American Civil War; Antietam/Sharpsburg; Bull Run, First/Manassas; Bull Run, Second/Manassas Junction; Cold Harbor, Battle of; Fredericksburg; Grant, Ulysses Simpson; Hill, Ambrose Powell; Hooker, Joseph; Lee, Robert Edward; Lincoln, Abraham; Longstreet, James; McClellan, George Brinton; Meade, George Gordon; Mexican-American War; Petersburg, Siege of; Pope, John; Sherman, William Tecumseh; Spotsylvania Court House; Wilderness

References and further reading:

Ballou, Daniel Ross. *The Military Services of Major-Gen. Ambrose Everett Burnside in the Civil War and Their Value as an Asset of His Country and Its History.* Providence: Rhode Island Soldiers and Sailors Historical Society, 1914.

Marvel, William. *Burnside.* Chapel Hill: University of North Carolina Press, 1991.

Woodbury, Augustus. *Ambrose Everett Burnside.* Providence, RI: N. B. Williams, 1882.

Bushy Run, Battle of (5–6 August 1763)

British forces' defeat of aboriginal Americans near Bushy Run (south and east of present-day Pittsburgh), said to have saved western Pennsylvania in one of the few major engagements of Pontiac's War.

Pontiac's War, the biggest and most far-reaching uprising of aboriginal Americans against Europeans in North America, broke out in the spring of 1763. By midsummer only two of the British posts west of Fort Niagara still held out, Fort Detroit and Fort Pitt, and both were under siege.

Colonel Henry Bouquet served as British commander for western Pennsylvania. The only troops he had available were the understrength remnants of the 42d (the Black Watch) and 78th (Montgomery's) Highlander, many of whom were ill with various fevers as they had just returned from the siege of Havana, and some fragments of his own 60th (Royal American) Regiment. Bouquet spent the early summer reinforcing the posts along the military road into western Pennsylvania, leaving garrisons of the most ill soldiers behind.

On 4 August 1763 Bouquet left Fort Ligonier for Fort Pitt, which had been besieged for about four months. Bouquet had a convoy of about 350 packhorses carrying sacks of flour, about 450 regular soldiers, some 15 colonial rangers, and an unknown number of packhorse drivers. Bouquet intended to relieve Fort Pitt, and he was also seeking battle with the aboriginal Americans. The next day he found it.

About a mile short of the way station of Bushy Run, Bouquet's advance guard encountered Shawnee, Delaware, Mingo, and Huron warriors on a ridgeline. The advance guard drove them off, but they returned and circled around to attack. That night, Bouquet had a small redoubt built with bags of flour to shelter the wounded. The first day's action had cost the British about 60 killed and wounded.

The next day, the aboriginal Americans tormented the British troops, who were short of water, for several hours by making brief rushes at the perimeter of the British defensive circle, firing a few shots, and then fading away back into the tree line when the British attempted to counterattack. After several hours of this, Bouquet devised a plan.

Two companies of light infantry moved over the crest of the hill, while the rest of the circular perimeter was shrunk. These movements were intended to simulate a retreat. Two other companies took up hidden positions within the perimeter. The aboriginal Americans fell for this ruse, and rushed the perimeter. As they did so, the first two companies sallied out and took them in their right flank. They were driven to their left, and into the fire of the other two companies. After this, the aboriginals withdrew.

Bouquet's forces suffered a total of 50 killed, 60 wounded, and five missing and he seems to have estimated that about 60 of his enemies had been killed.

Bouquet abandoned his supplies and moved on, with his wounded, to Fort Pitt, fighting a brief skirmish in the process. After this action the aboriginal American threat to Fort Pitt and to western Pennsylvania dissolved, and Pontiac's War began to dissolve as well. The victory at Bushy Run was widely celebrated at the time, and was seen as having saved western Pennsylvania.

Scott N. Hendrix

See also: Bouquet, Henry; French and Indian War; Pontiac's Rebellion; Seven Years' War

References and further reading:

Anderson, Fred. *Crucible of War: The Seven Years War and the Fate of Empire in British North America, 1754–1766.* New York: Alfred A. Knopf, 2000.

Nester, William R. *"Haughty Conquerors": Amherst and the Great Indian Uprising of 1763.* Westport, CT: Praeger, 2000.

Butler, Benjamin F. (1818–1893)

Controversial Civil War Union "political" general. Born in New Hampshire, Benjamin Franklin Butler had earned a prewar name as a controversial trial lawyer and Massachusetts politician. As a Democrat, and desiring to regain influence with the outbreak of the Civil War, Butler managed to parlay a brigadier's commission in the Massachusetts state militia into an active-duty role.

While generally successful as military governor in Baltimore and New Orleans, Butler kept order at the price of offending local elite opinion. This, and his reputation for corruption, overshadowed Butler's genuine achievements.

Butler also participated in the taking of New Orleans, the 1864 landing at Richmond (an operation of great potential value), and the assault on Fort Fisher. Having botched the latter two operations, Butler's reputation was so in ruins that his political talents could not save his military career, effecting his return to civilian life.

George R. Shaner

See also: Civil Affairs/Military Government

References and further reading:

Hearn, Chester G. *When the Devil Came Down to Dixie: Ben Butler in New Orleans.* Baton Rouge: Louisiana State University Press, 1997.

Sandler, Stanley. *Glad to See Them Come and Sorry to See Them Go: A History of U.S. Army Military Government and Civil Affairs, 1775–1992.* Fort Bragg, NC: U.S. Army Civil Affairs and Psychological Operations Command, 1998.

West, Richard S., Jr. *Lincoln's Scapegoat General: A Life of Benjamin F. Butler, 1818–1893.* Boston: Houghton Mifflin, 1965.

Byzantine Civil Wars (1322–1355)

Civil conflict erodes position of declining Byzantine Empire. In 1321, a family dispute within the ruling Byzantine created a rift between the reigning emperor, Andronicus II (r. 1282–1328), and his grandson, also named Andronicus. Subsequently, when Andronicus II attempted to expand the Byzantine armed forces, and imposed taxes upon the Byzantine nobility in order to pay for that expansion, he forfeited much popular support. As a result, Andronicus the younger proclaimed himself Emperor Andronicus III (r. 1328–1341). Although open violence was prevented, the empire was partitioned between the two claimants, and Andronicus III was excluded from Constantinople.

Open warfare began in 1328, when Andronicus III attempted to enter Constantinople and was refused. Andronicus III instead seized Salonika, the second largest city in the empire. Most of the Byzantine forces in the west declared for the younger Andronicus. In May, Andronicus III finally entered Constantinople, and forced the abdication of his grandfather.

Andronicus III fought successfully with Turks, Serbs, Genoese, and Bulgars, both adding territory to and occasionally losing imperial territory. He also finally defeated the Despotate of Epirus, and reintegrated it into the empire in 1339 and in 1341. Andronicus III also forced the submission of the Latin barons in the Peloponnesus and took steps to reattach the empire of Trebizond to the Byzantine Empire.

Andronicus III died suddenly in 1341. As his heir, he left behind his nine-year-old son, John V (r. 1341–1391). Andronicus had proposed in his will that John Cantacuzinus, a wealthy noble from Thrace, competent soldier, and trusted confidant, be made regent. Although Cantacuzinus was a friend of the late emperor, he had made some powerful enemies in the imperial court, including the empress Anna; John Kalekas, the patriarch of Constantinople; and the governor of Constantinople, Alexius Apocaucus. Cantacuzinus thus found it expedient to decline an offer made by Apocaucus to become coemperor with John V, but did assume the title of coregent.

Cantacuzinus led an army against the Serbs to recover territory lost by Andronicus III. He also prepared a campaign to force further concessions from the Peloponnesian barons. Before he could effectively undertake this, his opponents in Constantinople deposed him in September 1341 and incited anti-Cantacuzine and antiaristocratic riots in Constantinople and Salonika. A revolutionary government in Salonika was then recognized by the imperial court.

Cantacuzinus fled to Serbia and proclaimed himself Emperor John VI (r. 1347–1354). In 1341 and 1342, both Serbs and Bulgars raided the empire, and Cantacuzinus made little headway against the new regency.

By 1343, with Serb support, Cantacuzinus overran most of Epirus, and began to advance into Thrace, besieging Salonika. By 1344, he held most of Thrace. In 1345, Cantacuzinus took Adrianople. The Serbs, in the meantime, had ended their support for Cantacuzinus, who responded by reaching agreement with a group of Turks led by Umur, the emir of Aydin. Also in 1345, Apocaucus was murdered by a group of nobles in Constantinople. This encouraged Cantacuzinus to attempt to enter the city, but he failed. In an effort to further strengthen his position, Cantacuzinus allied himself with the Ottoman Turks in 1346.

In 1346, Cantacuzinus was formally crowned emperor in Adrianople, but only in February 1347 was he able to reach an agreement with the supporters of John V. John VI was to administer the empire as the senior emperor. John V was married to a daughter of John VI. John VI was not only able to recover his estates in Thrace, but was also able to distribute large tracts of land to be governed directly by his sons. Matthew Cantacuzinus, the eldest son of John VI, received land in Thrace, while Michael Cantacuzinus received land in the Morea.

In 1351 or 1352, John V demanded a greater role in the government, and received instead the lands in Thrace held by Matthew Cantacuzinus. Matthew, by way of compensation, received lands further east, around Adrianople. Border disputes between Matthew and John V led to renewed fighting between the Paleologoi and the Cantacuzines, and foreign powers were quickly drawn in. John V received help from the Serbs, while John VI received help from the Turks. Although John V enjoyed some initial success, his Serbian allies were defeated by the Turks at Demotika in October 1352.

John V escaped to the Island of Tenedos, held by Venice. From Tenedos, he attempted to seize Constantinople in March 1353. John VI declared John V deposed, and crowned Matthew coemperor in February 1354.

John V, meanwhile, sought aid from Genoa. Cantacuzinus, blamed for having brought the Turks into the empire, abdicated and became a monk. John V entered Constantinople in November 1354.

John V defeated Matthew in 1355, and forced him to relinquish his claims. He then sent Matthew to join his brother in the Morea. A subsequent attempt by John V to conquer the Morea failed.

As a result of the civil war, Byzantium lost much of its remaining territory. Serbia nearly doubled in size, and its ruler claimed the title of czar. The Ottoman Turks entered Europe and seized the Dardanelles. In the course of the war, the Byzantine economy and bureaucracy collapsed, and neither was ever properly restored.

Joseph M. Isenberg

See also: Byzantine-Ottoman Wars; Ottoman Empire

References and further reading:

Fine, John V. A. *The Early Medieval Balkans.* Ann Arbor: University of Michigan Press, 1983.

Treadgold, Warren. *A History of the Byzantine State and Society.* Stanford, CA: Stanford University Press, 1997.

Byzantine-Muslim Wars (633–1035)

A series of wars involving the Arab Muslim conquest of Byzantine territory in North Africa and the Near East, and the subsequent Byzantine recovery of some of that territory. In 629, the Byzantine Empire successfully concluded a long series of wars with Persia. As a result, the Byzantines recovered territory in Palestine, Lebanon, and Syria, but had little time to organize administration or defense there before the initial Moslem incursions.

An initial Muslim Arab invasion in 929 was defeated by Byzantine forces at Muta, but in 630, a Muslim expedition forced the submission of the town of Aqaba. A more earnest Arab effort at conquest began in 633 and 634, when the caliph Abu Bekr sent four armies, perhaps totaling 20,000 men, into Syria. By the end of 634, the Arabs had won a series of victories at Ayn Ghami, Dathin, Anyadayn, and Scythopolis. In 635, the Muslim armies occupied Damascus and Homs for the first time, but were forced to abandon the cities in the face of a Byzantine counteroffensive. This counteroffensive came to an end on 20 August 636, when Byzantine forces suffered a catastrophic defeat at Yarmuk.

As a result of the defeat, Muslim armies overran most of Palestine and Syria, save only for a portion of Syria granted a one-year truce. This allowed Christian Arabs to flee into Byzantine territory before the resumption of hostilities. After the expiration of the truce, Muslim forces quickly overran the remainder of Syria and Palestine.

In December 639, Arab forces began an invasion of Egypt, and raided Cilicia and Anatolia. The deaths in 641 of the Byzantine emperors Heraclius (r. 610–641) and Constantine III (r. 641), and then the removal, through a coup, of the emperor Heraclonas (r. 641), created political instability in Constantinople. As a result, no effective aid was given to isolated Byzantine forces in Egypt. Arab armies were able to complete their invasion of Egypt, and moved on Cyrenaica in 642.

After these initial disasters, the Byzantines turned to positional warfare, attempting to hold the major cities while letting interior areas go. While this policy slowed the Muslim advance by forcing Arab armies to take the time to seize fortified points, in the long term the strategy was doomed to failure. The cities could not be held without also securing the agricultural hinterland that supplied them.

In 642, the Arab invasion of Armenia began. Within 20 years, despite constant Byzantine efforts to control Armenia, the Muslims had successfully converted the region into a client state. In 650, the Arabs invaded Cyprus. It later became a Byzantine/Arab codominion by treaty.

In the 650s, the Arabs turned attention to the Persian Empire, and to further conquests in North Africa. While raids remained frequent, the Byzantine Empire was able to gain control of the Taurus mountain passes leading into Anatolia. Further Arab conquests in there were thus blocked.

In 648, an initial Arab invasion of the Byzantine province of Africa was bought off by local officials. The year 654 saw further Arab assaults on Cyprus, Crete, and Rhodes. The emperor Constans (r. 641–668) reorganized the Byzantine army between 659 and 662 in a further effort to stop Arab advances. He continued the process whereby the remaining mobile armies were associated with a new type of province, called *themes.*

In 670, the Arab invasion of Africa began in earnest, but met with firm resistance. That same year, Arab naval forces raided the Sea of Marmora for the first time. In response, the

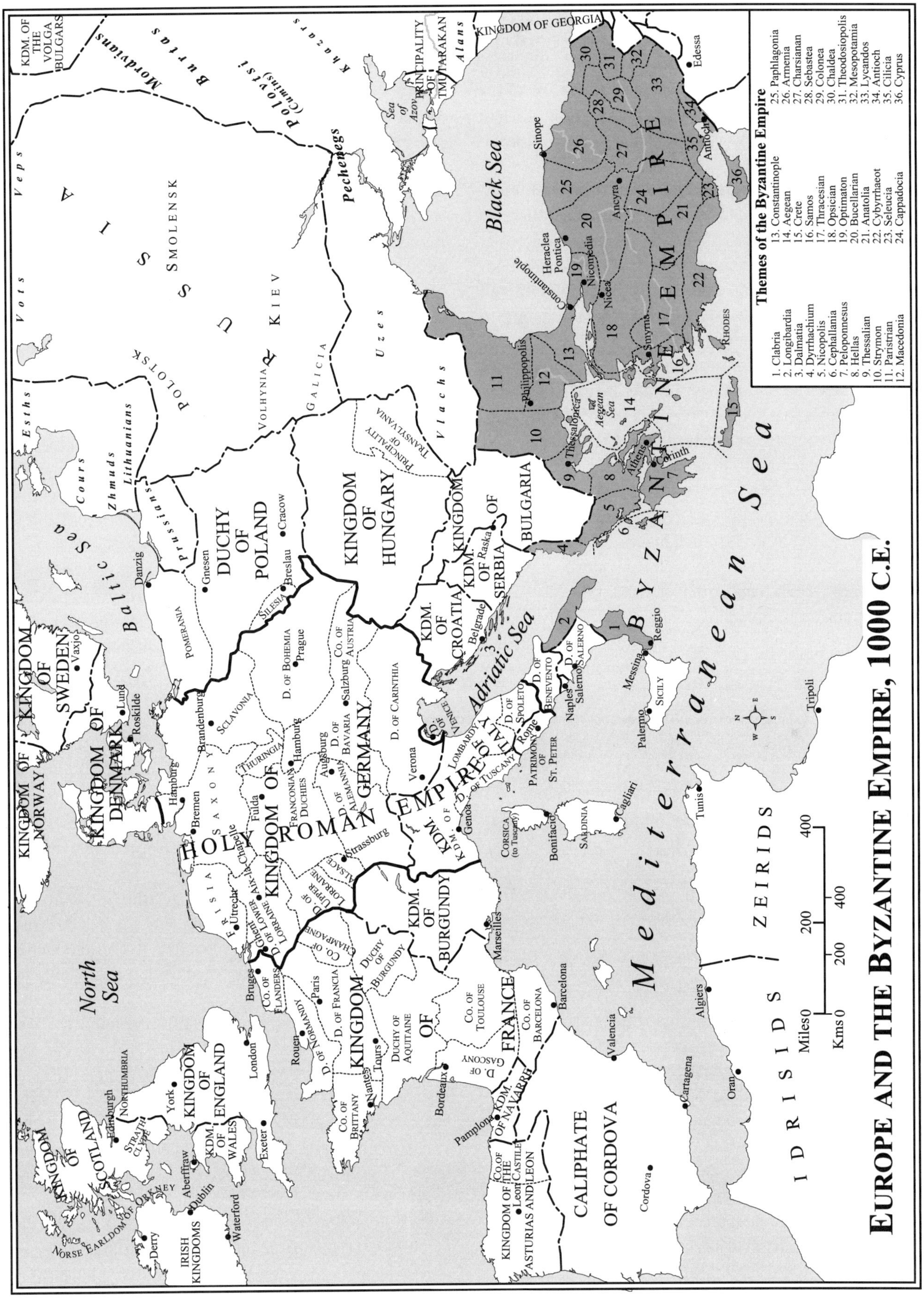

EUROPE AND THE BYZANTINE EMPIRE, 1000 C.E.

Byzantines prepared a fleet, which by 672 included the mysterious "Greek fire" in its arsenal. In 677, this improved Byzantine fleet won a major victory over the Arabs.

In 696, a second Arab expedition to Africa led to the seizure of Carthage. Byzantine efforts to recover the province proved unsuccessful, and this led to further political instability in Constantinople.

In 711, the Arabs breached the Taurus barrier and advanced into Anatolia. Unable to stop the Arab advance, the emperor Anastasius II (r. 713–715) began preparing the defenses of the capital. In 716, an Arab assault on Constantinople failed, but in 717 an Arab force of 120,000 men and 1,800 ships besieged the capital. The Bulgarians, hoping to take the city for themselves, attacked the Arabs, who were forced to build two sets of siege works, to contain the Byzantines on one side and to keep out the Bulgarians on the other. In September, the Arab fleet appeared, but was driven off by the Byzantines using Greek fire. The Arab army thus remained trapped in its siege works during an unusually harsh winter.

A fleet of 600 ships was sent to replenish the Arab forces. The ships landed near Chalcedon in order to avoid the Byzantine fleet. The crews of the Arab fleet, mostly Egyptian Christian, defected en masse to the Byzantines.

After an Arab reinforcing column was destroyed near Nicaea, and an epidemic had broken out among the Arab forces near Constantinople, the caliph Umar finally ordered a retreat on 15 August 718. The Arab retreat was not opposed, but surviving Arab ships were attacked, and the Arab fleet was further damaged by storms and by a volcanic eruption.

Between 718 and 741, a series of raids and counterraids ravaged Anatolia. In this period, Byzantine strategy embraced not only positional defense, but also a policy of intercepting Arab raids returning from plundering expeditions. These tactics proved moderately successful, and gave rise to a series of Byzantine epic poems and legends about border raiders and defenders. Of these the most famous was the epic poem *Digenes Akrites.*

In 741 Constantine V (r. 741–775) succeeded to the Byzantine throne. He gradually pushed the Byzantine frontier forward. This process was continued by his son, Leo the Khazar (r. 775–780).

Throughout the eighth century and the first part of the ninth century, Arab raids and conquests continued. In the 850s, the Arab Caliphate began to lose control of its border regions in the face of attacks by Paulician heretics in Anatolia and by local Byzantine governors.

In the tenth century, Byzantine defense efforts began to show real results. In 900, Leo VI (r. 886–912) invaded the emirate of Tarsus, defeated its army, and also invaded Armenia. Leo was able to recover much frontier territory.

In 926, Romanus Lecapenus (r. 920–944) renewed the attack upon the Arabs and sacked Melitene. In 927, the emir of Melitene submitted to the empire, and in 928 the city received a Byzantine garrison. This was initially driven out, but in 933 and 934 the Byzantines systematically occupied the fortresses around Melitene and around the city of Samosata. Melitene was taken and all non-Christians forced to leave. Samosata was taken and razed in 936.

In 942 and 943, John Curcuas, Grand Domestic, sacked a number of cities in Armenia and Mesopotamia, and besieged the city of Edessa, withdrawing only after the governor of the city had agreed to surrender the Mandilyon, a cloth said to bear the imprint of the face of Christ. The relic and Curcuas were accorded a triumphal entry into Constantinople.

Constantine VII (r. 913–959) assumed full power in 944, and, with his generals, Nicephorus Phocas, John Tzimisces, and Basil the Grand Chamberlain, conducted a successful raiding war against the Arab leader Sayf al Dawlah. Byzantine efforts to recapture Crete, Sicily, and Italy were less successful, and Arab forces repeatedly defeated forces sent to those places. Constantine's successor, Romanus II (r. 959–963), reorganized Byzantine forces in an effort to intensify the war against the Arabs. In 960, a massive expedition under Nicephorus Phocas landed on Crete. Phocas's army killed 40,000 Arabs and besieged the principal city of Chandax from 960 until the spring of 962, when it finally fell. A raid into Syria by Sayf in that year was suppressed by Byzantine forces.

After capturing Chandax, Nicephorus Phocas attacked and defeated the emir of Tarsus and took several towns in Anatolia and Syria. Phocas then marched against Sayf al Dawlah in Aleppo. Sayf sent his army north to meet the Byzantines, and quickly raised a force of militia to defend the city. Phocas evaded the main Arab force, and fell upon the poorly prepared defenses of Aleppo, which he overcame. Sayf fled, pursued by Tzimisces, and Phocas entered Aleppo after a three-day siege. Upon his return to Constantinople, Phocas found that Romanus II had died in a hunting accident, and assumed power.

Nicephorus Phocas (r. 963–969) continued the war against Sayf, taking the town of Mopsuestia, as well as Tarsus. Nicephorus also expanded Byzantine control of Armenia and cleared Cyprus of its Arab garrison. Sayf died in 967.

John Tzimisces (r. 969–976) seized power in a coup. During his reign, Aleppo ceded control of its coastal territory to Byzantium; this remained permanently under Byzantine control. Byzantium now bordered the Fatamid Caliphate of Egypt and the Hamdanid emirate of Mosul, both of which were willing to expand at the expense of Byzantine or Arab neighbors. Tzimisces thus found himself fighting against the Fatamids in 971 and in 975, and the Hamdanids from 972 to 974. Tzimisces was able, in the end, to subject most of Syria and Lebanon to either direct Byzantine government or tributary status. He died before these conquests could be followed up, and was succeeded by Basil II (r. 976–1025), the greatest ruler of medieval Byzantium.

Basil II faced civil disorder in the early part of his reign, but his Arab neighbors were too weak to take advantage of the situation. Basil was strong enough in the 990s to drive off Fatamid attacks upon the emirate of Aleppo, now a Byzantine client state.

Fatamid attempts at naval warfare were equally unavailing. An Egyptian fleet burned in 996 in a mysterious fire. A second fleet was defeated off of Syria by the Byzantines. The Fatamid caliph, al-Aziz, then died, and rebellions broke out over the succession.

Basil II turned his attention to Byzantine territories in Europe. He signed a 10-year truce with the Fatamids in 1001, which he renewed periodically for the rest of his reign. Basil refused to respond to Fatamid and Hamdanid attacks on Aleppo; the city fell to the Fatamids in 1015, only to be lost to them in a revolt in 1025.

At this point, Byzantine attentions became fixed elsewhere, with few exceptions. Romanus III (r. 1028–1034) forced the surrender of Edessa in 1031, and attempted to purchase the city of Aleppo. This offer was refused. The Fatamids renewed the 10-year truce in 1036, and the Byzantine frontier was secured by the presence of a client state, the Mirdanid emirate, in what is now modern Syria and Iraq. Though raids would occasionally continue, the initiative in Muslim expansion had passed from the Arabs to a group of Islamic mercenaries brought in by the Arabs, the Turks.

Joseph M. Isenberg

See also: Basil II Bulgaroctonus; Byzantine-Persian Wars; Constantine V; Constantinople, Siege of (717–718); John I Tzimisces; Muslim Civil War; Nicephorus II Phocas; Yarmuk, Battle of

References and further reading:

Kaegi, Walter. *Byzantium and the Early Islamic Conquests.* Cambridge, UK: Cambridge University Press, 1992.

Treadgold, Warren. *A History of the Byzantine State and Society.* Stanford, CA: Stanford University Press, 1997.

Whittrow, Mark. *The Making of Byzantium, 600–1025.* Berkeley: University of California Press, 1996.

Byzantine-Ottoman Wars (1302–1461)

A series of wars ending in the Ottoman conquest of the Byzantine Empire. Following the fall of the Latin Empire of Constantinople in 1261, the Byzantine Empire was a mere shadow of its former greatness, controlling only the city of Constantinople, the western coast of Asia Minor, northern Greece, and part of the Peloponnesus (the Morea). Two autonomous territories, Trebizond and Epirus, also were Byzantine in culture, but not politically dependent on Constantinople.

In this weakened state, the Ottoman Turks, having threatened the empire for more than two centuries, advanced. The Byzantine Empire often had to rely on outside help, sometimes from western Europe, to protect itself, and that help was often tied to a union of the Orthodox and Catholic churches, a union that was briefly achieved at the Council of Florence and Ferrara in 1439.

Among the western aid the Byzantines received were Roger de Flor's Catalan mercenaries, hired by Andronicus II in 1302. Roger was murdered three years later, but the Catalans went on to set up their own independent state in Athens in 1311. In 1396, Pope Boniface IX called a crusade to assist Constantinople, which ended in the defeat of Duke John the Fearless of Burgundy and Jean Bouciquaut, Marshal of France, at Nikopolis (Nicopol, Bulgaria). On 10 November 1444, the Last Crusade ended in defeat for King Ladislas of Hungary and Poland and his general, János Hunyadi, at Varna.

Western aid was to no avail, and in 1329 Sultan Orkhan I defeated Andronicus III at Maltepe and two years later captured the city of Nicaea. More than a decade later, Emperor John VI Cantacuzinus called on the Turks to assist him against the Serbian emperor Stephan Dushan, who threatened Constantinople. Turkish forces thus crossed into Europe in 1345, gaining a foothold in Gallipoli in 1354. In 1365, Sultan Murad I captured the important city of Adrianople and made it his capital. A year later crusades by Amadeus of Savoy and Louis of Hungary led to the temporary recapture of Gallipoli and the defeat of the Turks near Vidin, but these gains were short-lived.

In 1365, the Janissaries (literally "New Troops"), a corps of elite infantry composed of slaves, were established as a military corps by the sultan. This corps was to prove important in the remaining wars with the Byzantine Empire and in subsequent Ottoman politics into the nineteenth century.

The Turks continued reducing Byzantine power, taking Thessalonika in 1387 and again in 1430 after it had been recaptured by the Venetians and Byzantines. Epirus fell in 1430 also. Turkey invaded central Greece in 1397. In 1446, Constantine Paleologus (Constantine XI in 1448) attempted the recapture of central Greece but was repulsed.

Throughout the wars, the Ottomans besieged Constantinople itself. The first Ottoman siege lasted from 1391 to 1399. Sultan Murad II again besieged the city in 1422. Sultan

Mehmet II the Conqueror (*Fatih*) besieged the city again in April 1453; it fell on 29 May after bitter fighting, during which Constantine XI was killed. Mehmet then made Constantinople the Ottoman capital, renaming it Istanbul.

The Morea held out seven more years, falling in 1460. Trebizond, the last remnant of the Roman Empire, was conquered in 1461.

Michael C. Paul

See also: Constantinople, Siege of (1453); Hunyadi, János; Janissaries; Turkish Wars of European Expansion

References and further reading:

Nicol, Donald M. *The Last Centuries of Byzantium 1261–1453.* Cambridge, UK: Cambridge University Press, 1993.

Norwich, John J. *Byzantium: The Decline and Fall.* New York: Knopf, 1996.

Byzantine-Persian Wars (502)

A series of wars between the two great powers of western Eurasia defining the history of the sixth century. In 502, Sassanid shah Kavad I broke 60 years of peace with Byzantium in hopes of wringing money from Anastasius I (r. 491–518). Briefly dethroned during a period of domestic upheaval, and undergoing punishing raids by the White Huns on Persia's Asian borders, Kavad hoped to buy the raiders off. He felt that the wealthy Byzantines, at war in the Balkans and Arabia, would pay dearly to avoid a third front. But when Anastasius refused to contribute, Persian honor demanded action, so the shah invaded Byzantine Armenia and northern Mesopotamia. In response, the Byzantines assailed Persian Armenia. By 504, White Hun raids again distracted Kavad, and Anastasius bought peace for a token sum.

In 525, Emperor Justinian (r. 527–565) inattentively beguiled himself into an Armenian adventure. Christians in Iberia province, vassals of Persia, asked for a symbolic force to aid an independence scheme. Justinian's gesture provoked a massive Sassanid offensive. Byzantium sacrificed much blood, treasure, and three hard-fought years to restore the frontier.

In 530, Byzantine occupation of the Crimea gave Constantinople access to Asian nomads who might open an eastern front against Persia. Recognizing the risk, a new shah, Khusrauw I, proposed a truce in 531. Justinian, fixated on re-creating the Roman Mediterranean empire, obliged.

The peace of 532 required a heavy tribute from Justinian, but Khusrauw remained suspicious. Ostrogoth emissaries from Italy warned the shah that Justinian might turn eastward after subjugating them. Byzantine machinations routinely encouraged instability in Armenia. In 540, Khusrauw's army rolled en masse toward Antioch, the "Third City" of Byzantium, sacking forts, looting towns, and demanding an enormous tribute. When Justinian refused, Khusrauw viciously burned Antioch and took thousands hostage. Another weary stalemate followed, compounded by a horrendous epidemic.

Byzantine defeats in Armenia in 543 mirrored Persian failures before Edessa in 544. Increasingly preoccupied at home, the rivals renewed peace in 545. Although Armenia remained in a kind of proxy war, Justinian and Khusrauw extended their treaty for 50 years in 560, with Byzantium sending an annual stipend.

In 572, Justin II (r. 565–578) arrogantly broke the peace. Indignant, Khusrauw marched into Syria, capturing Dara, and into Armenia. In 578, Byzantine general Maurice staged counterattacks in Armenia and along the upper Tigris-Euphrates. In response, Hurmazd IV, Khusrauw's son, encouraged the Avars to ravage the Balkans to draw Byzantine forces from the east. In 582, Maurice became Byzantine emperor (r. 582–602), confronting wars in Italy, the Balkans, Armenia, and Iraq. Victories over Hurmazd in 586 and 587 failed to knock the Persians out. In 588, Byzantium's eastern army erupted in a terrifying mutiny. Yet Persia, also staggering from the war effort, lacked the initiative and energy to exploit a historic opportunity.

Two years later, the Persian commander Bahram mutinied, overthrew Hurmazd, and tried to enthrone Hurmazd's son, Khusrauw II. Khusrauw instead fled to the Byzantines and appealed for assistance from Maurice. Ignoring his timid advisers, the emperor threw his weight behind the young aspirant. Their combined forces overcame Bahram rapidly. In 591, Khusrauw and Maurice, now close friends, signed a peace that endured until 602.

Weston F. Cook Jr.

See also: Avars; Belisarius; Justinian I; Mauricius /Flavius Tiberius

References and further reading:

Norwich, John J. *Byzantium: The Early Centuries.* New York: Alfred A. Knopf, 1994.

Whitby, Michael. *The Emperor Maurice and His Historian.* Oxford, UK: Oxford University Press, 1988.

C

Caesar, Julius (Gaius Iulius Caesar 100–44 B.C.E.)

Shrewd political and military strategist. Caesar accurately assessed and energetically defeated domestic and foreign foes. He brilliantly inscribed for posterity his exceptional achievements. Nonetheless, his pride and resulting incapacity to cement lasting personal alliances paved the way for his downfall and, with it, the transformation from Roman republic to empire.

Caesar's family was firmly established in Marius's popular reform camp. Virtually exiled for opposing Sulla's party, Caesar served as a military tribune in Asia, 80–78 B.C.E., where he received Rome's highest decoration, the *corona civica.* In 65 B.C.E., as *curule aedile,* he spent profusely on games and public building projects, and subsequently became *pontifex maximus* and praetor. As governor of Further Spain (embracing modern Portugal and much of western, central, and southern Spain), he demonstrated his military prowess in highly successful policing actions against the tribes of the area in 61–60 B.C.E. In 61 he obtained the province of Hispania Ulterior and on his return was elected consul. He then wisely came to terms with Pompey and Crassus and they jointly constituted the First Triumvirate in 60 B.C.E. He was given a five-year governorship, extended to 10 years in 55, of the provinces of Illyria on the eastern shore of the Adriatic Sea and of both Transalpine and Cisalpine Gaul (present-day northern Italy, France, Belgium, part of Germany, and the southern Netherlands).

During his tenure as governor, Caesar conquered Gallic territory up to the River Rhine. His defeat of Vercingetorix at Alesia was notable for Caesar's use of innovative field engineering techniques. He suffered only two setbacks in this period: a detachment of 15 cohorts was annihilated in the winter of 54, and his attack on the Gallic fortress-town of Gergovia in 52 ended in a costly failure. His piecemeal conquest of Gaul is attributable to judicious tactics, excellent reconnaissance, vigorous exploitation, rapidity of decision, and deterrence by terror. When his governorship ended in 49, Caesar was immensely wealthy and the leader of a highly efficient and fanatically loyal army.

His military success aroused the jealousy of Crassus and conservative factions in the Senate who felt threatened by Caesar's ostensible populism. Crassus's defeat and death at Carrhae left Pompey to deal with civil unrest entailing violent clashes among rival gangs, including Caesar's popular reform alliance. Seeking to reestablish law and order, the Senate elected Pompey sole consul in 52, a position that he exploited to have Caesar sacked from governorship in Gaul with a demand that he disband his army in that province. When Caesar refused and marched his loyal legions across the Rubicon River into Italy proper, Pompey, believing his army too meager, withdrew to the east to recruit more forces there. After entering Rome in triumph, Caesar led troops into Spain to crush threatening pro-Pompey troops. Having been declared dictator, he led a force to Greece where he defeated Pompey's much larger army at Pharsalus. Siding with Cleopatra (later his mistress) against her brother in an Egyptian dynastic struggle, he defeated the deceased Pompey's naval expedition sent against him. Caesar, pausing in Asia Minor to defeat the upstart Pharnaces of Bosporus, returned to Rome where he continued his program of sweeping civic and legislative improvements. In the spring of 46 B.C.E. Caesar took an army to North Africa to eradicate Pompey's remnants there and thence to Spain for final mopping-up operations. He was assassinated upon his return to Rome by a cabal of patrician opponents.

Jim Bloom

See also: Alesia, Siege of; Gallic Wars; Pharsalus, Battle of; Pompey the Great; Roman Civil Wars (88–30 B.C.E.); Roman Republic, Wars of the; Sulla, Lucius Cornelius

References and further reading:
Dodge, T. A. *Caesar: A History of the Art of War among the Romans down to the End of the Roman Empire, with a Detailed Account of the Campaigns of Gaius Julius Caesar.* Boston: Houghton Mifflin, 1892. Reprint, New York: Da Capo Press, 1997.
Fuller, J. F. C. *Julius Caesar: Man, Soldier and Tyrant.* London: Eyre & Spottiswood Ltd., 1965.
Gelzer, Matthias. *Caesar—Politician and Statesman.* Cambridge, MA: Harvard University Press, 1968.
Meier, Christian. *Caesar: A Biography.* New York: Basic Books, 1995.

Calais, Siege of (1558)

The end of English rule over any part of the European continent. After the Spaniards, allied with the English, had routed the French army under Montmorency at St. Quentin on 10 August 1557, Paris was at the mercy of Phillip II of Spain, but his hesitation to attack the capital gave the French monarch, Henri II, time to organize the city's defenses. In October, François de Guise, returned from the Italian war theater, was created lieutenant of the realm, and gathered together the French troops in northern France. Both sides were desperately short of money, but while the Spanish king disbanded his troops for the winter, the French monarch boldly planned a winter campaign.

In November, Henri determined to attack Calais to avenge the humiliation of St. Quentin, and to seize an important pledge for future peace negotiations. At the direct order of Henri, de Guise agreed to lead the winter expedition, although he was not convinced of its prudence. The English had not the vaguest suspicion of the French plans until the end of December 1557.

On 1 January, 2,500 unprepared English, spread among a dozen forts and in Calais itself, faced the assault of 20,000–22,000 French foot soldiers, 4,000 cavalry, and 30 artillery pieces. Only one day later, the fort of Rysbank surrendered, and from there the French bombarded the castle across the harbor for two days. After a breach had been made, several French companies seized the castle and held it against two English counterattacks. On 8 January, the English commander, Lord Thomas Wentworth, asked for terms. Five hundred English troops along with some of the town's inhabitants were permitted to leave for the Flemish border. After holding Calais for 220 years, England had lost its last territory on the European continent.

Holger Th. Graef

See also: Anglo-French Wars; Boulogne, Siege of; Guise, François de Lorraine, Second Duke of; Hundred Years War; St. Quentin

References and further reading:
Baumgartner, Frederic J. *Henry II, King of France, 1547–1559.* Durham and London: Duke University Press, 1988.

The Cambodian Incursion (30 April–15 May 1970)

A U.S.–South Vietnamese incursion into Cambodia during the Vietnam War. President Richard Nixon had a three-part plan to end the interminable Vietnam conflict: American troop withdrawal, an end to the draft, and Vietnamization of the war effort. To help provide time for the government of South Vietnam to prepare to defend itself, Nixon in March 1969 had authorized a secret and possibly illegal bombing campaign across the border in neighboring Cambodia. The bombing did not significantly reduce North Vietnamese supply capacity, and it threatened the stability of the Cambodian government itself as the Communists moved away from the borders to avoid the bombing.

Therefore, the South Vietnamese and Americans engaged in a limited crossing of the border into the so-called Parrot's Beak and Fishhook regions to destroy base areas. Communist forces abandoned their bases, and the allies captured a vast amount of weapons and ammunition. In a smaller operation, U.S. Army units crossed the border in the Central Highlands also to attack base camps near the Laotian–Cambodian–South Vietnamese border.

In an immediate sense, the spoiling attacks were successful, and the North Vietnamese had to wait two more years to launch the Easter 1972 offensive. However, the U.S. Senate repealed the Gulf of Tonkin Resolution and, combined with the unrelated killings on the Kent State University campus, the offensive strengthened the antiwar movement in the United States, something carefully noted by Hanoi.

Charles M. Dobbs

See also: Abrams, Creighton William, Jr.; Ho Chi Minh; Tet Offensive; Vietnam Conflict; Vo Nguyen Giap; Westmoreland, William

References and further reading:
Herring, George C. *America's Longest War: The United States and Vietnam, 1950–1975.* 3d ed. New York: McGraw-Hill, 1996.
Karnow, Stanley. *Vietnam: A History.* New York: Viking Press, 1983.
Tran, Dinh Tho. *The Cambodian Incursion.* Washington, DC: U.S. Army Center of Military History, 1979.

Cambodian Wars (1970–1990s)

Although an indigenous Communist (Khmer Rouge) insurgency fought the central government from rural bases in the late 1960s, Cambodia remained on the fringes of the Indochina War until Prince Sihanouk was overthrown in an 18 March 1970 coup d'état. Both Communist and non-Communist forces had ignored Sihanouk's policy of neutrality when it suited them, but it was only with the establishment of a new regime (proclaimed the Khmer Republic in October 1970) under Premier Lon Nol that the country became a major Indochina battlefield.

Cambodia's wars can be divided into three periods: (1) 1970–1975, when Cambodian and Vietnamese Communist forces fought the Khmer Republic, which received massive military and economic aid from the United States; (2) 1975–1978, when Democratic Kampuchea (DK), the revolutionary regime headed by the Khmer Rouge leader, Pol Pot, waged brutal class warfare, resulting in the deaths of 1–2 million Cambodians out of a total population of 7 million, and provoked a war with Vietnam; and (3) 1979–1990s, which opened with the January 1979 establishment of a pro-Hanoi regime in Phnom Penh, the People's Republic of Kampuchea (PRK), and witnessed protracted fighting between the PRK and the Vietnamese army on one side and the Khmer Rouge based along the country's western borders on the other (joined by Prince Sihanouk and a rightist politician, Son Sann, in a largely symbolic coalition government in 1982). This last phase, characterized by Vietnamese dry season offensives against mobile Khmer Rouge guerrillas, continued until Hanoi agreed to withdraw its forces in 1989, and the signing of the Paris Peace Accords on 23 October 1991. But some observers date its termination with the late 1990s internal collapse of the Khmer Rouge and the apparent suicide of Pol Pot in April 1998.

After the coup d'état, Lon Nol's Khmer National Armed Forces received new American equipment and expanded rapidly in personnel, but lost control of the countryside to the North Vietnamese army and a revitalized Khmer Rouge, which gained wide popular support by promising to restore the popular Prince Sihanouk to power. Two government offensives in 1970–1971, Chenla I and II, attempted to recapture the agriculturally rich northwestern region, but were stopped by North Vietnamese regulars.

Washington launched a devastating air war against the Communists, peaking in summer 1973 (Operation ARC-LIGHT), in which the U.S. Air Force dropped, according to some accounts, more tonnage of explosives on the small country than on Japan during World War II. The bombing shattered rural society, driving hundreds of thousands of Cambodians into Phnom Penh and other towns (they also fled Khmer Rouge atrocities), but failed to bolster Lon Nol or stop the Communists. Most North Vietnamese troops withdrew from Cambodia by 1972–1973, and the Khmer Rouge built up their forces to 50,000 soldiers and captured Phnom Penh on 17 April 1975.

The Cambodia war was waged with untrammeled brutality. The worst perpetrators tended to be adolescent or preadolescent soldiers, the mainstay of Pol Pot's army. Most older soldiers had been killed off and the young Khmer Rouges, cut off from their families, proved to be zealous instruments of revolutionary terror, lacking the socialization in traditional Cambodian values that might have moderated their behavior.

Democratic Kampuchea was no puppet of Hanoi. The DK regime singled out Vietnamese living in the country for especially harsh treatment and murdered hundreds of Vietnamese peasants in cross-border raids. Pro-Hanoi elements in the Communist Party of Kampuchea were purged and executed, even before 1975. In May 1978, the commander of the Eastern Region bordering Vietnam rebelled, but was defeated, and an estimated 100,000 people, labeled as having "Khmer bodies but Vietnamese minds," were sent into the killing fields.

DK armed forces repeatedly provoked border clashes. Pol Pot apparently harbored the ambition of conquering the Mekong Delta, which before the seventeenth century had been Cambodian territory. In late 1977, Vietnamese forces counterattacked, occupying Cambodian territory, but withdrew in January 1978 (another motive for the incursion was DK's alliance with China, seen by Hanoi as a provocation).

In November 1978, the Vietnamese carved a "liberated area" out of Cambodian territory and proclaimed an anti–Pol Pol government in exile, the Khmer United Front for National Salvation (KUFNS). On Christmas Day, a 100,000-man infantry and armored force consisting of Vietnamese regulars and KUFNS auxiliaries launched a blitzkrieg, capturing Phnom Penh on 7 January 1979 and thrusting into the western provinces of Siem Reap and Battambang. Pol Pot and other DK leaders fled the capital just ahead of the Vietnamese, and established a guerrilla resistance in the hilly, thickly forested region around Cambodia's border with Thailand.

The third phase of the war (1979–1990s), like the 1970–1975 phase, was in large measure a proxy war between the great powers. Cold War geopolitics determined that the Soviet Union would back their Vietnamese allies and Hanoi's client state, the People's Republic of Kampuchea, while China and the Western powers refused to recognize the latter, denying the PRK a seat in the United Nations and much-needed economic aid. Beijing funneled military aid to Pol Pot's forces through a cooperative Thailand (this amounted to an estimated U.S.$60–100 million a year in the 1980s). Although the Paris Peace Accord of October 1991 opened the way to United Nations–sponsored political and economic reconstruction, Cambodia's tragic quarter century of war showed how big powers could take advantage of local conflicts, using small countries for their own ends, with scant regard for the welfare of their populations.

Donald M. Seekins

See also: Vietnam Conflict

References and further reading:

Chandler, David P. *A History of Cambodia.* 2d ed. Boulder, CO: Westview Press, 1992.

Shawcross, William. *Sideshow: Kissinger, Nixon, and the Destruction of Cambodia.* New York: Simon & Schuster, 1979.

Tatu, Frank. "National Security." In *Cambodia: A Country Study,* ed. Russell R. Ross. Area Handbook Series. Washington, DC: Library of Congress, 1990.

Cambrai, Battle of (20 November–8 December 1917)

British offensive of World War I, noted for the first significant use of tanks. The Cambrai attack was planned to break the Hindenburg Line. Timing was critical. The Allies hoped to collapse the western front before arrival of troops released by Germany's victory over Russia.

Tanks had been used before, in small numbers. Because bombardment always preceded attacks, giving the Germans time to reinforce, the advantage of surprise was usually lost. General Douglas Haig, the British commander, laid different plans: no preparatory bombardment.

Haig gathered 476 tanks, placing most in the assault columns. The staff perfected tactics to cross trenches with armor, as infantry followed closely, providing protection and holding ground. Cavalry poised to exploit any gaps.

The assault began on 20 November 1917 as the tank corps plowed into German lines. After the first day, the British held most objectives, driving the Germans four miles back along a six-mile front. While they inflicted heavy casualties and took thousands of prisoners, the infantry could not keep up with the tanks. Communications failed; soon tanks and tankers began to break down as fatigue grew. The Germans reacted vigorously, stabilizing the front. Arrival of troops from the eastern front proved decisive. On 28 November, Haig halted. On 30 November, the Germans counterattacked; by 8 December they had captured half of the lost ground.

The results were mixed. Each side lost more than 40,000 men. Tanks demonstrated potential, but tactics were unsatisfactory. The Hindenburg Line was broken—and restored. The British took ground, but exhausted their army. The Germans reassigned divisions along the front, shoring up defenses and preparing for offensives in 1918.

Michael S. Casey

See also: Armored Fighting Vehicles

References and further reading:

Cooper, Bryan. *The Battle of Cambrai.* New York: Stein & Day, 1968.

Horsfall, Jack, and Nigel Cave. *Cambrai: Hindenburg Line.* New York: Pen & Sword Books, 2000.

Camden, Battle of (15 August 1780)

One of the most thoroughgoing military defeats in U.S. history. By 1780, five years after the outbreak of conflict, Great Britain had devised a new strategy to win the American Revolution. General Henry Clinton's army would invade the southern colonies/states, build on loyalist support, and reconquer each colony as it marched north. The invasion got under way with the fall of Charleston, South Carolina, in May 1780. Clinton then returned to New York, leaving Lord Charles Cornwallis in command.

The loss of Charleston and its army was a disaster for the Americans unmatched until the surrender of the Philippines in 1942, and General George Washington quickly dispatched reinforcements to the South under General Horatio Gates. Marching south, Gates hoped quickly to retake the area. Cornwallis had reached Camden, in central South Carolina, and was preparing to meet Gates. The armies marched on the same road during the night of 15 August, and the advance parties collided around 2:30 A.M. Just a few miles north of Camden both sides halted and waited for daylight.

Gates's battlefield performance was poor: He did not reconnoiter the ground, develop a plan for battle, or even know how many men he had. At sunrise the armies faced each other across an open field broken by tall pines. About 3,700 Americans faced 2,179 English, yet the British were primarily regulars, while Gates commanded mostly militia. While Cornwallis put his units into line, Gates decided to send his militia to strike the British right. The English regulars charged with bayonets, routing these green troops and leaving the continentals on the American right unsupported. Under General Johan DeKalb they fought well despite being surrounded and outnumbered. Nonetheless, Gates fell back with the retreating militia, and eventually retreated to Hillsboro, North Carolina, supposedly setting some sort of record for a long-distance mounted dash. For another hour the Maryland and Delaware continentals fought until they were overwhelmed, when they too broke and fled.

Gates probably lost more than 1,000 men, while Cornwallis's casualties numbered 324. The rout firmly established British control over the colony, forcing American militia to flee to North Carolina until the Patriot victory on Kings Mountain in October began to turn the tide of battle.

Brian Dunkerly

See also: American Revolution; Marion, Francis

References and further reading:

Buchanan, John. *The Road to Guilford Courthouse.* New York: Wiley & Sons, 1997.

Flood, Charles B. *Rise, and Fight Again.* New York: Dodd, Mead, & Co., 1976.

Martin, James K., and Mark Lender. *A Respectable Army.* Abbington Heights, IL: Harland Davidson, 1982.

Campbell, Colin (1792–1863)

Perhaps the greatest British tactician and field commander of all time, noted especially for his deeds at Alma, Balaklava, and in the Indian Mutiny. Born the son of carpenter John M'Liver in Glasgow, Scotland, on 20 October 1792, Colin M'Liver enlisted in the British army in 1808 under his mother's maiden name, Campbell. He saw his first action in Portugal during the Peninsular War. From 1808 to 1813 he fought at Rolica, Vimeiro, Salamanca, Corunna, Walcheren, Gibraltar, Barossa, Tarifa, Vittoria, San Sebastian, San Bartholome, and Bidassoa.

In North America from 1813 to 1815, he fought under Sir Edward Pakenham in the British disasters at New Orleans on 8 January 1815. After serving in Gibraltar from 1816 to 1818, Barbados from 1819 to 1821, and British Guiana from 1821 to 1826, he remained in England as a major until 1832 and as a lieutenant colonel until 1842, when he led reinforcements to China for Sir Hugh Gough in the First Opium War. After Gough returned to India in 1843, Campbell remained in China, achieving the rank of brigadier general in 1844. Under Gough in India from 1846 to 1852, he fought at Ramnuggur, Chillianwallah, Gujerat, Kohat Pass, Panj Pao, and Iskakote.

As major general in command of the Highland Brigade during the Crimean War, he compensated for the errors of his superiors, almost single-handedly won at the Alma, prevented disaster at Balaklava, and led the reserves against the Redan in the siege of Sevastopol. Back in England, he became lieutenant general in 1856 and was appointed commander in chief of forces suppressing the Indian Mutiny in 1857. The tide turned after his arrival in Calcutta in August. Recognizing his value at Lucknow and Cawnpore, Queen Victoria created him First Baron Clyde of Clydesdale in 1858. He was promoted to field marshal in 1862, died on 14 August 1863 in Chatham, England, and is buried in Westminster Abbey.

Eric v. d. Luft

See also: Alma; Balaklava; Chillianwallah; Corunna, Battle of; Crimean War; Gujerat; Indian Mutiny; Napoleonic Wars; New Orleans, Battle of; Sevastopol, Siege of; War of 1812

References and further reading:

Forbes, Archibald. *Colin Campbell, Lord Clyde.* London: Macmillan, 1916.

Roy, James McLiver. *"Old Take Care": The Story of Field Marshall Sir Colin Campbell, Lord Clyde.* Glasgow: J. M. Roy, 1985.

Shadwell, Lawrence. *The Life of Colin Campbell, Lord Clyde.* Edinburgh: Blackwood, 1881.

Watson, Bruce. *The Great Indian Mutiny: Colin Campbell and the Campaign at Lucknow.* New York: Praeger, 1991.

Canadian Military

The military of Canada has been shaped throughout its history by the fact that the dominion is at one and the same time indefensible (because of its small population) and invincible (because of its vast landmass). It shares with Great Britain and the United States a dislike for professional military and has found it almost impossible to institute conscription even in time of war, let alone in peacetime. That English-speaking Canada's only experience of foreign invasion was the tragicomic U.S. incursion in the War of 1812 has also led to a refusal to take the military and national defense very seriously, except in time of war. In addition, the English-French divide has made it difficult in the past for the semiautonomous, officially French-speaking province of Quebec to see the necessity to "fight England's wars."

Yet Canadians themselves have long been interested in things military. Canada's military traditions are based on the militia system, citizen soldiers who volunteered to defend the nation in time of need. Loyalists during the American Revolution raised some of the first Canadian militia units. The militia later defended Canada from invasion by the United States during the War of 1812. Once the danger had passed, the militia became sedentary. In 1840 the Canadian militia rolls still carried 426 local battalions with 235,000 officers and men. An active militia, maintained at government expense, was authorized in the Militia Act of 1855. Successive parsimonious governments kept military appropriations low, but public interest in the militia remained high. The number of authorized units grew steadily throughout the nineteenth century. When Great Britain went to war with the South African Boer republics in 1898, public pressure forced the Canadian government to offer a volunteer field force as a symbol of imperial unity.

The South African war exposed weaknesses in the Canadian militia. Administrative reforms established departmental and service units, functions that were previously supervised by British officers seconded to the Canadian militia. Authorization for engineer and army service corps companies came in 1901, quickly followed by the establishment of ordnance, intelligence, signals, medical, and pay branches. These units provided the Canadian militia with a permanent support system to meet the needs of an independent army.

A large contingent of Canadian troops arriving in Great Britain on a transport ship, 1941. (Library of Congress)

In 1910, growing concern in Britain over the expansion of the imperial German navy resulted in the calling back to British home waters of a substantial portion of the Royal Navy's squadrons stationed in Canada. That, following the withdrawal of all but a handful of British troops, signaled the end of Britain's primary role in the defense of Canada.

When the British Empire went to war in August 1914, Canada had a permanent force of little more than 3,000 officers and men. On paper there was a reserve of more than 50,000 men, but most were poorly trained, poorly equipped, and at least half were unfit or unable to serve. The government was quickly inundated with volunteers. To compound an already chaotic situation, the minister of militia and defense scrapped the existing mobilization plans and created an overseas army of 260 newly formed battalions with no relation to existing regimental designation or organization.

Despite initial difficulties, the Canadian Expeditionary Force became an effective overseas army, serving with distinction on the western front and winning the respect and admiration of both allies and foes. Canadian troops even served in the misbegotten Allied intervention in post-Revolution Russia. During World War I the ranks of the Canadian military grew to an incredible 628,000 men, giving a sense of national unity. (One trooper, seeing entire Canadian divisions in formation preparatory to the Vimy Ridge operation, exclaimed that he had never seen so many Canadians together in any one place in his life, and didn't even realize that there were that many live Canadians.) But this sense of national unity was marred by the strong anticonscription sentiment and disorders in Quebec. In addition, the Royal Canadian Navy had over 5,000 serving personnel, and Canada's contribution to the British air services was in excess of 24,000 men, all from a country with a population that had just reached 8 million.

The end of the war signaled the return to the prewar regimental system. For Canada, the 1920s and early 1930s were described as a period of military economy but, more accurately, it was a period of military neglect. The Canadian militia existed, to a large extent, only on paper, with no tanks, no armored cars, no heavy or medium artillery, and no antiaircraft guns. In 1922 there were only two ships in the Royal Canadian Navy and few flightworthy aircraft in the Canadian Air Force. Politicians maintained that Canada was invulnerable to military conflicts, often comparing it to a "fireproof house." The government felt little need to concern itself with defense. As one Canadian prime minister asked, "Defense from whom?" (The Americans?) However, by the mid-1930s it was clear that not even Canada was isolated from disturbing events half a world away. Canadian defense budgets doubled from 1935 to 1939.

The year 1936 saw a dramatic change in the structure of the Canadian army. After an exhaustive review of Canada's military capabilities, resources, and shortcomings, major organizational changes were implemented: (1) the amalgamation (or uniting) of regiments; (2) the disbandment of regiments; and (3) the redesignation of regiments as artillery, armored (tank), motorized, or machine gun units.

The reorganization of 1936 reflected the changes in the way wars were fought. Before the reforms, the Canadian army, like most Western armies, was still geared for fighting World War I, and after 17 years of neglect was ill-prepared for modern warfare, as was being displayed in Spain and China. The reforms of 1936 and the increased military budgets provided Canada with the groundwork for an effective army and a command structure capable of putting that army in the field. Aside from a small Canadian contingent sacrificed at Hong Kong in the opening weeks of the war, and the botched Dieppe operation, Canadian troops did not go into action until 1943. But they fought their way up the "boot" of Italy, swarmed ashore on D day, serving in both the European and Pacific theaters of operation. By war's end more than 1,000,000 Canadian men and women were in uniform. Again, however, the divisive conscription controversy flared, and Canadians were not conscripted until so late in the war that none made it to a combat zone before hostilities had ceased. (These Canada-bound conscripts were known derisively as "zombies.")

When World War II ended in 1945, a swift demobilization was of prime concern to the government and service personnel alike. Canadians did not relish being a part of an occupation army. By December 1946 only a few hundred Canadians were left in Europe. Wartime units were disbanded and the remaining regiments were consolidated into a regular army. Canadian troops were dispatched to Korea in response to a call from the United Nations, and again recruiting offices were swamped; although some one-third had second thoughts and deserted before leaving Canada, the remainder gave a very good account of themselves.

The 1960s found Great Britain preoccupied with a rapidly disintegrating empire. Canadians looked to the United States as their chief ally and defense partner. The Canadian government launched a campaign to cut redundancies in the military by forming a single unified service. By 1968 this sweeping reorganization replaced the Royal Canadian Air Force, Royal Canadian Navy, and Canadian army with specialist branches of the newly created Canadian Armed Forces, much to the disgust of more traditionally minded officers.

The Canadian forces assumed the role of peace keeper, supervising the withdrawal of U.S. troops from the Vietnam War and providing troops for numerous United Nations operations. No nation has so large a proportion of its military on peace-keeping duties. In 1975 the three services were tentatively reconstituted, supplemented by a logistics branch and communications command. With the end of the Cold War, the Canadian government reduced its international treaty commitments and has increasingly relied on American military assets for defense, leaving many Canadians to question their future military needs. Canadian nationalists have always been at some pains to differentiate their nation from the colossus to the south, but it must be questioned as to just how sovereign a country is that, by default, leaves its defense primarily in the hands of another nation, however friendly or benign that latter power might be.

Eric Smylie

See also: Dieppe; Korean War; Normandy Landings; Russia, Allied Intervention in; Vimy Ridge; World War I; World War II

References and further reading:

Cuthbertson, Brian. *Canadian Military Independence in the Age of the Superpowers.* Toronto: Fitzhenry and Whiteside, 1977.

Douglas, W. A. B., and Brereton Greenhous. *Out of the Shadows: Canada in the Second World War.* Toronto: Oxford University Press, 1977.

Goodspeed, D. J. *The Armed Forces of Canada, 1867–1967: A Century of Achievement.* Ottawa: Queen's Printer, 1967.

Melady, John. *Canada's Forgotten War.* Toronto: Macmillan, 1983.
Morton, Desmond. *A Military History of Canada.* Edmonton, Alberta: Hurtig Publishers, 1990.
Nicholson, G. W. L. *Canadian Expeditionary Force 1914–1919: Official History of the Canadian Army in the First World War.* Ottawa: Queen's Printer, 1964.
Stanley, G. F. G. *Canada's Soldiers: The Military History of an Unmilitary People.* Toronto: Macmillan, 1974.

Cannae, Battle of (216 B.C.E.)

The Battle of Cannae was the third and last major defeat the Carthaginian general Hannibal inflicted on the Romans, and certainly the most terrible. The Romans had assembled an army of 80,000 foot and 6,000 horse, roughly half of them Italian allies. They greatly outnumbered Hannibal, who had only about 40,000 foot and 10,000 horse at his disposal. The Roman force was under the command of the consuls L. Aemilius Paullus and G. Terentius Varro. It met Hannibal's army near the citadel of Cannae in Apulia.

Hannibal had encamped on the north bank of the river Aufidius (Ofanto). The Romans set up camp about a mile downstream and established a minor camp on the south bank. When Hannibal challenged the Romans to fight in the wide plain north of the river they refused, probably because Hannibal could use his superior cavalry there to outflank the Romans. However, the next day they engaged Hannibal on the south bank. There the Roman left would be protected by a range of hills, the right by the river.

As usual, the Romans deployed their own infantry in the center and their allies on the wings. However, they assumed a formation that was deeper and more closely packed than they normally did. The formation was covered by a line of skirmishers. The cavalry took positions on the flanks, prepared to hold the line against the superior Carthaginian horse.

Hannibal deployed his men carefully. Behind a light infantry screen he positioned a convex crescent-shaped line of Celtic and Spanish infantry. Their formation increased in depth toward the middle of the line. This part would be nearest to the enemy and would bear the brunt of the Roman attack. The points of the crescent touched two deep columns of crack African troops that were positioned behind the cavalry positioned on the flanks.

At first the light infantry of both sides engaged. The Carthaginian horse on the right attacked the Roman cavalry and quickly the latter were in sore straits. Immediately the Roman infantry was ordered to attack. They had to break the enemy center before the Roman cavalry would yield. The Celts and Spanish were pushed back until the crescent line changed from convex to concave. Under the personal command of Hannibal they gradually and purposefully gave way, pulling their enemies into the crescent. If the Carthaginian center would break, the battle would be lost, but Hannibal's leadership prevented the gradual retreat from turning into a complete rout.

The Romans pushed themselves further and further into the sac until they had passed the columns of Africans on the wings. Then the latter attacked the Roman line in the flank and, having defeated the Roman cavalry on the left, the Carthaginian horse turned upon the Roman rear. The Roman army was completely encircled and a terrible slaughter ensued. The losses can be estimated at 50,000 men, almost two-thirds of the Roman force, among whom were the consul Paullus and some 80 senators. Hannibal's losses amounted to only some 7,000.

M. R. van der Werf

See also: Hannibal Barca; Punic Wars

References and further reading:

Bagnall, Nigel. *The Punic Wars. Rome, Carthage and the Struggle for the Mediterranean.* London: Pimlico, 1999.
Connolly, Peter. *Greece and Rome at War.* London: Greenhill Books; Mechanicsburg, PA: Stackpole Books, 1998.
Lazenby, J. F. *Hannibal's War. A Military History of the Second Punic War.* London: Aris & Phillips, 1978.

Cantigny (28–30 May 1918)

U.S. Army 1st Infantry Division attack in the Somme River region of northern France. In the midst of a series of German spring offensives designed to end the war, elements of the German Eighteenth Army held the key town of Cantigny. With the U.S. 28th Infantry Regiment in the lead, the division attacked the town and captured it on the first day, netting roughly 200 prisoners. The soldiers of the Big Red One then endured a total of seven German counterattacks but held fast. American artillery disrupted the first couple of German attacks, but the following attacks were blunted by a combination of artillery and infantry fire. The Americans suffered 1,603 casualties, including 199 killed in the battle. In the context of the overall situation on the western front, the victory at Cantigny was small, but it boosted Allied morale and served notice that American ground combat manpower would have a significant impact on the outcome of World War I.

John C. McManus

See also: Meuse-Argonne; Pershing, John J.; World War I

References and further reading:

Coffman, Edward M. *The War to End All Wars: The American Military*

Experience in World War I. New York: Oxford University Press, 1968.
Evarts, Jeremiah Maxwell. *Cantigny: A Corner of the War.* New York: Scribner Press, 1938.
Harbord, James G. *The American Army in France, 1917–1919.* Boston: Little, Brown, 1936.
Lyons, Michael J. *World War I: A Short History.* Upper Saddle River, NJ: Prentice-Hall, 2000.
Pershing, John J. *My Experiences in the World War.* 2 vols. New York: Stokes, 1931.

Cape-Xhosa Wars (1779–1878)

Nine wars between white colonists from the South African Cape Colony and the Xhosa people of the eastern coastland, leading eventually to the subjugation of "Xhosaland" by British South Africa.

In the 1770s, land-hungry Dutch settlers from the Cape Colony and the exploding Xhosa populations of the South African coast converged and clashed in the frontier called Zuurveld. Unlike other southern African peoples, the Xhosa were more numerous, better organized, and no less adept at commando-style combat than the trekboer Dutch. In 1779, the first of the Cape-Xhosa wars erupted in a series of reprisals between militias of both sides. After two years, the hostilities sputtered out. This 1779–1781 conflict would be the first of nine Cape-Xhosa wars spread over the next century.

Throughout the next decade, Prince Ndhambe sought to tighten control over the Zuurveld Xhosa. In 1793, Cape Boers, fearing Ndhambe's growing strength, tried to drive the Xhosa out of the Fish River zone. Instead, the Xhosa reacted strongly and drove out the white settlers. Cape governor H. C. Maynier rescued the settlers from total rout, but his 1795 treaty left the Xhosa in virtual control of the Zuurveld and let them keep plundered Boer farms and cattle. Infuriated, the Boers chased Maynier out of Cape Town but, before they could return to the Zuurveld, Great Britain seized and occupied the Cape. Catching the colony in turmoil, the Xhosa started the third war in 1799 and, by the truce of 1803, had proven their mettle to British forces. The Zuurveld remained under Xhosa domination.

In 1806, however, the British reinforced their position in the Cape. Among their goals was incorporation of the Zuurveld as an area for white settlement and the imposing of a fixed frontier between the colony and Xhosaland.

The 1811–1812 Fourth Cape-Xhosa War consisted of a direct invasion of the Zuurveld, driving Prince Ndhambe and his people east across the Fish River. Xhosa Zuurveld refugees packed into the overpopulated Xhosa lands, causing a near civil war between Ndhambe and another ruler, Ngqika. A religious visionary, Makana preached unity of the Xhosa in a divinely inspired war to drive out the whites, and his followers helped overthrow Ngqika, who had sought Cape help against his rival. When the Cape attempted to support Ngqika in 1818, Ndhambe and Makana besieged Grahamstown, the key Cape fort in Zuurveld. The attack turned into disaster and the British drove the Xhosa further eastward from Fish River past the Keiskama River. At the end of this Fifth War (1818–1819), Britain created an open buffer zone between the colony and the Xhosa. However, the zone itself became an arena of conflict among illegal British, Boer, and African squatters.

In 1834–1835, land desperation drove the Xhosa to revolt again, the Sixth Cape-Xhosa War. Impressed by Xhosa fighting capabilities, the British at first pledged to contain white infiltration. Nonetheless, colonists continued to settle in Xhosa territory, causing new communal frictions. Finally, British demands on Xhosa border communities set off the 1846–1847 War of the Axe, the Seventh War. Again victorious, Cape authorities directly occupied Xhosaland, making Xhosa chiefs agents of the crown in the new Kaffraria colony. Heavy-handed magistrates, intrusive settlers, and prohibitions against various Xhosa customs provoked a massive uprising from 1850 to 1852. Drained by another defeat and by starvation, the exhausted Xhosa staged their last, doomed uprising, the Ninth War, in 1877–1878.

Weston F. Cook

See also: Anglo-Zulu War

References and further reading:

Mostert, Noel. *Frontiers: The Epic of South Africa's Creation and the Tragedy of the Xhosa People.* London: Jonathan Cape, 1992.
Omer-Cooper, J. D. *History of Southern Africa.* 2d ed. Portsmouth, NH: Heinemann, 1994.
Thompson, Leonard. *A History of South Africa.* 2d ed. New Haven: Yale University Press, 1995.

Caporetto (24 October–9 November 1917)

Austro-German World War I breakthrough that exposed the demoralization of the Italian army and demonstrated the effectiveness of new assault tactics.

Having endured 11 Italian offensives along the Isonzo River between June 1915 and September 1917, the Austrians had enlisted German help in driving the Italians back beyond the Tagliamento River. To counter the 53 German and Austrian divisions massed against his own 34, General Luigi Cadorna, the Italian commander, concentrated his troops in forward trenches, almost stripped intermediate lines, and positioned his reserves too far back for effective use. His subordinate, General Luigi Capello, thinned his left wing from eight battalions per mile of front to two battalions so as to strengthen his right wing for a riposte on the flank of Austrian attackers. Cadorna countermanded this disposition and Capello's army was in transit when the attack came.

At 2 A.M. on 24 October, the Germans and Austrians began an intense two-and-a-half-hour gas and heavy mortar attack around Tolmino and Caporetto, horribly effective because of inadequate Italian gas masks and the crowding of troops in frontline trenches. At 6:30 A.M. the shelling resumed, targeting Italian command and communications, munitions, and artillery. Assault units jumped off at 7 A.M., bypassing strong points and infiltrating rear areas. German and Austrian divisions streamed through the 15-mile hole torn in the Italian defenses. Capello's army was split and retreated in confusion. Officers abandoned their troops and units surrendered en masse.

Cadorna improvised a defense on the Tagliamento, the original objective of the attack, but the Germans and Austrians, flushed with success, crossed the river on 2 November. A week later, their attack reached the Piave River, only 20 miles from Venice. In two weeks the Italians had lost 250,000 prisoners, 300,000 deserters, 40,000 killed and wounded, and had been pushed back 80 miles. And the Germans had gained valuable training and experience in breakthrough tactics that they would put to good use the following spring on the western front.

Joseph M. McCarthy

See also: World War I

References and further reading:

Cyril Falls. *Caporetto 1917.* London: Weidenfeld & Nicolson, 1966.

Ronald, Seth. *Caporetto.* London: Macdonald, 1965.

Carleton, Sir Guy (1724–1808)

British commander who laid the foundations of what eventually became the provinces of Upper and Lower Canada.

Guy Carleton was born at Strabane, Ireland, on 3 September 1724, and he joined the army in May 1742. In 1758 he was present at the successful siege of Louisbourg under Jeffrey Amherst, and the following year he fought under James Wolfe at Quebec. In light of his excellent service, Carleton gained appointment as governor of Quebec in 1766, and over the next decade he successfully reconciled the Catholic, French-speaking population with English rule.

When the American Revolution broke out in April 1775, Carleton ordered most of his garrison south to Boston. Ten months later he confronted and managed to defeat a much larger American invasion force during Benedict Arnold's assault on Quebec, on 31 December 1775. Reinforced the following spring, Carleton slowly began pushing the invaders back and built a fleet on Lake Champlain to invade New York. His forces again defeated Arnold at Valcour's Island in October 1776, but the season was too advanced to continue. Carleton was eventually superseded by General John Burgoyne, and after the war ended in 1783, he directed the evacuation of British forces in New York City. He spent the next 13 years ably administering Canadian affairs before dying in Ireland on 10 November 1808.

John C. Fredriksen

See also: American Revolution; Quebec, Battle of

References and further reading:

Nelson, Paul D. *General Sir Guy Carleton, Lord Dorchester: Soldier-Statesman of Early British Canada.* Madison, NJ: Fairleigh-Dickinson University Press, 2000.

Carlist Wars (1833–1876)

A struggle between absolutists and liberals but also a dynastic conflict. The First Carlist War, 1833–1839, commenced at King Ferdinand VII's death in 1833. He had revoked the Salic law prohibiting female inheritance of the Spanish throne so his daughter Isabella II (1830–1904) could become queen. Maria Cristina, Ferdinand's fourth wife and Isabella's mother, acted as regent and was supported by the Quadruple Alliance of Spain, Portugal, Great Britain, and France of 22 April 1834. This infuriated the former heir, Ferdinand's brother, Don Carlos, Maria Isidero de Bourbón (1788–1855), an extremely conservative Sandhurst-educated traditionalist planning to revive absolutist monarchical traditions. With secret support from the papacy, he was proclaimed Charles V by the Catalans and the Basques, who envisioned him as crusading for the faith but also as a useful tool in their long struggle to maintain their ancient traditions and rights.

Initial Carlists successes were due to the superb tactics of General Tomas Zumalacarregui. But General Baldomerio Espartero defeated Carlists at Luchana in 1836 with foreign aid. The 1837 Carlist offensive against Madrid failed, as did frequent attempts to attack Bilboa. Without Don Carlos's permission, commander Rafael Maroto compromised by signing the Convention of Vegara in 1839 in exchange for amnesty and liberal recognition of Basque legal privileges. No major city was won by the Carlists, despite brilliant guerrilla tactics led by Ramón Cabrera, who fought on until 1840. Executions, senseless reprisal killings, and unspeakable inhumanity characterized this civil war.

Don Carlos fled into exile in France and abdicated his pretensions in 1845 in favor of his son, Carlos Luis de Bourbón, whose abortive 1860 insurrection yielded nothing. His brother Juan became Carlist leader in 1867 but transferred his rights to his son Don Carlos III, Duke of Madrid. After inept and corrupt Isabella's deposition during the Spanish Revolution of 1868, Duke Amedeo of Aosta (1845–1890) became king. He abdicated in 1873, however, renewing Carlist military initiatives.

The Second Carlist War, 1873–1876, was a brutal conflict in which the Carlists seized Alcoy, Cadiz, Bilbao, Cartagena, Valencia, and Pamplona. By 1875 Carlists repeatedly lost savagely fought battles.

In 1876 Isabella's son Alfonso XII (1857–1885) became king and the fighting ultimately led to the Basques losing their independence. Don Carlos escaped to France and the Carlists were banished from Spain.

Carlists could not revive their movement until the 1930s. Then they slowly began to rebuild their organization because the church, their rallying point, was under great pressure from the new Spanish Republic. Carlists, naturally, joined the Nationalists Falange of General Francisco Franco, and fought with distinction in the Spanish Civil War (1936–1939). However, Franco made Bourbon prince Juan Carlos his successor in 1969, and in 1975 he ascended the throne. The Carlists, headed by Carlos Hugo, Duke of Bourbon-Parma, then withdrew to France.

Annette Richardson

See also: Spanish Civil War

References and further reading:

Castillo, José. *Recetas de Cocina de Abuelas Vascas: Alava-Navarra.* San Sebastién: Ttarttalo, 1998.

Kurlansky, Mark. *The Basque History of the World.* New York: Penguin, 2001.

Real Cuesta, Javier. *El Carlismo Vasco 1876–1900.* Madrid: Siglo Veintiuno de España, 1985.

Rubio, Antonio, and Manuel Cerdan. *Del Origen del GAL.* Madrid: Ediciones Tema de Hoy, 1997.

Carnatic Wars (1744–1754)

In the middle decades of the eighteenth century, the Carnatic, encompassing the southeastern extremity of India, was ripe for conflict among a variety of interests. In 1638, the English East India Company had founded a trading "factory" at Madras. A century later, France established its own factory less than 200 kilometers to the south, at Pondicherry. Yet, while the First Carnatic War (1744–1748) paralleled the War of Austrian Succession (1740–1748), the conflict on the subcontinent did not represent the expansion of the European struggle into a global war. Indeed, when word arrived of the outbreak of war in Europe, Joseph Dupleix, Governor of Pondicherry, sought to maintain peace between the French and British traders. Conflict proved inevitable, however, when Commodore Curtis Barnett's squadron arrived to prey on French shipping in 1745, a move countered by the French in June 1746. In the battle of Negapatam, 25 July 1746, Count de la Bourdonnais's squadron successfully drove the British naval forces from the Indian coast, and, together with land forces under the command of Dupleix, captured Madras in September 1746.

In the meantime, the conflict had not been limited to European combatants. Indeed, as primarily trading interests, the French and the British companies needed Indian military assistance. Both Madras and Pondicherry lay within the province of Anwar-ud-din, the Nabob of Arcot, and with the outbreak of hostilities between the Europeans the nabob allied himself to the British. In the event, this was unfortunate for the nabob and of little assistance to the citizens of Madras. On 21 September 1746, 11 days after the fall of Madras, the nabob's relieving army of 10,000 cavalry was defeated by a force of little more than 500 Frenchmen. Despite the capture of Madras and the defeat of Britain's indigenous allies, the laurels of victory were not to be France's alone. In November 1746, Dupleix commenced the siege of Fort St. David, south of Pondicherry at Cuddalore. Eighteen months later, in April 1748, the French were forced to raise their siege with the arrival of a British squadron under the command of Admiral Edward Boscawen. Boscawen, with land forces under the command of Major Stringer Lawrence, moved against Pondicherry in August 1748. However, the able defense of Dupleix and the onset of the monsoon season forced the British to raise their siege in October. At the end of 1748 word of peace between Britain and France arrived and, with unforeseen implications for the empires of both countries, the exchange of the French fortress of Louisbourg, captured by American provincials in 1745, for Madras.

If the First Carnatic War had seen indigenous forces coming to the aid of Europeans, however ineffectively, the Second witnessed the renewal of European conflict in India through the vehicle of indigenous power struggles. It is both fitting and ironic, therefore, that the dates of the Second Carnatic War (1749–1754) lie between those of the War of Austrian Succession and the Seven Years' War (1755–1763). The death in 1748 of Nizam ul-Mulk, the nabob of Hyderabad, precipitated the Second Carnatic War. Taking advantage of the confused political situation, Chanda Sahib moved against the pro-British Nabob of Arcot, and with the aid of French forces under the Marquis Charles de Bussy, easily overthrew him. Chanda was then challenged by Mohammed

Ali, the slain nabob's son, who was in turn supported by the British. Through 1749 and 1750, Ali was supported by Nasir Jang, who had succeeded his father as nabob of Hyderabad. For his part, Chanda received the aid of Muzaffar Jang, Nasir's son, who in 1750 succeeded as Nizam after the murder of his father.

As the bloodshed within the palaces seemed to settle, matters came to a climax on the battlefield. In September 1751, Chanda laid siege to Ali at Trichinopoly, supported by 1800 Frenchmen under de Bussy. It was clear that if Ali fell, British interest in the region went with him. Yet it was equally clear that the British lacked the resources to break the siege. Ali therefore urged that what forces were available be used to attack Chanda's capital at Arcot, thus forcing him to lift the siege on Trichinopoly. On 22 August 1751, 200 Europeans, 300 sepoys, and three cannon, under the command of Robert Clive, set out from Madras. Arriving on 1 September, they found Arcot deserted by its garrison. It was not until 22 September that Chanda's son, Raza Sahib, arrived with 4,000 men, plus 150 Frenchmen, and opened a 50-day siege that failed to drive Clive from the citadel.

The loss of his capital inflicted great damage to the prestige of Chanda and his French allies. It likewise encouraged the British to go on the offensive. On 3 December, Clive, commanding a force of European and native troops, defeated superior numbers under Raza at the hard-fought battle of Arni. Chanda's forces were not entirely broken, however, and in February 1752 Raza besieged Madras. Though the British succeeded in holding the city, at Kaveripak (28 February) Clive only narrowly averted annihilation when his forces were ambushed by Raza. Despite this, it was only a matter of time before Chanda's forces were forced to withdraw from Trichinopoly, and his French allies were forced to surrender to the British at Srirangam (4 June 1752).

Defeat at Srirangam meant the end for Chanda Sahib. Captured shortly after the battle, Chanda was summarily strangled and beheaded. Though fighting would continue intermittently for the next year, with the recall of Dupleix to France in August 1754 both companies quickly agreed to end the war. Although with the end of the Carnatic Wars the British had secured their candidate as nabob of Arcot, and thus secured their position within the region, death for their protégé did not mean defeat for French interests in India. Thanks largely to the efforts of de Bussy, the French had secured their candidate (Muzaffar Jang) for the superior nabob of Hyderabad. While both European powers thus profited greatly from the conclusion of the Carnatic Wars, given the stakes involved, future conflict between them was inevitable.

Adam Norman Lynde

See also: Clive, Robert

References and further reading:

Harvey, Robert Clive. *The Life and Death of a British Emperor.* London: Hodder & Stoughton, 1998.
Keay, John. *India: A History.* London: HarperCollins, 2000.
Mehra, Parshotam. *A Dictionary of Modern Indian History.* Madras: Oxford University Press, 1987.

Carnot, Lazare-Nicholas (1753–1823)

French soldier, engineer, scientist, and politician. Carnot, born in the French town of Nolay, entered the French army and the Royal Engineering School at Mézières in 1771. Upon completion of the course in 1773, he entered service as an engineer, and was posted to garrison towns in northeastern France. While on garrison duty, Carnot had ample opportunity for study, and also made the acquaintance of a rising lawyer, Maximilien Robespierre. Carnot wrote treatises on fortification and the motion of machines in this period.

The French Revolution of 1789 allowed him to become involved in politics. He was elected to the Legislative Assembly in 1791 and to the Convention in 1792. He voted for the execution of Louis XVI. In August 1793, he was brought on to the Committee of Public Safety by Robespierre to assist with military matters. He promoted talented officers, including Jourdan, Hoche, and Bonaparte, and set about reorganizing the French military, which had been greatly disordered by the Revolution. His scheme of reorganization included mass conscription, the amalgamation of untrained levies with former royal units, the introduction of systematic foraging into the army, and the development of permanent brigade, divisional, and corps structures. He also organized the creation of 14 armies to defend all parts of France. In 1796, he appointed Napoleon Bonaparte commander of the Army of Italy. Upon the fall of Robespierre, he was made a member of the Directory, and served until forced to flee to Switzerland as a result of the coup of the 18th Thermidor. Carnot remained in exile until Bonaparte's coup in 1799, when he was allowed to return.

Upon assuming power, Bonaparte made Carnot minister of war; Carnot resigned in 1800 after completing the organizational work for the Second Italian Campaign. As a dedicated antimonarchist, Carnot took no active part in the government of the French Empire. In 1814 he was appointed by Bonaparte as military governor of Antwerp and made *général de division.* He conducted an aggressive defense of the city, and surrendered it only after the Second Abdication of 1814 and the accession of Louis XVIII were confirmed.

During the Hundred Days, Carnot served as Bonaparte's minister of the interior, but had no time to bring an ambitious plan for public education to fruition. Having voted for the death of Louis XVI, he was obliged to go into exile upon the restoration of Louis XVIII in 1815. He spent his remaining years studying and writing on physics and mechanical engineering, and died at Magdeburg, Germany, in 1823.

Joseph M. Isenberg

See also: Conscription; Engineering, Military; French Revolutionary Wars; Marengo, Battle of; Napoleon I; Napoleonic Wars

References and further reading:

Dupre, Huntley. *Lazare Carnot.* Philadelphia, PA: Porcupine Press, 1975.

Watson, S. J. *Carnot.* London: Bodley Head, 1954.

Carolingian Empire

The zenith of the Frankish Empire. Charles (742–814), who became known as Charlemagne, gradually extended Frankish power in western Europe. After the occupation of practically all of Gaul he succeeded in pushing back the Muslims well behind the Pyrenees. In eastern Europe he defeated the Bohemians, Avars, Serbs, and Croats. To secure his southern flank Charles attacked and defeated the Langobards in northern Italy, which gave him access to Rome. There, on Christmas Day 800, he was crowned emperor of the restored Roman Empire by the pope. Through this act the Holy Roman Empire, which dominated the history of central Europe until 1806, was born.

Charlemagne concentrated the largest territory under a single ruler in Europe since the fall of the Roman Empire. However, the Carolingian Empire still was largely a conglomeration of Gauls and Germanic and Slavic tribes in which the local dukes held the balance of power. Charles tried to overcome the traditional social structures. Standardized education, the codification of tribal laws, and the creation of a common language were some components of a universal and Christian empire in which the Frankish king as emperor was the unifying force. Even more important, however, was the introduction of a completely new territorial constitution based on counties instead of tribal duchies. The counts would be installed as imperial officials and accountable exclusively to the emperor. The counties were the basic military territorial and organizational units and the most important judicial structures. The counties situated in the endangered border territories had wider military competences. Unfortunately this administrative reorganization was never completed. Instead of replacing the old tribal structures it was co-opted by local dukes when they added the count's titles and perquisites to their own as a feudal right.

With Charlemagne's death the empire broke apart, demonstrating just how much it was dependent on one strong man. With the Treaty of Verdun (843) Charles's heirs divided the empire into three independent kingdoms: the west to Charles II the Bald, the east to Louis II the German, and the central part to Lothair. Later this central part was shared out between the western and the eastern kingdoms, which eventually evolved into France and Germany respectively.

After the division, Frankish power withered away until in 887 Charles III the Fat was deposed as the last Frankish king. The concept of the Holy Roman Empire continued, however, taken over by Germanic Austrian kings and emperors.

Marcus Hanke

See also: Charlemagne

References and further reading:

Bullough, Donald A. *The Age of Charlemagne.* New York: Exeter Books, 1965.

Geary, P. J. *Before France and Germany: The Creation and Transformation of the Merovingian World.* New York and Oxford: Oxford University Press, 1988.

James, E. *The Franks.* Oxford and London: Blackwell, 1988.

Carrhae, Battle of (53 B.C.E.)

Parthians inflict a disastrous defeat on the attempted Roman invasion of Parthia headed by Crassus. Crassus commanded a force of seven legions: 28,000 heavy infantry, 4,000 cavalry, 1,000 Gallic auxiliary archers, and 4,000 peltasts (light infantry). He struck directly across the exposed arid plain of Mesopotamia, planning to reach Seleuciaon on the Tigris, by staging through Roman garrisons along the way.

Crassus was surprised by a force of 10,000 Parthian horsemen under Suren, chief general of the Parthian king Hyrodes. Crassus redeployed his force from the extended line to a heavy infantry square, allocating his inadequate cavalry and light infantry components equally among the four sides. Suren's Parthian cavalry archers, capitalizing on their excellent marksmanship with the powerful compound bow, incessantly inflicted casualties through coordinated high trajectory and direct fire.

In an effort to pin the elusive Parthians long enough for his heavy foot to engage, Crassus rested his men near the Balissus stream and sent his son Publius ahead with a combined force of 6,000 legionnaires, cavalry, and Gallic bowmen to where the enemy was thickest. A section of Parthian horse feigned retreat long enough to separate Publius from the main body. An attack of heavy lancers caused Publius to concentrate his men to receive cavalry, making a more compact target for the mounted archers, who decimated the Romans, killing Publius.

Crassus's calculation that the Parthians were nearly out of arrows was dashed when he observed a camel supply train regularly replenishing the Parthian quivers. Crassus attempted an orderly nighttime withdrawal to the garrison at Carrhae. When his broken remnant reached the town walls, they learned that supplies were inadequate to withstand a lengthy siege. While attempting to negotiate with Suren, Crassus was murdered. Crassus lost 24,000 killed and some 10,000 prisoners.

Jim Bloom

See also: Ancient Warfare; Caesar, Julius; Parthian Empire; Roman Civil Wars (88–30 B.C.E.); Roman Republic, Wars of the; Tactics

References and further reading:

Montagu, John Drogo. *Battles of the Greek and Roman Worlds.* London: Greenhill Books, Lionel Leventhal Limited, 2000.

Scullard, H. H. *From the Gracchi to Nero: A History of Rome, 133 BC to AD 68.* 5th ed. London: Methuen & Co., 1982.

Carus (Marcus Aurelius Carus) (r. 283–284)

Soldier emperor who marched on Ctesiphon. Carus was from Gaul and served as Praetorian prefect for Probus (276–282). Elevated by the troops after Probus's murder, Carus announced to the Senate that he was filling the vacant throne.

Leaving his eldest son, Carinus, as caesar in the west, Carus led troops against Persia with Numerianus. The Sasanid king, Varanes, was considered weak. Factions within his court were vying for power. When Persian ambassadors came to the Roman camp, they found Carus seated on the ground enjoying the same rations as his soldiers. Carus curtly told them that unless Varanes acknowledged Rome as his master, Carus's army would render Persia as bare of trees as the emperor's own bald head.

True to his threat, Carus took Ctesiphon, and then penetrated beyond the Tigris. There he died suddenly and mysteriously, allegedly struck by lightning. An ancient prophecy had declared the Tigris the eastern frontier of the empire. Roman conquest was to go no further. Apparently filled with superstition, the army pressured the new emperor, the young Numerianus, to march out of Persia with victory almost within his grasp.

Nic Fields

See also: Roman Civil Wars (235–284)

References and further reading:

Mattingly, H. "Carus and His Sons." In *The Cambridge Ancient History,* Vol. 12, 321–323. Cambridge: Cambridge University Press, 1939.

Cassino, Battle of (17 January–18 May 1944)

One of the most bitter and costly battles fought in Italy during World War II. The town of Cassino and Monte Cassino, site of the monastery founded in the sixth century by St. Benedict, anchored the German defensive Gustav Line that barred the Allied advance on Rome. German forces defending Cassino were under the command of General von Senger und Etterlin. In the first battle of Cassino the U.S. 34th Division fought bravely, only to be beaten back from the monastery walls. The Free French Expeditionary Corps, mostly colonial troops from North Africa, made some limited gains, at equally heavy cost. In the second battle, 15–18 February, it was hoped that the New Zealand Corps (2d New Zealand Division and 4th Indian Division) would be able to burst through the last few hundred yards. Its commander, General Bernard Freyberg, supported the requests of his divisional commanders to have Monte Cassino bombed, since all the troops were convinced that the Germans used the monastery as an observation post.

No event of the war caused more heated and lingering controversy than the Allied bombing of Monte Cassino on 15 February 1944. Critics have condemned the bombing as "a tactical mistake," "a criminal act," and "inexcusable." However, General von Senger himself had ordered that German defensive positions were to be made right up to the monastery walls if necessary. After the bombing, the elite 1st Parachute Division held the monastery ruins for three more months.

In May 1944 the Allies launched one of the great set-piece battles of the war that broke through the Gustav Line. Polish troops captured Monte Cassino on 18 May after taking almost 4,000 casualties. On 4 June the Allies liberated Rome.

Colin F. Baxter

See also: Infantry; World War II

References and further reading:

Jackson, W. G. F. *The Battle for Italy.* New York: Harper & Row, 1967.

Cassius (Gaius Cassius Longinus) (d. 42 B.C.E.)

Roman politician and soldier who participated in the conspiracy against Julius Caesar. Born into an illustrious senatorial family in the first half of the last century B.C.E., Cassius moved through the traditional *cursus honorum* of offices, serving as quaestor in 54 B.C.E., tribune in 49, legate for Caesar in 47, and praetor in 44 B.C.E. He served on Crassus's command staff during his disastrous campaign against the Parthians, surviving the massacre at Carrhae in 53 and successfully organizing the defense of Syria.

Fighting for Pompey during the civil war, Cassius held a naval command, submitting to Caesar only after Pompey's death in 48 B.C.E. However, privately he remained opposed to Caesar and, with his brother-in-law Brutus, assassinated him in 44. Although initially supported by the Senate and awarded imperial authority, the conspirators were declared outlaws in 43 and pursued by their adversaries Antony and Octavian (Augustus). Cassius and Brutus obtained the loyalty of most Roman troops in the eastern provinces and moved their armies into Greece, where they were routed at Philippi and committed suicide in 42 B.C.E.

Ian Janssen

See also: Caesar, Julius; Carrhae, Battle of; Philippi, Battle of; Pompey the Great; Roman Civil Wars (88–30 B.C.E.); Roman Republic, Wars of the

References and further reading:

Scullard, Howard. *From the Gracchi to Nero.* 5th ed. London: Routledge, 1982.

Castro Ruz, Fidel (1926–)

Cuban dictator. Born Fidel Castro Ruz on 13 August 1926 at the family plantation in Oriente province. Though his father was a Spanish immigrant from Galicia, he had amassed enough wealth to send Castro to the best Jesuit schools in Oriente and Havana. By 1945, Castro had enrolled in law school at the University of Havana. There he participated in the usual student activism, as well as gangsterism. The one political group with whom Castro associated was known as the Unión Insurreccional Revolucionaria. He was defeated in several student elections, but became known as a fine orator. In 1948, Castro became associated with and participated with those who made the Colombian *Bogotazo.* The Bogo-

Cuban president Fidel Castro speaking to the people of Camaguey, three days before taking action against the forces of dictator Fulgencio Batista, 4 January 1959. (Hulton/Archive)

tazo were a series of bloody riots in Bogota, ignited by the assassination of Liberal Party leader Jorge E. Gaitán. Though pursued by Colombian police, Castro escaped to the Cuban Embassy, whence he was able to effect his escape. Returning to the university, he became a follower of Eduardo Chibás, leader of the Partido Ortodoxo. The party's agenda was one that Castro favored—social justice, end to corruption, and economic independence. Castro was a candidate for the Cuban congress in 1952, when General Fulgencio Batista took power. Due to this coup, elections were never held.

Castro organized a group of his followers and on 26 July 1953 they abortively attacked the Moncada Barracks in Santiago de Cuba. For this he was sent to jail after uttering in his own defense that though he might be condemned here, "history will absolve me." Given amnesty by the Cuban Congress in 1955, he traveled to Mexico to organize an expeditionary force to overthrow Batista. On 2 December 1956 he embarked for Cuba aboard the yacht *Granma* with about 80 men. After landing, they took to the Sierra Maestra mountains, which offered good shelter, and commenced guerrilla warfare against the inept Cuban government and army. By December 1958, the Batista regime was on its way out. Finally, on New Year's Eve Batista fled and Castro took power the next day. Castro never held elections, although he claimed that the vast majority of the Cuban people supported him, and embarked on a program of socialization of the island. Lavish Soviet subsidies, the equivalent of $1 million per day, somewhat covered the numerous failures of the economy, as lines for everyday goods and food staples as well as rationing persisted. As in the Soviet Union, a privileged political class had access to those necessities either rationed or in very short supply to the bulk of the population. The collapse of the Soviet Union vastly exacerbated socialist Cuba's problems. In 1953, according to UN figures, Cuba had one of the highest per capita incomes in Latin America; after four decades of socialism, it had one of the lowest. Castro, of course, blamed these failures on the U.S. embargo clamped on the country since the early days of Castro's dictatorship. The aging dictator managed to retain some measure of popular support (as in the Bay of Pigs invasion early in his regime) by championing Cuban nationalism against the "bully to the north," and by constantly warning of an invasion at any moment by the *yanquis*. But, far from being "absolved by history," by the opening of the twenty-first century, history seems to have passed Castro's Cuba by.

Peter Carr

See also: Castro-Cuban Revolution; Cuban Missile Crisis; Guevara de la Serna, Ernesto "Che"

References and further reading:

Alvarez Díaz, José R. *The Road to Nowhere: Castro's Rise and Fall.* Miami: Editorial AIP, 1965.

Depestre, Rene, et al. *El Asalto al Moncada y la Revolución Latino Americana.* México: Editorial Nuestro Tiempo, 1975.

Castro-Cuban Revolution (1959–)

The result of public frustration with the failures of Cuba's first 60 years of independence. Sugar was the major component of Cuba's economy, and, with the development of the island's railway network, big landowners and mill owners gradually came to dominate the industry. This trend accelerated with increased American investment in sugar growing. Cuba's small independent farmers were gradually reduced to a marginal existence. Because of diminished public purchasing power, other industries failed to develop. From the 1920s on, income from sugar exports generally declined owing to growing foreign competition, despite some years of record production. Public disillusionment was compounded during several dictatorships led by Gerardo Machado (1925–1933) and Sergeant, later Colonel, Fulgencio Batista (1933–1944 and 1952–1958), which alternated with democratically elected but often corrupt political regimes.

American diplomatic meddling often complicated leadership struggles. Fidel Castro Ruz (born 1926), son of a wealthy farmer in northwestern Cuba, studied law at the University of Havana. From his midtwenties, he was involved in revolutionary activities against the Batista regime. He advocated land and educational reform, an end to government corruption and excessive arms expenditures, and a return to constitutional government, while hoping to avoid the mistakes made by previous administrations. His failed attack on the Santiago army barracks (July 1953) led to an 18-month prison term, during which publication of *History Will Absolve Me,* the speech Castro had given at his trial, brought him national recognition.

Freed in a general amnesty (1955), Castro landed in Oriente province in November 1956 with a small group of followers, some of whom were Communists. His supporters included Castro's brother Raul and Ernesto "Che" Guevara (1928–1967), an Argentinian-born physician and revolutionary. Although the operation was betrayed, the Castro brothers, Guevara, and a few others escaped to the Sierra Maestra mountains. There Castro's political strength grew, helped by a series of favorable *New York Times* articles by Herbert Matthews. Batista's increasingly brutal military repression brought Castro's guerrilla movement increased public support. By late 1958, Batista's unprofessional, poorly led, and unmotivated army was no match for Castro's much smaller force, led by Guevara. The United States ended its arms supply to Batista in March 1958. Lacking American support, Batista failed in an attempt to put a puppet president in place using preprinted ballots. With Castro's *barbudos* (bearded ones) in the city's outskirts, Batista and some cronies fled the Cuban capital on 1 January 1959. Castro's forces entered Havana two days later.

Keir B. Sterling

See also: Castro Ruz, Fidel; Guevara de la Serna, Ernesto "Che"

References and further reading:

Chilcote, Ronald H. *Cuba, 1953–1978: A Bibliographical Guide to the Literature.* 2 vols. White Plains, NY: Kraus International Publications, 1986.

Mazarr, Michael J. *Semper Fidel: America and Cuba, 1776–1988.* Baltimore, MD: Nautical and Aviation Publishing, 1988.

Ruiz, Ramon. *Cuba: The Making of a Revolution.* Amherst: University of Massachusetts, 1968.

Szulc, Tad, *Fidel: A Critical Portrait.* New York: Morrow, 1986.

Thomas, Hugh. *Cuba: The Pursuit of Freedom.* New York: Harper & Row, 1971.

Casualties, War in the Twentieth Century

War in the twentieth century has been characterized by appalling suffering and death, brought about by the weapons of mass destruction and the demands of "total war." One would have to go back to the seventeenth century's Wars of Religion to see parallels to the racial and ethnic hatreds that led to the wars of the twentieth century, and back even further to the times of the Mongol conquests to find comparable death totals.

The murderous ideologies of the twentieth century have made it fairly certain that the century will be remembered as the century of megadeath. The Nazi Holocaust claimed some 6 million Jews, as is well-known. But millions of Slavs, Gypsies, Poles, and other "subhumans" also died in German concentration camps during World War II. At the same time, Allied bombers killed hundreds of thousands of civilians in German and Japanese cities. (Some have calculated that a broad-based pyramid 100 feet high could have been constructed of the dead children alone immolated in the great Tokyo fire raid of March 1945.) The Japanese themselves slaughtered literally millions in their invasion of China between 1937 and 1945.

Yet many of the worst killings of the twentieth century were actually not the direct result of any wars or conflicts. The man-made famines in Ukraine in the early 1930s, China's 1960s' aberrant "Great Leap Forward," and the Ethiopian famines of the 1970s and 1980s were actually conducted in peacetime, to test academic agrarian reform theories. The theories did not work, and millions died. Stalin's incomprehensible blood purges are supposed to have cost several million Soviet lives in the 1930s and 1940s. But tiny Cambodia has the dubious distinction of having suffered, per capita, more violent deaths than any other nation or people in the violent twentieth century; some 1 million of a population of a little more than 4 million, inflicted by the Marxist Khmer Rouge after the close of the second Vietnamese War. The danger is that humanity in general will grow callous to megadeath, the legacy of the twentieth century. In the words of one who practiced megamurder, Joseph Stalin, "One person's death is a tragedy, a thousand is a statistic."

Totals:

Russo-Japanese War 1904–1905

Japanese: Killed in action (KIA) 47,152; Died of wounds (DoW) 11,424 (total 58,576); Disease 21,802 (total of these 80,378)

Russians: KIA 34,000; DoW 52,623; Disease 9,300; Missing in action/Prisoner of war (MIA/POW) 39,500; Wounded in action (WIA) 141,800

World War I

Russians: KIA 626,440; DoW 17,174; WIA 2,754,202 (total 3,397,816; POW 2,417,000

France: KIA 1,089,700; MIA 265,300; POW 477,800

UK and Empire: KIA 908,371 (Army 658,704); WIA 2,090,212; Civilians 30,633

Empire troops: Australia: KIA 54,431; WIA 156,173; POW/MIA 3,494 (total 214,098)

Canada: KIA 50,334; WIA 152,779: POW/MIA 8,245 (total 211,358)

India: KIA 21,642; WIA 46,969 (total 68,611)

New Zealand: KIA 15,000

South Africa: KIA 5,000

Italy: KIA 462,391 (incl. 98,000 WIA/DoW and 45,000 died as POW); WIA 953,886

Rumania: KIA 335,706; WIA 120,000; Civilians c. 275,000

USA: KIA 53,513 (or 36,931 + 13,673 DoW = 50,604); Other deaths 63,195 (or 62,668 – 23,583 abroad = 39,085 in USA); WIA 204,002 (or accident 4,503; WIA 189,059) (total 306,834)

Germany: KIA 1,611,104; WIA 3,683,143; MIA 772,522 (total 6,066,769)

Austria-Hungary: KIA 922,500; WIA 3,620,000; Civilians 300,000 (incl. 200,000 Poles)

Turkey: KIA 325,000; WIA 400,000; Civilians c. 2.5 million

Bulgaria: KIA 75,844; Disease, etc. 25,500; WIA 152,390; Civilians 275,000

Russian Civil War 1917–1920
No figures (est. 2,000,000)

Spanish Civil War 1936–1938
Spanish: KIA c. 285,000: Franco's troops 110,000; Spanish Republic 175,000; Civilians: all 125,000 (due to war direct causes) + 200,000 disease/starvation
Foreigners: KIA 2,725; WIA 10,558; MIA 222

Italian-Ethiopian War 1935–1936
Italians: KIA 4,539 (2,313 Italians, others allied)
Ethiopians: c. 275,000

Sino-Japanese War 1937–1945
Chinese: KIA 1,319,000; WIA 1,761,000; MIA 130,000; Civilians c. 3,000,000
Japanese: uncertain, but thought to be c. 500,000 KIA and died of wounds/disease

World War II
France: KIA 213,324; WIA 390,000; Civilians 390,000
Greece: KIA 17,024; WIA 47,290; Civilians 325,000
Belgium: KIA 8,460; WIA 55,513; Civilians 76,000
Norway: KIA 1,598; WIA 7,000; Civilians 364
Netherlands: KIA 6,344; WIA 2,860; Civilians 200,000
Denmark: KIA 1,800; WIA 2,000; Civilians unknown
Czechoslovakia (as ally): KIA 6,683; WIA 8,017; Civilians 215,000
Brazil: KIA 943; WIA 4,222; Civilians 972
Philippines: KIA 27,258; WIA: unknown; Civilians 91,000
Germany: KIA 2,850,000; WIA 7,250,000; Civilians 780,000; POW 3,400,000
Japan: KIA 1,506,000; WIA 326,000; Civilians 672,000; POW 810,000
Italy: KIA 110,823; WIA 225,000; Civilians 152,941; POW 350,000
Rumania: KIA 169,882; WIA unknown; Civilians 200,000; POW 100,000
Hungary: 147,435; WIA 89,313; Civilians 200,000; POW 170,000
Austria: WIA 350,117; other figures unknown

(Allies)
Russia: KIA 6,115,000; WIA 14,012,000; Civilians 14–20 million
China (1937–1945): KIA 1,319,958; WIA 1,761,335; Civilians 2,000,000; POW 124,470
Yugoslavia: KIA unknown; WIA 425,000; Civilians 1,200,000
Poland: KIA 123,178; WIA 236,606; Civilians 5,675,000 (plus at least 53,000 partisans); POW 420,760
UK: KIA 264,443; WIA 329,267; Civilians 92,673; POW 213,919
Australia: KIA 23,365; WIA 39,803; POW 32,393
Canada: KIA 37,476; WIA 53,174; POW 10,888
India: KIA 24,338; WIA 64,354; POW 12,430
New Zealand: KIA 10,033; WIA 19,314; POW 10,582
South Africa: KIA 6,840; WIA 14,363; POW 16,430
British Colonies: KIA 6,877; WIA 697; POW 22,323
USA: KIA 292,131; WIA 671,801; POW 139,709

French/Indochina War 1946–1954
French: KIA 75,867; WIA 65,125
Indochinese on the French side: KIA 18,714 (incl. MIA); WIA 13,002
Vietminh estimate: KIA 175,000; WIA 300,000; Civilians many 100s of thousands

Chinese Civil War 1945–1949
PLA: KIA and WIA 1,522,500
Nationalists: KIA 571,610; POW 6.8 million+; Defected to PLA 1,773,490

Israeli War of Independence 1947–1949
Egypt: KIA 1,500–2,000; WIA 4,231
Syria: KIA 1,000 plus 1,000 Arab legionnaires
Iraq: KIA 500
Lebanon: KIA 500
Saudi, Yemen, and Sudan: a few hundred KIA
Israel: military KIA 4,074; Civilians more than 2,000

Korean War 1950–1953
UK: KIA 710; WIA 2,278; MIA/POW 2,029
Turkey: KIA 717; WIA 2,246; MIA/POW 386
Canada: KIA 291; WIA 1,072 MIA/POW 33
Australia: KIA 291; WIA 1,240; MIA/POW 60
France: KIA 288; WIA 818; MIA/POW 29
Greece: KIA 169; WIA 543; MIA/POW 3
Thailand: KIA 114; WIA 794; MIA/POW 5
Ethiopia: KIA 120; WIA 536; MIA/POW 0
Colombia: KIA 140; WIA 452; MIA/POW 94
Belgium: KIA 97; WIA 350; MIA/POW 6
Philippines: KIA 92; WIA 299; MIA/POW 97
Netherlands: KIA 111, WIA 589; MIA/POW 4
New Zealand: KIA 34; WIA 80; MIA/POW 1
South Africa: KIA 20; WIA unknown; MIA/POW 22
USA: KIA 33,629; WIA 103,284; MIA/POW 10,218; noncombat dead 20,617

Vietnam 1965–1975
USA: battle dead 47,072; nonbattle dead 10,435; WIA 155,419 (plus nonhospitalized personnel total = 303,704); MIA/POW 932

South Vietnam: KIA 196,863; WIA 502,383
NVA-VC (est. 1961–1973): KIA 927,124 (or 731,000 by body count system, discounted by 30% to allow for many mistakes and murders)
Russians: KIA 13

6-Day War 1967
Israelis: KIA 983; WIA 4,517; MIA/POW 15
Arab forces: KIA 4,296; WIA 6,121; MIA/POW 7,550

Yom Kippur War 1973
Israelis: KIA 2,838; WIA 8,860; MIA 508; POW 301
Egypt: KIA 7,700; WIA 12,000; POW 8,031
Cubans: KIA 180; WIA 250
Plus other Arab casualties: KIA 8,258; WIA 19,459

Iran-Iraq War 1980–1988
Estimates only
Iran: KIA 450,000 (incl. 25, 000 by gas) up to 750,000; WIA more than 1.2 million; POW 45,000
Iraq: KIA 150,000 to 340,000; WIA 400,000 to 700,000; POW 70,000

Gulf War 1990–1991
USA: KIA 148 (of which 35 friendly fire); WIA 467 (incl. 72 friendly fire); POW 21; Accidents 118
Saudis: KIA 44
UK: KIA 24 (incl. 11 blue on blue/friendly fire); WIA 43
Egypt: KIA 14; WIA 120
France: KIA 2; WIA 25
Iraq: uncertain but suggested KIA 100,000–200,000 (most likely UK estimate 40,000); 5,000–15,000 civilians in bombing; 4,000–16,000 civilians starvation; 15,000–30,000 Kurds during Iraqi occupation and war; 2,000–5,000 Kuwaitis

Afghanistan Civil War and Soviet Intervention 1979–1990
Afghanistan: c. 1.3 million due to war
Russians: KIA 13,310; WIA 35,478; MIA 311

Falkland Islands War 1982
UK: KIA 218; Other causes 37; WIA 777
Argentina: KIA 746 (incl. 368 on *Belgrano*)

David Westwood

References and further reading:

Clodfelter, M. *Warfare and Armed Conflict: A Statistical Reference to Casualty and Other Figures.* 2 vols. Jefferson, NC: McFarland, 1994.

Dupuy, R. Ernest, and Trevor N. Dupuy. *The Harper Encyclopedia of Military History: From 3500 B.C. to the Present.* 4th ed. New York: HarperCollins, 1993.

Catapults

The artillery of ancient and medieval warfare. Catapults were among the earliest mechanically powered weapons. Based to some extent on the simple principle of a bow and arrow, initial examples were relatively straightforward attempts to increase the range and penetrating power of missiles by strengthening the bow and the stock that propelled them. The early *gastraphetes* of the Greek times, powered by the arm and stomach muscles of the warrior (hence the name), were in time replaced by the more powerful torsion catapults—the *euthytonon* (an arrow-shooting machine) and the *palintonon* (a stone-thrower). These made it possible to use larger and more lethal missiles over longer distances, aspects that were crucial to military expeditions, especially when cities were well fortified, and consisted of multiple sets of walls and moats.

The popular use of catapults by the Greeks and Romans is verified through the accounts of glorious war victories, including several by Alexander the Great, who employed these war machines even in his far-flung eastern expeditions. The art of building catapults survived through the Middle Ages, and was used efficiently at the Vikings, by Genghis Khan's massive Mongol armies, the crusader forces, and finally Tamerlane, who flung diseased carcasses of animals into fortified cities spreading death and destruction. By the sixteenth century the *trebuchet,* a modified catapult with a single pivoted arm, but with considerable force and a high trajectory, was the warring world's precursor to advanced gunpowder and ballistics.

Manu P. Sobti

References and further reading:

Hacker, Barton C. "Greek Catapults and Catapult Technology: Science, Technology and War in the Ancient World." In *Technology and the West: A Historical Anthology from Technology and Culture,* ed. Terry S. Reynolds and Stephen H. Cutcliffe. Chicago and London: University of Chicago Press, 1997.

Keegan, John. *A History of Warfare.* New York: Vintage Books, 1994.

Marsden, Eric W. *Greek and Roman Artillery: Historical Development.* Oxford, UK: Clarendon Press, 1969.

Parker, Geoffrey, ed. *The Cambridge Illustrated History of Warfare: The Triumph of the West.* Cambridge, UK/New York: Cambridge University Press, 1995.

Cavalry

The noble arm of battle.

The traditional horse-borne arm of mobility and shock from 2000 B.C.E., when chariots began to be used in organized warfare, till rapid-fire weaponry and mechanization swept the horse from the battlefield. While chariots were dominant in Bronze Age militaries across Eurasia, men rid-

A cavalry troop of the Red Army thundering to an attack on the German infantry on the eastern front, 1941. (Library of Congress)

ing directly on horses supplanted the inflexible chariot, beginning with the Assyrians in the seventh century B.C.E. This process reached its pinnacle in the army that Alexander the Great took to war. Integrating infantry and cavalry, using both missile weapons and shock action, to coordinated effect, this army remains a model for combined arms.

The tactical equation is that disciplined infantry can usually withstand cavalry seeking to break their ranks by main force and personal weapons ("shock action"), and Rome certainly epitomized this, despite the challenge of cavalry-dominated forces such as the Hellenistic armies of Hannibal and Pyrrhus or the Parthians. Due to many factors, though, Rome was overrun when its military machine broke down and it faced the challenge of the steppe horseman, starting with the Goths but epitomized by the Huns.

The Huns were not the first nomadic horsemen to make their mark on Western history, but they represented a factor that the Eurasian world would be contending with into the 1400s. With tough steppe-bred horses, the invention of the stirrup, and recurved compound bows, all the hordes needed were leaders capable of pulling together disciplined armies capable of defeating more sedentary societies.

While the medieval period was the epoch of the hordes, the armored horseman was also coming into his own. In another creation of the Near East, the Romans attempted to deploy heavy cavalry as early as the time of Hadrian. This began a trend by which the Roman army became a force of cavalry regiments, as opposed to infantry legions, as a premium was placed on mobility. This process eventually led to the creation of the feudal knight and a whole new social order.

Both forms of horseman, the knight and the steppe nomad, remained dominant into the thirteenth century, when a number of forces began to erode their predominance, such as the return of disciplined infantry and improved missile weapons, though this only brought a certain balance back to the battlefield. Further, the mounted man-at-arms was as likely to fight afoot as a heavy infantryman, where his weaponry and skill could bolster a battle line, despite the romantic image of the knight in single combat. It took the arrival of firearms simple enough to be used on horseback and capable of piercing all but the best armor to end the reign of the armored horseman. Once pistol-armed riders were capable of outmaneuvering the man-at-arms, the use of armor rapidly dwindled to near irrelevance by the end of the seventeenth century.

While firearms slowly eroded the cavalry's value as a shock arm on the battlefield worldwide, there was a compensatory trend in the creation of the dragoon in the seventeenth century, bodies of men capable of fighting as infantry or cavalry and useful for scouting, raiding, and "screening" purposes. Diverging over time into general-purpose cavalry and mounted infantry, such troops remained relevant into World War II, wherever terrain prevented the use of heavy forces and allowed freedom of maneuver. In fact, mounted troops have been used in antipartisan operations as late as the 1970s in Rhodesia and Portuguese Africa.

Having emphasized tactical considerations, it should never be forgotten that cavalry, be it the knight, the samurai, cavaliers, or sipahis, was the prestige arm of most armies. The expense of horses and the difficulty of mounted fighting generally reserved the role of cavalry for social elites, leaving a cachet that sustained that arm into the twentieth century. Even the tank was originally designed to break through trench positions—and allow the cavalry to get in among the enemy. Though now mounted in armored vehicles and helicopters, U.S. cavalry officers still wear spurs as part of their dress uniform.

George R. Shaner

See also: Alexander the Great; Ancient Warfare; Animals in War; Assyria; Attila the Hun; Avars; Cossacks; Firearms; Genghis Khan; Goths; Hannibal Barca; Magyars; Mamluks; Mongol Empire; Parthian Empire; Samurai; Scythians; Tamerlane

References and further reading:

Hall, Bert S. *Weapons and Warfare in the Renaissance: Gunpowder, Technology, and Tactics.* Baltimore, MD: Johns Hopkins University Press, 1997.

Hildinger, Erik. *Warriors of the Steppe: A Military History of Central Asia, 500 BC to 1700 AD.* New York: Sarpedon, 1997.

Hyland, Ann. *Equus: The Horse in the Roman World.* New Haven, CT: Yale University Press, 1990.

Jones, Archer. *The Art of War in the Western World.* Urbana: University of Illinois Press, 1987.

CEDAR FALLS, Operation (January 1967)

A major allied offensive in the Vietnam conflict. The American strategy for the Vietnam conflict was to kill, wound, capture, or cause to desert more enemy than the enemy could reasonably replace. At the point the attrition rate went negative, so the reasoning in Washington went, the enemy would end its aggression. Thus, the U.S. command favored so-called search-and-destroy missions, in which U.S., South Vietnamese, and/or allied forces typically would land to establish a strong point or anvil, and then strike forces would land and drive—or hammer—the enemy to the anvil and destroy it and its base/resources.

A classic example was Operation CEDAR FALLS, which sought to clear out a long-standing Vietcong base area known as the "Iron Triangle," northwest of Saigon. The operation began on 8 January 1967 with the anvil forces (several U.S. and Vietnamese divisions) moving into position; then the so-called hammer (more U.S. troops) began to descend on January 9.

Unexpectedly, the Vietcong main force units, the object of the operation, eluded the hammer group and escaped into the jungle. Nineteen days later the operation ended, and the U.S. claimed to have seized large quantities of arms and equipment and to have destroyed the base area, including an extensive and amazing tunnel complex. In fact, the base area was not destroyed, and the Vietcong returned to launch attacks on Saigon. The operation illustrated the problems with a strategy that conceded the strategic initiative—that is, whether to fight or not—to the enemy and then to abandon territory that was seized after hard fighting.

Charles M. Dobbs

See also: Ho Chi Minh; Tet Offensive; Vietnam Conflict; Vo Nguyen Giap; Westmoreland, William

References and further reading:

Rogers, Lieutenant General Bernard William. *Cedar Falls–Junction City: A Turning Point.* Washington, DC: Department of the Army, 1974.

Stanton, Shelby L. *The Rise and Fall of an American Army: U.S. Ground Forces in Vietnam, 1965–1973.* Novato, CA: Presidio Press, 1985.

Westmoreland, William C. *A Soldier Reports.* Garden City, NY: Doubleday, 1976.

Celts

The name applied by Greco-Roman writers (Keltai/Celtae, Galatai/Galatae, and Galli) generally to peoples occupying lands north of the Mediterranean, not all of whom actually spoke languages now identified as Celtic, and were thus culturally *Celtic.*

Greece and Rome mostly knew the Celts as invaders with a fearsome reputation for aggressiveness and, in the words of Strabo, "madly fond of war, high spirited, and quick to battle." There is no doubt that warfare played a central role in Celtic society. For the nobles, raiding offered an opportunity for wealth and prestige. Their retinues could only be maintained by actual fighting, and formed the nucleus for a tribal army primarily composed of all free tribesmen able to equip themselves. War bands based on clan, familiar, and settlement groupings made a man's relatives witnesses to his behavior.

Body armor was rare. The combination of shield, long slashing-sword, and spear(s) formed the equipment of most warriors. The appearance of the individual, his size, expressions, and war cries, added to the din of clashing weapons and of the war trumpet (*carnyx*), was clearly intended to intimidate any opposition. If an enemy could be persuaded that he was going to lose, then a Celtic charge would drive all before it. As was common in tribal armies, Celtic warriors were poorly disciplined, and had no training above the level of the individual.

The highest-quality troops were the cavalry. Recruited from the wealthier nobles and their retinues, equipment was of good quality. It consisted of a shield, spear, sword, and often a helmet and mail armor. Added to this was the four-horned saddle, later adopted by the Romans, which provided a secure seat. The morale of these horsemen was usually very high, but discipline was normally poor.

In early encounters between Celt and Roman, it was the two-horse war chariot that drew the latter's interest the most. The main use of the chariot was psychological, to cause panic. Charioteers would drive their vehicles in a rush against the enemy lines, the warriors throwing javelins. This, coupled with the speed and noise of the chariots, would be enough to unsettle the opposition. Once these preliminaries were finished, the warriors dismounted from their chariots and fought on foot, while the charioteers kept their chariots at the ready for a speedy retreat. Some 1,000 chariots took part in the battle of Sentinum (295 B.C.E.). At Telamon (225 B.C.E.), the chariots were stationed on the wings of the Gallic army. As the prowess and agility of their horsemen increased, the Gauls gradually gave up their chariots. They were no longer in fashion in the time of Caesar. He was surprised to find them still in use in Britain. Later, Tacitus mentions chariots in the Caledonian army at Mons Graupius.

Nic Fields

See also: Boudicca's Rebellion; Gallic Wars; Mons Graupius, Battle of

References and further reading:

James, Simon. *Exploring the World of the Celts.* London: Thames & Hudson, 1998.

Ritchie, W. F., and J. N. G. Ritchie. *Celtic Warriors.* Aylesbury, UK: Shire Publications; Cincinnati, OH: Seven Hills Books, 1985.

Central American Federation Civil Wars (1826–1840)

Series of civil wars that led to the dissolution of the Central American Federation. The United Provinces of Central America, consisting of the present-day nations of Guatemala, El Salvador, Honduras, Nicaragua, and Costa Rica, gained independence from Mexico on 1 July 1823. From the very beginning, the fledgling nation suffered from a lack of cohesiveness and from deep economic, political, and social divisions, which resulted in various periods of civil war leading to the eventual dissolution of the federation.

In 1826 hostilities broke out between the Conservatives, led by President Manuel José Arce of El Salvador, and the Liberals, whose most prominent early leader was the Honduran Francisco Morazán. Morazán led his army to victory over federal forces in Honduras and El Salvador in 1828 and pacified Guatemala in early 1829. When Morazán became president in 1830, he initiated a reform program that was particularly harsh on the Catholic Church. In Guatemala, the Liberal governor Mariano Gálvez, with the encouragement of the federal government, confiscated church property, ended some religious feast days, and legitimized divorce. When the Liberal Guatemalan government instituted the Livingston Codes on 1 January 1837, a peasant revolt quickly ensued. The popular uprising soon found a leader in José Rafael Carrera, who organized the peasants and issued a manifesto demanding abolition of the Livingston Codes, the restoration of religious orders, reinstatement of the archbishop, end of the head tax, and recognition of his authority. Gálvez resigned his position and Carrera's insurgents stormed into Guatemala City on 31 January 1838.

When Carrera's movement joined forces with the Conservatives, Morazán was determined to crush the insurgent movement and restore federal control of Guatemala. In addition, there were conservative movements against the federal government in the other four provinces. The battle lines were clearly demarcated in 1838. Carrera and the Conservatives stood for autonomy; Morazán and the Liberals wanted to preserve the federation. In 1839 Morazán gained the upper hand and briefly reestablished federal power in much of Guatemala. However, Carrera renewed his campaign to defeat Morazán and to rid Central America of the Liberals. He adroitly formed alliances with conservative groups in Honduras and Nicaragua. He drove the liberal army from Guatemala, El Salvador, and Honduras and subdued the rebellious area of Los Altos. The final blow was delivered in March 1840 when Carrera defeated Morazán's army at Guatemala City. Morazán evaded capture and fled to David in Panama. The United Provinces of Central America was dissolved and the five nations soon became independent republics. In 1842

Morazán made one last attempt to regain power with a revolt in El Salvador followed by an invasion of Costa Rica. His movement lacked popular support and he was captured, tried, and, on 15 September 1842, executed by firing squad. Carrera dominated politics in Guatemala, first holding the presidential chair from 1844 to 1848 and in 1854 declaring himself president for life. He died in 1865.

George M. Lauderbaugh

References and further reading:

Bethell, Lesli, ed. *Central America since Independence.* Cambridge, UK: Cambridge University Press, 1991.

Karnes, T. L. *The Failure of Union: Central America, 1824–1975.* Rev. ed. Tempe: Center for Latin American Studies, Arizona State University, 1976.

Woodward, Ralph Lee, Jr. *Central America: A Nation Divided.* New York: Oxford University Press, 1976.

Central Intelligence Agency

Principal intelligence and counterintelligence agency of the U.S. government, founded 18 September 1947. The Central Intelligence Agency (CIA) grew out of America's wartime experience with the Office of Strategic Services (OSS) and the need to centralize the collection and dissemination of strategic intelligence.

In 1947, Congress passed the National Security Act, creating the National Security Council (NSC) and, under its direction, the CIA, which was to advise the NSC on intelligence matters pertaining to national security. During the time of the CIA's creation, the idea of a single intelligence service had given way to the concept of an "intelligence community" comprising the Defense Intelligence Agency (DIA), State Department Intelligence, National Security Agency (NSA), and others, with the CIA as overall coordinator.

The CIA has come to perform three major assignments: (1) foreign intelligence collection, evaluation, and communication; (2) counterintelligence operations overseas; and (3) secret political intervention, psychological warfare, and paramilitary operations in foreign areas.

It is the third area of responsibility where the CIA has earned not only its greatest successes and failures, but also an aura of controversy. In Italy, the CIA's financial and technical assistance was successful in aiding the Christian Democrats in defeating communist parties in the 1948 parliamentary elections. Another major covert operation was the toppling of Mohammad Mosaddeq as premier and the restoration of the shah of Iran in 1953.

Among the failures of the CIA was the 1961 Bay of Pigs invasion of Cuba by CIA-supported anti-Castro dissidents. Finally, the CIA played a major role in the Kennedy administration's successful management of the Cuban Missile Crisis.

More recent years have given up convincing evidence of CIA law-breaking in the surveillance and manipulation of and experiments upon unwitting American citizens and the infiltration of dissident but legal domestic organizations. Because of these illegal activities and the failures noted above, the CIA has come under increasing scrutiny and control by the federal executive and Congress to the extent that some authorities now question its effectiveness.

Craig T. Cobane

See also: Bay of Pigs Invasion; Intelligence, Military; National Security Agency/Central Security Service

References and further reading:

Breckinridge, Scott D. *The CIA and the U.S. Intelligence System.* Boulder, CO: Westview Press, 1986.

Richelson, Jeffrey T. *The U.S. Intelligence Community.* Boulder, CO: Westview Press, 1995.

Cerisolles, Battle of (11 April 1544)

Tactical lesson of the Italian wars. By 1544, the Italian wars had turned into a generalized European conflict. France had to fight in Italy, but at the same time against the imperialist army on its eastern borders and along its northern coast to prevent English incursions. In Italy, war raged for more than 40 years without decisive result. During the new campaign along the Po, the French main army, commanded by the 24-year-old Duke d'Enghien, faced a veteran army of Spanish and German soldiers led by the Marquise Del Guasto.

Strategic planning by Henry VIII and Charles V assigned an important role to Charles's army in Italy: It was to cross the Alps and take Lyon; while it did this Charles would invade Champagne, and Henry VIII, Picardy.

To prevent an attack on the French southern borders, d'Enghien decided to besiege Carignano with 14,000 soldiers. In order for Del Guasto to relieve the besieged town with 20,000 infantry, he had to cross the Po using a bridge held by the French. D'Enghien chose to face the imperialists with his back to the river, making retreat all but impossible.

At dawn on 11 April 1544, the French deployed their troops in three "battles," mainly infantry. On their right were 800 skirmishers led by Blaise de Montluc, positioned to slow any imperialist advance. At 7 A.M., sporadic fire erupted into a continuous fight as the imperialists encountered Montluc's men. Skirmishing lasted for four hours with no winner, but by noon the imperialists were unsteady and retiring slowly. Noting this fact, Del Guasto launched his main force of 10,000 Landsknechten against the French center. With the help of artillery, the German mercenaries advanced without fear against their Swiss opponents, but the huge German column was taken in the flank by French fire and stalled. D'Enghien, unaware of the success of his left, ordered repeated charges against the Spanish, led personally by Del Guasto. Del Guasto then stopped the attack on the right, thereby leaving the left flank of the Germans unprotected. The French cavalry charged at once and routed the Landsknechten. The Spanish, alone now, held the French at bay as long as they had powders for their guns and then surrendered.

Imperialist casualties in the battle amounted to 12,000, with 3,000 prisoners, but the victory was without political consequence for France, which was still invaded in September, as previously agreed upon. But the invading forces had to face the victorious troops returning from Italy. The battle of Cerisolles foreshadowed the weakness of blocks of pikes and underscored the importance of flanking movement.

Gilles Boué

See also: Pavia, Battle of

References and further reading:

Cornette, Joël. *Chronique de la France Moderne: Le XVIème Siècle.* Paris: SEDES, 1995.

Corvisier, André. *A Dictionary of Military History and the Art of War.* Rev. ed. Trans. Chris Turner. Oxford: Blackwell, 1994.

Hardy, Etienne. *Batailles Françaises.* Paris: Dumaine, 1881.

———. *Origines de la Tactique Française.* Paris: Dumaine, 1881.

Cerro Gordo, Battle of (Mexican War, 17–18 April 1847)

A significant battle in the Mexican-American War that opened the way to an assault on Mexico City itself. Hoping to stop American general Winfield Scott as he passed through a long and narrow defile of the National Road, the Mexican commander, General Antonio López de Santa Anna, carefully placed artillery batteries and infantry to make the defile impassable, and assumed that no route around his position existed. Santa Anna established his strong defensive position at Cerro Gordo about three days' march inland from Veracruz.

Scott ordered a complete reconnaissance of the area, and several junior officers found a mule path, a mere trail, around and behind the Mexican left. Scott ordered the path widened to permit a large force to outflank the Mexican defenses.

The attack, beginning on 17 April, on the Mexican left and El Telegrafo hill was difficult. Scott had wanted a demonstration straight up the National Road to fix the attention of the defenders while he sent a strong column around the widened mule path, but the demonstration began rather late. In the meantime, American troops had to drag heavy artillery up steep hills, cross various barriers, and fight against brave troops defending the key position on the left, the hill of El Telegrafo. But by the next day the hill was seized, and the Americans surged down at the remaining Mexican defenders, who engaged in a general retreat.

The Americans thereupon continued on to Jalapa and rested at Puebla as Scott wrestled with issues of terms of enlistment, troop numbers, and supplies, and as Santa Anna prepared the defenses of Mexico City for the expected attack.

Charles M. Dobbs

See also: Mexican-American War; Santa Anna, Antonio López de; Scott, Winfield

References and further reading:

Eisenhower, John S. D. *Agent of Destiny: The Life and Times of General Winfield Scott.* Norman: University of Oklahoma Press, 1997.

Johnson, Timothy D. *Winfield Scott: The Quest for Military Glory.* Lawrence: University of Kansas Press, 1998.

Chaco War (1932–1935)

War between Bolivia and Paraguay for possession of a vast but desolate area, the Chaco Boreal, which both had claimed since gaining independence and saw as an important area for future development. Bolivia was also interested in gaining an outlet to the Atlantic by establishing ports on the Paraguay River. Both sides were ill-prepared and underestimated the formidable terrain where they had to wage war. The Paraguayans were able to mobilize faster and expanded their army from 3,000 to 60,000 men. Paraguay also found a competent leader in General José Estigarribia, who won the full backing of his government. Bolivia, on the other hand, failed to declare a general mobilization and the military command and civilian leadership bickered over strategy. Nevertheless, Bolivia advanced and quickly overran several small Paraguayan outposts. Estigarribia mounted a counterattack on Fort Boquerán on 29 September 1932 and drove the Bolivians back, but both sides sustained more than 3,000 casualties.

The Paraguayan victory sent shock waves throughout Bolivia and resulted in the recall of General Hans Kundt, former head of the German military mission to Bolivia, who had helped modernize the army before World War I. Kundt ordered a general offensive against a string of Paraguayan forts starting in December. Kundt's yearlong offensive failed to dislodge the Paraguayan defenders and yielded only heavy causalities, resulting in his being relieved of command. Having blunted Bolivia's attack, Estigarribia launched an offensive and gradually drove the Bolivians from many of their Chaco fortifications. By the summer of 1934, part of the Paraguayan army had driven into Bolivia proper. Estigarribia feared that his forces were overextended and ordered a gradual retreat. Bolivia pursued his army deep into the Chaco but in December 1934, a reinforced Paraguayan army encircled the Bolivians and took 6,000 prisoners. By early 1935, Paraguay controlled most of the disputed territory; however, Bolivia rallied and finally declared a general mobilization. The war had drained the national treasuries and manpower of both countries, and on 12 June 1935 a cease-fire was signed. In 1938 Bolivia and Paraguay signed the Treaty of Buenos Aires, formally ending hostilities. Paraguay gained three-fourths of the disputed territory, and Bolivia was granted access to the Paraguay River for use as an outlet to the Atlantic Ocean. The human toll was high: Bolivia lost 57,000 dead and Paraguay nearly 36,000.

George M. Lauderbaugh

References and further reading:

Fernandez, Carlos J. *Guerra del Chaco.* 6 vols. Buenos Aires: Imprenta Militar, 1956–1976.

Ynsfràn, Pablo Max, ed. *José Felix Estigarribia: The Epic of the Chaco.* Austin: University of Texas Press, 1952.

Zook, David H. *The Conduct of the Chaco War.* New Haven, CT: Bookman Associates, 1960.

Chadian Civil Wars (1960s–1984)

Chad, the fifth largest state in Africa, neglected by the French in colonial times, wedded an Arabic-speaking desert north to a French-speaking south, dependent on subsistence agriculture. In the 1960s, rebellions in the north left the inefficient national government, dominated by southerners, holding only four garrison towns in the north. A national liberation front, FROLINART, emerged to organize the rebellion. With French military and administrative assistance, the government stabilized the situation without improving it. A military coup in 1975 brought Félix Malloum to power. Unable to cope adequately with economic and social problems, he foolishly demanded the withdrawal of French troops even as FROLINART was growing more daring and powerful. With weapons and logistical support from Libya, the rebels were able to seize one of the northern garrison towns in 1977. Malloum took advantage of a split between rival FROLINART commanders, Goukouni Ouedeye and Hissene Habré, to create a National Union Government with himself as president and Habré as prime minister. Goukouni's forces advanced on the capital, N'Djamena, until stopped just short of their goal by French troops fighting alongside government forces.

The National Union Government collapsed early in 1979, with forces loyal to Habré seizing the capital and welcoming Goukouni's troops. The Lagos Accord on 21 August led to the establishment in November of a new government that included all linguistic, tribal, geographic, and ideological factions. By 1980 the two northerners, Goukouni and Habré, were fighting for control of the capital and government, with the south sitting on the fence. Libyan military intervention consolidated Goukouni's power, but an unwise announcement that Chad and Libya were working toward unity brought a demand by the Organization of African Unity that Libya withdraw. As soon as the Libyans left in November 1981, Habré returned to the attack and seized power on 7 June 1982. He was dictator by 1984.

Joseph M. McCarthy

References and further reading:

Buijtenhuijs, Robert. *Le Frolinart et les Guerres Civiles du Tchad (1977–1984): La Révolution Introuvable.* Paris: Karthala, 1987.

Lorell, Mark A. *Airpower in Peripheral Conflicts: The French Experience in Africa.* Santa Monica, CA: Rand, 1989.

Thomson, Virginia M. *Conflict in Chad.* Berkeley: University of California Press, 1981.

Chaeronea, Battle of (86 B.C.E.)

Roman legion use of field fortifications to defeat a numerically superior Greek force. Mithridates VI of Pontus, who had been steadily expanding his kingdom in Asia Minor, invaded the Roman client state of Bithynia in 89 B.C.E. He then overran the Roman provinces in Asia Minor and ignited a Greek revolt against Rome.

Rome dispatched Lucius Cornelius Sulla with an army of 30,000 to restore the situation. Delayed by the Roman Civil War (88–82 B.C.E.), Sulla's army did not arrive in Greece until 87 B.C.E. The Greek armies retreated into Athens, which Sulla invested and captured the following year. Sulla then marched his army north to seek battle with a new Greek army, recently reinforced by Mithridates.

Archelaus, who had escaped from Athens, had built this army in northern Greece to possibly 100,000 men, though many of them lacked significant military training or experience. Archelaus marched south, and the two armies met in Boeotia on the plain of Chaeronea. Faced with a force substantially larger than his own, Sulla had his troops erect palisades fronted by entrenchments to cover his flanks and awaited attack.

Archelaus began the battle with a cavalry charge, which Sulla's disciplined legions threw back in disorder from their secure position. Archelaus then launched his chariots at the Romans, and these fared even worse. Some of the chariots fled back through the Greek phalanx, throwing it into disorder. Sulla seized this opportunity and launched a counterattack that shattered the Greek phalanx and drove it from the field.

Mithridates sent Archelaus more reinforcements and the war continued. It took another defeat and another Roman army to force Mithridates to make peace.

Stephen Stein

See also: Mithradatic Wars; Roman Civil Wars (88–30 B.C.E.)

References and further reading:

Goldsworthy, Adrian. *The Roman Army at War.* Oxford, UK: Oxford University Press, 1996.

Keppie, Lawrence. *The Making of the Roman Army: From Republic to Empire.* Norman: University of Oklahoma Press, 1998.

Marsh, F. B. *Cambridge Ancient History.* Vol. 9, *The Roman World 146–30 B.C.* Revised by H. H. Scullard. Cambridge, UK: Cambridge University Press, 1994.

Chaeronea, Battle of (August 338 B.C.E.)

Battle in which the Macedonian king Philip II decisively defeated an alliance of Greek city-states, of whom the most important were Athens and Thebes.

When Philip had marched into central Greece to defeat the allies, the Greeks decided to fight on the defensive. They deployed their forces near the town of Chaeronea, the left flank protected by a range of hills, the right by the marshy bank of a stream. Their army, probably numbering between 30,000 and 35,000 men, consisted largely of heavy infantry (hoplites). The Athenian contingent was positioned on the left, the Theban on the right, and those of the minor allies in the center.

Philip's army consisted of 30,000 foot and 2,000 horse. Deploying his infantry in line and himself taking command over the guard infantry on the advanced right flank, he ordered his son Alexander (later called "the Great") to command the cavalry on the retarded left.

The Macedonians advanced in oblique order. Having approached the Greek right, Philip ordered his guard to fall back, feigning retreat. As the inexperienced Athenians enthusiastically pursued the Macedonians, the Greek line was disrupted and a gap appeared. Unhesitatingly Alexander charged ferociously at the head of the cavalry into the gap, rupturing the Greek line. Thereupon he turned upon the Greek right. In the meantime Philip halted the retreat of the guard and attacked the Greek left. Soon the allies broke and fled, the Athenians leaving 1,000 dead on the field of battle, 2,000 being taken prisoner. Theban dead numbered at least 300, all members of the Theban elite unit called the Sacred Band. They fell where they stood. Macedonian casualties were probably few. This victory earned Philip control over the whole of Greece, with the exception of Sparta in the Peloponnesus.

Maarten van der Werf

See also: Alexander the Great; Philip II of Macedon

References and further reading:

Cawkwell, George. *Philip of Macedon.* London: Faber & Faber, 1978.

Ellis, J. R. *Philip II and Macedonian Imperialism.* London: Thames & Hudson, 1976.

Hammond, N. G. L. *Philip of Macedon.* London: Duckworth, 1994.

Châlons, Battle of (Gaul, 20 June 451)

Coalition of Romans, Alans, and Visigoths under Aetius fought Attila's equally diverse army of Huns, Ostrogoths, and Gepids to a standstill. During his invasion of the western empire, Attila secured the Rhine, and then moved quickly into central Gaul. He laid siege to Aureliani (Orleans sur Loire), and had he prevailed would have been in a strong position to subdue the Visigoths of Aquitania Secunda, but Aetius had assembled a formidable coalition against him. The Visigoths and Alans, traditional foes of the Romans, joined Aetius's army out of common hatred for the Huns. Attila also had his friends, including the Vandal king Gaiseric, who had urged him to attack the Visigoths. The Visigoths, in turn, hated the Vandals. A generation earlier Gaiseric's son had married the daughter of Visigothic king Theodoric I, but in 442 the Roman emperor, Valentinian III, had agreed to a marriage between Gaiseric's son and his own daughter. The Visigoth princess was returned to her people, but with her nose and ears inhumanly mutilated. When Attila crossed the Rhine, the Visigoths joined Aetius. The Vandals stood aside.

Attila had not expected such vigorous action on the part of the Romans. Unwilling to allow his army to be trapped beneath the walls of Aureliani, he lifted the siege and withdrew north, to open country better suited to Hun horsemen.

The imperial army was drawn up with Romans on the left flank, and the Visigoths, under Theodoric I, on the right. The Alans, whose loyalty Aetius doubted, occupied the center. Attila stationed the bulk of his Germanic allies on the right wing of the Hun army, but the Ostrogoths took up a position opposite the Visigoths. The Huns themselves, the best of Attila's army, were positioned in the center.

The battlefield was a flat, open plain somewhere in what is now Champagne, northern France (perhaps closer to

Troyes than to Châlons). A large level area of some 300 acres, cut by a stream, rose sharply to a ridge. Precise details are lacking. The only surviving account is that of Romano-Gothic historian Jordanes. The battle apparently began when Theodoric dispatched his son Thorismund to occupy high ground overlooking the Hun left flank. Attila sent troops to drive away Thorismund, but these failed to gain the ridge, and were easily routed.

Attila then launched the rest of his forces straight at the imperial army, without first softening up the opposition with missile fire, the usual Hun tactic. According to Jordanes, Attila said the following to his men: "The Romans are poor soldiers, keeping together in rank and file. They are contemptible, the only worthy enemies are the Alans and the Visigoths."

The battle was hard fought. It lasted for most of the day with heavy casualties on both sides. Theodoric himself was among the slain. By nightfall the Romans were in possession of the field. Attila withdrew his exhausted and battered forces into a wagon laager, preparing for a fight to the death on the following day, but Aetius allowed him to withdraw unmolested. The Roman commander feared his Visigoth allies as much as the Huns, and was unwilling to remove a possible counterpoise to their power.

This battle has been reckoned as one of the most decisive in the world's history, but, even considering its violence, it decided very little. Attila retreated to his wooden capital in Pannonia. The next year he launched a major offensive into Italy.

Nic Fields

See also: Aëtius, Flavius; Attila the Hun; Huns

References and further reading:

Täeckholm, U. "Aetius and the Battle on the Catalaunian Fields." *Opuscula Romana* 7 (1969), 259–276.

Champlain, Samuel de (c. 1567–1635)

Established France as a power in North America, and founder of France's first permanent colony in Canada. Champlain was born around 1567, and saw service in the French Wars of Religion under Henry IV. In 1598, Champlain accompanied mercenary Spanish soldiers back to Spain, and unusually, served as a geographer on a Spanish voyage to Mexico, Havana, and Cartagena. Henry IV, his patron, had encouraged several trading settlements in France, but they had failed because of poor planning. In 1603, Champlain went to Canada under Grave du Pont and established a base at Tadoussac, taking care to establish friendly relations with the local tribes. Subsequent voyages explored the southern coast of Nova Scotia and established Port Royal, where Champlain spent three years from 1605 to 1608.

In 1608, Champlain returned with a concession to found a permanent settlement and selected Quebec, insisting on full storage cellars and defensive fortifications. Champlain ventured out to meet the Hurons, Algonquins, and Montagnais, promising them aid against the Iroquois, although he was disdainful of their lack of military discipline. In 1609 and again in 1610, Champlain personally led them against the Iroquois armed with muskets, at Ticonderoga and Soul, and although disgusted by the torture he witnessed, arranged an exchange program whereby he took one of the Huron braves to Paris, while his men lived amongst the tribes.

The problems of the new settlement required Champlain's utmost skill as an administrator. The Jesuit missionaries quarreled with the Recollets (strict Franciscan monks), traders challenged agricultural settlers, and the search for a northwest passage petered out after exaggerated native reports proved false. Champlain faced a major disappointment in 1615, when the Huron, fortified in a palisaded compound at Syracuse, could not be defeated by Hurons with muskets because they would not obey Champlain's orders for organization. This conflict, made deadly serious by the introduction of muskets, later led to the Iroquois's extermination of the Huron and their allies.

Champlain died after a stroke in December 1635, not before establishing French power in North America, buttressed by Native American allies, and providing the government with detailed maps allowing further colonization of New France.

Margaret Sankey

See also: American Indian Wars

References and further reading:

Berry, Gerald L. *Champlain.* Richmond Hill, ON: Pocket Books, 1967.

Heidenreich, Conrad. *Explorations and Mapping of Samuel de Champlain 1603–1632.* Toronto: B. V. Gutsell, 1976.

Morison, Samuel Eliot. *Samuel de Champlain, Father of New France.* Boston: Little, Brown, 1972.

Chan Chan, Battle of (Inca Empire, 1468)

Crowning Inca conquest of the most powerful coastal state, Chimu. As the Inca Empire developed and expanded under the 33-year reign of Emperor Pachacutec Yupanqui, it transitioned from highland conquests to the hot, coastal deserts. The Inca collided here with the powerful Chimu state on the north coast of modern Peru.

Pachacutec developed the tactic of cycling in 20,000-man armies every two months in the hotter realms. This allowed his troops to remain fresh and to mount constant pressure on the enemy. This tactic won over the Pachacamac region and then was used by Pachacutec's son Tupac to defeat the "Great Chimu" and capture the Chimu capital of Chan Chan after 1468.

The Inca defeated the superior metallurgical civilization of the Chimu by constant pressure, cutting off water supplies to the desert coast and removing tribute populations from the Chimu sphere of influence. Although the Inca never broke into the huge city of Chan Chan, they did manage to wear down the Chimu and cut off all resources. Eventually the Chimu leader was forced to surrender not because of battle losses but due to a lack of resource base, which had been denuded by the Inca strategy.

This use of economic warfare became the hallmark of Inca conquests as they used superior logistics to overcome military superiority by their enemies. The victory also marked the completion of the Inca transition from seasonal military systems under a single commander to a comprehensive grand strategy approach to empire with permanent armies, multiple campaigns, and numerous generals.

Christopher Howell

See also: Cuzco, Battles of; Inca Empire Imperial Wars; Pachacutec Yupanqui

References and further reading:

Bram, Joseph. *An Analysis of Inca Militarism.* New York: Augustin Publisher, 1941.

Garcilaso de la Vega, El Inka. *The Incas: Royal Commentaries.* Ed. Alain Gheerbrant. 3d ed. New York: Avon Books, 1971.

Chancellorsville, Battle of (30 April–6 May 1863)

One of the Confederacy's most brilliant, but costly, victories. In late April 1863, General Joseph Hooker conceived a brilliant idea for the newly revamped Army of the Potomac. Still facing Robert E. Lee and the Army of Northern Virginia at Fredericksburg, Hooker envisioned a gigantic turning movement: While John Sedgwick remained with up to 40,000 troops to occupy Lee, Hooker would take the bulk of the army north, cross the Rapidan and Rappahannock Rivers 40 miles upstream, and move behind Lee, threatening to cut his supply line and to trap him between two large Union forces. Either way, Lee would have to fight, retreat, or risk being trapped. At first, Hooker seized the initiative, moved out, crossed and secured several fords, and then proceeded east

Robert E. Lee with his soldiers in battle at Chancellorsville. (Library of Congress)

and south past Chancellorsville. Then, for reasons not entirely clear, Hooker seemingly lost his nerve and ordered his men to retreat to Chancellorsville.

At the same time, Hooker's advance clearly brought out the fighter in Lee. He divided his smaller command (he had about 60,000 compared to 125,000 for Hooker), leaving some 15,000 under Jubal Early at Fredericksburg, he moved some 45,000 to find Hooker. On the night of May 1–2, "Stonewall" Jackson proposed an audacious plan—to divide Lee's army yet again before a superior foe, and for Jackson to march around Hooker and attack his hanging flank from the rear. Lee agreed, and Jackson swung left with some 28,000 while Lee faced 75,000 Union troops with only 17,000 men.

Jackson's men were finally in position in late afternoon, and around 5:20 P.M., accompanied by a thunderous Rebel yell and noise from the startled wildlife, Jackson's men fell upon the Union rear guard commanded by General O. O. Howard. That night, while looking over the positions, Jackson was shot by his own men, and died the next day. Jeb Stuart assumed command and tried to continue the attack, and forced the Union position into a big bow, but superior Union numbers and artillery limited the Confederate victory. The cost was high to Lee—many casualties including his great right-hand man, Stonewall Jackson.

Charles M. Dobbs

See also: American Civil War; Hooker, Joseph; Jackson, Thomas "Stonewall"; Lee, Robert Edward

References and further reading:

Casdorph, Paul. *Lee and Jackson: Confederate Chieftains.* New York: Paragon House, 1992.

Gallagher, Gary W., ed. *Chancellorsville: The Battle and Its Aftermath.* Chapel Hill: University of North Carolina Press, 1996.

Stackpole, Edward J. *Chancellorsville: Lee's Greatest Battle.* Harrisburg, PA: Stackpole Company, 1958.

Chandragupta Maurya (r. c. 321–c. 298 B.C.E.)

Founder of the Gupta Dynasty. In the aftermath of Alexander the Great's invasion of northern India, Chandragupta Maurya, a regional landowner, about whose family and background little is known except that he came from a village of peacock tamers, began a war of "liberation" against the Greek rulers left behind to rule the farthest reaches of Alexander's conquest. Between 321 and 317 B.C.E., aided by Chanakya, the "Indian Machiavelli" and author of the political handbook *Arthsastra,* Chandragupta Maurya ousted the satraps of Punjab and Sindh and established himself as the founder of a new dynasty. To organize his empire, he created an elaborate centralized bureaucracy, including a state council, census takers, treasury department, and secret police force to administer the areas he ruled directly and indirectly through local rulers. Large public works projects improved the empire's agriculture, and also provided militarily strategic highways, hospitals, and dams. The empire was in diplomatic contact with the Alexandrian successor states, as well as its Indian neighbors. The state revenues supported an enormous army and navy of 650,000, including elephant cavalry, a logistical branch, chariots, and riverboats.

In 305 B.C.E., Seluceus I Nicator invaded the Gupta Empire, hoping to regain territory, but was defeated so badly by Chandragupta that he abandoned plans for further reconquest and gave up most of modern Afghanistan, including its capital, Kabul. His daughter was married to Chandragupta to seal the treaty. In exchange, Seluceus received 500 war elephants and safe passage home. With a solid northern frontier and a wealthy empire, Chandragupta retired from power and ended his life as a Jain monk, and probably starved himself to death in religious devotions, leaving his empire to his son Bindusara, father of perhaps the greatest Gupta emperor, Ashoka Maurya.

Margaret Sankey

See also: Alexander's Wars of Conquest

References and further reading:

Joshi, M. C., ed. *King Chandragupta and the Meharauli Pillar.* Meerut, India: Kusumanjali Prakashan, 1989.

Mookerji, Radha Kumud. *Chandragupta Maurya and His Times.* New Delhi: Sindar Lal Jain, 1961.

Sharma, L. P. *Ancient History of India.* New Delhi: Vikas Publishing House, 1981.

Chaplaincy, Military

Military officers who tend to the spiritual, moral, and physical needs of troops. The military chaplaincy has a tradition dating back through many centuries, during which time representatives of the cloth have held some of the highest offices at almost every administrative level.

A number of Old Testament Hebrew prophets led troops in battle in the name of God, but the first recorded involvement of clergy in postbiblical times occurred in 430 when Bishop Germanus, as general officer commanding the army in Wales, successfully led a campaign against a united force of Picts and Scots. This function of a priest as a military commander in time of national emergency was by no means an isolated occasion. Prelates and priests were engaged in warlike activities during and following the reign of William the Conqueror. When Scottish armies attempted to invade England, military service was required from every guild. The clergy were not exempt.

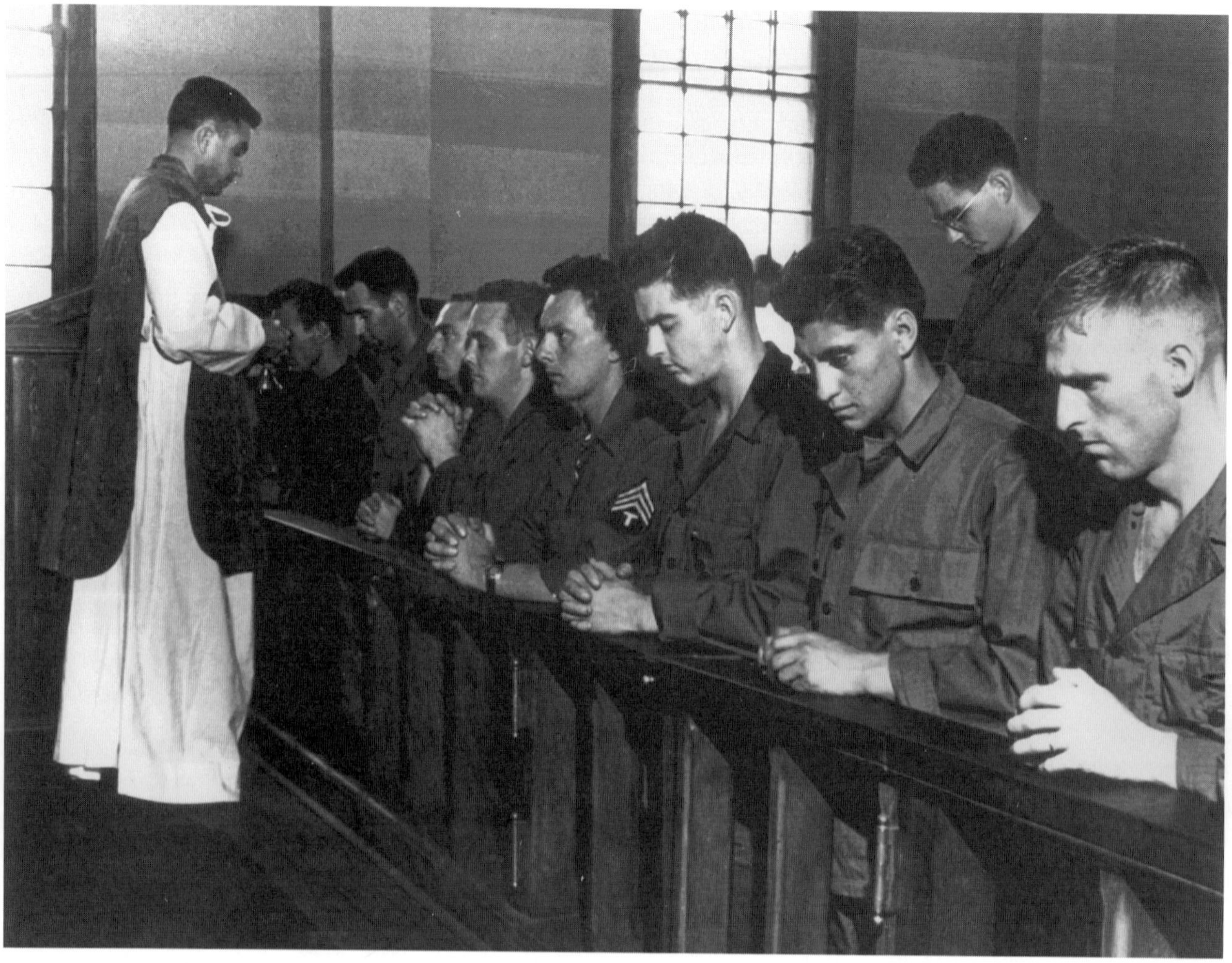

Army chaplain giving communion to soldiers at the New York Port of Embarkation during World War II, c. 1942. (Library of Congress)

The year 1272 marked the end of the predominantly warlike role of the clergy when, for what may have been the first time in history, a "chaplain on the Pay Roll" was recognized as a noncombatant member of the English armed forces. Modern chaplains are still called staff preacher, preacher, padre, or chaplain, and are drawn from all recognized denominations.

Chaplains' duties have altered over time and are dependent on a chaplain's location and proximity to conflict. In peacetime, their official duties include the conducting of religious services, bible classes, prayer meetings, hospital visitations, and burials. With the number of married soldiers increasing dramatically, military chaplains now find themselves very much involved in the marital problems of their troops. In rear areas during periods of conflict a chaplain is required to visit the men, and provide such things as coffee stalls and sporting activities, sit-down dinners, games or song nights, and personal consultations. Frontline chaplains are usually assigned to the battalion aid station, where they comfort the wounded and the dying and generally make themselves useful.

Chaplaincy can be dangerous work: 11 U.S. Army chaplains were killed in the line of duty during America's relatively brief combat involvement in World War I, and the four chaplains of different faiths who gave up their life belts and their lives when their army transport was torpedoed in the frigid North Atlantic early in World War II have become legends.

Margaret Hardy

See also: Religion and War

References and further reading:

Abbott, Douglas. *Australian Military Chaplaincy.* Unpublished history RAAChD, Russell Offices. Canberra: n.d.

Appelquist, A. Ray, ed. *Church, State, and Chaplaincy: Essays and Statements on the American Chaplaincy System.* Washington, DC: General Commission on Chaplains and Armed Forces Personnel, 1969.

Barish, Louis, ed. *Rabbis in Uniform: The Story of the American Jewish Military Chaplain.* New York: J. David, 1962.
Brandt, Hans Jurgen. *Priester in Uniform: Seelsorger, Ordensleute und Theologen als Soldaten im Zweiten Weltkrieg / Herausgegeben vom Katholischen Militäärbischofsamt und Hans Jüürgen Brandt.* Augsburg: Pattloch, 1994.
Brumwell, P. Middleton. *The Army Chaplain, The Royal Army Chaplains' Department. The Duties of Chaplains and Morale.* London: A & C Black, 1943.
Hardy, Margaret. "Military Chaplaincy: An Establishment of Chaplains 1901–1918." Honors thesis, Australian National University, Canberra ACT, 1996.
Hutcheson, Richard G., Jr. *The Churches and the Chaplaincy.* Atlanta, GA: John Knox Press, 1975.
Johnstone, Tom. *The Cross on the Sword—Catholic Chaplains in the Forces.* London: Continuum, 1996.

Charlemagne (742–814)

Frankish conqueror, first ruler of the Holy Roman Empire. Charlemagne inherited half of the Frankish kingdom from his father, Pepin, in 768. The initial years of his rule were occupied with a civil war between himself and his brother Carloman until Carloman's death in 771. Now sole ruler, Charlemagne embarked on a series of actions that resulted in the expansion of his kingdom into an empire, consolidation of the continental Christian kingdoms above the Pyrenees, a new administrative framework to govern his realm, establishment of a relationship between the papacy and German emperors that would continue for a millennium, and creation of an environment for learning that would be unrivaled in western Europe until the rise of the universities.

Frankish consolidation of the Christian kingdoms was the triumph of Orthodox Christianity, as defined by the pope, over Arian Christianity. As part of the process, all other Christian kingdoms were brought into line with the Orthodox Roman view of dogma. Forced conversion of Frisians and Saxons was part of the process. Charlemagne's final solution to Saxon disaffection was wholesale slaughter of Saxon males and the forced resettlement of the remaining Saxons. It should be noted that the absorption, conversion, and repelling of pagan invaders were part of what Charlemagne regarded as his obligation as defender of the faith. So too was the establishment of a papal state. In recognition of his actions, Pope Leo III crowned him, in 800, emperor of the Romans.

Administratively, Charlemagne created frontier border districts, called marches, and deeded them to powerful warlords. The areas buffered by the marches were divided into provinces, and were ruled by various nobles. Charlemagne institutionalized commissioners (*missi*), transforming them from periodic messengers into a corps representing the king's authority. Law was another area transformed and codified by Charlemagne. Throughout the empire, laws were divided into church administrative codes, local secular laws, and laws derived from the emperor.

In the end, Charlemagne's empire would not survive more than a generation and was divided by his grandsons. The strength of Charlemagne's vision and actions continues to the present. The intermingling in the West of religious and secular power in the person of the monarch, coupled with his ascension as Holy Roman Emperor, had far-reaching ramifications. First, it set up the basic constructs for dissension between the church and the state, which would later resolve itself into the disputes of the Guelfs and Ghibellines. Second, the question of investiture rights can be traced to Charlemagne's rule with two swords philosophy. Third, the resurrection of the Roman Empire and the unification of all Europe under a single government would be a recurring theme in European history.

Tamsin Hekala

See also: Roland

References and further reading:

Loyn, H. R., ed. *The Middle Ages: A Concise Encyclopaedia.* New York: Thames & Hudson, 1991.
Loyn, H. R., and J. Percival. *The Reign of Charlemagne: Documents on Carolingian Government and Administration.* London: Edward Arnold, 1975.

Charlemagne's Wars (771–814)

Charlemagne extends Frankish Empire to its greatest extent. The death of Pepin the Short in 768 left his two sons, Charles and Carloman II, in uneasy control of the western and eastern Franks respectively, until the death of Carloman in 771. Charles then launched his career as king of the Franks, conqueror of a Frankish Empire, and eventual protector of the Roman Catholic Church as Holy Roman Emperor.

Charles first waged war against his father-in-law, Desiderius, king of the Lombards in the Italian peninsula, at the request of the besieged pope Adrian I. Charles first divorced Desiderius's daughter, and then captured the family of Carloman II, eliminating potential rivals to his own position, and of Desiderius himself. Charles then reaffirmed the donation by Pepin, who had served as overlord of the Lombard Empire, of papal lands controlled by the Church. Thus, by 774, Charles had eliminated all internal rivals and allied himself firmly with the powerful Roman Catholic Church.

Charles also waged war on the Saxons (772–804), the last independent Germanic people who often raided the rich eastern Frank lands. A major Frank invasion in 775 was only partially successful, and Charles changed strategy, turning to annexation, fortifications, and missionaries. The Saxons got aid from Danes and Slavs, and turned to the leadership of the Westphalian Widukind. They caused Charles great difficulty from 777 to 785. In response, Charles mounted an increasingly cruel campaign against them, including annual invasions and replacement of Saxons with Franks. He once executed 4,500 Saxons after a Frankish defeat.

Charles fought his most famous war in his invasion of northern Spain (777–801). He led a large Frankish army to the city of Saragossa, where Umayyads were resisting the second Muslim Dynasty, the Abbasids. Charles destroyed several towns, including Pamplona, a Christian town apparently mistaken as Muslim. Charles was then recalled to deal with Saxon depredations, leaving behind a rear guard under his nephew Roland. The Christian Basques (Gascons), angered by the destruction of Pamplona, allied with Muslim forces to ambush the rear guard at Roncesvalles on 15 August 778, leading to the destruction of the entire Frankish force, including Roland.

Charles also carried out war (791–796) against the Avars, a Central Asian people who had settled in and plundered the Danube River area. Initially unable to pin down the mobile horsemen, his son Pepin finally defeated them in 796. The hidden treasury of the Avars was later captured and sent back to Charles at his capital in Aachen. They had been entirely subjugated by 805.

After Charles had reinvaded the Italian peninsula and put his son on the Lombard throne, he had himself crowned Holy Roman Emperor by the pope on Christmas Day in 800. Charles then waged a successful land and sea campaign against the Byzantines (803–810) to force them to recognize him as emperor of the west, and to gain control of Venice and Dalmatia on the Adriatic. The Byzantine emperor Nicephorus I, fully occupied with the Bulgarians, did so in 810.

Simultaneous with the Franco-Byzantine War, Charles fought other groups as well. He subdued the Saxons by 804, the Slavic Wends by 806, and fought the Danes as well. By 811, his capital at Aachen was the center of the only true medieval Christian empire, and Charles, then 62, set about passing on his legacy of a pan-European empire to his son Louis I, "the Pious."

Christopher Howell

See also: Avars; Charlemagne; Frankish-Moorish Wars; Roland

References and further reading:

Bullough, Donald. *The Age of Charlemagne.* New York: Putnam, 1966.

———. *Carolingian Renewal: Sources and Heritage.* Manchester, UK, and New York: Manchester University Press, 1991.

Burgess, Glyn, trans. *The Song of Roland.* London: Penguin Books, 1990.

Charles Martel (689–741)

After the death of the Merovingian king Dagobert in 638, a number of military-supported magnates, especially the mayors of the palaces in Neustria and Austrasia, took power in the northern Frankish kingdoms. Charles Martel (Martel meaning "the Hammer," for his military conquests) was the bastard son of Pepin II, the mayor of the palace in Austrasia. Shortly after Pepin's death, in 714, Charles fought for Pepin's title against another magnate, Ragamfred, and Ragamfred's allies. Charles had only a small war band; however, after a small victory over Ragamfred and the obtainment of his father, Pepin's, treasury, Charles had the resources ultimately to defeat his adversary. Charles, as the newest mayor of the palace in Austrasia, placed his own figurehead king on the throne, and spread his influence with military victories over the Saxons, Alamans, and Bavarians. As Charles continued his military successes he presented booty and confiscated lands, much of which had belonged to the Church, to his own supporters.

In 732, Charles came to the duke of Aquitaine with aid against the Arabs who had pushed north toward the towns of Poitiers and Tours. At Poitiers, Charles's army defeated that of Abd al-Rahman and thwarted the Arab Muslim threat. After his victory, Charles invaded Burgundy and led a naval expedition against the Frisians. He implemented siege warfare to capture the city of Avignon; and, on the banks of the Berre River, he defeated an Arab force for the second time. Before he died in 741, Charles divided control of the lands in his sway between his two sons, giving Austrasia, Swabia, and Thuringia to Carloman, and Burgundy, Neustria, and Provence to Pepin. In only a quarter century, Charles Martel had built up what would become under his grandson, Charlemagne, the Carolingian Empire.

Christopher P. Goedert

See also: Charlemagne; Frankish Civil Wars; Frankish-Moorish Wars; Merovingians; Tours

References and further reading:

Bachrach, Bernard S. *Early Carolingian Warfare: Prelude to Empire.* Philadelphia: University of Pennsylvania Press, 2000.

———. *Merovingian Military Organization, 481–751.* Minneapolis: University of Minnesota Press, 1972.

Riché, Pierre. *The Carolingians: A Family Who Forged Europe.* Trans. Michael I. Allen. Philadelphia: University of Pennsylvania Press, 1993.

Charles the Bold, Duke of Burgundy (1433–1477)

Last reigning duke of Burgundy and great military reformer. Charles was the son of Philippe the Good, and inherited Burgundy and the Low Countries from him in 1467. He opposed the growing power of King Louis XI of France, and tried to free his duchy from its dependence on France, and restore it as a kingdom.

Charles's ambition was to create a kingdom stretching from the mouth of the Rhine to the mouth of the Rhône. He ruled Burgundy, Flanders, Artois, Brabant, Hainaut, Luxembourg, Holland, Zeeland, and Friesland. He needed Alsace and Lorraine to achieve his territorial goal. His struggle with the Alsatian towns (1473) and occupation of Lorraine alienated the Swiss cantons allied with the French. War broke out between Charles and the Swiss. Charles's armies were routed in 1476 at Grandson and at Morat. He himself was killed at the Battle of Nancy in 1477. His heiress, Mary, married Maximilian of Habsburg, the future Emperor Maximilian I.

Charles was the chief military innovator of the late fifteenth century. The organization of his armies shows clearly the beginnings of a transition between feudal levies and independent mercenary bands, and standing armies. He improved on the military ordnance of the previous dukes. The new ordnances followed the pattern of the ordnances of the king of France of 1445 (that is, a system of tax-supported regular soldiers, with uniform equipment and tactics).

The Burgundian army was the fist in Europe to be placed on a largely regular footing. It included feudal levies (knights and men-at-arms), but also large numbers of mercenary specialists and a large field artillery park. Most of the mercenaries were grouped in units by weapon type (pikes, handguns, or crossbows).

Charles's military ordnances demanded strict discipline and regular drilling. The army that was to fight against the Swiss was composed of 15,000 regular soldiers from all over Europe. This fact provided a weakness, since Charles's multilingual army lacked the necessary cohesion. Charles also personally lacked the tactical skills needed to command against the ferocious Swiss.

Once powerful, Burgundy ceased to exist as a state after Charles's death. His impetuous gallantry earned him his surname, but his achievements were short-lived.

Gilles Boué

See also: Grandson and Morat, Battles of

References and further reading:

Nicholas, Michael. *Armies of Medieval Burgundy.* London: Osprey Publishers, 1983.

Soisson, Jean Pierre. *Charles le Téméraire.* Paris: Grasset, 1997.

Charles XII (1682–1718)

The most daring and most hazardous general of his times. On 13 April 1700, Charles XII left his native city of Stockholm, where he was born on 27 June 1682, to fight the Danish troops on Zealand; he never returned. The short war against Denmark ended with the Treaty of Travendal but marked the beginning of the Great Northern War (1700–1721). In this conflict among Sweden, Denmark, Russia, and Saxe-Poland the small Swedish army under Charles defeated Russian troops at Narva (1700) and Grodno (1706). Both victories, as all those that were to follow, were due to superior Swedish cavalry. After the triumph over the Saxon army near Fraustadt (1706), Charles dethroned King Augustus of Poland in the Peace of Altranstadt and led his 25,000–30,000 battle-weary troops into winter quarters in Saxony.

He left the country in autumn 1707 and marched with approximately 40,000 troops on Moscow, planning to link up with a further 30,000 Swedish soldiers under Loewenhaupt coming from Livonia. On 28 September 1708, a Russian army under Peter the Great defeated Loewenhaupt at Lesnaya and captured all the Swedish supplies. Nevertheless, in May 1709 Charles besieged Poltava with a battered remnant of 20,000 troops. Czar Peter arrived there with his army on 4 June and set up entrenchments. On 28 June, Charles attacked the Russian position but was heavily defeated. This time his horsemen were of no use. Charles fled to Turkey where he spent the next five years.

In 1715 he returned to Sweden but never regained the initiative. He lost Stralsund in that very year and his life during the siege of Frederikshall in Norway on 11 December 1718.

Juergen Luh

See also: Northern War, Great; Poltava

References and further reading:

Hatton, Ragnhild Marie. *Charles XII.* London: Historical Association, 1974.

Charleston, Siege of (April–May 1778)

One of the most severe reverses to American arms of the American Revolution. British general Henry Clinton decided to strike at the center of American and French strength in the American South with the siege of Charleston. In 1778, he withdrew the garrison from Newport, Rhode Island, and pulled back troops in New Jersey to New York City. The British navy commanded the American coast after the French withdrawal to the Caribbean, and Clinton could move his men at sea far faster than General George Washington could send troops across largely unsettled terrain. Thus, Clinton was able to mass 14,000 men and a powerful navy for the siege of Charleston.

The British army and navy cooperated efficiently in this operation. The navy landed men on John's Island to the south of the city, and then moved up the Ashley River (Charleston was bounded by the Ashley and Cooper Rivers). As the British moved to surround the city by land and sea, Charleston's leaders insisted that American general Benjamin Lincoln concentrate on defending the city proper, with little room for defense in depth or opportunity to save his army.

The British conducted the siege in exemplary fashion, first cutting off the city by sea, constructing siege lines ever closer to the city, having the navy sail past lightly defended Fort Moultrie, and finally beginning the bombardment of the town. By 9 May, the bombardment set many of the wooden houses in the city on fire, and city elders, who had insisted on Lincoln remaining to defend the city in the first place, asked to surrender to save the community from further destruction. On 12 May 1780, General Lincoln surrendered the city and his entire force of 5,466 men. The capitulation at Charleston was the worst American defeat until the surrender of the Philippines in 1942.

Charles M. Dobbs

See also: American Revolution

References and further reading:

Mattern, David B. *Benjamin Lincoln and the American Revolution.* Columbia: University of South Carolina Press, 1995.

Pancake, John S. *This Destructive War: The British Campaign in the Carolinas, 1780–1781.* Tuscaloosa: University of Alabama Press, 1985.

Chateau Thierry/Belleau Wood (1–26 June 1918)

American attack, usually associated with the United States Marine Corps, in a heavily forested region approximately 50 miles east of Paris. In the late spring of the decisive year of 1918, the German army was desperately pushing toward Paris in an effort to put an end to World War I on German terms. In the first few days of June, the German vanguard occupied and fortified Belleau Wood, just east of the town of Chateau Thierry along the Marne. The Germans attacked American positions on 4 June but failed to exploit a gap in the American line.

At the time two large American formations were serving as part of the French XXI Corps. One day after the German attack petered out, the French XXI Corps commander ordered the Americans to clear out this large patch of woods.

The 4th Brigade of the United States Marine Corps, attached to the army's 2d Infantry Division, was particularly eager to carry out this attack. With no reconnaissance and an almost complete lack of artillery support, the marines attacked the solidly dug-in Germans on 6 June 1918. The enemy inflicted massive casualties on the marines, who nonetheless pushed forward in rushes of small groups of survivors. The battle quickly degenerated into a maelstrom of hand-to-hand combat in the dense forest. By the end of the day the marines had suffered over 1,000 casualties but had held fast.

It took three more weeks for the marines and their comrades from the army's 7th Infantry Regiment to capture the woods. They did so without adequate artillery and logistical support. The American victory helped turn the tide of the war. Among the legacies of this battle was the rise of the Marine Corps in the American public's consciousness. Marine public relations officers did a masterful job of trumpeting their story to war correspondents, creating the impression that only marines fought in this battle, giving rise to a residue of interservice bitterness that would endure for many years.

John C. McManus

See also: Marne, Battle of the; World War I

References and further reading:

Asprey, Robert. *At Belleau Wood.* Denton: North Texas State University Press, 1996.

Toland, John. *No Man's Land.* Garden City, NJ: Doubleday, 1980.

Chattanooga, Battle of (23–25 November 1862)

A Union victory that opened the way for Union forces to the lower South. General Braxton Bragg and the Army of Tennessee continued the siege at Chattanooga from late September into late November 1862. Bragg divided his army, sending James Longstreet and his corps to Knoxville, although that action could not have a major effect on the outcome of fighting in the area.

In late October, President Abraham Lincoln appointed Ulysses S. Grant to command of the armies in the west, and Grant moved from Vicksburg to Chattanooga and began to reverse the situation.

Union troops were short of food, and Grant broke the blockade that Confederates had established with the so-called Cracker Line. With ample supplies for the men and horses, Grant moved General William Tecumseh Sherman and his four divisions in mid-November, and later General Joseph Hooker and his corps from the Army of the Potomac (using the superb northern-rebuilt railway system). Grant was ready for action.

First, Hooker's corps attacked Lookout Mountain in the "battle above the clouds" on 23 November. The outnumbered Confederates had little choice but to concede the position on their extreme left flank. Grant then wanted Sherman's units to turn the Confederate right, Tunnel Hill, but the attacks failed the next day. Grant had little respect for George Thomas's Army of the Cumberland, which had been beaten so badly at Chickamauga, but he ordered a demonstration in the middle to distract the Confederates so that Sherman could try to take the right one more time.

Thomas's men had something to prove, and began marching up Missionary Ridge against orders. Confederate rifle pits atop the hill had been poorly constructed and were not particularly useful. The spectacle of the moving line of Union blue broke the morale of the Confederate defenders, who fled the heights, and streamed into Georgia. As a result Grant had opened the gateway to the lower South.

Charles M. Dobbs

See also: American Civil War; Bragg, Braxton; Chickamauga, Battle of; Grant, Ulysses Simpson

References and further reading:

Bowers, John. *Chickamauga and Chattanooga: The Battles That Doomed the Confederacy.* New York: HarperCollins, 2000.

Cozzens, Peter. *The Shipwreck of Their Hopes: The Battles for Chattanooga.* Urbana: University of Illinois Press, 1994.

McDonough, James Lee. *Chattanooga: A Death Grip on the Confederacy.* Knoxville: University of Tennessee Press, 1984.

Cheka

Soviet secret police organs from 1917 to 1991. They included border troops and special troops that participated in the Bolshevik Revolution and World War II.

The Cheka, or Chrezvychainaya Komissariat po bor'be s kontrarevoliutsiei i sabotazhem (All-Russian Commission for Struggle against Counterrevolution and Sabotage; *cheka* is also the Russian word for "linchpin"), was established 20 December 1917 and numbered 30,000 members by the end of the civil war, with an additional 70,000 combat troops operating independently of the military chain of command. Cheka units crushed insurgent groups, referred to as *bandits,* with the aid of the Red Army, and took part in the suppression of the Kronstadt sailors' mutiny near Petrograd (St. Petersburg/Leningrad) in March 1921.

Under its head, Felix Dzerzhinskii (Iron Felix), the Cheka launched the Red Terror in the years immediately after the Revolution, executing perhaps 50,000 people. Less sanguinary Bolsheviks called repeatedly for an end to the terror, and the Cheka was finally disbanded by a decree of 6 February 1922. Its functions were transferred to the Narodnaya Komissariat Vnutrennikh Del (NKVD, People's Commissariat for Internal Affairs), under which an autonomous Gosudarstvennoe Politicheskoe Upravlenie (GPU, State Political Administration) fulfilled the functions of a political police organ.

The NKVD had been founded on 8 November 1917, and competed with the Cheka during the years after the Revolution, though Dzerzhinskii was a member of both bodies. Dzerzhinskii became head of both the NKVD and the GPU in 1919 and held these posts until his death in 1926. In 1923, the GPU became the OGPU (Unified State Political Administration) and was separated from the NKVD; the latter was abolished on 15 December 1930. The OGPU carried out the show trials of the 1920s and 1930s and operated the GULAG (Glavnoe Upravlenie Ispravitel'no-trudovykh Lagerei, Chief Administration of Corrective Labor Camps) in the early 1930s. Tens of millions of political prisoners were imprisoned in the Gulag and millions of them died. After an absence of several centuries, slavery had returned to Russia on a large scale.

The NKVD was reorganized on 10 July 1934 as an All-Union Commissariat, and the OGPU was subsumed into it. It was the NKVD that became infamous as the organ of repression at the height of the Great Terror of the late 1930s. In addition to its political police functions, it dealt with frontier security and the registration of civil acts (births, deaths, marriages, and so on). In 1941, it was divided into the NKVD and the Narodnaya Komissariat Gosudarstvennoi Bezopastnosti (NKGB, People's Commissariat for State Security) and in 1946 was reorganized yet again as the Ministerstvo Vnutrennikh Del (MVD, Ministry of Internal Affairs). Many of its security and secret police functions were taken over by the Komitet Gosudarstvennoi Bezopastnosti (KGB, Committee of State Security), when the latter was founded on 16 March 1956. (The reasons for all of these reorganizations and title changes are not always clear.)

During World War II, NKVD troops often took up positions behind Red Army troops and fired on them if they attempted to retreat or desert. It also operated filtration camps for Soviet POWs brought back from German captivity and Soviet citizens who had been in occupied territory. The NKVD also carried out the relocation of factories and wartime industries eastward ahead of the Nazis and exercised administrative control over partisan units.

The KGB was involved in multitudinous espionage and disinformation activities in the postwar decades.

Michael C. Paul

References and further reading:

Conquest, Robert. *The Great Terror: Stalin's Purges of the Thirties.* New York: Macmillan, 1968.

Ebon, Martin. *KGB: Death and Rebirth.* Westport, CT: Praeger, 1994.
Hingley, Ronald. *The Russian Secret Police.* London: Simon & Schuster, 1970.
Wolin, Simon, and Robert M. Slusser, eds. *The Soviet Secret Police.* New York: Praeger, 1957.

Chemical and Biological Warfare

The deliberate use of chemical or biological agents against an enemy has a long history. In the Hebrew Bible book of Judges, Abimelech conquered and destroyed a city and poisoned the soil with salt. Beginning about 300 B.C.E., Persian, Greek, and Roman literatures describe instances of animal or human corpses dumped into water supplies. This tactic was later used by the Confederates in the American Civil War, and apparently in nineteenth-century European wars and the South African Boer War. In the Battle of Eurymedon in 190 B.C.E., Hannibal's forces catapulted clay jars full of poisonous snakes onto the king of Pergamon's ship. The Tatars attacking Kaffa (now Feodossia in Ukraine) in the fourteenth century launched the bodies of plague victims over the city walls, a tactic apparently used in many other medieval battles.

In 1763 in North America, biological warfare using a specific agent was attempted. Sir Jeffrey Amherst, British commander in chief, approved a suggestion by his ranking officer on the Pennsylvania frontier, Colonel Henry Bouquet, to put blankets used by infected soldiers at Fort Pitt into the hands of Native Americans who were threatening the fort. Amherst approved the scheme. Debate over its effectiveness continues today, but the French tried a similar tactic in their conflicts with Native Americans. Clothing infected with plague was used in Tunisia in the 1780s by native tribes against Christian attackers. In 1863 a Confederate surgeon was arrested as he attempted to smuggle clothing with yellow fever infection into the North.

Chemical warfare took a conceptual leap forward in 1854 during the Crimean War. British admiral Lord Dundonald had noticed decades earlier the death of plant and animal life near sulfur kilns. In the war Dundonald suggested using sulfur fumes against the Russians at Sevastopol; he estimated that less than 500 tons of the chemical and 2,000 tons of coke would be needed. Various committees considered Dundonald's proposal and turned it down, deciding the tactic was dishonorable.

The modern era of chemical and biological warfare began in 1899 with the Hague Gas Declaration. This document was the first treaty to deal with poison gas. Twenty-seven nations signed it; the United States was not among them. The treaty banned delivery of gas by projectiles. The Germans ignored the declaration when they used chlorine from large canisters at the Battle of Ypres in 1915. Later in the war Germany used phosgene and mustard gas; these three agents killed or injured more than a million soldiers and civilians. German saboteurs allegedly infected mules, cattle, and horses in Argentina and the United States with anthrax and glanders; the worldwide flu epidemic of 1918 was even rumored to have been a deliberate German effort. But, overall, chemical warfare in world wars has been judged by historians to have been a clumsy affair, far less lethal than the machine gun.

The real and imagined acts of chemical and biological warfare during World War I led to the 1925 Geneva Protocol, which banned use of such agents in war. The United States did not sign this treaty until 1975. The Geneva treaty did not prevent nations from research on these agents or from developing systems to deliver them. In the late 1920s the Soviet Union began research into biological weapons. Before World War II, Japan began a massive effort in Manchuria to develop biological warfare agents such as plague, anthrax, typhus, and cholera. The Japanese built several factories that employed thousands of technicians and scientists, and eventually exposed more than 10,000 Chinese prisoners to numerous biological agents.

In 1941 the United States established the Chemical Warfare Service; the agency became the War Research Service the following year. During the war about $60 million was spent investigating biological and anticrop agents. During World War II, the Allied nations were concerned that Germany would use chemical and biological weapons against them; however, unbeknownst to them, Adolf Hitler had terminated research on such weapons in 1939. Both sides refrained from such weapons in this most total of wars for fear of retaliation by the other side. Such considerations did not, of course, inhibit Germany from the use of Zyclon B gas to kill millions of Jews and others in the concentration camps, and late in the war, research on biological weapons was revived in those camps.

Since the end of World War II, charges of chemical or biological warfare have surfaced several times. During the Korean War in the early 1950s both North Korea and China made detailed accusations that United Nations forces led by the United States were using a number of chemical and biological agents spread by insects. Both the United States and the United Nations denied the charges, and current documentation shows convincingly that the whole episode was an act of national theater by China and North Korea. On the other hand, we also now know that the United States (and possibly other Western nations) was conducting chemical and biological warfare research and experimentation with an abandon that would be deemed criminal today.

In the Second Vietnam War, 1965–1975, the United States charged North Vietnam with using a biological agent in the Laos campaign; like the Korean War allegations, these "yellow rain" claims were supported by weak evidence. Iraq began research into such warfare in 1974, and developed large stockpiles; chemical and nerve agents were used in 1982 and 1983 in its conflict with Iran. The Soviet Union is believed also to have used mustard gas in 1980 in its war in Afghanistan.

Between 1949 and 1960 the United States military conducted numerous tests of chemical and biological warfare agents on civilian populations in a number of locations in the country. These tests were designed to determine dispersal rates and population vulnerability. In 1968, some 6,000 sheep were killed in Utah in an accidental escape of nerve agents. The following year President Richard Nixon declared that the United States would stop research on offensive and biological weapons and destroy its stockpiles. In 1972 a multinational Biological Weapons Convention banned production of such agents.

In recent decades a new threat in this area has appeared: chemical or biological attacks by individuals or small groups. In 1982 seven people in the Chicago area died after ingesting Tylenol laced with cyanide. A salmonella outbreak in Oregon in 1984 was traced to members of the local Rajneesh commune. In March 1995 the Aum Shinrikyo cult in Japan released sarin gas in the Tokyo subway, with fatal results. That same year two members of a Minnesota militia were convicted for possession of ricin, which they had manufactured for use on local officials. Similar incidents have occurred around the world in the past decade.

Today, almost 20 nations are believed to have a biological weapons program. The ease of manufacture and use of these weapons makes them available to individuals and small groups as well as to nations. Research on chemical warfare no doubt continues in laboratories around the world. Over the centuries lack of capabilities or fear of severe reprisals has limited the wartime use of chemical and biological weapons, but as long as nations go to war, or there are disaffected groups or individuals, the temptation to develop them will continue.

A. J. Wright

References and further reading:

Barnaby, Wendy. *The Plague Makers: The Secret World of Biological Warfare.* New York: Continuum, 2000.

Drell, Sidney D., et al., eds. *The New Terror: Facing the Threat of Biological and Chemical Warfare.* Stanford, CA: Hoover Institution Press, 1999.

Haugen, David M., ed. *Biological and Chemical Weapons.* San Diego, CA: Greenhaven Press, 2001.

Chemin des Dames (16 April–3 June 1917)

The bloodiest French defeat of the Great War. After the Verdun and Somme bloodshed, the western front seemed set for a time of reorganization. The Germans dug deeper and wiser, and the French changed the high command: Joffre was sacked and Nivelle was appointed commander in chief on 12 December 1916. He was known as the brilliant organizer of a victorious attack at Verdun in April 1916. Nivelle gained the political battle to force the army's viewpoint on a breakthrough. The chosen part of the front was a line of wooded rocky ridges paralleling the River Aisnes between Rheims and Soissons, called Chemin des Dames. Nivelle had at his disposal more than a million soldiers, most of them young conscripts known as the *bleuets* (cornflowers). The 66 opposing German divisions knew that the French would attack there, thanks to the efficiency of their military intelligence and the incredible confidence of the French high command casually spreading information to the government (and German spies). The French plan was to assault the ridges after 14 days of artillery bombardment. But the Germans simply took cover in the numerous caves in this chalky area and emerged as the bombardment stopped. On 16 April, the offensive was launched all along the line, from Berry au Bac (first use of French tanks) to Laffaux (40 kilometers farther west). After five days, only the first German line had been reached at a tremendous human cost: more than 100,000 casualties (on average, 20 dead every minute). Contrary to what had been explained to the soldiers and the government (no breakthrough, no more attacks), Nivelle again and again ordered attacks that gained little ground. Desperate field officers led the cornflowers to certain death out of the trenches. Informed secretly by high commanding generals, the government tried to persuade Nivelle to put an end to this bloodbath immediately. As he refused to follow this wise advice, he was sacked and replaced by General Pétain on 11 May. The situation was critical: As early as late April, troops mutinied (half of the frontline regiments of the French army), accepting holding the trenches but refusing to attack anymore. Pétain faced the worst nightmare of a commander: defeat and mutiny. He decided to punish the mutineers with limited harshness and to forbid any large offensive. He said he was waiting "for the tanks and the Americans." The Chemin des Dames proved to be "a ridge too far" for the 271,000 fallen cornflowers.

Gilles Boué

See also: The Somme; Verdun; World War I

References and further reading:

Miquel, Pierre. *Les Poilus.* Paris: Plon, 2000.

Nobécourt, Roger. *Les Fantassins du Chemin des Dames.* Bertout: Luneray, 1983.

Pédroncini, Guy. *Les Mutineries de 1917.* Paris: Perrin, 1984.

Ch'i, Chi-kuang (Qi, Jiguang) (1528–1588)

One of the most innovative of Asian generals. In 1555, General Ch'i Chi-kuang was transferred to Chekiang, a coastal province and frequent target of Japanese pirates. To deal with the pirates, Ch'i organized a new army and developed new tactics. Among the tactics developed were fighting stances for the two-handed sword. He also developed a tactic called the Mandarin Duck Formation that was instrumental in defeating the pirates. The formation consisted of four soldiers wielding 12-foot lances, preceded by four soldiers, two carrying shields and two carrying bamboo trees complete with upper branches. The lancers were followed by two soldiers wielding tridentlike weapons from which arrows propelled by gunpowder could be fired. With support staff, the squad was made up of 12 men. Using such tactics, Ch'i inflicted a great defeat on the Japanese pirates *(wu-li)* at Taizhou in 1561, capturing the leader and 1,900 prisoners. By 1564, the pirate threat had ended.

In 1568 Ch'i became commander in chief at Chichou in North Chihli. This important command protected Peking and northern China against nomadic raiders from the steppe. Ch'i used this command to train conscripts. He demonstrated his skill in training troops in his book, *Chi-hsiao Hsin-shu* (1580). The book touched on recruitment procedures, pay, personnel assignments, combat formations, weaponry, military etiquette, and the general duties of officers and men, and other aspects of military life. In addition, Ch'i Chi-kuang designed banners for his armies, devised signals for use in combat, and issued orders on courts-martial.

At Chichou Ch'i also developed defensive tactics using "battle wagons"—oxcarts equipped with two pieces of light artillery and manned by 20 soldiers. He also built the first castlelike defensive towers along the Great Wall.

In 1583, Ch'i was appointed commander in chief of Kwangtung province and resigned a year later due to illness. He died 17 January 1588.

Michael C. Paul

References and further reading:

Huang, Ray. *1587, A Year of No Significance.* New Haven, CT: Yale University Press, 1981.

Millinger, James F. "Ch'i Chi-kuang, Chinese Military Official: A Study of Civil-Military Roles and Relations in the Career of a 16th Century Warrior, Reformer, and Hero." Ph.D. diss., Yale University, 1968.

Generalissimo Chiang Kai-shek and other officers standing in the field planning a military campaign, 1937–1943. (Library of Congress)

Chiang Kai-shek (1887–1975)

Nationalist Chinese general and political leader during the Chinese Civil War and World War II. Born 31 October 1887 in China's Chekiang province, Chiang Kai-shek, son of a salt merchant, received a military education in his country's Paoting Academy and Japan's Preparatory Military Academy (1907–1909). Serving briefly in the Japanese army, he returned to China to participate in the successful revolution against the Ch'ing (Manchu) Dynasty (1911–1912). He subsequently joined Dr. Sun Yat-sen's Kuomintang, heading the movement's Whampoa Military Academy (1924) and emerging as its military and political leader (1926) following Dr. Sun's death in 1925.

During the late 1920s, Chiang sought to unify a politically decentralized China. To this end, he launched the Northern Expedition (1926–1928), subduing many of the regional warlords who had dominated much of China since the revolution and establishing a new national government at Nanking. He then turned against the Chinese Communists, undertaking five campaigns of extermination (1930–1935) that culminated in their epic Long March of 1934–1935.

As a military leader, Chiang achieved his greatest international prominence as commander of nationalist forces during China's eight-year war against Japan (1937–1945). Under his guidance, China—despite much of the country's occupation by the militarily superior Japanese and fighting alone for more than four years—avoided defeat and ultimately emerged, along with the United States, Great Britain, and the Soviet Union, a major victor of World War II.

Chiang's military and political fortunes deteriorated rapidly in the late 1940s when China descended into civil war (1946–1949) between the Kuomintang and Mao Zedong's resurgent Communist Party. Although nationalist forces held the initiative early, and would receive some $2 billion in U.S. aid between 1945 and 1949, Chiang's inability to establish strategic priorities and his decision to attack in Manchuria where Mao's forces were strongest led to critical defeats in 1948 and Communist conquest of the entire mainland by 1949. The Nationalist Army and government were incompetent and corrupt, the nation racked by hyperinflation, and the Chinese masses seemed about as glad to welcome the Communists as to see off the discredited Nationalists.

Fleeing to the island of Taiwan with 2 million followers—and not at all welcomed by the Taiwanese, who remembered well that their 1947 protests against the Kuomintang had been harshly repressed—Chiang seemed doomed, but with the outbreak of the Korean War, the United States positioned a fleet between Taiwan and the mainland, and the Communists had to postpone their plans for an imminent invasion.

Chiang established the Republic of China, of which he was elected president five times. Chiang died in Taipei, 5 April 1975, having failed to fulfill his dream of reclaiming power on the mainland. Yet Taiwan today, bustling, emerging into democracy, and prosperous (at one time holding more gold reserves than the United States), might cause some historians to reevaluate Chiang's legacy.

Bruce J. DeHart

See also: Chinese Civil War; Mao Zedong; Sino-Japanese War (1937–1945); World War II

References and further reading:

Crozier, Brian. *The Man Who Lost China: The First Full Biography of Chiang Kai-shek.* New York: Scribner's, 1976.

Furuya, Keiji. *Chiang Kai-shek: His Life and Times.* New York: St. John's University Press, 1981.

Lattimore, Owen. *China Memoirs: Chiang Kai-shek and the War against Japan.* Tokyo: Tokyo University Press, 1991.

Chickamauga, Battle of (18–20 September 1863)

A drawn but bloody battle, typical of so many clashes in the American Civil War. In early September, Confederate commander Braxton Bragg had abandoned Chattanooga, Tennessee, gateway to the Deep South, after a brilliant campaign of maneuver by Union general William Rosecrans. Soon thereafter, General James Longstreet and his corps from Lee's Army of Northern Virginia joined Bragg's forces, and thus, unknown to Rosecrans, Bragg's Army of Tennessee would outnumber the opposing Army of the Cumberland—one of the very few instances in this war when the South had more men than the North.

The two armies faced one another at Chickamauga Creek in northern Georgia. After skirmishing on 18 September, Bragg wanted General (Episcopal Bishop) Leonidas Polk to turn the Union left, and he kept up the pressure the next day and then on to 20 September. Rosecrans kept moving units to his left to meet the strong attack from the Confederate right. Finally, unclear about dispositions, he ordered General Woods to move his division from the center-right side of his line almost at the same time as Longstreet's experienced corps attacked. With a loud Rebel yell and a great deal of energy, the Confederates drove through the opening, destroyed the Union right, and most of the remaining Union army fled, including Rosecrans, back to Chattanooga.

Bragg, rather than promptly following up the fleeing Union forces, instead sought to attack the remaining Union troops on Horseshoe Ridge at the Snodgrass House under General George "the Rock of Chickamauga" Thomas. Thomas and his men held the hill under fierce assault. That night, Thomas retreated in good order to Chattanooga, and Bragg appeared content to occupy the surrounding hills and wait.

Charles Dobbs

See also: American Civil War; Bragg, Braxton; Rosecrans, William Starke

References and further reading:

Abbazia, Patrick. *The Chickamauga Campaign, December 1862–November 1863.* New York: Gallery Books, 1988.

Cozzens, Peter. *This Terrible Sound: The Battle of Chickamauga.* Urbana: University of Illinois Press, 1992.

Woodworth, Steven E. *Six Armies in Tennessee: The Chickamauga and Chattanooga Campaigns.* Lincoln: University of Nebraska Press, 1998.

Children and War

Histories of war generally include the stories of battles, commanders, politicians, and military equipment. However, the effect of war on children has been woefully neglected by academics, professionals, and humanitarian agencies, none of whom consider children an integral component of war. The problems engendered by schooling disruption due to war, for example, have eluded historical inquiry, scarcely being acknowledged or recognized. Yet, obviously, wars have profoundly affected children. The disruption that war causes in their lives is often irrevocable and long-term. War causes

Angolan children in a military parade during the civil war in Angola, 1976. (Hulton/Archive)

family transformation, displacement, and ultimately a cessation of schooling that reverberates negatively throughout future generations.

According to UNICEF executive director Carol Bellamy, more children are involved in war and conflict now than at any other time in history. The most overlooked element of war is the role played by the loss of schooling. The term *schooling* defines prescribed years of formal academic and vocational programs taught by certified teachers through state-supported curricula. Education, on the other hand, is a lifelong experiential process. A child will be considered anyone up to age 18. One-sixth of the world's population consists of children and youths ages 10 to 19.

Throughout the twentieth century, but especially since 1945, political, religious, ethnic, and intrastate conflicts have become the catalyst for many educational problems. In the 1990s, more than 2 million children have died, nearly 5 million have been disabled, and over 1 million have lost or been separated from their parents; some 10 million children have been psychologically traumatized due to war.

Second only to separation from their parents is the loss or disruption of schooling.

First, the capture of the physical spaces of schooling takes place. Historically, school buildings are affected almost immediately. They are confiscated and used by the combatants as barracks, hospitals, or for administrative purposes. In Alsace Lorraine, for example, schools were used as barracks throughout World War I. And when the German High Command formally surrendered to the Allies in May 1945 in a Rheims, France, schoolhouse used as headquarters by the Supreme Allied Command, apparently no one wondered what the children of Rheims were using for schools. In most cases, however, school buildings are simply sabotaged or irreparably damaged, or are destroyed. In Mozambique's lengthy civil conflict more than 45 percent of children have been deprived of schooling because more than 50 percent of the schools were destroyed.

Teachers are removed, become forced laborers or soldiers, or are simply killed. In Posnan, Poland, during World War II, all the Jewish professors were removed and never heard from again. Teachers are often replaced by untrained appointees of the new regime so that children can be indoctrinated. As the conflict continues, teacher-training programs are targeted so that new teachers learn to espouse the prevailing ideology. As well, resource deprivation such as lack of school supplies or the destruction of whole libraries (as in the case of the University of Sarajevo) and laboratories invariably impairs the quality of education available to children living through war.

Second, a number of social and cultural factors are associated with schooling disruption. Family and parents play an important role in schooling. Often a child's schooling ends if a parent or guardian dies or is deported. Elder siblings often have to provide for younger ones. Over half a million children between ages 7 and 16 in Zambia do not attend school because they need to look after their families. The paucity of food, lack of health and dental care, and the subsequent malnutrition lead to prolonged school absences, as can lack of transportation to the school. Children are forced to walk but they need to traverse bombed roads and land mines. On their way to school, girls are raped and both boys and girls are abducted. Consequently some parents keep their children home.

Third, cultural disruptions, such as the imposition of a new language, curricular changes, and training programs, can be profound. In many instances teachers are not trained in the victor's language, or if they are, they then become indoctrinated by the victors. Displacement from their once secure environment results in children becoming victims or combatants. About half of the approximately more than 700,000 displaced Kosovar refugees in 1999 were children. Other children become sex slaves or child soldiers. Some boys as young as age seven are forcibly recruited into armies. One young boy was forced to burn his hut and decapitate his parents. This type of horrendous trauma ensures that children will learn to distrust adults.

Schooling disruption due to conflict situations can last for years. In Europe during World War II, schooling by 1943 was at a standstill. In Korea, schooling was denied to many children from 1950 to 1953. In Afghanistan, 3 million school-aged children are deprived of their schooling. Somalia's organized system of learning ceased entirely. Nearly half of Mozambique's children have been deprived of their schooling. Nigerian schooling ceased in 1966–1967 and 1969–1970. In Kuwait formal schooling ceased from July 1990 to 26 February 1991 during the Iraqi occupation. At the end of the twentieth century in Kosovo, schooling for refugee children was nearly nonexistent.

Postsecondary schooling also ceases during conflict situations. In the Netherlands, Delft University, despite having students registered, closed as early as 1941 during the German occupation, and remained so until World War II ended. Across the world, in Cambodia the Royal Phnom Penh University was closed for four years. The schooling disruption pattern thus is sadly obvious. The list of schools closed due to conflict is seemingly unending.

It takes at least two generations for an educational system to recover to where it stood before the conflict began. Some children never recover from their war experiences. Several generations of war-affected children have been written off by psychologists and psychiatrists as unstable, thus unable to lead normal lives.

One of the most destructive by-products of armed conflict, whether it is religious, ethnic, or political, is in the effects of involving generations of children in war either accidentally or purposely. The disruption of schooling and related violations of the rights of the child constitute a threat to global security. These actions need to be considered by military policy makers before a commitment is made to go to war. A proactive, child-centered approach should be in place to counter the disruption of schooling and the subsequent horrendous treatment of children. An addendum to the Geneva Convention, stating that disruption of schooling is a human rights violation and that schools should enjoy the same safety zone protection as hospitals, could be adopted by the United Nations. Such action might well be the single most important step to ameliorate the war on children.

Annette Richardson

References and further reading:

Abdi, Ali. "Education in Somalia: History, Destruction, and Calls for Reconstruction." *Comparative Education* (November 1998).

Ajdukovic, Marina, and Dean Ajdukovic. "Impact of Displacement on the Psychological Well-Being of Refugee Children." *International Review of Psychiatry* (August 1998).

Bellamy, Carol. "Saving Children, Stopping War." *Christian Science Monitor* (18 December 1995).

Harp, Stephen. "War's Eclipse of Primary Education in Alsace-Lorraine, 1914–1918." *Historian* (Spring 1995).

Machel, Graça. "Impact of Armed Conflict on Children." New York: United Nations Children's Fund, 1996.

Mazower, Mark. "Children and the Aftermath of War." *History Today* (June 1996).

Moreno, Zerka. "Healing the Hidden Wounds of War." *Journal of Group Psychotherapy, Psychodrama & Sociometry* 42: 4 (1990).

Richardson, Annette. "Introducing the Richardson Schooling Disruption Model." *Educational Practice and Theory* 21: 2 (1999).

———. "Waging War on Children and Youth through Schooling Disruption." *Paedagogica Historica* 35: 3 (1999).

———. "Social Injustice: The Irony of African Schooling Disruption." In *Cross Cultural Perspectives in Child Advocacy,* ed. Ilene R. Berson, Michael J. Berson, and Bárbara Cruz. Greenwich, CT: Information Age Publishing, 2001.

Wilhelm, Ron. "*Espane Neustra:* The Molding of Primary School Children for Fascist Spain." *Journal of Curriculum & Supervision* (Spring 1998).

Chillianwallah (1849)

The high point for the Sikhs in the second Sikh War of 1848–1849 when they fought the opposing British Army of the Punjab to stalemate. The local Sikh leader, Shere Singh, fought a series of blocking actions in late 1848 against the hastily formed British Army of the Punjab, commanded by Lord Gough, as it commenced crossing the Chenab River from the south. Gough delayed while waiting for reinforcements, but in January 1849 Sikh advances freed a force under Chuttar Singh to join Shere Singh. Gough then moved quickly northward toward the Jhelum River in pursuit of Shere Singh's army, hoping to force a speedy outcome to the campaign. Shere Singh positioned his force of 30,000 men and 62 guns in a line west of, and enveloping, the town of Chillianwallah, ready to surprise the British who he knew must halt at the wells there. On arrival, Gough was partly aware of the Sikh dispositions and intended to move his 12,000 troops and six batteries of guns into battle the next morning, 13 January 1849.

Some Sikh guns fired too early, before the British had fallen out, placing Gough in a very dangerous situation: either to advance immediately into an afternoon and night battle without reconnaissance, or attempt to withdraw in the face of an aggressive Sikh army four times his strength. He chose to attack the center of the Sikh line. His 12 infantry and two cavalry regiments lined up along a three-mile front but were still flanked by the Sikhs at both ends.

The British units attacked at a disadvantage through thick scrub. One infantry regiment was decimated, four others were repulsed, and a cavalry regiment routed, but by evening the Sikhs had withdrawn to the River Jhelum under pressure. Lord Gough had to withdraw to the wells for water, allowing the Sikhs to regain their ground during the night. However, heavy rainfall prevented further engagements, and Gough fortified his camp and awaited reinforcements.

Michael Hyde

See also: British-Indian Army; Gujerat

References and further reading:

Cook, H. *The Sikh Wars: The British Army in the Punjab, 1845–1849.* London: Leo Cooper, 1975.

Kar, H. C. *Military History of India.* Calcutta: Firma KLM Private Limited, 1980.

Chindits

British raiding forces in World War II northern Burma. Their name being a corruption of a Burmese term for lion, the Chindits were originated by Orde Wingate, a British officer with a flair for unconventional warfare. Wingate was convinced that a force could operate at will behind and in an enemy's perimeter while depending on radio links for coordination, and aviation for supply and transport, thus overthrowing the orthodoxy of the defending force enjoying interior lines of control.

This doctrine was put into action with LONGCLOTH, a brigade-sized operation intended as an accompaniment to a larger British counteroffensive against Japanese-occupied Burma. Though the larger operation was cancelled, LONGCLOTH went forward in February 1943, for morale-building purposes as much as anything else. Despite losing a third of his force outright, having to leave the wounded behind due to lack of transport, Wingate and his men were held up as an example of how British troops could defeat the Japanese on their own terms.

LONGCLOTH demonstrated enough success to mandate a corps-sized mission called THURSDAY. Mounted in 1944 with American and Chinese cooperation to reopen a land supply route to China, the so-called 3d Indian Division had its own air component in the form of the 1st Air Commando, providing more firepower and the ability to evacuate the wounded.

While THURSDAY succeeded in disrupting Japanese logistics during the Imphal-Kohima battles, the mission of reopening a road to China languished, as Allied strategy abandoned the idea of launching an invasion of Japan from the Asian mainland.

After Wingate's death in an air crash, the Chindits were placed under Joseph Stilwell's command, where excessive use as conventional infantry, while lacking the firepower of a regular infantry division, effectively destroyed the unit.

The Chindits' legacy is as a demonstration of air-ground cooperation granting operational flexibility, and an example of endurance in combat in some of the world's worst terrain. The actual contribution of the Chindits is still debated.

George Shaner

See also: Imphal and Kohima; Stilwell, Joseph Warren; Wingate, Orde

References and further reading:

Allen, Louis. *Burma: The Longest War, 1941–1945.* New York: St. Martin's Press, 1984.

Bidwell, Shelford. *The Chindit War: Stilwell, Wingate and the Campaign in Burma.* New York: Macmillan, 1980.

Chinese Civil War (1927–1949)

A long-running conflict that resulted in the victory of the Chinese Communist Party. On 4 May 1919, students in Beijing protested provisions of the Versailles Peace Treaty (to end World War I) that would award the former German concession in China to Japan; it was the beginning of modern

Chinese nationalism. The ensuing May 4th Movement helped revive the fortunes of the Nationalist People's Party, the Kuomintang, and helped start the Chinese Communist Party. For the next several years, the two groups pursued separate and not particularly successful paths. Several years later, Kuomintang leader Sun Yat-sen entered into an agreement with the Communist International (Comintern) through Michael Borodin. The two nationalist groups cooperated and received assistance from the Soviet Union and its agents.

In summer 1926, Chiang Kai-shek, who came to power in the Kuomintang after the death of Sun, led the Northern Expedition to smash warlords and reunify China. Chiang already was beginning to turn on the Chinese Communists while continuing to proclaim his support for the alliance with the Soviet Union. Then in April 1927, Chinese authorities raided the Soviet embassy, seized allegedly incriminating evidence, and on 12 April, Chiang and local criminal gangs in Shanghai turned on the so-called Urban Communists, and the resulting slaughter signaled the end to the alliance.

From 1927 to 1949, the Nationalists and Communists fought to control the mainland; there was something of a respite during the anti-Japanese resistance from 1937 to 1945, but the civil war began anew soon after Japan surrendered in September 1945.

There were four phases to this conflict.

Initially, Mao Zedong with Ju De established a refuge in the mountainous southeastern China. Mao established his Soviet *republic* on the border of provinces, recognizing that this permitted the governors of each province to blame the other for failing to attack the Communists. In late 1931, Chiang Kai-shek began the first of a number of extermination campaigns (the number varies since there were not always clear beginnings and endings to the campaigns). The first four campaigns did not succeed; Mao used sound guerrilla tactics to isolate Nationalist units, overwhelm them, escape around them, and generally frustrate their advance. In late 1933, Chiang, with the aid of German advisers, engaged in a *blockhouse* strategy: Before advancing, Nationalist troops would construct a series of blockhouses and connecting lines, all to limit Communist mobility, and gradually constrict the base area.

In October 1934, Mao and the Communists began the epic Long March. Chiang's blockhouse strategy worked, the Communist base area had shrunk too much to remain viable, and Mao determined to flee the southwest through warlord troops to escape the tightening net. For the next year, the Communists marched through southern and then southwestern China, through western and then northwestern China where few ethnic Chinese had walked before. It was an incredible story of human fortitude and determination, and in October 1935, Mao's greatly reduced force had marched some 6,000 miles and fought more than 150 battles and skirmishes, and settled into caves in Yan'an near the Great Wall and nearer to the Japanese whose aggression had carried them from Manchuria into Outer Mongolia and China proper. The Long March was an epic story of human endurance, but it also was a great defeat. Mao began with 80,000 troops and entered Yan'an with fewer than 10,000, although many, to be sure, remained behind to foment rebellion in Nationalist-controlled territory. Also, if one tracks the Long March, Mao and his colleagues sought to turn in to China, especially South China, and Chiang's resolute pursuit forced them almost literally "off the board." However, if one employs the strategies of the ancient Chinese game of go, rather than the more Western game of chess, one controls the board by being in a secure position on its side, and hence Mao may have inadvertently actually strengthened his situation.

The third phase was the long war against Japan. Both the Nationalists and Communists carefully watched one another rather than devoting their entire energies to anti-Japanese resistance. When Chiang tried to force several warlord generals to engage in a final extermination campaign, they captured him at Xian and forced him to enter into a united front against the Japanese, but Chiang's agreement was more fiction than fact. The Communists used the facade of a national war of resistance to revolutionize the peasantry and expand against both the Japanese and Nationalists; the Communists would later claim great areas of control based on these foci of peasant resistance.

When the war ended, the Nationalists were located mostly to the southwest; the Chinese Communists claimed base areas throughout North China, and the Soviets had occupied Manchuria in a series of lightning thrusts against the greatly weakened shell of the once strong Japanese Kwantung Army. Advantages seemingly resisted with the Nationalists: an army of more than 3 million with many American-equipped, American-trained divisions; an air force; and a small navy. But the troops were at best uninspired and Chiang was determined to occupy Manchuria before the Chinese Communists; he created his own grand strategic trap. American air and naval power helped transport the best troops in the Nationalist army over and around Communist positions in North China to Manchuria. Mao and his military commanders moved into Manchuria, and forced hesitant Nationalist generals to retreat to the cities. Isolated, they were soon forced to rely on resupply by air. In October 1948, Nationalist garrisons in Manchuria surrendered to the Communists, and Mao's forces gained modern weaponry and munitions. The Communists continued the propaganda war and took advantage of Nationalist corruption in the country's center. In January 1949, they won a grand victory at the Battle of Huai-hai, where they forced the Nationalist advance army to dig in and then went around that army to attack and destroy the slow-moving relief force, and then returned to the still-dug-in vanguard. Nationalist defeat was now inevitable. In January 1949 the Communists gained control of North China; in April they forced the crossing of the Yangtze River, which could and should have been better defended; and in May they entered Shanghai. Late in the year they had occupied southern, southwestern, and western China. Chiang and his remaining forces fled to the island redoubt of Taiwan.

The last act of the civil war has been delayed at least 50 years. As the Communists prepared to attack Taiwan in late spring 1950, North Korea invaded the south, the U.S. Navy entered the Taiwan Straits, and an independent Chinese government continues to exist on the island of Taiwan.

Charles Dobbs

See also: Chiang Kai-shek; Mao Zedong

References and further reading:

Bianco, Lucien. *Origins of the Chinese Revolution, 1915–1949.* Trans. Muriel Bell. Stanford, CA: Stanford University Press, 1971.

Chassin, Lionel Max. *The Communist Conquest of China: A History of the Civil War, 1945–1949.* Trans. Timothy Osato and Louis Gelas. London: Weidenfeld & Nicolson, 1965.

Wilson, Dick. *The Long March, 1935: The Epic of Chinese Communism's Survival.* New York: Penguin, 1982.

Chinese Imperial Wars (200 B.C.E. to 1800 C.E.)

Cycles of Chinese military power show alternating Chinese and barbarian control. During the 2,000-year history of imperial China prior to the coming of the West, there have been considerable differences in the ability of Chinese rulers to maintain an effective military and project power. While many of the stronger dynasties, including Qin (221–207 B.C.E.), Han (206 B.C.E.–220 C.E.), and Tang (618–906 C.E.), have not only maintained large and relatively effective armies, and advanced positions far beyond Chinese frontiers, such periods of Chinese power have alternated with periods of barbarian rule, and with periods in which even large and theoretically powerful Chinese dynasties have been forced to remain strictly on the defensive, usually for geopolitical reasons.

Chinese military power has thus fluctuated considerably over the centuries, but in general, strong dynasties and strong rulers have controlled the barbarians, while in other periods barbarians have controlled China. Overwhelmingly these barbarians were peoples of Central Asia, with southern groups rarely playing a significant role. Thus, for obvious reasons, much traditional Chinese military activity focused on China's northern frontiers, a fact reflected, among other things, in the existence of the Great Wall. This is a Ming Dynasty (1368–1644) successor to a series of frontier fortifications whose origins date back well before the imperial period. Originally, such walls delineated territory within China, but they also came to mark the border between settled China and the pastoral nomadism of most of China's neighbors to the north and northwest.

Classically, Chinese dynasties sought to control foreign peoples indirectly, through threats, bribery, and turning one against the other (*yi yi zhi yi* [using the barbarians to govern the barbarians]), with direct military intervention usually a final resort, since most Chinese rulers were only too aware of the limitations of their power when it came to campaigning in the deep steppe. While the main goal of this policy was a secure northern frontier, Chinese imperial dynasties were also interested in trade. From Han times on, they wanted to reach out to the extreme northwest, to control as much as possible of the so-called Silk Road, the trade route for Chinese silk and other precious commodities, extending from China to the world of the eastern Mediterranean, to Rome and Byzantium, the Arab caliphates, and later the trade controlled by the Italian cities.

After uniting China in 221 B.C.E., the first emperor of China, Qin Shi Huangdi, constructed extensive frontier fortifications and fought with the Xiongnu and other Inner Asian peoples outside the new barriers. The Qin did not last long, and were soon succeeded by the Han Dynasty, which roughly parallels the era of and the achievements of the Roman Republic and Roman Empire. The Han engaged in the wars typical of the stronger Chinese dynasties. After a period of diplomacy, the Han actively fought the Xiongnu, and sought to divide and conquer them. Later the Han extended their power as far west as the Ferghana valley.

As the Han Dynasty weakened, the yang of Chinese imperial power yielded to the yin of barbarian reaction. Although the northern state of (Cao) Wei (221–264), and later the Western Jin Dynasty (265–317), were at first able to maintain an effective defense, starting in the early fourth century, northern barbarians overran much of northern China. For three centuries, barbarians dominated almost completely, and many Chinese moved south, to new lands and new ways of life, to escape the warfare and deprivation. Most important of the barbarian groups ruling states in the north were the Xianbei, founders of the Toba Wei Dynasty (386–534), which effectively united the north under a single set of rulers and dominated southern states as well. The Toba were the originators of the *fubing* system of mixed Chinese and barbarian militias that later became the basis of early Tang (618–906) military power.

Almost inevitably, the pendulum swung back, and native Chinese power was restored under the Sui (589–618), which reunited China. It actively sought to control trade routes to the northwest and sent missions to Vietnam, Taiwan, and even reached out to Japan. It also invaded Korea, which led to its downfall.

In a pattern reminiscent of the transition from Qin to Han, in which a short, harsh, unifying dynasty fell quickly and was followed by a longer peaceful period, Sui was quickly succeeded by the Tang. This dynasty completely dominated its surroundings from its founding in 618 until the middle of the eighth century. While ultimately no more successful in Korea than the Sui, the Tang did make extensive conquests in Central Asia, and reached the Caspian Sea. In 751, the Tang suffered a major defeat by the Arabs on the Talas River in Central Asia, even as their supply routes were threatened by Tibetans and other barbarian groups in the rear. Shortly thereafter, the dynasty nearly collapsed in the wake of the great An Lushan rebellion.

An was a Sogdian general in Tang service and commanded an army largely composed of Turks and other barbarians. To suppress his rebellion other barbarian groups, including Uighurs and Tibetans, had to be mobilized, as the once strong *fubing* system inherited by the Tang disappeared entirely. Once again, the yang of Chinese power began to give way to the yin of barbarian power. By late Tang, the very survival of the dynasty was dependent upon one barbarian group, the Turkic Shato.

When the Tang finally collapsed in the early tenth century, there was another, but this time brief, era of disunity. Once again, China fell under the control of competing ruling houses, the majority of them of Turkic Shato extraction.

Although the Song (960–1279) managed to reunite China again in 960, the new dynasty was never able to establish effective control everywhere. The northeast, which had been ruled by separatist regimes during late Tang times, now fell under the control of the Khitan, who established their own dynasty, that of the Liao (907–1125). Almost simultaneously, the Tanguts, of uncertain ethnic affiliation, established their own separatist state in the northwest, later known as Xixia (destroyed 1227). These competing regimes robbed Song of defensible frontiers, to a large extent, and limited its access to urgently needed Central Asian warhorses. This meant that Song armies, often well armed technologically, including with the first gunpowder weapons, were overwhelmingly composed of relatively immobile foot soldiers, while its opponents mostly fielded cavalry.

Song economic, as opposed to military, power rested on the great medieval economic revolution, and the changed locus of Chinese power in the southeast where reliance on water and canals, a development from late Tang times on, frustrated the horse peoples of Inner Asia. Nonetheless, the Song lost control of the North China plain in the first third of the twelfth century, and retreated to the Yangtze basin. Henceforth the Tungus Jurchen, through their Jin Dynasty (1125–1234), controlled traditional northern China, and engaged in war and diplomacy with even more barbarian tribes to their north. As Jin power developed, Song China continued to rely on masses of foot soldiers and on fixed fortifications for its defense, but also developed an effective navy, operating both on the rivers and lakes of China and on the high seas.

In 1234 the great tide of Mongol power finally overwhelmed the Jin after conquering Xixia in 1227. Warfare between the Mongols and Song began almost immediately, and continued until 1279, when the last embers of Song resistance were crushed in the extreme south in the great naval battle of Yaishan, and China was finally reunited under Mongol rule. By this time Kublai Khan had established the Yuan Dynasty (1279–1368) in China, and in surrounding parts of Central Asia, including what is now Chinese Turkistan.

Under Kublai a Chinese dynasty, for the first time, conducted large-scale overseas expeditions, including two attacks on Japan, another into Vietnam, and another against Java. Such expeditions, carried out by fleets that were truly huge by the standards of the time, were unprecedented in Chinese history. Overland, the Mongols not only penetrated southern areas such as Yunnan, henceforth a Chinese province, and Burma, but united Tibet with China for the first time in its history, although the union was personal between the Mongol rulers of China and the principal Tibetan religious authority.

In 1368 Zhu Yuanzhang, a peasant from the south, finally defeated the Mongols and Chinese pretenders to power, and established the Ming Dynasty. Under Zhu and his immediate successors, once again the pendulum swung toward Chinese power, and the Ming regained control over all of China, as well as parts of Mongolia and Manchuria, although attempts to penetrate the deep steppe failed. Later one Ming emperor was even captured by the Mongols, who also threatened the later Ming capital of Beijing at one point. At sea, expansion continued as the eunuch admiral Zheng He explored the Indian Ocean all the way to East Africa and Southeast Asia.

Eventually the Ming weakened and internal rebellion threatened. At the same time a barbarian power, the Jurchen, who had once ruled the Jin Dynasty, later called the Manchus, sought to set up a Chinese-style regime in Manchuria. During this period China's response to its various foreign threats was one of the most massive efforts at the construction of frontier fortifications in world history, the Great Wall. It tied up huge resources and required an enormous garrison, but still proved quite ineffective in preventing foreign attack.

Taking advantage of internal breakdowns, in 1644 the Manchus, by then the Qing Dynasty (1644–1911), seized Beijing, and soon thereafter defeated remaining Ming loyalists, including mounting a massive overseas expedition with the help of the Dutch to conquer Taiwan. It became a part of China for the first time in its history.

The Qing restored the power of China, albeit under foreign rule, and under the Kangxi emperor (r. 1662–1722) China negotiated successfully with the advancing Russians. It neutralized the Mongols, conquering all of Mongolia in the mid-eighteenth century, seized control over Tibet, and fought with the Vietnamese. However, with the death of Kangxi's grandson, the Qianlong emperor, in 1799, and the end of the Napoleonic Wars soon thereafter, China would face a century of foreign, Western-based imperialism, several decades of costly invasion and civil war, and now five decades of Communist control. Nonetheless, the history of Chinese imperialism continues to be of concern to the peoples on China's borders.

Charles M. Dobbs

See also: Great Wall of China; Han Wudi; Kangxi; Kublai Khan; Li Shihmin; Manchu Expansion, Wars of; Mongol Empire; Qianlong; Qin Shi Huangdi; Talas River, Battle of; Yangzhou, Siege of; Yonglo

References and further reading:

Barfield, Thomas J. *The Perilous Frontier, Nomadic Empires and China.* Cambridge, MA, and Oxford, UK: Basil Blackwell, 1989.

Deng, Gang. *Maritime Sector, Institutions, and Sea Power of Premodern China.* Westport, CT, and London: Greenwood Press, 1999.

Elvin, Mark. *The Pattern of the Chinese Past.* Stanford, CA: Stanford University Press, 1973.

Fairbank, John King, and Frank A. Kierman, Jr., eds. *Chinese Ways in Warfare.* Cambridge, MA: Harvard University Press, 1974.

Hsu, C. Y. *The Rise of Modern China.* 6th ed. New York: Oxford University Press, 2000.

Hucker, Charles O. *China's Imperial Past.* Stanford, CA: Stanford University Press, 1975.

Waldron, Arthur. *The Great Wall of China: From History to Myth.* Cambridge, UK: Cambridge University Press, 1990.

Chinese Military (Twentieth Century: History, Organization/Structure)

The twenty-first century's dominant Asian military power? Twentieth-century Chinese military history is the story of three separate forces: the Western-style army of the Qing Dynasty, the Nationalist or Kuomintang army, and the People's Liberation Army (PLA) that has developed out of the Communist revolution.

Defeat in the First Sino-Japanese conflict (1894–1895) finally forced the Qing Dynasty to create up-to-date forces capable of standing up to the Japanese and European armies, though time was running out for imperial rule in China. The armies survived the Qing, and when central authority collapsed in 1911 and the new Chinese republic proved unable to rule effectively, regional commanders struck alliances with local civilian elites, leading to the so-called War Lord period.

While the strongest leaders of the old military might have reunified the country—and one of them, Yuan Shikai, did, if briefly and superficially, before his death in 1916—this avenue of Chinese history had definitively been closed not later than the early 1920s. By that time, a series of wars had ensured that the Manchurian warlord, Zhang Zuolin, who enjoyed Japanese support, had predominance in northern China, but little influence outside it, where a stalemate between various local forces and foreign powers continued.

This situation allowed the revival of the political fortunes of the founder of the Chinese republic, Sun Yat-sen, and the creation of the Central Army of his Kuomintang regime. Like the warlord armies, the Nationalist army was modeled on Western and Japanese examples. The difference was that the new army created an ideologically motivated and loyal officer cadre via the graduates of the Whampoa military academy, under the direction of Chiang Kai-shek. This was the army that defeated the weakened northern regional forces and pursued the nascent Communist guerrilla forces to near annihilation.

Structural problems remained, though, even after the political victory following the Northern Expedition of 1926. Many of the old regional forces were only partially assimilated into the new army, and China remained so impoverished that the 2 million–man army fielded by 1937 lacked the heavy equipment to fight the Japanese army on equal terms. Chiang was also primarily concerned with eliminating the Communists, appeasing Tokyo in hopes that the Western powers would eventually confront Japan over its aggression.

Eventually, China's internal political dynamics generated sufficient pressure to force Chiang to confront Japan, leading to full-scale war in 1937. As expected, the Nationalist regime could only inflict serious wounds on the Japanese, not a repulse, and eventually fell back to Chungking. This allowed a resuscitation of the Communist movement.

An ally of the Kuomintang in the mid-1920s, the 1930s saw the transformation of the Chinese Communist Party's military from an urban workers' militia into a rural guerrilla force under the direction of Mao Zedong and his circle. While the Kuomintang did develop an internal ideology, tending toward authoritarianism, even fascism, Mao made

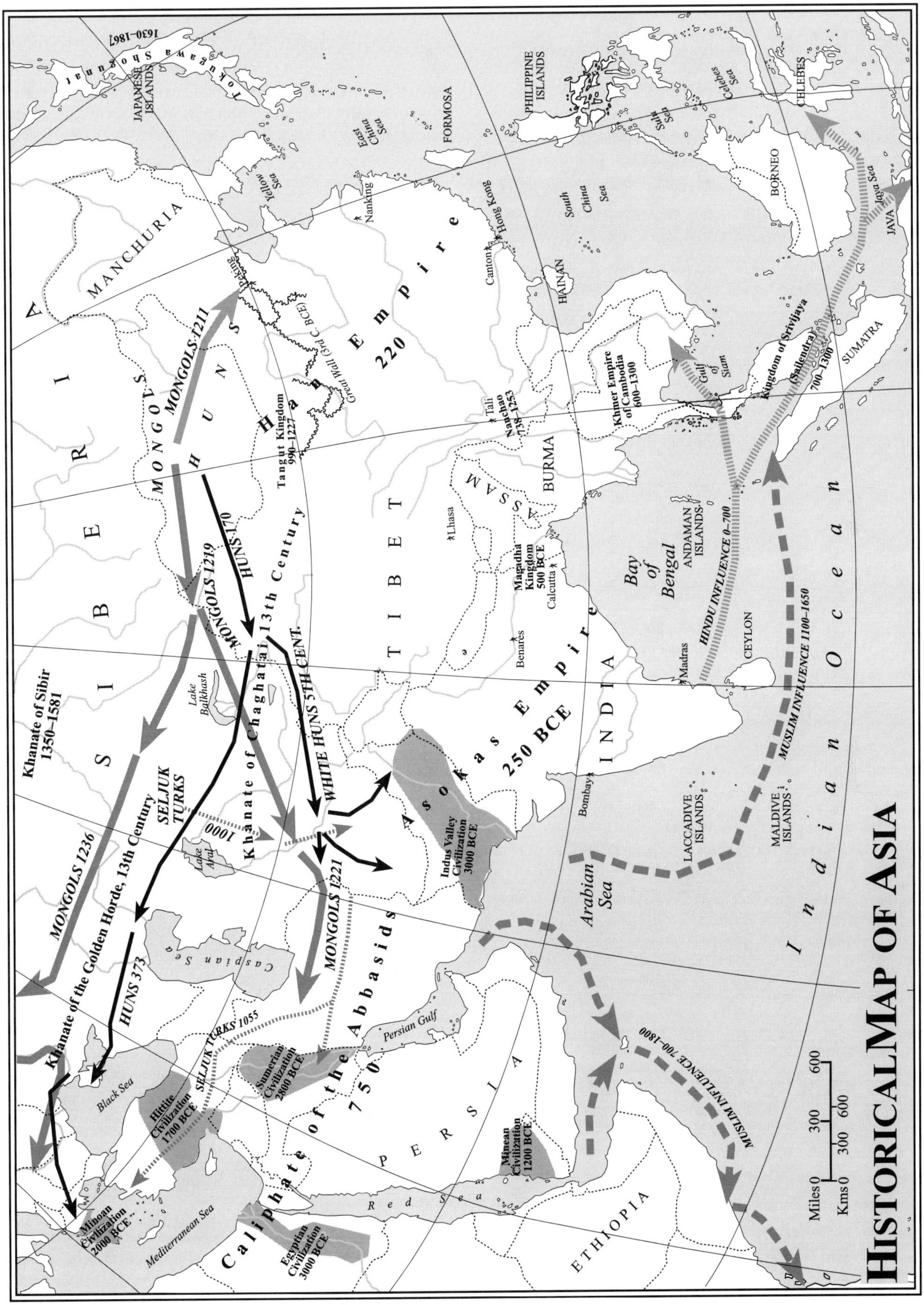

HISTORICAL MAP OF ASIA

ideological commitment the core of his military system, along with the belief that correct political and social practice could be a force multiplier. This was accompanied by the creation of an operational system conceptualizing how a revolutionary army could defeat a better-armed conventional force, based on locally raised part-time units.

As it was, neither the Kuomintang nor the Communists mustered the forces or the skill to score operational victories over the Japanese. Their armies remained forces-in-being while Anglo-American and Soviet forces defeated the empire.

With the return to civil war, Chiang's American-equipped forces were gradually defeated and exiled in 1950. Arguably, the PLA's victory stemmed from Chiang's rail-bound forces having reached a culminating point in their Manchurian offensive, and never regaining the initiative, as opposed to any inherent virtues of Mao's military system.

From there, the Chinese army committed to Korea was able to take devastating advantage of the United Nations force's operational overstretch. Still, quick tactical movement and strong morale were ultimately undone by serious limitations in logistics, professional expertise, and modern equipment. These faults and virtues remained in evidence until Mao's death, when serious reform of the PLA finally became possible.

The heritage of the PLA, like all the modern Chinese militaries, is to regain Chinese sovereignty and prestige. The common problem of all has been to overcome less-than-popular regimes, economic weakness, and the difficulty of confronting better-equipped and -trained opponents. This problem will persist into the foreseeable future.

George R. Shaner

See also: Boxer Rebellion; Chiang Kai-shek; Chinese Civil War; Chinese Revolution; Guerrilla/Partisan/Irregular Warfare; Japanese Colonial Wars; Korean War; Lin Biao; Mao Zedong; Peng Dehuai; Sino-Japanese War (1894–1895); Sino-Japanese War (1937–1945); Tibet, Chinese Occupation of; Yuan Shikai

References and further reading:

Dreyer, Edward L. *China at War, 1901–1949.* New York: Longman, 1995.

Mulvenon, James C., and Richard H. Yang. *The People's Liberation Army in the Information Age.* Santa Monica, CA: Rand Corporation, 1999.

Nafziger, George F. *The Growth and Organization of the Chinese Army, 1895–1945.* Westchester, OH: Nafziger Collection, 1999.

Zhang, Shu Guang. *Mao's Military Romanticism: China and the Korean War, 1950–1953.* Lawrence: University Press of Kansas, 1995.

Chinese Revolution (1911–1912)

The revolution that ended the Manchu Dynasty and paved the way for the establishment of the Chinese Nationalist government. The Manchu Dynasty that had ruled China since 1644 tottered to its end on 12 February 1912 when the young emperor, Hsuan T'ung, resigned. For some years there had been an increasing opposition to the Qing Dynasty; Chinese of all political persuasions were discouraged by Japan's apparent ability to leap to modernity and China's apparent inability to do the same, not to mention Japan's humiliating defeat of China in the war of 1894–1895. In particular, students and soldiers had organized a series of revolutionary study societies in central China, and Qing authorities discovered one of these groups of plotters on 9 October 1911. The next day, soldiers in Wuzhang, a city on the Yang-tzu River, mutinied and took control of the city. This mutiny, really caused by fear among the soldiers that they would be linked to the plotters and be executed as rebels, led to similar outbreaks across the country. Within a few weeks, provinces across China had declared their independence of the Manchus, now increasingly isolated in Beijing. Indeed, by early December, all of the southern, central, and even northwestern provinces had declared their independence; power was shared jointly by former government army officers and provincially selected political leaders. This decision to cast off the Manchu overlords was so broadly accepted that there was little need for fighting anywhere.

For a brief period of time it appeared there might be two contending centers of power or perhaps a conflict between them. The Qing turned to a senior military commander (and political adviser), Yuan Shi-kai. Yuan counseled them that the Manchu reign was over, and a hastily convened assembly elected him as prime minister of a provisional government. Yuan's troops, China's only real modern armed force, seized several cities from the rebels and ordered the rebellion to stop.

At about the same time, the real father of the Chinese Revolution, Sun Yat-sen, had been abroad in America and went to Great Britain seeking a loan or British help to prevent Japan from befriending the declining Qing Dynasty and thereby gaining control over China. After returning home, he was offered the position of provisional president and was inaugurated on 1 January 1912 at a provisional capital in Nanjing, farther away from Manchu and Yuan's power. However, Sun soon realized that, lacking an army and a secure base of support, he was vulnerable to Yuan and his dreams of power. So Sun offered to resign and give way to Yuan, assuming that Yuan would accept that he was subject to the power of the people through the provisional assembly.

Yuan soon gained control. While the rebels were stridently anti-Manchu, they were united on little else. They did fear rising Japanese imperialism, and did want a strong China to regain its stature in the world. And Chinese generally recognized Yuan's prestige and power. So Sun resigned,

Prisoners to be executed are escorted by armed guards during the first Chinese Revolution, 1912. (Hulton/Archive)

the last Qing emperor abdicated, and the newly established National Assembly elected Yuan as provisional president and nominally leader of China, although, to be sure, he really was the first among a series of warlords of varying power and control.

Charles Dobbs

See also: Yuan Shikai

References and further reading:

Chen, Jerome. *Yuan Shih-k'ai: Brutus Assumes the Purple.* Stanford, CA: Stanford University Press, 1971.

Liang, Ching-ch'un. *The Chinese Revolution of 1911.* Hong Kong: Joint Publishing Co., 1962.

Shinkichi, Eto, and Harold Z. Schiffrin, eds. *China's Republican Revolution.* Tokyo: University of Tokyo Press, 1994.

Wright, Mary Clabaugh, ed. *China in Revolution: The First Phase, 1900–1913.* New Haven, CT: Yale University Press, 1968.

Chippewa, Battle of (5 July 1814)

First American battle involving a majority of regular U.S. Army troops. British lieutenant general Sir Gordon Drummond commanded about 4,400 men around the western end of Lake Ontario in June 1814, defending the Niagara peninsula against the expected American invasion. On 3 July, about 3,500 Americans under Major General Jacob Brown, including 1,400 under Brigadier General Winfield Scott, crossed the Niagara River, easily and bloodlessly captured Fort Erie, and marched north. In response to this threat, British major general Phineas Riall deployed 2,100 regulars, militia, and Indians in a strong defensive position north of Street's Creek, with the river on his left and woods on his right.

The Americans camped just south of Street's Creek the night of 4 July. Because Brown considered a frontal attack across Street's Creek unwise, he ordered volunteers under Brigadier General Peter B. Porter to take the forest on the British right. Porter met little resistance and had accomplished his mission by 4:30 P.M. on 5 July, but then encountered a regiment of British regulars, which he successfully handled until British artillery drove him back. Scott, meanwhile, was parading 1,500 men in full sight of the British, perhaps to intimidate them, but suddenly wheeled, crossed the creek under fire, redeployed on the north bank, reinforced Porter, and attacked. Riall, shocked by the precision and deadliness of Scott's extraordinarily well-drilled brigade, is supposed to have exclaimed, "Those are regulars, by God!" Scott's charge forced the British to retreat, losing

148 killed and 321 wounded, compared with 60 Americans killed and 235 wounded. Brown rewarded Scott with a brevet major generalship.

Eric v. d. Luft

See also: Lundy's Lane, Battle of; Scott, Winfield; War of 1812

References and further reading:

Barbuto, Richard V. *Niagara 1814: America Invades Canada.* Lawrence: University Press of Kansas, 2000.

Davis, Paris M. *The Four Principal Battles of the Late War: Being a Full Detailed Account of the Battle of Chippeway, Fall and Destruction of the City of Washington, Battles of Baltimore and New Orleans.* Harrisburg, PA: Baab, 1832.

Graves, Donald E. *Red Coats and Grey Jackets: The Battle of Chippawa, 5 July 1814.* Toronto: Dundurn, 1994.

Green, Ernest. *Lincoln at Bay: A Sketch of 1814.* Welland, ON: Tribune-Telegraph Press, 1923.

Chosin/Changjin Reservoir (1950)

A reservoir in the far north of Korea, site of a series of bitter engagements between U.S. X Corps and Chinese People's Volunteers forces. Despite Chinese Communist attacks on advanced Republic of Korea (ROK) units nearing the Yalu River in late October 1950, United Nations (UN) Commander Douglas MacArthur continued his northward offensive, and American and other forces replaced ROK units for the final drive to eradicate communist military power on the peninsula.

However, there was little contact between U.S. Eighth Army on the west and combined X Corps on the east side of the Korean peninsula. Chinese Communist divisions—nearly 300,000 men—had infiltrated UN lines and prepared to attack, although that possibility had been heavily discounted by MacArthur and others caught up in the euphoria of the moment.

The Chinese Communists struck beginning 24 November 1950. They sought to cut UN lines in two, and roll each of the lines up against the coasts and thereby annihilate them. As the Chinese attacked, the U.S. 1st Marine Division continued to advance northwestward as part of MacArthur's final offensive thrust. Then, on the night of 27 November, they were surrounded by three Chinese divisions. Meanwhile, supporting ROK forces were falling back under Chinese pressure.

The 10,000 marines were able to break out and march to the port of Hungnam. The marines had never previously retreated in the long proud history of the Corps, so (as they said), they attacked "in another direction." Ultimately they fought their way through and past seven Chinese divisions. Meanwhile a relief column drove through Chinese lines. U.S. airpower kept the Chinese under continuous attack while dropping necessary supplies to keep the marines armed, fed, and clothed—indeed, airplanes even dropped bridging equipment. In the end, the marines escaped and later reentered the war farther south, but it was a retreat and a desperate one at that.

Charles M. Dobbs

See also: Korean War; MacArthur, Douglas; Peng Dehuai; Walker, Walton

References and further reading:

Appleman, Roy E. *East of Chosin, Entrapment and Breakout in Korea–1950.* College Station: Texas A & M University Press, 1987.

Hammel, Eric M. *Chosin: Heroic Ordeal of the Korean War.* Novato, CA: Presidio Press, 1981.

Spurr, Russell. *Enter the Dragon: China's Undeclared War against the U.S. in Korea, 1950–1951.* New York: Henry Holt, 1988.

Churchill, Sir Winston (1874–1965)

British statesman and wartime leader. Churchill came from a family with a long military tradition (the Marlboroughs), and he was a professional cavalry officer, serving in India (1895–1899) before he resigned to enter politics, though he fought again for six months in France (1915–1916). It is, however, for his passages at the head of the Admiralty (1911–1915 and September 1939–May 1940) and as prime minister of wartime Britain (May 1940–July 1945) that he ranks among the greatest war leaders in British history. This does not mean that he is above criticism in that field: On the contrary, the continuous flow of books on the two world wars, as well as the never-ending stream of biographies of him, always submit his actions to the closest critical scrutiny.

There is nearly unanimous agreement on at least one point: The Gallipoli fiasco of 1915 almost ruined his political career—he himself believed he was "finished." The incompetence displayed was indeed appalling, but Churchill is not alone to blame. In fact, the affair showed a constant trait of his character: He seldom knew where to draw the line between perseverance (a virtue) and obstinacy (a fault), a not unusual trait among great war leaders. Some historians argue that he never forgot the lesson, which explains his prudence over the *second front* and the Normandy landings of the next war.

The complexity of the British political scene between 1918 and 1940 explains both why his talents were sometimes recognized—indeed used—and why he remained a maverick in the eyes of his peers, and also in public opinion. By 1939 he had become the archetype of the warmonger as opposed to the appeaser, and the declaration of war made his inclusion in the Cabinet inevitable, to show that it meant business. "Winston was back" at the Admiralty, where disaster had again to be faced, in Norway in April 1940. One of the ironies of history is that this setback eventually gave Churchill the premiership, on 9 May 1940.

His obduracy ("We shall never surrender") was now an undoubted asset, which galvanized the British population ("their finest hour") and made it accept the full rigors of total war when Hitler's final peace feelers were unambiguously rejected by his War Cabinet in July 1940. This is the period described as "Alone" in his memoirs, the period on which he probably imprinted his greatest mark, and the period in which he benefited from the largest consensual agreement in the population, beyond the old party divides. The vital priority was resisting invasion, and in these particular circumstances he admirably led the country to success. When he had to compromise with the Soviets, with the Indians, and later the Americans, from 1941, his old imperialist, anti-Communist, Edwardian streak was an obvious liability, and the most severe criticism in current Churchillian historiography is that behind the florid rhetoric of the Big Three he could not ignore the fact that at Yalta he was agreeing to a postwar settlement that made Britain lose the peace despite its hard-won place among the victors.

In sum, Churchill's magnificent spirit braced the British people to their finest hour, for which most Britons would consider him to be their greatest leader of the twentieth century. But in hindsight, it can also be asserted that such spirit and such rhetoric masked Great Britain's decline from major power status.

A. Capet

See also: World War I; World War II

References and further reading:

Blake, Robert, and William Roger Louis, eds. *Churchill: A Major New Assessment of His Life in Peace and War.* Oxford, UK: Oxford University Press, 1993.

Charmley, John. *Churchill: The End of Glory—A Political Biography.* London: Hodder & Stoughton, 1993.

Gilbert, Martin. *Winston Churchill.* 8 vols. London: Heinemann, 1980–1988.

Manchester, William. *The Last Lion: Winston Spencer Churchill.* 3 vols. London: Michael Joseph, 1984–1987. (Two volumes are available: Vol. 1, *Visions of Glory, 1874–1932,* and Vol. 2, *Alone, 1932–1940.* Vol. 3, *Defender of the Realm, 1940–1965,* is forthcoming.)

Cimon (c. 510–451 B.C.E.)

Athenian statesman and general who played an active part in building up the Athenian Empire in the period following the Greco-Persian Wars.

Cimon was the son of Miltiades, the architect of the victory at the Battle of Marathon against the Persians in 490. His impressive performance in the victorious sea battle against the Persians at Salamis in 480 led to his election as *strategos,* one of Athens's 10 annual generals. In 478 he helped Aristides to secure the transference of the leadership of the Greek forces from Sparta to Athens and he became the principal commander of the Athenian-led alliance known as the Delian League.

Cimon's first task was to drive out the Spartan general Pausanias—who had been dismissed on suspicion of treason—from Byzantium. He then removed Eion in Thrace from Persian hands (476–475 B.C.E.) and soon after this he won the island of Skyros for Athenian settlers and returned to Athens the supposed remains of Theseus, Athens's legendary hero.

Cimon's greatest triumph took place in 466 when, as leader of an allied fleet of 200 ships, he crushed the much larger Persian fleet near the mouth of the River Eurymedon in Pamphylia and subsequently defeated the Persian king's forces on land. He then returned to the Aegean and drove the remaining Persians out of the Thracian Chersonese. When the rich island of Thasos seceded from the Delian League, Cimon besieged it and forced it to surrender (463).

In 461, Cimon was ostracized. On his return to Athens, he worked for peace with Sparta. When peace was achieved in 451, he once again mounted a big naval expedition against Persia in order to recapture Cyprus. During the siege of the city of Kition, however, he died of sickness or a wound.

Ioannis Georganas

See also: Marathon, Battle of

References and further reading:

Plutarch. *Life of Kimon.* Trans. A. Blamire. London: University of London, 1989.

Cincinnatus, Lucius Quinctius (c. 519–430 B.C.E.)

Roman citizen called upon to save the early Republic during the fifth century B.C.E. Cincinnatus served as one of the two Roman consuls in 460 B.C.E. and as a member of the Senate composed of 300 men representing Rome's aristocratic families. After his one-year term, during which he commanded the army and interpreted and enforced the law, he returned to his three-acre farm to plow the fields and provide for his family. Then in 458 B.C.E., the highland Aequi tribe, pushed by overpopulation down onto the slopes to the northeast of Rome, threatened the republic. The Senate appointed Cincinnatus as dictator with unlimited power to defeat the

Aequians. Upon hearing the news of the threat, Cincinnatus replied that he and his family might not have food the following winter but it was his duty to defend Rome. Leaving his plow behind, he led the Romans against the Aequi. Within 16 days he and his troops defeated the enemy. Hailed as a victor and still vested with unlimited powers, Cincinnatus might have abused his position, as the Roman plebeians feared. But after the victory he relinquished the office of dictator, putting their fears to rest. For his courage and leadership the Senate awarded Cincinnatus a golden wreath. After 60 days he turned over the reins of government and returned to his fields. Cincinnatus accepted the position of dictator again in 439 B.C.E., this time to deal with a plebeian uprising. After the immediate danger passed he once again relinquished the office. Cincinnatus remains a symbol of the virtuous farmer and exemplary citizen willing to defend his country.

Cynthia Clark Northrup

See also: Ancient Warfare

References and further reading:

Gruen, Erich S. *The Roman Republic.* Washington, DC: American Historical Association, 1972.

Scullard, H. H. *A History of the Roman Republic: 753–146 B.C.E.* New York: Routledge, 1991.

Civil Affairs/Military Government

Reconstruction after the battle. Military government is the installation of a martial administration over a region that has been subjected to forcible occupation. Ideally this occurs under the auspices of the Hague and Geneva Conventions, as opposed to an exercise in sanguinary force, until the restoration of civil authority. More complicated is the question of civic action as a military function, the application of military resources to the reconstruction of a population and region traumatized due to war or natural disaster.

Having first been subjected to military government as a client, and inheriting the British distaste for the military, American military government was on an ad hoc basis until World War II. American soldiers acquired on-the-job training by exercising authority in Mexico, the occupied South during the Civil War and Reconstruction, and in the Philippine Insurrection. The question was raised as to whether military occupation could be an avenue to social and political improvement, instead of simply keeping a hostile population in check.

The golden age of military government and civil affairs occurred during World War II and the immediate Cold War period. With the rise of total war ideologies mandating a role for all citizens and the threat of violent social revolution, even the United States began to try to "win the peace" by making over the fascist and authoritarian states defeated by the Allies in World War II into the liberal-democratic image of the victors. The Allies all had their own civil affairs/military governments, but only the United States and the Soviet Union had the resources to rebuild defeated societies in the victor's own image. (German and Japanese *civil affairs* would be almost a contradiction in terms.)

Civil affairs transitioned into counterinsurgency when Soviet premier Nikita Khrushchev made the region an arena of ideological conflict. Edward Lansdale in the Philippines and Gerald Templer in Malaysia became successful pioneers of counterinsurgency. Here, civil action became an integral portion of defeating guerrilla armies, the theory being that the provision of grassroots economic and social services could be a means of winning disputed populations while separating them from the threatened government's military opposition. While not totally original, the concept certainly received a higher profile, plus questioning as to whether this was simply another form of imperialism. Although the French enjoyed some success in Algeria, it was not sufficient to forestall independence for its former territory.

The climax of the strategy came during the Vietnam War where extensive (if shallowly applied) theories of civil action were administered by the American Office of Civil Operations of Rural Development Support (CORDS). Despite the ostensibly coordinated effort and some considerable success, the strategy could not be separated from the inadequacies of the Saigon government or the sheer destructive course of the war itself, making *nation building* a pejorative to the present day.

The irony is that in a world of failed states and transnational disaster, the requirement for military civil action seems as in demand as at the zenith of the Cold War. Civil affairs now revolve more around the mission of peace keeping as opposed to counterinsurgency or overt war. Civil affairs challenges of the early twentieth century include the calls for international coordination, including the role of nongovernmental organizations (NGOs), intense media focus, and the question of cultural imperialism.

George R. Shaner

See also: Hukbalahap Revolt; Malayan Emergency; Somalia, U.S. Military Operations in; Vietnam Conflict

References and further reading:

Carlton, Eric. *Occupation: The Policies and Practices of Military Conquerors.* Savage, MD: Barnes & Noble Books, 1992.

De Pauw, John W., and George A. Luz, eds. *Winning the Peace: The Strategic Implications of Military Civic Action.* New York: Praeger, 1992.

Kyre, Martin, and Joan Kyre. *Military Occupation and National Security.* Washington, DC: Public Affairs Press, 1968.

Sandler, Stanley. *Glad to See Them Come and Sorry to See Them Go: A History of the U.S. Army.* Fort Bragg, NC: U.S. Army Civil Affairs and Psychological Operations Command, 1998.
Weiss, Thomas G. *Military-Civil Interactions: Intervening in Humanitarian Crises.* Lanham, MD: Rowman & Littlefield, 1999.

Clark, General Mark Wayne (1896–1984)

Commander of the U.S. Fifth Army, the Fifteenth Army Group, in World War II and United Nations (UN) forces in Korea. A 1917 graduate of West Point, Clark was wounded as an infantry officer in World War I. As a decorated veteran of World War I and a keen military thinker, he caught the eye of General George C. Marshall in the interwar years and became one of a cadre of young officers—other notables were Omar Bradley and Dwight Eisenhower—slated for advancement at the outset of World War II.

After a stint with Army Ground Forces in Europe and as deputy chief of Allied forces in Northwest Africa, he was sent, clandestinely, into French North Africa in late October 1942 to negotiate a truce with Vichy French commanders in advance of the impending Allied invasion. Although he did not succeed in forestalling all resistance, he did eventually reach agreement with the French.

Clark assumed command of the Fifth Army in time for its invasion of Italy in September 1943. Here the trouble began. In hopes of achieving surprise, Clark elected to forego naval bombardment in advance of the Allied invasion of Salerno. The German defenders, surprised but unscathed from lack of bombardment, nearly pushed the Americans into the sea. Whether the fault of Clark or not, the ensuing Italian campaign was an arduous slugging match up the Italian peninsula, at great cost.

In January 1944, in the hope of relieving the embattled invasion force at Anzio, Clark ordered his troops to attack across the heavily defended Rapido River. He lost 2,100 men in the space of one day amid repeated assault orders. This bloody debacle provoked a congressional investigation. But Clark's greatest controversy came five months later after the Allied breakout from both Anzio and the Gustav Line. A brave but vainglorious man, Clark pushed the bulk of his forces toward Rome to liberate the first Axis capital, but at the cost of allowing the bulk of the German army in Italy to escape and survive to fight another day. They would hold out in northern Italy, and extract many Allied casualties, until the end of the war.

An adept self-promoter and bureaucratic infighter, Clark survived these controversies and rose to command United Nations forces in Korea in 1952–1953. After the war he retired from the army and served as president of The Citadel Military College in South Carolina until 1965. To his dying day, Clark defended his reputation against all critics.

John C. McManus

See also: World War II

References and further reading:

Blumenson, Martin. *Mark Clark.* New York: Congdon Weed/St. Martin's Press, 1984.
Clark, Mark W. *Calculated Risk.* New York: Harper, 1950.

Clark, George Rogers (1752–1818)

"Conqueror of the Old Northwest" during the American Revolution. A prominent explorer in the western frontier of Virginia, in 1774 Clark served as captain of the Virginia militia in Lord Dunmore's War against the Shawnee Indians. At the outbreak of the American Revolution, Clark commanded the frontier militia and organized the defense of the Kentucky country. In 1778, he captured Kaskaskia, Cahokia, and Vincennes, but lack of reinforcements forced him to abandon plans to take Detroit. Between 1780 and 1782, Clark, now a brigadier general, defended Virginia against General Benedict Arnold's invasion. He also defeated the Shawnee at Piqua and Chillicothe, defended Cahokia and St. Louis against a British-Indian expedition, and recaptured Vincennes. Clark's campaign against Detroit again failed to materialize, but his victories preserved American claims to the Old Northwest at the end of the war. Clark was not paid for his service and spent the remainder of his life deeply in debt. He remained in the West as Indian commissioner until 1786 when a military campaign against the Wabash ended in failure. Clark's involvement in foreign intrigues in the West further clouded his reputation. He tried to recoup his fortunes through several abortive Spanish and French colonization schemes. In 1793, he accepted a French major general's commission to lead an expedition of American frontiersmen against Spanish Louisiana. The scheme, part of the Genet Affair, also failed. In 1798, Clark joined the French army to lead another attempt to conquer Louisiana that never materialized. In 1812, in belated recognition for his Revolutionary War service, the Virginia legislature granted Clark an annual pension of $400.

Dean Fafoutis

See also: American Revolution; Arnold, Benedict

References and further reading:

Bakeless, John. *Background to Glory: The Life of George Rogers Clark.* Philadelphia, PA: J. B. Lippincott Company, 1957.
Bodley, Thomas. *George Rogers Clark: His Life and Public Services.* Boston: Houghton Mifflin, 1926.

Clausewitz, Karl Maria von (1780–1831)

Prussian general, one of the most influential military theorists of modern times. Born near Magdeburg, son of a civil servant and retired army officer, Clausewitz joined the Prussian army in 1792, was commissioned in 1793, and fought at Mainz (1793) and the Rhineland Campaign (1794) against France.

Clausewitz devoted time on garrison duties to self-education, graduated from the Berlin Military Academy, was appointed to General Staff (1803), and served as aide to Prussian ruler Prince Augustus (1804).

Clausewitz fought at Auerstädt and was captured at Prenzlau by the French in 1806. Repatriated with the Peace of Tilsit, he was employed on the Military Reorganization Commission from 1807 to 1811 under Gerhard von Scharnhorst, director of the Berlin Military Academy, working with August Wilhelm Gneisenau, Hans David Yorck, and Gebhard von Blücher, in modernizing the Prussian army to match that of Napoleon.

He resigned his army commission in 1812, protesting the Franco-Prussian alliance against Russia, and joined the Russian army in defensive campaigning, culminating in the Battle of Borodino. Clausewitz helped persuade Yorck to sign the Convention of Tauroggen, abandoning France for Russia.

He was Prusso-Russian liaison officer during the War of Liberation against Napoleon (1813) and corps chief of staff in Blücher' s army in Napoleon's final defeat (1815).

But his 1812 resignation had blighted his career. King Frederick William III, to whom Clausewitz was military tutor (1807–1811), would not readmit him into the Prussian army until Napoleon's first abdication (1814). He did serve as director of the Berlin War College (1818–1830), an administrative post, giving little input to teaching, army reformation, or future war planning.

Although Clausewitz never commanded in battle, he wrote of his war experiences, beginning in 1818. Ten volumes were published in unfinished form in 1832, the year after his death from cholera at Breslau (Wroclaw). His most important work was *Von Kriege* [On War]. Revising this volume from 1827, only the first chapter was satisfactorily completed, and inconsistencies therefore appear. Industrial and military developments were rapidly bypassed in favor of tactics, but this was the first attempt to analyze practically every concept of war, including unconventional or partisan war together—analyzing conflict, how future war may develop, drawn from his Napoleonic War experiences, knowledge of military history, and Kantian philosophy.

Von Kriege's best-known section deals with "War as a continuation of politics by other means." Clausewitz contended that "absolute" war, completely annihilating the enemy, gaining total victory, in reality could not always be achieved. Frictions—numerous diverse factors—limit war, the greatest friction being politics. Wars are fought for political goals, their scale determining the scale of war. Governments should oversee war, otherwise generals would set few if any limits. Political goals must be established at the onset of war.

Varying frictions make each war different, preventing the application of universal rules. Commanders must be quick to act and react, and be sure in their decisions, thus establishing good morale among the troops. The side with greatest morale (élan) will succeed. In that offensive action creates higher morale than defensive, decisive offensives should be mounted.

The enemy center of gravity should be identified: army, capital, army of stronger ally; these must be destroyed by concentration of superior forces in a decisive blow. This offense should be reduced to the least number of actions necessary to bring victory.

Thus, Clausewitz advocated the concentrated offensive to win. However, defensive war can be the stronger, in that it is easier to hold than to take. However, defense is negative in itself and should only be used to prepare for winning the counteroffensive.

Ignored until 30 years after his death, Clausewitz's writings were developed by von Moltke and used to great success in the Austro-Prussian War (1866) and the Franco-Prussian War (1870–1871), becoming seminal reading for mainland Europe's general staffs, but tended to be selectively employed.

Ignoring Clausewitz's argument that the defense was often stronger than the offense, but accepting his affirmation of élan and concentration of forces for decisive offensives, France and Germany adopted offensive doctrines that were thwarted by World War I defenses. Clausewitz also influenced "new" 1920s Soviet revolutionary doctrine.

Clausewitz's disregard of naval power produced an adverse reaction in Britain, with Basil Liddell-Hart insisting on the need for smaller, highly trained forces to be used sparingly with the major battle avoided by diplomacy—the "indirect approach."

Clausewitz's writings had much to do with the debate between absolute continental notions of mass armies and the limited naval-based power epitomized by Great Britain. Unfortunately, the major European military powers attempted to do both, while adding mass airpower. The result was the self-discarding of the French military, the collapse of the Soviet Union, the bankruptcy of Great Britain, and the complete defeat of two German empires in two world wars. It would seem apparent that for the foreseeable future only the United States, which emerged from the twentieth century's world wars far stronger than it entered, has the resources to deploy continental armies, dominate the sea lanes, and deploy war-winning airpower.

Neil Harvey Croll

See also: Napoleonic Wars

References and further reading:

Clausewitz, C. von. *Von Kriege* [On War]. 3 vols. Trans. J. J. Graham. London: Kegan, Paul, Trench, Trabuer, 1911. Rev. ed. F. N. Maude, abridged and with introduction by L. Willmot. Hertfordshire, UK: Wordsworth Editions Limited, 1997.

Howard, M. *Clausewitz.* Oxford, UK: Oxford University Press, 1983.

Parkinson, R. *Clausewitz: A Biography.* London: Wayland Publishers, 1970.

Clay, Lucius Dubignon (1897–1978)

U.S. military-political leader. Lucius Clay, youngest child of a U.S. senator from Georgia, entered the U.S. Military Academy at West Point and graduated twenty-seventh in his class in 1918. Assigned to the engineers (he would have preferred the artillery), he rose rapidly to captain, but reverted to first lieutenant in 1919, remaining in that grade for 17 years. With the coming of the F. D. Roosevelt administration, his career accelerated. Clay first worked capably with the Works Progress Administration and the Civilian Conservation Corps. He then served briefly as chief engineer in the Philippines and was then given responsibility for the Defense Airport Program for the Civil Aeronautics Authority. Though he desired combat duty during World War II, he was instead given charge of the logistics of war production for the army.

In June 1944, General Dwight D. Eisenhower asked that Clay come to Europe to resolve logistical problems at the vital French port of Cherbourg; Clay had it running smoothly within a month. Returning to Washington, he was named deputy to James Byrnes, director of War Mobilization and Reconversion. Returning to Europe, he oversaw problems of military government in Germany for several years, and was named both military governor and theater commander in the spring of 1947. Clay had to rely on his own judgment in great measure, having been given no precise guidance, and worked well with his State Department adviser Robert Murphy. His objectives, in the face of often difficult relations with Soviet and French authorities, were to restore a sound and denazified civil government, facilitating the creation of a new nation. He organized and oversaw the successful Operation VITTLES (Berlin Airlift) during 1948 and 1949 in the face of determined Soviet intransigence, and in this and other ways, earned the warm regard of the German people. Following his retirement in 1949, he was chairman and CEO of Continental Can and then a partner in the Lehman Brothers banking firm. He was also active in the Republican Party and various civic enterprises.

Keir B. Sterling

See also: Civil Affairs/Military Government

References and further reading:

Blacker, John H. *Winds of History: The German Years of Lucius DuBignon Clay.* New York: Van Nostrand Reinhold, 1983.

Clay, Lucius D. *Decision in Germany.* New York: Doubleday, 1950.

Smith, Jean E. *The Papers of Lucius D. Clay: Germany, 1945–49.* 2 vols. Bloomington: Indiana University Press, 1974.

———. *Lucius D. Clay: An American Life.* New York: Holt, 1990.

Clive, Robert (1725–1774)

Regarded as the founder of British India. Robert Clive was born in Shropshire in 1725. In 1743, at the age of 18, Clive became a writer in the East India Company, the lowest clerical position available. Arriving in Madras in 1744, Clive lived the life of a loner and, in a display of the depression that would eventually claim his life, attempted suicide. Opportunity for the excitement that suited Clive's nature was not long in coming, however, for soon after his arrival Clive found himself immersed in the First Carnatic War (1744–1754). Clive's initial taste of war was not propitious. In September 1746, French forces under Joseph Dupleix captured Madras, and Clive and his fellow company employees found themselves prisoners of war. Internment was not to Clive's liking, however, and he shortly escaped to Fort St. David. There, while assisting in the defense of the settlement during Dupleix's 18-month siege, Clive received his ensigncy in the army of the East India Company (May 1747). In 1748, Clive took part in Admiral Edward Boscawen's abortive attempt to capture the French station at Pondicherry.

Though the First Carnatic War ended in 1749, peace did not immediately end Clive's military experiences. In 1750, then ranked lieutenant, he was given command of a small force ordered to assault the Maratha town and fort of Devikota. In the subsequent action, Clive would demonstrate the daring verging on rashness that would be a hallmark of his career. After a short return to civil employment, Clive returned permanently to the military line with an appointment as commissary to the European troops in Madras. With the outbreak of the Second Carnatic War in 1751, Clive was promoted to captain and dispatched to Trichinopoly. Soon besieged by French and Indian forces, it was Clive who brought to the council at Madras Mohammed Ali's recommendation that a diversionary assault be launched against Arcot, in order to relieve pressure on Trichinopoly. Command of this expedition was given to Clive, with the rank of captain. The capture of the city and its subsequent defense through a 50-day siege meant that when Clive returned to England in 1753, he met a hero's welcome. Returning to India in 1755, Clive was appointed governor and commander of Fort St. David with the local rank of lieutenant colonel. Shortly after his arrival, he joined with Admiral Charles Watson successfully to attack the Maratha fortress at Gheria, which fell a day before Siraj-ud-daula, nabob of Bengal, captured the British station at Calcutta (21 June 1756). When word of "the Black Hole of Calcutta" atrocity reached Madras in August, Clive was given command of a force of more than 1,500 men, which, sailing on 16 October 1756, recaptured Calcutta on 2 January 1757.

Clive effectively drove the French from Bengal, leaving him free to deal with Siraj. Though the latter soon sued for peace, Clive intrigued with Siraj's commanders to replace him with Mir Jafar. It was the success of this intrigue that to a large extent accounted for Clive's subsequent victory at Plassey (23 June 1757), where the weaknesses of Siraj's ill-organized army were fatally compounded by the treason of his commanders, among them Mir Jafar. However ignobly it was won, Plassey secured the British position in the Bengal while Clive received a lifetime annuity of 30,000 pounds sterling, and the rank within the Mogul hierarchy of *mansabdar.*

Clive next turned his attention to the Carnatic, where French forces under Thomas Lally had captured Fort St. David (2 June 1758) and had since December 1758 laid siege to Madras. Remaining in Calcutta, Clive dispatched a relief force under Colonel Francis Forde, which defeated Lally at Masulipatam on 25 January 1759. Ill, Clive returned to England in February 1760. Received even more enthusiastically than seven years before, Clive was created Baron Clive of Plassey in 1762, and a knight of the Bath in 1764. In the latter year he was also returned to Parliament as member for Shrewsbury, and it was partly through his political connections that he returned to India as governor of Bengal in 1764. Foremost among Clive's efforts during this period was the establishment of Dual Government, whereby the East India Company became, in Clive's words, "nabobs in fact, if not in name, perhaps totally so without disguise." Clive also sought to suppress what he saw as corruption among company officials. Given that Clive had benefited so greatly from the venality that was part and parcel of the British system, his critics perceived something perverse in his newfound morality. Clive's success, and the wealth that accompanied it, as well as his often turbulent personality, had made him many enemies in India and at home, within the East India Company and the House of Commons. This in large part accounted for the public rancor that met his third and final return to England in 1767, and which resulted in a public inquiry into his conduct in India. Though in 1773 Clive was acquitted of charges of corruption and nepotism, the six-year ordeal left him a broken man. Physically ill and gripped by the depression that marked much of his life, Clive committed suicide on 22 November 1774.

Adam Norman Lynde

See also: Plassey, Battle of

References and further reading:

Harvey, Robert. *Clive. The Life and Death of a British Emperor.* London: Hodder & Stoughton, 1998.

Keay, John. *India: A History.* London: HarperCollins, 2000.

Mehra, Parshotam. *A Dictionary of Modern Indian History, 1707–1947.* Delhi: Oxford University Press, 1987.

Coastal Defense

Defending a nation's sea coast from an enemy sea invasion or blockade. While often viewed as forts and large-caliber guns, coastal defense systems can also include coastal ships, torpedoes, mines, and aircraft. The concept of coastal defense has been around for hundreds of years, with its greatest popularity in the late nineteenth and early twentieth centuries. In that period, coastal defenses often consisted of a series of forts that could provide mutual support. These forts were often quite substantial, and made up a large part of the defense planning (and budget) of many nations. They were strategically located, such as the Spanish forts in Florida or the U.S. forts guarding Charleston and Savannah.

With the development of large and accurate breach-loaded guns in the late nineteenth century, coastal defense systems became much larger and more complex, something of a counterpart to the land-based forts of Belgium, France, and Germany. Many countries at this time also built coastal defense ships, such as Norway's *Norge* (1900), which mounted four 8.2-inch guns in two turrets, or Sweden's *Sverige* (1915), with four 11-inch guns in two turrets.

While the coastal fortification may have seemed vulnerable, they were in fact able to stand up to heavy attack and still fulfill their role in defense. The Charleston, South Carolina, forts, for example, foiled the assault of Union monitors in 1863.

Later coastal fortification guns often used one of two designs. The first was the disappearing carriage, which mounted the gun on a folding trapeze; the gun could be aimed and loaded below a parapet, raised to be fired, and then would *disappear* after firing, as the trapeze was designed to lower from the recoil. The second design was more complex, and involved mounting the gun on a railroad carriage that was protected in a hardened bunker, often with blast-proof doors. The gun would be rolled out, fired, and then rolled back behind the heavy doors and reloaded in safety before counterfire could disable it.

The interwar period of 1919–1939 saw a great deal of coastal defense construction, such as the British complex at Singapore and the fortification of the Japanese islands in the Pacific. World War II was to see the decline of the traditional coastal fortification as sea-based airpower matured enough to allow large-scale attack from the sea. Singapore was outflanked by Japanese landings in the rear of Singapore, and Japanese fortifications, while tough to knock out, fell to a combination of ground troops, ship-based support fire, and ship-based aircraft. The Norwegian forts in the fjord near Oslo were able to sink the German cruiser *Blucher.* Although it was a shock to the Germans, the sinking did not repel the invasion.

World War II witnessed both the nadir and the apogee of coastal defense. The system was at its lowest when the Japanese basically marched into Singapore, whose great guns were designed to repel a seaborne assault. (It is an enduring myth that the guns "pointed the wrong way"; nonetheless, the British did not bother to fire their biggest pieces at the Japanese.) The American defenses of Manila Harbor, collectively called Corregidor after the main island forts, held out for five months, denying the harbor to the Japanese and upsetting their timetable of conquest. For those five months, the 14-inch guns of Fort Drum, the amazing army's Concrete Battleship were the largest guns in action against any enemy of America on any front.

Coastal defense today can be seen as part of a larger and more complex defensive network that includes patrol aircraft, light warships, and land-based radar stations. The static coastal fortifications of the turn of the century are too vulnerable to air weapons to serve in any defensive capacity. Many of the coastal forts of the nineteenth and early twentieth centuries are historical parks and museums from another time and place.

Drew Philip Halévy

See also: Artillery

References and further reading:

Coleman, James C. *Guardians on the Gulf: Pensacola Fortifications, 1698–1980.* Pensacola, FL: Pensacola Historical Society, 1982.

Floyd, Dale E. *Defending America's Coasts, 1775–1950: A Bibliography.* Alexandria, VA: Office of the Chief of History, U.S. Army Corps of Engineers, 1997.

Lewis, Emanuel Raymond. *Seacoast Fortifications of the United States: An Introductory History.* Annapolis, MD: Naval Institute Press, 1993.

Paloczi-Horvath, George. *From Monitor to Missile Boat: Coast Defense Ships and Coastal Defense since 1860.* Annapolis, MD: Naval Institute Press, 1996.

Sarty, Roger. *Coast Artillery, 1815–1914.* Ottawa: Service Publications, 1988.

Cochise (c. 1812–8 June 1874)

Chief of the Chirichua Apache in the southeastern area of present-day New Mexico and Arizona. He is best remembered for the war he waged against the United States from 1861 to 1872. The conflict began when Cochise and five other Apaches were questioned by Lieutenant George Bascom over a ranch hand's missing child and some abducted cattle. Maintaining their innocence, a struggle ensued after Bascom tried to arrest them. Cochise was able to escape with three bullet wounds after one Apache died. In order to free the Apaches that had been taken prisoner over the incident, Cochise seized some whites to exchange them for the captives. Bascom stupidly and brutally retaliated by hanging six tribe members. Cochise and Mangas Coloradas now waged war against any nearby Americans. Due to the ferocity of the campaign and the departure of many U.S. Army personnel because of the Civil War, most settlers abandoned the region.

In 1862, General James Carleton and 3,000 California volunteers engaged the Apache at Apache Pass. The military's howitzers gave the Americans the victory. Yet Cochise was able to escape with several hundred of his followers and eluded capture for more than a decade in the Dragoon Mountains. During this time, they used guerrilla tactics to raid travelers, ranches, miners, and homesteaders. Finally, General George Crook, through the use of Indian informants and scouts, was able to get Cochise to surrender in September 1871. He left the reservation in the spring of 1872 after a dispute over where his people would be located. He quickly surrendered again and spent his remaining years on the Chiricahua Reservation in Arizona.

T. Jason Soderstrum

See also: American Indian Wars; Custer, George Armstrong; Geronimo

References and further reading:

Roberts, Dave. *Once They Moved Like the Wind: Cochise, Geronimo, and the Apache Wars.* New York: Simon & Schuster, 1993.

Sweeney, Edward R. *Cochise: Chiricahua Apache Chief.* Norman: University of Oklahoma Press, 1991.

Coehoorn, Baron Menno van (1641–1704)

Dutch general, author, and engineer. Lieutenant General Menno van Coehoorn, a member of a prolific Swedish-Dutch family, was a general, civil engineer, and military engineer. His name is often linked with Vauban as one of the two great fortress designers who dominated the so-called wars of reason. This oversimplifies the lively world of fortification theory in the early 1700s, but his books, *Versterckinge de Viffhoeks met all syne Buylenwerken* (Leeuwarden, 1682), and *Nieuwe Vestingbouw op en natte of lage horizont* (Leeuwarden, 1685), were very successful and influential. The latter in particular appeared in many translated editions during the War of the Spanish Succession (1701–1714).

Coehoorn entered the Dutch army under the patronage of the Stadholder of Friesland, Casimir Hendrik, at the age of 16, taking part in several sieges. At Grave in 1673, he introduced his signature in attack, bombardment by multiple light trench mortars, called *coehorns* after him.

After the publication of his first book, Coehoorn was successively employed in reconstructing a series of Dutch fortresses. It was in this common endeavor that contemporary engineers made their reputations, and Coehoorn's redesigns impressed fellow professionals. In the subsequent War of the Grand Alliance (1688–1697), they proved very successful. Coehoorn served as a brigadier in the latter war, defending Namur against Vauban, and three years later retaking the fortress in the face of the improvements made by Vauban in the interim in a classic battle of talents. After the 1697 peace, Coehoorn became inspector general of fortresses.

In the War of the Spanish Succession (1701–1714) Coehoorn served as a corps commander and conducted several brilliant sieges, but given a field command, showed himself perhaps in over his head. He died of illness in March 1704.

Coehoorn was also involved in several large civil engineering projects, including redyking the lower Scheldt and designing new barrages for the mouth of the Zuider Zee.

Erik Lund

See also: Artillery; Engineering, Military; Grand Alliance, War of the; Louis XIV; Mortars; Vauban, Sébastien Le Prestre de; Spanish Succession, War of the

References and further reading:

Childs, John Charles Roger. *The Nine Years War and the British Army, 1688–1697: The Operations in the Low Countries.* Manchester, UK: Manchester University Press, 1991.

Duffy, Christopher. *Fire and Stone: The Science of Fortress Warfare 1660–1860.* Newton Abbot, UK: David & Charles, 1975.

Lynn, John. "Food, Funds, and Fortresses: Resource Mobilisation and Positional Warfare in the Campaigns of Louis XIV." In *Feeding Mars: Logistics in Western Warfare from the Middle Ages to the Present,* ed. John Lynn. Boulder, CO: Westview, 1993.

Coen, Jan Pieterszoon (1587–1629)

Founder of the colonial empire in the East Indies. Born on 8 January 1587, in Doorn, Holland (present-day Netherlands) and dying on 21 September 1629, in Batavia, Dutch East Indies (present-day Jakarta, Indonesia), he was the fourth governor-general of the Vereenigde Oost-Indische Compagnie (VOC, Dutch East India Company), who supplanted Portuguese interests, excluded English commercial penetration of the Indonesian Archipelago through extreme military measures, and ruthlessly established the Netherlands' commercial superiority in Asia. The VOC was founded by the States General (parliament) in the United Netherlands in 1602 for two reasons: to protect their eastern trade routes and to aid the lengthy war of independence against Spain.

Coen was brought up as a strict Calvinist and became a forceful personality, ruthless, punitive, efficient with a mean streak, intent on securing a trade monopoly for the VOC. Appointed governor-general in October 1617, Coen's first aggressive act was to conquer Jakarta. He razed, fortified, and renamed it Batavia in 1619 and used it as his base to conquer the surrounding area. In 1621 he set his sights on the Banda Island, a part of Jakarta regarded as a feudal benefice by the sultan of Bantam. Coen strongly abhorred lax rules in commerce and his conquest was based on a flimsy pretext: lack of close adherence to commercial agreements. Through extremely widespread slaughter, even for that time, Coen largely exterminated the indigenous Bandas, keeping those who survived as slaves or soldiers. He paid little attention to the reprimand he received from the VOC. He then laid claim to the kingdom of Jakarta, using it as his own personal domain. In 1622 he sent a large expedition to the Chinese coast and established a Dutch settlement on Formosa, which provided a base for lucrative trade with China and Japan. Coen unsuccessfully attempted to attract Dutch settlers. He died in Batavia in 1629 while the city was under siege by the Javanese.

Annette Richardson

See also: Dutch Colonial Wars

References and further reading:

Boxer, C. R. *The Dutch Seaborne Empire 1600–1800.* London. Hutchinson & Co., 1977.

Bruyn, J. R., F. S. Gaastra, and I. Schoffer, eds. "Dutch-Asiatic Shipping." *Rijks Geschiedkundige Publicatien* 165–167. Grote Serie. The Hague: 1979, 1987.

Israel, Jonathan I. *The Dutch Republic and the Hispanic World.* New York: Oxford University Press, 1986.

Cold Harbor, Battle of (31 May–12 June 1864)

After avoiding a potential Confederate trap at the North Anna River, Ulysses S. Grant continued the movement of the Army of the Potomac south and east to find the right flank of Robert E. Lee's Army of Northern Virginia.

Grant had sent cavalry under Phillip Sheridan with their new repeating carbine rifles to seize the crossroads at Cold Harbor. Sheridan's dismounted troopers withstood attacks

Grant's great campaign—Stevens's battery at Cold Harbor. Wood engraving, 1864. (Library of Congress)

from Confederate infantry rushing to retake the vital crossroads; other Confederates were entrenching as the bulk of Lee's army arrived. On 1 June, Union infantry assaults enjoyed some success, but two days later a much bigger attack resulted in wholesale slaughter among the Union II, XVIII, and IX Corps on the seven-mile front that reached from Bethesda Church to the Chickahominy River.

Grant regretted the assault, and the two armies faced one another until 12 June as Grant sought a way to break through Lee's lines or to turn his flank. The Union commander decided there were no reasonable opportunities for a breakthrough and determined to continue the advance toward Richmond. By 14 June, federal troops began crossing the James River, and Grant wisely decided to avoid the strong defenses of Richmond and rather threaten Petersburg to the south, through which communications to Richmond from the rest of the Confederacy passed. At this point, Grant and Lee began a long period of siege warfare around Petersburg and also Richmond; Lee's defenses were too strong for Grant to risk an attack, and he settled for stretching those defenses ever thinner by reaching around to cut rail and road communications to the south. By spring 1865, Grant would soon be able to cut off Richmond, and Lee had first tried a futile attack on federal positions and then the desperate flight that ended with surrender at Appomattox Court House.

Charles M. Dobbs

See also: American Civil War; Grant, Ulysses Simpson; Lee, Robert Edward

References and further reading:

Dowdey, Clifford. *Lee's Last Campaign: The Story of Lee and His Men against Grant—1864.* Boston: Little, Brown, 1960.

Furgurson, Ernest B. *Not War but Murder: Cold Harbor 1864.* New York: Knopf, 2000.

Miers, Earl Schenck. *The Last Campaign: Grant Saves the Union.* Philadelphia, PA: Lippincott, 1972.

Cold War (1946–1991)

The so-called Cold War began at the end of World War II when the victorious allies were unable to come to amicable settlements for governing a divided Europe. Joseph Stalin, leader of the Soviet Union, aggressively sought to improve

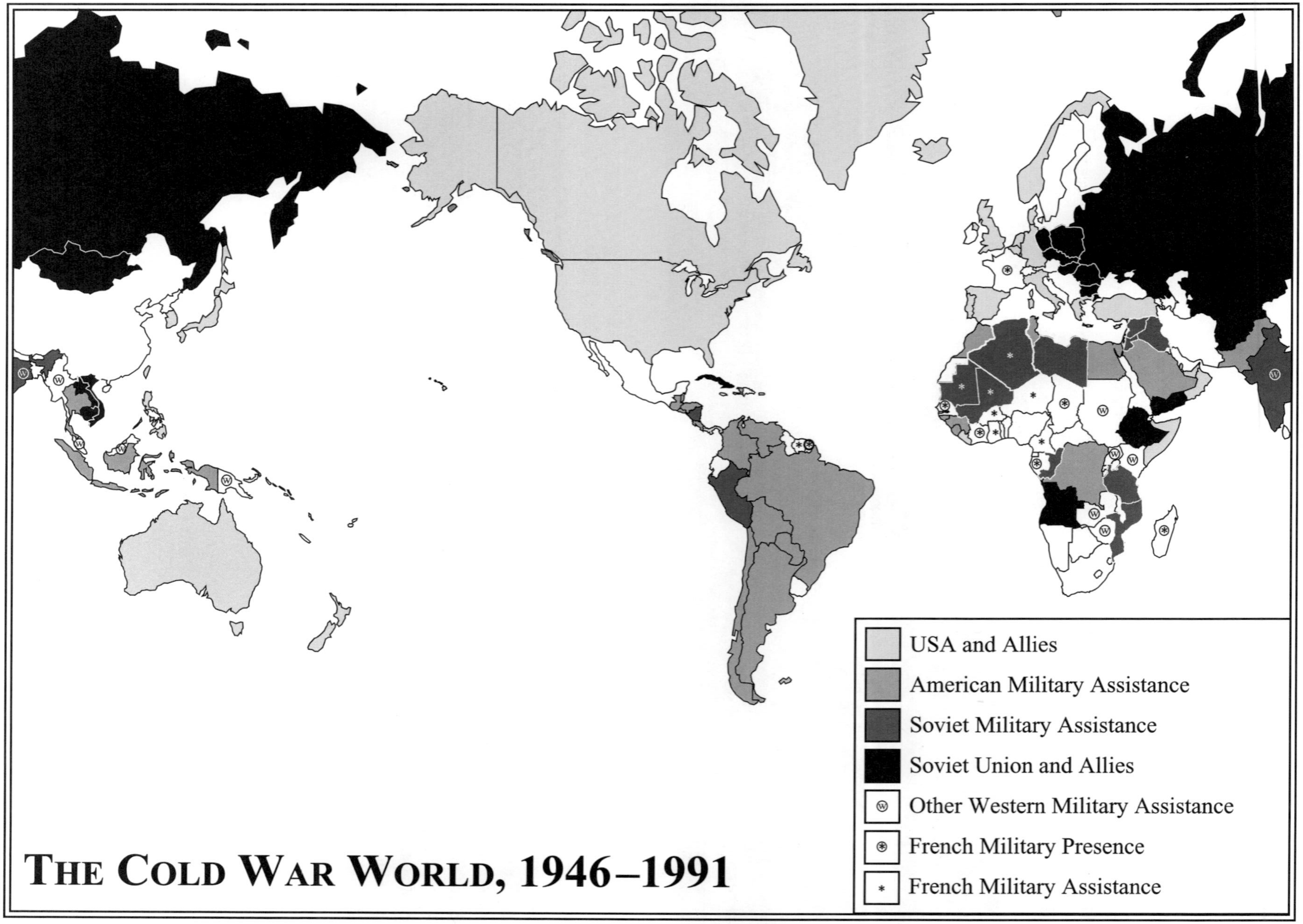
THE COLD WAR WORLD, 1946–1991
USA and Allies
American Military Assistance
Soviet Military Assistance
Soviet Union and Allies
Other Western Military Assistance
French Military Presence
French Military Assistance

the security of his nation by constructing a barrier of communist client states. The United States resented Stalin's blatant lies to allow elections, his anti-Western rhetoric, and the brusque manner in which the Soviets expanded their perimeter. Winston Churchill's famous "Iron Curtain" speech at Westminster College in Missouri in 1946 marked a recognition that Europe had become divided into two hostile camps, although many Americans, still mindful of the contribution and sacrifices of their Soviet ally in World War II, reacted unfavorably to the speech. But within a year the United States pledged with the Truman Doctrine to aid any country fighting communist aggression. It also spent billions of dollars through the Marshall Plan, a successful effort to bolster the European economy and undermine communist activities in France and Italy.

When the United States tried to reform the currency and economy of the British-American occupation German sector, the Soviets responded by blockading West Berlin in 1948. Unwilling to respond militarily, the United States conducted an airlift that supplied Berliners with food and fuel. Marking an emphatic end to joint occupation of Germany with the Soviets, the United States encouraged the creation of an armed and independent West Germany and welcomed the new nation into the newly formed North Atlantic Treaty Organization (NATO), a military alliance created in 1949 to prevent the Soviets from invading Western Europe. The Soviet construction of an atomic bomb in 1949 and enactment of a military draft by the United States further escalated tensions.

The fall of China to the Communist forces under Mao Zedong in 1949 marked a shift in the focus of the American-Soviet struggle from Europe to Asia. While Americans lashed out at each other in what became known as the Red Scare, Communist North Korea invaded South Korea on 25 June 1950. For two and a half years U.S. troops and their South Korean and United Nations allies battled the Communist Chinese and North Koreans to a standstill roughly along the 38th Parallel. By 1953 it was clear that the United States and the Soviet Union were locked in a bitter struggle that promised years of discord and conflict.

Throughout the 1950s and early 1960s the changing nature of global politics increased tensions as each side probed for some weakness in the other. In the United States, newly elected president Dwight Eisenhower enacted a policy of massive retaliation to any aggressive Soviet moves. Eisenhower hoped to reduce the military budget by building a cheaper atomic arsenal and airpower instead of conventional weapons. Eisenhower was also convinced that the United States could increase the allure of democracy and capitalism with a prosperous economy. Reducing the military budget was an important step in this plan.

Meanwhile, the Soviets increasingly exerted their domination of Eastern Europe. In 1955 they created the Warsaw Pact, a military alliance of their Eastern European client states. But in 1956 they brutally suppressed an uprising in Hungary that challenged the Soviet-backed Communist leadership. Finally, in 1961, Nikita Khrushchev, who succeeded Stalin as leader of the Soviet Union, ordered the construction of the Berlin Wall, one of the most provocative acts of the Cold War, primarily to keep skilled East German workers from escaping to the West. Not content to confine Soviet growth solely to Europe, Khrushchev aggressively backed and encouraged wars of national liberation in the decolonizing nations of Asia and Africa. The two most famous examples were Cuba and Vietnam. These two small nations dominated relations between the United States and the Soviet Union in the early to mid-1960s. Cuba, a mere 90 miles off the coast of Florida, enormously aggravated the American leadership, which regarded Communist presence in Cuba as a threat to national security. After a clumsy attempt by President John F. Kennedy to liberate Cuba by sponsoring the Bay of Pigs invasion by anti-Castro Cubans in 1961, Khrushchev authorized the deployment of Soviet nuclear missiles to the island. American U2 spy planes discovered the missiles in October 1962 and the world waited anxiously as the United States and the Soviet Union came the closest ever to nuclear war. The Soviets withdrew the missiles and Kennedy promised that America would not invade Cuba. Several months later the United States withdrew missiles from Turkey.

Ever since the French withdrawal from Vietnam in 1954, the United States had kept a wary eye on developments in Southeast Asia. President Dwight Eisenhower was suspicious of another American land war in Asia. Unlike Eisenhower's single approach to communist expansion, Kennedy adopted a policy of *flexible response,* meaning that each act of aggression would be met with a response in proportion to the threat. Accordingly, Kennedy incrementally increased assistance in both manpower and money to the pro-American government in South Vietnam and its efforts to fight rebels backed by communist North Vietnam. Fearful that Europeans would read any failure of the United States to honor its treaty commitments to South Vietnam as American unwillingness fully to support NATO, and seeing communist victory in Southeast Asia as a stepping-stone to the Philippines and Japan, the United States increased its commitment after the Gulf of Tonkin incident in August 1964. By 1968 the United States had the incredible number of more than 500,000 troops in South Vietnam. The war eventually proved unpopular at home, and protests spread throughout American college campuses. President Richard Nixon, elected in 1968, engaged in a policy of *Vietnamization,* or turning greater amounts of responsibility for fighting the land war to the South Vietnamese. Although Nixon briefly expanded the war into Cambodia in 1970, he decreased the commitment of the United States to Vietnam. In 1973 he negotiated an end to the war. In 1975, the year after Nixon resigned, the north overran the south and unified Vietnam under a communist government.

Despite the war in Vietnam, the Cold War entered a new phase, détente, or peaceful coexistence. Relations between the Soviet Union and the United States improved for a number of reasons. First, the new leadership of the Soviet Union, headed by Leonid Brezhnev, sought to hold on to its power by improving the quality of life for Soviet citizens. Less committed to communist ideology than to maintaining their own bureaucratic rule, they desired to cut military spending and increase the supply of consumer goods. Second, other powers were rising to question the supremacy of the United States and the Soviet Union. Economically, Germany and Japan challenged the dominance of the United States in the realm of global trade. Politically, growing nations such as India played the United States and the Soviet Union off one another to the best of their advantage while refusing to take sides in the Cold War. In Europe, Charles de Gaulle of France spearheaded a movement to create a continent that was politically, militarily, and economically independent of the Cold War. He even pulled France out of NATO's command structure. In Asia and Africa, China challenged the Soviet Union for the leadership of the ongoing communist revolution. President Richard Nixon recognized the importance of China with his astonishing state visit there in 1972, thus beginning the end of the more overt stage of Sino-American hostility dating from the Korean War. Finally, resentment in the United States against the war in Vietnam created a climate more open to discussion with the Soviets.

Agreements between the United States and the Soviet Union to limit the spread of nuclear weapons best exemplified the spirit of détente. Treaties signed between 1963 and 1979 limited atmospheric testing of weapons, pledged nations that possessed nuclear weapons not to spread the technology to nonnuclear, banned the building of defensive weapons, froze the land-based and submarine-launched nuclear arsenals for five years, and limited the construction of missiles and bombers.

Détente came to a rapid end when the Soviet Union invaded Afghanistan in 1979. And the election of Ronald Reagan as president of the United States in 1980 marked a more militant mood in the U.S. Believing that the Soviet Union could not survive a renewed arms race, Reagan increased defense spending, deployed missiles in Europe, and ratcheted up the rhetoric by dubbing the Soviet Union an "evil empire." The Soviet gerontocracy proved unable to deal effectively with Reagan. In 1985 Mikhail Gorbachev became leader of the Soviet Union. A younger, committed reformer, Gorbachev instituted two programs, Glasnost and Perestroika, that opened the infarcted Soviet society and traditionally state-run economy. Gorbachev and Reagan hammered out several arms-control agreements that not only limited construction of new weapons, but also eliminated some stockpiled arms. As economic conditions in the Soviet Union continued to decline (as did many of its vital statistics), Gorbachev lessened the Soviet commitment to upholding the communist regimes of Eastern Europe. One by one, beginning in 1989, the communist governments of Eastern Europe fell to a tide of democracy. The infamous Berlin Wall fell to peaceful crowds on both sides. Unable to stem the tide, the Soviet Union itself fragmented into its 15 constituent ethnic republics. In 1991 the Soviet Union no longer existed and the Cold War drew to a close—peacefully—a denouement almost totally unforeseen by experts and general public alike.

Gregory Dehler

See also: Arms Control; Atomic Bomb, Development of; Bay of Pigs Invasion; Castro Ruz, Fidel; Castro-Cuban Revolution; Clay, Lucius; Cuban Missile Crisis; Eisenhower, Dwight David; Greek Civil War; Guevara, Che; Hungarian Revolt; Hydrogen Bomb, Development of; Indochina Wars; Korean War; Laotian Civil War; Malayan Emergency; Mayaguez Operation; North Atlantic Treaty Organization; Soviet-Afghan War; Stalin; Vietnam Conflict; Mao Zedong

References and further reading:

Gaddis, John Lewis. *We Know Now: Rethinking Cold War History.* New York: Oxford University Press, 1998.

———. *The Long Peace: Inquiries into the History of the Cold War.* New York: Oxford University Press, 1987.

Hogan, Michael J., ed. *The End of the Cold War: Its Meanings and Implications.* New York: Cambridge University Press, 1992.

LeFeber, Walter. *America, Russia and the Cold War: 1945–1996.* New York: McGraw-Hill, 1996.

Miller, David. *The Cold War: A Military History.* New York: St. Martin's Press, 1998.

Painter, David. *The Cold War: An International History.* London: Routledge Press, 2000.

Powaski, Ronald. *The Cold War, the United States and the Soviet Union, 1917–1991.* New York: Oxford University Press, 1998.

Colenso, Battle of (15 December 1899)

British defeat during the Anglo-Boer War, 1899–1902. At the beginning of the war, General Sir Redvers Buller, British commander, began operations against the Boers with the initial goal of relieving Ladysmith. He launched his offensive near Colenso, on the Tugela River. His artillery fired into the Boer trenches along the river while his forces moved into position. The main attack was to come from the west, where

Major General Fitzroy Hart's 5th Brigade would strike the Boer flank. To the east, mounted troops under Lord Dundonald were to do the same. In the center, a force under Major General H. J. T. Hildyard was to make a general assault to keep the Boers from strengthening their flanks. Another unit was held in reserve. Also in the center, naval guns under Colonel L. J. Long were to fire into the Boer lines, keeping them down. Success depended upon each unit taking position and remaining out of Boer range until all were in place.

Before the action unfolded, General Hart to the west and Colonel Long in the center destroyed any hope of success, placing their men prematurely in range of Boer fire. Hart led his infantry in tight formation into a loop in the river, well to the east of their assigned position, where they received enfilade fire. Meanwhile, Long set up his guns too close, and they too came under fire. These mistakes spelled disaster. Hart's men were trapped, if not killed or wounded outright, and Long's men were also all down and the guns silenced. Sir Redvers quickly sent his reserve force to extract Hart, which they did.

Reaching Long proved impossible. All the gunners, including Long, were killed or wounded, and the guns had to be abandoned. Under such circumstances, Buller called off the entire attack on Colenso.

James B. Thomas

See also: Boer Wars; Buller, Sir Redvers Henry; Ladysmith, Siege of

References and further reading:

Pakenham, Thomas. *The Boer War.* New York: Random House, 1979.

Coligny, Gaspard II de (1519–1572)

Leader of the French Protestants during the Wars of Religion. Coligny was born 16 February 1519. As the nephew of Anne de Montmorency, constable of France, Coligny was brought to court as a childhood companion for the future king Henry II. Distinguishing himself in Francis I's Italian campaigns, particularly the Battle of Cessarole in 1544, Coligny was promoted quickly to colonel general of the French infantry, and in 1552, became admiral of France. During this period, Coligny became interested in French colonization and sponsored three expeditions, in 1552 to Brazil, in 1562 to Port Royal in Carolina, and in 1564, where French settlers established Fort Caroline on St. John's River. Unfortunately, none of these settlements survived, and Fort Caroline was wiped out by the Spanish in 1565. As one of the defenders of St. Quinten in 1557, he prevented the Spanish from moving into France, but was taken prisoner and held until 1559.

Shortly after his release, Henry II was killed in a jousting match, an accident that destabilized France as the Guises established control of the new king. Announcing his conversion to Protestantism through the influence of his brother, Francois d'Andelot, Coligny became the leader of the Huguenot forces, particularly after Conde was killed in action at Jarnac in 1569. Coligny became the military adviser to Henry of Navarre (later Henry IV of France), and although defeated at Moncontour, won at Arnay-le-Duc in 1570 and helped to negotiate the Treaty of St. Germain.

Once more acceptable at court, he became the tutor to the young king, Charles IX, from which position he advocated aid to the Netherlands against Spain, going so far as to give his daughter in marriage to William "the Silent" of Orange. Disliking Coligny and the Protestants, the Queen Mother, Catherine d'Medici, arranged for Coligny to be assassinated on 22 August 1572, but the attempt was botched. In the ensuing St. Bartholomew's Day Massacre, surrounding the wedding of Henry of Navarre and Margaret de Valois, Coligny was among the first located and killed by Henry de Guise, leading a party of German mercenaries.

Margaret Sankey

See also: French Wars of Religion

References and further reading:

Crete, Lilane. *Coligny.* Paris: Fayard, 1985.
Thompson, James Westfall. *The Wars of Religion in France 1559–1576.* New York: Frederick Ungar, 1958.
Whitehead, A. W. *Gaspard de Coligny, Admiral of France.* London: Methuen, 1904.

Collins, J. Lawton (1896–1987)

U.S. Army commander. General J. Lawton Collins was born 1 May 1896 in New Orleans, Louisiana. He graduated from the U.S. Military Academy at West Point in 1917 and served as an instructor or student during most of the interwar period. When World War II broke out, Collins moved through a succession of staff positions as chief of staff of VII Corps and the Hawaiian Department. In 1942, he assumed command of the 25th Infantry Division and took the unit to Guadalcanal. During the Guadalcanal and New Georgia campaigns, Collins earned a Distinguished Service Medal, Silver Star, and Legion of Merit, in addition to the nickname Lightning Joe, a reference to the insignia on 25th Division's shoulder patch. In 1944, Collins was transferred to Europe to prepare for the Normandy invasion. He led VII Corps ashore on D day–plus–1 and commanded the corps throughout the remainder of the fighting in Europe, from the capture of Cherbourg to the linkup with the Soviets along the Elbe River.

Following the close of World War II, Collins held a series of high-level staff positions, serving as army chief of staff from 1949 through 1953 during the Korean War, when he attempted to work with flamboyant Douglas MacArthur, Far East commander. Following this assignment, President Dwight Eisenhower asked him to stay on active duty to serve first as U.S. representative to NATO, and later as special representative in Vietnam with the rank of ambassador. In 1956, he retired from the army, and worked in the private sector and assisted numerous humanitarian causes. In his retirement years, he wrote a well-received history of the Korean War. Collins was known as an aggressive, confident, and enthusiastic commander—"a soldier's general."

William Hartley

See also: Korean War; World War II

References and further reading:

Collin, J. Lawton. *Lightning Joe: An Autobiography.* Baton Rouge: Louisiana State University Press, 1979.

———. *War in Peacetime: The History and Lessons of Korea.* Boston: Houghton Mifflin, 1969.

MacDonald, Charles B. *The Last Offensive: U.S. Army in World War II.* Washington, DC: U.S. Army Center of Military History, 1973.

Colombian Guerrilla War (1976–2000)

Continuing unrest and violence that threatened the very sovereignty of Colombia. In the mid-1960s three leftist groups began to organize in Colombia to stage low-intensity attacks on the government. The first group to emerge was the Fuerzas Armadas Revolucionarias de Colombia, or Colombian Revolutionary Armed Forces, commonly known by its Spanish acronym, FARC. Under the leadership of the insurgent Manuel Marulanda Vélez whose nom de guerre is Tirofijo, FARC took over areas of the Upper Magdalena Valley, drawing its support from peasants and settlers. A second group was the Ejercito de Liberación Nacional (ELN, Army of National Liberation). ELN operated in the middle Magdalena Valley and was inspired by Fidel Castro's Cuban Revolution. Unlike FARC, ELN drew its support mainly from middle-class youths, intellectuals, and priests discontented with Colombian government and society. The third and smallest of the rural guerrilla groups was the Ejercito Popular de Liberación (EPL, Army of Popular Liberation). In 1973, a fourth group, M-19 or Movimento 19 de Abril (19th of April Movement), began to make its presence known. M-19 was an urban-based terrorist group with vague goals and ideology. All four groups have waged war on the Colombian government for nearly 30 years, and FARC and ELN have gained control over portions of Colombia.

The M-19 guerrilla group achieved spectacular success in some of its early actions, including the theft of Simon Bolivar's sword from a Bogotá museum in 1974, the capture of a large cache of arms from the army in 1979, and takeover of the Dominican embassy during a diplomatic reception in 1980. In late 1985 M-19 seized the Palace of Justice seat of Colombia's Supreme Court. The group held the Supreme Court justices hostage. The army assaulted the palace, killing all of the terrorists including some of the key leaders of M-19. In 1990 the group demobilized after negotiations with the government and formed a legitimate political party.

The largest of Colombia's insurgent groups is FARC and it has always pursued a strategy to gain and hold portions of the countryside. In some regions and small towns it has achieved control for periods of two years or more. Although FARC has participated in several peace initiatives, it has always quickly withdrawn. While the Colombian army has struck some FARC-held strongholds, it has been unable to destroy the movement or force significant negotiations. Likewise ELN continues to be troublesome to the government with attacks on Colombia's oil pipelines. The EPL, on the other hand, negotiated with the government and demobilized in 1991.

An ominous development of the 1990s was the linkage between FARC and ELN and the drug cartels. In exchange for protection, narcotics traffickers provided large amounts of cash to the insurgent units, which enabled them to renew terrorist operations. Because of its interest in drug interdiction the United States has provided military assistance on a large scale to Colombia's armed forces.

George Lauderbaugh

References and further reading:

Busnell, David. *The Making of Modern Colombia: A Nation in Spite of Itself.* Berkeley: University of California Press, 1993.

Communications, Military

The application of various technologies to the transmission of military orders and intelligence. History's first known conqueror was King Sargon of Akkad (c. 2300 B.C.E.). We know this because he left inscriptions at various places that imperfectly record his conquests, for such were the limits of written communication in his era. Every advance in literacy made possible a like improvement in military communications, until fifteenth-century commanders such as Babur and Maximilian I could be known as sophisticated authors as well as conquerors. Yet up to 1800 the physical means of communication remained letters carried by muscle- and wind-powered means, and military needs strained to cross the twin barriers of space and time.

Distance was constant and particularly infuriating in the

sparsely settled regions and in overseas expeditions and colonial wars, but the time problem was related to administration and security. Letters had to be read and turned around, which depended on the size and efficiency of staffs, and secure letters had to be enciphered and then deciphered, something that could already be much delayed due to sophisticated codes as early as the Thirty Years' War.

The age of muscles ended about 1800 with the first advanced communications, the visual semaphore. By 1840 France had a nationwide semaphore network, and its general staff confronted for the first time the problem of administering a communications network. But the semaphore was subject to weather and visibility limitations.

Electrical telegraphs were already in operation by this time, but, based on ill-understood science, were underdeveloped and short-range. Yet they represented, for the first time, instantaneous communications not subject to weather conditions. In the latter half of the century the Maxwellians transformed the telegraph, even as railroads developed the capacity needed to support mass mobilizations. In the 1866–1872 German Wars of Unification, the telegraph transformed military communications, making it possible to fight a Napoleonic war of rapid movement in a shockingly short time. But even this early, commanders, particularly in the American Civil War, were heard to complain that they fought "tied to the end of a telegraph wire."

But in the 1880s and 1890s, electrical engineers were preparing a new kind of war. Mobile telephone equipment, linked by spooled cable or even earth conductance, introduced electrical communications on a tactical level.

The first intimation of the transformational importance of the new technologies was the use of reconnaissance balloons. Semaphore and other visual means were the prime link between air and ground well into World War I, but a new invention, telephones, disseminated their information ever more quickly. By the decade before 1914, artillerists realized that they would be able to fire on targets beyond visual range using telephoned information from air or advance ground observers. The next war would be a clash of fire, not of men, although men would die in unprecedented numbers.

Radio, invented in 1904, took a relatively minor place in World War I land warfare due to the crudeness of the technology. When it was used, there was little regard for security, even though electronic eavesdropping and jamming methods developed in 1904–1914. Russian radio security was noticeably poor, and may have contributed to the great loss at Tannenberg in 1914.

In the interwar period, automated encryption methods developed alongside much improved radios. By the end of World War II, radio communications extended down to platoon level, giving artillery and air support even greater tactical immediacy. Security was maintained by automated encryption and the use of narrow-beam microwave links. But as neither of these methods was perfect, cable-carried telephone and telegraph were used for greater security, while the dispatch rider and liaison officer survived, mounted on motorcycle and Jeep rather than horse or chariot.

In the postwar era, computer-mediated secure communications networks have proliferated at both the tactical and strategic levels. At one extreme, the modern Internet serves military uses, while at the other extreme tactical units developed into local area networks. Communications were becoming increasingly seen as a component of a larger entity, C3I, or command, control, communications, and intelligence. The enormous firepower of even the smallest units could be held back mainly by delayed information.

So important have communications become to early twenty-first-century military establishments that information warfare, that is, the jamming or disabling of the enemy's communications, is considered by military planners to be one of the most promising, or feared, future war scenarios.

Erik Lund

See also: Artillery; Electronic Warfare; Intelligence, Military; ULTRA

References and further reading:

Browne, J. P. R., and M. T. Thurbon. *Electronic Warfare.* Brassey's Air Power: Aircraft, Weapons Systems and Technology series. London: Brassey, 1998.

Munro, Neil. *The Quick and the Dead: Electronics Combat and Modern Warfare.* New York: St. Martin's Press, 1991.

Thomson, George R., et al. *The Signal Corps.* 3 vols. United States Army in WWII series. Washington, DC: Office of the Chief of Military History, 1957.

Condé, Louis II de Bourbon, Fourth Prince de (1621–1686)

The typical military prince of the Baroque era. The princes de Condé were the heads of an important branch of the House of Bourbon; their enormous fortune was second only to that of the king. Louis received a strict religious and military education. He was known as Duc D'Enghien till his father's death in 1646. He was, all his life, a cultivated man, a patron of the arts. At age 19, he first saw battle at the siege of Arras. His birth and titles pushed him to higher command. He won his first great victory at Rocroi in 1643 over the Spaniards, the greatest French victory of the first half of the seventeenth century. He followed his victory with successes along the Rhine, and with Turenne he was victorious at Fribourg (1644) and Nordlingen (1645). This charismatic and ardent leader was the complementary element of the more thoughtful Protestant Turenne. The Fronde civil war (1650–1659) was to change his destiny. After being loyal to the young king, he behaved with such arrogance as the government's savior that Mazarin (the king's main minister) had Condé arrested. The high nobility rebelled against the hated minister, and Condé was released. Condé then took the leadership of an open rebellion (1651) and allied himself with Spain. He made his way to Paris and was able to defy the royal troops under Turenne. But his position became politically and militarily untenable and he left Paris to become generalissimo of the Spaniards. With varying fortunes he opposed the royal troops but was finally soundly defeated by Turenne at the Battle of the Dunes (1658). After the Peace of the Pyrenees (1659) he reentered the king's good graces but, mistrusted by Louis XIV, he did not receive any command till 1668. During the short War of Devolution (1667–1668) he invaded the Spanish-held Franche-Comté in 15 days and added this province to the kingdom. Totally restored to Louis's favor, he was placed, with Turenne, in command of the main army. He won his last victory in 1674 at Seneffe in the Spanish Netherlands. He then retired to his Chantilly Palace, suffering from gout, and spent his late life surrounded by artists (Mignard, Le Brun) and writers (La Fontaine, Molière).

Gilles Boué

See also: Fronde, Wars of the; Louis XIV; Turenne, Henri de la Tour d'Auvergne, Vicomte de

References and further reading:

Bluche, François. *Dictionnaire du Grand Siècle.* Paris: Fayard, 1990.
Pujo, Bernard. *Le Grand Condé.* Paris: Albin Michel, 1995.

Conrad von Hötzendorf, Franz, Baron (1852–1925)

Chief of the Austro-Hungarian general staff during World War I. Conrad was born 11 November 1852 at Penzing near Vienna. Advancing rapidly in the monarchy's army and regarded as a brilliant strategist, he served as chief of general staff in 1906–1911 and 1912–1917. He mistrusted Austria's non-German nationalities and the expansionist tendencies of both Serbia and Austria's ally Italy, advocated preventive wars against them, and worked hard to strengthen Austria's military forces for the war that, in his opinion, was inevitably to come.

After the assassination of Archduke Franz Ferdinand at Sarajevo (28 June 1914), Conrad argued vigorously for war against the Serbs, holding them responsible for the murder as well as Slav agitation within the monarchy.

Lacking fighting experience, Conrad shared the short-war illusions of his contemporary generals. He underestimated warfare on two fronts against Serbia and Russia. When war started, Conrad shifted troops from the Balkans to the east and failed to concentrate on one enemy. As a result Austria suffered defeat on both fronts. Serbia was finally subdued by the end of 1915, whereas the situation in the east in 1914 turned disastrous with enormous casualties. Later offensives in 1915 and 1916 against Russia and Italy, carried out with German aid, were successful but no decisive victory was obtained. Conrad's strategy was sound but he lacked a sufficiently large army to carry it out. Austria became increasingly subordinated to Germany.

The new emperor, Charles I, dismissed Conrad on 1 March 1917. He took over an army group on the Italian front. Retiring after 1918, he wrote several volumes of his memoirs. He died on 25 August 1925 at Mergentheim (Germany).

Conrad was a gifted but unlucky soldier, a hardliner advocating ruthless military solutions for problems of domestic as well as foreign policy. Caught in his extreme right-wing, anti-Semitic, and anti-Slav ideology, his plans failed because of lack of resources and neglect of human and political realities.

Martin Moll

See also: Brusilov Offensive; Caporetto; Gorlice/Tarnow; Isonzo, Battle of the; World War I

References and further reading:

Rauchensteiner, Manfried. *Der Tod des Doppeladlers: Österreich-Ungarn und der Erste Weltkrieg.* Graz-Vienna-Cologne: Styria, 1993.

Conscription

Although the involuntary selection of men and women for military service has been practiced in various ages, large-scale modern conscription came into being with the *levée en masse* in Revolutionary France. (In tribal societies, of course, almost all young men are warriors, but they assume this occupation as a rite of manhood, and if any have doubts they are not likely to express them.)

The law of 23 August 1793 called all able-bodied men from the ages of 18 to 25 to the colors, instantly creating a military force sufficient to meet the pressures other powers were bringing to bear on France. Continued conscription on this scale provided 2.6 million men; France could keep the other powers at bay for more than two decades of almost continuous warfare. Despite this success, the other powers disdained conscription, except for Prussia, which adopted it in 1808. A comparatively small power, Prussia enacted the Defense Law of 3 September 1814 stipulating that every male at age 20 would be subject to three years of compulsory army service, two years with the active reserve, and then a lengthy commitment to the civilian-supervised long-term reserve, the Landwehr. In 1815, the restored French monarchy ended conscription, but low levels of recruitment weakened the army unacceptably. Universal conscription was unpopular with economists, conservatives, and the French people as a whole, so in 1818 France adopted a conscription system operated by lottery in which some males would get a life exemption while others were called to seven years of active service, unless they could hire a substitute or otherwise buy their way out. Those who could not evade seven years of active service found themselves at a disadvantage in competing for civilian careers and often reenlisted, thus creating a long-service professional army, distanced from the population as a whole. The Prussian system blurred class distinctions, promoted a responsible patriotism, and fostered hygiene and education alike. Friedrich Engels believed that "compulsory military service surpasses general franchise as a democratic agency." And the swift triumph of Prussia's mass army (led by highly professional officers) over the smaller French semiprofessional army in 1870 was ample vindication of the military efficacy of mass conscription. The new German Empire adopted the Prussian system, while France adopted a universal conscription system in 1872 and kept it until after World War I.

Some other nations, impressed by Prussia's success against France, themselves instituted mass conscription. In 1873, for example, Japan decided to raise a mass army by conscripting males for two years of active service, 17 years in the reserve National Army, and further service in the First National Army, a sort of home guard. But other nations declined to follow suit. Great Britain's centuries-long distrust of a standing army led it to avoid conscription until 27 January 1916, when the manpower stresses of World War I mandated a call-up of all males aged 18 to 41 who were not in occupations vital to the war effort. This wartime draft was reinstituted by the National Service Acts of 1939 and 1941, the latter maintained until 1960. The United States and the Confederate States of America both resorted during the Civil War to an iniquitous conscription so unpopular, with its provisions for substitutes and cash deferments, as to have sparked the New York Draft Riots of 1863 in which hundreds died, and it was quickly abandoned at war's end. The Confederate draft was, if anything, even more unpopular.

British conscripting German troops for service in America, eighteenth century. (Library of Congress)

The Selective Service Acts of 19 May 1917 and 16 September 1941 set up conscription for the two world wars, based, uniquely, upon unpaid local draft boards, which determined eligibility, and was generally accepted. After a hiatus in 1947–1948, a peacetime draft was adopted. It became highly unpopular and controversial during the Vietnam War for the same reason that the Civil War draft was detested: a general perception of unfairness because of exemptions. Students were exempt, as were those in certain occupations. And Americans reverted to their distrust and dislike of the military, questioning the necessity for a draft in the semiwar of the Korean and Vietnam conflicts. President Nixon made the abolition of this draft a campaign pledge, which he fulfilled in 1973. The dire predictions of those still favoring a peacetime draft—that an all-volunteer army would lead to a praetorianization of the U.S. Army, or that minorities would dominate the new force—never came true to any great extent, and the military triumph of the U.S. in the Gulf War (1990–1991) seemed to settle the question. A more intangible argument is that conscript national armies provide the one great common experience for young men, of whatever background of wealth, ethnicity, or race. Aside from the United States, the British Commonwealth, and Japan, most of the democracies have retained conscription.

Joseph M. McCarthy

See also: British Military, Twentieth-Century Organization and Structure; German Army

References and further reading:

Chambers, John Whiteclay, II, ed. *To Raise an Army: The Draft Comes to Modern America.* New York: Free Press, 1987.

Flynn, George Q. *America and the Draft, 1940–1973.* Lawrence: University of Kansas Press, 1993.

Kestnbaum, Meyer. "Partisans and Patriots: National Conscription and the Modern State in France, Germany and the United States." Ph.D. diss., Harvard, 1997.

Constantine V (718–775)

Byzantine emperor, fought Arabs and Bulgarians. The son of Emperor Leo III (r. 717–741), Constantine allegedly drew his name, Copronymus ("Named for Dung"), from an incident at his christening. Crowned coemperor by his father at age two, he became emperor in his own right in June 741, and began a campaign against the Arabs. His accession exacerbated divisions in the Byzantine military over personalities and over the role of icons within the Church. As a result, his first campaign ended immediately when his brother-in-law, Artabasdus, ambushed and defeated Constantine's army on 27 June 741. He fled to Amorium while Artabasdus marched to Constantinople and proclaimed himself emperor. In May 742, Constantine defeated Artabasdus at Sardis, and reentered Constantinople on 2 November 742.

In 746, Constantine invaded northern Syria and took the town of Germanica. He followed this up by defeating an Arab fleet from Alexandria. As a result of this campaign, he settled a group of Syrians in Thrace, and followed this up with a program of fortification building that would shortly have grave ramifications. In the meantime, the war against the Arabs continued until 750, when the defeat of the Umayyads of Damascus by the Abbasids of Baghdad allowed Constantine to turn his attention to the Bulgarian frontier of his empire.

In 756, the Bulgarians, offended by the resettlement of the Syrians near their territories and the construction of fortifications, invaded the Byzantine Empire. Constantine quickly defeated the invading force. This defeat sparked a civil war among the Bulgarians, and, against this backdrop, Constantine waged eight more campaigns against them. In 763 he won a major victory over Khan Teletz, and celebrated a triumphal entry into Constantinople. His campaigns in 773 and 774 were less decisive. During a campaign against the Bulgarians in 775, Constantine died on 14 September.

Constantine continued the policy of his father in strengthening troop units stationed near the capital, the so-called tagmatic forces; in the aftermath of the revolt of Artabasdus, he found it expedient also to reorganize and weaken the provincial or thematic troops. His continual victories over the Arabs and Bulgarians were sufficient to win the loyalty of his army, and the succession of his son was undisputed. At the same time, he was unable to prevent the loss of the Byzantine holdings in northern Italy to the Lombards.

Joseph M. Isenberg

See also: Byzantine-Muslim Wars

References and further reading:

Fine, John V. A. *The Early Medieval Balkans.* Ann Arbor: University of Michigan Press, 1991.

Kaegi, Walter. *Byzantine Military Unrest, 471–843.* Amsterdam: A. M. Hakkert, 1981.

Treadgold, Warren. *A History of the Byzantine State and Society.* Stanford, CA: Stanford University Press, 1997.

Whittow, Mark. *The Making of Byzantium, 600–1025.* Berkeley and Los Angeles: University of California Press, 1996.

Constantine the Great (280–337)

Reunified the Roman Empire, officially recognized Christianity. Born Flavius Valerius Constantinus, son of Constantius I Chlorus, Caesar to Augustus Maxentius in the West during the time of Diocletian, and his wife Helena. In 289 Constantine moved to the court of Diocletian with his father. In 305 Constantius rose to the position of coemperor. The following year, while both father and son were fighting in Britain, Constantius died. The troops, impressed with the military skills of Constantine, proclaimed him coemperor, but he was unable to eliminate potential rivals until 324. Defeating the forces of Maxentius at Milvian Bridge after a dream that Christ would lead him to victory, Constantine consolidated his power over the western empire, and ordered toleration for all Christians in the Roman Empire through the Edict of Milan in 313. A struggle between Constantine and Lucinius, emperor of the east, for supreme authority over the entire empire ended in 324 with Constantine as the victor. His reorganization of the government separated civil and military affairs, resulting in the restoration of the Senate to a position of prominence. The reunification of the empire secured the continuation of Roman control over the eastern region. Ecclesiastical changes implemented by Constantine also strengthened the empire.

Cynthia Northrup

See also: Milvian Bridge, Battle of

References and further reading:

Grant, Michael. *Constantine the Great: The Man and His Times.* New York: Scribner, 1994.

MacMullen, Ramsey. *Constantine.* New York: Croom Helm, 1987.

Pohlsander, Hans A. *The Emperor Constantine.* New York: Routledge, 1996.

Smith, John Holland. *Constantine the Great.* New York: Scribner, 1971.

Constantinople, Siege of (717–718)

One of the most spectacular medieval sieges. Byzantium's endurance and the Umayyad Caliphate's failure had historic consequences. As the eighth century began, the balance of power in the Mediterranean east had moved dramatically away from the Byzantine Empire in favor of the Umayyad caliphs of Islam. Exhausted by internal upheaval, Balkan strife, and coups, Byzantium seemed vulnerable and crumbling. Thus, around 715, the Umayyads, led by the ambitious House of Abd al-Malik, resolved to seize Constantinople at any cost.

As Caliph Süleyman and his brother Maslamah, his commander, assembled their army and a huge supporting navy, civil war again engulfed Byzantium, but the war brought to power Leo III, a brilliant officer especially experienced in fighting the Arabs. In July 717, as Maslamah's troops encircled Constantinople, Leo's ships converged from every harbor in Byzantium and devastated the Umayyad fleet. The Arab forces were bedeviled by Bulgarian guerrillas, Constantinople's towering walls, and the worst winter in centuries; casualties and disease shrank their numbers daily. Süleyman himself died unexpectedly, campaigning in Anatolia.

In spring 718, a relief fleet from Egypt reached the Straits. Here, the sailors, largely Christian, mutinied against the Muslims. Byzantine forces in Anatolia, meanwhile, savaged Umayyad reinforcements, forcing them back to Syria. Epidemics, storms, even a volcanic eruption at the port of Thera, pummeled the besiegers. Finally, the new caliph, Umar II, ordered the bedraggled survivors to withdraw home. Leo had broken the 13-month siege. Byzantium enjoyed a rebirth of power while, 30 years later, the discredited Umayyads collapsed in the face of the Abbasid revolution.

Weston F. Cook, Jr.

See also: Abbasid Revolution; Byzantine-Muslim Wars; Leo III

References and further reading:

Hawting, G. *The First Dynasty of Islam: The Umayyad Caliphate.* Carbondale: Southern Illinois University Press, 1987.

Treadgold, Warren. *A History of the Byzantine State and Society.* Stanford, CA: Stanford University Press, 1997.

Constantinople, Siege of (1453)

Siege that ended the Byzantine Empire. By 1450 the once-mighty Byzantine Empire had been reduced to parts of the Peloponnesus, a few islands, and the area around the ancient capital of Constantinople. After years of warfare, Ottoman forces controlled all the surrounding countryside. A new sultan, Mehmed II, ascended to the throne in 1451 at the age of 21. Young and ambitious, he focused his energy on capturing Constantinople. Diplomatically isolating the Byzantine emperor, Constantine XI Palaeologus (r. 1449–1453), Mehmed II signed treaties with the Venetians and the Hungarians. He then ordered the construction of fortress Rumeli Hisar on the European side of the Bosphorus. With his forces controlling access to the Black Sea from Rumeli Hisar, and another fortress, Anadolu Hisar, six miles south of the city, Mehmed II cut off the Byzantines from all European assistance. Although the emperor petitioned the pope to call another crusade for the purpose of defending the city, no help arrived.

After the completion of the Rumeli Hisar, the emperor ordered the gates of the city closed, and the siege of Constantinople began in July 1452. For the next six months the Ottomans made preparations for the battle, including the construction of the heavy cannons to bombard the walls. In the spring of 1453 the sultan ordered his troops onto the plain of Adrianople. Ottoman forces numbering close to 150,000 men assembled, and began moving slowly toward the city. The sultan, camping near the Military St. Romanus Gate, positioned the cannons, concentrating his forces on the land side of the city where the fortifications included a triple wall. The emperor gathered his 7,000 troops around the same gate, where he expected the main assault to occur.

After the sultan had demanded the surrender of the city, with the promise of sparing the lives of the inhabitants, the emperor refused and the Ottomans opened fire. The cannonade inflicted heavy damage, but the defenders repaired and protected the fortifications. On 20 April, four large European vessels carrying provisions for the city appeared on the Sea of Marmara and managed to fight their way past the Ottoman naval forces until they reached the Golden Horn. Mehmed II, after beheading his admiral, ordered the transport of several of his ships overland from the Bosphorus. Once the vessels reached the Golden Horn, his new admiral opened fire on the European ships, badly damaging them. Forty of the captured sailors were executed.

Having eliminated outside assistance, the Ottomans once again turned their attention to breaching the fortifications. On 7 and 12 May, Ottoman forces attacked heavily damaged sections of the walls. Then on 28 May they rested. The emperor, realizing that the final assault was imminent, ordered the religious icons and relics to be carried around the city in a processional as the church bells rang and the inhabitants prayed.

The final assault began just after midnight on 29 May. The first two waves consisted of the irregulars and the Anatolian troops of Ishak Pasha. The Byzantine emperor and his men successfully defended the narrow section of wall, but when the third wave, consisting of the Janissaries, attacked, the exhausted defenders collapsed. After the emperor was slain in battle, resistance faded quickly. Ottoman soldiers flooded into the city. Soldiers looted, massacred the inhabitants, burned buildings, raped the women, and enslaved the survivors. In the afternoon, the sultan finally entered the city and ordered the killing stopped. He then proceeded to Saint Sophia and decreed that it be turned into a mosque. As a result of the battle, Constantinople, later renamed Istanbul, was depopulated and the last remnant of the eastern empire destroyed.

Cynthia Clark Northrup

See also: Byzantine-Ottoman Wars; Constantinople, Sieges of; Turkish Wars of European Expansion

References and further reading:

Runciman, Steven. *The Fall of Constantinople, 1453.* New York: Cambridge University Press, 1990.

Constantinople, Sieges of (674–718)

Between 633 and 642 the eastern empire's eastern provinces were overrun by Islamic forces, and the southern provinces were lost, never to be recovered. With the Sassanid Empire destroyed in 641, the eastern empire was subjected to repeated Muslim invasions and also to Bulgar encroachment in the Haemus (Balkans).

The first Saracen assault on Constantinople began between 670 and 672, when their naval units seized various coastal towns, including Cyzicus on the Sea of Marmara. In 674 the Muslims initiated what proved to be a siege of Constantinople spread over five successive summers. The fully walled capital, however, proved invulnerable to the massive siege engines and catapults that the Saracens put into their ships, while the Muslims had no answer to the threat presented by a mysterious new weapon, Greek Fire, with which imperial ships devastated their opponents. After the siege was abandoned, what remained of the Saracen fleet was destroyed off Syllaecum, in southern Anatolia, in a freak storm in 678.

In 712 a Bulgar raid reached but could not take Constantinople, but three years later the capital fell after a siege. In two of the numerous civil wars that litter these years Theodosius III first besieged and took Constantinople and then in March 717 was forced to abdicate in favor of Leo III. It has been suggested that the Saracens made no move lest Leo, whom they trusted to surrender Constantinople, was discredited by association. For his part Leo may have sought Muslim passivity in order to oust Theodosius, but never contemplated any meaningful arrangement with the caliphate. Muslim raids on the empire continued for many decades, but the very existence of the empire was never seriously threatened thereafter.

H. P. Willmott

See also: Constantinople, Siege of

References and further reading:

Haldon, John. *The Byzantine Wars.* Stroud, UK: Tempus, 2001.

———. *Byzantium. A History.* Stroud, UK: Tempus, 2000.

———. *Warfare, State, and Society in the Byzantine. London World, 565–1204.* London: UCL Press, 1999.

Norwich, John Julius. *Byzantium: The Early Years.* London: Viking, 1988.

Obolensky, Dimitri. *The Byzantine Commonwealth: Eastern Europe, 500–1453.* London: Weidenfeld & Nicolson, 1971.

Córdoba, Fernandez de (1453–1515)

Prominent Spanish military commander. Fernandez de Córdoba was born on 1 September 1453 at the Castillo de Montilla in Córdoba. Upon his father's death, his mother sent him to Córdoba to be brought up by a relative. He served as page to the pretender to the Castillian throne, Don Alfonso, an infant. After Don Alfonso's premature death, de Córdoba was called to Segovia by Princess Isabel (Isabella) who had just married and was defending her claim to the throne.

After the death of the rightful monarch, Enrique IV, de Córdoba, at the start of the war of reconquest of Granada, led 120 lancers of his elder brother's unit in the army of King Ferdinand and Queen Isabella. He gained renown at the battles for the taking of Loja in 1486 and in the siege of Granada in 1492.

He further distinguished himself in battle in Italy. When Charles III of France invaded the Italian peninsula, King Ferdinand sent Fernandez de Córdoba to aid his nephew, King Fernando II of Naples. De Córdoba was instrumental in liberating Calabria and obtaining the surrender of the French commander, General Aubigny, along with his 7,000 troops.

The Spanish warrior also came to the aid of Pope Alexander VI, retaking papal lands seized by a pirate force. Soon after, at the request of King Ferdinand, he was sent to help defend Sicily against the Turks. He battled Turkish and French forces for several years, and his bravery and mastery of military tactics brought him much acclaim. He became known as "The Great Captain" and was revered by his troops. He died in Granada on 2 December 1515.

Peter Carr

References and further reading:

Martín Gómez, Antonio L. *El Gran Capitán: Las Campañas del Duque de Terranova y Santángelo.* Madrid: Almena Ediciones, 2000.

Cornwallis, Sir Charles (1738–1805)

British general in the American Revolution, governor-general of India, viceroy to Ireland. Some men are remembered for a single failure rather than for a general record of achievement. Cornwallis enjoyed a highly successful military career, but a key defeat in the American Revolution ended Britain's hopes for victory in that war and tarnished his reputation ever since.

Charles Cornwallis received his first army commission at the age of 17 after studying at Eton and Claire College. Military studies at Turin gave way to service in the Seven Years' War in Europe, where the young officer rapidly achieved the rank of lieutenant colonel. His fast-paced career continued with his succession to the House of Lords in 1762.

Although he opposed measures against the American colonies, he took the rank of major general in the British army in America in 1775. He proved an active and capable commander in the campaigns of 1776 and 1777, playing a key role in the Battle of Brandywine. As commander of an army in the southern colonies in 1780 and 1781, however, he led British forces on an arduous campaign through difficult country, which culminated in the surrender of his entire army at Yorktown, Virginia, a loss that made it impossible for Great Britain to continue the war.

The failure at Yorktown was the result of poor coordination between senior army and navy commanders; although Cornwallis must share in the blame, particularly for his troubled relationship with his superior in America, the loss cannot be attributed to bad generalship.

After serving as an envoy to Prussia, Cornwallis became governor-general of India in 1786, and personally led military campaigns that brought British victory in the Third Mysore War.

After assuming a cabinet appointment in 1795, he led forces that defeated the 1798 Irish rebellion; as viceroy to Ireland, he was instrumental in the passing of the Act of Union in 1802.

He was again appointed governor-general of India, where he died in 1805.

Don N. Hagist

See also: American Revolution; Guilford Court House; Irish Uprising; Long Island, Battle of; Yorktown

References and further reading:

Wickwire, Franklin, and Mary Wickwire. *Cornwallis: The American Adventure.* Boston: Houghton Mifflin, 1970.

———. *Cornwallis: The Imperial Years.* Chapel Hill: University of North Carolina Press, 1980.

Corregidor (December 1941–May 1942)

American coast defense fortifications in Manila Bay. Throughout the desperate American campaign to hold out in the Philippines in the spring of 1942, the island of Corregidor was the nerve center of resistance. The island served as a gateway to one of the finest natural harbors in the world, Manila Bay. Even though they had captured Manila by May 1942, the Japanese could not use its harbor until they neutralized Corregidor. After the collapse of Allied resistance on neighboring Bataan and most other locales in the Philippines, the Japanese redoubled their efforts to take Corregidor.

Garrisoned by about 14,000 American and Filipino troops, the tadpole-shaped island was dotted with forts, antiaircraft batteries, and tunnels, and was supported by "The Army's Concrete Battleship." Fort Drum, with its battleship guns, in Manila Bay was also a part of the Corregidor defenses.

By May the garrison had endured months of disease, hunger, and bombardments. In late April the Japanese began hurling artillery at Corregidor and stepping up their air raids. Each day the intensity of the attacks increased. Meanwhile much of the garrison holed up in the Malinta Tunnel, living a molelike existence amid rats and dust. The combat troops scattered about the island took the brunt of the bombardment, however. On 5 May the Japanese landed troops on the northern tip of the island. The ground fighting lasted little more than a day. Exhausted and with little other choice but to capitulate, Lieutenant General Jonathan Wainwright surrendered his forces to the Japanese. Just under three years later, in February 1945, United States Army Airborne troops assaulted Corregidor and recaptured it during the American invasion of the Philippines.

John C. McManus

See also: Philippines, U.S. Loss of; Philippines, U.S. Retaking of; World War II

References and further reading:

Morton, Louis. *The Fall of the Philippines.* Washington, DC: U.S. Army Office of the Chief of Military History, 1953.

Sandler, Stanley. "Defiant Fort Drum." *MHQ: The Quarterly Journal of Military History* (Winter 2000).

Wainwright, Jonathan M. *General Wainwright's Story: The Account of Four Years of Humiliating Defeat, Surrender and Captivity.* Westport, CT: Greenwood, 1946.

Cortez (Cortes), Hernando de (1485–1547)

Spanish conqueror and adventurer. Cortez was born at Medellín, the son of a retired soldier. He studied law at the University of Salamanca, but did not receive his degree. Cortez sailed to Santo Domingo in 1504 and worked as a farmer and notary public in Azua until 1511. He was secretary to Diego Velazquez during the latter's conquest of Cuba, and was appointed alcalde of Santiago. Cortez also mined gold and raised cattle in Cuba.

In 1518, Cortez persuaded Velazquez to let him lead an expedition to the Yucatán to rescue survivors of previous expeditions and to explore the land. He left Cuba on 19 February 1519 with 600 men, 16 horses, and 11 cannon. The expedition explored the coast of Yucatán, landing at Tabasco in March. Cortex defeated the Tabascans at Ceutla Plain and won them over as allies. Learning of the riches of the Aztec Empire, Cortez dismantled his ships and founded the town of Veracruz. Exceeding his original commission, Cortez led his expedition and Indian allies inland. Using a minimum of force, he formed alliances with local peoples, especially the Tlaxcalans, traditional enemies of the Aztecs. On 8 November 1519, the Aztec emperor Montezuma and his people welcomed Cortez as the incarnation of the god Quetzalcoatl to their capital, Tenochtitlán.

Cortez soon established himself as ruler of the empire. Meanwhile, Velazquez, concerned that Cortez would exceed his authority, sent an expedition under Panfilo de Narvaez to arrest him. Leaving 80 Spaniards and several hundred Indian allies in Tenochtitlán, Cortez returned to Veracruz and defeated Narvaez. Nearly all of Narvaez's soldiers then enlisted with Cortez. By the time Cortez returned to Tenochtitlán, the Aztecs rebelled against Spanish rule. Montezuma was killed. On 30 June 1520, the Spaniards battled their way out of the city, losing many men and horses and much of their booty. The Aztecs attacked Cortez on 7 July at Otumba. Although the Spanish won, they suffered heavy casualties. Cortez had to await reinforcements and could not attack Tenochtitlán until 26 May 1521. He captured the city on 13 August, virtually destroying it. The new capital of Mexico City was built on the ruins.

Cortez sent out further expeditions to Guatemala, Honduras, and the Pacific coast. He was accused of malfeasance by his enemies and defended himself in Spain in 1528. Cortez served as governor of New Spain until 1540, when he returned to Spain for good. He died at his estate near Seville in 1547. Cortez had conquered the Aztec Empire, the largest in Mesoamerica, with only a few hundred men. He was a capable soldier with a good eye for opportunity and the boldness to take advantage of his opponents' weaknesses.

Tim J. Watts

See also: Spanish Colonial Wars

References and further reading:

Abbot, John S. C. *History of Hernando Cortez.* New York: Harper, 1983.

Johnson, William Weber. *Captain Cortes Conquers Mexico.* New York: Random House, 1960.

Marshall, Edison. *Cortez and Marina.* Garden City, NY: Doubleday, 1963.

Corunna, Battle of (16 January 1809)

Costly British victory in the Peninsular War. Sir John Moore, in command of all 30,000 British troops in Portugal after September 1808, launched a 20,000-man offensive into northern Spain in December, marching northeast via Salamanca toward Burgos, where he hoped to engage Nicolas Jean de Dieu Soult. Napoleon, in Madrid since 4 December, learned of Moore's gamble and decided to lead 70,000 men across the icy Sierra de Guadarrama to cut Moore off from Lisbon and attack his rear. As soon as Moore heard about Napoleon's march, he retreated toward the nearest British fleet, 250 miles away at Corunna (La Coruña).

Moore fought successful rearguard actions against Napoleon at Sahagun on 21–22 December, Benavente on 26 December, and Castro Gonzalo on 28 December. Soult engaged Moore's rear guard at Astorga on 31 December. In Astorga on 2 January, Napoleon suddenly broke off pursuit and returned to Paris, disturbed by news of remobilization in Austria and intrigues against him in France.

The unusually cold and snowy weather hindered both

Moore and Soult. When Moore reached Corunna on 11 January, he had only 15,000 men left to face Soult's 24,000. Both armies were frozen and exhausted. Moore deployed 40 guns on high ground south of the city, near the village of Elviña, and began to embark his men. Confident in his superior numbers, even though he had only nine guns, Soult declined to maneuver. The morning of 16 January, he attacked frontally along the entire line. The battle seesawed across Elviña until evening, when the last of the British troops were rescued by sea. Each side lost about 900. Moore was among the British dead. Corunna was a British victory only in the sense that Moore was able to prevent Soult from annihilating his men before they could board ship.

Eric v. d. Luft

See also: Campbell, Colin; Napoleon I; Napoleonic Wars; Soult, Nicolas-Jean de Dieu

References and further reading:

Davies, David William. *Sir John Moore's Peninsular Campaign, 1808–1809.* The Hague: Martinus Nijhoff, 1974.

Fletcher, Ian. *Fields of Fire: Battlefields of the Peninsular War.* Staplehurst, UK: Spellmount, 1994.

Gordon, Alexander. *A Cavalry Officer in the Corunna Campaign, 1808–1809: The Journal of Captain Gordon of the 15th Hussars.* London: Murray, 1913.

Hibbert, Christopher. *Corunna.* London: Pan, 1967.

Cossacks

A social group with no single ethnicity, choosing a seminomadic lifestyle on the Eurasian steppe to escape slavery and poverty. At the end of the Mongol Golden Horde's occupation of the Rus' lands, 1237–1480, the nomadic peoples of the steppe gained the description Cossacks, rather than their specific ethnic or clan designation. Mongols, Tatars, Iranians, and Slavs formed communal groups and became identified according to their geographic location: Volga, Don, Zaporozhe, and so on.

Cossack societal organization exaggerated democratic structure and individual freedom. Exemption from formal inclusion within a state structure, institutions, or aspects of control and participation, such as taxation and conscription, enhances the myth that Cossack existence was carefree. Each Cossack region was ruled by a hetman.

Life on the steppe created equine proficiency and a reputation as tough fighters. During Ivan IV's expansion of Muscovy, Cossacks participated as allies of the Orthodox czar against the Ottoman Turks and Catholic Poles. They led peasant uprisings from the sixteenth to nineteenth centuries, but were also known for their brutality as czarist enforcement forces. The rebellions of Stepan "Stenka" Razin and the Don Cossacks in 1670–1671, and Emelian Pugachev and the Ural Cossacks in 1773–1774, were against czarist rule in reaction to the loss of traditional Cossack autonomy.

The population in Cossack regions swelled as serfs fled and as peasant settlers attempted to relocate beyond czarist administrative control. Often Cossacks relocated themselves and in so doing became an instrument of Russian control in Siberia by setting up outposts from Tomsk in 1604 to Okhotsk in 1649. Cossacks were used to protect and patrol the frontier. During the Russian Civil War, 1918–1921, Cossacks supported both White and Red armies. During World War II, Cossack formations fought as part of the Soviet Army.

Today, the Cossack tradition and reputation are preserved in their ethnic dancing and singing, rather than in military prowess.

Neville G. Panthaki

See also: Genghis Khan; Ivan III; Pugachev's Revolt; Razin's Revolt

References and further reading:

Feodoroff, Nicholas. *History of the Cossacks.* Commack, NY: Nova Science, 1999.

Gajecky, George. *The Cossack Administration of the Hetmanate.* Cambridge, MA: Harvard Ukrainian Research Institute, 1978.

Seaton, Albert. *The Horsemen of the Steppes.* London: Bodley Head, 1985.

Ure, John. *The Cossacks.* London: Constable, 1999.

Courtrai, Battle of (11 July 1302)

Feudal cavalry suffers decisive defeat at hands of militia. The Battle of Courtrai was a watershed in the history of western European warfare. A formed body of dismounted soldiers, acting as the principal rather than auxiliary force, utterly defeated a feudal mounted army. The outcome marked the beginning of the long road to irrelevance for feudalism and its mounted warrior elite.

In 1302, the Flemish townspeople revolted, threw off rule from Paris, and sent their militia to besiege Cassel and Courtrai, two towns occupied by the French. The French responded by sending a large relief army under the command of Count Artois, brother-in-law of King Philip IV. The Flemish Burghers resolved to meet the French at Courtrai; they would prevail or perish. A screen of crossbow marksmen stood in front of the militia phalanx, which was armed with spears and halberd-style weapons. Additionally, the Flemings occupied a strong position behind a small stream.

Count Artois recognized the strength of the Flemish posture, but he felt compelled to act before the Courtrai garrison surrendered, and therefore decided to attack. The Flemings at first recoiled from French javelins and bolts. Artois, hoping to gain an advantage by bringing his cavalry forward to cross the stream, pushed his cavalry forward through his own crossbow men. His heavy horse became disrupted and stalled midstream. With the French momentum broken, the Flemings took heart and advanced rapidly, thrusting and hacking at the French. Trapped between the stream and the enraged Flemings, Artois attempted to surrender. His enemies pretended not to understand French, and cut him down along with all his mounted men.

The Flemings owed their victory to unique circumstances: the topography certainly favored the foot soldiers; the French knights, impatient to get on with their business, failed to allow the missile troops to take their toll on the stationary Flemings; and the aggressive leadership of the Burghers ensured they would fight to victory or death. Circumstances like these did not recur for another hundred years.

Bryan R. Gibby

References and further reading:
Delbruck, Hans. *Medieval Warfare: History of the Art of War,* Vol. 3. Trans. Walter J. Renfoe, Jr. Lincoln: University of Nebraska Press, 1990.
Oman, Charles W. *A History of the Art of War in the Middle Ages.* London: Greenhill Press, 1991.

Cowpens (17 January 1781)

One of the most one-sided victories for American forces during the War for Independence. On 17 January 1781, American and British forces clashed at the Cowpens in western South Carolina. There were approximately 1,000 men on each side. General Daniel Morgan made brilliant use of his largely undertrained militia forces. He stationed some 50 sharpshooters on one hill and, taking advantage of the accuracy of their rifled muskets, he had them fire a round or two at the advancing British and then retire to their left. He had stationed about 400 better-trained militia on the next rolling hill 150 yards back with the same orders. On the third hill, he placed his 450 trained continental troops and his cavalry behind them and to the right.

The plan worked to perfection. The two lines of militia fired their rounds and retreated; the British took casualties but kept advancing. Serendipitously, as the British approached the line of Continentals, the militia had swung behind that line and reappeared with the cavalry on the right. As the British attacked, they were routed between the unplanned double envelopment.

In the end, virtually the entire British force was killed, wounded, or captured; the victory at Cowpens led to the battle at Guilford Court House, which reversed the tide of British victories in the South, leading later in the year to Lord Cornwallis's surrender at Yorktown.

Charles M. Dobbs

See also: American Revolution; Cornwallis, Sir Charles; Greene, Nathanael

References and further reading:
Davis, Burke. *The Cowpens–Guilford Court House Campaign.* Philadelphia, PA: Lippincott, 1962.
Higginbotham, Don. *Daniel Morgan, Revolutionary Rifleman.* Chapel Hill: University of North Carolina Press, 1961.
Treacy, M. F. *Prelude to Yorktown: The Southern Campaign of General Nathanael Greene.* Chapel Hill: University of North Carolina Press, 1963.

Crazy Horse (1840–1877)

Native American chief and war leader Tasunke Witko, or Crazy Horse, was born near Bear Butte in the Black Hills of South Dakota. By then, the Plains Indians, like the Sioux, had adopted horse and gun and were effective, if unorthodox, aboriginal cavalry.

In 1866 Crazy Horse achieved notoriety by luring 80 soldiers from Fort Kearny into a trap set by the Lakota Sioux, Arapaho, and Cheyenne. All 80 soldiers were killed in the Fetterman "Massacre." By 1876 Crazy Horse was Sioux war leader.

At the Battle of Rosebud, Crazy Horse and his warriors held George Custer and his Seventh Cavalry off and eight days later, the Lakota, Cheyenne, and Arapaho annihilated Custer and his Seventh Cavalry at the Battle of the Little Bighorn. This last Sioux victory caused the U.S. government to pursue a policy of starvation in the winter of 1877. In May 1877 Crazy Horse and his band surrendered at Fort Robinson in the Dakota Territory.

Later, a rumor suggested Crazy Horse was plotting against the government and he was to be arrested. He resisted and was stabbed to death with bayonets, probably by other Sioux serving as camp police, on 5 September 1877.

Chris Howell

See also: American Indian Wars; Little Bighorn; Sitting Bull

References and further reading:
Hogan. William T. *American Indians.* 3d ed. Chicago, IL: University of Chicago Press, 1993.
McGinnis, Anthony. *Counting Coup and Cutting Horses: Intertribal Warfare on the Northern Plains 1738–1889.* Evergreen, CO: Cordillera Press, 1990.
Secoy, Frank R. *Changing Military Patterns of the Great Plains Indians.* Lincoln: University of Nebraska Press; Bison Books, 1992 reprint of 1953 monograph.

Crazy Horse and his band of Indians on their way from Camp Sheridan to surrender to General Crook, 1877. (Library of Congress)

Starkey, Armstrong. *European and Native American Warfare 1675–1815.* Norman: University of Oklahoma Press, 1998.

Crécy, Battle of (25 August 1346)

Decisive English victory of Hundred Years War. The renewal of the Hundred Years War in 1345 saw the French invasion of Gascony in 1346, and an English landing near Cherbourg. It was quickly followed up by the capture of Caen (27 July). Advancing on Paris and devastating French territory as it went, the English army forced the French to abandon their effort in Gascony, in order to oppose the English advance. Knowing that the French had abandoned their siege of Aiguillon and were advancing north, Edward III decided to cross the Seine and open a line of communication, and retreat, to Flanders. In response, the French destroyed all the bridges over the Seine below Paris. This forced the English army to advance almost to the French capital until they found a bridge, at Poissy, that could be repaired. There the English army crossed the river, a numerically superior French army making no effort to prevent it. Once north of the Seine, the English army possessed the critical advantage of one day's lead over the French. This proved critical in enabling the English to ford the Somme just ahead of pursuing French forces.

Once north of the Somme, Edward III was prepared to give battle in Ponthieu. The position selected was at Crécy, where the English army would stand astride the road that the French would have to take as they came north. On the right flank was a river, and on the left, woods and the village of Wadicourt. Between the two, the English dug a series of ditches behind which they would fight in a double echelon. Estimates of English strength vary between 10,000 and 20,000 men, and of French strength between 28,000 and 60,000, the latter figure certainly being far too high. The main difference between the two armies, other than the English choice of ground on which to give battle, was that the English, who had Welsh and Irish in their ranks, were largely drawn from veterans of the Welsh and Scottish wars. The French, on the other hand, had no comparative experience of battle to draw upon, and did not possess the discipline within formations needed when battle was joined.

The French encountered the English late on the afternoon of 25 August, and immediately deployed their Genoese crossbowmen to the front to fire on the English line. The crossbowmen were decisively outranged by longbows, and the Genoese were first shattered by English fire, and then ridden down by French armored knights in the vanguard. Having to charge up a slope made slippery by rain, the French assault broke under a rain of arrows. Very few of the French reached the English line, and those who did were quickly repulsed. Thereafter a pattern imposed itself on the battle. Successive French formations, which did not deploy into line from their marching columns, and which did not support one another, sought to breach the English positions as night fell. It has been suggested that there were some 15 or 16 successive French assaults, all broken. The English stood at their positions throughout the night. The next day they found that the French army had abandoned the field. They left between 10,000 and 20,000 dead, estimated to have included over 1,500 members of the nobility.

Crécy is supposed to have been the first European battle in which gunpowder was used, but the real significance of the battle was twofold: it was a battle fought and won by infantry, and it was a decisive English victory that provided the psychological basis for English operations over the next 70 years. Nonetheless, the English lost any immediate advantage from the battle due to the outbreak of the Black Death, which after 1347 imposed a seven-year halt on proceedings.

H. P. Willmott

See also: Agincourt, Battle of

References and further reading:

Burne, Alfred H. *The Crecy War: A Military History of the Hundred Years War from 1337 to the Peace of Bretigny, 1360.* London: Eyre & Spottiswoode, 1955.

Seward, Desmond. *The Hundred Years' War: The English in France, 1347–1453.* London: Constable, 1978.

Sumption, Jonathan. *The Hundred Years War.* Vol. 1, *Trial by Battle.* Philadelphia: University of Pennsylvania Press, 1991.

Creek War (1813–1814)

Offshoot of the War of 1812, fought in Alabama by an alliance of Americans, Cherokees, and some Creeks against renegade Creeks armed by the British and supported by the Spanish. Inspired by stories of Tecumseh, several thousand young Creek warriors resisted assimilation into the agricultural lifestyle of the whites. One of their leaders, half-breed Peter McQueen, engaged Mississippi militia under Colonel James Caller at Burnt Corn Creek on 27 July 1813. Another half-breed, William Weatherford, or Red Eagle, leader of the extremist "Red Stick" faction, massacred all 400 whites at Fort Mims on 30 August, sparing only the black slaves and losing about 300 warriors.

Tennessee rapidly mobilized against the Creeks, sending about 5,000 regulars and militia under Andrew Jackson, John Coffee, James White, and John Cocke south in October. Jackson established headquarters at Fort Strother. Coffee attacked and destroyed the Creek village of Tallushatchee on 3 November. Jackson did the same to Talladega on 9 November. Cocke and White, against Jackson's orders, killed 60 friendly Creeks on 17 November. Mississippi militia under Frederick Claiborne and Georgia militia under John Floyd began in November to converge on the Creek "Holy Ground" near the juncture of the Coosa and Tallapoosa Rivers. Jackson fought at Emuckfau on 22 January 1814 and Enotachopco on 24 January. Floyd lost at Calibee Creek on 27 January. Hostilities were minimal after Jackson's victory at Horseshoe Bend on 27 March. By the Treaty of Fort Jackson on 9 August, the United States had seized 20 million acres of Creek land, regardless of whether the inhabitants of that land had fought for or against the Americans.

Eric v. d. Luft

See also: Horseshoe Bend, Battle of; Houston, Samuel; Jackson, Andrew; Tecumseh; War of 1812

References and further reading:

Martin, Joel W. *Sacred Revolt: The Muskogees' Struggle for a New World.* Boston: Beacon, 1991.

Rogin, Michael Paul. *Fathers and Children: Andrew Jackson and the Subjugation of the American Indian.* New Brunswick, NJ: Transaction, 1991.

Crete (1941)

After their completely successful Balkans campaign at the end of April 1941, the British troops that escaped from Greece were moved to Crete. They were joined by 12,000 reinforcements from Egypt and were also assisted by a 14,000-man Greek garrison. With the addition of the 15,500 men from Greece, the island boasted a defensive force of 41,500 men.

The strategic value of Crete meant that the Germans needed to capture it to secure their communications in the eastern Mediterranean, especially in view of the British presence elsewhere in the Middle East, particularly on the Suez Canal. Lieutenant General Kurt Student and 7th Airborne Division were placed with 5th Mountain Division in XI Airborne Corps, and ordered to capture the island.

The island defense was commanded by the New Zealander, Major General Bernard Freyberg, who was given advance ULTRA signal intercept intelligence of German intentions. Nevertheless, when the first wave of paratroops landed on 20 May, his forces concentrated on defense, rather than attacking with their superiority in numbers.

The Germans held on, and on 21 May the rest of 7th (German) Parachute Division was dropped. Heavy fighting took place around Maleme airfield, Rethymnon, and Herakleion. Despite British command of the sea, the Germans managed to land reinforcements of 5th Mountain Division and further glider-borne troops. Freyberg was forced back to the south coast of the island, and during the fighting for Sfakia the Royal Navy managed to evacuate a number of troops, but Crete was lost to the Germans.

German casualties were so heavy that Hitler never again used mass parachute attack. During the parachute landings the defenders had cut the German paratroops to pieces, mainly because the landing was made piecemeal against heavily defended areas instead of being concentrated in a quiet area to allow forming up before attacking.

David Westwood

See also: Airborne Operations; Balkans Campaign; Student, Kurt; Tactics

References and further reading:

Burdick, Charles, and Hans-Adolf Jacobsen. *The Halder War Diary 1939–1942.* London: Greenhill Books, 1988.

Reinhardt, Major General Hellmuth, et al. *Airborne Operations: A German Appraisal.* Washington, DC: U.S. Army Center of Military History, 1989.

Crimean War (1853–1856)

War fought against Russia by Britain, France, Turkey, and several small Italian states to check Russian aggression against Turkey. Because it involved a theater of operations beyond just the Crimean Peninsula, the Crimean War would more accurately be called the Anglo-Russian War. The British had a Baltic Sea strategy as well as a Black Sea strategy, and one could argue that the British naval and amphibious threat to Cronstadt and St. Petersburg in 1855–1856 was more decisive toward ending the war than was the fall of Sevastopol in 1855. The conflict is termed the Crimean War because its bloodiest and most famous engagements were the major land battles in the 1854–1855 Sevastopol campaign.

Despite French naval superiority over Russia, the two greatest world powers in the early 1850s were Great Britain and Imperial Russia. Queen Victoria and Czar Nicholas I were playing the so-called Great Game for the control of Asia and the Near East. Constantinople, the Bosporus, the Sea of Marmara, and the Dardanelles were together a key strategic area in the Great Game, because whoever held them could either allow or deny Russian naval access to the Mediterranean Sea. Whether Russia in fact had a Mediterranean strategy is a moot point, but Britain thought it did. Her Majesty's government believed that maintaining the sovereignty of Turkey against any Russian incursion was in the British national interest. France, under the maritime-minded Emperor Napoleon III, also did not want Russian warships in the Mediterranean.

In January 1853, the czar, believing he had Austrian support, perceiving Turkish weakness, and hoping to stalemate France with diplomacy, began to lay naval and military plans to seize from Turkey the passage to the Mediterranean. From 2 March to 21 May, Russian prince Alexandr Sergeevich Menshikov, backed by a large force, was in Constantinople trying to bully concessions from Sultan Abdul Mejid I. Menshikov failed to intimidate the Turks, and angrily departed when the sultan refused to yield.

Saber-rattling continued after the Turks refused Menshikov's ultimatums. On 2 July, four Russian infantry and two Russian cavalry divisions under Prince Mikhail Dmitriyevich Gorchakov crossed the Prut River, occupied the Turkish Christian principalities of Moldavia and Wallachia (modern Romania), and threatened, once reinforced, to cross the Danube. Skirmishes broke out on all Turko-Russian frontiers. Austria stayed neutral and hosted a futile peace conference in Vienna. Britain and France managed to restrain Turkey from declaring war on Russia until 4 October.

Turks under Omer Pasha crossed the Danube on 23 October and won a minor victory over Gorchakov at Oltenitza, Wallachia, on 4 November. Moving out from his established headquarters at Shumla (modern Shumen, Bulgaria) in December, Omer Pasha checked the advance of 30,000 Russians on Cetate, Wallachia, in January 1854 and on Calafat, Wallachia, in February.

On 30 November 1853, the Russian navy annihilated the Turkish fleet off Sinope, Turkey. Britain and France mobilized, but continued negotiating with Russia. When the czar rejected their final demands to withdraw from Moldavia and Wallachia, France declared war on 27 March 1854; Great Britain followed suit the next day.

Gorchakov crossed the Danube at Galatzi, Moldavia, on 23 March. Under Gorchakov's orders, Prince Ivan Feodorovich Paskevich laid siege to Silistra, in modern Bulgaria, on 4 May. Silistra would be the key to Russian flank defenses if the allies were to invade in the vicinity of Varna, Bulgaria. The siege was unsuccessful, the Russians lost a battle at Silistra on 9 June, and on 23 June Gorchakov obeyed the czar's order to retreat. Under pressure from Austria and frustrated by Austrian treaties with Prussia and Turkey, Russia withdrew from Moldavia and Wallachia on 8 August. Austria occupied the towns on 22 August.

A sharply divided British cabinet undertook to plan the overall strategy of the war. Some members wanted to secure the Turkish frontier in the Danube region and push the Russians back into Bessarabia and the Ukraine. Others wanted to fight an Asian land war to help the Turks push the Russians back into Georgia. Still others wanted primarily naval action in the Baltic and Black Seas. For reasons of homefront morale and propaganda rather than sound military logic, the cabinet decided on 28 June that their objective would be Sevastopol, jewel city of the Crimea and an important naval base.

When the British commander in chief, General Fitzroy James Henry Somerset, Baron Raglan, received the cabinet's dispatch a few weeks later at Varna, he knew that invading the Crimea would be a military blunder. The allies had insufficient intelligence, supplies, arms, and men to ensure victory. But Raglan also knew that he would be replaced with someone willing to attack Sevastopol if he did not treat the cabinet's advice as a direct order. Reluctantly, he persuaded the French to join the invasion.

The Allied landing began on 14 September at (the perhaps aptly named) Calamita Bay, just south of Eupatoria, 35 miles north of Sevastopol. Menshikov was prepared for land action around Varna or Odessa, but he did not expect the allies to invade the Crimea. Thus he was even less prepared than the allies. After routing the Russians at the Alma River on 20 September, the allies flank-marched to Balaklava to establish a supply base. The Russians attempted unsuccessfully on 24–25 October to recapture Balaklava, and made minor strategic gains.

Two battles were fought at Inkerman, a skirmish on 26 October and the major Russian offensive on 5 November, which resulted in total victory for the outnumbered British. Action around Sevastopol was then relatively quiet until April 1855. The French and British hoped to solidify their new friendship by taking Sevastopol on the fortieth anniversary of Waterloo, 18 June, but their massive combined assault that day failed. On 16 August the French lost only 1,800 as they inflicted 8,000 casualties on the Russians at the Tchernaya River.

After 11 months of brutal siege, the city of Sevastopol fell to the allies on 11 September. Thereafter the war was essentially over. The Russians captured Kars from the Turks on 28 November, but on 28 December, Austria threatened to enter the war on the side of the allies unless Russia made peace. The Treaty of Paris, 30 March 1856, ended the war. Russia returned what it had captured from Turkey, made other territorial concessions, and neutralized the Black Sea. On 15 April, Britain, France, and Austria signed a treaty regarding any Russian renavalization of the Black Sea or any threat to Turkey as casus belli. The Crimean War has gone down in history as one of the most unnecessary of modern conflicts, its cases intricate and remote, its course inept, its settlement almost a restatement of the *status quo ante bellum.*

Eric v. d. Luft

See also: Alma; Balaklava; Inkerman, Battle of the; Light Brigade, Charge of the; Sevastopol, Siege of

References and further reading:

Baumgart, Winfried. *The Crimean War 1853–1856.* London: Edward Arnold, 2000.

Coached, J. B. *Britain and the Crimea, 1855–56: Problems of War and Peace.* New York: St. Martin's Press, 1988.

Lambert, Andrew D. *The Crimean War: British Grand Strategy, 1853–56.* Manchester, UK: Manchester University Press, 1990.

Woodham-Smith, Cecil. *Reason Why.* New York: McGraw-Hill, 1954.

Croesus (fl. c. 560–546 B.C.E.)

Much of the information on Croesus, last king of Lydia and conqueror of Ionia, is shrouded in myth and fable. In the early sixth century, Croesus's father, King Alyattes, expanded the Lydian Empire to encompass most of western Asia Minor and Croesus served as a general during part of this time. After the death of Alyattes, Croesus won a short succession struggle over his stepbrother to gain the throne. He subjugated the coastal cities of mainland Ionia with the single exception of Miletus but he was forced to ally with the Ionian offshore islands because he lacked a navy capable of subduing them. He also made alliances with Egypt, Chaldea, and Sparta, hoping to form a bloc against the rising power of Persia under Cyrus the Great.

According to myth, Croesus often consulted the oracle at Delphi. On one occasion Croesus is supposed to have asked what would be the result of an invasion of Persia, and the oracle replied that it would destroy a great kingdom. Encouraged, Croesus proceeded with an invasion of Cappadocia, in eastern Anatolia in 547. After a battle with no clear victor at Pteria, Croesus retreated to his capital of Sardis to regroup. He was waiting for reinforcements from his Egyptian and Spartan allies when the smaller Persian army caught up and defeated him at Thymbra. Lydia became a province of Persia. This led to later conflict between the Greek cities of Ionia and Persia.

After his defeat, some accounts say that Croesus became attached to the Persian court and even became the governor of Barene. Others say Croesus was killed during the Persian conquest.

Harold Wise

See also: Cyrus II the Great; Persian Empire; Persian Wars of Expansion

References and further reading:

Herodotus, *The Histories*. Trans. Aubrey De Sélincourt. London and New York: Penguin, 1996.

Cromwell, Oliver (1599–1658)

English military and government leader who established a military protectorate. Cromwell was born on 25 April 1599 in Huntingdon, England, and educated at Cambridge. At 29, he was elected to Parliament, but was dismissed along with the other representatives when Charles I ruled as a despot for 11 years. Under financial pressure, Charles had to reconvene Parliament in 1640. In the power struggle between the king and Parliament, Cromwell, a strong Puritan or Roundhead, returned home to raise a cavalry unit. He first led troops at the Battle of Edgehill on 23 October 1642. Over the next several years, he rose to the rank of lieutenant general because his strong disciplinary methods allowed him to control his men in battle. Between 1643 and 1646, he fought in numerous battles, the greatest being the Battle of Marston Moor, outside of York, on 2 July 1644. His cavalry performed valiantly, but Cromwell criticized faulty leadership on the parliamentary side. Almost impeached in 1644, he proposed in December that in the future no member of Parliament should be allowed to hold a command. Parliament agreed to a new army with Sir Thomas Fairfax in command. In the summer of 1645, Fairfax insisted that Cromwell be made second-in-command. At the Battles of Naseby and Langport, the last of Charles I's two field armies were destroyed. In January 1646, Cromwell's commission was renewed and he was financially rewarded for his service. He then joined Fairfax at the siege of Oxford. Soon after, Charles surrendered to the Scots and was turned over to Parliament.

In the aftermath of the Civil War, the Parliament and army came into dispute. Many in Parliament felt the army could not be trusted, ordered it to be disbanded, and hired a Scottish army to protect them. Feeling his fellow soldiers had been disgracefully treated, Cromwell joined them on 4 June 1647. In November 1648, Cromwell's soldiers removed 110 members of Parliament and another 160 members refused to take their seats in opposition. This new Rump Parliament dismantled the government and consolidated power to an executive Council of State and themselves. Cromwell's army defeated the radical Levellers at Burford, slaughtered 40 percent of indigenous Irishmen, transported the remaining Irish to County Connaught with the Act of Settlement in 1653, and defeated the Scottish Presbyterians at Dunbar in 1650 and Worcester in 1651. Cromwell even dissolved the Rump government on 21 April 1653. Two years later, he again dissolved Parliament and established a military-religious dictatorship, with himself as Lord Protector of the Commonwealth. On 3 September 1658, he died in London.

With delusions of monarchy, Cromwell had appointed his son, Richard, as his successor. But "Tumbledown Dick" was no Oliver, and soon the house of Stuart, in the person of Charles II, was on the throne once again, to genuinely joyous public acclaim. The remains of Oliver Cromwell and other deceased regicides were dug up and publicly execrated (Cromwell's head being attached to a pike overlooking the Tower of London), while living antiroyalists suffered agonizing deaths at the hands of a vengeful monarch. Much of the English (and, later, American) suspicion of the military derives from the strict military reign of Oliver Cromwell.

T. Jason Soderstrum

See also: British Military, Twentieth Century Organization and Structure; English Civil War (1642–1649)

References and further reading:

Baldock, Thomas Stanford. *Cromwell as a Soldier.* London: K. Paul, Trench, Trubner, & Company, Ltd., 1899.
Coward, Barry. *Oliver Cromwell.* London: Longman, 1991.
Fraser, Antonia. *Cromwell.* New York: Grove Press, 2001.
Gillingham, John. *Cromwell: Portrait of a Soldier.* London: Weidenfeld & Nicolson, 1976.

Crusades (1095–1272)

Attempt to recover Christian territory from Muslim occupation in a series of military expeditions by western European Christians. The Crusades are often divided into seven distinct campaigns, but the crusading movement was actually an ongoing process with expeditions intermittently coming and going. Most were directed primarily toward what is now known as the Middle East, but some crusades also targeted lands in Europe occupied by Moors, other non-Christians, or even Christian heretics.

The motivations behind the Crusades were complex. Piety was certainly a factor, as crusaders sought to reclaim Christian territory held by infidels. Crusaders were also pilgrims who vowed to visit the Holy Sepulcher in Jerusalem, and the Church rewarded crusaders with remission from sin for participation. Also motivating factors were pursuit of wealth, thirst for adventure, and the desire to expand European influence eastward.

In the decades before the crusading movement began, Muslim armies had been encroaching on Byzantine territory. In 1071, the Byzantine army was crushed by invading Seljuq Turks led by Alp Arslan at the Battle of Manzikert, and stories of Turkish brutality against Christians began to appear in the West.

The crusading movement proper began in 1095 when Byzantine emperor Alexius Comnenus I, in a desperate attempt to stave off an approaching Muslim army, appealed to Pope Urban II for help. Urban in turn appealed to the European nobility in a famous speech at Clermont. He called on them to help their Byzantine brothers and regain Holy Lands that had been lost to the Muslims.

The resulting First Crusade (1096–1099) was the most successful of all crusades, culminating with a victory at Jerusalem in 1099. After this, the permanent outpost kingdoms of Edessa, Tripolis, Antioch, and Jerusalem (collectively known as the Outremer) were created.

Later crusades were not as successful as the First Crusade. On Christmas Eve 1144, Muslims led by Imad ad-Din Zengi captured Edessa, leading to renewed crusading fervor in the West and to the Second Crusade (1147–1148). The Second Crusade not only attempted to regain lands lost in the Middle East, but also strove to oust Moors from the Iberian Peninsula, and to convert non-Christian Wends in what is now northern Germany. The crusade to the Middle East failed miserably, but crusaders in Portugal were more successful and took Lisbon in 1147.

In 1187, Muslim forces led by a powerful new warrior named Saladin annihilated a crusader army at the Battle of Hattin. This resulted in the fall of Jerusalem, and the Third Crusade (1187–1193), which was led by Europe's most powerful monarchs, including King Richard I (better known as Richard the Lionheart) from England. Richard succeeded in regaining considerable portions of crusader territory, and defeated Saladin's forces at Acre, but ultimately failed to retake Jerusalem from the Muslims.

With Jerusalem still in Muslim hands, crusading zeal in the West continued. The Fourth Crusade (1201–1204) never reached the Holy Land at all. Diverted crusaders instead captured the Adriatic port of Zara on behalf of the Venetian government (for which they were summarily excommunicated by Pope Innocent III), and then took Constantinople itself, resulting in a short-lived Latin empire. Crusaders from the Fourth Crusade returned to Europe enormously wealthy and brought with them thousands of Holy Relics taken from the Byzantines.

The last crusades, the Fifth (1217–1221), Sixth (1248), and Seventh (1270), were directed toward Egypt, the center of Muslim power. Crusaders achieved negligible success, and ultimately returned to Europe as failures. In 1291, Mamluks led by Kalavun laid siege to Acre, the last bastion of Christian power in the Middle East. The crusader army was outnumbered seven to one, and Acre soon fell. This defeat led to a permanent end of the crusader presence in the Middle East.

In the Middle East, there were few long-term political consequences of the Crusades as all conquered territory was eventually lost, but the crusading movement had profound and lasting effects on European politics and culture. New military monastic orders, such as the Knights Hospitalers, Knights Templars, and Teutonic Knights, were born. Returning crusaders brought with them new military tactics and innovations, such as Greek Fire and advanced castle fortification techniques. Crusaders also imported Middle Eastern and Greek intellectual, mathematical, and philosophical ideas, which many believe helped open the door for an intellectual revival in the West.

Melanie Casey

See also: Alexius I Comnenus; Antioch, Battle of; Armor, Ancient and Medieval; Arsuf, Battle of; Byzantine-Muslim Wars; Dorylaeum, Battle of; Frederick I Barbarossa; Hattin, Battle of; Jerusalem, Siege of; Latin Empire–Byzantine Wars; Mansûrah, Battle of; Ramleh, Battle of; Richard I; Saladin; Seljuqs

References and further reading:

Cowley, Robert, and Geoffrey Parker, eds. *The Reader's Companion to Military History.* Boston: Houghton Mifflin, 1996.

Hollister, C. Warren. *Medieval Europe: A Short History.* 8th ed. Boston: McGraw-Hill, 1998.

Madden, Thomas F. *A Concise History of the Crusades.* Lanham, MD: Rowman & Littlefield, 1999.

Riley-Smith, Jonathan. *The Crusades: A Short History.* New Haven: Yale University Press, 1987.

Cuban Missile Crisis (October 1962)

The closest the world has yet come to nuclear war. After Castro seized power in Cuba on 1 January 1959, he set about reforming various facets of the Cuban economy and society. One of his major reforms was going to be in landownership. To this end the United States and Cuba began discussions on economic relations as early as April 1959. These did not go well, however, and Cuba seized American property in January 1960. This expropriation led to protests by the American ambassador and in October 1960 to an American embargo of all exports to Cuba except for medicine and food.

The situation went from bad to worse, and by January 1961 Cuba and the United States had severed diplomatic ties. In July 1962, Castro declared in one of his lengthy speeches that Cuba would soon have new defenses against American imperialism. On 29 August 1962, U-2 spy-plane photos verified surface-to-air defensive missiles (SAMs) in Cuba. After new reconnaissance photos revealed the possibility of offensive weapons in Cuba, President John F. Kennedy warned Soviet chairman Nikita Khrushchev not to place offensive nuclear missiles in Cuba. On 14 October 1962, U-2 spy-plane photos showed that launch pads for medium- and intermediate-range ballistic missiles were under construction in Cuba. On 22 October, President Kennedy announced to the American public that Soviet missiles were being placed in Cuba and that a "quarantine," basically a blockade of Cuba, was in effect. Between this speech and 24 October, Soviet ships acknowledged the American blockade, and turned back while still in international waters. By the 26th and 27th, the Soviet Union and the United States had exchanged letters addressing the resolution of this crisis. On 28 October, Chairman Khrushchev agreed to remove the Soviet missiles if President Kennedy would pledge not to invade Cuba. Kennedy agreed (informally) and the last Soviet missiles were removed from Cuba by 20 November 1962. This agreement basically ensured Castro's continuation in power as it also prevented Cubans from leading filibustering expeditions from the U.S. in any attempt to overthrow Castro.

The United States was able to get its way in the Cuban Missile Crisis because of its great lead in nuclear weaponry and naval power. The Soviet leadership was determined that the nation would never again be put in such a humiliating position. Consequently, Khrushchev was removed from office in 1964 and the USSR embarked on a massive buildup of Soviet arms, the costs of which, in turn, helped lead to the collapse of the Soviet Union three decades later.

Peter Carr

See also: Cold War

References and further reading:

Abel, Elie. *The Missile Crisis.* Philadelphia, PA: Lippincott, 1966.

Allison, Graham T., and Philip Zelikow. *Essence of Decision: Explaining the Cuban Missile Crisis.* 2d ed. New York: Longman, 1999.

Miller, David. *The Cold War: A Military History.* New York: St. Martin's Press, 1998.

Nash, Philip. *The Other Missiles of October: Eisenhower, Kennedy, and the Jupiters, 1957–1963.* Chapel Hill: University of North Carolina Press, 1997.

Painter, David. *The Cold War: An International History.* London: Routledge Press, 2000.

Powaski, Ronald. *The Cold War, the United States and the Soviet Union, 1917–1991.* New York: Oxford University Press, 1998.

Cuban Ten Years' War (1868–1878)

Failed guerrilla independence war. The Cuban Ten Years' War started on 10 October 1868 after Carlos Manuel de Cespedes called for the uprising to start at his plantation, La Demajagua, in the eastern province of Oriente. His proclamation is known as the Grito de Yara. Though he planned for the revolt to begin at a later date, word of betrayal precipitated its start.

Coincidentally de Cespedes also freed all of his slaves, although this war was not fought for the ending of slavery. Instead, after many years of failed attempts at getting political and economic reforms from the Spanish government in Cuba and in Madrid, Cuban Creole sugar planters decided it was time to take matters into their own hands. Denouncing heavy taxation, the independence movement promised a representative form of government and suffrage for all men. The abolition of slavery would be slowly implemented. In addition to de Cespedes, some of the other leaders were Salvador Cisneros Betancourt, Francisco Vicente Aguilera, Bartolome Masó, Pedro Figueredo, Ignacio Agramonte, Máximo Gómez from Santo Domingo, and the eventual Cuban army chief. More than 40,000 supporters joined the rebel forces by the start of the 1870s.

The revolt, for the most part, was an eastern Cuban phenomenon. Efforts by de Cespedes and his junta to gain support in the western part of the island were in vain. Cuban troops were composed largely of poor farmers and freed slaves. This smaller, ill-equipped, and poorly trained force kept the Spanish army at bay for 10 years. Spanish troops, on the other hand, numbered well over 100,000 during the mid- to later 1870s. In the province of Camagüey, a ditch or *trocha* was constructed by the Spanish army to prevent the easy passage by the Cuban insurrectionists. Shortly after the start of the war, some of the leaders, such as de Cespedes, Masó, Figueredo, and a few others, petitioned American secretary of state William Seward to accept Cuba as a state, but were denied. The United States was not about to take on another slave state. A year after the start of the war, a constitutional assembly was held at Guáimaro.

During the 1870s the war became one of attrition. The more Spanish troops that were killed, the more were sent from Spain. But in 1877, Spanish general Arsenio Martinez Campos arrived to command Spanish forces. He brought many more troops and promises for reforms. After negotiations in early 1878, the Pact of Zanjon was signed and the war was effectively brought to an end.

Though some factions continued to fight, the bulk of the separatists' army disbanded. The pact granted general amnesty to all insurrectionists as well as unconditional freedom to all slaves and other indentured servants. One of the leaders who denounced this pact was General Antonio Maceo, who promulgated the Protest of Baragua and who would be one of the major figures of the Cuban War of Independence, 1895–1898.

Peter Carr

See also: Cuban War of Independence; Spanish-American War

References and further reading:

Collazo, Enrique. *Desde Yara hasta el Zanjon, apuntaciones historicas.* 2d ed. Habana: Tip. de "La Lucha," 1893.

Cuban War of Independence (1895–1898)

After the end of the Cuban Ten Years' War of 1868–1878, many of the top military figures went into exile. Generals Máximo Gómez and Antonio Maceo, as well as the poet, author, and philosopher José Martí, lived throughout South and Central America as well as the United States, mostly in New York. Though various political and economic reforms were instituted by the Spanish government after the end of the Ten Years' War, these were not radical enough for those who wanted Cubans to be the masters of their own fate.

Starting in October 1868, to escape the ravages of war, many Cuban families went into exile in Key West and Tampa, Florida, and other major cities of the United States, such as New York, New Orleans, Philadelphia, and Baltimore. Many of these families had prospered during their years in the United States. They would be a major source of financial and moral support to the Cuban Revolutionary Party. Another source of financial support was the various bandits that roamed the Cuban countryside. Mostly concerned with lining their pockets, they too were interested in separating from Spain. The most significant of these was Manuel Garcia.

In April 1892, Martí officially established the Cuban Revolutionary Party in an attempt to unite all Cubans in the struggle for independence. On 24 February 1895, the "Grito de Baire," a "call" for armed insurrection against Spanish rule, was made. Though mostly in the eastern end of Cuba, the revolution quickly succeeded in fielding an army of more than 3,000 men. In April 1895, Martí, Gómez, and Maceo landed in Cuba. In the meantime, Spanish general Martinez Campos was sent to Cuba to try to work out some agreement to avoid conflict. In May 1895, Martí was killed in a small skirmish with the Spanish. Shortly thereafter, Maceo and Gómez were able to cross the Trocha into Las Villas province. The Cuban Revolutionary Army was divided into six corps, one for each province in existence then. Each corps was subdivided into divisions, regiments, and other smaller military divisions.

By 1896, the war again became one of attrition. Spain continued to send troops while Gómez and Maceo's troops killed them. In early 1896, Gómez initiated the policy of burning sugar plantations and their crops because he felt that the proceeds from the harvest provided funding for the counterrevolutionary operations by Spain. Of course, this policy also caused much misery among the peasant sugar harvesters. The revolutionaries recognized that they needed ships to break the Spanish blockade, bring in supplies and recruits, and conduct raids. Although many efforts were made to acquire vessels from the United States, that nation maintained an anti-filibustering policy.

Spanish general Valeriano Weyler y Nicolau then initiated the policy of *reconcentración,* by which Cubans in the outlying regions were concentrated in camps with barely enough food and water to stay alive. This policy, though somewhat effective in hindering the revolution, led to an even deeper hatred of Spain, and resulted in adverse publicity, fueled by exiled Cubans, in the United States.

In February 1898, the American battleship *Maine* arrived in Havana harbor on a goodwill visit. (By then, both Martí and Maceo had been killed in action.) Its visit was protested by both the Spanish and Cuban revolutionaries. Whether by Cuban revolutionaries or by the explosion of its own coal bunkers, the warship was destroyed by an explosion in Havana harbor. (Although the American yellow press screamed that the Spanish had blown up the warship, they would have had the least to gain by such a mad act, although it is possible that rogue Spanish nationalists, determined to avenge Spanish honor, were guilty, but no clue has ever emerged pointing in this direction.)

The mysterious explosion precipitated the entrance of the United States into the war, which ended with Spain's capitulation by August 1898. Unfortunately, the American entrance into the Cuban War of Independence helped to obscure the three years of Cuban fighting against Spain, and the war is remembered inaccurately as simply the Spanish-American War.

Peter E. Carr

See also: Cuban War of Independence; Martí y Pérez, José Julián; Weyler y Nicolau, Valeriano, Marquis of Tenerife

References and further reading:

Berner, Brad K. *The Spanish-American War: A Historical Dictionary.* Lanham, MD: Scarecrow Press, 1998.

Dressel, C. M. F. *The Cuban Spanish American War.* New York: Dressel Co. Publishers, c. 1898.

Foner, Philip Sheldon. *The Spanish-Cuban-American War and the Birth of American Imperialism, 1895–1902.* 2 vols. New York: Monthly Review Press, 1972.

Quesada, Gonzalo de, and Henry Davenport Northrop. *The War in Cuba: Being a Full Account of Her Great Struggle for Freedom: Containing a Complete Record of Spanish Tyranny and Oppression, Scenes of Violence and Bloodshed, Daring Deeds of Cuban Heroes and Patriots, Thrilling Incidents of the Conflict, American Aid for the Cause of Cuba, Secret Expeditions, Inside Facts of the War. . . .* Chicago: Wabash Publishing House, c. 1896.

Culloden, Battle of (1746)

The last land battle fought on British soil. Having retreated into Scotland after a war council at Derby determined that an attack on London, without French aid or local support, would be disastrous, the Jacobite armies of Charles Edward

British troops firing on fleeing and dying soldiers who are being tended to by their women, on Culloden Moor. (Hulton/Archive)

Stuart wintered close to the ground and steadily moved to the northwest of Scotland. The prince himself, disappointed by the failure of the march on London, pressed for a direct engagement with the Hanoverian forces, which had followed from Derby under the command of the Duke of Cumberland and General George Wade. Many of the prince's advisers, however, counseled a sustained guerrilla campaign, hoping to gain something out of a negotiated peace. Ultimately, the prince insisted on meeting the Hanoverian army at Culloden, east of Inverness, on a field with several significant obstacles, including marshy ground. When the battle began around 1 P.M. the afternoon of 16 October, the Jacobite army of about 5,000 immediately found their line thrown into disarray because of the obstacles, with second-line units being brought up to fill the gaps. Both sides fired about 15 minutes of artillery, an action that greatly demoralized the Jacobites, whose own artillery did little damage. When the Jacobite attack came, it was as a Highland charge, advancing at a run, into the Hanoverian lines, with many of the Jacobites drawing swords. The Hanoverians, numbering about 9,000, countered the charge by changing over from ball to canister shot, which devastated the charge, killing a disproportionate number of officers leading "from the front." The Jacobites were thrown into disarray as the survivors made contact with the front ranks of the enemy, who fired once, then used their bayonets.

While the Jacobites struggled to advance across boggy ground in the face of continuing fire, their right wings began to collapse and run, followed by many units from the left. At this moment, Cumberland sent the two troops of Cobham's 10th Dragoons in pursuit, followed by a general charge of the Hanoverian army, which turned the retreat into a rout. In a battle that lasted perhaps 40 minutes, the Jacobites had lost nearly 1,000 dead while the government lost only 50. Although the Jacobites claimed that their wounded were systematically butchered on the field, giving Cumberland his nickname "The Butcher," most survivors died in custody. The ruthlessness with which the government proceeded to end the rebellion, including the prohibitions of Highland traditions such as the bagpipes and plaid, as well as systematic road building and fortification, ended the Jacobite cause in all but romantic memory.

C. E. Wood

See also: British Military, Twentieth Century Organization and Structure

References and further reading:

Black, Jeremy. *Culloden and the '45.* New York: St. Martin's Press, 1990.

Prebble, John. *Culloden.* London: Secker & Warburg, 1966.

Reid, John. *1745: A Military History of the Last Jacobite Rising.* New York: Sarpedon, 1996.

Cunaxa, Battle of (401 B.C.E.)

A battle in Mesopotamia that served as prelude to Xenophon's March of the Ten Thousand. Cyrus the Younger, second son of Darius II and viceroy of Asia Minor, contracted a force of 13,000 Greek mercenaries to help him dispute the succession of his brother, Artaxerxes, to the Persian throne in 401 B.C.E. In addition to the Greek contingent, commanded by Clearchus—who was murdered by treachery after the battle—Cyrus's force contained an unknown number of Asiatic light troops. Cyrus met the host of Artaxerxes on the Euphrates River near the town of Cunaxa. The Greeks stood in their phalanx order on the right flank, against the river. Cyrus, with a bodyguard of 600 horses, held the center, and his remaining infantry deployed on the left, with no cavalry or terrain obstacle to cover this flank. Ominously, the Persian force under Artaxerxes possessed such length that the center position overlapped Cyrus's exposed flank. Taking advantage of this situation, the Persian right wing promptly marched forward, attacked, and enveloped the rebels' left flank, while on the right, the Greek phalanx simultaneously scattered the light horse facing it and crashed into the Persian left. The Greeks crushed and pursued their opposite number for some distance. Seeing the success of his right flank and impending disaster on his left, Cyrus launched his companion force of horsemen straight at Artaxerxes. In the melee that followed, Artaxerxes was wounded, but Cyrus was killed. The news of Cyrus's fall quickly spread through both armies. The rebel Asiatics broke and fled; the victorious Persians pursued and plundered their enemy's camp. Clearchus and his Greeks rallied and re-formed their phalanx with their back to the river, obliging Artaxerxes to reverse direction and file back to face them. The Greeks advanced and fought their way clear of the battlefield. Finding themselves unemployed, betrayed, and deep in hostile territory, the Greeks nominated leaders for their vanguard and rear guard. They then began the long walk home to Greece.

Bryan R. Gibby

See also: Xenophon

References and further reading:

Delbruck, Hans. *Warfare in Antiquity.* Trans. Walter J. Renfroe Jr. Lincoln: University of Nebraska Press, 1990.

Xenophon. *The Persian Expedition.* Trans. Rex Warner. Baltimore, MD: Penguin Books, 1949.

Custer, George Armstrong (1839–1876)

Nineteenth-century American commander. Born on 5 December 1839 in New Rumley, Ohio, Custer was educated at the United States Military Academy, from which he graduated last in the class of June 1861. Commissioned a second lieutenant of cavalry, Custer was assigned to the Army of the Potomac and, during the first two years of the war, he found himself serving as a captain on the staffs of Generals Philip Kearny, George B. McClellan, and Alfred Pleasonton. In June 1863, Custer was promoted to brigadier general of volunteers and given command of the 2d Brigade (Michigan). He commanded the brigade during the Gettysburg Campaign, and then took command of the 3d Cavalry Division during Major General Philip Sheridan's campaign in the Shenandoah Valley. After commanding the 3d Cavalry Division during the final battles of the war, Custer was promoted to major general of volunteers and brevetted major general in the Regular Army.

Custer remained in the U.S. Army following the war and became a lieutenant colonel in the 7th Cavalry Regiment. After campaigning against the Sioux and Cheyenne in 1867, he was court-martialed and suspended for leaving his regiment without permission to visit his wife. Custer was subsequently restored to duty and commanded the 7th Cavalry during the Washita Massacre of Cheyenne in 1868. In 1876, following years of clashes with the Sioux and Cheyenne, the U.S. Army decided to crush them once and for all with a three-columned attack. Custer led one of the support columns, Colonel John Gibbon led the second support column, and Brigadier General Alfred Terry both commanded the main force and served as overall commander. On 25 June 1876, the 7th Cavalry (numbering only 655 men) located a Sioux and Cheyenne encampment, and Custer, unaware that there were approximately 3,000 Indian warriors present, disregarded his orders to wait for the other two columns before attacking. Believing that he could surround the encampment, he divided his troops and personally led about 260 men in a frontal assault, while Major Marcus Reno and Captain Frederick Benteen were given command of the two supporting columns. Custer's column encountered the numerically superior Sioux and Cheyenne, and was quickly cut off and completely surrounded. Despite a desperate fight, Custer and his men were all killed. Along with the Alamo, Custer's Last Stand has passed into American mythology. In many nineteenth-century saloons a large painting of the battle hung over the bar.

Alexander M. Bielakowski

See also: American Indian Wars; Cochise

References and further reading:

Connell, Evan S. *Son of the Morning Star.* Norwalk, CT: Easton Press, 1984.

Custer, George A. *My Life on the Plains: Or, Personal Experience with Indians.* Norman: University of Oklahoma Press, 1977.

Utley, Robert M. *Cavalier in Buckskin: George Armstrong Custer and the Western Military Frontier.* Norman: University of Oklahoma Press, 1988.

George Custer (kneeling at left) at his Last Stand at Little Bighorn. (Library of Congress)

Custozza, Second Battle of (24 June 1866)

Second defeat at the same location that gave Venetia to Italy. After signing an alliance with Prussia on 12 May 1866, the Italian government (backed by the French) took advantage of Vienna's problems with Germany to declare war on 20 June. An Italian army under General Alphonso Ferrero, Marchese de la Marmora had gathered more than 200,000 soldiers, most of them untrained militiamen. The two Italian commanding officers were arguing about the forthcoming campaign: La Marmora favored an invasion from the River Mincio toward Verona; General Enrico Cialdini wanted a diversion on the Mincio and a main effort on the Po. Piedmont king Victor Emmanuel II chose to compromise; four corps (120,000) with La Marmora were to attack first, then Cialdini was to cross the Po with two corps to outflank the smaller Austrian army. The archduke Albrecht had only three corps (80,000) to defend Venetia, so he decided to deploy his troops in a central position. His cavalry was well aware of all the Italian moves and as soon as the Mincio was crossed (23 June), the main Austrian army marched to a line of steep hills north of Villafranca, covering their lines of communications and outflanking La Marmora's advance. The Italian general did not know where the Austrians had deployed, and on 24 June, the poorly led Italian columns encountered the awaiting Austrians. The Battle of Custozza did not follow any form or plan on either side. The Italians attacked piecemeal; La Marmora's right flank was halted near Sommacampagna by a spirited, old-fashioned charge of Austrian lancers that the Italian troops repulsed easily. But panic-stricken teamsters and deserters fled and prevented any reinforcements from crossing the Mincio bridges. In this hilly and cultivated area, the two headquarters were unable to see any action, and Albrecht and La Marmora spent the day riding their horses here and there, attempting to gather information. The conduct of the battle was left to corps or even division commanders. The Italians proved unable to advance up the hills, and the village of Custozza became the main point of Italian resistance against Austria's advance toward the Mincio. Slowly, the Italian wings were sent back to the river; by 4 P.M. Custozza was surrounded by cautiously advancing Austrian forces. The village was eventually evacuated by Italian Elite troops at 5 P.M. The Italians retreated in disorder across the Mincio having lost more than 8,000 men. The Austrian army did not pursue and was shortly afterward withdrawn to protect Vienna against the Prussians. Yet Albrecht's victory (and the Austrian triumph at sea in the

Battle of Lissa) proved useless; by the treaty of Vienna, Austria had to cede Venetia to the Italians.

Gilles Boué

See also: Bismarck, Otto von

References and further reading:

Perdezolli, Ippolito. *Da Custozza a Lissa.* Lugano: Einodi, 1866.
Wawro, Geoffrey. *The Austro-Prussian War.* Cambridge, UK: Cambridge University Press, 1996.

Cuzco, Battles of (Inca Empire, 1438, 1536)

The first battle established the Inca as the dominant Andean highland power. The second battle marked the final eclipse of the Inca Empire by the Spanish.

By the time of the seventh Inca emperor, Viracocha, three groups vied for control of the Andean highlands: the allied Quechua and Inca, and the Chanca. The aggressive Chanca were led by Hancohuallu and marched first on the Quechua capital of Antahuailla, capturing it around 1437. The Chanca then marched on the Inca capital of Cuzco in 1438 with an army of 30,000, forcing a showdown for control of the Andes highlands.

Viracocha's son, Yupanqui, was now in command of the Inca resistance, numbering 8,000 ethnic Inca. As the Chanca crossed the Apurimac River, Yupanqui decided to ambush them in the mountain passes between the river and the plains outside of Cuzco known as Yahuarpampa. Yupanqui, who had sent for help from Inca allies, then received word that 20,000 Quechua and other Inca allies were on the way. He changed plans and hid the growing reinforcements in the highlands next to the Yahuarpampa (blood-field), determined to engage the Chanca army with his Inca forces and then hit them with allied reinforcements as they arrived.

The Chanca and Inca armies joined battle two days later on the plain of Yahuarpampa within sight of Cuzco. After intensive morning fighting, Yupanqui hit the Chanca right flank with 5,000 allies hidden in the highlands. The Chanca recovered within two hours, only to find more Inca allies pouring onto the battlefield. In the late afternoon the Chanca found themselves surrounded in enemy territory, taking heavy losses, and Hancohuallu fled with the remnant Chanca forces.

Chanca losses numbered in the thousands, and many more were probably lost as they struggled home through enemy Quechua territory. The battle left the Inca in control of the Andean highlands, having played their more powerful neighbors against each other. Yupanqui was crowned Inca emperor Pachacutec, and developed the Inca Empire, conquering most of the Andean highlands and the powerful coastal Chimu kingdom, and establishing a regularized military system led by ethnic Inca trained in warfare.

The second battle of Cuzco at the start of 1536 marked the final attempt of the Inca to expel the Spanish from the Inca Empire lands. From 1535 to 540, the rebel Inca emperor Mancos, who studied and employed Spanish battle tactics, including the use of horse and pike, attempted to drive the Spanish out of the Inca Empire. Hundreds of Spanish soldiers were killed in empire-wide engagements, often ambushed in terrain unfavorable to Spanish horses.

Mancos launched his rebellion from his highland jungle–fortified city of Vilcambambos, with his grand strategy centered on regaining the highland Inca capital at Cuzco. At the start of 1536, he gathered a 100,000-man force to besiege Cuzco, protected by the brothers of Francisco Pizarro and a force of several hundred Spanish and Indian allies. Although the Spanish were sorely pressed in a series of pitched battles that lasted into the spring rains, they did not succumb. Much of Cuzco was burned to the ground but Mancos and his Inca-led forces failed to hold the huge fortification of Sacsahuaman and this provided refuge for the Spanish over the harvest months. Spanish reinforcements in 1537 and 1538, from as far away as Cuba and Spain, finally put Mancos on the defensive, and he retired to Vilcambambos, never again truly to threaten Spanish control of the Inca Empire.

Christopher Howell

See also: Chan Chan, Battle of; Inca Empire Imperial Wars; Pachacutec Yupanqui

References and further reading:

Bram, Joseph. *An Analysis of Inca Militarism.* New York: Augustin Publisher, 1941.
Garcilaso de la Vega, El Inka. *The Incas: Royal Commentaries.* 3d ed. Ed. Alain Gheerbrant. New York: Avon Books, 1971.
Hemming, John. *The Conquest of the Incas.* New York: Harvest-HBJ Inc., 1970.

Cynoscephalae, Battle of (197 B.C.E.)

Following the inconclusive First Macedonian War (215–205 B.C.E.), Philip V of Macedon renewed his effort to dominate Greece. Responding to pleas from its Greek allies, Rome declared war on Philip in 200 B.C.E. and dispatched two legions under Lucius Quinctius Flamininus to Greece. Flamininus convinced many Greek states to ally with Rome, built up his army to almost 25,000 men, and advanced into Thessaly against Philip's roughly equally sized army. At Cynoscephalae the legion finally triumphed unambiguously over the phalanx.

The two armies maneuvered against one another across

the rugged countryside for several days, often in pouring rain. When their patrols blundered into one another in a dense fog, both commanders dispatched reinforcements and then committed their whole armies to the developing battle. The hilly, rocky terrain was hardly suited to Philip's phalanx, but he sensed he had a chance to destroy the Roman army before it had properly deployed for battle. Leading his right wing in person, he drove back the Roman left. For a time it looked like he might win the battle, but his successful attack had opened a gap in his line.

The Macedonian phalanx, virtually unstoppable from the front with its rows of serried pikes, was particularly vulnerable to flank attacks. The Macedonian left wing, still deploying for battle from march column, was driven back in confusion by Flamininus. Some of the advancing legionnaires from the Roman right poured through the gap in the Macedonian line, fell on the flank of Philip's advancing wing, and routed it from the field. The tactical flexibility of the maniple had proven superior to the vaunted Macedonian phalanx. The following year Philip renounced his claim to Greece.

Stephen Stein

See also: Macedonian Wars

References and further reading:

Goldsworthy, Adrian. *The Roman Army at War.* Oxford: Clarendon, 1996.

Hackett, John. *Warfare in the Ancient World.* New York: Facts on File, 1989.

Polybius. *The Histories of Polybius.* Trans. from the F. Hultsch text by Evelyn S. Shockburgh. Westport, CT: Greenwood Press, 1974.

Cypriot Wars (1955–1977)

Struggles between the incompatible desires of the Greek-Cypriot majority to reunite with Greece versus Turkish-Cypriot desires for security through partition. Although Great Britain annexed Cyprus from Turkey in 1914, the ethnic Greek drive for *enosis* (homeland reunion) ensures continued tensions. The National Organization of Greek Fighters (EOKA) was formed on 1 April 1955 under the guidance of Greek army colonel Georgios Grivas (1898–1974) to eliminate British rule and reunite Cyprus with Greece. A Turkish-Cypriot countergroup, Vulkan ("volcano"), appeared in response, with a goal of *taksim* (partition). Although Greek-Cypriot community leader Archbishop Makarios III (Michael Mouskos, 1913–1977) called for passive resistance, EOKA attacks on British soldiers increased.

The British distanced themselves from hostilities while retaining basing rights through an April 1960 treaty; Britain, Turkey, and Greece were further authorized to intervene, unilaterally if necessary. The treaty pledged three of seven Cypriot cabinet positions (Defense, Health, Agriculture) and 30 percent of government positions to Turkish Cypriots, aggravating the Greek Cypriots.

The constitutional debates continued until 1963, stumbling over each faction's prejudices or insecurities—protection of minority rights, for example, or the presence of still-armed EOKA elements. In December 1963, Makarios introduced constitutional changes reducing Turkish-Cypriot parliamentary presence. The resulting reigniting of communal fighting and paramilitary reprisals triggered United Nations Security Council Resolution 186 (4 March 1964, extended repeatedly) authorizing 7,000 peace-keeping troops.

Canadian troops arrived on 14 March 1964, with soldiers from Ireland, Sweden, Denmark, Finland, and Austria following over the next three weeks. Grivas's use of the Greek Cypriot National Guard (GCNG) to eliminate suspected terrorists ensured that violence continued. In August, for example, Turkish aircraft conducted rocket and napalm attacks against the GCNG to protect a Turkish enclave. Grivas was recalled to Greece following the April 1967 Colonel's Coup, but returned secretly in 1971 to form EOKA-B, a paramilitary group additionally targeting Greek Cypriots who were not sufficiently pro-enosis.

In an effort to break the political deadlock in July 1974, Makarios ordered the expulsion of 600 Greek military personnel, hoping that the Turkish side would reciprocate. However, the expelled Greek soldiers overthrew the Cypriot government on July 15. The immediate aftermath saw several days of atrocities committed against both Turkish and pro-Makarios Cypriots.

The Turkish military intervened on 20 July 1974 with air attacks against Cypriot shipping, an amphibious offensive along the northern coast, and a parachute assault, followed by heliborne commandos, onto the plains north of the capital, Nicosia. The beachhead and linkup through the Kyrenia Mountains was established within 48 hours. The Greek response was limited to 14 transport aircraft with soldiers, several of which were shot down by the GCNG, who had not been informed of their arrival, causing the remaining aircraft to return to Greece.

Through tenuous cease-fires, UN troops established a buffer zone between the Turkish-occupied northern 30 percent of the island and the remaining Greek-Cypriot south. Makarios resumed the presidency of the divided island until his death in December 1977. The Cyprus troubles, fueled by traditional Greek-Turkish enmity, show few signs of abating.

Robert Martyn

References and further reading:

Borowiec, Andrew. *Cyprus: A Troubled Island.* Westport, CT: Praeger, 2000.

Cyrus II the Great (c. 600–530 B.C.E.)

Founder of the Achaemenid Dynasty of Persia, which lasted two centuries, and conqueror of a wide expanse from Central Asia to the Mediterranean Sea.

The sources for Cyrus's reign are diverse, ranging from his own inscriptions (e.g., the Cyrus Cylinder), contemporary Babylonian and biblical historical texts, and later classical writers, such as Herodotus, Xenophon, and Ctesias. Little is known about Cyrus's youth, although Xenophon wrote a largely unhistorical account of Cyrus's childhood and education.

Early in his reign (c. 550 B.C.E.) Cyrus defeated the rival Iranian king Astyages of Media, seized the Median capital of Ecbatana, and thereby became the king of the Medes and Persians, with his capital at Pasargade. He promptly began expansion in the west, where he met Croesus of Lydia at the Halys River in Anatolia, and fought to a draw. Soon thereafter, the Achaemenid king attacked Sardis, the Lydian capital, took the city, and became the undisputed master of nearly all of Anatolia, including a number of the Greek city-states on the western coast.

In 539 B.C.E. Cyrus crossed the Diyala River in Iraq and attacked the Chaldean Empire of Nabonidus. He defeated a Babylonian force at Opis, and sent a Persian army to capture Babylon itself. He declared his son, Cambyses II, king of Babylon, and then issued a decree allowing deported Jews to return to Judea to rebuild their temple.

Cyrus continued to campaign east of Iran in Central Asia, successfully subduing most of Afghanistan and southern Central Asia. However, he apparently died in battle in Central Asia against the Massagetae in 530 B.C.E. and was succeeded by Cambyses II. The Roman historian Arrian claimed that Alexander the Great restored Cyrus's tomb at Pasargade two centuries later.

Mark W. Chavalas

See also: Alexander the Great; Croesus; Persian Empire

Cyrus II the Great and his forces conquering Babylon, c. 538 B.C.E. (Hulton/Archive)

References and further reading:

Briant, P. *From Cyrus to Alexander: A History of the Persian Empire.* 2 vols. Winona Lake, IN: Eisenbrauns, 1999.

Kuhrt, A. "The Cyrus Cylinder and Achaemenid Imperial Policy." *Journal for the Study of the Old Testament* 25 (1983), 83–97.

Sekunda, N., and S. Chew. *The Persian Army 560–330 BC.* London: Osprey, 1992.

Stronach, D. "Cyrus the Great." *Revue d'archéologique et d'art iraniens* 7–8 (1971), 4–21.

D

Danish Wars with the Hanseatic League (1361–1370)

Hanseatic League deposes King Valdemar III of Denmark. The prelude to the Danish Wars with the Hanseatic League was a century of Danish civil war and the Black Death. The result was a complete breakdown of the Danish monarchy, and control of much of Denmark by the Holstein counts. Civil unrest continued until the counts were forced by a consortium of trading cities known as the Hanseatic League to impose peace in the Baltic. Valdemar III became king in 1340.

Valdemar wished to reestablish the control of the Danish crown over the territories it had once held. From the late 1340s to 1361, he pursued his policy of reacquisition with some success, until he attacked and conquered Gotland in 1361. Gotland's main town was the Hanseatic League member Visby. In response, the Hanseatic League entered into an alliance with Valdemar's enemy, Sweden. This alliance produced escalated hostilities, resulting in Valdemar's flight from Denmark, the conquest of Copenhagen in 1368 by the Hanseatic League, and the Treaty of Stralsund in 1370. The treaty gave the league control for 15 years of the Skaanor fisheries and three coastal forts, plus a formal veto over the Danish choice of king.

Tamsin Hekala

References and further reading:

Derry, T. K. *A History of Scandinavia: Norway, Sweden, Denmark, Finland, and Iceland.* Minneapolis: University of Minnesota Press, 1979.

Pulisiano, P., ed. *Medieval Scandinavia: An Encyclopedia.* New York: Garland Publishers, 1993.

David (r. c. 1000–960 B.C.E.)

King of a unified Israelite kingdom, established Israel as a major power, and became the standard for all later Israelite kings. As there is no specific mention of David in extrabiblical sources, historical details concerning his rise to power are derived almost entirely from biblical sources.

David, apparently a high officer in Saul's army, was forced to flee to the Judean hills, heading a band of fugitives, mercenaries, and outlaws that raided Saul's kingdom, undermining Saul's authority, and hired out as mercenaries to the Philistines. At Saul's death David managed to wrest control of the southern tribe of Judah from Ishbaal, Saul's heir, and created a rival state. After Ishbaal's death, David claimed the throne of Israel, partly by virtue of the fact that he had married into Saul's family. He subsequently captured the Canaanite fortress of Jerusalem, making it his capital.

David triumphed in a series of bitter foreign wars against the Philistines, three states to the south and east of the Jordan River (Edom, Moab, and Ammon), and the powerful Aramean states to the north. He incorporated much of Philistine territory and all of the Transjordanian states into his kingdom. He subdued the Arameans to the north and allied himself with the coastal city-states of Phoenicia, which provided a lucrative economic relationship that continued during the reign of his successor, Solomon.

Late in David's reign there were a number of internal revolts, including an attempted coup d'état by one of his sons, Absalom, and one by a Saul loyalist, Sheba, all unsuccessful. The line of succession to his son Solomon was also disputed by another son, Adonijah. David's united kingdom, which is reputed to have ranged from the border of Egypt to the Euphrates River, lasted only until the death of his successor, Solomon (c. 960 B.C.E.).

If biblical accounts are to be relied on, David apparently oversaw the transformation and reorganization of the Israelite military from a militia-based organization to a standing army. Battles of volunteer champions, a feature of pre-Davidic warfare, were replaced by organized tactical mass charges, and logistics were regularized.

Mark W. Chavalas

See also: Ancient Warfare

References and further reading:

Gunn, D. *The Story of King David.* Sheffield: *Journal for the Study of the Old Testament,* 1978.

Ishida, T., ed. *Studies in the Period of David and Solomon and Other Essays.* Winona Lake, MN: Eisenbrauns, 1982.

Mazar, B. "The Military Elite of King David." *Vetus Testamentum* 13 (1963), 310–320.

Na'aman, N. "The List of David's Officers." *Vetus Testamentum* 38 (1988), 71–79.

Davout, Louis-Nicolas, Duke of Auerstädt, Prince of Eckmühl (1770–1823)

One of Napoleon's most reliable field commanders, noted for his rigorous training, strategy, and reliability at crucial moments. Davout was born on 10 May 1770 in Annoux, Burgundy. He was commissioned in February 1788, but his devotion to revolutionary ideas led to his dismissal (15 September 1791). But a volunteer regiment elected him its lieutenant colonel. He fought at Neerwinden (18 March 1793) and rose to division general in July.

Cashiered as a former noble, Davout petitioned and was restored as a brigadier general of cavalry (21 September 1794). He served in the Rhineland (1795) under Charles Louis Desaix, who introduced him to Napoleon, and fought in Egypt with Desaix. Both escaped from Egypt and returned to France on 6 May 1800.

Named a marshal (19 May 1804), Davout held a virtually independent command during the march to Ulm. Then, after an incredible forced march, his corps played a major role at Austerlitz (2 December 1805). At Auerstädt (14 October 1806) Davout, heavily outnumbered, achieved a significant victory. Napoleon belatedly named him Duke of Auerstädt (28 March 1808). At Eylau (8 February 1807), Davout's right wing prevented a French defeat. He subsequently served as governor-general of the Grand Duchy of Warsaw.

In the campaign of 1809 Davout played a crucial role at Eckmühl (22 April). At Wagram (6 July) his leadership of the vital right wing secured victory. Napoleon then named him Prince of Eckmühl (15 August).

In the Russian campaign Davout led I Corps and later Prince Jerome's army too. At Borodino (7 September 1812) he unsuccessfully advocated a flanking attack instead of the bloody frontal assault Napoleon employed. Retreating from Moscow, Davout commanded the rear guard (26 October–3 November) but was relieved for lack of speed and failure to halt Cossack attacks.

Davout fought at Dresden (9–13 March 1813) and took and held Hamburg (30 May 1813–27 May 1814). During the Hundred Days he served Napoleon as war minister and governor of Paris. In surrendering he demanded and obtained an amnesty to protect his colleagues of the Hundred Days. Davout died in Paris on 1 June 1823.

James K. Kieswetter

See also: Aboukir; Austerlitz, Battle of; Borodino; Brunswick, Frederick William, Duke of; Carnot, Lazare-Nicholas; Dresden, Battle of; Frederick William, Elector of Brandenburg; French Revolutionary Wars; Jena and Auerstädt; Moscow, Retreat from; Napoleonic Wars; Pyramids; Wagram

References and further reading:

Chandler, David G. *The Campaigns of Napoleon.* New York: Macmillan, 1966.

———. "The Iron Marshal—Davout." In *Napoleon's Marshals,* ed. David G. Chandler. New York: Macmillan, 1987.

Gallaher, John G. *The Iron Marshal: A Biography of Louis N. Davout.* Carbondale: Southern Illinois University Press, 1976.

Macdonnell, Archibald Gordon. *Napoleon and His Marshals.* New York: Macmillan, 1934.

De Wet, Christiaan Rudolph (1854–1922)

Boer general who inaugurated the guerrilla campaign during the Second Anglo-Boer War (1899–1902). Although a citizen of the Orange Free State (OFS) republic, De Wet joined the Transvaalers in their war of independence against Britain (1880–1881), and in the process revealed his courage and natural military talents.

When the Second Anglo-Boer War broke out in October 1899, De Wet joined as an ordinary burgher. As an acting commandant, De Wet defeated the British at Nicholsonsnek (30 October 1899) and in December 1899 was promoted to combat general on the Kimberley front. He was unable to stop Lord Roberts's advance and on 13 March 1900 the British forces occupied Bloemfontein unopposed. De Wet then disbanded the commandos, ordering them to reassemble on 25 March 1900. Those who returned truly believed in the Boer cause and were prepared to fight with renewed commitment. In the meantime, at a Boer council of war meeting at Kroonstad, it was decided to abandon the large wagon laagers, and to concentrate on destroying the British lines of supply and communication.

De Wet, now in command of all the Orange Free State forces, inaugurated the guerrilla phase of the war by defeating a British force at Sannaspos (31 March 1900). He followed this up with victories at Mostertshoek (3 April 1900) and Rooiwal (7 June 1900). He succeeded in escaping the British encircling movement at the Brandwater Basin, falling back into the Transvaal (the first so-called De Wet Hunt, July–August 1900). His first attempt to invade the Cape Colony (November–December 1900) was thwarted (second De Wet Hunt), but in January 1901 he succeeded in crossing into that colony. What followed was the third (or great) De Wet Hunt, with 15,000 British troops chasing the Boers for 800 miles before De Wet broke back to the Orange Free State. On Christmas Day 1901 De Wet defeated the British at Groenkop. From January to May 1902 he successfully evaded Lord Kitchener's "new model drives." For most of the guerrilla phase, OFS president M. T. Steyn accompanied De Wet's commandos, and together they were the driving force behind the Boers' continued resistance. De Wet was a brilliant tactician, as well as a strict disciplinarian who sometimes used a whip against the weak-kneed.

Although De Wet favored the continuation of the struggle, he eventually bowed to pressure and signed the conditions of surrender as acting president of the Orange Free State. After the war, De Wet once again took up farming, was elected to the OFS Parliament in 1907, but in 1914 took up arms against the Union government in an effort to restore republican independence. He was captured and spent a few months in jail.

André Wessels

See also: Boer Wars; Kitchener, Horatio Herbert; Roberts, Frederick Sleigh, First Earl, Viscount St. Pierre of Kandahar

References and further reading:

De Wet, C. R. *Three Years War (October 1899–June 1902)*. London: Constable, 1902.

Rosenthal, E. *General de Wet: A Biography*. Cape Town: Unie-Volkspers Beperk, 1946.

Scholtz, W. L. von R. "Generaal Christiaan de Wet as Veldheer." D. Litt. thesis, University of Leiden, 1978.

Van Schoor, M. C. E. "De Wet, Christiaan Rudolph." In *Dictionary of South African Biography*, Vol. 1, ed. W. J. de Kock. Pretoria: National Council for Research, 1968.

Death Squads

Clandestine and usually irregular organizations, often paramilitary in nature, that carry out extrajudicial executions and other violent acts against clearly defined individuals or groups of people. Murder of this sort is their primary or even sole activity. They differ from other tools of repression in a number of significant aspects, notably in the way they mix state and private interests, and in the way they call into question the very legitimacy and substance of the state: Except in the rare case where an insurgent group forms them, death squads operate with the overt support, complicity, or acquiescence of government, or at least some parts of it. In many cases, government security forces have participated directly in the killing. Yet at the same time, death squads almost always involve the support and participation of elements outside of government, and develop considerable independence from their backers. Death squads are found all over the world today, and in just the last 30 years have been responsible for hundreds of thousands, perhaps millions, of deaths. Major examples of their use include El Salvador or Guatemala in the 1980s and 1990s, South Africa in the 1970s and 1980s, and Bosnia in the 1990s.

Bruce Campbell

See also: Paramilitary Organizations

References and further reading:

Campbell, Bruce, and Arthur Brenner, eds. *Death Squads in Global Perspective: Murder with Deniability*. New York: St. Martin's Press, 2000.

Mason, T. David, and Dale A. Krane. "The Political Economy of Death Squads: Toward a Theory of the Impact of State-Sanctioned Terror." *International Studies Quarterly* 33 (1989).

Sluka, Jeffrey, ed. *Death Squad: The Anthropology of State Terror*. Philadelphia: University of Pennsylvania Press, 1999.

Wolpin, Miles D. *State Terrorism and Death Squads in the New World Order. Peace Research Reviews* 12(3). Dundas, ON: Peace Research Institute, 1992.

Delhi Sultanate, Wars of (c. 1200–1556)

Campaigns that made the Delhi Sultanate the principal Muslim regime in north India between the thirteenth and sixteenth centuries. This hegemony emerged as a result of the military campaigns of Muhammad of Ghur and his commander Qutb-ud-Din Aybak between 1175 and 1206, and in particular their decisive victories at the Battles of Taraori in 1192 and Chandawar in 1194.

For the first few years, until the death of Aibak in 1210, Lahore remained the capital of Muslim India, though the sphere of Ghurid control extended to include Delhi and its environs. This changed with Sultan Iltutmish (r. 1211–1236), who took over the kingdom after Aibak, seizing political power in a coup at Delhi. His first political move was to establish a permanent capital at Delhi, causing the Ghurid soldiers of fortune finally to sever their ties with Ghur. Aibak also withdrew his forces from contact with the Mongol armies, now concentrating his resources toward consolidating control over the several urban centers of north India, and establishing control over the activities of the Hindu and Rajput kings of the region. It appears that beginning with Iltutmish, the center of the Indo-Muslim polity was no longer at Lahore, but moved to Delhi. This was partly to compensate for the growing threat of Mongol attacks on the northwestern territories of the empire. Delhi was a safer, more interior location, in comparison to Lahore, which was sacked in 1241. Iltutmish's death was followed by a period of factional strife and then stability under Balban (r. 1266–1267). Meanwhile, the resurgence of the widespread threat of Mongol invasions from the north between 1221 and 1258 was reason enough especially to pare the defenses of Delhi, leading to few campaigns in other areas of the country. The Khalji Dynasty (1290–1320), which followed Balban, provided relative stability to the Delhi Sultanate and made it an imperial power. The prolific Khalji king, Ala-ud-Din (1296–1316), greatly expanded the frontiers of the empire. He conquered the regions of Gujarat (1297) and Rajasthan (1301–1312), and created vassals out of some of the principal Hindu kingdoms in southern India (1307–1302). His forces also prevented strong Mongol onslaughts by the Chagatais of Transoxania into north India (1297–1306).

With the Mongol invasions now an event of the past, in the succeeding Tughluq Dynasty, Muhammad ibn Tughluq (r. 1325–1351) attempted to extend the Delhi sultanate's sphere of influence to the Deccan region of central India. His second capital at Daulatabad did little to create the much-desired new Muslim military, administrative, and cultural elite, save for a defiant Deccani military aristocracy—the remaining vestiges of the ambitious capital transfer. This aristocracy severed themselves from the overlordship of Delhi and set up a new Bahamani sultanate in 1347. Meanwhile the Central Asian powers had begun to show a renewed interest in the prosperous region of north India, and Muhammad's successor, Firuz Shah (r. 1351–1388), was so concerned about these invasions from the north that he paid little attention to reconquering the Deccan from the Bahamanis. In any case, despite his preparedness, Timur's sack of Delhi in 1398–1399 left the sultanate utterly devastated. During the reign of the Sayyids (1414–1451), who immediately followed this disaster, the Delhi Sultanate was a shadow of its former self, little more than a country power competing against petty Hindu and Muslim principalities. There was a relative recovery under the Lodi Dynasty (1451–1526), which encouraged significant immigrations from Afghanistan, mainly to swell recruitments into their large armies. The Delhi Sultanate also partly recovered its hegemony, only to be conquered by Babur—the last of the Central Asian Timurids—at the First Battle of Panipat in 1526. The Afghan leader Sher Shah Suri briefly reestablished the sultanate at Delhi between 1537 and 1555, when it fell again to Babur's son and successor, Humayun, who had returned from exile in Persia. Humayun's death in 1556 coincided with the definitive victory by his son, Akbar, at the second Battle of Panipat. This decisive battle finally encompassed the remains of the Delhi Sultanate into the Mogul Empire.

Manu P. Sobti

See also: Muhammad of Ghur, Conquests of

References and further reading:

Jackson, Peter. "Delhi: The Problem of a Vast Military Encampment"; Ali, M. Athar. "Capital of the Sultans: Delhi during the Thirteenth and Fourteenth Centuries"; and Hambly, Gavin R. "The Twilight of Tughluquid Delhi: Conflicting Strategies in a Disintegrating Imperium." In *Delhi through the Ages: Essays in Urban History, Culture and Society,* ed. R. E. Frykenberg. Delhi: Oxford University Press, 1986.

Sarkar, Jagish N. *The Art of War in Medieval India.* New Delhi: Munshiram Manoharlal, 1984.

Denain, Battle of (24 July 1712)

The battle that saved France and ended the "war of succession." By 1712, the French northern border was wide open to invasion. Prince Eugene with 130,000 soldiers (Dutch and imperials) had divided his army in two parts, one besieging Landrecies since 12 July, the other in a fortified camp near Marchiennes, which was linked by a fortification to the town of Denain, 12 kilometers south on the River Escaut. The second army was in charge of protecting the siege, and sending all the supplies through a fortified road called *le chemin de Paris,* as Landrecies was the last French fortified place to protect the capital. The French commanded by Villars had no less than 70,000 troops in a camp observing the siege.

The idea of the bold move that gave victory came from a judge, Le Febvre d'Orval, and was adopted by Villars. First, the French army advanced toward Landrecies to draw Eugene's reserve. Then, on the night of 23–24 July, with a forced march and a flanking movement, Villars crossed the Escaut and appeared before Denain. The Count of Albermarle commanding the Dutch in Denain immediately sent a message to Eugene calling for more soldiers. Thinking that the fortifications would be enough to stop the assaults and give time for reinforcements, Eugene said, "It is time for dinner" and went back to his headquarters. Around 2 P.M., he came back with troops to attack the French right wing. Albermarle had been left in inferior numbers behind Denain's walls. Villars formed an attack column of 40 battalions to storm the fortifications and deployed the rest of his army to prevent the crossing of the river by Eugene's troops. In less than an hour, the town was taken and Albermarle surrendered. Eugene, with exhausted reinforcements, tried uselessly to prevent the French advance. Eventually he retired after the French commander destroyed the last bridge on the Escaut.

This French victory had enormous consequences. Eugene had to retire and give up all the fortified places previously taken. Moreover, all the opponents were exhausted by 10 years of war and looked for an honorable way to end it. Denain's victory gave the French a new pride and the Dutch a good reason to negotiate.

Gilles Boué

See also: Eugene of Savoy; Spanish Succession, War of the

References and further reading:

Bluche, François. *Louis XIV.* Paris: Fayard, 1986.
Corvisier, André. *Dictionnaire d'art et d'histoire militaire.* Paris: PUF, 1988.
Lesage, Gérard. *Denain 1712.* Paris: Economica, 1992.

Denikin, Anton Ivanovich (16 December 1872–8 August 1947)

Czarist army general; White Russian civil war commander. Born near Warsaw, the son of a retired Russian army major, Denikin graduated from the Kiev Infantry Academy (1892), served as artillery brigade second lieutenant, graduated from the General Staff Academy (1902), and became colonel with the Transbaikal Cossack Division during the Russo-Japanese War.

In World War I, Denikin served in the Eighth Army, participated in "Brusilov Breakthrough" (June 1916), was appointed Eighth Army Corps Commander on Rumanian Front (September), and was promoted to lieutenant general.

After the February Revolution, he served as the provisional government's chief of staff (April–May 1917) and was Western Front commander in the disastrous June offensive, which hastened the czarist army's collapse.

Imprisoned after supporting General Kornilov's abortive rightist coup in August, he escaped in November, commanding the "volunteer army" from April 1918 upon Kornilov's death. He retreated to Caucasus strengthening his forces.

In January 1919, he became commander in chief of Armed Forces of South Russia, with Allied backing, and became the most effective White force of the Russian civil war, attacking in June 1919 with 100,000 well-equipped, well-armed troops, including 56,000 cavalry (Don and Kuban Cossacks). Despite capturing Orel (October 1919), 200 miles from Moscow, Denikin failed—in part because the Red Army matched him, especially in cavalry, but primarily through his own errors.

He ignored opportunities to combine with other anti-Soviet forces—Komuch, the Czechs, Krasnov in 1918; Kolchak in 1919—to launch combined attacks, instead seeking glory himself in his drive for Moscow. His vision of "Russia, One and Indivisible" alienated independence-seeking national minorities, and his returning of land to nobility alienated the peasant masses. Thus partisan activity, notably Nestor Makhno's Anarchist army, destroyed his rear.

Becoming Supreme White Commander (January 1920), Denikin retreated to Crimea under Red offensives (1919 and 1920). He departed Russia in April 1920, leaving Baron Wrangel in command.

Denikin wrote his memoirs in exile, in France until 1945, but he did not support the Nazis against the Soviet Union, in contrast to many White Russian émigrés. He moved to America and died in Michigan. He is buried in Detroit.

Neil Harvey Croll

See also: Russian Civil War (1918–1921); Trotsky, Leon

References and further reading:

Cherkasov-Georgievskii, V. *General Denikin.* Smolensk: Rusich, 1999.
Denikin, A. I. *Ocherki Russkoi smuty (Studies of the Russian Turmoil) v 5-kh tomakh.* Paris, Berlin, and Brussels: Chekova, 1921–1926.
———. *Put' Russkogo ofitsera (The Road of a Russian Officer).* New York: Chekova, 1953.
———. *The White Army.* Trans. C. Zregintsov. London: Jonathan Cape, 1930.

Dien Bien Phu (December 1953–7 May 1954)

The penultimate siege that ended French rule in Indochina. The war against the Vietminh in Indochina had been going badly for the French. By summer 1953, they controlled only the southern third of Vietnam, the coastline in the country's center, and the Red River delta in the north. The French people were tiring of the war, and depending on how one evaluated Vietminh irregular forces (compared to Vietminh main force units), the French and their Vietnamese allies were outnumbered and losing control.

General Henri Navarre, who had just taken command, felt he could find a solution, perhaps force a negotiated settlement, if he could lure the Vietminh into an open battle where he could defeat them. To tempt the Vietminh, Navarre eventually placed 12 battalions on a plateau—some 180 miles from French bases at Hanoi—in extreme northwestern Vietnam astride enemy supply lines to Laos. Vietminh general Nguyen Vo Giap took the bait and began concentrating more than 50,000 men—33 infantry and artillery battalions—to surround and isolate the French paratroopers at Dien Bien Phu. Giap, a logistical genius, managed to bring artillery where the French thought it difficult by breaking the big

French soldiers on patrol crouched in a thicket outside Dien Bien Phu, Vietnam, 1954. (Library of Congress)

guns down and dragging them up steep hillsides and reassembling them in position, his men often using bicycles. As a consequence, he was able unexpectedly to besiege the French position.

The siege began in December. Giap surrounded the French position and spent several months bringing more troops, equipment, and supplies to the area. Finally, on 13 March, and after an intense artillery barrage, the Vietminh sent two battalions each against the French companies defending the outposts of Gabrielle to the north, Beatrice to the northeast, and Anne Marie to the northwest. Thereafter Vietminh artillery closed the airfield; the French would have to be resupplied by parachute drops, which as often fell to the enemy as to the increasingly desperate French defenders.

After a two-week pause to consolidate, Giap resumed the offensive on 30 March, and soon overran the outposts of Eliane and Dominique to the east and Hugette and Claudine to the west. After a month of siege tactics—digging mines, exploding bombs under the French defenses, and constructing approach trenches ever closer to French lines—Giap launched the final attack on 1 May.

On 7 May 1954, the French surrendered. Casualties in this climactic battle were high: The Vietminh have never released casualty figures; the French admitted to 2,000 killed, 6,500 wounded, and 10,000 captured. The Geneva Conference on Indochina coincidentally began the same day and marked the end of the French empire in Southeast Asia.

Charles M. Dobbs

See also: Ho Chi Minh; Vietnam Conflict

References and further reading:

Devillers, Philippe, and Jean Lacouture. *End of a War: Indochina, 1954.* Trans. Alexander Lieven and Adam Roberts. New York: Praeger Publishers, 1969.

Fall, Bernard B. *Hell in a Very Small Place: The Siege of Dien Bien Phu.* Philadelphia, PA: Lippincott, 1967.
Roy, Jules. *The Battle of Dienbienphu.* Trans. Robert Baldick. New York: Harper & Row, 1965.

Dieppe (19 August 1942)

A disastrous landing that nonetheless provided valuable experience for the Allies in amphibious landing techniques for their main landings in the invasion of northeast Europe. Operation JUBILEE was launched on 19 August 1942 with the declared objective of seizing a strongly defended French port. The majority of the 6,000 troops involved were Canadian (5,000), the rest being British, with 50 American rangers, supported by an inadequate number of destroyers and fighter squadrons. No preliminary bombing was provided because Bomber Command believed that it could not spare aircraft allocated to the bombing of Germany.

The frontal attack by Canadians on Dieppe proper was to be facilitated by the destruction of German peripheral defenses in surrounding villages, entrusted to the commandos. But a combination of bad previous reconnaissance, bad luck (which ruined the surprise effect), and poor timing and improper location of the initial landings, more than a 10-mile zone left German firepower largely intact. The main thrust on Dieppe beach at 5:20 A.M. led to heavy casualties. Only a few actually reached the town's streets, protected by only 15 tanks, and were almost immediately killed or taken prisoner. Retreat was ordered at 11:00 A.M. and by 1:00 P.M. all fighting had ceased. Almost 1,000 Canadians lost their lives in what remains one of the most controversial operations of World War II.

Very few historians would now support Churchill's argument in his memoirs that the heavy toll was worth paying for its "mine of experience." Common sense could have suggested the obvious: the need for a near-exact knowledge of enemy strength in the area, the decisive role of advance bombardment by sea and air, and the necessity of massive amphibious equipment. Possibly the only advantage of the operation was to make the Germans more alert to the possibility of cross-Channel invasion: although JUBILEE immobilized a number of German divisions in the west, it also ensured that they would not be caught completely unprepared on D day.

It could also be argued that Dieppe was too large to be a successful raid, and of course, far too small for an invasion force.

A. Capet

See also: World War II

References and further reading:

Mordal, Jacques. *Dieppe: The Dawn of Decision.* London: Souvenir Press, 1963.
Robertson, Terence. *Dieppe: The Shame and the Glory.* London: Hutchinson, 1962.
Souster, Raymond. *Jubilee of Death: The Raid on Dieppe—The Moving Story of the Suffering and Achievement of the Canadians at Dieppe on August 19, 1942 Told by the Men Who Were There, in the Words of One of Canada's Greatest Living Poets.* Ottawa, ON: Oberon Press, 1984.
Villa, Brian Loring. *Unauthorized Action: Mountbatten and the Dieppe Raid.* Oxford, UK: Oxford University Press, 1989.

Diocletian (245–316)

Roman emperor (r. 284–305) responsible for the reorganization of the military, administration, and finances of the Roman Empire. Born in Dalmatia as Gaius Aurelius Valerius Diocletianus, he rose to power through the army fighting with Carinus against the Persians. During that campaign the murder of Carinus's brother and coemperor raised suspicions against the Praetorian prefect and provided an opportunity for Diocletian's troops to proclaim him emperor in 284. He exercised power in Asia Minor and Syria, where his troops dominated while Carinus continued to rule in the west. As the two armies converged against each other in the area around present-day Belgrade, soldiers assassinated Carinus, leaving Diocletian as the sole ruler of the empire. Recognizing that the vastness of the territory required an administrator as well as a military commander, Diocletian divided his power between himself and Maximian, an Illyrian, who suppressed revolts primarily along the German borders. Both men assumed the title of Augustus in 293. Responsibility was further divided when Diocletian decreed that Galerius and Constantius I Chlorus would be caesars, each one of them assigned to either Diocletian or Maximian. Militarily, during the reign of Diocletian the Roman legions suppressed revolts in Britain, Persia, Syria, Egypt, and along the Danube. The effectiveness of the military ushered in an era of peace. Domestic reforms initiated by Diocletian included the strongest persecution of Christians by any emperor. He also reorganized the offices of government by building a professional bureaucracy while reducing the power of the prefects, senators, and consuls.

Cynthia Clark Northrup

See also: Constantine the Great

References and further reading:

Barnes, Timothy David. *The New Empire of Diocletian and Constantine.* Cambridge, MA: Harvard University Press, 1982.
Williams, Stephen. *Diocletian and the Roman Recovery.* New York: Routledge, 1997.

Dionysian Wars (398–367 B.C.E.)

A series of four wars waged by Dionysius the Elder against Carthage for control of Sicily. Throughout the last half of the fifth century B.C.E., relations between the Carthaginians ensconced in western Sicily and the inhabitants of the Greek colonies throughout the island deteriorated as both sought to protect their commercial interests and expand their territories. Carthaginian armies mounted a series of bloody campaigns during the century's final decades, seizing important cities and allowing Athens to attack, albeit unsuccessfully, Syracuse (Siracusa, Italy), the foremost challenger to Carthaginian supremacy. In 406, a Carthaginian force invaded and threatened the few independent areas of eastern Sicily. Concurrently, Dionysius seized power at Syracuse and negotiated an armistice; its terms relegated free Greek possessions to the easternmost fringes of Sicily.

Fought between 398 and 392 B.C.E., the First Dionysian War marked the new tyrant's first attempt to wrest Sicily away from Carthage. Violating the truce, Dionysius sent troops in 398 to attack the seemingly impregnable stronghold at Motya (Mozia, Italy), which resisted its assailants until 397. Carthage retaliated, routing the Syracusan navy near Catana (Catania, Italy) and landing a mercenary force led by Himilco, which easily occupied the suburbs of Syracuse. Plague struck the Carthaginians in 395, permitting an opportunity for a successful counterattack by Dionysius, whose forces were reinforced with Spartan and mercenary contingents. Himilco escaped with a fraction of his army and committed suicide.

Carthage dispatched another invasion force under Mago in 393, starting the Second Dionysian War. Dionysius defeated Mago near Abacaenum (Tripi, Italy), but hurriedly demanded a treaty when his own soldiers mutinied unexpectedly in 392. The accord recognized Syracusan suzerainty over Sicily up to the Mazarus River, leaving only the westernmost sliver as a Carthaginian province.

In the Third Dionysian War, fought from 383 to c. 375 B.C.E., Dionysius took the offensive against Carthage, which formed an alliance with the remnants of the Italiote League in Calabria. On the latter front, a storm savaged his fleet while it attacked Thurii (Sibari, Italy), but his troops managed to take Croton (Crotone, Italy), consolidating Syracusan control of Calabria by 377. In Sicily that same year, his forces heavily defeated the Carthaginians at Cabala (location unknown), killing their leader Mago. However, 376 witnessed a reversal of Dionysius's fortunes, when Dionysius lost his army and his brother Leptines to the Carthaginians at Cronium (location unknown). The conflict ended with formal demarcation of Sicily into Carthaginian and Syracusan zones separated by the Platani River.

Dionysius made a final bid to dominate Sicily during the Fourth Dionysian War. In 368, he besieged Lilybaeum (Marsala, Italy), which replaced Motya as the Carthaginians' main western garrison. However, the Carthaginians caught his fleet at Drepana (Trepani, Italy) and Dionysius ended the campaign, dying shortly thereafter. Dionysius II abandoned the strategic ambitions of his father and relations between Syracuse and Carthage remained quiet until the outbreak of Timoleon's War decades later.

Ian Janssen

See also: Agathocles; Dionysius the Elder

References and further reading:

Caven, Brian. *Dionysius I, War-Lord of Sicily.* New Haven, CT: Yale University Press, 1990.

Talbert, Richard. "The Greeks in Sicily and South Italy." In *The Greek World in the Fourth Century,* ed. Lawrence Tritle, 137–165. London: Routledge, 1997.

Dionysius the Elder (c. 430–367 B.C.E.)

Greek ruler who briefly conquered most of Sicily. Born c. 430 B.C.E. into a wealthy family of Syracuse (Siracusa, Italy), Dionysius had a career that combined ruthless political opportunism and autocracy with brilliant, innovative soldiery. He entered public life through both intrigue and military service, participating in Hermocrates's abortive coup in 408 and fighting Carthaginian invaders in 406. After perfidiously convincing the Syracusan assembly that their generals were traitors, Dionysius became head of the Syracusan army and virtual ruler of the city in the next year. Immediately, he concluded a truce with the Carthaginians and brutally suppressed an aristocratic insurrection.

Three main opponents resisted Dionysius's expansionist policies: Greek colonists, Carthaginians, and the native Sicel population. His mercenaries vanquished the Greeks and Sicels of eastern Sicily in 399. During this campaign, Dionysius introduced the catapult and the quinquereme, a five-banked, oared warship, to Sicilian military practice. With little delay, he started the first of four Dionysian wars (398–367) with Carthage. By the close of the Second Dionysian War in 392, Sicily was his except for the westernmost segment. Dionysius then led his troops into mainland Italy in 390, crushing the Italiote League and gaining mastery of Calabria as far north as Thurii (Sibari, Italy) by 379. He founded colonies on the Adriatic coast of Italy and raided the Etruscan port of Pyrgi (Santa Severa, Italy).

After the mid-380s, Dionysius's fortunes declined somewhat. He became entangled in mainland Greek affairs, shifting his support from Sparta to Athens, the erstwhile enemy of Syracuse. Carthage defeated his armies during the final two Dionysian wars, resulting in territorial losses and military stalemate in western Sicily. His son Dionysius II succeeded him upon his death in 367. In subsequent classical literature, the elder Dionysius epitomized tyrannical excess and egotism.

Ian Janssen

See also: Dionysian Wars

References and further reading:

Caven, Brian. *Dionysius I, War-Lord of Sicily.* New Haven, CT: Yale University Press, 1990.

Disarmament

Just before dawn on the morning of 16 July 1945, the world's first atomic bomb was detonated over the desert sands (turning some of that sand to glass) of southern New Mexico. A few short weeks later, atomic bombs exploded over the Japanese cities of Hiroshima and Nagasaki, ending World War II. Less than a month before the first atomic test, the charter for the United Nations was signed in San Francisco. Recognizing the destructive power of atomic bombs, as evidenced by the devastation at Hiroshima and Nagasaki, the very first resolution adopted by this new international body created the United Nations Atomic Energy Commission. The commission was created to make recommendations for dealing with weapons of mass destruction, specifically to promote exchange of basic scientific information about atomic energy, advance ideas on the control of atomic energy, work toward the elimination of atomic weapons, and develop effective methods of inspection to monitor compliance with disarmament.

Disarmament is a generally accepted term that includes all aspects of arms regulation such as arms control, or arms limitation, as well as the elimination of nuclear weapons. However, disarmament is more often used to describe those efforts aimed at entirely eliminating weapons of mass destruction, whereas arms control is used to define activities related to treaties that reduce the number of weapons.

There have been several comprehensive attempts to limit or even eliminate certain types of weapons. But the only truly successful disarmament treaty was the Washington Treaty of 1922, which eliminated large numbers of battleships. The United Nations, where most disarmament work takes place, adheres to the concept that nuclear weapons disarmament is the long-term solution to international peace and security. In 1959, the United Nations declared, formally, that its ultimate goal was to achieve general and complete disarmament under international control.

The United Nations promotes two general activities in working toward disarmament—education and the reduction and eventual elimination of nuclear testing. Education efforts center on such activities as expert studies to facilitate understanding of the many complex issues underlying disarmament. President Eisenhower's Atoms for Peace proposal was both a disarmament and an educational effort that planned to give more rather than fewer nations access to atomic energy. Education programs benefit smaller nations, who might not normally be allowed to participate in disarmament negotiations, but who have a stake in the world political process.

The majority of the United Nations's disarmament efforts center on promoting an end to weapon testing. Cessation of nuclear testing, it is believed, will prevent continued improvements of existing nuclear weapons and retard the development of new weapons. To this end, the United Nations facilitated a number of treaties to limit and, in some cases, ban nuclear testing in specific areas. These treaties include the 1963 Partial Test Ban, 1967 Outer Space Treaty, 1978 Non-Proliferation Treaty, 1971 Sea Bed Treaty, 1977 Environmental Modification Convention, and the 1979 Agreement on the Moon and Other Celestial Bodies. Of these treaties, the 1963 Partial Test Ban and the 1978 Non-Proliferation Treaty have been the most important vehicles for reducing both testing and proliferation of weapons. Other treaties, such as the Strategic Arms Limitation Talks (SALT) and Strategic Arms Reduction Talks (START) agreements, although more properly categorized as arms control efforts, have contributed to disarmament by the very act of reducing the number of total weapons in existence.

Despite the number of disarmament treaties that have been adopted in the international arena, disarmament proponents, including the United Nations, have not been able to get universal approval of the Comprehensive Test Ban Treaty (CTBT). Although now generally prepared to accept the treaty, India, Pakistan, France, and China tested nuclear weapons in the 1990s to improve their weapons, something the CTBT is trying to stop. The United States declined to ratify the CTBT in the late 1990s, although it has tacitly abided by the treaty and has not conducted a nuclear test since the early 1990s. The failure to achieve universal acceptance of the CTBT highlights the importance of national politics in any international effort that might curb national self-determination.

Disarmament activities have had limited success in the almost 60 years since Hiroshima and Nagasaki. Unlike arms control efforts, which can be counted, measured, and safeguarded to protect national security, disarmament is an ideal and cannot be quantified. Although nations support the ideal of disarmament, they find it difficult to accept in practice. By the early 1960s, it was widely recognized that disarmament would be a very long-term goal at best. No satisfactory answer has been found to the basic question of how international relations would be governed in a disarmed world or how disputes would be settled if no agreement could be negotiated. The Cold War and its highly sensitive national security concerns, as an example, prevented significant movement to a disarmed world.

The end of the Cold War has rekindled hopes for general disarmament, despite the setback of not achieving universal acceptance of the CTBT. Reduced political tensions have allowed treaties, such as START I and II and the INF (intermediate range nuclear forces), to make deep cuts in nuclear stockpiles. The early commitment to general disarmament by the United Nations, although still the goal, has been replaced by an effort to approach the problem in stages. This incremental approach does not abandon disarmament as an ideal. Rather, it has been modified to emphasize work in smaller areas such as reducing the number of total weapons and preventing proliferation of fissionable material. With the breakup of the Soviet Union and the increase in terrorism, the disposition of fissionable material is of immediate and serious concern. Access to fissionable material is the most important prerequisite to building a nuclear weapon.

The key to any successful disarmament will be the continued improvement of the international political and economic systems. The end of the Cold War removed the most significant impediment to nuclear weapons disarmament—continual superpower confrontation. However, the end of the Cold War did not bring to an end national security interests. Nations still use arms to promote national self-interest. The number of nuclear nations and the number of near-nuclear nations have not yet diminished, although their increase in numbers has been far less than predicted. If the double threat of terrorism and the possible proliferation of fissionable material are not solved, disarmament will continue to be an ideal.

Roger A. Meade

See also: Arms Control

References and further reading:

United Nations Department for Disarmament Affairs. *The United Nations and Disarmament: 1945–1985.* New York: United Nations, 1985.

———. *The United Nations Disarmament Yearbook.* Vol. 14, *1989.* New York: United Nations, 1989.

Dominican Civil War (1965–1966)

Fraternal conflict that brought in U.S. armed intervention. In September 1963, a military coup removed freely elected president Juan Bosch after only seven months in office and replaced him with a three-man junta headed by Donald Reid Cabral, a prominent Dominican businessman. In less than two years the Cabral government lost popular support, due largely to widespread corruption. On 24 April 1965, disgruntled factions of the armed forces, led by Colonel Francisco Caamaño Deñó and supported by the Dominican Revolutionary Party (PRD) and some elements of the Christian Socialist Party, took over the government. Bosch's supporters, the Constitutionalists, called for his return to the presidential palace, a move that had some popular support. However, the U.S. ambassador, William Tipple Bennett, opposed the return of Bosch, as did some factions of the Dominican armed forces, especially the air force.

General Elías Wessín y Wessín ordered the bombing of Santo Domingo by the Dominican air force and attempted to take control of the presidential palace in the colonial part of the capital. The Constitutionalists resisted fiercely and soon appeared to have the upper hand. On 28 April U.S. president Lyndon Johnson ordered an intervention by 500 marines to protect American citizens. The marines reported that the country was on the verge of collapse and that the level of violence was higher than anticipated. Johnson then ordered 23,000 additional troops into the country to keep Bosch from returning and to prevent a possible communist takeover as had happened in Cuba.

Despite the U.S. intervention, the conflict between the Constitutionalists and the Loyalists (supporters of Cabral) raged on. Johnson's actions and policies drew heavy criticism from Mexico and Chile (not to mention opposition in the U.S. itself). By July Johnson realized that Bosch had little chance of returning and he called for a peace-keeping force from the Organization of American States to enforce a cease-fire. Nicaragua, Honduras, Costa Rica, and Brazil sent troops to serve as peace keepers.

The Act of Dominican Reconciliation was agreed to on 31 August 1965, and an interim government was established. In 1966 Dr. Joaquin Balaguer was elected president. Balaguer brought a measure of stability to the island republic during two consecutive terms as president.

George M. Lauderbaugh

See also: Banana Wars

References and further reading:

Gleijeses, Piero. *The Dominican Crisis: The 1965 Constitutionalist Revolt and the American Intervention.* Baltimore: Johns Hopkins University Press, 1978.

Lowenthal, Abraham F. *The Dominican Intervention.* Baltimore: Johns Hopkins University Press, 1972.

Slater, Jerome. *Intervention and Negotiation: The United States and the Dominican Revolution.* New York: Harper & Row, 1970.

Don Juan de Austria (1547–1578)

Spanish general, admiral, and statesman. Don Juan de Austria commanded Christian forces at the last great galley battle of Lepanto. He was also Phillip II's regent in the Netherlands and made significant progress in putting down the Dutch revolt.

Don Juan was the natural son of Emperor Charles V and Barbara Blomberg. He was raised first by a peasant, then brought to Spain to live with a noble family under the name Geronimo. Charles V's will acknowledged Don Juan as his son, and he entered Phillip II's court. Don Juan received his first military command in 1568, when he commanded a squadron of galleys against Barbary corsairs. When Moriscos in Granada rose in revolt, Phillip named Don Juan to head the campaign against them in March 1569. In a small-scale and irregular war, Don Juan developed and executed a competent campaign against their isolated mountain villages. He successfully closed the war in 1570.

Don Juan's next command was as head of the Christian Holy League forces against Ali Pasha and the Turkish fleet in 1571. Spain, Venice, and other powers provided men and galleys to try to stop the Turks' expansion and reconquer Cyprus. The opposing forces met at Lepanto, in Greece, on 7 October 1571. It was the last great galley battle, which assumed many characteristics of land battles. The greater number of Christian soldiers, as well as the heavier construction of their galleys, won the day. More than 200 Turkish galleys were taken or destroyed and the Turks were never again able to pose such a threat to Christendom. As a follow-up, Don Juan recaptured Tunis from the Turks.

In November 1576, Phillip named his brother governor of the Netherlands. Don Juan's natural charisma and energy won back some Dutch, but negotiations proved fruitless. When the states-general raised an army of 20,000 men, Don Juan mobilized Spanish forces. He captured the town of Namur in the fall of 1577. On 31 January 1578, Don Juan routed and nearly destroyed the rebel army at Gemblours. Lack of resources prevented him from exploiting his victory, allowing the Dutch to rebuild. Don Juan died on 1 October 1578, at Bouges, after a brief illness. Rumors abounded that he had been poisoned, possibly by a paranoid Phillip II.

Tim J. Watts

See also: Dutch War of Independence

References and further reading:

Petrie, Charles. *Don John of Austria.* New York: W. W. Norton, 1967.

Slocombe, George Edward. *Don John of Austria, the Victor of Lepanto (1547–1578).* New York: Houghton Mifflin, 1936.

Yeo, Margaret Routledge. *Don John of Austria.* New York: Sheed & Ward, 1934.

Dorian Invasion (c. 1200 B.C.E.)

The movement of peoples from the north of Greece into the Peloponnese, c. 1200 B.C.E. This has been considered by some scholars to be responsible for the destruction of the Mycenaean world. This migration is also called the Return of the Heraclids. According to myth, the sons of Hercules were expelled from the Peloponnese by Eurystheus, and after an unsuccessful attempt to come back, they settled in central Greece, until the Delphic Oracle informed them that the time had come for their return, whereupon they were divided into three groups, descended to the Peloponnese, and conquered most parts of it. Another version has them wandering until reaching Doris, a small region in the mountains of central Greece, where they settled on land given by the king, thus establishing a friendship with the Dorian people, who would later help them invade the Peloponnese. According to others and ignoring the myth, Doris was erroneously thought to be the Dorians' original homeland, but rather they emerged from Epirus in northwestern Greece and migrated south.

None of the ancient writers ever suggested that the Dorian migration was a destructive affair, and it was only after Schliemann's excavations in the nineteenth century that the picture changed. Schliemann found that the Mycenaean centers were suddenly and catastrophically destroyed, most probably by human hands, and immediately the Dorians were held responsible. Thus, in the nineteenth century the Dorian migration became the Dorian Invasion, and its date was synchronized with the fall of Mycenae, sometime between the end of the thirteenth and the end of the twelfth century B.C.E. The evidence, which suggests that a new population entered the Peloponnese, is mostly the appearance later on of the Doric dialect. In terms of archaeological evidence, the invasion of a foreign people can be attested by the appearance of new metal types: the bronze flange-hilted sword with a straight blade known as the Griffzungenschwert sword; the spearhead with flame-shaped blade and a complete cast socket; a type of flange-hilted dagger called the Peschiera dagger; a single-edged knife, with or without a curved blade; a type of lugged ax called the Armchenbeil; and the so-called violin-bow fibula (clasp), which was related to the use of thicker clothes. However, the evidence so far does not support the theory of an invasion, and most scholars do not accept it anymore. There is no distinctive material culture that would define the invaders, and most of what were thought to be new metal types actually appear long before the destruction. In addition, the destruction in the Mycenaean world occurred over quite a long period, and according to no geographical pattern that would imply an invasion. Innovations in burial practices (namely the frequent use of cyst graves) could imply the arrival of a new population, but they appeared well after the destruction of the Mycenaean world. Therefore, there is a possibility that the "Dorian Invasion" was actually an infiltration of peoples into a half-empty land rather than a military invasion.

Christina Aamont

See also: Ancient Warfare

References and further reading:

Drews, Robert. *The Coming of the Greeks.* Princeton, NJ: Princeton University Press, 1988.

Snodgrass, Antony M. *The Dark Age of Greece.* Edinburgh, Scotland: Edinburgh University Press, 1971.

Thomas, Carol G., and Craig Conant. *Citadel to City State: The Transformation of Greece, 1200–700 BCE.* Bloomington: Indiana University Press, 1999.

Dorylaeum (Eske Shehr), Battle of (Turkey, 1 July 1097)

First significant field battle between crusader and Muslim armies. After the forces of the First Crusade (1095–1099) had captured Nicea from the Seljuq Turks, they pressed on for the Holy Land. The route traveled required them to pass through the valley of Dorylaeum. There the Seljuq sultan, Kilij Arslan, awaited them.

The appearance of the crusaders had alarmed Kilij Arslan, forcing him to form an alliance with his rivals, the Danishmend Turks. The Turks, primarily light horse-archers, felt confident that they would defeat the crusaders in open battle.

The crusaders were marching in two divisions. The first, under Bohemund of Taranto, camped in a field not far from Dorylaeum. The following day at sunrise, the Seljuqs charged from the hills. Under direction by Bohemund, the knights dismounted and formed a protective barrier, with the noncombatants in their middle, where springs of water existed. Bohemund also sent a courier to the second crusader division.

The Turks rained a steady hail of arrows down on the crusaders. Surrounded, they faced destruction or slavery. Despite the constant barrage, the crusaders held out until midday, when the second crusader division arrived under Raymond of Toulouse. This second force surprised the Turks, who thought the crusaders were only a single body. As the crusader armies merged, they began to counterattack. The Turks struggled to withstand them, but their surprise turned to panic as a third crusader force appeared in their rear.

This force, led by Bishop Adhemar of Le Puy, had been detached by Raymond for this very purpose. The Turks were routed and the camp of Kilij Arslan fell to the crusaders. Afterwards, the crusaders were able to continue to the Holy Land with minimum difficulty.

Timothy May

See also: Antioch, Battle of; Armor, Ancient and Medieval; Byzantine-Muslim Wars; Crusades; Jerusalem, Siege of; John II Comnenus; Norman-Byzantine Wars; Religion and War; Seljuqs

References and further reading:

Holt, P. M. *The Age of the Crusades.* London: Longman Group, 1990.

Mayer, Hans Eberhard. *The Crusades.* Oxford, UK: Oxford University Press, 1988.

Riley-Smith, Jonathan. *The Crusades: A Short History.* New Haven, CT: Yale University Press, 1987.

Runciman, Steven. *A History of the Crusades.* 3 vols. Cambridge, UK: Cambridge University Press, 1951–1954.

DOWNFALL, Operation (1945–1946)

Code name for the planned Allied invasion of Japan during World War II. Assuming that only the conquest of their home islands could compel Japanese leaders to surrender, Allied planners began creating invasion plans in 1943. By mid-1945 they envisioned two massive amphibious assaults, each commanded by Douglas MacArthur. The first, code-named OLYMPIC, was scheduled for 1 November 1945 against the island of Kyushu; the second, code-named CORONET, was slated for 1 March 1946 against the island of Honshu. Collectively code-named DOWNFALL, these attacks were designed to end the war by late 1946.

The Kyushu plan called for a 10-division assault landing supported by thousands of aircraft from the Marianas, Okinawa, and Iwo Jima, and by 32 aircraft carriers and thousands of support ships. U.S. forces would battle an estimated 350,000 Japanese troops in a campaign that planners believed could last well into 1946, and would be the largest military campaign in history.

After conquering Kyushu, Allied leaders hoped to launch CORONET on 1 March 1946. The largest amphibious landing in history, CORONET called for 17 assault divisions to drive on Tokyo, battling more than 1.2 million Japanese soldiers and at least 1,600 enemy planes along the way.

Casualty estimates for OLYMPIC and CORONET ranged from 250,000 to 1 million American dead, and proved decisive in President Harry Truman's decision to drop atomic bombs on Japan. The destruction of Hiroshima and Nagasaki and the entry of the Soviet Union into the war against Japan compelled the Japanese to surrender, thus making DOWNFALL's planned bloody assaults unnecessary.

Lance Janda

See also: MacArthur, Douglas; World War II

References and further reading:

Allen, Thomas B., and Norman Polmar. *Code-Name Downfall: The Secret Plan to Invade Japan—And Why Truman Dropped the Bomb.* New York: Simon & Schuster, 1995.

Frank, Richard B. *Downfall: The End of the Imperial Japanese Empire.* New York: Random House, 1999.

Dresden, Battle of (26–27 August 1813)

French victory over Austria and Russia in the Leipzig Campaign of the Napoleonic Wars. The Austrian, Prussian, and Russian allies of the Sixth Coalition fought the French five times in the last week of August 1813, trying to liberate Saxony. On 23 August, Jean Baptiste Jules Bernadotte and Friedrich Wilhelm von Bülow defeated Nicolas Charles Oudinot at Grossbeeren. On 26 August, Gebhard Leberecht von Blücher defeated Jacques Étienne Joseph Alexandre MacDonald at the Katzbach and Dominique Joseph René Vandamme defeated Eugen von Württemberg at Pirna.

Three armies, led by Bernadotte in the north, Blücher in the center, and Karl von Schwarzenberg in the south, formed a semicircle around Napoleon's east, each independently threatening to push him toward the Rhine. Oudinot guarded the north, Michel Ney the center, while Napoleon patrolled both the south and the center. On 22 August, in order to prevent Schwarzenberg from capturing the French headquarters at Dresden, Napoleon abandoned his pursuit of Blücher and force-marched 100,000 men 120 miles in four days to defend Dresden. When he arrived, Schwarzenberg's 200,000 Austrians and Russians already held most of the high ground. The French attack prevailed by sheer tenacity, audacity, and desperation, not superior tactics. Napoleon himself commanded field artillery in the thick of battle and Joachim Murat used his cavalry expertly but recklessly. Schwarzenberg, with 6,000 killed, 8,000 wounded, and 24,000 missing or captured, ordered retreat after two full days of fighting. Napoleon lost 10,000 killed and wounded.

Dresden weakened Napoleon severely. Among the reasons that Schwarzenberg was able to beat him decisively in the Battle of the Nations at Leipzig on 16–19 October were Napoleon's general failure to profit from his victory at Dresden, his surrender of 7,000 trapped men to Schwarzenberg on 28 August, Vandamme's loss to Alexeii Ivanovich Ostermann-Tolstoy and Friedrich von Kleist at Kulm-Priesten and Nollendorf on 29–30 August, and especially Ney's 15,000 casualties in his loss to Bernadotte at Dennewitz on 6 September.

Eric v. d. Luft

See also: Bernadotte, Jean Baptiste Jules; Blücher, Gebhard Leberecht von; Leipzig, Battle of; Murat, Joachim, Grand Duke of Cleves-Berg, King of Naples; Napoleon I; Napoleonic Wars; Ney, Michel; Oudinot, Nicholas-Charles, Duc de Reggio; Schwarzenberg, Karl Philipp zu

References and further reading:

Brett-James, Antony, ed. *Europe against Napoleon: The Leipzig Campaign, 1813.* London: Macmillan, 1970.

Lawford, James Philip. *Napoleon: The Last Campaigns, 1813–15.* New York: Crown, 1977.

Nafziger, George. *Napoleon at Dresden: The Battles of August 1813.* Chicago: Emperor's, 1994.

Petre, Francis Loraine. *Napoleon's Last Campaign in Germany, 1813.* London: Greenhill, 1992.

Dreyfus Affair (1894–1906)

A uniquely French complex miscarriage of military justice. Captain Alfred Dreyfus (1859–1935), an independently wealthy Jew of German origins, a member of the General Staff since 1 January 1893, was charged with spying on the basis of a document known as the *bordereau* found at the German embassy. The staunchly anti-Semitic General Staff refused to provide Dreyfus access to the documentation used to try him in a court-martial in October 1894. He was found guilty and sentenced to degradation and imprisonment on Devil's Island. Lieutenant Colonel Georges Picquart realized that the *bordereau,* a memorandum, had been written by Major C. Ferdinand Walsin-Esterhazy, whose handwriting corresponded to other incriminating evidence. Picquart was quickly replaced by Commandant Joseph-Hubert Henry, who forged documents and upon discovery committed suicide. Esterhazy's court-martial completely absolved him of all charges.

Emile Zola's public letter "J'Accuse" on 13 January 1898 was partially responsible for a new trial at Rennes on 3 June 1899, where Dreyfus again was declared guilty but with extenuating circumstances. In 1903 the Supreme Court declared that the court-martial had been erroneous. Prime Minister Waldeck Rousseau offered Dreyfus a presidential pardon and he was reinstated on 22 July 1906 as a major and named Chevalier de la Légion d'Honneur. He served in World War I as a lieutenant colonel and retired in 1918.

The Dreyfus Affair was significant because the army knowingly allowed one of its own to be a sacrificial scapegoat because he was Jewish and partly because it would not admit its own duplicity. The mass hysteria engendered by the divisive affair reflected the current obsession in many quarters with Jewish and Protestant conspiracies, by those who put the "honor" of the French army above all else. On 7

Captain Alfred Dreyfus (second from right) with three other French military officers. (Library of Congress)

September 1995 the French army finally admitted it had been wrong from the beginning in the Alfred Dreyfus affair.

Annette Richardson

See also: French Army

References and further reading:

Chapman, Guy. *The Dreyfus Case: A Reassessment.* New York: Reynal, 1955, 1972.

Dreyfus, Pierre. *Souvenirs et correspondence:* Paris: Grasset, 1936.

Reinach, Joseph. *Histoire de l'affaire Dreyfus.* 7 vols. Paris: Editions de la Revue blanche, 1901–1911.

Dudley, John, Duke of Northumberland (1502–1553)

Son of Henry VII's hated tax collector, Edmund Dudley, who was executed by Henry VIII as a public relations move. John Dudley quickly set about restoring the family's favor. During the pilgrimage of Grace, a rebellion against the Reformation Parliament, Dudley pacified the north of England. Appointed a vice admiral in 1537, he dispatched a small fleet against Flemish pirates and successfully cleared the channel, receiving the title Lord Lisle in return for his service. As warden of the Scottish Marches and commissioner to Berwick, Dudley improved fortifications along the border and played a key role in the Battle of Pinkie in 1547 against a larger Scottish army, rallying his men to attack the Scottish right. Advancing to the title Earl of Warwick upon the accession of Edward VI, he joined the Privy Council and immediately clashed with the young king's guardian and uncle, the duke of Somerset, Edward Seymour.

Never a committed religious man, Dudley embraced the Reformation in order to strengthen his position at court, and invited criticism against Seymour, the Protector. When rebellion broke out in Norwich, started by peasants provoked by enclosure, Warwick was appointed to put it down, using an army of 7,500 men, with whom he seized the city of Norwich and defeated the rebel's leader, Robert Kett, at Dessingdale in August 1549. Meanwhile, Dudley, who became duke of Northumberland, engineered the disgrace and fall of Edward Seymour and took his place as Protector. The young king, without an heir and in poor health, would be succeeded by his sister Mary Tudor, a Roman Catholic, a situation that prompted Dudley to convince Edward VI to alter the line of succession and leave his crown to his cousin Jane Grey, whom Dudley had married to one of his sons. On Edward's death, Dudley proclaimed Jane queen and marched with an armed force to seize Mary, who eluded him and marched on London herself. Rallying the people, Mary deposed Jane and arrested Dudley. Judged guilty of treason for his unsuccessful attempt to seize the throne for his family, he was executed at the Tower of London in 1553.

Margaret Sankey

References and further reading:

Beer, Barrett L. *Northumberland.* Kent, OH: Kent State University Press, 1973.

Jordan, W. K. *Edward VI: The Threshold of Power.* Cambridge, MA: Belknap Press, 1970.

Loades, D. M. *John Dudley, Duke of Northumberland.* Oxford, UK: Clarendon Press, 1996.

Dunbar, Battle of (3 September 1650)

Victory over Scots demonstrates Cromwell's tactical skills. Following the 1649 execution of Charles I, his son Charles II landed in Scotland in June 1650, and raised his standard. In response, an English expeditionary force, consisting of 5,000 horse and 10,000 foot under the command of Oliver Cromwell, crossed the border on 22 July.

The Scottish commander, David Leslie, had an inexperienced, albeit larger, force, and avoided a direct confrontation with the invaders. Harassed, and with long supply lines, Cromwell was forced to halt his advance by late August, and pulled back to the port at Dunbar, which he fortified. On 2 September Leslie moved a force of approximately 22,000 men to the south of Dunbar, threatening Cromwell's positions. The rough ground did not favor their lines, but it appears possible that the movement was forced upon Leslie by political advisers.

Heavily outnumbered (disease had cut the size of his army to fewer than 12,000), Cromwell decided upon a surprise attack, and at 4 A.M. on 3 September his regiments launched their assault. The Scottish forces were caught unprepared, and a concerted effort on their right forced that wing to collapse. The Scottish center attempted to withdraw, but the rough ground prevented a safe escape, and it quickly became a rout.

It is estimated that Scottish casualties may have exceeded 3,000, with a further 10,000 taken prisoner; the English losses were negligible. This was perhaps Cromwell's most notable victory as it demonstrated his ability as a field commander. It also served as precursor to his final victory over the Scots, and allowed him to secure Glasgow and Edinburgh, as Leslie was forced to retreat.

Daniel German

See also: Cromwell, Oliver; English Civil War (1642–1649)

References and further reading:

Kenyon, John P. *The Civil Wars of England.* London: Weidenfeld & Nicolson, 1988.

Young, Peter, and Richard Holmes. *The English Civil War.* Ware, UK: Wordsworth Editions (Wordsworth Military Library), 2000.

Dunes (14 June 1658)

Climactic battle of the French-Spanish conflict over the Lowlands, which helped to create the northern boundaries of modern France. In 1658, a polyglot Spanish army of 16,000 (including 2,000 followers of the exiled Charles II of England, Condé's French rebels, and a few regiments of Spanish regulars) marched to raise a siege of the coastal town of Dunkirk, which was surrounded by Turenne's Anglo-French expeditionary force of perhaps 21,000. Turenne's force was somewhat more homogenous but did include a leavening of 6,000 English infantry sent by Oliver Cromwell's Protectorate.

On 13 June, the Spanish force drew up their lines on the dunes north of the town, between the strand and a canal. Turenne met them the next day with 9,000 cavalry and 6,000 infantry, leaving the remainder of his forces to maintain the siege. The guns of a small Protectorate fleet covered the strand.

The principal action took place between the right of the Spanish line and the left of the French. There both wings were composed of British troops, the Royalists atop a tall steep dune, while below them Cromwell's Protectorate infantry struggled to storm the dunes. After much bloody fighting, the weight of the Protectorate arms prevailed, and the right flank of the Spanish army buckled. Although the Spanish cavalry of the left had some success, eventually they too were forced to withdraw, and a rout ensued. It is estimated that the Spanish army suffered as many as 6,000 casualties, many of them prisoners, while the Anglo-French alliance lost only 400.

This decisive victory ensured the French domination over the region. Although Britain gained Dunkirk as part of the alliance, it was sold to France in 1662. In addition, France captured a number of other towns in this campaign and forced Spain to sign the Treaty of the Pyrenees, all of which marked the end of a serious Spanish threat to northern France.

Daniel German

See also: Cromwell, Oliver; English Civil War (1642–1649); Turenne, Henri de la Tour d'Auvergne, Vicomte de

References and further reading:

Firth, C. H. "Royalist and Cromwellian Armies in Flanders, 1657–1662." *Transactions of the Royal Historical Society* NS 17 (1903).

Marichal, Paul, ed. *Memoires de Marechal de Turenne.* Paris: Librairie Renouard, 1914.

Sells, A. Lytton, ed. *The Memoirs of James II: His Campaigns as Duke of York 1652–1660.* Bloomington: Indiana University Press, 1961.

Young, P., and R. Holmes. *The English Civil War 1642–51.* Ware, UK: 2000.

Dutch Colonial Wars (c. 1620–1949)

Colonial wars focused on the obtaining, retention, and loss of power throughout Dutch areas of economic and military power and influence. The Dutch fought with the Portuguese, the British, the Indians, and the Indonesians. The fighting mainly occurred through the auspices of the Dutch East India Company, founded in 1602, and the Dutch West India Company, founded in 1621.

The Dutch and Portuguese Colonial Wars from c. 1620–1655 were economically based and spread from Africa to Brazil. The Dutch West India Company seized Portuguese commerce in Brazil and Africa after successfully removing the Portuguese from India. Piet Heyn (1577–1629), who commanded 26 Dutch ships, seized Baha in 1624. Spain, which had annexed Portugal, retaliated with 52 ships, with 12,000 men led by Fadrique de Toledo, that removed Heyn, who, however, returned in 1627. Heyn's capture of a huge Spanish treasure in Cuba in 1628 resulted in his promotion to admiral. The newly wealthy Dutch West India Company sent 67 ships and 7,000 men, who captured Pernambuco and Recife. The Portuguese inhabitants revolted in 1654, causing the Dutch to withdraw. However, Portuguese forts in Ghana, Luanda, and Brazil were attacked. The Dutch gained control over Ghana and the Gold Coast, and the Portuguese were also expelled from Molucca and Sri Lanka.

From 1655 to 1664 the Dutch had negative relations with the Manhattan and Algonquin Indians in present-day New York State. A dispute led to Pieter Stuyvesant (1592–1672) applying rules restricting movements of the Indians. However, Indian raids on Long Island led to Stuyvesant's intervention at the request of settlers. A palisade was built. Indian attacks occurred again in August 1658, leading to Stuyvesant's victory over the Indians. Renewed fighting occurred in 1663. The following year the Indians were forced to surrender.

The Dutch also fought three wars with England, from 1652 to 1654, 1664 to 1666, and 1672 to 1674. These wars were based on commercial rivalry but extended to the colonies. The English 1651 Navigation Act restricted non-British crews and ships from trading, severely harming Dutch maritime profits. Nine naval battles culminating in the Battle of Scheveningen on 31 July 1652 resulted in the death of Dutch commander Maarten Tromp (1597–1652). The 1654 Treaty of Westminster ratified the English victory after they had blockaded Holland in 1653. The Dutch West African ports that handled the profitable slave trade were also attacked by the English.

New Amsterdam was seized by the English in 1664. Prince James (1622–1701), brother of King Charles II, defeated Jacob Opdam's 100-ship fleet in 1665. But Admiral Michiel de Ruyter (1607–1676) raided the Thames River in June 1667, destroying 16 ships and causing the English to sue for peace. The Treaty of Breda granted Surinam to the Dutch but gave the English control over present-day New Jersey, Delaware, and New York. As the Dutch said of their final defeat by the British, "The mountain of iron [England] defeated the mountain of gold [Holland]."

The final war involved France's king Louis XIV (1638–1715) and his alliance with King Charles II (1630–1685) of England and their attempt to blockade Holland. Although the French were generally victorious, in order to avoid financial ruin Louis made peace with the 1678 Treaty of Nijmegen.

In 1652 the Dutch founded a colony at Cape Town in present-day South Africa. By the 1680s French Huguenots and Dutch Reformed had families settled in the area. They became known as Boers and came to speak Afrikaans, a variant of the Dutch language. The British seized the area in 1806; consequently many Boers participated in the Great Trek north where they founded their own republics. The diamond discoveries in 1867 and the 1886 gold discoveries led to British encroachment in Boer territory. The Boers were defeated in the Boer War of 1899–1902, after a heroic resistance.

The final Dutch colonial war took place in the Dutch East Indies (contemporary Indonesia). The Dutch had governed their East Indies for nearly 300 years in a corrupt and ruthless manner that had little respect for the natives. The focus on trade and profit overwhelmed all other considerations. It is not surprising then that the Sarekat Islam (Islamic Union) was established in 1912 as a response to the Dutch Ethical Policy that divided Indonesian society into western educated haves and local have-nots. Some 2 million merchants created the Volksraad (People's Council). Although initially conciliatory in its approach, after the failed Communist-led insurrection in 1926 and 1927, the Dutch embraced a repressive policy. In response, the Partai Nasional Indonesia (Indonesian Nationalist Party, PNI) was established by Achmed Sukarno (1901–1970) and Muhammad Hatta (1902–1980), both of whom advocated complete independence. The PNI was banned and its leaders repeatedly arrested and exiled.

The Japanese occupation of Indonesia during World War II granted Sukarno and his followers political freedom. Although a repressive system, the occupation trained young men militarily throughout the Indonesian islands, a training that eventually created a postwar army of independence. The Japanese offered independence in October 1944 in exchange for support against the Allies. However, economic deprivations and the aggressive behavior of the Japanese produced considerable ill will toward the occupation. In short, the Japanese were behaving just like the former colonial masters, down to assertions of cultural and even racial superiority. But this ill will did not translate into support for the Dutch when they returned to the islands in the wake of the Japanese defeat in 1945. The surrender of the Dutch colonial forces and the general capitulation of all European authorities' empires in the Pacific and Asia to the Japanese early in World War II in the Pacific had led to a catastrophic loss of prestige and native deference to those regimes. Between 1945 and 1949, with the diplomatic aid of the United Nations, the Indonesians were able to resist the restoration of Dutch authority and to achieve independence.

By the middle of the twentieth century the Dutch had lost all of their colonial wars. This loss was due in part to reduced military power, the widespread independence movements, and the international condemnation of empires. Presently, only the Netherlands Antilles and Aruba are joined to the kingdom of the Netherlands.

Annette Richardson

See also: Guerrilla/Partisan/Irregular Warfare

References and further reading:

Boxer, C. R. *The Dutch Seaborne Empire 1600–1800.* New York: Knopf, 1965.

De Jong, Louis. *Het koningkrijk der Nederlanden in de tweede wereldoorlog.* 'S Gravenhage, Netherlands: Staatsuitgeverij, 1969.

Goonewardena, K. W. *The Foundation of Dutch Power in Ceylon, 1638–1658.* Amsterdam: Djambatan, 1958.

Hyma, Albert. *The Dutch in the Far East.* Ann Arbor, MI: G. Wahr, 1942.

Mak, Geert. *Amsterdam.* Cambridge, MA: Harvard University Press, 2000.

Reid, Anthony. *Southeast Asia in the Age of Commerce 1450–1680.* Vol. 1, *The Lands below the Winds?* New Haven, CT: Yale University Press, 1988.

———. *Southeast Asia in the Age of Commerce 1450–1680.* Vol. 2, *Expansion and Crisis.* New Haven, CT: Yale University Press, 1993.

Schama, Simon. *The Embarrassment of Riches.* London: William Collins & Sons, 1987.

Dutch War of Independence (1567–1648)

Better known as the Eighty Years' War or the Dutch Revolt, one of the longest struggles for national independence in history. The conflict began as a revolt by present-day Belgium and the Netherlands against the severe economic and religious oppression of the Spanish monarchy. The revolt was a bitter clash between ancient freedoms, absolutist tyranny, and differing ideologies, and converged into a religious and social revolution. The revolt further became entangled in the Thirty Years' War and ended with complete independence of the Netherlands from its Spanish overlords.

Burgundian rule over the present-day Netherlands and Belgium ended when in 1506 Charles V (1500–1558) inherited various territories from his grandparents. Serious problems emerged upon his abdication on 25 October 1555 when the prematurely worn-out, stately, but well-loved monarch bequeathed his empire to his son Philip II (1527–1598), an absolutist, narrow-minded fanatic Catholic. Philip appointed his conciliatory half sister Margaret of Parma (1522–1586) as regent in 1557. She governed through a reckless Council of Regency that excluded the local nobility and persecuted the sizable Protestant Calvinist faction that had arisen during the Reformation. In his bigoted obstinancy, Philip contravened many ancient privileges—one such was that no Spanish soldiers would ever set foot on Netherlands soil. Philip not only revoked that privilege but also championed Catholicism by increasing the number of Holy Sees. Through Antoine Perrenot Cardinal de Granvelle, Philip brought the Inquisition to the Netherlands. Properties were confiscated, titles removed, and thousands of people charged with treason for minor infractions. The barbarous cruelty against the Calvinists aroused considerable hatred against Spanish rule, which Margaret was powerless to stop. Granvelle left the Netherlands in 1564 but the damage was done.

Anti-Catholic riots spread throughout the Netherlands in 1566. Calvinists responded to Catholic terror by ransacking churches and by destroying religious relics and art. Outraged, Philip sent the notorious Fernando Alvarez de Toledo, Duke of Alba, to crush the Calvinists, not only with 20,000 troops but also through the Council of Blood. His job was to return the Netherlands to the Catholic fold. Margaret resigned her position. Alva's dictatorial rule and punitive military campaigns from 1567 to 1573 were conducted with unusual carnage and terror, added to which a retaliatory tax of 10 percent on each transaction to pay for the occupation and the war crippled the mercantile sector.

Meanwhile, William of Orange (1533–1584), a German-born nobleman who had been raised in the court of Charles V, and who had converted to Catholicism to please Charles, was commander of his troops near the French border. He had served Philip as a diplomat, objected strongly to Spanish tyranny, and had reconverted back to Calvinism. His diplomatic skills and cautious demeanor earned the sobriquet William the Silent, and he gained the firm allegiance of many beleaguered nobles and the common people.

The Geux, a rebel group of converted Calvinist adventurers, pirates and guerrillas, supported by William from 1568, raided Spanish-held territories. They were crucially important to the advances of the revolt. Although losing at Heiligerlee and Jemmingen in 1568, the Geux still successfully blockaded the sea outlet to Brussels. In 1572 they seized Den Briel, which proved fatal to the Spanish. The Geux also gained control over the entrances to the provinces of Zeeland and Holland.

William's tactic was to avoid Alva's well-prepared forces. Instead, he exploited his knowledge of the landscape by aiding besieged cities. In Leiden in 1574, for example, he opened the dikes, scattering the Spaniards. William soon had the majority of the northern Protestant areas under his control and formed the Union of Utrecht in 1579, which melded the various provinces into one entity. These seven provinces later became the Dutch Republic.

Then in 1578, Alexander Farnese of Parma (1578–1592), son of Margaret, became governor of the Spanish Netherlands (Belgium). He defeated the Dutch in 1578 at the Battle of Gembloux and returned the Belgian provinces to the Catholic fold. With the 1579 Union of Arras the Belgians swore allegiance to Spain. The revolt had by now become religious. The Union of Utrecht proclaimed its independence from Spain in 1581. William governed judiciously, but was assassinated in 1584.

Maurice of Nassau (1567–1625), William's son from his second marriage, succeeded his father and proved to be the savior of the revolt. This brilliant military reformer realized the need for a strong, effective counteroffensive. He professionalized the officer corps by establishing the first military academy and a system whereby troops were paid regularly. He devised 500-man battalions whose units could move en masse or independently for greater maneuverability. He insisted on the best artillery and was brilliant at siege warfare. This strategic genius built a defensive zone of fortified towns that seriously obstructed the Spanish. Maurice nearly single-handedly revolutionized warfare through improved training, sound economic policies, and logistics; his was the most modern army in Europe.

Maurice was appointed captain general in 1588 and admiral of the United Netherlands. In 1589 he was recognized as stadtholder (governor) of the provinces of Utrecht, Gelderland, and Overijssel. He captured Breda in 1590; Zutphen, Deventer, and Hulst in 1591; and won the battles of Turnhout in 1597 and Nieuport in 1600, among others. Maurice's military success resulted in the Twelve Years' Truce, signed in 1609. Fighting resumed in 1621 when Maurice would not extend the truce. The Spaniards by this time were commanded by Ambroglio de Spinola (1569–1630) who bested Maurice with his numerically superior forces at the siege of Breda. Maurice died after a five-month illness on 23 April 1625, leaving no heirs.

Under Maurice's half brother Frederick Henry (1584–1647), a charming, tactful leader, the Dutch quest for independence became enmeshed in the Thirty Years' War that began in 1618. With financial support from France he accomplished major and significant victories against the Spaniards at 's-Hertogenbosch in 1629, Maastricht in 1632, and Breda in 1637. His 1635 alliance with France and Sweden against Spain solidified his position, as did the marriage in 1641 of his son William (1626–1650) to Mary Stuart (1631–1660), daughter of King Charles I (1600–1649) of England.

A grateful Netherlands granted Frederick Henry's family hereditary rights to the stadtholder position, and they eventually became the royal family of the Netherlands. Upon his death in 1647 Frederick Henry was succeeded by his son William II, whose son William III (1650–1702), born eight days after his father's death, became joint monarch of England in 1688 along with his wife Mary Stuart, daughter of the deposed King James II.

The Spaniards were severely weakened by this time, and were forced to recognize the independence of the Netherlands at the Peace of Munster in 1648. The Eighty Years' War was the first European war in which independence was the desired objective. It would not be the last.

Annette Richardson, Ph.D.

See also: Alba, Fernando Alvarez de Toledo, Duque de; Maurice of Nassau; Thirty Years' War

References and further reading:

Geyl, Peter. *The Revolt of the Netherlands, 1559–1609.* 2d ed. London: Ernest Benn, 1958.

Israel, Jonathan I. *The Dutch Republic: Its Rise, Greatness, and Fall, 1477–1806.* Oxford, UK: Clarendon Press, 1998.

Parker, Geoffrey. *The Dutch Revolt.* Harmondsworth, UK: Penguin Books Ltd., 1981.

Wedgwood, C. V. *William the Silent: William of Nassau, Prince of Orange 1533–1584.* New York: W. W. Norton, 1968.

E

Economic Warfare

Compelling an enemy to capitulate either by direct action against its economic base or indirectly through blockade, boycotts, and so on. Joseph Goebbels, minister for Propaganda of the Third Reich, asked a group of German area leaders in 1943 whether they wanted total war. Their response was a resounding *Ja!* What lay behind the question was the fact that the German economy in 1943 was not yet fully mobilized for war, although it was at war with Russia and the Western Allies simultaneously.

Economic warfare can be tactical—denying a water hole by poisoning; burning a field of wheat—or strategic—destroying lines of communication necessary for the movement of raw materials and finished military and other products. Navies have often engaged in blockade to deny maritime trade, and in the twentieth century air forces have also been pressed into the battle.

In the siege warfare of the Middle Ages it was common to attempt to poison water supplies and to try to spread disease among the besieged inhabitants. In the Indian Wars in America, salting water holes or even leaving animal carcasses in water sources was a form of tactical economic warfare. On a larger scale, Stalin's order to burn all of Russia before the Germans was mainly a tactical, defensive form of this type of warfare, for the strategic factories had already been moved east to the Urals.

In World War I, the blockade of Germany eventually caused near-starvation, but did not destroy the German will to fight—the ultimate aim of economic warfare, coupled with the denial to the enemy of the means to fight. Economic warfare attacks the soldier in the field only indirectly, attacking instead his civilian counterpart, the maker of the equipment the soldier needs, as well as denying the civilian population the means to make that equipment, and causing morale to fall to such a level that continuation of the war is made impossible.

The advocates of strategic air bombing before World War II were sure that their bombers would always penetrate the defenses, and that the bombing would be so effective that countries would be obliged to sue for peace without land invasion. That this was a misconceived concept was made clear in the air campaign against Germany in World War II. The Royal Air Force switched to night bombing in the face of German antiaircraft defenses and the Luftwaffe fighter patrols, and then tried to justify random area bombing as economic warfare.

Heavy bomber advocates in Britain argued that bombing alone would bring Germany to its knees, but, basically, the bombing raids produced a stiffening of resistance in Germany—exactly as had happened in London during German air raids in 1940 and 1941.

The U.S. Army Air Forces made every attempt to carry out an economic air offensive against Germany. They attacked in daylight, with extreme courage and fortitude, dropping bombs on specific strategic economic targets such as oil refineries, U-boat yards, munitions factories, and aircraft production sites. The results only became clear in 1945, showing that despite their sacrificial efforts (and those of the RAF), air bombing could not win a war; it could only make it easier in the end for land forces to penetrate land defenses.

At sea, an extremely effective method of blockade is to sink the ships carrying materials needed to continue the war. In both of the world wars of the twentieth century, Germany used submarine warfare as its most effective economic assault on Britain. Supplies from the United States under the Lease-Lend agreement were fundamental to the continuance of British resistance. Admiral Karl Doenitz knew how effective the U-boats had been in World War I, and to make them even more so he created the wolf-pack. A group of U-boats would attack a convoy (previously the convoy system had defeated the U-boat campaign in World War

I), and by weight of numbers eventually render the Atlantic Ocean impassable.

Had Doenitz succeeded, there is little doubt that Britain would have had to sue for peace, but with the "Germany First" policy of President Franklin Roosevelt and the entry of the United States into the war, it was just a matter of time before the U-boats were rendered ineffective, as they were in May 1943.

The war in the Atlantic in both world wars is perhaps the single most significant example of economic warfare. It should be remembered that only by cooperation between the Allies was the U-boat menace defeated, despite every effort by British Bomber Command to ensure that RAF aircraft were not released for antisubmarine duties at a time when they were unable to find targets in Germany, let alone bomb accurately.

Economic warfare was used ineffectively in the Vietnam conflilct because it was based on Western living standards; the North Vietnamese were capable of withstanding deprivation on a scale inconceivable to the affluent Americans. Further, it seems that the Agent Orange defoliant caused as much damage to U.S. troops as it did to the Vietnamese. In addition, bombing an enemy "back to the Stone Age" is only effective against an enemy living well above Stone Age levels.

Air attacks and blockades can severely damage the infrastructure of a warring nation, but they cannot defeat that nation, and economic warfare can only be effective in the longer term. In short-term campaigns such as the Gulf War, the solution is to use one's air and sea assets against military targets.

Economic warfare thus has as its aim the denial to the enemy of the will and means to fight, but it is a two-edged sword that, unless planned meticulously and constantly evaluated, can lead to wasted effort and lives attacking targets that have no great value to the enemy.

David Westwood

See also: Nuclear and Atomic Weapons; Theory, Military

References and further reading:

Churchill, Winston S. *The Second World War.* London: Cassell, 1950.

Doenitz, Admiral Karl. *Ten Years and Twenty Days.* London: Weidenfeld & Nicolson, 1959.

Raeder, Grand Admiral Karl. *Struggle for the Sea.* London: Willam Kimber, 1959.

United States Strategic Bombing Survey. New York: Garland Publishing, 1976.

Edgehill, Battle of (23 October 1642)

The first major battle of the English Civil War. King Charles I raised his standard and began recruiting soldiers on 22 August 1642 on Castle Hill at Nottingham. Seeking more recruits, he marched west to the royalist region around Shrewsbury. By October, Charles felt strong enough to march on London, but his path was blocked by a parliamentary army under Robert Devereux, Earl of Essex.

Essex had about 20,000 men, Charles around 13,000. The royalists were poorly equipped, but superior in morale, especially among the cavalry. Essex's soldiers were mostly members of the prewar "trained bands" of militia. Their morale was low, due to lack of training and pay. Many officers on both sides had military experience in the Dutch wars against Spain and the Thirty Years' War. Charles's chief military adviser was his nephew, the dashing Prince Rupert.

The two armies blundered into each other during the afternoon of 22 October. They formed for battle the next morning, with the royalists between Essex and London. Both sides placed their infantry in the center, with cavalry on both wings. The royalists took six hours to form their line. When Prince Rupert, the king's nephew, objected to the earl of Lindsay's dispositions, the angry earl resigned as royalist commander just before the battle. After skirmishing, Rupert's cavalry on the right routed the opposing cavalry. The royalist troopers on the left quickly did the same, but both groups pursued their foes off the battlefield, a common battlefield error. The infantry then closed. After exchanging musket fire, the battle became a contest between pikemen. The eventual return of the exhausted royalist cavalry helped force a stalemate, but both sides had been roughly handled. They spent the night on the field. Both sides proclaimed victory, but Essex retreated toward reinforcements the next day. Charles declined a rapid advance on London and established his headquarters at Oxford. Essex was able to take up a blocking position and was reinforced by militia from the capital. When Charles finally marched on London in November, he was stopped at Turnham Green, ensuring that the war would not end quickly.

Tim J. Watts

See also: English Civil War (1642–1649); Rupert, Prince

References and further reading:

Fletcher, Anthony. *The Outbreak of the English Civil War.* London: Arnold, 1981.

Tincey, John. *Edgehill 1642: The English Civil War.* Oxford: Osprey Military, 2001.

Young, Peter. *Edgehill 1642: The Campaign and the Battle.* Moreton-in-Marsh, UK: Windrush, 1995.

Edington (Wessex, May 878)

Defeat of Danes results in establishment of the Danelaw. In 876, the Great Danish Army, which had arrived 11 years before, advanced on Wessex. Unlike previous years, it did not leave Wessex after collecting tribute, but wintered there. Although it retired again in August 877, the army was back again in January 878 in a surprise attack, demonstrating the Viking ability to raid even in the dead of winter. A serious attempt was now made to subjugate Wessex.

The army established itself at Chippenham (Wiltshire), and the *Anglo-Saxon Chronicle* records that the West Saxons submitted, "except King Alfred." By March of that year he and his retainers had been forced into hiding, and the hope of the West Saxons seemed to be fading. Alfred continued to harass the Danes from a fort at Athelney in the Somerset fens as he secretly assembled an army. He met the Danes at Edington. "There," according to the *Anglo-Saxon Chronicle*, "he fought against the entire host, and put it to flight, and pursued it up to the fortification [Chippenham], and laid siege there a fortnight."

The Danes surrendered and their leader, Guthrum, agreed to vacate Wessex and to accept baptism. Alfred himself stood as sponsor. The baptism took place at Wedmore (Somerset) some weeks later. It eventually led to an uneasy peace and establishment of the Danelaw (Treaty of Wedmore). The following year the Danes settled in East Anglia, where Guthrum reigned (880–890) as their king.

Nic Fields

See also: Alfred the Great; Viking Raids; Vikings

References and further reading:

Abels, R. P. *Alfred the Great: War, Kingship and Culture in Anglo-Saxon England.* London: Longman, 1998.

Peddie, J. *Alfred: Warrior King.* Stroud, UK: Sutton, 1999.

Edward, the Black Prince (1330–1376)

Leading English general of Hundred Years War. Edward was born 15 June 1330 at Woodstock, Oxfordshire, England; died 8 June 1376 at Westminster near London; and was buried at Canterbury. By the time of his death he was Edward of Woodstock, Prince of Aquitaine, Prince of Wales, Duke of Cornwall, and Earl of Chester. Edward was the son and heir apparent to Edward III, although he never assumed the throne. He is widely considered one of the principal players and exceptional generals of the Hundred Years War, much of his reputation being based upon the lopsided victory in the Battle of Poitiers in 1356. His sobriquet is a reference not to brutality or tyranny, as is often assumed, but rather to the black armor that was his signature.

Edward's first campaign was under his father in northern France, where he fought at Crécy. He was granted an independent command in 1355, and used it to win the Battle of Poitiers the following year. After he was created prince of Aquitaine in July 1362, he attempted to restore Peter the Cruel of Castile to his throne in 1367. He won a famous victory at Najera on 3 April 1367, but at a high cost. This weakened Edward's position in Aquitaine, and he was unable to establish rule there after the nobles and prelates of the area tried to establish Charles V of France as suzerain in 1368. In an attempt to quell this revolt, Edward was forced to rely on mercenaries he could not pay, and eventually had to surrender his principality to his father in January 1371. He had no successor as the prince of Aquitaine.

David J. Tietge

See also: Crécy, Battle of; Hundred Years War; Poitiers, Battle of

References and further reading:

Barber, Richard W. *Edward, Prince of Wales and Aquitaine: A Biography of the Black Prince.* New York: Scribner, 1978.

———, ed. and trans. *The Life and Campaigns of the Black Prince: From Contemporary Letters, Diaries and Chronicles, Including Chandos Herald's Life of the Black Prince.* New York: St. Martin's Press, 1986.

Harvey, John Hooper. *The Black Prince and His Age.* Totowa, NJ: Rowman & Littlefield, 1976.

Edward I (1239–1307)

Led unsuccessful effort to conquer Scotland, reformed England. Edward Longshanks was born 17 June 1239 in Westminster, Middlesex, England, and died 7 July 1307 at Burgh by Sands, Carlisle, Cumberland. He was the son of Henry III and king of England from 1272 to1307.

England during Edward's reign was characterized by an increasing sense of nationalism, and one of his major goals as king was the unification of Britain under one rule. To that end, Edward managed to bring Wales under English authority, but was not able to achieve the same results with Scotland. The campaigns involved, along with regular battles with France and a crusade during his early career, marked his rule as one of constant fighting and border skirmishes, most of them successful. This made Edward I one of the most historically influential rulers in the formation of modern England.

Edward I was sometimes referred to as the Lawgiver and the Father of the Mother of Parliaments, since it was through his efforts that arbitrary abuse of power under fiefdoms began to crumble and a more stable and uniform system of administration was established. Edward experienced his first taste of combat during the 1254 uprisings in Wales led by Llywelyn ap Gruffydd. In 1270, Edward embarked on a crusade to the Holy Land, and was nearly killed by a poisonous dagger. He did not have a coronation until 2 August 1274, and after he officially came to power, he began systematically to punish feudal lords who abused their power. He expelled the Jews from England in 1290 on the grounds of usury, and set off on a conquest of Scotland. This last attempt is perhaps the most well known, being the subject of the legends surrounding Robert the Bruce and William Wallace. The Scottish rebellions that began in 1295 were never fully resolved in Edward's favor.

David J. Tietge

See also: Anglo-Scots Wars (1290–1388); Anglo-Scots Wars (1513–1560); Bannockburn, Battle of; Falkirk, Battle of; Flodden, Battle of

References and further reading:

Jenks, Edward. *Edward Plantagenet: The English Justinian.* Freeport, NY: Books for Libraries Press, 1969.

Prestwich, Michael. *The Three Edwards: War and State in England, 1272–1377.* New York: St. Martin's Press, 1980.

Salzman, L. F. *Edward I.* Boston: Praeger Press, 1968.

Edward III (1312–1377)

Began Hundred Years War. Edward of Windsor was born 13 November 1312 at Windsor, Berkshire, England, and died 21 June 1377 at Sheen, Surrey. He was the eldest son of Edward II and Isabella of France, and was king of England from 1327 to 1377.

Early in his reign, Edward III tolerated the rule of his overbearing mother, Isabella, and his father's murderer, Mortimer, until he had the latter killed in 1330. In 1327, Edward led an army into northern England in response to continuing attacks by the Scots, but was not successful in quelling the Scottish rebels. In May 1328, under the Treaty of Northampton, the Scots were granted independence. Later, in 1332, Edward overthrew and later captured the Scottish king, David II, and won a major victory at Halidon Hill on 19 July 1333, restoring the Scottish throne to Edward Balliol, whom Edward supported.

Like his grandfather Edward I, Edward III was determined to claim the French throne, and revived his claim to it in 1337, marking the beginning of the Hundred Years War. After several years of minor skirmishing against the French king, Philippe VI, Edward won a major sea battle off the port of Sluys in Holland, a victory that prompted Edward to declare himself king of France. He finally reached a truce with Philippe in 1343, but this lasted only two years. The culminating battle between Edward and Philippe occurred at Crécy, near the Somme, on 26 August 1346, and Edward emerged victorious. After Philippe's death in August 1350, his son, Jean II, refused to acknowledge Edward's authority as king, initiating yet another series of battles that effectively ended with Edward's victory at Poitiers in September 1356.

Despite the glory of the victory, the pinnacle of Edward's reign was stained by the Black Death; the last half of his reign was financially crippled by still more war with France under Charles V; and Edward's very capable son, the soldier Edward the Black Prince, turned against his father's policies. Edward died of stroke at Sheen Palace in June 1377.

David J. Tietge

See also: Anglo-Scots Wars (1290–1388); Crécy, Battle of; Edward, the Black Prince; Edward I; Hundred Years War; Poitiers, Battle of

References and further reading:

Longman, William. *The History of the Life and Times of Edward III.* London: Longmans, Green, 1869.

Ormond, W. M. *The Reign of Edward III: Crown and Political Society in England, 1327–1377.* New Haven: Yale University Press, 1990.

Prestwich, Michael. *The Three Edwards: War and State in England, 1272–1377.* New York: St. Martin's Press, 1980.

Eichelberger, Robert L. (1886–1961)

Division, corps, and army commander under General Douglas MacArthur in the Pacific theater in World War II. Eichelberger knew MacArthur from his days serving on the latter's staff in the 1930s. Both of them had also been superintendents at West Point. In late 1942, when MacArthur's New Guinea campaign bogged down, he appointed Eichelberger to command the 32d Infantry Division, which was assigned the job of taking Buna, a key town on the northeast coast of New Guinea.

With MacArthur's order to "take Buna or don't come back alive" ringing in his ears, Eichelberger set about the task of restoring his division's morale and fighting spirit. The men suffered from lack of proper food, clothing, ammunition, and medical supplies. At any given time, almost one in three was stricken with malaria. In spite of these obstacles, Eichelberger's soldiers took Buna after bitter, close combat on 2 January 1943. Promoted to command I Corps, Eichelberger carried out a successful leapfrogging campaign on the north New Guinea coast for the next year and a half. He also functioned as MacArthur's unofficial troubleshooter. For example, when the Biak campaign stalled in June 1944, Eichelberger was brought in to turn the situation around, and did so.

In November 1944 he was promoted to lieutenant general and given command of the Eighth Army, the force spearheading the invasion of the Philippines. His troops carried out dozens of landings throughout the archipelago, and continued fighting until the end of hostilities. Eichelberger is remembered as an excellent, resourceful commander but probably never got his full due because he worked in the shadow of MacArthur, who went to great lengths to make sure that the spotlight shone only on his person.

John C. McManus

See also: MacArthur, Douglas; World War II

References and further reading:

Eichelberger, Robert. *Our Jungle Road to Tokyo.* New York: Viking, 1950.
Milner, Samuel. *Victory in Papua.* Washington, DC: U.S. Army Office of the Chief of Military History, 1957.
Smith, Robert. *Triumph in the Philippines.* Washington, DC: U.S. Army Office of the Chief of Military History, 1963.

Eisenhower, Dwight David (1890–1969)

American World War II military commander and postwar president. He is remembered for his ability to build a coalition among Allied forces in Europe during World War II, and historians in more recent years have come to appreciate the ability of the thirty-fourth president of the United States to use the same skills during the Cold War. His sometimes mangled and vague syntax is, somewhat improbably, seen as a very clever means of confusing his political enemies and of buying time until an unwelcome problem went away naturally.

Dwight David Eisenhower was born 14 October 1890 in Denison, Texas. The next year his family moved to Abilene, Kansas. Upon graduation from the U.S. Military Academy in 1915 ("the class the stars fell on") in the middle ranking of his class, he was assigned to Fort Sam Houston, Texas, and commissioned a second lieutenant. Promoted to captain and the brevet rank of lieutenant colonel by World War I, he was a training instructor at various military bases during the war. In 1922, he was assigned to the Panama Canal Zone for the next three years. After attending Command and General Staff School (1925–1926) and Army War College (1928–1929), where he compiled outstanding academic records, he was named assistant to the assistant secretary of War. In 1932 he was named aide to General Douglas MacArthur, holding that post until 1939. (Much later, he is reliably re-

President-elect Dwight D. Eisenhower (left) and commander of the 2d Infantry Division Major General James C. Fry (right) tour United Nations Command units on 4 December 1952, during Eisenhower's visit to Korea. (National Archives)

ported to have remarked that he "studied dramatics under MacArthur for seven years.") The Eisenhowers moved around between military postings until he was promoted to brigadier general in 1941.

After Pearl Harbor, Eisenhower has named assistant chief of staff in charge of war plans in Washington, D.C. In March 1942, he was given the rank of major general and named commander of U.S. forces in Europe. A month later, he was promoted to lieutenant general and was Allied commander in chief for the invasion of North Africa, Sicily, and Italy. Given the rank of general in February 1943, he was named supreme Allied commander in December. He oversaw Operation OVERLORD, the D day invasion of Normandy on 6 June 1944. In perhaps the greatest professional challenge of his career, Eisenhower had to make the decision as to the exact day to commence the invasion. His meteorologist had informed him of a window of a little more than 24 hours, after which the Channel would once again be subjected to tempestuous weather. With the tens of thousands of troops already loaded in their invasion craft and the paratroopers already taking off, "Ike" alone made the decision: "OK, let's go." He actually had a message in his pocket apologizing to the American public had the invasion failed. But his nerve never failed, and the invasion was a success. Promoted to five-star general in December, he directed the Allied campaign into the heart of Germany and accepted the unconditional surrender of the enemy high command at his headquarters, Rheims, France, at 2:41 A.M. local time on 7 May 1945.

In November 1945, he succeeded George Marshall as army chief of staff and resigned from the army in February 1948. Enormously popular in the postwar years, Eisenhower could have had the presidential nomination of either American political party for the asking in 1948. He declared himself a Republican in 1952, and was elected president in a landslide. "I Like Ike" was one of the most effective slogans in American political history.

Eisenhower's administration revived peace talks and ended the Korean War the next year. Acting as a moderating influence, making peace even with his opponent for the nomination, the conservative senator Robert A. Taft, Eisenhower tried to forward his brand of "Modern Republicanism" and seemed more frustrated by the conservative wing of the GOP than by the Democrats, who won control of Congress in the 1954 elections and retained that control for the rest of his administration. He took satisfaction, however, in his administration's building up of the nation's nuclear deterrent, in a "New Look" that downplayed the role of the army, and which critics charged left no middle ground between inaction and nuclear Armageddon, leaving the Soviets free to nibble at the perimeters of America's security interests.

Facing foreign crises in Hungary, Berlin, and Cuba, Eisenhower never came close to any overt clash with the Soviets, something he looked back upon with pride at the close of his administrations. To counter Soviet influence, his administration gave aid to Pakistan, Thailand, and the Philippines, and provided military advisers to South Vietnam.

He also promoted the Eisenhower Doctrine, which asserted that America would aid any country threatened by communism. Under this policy, he sent marines to Beirut, Lebanon, in 1958.

Another landslide confirmed Ike in office in 1956, but his last years in office seemed stale, a perception played upon by the youthful John F. Kennedy, his successor. In the last year of his administration, 1960, Eisenhower fumbled the U-2 spy plane incident, insisting that there had been no intentional violation of Soviet airspace, having been assured by his security people that the pilot of the offending spy plane had been killed—until Soviet prime minister Nikita Khrushchev produced a very much alive Francis Gary Powers in probably the nadir of the Eisenhower years.

Nonetheless, Ike remained one of the most popular American presidents to the day that he left office. His farewell speech is his most remembered, warning Americans about the growing "military-industrial complex" that he himself had done so much to shape. Dwight Eisenhower, one of the most acclaimed Americans of the twentieth century, in war and in peace, died of natural causes on 28 March 1969 at Walter Reed Medical Center, Washington, D.C.

T. Jason Soderstrum

See also: Cold War; MacArthur, Douglas; Normandy Landings; World War II

References and further reading:

Ambrose, Stephen E. *Eisenhower.* 2 vols. New York: Simon & Schuster, 1983, 1984.

Lyon, Peter. *Eisenhower: Portrait of the Hero.* Boston: Little, Brown, 1974.

Neal, Steve. *The Eisenhowers: Reluctant Dynasty.* Garden City: Doubleday, 1978.

El Alamein (July–November 1942)

The defeat that spelled the end of Axis hopes in North Africa, and one of the turning points of World War II. In late January 1942 General Erwin Rommel launched his German and Italian divisions eastward toward Egypt for the second time. By June they had reached Mersa Matruh, inside Egypt. Mussolini flew to Libya to prepare for a triumphant entry into Cairo; Roosevelt was briefed that the Suez Canal could fall within two weeks. Nonetheless, General Sir Claude Auchinleck skillfully husbanded the Allied forces into a mobile defense line at El Alamein, only 60 miles west of Alexandria.

General Rommel in the desert El Alamein, with his troops, 1942. (Library of Congress)

On 1 July 1942 the German and Italian forces made their first attack on the El Alamein line, but found it too strong and reverted to a defensive stance. Auchinleck was replaced by General Harold Alexander, who had orders to commence offensive operations, and General Bernard Montgomery was appointed to command Eighth Army. The El Alamein offensive finally started the day after the first convoy of Operation TORCH, the American invasion of western North Africa, sailed directly from the United States for Morocco.

A heavy artillery barrage opened at 9:40 P.M. on 28 October, followed by infantry attacks by XXX Corps (9th Australian, 51st Highland, 2d New Zealand, 1st South African, and 4th Indian Divisions). X Corps (1st and 10th Armored Divisions) moved up to exploit a breakthrough. Facing them were Italian Bersaglieri, Trento, Bologna, and Brescia Divisions, bolstered with German 164th Light Division and Ramke Parachute Brigade. In reserve were the 15th and 21st Panzer, 90th Light, Trieste, Littorio, and Ariete Divisions.

A vicious war of attrition was fought for 10 days, but fierce attacks by 9th Australian Division gradually drew the Panzer divisions north, allowing 2d New Zealand Division to force a salient between the German and Italian fronts by 2 November. Supported by 4th Indian Division, British armor raced into this gap and westward, and on 4 November Rommel ordered retreat.

The Allied cost was high—13,500 casualties, 500 tanks, and 110 guns—but the El Alamein victory directly contributed to the Axis capitulation in North Africa only 24 weeks later.

Michael Hyde

See also: Churchill, Sir Winston; Montgomery, Bernard Law; Rommel, Erwin Johannes Eugen; Sidi Barrani; Tobruk, Battle of; TORCH, Operation; Wavell, Archibald Percival, First Earl

References and further reading:

Liddell Hart, B., ed. *The Rommel Papers.* London: Collins, 1953.

Lucas Phillips, C. E. *Alamein.* London: Heinemann, 1962.

Maughan, Barton. *Australia in the War of 1939–1945.* Ser. One, *Army.* Vol. 3, *Tobruk and El Alamein.* Canberra: Australian War Memorial, 1966.

Playfair, I. S. O., and C. J. C. Molony. *History of the Second World War: The Mediterranean and Middle East.* Vol. 4, *The Destruction of the Axis Forces in Africa.* London: Her Majesty's Stationery Office, 1966.

El Cid, Rodrigo Diaz de Vivar (1040–1099)

Spanish general. Raised at the court of Fernando I of Castile as a companion to the king's son Sancho, Rodrigo Diaz entered Sancho's service when Sancho became king of Castile. Until 1072, when Sancho was killed at the siege of Zamora, Diaz acted as the king's constable in Sancho's wars against his siblings, especially Alfonso, who had also received lands at Fernando's death. Now king, Alfonso demoted Diaz, and although he was allowed to marry into the royal family and campaign with the king against Navarre in 1077, Diaz never recovered his status at court. Falling out with Alfonso over tributes from the Muslim city of Seville, Diaz was banished from Castile and took service as a mercenary in Zaragoza from 1081 to 1086, serving under a Muslim king. Far from being a great Christian hero, as portrayed in later chronicles, Diaz was an opportunistic exploiter of the complex world of a Spain divided into warring factions of Christian nobles and Islamic rulers, who warred both with and against one another.

The arrival of an Almoravid army from North Africa, and its defeat of Alfonso at Sagrajas in 1086, brought Diaz back to Castile with the extraordinary privilege of keeping any Muslim land he conquered. Immediately falling out with Alfonso over their failure to meet up en route to the siege of Aleto, Diaz was again banished and embarked on a tour to collect protection money from Muslim cities allied to the count of Barcelona. In an extraordinary battle against the army of Barcelona at Tevar, Diaz, in the mountains, defeated not only the troops advancing up the hillside, but also those who had secretly advanced into the hills above Diaz's lines, and secured a treaty giving him overlordship of the eastern coast of Spain, the Levant. After a raid into Castile to warn off Alfonso, Diaz began conquering the Levant, and in 1093 captured the city of Valencia, which he established as his own kingdom, securing it by defeating the Almoravid army at Curate and seizing nearby fortresses, like Murviedo in 1098.

Having established himself as an independent ruler, Diaz married his daughters into the dynasties of Barcelona and Navarre, and died in bed in July 1099. Unfortunately, his widow was unable to hold Valencia against the Almoravids' return, and evacuated the city in 1102.

Margaret Sankey

See also: Reconquest of Spain

References and further reading:

Fletcher, Richard. *Quest for El Cid.* New York: Alfred Knopf, 1990.

Fletcher, Richard, and Simon Barton, ann. and trans. *The World of El Cid: Chronicles of the Spanish Reconquest.* Manchester, UK: Manchester University Press, 2000.

Pidal, Ramon Menendez. *Cid and His Spain.* Trans. Harold Sunderland. London: Frank Cass, 1971.

Electronic Warfare

The use of the electromagnetic spectrum to gain knowledge of the presence and movement of an opposing force, and also to deny any opposing force the use of that spectrum.

Electronic warfare first emerged in World War II when the radio became an important form of communication. The emission of radio waves became a method of detecting enemy movement, based on Nikola Tesla's "electrical effect."

In the World War II naval theater, the Western Allies employed a networked system of Radio Direction Finding (RDF) devices to track German U-boats. The Allies developed a system that employed high-frequency radio direction finders called Huff-Duff (HF/DF). HF/DF devices were placed on Allied ships in 1942 in order to locate the German U-boats that had become a serious threat to the British Royal Navy, the British Merchant Navy, and allied shipping. Using triangulation (both sea- and land-based) the Allies were able to pinpoint the location of surfaced German U-boats by intercepting their radio communication. Unfortunately for the Allies, submerged submarines could not transmit radio signals, and consequently could not be traced with HF/DF. Another system was needed to track enemy objects moving below the waterline.

The British invented the first underwater listening device, called ASDIC (named after the Anti-Submarine Detection Investigation Committee), to detect submerged objects. From this evolved the more modern and commonly known American system of Sound Navigation and Ranging (SONAR). These listening devices, know as hydrophones, could either listen for underwater low-frequency noises (passive) such as a submarine propeller, or they could transmit pulses of sound energy through the water (active) and interpret the sounds and echoes that bounced back. World War II marked the beginning of electronic antisubmarine warfare (ASW).

Electronic warfare on land and in air also made its debut in World War II. Radio Detecting and Ranging (RADAR) had been under development prior to World War II in both Germany and Britain. By 1939 all of the major nations that would be involved in World War II—France, Italy, Japan, the Netherlands, Russia, and the United States—were working on their own radar research programs. The British program, however, was under greater pressure to develop a workable radar system, and their program was accelerated by the threat of Luftwaffe air raids.

The concept of radar is much like that of active sonar, which is to bounce radio waves off of objects to determine their shape and location. The British scientists most closely involved in the development of radar were P. M. S. Blackett, A. V. Hill, Professor Henry Tizard, H. E. Wimperis (a civil engineer at the British Air Ministry), and Robert Watson-Watt of the Radio Research lab. In the face of sustained German bombardment Britain established a special committee under Sir Henry Tizard to develop and maintain a network of radar stations. This would later be called the Chain Home, which was an effective air defense Command and Control (C2) system composed of 51 linked radar stations built around the British coast to detect incoming German bombers. A second chain to detect low-flying aircraft was later incorporated to supplement the Chain Home radar net.

Germany also had its own radar net, which meant that World War II also saw the use of numerous radio countermeasures to avoid radar detection. These countermeasures involved the use of chaff, airborne active self-protection jammers (aircraft that used electronics to suppress enemy air defenses), and ground-based systems aimed at radio and radar bombing and navigation aids. Further developments were made during the war, such as the incorporation of Identification Friend-or-Foe (IFF) transponders. The transponder could identify a radar signal as originating from a friendly or hostile aircraft.

The advancement of radar was crucial to an Allied victory in both the air and sea battles of World War II. The developments of electronic warfare played a critical role in the survival of Britain as the sole Allied power in Europe by arming it against the crippling German U-boat campaign of unrestricted warfare and the destructive Luftwaffe bombing of Britain. The electronic warfare inventions of World War II would affect the future of warfare, as well as having many useful civilian applications (such as in civil aviation and in the sailing of ocean-going ships).

The major advancements that took place in electronic warfare after World War II came primarily from the addition of computer-processing power to radar. The basics of radar have not evolved since World War II; the signal received by World War II operators is similar to that received by modern radars. What the addition of digital processing to radar means is that the signals can be much more accurately interpreted, giving radar operators a much clearer image of what they are tracking.

The Cold War saw a rapid advancement in electronic warfare. Americans were shocked when the Soviet air defense system tracked and shot down Gary Power's U-2 high-altitude spy plane. The electronic warfare in the Vietnam conflict called for fighter-bombers to be individually equipped with radar warning receivers, chaff dispensers, and active radar jammers (these were known as Wild Weasels). New aircraft were developed, such as the Boeing E-3 Airborne Warning and Control System (AWACS) that provides high-and low-altitude surveillance from the air, and the 2-C Hawkeye Airborne Early Warning (AEW) command-and-control systems that provide situational awareness to coalition aircraft.

The 1991 Gulf War offered the first glimpse of what electronic warfare may resemble in the twenty-first century. Stealth aircraft and precision-guided weapons played an important role in that conflict. One of the main threats to the allies in the Gulf War was infrared (IR) radar and weapons, which first emerged in the 1950s. These have called for the development of a new breed of countermeasures in electronic warfare. It is most likely that electronic warfare will continue to develop rapidly through the foreseeable future.

Matthieu J-C. Moss

See also: Intelligence, Military

References and further reading:

Dickson, Paul. *The Electronic Battlefield.* Bloomington: Indiana University Press, 1976.

Price, Alfred. *Instruments of Darkness: The History of Electronic Warfare.* New York: Scribner, 1978.

Emilia Plater, Independent Women's Battalion (1943–1945)

First Polish cohesive, all-female combat unit, consisting of volunteers forcibly resettled in the USSR. Ironically named after a leader of insurrection (1830) against Russia.

Initially attached to lst Tadeusz Kosciuszko Division, the battalion swore the oath of allegiance on 15 July 1943. It became directly subordinated to I Polish Corps on 19 August and to First Polish Army on 17 July 1944. Unlike women's auxiliary units in the West, it was not subjected to special military regulations. Its command personnel were men, its political officers women.

On 18 August 1943 the battalion consisted of command element; five companies (two infantry and one each of fusiliers, machine guns, and handheld antitank grenade launchers); and six platoons (mortar, reconnaissance, signals, medical, engineer, and logistics). In late 1943 a transport platoon was added. Personnel strength fluctuated as the battalion provided basic training to women assigned elsewhere.

The inability of some members to cope with very intensive training resulted in gradual transformation from a first-line combat unit to one assigned mainly sentry and military police duties. The organizational structure and training standards were maintained. The battalion's changed status likely reflected reluctance of senior commanders to expose women soldiers to the heavy personnel losses suffered alongside the Russians.

About 70 servicewomen were killed. In May 1945 the battalion's strength of roughly 500 members represented a small percentage of the total number of Polish women serving in the two Polish armies formed in the USSR, with estimates ranging from 8,500 to 14,000, including former members who had commanded all-male units.

Kazimiera J. Cottam

See also: Gierczak, Emilia

References and further reading:

Drzewicka, S. *We Came from Oka River Shores* (in Polish). 2d ed. Warsaw: MON, 1985.

Pawlowski, Edward. "Platerówki" (in Polish). *Wojsko Ludowe* 6 (June 1985).

Przeciszewski, Roman. "Nationwide Reunion of 'Platerówki" (in Polish). *Zwiazek Wolnosci* (8 September 1986).

Stasinski, Adam. "Platerówki" (in Polish). *Polska Zbrojna* (18–20 June 1993); "Honoring Platerówki" (in Polish). *Polska Zbrojna* (21 June 1993).

Engineering, Military

The application of science and technology for military purposes, historically primarily civil engineering.

Since at least the beginning of the Neolithic era, human beings have manipulated their environment for economic and religious purposes. City walls are extant from a very early date in the Middle East, while very large-scale permanent field fortifications dating to the archaic period still exist in north China, where conditions favored their construction. It is questionable whether these works can be called the products of military engineering, but they do demonstrate the parallel development of civil engineering technology and agriculture. Today, it is often difficult to determine whether a historic moat, canal, railway, or even subway line was built for the reason of agricultural and civil improvement, or for military purposes.

The true history of military engineering may be said to begin with classical times, when the archaeological and written record of warfare was supplemented by a theoretical literature that tells us of the role of military engineers in siege warfare, artillery engine construction, and bridging operations. Persian, Greek, Hellenistic, and Roman armies all employed engineers in Middle East campaigning. In this dry climate, armies rarely had difficulties with narrow water barriers or weather damage to roads, but we hear of major pontoon bridges across the Hellespont and the Indus, sometimes built under the supervision of naval officers. In sieges, the major targets of attack tended to be masonry citadels on local eminences, and the usual mode of attack was escalade with ladders or more sophisticated devices supported by muscle-powered artillery. Engineers were so closely associated with this technology that the very word is derived from the Greek for "machine." The fortresses that they attacked, however, were not of a very sophisticated engineering plan, their integrity depending more on geography than on design.

By the time the Roman legions left their own peninsula, they had formulated a particular national tradition of military engineering. Here we find the engineers combined with everyday tradesmen and specialists, particularly carpenters, not surprising given that Italy was significantly more heavily wooded than the Middle East and even Greece. The use of standing timber for military purposes became basic to Roman military practice. The lightly laden classical army, often accompanied by no heavier a burden than a pack mule, could easily employ a timber bridge or log road for rapid movement through wet areas, while fortifications often took the form of palisades or abatis. While Roman armies naturally adapted to more appropriate methods in the Middle East, when campaigning beyond the Alps in northern and western Europe, they found an even more heavily wooded and wetter region with heavy clay soils that resisted traditional Middle Eastern methods of earthwork construction and required the application of animal power along with the ax.

In China, the loess zone continued to present many opportunities for the use of extended field fortifications to hem in the largely cavalry armies that operated there. In the southwest, particularly in the Tang struggle with Tibet in 600–800, fighting in upland country drew out a different set of skills, while masonry fortifications began to replace earthworks around cities in the middle of the first millennium. Chinese imperial warfare was marked by a bifurcation between the light cavalry armies of the north, with relatively minor engineering requirements, and virtually amphibious forces operating in the southern river valleys, where engineering blended with naval architecture to provide mobility across and along river barriers. However, this engineering-intensive "southern" warfare was intermittent, for stable dynasties rarely faced military threats from the south, and defense toward the north took precedence. The southern Song Dynasty's war with the Mongols in the 1200s was the great exception, and it is perhaps not surprising that the Yuan-Song war supplies the first documentary evidence of gunpowder artillery, used to supply additional firepower for both besieger and besieged. Once peace was restored, there was little pressure to develop this technology in China, but Mongol armies may have helped disseminate it to the west and southwest. In China, military engineering technique tended to stagnate again until the Qing-Ming conflict of the seventeenth century.

In India and Southeast Asia, the early domestication of the elephant gave army engineers a draft animal of unique energy density. In the forest zone elephants allowed large armies to overcome jungle conditions, while in the drier east, elephant and war interacted to produce a landscape of cyclopean fortification, although the very low cost of gunpowder in the Indian subcontinent introduced a countervailing tendency to very artillery-rich armies.

The classical era in the Middle East ended with the widespread domestication of the camel around 400, freeing trade and armies from the roads. Subsequent armies were built up around cavalry and archery that made less use of engineering in siege warfare, and the art of fortification fell into relative neglect in the land of its birth.

In western Europe, in contrast, Roman-style urban civilization led to a proliferation of stockaded and diked cities and castles by c. 1000 Europe's relatively wet, flood-prone climate dictated fortifications (particularly lowland fortifications in contradistinction to previous fortified eminences) to protect from floods as much as from men. Fortification proliferated in Europe as nowhere else save southern China, and command pressure was heavy on military engineers, who soon adopted the Chinese innovation of gunpowder.

Responding to the increasing sophistication of artillery, gunpowder or not, European fortification achieved great intricacy in 1300–1700, with depth and geometric design superseding simple masonry walls. By 1700 a fortress city's defensive was often 100 meters deep or more, with fire zones stretching 300 meters farther out. This emphasis on control of territory in depth with earthworks and water barriers linked the revolution in fortification to contemporary improvements in agriculture.

Artillery also had significant implications for marching armies. European armies required more draft animals and carts. With artillery live loads reaching 12 tons, unprecedented pressures on roads and bridges soon brought the ancient technology of the timber footbridge and roads to their limits, leading armies to learn a whole technology of bridging and road building. Moreover, larger and heavier armies required careful surveying and planning. Some Venetian military engineers specialized as cartographic engineers in 1550–1650, mapping regions of real or potential conflict and producing march, quartering, and magazine plans. This activity formed one of the bases of general staff organizations, and military and state civil engineers tended to become a technocratic government arm charged with general responsibility for massive environmental intervention for economic purposes, notably in the taming of the Mississippi, the Po, and the rivers of Vienna.

Finally, the increasing importance and specialization of naval warfare led to the creation of professional permanent navies in most European states after 1600. In western Europe this bifurcation had few consequences for military engineers, but in central Europe, navies and army pontoon (bridge laying) forces sometimes merged in the mid-eighteenth-century European continental wars. The result in at least the Austro-Hungarian case was the rise of a pioneer corps responsible for roads and bridges alongside the older fortress engineering corps.

In China, the classical impressively and efficiently fortified interior city was, by the 1700–1800 period, in profound neglect. When this trend was combined with lack of experience with the rapid evolution of artillery technique in Europe, China was left vulnerable to attack by imperialist armies in the 1800–1900 period. In the Middle East and India, meanwhile, the appearance of European armies, beginning with the Turks in 1500 and accelerating thereafter, led to regional traditions in military engineering merging with worldwide trends, although expensive timber continued to impose heavy hidden costs on fortification.

The nineteenth century saw military engineers assume responsibility for railroads, telegraph, and powered traction, but these separate disciplines of engineering soon formed independent corps, and the mechanical and electrical engineers are not now considered part of the continuity in military engineering. This tendency to split off new branches from the military engineers reached its culmination when air forces followed navies into independence, ensuring that technological change within the military engineering corps during the twentieth century was less profound than the general development of science and technology might lead one to expect.

Modern military engineers can install a bridge capable of supporting a live load of 70 tons in five minutes and bridge a river the width of the Euphrates in a half hour, but the general nature of their difficulties and the means used to overcome them would be recognizable to Xerxes.

Erik Lund

See also: Academies, Military; Antwerp, Siege of; Armored Fighting Vehicles; Artillery; Atomic Bomb, Development of; Breda, Siege of; Carnot, Lazare-Nicholas; Catapults; Chemical and Biological Warfare; Coehoorn, Baron Menno van; Constantinople; Denain, Battle of; Dien Bien Phu; Electronic Warfare; Firearms; Gibraltar, Siege of; Great Wall of China; Jericho, Siege of; Jerusalem, Siege of; Kublai Khan; Kut-al-Amara; La Rochelle, Siege of; Leningrad, Siege of; Logistics; Louisbourg, Expedition against; Luxembourg, Siege of; Maginot Line; Malta, Siege of (May–September 1565); Masada, Siege of; Metz, Siege of; Mortars; Ostende, Siege of; Parma and Piacenza, Alessandro Farnese, Duke of; Pleven/Plevna, Siege of; Port Arthur, Siege of; Propellants; Rhodes, Sieges of; Savannah, Siege and Taking of; Sevastapol, Siege of; Stalingrad; Süleyman I; Syracuse, Siege of; Tobruk, Battle of; Tyre, Siege of; Vauban, Sébastien Le Prestre de; Vicksburg, Siege of; Vienna, Sieges of; Xerxes I

References and further reading:

Abbot, Robert W. *American Civil Engineering Practice.* 3 vols. New York: J. Wiley and Sons, 1956; London: Chapman & Hill, 1956.

MacAulay, J. S. *A Treatise on Field Fortifications, and Other Subjects Connected with the Duties of the Field Engineer.* London: Fraser, 1834.

Vegetius. *Vegetius: Epitome of Military Science.* Trans. with notes by N. P. Milner. Liverpool, UK: Liverpool University Press, 1993.

English Civil War (1215–1217)

Barons force King John to sign Magna Carta. King John was neither loved nor respected by the majority of his countrymen during his 17-year reign (1199–1216). His military endeavors, which included campaigns in France, Scotland, Wales, and Ireland, necessitated high taxes. These financial burdens, combined with John's dictatorial and often unpredictable style of ruling, led to widespread disenchantment among many of his barons. During the summer of 1214 John made a determined but unsuccessful effort to regain his French territories, only to return to England, short of finances, to face a baronial rebellion.

Refusing John's offer of arbitration, the barons captured London in April 1215 without resistance, and John had to evade the baronial force of 2,000 knights for two months as it followed him around the south of England. On 15 June, the two sides met at Runnymede and negotiated the Magna Carta, a document that conceded most of the baronial demands and limited the king's exclusive right to govern. John had no intention of abiding by its terms. He viewed the charter as providing a breathing space while he assembled a more formidable military force. Heavily dependent on borrowed money and hired mercenaries, John resumed hostilities in September 1215.

John's problems were exacerbated when Louis, the French king's son and heir, arrived in England (21 May 1216) at the invitation of the rebel barons. In what must count for one of the least known events in English history, Louis was welcomed by the citizenry of London, and crowned king of England. Emboldened by such obvious domestic support, his large army, which is believed to have numbered 35,000 at its peak, began driving the royal forces westward.

Shortly afterwards, on 23 October 1216, John died at Newark and was succeeded by his nine-year-old son, Henry III, who proved a more popular figurehead than his father. The primary responsibility for conducting the war now fell on the two regents, William Marshal and Hubert de Burgh. Marshal proved himself a shrewd diplomatist and strategist, winning the allegiance of many of the dissident barons. When Lord Fitzwalter raised a rebel army of 600 knights and 20,000 French soldiers, Marshal astutely retreated to Nottingham and waited for reinforcements. When he enjoyed numerical superiority, he laid siege to the rebel stronghold of Lincoln Castle, and secured a decisive royalist victory (23 May). This defeat, combined with the repulsion of the French naval fleet off Sandwich in Kent, forced Louis to accept peace terms. In return for a payment of 10,000 marks, the French dauphin agreed to waive his claim to the English throne and to restore Normandy. Neither promise was fulfilled.

Although Henry went on to rule England until 1272, the issue of regal versus baronial rights was never satisfactorily resolved during his reign. The most enduring legacy of the civil war was undoubtedly the Magna Carta, which inspired supporters of individual freedom long after the memory of the rebel barons had faded.

Donnacha Óbeacháin

See also: Barons' War; Philip II Augustus

References and further reading:

Bartlett, Robert. *England under the Norman and Angevin Kings, 1075–1225.* Oxford, UK: Oxford University Press, 2000.

Warren, W. L. *King John.* New Haven and London: Yale University Press, 1997.

English Civil War (1642–1649)

Also known as the Great Rebellion or the Puritan Revolution, the English Civil War was a conflict between King Charles I of England and his loyalists, known as Cavaliers, and his opponents, known as parliamentarians or Roundheads, that ended with the defeat and execution of the king and the establishment of a republican Commonwealth and Protectorate. The war centered on the constitutional issue of the divine right of a king and the privileges of Parliament independent of the crown.

The roots of the conflict lay in the king's need to raise money from Parliament. In 1638, Charles I attempted to enforce Anglican formulations on the Scottish church, which led to a march on Scotland in 1639, beginning what was called the Bishops' Wars (1639–1640). The king's forces were forced to leave Scotland without a battle due to lack of funds. Fearing French involvement, Charles mounted another expedition and convened Parliament to raise funds. An angry king dismissed Parliament after realizing that they were opposed to the invasion and wanted to argue grievances. Foiled by the Scots, the king was forced to call Parliament again in November 1640. On 4 January 1642, Charles failed to arrest five members of that body. The king left London and both parties began to stockpile military resources. Each had approximately 13,000 men, with the king's support mainly being found in the north, west, and Wales, and Parliament finding support in the more economically developed south and east.

The Royalists tried to end the war by marching on London but were met by Roundheads at Edgehill near Essex on 23 October. Although the battle was inconclusive, Charles was not able to march on London after meeting another force at Turnham Green, causing the king to withdraw to Oxford. In January 1643, parliamentary attempts to negotiate a peace were rejected by Charles. Over the next year, both sides seized each other's estates and fought several battles across England, including Cavalier victories at Adwalton Moor, Lansdown, and Roundway Down in June and July; an inconclusive battle at Newbury in September; and a Roundhead victory at Winceby on October 11.

The tide turned when Parliament drew up the Solemn League & Covenant, from which aid from the Scottish Presbyterians was obtained. In 1644, Parliament was able to take York with a major victory at Marston Moor on 2 July. Still, they lost at Lostwithiel and had to withdraw from Newbury after an inclusive battle. Led by Oliver Cromwell, Parliament reorganized the army into the New Model Army under the leadership of Thomas Fairfax. After another defeat at the hands of the Roundheads at Naseby on 14 June 1645 and unable to gain Irish aid, Charles surrendered to the Scots at Oxford in June 1646.

Charles was delivered to Parliament in 1647. Yet the army, alienated by Presbyterian rule, refused to disband. Fearing that the Parliament would negotiate with the king behind its back, Charles was removed to Hampton Court. When the army demanded 11 members of Parliament be arrested for secretly negotiating with the king, Parliament refused, and the army marched on London. As army discontent grew, Charles escaped to the Isle of Wight in November, where he tried to negotiate with both Parliament and the Scots. In December, he came to an agreement, known as the Engagement, with the Scots in which he would accept Presbyterianism in return for military support. The army led by Cromwell was able to defeat them at Preston on 17 August 1648. Parliament again tried to come to an agreement with Charles, but the army disposed of its enemies in Parliament in Pride's Purge in December. The remaining members, known as the Rump Parliament, tried the king for treason, and beheaded him on 30 January 1649.

The Rump Parliament established a Commonwealth, but in 1653 Cromwell and his army set him up as Lord Protector, with rights of succession.

Charles II, the king's son, was recognized as monarch in most of Ireland and in parts of Scotland. This uprising was crushed at the Battle of Worcester in 1651. The future king of England fled to the Continent, remaining there until 1660, when the Protectorate ended with Cromwell's death and the monarchy was reestablished.

T. Jason Soderstrum

See also: Cromwell, Oliver; Edgehill, Battle of; Marston Moor; Worcester, Battle of

References and further reading:

Ashley, Maurice. *The English Civil War: A Concise History.* London: Thames & Hudson, 1974.

Bennett, Martyn. *The English Civil War, 1640–1649.* London: Longman, 1995.

Burne, Alfred Higgins. *The Great Civil War: A Military History of the First Civil War, 1642–1646.* London: Eyre & Spottiswoode, 1959.

Hibbert, Christopher. *Cavaliers & Roundheads: The English Civil War, 1642–1649.* New York: C. Scribners Sons; Maxwell Macmillan International, 1993.

English Wars in Ireland (1688–1691)

War had broken out in Europe in 1688 between the French king, Louis XIV, and a league of European states led by William of Orange. Under King James, England had supported Louis, but when William ascended the throne, the state changed sides. This was a serious blow to France, but Louis hoped to offset it by keeping William and his army occupied in Ireland. In March 1689, James landed in Kinsale with 400 French officers and arms and ammunition for 10,000 men.

At first things went well for James and, with the exception of parts of Ulster, Ireland declared for him. Panic spread among Ulster Protestants. Thousands fled to Scotland, or to the only walled cities in Protestant hands, Derry and Enniskillen. In April, James marched to Derry to demand its surrender. The Jacobite army had no guns to batter down the walls, so they hoped to starve the city into submission. On 30 July, however, one ship, the *Mountjoy,* forced its way through, bringing the 105-day siege to an end. Soon after, Williamite Marshall Herman Schomberg landed near Belfast and took Carrickfergus, and gradually most of Ulster fell into Williamite hands.

Back in Dublin, James had summoned a Parliament, which removed all discriminatory laws against Catholics and restored lands confiscated by the crown. In June 1690, William landed at Carrickfergus and, on 12 July, his army of 34,000 defeated James's army of 22,000 at the Battle of the Boyne. The Irish commanders left in control of the Jacobite army fell back on the Shannon, and after a successful defense of Limerick a treaty was signed between the Irish leader, Patrick Sarsfield, and the Williamite commander, General Godard Van Ginkel. Under the terms of the Treaty of Limerick, Jacobite soldiers still in arms against William might return to their homes, join William's army, or go to France. Sarsfield was anxious to protect the interests of Irish Catholics and secured for them a promise of religious toleration. King William was prepared to abide by the terms agreed upon in Limerick, but most of his supporters considered them too lenient, and as soon as the Irish were demobilized the provisions of the treaty were ignored.

Donnacha Óbeacháin

See also: Boyne

References and further reading:

Doherty, Richard. *The Williamite War in Ireland, 1689–91.* Dublin: Four Courts Press, 1998.

Wauchope, Piers. *Patrick Sarsfield and the Williamite War.* Dublin: Irish Academic Press, 1992.

Entebbe Rescue Raid (4 July 1976)

Legendary commando raid and hostage rescue. On 27 June 1976, four Palestine Liberation Organization terrorists hijacked 257 passengers and crew aboard Air France Flight 139 en route from Athens to Paris. After refueling in Libya, the hijackers directed the plane to Entebbe Airport in Uganda, where they demanded the release of 53 convicted terrorists lest they execute their hostages.

Negotiation led to the release of 149 passengers, but leaving more than 100 Israelis and the French flight crew captive in Uganda. Further diplomatic efforts seemed hopeless because the increasingly unstable Ugandan dictator Idi Amin Dada supported the terrorists. Israeli officials therefore launched Operation THUNDERBOLT to free the hostages.

Under cover of darkness, Israeli paratroopers in four C-130 transports escorted by fighters and two Boeing 707 support aircraft surreptitiously landed at Entebbe on 4 July 1976. Utilizing half-tracks and jeeps armed with recoilless rifles and machine guns, they took the airport completely by surprise and freed all of the surviving hostages. The 90-minute operation included the destruction of airport communication and fuel equipment as well as Soviet-built Mig 17 and Mig 21 fighter planes, at a cost of three hostages and one paratrooper killed. In exchange, Israeli forces claimed all the terrorists and approximately 45 Ugandan soldiers dead. After reboarding their aircraft with five hostages and one wounded soldier, the strike force flew to Nairobi for fuel before returning safely to Israel.

Crowd lifting the squadron leader of the rescue planes on their return to Israel from Uganda, 1976. (David Rubinger/Corbis)

International acclaim followed the surgical strike, which astounded military analysts by covering more than 5,000 miles from start to finish. The raid cemented Israel's reputation as a world leader in the fight against terrorism.

Lance Janda

References and further reading:

Hastings, Max. *Yoni: Hero of Entebbe.* New York: Dial Press, 1979.

Stevenson, William. *90 Minutes at Entebbe.* New York: Bantam Books, 1976.

Yesha Yahu, Ben Porat. *Entebbe Rescue.* New York: Delacorte, 1977.

Enver Pasha (1881–1922)

Turkish general and politician who reformed the Ottoman Empire and was virtual dictator during World War I. Ismail Enver was born in Constantinople (Istanbul) in 1881 and received his commission as a lieutenant in the Turkish army in 1903 and additional training in the German army. He rose quickly in rank, becoming lieutenant colonel and then full colonel in 1913, playing a significant role in recapturing Edirne (Adrianople) from the Bulgarians. By the end of the year he attained the rank of brigadier general and, in 1914, the title Pasha.

In 1908, he was one of the three main leaders of the Young Turks, a group of army officers who revolted against the misrule of Sultan Abdul Hamid. Enver fought in Libya during the Italo-Turkish War (1911–1912), and in the Balkan Wars (1912–1913), and his dissatisfaction with Turkish losses during the Balkan Wars led him to bring Turkey into World War I on the side of the Central Powers.

In early 1914, Enver was named minister of War and commander in chief. During the war, he led several unsuccessful campaigns against the Russians on the Caucasian front. When the armistice was signed, Enver fled to Odessa, then Berlin, and finally to Turkestan (Russian Central Asia) where he was killed on 4 August 1922 leading an anti-Soviet revolt at Bukhara. His body was repatriated to Turkey in 1996.

Michael C. Paul

See also: Ataturk, (Mustafa) Kemal; Balkan War, First; Italo-Turkish War; World War I

References and further reading:

Swanson, G. W. "Enver Pasha: The Formative Years." *Middle Eastern Studies* 16: 3 (1980).

Turfan, M. Naim. *Rise of the Young Turks: Politics, the Military and Ottoman Collapse.* London: I. B. Tauris, 2000.

Yamauchi, Masayuki. *The Green Crescent under the Red Star: Enver Pasha in Soviet Russia 1919–1922.* Tokyo: Institute for the Study of Languages and Cultures of Asia and Africa, 1991.

Epaminondas (c. 410–362 B.C.E.)

Theban statesman and general of the fourth century B.C.E. Epaminondas was born to a noble Theban family and during his early adult life participated in the liberation of Thebes from Spartan occupation. In 371, when the Spartans invaded Boeotia once again, Epaminondas commanded the Theban army. At the Battle of Leuktra he showed his tactical brilliance by massing his fighting elite on the left of the Theban battle line, and not on the right as usual, in a great block 50 deep and possibly 80 wide. The Spartans attempted to change their formation but with no success, as the Theban Hieros Lochos (Sacred Band) crushed the Spartan elite. The battle ended with the Spartan king Cleombrotus dead and Sparta's supremacy in ruins. In its place, Epaminondas established the Theban hegemony, which was to last until 362.

In 370, Epaminondas led a massive invasion in Laconia. Although he failed to capture Sparta, he liberated Messenia and in 369 directed the foundation of Messene, its new capital. After a brief stay at Thebes, Epaminondas returned to the Peloponnese and attacked Corinth and its neighboring polis, leaving Sparta with no allies in the northeastern Peloponnese. However, due to subsequent diplomatic mistakes, Thebes gradually lost all its allies in the Peloponnese, giving the opportunity to Sparta to rise again.

Realizing that Thebes had lost control of the situation, Epaminondas led an army of about 30,000 men against a coalition of Spartan, Athenian, Elian, and Arcadian forces, numbering some 22,000 soldiers. The crucial battle was to be fought south of Mantinea in 362. The Theban army charged and cut through the Spartan line easily, but at the moment of triumph, Epaminondas fell. His death marked the beginning of the decline of the Theban hegemony.

Ioannis Georganas

See also: Mantinea, Battle of

References and further reading:

Buckler, John. "Plutarch on Leuktra." *Symbolae Osloenses* 55 (1980), 75–93.

———. "Epaminondas." In *Encyclopedia of Greece and the Hellenic Tradition,* ed. Graham Speake, 554–555. London: Fitzroy Dearborn, 2000.

Hanson, Victor Davis. *The Wars of the Ancient Greeks.* London: Cassell, 1999.

Ethics of Warfare

Commentators on ethics note that its subject matter includes not only rules or principles but also the virtues, the characteristics of perfection or excellence. Opposed to these notions is the usual caricature of warfare that war is the realm of necessity and horror in which anything is acceptable. "War is hell" and there is no good of any kind in hell. Some call this view of war *Realism.*

Various writings have survived from China and India of the sixth and fifth centuries B.C.E. in which Chinese writers such as Lao-tse and Sun-tzu offered moral guidance for generals regarding war and warfare. The Hindu *Book of Manu* also provided rules limiting violence similar to those that humanitarians support today.

Writings from the Jews, Greeks, and Romans established the foundational ideas of what is known as the Just War Tradition. For example, Deuteronomy 10:10–20 listed a variety of rules beginning with the requirement of offering peace to a city before fighting its inhabitants. Plato argued in *Laws* that war is necessary but that it is fought to obtain peace. In both *Nichomachean Ethics* (1277b6) and *Politics* (1256b25), Aristotle followed Plato's view and first used the term *just war.* Rome had explicit laws governing its commanders engaging in war. Cicero's *De Re Republica* 3, XXIII, provided a full discussion of the essentials.

The just war tradition became the work over the centuries of Christian thinkers, including Augustine, Aquinas, and Vitoria. Augustine (354–430) wrote no treatise on war or warfare, but did express in many places both remarks on war in general and specific recommendations to Roman leaders about their duties in war. Augustine's teachings served as the foundation for Western Christianity until the synthesis achieved by Thomas Aquinas (1225–1274) in *Summa Theologicae.* In *ST* II-II, Q. 40, A. 1, Aquinas established the basic principles of the just war. There must be a just cause. There must be proper authority. There must be a right intention. Centuries later, Spaniard Francisco de Vitoria (d. 1546) developed just war thinking in his *De Indis* and *De Jure Belli* as he criticized the Spanish conquistadors' treatment of the indigenous populations of South and Central America. He argued that the monarch's justification for war must meet objective standards held by wise and upright men. Also it is never lawful to deliberately slaughter innocents, that is, those not directly involved in the fighting.

Thus, the just war tradition proposes two sets of principles that specify the ethics of warfare. One set of principles, known as the *jus ad bellum,* answers the question: When is recourse to war morally licit? The second set, known as the *jus in bello,* answers the subsequent question: What means are morally allowed in warfare?

There are differing opinions within the tradition about which principles are to be included as the *jus ad bellum.* Indeed, there are ranges of understanding and questions of application in all these matters about which individuals may honestly disagree in their judgments despite agreeing on the principles in question.

All agree that there must be a *just cause.* States are entitled to defend themselves against armed aggression. Thus defensive wars are allowed. Similarly, if a great injustice has been perpetrated against a country, then it may be necessary to take up arms to rectify the injustice. There is also the debated view about whether a state may preemptively initiate a war when it has persuasive evidence that its enemies may attack soon.

There is general agreement over *proper authority.* Only the highest political authority may declare war. Thus, no one thinks that California could declare war on Nevada as that usurps the proper authority of the government of the United States of America. But this same principle has difficulties when a state is breaking up. What constitutes proper authority? Consider what happened recently in Yugoslavia as Slovenia, Croatia, and then Bosnia-Herzegovina became independent states, recognized by other European states and even the United Nations during 1991 and 1992.

Unlike just cause and proper authority, *right intention* has come under much scrutiny and debate. When proper authority existed in times of emperors and kings, right intention could focus on the leader and his intention. Proponents of this principle realize that a person can do something that is objectively good, that is, in all outward appearance seems to be unobjectionable; yet, that person could have an ulterior motive that subverts good into evil. At a later time, people could come to understand that what seemed good could in fact be part of a nefarious scheme and thus really evil.

So the judgment is about the war and the leader who declared war. It is not about those who in good faith carried out the war, as they were obliged to do. Soldiers in war do not bear the burden of moral blame in such matters. The purpose of this principle is to assign blame to those who deserve it, those who had the evil intention.

Thinkers who object to this principle of right intention argue that in times like today, dominated by nation-states, it is not clear whether "states" as an entity can have an intention. Moreover, whether a state or a leader has an intention, how can others ever claim to know what another's intention is? Thus some refuse to apply this principle, thinking that the real question is whether there was a just cause.

Certain thinkers address these issues by stipulating another principle, *declaration of war aims.* There is ordinarily a need to issue a formal declaration of war, or in the case of the United States, which has not issued a declaration of war since World War II, the need to state to its citizens the aims that it seeks in committing its military to battle.

Finally, there are several other principles that certain thinkers use in evaluating whether a war is just. They include *last resort, comparable justice, consideration of the costs,* and *probability of success.* All of these considerations require some fair-minded judgment that there are additional good reasons why war is just in this case or unjust in that case. Perhaps there are other solutions to deal with the injustice. Or the enemy may have valid reasons that justify its position. Or the costs of the war, in both material and human/moral considerations, may outweigh the good that could be achieved by war. Or last, there may be a good reason not to go to war over a just cause, because there is little likelihood that ultimately the military will succeed in overturning the injustice.

If we turn to the *jus in bello,* two principles stand out: *discrimination* and *proportionality.* Discrimination is the principle of noncombatant immunity. It is wrong intentionally to target civilians, noncombatants. The principle of proportionality brings a moral weighing to particular acts of war: Will this act of war cause more evil than the good it accomplishes? (Does destroying a village to save it make moral sense?) Many just war thinkers find the principle of discrimination to be primary and prior to the application of the principle of proportionality, though others would allow judgments of proportionality to overcome that moral limit in times of military necessity.

These ethical considerations have become part and parcel of the training that military professionals receive in the United States and in many other developed countries. The military is a profession and has great concern to develop the professional ethics of its leaders. At one level, the United States has entered into international agreements, signed the various conventions, and established its Uniform Code of Military Justice and Rules of Engagement. Soldiers must follow these tenets. At another level, the military in the United States is entirely composed of citizen-soldiers. Officers who lead men into battle must demonstrate the characteristics of leadership, the military virtues, to develop not only their own morale and confidence, but also that of their subordinates. Finally, both the government and the military must obtain and retain popular support. Since the Vietnam conflict, the matters of the just war tradition have become slowly, but surely, the terms of presentation and approval of the policies regarding the committing of United States troops. For instance, on 28 January 1991, President George H. W. Bush explicitly used these principles of the just war to explain and defend the Persian Gulf War.

John R. Popiden

See also: Laws of War; Pacifism/War Resistance; Prisoners of War; Refugees and Victims of Ethnic Cleansing; War Crimes

References and further reading:

Christopher, Paul. *The Ethics of War and Peace.* 2d ed. NJ: Prentice Hall, 1999.

Coates, A. J. *The Ethics of War.* Manchester and New York: Manchester University Press, 1997.

Johnson, James Turner. *Morality and Contemporary Warfare.* New Haven: Yale University Press, 1999.

Walzer, Michael. *Just and Unjust Wars.* 2d ed. New York: Basic Books, 1992.

Eugene of Savoy (Eugene Prince of Savoy-Carignan) (1663–1736)

Austrian general, noted for his courage and strategic skill. Born in Paris, Eugene spent his childhood at the court of Louis XIV. In 1683 Eugene joined the Austrian army, distinguishing himself in the relief of Vienna, and participated in the capture of Buda (1686) and Belgrade (1688). Promoted to lieutenant general, Eugene was sent to Italy in 1689 to assist Savoy in the War of the Grand Alliance. After the defection of Duke Victor Amadeus, Eugene returned to Hungary to command the imperial army. His victory over the Ottomans at Zenta (1697) led to the Treaty of Karlowitz (1699).

In the War of the Spanish Succession, Eugene initially directed the Italian front, winning victories at Chiari (1701) and Cremona (1702), and inducing Savoy to abandon the French side. In 1703 he was made president of the Imperial Council of War. Eugene cooperated closely with the Anglo-Dutch commander John Churchill, First Duke of Marlborough, in planning and executing the Blenheim campaign (1704). Sent to Lombardy in 1706, Eugene outmaneuvered the French army, raised the siege of Turin, and drove the

Undated portrait of Eugene of Savoy. (Library of Congress)

French out of northern Italy. Returning to the Low Countries in 1708, he led the German forces at Oudenaarde, directed the siege of Lille, and commanded the right wing at Malplaquet (1709). After Marlborough's recall in 1711, Eugene assumed command of the allied army in the Netherlands. Defeated at Denain (1712) and without Anglo-Dutch support after the Treaty of Utrecht (1713), Eugene negotiated the Treaty of Rastatt (1714) with France, ending the war.

In the renewed Austro-Turkish War, Eugene defeated the Ottomans at Peterwardein and Temesvár (1716) and recaptured Belgrade (1717). Eugene took his last field command in the War of the Polish Succession (1733–1738). Greatly outnumbered by the French, Eugene failed to prevent the capture of Philippsburg but managed to protect his own forces by skillful maneuvering. He died in Vienna on 20 April 1736.

Brian Hodson

See also: Austro-Turk Wars; Grand Alliance, War of the; Spanish Succession, War of the

References and further reading:

Braubach, Max. *Prinz Eugen von Savoyen: Eine Biographie.* 5 vols. Munich: Oldenbourg, 1963–1965.

McKay, Derek. *Prince Eugene of Savoy.* London: Thames & Hudson, 1979.

F

Fabius Maximus Verrucosus "Cunctator" (c. 285–203 B.C.E.)

Fabius Maximus became famous through his delaying tactics in the Second Punic War. When this war began, Fabius was already of respectable age, having an impressive career behind him. He had won a triumph by defeating the Ligurians (233 B.C.E.) and in 222 B.C.E. he had been dictator. He may have been part of the delegation that declared war on the Carthaginians in 218 B.C.E., starting the Second Punic War.

Following the terrible Roman defeat at the hands of Hannibal at Lake Trasimene (217 B.C.E.), Fabius was appointed dictator again. Knowing that Rome was as yet incapable of defeating Hannibal in the field, he initiated a tactic of wearing Hannibal out, refusing him battle, and harassing him and undoing his political successes. At first Fabius's strategy met with strong opposition, but after the terrible defeat at Cannae (216 B.C.E.) the Romans generally stuck to it. Avoiding battle with Hannibal, the Romans remained in his vicinity, obstructing supply, threatening potential renegade allies, and frustrating Hannibal's attempts to gain widespread support in the Italian Peninsula.

Fabius had several commands during the war, during which he kept rigidly to his own tactics. In 215 B.C.E. Fabius was involved in harassing Capua, a renegade ally that had welcomed Hannibal, who encamped in its vicinity. Fabius pillaged the countryside and gave orders to occupy neighboring towns to defend them and keep them from defection. In 209 B.C.E. Fabius took Tarentum (Tarente).

Fabius's strategy denied Hannibal the opportunity to win decisively in the first years of the war, giving Rome precious time to overcome its defeats. Therefore he was called the Shield of Rome, and in time his nickname Cunctator (the Delayer) became a title of honor. More aggressive generals like Scipio Africanus eventually brought decisive victory, yet without Fabius's conservative tactics Rome might not have survived to see it.

M. R. van der Werf

See also: Cannae, Battle of; Hannibal Barca; Mantinea, Battle of; Punic Wars

References and further reading:

Bagnall, Nigel. *The Punic Wars: Rome, Carthage and the Struggle for the Mediterranean.* London: Pimlico, 1999.

Caven, Brian. *The Punic Wars.* London: Weidenfeld & Nicolson, 1981.

Lazenby, J. F. *Hannibal's War. A Military History of the Second Punic War.* London: Aris & Phillips, 1978.

Falaise-Argentan Pocket (August 1944)

The decisive moment of the Battle of Normandy. After being penned into Normandy for the better part of two months, the Allies finally started to make solid gains, beginning in late July. Soon their advances developed into a full-fledged breakout. American forces under Lieutenant General George S. Patton swung to the west of the German armies in Normandy and then turned eastward, racing for the Seine River. After a German counterattack at Mortain failed by 12 August 1944, Hitler's forces now faced the possibility of annihilation. At Mortain the Germans had hoped to punch through American lines, thus cutting off Patton's lead elements and dealing the Americans a major defeat. Instead, the Germans now had little choice but to retreat eastward in a fight for their lives.

From the north, Canadian and British troops attacked southward, capturing the Norman town of Falaise by 16 August. In the south, Patton's troops entered the town of Argentan. At this point only 14 miles separated the Americans and the Canadians. Closing the jaws of this giant pincer would destroy roughly 21 German divisions desperately trying to escape to the east. The pocket for them was a hellish nightmare of Allied artillery and air attacks. In one of the war's most controversial decisions, Lieutenant General Omar Bradley, commander of the American Twelfth Army Group, ordered a halt on further northward movement. He cited concerns that American and Canadian troops might accidentally clash, but he may well have also worried about the possibility of a large German counterattack. The decision allowed some 20,000 of the German troops to escape the pocket. By the time the sack was officially closed on 21 August, the Germans had lost close to 100,000 men. The fact that any Germans had been allowed to escape, though, has dominated the postwar literature about the battle.

John C. McManus

See also: Bradley, Omar Nelson; Normandy Landings; Patton, George Smith, Jr.; World War II

References and further reading:

Blumenson, Martin. *Breakout and Pursuit: United States Army in World War II.* Washington, DC: Office of the Chief of Military History, 1961, 1984.

D'Este, Carlo. *Decision in Normandy.* New York: E. P. Dutton, 1988.

Hastings, Max. *Overlord: D-Day and the Battle for Normandy.* New York: Touchstone, 1984.

Falkenhayn, Erich von (1861–1922)

Prussian minister of War and chief of the German general staff. Falkenhayn was born on 11 September 1861, near Thorn in West Prussia. After attending cadet school, he enrolled in the War Academy, graduating third in his class in 1880. He joined the Prussian general staff in 1890. His prewar service included a tour in China, where he taught at the Chinese Military School in Nankow and participated in the suppression of the Boxer Rebellion. In July 1913, he was promoted to major general and appointed Prussian minister of War, a post he still held at the outbreak of World War I. Falkenhayn's appointment as chief of the general staff coincided with the First Battle of the Marne and the failure of Germany's attempt to achieve a quick and decisive victory on the western front. For the next two years Falkenhayn oversaw German attempts to break the strategic stalemate in the war, which brought him into increasing conflict with the heroes of the eastern front, Hindenburg and Ludendorff.

Convinced that a strategy of attrition offered the best chance for German victory, Falkenhayn launched the German assault on Verdun in February 1916, the purpose of which was to "bleed France white," as he cold-bloodedly put it. Unfortunately the German army was also bled white at Verdun. By the summer of 1916, his failure to achieve victory, coupled with political intrigue, sealed Falkenhayn's fate and he was replaced as chief of the general staff by Hindenburg and Ludendorff. In September 1916 Falkenhayn took command of the Ninth Army and oversaw the conquest of Rumania. In July 1917 he was sent to oversee operations in Palestine and Mesopotamia. In March 1918 he assumed command of the Tenth Army in Lithuania, where he ended the war. Falkenhayn retired in 1919 and died near Potsdam on 8 April 1922.

J. David Cameron

See also: Hindenburg, Paul von Beneckendorf und von; Ludendorff, Erich Friedrich Wilhelm; Moltke, Graf Helmuth Johannes Ludwig von; William II

References and further reading:

Afflerbach, Holger. *Falkenhayn: Politisches Denken und Handeln im Kaiserreich.* Munich: Oldenbourg, 1994.

Falkenhayn, Erich von. *Die Oberste Heeresleitung 1914–1916 in ihren Entschliessungen.* Berlin: E. S. Mittler und Sohn, 1920.

Janssen, Karl-Heinz. *Der Kanzler und der General: Die Führungskrise um Bethmann-Hollweg und Falkenhayn, 1914–1916.* Göttingen: Musterschmidt, 1967.

Falkirk, Battle of (22 July 1298)

English victory paves way for conquest of Scotland. After the Scottish uprising and victory over the forces of Lord Warenne at Stirling Bridge in 1297, and then the Scottish invasion of northeast England, Edward I invaded Scotland with what was "probably the largest single army that had been raised up until that time by an English government." It was an army that had been raised without the issuing of a feudal summons. The majority of the cavalry was not paid: The army appears to have been raised on the basis of personal loyalty to the king, and of acquisition due to right of conquest.

Edward I intended to invade Scotland in such numbers that the Scots would be forced to give battle, where they would then lose. The exact size of his army is disputed, but it would seem that both armies numbered between 25,000 and 30,000 men, with the English having some 10,000 cavalry to the 3,500 of the Scots. It was this advantage of numbers that enabled the English cavalry to drive the Scottish horse from the field. With the protection thus conferred, English longbowmen, employed for the first time outside Wales, then closed on the massed schiltrons of Scottish pikemen and systematically shredded them. With these formations gravely weakened, an English cavalry charge completed the annihilation of the Scottish army.

The defeat in effect cost Wallace his claims on the leadership of Scotland. Nonetheless, it was not until 1305 that he was captured and executed.

H. P. Willmott

See also: Anglo-Scots Wars (1290–1388); Anglo-Scots Wars (1513–1560); Bannockburn, Battle of; Flodden, Battle of

References and further reading:
Prestwich, Michael. *The Three Edwards: War and State in England, 1272–1377.* New York: St. Martin's Press, 1980.
———. *Edward I.* Berkeley: University of California Press, 1988.

Falkland Islands War (2 April–20 June 1982)

Brief undeclared war between Argentina and Great Britain over a group of islands 300 miles east of Argentina; also called the Falklands War, Malvinas War, or the South Atlantic War. (The British call the islands the Falklands and the Argentineans call them the Malvinas.)

Conflicting claims to the islands go back to the eighteenth century. Armed conflict between the two broke out following the occupation of the Falklands by Argentine forces on 2 April 1982. The British government, under Prime Minister Margaret Thatcher, declared a 200-mile war zone around the Falklands and immediately began assembling a naval task force.

On 25 April, the British task force was steaming the 8,000 miles to the war zone. During this time the Argentine cruiser *General Belgrano,* just outside the war zone, was torpedoed and sunk by the British nuclear-powered submarine HMS *Conqueror* (the only use of a nuclear weapon since Nagasaki). Further battles led to the downing of a number of the land-based Argentine air force jets and the sinking of several British ships with French Exocet missiles.

While the naval-air battles were being fought, the British were utilizing their excellent Special Forces units to gather intelligence on the Argentine troops, reconnoiter for a landing site, and make life difficult for the Argentinean forces. As early as 18 April, members of the Special Air Service (SAS) and Special Boat Service (SBS) were in service and by the invasion, about 300 of them would be operating ashore.

One such operation was a raid against an airstrip on Pebble Island where several ground-attack aircraft were housed, along with large stores of ammunition, fuel, and a strategically important radar station. On 15 May 48 SAS men flew from HMS *Hermes* and began the first British land attack since the occupation. The 30-minute attack was a complete success, destroying 11 aircraft, the radar station, and the fuel and ammunition dumps.

The raids were a precursor to the amphibious assault at Port San Carlos. The site was adequate, but not ideal; it was small enough to offer protection from Exocet missiles (though too small to allow ships to maneuver), it possessed a deep anchorage, it had a perimeter of hills for a safe bridgehead, and it provided three possible routes toward the capital of Port Stanley.

The British amphibious assault was very successful; except for the loss of 21 SAS personnel who drowned when a Sea King helicopter ferrying troops was struck by a rearing ship during takeoff and fell into the ocean. Other than the Sea King crash, most of the British casualties were a result of the Argentine air attacks on the vulnerable British ships in the confined waters. Over the next week, the British lost three ships, with several others damaged, and the Argentine air force lost more than 40 aircraft.

With the bridgehead secure, the focus of the conflict shifted to the army. The British approach to Port Stanley utilized the classic pincer movement, one hook going north and then east, the other going south and then east via Goose Green.

The 27 May assault on Goose Green involved 600 members of the 2d Battalion of the Parachute Regiment attacking 1,400 Argentine defenders. The British suffered 17 dead and 31 wounded while the Argentineans lost 250 dead and 121 wounded, with approximately 1,200 taken prisoner. It was a critical victory for the British, demonstrating the effectiveness of highly trained professional troops over a larger force of inadequately trained soldiers.

The siege of Port Stanley was short but fierce. Starting on 11 June, various units of British forces pressed the Argentine defenders. On the night of 13–14 June, heavy fighting took place at Mount Longdon (3d Battalion Paratroopers), Mount William (7th Gurkhas), Sapper Hill (Welsh Guards), and Mount Tumbledown (Scots Guards). The 3d Para was involved in what has been described as the heaviest fighting of the campaign against Argentine commandos. By midday, taking heavy casualties and retreating, Argentinean troops started hoisting white flags. By evening on 14 June, General Moore, overall commander of British land forces, was able to report that all Argentine armed forces on the island had surrendered.

On 20 June, Britain formally declared an end to hostilities. The British had captured more than 10,000 Argentine prisoners during the war. Argentina sustained 655 men killed, while the British lost 236.

Politically, the war helped to rejuvenate the flagging political fortunes of Margaret Thatcher (and led to her reelection) and brought down General Leopoldo Galtieri, who had cold-bloodedly hoped to improve his political situation by starting the war. Galtieri's quick resignation paved the road to the restoration of civilian rule in 1983.

Craig T. Cobane

See also: Goose Green, Battle for

References and further reading:
Hastings, Max, and Simon Jenkins. *The Battle for the Falklands.* New York: W. W. Norton, 1984.
Laffin, John. *Fight for the Falklands!* New York: St. Martin's Press, 1982.

Fallen Timbers (1794)

An American victory over a confederation of Indian tribes. In the years leading up to the battle, President George Washington and the Indian tribes of the Ohio region disputed the boundary between Indian lands and the new republic. Tensions simmered between American settlers and the Indian inhabitants for a decade until 1791, when General Arthur St. Clair led a disastrous and ultimately failed effort to "pacify" the Indians near Fort Recovery. St. Clair lost nearly 600 troops to the Indians, which produced a public outcry for the new administration to take more decisive steps to resolve the frontier question. Washington responded by placing General "Mad" Anthony Wayne in command of an expedition to the Ohio and Miami Rivers to establish the American claim on the region and to discourage British officials stationed in the Old Northwest from aiding and abetting the Indians.

In early 1794, Wayne marched north along the Miami and Wabash Rivers, against resistance from bands led by Chief Little Turtle of the Miamis. Retreating northward, Little Turtle organized a confederation that included Shawnee, Ottawa, Wyandot, and Delaware warriors, who determined to meet the American advance along the Maumee River in northern Ohio. Some accounts placed the legendary Chief Tecumseh in Little Turtle's camp as well. Wayne's advance guards discovered Little Turtle's position near a grove of trees felled in a recent tornado, at a site along the Maumee River near present-day Toledo, and Wayne prepared for battle.

On 20 August, Wayne's troops, some 2,600 men, including riflemen from Kentucky, formed lines before the "fallen timbers" and a firefight ensued. Fixing bayonets, the Americans charged the Indian warriors, who scattered amid chaotic, hand-to-hand combat. Both sides lost nearly 100 men, but the result was a decisive victory for Wayne and the United States forces. This military success stimulated negotiations that led to the Treaty of Greenville (1795), which composed affairs between the United States and key Indian nations of the Old Northwest, bringing a large portion of the Ohio River valley under American control. Washington's administration gained much-needed credibility among American citizens and was better positioned to extract favorable terms from Indian tribes of the Old Northwest through the remainder of the 1790s.

General Anthony Wayne's American troops defeat the Miami Indians at the Battle of Fallen Timbers in August 1794. (Library of Congress)

Jeffrey B. Webb

See also: American Revolution; Washington, George; Wayne, Anthony

References and further reading:
Carter, Harvey Lewis. *The Life and Times of Little Turtle: First Sagamore of the Wabash.* Urbana: University of Illinois Press, 1986.
Nelson, Paul David. *Anthony Wayne, Soldier of the Early Republic.* Bloomington: Indiana University Press, 1985.
Sword, Wiley. *President Washington's Indian War: The Struggle for the Old Northwest, 1790–1795.* Norman: University of Oklahoma Press, 1985.

Ferdinand, Duke of Brunswick (1721–1792)

Commander in chief of the allied army during the Seven Years' War (1757–1762). When appointed commander in chief of the combined Hanoverian, Hessian, Brunswick, English, Saxe-Gotha, and Schaumburg-Lippe army in November 1757, the 36-year-old prince (born 12 January 1721 in Brunswick) had been serving with distinction in Prussian service under Frederick the Great. In his new position he showed both military leadership and political responsibility. Despite the numerical inferiority of his troops, he masterfully kept two French armies in check on the western theater of operations (Westphalia, Hesse, Hanover) year in and year out between 1758 and 1762. He did so by either occupying impregnable positions or giving battle, or using his famous light troops (under the command of Luckner, Freytag, and Scheither) in a very sophisticated manner. He led his army to victories over the French at Krefeld (1758), Minden (1759), Warburg (1760), Vellinghausen (1761), and Lutterberg (1762), each action or battle skillfully planned and carefully carried out. Minden particularly was a great success for the allies but it could have been an even more impressive one if Lord Sackville, commander of the cavalry, had not refused to advance his troops despite repeated orders from the prince.

Portrait of the duke of Brunswick. (Library of Congress)

After the war, Ferdinand became governor of the Prussian fortress of Magdeburg only to retire three years later, in 1766, because of severe differences of opinion with Frederick the Great. The Austrian queen, Maria Theresa, offered him the position of field marshal in her army. But the prince declined the offer as well as one from George III actually to lead the British forces in North America during the War of Independence. Ferdinand died at his country seat in Vechelde (Lower Saxony, Germany) on 3 July 1792.

Juergen Luh

See also: Minden; Seven Years' War

References and further reading:

Luh, Juergen. *Ancien Régime Warfare and the Military Revolution: A Study.* Groningen: INOS, 2000.

Savory, Reginald. *His Britannic Majesty's Army in Germany during the Seven Years War.* Oxford, UK: Oxford University Press, 1966.

Ferdinand, Karl Wilhelm, Duke of Brunswick (1735–1806)

German commander known chiefly for his command of the armies at both Valmy in 1792 and Jena-Auerstädt in 1806. Born on 9 October 1735 in the ducal palace at Wolfenbuttel, Karl Wilhelm was the eldest son of Duke Karl I of Braunschweig. He earned acclaim early in his military career, serving in the Prussian service in the western theater of the Seven Years' War. Acceding to the throne of Braunschweig in 1780, he retained his position in the Prussian army. Raised to the rank of field marshal, he commanded the Prussian forces that subdued the Dutch uprising in 1787. In 1792 Karl Wilhelm declined a commission offered him by the French to lead their new revolutionary army, choosing instead to command the counterrevolutionary forces sent by Prussia and Austria to invade France later that year. After initial successes against the revolutionary citizen armies, Brunswick's advance was stopped at the Battle of Valmy in 1792. Though the allied forces were subsequently driven back, surrendering much of western Germany to the French, by December 1793 Brunswick's troops had succeeded in recapturing this lost territory. Frustrated by incompetent political and military intervention by King Frederick William II, however, the duke resigned his commission in January 1794, only returning under Frederick William III in 1797 to command an army of observation. With the outbreak of war against France again in 1806, Brunswick was named commander in chief and was charged with remobilizing and commanding the ill-prepared Prussian armies, a task for which he was not well suited. During the Battle of Jena-Auerstädt on 14 October, while leading a body of grenadiers in an assault on the latter village, he was shot through both eyes and later died of his wounds in Ottensee on 10 November.

Christopher C. W. Bauermeister

See also: Frederick the Great, King of Prussia; French Revolutionary Wars; Jena and Auerstädt; Minden; Seven Years' War; Valmy

References and further reading:

Chandler, David G. *The Campaigns of Napoleon.* New York: Macmillan Publishing Company, 1966.

Fitzmaurice, Lord Edmond. *Charles William Ferdinand, Duke of Brunswick: An Historical Study, 1735–1806.* London and New York: Longmans, Green, 1901.

Schroeder, Paul W. *The Transformation of European Politics 1763–1848.* Oxford, UK: Oxford University Press, 1994.

Film and War

The images of Errol Flynn as George Custer, dressed in buckskin, standing bravely in the face of an avalanche of Indians; of Steve McQueen on a motorcycle in the midst of his great escape; and of Gary Cooper in his olive-drab World War I uniform as Sergeant York are some of the most ingrained images in American contemporary culture. Movies and films have influenced how society views and thinks about warfare. They impart cultural values and dispositions and provide propaganda for national causes. They sum up how certain segments of the public feel about internal events and can sway the larger population more efficiently than rational arguments. War movies led activist Ron Kovic and other young men during the Vietnam conflict era to join the services to "play John Wayne." Richard Nixon insisted that his top foreign policy advisers watch *Patton* during a critical time in his Vietnam conflict policy, and Ronald Reagan imparted to moved audiences the dying words of a cinematic flyer who went down with his plane as if he were actually flesh and blood.

The power of the image to affect how a country views conflicts and combatants occurred long before film. Leaders like Napoleon commissioned the production of artwork to celebrate the completion of certain campaigns and victories to influence attitudes at home. Matthew Brady and other Civil War photographers took photographs that altered how the war was remembered and may, in fact, in their gruesome reality have influenced at least the American public away from a glorification of the battlefield. Yet films provided a unique forum in which images and stories could be related.

Film, by its very nature, tells stories and some of the most powerful are about war. In 1897, correspondent Frederic Villiers filmed the first real images of combatants in the Greco-Turkish War and a number of cameramen recorded the Boer War and the Spanish-American War the next year. The cumbersome nature of the camera equipment, technological improvements in weapons, and the attitude of countries involved made filming of actual combat nearly impossible. This gave rise to the practice of faking war footage. The practice was so common that a 1900 press cartoon of the Boxer Rebellion shows it being stage-managed for the camera. Whether fake or real, war films drew large audiences. These images, coupled with sound effects, music, and commentary, were powerful vehicles of patriotism during the Spanish-American War, as in *Tearing down the Spanish Flag* (1898). A number of newsreels from companies like Pathe (France) and Rosie (England) supplied the world with footage during the first decade and a half of the twentieth century. In 1913–1914, Major T. J. Dickson lectured throughout the eastern United States showing audiences film on the mobile army. During this time the U.S. military began to take motion pictures as a serious means of instructing and entertaining troops. By 1915 more than 60 projectors could be found at the U.S. military and naval academies.

Due to limited technology and the high cost of production, war films as a genre did not begin until just before World War I. Although there were a number of smaller films, including *Washington at Valley Forge, France Marion: Swamp Fox, Soldiers of Fortune,* and *The Battle of Shiloh,* it was D. W. Griffith's epic *Birth of a Nation* (1915) that brought war films to a national audience. Centering on the Civil War and its aftermath, Griffith showed a hooded Ku Klux Klan army defending the South against mulattoes and blacks. This masterpiece was followed by Cecil B. DeMille's *Joan the Woman* (1916), a movie on Joan of Arc that served as a rallying point for the Allies. Even though Griffith's later epic *Intolerance* (1916) urged pacifism, nonetheless, facing a looming conflict in Europe, Hollywood released *Hearts of the World* and *To Hell with the Kaiser* (1918), which portrayed Germans as brutish thugs. Studios began gearing their advertising campaigns and films to support the war effort. The nation's greatest star, Mary Pickford, in *The Little American* (1917), was in the forefront of propaganda efforts. Other pictures like *The Unbeliever* (1918) made with the cooperation of the marines showed the gallantry of American troops. As the war progressed, the War Department increased its use of film for training and instructional purposes.

With the armistice and the American public's weariness with war, combat films entered a lull and bottomed in 1921, when only nine such feature movies were released. Starting in 1925, three films would revive cinematic battle stories. Reflecting the feeling of the "lost generation," *The Big Parade* (1925) offered the viewing public a more realistic account of conditions in the trenches of France. The next year, another, Laurence Stalling's story *What Price Glory?,* was screened.

The depiction of the Battle of Belleau Wood was extremely realistic for this time. Finally, *Wings* (1927), the film to win an academy award, offered audiences their first glimpse of aerial combat.

Hollywood would continue to draw on War World I with the advent of sound. Erich Maria Remarque's novel *All Quiet on the Western Front* was screened in 1930. This antiwar film recounted the griminess of war from the German perspective. Millionaire Howard Hughes, using new camera techniques, shot realistic aerial dogfights in *Hell's Angels* (1930). Yet not all the Great War movies were so somber. *The Dawn Patrol* (1930) was an action film that told the story of the British Royal Flying Corps. While several military action adventure movies were film, due to the Great Depression Hollywood turned to comedies and movies that were more light and airy.

The coming of World War II marked a new era for war films. From 1939 to 1947, American studios responded to the larger political environment as never before. They played a major part in the war effort and often worked hand-in-hand with the government in the form of the Office of War Information and the War Production Board. While much of the nation was isolationist in 1939, Hollywood's *Confessions of a Nazi Spy* (1939) is considered by film historians to be the first of many anti-Nazi films. The horrors of life in Nazi Germany were depicted in *Beasts of Berlin* (also 1939). Charlie Chaplin portrayed a buffoonish Adolf Hitler–like character in *The Great Dictator* (1940). Yet not only was Hollywood trying to alarm the nation to outside threats, but it had rallied around the Roosevelt administration and produced many patriotic films, such as *Sergeant York* (1941). Gary Cooper played the heroic Alvin York as the personification of national values. He is a peace-loving man who is religious, moral, and just, even to the point of refusing to cash in on his battlefield heroics. Hollywood's advocacy in this prewar era is shown clearest in *A Yank in the R.A.F.* (1941), an exciting story starring Tyrone Power as an American who became a flyer in England during the early stages of the war.

Patriotic musicals, such as the slam-bang classic *Yankee Doodle Dandy* (1942), *Cover Girl* (1944), *Pin-Up Girl* (1944), and *Anchors Aweigh* (1945), kept the American public encouraged and lightened spirits during World War II. Love was the cost for patriotism in the classic *Casablanca* (1942) and the audiences were reminded of the importance of sacrifice on the home front in movies like *The Human Comedy* (1943) and *Since You Went Away* (1944). *Wake Island* (1942), *Gung Ho!* (1943), *Action in the North Atlantic* (1943), *Bataan* (1943), *Guadalcanal Diary* (1943), *Destination Tokyo* (1943), *Sahara* (1943), *Immortal Sergeant* (1943), and *Thirty Seconds over Tokyo* (1944) depicted for the American people the heroic efforts of servicemen in the war. Grimmer and more realistic war films like *They Were Expendable* (1945), *The Story of GI Joe* (1945), *Back to Bataan* (1945), and *A Walk in the Sun* (1946) made their way into theaters toward the end of the war.

After World War II, war films again went out of fashion. The films *The Best Years of Our Lives* (1946) and *The Men* (1950) depicted the struggles of servicemen adjusting to civilian life, while *Gentleman's Agreement* (1947), dealing with anti-Semitism, typically used returned servicemen as its protagonists. With the advent of the Cold War, Americans were again reminded of their patriotic duty and the bravery of servicemen in films like *Battleground, Sands of Iwo Jima,* and *Twelve O'Clock High* (all 1949). Yet for the most part, in the 1950s Hollywood released movies portraying the folly of war and the growing threat of nuclear war. The madness of the Korean War formed the backdrop for two of Samuel Fuller's films, *Fixed Bayonets* and *Steel Helmet;* the trauma and hardships of combat are detailed in John Huston's civil war drama *The Red Badge of Courage* (all 1951). The antiwar message was further spread by Stanley Kubrick's *Paths of Glory,* which recounted court-martialing of French soldiers during World War I, and David Lean's *The Bridge on the River Kwai* (both 1957). Gregory Peck, playing an army officer ordered to assault a strategically insignificant hill, shows the insanity of war in Lewis Milestone's *Pork Chop Hill* (1959). *The Burmese Harp* (1959) examines the horrors of World War II from the Japanese perspective. Other notable war movies during the 1950s include Billy Wilder's *Stalag 17,* which detailed life in a Nazi POW camp, and *The Man Who Never Was* (1955), about British intelligence use of a dead body to fool Germany about the location of the Allied invasion.

The 1960s brought a number of epic historical war dramas to the screen, including *The Longest Day* (1962), *In Harm's Way, The Battle of the Bulge* (both 1965), *Patton,* and *Tora! Tora! Tora!* (both 1970). Other action war movies like *The Great Escape* (1963), *The Dirty Dozen* (1967), and *Where Eagles Dare* (1969) were greeted favorably by audiences. Stanley Kubrick satirized the insanity of nuclear war in his dark comedy *Dr. Stangelove or: How I Learned to Stop Worrying and Love the Bomb* (1964). *M*A*S*H** and *Catch-22* (both 1970) were both comedic commentaries on the folly of warfare. With the exception of *The Green Berets* (1968), which was savaged by critics, patriotic combat films fell out of favor once the Vietnam conflict escalated.

On the other hand, the popular reenactment movement provided a ready and inexpensive background for literally scores of documentary war films, most made for television. Reenactors meticulously re-created in their persons every type of soldier from Roman centurian, to crusader, to English and American Civil War troopers, to World War I doughboys.

(One of the more improbable was a group portraying U.S. World War II GIs in England, reenacted by Englishmen who rode about in restored Jeeps and half-tracks, their pockets stuffed with Wrigley's chewing gum and Life Savers!)

By the late 1970s, Hollywood was finally able to address lingering feelings and attitudes about the Vietnam conflict in such films as *Coming Home, The Deer Hunter* (both 1978), and *Apocalypse Now* (1979). *The Killing Fields* (1984) showed the social upheaval caused by conflict. Director Oliver Stone struggled to make sense of the war in *Platoon* (1986). *Full Metal Jacket, Hamburger Hill* (both 1987), *84 Charlie Mopic, Casualties of War, Born on the Fourth of July* (all 1989), and *Heaven and Earth* (1993) also offered harsh commentary on the war. The unresolved issue of POWs received attention in *Uncommon Valor* (1983), *Missing in Action* (1984), and *Rambo* (1985), while Clint Eastwood's *Heartbreak Ridge* (1986) tried to return glory to the military with its tale of a green platoon in Grenada. But even epic films like Samuel Fuller's *The Big Red One* (1980) and *Gallipoli* (1981) spoke to the terror of war.

Yet war films have also been influential throughout the rest of the world. Though they have to deal with governmental oversight, some of the greatest epic war films have come from communist- or fascist-controlled countries. Even when these governments have been overthrown, the imprint can still be seen in filmmakers. Out of Russia, Grigori Chuckrai shows warfare from the viewpoint of a young soldier movingly in *Ballad of a Soldier* (1960). Elem Klimov, who had gotten himself censored for being the first Russian filmmaker to feature Rasputin, on regaining his artistic freedom detailed the evils of ethnic cleansing in the 1943 German invasion of Belorussia in *Come and See* (1986). Alexandr Askoldov lashes at anti-Semitism in *Commissar* (1967), his first and last film, a movie about a Red Army commissar who finds herself pregnant. Mikhail Kalatozov's *The Cranes Are Flying* (1957) shows the radical nature of Soviet cinema with its tale of a beautiful young woman confronted by the horrors of war. Famous poet Yevgeny Yevtushenko re-creates his childhood memories of the Nazis trying to seize Moscow during World War II in *Kindergarten* (1983). The epic *War and Peace* (1968) was directed by Sergei Bondarchuk and remains faithful to the Leo Tolstoy novel. Sergei Bodrov takes up another Tolstoy story in *Prisoner of the Mountains* (1996). Finally, Andrei Tarkovsky's powerful *The Mirror* (1975) uses documentary footage from the Spanish Civil War and the Soviet-Chinese confrontation.

From Bosnia comes Srdijian Dragojevic's provocative *Pretty Village, Pretty Fame,* which looks at the beginnings of the war in Bosnia in 1992. Czechoslovakian director Jiri Weiss comments on heroism in World War II in *The Coward* (1962). Fellow countryman Jan Sverak also comments on human weakness during war from a child's point of view in *Elementary School* (1991). Hungarian cinema has produced the powerful and moving *Budapest Tales* (1976), *Cold Days* (1966), and *The Red and White* (1968). Also of interest is the Serbo-Croatian movie *Vukovar* (1994). Yet, surprisingly, it is Poland that has been the most prolific and offered the most powerful commentaries on war. Jerzy Hoffman looks at the seventeenth-century Polish-Swedish War in *The Deluge* (1973), which was nominated for Best Foreign Language Film in 1974. Andrzej Munk looks at the aftermath of Auschwitz in *The Passenger* (1963). Janusz Zaroski's *Mother of Kings* (1976) examines the worst moment of Poland's history from World War II and Stalinist times. Andrei Tarkovsky won the Golden Lion at the Venice Film Festival for his brutal depiction of war in *My Name Is Ivan* (1962). Yet it is Andrej Wajda who has directed the most renowned images of World War II in *Kanal* (1956), *A Generation* (1954), and *Ashes and Diamonds* (1958).

Losing World War II had a lasting effect on Japanese cinema. Shohei Imamura's *Black Rain* (1988) won a Cannes Film Festival award for his recounting of the bombing of Hiroshima. Similarly, Kon Ichikawa's *The Burmese Harp* (1956) and *Fires of the Plains* (1959) condemned warfare. Akira Kurosawa directed his most personal film on the fears of nuclear war in *Dreams* (1973) and covers similar ground in *Rhapsody in August* (1991). Masaki Kobayashi's *The Human Condition* Part I (*No Greater Love*) (1958) and Euthana Mukdasanit's *Sunset at Chapraya* (1996) are also pacifistic in tone. The most overlooked film for those interested in World War II aircraft movies is *I Bombed Pearl Harbor* (1960).

Another country dealing with the tragedy of its military past is Cambodia in H. M. Norodom Sihanouk's *Peasants in Distress* (1994). Three Australian productions examine the horrors of war in *Breaker Morant* (1980), *Gallipoli* (1981), and *Stories Our Mothers Never Told Us* (1995). German filmmakers try to make sense of their military losses in the two world wars in *Das Boot* (1981), *Germany, Pale Mother* (1979), *Gorilla Bathes at Noon* (1992), *Jacob the Liar* (1974), *Songs of Life* (1968), *Stalingrad* (1992), and *Westernfront 1918* (1930).

The home of many of the great epic war films is France. Abel Gance made the antiwar story *J'Accuse* (1918), which many critics place on par with D. W. Griffith's *Birth of a Nation.* Bertrand Tavernier examines the last days of World War I in *Capitaine Conan* (1996). Jean Renoir tells one man's story in a POW camp in *Elusive Corporal* (1962). Jean-Pierre Denis examines France's Napoleonic roots in *Field of Honor* (1987). Other French films of note include *The Green House* (1996), *Is Paris Burning?* (1968), *Grand Illusion* (1937), *For Ever Mozart* (1996), *Here and Elsewhere* (1974), *The Ogre* (1996), *Queen Margot* (1994), *La Silence de la Mer* (1947),

Underground (1995), *Uranus* (1990), and *Interred in a Prison*.

Finally, other war films of interest include the Welsh *Hedd Wyn* (1996), Spain's *The Hunt* (1966), Israel's *Siege* (1970), Italy's *Man with a Cross* (1943), *Night of the Shooting Stars* (1982), and *Le Soldatesse* (1964). Also noteworthy are Greece's *Ulysses Gaze* (1995), Sweden's *The Shame* (1968), Argentina's comedy *Funny Dirty Little War* (1983), a powerful commentary on the Falkland War in *Veronico Cruz* (1989), and Colombia's *Confessing to Laura* (1990).

In the late 1990s, Hollywood war movies made a comeback. Terrence Malick's epic *The Thin Red Line* (1998) provided thought-provoking images of the attack on Guadalcanal in World War II. Steven Spielberg won the Academy Award for best director for *Saving Private Ryan* (1998). Mel Gibson made three movies about the glory of war with *The Patriot* (2000), *Braveheart* (1995), and *We Were Soldiers* (2001). Audiences were even given a Soviet sniper as a hero in the World War II epic about the battle for Stalingrad in *Enemy at the Gates* (2001). With the opening of the twenty-first century, Hollywood is again recognizing the box office appeal of war stories. Given the historical record, feature films on the subject of war will remain popular with audiences worldwide for the foreseeable future.

T. Jason Soderstrum

See also: Art in War; Military and Society; Music, Military

References and further reading:

Butler, Ivan. *The War Film.* South Brunswick: A. S. Barnes, 1974.
Campbell, Craig. W. *Reel America and World War I: A Comprehensive Filmography and History of Motion Pictures in the United States, 1914–1920.* Jefferson, NC: McFarland & Company, 1985.
Dick, Bernard F. *The Star-Spangled Screen: The American World War II Film.* Lexington: University of Kentucky Press, 1985.
Dittmar, Linda, and Gene Michaud, eds. *From Hanoi to Hollywood: The Vietnam War in American Film.* New Brunswick, NJ: Rutgers University Press, 1990.
Evans, Alun. *Brassey's Guide to War Films.* Washington, DC: Brassey's Publishing Company, 2000.
Parish, James Robert. *The Great Combat Pictures: Twentieth-Century Warfare on the Screen.* Metuchen, NJ: Scarecrow, 1990.
Suid, Lawrence H. *Guts & Glory: Great American War Movies.* Reading, MA: Addison-Wesley Publishing, 1978.

Finances, Military

For most civilizations, the financing of military exertions has proved the crucial problem of governance. Among the river valley civilizations, financial administration and the government were inseparable. Governments were organized on hierarchical bases, with the rulers possessing control over military decisions. Taxes were paid in kind, which made it more difficult to utilize the revenue for military campaigns over great distances. For these agricultural economies, supporting a large standing army was the principal expenditure item. The optimal size of an empire was determined by the efficiency of taxation, resource extraction, and its transportation system. The supply of metal and weaponry was seldom a crucial variable in its military success. Some changes were brought, however, by the use of cheap iron weaponry after about 1200 B.C.E.

These civilizations, nonetheless, paled in comparison with the military efficiency and economy of the Roman Empire. Military spending was the largest item of public spending throughout Roman history. During the first two centuries of the empire, the Roman army had about 150,000 to 160,000 men, in addition to 150,000 other troops, with soldiers' wages increasing rapidly. For example, in republican and imperial Rome military wages consumed more than half of the revenue.

During the Middle Ages, following barbarian invasions, a varied system of European feudalism emerged in which feudal, aristocratic lords provided protection for their communities. Prior to 1000, the command system was still preeminent in the mobilization of military resources, mostly on a contingency basis. It was not until the twelfth century and the Crusades that the feudal kings needed to supplement their ordinary revenues to finance their armies. Internal discontent in the Middle Ages often led to an expansionary drive as the spoils of war helped calm the elite. The political ambitions of medieval kings, however, relied on revenue strategies that catered to short-term deficits and military campaigns. Innovations in the ways of waging war, aided by the gunpowder revolution of the fifteenth century, permitted armies to attack and defend larger territories. These developments also made possible the commercialization of warfare in the fourteenth and fifteenth centuries as volunteer feudal forces had to give way to professional mercenary armies. Accordingly, medieval states had to increase their taxes to support the growing costs of warfare. The age of commercialization of warfare was accompanied by the rising importance of sea power, as European states began to build their overseas empires. States such as Portugal, the Netherlands, and England, respectively, became the "systemic leaders" due to their extensive fleets and commercial expansion. The early winners in the fight for European leadership were usually supported by the availability of inexpensive credit. This pattern can be discerned in the English case. In the period 1535–1547, the English defense share (military expenditures of central government expenditures) averaged 29.4 percent, with large yearly fluctuations. However, in the period 1685–1813, its mean defense share increased to 74.6 percent, never falling below 55 percent. The newly emerging nation-states began to develop more centralized and productive revenue-expenditure systems, especially in the absolutist era. These systems also reflected the growing cost and scale of warfare: During the Thirty Years' War between 100,000 and 200,000 men fought under arms, whereas 50 years later 450,000 to 500,000 men fought in the War of the Spanish Succession. The participants of these wars became even more dependent on access to long-term credit. For example, Spain's decline in the seventeenth century can be linked to the lack of long-term credit and poor management of government finances. The new style of warfare, prominent in the Revolutionary Wars, and rapid population growth increased the European manpower and spending requirements in the eighteenth century. For example, the French army grew more than 3.5 times in size from 1789 to 1793, up to 650,000 men. Similarly, the British army grew from 57,000 men in 1783 to 255,000 men in 1816. The Russian army reached a massive size of 800,000 men in 1816.

This kind of mobilization, which became more or less permanent in the nineteenth century, required new sources of financing. The nineteenth century saw just such reforms: centralized public administration, reliance on balanced budgets, innovations in public debt management, and direct taxation. These reforms were also supported by the spread of industrialization and rising productivity, so that the burdens of military spending could be borne relatively easily. The nineteenth century, moreover, made possible the industrialization of war and armaments production, starting in the midcentury and gathering speed quickly. The economic consequences posed by these changes differed. In the French case, the mean defense share remained roughly the same in 1870–1913, around 35 percent, whereas its military burden (military expenditures of gross domestic product [GDP]) increased about 1 percent to 4.2 percent. In the British case, the mean defense share declined 2 percent to 36.7 percent in 1870–1913 compared to the early nineteenth century. However, the strength of the British economy actually enabled a slight military burden decline to 2.6 percent, a similar figure incurred by Germany in the same period. For most countries the period leading to World War I meant comparatively higher military burdens, easily borne. The United States, the new economic leader, had a meager 0.7 percent average military burden.

In World War I this military potential was unleashed in Europe, with a war of attrition causing, along with millions of casualties, property damage amounting to perhaps U.S.$36 billion. In the interwar period, especially in the 1920s, public spending was static, plagued by budgetary immobility. However, although among democracies defense shares dropped noticeably, their respective military burdens stayed either at similar levels to before the war or even increased—for example, the French military burden rose to a mean level of 7.2 percent. Also in Great Britain, the defense share mean dropped to 18.0 percent, yet the military burden mean actually increased, despite the Ten-Year Rule (no major war to be anticipated for the next decade). For these countries, the mid-1930s marked the beginning of rearmament. Nazi Germany increased its military burden from 1.6 percent in 1933 to 18.9 percent in 1938, a rearmament program aided by creative financing and promising both "guns and butter." Mussolini was not quite as successful in his efforts, with military burdens fluctuating between 4 and 5 percent in the 1930s, but not giving the Italians much for their money. The Japanese rearmament drive was perhaps the most impressive, amassing a military burden as high as 22.7 percent in 1938.

In World War II, the initial phase from 1939 to early 1942 favored the Axis as far as their strategic and economic potential was concerned. After that, the war of attrition, with the United States and the USSR joining the Allies, made it impossible for the Axis to prevail. For example, in 1943 the Allied total GDP was U.S.$2,223 billion (1990 prices), whereas the Axis accounted for only U.S.$895 billion. It was sheer madness for the Axis to enter war with the Allies. The economic demands of the war were unprecedented, as for example Great Britain's maximum military burden of around 27 percent in 1918 was dwarfed by the over 50 percent level maintained throughout World War II. This war also brought with it a new military-political leadership role for the United States, especially within NATO, a formidable defense alliance formed among democratic countries in 1949. The USSR, rising to new prominence in the war, established the communist Warsaw Pact in 1955 to counter these efforts. The war also gave impetus to welfare states that brought the European social democracies' government expenditure average from under 30 percent of GDP in the 1950s to more than 40 percent in the 1970s. Military spending levels followed suit and peacetime spending peaked during the early Cold War. The American military burden increased above 10 percent in 1952–1954, and the United States retained a high mean value of 6.7 percent for the postwar period. Great Britain and France followed the American example. The Cold War embodied a relentless armaments race between the two superpowers, with the USSR, according to some figures, spending as much as 60 to 70 percent of the American equivalent in the 1950s. Other figures suggest an enduring yet dwindling lead for the U.S. even in the 1970s. Furthermore, the same figures point to a two-to-one lead in favor of the NATO countries over the Warsaw Pact. Part of this armaments race was due to technological advances that produced a mean annual increase in real costs of around 5.5 percent. Yet, spending on personnel has remained the biggest fiscal item, as soldiers, at least in the democracies, demanded living standards comparable to those in civilian life.

The world military spending levels began a slow decline from the 1970s onward, with the exception of the Reagan years. In 1986, the American military burden was 6.5 percent, whereas in 1999 it had dropped to 3.0 percent, and conscription had been dropped in 1973. In France (1977–1999), the military burden has declined from the postwar period to an average of 3.6 percent. This reduction has mostly been triggered by the reduction of tensions between the rival groups and the subsequent collapse of the Soviet Union. The USSR continued to challenge the American military dominance until the mid-1980s, and the Soviet military burden remained high at 12.3 percent in 1990. In the successor Russian Federation, with a declining GDP, this level has dropped rapidly to 3.2 percent in 1998. Similarly, other nations have downscaled their military spending in this period. For example, German military spending in 1991 was more than U.S.$52 billion, whereas in 1999 it had declined to less than U.S.$40 billion. In the French case, the decline was more modest, with its military burden decreasing from 3.6 percent to 2.8 percent. Overall, there was a reduction of approximately one-third in real terms in world military spending in 1989–1996, with small gains again in 1999. In the global scheme, world military expenditure remains highly concentrated, with the 15 major spenders accounting for 80 percent of the world total in 1999. If there is one constant in military expenditures through the ages, it is that such spending is invariably the largest single item in the budget of any nation.

Jari Eloranta

See also: Arms Control; Babylonian Empire; Conscription; Crusades; Disarmament; Economic Warfare; French Revolutionary Wars; Korean War; Spanish Succession, War of the; Thirty Years' War; World War I; World War II

References and further reading:

Bonney, Richard, ed. *The Rise of the Fiscal State in Europe c. 1200–1815.* Oxford, UK: Oxford University Press, 1999.

European State Finance Database (ESFDB). Maintained by Richard Bonney. Available at: http://www.le.ac.uk/hi/bon/ESFDB/dir.htm.

Harrison, Mark, ed. *The Economics of World War II: Six Great Powers in International Comparison.* Cambridge, UK: Cambridge University Press, 1998.

Hobson, J. M. "The Military-Extraction Gap and the Wary Titan: The Fiscal-Sociology of British Defence Policy 1870–1913." *The Journal of European Economic History* 22: 3 (1993).

Kennedy, Paul. *The Rise and Fall of the Great Powers: Economic Change and Military Conflict from 1500 to 2000.* London: Unwin Hyman, 1988.

McNeill, William H. *The Pursuit of Power: Technology, Armed Force, and Society since A.D. 1000.* Chicago, IL: University of Chicago Press, 1982.

Singer, J. David, and Melvin Small. *National Material Capabilities Data, 1816–1985.* Computer file. Ann Arbor, MI: Inter-university Consortium for Political and Social Research (distributor), 1993.

SIPRI Yearbooks of World Armaments and Disarmament. Stockholm: Almqvist & Wiksell; Oxford, UK: Oxford University Press, 1969/70–2000.

Webber, Caroly, and Aaron Wildavsky. *A History of Taxation and Expenditure in the Western World.* New York: Simon & Schuster, 1986.

Finnish Civil War (1918)

A conflict that lasted three and a half months and claimed some 6,500 military casualties, and which established conservative supremacy in Finland. The explosive situation in newly independent Finland in January 1918 can be seen in the concentration of troops: up to 70,000 Russian soldiers were situated across Finland, as well as 40,000 White Civic Guards; the Red Guards numbered about 30,000 men. The first clashes occurred in the town of Viborg in late January. The headquarters of the White forces were assembled in the town of Vasa from mid-January on under the leadership of C. G. E. Mannerheim. The disarming of Russian soldiers in southern Ostrobothnia by the White forces and the revolution in Helsinki by the Reds occurred simultaneously on 27 January 1918. The Red Peoples Delegation concentrated its leadership in Helsinki.

The war started slowly and was uncoordinated on both sides. The frontier lines were more or less stabilized by the beginning of February, with the Reds controlling most of the industrialized south and Finland proper, whereas the Whites held the northern and central regions. The Reds achieved their first significant victory on 21 February, yet the initiative soon shifted to the White forces in the west. The Reds' attack on the western front was stopped on 25 February. At the beginning of March, the Reds suffered a major setback, as Soviet Russia was forced to sign the peace of Brest-Litovsk and to evacuate Finland.

On 12 March Mannerheim mounted a major offensive toward Tampere, Finland's biggest industrial center. The Reds' counteroffensives were largely unsuccessful, as were also the Whites' efforts to round up the retreating Red forces. The White forces also suffered heavy losses.

The final offensive began on 3 April, against heavy Red resistance. After bitter fighting, Tampere finally surrendered on 6 April, a crucial victory for the Whites. Afterwards, Mannerheim divided his forces into two operational units, the Eastern and the Western Armies.

German-trained Finnish troops, Jaegers, and German units were essential for the Whites. The Germans themselves began landing early in April, signaling the beginning of the end for the revolutionary forces.

On 8 April, the Peoples Delegation was forced to flee to Viborg, and Helsinki surrendered on 14 April. With the Reds retreating but fighting fiercely, the White army began its eastern offensive on 23 April. Viborg fell into White hands, and the remainder of Red resistance withered quickly, with the last Red group surrendering on 5 May.

Jari Eloranta

See also: Bolshevik Revolution; Mannerheim, Carl Gustaf Emil; Russian Civil War (1918–1921); World War I

References and further reading:

Kronlund, Jarl. *Suomen puolustuslaitos 1918–1939: Puolustusvoimien rauhan ajan historia.* Porvoo: WSOY, 1990.

Singleton, Fred. *A Short History of Finland.* Cambridge, UK: Cambridge University Press, 1989.

Suomen historian pikkujättiläinen. Porvoo: WSOY, 1987.

Ylikangas, Heikki. *Tie Tampereelle: Dokumentoitu kuvaus Tampereen antautumiseen johtaneista sotatapahtumista Suomen sisällissodassa 1918.* Helsinki: WSOY, 1993.

Firearms

The basic principle of the firearm is simple: a tube, closed at one end, has in it an explosive with a projectile above it, nearer to the open end of the tube. An ignition system fires the explosive charge, which forces the projectile along the tube by means of the gases from the explosion. Possibly one of the earliest examples is held in the Bernisches Historisches Museum in Switzerland, and consists of a length of wood with a short tube fixed at one end. Just behind the rear, closed end of the tube, is a spike. The spike was used to lessen the recoil of the weapon, because of its large caliber, by being pushed into the ground.

Handheld, man-portable weapons date from the discovery of gunpowder, the first propellant known to man. Gunpowder is a coarse mixture of saltpeter, charcoal, and sulfur. (Modern propellants are dealt with elsewhere, as is artillery, defined as not man-portable weaponry.) The earliest firearms were muzzle-loaded. The barrel of the weapon was a simple, smooth tube, with a small aperture near the closed end to allow the firing spark access to the gunpowder charge. It seems certain that firearms as light weapons developed from cannon. Cannon were available in the early part of the thirteenth century, but firearms were not common until about 1375. The earliest firing method was by means of the matchlock: simply a length of fuse held in a clamp that could be moved into the firing position by the musketeer when he chose. Naturally such a firing method required the musketeer to keep his fuse alight, not an easy task in rain or wind.

The next developed firing mechanism was the wheel lock, followed by the snaphaunce, the flintlock, and the miquelet. All worked on the principle of causing sparks to be generated by a flint that lit a powder train, the train in turn firing the main charge. In terms of time, from the development of the wheel lock to the development of the next stage of ignition—the percussion system—a period of 300 years elapsed.

Firearms were inevitably rough weapons until the middle of the nineteenth century, simply because every maker had his own standards and methods of measurement. This meant that calibers were erratic, even when makers were in the same town or even street. Further, bullet makers were equally nonstandardized, so it was lucky that the main projectile was the lead ball, which could be adjusted reasonably easily in the field. Private firearms, as opposed to military weapons, always exhibit a much higher standard of workmanship and finish, simply because the purchaser is able to pay for such refinement in a single weapon.

The nineteenth century saw a complete change in the technical aspects of firearms. The Rev. Dr. Alexander Forsyth patented the percussion system in 1807; this system still operated in a muzzle loader, but the ignition charge was sealed in a percussion cap that the firer placed over a nipple, which was connected to the main charge. When the trigger is pressed a hammer hits the cap, causing a high-pressure explosion that ignites the propellant.

Pistols developed in exactly the same way as muskets, and the caliber of all weapons up to the mid-nineteenth century were large in comparison with today's firearms. This was due to problems in barrel making, for no effective method had been found to make barrel steel capable of withstanding the propellant explosion except by thickness of the barrel; further, gunpowder was so weak in its explosive effect that large amounts were needed to propel the ball down the barrel and on to the target. This insoluble circle was broken with the invention of nitrocellulose powders in the nineteenth century, together with Whitworths standards and gauges and the invention of the cartridge and rifling.

The rate of fire of standard military firearms did not increase fundamentally until the same time. In 1812 Dean Pauly invented the first cartridge-loaded breech-loading rifle, then Emile Lefauchaux invented the pin-fire cartridge. At last the igniter, the propellant, and the bullet were combined into one unit. This was of fundamental significance to Samuel Colt when he invented the first repeating firearm, which enjoys great status today as the first true modern firearm.

So, by the second half of the nineteenth century, firearms were beginning to operate in the same way as they do today. The breech-loaded weapons were firing a unitary cartridge, and the barrels were rifled, enabling sights to become things of accuracy rather than indicators of the possible final resting place of the projectile.

The next development was even more significant: single-shot rifles were standard issue until the bolt action was invented. Single-shot weapons were loaded by means of rolling blocks, falling blocks, or lever or breaking action. This meant that only a slow rate of fire was possible, although this was a much higher rate than that ever achievable by the muzzle loader. The bolt action, however, allowed the rifleman to reload his weapon by means of a mechanical device that would eject the spent cartridge from his last shot and load a fresh cartridge into the chamber from the magazine. In no time at all military rifles were capable of firing from five to fifteen aimed rounds per minute. That this was effective is shown by the German conclusion that the British army at Mons in 1914 must have had a lot of machine guns, so dense and effective was the fire directed at them. In fact British battalions at the time had only two machine guns; it was the British standard of 15 aimed rounds per minute per man that caused the misunderstanding.

Machine guns were another logical step; magazines provided rifles with increased rates of fire, but a continuous belt of ammunition and an automatic reloading mechanism brought about the next step. By the end of World War I it was well known that artillery and machine-gun fire had killed a generation by their effectiveness on the battlefield. A further development in World War I was the use of the shotgun firing solid shot for trench clearing, a use now continued by special forces for opening doors.

One of the earliest firearms that has seen ubiquitous use is the pistol. The small, often concealed, hand weapon is for local protection only, but it has a long history. Its technical development has mirrored that of the rifle to a very great extent, and so flintlock pistols marched side by side with flintlock muskets. By the early part of the twentieth century there were automatic pistols, however, that were well in advance of rifle technology of the time. It is an interesting fact that the machine gun also appeared at about the same time, yet semiautomatic and automatic rifles did not make a practical appearance until the era of the American Garand rifle and the Springfield carbine. World War II also saw the Germans issuing (in small numbers only) a very effective assault rifle that was semiautomatic. The reasoning behind the apparent failings of the semiautomatic rifle was simple: Infantrymen could not be trusted with such a weapon, and would fire all their ammunition ineffectively. There has always been a certain amount of truth in this argument, and many television images show that the opportunity to make a lot of noise with an automatic rifle is very tempting, even to Westernized infantry, especially if it can be done without showing any body parts to the enemy.

One aspect of the firearm that leads the way in development is the cartridge. Since all-in-one cartridges were first invented, the caliber of the military round has diminished considerably. Snider-rifled muskets fired rounds of caliber 0.577 inch, the Martini rifle fired 0.45 inch. World War I saw weapons of calibers around 8 mm or about 0.3 inch. Nowadays the common cartridge is 5.56 mm, a caliber developed in America and now in common use by many European armies. The cartridge is small, the round seems almost too small, but the muzzle velocity is such that a single round will do just as much damage as the bigger .30 caliber of World War II.

Firearms development underwent a radical change in the twentieth century with the appearance of the submachine gun. This weapon was an offshoot of both pistol and machine-gun technology, and by 1930 formed a separate branch of firearm technology. One of the most famous sMGs was the Thompson, made in America, and favored supposedly by Chicago gangsters. Although bulky, it fired the 0.45-inch automatic cartridge pistol (ACP) round, which was undoubtedly a man stopper. Developments in Europe, however, settled on the alternative pistol round, the 9 mm parabellum.

Submachine guns have two advantages in warfare: They are light and easy to carry, and they are extremely valuable in close-quarter battle. The Germans issued many, of which the most famous were the MPs 38 and 40; the Americans stuck to 0.45-inch caliber with the slimmed-down Thompson (with a 20-round box magazine) and the truly utility sMG, the M3. The Russians, however, always willing to learn from the enemy, adopted the sMG as their own, issuing millions to infantry, cavalry, tank crews, vehicle crews, and rear area troops.

The combination of pistol, rifle, and machine gun is still available to all armed forces today. Pistols are becoming more rare, however, as are sMGs, for the reason that rifles are becoming smaller. Modern rifles are built on the bull-pup principle; this means that the magazine is behind the pistol grip of the weapon. This in turn means that the barrel length of the weapon is not severely shortened, but the overall length of the weapon is reduced.

The modern rifle is supported by the machine gun still; tactical field requirements are such that there is still a need for a rapid-firing weapon that will keep the enemy's heads down while the riflemen maneuver for assault positions. The machine gun itself is the subject of much debate. In Vietnam, although riflemen were issued with the M16 in 5.56 mm caliber, their machine gunners still carried the M60 machine gun in 7.62 mm caliber. Again, in Europe, although the British army has the light support weapon in 5.56 mm, it has not completely relinquished the general-purpose machine gun, also in 7.62 mm.

The ability of infantry to carry large amounts of ammunition has always been a military goal, but questions seem to arise as to whether the 5.56 mm cartridge is sufficient to the general infantry task. It wounds and kills very effectively, but it is an extremely unmilitary round if it meets obstacles on its way to the target. The 7.62 mm NATO round will punch through a wall and still be lethal on the other side; the 5.56 mm round, however, has a tendency to disintegrate or deflect if it catches anything of substance on its way to the target.

But firearms are not just handguns. Artillery figures significantly in the field, as does rocketry. Artillery is merely a larger-caliber weapon, firing a variety of shells according to the task. Rocketry often has even greater range, but it has the same role as gun artillery.

Military firearms have always been influenced to a degree by the civilian products, but this is now less marked than before. The civilian firearm for hunting still demonstrates the classic arrangement of stock, receiver, and barrel, with often elegant furniture and delicate scrollwork decoration. Military firearms, however, are more functional, and recognize the need for ease of handling within vehicles, for troops are nowadays delivered to the battlefield in armored personnel carriers, from which they may have to fight. Further, military firearms need to be able to produce a heavier weight of fire, and so are often automatic or burst-firing weapons.

The firearm of the future will undoubtedly benefit from firing consumable cartridges—cartridges that are burned up in the firing process, removing the need for extraction.

David Westwood

See also: Artillery; Machine Gun; Mortars

References and further reading:

Allen, Major W. G. B. *Pistols, Rifles and Machineguns.* London: EUP, 1953.

Ezell, Edward C. *Small Arms of the World: A Basic Manual of Small Arms.* Harrisburg, PA: Stackpole; London: Guns and Armor Press, 1983.

Jane's Infantry Weapons. London: Jane's, 1975.

Fleurus, Battle of (26 June 1794)

French defeat of an allied army that contributed to the end of the Reign of Terror. In the spring of 1794, as the domestic terror accelerated, members of the antirevolutionary coalition prepared to invade France from the east and northeast. Foreseeing this, the French prepared drives of their own toward the Rhineland and into Belgium.

The goal of Jean Baptiste Jourdan, commander of the 75,000-man Army of Sambre and Meuse, was to cross the Sambre River and seize Charleroi, which they captured on 25 June. An Austro-Dutch force of 52,000, commanded by the prince of Saxe-Coburg, moved against Jourdan. Believing himself outnumbered, on 25 June Jourdan entrenched in a large arc to cover Charleroi from the north.

Coburg actually began the battle about 3:00 A.M. on 26 June. It evolved into a series of relatively disconnected engagements. On the French left Jean-Baptiste Kléber held the allies and in the afternoon drove them back. The French right yielded, and the Austrian archduke Charles and Johann Beaulieu almost succeeded until stopped by François Lefebvre. In the center Jourdan began a retreat but then counterattacked and drove the Austrians back. The battle ended about 7:00 P.M. The French had suffered approximately 7,000 casualties, the allies 10,000. The Austrians, who could have continued the battle the next day, instead withdrew. A factor in the French victory was the use of an artillery observation balloon, the first use of aviation in warfare.

By the end of September the French had occupied Belgium, conquered Holland, crossed the Rhine, and were besieging Mainz. No threat of invasion from this quarter existed again until 1814.

Politically Fleurus had even greater consequences. By early summer of 1794 many French had wearied of the domestic terror. Its rationale of defeating the enemy at home was no longer credible. Now Fleurus shattered the enemy abroad and destroyed the military rationale. On 27 July (9 Thermidor) Maximillian Robespierre was overthrown and the terror declined.

Fleurus was the military fulcrum for this change in French politics.

James K. Kieswetter

See also: French Revolutionary Wars; Kléber, Jean-Baptiste

References and further reading:

Dupuis, Victor. *Les Opérations Militaires sur la Sambre en 1794: La Bataille de Fleurus.* Paris: R. Chapelot, 1907.

Lynn, John A. *The Bayonets of the Republic: Motivation and Tactics in the Army of Revolutionary France, 1791–1794.* Chicago: University of Illinois Press, 1984.

Phipps, Ramsay Weston. *The Armies of the First French Republic and the Rise of the Marshals of Napoleon I.* 5 vols. Oxford, UK: Oxford University Press, 1926–1939.

Flipper, Henry Ossian (21 March 1856–3 May 1940)

First African-American graduate of the U.S. Military Academy at West Point. Although other African Americans had previously entered West Point, he was the first to complete his studies. The son of slaves, Flipper was educated in American Missionary Association schools after the Civil War. He received an appointment to West Point in 1873. Despite racism from other cadets, Flipper remained dignified and hardworking. He graduated fiftieth in a class of 64 on 14 June 1877.

Flipper was commissioned a second lieutenant in the

Tenth Cavalry, a black unit. He served in various frontier posts in Texas and the Indian Territory. Most of his time was spent on engineering projects, such as drainage of swamps and installation of telegraph lines. He earned a commendation for service in the 1880 campaign against the Apache Victorio.

In November 1881, Flipper was accused by Colonel William R. Shafter of embezzling commissary funds. A court-martial found Flipper innocent, but he was dismissed from the service on 30 June 1882 for conduct unbecoming an officer. He believed racism was the real cause of his dismissal and spent the remainder of his life trying to prove his innocence and seeking readmission to the army.

Flipper worked as an engineer in the West. He developed a fluency in Spanish and a knowledge of local land use and mining laws. His translations of Mexican and Venezuelan mining laws were regarded as standards for many years. During the 1920s, Flipper was an assistant to the secretary of the Interior. After his death, Flipper was granted an honorable discharge by the army in 1976. A bust of him was unveiled at West Point in 1977, and an award in his name was instituted to honor the cadet who best exemplifies the high personal standards displayed by Flipper.

Tim J. Watts

References and further reading:

Black, Lowell D. *An Officer and a Gentleman: The Military Career of Lieutenant Henry O. Flipper.* Dayton, OH: Lora Co., 1985.

Flipper, Henry Ossian. *The Colored Cadet at West Point.* Lincoln: University of Nebraska Press, 1998.

Robinson, Charles M. *The Court-Martial of Lieutenant Henry Flipper.* El Paso, TX: Western Press, 1994.

Flodden, Battle of (9 September 1513)

Disastrous early sixteenth-century defeat for the Scots at the hands of the English. By August 1513, with Henry VIII engaged in war with France, King James IV of Scotland had seized control of the region south of the Rivers Tweed and Till from the English. Henry's commander in the north, Thomas Howard, Earl of Surrey, led an English army into the region to keep James from retreating back into Scotland. Surrey engaged in diplomatic exchanges with James that ended on 6 September when James agreed to a battle at noon three days later. Surrey's main army crossed the River Till, five miles north of the Scottish encampment, on the morning of 9 September, cutting off James's retreat. James's army settled at the edge of Branxton Hill near Flodden Field, maintaining a strategic advantage.

The brash James immediately engaged the English, and an artillery duel ensued. The Scottish army failed to find an effective range. After four hours of being bombarded by gunstones and heavy rain, the Scots advanced. The Scottish 17.5-foot pikes and long swords were no match for the English 9-foot halberds or bills, which combined an axe-head and spearhead. When night fell, the battle came to a merciful end for the Scots. The English army lost about 1,700 of its near 26,000 troops. The Scottish army lost more than 5,000 of its 28,000-plus troops, including most of its nobility, and King James himself.

Christopher P. Goedert

See also: Anglo-Scots Wars (1513–1560)

References and further reading:

Kightly, Charles. *Flodden: The Anglo-Scottish War of 1513.* London: Almark Publishing Co. Ltd., 1975.

Mackenzie, W. M. *The Secret of Flodden.* Edinburgh: Grant & Murray, 1931.

Mackie, R. L. *King James IV of Scotland.* Edinburgh: Oliver and Boyd, 1958.

Foch, Ferdinand (1851–1929)

Marshal of France and World War I commander. Born in Tarbes, southwestern France, on 2 October 1851, Ferdinand Foch interrupted his education during the Franco-Prussian War to enlist in the army but did not see action and entered the École Polytechnique in 1871. Three years later he was commissioned a lieutenant in the artillery (1883). Foch attended the War College (1885–1887) and was assigned to the general staff.

In 1894 Foch taught at the War College, where he stressed the importance of the commander's will and the offensive. Promoted to brigadier general (1907), he then headed the War College (1908). Promoted to major general (1912), he commanded a division. He then commanded XX Corps (1913).

Foch distinguished himself early in World War I in the Battle of the Frontiers (August 1914), and French army commander General Joseph Joffre assigned him command of the new Ninth Army, which held a crucial sector during the Battle of the Marne (September). Joffre then appointed Foch his deputy commander in chief to coordinate operations in the northern wing of armies (October). He next commanded Northern Army Group (January 1915).

When Joffre was dismissed in favor of General Robert Nivelle (December 1916), Foch was transferred to an unimportant advisory post. But when General Henri Pétain became commander of the French army (May 1917), Foch returned to prominence as chief of the general staff, although command authority rested with Pétain. Premier Georges Clemenceau then appointed Foch to the military committee of the new Supreme Allied War Council (November 1917).

Foch's opportunity came in spring 1918 when the Allies, close to defeat during the Ludendorff Offensive, adopted unity of command. Clemenceau selected Foch to command the Allied armies, largely for his fighting spirit. Originally, he had power only to coordinate Allied operations, but soon he gained full control, and on 14 April he was named commander in chief of the Allied Armies, first on the western front, then over all Allied military operations.

As generalissimo of the Allied armies, Foch directed the defeat of the German offensive. When the German drive had spent itself, on 18 July he launched his own succession of offensives over the whole of the front to give the Germans no time to regroup. This continued until victory. In gratitude for that achievement, Clemenceau awarded Foch the rank of marshal of France in August 1918.

Foch headed up the Allied side in armistice negotiations with the Germans in early November, insisting on terms that would make it impossible for Germany to renew the war. During the 1919 Paris Peace Conference, he advocated a tough stance toward Germany. He later denounced the Treaty of Versailles, predicting that the Germans would invade France again within a generation.

Foch was elected to the Académie Française and was the only French general named a British field marshal. He also was made a field marshal of Poland. He died in Paris on 20 March 1929.

Spencer C. Tucker

See also: French Army; Joffre, Joseph Jacques Césaire; Pétain, Henri-Philippe; World War I

References and further reading:

Foch, Ferdinand. *The Memoirs of Marshal Foch.* Trans. T. Bentley Mott. Garden City, NY: Doubleday, Doran, 1931.

Hunter, T. M. *Marshal Foch: A Study in Leadership.* Ottawa: Directorate of Military Training, Army Headquarters, 1961.

King, Jere Clemens. *Foch versus Clemenceau: France and German Dismemberment, 1918–1919.* Cambridge, MA: Harvard University Press, 1960.

———. *Generals and Politicians: Conflict between France's High Command, Parliament, and Government, 1914–1918.* Berkeley: University of California Press, 1951.

Marshall-Cornwall, James H. *Foch as Military Commander.* New York: Batsford, 1972.

Fontenoy (1745)

Major engagement on 11 May 1745 between the French and a coalition of English, Hanoverian, Dutch, and Austrian forces during the War of Austrian Succession (1740–1748). The French, under the command of Maurice, Comte de Saxe, with about 50,000 men, had laid siege to the fortified city of Tournai in the hope of luring the Allied army into a position where it could be successfully attacked. Marching to relieve the siege was the second son of King George II of England, the duke of Cumberland, with approximately 46,000 Allied troops. Responding to Cumberland's approach, Saxe placed his forces on a low plateau near the present-day Belgian town of Fontenoy. In addition, the experienced Saxe ordered defensive redoubts to be constructed along his lines.

To begin the battle, the inexperienced Cumberland chose a frontal attack. During this assault, the Allied forces were severely mauled by the French-held redoubts, greatly depleting Cumberland's available force. Although the initial Allied attack was able to break the French line, it did so at great cost. Furthermore, the Allied cavalry failed to appear in order to capitalize on the infantry's hard-earned success. While French cavalry squadrons maintained pressure on the Allied formations, Saxe reorganized his battered forces and launched a counterattack spearheaded by a great concentrated artillery barrage and supported by fresh troops and cavalry. In the end, the Allied forces under Cumberland were defeated, illustrating the advisability of a well-positioned defensive line, superior artillery, and the successful use of field fortifications and light infantry. The defeat of the Allied forces at Fontenoy allowed Saxe and the French to seize the key cities of Flanders—including Ghent, Antwerp, Brussels, Mons, and Namur.

Andrew G. Wilson

See also: Austrian Succession, War of the

References and further reading:

Boudet, Jacques. "Fontenoy." In *Great Military Battles,* ed. Cyril Falls. New York: Macmillan Company, 1964.

Weigley, Russell F. *The Age of Battles: The Quest for Decisive Warfare from Breitenfeld to Waterloo.* Bloomington: Indiana University Press, 1991.

Fontenoy en Puisaye, Battle of (France, 25 June 841)

Carolingian Empire permanently divided. Charlemagne's heir, Louis the Pious (814–840), had three sons. To avoid a succession dispute, Lothair was to receive the title of emperor, while Louis and Pepin (later, Pepin's son Charles) were to become kings. A civil war quickly erupted between the three brothers while Louis the Pious was still alive, and intensified after his death.

Lothair's plan was to defeat the forces of his brothers separately, but he forced them into an alliance against him instead. His forces now outnumbered, Lothair had to recognize his nephew, Pepin, as independent ruler of Aquitaine to gain support, and, with Pepin's help, field an army powerful enough to defeat his two rivals. Marching to join forces with Pepin, who was advancing north from the Loire, Lothair was met by Louis and Charles. They offered battle in the Yonne Valley, near Fontenoy.

Charles's Franks and Louis's Saxons and Bavarians stood on a hill facing Lothair. Little else is known about the battle, but reinforcements of Burgundian troops under Count Warin may have helped repulse and then slaughter Lothair's right wing. Lothair's army was routed. Lothair fled, and Pepin returned to Aquitaine, bringing the civil war to an end.

Outstanding issues were settled in the Strasbourg Oath of 842, between Louis and Charles, and in the Treaty of Verdun of 843. The treaty officially divided the territories of Louis the Pious into three parts: Lothair was given Italy, Burgundy, and Lotharingia (a strip of land extending from the Swiss mountains to the North Sea, along the Rhine valley); Louis became king of Eastern Francia, the later Germany, and Charles of Western Francia, the later France.

Gilles Boué

See also: Charlemagne; Frankish Civil Wars

References and further reading:

Halsall, Guy, ed. *Violence and Society in the Early Medieval West.* Rochester, NY: Boydell Press, 1998.

Heath, Ian. *Armies of the Dark Ages.* Worthing, UK: WRG Publication, 1980.

Forrest, Nathan Bedford (1821–1877)

The most successful Confederate cavalry field commander in the western theater of the American Civil War. Forrest was born in Chapel Hill, Bedford County, Tennessee, on 13 July 1821. A self-made millionaire through planting, land deals, and slave trading, he enlisted as a cavalry private under Josiah H. White on 14 June 1861. Although Forrest had no previous military experience or training, Tennessee governor Isham Green Harris commissioned him lieutenant colonel in October and requested him to create a cavalry battalion. Forrest did this at his own expense.

Forrest's natural aptitude for tactics soon became apparent in his daring escape from Ulysses S. Grant at Fort Donelson on the night of 15–16 February 1862. He was wounded while commanding the rear guard on the retreat from Shiloh, Tennessee, to Corinth, Mississippi, in April. After his promotion to brigadier general on 21 July, his service was mostly as a raider. His attacks on federal supplies and communication lines gained territory in Tennessee, prevented a summer assault on Chattanooga, and stifled Grant's initial advance toward Vicksburg, Mississippi. He fought under Braxton Bragg at Murfreesboro and Chickamauga.

Promoted to major general on 4 December, he thereafter commanded all Confederate cavalry in the west. After capturing Fort Pillow, Tennessee, on 12 April 1864, and finding the surrendered garrison about half white and half black, he massacred in cold blood about 260 troops of the 11th Colored Troops and the 4th Colored Light Artillery.

One of Forrest's favorite tactics was to use cavalry as mounted infantry, thus combining the speed and mobility of horse with the power of foot. Although outnumbered two to one by Samuel D. Sturgis at Brice's Cross Roads, Mississippi, on 10 June, his dismounted soldiers achieved a decisive victory.

Forrest harassed William T. Sherman for most of the rest of the war. Promoted to lieutenant general on 28 February 1865, but outnumbered, he was defeated at Selma, Alabama, in April and surrendered in May.

After the war, Forrest helped to create the Ku Klux Klan and may have been its first Grand Wizard. He died in Memphis, Tennessee, on 29 October 1877.

Eric v. d. Luft

See also: American Civil War; Bragg, Braxton; Chickamauga, Battle of; Fort Donelson; Grant, Ulysses Simpson; Hood, John Bell; Murfreesboro; Sherman, William Tecumseh; Shiloh

References and further reading:

Hurst, Jack. *Nathan Bedford Forrest: A Biography.* New York: Vintage, 1994.

Jordan, Thomas, and J. P. Pryor. *The Campaigns of Lieut.-Gen. N. B. Forrest and of Forrest's Cavalry.* New York: Da Capo, 1996.

Maness, Lonnie E. *An Untutored Genius: The Military Career of General Nathan Bedford Forrest.* Oxford, MS: Guild Bindery, 1990.

Wills, Brian Steel. *The Confederacy's Greatest Cavalryman: Nathan Bedford Forrest.* Lawrence: University Press of Kansas, 1992.

Fort Donelson (11–16 February 1862)

The military operation that first brought U. S. Grant to public attention. During the winter of 1861–1862, Confederate general Albert Sidney Johnston had established a defensive line stretching across neutral Kentucky from the Appalachians to the Mississippi River; he had only about 45,000 men to maintain the cordon. One key point was at Fort Henry on the Tennessee and Fort Donelson on the Cumberland River in western Tennessee not far from the Ohio River.

Union general Ulysses S. Grant, working well with his naval counterpart, Flag Officer Andrew Foote, and his flotilla, began his offensive in early February. On 6 February, most of the defending Confederates abandoned Fort Henry (it sat on a poorly chosen site for defense—easily flooded and easily bombarded from higher ground across the river) and retreated to Fort Donelson. Several days later, Grant sur-

Battle of Fort Donelson: capture of General S. B. Buckner and his army, 16 February 1862. (Library of Congress)

rounded the fort, but Foote's flotilla could not bombard the fort into submission.

The Confederate commanders decided to break out, attacking south, to roll up Grant's besiegers and either threaten him or open the way to retreat safely. The unexpected attack worked well, but Grant's coolness under fire—he responded to an attack on his right by ordering an attack from his left against Confederate defenses—perhaps caused his opponents to lose their nerve. The Confederates retreated into the fort with its outer defenses compromised, and the two senior commanders fled. When Simon Buckner asked Grant for his surrender terms, Grant replied that "no terms except an unconditional and immediate surrender can be accepted" and at least for the moment a northern hero was born—"Unconditional Surrender" Grant.

Defeat at Henry and Donelson caused Johnston to concede Kentucky and northern Tennessee and to regroup further south, seriously compromising the Confederate position in this western theater of war.

Charles M. Dobbs

See also: American Civil War; Grant, Ulysses Simpson; Johnston, Albert Sidney

References and further reading:

Cooling, Benjamin Franklin. *Forts Henry and Donelson: The Key to the Confederate Heartland.* Knoxville: University of Tennessee Press, 1987.

Williams, Kenneth P. *Grant Rises in the West: The First Year, 1861–1862.* Lincoln: University of Nebraska Press, 1997.

Fort Duquesne, Seizure of (1758)

A significant British military success in the Seven Years' War (French and Indian War). In 1758, British prime minister William Pitt charged Brigadier General John Forbes with an objective that had eluded General Edward Braddock several years earlier. He was ordered to seize Fort Duquesne, and thereby gain control of the upper Ohio River valley and cut French Canada from French Louisiana. Forbes initially intended to follow the same route of advance as Braddock, but decided to move westerly through Pennsylvania and therefore to have to hack yet another route through the wilderness. The advance went slowly.

By early September, the British army—about 25 percent regular troops and about 75 percent colonial militia—had moved to within 40 miles of Fort Duquesne. Several weeks later, an advance party ineptly attacked the fort, and most of the troops were captured. After a pause, Forbes decided to attack in early November. He disregarded advice of some subordinates who counseled going into winter quarters, and he ignored his own illness, which would kill him soon after he seized the fort. The French commander had sent away his Canadian militia and Native American allies because he did not have food and supplies for all of them for winter, and he assumed the British would retire to winter quarters as well. Faced with a larger foe, he decided to save his army, evacuated the fort, and retreated north toward the lakes and Canada. Control over the upper Ohio valley passed to the British, furthering the isolation of the French in Quebec.

Charles M. Dobbs

See also: French and Indian War

References and further reading:

Collier, Christopher, and James Lincoln Collier. *The French and Indian War: 1660–1763.* New York: Benchmark Books, 1998.

Fregault, Guy. *Canada: The War of Conquest.* Trans. Margaret M. Cameron. Toronto: Oxford University Press, 1969.

Fort Sumter (12–14 April 1861)

Scene of the opening shots of the American Civil War. In the aftermath of the secession of seven Deep South states from the Union, most federal facilities in the South went over to the new Confederate States of America. One very visible facility held out—Fort Sumter in the middle of Charleston, South Carolina, harbor.

It was an intriguing situation. If the Confederate States of America could not gain control over the fort, it could hardly maintain its claim to be an independent nation; on the other hand, if President Abraham Lincoln tried to bring in reinforcements, the North might appear the aggressor and thus induce some or all of the remaining eight southern slave states to secede.

Lincoln sent a provisioning ship to Charleston to continue the stalemate. Southern gunners drove the ship away. On 10 April 1861, Brigadier General Pierre G. T. Beauregard called for surrender of the fort; the next day, negotiators rowed to Sumter but failed to reach agreement on the fort's evacuation. On 12 April at 4:30 A.M., bombardment of the fort began; by midafternoon the next day Major Robert Anderson surrendered, being unable to resist the Confederate batteries along the harbor. Union forces evacuated the fort the next day.

As a result, Lincoln was able to argue that the South had attacked first; he kept four of the remaining eight slave states in the Union. The South, in celebrating its victory, may have drawn the conclusion that the war would be too easily won and the defense too easily conducted.

Regardless, Fort Sumter was a key to an impressive arrangement of forts and batteries protecting Charleston harbor that helped keep the port under Confederate control until February 1865.

Charles M. Dobbs

See also: American Civil War; Anaconda Plan; Lincoln, Abraham

References and further reading:

Meredith, Roy. *Storm over Sumter.* New York: Simon & Schuster, 1957.

Swanberg, W. A. *First Blood: The Story of Fort Sumter.* New York: Scribner, 1957.

Fort Ticonderoga

A fortress guarding the traditional invasion route either to or from Canada and the southern colonies and the United States.

To help protect approaches to the lakes leading to Montreal and the heart of Quebec colony, the French in 1755 constructed a fort 10 miles south of Crown Point by Lake George and called it by its Indian name of Ticonderoga. It was to serve as a Gibraltar of North America, and the French initially and the British during the American Revolution invested greatly in this fortress.

As part of the broader offensive of 1758 during the French and Indian War, Major General James Abercromby with a mixed force of regulars, militia, and some colonial rangers moved to attack Ticonderoga. On the night of 5–6 July, they landed on the west side of Lake George a few miles from their objective. Meanwhile, the Marquis de Montcalm had sent some reinforcements and supplies to the fort, and the French sought to strengthen defenses. An early battle on 6 July between advance forces on both sides led to high casualties and lowered morale among the attackers. Abercromby believed reports of large numbers of defenders—doubtful given the size of the fort—and left his artillery behind, which would have permitted him to engage in a punishing siege. Instead, on 8 July he attacked, and many intrepid troops died as they sought to climb over the barriers the French had constructed on the fort's approaches. The French assumed another attack the next day, but when the sun rose,

Capture of Fort Ticonderoga: Ethan Allen and Captain de la Place. (National Archives)

the attackers had left and Ticonderoga survived in French hands.

In July 1759, Major General Jeffrey Amherst renewed the offensive. He brought cannon and prepared for a siege. Unknown to him, the French were prepared to retreat to an inner defense ring to protect Montreal and the colony's heart. So the several hundred defenders blew up the stone fort and retreated north to Fort Saint Frederic. Amherst then spent two months rebuilding the stone fortress before continuing the offensive.

During the siege of Boston at the opening of the American Revolution, Americans had attacked and seized Fort Ticonderoga on 10 May 1775, capturing some 60 cannon, which they laboriously transported to Boston to strengthen their position. The introduction of these cannon caused British general Thomas Gage to withdraw from Boston to Halifax, Nova Scotia. And in the aftermath of the failed invasion of Quebec, the retreating Americans returned to Ticonderoga to recuperate, conceding Canada thereafter to the British.

Finally, in 1777 the British under Major General John Burgoyne once again seized Ticonderoga. Having constructed a small fleet for passage down the lakes, the British took Ticonderoga and then stumbled as they approached Saratoga, leading eventually to Burgoyne's surrender and a turning point in the conflict. This was the end of Fort Ticonderoga's crowded military history.

Charles M. Dobbs

See also: American Revolution; Burgoyne, John; French and Indian War; Montcalm-Gozon, Louis-Joseph de, Marquis de

References and further reading:

Bird, Harrison. *March to Saratoga: General Burgoyne and the American Campaign, 1977.* New York: Oxford University Press, 1963.

Furneaux, Rupert. *The Seven Years War.* London: Hart-Davis MacGibbon, 1973.

———. *The Battle of Saratoga.* New York: Stein & Day, 1971.

Peckham, Howard H. *The Colonial Wars, 1689–1762.* Chicago: University of Chicago Press, 1964.

France (1940)

One of the strangest defeats of modern times. When France mobilized in September 1939, it fielded some 5 million soldiers, 2.7 million of whom were stationed on the front lines. However, budgetary restrictions caused in part by construction of the Maginot line led to a limited supply of infantry rifles and antiaircraft. On the other hand, French tanks outnumbered and even outclassed German tanks in the northern area, but lacked speed. Hence began what became known as the Phoney War (Drole de Guerre) characterized by sporadic air incursions and artillery fire between the Maginot line and the so-called Siegfried line. The Saar offensive of September 1939 had ended without any follow-up, thus leaving German troops to focus on Poland's demise. During that time, seeking to reinforce Allied commitments, France and Great Britain signed an agreement stipulating that neither side would seek a separate peace without consultation.

Heading the French general staff was the aging and monkish Maurice Gamelin, isolated in his headquarters, without even radio or wire communications, with his subordinate commanders. His instructions to his commanders remained vague and frustrated any initiative for action. Furthermore, he and the chief of the French government, Paul Reynaud, disagreed on what actions to take to check the German threat. This eventually led to the fall of the Daladier cabinet shortly before the German offensive.

Hence, the German offensive on 10 May 1940 took the Allies by surprise, as usual. Instead of invading through Belgium, which the French general staff was prepared to counter with Escaut and Dyle-Breda lines, the German army broke through the Ardennes. This rough mountainous and heavily wooded terrain had been deemed impenetrable by both sides, which is why the French Ninth Army, composed of mostly aged reservists, defended the northern sector. The French Second Army, covering the eastern area, failed to send reinforcements. Within days, General Alphonse-Joseph Georges resigned under the mental strain of the war. Paul Reynaud was recalled as prime minister. For several days, the high command showed no sign of decision, although Gamelin did offer a counteroffensive plan. On 19 May he was replaced by Maxime Weygand, who offered nothing new in terms of strategy. The Belgian army capitulated, and British troops, along with some French units, held off the Germans at Dunkerque long enough to board ships under intense German aerial attack and leave the battle. Furthermore, as conditions changed and German invaders cut deeper into French positions, the French high command was often unable to dispatch new orders due to the lack of radio communication with field commanders. The German blitzkrieg (lightning war) made it a point to attack communications and mobilization points by air and armor, making it almost impossible for the French high command to mount a coordinated counterattack.

As for the French air force, although offering impressive numbers on paper, it struggled to make up for delayed aircraft deliveries. It thus found itself unable to counter such threats as the tactics of German Stuka dive bombers. Many French pilots did successfully engage and shoot down German fighters, but the exact number is hard to determine. Indeed, the official number of over 1,000 is likely a myth due to a combination of victory confirmation methods (confirmed participants in a joint victory were each credited with a full victory), the loss of records, and the spread of the strange notion of "defeat with honor" through the Vichy propaganda services as a means to motivate the remnants of a demoralized army.

On 10 June, Italy opportunistically launched its own attack against France. Even with the French *in extremis* militarily, Italian border penetrations were held to a few miles.

Weygand asked for permission to break the 28 March 1940 Franco-British agreement, which forbade negotiating a separate peace with Germany. He was overruled and refused to consider continuing fighting from North Africa. On 16 June, Paul Reynaud resigned and retired. The defeatist Marshall Philippe Pétain, the aged hero of Verdun, replaced him and wasted no time in calling on Germany for armistice terms. The armistice was signed on 22 June in the same railway sleeping car in which the Germans were forced to sign their armistice in 1918—a Hitler touch—and went into effect on 25 June. By then 120,000 French soldiers had died, another 250,000 were wounded (remarkably high numbers considering the briefness of the campaign), and close to 1.5 million were prisoners of war.

Meanwhile, General Charles de Gaulle, in exile in London, issued a call to resistance on 18 June that signaled a split of the French army between the future Vichy government forces and those that would join the Allies, some at once, others only in 1942.

On 10 July 1940, the French Third Republic was dissolved and a new government headed by Pétain, headquartered at Vichy and in control of southern France (also known as the Free Zone), began operations. Vichy was allowed to continue operating a small defense force of some 100,000 soldiers and 60,000 sailors, along with a small air force to help control French colonies. The French themselves described the armistice terms as "hard, but not unbearable." France was not dismembered, although Alsace and Lorraine went back to Germany, and, as noted, Vichy could keep a small defense force. But Hitler vowed privately that after the war he would "take care of France."

Guillaume de Syon

See also: Armored Fighting Vehicles; French Army; Gamelin, Maurice; German Army; World War II

References and further reading:

Bankwitz, P. C. F. *Maxime Weygand and Civil-Military Relations in Modern France.* Cambridge, MA: Harvard University Press, 1967.

Gunsburg, J. A. *Divided and Conquered.* Westport, CT: Greenwood Press, 1979.

May, Ernest R. *Hitler's Conquest of France.* New York: Hill & Wang, 2000.

Young, Robert J. *In Command of France.* Cambridge, MA: Harvard University Press, 1978.

France and the American Revolution

Next to George Washington, the single most valuable asset of the Americans during the Revolutionary War. It is difficult to see how the United States could have won its independence without the help of France.

After its defeat in the Seven Years' War in 1763, France looked for revanche against Great Britain, and it carefully followed events of the American Revolution. The French foreign minister, the Comte de Vergennes, believed British power was based on trade and the loss of the American colonies would serious weaken France's longtime foe.

At first, France carefully sought a balance between giving the rebellious Americans enough aid to keep fighting and not provoking the British into an attack on France or its remaining overseas possessions. Through the playwright Baron de Beaumarchais, the French arranged for clothing, arms, powder, and medicine for the new republic. Vergennes also permitted American privateers, which preyed on British commerce, to use French ports. And the large French fleet forced Britain to keep part of its fleet in home waters for defense of the British Isles.

After General John Burgoyne surrendered at Saratoga and in the wake of Washington's generally impressive performance in defeat at Germantown outside Philadelphia, Vergennes urged his king to enter into an outright alliance with the new United States to expand the war of attrition against Britain.

After Saratoga, the Franco-American alliance (and the later entry of Spain and the Netherlands into the conflict) ensured an American victory over Great Britain. The drain on British imperial resources was too great.

There were, to be sure, awkward moments in the alliance. The French did not necessarily respect their American allies. There also were difficulties in coordinating the movement of American troops across a largely primitive and undeveloped terrain with that of French naval forces fearing changes in the seasons and the sudden appearance of British fleets. Thus, a planned counterattack to retake Savannah was unsuccessful, and George Washington was never able to use French naval power to bottle up the British in New York and retake the city.

However, there was one moment of supreme triumph. In fall 1781, Lord Charles Cornwallis retreated down the Virginia tidewater to the York River, where he waited for clear orders and/or the appearance of a British fleet to transport his men elsewhere. A French fleet drove off a British relief force; another French naval force helped transport French troops and a siege train from around New York City; and the Americans marched from New Jersey to Virginia. The result was the classic eighteenth-century European-style siege of Yorktown, the successful break through British defenses, and the subsequent British surrender, which ended fighting in North America and led to the Treaty of Paris in 1783, proclaiming the new United States of America. French sea power, along with French gold and French troops, had given the margin of victory.

Charles M. Dobbs

See also: American Revolution; Cornwallis, Sir Charles; Rochambeau, Jean-Baptiste Donatien de Vimeur, Comte de; Washington, George

References and further reading:

Dull, Jonathan. *The French Navy and American Independence.* Princeton, NJ: Princeton University Press, 1975.

Murphy, Orville Theodore. *Charles Gravier, Comte de Vergennes: French Diplomacy in the Age of Revolution, 1719–1787.* Albany: State University of New York, 1982.

Stinchcombe, William C. *The American Revolution and the French Alliance.* Syracuse, NY: Syracuse University Press, 1969.

Franco, Francisco (1892–1975)

Spanish soldier and dictator. Franco ruled Spain for nearly 40 years after defeating the liberals, socialists, and communists in the Spanish Civil War. Although his victory was made possible by the fascist states, he did not join them in World War II.

Franco was born at El Ferrol, the son of a naval officer. He attended the Infantry Academy in Toledo, graduating in 1910. He served against the Riff in Morocco from 1912 to 1916. Franco earned a reputation as a courageous officer and was quickly promoted. After being wounded in 1916, Franco became the youngest major in the Spanish army. He returned to Morocco in 1920 and helped organize the Spanish Foreign Legion. By 1923, he commanded the legion and was a lieutenant colonel. He defied government orders to withdraw from Morocco and led a brilliant assault on Alhucemas Bay in 1925.

Franco was Spain's youngest brigadier general ever at age 33 in 1926. He directed the Military Academy at Zaragosa until 1931, when the Republican government transferred him because of his monarchist sympathies. He served as governor of the Balaeric Islands for three years, then led the legion and Moorish soldiers in repressing a miners' revolt in Asturias. Franco's actions earned him the nickname "Butcher" and the enmity of the Left, but he became the conservatives' hero. After a year as commander in chief of the army, Franco was posted to the Canary Islands by the new Republican government in February 1936.

Franco soon joined other officers in plotting to overthrow the government. He secretly traveled to Morocco, where he led the legion and Moorish troops in revolt on 17 July 1936. Securing his base, Franco used Italian planes to fly large numbers of soldiers to Spain by August. On 29 September, the rebels named Franco generalissimo and head of state. Franco's two fellow conspirators, Generals Sanjurjo and Mola, were killed in convenient airplane crashes, leaving Franco as the sole leader of the revolt. His advance on Madrid was stopped just short of the capital. The Spanish agony continued for another two years, as Spaniards killed some 600,000 of each other. He accepted large amounts of aid from Italy and Germany, but retained control over the war. Franco was a methodical commander. Although he used motorized units, his carefully paced campaigns against his enemies were more reminiscent of World War I. In March 1939, the Republicans surrendered, and Franco became caudillo, or absolute dictator, of Spain. Thousands of opponents were eliminated after the civil war, and Franco allowed no dissent. However, he gradually allowed greater freedoms to the Spanish people.

Franco sympathized with the fascists, but carefully remained neutral during World War II. After the war, the United Nations isolated Spain, but Franco successfully built a reputation as a strong anticommunist. In 1950, the United States reestablished diplomatic relations, and beginning in 1953 provided massive military and economic assistance in exchange for air bases in Spain. Franco promulgated a Law of Succession in 1947, making Spain a monarchy, but reserving the power to name a king to himself. In 1969, he named Don Juan Carlos de Bourbon future king of Spain. Franco died after a long illness in 1975, leaving a Spain that soon moved on to democracy.

Tim J. Watts

See also: Guadalajara; Guernica, Bombing of; Spanish Civil War

References and further reading:

Coles, S. F. A. *Franco of Spain.* London: Neville Spearman, 1955.
Crozier, Brian. *Franco: A Biographical History.* London: Eyre & Spottiswoode, 1967.
Ellwood, Sheelagh. *Franco.* London: Longman, 1994.

Franco-German War (978–980)

King of Francia fights for his independence against Ottonians. Lothar IV (r. 954–986) was dominated first by Hugo the Great, and then by his uncle Brun, archbishop of Cologne and duke of Lotharingia (Loraine). Brun's support was invaluable, but he also used his influence in the interests of his brother, Otto I, "the Great." After Brun's death (965), Lothar's position deteriorated even further. Although his relations with Hugo Capet, the son of Hugo the Great and nephew of Brun, were generally amenable, he had only a tiny domain much distracted by feudal conflict. Also, a persistent desire to win back Lotharingia from the German allegiance was to bring disastrous consequences.

On Otto's death in 973, his son Otto II (955–983) was accepted without opposition as successor; already in 961 he had been crowned coregent king of Germany and Italy with his father. Six years later he was made coregent emperor as well. Although revolts in the realm were to occupy the early years of his reign, Otto was to continue his father's policy of promoting a strong monarchy in Germany, and of extending the Ottonian influence in Italy.

Lothar's backing a revolt in Lotharingia impelled Otto, who was also hampered by a rebellion in Bavaria at the time, to give lower Lotharingia to Lothar's refractory brother, Charles, in exchange for a pledge of fealty (976). A second invasion was mounted, and Lothair pillaged Aix, but advanced no further (978). It was during this campaign that Lothar also made a surprise attack on the imperial palace of Aachen, and nearly captured Otto in person. His troops occupied the palace, and as a symbol of their victory, they turned the bronze eagle fixed to the roof so that it faced eastward. Yet only three days later they withdrew.

Otto immediately responded with a punitive thrust into Francia that reached as far as the gates of Paris, but there it came to a halt. Otto, with an army reckoned at 60,000 strong, was forced to retreat because of a shortage of supplies. Otto's retaliatory attack on Paris satisfied honor, but little more. Lothar's renunciation of Lotharingia at Margut, near Sedan, was no more binding than those his predecessors had made (980). Still, the third invasion of the duchy, despite the sacking of Verdun, not only failed in its purpose, but determined the powerful Archbishop Adalbero of Reims to support Hugo Capet against Lothar (985). The following year Lothar was planning yet another expedition into Lotharingia when he died.

Nic Fields

See also: Holy Roman Empire; Otto I, the "Great"

References and further reading:

Reuter, T. *Germany in the Early Medieval Age, 800–1056.* London: Longman, 1993.

Franco-Prussian War (1870–1871)

The war that made Germany the major European power. As soon as the Austro-Prussian war ended in 1866, Napoleon

III worried about the growing power of Prussia and Bismarck's policy of German unification. Oddly enough, the spark that would start the war began in Spain. The queen had been deposed and the new monarch should have been a relative to the king of Prussia; France couldn't accept the possibility of a two-front war. French diplomacy persuaded King William to withdraw his support but Bismarck reworded a French official dispatch (Ems Telegram) in a way deemed to be insulting to the king. This machiavellian maneuver pushed France to declare war on 15 July 1870.

The German forces were made of Prussian and German state troops (Hanover, Bavaria, and smaller states); 400,000 men in three armies were to invade France following a well-directed plan of mobilization and troop concentration, using the railway net. German high command was given to King William but in fact to his expert chief of staff, Helmuth von Moltke.

The French army seemed to be a fierce opponent, but Napoleon had impetuously declared war before preparations had been completed, and confusion reigned. The eight army corps (225,000) remained behind the frontier without a war plan, transport was unorganized, munitions for the outstanding chassepot rifle (twice the Prussian gun range) were scanty, and units were understrength. Tactics were obsolete and relied on columns attacking with fixed bayonets and heroic cavalry charges à la the first Napoleon. French artillery consisted of muzzle-loaded cannon outranged by the breech-loaded Krupp German pieces. Moltke's carefully planned campaign against France had been drawn up years before and seemed to run like clockwork. The French were beaten at Spicheren and Froeschwiller on 6 August; in both cases every German division marched to battle as French corps waited for orders that never materialized, ensuring their defeat. The two main French armies were defeated one after the other: Bazaine was trapped in Metz after several drawn battles (Mars la Tour, Saint-Privat). Napoleon III's own army was crushed at the battle of Sedan on 1–2 September; the army surrendered and Napoleon himself was taken prisoner.

The capture of Emperor Napoleon was the end of the Second Empire; on 4 September a provisional "government of National Defense" declared France a republic. The German invasion was directed toward Paris but as it penetrated inland, all France flamed up in a patriotic fervor. Paris, already heavily fortified, was hastily garrisoned by untrained newly raised troops. Moltke besieged Paris from 19 September. Republican armies were raised but were no match for the seasoned German troops. Fighting continued through the winter, with pitched battles around Paris and Orleans, but also large-scale skirmishing against German lines of communication. Paris, its citizens almost starving, surrendered on 28 January 1871, a capitulation that put an end to the war.

The main result of the war was the founding of the German Second Empire at Versailles on 18 January 1871. Germany was united under Prussian domination; the king of Prussia became the emperor of Germany. France, humiliated, losing Alsace-Lorraine, burned with a demand for revanche well into the twentieth century.

Gilles Boué

See also: Bismarck, Otto von; Metz, Siege of; Moltke, Graf Helmuth Karl Bernhard von; Sedan; Paris, Siege of

References and further reading:

Bucholz, Arden. *Moltke and the German Wars (1864–1871).* New York: St. Martin's Press, 2000.

De Lonlay, Dick. *Français et Allemands.* Paris: Garnier Frères, 1888.

Roth, François. *La Guerre de 1870.* Paris: Fayard, 1990.

Franco-Spanish War (1648–1659)

War within a war, part of the Thirty Years' War. Actually lasting about 50 years, the various wars were part of the European struggle for balance of power as well as religious freedom and hegemony. In 1648, Spain signed various treaties with the United Provinces of Westphalia that became known as the Peace of Westphalia. With this part of the struggle over, Spain was able to concentrate on its struggle against France. However, with the outbreak of the rebellion in France, known as the Fronde, the Spanish slowly began to lose ground. The Fronde rebellion first came about as a movement in the parliament of Paris against absolute royal authority, later spreading to encompass other areas of society. During this period, many of those in rebellion made separate peace treaties with Spain even as the Fronde took place.

However, in 1655 France began to enjoy success against the Spain. Additionally, France and England made various treaties of friendship from 1656 to 1657, which allowed France to concentrate on its struggle with Spain. On 7 November 1659 France and Spain signed the Peace of the Pyrenees, through which France obtained the city of Roussillon and was able to establish its borders with Spain along the Pyrenees. In effect, the Franco-Spanish War brought France to the forefront as a major power in Europe and in the newly discovered Americas, while Spain lost much of its worldwide hegemony.

Peter Carr

See also: Breda, Siege of; Breitenfeld; Gustavus II Adolphus; Holy Roman Empire; Lützen, Battle of; Magdeburg, Siege of; Maurice of Nassau; Rupert, Prince; Tilly, Johann Tserclaes Graf von; Turenne, Henri de la Tour d'Auvergne, Vicomte de; Wallenstein, Albrecht von

References and further reading:
Asch, Ronald G. *The Thirty Years War: The Holy Roman Empire and Europe, 1618–48.* New York: St. Martin's Press, 1997.
Sanabre Sanromá, José. *La acción de Francia en Cataluña en la pugna por la hegemonía de Europa, 1640–1659.* Barcelona: Distribuidor Librería J. Sala Badal, 1956.

Frankish Civil Wars (670–719)

Carolingians seize power from Merovingians. By the late seventh century, the Frankish Merovingians were in decline. Real power had passed to three mayors of the palace (in Neustria, Austrasia, and Burgundy), who had been able to acquire large territories for themselves. Civil war resulted as competing aristocratic clans, each claiming autonomy, tried to establish hegemony.

Ebroïn, mayor of the palace in Neustria, attempted to unify the kingdom, but met violent opposition, especially from Austrasia. Bishop Leodegar, eventually assassinated (c. 679), led resistance in Burgundy. In Aquitaine, Duke Lupus made his duchy an independent principality.

The murder of Ebroïn (c. 683) favored Austrasia and the Pepinids (or Carolingians). Pepin II defeated the Neustrians at Tertry (687), and made himself lord of the land between the Loire and Meuse. Austrasia and Neustria were reunited under a Merovingian figurehead, but Pepin II actually governed as mayor. Simultaneously, Pepin II partially reconsolidated the frontiers of northern Francia. He drove the Frisians north of the Rhine, and restored Frankish suzerainty over the Alamanni. Control over the south continued to elude him and his supporters. Aquitaine remained an autonomous duchy, while power in Burgundy was divided.

Pepin II's death in 714 jeopardized Carolingian hegemony. His heir was a grandchild, with his widow, Plectrude, regent. Neustria began the revolt. Eudes, Duke of Aquitaine, took advantage to increase his holdings, and allied himself with the Neustrians. Pepin's illegitimate son, Charles Martel ("the Hammer"), responded by seizing what he considered his birthright. Defeating the Neustrians at Amblève (716), Vincy (717), and Soissons (719), he became master of northern Francia, like his father before him. To gain legitimacy, he declared the Merovingian Chlotar IV king of Austrasia, keeping the office of mayor of the palace for himself.

The Merovingians had discouraged the growth of nobility within their ranks. They preferred to delegate administrative and military duties to directly appointed counts or dukes. This entailed honor, but few material rewards. It also failed to curb the ambitions of an unscrupulous subject and of his clan. Out of the confusion, a firm military and administrative hierarchy was gradually forged. The process began with Charles Martel; continued with his son, Pepin the Short; and finally came to fruition under his grandson, Charlemagne.

Nic Fields

See also: Charlemagne; Charles Martel; Merovingians

References and further reading:
Wallace-Hadrill, J. M. *The Barbarian West, 400–1000.* London: Hutchinson, 1957.

Frankish-Moorish Wars (718–759)

The First Franco-Moorish War (718–732) saw the Moors conquer the Visigoths in Spain, and then invade the Frankish kingdoms north of the Pyrenees Mountains. The Moors captured Narbone in 719, but were defeated at Toulouse in 721 and pushed back into Spain by Eudes. The Moors returned in 725 and conquered the Septimania region. A new Moorish offensive spearheaded by Abd-ar-Rhaman, governor of Spain, defeated Eudes at Garonne and laid waste to Bordeaux and Aquitaine. The Moors advanced north past Poitiers in west-central France, forcing Eudes and other independent Frankish rulers to ask for help from Charles Martel of the Carolingian family. Charles utilized the army developed by his father, Pepin, to defeat the Moors at the Battle of Tours in 732. The Moorish advance into Europe ended, and Charles became known as "the Hammer." Eudes, now effectively subject to Charles, but free of the Moorish threat, kept the Moors in check across the Pyrenees.

The Second Franco-Moorish War (734–759) began with the death of Eudes and the rebellion of his sons against Charles Martel. Moorish raids increased up the Rhone River valley, but these were checked at Valence in 737 and Lyon in 739. After Charles died, his sons Carloman and Pepin the Short ruled jointly as mayors of the palace, until 747 when Carloman joined a monastery. Pepin the Short drove the Moorish raiders back into Spain as he annexed Septimania. Moorish raids north of the Pyrenees became sparse with Abbasid-Umayyad rivalries in Islam. This dynastic rivalry in the Islamic world also aided the Franks in establishing Christian kingdoms in the Iberian Peninsula of Spain. The kingdoms of Navarre and Asturias formed a buffer zone and would provide a launching point for Charlemagne's invasion of the peninsula.

Christopher Howell

See also: Charlemagne's Wars; Charles Martel; Tours

References and further reading:
Beeler, John. *Warfare in Feudal Europe.* Ithaca, NY: Cornell University Press, 1971.

Burgess, Glyn, trans. *Song of Roland.* New York: Penguin USA, 1990.
Nicolle, David. *The Age of Charlemagne.* London: Osprey Press, 2000.
———. *Medieval Warfare Source Book.* London: Brockhampton Press, Arms & Armour, Cassell Group, 1996.

Franklin, Battle of (30 November 1864)

Seeking to lure Union general William Sherman from Atlanta, Confederate general John Hood continued his invasion through Alabama into Tennessee. At Franklin he attacked a smaller army commanded by General John Schofield. Schofield wanted to hold his position on the southern edge of town to permit his supply wagons to retreat unhindered to the north. His men improved some defense works constructed in spring 1863 by northern troops.

Hood was convinced that timidity among his subordinates and troops had prevented a major victory a few days earlier at Spring Hill. He believed in seizing the initiative, and was determined to demonstrate its benefits. However, he also promised his men, who were war-weary, that he would not commit them to any more deadly frontal assaults against prepared positions.

Around 4:00 P.M. on 30 November, Hood launched an attack. Confederates assaulted the hastily constructed northern defenses before their artillery arrived at the battlefield—that is, they did what Hood had promised his men he would not order them to do. They charged 20,000 well-dug-in Union troops who held their fire until the southerners were well within firing range. The battle became a deadly clash that lasted well into the night, and Hood's Army of Tennessee lost about 25 percent of its effective strength. Schofield did have to retreat to Nashville, but he had bought time for Thomas's defense of Nashville.

Charles M. Dobbs

See also: American Civil War; Hood, John Bell; Nashville, Battle of

References and further reading:

McMurry, Richard M. *John Bell Hood and the War for Southern Independence.* Lexington: University of Kentucky Press, 1992.
Sword, Wiley. *Embrace an Angry Wind: The Confederacy's Last Hurrah: Spring Hill, Franklin, and Nashville.* New York: HarperCollins, 1993.

Franks

Germanic conquerors of the former Roman province of Gaul. The Franks created the strongest and most stable barbarian kingdom in the west. The name Franci (bold, fierce) suggests an association with a coalition of Germanic tribes of the middle and lower Rhine. It caused the Roman Empire much trouble, from around 260 until the end of the fourth century.

The Franks did not fully emerge as a people until the establishment of the Merovingian Dynasty by Clovis (r. 481–511). Clovis drove the Gallic Visigoths into Spain. He also absorbed much of the Burgundian kingdom and many Alamanni into his kingdom. Clovis converted to orthodoxy, an act that made him king of the Franks in the eyes of the Roman Church.

Agathias, the sixth-century chronicler, provides a graphic description of Frankish warriors. They had few horsemen, he says, but the Frankish infantry were bold and experienced in war. Disdaining body armor and rarely wearing a helmet, they preferred to fight bare-headed and with their chests exposed. Their weapons were a javelin with a long iron shank (*angon,* a throwing weapon similar to the Roman *pilum*) and a throwing-axe with a sharply swept head (*francisca*). Both could split a shield or penetrate armor at short range. Frankish shields were round, solidly constructed of wooden planks held by a single grip behind a heavy iron boss. Sidonius Apollinaris, fifth-century Gallo-Roman eyewitness to the Frankish invasion of Gaul, mentions the characteristic hairstyle of Frankish warriors, with side-braids, topknot, and the rear of the skull shaved. He also says they "hurl their axes and cast their spears with great force, never missing their aim . . . and rush on their enemy with such speed, they seem to fly more swiftly than their spears."

Nic Fields

See also: Burgundians

References and further reading:

James, E. *The Franks.* Oxford: Blackwell, 1988.
Lasko, P. E. *The Kingdom of the Franks.* London: Thames & Hudson, 1971.

Frederick I Barbarossa (1152–1190)

German king and Holy Roman Emperor. Frederick Barbarossa's reign as German king and Holy Roman Emperor was one of conquest and consolidation. He succeeded Conrad III as the elected king of Germany in 1152. Unlike Conrad, he was also crowned Holy Roman Emperor, as Frederick I, in 1155. Frederick had strengthened ties with both his Staufen and Welf lineages to gain their support for his election. In 1153, he campaigned in Italy and secured the papacy as an ally. In 1158, after his election, Frederick campaigned against the Normans of Sicily and the Byzantines in imperial northern Italy. He seized Milan and established his own direct rule over the region.

After the death of Pope Adrian IV in 1159, Frederick supported pro-imperial Victor IV against the anti-imperial

Alexander III as the new pope. After 1167, Frederick was opposed by the Lombard League formed by Alexander III, the Norman king, the Byzantine emperor, and the anti-imperial resistance in northern Italy.

In 1169, he secured the succession of the Staufen line with his son, Henry. In July 1177, to free himself for further consolidation in Germany, he negotiated a peace settlement in Venice. It gave part of northern Italy to Alexander III, and established a truce with the Lombard League and with the Norman king. In 1183, in the Peace of Constance, Frederick finally settled matters with the Lombard League, led by Milan. In his sixth and final Italian campaign, he strengthened his influence in Lombardy, and in the 1180s, Frederick again consolidated power in Germany. In May 1189, he left for the Holy Land on crusade, leaving his son, Henry VI, in control. After a victory over the Turks at Iconium, Frederick drowned in the River Salef on 10 July 1190.

Christopher Goedert

See also: Holy Roman Empire; Holy Roman Empire—Papacy Wars

References and further reading:

Haverkamp, Alfred. *Medieval Germany, 1056–1273.* Trans. Helga Brown and Richard Mortimer. Oxford, UK: Oxford University Press, 1992.

Munz, Peter. *Frederick Barbarossa: A Study in Medieval Politics.* London: Eyre & Spottiswoode, 1969.

Otto of Freising and his continuator, Rahewin. *The Deeds of Frederick Barbarossa.* Trans. Charles C. Mierow and Richard Emery. New York: Columbia University Press, 1953.

Frederick II (1194–1250)

King of Sicily and of Germany, the last strong Holy Roman Emperor to clash with the papacy during the Middle Ages. Frederick expanded his authority through crusade to the Holy Land and threatening the conquest of the whole of Italy.

Frederick was born on 26 December 1194, at Jesi in the kingdom of Sicily. Frederick was the son of Holy Roman Emperor Henry VI of Germany and Constance of Sicily, attaining the latter crown by inheritance. After a lengthy struggle with his father's immediate successor, Otto IV, Frederick officially became the German king in 1211 and Holy Roman Emperor in 1220. In 1228, he led a crusade against the Muslims in the Middle East, acquiring Jerusalem from Sultan al-Kamil through diplomacy instead of combat.

After returning to Sicily, Frederick suppressed a rebellion by his oldest son Henry, whom he had made the new German king. Frederick created a bodyguard of loyal Saracens, locating them near his new arsenal at Lucera. He consolidated power in Germany and Sicily and invaded northern Italy. At the height of his power, Frederick defeated a Milanese army at Cortenuova in 1237, threatening the power of Pope Gregory IX. Frederick was excommunicated (for a second time), but he continued to press toward Rome despite multiple defeats. Pope Innocent IV deposed Frederick in 1245. After numerous campaigns and attempts at his life, Frederick died not in battle but of dysentery at Castel Fiorentino on 13 December 1250. He left the weakened empire to his second son, Conrad; Arles and Jerusalem to his third son, Henry; and Austria to his grandson, Frederick.

Christopher P. Goedert

See also: Crusades; Holy Roman Empire; Holy Roman Empire—Papacy Wars

References and further reading:

Abulafia, David. *Frederick II: Medieval Emperor.* London: Allen Lane/Penguin Press, 1988.

Cleve, Thomas Curtis van. *The Emperor Frederick II of Hohenstaufen: Immutator Mundi.* Oxford, UK: Clarendon Press, 1972

Frederick the Great, King of Prussia (1712–1786)

Brilliant Prussian king, statesman, and field commander. Frederick was born in Berlin on 24 January 1712 as Friedrich von Hohenzollern, son and heir of King Frederick William I of Prussia. The king systematically tormented and abused his son until 1732, when the two reconciled and Frederick became colonel of the Ruppin regiment of infantry. Frederick fought in the War of the Polish Succession in 1734 and ascended the throne as Frederick II on 31 May 1740. He immediately began to use the formidable army that his father and great-grandfather had built.

In the War of the Austrian Succession, Frederick captured Silesia in a swift December 1740 to January 1741 campaign, defeated the Austrians at Mollwitz on 10 April, invaded Bohemia on 5 February 1742, won at Chotusitz on 17 May, and achieved favorable terms in the Peace of Breslau on 11 June. He invaded Bohemia again on 17 August 1744; took Prague on 16 September; defeated Austria at Hohen-Friedberg on 4 June 1745, Soor on 30 September, Hennersdorf on 23 November, and Görlitz on 24 November; and successfully exited the war by the Treaty of Dresden on 25 December. Back home, he published his philosophy of war in 1747 and continued to enlarge his army.

Frederick's opening campaign in the Seven Years' War was an invasion of Saxony in August 1756. Usually outnumbered in battle, he compensated with clever tactics and precise maneuvers. He was thwarted at Pirna in September, but won at Lobositz on 1 October. Attacking Bohemia in April 1757, he won at Prague on 6 May, but lost at Kolin on 18 June and retreated to Silesia. His victories at Rossbach on 5 November and Leuthen on 6 December were decisive. He besieged Olmütz in May 1758, won at Zorndorf on 25 August, but lost at Hochkirch on 14 October, Kunersdorf on 12 August 1759, and Maxen on 20–21 November. Regrouping his exhausted and overextended forces, he won at Liegnitz on 15 August 1760, Berlin on 12 October, and Torgau on 3 November; retreated to Bunzelwitz in August 1761; and won at Burkersdorf on 21 July 1762.

Frederick spent most of the rest of his life restoring Prussia from the wreckage of the Seven Years' War, solidifying his territorial gains, adding to them through the 1772 partition of Poland, and making Berlin a European cultural center. He died at the Palace of Sans Souci in Potsdam on 17 August 1786.

Eric v. d. Luft

See also: Austrian Succession, War of the; Blücher, Gebhard Leberecht von; Brunswick, Frederick William, Duke of; Frederick William I, King of Prussia; Hochkirch, Battle of; Leuthen, Battle of; Rossbach; Seven Years' War; Steuben, Friedrich Wilhelm Augustin, Freiherr von; Torgau

References and further reading:

Duffy, Christopher. *The Military Life of Frederick the Great.* New York: Atheneum, 1986.

Frederick II, King of Prussia. *Frederick the Great on the Art of War,* ed. and trans. Jay Luvaas. New York: Da Capo, 1999.

Showalter, Dennis E. *The Wars of Frederick the Great.* London: Longman, 1996.

Frederick William, Elector of Brandenburg (1620–1688)

German prince who created the first standing army in Europe. Born as Friedrich Wilhelm von Hohenzollern, Kurfürst von Brandenburg, on 16 February 1620 in Berlin, he attended the University of Leiden, Netherlands, from 1634 to 1637 and succeeded his father as elector of Brandenburg on 1 December 1640, in the midst of the Thirty Years' War.

Frederick William's greatest achievement was building Brandenburg into a European power after the devastation of the Thirty Years' War. Brandenburg gained some territory by the 1648 Peace of Westphalia, but its population was decimated, its farmland was wasted, and it was under constant threat of attack by Sweden and Poland. Frederick William realized that the only way to overcome this vulnerability was with a standing army. He created the noble Junker class of military officers, established military schools, and within 20 years had 30,000 rigorously disciplined and highly trained professional troops at his command. During this period of development, he had to be diplomatically careful to avoid preemptive strikes from his neighbors. He skillfully played them off against each other in the Anglo-Dutch Wars of 1652–1654 and 1665–1667 and the first Northern War of 1655–1660.

During the 1672–1679 Franco-Dutch War, France prompted Sweden to invade Brandenburg in December 1674. Brandenburg lost to Viscount Turenne at Turckheim on 5 January 1675; but at Fehrbellin, just northwest of Berlin, on 28 June, 5,600 Brandenburger cavalry with 13 guns defeated Sweden's 7,000 infantry, 4,000 cavalry, and 38 guns. This news stunned all Europe, and turned France permanently against Brandenburg. Although still outnumbered by the invading army, Brandenburg defeated the Swedes at Stettin and Rügen in 1677; Stralsund in 1678; and Tilsit, Splitter, and Heydekrug in January 1679.

In 1685, Frederick William countered Louis XIV's revocation of the Edict of Nantes with the Edict of Potsdam, shrewdly welcoming hardworking Huguenots to Brandenburg. During his reign, Brandenburg's population rose from 600,000 to 1,500,000 and its territory increased by 40 percent.

"The Great Elector" died on 9 May 1688 in Potsdam. His creation of the first military secular state in modern Europe enabled his son, Frederick I, to become the first king of Prussia in 1701, laid the foundation of the military power of eighteenth-century Prussia, and eventually led to the unification of Germany in 1871.

Eric v. d. Luft

See also: Bismarck, Otto von; Frederick the Great, King of Prussia; Frederick William I, King of Prussia; German Wars of Unification; Louis XIV; Northern War, Second; Thirty Years' War; Turenne, Henri de la Tour d'Auvergne, Vicomte de

References and further reading:

Maurice, C. Edmund. *Life of Frederick William, the Great Elector of Brandenburg.* Westport, CT: Greenwood, 1981.

Mitchell, Otis C. *A Concise History of Brandenburg-Prussia to 1786.* Washington, DC: University Press of America, 1980.

Schevill, Ferdinand. *The Great Elector.* Chicago, IL: University of Chicago Press, 1947.

Shennan, Margaret. *The Rise of Brandenburg-Prussia.* London: Routledge, 1995.

Wilson, Peter H. *German Armies: War and German Politics, 1648–1806.* London: University College of London Press, 1998.

Frederick William I, King of Prussia (1688–1740)

The "Soldier King" of Prussia, architect of a mighty military power in central Europe. Born in Berlin on 14 August 1688 as Friedrich Wilhelm von Hohenzollern, son and heir of Elector Frederick III of Brandenburg who became King

Frederick I of Prussia in 1701, Frederick William was always devoted to military life. He served under the duke of Marlborough in the War of the Spanish Succession and genuinely enjoyed fighting at Malplaquet. As soon as he ascended the Prussian throne on 25 February 1713, he resumed the militaristic policies of his namesake grandfather, the Great Elector of Brandenburg, which his father had allowed to fall into disarray. The treasury was empty, military manpower was about 25,000, and national morale was low.

Frederick William immediately ordered national austerity and poured all available resources into the military. He conscripted vagabonds; invited immigrant Protestants to enlist; reformed military and governmental bureaucracy; made sophisticated military education available to all male Prussians, even peasants; and paid special bonuses for tall or physically gifted recruits. The feudal nobility became a military hierarchy, as Frederick William encouraged noble sons to military careers. An absolute monarch, he enforced his personal Calvinist values of frugality, piety, severity, discipline, diligence, punctuality, efficiency, and obedience upon the whole nation. He was brutal, unforgiving, coarse, and often cruel. As Thomas Carlyle put it, he ruled like a "drill sergeant."

Frederick William used his army only once during his entire reign, toward the end of the Great Northern War. He forced Sweden to cede Stettin to Prussia in 1715, and annexed Pomerania by the Treaty of Stockholm in 1719, with a small additional payment in 1720. He successfully disguised his growing military might by pretending to neighbor nations that he merely loved military parades. Austria, especially, was convinced that these frequent displays would have no relevance on the battlefield.

Frederick William I died in Potsdam on 31 May 1740 and bequeathed to his son, who became Frederick the Great, a bulging treasury, a magnificent army of 85,000, and a proud nation.

Eric v. d. Luft

See also: Charles XII; Frederick the Great, King of Prussia; Frederick William, Elector of Brandenburg; Malplaquet, Battle of; Marlborough, John Churchill, First Duke of; Northern War, Great; Quadruple Alliance, War of the; Spanish Succession, War of the

References and further reading:

Büsch, Otto. *Military System and Social Life in Old Regime Prussia, 1713–1807: The Beginning of the Social Militarization of Prusso-German Society.* Atlantic Highlands, NJ: Humanities, 1997.

Carsten, Francis Ludwig. *The Origins of Prussia.* Westport, CT: Greenwood, 1981.

Dorwart, Reinhold August. *The Administrative Reforms of Frederick William I of Prussia.* Westport, CT: Greenwood, 1971.

Ergang, Robert Reinhold. *The Potsdam Führer: Frederick William I, Father of Prussian Militarism.* New York: Octagon, 1972.

Thadden, Rudolf von. *Prussia: The History of a Lost State.* Cambridge, UK: Cambridge University Press, 1987.

Fredericksburg (11–15 December 1862)

A bloody repulse for the Union. In early November 1862, President Abraham Lincoln had replaced George McClellan with Ambrose Burnside as commander of the Army of the Potomac; McClellan was too cautious and Lincoln understood that to win the North had to fight.

Burnside in mid-November moved the army quickly from camp near Warrenton, Virginia, to cross the Rappahannock River at Fredericksburg; he soon had his entire command at the rain-swollen river. He intended a quick series of marches to force the Confederates to fight for their line of communications to Richmond, but the lack of pontooning material caused the Union army to sit while Robert E. Lee concentrated the Army of Northern Virginia south of the town. Lee had placed his army in very strong defenses, anchored by a sunken road and a stone wall along Marye's Heights, behind stone walls and wood fences reaching down to Jeb Stuart's cavalry who anchored the Confederate right. Union artillery on Stafford's Heights could reach the town but not the Confederate emplacements, and vice versa. Any assault would be near suicidal.

On 11 December, Union engineers threw five pontoon bridges across the river under fire (Lee had some sharpshooters in the town to disrupt the engineers). Union troops crossed the river the next day and on 13 December charged the well-dug-in Confederates, with very high casualties—13,000—nearly three times as many as the Confederates suffered. Indeed, in some places, Lee had his men standing five deep to keep up a continuous, murderous volley of fire. Two days later, a shaken Burnside cancelled the offensive, and the men returned to the heights on the north side of the river.

Charles M. Dobbs

See also: American Civil War; Burnside, Ambrose Everett; Lee, Robert Edward

References and further reading:

Gallagher, Gary W. *The Fredericksburg Campaign: Decision on the Rappahannock.* Chapel Hill: University of North Carolina Press, 1995.

Sutherland, Daniel E. *Fredericksburg and Chancellorsville: The Dare Mark Campaign.* Lincoln: University of Nebraska Press, 1998.

French, John Denton Pinkstone, First Earl of Ypres (1852–1925)

British field marshal, first commander of the British Expeditionary Force (BEF) during World War I. French was born on 28 September 1852 at Ripple, Kent. He spent two years in the Royal Navy before entering the militia in 1870. Four years later French transferred to the army. After serving in Britain,

India, and Egypt, he achieved prominence with his successful performance as a cavalry commander during the Boer War. In 1912 French was appointed chief of the imperial general staff.

With the outbreak of World War I, French was selected to command the BEF. After being defeated at Mons in August 1914 he ordered a retreat. During these early operations French failed to coordinate closely the movements of the two corps under his command. Only General Sir Horace Smith-Dorrien's decision to fight a rearguard action at Le Cateau slowed the German advance. During the retreat French lost his nerve and decided to withdraw to the French port of Saint Nazaire. The intervention of the secretary of War, Lord Kitchener, was required before French agreed to participate in the Allied counteroffensive.

After the BEF was transferred to Flanders, French stalemated the German attacks at Ypres. He in turn launched several offensives designed to breach the German lines at Neuve Chapelle, Auber Ridge, Festubert, and Loos; all ended in failure.

Possessed of a mercurial personality, French blamed others for the setbacks in France and constantly quarreled with his subordinates. Amid concerns for his competence, French was forced to resign at the end of 1915.

Following his resignation, and to keep him out of trouble, French was assigned as commander in chief, Home Forces (1916–1919), and as lord lieutenant of Ireland (1919–1921). He died on 22 May 1925 at Deal Castle, Kent.

Bradley P. Tolppanen

See also: Kitchener, Horatio Herbert; World War I; Ypres, Battles of

References and further reading:

Cassar, George. *The Tragedy of Sir John French.* London: Associated University Presses, 1985.

Holmes, Richard. *The Little Field Marshal.* London: Jonathan Cape, 1981.

French and Indian War (1759–1763)

Known in Europe as the Seven Years' War, this conflict destroyed French power in North America. Three times previously the American colonists and the French in Quebec had fought as part of larger conflicts between the British and French Empires. And three times the gains and losses in North America were bargained away because of more important possessions in the West Indies, Africa, and India.

The last of these wars, the French and Indian War, would be different. The British prime minister, Sir William Pitt, chose to hire mercenaries to contain the French on the European continent while he concentrated British power on conquering French colonial possessions. Despite some fits and starts and perhaps less than ideal cooperation by American colonial governments and militia, the British ultimately secured a great victory over the French in North America, adding greatly to the British Empire, and perhaps inadvertently beginning the process that would lead to the American Revolution a decade later.

Pitt put the wealth of Britain into the war effort. He sent large armies transported by huge navies to the New World; he purchased supplies from colonial purveyors; he engaged colonial militia and rangers. And he kept his focus on the objective—winning Quebec.

The conflict started badly for the British in 1755. General Edward Braddock refused to compromise with the colonial terrain, and marched in rather traditional style from western Virginia toward Fort Duquesne at the junction of the Ohio, Allegheny, and Monongahela Rivers. His advance force was surprised by the French and their Native American allies, and Braddock's attack turned into a disastrous retreat. Braddock's defeat (and death) was compounded by failure to take Fort Niagara; New Yorkers more welcomed trade with the French and Iroquois than they favored supporting General William Shirley's plan to attack Niagara. The French continued their successes on Lake George, at Fort Oswego, and elsewhere.

Several years later, in 1758, the weight of British (and colonial) power began to be felt in the conflict. Pitt became prime minister, and he brought focus and determination to the conflict. The British under General John Forbes once again attacked Fort Duquesne, this time advancing from western Pennsylvania in the fall. The attack succeeded, the French retreated, and the British cut off Quebec from French Louisiana. Also that year, another offensive commanded by General Jeffrey Amherst sailed from Halifax, Nova Scotia, to attack Louisbourg, a French fort on Cape Breton Island, guarding the entrance to the Gulf of St. Lawrence and the St. Lawrence River, the route connecting Quebec and France. Amherst landed west of Louisbourg, brought up siege weapons, and the fort surrendered. However, an attack against Fort Ticonderoga on Lake George did not fare as well, and the British rather unexpectedly retreated whereas the French commander expected a continuation the next day and he, in turn, contemplated retreat.

The victories of 1758 were completed by greater victories in 1759. General Amherst gained control over Ticonderoga, which the French had blown up. Meanwhile, General James Wolfe sailed to Quebec, approached the city from the western side, engaged in the penultimate battle on the Plains of Abraham against the French commander, the Marquis de Montcalm, and the British compelled the surrender of the citadel. (Both Montcalm and Wolfe died in the battle.)

In 1760, Amherst compelled the surrender of Montreal,

and the French had lost their colony in Quebec. The war would drag on for another several years as the conflict expanded to include the Spanish, which made Cuba and Gibraltar areas of operation. The French and Indian War—the Seven Years' War elsewhere—was a great, but ultimately a costly, British victory.

Charles M. Dobbs

See also: Braddock's Defeat; Fort Duquesne, Seizure of; Fort Ticonderoga; Montcalm-Gozon, Louis-Joseph de, Marquis de Montcalm de Saint-Véran; Pitt, William, the Elder; Plains of Abraham; Washington, George; Wolfe, James

References and further reading:

Brecher, Frank W. *Losing a Continent: France's North American Policy, 1753, 1763.* Westport, CT: Greenwood Press, 1998.

Ferling, John. *Struggle for a Continent: The Wars of Early America.* Arlington Heights, IL: Harlan Davidson, 1993.

Furneaux, Rupert. *The Seven Years War.* London: Hart-Davis MacGibbon, 1973.

Peckham, Howard. *The Colonial Wars, 1689–1762.* Chicago: University of Chicago Press, 1964.

French Army

The army of the French king was respected and feared in early modern Europe. It became the French army when the revolutionary government transformed it into a mass organization based on universal conscription and replaced most of the aristocrats in the officer corps with experienced noncommissioned officers. This revolutionary army was organized into divisions, each with infantry, artillery, and cavalry, to provide better command and control. Napoleon soon organized the divisions into corps to support more flexible tactics, ease problems of logistics, and speed marches by utilizing several simultaneous lines of advance. Combining these organizational changes with significant tactical advances, Napoleon harnessed the best ideas of military innovators to create the first modern army. The restored monarchy introduced limited conscription, which, combined with a seven-year term of service, produced a long-service army with high rates of reenlistment. This army fought numerous colonial campaigns, muddled through the Crimean War, and acquitted itself well in the bloody bayonet charges of the 1859 war against Austria, but was no match in 1870 for the Prussian mass army created with universal conscription and a carefully organized reserve system.

The French army of 1914, with 48 active and 37 reserve divisions, performed well against the German army whose 51 active and 67 reserve divisions had to defend also against Russia. Eventually, France mobilized 7.5 million troops, of which about 1.4 million were killed and up to 4.5 million wounded or missing. Such heavy losses led some divisions to mutiny in 1917, and postwar military plans emphasized the defensive. After quick defeat at thc hands of the Germans in 1940, the greater part of the army remained loyal to the collaborationist Vichy regime while some personnel and units joined the Free French forces on the Allied side.

After 1945, the French army bore the burden of homeland defense (in cooperation with the North Atlantic Treaty Organization nations after 1950) and the quashing of Indochinese and Algerian liberation movements. Failure in this latter enterprise ate at army morale, and in the late 1950s the rightist Secret Army Organization (OAS) attempted to force the government's hand in Algeria but the situation was defused by General Charles de Gaulle. In 1966, he withdrew France from military participation in NATO.

France's defense reorganization of 1997, aimed at more rapid deployment and more efficient operations outside France, created an army of nine brigades composed of four to seven battalions of infantry, artillery, and engineers; two logistics brigades; and one air mobile brigade. The force-level target for the year 2002 was 16,000 officers, 50,000 noncoms, 66,500 enlistees, 5,500 volunteers, 30,000 reservists, and 34,000 civilian employees.

As was the case with the British army, the French army had to undergo a wrenching adjustment to the postwar reality that France was no longer a major power. Perhaps because of the catastrophic 1940 defeat, the French army fought to hold French colonial possessions longer than did the British. But by 1962, with Algerian independence, France was at peace, its global *mission civilitrice* limited to economic and cultural aid to former colonies. In addition to various United Nations peace-keeping duties, the French army gave good service in the Gulf and Balkan Wars.

Joseph McCarthy

See also: Algiers, Battle of; Dien Bien Phu; Franco-Prussian War; Marne, Battle of the; Napoleonic Wars; Sedan; World War I; World War II

References and further reading:

Haythornthwaite, Philip J. *Napoleon's Military Machine.* New York: Sarpedon, 1995.

Horne, Alistair. *The French Army and Politics, 1870–1970.* New York: Harper & Row, 1984.

Masson, Philippe. *Histoire de l'armée française.* Paris: Perrin, 1999.

Porch, Douglas. *March to the Marne: The French Army, 1871–1914.* New York: Cambridge University Press, 1981.

French Colonial Wars (1800–1939)

During the nineteenth century and the first part of the twentieth, the French waged small wars to acquire additional colonies. In 1830, they occupied Algiers and deposed the bey, ostensibly to solve the piracy problem. Resistance in the interior led to the appointment of General Thomas Bugeaud commanding 59,000 French troops. Seeing static blockhouse as ineffective in the face of Algerian raiding, Bugeaud developed light infantry and cavalry units, *voltigeurs, chasseurs, zouaves,* and *turcos,* and adopted an attacking infantry-square formation, the *bull's head,* to be employed with cavalry, the *bull's horns.* Using flying columns to destroy infrastructure, Bugeaud drove the opposition leader, Abd-el-Kadr, into Morocco and destroyed his army at Isly on 15 August 1844. The main resistance eliminated, intermittent pacification campaigns continued for nearly 40 years. This experience, from intervention on a pretext, to use of superior firepower by small swift units, to lengthy pacification efforts, provided the blueprint for subsequent conquests.

From trading posts developed in Senegal in the eighteenth century, France seized posts on the Ivory Coast in 1842–1843, and from them fought three wars against the Mandingos between 1885 and 1898 to establish a French protectorate over the area. Between 1854 and 1864, French troops fought their way gradually from Senegal to the upper Niger. From there, they launched operations between 1883 and 1890 along the Dahomey frontier and in two short wars in 1889 and 1892 annexed Dahomey. Columns from Algeria and Congo converged to conquer Chad in 1899–1900, with eastern Chad finally succumbing in a campaign lasting from 1909 to 1911. France exercised a protectorate over coastal areas of Madagascar from 1859 and a war from 1883 to 1885 extended the protectorate to the whole island. In 1895–1896, the capital was taken and the island was made a French colony. From bombarding Casablanca and occupying the Moroccan coast in 1907, the French proceeded to claim a protectorate over Morocco in 1911. Only after a joint French-Spanish force put down the Riff rebellion of 1925–1926 was the country fully in French control. In 1908–1909, the French conquered Mauretania.

In Southeast Asia, half a dozen years of clashes with the king of Cochin China led to French occupation of Tourane in 1858. After moving to Saigon, the French were besieged there while other French forces in the region joined the British in attacking Peking, from 13 May to 18 October 1860, in the Second Opium War. When the siege was lifted in February 1861, the French speedily conquered the three eastern provinces. In 1863 they claimed a protectorate over Cambodia, despite Siamese protests, and suppressed a Cambodian rebellion in 1866–1867. By 1873, French pacification of rebellions in Cochin China led to the conquest of the three southwestern provinces. Ten years later, the capture of Hanoi and Hué led to a French protectorate over all of Vietnam, precipitating a brief war between France and China. Uprisings in Vietnam, Cochin China, and Tongking occupied the French from 1885 to 1895. In 1893, French gunboats threatening Bangkok compelled acceptance of a French protectorate over Laos and cession of western Cambodia.

The most intriguing episode of the French colonial wars was the attempt by Emperor Napoleon II to set the Austrian archduke Maximilian on the throne of Mexico. Intervening in a civil war in 1862, French troops marched on Mexico City. Maximilian wore the crown briefly before the project was abandoned in 1867 in the face of opposition from the United States.

In pursuit of its League of Nations mandates in the Middle East after World War I, French forces ejected the Emir Facial out of Syria in 1920 and fought an insurrection by the Douses in Lebanon in 1925–1926.

In the post–World War II era of violent anticolonialism in the colonies and growing resistance to imperialism in the mother countries everywhere, the French fought a series of losing actions to retain something of their overseas empire. They progressively lost their holdings in northern Vietnam, Africa, and, finally, Algeria (the latter was actually considered a part of Metropolitan France). Although France enjoys strong cultural ties with some of its former colonies, particularly in Africa, its actual overseas holdings today consist of two tiny islands in the mouth of the St. Lawrence River.

Joseph M. McCarthy

See also: French Army

References and further reading:

Aldrich, Robert. *Greater France: A History of French Overseas Expansion.* Basingstoke, UK: Macmillan, 1996.

Quinn, Frederick. *The French Colonial Empire.* Westport, CT: Praeger, 2000.

French Foreign Legion

Legendary fighting force. King Louis-Philippe created the French Foreign Legion on 10 March 1831 to enlist foreign nationals who flocked to France during the 1830 revolution. Legionnaires served only *outside* France, which had the double advantage of providing France with troops and taking foreigners who might join revolutionary groups out of the country.

Over time the legion grew into a multinational, polyglot force whose composition mirrored instability around the world. Men joined to escape economic, social, or criminal problems in their own countries, their families, or simply to pursue adventure. They served under predominantly French officers in all of France's colonial wars, from Tonkin, Da-

Three members of the Free French Foreign Legion who distinguished themselves in the Battle at Bir Hacheim, 1942. (Library of Congress)

homey, Madagascar, Morocco, and Mexico to service in France itself during World War I.

Along the way legionnaires became legendary for their bravery, and the legion developed an almost mythical reputation as a haven where men could reinvent themselves with new identities. Newspapers, magazines, and eventually films capitalized on these legends and myths, portraying legionnaires as colorful and violent characters with an exotic reputation for adventure.

Since World War I, the legion has added to its reputation for bravery, fighting in Algeria, Indochina, the Persian Gulf, Egypt, Chad, World War II, and throughout the French Empire. Legionnaires were involved in an abortive coup in 1961, but for the most part have served France with distinction by fighting the unpopular, dirty wars that French nationals are less willing to fight themselves.

Lance Janda

References and further reading:

McGorman, Evan. *Life in the French Foreign Legion: How to Join and What to Expect When You Get There.* Central Point, OR: Hellgate Press, 2000.

Porch, Douglas. *The French Foreign Legion: A Complete History of the Legendary Fighting Force.* New York: HarperCollins, 1991.

Simpson, Howard R. *The Paratroopers of the French Foreign Legion: From Vietnam to Bosnia.* Washington, DC: Brassey's Defense Publishers, Inc., 1999.

French Revolutionary Wars (1792–1802)

A series of military conflicts between France and the European powers from the French declaration of war on Austria in April 1792 to the conclusion of the peace treaty of Amiens between France and Britain in March 1802, which ended the Revolutionary period and ushered in the Napoleonic Wars.

From the outset, the French Revolution aroused the hostility of all European crowned heads who feared that the new ideas of liberty, equality, and fraternity could spread beyond the French frontiers and stir up upheavals among the people in their own countries. Likewise, the French émigré nobles abroad added fuel to the war fervor as they had long been trying to goad the European rulers to intervene and suppress the revolution. Friction with Austria intensified when King Louis XVI and his wife, Marie Antoinette, tried to flee abroad in June 1791 and join the enemies of the Revolution. Finally, the Pillnitz Declaration issued by the Austrian and Prussian rulers on 27 August 1791 made it clear that they would intervene in France and restore the power of King Louis if the rest of the European powers would join them. The French interpreted the declaration as an act of interference in the internal affairs of their country.

The push for war came from inside France. When Austria and Prussia concluded a military convention in February 1792, the Legislative Assembly responded by declaring war on Austria on 20 April 1792.

But soon the French troops suffered a series of military setbacks. Although the army had numbers and enthusiasm in 1792, it was nevertheless in a state of confusion and lacked discipline. The majority of the officer corps had either resigned or emigrated. Regulars and volunteers were amalgamated, but could not face a well-trained and disciplined allied army. An attempt to invade Belgium ended in complete disaster. Demoralized, the French soldiers turned against their officers, murdering them, while entire army units deserted to the enemy.

Meanwhile, the Austro-Prussian forces under the duke of Brunswick crossed the border and advanced toward Paris. On 25 July 1792, the duke issued the famous Brunswick Manifesto threatening to raze the capital if the royal family were endangered. The Legislative Assembly proclaimed France in danger. In Paris the revolutionaries accused King Louis of conspiring with the Austrians. Tension grew in the capital when a crowd of sans-culottes stormed the Tuileries palace on 10 August 1792 and arrested and imprisoned the royal family. In September, the new legislature, the National Convention, abolished the monarchy and proclaimed France a republic.

Two political parties dominated the Convention: the Girondins, a moderate republican group, and the Jacobins, a radical revolutionary faction. The immediate task of the Convention was the continuation of the war, which was going badly for France. Verdun surrendered on 2 September 1792, and Lafayette with his troops had already deserted to the Austrians.

The Convention decreed the enlistment of new volunteers. It reorganized and equipped the army and placed it under the command of Generals Charles Dumouriez and François Kellemann. Soon the republican army stopped the invaders and on 20 September 1792 won the first victory over the army of the duke of Brunswick at Valmy. The engagement, far from being an important military event, was nevertheless a victory of the revolutionary army and a turning point in the war against Austria and Prussia.

In the autumn of 1792 the French continued the offensive. The Prussians evacuated Verdun and General Adam Custine actually occupied Maintz and Frankfurt. Dumouriez defeated the Austrians at Jemappes, took Brussels, and occupied the Austrian Netherlands, roughly modern Belgium. French troops entered Savoy and Nice and their patriots requested annexation to France.

But there were sharp disagreements on the question of annexation. In the end, the Convention abandoned the *no conquest* formula and decided that France should achieve its "natural frontiers." Far from acceding to the wishes of the liberated people, the Convention—which had proclaimed the right to intervene anywhere where the people aspired to regain their liberty—now took the first step toward conquest and annexation.

Meanwhile, the execution of King Louis in January 1793 and the annexation of Belgium and most of the Rhineland forced Britain to form the First Coalition against France. Russia, Austria, Prussia Spain, Holland, and Sardinia joined in.

France was now confronted by an almost universally hostile Europe. In response, the Convention boldly declared war on Britain, Holland, and Spain in February 1793. The Austrians pushed the French out of Belgium, defeated General Dumouriez at the Battle of Neerwinden, and a few weeks later he abandoned his troops and also defected to the Austrians. In southeastern France, the British besieged and captured the port of Toulon. The troops of the republic had suddenly once again been pushed back on all fronts.

In Paris, the Jacobins and a large crowd stormed the Convention on 2 June 1793, expelled the Girondin deputies, and took control of the government.

Once in power, the Jacobins faced a dangerous war situation and widespread antigovernment outbreaks throughout the country. To deal with the internal and foreign crises, the Jacobins set up the Committee of Public Safety and instituted the Reign of Terror to win the war efforts and to suppress the revolts of the Girondins, royalists, and Catholic clergy.

On 23 August 1793, the Convention issued a *levée en masse*—universal conscription of all single men, age 18 to 25, to defend France. The revolutionary army employed new military tactics, with heavy emphasis on the use of artillery and rapid mobility of the cavalry. Promotion was based on merit. Patriotism animated the new citizen army. In December 1793 the republican army took the offensive against the Coalition armies, forcing them to retreat across the Rhine, while the young colonel of Artillery, Napoleon Bonaparte, helped wrest the city of Toulon from the British.

In Paris, control of the government passed from the realist Danton, the great patriot who had inspired the events of 10 August 1792, to the idealist and revolutionary Maximilien Robespierrre, who would dominate the Convention until July 1794.

Robespierre was a fanatical follower of Rousseau, who dreamed of establishing the Republic of Virtue through the Reign of Terror. He cracked down on the Hebertists, the Paris group who wanted to carry the economic revolution further, and guillotined the Dantonists, who called for an end to the Terror. After General Jean Jourdan's victory at Fleurus against the armies of the First Coalition in June 1794, when the initiative on the Continent favored France, terror seemed increasingly unnecessary. Finally, reacting to the excesses of Terror, the moderate faction in the Convention and Robespierre's opponents in the Committee of Public Safety brought him down in July 1794 and beheaded him.

The rule of the Thermidorians, as the period from July 1794 to the autumn of 1795 was known, brought the dismantling of the apparatus of Terror, the abolition of the Paris Commune, and the closing of the Jacobin clubs. In 1795 the Thermidorians drafted a new constitution and set up the Directory (1795–1799)—a political body of five directors jointly responsible for conducting France's domestic and foreign affairs.

In the war against the European Coalition, the army of the Directory under Charles Pichegru and Jourdan drove the Austrians out of Belgium and occupied Holland in December 1794. Prussia, financially exhausted, withdrew from the war and concluded the Treaty of Basel in March 1795. Soon Spain and Holland made peace. The Polish revolt forced Russia to leave the Coalition. Only Britain and Austria remained at war with France.

The war against Austria continued during 1796–1797. Jordan invaded southern Germany, while Napoleon attacked the Austrians in Italy. The campaign in Germany, however, ended in failure, and only that of Italy succeeded. In the Italian campaign, Napoleon defeated the Austrians at the Battles of Millesino, Lodi, Castiglioni, Rovedo, Bassano, and Arcola in 1796. He struck at the Sardinian army at Mondovi, knocked it out of the war, and occupied the fortresses of Mantua and Milan in January 1797. He advanced toward Rome, but the pope hastily signed the Treaty of Tolentino.

In April 1797, Napoleon crossed the Alps and forced the Austrians to conclude the armistice of Leoben. Six months later, they signed the Treaty of Campo Formio. Austria ceded Belgium and the Rhineland to France, but retained Venice, Istria, and Dalmatia. The Italian campaign effectively ended the First Coalition, and Britain alone still remained at war with France.

Next Napoleon decided to strike back at Britain. He organized the expedition to Egypt, ultimately hoping to invade India and deprive the British of their most precious "pearl of the English crown"—a most unlikely prospect from the start in view of British sea power. In May 1798, he sailed for Egypt with 350 ships, more than 30,000 troops, and a corps of scientists. He defeated the native Mamluk cavalry at the Battle of the Pyramids near Cairo in July. Admiral Horatio Nelson, however, destroyed his fleet at the Battle of the Nile, cutting him off from Europe. In early 1799, Napoleon set out for Palestine and Syria, but failed to capture them. He returned to Egypt in May and smashed a Turkish force at the Battle of Abukir. This victory left him free to consider his return to France. In August he sailed from Egypt, leaving General Kléber in command of the French army.

In the European theater, Britain formed the Second Coalition (December 1798), which included Austria, Russia, Naples, Portugal, and Turkey. During 1799, the Coalition armies defeated the French at the Battle of Magnano, while at the Battle of Cassano, the Austrian general Melas and the Russian army under Marshal A. Suvorov won a major victory over General Jean Moreau. Suvorov overcame another French army under MacDonald at the Battle of Tribia in June, while at the Battle of Novi, he and Melas routed the French under Joubert. Only General André Masséna was successful over the Russians under General Aleksandr Korsakov, driving them out of Zurich, while General Guillaume Brune defeated an Anglo-Russian force in Holland.

But soon the Second Coalition began to fall apart. In October 1799, Russia withdrew from it and made peace with France when Czar Paul I changed sides and revived the Armed Neutrality against Britain.

Back from Egypt, Napoleon staged the coup d'état on 9 November 1799, toppled the Directory, set up the Consulate, and proclaimed himself First Consul. He offered peace to the Allies, but they rejected his overture.

At the beginning of 1800 Napoleon renewed his campaign in northern Italy against the Austrians. Marching through the most difficult snow-covered terrain, he crossed the St. Bernard Pass riding on a mule, attacked the Austrians, and took Milan. On 14 June, he faced the main Austrian army under General Melas. With a force of only 22,000 men against 30,000 Austrians, he fought an obstinate battle at Marengo, turning a likely French disaster into an Austrian rout. The Austrians then asked for an armistice, but it was only after Moreau's victory at Hohenlinden (December 1800) and MacDonald's and Brune's invasion of Austria (January 1801) that the emperor negotiated the Treaty of Lunéville on 9 February 1801. By its terms, France annexed Belgium and all southwestern German territory, while Austria recognized the Helvetian (Switzerland), Batavian (Holland), Liqurian (Genoa), and Cisalpine (Lombardia) *republics,* which in practice became French dependencies.

Britain still remained at war with France until 1802. But both countries were now eager for a breathing spell. On 27 March 1802, they negotiated the Treaty of Amiens, which ended Anglo-French hostilities.

For the first time after 10 years of costly and bloody wars, Europe was at peace. Yet the peace of Amiens proved to be merely a brief truce. A year later, the second round of wars, the Napoleonic Wars, would begin and would last until 1815.

James Farsolas

See also: Lodi; Marengo, Battle of; Pyramids; Rivoli; Toulon, Siege of; Valmy

References and further reading:

Bertaud, Jean-Paul. *The Army in the French Revolution.* Princeton, NJ: Princeton University Press, 1988.

Doyle, William. *The Oxford History of the French Revolution.* New York: Oxford University Press, 1990.

Schama, Simon. *Citizens: A Chronicle of the French Revolution.* New York: Random House, 1990.

French Wars of Religion (1562–1598)

During the first half of the sixteenth century, the French monarchy struggled to gain control of France and its nobles. The eight distinct civil wars known as the French Wars of Religion constituted part of the greater religious conflict engulfing Europe following the Protestant Reformation. However, the age-old struggle between kings and nobles over economics, dynastic power, and foreign expeditions now included a new religious division between Catholics and Protestants. Some of the major noble houses converted to Protestantism while others remained Catholic. Opposition to centralized royal power began under Henry II when three aristocratic families—the Guise (Catholic) and Montmorency and Bourbon (both Protestant)—maneuvered for control of royal policy. The fanaticism and brutality of the soldiers made it a struggle in which pillage, cruelty, and atrocities were normal.

The Valois king Henry II faced considerable opposition to his attempts to centralize royal authority. The death of Henry II in a jousting accident in 1559 began a long period of royal weakness that would not end until the Bourbon Henry of Navarre took the throne in 1589. Henry II's death brought to the throne his sickly 15-year-old son, Francis II, who reigned only two years. His successor was his 13-year-old brother, Charles IX. The queen mother, Catherine de Médicis, served as regent during both reigns and continued to be influential in the reign of her third son, Henry III (1574–1589).

The duke of Montmorency, constable of France, had immense landholdings and a personal following of several hundred vassals. When the Huguenot Montmorency converted other major nobles of France to the new religion, they became a truly dangerous political threat to the weak Catholic Valois kings who were not strong enough to eliminate Protestantism from France the way the Spanish had—with blood and fire. However, state institutions like the law courts and the royal family remained resolutely Catholic.

In 1562, François de Guise attacked, butchered, and burned hundreds of Protestants worshipping at Vassy, sparking the first of the Wars of Religion. Between 1562 and 1598, eight religious wars split France (1562–1563, 1567–1568, 1569–1570, 1573–1574, 1576, 1577, 1579–1580, 1585–1598). Both sides committed atrocities in an effort to rid their communities of what each side considered a cancer. When Catholics and Protestants gathered in the capital to celebrate the marriage of the king's sister to the recently converted Henry of Navarre, the festivities ended in violence. The 1572 St. Bartholomew's Day Massacre of 3,000 Protestants in Paris, including Protestant leader Admiral Gaspard de Coligny, was only part of the larger pattern of violence in the period, but it did spark a new fanaticism in France. Henry of Navarre renounced his conversion and the wars continued.

Catholic extremists, led by the Guise family, formed the Catholic League in Paris, bound and determined to control the monarchy. When Henry III had the Guise brothers assassinated in 1588 at Blois, the Council of Sixteen, which governed Paris, called for tyrannicide. In desperation, Henry III allied with Henry of Bourbon, descendent of Louis IX and the Protestant king of Navarre, and made him his heir. Following Henry's assassination in 1589, Henry of Navarre judged Paris "worth a Mass" and reconverted to Catholicism, becoming Henry IV and establishing the Bourbon Dynasty on the French throne, which would reign until 1792. He allied with a moderate group of Catholics called the *politiques*, and together they put France's needs ahead of religious differences. Henry IV issued the Edict of Nantes in 1598, which allowed freedom of conscience and equality of rights, but allowed Protestants to worship only in locations where they had already established strong communities.

The long-term impact of the Wars of Religion is certainly as important as the wars themselves. Population losses were severe in areas where fighting had been fierce. However, Henry IV came to realize that he required absolutism to maintain social order in France and keep all the nobility in check, contributing to the rise of a larger state. Further, French national consciousness began to take shape in this period of civil war in a form of a proto-nationalism tied to the person of the king.

David C. Arnold

References and further reading:

Diefendorf, Barbara. *Beneath the Cross: Catholics and Huguenots in Sixteenth-Century Paris.* Oxford, UK: Oxford University Press, 1991.

Dunn, Richard S. *The Age of Religious Wars, 1559–1715.* 2d ed. New York: W. W. Norton, 1979.

Holt, Mack P. *The French Wars of Religion, 1562–1629.* Cambridge, UK: Cambridge University Press, 1995.

Friedland (14 June 1807)

A battle in East Prussia between French forces, ultimately numbering 80,000, commanded by Napoleon Bonaparte, and Russian forces, numbering about 46,000 under Levin, Count Bennigsen. In June 1807, the Russian army under Count Bennigsen began operations designed to assist their Prussian allies, besieged in Danzig and threatened in Konigsberg. Although Danzig fell before any practical relief could be given by the Russians, operations continued. Napoleon, in an effort to keep the Russians from Konigsberg, attempted to move between the Russians and the city. This led to a desultory battle at Heilsburg on 10 June and a Russian retreat. In a further effort to prevent the Russians from reaching Konigsberg along an indirect route, Napoleon sent the corps of Jean Lannes to occupy Friedland, along the path of the Russian advance. Most of the French army moved toward Konigsberg. Bennigsen, learning this, resolved to attack and defeat Lannes before he could be supported by the French.

Friedland sits in an oxbow formed by the River Alle. The town is further divided in two by a stream and millpond. Russian forces arrived late on 13 June and drove off French reconnaissance parties. Bennigsen then undertook the construction of three pontoon bridges across the Alle to supplement the existing bridge over the river. These completed, the Russians deployed on both sides of the millstream.

Early on the 14th, the battle began in earnest as the Russians sought to drive Lannes's French out of two villages near Friedland. Bennigsen failed to press his early superiority in numbers, and Lannes, realizing his danger, called for and received reinforcements. Napoleon arrived around noon, and ordered an attempt to destroy the Russian army, now fighting with its back to a river. By 1700 Napoleon felt confident enough to attack the Russians, first on the southern side of the millstream, then on the north. Napoleon thus ordered Ney and Victor to attack in the south, and to destroy at least two of the pontoon bridges. The French attack, initially hidden by woods, enjoyed considerable success. Russian efforts at a counterattack failed as the French brought up artillery to bombard Russian troops being driven into an increasingly constricted area. In an effort to relieve the pressure on his southern flank, Bennigsen ordered an attack on the north side of the millstream. This failed and only diverted Russian troops from the fight in the south.

By 8.30 P.M. Ney had entered Friedland, and the retreating Russians set fire to the town. This only increased the danger faced by Bennigsen's army as the fire burned down the pontoon bridges. Complete disaster was avoided only by the Russian discovery of a ford over the river. Russian losses were between 18,000 and 20,000 men. The French lost about 8,000. As a result of the Battle of Friedland, Czar Alexander I of Russia found it prudent to open negotiations with the French. These led to the Treaty of Tilsit and an end to the 1806–1807 war between France and Russia and Prussia.

Joseph M. Isenberg

See also: Jena and Auerstädt; Lannes, Jean, Duke of Montebello; Lefebvre, Pierre-François-Joseph, Duke of Danzig; Napoleon I; Napoleonic Wars; Ney, Michel, Duc d'Elchingen, Prince de la Moskova; Oudinot, Nicholas-Charles, Duc de Reggio

References and further reading:

Chandler, David G. *Campaigns of Napoleon.* New York: Scribner, 1966.
Connelly, Owen. *Blundering to Glory.* Wilmington, DE: Scholarly Resources, 1999.
Gates, David. *The Napoleonic Wars, 1803–1815.* London: Arnold, 1997.

Fronde, Wars of the (1648–1653)

Series of unsuccessful attempts by French judges, nobles, provincial parliaments, and workers to limit royal power. The several insurrectionist parties were collectively called the Fronde, from a French word meaning "slingshot," a common weapon of the time among poor people. When Louis XIV ascended the throne in 1643 at the age of five, his mother, Anne of Austria, ruled nominally as regent, but power actually rested with her prime minister, Jules Cardinal Mazarin, an Italian. Anne and Mazarin withheld judges' salaries and imposed oppressive taxes on the working poor in order to finance French involvement in the Thirty Years' War in Germany. The Parlement de Paris rejected these decrees in July 1648, ruling that they were illegal during a regency, especially in that the regent and her minister were both foreigners. Anne and Mazarin retaliated by arbitrarily arresting Parisian judges, including the popular Pierre Broussel, on 26 August, and placing Paris under martial law. The Parisian public openly supported its parliament and erected barricades on 27 August. Anne surrounded Paris with troops, intending to starve the dissidents into submission.

Revolt did not abate when the Peace of Westphalia ended the Thirty Years' War in October. The mob gradually gained the upper hand in Paris, kept Anne and Louis under virtual house arrest in the Louvre, and frightened her into fleeing to Saint-Germain-en-Laye on 5 January 1649. News of the execution of English king Charles I hardened her resolve to centralize power and establish her son's invulnerability to public discontent. She ordered Louis II de Bourbon, Duke of Enghien, known as Condé the Great, to invade Paris and put down the revolt by any means necessary. He obeyed with brutal efficiency and was entirely successful. The pro-royalist Peace of Reuil was concluded on 1 April.

Mazarin grew fearful of Condé's ambition and had him arrested on 18 January 1650. This blunder only galvanized anti-Mazarin sentiment among the nobles, who were already jealously trying to protect their traditional feudal privileges from central authority. Sparked by Condé's sister, the duchess of Longueville, the nobles rebelled, first in Normandy and soon all over France. Chaotic civil war raged throughout 1650. As the advantage seesawed, Mazarin released Condé in February 1651 and went into brief exile.

Eager for revenge, Condé played on the general hatred of Mazarin among all classes. With some help from Habsburg Spain and Cromwellian England, by September he had placed himself at the head of the combined armies of the Fronde. The duke of Luxembourg was among his lieutenants. Condé captured Bordeaux, won the allegiance of Orleans, lost to the Viscount of Turenne's royalist forces on 6–7 April 1652 at Bléneau, then marched on Paris. On 2 July he and the duchess of Montpensier took Paris by defeating Turenne at the Bastille. Mazarin went into exile again. Anne and Louis returned safely to Paris in October after Condé's arrogance had alienated the Parisians. Condé and Turenne faced each other on many battlefields before Turenne finally wore Condé down late in 1652. Condé defected to Spain. Mazarin returned on 3 February 1653 as the insurrection fizzled, though various protests continued until his death in 1661.

The defeat of the Fronde paved the way for Louis XIV and Louis XV to rule as absolute monarchs. The dictatorial power of French kings was not again seriously challenged until the excesses of Louis XVI and Marie Antoinette incited the French Revolution.

Eric v. d. Luft

See also: Condé, Louis II de Bourbon, Fourth Prince de; Cromwell, Oliver; English Civil War (1642–1649); Franco-Spanish War; Louis

XIV; Luxembourg, François Henri de Montmorency-Bouteville, Duc de Piney; Thirty Years' War; Turenne, Henri de la Tour d'Auvergne, Vicomte de

References and further reading:

Bercé, Yves Marie. *The Birth of Absolutism: A History of France, 1598–1661.* New York: St. Martin's Press, 1996.

Knachel, Philip A. *England and the Fronde: The Impact of the English Civil War and Revolution on France.* Ithaca, NY: Cornell University Press, 1967.

Moote, Alanson Lloyd. *The Revolt of the Judges: The Parlement of Paris and the Fronde, 1643–1652.* Princeton, NJ: Princeton University Press, 1972.

Ranum, Orest A. *The Fronde: A French Revolution, 1648–1652.* New York: W. W. Norton, 1993.

Treasure, Geoffrey Russell Richards. *Mazarin: The Crisis of Absolutism in France.* London: Routledge, 1995.

Frunze, Mikhail Vasil'evich (1885–1925)

Revolutionary, Russian Civil War commander, political actor, military theoretician. An army doctor's son, Frunze was born in Pishpek (later renamed Frunze), Kirghizstan. He attended St. Petersburg Polytechnic Institute (1904–1907), but was expelled for poor class attendance. He joined the Social Democratic Party in time to be active in the 1905–1907 Revolution, leading a militia uprising in Ivanovo-Voznesensk in December 1905. He was imprisoned on political grounds in 1907 and exiled to Siberia in 1914.

In 1916 Frunze headed the Bolshevik underground in Minsk, agitating among the Third and Tenth Russian Armies, before heading the militia of the same city after the February Revolution. In October 1917, he led a workers' detachment in the Moscow uprising. In 1918, he held Ivanovo-Voznesensk Party and Military Commissar posts before participating in suppression of Left SR uprisings in Moscow (July) and Iaroslavl (August); he then served as Iaroslavl Military District's military commissar between August and December 1918.

Mikhail Frunze made his military reputation during the Russian Civil War, on the eastern front in 1919, serving as Fourth Red Army commander (January–March), then Southern Group commander (March–July) and eastern front commander (July–August), where he coordinated the Red offensives against Kolchak, splitting Kolchak's forces and capturing the North and Central Urals. As Turkestan Front commander (August 1919–September 1920), he planned the defeat of Kolchak's Southern Group and captured the Southern Urals before securing the Soviet grip in Central Asia, taking Turkestan and Kazakhstan. He then served in political posts there before appointment as Southern Front commander (September–November 1920), where he defeated the final White force under Wrangel' in the Crimea.

From December 1920 to March 1924 Frunze held Ukrainian and Crimean military and political positions, also heading the Ukrainian diplomatic delegation to Turkey (November 1921–January 1922).

Frunze held a succession of high military posts after the Civil War. He was appointed deputy chairman of Revvoensovet (RVS, Revolutionary Military Council) of the USSR, and deputy peoples commissar for Military and Naval Affairs (March 1924); then Red Army chief of staff, head of the Red Army Military Academy (renamed Frunze Military Academy, 1925) in April 1924; chairman RVS USSR and peoples commissar for Military, Naval Affairs (January 1925).

Mikhail Frunze died in Moscow while reluctantly undergoing an operation for stomach ulcers. The surgery was on Stalin's orders, lending suspicion as to the actual nature of his death. Later, in the following decade, Stalin would not need to resort to such a subterfuge to eliminate popular army commanders whom he deemed as somehow a threat.

Although lacking formal military training, Frunze was an immense contributor to Soviet military development, combining his Marxist revolutionary convictions with military knowledge acquired during World War I and the Russian Civil War maneuver warfare. He wrote military works, inventing the idea of a distinct Soviet unified military doctrine, developed by Marshal M. Tukhachevsky (later killed by Stalin) and others after his death.

Neil Harvey Croll

See also: Russian Civil War (1918–1921)

References and further reading:

Frunze, M. V. *Izbrannye proizvedeniia v 2-kh tomakh [Selected Works].* 2 vols. Moscow: Voenizdat, 1957, 1971.

Frunze, M. V. *Vospominaniia druzei i soratnikov [Recollections of Friends and Comrades-in-Arms].* Moscow: Voenizdat, 1965.

Jacobs, Walter Darnell. *Frunze: The Soviet Clausewitz, 1885–1925.* The Hague: Martinus Nijhoff, 1969.

Vladimirov, M. I. *O Mikhail Frunze: Vospominaniia, ocherki, stat'i sovremennikov [About Mikhail Frunze: Recollections, Essays, Articles of Contemporaries].* Moscow: Politizdat, 1985.

Fuller, John Frederick Charles (1878–1966)

One of the most important interwar military theorists; highly influential writer on the tactics and operational employment of tanks. Born on 1 September 1878 in Chichester, England, Fuller entered Sandhurst as a cadet in 1897. After being commissioned in 1878, he served in Ireland, India, and in the Boer War. The event most significant for his career as a writer and theorist was his transfer to the Heavy Branch (later Tank Corps) HQ at Bermicourt, France, in January 1917. Fuller realized that the tank was the solution to the deadlock on the western front; the success of the Tank Corps was in great measure due to Fuller's ingenious schemes and tactical concepts, principally as operations officer from April 1917 to August 1918. He was largely responsible for the planning of the first mass tank attack at Cambrai in November 1917; he wrote a memorandum, Plan 1919 (24 May 1918), now regarded as the first blueprint for the blitzkrieg-style operations of World War II.

From 1918 to 1922 he served in the War Office, and from 1923 to 1925 as an instructor at the Staff College, Camberley. During the 1920s he published widely on mechanized warfare in military journals. In books such as *The Reformation of War* (1923) and *The Foundations of the Science of War* (1926) he advocated that war be treated as a science, discussing also future warfare and proposing wide-ranging military reforms. Fuller summarized his thinking on tank warfare in *Lectures on F.S.R. III (Operations between Mechanized Forces)* (1932), a work that indicates that he saw the tank as an operational weapon. There is little doubt he influenced German and Russian armor theory and doctrine in the 1920s. However, he enraged conservative military opinion in Britain and was retired in 1933, albeit with the rank of major general.

In 1934 Fuller joined the British Union of Fascists and turned increasingly to journalism. Playing no military role in World War II, he was a strong critic of strategic bombing against Germany. After 1945 he devoted himself almost exclusively to military history. He died in Cornwall on 10 February 1966, the author of nearly 50 books.

Alaric Searle

See also: Cambrai, Battle of; Theory, Military

References and further reading:

Holden Reid, Brian: *J. F. C. Fuller: Military Thinker.* New York: Macmillan, 1987.

Searle, Alaric. "J. F. C. Fuller, Tukhachevsky and the Red Army, 1923–1941." *Journal of Slavic Military Studies* 9 (December 1996).

Trythall, A. J. *"Boney" Fuller: The Intellectual General 1878–1966.* London: Cassell, 1977.

G

Gage, Thomas (1721–1787)

British general and governor whose policies toward and critical reports about colonists helped lead to the American Revolution. Born in Firle, Sussex, in 1721, he joined the army in 1741 and was aide to the duke of Albemarle in Flanders and Scotland. He went to Canada in 1754 under General Edward Braddock at the disastrous attack on Fort Duquesne. He later served under James Abercromby at Fort Ticonderoga, was promoted to brigadier general in 1758, and later found success in operations in Quebec in 1759–1760.

Made governor of Montreal in 1760, he was promoted to commander in chief of all British forces in North America in 1763. Stationed in New York, he was in charge of enforcement of the Intolerable (Coercive) Acts in 1774 after the Boston Tea Party. Instead of quieting their rebellious activity, American colonialists became more agitated with the quartering of British troops in private homes, the closure of Boston Harbor, and the loss of some political representation. Realizing the tense situation, Gage unsuccessfully urged King George to pressure Parliament to repeal some of the Coercive laws.

In 1774, Gage was named the military governor of Massachusetts and ordered 700 British troops to march on Lexington and Concord on the night of 18–19 April 1775 to seize ammunition stores and arrest Samuel Adams and John Hancock. Local militia met British forces at Lexington and eight rebels were killed. In Concord, the redcoats encountered 400 patriots, with similar results. On the retreat to Boston, ambushed along the way by more than 1,000 patriots, 73 British soldiers were killed and 174 wounded. Instead of this open show of force ending hostilities, it sparked the American Revolution. On 17 June, Gage ordered a frontal attack on rebel forces occupying Breed's Hill (Battle of Bunker Hill). After heavy criticism for casualties, he resigned and was succeeded by General William Howe in October 1775. Returning to England, he died on 2 April 1787, five years after being appointed a full general.

T. Jason Soderstrum

See also: American Revolution

References and further reading:

Alden, John Richard. *General Gage in America: Being Principally a History of His Role in the American Revolution.* New York: Greenwood Press, 1969.

Billias, George Athan. *George Washington's Opponents: British Generals and Admirals in the American Revolution.* New York: Morrow, 1969.

Gage, Thomas. *The Correspondence of General Thomas Gage.* Ed. Clarence Edwin Carter. Hamden, CT: Archon Books, 1969.

Gallic Wars (58–51 B.C.E.)

From 58 to 51 B.C.E., Julius Caesar waged a series of campaigns to consolidate Roman power over Transalpine Gaul. These began when the Helvetii, pushed from their lands north of Lake Geneva by the Suebi, sought to cross Roman territory in Provincia (Provence) en route to a new home in southwest Gaul. While negotiating, Caesar threw up an earthwork from Lake Geneva to the Jura Mountains and brought five legions into northern Italy. Orgetorix nonetheless led the Helvetii west, so Caesar drove them northward, destroyed their rear guard while they were crossing the Arar (Saône) River, and defeated them at Bibracte (Mont Beuvray). He sent them back to their homelands, then marched northeast and pushed the Suebi back across the Rhine, clearing Gaul of Germans. The tribes of Belgic Gaul then combined to attack Caesar's camp on the Axona (Aisne) River in the summer of 57 B.C.E. When the attack failed, they tried to withdraw and were defeated piecemeal by the Romans. The strongest tribe, the Nervii, were subdued in a closely fought battle on the Sabis (Sambre) River. Though the maritime Gauls made their submission at this time, they rebelled in 56 B.C.E. under the leadership of the Veneti Tribe in Brittany. This rebellion was put down only after a difficult campaign in which Roman galleys had to be used against stout Gallic sailing ships with crews thoroughly familiar with the rocky coastal area. Caesar decided to consolidate his gains in the north and west of Gaul and to romanize the region. In 55 B.C.E. he destroyed a group of Germans crossing the Rhine near the Meuse, then bridged the Rhine and menaced the Suebi with a reconnaissance in force to end German infiltration. Late in the summer, he transported two legions to Britain in 80 vessels and engaged the Britons by both arms and diplomacy. The following year, he returned with five legions plus cavalry and defeated King Cassivellaunus in the field. At this point his conquest of Gaul was complete except for the necessity of putting down several rebellions. The most serious of these was an uprising in central Gaul led by Vercingetorix that was crushed by the Roman capture of Alesia (Alise-Ste-Reine) after a bloody siege in 52 B.C.E., though the mopping up took another year.

Joseph M. McCarthy

References and further reading:

Caesar, C. Julius. *The Battle for Gaul.* Boston: David R. Godine, 1980.
Goudineau, Christian. *César et le Gaule.* Paris: Errance, 1980.

Galliéni, Joseph Simon (1849–1916)

French field marshal. Born 24 April 1849, at St. Beat, Haute Garonne, Joseph Galliéni attended St.-Cyr, graduating from there in 1870. Commissioned in the French Naval Infantry (marines), he fought with distinction in the Franco-Prussian War and then served in the French Empire, most notably in Niger, Senegal, Tonkin, and Madagascar. Between 1885 and 1905 he wrote five books describing his colonial experiences.

Galliéni returned to France in 1905 to command first XIII, then XIV Corps. He joined the general staff in 1908 and served there until his retirement in 1914.

At the outbreak of World War I, Galliéni was recalled to active service. On 28 August 1914, he became military governor of Paris and designated successor to French army commander General Joseph Joffre. When Galliéni learned that the German First Army had turned southeast of Paris, making it vulnerable to a thrust by his Sixth Army and Paris garrison, he pressed upon Joffre the need for an immediate counterattack. Although Joffre had overall responsibility for this decision, Galliéni dictated both the site and timing of the action. On the afternoon of 5 September, Sixth Army opened the Battle of the Marne. Certainly, Galliéni's efforts helped prevent a German victory in 1914.

Galliéni was minister of war from October 1915 to March 1916, until poor health and disputes with Joffre over the defense of Verdun forced his retirement. He died at Versailles on 27 May 1916. Galliéni, perhaps the only French commander in 1914 to have escaped severe criticism, was posthumously promoted marshal of France in May 1921.

Spencer C. Tucker

See also: Marne, Battle of the

References and further reading:

Bernhard, Jacques. *Galliéni.* Vagney, France: G. Louis, 1991.
Galliéni, Joseph Simon. *Les Mémoires du Maréchal Galliéni, Défense de Paris.* Paris: Payot, 1926.
Gheusi, Pierre B. *Galliéni.* Paris: Charpentier, 1922.
Palmer, Alan. "The Marne." In *The March to the Marne: The French Army, 1871–1914,* eds. Noble Frankland and Douglas Porch. Cambridge, UK: Cambridge University Press, 1981.

Gallipoli (1915–1916)

Abortive Franco–British Empire attempt to force an Allied path to Istanbul. The battle, which consisted of two main offensives from April 1915 to January 1916, was fought between the Allies and the Turks under General Otto von Sanders along six beaches on Gallipoli Peninsula. The Allied forces, which initially consisted of British, French, Australian and New Zealand (ANZAC), and Indian divisions, were commanded by General Sir Ian Hamilton and later General Sir Charles Monro.

The aim of the campaign was to land ground troops on the Gallipoli Peninsula and destroy the Turkish batteries, thereby making way for the navy to capture Constantinople. The Turks would then be unable to prevent the Allies from joining the Russians in their fight against Austria, Hungary, and Turkey. This objective was to be achieved by landing the main British force of 35,000 troops on five beaches at Cape Helles, while a smaller ANZAC corps of 17,000 men were to land 13 miles farther north. The landings were accomplished, but at heavy cost.

Fighting began on 25 April 1915. The British forces at Cape Helles tried to take the high position at Achi Baba, yet without a divisional commander on hand—he was safely off shore on board his command ship. The troops were disorganized and ill prepared to meet heavy resistance from the Turks. By the end of the first day's fighting they had gained little ground. The ANZACs who landed at Ari Birun (now known as Anzac Cove), a mile north of their objective, moved up the rocky slopes toward the high position of Chunuk Bair, but Turkish resistance under the command of

Picture Section: Part 4, Shakespeare Section.

The New York Times

Sunday, March 5, 1916

BRITISH STORES WHICH COULD NOT BE REMOVED FROM GALLIPOLI BURNING ON THE SHORES OF SUVLA BAY.
From a British Official Photograph Taken from the Deck of H. M. S. Cornwallis, the Last Ship to Leave the Dardanelles.
(Photo © by American Press Assn.)

A photograph taken from the deck of H.M.S. Cornwallis showing British stores from Gallipoli burning in the background, 1916. (Library of Congress)

Mustapha Kemel drove them back to the beaches by nightfall. Without these two critical heights the Allies remained pinned down on the beaches.

Hoping to break the deadlock, Hamilton mounted a second assault on 6 August 1915 during which the newly reinforced ANZACs were to embark upon a night attack to take the heights of Chunuk Bair and Sari Bair ridge. British troops at Helles were to pin down the Turks, while three new British reinforcement divisions were to be landed at Suvla Bay to the north. While the holding attack at Helles and the Suvla Bay landing were initially successful, failure by the British to move inland and pursue these initial gains gave the Turks valuable time to get reinforcements up to the front lines. The failure of the Suvla Bay landing was made the more tragic because of the heavy losses sustained by the ANZACs as they became bogged down in the dark. For months the Allies remained on the beaches, unable to drive the Turks from their hillside positions.

The stalemate was relieved only by the appearance of Lord Kitchener, who, after having inspected the battlefields, was so appalled with the situation that he recommended the complete withdrawal of all Allied troops. On 20 December 1915 Suvla Bay and Anzac Cove were evacuated, as too was Helles on 9 January 1916. The evacuation was completed without loss of life. In the nine months of fighting the Allies gained a few hundred yards at a cost of 41,000 British killed, 78,500 wounded; 9,000 French soldiers killed, 13,000 wounded; and 100,000 French, British, and ANZAC soldiers ill. The Turks lost 66,000 men who were killed in action and 152,000 wounded. The western front itself would be hard pressed to produce so little gain at so high a cost.

Margaret Hardy

See also: Churchill, Sir Winston; Enver Pasha; World War I

References and further reading:

Laffin, John. *Damn the Dardanelles! The Story of Gallipoli.* London: Osprey, 1980.

Murray, Joseph. *Call to Arms: From Gallipoli to the Western Front.* London: W. Kimber, 1980.

Reid, Richard, Courtney Page, and Robert Pounds. *A "Duty Clear Before Us": North Beach and the Sari Bair Range, Gallipoli Peninsula: 25 April–20 December 1915.* Canberra, A.C.T. : Dept. of Veterans' Affairs, 2000.

Steel, Nigel. *Gallipoli.* Barnsley, UK : Leo Cooper, 1999.

Gamelin, Maurice (1872–1958)

The general who calmly led France to disaster in 1940. Born into a military family, Maurice Gamelin was the top cadet of St.-Cyr (France's preeminent military academy) in 1893. He served in Africa before going back to the prestigious Ecole de Guerre, where he proved to be one of the brightest students: clever, educated, and hardworking. In 1906 he was appointed aide-de-camp to General Joffre, who would be his mentor. Gamelin became Joffre's aide during the Battle of the Marne before being sent to the front as a brigade commander. As a general in 1916, he commanded an infantry division through 1918. After the war, he made his way, slowly but surely, up the military hierarchy. He was made chief of the French general staff (1931) and eventually chief of staff of National Defense (1938). As the highest commanding officer, he was responsible for global French strategy and the main military adviser of the numerous French governments of the 1930s. Gamelin was seen as a fine diplomat and a cautious general, always preferring playing for time, and treading carefully where government was concerned. He promoted the overly scrupulous French diplomacy of appeasing Hitler. During the initial period of inactivity in World War II, Gamelin was commander in chief of the French and British armies on the western front. He was responsible for the Wait and See military strategy that produced the Phony War of the 1939–1940 winter. Fearing reprisals, he forbade any offensive action against Germany (bombing of the Ruhr or mining of the Rhine river). Gamelin located his headquarters in the Chateau de Vincennes in a Paris suburb to stay as close as possible to the government. He rarely visited front units and engendered little popularity.

From the beginning of the German invasion (10 May 1940), he was overtaken by events and took cover behind front commanders. His most famous written order of the day began: "without being willing to intervene in the battle in progress." He was sacked on 19 May 1940. In 1942 Gamelin was court-martialed for his failure and was turned over to the Germans. Set free in 1945, he spent the rest of his life trying to justify his actions. Gamelin was an outstanding field officer but a wretched commander in chief.

Gilles Boué

See also: France; World War II

References and further reading:

Masson, Philippe. *Histoire de l'armée française de 1914 à nos jours.* Paris: Perrin, 1999.

Porch, Douglas. "Why Did France Fall?" *MHQ* 2 (Spring 1990).

Vanwelkenhuezen, Jean. *1940 Pleins feux sur un désastre.* Bruxelles: Racine, 1995.

Garibaldi, Giuseppe (1807–1882)

The most prominent figure of the Italian Risorgimento (resurgence), Italy's unification during the mid–nineteenth century. Garibaldi was born in Nice, in Napoleonic France, on 4 July 1807 to a family of sailors and grew up on the Mediterranean. He enrolled in Giuseppe Mazzini's Young Italy in 1833 and joined the Piedmontese navy in a failed nationalist insurrection a year later. In 1835, Garibaldi fled to South America, where he led independence movements for, first, Rio Grande do Sul from Brazil and then Uruguay from the Argentine. Upon his return to Italy in 1848, Garibaldi joined uprisings against Austria for Italian unity and established a brief Roman Republic, which was ended by the French, who supported the pope. By 1860, Count Camillo di Cavour had organized the unification of the northern Italian states into a kingdom ruled by Victor Emmanuel II of Piedmont. Garibaldi made it his goal to unify southern Italy with the northern states.

In 1860, Garibaldi led just over 1,000 nationalists, known as The Thousand, into Sicily, where he defeated a Neapolitan army at Calatafimi, near Palermo. He captured Naples and won a major victory in October on the Volturno River, only to be bypassed by Cavour, who opposed a republic and united northern and southern Italy himself in November. Although Garibaldi did not want a kingdom, his nationalism won out over his republicanism. In March 1861, the kingdom of Italy, excluding Austrian Venetia and the region around Rome, was declared.

In 1862, Garibaldi led another failed uprising to annex Rome into Italy and was wounded at Aspromonte. After helping to obtain Venetia in 1866, Garibaldi again sought Rome but was ultimately defeated at Mentana in 1867. He nonetheless lived to see the final unification of the Italian peninsula in 1871. Garibaldi spent the rest of his life supporting political reform in Italy and died in Caprera on 2 June 1882.

Christopher P. Goedert

See also: Revolutions of 1848

References and further reading:
Mack Smith, Denis. *Garibaldi: A Great Life in Brief.* New York: Alfred A. Knopf, 1956.
Ridley, Jasper. *Garibaldi.* New York: Viking Press, 1976.
Trevelyan, George M. *Garibaldi and the Making of Italy.* New York: Longmans, Green, and Co., 1911.
Viotti, Andrea. *Garibaldi: The Revolutionary and His Men.* Dorset, UK: Blandford Press, 1979.

Gates, Horatio (1728–1806)

Ineffectual American field commander in the Revolutionary War and supposed victor at Saratoga. Gates was born in England, probably in April 1728. Intending a career in the British army, he was commissioned in 1749, stationed in North America, and soon made captain. Serving under Edward Braddock in 1755, he made friends with George Washington and was wounded at the Monongahela on 9 July. He participated in Robert Monckton's 1761 invasion of Martinique, became a major in 1762, resigned in 1772, and, with Washington's help, settled down as a Virginia planter in 1773.

Washington recalled Gates to active duty on 17 June 1775 as a brigadier general. For distinguished service in the Boston campaign, he was promoted to major general on 16 May 1776. From June 1776 until 4 August 1777, Gates and Philip Schuyler jockeyed for command of American forces in northern New York, but after Schuyler lost Fort Ticonderoga to John Burgoyne, Congress decided in favor of Gates.

Against Burgoyne's march toward Albany from Canada, Gates fortified Bemis Heights. When the British attacked at Freeman's Farm on 19 September, the tactics and courage of Benedict Arnold and Daniel Morgan won the decisive victory, despite Gates's overcautiousness. When Gates failed to mention Arnold in his official report, Arnold protested, and Gates relieved him of command. During the second battle of Saratoga at Bemis Heights on 7 October, Arnold simply seized command, again won the day, and again Gates took credit for Arnold's heroism. Gates received Burgoyne's surrender on 17 October.

Gates may or may not have been involved in the Conway Cabal against Washington, but he certainly stood to gain if it succeeded. In July 1780 he took command of the American troops in the Carolinas. A laughingstock and a disgrace after he turned tail and ran from the field at Camden, he was replaced by Nathanael Greene in October. He served in the New York state legislature in 1800–1801 and died in New York City on 10 April 1806.

Eric v. d. Luft

See also: American Revolution; Arnold, Benedict; Braddock's Defeat; Burgoyne, John; Camden, Battle of; Fort Ticonderoga; French and Indian War; Greene, Nathanael; Saratoga; Washington, George

References and further reading:
Mintz, Max M. *The Generals of Saratoga: John Burgoyne and Horatio Gates.* New Haven, CT: Yale University Press, 1990.
Nelson, Paul David. *General Horatio Gates: A Biography.* Baton Rouge: Louisiana State University Press, 1976.
Patterson, Samuel White. *Horatio Gates: Defender of American Liberties.* New York: Columbia University Press, 1941.

Gaugamela, Battle of (1 October 331 B.C.E.)

The third and decisive victory of the Macedonian king Alexander III the Great over the Persians, led by Darius III, took place near Gaugamela in present-day northern Iraq. Although we know that the Persian force was huge, mustered from all over the empire, their exact numbers are difficult to assess. However, it severely outnumbered Alexander, especially in cavalry. Relying on this advantage, Darius positioned a line of some 40,000 horse and his 200 scythed chariots in the front. In the middle of this line Darius himself took position with his guard infantry (2,000) and his Greek mercenaries (2,000). The rear line consisted of an unknown number of Persian infantry.

Alexander had some 40,000 infantry and 7,000 cavalry at his disposal. In the center he positioned his Macedonian pikemen, flanked on the left by cavalry and light troops, and on the right by the guard infantry, Alexander with the guard cavalry and additional horse and skirmishers. In the rear, Alexander positioned a second line of allied Greek infantry as a reserve.

Darius planned to use his numerical superiority in cavalry to outflank the Macedonians on both sides. The Macedonians marched in oblique order, the right wing taking the more advanced position. The Persian cavalry attacked this part of the Macedonian army first. They were repelled and Alexander, leading his guard, attacked the Persian line and broke through. Thereupon he rode immediately toward the Persian center, ferociously attacking Darius himself. As at Issus, the Persian king turned and fled.

In the meantime the Persians had not been without success. A detachment of horse circumvented the Macedonian left and attacked the Macedonian camp. Persian cavalry in the center took advantage of a gap in the Macedonian line and fought their way through, only to be repelled by the second line. However, Persian resistance did not survive the flight of the king. Following his example, the army routed, suffering heavy losses during the Macedonian pursuit. Macedonian losses were limited, perhaps a few hundred dead.

Maarten van der Werf

See also: Alexander the Great; Alexander's Wars of Conquest

References and further reading:

Fuller, J. F. C. *The Generalship of Alexander the Great.* London: Eyre & Spottiswoode, 1958.

Hammond, N. G. L. *The Genius of Alexander the Great.* London: Duckworth, 1997.

Marsden, E. W. *The Campaign of Gaugamela.* Liverpool, UK: Liverpool University Press, 1964.

Tarn, W. W. *Alexander the Great.* Cambridge, UK: Cambridge University Press, 1948.

de Gaulle, General Charles (1890–1970)

The personification of twentieth-century France. After attending St.-Cyr military school, where he distinguished himself in his knowledge of tactics, history, and geography, de Gaulle graduated in 1912, and was assigned to the 33d Infantry Regiment under the leadership of then Colonel Philippe Pétain. Entering World War I as captain, de Gaulle fought at the front but was captured at Douaumont on 2 March 1916. He made several escape attempts (five or seven depending on the accounts), all of which failed.

Returning from the war, Captain de Gaulle was assigned, at his request, to face off against the Red Army in Poland by helping train Polish officers, in 1921. He then became a history professor at St.-Cyr. Between 1925 and 1931, de Gaulle was assigned various tasks, from working as a member of Pétain's war cabinet, where he stunned his colleagues with his amazing memory, to serving in Syria before returning to work with the National Superior Defense Council. During the interwar years, de Gaulle published no fewer than four books on the military arts, preaching the importance of tank units as offensive weapons. These works had a limited impact, for they criticized the notion of defensive strategies, which the French high command swore by, thanks to their success in World War I. De Gaulle's 1938 volume, *La France et son armée,* caused dissension between him and his mentor, Pétain, and prompted other opponents to nickname him "Colonel Motor," after his obsession with mechanized units.

At the beginning of World War II, de Gaulle was assigned to head the 4th Armored Division. Local successes in staving off German advances in 1940 granted him a temporary field promotion to the rank of general and he was named assistant secretary of war in the last cabinet of the French Republic on 6 June 1940.

De Gaulle's call of destiny came when he rejected the armistice France had asked of Germany and left for London, where on 18 June he broadcast on the BBC a call to all French to carry on the fight. He also formed a Committee for Free France, which relied on support from exiled French and some of the French Empire's possessions. However, distrust between him and President Franklin Roosevelt prompted the United States to first deal with the collaborationist Vichy government under the aged Pétain, Admiral Jean-Louis Darlan, and General Henri Giraud after the American landings in North Africa. De Gaulle then moved to negotiate an alliance with the head of the interior French resistance, Jean Moulin, and used his own growing prestige to push Giraud aside. By 1944, de Gaulle was able to assume leadership of the provisional French government, in the teeth of American and British resistance to him, following the liberation of Paris in August.

Dissatisfied with political infighting, de Gaulle resigned his position in 1946 and formed the Assembly of the French People Party (RPF) which called for a reorganization of the Fourth French Republic and preparation for a third world war, which de Gaulle thought imminent. In the 1948 elections, the party collected some 40 percent of the vote, but three years later fell back to less than 20 percent, with the republic still alive and no world war.

By 1953, de Gaulle had retired from political life, but five years later, during a crisis over the French war in Algeria, a consensus was reached between both colonial and anticolonial parties that de Gaulle alone could resolve the troubles. De Gaulle then used his temporary full powers to force through a new constitution, approved in September 1958. He was then elected first president of France's Fifth Republic in December. Over the following four years, he initiated a series of economic measures to assist the country's recovery and reaped the benefits of several projects begun under the fourth republic, including the acquisition of the atomic bomb in 1960. De Gaulle also established new military programs to assert further French independence from NATO (which he termed the "machin" or "watchmacallit"). First removing the nation from NATO's Mediterranean command, de Gaulle then refused to stockpile NATO nuclear weapons. When the U.S. proposal of a multilateral force involving Germany in nuclear weapons decisions failed in 1966, de Gaulle seized the initiative and took France out of NATO's integrated command in 1966, keeping it in the Atlantic Alliance only in name. Following the student and worker movements of 1968 de Gaulle held a referendum on political reform the following year. The rejection of his proposals prompted him to resign from office and retire to write his memoirs. He did not complete them, dying of a ruptured artery in his home in 1969.

It would be difficult to imagine any Frenchman more French than Charles de Gaulle (down to his very name). An aide responded to several foreigners that they should not take the general's hauteur personally; de Gaulle himself didn't like the French either, and the French didn't like themselves!

Guillaume de Syon

See also: Armored Fighting Vehicles; Verdun; World War I; World War II

References and further reading:

Cogan, Charles G. *Charles de Gaulle: A Brief Biography with Documents.* New York: St. Martin's, 1996.

De Gaulle, Charles. *The Army of the Future.* Philadelphia, PA: Lippincott, 1941.

———. *The Complete War Memoirs of Charles de Gaulle.* 3 vols. New York: Simon & Shuster, 1964.

Gordon, Philip H. *A Certain Idea of France: French Security Policy and the Gaullist Legacy.* Princeton, NJ: Princeton University Press, 1993.

Lacouture, Jean. *De Gaulle.* Vols. 1 and 2. New York: Simon & Shuster, 1991, 1992.

Gempei War (1180–1185)

War fought between the Minamoto and the Taira clans for control of Japan. It ended in the defeat of the Taira, and the formal establishment of the first shogunate under Minamoto Yoritomo in 1192 at Kamakura. The name is derived from the Chinese reading of the names of the two clans (Genji, and Heike or Heiji).

In the mid–twelfth century, the Taira supplanted the Fujiwara as the leading noble family at court in Kyoto. In 1159, conflict between the Taira and the Minamoto clans led to the defeat of the Minamoto, during what is called the Heiji Disturbance (Heiji-no-ran), and the murder of the head of the Minamoto, Yoshitomo (1123–1160). The head of the Taira, Kiyomori Taira (1118–1181), was named chancellor (*daijo-daijin*), and the Taira came to dominate the political life of Japan. In 1180, Yoshitomo's son, Yoritomo, revolted against the Taira, arguing that they had usurped the power of the emperor, and that the Minamoto would restore the proper order. Prince Mochihito, second son of the cloistered (or retired) emperor Go-Shirakawa, issued a decree outlawing the Taira and authorizing the Minamoto to destroy them. This decree provided moral and legal justification for the Minamoto cause, though it was later rescinded by Cloistered Emperor Go-Shirakawa himself, who issued a counterdecree outlawing the Minamoto.

Yoritomo was initially defeated at the Battle of Ishibashi Mountain in Sagami Prefecture and fled into the Hakone Mountains, narrowly escaping capture. Making his way to Kamakura, he marched almost immediately against the Taira in Suruga Prefecture. There the Taira and Minamoto armies encamped across from one another along the Fuji River near Kajima. In the night, a member of the Minamoto army startled flocks of birds nesting in the marshes. The Taira, mistaking the commotion for a night attack by a vastly superior foe, fled back to Kyoto and abandoned the campaign in eastern Japan, preferring instead to maintain control of the capital and western Japan. The Minamoto did not pursue, remaining instead in eastern Japan to build up an army of supporters against the Taira.

In 1183, Yoritomo's cousin and vassal, Yoshinaka, on his own initiative, attacked and defeated the Taira in Etchu Prefecture, pursued them back to Kyoto, and captured the capital. There another decree was issued giving Yoshinaka authority to attack the now-outlawed Taira. He also sought a decree outlawing Yoritomo. Yoritomo's brother, Yoshitsune, was sent to subdue Yoshinaka, whose independent actions now threatened Yoritomo's position. Yoshitsune broke through Yoshinaka's defenses and took Kyoto with surprising speed; Yoshinaka was killed attempting to flee.

Yoshitsune then launched an attack against the Taira, winning decisive victories in the battles of Ichinotani in Settsu Prefecture, at Yashima in Sanuki Prefecture (on Shikoku), and at the naval battle of Dannoura off the western tip of Honshu. This final battle ended in a crushing defeat for the Taira, brought an end to the war, and gave the Minamoto control of Japan.

Michael C. Paul

See also: Minamoto, Yoshitsune; Samurai

References and further reading:

Kitagawa, Hiroshi, and Bruce T. Tsuchida, trans. *The Tale of the Heike (Heike monogatari).* Tokyo: University of Tokyo Press, 1975.

Morris, Ivan. *The Nobility of Failure.* New York: Holt, Rinehart & Winston, 1975.

Shinoda, Minoru. *The Founding of the Kamakura Shogunate 1180–1185.* New York: Columbia University Press, 1960.

General Order No. 100 (24 April 1863)

The first codification of the law of war. Issued at the direction of President Abraham Lincoln, and at the request of General in Chief Henry Halleck, it was the enactment of a report by Francis Lieber, a German-American law professor at Columbia College in New York.

Titled *Instructions for the Government of Armies of the United States in the Field,* it augmented the U.S. Army's Articles of War, written (with few modifications) in 1775 to maintain discipline among the troops. The new order went much further, setting uniform and humane standards for the treatment and exchange of prisoners, the freeing of slaves, the treatment of people and property in occupied territory, and the treatment of opposing combatants.

The Order was a success in America. With one exception —Sherman's bombardment of Atlanta without warning— the Union army generally followed its strictures. The U.S. Supreme Court adopted it in *ex Parte Vallandigham* (1863).

GO 100 also influenced international law. Translated into German by Johann Bluntschli, it formed the basis of his *Das Moderne Kriegsrecht* (1866). Indeed the Lieber Code was reprinted intact in most international law texts for the next 50 years. Its terms and ideas influenced the European powers at the first Geneva Convention (1864) to agree to standards of treatment for wounded prisoners of war. International conferences at the Hague in 1899 and 1907 codified much of the Code regarding the definition of combatants and treatment of neutrals into the international laws of war.

Steve Sheppard

See also: American Civil War; Laws of War

References and further reading:

Friedel, Frank. *Francis Lieber, Nineteenth Century Liberal.* Baton Rouge: Louisiana State University Press, 1947.

General Order No. 100—Adjutant General's Office: Instruction for the Government of Armies of the United States in the Field, Revised by a Board of Officers. Washington, DC: War Department, 24 April 1863.

Geneva Conventions (1864–1949)

A series of international conventions that laid out mutually recognized rules of modern warfare. The horrors of the Crimean War (1854–1856) highlighted the need for some sort of rules protecting the wounded and those who tend them. This sentiment culminated in the Geneva Conventions as first established in 1864 and then modified and extended three more times through 1949. Although idealistic in nature, the conventions themselves remained grounded in reality. They never ask that a nation do more than it is willing to do in the light of military necessity.

The first convention stipulated that belligerents treat hospitals, ambulances, and all medical personnel as neutrals in war. It also mandated that wounded and sick soldiers were to be cared for regardless of nationality, and that those giving such care, including civilians, would be protected from harm. Furthermore, those too wounded to continue fighting might return to their nation of origin as soon as practicable. This convention, however, did not extend to maritime conflict.

The 1906 Convention extended earlier protections to those at sea. The neutrality provision became "respect and protection" but extended such protection to nonduty hours also. It also provided for the burial of the dead and notification of such to the family.

The 1929 Convention extended the protection clause to medical aircraft and authorized the use of the Red Cross emblem in peacetime activities. It also boldly declared the conventions binding, even if all the belligerents in a war were not signatories. Most noticeably, it called for the humane treatment of all prisoners of war, not just the wounded.

The final convention in 1949 watered down the earlier protection clause to allow belligerents to hold medical personnel as POWs rather than return them to their nation of origin. Nevertheless, this convention extended the earlier agreements regarding treatment of wounded and sick to civilians. It stated that any victims of war, military or civilian, should receive the widest possible safeguards consistent with wartime conditions.

At least in open societies, with a vigorous political opposition, the Geneva Conventions act as a line that military commanders or their troops must consider before crossing.

Elizabeth Pugliese

See also: Ethics of Warfare; Laws of War; Military and Society; War Crimes

References and further reading:

Draper, G. I. *The Red Cross Conventions.* New York: Frederick A. Praeger, 1958.

Kalshoven, Firits. *Constraints on Waging War.* Geneva: International Committee of the Red Cross, 1987.

Pictet, Jeans S., ed. *The Geneva Conventions of 12 August 1949 Commentary.* Vols. 1–4. Geneva: International Committee of the Red Cross, 1952.

Genghis Khan (c. 1162–1227)

Founder of the Mongol Empire. Temüjin, later Cinggis-qan, a title assumed by him after his election as Mongolian supreme *qan* in 1206, and familiar to us in the Arabic form Genghis Khan, was born some time toward the middle of the twelfth century, son of a minor Mongolian chieftain. He grew up under most difficult circumstances, including the poisoning of his father by Tartar enemies and the abandonment of himself, his brothers, and his mother on the open steppe by his father's former retainers, a virtual death sentence. Despite these setbacks, young Temüjin, in part drawing upon the resources of his father-in-law in a marriage arranged shortly before his father's death, not only survived but prospered and eventually built up a following. This was possibly as an ally of the Chinese Jin Dynasty in the steppe, but this aspect of his career is obscure in our sources, perhaps intentionally so since the Jin Dynasty was later the mortal enemy of the Mongols.

By 1205, Temüjin had defeated his rivals in years of brutal conflict and his movement had begun to penetrate outside Mongolia. Initially this was in the form of small raids into sedentary domains closest to Mongolia, but as time

Portrait of Genghis Khan.
(Print and Picture Collection, The Free Library of Philadelphia)

passed, and as the experience of the already militarized Mongols grew, more ambitious advances were attempted. The most ambitious of these, ultimately taking the Mongols as far as Russia, first began in a pursuit of defeated enemies and then continued as a vast reconnaissance in force into what is now Turkistan and then north into what is now northern Kazakhstan. These advances laid the groundwork for a still more serious assault beginning in 1217 led by the *qan* himself. It quickly overran the entire Khwarezmian Empire. Further advances continued into Iran, from where a Mongol detachment led by Jebe and Sübötei (1176–1248) advanced into Armenia and Georgia, and then north into Russia before returning to Mongolia, through the steppes north and east of the Caspian. In the meantime, other Mongol armies, in some years with the personal participation of the great *qan* himself, had conquered most of north China. Genghis Khan died in 1227 while completing the conquest of the Xixia Empire. His descendants consolidated his conquests and carried his empire to the suburbs of Vienna, into the Sinai, as far south as Java, and to points in between.

Paul D. Buell

See also: Kublai Khan; Mongol Empire; Ögödei

References and further reading:

Buell, Paul D. "Sübötei-ba'atur." In *In the Service of the Khan, Eminent Personalities of the Early Mongol-Yuan Period (1200–1300)*, ed. Igor de Rachewiltz, Chan Hok-lam, Hsiao Ch'i-ch'ing, and Peter W. Geier, 13–26. Wiesbaden: Otto Harrassowitz, 1993.

———. "Early Mongol Expansion in Western Siberia and Turkestan (1207–1219): a Reconstruction." *Central Asiatic Journal* 36: 1–2 (1992), 1–32.

Ratchnevsky, Paul. *Genghis Khan, His Life and Legacy.* Trans. Thomas Nivison Haining. Oxford, UK, and Cambridge, USA: Blackwell, 1991.

German Army

Its history has been influenced by the role of Germany as a predominantly land power, the need to defend long and vulnerable borders, innovation in fighting methods and military planning, but also by a failure to grasp the political dimension of strategy and by a persistent underestimation of its enemies' power in alliance.

The real origins of the German army lie in the military policies of Frederick William, the Great Elector of Brandenburg, who by 1678 had an army of 45,000 men, made up of mercenaries, socially disadvantaged members of his state, and some Junker (landowner) officers. Frederick William I later completed the process of creating an officer corps in which the aristocracy dominated; in 1739 all 34 generals were aristocrats. In 1740, Frederick II inherited a Prussian army of 83,000, an indication of the economic priority that the army had been accorded, given the population of 2,200,000.

The defeat of Prussia at Jena and Auerstädt in 1806 led to the introduction of the military reforms advocated by Gerhard Scharnhorst, August Niedhardt von Gneisenau, and Hermann von Boyen, most notably compulsory military service and the end to exemptions for the middle class; by midcentury the other states in the German confederation had followed the Prussian example. Prussia also led the way with its general staff system, exploiting in particular the military applications of railways. The fruit of these reforms was obviously seen in Prussia's victorious wars of 1863, 1866, and 1870–1871. Its dominance was confirmed when, after the creation of the German army, consisting of 18 army corps (13 in Prussia, two in Bavaria, one in Württemberg, one in Saxony, one in Alsace-Lorraine), the king of Prussia and German emperor was given the supreme command of all army corps in time of war. On 27 April 1871, the first memorandum by the Prussian-German general staff on a two-front war against France and Russia was issued. Following the Franco-Russian Alliance of 1894, military planning came to be dominated by fear of a two-front war, leading to the Schlieffen Plan of 1905.

During the course of World War I, the German army performed consistently more effectively than its opponents at both the tactical and operational levels, achieving some notable innovations in infantry and artillery tactics. However, the army leadership failed at the strategic level due to its unwillingness to give up exaggerated territorial goals in the East with a compromise peace.

Following the defeat in 1918, left with 100,000 men as a result of the Versailles Treaty, the German army (Reichsheer) was forced to conduct much of its general staff and field training in secret; air and armor training was conducted in Russia, another pariah nation of the time. During the 1930s, sections of the army were impressed by Hitler because he favored mechanization, introducing an ideological element into military reform and a split between younger officers and older, more traditional generals.

In 1939 the German army was not entirely ready for war, having been subjected to a rapid program of expansion during the 1930s. However, it studied the lessons of the Polish campaign closely, which contributed to its remarkable defeat of France in 1940. The successful campaigns of 1939–1942 were marked by innovative new tactics, particularly in the use of armored forces and aircraft in a ground-support role (blitzkrieg, lightning war). But the influence of the chief of the army general staff had been reduced by the creation of the Wehrmacht high command, and the army lost influence on the direction of the war at the strategic level. All major decisions came to be made by Adolf Hitler, and the German army was decisively overwhelmed on all fronts in 1944–1945. No better army ever fought in a worse cause.

The victorious Allies completely demilitarized Germany after the unconditional surrender of the Wehrmacht on 8 May 1945. (Its two highest commanders, Wilhelm Keitel and Alfred Jodl, were executed in 1946.)

But eventually two new German armies were founded in postwar Germany, reflecting the Cold War division of the nation. The Federal Republic of (West) Germany's armed forces (Bundeswehr) were activated on 11 November (of all dates!) of 1955. The German Democratic Republic's People's Army (Nationale Volksarmee) followed on 18 January 1956. In both armed forces, the army was numerically dominant; in the East more emphasis was placed on Prussian army traditions (including the odious goose-step). With the fall of the German Democratic Republic (GDR) on 3 October 1990, the People's Army was officially dissolved and its personnel and equipment taken over by the Bundeswehr.

Limited German forces were deployed in the Gulf War (1990–1991), but their presence was kept as almost a military secret, in view of the memories still held by older generations about the German army.

Alaric Searle

See also: Austerlitz, Battle of; Franco-Prussian War; Frederick the Great, King of Prussia; Gneisenau, August Neidhart, von; Hindenberg, Paul von Beneckendorf und von; Jena and Auerstädt; Jodl, Alfred; Keitel, Wilhelm; Ludendorff, Erich; Moltke, Graf Helmut Johannes Ludwig von; Moltke, Graf Helmut Karl Bernhard von; Scharnhorst, Gerhard Johann von; Schlieffen, Graf Alfred von; Seeckt, Hans von; World War I; World War II

References and further reading:

Dupuy, Trevor N. *A Genius for War: The German Army and General Staff, 1807–1945.* Englewood Cliffs, NJ: Prentice-Hall, c. 1977.

Rosinski, Herbert, ed. *The German Army.* London and New York: Praeger, 1966.

German Colonial Wars (1884–1919)

A short-lived "place in the sun." The young German Empire did not establish colonies until 1884, and then mostly in areas of little economic or strategic value. It took the expenditure of military resources both to found and to maintain Germany's place in the sun.

Beginning in 1884 the German flag was raised over several territories in Africa: Togo, Cameroon, German Southwest Africa (Namibia), and German East Africa (Tanzania, Rwanda, and Burundi). In the Pacific, Germany also claimed possession of northeast New Guinea and the Marshall and Solomon Islands, with the close support of the German navy.

The first military action was in Cameroon in December 1884, when a naval landing party of 350 men defeated the forces of a pro-British chieftain. One of the leaders of this party was Lieutenant Reinhard Scheer, later commander of the High Seas Fleet during World War I.

The creation of German East Africa brought about more conflict, both with the sultan of Zanzibar and with Arab slave traders. A blockade of Zanzibar harbor in 1885 established the German presence. A subsequent blockade of the entire coast, this time in conjunction with the British Royal Navy in 1888, was directed against the slave trade.

As control of German territories extended further inland, it became necessary for the Germans to create a permanent protection force, the Schutztruppen. The Schutztruppen was composed mainly of native troops led by German officers.

In 1904, a large revolt by the Herero tribe broke out in southwest Africa. The Hereros were upset at losing their traditional pastures, and German traders had forced many Hereros into debt through unscrupulous practices. The impassable wastes between the Atlantic coast and the settled regions made transportation difficult: the Germans relied on a single narrow-gauge railway line that could support only one train per day. Through quick raids the Hereros were able to keep the Germans off balance.

The German public was shocked to hear of the revolt and were angry at the army's inability to defeat the natives. This outcry led to 10,000 troops being shipped to southwest Africa. However, most of these troops were used just in pacifying the countryside. The only real battle of the revolt, at Waterberg on 11 August 1904, was fought before they arrived. A subsequent revolt in late 1904 by the Hottentots took nearly three years to quell simply because the vastly outnumbered natives used guerrilla tactics and avoided pitched battles.

Meanwhile, another revolt in East Africa was subdued with far fewer German soldiers. A number of tribes were persuaded of the magical properties of a potion of water, castor oil, and millet seeds. This "maji-maji" was said to turn German bullets to water. The uprising broke out in central East Africa in late July 1905. However, without any coordination between the various tribes, no serious threat was posed to German settlers. When the maji-maji was proven to be no defense against guns, the rebels quickly gave up the fight.

An unfortunate feature common to both uprisings was the harshness of German reprisals. In southwest Africa, the Herero population plummeted from 80,000 in 1904 to 15,000 in 1911. In East Africa, the rebelling tribes were punished by having their crops destroyed; the resulting famine killed between 250,000 and 300,000 people, more than 10 times the number of natives who had taken up arms.

The empire was in no position to defend itself against Allied assault when World War I began in 1914. In Oceania, the German island chains were unprotected and easily taken by Australian and Japanese forces. The only resistance of note was by the port of Tsingtao, taken from China in 1897. Although a Japanese force laid siege at the outbreak of war, Tsingtao was not taken until 7 November.

In Africa, Togo was quickly overrun, Cameroon offered little resistance, and German Southwest Africa was taken by South African units. However, German East Africa posed an unexpected obstacle. The German commander, Colonel Paul von Lettow-Vorbeck, was a skilled jungle fighter. His Schutztruppen launched offenses against British-held Kenya. As well, the six-inch guns from a stranded German light cruiser were put to good use. The Schutztruppen in East Africa did not surrender until late November 1918 and the Armistice.

The German colonial empire produced no economic or political gain for Germany, rather the opposite when the costs of defense are weighed. Nevertheless, matters of honor dictated that Germans abroad should defend their newly acquired territories against both insurrection and foreign invasion. In the end, the German legacy in these areas vanished.

David H. Olivier

See also: Lettow-Vorbeck, Paul Emil von

References and further reading:

Bridgman, Jon M. *The Revolt of the Hereros.* Berkeley: University of California Press, 1981.

Farwell, Byron. *The Great War in Africa, 1914–1918.* London: W. W. Norton, 1986.

Pakenham, Thomas. *The Scramble for Africa, 1876–1912.* London: Weidenfeld & Nicolson, 1991.

German Wars of Unification (1864–1871)

Three Prussian wars against Denmark, Austria, and France, engineered by Prussian chancellor Otto von Bismarck, through which Germany was unified. Denmark's attempt to assert constitutional control over the duchies of Schlesswig and Holstein in 1863, in violation of the London Treaty of 1851–1852, sparked a war with Austria and Prussia. An Austro-Prussian force of 57, 000 under the command of General Wrangel, occupied Holstein without incident on 21 January 1864. However, the Danish force under General Christian de Meza refused to evacuate Schlesswig. On 1 February, Wrangel's forces crossed the Eider River, stormed the Dannevirke fortifications, and forced Meza and 44,000 Danes to retreat. After a siege from 2–18 April, the Dybbol fortifications capitulated.

The war resumed on 26 June, after an armistice during the London Conference, and 24,000 Prussians landed on Als Island. The war ended 20 July, and Denmark relinquished claims to the duchies by the Treaty of Vienna on 1 October. According to the Treaty of Gastein, August 1865, Austria would administer Holstein while Prussia would administer Schlesswig. At the conclusion of this, what became known as the first war of German unification, Prussia became the dominant regional power.

Prussian chancellor Otto von Bismarck realized that Austria would contest Prussia's growing dominance. Bismarck made diplomatic gestures to gain Europe's benevolent neutrality in case of an Austro-Prussian conflict. To this end, he prevented Prussian Poles from aiding Russian Poles in revolt during 1863.

In 1865, Napoleon III was given tacit Prussian support for French ambitions in Belgium and Luxembourg. A Prussian alliance with Italy was concluded in April 1866. In response, Austria mobilized and brought problems of joint administration of Schlesswig-Holstein before the Frankfurt Diet. Bismarck took this opportunity to accuse Austria of breaching the Gastein Convention and marched Prussian troops into Holstein. Hostilities began on 16 June when Prussia invaded Hannover, Saxony, and Hesse.

Although a quarter of the Austrian army was stationed along the Italian border, the remaining Austrian and southern German forces were still numerically superior to the Prussian army. However, the former lacked a coordinated strategy.

The expansion of railways 1840–1870 enabled the Prussian general staff to transport a mass army in a short period of time. The railway also reduced the loss of manpower and supply due to exhaustion from forced marches to the battlefield. The Prussian infantry was empowered by the Dreyse needle-gun, a breech-loading rifle that increased the rate of fire and made it possible for the soldier to load on the move. Firepower was also made more accurate by the advent of conical bullets. These two innovations forced enemy artillery to reposition to the rear.

Count Helmuth von Moltke, chief of the Prussian general staff 1858–1891, rapidly marched two armies into Bohemia; the army commanded by Prince Frederick Charles marched toward Muchengratz, and the army commanded by Crown Prince Frederick William, toward Trautenau-Nachod. Moltke's Austrian counterpart, Marshal Ludwig von Benedek, ordered a retreat to Sadowa after his generals Eduard Graf Clam-Gallas and Wilhelm Ramming were defeated. The Prussian armies linked on 30 June 1866. On 3 July, 280,000 Prussians with 900 guns defeated 245,000 Austrians and 25,000 Saxons with 600 guns at Königgrätz/Sadowa. Rather than numerical superiority, it was the Prussian technique of organization and mobilization as well as flanking attacks that won the battle.

However, Benedek and 150,000 troops escaped encirclement due to Prussian error, and Moltke could not resume the advance to Wagram until 18 July. The Austrians asked for an armistice on 21 July, and the Peace of Prague was signed on 23 August. Bismarck forbade a war indemnity, Prussian occupation, or a victory parade through Vienna, thus wisely avoiding Austrian revanchist feeling. The North German Confederation was formed between Prussia and the German states north of the Main River. The result of this, the second German war of unification, was that Prussia emerged as the dominant German power.

The prospect of a unified German entity was against French policy from the time of Cardinal Richelieu. However, Bismarck realized that only a common German patriotism would convince the independent southern German states to ally under Prussia's political and military leadership. Bismarck's opportunity arose when Prince Leopold von Hohenzollern-Sigmaringen was offered the Spanish throne after a military coup ousted Queen Isabella II in September 1868. French ambassador Count Benedetti protested to Prussian king Wilhelm I on 9 July 1870, while that monarch was at spa in Bad Ems, and received assurances that Leopold would retract his candidacy, which indeed occurred on 12 July.

Seeking guarantees, Benedetti revisited Wilhelm, demanding that Leopold never renew his candidacy. Having already given his word, Wilhelm considered the matter closed and refused to discuss the matter. Bismarck received a telegram of this conversation on 13 July and released it to the press. The reaction of the German public was to condemn French snobbery, while the outcry of the indignant French forced their legislature to declare war on 19 July 1870.

The Prussians possessed a two to one advantage over French forces, and since 1866, Prussian mobilization time had improved from five weeks to eighteen days. The Prussians also possessed a better order of battle, their peacetime army being organized into divisions, corps, and armies; the French were not organized beyond regiment level. The French railway system was efficient, but the military (still dreaming of Napoleonic glories) did not spend much time with such modern impedimenta. The only strategic plan was a series of arrangements to form armies around Paris, Lyons, and Toulouse. In comparison, the Prussian general staff produced a war plan in 1867, with regular revisions.

While the French chassepot was accurate at 1,600 yards compared to the needle-gun's 600 yards, the Prussian Krupp breech-loading cannon was a superior and more widely distributed artillery piece than the French *mitrailleuse* (machine gun).

Moltke reasoned that the French would be unable to assemble more than 250,000 troops or interfere with the concentration of Prussian forces west of the Rhine at the mouth of the Lorraine River. This was the only path for either army, besides a route through Belgium, because the Belfort Gap had been fortified by the French. Assembling their troops around Metz and Strasbourg, the French forces were separated by the Vosges Mountains.

On 1 August, Marshal François Bazaine advanced 7 kilometers into the Saar, but fell back on 2 August. The same day Prussian armies attacked: General Karl Friedrich von Steinmetz and 60,000 troops marched through the Lorraine Gap, Prince Frederick Charles and his 131,000 troops advanced on the right flank, the Crown Prince and 130,000 advanced on the left flank. The French force of Marshal Patrice MacMahon encountered at Wört 5–6 August, retreated through the Vosges, reaching Neufchateau on 14 August. The main French army was pushed beyond Spicheren on 6 August; however Steinmetz and Charles had advanced too far too fast and were unable to effect encirclement.

The Prussians sustained greater casualties but advanced due to their well-organized artillery and Moltke's empowerment of battlefield commanders to make tactical decisions. The French imperial headquarters at Metz began to issue contradictory orders, further complicating their tactical situation. On 9 August, a decision was reached that all French troops east of Metz should halt their retreat and repair to Metz.

However, while Bazaine received the order and halted, MacMahon continued toward Paris. On 12 August, Napoleon yielded as supreme commander, to Bazaine. The Prussians exploited the chaos, dividing the French forces and encircling the larger contingent commanded by Bazaine, at Metz on 16–18 August. This occurred after the battle at Gravelotte/St. Privat, while Bazaine attempted to retreat from the Moselle to the Meuse River.

MacMahon gathered remaining French forces at Châlons and, accompanied by Napoleon III, left to relieve Metz on 24 August. MacMahon effectively outflanked himself and had to abort the relief effort on 27 August, attempting instead to retreat to Mezieres. French communications broke down completely, and MacMahon reissued orders on 28 August for resistance to organize at Sedan.

The battle began 1 September, and MacMahon was immediately replaced by General Auguste Alexandre Ducrot after being wounded. Ducrot issued an order to retreat to Illy, but before it could be carried out, he was reprimanded and superseded as commander in chief by General Comte Felix von Wimpffen. Later that evening, Wimpffen, Napoleon III, and 104,000 troops capitulated at Sedan.

In Paris, the Third Republic was proclaimed by General Louis Jules Trochu and Minister of the Interior Léon Gambetta. Paris mobilized its National Guard as well as 72,000 sailors, for a total force of 300,000 to defend the capital to which the Prussians laid siege on 17 September. The Prussian force numbered 147,000 men along 15 miles of siege lines.

On 29 September, Bazaine and 154,000 troops surrendered at Metz. On 19 January 1871, 90,000 French troops failed to break out of Paris, which capitulated on 28 January.

Meanwhile, William had been crowned German emperor in the Hall of Mirrors at Versailles on 18 January 1871. The Peace of Frankfurt concluded the Franco-Prussian War on 10 May. Germany required France to pay an indemnity of 5 billion francs, and northern France remained occupied until the sum was paid in 1873. France also was forced to cede Alsace and eastern Lorraine. The result of this, the third and final war of German unification, was the creation of the German Second Reich. But the Franco-Prussian War also caused a lasting French bitterness that led to the European alliance systems that, 43 years later, were a major cause of World War I.

Neville G. Panthaki

See also: Bismarck, Otto von; Franco-Prussian War; Königgrätz, Battle of; Metz, Siege of; Moltke, Graf Helmuth Karl Bernhard von; Paris, Siege of; Revolutions of 1848; Sedan

References and further reading:

Carr, William. *The Origins of the Wars of German Unification.* New York: Longman, 1991.
Dupuy, Trevor Nevitt. *A Genius for War.* Englewood Cliffs, NJ: Prentice-Hall, 1977.
Friedrich, Otto. *Blood and Iron.* New York: HarperCollins, 1995.
Showalter, Dennis. *Railroads and Rifles.* Hamden, CT: Archon Books, 1975.

Germantown (1777)

Drawn battle of the American Revolution. On 4 October 1777, General George Washington's American troops engaged British general William Howe at Germantown, west of Philadelphia, producing an indecisive outcome and a continuation of the American Revolution. In October 1777, General Howe marched through eastern Pennsylvania in his drive toward the rebel capital, Philadelphia, attempting along the way to draw George Washington's army into a general battle. Failing to entice Washington, Howe settled into Philadelphia and ordered the main part of his force to camp in the vicinity of Germantown.

Pressured by the Continental Congress to dislodge Howe, and infuriated by the Paoli Massacre, Washington planned a complex, four-column operation against the British encampment. His forces advanced on the British in a dense fog, enjoying early successes, but many troops lost contact with their commanding officers, and several units exchanged friendly fire. Meanwhile, British troops found their way to a stone house in the American rear, owned by loyalist Benjamin Chew, which diverted attention from the point of attack. Both sides suffered heavy casualties, with the Americans losing more than 1,000 and the British 534. Washington called on General Nathanael Greene to organize a retreat, which was performed with great skill. Given his losses, Howe declined to follow, enabling the Continental army to continue the rebellion, but only after Washington's harrowing winter at Valley Forge in 1777–1778.

Jeffrey B. Webb

See also: American Revolution; Greene, Nathanael; Valley Forge; Washington, George

References and further reading:

Fleming, Thomas. *Liberty! The American Revolution.* New York: Viking, 1997.
Jensen, Merrill. *The Founding of a Nation: A History of the American Revolution, 1763–1776.* New York: Oxford University Press, 1968.
Robson, Eric. *The American Revolution in Its Political and Military Aspects, 1763–1783.* New York: Norton, 1966.

Portrait of Geronimo, Apache chief, c. 1907. (Library of Congress)

Geronimo (c. 1827–1909)

Apache medicine man and war leader. Geronimo, born around 1827, was originally named Goyathlay (He Who Yawns) by his Chiricahua Apache relatives. When his family was killed by Mexican troops during the 1850s, he set out to avenge their deaths, raiding north and south of the border.

Killing and stealing, he honed his skills as a combat leader. His fierceness as a warrior struck fear into the Mexicans upon whom his band usually preyed. They named him "Geronimo."

Never a true chieftain, Geronimo initially fought under the great Cochise, building a reputation among his people as the most skillful Apache tactician. Forced onto a reservation in 1877, Geronimo broke out in 1881 and again in 1885. General George Crook with thousands of troops corralled him the first time, General Nelson Miles and thousands more troops the second time. Geronimo and his small band of warriors, accompanied by their women and children, led the army on wild chases back and forth across the desert Southwest. In both cases, Geronimo was eventually coerced into returning to the reservation once it became clear that his exhausted people simply could not forever outrun, outwit, and outfight the U.S. Army.

Geronimo's surrender to Miles in September 1886 essentially ended the Apache Wars. Geronimo and others from his tribe were transported to military incarceration in Florida, and later, in Oklahoma. Geronimo, who had captured the imagination of the American people, appeared at the 1904 World's Fair and rode in President Teddy Roosevelt's inaugural parade. His exploits as a warrior were legendary, but his military efforts were ineffective in turning back American control of the region. Geronimo died of pneumonia in 1909 on the Fort Sill reservation in Oklahoma, as widely admired by white Americans as by his fellow aboriginal peoples.

Michael S. Casey

See also: American Indian Wars

References and further reading:

Adams, Alexander B. *Geronimo: A Biography.* New York: Putnam, 1971.

Debo, Angie. *Geronimo: The Man, His Time, His Place.* Norman: University of Oklahoma Press, 1976.

Faulk, Odie B. *The Geronimo Campaign.* New York: Oxford University Press, 1969.

Gettysburg (American Civil War, 1–3 July 1863)

The greatest battle ever fought in the Americas and the turning point of the American Civil War. The situation for the Confederacy in summer 1863 had worsened from the previous summer. Ulysses Grant had a bear-hold on Vicksburg, which, if it fell, would cut the Confederacy in half; the chances for foreign recognition had lessened; and total war was extracting a terrible price in the South.

After the Battle of Chancellorsville, Confederate general Robert E. Lee felt he had no choice but to invade the North. He believed another Union general replacing John Hooker eventually would commence another offensive, further destroying the Virginia countryside and forcing Lee's army to depend ever more on the agricultural bounty of the Shenandoah Valley. He was committed to the idea of the offensive/defensive. He hoped that a successful offensive into Maryland and southern Pennsylvania might cut the railroad lines connecting the East and Midwest, revive the northern peace movement, or cause U.S. president Abraham Lincoln to recall Grant and his army from the siege at Vicksburg and relieve the threat there.

On 3 June, Lee's Army of Northern Virginia began moving from camps near Fredericksburg and retraced ground from its invasion of the previous summer. Lee used the

Confederate general James Longstreet's attack on Blue Ridge during the Battle of Gettysburg on 3 July 1863. (Library of Congress)

mountains to shield his movement in a broad cartwheel around Washington, D.C. In late June, Lincoln replaced Hooker and appointed George Meade as commander of the Army of the Potomac. By mid-June, Lee's units were nearing the Potomac River, and soon thereafter began to move into Maryland and then into southern Pennsylvania.

On 1 July, the vanguard of Lee's army was approaching the quiet Pennsylvania market town of Gettysburg from the northwest, where it ran unexpectedly into the vanguard of Meade's army coming from Frederick, Maryland; the southerners had heard rumors of a supply of shoes, always a problem for the South. First Ewell's men and later A. P. Hill's drove Union troops back through the town and eventually the Union army settled into a strong defensive position atop Cemetery Ridge to the southeast of Gettysburg while Confederates occupied Seminary Ridge as reinforcements for both sides streamed to the emerging battlefield.

On 2 July, Lee decided to turn the flanks of Meade's army. He had agreed to a flamboyant cavalry raid by Jeb Stuart, thus denying himself the intelligence about the Union army he so desperately needed. And he seemingly forgot his earlier promise to his senior subordinate, James Longstreet, that he would avoid a battle—particularly an attack—that would result in great casualties for the South. He felt the Union left might be "hanging" and Longstreet engaged in a roundabout approach to preserve secrecy, but he could not attack until about 4:00 P.M.; although the Union position was precarious, by nightfall Longstreet was unable to break the flank. Meanwhile, Richard Ewell attacked the Union right at Culp's Hill about the same time; again, when nightfall came, the Union position largely had held, for Meade had moved his reserves around well to meet the threats to his flanks.

The climax of the battle was on 3 July. Lee had decided on an artillery barrage by 160 cannon and then an attack by some 12,000 to 15,000 men right at the Union center. It was an awesome sight late in the afternoon as Confederate cannon fired—largely over Union positions—for two hours. Then troops from George Pickett's and James Pettigrew's commands crossed the 1,400 yards to the Union defenses. Union artillerymen and Union infantry poured shot and shell into the advancing Confederates. Lee lost more than 50 percent of his force, thus demonstrating the power of de-

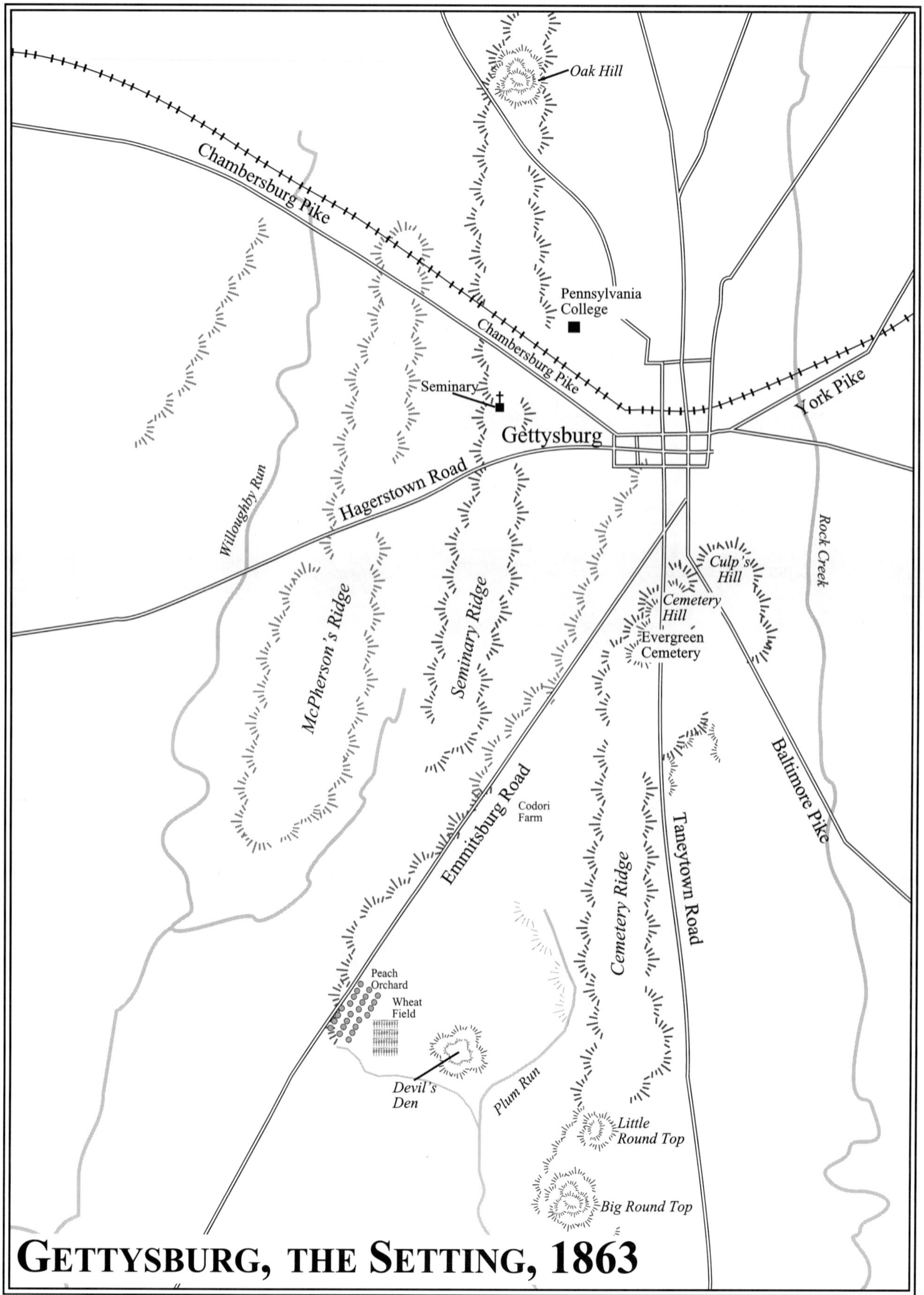

Gettysburg, the Setting, 1863

fense, and as he watched his attack crumble, he turned to his subordinates, took complete responsibility for the failure, and asked them to help him save the army.

Fortunately for Lee, Meade did not follow up his victory by pursuing the retreating Confederates, and Lee was able to return to Virginia without much interference. The next day Pemberton surrendered to Grant at Vicksburg, and the war greatly worsened for the Confederacy.

The carnage at Gettysburg has never been surpassed in the Western Hemisphere: Of the 163,000 troops engaged, more than 6,000 died in battle, 22,000 were wounded, and an incredible 10,000-plus went missing.

Charles M. Dobbs

See also: American Civil War; Lee, Robert Edward; Meade, George Gordon

References and further reading:

Gallagher, Gary W., ed. *Three Days at Gettysburg.* Kent, OH: Kent State University Press, 1999.

Gordon, Lesley J. *General George E. Pickett in Life and Legend.* Chapel Hill: University of North Carolina Press, 1998.

Gramm, Kent. *Gettysburg.* Bloomington: Indiana University Press, 1994.

Ghaznavid Empire (977–1180)

The Ghaznavids originated as a Turkish military force stationed by the Persian Samanid Dynasty in Ghazna on the eastern frontiers of Iran. Sebuktigin, their first notable leader, increased his autonomy through wealth raised by looting the Hindu cities of the Indus Valley. As the Samanid state declined, he put more lands under his authority, including Bactria and Baluchistan. A Sunni Muslim, he also raided the lands of the Ismaili Shia of Multan and Sindh.

His son, Mahmud, was probably the most significant figure of the dynasty. Taking control in 999, he spent much of his life campaigning on both his eastern and western frontiers. His forces expanded into Iran and into areas along the Oxus River. He took control of the Punjab and most of the lower Indus, founding mosques, schools, and other centers for propagating Islam, and raided into Sindh and the Ganges plain. He pounced on the temple city of Somnath in 1025–1026 and made off with 2 million dinars of gold and jewels, putting over 50,000 Hindus to the sword. Despite his distaste for those he called infidels and idolaters, Mahmud added Hindi infantry and elephant units to his Turkish cavalry, lancers, and bowmen. At his death in 1030, a number of Hindu rajahs in the Ganges had become his vassals.

The Ghaznavid Empire fell eventually to a combination of strong neighbors and internal disputes. The Ghaznavid western possessions in Iran fell in 1040 to the Seljuqs, and Sultan Ibrahim, who came to power in 1059 after a civil war, recognized Seljuq control of the west. Ibrahim reigned in Lahore for 40 years of relative peace. His son, Mas'ud III, revived the wars, seizing the Hindu town of Kanawj, but an accession dispute between his three sons after his death tore the empire apart. In 1150, the Ghurrids, another Turkish confederacy, captured Ghazna, virtually destroyed the city, and massacred or deported most of its people. Bahram-Shah, Mas'ud's heir, died in the wreck of his capital in 1157.

In 1163, the last of the Ghaznavid sultans, Khusraw Khan, was evicted from Ghazna and settled in Lahore as a Ghurrid vassal, but he and his sons were executed by the Ghurrids in 1190 after several sieges of the city.

Weston F. Cook Jr.

See also: Mahmud of Ghazna

References and further reading:

Bosworth, C. E. *The Ghaznavids.* Edinburgh: Cambridge University Press, 1963.

Wink, Andre. *Al-Hind: The Making of the Indo-Islamic World.* Vol. 1. Leiden: E. J. Brill, 1991.

Gibraltar, Siege of (1779–1783)

Abortive attempt by French and Spanish to take the British stronghold. When France declared war on Great Britain in 1778, what had been a rebellion in the American colonies became a global war. France joined forces with Spain to attempt to unseat the British from one of their smallest but most strategic possessions, the fortress of Gibraltar at the mouth of the Mediterranean Sea.

This small but naturally strong bit of land joined to the southwestern tip of Spain was home to a well-fortified garrison; lacking natural resources, however, it had to be sustained from the sea. The French and Spanish allies recognized that it would be a straightforward operation to lay siege to the place and starve out the garrison, if they could prevent relief by Britain's superior navy.

Operations commenced in June 1779, with Spanish forces raising siege works on the landward side, and French and Spanish ships blockading the seaward approaches. The small garrison, commanded by General George Elliot, faced the cannonading and deprivation of supplies with dogged determination, and the besiegers themselves ran into many difficulties coordinating their forces. Twice, British fleets carrying provisions and reinforcements broke through the naval blockade, and in 1781 the garrison sallied out and destroyed many of the gun batteries on the landward side.

The attackers turned to technology to assail the harbor at Gibraltar. Knowing that the fortress could outgun normal warships, they constructed armored floating batteries, precursors to the ironclads of the next century. These gun platforms included elaborate water systems to douse heated shot from the enemy. Their complexity, however, caused lengthy construction delays. When they were finally hurried into service in a grand assault in September 1782, they were no match for the defenders' heavy guns; the floating batteries exploded with a fury that signaled the collapse of the will of the attackers. Soon after, the siege was broken by a large British naval force; possession of the Rock of Gibraltar was never again challenged by force.

Don N. Hagist

References and further reading:

Drinkwater, John. *A History of the Late Siege of Gibraltar.* London: J. Johnson, T. & J. Edgerton, 1785. Widely reprinted.

Russell, Jack. *Gibraltar Besieged 1779–1783.* London: William Heinemann Ltd., 1965.

Gierczak, Emilia (1925–1945)

Second lieutenant, platoon commander, 10th Infantry Regiment, Poland, World War II. Polish patriotism was an important element in Gierczak's upbringing. Her name, Emilia, was made famous by the legendary Emilia Plater, a leader of the 1830 Polish uprising against Russian rule.

She at first served in the Emilia Plater Women's Battalion, initially subordinated to Tadeusz Kosciuszko 1st Division created in the USSR in 1943. She was among the battalion soldiers trained at Infantry Officers' School in Riazan to command companies and platoons of the new Polish army, due to a drastic shortage of Polish male officers, many of whom had been killed by the Soviets at the Katyn Massacre, 1939–1940.

Gierczak distinguished herself during the fighting on the Pomeranian Rampart and fell on 17 March 1945 while leading an assault group in Kołobrzeg (former Kolberg).

Kazimiera J. Cottam

See also: Emilia Plater, Independent Women's Battalion

References and further reading:

Cottam, K. Jean. "Veterans of Polish Women's Combat Battalion Hold a Reunion." *Minerva* 4 (Winter 1986).

Jagielski, B. "This Happened in Kołobrzeg." *Żołnierz Polski* 10 (11 March 1979).

Kuliński, Rajmund. "Legend Beyond History" *Żołnierz Polski* 10 (11 March 1979).

Rychliński, Czesław. "They Came from Oka River Shores" *Żołnierz Polski* 11 (17 March 1985).

Glendower's Revolt (1400–1413)

Revolt that seriously threatened English rule in Wales. The century after the defeat of the last Prince of Wales, Llewelyn the Great, by Edward I had not been kind to Wales. The black plague, combined with its attendant economic and social dislocation, only exaggerated tensions between the Welsh gentry and English neighbors in the Welsh marchland. Owain Glyndwr, descended from Welsh princely families on both sides, and with an English legal education, was only a well-to-do landowner with a fondness for bardic predictions of Welsh resurgence. A crisis within the English monarchy, the deposition of Richard II, who enjoyed support in north Wales, and his replacement by his cousin Henry IV, as well as Henry's heavy tax demands at his accession, stirred resistance.

Glyndwr's own participation stemmed from a longtime feud with Lord Grey of Ruthin over the borderland of their estates. Parliament, petitioned in 1400 by Glyndwr to hear the case, ruled for Ruthin. To make matters worse, Henry IV's eldest son had been installed as Prince of Wales, with authority vested in the hands of the Percy family of Northumberland, a slight to the Welsh lords. As early as September 1400, Glyndwr, supported by his retainers and family, proclaimed himself Prince of Wales and plundered Ruthin before being scattered by an English force. Henry IV, who might have ended the problem once and for all by redressing Welsh grievances, instead allowed Parliament to enact harsh anti-Welsh policies while focusing his attention on threats from France and Scotland. Spurred by their new grievances, Glyndwr and his followers began guerrilla attacks in the spring of 1401, evading an invasion into south Wales launched by the king in the fall of that year and capturing Caernarfon Castle. Glyndwr also defeated an English force at Bryn Glass in June, capturing Edmund Mortimer.

Glyndwr, needing help to continue his rebellion, played upon Mortimer's resentment that he had not been ransomed and his overlooked claim to the throne. Mortimer allied himself to Glyndwr and married his daughter to seal the agreement. Through this connection, Glyndwr negotiated with Henry Percy (Hotspur), Mortimer's brother-in-law, who declared his support in 1402. The king, although still engaged in Wales and chronically overstretched confronting the Scots and the French, forced Hotspur to a fight at Shrewsbury, which ended in Hotspur's death. Glyndwr attempted to organize a Welsh government and use Wales's difficult terrain and bad weather to hold off the English. He also drew on the support of the French to supply him by sea and the ships of Brittany-based pirates to harass the English coast.

In 1405, Henry IV captured the Scottish king. He believed that internal French politics would hinder intervention in

Wales, and with more generous financial resources granted by Parliament, he began a long-term campaign against the rebellion with his son at his side and with pressure on the Marcher lords to keep order in their lands. Slowly retaking Wales, the king captured Glyndwr's wife, two daughters, and three granddaughters in 1408 at Aberystwyth Castle. Although still a threat as raiders, Glyndwr and his followers were on the run, and in 1413 most of the rebels succumbed to Henry V's offer of a pardon and employment in his wars in France. Glyndwr himself retired to his daughter's lands and died by 1416.

Margaret Sankey

See also: Henry V, King of England

References and further reading:

Allday, D. Helen. *Insurrection in Wales.* Suffolk, UK: Lavenham, 1981.

Davies, R. R. *Revolt of Owain Glendwr.* Oxford, UK: Oxford University Press, 1995.

Williams, Glanmore. *Owain Glyndwr.* Cardiff: University of Wales Press, 1993.

Gneisenau, August Neidhart von (1760–1831)

Natural Prussian hero, founder of the modern general staff. Gneisenau was born at Schilda bei Torgau on 27 October to a lieutenant in the imperial army battling against Frederick the Great. He was raised in Würzburg and educated by Jesuits before returning to his father in Erfurt at age 13. After attending the university, he joined a mercenary Ansbach regiment, and arrived in Canada in 1783 just as the American colonists won their independence. Back in Europe, he applied in person to Frederick II to enter Prussian service. The poise and engineering skills of the handsome, brown-haired, blue-eyed lieutenant pleased the king. His battalion garrisoned Polish territory in 1793, and following a promotion to captain in 1795, he performed staff duties. After Prussia joined Russia against Napoleon in 1806, Gneisenau was wounded at Saalfeld, but he continued to command his battalion at Jena before the army's disastrous retreat. In early 1807, the major oversaw the defenses of Kolberg in Pomerania, where he gained fame for his heroic stand against the French.

The Prussian king promoted him to lieutenant colonel, and he served with Gerhard Scharnhorst, Karl Grolman, WIlhelm von Boylen, Carl von Clausewitz, and other reformers on the Military Reorganization Commission. These well-educated, liberal intellectuals envisioned a people's army to liberate Germany but foresaw the need for radical political social reforms. Napoleon watched these men closely until 1812, when he demanded a Prussian contingent for the invasion of Russia. Meanwhile, Gneisenau performed secret missions in Vienna, Riga, and Stockholm. In England, he procured aid—gold and weapons—from the prince regent. In Germany's War of Liberation in 1813, Gneisenau served as Gebhard Blücher's quartermaster general at Gross-Görschen and Bautzen, where he led a masterful retreat. During the armistice from June to August, he organized the Landswehr in Silesia. He was instrumental to Blücher's victory on the Katzbach and supported his superior's bold decisions, which precipitated the Battle of Nations at Leipzig in October. He supported Blücher's energetic offensive into France in 1814 and took personal command for two weeks after the defeat at Craonne.

In 1815, he and Blücher resumed their partnership against Napoleon. When the field marshal was incapacitated at Ligny on 16 June, Gneisenau ordered the army's retreat toward Wavre, which allowed the Prussians to join Wellington's left flank at Waterloo two days later. Gneisenau was named field marshal in 1825 and commanded during the Polish insurrection in 1831, but fell ill and died on 24 August.

Llewellyn Cook

See also: Blücher, Gebhard Leberecht von; Leipzig, Battle of; Scharnhorst, Gerhard Johann von; Waterloo

References and further reading:

Brett-James, Anthony. *Europe Against Napoleon: The Leipzig Campaign, 1813 from Eyewitness Accounts.* London: Macmillan and Co., 1970.

Craig, Gordon. *The Politics of the Prussian Army, 1640–1945.* New York: Oxford University Press, 1956.

Dupuy, T. N. *A Genius for War: The German Army and General Staff, 1807–1945.* Englewood Cliffs, NJ: Prentice-Hall, 1977.

Paret, Peter. *Clausewitz and the State.* Oxford, UK: Clarendon Press, 1976.

Goethals, George Washington (1858–1928)

American engineer and builder. Born in Brooklyn, New York, Goethals attended the City College of New York before his admittance to the U.S. Military Academy at West Point. He graduated second in his class in 1880. As an engineer officer from 1882 until 1905, Goethals capably filled many assignments, including two tours as engineering instructor at West Point. In 1907, he retired as a lieutenant colonel when appointed third chief engineer for the Panama Canal project by President Theodore Roosevelt. Goethals supervised 44,000 men and resolved many critical engineering problems, including the construction of the lock system and recurrent landslides. He also confronted innumerable logistical and sanitation challenges. The 40-mile canal was completed in August 1914. Goethals was appointed major general and first civil governor of the Canal Zone (1914–1916). Retiring in 1916, he was recalled to active duty as acting army quartermaster general in 1917. He worked to reorganize the corps as a purchasing organization, streamlining the supply and transport of troops. In April 1918, he was appointed director of the Purchase, Storage, and Traffic division of the army general staff. His efforts to impose vertically integrated logistical procedures were only partially successful, owing to staunch bureaucratic resistance. Awarded the Distinguished Service Medal, he again retired in 1919, and headed a private engineering firm. Despite his other considerable achievements, Goethals will always be associated with his greatest engineering triumph, the construction of the Panama Canal.

Keir B. Sterling

References and further reading:

Bishop, Joseph B. *Goethals: Genius of the Panama Canal, A Biography.* New York: Harper, 1930.

Goethals, George W., ed. *The Panama Canal: An Engineering Treatise.* 2 vols. New York: McGraw-Hill, 1916.

McCullogh, David. *The Path Between the Seas.* New York: Simon & Schuster, 1977.

Zimmerman, Phyllis. *The Neck of the Bottle: George W. Goethals and the Reorganization of the U.S. Army Supply System, 1917–1918.* College Station: Texas A&M Press, 1992.

Goose Green, Battle for (28–29 May 1982)

First major land battle of the Falkland Islands War. Considered the decisive battle of the conflict, it was fought on the isthmus connecting northern and southern East Falklands Island.

The assault on Goose Green, by the 600-man British 2d Battalion of the Paratroops (2 Paras), was the first step in moving away from the Port San Carlos beachhead toward the capital of Port Stanley.

The British assault on Goose Green was plagued with intelligence errors, lack of artillery support, and supply problems. British intelligence reported that a weak battalion (less than 600 men) defended Goose Green. Instead, they found heavily fortified, reinforced positions. The British attacked at night, advancing rapidly until daylight when they were caught without cover and pinned down by Argentine troops.

However, through superior fighting ability, personal heroics, and timely air support the British were able to break Argentine resistance. The 2 Paras captured more than 1,200 Argentines. Casualties were heavy: the Argentines counted 250 dead and 121 wounded; the British 17 dead and 35 wounded. In the end, the British mistakes were more than countered by Argentine strategic and operational ground failures.

Craig T. Cobane

See also: Falkland Islands War

References and further reading:

Adkin, Mark. *Goose Green: A Battle Is Fought to Be Won.* London: Cassell, 1992.

Laffin, John. *Fight for the Falklands!* New York: St. Martin's Press, 1982.

Hastings, Max, and Simon Jenkins. *The Battle for the Falklands.* New York: W. W. Norton & Co., 1984.

Gordon, Charles George ("Chinese" Gordon) (1833–1885)

British field commander. Gordon was born in Woolwich, England, on 28 January 1833, the son of a general. After four years at the Royal Military Academy, Woolwich, he was commissioned second lieutenant in 1852. Ordered to the Crimea in 1854 and arriving there in 1855, he was a trench warrior in the siege of Sebastopol. Bored by military surveying on the Russian/Turkish border from 1856 to 1858 and by the home front in 1859, he requested transfer to a combat zone. In China during the Second Opium War, sometimes called the Third Anglo-Chinese War, he saw action at Beitang on 1 August 1860, the Dagu forts on 21 August, and Beijing on 6 October. With permission from his British superiors, he remained in China as military adviser to the imperial Qing Dynasty, to help suppress the bloody Taiping Rebellion. In April 1863 Emperor Tong Zhi gave him command of the Ever Victorious Army (EVA), the Qing mercenary unit founded by the American adventurer Frederick Townsend Ward (1831–1862). He drilled the EVA into a crack fighting outfit, led it into battle with only a cane, won more than 30 battles, and became a legend in Britain as "Chinese" Gordon.

The British War Office acknowledged that Gordon was an excellent soldier, but suspicious of his eccentricities, Christian fanaticism, and alleged pedophilia, they were reluctant to place him in command of British combat troops. As a lieutenant colonel from 1865 to 1871, he commanded engineers in Gravesend, England. After a brief assignment in Romania, he was appointed governor of the southern Sudan (Equatoria) in 1873, then, under the khedive of Egypt, governor of Sudan in 1877. He resigned in 1880, accepted missions in England, India, China, Mauritius, and South Africa, then spent most of 1883 in Palestine studying the Bible. Early in 1884, he planned to become governor of the Belgian

Death of General Gordon at Khartoum. (Library of Congress)

Congo under King Leopold I, but instead, as a major general, returned to the Sudan at Prime Minister William Gladstone's request to quell the Mahdist uprising. At Khartoum, he and his Egyptian garrison were besieged for 10 months and were all killed during the Mahdi's final assault on 26 January 1885. Gordon's life resembles that of at least two other well-known British eccentric military commanders, Orde Wingate and Horatio Herbert Kitchener.

Eric v. d. Luft

See also: Crimean War; Hong Xiuquan; Khartoum, Siege of; Kitchener, Horatio Herbert; Li Hongzhang; Muhammad Ahmad; Religion and War; Sevastopol, Siege of; Taiping Rebellion; Wolseley, Garnet Joseph, Viscount; Yang Xiuqing; Zuo Zongtang

References and further reading:

Johnson, Peter. *Gordon of Khartoum.* New York: Sterling, 1985.

MacGregor-Hastie, Roy. *Never to Be Taken Alive: A Biography of General Gordon.* New York: St. Martin's Press, 1985.

Pollock, John Charles. *Gordon: The Man behind the Legend.* London: Constable, 1993.

Waller, John H. *Gordon of Khartoum: The Saga of a Victorian Hero.* New York: Atheneum, 1988.

Goring, Hermann Wilhelm (1893–1946)

Nazi political and military leader. Goring was born in Bavaria on 12 January 1893. Educated at the cadet college in Karlsruhe, he served as an infantry lieutenant before transferring to the air force. In 1918 he assumed command of the air fighter squadron Jagdgeschwader Richthofen after the death of Manfred von Richthofen. He was awarded Germany's highest military decoration, the Pour le Mérite.

For some years after the war he flew as a civilian, and met Hitler in autumn 1922. He was wounded in the Beer Hall Putsch in 1923 and became addicted to morphine; there is some controversy as to whether he was able to shake off this habit.

In 1932 he became president of the Reichstag and was the Nazis' war hero, their man to be put on show. Rapidly promoted, he was given command of the new Luftwaffe in 1935. His life was henceforth one of luxury and decadence, and he grew obese. Like other Nazis, he was not averse to killing, nor to despoiling art treasures from the conquered countries.

In 1936 he was made a full general and took over economic control of the German War Plans. Simultaneously he made a vast fortune out of state-owned businesses.

In 1940 he became a Reichsmarschall, but his failure to defeat the Royal Air Force in the Battle of Britain resulted in his losing influence with Hitler and was the first of a string of blunders and miscalls. He had foolishly boasted that if one British bomb fell on Berlin, "then you can call me Meyer"—a common Jewish name. He also promised—and failed to make good on his promise—to supply by air the surrounded Sixth Army at Stalingrad. Later, he absolutely denied that any Allied fighter aircraft could fly to Berlin, and when one was reliably reported to have done just that, he asserted that it must have been "blown" there by high winds!

As Hitler's nominated successor, he made a naïve attempt to take control as the Third Reich disintegrated in April 1945, a final gaffe for which Hitler stripped him of all rank and titles. Captured by the Americans, his braggadocio failed to impress the judges of the International Military Tribunal at Nuremberg, and he was condemned to death. A trickster to the end, he cheated the hangman by ingesting poison on 15 October 1946, two hours before he was due to take what he had blithely termed "the high jump."

Guillaume de Syon

See also: Hitler, Adolf; World War II

References and further reading:

Burleigh, Michael. *The Third Reich: A New History.* 4th ed. New York: Hill & Wang, 2001.

Overy, Richard. *Goering.* London: Phoenix Press, 2000.

Snyder, Louis L. *Encyclopedia of the Third Reich.* New York: McGraw-Hill, 1976.

Hermann Goring shaking hands with Adolf Hitler, 1941. (Library of Congress)

Gorlice/Tarnow (May 1915)

Led to the Austro-German conquest of Russian Poland. In order to relieve the beleaguered armies of Austria-Hungary, the German commander in chief, Erich von Falkenhayn, conceived a plan designed to cripple the Russian field armies. In a best-case scenario it could force them out of the war. He placed General August von Mackensen in command of the German Eleventh Army and the Austro-Hungarian Third and Fourth Armies. It would be the first time in the war that the Austro-Germans fought under a unified command. The attack was scheduled to begin along a 30-mile front between the Hungarian towns of Gorlice and Tarnow, where the Austro-Germans had a slight numerical advantage over the Russians. The Austrian armies served on the flanks of the German drive.

The preparatory battles began on 28 April, with the Austro-Germans assaulting Russian positions along the Biala River, the main Russian line of defense. The actual battle began on 2 May 1915. A massive artillery bombardment, which included Austrian 30.5-centimeter howitzers, destroyed the Russian trenches and wire entanglements. The Russians were unable to halt the Austro-German advance and were forced to retreat from the Biala River and fall back to vulnerable interior lines. Tactically and strategically the offensive was a success. The Austro-German armies ruptured the Russian front within the first two days of the drive, and the Russian Third Army was destroyed. The Austro-Germans continued their penetrating drive into the Russian lines, eventually threatening the flanks and rear of the Russian armies in western Poland and the Carpathian mountains. By June the Russian front showed signs of disintegrating, the Russians having suffered more than 400,000 casualties.

The victories around Gorlice/Tarnow led to a general Austro-German offensive along the entire eastern front, which ended in a 300-mile advance into Poland.

Stephen Chenault

See also: Falkenhayn, Erich von; Hindenburg, Paul von

Beneckendorf und von; Ludendorff, Erich Friedrich Wilheim; Mackensen, August von; Tannenberg, Battle of; World War I

References and further reading:

Asprey, Robert B. *The German High Command at War: Hindenburg and Ludendorff Conduct World War I.* New York: William Morrow and Company, 1991.

Herwing, Holger H. *The First World War, Germany and Austria-Hungary, 1914–1918.* London: Arnold, 1997.

Gothic War (534–554)

Part of Justinian's opportunistic scheme to conquer the western Mediterranean. The Gothic War demonstrated the strengths and weaknesses of the early Byzantine military system. Having conquered Vandal Africa, Belisarius took Ostrogothic Sicily (535), then Naples and Rome (536). Successfully defending Rome for three years against King Vittigis (r. 536–539), Belisarius next fought his way up the peninsula and occupied the former imperial capital of Ravenna (539). Unfortunately, this rapid conquest proved ephemeral. From the main Ostrogothic settlements north of the Po, King Totila (r. 542–552) quickly reconquered Italy and Sicily, except for a few coastal strongholds. Elsewhere, Persian attacks in the east left imperial armies in Italy starved of men and matériel.

Not until 550 were sufficient resources available for Italy. Narses's demands for sole command and full financial backing were met, and his army, 20,000 strong, became the largest Justinian sent west. It was primarily comprised of Narses's household troops (*bucellarii*), with a high proportion of Huns and Lombards.

Having destroyed the Ostrogothic fleet, Narses marched out of Illyria into northern Italy, using his own fleet to outflank the Ostrogothic river defense system (551). To cover Rome, Totila deployed at Taginae (552). Here the Ostrogothic cavalry failed to break the Roman center. Narses had strengthened his infantry line with dismounted cavalrymen. As a result, Totila's army was enveloped and overwhelmed by Narses's armored horse-archers. Totila was killed and his army shattered.

Ostrogothic resistance continued north of the Po. At Casilinum (554), Narses again used an enveloping tactic to destroy a supporting Frankish-Alamanni force, reputedly 80,000 strong. The Roman infantry in the center absorbed the ferocious Frankish charge, while the armored horse-archers on the wings closed the trap. Despite this success, it was not until 561 that Narses reduced the last Ostrogothic garrisons in northern Italy.

Infantry played only a minor role in the campaigns of Belisarius and Narses. Imperial infantry might be tasked to garrison or lay siege to a city, but whenever serious fighting was called for it was up to the armored cavalry, who wielded both lance and composite-bow. They were equally well equipped for shock action or for fighting from a distance, as the situation required.

Although the campaigns of the Gothic War were ultimately successful, in the long term it proved difficult for the Romans to retain effective control of the old western provinces. Much of what had been gained was quickly lost to the invading Lombards only a few years after Justinian's death, and the imperials were never able to recover the lost ground. In the end, imperial resources proved sufficient for defense but utterly inadequate for sustained overseas conquest.

Nic Fields

See also: Belisarius; Justinian I; Narses; Ostrogoths; Taginae, Battle of

References and further reading:

Bury, J. B. *A History of the Later Roman Empire from the Death of Theodosius I to the Death of Justinian (A.D. 395 to 565).* 2 vols. London: Macmillan, 1923.

Teall, J. "The Barbarians in Justinian's Armies." *Speculum* 40 (1965), 294–322.

Goths

A Germanic peoples of ultimately Scandinavian origin, according to Jordanes. By the end of the second century, the Goths controlled large tracts of steppe north of the Black Sea. These were territories once belonging to the nomadic Roxolani, and it was from the Roxolani that the Goths learned the use of heavy cavalry. Invading Thrace and Greece in the mid–third century, they were evicted by Claudius II "Gothicus," but Aurelian, his successor, allowed some to settle in the former Roman province of Dacia, north of the Danube.

Toward the end of the fourth century, hard-pressed Goths sought the right to settle within the Roman Empire. They promised to serve in its armies to escape the Huns, who had overrun former Gothic steppe territories and were now pursuing the Goths themselves. Emperor Valens, preoccupied with Persia, tried to disarm the refugees before admitting them into Thrace. His officials, taking advantage of the situation, shamelessly exploited the starving refugees. The Goths rose in revolt, and Valens was obliged to take command of the situation himself. He died at Adrianople (9 August 378), along with most of his army, but the Goths, lacking both the skill and resources to storm fortifications, failed to overrun the empire. Valens's successor, Theodosius I, recruited Christian Goths to resist the newcomers. A treaty was finally made in 382. The Goths were allowed to settle south of the lower Danube as a confederate people under their own chieftains (*foederati*) in return for military service.

Long before the arrival of the Huns, according to Jordanes, the Goths were already divided into two peoples, the Ostrogoths and Visigoths, but at the time that the Huns attacked, there were actually six or more Gothic tribes. It was the Hunnic attack itself that gave rise to the division of the Goths into two peoples. Those who entered the Roman Empire became the Visigoths. Those who remained behind became subjects of the Huns and were called Ostrogoths.

Nic Fields

See also: Adrianople, Battle of; Aurelian, Lucius Domitus; Huns; Ostrogoths; Visigoths

References and further reading:

Heather, P. J. *Goths and Romans, 332–489.* Oxford, UK: Clarendon Press, 1991.

———. *The Goths.* Oxford, UK: Blackwell, 1996.

Wolfram, H. *History of the Goths.* Berkeley: University of California Press, 1988.

Gotthard Abbey (1664)

Conclusive battle of the Austro-Turkish War (1663–1664). In 1658 Ottoman grand vizier Ahmed Köprülü deposed Transylvanian prince György II Rákóczi, replacing him with Michael I. Apafi. When Habsburg emperor Leopold I was slow to recognize Apafi, Köprülü attacked Habsburg-controlled Hungary in 1663. Leopold's army under Count Raimundo Montecuccoli was unprepared for a major war and retreated before the invading Ottomans.

Köprülü took the fortress of Neuhausel (Nové Zamky) in September after a six-week siege and withdrew into winter quarters. In November 1663, Croatian Ban Miklós Zrinyi replaced Montecuccoli as Habsburg field commander. Throughout the winter Zrinyi conducted bold mounted operations in the Ottoman rear areas along the Mura River, disrupting supply lines but failing to recover any territory.

The following summer, Montecuccoli was again given command of the imperial armies, reinforced by a French contingent under the duke of Lorraine. When the Ottomans invaded Croatia, Montecuccoli refused Zrinyi's demands for reinforcements to relieve the siege of Zrinyi-jvr, which fell on 30 June. Instead, he withdrew to a position on the Raab River, where on 1 August he caught and defeated Köprülü's army at Szént Gotthard/Mogersdorf as it attempted to cross into Austria. Leopold used the opportunity of the victory to conclude quickly the Treaty of Vasvár, which established a 20-year peace and awarded the Ottomans an indemnity and possession of Neuhausel. Leopold's neglect of Hungarian interests in securing the peace provoked a magnates' conspiracy against Habsburg rule, which was exposed and suppressed in 1670. Refugees from Habsburg reprisals fled to Transylvania, from which they launched *kuruc* raids into Hungary for the next decade.

The Battle of St. Gotthard was the only major confrontation of the Habsburg and Ottoman armies between the Fifteen Years' War (1591–1606) and the Second Siege of Vienna (1683).

Brian Hodson

See also: Austro-Turk Wars

References and further reading:

Peball, Kurt. *Die Schlacht bei St. Gotthard/Mogersdorf, 1664.* Vienna: Österreichische, Bundesverlag, 1964.

Wagner, Georg. *Das Türkenjahr 1664.* Eisenburg: Druckerei Rötzer, 1964.

Grand Alliance, War of the (1688–1697)

An indecisive European struggle over hegemony. The conflict was also known as Nine Years' War or War of the League of Augsburg. After reaching the top of its hegemonial power in 1680, the France of Louis XIV feared a counterstroke by other European states, especially after Austrian successes against the Turks, which led to the rise of Habsburg power. So Louis began a preventive war to secure his conquests of the past. In the autumn of 1688, the war began with a sudden French attack on southwest Germany, where serious atrocities were committed against the civilian population in the so-called devastation of the Palatinate. In 1689, a Grand Alliance between the emperor, the Netherlands, England, Spain, Savoy, and some minor German states, and later also some imperial circles, was formed to fight against the French predominance. The fighting took place mainly in Flanders, where the battles at Fleurus (1690), Steenkerke (1692), and Neerwinden (1693) ended in French victories. William III of Orange was preoccupied with securing the British Isles until 1690. After this, he could pay full attention to the war on the Continent.

The Upper Rhine region was successfully defended by the joint forces of the Swabian and the Franconian imperial circles under the command of Margrave Ludwig Wilhelm of Baden. Fighting also took place in Catalonia, in Italy, and in some colonial areas. The naval war saw the French victorious at Beachy Head (1690) and the English off La Hogue (1692). French privateers enjoyed some success against English trade. But none of these victories or defeats was decisive. All belligerent parties were exhausted after years of fighting. The treaty of Rijswijk brought the war to an end in late 1697. The threat of French hegemony was successfully checked for the moment, but France was not beaten. It had to give some earlier conquests back to the former owners, but it secured the city and fortress of Strasbourg, which stabilized its eastern frontier.

Max Plassmann

See also: Boyne; English Wars in Ireland; Jacobite Rebellions; Killiecrankie

References and further reading:

Childs, John. *The Nine Years' War and the British Army 1688–1697: The Operations in the Low Countries.* Manchester, UK: Manchester University Press, 1991.

Plassmann, Max. *Krieg und Defension am Oberrhein.* Berlin: Duncker & Humblot, 2000.

Grandson and Morat, Battles of (Switzerland, 2 March and 22 June 1476)

Reemergence of "pike power" against cavalry. Austria needed a powerful ally against the Swiss Confederacy in the fifteenth century. To this end, Austria signed the Treaty of Saint Omer in 1474 with Charles the Bold. Charles received possessions in Alsace, bordering Switzerland, and the Burgundian war was the outcome of Charles's subsequent attempt to establish a Burgundian Empire stretching from Netherlands to Italy.

In a first clash at Héricourt (13 November 1474), a Burgundian army of 12,000 was forced to retreat to avoid being overrun by Swiss pikemen. In 1476, Charles invaded Switzerland and advanced on Berne, intending to cut a path through the Confederation. Berne, well aware of the Duke's plan and approach routes, had reinforced the garrison at Grandson. After a short siege, the Burgundians slaughtered all Swiss survivors. The massacre infuriated the regrouping Swiss.

On 2 March 1476, the armies blundered toward one another and Swiss leading elements came upon the Burgundian vanguard pitching camp. A short skirmish between Swiss handgunners and Burgundian archers gave time for the first Swiss battle corps to emerge from the wooded high ground. It deployed in a characteristic square of 10,000, with halberdiers and ensigns in the center, surrounded by a forest of 6-meter-long pikes. Charles intended to bombard the square with artillery and to finish it with cavalry charges. The Burgundians easily dispersed the Swiss skirmishers but were repeatedly repulsed by the impenetrable wall of pikes.

Charles tried to draw the Swiss square further forward into the low ground to expose its flanks. He ordered his troops to withdraw a short distance and moved archers and artillery to the flanks. During this complex maneuver, the main Swiss column appeared and advanced on the disorganized Burgundians. A rout ensued and Charles's panic-stricken army abandoned 420 pieces of artillery and his fabulously rich camp to the Swiss.

Casualties were incredibly low: 200 Swiss and fewer than 500 Burgundians. The Swiss failed to press their advantage and Charles's army was soon ready again. He sent for reinforcements. Mercenaries were hired and a new artillery train assembled. The Swiss found new allies from Lorraine, Alsace, and Swabia, and their army was reinforced to a total of 25,000 foot and 2,000 cavalry. The Burgundians numbered only 12,000, most survivors of the Grandson disaster.

Charles decided to attack Berne from the west and besieged the small town of Morat on the shores of the lake of the same name. The Burgundian camp was protected by a wooden palisade called the Grünhag, one mile from the main camp. The Swiss divided their army into three pike blocks. The allied cavalry was to outflank the palisade one mile north. The Burgundians were completely unaware of Swiss preparations and had left only 3,000 men to guard their stronghold. The Swiss advanced in echelon toward the Burgundian camp on the afternoon of 22 June 1476. The northern attack was checked by artillery and archers, but the Swiss center column reached and overran the wooden palisade. Burgundian reinforcements were defeated in detail by the advancing Swiss while outflanking cavalry cut off the northern escape route. The Burgundians were slaughtered or drowned in the lake by the enraged Swiss. Their army suffered more than 8,000 casualties, to 410 Swiss. Charles the Bold lost his army and barely escaped with his life, only to lose it in his next encounter with the Swiss (Nancy, 5 January 1477). The fierce reputation gained by the Swiss in the battle gained them offers of foreign employment. Swiss infantry would be the "queen of battles" for more than a century.

Gilles Boué

See also: Charles the Bold, Duke of Burgundy

References and further reading:

Miller, Douglas. *The Swiss at War.* London: Osprey, 1979.

Kurz, Hans. *Schweizerschlachten.* Berne: Francke, 1977.

Granicus, Battle of the (May/June 334 B.C.E.)

The first battle the Macedonian king Alexander III the Great fought after he had invaded the Persian Empire. It took place at the river Granicus (Kocabas) in northwestern Turkey. After Alexander had crossed the Hellespont, the Persian satraps (governors) of Asia Minor assembled an army to fight him, consisting of 20,000 horse and 20,000 foot, the lat-

Alexander the Great defeating the Persians at the Battle of Granicus. (Bettmann/Corbis)

ter largely Greek mercenaries. As Alexander was on the west bank of the river, the Persians awaited him on the other side in two lines. The first consisted of the cavalry positioned at the riverside. The Greeks formed the second line on a height some distance behind.

Although Alexander had an army of 37,000 men, he brought only part of it to the field at the Granicus, probably 13,000 infantry and 5,100 cavalry. He deployed his Macedonian pikemen in the center. The left flank was covered by cavalry. On the right wing the Macedonian guard infantry and the guard heavy cavalry took position, reinforced by light cavalry and skirmishers.

The Macedonians advanced to the riverside and an assault force crossed the river and attacked the Persian horse. Thereupon Alexander himself followed, leading his guard cavalry through the river into the Persian line. Slowly the Macedonians fought their way up the riverbank. In the meantime the Macedonian infantry also crossed the river. The Persian cavalry broke and fled under the pressure of the cavalry and the pikemen, whereupon the Macedonians turned upon the Greek mercenaries. Most of them perished; 2,000 were taken prisoner and enslaved. The Persian losses in cavalry were probably limited to about a thousand, but several important commanders were killed. The Macedonians counted only 120 dead.

Maarten van der Werf

See also: Alexander the Great; Alexander's Wars of Conquest

References and further reading:

Green, Peter. *Alexander of Macedon, 356–323 BC.* Harmondsworth, UK: Penguin, 1974. Reprinted Berkeley and Los Angeles: University of California Press, 1991.

Fuller, J. F. C. *The Generalship of Alexander the Great.* London: Eyre & Spottiswoode, 1958.

Hammond, N. G. L. *The Genius of Alexander the Great.* London: Duckworth, 1997.

Tarn, W. W. *Alexander the Great.* Cambridge, UK: Cambridge University Press, 1948.

Grant, Ulysses Simpson (1822–1885)

The most successful Union field commander in the American Civil War, twice elected president of the United States. Grant was born as Hiram Ulysses Grant in Point Pleasant, Ohio, on 27 April 1822. Through a transcription error, he was enrolled in the U.S. Military Academy at West Point in 1839 as "Ulysses Hiram Grant." When "Hiram" was misread as "Simpson," Grant accepted it. Nicknamed "Uncle Sam" and "United States" by his fellow cadets, he later became known as "Unconditional Surrender" after the terms he dictated at Fort Donelson. His friends called him "Sam."

Graduated from West Point in 1843, Grant was commissioned in the 4th Infantry. He served in the Mexican-American War under Zachary Taylor at Palo Alto, Resaca de la Palma, and Monterrey, then under Winfield Scott at Veracruz, Cerro Gordo, Churubusco, Molino del Rey, and Chapultepec. He resigned as captain in 1854 and tried farming and several small business ventures, failing at all of them.

When the Civil War began, Grant volunteered for the militia in Galena, Illinois. He was ordered to the state adjutant general's staff, then promoted to colonel of the 21st Illinois Volunteer Infantry in June 1861 and brigadier general of volunteers in August, commanding the District of Southeast Missouri.

Eager to fight, he took the offensive on his own initiative. He captured Paducah, Kentucky, on 6 September, but lost at Belmont, Missouri, on 7 November. At odds with his superior officer, Henry Halleck, over both strategy and temperament, he acted on his own whenever he could. He took Fort Henry, Tennessee, on 6 February 1862 and won the war's first major Union victory when Simon Bolivar Buckner surrendered 14,000 men to him at Fort Donelson, Tennessee, on

16 February. These actions earned him promotion to major general in command of the District of West Tennessee on 21 February.

Albert Sidney Johnston surprised Grant in an unfortified position the first day at Shiloh while Grant was awaiting Don Carlos Buell's reinforcements. Union casualties were heavy as Grant hastily built a defensive line. He gained back all the lost ground through sheer doggedness on the second day. Halleck wanted to dismiss him for recklessness, but Abraham Lincoln would not allow it.

Lincoln replaced Halleck with Grant that summer. After Grant's new subordinate, William S. Rosecrans, won at Iuka and Corinth, Grant felt free to launch his Vicksburg campaign in November. It was Grant's only brilliant campaign. By combined amphibious and overland assault, victories at Jackson on 14 May and Champion's Hill on 16 May 1863, and a six-week siege, he captured Vicksburg on 4 July, coincident with George Gordon Meade's victory at Gettysburg.

Grant broke Braxton Bragg's siege of Chattanooga on 28 October and defeated Bragg at Lookout Mountain on 24 November and Missionary Ridge on 25 November. Lincoln promoted him to lieutenant general in command of all Union armies on 12 March 1864. Thereafter he made his headquarters with Meade's Army of the Potomac.

Portrait of Lieutenant General Ulysses S. Grant, wearing uniform, c. 1865. (Library of Congress)

Grant waged an unimaginative, horribly bloody war of attrition against Robert E. Lee. Staggering Union losses in the Wilderness and at Spotsylvania provoked the northern press to call him "Butcher." Grant's worst tactical mistake of this campaign was at Cold Harbor, where Lee's entrenched 60,000 massacred Grant's frontally attacking 110,000. Nevertheless, Grant's superior numbers eventually pushed Lee away from Richmond and Petersburg, outflanked him, and forced the Confederate surrender at Appomattox on 9 April 1865.

Elected president easily as a Republican in 1868 on the strength of his war record, Grant served two terms. Having mismanaged his money, betrayed by friends, he spent most of his retirement in poverty. In perhaps the most heroic campaign of his life, Grant raced to complete his memoirs before the cancer gnawing at his throat took him. He completed his memoirs just weeks before his death on 23 July 1885 in Mount McGregor, New York; they were among the finest of autobiographical works in American literary history and freed Grant's family from financial worry. Grant was buried in a mausoleum that soon became a New York City landmark.

Eric v. d. Luft

See also: American Civil War; Chattanooga, Battle of; Cold Harbor, Battle of; Fort Donelson; Halleck, Henry Wager; Johnston, Albert Sidney; Johnston, Joseph Eggleston; Lee, Robert Edward; Lincoln, Abraham; McClellan, George Brinton; Meade, George Gordon; Mexican-American War; Monterrey; Petersburg, Siege of; Resaca de la Palma; Scott, Winfield; Sherman, William Tecumseh; Shiloh; Sioux Wars; Spotsylvania Court House; Taylor, Zachary; Vicksburg, Siege of; Wilderness

References and further reading:

Anderson, Nancy Scott, and Dwight Anderson. *The Generals: Ulysses S. Grant and Robert E. Lee.* New York: Wings, 1994.

Arnold, James R. *Grant Wins the War: Decision at Vicksburg.* New York: Wiley, 1997.

Conger, Arthur Latham. *The Rise of U. S. Grant.* New York: Da Capo, 1996.

Perret, Geoffrey. *Ulysses S. Grant: Soldier & President.* New York: Modern Library, 1999.

Great Wall of China (16th Century)

An extensive defensive wall defining China's northern frontier. Chinese states and dynasties had built frontier walls from time immemorial. Such walls delineated frontiers for political and economic purposes. They could also serve as military barriers, although the costs of garrisoning a long wall along with the supporting bases and reserves necessary

View of the Great Wall of China at Nankow Pass. (Hulton/Archive)

to make it viable could prove prohibitive. Much of the foundation for the Ming wall was laid by Qin Shi Huangdi, the first emperor of China in the third century B.C.E.

The Great Wall of Ming times was built along what was perceived at the time as a fault line between China and its northern neighbors: Mongol groups as well as the Manchus that later were to create their own "Chinese" dynasty. The wall also governed movement between China and the north and funneled trade through barriers so that it could easily be taxed. Finally the wall was a military barrier, but only against the low-key raids that had troubled north China for centuries. Any serious attack was bound to get through unless met by a major mobilization, a lesson that the Ming ultimately forgot.

The present wall dates to that era after the capital had been moved to Beijing (1421) and the Ming government found itself increasingly unable to deal with the Mongols, who captured a Ming emperor in 1449. Gradually a linear complex of frontier fortifications was seen as the only solution and a massive effort was made in the sixteenth century, resulting in the wall as we now know it. The wall continued to expand right down to the end of the dynasty but never provided effective resistance to any attack, not the Mongol Altan-qan in 1550, nor the Manchus in 1644.

Paul D. Buell

References and further reading:

Barfield, Thomas J. *The Perilous Frontier, Nomadic Empires and China.* Cambridge, MA, and Oxford, UK: Basil Blackwell, 1989.

Waldron, Arthur. *The Great Wall of China, From History to Myth.* Cambridge, UK: Cambridge University Press, 1990.

Greco-Turkish War (1920–1922)

Conflict following World War I that established the republic of Turkey. At the end of World War I, the victorious Allies occupied much of the Ottoman Empire and began the process of dismembering that empire. Certain provinces, such as Palestine, were made into protectorates under the Allies. The Smyrna area of Anatolia, with a large Greek population, was awarded to Greece. In May 1919, a Greek army occupied that city and its hinterland, resulting in conflicts with the Turks. The Greeks occupied a large area, up to the so-called Milne line. The Greek government was encouraged by their success, and with the approval of the British, they expanded their control further. Beginning on 22 June 1920, the Greeks marched on Panderma on the Sea of Marmara, and Bursa and Usak. Resistance by the Turks was limited. The Ottoman government of Sultan Muhammad VI Vahideddin in Istanbul was convinced that resistance was futile. The sultan signed the Peace of Sevres, which surrendered eastern Thrace to Greece, as well as control over the Smyrna area for five years, to be followed by a plebiscite.

Turkish resistance to such national amputation was led by Mustafa Kemal, later known as Kemal Ataturk. In 1919, Kemal became the leader of the Association of the Rights of Anatolia and Rumelia. He established a provisional government in Ankara and set up a nationalist movement to oppose both the sultan and the Greeks. He declared that Turkish borders were those existing on 11 November 1918.

In January 1921, the Greek commander in Anatolia, General Anastasios Papoulas, launched a reconnaissance in force against the important Turkish rail center of Eskisehir. Turkish troops repulsed them. The allies voiced growing dissatisfaction with Greek ambitions, but Papoulas attacked Eskisehir in greater force in March 1921 and suffered another defeat. Additional Greek troops were brought to Anatolia. They suffered from a lack of proper equipment and a poor supply situation. Still, by June 1921, Papoulas had more than 200,000 men. He began a general advance in June. His army quickly took Eskisehir and Afyon, another important rail center. The Greeks slowly pushed eastward toward Ankara. The Turks, commanded by Kemal in person, conducted a fighting retreat until they reached the Sakarya River, 70 miles west of Ankara. A three-week battle began on 24 August and ended on 16 September. The Greeks were defeated by a lack of supplies and equipment, as well as by the Turks. They retreated to a partially fortified line stretching from the Sea of Marmara, then east of Eskisehir and Afyon, and on to the Menderes River. Despite poor supply, the Greeks stayed in that position for the next 11 months. Kemal used the time to build up his forces. He was assisted by Soviet Russia, which provided military equipment left from its own civil war.

On 26 August 1922, Kemal launched a series of major attacks against the central and southern parts of the Greek front. Despite strong resistance, the Turks broke through. By 5 September, they had captured Bursa. On 9 September, they captured Smyrna. The collapse of the Greek army in Anatolia was followed by the slaughter of thousands of Greek soldiers and civilians by the enraged Turks.

A revolution broke out in Greece, which overthrew the government. Two prime ministers, three cabinet ministers, and the last commander in Anatolia, General Georgios Hatzianestis, were all executed. The Peace of Sevres was replaced by the Treaty of Lausanne, which included everything Kemal had wanted. Eastern Thrace and all of Anatolia were returned to the Turks. Greeks living in areas controlled by the Turks were forced from their homes and exiled to Greece, and vice versa in Greece. Kemal abolished the sultanate, established a republican form of government, and served as the country's first president.

Tim J. Watts

See also: Ataturk, Kemal; World War II

References and further reading:

Kinross, Patrick Balfour. *Ataturk: A Biography of Mustafa Kemal, Father of Modern Turkey.* New York: Morrow, 1964.

Mango, Andrew. *Ataturk: The Biography of the Founder of Modern Turkey.* Woodstock, NY: Overlook Press, 1999.

Pope, Nicole, and Hugh Pope. *Turkey Unveiled: Ataturk and After.* London: John Murray, 1997.

Greek Civil War (1944–1949)

A two-phase civil war (1944–1945 and 1946–1949) that had its beginning in the German occupation of Greece in 1941. In both phases, Communist forces staged an unsuccessful attempt to take over the country.

During the German occupation, two resistance forces emerged to oppose the Germans. One, the National Liberation Front–National Popular Liberation Army (known as the EAM-ELAS in Greek) was Communist controlled and directed; the other, the Greek Democratic National Army (known as the EDES in Greek), was Royalist.

During the German occupation, both groups fought against the Germans and, on some occasions, worked together. Due to the brutality of German reprisals, both sides worked to bide their time until the Germans left and the inevitable postwar struggle for power would begin. By 1944, when the Germans withdrew from Greece, the EAM-ELAS had removed all political and military opposition in Greece except for the EDES. They established a provisional government in the mountains of Greece and planned for a regime that would not include the king of Greece nor the government in exile in London.

The British had a presence in Athens and were able to cobble together a fragile coalition government between the EDES and the EAM-ELAS in October 1944. The agreement between the two sides failed over the issue of disarming the Communist forces in the field. On 3 December 1944, the EAM-ELAS launched an uprising in Athens, which the British suppressed only after great effort. The Communist forces had been able to capture almost all of Greece but were finally defeated.

With their defeat, the Communists entered into talks and agreed to disarm at a meeting in February 1945. Part of the agreement was that in March 1946 a general election was to be held across Greece. When the elections came, the Communists sat it out and saw a Royalist government placed into power. In September 1946, the king of Greece was restored to the throne via a plebiscite.

All of this was rejected by the Communists, and full-scale civil war broke out in 1946. Britain, which since 1944 had been responsible for the defense of Greece, found that it was a burden it could no longer bear and transferred the task to the United States in 1947. By this time, the EAM-ELAS had established a provisional government in the mountains of the north of Greece. The United States issued the Truman Doctrine, stating that the U.S. would aid those nations struggling against outside-supported insurgency, and started to pour military and economic aid into Greece.

Over the next two years, supported by massive amounts of U.S. aid and military advisers, the Greek government was finally able to push the Communists back, and they abandoned the armed struggle, with many Greek Communists fleeing to Albania. The two phases of the Greek Civil War led to more than 50,000 battle deaths and the creation of over half a million internal refugees.

Drew Philip Halévy

See also: Cold War

References and further reading:

Weinberg, Gerhard L. *A World at Arms: A Global History of World War II.* New York: Cambridge University Press, 1995.

Greek War of Independence (1821–1832)

On 2 April 1821, lingering resentment against continuing Ottoman rule resulted in outright rebellion when Greek Orthodox cardinal Germanos led an uprising at Kalaviyta. The Turkish garrison and a large part of the Muslim population was massacred, which led to counteratrocities by occupation forces. Nevertheless, within months Greek forces had effective control of the peninsula. In the spring of 1822 Ottoman troops counterattacked by seizing the island of Chios and enslaving the population. The Greek extracted a measure of revenge on the night of 18–19 June 1822, when fire ships under Captain Constantine Canaries sank a Turkish flagship, killing an estimated 3,000 Turks. By January 1823, Turkish forces abandoned the siege of Missolonghi and it appeared that Greek independence had been won.

The news was certainly applauded by the governments of Western Europe, then undergoing a renaissance of classical Greek culture among its elite. Popular and political sentiment favoring Greek independence would prove decisive. *Greece* and *Athens* meant Homer, Aristotle, the highest culture, science, philosophy; the *Terrible Turk* seemed to stand for harems and galley slaves.

Unfortunately for the Greeks, they had no sooner expelled occupying forces than they quarreled amongst themselves and a period of civil war ensued. Sultan Mahmud II took advantage of this impasse by appealing to Mohammed Ali of Egypt for help, who dispatched his son Ibrahim and a large squadron of ships and troops. This infusion of Muslim strength caught the Greeks unprepared and they were slowly rolled back. Missolonghi was finally captured after a long siege, as was the Acropolis in Athens. Greek attempts to recapture this cultural landmark ended in failure, and by June 1827 nearly the entire peninsula was back under Turkish/Egyptian control. At this juncture fate intervened when the governments of Great Britain, France, and Russia demanded a Muslim withdrawal from Greece and de facto independence. When the sultan refused, all three nations dispatched heavily armed squadrons into the region.

On 20 October 1827, a combined allied fleet under Admiral Sir Edward Codrington entered the harbor of Navarino and anchored in full view of the Egyptian fleet. Their leader, Tahir Pasha, arranged his ships defensively in a large crescent formation and awaited events. A tense standoff ensued and negotiations were under way when a Turkish vessel suddenly fired upon a British dispatch boat, precipitating an all-out battle. In this duel between long lines of anchored warships, the allies, with larger vessels and heavier guns, completely defeated their Muslim opponents. They lost slightly over 500 casualties in this, the last pitched engagement between wooden warships, while Muslim losses totaled several thousand. Furthermore, this defeat, and the declaration of war by Russia against Turkey, finally induced the Ottomans to relinquish their grasp of Greece. After four centuries of Muslim dominance, Greek independence had been achieved and was confirmed by the 1832 Treaty of London.

John C. Fredriksen

See also: Turkish Wars of European Expansion

References and further reading:

Clogg, Richard, ed. *The Struggle for Greek Independence.* London: Macmillan, 1973.

Crawley, C. W. *The Question of Greek Independence: A Study of British Policy in the Near East, 1821–1833.* New York: H. Fertig, 1973.

Phillips, W. Alison. *The War of Greek Independence, 1821 to 1833.* New York: C. Scribner's Sons, 1897.

Greek-Persian Wars (499–448 B.C.E.)

Series of unsuccessful attempts by the Persian Empire to subdue a loose coalition of Greek city-states. The Persian province, or satrapy, of Ionia, on the Aegean coast of Turkey, revolted against Persian king Darius the Great in 500 B.C.E. Aristagoras, the Greek tyrant of Miletus, led the revolt. Athens, Eretria, and a few other Greek city-states supported it. After Darius ruthlessly counterattacked in 499, Aristagoras fled to Thrace and died there in 497. Darius destroyed Miletus and deported its inhabitants to Mesopotamia in 494. With the rebellion crushed, Darius vowed to add all Greece to his empire to punish Athens for helping Ionia. In 492 he invaded from Thrace, conquered Macedonia, but had to abandon his plans when a storm wrecked his fleet.

In 490 Datis and Artaphernes jointly led Darius's second invasion force, estimated to be between 20,000 and 200,000, into Greece. Crossing the Aegean in 600 ships from Samos, Ionia, they easily subdued Euboea and moved into Attica. Athenian general Miltiades the Younger chose to fight them on the open plain at Marathon and there deployed his 20,000 Athenians and Plataeans to await the invaders. Hippias, a Greek, told the Persians that they could use their cavalry effectively on the plain and guided them to Marathon.

As Darius's army marched toward Marathon, the Persian Empire was at its height and included Libya, Egypt, Palestine, Turkey, Cyprus, Thrace, and Macedonia. With the Carthaginians friendly to the Persians, Greece was isolated. For Greece, everything was at stake.

As the Persians assembled in the center of the plain in September 490, Miltiades lured them toward his weak center and between his strong flanks. When the Persians had advanced far enough, both Greek flanks charged simultaneously at a dead run and soon surrounded their enemy. The mobility and versatility of hoplites in phalanx formation, enclosed in networks of shields to protect them from arrows and using their long spears like lances, completely frustrated the inflexible tactics of the Persians and won the day. Casualties were reported as 6,400 Persian and 190 Greek. The Persians retreated to Turkey.

Darius's son and successor, Xerxes the Great, was determined to avenge his father's loss in Greece. After suppressing a revolt in Egypt, he began in 483 to prepare an unprecedentedly massive assault against Greece. In 480 he built a pontoon bridge across the Hellespont with planks laid over 674 ships lashed together in two rows and thus brought his men, animals, wagons, and war machines from Asia to Europe. The Persian navy guarded the army's left flank.

The nearly contemporary Greek historian Herodotus is the main source of information about the Greek-Persian Wars, but scholars know that he exaggerated and sometimes even lied. We must therefore be skeptical about some of his fantastic claims and huge numbers. Nevertheless, the size of Xerxes's invasion force has been reasonably estimated at between 2 and 5 million, including noncombatants and camp followers. The number of actual combatants may have been as low as 100,000. The sheer size of this unwieldy mass contributed to its undoing.

When the gigantic Persian host tried to squeeze through the small pass at Thermopylae in August 480, 300 Spartans and 700 Thespians under Leonidas held them for three days until a traitor, Ephialtes, showed the Persians a secret mountain path from which they could use archers against Leonidas's rear. The defenders were annihilated and the Persian march southward continued. After burning Athens, Xerxes returned home and left his army under the command of Mardonius.

Themistocles lured 800 large and clumsy Persian ships into the shallow bay at Salamis in September 480, then trapped and overwhelmed them with his 380 small, speedy, shallow-draft vessels. This decisive naval victory inspired the Greek land forces. They immediately massacred the Persian garrison on the island of Psyttaleia. The Greek army, led by Sparta, finally beat back the third Persian invasion at Plataea in 479. Mardonius was among the Persian dead. The same day, Athens won a decisive naval battle against Persia at Mycale.

About 478, Athens created the Delian League, or Attic Naval Alliance, a mutual defense bloc that relied upon Athenian sea power to prevent any future Persian aggression. Gradually the Delian League wrested Aegean coastal settlements from Persia and forced them to join the alliance. Athens was establishing its own empire. Both Sparta and Persia grew anxious about this Athenian expansion.

For the next 30 years most of the combat between Greece and Persia was naval, with Greece usually getting the best of it. The reign of Xerxes's son, Artaxerxes I, was generally peaceful except for sporadic minor revolts in far-flung provinces. Yet his empire eroded and he presided over its decline. Less and less able to resist Greek encroachment, Artaxerxes sought peace. Sparta concluded a five-year truce with Persia in 451. The Treaty of Callias ended conflict between Persia and Athens in 448.

Eric v. d. Luft

See also: Alexander the Great; Ancient Warfare; Cimon; Marathon, Battle of; Peloponnesian Wars; Persian Empire; Persian Wars of Expansion; Plataea, Battle of; Thermopylae, Battle of; Xerxes I

References and further reading:

Bradford, Ernle Dusgate Selby. *Thermopylae: The Battle for the West.* New York: Da Capo, 1993.

Green, Peter. *The Greco-Persian Wars.* Berkeley: University of California Press, 1996.

Hanson, Victor Davis. *The Wars of the Ancient Greeks and Their Invention of Western Military Culture.* London: Cassell, 1999.
Hignett, Charles. *Xerxes' Invasion of Greece.* Oxford, UK: Clarendon, 1963.

Greene, Nathanael (1742–1786)

American field commander, perhaps the best strategist on either side in the Revolutionary War. Born a Quaker in Potowomut, Rhode Island, on 7 August 1742, Greene went into his father's business as an ironworker and from 1770 to 1772 served in the General Assembly of the Colony of Rhode Island. Devoted to the defense of liberty, he disappointed his family in October 1774 by rejecting the pacifist ideals of his religion and raising a company of militia. He became brigadier general of militia in May 1775 and brigadier general in the regular Continental army in June.

During the siege of Boston, Greene impressed George Washington with his strategic, tactical, and logistical plans. He garrisoned Long Island after 1 April 1776 and was promoted to major general on 9 August. He was ill and out of action for most of William Howe's invasion of New York but attacked Staten Island on 12 September and participated in planning the final battle of that campaign, the loss of Fort Washington to the British on 14 November. He fought with distinction at Trenton, Brandywine, and Germantown and was with Washington at Valley Forge. Following the demise of the Conway Cabal against Washington, the commander in chief put his most trusted officers in key positions. Thus Greene, an expert field commander, became quartermaster of the Continental army on 2 March 1778. Commanding the right flank at Monmouth, he valiantly tried to retrieve Charles Lee's blunders. He further distinguished himself at (fittingly) Quaker Hill, Rhode Island, on 29 August.

Portrait of General Nathanael Greene. (Library of Congress)

In September 1780 he was assigned to replace Benedict Arnold in the Hudson valley after the latter's treason and defection, but after Horatio Gates failed at Camden, Washington appointed him commander of the Southern Department of the Continental army on 14 October.

Greene relieved Gates on 2 December and collaborated with Daniel Morgan, Francis Marion, and other partisan leaders, unlike many Continental commanders. He induced Charles Cornwallis to chase him through the backwoods of North Carolina toward Virginia, losing tactically at Guilford Courthouse but achieving strategic victory by overextending the British supply lines. Using a similar strategy, he lost at Hobkirk's Hill on 25 April, won at Augusta on 5 June, and lost at Ninety-Six on 19 June and Eutaw Springs on 8 September; but by the end of that campaign, the British could hold only Savannah and Charleston. Even after Yorktown, Greene continued to harass the British in the South until August 1783. He died at his estate given to him by a grateful state, near Savannah on 19 June 1786. Greene is remembered in scores of "Greene Counties," "Greenevilles," and "Greene Streets" throughout the eastern United States.

Eric v. d. Luft

See also: American Revolution; Arnold, Benedict; Brandywine; Camden, Battle of; Cornwallis, Sir Charles; Cowpens; Gates, Horatio; Germantown; Guilford Court House; Lee, Henry; Long Island, Battle of; Monmouth; Trenton; Washington, George; Yorktown

References and further reading:

Greene, George Washington. *The Life of Nathanael Greene: Major-General in the Army of the Revolution.* Freeport, NY: Books for Libraries Press, 1972.
Johnson, William. *Sketches of the Life and Correspondence of Nathanael Greene.* New York: Da Capo Press, 1973.
Thane, Elswyth. *The Fighting Quaker: Nathanael Greene.* New York: Hawthorn Books, 1972.
Thayer, Theodore George. *Nathanael Greene: Strategist of the American Revolution.* New York: Twayne, 1960.
Treacy, M. F. *Prelude to Yorktown: The Southern Campaign of Nathanael Greene, 1780–1781.* Chapel Hill: University of North Carolina Press, 1963.

Grenada (October 1983)

The American invasion of the island nation of Grenada to rescue American students and end the Marxist regime. Grenada experienced a coup in March 1979 when Maurice Bishop, leader of the New Jewel movement, took advantage of Prime Minister Sir Eric Gairy's absence from the country. By 1983 the leaders of JEWEL, feeling that Bishop was not moving fast enough in implementing a Marxist ideology, staged their own coup. On 19 March 1983 the People's Revolutionary Army (PRA) killed Bishop and some 50 of his followers during the attack on Fort Rupert. By 23 October 1983 the U.S. had established Joint Task Force 120 under Vice Admiral Joseph Metcalf III to begin planning operations in Grenada to evacuate the 1,000 American citizens and other foreign nationals there. By 21 October the U.S. received a formal request by the Organization of Eastern Caribbean States (OECS) to help restore order. On 22 October 1983 American president Ronald Reagan approved the plan that was to commence by dawn on 25 October 1983.

Operation URGENT FURY took place on 25 October 1983 with simultaneous operations by the Army and Marine Corps. The airborne assault was carried out by members of the 75th Ranger Regiment and the 82d Airborne Division to secure the airport at Port Salines. After encountering some antiaircraft fire, the flight of C-130s carrying the rangers prepared for a low-level jump to secure the airport. Immediately after landing, the rangers cleared the runway of debris placed there to prevent its use. One ranger, under fire, used a commandeered steam roller to drive spikes into the runway. After the runway was secured by the rangers, the 82d landed to relieve them, allowing the rangers to proceed to St. Georges to release the American students there. Later that day the Caribbean Peace-keeping Force landed to assist the American forces on the island in keeping order after combat operations ceased.

On the other side of the island the U.S. Navy and U.S. Marine Corps conducted combat operations on Pearls Airport and the surrounding area. This operation began early on 24 October 1983 when a sea, air, and land (SEAL) team was inserted to check the feasibility of an amphibious assault. The SEALs' reconnaissance of the beach showed that the reefs would obstruct landing craft. The SEALs also determined the defenses of the Pearls Airport. The marines' helicopter assault on Pearls Airport encountered a resistance from a Peoples' Revolutionary Army militia platoon, which ceased when marine AH-1 helicopters fired on the PRA.

Another operation was the Delta Force's failure to secure the Richmond Hill Prison when the helicopters transporting them encountered heavy antiaircraft fire that downed one 101st Airborne helicopter. It was decided that any further attempt to take the prison was futile. The navy SEALs, who were to rescue the governor-general, came under heavy fire. The siege of the SEALs was finally broken when Metcalf ordered aircraft from the USS *Independence* to bomb the area around the governor-general's residence. The ensuing attack led to the accidental bombing of a mental hospital. The hospital had been a military base known as Fort Matthew; the Americans did not know it had been converted to civilian use.

The second day of the operation saw the rangers secure the students at Grand Anse. This part of the operation was improvised because it was originally believed that all of the students were at the St. George's campus. By this time most of the combat operations involved mopping up the PRA's remaining resistance. The U.S. suffered 18 dead and 116 wounded; the Grenadian cost was 45 dead and 350 wounded. By December 1983 the U.S. forces were withdrawn and the Caribbean Peace-keeping Force took over.

Michael Mulligan

See also: Airborne Operations; Special Operations Forces

References and further reading:

Adkin, Major Mark. *Urgent Fury: The Battle for Grenada.* Lexington, MA: Lexington Books, 1989.

Russell, Lee E., and Albert Mendiz. *Grenada 1983.* Men-at-Arms Series. London: Osprey, 1985.

Gribeauval, Jean Baptiste Vaquette de (1715–1789)

French military engineer whose developments of artillery contributed to Napoleon's successes. Born in a newly noble family, Gribeauval is the best example of promotion by merit. He entered the king's service at 17 years of age at the artillery school of La Fère. He was appointed artillery officer two years later and took part in the war of Austrian Succession (1740–1748). Gribeauval was sent to Prussia and Austria to study their artillery and fortification systems. He was lent to the Austrian army during the Seven Years' War (1756–1763) and distinguished himself in the siege of Schweidnitz, a fortified Silesian town he defended against Frederic II. Back in France he became general inspector of the artillery, and with Secretary of War Choiseul's support, he began the reform of the artillery. He separated the artillery into three categories: field artillery, to be used on campaign; siege artillery, for the attack on or defense of strongholds; and coastal artillery, to defend the numerous French ports. Gribeauval reduced the multiplicity of different-caliber field pieces to three: according to the weight of the cannon ball, 4 pounds, 8 pounds, and 12 pounds. He reduced the length and weight of the field artillery. To obtain greater ranges with less powder, he used prefabricated, standardized powder and shot combinations. Gribeauval introduced interchangeable wheels and parts on lighter gun carriages and also an iron elevating screw for easier raising of the barrel. He improved the disposition of the draft horses, proposing double files of the animals instead of single file. This change improved greatly the pulling capacity and reduced the numbers of horses necessary in a team. The mobility and effectiveness of the French artillery was the best of the European armies at the end of the eighteenth century. The standardization of the Gribeauval system was adopted in 1785 after some resistance from the older artillery officers. The Gribeauval system was intensively used during the Wars of the Revolution and the Napoleonic Wars and was not replaced until 1825.

Gilles Boué

See also: Artillery; Napoleonic Wars

References and further reading:

Corvisier, André. *Dictionnaire d'Art et d'histoire militaire.* Paris: PUF, 1988.

Nardin, Pierre, *Gribeauval.* Paris. *Cahiers de la Fondation pour les études de Defense nationale* 24 (1982).

Grotius, Hugo (1583–1645)

Through his writings on natural and international law, established a framework for relations between states. Hugo Grotius (Huig de Groot) was born on 10 April 1583 in Delft, Holland, the son of a lawyer. He was a child prodigy, receiving his doctorate at age 15. Grotius served as *advocaat-fiscaal* (attorney-general) for the province of Holland and held a seat in both the provincial and United Netherlands legislatures but was imprisoned in 1618 during a religious controversy. Through the ingenuity of his wife, he escaped in 1621 to Paris. From 1634 until 1644 he was the Swedish ambassador to France.

Grotius was a prolific author, but his greatest contribution was his works on natural and international law. Grotius compiled the writings of ancient and medieval authorities and produced a series of principles about the *just war* rooted in the law of nature. In *De Jure Belli ac Pacis* (On the Law of War and Peace), written in 1625, Grotius argued that war was a legitimate means of settling one state's grievance with another, provided there was just cause. However, a just war must be waged within limits. This was a reaction to the lawless nature of wars in his era.

Grotius also argued that, as killing in wartime was not murder, the capture of private property was not robbery. This served as a legal justification for privateering on the high seas.

Grotius died on 28 August 1645 in Rostock, after falling ill on a sea voyage.

David H. Olivier

See also: Ethics of Warfare; Laws of War; Military and Society

References and further reading:

Bull, Hedley, Benedict Kingsbury, and Adam Roberts, eds. *Hugo Grotius and International Relations.* Oxford, UK: Clarendon Press, 1990.

Dumbauld, Edward. *The Life and Legal Writings of Hugo Grotius.* Norman: University of Oklahoma Press, 1969.

Edwards, Charles S. *Hugo Grotius: The Miracle of Holland. A Study in Political and Legal Thought.* Chicago: Nelson-Hall, 1981.

Guadalajara (8–18 March 1937)

Italian defeat during early stages of the Spanish Civil War (1936–1939). Early in March 1937, a division of 30,000 Italian troops under Fascist general Mario Roatta, supported by a mixed group of 22,000 Spanish Nationalists, legionnaires, and Moroccan troops, under General Jose Moscardo, moved south from their positions in northern Guadalajara Province. Their objective was the provincial city of Guadalajara, 34 miles northeast of Madrid. Both cities were held by the Loyalists. Accompanied by 2,400 Italian trucks, 80 light tanks, and 50 warplanes, Roatta's intention was to seize Guadalajara and extend the Nationalist encirclement of Madrid. Road conditions were wet and muddy. Roatta's less-experienced but more numerous Loyalist opponents (approximately 100,000 men) initially fell back. The town of Brihuega, halfway to Guadalajara, fell to the Italians on 10 March. Within several days, however, the Loyalist forces, made up of Spaniards, anti-fascist Italians, and many international volunteers, supported by some Russian tanks and aircraft, unexpectedly counterattacked. Surprised, Roatta's men retreated. The retreat turned into a rout, which carried along Moscardo's men. Brihuega was lost, together with many supplies. Roatta's casualties totaled some 6,500 men, of whom 2,200 were killed. Loyalist losses were roughly the same. Rotatta was relieved by General Ettore Bastico. Though the Nationalists had gained some ground, their efforts to take Guadalajara were foiled, and the Loyalist cause heartened.

Keir B. Sterling

See also: Franco, Francisco; Spanish Civil War

References and further reading:

Bolleten, Burnet. *The Spanish Civil War.* Chapel Hill: University of North Carolina Press, 1979.

Thomas, Hugh. *The Spanish Civil War.* Revised edition. New York: Harper & Row, 1977.

Wyden, Peter. *The Passionate War.* New York: Simon & Schuster, 1983.

Guadalcanal (August 1942–February 1943)

One of the most bitter land, sea, and air campaigns of World War II in the Pacific. The campaign for Guadalcanal, one of the Solomon Islands, began in the aftermath of the naval battles of the Coral Sea and Midway. The Japanese wanted to continue their plans to take Port Moresby on the tip of New

Guinea and threaten the most direct shipping routes to Australia; their naval defeats placed increased importance on airfields in the Solomon Islands, including Guadalcanal. The United States decided to take the offensive and keep the Japanese off balance.

On 7 August 1942, navy forces began landing marines (1st Marine Division and a regiment of the 2d Marine Division) on Guadalcanal and nearby islands as part of Operation WATCHTOWER. Japanese forces on the island retreated, and the marines seized the airfield, which was under construction, and renamed it Henderson Field. Fearing exposure, Admiral Frank Jack Fletcher withdrew naval support forces, and the supply ships soon left as well. There were a series of night naval battles in the waters to the north, the so-called Slot near Savo Island, Cape Esperance, and nearby waters, and for many months these battles tested the night-fighting capabilities of the U.S. Navy, which had a lot to learn.

For the marines and U.S. Army forces that reinforced them, the fighting was difficult and the tropical conditions caused many to contract various diseases. For nine days after securing their positions, the marines prepared, and on 18 August Japanese troop reinforcements arrived. The Tokyo Express ran nightly, and beginning on 21 August there were fierce nighttime battles, sometimes ending in suicidal mass attacks on American positions. At night, a Japanese bomber with its engines out of synchronization, "Washing Machine Charlie," regularly flew over American positions, to deny U.S. troops needed rest. There were several noted battles—the Tenaru River on 21 August, Edson's Ridge on 12 September, and Henderson Field on 24 October.

By December, the Japanese effort lessened; their losses in troops, ships, airplanes, and pilots had all become too great. Indeed, a major effort to bring reinforcements in November resulted in the sinking of six troop transports. Meanwhile, American forces were strengthened. Finally, on 7 February 1943, the Japanese secretly completed the removal of their troops before U.S. forces realized what had occurred. More than 25,000 Japanese died on Guadalcanal, including about 9,000 from disease and primitive conditions; about 1,500 marines and soldiers died and about 4,800 were wounded or incapacitated by disease. One American veteran was supposed to have remarked that "Guadalcanal wasn't a battle, it was an emotion." Japanese survivors would undoubtedly have agreed.

Charles M. Dobbs

See also: Kokoda Trail; Pearl Harbor Attack; Philippines, U.S. Loss of; Tarawa

References and further reading:

Griffith, Samuel B. *The Battle for Guadalcanal.* Urbana: University of Illinois Press, 2000.

Morison, Samuel Eliot. *The Two Ocean War, A Short History of the United States Navy in the Second World War.* Boston: Little & Brown, 1963.

Tregaskis, Richard. *Guadalcanal Diary.* New York: Random House, 1943.

Guatemalan Civil War (1954)

In 1950 Jacabo Arbenz, a 36-year-old army officer, was elected president of Guatemala and, soon after assuming office, moved Guatemala into a friendly relationship with the Soviet Union, much to the consternation of the United States. In addition, the Guatemalan Communist Party, known after 1952 as the Labor Party, gained greater influence. In 1952 the Agrarian Reform Law was passed and provided for the expropriation and redistribution of large land holdings in the country. The potential threat of expropriation of lands owned by the United Fruit Company and Guatemala's pro-Soviet position greatly alarmed John Foster Dulles, secretary of state in the Eisenhower administration. The U.S. ambassador to Guatemala, John E. Puerifoy, reported Arbenz's pro-Soviet actions to Washington and plotted with the Guatemalan armed forces to have him overthrown.

When the Arbenz regime received a shipment of small arms from Communist Czechoslovakia the United States decided to act by supporting Lieutenant Colonel Carlos Castillo Armas, a Guatemalan exile who had been organizing a guerrilla group in neighboring Honduras. The Central Intelligence Agency (CIA) oversaw Operation EL DIABLO, which consisted of shipping arms to Castillo Armas, providing air support, and equipping the rebels with radio transmitters to broadcast propaganda to Guatemala. Arbenz had lost some domestic support when he cut the military budget and allowed the Labor Party to assume too much power. The invasion was launched on 18 June 1954 when Castillo Armas and a band of around 200 guerrillas entered the country. CIA bombers attacked Guatemala City and, while doing little physical damage, caused panic in the Guatemalan high command, which ordered the armed forces not to resist the invasion. On 27 June Arbenz resigned and fled the country. In October, Castillo Armas became president of Guatemala and served until he was assassinated in 1957. Guatemala would be plagued by civil strife for the next four decades, resulting in more that 100,000 deaths.

For many, the U.S. intervention in Guatemala's internal affairs was a near-perfect example of the Colossus of the North's manipulating affairs in Latin America with scant regard for the people who live there, showing hostility to agrarian reform groups, but supporting right-wing circles that represented the most exploitative elements in those nations. On the other hand, the Eisenhower administration argued that anything, including direct American involvement in the affairs of a sovereign nation, was preferable to a communist regime, and that, at any rate, the Soviets were certainly involving themselves across the board in Latin American affairs, and that many of those "agrarian reformers" were either communists or their dupes.

George Lauderbaugh

References and further reading:

Gleijeses, Piero. *Shattered Hope: The Guatemalan Revolution and the United States, 1944–1954.* Princeton, NJ: Princeton University Press, 1991.

Schlesinger, Stephen C. *Bitter Fruit: the Untold Story of the American Coup in Guatemala.* Garden City, NY: Doubleday, 1982.

Guderian, Heinz (17 June 1888–14 May 1954)

Recognized as one of the military theorists and practitioners of blitzkrieg, involved in the defeat of Poland, France, and Operation BARBAROSSA during World War II. Guderian was born on 17 June 1888 in Kulm. He attended cadet schools and was commissioned second lieutenant on 27 January 1908. During World War I, Guderian served in technical and staff positions. On 1 April 1922, General Staff officer Guderian was transferred to the Motorized Troops Department of the Defense Ministry. Involved in tank development, he advocated the creation of independent armored formations with air and motorized infantry support.

Guderian's intention was to increase battlefield mobility by rapid penetration of enemy lines, resulting in encircling movements. He requested that the army reorganize into divisions built around armor, publishing these views in *Actung! Panzer* in 1937. Hitler sanctioned the creation of three Panzer divisions to test Guderian's ideas, and on 20 November 1938, Guderian was made general of Panzer Troops and chief of Mobile Troops.

During the Polish and French campaigns of World War II, Guderian commanded the Nineteenth Army Corps. In France, his success was more striking in that British and French tanks were in many respects superior to those of the Wehrmacht, but theirs were basically deployed as infantry support rather than in the Blitzkrieg format of independently acting armored divisions.

He led the Second Panzer Group (renamed Second Panzer Army) in Operation BARBAROSSA but was relieved of command on 26 December 1941 and cited for insubordination to General von Kluge of the Fourth Army.

On 1 March 1943, Hitler recalled Guderian, making him inspector general of Panzer Troops with the authority to establish priorities for armored vehicle production and their employment within a revised force structure.

Guderian became army chief of staff on 21 July 1944. He was dismissed on 28 March 1945 for contradicting Hitler's idea of how to organize the defense of the Reich.

Guderian was not compelled to appear before the Nuremberg tribunal. He died on 14 May 1954.

Neville G. Panthaki

See also: France; Fuller, John Frederick Charles; Gamelin, Maurice; Hitler, Adolf; Keitel, Wilhelm; Kiev; Liddell Hart, Sir Basil Henry; Moscow; Seeckt, Hans von; Smolensk; World War II; Zhukov, Georgy Konstantinovich

References and further reading:

Guderian, Heinz. *Panzer Leader.* London: M. Joseph, 1952.

Macksey, Kenneth. *Guderian, Creator of the Blitzkrieg.* London: Macdonald & Jane's, 1976.

Rothbrust, Florian. *Guderian's XIXth Panzer Corps and the Battle of France.* New York: Praeger, 1990.

Walde, Karl. *Guderian.* Frankfurt: Ullstein, 1977.

Guernica, Bombing of (April 1937)

The most notorious atrocity of the Spanish Civil War. Hitler's air force commander, General Hermann Goring, from July to November 1936 had built up gradually, and secretly, a 5,000-man air unit known as the Legion Condor to assist General Francisco Franco to win his civil war in Spain. The chief of staff of the Legion Condor was a career aviator and a cousin of the famous Red Baron of World War I fame, Lieutenant Colonel Manfred von Richthofen.

On 26 April 1937, about 18 bombers accompanied by 16 fighters struck the Basque town in seven waves. Three Savoia 79s with 36 50-kilogram bombs from the Italian Aviation Legion joined one Donier 17 with 12 250-kilogram high-explosive bombs. In the fourth through sixth waves, nine to twelve Junker 52s, the workhorse bombers of the legion, dropped their loads of incendiary bombs, inflicting the most destruction on the wood-built city. Fires burned for three days in unprepared Guernica, killing from 200 to 1,600 civilians. Later research showed that the attack damaged 721 buildings, of which nearly 500 were destroyed.

The reason for the disparity in the numbers killed is that news media and historians debated for 40 years the political, ideological, and strategic results of a raid that General Franco never admitted. The three officers most responsible for the experimental attack were Richthofen, working with Colonel Juan Vign, chief of staff to the commander of the Army of the North, General Emilio Mola. General Franco did not repeat the technically successful experiment in terror-bombing an entire town or city. Mola's infantry captured Guernica from the retreating Basques on 29 April.

The bombing constituted an atrocity according to certain principles of international law enunciated at Geneva in 1925, which had attempted to outlaw the bombing of non-combatant civilians. Nonetheless, Guernica might have remained just another unhappy episode in the bloody Spanish Civil War were it not for Pablo Picasso's surreal depiction, drawn up in white heat, of the scene, the most famous painting of the twentieth century.

Robert Whealey

See also: Franco, Francisco; Spanish Civil War

References and further reading:

Cava Mesa, Mara Jess. *Memoria Colectiva del Bombardeo de Gernika.* Bilbao: Bakeaz/Gernika Gogoratuz, 1997.

Chipp, Herschel Browning. *Picasso's Guernica: History, Transformations, Meanings.* Berkeley: University of California Press, 1988.

Corum, James S. "The Luftwaffe and the Coalition Air War in Spain, 1936–1939." *Airpower: Theory and Practice,* ed. John Gooch. London: Frank Cass, 1993.

Fitzmaurice, G. G. Unpublished memos on the bombing issue, 1937–1938. FO/371, Public Record Office, London.

Guernica 26. 4. 1937: Die deutsche Intervention in Spanien und der "Fall Guernica," ed. Militrgeschichtlichen Forschungsamt. Freiburg: Rombach, 1975.

Maier, Klaus A. "Guernica, Fakten und Mythen." *German Studies Review* 18 (October 1995).

Montheath, Peter. "Guernica Reconsidered: Fifty Years of Evidence." *War & Society* (May 1987).

Guerrilla/Partisan/Irregular Warfare

Unconventional warfare is fought by relatively weaker combatants or through proxies by nations seeking to deny responsibility. Unconventional warfare has been conducted by guerrilla, partisan, and irregular formations throughout recorded history. In the west, at about the time Persian king Darius was being defeated by Scythian guerrillas (512 B.C.E.), the eastern warrior-philosopher Sun-Tzu was expounding the virtues of this indirect form of warfare (c. 500 B.C.E.). Thus, the utility of unconventional warfare developed independent of military or social cultures. This form of combat is usually considered unconventional by the norms and style of battle practiced by the ruling group's military doctrine. That is, a soldier sniping from a concealed position in a flanking wood line is deemed irregular if the contemporary tradition sees conventional fighting fought by tightly packed ranks of red tunic–clad soldiers firing volleys.

Guerrilla warfare may be defined as hostilities conducted by small, independent groups, often against larger regular forces. Guerrillas often remain in the countryside, either in hiding or in relatively secure areas, until they attack. Usually the guerrillas will avoid committing themselves to large set-piece battles, where they tend to be at a disadvantage against the army forces. The term *guerrilla* (small war) was coined during Napoleon's Peninsula campaign due to the small-scale harassing tactics of the Spanish in support of Wellington's operations. The Vietcong conducted similar guerrilla warfare as part of the North Vietnamese campaign against the United States in Vietnam (1966–1975).

Partisans are named for the loyalty displayed toward their cause. For example, the French resistance "Gaullist Maquis" were General de Gaulle's partisans (1940–1945). Partisans are seldom uniformed, allowing them to slip back into regular society when not conducting operations. Some distinguishing mark, such as a particular armband or color, may be displayed, becoming more overt as the enemy weakens and the partisans gain strength.

Irregular warfare consists of uniformed regular military personnel utilizing nonstandard tactics to conduct irregular warfare against the enemy. This was seen in the airfield and logistics raiding of the British Special Air Service during the Desert Campaign (1941–1943). Irregular warfare also features regular military personnel assisting guerrilla and partisan operations, providing tactical expertise and logistics support. This is a primary role of the U.S. Special Forces.

Two often-conflicting roles of unconventional forces have been intelligence gathering and raiding enemy rear areas. Because an adversary denies access to their territory, information may often be gathered only through indigenous partisans or irregular "stay-behind parties" operating once the forward edge of the battle has passed, leaving them in the enemy's rear area. While higher headquarters use this information, the guerrilla's operational effectiveness, or even survival, depends upon awareness of the enemy's activities. The U.S. Confederate raider Colonel Mosby kept a force of Union soldiers at least twice his number occupied utilizing only locally acquired information during the early portion of the American Civil War (1861–1865). Information is easier to obtain when the enemy feels secure and becomes complacent. Naturally, this condition is difficult to obtain if the raiding threat remains high. However, raiding degrades enemy morale and forces him to assign combat troops for rear area security. This operational divergence became so great between the British Special Air Service (SAS) raids and the Long Range Desert Group's intelligence gathering that an arbitrary line had to be chosen in the North African desert dividing their respective areas of operations (1941–1943).

Fighting from a position of military or political weakness usually necessitates guerrilla, partisan, or irregular warfare. This is a common strategy during conflicts involving a foreign occupation or during independence movements. However, it also contributes to conventional warfare by providing either the initial response upon the outbreak of hostilities until regular forces can be raised or in conjunction with an ongoing conventional campaign. The British Layforce commandos marauding in post-Dunkirk France (1940) are an example of the former, while Colonel Mosby's raiders illustrate the latter. Guerrilla tactics are most successful when used in conjunction with a conventional military campaign. While there are exceptions, such as Castro's victory in Cuba (1956–1958), the North Vietnamese (1964–1975) and American Revolution's (1775–1783) guerrilla operations provide a common model. Fighting in conjunction with a regular force prevents the counterinsurgent elements from both forming smaller, more effective guerrilla-hunting units and focusing exclusively on defeating the conventional threat. Also, a regular army may bolster guerrilla forces with fire support or logistics, as seen when a complete Soviet airborne brigade moved by foot through the German lines to assist partisan operations after Smolensk had been overrun (1941). Conventional and unconventional forces benefit from cooperation regardless of motivation. From 1916 to 1918 British Lieutenant Colonel T. E. Lawrence's Arab force fought in the Palestine campaign. While Britain battled German ally Turkey, Arabs were fighting Ottoman rule. Nonetheless, Arab tactics played an integral part in General Allenby's conventional strategy of breaking through Turkish defenses through standard, methodical combat.

In countering guerrilla warfare, conventional armies face numerous constraints. Often it is simply a matter of relative size, in that the partisans may be only two or three individuals from a community of several thousands, or they may be operating within several hundred square miles of forested mountains. In either case, security elements face difficulty isolating these individuals, particularly given political constraints. The ruling faction may hesitate to acknowledge a rebellion, believing this may weaken their credibility or control. However, this allows the guerrillas to increase strength as their unchecked actions draw further popular support. Wartime constraints may be eliminated, as seen by the Gestapo hunting suspected partisans in occupied Europe. However, coercive military power may prove irrelevant, often counterproductive, in a strategy of "winning hearts and minds."

Western military results in countering unconventional warfare have varied, often due to conventionally educated commanders having difficulty adapting to nonstandard warfare. In 1942, for example, Japanese jungle warfare tactics forced the thoroughly unprepared British to abandon Burma and Malaya. English leadership held that the jungle was unsuited for war, and hence limited prewar training was conducted. Not only had the Japanese trained in jungle guerrilla warfare, they also exploited the Burmese dislike of the colonial British for intelligence purposes, a requirement when fighting at a numerical disadvantage. Some, however, such as British Major General William Slim, learned quickly. He insisted that all units, including medical sections, emphasize patrolling for security and intelligence. Commanders minimized long communications lines and frontal attacks; basically, a counterguerrilla force had to forget staff college training and fight as the enemy fought. A conventional war-planning view often focuses upon equipment and materièl strength, whereas guerrilla struggles emphasize military and civilian personnel. Countering guerrillas requires addressing the perceived injustices that provide recruits. Unconventional revolutionary warfare is seen as an instrument for realizing social, political, and economic aspirations of the underprivileged people. For a group to evolve from minor terrorist acts into a popular guerrilla movement requires sufficient oppression of the people to provide recruits. While successful counterguerrilla operations often include social change, a substantial commitment of troops is necessary to deny guerrilla victory while these changes occur. The successful French campaign in Algeria (1956–1962) reached a ratio of 18 soldiers/security personnel for each guerrilla, while in France's failed Indochina struggle (1946–1954), it seldom deployed in excess of 1.7 to 1. Thus, commanders must find an elusive balance between actions providing the guerrillas with more credibility and providing security necessary to govern effectively. These political factors became more significant during the twentieth century as several guerrilla campaigns, inspired directly or indirectly by the Soviet and Maoist Chinese examples, dominated the Cold War period. While many of the sabotage and subversion tactics remained unchanged, a new feature was the prevalence of political indoctrination. This provided increased motivation, especially necessary during protracted campaigns where simple tactics prove effective provided they are repeated sufficiently to wear down the opponent. Cold War politics provided increased awareness of these conflicts. A domestic struggle with little international significance would attract support because of the global powers' ideology. Covertly supporting guerrilla movements, fighting by proxy provides deniability to the patron nation. This has been further enhanced by the increased pervasiveness of media coverage, variously portraying an enemy as more oppressive and/or vulnerable depending on reportage sympathies. Thus the postcolonial struggles in Angola, for example, drew soldiers and weapons from the Soviet Union, China, Cuba, the United States, and South Africa, to name but a few of the protagonists (1961–1976).

Superpower nuclear constraints exacerbated by decolonizing and revolutionary struggles increased guerrilla warfare prominence in the second half of the twentieth century.

Standing armies established unconventional warfare doctrine and formations only since 1945. While lines of communications, being typical guerrilla targets, have usually been relatively exposed, the increased industrialized nature of modern conflict has increased the utility of striking these targets. Although the end of the Cold War has cast doubt upon various ideologies, ethnic tensions and socioeconomic desperation will ensure that unconventional warfare remains a viable method of conflict.

Robert Martyn

See also: Special Operations Forces; Theory, Military

References and further reading:

Asprey, Robert. *War in the Shadows: The Guerrilla in History.* 2d ed. New York: Morrow, 1994.

Heilbrunn, Otto. *Warfare in the Enemy's Rear.* London: George Allen & Unwin, 1963.

Guevara de la Serna, Ernesto "Che" (1928–1967)

Argentine-born participant in the Cuban Revolution of 1956–1959. Guevara was born on 14 June 1928 in Rosario, Argentina. Since he suffered from asthma as a child, he spent much of his youth in the mountains near Rosario. He was influenced in his early youth by the writings of Pablo Neruda, a Chilean communist. At 19 years of age, he entered medical school at the University of Buenos Aires. He cut his studies short to go on an adventurous trip across South America and then briefly to Florida. Afterwards, he returned to medical school. In 1953, he graduated with a degree of doctor of medicine and surgery. In 1954, after the overthrow of the Jacobo Arbenz regime, he fled Argentina and went to Mexico. He met Fidel and Raul Castro in Mexico and joined them in their expedition to Cuba in 1956. Supposedly an expert in guerrilla warfare, he assisted Castro in the Sierra Maestra. After Castro took power, Che became president of the National Bank as well as minister of industry. In 1960, Guevara traveled extensively throughout the Soviet bloc to seek economic aid.

Che Guevara smoking a cigar, 1964. (Library of Congress)

Castro then dispatched Che to foment communist revolution throughout Latin America. The supposed "expert" in rural guerrilla warfare failed miserably and was reduced to haranguing uncomprehending indigenous peoples on their woes in his Argentinian Spanish. He was captured in Bolivia by Bolivian Army Rangers assisted by U.S. Army Special Forces and was executed by the former on 8 October 1967. With his death (which he apparently met bravely), Castro's dream of a Latin America ablaze with indigenous insurgencies foundered.

Peter Carr

See also: Castro Ruz, Fidel

References and further reading:

Alvarez García, John. *Che Guevara; datos biográficos, cartas, versiones sobre sumuerte, discursos, mensaje a la Tricontinental, Guerra de guerrillas, mensajes de escritores de América y Europe.* Medellín: 1968.

Gadea, Hilda. *Ernesto: A Memoir of Che Guevara.* Trans. Carmen Molina and Walter I. Bradbury. Garden City, NY: Doubleday, 1972.

Guilford Court House (15 March 1781)

A drawn battle in the American Revolution that had much to do with the British evacuation of the Carolinas and retreat to Virginia.

After the victory at Cowpens, American generals Nathanael Greene and Daniel Morgan began a retreat across North Carolina to the Dan River into Virginia. Sir Charles Cornwallis destroyed much of his baggage and supplies and hurried his troops after them. Greene arranged to transport his men across rain-, sleet-, and snow-swollen rivers; Cornwallis arrived frequently to find boats destroyed. The further Cornwallis advanced, the more extended and weaker his forces became.

Finally, Greene fought at a field he had previously chosen at Guilford Court House. He sought to arrange his troops like Morgan at Cowpens, but the distances were too great to have the same effect. He placed two lines of militia in front—one behind a log fence and another behind a line of trees—and then his battle-tested Continentals with cavalry guarding their flanks. American casualties in the resulting battle were light and British losses were heavy—more than 630 out of 2,200 effective. Even though Greene abandoned the field,

Cornwallis soon retreated to Hillsborough, then Wilmington for supplies, and finally moved into Virginia, conceding the Carolinas.

Charles M. Dobbs

See also: American Revolution; Cornwallis, Sir Charles; Greene, Nathanael

References and further reading:

Hatch, Charles E., Jr. *The Battle of Guilford Court House.* Washington, DC: Office of History and Historic Architecture, Eastern Services Center, 1971.

Thayer, Theodore. *Nathanael Greene: Strategist of the American Revolution.* New York: Twayne, 1960.

Wickwire, Franklin, and Mary Wickwire. *Cornwallis: The American Adventure.* Boston: Houghton Mifflin, 1970.

Guinea-Bissauan War of Independence (1961–1975)

Conflict that led to the independence of Guinea-Bissau. Though conquered by Portugal in 1446, the inhospitable terrain of Guinea-Bissau in West Africa ensured that it never became a major center for European settlers, and it was only after World War I that Portugal established the territory's boundaries. In 1952, a constitutional amendment was introduced whereby Portugal's colonies were renamed overseas provinces, a thinly disguised device to evade international scrutiny.

In 1956, the African Party for the Independence of Guinea and Cape Verde (PAIGC) was founded by the nationalist and Marxist revolutionary Amilcar Cabral (1924–1973). Five years later, the PAIGC embarked on a protracted guerrilla war against the Portuguese colonial administration. The PAIGC succeeded in recruiting and training an armed force of some 7,000–10,000 fighters while the Portuguese forces reached a peak of 35,000 (a figure that includes 10,000 conscripted African troops). Despite its numerical inferiority, the PAIGC enjoyed many advantages, not least an intimate knowledge of the difficult terrain, which was ill-suited for conventional warfare. It also received crucial support from the Soviet Union and Cuba, which provided the PAIGC with money, arms, training, and practical assistance.

By 1965 the PAIGC controlled more than half of the countryside and as rural areas were liberated an alternative government was established. Eventually, Portuguese control was confined to the refugee-swollen capital, Bissau, a few small towns, and pockets in the northeast where the colonial authorities received cooperation from the more prominent Muslim Fula in an attempt to preserve traditional privileges.

The assassination of Cabral in Conakry, Guinea (23 January 1973) by PAIGC dissidents working with Portuguese intelligence officers did not alter the fortunes of the Portuguese. The intensification of its colonial wars had forced Portugal to take extraordinary measures, including conscription, and the government was forced to allot half of the national budget (6 percent of GNP) to military spending. Disillusioned, the commander in chief of the Portuguese forces in Guinea-Bissau, Antonio Spínola, published a radical manifesto, *Portugal and the Future,* which gave conscripted soldiers a moral justification for refusing to fight in colonial wars. A successful coup d'état by professional soldiers (25 April 1974) dislodged the 48-year-old military dictatorship in Lisbon whose credibility had depended greatly on its ability to preserve the Portuguese Empire.

On 10 September 1974, the new Portuguese administration formally recognized the independence of Guinea-Bissau, although it delayed conferring the same status on the Cape Verde Islands until the following year. Despite initial moves toward unity under PAIGC guidance, Guinea-Bissau and Cape Verde remain separate states.

The PAIGC had been able to inflict about 300 Portuguese army casualties annually, while toward the end of the war, PAIGC fatalities were running at more than 1,000 a year. An estimated 15,000 people lost their lives during the conflict, of which about 10,000 were combatants, the remainder civilians.

Donnacha Óbeachdáin

See also: Angolan War of Independence; Mozambican War of Independence

References and further reading:

Mustafah Dhada. *Warriors at Work: How Guinea Was Really Set Free.* Boulder: University Press of Colorado, 1993.

Chaba, Patrick. *Amilcar Cabral.* London: C. Hurst and Co., 2000.

Lobban, Richard Andrew, Jr., and Peter Karibe Mendy. *Historical Dictionary of the Republic of Guinea-Bissau.* Kent, UK: Scarecrow Press, 1997.

Guiscard, Robert (1016–1085)

Norman adventurer in the south. The eldest son of Tancred de Hauteville, a minor Norman baron, by his second marriage, Robert had little chance of inheriting land or of making his name in northern Europe. Instead, he followed his elder brothers to southern Italy, where they had established themselves first as mercenaries, and then, in the competing spheres of Lombard, Muslim, and Byzantine cities, as warlords with fortified cities.

Robert arrived in 1046, and quickly succeeded as a freebooter in Calabria, earning the nickname "Guiscard" (clever) for tricking the monks inhabiting a strategic mountain monastery into letting him in the gates by faking a funeral procession. In 1053, by then a member of the Norman establishment in southern Italy, he commanded the left wing of the victorious Norman forces at Civitate, fighting against Pope Leo IX's claims on the city of Benevento and papal power in the south of Italy.

In 1057, he inherited his brother Humphrey's city of Melfi and married Seichelgaita of Salerno, a Lombard noblewoman who frequently fought at his side in battle. When Robert and the new pope, Nicholas, reconciled in 1059, the pope granted Robert claim to Apuila, Calabria, and Sicily, which was still in Byzantine and Muslim hands. Aided by his brother Roger, newly arrived from Normandy, Robert invaded Sicily.

Always forced to leave Roger, ultimately Great Count of Sicily, in charge of the Sicilian invasion because of the constant rebellions and Byzantine invasions of his Italian possessions, Robert expended most of his resources subduing his neighbors and taking advantage of the contest between the pope and the Holy Roman Emperor to seize territory. In 1080, using the excuse of his daughter Helena's marriage to a son of the deposed Byzantine emperor Constantine X Ducas, Robert attacked Durazzo, using an ingenious solid blockade of ships, which defeated a Venetian fleet. He also defeated Alexius Comnenus, and the Byzantine army, on land.

Returning to quell a disturbance in Apuila, Robert aided the pope, Gregory VII, who had been evicted from Rome by an imperial army, by retaking the city, but then unfortunately he allowed his army to sack it. With the pope's authority reestablished, Robert returned to attack Greece, but after taking Corfu, he died in 1085 in an epidemic, probably typhoid, that was raging in his camps. Succeeded by his and Seichelgaita's son Roger Borsa, Robert also fathered Bohemond, one of the first and most successful crusaders, while his brother Roger founded a Sicilian dynasty of his own. Together they created the powerful Norman kingdoms in southern Europe.

Margaret Sankey

See also: Norman-Byzantine Wars

References and further reading:

Loud, G. A. *The Age of Robert Guiscard.* New York: Longmans, 2000.

Norwich, John Julian. *Normans in the South.* London: Longmans, 1967.

———. *Kingdom in the Sun.* Harlow: Longmans, 1970.

Guise, François de Lorraine, Second Duke of (1519–1563)

French military commander. François de Lorraine, second duke of Guise, distinguished himself in various battles of the wars of France against Charles V, especially at Metz in 1552. In 1557 he fought the Spanish in Italy and in 1558 he took Calais from the English. At court, he was the rival for power of Anne, duke of Montmorency. With the accession in 1559 of the youthful Francis II, who was married to the duke's niece, Mary Stuart, Guise and his brother Charles, Cardinal de Lorraine, were given control of the government. Guise and his brother gained great influence over the youthful Francis II and almost completely controlled the government during Francis's two-year reign. In that period the Guises directed the persecution of the Huguenots, becoming widely detested for their violent suppression of the Huguenot conspiracy of Amboise in 1560. Shortly afterward, however, the death of Francis II deprived the Guises of power. Catherine de Médici, mother and regent for the succeeding Charles IX, ousted the Guises from their position of influence. François subsequently joined the leadership of the Catholic party, opposing both the Huguenots and the tolerance of the regency. The massacre of the Huguenots at Wassy by Guise's soldiers in 1562 led to the French Wars of Religion (1562–1598) between Catholics and Protestants. He was assassinated while preparing to attack Orléans the following February. Though Guise has been charged with cruelty, his soldiers considered him a generous man and respected him for his military skill.

David C. Arnold

See also: French Wars of Religion

References and further reading:

Cuisiat, Daniel. *Lettres du Cardinal Charles de Lorraine (1525–1574).* Geneva: Librairie Droz S.A., 1998.

Sedgewick, Henry Dwight. *The House of Guise.* London: Lindsay & Drummond, 1938.

Gujerat (1849)

The final battle of the Second Sikh War (1848–1849). After the Chillianwallah battle on 13 January 1849, in which the British Army of the Punjab had been severely mauled by the Sikhs, the latter withdrew into hills along the Jhelum River. The British force, commanded by Lord Gough, remained in camp at Chillianwallah, awaiting further Sikh movement. On 2 February Shere Singh's army headed southeast into the fertile district of Gujerat to obtain supplies. Gough had anticipated such a move and, once alerted, ordered his force southward in pursuit. However Shere Singh, preferring to fight defensively, had halted immediately south of Gujerat and prepared to fight the next battle from a static position.

The Sikh position faced south and straddled two small gullies. Their cavalry was concentrated at each end of the line on the far sides of the gullies, with the infantry positioned between the two. No breastworks had been prepared for the 35,000 Sikh troops.

Gough had been reinforced by about 15 regiments so was able to field some 20,000 men with 96 guns. He planned a frontal assault on the left and center of the Sikh line and positioned half his force between the two gullies and half to the left of the left-hand one.

After a fierce and effective artillery barrage, British troops advanced according to plan. The Sikhs attempted a counterattack into a break in the British line along the dry left-hand gully but were broken up by artillery. By midday, although fighting strongly and maintaining order, they were in retreat. The British infantry advanced two miles, then left off the pursuit to the cavalry, who harassed the Sikhs 15 miles further.

The Sikh army was unable to organize or resupply effectively after the battle and finally surrendered unconditionally at Rawalpindi on 14 March 1849, ending the second Sikh War. But of all the indigenous peoples resisting British colonial incursion, they had put up the stiffest opposition.

Michael Hyde

See also: Anglo-Sikh Wars; Chillianwallah

References and further reading:

Cook, H. *The Sikh War: The British Army in the Punjab, 1845–1849.* London: Leo Cooper, 1975.

Kar, H. C. *Military History of India.* Calcutta: Firma KLM Private Limited, 1980.

Gulf War (2 August 1990–28 February 1991)

First major conflict of the post–Cold War period, fought between Iraq and an American-led international coalition. Known as one of the most decisively one-sided victories in military history, it has been described as a battle of Soviet equipment and tactics versus Western equipment and tactics. The conflict can be divided into two parts, DESERT SHIELD and DESERT STORM.

DESERT SHIELD was the U.S. reaction to Iraq's invasion of Kuwait, to deter Iraqi aggression into Saudi Arabia. In the process, it became the first time in history that a major com-

British troops and units from other coalition countries gather for review by King Fahd of Saudi Arabia. The units took part in an assembly of international coalition forces united against Saddam Hussein during Operation DESERT STORM. *(Department of Defense)*

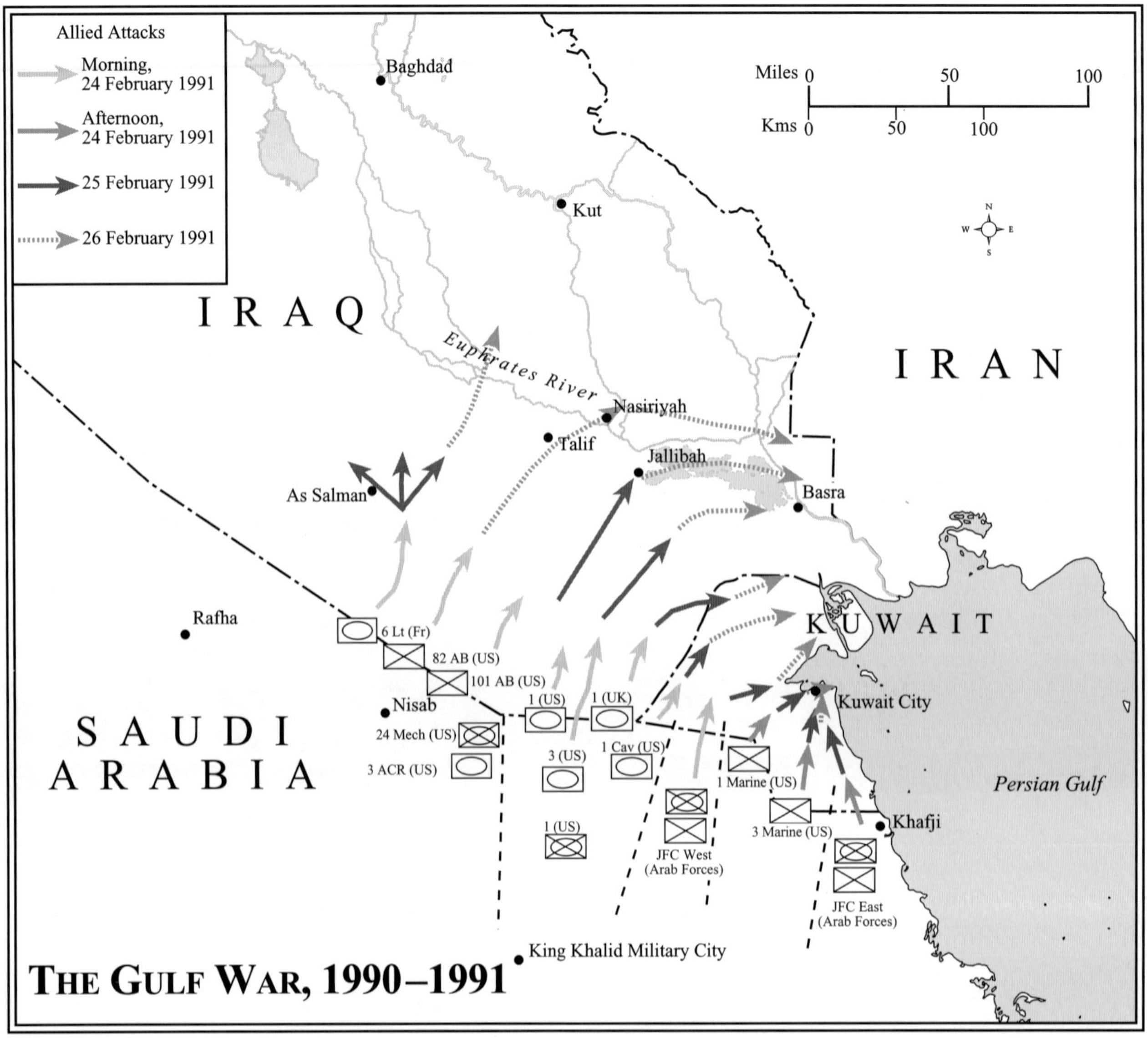

bat army, consisting of coalition forces, was allowed time to build up and train, then bomb an enemy, without threat to its own ground forces.

Once the coalition bombs starting falling, on 17 January 1991, part two of the Gulf War had begun—DESERT STORM. There were two phases to DESERT STORM: first, the devastating five-week-long bombing campaign, and second, the ground assault. The coalition's air campaign cut the Iraqi lines of communication and reinforcement, seriously degraded their military capacity, and vastly reduced morale. In short, the air assault prepared the Iraqis for the coup de grace—the ground war.

The coalition strategy, called Air Land Battle (ALB) Doctrine, was developed by NATO to fight a major land war against the Soviet Union. The crux of the combined-arms doctrine is to "look deep," a hundred miles or more behind the front. The philosophy celebrates deception and maneuver in the spirit of Robert E. Lee and Stonewall Jackson. The key to ALB is a combination of speed, surprise, and mobility, which protects one's own forces and keeps the enemy off balance.

The strategy was ideally suited for the region's terrain, the enemy's strategy, and the military equipment available. The "left hook" took coalition troops 300 miles west of Kuwait to swing around Iraqi defenses and cut off reinforcement and retreat. The assault by the 101st Airborne is considered the largest air assault operation in military history.

The right flank of the coalition was intended to push up through the teeth of Iraqi defenses in Kuwait and draw the elite Republican Guard reserves down into the fray. With the Iraqis being pulled in two directions, the center, or "mailed fist," was to slash through the gap. The mailed fist was made up of American and British heavy armored units whose task was to find, engage, and destroy the Republican Guard formations.

Although political pressure ended the ground operation before the mailed fist could finish its job, the Iraqis lost more than 5,000 tanks and armored vehicles compared to 13 lost by the coalition. The Iraqi military lost some dead and wounded, compared to the coalition's dead and wounded. In spite of the experts' tales of a battle-hardened Iraqi army, no other twentieth-century large-scale war had proved as one-sided.

Craig T. Cobane

See also: Armored Fighting Vehicles; Iran-Iraq War; Jackson, Thomas "Stonewall"; Lee, Robert Edward

References and further reading:

Atkinson, Rick. *Crusade: The Untold Story of the Persian Gulf War.* New York: Houghton Mifflin Company, 1993.

Gordon, Michael. R., and Bernard E. Trainor. *The Generals' War.* New York: Little, Brown and Company, 1995.

Scales, Robert H. *Certain Victory: The U.S. Army in the Gulf War.* Washington, DC: Brassey, 1997.

Gustavus II Adolphus (1594–1632)

One of the great captains, who advanced the institution of disciplined and professional standing armies. Gustavus Adolphus, known as the Lion of the North, established an enduring personal and military legacy during the seventeenth century. A devout and pious Lutheran, he was also an exceptional nationalist who took advantage of international instability to advance the cause of Sweden as a major power.

Born into the House of Vasa on 19 December 1594, he inherited his family's impetuousness and daring. In 1611, during a war with Denmark, Gustavus led a 500-man detachment that successfully bluffed its way into a Danish garrison and captured it without loss. In October that same year, Gustavus became heir to the throne following the unexpected death of his father, Charles IX. In an unprecedented move by the regency, who so regarded the young prince's intellect and boldness, 16-year-old Gustavus was formally recognized as king in December 1611.

Gustavus had the good fortune to ascend to the throne at a time of Swedish prosperity and strengthening nationalism. Sweden clashed with its neighbors Russia, Poland, and Denmark repeatedly during the early years of his reign. In these wars, the king learned invaluable lessons from his own direct experience as well as from soldiers who came into his service from the suspended conflict in the Netherlands. From these veterans he learned the revolutionary advances in tactics, army organization, and finances pioneered by Maurice of Nassau.

Gustavus carefully crafted his military instrument, which he raised predominantly through national conscription. Upon this core of stout soldiery, Gustavus imposed strict discipline, a sense of uniformity, and meticulous drill. In fact, Gustavus most thoroughly completed the reforms envisioned by Maurice. His cavalry discarded the pistol and trained to charge with the sword; infantry learned rapid musket fire combined with articulated maneuver in smaller linear formations; and the Swedish artillery—light, mobile, and plentiful—became the bane of all of Sweden's enemies.

Gustavus spent the early years of the Thirty Years' War fighting against his Catholic uncle Sigismund, king of Poland. In 1630, Gustavus recognized an opportunity to intervene in the massive German conflict on the side of the Protestants, who had suffered terrible military and political setbacks. Subsequently, he landed his force on the Baltic shores of Pomerania and commenced a campaign that lasted nearly three years. During the first year of his campaign, Swedish arms conquered the Oder valley as far as the Catholic-held city of Frankfort am Oder. Gustavus won his first major field victory against the Hapsburg imperial army under Count Tilly at Breitenfeld on 17 September 1631. From that victory, Protestant hopes and resolve soared and the Swedes marched throughout central and southern Germany. After some inconclusive skirmishing in Bavaria, Gustavus felt compelled to return to Saxony to confront a second imperial army, this time commanded by Count Wallenstein. The two forces met at Lützen on 16 November 1632. The Swedes won the desperate battle, but only after the Swedish king fell from his charger, mortally wounded by at least three musket balls, while leading a cavalry attack. Gustavus earned his lasting rank as a great captain due to his relentless energy, keen tactical insight, and extraordinary leadership. He fully recognized the elements of early modern warfare: firepower, maneuver, and decisive shock action. He honed these elements to near perfection, and when married with his well-administrated army, he acquired the praise and fearful respect of ally and enemy alike.

Bryan R. Gibby

See also: Breitenfeld; Lützen, Battle of; Thirty Years' War

References and further reading:

Dodge, Theodore A. *Gustavus Adolphus.* New York: Da Capo Press, 1998.

Rothenberg, Gunther. "Maurice of Nassau, Gustavus Adolphus, Raimondo Montecuccoli, and the 'Military Revolution' of the Seventeenth Century." In *Makers of Modern Strategy,* ed. Peter Paret. Princeton, NJ: Princeton University Press, 1986.

GROUND WARFARE

An International Encyclopedia

INTERNATIONAL WARFARE ENCYCLOPEDIAS FROM ABC-CLIO

Spencer C. Tucker, General Editor

Air Warfare: An International Encyclopedia, Walter J. Boyne, Editor

Naval Warfare: An International Encyclopedia, Spencer C. Tucker, Editor

Ground Warfare: An International Encyclopedia, Stanley Sandler, Editor

GROUND WARFARE

An International Encyclopedia

VOLUME TWO, H–Q

EDITED BY
Stanley Sandler

ASSOCIATE EDITORS
Michael Ashkenazi
Paul D. Buell

FOREWORD BY
General Henry H. Shelton, USA (R)

ABC CLIO

Santa Barbara, California Denver, Colorado Oxford, England

Library of Congress Cataloging-in-Publication Data
Sandler, Stanley, 1937–
Ground warfare : an international encyclopedia / Stanley Sandler ;
foreword by Henry H. Shelton.
p. cm
Includes bibliographical references and index.
ISBN 1-57607-344-0 (Hardcover: alk. paper) — Ebook ISBN 1-57607-733-0
1. Battles—Encyclopedias. 2. Military history—Encyclopedias.
3. Military biography—Encyclopedias. I. Title.
D25.A2 S26 2002
355'.003—dc21

2002004568

07 06 05 04 03 02 10 9 8 7 6 5 4 3 2 1

This book is also available on the World Wide Web as an e-book. Visit abc-clio.com for details.

ABC-CLIO, Inc.
130 Cremona Drive, P.O. Box 1911
Santa Barbara, California 93116-1911

This book is printed on acid-free paper.
Manufactured in the United States of America

CONTENTS

A-TO-Z LIST OF ENTRIES

Tecumseh (1768–1813)
Tel-el-Kebir (13 September 1882)
Terauchi, Hisaichi (1879–1946)
Terrorism
Tet Offensive (January–March 1968)
Teutoburger Wald, Battle of (9)
Teutonic Knights
Teutonic Tribes
Tewkesbury (4 May 1471)
Texas War of Independence (1835–1836)
Thai (Tai) Wars (c. 1300–1569)
Thames (5 October 1813)
Thayer, Sylvanus (1785–1872)
Theory, Military
Thermopylae, Battle of (480 B.C.E.)
Thirty Years' War (1618–1648)
Thomas, George Henry (1816–1870)
Thutmose III (d. 1450 B.C.E.)
Tiberius (42 B.C.E.–37 C.E.)
Tibet, Chinese Occupation of (1949–)
Tiglath-Pileser I (r. 1115–1077 B.C.E.)
Tiglath-Pileser III (r. c. 745–727 B.C.E.)
Tigranes the Great (c. 140–c. 55 B.C.E.)
Tilly, Johann Tserclaes, Graf von (1559–1632)
Timoshenko, Semen Konstantinovich (1895–1970)
Tinian (24 July–1 August 1945)
Tippecanoe, Battle of (7 November 1811)
Tito (Josef Broz) (1892–1980)
Tobruk, Battle of (April 1941)
Tokugawa, Ieyasu (b. Matsudaira Takechiyo) (1543–1616)
Tondibi (1591)
TORCH, Operation (1942)
Torgau (3 November 1760)
Toulon, Siege of (September–December 1793)
Tours (October 732)
Toussaint L'Overture, Wars of (1793–1803)
Towton, Battle of (29 March 1461)
Trajan, Marcus Ulpius (53–117)
Trebia, Battle of the (22–23 December 218 B.C.E.)
Trenton (26 December 1776)
Trinh-Nguyen Dynastic Struggles (1620–1673)
Triple Alliance, War of the (1864–1870)
Trojan War (12th or 13th Century B.C.E.)
Trotsky, Leon (1879–1940)
Trumpeldor, Yosef (1880–1920)
Trung Sisters, Rebellion of (39–43)
Tukhachevsky, Mikhail Nikolayevich (1893–1937)
Turenne, Henri de la Tour d'Auvergne, Vicomte de (1611–1675)
Turkish Wars of European Expansion (1413–1699)
Tyre, Siege of (January–July 332 B.C.E.)

ULTRA
Unarmored Fighting Vehicles
Uniforms
United Nations and Conflict Resolution
U.S. Army
U.S. Marines
U.S. Militia (1603–1815)
Utah War (1857–1858)

Vacietis, Jukums (1873–1938)
Valley Campaign (23 March–9 June 1862)
Valley Forge (1777–1778)
Valmy (20 September 1792)
Valois-Habsburg Wars (1521–1559)
Van Fleet, James A. (1892–1992)
Vandals
Vauban, Sébastien Le Prestre de (1633–1707)
Vegetius Renatus, Flavius (fl. late 300s)
Venetian-Genoese War (1255–1381)
Venezuelan Civil Wars (1858–1870)
Veracruz, Siege of (9–28 March 1847)
Veracruz, U.S. Landings at (1914)
Vercingetorix (d. c. 45 B.C.E.)
Verdun (21 February–18 December 1916)
Vespasian (9–79)
Vicksburg, Siege of (18 May–4 July 1863)
Vienna, Sieges of (1529, 1683)
Vietnam Conflict (1961–1975)
Vietnamese Civil War (Tayson Rebellion, 1773–1802)
Viking Raids (c. 800–1016)
Vikings
Villa, Francisco "Pancho" (Doroteo Arango) (1878–1923)
Vimy Ridge (9 April 1917)
Visigoths
Vo Nguyen Giap (1911–)
Vouillé, Battle of (spring of 507)

Waffen SS (1934–1945)
Wagram (5–6 July 1809)
Wainwright, Jonathan Mayhew, IV (1883–1953)
Wake Island (8–23 December 1941)
Walker, Walton (1889–1950)
Walker, William (1824–1860)
Wallenstein, Albrecht von (1583–1634)
War Crimes
War of 1812 (1812–1815)
War Plan Orange (1907–1940)
Wars of the Roses (1455–1464, 1467–1471, and 1483–1485)
Warsaw/Vistula (August 1920)
Washington, George (1732–1799)
Washington, Burning of (24–25 August 1814)

MAPS

ACRONYMS

ABM	Antiballistic Missile Treaty
ACP	automatic cartridge pistol
AEW	Airborne Early Warning
AFVs	armored fighting vehicles
AIF	Australian Imperial Force
ALB	Air Land Battle Doctrine
ANZAC	Australian and New Zealand Army Corps
ARVN	Army of the Republic of Vietnam
ASDIC	Anti-Submarine Detection Investigation Committee
ASW	antisubmarine warfare
ATC	American Tobacco Company
AWACS	airborne warning and control system
BEF	British Expeditionary Force
C2	Command and Control system
C3I	command, control, communications, and intelligence
Cheka	Chrezvychainaya Komissariat po bor'be s kontrarevoliutsiei i sabotazhem (All-Russian Commission for Struggle against Counterrevolution and Sabotage)
CI	counterintelligence
CIA	Central Intelligence Agency
CIGS	chief of the Imperial General Staff
CORDS	American Office of Civil Operations of Rural Development Support
CPB	Communist Party of Burma
CTBT	Comprehensive Test Ban Treaty
DIA	Defense Intelligence Agency
DK	Democratic Kampuchea
DPRK	Democratic People's Republic of Korea
DSM	Distinguished Service Medal
EAM-ELAS	National Liberation Front–National Popular Liberation Army
EDES	Greek Democratic National Army
ELN	Ejercito de Liberación Nacional (Army of National Liberation)
EOKA	National Organization of Greek Fighters
EPL	Ejercito Popular de Liberaci (Army of Popular Liberation)
EVA	Ever Victorious Army
FARC	Fuerzas Armadas Revolucionarias de Colombia (Colombian Revolutionary Armed Forces)
FLEC	Front for the Liberation of Cabinda
FLN	National Liberation Front
FNLA	National Front for the Liberation of Angola
GCNG	Greek Cypriot National Guard
GDP	gross domestic product
GDR	German Democratic Republic
GPU	Gosudarstvennoe Politicheskoe Upravlenie (State Political Administration)
GULAG	Glavnoe Upravlenie Ispravitel'no-trudovykh Lagerei (Chief Administration of Corrective Labor Camps)
HUMINT	human intelligence
I&W	Indicators and Warnings
IDF	Israeli Defense Forces
IFF	Identification Friend-or-Foe transponders
IMINT	(aerial and satellite photo and radar) imagery intelligence
INA	Indian National Army (Azad Hind Fauj)
INF	intermediate-range nuclear forces
IR	infrared (radar)
IRA	Irish Republican Army
IRB	Irish Republican Brotherhood
JDA	Japanese Defense Academy
JTF	Joint Task Force
KGB	Komitet Gosudarstvennoi Bezopastnosti (Committee of State Security)
KNU	Karen National Union

KPA Korean Peoples Army
KUFNS Khmer United Front for National Salvation
M-19 Movimento 19 de Abril (19th of April Movement)
MACV U.S. Military Assistance Command
MAUD Military Application of Uranium Detonation
MED Manhattan Engineer District
MFA armed forces movement
MID Military Intelligence Division of the Army General Staff
MLRS Multiple Launch Rocket System
MPLA Popular Movement for the Liberation of Angola
MVD Ministerstvo Vnutrennikh Del (Ministry of Internal Affairs)
NATO North Atlantic Treaty Organization
NF Intermediate Nuclear Forces Treaty
NKGB Narodnaya Komissariat Gosudarstvennoi Bezopastnosti (People's Commissariat for State Security)
NKVD Narodnaya Komissariat Vnutrennikh Del
NSA National Security Agency
NSC National Security Council
NVA North Vietnamese Army
OAS Secret Army Organization
OECS Organization of Eastern Caribbean States
OFS Orange Free State
OGPU Unified State Political Administration
OKW Wehrmacht High Command
OPEC Organization of Petroleum Exporting Countries
ORBAT Order of Battle information
OSRD Office of Scientific Research and Development
OSS Office of Strategic Services
PAIGC African Party for the Independence of Guinea and Cape Verde
PG Provisional Government
PLO Palestinian Liberation Organization
PNI Partai Nasional Indonesia (Indonesian Nationalist Party)
PPA Planters Protective Association
PRA People's Revolutionary Army
PRD Dominican Revolutionary Party
PRK People's Republic of Kampuchea
RADAR Radio Detecting and Ranging
RAR Royal Australian Regiment
RDF Radio Direction Finding
RKKA Workers and Peasants Red Army
ROK Republic of Korea
RPF Assembly of the French People Party
SALT Strategic Arms Limitation I Interim Agreement
SAMS surface-to-air missile defenses
SAS Special Air Service
SBS Special Boat Service
SEAL sea, air, and land team
SIGINT signals intelligence
SLORC State Law and Order Restoration Council
SONAR Sound Navigation and Ranging
START Strategic Arms Reduction Talks I and II
TECHINT technical intelligence
TOW target on wire
UN United Nations
UNC United Nations Command
UNEF United Nations Emergency Force
UNITA National Union for the Total Independence of Angola
VOC Vereenigde Oost-Indische Compagnie (Dutch East India Company)

GLOSSARY

(Recurring topics. Each listing in this glossary has its own article in the text.)

Airborne Operations The insertion of troops and equipment on the battlefield by means of parachute, helicopter, or glider.

Armor Body protection for soldiers.

Armored Fighting Vehicles Protected vehicles used on the battlefield.

Arms Control Limits, usually set by treaty, on the number and types of weapons; usually refers to nuclear weapons.

Artillery Basically, a heavy metal tube from which a missile is discharged violently by explosive force.

Awards and Honors Recognition granted by authorities for meritorious service by soldiers.

Ballistics The science of projectiles, divided into interior and exterior ballistics. Its aim is to improve the design of shells/projectiles so that increased accuracy and predictability are the result. It deals also with rockets and ballistic missiles.

Bayonet Metal blade or spike that, when fixed to a musket or rifle, facilitates its use in hand-to-hand combat. In recent decades, more likely to be used for opening ration tins.

Bazooka Shaped-charge, smoothbore, man-portable, antitank, and pillbox weapon.

Buffalo Soldiers African-American troops of the late-nineteenth-century U.S. regular army.

Catapult Engine for throwing a heavy weight, using an arm released from tension; the artillery of the ancient world.

Cavalry The noble, mobile arm of battle; the traditional horse-borne arm of mobility; can now refer to a motor vehicle–mounted unit.

Chaplains Military officers who tend to the spiritual, moral, and physical needs of troops in the field and in camp; pastors in uniform.

Chemical and Biological Warfare The deliberate use of chemical or biological agents against an enemy.

Civil Affairs/Military Government Those activities of a commander that embrace the relationship between the military forces and civil authorities and people in a friendly country or area (Civil Affairs) or occupied country or area (Military Government).

Coastal Defense The defense of a nation's coast from an enemy sea invasion or blockade, accomplished with heavy artillery, mines, small warships, and nets.

Cold War (1946–1989) Period of tension between the Soviet Union and the People's Republic of China and their allies, representing communism (the East); and the United States and its allies, representing capitalism and democratic socialism (the West); punctuated with several "hot" wars, the most significant being the Korean War (1950–1953) and the Vietnam War (c. 1955–1975). The Cold War ended unexpectedly with the fall of the Berlin Wall and the bloodless victory of the West.

Communications, Military The application of technology to the transmission of military orders and intelligence.

Conscription The selection of persons for involuntary military service.

Death Squads Clandestine and usually irregular organizations, often paramilitary in nature, that carry out extrajudicial executions and other violent acts against clearly defined individuals or groups of people.

Disarmament The removal or drastic reduction by nation-states of major weapons.

Economic Warfare Compelling an enemy to submit either by direct action against its economic basis or indirectly through blockade or boycott.

Electronic Warfare The use of the electromagnetic

spectrum to gain knowledge of the presence and movement of an opposing force and also to deny any opposing force the use of that spectrum.

Engineering, Military The application of science and technology for military purposes, primarily through the use of civil engineering.

Ethics of War Rules, principles, or virtues applied to warfare.

Firearm A tube, closed at one end, that has in it an explosive with a projectile above it, nearer to the open end of the tube. An ignition system fires the explosive charge, which forces the projectile along the tube by means of the gases from the explosion; can include both artillery and small arms.

History, Military History dealing with the use of organized armed force, either on behalf of some form of recognized state authority or against it.

Infantry Lightly armed ground troops; the backbone of any army; not only the most numerous of the fighting arms but the only one that can actually take and hold ground.

Intelligence, Military Military specialty that provides a commander and staff with the knowledge of the enemy and of weather and terrain required for the planning and conduct of operations. (There is no truth to the assertion that "military intelligence" is an oxymoron.)

Laws of War International laws, enforced sometimes by nations after war and sometimes by commanders in battle, governing both the decision to engage in war and the manner of its conduct, particularly the forms of violence used, the definition of combatants, the treatment of prisoners, and the treatment of neutrals and noncombatants.

Logistics Largely an American usage, encompassing military supply, transportation, medical service, and construction-maintenance.

Machine Gun Rapid-firing small arm that can maintain a high rate of fire without the requirement of reloading after each round, which today means a fully automatic weapon; either man-portable ("machine pistol") or heavier.

Maps and Cartography The result of the utilization of cartography and topographical reproduction for military strategy and operations.

Medals and Decorations Tangible recognition of faithful military service or success awarded to individual soldiers.

Medicine, Military The medical and surgical specialty concerned with the ailments of soldiers and sailors.

Mercenaries Hired professional soldiers who fight for a state or entity without regard to political interests or issues.

Military-Industrial Complex The institutions and people that plan, procure, and fight a war and that supposedly shape the economy, the political realm, and the wider society, even in peacetime. Term first used by outgoing U.S. President Dwight D. Eisenhower in 1961.

Mortar Muzzle-loaded weapon firing its bomb at a high angle to attack protected positions and trenches.

Pacifism/War Resistance The organized opposition to war, killing, or violence; usually divided into two segments: religious (e.g., Quakers, Mennonites) and secular.

Paramilitary Organizations Unofficial groups organized along military lines yet lacking the traditional role or legitimization of conventional or "genuine" military organizations.

Propellants Compounds used to move a projectile from the firing device to the target.

Psychological Operations The use of psychology and propaganda by military units to persuade target audiences to adopt at least some of their views and possibly to modify their behavior.

Rank, Military Official indication of a soldier's length and quality of service in organized militaries.

Rifles Firearms designed with barrel grooves to impart a spin and thus far greater accuracy; either small arms or artillery.

Tactics The theory and practice of using military forces in combat.

Terrorism Acts of violence intended for a wide audience in order to create an environment of fear for political reasons.

Theory, Military That body of knowledge usually published in books and journals that examines the nature of wars and the art of war on an abstract level.

Unarmored Fighting Vehicles Unprotected military vehicles, either specially built for the military or adapted from commercial models and used in combat-support roles (e.g., Jeeps, trucks, ambulances).

Uniforms Military clothing worn by organized bodies of troops to distinguish them from the uniformed personnel of other armed forces and to strengthen morale.

War Crimes Actions in wartime that violate the laws or usages of war.

H

Hadrian (Publius Aelius Hadrianus) (76–138)

Military reformer, Roman emperor from 117 to 138. Before becoming emperor, Hadrian had a long military career, serving in Spain, Pannonia, Moesia, Germany (with the XXII Legion, the Primigenia Pia Fidelis), and Parthia. During the First and Second Dacian Wars (102–103 and 105–106), Hadrian served variously as quaestor, legate, praetor, and commander of a legion. He became governor of Pannonia in 107, consul in 108, and governor of Syria in 114. Hadrian was in line to receive a second consulship in 118 when, upon hearing of the death of Emperor Trajan (d. 8 August 117), the armies of Syria proclaimed Hadrian to be emperor of Rome. Deeming Trajan's wars of expansion a waste of blood and treasure, Hadrian pursued a policy of imperial consolidation. In 122, preferring peace to war, Hadrian negotiated an armistice with the Parthians. Touring the empire and inspecting the provinces in order to make reforms, Hadrian established large-scale border fortifications, which not only protected the frontiers from Barbarian attacks, but also served as checkpoints for trade. The most famous of these fortifications, Hadrian's Wall, was erected between Tyne and the Solway Firth in Britain. Although he demanded rigid discipline, Hadrian's military reforms and personal inspection tours throughout the empire won the intense loyalty of his legions. Perhaps one of Hadrian's most significant reforms was the elimination of distinctions between the legions and the auxiliary corps, which meant that Roman citizens and noncitizens now served in the same units. Hadrian also reintroduced the tactical modification of the Macedonian phalanx, in which auxiliary troops led an attack followed later by a reserve of legionnaires. Hadrian's reign was generally marked by peace and sensible policies, except in Judea, where Hadrian's insensitivity provoked the Second Jewish Revolt (132–135).

Eric D. Pullin

See also: Jewish Revolts

References and further reading:

Birley, Anthony R. *Hadrian: The Restless Emperor.* London: Routledge, 1997.

Goldsworthy, Adrian Keith. *The Roman Army at War, 100 B.C.–A.D. 200.* New York: Oxford University Press, 1996.

Webster, Graham. *The Roman Imperial Army of the First and Second Centuries C.E.* Norman: University of Oklahoma Press, 1998.

Haig, Douglas (1861–1926)

British field commander in World War I. Haig was born into the famous family of distillers in Edinburgh, Scotland, on 19 June 1861. Educated at Oxford and graduating first in his class at Sandhurst in 1885, Haig was commissioned in the 7th Hussars and saw service in India, Africa, and on the home front. In the Nile Valley during the Mahdist War, he fought at Atbara and Omdurman. In the Second Boer War, with his superior officer, John French, and ambulance volunteer Mohandas K. Gandhi, Haig escaped Ladysmith by train just before the Boers surrounded it. In 1901–1902 he fought Jan Smuts's guerrillas in the Cape Province. He was promoted to major general in 1905 and to lieutenant general in 1910.

Again under French in 1914, he commanded the I Corps of the British Expeditionary Force at Mons and the Marne and in Picardy and Artois. He became commander of the First Army in February 1915. The Germans stopped him at Neuve-Chapelle from 10 to 13 March, at Festubert from 9 to 26 May, and at Loos from 26 September to 14 October. Haig replaced French in December and began to plan the great offensive of the Somme. After this offensive succeeded (at enormous cost), late in 1916, in reducing threats to the poilus at Verdun, Haig was promoted to field marshal.

Operating under Robert Nivelle, Haig made gains at Arras from 9 to 15 April 1917. After Henri Pétain replaced Nivelle on 15 May, Haig improved the French position by attacking at Passchendaele from 31 July to 10 November. He contained Erich Ludendorff's offensives at the Somme and the Lys in March and April 1918, then counterattacked at Amiens on 8 August. He commanded the Flanders operation in Ferdinand Foch's final offensive from 26 September to 11 November. Created earl in 1919, he died in London on 29 January 1926.

Haig has come to symbolize the unimaginative head-on tactics on the western front that nearly destroyed a British generation in World War I.

Eric v. d. Luft

See also: Amiens; Boer Wars; Foch, Ferdinand; French, John Denton Pinkstone, First Earl of Ypres; Ladysmith, Siege of; Ludendorff, Erich Friedrich Wilheim; Marne, Battle of the; Nivelle, Robert; Omdurman; Pétain, Henri-Philippe; The Somme; Verdun; World War I

References and further reading:

Bond, Brian, and Nigel Cave, eds. *Haig: A Reappraisal 70 Years On.* London: Leo Cooper, 1999.

Terraine, John. *Douglas Haig: The Educated Soldier.* London: Leo Cooper, 1990.

Warner, Philip. *Field Marshal Earl Haig.* London: Bodley Head, 1991.

Winter, Denis. *Haig's Command: A Reassessment.* London: Viking, 1991.

Haitian Civil War (1806)

After independence Haiti was ruled by the despotic Jean-Jacques Dessalines, who proclaimed himself emperor. Dessalines, who had been born in Africa, discriminated against the mulattoes. Moreover, his regime was corrupt and stole most of the national treasury. In 1806 Dessalines was murdered and Haiti plunged into civil war. The civil war centered around two strongmen who vied to be master of Haiti. The first was Henri Christophe, a black man who had been one of Tousaint L'Overture's lieutenants in the war for independence from France. The other was Alexandre Pétion, a prominent mulatto who had led that segment of Haitian society in battles with the French and Haitian slaves. An assembly attempted to establish a national government with Christophe as president and Pétion as head of the legislature. Under this arrangement the mulattoes would dominate and Christophe would be a mere figurehead.

Christophe rejected the plan and attempted to seize power by marching on Port-au-Prince, but he was thwarted by Pétion's superior army, which was equipped with artillery. Christophe retreated and established a state in the northern portion of Haiti with the capital at Cap-Haitien. In 1811 he proclaimed himself King Henry I and built a magnificent palace. In order to maintain his personal power, Christophe brought African warriors who formed his palace guard and were called Royal Dahomets.

Pétion established a republic in the southern portion of Haiti with himself as president for life. The division of Haiti into two distinct states resulted in clashes, but for many years neither section had the strength to defeat the other. Pétion pursued economic policies that rewarded the mulatto elite, but his racial policies were less discriminatory than those in Christophe's kingdom. Both states showed a marked contrast between the small wealthy ruling class and the balance of the population, which was impoverished.

When Pétion died in 1818 Christophe sought to unify the country under his leadership, a move that was rejected by the southern elite, who did not want a black leader. Instead, General Jean-Pierre Boyer was selected by the republican senate to be president. In 1820 Christophe suffered a severe stroke and later took his own life. Boyer then united the country and ruled until 1843.

George M. Lauderbaugh

See also: Toussaint L'Overture, Wars of

References and further reading:

Heinl, Robert Debs, Jr., and Nancy Gordon Heinl. *Written in Blood: The Story of the Haitian People, 1492–1971.* Boston: Houghton Mifflin, 1978.

Moran, Charles. *Black Triumvirate: A Study of Louverture, Dessalines, Christophe—The Men Who Made Haiti.* New York: Exposition Press, 1957.

Nicholls, David. *From Dessalines to Duvalier: Race, Colour, and National Independence in Haiti.* London: Cambridge University Press, 1979.

Halleck, Henry Wager (1815–1872)

Union field commander and administrative officer in the American Civil War, known (and not with affection) as "Old Brains." Halleck was born in Westernville, New York, on 16 January 1815. Assigned to the engineers after graduating from West Point in 1839, he studied fortifications in Europe and wrote a book on military science, published in 1846. As brevet captain in the Mexican-American War, he saw little battle action but excelled at engineering. He resigned his commission in 1854 and became a lawyer in San Francisco.

Recalled to active duty upon the recommendation of Winfield Scott, Halleck was commissioned major general on 19 August 1861. He replaced John C. Frémont in command of the Department of Missouri on 19 November and straightened out the administrative mess that Frémont had made. After Ulysses S. Grant captured Forts Henry and Donelson in February 1862, Halleck, as Grant's commanding officer, was rewarded on 13 March with command of all Union forces in the western theater.

Halleck's only field campaign was the march on the Confederate supply base at Corinth, Mississippi, in May and June. He was severely criticized for his creeping, mile-a-day advance, which allowed P. T. Beauregard and Braxton Bragg to escape and regroup after Shiloh. Abraham Lincoln relieved him of field command on 19 September and brought him to Washington, D.C. As general in chief from 11 July 1862 to 12 March 1864, and as chief of staff until 16 April 1865, he performed much better as an administrator than he had as a field commander.

After the Civil War, Halleck held commands in Virginia, the Pacific, and Kentucky. He died in Louisville, Kentucky, on 9 January 1872.

Eric v. d. Luft

See also: American Civil War; Beauregard, Pierre Gustave Toutant; Bragg, Braxton; Engineering, Military; Fort Donelson; Grant, Ulysses Simpson; Mexican-American War; Scott, Winfield; Shiloh

References and further reading:

Ambrose, Stephen E. *Halleck: Lincoln's Chief of Staff.* Baton Rouge: Louisiana State University Press, 1962.

Simon, John Y. *Grant and Halleck: Contrasts in Command.* Milwaukee: Marquette University Press, 1996.

Hamilcar Barca (c. 270–228 or 229 B.C.E.)

Carthaginian general during and after the First Punic War, and father of Hannibal and Hasdrubal. In 247 B.C.E. Hamilcar was sent to Sicily to take over command of the fleet. He raided the coast of Italy, hoping both to bring the Italian population to revolt against the Romans and to keep Roman forces occupied. When this remained ineffective he landed in Sicily with a force of mercenaries and started a guerrilla war. After he had captured the town of Eryx (244 B.C.E.) he succeeded in prolonging the war in Sicily until 241 B.C.E. The Carthaginians were defeated at sea near the Aegeates islands. Hamilcar was given full authority to negotiate a peace treaty.

Due to the cost of the war effort Carthage was not able to pay its mercenaries, which were the bulk of its army. They revolted and besieged Carthage. Hamilcar assumed command of the Carthaginian army. As in Sicily he preferred a war of mobility and small-scale action rather than one of large battles. By cutting off the supply lines of the insurgents he forced them to raise the siege. The war was brought to an end in 237 B.C.E. by Hamilcar, sharing command with his political rival Hanno, in a battle near Leptis Minor.

After the revolt Hamilcar went to southern Spain to recover the territories Carthage had lost during the war with Rome. He campaigned in the peninsula until he drowned during the siege of Helice (near Alicante) on the east coast of Spain, where he was treacherously defeated by a local king. By that time he had reestablished Carthage's Iberian empire. As a result the city regained its position as a major power in the western Mediterranean. Hamilcar became the effective ruler of the province, a power base that he passed on to his house. With its resources his son Hannibal almost brought Rome to its knees.

M. R. van der Werf

See also: Hannibal Barca; Punic Wars

References and further reading:

Bagnall, Nigel. *The Punic Wars. Rome, Carthage and the Struggle for the Mediterranean.* London: Pimlico, 1999.

Caven, Brian. *The Punic Wars.* London: Weidenfeld & Nicolson, 1980.

Lazenby, J. F. *The First Punic War.* London: University College London Press, 1996.

Hamilton, General Ian Standish Monteith (1853–1947)

British soldier and author. Ian Hamilton was first noticed by Lord Roberts during the Second Anglo-Afghan War (1878–1880) and became his aide-de-camp in 1882. He saw action in South Africa during the Transvaal War of Independence (1880–1881) and was severely wounded and captured during the fateful battle at Amajuba (27 February 1881). He then took part in the Gordon Relief Expedition in the Sudan (1884–1885). At the time of the outbreak of the Anglo-Boer War (1899–1902) he was General George White's assistant adjutant general in Natal and was besieged with White in Ladysmith, playing a pivotal role in several battles, including Platrand (Wagon Hill and Caesar's Camp, 6 January 1900). After the relief of Ladysmith, Hamilton commanded Roberts's Mounted Infantry Division. He took part in Roberts's advance from Bloemfontein to Pretoria, saw action at Donkerhoek (Diamond Hill, 11–12 June 1900), but was unable to corner General Christiaan De Wet (August 1900). At the end of 1900, Hamilton accompanied Roberts back to England and became his military secretary at the War Office. He returned to South Africa in November 1901 to become Lord Kitchener's chief of staff but was soon ordered to command mobile columns in the western Transvaal during the final weeks of the antiguerrilla campaign. He defeated the Boers at Roodewal (11 April 1902) and was one of the few British commanders who emerged from the war with an enhanced military reputation.

During the Russo-Japanese War (1904–1905), Hamilton was a British observer in the field. He was adjutant general, 1909–1910, and became commander in chief of the Mediterranean Command in 1910.

When World War I broke out, he was given command of the Central Force for the defense of the United Kingdom and in March 1915 he was placed in command of the Dardanelles operation. However, lack of success (and heavy casualties) led to his recall in October 1915. He was not given another command.

Hamilton was the gifted author of several books, including *The Fighting of the Future* (1885), *A Staff Officer's Scrap Book* (2 volumes, 1905–1907), *Gallipoli Diary* (2 volumes, 1920), *Anti-Commando* (with A. Wools-Sampson, 1931), *When I Was a Boy* (1939), and *Listening for the Drums* (1944). He also published novels and poems. Hamilton was a confident and resourceful officer, but "The Happy Warrior" lacked the ruthless drive and single-mindedness that are prerequisites for a truly great commander. Nevertheless he was a rare phenomenon, an intellectual professional soldier with a keen interest in all the arts.

André Wessels

See also: Boer Wars; De Wet, Christiaan Rudolph; Gallipoli; Kitchener, Horatio Herbert; Roberts, Frederick Sleigh, First Earl, Viscount St. Pierre of Kandahar; World War I

References and further reading:

Aspinall-Oglander, C. F. "Hamilton, Ian Standish Monteith." In *The Dictionary of National Biography 1941–1950.* Eds. L. G. Wickham Legg and E. T. Williams. London: Oxford University Press, 1959.

Churchill, W. S. *Ian Hamilton's March.* London: Longmans, Green, 1900.

Hamilton, I. S. M. *The Happy Warrior: A Life of General Sir Ian Hamilton.* London: Cassell, 1900.

Spies, S. B. "Hamilton, Ian Standish Monteith." In *Dictionary of South African Biography,* vol. 2, eds. W. J. de Kock and D. W. Krüger. Pretoria: Human Sciences Research Council, 1972.

Han Wudi (r. 141–87 B.C.E.)

Chinese emperor of the Han Dynasty who maintained strong, forward positions against the Xiongnu Rising in the north. More or less contemporaneous with the unified Chinese empires of Qin and Han was the steppe state of the Xiongnu (possibly the ancestors of the Huns), which not only seriously threatened Chinese rule in north China but also prevented direct Chinese contacts with the wealthy Greek world of the distant west. Han rulers responded to the Xiongnu threat in various ways, but under Wudi a "modernist" school of administration insisted upon direct confrontation with their steppe enemies and the maintenance of a forward position of military bases and border colonies to keep the Xiongnu as far away from China as possible.

Wudi's policy was a concerted attempt to "use the barbarians to control the barbarians." This involved the encouragement of subversion within the Xiongnu, above all through substantial bribes to the right parties, a carefully controlled marriage policy, and a search for allies to support the Han cause. It was in support of this latter goal that the courtier Zhang Qian was sent west to establish contact with the Yuezhi, "moon clan," traditional enemies of the Xiongnu. After harrowing adventures that brought the Chinese explorer as far as Sogdia and Bactria, whence the Yuezhi had moved, he returned with no alliance but with abundant information about the west and the roads leading there. Armed with this intelligence, Wudi's armies, which had already begun an advance into what is now Chinese Turkistan, quickly conquered the entire area as far as Ferghana in what is now western Turkistan. This advance not only outflanked the Xiongnu but brought China, for the first time, into direct contact with the west. This was the real beginning of the famous Silk Route.

Paul D. Buell

See also: Ban Chao; Huns

References and further reading:

Barfield, Thomas J. *The Perilous Frontier, Nomadic Empires and China.* Cambridge, MA, and Oxford, UK: Basil Blackwell, 1989.

Loewe, Michael. "The Campaigns of Han Wu-ti." In *Chinese Ways in Warfare,* eds. Frank A. Kierman and John K. Fairbank, 67–122. Cambridge, MA: Harvard University Press, 1974.

Hancock, Winfield Scott (1824–1886)

Impeccable Union field commander in the American Civil War. Hancock was born in Montgomery Square, Pennsylvania, on 14 February 1824. After graduating from the U.S. Military Academy at West Point in 1844, he served under his namesake in the Mexican-American War and was brevetted first lieutenant at Churubusco. In the 1850s, he fought the Seminoles in Florida, the factions in Kansas, and the Mormons in Utah.

Stationed in California when the Civil War broke out, he returned east and was promoted to brigadier general of volunteers on 23 September 1861. After leading a masterful flank attack at Williamsburg, Virginia, on 5 May 1862, he was called "Hancock the Superb." He fought at Seven Pines and Fair Oaks from 31 May to 1 June, at Frayser's Farm on 30 June, commanded a division at Antietam, and became major general of volunteers on 29 November. He attacked Marye's Heights at Fredericksburg and performed expert rearguard maneuvers at Chancellorsville.

On the first day at Gettysburg, Confederates under Henry Heth and A. P. Hill pushed him southeast to defensive positions on Cemetery Ridge. He held the Federal center left on the second day and the center the third day but was critically wounded on Cemetery Ridge as his forces brought George Pickett's famous charge to a halt. Returning to action six months later, he distinguished himself at the Wilderness and was brevetted major general in the regular army for his service at Spotsylvania. He failed at Cold Harbor only because of Ulysses S. Grant's error. Troubled by his Gettysburg wound, he deferred command at Petersburg and went on furlough. Hill and Wade Hampton dealt him an embarrassing defeat at Reams' Station, Virginia, on 25 August. Relieved of field command on 27 November, he finished the war commanding garrisons around Washington, D.C.

Hancock was the Democratic candidate for president in 1880. After losing to James A. Garfield, he returned to active military duty and died at his headquarters on Governor's Island, New York, on 9 February 1886.

Eric v. d. Luft

See also: American Civil War; Antietam/Sharpsburg; Chancellorsville, Battle of; Cold Harbor, Battle of ; Fredericksburg; Gettysburg; Grant, Ulysses Simpson; Hill, Ambrose Powell; McClellan, George Brinton; Meade, George Gordon; Mexican-American War; Mexico City, Battles for; Mormon War; Petersburg, Siege of; Pickett, George Edward; Seven Days' Battles; Sioux Wars; Spotsylvania Court House; Utah War; Wilderness

References and further reading:

Coates, Isaac Taylor. *On the Plains with Custer and Hancock: The Journal of Isaac Coates, Army Surgeon.* Boulder, CO: Johnson, 1997.

Gambone, A. M. *Hancock at Gettysburg and Beyond.* Baltimore: Butternut and Blue, 1997.

Jordan, David M. *Winfield Scott Hancock: A Soldier's Life.* Bloomington: Indiana University Press, 1988.

Tucker, Glenn. *Hancock the Superb.* Indianapolis, IN: Bobbs-Merrill, 1960.

Hannibal Barca (247–188 B.C.E.)

The greatest Carthaginian general during the Second Punic War. Hannibal was the son of the famous general Hamilcar Barca, who took him to Spain in 237 B.C.E. According to Roman tradition Hannibal was raised to hate Rome, swearing an oath forever to be its enemy. Although Hannibal became an implacable enemy of Rome, his reputation for cruelty, malice, and greed, attributed to him by biased Roman writers, is not backed by evidence.

In 220 B.C.E. Hannibal became commander of the Carthaginian army in Spain. After subduing the north of the peninsula he attacked the city of Saguntum (Sagunto) in 219 B.C.E. The Romans protested, claiming that the city was an ally, and declared war. Hannibal immediately prepared for an invasion of Italy. His plan was based upon the assumption that if he attacked the Romans in Italy, many of Rome's allies would change sides. Having set out with an army of 50,000 foot, 9,000 horse, and some 40 elephants, Hannibal marched through the Pyrenees and southern Gaul. After a difficult march through the Alps he reached the Po valley and defeated the Romans at the Trebia (218 B.C.E.).

The following year Hannibal marched over the Apennines into Etruria (Tuscany), where he destroyed a Roman army at Lake Trasimene (217 B.C.E.). Again a Roman force marched to fight Hannibal. At Cannae (216 B.C.E.) two Roman armies were totally destroyed. Hannibal showed superior generalship and complete control over his troops.

Although Hannibal had won three major victories, things did not turn out as he expected. The Romans showed remarkable tenacity. Following the advice of Fabius Maximus, they dogged Hannibal's footsteps, denying him battle but undoing his successes behind his back. The delaying tactics wore Hannibal down, denying him the opportunity to win the war decisively.

Rome's allies also proved far more loyal than Hannibal had expected. Few of Rome's major allies came over to him. Moreover, as abandoning his newfound allies would negate any chances of more Italians joining him, Hannibal was forced to protect them. As the Romans set in on pushing renegade allies into line, Hannibal had to march regularly to their relief and had to use up valuable manpower on garrisons.

Despite Hannibal's initial successes, he received few reinforcements from Carthage partly because of significant political opposition. Moreover the Carthaginians deemed the protection of their empire in Spain more important than the war in Italy. An attempt by Hannibal's brother Hasdrubal to reinforce him with an army from Spain was foiled at the Metaurus in northern Italy (207 B.C.E.).

Hannibal campaigned in Italy until he was recalled to Africa in 202 B.C.E., when Carthage itself was threatened by Scipio Africanus. At Zama (202 B.C.E.), Hannibal was defeated decisively. Knowing that Carthage had lost the war, he advocated peace. For a while Hannibal was allowed to follow a political career in Carthage, but eventually he was forced to flee to King Prusias of Bythinia in Asia Minor by Roman machinations. To avoid being extradited to the Romans, Hannibal committed suicide.

M. R. van der Werf

See also: Animals in War; Cannae, Battle of; Fabius Maximus Verrucosus "Cunctator"; Hamilcar Barca; Marcellus, Marcus Claudius; Punic Wars; Scipio Africanus Major, Publius Cornelius; Trebia, Battle of the; Zama, Battle of

References and further reading:

Bagnall, Nigel. *The Punic Wars: Rome, Carthage and the Struggle for the Mediterranean.* London: Pimlico, 1999.

Carthaginian General Hannibal crossing the Alps into Italy with elephants during the Second Punic War between Carthage and Rome 218-202 B.C.E. (Hulton/Archive)

Beer, Sir Gavin. *Hannibal. The Struggle for Power in the Mediterranean.* London: Thames & Hudson, 1969.
Bradford, Ernle. *Hannibal.* London: Macmillan, 1981.
Lazenby, J. F. *Hannibal's War. A Military History of the Second Punic War.* London: Aris & Phillips, 1978.

Harpers Ferry (American Civil War, 12–15 September 1862)

One of the greatest capitulations in American military history. After the Battle of Second Bull Run/Manassas Junction, Robert E. Lee and the Army of Northern Virginia "invaded" Maryland with plans to move into Pennsylvania. Using cavalry along the fall line of the mountains, Lee screened his movements from George McClellan, who moved cautiously.

When Lee learned that the Union garrison at Harpers Ferry, (West) Virginia, had not abandoned the munitions and supply depot there, he ordered General Thomas J. "Stonewall" Jackson and three of Lee's four columns to surround the town and compel its surrender.

The Union garrison under Colonel Dixon Miles held out until the Confederates placed artillery on the heights overlooking the town, which was in the valley at the confluence of three rivers. Miles surrendered his garrison of more than 12,000 men and the vital supplies to Jackson, who meanwhile sent his columns hurrying to join Lee at Antietam, where Lee had decided to make a stand against McClellan. The last of Jackson's columns, under General A. P. Hill, was the last to leave and therefore the last to arrive at Antietam, blunting Burnside's late afternoon assault.

Charles M. Dobbs

See also: American Civil War; Antietam/Sharpsburg; Jackson, Thomas "Stonewall"

References and further reading:

Farwell, Byron. *Stonewall: A Biography of General Thomas J. Jackson.* New York: W. W. Norton, 1993.

Gallagher, Gary W., ed. *The Antietam Campaign.* Chapel Hill: University of North Carolina Press, 1999.
Robertson, James I., Jr. *Stonewall Jackson: The Man, the Soldier, the Legend.* New York: Macmillan, 1997.

Harrison, William Henry (1773–1841)

American frontiersman, field commander, politician, and president. Born on his father's plantation, Berkeley, in Charles City County, Virginia, on 9 February 1773, Harrison graduated from Hampden-Sidney College in 1790. His father, Benjamin Harrison, a signer of the Declaration of Independence and the governor of Virginia from 1782 to 1784, sent him to Philadelphia to study medicine under Benjamin Rush. When his father died in 1791, Harrison immediately abandoned medicine and received an ensign's commission in the 1st Infantry, which was stationed in Ohio with standing orders to patrol the Northwest Territory.

As lieutenant and aide-de-camp to "Mad Anthony" Wayne, Harrison fought with distinction at Fallen Timbers on 20 August 1794. Promoted to captain in 1797, he commanded Fort Washington, near Cincinnati, until he resigned in 1798 to become territorial secretary for the Northwest. He was territorial delegate to Congress in 1799 and governor of the newly created Indiana Territory from 1800 to 1812. A significant part of his duties was to maintain good relations with the natives of the territory and thus ensure safety for white settlers. He undercut his own efforts at peace by forcing a series of land-grabbing treaties on the Shawnee and other indigenous nations. Tecumseh and his brother, Tenskwatawa, organized native resistance against Harrison's policies.

On 7 November 1811, at the confluence of Tippecanoe Creek and the Wabash River, Harrison defeated Tenskwatawa and burned his village, losing about 180 of the 950 regulars and militia under his command. This battle destroyed the fragile coalition of natives in Indiana and made Harrison a national military hero with the nickname "Tippecanoe." Tecumseh led the remnants of the tribes into Canada, where they became staunch allies of the British during the War of 1812.

Appointed major general of the Kentucky militia in August 1812, Harrison relieved Fort Wayne and was commissioned brigadier general of regulars in September. Seeking to recoup the losses suffered by James Winchester in the west, Harrison built Fort Meigs and Fort Stephenson in Ohio and waited for reinforcements. Promoted to major general of regulars in March 1813, he marched north that autumn, recapturing Detroit on 29 September and decisively defeating the British and Indians at the Thames on 5 October. His nemesis, Tecumseh, was killed in that battle. He resigned his commission in May 1814 and returned to Ohio.

Harrison represented Ohio in Congress from 1816 to 1819 and in the Senate from 1825 to 1828. He lost a four-way election for president to Martin Van Buren in 1836, but won the White House as a Whig in 1840. He died on 4 April 1841 from the pneumonia he caught while delivering his inaugural address a month earlier, the first American president to die in office.

Eric v. d. Luft

See also: American Indian Wars; Fallen Timbers; Taylor, Zachary; Tecumseh; Thames; Tippecanoe, Battle of ; War of 1812; Wayne, Anthony

References and further reading:

Cleaves, Freeman. *Old Tippecanoe: William Henry Harrison and His Time.* Newtown, CT. American Political Biography Press, 1990.
Goebel, Dorothy Burne. *William Henry Harrison: A Political Biography.* Philadelphia, PA: Porcupine, 1974.
Green, James A. *William Henry Harrison: His Life and Times.* Richmond, VA: Garrett & Massie, 1941.
Todd, Charles Stewart, and Benjamin Drake. *Sketches of the Civil and Military Services of William Henry Harrison.* New York: Arno, 1975.
Young, Stanley. *Tippecanoe and Tyler Too!* New York: Random House, 1957.

General William H. Harrison at the Battle of Tippecanoe in 1811. (Library of Congress)

Harsha (c. 590–c. 647)

Starting as the teenage ruler of a small Indian state, Harsha came close to recreating the mighty Gupta Empire. In 606, Harsha Vardhana became ruler of Thaneswar, a small state located in the Punjab between the northern Indus and northwestern Ganges Valley. He married into the neighboring Maukhar kingdom and relocated to the capital at Kanauj. In 618, Harsha invaded the Gauda kingdom with 50,000 troops, 20,000 horsemen, and 5,000 elephants. Centered in modern Bangladesh, Gauda was responsible for the death of Harsha's older brother. "The elephants were never unharnessed and the soldiers never unhelmeted" until Harsha made Gauda his tributary.

Generally, however, Harsha preferred to negotiate his neighbors into his empire as allies rather than use force. These tactics made the rulers of Sindh, Ghujarat, and Valabhi become vassals (Samantas), placing much of the Indus Valley and the Arabian Sea coast under his sway. However, Pulakeshin II, ruler of the Decca plains south of the Narmada River, spurned Harsha's diplomacy. In 633, when Harsha attempted to march into Pulakeshin's territory, the Deccans forced him back across the Narmada. Three years later, he expanded his empire to the east, moving down from Gauda to annex more of Bengali coast.

Later foreign sources comment on the extent of Harsha's cavalry (100,000 horse and 60,000 elephants). Despite this great host, Harsha preferred a feudal-confederal decentralism over militaristic despotism. Patron of arts, culture, public charities, and scholarship, historical sources are kind to his memory. Nonetheless, Harsha was assassinated in 647 and his empire disintegrated almost immediately.

Weston F. Cook Jr.

References and further reading:

Devaluti, D. *Harsha: A Political Study.* London: Oxford University Press, 1970.

Smith, Vincent A. *The Oxford History of India.* 4th ed. Oxford, UK: Oxford University Press, 1981.

Harun al-Raschid (766–809)

Sultan of the Abbasid Empire. Member of the dynasty that had overthrown the Umayyads, and established themselves as the successors of Muhammad and rulers of a rich empire centered on Baghdad, Harun al-Raschid is primarily known as a patron of the arts and sciences. With extraordinary revenues from trade, Harun al-Raschid supported a spectacular court that inspired the *Thousand and One Nights* and perfected the pursuits of falconry, polo, and chess. He was one of the first sultans to make diplomatic contact with western Europe, sending Charlemagne an elephant as a gift and establishing himself as the protector of Christian pilgrims to the holy land.

Nonetheless, Harun al-Raschid's rule was far from peaceful. Constantly on the alert against internal threats, he maintained an elaborate secret police network. His system of regional emirs contributed to constant revolts of Berbers, and Egyptians, as well as Khazars, who collaborated with the Byzantines. Public works projects such as hospitals and universities were offset by large-scale military fortress building on the borders, garrisoned by fanatical ghazis. The Byzantines were Harun al-Raschid's greatest enemy, although despite the use of Greek fire, there were few permanent gains for either side because of prisoner exchanges and truces. Unfortunately, the excesses of his court and the internal dynamics of his family guaranteed that the empire was plunged into destructive rounds of fratricide and civil war when Harun al-Rashid died.

Margaret Sankey

See also: Abbasid Revolution

References and further reading:

Audisio, Gabriel. *Harun al-Raschid, Caliph of Baghdad.* New York: McBride & Company, 1931.

Bekrine, Mustapha. *Haroun al-Raschid.* Algir: SNED, 1971.

Hasegawa, Yoshimichi (1850–1924)

Japanese field marshal who earned worldwide attention with his accomplishments during the Russo-Japanese War. Hasegawa was born in the Iwakuni subfief of the Choshu clan's territory, now the prefecture of Yamaguchi. When the Choshu joined with the Satsuma clan to overthrow the shogun and restore the emperor to power in 1868, Hasegawa participated in the fighting. When the new government established an army to replace the old clan armies, he joined as a captain in 1871. By the time of the Satsuma Rebellion in February 1877, he was a major and commanded a regiment. Hasegawa's regiment was among those sent to relieve Kumamoto Castle in April 1877, and he distinguished himself in the fighting. After the war, Hasegawa was sent to France during 1885–1886, to review military developments in that nation. When he returned, the army promoted him to major general.

Hasegawa's next war was the first Sino-Japanese War of 1894–1895. He commanded a brigade during the fighting in Korea. At the battle of Pyongyang on 15 September 1894, Hasegawa won distinction for his valor and for the performance of his brigade. During the fighting at Haicheng in December 1894 and January 1895, his unit once again was recognized. As a reward, Hasegawa was promoted to command of the Guards Division in General Kuroki's First Army during the Russo-Japanese War of 1904–1905. He led his division from its landing in Korea to the Yalu River. During the Battle of the Yalu (30 April–May 1904) Hasegawa was recognized for his aggressive drive against the Russian defenders; crossing the river, he forced them to retreat. In June 1904, he was promoted to general.

Hasegawa served as commander of the Korean Garrison Army from September 1904 to December 1908. He tried to exclude civilian authorities from his area of responsibility as much as possible. In 1912, he was promoted to chief of staff of the army and served until 1915. His disdain for civilians was made obvious in 1913, when he protested directly to the emperor regarding a plan to allow reserve officers to hold positions as service ministers in the government. Hasegawa was promoted to field general when his term as chief of staff ended.

Tim J. Watts

See also: Kuropatkin, Aleksey Nikolaevich; Mukden, Battle of; Nogi, Maresuke; Russo-Japanese War; Sino-Japanese War (1894–1895); Yalu River

References and further reading:

Connaughton, R. M. *The War of the Rising Sun and Tumbling Bear: A Military History of the Russo-Japanese War, 1904–5.* New York: Routledge, 1988.

Walder, David. *The Short Victorious War: The Russo-Japanese Conflict, 1904–5.* London: Hutchinson, 1973.

Warner, Denis Ashton. *The Tide at Sunrise: A History of the Russo-Japanese War, 1904–1905.* New York: Charterhouse, 1974.

Hastings, Battle of (14 October 1066)

Decisive battle between William of Normandy and Harald II Godwinsson. Hastings provided the final resolution to the Danish Wars of Succession and marked the end of the Anglo-Saxon kingdoms. It resulted in the administrative, legal, and social restructuring of England by William I, the Conqueror.

The events leading to the battle were the death of Edward the Confessor with no clear heir to the throne; election by the Witan of one of the claimants, Harald II Godwinsson, brother-in-law to Edward; and the battle of Stamford Bridge between two of the claimants, Harald Sigurdsson of Norway and Harald II Godwinsson of England. Immediately after Harald II Godwinsson's victory at Stamford Bridge (25 September 1066), word came of the landing of William's invasion force at Pevensey on 29 September 1066. Hoping to duplicate the tactics that had worked so well at Stamford Bridge—a surprise attack cutting the enemy off from his ships—Harald quickly returned to the south.

The speed of his forced march coupled with losses at Stamford Bridge meant that Harald arrived with depleted resources. Many of Harald's foot troops and archers were left behind. That, coupled with insufficient time to regroup and call up fresh reinforcements, played a significant role in Harald's subsequent loss of the battle to the Normans.

The opponents were positioned on two hills with an intervening valley. At 9 A.M. William's attack surprised Harald, forcing him to fight a defensive battle with largely unseasoned levies. Harald's housecarls took the front and flank positions of the tightly grouped Anglo-Saxon army. This was a good defensive stand. It also blocked the road to London.

William deployed his army into three main groups: Breton auxiliaries on the left, the bulk of Normans in the center, and a mixed group on the right. The van was comprised of light foot soldiers and archers, who were followed by more heavily armed infantry, and finally, squadrons of mounted knights. Initially, William's battle strategy was an uncoordinated series of attacks by infantry and cavalry. After the first waves were repulsed by Harald, William altered his strategy to combined attacks. Archers shot high to disable and occupy the defenders while the knights attacked to break the defense line. William was successful. Harald died in the blizzard of arrows, and the Anglo-Saxon line broke.

Tamsin Hekala

See also: Norman Conquest; Stamford Bridge, Battle of; William the Conqueror

References and further reading:

Brown, R. A. "The Battle of Hastings." In *Proceedings of the Battle Conference on Anglo-Norman Studies III 1980,* ed. R. A. Brown, 1–21. Woodbridge, UK: Boydell Press, 1981.

Douglas, David C. *William the Conqueror: The Norman Impact upon England.* Berkeley: University of California Press, 1964.

Walker, Ian W. *Harold: The Last Anglo-Saxon King.* Thrupp Stroud, Glouchestershire, UK: Sutton Publishing, 1997.

Hattin, Battle of (4 July 1187)

A severe defeat of the crusaders by the Muslims under Saladin. Sultan Saladin (Salah al-Din) organized forces to retake the Holy Land from the Christian kingdoms established in the Levant after the First Crusade. To counter this, King Guy of Lusignan organized an army of 15,000 at Sepphoris, near Nazareth. Against the advice of his advisers, Guy's army began their march into the arid region of the eastern Galilee on 3 July 1187 to relieve the Muslim siege of Tiberias.

The Christian army consisted of about 1,500 knights and 4,000 cavalry, the remainder being infantry. The march proceeded slowly as Muslim cavalry harassed the advanced units. By the afternoon the Christians were out of water and growing weary. Guy decided to turn toward the springs at Hattin, overlooking the sea of Galilee. Two rocky outcrops, known as the Horns of Hattin, dominated the barren plain. As the Christian army camped that night Saladin encircled their position.

At dawn on 4 July the Christians resumed the march. Saladin's infantry lit brush fires that blew smoke toward the Christians. With the two armies arrayed, the Christian knights charged, driving back the Muslims. The Christians were unable to drive them off, however, and became more fatigued from lack of water, heavy mail, and the smoke. Some of the Christian cavalry broke through and escaped, the remainder of the army was mired near the rocky horns.

Saladin's forces attacked on all sides, and the exhausted Christians surrendered, with a remnant of the True Cross falling into Saladin's hands. The numerous bishops and noblemen captured along with King Guy were ransomed, while the foot soldiers were sold into slavery. Saladin took the rest of the region, eventually capturing Jerusalem.

Robert Dunkerly

See also: Crusades; Saladin

References and further reading:

Nicolle, David. *Hattin: Saladin's Greatest Victory.* London: Osprey Publishing Co., 1993.

Hawaiian Wars (1782–1810)

The three decades from about 1780 to 1810 that saw the Hawaiian Islands brought together into a unified kingdom for the first time by King Kamehameha "the Great" (c. 1752–1819). As in other parts of the world, this consolidation was made possible in the Hawaiian Islands in great part through the introduction of firearms.

When Captain James Cook was killed on the Big Island of Hawaii in 1778 by armed warriors of that island's primary chief, Kalaniopuu, the islands of Hawaii were far from a unified polity. Political power and control varied from island to island, with even the Big Island divided among rival chieftains. Yet within a generation the armaments and technology that Cook and other Western traders and explorers introduced would become decisive in that archipelago's unification. Soon after Chief Kalaniopuu's death in 1782 a rivalry ensued between Kalaniopuu's relations, including his sons Kiwalao and Keoua and his nephew Kamehameha, for control of the Big Island. But the rival chieftains and their bands of warriors were of relatively equal strength, and as a result their struggle persisted throughout the 1780s without conclusive results.

In 1790 an American trading vessel, the *Fair American,* along with its guns and two English crewmen, fell into the hands of Kamehameha after it was attacked and seized as retaliation for losses suffered in an encounter with an earlier Western ship. Such trading vessels had begun to appear with increasing frequency in the islands, a convenient watering hole between China and the West Coast of the Americas. Kamehameha would use the two foreigners to manufacture Western handguns and train his men in Western fighting tactics.

Even before establishing his power on the Big Island, Kamehameha decided to attack the neighboring island of Maui, then under the control of the most powerful chief in the islands, Kahekili. In the narrow valley of Iao on Maui, Kamehameha, employing his two Englishmen and newly acquired guns, inflicted a decisive defeat upon an army led by Kahekili's son. Despite this victory Kamehameha returned to the Big Island, where fighting had erupted again in his absence. The renewed struggle on the Big Island was again indecisive until Kamehameha ambushed and killed his chief rival, Keoua, along with his retinue of warriors, after inviting him to meet at a newly constructed *heiau* (temple), dedicated tellingly to the god of war. With this death Kamehameha established himself as master of the Big Island of Hawaii.

Soon thereafter Kahekili sent a fleet of native canoes and special bands of warriors, along with his own Western vessel, to harass Kamehameha on his own turf. A sea battle was fought off the Big Island between the two rival chieftains' vessels, which proved sanguinary but indecisive. Kahekili died on his home island of Oahu soon afterward, his domains, like those of Kalaniopuu previously, falling into dispute between his various heirs. Only in late 1794 did Kahekili's son Kalanikupule emerge as victor, following the defeat on Oahu of his half brother, and primary foe, with the help of guns supplied by an English merchant. In January 1795 the victorious Kalanikupule decided to take his campaigns to the Big Island of Hawaii, hoping to defeat his father's rival Kamehameha. Now equipped with a plentiful supply of firearms and several Western vessels, his hopes of bringing the Big Island under his control were not farfetched. His luck did not hold, however, and the foreign crews of his ships, pressed into his service, mutinied and succeeded in driving Kalanikupule and his warriors overboard and back to Oahu in humiliation.

Kamehameha meanwhile had been colluding with the English. In 1794 he agreed to "cede" the Big Island of Hawaii to Great Britain and in return received English help in building a fighting ship. Eyeing his strategic opportunity, Kamehameha decided to move and in early 1795 seized Maui and the narrow island of Molokai, which lay just to its north. Despite the defection of one of his primary chiefs to Kalanikupule, Kamehameha proceeded with plans to attack Oahu and landed on that island's southern coast near modern Waikiki. Kamehameha scattered his foe, driving many over the high cliffs of the pass, and with his victory, and the death of Kalanikupule, secured his control over Oahu.

The only island remaining outside Kamehameha's control was the far western island of Kauai. On Oahu Kamehameha received further British help in building a 40-ton ship with which to attack Kauai. Kamehameha and his forces set sail for Kauai in summer 1796, only to have his plans postponed at the last moment by an uprising on the Big Island. Perhaps the delay was fortunate. The uprising was soon subdued but plans for the invasion of Kauai were put on hold. The interval allowed Kamehameha time to consolidate his newly won domains and to set up efficient means of administration and communication. He set up governors on each of the islands, and like resourceful rulers before him, such as France's Louis XIV or Toyotomi Hideyoshi in Japan, he invited potential rivals to dwell with him in his capital. He also set about building a stronger navy, switching to innovative twin-hulled canoes rather than the traditional and less stable single-hulled ones. From the foreigners arriving in increasing numbers and with increasing frequency in the islands Kamehameha procured yet more armaments and foreign vessels.

In 1802 Kamehameha finally sailed again for Kauai, then ruled by the chief Kaumualii, with a fleet of nearly 800 vessels and an armed force of thousands. Kamehameha and his fleet tarried for some time on Maui, hoping unsuccessfully to threaten Kaumualii into submission, before continuing westward to Oahu. On Oahu in 1804 Kamehameha's efforts were struck an almost fatal blow, in the form of an epidemic that wiped out many of his troops, though it spared him. For several more years Kamehameha stayed on in Oahu, which was yearly growing in population and prosperity. At this point Kamehameha let it be known that he would be satisfied with the outward submission of his rival on Kauai, and gaining it would allow him to rule on there as his governor. The two rival chieftains were finally brought together in Honolulu in early 1810. The result was the formal inclusion of Kauai as a tributary island to Kamehameha with Kaumualii as its leader. It was a diplomatic terminus to almost two decades of conflict, and with it Kamehameha secured his control over all of Hawaii and effected the first unification of the islands in their history.

Daniel Kane

References and further reading:

Cahill, Emmett. *The Life and Times of John Young: Confidant and Advisor to Kamehameha the Great.* Honolulu: Island Heritage Publishers, 1999.

Daws, Gavan. *Shoal of Time: A History of the Hawaiian Islands.* Honolulu: University of Hawaii Press, 1968.

Kuykendall, Ralph S. *The Hawaiian Kingdom.* 3 vols. Honolulu: University of Hawaii Press, 1967.

Hawkwood, John, Sir (c. 1321–1394)

English soldier and mercenary captain-general in Italy. The son of Gilbert Hawkwood of Essex, he fought under King Edward III in the Hundred Years War at the battles of Crécy (1346) and Poitiers (1356). Knighted after Poitiers, and unemployed after the Treaty of Bretigny (1360), he joined the mercenary White Company, so named for their brilliantly polished armor. While serving the Italian city-state of Pisa, Hawkwood was elected captain-general of the White Company in January 1354. He led the Pisan forces against Florence (1364), before committing his company to the service of Bernabo Visconti of Milan against Emperor Charles IV. Hawkwood was captured near Arezzo and held until ransomed in 1369. Entering the service of Pope Gregory XI in 1372, he fought in an indecisive war against Milan in 1374 and led papal forces during the War of the Eight Saints (1375). Leaving papal service, Hawkwood joined the antipapal alliances and served as captain-general of Florence from 1378 to 1381. He led Padua, an ally of Florence, to a decisive victory over Verona in the Battle of Castagnaro (11 March 1387). Hawkwood led the Florentine army in his last campaign during an inconclusive war against Milan in 1390–1392.

Hawkwood utilized the English longbow and tactics developed in the French war during his Italian service. He was renowned for his infantry tactics, unit discipline, and utilization of lighter armor and equipment to improve the rapidity of troop movements.

Hawkwood died on 16 March 1394 in Florence. King Richard II later had his body returned and reinterred in Hawkwood's native village.

Brigitte F. Cole

References and further reading:

Temple-Leader, John, and Guiseppe Marcotti. *Sir John Hawkwood.* London: T. F. Unwin, 1849.

Trease, Geoffrey. *The Condottieri.* New York: Holt, 1971.

Henry II, King of England (1133–1189)

Henry Plantagenet, king of England, conqueror, reformer. Henry was duke of Normandy from perhaps 1149, count of Anjou from 1151, and duke of Aquitaine through his marriage to Eleanor in 1152. He is best remembered for his quarrel with Archbishop Becket, for his troubled relations with his wife and sons, among them the future kings Richard I and John, and for his sweeping constitutional reforms in England. His contemporaries recognized him as a military leader without peer. On behalf of his mother, Mathilda, he organized the forces against Stephen of Blois in the civil war for the crown that followed the death of Mathilda's father, Henry I. On the Continent, he regularly subdued rebellious vassals, later including those supporting his sons, and established dominance over virtually all of the princes of northern France, as well as a cowed King Louis VII. While he marshaled the full feudal resources of his realms for major campaigns in Toulouse (1159), Wales (1165), and Ireland (1171), most of his career was spent besieging castles. Castles were both a symbol and a consequence of baronial power at the expense of overlords in the twelfth century.

Henry first besieged and leveled the castles of those refusing to accept his lordship, then strengthened and rebuilt his own castles to preserve order in his realms. His successful castle strategy avoided the necessity of expensive pitched battles. To accomplish this strategy, he increasingly relied on mercenary footmen trained and equipped for castle siege and defense, and not on his feudal levies.

Despite this, Henry was still a feudal king and his administrative reforms in England were marked by two significant initiatives in this capacity: the 1166 Cartae Baronum (Baronial Charters) and the 1181 Assize of Arms. The Baronial Charters were written statements from all of Henry's tenants-in-chief identifying feudal obligations to knight service, thus allowing the king to discover the full extent of his feudal military resources. The Assize of Arms identified and classified each vassal's obligation according to wealth, establishing a hierarchy of military obligation based on a single recruitment system. It would be the first of a series of such attempts that continued into the thirteenth century.

Robert Babcock

See also: Philip II Augustus; Richard I

References and further reading:

Keefe, Thomas K. *Feudal Assessments and the Political Community under Henry II and His Sons.* Berkeley: University of California Press, 1983.

Warren, W. L. *Henry II.* 2d ed. New Haven, CT: Yale University Press, 2000.

Henry V, King of England (1387–1422)

English king, victor at Agincourt. Henry V was born 16 September 1387, at Monmouth, Monmouthshire, Wales, and died 31 August 1422 at Bois de Vincennes, France, as Prince of Wales, duke of Cornwall, earl of Chester, prince of Aquitaine, duke of Aquitaine, and duke of Lancaster. Henry was the eldest son of Henry IV, and he first fought alongside his father during the Welsh (against Owain Glyn Dwr) and English (against Henry Hotspur Percy and Edmund Mortimer) rebellions. Like his father, Henry wished to expand English influence in French territories, and his most famous victory at Agincourt on 25 October 1415 confirmed his place in British history. This battle, which resulted in the destruction of many of the most powerful French nobles, saw the loss of some 6,000 Frenchmen, but only 400 English. Not a decisive victory, Henry was forced to continue to push his way inland, finally capturing Normandy in the spring of 1419. Eventually, Henry signed the Treaty of Troyes with the Burgundians in May 1420, and this solidified his claim to the French throne.

Henry married Katherine, the daughter of the French king, and after a brief tour of the English countryside with his new bride, he returned to France and defeated the stronghold at Meaux in May 1422. It was during this time that Henry's health began to fail him, and he died prematurely of dysentery on 31 August 1422, at the young age of 34. His death was a major blow to the English, who had become quite loyal to him, and while he never saw his goal of conquering France come to fruition, he did leave his kingdom to a nine-month-old successor, Henry VI.

David J. Tietge

See also: Agincourt, Battle of

References and further reading:

Allmand, C. T. *Lancastrian Normandy, 1415–1450: The History of Medieval Occupation.* Oxford, UK: Oxford University Press, 1983.

Jacob, Ernest Fraser. *Henry V and the Invasion of France.* London: English Universities Press, 1947.

Wylie, James Hamilton. *The Reign of Henry the Fifth.* Westport, CT: Greenwood Press, 1968.

Heraclius (c. 575–641)

Byzantine emperor (r. 610–641) who managed to save the empire from defeat at the hands of the Persians and Avars. During the eight-year reign of Phocas I, responsible for a rule of mindless cruelty and terror, the eastern empire all but ceased to exist. Its position in the Haemus collapsed and Persian armies reached the Bosphorus in 608. With the empire in extremis, the exarch of Africa, Heraclius the Elder, raised the standard of revolt and a fleet commanded by his son made its way to Constantinople. There Heraclius the Younger overthrew Phocas and was installed as emperor.

The change of emperor brought no immediate improvement of the empire's position. Damascus fell in 613, Jerusalem in 614, Chalcedon, on the Marmara, in 616. The loss of Egypt (616–619) followed. With Avar raids reaching the walls of Constantinople (617–619), Heraclius determined to abandon Constantinople in favor of the African provinces but was prevented by popular protests and a pledge by Church and people to defend the empire (619). From this time dated the militarization of the state: It underwent fundamental reform, which, over decades, gave rise to the creation of an Anatolian peasantry that held land in return for military service.

With these various changes in hand, in 622 Heraclius left Constantinople with an army to begin operations in Cilicia and Syria. The war was taken into Armenia and thence into Media and Mesopotamia: In 624 Istfahan was occupied. By 625 the Persian Empire, under obvious threat, allied itself with the Avars. Heraclius, returning to Constantinople, resorted to alliance with the Khazars in an attempt to divide Persian attention. Then, leaving Constantinople under the command of the patriarch, Heraclius returned to Armenia and the Caucasus.

Another Persian invasion of Anatolia reached the Bosphorus on 29 June 626 at the same time as an Avar army, of about 80,000 men, reached Constantinople's walls. There followed a climactic 10-day battle on land and at sea as the Avars sought to land forces inside the Golden Horn. The shattering of this attempt forced the Avars to abandon their effort: The Persians were left in Anatolia with a double threat in the mountains and Mesopotamia. Over the next two years the war was carried into the Persian heartland once more. In 628 a peace agreement restored all the eastern empire's lost possessions.

To have been responsible for such a feat of survival would have assured Heraclius of a place in history, but in the 13 years after his return to Constantinople in May 628 the eastern empire lost one-third of its territory. Between 634 and 640 Palestine and Syria were overrun once again; Egypt was lost for good between 639 and 640. Within 25 years of the death of Heraclius the empire had lost half of its territory and faced threats as grave as those that it had faced in 626.

H. P. Willmott

See also: Byzantine-Persian Wars

References and further reading:

Haldon, John. *Warfare, State and Society in the Byzantine World, 565–1204.* London: University College London Press, 1999.

Norwich, John Julius. *Byzantium: The Early Years.* London: Viking, 1988.

Hideyoshi, Toyotomi (1537–1598)

Hideyoshi began his career as a peasant's son in Nakamura, Japan, but rose through the ranks of Oda Nobunaga's army until he became a general. When Akechi Mitsuhide assassinated Nobunaga, Hideyoshi halted his attacks, returned to Kyoto, and executed Akechi. Hideyoshi then served as one of four regents to the grandson of Nobunaga in 1582. By 1585, Hideyoshi had established himself as the successor to Nobunaga's legacy and concluded several alliances, including one with Tokugawa Ieyasu. He then proceeded to unify Japan. Hideyoshi amassed an army of 200,000 men and invaded Chosokase in 1585. Then, in 1587, he assembled an even larger army of 280,000 men and took Kyushu.

Portrait of Hideyoshi, "the Napoleon of Japan." (Bettmann/Corbis)

By 1590, Hideyoshi, after the battle of Odawara, had successfully unified Japan. The key to his success was due more to his political acumen in arranging alliances than to military force. He did, however, possess the military might to enforce his alliances. Furthermore, in order to prevent rebellion, he arranged national sword hunts and disarmed all but the bushi, or samurai class.

Hideyoshi was not satisfied with the conquest of all of Japan but envisioned the conquest of China. He intended to march through Korea and then invade China from the north. After the kingdom of Korea refused him free passage, he invaded it in 1592. His army of 200,000 overran most of Korea but encountered the Ming Chinese army at Pyongyang. Negotiations began, but they eventually broke down. So another invasion occurred in 1597–1598 with another massive army but in the process Hideyoshi died.

Hideyoshi's greatest contribution was the unification of Japan. Furthermore, he was able to harness its nascent military ability and focus it away from deleterious civil war and into foreign ventures. The sheer size of his armies are testaments to Hideyoshi's organizational and logistical abilities.

Timothy May

See also: Japanese Invasion of Korea; Japanese Wars of Unification; Oda, Nobunaga; Samurai; Sekigahara; Tokugawa, Ieyasu

References and further reading:
Berry, Mary Elizabeth. *Hideyoshi.* Cambridge, MA: Harvard University Press, 1982.
Dening, Walter. *The Life of Toyotomi Hideyoshi.* New York: AMS Press, 1971.
Toyotomi Hideyoshi. *101 Letters of Hideyoshi: The Private Correspondence of Toyotomi Hideyoshi.* Ed. Adriana Boscaro. Tokyo: Sophia University Press, 1975.

Hill, Ambrose Powell (1825–1865)

Confederate general, tenacious fighter, and one of Robert E. Lee's most valued subordinates. A. P. Hill was born in Culpeper, Virginia, on 9 November 1825. At the U.S. Military Academy at West Point, he was a member of the class of 1847 and the roommate of future Union commander George B. McClellan. He fought as an artilleryman in the Mexican-American War and against the Seminoles. On 31 March 1861, he resigned from the U.S. Army as first lieutenant.

Commissioned colonel of the 13th Virginia Regiment in May, Hill was in reserve at First Bull Run. Promoted to brigadier general on 26 February 1862, he distinguished himself at Yorktown, Williamsburg, and Hanover Court House, Virginia, in May. He was made major general on 26 May and quickly assembled the famous Hill's Light Division, which became a key element in the peninsular campaign. Hill's setbacks during the Seven Days Battles were not through lack of either courage or tactical skill, but from the absence of the support he expected from Stonewall Jackson.

Hill and Jackson subsequently worked better together. At Cedar Mountain on 9 August they combined to rout Nathaniel Banks. Their cooperation helped the Confederates at Second Bull Run, Harpers Ferry, Antietam, and Fredericksburg. Hill accompanied Jackson around the Union right flank at Chancellorsville and took over command when Jackson was hit. Robert E. Lee promoted Hill to lieutenant general on 24 May 1863, hoping to replace Jackson.

Hill proved inadequate for corps command. His accomplishments after Chancellorsville never matched his earlier work. His III Corps was the first to attack at Gettysburg, but it did not perform well. His hasty attack on a superior Union force under Gouverneur Kemble Warren at Bristoe Station, Virginia, on 14 October without first gathering the necessary reconnaissance devastated his troops and thwarted Lee's offensive. He likewise failed to meet Lee's expectations at the Wilderness, North Anna, and Cold Harbor. He was killed in action at Petersburg on 2 April 1865.

Eric v. d. Luft

See also: American Civil War; Antietam/Sharpsburg; Bull Run, First/Manassas; Bull Run, Second/Manassas Junction; Chancellorsville, Battle of; Cold Harbor, Battle of; Fredericksburg; Gettysburg; Harpers Ferry; Jackson, Thomas "Stonewall"; Lee, Robert Edward; Longstreet, James; McClellan, George Brinton; Petersburg, Siege of; Seven Days Battles; Wilderness

References and further reading:
Hassler, William Woods. *A. P. Hill, Lee's Forgotten General.* Richmond: Garrett & Massie, 1962.
Schenck, Martin. *Up Came Hill: The Story of the Light Division and Its Leaders.* Harrisburg, PA: Stackpole, 1958.

Hindenburg, Paul von Beneckendorf und von (1847–1934)

World War I commander and president of the Weimar Republic. Hindenburg was born on 2 October 1847 in Posen. He entered the Prussian army in 1866 and fought in the last two wars of German Unification, the Austro-Prussian War (1866) and the Franco-Prussian War (1870–1871). From 1873 to 1876 he studied at the Prussian War Academy before entering the Prussian General Staff with the rank of captain. A solid officer, Hindenburg steadily rose through the ranks of the Prussian army and was even considered for the posts of chief of the general staff and Prussian war minister. Hindenburg commanded an army corps from 1903 to 1911, when, having achieved the rank of lieutenant general, he retired after a successful if not brilliant military career. World War I changed that.

On 22 August 1914, Hindenburg was recalled to active duty and given command of the Eighth Army in East Prussia. His chief of staff was Erich von Ludendorff. The careers and fortunes of both men would henceforth be inexorably linked. Sent to East Prussia to deal with the Russian threat, Hindenburg and Ludendorff inflicted two major defeats on the Russians at the battles of Tannenberg and the Masurian Lakes (August–September 1914). As a result of these spectacular victories, Hindenburg became the most famous and most popular general in Germany, a status that he was to keep for the remainder of his long life. In the wake of Tannenberg and the Masurian Lakes, Hindenburg was promoted to the rank of field marshal and appointed commander in chief on the eastern front in November 1914. In this position Hindenburg and his chief of staff Ludendorff came into increasing conflict with the commander in chief of the general staff, Erich von Falkenhayn, over the question of German military strategy. Moreover, Falkenhayn's inability to break the military stalemate redounded to Hindenburg's benefit, as many ordinary Germans now looked to the hero of the eastern front as Germany's savior. On 28 August 1916, Hindenburg replaced Falkenhayn as chief of the general staff, with Ludendorff serving as first quartermaster general. In addition to their responsibilities for German military strategy, Hindenburg and Ludendorff were increasingly in charge of German domestic policy as well, which led to the creation of a "silent dictatorship."

Domestically, Hindenburg and Ludendorff sought to place the German economy on a total war footing by enacting a massive munitions program, accompanied by an Auxiliary Service Law aimed at mobilizing German manpower. Yet they could do little about the food situation, brought about by the Allied naval blockade, and for several winters, turnips seemed to be about the only food available. This near famine is all the more remarkable considering that after 1917, Germany had control of the Ukraine, one of the world's greatest food producers, as well as one of Europe's most efficient railway systems to transport the wealth of the Ukraine to the Reich. Germany did exact huge reparations in specie from the defeated Russians, but the German people could not eat gold.

Hindenburg and Ludendorff were also instrumental in the downfall of Chancellor Theobald von Bethmann-Hollweg in July 1917 and his replacement by a string of weak chancellors who could be bent to the military's will. In terms of military strategy Hindenburg and Ludendorff were responsible for the reintroduction of unrestricted submarine warfare in April 1917, which brought the Americans into the war; the annexationist peace treaties imposed on Rumania and Russia in 1918; and the final German effort on the western front in 1918. As grand strategists both left much to be desired.

Hindenburg retired from the army for the second time in 1919 but remained a popular figure in a postwar Germany bitterly resentful of the Versailles diktat. (Of course, the terms imposed by Germany on her defeated enemies were certainly diktats by any definition of the word.) In 1925 Hindenburg again emerged from retirement and was elected president of the Weimar Republic. Although a monarchist at heart, he was initially loyal to the republic. However with the onset of the Depression in 1929, the rise of Nazism after 1930, and his advancing age, he became increasingly dependent on a rightist camarilla determined to destroy German democracy. Under the influence of this rather unsavory circle, Hindenburg appointed Adolph Hitler chancellor on 30 January 1933. Relegated to the sidelines as a senile figurehead, Hindenburg died on 2 August 1934, at his estate in Neudeck, East Prussia. It was an ill day for Germany when Hindenburg was translated from his field command to political power.

J. David Cameron

See also: Falkenhayn, Eric von; Ludendorff, Erich Friedrich Wilheim; William II

References and further reading:

von Hindenburg, Paul. *Aus meinem Leben.* Leipzig: S. Hirzel, 1920.

Kitchen, Martin. *The Silent Dictatorship: The Politics of the German High Command under Hindenburg and Ludendorff, 1916–1918.* New York: Holmes & Meier, 1967.

Wheeler-Bennett, John W. *Hindenburg: The Wooden Titan.* New York: Macmillan, 1967.

Hiroshima and Nagasaki, Atomic Bombings of (1945)

Ended World War II and began the nuclear age. The Japanese cities of Hiroshima and Nagasaki were devastated by atomic bombs in August 1945. Their destruction culminated years of intensive research and the investment of $2 billion into the Manhattan Project, the top-secret American effort to develop atomic weapons.

By 1944, American scientists believed they could produce an atomic chain reaction sufficient to generate an unprecedented explosion, and army leaders created the 509th Composite Bomb Group to develop tactics for utilizing the new weapon as a bomb. Colonel Paul Tibbets's 393d Bombardment Squadron pioneered these tactics in the 509th, using modified B-29 Superfortresses to drop single bombs accurately from high altitudes and then climb, turn, and dive away at high speed to escape the shockwave generated at detonation. After training in Utah, Tibbets and his men flew to the Pacific island of Tinian and awaited delivery of atomic bombs for use against Japan.

Their delivery came closer on 15 July 1945, when scientists detonated the first atomic device at the Trinity test site, near Alamogordo, New Mexico. With an estimated yield of 15,000–20,000 tons of TNT, the explosion exceeded even the wildest expectations of Manhattan Project scientists, and President Harry Truman quickly approved the use of atomic weapons against Japanese cities. Hiroshima, Kokura, Niigata, and Nagasaki were the possible targets, chosen for their military value and because they had been relatively undamaged by previous raids.

Tibbets and crew began their mission to Hiroshima by boarding the *Enola Gay,* a B-29 named for Tibbets's mother, on 6 August 1945. They carried a bomb nicknamed Little Boy, in honor of Franklin Roosevelt, which utilized a uranium 235 core to generate an atomic explosion. Tibbets dropped the device at 8:15 A.M. and it detonated 1,900 feet above Hiroshima with a force of 15,000 tons of TNT. It destroyed five square miles of the city, along with 140,000 people, who died from the initial explosion or from radiation and blast burns over the next several months.

On 9 August, the United States launched a second atomic strike, sending a B-29 named *Bock's Car* commanded by Ma-

The atomic bombing of Nagasaki, Japan. (Library of Congress)

jor Charles Sweeney to attack the city of Kokura. When clouds obscured the city, Sweeney switched targets and bombed Nagasaki at 11:02 A.M. with a plutonium bomb named Fat Man (after Winston Churchill). Though more powerful than Little Boy, Fat Man caused less damage because hills around Nagasaki contained the explosion. The bomb killed 73,884 people outright, injured 74,909, and destroyed 2.6 square miles of the city.

Historians and many concerned citizens still vehemently debate the wisdom and morality of dropping the atomic bombs and their role in compelling the Japanese surrender on 14 August 1945. The surrender came five days after the destruction of Nagasaki, but Japanese leaders were influenced by the entry of the Soviet Union into the war against them perhaps as much as by the power of atomic bombs. Their surrender made the invasion of Japan, the greatest planned military operation in history, completely unnecessary.

Lance Janda

See also: Atomic Bomb, Development of; World War II

References and further reading:

Alperovitz, Gar. *The Decision to Use the Atomic Bomb.* New York: Vintage Books, 1995.

Rhodes, Richard. *The Making of the Atomic Bomb.* New York: Touchstone Books, 1995.

History, Military

The study of the planning for and the use of organized armed force, either on behalf of some form of recognized state authority or against it. It differs from other branches of history insofar as it calls upon knowledge of certain specialized areas, such as strategy and tactics.

Military history can be said to have existed since men began to write about armed conflict: The Hebrew Scriptures contain many accounts of battles and can be considered a military historical source. The Athenian Thucydides, whose history of the Peloponnesian War is often cited as an early example of military history, provided accounts of sea operations, clashes between armies, and the strategies of the opposing sides and did not demonize the opposing forces. But he also examined the nature of state power, the effects of war, and offers of peace, so much so that one of the first translators of the work, Thomas Hobbes, described its author as "the most Politick Historiographer that ever writ," thereby highlighting a perennial problem of military history: deciding the point at which it overlaps so much with other areas that it becomes political or general history.

Nevertheless, throughout its development, military history has been marked by certain recurring characteristics. In particular, the motivation of authors and the functions that it has served can be identified as follows: to provide a record of heroic or sacrificial deeds for coming generations; to pay tribute to the achievements of those who fought soon after the conflict; to assist in the discovery of lessons for future wars; to satisfy the demand of the public for accounts of battles; to cover up mistakes by commanders or leaders by falsifying, omitting, or distorting events; to provide inspiration to soldiers and future soldiers; to act as a teaching tool for military instructors; to encourage feelings of identification within a group, community, or state; and to satisfy scholarly interest. Trends in military historiography have been largely defined by the interplay of these motivations.

While in many early accounts enjoyment in the destruction and slaughter can be clearly detected, before and during the early modern period the *genius* of the warlord played a central role in the writing of military history. Although the organization, equipment, and movement of armies were described, the commander was usually the monarch or nobleman and hence was honored through accounts of his campaigns. The function of this military history, though, was not simply to generate propaganda on behalf of the ruler: It was assumed that the performance of the commander was normally the factor that decided the outcome. It should also not be forgotten that in times of high illiteracy rates military history should also be considered to have included visual records, such as the Bayeux Tapestry, or the woodcut illustrations of the sixteenth and seventeenth centuries.

Two parallel developments at the beginning of the nineteenth century, however, ushered in important changes: the professionalization of historical research and writing at universities and of armies through the establishment of military academies and general staffs.

Even before the Franco-Prussian War, military history began to be systematized, with the use of battles to illustrate certain points of military operations and tactics. In Britain the first example was probably Major General Patrick MacDougall's *Theory of War Illustrated by Numerous Examples from Military History,* published in 1856, followed by the more influential work by General Sir Edward Hamley, *The Operations of War Explained and Illustrated,* first published in 1866 and continually republished until 1922. Hamley's work is interesting for the way in which it shows the blurred dividing line between military theory and military history. Hamley's starting point was to complain that military history was written very much like novels, with the reader accepting uncritically the opinions of the writer. In fact, in the nineteenth century mainstream military history was dominated by the assumption that turning points in history were epitomized by decisive battles, an approach enshrined in Sir Edward Creasy's *The Fifteen Decisive Battles of the World* (1851).

After the conclusion of the Franco-Prussian War, military history enjoyed a new period of popularity in Europe, with numerous illustrated and well-documented histories of that war appearing. The most significant development was arguably the emergence of official histories, the American Civil War being most memorably commemorated by the *War of the Rebellion: A Compilation of the Official Records of the Union and Confederate Armies,* consisting of an astonishing 128 volumes, published between 1881 and 1901. But most professional of the official military histories was the more manageable eight-volume history of *The German-French War of 1870–71,* undertaken by the Military History Section of the German General Staff; the first volume was published in 1874, the last in 1881. This work was the first official military history to be researched and written in conformity with the standards of the emerging history profession and was characterized by detail and analysis down to company level and up to panoramic descriptions of all the engagements. The first volumes won the prize for German History awarded by the Prussian Academy of Science in 1878, and their publication opened the era of professionally respectable military history.

With the expansion of staff work, so began an increase in the volume of documents left behind by wars; and as the official archives grew, so did the desire of armed forces to keep them under their own control. One of the few historians in Germany to challenge this monopoly was Hans Delbrück, an academic at Berlin University, who argued that the subject needed to broaden its focus, taking into account the political intentions of national leaders. His view that military history ought to be conducted outside the general staff provoked bitter opposition.

The writing of the history of World War I saw the apogee of the dominance of official military history. Although the Red Army did not produce an official history of Russian operations 1914–1917, and the Americans contented themselves with the publication of documents, the voluminous British, German, and French historical undertakings were severely hampered by the determination of the military establishments to prevent damage to the reputation of commanders and their armies. The German official history, although published by the state archives, was written mainly by former officers who had not been accepted by the Reichswehr, the intention being to counteract the defamation of the military that followed the defeat in 1918. The British official history demonstrated a further problem: With the first volume appearing in 1922, and the final one in 1947, the slow progress put into question the value of such works for instructional purposes.

In Britain, a reaction occurred against self-serving military history in the 1930s, Basil Liddell Hart and J. F. C. Fuller in particular making influential contributions. Although their works on the American Civil War and World War I did not meet the standards of later academic monographs, they encouraged a more critical attitude to the use of sources and the official version of events.

As it turned out, the official British history of World War II far exceeded the quality of its predecessor. The U.S. histories were likewise of a high standard, the range of publications being particularly noteworthy. Only the Russian official histories of the "Great Patriotic War" can be criticized for omitting significant facts and for its propaganda. The history the Wehrmacht remains, however, the most problematic case study in the writing of the history of World War II. Many of the surviving German records were captured by the Western Allies, and the U.S. Army established an Operational History (German) Section in January 1946, employing 328 ex-Wehrmacht officers by June 1946. Its successor, the Control Group, survived in skeleton form until 1961. The studies conducted under American auspices assessed the war only within the narrow framework of "lessons learned," yet they were not only the basis for many later published works, the results flowed into official U.S. military thinking. Mistakes and atrocities were almost always blamed on Hitler. No doubt as a reaction to this traditional concept, the first volumes of the Federal German official history of World War II (volume 1 was published in 1979) undertaken by the armed forces Military History Research Office adopted a wider and more critical approach. But by the close of the twentieth century the work was still uncompleted, illustrating that some of the problems of official history have remained.

Despite the dominance of World War II as the main subject for military historians throughout the 1950s and 1960s, the postwar period has seen a revolution in the writing of military history. The increasing numbers of military officers who take doctorates before assuming posts as historians has led to a remarkable improvement in research standards. The U.S. Army Command and General Staff College at Fort Leavenworth has been influential in promoting an expansion in the scope of combat studies, geographically and thematically. Equally important is the emergence of a new concern with the experience of battle, a trend confirmed by John Keegan's innovative *The Face of Battle* (1973). Keegan's book inspired many more studies of this kind, marking a decisive break with the operational narrative tradition, which usually excluded the horrors of battle. The new concern for the "ordinary soldier" (undoubtedly a reflection of the prevailing interest in the academic history establishment of "history from the bottom up") expressed itself particularly effectively in studies of war in the trenches, such as Tony Ashworth's *Trench Warfare 1914–1918: The Live and Let Live System* (1980). These works opened up new possibilities for historians to use microstudies to put official accounts to the test, showing how *the system* actually worked.

Beyond the experience of combat, other types of military history have been developing, most notably studies of armies in peacetime. Building on insights from sociology and organization theory and pioneering sociological studies of various officer corps, major studies have been published on military reform, military debates, and the dynamics of the introduction of new weapons. Well-researched monographs on the development of armored forces, 1919–1939, for instance, have not simply added to gaps in knowledge, they have improved awareness of the fact that poor battlefield performance can have doctrinal causes that stem from decisions taken during peacetime. They have also contributed to the development of methodologically pioneering works on military effectiveness, innovation, doctrine, and culture, a recent example being *Military Innovation in the Interwar Period* (1996), edited by Williamson Murray and Allan Millett.

The organizational approach in studies of armies has likewise led to significant improvements in understanding such subjects as war planning, mobilization, surprise attack, and failures in intelligence assessment. Indeed, the role of intelligence in both peacetime and wartime has emerged of late as a subject in its own right.

Perhaps the most important postwar trend in the writing of military history has been the body of literature that falls under the rubric of "war and society." This school of thought, which became popular in the 1970s, seeks to examine the effects of warfare on political institutions, economies, popular culture, and postwar societies. It has assisted, for example, in encouraging study of the home fronts during both world wars and militarism in Wilhelmine Germany. The problem with this approach is that it lacks any clear theoretical or methodological underpinning, leading often to imprecise generalizations. Perhaps more stimulating have been those works that have examined specific aspects of warfare over several centuries, original studies such as Martin Van Creveld's path-breaking *Supplying War* (1977), which examined logistics since Gustav Adolphus, or William McNeill's *The Pursuit of Power* (1983), discussing the interaction of war, technology, finance, and society. One criticism that can be leveled against such works is that they tend to be bloodless accounts of a very bloody process: men with lice crawling over their testicles dying and being mutilated in very unpleasant ways.

In short, since 1945 the thematic and methodological scope of military history has widened dramatically, producing a concomitant trend of specialization within an increasing number of subdisciplines. The use of military history for instructional purposes has continued, but it has become more subject to academic scholarly standards than ever before, suggesting perhaps the fulfillment of Hans Delbrück's original demands.

Alaric Searle

See also: Theory, Military

References and further reading:

Fortescue, J. W. *Military History.* Cambridge, UK: Cambridge University Press, 1914.

French, David. "Sir James Edmonds and the Official History of the Great War." *RUSI Journal* 131 (March 1986).

Liddell Hart, Basil: *Why Don't We Learn from History?* London: Allen & Unwin, 1944.

Kühne, Thomas, and Benjamin Ziemann, eds. *Was ist Militärgeschichte?* Paderborn: Schöningh, 1999.

Hitler, Adolf (1889–1945)

Austro-German statesman, classic and supreme symbol of evil for the twentieth century. Hitler was born on 20 April 1889 to Alois, a focused, able civil servant of the multiethnic Austro-Hungarian Empire, and his much younger wife, Klara. Alois died in January 1903, after years of often-violent conflict with Adolf that apparently only taught the son to defy the father by rejecting everything that Alois stood for: thrift, discipline, work, a family life, and a measure of political tolerance. Hitler fully justified his father's fears when his mother allowed him to lead an idle life in the family home in Linz.

In September 1907, Hitler went to Vienna, seeking admittance to the Academy of Fine Arts. Rejected for inadequate

Hitler with Nazi troops. (Library of Congress)

preparation, he returned to Linz to care for his mother during her long, painful death from breast cancer, moving to Vienna in February 1908. There, he wasted his mother's inheritance in lazy fantasy at the opera and reading a gutter press that intensified the pan-German nationalism and anti-Semitism he had absorbed as a child. By the winter of 1909–1910, he was homeless, seeking shelter in a men's hostel. Only then did he paint hack work routinely. In May 1913, Hitler received his father's inheritance and moved to Munich immediately afterwards. He refused service in the multiethnic Austro-Hungarian army during the Balkan War only to volunteer for service in August 1914 in the Bavarian army, eventually serving as a dispatch runner in the 16th Reserve Infantry Regiment.

A committed soldier, he was highly decorated, respected, and very odd: He seems to have had no lovers nor close friends even though his regiment may have been his first home since his mother's death. He vehemently disapproved of humanitarian considerations influencing military policy and reacted violently to defeatist comments. The mass killing of World War I apparently deadened Hitler's already numb emotions. In an era tainted by a pseudo-Darwinian emphasis upon combat as the ultimate test of nations, as well as of men, he thought that the Jews had grown rich upon German suffering and defeat. Selected for army propagandist training in 1919, he created a theoretical framework for his obsessions; in September of that year, he first wrote of annihilating the Jews.

Hitler considered himself the drummer of Germany's *völkisch* right when he led an almost comical putsch in Munich on 8 November 1923. His trial for high treason transformed him into the right's national leader. Sentenced to five years' fortress confinement in Landsberg prison, he served only nine months, during which time he dictated *Mein Kampf* to his epigone, Rudolph Hess.

Like his father, Hitler preferred young, dependent women who could be manipulated and controlled. In 1929, he took as his lover his niece Geli Raubal, 19 years his junior. His insecurity permitted her no independence: the evidence points to her suicide, rather than murder, on 18 September 1931. Hitler claimed to be married to Germany and to every German woman. In reality, he simply could not permit a woman any standing of her own, even in the domestic sphere national socialism prescribed for women. Nevertheless, it is a mistake to stress Hitler's very real misogyny too much: However dim his view of German womanhood, he was also a catastrophe for German manhood.

Hitler and the Nazi Party did not seize power but received a majority of the votes cast in the Reichstag elections of July 1932. Despite defeat in the November 1932 elections, Hitler had mobilized the masses and so he was brokered into the chancellorship in January 1933 by conservative allies. However antidemocratic and authoritarian they were, they were also deeply divided amongst themselves about how to control him. On 23 March 1933, the Reichstag granted Hitler an Enabling Act, figuratively voting itself out of existence; on 30 June 1934, the Night of the Long Knives, Hitler crushed any internal Nazi opposition with the blessing of the German army.

By 1939, Hitler had stamped Germany with his personal insecurity and pseudo-Darwinian preference for conflict, often duplicating, even triplicating, traditional institutions with competing Nazi ones. The result was that even quite trivial decisions were referred to Hitler, which may have salved his ego but as policy was folly in peacetime and calamity in wartime. This subversion of bureaucratic power bases ensured confidants who would tell Hitler what he wanted to hear. Many Germans shared his racial views of Nordic superiority, Slavic inferiority, the necessity of exterminating the Jews, and cleansing the Germanic race. However, the increasing violence of Nazi rule was an extremely powerful motivation for *all* Germans to "work toward the Führer," on these matters. It was thus unnecessary for Hitler personally to direct the extermination of Europe's Jews, the Final Solution, chosen during the Wannsee Conference in January 1942.

By nature a gambler, Hitler actively pursued war with Britain and France during the Polish crisis of August 1939. An unhealthy hypochondriac, he was convinced he would die young. He thought no one else could lead Germany down his path of annihilatory racism, endemic though racism and anti-Semitism were in German society. He was probably right: Hitler was a revolutionary, not a reactionary.

Insecure and undisciplined, Hitler surrounded himself with advisers who craved his favor, shared his racism, and could not contradict his flood of facts, half-truths, and outright lies masquerading as knowledge. Hitler deliberately created an environment for himself that made it impossible for him accurately to assess risks, strengths, and weaknesses in any given situation. His successes before 1941 made him utterly resistant to criticism. The pivotal events of World War II show this clearly.

Despite the advice of his armor experts, Hitler regarded the area around Dunkirk as unsuited to armored operations, and so he entrusted the destruction of Britain's only field army to the Luftwaffe in May and June 1940. Wary of the British navy until he had a blue-water fleet capable of defeating it, Hitler postponed the invasion of Britain, leaving it an unreduced fortress in his rear while he turned east.

Supported this time by his military staff, Hitler insisted that Germany could invade the Soviet Union on 21 June 1941 and defeat the Red Army in six weeks, a tacit admission that German victory depended upon defeating the Soviet Union within that time frame. On 10 December 1941, as German troops retreated from Moscow, Hitler, in an act of supreme folly, declared war upon the United States of America in the aftermath of Pearl Harbor. Once the Soviet Union was defeated, Hitler intended to turn upon Britain and crush its resistance. The Reich would then fight and defeat the U.S., which Hitler despised for its racial "impurity" and "weaknesses," despite its immense industrial resources.

As the war progressed, Hitler intervened at increasingly lower levels of military decision making. He had always mistrusted the army leadership, which accepted his bribes and decorations to keep faith with him during the war, only to blame him for their mistakes as well as his own afterwards. From the army's middle ranks came a series of assassination attempts motivated by compassion toward the conquered, a horror of retribution for German crimes, especially against the Soviets, and shame for what Germany had become. Rather than await the end of the war in relative safety, these men made a final attempt to kill Hitler and overthrow his regime on 20 July 1944. Few survived. Their gesture of atonement only strengthened Hitler's belief that he was appointed by Providence to lead Germany to victory or doom. He would not countenance surrender.

Despite looming catastrophe, the German people greeted Hitler's survival with public expressions of relief. Although his personal popularity declined with approaching defeat, Hitler retained an amazing degree of loyalty from the common German, civilian or soldier, who often attributed misdeeds to party functionaries. To a great extent, Hitler was never a tyrant. It is true that the mechanisms of repression, selective breeding, and genocide were created for and first practiced upon the German people. However, many Germans supported Hitler's most extreme plans not only for the conquered, but also for Germany, including scorching German earth before the Allied onslaught.

On 29 April 1945, Adolf Hitler married his longtime mistress, Eva Braun, in Berlin. Although for many years the greatest desire of this unsophisticated young woman had been to become Frau Hitler, her lover obliged her only when it no longer mattered. They committed suicide the next day, rather than face their people and the enemy or assume responsibility by fighting to the death in the ruins of what had once been an imperial capital. Hitler left Germany utterly defeated, ground like grain between the Soviets and the Western Allies, in moral and physical ruins, soon to be partitioned, its rich cultural heritage almost effaced by twelve years of Nazism.

The profound cruelty Hitler inflicted on the world is not reducible to his anti-Semitism. The destruction of the Sixth Army at Stalingrad, the militarily useless sacrifice of poorly armed and trained troops, and the execution of thousands of German soldiers by their comrades are not comparable to the Holocaust, the vicious antipartisan warfare in Serbia, or the destruction of Warsaw. But all of these German crimes were part of a whole, rooted in German history and culture, as indeed any nation's actions must be, just as all serious German resistance to Hitler was fiercely reactionary, based on older, more humane values that were also authentically German. Hitler brilliantly understood his adopted country, and he shared its hopes and fears to an extraordinary degree, tremendously exaggerating them in his slaughterhouse heart.

Erin E. Solaro

See also: Bolshevik Revolution; Churchill, Sir Winston; Ethics of Warfare; German Wars of Unification; Guderian, Heinz; Rommel, Erwin Johannes Eugen; Roosevelt, Franklin D.; Stalin; World War I; World War II; Ypres, Battles of

References and further reading:

Bullock, Alan. *Hitler and Stalin: Parallel Lives.* New York: Alfred A. Knopf, 1992.

Burleigh, Michael. *The Third Reich: A New History.* 4th ed. New York: Hill & Wang, 2001.

Fest, Joachim. *Hitler.* New York: Harcourt, Brace, Jovanovich, 1974.

Kershaw, Ian. *Hitler, 1889–1936: Hubris.* New York: W. W. Norton & Co., 1999.

———. *Hitler, 1936–1945: Nemesis.* New York: W. W. Norton & Co., 2000.

Lukacs, John. *The Hitler of History.* New York: Alfred A. Knopf, 1998.

Weinberg, Gerhard L. *A World at Arms: A Global History of World War II.* London: Cambridge University Press, 1994.

Hittites (c. 2000–1100 B.C.E.)

A major Anatolian state during the late Bronze Age, rivaling New Kingdom Egypt in power. Hittite political and military history is known from the thousands of cuneiform texts from Anatolia in the late Bronze Age. Beginning with Hattushili I, approximately 15 kings are known to have ruled from Hattusha in the Old Kingdom (c. 1750–1600 B.C.E.). During the reign of Hattushili I (c. 1700 B.C.E.), the Hittites expanded into northern Syria and west into the land of Arzawa. Mursili I (c. 1600 B.C.E.) raided the city of Babylon (c. 1595 B.C.E.) and ended the First Dynasty of Babylon. However, the Hittites were unable to permanently expand into Mesopotamia, and Hittite control of eastern territories seems to have collapsed soon thereafter.

Hittite influence in western Asia Minor and northern Syria was reasserted by Tudhaliya II (c. 1420–1370 B.C.E.). The greatest expansion took place during the reign of Shuppiliuiluma I and his immediate successors (c. 1350–1250 B.C.E.). The Hittites conquered the powerful Hurrian state of Mitanni, controlled all of Syria north of Damascus, and fought with the Egyptians in Syro-Canaan. Hattushili III (c. 1250 B.C.E.) made a treaty with Rameses II following Kadesh and gave him a Hittite princess in marriage. This treaty stayed in effect until the fall of Hittite power in 1180 B.C.E. Due to a number of factors, Hittite power began to decline during the reigns of the three monarchs following Hattushili III: Tudahiyah IV, Arnuwanda III, and Shuppiluiluma II. The rising power of Assyria in northern Iraq severely truncated Hittite power in Syria. Ahhiyawa (possibly the Hittite term for the Achaeans), a powerful kingdom to the west, threatened Hittite power in western Anatolia. The Hittites also had serious troubles with the rival Hittite kingdom of Tarhuntassha in the south.

What is not certain, however, is what brought about the fall of the Hittite capital, Hattusha. Invaders from the west, usually identified with the Sea Peoples in Egyptian sources, may have been the catalyst. Contrary to popular scholarly tradition, Hittite power did not end with the fall of Hattusha; successor dynasties continued at Tarhuntassha and southeast at Carchemish on the Upper Euphrates. Smaller Neo-Hittite states continued in southeast Anatolia and Syria for at least the next 500 years (to c. 700 B.C.E.). These states were often in conflict with rival Aramean dynasties. Both Aramean and Neo-Hittite states were absorbed into the Assyrian world state. Passages from 1 Samuel and 1 and 2 Kings that mention the Hittites most likely refer to the Neo-Hittite states of Syria.

The Hittite king was the supreme war commander, and the Hittite annals imply that all kings were required to campaign on a regular basis. Their wars were a major source of tribute income and manpower that was used to maintain the agricultural base of the Hittite state. A Hittite king successful in war signaled the favor of the gods. The Hittites were also known to have performed purification and scapegoat rituals directed toward the enemy army before important pitched battles such as the Battle of Kadesh.

Mark W. Chavalas

See also: Kadesh, Battle of

References and further reading:

Beal, R. *The Organization of the Hittite Military.* Heidelberg: Carl Winter Universitätsverlag, 1992.

Bryce, T. *The Kingdom of the Hittites.* Oxford, UK: Clarendon Press, 1998.

Goetze, A. "Warfare in Asia Minor." *Iraq* 25 (1963), 124–130.

Howinckten Cate, P. "The History of Warfare according to Hittite Sources: The Annals of Hattusilis I." *Anatolica* 11 (1984), 147–183.

Ho Chi Minh (1890–1969)

Ho Chi Minh, who used at least 20 pseudonyms during his long and adventurous career (adopting his final most famous name only in 1944), drew upon Vietnam's millennia-old tradition of nationalist resistance to foreign rule and combined it with Marxist-Leninist doctrine to lead successful resistance against the Japanese, French, and Americans during the Pacific War and the two Indochina Wars. Originally named Nguyen Sinh Cung, he was born into a mandarin family in Nghe An Province in north central Vietnam, traditionally a center of rural unrest. His father and siblings were bitterly opposed to French colonial rule. After studying at the elite Quoc Hoc Secondary School in Hue, he began a career overseas, working first as a sailor aboard a French liner and not returning to his own country until 1941. From 1917 to 1923 he lived in France and became one of the founding members of the French Communist Party, established in December 1920.

In 1923, Ho journeyed to Moscow and the following year attended the Fifth Congress of the Communist International (Comintern). Thus began his Bolshevik phase from 1924 to 1941, when he lived and worked in Soviet Russia, Europe, China, and Thailand. He organized revolutionary groups among overseas Vietnamese, of which the most important were the Revolutionary Youth League (1925) and the Indochina Communist Party (1930).

After the Japanese military occupation of French Indochina in late 1940, Ho Chi Minh's career took a fundamentally new turn, the "people's war" phase. In May 1941, he established the Vietnam Doc Lap Dong Minh Hoi (League for the Independence of Vietnam), or Vietminh, which made village-based guerrilla warfare an extension of revolutionary politics, abandoning urban-based Bolshevik methods, which were of limited relevance in a largely rural society. Influenced by Mao Zedong's military doctrines, the Vietminh operated in the mountainous China-Vietnam border region.

Ho and his comrades, including Vo Nguyen Giap, commander of the Vietminh's armed forces, established village-level networks of National Salvation Associations and Vietminh Committees, which assumed local governmental

Portrait of Ho Chi Minh with two young girls, 1954. (Library of Congress)

functions and served as the "infrastructure" for guerrilla resistance. Ho's united front tactics, his emphasis on peasant mobilization, diplomatic skills in winning Chinese and American support, and highly popular appeals to nationalism and anti-imperialism at a time when Vietnam suffered greatly from the Japanese occupation (including a famine that killed between 1 and 2 million people in the north) contributed to the rapid expansion of Vietminh "liberated areas" north of the Red River. On 19 August 1945, the revolutionaries entered Hanoi, and on 2 September Ho Chi Minh declared Vietnam's independence. The Vietminh had to fight the American-backed French for nine more years before the independence of the North, following the brilliant victory at Dien Bien Phu, became a fact. Following the 1954 Geneva Conference, Ho became leader of the Democratic Republic of Vietnam (North Vietnam).

By 1955 Ho had embarked upon guerrilla warfare against the American-backed government of the Republic of (South) Vietnam. Again, it would be a long struggle, not ending for another two decades with the complete surrender of the South in 1975. The two Indochina Wars, against France (1946–1954) and against South Vietnam and its American backers (1959–1975), validated the "people's war" strategy. Ho did not live to see the reunification of Vietnam in 1975. North Vietnamese regulars rather than guerrillas won the final victory in the South. It can be argued that the United States and the Republic of Vietnam had indeed won the war against the North after the disastrous Tet Offensive in early 1968, but by then Americans were sick of the war and gradually withdrew. The forces of the South, with some exceptions, could not match the discipline, organization, and dedication of Ho's armies.

Donald M. Seekins

See also: Vietnam Conflict; Vo Nguyen Giap

References and further reading:

Harrison, James P. *The Endless War: Vietnam's Struggle for Independence.* New York: Columbia University Press, 1989.

Lacoutre, Jean. *Ho Chi Minh: A Political Biography.* New York: Random House, 1968.

Tin, Bui. *From Cadre to Exile: The Memoirs of a North Vietnamese Journalist.* Chiang Mai, Thailand: Silkworm Books, 1995.

Hochkirch, Battle of (14 October 1758)

Rare defeat of Frederick the Great by the Austrians. Frederick the Great marched from East Prussia to defend Saxony from an Austrian advance. Weakened by the battle of Zorndorff (25 August), he could muster only 36,000 men compared to 80,000 Austrians under Field Marshal Leopold Daun, which stood east of Dresden in broken country suitable for the defense. He hoped Frederick would attack his larger force. Frederick decided to maneuver Daun's army from its position. On 9 October, after nearly a month of maneuver, Frederick encamped near the village of Hochkirch to resupply. His forces stretched from north to south with the center and right fortified by redoubts around Hochkirch and its prominent walled church.

Frederick assumed incorrectly that Daun would remain inactive. Instead, Daun, urged by his chief of staff Franz Moritz Lacy, attacked Frederick's camp. At 5:00 A.M. on 14 October their army advanced in columns and completely surprised the Prussian right flank, overrunning tents with still-sleeping soldiers. As his army disintegrated at first Frederick dismissed the shouts of his men. Although Prussian cavalry attacks failed to stem the Austrian advance, it was halted by the walled churchyard. This delay allowed James Keith and Prince Moritz of Dessau to reform the Prussian center and counterattack. Both Keith and Moritz died as their desperate charge collapsed under the weight of Austrian numerical superiority. Frederick arrived on the scene, rallied his men, but knew the battle was lost when the his left flank collapsed. By 10:00 A.M. he had withdrawn his army toward the northwest.

Hochkirch resulted from Frederick's underestimation of his enemy. It had little impact on the war. Nearly 9,000 Prussians were killed or wounded while Austria lost 7,000 dead and wounded, but Daun failed to derive any strategic advantage from his victory due to Frederick's skillful retreat.

Patrick J. Speelman

See also: Frederick the Great, King of Prussia; Seven Years' War

References and further reading:

Duffy, Christopher. *Frederick the Great: A Military Life.* London: Routledge & Kegan Paul, 1985.

Showalter, Dennis. *The Wars of Frederick the Great.* London and New York: Longman, 1996.

Weigley, Russell F. *The Age of Battles: The Quest for Decisive Warfare from Breitenfeld to Waterloo.* Bloomington and Indianapolis: Indiana University Press, 1991.

Holy Roman Empire (800–1806)

Greatest power in central Europe until the eighteenth century. Though unified under a single emperor, the empire was composed of lesser powers. As a result, aside from defense and an initial period of expansion, warfare served chiefly as a means for resolving frequent conflicts over territory and authority among them, and between them and the emperors.

Charlemagne's empire (800–814) required annual campaigns to provide land and wealth to secure the allegiance of lesser lords whose forces made up his army. This concentration of military strength in the hands of powerful subjects contributed to the fragmentation of the empire but provided a basis for the development of localized feudal hierarchies. These later proved effective against Viking and Magyar incursions of the ninth and tenth centuries.

The empire's initial avenue of military expansion was southwards into Italy. The papacy initially welcomed the imperial presence as a safeguard, but papal attitudes changed as the popes sought to establish their own authority against the emperors. The ensuing conflict contributed to the dissolution of central authority within the empire between the tenth and thirteenth centuries. Both lesser princes and rival emperors could justify their opposition to an excommunicated emperor.

The greatest expansion occurred in the east between the twelfth and fourteenth centuries into areas inhabited by Slavic peoples. This expansion included the conquest of Prussia and the Baltic provinces by the Teutonic Knights, but these territories did not come under imperial jurisdiction. By the end of the fourteenth century the empire had reached the limits of its growth.

Although the medieval empire had engaged in frequent wars with France over claims to territories lying between them, the most serious rivalry between the two came with the accession of Charles V (1516–1556). Charles was not only emperor, but also king of Spain, and France was now surrounded by Habsburg territories. French kings from Francis I (1515–1547) onward sought to undermine Habsburg power to prevent the encirclement of France.

Conflict with France further complicated the struggle for authority within the empire, particularly after the Protestant Reformation. Many territorial princes adopted Lutheranism and denied the authority of the Catholic Charles. Tensions culminated in the Schmalkaldic War between the Protestant princes (supported by France) and the emperor. Unable to confront both internal and external enemies simultaneously, Charles was forced to concede the existence of a rival Protestant religious power within the empire in the Peace of Augsburg (1555).

The tensions continued and later erupted into the Thirty Years' War (1618–1648). Attempts to consolidate imperial authority were finally defeated due to the support of resisting princes by Denmark, Sweden, and France. The Peace of Westphalia (1648) not only granted the princes autonomy and the ability to conclude military allegiances with foreign powers, but foreign intervention within the empire was constitutionally recognized. France and Sweden became guarantors of the peace and of religious equality.

From the end of the seventeenth century until the middle of the eighteenth, the empire had to confront the Ottoman Empire as well as the France of Louis XIV. The Ottomans had controlled a large part of Hungary since the sixteenth century, but the high-water mark of Ottoman expansion into the empire was reached in the Siege of Vienna in 1683. A successful war of reconquest followed under the emperor Leopold I (1658–1705), leading the Holy League. This army combined imperial forces with those of other European nations in a new crusade declared by the pope, but tension between the emperor and the princes resurfaced. The latter resented the use of the imperial army to regain purely Habsburg territory (Hungary, like Prussia, was beyond imperial jurisdiction). When the French attacked the Rhineland in 1688, most of the princes withdrew to defend Germany. Further conflict in the east became a purely Habsburg affair.

Some of the larger territories, including Prussia, succeeded in establishing themselves as rivals to the Habsburg emperors. Frederick the Great's (1740–1788) victories in the War of Austrian Succession (1740–1748) and in the Seven Years' War (1756–1762) turned Prussia into a great power and a chief rival of the Austrian Habsburgs. The empire's final half century was marked by Austro-Prussian dualism as each sought to consolidate the allegiance of the lesser princes or gain control over them. Despite this, the princes resisted assimilation and clung to imperial traditions to defend their autonomy. As a consequence, it was only the conquest and consolidation of Germany by Napoleon that finally spelled the end of the empire.

Christopher C. W. Bauermeister

See also: Austrian Succession, War of the; Austro-Swiss Wars; Austro-Turk Wars; Charlemagne; Charlemagne's Wars; Franco-German War; Frederick I Barbarossa; Frederick II; Frederick William, Elector of Brandenburg; French Revolutionary Wars; Grand Alliance, War of the; Holy Roman Empire–Papacy Wars; Hungarian War with the Holy Roman Empire; Hussite Wars; Magyars; Napoleonic Wars; Otto I, the "Great"; Schmalkaldic War; Seven Years' War; Spanish Succession, War of the; Teutonic Knights; Thirty Years' War; Valois-Habsburg Wars; Vienna, Sieges of; Viking Raids

References and further reading:

Arnold, Benjamin. *Princes and Territories in Medieval Germany.* Cambridge, UK, and New York: Cambridge University Press, 1991.

Barraclough, Geoffrey. *The Origins of Modern Germany.* New York and London: W. W. Norton and Company, 1984.

Gagliardo, John. *Germany under the Old Regime.* London and New York: Longman, 1991.

Wilson, Peter H. *German Armies: War and German Politics 1648–1806.* London: University College London Press, 1998.

Holy Roman Empire–Papacy Wars (1077–1250)

The conflicts between the Holy Roman Empire and the papacy were defining events during the eleventh through thirteenth centuries as the papacy became a formidable power in Europe. As part of these conflicts, Holy Roman emperors and popes, and their respective supporters, wrote countless documents debating such topics as the source of the emperor's power: Was it from election by the German princes or from the pope who crowned him? This war of words carried over into physical altercation. The emperors wished to consolidate power in northern Italy and, later, attain the kingdom of Sicily. The papacy, as the authority in central Italy, did all it could to repel this incursion.

The conflict began in the ambitions of Emperor Henry IV (1071–1106) and Pope Gregory VII (1073–1085). Gregory had continued the reforms of Pope Leo IX, influenced by monastic movements, which focused on such corruptions as lay investiture. Henry invested—that is, appointed—his own bishops in Germany, and Gregory excommunicated him. Henry begged for and received reconciliation from Gregory at Canossa in January 1077, but in March a number of German dukes rebelled against the emperor and elected another king. Henry, with military victories at Mellrichstedt, Flarcheim, and the river Elster, reasserted his claim to the throne.

Gregory excommunicated Henry a second time, only to be deposed himself by Henry in 1080. Henry placed his own pope in power, but neither that antipope nor his successors posed any major threat to the cardinal-elected popes. Between 1090 and 1092, Henry fought Matilda of Tuscany and Duke Welf of Bavaria, allies of Pope Urban II. On 31 December 1105, Henry's son, Henry V, forced his father to abdicate the throne. In 1122, Henry V and Pope Clement II reached an agreement concerning lay investiture at the Concordat of Worms, nominally ending lay investiture but allowing the emperor to veto the election of a bishop by refusing to accept feudal homage from him.

The larger papal-imperial conflict escalated with Frederick I (Barbarossa, 1152–1190). In 1159, Frederick tried to consolidate his power in northern Italy and to place his own pope in power. The cardinal-elected pope, Alexander III (1159–1181), the Norman king of Sicily, the Byzantine emperor, and the northern Italian resistance formed the Lombard League to keep Frederick at bay.

After six separate campaigns in Italy, Frederick finally withdrew, maintaining some influence in Lombardy. Frederick's son, Henry VI, succeeded him and conquered the whole of Italy, including Sicily and the Papal States. When Henry died in 1197, he failed to make the empire hereditary as the German princes reclaimed the right to elect the German king. Pope Innocent III (1198–1216) took advantage of the situation and chose another man as successor. Henry's son Frederick was too young; and when the German-elected king died in 1208, Otto IV, supported by the papacy, became Holy Roman Emperor.

Otto defied Innocent and revived imperial authority in northern Italy. Innocent gained support from King Philip Augustus of France, accepted Henry's son, Frederick II (1215–1250), whom he had earlier bypassed as emperor, and, ultimately, stopped Otto. As emperor and as king of Sicily, Frederick II also turned against the papacy and continued the policy of invading northern Italy. He was excommunicated, first by Pope Gregory IX and then by Innocent IV as the conflict reached new heights; but he could not vanquish the papacy. After Frederick's death in 1250, subsequent Holy Roman emperors could not gain enough power to threaten the papacy, and the papal-imperial conflict faded away.

Christopher P. Goedert

See also: Frederick I Barbarossa; Frederick II

References and further reading:

Jedin, Hubert, and John Dolan, eds. *Handbook of Church History.* Vols. 3 and 4. Trans. Anslem Biggs. New York: Herder & Herder, 1969.

Morris, Colin. *The Papal Monarchy: The Western Church from 1050 to 1250.* Oxford, UK: Clarendon Press, 1989.

Tierney, Brian. *The Crisis of Church and State, 1050–1300.* Englewood, NJ: Prentice-Hall, 1964.

Tout, T. F. *The Empire and the Papacy, 918–1273.* London: Rivingtons, 1965.

Honduran-Nicaraguan War (1907)

Honduras invaded Nicaragua in January 1907 on the pretext that Nicaragua was aiding Honduran rebels who were trying to overthrow President Policarpo Bonilla. Nicaraguan dictator Jose S. Zelaya denied involvement in the affair and offered to submit the dispute to an international tribunal. Negotiations were unsuccessful and both sides continued fighting until the Nicaraguan army defeated the invaders. El Salvador and Guatemala threatened to intervene and place former Honduran president, Terencio Sierra, in office. Mexico and the United States persuaded all five Central American nations to meet in Washington to resolve their differences.

The conference met from 14 November until 20 December and was noteworthy for the establishment of the Central America Court of Justice to resolve disputes between the five member states. In the summer of 1908, Nicaragua charged that Guatemala and El Salvador were supporting another attempt from Honduras to overthrow Zelaya and presented its case to the court. After an investigation, the court ordered all four nations to cut the size of their armies and to agree not to intervene in each other's internal affairs. All parties agreed and peace was restored, making this one of the first successful resolutions of war by an international tribunal.

George M. Lauderbaugh

References and further reading:

Woodward, Ralph Lee, Jr. *Central America: A Nation Divided.* New York: Oxford University Press, 1976.

Hong Xiuquan (1814–1864)

Charismatic founder of the Taiping movement and leader of the Taiping Rebellion. Hong was born as Hong Fuoxiu in Huaxian, Guangdong Province, on 1 January 1814, of a Hakka farming family.

At fifteen Hong passed the preliminary level of imperial examinations; he taught school from 1830 to 1843 but in four tries did not manage to pass the advanced civil-service examination. He received Christian pamphlets and, after falling dangerously ill, dreamt that he had been approached by God and by Elder Brother Jesus and exhorted to fight evil. After a brief period of Bible study in 1847, Hong joined Feng Yünshan, founder of the Society of God Worshippers, at Thistle Mountain in Guangxi Province and declared himself the Second Son of God. By 1850 he had about 20,000 militant anti-Manchu followers, mostly Hakkas and local miners. They rebelled in summer 1850. On 11 January 1851 Hong declared the new dynasty of Taiping Tianguo, with himself as the Heavenly King (Tian Wang). He achieved initial successes, but after 1856, mostly reverses. The military ineptitude of his own leadership was exceeded only by the fragmentation and unpreparedness of the imperial forces.

Increasingly debauched, withdrawn, and irrational, Hong continued to lead the revolution until April 1864 when he mysteriously fell ill. Trusting in divine providence rather than sound military policy as the Qing army besieged the Taiping capital, Nanjing, he died in his palace on 1 June 1864, possibly by suicide but more likely from disease.

Eric v. d. Luft and Sarah Luft

See also: Chinese Imperial Wars; Gordon, Charles George; Li Hongzhang; Religion and War; Taiping Rebellion; Wolseley, Garnet Joseph, Viscount; Yang Xiuqing; Zuo Zongtang

References and further reading:

Anderson, Flavia Giffard. *The Rebel Emperor.* Garden City, NY: Doubleday, 1959.

Boardman, Eugene Powers. *Christian Influence upon the Ideology of the Taiping Rebellion, 1851–1864.* New York: Octagon, 1972.

Hamberg, Theodore. *The Visions of Hung-Siu-tshuen, and Origin of the Kwang-si Insurrection.* New York: Praeger, 1969.

Spence, Jonathan D. *God's Chinese Son: The Taiping Heavenly Kingdom of Hong Xiuquan.* New York: W. W. Norton, 1996.

Honors and Awards, Military

From earliest times honors have been bestowed on victorious warriors. Many of the earliest known pictographs illustrate the exploits of men in battle. Ancient grave sites reveal ritually buried warriors complete with their weapons and armor. Successful military leaders were showered with lands, riches, and the spoils of war. Even simple soldiers shared in the treasure of conquest. There are stunning examples of Greek, Roman, Chinese, Egyptian, and Persian art displaying returning armies bearing captured arms and slaves from their vanquished enemies.

The ancient Greeks awarded arms and armor to victorious commanders. The Romans, who also crowned their successful generals with laurel wreaths of silver and gold, adopted this practice. Later they awarded large circular medallions, or phalerae, that were mounted on the breastplate of the recipient. These medallions were also awarded to entire Roman legions that distinguished themselves in battle. The phalerae, mounted on the legion flagstaff, served as a public symbol of distinction and a symbol of honor to the legion.

In time, medals, orders, and decorations were authorized to honor the heroism of individual soldiers. For military formations the banner or flag displayed the unit's collective honors. Flags served an important purpose in early warfare, providing the soldier with a point to rally in the confusion of battle. The first flags bore heraldic devices of the leaders for whom the soldiers fought. With the advent of body and horse armor in the Middle Ages, soldiers themselves wore these devices.

The armies of Spain under the royal flag and the banners of the church carried out the conquest of the Americas. The armies of Pizarro and Cortes had their own standards, as did the native armies they faced. At the Battle of Otumba (1520) the Aztec standard, known as Quetzalteopamil, composed of a dazzling representation of the sun surrounded by quetzal plumes, was lashed to the back of the Aztec general. The capture of this standard by a young conquistador, Juan de Salamanca, demoralized the Aztec troops and led to an overwhelming Spanish victory. The Aztec banner was incorporated into Salamanca's coat of arms.

The creation of modern, state-organized and supported European armies in the sixteenth and seventeenth centuries saw the size and function of military units stabilize. King Gustavus II Adolphus of Sweden recognized the value of experienced and cohesive military formation. His administrative and tactical innovations led to the establishment of the modern regiment. One of the key elements in creating a collective identity for a regiment was the regimental standard.

In England, the regimental system was introduced with the restoration of the monarchy in 1660. The years that followed were filled with both domestic and international strife. Between 1660 and 1747, British noblemen received royal warrants to raise and maintain regiments for use by the monarch. Some distinguishing badge was allowed on the regimental standard acknowledging the sponsor. In 1747, the crown forbade the use of these private badges on the standard. Regulations also stipulated that two flags would be used by the regiment, the King's (or Queen's) First Colour, or the Great Union, and the regimental color. The regimental flag had as its background the same hue as the facings (collars, cuffs, and piping) on the regimental uniform.

King William III granted the 18th Regiment of Foot (the Royal Irish) the right to have the Lion of Nassau and the motto VIRTUTIS NAMURCENSIS PROEMIUM emblazoned on its regimental color in recognition of its valor at the Battle of Namur in 1695. This was the first battle honor granted by an English monarch.

In Britain, battle honors are the gift of the sovereign to particular regiments for distinguished service in a specific engagement. Honors may take the form of a badge, a motto, or, more often, a name embroidered on a ribbon sewn on the field of the regimental color. Only battles that are deemed victories are eligible as honors. Battles fought in civil wars, like the English Civil War and the American War of Independence, are not carried on the color. With well over 100 battle honors authorized for World War I, honors on a regimental color were limited to 10 from any single war. Army units that participate in almost every engagement, like the Royal Artillery, do not have individual battle honors. However, the badge and colors of the Royal Artillery both bear witness to its wide service through its motto, UBIQUE (Everywhere).

In the United States, militia regiments had their own standards or color before the establishment of the United States Army in 1775. After the revolution, two flags were authorized for each infantry regiment, a regimental standard of a blue field with the seal of the United States in the center, below which was a riband bearing the regiment's designation and a national color, the Stars and Stripes. Cavalry regiments had regimental color but did not carry the national color. They used a smaller flag, a guidon, similar to the national color, with 13 stripes and a blue field and no set pattern for the stars but cut like a swallowtail.

Shortly after the opening of the American Civil War, the War Department authorized the inscribing on its national color the names of engagements in which a regiment distinguished itself. It was hoped that units would regard their colors as representing the regiment's honor and rally to protect the colors in battle. To underscore the importance of the colors, the War Department also encouraged the award of the Medal of Honor to flag bearers who saved a regiment's flags from capture.

As battle followed battle, there was little room left on most regimental national colors for additional honors. The War Department authorized silver rings engraved with battle honors affixed to the staff bearing the regimental color.

Initially, the regiments themselves chose battles commemorated by honors. An inventory of honors was undertaken by the army in 1878 to verify and standardize the criteria for their use. This process was not completed until 1919, when 76 battles were deemed suitable for recognition. The next year a new method of displaying honors was introduced: ribbon streamers in the colors of the appropriate campaign medal, embroidered with the name of the engagement and suspended from the finial of the staff bearing the regimental color. Today, the United States Army recognizes more than 172 battle honors.

A regiment's participation in a particular battle can also be incorporated into its regimental badge. The British Army's Royal Regiment of Wales was created through the amalgamation of several regiments, one of which was the South Wales Borderers. In 1879, that regiment, then known as the 24th Regiment of Foot, lost 599 officers and men in the Battle of Isandhlwana to Zulu regiments. Queen Victoria placed a wreath of immortelles on the regimental Queen's Colour and commanded that a silver replica of the wreath be carried on the flagstaff thereafter. The cap badge of the South Wales Borderers, and later the collar badge of the Royal Regiment of Wales, is surrounded by an unbroken wreath of immortelles to commemorate the queen's gesture.

It is interesting to note that two of the three Victoria Crosses awarded for Isandhlwana went to Lieutenants Melville and Coghill of the 24th Regiment, ordered to take the Queen's Colour to safety when it was obvious the column was doomed. Their bodies were later found in a nearby river alongside the Colour.

The distinctive insignia used by the regiments of the United States Army may also incorporate elements of a regiment's history and battle honors into their design. For example, the distinctive insignia of the 19th Infantry is a shield of azure with an infantry bugle of 1861 enclosing the numeral 19 with three white stars above the bugle. Below the shield is a riband bearing the unit motto. The bugle with the number is a reproduction of the 19th Infantry's Civil War insignia when it received the nickname "The Rock of Chickamauga," which also serves as its motto. The three stars represent service in the Civil War, the war with Spain, and the Philippine Insurrection.

In many of the world's armies, units may be awarded decorations for particularly meritorious service. The United States, the Philippines, and the Republic of Korea each have a Presidential Unit Citation, designed to honor entire units. Several regiments of the United States Army were awarded the French Croix de Guerre and Medaille Militaire in recognition of their service in the first and second world wars. The Soviet government often awarded orders and decorations to outstanding regiments during World War II.

Honors and awards are intended to recognize exceptional service and build loyalty. This is true of medals and decorations awarded to individual soldiers as well as battle honors, flags, and heraldic devices bestowed upon military units. They bind the recipient to the government that grants the honor. In the case of military units, awards and honors create a bond between the soldiers in those units by paying tribute their common history, sacrifice, and spirit.

The greatest threat to the value of such devices and emblems is their occasional wholesale awarding and subsequent cheapening, as was the case when the U.S. Army after the Grenada incursion awarded more medals than there were troops involved! There is also the danger that they may be awarded to the unworthy, as in one-party states, where commanders can cover their chests with medals that bear little correlation with their actual battlefield experience. Wise government avoids this cheapening of their military awards and honors, and even the Soviet Union and Nazi Germany reserved special recognition devices that, at least in wartime, had nothing to do with a soldier's politics.

The physical cost of producing these bits of cloth and metal is infinitesimal. But such awards and honors repay their cost many times over in troop and unit efficiency, motivation, and morale.

Eric Smylie

See also: British Military; Cortes, Hernando; German Army; Gustavus II Adolphus; Pizarro, Francisco; U.S. Army

References and further reading:

Baker, Anthony. *Battle Honours of the British and Commonwealth Armies.* London: Ian Allen, 1986.

Edwards, T. J. *Regimental Badges.* London: Charles Knight & Co., 1974.

Farwell, Byron. *Mr. Kipling's Army.* New York: Norton, 1981.

Joslin, E. C., A. R. Litherland, and B. T. Simpkin. *British Battles and Medals.* London: Spink & Son, 1988.

Lemonofides, Dino. *British Infantry Colours.* Edgeware, Middlesex, UK: Altmark Publishing, 1971.

Mahon, John K., and Romana Danysh. *Army Lineage Series, Infantry.* Part I, *Regular Army.* Washington, DC: Office of the Chief of Military History, United States Army, 1972.

Taylor, Arthur. *Discovering Military Traditions.* Tring, Hertsfordshire, UK: Shire Publications, 1969.

Wilson, John B. *Campaign Streamers of the United States Army.* Arlington, VA: Association of the United States Army, 1995.

Hood, John Bell (1831–1879)

Confederate general, capable at the brigade or division level, but unqualified to command larger forces. Hood was born in Owingsville, Kentucky, on 1 June 1831. After graduating from the U.S. Military Academy at West Point in 1853, he served with the infantry in Missouri and California until assigned to the Second Cavalry in Texas in 1855. He took a Comanche arrow in the left hand at Devils River, Texas, on 20 July 1857.

Hood resigned as first lieutenant from the U.S. Army on 16 April 1861 and was immediately commissioned captain in the Confederate cavalry. On 30 September he was named colonel of the 4th Texas Infantry and on 2 March 1862 brigadier general of the Texas Brigade, which soon became one of the Confederacy's toughest fighting forces. It excelled during the Seven Days' Battles, especially at Gaines's Mill on 27 June. Hood commanded a division of the Army of Northern Virginia, including his Texas Brigade, at Second Bull Run, Antietam, and Fredericksburg. His stellar performances earned his promotion to major general on 10 October.

His left arm was crippled near Little Round Top on the second day at Gettysburg, and he lost his right leg at Chickamauga. While recovering from the amputation, he was promoted to lieutenant general on 1 February 1864.

Hood served under Joseph Johnston trying to stop William T. Sherman's march to the sea. Jefferson Davis, dissatisfied with Johnston in this campaign, brevetted Hood full general on 18 July and placed him in command. Hood's immediate and reckless attacks on Sherman showed that Davis was mistaken. Beaten by Sherman, Hood retreated into Tennessee to disrupt Union supply lines. Crushed by John McAllister Schofield at Franklin, Tennessee, on 30 November and almost annihilated by George H. Thomas at Nashville on 15–16 December, he was relieved of command at his own request on 23 January 1865. Fleeing toward Texas, he surrendered in Natchez, Mississippi, on 31 May. He died in New Orleans on 30 August 1879.

Eric v. d. Luft

See also: American Civil War; Antietam/Sharpsburg; Atlanta, Battles Around; Bull Run, Second/Manassas Junction; Chickamauga, Battle of; Fredericksburg; Gettysburg; Johnston, Joseph Eggleston; Lee, Robert Edward; Longstreet, James; Nashville, Battle of; Seven Days' Battles; Sherman, William Tecumseh; Thomas, George Henry

References and further reading:

Bailey, Anne J. *The Chessboard of War: Sherman and Hood in the Autumn Campaigns of 1864.* Lincoln: University of Nebraska Press, 2000.

Dyer, John Percy. *The Gallant Hood.* Indianapolis, IN: Bobbs-Merrill, 1950.

Hay, Thomas Robson. *Hood's Tennessee Campaign.* Dayton, OH: Morningside, 1976.

McMurry, Richard M. *John Bell Hood and the War for Southern Independence.* Lincoln: University of Nebraska Press, 1992.

Hooker, Joseph (1814–1879)

Aggressive, popular, and sometimes reckless Union field commander in the American Civil War. Hooker was born in Hadley, Massachusetts, on 13 November 1814. Assigned to artillery after graduating from the U.S. Military Academy at West Point in 1837, he fought in the Seminole Wars and, under both Zachary Taylor and Winfield Scott, in the Mexican-American War. When the Civil War began he was colonel of the California militia.

After witnessing the Union loss at First Bull Run as a civilian and thinking that he could do better, he raised a brigade of volunteers and commanded a division at Williamsburg, Virginia, on 5 May 1862. Promoted to major general of volunteers, he fought well at Seven Pines, in the Seven Days' Battles, and at Second Bull Run and South Mountain. Wounded at Antietam, he was promoted to regular brigadier general on 20 September, led a corps at Fredericksburg, and replaced Ambrose Burnside as commander of the Army of the Potomac on 26 January 1863.

Uncharacteristic hesitancy made "Fighting Joe" the loser at Chancellorsville. His usual cockiness might have beaten Robert E. Lee, whom he outnumbered about 75,000 to 50,000. Abraham Lincoln replaced Hooker with George G. Meade three days before Gettysburg.

Serving under William S. Rosecrans and Ulysses S. Grant, Hooker achieved an important victory with the capture of Lookout Mountain, Tennessee, on 24 November. For this he was brevetted major general. Reassigned under William T. Sherman, he distinguished himself during the siege of Atlanta. Disgruntled when passed over for command of the Army of the Tennessee after James Birdseye McPherson's death on 24 July 1864, he asked to be relieved of field service and spent the rest of the war behind a desk.

Hooker was a man of intemperate morals and routinely allowed prostitutes to follow his camp. There is no truth to the prevalent notion that the slang term *hooker* for prostitute derives from his name, but the coincidence of his name with the preexistent word certainly helped to popularize it.

Hooker retired in 1868 when he suffered a stroke. He died in Garden City, New York, on 31 October 1879.

Eric v. d. Luft

See also: American Civil War; Antietam/Sharpsburg; Atlanta, Battles Around; Bull Run, First/Manassas; Bull Run, Second/Manassas Junction; Burnside, Ambrose Everett; Chancellorsville, Battle of; Chattanooga, Battle of; Fredericksburg; Gettysburg; Grant, Ulysses Simpson; Jackson, Thomas "Stonewall"; Lee, Robert Edward; Lincoln, Abraham; McClellan, George Brinton; Meade, George Gordon; Mexican-American War; Pope, John; Rosecrans, William Starke; Scott, Winfield; Seven Days' Battles; Sherman, William Tecumseh; Thomas, George Henry

References and further reading:

Hebert, Walter H. *Fighting Joe Hooker.* Lincoln: University of Nebraska Press, 1999.

Horseshoe Bend, Battle of (27 March 1814)

Decisive battle in the Creek War. Andrew Jackson marched about 4,000 regulars, volunteers, and Cherokees south from Fort Strother, Alabama, in March 1814 to seek and destroy the renegade "Red Stick" Creeks and their Spanish allies. Under Chief Red Eagle, alias William Weatherford, about 1,200 Creeks had fortified a small peninsula formed by a "horseshoe bend" in the Tallapoosa River northeast of Montgomery, Alabama. With earthworks across the neck of the thickly wooded peninsula, the Creeks were well equipped to withstand bombardment, assault, or siege.

The morning of 27 March, Jackson surrounded the Creek position and dispatched the Cherokees to swim the river and steal the Creek canoes. That done, he fired his only gun, a six-pounder, at the earthworks to signal attack. First, John Coffee's Tennessee militia attacked amphibiously on three sides in the stolen canoes while the six-pounder provided covering fire as rapidly as possible, then Jackson led about 600 regulars in a frontal assault against the earthworks. Because Coffee had already engaged the barricade's defenders from the rear, Jackson's infantry met little resistance.

About 700 Creek warriors died. Jackson's casualties were 111 wounded, including Sam Houston, and 26 killed. Red Eagle was not present. Jackson marched downstream toward the Creek "Holy Ground" near the confluence of the Coosa and Tallapoosa Rivers. When Red Eagle saw that his headquarters were doomed, he offered unconditional surrender and Jackson gave him a generous settlement. Living thereafter as Weatherford, Red Eagle became a peaceful Alabama planter.

Eric v. d. Luft

See also: Creek War; Houston, Samuel; Jackson, Andrew; War of 1812

References and further reading:

Brantley, William Henderson. *Battle of Horseshoe Bend in Tallapoosa County, Alabama, March 27, 1814.* Birmingham, AL: Southern University Press, 1969.

Holland, James Wendell. *Andrew Jackson and the Creek War: Victory at the Horseshoe.* Tuscaloosa: University of Alabama Press, 1968.

Mackenzie, George C. *The Indian Breastwork in the Battle of Horseshoe Bend: Its Size, Location, and Construction.* Washington, DC: Division of History, U.S. Office of Archeology and Historic Preservation, 1969.

Houston, Samuel (1793–1863)

American and Texan field commander and politician. Born near Lexington, Virginia, on 2 March 1793, Houston grew up in Blount County, Tennessee, and lived with the Cherokees from 1808 to 1811. Commissioned ensign in the regular army in 1813, he served under Andrew Jackson in the Creek War and was wounded in the left thigh and right shoulder during the assault on the earthworks at Horseshoe Bend. From 1815 until he resigned in May 1818, he was a military agent to the Cherokees, usually dressing like them and tak-

Photograph of Samuel Houston, c. 1860. (Library of Congress)

ing their side on almost all issues. He began practicing law in Nashville in 1818, became major general of the Tennessee militia in 1821, a member of Congress from Tennessee in 1823, and governor of Tennessee in 1827, but suddenly resigned in 1829, despondent when his wife, Eliza Allen, fled back to her father.

Houston went west and returned to the Cherokees, who called him "Big Drunk." He lived in despair for three years, but in 1832, with his sense of purpose restored, he traveled to Washington to lobby for the Cherokees with the Jackson administration. He impressed Jackson by beating up an anti-Jackson congressman. Jackson pardoned him immediately after his conviction and sent him on a fact-finding mission to Texas late in 1832. Houston settled there and soon became an activist for Texan rights and separate statehood within the nation of Mexico.

In November 1835 Houston became commander in chief of the Texan army and on 2 March 1836 he cosigned the Texas Declaration of Independence. His battle plans and tactics crushed the Mexicans at San Jacinto and won independence for Texas. He was president of the Republic of Texas from 1836 to 1838 and from 1841 to 1844. After the United States annexed Texas in 1845, Houston represented Texas as a senator from 1846 to 1858 and served as governor from 1859 until he was overthrown in March 1861 for refusing to support secession. He died in Huntsville, Texas, on 26 July 1863.

Eric v. d. Luft

See also: Alamo; Creek War; Horseshoe Bend, Battle of; Jackson, Andrew; San Jacinto; Santa Anna, Antonio López de; Texas War of Independence; War of 1812

References and further reading:

Braider, Donald. *Solitary Star: A Biography of Sam Houston.* New York: Putnam, 1974.

De Bruhl, Marshall. *Sword of San Jacinto: A Life of Sam Houston.* New York: Random House, 1993.

James, Marquis. *The Raven: A Biography of Sam Houston.* Norwalk, CT: Easton, 1988.

Williams, John Hoyt. *Sam Houston: A Biography of the Father of Texas.* New York: Simon & Schuster, 1993.

Hue, Battle of (31 January–2 March 1968)

Major battle of the Tet Offensive during the Vietnam conflict. As part of the Tet Offensive, North Vietnamese and Vietcong main force units attacked the old imperial capital of Hue in northern South Vietnam. The effort to retake the city soon became the longest and bloodiest of all of the Tet Offensive fighting. To appease Vietnamese sensibilities, the old imperial citadel was off-limits to U.S. combat units, and the enemy quickly seized control. During the period of Communist control, cadres rounded up various "enemies of the people," apparently conducted "people's trials," and murdered them in a bloodbath.

South Vietnamese and American forces quickly responded. The South Vietnamese 1st Division immediately counterattacked and requested U.S. support. Elements of the U.S. 1st Cavalry Division and 101st Airborne Division blocked off the city on the west to prevent possible enemy reinforcements. Still, General William Westmoreland limited reinforcements, believing that the siege at Khe Sanh potentially was a more critical situation and wanting to preserve his strategic reserve in the region. Meanwhile Army of the Republic of Vietnam (ARVN) troops and U.S. Marines engaged in block by block, house by house fighting to regain the old capital. The U.S. Marines took one casualty for every yard gained, and one combat reporter, Don Oberdorfer of the *Washington Post,* called it "quite possibly, the longest and bloodiest single action of the Second Indochina War."

On 2 March, the Military Assistance Command, Vietnam declared the battle officially over. More than half of the city had been destroyed in the fighting to retake it; Communist military units had suffered more than 5,000 casualties, while

A U.S. Marine carries a Vietnamese woman to safety during the Battle of Hue, the longest and bloodiest of all the Tet Offensive battles. (National Archives)

Vietnamese, U.S. Marine, and U.S. Army casualties combined exceeded 3,500 killed and wounded.

Charles Dobbs

See also: Tet Offensive; Vietnam Conflict; Westmoreland, William

References and further reading:

Hammel, Eric M. *Fire in the Streets: The Battle for Hue, Tet 1968.* Chicago, IL: Contemporary Books, 1991.

Nolan, Keith William. *Battle for Hue: Tet 1968.* Novato, CA: Presidio Press, 1983.

Smith, George. *The Siege at Hue.* Boulder, CO: Lynne Rienner, 1999.

Hukbalahap Revolt (1945–1959)

One of the Philippines' periodic agrarian rebellions. The Hukbalahap revolt's roots were in the conditions of peasant distress before World War II, though the Hukbalahap (the term essentially meaning People's Army) came into being in 1942, when fragments of left-wing guerrilla units fighting the Japanese came together under Luis Taruc as their military commander. Centered in Luzon, these remnants grew into an army of 15,000 by the end of the war, despite competition with partisan forces organized by the U.S. Army.

Seen as little more than bandits by the newly established government of the Philippines, the Huks reformed their political organization and established plans to create a people's republic. This course of action was greatly assisted by the distressed state of the country in the wake of World War II and further abetted by the counterproductive fashion in which Manila fought. Against the backdrop of a kleptocratic government, occasional infantry sweeps of battalion strength were conducted with indiscriminate violence and looting, driving the population into the arms of the rebels. By 1949, 37,000 government soldiers faced a movement with a population base of a million supporters dominating half of Luzon. Very little held the Huks back except their own material weaknesses, and the occasional overreaching atrocity, such as the assassination of the revered widow of wartime president Manuel Quezon in a convoy ambush near Manila.

In 1950, the ascension of Ramón Magsaysay to the post of defense chief and the outbreak of the Korean War led to a change of fortune. It became possible to finance an expansion of the war while the necessary leadership was available to reform the Filipino military.

By increasing troops levels from 10 to 26 battalions, and providing adequate logistical and air support, enough forces were available to cover the affected region. These troops mounted continuous patrols so as to give the guerrillas no sanctuary, while specialized units mounted deep-penetration missions to further add to the disruption. This went hand-in-hand with a civic action program designed to provide training and land for those in the areas most dominated by the Huks, beginning the process of winning the population back to Manila. Sophisticated psychological warfare also alienated the population from the Huks and made the rebels themselves doubt their cause. By 1953, the Huks had been reduced to 2,300 active fighters in uncoordinated bands, while Luis Taruc himself surrendered in 1954—and became a born-again Christian.

The suppression of the Huks was seen as a model of war in the postcolonial world. The essential problem was that the course of this campaign depended on the existence of exemplary leadership—Magsaysay and U.S. intelligence adviser Edward Lansdale were lionized; Magsaysay's death in a 1957 airplane crash was a near-catastrophe for the Philippines—a condition that did not obtain in the Vietnam conflict or in future Filipino insurgencies.

George R. Shaner

See also: Cold War; Guerrilla/Partisan/Irregular Warfare; Indochina Wars; Magsaysay, Ramón; Malayan Emergency

References and further reading:

Currey, Cecil B. *Edward Lansdale: The Unquiet American.* Boston: Houghton Mifflin, 1988.

Greenberg, Lawrence M. *The Hukbalahap Insurrection: A Case Study of a Successful Anti-Insurgency Operation in the Philippines, 1946–1955.* Washington, DC: U.S. Army Center of Military History, 1987.

Kerkvliet, Benedict J. *The Huk Rebellion: A Study of Peasant Revolt in the Philippines.* Berkeley: University of California Press, 1977.

Hundred Years War (1337–1453)

Most important Anglo-French conflict of the Middle Ages. The war had complex roots. It was the last and more violent stage of an old conflict arising due to the fact that the English king was, at the same time that he was king of England, a vassal of the French king for Guyenne. By the early fourteenth century, there were many new areas of confrontation as well: who was to control Flanders (vital for the English wool market); the succession war of Brittany; the revival of the "Auld Alliance" between the Scots and the French that caught the English in a vise; who was to control the Channel and North Sea; and, from 1328, the French dynastic conflict.

The last son of King Philippe IV ("the Fair") had died without heir. The closest male in a collateral line was Edward III of England, Philippe's grandson through Philippe's daughter (Isabelle of France, who had married King Edward II of England). Edward III thus had a valid claim on the French crown, but French lords were unwilling to accept the idea that Edward might become their king. In order to reject his claim, French lawyers drew on old Frankish law (Salic law), which stated that property could not descend through a female.

Philippe VI de Valois, Philippe's IV nephew, was consequently chosen as king. Despite the disparity in power (France was the most populous and wealthiest country in western Europe), England was ready to support the claim of its popular king. In 1337, Philippe VI seized Edward's fiefs in France, marking the beginning of the war. The conflict can be divided in four phases.

During the first phase, 1337–1360, France suffered repeated military disaster (Sluys in 1340, Crécy in 1346, Poitiers in 1356). The 1360 Treaty of Brétigny gave England a third of French territory.

During phase two, 1360–1415, French armies gradually recovered lost territory. As a consequence, at the end of Charles V's reign (1380), England had lost most of its holdings in the French interior, ending up controlling only a few ports. Neither France nor England was able to prosecute the war to its end due to internal difficulties. In France, the intermittent insanity of Charles VI ("the Mad," 1380–1422) prompted a power struggle between the dukedoms of Orleans and Burgundy. This grew into a full-scale civil war, considerably weakening the kingdom. England also had to face internal instability, including the Wat Tyler uprising in 1381, but central authority was reestablished under Kings Henry IV (1399–1413), and Henry V (1413–1422).

During phase three, 1415–1429, Henry V seized the opportunity presented by French anarchy. Henry crushed the French royal army at Agincourt (1415). He conquered Normandy and used diplomacy (an alliance with the Burgundian party) to force Charles VI to sign the Treaty of Troyes (1420), which disinherited Charles's son. Henry then married Charles's daughter. His infant Henry VI was declared king of England, as well as of France (1422). But most of southern France remained loyal to Charles VII, the Dauphin, nicknamed "King of Bourges" (his tiny capital). England controlled northern France and had nearly rid itself of Charles VII when Jeanne d'Arc raised France's spirits again.

During the years 1430–1453, the final stage of the Hundred Years War, English troops were regularly defeated by French armies (Orleans and Patay, 1429; Formigny, 1450; Castillon, 1453). In 1435, Charles VII signed the Peace of Arras with Burgundy, destroying the Burgundian-English alliance. France had rallied around the idea that it was a united nation and that the *goddons* (English) were foreigners to be swept away. Between 1449 and 1453, England lost all his territories save for the tiny foothold of Calais. The Hundred Years War had definitively come to an end by 1475, when Louis XI of France prevented an invasion by bribing Edward IV into returning to England.

This Hundred Years War was a chaotic conflict between princes, with no definitive peace at its end. It led to political unrest (War of the Roses in England, War of the Public Good in France) but also gave birth to a protonationalism. Fewer than 20 major battles occurred. The real war was comprised of innumerable skirmishes. Towns were more often taken by surprise, or treason, than by sieges. English raids (five between 1339 and 1360) often had a greater psychological than physical impact. The employment of a scorched earth policy by French kings proved a useful deterrent.

The conflict had a disastrous impact on French wealth. The defeats of the fourteenth century, combined with the Black Death, paralyzed the economy. It was the middle of the fifteenth century before trade returned to the levels of 1300.

The war proved to be too costly for England's 3 million inhabitants as well. As long as the booty was flowing from France, English kings enjoyed the support of Parliament, but any setback could provoke a change of dynasty (Lancaster, in 1400; then York, in 1455). England failed to foresee the military vicissitudes of the fifteenth century.

During the entire war, the English relied on mobility and a powerful archery, but English mobility was checked by the hundreds of fortified places that a wealthy France could afford to build. Archery won battles as long as French knights looked for glory and fame, but the emergence of professional soldiers (1445), in a standing French royal army, resulted in tactical and technical superiority for the French.

French armies of the late war could choose when to give battle (Formigny or Castillon). The French standing army was paid by the first modern tax system. The French king no longer had to live from the proceeds of his estates. A permanent tax gave him the power to make policy. The idea of the nation and of a modern, centralized monarchy is the principal legacy of the Hundred Years War.

Gilles Boué

See also: Agincourt, Battle of; Crécy, Battle of; Edward, the Black Prince; Edward III; Henry V, King of England; Joan of Arc; Orleans, Siege of; Poitiers, Battle of

References and further reading:

Allmand, Christopher. *The Hundred Years War.* Cambridge, UK: Cambridge University Press, 1988.

Contamine, Philippe. *War in the Middle Ages.* Trans. Michael Jones. New York: Blackwell, 1984.

———. *La Guerre de Cent Ans.* 7th ed. Paris: PUF, 1992.

Sumption, Jonathan. *The Hundred Years War.* 2 vols. Philadelphia: University of Pennsylvania Press, 1990–1999.

Hungarian Civil Wars (1526–1547)

Wars between rival claimants of the Hungarian throne, ending in the division of the country between the Habsburg and Ottoman empires and the dependent Principality of Transylvania. The death of Louis II at the Battle of Mohács (1526) left the Hungarian throne vacant. A majority of the nobles elected the Transylvanian Vajda János Zápolyai king in October 1526; a smaller number, joined by the chancellor, supported Austrian Archduke Ferdinand Habsburg, brother of Emperor Charles V. Reinforced by German mercenaries after Charles's conquest of Rome (1527), Ferdinand quickly drove Zápolyai out of the country.

Rather than abdicate, Zápolyai appealed to Sultan Süleyman I for aid. Süleyman recognized Zápolyai as the legitimate king and led an army into Hungary to reestablish his position. The Ottoman army easily pushed Ferdinand's forces out of central Hungary but failed to capture Vienna (1529). A second campaign by Süleyman against Ferdinand's capital in 1532 was stopped by the determined resistance of the town of Köszeg (Gün).

In the following years, while the two kings' armies competed for control of the country, the Ottomans expanded their base in Hungary by occupying Slavonia and placing a garrison near Buda. As it became evident that the Ottomans alone stood to profit from the continued division of the kingdom, Ferdinand and Zápolyai worked to negotiate a settlement. By the Treaty of Várad (1538), Ferdinand recognized Zápolyai's claim and pledged to support him with imperial forces; in return, Zápolyai named Ferdinand his heir to the throne. At Zápolyai's death in 1540, however, his treasurer György Martinuzzi, bishop of Várad, refused to honor the agreement and had Zápolyai's infant son elected King János II. Ferdinand's forces were too small to occupy the kingdom and failed in two attempts to capture Buda.

In August 1541 Süleyman marched to the capital, declared himself János's guardian, and occupied the castle. He made Buda the administrative center of a new Ottoman pashalik and gave Transylvania and the lands east of the Tisza River to János to hold as a dependent principality.

After a failed attempt by Ferdinand to recapture Buda in 1542, Süleyman carried out another campaign in Hungary, conquering Siklós, Székesfehérvár, Esztergom, and Szeged (1543). Unable to break Süleyman's hold on the country, Ferdinand and Charles V, in the Treaty of Edirne (1547), finally extended de facto recognition of the Ottoman conquest of Hungary by agreeing to pay Süleyman an annual gift of 30,000 gold florins for possession of the northern and western portions of Hungary still in Habsburg control.

Brian Hodson

See also: Austro-Turk Wars; Mohács, Battles of; Vienna, Sieges of

References and further reading:

Perjes, Géza. *The Fall of the Medieval Kingdom of Hungary: Mohács 1526-Buda 1541.* Ed. Mario Fenyö. Boulder, CO: East European Monographs, 1989.

Hungarian Revolt (1956)

Popular uprising against the Soviet-backed Communist regime. Following the harsh imposition of Communist rule after World War II, the New Course of de-Stalinization, beginning in 1953, split the Hungarian Communist Party between conservatives and reformers and led to a loosening of political controls over public discourse. In the summer of 1956, conservatives attempted to placate the reformers by removing the unpopular party secretary, Mátyás Rákosi, and rehabilitating a number of victims of Stalinist-era purges.

Encouraged by this and the success of similar reform efforts in Poland, Budapest University students, joined by thousands of workers, marched to the Parliament building on 23 October with a list of demands for further change. Reform leader Imre Nagy addressed the crowd, which dispersed, but later in the evening protestors toppled the Stalin monument in the city park and seized the radio station. In response, the party appointed Nagy prime minister on 24 October, simultaneously announcing that it had requested Soviet military assistance to maintain order. The announcement provoked a general strike, followed by armed clashes between the Soviet troops and freedom fighters, who seized key positions in the capital.

Relative peace was restored after 28 October, when Nagy announced a cease-fire, amnesty, reorganization of the party, and the withdrawal of Soviet forces. Though willing to cooperate with the new government, the freedom fighters hesitated to disarm. A confrontation with security troops on 30 October led to a siege of the Communist Party headquarters and the lynching of a number of security personnel. The following day the Soviet army began to reoccupy strategic points. After receiving evasive replies from the Soviet government about their intentions, Nagy declared Hungarian neutrality and its withdrawal from the Warsaw Pact. To ward off a Soviet attack, the Nagy government attempted to quiet the revolutionary situation in the country, arranging the end of the general strike on 3 November.

On the night of 3–4 November, 16 Soviet divisions invaded Hungary and installed a new government under János Kádár. The Hungarian army was quickly disarmed and most of the country occupied within days. Fighting continued in Budapest until 11 November, while 200,000 refugees fled the country. As many as 25,000 Hungarians were killed in the fighting between 23 October and 11 November; officially, Soviet losses were 700.

The blatant suppression by the Red Army of the relatively modest Hungarian demands led to worldwide condemnation, and communists left their national parties in numbers not seen since the Hitler-Stalin pact of 1939; many fellow travelers also were alienated. The Kádár government embarked on a program of economic reform but also brutally hanged Nagy and Hungarian army general Paul Maleter in 1958.

Brian Hodson

References and further reading:

Hodson, Brian. *The Hungarian Revolution of 1956: Reform, Revolt and Repression 1953–1963.* Ed. György Litván. Trans. János Bak and Lyman Legters. London and New York: Longman, 1996.

Zinner, Paul. *Revolution in Hungary.* New York: Columbia University Press, 1962.

Hungarian War with the Holy Roman Empire (1477–1485)

War fought between Hungarian king Matthias Corvinus, and Holy Roman Emperor Frederick III, over possession of Lower Austria. In the early spring of 1477, Matthias sent his light cavalry on raids to distract and occupy Frederick's army, while his main force operated against the towns. Though he failed to capture Vienna, by the end of the year Matthias had gained possession of Styria and most of Lower Austria without fighting a single major battle, forcing Frederick to sue for peace. For the next four years, Matthias's Hungarian forces were occupied by the Hungarian-Turkish Wars, allowing Frederick to recover most of his losses.

After securing peace with Mehmed II of Turkey, Matthias again declared war against Frederick in 1482. To break Frederick's hold on Lower Austria, Matthias conducted a systematic campaign against the towns surrounding Vienna and Wiener Neustadt, capturing Hainburg (1482), Klosterneuburg (1483), and Bruck an der Leitha (1484), using light cavalry as before to harass Frederick's less mobile forces. Frederick's attempt to raise the siege of Kornenburg resulted in a disastrous defeat at Leitzersdorf (November 1484), destroying his field army and leaving the rest of his forces isolated in the towns without a chance for relief. The subsequent capture of Kornenburg by Matthias closed the ring around Vienna, which surrendered in May 1485.

The following year, Matthias completed his conquest of Lower Austria by taking Wiener Neustadt and forcing Frederick to recognize his claims. Matthias's success left him isolated. He was unable to prevent the election of Frederick's son Maximilian I as Holy Roman Emperor (1486), nor was he able to consolidate his hold on Austria, which reverted to Maximilian at Matthias's death in 1490.

Brian Hodson

See also: Bohemian Civil Wars; Matthias I

References and further reading:

Rázsö, Gyula. *Die Feldzüge des Königs Mathias Corvinus in Niederösterreich, 1477–1490.* Militärhistorische Schriftenreihe 24. Vienna: Heeresgeschichtliches Museum, Militärwissenschaftliches Institut, 1981.

Hungarian-Turkish Wars (1437–1526)

Series of wars between the Kingdom of Hungary and the Ottoman Empire, beginning with the Ottoman occupation of Serbia (1438–1439) and ending with the collapse of Hungary in the Hungarian Civil Wars (1526–1547). The failure of Hungarian king Albrecht's crusade in 1437 introduced a new phase of the Ottoman wars of European expansion in the Balkans, which were now waged up to and across the borders of Hungary. To support deposed Serbian despot George Brankoviæ, Hungarian general János Hunyadi counterattacked into Wallachia in 1442. In the winter of 1443–1444 Hunyadi invaded Bulgaria, forcing Sultan Murad II to agree to the restoration of Serbia to Brankoviæ. Assured by the pope that promises made to infidels need not be honored, Hungarian King Ulászlo I broke the peace and launched another crusade in 1444. The crusading army was cut off and destroyed by Murad at Varna, where Ulászlo was killed. Hunyadi escaped but was defeated again at Kosovo Polje in 1448. A continuing succession crisis left Hungary too weak to intervene in the Ottoman conquest of Constantinople (1453). Hunyadi gathered sufficient forces to break the siege of Belgrade (1456), but the Hungarians were unable after his death to prevent the Ottoman conquest of Serbia (1457–1458).

Though Hunyadi's campaigns against the Ottomans ultimately failed to recover any territory, they did revitalize and provide leadership for the resistance of the Balkan peoples fighting against the Turks, encouraging Skander Beg (George Kastriota) to renounce Ottoman suzerainty and launch the Albanian-Turkish wars for independence.

In 1463 Mehmed II invaded and occupied Bosnia, prompting a winter counterattack by Hunyadi's son, Matthias Corvinus, who recaptured the strategic fortress of Jajce. From 1464–1466 the Hungarians and Ottomans fought ineffectually in Bosnia, eventually dividing the kingdom between themselves.

Subsequently, Matthias focused on strengthening the line of fortresses established by King Sigismund along the southern borders of Transylvania and Slavonia through Bosnia to the Adriatic while the Ottomans consolidated their Balkan conquests. The following 50 years were marked by repeated border incursions and raids from both sides, over time weakening the fortress system. A large raid by Ali Beg of Smederevo in 1479 was followed by a campaign by Matthias into Wallachia, Serbia, and eastern Bosnia in 1480, capturing Srebrenica and briefly restoring the frontier defenses.

After Matthias's death, the Hungarians successfully repulsed an attack on Belgrade in 1494, but by the first decades of the sixteenth century Ottoman raiders were penetrating deeper into the frontier zone and inflicting defeats on Hungarian counterattacks inside Croatia and Hungary, notably at Sinj (1508), Knin (1511), and Dubica (1520). The recurrent raids devastated the frontier regions, leaving the fortresses isolated and unsupported in the deserted land. Srebrenica was recaptured by the Ottomans in 1512, completing the Turkish conquest of Bosnia. The border defenses were fatally breached with the capture of Belgrade by Süleyman I in 1521 and the fall of Orsova and Knin the following year.

With the lower Danube firmly in his control, Süleyman invaded Hungary in force, defeating the Hungarian army in the Battle of Mohács (1526), at which King Louis II was killed. Louis's death marked the end of the medieval Hungarian kingdom, which was subsequently divided among the Ottomans, Austrian Habsburgs, and the dependent principality of Transylvania.

Brian Hodson

See also: Matthias I; Hunyadi, János; Mohács, Battles of; Süleyman I; Turkish Wars of European Expansion

References and further reading:

Sugar, Peter. *Southeastern Europe under Ottoman Rule: 1389–1814.* Seattle: University of Washington Press, 1977.

Szakály, Ferenc. "Phases of Turco-Hungarian Warfare before the Battle of Mohács." *Acta Orientalia Academiae Scientiarum Hungaricae* 33 (1979): 65–111.

Hungarian-Venetian Wars (1345–1381)

Three wars between Hungary and Venice that took place during the reign of Hungarian king Louis I ("The Great," 1342–1382). At that time both the Hungarian kingdom and the Venetian republic were prosperous and major powers in southern Europe. Dalmatia was the major flash point between them. Louis sought to extend Hungarian power into the Balkans and, on the basis of dynastic right, to secure the throne of Naples. Louis's aspirations in Italy were unacceptable to Venice and to the papacy.

The first war was fought during 1345–1348, with Genoa on Louis's side. Doge of Venice Andrew Dandolo ordered the siege of Zara, under Hungarian rule, and a rival trading city to Venice on the Adriatic. The city was well fortified and resisted. In July 1348 Louis attacked the Venetian siege positions but was defeated. His army, which was principally of mounted knights, was unable to achieve any significant success against the Venetian fortified positions, and Louis decided to return home. The decision was also forced by the situation in Naples, where the king, Louis's younger brother, had been murdered. Zara continued to resist the Venetians but finally had to surrender in December. The first war ended with a cease-fire concluded on 5 August 1348, under the terms of which Hungary accepted Venetian rule in Dalmatia in return for which the trade and logistics routes between Hungary and Naples would remain open.

The second war occurred during 1356–1358. Peace talks started in Buda in 1349, but no agreement was possible because both parties wanted to rule Dalmatia. Louis began this second war in 1356. He divided his forces near Zagreb and sent the viceroy of Croatia to besiege Zara, while he led the main forces against the Venetian continental territories known as Terra Ferma. Following a few successful minor battles and sieges, in July Louis initiated a siege of Treviso. The city was close to capitulation when in November papal mediation brought about a cease-fire.

Louis's plan to attack Venice's mainland territories, where it could not use its formidable fleet, was brilliant, but the Hungarian forces lacked adequate siege equipment. Treviso pinned down the stronger Hungarian land forces, and they were unable to engage the Venetians in open battle.

Peace talks were again unsuccessful, and Louis changed his strategy. Instead of a large heavy force, he sent smaller cavalry units to raid the Venetian hinterland. He also ordered the viceroy of Croatia John Csuzi to carry out the same tactics in Dalmatia. These were successful, as the exhausted Dalmatian cities changed sides. The war ended with a peace agreement on 18 February 1358, with Venice giving up its claims to Dalmatia; Louis's forces departed Terra Ferma and freedom of the trade on the Adriatic was established.

In 1370 Louis acquired Poland from his uncle. The third war with Venice occurred during 1372–1381. Venice attacked Louis's ally Prince Francis Carrara of Padua in 1371, and the Hungarians came to his rescue in 1372. Carrara and the Hungarians defeated the Venetian forces at the Battle of Piave di Sacco but failed to take Treviso. This led to a disadvantageous peace for Padua.

Hungary, Genoa, and Padua formed a new alliance against Venice in 1378, but after a Venetian victory at sea, the allies turned to the old strategy of raiding the Venetian land-supply lines. The year 1379 brought successes to the allies. They defeated the Venetian fleet at Pola, and Charles of Durazzo, the new Hungarian commander in chief, besieged Treviso. Finally, the allied forces occupied Choggia, the "gateway to Venice." An exhausted Venice sued for peace, but the alliance broke up and Venice was able to retake Choggia.

The wars between Hungary and Venice ended by the peace of Turin on 8 August 1381. Its terms were advantageous to Hungary. Venice had to pay an annual tribute to Hungary and lost Dalmatia. But it had survived the war and maintained a leading role in Adriatic trade.

Ákos Tajti

References and further reading:

Bánlaki, József, of Doberdo. *A magyar nemzet hadtörténelme* (The Military History of the Hungarian Nation). Vol. 8. Budapest: Grill, 1934.

Kristó, Gyula. *Az Anjou kor háborúi* (The Wars of the Anjou Era). Budapest: Zrínyi, 1988.

Molnár, Erik, ed. *Magyarország története* (History of Hungary). Vol. 1. 2d ed. Budapest: Gondolat, 1967.

Ráth, Károly. *Magyar királyok hadjáratai utazásai és tartózkodási helyei* (Campaigns, Travels, and Residences of the Hungarian Kings). Gyor, Hungary: Sauervein, 1861.

Huns

Pastoral nomads of uncertain origin threatening the Roman world. Appearing unexpectedly in the steppes north of the Black Sea around 370, the Huns pushed the Goths west, across the Danube into Thrace. In 395, the Huns themselves crossed the Danube, thereby coming into direct conflict with the Roman world.

Superb horsemen—Ammianus Marcellinus and Zosimus describe them eating, sleeping, and even performing bodily functions while on horseback—they rode the ill-shaped but hardy steppe breed of horse. Thus mounted the Huns fought with a reflex composite bow—with which they were highly skilled even when drawn from a galloping or wheeling horse—a small shield, and a spear. The stave of the composite bow was constructed of laminated materials, usually wood (core), sinew (back), and horn (front). When strung, the bow was opened back against its natural curve, and held that way by the bowstring. More powerful than the longbow, it could penetrate armor at 100 meters.

A Hunnic army is described by Ammianus as forming up, with much disorderly movement and savage noise, into wedge-shaped masses. Some of these would break up into scattered bands, which would rush around with surprising speed and apparent chaos, inflicting casualties with their shooting. Other bands would relieve the first to maintain an incessant barrage, until the enemy was sufficiently weakened or demoralized. Each warrior would then charge at the gallop, regardless of risk to his own safety, to fight at close quarters with sword and spear. A Hun charge was executed with such speed and suddenness that it usually overwhelmed everyone and everything in its path. Huns terrified people by their outlandish appearance, but it was their very name that soon came to symbolize the epitome of swift, merciless destruction.

Nic Fields

See also: Attila the Hun; Châlons, Battle of; Goths

References and further reading:

Lindner, R. P. "Nomadism, Horses and Huns." *Past and Present* 92 (1981), 1–19.

Maenchen-Helfen, O. J. *The World of the Huns.* Berkeley: University of California Press, 1973.

Hunyadi, János (c. 1407–1456)

Hungarian general and regent. The son of a Romanian noble granted the estate of Hunyad (Hunedoara) in Transylvania by King Sigismund, Hunyadi began his military career serving the king in Italy and in the Hussite Wars in Bohemia. Originally a supporter of Albert of Habsburg following Sigismund's death, Hunyadi later opposed the succession of Albert's infant son, Ladislas Posthumus. In 1441, Hunyadi led the army of Polish king Wladyslaw II Jagiellon to victory over Ladislas's guardian, Frederick III at Bátaszék, winning the throne for Wladyslaw, crowned Ulászló I of Hungary.

In 1443 Hunyadi led Ulászló's Hungarian-Polish crusade against the Ottomans, driving the Turks out of Serbia, Bosnia, and Bulgaria in a winter campaign. A second crusade in 1444 ended in disaster at Varna on the Black Sea coast, where Sultan Murad II surprised and destroyed the Christian army and Ulászló was killed. Hunyadi escaped to Hungary and headed a regency council that opened negotiations with Frederick for the return of Ladislas. From 1446 to 1453, Hunyadi served as governor of Hungary in the name of the still-absent king. His third crusade against the Ottomans resulted in another defeat at the Second Battle of Kosovo Polje (1448).

After Ladislas's return to the throne, Hunyadi continued to serve the king as chief captain and administrator of royal revenues. In 1456, Hunyadi and the Minorite monk Giovanni di Capestrano organized a popular crusade in Austria and Hungary for the relief of Belgrade, besieged by Ottoman sultan Mehmed II. Hunyadi's small army, consisting largely of peasants and townsfolk, cut the Ottoman supply lines, repelled a Turkish assault, and impetuously attacked and broke the Ottoman army. On 11 August 1456, two weeks after his greatest victory, Hunyadi died in an epidemic that broke out in the Christian camp.

Brian Hodson

See also: Hungarian-Turkish Wars

References and further reading:

Bak, János. "The Late Medieval Period, 1382–1526." In *A History of Hungary,* ed. Peter Sugar. Bloomington: University of Indiana Press, 1990.

Held, Joseph. *Hunyadi: Legend and Reality.* Boulder, CO: East European Monographs, 1985.

Hurrians (c. 2300–1100 B.C.E.)

An Upper Mesopotamian people who were the dominant ethnic group of the Mitanni state (c. 1600–1100 B.C.E.) that ruled much of the ancient Near East in the late first millennium B.C.E. The earliest attestation of the Hurrians are in Sumero-Akkadian sources from the reign of Naram-Sin in the late third millennium B.C.E. They are described as inhabiting the land of Subartu—the Upper Mesopotamian regions of the Khabur and Balikh river basins in Syria, as well as the Tigris River basin in northern Iraq. By the end of the third millennium B.C.E., north Mesopotamia was thoroughly Hurrianized, with well-established Hurrian states that continued until the rise of a powerful Hurrian-based kingdom of Mitanni (c. 1600 B.C.E.).

The kingdom of Mitanni was a confederation of Hurrian states in Upper Mesopotamia in the late second millennium B.C.E. Its capital, Washukanni, has not been located for certain; it may have been Tell Fakhariyah, a mound located near the headwaters of the Khabur River in Syria. By 1450 B.C.E., Mitanni was the most powerful state in the Tigris-Euphrates region, composed of confederate and vassal city-states, each with its own king. Although the state of Mitanni was dominated by Hurrians, there was a significant percentage of Indo-European West Semitic–speaking peoples, Hittites, and Assyrians.

Because of our fragmented sources, the military history of the Mitanni state can only be partially reconstructed. It appears that by 1500 B.C.E. Mitanni had expanded into most of Syria under the reigns of Paratarna and Saustatar. This newly formed confederation was likely in conflict with the expansionist policies of Thutmose III of Egypt (c. 1504–1450 B.C.E.). Later Mitanni kings are known primarily through the Amarna letters from Egypt (c. 1411–1350 B.C.E.), where the Mitanni kings engaged in diplomatic relations with the kings of Egypt. The marriage alliances between the two states may have been due to the rise of Assyria in northern Iraq and the Hittites in Anatolia, which severely threatened the Mitanni state. Under Tushratta Mittani became somewhat fragmented and suffered defeat at the hands of the Hittite king Shuppiliuiluma. Thus, after c. 1350 B.C.E., the Hurrian state ceased to be a major role player in ancient Near Eastern politics. Mitanni continued to be a buffer between the Hittites and Assyria for at least the next two centuries, until the area was absorbed into the Assyrian Empire.

Mark W. Chavalas

See also: Assyria; Babylonian Empire; Hittites

References and further reading:

Gelb, I. *Hurrians and Subarians.* Chicago: Oriental Institute, 1944.

Morrison, M. A., and D. I. Owen, et al. *Studies on the Civilization and Culture of Nuzi and the Hurrians.* 9 vols. Winona Lake, WI: Eisenbrauns, 1981.

Wilhelm, G. *The Hurrians.* Westminster, UK: Aris & Phillips, 1989.

Wiseman, D. *The Alalakh Tablets.* London: British School of Archaeology in Iraq, 1953.

Hussein, Saddam al-Tikriti (1937–)

Iraqi dictator. Born 28 April 1937 near Tikrit, Saddam Hussein never knew his father and has spent his life seeking to become the father of a new empire centered on Baghdad. He became involved with the Ba'athist Party and participated in a failed coup attempt in 1959 against Abdul-Karim Qassem. He fled to Egypt, where he was educated until he returned to Iraq. Following a coup in 1968, his cousin Ahmad Hassam al-Bakr became president. Saddam Hussein became the vice president of the Revolutionary Command Council, commanding internal security. Replacing his cousin, he took over Iraq in 1979.

Aiming to reestablish the dominance of Baghdad in the Arab world, Saddam Hussein studied and patterned himself after Joseph Stalin. Indeed, his regime continues to exist in large measure due to repeated purges of any whom Saddam Hussein suspects of harboring even thoughts of rebellion.

Iraq is an oil-producing country, but Saddam Hussein sought greatness through military adventures. He pursued the acquisition of nuclear, biological, and chemical weapons but was thwarted in developing a nuclear capacity initially by an Israeli air strike in 1981. His eight-year war with Iran failed to achieve its goals. Iraq emerged heavily in debt from that war with the fourth largest military in the world, numbering 1 million.

Saddam wasted little time before embarking on his sec-

Iraqi propaganda art glorifying Saddam Hussein's supposed military prowess. (National Archives)

ond war, the invasion of Kuwait. His defeat, in one of the most one-sided conflicts of modern times, resulted in the decimation of Iraq's military, economy, and infrastructure. One of the world's most unsuccessful military leaders, he remains the president of Iraq.

John R. Popiden

See also: Gulf War; Iran-Iraq War

References and further reading:

Makiya, Kanan. *Republic of Fear: The Politics of Modern Iraq.* Los Angeles and Berkeley: University of California Press, 1989, 1998.

Miller, Judith, and Laurie Mylroie. *Saddam Hussein and the Crisis in the Gulf.* New York: Times Books/Random House, 1990.

Hussite Wars (1419–1436)

Religious wars and revolt of Bohemia against the Holy Roman Emperor. The Hussite revolt began as a religious reform movement with strong popular support. In 1415, the movement's doctrine was declared heretical and its leader, Ján Hus, was burned at the stake at the Council of Constance. King Václav (Wenceslas) IV failed to control the resulting unrest in Bohemia, which spread along both national and class lines.

The Hussites were predominantly Czech and included both a moderate, mostly bourgeoisie, and noble Utraquist party and the more radical Taborites, largely peasants and villagers. Their opponents drew support from the German-speaking inhabitants of the kingdom and the upper nobility.

When Václav died in 1419, his brother, Holy Roman Emperor Sigismund, claimed the throne over the objection of the Hussites, who held him responsible for Hus's death. With the support of Pope Martin V, Sigismund entered Bohemia with an army of German and Hungarian crusaders in 1420 but was defeated by the Taborite general Ján Žižka outside Prague and forced to retire. A Second Crusade by German princes in 1421 ended in the failed siege of Zatec. Separately, Sigismund reentered Bohemia from Hungary and captured Kutna Hora but failed to trap Žižka's army.

Žižka returned to attack Sigismund's forces in their winter quarters, destroying his army in a running battle from Nebovidy to Německý Brod (6–10 January 1422). After defeating a Third Crusade led by Frederick I of Brandenburg, the Hussites were split by tensions between the Utraquists, who sought a settlement with the Catholic Church, and the Taborites, who pressed for continued reform and resistance. Taborite victories under Žižka's command over the Utraquists at Strachov (August 1423) and at Malešov (June 1424) shortly before his death ensured the continuation of the war.

Under Žižka's successor, the Taborite priest Prokop the Bald, the Hussites defeated a German army at Usti and launched raids into Silesia and Austria, burning Landshut (October 1426) and defeating an Austrian army near Linz (March 1427). A Fourth Crusade against the Hussites ended with the aborted siege of Støíbro and Prokop's victory over the crusaders at Tachau (1427).

In the following years, Hussite armies continued their raids into Germany and Hungary and briefly aided the Poles in their war with the Teutonic Knights. The defeat of a Fifth Crusade at Domazlice (1431) led the Council of Basel to open negotiations with the Hussites, resulting in a settlement with the Utraquists (1433), which the Taborites rejected. Allied now with the Catholics and barons, the Utraquists defeated the Taborites at the Battle of Lipany (1434), in which Prokop was killed. A final agreement between the two sides was reached in the Compacts of Jihlava (1436), which permitted the Hussites the use of their own communion rite in return for recognition of the authority of the Catholic Church and Sigismund's kingship.

Brian Hodson

See also: Německý Brod, Battle of; Prague, Siege of; Žižka, Ján

References and further reading:

Bartoš, František. *The Hussite Revolution, 1424–1437.* Trans. John Klassen. Boulder, CO, and New York: East European Monographs, 1986.

Heymann, Frederick. *John Zizka and the Hussite Revolution.* Princeton, NJ: Princeton University Press, 1955.

Hydaspes, Battle of the (May 326 B.C.E.)

Alexander's final major victory in Asia, in the Indus River valley. After the Macedonian king Alexander III the Great had conquered the Persian Empire, he crossed the Hindu Kush into India. A local king called Poros opposed him, holding the far bank of the river Hydaspes (Jhelum).

Poros's army consisted of 30,000 foot, 3,600 horse, 200 elephants, and 180 chariots. The elephants were deployed in a line in the center, with the infantry positioned behind them, opposing Alexander's cavalry, whose horses were terrified of the elephants. Poros's flanks were covered by cavalry and chariots. Alexander's force consisted of 5,000 cavalry and 10,000 infantry. The infantry were deployed in line, with the right flank covered by the majority of the cavalry. The remainder of the cavalry were deployed on the left, possibly shielded behind the infantry line.

Alexander left a holding force to keep Poros's army in place, while Alexander with the rest of his forces forded the river upstream and attacked the Indian king. Alexander initiated the battle by attacking the enemy cavalry with his own cavalry force on the right. As the numerically inferior Indian horse were under severe pressure, Poros ordered the cavalry on the other flank to reinforce them. Thereupon the Macedonian cavalry detachment on the left crossed the battlefield and attacked the Indian horse in the flank. The Indian horse were surrounded and destroyed.

Meanwhile the infantry lines had clashed. At first the Macedonians had a hard time with the elephants, but soon they opened gaps to let the panicked beasts through or turned them with their pikes, driving them back into the Indian line, then charged the disordered Indians. Twelve thousand Indians were killed, 9,000 captured. Their king fought heroically until he was persuaded to surrender. The Macedonians lost 1,000 men, a heavy toll for Alexander's depleted and weary troops.

Maarten van der Werf

See also: Alexander the Great; Alexander's Wars of Conquest

References and further reading:

Fuller, J. F. C. *The Generalship of Alexander the Great.* London: Eyre & Spottiswoode, 1958.

Green, Peter. *Alexander of Macedon, 356–323 B.C.E. A Historical Biography.* Harmondsworth, UK: Penguin, 1974. Reprinted Berkeley and Los Angeles: University of California Press, 1991.

Hammond, N. G. L. *The Genius of Alexander the Great.* London: Duckworth, 1997.

Tarn, W. W. *Alexander the Great.* Cambridge, UK: Cambridge University Press, 1948.

Hydrogen Bomb, Development of (1942–1952)

The United States detonated the world's first hydrogen, or thermonuclear, bomb on 31 October 1952. This bomb, codenamed Mike, stood three stories high and exploded with a force of 10.4 megatons, or over 10 million tons of TNT. The mushroom cloud from the explosion rose to a height of more than 100,000 feet and could be seen from a distance of 50 miles. A new era in nuclear weaponry began as the destructive power of such weapons increased from thousands of tons to millions of tons of TNT equivalence.

Nuclear weapons are of two basic types. Atomic bombs, such as those used in World War II combat against Hiroshima and Nagasaki, split atoms of the elements uranium or plutonium. Hydrogen bombs fuse atoms of the hydrogen isotopes deuterium and tritium. In both types of weapons, large amounts of energy are released in the form of blast waves and radiation. The very sudden release of this energy causes the enormous destructive power of nuclear weapons.

Edward Teller and Enrico Fermi first studied the possibility of a hydrogen bomb in 1942. Based on the research of Hans Bethe, who pioneered studies of thermonuclear reactions in stars, Teller and Fermi believed that atoms of deuterium, an isotope of hydrogen, could be fused into helium with a simultaneous release of energy. Because such a process required stellar temperatures, then unobtainable on earth, Teller originally doubted that a hydrogen bomb could be built.

Despite his original doubts, Teller continued to study thermonuclear reactions and became convinced that a hydrogen bomb was possible. By this time, however, it had been decided that the United States would concentrate its World War II nuclear efforts on building a fission device at the Los Alamos Laboratory in New Mexico. A thermonuclear weapon would be too difficult to develop in time to assist the war effort. Despite the wartime concentration on the fission weapons, Fat Man and Little Boy, Teller and a small group of scientists did conduct some elementary research on the hydrogen bomb. In particular, Teller and his colleagues found that much less deuterium and tritium would be required than originally thought, thereby making a hydrogen bomb more realistic.

With the end of World War II, the United States demobilized. The wartime weapons laboratories, including Los Alamos, faced severe shortages of manpower as senior scientists returned to their prewar university positions and younger staff left to enter graduate school. As a result, very little research and development of the hydrogen bomb took place. Between 1946 through 1949, for instance, fewer than 12 theoretical physicists worked full time on the hydrogen bomb. Despite the shortage of scientist staff, work continued at Los Alamos during the late 1940s on the hydrogen bomb. Among the technical accomplishments during the late 1940s was the improvement of fission devices, the use of computers and computational modeling, and the development of cryogenic technology to produce the liquid deuterium required to make a hydrogen bomb.

The ultimate success of the United States' thermonuclear program rested on five accomplishments. First was the discovery of how to make such a device work. This discovery had to overcome a fundamental problem uncovered during early thermonuclear research. Thermonuclear systems lose as much energy as they create. Second, Los Alamos had to increase the size of its scientific staff significantly. The hydrogen bomb problem required complex interactions among the entire range of physicists, chemists, and metallurgists and their respective skills. Third, smaller and more efficient fission bombs were needed to start a thermonuclear fire. Fourth, computational ability had to be greatly enhanced. Fifth, the political will had to exist to marshal the resources necessary to accomplish the task in a complex technical and political environment.

The design adopted for the first hydrogen bomb did not come easily or quickly. Unlike fission weapons that are one of two types, scientists did not have a clear idea of the range of physical constraints governing thermonuclear weapon design. Extensive mathematical modeling and simulation were required before any kind of scientific judgment could be made. It took five years to discover the ultimate design.

The shortage of scientific staff, particularly theoretical and experimental physicists, also took five years to overcome. In the short term, the Los Alamos Laboratory used consultants such as university professors on leave. Many of these consultants had worked at Los Alamos during World War II and knew the nature of problems facing thermonuclear development. Eventually the pool of scientists grew as postwar university graduation rates increased. However, not enough can be said for the small cadre of scientists who stayed at Los Alamos after the war and advanced thermonuclear work under less than ideal conditions.

Because a hydrogen bomb requires extremely high temperatures to ignite the thermonuclear fuel, temperatures found only in stars, the only possible way to ignite a hydrogen bomb was by using an atomic bomb. For the development of the hydrogen bomb to go forward, fission weapons had to be improved. The two wartime fission devices, Fat Man and Little Boy, were crude prototypes and not capable of being adapted for use in a hydrogen bomb. A new class of fission bombs had to be designed, built, and tested. This process took years. The first testing of new fission devices did not take place until 1948, with more design improvements following in 1950.

Because all of the design work on a hydrogen bomb involved complex mathematical modeling and simulation, the need for better and better computers was compelling. During World War II, all computing at Los Alamos was done with desktop calculators and a variety of IBM business machines. Such machines were not capable of handling the complex modeling required for developing the hydrogen bomb. Beginning shortly after the war, true computers started to become available, beginning with the ENIAC, IBM's SSEC, and the National Bureau of Standards' SEAC. Because these machines were on the East Coast, many of the thermonuclear calculations actually took place far from Los Alamos. Although the first hydrogen bomb could have been developed without modern computers, such development would have been substantially delayed.

Finally, a political mandate was necessary to concentrate and focus the final development of the hydrogen bomb. By the late 1940s the Cold War had begun in earnest, and in 1949 the Soviet Union detonated its first atomic bomb. Concern arose in the national security establishment that the United States could be overtaken militarily by the USSR if the U.S. did not develop thermonuclear devices. Fission bombs have a natural upper limit to the explosive power they could produce. Hydrogen bombs do not. By adding more fuel, thermonuclear bombs can be made ever larger. A Soviet Union with unchallenged thermonuclear capability was a sobering thought. Taking all of this into account, President Harry Truman directed an accelerated effort to develop the hydrogen bomb in January 1950.

Shortly after Truman's directive, hydrogen-bomb research began to bear fruit. Edward Teller and Stanislaw Ulam came up with a promising design, radiation implosion, which was translated by Richard Garwin into a working design. Once the design concept was reviewed and approved, work began on constructing the Mike device and planning for a full-scale test at the Pacific Proving Ground in the Marshall Islands. Every atmospheric nuclear test was a major undertaking. The first hydrogen bomb test was even more so, involving thousands of people, millions of dollars, and countless pieces of equipment and conducted in the remote reaches of the Pacific Ocean.

Mike was not a bomb in the combat sense of the word. Weighing more than 1 million pounds, standing three stories high, and using a cryogenic fuel, it was a stationary device. Work began immediately to make hydrogen bombs deliverable, first by aircraft and later by missiles. It took some two years after the Mike shot to reach this goal. This work involved moving away from the constraints of cryogenic to dry fuel and building devices based on the demands of air force and navy aircraft and the nose cones of ballistic missiles.

Atomic bombs ended World War II. As destructive as Fat Man and Little Boy proved to be, some governments thought a nuclear war might be won—albeit at great cost. Thermonuclear weapons destroyed this concept of survivability. In 1954, the United States tested its largest ever hydrogen bomb, Bravo, which exploded with the energy of 15 megatons. The Soviet Union, in the early 1960s, exploded a device of 50 megatons—the world's largest explosion. Capable of theoretically unlimited destruction, thermonuclear weapons could destroy not only entire cities, but civilizations as well. The testing of thermonuclear devices brought the United States and the Soviet Union to the bargaining table and a test moratorium. The consequences of a thermonuclear war were too terrible to ignore.

Roger A. Meade

See also: Cold War

References and further reading:

Hewlett, Richard G., and Francis Duncan. *Atomic Shield: A History of the United States Atomic Energy Commission,* Vol. 2, *1947–1952.* Berkeley: University of California Press, 1990.

Rhodes, Richard. *Dark Sun: The Making of the Atomic Bomb.* New York: Simon & Schuster, 1995.

I

Ia Drang Valley (October–November 1965)

The first real test of North Vietnamese and U.S. regular armed forces in the deepening conflict in Vietnam. Although the Americans won, both sides drew important conclusions about strategy and tactics that would influence the course of the war.

In early 1964, the government in Hanoi had begun sending regular North Vietnamese Army main force units to augment Vietcong guerrillas in South Vietnam; a year later, the United States also began committing regular army and Marine Corps forces.

In October 1965, the North Vietnamese began attacking a U.S. Army Special Forces camp at Plei Me in the central highlands of South Vietnam. They were planning to destroy an expected South Vietnamese Army relief column. However, the U.S. Army's 1st Cavalry Division had arrived at Pleiku and was ready to test a new type of warfare, using helicopters to fly over enemy positions to deliver troops and firepower where needed. On 14 November, U.S. troops landed at a site some 14 miles east of Plei Me, which to their surprise was in the middle of a North Vietnamese Army (NVA) regiment, and a major two-day battle began. The Americans were supported by massive artillery and air strikes, including the first use of B-52 bombers in Vietnam, flying from Guam, each with payloads of 200 tons of 500-pound bombs.

The North Vietnamese eventually broke off contact and retreated across the Cambodian border. They had lost 2,000 killed and wounded; the Americans, 79 killed and 121 wounded. The NVA leadership concluded they could not fight another such battle; either they would have to select situations where they could overwhelm smaller and more isolated American units or else they would have to "hug" American forces, getting so close that the U.S. troops would not call in air strikes and artillery for fear of being hit themselves. General William Westmoreland concluded that the so-called airmobile concept and his seizing of the tactical initiative (while conceding the strategic initiative) would work, and so the American phase of the protracted conflict proceeded.

Charles M. Dobbs

See also: Vietnam Conflict; Vo Nguyen Giap; Westmoreland, William

References and further reading:

Moore, Harold G., and Joseph L. Galloway. *We Were Soldiers Once and Young: Ia Drang, the Battle that Changed the War in Vietnam.* New York: Random House, 1992.

Pimlott, John. *Vietnam: The Decisive Battles.* New York: Macmillan, 1990.

Turley, William S. *The Second Indochina War: A Short Political and Military History, 1954–1975.* Boulder, CO: Westview Press, 1986.

Illyrian Wars (229–219 B.C.E.)

Two wars fought between Rome and the coastal kingdom of the Illyrians (modern Croatia). In the First Illyrian War (229–228 B.C.E.) the expanding Roman Republic clashed with Queen Teuta and her Illyrian pirates. The Illyrians had raided Greco-Roman shipping lanes, besieged the Greek port of Corcyra (Corfu), and killed Roman ambassadors in an ambush after attempts to negotiate with the aggressive Illyrians had failed. The Roman senate, seeking revenge, sent a Roman land and sea expedition to relieve the siege at Corcyra. Queen Teuta capitulated and agreed to give up land claims and pay reparations. Macedonia allied with Rome to form an additional counter to Illyrian aggression.

In the Second Illyrian War (219 B.C.E.) Demetrius of Pharos succeeded Queen Teuta as the Illyrian leader and renewed land intrusions into Roman protectorates and piracy against Roman shipping. The Romans acted quickly, perhaps because war with Carthage was imminent. Two Roman

Men from the 25th Infantry Division take cover during the bitter fight at the Imjin River on April 23, 1951. (National Archives)

armies besieged the fortified Illyrian cities of Dinale and Pharos, with the former falling in seven days and the latter in one day. Demetrius fled and the conditions of the first war were reimposed on the Illyrians.

Christopher Howell

See also: Macedonian Wars; Punic Wars; Roman Republic, Wars of the

References and further reading:

Adcock, Frank E. *The Roman Art of War under the Republic.* Revised edition. New York: Barnes & Noble, 1963.

Harris, W. V. *War and Imperialism in Republican Rome.* Oxford: Clarendon, 1979.

Sherwin-White, A. N. *Roman Foreign Policy in the East.* Norman: University of Oklahoma Press, 1984.

Imjin River (April 1951)

Prominent river in Korea, scene of heavy fighting numerous times during the Korean War (1950–1953), most notably in the spring of 1951 during the Communist Fifth Phase Offensive. The offensive, primarily carried out by the Chinese People's Volunteers, was designed to capture Seoul, the South Korean capital, which had already changed hands several times since the commencement of hostilities in June 1950. The Chinese commanding general, Teh-huai Peng, hoped to present Seoul to Premier Mao Zedong as a gift for the 1 May communist holiday.

Beginning on 22 April 1951, the Communists pushed southward across the entire expanse of Korea. Particularly hard hit were the positions of the British 29th brigade, especially the Gloucestershire Battalion, usually referred to as the Glosters. The Glosters held positions immediately south of the river. Troops from the Chinese Sixty-third Army fought desperately and repeatedly to ford the river in the area defended by the Glosters. Finding it impossible to breach the river in the face of the Glosters' deadly firepower, the Chinese pressed across in weaker, adjacent sectors. By 25 April, the Glosters were threatened with encirclement. Efforts to relieve them failed in the face of strong Chinese resistance. By the end of the month, almost all of the battalion had been killed, wounded, or captured. These were the kind of losses the hard-pressed British could ill afford.

The annihilation of the Glosters had political, not just military, consequences. Striving to hold together its diverse coalition of United Nations forces, the United States sought to maintain the continued support of its British ally. General Mathew Ridgway, commander of all UN forces, even demanded a formal report on the loss of the Glosters. Although the destruction of the unit was blamed primarily on the size and ferocity of Chinese opposition, Ridgway took great care from that point forward to make sure that no similarly disastrous situations confronted British forces in Korea. The Communist offensive put a major dent in allied lines but came nowhere near capturing Seoul. This campaign was the last Communist offensive against non-ROK (Republic of Korea) United Nations Command (UNC) forces. The Communists had suffered terrible losses and were not willing to try matters again against the Americans, the Commonwealth Division, and so on, on a large scale. The Americans, for their part, were unwilling to invest the resources needed for victory in Korea while Europe, where in their eyes the main threat lay, remained basically unprepared for defense against Communist designs. Two months later, the Communist side called for negotiations and an armistice.

John C. McManus

See also: Korean War; Ridgway, Mathew B.; Van Fleet, James A.

References and further reading:

Alexander, Bevin. *Korea: The First War We Lost.* New York: Hippocrene, 1993.

Appleman, Roy E. *Ridgway Duels for Korea.* University Station: Texas A&M Press, 1990.

Sandler, Stanley. *The Korean War: No Victors, No Vanquished.* Lexington: University of Kentucky Press; and London: Baker & Taylor, 1999.

Whelan, Richard. *Drawing the Line: The Korean War, 1950–1953.* Boston: Little, Brown, 1993.

Imphal and Kohima (8 March–22 June 1944)

Critical World War II battles in the China-Burma-India theater, arguably the worst defeat suffered by Japanese ground forces in the war. Located in northeast India, Imphal and Kohima became focal points of one of World War II's most desperate campaigns when Japan's Fifteenth Army, commanded by General Mutaguchi Renya, invaded India (Operation U-GO) in late winter 1944. Mutaguchi's objectives included preempting a British invasion of Burma, establishing a foothold for the collaborationist Indian National Army (INA) in hopes of provoking an India-wide revolt against British rule, and capturing large portions of Manipour and Assam provinces, both essential for the Allies' hump air route to China.

During the campaign's opening month, Mutaguchi's forces—three Japanese divisions and 7,000 troops from the INA—besieged both Imphal and Kohima, key supply bases for General William Slim's British Fourteenth Army. Bitter, attritional fighting occurred at both towns, the British demonstrating a resolve and skill not evident during their ignominious retreat from Burma just two years earlier or their aborted Assam offensive the year earlier. Ultimately a combination of combat skill, determination, quantitative superiority in men and equipment, air power, and Slim's command decisions proved decisive. The Fourteenth Army relieved both Imphal and Kohima in June and then drove the Japanese back into Burma in July.

At a cost of 17,000 casualties, British Fourteenth Army inflicted more than 50,000 casualties, destroying Japan's Fifteenth Army, eliminating the Japanese threat to India, and paving the way for the reconquest of Burma that followed between August 1944 and May 1945.

Bruce J. DeHart

See also: Chindits; Merrill's Marauders; Slim, William Joseph, First Viscount

References and further reading:

Allen, Louis. *Burma: The Longest War, 1941–1945.* London: J. M. Dent, 1984.

Grant, Ian Lyall. *Burma: The Turning Point.* Chichester: Zampi, 1993.

Rooney, David. *Burma Victory: Imphal and Kohima, March 1944 to May 1945.* New York: Continuum, 2000.

Inca Civilization

The Andean valleys of northern and western South America were populated from approximately 1200 B.C.E. by a succession of societies. The Chavin culture, first of the great civilizations in the region, was noted for its pottery, architecture, and sculpture, especially bas relief, and for their introduction of maize as a major food crop prior to the end of the eighth century C.E. Later societies added their own distinct contributions, many of which were later utilized by the Incas. The Nazca people were noted for their colorful ceramics and for developing irrigation systems for agriculture. The Mochica, a theocratic people, built temples, pyramids, and roads, engaged in metal work, and made ceramic portraiture, which explicated their life activities. The Tiahuanaco and Huari peoples developed a military state. Under the Chimu, urbanization increased. For several millennia, the llama was the principal beast of burden for all peoples. The polygamous Quechua-speaking Incas, originally humble mountain people, arose near present-day Cuzco and borrowed heavily from the cultural, administrative, and military precedents set by their predecessors. They eliminated many elements of earlier civilizations and developed myths demonstrating the divinity of the Inca, their ruler, from which their society took its name. The *coya* (queen) was chief wife of the Inca ruler. The Inca (ruler) owned all land, and individual citizens through their local communes, the *ayllu,* merely had the use of it. Major Inca expansion began under Pachacuti Inca Yupanqui (r. 1438–1471), and their centralized feudal society, with its administrative districts and advanced social welfare system, exercised effective control over a widely dispersed population. Public buildings were constructed of stone, carefully cut and fitted so as to preclude any need for mortar. Wide, well-maintained roads, facilitating wars of conquest, stretched from Peru south to the town of Constitucion in modern central Chile. The Inca were almost unique in being a civilization that had no written language, but government runners, carrying *quipu,* a message system employing knotted strings, communicated essential intelligence throughout their far-flung empire. Their army operated as a form of agricultural militia. The government provided needed arms and armament. Tightly disciplined, Inca warriors fought en masse in battle. Principal weapons included slings, a six-foot-long spear and spear throwers, stone or metal-headed clubs, war axes, and double-edged wooden swords. Helmets were made of cane or wood, and their square shields were made of wood and animal hides. Body armor consisted of quilted cotton jackets. Under Pachacuti's successors, Topa Inca Yupanqui (1471–1493) and Huayna Capac (1493–1527), Inca power was consolidated. In 1527, the Inca leader Huascar succeeded, but during much of his six-year tenure, he was engaged in civil war with his illegitimate half brother Atahualpa. Huascar's forces were ultimately defeated by Atahualpa in 1532. Spanish conquistadors under Francisco Pizarro arrived shortly thereafter. Atahualpa met Pizarro at Cajamarca on 16 November 1532. His unarmed troops were slaughtered and Atahualpa captured. Despite having met Spanish demands for a substantial ransom, in return for which he had been promised his freedom, Atahualpa was publicly garroted at Pizarro's order on 29 August 1533. The Spanish conquest was completed by 1535.

Keir B. Sterling

See also: Pizarro, Francisco

References and further reading:

Garcilaso de la Vega. *El Inca. Obras Completas.* Madrid: 1960.

Hemming, John. *The Conquest of the Incas.* London: Macmillan, 1970.

Prescott, William H. *The History of the Conquest of Mexico and the History of the Conquest of Peru.* New York: Modern Library, n.d.

Stern, Steve J. *Peru's Indian Peoples and the Challenge of Spanish Conquest.* Madison: University of Wisconsin Press, 1982.

Inca Empire Imperial Wars (1438–1540)

Wars that led to the Inca Empire becoming the largest of its day. The Inca Empire eventually dominated much of western South America, including the highland and coastal Andes areas. At the heart of this empire was its ability to use grand strategy to wage imperial wars against less organized, though often militarily superior, opponents. Although Spanish and native sources differ with archaeological data on exact dates, it is clear that Pachacutec Yupanqui deserves much of the credit. He ruled from 1438 to 1471 and developed a permanent military system and imperial grand strategy.

Significant wars were mounted against highland Andes chiefdoms like the Chanca and Lupaca after the Battle of Cuzco in 1438. Highland strategies involved the use of alliances, a complicated road network and supply system, fortifications, and llamas as backpack animals. Inca logistics was far superior and the empire infrastructure, based on labor taxes and movement of people, became the rival of any in the world. Considering the lack of draft animals, wheels, and iron metallurgy, the gain and consolidation of territory are even more astounding.

Coastal lowland wars were fought using rotating 20,000-men armies based on decimal units. During the war with the Chimu state, bronze weaponry became more common. Military schools trained ethnic Inca to lead loyal allied troops into battle. The Sapa Inca, or emperor, often appointed kin to command the rotating armies and eventually stayed in Cuzco to oversee all operations. By the time of Spanish invasions in 1531, the Inca had permanent standing armies, excellent logistics, and a tested grand strategy based both on military might and on management of human and natural resources that could be used to wage economic warfare.

After Pachacutec had established the empire, his son Topa Yupanqui continued conquests in the late fifteenth century and passed on the title of Sapa Inca to his son Hauna Capac, who conquered much of Ecuador and Columbia in the early sixteenth century.

Capac died of smallpox along with many other Andean Indians, and his sons Atahualpa and Hauscar fought a civil war in the early 1530s that devastated the empire further. At this point conquistador Francisco Pizarro and about 150 fellow Castillians had boldly invaded and taken control of the Inca Empire, executing Atahualpa, taking the Inca capital of Cuzco, and establishing a Spanish port capital at Lima. The royal Inca Paullu was established as a puppet ruler at Cuzco while his brother Manco carried on against the Spanish from the highland jungle fortifications of Vitcos and Vilcambamba, the legendary lost city of the Inca.

Between 1535–1540, Mancos, who studied and employed Spanish battle tactics—even riding a captured horse himself—organized huge rebellions that killed hundreds of

Spanish, mainly by ambushes in battles, and threatened to push the Spanish back into the Pacific. In 1536, he gathered a 100,000-man force to besiege Cuzco but failed to kill the brothers of Francisco Pizarro and his forces there. The battle centered around the monstrous Inca fort of Sacsahuaman, which was eventually retaken by the Spanish.

Although Manco's son, Tito Cusi, carried on the defense against the Spanish into the 1560s from Vilcambamba, the chance for regaining the empire had passed. A devastating epidemic in the 1570s killed millions more Andean Indians and ended all hope of future empire-wide rebellions.

Christopher Howell

See also: Chan Chan, Battle of; Cuzco, Battles of; Pachacutec Yupanqui

References and further reading:

Bram, Joseph. *An Analysis of Inca Militarism.* New York: Augustin Publisher, 1941.

D'Altroy, Terrence, ed. *Provincial Power in the Inca Empire.* Washington, DC: Smithsonian Institution Press, 1992.

Garcilaso de la Vega, El Inka. *The Incas: Royal Commentaries.* 3d ed. Ed. Alain Gheerbrant. New York: Avon Books, 1971.

Hemming, John. *The Conquest of the Incas.* New York: Harvest-Harcourt Brace Jovanovich, 1970.

Inchon Landings (15 September 1950)

The U.S. landings that changed the course of the Korean War, temporarily. For several weeks, Eighth U.S. Army and Republic of Korea forces had stubbornly held a perimeter around Pusan in extreme southeastern Korea. The Korean People's Army of North Korea had committed the bulk of its forces there and allied air attacks had limited supplies to the front. The North Koreans were stretched dangerously thin.

MacArthur overrode concerns by U.S. Navy and Marine Corps strategists about the dangers of assaulting Inchon by sea—Operation CHROMITE. High tides, a narrow shipping channel, and a very tall sea wall worried virtually every senior officer who reviewed the plans. MacArthur believed that the advantages, especially an attack along a line of least expectation, outweighed the concerns. And he ordered decoy attacks farther south along both coasts to distract the North Korean command.

The attack at Inchon went spectacularly well. U.S. Marines landed first and secured the port; U.S. Army units provided added punching power. After securing the port, the invasion force moved the short distance to Seoul by September 29. Meanwhile, United Nations (UN) forces in the Pusan perimeter scheduled a simultaneous breakout that linked up with the Inchon invading force on September 26 near Osan, site of the first American resistance to and defeat by the North Koreans. The North Koreans were cut off and, by September's end, that army largely had ceased to exist as an organized force south of the thirty-eighth parallel. The Inchon landings have been held up through the subsequent decades as a near-perfect example of amphibious operations. Inchon was not the "desperate gamble" of legend, however; American forces had complete control of the air and seas and a substantial superiority in numbers. Still it was MacArthur's last victory and a most impressive operation.

Charles M. Dobbs

See also: Korean War; MacArthur, Douglas

References and further reading:

Heinl, Robert Debs, Jr. *Victory at High Tide: The Inchon-Seoul Campaign.* Annapolis: Nautical and Aviation Publishing Company of America, 1980.

Langley, Michael. *The Inchon Landing: MacArthur's Last Triumph.* New York: Times Books, 1979.

Sandler, Stanley. *The Korean War: No Victors, No Vanquished.* Lexington: University of Kentucky Press; London: Routledge, 1999.

Sheldon, Walt. *Hell or High Water. MacArthur's Landing at Inchon.* New York: Macmillan, 1968.

Indian Border Conflicts (1962–1971)

The granting of Indian independence in August 1947 occasioned a series of long-standing border disputes with neighboring countries. Foremost of these was the Indian annexation of Kashmir to the north, despite its predominately Muslim population. Beyond Kashmir was the Jammu region adjacent to Tibet, then jointly claimed by both India and China. Prime Minister Jawaharal Nehru was intent upon maintaining friendly relations with the People's Republic, but as early as August 1958 he complained of Chinese border violations in the Longju and Ladakh regions. Troops were rushed in to enforce Indian claims, but in October 1962, the government of Chairman Mao Zedong launched a massive surprise attack to evict them. Through a series of lightning thrusts, Indian forces were handily repulsed from the disputed territories, at which point China declared a unilateral cease-fire. India, soundly trounced, had no recourse but to accept a truce.

In the spring of 1965 violence sprang up along the border with Muslim Pakistan over the issue of Kashmir, which also spilled over into the Indian state of Punjab. In August 1965 India commenced formal hostilities with a large-scale raid into Kashmir, which was followed by a major Pakistani thrust into the same region. The fighting was particularly intense and featured widespread use of aircraft against cities and large-scale tank battles. The result, however, was a bloody stalemate and by September 1965 a United Nations–brokered cease-fire was enacted. Further border tensions with China also induced the Indian government to enter a 20-year Treaty of Friendship with the Soviet Union in exchange for large shipments of modern weapons. The Indian governments resolved to be fully prepared for the next round of fighting, when it occurred.

In December 1971, civil war erupted in the nominally Muslim state of East Pakistan, which was separated from West Pakistan by 1,000 miles of Indian territory. Rebels in East Pakistan were seeking autonomy and readily turned to Hindu India for help. The result was an increase of border tensions with West Pakistan, which goaded that country into launching a preemptive air strike against India on 3 December 1971. The Indians, however, were ready and easily parried the blow. Moreover, enjoying a threefold manpower advantage, their forces rolled into East Pakistan. The West Pakistani garrison there was hard-pressed, so as a diversion Pakistani forces also launched an unsuccessful diversionary attack into Kashmir. However, Indian numbers prevailed, and on 16 December 1971 they captured the East Pakistani capital of Dacca, which signaled the general collapse of resistance. Pakistan had been badly defeated, with losses of 4,000 dead and 10,000 wounded; Indian losses were half that. Moreover, a new state, Bangladesh, was created from the former East Pakistan, which effectively reduced West Pakistan's influence throughout the region. The Kashmir issue remains unresolved, and Indian and Chinese forces still dispute some of the most remote and elevated regions on earth.

John C. Fredriksen

References and further reading:

Ganguly, Sumit. *The Origin of War in South Asia: Indo-Pakistani Conflicts since 1947.* Boulder, CO: Westview Press, 1986.

Palit, D. K. *War in High Himalaya: The Indian Army in Crisis, 1962.* New York: St. Martin's Press, 1991.

Indian Mutiny (1857)

Violent uprising of the Indian Bengal Army, hailed by many Indians as a nearly successful revolution. The Indian army of the time was a mixture of 39,750 British officers and troops supervising 226,400 trained Indian soldiers. During the 1850s the British Indian Officer Corps had declined in numbers, quality, and experience and had begun to cause bitter resentment among Indian troops by widespread disrespect for their religious and ethnic beliefs.

In 1856 an Indian revolutionary movement began targeting Indian troops with stories of religious persecutions and blasphemies planned by the British. Although the movement was not highly organized, the situation was aggravated by continued British acts of ignorance, and by 1857 an uprising was inevitable.

Rumors had been spread that cartridges for the new Enfield rifle, which required handling during loading, contained pig and beef fat, which was highly offensive to both Muslims and Hindus. On 8 May 1857, the mutiny was triggered when some Indian troops of the 3d Light Cavalry stationed in Meerut were forced to declare their individual position regarding the new cartridges; those renouncing use of them were court-martialed and jailed. The Indian garrison in Meerut mutinied and freed the prisoners on Sunday, 9 May, and with a growing civilian mob took over the town, which led to a massacre of local Europeans.

The mutineers then moved to Delhi, 40 miles distant. Delhi had symbolic importance as the former center of the Mogul Empire and was also the focus of British administration for the whole of northern India. The 54th Native Infantry Regiment in Delhi turned on its English officers and joined the 3d Light Cavalry in capturing the city, and most resident Europeans were slaughtered. Bahadur Shar, last of the line of Mogul rulers, was proclaimed head of a new Mogul Empire. Unfortunately for the rebels, he was ineffectual; a strong leader might have driven the British out of India entirely. (The Indians had a saying, "If we would all only spit together, the British would drown!")

British garrisons and civilian enclaves throughout central and northern India were aware of the danger but could only await events during May 1857. Regiments disarmed their Indian troops or retreated into semifortified positions. However, word of events in Delhi spread among the Indians and many local revolts occurred rapidly. Cawnpore was captured after a three-week battle, and Lucknow besieged. In most smaller towns Europeans were massacred before rebel troops moved on to join larger actions.

It took a series of campaigns lasting until March 1858 (when Lucknow was finally recaptured) to end the mutiny. Sir Henry Havelock led the first relief column from Allahabad to Lucknow via Cawnpore but was besieged in turn, and not until November were the surviving Europeans evacuated. Delhi was recaptured in September 1857 by British troops with help from Sikhs under John Lawrence. Sir Hugh Rose suppressed the mutiny in the central districts of India in a series of operations. As the details of the Cawnpore massacre became widely known, British treatment of rebels was brutal, and executions were indiscriminate—many captured rebels being blown from cannons.

After the mutiny, the British army replaced East India Company garrisons throughout India, and native troops were not trained in artillery specialities.

Michael Hyde

See also: British-Indian Army

References and further reading:

Forbes-Mitchell W. *Reminiscences of the Great Mutiny 1857–59.* London: Macmillan and Co., 1893.

Haq, S. Moinul. *The Great Revolution of 1857.* Karachi: Pakistan Historical Society, 1968.

Harris, J. *The Indian Mutiny.* London: Hart-Davis MacGibbon, 1973.

Holmes, T. R. E. *A History of the Indian Mutiny, and the Disturbances which Accompanied It among the Civil Population.* London: W. H. Allen and Co., 1883.

Indian National Army (1943–1945)

An "independent" Indian army established under Japanese tutelage in 1943 to overthrow British rule in India. Started by the Japanese, it came into its own with Subhas Chandra Bose as its leader. Bose had been elected president of the Indian National Congress in 1938 and was a political and ideological rival to Gandhi. He resigned when he was unable to gain Gandhi's support after a bitter reelection. Arrested by the British in 1940, he went on a hunger strike that secured his release from prison. He evaded the British and fled in January 1941 to Berlin, via Moscow, to work for the Nazi's Special Bureau for India in 1942. A year later, after Japanese advances into Southeast Asia, he traveled to Tokyo and proclaimed a provisional Indian government. He became leader of the Indian Independence Movement and set about building up a Indian National Army (INA, also known as the Azad Hind Fauj), many of whose troops were disaffected Indian POWs from the ignominious fall of Singapore to the Japanese.

In 1944, his army took to the field with Japanese troops, advancing through Rangoon, and then into India, across the national frontier on 18 March 1944. In rough fighting, the Indian National Army was unable to occupy Kohima and advance toward Imphal. The British under General William Slim were able to resist and the joint Indian National/Japanese Army force was driven back, due in large part to a lack of air support. The INA was able to maintain some semblance of legitimacy, of carrying the banner of Indian liberation, but waned as the Japanese were driven back by the Allies, and many troops deserted back to the British-Indian forces. When Japan surrendered, so did the Indian National Army. Bose died in a mysterious aircraft crash sometime in August 1945.

Drew Philip Halévy

See also: Slim, William Joseph, First Viscount; World War II

References and further reading:

Calvocoressi, Peter, and Guym Pritchard. *Total War: Causes and Courses of the Second World War.* New York: Pantheon Books (1989).

Indochina Wars (1945–1954)

When the Pacific War ended suddenly in September 1945, Great Britain arranged for the return of the French to their colonies in Indochina. During the war, the Japanese had taken advantage of the regime in Vichy to occupy Indochina, and later, in March 1945, to disarm and imprison French troops. When Chiang Kai-chek's Nationalist forces left northern Indochina, it appeared the French had successfully reestablished that part of their empire.

It was not to be. During the long years of war a Vietnamese independence movement, dominated and led by Communist cadres, had arisen. Led by Ho Chi Minh, a charismatic leader, the so-called Vietminh had proclaimed Vietnamese independence in fall 1945 (the French had divided Vietnam into three units—Annam, Tonkin, and Cochin China—and thus Indochina into five states). Nonetheless, the French had little difficulty in reestablishing control. They drove Ho and his colleagues out of Hanoi and established a nominally independent regime under the emperor Bao Dai, although all real power remained in French hands.

Until the Communists conquered southern China in late 1949, the French were able to maintain control. But with the Communists on Vietnam's border, Ho's guerrillas would benefit from more secure access to supplies and advice, and the tide of war began to turn slowly but increasingly against France. Within a couple of years the French largely controlled only the two major river deltas, the Red River in the north (centering on the port of Haiphong and the capital city of Hanoi) and the Mekong River in the south (centering on the one-time fishing village of Saigon), as well as a coastal strip connecting the two. Control over much of the countryside and, at night, even major transportation routes was ceded to the Vietminh.

By 1953, the Vietminh had the upper hand. The French commander, Henri Navarre, decided to reverse policy and establish a strong base in northwestern Vietnam at Dien Bien Phu, about 180 miles from French bases around Hanoi, to interdict Vietminh forces moving onto the strategic Plaine des Jarres in Laos. Dien Bien Phu is an isolated plateau, surrounded by mountains; the French gambled that they could bring in a large garrison, occupy the surrounding hills, and destroy the Vietminh when they came out onto the plains to attack the entrenched French.

The French had seriously miscalculated. The Vietminh commander, Nguyen Vo Giap, decided to concentrate against the French garrison. He moved at least 50,000 men into the surrounding hills; using gang labor, he brought up artillery and weapons the French believed they could not transport. Soon the French were caught in a trap of their own making. The siege began in December 1953. On 13 March, after an intense artillery barrage, the Vietminh sent two battalions each against French outposts, and thereafter Vietminh artillery closed the airfield. Air dropping of supplies was haphazard and only added to the desperation of the French defenders. In late March, Giap's forces overran the other French outposts, and then a month of siege tactics ensued, followed by an attack on 1 May that led to the garrison's surrender on 7 May 1954. When the garrison surrendered, it seemed to mark an end to the French effort; the Geneva Conference on Indochina had begun, and the French withdrew. It would be America's turn next.

Charles M. Dobbs

See also: Chinese Civil War; Dien Bien Phu; Indonesian War of Independence; Vietnam Conflict

References and further reading:

Devillers, Philippe, and Jean Lacouture. *End of a War: Indochina, 1954.* Trans. Alexander Lieven and Adam Roberts. New York: Praeger, 1969.

Fall, Bernard. *Street without Joy: Indochina at War, 1946–1954.* Harrisburg, PA: Schoken Books, 1961.

Hammer, Ellen J. *The Struggle for Indochina, 1940–1955.* Stanford, CA: Stanford University Press, 1966.

Pike, Douglas. *History of Vietnamese Communism, 1925–1976.* Stanford, CA: Hoover Institution Press, 1978.

Indonesian War of Independence (1945–1949)

An internal conflict, that, with United Nations and United States aid, brought about the independence of Indonesia. As the Japanese were surrendering in World War II, Indonesian nationalists developed the Jakarta Charter, a plan for an independent, secular Indonesian state, which was proclaimed on 17 August 1945. Before Allied forces could intervene, Indonesian armed forces were formed and put to work crushing Muslim and Marxist revolutions in the north. British Indian troops finally arrived at Jakarta on 30 September. Violence erupted at Surabaya in East Java on 28 October. Hundreds of Indian troops were killed and thousands of Indonesians perished in British reprisals and Indonesian counterattacks. Convinced that they could not manage a military solution, the British urged talks between the Indonesians and Dutch. The Linggajati Agreement of 25 May 1947 provided for a Netherlands-Indonesian union, to which neither side was truly committed. Dutch forces, now numbering about 92,000, launched what they termed a "police action" on 21 July 1947, extending Dutch control to all of Java and Sumatra except the area around Jakarta. On 17 January 1948, the Renville Agreement recognized Dutch control of these areas but mandated plebiscites. Despite this, all Dutch-held areas were organized into a system of federated republics. Outrage at the situation sparked a communist uprising in September, whose speedy suppression by loyal Indonesian forces led the United States to view a prospective Indonesian republic as a bulwark against communism. The U.S. could exert enormous pressure on the Netherlands because of the Dutch need for American economic aid in the wake of World War II. But the fact that the communist uprising had happened at all led the Dutch to think that the republican movement was about to fracture, and on 19 December they occupied Jakarta and imprisoned the republican leaders. Guerrilla activity intensified, and the United Nations Security Council demanded restoration of the republican leadership and a complete transfer of authority. On 27 December 1949 full sovereignty was transferred to the Indonesian Republic and the 15 Dutch-established states. By November 1950 the 15 had been absorbed by the republic, and Indonesian sovereignty was complete.

Joseph M. McCarthy

See also: Dutch Colonial Wars

References and further reading:

Cribb, Robert E. *Gangsters and Revolutionaries: The Jakarta People's Militia and the Indonesian Revolution, 1945–1949.* Honolulu: University of Hawaii Press, 1991.

Frederick, William H. *Visions and Heat: The Making of the Indonesian Revolution.* Athens: Ohio University Press, 1989.

Wild, Colin, and Peter R. B. Cary, eds. *Born in Fire: The Indonesian Struggle for Independence, an Anthology.* Athens: Ohio University Press, 1988.

Infantry

Lightly armed ground troops. The backbone of any army is the infantry. Not only the most numerous of the fighting arms, but the only fighting arm that can take and hold ground. It has long been recognized that well-trained infantry would always beat mere masses of men. Further, if the well-trained men were also well armed, well led, and well equipped, they would be second to none.

The armies of Greece boasted the men of the phalanx; of Rome, the men of the legions. When led by great generals such as Philip of Macedon and Marius, Caesar, and Pompey, these infantry were capable of great feats of arms and endurance, achieved almost always by maneuver rather than face-to-face slogging. These were the forebears of all the infantry who have come onto the battlefield since, and their tradition of fighting with honor is perpetuated today.

Traditionally the infantryman has always been regarded as the "footslogger," the soldier who gets about on foot, and this has been true until the twentieth century. The rate of advance of the Greek phalanx and the German army of 1914 were virtually identical. The need for mobility and maneuver was foreseen by Napoleon, and his troops were able to move quickly by forced march, but little could be done to bring true mobility to the infantry until the advent of the internal combustion engine. During the wars of the nineteenth and early twentieth centuries, just as before, the infantry had the task of advancing toward the enemy until contact was established. Once that had occurred, with or without support from artillery, and later armor, it was the infantry's task to bring the enemy to battle and to kill him. The great generals, of course, were those who outmaneuvered the enemy in these advances, catching the enemy in the flank or the rear rather than head-on.

World War I saw more massed infantry involvement than in any previous conflict. In the east, German infantry fought and beat Russian infantry after three years of war. In the west the so-called war of attrition lasted for more than four years. Despite every indication that frontal attack was not viable against barbed wire, entrenched defenders, machine guns, artillery, and gas, both sides continued to make frontal assaults at a cost in men that presents a sickening picture of classic military ineptitude at command and staff level.

The arrival of effective armor and air support, heralded by the German blitzkriegs of 1939 and 1940, preempted the expected stalemate, in both the west and the east. Hitler's highly mobile Panzer formations, supported in the main by infantry on foot, defeated the forces of Poland, France (and the static Maginot line), Great Britain, Belgium, Holland, Norway, Yugoslavia, and Greece.

Hitler then sent his army into Soviet Russia. Initially successful, the German infantry got to the gates of Moscow. Then, however, the Hitlerian exhortations to hold at all costs, avoid retreat, and fight to the last man and bullet led to ultimate defeat. Despite the many tanks and aircraft at their disposal, it was infantry who formed the heart of both the German and Russian armies.

Masses of infantry moving on foot were standard until motorization. Aside from the American and British forces, armies have only truly become motorized since World War II. The German army between 1939 and 1945 was still heavily dependent upon the horse, in both offense and defense. However, at the end of the twentieth century, most developed nations have equipped their infantry with armored transport, either wheeled or tracked, the aim of which is to deliver them as far forward in the battle zone as possible under protection and unfatigued. Infantry, it has finally been recognized, need to be fresh and unharmed when they go into battle.

Modern infantry still have the prime role in land warfare—taking and holding ground. However, the modern infantryman has become a specialist. Regular infantry must be expert in small arms, antitank weapons, signals, and field engineering, as well as able to cope with threats from nuclear, chemical, and biological weapons.

In the last decade of the twentieth century many infantrymen have become peace keepers, especially in the former Yugoslavia, as well as in some areas of Africa.

Today all entrants to land forces are trained initially as infantrymen. Infantry skills are indispensable to all soldiers, and every soldier, whether truck driver, cook, bandsman, artilleryman, and so on, must be prepared to fight as an infantry soldier. Skills taught include group training, such as drill, battle PT, and first aid. Further infantry skills are added and exercised throughout every man's career. Expertise in using rifles, machine guns, mortars, and antitank and antiaircraft weapons; practice in field craft, battle craft, river crossing, mountain and arctic warfare, and warfare in wooded and built-up areas; and signals communications, infantry-tank, infantry-air, and infantry-artillery cooperation are all taught, practiced, and used. No longer is the infantryman just a body on the battlefield; he is, today, as skilled as any other specialist, and furthermore he is always at the sharp end.

Further infantry specialization has led to the formation of airborne troops (both parachute and air-mobile) and special forces. Airborne forces are projected behind enemy lines to capture and hold strategic points, where they hold on until relieved by troops advancing to relieve them. Special operations forces, such as Rangers or Special Forces, perform specialized tasks, often in aid of other governmental authorities. In war they destroy headquarters and other strategic targets by coup de main, raise insurgency warfare against unfriendly powers, or train friendly forces in counterinsurgency.

Infantry have benefited greatly from modern concepts and are rarely committed without armor, artillery, air, and logistic support. The infantry may still march on its stomach, but it fights as a combined team. The use of infantry is now part of the maneuverist approach, whether by land, sea, or air. Every modern generation has its experts predicting their war of tomorrow, push-button warfare, and so on, but almost all wars tend, finally, to come down to the infantryman in the mud.

David Westwood

See also: Airborne Operations; Armor, Ancient and Medieval; Artillery

References and further reading:

Eady, Major H. G. *Historical Illustrations to Field Service Regulations.* Vol 2. London: Sifton Praed and Co, 1927.

Chandler, David, ed. *The Oxford History of the British Army.* Oxford and New York: Oxford University Press, 1994.

Dupur, R. Ernest, and Trevor N. Dupuy. *The Collins Encyclopedia of Military History.* London: BCA, 1993.

Inkerman, Battle of the (5 November 1854)

Allied victory over Russia, regarded by Britain as a moment of glorious triumph in the Sebastopol campaign of the Crimean War. The British 2d Division under General Sir George De Lacy Evans, 3,000 strong, defended 18 field pieces on the rugged plateau east of Sevastopol between the Tchernaya River and Careening Ravine. It was a weak point in the allied lines because the position allowed concealed approaches through Careening Ravine to the west, Volovia Ravine to the north-northwest, Quarry Ravine to the north, St. Clements' Ravine to the north-northeast, and several adjoining gullies. Also, many of its key defenders, including General Sir Colin Campbell's 93d Highlanders, had been transferred south to defend Balaklava. The Light Division, 1,200 under General Sir George Brown, and the Guards, 1,300 under General H. J. Bentinck, were camped about a mile south. French and Turkish reserves, 23,000 under General Pierre Jean François Bosquet, were camped a few miles east and southeast.

On 5 November 1854 Prince Alexandr Sergeevich Menshikov ordered 19,000 men with 38 guns under Lieutenant General F. I. Soimonov to split the British army from the west, 16,000 men with 96 guns under Lieutenant General P. I. Paulov to attack from the north, 22,000 men with 88 guns under Prince Mikhail Dmitriyevich Gorchakov to prevent Bosquet from providing reinforcements, and 20,000 men under General P. A. Dannenberg to wait in reserve to the northwest. The Russians had to attack uphill and the British Minié rifles had twice the range of the Russian muskets, but the main reason that the British defense succeeded was because the Russian generals did not communicate effectively. British reinforcements broke through early in the day. Bosquet joined the battle much later.

In total, about 22,000 British and 20,000 French and Turks took part in the engagement. After eight hours of fighting, Russian casualties were more than 12,000, British about 2,500, and French and Turkish about 1,700. Inkerman was the last serious Russian attempt to destroy the allies in the Crimea. From then on, attrition and logistics decided the Sebastopol campaign.

Eric v. d. Luft

See also: Alma; Balaklava; Campbell, Colin; Crimean War; Light Brigade, Charge of the; Sevastopol, Siege of

References and further reading:

Barker, A. J. *The War Against Russia, 1854–1856.* New York: Holt, Rinehart & Winston, 1971.

Barthorp, Michael. *Heroes of the Crimea: The Battles of Balaclava and Inkerman.* London: Blandford, 1991.

Royle, Trevor. *Crimea: The Great Crimean War, 1854–1856.* New York: St. Martin's, 2000.

Seaton, Albert. *The Crimean War: A Russian Chronicle.* London: Batsford, 1977.

Intelligence, Military

Military speciality that provides a commander and staff with the knowledge of enemy weather and terrain required for the planning and conduct of operations.

The requirement for information on an opponent has existed throughout recorded history. The Hebrew Scriptures and Sun Tzu's *Art of War* (c. 500 B.C.E.) refer repeatedly to aspects of intelligence, reconnaissance, counterintelligence, and awareness of weather and terrain. Although military espionage existed in the intervening period, it was not until the twentieth century that the importance of military intelligence was firmly established, with increased efforts made to formalize its processes. This change began during the industrialized stalemate of World War I. Sheer weight of arms often proved inadequate to force a military decision, thus providing impetus to fight more wisely. World War II's technological improvements, such as the interception and analysis of strategic communications, brought international consensus on the growing significance of intelligence. The threat of nuclear warfare by the five declared atomic powers, coupled with development programs in other countries, keeps postwar intelligence efforts robust. The main focus has been on acquiring information on weapons characteristics and seeking warnings of potential nuclear attacks.

The emphasis on military intelligence operations continued beyond the end of the Cold War as targeting shifted to other concerns, such as terrorism and narcotics. Concurrently, traditional war fighting interests of military intelligence have been expanded by the concept of *information operations,* which envisages attacking an adversary through electronic and information systems technology. The span of options include traditional psychological operations or tactical radio jamming as well as attacking an electronic banking system or those computers that control the dams that keep an enemy's country from flooding. In all of these, intelligence provides the knowledge of the adversary's vulnerabilities and a follow-on assessment of the operation's success. There is naturally a concomitant requirement to defend one's own exposed infrastructure.

Intelligence will often occupy a separate appointment among the headquarters staff. Within Western armies the four key appointments are G1-Personnel Administration, G2-Intelligence, G3-Operations, and G4-Logistics, although the letter may vary; for example, J2 for Joint Staff Intelligence or A2 for Air Staff Intelligence. The Prussian General Staff popularized this practice of separate staff appointments in the nineteenth century; hence, G for General Staff. Specific appointments occurred as the functions became sufficiently specialized that the tasks required dedicated expert personnel. An additional benefit is that the division between operations and intelligence reduces the likelihood that the intelligence product will be skewed to fit a preconceived operational plan. During Operation MARKET GARDEN in 1944 Holland, repeated cancellations of airborne operations due to the rapid Allied advances proved frustrating to the senior paratroop officers. Thus significant indications of German tank formations in the vicinity of the farthest objective, Arnhem, were ignored in order to ensure that the operation was approved.

The general focus on enemy, weather, and terrain provides awareness sufficient to conduct most war-fighting operations. The elements of strategic intelligence are generally accepted as collection, analysis, counterintelligence, and covert action. However, in a military context the latter is a Special Operations Forces responsibility, notwithstanding their distinct military intelligence requirements.

Collection refers to the gathering of information for intelligence staff assessment. Information sources can be divided roughly between technical and human means. The former will include signals intelligence (SIGINT), gained from intercepting radio and radar emissions; aerial and satellite photo and radar imagery (IMINT); and capabilities and weaknesses determined from examining captured equipment or intercepted telemetry (TECHINT). Human intelligence (HUMINT) traditionally meant spies. Within a military intelligence context, however, reconnaissance troops and questioning prisoners of war or refugees provide HUMINT. Humanitarian operations with the attendant nongovernmental organizations throughout a theater of operations are further potential information sources. Finally, open source information from readily accessible publications and broadcasts provides much data.

Analysis is the process of turning information into intelligence. Interpreting often fragmentary or ambiguous information requires reasoned assessment through comparison with other details, either known facts or previous assessments of how an adversary traditionally operates. This process generates three types of intelligence: basic intelligence, such as the enemy force strength and composition (order of battle, or ORBAT, information); current intelligence, providing awareness of an adversary's current activities and specific indicators and warnings (I&W); and estimates or forecasts of the enemy's future activity. Military intelligence estimates are often prepared with a view toward the enemy's most likely and most dangerous courses of action. These intelligence types are interrelated. For example, IMINT and national-level ORBATs indicate that the enemy has a tank division on the far side of a river (basic intelligence). A captured soldier and intercepted radio communications indicate that the enemy division is preparing to attack (current intelligence). The intelligence staff assesses that the enemy will likely attack where the river is narrowest, but could attack downstream where the terrain would provide the defenders with difficulty counterattacking. The commander may then choose to reinforce the downstream site before the attack, but concentrate on defending at the river's narrowest point.

The focus of counterintelligence (CI) is different from, but complementary to, traditional intelligence. CI protects information and a military's intelligence system rather than aiming to collect and analyze information, thus providing similarities with a police function. The two subsets of CI are security and counterespionage. Because security aims to keep information from those not authorized access, coverage often falls upon physical security: locks, fences, and security sweeps. In a tactical situation, field security would be concerned with keeping soldiers from carrying unit identifiers or operational plans into battle. Counterespionage previously emphasized apprehension or neutralization of an enemy's information-collection agents. However, in a technical environment, human spies are no longer an overarching counterespionage concern. Much information previously acquired through spying can now be gathered through technical means, such as monitoring the electronic emanations from computers rather than stealing their documentary product. Additionally, neutralizing an enemy's collection efforts can be accomplished through deception operations. These aim to provide an enemy intelligence service, and hence their commander, with false or misleading information. Deception operations will be more effective if they are fed through multiple channels, thus appearing to provide confirmation, and if these messages reinforce an enemy's preconceived notions. This was the case during the World War II Normandy invasion when false radio traffic, mock-up vehicles, and the physical isolation of the southeast coast of England from German espionage all indicated that a First U.S. Army under General Patton was poised to strike at Calais, a point already believed by Hitler to be the assault objective.

A number of factors make intelligence failure or surprise practically inevitable. In the previous example an attack is expected, but because of active deception and inclement weather, the intelligence picture indicated a different location and time. A more damaging surprise occurs when intelligence provides no warning of enemy assault, as seen with the Japanese naval attack at Pearl Harbor. Surprise may also result from having adequate information misinterpreted through faulty understanding of an adversary's doctrine. During the Korean War (1950–1953), Chinese light infantry dispersed away from the main highways, unlike North Korea's mechanized army. This differing doctrine negated the American air superiority, allowing the Chinese initially to rout the allied forces. Intelligence failures are therefore a problem of active deception (Normandy), information not being available in a clear and timely manner (Pearl Harbor), or mirror-imaging (blinding) analysts to the significance of different enemy doctrine (North Korea).

The breadth and depth of information requirements will vary with the strategic versus tactical level of operations. For example, in a conventional war, knowing an enemy's strategic oil reserves is of little interest to a platoon commander holding a ridgeline against enemy attack. Conversely, the difference between that attacking enemy possessing three tanks or twenty is of little interest to the National Command Authority. However, nontraditional conflicts falling outside of conventional war, such as humanitarian relief operations or those missions conducted by Special Operations Forces, pose different demands upon military intelligence. An on-scene commander may need intelligence on the religious beliefs of a particular village or cleared routes to a neonatal care facility rather than the range of a particular antitank weapon system. In these operations, political and economic concerns, previously the purview of strategic-level staffs, are required at the tactical level.

Military intelligence will remain focused upon the basics of enemy, weather, and terrain for the foreseeable future. However, changes in technology, doctrine, and forms of conflict will continue to force military intelligence evolution.

Robert Martyn

See also: Psychological Operations; Special Operations Forces

References and further reading:

Gudgin, Peter. *Military Intelligence: A History.* New York: Sutton Publishing, 2000.

Handel, Michael. *Intelligence and Military Operations.* London: Frank Cass, 1990.

Shulsky, Abram. *Silent Warfare: Understanding the World of Intelligence.* 2d ed. Washington: Brassey, 1993.

Interventions in Civil Unrest, Strikes, Military

Around the world, there remain many regimes in power only by the intervention of the military within that country. In China, the picture of the man and the tank from Tiananmen Square speaks volumes. The demise of the Soviet Empire and the Warsaw Pact began when Prime Minister Mikhail Gorbachev refused to use troops in the Baltic countries or to support regimes such as East Germany.

In the United States, military intervention in civil unrest is the exception, not the rule. However, the drawing up of the Constitution of the United States drew impetus from the failure of the national government under the Articles of Confederation to aid Massachusetts in suppressing Shays's Rebellion in 1786. Thus the nation ratified a constitution that authorized the federal government to use military force in civil disorders, first to enforce federal authority, and second to assist states when they were unable to restore domestic tranquility.

The first use of military intervention under the Constitution occurred in response to the Whiskey Rebellion of 1794. President George Washington called out militia from four states to march on western Pennsylvania to suppress the tax revolt. President Washington set a precedent for presidents acting with great restraint in matters regarding military intervention into civilian life.

For the next 80 years, slavery and abolition provided most of the occasions for military intervention. Federal troops were sent to Boston in the 1850s to quell demonstrations against the enforcement of the Fugitive Slave Law. Similarly they were dispatched to "Bloody Kansas" in the late 1850s. Of course, federal troops quelled the "rebellion" in the southern states, 1861–1865.

One exceptional case in 1857–1858 saw the use of 2,500 regulars under the command of A. S. Johnson to enforce federal law and court orders in Utah Territory regarding the Mormon Church.

After the Civil War, the main use of U.S. troops in domestic affairs was in the widespread activities of the Freedmen's Bureau, established to aid the newly freed blacks in the South, headed by an army general whose policies were carried out throughout the South by regular troops and black state militias. Although the Freedmen's Bureau carried out many good works (providing the South with its first tax-supported public school system, for example), it was detested by white southerners, who liked military government no more than Americans from any other section of the nation.

The U.S. military was also involved in industrial strikes and disturbances in the postwar decades, and often reserve or state militia troops were involved, rather than regulars. Five times in 1877, President Hayes sent in troops to assist states, although he denied four other requests. In 1894 President Grover Cleveland sent regular troops to Chicago during the railroad strike. At the beginning of the twentieth century, President Theodore Roosevelt exercised greater reticence in responding to similar requests. However, both Presidents Woodrow Wilson and Warren Harding sent in troops in 1914 and 1921, respectively.

In most of these cases, when federal troops arrived, calm was restored quickly with no loss of life. Aside from the Civil War, one of the worst incidents was the Ludlow Massacre, in which 13 women and children died in 1914.

Since World War II, urban and racial disturbances have occasioned most military interventions. Detroit (1943, 1967), Los Angeles (1965, 1997), and the numerous riots that occurred upon the assassination of Martin Luther King Jr. (1968) were the most prominent. Regular troops and National Guardsmen were also deployed to enforce federal court desegregation orders in the 1950s and 1960s.

In general, the United States military has had little desire or preparation for military inventions in civil disorder. During most of the history of the United States, the military operated only under direct orders of the commander in chief. The one exceptional period occurred during 1917–1920, when the newly created Military Intelligence Division (MID) of the Army General Staff engaged in its own domestic surveillance and even arrest of various labor radicals deemed Bolsheviks.

At present, guided by their oath "to support and defend the Constitution of the United States against all enemies, foreign and domestic," the military is much more interested in confronting foreign enemies. It is glad to leave the decision to intervene militarily in any civilian turmoil in the hands of the civilian leaders. The marked American distaste for the military, despite the nation's propensity for individual firearm violence, has made military intervention in civil life all the more rare.

A few other democratic states have had to use the military in a far more direct role than was the case in the United States. Canadian military forces deployed into the streets of Quebec's cities in 1970, arresting and holding citizens without warrant in the wake of Separatist violence, and British troops have policed Northern Ireland since the early 1970s. The German, Japanese, Italian, French, and Scandinavian governments have, on the other hand, refrained since the end of World War II from using their militaries to intervene in civil disturbances.

John R. Popiden

See also: Civil Affairs/Military Government; Military and Society

References and further reading:

Coakley, Robert W. *The Role of Federal Military Forces in Domestic Disorders, 1789–1878.* Washington, DC: Government Printing Office, 1988.

Cooper, Jerry M. "Federal Military Intervention in Domestic Disorders." In *The United States Military under the Constitution of the United States, 1789–1989,* ed. Richard H. Kohn, 120–150. New York: New York University Press, 1991.

Engdahl, David E. "Soldiers, Riots and Revolution." *Iowa Law Review* 57 (October 1971), 35–42.

Sandler, Stanley. *Glad to See Them Come and Sorry to See Them Go: A History of U.S. Army Tactical Civil Affairs/Military Government, 1775–1991.* Fort Bragg, NC: U.S. Army Special Operations Command, 1998.

Iran Hostage Rescue Attempt (24–26 April 1980)

Failed attempt by the U.S. Armed Forces to rescue American hostages being held in the American embassy in Teheran. The rescue attempt was made after Iranian "students" seized the embassy on 4 November 1979 and seized most of its staff. Shortly after the seizure a joint task force (JTF) was established to begin planning a rescue. The JTF's plan was complex, involving members of all of the armed services and calling for several C-130s to land at a site in the Iranian desert code-named Desert One. The aircraft would transport the highly secret special operations unit, Delta Force, and other support troops to the site and refuel the RH-53 helicopters that would transport the commandos to Teheran to rescue the hostages. Almost immediately after landing at the site, trouble occurred when a bus full of Iranians was detained on what was supposed to be a remote road. The helicopters encountered the weather phenomenon known as a *haboob,* a dust cloud, which delayed their arrival at Desert One. During refueling at the site one of the helicopters suffered a hydraulic failure, leaving the mission with five helicopters when six had been determined as the mission minimum. The decision had already been made to abort the mission when one of the helicopters collided into a C-130. The ensuing explosion killed eight flight crew members.

The failure of this mission led to the establishment of the Special Operations Command and the U.S. 160th Special Operations Aviation Regiment and to an upgrading of all American special operations assets by the incoming Reagan administration.

Michael Mulligan

See also: Entebbe Rescue Raid; Special Operations Forces

References and further reading:

Beckwith, Charlie A. *Delta Force.* New York: Harcourt Brace Jovanovich, 1983.

Kyle, James H. *The Guts to Try.* New York: Orion Books, 1990.

Iran-Iraq War (1980–1988)

The longest war between Third World countries since the end of World War II, and Saddam Hussein's first war. On 22 September 1980, Iraqi divisions crossed the Iranian border. In eight days Iraqi troops reached six to twenty-five miles deep into Iran, including the outskirts of the oil port cities Khorramshahr and Abadan on the Shatt al Arab.

The causes of this war were political, economic, and religious. In 1975 Iran and Iraq agreed over their border along the Shatt al Arab. In 1980, now ruling Iraq, Saddam Hussein claimed complete control of the Shatt al Arab to gain clear access to the Persian Gulf for oil export. Similarly, Iran, now under the rule of the Ayatollah Khomeini, had sought to undermine the Iraqi regime by instigating rebellion by both the Shiites in the south and the Kurds in the north. Also random Iranian artillery fire and air strikes struck towns and oil facilities in Iraq.

The initial success led Saddam Hussein to call for negotiations, which Khomeini rebuffed. In November 1980, Iraqi troops took Khorramshahr.

However, from then on, Iran gained the military advantage. Iranian volunteers, many of them teenagers, formed the Revolutionary Guards Corps. Also, Iran had the better air force, striking Iraqi air bases and oil facilities with impunity. By the end of the first year, losses were estimated to be 38,000 Iranians and 22,000 Iraqis.

Military tactics stressed entrenchments and the husbanding of valuable equipment. Iranian tactics relied on World War I–style frontal assaults. Iraq introduced chemical weapons on occasion against Iranian troop concentrations, Iranian cities, and Kurd civilians. Both sides used surface-to-surface missiles (SAMs) to attack each other's cities. Eventually, Iran regained its lost territory and entered Iraq, threatening Basra.

In early 1984, both sides escalated the war into the Persian Gulf (the Tanker War). Iraq's best weapon was the Super-Etendard jet using Exocet missiles against tankers carrying Iranian oil. As the flow of oil became hampered, the United States sent its naval forces to protect shipping. In 1987 the United States reflagged Kuwaiti tankers to provide direct protection. Although U.S. activities tended to favor Iraq, the U.S. surreptitiously sold 4,000 target on wire (TOW) and additional spare parts to Iran (Iran-Contra Affair).

In 1987 two Iraqi Exocet missiles struck USS *Stark*, killing 37 U.S. sailors. In 1988, USS *Vincennes* inadvertently shot down an Iranian airliner, killing 290 civilians.

An Iraqi offensive with the greatly expanded Republican Guard recaptured lost territory, returning the frontlines to nearly the original borders. Both sides then agreed to a cease-fire 20 August 1988, after some 262,000 Iranians and 105,000 Iraqis had died. As the interminable war finally ended, Iraq was militarily strong but economically crippled and in massive debt to Kuwait and Saudi Arabia.

John R. Popiden

See also: Hussein, Saddam al-Tikriti

References and further reading:

El-Shazly, Nadia El-Sayed. *The Gulf Tanker War.* New York: St. Martin's, 1998.

Hiro, Dilip. *The Longest War.* New York: Routledge, 1991.

Rajaee, Farhang, ed. *The Iran-Iraq War.* Gainesville: University Press of Florida, 1993.

Irish Easter Uprising, War for Independence, and Civil War; The Easter Rebellion (1916–1923)

The first of a series of events that culminated in the independence of 26 of Ireland's 32 counties. The Irish Republican Brotherhood (IRB) had patiently waited for a chance to stage a rebellion against British rule, and World War I provided an opportunity to achieve their stated goal of an independent Irish republic. Confused by the contradictory orders emanating from the leadership, most of the rural volunteers failed to rise and only 1,600 republicans were engaged in the fighting in the capital, Dublin, throughout Easter week. Without the expected German arms what little hopes the conspirators may have entertained of victory evaporated: 550 were killed, 2,000 wounded, and £2.5 million of damage caused. Initially, the Irish response to the rising was hostile, but the mindless executions of 15 leaders fundamentally altered the political landscape.

The War for Independence

The conflict with Great Britain that won the independence of the 26 counties in the south of Ireland. Sinn Fein, the political arm of the Irish Republican Army (IRA), won the vast majority of seats in Ireland during the British general election of November 1918 and proceeded to establish an independent parliament in Dublin, which Britain refused to recognize. At the same time the IRA began a guerrilla war against British forces. According to IRA leader Michael Collins, the organization's effective fighting strength never exceeded 3,000. The British, however, used enough force to alienate Irish and international (particularly American) opinion, but never enough to suppress the IRA. On 11 July a truce between British forces and the IRA came into force, the terms of which allowed both sides to retain their arms. Negotiations followed, during which the British delegation, led by Prime Minister David Lloyd George, employed the threat of war to ensure that the Sinn Fein delegation accepted a treaty that fell well short of its demands.

Civil War

The treaty, which left the issue of Northern Ireland unresolved, led directly to civil war. An oath of fidelity to the British monarch, the constitutional status of the new Irish Free State, and the partition of the country were among the

Soldiers and civilians shoot at each other on a narrow, smoke-filled street as an overturned cart burns in the foreground; Dublin, April 1916. (Hulton/Archive)

issues that brought erstwhile comrades into mutual armed conflict. Large-scale engagements were confined to the early stages of the war. The assassination of pro-treaty IRA leader Michael Collins on 22 August 1922 strengthened the hand of those who sought a more rigorous campaign. Seventy-seven republicans were executed and 12,000 were interned. By April 1923, most of the IRA's original leadership was either dead or in prison, and during the following month it announced that it was dumping its arms. As a result, Northern Ireland (Ulster) remained an integral part of the British Isles, and the Irish Free State (Erie) an independent nation.

Donnacha Óbeachdáin

See also: Boyne; Irish Uprising; Northern Ireland, Civil War

References and further reading:

Augusteijn, Joos., *From Public Defence to Guerrilla War.* Dublin: Irish Academic Press, 1996.

Neeson, Eoin. *The Civil War 1922–23.* Dublin: Poolbeg, 1989.

O'Malley, Ernie. *The Singing Flame.* Dublin: Anvil, 1979.

Irish Rebellion, Great (1641–1649)

A free-for-all between disunited factions of Scots Protestants, Old English Catholics, Irish Catholics, English Protestants, and the forces of Charles II. Under the "thorough" and ruthless administration of Thomas Wentworth, Earl of Stafford, the Old England and Old Irish were increasingly alienated from the English Crown. After Stafford was executed in London at the wishes of Parliament in 1641, the Ulster Irish exploded in rebellion, attacking English and Scottish Protestant settlers. This attack, beginning October 1641, was chaotic and bloody and quickly focused on the Protestant towns such as Dundalk, Dungannon, and Newry, which fell to the Irish one by one. Because Wentworth's army had been disbanded at his death, the Protestant Earl of Ormond raised a defense force and waited for reinforcement from England, while the Old English, angry at being blamed for the rising alongside the Irish, confederated themselves at Kilkenny and declared their loyalty to the king and to the rebellion.

In the spring of 1642, a Scottish Covenanter army arrived and began to retake Antrim and Down. Faced with a complex and bizarre civil war, the beleaguered King Charles I demanded that Ormonde reach a truce with the rebels in 1642 so that Irish troops could be used in England against Parliament. Although Ormonde offered generous terms in 1646, the papal nuncio threatened to excommunicate Catholics who accepted, so the fighting continued. At Benburn, on the River Blackwater, Owen Roe O'Neill, a professional soldier who had served in Spain, led the Catholics to defeat Ormonde, but then retreated to Kilkenny and did not move on Dublin. To confuse matters further, Parliament sent a Roundhead army under Colonel Michael Jones later that year, while O'Neill campaigned in Munster and Ormonde briefly resurfaced as the leader of royalists in Cork and was defeated by parliamentarians in August 1649 at Rathmines.

With Charles I dead at the hands of Parliament, Oliver Cromwell turned his attention to the chaotic situation in Ireland and arrived with an army. At Drogheda, he set a pattern for the conquest of Ireland: surrender or be brutally sacked. Unlike the 1641 atrocities, which were disorganized and the result of personal animosity, Cromwell's were object lessons in the futility of defiance, and the major fortresses capitulated, including Wexford, Dundalk, and Trim. O'Neill, who escaped from Kilkenny before it surrendered to Cromwell, died in November 1648 trying to hold the Old English and Irish together, while Ormonde, who had failed to hold Drogheda, joined Charles II in Scotland. Satisfied that Ireland was under control, Cromwell returned to England, leaving behind his son-in-law, Henry Ireton, who oversaw the surrender of Limerick in 1651.

Cromwell's triumph over his enemies on behalf of Parliament led to the annexation of Ireland by the Protectorate and a harsh program of repression and land redistribution.

Margaret Sankey

See also: Cromwell, Oliver; English Civil War; English Wars in Ireland

References and further reading:

Bennett, Martin. *Civil Wars Experienced.* New York: Routledge, 2000.
O'Siochan, Michael. *Confederate Ireland 1642–49.* Dublin: Four Courts Press, 1999.

Irish Uprising (1798)

One of the very few Irish Protestant resistance movements. Heavily influenced by the ideals proclaimed in the American and French Revolutions, the United Irishmen was founded in 1791 by Irish Protestants who sought to ally themselves with the native Catholic majority in an effort to establish an Irish republic, independent of Britain, which had ruled the country with varying degrees of success since 1169. The United Irishmen was a secret oath-bound organization with a peak membership of about 300,000.

Bedeviled by informers, its leadership arrested or in exile, and bereft of expected French military assistance, the United Irishmen rebellion of 1798 ("The '98") was largely a collection of isolated battles. Though large in numbers, the Irish mainly fought with pikes against smaller contingents of well-trained British professional soldiers backed with heavy artillery. A tiny French force of 1,100 arrived on 22 August and with Irish peasant support enjoyed some initial victories before being defeated by a much larger British force at Ballinamuck. Approximately 30,000 died during the rebellion. Reflecting the marked imbalance in military might, only 3,000 of these were killed by Irish insurgents.

Having destroyed the most serious military threat to emerge in Ireland for more than a century, British rule was consolidated and institutionalized by the Act of Union (1801), which established the United Kingdom of Great Britain and Ireland.

Donnacha Óbeacháin

See also: Irish Easter Uprising, War for Independence, and Civil War; The Easter Rebellion

References and further reading:

Cullen, Mary, ed. *1798: 200 Years of Resonance.* Dublin: Irish Reporter, 1998.
Pakenham, Thomas. *The Year of Liberty.* London: Abacus, 2000.
Keogh, Daire, and Nicholas Furlong. *The Women of 1798.* Dublin: Four Courts Press, 1999.

Iroquois-French Wars (1609–1697)

A war over control of the fur trade by aboriginal tribes in North America, which soon led to a war between the Iroquois League and the French. Some historians trace its roots to when a Huron band of warriors led by Samuel de Champlain in July 1609 killed two Iroquois chiefs, but more likely it was an outgrowth of the ancient hostility between the tribe of the Iroquois League (Mohawk, Oneida, Onondaga, Cayuga, and Seneca) and the Algonquin tribe allied with the French. As the beaver trade flourished, the Iroquois depleted the populations in their area. The Iroquois in all likelihood did not want to fight the French but rather to displace tribes such as the Illini and Hurons as middlemen in the pelt trade and take over more plentiful lands. They were given the resources to do this when the Dutch traders in the Hudson Valley of New York in 1610 began to trade steel knives, tomahawks, powder, guns, and ammunition for pelts. While the French were not their primary target, they continued raids and engaged in hit-and-run warfare against them in the St. Lawrence Valley and southern Ohio.

The Iroquois began to expand their area to take over lands controlled by the Neutrals and Erie Indians south of the Great Lakes. By 1655, they had also defeated the Nipissings and Hurons, but they could not overcome the Ottawa, located in the western Great Lakes. They were also defeated the next year in warfare with the Chippewas and Illini. Not giving up, a war party of 500 Iroquois and 100 of their allies prepared to attack an Illini village of Kaskaskia, near what is today Peoria, in September 1680. While most of the Illini warriors and Chief Chassagoac were at Cahokia attending a religious ceremony, the remaining Illini warriors attempted to ambush the advancing Iroquois between the Illinois and Vermillion Rivers before they reached the village. They met with some success, and French representative Henri de Tonti tried to negotiate a peace treaty with them. Iroquois warriors stabbed him, causing him to retreat to Lac du Illinois. After eight days of fighting, the Iroquois laid siege to the village. Surviving Illini fled down the Illinois River while captives were killed and mutilated. The Kaskaskia, Cahokia, Chinkoa, Omouahoa, Coiracoentanon, Moingwena, Chepoussa, and Peoria all left the area, while the Iroquois killed or captured 700 members of the tribes remaining at the mouth of the Illinois.

The main body of Illini tribesmen came back to Kaskaskia but were soon confronted by the news that the Iroquois were about to return. They turned to the French Fort St. Louis, located near present-day La Salle, Illinois, under the command of de Tonti. Informed that the fort was too small to protect them, the Illini fled. The Iroquois, finding their enemy gone, attacked Fort St. Louis. After firing on the fort for several days, the Iroquois tried to assault the walls but were beaten back by cannon and musket fire, leading them to retreat. Shortly after this, Tonti and 200 warriors joined a large contingent of Canadian soldiers in an invasion of what is now upper New York, destroying several Iroquois villages along the Mohawk River. By 1696, the Iroquois, with the exception of northern Pennsylvania and eastern Ohio, had retreated to their traditional homelands. In 1687 the French attacked Seneca and Onodaga villages on Iroquois land, and in August 1689 a massive warrior party of 1,200 Iroquois attacked Lachine, near Montreal, killing 200 French settlers. The next year, the French and their Indian allies attacked Schenectady and in retaliation the Mohawk attacked the Sokoki at St. François in 1690 and 1692. The French then launched three campaigns under Louis Frontenac from Quebec between 1693 and 1696 on Iroquois villages. Yet it was smallpox that brought the Iroquois to the peace table. Although they were unable to work out a peace, the Treaty of Ryswick in 1697, which ended the war between England and France, also placed the Iroquois under British protection. Worrying that any conflict with the league would bring another conflict with the British, the French agreed to mediate any disputes between the Iroquois and the Algonquin.

T. Jason Soderstrum

See also: American Indian Wars; French and Indian War

References and further reading:

Eckert, Allan W. *The Wilderness War: A Narrative.* Boston: Little Brown, 1978.

Hunt, George T. *The Wars of the Iroquois.* Madison: University of Wisconsin Press, 1960.

Stewart, Alexander McGinn. *French Pioneers in North America.* 3 vols. Millwood, NY: Kraus Reprint Company, 1976.

Isandlwana (South Africa, 22 January 1879)

The greatest Zulu victory of the nineteenth century over the forces of colonialism and one of the heaviest defeats ever suffered by British troops during the small wars of the Victorian era. Paradoxically, Isandlwana sealed the fate of the independent Zulu kingdom, for in order to reassert their paramountcy in southern Africa, it became imperative for the British to avenge their defeat, crush the Zulu army in battle, and impose a peace settlement entirely on their own terms.

On 20 January 1879 the British Number 3 Column under Lieutenant General Lord Chelmsford encamped at the eastern base of Isandlwana hill. The next day Chelmsford despatched a reconnaissance-in-force under Major J. G. Dartnell toward the southeast to seek out the Zulu army and to establish the column's next campsite. During the night, Chelmsford strongly reinforced Dartnell, leaving the depleted garrison at Isandlwana under the command of Colonel H. B. Pulleine. For the whole of 22 January Chelmsford skirmished with Zulu irregulars, who, in accordance with Zulu strategy, steadily drew him away from his camp.

Colonel A. W. Durnford reinforced the camp midmorning of 22 January with men from Number 2 Column and immediately moved out northeastwards to intercept Zulu reportedly threatening Chelmsford's rear. One of his mounted pa-

Cetshwayo's Zulu warriors defeat the invading British forces at the Battle of Isandlwana in 1879. (Library of Congress)

trols stumbled upon the Zulu army concealed in the Ngwebeni valley only nine miles from the camp.

Undetected by the British, the Zulu army under Chief Ntshingwayo kaMahole Khoza and Chief Mavumengwana kaNdlela Ntuli had reached their bivouac by the early hours of 22 January. Their plan to divide the British forces had succeeded, and they were preparing to fall on the British camp held by 67 officers and 1,707 men, about half of whom were African auxiliaries.

At about 12:15 P.M. the British formed an extended skirmishing line about half a mile in advance of the camp, both to command the dead ground and to support Durnford's horsemen and other detached units as they fell back before nearly 20,000 Zulu. The Zulu center was pinned down by British fire, but the horns extended to outflank the British line and raced around to enter the rear of the camp. (Without their realizing it, the Zulu leaders had achieved every European commander's dream: a "Cannae," an encirclement of an enemy.) Realizing they were being enveloped, at 1:00 P.M. the British precipitately fell back on their camp, losing all cohesion in hand-to-hand fighting with the Zulu. Though harried, a few mounted men broke southwestwards through the Zulu encirclement to escape over the Mzinyathi (Buffalo) River at the aptly named Fugitives' Drift. Many of the infantry conducted a fighting retreat in the same direction but were all cut off and killed before they reached the Manzimyama stream one and a half miles away. The Zulu pillaged the camp and retired at nightfall when Chelmsford and his force finally marched back in battle order.

No less than 1,000 Zulu died in the encounter. The British lost 52 officers and 739 white troops, 67 white NCOs, and close to 500 of the Natal Native Contingent.

John Laband

See also: Anglo-Zulu War; Khambula; Rorke's Drift

References and further reading:

Knight, Ian. "The Battle of Isandlwana, 22 January 1879." In *Great Zulu Battles.* London: Arms & Armour Press, 1998.

Laband, John, ed. *Lord Chelmsford's Zululand Campaign 1878–1879.* Stroud, UK: Alan Sutton Publishing for the Army Records Society, 1994.

———. *The Rise and Fall of the Zulu Nation.* London: Arms & Armour Press, 1997.

Laband, John, and Paul Thompson. *The Illustrated Guide to the Anglo-Zulu War.* Pietermaritzburg: University of Natal Press, 2000.

Isonzo, Battles of the (1915–1917)

Series of no less than 12 battles fought between the Austro-Hungarians and the Italians along the Isonzo River. On 23 June 1915 the Italians attacked Austria across the Isonzo with superior numbers. The Italian generals failed to supply forward units with enough arms, ammunition, and artillery. As the Austrians rushed troops to defend the inadequately held frontier, the Italians made three more bids to break through. The second (July–August), third (October–September), and fourth (November–December) battles of the Isonzo were as inconclusive as the first, and the customary trench warfare ground into an even more bitter contest along the frozen Alpine mountain tops.

Here the soldiers of both sides endured the horrors of fighting in the trenches; however, dizzying heights, the constant cold, flash floods in the valleys, poor supplies, rudimentary medical attention and sanitation, and poor shelter compounded the suffering troops endured. The fifth bid by the Italians to break through began on 11 March 1916, but an Austrian offensive in the Trentino sector offset the small gains the Italians made. A sixth bid to break the river barrier began in August 1916; this time the Italians managed to seize the town of Gorizia but failed to achieve a breakthrough. The seventh, eighth, and ninth battles, September–November 1916, developed along the same patterns as the first six battles. In May 1917 the tenth Isonzo began as the Italians joined the overall allied spring and summer offensive. In particularly brutal fighting, 125,000 casualties achieved scant gains. The Italians followed this with the eleventh battle in August and at last achieved a limited breakthrough at Bainsizza, once again at a heavy price. The Battle of Caporetto (twelfth Isonzo), where the Germans led their Austrian allies in a major drive against Italy, offset all the summer gains, forcing the Italians off the frontier and back to the Piave River.

Stephen Chenault

See also: Caporetto; World War I

References and further reading:

Herwing, Holger H. *The First World War, Germany and Austria-Hungary, 1914–1918.* London: Arnold, 1997.

May, Arthur J. *The Passing of the Habsburg Monarchy, 1914–1918.* Philadelphia: University of Pennsylvania Press, 1966.

Israeli Military

Israel's armed forces have evolved from prestate vigilante bands to the paramount strike force in the Middle East. In 1908 Jewish settlers formed an ad hoc society of armed sentinels, called Ha-Shomer, or Guardsmen, to replace mercenary, untrustworthy hired Arab watchmen in thwarting robbery, cattle-rustling, and poaching. The Jewish farm laborers on the settlements (the Shomrim would not protect farms with an Arab or mixed workforce) comprised a reserve to assist the patrol cadres, who were equipped with a motley collection of obsolescent firearms. Another armed band, the Jaffa Group, provided security for the Jews of that community and Tel Aviv.

These organizations kept a low profile during World War I, as the Turks (and Palestine) were formally allied with Germany, and Jewish loyalties were suspect. Nevertheless, the Palestine-based spy ring NILI—an acronym for the Hebrew verse Netzah Yisrael Lo Yeshaker (The strength of Israel will not lie [1 Sam. 15:29]), which served as its password—comprised of a former Zionist youth corps, worked for British interests, while other Palestinian Jewish adolescents were conscripted into the Turkish army. Zionists expelled from Palestine by the Turks formed the Zionist Mule Corps, providing logistic support in the Gallipoli Campaign. Veterans of the Mule Corps created the nucleus of new all-Zionist battalions, the British 38th and 39th King's Fusiliers, which met the test of fire in Allenby's Palestine Campaign, 1917–1918.

All of these ventures provided valuable military experience for a proficient cadre, which returned to postwar Palestine, now under the British Mandate, whereby a Jewish homeland was to be established alongside an Arab nation. The British high commissioner encouraged the Zionist administration in Palestine to set up an executive framework under the auspices of the formally sanctioned Jewish Agency for Palestine. Independent of but allied to these was the Histadrut, nominally a populist labor society, but in reality the linchpin of Zionist industrialization and agricultural expansion in Palestine.

Established by the labor mainstream, Histadrut, after the murderous Arab riots in 1920, set up the Hagganah (self-defense force), which rapidly became a country-wide Jewish army as Jewish immigration swelled to a flood upon the rise of Hitler, and Palestinian Arab resistance hardened in response. Armed defense organizations were illegal under the mandate, so Hagganah had to organize, train, and procure weapons covertly. The organization even managed to establish a backyard armaments industry producing bullets and crude submachine guns and mortars.

After the Arab riots of 1929, Hagganah was transferred from the authority of Histadrut to the quasi-governmental Jewish Agency, which had formerly spurned any connection with outlawed clandestine groups. This progression entailed splitting Hagganah high command into left and right political factions. Although illegal, every Jewish town and neighborhood in Palestine was affiliated with a district command of Hagganah. Elements of the conservative wing split from the labor-left mainstream and allied themselves with Zeev Jabotinsky's Revisionist Party, the foundation of today's Likud, thereby forming the renegade National Military Party, or Etzel (alternately, Irgun). The latter, at most consisting of 250 firebrands, were ruthless in their attacks on Arabs, while Hagganah counseled restraint so as to forestall unwelcome British attention.

The Palestinian Arab Revolt of 1936–1939, largely directed at the British Mandatory government, brought a measure of unusual cooperation between the colonial counterterrorism effort and specially constituted Jewish Supernumerary Police, developing commando expertise under the able direction of the New Testament Zionist Orde Wingate. In 1938 and 1939, Hagganah benefited by the appointment of a nonpartisan nationwide commander and the establishment of a professional military general staff to coordinate the formerly diversified elements.

During World War II, Hagganah and segments of the Irgun collaborated with the British authorities in fighting the Nazi menace; however, a renegade sector of the Irgun, Lehi, informally the Stern Gang, rashly attacked British and Arab civilians. (The British high commissioner for Egypt was assassinated by Jewish terrorists in 1944.) Also during the war, the left wing of the Hagganah formed an elite strike force, or Palmach, sanctioned because it would embody a potential guerrilla resistance to Rommel's advancing Afrika Corps. In addition, 30,000 Palestinian Jews gained invaluable experience serving with the British armed forces, including a Jewish Brigade Group. Hagganah meanwhile developed a field corps, a medical service, a signals corps, an arms industry, and an intelligence section, the last assisting illegal immigration of Holocaust refugees. Intelligence became paramount in the postwar years when Hagganah conducted an effective insurgency campaign against British military and police logistics, while the Irgun and Stern groups focused on terrorizing British individuals.

During the 1948 Israeli War of Independence, Israeli political leader David Ben-Gurion remodeled Hagganah as the national defense force of the newborn state. Early in the war, Irgun was forcefully integrated into the Israeli Defense Forces (IDF), the new designation for the armed forces of Israel. In the course of desperate fighting, the patchwork geographical structure of company-sized lots amalgamated into battalions, then regiments, which nonetheless retained their regional identification. Makeshift aviation and naval assets were likewise "born in battle" and integrated into the whole. The IDF would win every one of its subsequent conflicts against its hostile Arab neighbors, and such success would be studied in some detail by military academies around the world (with the conspicuous exception of Arab military academies). But the reason for the IDF success, aside from the numerous Arab military deficiencies (lack of coordination, unprofessional officers, untrained troops, etc.), came down to a simple point: Israel had no alternative to victory. Defeat meant literal death.

Jim Bloom

See also: Gallipoli; Israeli-Arab Wars; Megiddo (September–October 1918); October War; Sinai-Suez Offensive; Six-Day War

References and further reading:

Allon, Yigal. *The Making of Israel's Army.* New York: Universe Books, 1971.

Dupuy, Trevor N. *Elusive Victory: The Arab-Israeli Wars, 1947–1974.* New York: Harper & Row, 1978.

Morris, Benny. *Righteous Victims: A History of the Zionist-Arab Conflict, 1881–1999.* New York: Alfred A. Knopf, 1999.

Schiff, Zeev. *A History of the Israeli Army (1870–1974).* San Francisco: Straight Arrow Books, 1974.

Van Creveld, Martin. *The Sword and the Olive: A Critical History of the Israeli Defense Force.* New York: Public Affairs, 1998.

Israeli-Arab Wars (1948–1999)

A series of wars between Israel and its Arab neighbors that saw the establishment of the Jewish state.

1948, War of Independence

After Britain announced it would withdraw from its mandate in Palestine, the United Nations (UN) passed a partition plan on 29 November 1947 that divided Palestine along existing settlement lines between the Arab and Jewish populations, which the Jewish authorities accepted and the Arab leaders rejected. Civil war soon broke out between the Arabs and Jews. The Jewish leadership declared the independent state of Israel on 14 May 1948, an announcement followed immediately by a declaration of war by the Arab states of Egypt, Iraq, Lebanon, Syria, and Transjordan. Coop-

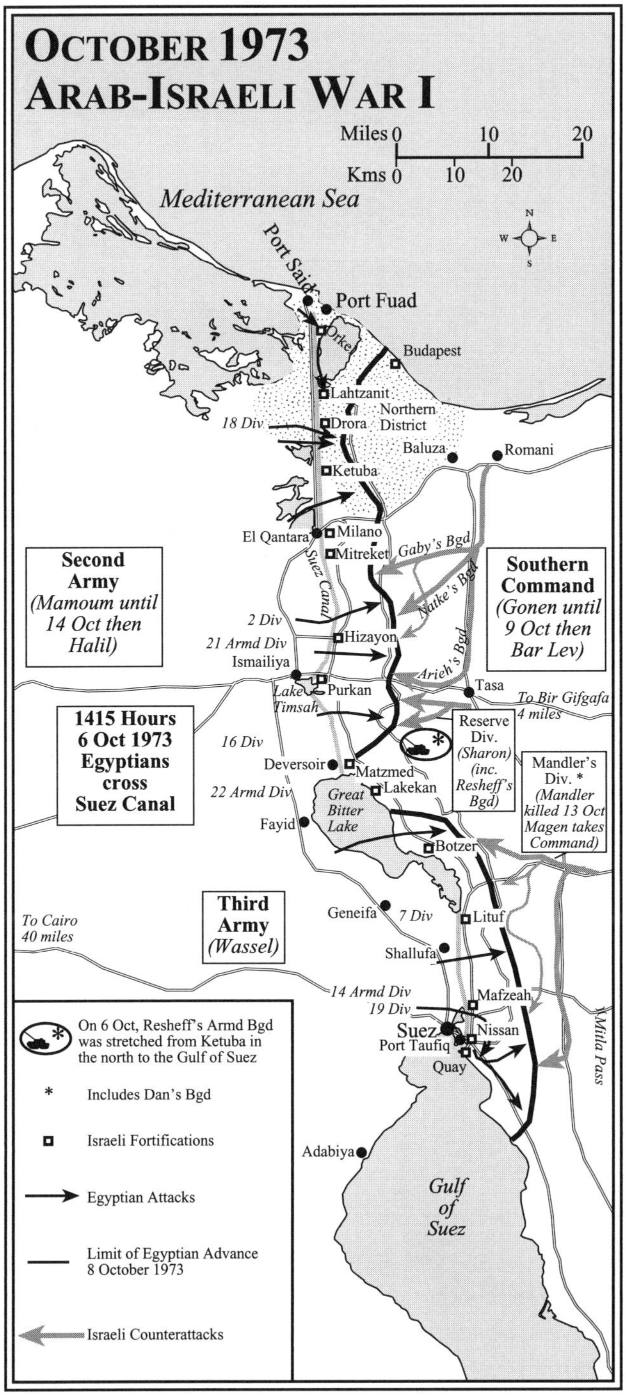

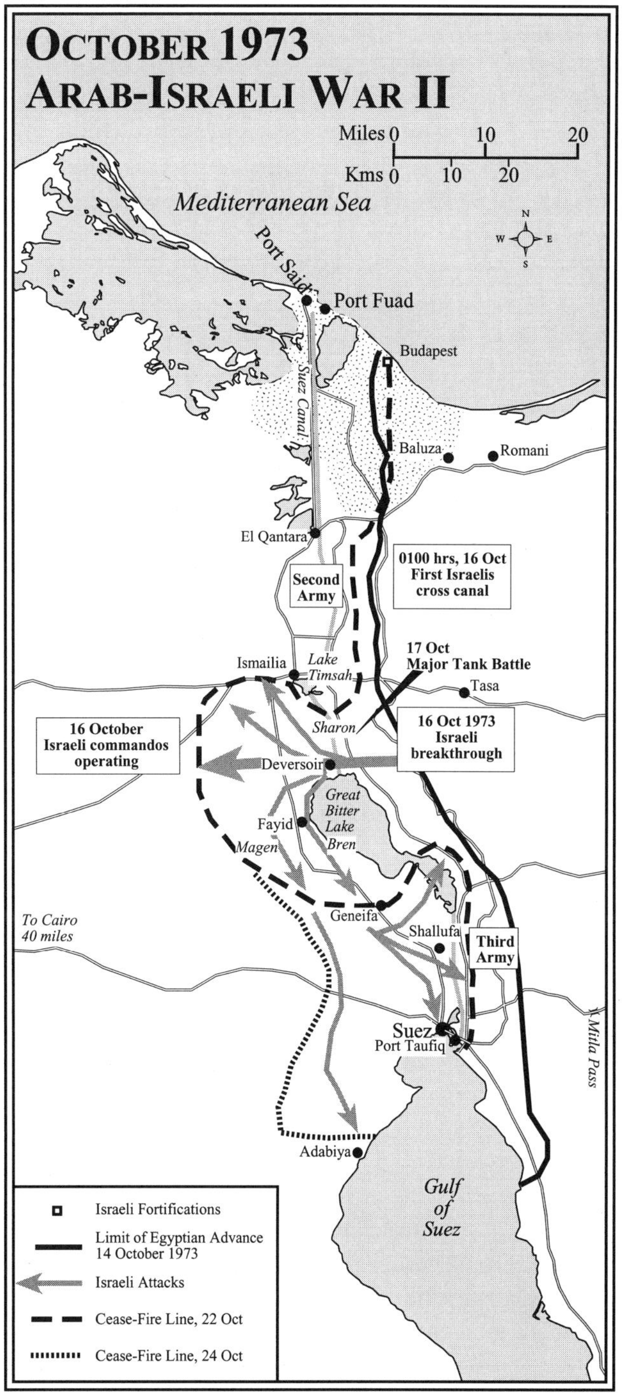

erating with local Arab militia, the invading Arab states enjoyed early success, choking the supply roads to Jewish population centers and driving Israelis from the Old City of Jerusalem. These successes were short-lived. The Arab states claimed they were liberating Palestine from the Jews but failed to coordinate military operations and often pursued conflicting territorial goals. Moreover, the Arab forces were poorly supplied, received inadequate training, and suffered from poor morale. The Israeli Defense Forces (IDF) did not fare much better during the first month of the war. The tide began to shift in Israel's favor after the IDF obtained weapons and supplies despite a UN embargo. Between June 1948 and January 1949, the IDF counterattacked and captured significantly more territory than it had been allotted by the UN partition plan, including the Negev and the Galilee. Four thousand Israeli soldiers and 2,000 civilians were killed during the war; the number of Arabs killed is unknown. Between 550,000–750,000 Arab refugees (estimates vary) fled Israel as a result of the war. Their fetid camps became a breeding ground for generations of young Arab fighters against the "Jewish entity."

1956, Sinai-Suez

During 1955–1956, Gaza and the Sinai Peninsula served as bases for numerous terrorist attacks against Israel. In September 1955, Egyptian prime minister Gammal Abdel Nasser ordered the closure of the Straits of Tiran, an international waterway, effectively blockading Israel's port at Eilat. Israel regarded the blockade as an act of war. At the same time, Britain and France had grown increasingly frustrated with Nasser. The French resented Nasser's support of Algerian nationalists, who were engaged in a war of independence from France, and the British opposed his nationalization of the Suez Canal on 26 July 1956. Thus Israeli, British, and French interests converged in the Suez-Sinai campaign. Their joint plan was divided into two distinct military operations, one led by the Israelis and the other by the British and French, and had the primary goal of deposing Nasser. Israel, ostensibly acting alone, launched its invasion of the Sinai on October 29. The Israelis, led by General Moshe Dayan, combined airborne and armored assaults to seize the Suez Canal by 31 October and to gain control of most of the Sinai by 3 November. As planned, the British and French then demanded that both Israel and Egypt pull back from the Suez Canal. When Nasser, as predicted, refused to withdraw, the British and French began a three-day air assault on 31 October, destroying the Egyptian air force on the ground. Within two days, British and French airborne troops had secured the canal. Militarily, the campaign was a success; politically, it was a disaster. U.S. president Dwight D. Eisenhower, preoccupied with the upcoming presidential election and Soviet repression in Hungary, opposed the Israeli-British-French collusion. On 2 November, the United Nations passed a U.S.-sponsored resolution calling for the immediate cessation of hostilities and the withdrawal of all foreign forces from Egyptian territory. France and Britain withdrew in humiliation, but Israel remained in the Sinai until March 1957. Nasser regained control of the Suez Canal and convinced the UN to dispatch a United Nations Emergency Force (UNEF) to the Sinai as a buffer between Israel and Egypt. The UNEF also prevented any further blockade of the Straits of Tiran. The Suez-Sinai campaign bolstered Israel's military reputation, enhanced Egypt's political influence in the Third World, and demonstrated that Britain and France were no longer world powers.

1967, Six-Day War

Tensions between Israel and its Arab neighbors significantly increased during 1966–1967. Israel's borders saw repeated Arab terrorist attacks and Syrian military activity. Throughout May, Nasser called for the destruction of Israel, and on 16 May demanded the withdrawal from the Sinai of the UNEF, which cravenly left three days later. On 22 May, Egypt closed the Straits of Tiran to Israeli shipping. Three days later, Egypt, Iraq, Jordan, Kuwait, Lebanon, Saudi Arabia, and Syria mobilized 547,000 troops, 2,504 tanks, and 957 aircraft along the Israeli border. Threatened by terrorist attacks, anti-Israel rhetoric, the closure of the Straits, and hostile troops on its borders, the Israelis launched their preemptive attack on 5 June 1967. On the first day, the Israeli Air Force, flying French Mirage fighter-bombers, attacked Arab airfields, destroyed the bulk of its enemies' aircraft on the ground. The IDF then struck into the West Bank and the Sinai. Lacking air cover, the Jordanian and Egyptian armies quickly retreated. On 7 June, the IDF seized the West Bank (Judea and Samaria), previously controlled by Jordan, and captured east Jerusalem. Suffering 5,000 casualties, King Hussein of Jordan agreed to a cease-fire that same day. In the Sinai, the IDF quickly raced across the Sinai and inflicted perhaps as many as 12,000 casualties upon the Egyptians. Nasser agreed to a cease-fire on 9 June. With Jordan and Egypt defeated, Israel turned its armor and aircraft against Syria. The Israelis, after only 27 hours, entered Kuneitra and captured the Golan Heights, inflicting more than 1,000 Syrian casualties.

The UN negotiated a cease-fire between Syria and Israel on 10 June. Israeli air power was crucial to the destruction of the Arab armies in the war. Israel gained control of the Sinai Peninsula, the West Bank with all of Jerusalem, and the Golan Heights. Israel suffered the loss of 766 soldiers, compared with over 18,000 Arab casualties during the war.

1973, Yom Kippur War

In an attempt to regain the territories lost in 1967, Egypt and Syria launched a surprise attack against Israel, on 6 October 1973, the Jewish Day of Atonement (Yom Kippur). Poor political and military intelligence and an easy contempt for Arab military ability caused Israel to discount Arab frustration with the results of the Six-Day War and also misinterpreted Arab mobilization for the attack as annual military maneuvers. The Israelis finally realized the threat eight hours before the attack, but Israeli prime minister Golda Meir ruled out a preemptive strike for fear of being perceived as the aggressor. When fighting began at 2:00 P.M., the Israelis had not yet fully mobilized and were outnumbered 12 to 1. Backed by $3.5 billion worth of aid from the Soviet Union, the Egyptians opened with an air and artillery assault into the Sinai, and the Syrians followed soon after with a thrust into the Golan Heights, taking Mount Hermon. Initial Arab gains were impressive, but the Israelis stopped the

Egyptians by 8 June, the Syrians by 11 June, and captured the Suez Canal on 18 October. Initially reluctant to offer assistance, the U.S. then funded Israel with $2.2 billion worth of aid, which proved vital to Israel's war effort. The Organization of Petroleum Exporting Countries (OPEC), because of U.S. support for Israel, imposed an oil embargo upon the U.S. that lasted until 1974. Despite the active involvement of the U.S. and the USSR, the risk of superpower confrontation was never a serious possibility, and both powers brokered a cease-fire between the Israelis and the Syrians and Egyptians. Israel's military reputation declined as a result of the war; the nation had suffered 3,000 killed and 8,000 wounded, compared with 8,500 Egyptians and Syrians killed and 20,000 Syrians wounded.

But this conflict yielded two positive consequences. Egyptian president Anwar as-Sadat could now treat with the Israelis as a military equal for peace and recognition of the state of Israel. American president Jimmy Carter brokered the agreements in 1977, and both shared Nobel Peace Prizes for their work. But Sadat was assassinated by Muslim extremists five years later. The end of the war also saw the emergence of the Israeli peace movement.

1982, Lebanon and Beyond

Between 1978 and 1982, the Palestinian Liberation Organization (PLO), using Lebanon as a base for terrorist attacks, created instability along Israel's border. In early 1982, Maronite Christians in Lebanon and the Israelis began discussing a joint punitive war against PLO targets in Lebanon. Using the attempted assassination of the Israeli ambassador to the United Kingdom as a pretext, Israeli defense minister Ariel Sharon devised an invasion plan of Lebanon with three goals: the elimination of the PLO in Lebanon; the creation of a stable Maronite government in Lebanon; and the expelling of Syria from Lebanon. On 6 June 1982, IDF armor struck into Lebanon, attacking the PLO along the coast toward Beirut and Syrian forces near the Bekaa Valley. The IDF linked up with Maronite Christian forces outside Beirut on 8 June, but the Maronites were unwilling to attack the PLO in west Beirut. Fighting alone, the Israelis besieged west Beirut on 1 July. The Israelis led air, artillery, and naval bombardments against PLO positions in west Beirut until 6 August, when the PLO agreed to a U.S.-sponsored withdrawal agreement. Fighting elsewhere in Lebanon continued until 21 August. On 16 September, members of the Lebanese Christian Kataib militia massacred between 700 and 2,000 Palestinians at the Sabra and Shatila refugee camps. The IDF claimed to be unaware of the massacre going on about them. On 28 September, the IDF pulled out of west Beirut. Although a military success, the Israelis achieved only one of their three goals, the elimination of the PLO in Lebanon. The Syrian military remained in Lebanon and the Maronites were unable to maintain a stable government friendly to Israel. The IDF withdrew from Lebanon altogether in 1985 but left a buffer zone maintained by Lebanese-Christian police north of its border until May 1999. Although this was the last overt Arab-Israeli conflict to date, cross-border raids, Israeli air strikes, and nearly continuous unrest between Palestinians and Israelis have kept the area in turmoil.

The Israelis were victorious in their Arab wars because they knew that, for them, defeat was not an option. (More than one Arab leader of the time criticized Adolf Hitler for not killing more Jews.) The Arabs knew, conversely, that they could be defeated in a particular war and come back to fight another day. Israeli society, with its many highly educated and technically skilled citizens and, usually, facing annihilation, fostered innovation and careers open to all talents, just what was needed in the fast-moving, combined-arms Middle East conflicts. The impoverished, authoritarian, underdeveloped Arab societies, with their formulaic, hierarchical militaries, at least until 1973, could hardly compete.

Eric D. Pullin

See also: Israeli Military; Refugees and Victims of Ethnic Cleansing; Sinai-Suez Offensive; Six-Day War; United Nations and Conflict Resolution

References and further reading:

Gilbert, Martin. *Atlas of the Arab-Israeli Conflict.* 6th ed. New York: Oxford, 1993.

Lacquer, Walter. *Confrontation: The Middle East and World Politics.* New York: New York Times Books, 1974.

Ovendale, Ritchie. *The Origins of the Arab-Israeli Wars.* 3d ed. London and New York: Longman, 1999.

Schiff, Ze'ev. *A History of the Israeli Army.* London: Sidgewick & Jackson, 1987.

Schulze, Kirsten E. *The Arab-Israeli Conflict.* London and New York: Longman, 1999.

Issus, Battle of (November 333 B.C.E.)

The second victory of the Macedonian king Alexander III the Great over the Persian king Darius III that took place on the eastern coastline of the gulf of Iskanderun in southern Turkey. The battlefield was divided by a river called the Pinarus.

After Alexander had marched through Asia Minor, he followed the Syrian coastline southwards. The Persian army occupied a position north of the Macedonian army, blocking the road to the rear. Alexander immediately turned to meet Darius in battle, the Macedonians deploying south of the Pinarus, the Persians to the north.

Alexander's army consisted of 26,000 infantry and 5,300 cavalry. He positioned his heavy infantry in the center, with a small force of cavalry on the left. The right wing consisted of the guard infantry and the bulk of the cavalry, including Alexander at the head of his guard-heavy cavalry, and light troops.

We know that the Persian army greatly outnumbered Alexander's, but it is impossible to assess its exact numbers. It contained a sizable cavalry force, a large contingent of Greek mercenaries, probably some 30,000 strong, and a large number of Asiatic infantry. Having reinforced the riverbank with battlements, the Persians took position on the riverbank with the Greeks in the center, some 30,000 Persian infantry on each wing, and an unknown number of local troops in the rear. The cavalry formed the right wing, closing the gap between the main battle line and the sea.

Alexander opened the battle by crossing the river and charging the Persian left, routing the Persian infantry. The Persian king fled immediately, though the Greek mercenaries had put severe pressure upon the Macedonian center, which was driven back into the river. However, as Alexander turned with his guard troops upon the Greek mercenaries from rear and flank, the Persian army was routed and broke, suffering heavy losses. The Macedonian losses numbered 450 dead and 4,500 wounded.

Maarten van der Werf

See also: Alexander the Great; Alexander's Wars of Conquest

References and further reading:

Fuller, J. F. C. *The Generalship of Alexander the Great.* London: Eyre & Spottiswoode, 1958.

Green, Peter. *Alexander of Macedon, 356–323 B.C.E. A Historical Biography.* Harmondsworth, UK: Penguin, 1974. Reprinted Berkeley and Los Angeles: University of California Press, 1991.

Hammond, N. G. L. *The Genius of Alexander the Great.* London: Duckworth, 1997.

Tarn, W. W. *Alexander the Great.* Cambridge, UK: Cambridge University Press, 1948.

Italian Colonial Wars (1882–1936)

The Italian quest for prestige. Though it would seem that the pursuit of an overseas empire would be a low priority for the newly unified Italian state, with its lack of internal integration, serious border disputes with Austria, and a general paucity of resources, Rome still sought this objective in competition with the other major powers of the nineteenth century. The objectives being international prestige, potential markets, and an outlet for Italy's excess population that would still be under Rome's political control. A particular impetus for Italian expansion was the resonance of the very name Rome with *empire.*

While allowed by the British government to gain a foothold in what is now Somalia, Italian efforts to create a protectorate over the Ethiopian Empire failed. Despite losing to feudal levies at Dogali in 1887, Rome signed the treaty of Wichale with Emperor Menelik II in 1889. The emperor believed he had signed a treaty recognizing his sovereignty. The Italian government felt they had cleverly bound the Ethiopians with an admission of overlordship. When Menelik came to understand the true meaning of the treaty, he repudiated the document in 1893 and went to war.

The climax of this campaign was the disastrous Battle of Adowa in 1896, though the Italians were also fought to a standstill at Amba Alagi (1895) and Macalle (1896). Essentially, the Italo-Ethiopian conflict had become a proxy fight between London and Paris over control of the Sudan, with the result that the Ethiopian military, while essentially a feudal horde, had access to modern French and Russian weapons. When added to the raw numbers and traditional warrior ferocity of Menelik's army, the result was a crushing Italian defeat. Rome was unable even to mount a retaliatory campaign to exact revenge for the worst humiliation ever visited by a traditional state on a modern Western army.

The second major effort by the Italians came in North Africa, as the Agadir Incident encouraged Rome to try to turn its area of influence in Cyrenica and Tripolitania (modern Libya) into a formal colony, mostly from fear of French aggrandizement. This move led to the Italo-Turco War (1910–1911), which, although a war between organized armies, was mostly about securing colonial possessions in compensation for other governments' gains.

Once the Turks had stepped back from their confrontation with the Italians, mostly to deal with the Balkan War, Rome found itself locked into a long-running guerrilla war with the Senussi, a culture of desert nomads with no intention of compromising their traditions for the sake of Rome's economic and political aspirations. With encouragement from Turkey and Germany, the Senussi (under their emir Idris) were able to fight the Italians to a standstill; by 1919 Rome was forced to grant the nomads autonomy.

These were circumstances that Benito Mussolini was not prepared to tolerate upon his accession to power, though it was not obvious that he would be interested in pursuing a formal empire. As a former socialist, Mussolini was nominally an anti-imperialist. There was also the more cynical consideration of whether adopting such a pose would better serve the ends of the new regime. In the end, Mussolini was further concerned with achieving victories for his regime so as to solidify his domestic power, in addition to the usual Italian colonial goals.

Mussolini assigned Emilio De Bono the task of bringing the Senussi to heel, thus beginning a campaign that lasted into the early 1930s. De Bono was chosen because he was the most eminent soldier to join the Fascist cause, and his success would reflect glory on the Blackshirt movement, but his lack of progress led to his replacement by the regular army generals Pietro Badoglio and Rodolfo Graziani.

Their strategy was the traditional anti-insurrection method of concentrating the noncombatant population in secured camps so as to separate them from the active fighters, there being no more than 1,000 active guerrillas at any one time. The anti-insurgency campaign was carried out with all of the expected Fascist brutality and much of the social and economic infrastructure of Libya's traditional peoples was destroyed; it was estimated that by 1932 some 100,000 persons had died in Cyrenica alone, roughly half the population of that region.

It was probably inevitable that Mussolini would revisit the question of exacting revenge from Ethiopia for the debacle of 1896, the rationale given to the Italian population. Though Rome had been able to exert more influence over Addis Ababa, Haile Selassi had continued to try to play off the major European powers against each other so as to maintain the sovereignty of his state. Believing himself to have a free hand from London and Paris, Mussolini began his second colonial war much as he had his campaign in North Africa, by dispatching De Bono with a large force of Blackshirt militia to march on the Ethiopian capital so as to monopolize all the glory for his regime.

As before, a larger than expected force (some 800,000 men were mobilized) under professional officers was required to bring the formal campaign to a conclusion, a campaign that nauseated the democracies with its use of chemical weapons, indiscriminately spread by air. Much of the war became desultory after the Ethiopian regular army was defeated.

However, formal military conquest did not lead to a pacified region and the Italians found themselves contending with a constant level of insurrection, a matter not helped by the fragmentary control that the Ethiopian central government had exerted over the country. Neither sanguinary violence nor relative benevolence was able to solidify the Italian position before the country's defeat at the hands of the British in 1941 and the total loss of empire. The entire Italian adventure in imperialism brings to mind the supposed quotation of Chancellor Otto von Bismarck, "The Italians have strong appetites but weak teeth."

George R. Shaner

See also: Adowa; Italo-Turkish War

References and further reading:

Gooch, John. *Army, State and Society in Italy, 1870–1915.* London: Macmillan, 1989.

Mack Smith, Denis. *Mussolini's Italian Empire.* New York: Viking Press, 1976.

Mockler, Anthony. *Haile Selassi's War: The Italian-Ethiopian Campaign, 1935–1941.* New York: Random House, 1984.

Tripodi, Paola. *The Colonial Legacy in Somalia: Rome and Mogadishu from Colonial Administration to Operation Restore Hope.* New York: St. Martin's Press, 1999.

Italian Wars of Unification (1848–1870)

Series of wars that resulted in the unification of Italy. The unification of Italy, or Risorgimento, was achieved in several wars fought between 1848 and 1870 by the various states that comprised the Italian peninsula. With French support these states fought against Austria and the pope. By 1870 Italian patriotism created a unified Italy under the Royal House of Savoy.

The unification wars stemmed from the inadequate 1814–1815 Congress of Vienna settlements that restored the status quo of the pre-Napoleonic era. The various states had different monarchs who aimed to increase their power. Unification also was spurred by the liberal ideas of Guiseppi Mazzini (1805–1872), Vincenzo Gioberti (1801–1852), Massimo d'Aseglio (1798–1866), and Giuseppe Garibaldi (1807–1882), who had stirred revolutionary agitation throughout Italy since the 1830s. Revolts in Parma, Romagna, Umbria, and the Marches were harshly put down by the Austrians.

Yet Camillo Benso di Cavour (1810–1861), Piedmont's prime minister from 1852 to 1859, was the chief architect of Italian unification. Piedmont king Charles Albert (1798–1849) had agreed to a constitutional monarchy in February 1848, but the first battle for unification occurred at Custozza on 23–25 July 1848. There the Austrians, under their master tactician Field Marshal Joseph Radetsky, defeated the Italians. The Austrians also outmaneuvered them at the 23 March 1849 Battle of Novara. These humiliating defeats led King Charles Albert to abdicate in favor of his son Victor Emmanuel II (1820–1878).

Pope Pius IX (1792–1878), elected in 1846, exacerbated the situation. He had initially seemed liberal with his welcome refreshing reforms. However, his proclamation on 20 April 1848 that Italians should not fight another Catholic country earned him great enmity from his compatriots and resulted in rioting. The pope's temporal powers were abolished on 9 February 1849; he fled from Rome on 29 November 1849. He was returned to Rome under French protection on 12 April 1850. French troops occupied Rome as a protective force until 1870.

Cavour cleverly persuaded French emperor Napoleon III to join Piedmont and Sardinia in July 1858. Success finally came when the combined French-Piedmontese army attacked the Austrian army on 20 May at Montebello. The Piedmontese won again on 30–31 May at Palestro. Then this combined army of 59,100 soldiers and 91 artillery pieces defeated the 58,000 troops and 176 pieces of the Austrians at the Battle of Magenta on 4 July 1859, where some 6,000 died.

The same forces defeated the Austrians at the Battle of Solferino on 24 June 1859, which resulted in 14,000 Austrians either killed or wounded with 8,000 missing or prisoners. The victors lost 15,000 killed or wounded and 2,000 missing or prisoners. The victory led to most of Lombardy being annexed to Sardinia-Piedmont. The horrific slaughter and bloodshed severely stretched the medical capabilities of both camps. Henry Dunant (1828–1910) succored the wounded for three days and nights and subsequently established the International Red Cross. Napoleon then signed an armistice with Emperor Franz-Joseph at Villafranca, allowing Austria to retain Venice, which annoyed Cavour, who resigned.

Cavour returned to power in January 1860 and secretly agreed to turn Nice and Savoy over to Napoleon. The remnants of the Papal States, except for Rome itself, were occupied. The Piedmontese Parliament proclaimed Victor Emanuel II king of Italy. Rome was to be the future capital of the Italian nation.

Although Cavour was forced to cede Nice and Savoy to France, a plebiscite held on 15 April 1860 resulted in Parma, Tuscany, Romagna, and Modena voting to join Piedmont.

On 6 February 1860, Victor Emmanuel annexed Umbria and the Marches. In 1860, Garibaldi invaded Sicily and defeated the Neapolitans at Calatafimi on 15 May 1860. He seized Palermo and Naples in a plebiscite, again resulting in a decision to join Italy. Victor Emmanuel not only defeated the pope's army at Castelfidardo on 18 September and Ancona on 30 September, but he also confiscated all of the papal possessions outside Rome. Although Italy lost several battles (including a second loss at Custozza), it was rewarded with Venetia in 1866 as a reward for Italian aid to Prussia against Austria. Finally, on 20 September 1870, with the guardian French troops being withdrawn to fight the Prussians in the Franco-Prussian War, Victor Emmanuel seized Rome and made it Italy's capital city, thereby ending the Italian wars of unification.

Annette Richardson

See also: Garibaldi, Giuseppi

References and further reading:

Coppa, Frank J. *Camillo di Cavour.* New York: Twayne Publishers, 1973.

———. *The Origins of the Italian Wars of Independence.* London and New York: Longman, 1992.

Garibaldi, Giuseppe. *Autobiography.* Trans. A. Werner. New York: Howard Fertig, 1971.

Katz, Robert. *The Fall of the House of Savoy.* New York: Macmillan, 1971.

Mack Smith, Denis. *Cavour.* New York: Knopf, 1985.

———. *Mazzini.* New Haven: Yale University Press, 1994.

Reader, Harry. *Italy in the Age of the Risorgimento.* New York: Longman, 1983.

Ridley, Joseph. *Garibaldi.* New York: Viking, 1976.

Italo-Turkish War (1911–1912)

Italy's preoccupation with national unification in the mid- to late nineteenth century meant that it invariably fell behind other European countries in the quest for overseas colonies. Imperial aspirations thus became inextricably linked with national pride, and the government focused upon nearby North Africa for territorial gains. Italian immigrants and merchants began arriving in Libya in 1880 and within two decades constituted the most numerous foreigners in that Turkish colony. The Ottoman Empire being in decline, on 29 September 1911 Italy suddenly declared war on the Turks, alleging mistreatment of its nationals in Libya. A force of 50,000 men was then dispatched overseas, which caught the Ottoman garrison completely unprepared. By October Italian forces under General Carlo Caneva were in complete control of Tripoli and Tobruk and had carved out several enclaves along the coast. Turkish resistance was either weak or ineffective. However, strong resentment and resistance from the Muslim population culminated in the Battle of Tripoli, 23–26 October 1911, where Turkish-leaning Senussi tribesman tried and failed to recapture the capital after heavy casualties. Italian losses had also been considerable and they were dissuaded from pushing further inland. Despite the Italians' enjoying every advantage in terms of modern weaponry, an embarrassing stalemate ensued.

As events in Libya unfolded, the Italians also dispatched naval forces to harass the Ottoman coast. In April 1912 Italian warships bombarded the Dardanelles while other forces seized the Dodecanese Islands and Rhodes. It was not until July 1912 that Caneva led his forces away from the coast, and several pitched encounters were waged, but the Italians finally emerged victorious. Furthermore, because events in the Balkans began spinning rapidly out of control, the Sublime Porte entered negotiations to cease hostilities altogether. The Italo-Turkish War was finally concluded by the Treaty of Ouchy, signed 15 October 1912, whereby Italy gained Libya, Rhodes, and the Dodecanese Islands. Ottoman losses amounted to 14,000 men, while the Italians, better armed and equipped, lost half that tally. It was hardly a glorious triumph for the Italians, and the conflict may prove of interest more for the fact that it was the first to employ aerial reconnaissance, aerial bombardment, and armored vehicles.

John C. Fredriksen

See also: Balkan War, First; Italian Colonial Wars

References and further reading:

Childs, Timothy W. *Italo-Turkish Diplomacy and the War over Libya, 1911–1912.* New York: E. J. Brill, 1990.

Simon, Rachel. *Libya between Ottomanism and Nationalism: The Ottoman Involvement in Libya during the War with Italy (1911–1919).* Berlin: K. Schwarz, 1987.

Ivan III (1440–1505)

Ruler who laid the foundation of the modern Russian state, expanded and recovered much of Russia's western "historical lands," and shook off Mongol, or Tartar, rule. Ivan was born in Moscow on 22 January 1440. During his long reign, he gathered, absorbed, and incorporated into the Muscovite state most of the Russian independent principalities and free cities through conquest, diplomacy, annexation, or voluntary surrender. He transformed the small and often contested role of the principality of Moscow into the political center of a unified Russian state.

Ivan's most important achievement was the ending of Mongol rule over the Russian people. Friction between Moscow and the Mongol khans of the Golden Horde came to a head when the Russian and Mongol armies met on the opposite banks of the Ugra River in 1480. For more than two months neither army attacked the other. The Mongols suddenly withdrew their troops without a battle. In this rather inglorious manner Ivan ended Mongol domination.

Ivan's next major objective was the recovery of the ancient territories of Kievan Russia from Lithuania. In 1500, Ivan's army invaded Lithuania, and during the next three years, his forces captured much of Russia's western lands and saved Kiev and Smolensk.

Ivan's successes and victories over Lithuania brought Russia into direct contact with Europe, whose sovereigns viewed him now as a powerful and independent ruler. To augment his authority, he added the title of Sovereign of All Russia to that of czar.

Ivan died in Moscow on 27 October 1505, leaving a much expanded and more powerful country than when he had ascended to power.

James J. Farsoals

See also: Ivan IV, ("the Terrible"); Novgorod, Muscovite Conquest of

References and further reading:

Fennel, J. L. *Ivan the Great of Moscow.* New York: St. Martin's, 1962.

Grey, Ian. *Ivan III and the Unification of Russia.* New York: Collier Books, 1964.

Soloviev, S. M. *The Reign of Ivan III.* Gulf Breeze, FL: Academic International Press, 1978.

Ivan IV ("the Terrible") (1530–1584)

Complex and violent monarch who strove to create a powerful national monarchy and to expand Muscovite territory beyond the Ural Mountains. Ivan was born in Moscow on 25 August 1530. He was orphaned at the age of eight and a group of aristocrats (boyars) assumed controlled of the regency. His childhood was marred by acts of violence and murders, often in his presence, leaving a deep psychological trauma on his personal life and character. It was perhaps this violent experience that contributed to his vindictive and brutal conduct, earning him the epithet "the Terrible."

Ivan was crowned Czar of Moscow and All Russia in 1547. During the first years of his reign he introduced a series of administrative reforms. He organized a new force of infantrymen (*streltsy*), the first Russian soldiers to carry firearms. With this new army, he invaded the Mongol, or Tartar, khanates of Kazan and Astrakhan to the east and southeast of Russia and annexed them to the Muscovite state in the 1550s.

But the most important military event during Ivan's reign was the Livonian War—an abortive attempt to gain an outlet to the Baltic. In 1558, Ivan opened hostilities against the Livonian Knights, who had ruled Livonia (Estonia) since the thirteenth century, and captured much of its territory. But in 1561, Sweden, Lithuania, and Poland entered the war, defeated the Russian army, and forced Ivan to give up all of his gains along the Baltic.

In 1581 the exploration of western Siberia began. By 1583, the entire region came under Russian control, thus opening the road to further expansion into Siberia.

On 18 March 1584, in the midst of a chess game, Ivan suddenly collapsed and died in Moscow.

James J. Farsolas

See also: Livonian War

References and further reading:

Bobrick, Benson. *Fearful Majesty: The Life and Reign of Ivan the Terrible.* New York: Putman, 1987.

Carr, Francis. *Ivan the Terrible.* New York: Barnes & Noble, 1980.

Troyat, Henri. *Ivan the Terrible.* New York: Dutton, 1984.

Ivry, Battle of (14 March 1590)

The decisive battle that ended the Wars of Religion in France. Henry de Navarre became king after the murder of Henry III in 1589; only the Catholic League supported by Spain dared to challenge him. Henry tried to take Paris but was repulsed by the rebels led by the Duc de Mayenne. He decided then to besiege the town of Dreux, 40 miles west of Paris. Mayenne, with the help of Spanish veteran troops, attacked the royal/Huguenot army.

The king's army was made of French loyal subjects but also of German *reiters* and Protestant Swiss. His 11,000 soldiers had to face more than 15,000 Catholics, including elite cavalry from Spanish Flanders and German Protestant mercenaries.

After some skirmishing, the two armies faced each other on a plain without any terrain features. They deployed like pawns on a chessboard. The royal army had its left wing slightly forward and the right wing some way back as a reserve. Mayenne, confident in his superior number, disposed his troops in a crescent around the royals. Henry IV gave his troops a rallying point: three pear trees he could see behind Mayenne's lines. At 12 P.M., the royal artillery fired nine volleys into the crowding Catholics before Mayenne ordered his German cavalry to charge. The Protestants advanced but refused to fire on their coreligionists and instead they rode at full speed through the waiting Catholic infantry. The infuriated Spanish cavalry charged home on the royal artillery. Henry, leading his gendarmes, stopped this dangerous move; the fight was now a tangle of intermingled cavalry. The royal reserve took the opportunity to charge into the fray, after seeing that all the Catholics had fired their pistols and had no more powder. The Spanish survivors left the battlefield around 1 P.M. The king, following them, led the pursuit to the pear trees and then turned his cavalry toward the three large blocks of Catholic infantry. After a few volleys of artillery, he offered them to surrender. As Mayenne fled the battlefield, they asked for mercy. More than 4,000 were taken prisoners; all the German *landsknechts* were slaughtered by the royal Swiss.

The battle was over by 2 P.M. The Catholics had lost their field army and had to rely more than ever on the Spanish troops. It would take four more years before Mayenne and the Catholic League chose to rally Henry IV.

Gilles Boué

See also: French Wars of Religion

References and further reading:

Bayrou, François. *Henry IV.* Paris: Fayard, 1996.
Cornette, Joel. *Le XVIème Siècle.* Paris: SEDES, 1995.
Hardy, Etienne. *Batailles Françaises.* Paris: Dumaine, 1881.

Marines raising the American flag on Mount Suribachi, Iwo Jima, 1945. (Library of Congress)

Iwo Jima, Battle of (19 February–15 March 1945)

One of the bloodiest battles of World War II in the Pacific. In order to provide American warplanes with an airfield only 600 miles from Japan and to shorten the attack route toward the Japanese home islands, the United States planned an invasion of Iwo Jima for February 1945. Iwo Jima, only 8 square miles in area, was defended by 21,000 soldiers who were literally dug into the island. The Japanese commander, General Kuribayashi, was determined to inflict maximum casualties on the American forces by fighting to the last man. His soldiers used Iwo's soft volcanic ash and extensive caves to build an effective network of underground tunnels, fortifications, and concrete pillboxes, and from these well-entrenched positions the Japanese prolonged the battle and exacted high American losses.

After a three-day naval bombardment rather than the 10 days originally requested by the marines, the main American assault led by the 4th and 5th Marine Divisions started on 19 February 1945. Although 30,000 U.S. soldiers were ashore by the end of the first day, the volcanic sand made mobility difficult. With the beach crowded with marines and equipment, the Japanese opened fire and inflicted heavy losses, and even marine veterans commented on the ferocity and violence of the first days. Moving to the south, one regiment of marines captured Mount Suribachi in the face of withering Japanese fire and decided to raise the American flag in order to increase the morale of the marines below and naval vessels offshore. The image of the flag raising, captured by war photographer Joe Rosenthal, provided one of the most famous images of World War II. Despite this symbolic victory, the battle raged for 30 more days, with reinforcements from the 3d Marines joining the 4th and 5th divisions driving north against difficult terrain and the Japanese tunnels throughout the island. The United States employed heavy firepower from the air and from artillery, grenades, mortars, and tanks with flame throwers and slowly began to move northward across the island. The carnage on both sides can be seen in the nickname the "Meat Grinder" for Japanese defensive positions in the reinforced hills. Finally the Japanese defenders were isolated in the northern corner and destroyed.

While nearly the entire Japanese force of 21,000 was killed, the Americans suffered 26,000 casualties, including more than 6,000 killed, an average of more than 700 dead for every square mile on the island and the first time that American casualties exceeded those of the Japanese. In the end, 27 marines were awarded the Medal of Honor. General Holland Smith praised General Kuribayashi's toughness and tenacity, while Admiral Chester Nimitz proclaimed: "Among the Americans who served on Iwo Island, uncommon valor was a common virtue."

Harold J. Goldberg

See also: U.S. Marines

References and further reading:

Bradley, James. *Flags of Our Fathers: Heroes of Iwo Jima.* New York: Delacourt Press, 2001.

Garand, George W., and Truman R. Strobridge. *Western Pacific Operations: History of the U.S. Marine Corps Operations in World War II.* Washington, DC: United States Marine Corps, 1971.

King, Ernest J. *U.S. Navy at War, 1941–1945. Official Reports to the Secretary of the Navy.* Washington, DC: United States Navy Department, 1946.

Spector, Ronald H. *Eagle against the Sun.* Norwalk, CT: Free Press, 1985.

J

Jackson, Andrew (1767–1845)

American field commander, politician, and seventh president, the first to be elected from the frontier. Jackson was born on 15 March 1767 in Waxhaw, South Carolina, the youngest of three sons of recent Scots-Irish immigrants. Because his father and namesake died before his birth, his mother, Elizabeth, moved the family into the home of her brother-in-law, James Crawford. Jackson received only rudimentary schooling.

When the British invaded the Carolinas in 1780, Jackson's eldest brother, Hugh, volunteered for the militia and was soon killed. After Colonel Banastre Tarleton defeated the militia near Waxhaw in May, Elizabeth, Andrew, and his brother Robert helped care for the American wounded. The brothers volunteered for the mounted militia and fought at Hanging Rock, South Carolina, on 1 August. After several months as guerrillas in the backwoods, they were captured early in 1781. Both received saber cuts for refusing to polish a British officer's boots. As prisoners of war in Camden, South Carolina, they contracted smallpox. Elizabeth negotiated their exchange but Robert died on the way home. After she nursed Andrew back to health, she nursed American prisoners on British ships in Charleston harbor, where she died of typhus. These incidents prompted Jackson's lifelong hatred of the British.

In 1786 Jackson began to practice law in Martinsville, North Carolina. Eager for the rough life, he moved to Tennessee in 1788 and became a prosecuting attorney for Nashville in 1789. He married Rachel Donelson Robards in 1791, both of whom believed in good faith that her divorce from Lewis Robards was final. Discovering later that she was not legally divorced until 1793, they remarried in 1794. These circumstances dogged Jackson the rest of his life and led to his fighting many duels to defend his wife's honor. In a duel with Charles Dickinson on 30 May 1806, he took a bullet in the chest, then deliberately aimed and "killed his man." Jackson carried his bullet to the grave.

Through Rachel's family's money and his own shrewd land deals, Jackson quickly became a wealthy man with a large plantation. He served in the U.S. House of Representatives from 1796 to 1797, in the U.S. Senate from 1797 to 1798, and on the Tennessee Supreme Court from 1798 to 1804. After 1802 he was a major general of the Tennessee State Militia. He volunteered his troops for service as soon as the War of 1812 began. Ordered to Natchez, Mississippi, to prepare to invade Florida, he earned the nickname "Old Hickory" by being tough enough to maintain discipline after Congress canceled the invasion. He did not see action until after the Fort Mims massacre on 30 August 1813. His victories at Talladega, Alabama, on 9 November and Horseshoe Bend on 27 March 1814 won him a commission as a major general in the regular army on 28 May.

Using Mobile, Alabama, as a base, he captured Pensacola, Florida, on 7 November and thoroughly defeated the British at New Orleans on 8 January 1815. Now a national hero, he commanded the Southern Division, headquartered in Nashville. From 1817 to 1819 he fought the Seminoles in Spanish Florida, hanging two British subjects as spies in April 1818 and capturing Pensacola again in May, nearly provoking war with both Britain and Spain. He resigned his commission and served briefly as the first governor of Florida Territory in 1821, then was a U.S. Senator from 1823 to 1825. After losing a controversial four-way presidential election in 1824, he won by landslides in 1828 and 1832. Throughout his presidency, he vigorously supported westward expansion and the forcible appropriation of Indian lands. It would be difficult to imagine anyone at the time who more exemplified the virtues and vices of mid-nineteenth-century Americans.

Eric v. d. Luft

See also: American Indian Wars; American Revolution; Creek War; Horseshoe Bend, Battle of; Houston, Samuel; New Orleans, Battle of; War of 1812

References and further reading:

Bassett, John Spencer. *The Life of Andrew Jackson.* Hamden, CT: Archon Books, 1967.

Davis, Burke. *Old Hickory: A Life of Andrew Jackson.* New York: Dial, 1977.

Heidler, David Stephen, and Heidler, Jeanne T. *Old Hickory's War: Andrew Jackson and the Quest for Empire.* Mechanicsburg, PA: Stackpole, 1996.

James, Marquis. *The Life of Andrew Jackson.* Indianapolis: Bobbs-Merrill, 1938.

Remini, Robert Vincent. *Andrew Jackson.* New York: HarperPerennial, 1999.

Jackson, Thomas "Stonewall" (1824–1863)

Confederate commander-hero and General Robert E. Lee's irreplaceable "right arm." Born on 21 January 1824 in Clarksburg, Virginia, and orphaned at seven, he was raised by an uncle. Despite a poor formal education, he was appointed to the U.S. Military Academy. Through determination, he improved his academic standing steadily, graduating in 1846 in the upper third of his class.

In the Mexican War Jackson served with distinction as an artillerist during General Winfield Scott's campaign to Mexico City. Jackson saw action at Veracruz, Cerro Gordo, Churubusco, and Chapultepec, twice earning brevet promotions. In 1851, Jackson resigned his commission for a position as a professor of artillery and natural philosophy at the Virginia Military Institute.

At the outbreak of the American Civil War, Jackson was commissioned a brigadier under General Joseph E. Johnston in the Confederate States Army. At First Bull Run, his steadiness anchored the Confederate line at the crucial moment, buying time for the tide of battle to turn. The nickname "Stonewall" was thereafter applied to both Jackson and his heroic brigade. Jackson was subsequently promoted to major general.

Commanding his own small army, Jackson received the task of preventing several Union armies from converging on Richmond. His subsequent "Valley Campaign" was masterful and is still studied today in war colleges around the world. Striking at first one then another of the disjointed Union formations, Jackson tied down superior forces under generals Nathaniel Banks, Irvin McDowell, and John C. Frémont, while General Lee dealt with the main Union advance on Richmond.

After a string of minor but timely victories, Jackson's troops made a key contribution to Confederate victory at Second Bull Run. Though Antietam was a strategic loss, Jackson's role in the invasion of Maryland had been laudable, earning him a promotion to lieutenant general and corps command in Lee's Army of Northern Virginia. In late 1862, at Fredericksburg, Jackson held down the vulnerable flank before executing an explosive counterattack. During Chancellorsville in May 1863, Jackson received a mortal wound from the fire of his own troops, died from his wounds on 10 May 1863 at Guiney's Station, and was buried in Lexington. Jackson's loss left an almost inconsolable General Lee without his most trusted and reliable lieutenant. It would be difficult to imagine a worse blow to the Confederacy.

Michael S. Casey

See also: American Civil War; Antietam/Sharpsburg; Chancellorsville, Battle of; Bull Run, First/Manassas; Bull Run, Second/Manassas Junction; Lee, Robert Edward

References and further reading:

Farwell, Byron. *Stonewall: A Biography of General Thomas J. Jackson.* New York: W.W. Norton, 1992.

Robertson, James I., Jr. *Stonewall Jackson: The Man, The Soldier, The Legend.* New York: Simon & Schuster, 1999.

Selby, John. *Jackson as Military Commander.* Princeton: Van Nostrand, 1968.

Tanner, Robert G. *Stonewall in the Valley: Thomas J. "Stonewall" Jackson's Shenandoah Valley Campaign, Spring, 1862.* Garden City, NY: Doubleday, 1976.

Jacobite Rebellions (1689–1746)

Uprisings that attempted to restore the Stuart Dynasty, and the last land battles on English and Scottish soil. Following the Revolution of 1688—which had been sparked by the birth of a son and potentially Catholic heir to James II of Great Britain, and the successful invasion of William of Orange, the king's Protestant son-in-law, who assumed the throne with his wife, Mary—James fled to France with his wife and son. For the next 57 years, Jacobitism, the support for the exiled branch of the Stuart family, would be a major tool of European foreign policy and spark four armed uprisings and numerous plots in England, Scotland, and Ireland.

Returning to Ireland in March 1689 with a contingent of French officers sent by Louis XIV, James began to muster an Irish army. Meanwhile, in Scotland, John Graham, Viscount Dundee, had also raised a Jacobite army and, after a successful series of raids, had defeated government forces at Killiecrankie. Although defeated at Dunkeld, Dundee's army, gaining support among the Highland Scots, fought on. In Scotland, the government was forced to an expensive policy of fort building and slow harassment of the clans, while in Ireland, the government fought using mercenaries hired

Charles Edward Stuart (also known as Bonnie Prince Charlie) departs Scotland in 1746, five months after the disastrous Jacobite defeat at the Battle of Culloden. (Library of Congress)

from Europe to augment English forces. William of Orange defeated James on 1 July 1690 at the Battle of the Boyne, although French naval support allowed the Jacobites to continue fighting until government reinforcement arrived and captured Limerick and Galway. With the Jacobites defeated, William turned his resources to fighting France until the treaty of Ryswick in 1697.

The next Jacobite rising occurred in 1715, following an abortive invasion scare in 1708. The Earl of Mar, unhappy with his prospects under the new king, George I, raised the banner of James II's son, James ("the Old Pretender"), in the highlands, expecting significant French assistance. A corresponding English rising, centered on northern Catholics, failed to accomplish much and was defeated and captured at Preston in November 1715. Mar, meanwhile, fought an indecisive battle at Sheriffmuir on 13 November against the Duke of Argyll. When James Stuart arrived in December, without French aid, the rebellion was fading, and most of the leaders had fled to France by February 1716.

In 1719, with Britain and Spain on hostile terms because of Spain's invasion of Sicily, Cardinal Alberoni, prime minister to Philip V of Spain, lent his support to a Jacobite invasion of Scotland. Again mustering highlanders to augment 250 Spanish regulars, the Jacobites, under the command of the earl Marischal and marquis of Tullibardine, quarreled among themselves and were caught at the pass of Glenshiel by government forces. The Scots fled, leaving the Spanish as prisoners of war until ransomed by their own government.

The final Jacobite campaign, "The '45," was first sponsored as a French diversion meant to draw British troops out of the War of the Austrian Succession. A 1744 invasion, to be led by the Marshal de Saxe, fell through after a great storm not only destroyed stockpiled supplies but disrupted the French fleet sent to gain control of the English Channel. Not to be dissuaded, the Stuart claimant, Charles Edward Stuart ("the Young Pretender" or "Bonnie Prince Charlie"), prepared an invasion on his own, borrowing money and counting on a mass uprising upon his arrival in Scotland. Again due to poor weather, the Jacobites arrived with only half the planned men and supplies. (Many began to mutter that "God is a Protestant!") The Jacobites did muster a number of highlanders, captured Edinburgh Castle, and defeated the local government forces at Prestonpans, before marching south into England with an army of about 4,500. The Jacobite army turned back at Derby, now convinced that there was no support in England or substantial French aid on the way, abandoning a strike at London in favor of a retreat back to Scotland.

Pursued by two Hanoverian armies under the Duke of Cumberland and George Wade, they collected a trickle of smuggled French supplies and, after successfully holding off the government troops at Falkirk, went to ground over the winter of 1745/6. Charles Edward Stuart, emerging from a fit of petulance over the retreat from Derby, insisted on a conventional action rather than continued evasion and in April 1746 met Cumberland at Culloden, where the Jacobites were decisively defeated.

Fleeing, Charles Edward Stuart dismissed the survivors of his army who had rallied after the battle and made his way in secret through the Highlands before reaching France. Ruthlessly punished by the government for their participation, the Scots, disenchanted by Jacobitism, abandoned the Stuarts to romantic nostalgia and the Jacobite threat ceased to exist.

Margaret Sankey

References and further reading:

Jarvis, R. C. *Collected Papers on the Jacobite Risings.* 2 vols. Manchester, UK: Manchester University Press, 1972.

Reid, Stuart. *1745: A Military History of the Last Jacobite Rising.* New York: Sarpedon, 1996.

Szechi, Daniel. *The Jacobites: Britain and Europe.* Manchester, UK: Manchester University Press, 1994.

Jan III Sobieski (1629–1696)

Polish military commander, savior of Vienna. Heir to three of Poland's wealthiest landowning families, Jan Sobieski received a cosmopolitan education and as a young man traveled western Europe on the Grand Tour. In 1648, he joined the Polish army, beginning his career with the suppression of the rebellion of Hetman Chmielnicki and his Cossacks. In protest against King John Casimir, Sobieski briefly served under Charles X of Sweden in his 1655 invasion of Poland but changed sides the following year to drive out the invader. With court patronage, Sobieski rose quickly to field hetman, then grand hetman of the Polish forces while conducting successful campaigns against the Tartars and Cossacks in the Ukraine.

In 1672, the Ottoman Empire, whose border was only 40 miles from Crakow, invaded Polish territory and Sobieski took the field against the invaders, annihilating an entire Ottoman army at the Battle of Chocim, on the Dnieper River, in 1673, only days after the death of King Michael Piast.

This victory, and Sobieski's reputation, won him the elective monarchy of Poland. Still at war with the Ottomans, he defended the fortresses of Lwow and Trembowla and reached a truce with the sultan in September 1676. This truce gave him breathing space to improve the army, increase the mobility of artillery, enlarge the dragoons, and reduce reliance on pikemen. However, still operating within the semifeudal politics of Poland, he was unable to centralize fiscal planning or logistical supply.

Sobieski's planned pro-French foreign policy crumbled against the realities of Poland's relationship with the Habsburg emperor and their mutual enmity to the Ottomans; and he willingly answered the appeal of Leopold II for troops to relieve Vienna from Ottoman siege. Acting as commander in chief of a 75,000-man relief force of Lithuanians, Poles, and Germans, Sobieski, particularly skillful in his transport of artillery into the Vienna Woods and the construction of a pontoon bridge across the Danube, defeated Kara Mustafa at Khalenberg and saved Vienna. Unfortunately, attempts to follow this victory with repeated campaigns in Moldavia between 1687 and 1691 failed and drained Polish resources. Suffering repeated heart attacks after 1691 and forced to deal with the rising power of Prussia and the increasing intransigence of Lithuania, Sobieski died in 1696.

Margaret Sankey

See also: Vienna, Sieges of; Turkish Wars of European Expansion

References and further reading:

Davies, Norman. *God's Playground: A History of Poland.* New York: Columbia University Press, 1982.

———. *Sobieski's Legacy.* London: Orbis, 1985.

Morton, J. B. *Sobieski, King of Poland.* London: Eyre & Spottiswoode, 1932.

Janissaries

An elite corps of the Ottoman Empire and perhaps the first standing army in Europe. It helped conquer the Byzantine Empire, Egypt, and much of the Balkans, eventually becoming the praetorian element in Ottoman palace intrigues until brutally suppressed in 1826.

The Janissaries, from the Turkish *yeni cheri* (new troops), were formed in 1330 as a bodyguard under the personal command of Sultan Orkhan. They were reformed as a military corps by Sultan Murad I in 1365, using captives from the Byzantine city of Adrianople (Eridne, Turkey). Originally recruits consisted of 1,000 men drawn from the fifth part (*besinci*) of the human booty, which, according to Ottoman law, belonged to the sultan. Replacements of Christian prisoners of war forced to accept Islam or Christian volunteers who later converted to Islam proved insufficient for meeting the demands of the growing corps, and captive Christian

Engraving of Janissaries and Spahis, 17th century. (Chris Hellier/Corbis)

children were recruited beginning in the second half of the fourteenth century. This too proved insufficient and Sultan Selim I imposed the *devsirme,* or forced levy of children, on subject Christian populations in the Balkans. The Janissaries numbered 16,000 by 1520 and 37,000 by 1609. Christians were recruited because Janissaries became slaves of the sultan, and the enslavement of Muslims is forbidden by the Koran. However, once recruited and enslaved, a Janissary was converted to Islam.

The Janissary corps consisted of cavalry, infantry, artillery, and sailors. They often were armed with firearms, taking part in all the major battles and wars of the Ottoman Empire for more than half a millennium, including the capture of Constantinople in 1453, the conquest of Egypt in 1517, the Battle of Mohács in 1526, and the siege of Vienna in 1683. The Janissaries' failure to suppress the Greek revolt in 1820, and their own numerous mutinies and stiff resistance to military reforms led Sultan Mahmud II ruthlessly to suppress the corps after its attempted revolt on 14–15 June 1826.

Michael C. Paul

See also: Byzantine-Ottoman Wars; Constantinople, Siege of (1453); Mamluks; Turkish Wars of European Expansion

References and further reading:

Goodwin, Godfrey. *The Janissaries.* London: Saqi, 1997.

Japanese Civil Wars (1450–1550)

The period from 1450 to 1550 in Japan known as the Epoch of the Warring Country, or Sengoku Period, that marked a time of continual civil war and unrest. It also marked the decline of central power and the rise of samurai and daimyo (who as great provincial lords were also samurai). Daimyo and their loyal retainers, samurai, battled other daimyo to take advantage of perceived weaknesses or to gain control over economically important areas. The older order, the powerful families in Kyoto, the land stewards, and so on, began to disappear and power diffused throughout the country to the new class of daimyo.

For many years, Japan was controlled by a series of families operating at the capital, Kyoto, who employed stewards throughout the country. While the emperor was weak, power was centralized among these families. The Onin War and the ensuing century of civil wars ended their control outside Kyoto and ushered in a new era that eventually would lead to the consolidation of power behind a new elite.

In addition to the level of destruction, there was the beginning of a new order, or a reorder. Daimyo sought to control their holdings and tried to categorize the quality of agricultural land and manufacturing production in towns. They engaged in *sword hunts* to disarm the peasantry and also sought to elevate the status of samurai, the only warriors allowed to wear the two swords, one long and one short. Samurai were encouraged to settle in fortified castle towns and leave the countryside, and thus the daimyo sought to increase economic production to expand their sources of wealth.

There were other issues, to be sure. The daimyo to the south and west benefited from trade with China and, toward the end of the Sengoku Period, contact with the West; daimyo received valuable goods from China for trade and received more accurate calendars, improved medical knowledge, and guns and gunpowder from the Portuguese, Spanish, and later Dutch traders. The fighting of samurai to samurai with swords, reflecting many years of training and fierce discipline, began to give way to fighting with muskets and cannon and favored those daimyo with greater economic resources or the good fortune to consider converting to Christianity. The samurai perfected the tea ceremony and looked to Buddhism to balance the carnage of continual fighting.

Toward the end of the period (for which there really is no exact date), several groupings of daimyo began to emerge who gathered great power. In time, three great daimyo arose—Oda Nobunaga, Toyotomi Hideyoshi, and Tokugawa Ieyasu—who eventually unified the country, made peace among the contending factions, tried to close the country to foreign influences, and took Japan into a period of 250-plus years of isolation and yet great change.

Charles M. Dobbs

References and further reading:

Elison, George, and Bardwell L. Smith, eds. *Warlords, Artists & Commoners: Japan in the Sixteenth Century.* Honolulu: University of Hawaii Press, 1981.

Hall, John Whitney, ed. *Japan before Tokugawa: Political Consolidation and Economic Growth, 1500–1650.* Princeton, NJ: Princeton University Press, 1981.

Totman, Conrad D. *Early Modern Japan.* Berkeley: University of California Press, 1993.

Japanese Colonial Wars (1874–1945)

Japan's 60-year period of seeking to establish a colonial empire for economic exploitation and to gain equality of a sort with the major European powers. The Meiji Restoration oligarchs had their differences, and one was over an early invasion of Korea as a prelude to expansion on the Asian continent. Realizing that Japan was still too weak, the leaders instead decided to bring the Ryukyu Islands firmly under Japanese control. The people of the Ryukyus were related to the Japanese and were controlled by the former daimyo of Satsuma but had long-established tributary relations with China. In 1872, the new government in Tokyo claimed control and, acting on behalf of the islanders, sent a punitive expedition to Taiwan to avenge the murder of some islanders by Taiwanese aborigines. This campaign took until 1874. The Chinese government in Beijing paid an indemnity, implying a recognition of Tokyo's new territory in the Ryukyus.

The next target, not unnaturally, was Korea. In 1875, the Japanese demonstrated several of the Western-built ships in their new navy off Korean waters and in 1876 gained the Treaty of Kanghwa, opening the former Hermit Kingdom to Japanese trade and claiming Korean independence (from China). Japanese interest in Korea eventually led to the Sino-Japanese War of 1894–1895, pitting the small modern army and navy of Japan against the larger but seriously outdated military forces of China; the Japanese won easily but saw the fruits of their victory—a leasehold over the Kwantung Peninsula—taken away in the Triple Intervention. But Japan did gain control over Taiwan and the nearby Pescadores Islands, reflecting Japan's increasing economic interest along China's coast. In 1902, the Anglo-Japanese Alliance brought British recognition of Japan's interest in Korea "in a peculiar degree politically as well as commercially and industrially."

Japan faced few difficulties in its occupation of Taiwan. The effort to develop Taiwan economically and to bring it within the emerging Japanese system really began in 1898. Japanese authorities suppressed bandits and established order; they also raised health standards, improved diets, and provided widespread basic education for all Taiwanese. However, to advance beyond basic education required Japanese language ability and a willingness to accept Japanese cultural norms. While the Japanese did tighten controls in the mid-1930s, they never faced any real threats of rebellion until the end of the Pacific War.

The stage was set for the Russo-Japanese War of 1904–1905. When the exhausted combatants agreed to the invitation by U.S. president Theodore Roosevelt for peace negotiations, the resulting Treaty of Portsmouth recognized Japan's "paramount interest" in Korea, implied that the Russians would vacate Manchuria (thereby leaving it available for Japanese exploitation), and turned over to Japan the Russian leasehold on the Kwantung Peninsula, ownership of the South Manchurian Railway, and the southern half of Sakhalin Island.

The Koreans did not accept the end to independence easily. Korea had a long history, perhaps longer than Japan's, and Koreans were proud of their past; they did not move easily into Japan's Inner Empire and give up their sense of nationhood. When Japan declared Korea a protectorate in 1907 the Koreans responded with more than 1,400 armed attacks and outbreaks of violence between 1908 and 1910, and a Korean patriot in late 1909 assassinated a key Japanese oligarch, Ito Hirobumi, who had appointed himself resident general in Korea. Japan exploited Korea's economy more than Taiwan's, and the result was a declining standard of living for many Koreans; the lack of political freedoms, omnipresent Japanese control, and economic deprivations combined to spark the great Mansei revolt in March 1919; more than a million demonstrators protested for Korean independence as the world was negotiating the Versailles Peace Treaty in France. In turn, the Japanese moved to convert the well-ordered system of Korean schooling into Japanese-intensive education and to control Korean life ever more tightly. But between a strongly developed sense of Korean nationalism, the continuing influence of American Christian missionaries, and the growing influence of Marxism, especially of the Chinese Communist variety, Koreans were most uneasy subjects in the Japanese empire.

World War I increased the growing Japanese colonial empire. In addition to Korea, the Ryukyus, the Pescadores, Taiwan, and also interests in southern Manchuria, Japan sided with the Allies (based on its treaty with Great Britain) and as a consequence gained at the Versailles peace negotiations. Japan received German concessions in the Pacific—a series of island chains—and German holdings in China, especially Kiaozhou Bay and the port of Qingtao in Shandong. (No one inquired of the Chinese as to these transferrals of what was originally their territory.)

The East Asian world was somewhat fluid during the 1920s, but in September 1931 the Japanese Kwantung Army used the Mukden Incident as an excuse to seize control over Manchuria the next year and then gradually to expand, first into that part of China outside the Great Wall and then by the mid-1930s to significant parts of North China inside the Great Wall. In December 1941 the Pacific War began in earnest and within months the Japanese had acquired a large colonial empire for its Greater East Asia Co-Prosperity Sphere; many Asian nationalists admired Japan's early victories over European nations and the United States. In fact, Japan declared the "independence" of a number of former Asian colonies that they had conquered, including Malaya, Burma, and the Philippines. But nationalists increasingly resented the substitution of Japanese colonial administration for that of Europe or the United States. In August 1945, Japan's surrender to the Allies ended World War II and Japan's overseas empire.

Charles M. Dobbs

See also: World War II

References and further reading:

Beasley, W. G. *Japanese Imperialism: 1894–1945.* Oxford and New York: Clarendon, 1987.

Giffard, Sydney. *Japan among the Powers, 1890–1990.* New Haven, CT: Yale University Press, 1994.

Montgomery, Michael. *Imperialist Japan: The Yen to Dominate.* New York: St. Martin's Press, 1987.

Nish, Ian Hill. *Japanese Foreign Policy, 1869–1942.* London and Boston: Routledge & Kegan Paul, 1977.

Japanese Invasion of Korea (1592–1598)

The Japanese invasions of Korea describe two campaigns launched against Korea by Toyotomi Hideyoshi (1537–1598), the warlord and hegemon of Japan. Though the Japanese enjoyed initial success on land, Japan's inability to maintain control of the seas, continuous harassment by Korean guerrilla forces, the intervention of Ming Chinese forces, and ultimately, the death of Hideyoshi doomed the invasion to failure.

After gaining control over Japan in the last quarter of the sixteenth century, Hideyoshi turned his sights toward the conquest of China. This had as much to do with a need to occupy Japan's warrior class as with Hideyoshi's own grandiose visions. Such a continental invasion (just as with the Mongol invasions of Japan in the thirteenth century) necessitated the active cooperation of Korea, the closest mainland coast to the Japanese archipelago. Korea refused to collude with Hideyoshi, for geopolitical reality as well as Confucian principles required Korea's loyalty to China. In 1591 Hideyoshi ordered the invasion of Korea as a preliminary to the proposed China conquest.

Hideyoshi delegated leadership to a handful of loyal generals under the command of the Christian Konishi Yukinaga. In spring 1592 a vanguard Japanese force of over 50,000 embarked for the Korean city of Pusan. Though the Korean court had made some preparations for the attack, having received alarming reports of Japanese preparations, the Korean armies were eminently ill prepared to deal with the invasion. After landing with overwhelming force at Pusan the Japanese forces advanced in three prongs northward, making quick progress toward the Korean capital of Seoul. King Sonjo (r. 1567–1608) fled with his royal entourage to the more northern city of Pyongyang. In late spring 1592 Japanese forces captured and burned Seoul. Soon thereafter reinforcements arrived from China via Manchuria, and the land conflict was pushed to stalemate, with Chinese and Korean troops holding around Pyongyang and the Japanese barricaded in and around Seoul.

Meanwhile, Korean guerrilla armies and the navy under admiral Yi Sun-shin (1545–1598) harried Japanese troops and supply lines. Yi defeated the Japanese fleet at Pusan in late 1592, seriously hampering Japan's ability to procure reinforcements and supplies. Konishi agreed to armistice talks, and by 1594 the Japanese had completed their withdrawal from Korea.

Hideyoshi ordered a second invasion in 1597. Once again Hideyoshi chose Konishi to lead the attack. As before, the Japanese armies, numbering almost 150,000, met with initial success, defeating a Chinese army near Ulsan in early 1598. In late 1597, however, the Korean Admiral Yi Sun-shin had again inflicted a heavy blow on the Japanese fleet in the Battle of Myongnyang Straits. The sudden death of Hideyoshi in late 1598 brought the second invasion to an abrupt conclusion.

The invasion inflicted serious damage to the Korean landscape, and the famines and epidemics that resulted were harbingers of popular unrest to come. The cultural damage inflicted by the invaders on temples, official structures, and artifacts was also severe. The expenses and losses suffered during the invasions by China's Ming Dynasty (1368–1644) contributed to that dynasty's overthrow by the Qing (1644–1911) less than 50 years later.

Daniel Kane

See also: Japanese Wars of Unification; Hideyoshi, Toyotomi

References and further reading:

Lum, Peter. *Six Centuries in East Asia: Japan and Korea from the 14th Century to 1912.* New York: S. G. Phillips, 1973.

Turnbull, S. R. "Hideyoshi's Korean War." In S. R. Turnbull, *The Samurai: A Military History.* New York: Macmillan, 1977.

———. *The Samurai: A Military History.* New York: Macmillan, 1977.

Japanese Military, Twentieth Century

The first modern Asian military power. Combining Prussian military theory and traditional warrior values, the Imperial Japanese military had two prime missions: defense of the home islands against Western powers and the unification of the local polities under the new central government. The first campaign of the new Japanese military was the suppression of the Satsuma revolt (1877–1878).

Having defeated the traditional horde fielded by China (1894–1895) and a Russian military in decline (1904–1905), the roots of ultimate Japanese military failure were nonetheless planted during World War I. Lacking direct experience of the fighting in Europe, there was no experience of the traumatic challenges of these campaigns to force the reassessment of received wisdom, the exception being Germany's vulnerability to economic blockade. The ultimate response was conversion of the informal zone of influence in China into a formal empire, leading to the Manchurian Incident (1931) and Japan's political isolation from the liberal West.

Besides deteriorating economic and geopolitical circumstances, the failings of the Meiji constitution were a factor. Active-duty officers held the positions of army and navy ministers and had open access to the emperor. Being a free agent under this system, the military could pull down any given cabinet and arrogated increasing political influence for themselves, culminating with General Hideki Tojo becoming prime minister in October 1941.

Another factor leading to the drive for military political supremacy were political cliques in the army itself, as the so-called Imperial Way and Control factions strove for predominance. The Imperial Way desired the demise of the liberal order of the 1920s and the imposition of direct imperial rule. Meanwhile, the Control faction sought to suppress the political extremists and build up the modern capabilities of the army, in expectation of a coming war with the Soviet Union. By the end of 1936, the Control faction had brought to heel its rival, but the unifying factor for the army remained the drive for a secure economic base on the Asian mainland, particularly as massive resources were poured into the Manchurian and northern Korean industrial base.

As all-out war with China opened in 1937, Japan finally had to face its limitations. Despite the imperial forces' many advantages, the Chinese were still able to muster superior numbers and inflict sobering losses. Further, the Japanese defeat by the Soviet army at Nomonhan taught a direct lesson as to how war had changed in the wake of World War I.

As a result, Tokyo signed a nonaggression pact with Moscow and cultivated a new interest in the Western colonial interests in Southeast Asia, a course of action that brought strategic congruence with the imperial navy.

In choosing war with the West rather than the USSR, Prime Minister Tojo and the army general staff allowed themselves few doubts, in that they believed it was better to lose than not to try. They also hoped that victory would buy the time to create the resources needed to fight the larger war.

Their belief had always been that a well-led and motivated force would be able to overcome either the mass army of a traditional state or the small constabulary armies of the Western colonial powers. In reality, the Japanese military lacked the resources to go up against the modern mass army of a determined state for any length of time, a lesson that could have been learned as early as 1905.

By 1943, every Japanese liability was being ruthlessly exposed. The veneer of modernity created by aircraft and light arms disguised that many structural basics were lacking in the composition of the force. Relative poverty had put an emphasis on doing more with less, but it also inculcated an indifference to questions of intelligence, logistics, and signals.

There were also basic failures to adjust to modern circumstances at the psychological level, as officers continued to consider themselves a social caste and cultivated a theatrical style of leadership. Failing to master the skills of modern military professionals, Japanese staff schools taught advanced infantry tactics instead of preparing Japanese officers to think in operational-level terms. The final result would be an empty force reduced to claiming that superior spirit would defeat the Allies in 1945.

The demise of a Japanese military system was a short-lived development, though. In spite of the new Japanese constitution forswearing war, the North Korean invasion of 1950 forced the creation of the Japanese "Self-Defense" Force.

Starting with the so-called National Police Reserve—created at the direction of General MacArthur's headquarters to replace the American occupation troops deployed to fight in Korea—American officers and Japanese enlisted personnel were formed into new units. Besides the steady infusion of Japanese company and field-grade officers from the former imperial army, the sign that the new force was more than an

Japanese soldiers in action, 1942. (Library of Congress)

emergency extemporization was the creation of the Japanese Defense Academy in 1953. The JDA was an all-services institution designed both to overcome the factionalism of the prewar military and to inculcate a respect for civilian authority and the rule of law in the new officers.

Despite frequent domestic criticism as to its role, the Japanese Self-Defense Force has functioned as a component of the Japanese-American security agreement. Mostly existing to deny an enemy an easy target, it represented a return to the original function of the first modern Japanese military.

George R. Shaner

See also: Hasegawa, Yoshimichi; Japanese Colonial Wars; Khalkin-Gol; Russo-Japanese War; Sino-Japanese War; Terauchi Hisaichi; Yamagata, Aritomo; Yamashita, Tomoyuki; World War II

References and further reading:

Drea, Edward J. *In the Service of the Emperor: Essays on the Imperial Japanese Army.* Lincoln: University of Nebraska Press, 1998.

Humphreys, Leonard A. *The Way of the Heavenly Sword: The Japanese Army in the 1920s.* Stanford, CA: Stanford University Press, 1995.

Maeda, Tetsuo. *The Hidden Army: The Untold Story of Japan's Military Forces.* Chicago: Edition Q, 1995.

Millett, Allan R., and Williamson Murray. *Military Effectiveness.* 3 vols. Boston: Allen & Unwin, 1988.

Japanese Wars of Unification (1550–1615)

The years from about 1550, with the introduction into Japan of Western firearms, to the siege of Osaka Castle in 1615 by the forces of the shogun Ieyasu Tokugawa, were the final act in an extended period of armed struggle in Japan, going back to the early fifteenth century. It was during these approximately 75 years that three military leaders would emerge to forge a unified Japan: Nobunaga Oda (1534–1582), Toyotomi Hideyoshi (1536–1598), and Ieyasu Tokugawa (1542–1616). It would end with the victory of Ieyasu and the unchallenged hegemony of the Tokugawa Shogunate he established and usher in a period of unprecedented centralized power in Japan.

The year 1549 saw the beginning of regular contact between foreign, which is to say Catholic, missionaries and the Japanese. The Japanese reaction to Western Christianity aside, it was the Western introduction of firearms, in the form of the arquebus and cannon, that was to prove revolutionary and instrumental in reshaping the Japanese political landscape. By 1560 they were being used in battle, and it was the richer and more powerful local lords, or daimyo, who could afford them who would benefit most.

The Ashikaga Shogunate (1338–1573) had from its establishment proved lacking in central authority. Its founder Ashikaga Takauji was but one among a group of powerful military leaders vying for control. Indeed, the Ashikaga government ruled through a system of delegating authority in the provinces to constabularies, usually powerful military leaders loyal to the Ashikaga, who in return for their formal recognition of Ashikaga supremacy were granted almost total control over provincial domains. The central authority of the Ashikaga house, already delicate, was further weakened in the fifteenth century by civil wars between varying claimants to the shogunate. The Onin War (1467–1477) ushered in a period of more vigorous and ruthless warfare between local military leaders, who with Ashikaga weakness had begun to consolidate their individual power and private loyalties and to expand territorially. These local military leaders were for the most part no longer related to the officially sanctioned constabulary of the earlier Ashikaga. They were local military men who had risen to the top by dint of their ruthlessness, cunning, and military abilities. By the late fifteenth century the Ashikaga shoguns were reduced to insignificant bystanders in a contest between these powerful and autonomous local military lords, or daimyo. This so-called Period of Warring States (Sengoku jidai), from approximately 1467 to 1568, may have been fought into a stalemate, but the introduction of Western firearms proved decisive.

In 1568 a young man named Nobunaga Oda, then the heir to a daimyo around Kyoto, seized Kyoto, the capital of the Ashikaga, eventually overthrowing the last Ashikaga shogun in 1573 (an act in itself that garnished little attention, such had the Ashikaga fortunes fallen). In a close working alliance with another regional daimyo, Ieyasu Tokugawa, Oda set about consolidating the central region of Japan under his control. This meant eliminating many of the powerful Buddhist sects that had come to play prominent political roles through their alliances with rival families. In a battle with the Takeda clan of central Honshu in 1575 (Nagashino), Oda became the first military leader to rely primarily on muskets in battle, soundly defeating his foe. Oda's assassination at the hands of one of his own vassals in 1582 put a premature end to his efforts at reunification.

Oda typifies the type of man who could rise to prominence in the chaos of the Warring States period. He owed much of his success to the sheer will of his ambition and the ruthless means he was prepared to employ to augment and preserve his power. Such were the characteristics of his successor as well.

One of Oda's most able generals was Toyotomi Hideyoshi, a man who had risen from the most humble origins as a peasant's son. It was while fighting the forces of the Mori family in western Honshu for Oda that he heard of his master's assassination. Hurrying back to Kyoto, Hideyoshi quickly eliminated his rivals (including Oda's own son) to establish himself as successor. Gaining recognition as Oda's successor within Oda's coalition of daimyo, Hideyoshi then set about expanding his territorial control. His most powerful rival was perhaps Ieyasu Tokugawa, established in central Honshu. But in 1584, after an indecisive battle, Tokugawa had opted to swear fealty to Hideyoshi. Through marriages the Hideyoshi and Tokugawa clans were brought closer together and in 1590 Tokugawa aided Hideyoshi in his campaigns against the Hojo clan in the Kanto region around Edo (modern Tokyo). In return for his support, with the defeat of the Hojo Tokugawa was granted an immense swath of their former domains, transferring Tokugawa's power base north to Edo.

By 1590 Hideyoshi had succeeded in subduing, through battle or otherwise, all the rival daimyo. Though he was now in the position to name himself shogun, Hideyoshi deferred, perhaps recognizing the aristocratic pedigree that position had customarily held. Rather, he satisfied himself with the titles of regent *(kampaku)* in 1585 and then chancellor *(dajo daijin)* in 1586. Hideyoshi's domestic policies went far in establishing central authority. To quell disturbances in the countryside, he issued orders forbidding peasants from leaving the land, and he had their weapons confiscated. To eliminate daimyo rivalry, he forbade alliances between them without his approval, whether marital, military, or political. Seeing Western Christianity as a potential threat to central authority, in 1587 he ordered all foreign missionaries out of Japan. Though only sporadically enforced under Hideyoshi, this would later become strict policy under Tokugawa.

When Hideyoshi died in 1598 he left only an infant son, Hideyori, as heir. To govern, Hideyoshi had arranged for a coregency of five powerful daimyo to rule during Hideyori's minority. The most powerful of these daimyo-regents was Tokugawa. When it became clear that Tokugawa was acting as nominal ruler above the other coregents a coalition soon rose up against him, led by one of Hideyoshi's former vassals Ishida Mitsunari (1560–1600). Gathering a coalition of other daimyo around him, primarily from the western regions of Japan, Mitsunari eventually met the forces of Tokugawa at the battle of Sekigahara in central Honshu in 1600. Tokugawa's victory secured his position as hegemon. In 1603 Tokugawa named himself shogun, establishing the Tokugawa Shogunate (1603–1868) based at the Tokugawa capital of Edo. A final rally by the forces still loyal to Hideyoshi, who had lost at Sekigahara, occurred in 1614, briefly threatening the young Tokugawa government. When thousands of masterless samurai rallied to the side of the young Hideyori, now ensconced in the family castle at Osaka, Tokugawa sent a large force against them, eventually overcoming the castle defenses in 1615, putting an end to the Hideyori threat.

The establishment of the Tokugawa government was to usher in a period of unprecedented peace and central authority that would last until well into the nineteenth century. Tokugawa went to great lengths to further the power of the Tokugawa government by expanding and augmenting the domestic policies of Hideyoshi. The Tokugawa systematically eliminated those daimyo who were thought untrustworthy by confiscating their holdings. An "alternate attendance system" (*sankinkotai*), whereby the daimyo were forced to reside part of the year at Edo and to leave family members there permanently as hostages, further ensured daimyo loyalty. Oaths of loyalty were also administered to all daimyo, in which they pledged to aid the shogun in times of trouble. As before, the daimyo were forbidden from concluding alliances, or even from building castles or bridges, without the express consent of the shogunal government. And the threat from abroad, which in many ways had ushered in the unification of Japan, was eliminated when Japan began to seal itself off from the outside from 1635. All foreign contact was soon reduced to the small island of Deshima in Nagasaki harbor.

Daniel Kane

See also: Hideyoshi, Toyotomi; Nagashino, Battle of; Oda, Nobunaga; Osaka Castle, Siege of; Samurai; Sekigahara; Tokugawa, Ieyasu

References and further reading:

Hall, John Whitney, et al., eds. *Cambridge History of Japan.* Volume 4: *Early Modern Japan.* Cambridge, UK: Cambridge University Press, 1988.

Jansen, Marius B. *The Making of Modern Japan.* Cambridge, MA: Harvard University Press, 2000.

Sadler, A. L. *The Maker of Modern Japan: The Life of Tokugawa Ieyasu.* Tokyo: Charles E. Tuttle, Co. 1937.

Java War (1825–1830)

The last instance of armed resistance to the Dutch colonial regime by the aristocrats of Java and an inspiration to future generations of Indonesian nationalists who celebrated it as a patriotic struggle uniting noble and commoner alike. Essentially a guerrilla conflict, villagers of central Java accounted for most of the 200,000 casualties due to combat, disease, and starvation (the island's total population in 1830 was around 7 million).

The British occupation of Java during 1811–1816 caused great unrest, particularly in the old, semiautonomous kingdoms of central Java, Yogyakarta, and Surakarta. Conditions were not improved by Dutch attempts to reassert their weakened authority after the British returned the East Indies to them in 1816. Their governor-general passed a decree, highly unpopular with the nobles, that prohibited them from leasing their land to European and Chinese planters. Rebellion broke out in July 1825. Its leader was the charismatic Pangeran Dipanagara (1785–1855), eldest son of the Yogyakarta sultan who had been promised, then denied, the throne.

Dipanagara united in himself diverse strands of Javanese culture: Islam (the theme of holy war against the Dutch infidels), Javanese mysticism (he meditated in sacred caves and claimed the Goddess of the Southern Ocean came to him in a vision), and the royal traditions of the Mataram Dynasty, which had fought against, then been subjugated by, the Dutch in the seventeenth and eighteenth centuries. Common people as well as nobles were attracted to these appeals, and the Dutch found themselves hard-pressed to reimpose control, especially in Yogyakarta.

By implementing the *bentengstelsel* (fortress system), the Dutch matched the speed and flexibility of Dipanagara's guerrillas through the construction of a network of fortified points linked by roads, where they posted mobile columns. These could strike quickly, before local resistance organized. By 1827–1828, the course of the war turned against Dipanagara. He was imprisoned by the Dutch while conducting negotiations with them in March 1830 and exiled to the eastern island of Sulawesi.

The Java War taught the Dutch the need for brutal, grassroots policing of the villages and the wisdom of co-opting the old nobility. Many central Java aristocrats, including the ruler of Surakarta, chose to back the Dutch against the rebels. After 1830, they evolved into a parasitic class who prospered while Javanese farmers were ground down by Holland's increasingly harsh policies of economic exploitation.

Donald M. Seekins

See also: Dutch Colonial Wars

Javanese Wars of Succession (1685–1755)

Three wars that increased the dominance of the Dutch East India Company. The Dutch, through the power of the Vereenigde Oost-Indische (VOC, Dutch East India Company), intruded on the affairs of Javanese rulers in the East Indies.

The first war began in 1685 with the organization of a rebellion against the Dutch by a former VOC soldier, Surapati (or Untung), a former slave and outlaw. His followers killed officials who had rebelled against the authority of the central Javanese kingdom of Mataram. The grateful king of Mataram, Susuhunan Amangkurat II (1677–1703), granted Surapati refuge. Upon Amangkurat II's death in 1703 the Dutch supported a rival claimant to the throne, Amangkurat's uncle Pakubuwono (1704–1719). Surapati eventually claimed kingship over the northeastern part of Java. War continued until Surapati died in 1706 of battle wounds. Susuhunan Amangkurat III (1703–1708) surrendered all of his possessions, became a prisoner of war, and was exiled to Ceylon.

The second war erupted when Susuhunan Pakubuwono I died in 1719. The claims of numerous princes instigated war. The Dutch again intervened and gave support to whoever upheld their endeavors. Four years of fighting led to all the rival claimants and Surapati's descendants being captured by VOC forces and sent into exile. The Dutch extended their control in Java.

During the third war Mataram became a vassal of the Dutch East India Company. Susuhunan Pakubuwono III (1749–1788) received Dutch military support against two challengers. In 1751 the Dutch were seriously defeated and lost their commander. The 13 February 1755 Gianti Agreement, agreed upon by one challenger, split Mataram in two. The eastern region of Pakubuwono had its capital at Surakarta with Susuhunan Pakubuwono III as king. The western region made Yogyakarta its capital with Sultan Hamengkubuwono I (1749–1792) as its ruler. The VOC retained control over the northern provinces. Ultimately, the VOC militarily expanded its power to gain commercial supremacy over the region.

Annette Richardson

References and further reading:

Boxer, C. R. *The Dutch Seaborne Empire, 1600–1800.* Harmondsworth, Middlesex, UK: Penguin Books, 1990.

Greig, Doreen. *The Reluctant Colonists. Netherlanders Abroad in the 17th and 18th Centuries.* Assen, Netherlands: Van Gorcum, 1987.

Israel, J. *The Dutch Primacy in the World Trade 1585–1740.* Oxford, UK: Clarendon Press, 1989.

Jayavarman VII (r. 1181–c. 1220)

One of the greatest kings of Cambodia's Angkor period (802–1431). Jayavarman VII is best known today for his construction of the temple city of Angkor Thom, the mysterious Bayon temple decorated with huge, enigmatic faces of the Bodhisattva Avalokitesvara, and the temple-monastery of Ta Prohm, dedicated to his mother. He is also remembered as a fervent patron of Mahayana Buddhism who broke with Angkor's Hindu traditions of aloof, godlike kingship and devoted himself to good works in order to relieve the suffering of his subjects. For their benefit, he built hundreds of hospitals, rest houses, and roads.

To his contemporaries, however, Jayavarman VII was probably best known as a military leader. In the late twelfth century, the Angkor empire was one of the strongest states in mainland Southeast Asia but was vulnerable to the aggression of its eastern neighbor, the coastal state of Champa (located in what is now central Vietnam). In 1177, a Cham naval expedition sailing up the Mekong and Tonle Sap Rivers to the Tonle Sap (Great Lake) captured the Cambodian capital (near the modern town of Siem Reap), putting the Cambodian king to death and carrying away huge amounts of booty and slaves. During 1178–1181, Jayavarman drove the Chams out of his homeland, becoming king in 1181, and invaded Champa in 1190, making it a province of the Angkor Empire in 1203–1220. His realm also encompassed much of modern Laos, the northern Malay Peninsula, and the Menam Valley in Thailand.

Bas-reliefs in the Bayon temple depict Cambodian and Cham armies fighting on land and water and grim vignettes such as Cambodian soldiers displaying the severed heads of their enemies, artistic testimony to the savagery of war.

Donald M. Seekins

References and further reading:

Chandler, David P. *A History of Cambodia.* 2d ed. Boulder, CO: Westview Press, 1992.

Coedes, G. *The Making of South East Asia.* Berkeley: University of California Press, 1969.

Freeman, Michael, and Claude Jacques. *Ancient Angkor.* Bangkok: Asia Books, 1999.

Jena and Auerstädt (13–14 October 1806)

Decisive simultaneous French victories that cleared Napoleon's way to Berlin. After Napoleon created the Confederation of the Rhine on 12 July 1806 to bring the German states under French control, and after Emperor Francis II dissolved the Holy Roman Empire on 6 August, only Prussia, some Saxon states, and Russia continued to resist Napoleon. In October Napoleon deployed 148,500 French against 111,500 Prussians and Saxons under King Frederick William III in the Saale Valley, Thuringia, Germany.

On 13 October François Joseph Lefebvre's 15,000 Imperial Guard infantry occupied Jena and Napoleon inspected the town. Jean Baptiste Jules Bernadotte's I Corps, with 20,000 men, was just south of Naumburg. Louis-Nicolas Davout's III Corps, with 26,000, was at Kösen, between Naumburg and Auerstädt. Nicolas-Jean de Dieu Soult's 18,000 in IV Corps were evenly divided between Jena and Eisenberg. Jean Lannes's V Corps had 20,500 a mile north of Jena. Michel Ney, southeast of Jena, sent 4,500 of his VI Corps to Jena and the remaining 15,000 to Roda. Pierre François Charles Augereau, southeast of Jena with 16,500, separated his VII Corps into thirds, northwest toward Jena, east-southeast toward the Saale, and west toward Magdala. Joachim Murat split his cavalry reserves, 6,000 east of Bernadotte, 7,000 southeast of Ney.

Karl William Ferdinand, Duke of Brunswick, led 63,500 Prussians east toward Auerstädt. Friedrich Ludwig Hohenlohe, commanding 35,000 Saxons and Prussians northwest of Jena, just beyond Lannes, deployed southwest and northeast. Ernst Friedrich Wilhelm Philipp von Rüchel kept 13,000 infantry in reserve at Weimar.

In the midmorning of 14 October, Brunswick and Davout engaged frontally halfway between Auerstädt and Kösen. Brunswick failed to exploit his superior numbers. Davout used artillery, then seized an opportunity to sweep one division around to crush Brunswick's right. Brunswick was mortally wounded. Soon the French were able to create the classic pincer attack, routing the Prussians back through Auerstädt and southwest.

Meanwhile, Bernadotte and his portion of Murat's cavalry hurried southwest to check Hohenlohe's left flank. Lannes held the center while Soult attacked Hohenlohe's left and Augereau his right. Murat made random assaults. Napoleon took full advantage of the various skills of his generals to deploy his troops in diverse, flexible formations to confuse and entrap the Germans. Hohenlohe's conservative tactics could not counter these swift maneuvers. Soult and Lannes had routed their enemy by midafternoon. Augereau mastered first Hohenlohe's Saxons, then Rüchel's reinforcements.

The French lost 6,000 at Jena and 7,100 at Auerstädt. German casualties were 12,000 killed or wounded and 15,000 taken prisoner at Jena, 10,000 killed or wounded and 3,000 taken prisoner at Auerstädt, plus 20,000 more taken prisoner within the week.

Eric v. d. Luft

See also: Bernadotte, Jean Baptiste Jules; Blücher, Gebhard Leberecht von; Brunswick, Frederick William, Duke of; Clausewitz, Karl Maria von; Davout, Louis-Nicolas, Duke of Auerstädt, Prince of Eckmühl; Lannes, Jean, Duke of Montebello; Murat, Joachim, Grand Duke of Cleves-Berg, King of Naples; Napoleon I; Napoleonic Wars; Ney, Michel, Duc d'Elchingen, Prince de La Moskova; Scharnhorst, Gerhard Johann von; Soult, Nicolas-Jean de Dieu

References and further reading:

Chandler, David G. *Jena 1806: Napoleon Destroys Prussia.* London: Osprey Military, 1993.

Hourtoulle, F. G. *Jena-Auerstaedt: The Triumph of the Eagle.* Trans. Alan McKay. Paris: Histoire & Collections, 1998.

Maude, F. N. *The Jena Campaign, 1806.* London: Greenhill, 1998.

Petre, F. Loraine. *Napoleon's Conquest of Prussia, 1806.* London: Greenhill, 1993.

Jericho, Siege of (1400? B.C.E.)

One of the oldest cities, and probably the most ancient walled settlement on earth, made famous by the biblical account of its siege and destruction. Emerging from their years of wandering in the Sinai, the tribes of Israel needed to establish a beachhead west of the Jordan River to facilitate their invasion and conquest of Canaanite territory. The most obvious site was Jericho. From this oasis, they could advance into the Judean hills. Men sent ahead to gather intelligence and reconnoiter the city stopped at the inn run by Rahab "the harlot" and learned that the people of Jericho were demoralized by the news of the Israelites' defeat of two Amorite kings.

Joshua, given supreme command by the leaders of the tribes, crossed the Jordan River with the help of a minor earthquake, which temporarily dammed the river, a phenomenon seen as recently as 1927. Although the biblical story attributes the fall of Jericho to the walls actually falling because of the blasts from Joshua's trumpet, the real story may be far simpler and militarily plausible. Encamped around the city, whose walls may have been damaged in the earthquake or in poor repair, the Israelite army in full battle dress marched around the city every day for six days, parading the Ark of the Covenant before their troops. This provoked the response of the city's defenders, who, by the seventh day, tired of the false alarms and relaxed their vigilant defense. Joshua led the troops in assaulting the walls, leading to the surrender of the city. Immediately settling the area with trustworthy tribesmen and their families, Joshua had secured a base for supply and further operations as he moved into the second stage of his conquest.

Margaret Sankey

See also: Ancient Warfare

References and further reading:
Herzog, Chaim, and Mordechai Gichon. *Battles of the Bible.* New York: Random House, 1978.
Kenyon, Kathleen Mary. *Digging Up Jericho.* New York: Praeger, 1957.

Jerusalem, Siege of (Palestine) (1099)

On 7 June 1099, the members of the First Crusade encamped before the walls of Jerusalem, the future capital of the Latin Kingdom of Jerusalem. After years of hardship they had reached their goal, although to enter the city the crusaders had to defeat the Fatamid governor, Iftikhar al-Dawla, and his garrison of Arab and Sudanese troops. To prevent treachery, Iftikhar al-Dawla sent the Christian population outside the city.

Due to Jerusalem's immense size, the crusaders could not invest it completely. They also faced a shortage of water and food. The Fatamid garrison, on the other hand, though numerous and well armed, could not man the entire wall. They hoped for the arrival of a relief army from Egypt.

On 13 June 1099, the crusaders attempted an assault. This quickly overran the outer walls, but due to insufficient siege weapons and ladders to continue the assault, they retreated. In order to build siege engines the crusaders endured skirmishes while securing lumber from as far away as Lebanon. Morale also declined as news of the approach of a Fatamid army circulated in their camp.

Raymond of Toulouse and the other leaders realized that time was short. With approximately 14,000 soldiers, they assaulted Jerusalem in the middle of the night on 13 July 1099. On 15 July 1099, the crusaders seized a portion of the wall. The crusaders stormed the city through this breach. Iftikhar al-Dawla realized he had lost Jerusalem and surrendered to Raymond. The rest of the Muslim and Jewish population was massacred, even those who took refuge in mosques and synagogues.

Despite petty jealousies throughout the leadership of the crusade, the capture of Jerusalem culminated in the establishment of the Kingdom of Jerusalem, which lasted as a Latin presence in the Holy Land for almost 200 years.

Timothy May

See also: Crusades; Religion and War

References and further reading:
Holt, P. M. *The Age of the Crusades.* London: Longman Group, 1990.
Mayer, Hans Eberhard. *The Crusades.* Oxford, UK: Oxford University Press, 1988.
Runciman, Steven. *A History of the Crusades.* Vol 1. Cambridge, UK: Cambridge University Press, 1951.
Riley-Smith, Jonathan. *The Crusades: A Short History.* New Haven, CT: Yale University Press, 1987.

Jewish Revolts (66–135)

Two major Jewish rebellions against the Romans.

66–73 Revolt (the Zealots Revolt)

Upon arriving as Roman procurator of Judea in 64, Gessius Florus immediately encountered political crises. In Caesarea Maritima, the major Roman seaport of Judea, Florus angered Jews by siding with Greeks during intercommunal rioting. In Jerusalem, he angered Jews by seizing the payment of a large fine from the Temple treasury because Jewish payment of taxes was in arrears. When rioting erupted there, Florus's reprisals were brutal. In order to defuse talk of rebellion, King Agrippa II, the Jewish puppet ruler of Judea, argued that subordination to Rome was no shame and warned that no foreign power, particularly Parthia, would intervene on behalf of the Jews. Indeed, King Vologases I of Parthia agreed to a pact of friendship with Rome in 66. And, despite widespread opposition to Roman rule, the Jews were not united.

Eleazar ben Ananias, guardian of the Temple, prohibited sacrifices by foreigners. This was, in effect, an act of rebellion because it rudely ended the daily sacrifices on behalf of Rome and the emperor. Eleazar then seized control of the Temple, the Lower City, and the Upper City. Soon Jewish forces led by Menachem ben Judas captured the Dead Sea fortress of Masada from a small Roman garrison. By September 66, the Romans had been pushed from most areas of Judea. Cestius Gallus, the Syrian governor, marched a legion and 6,000 auxiliaries into Judea and pacified the Galilee in October. He then marched on Jerusalem but was repulsed and harassed into a humiliating retreat through the area of Beit-Horon around Jerusalem. The Jews' early successes were impressive but short-lived. The Jewish rebels in Jerusalem soon became overconfident and divisive. Eleazar sanctioned the murder of Menachem, and Eleazar himself was soon deposed and sent to command a small force in Idumaea.

In the spring of 67, Emperor Nero dispatched Titus Flavius Vespasianus (Vespasian) to Judea with two legions. Joined by a third legion commanded by his son, Titus, Vespasian now commanded a force of 60,000 and laid siege to Jotapata in the Galilee. After 47 days, the Jewish commander Yosef ben Mattitias (Josephus) surrendered, becoming the Romans' official historian. Instead of advancing directly upon Jerusalem, Vespasian cautiously spent the next three years suppressing revolt in cities like Tiberias, Gischala, and Gamala. Vespasian could afford delay, because violent disunity was destroying Jews in Jerusalem as effectively as Rome's legions. Vespasian returned to Italy in order to become emperor, following Nero's assassination, and left Titus in command of the Judean legions. Titus marched on

Jerusalem in the spring of 70 and captured the Temple in early August. (The Jewish fast day on the Ninth of Av laments this catastrophe.) Roman victory was inevitable despite the resistance of Eleazar ben Ya'ir at the fortress of Masada. Flavius Silva led a siege of three years, which finally captured the fortress in 73, whereupon the defenders committed suicide. In June 71, Titus held a triumph in Rome, where the Arch of Titus was erected next to the Forum to commemorate his victory.

The 132–135 Revolt (Bar Kochba's Revolt)

After the Roman Emperor Hadrian (Publius Aelius Hadrianus) banned circumcision and changed the name of Jerusalem to Aelia Capitolina (named after Hadrian himself), Judea rebelled against Roman rule. The Second Jewish Revolt was well planned in secret and ably led by a messianic figure named Shimon bar Kosiva (later bar Kochba). In the fall of 132, bar Kochba seized Jerusalem from its small garrison, expelled Tineius Rufus, Judea's consular governor, and struck coins in commemoration of the victory. Although the exact course of the rebellion is obscure, it appears that bar Kochba's forces used guerrilla tactics rather than engage the Romans in open field combat. The Jews enjoyed a number of early successes. In 133, Hadrian sent eight legions, commanded by the able Sextus Julius Severus, who employed a strategy of attrition against the Jews. Unable to draw bar Kochba into open combat, Severus surrounded strongholds and starved out the rebels. Despite this strategy, the Romans suffered significant losses. For example, almost an entire legion was killed with poisoned wine, and many thousands more were killed by the Jews' guerrilla attacks. In the spring of 135, after a two-year-long siege, the Romans captured the fortress of Betar, the last refuge of bar Kochba, who was killed in the battle. As punishment for the rebellion, the Romans changed the name of Judea to Palestine and expelled most of the Jews into the diaspora, into what amounted to an exile of over 1,800 years.

Eric D. Pullin

See also: Hadrian; Josephus, Flavius; Masada, Siege of; Parthian Empire; Vespasian

References and further reading:

Grant, Michael. *The Jews in the Roman World.* New York: Scribner, 1973.

Josephus, Flavius. *The Jewish War.* New York: Penguin, 1981.

Tacitus, Cornelius. *The Complete Works of Tacitus.* New York: Modern Library, 1942.

Portrait of Joan of Arc. (Library of Congress)

Joan of Arc (Jeanne d'Arc) (1412–1430)

Patron saint of soldiers and probably most famous female military commander in history. Jeanne d'Arc was probably born on 6 January 1412 in Domremy, Champagne, France. At age 12, she claimed to hear voices and have visions informing her of her mission to restore the dauphin, Charles Valois, later Charles VII, to the French throne and rid the country of English occupation and their Burgundian allies. In February 1429, she met with the dauphin and his commander, Robert de Baudricourt, and somehow convinced them of her mandate.

Dressed in male garb and examined by clergy, Joan was awarded the rank of captain and given command of a small army at Blois. She led her army to the besieged city of Orleans in May. A few days after her arrival, while her army marched to the city by a northern route, citizens armed themselves and took the weakest of 10 Norman/English blockhouses by coup de main. On 7–8 May 1429, she led her troops on a series of successful sorties against the enemy, causing them to retreat. While other advisers wanted to attack Normandy, she made the decision to march upon Rheims.

In the Loire campaign, beginning with the fall of Jargeau on 10 June, she mastered the use and placement of artillery.

At the Battle of Patay, on 18 June, the French defeated the Norman/English under Sir John Fastolfe and Lord Talbot, with 1,800 English soldiers lost. Retreating to Paris, French forces had a clear road to Rheims. On 16 July, Charles entered the city and was crowned king the next day.

In the spring of 1430, Joan was captured by Burgundians at Compiegne, near Paris, in a failed sortie. Sold to the English, she was put on trial by the ecclesiastical court at Rouen. On 30 May 1431, she was burned at the stake in the marketplace after being convicted of heresy and witchcraft. Canonized in 1920, her moral importance to the French during the Hundred Years War has been stressed by historians, but her abilities as a military commander have often been overlooked. Nonetheless, Joan of Arc became almost immediately after her death a symbol, a veritable icon, of the spirit of France.

T. Jason Soderstrum

See also: Hundred Years War

References and further reading:

DeVries, Kelly. *Joan of Arc: A Military Leader.* Stroud, UK: Sutton, 1999.

Giles, France. *Joan of Arc: The Legend and the Reality.* New York: Harper & Row, 1981.

Smith, John Holland. *Joan of Arc.* London: Sidgewick & Jackson, 1973.

Jodl, Alfred (1890–1946)

Hitler's most important military adviser during World War II. Born 10 May 1890 at Würzburg (Bavaria), Jodl married Irma von Bullion (d. 1943) in 1913 and Luise von Benda in 1944 but had no children. He entered military service at the age of 13 and was promoted to lieutenant in 1912. He took part in World War I and joined Reichswehr in 1919, held several commands, and was promoted to general major in 1939.

Called back to Berlin he took over Wehrmachtführungsstab, the operation staff within Wehrmacht High Command (OKW). Nominally subordinated to Wilhelm Keitel as chief of OKW, he worked largely on his own, having permanent access to Hitler. After the invasion of Russia (June 1941) Jodl's staff was responsible for every theater of war except the eastern front. His duty was to prepare reports on strategic questions, to participate in the daily briefings (more than 5,000 during the war), and to transmit Hitler's directives. Working very hard, he hardly left the führer's headquarters to visit the front. Frictions with Hitler culminated in autumn 1942, but Jodl held his job. No successor was at hand and Hitler valued Jodl's competence, diligence, and admiration for him. He was promoted to general colonel in January 1944.

On 22 April 1945 Jodl left Berlin and joined Grand Admiral Karl Dönitz's staff in northern Germany. His final duty was to sign the German surrender at Reims, 7 May 1945.

Soviet demands led to his arraignment at the Nuremberg trial of the major German/Nazi leaders. Confronted with numerous criminal orders bearing his signature, he argued that as a German officer he had to obey Hitler's directives. The court did not believe that he, staying in daily contact with the dictator, could have failed to notice events and orders bearing on war crimes. Found guilty, he was hanged on 16 October 1946.

Questions as to whether Jodl deserved the death penalty arose among the Nuremberg judges and are still current among historians.

Martin Moll

See also: Hitler, Adolf; Keitel, Wilhelm; World War II

References and further reading:

Scheurig, Bodo. *Alfred Jodl. Gehorsam und Verhängnis.* Schnellbach: Bublies,1999.

Joffre, Joseph Jacques Césaire (1852–1931)

French marshal. Born on 12 January 1852, at Rivesaltes, Joseph Joffre interrupted his studies at the École Polytechnique to serve in the 1870–1871 Franco-Prussian War. He resumed his studies after the war and on graduation was stationed in the Far East and Madagascar. He led an expedition to Timbuktu (Africa) in 1893 and served as a fortifications specialist in Madagascar during 1900–1905. He joined the Supreme War Council in 1910 and became chief of the general staff and commander in chief designate the next year.

At the outbreak of World War I, Joffre's War Plan XVII, intended to secure Alsace-Lorraine, failed to anticipate the main German deployment through Belgium because he was convinced the Germans would not use their reserves on the front lines. Following the defeat of the French offensive and the development of the German threat to the north, he skillfully redeployed his assets and conducted a fighting retreat to the Marne.

Joffre orchestrated the Battle of the Marne, 5–11 September 1914. It ended in a German withdrawal and was the high point of his career. Joffre's major offensives of 1915 and 1916, including Champagne, Somme, and Artois, were failures. He justified these as serving to convince Italy to join the Entente and to take pressure off Russia.

Joffre fell under severe criticism, especially when the February 1916 German offensive at Verdun caught him by surprise and with French defenses unready. He was also blamed for the Romanian disaster later that year. Removed as commander of French forces on the western front in December 1916, he was made marshal of France and named technical adviser to the government. Joffre then headed the French military mission to the United States in 1917, retiring after the war. He died in Paris on 3 January 1931.

Spencer C. Tucker

See also: Foch, Ferdinand; French Army; Pétain, Henri-Philippe; World War I

References and further reading:

Porch, Douglas. *The March to the Marne. The French Army, 1871–1914.* Cambridge, UK: Cambridge University Press, 1981.

Varillon, Pierre. *Joffre.* Paris: Arthème Fayard, 1956.

Williamson, Samuel R., Jr. *The Politics of Grand Strategy; Britain and France Prepare for War, 1904–1914.* Cambridge, MA: Harvard University Press, 1969.

John I Tzimisces (924–976)

Byzantine general and emperor. Tzimisces began his career as an officer in the Byzantine forces commanded by his uncle, Nicephorus Phocas. In 956, he served as strategus of Mesopotamia, with little success. In 958, he led a raid into southern Armenia, and in 962 he accompanied his uncle in a very successful invasion of Syria. In 963, he encouraged Phocas to stage a coup against Joseph Bringas, an official in the court of the recently deceased Romanus II Porphyrogenitus. As a result of this coup, Nicephorus married the empress Theophano, widow of Romanus, and became coemperor with her two minor children.

Tzimisces was rewarded with the title Domestic of the Scholae, essentially commander of the Byzantine army in Anatolia. He held this position for less than two years before being dismissed and replaced with another nephew of the emperor, Peter Phocas the Eunuch, who was deemed to pose no danger to the regime.

In December 969 Tzimisces entered into a conspiracy to murder Nicephorus, having been encouraged in this by the empress Theophano. The conspiracy succeeded, and Tzimisces took part personally in the murder of his uncle. Crowned coemperor, Tzimisces exiled all of the members of the Phocas clan, except for Peter the Eunuch, whose services he retained. Tzimisces also sent the empress Theophano to a convent.

Shortly thereafter, Rus forces under Svyatoslav, Prince of Kiev, invaded Bulgaria, took Preslav, and captured the Bulgarian royal family. A further Rus advance, into Byzantine territory, was defeated by imperial troops sent by Tzimisces and commanded by Bardas Sclerus. It is possible that some sort of treaty was agreed to at this time between the empire and the Rus. Although Tzimisces was preparing a further expedition against the Rus in 970, he was diverted from his purpose by a further revolt of the Phocas clan, which he suppressed.

In spring 971, Tzimisces was finally ready to attack the Rus. He found the passes into Bulgaria unguarded. The Byzantine forces moved quickly to Preslav, where they destroyed most of the Rus army. A siege of Preslav allowed Tzimisces to capture the Bulgar royal family, whom he imprisoned in Constantinople. The Byzantines then discovered that Svyatolav had gone to Dristra, which the Byzantines besieged until 24 June, when the remaining Rus were captured after an attempt to break out of the town. Svyatoslav and his followers were released and, for the most part, were massacred by the Petchenegs while attempting to return home.

Tzimisces was thus free to turn his attention back to the Arab frontier. An expedition to Baghdad was contemplated in 972 or 973 but never launched. In 974, Byzantine forces marched through Armenia, securing that kingdom as a client state and a recruiting ground. In 975, the emperor marched into Syria, Lebanon, and Palestine, taking Homs, Damascus, Beirut, Sidon, Tiberias, and Nazareth. Pockets of opposition forced Tzimisces to turn back and prevented him from capturing Jerusalem. At the end of the campaign, the emperor fell ill and returned to Constantinople, dying on 10 January 976, shortly after his arrival.

Tzimisces recovered territory lost to the Byzantines and revitalized the army. At the same time, his failure to leave an heir, his suppression of the Phocas family, and his promotion of other families to supplant the Phocas clan created factions within the army, which would shortly cause a civil war over the succession.

Joseph M. Isenberg

See also: Byzantine-Muslim Wars; Nicephorus II Phocas

References and further reading:

Fine, John V. A. *The Early Medieval Balkans.* Ann Arbor: University of Michigan Press, 1991.

Norwich, John Julius. *Byzantium: The Apogee.* New York: Knopf, 1991.

Wittrow, Mark. *The Making of Byzantium, 600–1025.* Berkeley: University of California Press, 1996.

John II Comnenus (1088–1143)

Byzantine emperor, son of Alexius I Comnenus. John II succeeded to the throne upon the death of his father in 1118. He was initially forced to deal with efforts by his mother, sister, and brother-in-law to overthrow him, but by 1119 he felt secure enough to cancel the trading privileges of Venice in Byzantium, which provoked a war, and to attack Turkish strongholds in the Meander valley.

John II's campaigns against the Turks in 1119 and 1120 proved successful. He recaptured Laodicea and Sozopolis and separated the Turkish sultanate of Iconium from the Turkish settlements in Caria. A Pecheneg and Cuman invasion in 1121 forced John to divert his attention from the Anatolian Turks. He negotiated with the invaders in 1121 and in 1122 launched a surprise attack upon the nomads. He defeated the Petchenegs and Cumans, enlisting some, while enslaving others. The war with Venice proved less fortunate, as the Venetians quickly overran several Aegean islands. John, having disbanded the Byzantine navy as a cost-saving measure, settled the dispute with the restoration of Venetian trading privileges.

In the autumn of 1127, the Hungarian king, Istvan II, attacked the Byzantine Empire, which had given refuge to a rival claimant to the Hungarian throne. The Hungarians sacked Belgrade, Nish, Sardica, and Philippopolis (Plovdiv) before being forced to withdraw. In 1128, John attacked the Hungarians and defeated them at Sirmium. Nevertheless, the Hungarians reinvaded Byzantine territory and incited a rebellion among the Serbians of Raska. John defeated the Serb rebellion and was able to negotiate a peace with the Hungarians upon the death of the offending claimant.

In 1130, John began a series of campaigns against the Turks. The first campaign ended suddenly, when John's brother, Issac, defected to the Turks. A campaign in 1132 was more successful, with the Byzantine forces ravaging considerable territory. In 1135, John captured the fortresses of Castamon and Gangra. John then turned his attention to forcing the crusader principality of Antioch into vassalage, which he accomplished in 1137. An attempt to seize Aleppo in 1138 failed.

From 1139 to 1141, John recovered the Black Sea coast of Anatolia, from Sinope to Trebizond, for the empire but made little progress in reconquering the interior of Anatolia. In 1142, the emperor forced the count of Edessa to swear homage to him and made further demands upon the principality of Antioch. Before John could settle affairs with Antioch in 1143, though, he died, allegedly from cutting himself with a poisoned arrow while hunting.

John Comnenus continued the policy of his father in improving the army. He also developed a stronghold at Lopadium, to better defend Anatolia. At the same time, he demobilized the Byzantine navy, leaving the empire dependent upon the goodwill of the Genoese and Venetian governments. John also failed to evict the Turks from Central Anatolia, which, though poor, was a region crucial for the defense of the Anatolian frontier. This failure left the borders of the empire nearly impossible to defend.

Joseph M. Isenberg

See also: Alexius I Comnenus; Antioch, Battle of; Crusades; Seljuqs

References and further reading:

Comnena, Anna. *The Alexiad of Anna Comnena.* Trans. E. R. A. Sewter. New York: Penguin Books, 1969.

Norwich, John Julius. *Byzantium: The Apogee.* New York: Alfred A. Knopf, 1991.

Ostrogorsky, George. *History of the Byzantine State.* Trans. Joan Hussey. Rev. ed. New Brunswick, NJ: Rutgers University Press, 1969.

Johnston, Albert Sidney (1803–1862)

Confederate field commander whose death in battle on 6 April 1862 was a devastating loss to the Confederacy. A. S. Johnston was born in Washington, Kentucky, on 2 February 1803. After graduating from the U.S. Military Academy at West Point in 1826, he fought in the Black Hawk War of 1832. He resigned from the U.S. Army in 1834, moved to Texas, and enlisted as a private in Sam Houston's army in 1836. By January 1837 he was commander in chief of the Texas army, and from 1838 to 1840 he served as secretary of war for the Republic of Texas.

As brevet colonel of the 1st Texas Volunteer Rifles, he fought under Zachary Taylor in the Mexican-American War, distinguishing himself at Monterrey. He became a U.S. Army paymaster with the regular rank of major in 1849 and regular colonel of the Second Cavalry in 1855. As brevet brigadier general, he led a successful expedition against the Mormons in 1857 and was named commander of the Department of the Pacific in 1860.

Johnston resigned his commission in April 1861 and traveled by way of Texas to Richmond, Virginia, where Jefferson Davis made him a full general on 30 August and gave him command of all Confederate forces in the west. He ranked second among the five original Confederate generals, behind only Samuel Cooper, and first among field commanders, ahead of Robert E. Lee, Joseph Johnston, and P. T. Beauregard.

Outnumbered from the start, he raised the Army of the Mississippi to face Ulysses S. Grant in Kentucky and Tennessee. He retreated from Kentucky after George B. Crittenden lost to Don Carlos Buell and George H. Thomas at Mill Springs on 19 January 1862 and deeper into Tennessee after the fall of Fort Henry on 6 February and Fort Donelson on 16 February. With forces and supplies gathered at Corinth, Mississippi, and reinforced by Beauregard, he took the offensive against Grant in April. Hit in the hip the afternoon of the first day at Shiloh, he bled to death before surgeons could reach him.

Eric v. d. Luft

See also: American Civil War; Beauregard, Pierre Gustave Toutant; Bragg, Braxton; Fort Donelson; Grant, Ulysses Simpson; Houston, Samuel; Johnston, Joseph Eggleston; Lee, Robert Edward; Mexican-American War, Mormon War; Shiloh; Texas War of Independence; Thomas, George Henry

References and further reading:

Johnston, William Preston. *The Life of Gen. Albert Sidney Johnston, Embracing His Services in the Armies of the United States, the Republic of Texas, and the Confederate States.* New York: Da Capo, 1997.

Moore, Avery C. *Destiny's Soldier.* San Francisco: Fearon, 1958.

Roland, Charles Pierce. *Albert Sidney Johnston: Soldier of Three Republics.* Lexington: University Press of Kentucky, 2000.

———. *Jefferson Davis's Greatest General: Albert Sidney Johnston.* Abilene, TX: McWhiney Foundation, 2000.

Johnston, Joseph Eggleston (1807–1891)

Extraordinarily skillful Confederate field commander whose effectiveness was limited throughout the Civil War by his quarrels with President Jefferson Davis. Johnston was born in Prince Edward County, Virginia, on 3 February 1807. Assigned to artillery after graduating from the United Sates Military Academy at West Point in 1829, he fought in the Seminole Wars, the Mexican-American War under Winfield Scott, and the Utah War. Brevetted colonel and wounded at Cerro Gordo, he earned an excellent reputation. He was a brigadier general and U.S. Army quartermaster when he resigned to join the Confederacy on 22 April 1861. With Stonewall Jackson at Harpers Ferry, he organized Virginia volunteers into the Army of the Shenandoah.

The five original Confederate generals were, in order of seniority, Samuel Cooper, A. S. Johnston, Robert E. Lee, J. E. Johnston, and P. T. Beauregard. At First Bull Run, Beauregard served under J. E. Johnston, but they commanded as equals to defeat Irvin McDowell. Johnston was wounded at Fair Oaks on 31 May 1862. Commanding the Department of the West after 4 December, he opposed Ulysses S. Grant around Vicksburg in May 1863, losing his base at Jackson, Mississippi, on 14 May.

In a brilliant defensive campaign, Johnston defeated William Tecumseh Sherman at Kenesaw Mountain on 27 June 1864 but could not prevent Sherman's larger and better equipped force from entering Atlanta. Davis, unreasonably disappointed that Johnston could not hold Sherman, replaced Johnston with John Bell Hood on 18 July. Hood's command was a complete disaster, and Lee had Davis recall Johnston on 25 February 1865. Severely outnumbered but undaunted, he lost to Sherman at Bentonville, North Carolina, on 19–21 March and surrendered to him at Durham Station, North Carolina, on 26 April.

After the war, Johnston was a paragon of reconciliation with the North. A Democrat, he represented Virginia in Congress from 1879 to 1881 and was President Grover Cleveland's railroad commissioner from 1887 to 1891. He died on 21 March 1891 of pneumonia, caught while acting as pallbearer at Sherman's funeral in Washington, D.C.

Eric v. d. Luft

See also: American Civil War; Atlanta, Battles Around; Beauregard, Pierre Gustave Toutant; Bragg, Braxton; Bull Run, First/Manassas; Bull Run, Second/Manassas Junction; Cerro Gordo, Battle of; Chattanooga, Battle of; Grant, Ulysses Simpson; Harpers Ferry; Hood, John Bell; Johnston, Albert Sidney; Jackson, Thomas "Stonewall"; Lee, Robert Edward; McClellan, George Brinton; McDowell, Irvin; Mexican-American War; Mormon War; Scott, Winfield; Sherman, William Tecumseh; Utah War; Vicksburg, Siege of

References and further reading:

Davis, Stephen. *Atlanta Will Fall: Sherman, Joe Johnston, and the Yankee Heavy Battalions.* Wilmington, DE: Scholarly Resources, 2001.

Johnston, Joseph Eggleston. *Narrative of Military Operations during the Civil War.* New York: Da Capo, 1990.

Lash, Jeffrey Norman. *Destroyer of the Iron Horse: General Joseph E. Johnston and Confederate Rail Transport, 1861–1865.* Kent, OH: Kent State University Press, 1991.

Symonds, Craig L. *Joseph E. Johnston: A Civil War Biography.* New York: Norton, 1992.

Jomini, Antoine Henri, Baron de (1779–1869)

Military thinker and writer. Born Payerne, Swiss canton of Vaud, Jomini, son of town mayor. Banker in Basle (1796) and Paris (1798, 1801), becoming major in French-sponsored Swiss army in the Swiss Revolution (1800). Obtaining a French army staff position under Marshal Michel Ney (1800–1802) and serving as an aide-de-camp to Ney (1803), he wrote the first of his renowned works, *Traités des grandes operations militaires* (1801–1804), published in 1805. On Ney's staff during the Austerlitz campaign (1805) he served at Ulm (October) and Austerlitz (December). Napoleon (who had read Jomini's book) appointed him a colonel on his staff for the Auerstädt campaign (1806–1807). Jomini then served at Jena (October 1806) and Eylau (February 1807). He was created a baron (1807) and was appointed chief of staff in Ney's corps in the Peninsular War (1808–1811), becoming brigadier general (1810) and director of the pioneering French General Staff's historical section (1811). In the Russian campaign (1812) he was governor of

Vilno and Smolensk but rejoined Ney for the battles of Lützen and Bautzen (1813).

Jomini failed to achieve further promotion, possibly because of his disconcerting tendency to resign posts on impulse, but also likely because he was jealously disliked by Berthier, Napoleon's imperial chief of staff, who had him arrested for a minor misdemeanor in 1813.

Jomini responded by defecting to the Russian army, gaining the rank of lieutenant general, and serving as an aide-de-camp to Czar Aleksandr I (1813–1814) during the War of Liberation, although he refused to enter Paris with allies in 1814.

Living between Russia and Brussels, Jomini wrote more books, serving as military tutor to the Russian Grand Dukes Michael and Nicholas (1816–1826), becoming the latter's aide-de-camp and general in chief on his ascension as Czar Nicholas I (1825), and fighting at the Siege of Varna during the Turkish War (1828–1829). He then played a major role in establishing the Russian Nikolaevskii General Staff Academy in 1832. Four years later he published his most influential work, *Précis de l'art de guerre.* He remained an adviser to Nicholas I, fulfilling this role in the Crimean War (1853–1856). He then retired to Passy, near Chamonix in 1859, dying there 10 years later.

Although Jomini lacked a formal military education and never commanded in battle, his theories, along with those of Karl von Clausewitz, formed the basis of modern military thought. His contributions were more quickly utilized than those of Clausewitz, strongly influencing European and American military leaders alike. Jomini's conclusions rested on his knowledge of the wars of Frederick the Great, French Revolutionary Wars, eighteenth-century military theoretical study, and his personal Napoleonic Wars experience.

Far from proclaiming Napoleon an original commander, Jomini explained his rise and fall through his adherence to or straying from historic military principles, which Jomini reduced to several simple codified laws, which supposedly explained the "art" of warfare.

His greatest contribution was his identifying of the fundamental principle of war as the concentration of the mass of troops against an enemy's weak point, thus allowing breakthrough and victory. Maneuver was vital, positioning the attacking force in the most advantageous position; concealment, surprise, and intelligence—gathering essential information to determine the best point to strike and when. Jomini stressed enveloping one or both enemy flanks (if numbers allowed) and recommended cutting enemy supply lines and communications. Fronts should be kept small, allowing swift maneuver to any weak point, with feints used to break up enemy forces. Commanders must strike decisively to forestall enemy countermaneuvers. Internal operational lines were preferable, as these allow swift movement and concentration of troops, whereas external lines split the force, creating the potential for defeat. These principles allow coordination of offensive actions, leading to total victory. Viewing the battlefield as a mathematical square, territory was the important factor for victory, not destruction of the enemy army. Jomini abhorred total war on the Napoleonic scale; eighteenth-century warfare was his ideal.

Limiting his writings to commanding armies in the field, always recommending limited wars along pre-Napoleonic dimensions, preventing large-scale endless bloodshed to retain status quo, Jomini was a prisoner of his times and experiences. Much of his writing was overtaken by industrialization and technology, but his basic principles remain valid.

Neil Croll

See also: Austerlitz, Battle of; Berthier, Louis-Alexandre, Prince of Neuchatel and Valangin, Prince of Wagram; Clausewitz, Karl Maria von; Crimean War; Friedland; Jena and Auerstädt; Napoleon I; Napoleonic Wars; Ney, Michel

References and further reading:

Brinton, C., G. A. Craig, and F. Gilbert. "Jomini." In *Makers of Modern Strategy.* Princeton, NJ: Princeton University Press, 1943.

Jomini, Baron A. H. de. *Précis de l'art de guerre. [Summary of the Art of War].* Reprint of 1862 translation, with introduction by C. Messenger. Philadelphia: Greenhill Books, 1992; London: Stackpole Books, 1996.

Shy, J. "Jomini." In *Makers of Modern Strategy from Machiavelli to the Nuclear Age,* ed. P. Paret. Oxford, UK: Clarendon Press, 1986.

Strachan, H. *European Armies and the Conduct of War.* London: Routledge, 1991.

Joseph the Younger, Chief (Hinmaton Yalatkit, Heinmot) (1840–1900)

Aboriginal American chief and military leader. Joseph was born Hin-mah-too-yah-lat-kekht, or Thunder Rolling Down the Mountain, in the Wallowa Valley in northeast Oregon in 1840. His father, who also had the Christian name Joseph, was leader of the Nez Percé, a tribe that had peaceably coexisted with the United States and had willingly settled on a reservation in Idaho. In 1863, the federal government renegotiated its treaty with the tribe and reduced the size of the reservation by 90 percent, seizing 6 million acres. The Nez Percé refused to honor this agreement because the new location of the reservation was difficult to survive in and they wanted to stay on their traditional homelands. Thunder Rolling Down the Mountain became chief in 1871 when his father died and took up the name Joseph. In 1877, Joseph was given an ultimatum by General Oliver Otis Howard to return to the boundaries of the new reservation or risk military consequences. Resisting, Joseph moved his followers

Painted portrait of Chief Joseph, Nez Percé chief. (Library of Congress)

slowly to their new home. Before reaching their destination, a small band of young warriors killed some settlers in anger.

Joseph and 700 of his followers decided to flee to the safety of Canada. He was able to elude and foil the army on several occasions for more than three months and cover more than 1,600 miles using rearguard tactics, skirmish lines, and field fortifications. But while camping near the Chinook on September 30, the army surprised the Nez Percé. After five days of fierce fighting and the loss of three chiefs, more than 30 warriors and a number of women, children, and horses, Joseph decided to surrender. The government dispersed the remaining Nez Percé to several different reservations. Joseph died on the Colville Reservation in Washington State in 1900.

T. Jason Soderstrum

See also: American Indian Wars

References and further reading:

Beal, Merrill D. *I Will Fight No More Forever: Chief Joseph and the Nez Perce War.* Seattle: University of Washington Press, 1996.

Josephy, Alvin M. *The Nez Perce Indians and the Opening of the Northwest.* New Haven and London: Yale University Press, 1965.

Lavender, David. *Let Me Be Free: The Nez Perce Tragedy.* New York: HarperCollins Publishers, 1992.

Josephus, Flavius (c. 37–c. 100)

Jewish historian and military leader who chronicled the fall of Jerusalem during the First Jewish Revolt. Born the child of aristocrats in Jerusalem circa 37, as a young man he was a member of the Pharisees, a sect that wanted religious freedom but did not necessarily desire independence from Roman rule. In 64, Josephus traveled to Rome on a diplomatic mission and was awed by the seemingly invincible strength of the empire. In 66 Judea revolted against Rome, and despite his misgivings, Josephus became a military leader in the Galilee. Though sometimes portrayed as a collaborator, he withstood a Roman army led by future emperor Vespasian for 47 days at the fortress of Jotapata before being forced to surrender. As a prisoner, he curried favor with Vespasian and regained his freedom in 69. Convinced the Romans were too powerful to be resisted, he served with the Roman army during the siege of Jerusalem in 70 and thus was seen as a traitor by the Judeans. His attempts to act as a neutral arbiter failed. After Jerusalem fell and the revolt was put down, Josephus retired to Rome to write. He is remembered mainly for his histories of the revolts entitled *History of the Jewish War,* and for a history of the Jewish people, *The Antiquities of the Jews.*

Harold Wise

See also: Jewish Revolts; Masada, Siege of; Vespasian

References and further reading:

Schwartz, Seth. *Josephus and Judean Politics.* Boston: Brill Academic Publishers, 1997.

Thackery, Henry. *Josephus, the Man and the Historian.* Hoboken, NJ: Ktav Publishing House, 1968.

Joubert, Petrus Jacobus ("Piet") (1831–1900)

Charismatic Boer leader against the British. Joubert was born on the family farm near Prince Albert, South Africa, on 20 January 1831. His parents took him on the Great Trek to Natal when he was six. Settling later in Transvaal, he was always a farmer at heart, with a kindly, diplomatic nature, but was also successful in law, business, and politics. He was a popular civilian leader with almost no military experience when he was elected commandant general at the start of the First Boer War in December 1880. He defeated General Sir George Pomeroy Colley at Laing's Nek on 28 January 1881, Ingogo on 8 February, and Majuba Hill on 27 February, thus liberating Transvaal and becoming a national military hero. He nonetheless lost four presidential elections to Paul Kruger in the 1880s and 1890s but at the same time remained in power to enlarge and improve the Boer army, especially the artillery. He brought in the famous 155 mm

"Long Tom" siege gun from Schneider-Creusot and the 120 mm rapid-fire howitzer from Krupp.

When the Second Boer War erupted in October 1899, Joubert defeated the British again at Laing's Nek on 12 October, at Talana on 15 October, Elandslaagte on 21 October, and Nicholson's Nek and Modderspruit on 30 October. Having forced the British to retreat into Ladysmith, he began the siege of that town on 2 November. Joubert's victories were mostly due to British mistakes, which offset Joubert's shortsighted preparations, overly defensive tactics, and natural aversion to violence.

The gentle Joubert was especially ineffective while besieging Ladysmith; his main concern was to make peace. Disabled on 23 November by a fall from his horse, he resigned his commission and was replaced as commandant general by Louis Botha on 25 November. He fell ill and died on 27 March 1900 in Pretoria.

Eric v. d. Luft

See also: Boer Wars; Botha, Louis; Buller, Sir Redvers Henry; French, John Denton Pinkstone, First Earl of Ypres; Haig, Douglas; Kruger, Paul-Stephanus Johannes Paulus; Ladysmith, Siege of; Smuts, Jan Christiaan

References and further reading:

Bateman, Philip. *Generals of the Anglo-Boer War.* Cape Town: South African Historical Mint, 1977.

Coetzer, Owen. *The Anglo-Boer War: The Road to Infamy, 1899–1900.* London: Arms & Armour, 1996.

Meintjes, Johannes. *The Commandant-General: The Life and Times of Petrus Jacobus Joubert of the South African Republic, 1831–1900.* Cape Town: Tafelberg-Uitgewers, 1971.

Ransford, Oliver. *The Battle of Majuba Hill: The First Boer War.* New York: Crowell, 1968.

Julian (Flavius Claudius Julianus "The Apostate") (332–363)

An outstanding military leader of the later Roman Empire. Flavius Claudius Julianus was born at Constantinople in 332, a relative of Emperor Constantine the Great. He survived a massacre of his family by political rivals and spent most of his early youth in exile. At the age of 20 he secretly renounced Christianity and embraced both paganism and Hellenic philosophies. In 356 Julian was regarded as safe enough for public service, so his cousin, Emperor Constantius II, appointed him caesar of Gaul.

Gaul was then besieged by German tribes, who swarmed across the Rhine River in large numbers. Despite his lack of formal military training, Julian rebuffed the Alamanni in several small encounters, and in August 357 his 13,000 Romans soundly defeated 30,000 warriors at Strasbourg and subsequently drove the Franks from Gaul. For the first time in many years, Roman military authority was reasserted along the Rhine frontier. Julian also effectively overhauled provincial administration, lowered taxes, and won the affection of his troops.

Constantius II, fearing a potential rival, ordered Julian to transfer the best parts of his army to the east for a war against Persia. The troops refused and proclaimed Julian emperor. Constantius II died of a fever as both men's armies advanced toward Constantinople, and Julian was proclaimed emperor in December 361. He was nicknamed "the Apostate" by later Church historians for officially embracing paganism and shunning Christianity, although he never initiated any persecution.

Over the next year, the Romans made feverish preparations for a showdown against the Persian Empire under Shapur II. Julian led 65,000 well-trained men through Mesopotamia and across the Tigris River. Supplied by a large fleet of warships accompanying him downstream, Julian easily defeated Shapur II in several sharp engagements and at length stood before the winter capital at Ctesiphon. He then waited for reinforcements from Armenia that were not forthcoming. Julian commenced a fighting withdrawal northward, defeating the Persians in several battles. In a minor skirmish on 26 June 363, he was mortally wounded. His successor, Jovian, concluded a peace accord by ceding considerable Roman territory to Persia.

John C. Fredriksen

See also: Shapur II

References and further reading:

Bowersock, Glen W. *Julian the Apostate.* Cambridge, MA: Harvard University Press, 1997.

Browning, Robert. *The Emperor Julian.* Berkeley: University of California Press, 1976.

Nicasie, Martinus J. *Twilight of Empire: The Roman Army from the Reign of Diocletian until the Battle of Adrianople.* Amsterdam: J. C. Gieben, 1998.

Justinian I (482–565)

Ruler of the eastern Roman Empire, sought to restore the unity of the Roman Mediterranean through military, diplomatic, religious, economic, and cultural initiatives that instead created Byzantium. When Justinian became emperor in 527, he inherited a strategic orientation founded on four principles: (1) keeping Constantinople, the capital, impregnable; (2) guarding against the Persian Empire to the east, Byzantium's primary military threat; (3) managing the ill-organized German kingdoms of western Europe, and the Maghrib as a "second front"; and (4) supervising the north-

A mosaic from San Vitale, Ravenna, depicting "Justinian the Great." *(Library of Congress)*

ern frontier created by the Danube River and the Balkans, that unstable territory between Constantinople and the river. Instead, he resolved to alter this approach permanently by reclaiming the lost provinces of the west.

A diplomatic quarrel led Justinian into an unwanted war with Persia in 530, but a new shah, Khusrauw, ended the hostilities in 532. Justinian used peace with Persia to launch an invasion of the Vandal kingdom of Carthage. Belisarius, the Byzantine general, made unexpectedly quick work of the Vandals, and in 533 the lands of Libya and Tunisia became Byzantine provinces. Consolidating his hold in Africa, a crisis in the Ostrogothic state of Italy offered Justinian another opportunity. In 535, backed by the navy, Belisarius invaded Sicily, crossed into Italy, and entered Rome in December 536. Although the Ostrogoth king Vitigis besieged Rome for over a year, the eternal city held, and he withdrew to his capital of Ravenna. Three years later, Vitigis surrendered the city to Belisarius.

Despite these successes, Justinian's fortunes reversed dramatically in the 540s. Shah Khusrauw, prompted by Arab and Armenian allies and fearing that the balance of power was tipping against Persia, stormed through Syria and sacked Antioch. Five years of inconclusive invasion and counter-invasion followed. While Persia preoccupied Justinian, wandering Slavic, Bulgarian, and Germanic peoples began to infiltrate the Balkans. In addition, Totila, a new Goth ruler, recaptured most of Italy and in 549 took Rome.

In 552, Justinian dispatched General Narses to Italy with 30,000 men, including Lombard mercenaries. Narses crushed Totila at Busta Gallorum, occupied Rome, and permanently ended Gothic power in Italy. He also sent his general Solomon to seize a slice of coastal Spain as a weight against the Visigothic kings there. In the Balkans, the crisis continued unabated as Zabregan, chief of the Kutrigurs, raided with virtual impunity, and even threatened Constantinople in 557. Lacking local resources, Justinian simply bribed Zabregan to leave, and then bribed another barbarian king to make war on the Kutrigur. Byzantium adopted this bribe and divide tactic repeatedly to manage barbarian threats in the Balkans. It could backfire. In 562, for example, impervious to payoffs, the Huns sacked the city of Anastasiopolis.

Justinian's successors soon lost Spain and, within a few centuries, Italy and most of his other acquisitions. Historians continue to debate whether his military legacy to Byzantium was one of short-lived grandeur or debilitating overextension.

Weston F. Cook, Jr.

See also: Belisarius; Byzantine-Persian Wars; Gothic War; Huns; Narses

References and further reading:

Moorhead, John. *Justinian.* London and New York: Longman Press, 1994.

Norwich, John J. *Byzantium: The Early Centuries.* New York: Alfred A. Knopf, 1994.

K

Kadesh, Battle of (1274 B.C.E.)

The Hittite and Egyptian empires clashed for control of the Levant at Kadesh in the largest, best-documented Bronze Age battle. In 1275–1274 B.C.E. Egyptian pharaoh Rameses II ("the Great") invaded the Levant, sending his Na'arm division along the coast while he led the Amun, Ra, Ptah, and Seth divisions inland. Egyptian forces included 20,000 Nubian and Egyptian infantry, Libyan mercenaries, and Egyptian chariots. The Hittite king, Muwatallis, responded with 35,000 allies and 2,500 chariots. Egyptian forces utilized bows and light chariots while Hittites were known for iron weapons and heavy chariots.

At Kadesh, Muwatallis set a trap, using spies to suggest that the Hittite army was still far to the north. Rameses fell for the ruse and raced ahead with the Amun division to seize Kadesh. At this time, the Hittites emerged from behind Kadesh, crossed the Orontes River and hit the Ra division squarely on its right flank. The Ra division broke and fled toward the Amun division. Fortunately, the Egyptian Na'arm division arrived just in time to stop a rout.

Rameses rallied the fast-moving Egyptian forces and rolled back the Hittites with swift chariot attacks. The Hittites were in danger of being crushed between the northerly Na'arm, Amun, and Ra divisions and the southerly Ptah and Seth divisions. Muwatallis decided to withdraw east across the Orontes River and occupy Kadesh. With no siege equipment Rameses could not take Kadesh. Both sides eventually agreed to a peace treaty (each loudly proclaiming victory), with Rameses marrying a Hittite princess.

Christopher Howell

See also: Ancient Warfare; Hittites

References and further reading:

Newby, P. *Warrior Pharaohs: The Rise and Fall of the Egyptian Empire.* New York: Faber Press, 1980.

Shaw, Ian. *Egyptian Warfare and Weapons.* Buckinghamshire, UK: Shire Egyptology Publications, 1991.

Yadin, Y. *The Art of Warfare in Biblical Lands.* New York: McGraw-Hill, 1963.

Kamenev, Sergei Sergeevich (1881–1936)

Czarist officer, Red Army commander in chief, military theoretician. Born in Kiev, son of a military engineer, Kamenev graduated from the Aleksandrovskii Military Academy (1900), joining the 165th Lutskii Regiment; he graduated from the General Staff Academy in 1907.

In World War I, Kamenev was a semi-adjutant in the Operational Department, First Army, 30th Poltavskii Infantry Regiment; a commander of a rifle corps; and a chief of staff. He emerged a colonel.

After the October Revolution, he sided with the Bolsheviks, was elected chief of staff, XV Rifle Corps, then Third Army, before demobilization.

During the Russian Civil War, Kamenev was a Red Army volunteer; a military head of the Nevel'skii District, western screens (April 1918); and Smolensk District military commander (August).

From September 1918 to July 1919, he was eastern front commander and oversaw the counteroffensive against Kolchak (April–July), taking the north and central Urals. But he was stripped of command in July by Trotsky after strategic disagreements with Vacietis, who wished the eastern front to dig in, allowing troop transfers southward to face Denikin. Eastern front commanders' complaints persuaded Lenin to reinstate Kamenev.

Sponsored by Stalin, Kamenev replaced Vacietis (Trotsky's candidate) as Red Army commander in chief (July 1919–April 1924). He oversaw the Red Army campaigns that defeated Kolchak (July–December 1919), Denikin (August 1919–April 1920), and Wrangel (April–November 1920), but was partially to blame for the defeat in Soviet-Polish War (April–October 1920). He failed to coordinate or control the western (Tukhachevskii) and southwestern (Egorov) fronts' advances into Poland, allowing the Polish counteroffensive at Warsaw (August 1920). He then oversaw the clearing of anti-Soviet forces in Ukraine, Belorussia (Makhno, Bulak-Balakhovich), and Turkestan (Basmachi).

Kamenev supported Frunze's military doctrinal ideas and became the principal tactics lecturer at the Red Army Military Academy; inspector, Red Army (April 1924); chief of staff (March 1925); head, Main Administration (November 1925), CEC member.

As deputy peoples commissar, Naval, Military Affairs, deputy chairman RVS USSR (May 1927), Kamenev finally joined the Communist Party in 1930. He was appointed head of administration, Red Army antiair defense, then a member of the Military Soviet Under-Defense Commissariat (June 1934). He also found time to lead Arctic exploration efforts. Kamenev wrote military works assessing civil war operations, changing conditions, and modern warfare developments.

Kamenev died, supposedly of heart failure, in Moscow when Stalin's purges of the military had gathered strength. He was branded a conspirator in the Tukhachevskii Plot (1937) but was later rehabilitated in the post-Stalin era. (Note: S. S. Kamenev should not be confused with L. Kamenev, the party leader executed by Stalin in the same year that S. S. Kamenev supposedly died of a heart attack.)

Neil Harvey Croll

See also: Russian Civil War (1918–1921); Russo-Polish War; Stalin; World War I

References and further reading:

Bubnov, A., S. S. Kamenev, R. Eidemanis, and M. N. Tukhachevsky. *Grazhdanskaia voina, 1918–1921* (Civil War, 1918–1921). Moscow: Izdat. Voennye vestnik, 1928, 1930, 1972, 1974, and 1978.

Bystrov, V. E. *Sovetskie polkovodtsy i voenachal'niki. Sbornik* (Soviet Leaders and Military Chiefs Collection). Moscow: Molodaia gvardiia, 1988.

Kamenev, S. S., L. M. Spirin, and P. P. Chernushkov. *Zapiski o grazhdanskoi voine i voennom stroitel'stve. Izbrannye stat'i.* (Notes about the Civil War and Military Construction. Selected Essays). Moscow: Voenizdat, 1963.

Kameneva, N. S., *Put' polkovodtsa: vospominaniia ob ottse.* (The Path of a Leader: Recollections about My Father). Kiev: Politizdat Ukrainy, 1982.

Kandahar (31 August–1 September 1880)

Final battle of the Second Anglo-Afghan War (1878–1880), the decisive British victory that made a hero of Frederick Sleigh Roberts. In the context of the so-called Great Game or Tournament of Shadows between Russia and Great Britain throughout most of the nineteenth century for control of southern Asia, Britain kept close watch on the northwest Indian frontier. Afghanistan was an important buffer state, and Britain sought to extend its influence there, despite the Afghans' dangerously intense and implacable hatred of all foreigners.

Roberts won the battle at Charasiah on 6 October 1879; occupied the capital, Kabul, on 12 October; deposed Amir Yaqub Khan; and ruled Afghanistan by strict martial law for the next eight months. In June 1880 Yaqub Khan's younger brother, Ayub Khan, began a jihad, or holy war, against the British occupation. At Maiwand on 27 July, 10,000 Afghans under Ayub Khan ambushed a British and Indian brigade of 2,500 under George Reynolds Scott Burrows, inflicting 40 percent casualties while suffering only 25 percent of their own. Ayub Khan then besieged the garrison of 4,000 under James Primrose at Kandahar.

Roberts immediately mobilized 10,000 men, mostly Indians, and marched them across 313 miles of treacherous mountain paths and burning deserts in just 23 days to relieve Primrose. Roberts's logistical and psychological genius ensured the success of this march and earned him an honored place in military history. Arriving north of Kandahar on 31 August, he attacked Ayub Khan's camp near Baba Wali Pass the next day and overwhelmed him, thus ending the war. The British and Indians lost 40 killed, but more than 600 Afghans died.

Satisfied with both the military and political outcome, the British withdrew from Afghanistan in April 1881. The new emir, Abdul Rahman Khan, Yaqub Khan's nephew, then ruled Afghanistan as a neutral state.

Eric v. d. Luft

See also: Roberts, Frederick Sleigh, First Earl, Viscount St. Pierre of Kandahar

References and further reading:

Hannah, W. H. *Bobs: Kipling's General: The Life of Field-Marshal Roberts of Kandahar, VC.* Hamden, CT: Archon, 1972.

O'Ballance, Edgar. *Afghan Wars, 1839–1992: What Britain Gave Up and the Soviet Union Lost.* New York: Brassey's, 1993.

Roberts, Frederick Sleigh. *Roberts in India: The Military Papers of Field Marshal Lord Roberts, 1876–1893.* Ed. Brian Robson. Dover, NH: Alan Sutton for the Army Records Society, 1993.

Robson, Brian. *The Road to Kabul: The Second Afghan War, 1878–1881.* London: Arms & Armour, 1986.

Kangxi (K'ang-his) (1662–1722)

Reign name of a Chinese Manchu emperor born with the name Hsüan-yeh. He survived smallpox as a youth and ruled the Manchu Empire for 60 years. His reign flourished in all respects, but in the realm of military affairs, Kangxi demonstrated particular acumen.

The completion of the conquest of China marked Kangxi's greatest accomplishment. Several Chinese generals had assisted the Manchu in their conquest of China; three of these renegade generals had established quasi-independent states known as the Three Feudatories. From 1673 to 1681, Kangxi's armies battled the Three Feudatories until he achieved victory. The complete domination of China came when Kangxi's army finally seized Taiwan from the heirs of Koxinga (Zheng Chenggong) in 1683.

Kangxi also expanded the empire into inner Asia. Freed from the threat of civil war and rebellion with the defeat of the Three Feudatories, Kangxi turned to meet the threat of the Russians and the Oirat Mongols. Russian Cossacks had made inroads on the Manchu northern frontier and even exacted tribute from his subjects. After destroying the Russian outpost of Albazin, Kangxi offered a peace treaty to the Russians at Nerchinsk in 1689. Thus, Kangxi could now bring the full weight of his military against the Oirats, without fear that the Czar would ally with them. The Oirats threatened to control all of Mongolia, Xinjiang, and part of Kazakhstan. The Khalkha Mongols, of modern Mongolia, turned to Kangxi and offered their submission in return for protection from the Oirats. Kangxi consented at the treaty of Dolon Nur in 1696. At the battle of Jao Modo, Kangxi defeated Galdan Khan, ruler of the Oirats. With this victory, Kangxi secured Mongolia and extended his influence into Xinjiang.

Timothy May

See also: Koxinga; Manchu Expansion, Wars of; Qianlong

References and further reading:

Crossley, Pamela Kyle. *The Manchus.* Cambridge, UK: Blackwell Publishers, 1997.

———. *A Translucent Mirror: History and Ideology in Qing Imperial Ideology.* Berkeley: University of California Press, 1999.

Kessler, Lawrence D. *K'ang-Hsi and the Consolidation of Ch'ing Rule, 1661–1684.* Chicago: University of Chicago Press, 1976.

Rossabi, Morris. *China and Inner Asia.* New York: Pica Press, 1975.

Kars, Battle of (16 November 1877)

One of the battles of the Russo-Turkish War (1877–1878), resulting in a Russian victory and the aggrandizement of Russian territory in the Caucasus. In support of Balkan rebellions against Ottoman rule, Russian czar Alexander II declared war on 24 April 1877. Russian forces were divided into two main armies: Danube and Caucasus.

The Russian army of the Caucasus was commanded by Grand Duke Michael and numbered 60,000, with a reserve force of approximately 50,000. It was divided into four operational groups: Kobulety, Akhaltsykh, Aleksandropol, Erevan. Their mission was to overcome the Turkish fortresses of Batumi, Arduhan, Kars, Bayazid, and Erzerum. Opposing Turkish forces were divided into two groups: 60,000 commanded by Mukhtiar Pasha and an army of 40,000 near Erzerum.

By the end of April, Michael's forces had occupied Bayazid. By mid-May, Russian forces had seized Ardahan and were besieging Kars. However, by the end of August, Turkish counterattacks lifted the siege of Kars.

Kars was a stone citadel upon a gorge, composed of 12 detached forts for a defensive line of 17 kilometers. Mukhtiar Pasha's army was encircled and destroyed by a two-prong Russian pincer in October. However, Khalil Pasha refused to surrender Kars and his force of 25,000 men.

The Russians divided their 50,000 troops into seven detachments, two of which were diversionary. Kars was captured on 16 November. The main fort (Kamli) was seized by three Russian detachments. The Turks lost 70,000 men and 17,805 were taken prisoner. The number of Russians killed and wounded was 2,270.

Adrianople/Edirne was captured on 20 January 1878, and the war was concluded by armistice on 31 January, with Russian forces outside Constantinople.

Kars remained part of the Russian Empire until the Treaty of Brest-Litovsk in March 1918 returned it to Turkey.

Neville G. Panthaki

See also: Balkan War, First; Poltava; Russo-Turkish Wars (1676–1878)

References and further reading:

Anderson, M. S. *The Eastern Question.* London: Macmillan, 1970.

Menning, Bruce. *Bayonets before Bullets: The Imperial Russian Army, 1861–1914.* Bloomington: Indiana University Press, 1992.

Rupp, George. *A Wavering Friendship: Russia and Austria, 1876–1878.* Philadelphia, PA: Porcupine Press, 1976.

Seaton-Watson, Hugh. *The Decline of Imperial Russia.* London: Methuen, 1966.

Kasserine Pass (14–23 February 1943)

A gap in Tunisia's western Dorsal mountains where American troops met defeat in their first major engagement with the Germans in World War II. The two-mile-wide pass, situated between two 4,000-foot mountains, was considered the gateway to Tunis, a city the Allies hoped to capture in early 1943. To prevent the capture of Tunis, and with the hope of splitting Allied forces in two and reaching the north coast of Tunisia, German commander Erwin Rommel launched an armored offensive against thinly held Allied lines near the pass. Because they misinterpreted ULTRA intelligence, the

Allies had made the mistake of placing most of their reserves too far north, leaving the troops guarding the pass extremely vulnerable to the German attack.

On 19 February Rommel's 10th Panzer Division, along with Italian armor and infantry, attacked northwestward and forced a breakthrough in the American lines. The inexperienced American troops gave way too easily, leaving behind many of their weapons and much of their heavy equipment. More than 1,000 were killed and many hundreds taken prisoner. The American commander on the scene, the blowhard General Lloyd Fredendall, proved inadequate to the crisis created by Rommel's attack. After the battle he would be relieved in favor of Lieutenant George S. Patton.

Instead of continuing westward to exploit his breakthrough, Rommel was ordered to turn north toward the port of Le Kef. In so doing his spearheads ran into the teeth of the Allied reserves, who fought the Germans to a standstill amid heavy rains. In the process the tide of battle turned and the pass was recaptured on 24 February. Kasserine Pass served as a hard-learned lesson to the United States Army as to the kind of training, toughness, command, and coordination necessary to defeat the Germans in World War II. Rommel himself warned the German military against making too much of their easy victory; the Americans would learn, he asserted, and more quickly than their allies.

John C. McManus

See also: Eisenhower, Dwight David; Montgomery, Bernard Law; Rommel, Erwin Johannes Eugen; World War II

References and further reading:

Blumenson, Martin. *Kasserine Pass.* New York: Tower Books, 1966.

Eisenhower, Dwight D. *Crusade in Europe.* Garden City, NY: Doubleday, 1948.

Howe, George F. *North Africa: Seizing the Initiative in the West.* Washington: U.S. Army Office of the Chief of Military History, 1957.

Kearny, Philip (1814–1862)

Internationally renowned cavalry officer and Union general. Philip Kearny was born into a wealthy family in New York City on 1 June 1814. His uncle was the famous soldier Stephen Watts Kearny. Philip attended Columbia University in 1833, graduating in 1837 with a law degree. He soon accepted a commission in the 1st Dragoons. In 1839 he was sent to Europe to study cavalry tactics. After attending the French cavalry academy, he served as a volunteer in Algiers with the Chasseurs d'Afrique, seeing action across North Africa and building a wide reputation as a fearless tactician.

Returning to the United States in 1840, he served in administrative billets until he secured a position on the staff of General Winfield Scott. Kearny commanded a cavalry company during the Mexican-American War. In engagements at Contreras, Churubusco, and Mexico City, he increased his reputation but lost his left arm. In 1851 he resigned his commission and traveled widely and in 1859 rejoined the chasseurs to participate in the Italian war. At Solferino, his charge essentially won the day and earned him the cross of the French Legion of Honor, the first American so honored.

Upon the outbreak of the American Civil War, Kearny offered his services to the Union. As a brigadier with the New Jersey Volunteers in the Army of the Potomac, he saw action at Williamsburg and Seven Pines during the Peninsular campaign. In May 1862, at Second Bull Run, his troops turned back the assault by General Thomas "Stonewall" Jackson's troops. Nearby, at Chantilly, on 1 September, he mistakenly rode into rebel lines and was shot down. The Confederates graciously returned his body for burial. At the time of his death, Kearny was considered perhaps the bravest man in an era of brave soldiers.

Michael S. Casey

See also: American Civil War; Mexican-American War

References and further reading:

Kearny, Thomas. *General Philip Kearny.* New York: G. P. Putnam's Sons, 1937.

Werstein, Irving. *Kearny the Magnificent: The Story of General Philip Kearny, 1815–1862.* New York: John Day, 1962.

Kearny, Stephen Watts (1794–1848)

American commander. Kearny was born on 30 August 1794 in Newark, New Jersey, into a wealthy family. He served as an ensign in the New York State Militia while he attended Columbia College in New York City. At the outbreak of the War of 1812 Kearny entered the United States Army and was commissioned a second lieutenant. At the Battle of Queenston Heights Kearny was wounded and captured.

Kearny spent most of the next 35 years serving in the American West, where he took part in many exploration expeditions. In 1836 Colonel Kearny succeeded to the command of a dragoon regiment. In 1837 he wrote a manual for the dragoons, earning him the nickname "Father of the Cavalry." Upon the declaration of war against Mexico in May 1846 he was appointed commander of the Army of the West and given orders to secure New Mexico and California. He entered Santa Fe in August and, acting as military governor, established a government. In September Kearny marched on California. Having received word that Commodore Richard Stockton had pacified California, Kearny brought only 110 dragoons. Upon arrival, however, he discovered that the Cali-

Field Marshal Wilhelm Keitel (right) standing in an open car in a victory parade in Bucharest, upon the return of the Romanian troops from Odessa, 1941. (Library of Congress)

fornians had reestablished themselves. Kearny led his weary men into action and obtained a costly victory at the Battle at San Pascual on 6 December 1846. By January he had secured California for the United States. A dispute between Kearny and Stockton developed over who should govern California. When the authorities in Washington, D.C., sustained Kearny, he had Stockton's choice of governor, Lieutenant Colonel John C. Frémont, arrested for insubordination.

In the spring of 1848 Kearny was ordered to Mexico, where he served as military governor of Veracruz, and later Mexico City, where he contracted yellow fever and took ill. He died 31 October 1848 in St. Louis, Missouri.

Gregory Dehler

See also: Cavalry; Mexican-American War; Queenston Heights; Scott, Winfield; War of 1812

References and further reading:

Clarke, Dwight L. *Stephen Watts Kearny: Soldier of the West.* Norman: University of Oklahoma Press, 1961.

Goetzmann, William H. *Army Exploration in the American West, 1803–1863.* New Haven, CT: Yale University Press, 1959.

Harlow, Neal. *California Conquered: The Annexation of an American Province.* Berkeley: University of California Press, 1982.

Keitel, Wilhelm (1882–1946)

Hitler's chief of Wehrmacht High Command (OKW). Keitel was born 22 September 1882 at Helmscherode near Hannover and joined the Prussian army in 1901, becoming a lieutenant the following year. He married Lisa Fontaine in 1909 and had six children.

Keitel fought during World War I and entered the postwar Reichswehr as a captain in 1919. From 1925 on he held administrative positions within the then-camouflaged general staff responsible for organizational matters and plans for rebuilding a strong army. As a general major he led the armed forces department within the War Ministry from 1935.

After the dismissal of War Minister Werner von Blomberg and Army Commander Werner von Fritsch in February 1938, Hitler made himself Wehrmacht's supreme commander and created OKW under Keitel as his personal staff. Keitel was without executive power toward the army, navy, or air force because Hitler kept command authority to himself, using Keitel as a high-ranking secretary. Because of his weak position and submissiveness toward Hitler, Keitel, promoted to field marshal on 19 July 1940, was called Lakeitel (lackey) by those who knew him. He occasionally disputed matters with the dictator, but Hitler kept the busy and experienced field marshal in office and Keitel progressively abandoned any critical attitude. Without prestige among the officer corps but backed by Hitler, Keitel was allowed to sign orders in Hitler's name ("Der Führer—by authority Keitel"), among them the infamous commissar order and many other criminal directives.

Little research exists about his activities in wartime industry and domestic activities as member of the Reich Defense Council. Leaving operational matters to Jodl he participated in many conferences on coordinating warfare, negotiated with German allies, and held important administrative functions, which indicates that his position as near–war minister is still underestimated.

Keitel left Berlin a week before Hitler's suicide but returned to sign the surrender to the Soviets, 9 May 1945. He was found guilty of war crimes at Nuremberg and hanged 16 October 1946.

Martin Moll

See also: Hitler, Adolf; Jodl, Alfred; World War II

References and further reading:

Görlitz, Walter, ed. *Generalfeldmarschall Keitel.* Verbrecher oder Offizier. Göttingen-Berlin-Frankfurt/Main: Musterschmidt, 1961.

Kellogg-Briand Pact (27 August 1928)

A multinational treaty actually outlawing war. The Kellogg-Briand Pact grew out of two trends that dominated diplomacy in the 1920s. First, the French search for security, and second, the intense desire in the United States for peace in isolation after World War I.

Aristide Briand, the French foreign minister, sought to protect France through a series of treaties known as the Locarno agreements. After solidifying alliances with potential enemies of Germany, he looked to secure improved Franco-American relations, chilled by America's insistence on collecting war debts for World War I. But Briand hoped at least to sign a pact with the United States that proposed to ban war as an instrument of national policy. United States secretary of state Frank Kellogg feared that France was attempting to draw the United States into a negative alliance and preventing the United States from ever declaring war against France, no matter the circumstances. War with France seemed highly unlikely, but such a treaty might prevent the United States from retaliating against French interference with American flagships trading with Germany during a time of war.

Briand communicated his plan directly to the American people in April 1927, ignoring the usual diplomatic channels and infuriating President Calvin Coolidge. Moreover, the wily Frenchman met with leaders of the peace movement in the United States in an effort to encourage them to spur on their reluctant government. The trans-Atlantic flight of Charles Lindbergh in May 1927 seemed to draw the two nations closer.

Throughout the fall of 1927 Kellogg stalled negotiations as he considered options to outmaneuver Briand. In November he proposed that the treaty should be multilateral, with an invitation extended to all nations to sign the agreement. Although Kellogg at first hoped this would scuttle talk of a concept he considered foolish, he later became convinced of its practicality and its rewards (a possible Nobel Prize) for himself. Unable to counter Kellogg's move, Briand assented. On 27 August 1928, 15 nations signed the agreement. Within five years, a total of 64 countries signed the treaty to renounce war as an instrument of national policy.

The treaty failed to include any enforcement mechanisms or even of any method to judge if a nation did in fact use war as an instrument of foreign policy. There was no mention of such issues as defense or alliances. Despite these shortcomings, the Kellogg-Briand Pact was, in fact, the legal basis for bringing Germany to account at the Nuremberg tribunal following the end of World War II.

Gregory Dehler

See also Disarmament; Laws of War; Pacifism/War Resistance

References and further reading:

Ellis, L. Ethan. *Frank B. Kellogg and American Foreign Policy, 1925–1929.* New Brunswick, NJ: Rutgers University Press, 1961.

Ferrell, Robert H. *Peace in Their Time: The Origins of the Kellogg-Briand Pact.* New Haven, CT: Yale University Press, 1952.

Kesselring, Albert (1885–1960)

German field marshal most famous for his defense of Italy, 1943–1944. Born at Marktsteft, near Bayreuth, 30 November 1885, Kesselring enlisted in the 2nd Bavarian Foot Artillery in 1904. During World War I he served in artillery staff positions on the western front, becoming a general staff officer in late 1917, then was assigned to a division on the eastern front. Promoted to colonel in 1930, and having already proved his ability as an administrator, he was transferred to the Air Ministry in 1933, promoted to general major in 1936, and became the air force chief of staff in the same year.

During the Polish campaign, Kesselring commanded the First Air Fleet and succeeded in breaking the defenses in Warsaw. He also played an important role in the French campaign in 1940 as the commander of the Second Air Fleet, supporting the actions of Army Group B. In the first three months of Operation BARBAROSSA (invasion of the Soviet Union) Kesselring commanded the air fleet assigned to Army Group Center but was transferred to Italy in November 1941, becoming commander in chief south, nominally the direct superior of Rommel. On 21 November 1943 he became commander in chief of Army Group C, the commander of German forces in Italy. He conducted defensive operations against the Allies skillfully until he suffered a severe road accident in October 1944. After returning to Italy for a few weeks, Kesselring became commander in chief west on 9 March 1945.

In February 1947 Kesselring was charged with the shooting of civilians in Italy in March 1944 and sentenced to death, subsequently reprieved, then released on 23 October 1952 on the grounds of ill health. Kesselring died on 16 July 1960 at Bad Nauheim.

Alaric Searle

See also: Anzio, Battle of; Salerno

References and further reading:
Kesselring, Albert: *Memoirs.* London: Kimber, 1953.
———. *Gedanken zum Zweiten Weltkrieg.* Bonn: Athenäum 1955.
Macksey, Kenneth: *Kesselring.* New York: David MacKay, 1978

Kett's Rebellion (1549)

In 1548, agrarian disturbances, spurred by rising inflation, enclosure of common lands, and debasement of the coinage, began to break out all over England. Most were handled firmly but leniently by the local authorities, urged by Protector Somerset, regent for Edward VI, to address grievances and disperse the rebels without using force, especially because English troops were involved in fighting Scotland. Unfortunately, by 1549, Norfolk was involved in a serious uprising that no threats of forfeitures or martial law could stop. Norfolk was without a great noble family the people respected and obeyed after the purges of the Tudors' Plantagenet relatives, and its citizens had little respect for their unpopular bishop, William Rugge. The ensuing rebellion centered on Robert Kett, a 57-year-old gentleman landowner, who assumed leadership of the angry mob of yeomen that had come to tear down his own fences and kill his sheep.

Kett led the increasingly large group to march on Norwich, where it camped outside the city, on Mousehead Heath, en route tearing down fences and destroying the hated dovecotes of landlords, whom they took as hostages and used as a human shield as they marched. On July 22, the rebels attacked the walled city with help from disaffected city dwellers and broke in through one of the gates, seizing all of the arsenal before retreating back to the heath, where Kett had set up an elaborate administration, with clerks, legal courts, and a provisioning system. Kett and his men issued a declaration to the crown, "Twenty-Nine Demands," most of which called for the preservation of traditional, conservative rights against recently corrupt practices of enclosure and economic repression.

The government first sent William Parr, Marquis of Northampton, who allowed himself to be trapped inside the city of Norwich, attacked, and forced to retreat, then John Dudley, Duke of Northumberland, who arrived on 24 August and took possession of the city. Dudley declared that he would stay there to the last man in his army, which included Italian mercenaries. Kett, who had sent out agents to try to enlarge the rebellion by contacting other agrarian rebels engaged in minor revolts nearby, failed to rally more men, and also failed to repeat his previous success of attacking Norwich, once Dudley was inside. Street fighting quickly fizzled, and Kett was forced to engage the Crown's army in an open field battle northeast of the city, where the professional soldiers decimated the rebels, killing perhaps 3,000 of them in battle and executing a number of men identified as ringleaders, before offering a pardon to those who surrendered.

Kett himself, and his brother, William, were captured and taken to the Tower, where they were tried for treason. Returned to Norwich for execution, they died on 9 December 1550. Ironically, the rebellion brought about the collapse of the government of Protector Somerset, who was sympathetic to the grievances of the rebels, and led to the rise of Robert Dudley, the Duke of Northumberland.

Margaret Sankey

References and further reading:
Clayton, Joseph. *Robert Kett and the Norfolk Rising.* London: M. Seker, 1912.
Cornwall, Julian. *Revolt of the Peasantry 1549.* London: Routledge & Kegan Paul, 1977.
Land, Stephen K. *Kett's Rebellion: The Norfolk Rising of 1549.* Totowa, NJ: Rowman & Littlefield, 1977.

Khalid ibn al-Walid (d. 642)

Primary Arab general during the first phase of the Arab conquests. Khalid fought against the Prophet Muhammad at the Battle of Uhud but later converted to Islam in 627 (or 629; historians are unsure). Khalid took part in the conquest of Mecca in 629, and then led several other expeditions.

After the death of the Prophet in 632, Abu Bakr sent Khalid on several missions to quell rebellion. As Khalid demonstrated exceptional prowess, he was placed in charge of an army to invade Iraq. In Iraq, Khalid conquered several locations before leading his army into Syria, by crossing the desert, although there is disagreement in sources as to whether Khalid went first to Syria, and then marched to Iraq to take the strategic location of Dumat al-Djandal before returning to Syria. Once in Syria, Khalid assisted the Arab armies already fighting the Byzantines there.

Eventually, Khalid's army returned to Iraq to resume its duties there, but Khalid himself remained in Syria. Khalid's success by this time was enormous and he eventually gained the sobriquet of Sayf Allah, or the Sword of God, although later sources refer to him with the less prestigious title of Sayf Rasul Allah, or the Sword of the Messenger of God (that is, of Muhammad).

Khalid rose to be the commander of the armies in Syria, but after the death of Caliph Abu Bakr in 634, his fortunes declined momentarily. Caliph 'Umar ibn al-Khattab demoted Khalid from his position, although he was not removed from Syria. He continued to lead troops in northern Syria as a lieutenant of Abu 'Ubayda. Khalid led numerous raids on the Byzantine border until his death in 642. There is some speculation that Khalid was assassinated by the future Caliph Mu'awiya, who envied Khalid's status.

Timothy May

See also: 'Amr ibn al-'As; Byzantine-Muslim Wars; Charles Martel; Heraclius; Musa ibn Nusayr; Muslim Civil War; Muslim Conquests; Religion and War; Sassanid Empire; Tariq ibn Ziyad; Yarmuk, Battle of

References and further reading:

Belyaev, E. A. *Arabs, Islam, and the Arab Caliphate in the Early Middle Ages.* London: Pall Mall Press, 1969.

Donner, Fred M. *The Early Islamic Conquests.* Princeton, NJ: Princeton University Press, 1981.

Shaban, M. A. *Islamic History: A New Interpretation.* London: Cambridge University Press, 1971.

Khalkin-Gol (Battle of Nomonhan, May–September 1939)

A series of border battles between Soviet and Japanese forces, resulting in Japan's defeat. In the summer of 1939, the Japanese Kwantung Army in Manchuria fought several battles against Soviet forces in Siberia over the exact border along the Khalkin-Gol River at a town named Nomonhan. There were several stages to the fighting.

First, Soviet troops occupied the disputed territory to a depth of about 12 miles to the east in May 1939.

Second, the Japanese attacked with a reinforced division and were initially successful. They had expected to fight Mongolian troops and were surprised at the appearance of regular Soviet Red Army forces. After the initial success in driving the Soviets out of disputed territory, Japanese troops crossed into Mongolia, where despite reinforcements of artillery and tanks, they were stopped.

Third, Soviet forces, now commanded by General Georgy K. Zhukov, drove back the badly outnumbered Japanese across the border and back to Nomonhan. Zhukov threw 65,000 troops against 28,000 Japanese and soon pushed them 20 miles into Manchuria.

Finally, before the Japanese Kwantung Army commander could employ the three divisions he had concentrated for a counterattack, the Japanese high command took control from the nearly autonomous Kwantung Army and called a halt to the fighting in mid-September 1939.

The Japanese learned little from the Soviet attack. Supposedly, the Japanese experience in fighting the Soviets and the situation of the European empires in Southeast Asia after the German victories in the West in the spring of 1940 caused the Japanese to look south to expand into the resource-rich southern Asia-Pacific region and hence to consider a preemptive strike against the American Pacific Fleet in its anchorage at Pearl Harbor, Honolulu, Hawaii.

Charles M. Dobbs

See also: Zhukov, Georgy Konstantinovich

References and further reading:

Coox, Alvin D. *Nomonhan. Japan against Russia, 1939.* Stanford, CA: Stanford University Press, 1985.

Drea, Edward J. *Nomonhan: Japanese-Soviet Tactical Combat, 1939.* Ft. Leavenworth, TX: United States Army Command and General Staff College, Combat Studies Institute, 1981.

Zhukov, Georgi K. *Marshal Zhukov's Greatest Battles.* Trans. Theodore Shabad. New York: Harper & Row, 1969.

Khambula (29 March 1879)

The turning point of the Anglo-Zulu War, breaking the morale of the Zulu army. Colonel Evelyn Wood's number 4 (left) column of 2,086 officers and men maintained an active presence in northwestern Zululand after the other invading British columns were thrown onto the defensive by defeat at Isandlwana. Wood's fortified position at Khambula consisted of a wagon laager, connected to an earthwork redoubt and a smaller cattle laager. At midday on 29 March the Zulu army of about 20,000 men, under the command of Chief Mnyamana kaNgqengelele, halted four miles southeast of the camp. Aware of the danger of attacking entrenched positions, King Cetshwayo had instructed Mnyamana to draw the British into the open by threatening their line of supply. But Mnyamana was overborne by the younger warriors, who insisted on a direct assault.

The Zulu army deployed with the intention of enveloping Khambula, but at 1:30 P.M. the right horn began an unsupported advance from the north, drawn on by mounted troops sent forward by Wood, and was repulsed. The Zulu were consequently unable to complete their envelopment of the camp, whose northern and western salients remained unthreatened, thus enabling the British to concentrate against the main Zulu attack, which unfolded at 2:15 P.M. from the south.

The Zulu drove the British from the cattle laager and threatened the main laager. Several British companies then sortied and drove the Zulu back at bayonet point, and by 3:00 P.M. the Zulu had abandoned their assault from the south. Over the next two hours they renewed the attack, first from the east, and then from the northeast. At about 5:00 P.M., when the Zulu attack slackened off, British infantry sortied once more, supported by the mounted troops. The exhausted Zulu were unable to rally, and their retirement turned into a rout. The mounted troops relentlessly pursued them eastwards until night fell.

The British lost 29 killed, the Zulu more than 1,000. The fighting spirit of the Zulu army never recovered from this crushing defeat, and the Zulu had lost the initiative in the war.

John Laband

See also: Anglo-Zulu War; Isandlwana; Rorke's Drift

References and further reading:

Jones, Huw M. "Why Khambula?" *Soldiers of the Queen* (September 1993).

Laband, John. "The Battle of Khambula, 29 March 1879: A Reexamination from the Zulu Perspective." In *Kingdom and Colony at War: Sixteen Studies on the Anglo-Zulu War of 1879,* John Laband and P. Thompson. Pietermaritzburg and Cape Town: University of Natal Press and N & S Press, 1990.

———. *Kingdom in Crisis: The Zulu Response to the British Invasion of 1879.* Manchester and New York: Manchester University Press and St. Martin's Press, 1992.

Laband, John, and P. Thompson. *The Illustrated Guide to the Anglo-Zulu War.* Pietermaritzburg: University of Natal Press, 2000.

Kharkov (12–28 May 1942)

World War II battle on eastern front, catastrophic defeat for Red Army. The second largest city in Ukraine, Kharkov fell to German forces 24 October 1941, four months into Hitler's invasion of the Soviet Union. Seven months later, in May 1942, Kharkov became the site of a major defeat for the Red Army.

The so-called Second Battle of Kharkov (12–28 May 1942) developed when Marshal S. K. Timoshenko's Southwestern Front, supported by Marshal R. Ia. Malinovsky's Southern Front, undertook an offensive to recapture the city and destroy the German forced deployed in its defense. Attempting a pincer operation, Soviet Twenty-eighth, Twenty-first, and Thirty-eighth Armies attacked from Volchansk in the northeast, while Soviet Sixth and Bobkin Army Group struck from the Izyum Bulge in the southeast.

Both arms of the Soviet pincer made substantial advances until 16 May. However, Timoshenko's failure to commit armored forces fast enough allowed the Germans to blunt the Soviet drives and mount a counterstrike. Codenamed Operation FREDERICUS, the German counteroffensive of 17–18 May saw the southern prong of the Soviet offensive encircled by General Friedrich Paulus's Sixth Army, attacking from the north, and General Ewald von Kleist's Army Group (Fourth Panzer and Seventeenth Army), attacking from the south.

Although a few Soviet units managed to escape, the Second Battle of Kharkov cost the Red Army an estimated 277,000 men, 4,900 guns and mortars, and 652 tanks.

Bruce J. DeHart

See also: Timoshenko, Semen Konstantinovich; World War II

References and further reading:

Erickson, John. *The Road to Stalingrad: Stalin's War with Germany.* New Haven, CT: Yale University Press, 1999.

Glantz, David, and Jonathan House. *When Titans Clashed: How the Red Army Stopped Hitler.* Lawrence: University of Kansas Press, 1995.

Ziemke, Earl, and Magna E. Bauer. *Moscow to Stalingrad: Decision in the East.* Washington, DC: U.S. Army Center of Military History, 1987.

Khartoum, Siege of (13 March 1884–26 January 1885)

Evacuation, defense, and massacre in the First Mahdist War. In the early 1880s the Sudan was ruled by Egypt, which was itself a puppet state of Great Britain and nominally part of the Ottoman Empire. Native Sudanese resented foreign domination and grumbled for home rule. Sudanese nationalist/theocratic uprisings led by the Mahdi, a charismatic Nubian Sufi fundamentalist mystic, began in 1881. British prime minister William Gladstone tried to remain aloof from what he perceived as an internal Egyptian affair, but the British became embarrassed by their inability to safeguard Egypt's control of the Sudan. The khedive of Egypt pleaded unsuccessfully for British support.

Finally bowing to public pressure in January 1884, Gladstone reluctantly sent a popular hero, Charles George "Chinese" Gordon, to the Sudan with orders first to evacuate Egyptian troops from Khartoum, then, if possible, regroup and retaliate against the Mahdi. Gordon left London on 18 January, arrived in Khartoum on 18 February, judged immediately that transport was insufficient for a well-managed withdrawal of troops, and set about to evacuate women, children, the sick, and the disabled. Gordon had evacuated only about 2,000 of these civilians when the Mahdi besieged Khartoum on 13 March. Gordon's garrison was about 8,000 men. Besides an unknown number of Mahdist troops, Gordon estimated that two-thirds of Khartoum's native population of 40,000 was against him.

Gladstone, annoyed at Gordon's apparent insubordination, did not authorize relief until October, then sent a rescue force under Garnet Joseph Wolseley. Advance British gunboats under Lord Charles Beresford arrived on 28 January 1885, two days too late, as Mahdists had already breached the walls and killed the entire garrison. Wolseley withdrew, leaving the Mahdi free to govern. The Mahdi proclaimed an Islamic state from the Red Sea to central Africa with its capital at Omdurman but died of natural causes in June. Gladstone's weak handling of the Sudan crisis contributed to his replacement as prime minister by the marquis of Salisbury on 23 June; many Britons blamed him personally for Gordon's death and Khartoum's fall.

Eric v. d. Luft

See also: Buller, Sir Redvers Henry; Gordon, Charles George; Kitchener, Horatio Herbert; Muhammad Ahmad; Wolseley, Garnet Joseph, Viscount

References and further reading:

Compton, Piers. *The Last Days of General Gordon.* London: Hale, 1974.

Hake, Alfred Egmont. *Gordon in China and the Soudan.* London: Darf, 1987.

Neillands, Robin. *The Dervish Wars: Gordon and Kitchener in the Sudan, 1880–1898.* London: Murray, 1996.

Nushi, Muhammad. *General Report on the Siege and Fall of Khartum by Mohammed Nushi Pasha and Several Native Officers Who Were in Khartum or Its Vicinity at the Time of the Siege and Assault.* Khartoum, Sudan: 1970.

Khe Sanh, Siege of (21 January–8 April 1968)

A major siege operation against a U.S. Marine combat base. At the time the siege of Khe Sanh, an isolated marine base along Route 9 in extreme northern South Vietnam, caused some to recall the French debacle at Dien Bien Phu in 1954. President Lyndon Johnson emphasized to the chairman of the Joint Chiefs of Staff, Earle Wheeler, "I don't want any damn Dinbinfoo," and Johnson received assurances that, unlike the French, the United States could sustain the marines.

Unlike the French, the marines held the hills surrounding their base camp and prepared for the attack. The North Vietnamese set up positions around the marines' perimeter, moved in thousands of troops, and engaged in classic siege measures. To upset North Vietnamese plans, U.S. air power saturated the area: B-52 bombers, marine and naval aviators, and army long-range artillery dropped more than 100,000 tons of bombs and shells on suspected enemy positions. Finally, in early April, the U.S. 1st Cavalry Division moved up Route 9 and linked up with the marines on 8 April, ending the siege. Later, to some bitter reaction, Khe Sanh was abandoned.

However, there will always be a question as to the goals of the North Vietnamese. Was this a decoy, to draw U.S. attention away from the cities and the coast immediately prior to the Tet Offensive? Was it simply a probe to test American resolve? Or was it preparation for a cross-parallel invasion to follow up a supposedly successful Tet Offensive and popular, antigovernment uprising in the South? The answers remain, to this day, unclear.

Charles M. Dobbs

See also: Tet Offensive; Vietnam Conflict; Vo Nguyen Giap; Westmoreland, William

References and further reading:

Pimlott, John. *Vietnam: The Decisive Battles.* New York: Macmillan, 1990.

Pisor, Robert. *The End of the Line: The Siege of Khe Sanh.* New York: W. W. Norton, 1982.

Prados, John. *Valley of Decision: The Siege of Khe Sanh.* Boston: Houghton Mifflin Co., 1991.

Khmer-Cham Wars (1050–1203)

A series of wars between the Khmer Empire and a resurgent Champa state for control of Indochina. In Southeast Asia, Hindu and Buddhist kingdoms from India and China had developed into independent polities centered around ethnic groups such as the Annam in north Vietnam, the Cham in central Vietnam, the Khmer in Cambodia, and the Tai in Siam. These polities had thrown off the yoke of Chinese and Indian overlordship and then turned on each other. The Khmer-Cham Wars, Khmer-Thai Wars, Vietnamese-Cham Wars, and the Vietnamese-Khmer Wars all stemmed from this struggle to control Southeast Asia and the rich trade that flowed between China and India.

The first Khmer-Cham War (1050–1051) involved revolts in Khmer territories. The Cham king Jaya Paramesvarman and his son Yavuraja then invaded the northern Khmer territory of Sambhupura while supporting a revolt in the Khmer south. Khmer generals suppressed the revolt and crushed Cham forces in the north.

The second Khmer-Cham War (1144–1150) followed the Vietnamese-Khmer War of 1123–1136. The Khmer king Suryavarman II controlled modern Cambodia and Laos but sought all of Vietnam by allying with Champa and invading Annam. Instead, a Cham-Annam alliance formed and opposed the Khmers. Khmer armies invaded Cham and captured its capital of Vijaya. Cham forces still held out under Harivarman I at Chaklyang and destroyed Khmer and Vijayan armies in two Phanrang Valley battles. Champa troops then marched north, captured Vijaya, and defeated the Khmer forces at Mahisa. A final Khmer force sent by Suryavarman II was also crushed.

The third Khmer-Cham War of 1167–1190 saw the Cham king Jaya Indravarman IV invade the weakened Khmer Empire using cavalry and naval forces, instead of the traditionally slow elephants and water buffalo. These concepts, learned from the Chinese, led to startling initial victories over the Khmers. In 1177 Cham naval forces sailed up the Tonle Sap (great lake) of the Khmer and destroyed the Khmer capital at Angkor (Yasodharupura).

Cham now threatened to overrun Southeast Asia but a wealthy Thai-Khmer alliance under the Khmer Jayavarman VII arose against it. Both sides became dependent on China for naval aid and horses, with most going to the wealthy Khmer Empire. An 1181 sea victory over Champa allowed the rebuilding of Angkor Thom, north of the original Khmer capital. Champa was invaded in 1190 and its capital at Vijayana sacked.

The fourth Khmer-Cham War of 1191–1203 saw Champa rebel against Khmer rule. Two Khmer armies were defeated, but the Khmer king Jayavarman VII used Cham traitors to regain control of the Champa throne. The renewed Khmer Empire remained dominant in Southeast Asia until Thai-Cham forces sacked Angkor-Thom after a seven-month siege in 1430 and destroyed vital irrigation systems in 1444, signaling the end of the mighty Khmer Empire.

Christopher Howell

See also: Jayavarman VII

References and further reading:

Briggs, Lawrence. *The Ancient Khmer Empire.* New York: White Lotus, 1999.

Embrick, A., and C. Gluck. *Asia in Western and World History.* London: M. E. Sharpe, 1997.

Kiev (16–26 September 1941)

The greatest single military victory in modern history, and the worst defeat in Soviet military history. On 16 September 1941, 680,000 Soviet troops of General M. P. Kirpanov's Southwestern Front suffered encirclement in the Kiev pocket (130 miles in width and depth) when the German Second Panzer Group, commanded by General Heinz Guderian, advancing from Smolensk in the north, linked up with the German First Panzer Group, commanded by General Ewald von Kleist, advancing from Kremenchug in the south at Lokhvitsa (125 miles east of Kiev). For the next 10 days, six trapped Soviet armies, the entire strength of the southwestern front, struggled to break out, while German forces, their movements coordinated by Field Marshal Gerd von Rundstedt, commander of Army Group South, fought to reduce the pocket.

Although 15,000 Soviet troops ultimately escaped, Kirpanov's armies lacked sufficient power to achieve a large-scale breakout against a numerically superior enemy who also enjoyed command of the skies. Kiev itself fell 20 September, and six days later the last Soviet resistance inside the pocket ended. Four entire Soviet armies were destroyed, and two others severely emasculated. German statistics revealed that the Battle of Kiev cost the Soviets 665,000 prisoners, 3,018 guns, and 418 antitank guns. General Kirpanos himself was killed attempting to break out of the pocket.

The annihilation of the Soviet Southwestern Front opened the door for Army Group South to capture central and eastern Ukraine and most of the Crimea in the last months of 1941.

And yet Hitler's diversion of additional German forces to the Kiev encirclement delayed the drive on what should have been the ultimate goal of the German invasion—Moscow. By the time German forces finally approached the Soviet capital in early December, it was winter and too late. Germany had won a great operational victory at Kiev but had lost any chance of destroying the Soviet union.

Bruce J. DeHart

See also Guderian, Heinz; World War II

References and further reading:

Boog, Horst, et.al. *Germany and the Second World War.* Vol. 4, *The Attack on the Soviet Union.* Oxford, UK: Clarendon Press, 1998.

Erickson, John. *The Road to Stalingrad: Stalin's War with Germany.* New Haven, CT: Yale University Press, 1999.

Glantz, David, and Jonathan House. *When Titans Clashed: How the Red Army Stopped Hitler.* Lawrence: University of Kansas Press, 1995.

Killiecrankie (27 July 1689)

A major defeat of the British at the hands of the Highland Scots, and the last true Scottish battle. The overthrow of James II by the English Parliament and the subsequent ascension to the throne by William III and Mary II led to an uprising by Scottish clans supporting the Stuart claim. In 1689, John Graham of Claverhouse, Viscount Dundee, unfurled the Stuart standard in the Highlands. He quickly raised an army of 1,800 soldiers, including men from the various clans, along with 500 Irish mercenaries. Dundee marched his force to secure Blair Castle (Atholl), an important strategic site in the Grampian Highlands near Perth.

To suppress this revolt, General Hugh Mackay marched north with a force of approximately 3,500 royalist troops. Mackay's force consisted mostly of Lowland Scots, many of whom were recent recruits. The most direct route to Blair Castle was to traverse the Pass of Killiecrankie, a narrow rugged area that was easily defensible. Furthermore, Dundee knew this was the likely avenue of march and moved his army to defend the pass. On 27 July, the advance guard of the royalist army entered the pass and found it undefended. Here Mackay made a critical error. Instead of quickly moving his entire force through the pass, he ordered his men to take up positions in the nearby grain fields to allow the artillery and baggage train to traverse the steep hills. This allowed Dundee to consolidate his forces on the battlefield. Mackay could have ordered an assault on the enemy, but lacking intelligence about the size of his opponent, he organized his units in defensive positions. He divided each battalion into two units, three men deep with intervals between each group. Mackay unlimbered his three artillery pieces and began a harassing fire on the Jacobites.

Dundee arranged his smaller army by clans with large intervals between units. He also shifted his line toward the enemy's right flank to prevent his own flank from being enveloped. At 7 P.M., the weight of the royalist cannon fire prompted Dundee to order an assault. The Highlanders launched a fierce charge, with each man bent forward to provide a smaller target, using his shields to protect the upper body. Mackay's troops fired three volleys at the advancing Jacobites, with little effect because each unit fired by platoon, instead of withholding fire for a concentrated volley. Some units did not fire at all. This allowed the Highlanders to fall upon Mackay's unnerved troops with broadswords and pole-axes before they could insert their socket bayonets. Hundreds were hacked down and the royalist infantry broke into rout. The entire force would have been destroyed had Dundee not been killed at the moment of victory and the Jacobites stopped to plunder the royalist baggage train. After suffering nearly 2,000 casualties, the remnants of Mackay's army retreated to Perth. Jacobite losses were approximately 500 men.

Killiecrankie is touted as the last of the true Scottish battles, as most of the participants were Scots. It was a tactical masterpiece by Dundee, but his death prevented the Jacobites from exploiting the victory. Colonel Cannon, Dundee's successor, was repulsed by the royalists at Dunkeld and the uprising soon lost momentum.

Barry P. Neville

See also: Anglo-Scots Wars (1513–1560)

References and further reading:

Brown, Peter H. *History of Scotland.* Vol. 3. New York: Octagon Books, 1971.

Reese, Peter. *The Scottish Commander: Scotland's Greatest Military Leaders from Wallace to World War II.* Edinburgh: Canongate Books, 1999.

Kim Il-sung (1912–1994)

Founding leader of the Democratic People's Republic of (North) Korea. Kim Il-sung was born in 1912 near Pyongyang in Japanese-occupied Korea and began his revolutionary career in 1929 when he was jailed for student activism. The next year he received the nom de guerre of Kim Il-sung—a famous former revolutionary—and for the next decade probably fought with Chinese guerrillas in Manchuria. In 1939, he likely reentered Korea to fight the Japanese occupiers and two years later he retreated into the Soviet Union. Little is known of his time in the USSR; there is no firm evidence to support the rumor that he fought at Stalingrad.

At the end of World War II, in September 1945, Soviet authorities brought Kim back to Korea and later presented him as the leader of the Soviet-imposed regime north of the thirty-eighth parallel. There probably was a struggle for power between Kim, an expatriate, and local Korean communists. In 1948 the Soviets established the Democratic People's Republic of Korea with Kim as its leader and set up the Korean People's Army. By the next year, American intelligence experts were expecting warfare across the parallel in the near future.

That attack began on 25 June 1950, and North Korean troops seemingly overwhelmed the South Korean defenders. Eventually, U.S. air power inhibited North Korean supply lines and General Walton Walker held the Pusan perimeter; on 15 September General Douglas MacArthur launched the Inchon offensive and North Korea nearly collapsed. Had the Chinese not intervened, Kim would have lost everything; presumably in the aftermath of Chinese intervention, he owed the Chinese a huge debt of gratitude, which probably was difficult for a committed though Communist Korean nationalist.

Kim retained power after the armistice with the United Nations forces defending South Korea and served as general secretary of the party as well as chairman of the party's military commission and president of the Democratic People's Republic of Korea (DPRK). He increasingly almost deified himself in a cult of personality that seemed to out-Stalin Stalin. To the consternation of the more credulous DPRK citizens, he died on 8 July 1994.

Charles M. Dobbs

See also: Chosin/Changjin Reservoir; Korean War; MacArthur, Douglas; Pusan Perimeter; Ridgway, Mathew B.

References and further reading:

Kim, Doug Joong, ed. *Foreign Relations of North Korea during Kim Il Sung's Last Days.* Seoul: Sejong Institute, 1994.
Sandler, Stanley. *The Korean War: No Victors, No Vanquished.* Lexington: University of Kentucky Press, 1999.
Scalapino, Robert, and C. Lee. *Communism in Korea.* Berkeley: University of California Press, 1972.
Suh, Dae-sook. *Kim Il-sung: The North Korean Leader.* New York: Columbia University Press, 1988.

Kim Yu-sin (595–673)

General and statesman of the Silla Kingdom (c. 200–905), one of the primary architects of Silla's unification of the Korean peninsula in 668. Kim Yu-sin was born in an aristocratic family in 595, at a time of increasing rivalry between Korea's three dominant kingdoms—Paekche in the southwest, Koguryo in the north, and Silla in the southeast—and was brought up in the strict military and Buddhist discipline common to the youths of the Silla nobility. Later legends relate how he was visited by a mountain spirit that predicted his eventual defeat of Silla's rivals.

For almost two centuries before Kim's birth, the three Korean kingdoms had fought an ever-shifting, three-way struggle for territorial expansion. By the early seventh century, the issue had become one of peninsular hegemony. After Silla secured an alliance with the Tang Dynasty (618–907) to overcome both Koguryo and Paekche, Kim Yu-sin led Silla forces in 661 in a combined Silla-Tang attack upon Paekche that culminated in that kingdom's utter defeat. The defeat of Koguryo, again through a Silla-Tang alliance, soon followed in 668, but Kim Yu-sin would die before witnessing the final Silla unification. This came in 676 with Silla's defeat of its erstwhile ally and would-be overlord, Tang China. Regardless, Kim Yu-sin is viewed as the father of Korean unification, and legends have continued to be built around him since his death. His tomb can still be seen in the historic city of Kyongju, the former capital of the Silla Kingdom.

Daniel Kane

See also: Chinese Imperial Wars; Koguryo; Paekche; Silla Kingdom; Sino-Korean Wars and the Wars of Korean Unification

References and further reading:

Adams, Edward B. *Korea's Golden Age: Cultural Spirit of Silla in Kyongju.* Revised edition. Seoul: Seoul International Publishing House, 1991.
Chu, Yo-sop. *Kim Yusin :The Romances of a Korean Warrior of the 7th Century.* Seoul: Mutual Publishers, 1947.
Yi, Ki-baek. *A New History of Korea.* Trans. Edward W. Wagner with Edward J. Shultz. Seoul: Ilchokak Publishers, 1984.

Kimberley, Siege of (14 October 1899–15 February 1900)

Unsuccessful siege in the Second Boer War. The diamond-mining frontier town of Kimberley was a key British outpost in the Cape Colony of South Africa in the late nineteenth century. Just days after the onset of hostilities, 4,000 Boers under Cornelius J. Weasels invested 600 British regulars and 4,000 local police and militia inside the town on 14–15 October 1899. The British command was uneasily divided between Lieutenant Colonel Robert G. Kekewich and a domineering civilian, Cecil John Rhodes, the founder of De Beers Consolidated Mines and a former prime minister of the Cape Colony, who usurped Kekewich's authority at every turn and generally harassed his efforts to run an efficient military operation. The Boers never attacked Kimberley, but surrounded it, cut communication, and tried to starve the garrison and inhabitants into submission. There was sporadic artillery fire from both sides. On 4 November Weasels issued a written ultimatum to surrender, which Kekewich immediately rejected.

In mid-November British commander in chief Redvers Buller ordered Lord Paul Sanford Methuen to reopen the railroad to Kimberley and rescue the garrison. Boer leaders Jacobus Hercules De La Rey and Jacobus Prinsloo were more than a match for Methuen's unimaginative tactics at Belmont on 23 November, Graspan on 25 November, and the Modder River on 28 November. Methuen's frontal assault withered against small arms fire from Piet Cronjé's skillfully entrenched Boer positions at Magersfontein on 11 December. Thus the first significant British effort to break the siege failed.

In January 1900 the new British commander in chief, Frederick Sleigh Roberts, demoted Methuen and ordered John French to relieve Kimberley. French led an exhausted division of cavalry into the town on 15 February. Only 21 inside the town had died from enemy action, but more than 1,500 had expired from disease. With the siege of Kimberley lifted, Roberts was able to defeat Cronjé at nearby Paardeberg on 27 February.

Eric v. d. Luft

See also: Boer Wars; Buller, Sir Redvers Henry; French, John Denton Pinkstone, First Earl of Ypres; Roberts, Frederick Sleigh, First Earl, Viscount St. Pierre of Kandahar

References and further reading:

Ashe, Evelyn Oliver. *Besieged by the Boers: A Diary of Life and Events in Kimberley during the Siege.* New York: Doubleday, Page & Co., 1900.

Duminy, Kokkie. *Summer of 1899: The Siege of Kimberley, 14 October 1899 to 15 February 1900.* Ed. Steve Lunderstedt. Kimberley: Kimberley Africana Library, 1999.

Gardner, Brian. *The Lion's Cage.* London: Barker, 1969.

Roberts, Brian. *Kimberley: Turbulent City.* Cape Town: D. Philip in Association with the Historical Society of Kimberley and the Northern Cape, 1976.

King Philip's War (1675–1676)

Proportionally in terms of population, both aboriginal and settler, the most devastating war in American history. The conflict basically erupted over differing concepts of land ownership. Indians viewed land as a shared environment; however, the English perspective was based on boundaries and exclusive ownership. The Plymouth settlement had expanded rapidly to almost 50,000 colonists by the mid-1670s. This development combined with religious zeal to threaten the stability of Indian culture, leading to resentment and eventually violence.

In the midst of this tension Philip assumed leadership as sachem of the Wampanoag in 1662. Philip sought to demonstrate leadership through shows of force. However he was coerced into surrendering all weapons and recognizing English sovereignty.

When Christian Indian John Sassamon was found murdered, the colonists accused and executed three Wampanoags. Philip's perceived weakness led stronger and younger warriors to seek direct conflict with the colonists. The Wampanoag began to ambush settlers in outlying areas while using the swamps for protection. They destroyed 12 towns, including Deerfield, Massachusetts, and killed thousands of settlers. Other tribes, such as the Narraganset and Nipmuc, joined with the Wampanoag to attack settlements across New England.

The colonists soon raised armies adapted to wilderness warfare that destroyed Indian crops, captured their families, and offered protection for those who rejected Philip's leadership.

The Narraganset Indians were decisively defeated during the Great Swamp Fight of December 1675, in which colonists overran a defensive stockade and killed more than 600 Indian warriors and their families. The colonists' superior

English troops taking King Philip's fort at South Kingston, Rhode Island, December 1675. (Library of Congress)

numbers and firepower, combined with Indian allies from the Iroquois tribes, spelled doom for Philip's efforts. Philip was killed during an ambush in August 1676 and his head put on display in Plymouth.

The war effectively ended with Philip's death, but not the violence as the colonists sought to exterminate or enslave the Indians. The death of almost 5,000 Indians from war, starvation, and disease destroyed the tribal societies of New England and removed barriers to further white settlement. The English paid a heavy price as well, with half their towns damaged and thousands of colonists killed during the merciless and brutal struggle. Most significantly, King Philip's War reflected a pattern of cultural conflict that would repeat itself over the next 200 years.

Steven J. Rauch

References and further reading:

Ferling, John. *Struggle for a Continent: The Wars of Early America.* Arlington Heights, IL: Harlan Davidson, 1993.

Lepore, Jill. *The Name of War: King Philip's War and the Origins of American Identity.* New York: Alfred A. Knopf, 1998.

Leach, Douglas Edward. *Flintlock and Tomahawk: New England in King Philip's War.* 2d ed. East Orleans, MA: Parnassus Imprints, 1995.

King's Mountain (7 October 1780)

A small but sanguinary conflict between civilians on both sides of the American Revolution. British Major Patrick Ferguson commanded approximately 1,000 loyalist troops moving through the western Carolinas guarding the left flank of Charles Lord Cornwallis's advance. On 7 October 1780, along the South Carolina/North Carolina border, Ferguson and his men occupied heavily wooded King's Mountain, about 60 feet above the surrounding plain. The plateau was about 600 yards long and varied in width from 70 feet to 120 feet; Major Ferguson assumed it was too steep to be scaled. In fact, he felt sufficiently safe to ask Lord Cornwallis for reinforcements to reach out and secure the backcountry.

Patriot forces from the surrounding Appalachian Mountains, aroused by Ferguson's threats against them if they did not submit to the Crown, arrived at the mountain, tied up their horses, and moved through the woods. These militiamen also sought revenge for an earlier battle (and massacre) at Waxhaws Creek.

At this point, the weakness of the loyalist position became clear. The thick woods prevented the kind of advance by column and deployment into line that was the basis of late-eighteenth-century European warfare; meanwhile, to defeat the patriots, the loyalists had to attack them. Loyalist units sought to descend the plateau and come down the hillsides to attack the patriots. Exposed to sharpshooters hiding in the trees and brush, the loyalists suffered many casualties; Ferguson, easily recognizable in a red hunting shirt, was wounded several times and died.

Soon thereafter the British surrendered, and several of the worst offenders in the hanging of captured patriot militiamen were themselves hanged after brief courts-martial. Fearing that Cornwallis would seek revenge, most of the patriot militia went home, never really to appear again in the conflict, but they did set the stage for the ensuing victory at the Cowpens, the skillful retreat to the Dan River, the key battle at Guilford Court House, and Cornwallis's decision to abandon the Carolinas and retreat to Virginia—and Yorktown.

Charles M. Dobbs

See also: American Revolution; Greene, Nathanael

References and further reading:

Lumpkin, Henry. *From Savannah to Yorktown: The American Revolution in the South.* New York: Paragon Publishers, 1987.

Messick, Hank. *King's Mountain.* Boston: Little, Brown, 1976.

Weigley, Russell. *The Partisan War: The South Carolina Campaign of 1780–1782.* Columbia: University of South Carolina Press, 1970.

Kinsale, Siege of (1601)

The climactic action of the Nine Years' War in Ireland. Hugh O'Neill, Earl of Tyrone, who had been in rebellion since 1595, had long awaited Spanish aid, but the arrival of Don Juan Aguila's 3,500 men at Kinsale on 22 September 1601 presented problems for both sides. Kinsale was in southwestern Ireland, while O'Neill's main forces were in Ulster, so joining the Spanish required a march across the island. Meanwhile, Charles Blount, Lord Mountjoy, the English lord deputy, faced the logistical problem of moving his own forces, including his siege train, from Dublin. Mountjoy successfully concentrated 7,000 men outside Kinsale, and English control of the sea allowed delivery of heavy guns and supplies needed to initiate a siege. However, Mountjoy's troops were poorly supplied, and as the siege continued through November, sickness began to reduce his numbers.

O'Neill also experienced difficulty moving his army in that season. His advanced guard, commanded by Hugh O'Donnell, arrived near Kinsale in November, but O'Neill did not arrive until December. With 6,000 men, his plan was to blockade Mountjoy's besieging forces until disease and hunger forced their withdrawal. O'Donnell and Spanish representatives, however, insisted on a relief attempt. This operation was mounted on 24 December. The English had the advantage in the ensuing set piece battle, and their cavalry routed O'Neill's advanced guard as it passed through a bog. Its retreat threw the army into disorder, and a charge by Mountjoy's reserves drove the Irish from the field with more than 1,200 casualties. During the action Aguila made no move to break out of Kinsale, and O'Neill's defeat left him with no hope of relief. The Spanish surrendered on 2 January 1602.

John S. Nolan

See also: Mountjoy, Charles Blount, Lord; Nine Years' War

References and further reading:

Falls, Cyril. *Elizabeth's Irish Wars.* Syracuse, NY: Syracuse University Press, 1997.

Silke, John J. *Kinsale: The Spanish Intervention in Ireland.* Liverpool: Liverpool University Press, 1970.

Wernham, R. B. *The Return of the Armadas: The Last Years of the Elizabethan War against Spain.* New York, Oxford: Clarendon, 1994.

Kitchener, Horatio Herbert (1850–1916)

British field commander, staff officer, and politician. The son of a career army officer, Kitchener was born in County Kerry, Ireland, on 24 June 1850. After graduating from the Royal Military Academy in Woolwich, he was commissioned in the Royal Engineers in January 1871 and saw action as a volunteer on the French side in the Franco-Prussian War. After extensive service as a British military surveyor and intelligence officer around the eastern Mediterranean, he was assigned to cavalry in Cairo, Egypt, in 1882, and was Sir Garnet Wolseley's intelligence officer from October 1884 to March 1885 during the futile expedition to rescue "Chinese" Gordon.

As governor of British Red Sea Territories after 1886, Kitchener faced significant Mahdist resistance. He was wounded and defeated by Mahdist leader Osman Digna at Suakin on 17 January 1888. While still a colonel in the British army, he became sirdar, or commander in chief, of the Egyptian army in 1892. After training this army in modern methods for four years, he invaded the Sudan to avenge Gordon, winning at Dongola on 21 September 1896, Abu Hamed on 7 August 1897, Atbara River on 7 April 1898, and Omdurman on 2 September. He drove Jean-Baptiste Marchand's French army from Fashoda, Sudan, on 18 September 1898.

In the Second Boer War, Kitchener was assigned to Frederick Sleigh Roberts as chief of staff on 18 December 1899. His tactics defeated Piet Cronjé at Paardeberg on 18–27 February 1900, frustrated Boer guerrillas, and resulted in the capture of Bloemfontein on 13 March, Johannesburg on 31 May, and Pretoria on 5 June. He replaced Roberts as com-

Horatio Herbert Kitchener. (Library of Congress)

mander in chief of British forces in South Africa on 29 November. Continuing his offensive against guerrillas, he invented the concentration camp, in which he imprisoned guerrillas, their families, and their supporters.

Appointed secretary of war in July 1914, Kitchener, almost uniquely, foresaw a long war against Germany and urged massive British mobilization. But he was among those lost when HMS *Hampshire* struck a German mine near the Orkneys and sank on 5 June 1916.

Eric v. d. Luft

See also: Boer Wars; Botha, Louis; Gordon, Charles George; Khartoum, Siege of; Omdurman; Roberts, Frederick Sleigh, First Earl, Viscount St. Pierre, of Kandahar; Wolseley, Garnet Joseph, Viscount; World War I

References and further reading:

Cassar, George H. *Kitchener: Architect of Victory.* London: Kimber, 1977.

Royle, Trevor. *The Kitchener Enigma.* London: Joseph, 1985.

Smithers, A. J. *The Fighting Nation: Lord Kitchener and His Armies.* London: Cooper, 1994.

Warner, Philip. *Kitchener: The Man Behind the Legend.* New York: Atheneum, 1986.

Kléber, Jean-Baptiste (1753–1800)

One of the most capable field commanders of the French Revolutionary and early Napoleonic wars. Kléber was born in Strasbourg on 9 March 1753. He served in the imperial army (1777–1785). Influenced by French revolutionary ideas, he joined the national guard at Belfort and rose to lieutenant colonel in a volunteer battalion. His aggressive defense of Mainz earned him a promotion to brigadier general (17 August 1793). He fought the Vendean rebels, and his achievements, especially at Cholet, earned him a battlefield promotion to division general (17 October). He captured Le Mans in December.

As a division commander Kléber played a crucial role at Charleroi (25 June 1794). At Fleurus (26 June) his left wing drove the Austrians into retreat. He captured Maestricht (4 November) and participated in the invasion of the Rhineland. But in spite of a series of victories Kléber, perhaps feeling inadequately rewarded, resigned all active field command on 21 December 1796.

Kléber by now had earned a reputation for great concern for his men's welfare. He was known for great tactical flexibility. He employed both line and column formations but preferred columns for attack.

Kléber returned to active service in 1798. Seriously wounded leading an attack at Alexandria, Egypt (21 July 1798), he also fought in Syria, distinguishing himself in the capture of El Arish (20 February 1799) and at Gaza, Jaffa, and Acre. At Mount Tabor (16 April) his troops held off the main Turkish army until relief arrived. He commanded the rear guard on the retreat into Egypt. Kléber, who had opposed the Egyptian campaign, now advocated a French evacuation. Ironically Napoleon appointed him to command the French forces left in Egypt.

On 24 January 1800 Kléber signed the Armistice of El Arish, providing for a French evacuation. The British disavowed it. With growing enemy opposition, Kléber audaciously resumed hostilities. He defeated the Turks at Heliopolis (20 March) and reconquered Cairo (25 April) and lower Egypt. On 14 June 1800 a Muslim nationalist assassinated Kléber in Cairo. His remains were returned to France in 1801.

James K. Kieswetter

See also: Aboukir; Fleurus, Battle of; French Revolutionary Wars; Napoleon I; Pyramids

References and further reading:

Chandler, David G. *The Campaigns of Napoleon.* New York: Macmillan, 1966.

Lynn, John A. *The Bayonets of the Republic Motivation and Tactics in the Army of Revolutionary France, 1791–1794.* Chicago: University of Illinois Press, 1984.

Herold, Jean Christopher. *Bonaparte in Egypt.* New York: Harper & Row, 1962.

Lucas-Dubreton, Jean. *Kléber 1753–1800.* Paris: Paul Hartmann, 1937.

Knox, Henry (1750–1806)

Continental army officer and secretary of war. At the beginning of the American Revolution, Knox constructed American defenses near Boston. Impressed by Knox's work and his knowledge of artillery, General George Washington commissioned him colonel of the Continental Artillery Regiment. Knox remained a trusted adviser to Washington throughout the war.

During the winter of 1775–1776, Knox directed the transfer of British ordnance captured at Fort Ticonderoga to Boston, which compelled the British to abandon the city. He participated in almost every major battle of the northern campaigns and Yorktown, rising in rank to major general. In 1783, Knox founded the Society of Cincinnati.

As secretary of war (1785–1794), he battled persistent distrust of standing armies yet organized Anthony Wayne's successful Indian campaign, reestablished the U.S. Navy, and secured congressional support for the Uniform Militia Act of 1792. In 1798, Knox was appointed major general in the Provisional Army in anticipation of war with France but refused to serve under Alexander Hamilton, his subordinate during the American Revolution.

Dean Fafoutis

See also: American Revolution; Brandywine; Fallen Timbers; Germantown; Monmouth; Princeton, Battle of; Trenton; Wayne, Anthony; Washington, George; Yorktown

References and further reading:

Callahan, North. *Henry Knox: General Washington's General.* New York: Rinehart, 1958.

Ward, Harry M. *The Department of War, 1781–1795.* Pittsburgh, PA: University of Pittsburgh Press, 1962.

Koguryo (attributed 37 B.C.E.–668 C.E.)

Northernmost of three kingdoms dominating the Korean peninsula. Although the oldest extant in Korean history, the *Samguk sagi* (*History of the Three Kingdoms,* 1145), dates the foundation of Koguryo to 37 B.C.E., there is no evidence for its existence prior to the first century C.E. At that time it emerged in the mountainous region now separating North Korea from China, expanding at the expense of lowland groups.

Following the fall of the Han (206 B.C.E.–220 C.E.), the various small states of northeast China entered a period of intensive competition. Koguryo's power was briefly eclipsed by that of the state of Wei (225–265), which overran its capital in 244, but it soon recovered. In the late fourth century, primarily under the inspired leadership of young king Kwanggaet'o (r. 391–413), a series of successful military campaigns brought Koguryo hegemony over a good part of Manchuria and northern Korea. Early in the fifth century, Koguryo moved its capital south to modern Pyongyang in North Korea, leaving it better positioned for competition with other Korean states.

The next 250 years were ones of growing competition between Koguryo and the other two major states on the peninsula, Paekche and Silla. In 668, Koguryo was conquered by the southern state of Silla, then in alliance with Tang (618–906) China. Although Silla was able to incorporate Koguryo territories as far as the Yalu River (separating modern North Korea from China), Koguryo possessions beyond permanently passed from Korean control with Koguryo's defeat.

Because of its positioning, Koguryo faced frequent and formidable military threats from nomadic tribes and other groups living in Manchuria, and from more centralized Chinese dynasties, Sui (587–618) and Tang in particular. For this reason it gained a reputation for martial spirit and strength of arms. Koguryo is often lauded by modern Koreans for its heroic defense of the peninsula from would-be foreign conquerors.

Daniel Kane

See also: Paekche; Silla Kingdom; Sino-Korean Wars and the Wars of Korean Unification

References and further reading:

Gardiner, K. H. J. *The Early History of Korea: The Historical Development of the Peninsula Up to the Introduction of Buddhism in the Fourth Century A.D.* Honolulu: University of Hawaii Press, 1969.

Iryon. *Samguk Yusa: Legends and History of the Three Kingdoms of Ancient Korea.* Trans. Tae-Hung Ha and Grafton K. Mintz. Seoul: Yonsei University Press, 1972.

Lee, Ki-baik. *A New History of Korea.* Trans. Edward W. Wagner with Edward J. Shultz. Cambridge, MA: Harvard University Press, 1984.

Kokoda Trail (1942)

The first substantial land defeat of Japanese forces in World War II. In early 1942 the Japanese were consolidating the southern boundaries of their Southern Economic Zone along a line through Timor, western New Guinea, Rabaul, and Micronesia, when Australian and U.S. air forces commenced an aggressive bombing campaign from Australia and Port Moresby in New Guinea. In response, the Japanese commenced operations to capture the whole of New Guinea. The first attempt was turned back during the Battle of the Coral Sea in May 1942.

On 22 July 1942 the Japanese landed a 2,000-man advance unit of General Horii's South Seas Force at Buna on the north coast of New Guinea with orders to commence an overland assault on Port Moresby. Their route was to be a walking track leading over 10,000-foot-high passes in the Owen Stanley Mountains—the Kokoda Trail.

A component of the Australian 39th Militia Battalion made an orderly fighting withdrawal without slowing the Japanese. By 21 August, 13,500 Japanese troops had been committed and had captured the town of Kokoda near the summit of the Owen Stanley Mountains. The Australian 53d battalion was ordered northwards up the Kokoda Trail to reinforce the 39th and was followed by the 21st brigade, which reached Isuvura at the southern side of the summit on 23 August.

Conditions were horrific. All supplies, including dismantled artillery and mortars, had to be carried by hand. The muddy, narrow trail crossed a series of very steep valleys, up to 3,000 feet deep, as it climbed either side of the mountains. The Japanese were now two full brigades in strength and continued to push the Australians, who had been reinforced by elements of the 25th brigade, back in fierce and continuous fighting. By 16 September they reached Imita Ridge, only 26 miles from Port Moresby. However, because of the situation at Guadalcanal, Japanese Imperial Army Headquarters ordered a withdrawal on 20 September, and with all supplies exhausted, the retreat commenced. It rapidly degenerated into a rout for the Japanese, and starvation, total breakdown in discipline, killing of sick and wounded, and cannibalism were evident. However, savage firefights did still occur frequently. Of 14,500 Japanese troops committed, only 5,000 survived to reach Buna by mid-November 1942.

Michael Hyde

See also: Guadalcanal; Milne Bay; World War II

References and further reading:

Japanese Monograph No. 37. "South-east Asian Operations Record: 18th Army Operations on New Guinea and Rabaul (January 1942–June 1943), The Southern Area (Part II)." In *War in Asia and the Pacific,* Vol. 7, ed. D. Detwiler. New York: Garland Publishing Co., 1980.

McCarthy, D. *Australia in the War of 1939–1945.* Series One, *Army.* Vol. 5, *South West Pacific Area—First Year Kokoda to Wau.* Canberra: Australian War Memorial, 1959.

Odgers, George. *Army Australia—An Illustrated History.* New South Wales: Child and Associates, 1988.

"Aleksandr Vasilyevich Kolchak, supreme ruler of all Russian government." Poster, 1918. (Library of Congress)

Kolchak, Aleksandr Vasil'evich (1874–1920)

Russian admiral, polar explorer, White Civil War leader. Born in St. Petersburg, son of a naval officer, Kolchak graduated from St. Petersburg Naval Academy (1894) and served in the Pacific and Indian Oceans, becoming a naval lieutenant in 1899. A specialist in oceanography and hydrology, he completed two Arctic exploratory trips (1900–1904) and was prestigiously awarded for his published results (1909).

With the outbreak of the Russo-Japanese War (1904–1905) he rejoined the navy, commanding a destroyer, and was decorated for mine-laying work and sinking an enemy cruiser. Forced ashore by ill health, he commanded a naval battery during the siege of Port Arthur, was captured, but repatriated in 1906.

A founder of the Naval General Staff, he served in this, becoming chief of the organization section for the Baltic Fleet, pressing for naval modernization and reforms. Between postings, he organized and planned another Arctic expedition to chart the northern Siberian coast.

During World War I he enjoyed a distinguished active service: Baltic Fleet (1914–1916) and commander of the Black Sea Fleet (July 1916–July 1917). Becoming vice admiral in August 1916, he won further decorations for mine-laying and naval defense work.

After the February Revolution, he supported the provisional government but resigned in June 1917 with his command disintegrating. Traveling to America via London, he made links with the American and British navies, joining the latter in December 1917 after the Bolshevik Revolution. In Japan December 1917–April 1918 and July–September 1918, he also served General Horvath in Harbin. Coveted as an anti-Bolshevik figurehead after his impressive military career and foreign connections, he was summoned back to Russia and installed as supreme ruler of Russia after the Omsk coup in November 1918.

During the Russian Civil War, Kolchak established a military dictatorship, but his initial support deteriorated amidst political corruption, repression, and misuse of Allied aid to create vast discontented partisan networks in his rear, which came to favor the Red Army. Kolchak launched an initially successful offensive, taking Perm (December 1918), but was pushed back by Red eastern front counteroffensives (April–December 1919). His capital, Omsk, fell on 14 November; he resigned on 4 January 1920 as supreme ruler of Russia, naming Denikin as his successor.

Fleeing eastwards, he was captured by Czechs and handed over to social revolutionaries in Irkutsk. Tried by a hastily assembled Communist revolutionary tribunal, he was executed on 7 February.

Neil Harvey Croll

See also: Denikin, Anton Ivanovich; Russian Civil War (1918–1921); Tukhachevsky, Mikhail Nikolayevich; World War I

References and further reading:

Collins, D., and J. Smele. *Kolckak i Sibir': Issledovaniia* (Kolchak and Siberia: Documents and Studies), *1919–1926.* Vols. 1 and 2. New York: Kraus International Publishers, 1988.

Krasnov, V. *Kolchak i zhizn', i smert' za Rossiiu, knigi I & II.* Moscow: Olma-Press, 2000.

Smele, J. *Civil War in Siberia: The Anti-Bolshevik Government of Admiral Kolchak, 1918–1920.* Cambridge, UK: Cambridge University Press, 1996.

Varneck, E., and H. H. Fisher, eds. *The Testimony of Kolchak and Other Siberian Materials.* Stanford, CA: Stanford University Press, 1935.

Konev, Ivan Stepanovich (1897–1973)

Soviet military commander marshal of the Soviet Union (1944). Of peasant origins, from northern Dvina Province,

Konev joined the Russian army (1916), saw little action, but emerged an NCO. A Bolshevik supporter, he returned home after the October Revolution, becoming Nikol'sk District military commissar. He ended the Russian Civil War as an armored-train political commissar, serving in the east against Kolchak, Semenov, and Japanese forces until 1922, and in suppression of the Kronstadt Revolt (March 1921).

He graduated from Frunze Military Academy staff training courses (1926, 1934) and somehow survived Stalin's purges to gain rapid promotion. He commanded 57th Special Corps in Mongolia (1937) and Second Separate Red Banner Army in the Far Eastern, Transbaikal, and Northern Caucasus Military Districts.

After mixed fortunes, Konev emerged as one of the most original, capable Soviet World War II commanders. As Nineteenth Army commander, he counterattacked at Smolensk to delay the Germans, but as Western Front commander in September 1941, he was partly responsible for the Viazma-Briiansk encirclement, wherein 500,000 Russians were taken prisoner. Escaping trial for this, he was given the Kalinin Front, counterattacking with Zhukov in December to halt Operation BARBAROSSA before Moscow.

He commanded the Western Front from August 1942, Northwestern Front from March 1943, the Steppe (later 2d Ukrainian) Front from June 1943, and the 1st Ukrainian Front (May 1944–May 1945), playing a leading role in Soviet Operations at Kursk, Korsun'-Cherkassy, Vistula-Oder, Berlin, and Prague.

Postwar, Konev served as Soviet commander in Austria (1945–1946), chief inspector of the Soviet Army (1950–1951), commander in chief of Soviet Land Forces and deputy minister for war (1946–1950, 1955–1956), and Carpathian Military District commander (1951–1955). Benefiting under Khrushchev at Zhukov's expense, Konev became first deputy minister for defense and commander in chief of Warsaw Pact forces (1956–1960). He commanded Soviet forces in suppressing the 1956 Hungarian uprising and in Germany during the construction of the Berlin Wall, all actions which Stalin would have approved.

Konev remained an adviser in retirement, wrote his memoirs, and died of cancer in Moscow.

Neil Harvey Croll

See also: Hungarian Revolt; Kursk, Battle of; Moscow; Stalin; World War II; Zhukov, Georgy Konstantinovich

References and further reading:

Bystrov, V. E. *Sovetskie polkovodtsy I voenachal'niki. Sbornik* (Soviet Leaders and Military Chiefs. Collection). Moscow: Molodaia gvardiia, 1988.

Kolesnikov, A. A. *Marshaly Rossii* (Marshals of Russia). Iaroslavl': Izdat. Niuans, 1999.

Konev, I. S. *Zapiski komanduiushchego frontom 1943–45* (Notes of a Front Commander, 1943–45). Moscow: Voenizdat, 1982 and 1991.

Shukman, H., ed. *Stalin's Generals.* London: Weidenfeld & Nicolson, 1993.

Kongo, Kingdom of the (14th–17th Centuries)

An ally of Portugal, adopting European military technology, but eventually undone by internal conflict, marauding neighbors, and the intrigues of foreign slave traders. The Kingdom of the Kongo arose in the 1300s along the mouth of the Congo River in western central Africa. The ruler, the Mani-Kongo, functioned as an absolute monarch, supervisor of an intricate administrative organization, and a semidivine religious authority in an intricate system that extended from the capital through all cities, provinces, and villages. An agrarian people, the Bakongo people were also active in mining, smelting, and in woven raffia products. Advised by a council of landed nobles, the king used trade, tribute, and occasional demonstrations of force to ensure peace with neighboring peoples.

In 1483, a Portuguese vessel arrived at the mouth of the Congo River and began a long and complex relationship with Kongo. By 1506, the Mani-Kongo, many nobles, and ordinary people had converted to Catholicism. King Afonso I (1509–1543) used the church, Portuguese mercenaries and advisers, and trade revenues to centralize his power. He mixed traditional Kongolese light infantry, shield-bearing "heavy" infantry, and Portuguese mercenaries armed with pikes and arquebus rifles. He campaigned against his neighbors, the kingdoms of the Angola, Banguela, and Nziki as well as more rustic peoples like the Mbundu, the Teke, and the Jaga. To support these efforts, Afonso engaged in slave trading with Portuguese merchants, operating out of São Tomé and other offshore islands. The demand for slaves generated a pervasive, pernicious, and hugely profitable commerce that neither the Portuguese authorities nor Afonso could control. Slavers relentlessly thwarted all Afonso's work by allying with other kingdoms, funding slaver gangs and rebellious nobles, and draining resources from the monarchy.

Future kings had to balance their dependency on Portugal for arms, assistance, and trade with the intrigues of Portuguese merchants, slave dealers, and mercenaries to promote rivalries, wars, and instability throughout the region. In 1569, for example, King Alvaro confronted a massive invasion by the Jaga. The Jaga sacked the capital at São Salvador, nearly drove Alvaro to the coast, and sent thousands of captives to the eager slavers at São Tomé. Lisbon came to Alvaro's rescue in 1571 and drove the Jaga out. Shortly thereafter, Portugal established another great enslavement center south of Kongo at Luanda in Angola territory. Thus, relations between Kongo and Portugal would remain active, but the slave trade also ensured that these relations would be brittle and unstable.

In the 1640s, Dutch efforts to seize Angola encouraged the Kongo kings to reduce Portugal's influence. In 1665, a boundary dispute led to war between the Kongo and Portuguese forces protecting Angola. At the Battle of Mbwila, the Kongo army shattered on the Portuguese formations and King Antonio died of a bullet wound. The Kongo fragmented into warring components.

In 1670, another Portuguese force from Luanda tried to place a candidate on Kongo's throne, only to be annihilated at the Battle of Colombo by troops from Nsoyo Province. São Salvador was destroyed and abandoned in 1678. The Kongo knew no peace until 1709, when Pedro IV finally emerged as the victor.

Pedro's Kongo was not the Kongo envisioned by Afonso 200 years before. Decentralized, unstable, plagued by rebel nobles and outside incursions, the proliferation of firearms in the eighteenth century only sharpened the turbulence. The Kongo had, in effect, imploded.

Weston F. Cook, Jr.

References and further reading:

Thornton, John K. *The Kingdom of Kongo.* Madison: University of Wisconsin, 1983.

———. *Warfare in Atlantic Africa, 1500–1800.* London: University College Press, 1999.

Königgrätz, Battle of (1866)

Battle that secured Prussian dominance of central Europe for Otto von Bismarck. The battle began in the morning of 3 July 1866 with a Prussian enveloping attempt, but the maneuver failed when two Prussian armies failed to connect and a third failed to advance because of ground fog. Prussia was forced to engage all its reserves, while Austria did not even have to commit its cavalry. When the Prussian Second Army finally arrived on the Austro-Saxon flank in the late afternoon, they forced an Austrian retreat. Some 250,000 Prussians had defeated a comparable number of Austrians and Saxons, ending the Seven Weeks' War. The Austrians had lost 45,000 casualties and 20,000 captured, while Prussian forces sustained only 10,000 casualties. Vienna lay open to Prussian attack, forcing Austria to sue for peace.

Prussian soldiers had several important advantages in the battle. First, their breech-loading needle-guns (so-called because of their long firing pins) were far superior rifles to the Austrian muzzle-loaders. The Austrian rifle had better range and accuracy, but men lying down and presenting a smaller target could use the Prussian gun.

Second, the superior organization and planning of the Prussian army, led by the general staff under Helmuth Karl von Moltke, quickly recovered from the mistakes of the morning and saved the day.

Finally, the Prussians moved large numbers of forces rapidly using the extensive Prussian railroad system. In contrast, Austrian movements were clumsy and slow. The Prussians saved their troops for battle rather than losing them to fatigue on march and earned an element of surprise.

The battle, which took place near the Czech town of Sadowa, was the largest European battle until 1914.

David C. Arnold

See also: Bismarck, Otto von; Custozza, Second Battle of

References and further reading:

Blackbourn, David. *The Long Nineteenth Century: A History of Germany, 1780–1918.* Oxford, UK: Oxford University Press, 1998.

Goerlitz, Walter. *History of the German General Staff, 1657–1945.* Trans. Briand Battershaw. New York: Praeger, 1962.

Showalter, Dennis E. *Railroads and Rifles: Soldiers, Technology and the Unification of Germany.* Hamden, CT: Archon Books, 1975.

Korean War (1950–1953)

Korea had been a part of Japan's Inner Empire until Japan surrendered to end World War II in the Pacific. At the Yalta Conference, Soviet leader Joseph Stalin promised Soviet intervention in the war against Japan soon after the war in Europe ended. Meanwhile, on the night of 10–11 August 1945, two American army colonels, Dean Rusk and Charles Bonesteel, sat down with a map to determine an American zone of occupation to accept the Japanese surrender in Korea; they chose a temporary demarcation line at the thirty-eighth parallel. Thus, the most homogeneous nation on earth was divided as an administrative convenience.

By 1950, the situation had worsened for the U.S. government. The Soviets had isolated their sector and had set about almost immediately to organize life in the north along People's Republic lines. Meanwhile, the southern zone needed continuing American economic assistance, which was difficult in those days of limited foreign aid budgets and many more deserving or perhaps more important countries.

In a notorious speech in January 1950, Secretary of State Dean Acheson outlined the U.S. defense perimeter in the western Pacific and Asia—and explicitly excluded southern Korea. Perhaps the North Korean invasion was inevitable; perhaps the government in Pyongyang assumed American noninterference given Acheson's remarks and America's

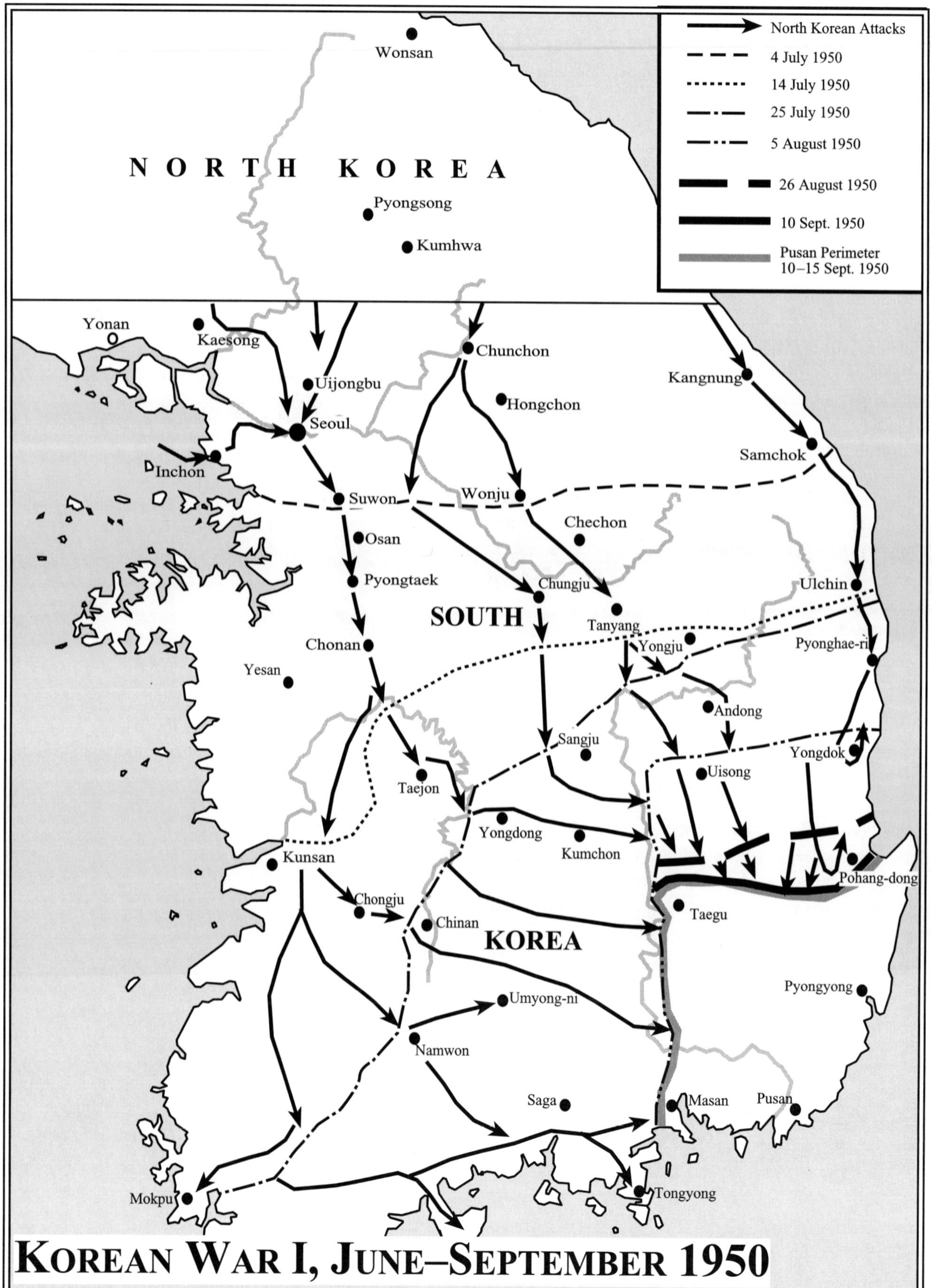

Korean War I, June–September 1950

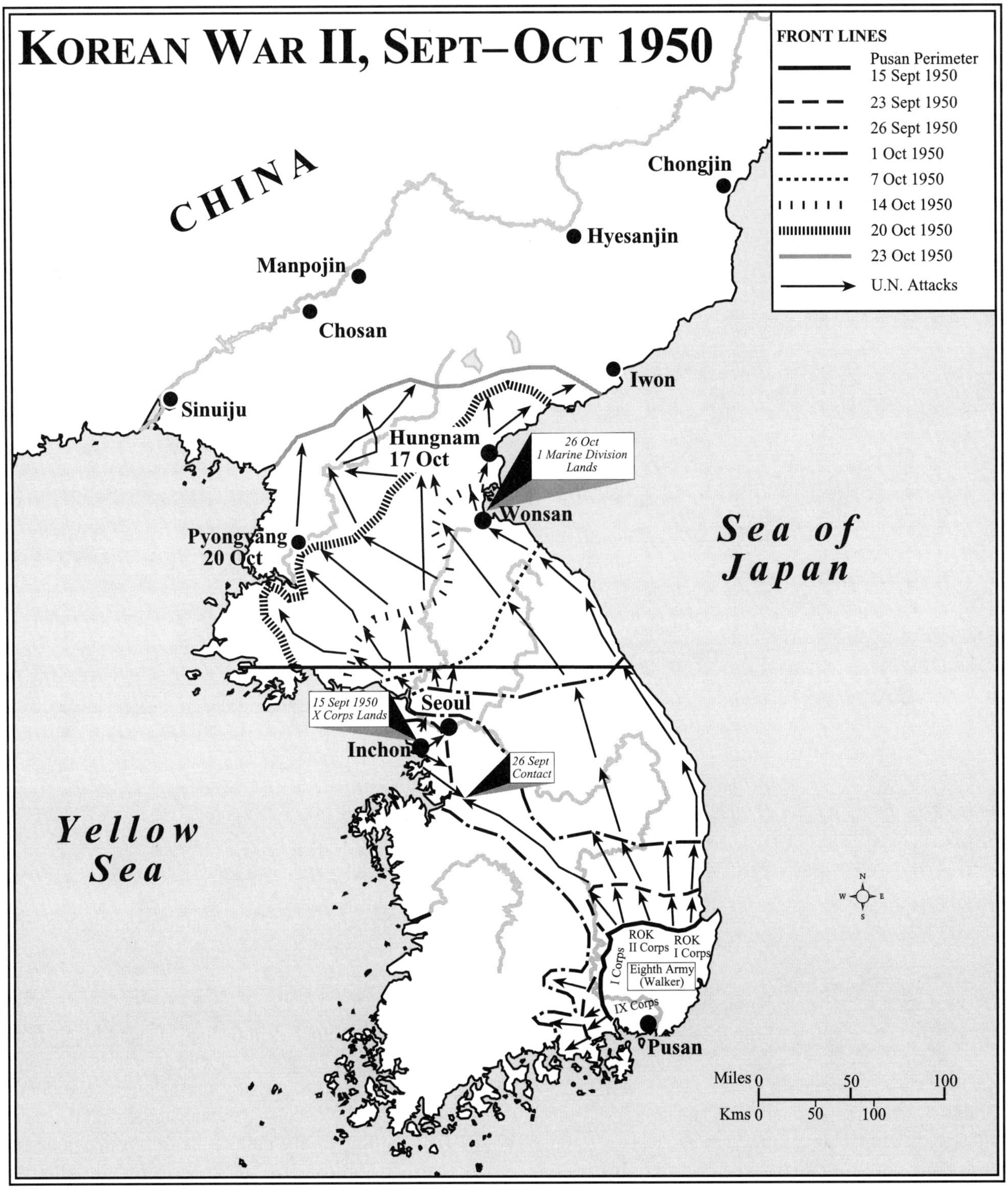
KOREAN WAR II, SEPT–OCT 1950
FRONT LINES
Pusan Perimeter 15 Sept 1950
23 Sept 1950
26 Sept 1950
1 Oct 1950
7 Oct 1950
14 Oct 1950
20 Oct 1950
23 Oct 1950
U.N. Attacks
CHINA
Chongjin
Hyesanjin
Manpojin
Chosan
Iwon
Sinuiju
Hungnam 17 Oct
26 Oct 1 Marine Division Lands
Wonsan
Pyongyang 20 Oct
Sea of Japan
15 Sept 1950 X Corps Lands
Seoul
Inchon
26 Sept Contact
Yellow Sea
ROK II Corps
ROK I Corps
I Corps
Eighth Army (Walker)
IX Corps
Pusan
Miles 0 50 100
Kms 0 50 100

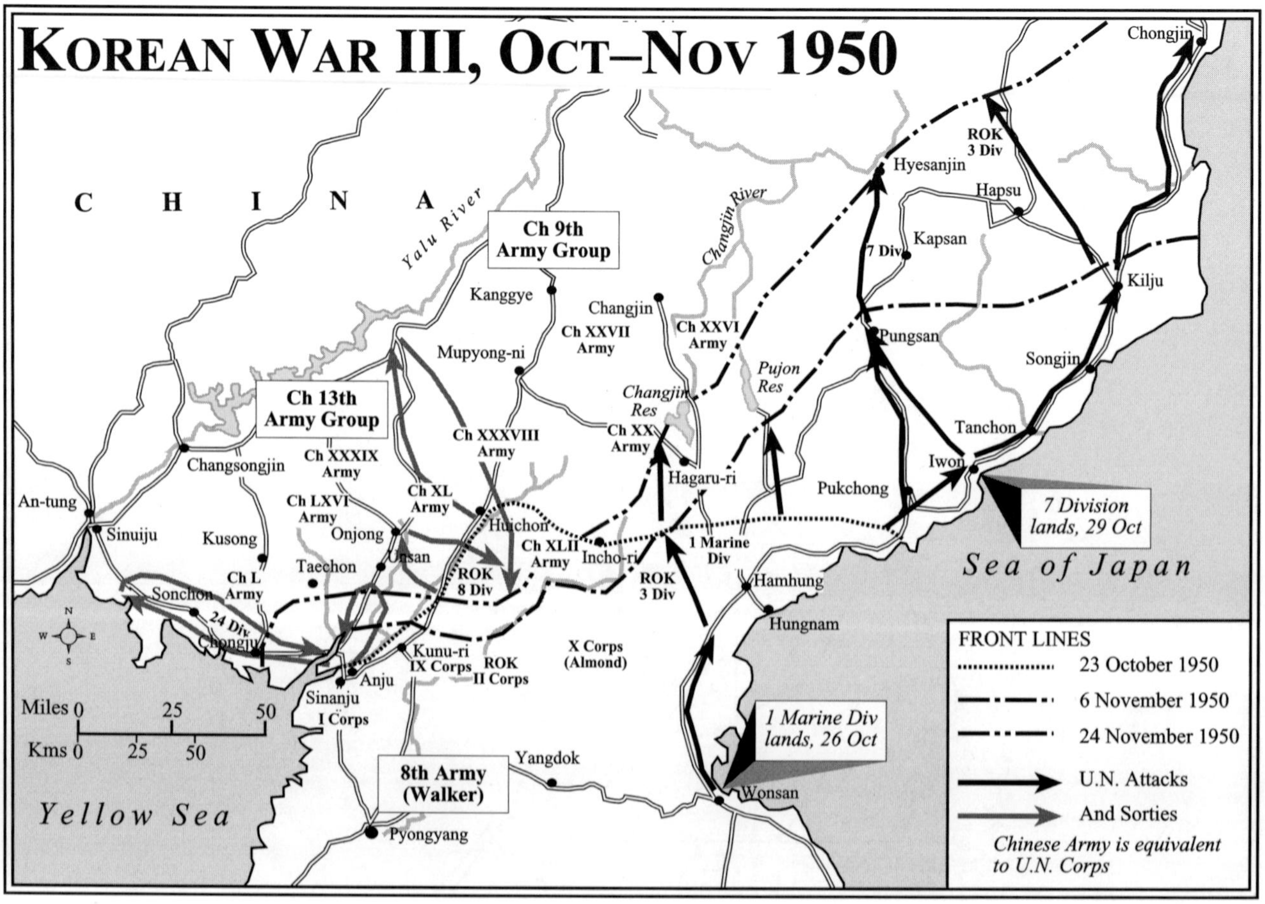

nonreaction to the "loss" of China. Regardless, on 25 June 1950, North Korea launched a major invasion of the south.

The Korean War can be seen in five stages. Initially, North Korea's Korean People's Army (KPA) unleashed a surprise attack against the lightly armed South on 25 June 1950. The South Korean army was really a constabulary with mostly antiguerrilla warfare, few heavy weapons, and practically no air power. Some of its higher-ranking officers did have military experience—fighting in the Japanese army in World War II. Many North Korean soldiers had experience fighting with the Chinese Communists in World War II or in the long Chinese civil war; more tellingly, they had Soviet T-34 tanks, artillery, and antitank weapons. They easily broke through the defenses at the thirty-eighth parallel and, after pausing for several days to cross the Han River before Seoul, they poured southwards. President Harry Truman committed U.S. forces to resist this aggression, and available infantry units sought to delay the Communist drive down the west side of the peninsula. But the Americans, pulled from their easy Japanese occupation duties, were too few in number, had too little artillery and tanks, and were kept off-balance by the North Koreans, who were able to move faster and punch harder. By early August the Communists had pinned the South Koreans and their American allies to a small perimeter around Pusan in southeast Korea. But they had a long supply line and the U.S. was committing more ground, sea, and air power to hold the perimeter and to interdict the flow of supplies to the aggressors.

Then came the American counterattack—the second stage. On 15 September 1950 General Douglas MacArthur mounted a surprise amphibious assault deep behind enemy lines at Inchon on the west coast near the capital, Seoul, and UN forces in the Pusan perimeter soon linked with the invasion force, cutting off the bulk of the North Korean army. U.S. Navy and Marine Corps commanders had great doubts about the wisdom of invading Inchon, but events proved that MacArthur was correct to follow the line of least expectation. Within a few weeks, the North Korean army in the south ceased to exist as an organized fighting unit, although many North Korean soldiers escaped north of the thirty-eighth parallel but without their equipment and armaments. After considering the alternatives, including a halt to the

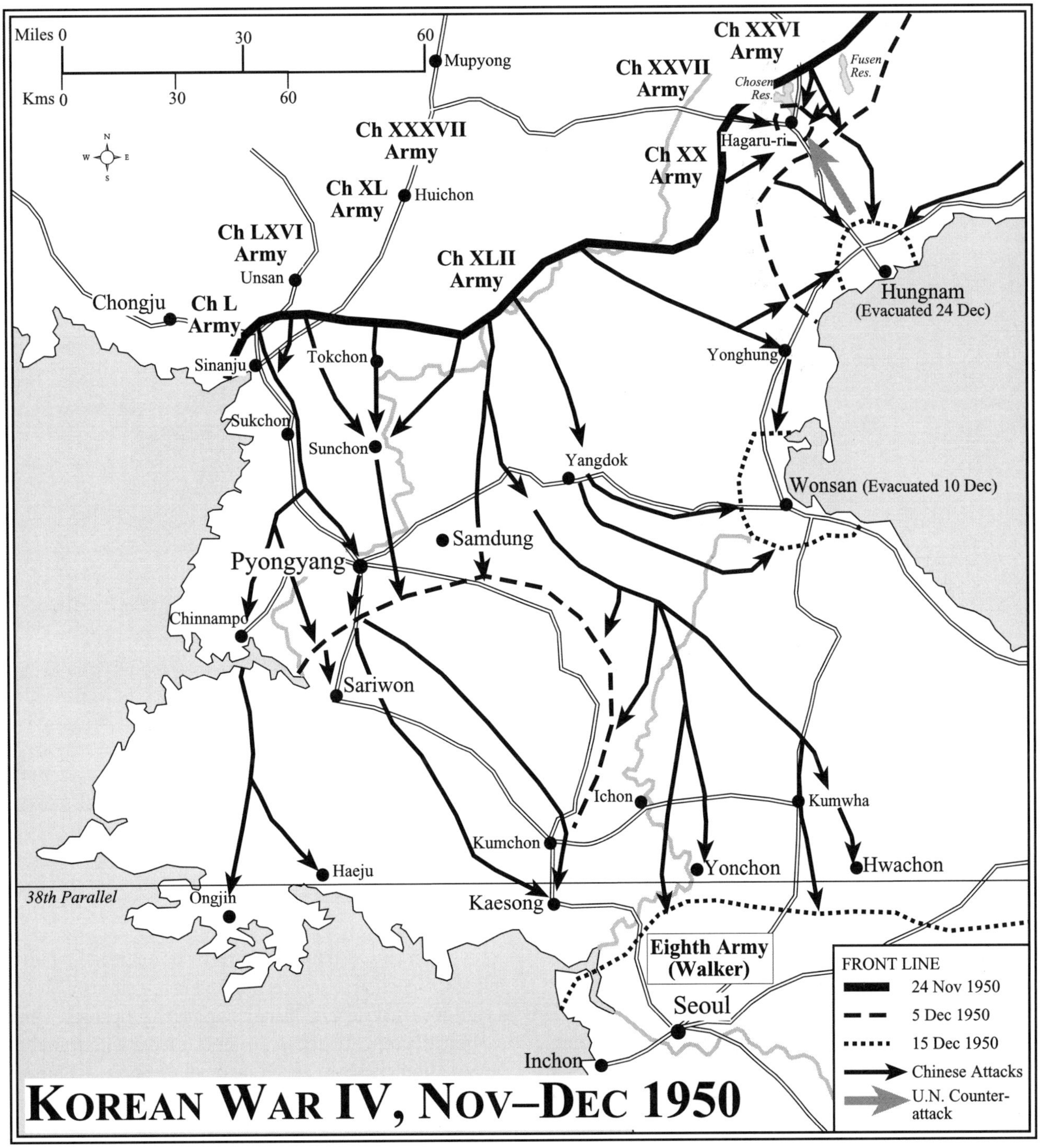

fighting at the thirty-eighth parallel, President Truman gave permission for MacArthur to cross the thirty-eighth parallel and to proceed north to "free" the entire peninsula from Communist control unless UN forces ran into Chinese or Soviet troops operating in Korea. Only South Korean units were to approach the sensitive borders with China and the Soviet Union. But MacArthur in Tokyo and many American officials in Washington, D.C., discounted warnings from the Chinese Communist leaders in Beijing, relayed through the Indian government, that the Chinese would intervene unless the drive northwards was halted; similarly, U.S. intelligence discounted evidence of hundreds of thousands of Chinese soldiers moving north to Manchuria from the provinces opposite Taiwan and in south China. (Recent evidence suggests

U.S. Marines move forward after effective close-air support flushed out the enemy from their hillside entrenchments on December 26, 1950. (National Archives)

that the Chinese Communist leadership seriously considered intervention in the Korean War from as early as late July 1950.)

Once again, there was a surprise attack—the third stage. In late October, Chinese Communist "volunteers" struck at Republic of Korea (ROK) units that had neared the Yalu River. They bloodied the South Koreans and then seemingly disappeared. MacArthur stubbornly continued to discount the possibility of a major Chinese Communist intervention. Then, in November, the Chinese Communists, who had infiltrated more than 300,000 men, struck hard and suddenly, and broke the United Nations Command (UNC) drive to the Yalu River. As MacArthur noted, "It was an entirely different war." Within a few weeks, American and ROK troops were retreating south of the thirty-eighth parallel and abandoning Seoul for the second time. The Chinese recognized their limitations in terms of firepower and logistics, and sought to set up positions behind the Americans and South Koreans, and cut off their expected lines of retreat; this tactic was, at least initially, incredibly disconcerting and hence effective. It was so effective that General MacArthur may have panicked and suggested a retreat back to Pusan or even a withdrawal to Japan. Although there would be media reports of Chinese hordes, the truth was that both sides had approximately equal numbers of troops, and UN forces clearly had more artillery, logistical support, and air power than the Chinese and North Koreans. The Communists were more willing to fight close up and were clearly less reliant on motorized transport and tank and artillery support, and it would take the American infantry time to relearn how to fight hard.

After General Walton Walker was killed in a jeep accident, General Mathew Ridgway was appointed to command the U.S. Eighth Army and later to overall command; he managed to stabilize the lines near the original border marking northern and southern Korea. He understood the limitations of the enemy and designed his strategy to take advantage of their weaknesses in logistics. He realized that American morale was low, and troops had become too accustomed to motorized transport, artillery support, and air power. He raised their fighting spirit, prepared them better for the changed nature of the conflict, and took advantage of the Chinese dependence on porters carrying supplies from Manchuria to the front lines. He realized the Chinese could not sustain an offensive for more than two weeks without running short of supplies. Operation RIPPER (called Operation KILLER by the press) accepted the initial Communist attack and then, as Communist supplies of ammunition decreased, prepared for a powerful counterattack. By summer 1951, the lines had largely stabilized along the thirty-eighth parallel, a little to the south (on the west)—Line Kansas—and a little more to the north (on the eastern side of the peninsula)—Line Wyoming.

Then came the final phase—not peace, not war—for another two years, during which Dwight Eisenhower became U.S. president and Soviet leader Joseph Stalin died. Suddenly the armistice talks, basically deadlocked for almost two years over the question of the forced repatriation of POWs, moved to closure and the guns fell silent in late July 1953, although the Chinese did threaten along the mostly American sector to the west and engaged in several serious efforts to destroy South Korean army units, perhaps to indicate they were still capable of fighting.

To the rest of the world, Korea was an example of a kind of limited war, fought after 1950 for limited goals. For the troops who fought in Korea, it was an intense conflict, with bloody hand-to-hand combat and incredible artillery duels reminiscent of World War I. And certainly the troops had to be motivated, prepared, supplied, and led as in any previous intense conflict—which was not easy to do, given the limited goals of the fighting.

But the conflict was limited because the major power combatants had other obligations and tasks that prevented them from investing the manpower and logistics of a major war in the Korean peninsula. The United States had to retain enough strength to meet an imagined Soviet threat to Western Europe; the Chinese Communists wanted to remake a society and rebuild a country devastated by years of political decline and corruption, external invasion and exploitation, and war. The Soviet Union was still rebuilding from the devastation of World War II. Thus neither side would invest what it would take to achieve victory in the sense of destroying one's enemy; instead, each side became content to demonstrate that it could resist and make the cost of an advance too great to be worth the expense.

Equally important, neither the United States nor the Soviet Union turned to nuclear weapons to force a decision. Although such weapons appeared to be the ultimate weapon, fortunately for the world the bombs remained safely away from the front, although there have been unsubstantiated claims that Dwight Eisenhower threatened their use to force the Chinese to hold real negotiations for an armistice after he became president in January 1953.

Charles M. Dobbs

See also: MacArthur, Douglas; Peng Dehuai; Ridgway, Mathew B.; Van Fleet, James A.; Walker, Walton

References and further reading:

Alexander, Bevin. *Korea: The First War We Lost.* New York: Hippocrene Books, 1997.

Rees, David. *Korea: The Limited War.* London: Macmillan; New York: St. Martin's Press, 1964.

Ridgway, Matthew B. *The Korean War.* New York: Da Capo, 1988.

Sandler, Stanley. *The Korean War: An Encyclopedia.* New York, London: Garland Publishing, 1995.

———. *The Korean War: No Victors, No Vanquished,* London: Routledge; Lexington: University of Kentucky Press, 1999.

Whelan, Richard. *Drawing the Line: The Korean War, 1950–1953.* New York: Little, Brown, 1990.

Kosciuszko, Tadeusz Andrezj Bonawentura (1746–1817)

Polish officer who fought against the British in the American Revolution. Born on 4 February 1746, in Mereczowszczyzna (in present-day Belarus), Kosciuszko was educated in military engineering in Warsaw and in military and civil architecture in Paris. Influenced by French ideas of liberalism, he went to America in 1776 to serve alongside the colonists in the American Revolution. Appointed a colonel of engineers under Major General Horatio Gates, Kosciuszko selected the defensive position that contributed to the American victory at the Battle of Saratoga in 1777. After spending two years directing the construction of fortifications at West Point, New York, he served under Major General Nathanael Greene in South Carolina. In 1783, as a reward for his services, Kosciuszko was granted U.S. citizenship and promoted to brigadier general.

In 1784, Kosciuszko returned to Poland and became a major general in the Polish army. Following the second partition of Poland in 1793, he led a rebellion against Russia and Prussian rule. Proclaimed supreme commander and given dictatorial powers, Kosciuszko defeated the Russians at Raclawice in April 1794 but was defeated in June by a combined Russian and Prussian force. After successfully defending Warsaw against both the Russians and Prussians (July–September 1794), he was defeated and wounded at the Battle of Maciejowice in October, which marked the end of the Polish nation. Kosciuszko was held prisoner in Russia

Engraved portrait of Thaddeus Kosciuszko. (Library of Congress)

until 1796, when he was released on the promise that he never again take up arms against Russia. In 1797 he visited America, where he was awarded a $15,000 pension and a grant of 500 acres of land in Ohio. After 1798, Kosciuszko lived in France and Switzerland, where he unsuccessfully sought Polish independence, while refusing to break his promise by joining Napoleon's invasion of Russia. He died 15 October 1817, in Solothurn, Switzerland. Kosciuszko's remains were carried to Krakow and were buried among the Polish kings in the Royal crypt of Wawel Cathedral. In accordance with his will, his Ohio property was sold and the money used to establish a school to educate freed African Americans at Newark, New Jersey.

Alexander Bielakowski

See also: American Revolution

References and further reading:

Haiman, Miecislaus. *Kosciuszko in the American Revolution.* New York: Kosciuszko Foundation, 1975.

Kajencki, Francis C. *Thaddeus Kosciuszko: Military Engineer of the American Revolution.* El Paso, TX: Southwest Polonia Press, 1998.

Pula, James S. *Thaddeus Kosciuszko: The Purest Son of Liberty.* New York: Hippocrene Books, 1999.

Kosovo, Battles of (20 June 1389, 17 October 1448)

Ottoman Turkish victories in the Balkans. The First Battle of Kosovo was fought on 20 June 1389 as part of the Ottoman Wars of Expansion. King Lazar of Serbia had taken advantage of issues of succession to the sultanate throne caused by the death of Süleyman the Magnificent and by Byzantine intrigues. Lazar formed a Christian coalition army that defeated the Turks at the River Topolica in 1387. In response the Turks, led by Murad I, advanced into Bulgaria, crushing the rebellion there. Murad I then turned his 60,000 Turks to face Lazar's 100,000-man league of Serbs, Bosnians, Bulgars, Poles, and Albanians on the plain of Kosovo (plain of Blackbirds), 60 miles north of Uskub.

Murad's army was composed of Janissaries and *sipahis.* The Janissaries—the name comes from the Turkish for new army (*yeniçeri*)—was an army raised from Christian slave recruits from all over Europe, most of whom were raised by the Turks themselves as foster children. The *sipahis,* developed by Süleyman, were heavy cavalry and proved effective in the battle to come.

The Turks won a hard-fought key victory, but Murad was mortally wounded by a Serbian aristocrat posing as a deserter. Murad's brother, Bayazid, led the Turks to victory by encouraging the 12,000 men under Vuk Brancovic to switch to the Turkish side late in the day when it appeared Lazar's forces had the upper hand. Lazar was captured and beheaded.

The Turks then turned their attention to a siege of Constantinople, having taken almost all other Byzantine territory. Only the arrival of mercenaries looking for payment from the last vestiges of the Byzantine treasury saved the city, albeit temporarily. After the first Battle of Kosovo, the Byzantines had no allies within territories nominally controlled by them. Instead they had to rely upon tenuous tricks to survive, including encouraging pretenders to the sultan's throne.

The Second Battle of Kosovo, 17 October 1448, was much more closely fought than the first. Eighty thousand Hungarians and Wallachians under John Hunyadi constructed trenchworks against a much larger Turkish army. On the 17th, the Christian forces, attempting to relieve Constantinople, surprised the Turks by leaving their trenches and attacking and holding them at bay for the entire day. On the 18th, the Turks, wily as always, enticed the Wallachians to switch sides, causing the Hungarians to retire from the field.

The Hungarians lost at least 17,000, while the Turks lost perhaps 40,000. The Turks, led by Sultan Muhammed II, "the Conqueror," could now turn their attentions entirely to the siege and capture of Constantinople, which they finally took in 1453.

Christopher Howell

See also: Byzantine-Ottoman Wars; Constantinople, Siege of (1453); Hunyadi, János; Ottoman Empire; Turkish Wars of European Expansion

References and further reading:

Turfan, Naim. *Rise of the Young Turks: Politics, the Military and Ottoman Collapse.* Istanbul, London: I. B. Tauris & Co. Ltd., 2000.

Vaughan, Dorothy. *Europe and the Turk: A Pattern of Alliances, 1350–1700.* Liverpool, UK: Liverpool University Press, 1954.

Koxinga (Zheng Chenggong) (1662–1722)

Leader of the last effective Ming resistance against the Manchus. Zheng Chenggong, who was half Japanese, is best known by the nickname given him by the Portuguese, Koxinga. This was based upon his honorary title, Guoxing Ye (Gentleman with the Dynastic Surname), which he acquired after being granted the imperial surname, Zhu, by the Ming pretender, the prince of Tang, as a reward for his loyalty.

Koxinga's military career began soon after the Manchus, founders of the Qing Dynasty (1644–1911), invaded China. As the son of a Ming admiral, Koxinga was quickly thrust into service. At first Ming loyalist forces, who had lost the north and the capital of Beijing to the rapidly advancing Manchus, hoped to make a stand along the Yangtse, but further Manchu advances, and confusion within loyalist ranks, soon made this impossible.

Hoping to repeat the success of the founders Southern Song, who when faced with an invasion by the Jurchen, the ancestors of the Manchus, in the mid–twelfth century adopted a flexible maritime strategy based on the Chinese southeast, Koxinga took to the seas. He and a band of men rallied on Nanao island, in eastern Guangdong. His forces steadily grew, and in 1647 Koxinga led an army unsuccessfully against the city of Quanzhou in Fujian. Although repulsed, Koxinga continued to raid the coast of China with growing success.

In 1650, Koxinga annexed the trading port of Amoy in Fujian and Quemoy, an island guarding its approaches. From these bases, Koxinga became involved in Chinese overseas trade to finance his operations. During this period, although Koxinga enjoyed success against the Qing armies on land, it was not universal; however he did control the sea, and the entire coastline of China was fair game for his operations.

In 1659, Koxinga attempted a major invasion into the lower Yangtze River valley. He sailed directly to Nanking, sweeping opposition before him, but his army was routed at the gates of the former Ming capital. Koxinga's overconfidence and lack of military strategy had undermined his efforts. This invasion effectively ended Koxinga's ability to restore the Ming and reclaim China, although he remained a thorn in the side of the Qing Dynasty.

By 1660, Koxinga was the sole remaining resistance to the Qing armies. In need of a secure food supply, he invaded Taiwan in 1661. Despite stiff resistance from the Dutch, who held the island with Qing consent, Koxinga eventually succeeded. Thereafter, Koxinga established an administration and governed the island, but this was his last adventure, although he did threaten the Spanish governor of the Philippines with invasion. In 1662, Koxinga died of illness. His successors continued to hold Taiwan until 1683, when the island was surrendered to the invading Qing.

Timothy May

See also: Chinese Imperial Wars; Kangxi; Manchu Expansion, Wars of; Yangzhou, Siege of

References and further reading:

Crozier, Ralph. *Koxinga and Chinese Nationalism: History, Myth, and the Hero.* Cambridge, MA: Harvard University, East Asian Research Center, 1977.

Hsu, Immanuel. *The Rise of Modern China.* New York: Oxford University Press, 1970.

Kruger, Paul-Stephanus Johannes Paulus (1825–1904)

South African statesman and military leader. Bush-trained trekboer, soldier, pioneer, farmer, politician, and South African statesman, Kruger took part with his family in the Great Trek from Cape Colony to Transvaal. Strictly reared in Dutch Calvinism, with scant formal education, he was profoundly affected by tribal opposition and his lifelong struggle for independence from the despised British, which he countered with fearlessness and cunning. While a teenager Kruger simultaneously farmed and held military and civil posts. In 1864 he became Transvaal commandant general and worked toward establishing constitutional authority. Lucrative minerals were found in 1868, leading the British to annex Transvaal in 1877.

In 1880 he traveled to England for talks to regain independence but was thwarted, leading Kruger, Piet Joubert, and Martinius Pretorius to oppose a British-inspired federation.

The Transvaal War saw numerous Kruger-led victories, climaxed by the Battle of Majuba Hill (27 February 1881), which restored independence to the Afrikaners at the Pretoria Convention of 1881. Once president in 1883, a position to which he was reelected in 1888, 1893, and 1898, Kruger stringently opposed political equality for *uitlanders* (mostly British and German foreigners) who arrived with the gold rush of Witwatersrand in 1886 and doubly outnumbered Afrikaners.

The Cecil Rhodes–inspired 1895 Jameson Raid to overthrow Kruger's government failed. Kaiser Wilhelm II's telegram congratulating Kruger for the victory was widely interpreted as German support. This struggle led to the Boer War of 1899–1902, which resulted in the 31 May 1902 Peace of Vereeniging, making Transvaal a British crown colony.

During the Second Boer War Kruger did not take to the field, but traveled to Europe to obtain aid. He was unsuccessful and never returned to Transvaal. The "old lion of Transvaal" died in Switzerland in 1904.

Annette Richardson

See also: Boer Wars; Buller, Sir Redvers Henry; Joubert, Petrus Jacobus; Kimberly; Ladysmith, Siege of

References and further reading:

Fisher, John. *Paul Kruger.* London: Secker & Warburg, 1969.

Gordon, C. T. *The Growth of Boer Opposition to Kruger, 1890–1895.* Cape Town and New York: Oxford University Press, 1970.

Marais, J. S. *The Fall of Kruger's Republic.* Oxford, UK: Clarendon Press, 1962.

Nathan, Manfred. *Paul Kruger: His Life and Times.* Durban: Knox Publishing Company, 1941.

Kublai Khan (1215–1294)

Founder of the Mongol-Chinese Yuan Dynasty. Kublai Khan, or in Mongolian, Qubilai-qan, was the second of four sons of Tolui-noyan (c. 1190–1231/32), who was the youngest son of Cinggis-qan (Genghis Khan). Kublai was at first just one of many Mongolian princes holding appanages near China. This changed with the events of 1251 that brought his elder brother, Möngke (Mongke), to the throne. Kublai became his viceroy in northern China and went on to build up an independent power base there that stood him in good stead after Möngke's death in 1259 and the end of a unified Mongolian empire.

Kublai rushed back to north China from Yunnan, where he had been campaigning, convened a *quriltai* of supporters, and had himself elected *qan.* He then set about establishing a regime largely using savvy local advisers he had acquired as Möngke's viceroy.

Kublai's claim as Möngke's successor was challenged by his younger brother, Arigh Böke (d. 1266). The latter enjoyed wide support and controlled Qaraqorum, the capital. Also an opponent of Kublai and his house from domains in Siberia was Qaidu, representing the line of *qan* Ögödei (Ogadei), excluded from the succession in 1251.

The war with Arigh Böke lasted until 1264 and ended with his rival's defeat, thanks to Kublai's superior resources. The war with Qaidu continued until 1303. At one point it seriously threatened the survival of Kublai's *qanate* (after 1271, Yuan Dynasty). Such central Asian concerns remained critical for Kublai and his house until the end.

Once relatively secure in Mongol China, that is, the north and parts of the southwest, Kublai set about expanding his power. His most important line of advance was due south, into Southern Song domains, definitively conquered by the Mongolian general Bayan in 1276, although the mop-up continued until 1279. Other campaigns were two failed invasions of Japan (1274 and 1281) and successful campaigns in Vietnam, Burma, and across the sea to Java, as, under Kublai, China became a base for a most aggressive sea power for the first time in its history.

Kublai died in 1294, nearly 80. No other ruler of Mongol China ever rose to his stature. Thanks to Marco Polo, Kublai has remained the very symbol of the Oriental potentate. It was his world, by then long lost, that the Portuguese and others went looking for in the fifteenth and sixteenth centuries, marking the beginning of our own era.

Paul D. Buell

See also: Genghis Khan; Mongol Empire; Ögödei

References and further reading:

Allsen, Thomas T., *Mongol Imperialism.* Berkeley: University of California Press, 1987.

Rossabi, Morris. *Khubilai Khan: His Life and Times.* Berkeley: University of California Press, 1988.

Kuropatkin, Aleksey Nikolaevich (1848–1925)

Disastrous Russian commander. Of noble origins, Kuropatkin was born in Pskov Province, graduated from Pskov Military Academy (1866), General Staff Academy (1874), and served in Turkestan (1866–1871, 1875–1877, 1879–1883). He participated in the capture of Samarkand (1868) and Kokand (1876) and was an infantry division chief of staff under General Skobelev in the Russo-Turkish War of 1877–1878. He fought at the third Battle of Plevna (1877) and Senova (1878), participated in the attack on Geok Tepe (1880–1881), and became a major general (1882), serving on the general staff (1883–90). He also published *Deistvia otriadov generala Skobeleva* (1885), on the Balkan and

Russo-Turkish War campaigns, and became head of the Transbaikal District (1890–1898).

His impressive military record and diplomatic experience in the Far East saw him appointed minister for war from 1898 to 1904. Kuropatkin supported expansion into the Far East but failed to negotiate the corruption of czarist court politics to modernize and professionalize the army or keep pace with technological developments. His failure was brutally exposed in the Russo-Japanese War (1904–1905). Kuropatkin's underestimation of Japan, his indecisiveness in planning and in battle as Russian Land Forces commander in chief were similarly revealed. Pressured to relieve Port Arthur, he suffered reverses at Laioyang, Sha-Ho River, and Sandepu before defeat at Muckden saw him demoted to First Army commander.

While a State Duma member Kuropatkin wrote "The Russian Army in the Japanese War" (1909), an attempt to justify his performance.

Kuropatkin's shortcomings in the Russo-Japanese War did not preclude him from commanding a corps and the Fifth Army (1915), and on the northern front (February–July 1916). He also seemed not to have learned much, compiling a similarly poor command in this most disastrous of czarist wars. He initially doubted the possibility of the Brusilov Breakthrough, then failed to support it, contributing to its loss of impetus.

He was finally removed from military command and appointed Turkestan governor-general, until February 1917. He was arrested in April by the Tashkent Soviet, transferred to Petrograd, but freed by the provisional government. Kuropatkin rejected White overtures to fight in the Civil War. French offers to emigrate probably enabled him to live out his life back in Pskov Province, teaching at a middle school and an agricultural school that he founded.

Neil Harvey Croll

See also: Russo-Japanese War

References and further reading:

Jones, N. F. *The First and Last Man of the War: The Far Eastern Crisis and the Career of General Kuropatkin, 1895–1905.* M.A. diss., University of North Carolina at Greensboro, 1991.

Kuropatkin, A. N., A. B. Lindsay, and E. D. Swinton. *The Russian Army and the Japanese War: Being Historical and Critical Comments on the Military Policy and Power of Russia and on the Campaign in the Far East by General Kuropatkin.* London: J. Murray, 1909.

———. *Russko-iaponskaia voina* (The Russo-Japanese War). Leningrad: Gosizdat, 1925.

Story, Douglas. *The Campaign with Kuropatkin.* London: T. W. Laine, 1904; Philadelphia: J. B. Lippincott Co., 1905.

Aleksey Nikolaevich Kuropatkin. (Library of Congress)

Kursk, Battle of (1943)

The greatest tank battle in history. After the disastrous loss at Stalingrad in World War II, German army commanders, with Adolf Hitler's enthusiastic approval, planned a massive offensive in 1943. In March, German field marshal Fritz von Manstein had achieved a great victory at Kharkov for Army Group South, leaving a bulge of 100 miles north and south and 75 westward into the German lines of Army Group Central at Kursk. Bad weather and German indecision interfered with von Manstein's plans for a pincer movement to encircle the Soviet army in the salient, allowing the Red Army under General Nikolai Vatutin in the south and General Konstantin Rokossovski in the north time to build defenses and prepare for a large counterattack.

On 10 May, Hitler consented to the plan called Operation ZITADELL. Colonel General Model's Ninth Army, with seven Panzer, two Panazergrenadier, and nine infantry divisions, were to attack from the north and Colonel General Hoth's Fourth Panzer Army, with 10 Panzer, one Panzergrenadier, and seven infantry, would advance from the south—roughly 570,000 men.

Delayed until 4 July, the Germans found themselves confronted by 11 Russian armies (alerted by Allied ULTRA intelligence) of 977,000 men. Soviet defenses in the north corner of the salient were particularly dense with 2,200 antitank and 2,500,000 antipersonnel mines and 20,000 guns of various kinds.

For the next eight days, the Germans tried to advance in the face of bitter fighting but Soviet artillery knocked out 40 percent of German armor. On 12 July, at the battle of Prokhorovka, 600 German tanks clashed with 850 Soviet tanks. The battle became a war of attrition with both sides calling for reinforcements.

On 13 July, due to the American invasion of Sicily and fears of a landing in Italy, Hitler ordered German units to disengage. Retreating on 17 July, the Nazis left 70,000 dead and 2,950 tanks on the battlefield. The Soviets following the German retreat ordered an immediate counteroffensive. The lack of victory at Kursk by the German army spelled an end to any more major offenses on the Russian front to the end of the war.

T. Jason Soderstrum

See also: Armored Fighting Vehicles; World War II

References and further reading:

Dunn, Walter S. *Kursk: Hitler's Gamble.* Westport, CT: Praeger, 1997.
Glantz, David M. *The Battle of Kursk.* Lawrence: University Press of Kansas, 1999.
Solov'ev, Boris G. *The Battle of Kursk, 1943.* Moscow: Novsti Press Agency Publishing House, 1988.

Kut-al-Amara (1915–1916)

One of the very few capitulations in British military history. In December 1915, after an unexpected defeat at Ctesiphon on the drive to capture Baghdad from the Turks, British troops fell back to the town of Kut-al-Amara, located on the Tigris River. Lacking the troops to relieve them immediately, the British High Command ordered them to retire further south. The order came too late, as the British force of about 16,000 was now surrounded by the Turks. When General Charles Townsend, in command of the British forces, suggested a breakout, he was told to stay in Kut-al-Amara in order to tie down as many Turkish troops as possible. The Turks immediately ordered a number of failed assaults on the town, with heavy losses on both sides. While the troops in Kut-al-Amara held out, a relief expedition was formed in order to try to break through the British lines. It failed, with the British suffering 23,000 casualties, and the Turks about 10,000. Attempts to resupply by river boat failed, and in April 1916, with food supplies dwindling and threats of epidemics looming, General Townsend asked the Turks for a six-day armistice in order to discuss surrender. The Turks, wanting to add to the British humiliation at Gallipoli, replied that they would only accept unconditional surrender. On 29 April 1916, British forces surrendered to the Turks. It was one of the largest capitulations for the British army up to that time.

For the British troops who had endured the siege, the worst came after the surrender. The surviving troops, weakened by disease and near starvation, were taken to POW camps under a brutal forced march by the Turks. More than 3,000 perished on the way, or about 2 percent of all those who surrendered. Conditions in the POW camps were little better, and those who survived the inhuman conditions came out of the camps in 1918 as little more than skeletons.

Drew Philip Halévy

See also: Gallipoli

References and further reading:

Braddon, Russell. *The Siege: The Forgotten Siege of Kut El Amarah, Mesopotamia, 1916.* Viking: New York, 1970.
Miller, Ronald. *Death of an Army: The Siege of Kut, 1915–1916.* Boston: Houghton Mifflin Company, 1970.

Kutuzov, Prince Mikhail Illarionovich Golenishchev (1745–1813)

Credited for defeating Napoleon's Grande Armée in the Russian campaign. Kutuzov was born in St. Petersburg and enrolled in a military engineering school at age 12. A corporal at age 14, he lost the sight in one eye during the Russo-Turkish War (1768–1774). Kutuzov served six years under the command of Aleksandr Suvorov in the Crimea and was promoted to colonel in 1777 and major general in 1784.

During the War of the Third Coalition, Kutuzov was appointed commander of Austro-Russian forces at Vienna and attempted to link up with General Mack at Ulm. However, the latter was defeated by Napoleon on 19 October 1805, before Kutuzov arrived. Kutuzov avoided Napoleon's pursuit, but Czar Alexander I demanded Kutuzov engage the French. At Austerlitz on 2 December, 90,000 Austro-Russian troops were defeated.

Kutuzov was retired and reappointed as supreme commander in 1811 during the Russo-Turkish War (1806–1812).

After Napoleon's 600,000 Grande Armée invaded Russia and reached Smolensk, Kutuzov was made a prince and appointed supreme commander on 20 August 1812. Kutuzov withdrew Russian forces, ordering the destruction of all unevacuated supplies and lodgings. To further weaken the

French, Kutuzov fought minor engagements. However, Alexander ordered Kutuzov to stand at Borodino on 7 September, where Kutuzov lost 42,000 of his 112,000 troops and withdrew southeast, allowing the remaining 58,000 of 130,000 French forces to enter Moscow. Failing to defeat the Russians and unwilling to winter in a Moscow that had been burned out, Napoleon retreated.

By giving battle at Maloiaroslavets on 24 October, Kutuzov forced Napoleon to retrace the path of the French advance. Kutuzov's troops continuously molested the retreating French, engaging them at Viazma and Krasnoie. The remnants of Napoleon's army narrowly escaped annihilation during the crossing of the Berezina River on 27–28 November. Kutuzov died while in pursuit of the French, at Bunzlau in Silesia.

Neville G. Panthaki

See also: Austerlitz, Battle of; Berezina River, Battle of; Borodino; Moscow, Retreat from; Napoleon I; Napoleonic Wars; Russo-Turkish Wars

References and further reading:

Austin, Paul Britten. *1812 Napoleon's Invasion of Russia.* London: Greenhill Books, 2000.

Cate, Curtis. *The War of the Two Emperors.* New York: Random House, 1985.

Nafziger, George F. *Napoleon's Invasion of Russia.* Novato, CA: Presidio Press, 1988.

Nicolson, Nigel. *Napoleon 1812.* New York: Harper & Row, 1985.

L

Ladysmith, Siege of (1899–1900)

Major siege of the second Boer War. The Boer independence movement conflicted with British governance in Natal colony, located on the route around Africa. Ladysmith was strategically sited where the main rail line from coastal Durban divided, providing communications with Transvaal and the Orange Free State. General Piet Cronjé's Boers surrounded Ladysmith, severing the rail line on 2 November 1900. Lieutenant General Sir George White, Victoria Cross, commanded 12,500 British soldiers in this town of 21,300 persons.

Similar to the French at Dien Bien Phu in Vietnam (1953–1954), the British garrison was situated in low ground and dominated by enemy observation and artillery fire from surrounding hills. The Boer's rapid-firing German and French artillery was superior to that of the British, with the exception of the two 4.7-inch naval guns and four long-range 12-pounder guns from the *Powerful.* Although the naval guns were more effective, the navy provided merely 500 rounds of ammunition. The Boer advantage was negated by poor gunnery. Thus, the battle reverted to the centuries-old siege technique of starvation and disease.

Ladysmith, unlike the other major sieges at Kimberly and Mafiking, was noteworthy for the level of sickness and disease. The garrison held sufficient rations for two months, with the perimeter holding one month's forage for the horses and pack animals. Fever cases peaked at 1,314 on 27 January 1900, when eight deaths per day became the average.

Field Marshal Lord Roberts commanded a relief column of 30,000 troops, having replaced General Sir Redvers Bullers, Victoria Cross, following his defeat at Colenso. The lead elements under Lord Dundonald arrived on 28 February 1900, ending the four-month siege.

Robert Martyn

See also: Boer Wars

References and further reading:

Sharp, Gerald. *The Siege of Ladysmith.* London: MacDonald and Janes, 1976.

Lafayette, Marie Joseph Paul Yves Roch Gilbert du Motier, Marquis de (1757–1834)

French field commander and politician and hero of the American Revolution. Lafayette was born to wealth in Chavaniac, Auvergne, France, on 6 September 1757. He joined the infantry in 1771 and the dragoons in 1773 and was promoted to captain in 1774. Inspired by the American cause and intending a jab at Britain for the sake of France, Lafayette arrived in America in 1777 with Baron Johann de Kalb and offered his services to the Continental Congress. He was commissioned a major general in the Continental army on 31 July.

Lafayette distinguished himself at Brandywine, where he was wounded, spent the winter at Valley Forge, and excelled at Barren Hill on 18 May 1778, Monmouth on 28 June, and in the Rhode Island campaign of July and August. He returned to France in 1779 to secure more French help for the Americans, was promoted to colonel in the French army, and returned to his command in Virginia in April 1780. He sat as a member of the court-martial that convicted Major John André in September, engaged Benedict Arnold's British forces several times in Virginia in 1781, supported Anthony Wayne at Green Spring, and was an important factor in the defeat of Charles Cornwallis at Yorktown. Upon his return to France in December, he became a major general in the French army.

Lafayette entered French politics in 1787, represented Auvergne in the Estates General in 1789, assumed command of the National Guard on 26 July, and tried to mediate among the various revolutionary factions and the crown. Promoted to lieutenant general and briefly in command of the Army of the Center, he was persecuted by the Jacobins and fled to Belgium in 1792. Captured, he was held as a prisoner of war, first by Austria and then by Prussia. After his release on 23 September 1797, he returned to France, avoided Napoleon, and kept a low profile, but he reentered politics during the Hundred Days, helping to ensure Napoleon's second abdication. Thereafter, he served liberalism in France until his death in Paris on 20 May 1834. Lafayette was lionized in the United States, with numerous towns and cities (e.g., Fayetteville, North Carolina and Arkansas; Lafayette, Indiana) and counties named in his honor.

Eric v. d. Luft

See also: American Revolution; Arnold, Benedict; Brandywine; Cornwallis, Sir Charles; France and the American Revolution; French Revolutionary Wars; Monmouth; Napoleon I; Napoleonic Wars; Revolutions of 1830; Valley Forge; Washington, George; Wayne, Anthony; Yorktown

References and further reading:

Bernier, Olivier. *Lafayette: Hero of Two Worlds.* New York: Dutton, 1983.

Buckman, Peter. *Lafayette: A Biography.* New York: Paddington, 1977.

Criss, Mildred. *La Fayette: On the Heights of Freedom.* New York: Dodd, Mead, 1954.

Gerson, Noel Bertram. *Statue in Search of a Pedestal: A Biography of the Marquis de Lafayette.* New York: Dodd, Mead, 1976.

Lake Trasimene, Battle of (2 June 217 B.C.E.)

The second of three disastrous defeats the Romans suffered at the hands of the Carthaginian general Hannibal. The year after his victory at the Trebia River, Hannibal crossed the Apennines into Etruria (Tuscany) with an army of 55,000 men. A Roman army under the command of the consul Flaminius tried to intercept him.

Hannibal directed his army to the northern shore of Lake Trasimene (Lago di Trasimeno), a perfect place for an ambush. The road passed through a narrow valley that was bordered by the lake to the south and by hills to the north. There was only a narrow path westward. An easily defensible hill blocked the eastern side. Because the Romans would come from the west, Hannibal positioned his Spanish troops and his light infantry in the east, on the hill at the end of the valley. Both Hannibal's cavalry and his Celtic troops took a position in hiding along the hills to the north. A large force of Celts was positioned near the western passage.

The Roman force, 30,000 strong, entered the valley in marching order. A dense mist obscured the hills, and the Romans failed to send out reconnaissance parties. When the entire Roman column had marched into the valley, the Celts attacked the rear end of the column, blocking the retreat. The cavalry charged downhill upon the Roman troops. The Romans were totally surprised.

What followed was more a slaughter than a battle. Thousands were killed on the shore or driven into the lake. Only the Roman vanguard, 6,000 men, succeeded in fighting through the Carthaginian line, though they too were captured the following day. The Roman army was crushed, with 15,000 troops taken prisoner and 15,000 killed, among them the consul himself. Hannibal lost only 1,500 men.

M. R. van der Werf

See also: Cannae, Battle of; Hannibal Barca; Punic Wars; Trebia, Battle of the

References and further reading:

Bagnall, Nigel. *The Punic Wars. Rome, Carthage and the Struggle for the Mediterranean.* London: Pimlico, 1999.

Conolly, Peter. *Greece and Rome at War.* London: Greenhill Books; Harrisburg, PA: Stackpole Books, 1998.

Lazenby, J. F. *Hannibal's War. A Military History of the Second Punic War.* London: Aris & Phillips, 1978.

Land Mines

Explosive charges that are usually concealed by their small size or by being buried just under the surface of land; pressure sensors detonate land mines when they are passed over by troops or vehicles. Their primary objective was originally to slow down large-scale troop advancement, but as they evolved, they came to be used in all phases of war, for purposes ranging from tactical defense to strategic offense.

Land mines were first used in World War I, when Germans troops buried live artillery shells and left only their fuses exposed. They were meant to act as a defense against advancing French and British tanks, but these land mines could easily be dug up, removed, and stolen by enemy troops on foot. This practice gave rise to the development of smaller mines, which were placed alongside the bigger mines to prevent them from being moved. Since World War I, technology has made land mines smaller and more destructive.

The two main classes of mines are antitank (AT) and antipersonnel (AP) mines. Antitank mines employ a large amount of explosives, weigh more than 5 kilograms, are detonated by a pressure of more than 120 kilograms, and are used primarily to immobilize tanks. It can be done by either a blast explosion that cuts the vehicle's tracks or a killer explosion that sends a plate through the tank and destroys both the tank and crew. Antipersonnel mines employ a smaller explosive charge and are detonated by pressure of more than 5 kilograms. Antipersonnel mines are intended to disable or kill personnel. Blast mines cause injury by shattering extremities or otherwise disabling personnel. Fragmentation mines are designed to kill personnel by exploding into the air and producing fragments. Offensively, land mines are used for protection along flanks and to cut off an enemy's withdrawal. Defensively, they are used to slow the advance of an enemy.

Since the end of World War II, antipersonnel land mines have become a lingering threat in many areas of the world, mainly to civilians. The United Nation estimates that there are more than 110 million buried mines left from the numerous wars of the twentieth century in more than 62 countries. Some 20,000 antipersonnel land mines detonate a year, taking with them a limb or the life of a civilian adult or child. The difficulty in avoiding them is that many minefields are uncharted because an accurate track of their location has not been kept. To add to the problem, land mines are mainly used by Third World nations because they are cheap and can be easily dispersed over a large area. Although easy to deploy, the recovery cost of land mines can be upwards of U.S.$1,000 per mine. There is some hope for the future in mines than can be remotely exploded when no longer required. But these weapons are expensive and so far are mainly used by wealthy nations.

It is only recently that nations have come to acknowledge that the strategic and tactical value of land mines does not outweigh humanitarian concerns. In May 1997, a U.S.-led United Nations resolution won the support of more than 150 counties in calling for a global ban on land mines. The United States was following a different path from that taken in Canada. On 3 December 1997, the Convention on the Prohibition of the Use, Stockpiling, Production, and Transfer of Anti-Personnel Mines and on Their Destruction (informally known as the Ottawa Convention), a Canadian-led initiative, was signed by 122 nations (11 nations subsequently added their signatures) and came into legal force in March 1999. However, two of the top producers and sellers of land mines, Russia and China, have not signed the Ottawa Convention; nor has the United States.

Matthieu J.-C. Moss

References and further reading:

Croll, Mike. *The History of Landmines.* Barnsley, UK: Leo Cooper, 1998.

Lannes, Jean, Duke of Montebello (1769–1809)

One of the most audacious field commanders of the French Revolutionary and Napoleonic Wars. Born on 10 April 1769 at Lectours (Gers), Lannes was commissioned on 10 June 1792, earning rapid promotion and the reputation of leading from up front. As a colonel he was the first across the Po River at Piacenza (7 May 1796) and was the first to reach the Austrian positions at Lodi (10 May) in Napoleon's audacious and decisive Italian campaign. This campaign brought him Napoleon's attention and a promotion to brigadier general.

Lannes participated in the capture of Alexandria, Egypt (2 July 1798). He distinguished himself at El-Arish (8–19 February 1799) and at Acre (8 May). Appointed a division general (10 May), Lannes left Egypt with Napoleon (22 August). He provided vital support for Napoleon at the coup of 18 Brumaire, which made Napoleon first consul of France.

Lannes led the advance guard through the Great Saint Bernard Pass (16–21 May 1800) and defeated the Austrians at Montebello (9 June). At Marengo (14 June), his badly outnumbered troops held the Austrians until Louis Desaix arrived. Appointed a marshal (19 May 1804), Lannes helped trap Karl Mack at Ulm (20 October 1805). At Austerlitz (2 December), he blocked Austro-Russian attempts to outflank the French left. Lannes drove the Prussians and Saxons back at Jena (14 October 1806). His decisive command of the center at Friedland brought victory and rewards, including the title Duke of Montebello (15 June 1808).

Sent to Spain, Lannes was victorious at Tudela (23 November 1808) and captured Saragossa (21 February 1809). Napoleon, however, recalled him for the 1809 campaign against Austria. At Ratisbon, when his troops balked at assaulting the walls, Lannes himself shouldered a scaling ladder. His troops took the city (23 April 1809). On 21 May Lannes and André Masséna spearheaded the ill-fated attempt to cross the Danube River. Then Lannes seized and held Essling against overwhelming numbers while the French evacuated. On 22 May, a cannonball shattered his legs, necessitating the amputation of the right leg. Gangrene developed, and Lannes died on 31 May at Ebersdorff, Austria. The first marshal of France to die of battle wounds, Lannes was Napoleon's friend among the marshals and one of his most aggressive commanders.

James K. Kieswetter

See also: Aboukir; Austerlitz, Battle of; French Revolutionary Wars; Friedland; Lodi; Marengo, Battle of; Masséna, André, Duc de Rivoli, Prince d'Essling; Murat, Joachim, Grand Duke of Cleves-Berg, King of Naples; Napoleonic Wars; Rivoli

References and further reading:

Chrisawn, Margaret Scot. *The Emperor's Friend: Marshal Jean Lannes.* Westport, CT: Greenwood Press, 2000.

Damamme, Jean-Claude. *Lannes: Maréchal d'empire.* Paris: Payot, 1987.

Laffargue, André. *Jean Lannes, maréchal de France.* Paris: Bouquet, 1975.

Macdonnell, Archibald Gordon. *Napoleon and His Marshals.* New York: Macmillan, 1934.

Laotian Civil War (1954–1973)

Laos declared its independence from France in July 1949 and was finally recognized in 1954. A three-sided conflict immediately erupted, involving royalist, neutralist, and Communist forces. In 1959 General Phoumi led a coup against the indigenous Pathet Lao (Communist) forces of Prince Souphanouvong. The Communists were driven into the bush, but aided and abetted by North Vietnam, China, and the Soviet Union, they retained military viability. Over the next decade and a half, the government alternated between various neutralist and royalist factions, which invariably brought on Pathet Lao offensives exploiting the chaos. By 1965, the Communists were poised to overrun the entire country before the United States intervened with direct air strikes against Pathet Lao positions and the Ho Chi Minh trail. Military advisers were also introduced, along with large-scale recruitment of Meo tribesmen as mercenaries.

For nearly a decade, the political and military situation in Laos revolved around control of the central highland region known as Plain of Jars. A seasonal succession of offensives and counteroffensives by both sides secured and lost this vital area with little diminution of fighting. At length, the Pathet Lao gained the upper hand, thanks to direct intervention by North Vietnamese forces. A cease-fire was agreed to in 1973, but two years later, with the retreat of U.S. forces from Indochina, the Pathet Lao took formal control of the country. Within two years, anti-Communist forces began a low-intensity war against the regime that lasted until 1990 and also involved incursions by neighboring Thai forces. Sporadic fighting lasted until 1991, when a final political accommodation was reached.

John C. Fredriksen

References and further reading:

Adams, Nina S., and Alfred W. McCoy, eds. *Laos: War and Revolution.* New York: Harper & Row, 1970.

Langer, Paul F. *Revolution in Laos: The North Vietnamese and the Pathet Lao.* Santa Monica, CA: RAND Corporation, 1969.

Larrey, Dominique Jean (1766–1842)

French military surgeon of the French Revolutionary and Napoleonic eras; renowned for pioneering rapid evacuation and treatment of wounded soldiers. Born at Beaudéan (Hautes-Pyrénées) on 8 July 1766, Larrey studied medicine at Toulouse. While serving with the Army of the Rhine in 1792, Larrey was appalled at the delay in treating the wounded. He helped organize mobile field hospitals, which closely followed the troops. He also recognized that rapid evacuation of wounded from the battlefield was crucial. Therefore he developed a light, sprung, horse-drawn, two-wheeled cart—the "flying ambulance"—to pick up wounded even under fire. A four-wheeled version to carry more wounded came later. In Egypt in 1798, he designed camel-mounted pannier baskets to transport the wounded.

Larrey served in every Napoleonic campaign after Egypt and was appointed chief surgeon of the Imperial Guard. Skilled in the accepted practice of rapid amputation of any severely injured limb, Larrey unwittingly pioneered the use of antiseptic dressings. He could be quite unorthodox. After the Battle of Aspern and Essling (20–22 May 1809), Larrey ordered the slaughter of horses, including officers' mounts, to provide broth for the wounded who were stranded without food.

Noted for his boundless energy in treating wounded of all ranks, Larrey distinguished himself by his courage, efficiency, ingenuity, and medical observations and writings. As surgeon general of the Grand Army in Russia in 1812, Larrey labored prodigiously to save the wounded while simultaneously making extensive observations of the effects of cold on them. He served at Waterloo, where the Duke of Wellington, noticing the brave surgeon at work under fire, ordered his gunners not to fire on him.

After Waterloo, Larrey devoted himself to a distinguished career in medicine and wrote many works dealing with military medicine. He died in Paris on 1 August 1842. In his "Testament" (15 April 1821), Napoleon described Larrey as "the most virtuous man" he had known.

James K. Kieswetter

See also: French Revolutionary Wars; Napoleonic Wars

References and further reading:

Haythornthwaite, Philip J. *Napoleon's Military Machine.* New York: Hippocrene Books, 1988.

Larrey, Dominique Jean. *Memoirs of Military Surgery and Campaigns of the French Armies on the Rhine, in Corsica, Catalonia, Egypt and Syria; at Boulogne, Ulm, Austerlitz; in Saxony, Prussia, Poland, Spain and Austria.* Trans. Richard Willmott Hall. 4 vols. Baltimore: Joseph Cushing, 1814.

Soubiran, André. *Le baron Larrey, chirurgien de Napoléon.* Paris: Fayard, 1967.

Vess, David M. *Medical Revolution in France 1789–1796.* Gainesville: University Presses of Florida, 1975.

Latin Empire–Byzantine Wars (1204–1267)

The Fourth Crusade shattered Byzantium and installed a Latin regime. In 1203, the knights of the Fourth Crusade captured Constantinople and drove out Emperor Alexius III (r. 1195–1203). The following year, after attempting to rule through several Byzantine puppets, the crusaders again took the city to suppress the anti-Latin regime of Alexius V Ducas Murtzuphlus (r. 1204). This time they thoroughly looted it and destroyed the Byzantine administration. Churches were converted from orthodox to Catholic (Latin) Christianity, and Count Henry of Flanders was crowned as the "Latin" emperor of Constantinople. Henry then parceled out provinces to his vassals, assuming that they could conquer them. Even as he did so, opposition was forming. In 1205, Alexius III's son-in-law, Theodore Lascaris (r. 1204–1222), created a Byzantine government in exile at Nicaea. The Latins also had to confront rivals in Epirus (present-day Albania and northwestern Greece) and in Trebizond, as well as predatory Turks and Bulgarians.

To challenge Lascaris, the Ducas family of Epirus began to expand into the Balkans. In 1216, Count Henry died, and his heir Peter fell in an Epiran ambush. With the Latins in disarray, Theodore Ducas moved in and captured Thessalonica in 1224, a great victory. Meanwhile, Nicaea, now under John Vatatzes (r. 1222–1254), was consolidating its hold on Anatolia after defeating both the Comneni of Trebizond and the Turks. With Ducas now claiming to be emperor of Byzantium, John Vatatzes, John Asen of Bulgaria, and even the Constantinople Latins united against him. In 1230, Ducas decided first to punish the Bulgarians, but John Asen crushed his army at the Battle of Klokotnitsa and annexed half of Ducas's former conquests.

With the Ducas family in eclipse, Vatatzes and the Bulgarians agreed in 1235 to divide the Balkans and besiege Constantinople together. The siege failed, and by 1237, the alliance had soured. Plagues and Mongol incursions delayed new operations, but circumstances still benefited Vatatzes. The Mongols crushed the Seljuqs, securing Nicaea's eastern flank. In 1246, Vatatzes invaded an enfeebled Bulgaria, securing a European base. In 1251, Michael II of Epirus invaded Nicaean holdings in Greece. Vatatzes intercepted the Epirans, pulverized their army, and forced Ducas into vassalage. Nicaea was now ascendant.

Vatatzes' sickly heir, Theodore II (r. 1254–1258), soon confronted a new Ducas plot, an alliance of Serbs, Bulgarians, Peloponnesian Latin knights, and the bellicose Manfred of Sicily. When Theodore II died, he left Nicaea to a child, John Lascaris (r. 1258–1261), and to his grand commander, Michael Paleologus. Paleologus immediately went on the attack. Although he inflicted severe defeats on the coalition, these campaigns were actually holding actions. Nicaea's real objective was not Epirus but Constantinople. Risking another invasion from Albania, Paleologus concentrated forces against the city in 1260 and took it, ironically, by stealth the next year. Now Emperor Michael VIII (r. 1259–1282), he played divide-and-conquer to force treaties on Bulgarians, Serbs, Venetians, the Latin knights in Greece, and the Ducas family. The death of Manfred of Sicily temporarily halted the menace from western Europe. By 1267, the "Latin Empire" of Constantinople was officially dead. It had taken the Byzantines over 60 years of convoluted war and diplomacy to overthrow it.

Weston F. Cook Jr.

See also: Seljuqs

References and further reading:

Bartutis, Mark. *The Late Byzantine Army.* Philadelphia: University of Pennsylvania Press, 1992.

Fine, John. *The Late Medieval Balkans.* Ann Arbor: University of Michigan Press, 1987

Laupen, Battle of (21 June 1339)

Swiss shock action that defeated mounted heavy cavalry. The Battle of Laupen pitted the Swiss city of Bern against a coalition of Burgundian cities in what boiled down to a land dispute. The city of Bern had aggressively acquired two Burgundian towns, Thurn and Laupen. In 1339, a Burgundian coalition formed to take back the disputed regions. In June 1339, the Burgundians (1,200 mounted knights and an unknown number of infantry auxiliaries) surrounded the town of Laupen, which was defended by 600 Bernese militia. The Swiss from the forest cantons came to the aid of the Bernese. On 21 June, the Bernese and their allies (about 6,000 men) moved to relieve the town.

The Burgundians, led by Count Gerard of Vallangin, intending to crush the Swiss by a strong attack against their left, deployed their horse on the right, with foot soldiers holding the center and left positions. The Swiss formed a staggered line, with the Bernese, armed with the pike, thrust forward onto the crest of a hill. Two other squares of Swiss halberdiers followed behind and were staggered to the left of the Bernese. There they awaited the Burgundian attack. After some minor skirmishing, the Swiss swept forward. Their right and center squares immediately drove into the enemy foot soldiers and dispersed them with great slaughter and panic. They then wheeled to the left to take the knights, who were engaged by the third Swiss square, in their flank. The knights had been launching successive charges against the Swiss with some effect, but they could not break the square, which bristled with halberds. With the field entirely in the hands of the Swiss, the knights were soon surrounded and forced to flee, leaving behind many of their fallen nobility.

Laupen marked the beginning of the end of the unquestioned dominance of the mounted warrior. Laupen was the first victory won by infantry attacking heavy cavalry. For nearly two hundred years afterward, the Swiss maintained the reputation for fearless and resolute offensive action.

Bryan R. Gibby

See also: Sempach, Battle of

References and further reading:

Delbruck, Hans. *Medieval Warfare.* Trans. Walter J. Renfroe, Jr. Lincoln: University of Nebraska Press, 1990.

Oman, Charles W. *The Art of War in the Middle Ages.* Ed. John H. Beeler. Ithaca, NY: Cornell University Press, 1953.

Lawrence, Thomas Edward (T. E.) (1888–1935)

Eccentric British scholar-adventurer, better known as "Lawrence of Arabia," who helped lead the Arab revolt against Turkey during World War I. Lawrence was born on 15 August 1888 in the Welsh town of Tremadoc. In 1910, he took a degree in modern history at Jesus College, Oxford University, while traveling regularly to the Middle East to study crusader castles. His reputation as an archaeologist and a linguist brought him to the attention of the War Office in 1914.

The writer and soldier T. E. Lawrence ("Lawrence of Arabia") c. 1918. (Hulton/Archive)

Lawrence was commissioned and went to work in the War Office's Geographical Department in London. In December 1914, he went to Cairo to help instigate a revolt on the Arabian peninsula.

In late 1916, Lawrence, now a captain, was part of a British delegation that met with Sharif Hussain of Mecca, whose son Feisal led a small insurgent army in southern Arabia. In November, Lawrence joined Feisal's army as an adviser and liaison. With British support, the Arab army launched attacks on the vital Hadjis Railroad. Feisal and Lawrence seized the Red Sea ports of Wejh (24 January 1917) and Aqaba (6 July 1917). Lawrence was promoted to major, and the army turned north in support of Edmund Allenby's offensive in Palestine. The Arabs played an important role in the fighting in 1918, taking Der'aa (27 September 1918) and Damascus (1 October 1918).

An outspoken supporter of Arab nationalism, Lawrence attended the Versailles Conference but was bitterly disappointed by continued Anglo-French influence in the region. Lawrence, now a colonel and uncomfortable with the publicity he received in Britain and the United States, resigned in 1922. He enlisted in the Royal Tank Corps and the Royal Air Force (RAF) under assumed names but was unmasked by the press. In 1926 he published his memoir, *The Seven Pillars of Wisdom.* On 19 May 1935, shortly after leaving the RAF, Lawrence died of injuries sustained in a motorcycle accident.

Adam R. Seipp

See also: Allenby, Edmund Henry Hyman, Viscount; World War I

References and further reading:

Graves, Robert. *Lawrence and the Arabs.* London: J. Cape, 1927.

Mack, John E. *A Prince of Our Disorder: The Life of T. E. Lawrence.* Boston: Little, Brown, 1976.

Wilson, Jeremy. *Lawrence of Arabia: The Authorised Biography of T. E. Lawrence.* London: Heinemann, 1989.

Laws of War

International laws, enforced sometimes by nations after war and sometimes by commanders in battle, governing both the decision to engage in war and the manner of its conduct, particularly the forms of violence used, the definition of combatants, the treatment of prisoners, and the treatment of neutrals and noncombatants. Throughout recorded military history, there have been limits on the conduct of war. These limits have been often breached, in the same manner that almost all laws are broken, but they have not only persisted but have grown more comprehensive, and whether a nation's armies and soldiers comply with them is a hallmark distinguishing that nation's military as professional.

Laws of war arose in ancient Greece from the custom of commanders as well as from the specific treaties concluded among cities. Thucydides recorded a high level of compliance with truces, armistices, alliances, and peace treaties between city-states. Further, commanders honored flags of truce and heralds, truces to bury the dead, conditions of surrender, and the sanctity of triumphant monuments erected to celebrate victory. The rights of neutrals were usually honored, as was the neutrality of religious temples and the right to travel to Olympic games. Soldiers and commanders obeyed these rules for three different reasons: they were defining habits of civilized Greeks, a status they highly valued; they were religious obligations; and they expected reciprocal treatment from their opponents.

Roman conquest was very little constrained by laws, although the army did maintain strict rules for discipline within its ranks, and there were detailed regulations for the treatment of conquered lands and people. Further, Rome concluded many treaties for truces and peace with both allies and opponents that were honored, at least for the time specified in the treaties. The most essential characteristics of the Roman *iustum bellum,* though, were that war be conducted *pium* (according to religious prescriptions) and only after unsatisfied demands had been made of the opponent or a declaration of war had been sent.

Medieval warfare was regulated only by religious doctrine. Christians were constrained in the treatment of opposing Christians, just as Muslims were constrained in their treatment of fellow Muslims. Neither, for instance, could enslave a captured enemy of the same religion. In both cultures, forms of chivalry emerged by the late Middle Ages that specified limits on the forms of battle and the treatment of prisoners, although these rules usually mandated such good treatment only of those with high rank.

Christian scholars from Ambrose and Augustine to Thomas Aquinas, Francisco de Vitoria, and Francisco Suárez argued that only some wars were just and that there was no justification in killing noncombatants. Despite their reflection in the Gratian decretals of canon law, these arguments for just wars were predicated on theology and found little adherence in national legal systems.

The modern law of war is the result of the development of the modern nation-state. Both were foreseen and described in the great text of Dutch scholar Hugo Grotius, *De Jure Belli ac Pacis* (1625), which argued (based in part on the system of Alberico Gentili) for national obligations of rational conduct through law, in peace and in war. Although initially few attempts were made at international agreement or enforcement, custom followed many of Grotius's observations and edits, such as the use of declarations of war and the honoring of pledges not to destroy surrendered cities.

With the balance of power created by the Treaty of Westphalia in 1648, states were subject to no obligations but those they voluntarily accepted, usually as treaties. There was no enforcement of a breached obligation other than by very powerful states or alliances. This process of "horizontal enforcement" has persisted for 350 years. As armies grew larger with massed men and as weapons became more capable of destruction, the identification of combatants, limitations of weaponry, and treatment of prisoners became more acute problems for all nations, and both internal military culture and international treaties began to set new standards.

The Declaration of Paris (1856) abolished privateering, leaving naval engagements solely to professional combatants and banning all others as pirates. In 1863, President Abraham Lincoln issued General Order No. 100, setting standards for the identity of combatants and treatment of prisoners of war. The first Geneva Convention (1864) set standards for the treatment of enemy wounded by the great states of Europe. Further international conventions, at The Hague in 1899 and 1907 and in Geneva in 1906, 1929, and 1949, refined the law of war as to battlefield conduct and weapons, civilians, prisoners of war, and wounded and sick military personnel. Other conventions banned particular weapons, beginning with the Geneva Protocol on Gas Warfare of 1925.

The Bryan Treaties of 1913 and 1914 and, perhaps most controversially, the Versailles Treaty of 1919 promoted limits on the grounds for commencing war, including requirements for investigation, arbitration, and peaceful settlement of disputes, and the Kellogg-Briand Pact (1928) condemned recourse to war in all cases but self-defense. None of these instruments prevented the horrors of World War II, and at the conclusion of the war, the victorious allies established tribunals at Nuremberg and Tokyo to try individuals accused of war crimes, particularly the new category of crimes against peace, which included planning, initiating, and waging wars of aggression in violation of international treaties and agreements; crimes against humanity, which included exterminations, deportations, and genocide; war crimes, or the violations of the laws of war on the battlefield; and conspiring to commit the criminal acts listed in the first three counts. In Germany, of 22 defendants, 3 were acquitted, 4 imprisoned from 10 to 20 years, 3 imprisoned for life, and 12 sentenced to hang. In Japan, of 25 defendants, 2 were sentenced to prison terms, 16 to life imprisonment, and 7 to hang. The most important change in law from these trials was the Nuremberg principle, holding the individual and not just the state accountable for violations of the laws of war.

With the adoption of the United Nations Charter in 1945, almost all nations of the world committed to the peaceful settlement of disputes and the renunciation of war except in self-defense. Among the signatory regimes, very effective conventions have been adopted outlawing genocide and crimes against humanity; further limiting the use of weapons of mass destruction, such as nuclear and biological weapons, and certain weapons of particular inhumanity, such as exploding bullets; and further refining standards for the treatment of prisoners and the wounded.

The UN Security Council created the International Criminal Tribunal for the Former Yugoslavia on 25 May 1993 and the International Criminal Tribunal for Rwanda on 8 November 1994. Both ad hoc tribunals are charged with the investigation of war crimes, and both have actively investigated and convicted individuals for violations of the Geneva Convention of 1949; genocide; violations of the laws and customs of war; and crimes against humanity. The specific articles for violations of the laws and customs of war include the use of poisonous weapons or other weapons calculated to cause unnecessary suffering; wanton destruction of civilian areas not justified by military necessity; attack or bombardment of undefended towns; seizing or harming buildings dedicated to religion, charity, education, and the arts and sciences, as well as historic monuments and works of art and science; and plundering public or private property.

The UN Rome conference (opened on 17 July 1998) drafted a treaty establishing an International Criminal Court with global jurisdiction to try individuals accused of crimes similar to those prosecuted by the tribunals for Yugoslavia and Rwanda and whose governments will not try them. The treaty will become effective when 60 states have ratified it; by December 2000, 120 states had signed it, and 25 had ratified it.

Steve Sheppard

See also: General Order No. 100; Geneva Conventions; Nuremberg Principle

References and further reading:

Best, Geoffrey. *War and Law since 1945.* Oxford: Clarendon Press, 1994.

Brownlie, Ian. *International Law and the Use of Force by States.* Oxford: Clarendon Press, 1963.

Holland, Thomas Erskine. *The Laws of War on Land (Written and Unwritten).* Oxford: Clarendon Press, 1908.

Lebanese Civil Wars (1958, 1975–2000)

Two civil wars that have devastated the Lebanese "nation" created by ill-conceived World War II peace settlements. The 1958 civil war was due to pro-Western president Camille Caiman's (1900–1987) wish to serve a second term, contrary to the constitution, and Lebanese Muslims' desire for a stronger pro-Arab relationship. To quell the riots of May 1958, Caiman asked the United States for help, and it responded by sending the Sixth Fleet with 14,000 U.S. troops. On 23 September 1958, Caiman was succeeded by General Fuad Chehab (1902–1973). The U.S. troops withdrew in October 1958.

The Lebanese civil war of 1975 to 2000 focused on the discontent that evolved from the unofficial, inequitable National Pact of 1943. The Maronite Christians governed Lebanon at the expense of the Shiite, Sunni, and Druze sects, which constituted 50 percent of the population by 1975. Disgruntled Palestinian Muslims supported the Palestine Liberation Organization's (PLO) guerrilla attacks on Israel.

Civil war was sparked on 13 April 1975, when some Christian Phalangists were attacked in a church; they retaliated by killing 27 Palestinian bus passengers. The PLO joined other Muslim groups in raids. Israeli became involved by supplying arms to the Christians. The League of Arab States, led by Syria, sent 30,000 troops to implement a peace plan in 1976. Beirut was divided into political and religious enclaves in 1977, with most groups fighting each other. Israeli troops invaded Lebanon on 14 March 1978 to eliminate PLO bases. Some 6,000 UN peacekeepers replaced the Israelis, who left later that year.

By 1981, Syria had intervened militarily and placed armed forces in Lebanon's Bekaa Valley, along with Soviet missiles. The Phalangists responded by occupying the area around the Beirut-Damascus highway, which sparked a Syrian offensive. The Israelis attacked the Syrians and bombed some Beirut sectors in retaliation for PLO rocket attacks into northern Israel. A cease-fire was called on 24 July 1981.

In June 1982, the Israelis reinvaded Lebanon, killing 18,000 Lebanese. Israel forced the PLO guerrillas to evacuate Beirut. On 14 September, President-elect Bashir Gemayel (1947–1982), a Phalangist, was assassinated, but he was succeeded by his brother Amin Gemayel (b. 1942). Then 328 Palestinian were killed by Phalangists in Beirut refugee camps from 16 to 18 September. Peace-keeping forces from the United States, the United Kingdom, France, and Italy arrived.

The U.S. Embassy was bombed on 18 April 1983, killing 50 people. Suicide bomb attacks on French and U.S. military headquarters in Beirut led to more than 300 deaths on 23 October 1983 and caused the international forces to withdraw. Israeli forces left the Shuf Mountains, which were then occupied by the Druze, who fought both the Lebanese army and the Christians. PLO Leader Yasir Arafat (1929–) was forced to leave Beirut on 20 December 1983 after some PLO dissidents supported by Syria attacked and besieged his stronghold.

By February 1984, Lebanon was occupied by both Israel

and Syria. Fierce fighting in Beirut between Druze and Shiite militias occurred. The extremist group Hezbollah (Party of God) emerged in 1985. Sheer anarchy reigned when Christian general Michel Aoun (b. 1935) became president in 1988, while Muslim prime minister Selim al Hoss (b. 1930) established a competing government. Their respective militias physically destroyed Beirut.

The League of Arab States drafted the National Reconciliation Charter for Lebanon in 1990, providing greater equality and diminished presidential power. It also called for the withdrawal of the militias and the destruction of the "'green line" that bisected Beirut; thus the civil war supposedly ended. However, more attacks and counterattacks continued to plague desolate Lebanon. Nearly 150,000 lives, civilian and military, have been lost in the civil war, and nearly 200,000 people have been wounded.

Annette Richardson

References and further reading:

Brogan, Patrick. *The Fighting Never Stopped.* New York: First Vintage Books, 1990.

Salibi, Kamal. *A House of Many Mansions: The History of Lebanon Reconsidered.* Berkeley: University of California Press, 1988.

Winslow, Charles. *Lebanon: War and Politics in a Fragmented Society.* London: Routledge, 1996.

Lechfeld (10 August 955)

Decisive defeat of Magyar raiders by Otto I. The Magyars, emerging from the steppes of Russia, migrated to the Danube River basin in the ninth century and settled in what is now Hungary. Fighting primarily as horse-archers, they sought to avoid close-quarter action, preferring to encircle their foes and engage them from afar.

As king of Germany, Henry the Fowler (r. 919–936) had transformed the Saxon nobility, hitherto used to fighting as light cavalry, into a disciplined force of armored horsemen. This "military revolution" gave the Ottonian dynasty the edge not only over its rivals in Germany but also over the Magyars. When Henry fought them at the Battle of Riade (933), he reminded his "men of iron" of the need to maintain their line, to use their shields to deflect the first discharge of arrows, and only then to spur their mounts to close contact, but it was left to his son, Otto I the Great, to banish the threat of the Magyars.

The Magyars resumed their attacks in 954 with a major invasion, followed by a second one in 955. Their numbers are unknown, but the chroniclers report that they had never been seen in such force before. Crossing Bavaria, they laid siege to Augsburg. Otto moved rapidly from Saxony, with an army amounting to some 8,000 horsemen, its various contingents coming from Saxony, Bavaria, Swabia, Franconia, and Bohemia. When Otto approached the Lech River, the Magyars broke camp. While their main body engaged the king, a smaller detachment, outflanking the German host, turned up in the rear of the Swabians and wreaked havoc among them. The Franconians under Conrad the Red, duke of Lotharingia, had to turn so as to face this surprise attack, but when Otto ordered a general charge against the main body of the enemy, everything favored his host. The Magyars, on their smaller mounts and lacking armor, discharged a volley of arrows before they were overtaken and slain, many as they attempted to negotiate the Lech River.

The Germans had suffered heavily—Conrad was killed the moment he lifted his helmet to get some air—but what turned defeat into irreparable disaster for the Magyars was the relentless pursuit of the vanquished by Otto. The remnants of the Magyar army were decimated and its captured leaders executed.

Nic Fields

See also: Otto I, the "Great"

References and further reading:

Leyser, K. J. *Medieval Germany and Its Neighbours, 900–1250.* London: Hambledon Press, 1982.

Lee, Henry ("Light Horse Harry") (1756–1818)

American cavalry commander and politician. Lee was born into a wealthy Virginia family in Prince William County on 29 January 1756. After graduating in 1773 from the College of New Jersey (later Princeton University) he volunteered for Captain Theodorick Bland's Virginia Cavalry in June 1776. This unit, with Bland as colonel and Lee as captain, was incorporated into the First Continental Dragoons on 31 March 1777 and saw action at Brandywine. Lee blamed Bland's poor reconnaissance of the northern fords for George Washington's being outflanked on the right and losing that battle.

Promoted to major on 7 April 1778 by virtue of distinctive service against Banastre Tarleton at Spread Eagle Tavern, Pennsylvania, on 20 January, he drilled his battalion into an effective force and led it to victory at Paulus Hook, New Jersey, on 19 August 1779. Promoted to lieutenant colonel on 6 November 1780, he named his regiment "Lee's Legion" and reinforced Nathanael Greene in South Carolina on 13 January 1781. Supporting the guerrilla warfare of Francis Marion, the "Swamp Fox," Lee excelled throughout the Carolinas in 1781 at Georgetown, Haw River, Guilford Court House, Fort Watson, Fort Motte, Fort Granby, Augusta, Ninety-Six, and Eutaw Springs. His raids on Charles Cornwallis's outposts were an important factor in the success of the Continental army's southern strategy.

After the war, Lee held several high political offices, including Virginia state legislator, Virginia governor, and U.S. congressman. In 1794 he commanded troops against the Whiskey Rebellion in western Pennsylvania. In 1799 he eulogized the just-deceased Washington as "First in war, first in peace, and first in the hearts of his countrymen." To recover from injuries he suffered while trying to impede an antiwar riot in Baltimore in July 1812, he retired to the Caribbean in 1813. He died on Cumberland Island, Georgia, on 25 March 1818, during the return voyage to Virginia. His third son was Robert E. Lee, the great Confederate general.

Eric v. d. Luft

See also: American Revolution; Brandywine; Greene, Nathanael; Guilford Court House; Lee, Robert Edward; War of 1812; Washington, George; Whiskey Rebellion

References and further reading:

Hartmann, John W. *The American Partisan: Henry Lee and the Struggle for Independence, 1776–1780.* Shippensburg, PA: Burd Street, 1999.

Lee, Henry. *The Campaign of 1781 in the Carolinas: Memoirs of the War in the Southern Department of the United States.* Reprint, Spartanburg, SC: 1975.

———. *The Revolutionary War Memoirs of General Henry Lee.* Edited with a biography of the author by Robert E. Lee. New introduction by Charles Royster. New York: Da Capo, 1998.

Royster, Charles. *Light-Horse Harry Lee and the Legacy of the American Revolution.* Baton Rouge: Louisiana State University Press, 1994.

Lee, Robert Edward (1807–1870)

Commander in chief of the Confederate armies in the American Civil War. Lee was born in Stratford, Virginia, on 19 January 1807, the third son of "Light Horse Harry" Lee. Ranking second in the U.S. Military Academy at West Point, class of 1829, and with no demerits, he was commissioned in the engineers and stationed in the southeastern states. Until 1846 he worked on bridges and fortifications in Georgia, Virginia, New York, and Missouri. In the Mexican-American War, he served first under John Ellis Wool and then under Winfield Scott, building bridges and securing important reconnaissance for the Battles of Saltillo, Veracruz, Cerro Gordo, Contreras, Churubusco, and Chapultepec. From 1852 to 1855, he was superintendent of the U.S. Military Academy.

Promoted to lieutenant colonel in the Second Cavalry, Lee served in Texas from 1855 to 1857. In October 1859 he led the expedition to capture John Brown at Harpers Ferry. He served again in Texas until February 1861, when Scott recalled him to Washington. When the Civil War broke out, both Scott and President Abraham Lincoln tried to persuade him to accept command of the Union armies, but choosing instead to support Virginia, he became major general of Virginia forces on 23 April. Confederate president Jefferson Davis promoted him to full general on 14 June.

Lee's debut as a Confederate field commander at Cheat Mountain, West Virginia, on 12–13 September was a failure caused by the insubordination of his junior officers. Davis reassigned him to the Carolinas and Georgia to oversee coastal defenses but recalled him to Richmond as a military adviser in March 1862. After Joseph E. Johnston was seriously wounded at Fair Oaks on 31 May, Davis ordered Lee to take over Johnston's command and renamed it the Army of Northern Virginia on 1 June. It was Lee's command for the remainder of the war.

Lee's brilliance as a field commander first became apparent in the Seven Days' Battles. Leaving only a small garrison in Richmond, Lee boldly attacked George B. McClellan north of the Chickahominy River. This bloody series of tactical defeats was in fact a strategic victory because it pushed the Union army away from Richmond. After a stunning victory over John Pope at Second Bull Run, Lee invaded Maryland, but McClellan eventually stopped him at Antietam. Lee tried a second invasion of the north after beating Ambrose Burn-

Photographic portrait of Robert E. Lee, 1869. (Library of Congress)

side at Fredericksburg and Joseph Hooker at Chancellorsville. He worsened his losses at Gettysburg on the third day by ordering "Pickett's Charge," a massive frontal infantry attack across a three-quarter-mile-wide plain under constant fire from the Union center.

In the wake of Gettysburg, the Army of Northern Virginia could not resume the offensive. Lee made full use of delaying tactics and entrenchments, forcing Ulysses S. Grant to waste huge numbers of Union troops assaulting his positions in the Wilderness, at Petersburg, and before Richmond. Davis named Lee commander in chief on 23 January 1865. In the end, Grant gave Lee generous terms of surrender at Appomattox on 9 April 1865.

Lee spent the rest of his life as a paroled prisoner of war. In October 1865, he became president of the nearly defunct Washington College, now Washington and Lee University in Lexington, Virginia. Unlike most Confederates, he was never able to regain his U.S. citizenship, even though he had opposed slavery before the war and epitomized reconciliation after the war. His greatest gift to his reunited nation, in fact, may have been his strong discouragement of any ideas of continuing the battle by guerrilla warfare, thus sparing the country the agony of prolonged insurgency and counterinsurgency conflict. He died in Lexington on 12 October 1870. A person of seemingly absolute integrity, Lee was revered throughout the South and honored in the North by the time of his death.

Eric v. d. Luft

See also: American Civil War; Antietam/Sharpsburg; Second Bull Run/Manassas Junction; Cerro Gordo, Battle of; Chancellorsville, Battle of; Fredericksburg; Gettysburg; Grant, Ulysses Simpson; Hill, Ambrose Powell; Jackson, Thomas "Stonewall"; Johnston, Joseph Eggleston; Lee, Henry; Lincoln, Abraham; Longstreet, James; McClellan, George Brinton; Meade, George Gordon; Mexican-American War; Petersburg, Siege of; Pope, John; Scott, Winfield; Seven Days' Battles; Veracruz, Siege of; Wilderness

References and further reading:

Chaney, William Franklin. *Duty Most Sublime: The Life of Robert E. Lee as Told through the Carter Letters.* Baltimore: Gateway, 1996.

Davis, Burke. *Gray Fox: Robert E. Lee and the Civil War.* Short Hills, NJ: Burford, 1998.

Fellman, Michael. *The Making of Robert E. Lee.* New York: Random House, 2000.

Gallagher, Gary W., ed. *Lee the Soldier.* Lincoln: University of Nebraska Press, 1996.

LeFebvre, Pierre-François-Joseph, Duke of Danzig (1755–1820)

French marshal. Pierre-François-Joseph LeFebvre was born at Rouffach on 25 October 1755. He enlisted in the Gardes Françaises in 1773 and rose to the rank of sergeant by 1789. During the French Revolution, he was promoted rapidly and served in the Army of the Moiselle. By December 1793, he was promoted to *general de brigade;* in 1794, he was appointed to command a division. From September to December 1797, he served as commander of the Army of the Sambre-et-Meuse after the death of Lazare Hoche. In March 1799, he was appointed to command the S17th Military Division, the district containing Paris. He remained quiet during Napoleon's coup of 18 Brumaire, thus ensuring its success. During the consulate, he served as president of the Senate.

Pierre-François-Joseph LeFebvre. (Library of Congress)

In May 1804, LeFebvre was made a marshal of the empire. Two year later, he commanded the V Corps and in October 1806 was appointed to command the infantry of the Imperial Guard, which he led at Jena. During the closing months of the campaign against Prussia, he was sent to besiege Danzig, in command of the X Corps. After the siege, lasting from January to May 1807, and the surrender of the city, he was created duke of Danzig.

In 1808, LeFebvre commanded the Fourth Corps in Spain, defeating Spanish armies led by Joaquin Blake and the marquis de la Romana. Recalled to Germany during the 1809 campaign, he commanded the VII Corps, composed mostly of Bavarians. Later, he was appointed to command the Army of the Tyrol, operating against pro-Austrian insurgents. He defeated the insurrection and restored order.

In 1812, LeFebvre again commanded the infantry of the Imperial Guard during the Russian campaign. He was thus present at most of the major battles of that invasion but figured little in the fighting. He did not participate in the 1813 campaign. In the 1814 campaign, he fought at Champaubert, Montmirail, and Monteneau, where, according to the diarist Coignet, he "fought so hard that he foamed at the mouth." At the end of the 1814 campaign, he was part of the group of marshals, including Michel Ney, MacDonald, and Nicholas-Charles Oudinot, who secured the first abdication from Napoleon, on 6 April 1814.

During the Hundred Days, he played no active role on behalf of the Bonapartists, pleading old age and thus avoiding exile or death. He lived in retirement until his death in 1820.

Joseph Isenberg

See also: Berezina River, Battle of; Borodino; French Revolutionary Wars; Jena and Auerstädt; Moscow, Retreat from; Napoleon I; Napoleonic Wars

References and further reading:

Chandler, David. *Napoleon's Marshals.* London: Weidenfeld and Nicholson, 1987.

Young, Peter. *Napoleon's Marshals.* Reading, Berkshire, UK: Osprey, 1973.

Leipzig, Battle of (16–19 October 1813)

Decisive victory for the allies over Napoleon, ending his last German campaign. When Austria declared war against France on 12 August 1813, Napoleon had about 443,000 field troops in Germany south of Berlin. Trying to surround him were three separate multinational armies, a total of about 512,000 men under Karl Philipp zu Schwarzenberg in the south, Gebhard Leberecht von Blücher in the east, and Jean Baptiste Jules Bernadotte, crown prince of Sweden, in the north. From a skirmish on 16 August to the great cavalry battle at Liebertwolkwitz on 14 October, they fought many engagements, and both sides suffered enormous losses.

Napoleon's main force on 16 October was deployed in and around the city of Leipzig. Schwarzenberg attacked the southern and eastern outskirts at 8:00 A.M. Meanwhile, Blücher tried to sever Napoleon's communication lines in the west. Inconclusive fighting raged all day. Little happened on 17 October; the crucial day was 18 October. At 9:00 A.M., Schwarzenberg advanced his left in the south and soon was able to send his center against Napoleon's diverted forces in the southeast. Napoleon's situation was precarious by early afternoon. Bernadotte attacked from the northeast and Blücher from the north, completing a three-quarters' circle around Napoleon's entire army, compressing it, and pushing it into the city. Schwarzenberg's final assault of the day came from the east, dramatically reducing Napoleon's perimeter and inflicting heavy casualties. Napoleon made a defensive stand inside the city on 19 October. The allies commenced urban warfare in the northwest, east, and southeast. Napoleon's main concern was to organize an orderly retreat to France without being cut off. His defeated army escaped through a swamp and over a causeway west of the city.

Leipzig is frequently called the Battle of the Nations because of the great numbers of troops actually engaged: about 300,000 Russians, Prussians, Austrians, Swedes, British, and Germans united against Napoleon's 190,000 French, Italians, Germans, and Poles. Napoleon's losses were about 40,000 and those of allied forces about 54,000. In terms of numbers of troops engaged, Leipzig was the greatest battle of the nineteenth century, eclipsing even the massive engagements of the American Civil War half a century later.

Eric v. d. Luft

See also: Bernadotte, Jean Baptiste Jules; Berthier, Louis-Alexandre, Prince of Neuchâtel and Valangin, Prince of Wagram; Murat, Joachim, Grand Duke of Cleves-Berg, King of Naples; Napoleon I; Napoleonic Wars; Ney, Michel, Duc d'Enchingen, Prince de La Moskova; Oudinot, Nicholas-Charles, Duc de Reggio; Schwarzenberg, Karl Philipp zu

References and further reading:

Hofschroer, Peter. *Leipzig 1813: The Battle of the Nations.* London: Osprey, 1993.

Jansson, Per-Eric. *Leipzig.* London: Almark, 1975.

Nafziger, George F. *Napoleon at Leipzig: The Battle of Nations, 1813.* Chicago: Emperor's Press, 1996.

Petre, Francis Loraine. *Napoleon's Last Campaign in Germany, 1813.* London: Greenhill, 1992.

Lend-Lease (1940–1945)

A program devised by President Franklin D. Roosevelt to provide material aid to England (and later the other Allied powers) during World War II. By late 1940, the United Kingdom had run out of gold and cash reserves and thus informed the United States that it would be unable to pay for any more war matériel. In response, President Roosevelt proposed the Lend-Lease Act, which Congress passed in March 1941. Under the act, the president was empowered to sell, transfer, or exchange matériel to any country deemed vital to the defense of the United States. The United States was still neutral, and Roosevelt had to deal with many Americans who took an isolationist view of the war in Europe. To get support for the act, Roosevelt appealed directly to the American people, using a water hose analogy: if your neighbor's house is burning down, you do not worry about selling him the hose, but you give him the hose to deal with the cri-

Russian militia during the Siege of Leningrad c. 1942. (Hulton/Archive)

sis at hand and worry about the cost later. In presenting the idea of Lend-Lease to Congress in January 1941, Roosevelt first conceived of the idea of the Four Freedoms (freedom of speech and expression, freedom of worship, freedom from want, and freedom from fear) that later became part of the Atlantic Charter.

One of the foundations for the Lend-Lease Act was the transfer, in September 1940, of 50 obsolete destroyers to the United Kingdom in exchange for 99-year leases on a number of bases in the Caribbean and Newfoundland. The initial act included $1 million for aid to the Allies. By the fall of 1941, China and the Soviet Union had been added to the list of recipients. In October 1941, the Office of Lend-Lease Administration was developed to handle the growth of the Lend-Lease program. By 1943, Lend-Lease had been taken over by the State Department. Although the United Kingdom was the first (and main) recipient of Lend-Lease, the program was later expanded to help more than 40 counties. By August 1945, when Lend-Lease was abruptly terminated, more than $49 billion in aid, ranging from fighter planes to canned meat, had been extended to Allied nations. The countries of the British Commonwealth received about 63 percent and the Soviet Union about 22 percent. The United States received about $8 billion in "Reverse Lend-Lease," in which host countries helped pay for U.S. troops overseas during the war.

Drew Philip Halévy

See also: World War II

References and further reading:

Dobson, Alan P. *U.S. Wartime Aid to Britain, 1940–1946.* New York: St. Martin's Press, 1986.

Leningrad, Siege of (1941–1944)

Monumental siege during World War II. Leningrad's 900-day ordeal began on 8 July 1941, when the German Fourth Panzer Army severed the city's land contact with the Soviet Union less than three weeks after invading Russia. Hemmed in by Finnish troops to the north and German forces to the south and west, the 2.5 million inhabitants of Leningrad depended on meager supplies brought across Lake Lagoda from the east. They were not prepared for a siege, the authorities having dismissed calls to stockpile supplies and prepare defenses as "defeatist."

The "siege" itself was really an extended blockade ordered by Adolf Hitler to annihilate the city. German forces remained within artillery range but were ordered not to accept any form of surrender. They dug in permanently after the diversion of troops for attacks against Moscow in 1941 made offensive schemes impossible, and their Finnish allies, reluctant to cross their pre-1940 border with the Soviet Union, refused to attack from the north.

For the next two and a half years, German artillery and aircraft pounded the city, and the Wehrmacht denied all Soviet attempts to break the blockade. Leningrad's inhabitants froze to death in the winter, died under the incessant shelling, or starved when no more animals or glue could be found for consumption. More than 850,000 were evacuated in 1942, but at least 1 million are estimated to have died by the time a Soviet offensive relieved the city in January 1944. No city in modern history has suffered such losses without surrendering.

Lance Janda

See also: Stalin; World War II

References and further reading:

Glantz, David M. *The Siege of Leningrad, 1941–44: 900 Days of Terror.* Staplehurst: Spellmount, 2001.

Salisbury, Harrison Evans. *The 900 Days: The Siege of Leningrad.* New York: Da Capo Press, 1985.

Wayne, Kyra Petroskaya, and Harrison E. Salisbury. *Shurik: A WWII Saga of the Siege of Leningrad.* New York: Lyons & Burford, 2000.

Leo III (c. 675–741)

Founder of the Isaurian dynasty, iconoclast, energetic military leader, and adept strategist. Leo III took power in 717, at a time when Byzantium seemed on the verge of disintegration and conquest. Seven emperors had reigned in two decades of coups, revolts, and murderous attacks by Arabs, Slavs, and Bulgarians (697–717). As Leo deposed the reluctant emperor, Theodosius III, and entered the capital, massive Arab armies converged behind him, determined to bring Constantinople under Islam.

Fortunately, Leo was a wily and experienced commander. His successes in the Caucasus had earned him command of the Anatolian-Syrian frontier. In that office, Leo had courted the Bulgars and Armenians and employed his fluent Arabic. He defended Constantinople against an overwhelming force in a masterful economy of force strategy, using fire ships, encouraging guerrilla raiding and ambushes by the Slavs and Anatolians, and provoking revolts in the enemy camp. The Umayyad caliph ended the siege in 718, and Leo quickly shifted forces to reclaim control of Sicily and southern Italy.

Although Arab raids into Anatolia continued throughout his reign, Leo had rescued the empire and restored the frontiers. He also reinforced the defense of many cities and updated the law codes of Emperor Justinian, but Leo's ban on the use of icons in Orthodox Christian worship ignited the iconoclastic struggle. It inflicted decades of internal strife upon Byzantium. His policies also fueled the growing alienation between eastern and western Christendom, when he attempted an invasion of northern Italy in 733.

Weston F. Cook, Jr.

See also: Constantinople, Siege of (717–718)

References and further reading:

Ostrogorsky, George. *History of the Byzantine State.* New Brunswick, NJ: Rutgers University Press, 1991.

Treadgold, Warren. *A History of the Byzantine State and Society.* Stanford, CA: Stanford University Press, 1997.

Lettow-Vorbeck, Paul Emil von (1870–1964)

Successful German guerrilla warfare leader in East Africa during World War I. Paul von Lettow-Vorbeck was trained as an officer at the Military Academy at Potsdam and served in the German expeditionary force that quelled the Boxer Rebellion in China (1899). As a captain in the Schutztruppe (German colonial troops), he saw action against local tribes in German South-West Africa (now Namibia) from 1904 to 1907.

In 1913, Lettow-Vorbeck was placed in command of the German forces in German East Africa (today Tanzania). When World War I broke out, he defeated the invading Allied forces on several occasions, for example, at Tanga harbor (November 1914); Jassini (January 1915); Kilbata (1916); and Narungombe, Mahiwa, and Lukuledi (1917). He led by example and was wounded on three occasions. At most, he commanded only 3,007 white soldiers and 12,100 Askaris (black soldiers). Lettow-Vorbeck succeeded in evading the vastly superior Allied forces (of approximately 150,000 men) commanded by the South African generals Jan Christiaan Smuts (1916) and J. L. van Deventer (1917–1918), both of whom had waged successful guerrilla campaigns against the British during the Boer War (1899–1902) but were incapable of implementing successful antiguerrilla warfare against the Germans. In fact, Lettow-Vorbeck was able to recruit a number of former Boer guerrilla fighters into his ranks, who assisted him in developing guerrilla tactics.

The German command attempted to resupply Lettow-Vorbeck by the only means possible in the face of British control of the seas—by dirigible. A German airship was dispatched from Bulgaria in 1916 and almost reached the guerrilla commander when it had to turn back because of adverse weather. It was the longest unrefueled flight in history until 1986.

In November 1917, Lettow-Vorbeck led his force across the border into Portuguese East Africa (Mozambique), defeated the Allies at Ngomano and Namakurra, and returned to German East Africa by September 1918. In November 1918, shortly before the armistice was signed, he occupied a strong position in Northern Rhodesia (now Zambia), but soon afterward he was ordered by his superiors in Germany to lay down his arms, which he and his remaining 155 white soldiers and 1,156 Askaris did reluctantly.

Lettow-Vorbeck's force contributed a brilliant chapter to the military history of irregular operations. His courageous resistance earned him worldwide fame, and he was honored and respected by both friend and foe. After the war, he became a symbol of inspiration for his humiliated fellow Germans. In 1919 he was promoted to brigadier general but was dismissed from the army in 1920 for taking part in the abortive right-wing Kapp Putsch. He was a member of the Reichstag (German parliament) from 1928 to 1933 and published three books on his war experiences in East Africa, two books on postwar visits to Africa, and his autobiography in 1957. Lettow-Vorbeck was a fervent patriot and strict disciplinarian but had good relations with his subordinates and was compassionate toward captured or wounded enemies. His Askaris called him *Bwana mkubwa ya akili mingi* (the big man who can do everything). His campaigns are still a cornerstone of unconventional warfare studies.

André Wessels

See also: Guerrilla/Partisan/Irregular Warfare; Smuts, Jan Christian

References and further reading:

von Lettow-Vorbeck, P. E. *Mein Leben.* Biberbach an der Riss: Koehlers, 1957.

———. *My Reminiscences of East Africa.* London: Hurst and Blackett, n.d.

Schnee, A. H. H. Deutsch. *Ostafrika im Weltkriege. Wie wir lebten und kämpften.* Leipzig: Verlag Quelle & Meyer, 1919.

Trümpelmann, G. P. J. "Von Lettow-Vorbeck, Paul Emil." In *Dictionary of South African Biography,* ed. C. J. Beyers. Vol. 4. Pretoria: Human Sciences Research Council, 1981.

Leuctra, Battle of (371 B.C.E.)

Epaminondas, with a force of approximately 6,000 soldiers, defeated the Spartan army, which numbered about 10,000 men, at Leuctra in 371 B.C.E. Epaminondas arrayed his force to meet the Spartan right in oblique order, with a reinforced left consisting of a deep column that was 48 ranks deep, compared to the customary eight. He may have protected this column's left flank with a terrain obstacle or with a detachment of the Theban Sacred Band. He held back the forces remaining to his right and assigned them the task of holding in place the numerically superior Spartan left. In standard phalanx warfare, the right flank tended to overlap the enemy's left because soldiers unconsciously inched to the right to find cover for their exposed side from their neighbor's shield. The resulting rightward drift of both sides often resulted in a quarter rotation of the two lines as each right wing bore against the enemy's less steady left. Epaminondas's oblique order allowed an overwhelming concentration to overthrow the enemy's strongest wing while immobilizing the remainder of the enemy force.

The battle opened, typically, with a cavalry skirmish, in which the Thebans had more success than the Spartans. The climax occurred soon after, when Epaminondas led his heavy column against the Spartan line. Following the impact there was considerable fighting, but the Spartan right could not resist the Theban mass. The Spartan center and left, held to their front by the remaining Theban line and the cavalry, took no part in the fight. When Epaminondas completed the destruction of the Spartan right, his column wheeled to the right and advanced on the flank of the remaining Spartan line. In short order, the Spartan line broke and fled, leaving Epaminondas master of the field. Although this new battle tactic had some shortcomings, it changed military tactics in ancient Greece and brought Sparta down from its preeminent military position.

Bryan R. Gibby

See also: Epaminondas; Mantinea, Battle of

References and further reading:

Delbruck, Hans. *Warfare in Antiquity.* Trans. Walter J. Renfroe, Jr. Lincoln: University of Nebraska Press, 1990.

Dodge, Theodore A. *Alexander.* New York: Da Capo Press, 1996.

Jones, Archer. *The Art of Warfare in the Western World.* Urbana: University of Illinois Press, 1987.

Leuthen, Battle of (5 December 1757)

Major battle of the Seven Years' War (1756–1763) between Prussia and Austria. The Austrian army occupied Silesia as Frederick the Great defeated the Austrian army at Rossbach on 5 November. Prince Charles of Lorraine's army overwhelmed the Prussian detachment guarding the province and captured both Schweidnitz and Breslau. On 2 December, Frederick returned to Silesia and marched toward Breslau with 39,000 troops and 170 guns. He planned to lure the

Austrians into a decisive battle that would end the war. Opposing him were 66,000 men and 210 guns.

The Austrians stood near the small village of Leuthen. Charles and Field Marshal Leopold Daun wanted to entice Frederick into making a frontal assault along their 4.5-mile line and lock his army in a battle of attrition. On the morning of 5 December, Frederick advanced his army directly toward Leuthen and discovered the Austrian left flank exposed and unguarded. He altered his army's angle of advance so that it would be oblique to the Austrian left flank. Feigning an advance against the Austrian main line, Frederick's army redeployed without the Austrians' knowledge. Around noon, Prussian forward units attacked and broke through the Austrian positions, followed by a large-scale cavalry charge that put the Austrian command in chaos. Charles's attempt to shift his front only contributed to a growing panic within the Austrian ranks. At 3:30 P.M., Frederick unleashed a final assault of 40 cavalry squadrons that resulted in a general Austrian retreat but failed to organize a general pursuit to annihilate the remnants of Charles's army.

Leuthen ranks as Frederick's greatest tactical masterpiece. Nearly 6,000 Prussians were killed or wounded. Austria losses included 3,000 dead, 7,000 wounded, 12,000 prisoners, and 130 guns. The battle did not decide the war, but it did reinvigorate Frederick's war effort and ensured the survival of Prussia.

Patrick J. Speelman

See also: Frederick the Great, King of Prussia; Rossbach; Seven Years' War

References and further reading:

Duffy, Christopher. *Frederick the Great: A Military Life.* London: Routledge & Kegan Paul, 1985.

Showalter, Dennis. *The Wars of Frederick the Great.* London: Longman, 1996.

Weigley, Russell F. *The Age of Battles: The Quest for Decisive Warfare from Breitenfeld to Waterloo.* Bloomington: Indiana University Press, 1991.

Lewis, Meriwether (1774–1809)

American explorer, leader of the Lewis and Clark expedition. Born near Charlottesville, Virginia, on 18 August 1774, the son of an American Revolutionary War officer, Lewis was an avid outdoorsman. He served during the 1794 Whiskey Rebellion in the Virginia Militia. Later, in the regular army, he served under Captain William Clark in the Chosen Rifle Company. Between 1796 and 1801, Lewis was assigned to the 1st U.S. Infantry Regiment on the Ohio and Tennessee frontier and was promoted to lieutenant in 1798 and captain in 1800. In 1801, Lewis became private secretary to President Thomas Jefferson, an old family friend.

In 1803, the United States acquired a enormous land tract, the Louisiana Purchase, from France, and Jefferson assigned Lewis and his old captain, William Clark, to explore the new territory with Lewis in command. Along with mapping and exploring, Jefferson wanted the team to look into the feasibility of a route to the Pacific Ocean and to bolster U.S. claims to western lands not included in the purchase.

The expedition began on 14 May 1804 from St. Louis, Missouri Territory, with a total of 50 men. The explorers went up the Missouri River to North Dakota, where they spent the winter before continuing across the Rocky Mountains toward Oregon, arriving at the Pacific Ocean in early November 1805. Along the way, they received valuable assistance from American Indians, most notably the Shoshone guide, Sacagawea. They wintered on the coast, waiting for a transport ship that never came.

In March 1806, they began the return journey and were back in St. Louis on 23 September 1806. Hailed as a hero, Lewis was nominated by Jefferson as governor of the Louisiana Territory in 1807. Lewis was a poor administrator and died during a trip from St. Louis to Washington, D.C., in a tavern 70 miles southwest of Nashville, Tennessee, on 11 October 1809. The cause of his death remains a mystery; most modern historians consider it a suicide.

Harold Wise

References and further reading:

Ambrose, Stephen. *Undaunted Courage: Meriwether Lewis, Thomas Jefferson and the Opening of the American West.* New York: Simon & Schuster, 1996.

Dillon, Richard. *H. Meriwether Lewis: A Biography.* New York: Coward-McCann, 1965.

Lexington and Concord (1775)

Traditionally, the first military engagement of the American Revolution. The clashes at Lexington and Concord on 19 April 1775 produced new pressures for independence within the British North American colonies and a new resolve to suppress colonial resistance on the part of the British government.

General Thomas Gage, commander of British forces in Massachusetts, received orders to act decisively against the leaders of colonial American resistance. Gage devised a plan to strike against the colonial militias and ordered 700 troops under Colonel Francis Smith and Major John Pitcairn to seize gunpowder and cannon reportedly hidden at Concord. News of this plan reached leaders of the resistance, including Paul Revere, who subsequently spread word of the planned British raid along the road from Boston to Concord.

Revere's ride brought him through the town of Lexington

Painting of the Battle of Lexington. (Library of Congress)

(and into immortality), where colonial militiamen dedicated to the resistance movement—the Minutemen—turned out in force under Captain John Parker to meet Smith and Pitcairn. On the village square, British and colonial American forces exchanged fire following a shot whose source remains a mystery, leaving 1 British and 18 American casualties.

Pressing his mission, Smith ordered his men to Concord, where they encountered a much larger and more organized force of Minutemen under Colonel James Barrett arranged on the heights above the settlement. Barrett moved his force toward the town center when he learned that British troops had began to burn houses and goods in their search for the gunpowder. Barrett's colonials met a British detachment at North Bridge on the outskirts of town, leaving a few casualties on both sides. Smith ended his futile search for the gunpowder and determined to withdraw, but the shots and alarm bells had drawn hundreds of new Minutemen from surrounding areas into his return path. Before their arrival back in Boston, the British withstood repeated guerrilla attacks, their retreat almost degenerating into a rout, and engaged a concentrated force at Menotomy, ultimately suffering 273 casualties in the entire operation. As a consequence, the British government reinforced Gage's army with thousands of new troops, while the American resistance movement recruited new adherents with their supposed evidence of British tyranny. Given the wider conflict that ensued, the exchange at Lexington and Concord eventually became known as "the shot heard round the world."

The brave and defiant performance of the citizen-soldiers at Lexington and Concord entered American mythology: the yeoman citizen-soldier, ready to leave his plow in its furrow and shoulder his musket to defend hearth and home. This ideal, given voice by Thomas Jefferson in particular and strengthened by the British heritage of distrust of a standing army and the Roman Republic's "Cinncinnatus at the Plow" for the classically literate British American elite, saw to it that the United States as late as 1939 had a standing army about the size of Romania's.

Jeffrey Webb

References and further reading:

Fischer, David Hackett. *Paul Revere's Ride.* New York: Oxford University Press, 1994.

Fleming, Thomas. *Liberty! The American Revolution.* New York: Viking, 1997.

Gross, Robert A. *The Minutemen and Their World.* New York: Hill and Wang, 1976.

Li Hongzhang (1823–1901)

Chinese imperial politician and military administrator, instrumental in modernizing the military infrastructure of China. Li was born into a Confucian scholarly family on 15 February 1823 in Hefei, Anhui Province, China. In 1844 he became a government official in Beijing under his mentor,

Zeng Guofan. Li's father, Zeng, and in 1847 Li himself both earned the terminal *jinshi* degree in the rigorous Confucian examination system.

Li raised militias to support the Qing Dynasty during the Taiping Rebellion. He was appointed a judge in 1856 and governor of Jiangsu Province in 1862. Militarily, he was both an independent field commander and a member of Zeng's staff. Li's negotiations brought Western personnel and weapons to the Qing side against the Taipings. After the death of Frederick Townsend Ward in 1862, Li held administrative command of the "Ever-Victorious Army," while Charles "Chinese" Gordon commanded this crack unit of foreign mercenaries in the field.

Convinced that China could never compete with the West, either militarily or economically, unless it adopted Western technology, Li dedicated the rest of his life to what he called after 1872 the "Self-Strengthening Movement." Under his leadership, China built railroads, shipyards, arsenals, factories, military academies, technological institutes, and communications systems. He grew increasingly wary of Japan, which was rapidly becoming a military threat to China. Because so few of his colleagues perceived this danger as accurately as he did, he was unable to persuade them to develop fleets or improve standing armies. The result was that Japan humiliated China in the 1894–1895 Sino-Japanese War, ending with the Treaty of Shimonoseki.

Li died on 7 November 1901 in Tianjin, brokenhearted by the Boxer Protocols signed on 12 September. Even though he had foreseen and tried to prevent both the Sino-Japanese War and the Boxer Rebellion, Li was ashamed that foreigners had twice so severely humbled China on his watch.

Eric v. d. Luft and Sarah Luft

See also: Boxer Rebellion; Chinese Imperial Wars; Gordon, Charles George; Hong Xiuquan; Religion and War; Sino-Japanese War; Taiping Rebellion; Wolseley, Garnet Joseph, Viscount; Yang Xiuqing; Yuan Shikai; Zuo Zongtang

References and further reading:

Cahill, Holger. *A Yankee Adventurer: The Story of Ward and the Taiping Rebellion.* New York: Macaulay, 1930.

Chu, Samuel C., and Kwang-Ching Liu, eds. *Li Hung-Chang and China's Early Modernization.* Armonk, NY: M. E. Sharpe, 1994.

Folsom, Kenneth E. *Friends, Guests, and Colleagues: The Mu-Fu System in the Late Ch'ing Period.* Berkeley: University of California Press, 1968.

Kuhn, Philip. *Rebellion and Its Enemies in Late Imperial China.* Cambridge, MA: Harvard University Press, 1971.

Li Shihmin (600–649)

Grand strategist, second emperor of the Tang Dynasty (618–907), and son of Tang Dynasty founder Li Yuan. Historians continue to debate how much Li Shihmin contributed to his father's rise and the precise nature of the coup that forced the latter's abdication in his favor in 626. Clearly he made contributions to the rise of Tang, but they may not have been as great as the historical tradition indicates, and equally clearly, Shihmin was no innocent party in 626.

In 630, Shihmin, later known as Emperor Tai Zong, defeated the eastern Turks and in several great campaigns brought the Tarim Basin and its Silk Road under Tang control. He eventually pushed Chinese power well beyond the Pamirs, its greatest extension in history. He also established a short-lived Chinese protectorate over Tibet and tried to annex Korea, a process completed in 668 by his successor, Emperor Gao Zong (643–683).

To maintain control over China's extended borders, Li Shihmin used permanent garrisons of soldiers who grew their own food and were otherwise self-supporting. He did so because the Tang tax base was still limited, and he wished to reduce the strain on government resources. The practice also drew upon a long northern tradition of similar forces (*fubing*) predating Tang.

Such garrisons tended to become more settled as they established families and grew more and more integrated into the local population over time. Thus they were less available as mobile forces. Mobilizing such local forces for distant campaigns had never been easy in any case. Later, to solve this problem, the Tang had to turn to expensive mercenaries, who often had their own agendas.

Among his many other contributions, Li Shimin restored local government after years of neglect. He continued the commitment of the Sui Dynasty (581–618) to a bureaucracy based on merit and primarily selected through a written examination system requiring mastery of Confucian philosophy and current topics. The energies of potential opponents were thus directed to passing the examinations and to moving ahead in the bureaucracy rather than to rebellion. Individual emperors (Empress Wu, for example) also used the examination system to recruit "new men," ones not associated with the traditional northern elite. It was largely thanks to Li Shimin's conquests and the institutions that he established that the Tang Dynasty lasted for nearly another three centuries, in spite of its near collapse in the aftermath of the An Lushan rebellion of 751 and the need for late Tang to reinvent itself.

Charles M. Dobbs

See also: Chinese Imperial Wars; Sino-Korean Wars and the Wars of Korean Unification; Talas River, Battle of

References and further reading:

Bingham, Woodbridge. *The Founding of the Tang Dynasty; the Fall of Sui and Rise of Tang: A Preliminary Survey.* New York: Octagon Books, 1975.

Fitzgerald, C. P. *Son of Heaven (A Biography of Li Shih-min, Founder of the Tang Dynasty)*. New York: AMS Press, 1971.
Pulleyblank, Edwin G. *The Background of the Rebellion of An Lu-shan.* New York: Oxford University Press, 1955.

Liberia (1989–1997)

Founded in 1822 by freed slaves from the United States, the Republic of Liberia quickly became an autocracy in which the freed slaves (Americo-Liberians) held power over the indigenous peoples. This, along with the growing unrest of the 1970s and a faltering economy, sowed the seeds of the Liberian civil war.

In 1980 a group of enlisted soldiers led by Samuel Doe entered the presidential palace and executed President William Tolbert and his ministers. Doe declared himself president; his 9-year rule was marked by corruption, brutality, and nepotism. Doe both showed favor to his own ethnic group, the Krahn, and violently suppressed rival ethnic groups to the north.

In December 1989, Charles Taylor, a former Doe minister, launched an invasion of Liberia with his rebel group, the National Patriotic Front of Liberia (NPFL), from bases in Ivory Coast. Taylor and his soldiers had received training and arms from Libya as well as Burkina Faso. Once the invasion began, the NPFL advanced rapidly, driving the impotent Armed Forces of Liberia (ALF) before it.

Even before the invasion, there were numerous factions vying for power and position within the NPFL. Once the invasion had begun, the warring factions quickly became more hostile. Prince Yourmie Johnson, who split with Taylor and formed the Independent National Patriotic Front (INPFL), was the first of many defectors.

Within seven months, the NPFL was on the outskirts of the capital, Monrovia, and in control of 95 percent of the country. With the international community obviously not willing to intervene, the surrounding West African states initiated a historic first: an all-African peace-keeping force for service in Africa. The Economic Community of West African States (ECOWAS) deployed ECOMOG (the Economic Community Monitoring Group) to Monrovia in early 1990.

The situation in Monrovia quickly became a stalemate and turned into a West African Beirut, with different factions controlling specific areas of the city. Doe was killed in an ambush on his way to peace talks with NPFL, ECOMOG, and INFLP in September 1990. After President Doe's death, the remnants of the AFL (which was mainly Krahn) regrouped under the banner of the United Liberation Movement for Democracy in Liberia (ULIMO).

The situation remained relatively static until elections were held in 1997. These elections, monitored by international observers, drew 90 percent of the 1 million registered Liberians to the ballots. Taylor won the presidency in a landslide and was installed as the president of the State of Liberia, and a measure of stability returned to the wounded nation.

James Corbin

References and further reading:

Beyan, Amos J., Carl P. Burrowes, and D. Elwood Dunn. *Historical Dictionary of Liberia.* 2d ed. Lanham, MD: Scarecrow Press, 2001.
Dunn, D. Elwood, and S. Byron Tarr. *Liberia: A National Polity in Transition.* Metuchen, NJ: Scarecrow Press, 1988.
Ellis, Stephen. *The Mask of Anarchy: The Destruction of Liberia and the Religious Dimension of an African Civil War.* New York: New York University Press, 1999.
Harold D. Nelson, ed. *Liberia: A Country Study.* Washington, DC: U.S. Government Printing Office, 1985.
Liebenow, J. Gus. *Liberia: The Quest for Democracy.* Indianapolis: Indiana University Press, 1987.

Liddell Hart, Sir Basil Henry (1895–1970)

British defense analyst. An infantry company commander when he was gassed in the Somme offensive, Liddell Hart became deeply involved in the development of infantry tactics in an attempt to restore mobility to the battlefield. His analyses caused senior officers to treat him with great courtesy, and he was unflinchingly proud of the ordinary British soldier, Britain's military leadership, and war aims.

In 1920, he made John Frederick Charles Fuller's acquaintance, and they sharpened each other's insights, writing for a wide audience about armored warfare. Both believed armor would profoundly change the nature of warfare, but Liddell Hart came close to advocating all-tank armies rather than Fuller's combined arms approach. In 1924, Liddell Hart was discharged for medical reasons from the service he loved, and his views of British military leadership deteriorated accordingly. As Adolf Hitler posed an increasing threat to Britain, Liddell Hart became convinced that British involvement in World War I had been a mistake. He therefore changed his policy positions to stress the primacy of the defense and became an appeaser as the Chamberlain government moved toward war.

His reputation shattered by the conquest of France, Liddell Hart lived quietly during World War II. After Germany's defeat, he began interviewing captured German officers, seeking to resurrect his reputation. Eventually, Generals Fritz Bayerlein and Heinz Guderian credited his writings with influencing German *offensive* armored doctrine at a time when he had stressed the *defensive* potential of armor. In turn, Liddell Hart vouched for the Wehrmacht leadership's blackened honor.

Israeli army commanders, thinking it wise to link the Israeli Defense Forces to the Wehrmacht's efficiency, proclaimed themselves his other best pupils, further enhancing his status. By the time of his death on 29 January 1970, the development of German and Israeli blitzkrieg doctrine was firmly associated with Liddell Hart's interwar writings, in defiance of the historical record.

Erin E. Solaro

See also: Academies, Military; Armored Fighting Vehicles; British Military, Twentieth Century Organization and Structure; France; Fuller, John Frederick Charles; Guderian, Heinz; Haig, Douglas; History, Military; Rommel, Erwin Johannes Eugen; The Somme; World War I; World War II

References and further reading:

Bond, Brian. *Liddell Hart: A Study of His Military Thought.* New Brunswick, NJ: Rutgers University Press, 1977.

Corum, James S. *The Roots of Blitzkrieg: Hans von Seeckt and German Military Reform.* Lawrence: University Press of Kansas, 1992.

Mearshheimer, John J. *Liddell Hart and the Weight of History.* New York: Cornell University Press, 1988.

Light Brigade, Charge of the (25 October 1854)

Near-suicidal British cavalry charge against Russian artillery in the Battle of Balaklava in the Crimean War. The British commander at Balaklava, General Fitzroy James Henry Somerset, Baron Raglan, saw from his headquarters on Sapoune Heights that the Russians on Causeway Heights were removing allied guns from the redoubts they had captured from the Turks that morning. Raglan's perhaps erroneous recollection that his hero, the Duke of Wellington, had never lost a gun roused him to action. He immediately tried to prevent the Russians from adding these guns to their own artillery.

Raglan routinely delegated tactical decisions to his staff and field officers, even when communication among them was poor. He chose to defer to General George Charles Bingham, Third Earl Lucan, nominal commander of the British cavalry, who was then with his troops east of Sapoune at the western end of the North Valley. From that vantage point, Lucan could not see the redoubts.

Raglan dictated the following order to his quartermaster general, Colonel Richard Airey: "Lord Raglan wishes the Cavalry to advance rapidly to the front, follow the Enemy & try to prevent the Enemy carrying away the guns. Troop Horse Attily may accompany. French Cavalry is on yr left. Immediate." The obscurity of this order has prompted much controversy among historians over the years, but most of them blame Raglan for the disaster.

Airey's aide-de-camp, Captain Lewis Edward Nolan, galloped the order downhill to Lucan. Nolan attempted to indicate which guns Raglan meant, but Lucan could see only the three main Russian batteries, aimed south from Fedioukine Heights, north from Causeway Heights, and west from the eastern end of the North Valley. Lucan ordered Brigadier General James Thomas Brudenell, Seventh Earl Cardigan, to charge the easternmost battery, about a mile and a quarter away. Cardigan simply obeyed. Disgusted with Lucan, Nolan voluntarily took part in the charge and was among the first to die. Of the 675 cavalrymen in the Light Brigade, only about 400 reached the guns, where they avenged their comrades by furiously sabering the Russian cannoneers.

"The Charge of the Light Brigade" soon entered legend as a symbol of bravery, perhaps blind bravery, against overwhelming odds. Poet Laureate Alfred Lord Tennyson's poem of the same title even included the line, "Someone had blundered," but the offending words were excised, supposedly at the behest of Queen Victoria.

Eric v. d. Luft

See also: Balaklava; Crimean War; Sevastopol, Siege of

References and further reading:

Adkin, Mark. *The Charge: Why the Light Brigade Was Lost.* London: Leo Cooper, 1996.

Bachrach, Deborah. *The Charge of the Light Brigade.* San Diego: Lucent, 1997.

Harris, John. *The Gallant Six Hundred: A Tragedy of Obsessions.* London: Hutchinson, 1973.

Lummis, William Murrell. *Honour the Light Brigade.* London: J. B. Hayward and Son, 1973.

Lin Biao (1907–1971)

Chinese Communist military leader. Lin Biao was, for many years, a close associate of Mao Zedong and a key figure behind the successful Communist effort to defeat the Nationalists and gain control of China and the more mixed effort to face American intervention in Korea. His downfall under mysterious circumstances only added to his image.

At age 18 years, he entered the famous Whampoa Military Academy in Guangzhou, and soon thereafter he served as a platoon leader and later battalion commander in the famous Northern Expedition in 1927–1928 to "unify" China under Chiang Kai-shek's Nationalist regime. When Chiang turned against the Communists, Lin defected to Mao's forces, and several years later, his corps was the vanguard in the Long March of 1934–1935, the long defeat—yet ultimately victory—as the Communists moved from southeastern to northwestern China.

When the anti-Japanese war (in which the Communist were little more effective than their Nationalist rivals) ended in 1945, Lin took command of Communist forces in Manchuria, and putting Mao's theories into practice, he abandoned the cities, gained the support of peasants, and took control of the countryside. By 1948, the Nationalists had suffered irreversible defeat in the northeast, despite massive American aid, and within a year had lost the battle for China.

In the late 1950s, Lin took over the army. He became a close ally of Mao during the Cultural Revolution of the mid-1960s that threatened to tear China apart. Lin compiled Mao's quotations into the famous book, *Quotations of Chairman Mao.* In 1966, Mao named Lin as his successor.

Then in 1971, it appeared—proof is sketchy—that Lin helped organize a plot against Mao (perhaps in opposition to the latter's opening to the Americans). When discovered, he tried to flee on an airplane to the Soviet Union; his plane supposedly crashed. But the reasons for and exact means of his demise remain shrouded in mystery.

Charles M. Dobbs

See also: Chinese Civil War; Korean War; Mao Zedong; Vietnam Conflict

References and further reading:

Chassin, Lionel Max. *The Communist Conquest of China: A History of the Civil War, 1945–1949.* Cambridge, MA: Harvard University Press, 1965.

Yao Ming-le. *The Conspiracy and Death of Lin Biao.* New York: Alfred A. Knopf, 1983.

Lincoln, Abraham (1809–1865)

Sixteenth president of the United States and for many the model of how a commander in chief should conduct a war. The American Civil War dominated the Lincoln administration and consumed the president's daily thoughts and actions. Although it can be claimed that the South had better generals, at least at the beginning of the war, Lincoln's leadership was the decisive factor in the North's favor, and historians have often listed Abraham Lincoln as the Union's greatest "asset."

On a practical level, Lincoln made several strategic decisions that altered the course of the war. He visited the army in the field 11 times, spending 42 days with them, and had voluminous correspondence with commanders. He studied military strategy and spent long hours at the War Department telegraph office. In May 1862, he personally issued or-

Detail of a photograph by Alexander Gardner of Abraham Lincoln with General McClellan (facing Lincoln) at Antietam, 1862. (Library of Congress)

ders for the occupation of Norfolk and later tried to organize Union attempts to defeat Stonewall Jackson in the Shenandoah Valley. Fearing for Washington's safety, he transferred the Army of the Potomac from southeast of Richmond to northern Virginia. In September 1863, he later decided to transfer four divisions of that army to Chattanooga under General William Rosecrans after the loss of the Battle of Chickamauga. It was not uncommon for the president to study maps and make suggestions to his generals. Yet Lincoln left most military decisions to his commanders; he was not what a later generation would call a "micromanager." He insisted that commanders act promptly, not move slowly, and take advantage of numeric superiority. Such considerations led him to remove or demote generals such as George McClellan and remain fearlessly loyal in the face of criticism of commanders like Ulysses Grant.

Lincoln's greatest impact on the war was in the political arena, where he was able to promote and sustain northern participation. He strengthened and unified the Republican Party through the use of political and military appointments. He kept many of the border states in the Union and worked with Democrats who opposed emancipation and "subjugation" of the Confederate states. He was also able to dominate the Peace Democrats and their 1864 platform,

The Battle of the Little Bighorn. Painting by C. M. Russell, 1903. (Library of Congress)

which called for an armistice and a negotiated end to the fighting. Lincoln issued directives and urged the passage of laws that harnessed the industrial capacity of the Union for war production, promoted volunteerism, and stifled criticism. His declaration of martial law is perhaps the most controversial aspect of his war leadership. Through speeches, letters, and the Emancipation Proclamation, the president was able to bring a moral clarity and authority to the cause and conduct of the war. The conflict was transformed from one over states' rights and secession to a crusade for freedom and justice. For many, he had come to personify the northern cause, and his passing is often referred to as the last death of the Civil War. Abraham Lincoln was the only American president whose entire administration was conducted in time of war.

T. Jason Soderstrum

See also: Chickamauga, Battle of; Grant, Ulysses Simpson; McClellan, George Brinton; Rosecrans, William Starke

References and further reading:

Borritt, Gabor S., ed. *Lincoln's Generals.* New York: Oxford University Press, 1994.

Davis, William C. *Lincoln's Men: How President Lincoln Became Father to an Army and a Nation.* New York: Free Press, 1999.

Williams, T. Harry. *Lincoln and His Generals.* New York: Alfred A. Knopf, 1952.

Little Bighorn (25–26 June 1876)

Perhaps the most written about and hotly debated battle in U.S. military history. The deeper symbolism and mythology linked to what is popularly known as "Custer's Last Stand" have often outweighed in the minds of Americans what actually occurred on 25–26 June 1876. Even Americans severely ignorant of their own history know something of "Custer's Last Stand" and of that other "They Died with Their Boots On" epic—the Alamo.

In December 1875, under the direction of President Ulysses S. Grant, the commissioner of Indian affairs ordered all northern Plains Indians to report to their agencies by 31 January 1876 or be forced to do so by the army. General William Sherman and General Philip Sheridan had planned a winter campaign to break the hostiles, but General George Armstrong Custer had been delayed in reporting to his command because of a conflict with the president. On 17 May, General Alfred Terry, Custer, and the Dakota column left Fort Abraham Lincoln, in a coordinated effort with troops under the command of General George Crook, to round up renegade Sioux and Cheyenne warriors. On 22 June, Terry ordered Colonel John Gibbons and Custer to lead two columns into the valley of Little Bighorn, trapping any Indians found in the valley between them.

The nearly 600 soldiers of Custer's Seventh Cavalry and 35 Crow scouts marched south along Rosebud Creek. On the night of 24 June, from a tall peak in the Wolf Mountains, scouts saw the unmistakable signs of a large village in the Little Bighorn Valley. Believing they would scatter, Custer ordered a forced march to engage them before they could escape. A few miles from the camp, Custer ordered Major Marcus A. Reno to led three companies directly into the Little Bighorn and attack the southern end of the village. Custer would take the remaining five companies east of the river and attack the village's northern end. Captain Frederick Benteen would take three companies to the south, along the Wolf Mountains, to make sure no one escaped.

Although the exact size of the encampment is debated, it was the largest Indian village ever to congregate on the Plains. On the morning of the 25th, Reno's troops crossed the Little Bighorn River 2 miles south of the village and then advanced on the lodges. Just short of the encampment, Hunkpapa warriors met the oncoming soldiers. Able to hold his position for just 15 minutes, Reno ordered his men to retreat to a grove of cottonwood trees along the river. Feeling surrounded, Reno ordered his command to withdraw to the steep bluffs on the other side of the river. Of his 140 men, Reno had 40 killed, 13 wounded, and 17 stranded in the trees below. Benteen's troops arrived on the scene and dug in with Reno. Custer's companies attacked the village from a broad coulee know as Medicine Tail. As the troopers tried to cross the river, they received heavy fire from warriors. After they were driven back to the bluffs, it is debatable what exactly occurred. Whether they were overwhelmed by Indians led by the warrior Gall or encircled from the rear by Crazy Horse's Oglala, every soldier under Custer's command was killed. Although the exact number of Sioux and Cheyenne who died is unknown, 263 soldiers were killed and 60 wounded (of Reno's men). It is said that for the next four decades, every saloon in the United States had mounted over the bar either a full-length painting of a nude or a depiction of Custer's Last Stand.

T. Jason Soderstrum

See also: American Indian Wars; Crazy Horse; Custer, George Armstrong; Sitting Bull

References and further reading:

Gray, John S. *Custer's Last Campaign: Mitch Boyer and the Little Bighorn Reconstructed.* Lincoln: University of Nebraska Press, 1991.

Magnussen, Daniel O. *Peter Thompson's Narrative of the Little Bighorn Campaign 1876: A Critical Analysis of an Eyewitness Account of the Custer Debacle.* Glendale, CA: Arthur H. Clark Company, 1974.

Utley, Robert M. *Cavalier in Buckskin: George Armstrong Custer and the Western Military Frontier.* Norman: University of Oklahoma Press, 1988.

Livonian War (1558–1583)

A military conflict initiated by Ivan IV of the Muscovite state against Livonia in an attempt to gain an outlet to the Baltic Sea for trade with the West. Hostilities broke out in January 1558. Ivan invaded Livonia, which the grand master of the Livonian Order of Knights had occupied and ruled since the thirteenth century. His forces seized much of its territory, including the commercial port of Narva (Dorpat).

The Livonian knights, unable to face Ivan's offensive, placed themselves under the protection of the king of Poland. Subsequently, Livonia was partitioned among Poland, Denmark, and Sweden. This development precipitated the war between Lithuania and the Muscovite state. Ivan's army captured several towns, including the fortified city of Polotsk. Faced with complete collapse, Lithuania formed a political union with Poland in 1569, and Stepan Bathory was elected king of the Polish-Lithuanian kingdom. An able military commander, Bathory led his well-trained army against the Russians, defeated them repeatedly, and captured Polotsk and several border towns. His advance was finally halted at Pskov. The Swedes, meanwhile, took advantage of Ivan's unfavorable position, invaded the Baltic, seized Narva, and occupied the entire coast of the Gulf of Finland.

Finally, Ivan appealed to Pope Gregory XIII to mediate the conflict. Hoping to bring Muscovy into the Catholic fold, the pope dispatched the Jesuit Antonio Possevino, who arranged an armistice between Ivan and Bathory in 1582. The following year, Ivan ceded Livonia and Polotsk but kept his former possessions along the Lithuanian border. A year later, he signed a less favorable armistice with Sweden, surrendering most of the Baltic coastline. After 25 years of intermittent warfare, Ivan IV was no closer to gaining a window to the West, and Moscow had to wait another century to achieve this objective.

James J. Farsolas

See also: Ivan IV

References and further reading:

Attman, Artur. *The Struggle for Baltic Markets: Powers in Conflict, 1558–1618.* Goteborg: Vetenskaps, 1979.

Urban, William L. *The Livonian Crusade.* Washington, DC: University Press of America, 1981.

Lobengula (a.k.a. Lopenule, Nobengulu, or Ulopengule) (c. 1830–1894)

King of the Ndebele (Matabele), in what is today Zimbabwe. At the beginning of the 1840s, Lobengula's father, the Ndebele king Mzilikazi, ordered many of his subjects, including his sons, to be killed in an effort to stamp out possible opposition. However, Lobengula was hidden by his mother and lived in seclusion for several years.

Consequently, little is known about Lobengula until the death of Mzilikazi in 1868. Two years later, Lobengula was installed as king at the insistence of the witch doctors and *indunas.* However, not all the Ndebele tribes recognized Lobengula as their king, and he took action against his opponents, routing them in battle. In 1881 he moved to the new town he had built for himself at Gibexhegu, later renamed Bulawayo. Aggressive raiding against other tribes was an integral part of Ndebele life, which for several years ensured dominance in his sphere of influence. However, in due course many European hunters, traders, and concession seekers entered Matabeleland and Mashonaland. Against the background of Cecil John Rhodes's plans for British expansion northward from the Cape Colony, Lobengula first signed a treaty with the Boer Transvaal Republic (1887) and then with John Smith Moffat, a British representative (1888).

Also in 1888, Lobengula ceded his country's mineral rights to a group representing Rhodes. Rhodes then formed the British South Africa Company (BSAC), and soon British settlers started to occupy portions of what became known as Southern Rhodesia (now Zimbabwe). Soon clashes between white and black ensued, and the BSAC's forces were ordered to occupy Matabeleland. Lobengula set fire to his capital, Bulawayo, on 3 November 1893, the day before the company's forces arrived. Lobengula fled with several of his regiments, pursued by the company's men. A patrol under Major Allan Wilson caught up with Lobengula's forces on the Shangani River, but all 33 members of the "Shangani Patrol" were killed. Soon after, Lobengula reached Pashu's country, but there he died—of self-administered poison, smallpox, or arthritis. Lobengula was a man of high intelligence, but he lacked the military genius and ruthless cruelty of his father. He was not inclined to be a warrior, but his suspicious nature and the influence of witch doctors and *indunas* led him to commit acts of cruelty.

André Wessels

References and further reading:

Burke, E. E. "Lobengula." In *Dictionary of South African Biography.* Vol. 3. Eds. D. W. Krüger and C. J. Beyers. Pretoria: Human Sciences Research Council, 1977.

Hole, H. M. *Lobengula.* London: Philip Allen & Company, 1929.

Preller, G. S. *Lobengula: The Tragedy of a Matabele King.* Johannesburg: Afrikaanse Pers-Boekhandel, 1963.

Van Zyl, M. C. "Lobengula." In *Standard Encyclopaedia of Southern Africa,* ed. D. J. Potgieter. Cape Town: Nasou Limited, 1947.

Lodi (10 May 1796)

A battle fought in northern Italy between the French Army of Italy, commanded by Napoleon Bonaparte, and an Austrian army commanded by J. P. Beaulieu. Bonaparte opened the first Italian campaign in April 1796 with an attack on the Piedmontese army, commanded by General Colli, and a supporting Austrian force, commanded by Generals Provera and Argentau. The French assault led to several inconclusive actions and the withdrawal of the Piedmontese army.

By 24 April 1796, Bonaparte was threatening the Piedmontese capital, Turin. The king of Piedmont asked for an armistice, which Bonaparte granted. The effective neutralization of the Piedmontese allowed Bonaparte to attempt to trap the main Austrian army, which had been left in a dangerously exposed position southwest of Milan as a result of the Piedmontese negotiations.

The French army, 30,000 strong, thus began a 50-mile forced march in an attempt to seize crossings over the River Adda, including the town of Lodi, in order to trap the Austrian army. Beaulieu, aware of his danger, was in full retreat, and with a vigor unusual for Austrian generals of the period, managed to escape with most of his army across the Adda River, leaving only a 10,000-strong rear guard by the time the French reached Lodi on 10 May.

The Battle of Lodi itself consisted simply of determined French assaults upon a bridge over the River Adda. The first of these failed. The second, however, prevailed, and the French main column, assisted by a body of cavalry that had forded upriver, routed the Austrians. The Austrians lost about 150 men, and 1,700 were taken prisoner. The French lost 350 men.

Despite the relatively small scale of the action, the French victory at Lodi opened the way for the capture of Milan. As a result, the Kingdom of Savoy and the Duchies of Parma and Modena were forced to make peace with the French, and the Kingdom of Piedmont found it expedient to complete peace negotiations with the French. The French were now masters of the northwestern Italian peninsula, and Bonaparte was free to begin efforts to force the Austrians from Mantua, which would culminate in the Battle of Rivoli.

Joseph Isenberg

See also: French Revolutionary Wars; Napoleon I

References and further reading:

Chandler, David G. *Campaigns of Napoleon.* New York: Scribner, 1966.

Connelley, Owen. *Blundering to Glory.* Wilmington, DE: Scholarly Resources, 1999.

Logistics

Umbrella term for military activities other than strategy and tactics that emerged about two centuries ago. Today it is largely an American usage crystallized in World War II and encompassing military supply, transportation, medical service, and construction-maintenance.

Historically, there have been three solutions to the problem of supplying an army with the least sacrifice of power, mobility, and range: (1) self-containment, in which the force carries all its supplies, mortgaging power and range to mobility; (2) local supply, a forage system that narrows military activity to growing seasons and fertile areas; and (3) supply from bases, favoring power and range over mobility. Alexander the Great's army marched, self-contained, from Macedonia to India. Napoleon's armies foraged their way across Europe for a dozen years, but his Grand Army starved and froze while retreating from Moscow through country it had already denuded. The arrival of mass armies and greatly increased firepower after the Industrial Revolution, combined with highly developed systems of communication and economic support, brought widespread reliance on supply from bases. Generals continued to be more interested in strategy and tactics than in logistics, however, and thus took less advantage of manpower and munitions than they might have done. In 1870 the Germans used staged, continuous resupply by railroad to support their invasion of France by a mass army and found that even constant forward movement of supply depots was not enough to forestall the necessity of foraging. The long-planned envelopment of Paris by German armies was frustrated in 1914 in part because their plans called for more troops than the road system could bear and in part because rail lines used by the French to bring reinforcements from the Alsace-Lorraine front had not been targeted by German planners.

World War II was a breathtaking exercise in the logistics of total war. The Allies produced seven times as many planes as the Axis powers, five times as many trucks and artillery pieces, and more than four times as many machine guns and tanks. All these and millions of men and women had to be transported to fronts all over the world in the face of enemy resistance. But the inability of the Luftwaffe to resupply the German army trapped at Stalingrad, like the halting of General George Patton's offensive late in 1944, indicated the limits on logistical problem solving in total war. Nonetheless, the United States was able not only to lavishly supply (by the standards of the other belligerents) its own forces but also to provide millions of tons of equipment for the Nationalist Chinese, the Free French, the Soviets, and even the British forces.

At war's end, the advent of nuclear weaponry threatened to make conventional logistics irrelevant, and superpower rivalry raised new problems in the organization, deployment, and supply of mass armies. At the same time, Third World conflicts increasingly demanded rapid deployment of special forces. The United States solved these problems in the post-Vietnam era by elaborating AirLand, mechanized combined operations that could stop a Soviet offensive in Europe and be adapted to more limited scenarios, such as the deployment of 527,000 personnel and 3,500 aircraft in the Gulf War.

Joseph M. McCarthy

See also: Patton, George Smith, Jr.; World War II

References and further reading:

Lynn, John A., ed. *Feeding Mars: Logistics in Western Warfare from the Middle Ages to the Present.* Boulder, CO: Westview, 1993.

Shrader, Charles R. *United States Army Logistics, 1775–1992: An Anthology.* 3 vols. Washington, DC: Center of Military History, United States Army, 1997.

Van Creveld, Martin L. *Supplying War: Logistics from Wallenstein to Patton.* New York: Cambridge University Press, 1977.

Long Island, Battle of (22 August 1776)

A major defeat of the patriot American forces early in the Revolutionary War. In summer 1776, a huge British fleet with a large army arrived off New York City. The British occupied Staten Island, and the Americans, led by George Washington, defended Long Island along Brooklyn Heights as well as the Battery, Fort Washington, and other places in and around New York City. For seven weeks, the British strengthened their positions.

British general William Howe and 10,000 troops landed on Long Island on 22 August and proceeded to formulate a plan. The American defenders were at best untrained in warfare and inexperienced in battle; at worst, a mob without artillery. While some British and German troops demonstrated in front of the American position, Howe and about half his men marched around and behind the American left flank. They gained complete surprise, General John Sullivan's left wing was crushed, and the men fled. Washington ordered a retreat.

After considering a defense of Long Island at Gowanus Bay and realizing the British fleet could trap him, Washington had his men rowed across the East River to Manhattan, where after a battle at Harlem Heights and some fighting at Fort Lee and Fort Washington, he retreated to White Plains and later across the Hudson to New Jersey. General William Howe and his brother, Admiral Richard Howe, not wanting a repeat of the slaughter at Breed's Hill, pursued rather slowly. Howe also feared exasperating the Americans and hoped for some sort of negotiated reconciliation. Although the brothers Howe may have felt they had demonstrated the Americans' inability to withstand a proper British attack, they also had given Washington time to withdraw and save his army to fight another day.

Charles M. Dobbs

See also: American Revolution; Washington, George

Washington's retreat at Long Island, 1776. (Library of Congress)

References and further reading:
Bliven, Bruce. *Battle for Manhattan.* New York: Holt, 1956.
Gruber, Ira D. *The Howe Brothers and the American Revolution.* New York: Atheneum, 1972.

Longstreet, James (1821–1904)

Confederate field commander in the American Civil War, a skillful subordinate but unsuited for independent command. Longstreet was born in Edgefield District, South Carolina, on 8 January 1821. As a member of the West Point class of 1842, his roommate was William S. Rosecrans. An infantry lieutenant in the Mexican-American War, he won brevets to captain at Churubusco on 20 August 1847 and major at Molino del Rey on 8 September and was severely wounded at Chapultepec on 13 September. Promoted to captain in 1852 and major in 1858, he resigned on 1 June 1861 to join the Confederacy, which commissioned him brigadier general on 17 June.

Longstreet defeated Irvin McDowell's vanguard at Blackburn's Ford, Virginia, on 18 July and distinguished himself at First Bull Run. Promoted to major general on 7 October, he served under Joseph Johnston in the Peninsula campaign, fighting at Yorktown, Williamsburg, Seven Pines, and Fair Oaks. He excelled at Second Bull Run; fought at South Mountain, Maryland; commanded Robert E. Lee's right flank at Antietam; and held Marye's Heights at Fredericksburg. He made lieutenant general on 9 October 1862.

On the way to Gettysburg, Longstreet counseled Lee to maneuver between George Meade's army and Washington, D.C., thus forcing Meade to attack. Nevertheless, Lee attacked, with Longstreet commanding the right. By the third day, Longstreet was so disheartened that he had difficulty bringing himself to convey Lee's order to George Pickett to charge the Union center. Historians analyzing Gettysburg from both tactical and strategic viewpoints usually agree that Longstreet's plans were more reasonable than Lee's.

After Gettysburg, Lee detached Longstreet to Georgia to reinforce Braxton Bragg. Longstreet's arrival in time to command the left on the second day at Chickamauga ensured the Confederate victory. His siege of Knoxville, however, deprived the South of needed manpower at Chattanooga. Wounded at the Wilderness, he recovered for the final Petersburg and Richmond campaigns and was with Lee at Appomattox.

After the war, Longstreet became a Republican and held several federal offices in the Grant, Hayes, Garfield, Arthur, McKinley, and Theodore Roosevelt administrations. He died in Gainesville, Georgia, on 2 January 1904.

Eric v. d. Luft

See also: American Civil War; Antietam/Sharpsburg; Bragg, Braxton; Bull Run, First/Manassas; Bull Run, Second/Manassas Junction; Chickamauga, Battle of; Fredericksburg; Gettysburg; Grant, Ulysses Simpson; Hill, Ambrose Powell; Johnston, Joseph Eggleston; Lee, Robert Edward; Wilderness

References and further reading:

Connelly, Thomas Lawrence, and Barbara L. Bellows. *God and General Longstreet: The Lost Cause and the Southern Mind.* Baton Rouge: Louisiana State University Press, 1982.

Eckenrode, Hamilton James, and Bryan Conrad. *James Longstreet: Lee's War Horse.* Chapel Hill: University of North Carolina Press, 1986.

Piston, William Garrett. *Lee's Tarnished Lieutenant: James Longstreet and His Place in Southern History.* Athens: University of Georgia Press, 1987.

Wert, Jeffry D. *General James Longstreet: The Confederacy's Most Controversial Soldier.* New York: Simon & Schuster, 1993.

Louis XIV (1638–1715)

King of France and warlord for more than 70 years. In the eyes of a younger generation that had forgotten the former power of Spain, Louis XIV was a vain warmonger and enduring threat to the European balance of power, largely because of a policy directed at the destruction of Spain's power that continued long after Spain had apparently lost its influence on the international stage.

Louis led his armies in his youth but never fought a major battle. In later years, he sometimes assumed personal command at sieges because this form of war permitted him to establish permanent camps where he could carry out his administrative responsibilities. His letters reveal a lively understanding of siege craft, and he certainly had a gift for finding able commanders.

Louis inherited the Thirty Years' War (1618–1648), a long-drawn-out conflict with Spain, and the civil wars of the Fronde. In 1661 he launched the War of Devolution with Spain, only two years after the signing of the Peace of the Pyrenees. A Dutch War (1672–1678) soon followed, in which Franche-Comté fell into his hands. Between 1678 and 1688, Louis used armed force to intimidate his neighbors, annexing Strasbourg and other territory in Germany. This policy shaded into a general European conflict in 1688, the War of the Grand Alliance or War of the League of Augsburg (1688–1697). Primarily defensive in Europe, it ended with significant French gains in the Americas as Britain was driven to unfortunate terms by near-bankruptcy. In the long War of the Spanish Succession (1701–1714), Louis devastated France's economy while establishing a king of his house in Spain (although not in many of Spain's former European territories) and ending the Spanish menace, but in the process he sacrificed his earlier gains in the Americas. He died in 1715, admitting on his deathbed that he had "loved war too much."

Erik A. Lund

See also: Blenheim-Höchstädt, Battle of; Condé, Louis II de Bourbon, Fourth Prince de; Denain, Battle of; Franco-Spanish War; Fronde, Wars of the; Grand Alliance, War of the; Louvois, François-Michel Le Tellier, Marquis de; Luxembourg, François Henri de Montmorency-Bouteville, Duc de Piney; Luxembourg, Siege of; Malplaquet, Battle of; Oudenaarde, Battle of; Queen Anne's War; Spanish Succession, War of the; St. Gotthard Abbey; Turenne, Henri de la Tour d'Auvergne, Vicomte de; Vauban, Sébastien Le Prestre de

References and further reading:

Bluche, François. *Louis XIV.* Trans. Mark Greenglass. French edition, 1984. Oxford: Blackwell, 1990.

Chandler, David G. *The Art of Warfare in the Age of Marlborough.* London: Batsford, 1976.

Rowen, Herbert Harvey. *The King's State: Proprietary Dynasticism in Early Modern France.* New Brunswick, NJ: Rutgers University Press, 1980.

Vault, Françoise Eugène de, and [J. J.] Pelet. *Mémoires militaire relatifs à la Succession d'Espagne sous Louis XIV.* 11 vols. Atlas. Paris: Imprimérie Royale, 1835–1862.

Louisbourg, Expedition against (May–June 1758)

Successful British and colonial operation that forced the surrender of the linchpin of the defense of New France. Early in 1757, the new British prime minister, William Pitt, ordered the Earl of Loudoun to seize Louisbourg, a fort on Cape Breton Island guarding the entrance to the Gulf of St. Lawrence. This effort failed miserably. By the time 10 regiments arrived from Ireland to join forces already gathered in Halifax, Nova Scotia, a large French fleet had reached Louisbourg and brought 2,500 reinforcements. Stormy weather halted British operations for the rest of July and August. The British naval commander, Vice Admiral Francis Holbourne, believed offensive action was hopeless, and Loudoun agreed and abandoned the attack and retreated to New York City.

General William Pepperrell at the siege of Louisburg in 1745. (Library of Congress)

The next year, Pitt wisely promoted three junior officers who all would play major roles in the eventual British victory over the French in North America. One of these officers, Jeffrey Amherst, commanded the attack on Louisbourg. A British fleet sailed to Halifax in winter 1758, and reinforcements arrived several months later in early May. On 28 May, the expedition set out. Amherst followed the plan New Englanders had used in their successful attack in 1745. Led by James Wolfe, men landed at Gabarus Bay, west of Louisbourg, and soon established a siege that greatly damaged the town. On 26 July 1758, the governor raised the white flag of surrender.

The loss of Louisbourg imperiled the French colony in Quebec, which depended on seaborne transportation through the Gulf and past the fort to France. Defeat at Louisbourg foreshadowed France's defeat in North America several years later.

Charles M. Dobbs

References and further reading:

Anderson, Fred. *Crucible of War: The Seven Years' War and the Fate of Empire in British North America, 1754–1766.* New York: Alfred A. Knopf, 2000.

Schwartz, Seymour I. *The French and Indian War, 1754–1763: The Imperial Struggle for North America.* New York: Simon & Schuster, 1995.

Louvois, François-Michel Le Tellier, Marquis de (1639–1691)

The most outstanding military organizer of the seventeenth century. Louvois was the son of one of the most powerful official of Louis's XIV early reign. In 1655, the senior Le Tellier brought his dissolute teenaged son into the war department to teach him military administration. He emerged as a hard worker, supremely confident in his own ability to be a brilliant administrator. His administrative career is exceptional; as early as 1661 he was working with his father in the foreign affairs council and was appointed state minister in 1672, in charge of military affairs and administration, after his father's retirement. With Louis's backing, Louvois transformed the French army from a feudal semi-independent force to the first modern army. The reform of the financial and tax system gave Louvois the money he needed to increase the king's troops both in quantity and quality.

His first task was to build a centralized army control: he used civilian inspectors as war commissioners or army intendants whose job was to reduce corruption by regular inspections of troops. Officers, whatever their rank of birth or their commission, were responsible for obedience to orders and regulations, which were prescribed by Louvois. Corrupt or insubordinate officers were dismissed. Officers were also expected to be conversant with current military theories and practices. The purchasing of officers' commissions was fought by Louvois. He instituted ranks that were appointed by the king rather than bought: from 1661, lieutenant colonels were promoted solely on merit, the rank of brigadier general (given by the king) opened the way to high command to poorer nobility (Sébastien Le Prestre de Vauban's career is the best example). Louvois's work included the reform of various administrative branches: creating militia for second-line duties, which was the first attempt at a national conscription system; constructing barracks for the troops instead of quartering then in citizens' houses; providing regular pay and a uniform; implementing tighter discipline and a military justice code; and improving standardized weapons.

All these reforms were conducted to improve the soldier's lot and to ensure his loyalty to the king, as well as to heighten morale and corps pride. Louvois supported the first veterans' hospital, Les Invalides. The French army reached the incredible number of 450,000 soldiers by Louvois's death. He had constructed the military tool needed by Louis XIV for his aggressive foreign policy.

Gilles Boué

See also: Grand Alliance, War of the; Louis XIV; Vauban, Sébastien Le Prestre de

References and further reading:

Chartrand René. *Louis XIV's Army.* London: Osprey, 1988.

Corvisier, André. *Louvois.* Paris: Fayard, 1983.

Ludendorff, Erich Friedrich Wilhelm (1865–1937)

World War I general and postwar politician. Ludendorff was born on 9 April 1865, near Posen. He entered the army in 1883 and was appointed to the general staff in 1895. On 2 August 1914, he was appointed deputy chief of staff for the Second Army. He saw action on the western front and immediately won fame through his contribution to the capture of Liège. On 22 August 1914, he was appointed chief of staff of the Eighth Army, serving under Paul von Hindenburg. Their spectacular victories over the Russians at Tannenberg and the Masurian Lakes (August–September 1914) made them the most popular generals in Germany. In November 1914, Ludendorff was made chief of staff of the Supreme Command on the eastern front, again serving under Hindenburg. Convinced that the war could be won on the eastern front, Ludendorff came into increasing conflict over strategy with Erich von Falkenhayn, chief of the general staff. Falkenhayn's dismissal in August 1916 resulted in his replacement by Hindenburg, with Ludendorff serving as first quartermaster general. Ludendorff and Hindenburg were now in charge of German military strategy and increasingly of German domestic policy, a situation that by 1917 had led to the creation of a "silent dictatorship." Domestically, Ludendorff and Hindenburg sought to place the German economy on a total war footing. Militarily, they were responsible for the reintroduction of unrestricted submarine warfare in April 1917 and the annexationist peace treaties imposed on Russia and Romania in 1918. Following victory in the East, Ludendorff oversaw Germany's final attempt to break the stalemate on the western front in 1918. After the war, Ludendorff became active in right-wing politics. He participated in Hitler's failed Beer Hall Putsch in 1923 and served in the Reichstag from 1924 to 1928. Ludendorff died on 20 December 1937 in Bavaria.

J. David Cameron

See also: Falkenhayn, Erich von; Hindenburg; Paul von Beneckendorf und von; William II

References and further reading:

Ludendorff, Erich. *Meine Kriegserinnerungen, 1914–1918.* Berlin: E. S. Mittler und Sohn, 1921.

Kitchen, Martin. *The Silent Dictatorship: The Politics of the German High Command under Hindenburg and Ludendorff, 1916–1918.* New York: Holmes and Meier, 1967.

Lundy's Lane, Battle of (25–26 June 1814)

The most sanguinary battle of the War of 1812, a draw. The Americans sought to follow up their victory several weeks earlier at Chippewa on the Niagara frontier. General Winfield Scott and about 1,000 troops were in the vanguard, with General Jacob Brown in command of the main unit. Scott was aggressive and unexpectedly encountered a British force of 1,600 to 1,800 troops. Although outnumbered, he believed he had to attack because a retreat might result in panic among his main body of troops.

So, in the early evening of 25 July 1814, Scott ordered an attack uphill against British troops and a gun battery. The attack failed, for British artillery was extremely effective, and Scott had to retreat. As more and more Americans arrived, they once again assumed the offensive as dusk turned into night. The British too received reinforcements, and the fighting became intense with heavy casualties—more than 850—on each side.

After dark, the Americans began to run out of ammunition, and General Brown ordered them to retreat. The British and Canadians held the field but were too exhausted to give chase as the Americans retreated to Fort Erie. The earlier American victory at Chippewa was undone, and bloody Lundy's Lane ended this last U.S. effort to invade Canada.

Charles M. Dobbs

See also: War of 1812; Scott, Winfield

References and further reading:

Graves, Donald E. *The Battle of Lundy's Lane: On the Niagara in 1814.* Baltimore: Nautical and Aviation Publishing Company of America, 1993.

Hitsman, J. Mackay. *The Incredible War of 1812.* Toronto: University of Toronto Press, 1965.

Stanley, George F. G. *The War of 1812: Land Operations.* 1983.

Lützen, Battle of (16 November 1632)

The high tide of Swedish strategic dominance under Gustavus Adolphus during the Thirty Years' War. Following the Swedish victory at Breitenfeld in the fall of 1631, the Swedes under Gustavus marched into Bavaria to attack the senior member of the Catholic League. The Hapsburg general Albrecht von Wallenstein skillfully rallied the imperial troops and wore down the Swedish force. After the imperial victory at Nuremberg on 3 September 1632, a series of inconclusive engagements brought the combatants back into Saxony.

The two forces clashed again at Lützen on 16 November 1632. Eighteen thousand Swedes faced 25,000 imperial troops. Wallenstein occupied a strong position with cavalry on the wings, four *tercios* composed of pikemen and musketeers in the center, and artillery emplaced in the center and on the right flank. Additionally, a ditch defended by musketeers extended across his front. The Swedes took up their usual linear formation of two lines with infantry in the center and cavalry-infantry combinations on the wings. Gustavus, commanding his right wing of cavalry, planned to attack Wallenstein's left flank to push him back and away from Lützen. After an intense artillery barrage, Gustavus led his cavalry to penetrate the imperial musketeers and crash into Wallenstein's cavalry. Swedish infantry likewise advanced and captured the imperial artillery in the center. This success was temporary, however. Imperial cavalry charged into the flanks of the Swedish infantry, throwing them back. At this point, Gustavus moved across the field to rally his infantry and personally lead his final charge against the enemy. He died with musket balls in his arm, back, and head.

The king's subordinates quickly marshaled all available forces and plunged ahead. The battle seesawed as infantry formations pushed with pikes and cavalry repeatedly charged and countercharged. The Swedes finally rallied and swept the imperial forces off the field, capturing by nightfall all their artillery and killing upward of 12,000 men. Swedish losses were about 6,000. Though a technical victory for the Swedes, the loss of their king brought the military balance into equilibrium. The Swedes lost much of the strategic initiative and direction Gustavus provided. The Battle of Nordlingen (1634) completed the shift of fortunes back to the Catholic cause.

Bryan R. Gibby

See also: Breitenfeld; Gustavus II Adolphus; Nordlingen; Thirty Years' War

References and further reading:

Dodge, Theodore A. *Gustavus Adolphus.* New York: Da Capo Press, 1998.

Parker, Geoffrey, ed. *The Thirty Years War.* London: Routledge, 1997.

Luxembourg, François Henri de Montmorency-Bouteville, Duc de Piney (1628–1695)

The most brilliant tactical commander of Louis XIV's early reign. The son of a duelist beheaded in 1627, he was educated at a prince's court and became a close friend of the Great Condé. At the age of 20, he was appointed brigadier general after the Battle of Lens (1648). Being loyal to Condé, he took part in the Wars of the Fronde on the rebels' side. His choice leading him straight to treason against the young king, he commanded Spanish troops during the Battles of Arras and the Dunes (1658). When the peace treaty of the Pyrenees was concluded, he was allowed to return to Paris in poverty. Once again, Condé helped him in finding a rich, noble young lady to marry. He claimed for his wife the restoration of the peerage of Luxembourg and eventually added this title to his own. He became the duke of Luxembourg and had to take legal action to be accepted as such by the king's courtiers.

Luxembourg's fortune had to wait 10 years before Louis XIV again allowed him to command French troops. The Dutch War of 1672–1678 gave him the opportunity to distinguish himself. Commanding the army in 1672 after the king's departure, he won the Battle of Senef with Condé in 1674 and was made marshal of France in July 1675. The victories accumulated in the following years: Valenciennes and Cassel in 1677 and Ypres and Saint Denis in 1678. His cleverness and his composure were the cornerstones of his tactical skill, but he took a hard line with those who didn't share his views and made numerous personal enemies, including François-Michel Le Tellier, Marquis de Louvois.

In 1679, Luxembourg was deeply involved in the famous poisoning of a king's mistress and spent 14 months in the Bastille jail. After being discharged the following year, he returned to court. The war of the Augsburg League gave him a further opportunity to add new victories to the king's glory: Fleurus in 1690, Leuze in 1691, Steinkeerk in 1692, and Neerwinden in 1693. This last battle crowned him as an outstanding tactical general. After besieging Huy, Luxembourg lured William of Orange from a very strong camp into a less sound position, brought him to battle, and successfully routed the allied army. His death in January 1695 left France without anyone of his quality to take his place. His nickname was "le tapissier de Notre Dame"—Notre Dame's decorator—in token of the dozens of captured standards displayed in that cathedral.

Gilles Boué

See also: Condé, Louis II de Bourbon, Fourth Prince de; Louis XIV; Louvois, François-Michel Le Tellier, Marquis de

References and further reading:

Bluche, François. *Dictionnaire du Grand Siècle.* Paris: Fayard, 1990.

Dussieux, Louis. *Les grands généraux de Louis XIV.* Paris: Lacoffre, 1888.

Luxembourg, Siege of (April–June 1684)

Probably the best example of siege warfare in the seventeenth century. During the seventeenth century, sieges were the rule and battle the exception. Fortified cities dominated communication lines and the most fertile parts of land. A field army could not allow harassing garrisons in its rear. Luxembourg was one of the most feared and well-defended fortified towns of this era. The five-stage siege of 1684 is archetypal.

The first stage was in January, when a French army under the Marechal de Créqui isolated Luxembourg from the Spanish main army. Then, a covering force of 20,000 took position between Brussels and Luxembourg to mislead the Spanish commanders. The main army approached Luxembourg with Marechal Sébastien Le Prestre de Vauban as commanding officer. A siege had to be conducted by specialist officers, who sometime overruled superiors, including the king. The besieging army was composed of more than 25,000 soldiers, including 40 royal engineers, and more than 70 guns. The besieged Spanish troops numbered no more than 3,000 under the governor, the prince de Chimay, and the comte de Tille. From 28 April to 8 May, defending lines were dug a few miles away from the town to protect the besieging troops. Vauban had to use 12,000 workers, forcing unwilling peasants into service. The inner lines were dug by troops under artillery fire.

The fortified town of Luxembourg was protected by high cliffs, a river, and modern fortifications made of a glacis-covered tunnel in front of ditches and bastions protecting the curtain walls. From early May, parallels were dug, and then zigzag trenches were constructed. Meanwhile, sappers had dug explosive mines. The explosion on 27 May was the sign of the assault on this first line by elite troopers. The French took the covered way and had to begin a new siege on the second defense line by a bastion. Hundreds of gabions were made to protect the head of the approaching trench. During the siege operations, the French artillery fired more than 55,000 rounds, night and day. The governor of Luxembourg, not expecting any relieving army and fearing the plunder and massacre that would follow a general assault, asked to surrender on 3 June. Four days later, he left the town with 2,000 survivors. The siege had cost more than 373,000 French livres, less than the million needed to rebuild new fortifications with Vauban as the main architect.

Gilles Boué

See also: Malplaquet, Battle of; Oudenaarde, Battle of; Ramillies, Battle of; Spanish Succession, War of the; Vauban, Sébastien Le Prestre de

References and further reading:

Dollar, Jacques. *Vauban à Luxembourg.* Luxembourg: RTL Editions, 1983.

Rorive, J. P. *La guerre de siège sous Louis XIV en Europe et à Huy.* Brussels: Racine, 1998.

Lyautey, Louis-Hubert-Gonzalve (1854–1934)

French colonial administrator and soldier. Born 17 November 1854 in Nancy, France, Lyautey studied at the Saint-Cyr Military Academy, spent time in a cavalry regiment, and served in Algeria from 1880 to 1882. He served in Indochina under the influential Joseph Simon Gallieni in 1894 and transferred to Madagascar in 1896. After successfully subduing the southern portion of the island, he returned to France to command the 14th Hussars at Alençon (1902–1903). Transferred to Algeria, he commanded the Ain Sefra region and gradually began expanding French territory into Morocco. In 1906, he became commandant of Oran and squelched Moroccan resistance to his expansion, particularly the rebellion of the Beni Snassen in late 1907. Lyautey returned to France and commanded the X Corps at Rennes from December 1910 to March 1912. He returned to Africa as resident general of the protectorate of Morocco in April

1912 and stepped into the middle of a revolt. He managed to reassert French rule and restore order fairly quickly and again was able to increase French holdings.

At the start of World War I, most of his troops left for France, but Lyautey kept a potentially explosive situation in hand by working within existing tribal institutions. From December 1916 to March 1917, he served as minister of war in Paris. Afterward, he assumed his former post in Morocco and successfully defended against a rebel force led by Abd-El Krim at Taza in spring 1925. Lyautey resigned on 5 September of that year, partly as a result of the appointment of Marshal Henri-Philippe Pétain as commander in chief in Morocco. He retired at Thorey, where he died on 21 July 1934.

Harold Wise

See also: French Colonial Wars; Galliéni, Joseph Simon

References and further reading:

Hoisington, William A. *Lyautey and the French Conquest of Morocco.* New York: St. Martin's Press, 1995.

Maurois, André. *Marshal Lyautey.* Trans. Hamish Miles. London: John Lane, 1931.

Lysander (d. 395 B.C.E.)

Spartan commander who combined land and sea strategies to defeat the Athenians in the final phase of the Second Peloponnesian War. In 406 B.C.E., Lysander, with a fleet of 140 ships, engaged the Athenian navy at Lesbos, destroying 30 of its 70 ships. The Athenians, in desperation, melted down the dedications to the gods, offered freedom to slaves and citizenship to the metics (resident foreigners), and built 150 additional ships. They engaged the Spartan navy at Arginusae, where they won their last naval victory of the war. The following year, Lysander ordered his ships to the Hellespont, where he observed the Athenian fleet. The Athenians attempted to draw the Spartan ships into battle, but Lysander ordered them to wait. Frustrated by the delays and needing fresh supplies, the Athenians anchored their ships and went ashore for provisions. Lysander seized this opportunity to dart across the straits separating the two fleets and captured 160 of the 180 Athenian ships. He then pushed all Athenians living outside Athens back to the city. After an eight-month siege, the people surrendered, and Lysander installed a new oligarchic government of the Thirty. In 403 B.C.E., Lysander put down an Athenian revolt intended to restore democracy.

After the death of the Spartan king Agis II, Lysander pushed the claim of Agis's brother Agesilaus as the rightful heir to the throne. Confident that he could exercise control over Agesilaus because of a personal relationship between the two men, Lysander received an appointment as the head of the board of 30 advisers for the new king. Agesilaus often rejected Lysander's advice, and at the beginning of the Corinthian War in 395 B.C.E., Lysander returned to Greece and was slain in the first battle of the war.

Cynthia Clark Northrup

See also: Peloponnesian Wars

References and further reading:

Mattingly, Harold B. *The Athenian Empire Restored: Epigraphic and Historical Studies.* Ann Arbor: University of Michigan Press, 1996.

M

MacArthur, Arthur, Jr. (1845–1912)

Prominent U.S. Army commander. Born in Springfield, Massachusetts, on 2 June 1845, Arthur MacArthur Jr. moved with his family to Milwaukee, Wisconsin, in 1849. Following the outbreak of the Civil War, MacArthur attempted to get an appointment to West Point, but when he was unable to do so, his politically influential father, Judge Arthur MacArthur Sr., managed to have him commissioned a second lieutenant and named the adjutant of the 24th Wisconsin Infantry Regiment on 4 August 1862. At the Battle of Missionary Ridge (25 November 1863), he seized the regimental colors at a critical moment and led his regiment to the crest of the ridge. The 24th Wisconsin's colors were the first to be planted on the enemy's breastworks, and for his daring action, MacArthur was awarded the Medal of Honor on 30 June 1890. By the end of the Civil War, MacArthur had been promoted to lieutenant colonel and brevetted to full colonel.

After being mustered out of the volunteers in June 1865, MacArthur was commissioned a first lieutenant in the regular army in February 1866. For the next 30 years, he served on the frontier, eventually being promoted to lieutenant colonel in May 1896. After the outbreak of the Spanish-American War, MacArthur was promoted to brigadier general of volunteers in May 1898 and commanded a brigade during the capture of Manila. Promoted to major general of volunteers in August 1898, he was given command of the U.S. forces in the Philippines, which were fighting against the insurrectionist army of Emilio Aguinaldo. MacArthur was promoted to brigadier general in the regular army in January 1901 and major general in February, which was followed by an appointment as the military governor of the Philippines in May 1901. In September 1906, MacArthur was promoted to lieutenant general and given command of all U.S. Army personnel in the Pacific. Despite being the most senior officer in the U.S. Army, he was passed over for the position of chief of staff and retired in June 1909. MacArthur died when a blood vessel in his brain burst while he was giving a speech at the 24th Wisconsin's fiftieth reunion in Milwaukee on 5 September 1912.

Alexander M. Bielakowski

See also: Aguinaldo, Emilio; Spanish-American War

References and further reading:

James, D. Clayton. *Years of MacArthur.* Vol. 1: *1880–1941.* Boston: Houghton Mifflin, 1970.

MacArthur, Douglas. *Reminiscences.* New York: Time, 1964.

Young, Kenneth R. *General's General: The Life and Times of Arthur MacArthur.* Boulder, CO: Westview Press, 1994.

MacArthur, Douglas (1884–1964)

The most successful and most controversial of any American general. His real and unchallenged accomplishments included the development of modern staff procedures, the foundation for American military expansion in the face of World War II, and the stunning employment of amphibious force in World War II and at Inchon in 1950. Still, his record has been marred by his own arrogance and by his refusal fully to disclose the reasons for many of the high command decisions.

MacArthur was born on 26 January 1880 into a military family. His father was a Civil War hero who was later to serve as the first U.S. governor of the Philippines and who would be rewarded by promotion to lieutenant general, then a rare rank in the American service. His older brother was a Naval Academy graduate who was himself a distinguished officer and whose death at a relatively early age was much lamented by his peers.

As is the case with most complex people, MacArthur was

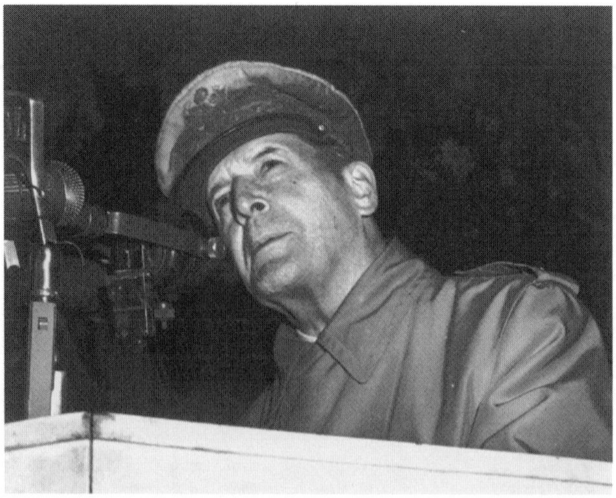

Douglas MacArthur, after being fired by President Truman, making a speech, 1951. (Library of Congress)

a morass of contradictions. He was not an innate modernizer, but when the need for modernization was proven to him, he demanded it from his subordinates and from the military system. He detested staff work but, in World War I, developed and ran what might have been the finest divisional staff in that war. In World War II, he allowed his own staff to become rather slack at conducting routine chores while he developed elaborate plans on his own that emerged fully hatched and capable of implementation. His faith in the American fighting man—army, navy, Marine Corps, and air force was immense—yet he spent most of his career with the navy, air force, and Marine Corps distrusting him to the point at which he had to challenge them with barbs to achieve his ends. He sought closeness with political leaders but only achieved it either with those whose grip on power was slight, such as Newton Baker or Herbert Hoover, or those who failed fully to understand his capacities, such as Franklin D. Roosevelt. At the same time, he alienated the political leaders who could have worked with him for even greater accomplishments than were to be his due, such as Henry Stimson or Harry Truman. MacArthur's entire life is one of contradiction.

Although MacArthur did not develop the idea, he was the first major U.S. commander to accept the necessity of bypassing Japanese strong points in the Pacific. He was late in comprehending the significance of tactical air power, yet, once convinced, no U.S. commander has ever understood or employed air power to better advantage than did MacArthur. He had been an early advocate of strategic air power, but his advocacy was so shielded that it seemed nonexistent to many of the leaders of the army and air force in the buildup to World War II. In a similar vein, MacArthur alienated the navy, though his relationship with the chief of naval operations during MacArthur's tenure as army chief of staff marked the high-water level of army-navy relations for almost the first half of the twentieth century.

MacArthur committed few grave errors, but the arrogance of his personality caused him to be blamed with many. MacArthur was held to account by the public for perceived harshness in disbursing the Bonus Marchers in 1932, for laxness in allowing half of his heavy bomber force to be wiped out on the ground when World War II erupted over the Philippines in 1941, for the supposed cowardice of "Dugout Doug" in facing combat, for not recognizing the real threat of Chinese Communist intervention in the Korean War, and for treating the president of the United States in a dismissive and belittling matter. Yet, at the same time, there is far more to each of these tales than these bare outlines, and the complexity of MacArthur's character is matched closely by the intricacies of the crises in his life.

MacArthur was an outstanding field commander and staff officer in World War I. He was an innovative superintendent at the U.S. Military Academy at West Point. He was a successful corps commander and department commander in the Philippines and was noted for the leadership he projected in the military doldrums of the 1920s. He headed the successful U.S. Olympic Team in 1928, arguably the first occasion in which the nation mounted a major effort at these international games. His term as chief of staff was outstanding and lasted longer than that of any other officer in recent times save for George C. Marshall; during his term, MacArthur laid the foundations for the rapid mobilization that the United States was able to mount in the late 1930s and early 1940s and modernized military organizations and structures. He displayed great ability in establishing the Civilian Conservation Corps and established the beginnings of the army-navy cooperation that flourished in the decade after MacArthur ended his term as chief of staff in 1935.

MacArthur, following his withdrawal to Australia from the doomed Philippines, rapidly learned the application of modern warfare and proved a master of combined arms operations, integrating air, land, and sea power into a potent force that enabled those in the Pacific theater to handle immense distances, prepared enemy defenses, and meager resources and still advance from the Stanley Owens Mountains in New Guinea in late 1942 to the invasion of the Philippines barely two years later. His ability to seize the initiative and to combine overwhelming force with only short windows of opportunity marked him as perhaps the best-balanced of American senior commanders during World War II.

MacArthur, after having been designated to command DOWNFALL, the invasion of Japan, found himself instead designated to command the Allied occupation of the former enemy and was charged with converting a warlike state to a nation of peace. In this, he was successful beyond all expectation, balancing the demands of competing Allied requirements and often ambiguous guidance from Washington into a sure course of direction that rapidly turned Japan into both a democracy and the first of the economic "Asian Tigers."

MacArthur's final command ended in frustration and embarrassment, in some measure caused by his own failure to recognize the changed attitudes in both the United States and Europe toward warfare in Asia. From his initial chagrin at U.S. troops proving themselves to be ill-trained and ill-equipped to the magnificent stroke of CHROMITE (the Inchon Landings)—certainly one of the finest amphibious operations ever mounted—and then to the hectic advance north and the dismal retreat south, MacArthur made this war his, with his frequent appearances at the front and his even more frequent press releases. But the intervention of the Chinese Communists caused MacArthur to become a changed man, negative and complaining, and this attitude, in turn, seems to have led to his refusal to listen more carefully to how the winds were blowing in Washington. At the end, MacArthur was perceived by his superiors as being out of control, a conclusion with some basis in truth but also one tainted with personal envy from some of those involved. In any event, despite MacArthur's unrivaled seniority—he had been the senior eligible officer for the position of chief of staff in 1930, for instance, and both of his successors, and all of his rivals, had long ago retired—he was relieved of the far eastern command in a botched episode embarrassing to both Washington and Tokyo in April 1951.

MacArthur returned home to vast public accolades, and his address to Congress flooded the media. He was the keynote speaker at the Republican National Convention in 1952, but his political aspirations were blocked by the ready nomination of Dwight D. Eisenhower for the presidency.

MacArthur then moved to New York, where at the instigation of former president Herbert Hoover he lived at the Waldorf-Astoria. His birthday celebrations were events of great interest to the U.S. military community, as his former soldiers, many of them now risen to military power, paid homage to the man they credited with forming their careers. MacArthur did serve as a corporate officer with some success but otherwise avoided the limelight.

Although President Eisenhower declined contact with his former commander, President John F. Kennedy frequently consulted MacArthur as the United States became drawn into the Vietnam conflict, though MacArthur's analysis was that the United States should avoid military support to the South Vietnamese government, advice unwelcome to Kennedy. MacArthur enjoyed a tumultuous return visit to the Philippines in 1961 and completed his military duties as the graduation speaker at West Point in 1962. He died following a brief illness at Walter Reed Hospital in 1964 and is buried in his mother's home city of Norfolk, Virginia.

MacArthur was a soldier marred by an arrogant personality and one who failed properly to protect his record by explaining in detail the reasoning behind his actions but nevertheless a soldier first, last, and always.

Marc Small

See also: Philippines, U.S. Loss of; Philippines, U.S. Retaking of; World War I; World War II

References and further reading:

Appleman, Roy E. *South to the Naktong: North to the Yalu.* Washington, DC: Office of the Chief of Military History, 1961.

James, D. Clayton. *The Years of MacArthur.* Vol. 1, *1880–1941.* Boston: Houghton Mifflin, 1970.

———. *The Years of MacArthur.* Vol. 2, *1941–1945.* Boston: Houghton Mifflin, 1975.

———. *The Years of MacArthur.* Vol. 3, *Triumph and Disaster, 1945–1964.* Boston: Houghton Mifflin, 1985.

Linn, Brian McAllister. *Guardians of Empire: The U.S. Army and the Pacific, 1902–1949.* Chapel Hill: University of North Carolina Press, 1997.

MacArthur, Douglas. *Reminiscences.* New York: McGraw-Hill, 1964.

Morton, Louis. *The Fall of the Philippines.* Washington, DC: Center for Military History, 1953.

Perret, Geoffrey. *Old Soldiers Never Die: The Life of Douglas MacArthur.* New York: Random House, 1996.

Petillo, Carol Morris. *Douglas MacArthur: The Philippine Years.* Peru: Indiana University Press, 1981.

Maccabees, Revolt of the (168–143 B.C.E.)

The Jewish-dominated region of Judea rebelled against the religious oppression of the Seleucid king, Antiochus IV, and forged an independent nation-state in one of the world's first successful guerrilla campaigns. For about a century after Alexander the Great had conquered the Persian Empire, the coastal zone southwest of Syria became the pivot point of the rivalry between the Ptolemaic (Egypt-based) and Seleucid (Syria–Asia Minor) successor regimes. By 170 B.C.E., Antiochus III and his sons Seleucis IV and Antiochus IV Epiphanes, rulers of the Seleucid Empire, understood that they would have to confront the expanding influence of Rome. Antiochus IV felt that it was crucial to forge religious unity throughout his dominion, integrating all the various gods with Olympian Zeus at their head. Especially significant was Jewish-dominated Judea, the strategic fulcrum covering southern invasion routes. As a result, he violated the promise of his father Antiochus III to respect the religious autonomy of the Jews. To this end, Antiochus took sides in a power struggle between rival high priests in Jerusalem, as a pretext to intervene openly in Jewish religious matters. This confrontation culminated when Antiochus Epiphanes occupied Jerusalem. He destroyed the city walls, raided the Temple treasury to fund his dwindling war chest, and decreed the abolition of Jewish separatism. He next converted the temple into a pagan shrine, setting up a fortress opposite, and forbade the hallowed practices of circumcision and Sabbath observance.

In 168 B.C.E., Seleucid troops set up pagan altars in the countryside and, in the village of Modi'in, ordered a lower-ranking priest, Mattathias, to ritually eat pig's flesh. He refused and killed another villager who complied, leading the townsfolk in a massacre of the Greek garrison, thereby starting the revolt. Mattathias withdrew to a concealed and well-guarded training camp for irregular forces in the foothills near Gophna, northwest of modern Ramallah. The training went on for a year prior to conducting operations and included politico-religious indoctrination and the establishment of friendly ties with the populace in key localities, where they established logistical bases, intelligence networks, safe houses, and weapons hoards. The moribund Mattathias designated one of his sons, Judas (who became known as Maccabee, or "the Hammer") to take over for him.

In a series of brilliant guerrilla actions, Judas defeated a succession of Syrian generals. His most renowned victories occurred at Beth Horon Pass (166 B.C.E.), Emmaus (166 B.C.E.), and Beth Zur, in the vicinity of Hebron (165 B.C.E.). His success is attributed to the first recorded instances of successful irregular warfare: hit-and-run night raids, ambushes at defiles, and attacks on rear-echelon units and individuals. After Beth-Zur, Judas captured Jerusalem, liberating the temple, though a Seleucid garrison held out in the citadel.

In 165–164 B.C.E., Judas extended his control over most of Judea, maintaining a close siege of the Syrian troops in the citadel. Since Antiochus was preoccupied with a triumphant campaign in the East, it was left to the Syrian regent Lysias to lead an invasion of Judea to recapture Jerusalem. After defeating the Jews at Beth Zacharia, he had to cut the campaign short to suppress a revolt in Syria in 164 B.C.E..

In 164 B.C.E., Bacchides, in charge of Seleucid forces in Judea, defeated Judas, driving him from Jerusalem. Quickly rebounding from this reverse, Judas took the offensive and, in 160 B.C.E., routed and killed the Syrian general Nicanor at Adasa, close to his earlier victory at Beth Horon. Judas himself was killed in battle by Bacchides at Elasa later that year.

Leadership of the Maccabees passed to Judas's brother, Jonathan, who continued guerrilla campaigns against the Syrians. In 143 B.C.E., Syrian troops, in league with alienated Jews, captured and eventually executed Jonathan at Ptolemais (Acre). Subsequently, the Seleucids recognized another of Judas's brothers, Simon, as king of Judea, establishing the Hasmonean dynasty, which ended with the accession of Herod the Great, after his marriage to the last Hasmonean queen, Miriam.

James Bloom

See also: Syrian-Egyptian Wars

References and further reading:

Bar-Kochva, Bezalel. *The Seleucid Army: Organization and Tactics in the Great Campaigns.* Cambridge, UK: Cambridge University Press, 1976.

———. *Judas Maccabeus. The Jewish Struggle against the Seleucids.* Cambridge, UK: Cambridge University Press, 1989.

Farmer, William R. *Maccabees, Zealots and Josephus: An Inquiry into Jewish Nationalism in the Greco-Roman Period.* New York: Columbia University Press, 1956.

Gichon, Mordechai, and Chaim Herzog. *Battles of the Bible.* 2d ed. London, Greenhill Books, Lionel Leventhal, 1997.

Macedonian Wars (215–146 B.C.E.)

A series of wars during which the Romans gained control over Greece and destroyed the Macedonian kingdom.

The First Macedonian War (215–205 B.C.E.)

In 215 B.C.E. the Macedonian king Philip V signed a treaty with the Carthaginian general Hannibal, who had invaded Italy. Thereupon Philip invaded Rome's possessions in Illyria (Albania). Although the Romans succeeded in keeping the important city of Apollonia out of Philip's hands, the war effort in Italy prevented them from intervening until 211 B.C.E., when the Romans allied with the Greek confederacy of Aetolia.

The Romans left land warfare to the Greeks, confining themselves largely to naval support. In the following years, Rome's successes were mainly diplomatic. Several Greek states chose Rome's side. Philip was driven out of Greece by diplomatic means. Then, during a lightning campaign, Philip defeated the Greeks and their Balkan tribal allies, recovered his position in Greece, and attacked Aetolia. The Aeolians sued for peace (206 B.C.E.) after their Spartan allies were defeated by the Achaeans. After an unsuccessful campaign in Illyria, the Romans did so as well. The Peace of Phoenice (205 B.C.E.) left Philip in possession of his conquests in Illyria.

The Second Macedonian War (200–197 B.C.E.)

In 200 B.C.E., war broke out again. The Romans landed in Illyria with two legions and marched inland. They failed to push through into Macedonia but did succeed in coercing several states to join the many Greek states that had already joined them.

In 199 B.C.E., an army of Greek allies raided Thessaly and

southern Macedonia, but during a lightning campaign, Philip succeeded in fighting off both these invaders and Rome's tribal allies on the Balkan frontier. In the following year, Philip took the initiative and moved his army into a strategic position, where he threatened the lines of communication of the Roman army in Illyria. The Romans assaulted Philip's position, a costly but eventually successful campaign. Thereupon Philip retreated into Macedonia. In the meantime, the allies of Rome were successful at sea, and even more Greek states joined the Romans.

Philip advanced into Thessaly but was engaged by the Roman army before he had reached his objective. He was forced to do battle at Cynoscephalae and was defeated. He had to abandon all territories outside Macedonia and respect the independence of all Greek cities. The Romans assumed Macedonia's role of dominant power in Greece.

The Third Macedonian War (171–168 B.C.E.)

The Romans felt threatened in their hegemony when King Perseus, the son of Philip V, again started to acquire influence in Greece. Unscrupulously taking advantage of Perseus's diplomatic advances to avoid hostilities, they brought an army into Illyria and Greece. Perseus reacted with speed and outmaneuvered the Romans in Thessaly, cutting off their line of supplies.

In the meantime, the Romans alienated themselves from the Greeks by their brutality, heavy-handedness, and greed. Perseus, on the contrary, became increasingly popular. Moreover, he was successful, while the Romans suffered from bad discipline and command. Two invasions of Macedonia failed, and Perseus counterattacked, regaining territory and defeating Rome's Balkan tribal allies. During the following winter (169 B.C.E.), he campaigned successfully against the Romans on Macedonia's northwestern frontier and in Greece and Epirus.

In 168 B.C.E., the consul Lucius Aemilius Paullus assumed command of the Roman army. The Romans decided to attack on three fronts: a naval offensive in the Aegean Sea, an offensive from the west from Illyria, and an offensive from Thessaly. After initial Macedonian success, Perseus met the Romans at Pydna. The well-deployed Macedonian phalanx attacked the unprepared Romans, but the Macedonian line became disrupted. The Romans counterattacked and broke the Macedonians. Perseus was captured and brought to Italy. Macedonia was divided into four republics, tributary to Rome.

The Fourth Macedonian War (146 B.C.E.)

The so-called Fourth Macedonian War was in fact an insurrection. The Macedonians had always been very loyal to their royal house, and in 152 B.C.E., a pretender to the throne named Andriscus aroused the Macedonians into a rebellion to reinstate the royal dynasty. The insurgents initially succeeded in defeating an army consisting of a Roman legion and local militia, though another Roman army soon crushed the revolt.

Maarten van der Werf

See also: Cynoscephalae, Battle of; Hannibal Barca

References and further reading:

Errington, Robert Malcom. *A History of Macedonia.* Berkeley: University of California Press, 1990.

Hammond, Nicholas Geoffrey Lem Prière, and F. W. Walbank. *A History of Macedonia.* Vol. 3, *336–167 B.C.E.* Oxford: Clarendon Press, 1988.

Walbank, Frank William. *Philip V of Macedon.* Cambridge, UK: Cambridge University Press, 1940. Reprint, Hamden, CT: Archon Books, 1967.

Maceo y Grajales, Antonio (1845–1896)

Noted black general in the struggle for Cuban independence. Maceo y Grajales was born on 14 June 1845 in Majaguabo, San Luis, to a farming family of free blacks. At the outbreak of the Ten Years' War in 1868, he joined the revolutionary forces and rose to the rank of general. Although the war ended with the Pact of Zanjon, in which Spain promised various social and economic reforms, Maceo was one of a number of Cuban revolutionaries who refused to accept the pact's terms.

By 1880, Maceo was traveling outside Cuba attempting to acquire arms, munitions, and men to return to the battle. Though it would be 1895 before he returned to Cuba to fight, he worked tirelessly for independence. He met with Máximo Gómez and José Martí various times to plan the continued war for independence. During these meetings, Martí would have to mediate at various times because Maceo and Gómez were rivals.

With the renewal of the war, Maceo was wounded more than 200 times. He was emphatically against annexation of Cuba by the United States, a course of action considered by some revolutionaries. But Maceo, appalled by the increasingly restrictive "color line" being drawn in the post-Reconstruction United States, stated that if annexation were attempted, he would be forced to switch allegiance and fight on the Spanish side.

During a fierce battle at San Pedro in Havana province on 7 December 1896, Maceo and his aide, a son of Máximo Gómez, fell in battle.

Peter Carr

See also: Cuban War of Independence; Cuban Ten Years' War; Martí y Pérez, José Julián

References and further reading:
Carbonell y Rivero, Miguel Angel. *Antonio Maceo.* Havana: Imprenta "La Prueba," 1924.
Costa, Octavio R. *Antonio Maceo, el héroe.* Havana: Academia de la Historia de Cuba, 1947.

Machiavelli, Niccolò (1469–1527)

Secretary to the chancellor on diplomatic relations for the Republic of Florence, better known as the first great political philosopher of the Renaissance. Born in Florence, Italy, on 3 May 1469, Machiavelli was the son of a lawyer of modest means who provided his son with a strong background in the humanities. Over his remains stands a monument bearing, in Latin, the phrase, "No eulogy would do justice to so great a name."

In 1512, the Medici family overthrew the republic, and Machiavelli lost his position. He spent his forced retirement reflecting upon events, reading history, and writing political philosophy. It was during this time he wrote *The Prince* (1513), *Discourses on the First Ten Books of Titus Livius* (1513–1517), and *The Art of War* (1520).

Portrait of Niccolò Machiavelli. (Library of Congress)

In *The Prince,* his most famous work, Machiavelli discusses the political necessity of "vices" over traditional "virtues," the political benefits of deceitfulness and miserliness, and how a prince should prefer being feared to being loved while avoiding being hated by his subjects. The book is a technical manual on how a prince was to grasp and hold power.

Most of Machiavelli's later work focused on his love of republican values. In *Discourses,* he examines the elements of the ancient Roman Republic that led to its success and draws lessons for republican governments. In *The Art of War,* Machiavelli advocates replacing unreliable mercenaries with a patriotic militia imbued with civic virtue and possessing an intense desire to protect their republican rights.

Many of Machiavelli's works, especially *The Prince,* were favorite reading of history's great military minds, such as Frederick the Great, Napoleon, and Clausewitz.

Craig T. Cobane

See also: Mercenaries

References and further reading:
Pocock, J. G. A. *The Machiavellian Moment.* Princeton, NJ: Princeton University Press, 1975.
Skinner, Quintin. *Machiavelli.* New York: Farrar, 1981.

Machine Gun

A rapid-firing small arm, which today means a fully automatic weapon. Muzzle-loading firearms were slow to reload, inaccurate, and effective only at very short ranges. Since their first appearance, those who make and use them have striven to increase the guns' rate of fire and combine great rate of fire with accuracy. Muzzle-loading separate ball and powder prevented any development along these lines. Breech-loading weapons and the complete round (bullet and cartridge in one unit) made machine guns possible. The machine gun is the easiest method of putting heavy fire down quickly on unprotected troops and areas.

One of the earliest successful machine guns was the Gatling (1862), a hand-cranked gun, which had between 6 and 20 rotating barrels. By the end of the nineteenth century, machine guns were fully automatic, in that once the trigger was depressed the gun continued to fire until it either ran out of ammunition or the trigger was released. This effect was achieved by tapping some of the gases used to propel the bullet down the barrel or by using the recoil of the cartridge case to work the mechanism of the gun. Ammunition is fed to the gun from magazines or belts.

The machine gun has a purpose in attack (to suppress the enemy) and in defense (to mow down attackers who do not use practical anti–machine gun field craft and tactics). Its greatest effect was in World War I, when the machine gun took second place only to artillery in killing and wounding on both sides.

The machine gun is an infantry weapon and can be used for direct or indirect fire. In the direct fire role, the machine gun is a very effective defensive weapon and gives very good fire support for moving troops. In the indirect role, a machine gun or a battery of guns can deny the enemy unprotected movement well behind the forward battle lines.

Light machine guns are weapons carried by the infantry, and every infantry section has one or more light machine guns. Medium machine guns (used mainly in the indirect fire role) are concentrated normally at battalion level, as are heavy machine guns. All armored fighting vehicles carry a machine gun for local support. Machine guns have also been used in aircraft for air warfare and for ground strafing, but these roles have now mainly been taken on by missiles. But even in an age of high-tech electronic warfare, the machine gun, by general agreement, is still a most important weapon in any army's armory.

David Westwood

References and further reading:

Allen, W. G. B. *Pistols Rifles and Machine Guns.* London: English Universities Press, 1953.

Hobart, F. W. A., ed. *Jane's Infantry Weapons.* London: MacDonald and Jane's, 1975.

Mackensen, August von (1849–1945)

German field marshal who achieved a string of spectacular victories on the eastern front in 1914–1916. Born on 6 December 1849 in Leipnitz, near Wittenberg, Mackensen joined the Death's Head Hussars in 1869 as a cadet. Serving in the Franco-Prussian War as a junior officer, he was appointed to the general staff in 1882. Although not from the nobility, he rose rapidly through the ranks and was promoted to cavalry general in 1908.

As commander of XVII Corps on the eastern front in 1914, he suffered defeat at Gumbinnen (20 August) but contributed to victory at Tannenberg and in the first Battle of the Masurian Lakes. As commander of the Ninth Army, he conducted the successful offensive at Lodz (11–21 November). The following year, he succeeded in breaking through the Russian lines at Gorlice-Tarnów as commander of the Austro-German Eleventh Army, his advancing forces taking 120,000 prisoners, destroying the Russian Third Army, and capturing Lemberg and Brest-Litovsk. Promoted to field marshal in recognition of his achievements, he was ordered to renew the attack on Serbia and captured Belgrade on 9 October 1915 as commander of German, Austro-German, and Bulgarian armies. On 1 September 1916, he launched a successful attack on Romania with combined Bulgarian, Turkish, and German forces, entering Bucharest on 6 December. Mackensen was a modest man, and his battlefield successes have been attributed to his ability to work harmoniously with different chiefs of staff.

During the 1930s, he was used by Adolf Hitler at various public displays to symbolize the continuity of military tradition in the Third Reich. He died on 8 November 1945, at Burghorn, near Celle, having lived just long enough to witness Germany's utter defeat in World War II.

Alaric Searle

See also: Gorlice/Tarnow; Tannenberg, Battle of

References and further reading:

Mackensen, August von. *Briefe und Aufzeichnungen aus Krieg und Frieden.* Leipzig: Bibliographisches Institut, 1938.

Schwarzmüller, Theo. *Zwischen Kaiser und "Führer." Generalfeldmarschall August von Mackensen.* Paderborn: Schöningh, 1996.

Mactan, Battle of (1521)

An avoidable defeat of Spanish forces led by the Portuguese Fernao de Magalhaes (Ferdinand Magellan). The Battle of Mactan delayed Spain's colonization of the archipelago later named Las Islas Felipinas, or the Philippines.

With Pope Alexander VI's bull (decree) of 1493 and the Treaty of Tordesillas of 1494 (later approved by Pope Julius II), conflict between Catholic Spain and Portugal over prospective colonies in the Western Hemisphere was adjudicated. Spain then sought a direct water route to the Spice Islands in Southeast Asia.

Captain General Magellan commanded three ships sailing from Spain in 1519, intent on reaching the Spice Islands. He sailed from the Atlantic to the Pacific and thence through insular Southeast Asia. Magellan baptized the Muslim Rajah Humabon of Zebu (Cebu), whose gold and ginger attracted him, obtaining a pledge of loyalty to Spain. Although Rajah Lapulapu of Mauthan (Mactan) Island, 2 kilometers east, was refusing to submit, Magellan overconfidently declined military support from Humabon.

Magellan's three galleons anchored at low tide in the northern gulf of Mactan Island early on Sunday morning, 27 April 1521, leaving inland Mactan beyond the firing range of cannons on deck. In lieu of a surprise attack, Magellan sent an envoy on a fruitless mission seeking Lapulapu's obeisance. At daybreak, the Spanish contingent of less than 60 mostly inexperienced volunteers slogged shoreward over reefs. They confronted a crescent formation of 3,000 fighters. Invoking military analogies from Mexico, Magellan injudiciously ordered his forces inland. However, crossbows, muskets, and spears were ineffective against bolos, shields, longer iron-tipped bamboo lances, and poisoned arrows. Squeezed on two sides, the volunteer force retreated—first with discipline and then in disorder. In the hour-long rout, 7 Europeans and 15 Mactan Islanders were killed. Covering his unit's retreat in hand-to-hand fighting, Magellan perished.

Circumnavigating the globe, 18 survivors returned to Spain in 1522. Lapulapu's victory simply delayed Spanish colonization. In 1564, Spain returned to stay for 334 years.

Vincent Kelly Pollard

See also: Spanish Colonial Wars; Spanish-Portuguese Wars

References and further reading:

Joyner, Jim. *Magellan.* Camden, ME: International Marine, 1992.

Pigafetta, Antonio. *The Voyage of Magellan: The Journal of Antonio Pigafetta.* A translation by Paula Spurlin Paige from the edition in the William L. Clements Library, University of Michigan, Ann Arbor. First published, 1536. Englewood Cliffs, NJ: Prentice-Hall, 1969.

Spate, Oskar Hermann Khristian. *The Spanish Lake. The Pacific since Magellan.* Vol. 1. Canberra: Griffin Press for the Australian National University Press, 1979.

Transylvanus, Maximilianus. *De Moluccis Insulis* (Of the Moluccas Islands). Cologne, 1523. In *Documentary Sources of Philippine History,* comp. and ed. Gregorio F. Zaide. Manila: National Book Store, 1990, Vol. 1, Doc. 20.

Maczek, Stanislaw (1892–1994)

Polish World War II commander and the last surviving senior Allied general officer of World War II. Born in Szczerzec in Austro-Hungarian-occupied Poland on 31 March 1892, Maczek served in the Austro-Hungarian army during World War I. After active service during the Russo-Polish War of 1919–1921 and as an infantry officer during most of the interwar period, he commanded the 10th Motorized Mounted Rifle Regiment during the Polish campaign of 1939. Though his regiment, which was one of only two tank regiments in the Polish army, particularly distinguished itself against the Germans, it was overwhelmed by the vastly superior quantity and quality of German armor and was forced to seek refuge in Hungary. Promoted to major general, Maczek was reunited with many of his men in France, where he commanded the 10th Polish Mechanized Cavalry Brigade during the French campaign of 1940. Again forced into exile, he and his men eventually found their way to Scotland, where they formed the 1st Polish Armored Division. As part of the First Canadian Army, the 1st Armored fought its way across northern Europe after the Normandy invasion. During the Battle of the Falaise Gap in August 1944, it was Maczek and the 1st Armored that closed the "pocket" and prevented tens of thousands of German troops from escaping. Promoted to lieutenant general in May 1945, Maczek refused to return to Soviet-dominated Poland after the war and lived out his exile in Scotland. He was promoted to full general by Polish president Lech Walesa in March 1994 and died in Edinburgh on 11 December 1994, the only senior Polish World War II commander to witness the fall of the Soviet Union and of European communism.

Alexander M. Bielakowski

See also: Polish Campaign of 1939; Russo-Polish War; Sikorski, Wladyslaw Eugeniusz; World War II

References and further reading:

Keegan, John. *Six Armies in Normandy: From D-Day to the Liberation of Paris, June 6th–August 25th, 1944.* New York: Viking Press, 1982.

Stachura, Peter D., ed. *Themes of Modern Polish History—Proceedings of a Symposium on 28 March 1992 in Honor of the Centenary of General Stanislaw Maczek.* Glasgow: Polish Social and Educational Society, 1992.

Magdeburg, Siege of (1630–1631)

A symbol of the fury of the Thirty Years' War. Magdeburg was the first imperial town (Reichsstadt) to enter an alliance with the Swedish king on 1 August 1630. Thus the town became an enemy of the Habsburg emperor, but it did not become the center of the operations at that time because the main imperial army, commanded by Johann Tserclaes, Graf von Tilly, advanced to Pomerania to fight Gustavus Adolphus. As the Swedish king avoided a battle, Tilly turned his whole army against Magdeburg at the end of March 1631.

Now the siege became part of a war of diversion. Tilly intended to force Gustavus Adolphus to relieve his ally, who himself tried to distract the imperial army by advancing along the Oder River and posing a threat to the Habsburgian provinces of Silesia and even Bohemia. Ignoring this advance, Tilly reinforced his efforts in besieging the town instead, but the Magdeburgians relied on the Swedish promise to relieve the town and did not surrender. After several attempts, a general assault on 20 May 1631 was successful. In the course of fighting, a fire broke out and reduced almost the whole town to ashes, killing not only most of its inhabitants but also many of the invading imperial soldiers.

The question of who started the fire still remains unclear. Undoubtedly, the devastated town was of less worth for Tilly, who failed to gain a logistic stronghold for his campaign against Gustavus Adolphus. To the Swedish king, the fall of

Magdeburg served to damage his reputation as protector of the German Protestants, so massive Protestant propaganda (mostly by broadsheets) put the blame on the victorious Tilly.

Michael Kaiser

See also: Gustavus II Adolphus; Thirty Years' War; Tilly, Johann Tserclaes, Graf von

References and further reading:

Kaiser, Michael. *Politik und Kriegführung: Maximilian von Bayern, Tilly und die Katholische Liga im Dreißigjährigen Krieg.* Münster: Aschendorff, 1999.

Roberts, Michael. *Gustavus Adolphus: A History of Sweden 1611–1632.* 2 vols. London: Longmans, Green, 1953, 1958.

Magersfontein, Battle of (11 December 1899)

Second British defeat of "Black Week." In the opening months of the Boer War, Redvers Buller ordered Lord Paul Sanford Methuen to relieve the beleaguered garrison at Kimberley and remove all noncombatants. Methuen proceeded northward along the Western Railway to safeguard his supply line and to ensure the integrity of his only means of evacuating the civilian population. Anticipating the British strategy, the Boers positioned themselves along the railway. In late November, Methuen's force defeated the Boers in three successive battles at Belmont, Graspan, and Modder River.

On 10 December, British artillery opened fire upon Boer entrenchments at Magersfontein. Methuen ordered a night march, to be followed by a frontal attack. Despite heavy rain and poor reconnaissance, British troops managed to reach their destination, a few hundred yards from the Boer trenches. Orders to extend, however, came too late. Daybreak caught the advancing infantry in tight formation. Boer firepower left the brigade in a confused mass scrambling for cover. All attempts to reform and renew the advance failed. The British were forced to withdraw by late afternoon.

In 1899, no British soldier had ever come across such an elaborate design of trenches as those constructed at Magersfontein. British artillery failed to disturb them. British casualties approached 1,000; Boer casualties, 275. The failure at Magersfontein, one of three British defeats that week, demonstrated that the British army was not ready for modern warfare and led to major changes in leadership, mobilization, and organization.

Stephen M. Miller

See also: Boer Wars; Buller, Sir Redvers Henry; Kimberley, Siege of

References and further reading:

Duxbury, George. *The Battle of Magersfontein.* Johannesburg: National Museum of Military History, 1979.

Miller, Stephen M. *Lord Methuen and the British Army.* London: Cass, 1999.

Pemberton, W. Baring. *Battles of the Boer War.* Philadelphia: Dufour Editions, 1964.

Maginot Line

An extensive series of border fortifications that failed France in 1940. At the end of World War I, the dominant school of thought in the French army emphasized the value of a defensive stance in warfare. Based on experience from the early phase of the war, when frontal attacks against machine guns caused massive casualties, and the experience acquired at Verdun, the French general staff favored the option of "digging in" as a means to hold off the enemy while limiting losses. In addition, in the event of surprise attack, a massive fortification would gain the French army some valuable time as it assembled its forces. Throughout the 1920s, the French General Staff argued about what the best defensive attitude was and where to build fortifications. Eventually, the focus fell on the relatively flat area that extended from the Ardennes to Alsace. In this context, politician André Maginot came to oversee the design and construction of a defensive line that bore his name. As war minister from 1922 to 1924 (and again in 1929), he remained involved in the planning as the head of the Parliamentary Armament Committee. In 1927, the basic design was approved, and Paul Painlevé, then minister of war, authorized the call for bids. Private contractors built the Maginot line. The reason for naming it Maginot rather than Painlevé goes back to the level of involvement of the former and his successful lobbying of the French parliament to allocate funds at a time when Germany, prior to Adolf Hitler's takeover, was not considered an immediate threat to French security. By the time the Maginot line was built, it had cost twice the original estimate, exceeding 6 billion French francs at the time.

The design of the Maginot line posed several challenges to engineers, and no two forts (also known as *ouvrages*) were the same because of terrain and communication constraints. In addition, drainage was a constant worry, and several modifications were required in the 1930s to make the installations livable. Common to all forts were the garrison personnel, ranging from 200 to 1,200 men and divided into infantry, artillery, and engineering sections. The layout followed the same structure everywhere, with living quarters at the bottom of the installation, a railroad that brought supplies to a narrow-gauge network that linked forts, and a diesel-powered generating facility to supply electricity in times of war (in peacetime, the French national power grid fed the entire Maginot line). Armament was also standard to all installations and involved modern cannons, some specially modified to fit in the forts alone. Construction began in late 1929, and in 1936 the line was first used on the occasion of Hitler's military occupation of the Rhineland. Some installations were added later, but limited funding and wavering political will made them no more than perfunctory complements, especially near the English Channel.

Divided into sectors, the Maginot line was placed on war alert on 24 August 1939, nine days ahead of the general mobilization order. However, the state of war was not confirmed until 6 September, and garrison commanders were ordered to fire on the enemy only in case of incursion on French territory. The subsequent "Phony War" lasted until May 1940, when the German army invaded Belgium and circumvented the Maginot line. Germany's striking success encouraged the implementation of Operation ROT (red), whereby the Germans began testing the Maginot line's defenses. Although some sections held up extremely well against 88 mm cannon, turrets and other mobile steel gear were often damaged. The greatest weakness of the line was that even though the heavy cement foundations facing the border were further protected by earth, the rear was not, which made rear assaults easier. Several forts, however, held until the Armistice. Morale all along the line remained excellent by most accounts, and the Maginot cannons ceased firing only on 24 June 1940. Five days later, the men left their quarters and were made prisoners of war, even though they should have, in theory, been allowed to join French troops in the Free (Vichy) zone under the terms of the cease-fire.

German and American troops made sporadic use of the installations until 1944, after which the French army took over and maintained part of the installations for another 20 years. Since then, much of the Maginot line has been auctioned off to civilian businesses, from mushroom farms to dance clubs. The Maginot Line remains as a literal monument to unimaginative military thinking.

Guillaume de Syon

See also: France; French Army, Verdun, World War I, World War II

References and further reading:

Alexander, Martin. *The Republic in Danger: Gen. Maurice Gamelin and the Politics of French Defense.* 1992.

Chelminski, R. "The Maginot Line." *Smithsonian* (June 1997).

Hughes, Judith M. *To the Maginot Line: The Politics of French Military Preparation in the 1920s.* Cambridge: Harvard University Press, 1971.

Kemp, Anthony. *The Maginot Line: Myth and Reality.* New York: Stein and Day, 1982.

Mary, Jean-Yves, and Alain Hohnadel. *Hommes et ouvrages de la Ligne Maginot.* Vol. 1. Paris: Histoire & Collections, 2000.

Young, Robert M. *In Command of France: French Foreign Policy and Military Planning, 1933–1940.* Cambridge: Harvard University Press, 1978.

Magsaysay, Ramón (1907–1957)

Philippine guerrilla, counterinsurgency leader, and later president. Magsaysay was born into a poor rural family in 1907 in Zambales Province. After a brief career in business, he became a noted guerrilla leader when the Japanese invaded the Philippines in 1941. Later, appointed secretary of national defense, he was faced from the outset with the problem of putting down the Hukbalahap, a peasant guerrilla movement that by 1953 had been penetrated by left-wing cadres. He initiated a three-pronged program. Corrupt and inefficient officers and soldiers were removed; a ground-up development program in Hukbalahap areas was initiated to deny the guerrillas local support, and guerrillas who surrendered were given land grants (land hunger was supposedly the main reason for the Huk revolt) and an amnesty. By 1955, when Magsaysay had been elected president, the Huk rebellion had faded to a small hard-core group of communist cadres. Magsaysay himself died in office on 17 March 1957 in a plane crash.

Portrait of Ramón Magsaysay. (Library of Congress)

Magsaysay's success can be attributed to two main factors. In an environment rife with corruption, he was one of the few seen as uncorrupt. Second, as both secretary of national defense and later as president, he managed to inspire support from line soldiers and civilians by leading from the front, personally supervising operations, and demonstrating concern and an aptitude for leadership.

Magsaysay also oversaw a number of military innovations in his counterinsurgency campaign. Small units with overwhelming firepower were used to track down guerrilla groups. Heightened real-time intelligence identified and tracked guerrilla leaders and groups, as did the extensive use of bribed and "turned" informants from among guerrilla ranks themselves. Psychological warfare was deployed extensively to convince the Huks that they were on the losing side. Many of these tactics were absorbed by the U.S. Army in Vietnam. Magsaysay's death was a disaster for the Philippines, for much of his good work was undone, and persisting Marxist guerrilla war sprang up anew beginning in the 1960s.

Michael Ashkenazi

See also: Guerrilla/Partisan/Irregular Warfare; Vietnam Conflict

References and further reading:

Abueva, Jose Veloso. *Ramon Magsaysay. A Political Biography.* Manila: Solidaridad Publishing House, 1971.

Starner, Frances Lucille. *Magsaysay and the Philippine Peasantry. The Agrarian Impact on Philippine Politics, 1953–1956.* Berkeley: University of California Press, 1961.

Magyars

Seminomadic tribes, originally from western Siberia, that conquered the middle Danube and raided western Europe in the ninth and tenth centuries. The Magyars were the last invading people to establish a permanent presence in central Europe.

After destroying the Moravian state (902) and defeating the Bavarians at Breclavsburg (907), the Magyars established themselves in Pannonia. From this base, they raided neighboring lands, extorting tribute. Their armies, consisting mostly of light cavalry, were highly mobile. Attacking without warning, they quickly plundered the countryside and departed before any defensive force could be organized. If forced to fight, they would harass their enemies with arrows or suddenly retreat, tempting their opponents to break rank and pursue, after which the Magyars would turn to fight them singly. After 937, the Magyars began to range ever farther, across Germany and France and into Italy. The raids finally ended after emperor Otto I defeated the Magyars at the Lechfeld, near Augsburg (955). In the following centuries, the Magyars settled down in Hungary and adopted western European forms of feudal military organization, including the predominant use of heavy armored cavalry. The light cavalry tradition was partly revived through the settlement of nomadic Cuman tribes in southern Hungary in the thirteenth century, the employment of Serb mercenaries in the fifteenth century, and finally in the evolution of Hungarian *huszar* light cavalry in the sixteenth and seventeenth centuries, which served as the model for the later appearance of hussar units in western European armies.

Brian Hodson

See also: Lechfeld

References and further reading:

Macartney, C. A. *The Magyars of the IXth Century.* Cambridge, UK: Cambridge University Press, 1930.

Makkai, László. "The Hungarians' Prehistory, Their Conquest of Hungary, and Their Raids to the West to 955." In *A History of Hungary,* ed. Peter Sugar. Bloomington: University of Indiana Press, 1990.

Mahan, Dennis Hart (1802–1871)

American educator whose theories and insights had a great influence on American military strategists in both the Mexican War and the Civil War. Born in New York City on 2 April 1802, Mahan spent his childhood in Norfolk, Virginia. A gifted student and protégé of Sylvanus Thayer, he graduated from the top of his class at the U.S. Military Academy at West Point in 1824 and was named assistant professor of mathematics in his third year at the institution. A year after graduation, Second Lieutenant Mahan was transferred to the position of assistant professor of engineering. He then went abroad to study public engineering work and military institutions at the Military School of Application for Engineers and Artillerists in Metz, France, and was befriended by the Marquis de Lafayette. Returning to West Point in 1830, he was named professor of engineering and was promoted to chairman of the department two years later. In 1838, he was also named to the position of dean. He committed suicide on 16 September 1871 in Stony Point, New York, when he learned that he was going to be forced into retirement by an overseeing board. He was the father of Alfred Thayer Mahan.

Among Mahan's most important works are *Complete Treatise on Field Fortifications* (1836), *Summary on the Cause of Permanent Fortifications and the Attack and Defense of Permanent Works* (1850), and *An Elementary Course of Military Engineering* (2 vols., 1866–1867). He exposed cadets to the principles of Antoine Henri, Baron de Jomini.

According to student Henry Wager Halleck, Mahan stressed that war should be looked at as both an art and a science; a science as it analyzes general principles and military operations and an art when referring to the practical rules of conducting campaigns, sieges, battles, and so on.

T. Jason Soderstrum

See also: Jomini, Antoine Henri, Baron de; Thayer, Sylvanus

References and further reading:

Mahan, Dennis Hart. *Descriptive Geometry: As Applied to the Drawing of Fortification and Stereotomy: For the Use of the Cadets of the U.S. Military.* New York: J. Wiley & Sons, 1871.
———. *An Elementary Course of Civil Engineering.* Edinburgh: A. Fullarton, 1845.
Weigley, Russell F. *The American Way of War: A History of United States Military Strategy and Policy.* New York: Macmillan, 1973.

Mahmud of Ghazna (Yamin al-Daula Abu'l-Qasim Mahmud ibn Sebuktigin) (971–1030)

Founder of the medieval Ghaznavid dynasty, his conquests made the Turkish people and the Islamic faith a part of Indian civilization. Sebuktigin, Mahmud's father, held command of Ghazna, a military camp town in eastern Afghanistan. By the time Mahmud reached maturity, Ghazna had become virtually independent of Iran. Quickly proving himself an adept commander, he established himself as emir in 998 after a brief war with his brother Isma'il. At that time, the Samanid government of Iran was crumbling, and in the scramble for the pieces, Mahmud also annexed the northern territories of Khurasan. A rich and desirable province, Mahmud had to return repeatedly to Khurasan to defend his northern frontier from Turkish interlopers like the Kharakhanids.

Mahmud inherited from his father the practice of conducting annual raids into the Indus valley. A strict Sunni Muslim, he believed that these raids not only filled his treasury with loot but also served as jihad against infidel Hindus, Buddhists, and Shiite Muslim cities. Mahmud made his capital a rich center of Islamic culture but also invested large sums into organizing his state and army. He supplemented his Arab, Turkish, and Afghan units with Hindu infantry and elephants. The sultan's wars, wealth, and militancy also magnetized thousands of Turkish plainsmen to his standard.

Mahmud conducted over 17 campaigns in northeastern and central India and jihads against the Shiite Buyid state in central Iran. He imposed a unified Islamic regime over most of the northern Indus valley. In 1024, Mahmud marched across the Thar wastelands and plundered the great Hindu center at Somnath. Under Mahmud, Islam penetrated for the first time onto the Ganges plains. At his death, the sultan was contemplating the annexation of Iraq and a jihad against the Fatimid Shiites in Syria.

Weston F. Cook, Jr.

See also: Ghaznavid Empire

References and further reading:

Bosworth, C. E. *The Ghaznavids.* Edinburgh: Cambridge University Press, 1963.
Wink, Andre. *Al-Hind: The Making of the Indo-Islamic World.* Vol. 1. Leiden: E. J. Brill, 1991.

Majorian (Julius Valerius Majorianus) (d. 461)

Emperor of the western Roman Empire from 1 April 457 to 2 August 461. The date of his birth is unknown; he was executed on 7 August 461 at Dertona (Tortona, Italy). After a distinguished early military career, Majorian was appointed *comes domesticorum* (roughly, commander of the imperial bodyguard). In league with the Suevian commander Ricimer, he defeated the western emperor Avitus at Placentia (Piacenza, Italy) on 17 October 456. Subsequently promoted to *magister militum,* Majorian was acclaimed emperor by his troops; formal recognition from the eastern empire came on 1 December 457.

As emperor, Majorian ambitiously sought to strengthen the western empire by reintegrating with it former Roman territories in Gaul and North Africa. In Europe, he campaigned successfully against Huns, Vandals, Goths, and Alamanni and subdued Gallic rebels. In 460 and 461, he launched two expeditions intended to expel the Vandals from Africa but in both cases was defeated, the last before his transports left Nova Carthago (Cartagena, Spain). He initiated tax reforms and was admired greatly by Sidonius Apollinaris, bishop of Lyons, France. In 461, Ricimer, his erstwhile ally, deposed Majorian.

Ian Janssen

See also: Goths; Huns; Vandals

References and further reading:

Jones, A. H. M. *The Later Roman Empire, 284–602.* 2 vols. Baltimore: Johns Hopkins University Press, 1986.

Malayan Emergency (1948–1960)

An insurgency inspired by and modeled on Mao Zedong's success in China. The dates are misleading: emergency restrictions remained in several areas after 1960, and the communists remained in the field in southern Thailand until 1990. Nonetheless, by 1960 communist guerrillas had been reduced to a futile and ineffective existence.

British success in Malaya was largely the result of four factors. Perhaps the most important was the fact that at no stage did the government collapse in any part of Malaya: the British never faced the problem of creating government where none existed. Moreover, Malayan society was racially divided. With the communist movement overwhelmingly drawn from the Chinese population (physically, readily identifiable) the counterinsurgency effort could be focused on just one section of society. Furthermore, physical geography limited insurgency to no more than one-tenth of the country. The interior was all but uninhabited, and the Chinese were scattered along the jungle fringe where it met the main north-south roads. Britain never had a major numbers-to-space problem, and the fact that Malaya was a peninsula ensured the communists' isolation from outside support.

There were other factors. The uprising of 1947 miscarried and cost the communist movement some 6–12 months as it tried to recover from failure and initiate rural-based insurgency. Failure provided the British administration with breathing space, and over time, four strands of policy were bound together: the streamlining of all levels of government in order to ensure speed and implement decisions; the principle of civilian supremacy, specifically the primacy of the police, with the military assigned the supporting role; a comprehensive resettlement program as the means of isolating the insurgents from all sources of supply; and the policy of tackling the least affected areas first. Results were slow in manifesting themselves, and resettlement made the communist task of infiltration easier, but the long-term effectiveness of resettlement can be seen by the fact that only 2 of some 400 new villages were abandoned after 1960. The policy of providing the population with something to lose, which included independence, with the communists portrayed as the obstacle to its being granted, was underwritten by boom conditions in tin and rubber prices, but the basic point was that the main features of the Briggs Plan—integrated government structure, civilian primacy, and the new villages—worked in the very special conditions of Malaya at this time. It should also be noted that the British did not send National Servicemen (conscripts) to Malaya; all were professionals who either knew their jobs or were willing to learn them in good time.

Of course, British strategy and policy may seem obvious, even easy in hindsight, but at the time it was felt that the struggle could go either way, particularly after the assassination of the high commissioner, Sir Henry Gurney, in 1951.

H. P. Willmott

See also: Guerrilla/Partisan/Irregular Warfare

References and further reading:

Coates, John. *Suppressing Insurgency: An Analysis of the Malayan Emergency.* Boulder, CO: Westview, 1993.

Jackson, Robert, *The Malayan Emergency: The Commonwealth Wars 1948–1966,* London: Routledge, 1991.

MacKay, Donald. *The Malayan Emergency, 1948–1960. The Domino That Stood.* London: Brassey's, 1997.

Stubbs, Richard. *Hearts and Minds in Guerrilla Warfare.* Oxford, UK: Oxford University Press, 1990.

Maldon, Battle of (10–11 August 991)

Brave Saxon defense that failed to prevent Viking inroads. After nearly half a century of relative stability, Anglo-Saxon England was unprepared for a new wave of Viking invasions beginning in 991. Danish king Swein Forkbeard led several expeditions along the southern half of England; the English king Ethelred, consequently nicknamed "the Unready," was on the defensive. Maldon, located about 12 miles north of the mouth of the Thames River, near the east coast of England, was the site of one of the initial battles between the Anglo-Saxons and the invading Danes. We know about the Battle of Maldon because of the near-contemporary poem, of which most has been preserved.

The Danes had ravaged the town of Ipswich, and the Anglo-Saxon ealdorman Byrhtnoth prepared to engage them along the coast of Essex in August 991. Byrhtnoth and his local shire force, the *fyrd,* allowed the Danes to land near Maldon in order to incite them to pitched battle. The Anglo-Saxons formed "shield-walls," cohesive though mobile formations of troops, and used spears and shields to hold the Danes in check. Byrhtnoth, an ealdorman for 35 years, was portrayed as the brave leader in the poem, rallying his troops for battle. Only after his death did many of the Anglo-Saxons retreat. Three years after the battle, Ethelred was forced to make payments to the Danes as appeasement. In 1016, new Viking invasions culminated with the Danish king Cnut succeeding Ethelred's son, Edmund, as king of England.

Christopher P. Goedert

See also: British Dynastic Wars; Viking Raids; Vikings; William the Conqueror

References and further reading:

Cooper, Janet, ed. *The Battle of Maldon: Fiction and Fact.* London: Hambledon Press, 1993.

Laborde, Edward D. *Byrhtnoth and Maldon.* London: Heinemann, 1936.

Scragg, Donald, ed. *The Battle of Maldon, A.D. 991.* Oxford: Basil Blackwell in association with the Manchester Centre for Anglo-Saxon Studies, 1991.

Malplaquet, Battle of (11 September 1709)

The duke of Marlborough's last and most useless battle. After a long series of victories (Ramillies, Oudenaarde), the allied armies intended to destroy the French army and then invade France. The French main army was south of Mons, waiting behind field fortifications near the woods of Malplaquet. Marlborough and Prince Eugene, with their 7,500 men, intended to use the same tactical dispositions they had used formerly for Bleinheim and Ramillies, a straightforward advance on the enemy lines. The duc de Villars, commanding the French, had taken a defensive position, protected on his two wings by occupied forests forming a funnel-shaped line. The center was on higher grounds and the fortifications were defended by his elite regiments.

The first shots were fired at 7:30 A.M. on the French left wing by Eugene's troops; the wood of Sars was taken at 11:00, but the winning troops were too exhausted to follow the reforming French. The French right was attacked by Dutch troops, who were repulsed; the French commander was too cautious to give orders to pursue the hesitating Dutch. Villars had reinforced the wings from his well-defended center when no attack had been launched by noon. He wanted now to attack and gathered 50 battalions, but the British artillery covered the French front with a deadly fire. Two events turned the tide of the battle: Villars was wounded and taken away from the battlefield, and Marlborough and Eugene, seeing the weakened French center, launched a decisive attack. The fortifications were assaulted at a terrible human cost. Boufllers, now in command, ordered his cavalry to charge against the emerging infantry around 1:00 P.M., and six futile charges ensued. The French then decided to retire in good order, while the allies were too exhausted to pursue.

Malplaquet was the bloodiest battle of the eighteenth century, with no fewer than 11,000 killed and 22,000 wounded. The allied casualties were higher than those the French suffered; they were unable to follow their invasion plan. Malplaquet was seen as a glorious defeat in France and raised the fighting spirit of the army. Conversely, this dubious victory gave Marlborough's court enemies an argument to undermine Marlborough's position, and he was recalled in 1711.

Gilles Boué

See also: Denain, Battle of; Marlborough, John Churchill, First Duke of; Spanish Succession, War of the

References and further reading:

Corvisier, André. *La bataille de Malplaquet.* Paris: Economica, 1997.
Wijn, Jan W. "Les troupes hollandaises à la bataille de Malplaquet." *Revue Internationale d'Histoire Militaire* 19 (1957), 334–379.

Malta, Siege of (May–September 1565)

A defeat that blocked Turkish expansion to the west. The Ottoman emperor Süleyman, determined to seize Malta, the strategic gateway to the West, assembled nearly 30,000 men, including Janissaries and Spahis. The formidable Turkish artillery included specially made heavy artillery. One piece weighed nearly 40 tons and fired 200-pound balls. Two other pieces weighing nearly 20 tons each could fire 90-pound iron balls. The Turks also brought with them 100,000 cannonballs and 170,000 tons of powder.

These overwhelming odds could hardly be matched by Malta's 500 knights, around 5,000 Spanish soldiers, and approximately 4,000 other troops. The knights were led by Grand Master Giovanni Parisot de la Valette, who was able to strengthen Malta's fortifications before the Turkish onslaught. The knights also created a new weapon. After wrapping a circular iron band in tow, placing it in boiling pitch, and wrapping it with tow again, they lit the device and threw it at Turks scaling the fortifications. The knights also used entrenchments to stop the Turks.

The Turks began the bombardment of Saint Elmo on 10 June. Despite the death of their leader, Dragut, the governor of Tripoli, the Turks took the fortress. During the course of the siege, the fierce resistance by the knights caused the deaths of more than 24,000 Turks. Süleyman sent an additional 20,000 reinforcements, and by August Malta was nearing the end of its resistance. But the "Great Relief Force" of around 9,000 men led by Garcia de Toledo reached Mellieha Bay and refortified Malta's defenses on 7 September. Frustrated, the Turks began to evacuate Malta the following day.

Annette Richardson

See also: Süleyman I

References and further reading:

Blouet, Brian. *The Story of Malta.* Malta: Progress Press, 1993.
Ellul, Joseph. *The Great Siege of Malta.* Siggiewi, Malta: Ellul, 1992.
Sire, H. J. A. *The Knights of Malta.* New Haven: Yale University Press, 1994.

Malta, Siege of (June 1940–November 1942)

German siege of British-held island, the failure of which kept open the British supply line to the Middle and Far East. Malta's strategic position in the Mediterranean between Sicily and Libya led to the island's second great siege during World War II.

Italy's declaration of war against Britain brought numerous air raids against the island, for a period defended by only three Gloster Gladiator biplanes and a small assortment of antiaircraft guns. The arrival of German troops in North Africa (February 1941) was accompanied by frequent Luftwaffe attacks against the island and its vital supply convoys. Despite the Axis blockade, Malta's defenses steadily improved throughout the siege, with it becoming a crucial base for surface vessels, submarines, and aircraft to operate against Field Marshall Erwin Rommel's supply lines. An Italian naval attack against Malta by explosive motorboats and "pigs," two-man piloted torpedoes, was repulsed on 15 July 1941, while a renewed Axis air offensive commenced in January of the following year. The sinking of numerous Allied supply ships increased the threat of starvation during the summer of 1942; between March and August, only 7 of 35 merchant ships bound for Malta arrived, with the most urgently needed supplies delivered by submarine.

In recognition of its heroic struggle, Malta was awarded the George Cross in April 1942. A combined German-Italian invasion, Operation HERCULES, was proposed for late June 1942, once Tobruk had been captured, but was postponed after Rommel's decision to strike against Egypt instead. However, the subsequent Axis defeat at El Alamein and lengthy retreat to Tunisia removed the threat of invasion to Malta. Following a final effort to neutralize the embattled island by air in October, Axis bombing raids progressively decreased. An ironic postscript to the siege was the surrender of the Italian fleet in Malta's Grand Harbor on 8 September 1943 (the anniversary of the ending of the Great Siege of 1565) and the awarding of a presidential citation by U.S. president Franklin D. Roosevelt.

David Green

See also: World War II

References and further reading:

Bradford, E. *Siege: Malta 1940–1943.* New York: William Morrow, 1986.

Belfield, E. *Defy and Endure: Great Sieges of Modern History.* New York: Crowell-Collier, 1967.

Mamluks (1000–1600)

Soldiers who controlled the Abbasid Caliphate in Cairo between 1250 and 1517. The origins of the Mamluks lie in the famous regiment of the Abbasid caliph al-Mu'tasim (r. 833–841). They were the backbone of caliphate troops until the destruction of the caliphate at Baghdad in 1258 by the Mongols.

The Mamluks were essentially slaves recruited from the regions of Central Asia and Khorasan. They were a one-generation nobility, and their descendants were not allowed to join the same military aristocracy as their fathers, leading to rampant conflict and unrest. Therefore, in order to ensure the continuity of this aristocracy and to safeguard nomadic vitality, a constant supply of nomadic children was maintained. Furthermore, the persistent effort at investing this aristocracy with a superior status and segregating them from the remaining urban classes had important physical implications. In earlier times, a similar situation had caused the transfer of the Abbasid capital from Baghdad to Samarra and the development of an exclusive residential district. In the era of the Mamluk sultanate's residence in Cairo, it was responsible for the use of the Cairo citadel as segregated residential quarters.

In the complex political history of the Mamluk state, as many as 45 sultans ruled for varying periods of time. In the absence of a system of legitimacy, a sultan's son succeeded him only until another Mamluk gathered enough support to seize the throne. However, some rulers, such as Qalaun (r. 1279–1290), were still able to establish dynasties of continuing rulers. Several others were also able to provide a degree of internal stability and initiate foreign conquests in the course of their short reigns. The success of this system is attested by the important Mamluk victory over the Mongols at 'Ayn Jalut in 1260, which brought Baybars I (r. 1260–1277) into power. Under his leadership, the Mamluks campaigned successfully against the remaining crusader possessions in Palestine and Syria and concluded a truce with the Mongols in 1323 during the long reign of al-Malik an-Nasir (1293–1341).

Egypt continued to dominate eastern Arabdom after Nasir's death in 1341, though the first signs of political and economic decline had already set in. Beginning in 1348, the Black Death repeatedly struck Egypt with large losses. So did Timur's Syrian victory in 1400 and Egypt's loss of control over its Indian trade routes to the Portuguese. The final blow was dealt by the actions of unruly Mamluk corps, which the sultans failed to effectively control. The heyday of the Mamluk sultanate was hence long past when Qait Bay (r. 1468–1496) lost the Syrian Empire to devastating raids by the Turkoman states of Anatolia and Azerbaijan and the campaigns of the Ottoman Turks.

Manu P. Sobti

See also: 'Ayn Jalut, Battle of

References and further reading:

Ayalon, David. "The Muslim City and Mamluk Military Aristocracy." *Journal of Middle Eastern Studies* (1986), 311–329.

———. "Studies on the Structure of the Mamluk Army." *Bulletin of the School of Oriental and African Studies* 15 (1953), 208–228, 448–476; 16 (1954), 57–90.

Rogers, J. M. "Samarra: A Study in Medieval Town Planning," in *The Islamic City: A Colloquium,* ed. R. Stern and A. Hourani. Philadelphia: University of Pennsylvania Press, 1970.

Manchu Expansion, Wars of (1600–1681)

In the sixteenth century, the leader of one of the Jurchen tribes that lived outside the Great Wall in what is now Manchuria, Nurhaci, began to unite the Jurchen people, take on the trappings of a Chinese state, and gather talented Chinese political and military officials around him. In time, he became sufficiently powerful to threaten Ming control over the area outside the Great Wall in the northeast but not to take over China.

Nurhaci created mechanisms of government midway between nomadic rule and the bureaucratic Chinese system. One of his foremost creations was the banner system, in which companies of 300 warriors were grouped under four banners, colored either yellow, white, blue, or red; later, four more banners were added, the first three bordered in red and the last bordered in white. He appointed officers, and the system organized his people for peace and war. By 1644, when the Manchus occupied Beijing, there were 278 Manchu companies, 120 Mongol, and 165 Chinese—making an army of 169,000. Although this was a formidable striking force, it was not large enough to conquer China by itself.

Nurhaci's son, Abahai, sought to build greater power and to threaten the Ming in the north. He attacked Korea in 1627, and in 1636–1637, a renewed attack made Korea a vassal state of the Manchus. Abahai led his warriors through the Great Wall at least on three major occasions (1629, 1632, and 1634) and gained control of nearly all the areas to the north and northeast of the wall. To strengthen his appeal to those Chinese discouraged by the weaknesses and excesses of the Ming regime, he renamed his dynasty Qing, or "pure," and encouraged Chinese "defectors" where it would aid his effort.

The Qing victory owed a great deal to the efforts of a Ming general, Wu San-gui. Wu was called to the capital to help defend the dynasty against a Chinese rebel but arrived too late and turned to the Manchus for help in restoring control over these internal rebels. Wu allowed the Manchu banners to pass through the Great Wall unhindered and to gain control of Beijing; his armies worked with Manchu armies to defeat various rebel groups and local warlords. For more than 30 years, Wu worked with the Manchus to extend their control over the whole of China.

It appeared that Wu accumulated personal power in southern China as he helped the Manchus gain control over the whole country, and in time he revolted. He set himself up as a regional warlord of a sort, and with two fellow warlords, he rebelled in 1673. It would take the great-grandson of Nurhaci, the Kang-xi emperor, eight years to end this rebellion and to gain control over continental China proper.

However, the new Qing dynasty would continue to battle to extend its power over territory typically controlled by other expansive Chinese dynasties. Thus, the Manchus seized control of Chinese coastal areas opposite Taiwan, and the local warlord, Guo Xing-ye, fled to Taiwan, which he temporarily seized from the Dutch. The Dutch helped the Manchus gain control over Taiwan in 1683. Meanwhile, the Qing dynasty would take well into the eighteenth century to establish control over Tibet and over what is now western China and independent Mongolia, an area of traditional Chinese interest.

Charles M. Dobbs

See also: Nurhaci

References and further reading:

Chan, Albert. *The Glory and Fall of the Ming Dynasty.* Norman: University of Oklahoma Press, 1982.

Parson, James Bunyan. *The Peasant Rebellions of the Late Ming Dynasty.* Tucson: University of Arizona Press, 1970.

Rossabi, Morris. *The Jurchens in the Yuan and Ming.* Ithaca, NY: Cornell University Press, 1982.

Spence, Jonathan D., and John E. Wills Jr., eds. *From Ming to Ch'ing: Conquest, Region and Continuity in Seventeenth Century China.* New Haven: Yale University Press, 1979.

Mannerheim, Carl Gustaf Emil (1867–1951)

Finnish military leader and statesman. Mannerheim was born on 4 June 1867 in Louhisaari, Finland, to a noble family. He first studied unsuccessfully at the Hamina Cadet School (1882–1886) and then graduated from a private school in 1887. He completed his military training at the Nicholas Cavalry School in St. Petersburg (1887–1889). Mannerheim served in the Russian imperial army, making his way through the officer ranks and participating in the Russo-Japanese War (1904–1905).

During the initial years of World War I, Mannerheim was initially commander of a cavalry brigade and of a cavalry division in Poland and Galicia and then commander of a cavalry corps in 1917. He resigned his commission on 1 January 1918. In mid-January 1918, as a firm anticommunist, he took over the leadership of the Finnish White Guard. He was awarded the rank of general during his victorious command in the Finnish Civil War but was compelled to resign in May 1918. After Germany's defeat in November 1918, he was elected regent of Finland (1918–1919). He returned to Finnish military decisionmaking in the 1930s as the chairman of the Defense Council (1931–1939) and was promoted to the rank of field marshal in 1933.

In the Winter War with the USSR (1939–1940) and the Continuation War with Germany (1941–1944), Mannerheim was the commander in chief of the Finnish armed forces. He was awarded the title of marshal of Finland in 1942. It may well be that the stout Finnish resistance to Soviet demands induced Stalin, who respected only force and power, to refrain from making Finland another Soviet satellite. Finland would enjoy national independence, although it was always wary of offending its giant neighbor. Much of the credit for Finnish national survival therefore must go to Mannerheim.

After the wars, with Finland having to agree to harsh peace terms with the Soviet Union, Mannerheim was elected president (1944–1946) to steer the country toward peace. During the war crimes trials (1945–1946), which were meant to satisfy the Allies' demands to assign responsibility for the Continuation War, the possibility of his prosecution undermined his position. He resigned in March 1946, exhausted by illness. Mannerheim died on 28 January1951 in Lausanne, Switzerland.

Jari Eloranta

See also: Finnish Civil War; Mukden, Battle of; Russo-Finnish Wars; Russo-Japanese War; World War I

References and further reading:

"C. G. E. Mannerheim," http://www.mannerheim.fi.

Jägerskiöld, Stig. *Mannerheim: Marshal of Finland.* London: Hurst, 1986.

Mannerheim, Carl Gustaf Emil. *Muistelmat,* Part 1. Helsinki: Otava, 1951.

———. *Muistelmat,* Part 2. Helsinki: Otava, 1952.

Field Marshal von Manstein on an inspection tour in Russia, c. 1941. (Library of Congress)

Manstein, Fritz Erich von (1887–1973)

The most brilliant German strategist of World War II was born 24 November 1887 at Berlin as "von Lewinski" and was adopted by his aunt. Following family tradition, he joined a military college, becoming a lieutenant in 1907. After service in staff positions from 1914 to 1918, he entered the post-Armistice Reichswehr and made his way up (deputy chief of the general staff 1937–1938) but was removed to take over a division. When during the winter of 1939–1940 Germany prepared to attack France, Manstein managed to impress Adolf Hitler with his own plan of operations (attack through the Ardennes mountains). Hitler took Manstein's suggestions, and France surrendered within six weeks.

Considered a military genius, Manstein received higher commands in the Russian campaign (Eleventh Army from September 1941 onward). Being fully aware of the mass murders carried out by SS units, Manstein issued an order excusing harsh measures against Jews. After the conquest of the Crimean peninsula, he was promoted to field marshal on 1 July 1942. Beginning in the autumn of 1942, he commanded Army Group Don, trying in vain to relieve the Sixth Army at Stalingrad. He refused to order a withdrawal without Hitler's consent. After the disaster, Manstein managed to stabilize the southern part of the eastern front, but Hitler ignored his suggestions for a reorganization of defenses to allow flexible tactics under Manstein as supreme commander in the east. After several quarrels and Manstein's disregard of Hitler's "stand fast" orders, Manstein was dismissed 30 March 1944 but continued to hope for a comeback until the final surrender.

Manstein repeatedly refused to join military resistance circles but kept silent about contacts with conspirators. A witness during the Nuremberg trials, he stood accused in 1949 of the killing of civilians and commissars. His trial received worldwide attention: Winston Churchill, Bernard Montgomery, and others intervened for Manstein, who nonetheless was sentenced to 18 years' imprisonment but was released four years later. He acted as an adviser for the reestablished German army and in his memoirs blamed Hitler for "lost victories." He died 10 June 1973 at Irschenhausen, near Munich.

Although his strategic abilities are undisputed, Manstein's critics focused on his tolerance for, if not support of, war crimes in Russia and his unwillingness to act against Hitler. He despised the dictator but stayed loyal to him, refusing to draw political conclusions from military developments.

Martin Moll

See also: France; Hitler, Adolf; Stalingrad

References and further reading:

Manstein, Erich von. *Verlorene Siege.* Bonn: Athenäum, 1958.

Smelser, Ronald, Enrico Syring, eds. *Die Militärelite des Dritten Reiches: 27 biographische Skizzen.* Frankfurt am Main: Ullstein, 1995.

Mansûrah, Battle of (November 1249)

Pivotal battle of King Louis IX's (Saint Louis) crusade in Egypt. After wintering in Cyprus, King Louis marshaled his army of approximately 20,000 men and sailed for Egypt in May 1249. Despite attempts by the army of Sultan al-Salih to prevent a landing, the crusaders successfully captured Damietta. On 20 November 1249, King Louis began an advance southward along the Nile toward Mansûrah.

Arriving before Mansûrah in December, the crusaders found their army wedged between the Nile and a tributary. Louis intended to cross the tributary and strike Mansûrah from the rear, but it was not until 8 February 1250 that the army's vanguard successfully forced a crossing. Led by Robert d'Artois, Louis's brother, the crusaders crossed the river, attacked the Egyptian camp, and killed the Muslim commander, Fakhr al-Dîn ibn al-Shaykh. Emboldened by his success, Robert d'Artois then advanced into the city of Mansûrah rather than waiting for reinforcements.

The narrow streets of Mansûrah became a gauntlet for Robert's knights as arrows, stones, and tiles hailed down upon them from the rooftops. An Egyptian counterattack led by the Mamluks, the sultan's military slaves, annihilated the survivors. When King Louis and the main army later attempted to cross the river, they met stiff resistance from the reinvigorated Egyptians.

With the crusaders' forward advance thwarted, Muslim forces increased pressure on King Louis's army by harrying its supply lines. Disease also struck the Christian camp, leaving King Louis himself with dysentery. In March, the crusaders had no choice but to fall back on Damietta but were subject to constant attack along the way. Finally, on 6 April 1250, King Louis surrendered and offered himself as a hostage. The crusaders were too enfeebled to resist or renew the offensive. The invasion of Egypt ended with the exchange of King Louis for the city of Damietta and the ransoming of his army.

Timothy May

See also: Arsuf, Battle of; 'Ayn Jalut, Battle of; Crusades; Hattin, Battle of; Mamluk; Saladin

References and further reading:

Joinville, John de, and Geoffroy de Villehardouin. *Chronicles of the Crusades.* Translated by M. R. B. Shaw. London: Penguin Books, 1963.

Mayer, H. E. Desmond. *The Crusades.* Trans. John Gillingham. London: Oxford University Press, 1972.

Riley-Smith, Jonathan. *The Crusades: A Short History.* New Haven: Yale University Press, 1987.

Runciman, Steven. *A History of the Crusades.* Vol. 3. Cambridge, UK: Cambridge University Press, 1954.

Mantinea, Battle of (362 B.C.E.)

Thebes repeatedly clashed with Sparta over the Spartans' tyrannical rule following the conclusion of the Peloponnesian Wars. Epaminondas led the Theban army on four separate invasions. His final invasion occurred in 362 B.C.E. and culminated in the Battle of Mantinea, in which he managed to repeat his tactical victory at Leuctra against the allied army of Spartans, Mantineans, Arcadians, and others.

Epaminondas deceived the Spartan allies by the manner of his march. His best troops, the Theban Sacred Band, led the march column northward toward the allies but deftly maneuvered to the left, forcing the allies to rotate at right angles. Epaminondas pretended to set up camp without giving battle. Consequently, his enemies began to remove armor and break ranks. Epaminondas maintained the integrity of his formation and posted his cavalry to protect his flanks; he made special provisions for his left flank to prevent it from envelopment.

At the signal, the Theban line and its auxiliary cavalry began a rapid advance against the surprised allies. They managed to reform, but their haste ensured a disorganized and loosely formed line. In keeping with his tactical device at Leuctra, Epaminondas withheld the right portion of his line and vigorously pushed forward his left, which he reinforced in depth at the expense of breadth. The strong impetus of the Thebans smashed into the Spartan phalanx and broke its moral and physical cohesion. After a bloody struggle on the left, the Theban center and right advanced in good order against the weaker portion of the allied line, which quickly gave way. Epaminondas won but at the highest personal cost. Wounded in the chest by a spear, he died shortly thereafter. Although not as decisive as it would have been had he lived, Epaminondas's tactical genius won peace but not hegemony for Thebes.

Bryan R. Gibby

See also: Epaminondas; Leuctra, Battle of

References and further reading:

Delbruck, Hans. *Warfare in Antiquity.* Trans. Walter J. Renfroe Jr. Lincoln: University of Nebraska Press, 1990.

Dodge, Theodore A. *Alexander.* New York: Da Capo Press, 1996.

Mao Zedong (1893–1976)

Leader of an agrarian-based, Communist movement in China during the long period of anti-Japanese resistance and civil war against Chiang Kai-shek's Nationalist regime in the 1930s and 1940s who ultimately came close to destroying Chinese society in the Great Proletarian Cultural Revolution in the 1960s. Mao was born in Hunan Province in central China and moved to Beijing where, in the aftermath of World War I, he became active in student and revolutionary causes, joined the Communist Party, and, during the period of alliance with Chiang's Kuomintang, studied agrarian conditions in China's poverty-stricken countryside. He concluded that peasants had a great anger that could be harnessed to a revolutionary movement, and thereby he broke with the more orthodox Communists who sought the revolution among the limited number of urban proletariat in China's Western-dominated cities along the coast and the Yangtze River.

After Chiang crushed the urban Communists in 1927–1928, Mao gained more power; his "soviet" (revolutionary area) in the rural southeast became a haven for Communists. But in October 1934, after a great many Nationalist "extermination campaigns," Mao and his followers were forced to abandon the soviet and undertake a 12-month "Long March" from southeastern to western China and eventually to the northwest by the Great Wall, near the imperialist Japanese.

Portrait of Mao Zedong. (Library of Congress)

Mao espoused a philosophy that both revolutionized the peasantry and helped guide them in the desperate conflict against Japanese and Nationalist power. He helped them identify entrenched interests—landlords, government officials, businesspeople, the privileged—as being in league with one another and with foreign imperialists and thereby helped peasants identify them as enemies. He also helped them understand how to fight—to harass when weak, to combat when stronger, and always to propagandize the peasantry.

Although the Communists did not play a major role in the defeat of Japan (however they later claimed to have defeated the Japanese almost single-handedly), Mao helped direct the successful campaign in Manchuria that first tied Nationalist troops to big cities, then isolated them, and finally forced them to surrender. Thereafter, as many Nationalist troops deserted or changed sides as were beaten in battle.

Before the final battle of the long civil war could take place—the invasion of Taiwan and the destruction of Chiang's remaining forces and resources—Mao felt forced to intervene in the Korean conflict as Republic of Korea troops and United Nations forces neared the sensitive Yalu River border with Manchuria. For nearly three years, Chinese troops fought better-armed and better-supplied UN forces to a draw, reestablishing the North Korean regime and the thirty-eighth parallel, but at a fearsome human cost.

During the late 1950s, Mao turned his attention inward to remake China. Frustrated with what he viewed as entrenched interests in the government and party bureaucracies, he unleashed the power of youth, backed by a highly politicized army, in the destructive Great Proletarian Cultural Revolution of the 1960s. His later Great Leap Forward was an economic disaster (in which peasants were "encouraged," for example, to build backyard blast furnaces, producing useless metal), and millions died of starvation. Eventually, Mao had to concede the breakdown in power, and he most likely died frustrated with his failure to secure his "Continuing Revolution."

Charles M. Dobbs

See also: Chinese Civil War; Lin Biao

References and further reading:

Schram, Stuart R. *Mao Zedong: A Preliminary Reassessment.* New York: St. Martin's Press, 1983.

Spence, Jonathan D. *Mao Zedong.* New York: Viking Press, 1999.

Terrill, Ross. *Mao: A Biography.* Stanford: Stanford University Press, 1999.

Wilson, Dick, ed. *Mao Tse-tung in the Scales of History: A Preliminary Assessment.* New York: Cambridge University Press, 1977.

Maps and Cartography

The utilization of cartography and topographical mapping for military strategy and operations. Throughout history, considerations of terrain have influenced land warfare. Basic knowledge or ignorance of the lands in which armies traveled and fought sometimes determined their victory or defeat. However, until the invention of the scientific tools and methodologies necessary for composing precise maps, cartography held limited value for soldiers. Scouts, spies, and travelers fulfilled their ever-pressing needs for information about enemy landscapes. Because of the small scale of most campaigns in the premodern world, reliance on such non-graphically expressed intelligence was neither surprising nor particularly injurious to the conduct of war. Even the exceptionally long-distance land campaign of Alexander the Great against the vast Achaemenid Persian Empire or the Mongol incursions into Russia and eastern Europe were not expedited greatly by any formal cartographic organization of knowledge about their enemies' territories and deployments, but rather were facilitated through intelligence obtained from mounted scouts and local informants in a gradual, unfolding fashion. Likewise, the successful administration and defense of small premodern polities did not require exceptionally strong cartographic skills.

These conditions do not mean that the strategic and tactical values of landscape went undetected and unexploited by ancient and medieval commanders. As small chiefdoms and city-states expanded into kingdoms and empires with diplomatic and commercial connections of continental scope, the need for graphic representations of geographic space acquired a heightened significance. The earliest known examples of methodical collations of geographic information occurred soon after the advent of writing among the peoples of ancient Mesopotamia, where cuneiform lists of towns, rivers, and mountains appear as early as the third millennium B.C.E. Simple maps, building plans, and property surveys emerged in the next millennium, both in Mesopotamia and Egypt. Whether the traditions of the ancient Near East and Egypt considerably influenced the evolution of Mediterranean cartography presents a contentious issue, but mapping certainly reached its pinnacle in the premodern West with the Greeks and Romans, especially with the massive influx of geographic data produced by Alexander's conquests and the proliferation of long-distance trading routes connecting Europe with Central Asia, Arabia, India, and the coasts of sub-Saharan Africa in the last four centuries B.C.E. Global mapping, at least in terms of the world known to them, and mathematically grounded cartography emerged in this period, as seen in the works of Eratosthenes, followed by Marinus of Tyre and Ptolemy. The latter's *Geographike Hyphegesis* (*Manual of Geography*) served as the standard for centuries and strongly influenced both European and Arab cartographers, the latter translating his work into Arabic as early as the ninth century.

However, these achievements existed predominantly within the civilian realm; intelligence and maps typically followed conquest in classical Europe. When produced, maps rarely incorporated significant amounts of expressly military data, as seen in the Peutinger Table, a medieval copy of a third-century map of the Roman Empire that lacks any references to military installations. Geographic and ethnographic literature was written after campaigns and oriented toward civilian audiences, such as the fantastic descriptions of India composed by Alexander's admiral Nearchus. Land surveying, an art practiced by every literate culture with increasing precision since the Sumerians, was a vital aspect of Greek and Roman cartography but was employed primarily for civilian or civil administrative purposes, such as the planning of cities and land apportionments. *Agrimensores* (land measurers) accompanied Roman legions, but their tasks principally consisted of laying out fortified camps and buildings, constructing roads, and demarcating lands for veterans, not producing maps; intelligence about territory and terrain was acquired by *exploratores* (scouts). Roman road itineraries constituted the main cartographic tool exploited for military purposes in premodern Europe. Itineraries were written descriptions of roads accompanied by enumerations of the settlements, way stations, and other important features situated along them, along with their intervening mileages, thus allowing for some marching-rate calculation. They are clearly referred to in the military textbook of the late Roman author Vegetius and sometimes were paired with rough graphic illustrations.

With Rome's demise, most of these cartographic developments were lost and replaced by unrealistic, fanciful depictions of the world, shaped more by Christian theological considerations than actual geographic data. This trend is seen most plainly in the *mappaemundi* (literally, "table-cloth worlds" for the size of their media) manufactured in monasteries as didactic tools. Itineraries retained some importance, but systematic, mathematically based cartography, military or otherwise, largely was unpracticed and ignored. Production of maps increased slightly during the Crusades, but again the chief motivation was religious, as demonstrated by the maps of holy sites and pilgrimage destinations in Palestine created by the English monk Matthew Paris.

Mapping reached a sophisticated level very early in China and was applied with considerably greater frequency to military affairs than in contemporary Europe. Dating to 168 B.C.E., silk maps clearly depicting topography, fortifications, depots, and other military sites have been unearthed by archaeologists from a tomb of an officer at Mawangdui (near Changsha, China). There are several unambiguous references to military maps contained within the historical literature of the Han dynasty, such as passages in the *Guanzi* (*Book of Master Guan*), a military text of the third century B.C.E., that urge their use for planning marches and exploiting terrain for strategic advantage. Three-dimensional military maps dating to the early first century and constructed of wood and molded rice are known, and precise, mathematically based topographical maps were drawn up as early as the third century by the imperial cartographer Pei Xiu, whose methods possibly are reflected in extremely accurate gridded maps of the entire Chinese coastline that survive in stone copies dating to the twelfth century. Throughout Chinese history, maps provided by tributary states and peoples signified their submission and allowed Chinese military leaders to amass a substantial amount of strategic geographic intelligence. Their precocious cartographic methods spread to Korea and Japan, where military administrators also adopted them. Beginning in the mid-1400s, Korean cartographers produced particularly high-quality maps for arranging the defense of their Manchurian frontier against potential Manchu invasion.

The history of medieval Islamic cartography is lengthy and rich, but its military relevance remained unrealized until the rise of the Ottoman Empire. One of the few early examples of military Islamic cartography is a literary reference to maps drafted for al-Hallaj ibn Yusuf, an eighth-century administrator who relied on them to coordinate military activities near the Caspian Sea from his palace in Iraq. Arab geographers, cosmographers, and mathematicians, although certainly gifted, drew maps from illustrative rather than strategic intent. However, the Ottoman army definitely exploited small-scale military mapping in its European operations. The *kulaguz* (reconnoiterer) figured prominently in Ottoman military practice prior to the adoption of European cartographic methodology in the nineteenth century, drafting maps and plans of enemy fortifications for use in future conflicts. The earliest example is a late fifteenth-century map of Kiev and its suburbs prepared for Sultan Bayezid II; it proposed a never-executed naval assault on the city, thus revealing its intended role in the planning of campaigns. Süleyman I probably relied upon a still-extant plan of Belgrade, drafted by an army reconnoiterer, during his successful siege of the city in 1521; similar plans for operations in Malta and Szigetvár (Hungary) also exist, as well as a map drawn much later for the ill-fated second siege of Vienna in 1683. For the most part, these "maps" were impressionistic artistic views of the targeted cities and fortresses but stand as concrete illustrations of cartographic-based advance military planning.

Mapping traditions also flourished in the Americas prior to European contact, although most of these materials were lost and what survives is known mainly from European copies of varying quality and frequently dubious veracity. The best examples of pre-Columbian military cartography are maps allegedly utilized and drawn by Aztec *pochteca* (long-distance traders), sent as spies into adjacent towns and regions in order to collect geographic intelligence under cover of their commercial endeavors. According to an illustration contained in the *Codex Florentine,* a Spanish-Nahuatl "encyclopedia" of preconquest Mexico written about 1570, it seems that the Spanish integrated Nahuatl information into new maps, which then guided the troops to their objectives; unfortunately, no actual *pochteca* maps exist today. Also, the conquistador Hernando de Cortez mentioned Aztec ambassadors who brought to him cloth maps detailing the Gulf of Mexico coast, whereupon he ordered his staff to make copies in European form and used them to plot further expeditions.

Comprehensive and accurate cartography useful for military applications in the modern sense emerged only toward the close of the medieval period in Europe. New technologies and social forces coincided to produce the enormous upsurge in map production during the Renaissance. Directional compasses appeared in China around 1150 and passed to Europe about a century later, eventually revolutionizing Western navigation and cartography. The revived study of classical urban planning and military architecture intensified interest in surveying and mapping generally, especially in Italy, where district maps and diagrams of fortifications prepared by military engineers, including the famous artist Leonardo da Vinci, appeared throughout the fifteenth century. This is a trend reflected most clearly in the military district maps commissioned in 1460 by the Council of Ten of Venice and the large regional maps of Lombardy produced as early as the 1440s. Triangulation surveying techniques, introduced by Gemma Frisius in 1533, allowed for considerably more precise measurements of distances than previously available, enabling the development of scale mapping. True-to-scale local and property maps began to be drawn in Italy, the Netherlands, and England, and English engineers produced accurately scaled maps of their coastal fortifications by the mid-1500s after French raids along the Channel coast. From the fifteenth to the seventeenth centuries, the escalation of border and frontier disputes between the emergent absolutist nation-states, extensive nautical exploration and the construction of transoceanic empires, massive religious warfare, and insurrections like the Fronde in France emphasized the need for reliable maps for the rulers and military leaders of Europe, who then enthusiastically employed surveyors to map their territories to meet such dire military and diplomatic concerns. The first known official military cartographic project in France began in 1495 with Charles VIII's order to Jacques Signot to map the mountain passes by which the French army could traverse the Alps and invade Italy. In Habsburg Spain, where military cartography evolved slowly and to an arguably lesser extent than in other European realms, Philip II nonetheless often utilized maps to coordinate information about his far-flung empire and charted the route of his doomed armada with maps and nautical charts, and his subordinate, Fernando Alvarez de Toledo, the Duke of Alba, relied extensively on rough maps of the Netherlands to plan marches and attacks during his attempts to subdue the Dutch insurgents.

Despite these advances, before 1700 most military maps were simplistic and schematic, drafted more frequently by and for civilians and bureaucrats than soldiers, and not exploited regularly by commanders in the field. Pamphlets published in the 1750s urged British officers in India to use their maps and avoid overreliance on native guides. Frederick II constantly complained about the dearth of good military maps and was compelled to rely on captured Austrian maps for his campaigns in Silesia.

The eighteenth century brought the formalization of military cartography requisite for the production of maps serviceable for strategic planning and combat operations. Surveyors and engineers increasingly were incorporated into formal army structures, beginning with the organization of the French army corps of surveyors in 1696. Improvements in road construction, the desire for better control over larger numbers of troops, which required detailed advance logistical and operational planning, and the realization that successful defense from foreign invasion rested significantly upon geographic intelligence led to the first truly scientific efforts to map entire countries. Important examples include the topographical surveys of France undertaken by the Cassini family between 1733 and 1788, General William Roy's mapping of Scotland for the British army in the wake of the Jacobite Rebellion, F. W. Schetten's survey of Prussia in the 1780s, the mapping of Austrian crown lands ordered by Joseph II in the same decade, and the initiation of the Board of Ordnance surveys in Britain in 1791. Topographic mapping, a necessary component of modern military cartography, began to be standardized during the 1700s. First proposed in 1777 by Jean-Baptiste Meusnier, a French lieutenant of engineers, contour lines dramatically increased the military efficacy of topographical maps, especially at the tactical level; however, they did not replace entirely older topographical maps drawn with hachure lines until the next century. Thoroughly map-based advanced military planning emerged toward the end of the eighteenth century, noticeable as early as 1775 with General Pierre-Joseph de Bourcet's work, *Les principes de la guerre de montagne.* Based on his detailed topographical surveys in the Savoy and Piedmont border regions of France, the work presented a carefully organized system by which all rates of march and supply issues for a campaign could be anticipated through reliance on detailed topographical maps. Coupled with the creation of divisional units, modern topographical maps contributed to the expansion of commanders' abilities to control significantly larger armies than existed in previous ages, thus transforming European warfare.

This process of formalization and large-scale strategic mapping continued throughout the nineteenth century. Napoleon I organized his Imperial Corps of Surveyors in 1809, whose efforts facilitated the speed with which his armies penetrated Russia a few years later. The implementation of general army staffs, which brought the art of preparing for future wars during peacetime to an unprecedented level, sustained interest in military cartography throughout the century. The predecessor of these bodies was the Great General Staff of the Prussian army, created during the Napoleonic Wars to coordinate technical and logistical information for field commanders; these goals required the systematic compilation of immense amounts of topographical data collected well before hostilities began. By the 1840s, general staff maps were available both in France and Prussia, where they exercised a strong influence on military strategy and planning, seen most clearly with the formulation of the Schlieffen Plan in Germany prior to World War I. Military cartography disseminated outside Europe, replacing older traditions in China, Japan, and the Ottoman state.

Army surveyors proved invaluable in exploring and mapping unsettled regions in the western United States throughout the nineteenth century, beginning with the famous expeditions led by Captains Meriwether Lewis and William Clark, Lieutenant Zebulon Pike, and Lieutenant John Frémont, the latter an officer with the Army Corps of Topographical Engineers, formed in 1838. The corps' work, as well as numerous surveys undertaken by the War Department, allowed for the rapid settlement and railroad construction that occurred in the American West after the Civil War. Although the western and "Indian" territories were mapped systematically, that conflict also highlighted the lack of scientific surveys of the country's eastern and southern sectors, further stimulating the development of American military cartography, so that by the Civil War's end, the Coast Survey and Army Corps of Engineers supplied Union forces with approximately 43,000 printed maps per year. Similar increases in the scale of map production took place elsewhere, especially in connection with trench warfare during World War I, which necessitated highly accurate large-scale maps for accurate, long-range artillery fire. For example, the British Expeditionary Force (BEF) originally landed in France with two cartographers, an officer, and a clerk responsible for the provision of all maps; by 1918, the BEF contained 5,000 cartographic personnel who produced over 35 million maps throughout the course of the war.

It was in the twentieth century that the most substantial advances in scientific military cartography were made, as mapping moved into the skies. Although surveyors and artillery observers working from balloons participated effectively in nineteenth-century wars, notably the American Civil War, the invention of the airplane utterly transformed their work. Significantly more mobile than balloons and dirigibles, aircraft were acknowledged to have an advantage quite early, with the first examples of aircraft reconnaissance occurring in 1911 during the Italian campaign in Libya. The armies of all participating countries employed them during World War I, eventually equipping them with machine guns and primitive bombs and transforming them into the first fighter and bomber aircraft; the American army flew reconnaissance missions as early as 1916 during the Mexican Punitive Expedition directed against Pancho Villa.

By the outbreak of World War II, aerial operations were regarded everywhere as vital components of land warfare, leading to a demand for detailed maps for strategic bombing and aerial support of land combat and, when coupled with photography, allowing for the composition of extremely accurate maps. Two examples from the experience of the German Luftwaffe clearly illustrate this double facet of modern warfare: Nazi bomber crews relied on British Ordnance Survey maps to locate targets in England and then in preparation for the invasion of the Soviet Union flew modified high-altitude bombers over Soviet territory to collect intelligence about geography and troop deployment. Similar photo-reconnaissance missions enabled the amphibious landing of their Allied adversaries at Normandy a few years later.

During the Cold War, aerial reconnaissance and mapping saw extensive application as nuclear threats emerged. The United States conducted high-altitude reconnaissance flights over Soviet territory beginning in 1946, successfully charting their bomber bases and revealing their nuclear capabilities, although the constant emphasis that Strategic Air Command placed on using aircraft instead of missiles or rockets throughout the 1950s hampered the development of American space and satellite programs. Between the Apollo moon landing program of the 1960s and the space shuttle program of the 1980s, this disadvantage was rectified, and American satellites now provide the most accurate military cartographic services ever known. Without global positioning system (GPS) and geographic information system (GIS) satellites and computer software that produces precise three-dimensional maps of enemy terrain, the pinpoint accuracy of armor fire achieved by U.S. forces during the Gulf War or in later cruise missile strikes against antiterrorist objectives in Sudan and Afghanistan would have been unachievable. Recent space cartography projects, such as the hyperaccurate digital terrain maps generated by the new Shuttle Radar Topography Mission, certainly hold potentially enormous consequences for the nature of land warfare in the twenty-first century.

Ian Janssen

References and further reading:

Black, Jeremy. *Maps and Politics.* Chicago: University of Chicago Press, 1997.

Buisseret, David, ed. *Monarchs, Ministers, and Maps: The Emergence of Cartography as a Tool of Government in Early Modern Europe.* Chicago: University of Chicago Press, 1992.

Delano-Smith, Catherine, and Roger Kain. *English Maps: A History.* Toronto: University of Toronto Press, 1999.

Harley, J. Brian, et al., eds. *The History of Cartography.* Vols. 1–2. Chicago: University of Chicago Press, 1987–1998.

Harvey, P. *The History of Topographical Maps.* London: Thames and Hudson, 1980.

Konvitz, Josef. *Cartography in France, 1660–1848.* Chicago: University of Chicago Press, 1987.

Luebke, Frederick, et al., eds. *Mapping the North American Plains.* Norman: University of Oklahoma Press, 1987.

Maratha Wars (1775–1818)

Three wars that doomed the Maratha Confederacy. By the late eighteenth century, the Maratha Confederacy was one of the greatest powers in the central Indian subcontinent. During the reign of Shivaji (d. 1680), the Marathas erupted from their mountain domain of the Western Ghats on the Arabian Sea and were transformed from Hindu mercenaries into warlords in their own right. Significant as was their military proficiency in the rise of the Marathas, equally important were the dissensions that afflicted Mogul rule of India. The reign of Shivaji paralleled that of the Mogul emperor Aurangzeb, whose vigorous persecution of non-Muslims in the last half of the seventeenth century served to encourage revolt among not merely the Marathas but the Rajputs and Sikhs as well. The resulting overextension of Mogul forces meant that, by the time of Aurangzeb's death in 1707, the Marathas had established themselves as the dominant power in the Deccan. In the succession crises that followed the death of the emperor, the Marathas were able to extend their power still further, until by 1740, their ruler, the Peshwa of Pune, governed from the Arabian Sea to the Bay of Bengal.

Given the nature of the Maratha rise to power, it was fitting, if not ironic, that the ultimate British victory in the three Maratha Wars had as much to do with dissension among the Marathas as to particular proficiency on the part of their European enemy. As the Marathas extended their power across India, governance of lands seized from the Mogul emperor was left in the hands of Maratha commanders. Consequently, rather than governing a centralized state, the Peshwa of Pune oversaw what the nineteenth-century historian James Duff described as a "communion of interests." The same tendency to internal dissension that so afflicted the Mogul empire and that largely accounted for the rise of the Marathas soon became a feature of Maratha political life.

Matters took a serious turn in the early 1770s, when the British East India Company's Bombay presidency saw in the succession crises of that time an opportunity to expand its interests in the region. In 1775, Bombay recognized Raghunath Rao as the legitimate claimant to the Peshwaship in exchange for territorial gains and in so doing set the scene for the First Maratha War (1775–1782). It was a conflict that ran in fits and starts. At Aras (18 May 1775), Raghunath and his British allies were defeated by the Maratha commander, Hari Pant. Facing war with France and its primary Indian ally, the Sultan of Mysore, Governor of India Warren Hastings took the opportunity of this defeat to condemn Bombay's actions and ordered the company troops back to their quarters. Subsequent negotiations with the Maratha Regency Council failed—despite British offers to abandon Raghunath Rao for territorial concessions—and in 1778 troops from Bombay again marched against the Marathas. Though their defeat at Talegaon (11–17 January 1779) represented a serious setback to British interests in the region, 1779 also marked an important turning point in British strategic interests in India. Concerned that Bombay's actions could increase French interest in the region, Hastings ordered six battalions to march from Bengal. When the commander of this force, Colonel Thomas Goddard, heard of the defeat at Talegaon, he did not turn back but pushed on, covering 300 miles in 19 days. When negotiations with the Regency Council failed, Goddard was joined by the Gaikar, Maratha princes from Baroda, and successfully stormed Ahmadabad. Meanwhile, in 1780, a second Bengal force under Captain Thomas Popham captured the Marathas' mountaintop fortress at Gwalior (November 1780). Just as British victory over the Marathas seemed to be assured, however, Haidar Ali, Sultan of Mysore, invaded the Carnatic, opening the Second Mysore War. As a result of the very real threat to Madras and the British position on the Carnatic, the East India Company opened negotiations with the Marathas in 1781 that would end with the Treaty of Salbai (1783). Although the latter forced the British to relinquish their support of Raghunath Rao, the expeditions of Goddard and Popham nonetheless demonstrated the British ability to strike at will anywhere on the subcontinent.

Peace between the British and the Marathas would last for the next two decades, until a succession crisis shook the confederacy in 1803. Defeated by his rival, Holkar of Indore, Baji Rao II entered into the Bassein Treaty (1803), by which the British agreed to restore Baji as Peshwa in return for the Marathas accepting and paying for British troops in their capital, together with other obligations. To restore Baji Rao and bring those princes who rejected the Bassein Treaty to heel, Governor General Sir Richard Wellesley planned a twofold campaign against the Marathas. First, Wellesley's brother, Arthur, would lead a force of 9,000 Europeans and 5,000 Indian troops into the Maratha homeland. A second force under Gerard Lake invaded Hindustan. In March 1803, Wellesley's force captured Pune and restored Baji as Peshwa. On 23 September 1803, Wellesley's army met and defeated the Marathas under Doulut Rao Sindhia at Assaye. Though victorious, Wellesley would later recall this campaign as the hardest-fought action of his long career. Meanwhile, Lake captured Delhi on 16 September 1803 and in the Battle of Laswari (1 November) finally destroyed the forces of the Maratha prince, Sindhia. In the meantime, the British government had grown concerned with the extent of these operations. In particular, the siege of Bhurtpore (January–April 1805) had claimed 3,100 men before the British were victorious. Accordingly, Lord Wellesley was recalled, and with the capitulation of Holkar at Amritsar (December 1805), the Second Maratha War came to an uneasy close.

The Third Maratha War (1817–1818) was in large part the consequence of the turmoil that gripped India as Maratha power finally crumbled. In the aftermath of the Second Maratha War, a vast horde of former Maratha soldiers known as the Pendaris spread out across central and southern India in an organized campaign of violence and depredation. When the governor of India, Francis Hastings, Lord Moira, decided to move against the Pendaris with two large armies, he expected Maratha support. Instead, the Peshwa Baji Rao II, justifiably resentful of the conditions imposed on him by the British in return for their support in the 1803 war, turned on the British and attacked and destroyed their residency at Pune with a force of 27,000 men (5 November 1817). Holkar likewise took to the field but suffered defeat at the hands of Sir Thomas Hyslop at Mahidput (21 December 1817) before being forced to surrender. Nor did the Peshwa fare any better, for on the same day his forces destroyed the British residency at Pune, they were defeated at Kirkee by a mere 2,800 British soldiers. Defeated again at Koregaon (1 January 1818) and Ashti (20 February), the Peshwa finally surrendered to Hastings's army on 2 June 1818. With Baji Rao's surrender, the hereditary office of Peshwa was abolished by the victorious British, and with it the political and military power of the Maratha Confederacy ceased to exist.

Adam Lynde

See also: Mysore Wars

References and further reading:

Duff, James G. *History of the Mahrattas.* Ed. J. P. Guha. New Delhi: Associated Publishing House, 1971.

Gordon, Stewart. *The Marathas, 1600–1818.* Cambridge, UK: Cambridge University Press, 1993.

Keay, John. *India. A History.* London: Harper Collins, 2000.

Mehra, Parshotam. *A Dictionary of Modern Indian History, 1707–1947.* Delhi: Oxford University Press, 1987.

Marathon, Battle of (490 B.C.E.)

In August 490, a Persian force of some 30,000 men under the command of Darius landed at Marathon in Attica. Nine thousand Athenians and 1,000 Plataean allies, all under Miltiades, defeated the enemy, despite having been refused immediate help from Sparta.

Who took the initiative for battle is still under debate, though it seems that the Greeks decided to attack first after they heard a rumor saying that the Persian cavalry was withdrawing. Miltiades, wanting to make his army look equal in length to the Persian army, made its center only a few ranks deep but kept the wings deeper and stronger. The Athenians occupied the center and the right side of the line, and the Plataeans were on the left. Little is known of the Persian dispositions, except that the Persians and Sacae (a warlike people from Central Asia) formed the very strong center of the line.

After the usual sacrifice, the Greeks charged at the double (according to Herodotus, the first Greeks to have done so) across no-man's-land, taking the Persians by surprise and thus reducing the effectiveness of their archers. The fighting, according again to Herodotus, was severe and lasted a long time. The Persians broke the weak Athenian center and pursued the survivors inland. Meanwhile, the stronger Greek wings, which had already managed to route their opponents, reformed (most probably joining forces) and attacked the Persian troops who had broken through their center. The Persians started to flee toward their ships, with the Greeks in close pursuit. Large numbers of Persians perished in a nearby marsh, and even more were killed by the pursuing

Greeks attacking Persian ships during the Battle of Marathon. (Bettmann/Corbis)

Greeks. Most of the Persian navy was already at open sea, but seven Phoenician ships that were still lying close inshore, waiting for the last fugitives, were captured by the Greeks. Herodotus gives the losses as 6,400 Persians and 192 Athenians dead. The former figure may seem exaggerated, but the latter can surely be accepted as precise.

Despite the Greek victory, the Persian fleet laid course for Sounion in order to attack Athens. Miltiades immediately ordered his troops to march post-haste to Athens and managed to reach the city before the enemy fleet. When the Persians arrived soon afterward and saw an army waiting for them, they set off toward Asia.

This amazing victory of the Athenians and their Plataean allies had a huge impact on Athens's later history, and many scholars have praised the strategy and tactical brilliance of Miltiades. It should be noted, however, that although all these tactics appear to be quite sophisticated and planned well ahead (even down to what becomes, according to some views, the "falling back of the Athenian center"), it is more likely that what happened was almost accidental. If we are to believe Herodotus, the thinning of the Greek center was a purely defensive move rather than a specific plan for the wings to first crush the enemy wings and then move on to smash their center by a "double envelopment." In addition, there is no reason to believe that the victory was due to the superiority of Greek discipline because most of the Greek *hoplitai* were just common citizens and not professional soldiers. All these caveats do not decrease the importance of the victory but on the contrary enhance it and make it more astonishing.

Ioannis Georganas

References and further reading:

Hackett, John, ed. *Warfare in the Ancient World.* London: Sidgwick & Jackson, 1989.

Hammond, Nicholas. "The Campaign and Battle of Marathon." *Journal of Hellenic Studies* 88 (1968), 13–57.

Lazenby, John. *The Defence of Greece 490–479 BC.* Warminster: Aris & Phillips, 1993.

Marcellus, Marcus Claudius (c. 275–208 B.C.E.)

Probably the most popular commander in Rome during the Second Punic War. Middle-aged when the Second War began, Marcellus had already won his spurs. While serving in the First Punic War, he saved the life of his adopted brother. As consul, he campaigned against the Gallic tribe of the Insubres in 222 B.C.E. He relieved the city of Clastidium and killed the Gallic chief Viridomarus in single combat, by which he won the so-called Spolia Opima (the spoils of honor), that is, the right to offer his vanquished opponent's personal spoils to Jupiter. This honor was given only twice before in Roman history.

When in 216 B.C.E., Rome was disastrously defeated by Hannibal at Cannae, Marcellus was sent to take command of the remnants of the defeated army. He reformed it into two legions and followed Hannibal into Campania. Using the delaying tactics advised by the general Fabius Maximus, Marcellus dug in on a strategically important position, afterward known after him as "Castra Claudiana." From there he could cover a large part of Campania, threaten the defecting town of Capua, and secure important lines of communication without risking battle with Hannibal. During the following years, Marcellus frustrated Hannibal's attempts to extend his foothold in Campania.

In 213 B.C.E., Marcellus was dispatched to Sicily and assaulted Syracuse. Because of the defensive machinery created by the scientist Archimedes, the Romans were repelled, and Marcellus laid siege to the city. Although the Syracusans held out for three years, in 211 the city was taken and sacked. In later years, Marcellus remained faithful to the Fabian method. He doggedly followed Hannibal's footsteps on the Italian mainland without engaging his army. However, in 208 B.C.E., Marcellus was attacked and killed reconnoitering the battlefield. Hannibal gave him a hero's funeral, a fitting end for the commander who was called "the Sword of Rome."

M. R. van der Werf

See also: Cannae, Battle of; Fabius Maximus Verrucosus "Cunctator"; Hannibal Barca; Punic Wars

References and further reading:

Bagnall, Nigel. *The Punic Wars. Rome, Carthage and the Struggle for the Mediterranean.* London: Pimlico, 1999.

Caven, Brian. *The Punic Wars.* London: Weidenfeld and Nicholson, 1981.

Lazenby, J. F. *Hannibal's War. A Military History of the Second Punic War.* London: Aris & Phillips, 1978.

March, Peyton (1864–1955)

American general of World War I, sometimes called the "Father of the Modern U.S. Army." Born 27 December 1864 in Easton, Pennsylvania, March graduated from the U.S. Military Academy at West Point in 1888 and served in the Philippine Insurrection of 1899–1902, having field command in the Battle of the Clouds at Tilad Pass on 2 December 1899. He served on the general staff from 1903 to 1907, acted as an observer during the Russo-Japanese War of 1904–1905, and served as commander of the 1st Battalion, 6th Field Artillery

Regiment in Fort Riley, Kansas, from 1911 to 1916. As a colonel, March commanded the 8th Field Artillery on the Mexican border in August 1916. In June 1917, after the United States entered World War I, March was promoted to brigadier general and led the 1st Field Artillery Brigade in the American Expeditionary Force in France. He became acting chief of staff of the army in March 1918 and chief of staff on 19 May of that year and held the position until 30 June 1921. During these crucial years, he presided over the buildup of U.S. forces in the closing months of the war, as well as the demobilization of those forces. Although today a little-remembered figure, Peyton March is credited with creating the Air Service, Tank Corps, and Chemical Warfare Service during the war. He died 13 April 1955 in Washington, D.C.

Harold Wise

See also: Philippine Insurrection; World War I

References and further reading:

March, Peyton C. *The Nation at War.* Garden City, NY: Doubleday, Doran and Company, Inc., 1932.

Statue of Emperor Marcus Aurelius. (Library of Congress)

Marcus Aurelius (Antoninus) (121–180)

Roman emperor, philosopher, and soldier. Born Marcus Annius Verus, Marcus Aurelius succeeded Antoninus Pius as Roman emperor (Augustus) in 161. Marcus appointed Lucius Verus coemperor to help rule and consolidate the large empire, reserving the title of chief priest (*pontifex maximus*) for himself. In 162, Marcus sent Lucius to the eastern part of the empire to campaign against the Parthian ruler Vologases III, who had placed his brother on the throne of Armenia and had attacked Roman garrisons along the Armenian border. Lucius defeated Vologases and reinforced the border along the Euphrates River before returning to Rome in 166. Marcus and Lucius suppressed rebellions in Britain and Africa and impeded an invasion of Italy by a Germanic group, the Marcomanni, who under their leader, Ballomar, besieged the strategic frontier town of Aquileia. Marcus and Lucius retook Aquileia and pushed the Marcomanni back across the Danube River among the other Germanic tribes. Lucius died on the journey north in 169, and Marcus was left to carry on the Danubian campaigns for another 11 years.

Marcus temporarily left these campaigns when Avidius Cassius, a once loyal governor of Syria, claimed for himself the title of Augustus in 175. Although Avidius was killed by his own soldiers before Marcus could reach Syria, Marcus used the opportunity to secure his power in the east and returned to the Germanic wars in 176. He campaigned separately against the Marcomanni, Quadi, and Sarmati tribes in order to break up any Germanic unity. Marcus planned to extend the empire past the Danube River and to place these tribes into the new provinces, thus creating a larger buffer zone between the tribes and Italy. Only the last step in his project remained when Marcus died in his camp at Vindobona on 17 March 180.

Christopher P. Goedert

References and further reading:

Garzetti, Albino. *From Tiberius to the Antonines: A History of the Roman Empire,* C.E. *14–192.* Trans. J. R. Foster. London: Methuen, 1974.

Grant, Michael. *The Antonines: The Roman Empire in Transition.* London: Routledge, 1994.

Watson, Paul Barron. *Marcus Aurelius Antoninus.* 1884. Reprint, Freeport, NY: Books for Libraries Press, 1971.

Marengo, Battle of (14 June 1800)

French victory over Austria in northern Italy. In early 1800, Napoleon sought to defeat Austria, his one active enemy. Crossing the Alps into Lombardy, he believed the Austrians would flee. Therefore he detached units of his army to block their escape. However, Austrian commander Michael Melas decided to fight and deployed around Alessandria. On 13 June Napoleon, increasingly convinced Melas would retreat, sent one division north to the Po River and ordered another, under Louis Charles Desaix, south to block the road to Genoa. Thus the Austrians had concentrated 31,000–34,000 troops around Alessandria, whereas Napoleon's were reduced to less than 28,000.

Early on 14 June, three strong Austrian columns surprised Napoleon by marching from Alessandria to attack. Battle was joined just west of the village of Marengo, 3.5 miles east of Alessandria. Not until late morning did Napoleon realize he confronted the bulk of Melas's army. He then ordered the recall of units he had previously detached. But the situation deteriorated rapidly for the French. Retreating eastward, by noon they had no reserves, were short of ammunition, and faced being outflanked on their right.

The Austrians, confident of victory, paused to regroup. They resumed the attack about 1:00 P.M., pushing the French back to San Giuliano, nearly 5 miles from Marengo. About 3:00 P.M. a confident Melas turned over command to his chief of staff, Anton Zach. Then Desaix arrived, having marched directly to the sound of the guns. Zach did not press the attack until approximately 4:30. Desaix led a counterattack in which he was killed, but Auguste Marmont and François Kellermann (the younger) rallied the French, who then routed the Austrians. The battle ended about 10:00 P.M. The French suffered some 4,700 killed and wounded, the Austrians 6,500.

Marengo was a very close call because Napoleon had badly miscalculated. Nevertheless, Marengo enhanced his image, strengthened his position as first consul, and quelled possible political opposition at home.

James K. Kieswetter

See also: French Revolutionary Wars; Masséna, André, Duc de Rivoli, Prince d'Essling; Napoleon I; Napoleonic Wars

References and further reading:

Chandler, David G. *The Campaigns of Napoleon.* New York: Macmillan, 1966.

Connelly, Owen. *Blundering to Glory: Napoleon's Military Campaigns.* Wilmington, DE: Scholarly Resources, 1987.

Rodger, Alexander Bankier. *The War of the Second Coalition: A Strategic Commentary.* Oxford: Clarendon Press, 1964.

Sargent, Herbert Howland. *The Campaign of Marengo.* 6th ed. Chicago: A. C. McClurg, 1914.

Marignano, Battle of (13–14 September 1515)

Marignano ended the myth of Swiss invincibility. At 20 years of age, François I became king on New Year's Eve, 1515. He was determined to bring the Duchy of Milan back to his crown. Gathering his army at Lyons the following spring, he faced an alliance of Massimiliano Sforza, Duke of Milan, the Habsburg emperor, and the cardinal of Sion. Their army, mainly Swiss mercenaries, occupied the easiest pass to cross the Alps. Francis I chose a smaller and more difficult draw to bypass the waiting Swiss. The two sides now looked for a financial agreement to avoid battle. The Swiss, followed by the French troops, marched to Milan. The king joined a Venetian army, and on 8 September, the agreement was to be signed. But the cardinal of Sion persuaded the Swiss to attack the king in his camp at Marignano on 13 September. With the Venetian army encamped at Lodi, the French were left on their own. The battlefield was difficult: a large plain cut by hedges and small irrigation ditches that would disrupt any French cavalry charge.

The surprised French placed their artillery in front of the large blocks of Swiss pikemen. The French king charged at the head of the royal gendarmerie more than 30 times, only to be repulsed by the wall of pikes of the Swiss phalanx. The falling night stopped the fight, with the armies resting less than 50 yards from each other. During the night, Francis I sent a messenger to Alviano's Venetian troops to join as soon as possible.

The dawn of the 14th saw a new battle, as the 20,000 Swiss attacked, again facing deadly artillery fire. The German mercenaries on the French side were overwhelmed by the Swiss, some of the latter reaching the French artillery only to be cut to pieces by the gendarmerie's heavy knights. The first Venetians reached the battlefield by 8:00 A.M., and the main Swiss pike block was assaulted on its front by the king's cavalry and on its left flank by Venetian infantrymen. By 1:00 P.M., the surviving Swiss retired to Milan, leaving 14,000 dead on the battlefield. Milan surrendered the following day, giving the duchy to the king. This battle was won by the superior French artillery and the king's stubborn defense. The most important result was the Perpetual Peace, signed in 1516, which gave France the exclusive use of Swiss soldiers. The treaty lasted until 1792.

Gilles Boué

See also: Cerisolles, Battle of; Pavia, Battle of

References and further reading:

Contamine, Philippe. *Histoire militaire de la France.* Vol. 1. Paris: PUF, 1992.

Oman, Charles. *A History of the Art of War in the Sixteenth Century.* London: Greenhill, 1997.

Marion, Francis (1732–1795)

American partisan leader in the Revolutionary War. Born in Winyah, South Carolina, and by occupation a planter, Francis Marion first saw military action as a lieutenant of South Carolina militia in campaigns against the Cherokees in 1759 and 1761. In 1775, he took a captaincy in the 2d South Carolina Regiment of the Continental army.

After being promoted to major the following year, he helped defend Fort Sullivan during the British attack on Charleston, firing the last shot at their departing fleet. As lieutenant colonel, he commanded the regiment until he was injured in the spring of 1780.

After the disastrous American defeat at Camden in August had cleared the Southern Department of practically all regular U.S. troops, Marion began a partisan campaign. Commanding about 50 militiamen, he defeated a Loyalist militia force five times larger in September and then squelched a Tory uprising. British lieutenant colonel Banastre Tarleton fruitlessly chased his guerrillas fighters through the swamps in November and commented disgustedly, "but as for this old fox, the devil himself could not catch him," thus giving Marion his famous nickname, the "Swamp Fox." Through 1781, now a brigadier general of militia, he harassed the British from bases in the swamps and woods, often in tandem with "Light Horse Harry" Lee, capturing Forts Watson and Motte. His brilliant and ceaseless guerrilla campaign distracted the British, while General Nathanael Greene's Continentals reestablished control of the area. In September, Marion joined Greene and commanded the militias of North and South Carolina in the Battle of Eutaw Springs. An abstemious and nonviolent man in an age of hard-drinking fighters, Marion is supposed to have clambered out of a second-story window to avoid a round of alcoholic toasts and, when forced to draw his sword (presumably in battle), found it rusted into its scabbard! After this bitter war, Marion was a leader in the reconciliation between patriot and Tory.

Elected to the South Carolina Senate in December, he was subsequently reelected twice. From 1784 to 1794, he was militia general and commandant of Fort Johnson in Charleston Harbor.

Joseph McCarthy

See also: American Revolution; Camden, Battle of; Cowpens; Greene, Nathanael

References and further reading:

Bass, Robert D. *Swamp Fox: The Life and Campaigns of General Francis Marion.* New York: Henry Holt, 1959.

Rankin, Hugh F. *Francis Marion, the Swamp Fox.* New York: Thomas Y. Crowell, 1973.

Weigley, Russell F. *The Partisan War: The South Carolina Campaign of 1780–1782.* Columbia: University of South Carolina Press, 1970.

Marius, Gaius (157–86 B.C.E.)

Roman general, statesman, and military reformer who led the Populares Party during the civil war. Born in 157 B.C.E. at Arpinum, Marius—a "new man" since he achieved his position through ability and talent instead of by birth—represented the interests of the plebeian class. In 116 B.C.E., he administered Further Spain, suppressing bandits and establishing his personal wealth through mining investments. During his consulship, Marius initiated numerous military reforms, including the recruitment of troops through a headcount system in the cities, which replaced the old volunteer plebeian army of farmers. He developed a strict training program and required his men to carry their own equipment and supplies. He is credited with redesigning the *pilum* (Roman spear) so that the shaft bent after hitting a target, making it impossible to throw back on his own men. In 107 B.C.E., as consul, he led the Roman forces against Jugurtha, King of Numidia (present-day Algeria), but his rival Sulla received credit for the victory after arranging for the capture of the king. A rivalry developed between Marius and Sulla that resulted in years of bloodshed within Rome.

At the conclusion of hostilities in North Africa, Marius and Sulla fought together against the Germanic tribes. Their rivalry continued to intensify until civil war broke out in 88 B.C.E., when Marius arranged to have Sulla's command, to lead the Roman army against the Asian king Mithradates VI, transferred to himself. Sulla returned to Rome with his army, and Marius lost his status as consul and was forced to flee. Many of his supporters lost their property and lives. After Sulla departed for Asia, Marius returned to Rome, murdering Sulla's supporters and ruling the city. In 86 B.C.E., Marius declared himself consul but died a few days later.

Cynthia Clark Northrup

See also: Sulla, Lucius Cornelius

References and further reading:

Kildahl, Phillip A. *Caius Marius.* New York: Twayne Publishers, 1968.

MARKET GARDEN (10–24 September 1944)

The largest airdrop in history and the last defeat for British forces in World War II. In September, Allied forces had liberated most of France and Belgium. Supreme Allied Commander General Dwight Eisenhower wanted to continue along a broad front, but Field Marshal Bernard Montgomery and other commanders wanted a quick thrust along a narrow front through Holland, outflanking the Siegfried Line, breaking the paper-thin German defenses, and then turning eastward into Germany's Ruhr Valley. Such a move would cut the heart out of Germany's industrial capacity.

The success of such an operation depended on securing bridges on the Rhine River at Grave, the Waal River at Nijmegen, and the Mass River at Arnhem. Montgomery decided that airborne troops would seize and hold each bridge, while the Second British Army moved northward along what was called Hell's Highway. In the largest airborne operation in history, the British 1st Airborne Division was to capture the bridge at Arnhem, while the 82nd American Airborne Division was to capture the crossings at Nijmegen and Grave and the 101st Airborne Division was to secure crossings between Grave and Eindhoven.

On 10 September, the first part of the operation, called Operation MARKET and involving the seizure of bridges at Eindhoven and Nijmegen, was successful. Operation MARKET GARDEN, the British seizure of the Mass River bridge at Arnhem, turned into a logical and strategic nightmare. British forces found themselves confronted by 9th and 10th SS Panzer Divisions, which were refitting in the area. The British 1st Airborne Division was able to seize and hold one of the two bridges until 20 September, when they surrendered after waiting for the British 2nd Army, which was not able to keep to the schedule. The 1st Airborne Division was effectively wiped out, with 1,000 killed and 6,000 taken prisoner. With German reinforcements gathering around Arnhem, Montgomery ordered British forces to withdraw on 24 September. For students of military strategy, Operation MARKET GARDEN provides numerous examples of faulty decision-making on the part of commanding officers but also of simple bad luck.

Jason Soderstrum

See also: Airborne Operations; Montgomery, Bernard Law; World War II

References and further reading:

Harclerode, Peter. *Arnhem: A Tragedy of Errors.* London: Arms and Armour, 1994.

Hibbert, Christopher. *The Battle of Arnhem.* London: B. T. Batsford, 1962.

Middlebrook, Martin. *Arnhem 1944: The Airbone Battle, 17–26 September.* Boulder, CO: Westview Press, 1994.

Marlborough, John Churchill, First Duke of (1650–1722)

Led the British armies in the War of the Spanish Succession with great success and made the fortune of the house of Churchill in British affairs. John Churchill, born to a wealthy but far from prominent family, rose to prominence thanks to the patronage that the Duke of York, the heir to the British

Painting of John Churchill, First Duke of Marlborough. (Library of Congress)

throne and subsequent king, and later Queen Anne conferred upon his family. On the duke's accession as James II (1633–1701, r. 1685–1688), Churchill, a prominent member of James's clique of young officers, received a senior command in the king's large new army and took part in the campaign leading up to Sedgemoor. Nevertheless, he betrayed James II in the revolution of 1688 and then turned coat again and conspired with the deposed pretender during the reign of his successor, William III. These political efforts kept Churchill from much active service in 1689–1702, but Queen Anne made him her captain general upon Britain's entry into the War of the Spanish Succession.

As commander on the continent, Marlborough conquered the Lower Palatinate at the head of an Anglo-Dutch army in 1702 and won the Battle of Blenheim-Höchstädt (1704) in combination with Prince Eugene. For this, Queen Anne conferred upon him the title duke of Marlborough. The years 1703, 1705, and 1707 were fallow periods of failed campaigns, but in 1706 Marlborough led the Anglo-Dutch army into Brabant in a brilliant campaign of maneuver that made possible the victories of Ramillies and Turin. In 1708, the Anglo-Dutch forces and the imperial Habsburg army, at last combined in a single theater under the joint command of Marlborough and Eugene, won the Battle of Oudenaarde, and, showing great strategic and logistical daring, took Lille. The year 1709 saw the more ambiguous victory of Malplaquet and the fall of Mons. In subsequent years, Eugene and Marlborough took many fortresses and devastated northern France, conducting an economic warfare that strained the French economy yet could not force peace upon the enemy. By 1712, Queen Anne faced state bankruptcy. She maneuvered her way out of the war, and Marlborough and his aggressive allies fell from office in the process.

Some historians have exaggerated Marlborough's abilities, portraying him as a prophet rising above the limits of his age and making his Dutch allies scapegoats for his failure to execute a presumed "Napoleonic" vision. Although one of history's great captains, Marlborough's genius was in keeping with the spirit of eighteenth-century warfare, and it was nowhere better exhibited than in his operational masterwork, the passage of the determinedly defended ne plus ultra French lines (1711).

Erik A. Lund

See also: Blenheim-Höchstädt, Battle of; Coehoorn, Baron Menno van; English Wars in Ireland; Eugene of Savoy; Jacobite Rebellions; Malplaquet, Battle of; Northern War, Great; Oudenaarde, Battle of; Ramillies, Battle of; Sedgemoor; Spanish Succession, War of the; Turenne, Henri de la Tour d'Auvergne, Vicomte de

References and further reading:

Chandler, David. *Marlborough as Military Commander.* London: Batsford, 1977.

Churchill, Winston. *Marlborough, His Life and Times.* 4 vols. London: Harrap, 1933–1938.

Marlborough, John Churchill, Duke of. *Letters and Dispatches from 1702–1712.* Ed. George Murray. 5 vols. London: J. Murray, 1845.

———. *The Marlborough-Godolphin Correspondence.* Ed. Henry L. Snyder. 3 vols. Oxford: Clarendon, 1975.

Marne, Battle of the (5–10 September 1914)

Crucial battle in the opening days of World War I. Heinrich von Kluck, commander of the German First Army and right wing of the Schlieffen Plan, advanced through Belgium and northern France in August 1914. Approaching Paris, von Kluck lost contact with the German Second Army on his left and opened a gap between the two advancing armies. He also exposed his right flank when he turned southeast to roll up what he believed to be the French left flank in front of Paris. Taxicabs transported French troops from Paris and caught von Kluck unaware. He quickly transferred two corps from his left flank to the west, further opening the gap north of the Marne River between his army and the German Second Army.

The British Expeditionary Force, under Sir John French, and the French Fifth Army crossed the Marne River and surged into the breach between the two German armies. The German Second Army tenaciously attacked the French Ninth Army under Ferdinand Foch near the marshes of St. Gond. These attacks failed, and the German chief of the general staff, Helmuth von Moltke the younger, ordered a withdrawal of the First and Second Armies to the Aisne River on 10 September 1914. The Germans held this line in the face of stiff attacks, and the "race to the sea" ensued when each side attempted to outflank the opponent. Both sides established a network of linked trenches that would characterize the next four years of warfare on the western front.

The German defeat at the Marne proved a decisive Anglo-French strategic victory and has been termed "the miracle of the Marne." Despite a string of costly defeats on the frontiers, the Anglo-French forces seized the strategic initiative with the victory. As a result, Germany had to fight on two fronts, which ultimately cost it the war.

Mark A. Mengerink

See also: Foch, Ferdinand; French, John Denton Pinkstone, First Earl of Ypres; Joffre, Joseph Jacques Césaire; Moltke, Graf Helmuth Johannes Ludwig von; Schlieffen, Graf Alfred von; World War I

References and further reading:

Asprey, Robert B. *The First Battle of the Marne.* Westport, CT: Greenwood Press, 1962.

Keegan, John. *The First World War.* New York: Alfred A. Knopf/Random House, 1999.

Tuchman, Barbara. *The Guns of August.* New York: Macmillan, 1962.

Marne Counteroffensive (15 July–16 September 1918)

Allied response to the final German offensive push of the war, also known as the Second Battle of the Marne. This battle marked the final turning of the tide for the Allies in World War I. In early July 1918, a German offensive menaced Paris. Many residents of the French capital prepared for evacuation in a situation that bordered on panic. Worse, German long-range artillery sporadically bombarded vital war industries in and around the capital. In the middle of July, the offensive ran out of energy 50 miles east of Paris, along the Marne River.

On 15 July, the U.S. Army's 3d Infantry Division held staunchly against strong local attacks, even though the American troops were nearly surrounded. British, French, Italian, and American reinforcements were rushed to the area and thrown into a large-scale attack. Now began a multinational battle, the outcome of which could very well decide the war.

The Germans, faced with the choice of imminent retreat versus reinforcing their troops along the Marne River at the expense of their impending Flanders offensive, chose the latter. For the first month of the battle, Allied soldiers advanced eastward against the entrenched, reinforced German defenders. Some of the fiercest fighting took place near Soissons, where the Americans encountered fanatical enemy resistance. For much of August, the Germans dug in along the Aisne and Vesle Rivers and held fast against American and French attacks. A break for the Allies came on 2 September, when the American 32d Infantry Division captured the key town of Juvigny, effectively cutting off a major German supply route. Meanwhile, French troops in the south and British troops in the north steadily pushed the Germans eastward.

Day by day, Allied soldiers fought and died to reduce the German salient and eliminate the threat to Paris. Eventually, they succeeded in reducing that salient to such an extent that Paris was no longer in range of German artillery. By mid-September the question was no longer if the Germans would take Paris, but rather if they could stave off the Allies from pushing any further eastward.

The ultimate significance of this battle is that the initiative passed from the Germans to the Allies for the rest of the war. In the spring and early summer of 1918, the Germans had hoped to launch one last great offensive to win the war before fresh American manpower could have any impact on the battle fronts. At the Marne, in the summer of 1918, the Germans ran out of time. From this point forward, the Allies would launch a series of offensives designed to drive the Germans from France, a goal never completely fulfilled, even though World War I would conclude in November 1918 on Allied terms.

John C. McManus

See also: Marne, Battle of the; World War I

References and further reading:

Coffman, Edward. *The War to End All Wars.* Lexington: University Press of Kentucky, 1998.

Toland, John. *No Man's Land.* Garden City: Doubleday, 1980.

General George Marshall, c. 1944. (Library of Congress)

Marshall, George Catlett (1880–1959)

U.S. Army chief of staff and one of the foremost soldier-statesmen of the twentieth century. Born in Uniontown, Pennsylvania, on 31 December 1880, George Marshall graduated from the Virginia Military Institute in 1901 and was commissioned in the infantry in 1902. He held a variety of assignments, including in the Philippines and Oklahoma. He attended the Infantry and Cavalry School in Fort Leavenworth (1906–1907) and in 1908 graduated from the Army Staff College there. After working as an instructor at the Staff College (1908–1910), he was an instructor-inspector for the Illinois National Guard and an infantry company commander. He again served in the Philippines until 1916.

Marshall established his reputation as a brilliant staff officer following U.S. entry into World War I. Sent to France in June 1917, he served on the staff of American Expeditionary Force commander General John J. Pershing and planned the September 1918 American offensive at St. Mihiel. Made operations officer for the First Army, Colonel Marshall planned the transfer of some 400,000 U.S. troops, carried out in just six days, for the September–November Meuse-Argonne offensive.

After the war, Marshall was aide to then army commander General Pershing (1919–1924). He served in China (1924–1927) and afterward was deputy commander of the Infantry School at Fort Benning (1927–1932). He held several command positions before he was promoted to brigadier general (1936). Marshall became deputy chief of staff of the army in 1938, won a promotion to major general that July, and became chief of staff on 1 September 1939.

President Franklin D. Roosevelt advanced Marshall over many more senior officers to appoint him chief of staff of the army with the rank of temporary general in September

1938. Promoted to general of the army in November 1944, he continued in that capacity, earning the unofficial title of "Organizer of Victory" for his masterful leadership, until his retirement in November 1945.

President Harry S. Truman recalled Marshall to serve as special envoy to China (1945–1947) and secretary of state (1947–1949). He was serving as president of the American Red Cross when Truman again recalled him in September 1950 as secretary of defense to preside over the 1950–1953 Korean War military buildup. He held that post until he retired altogether in September 1951. He was awarded the Nobel Prize for peace in 1953, the first soldier so honored. Marshall died in Washington, D.C., on 16 October 1959.

Spencer C. Tucker

See also: Meuse-Argonne; Pershing, John J.; Roosevelt, Franklin D.; St. Mihiel; Truman, Harry S.

References and further reading:

Pogue, Forrest C. *George C. Marshall.* 4 vols. New York: Viking Press, 1963–1987.

Marston Moor (2 July 1644)

Largest battle of the English Civil War. By the middle of 1644, the Scots had joined Parliament against King Charles I, threatening decisively to tip the scales against the royalists. Scottish forces joined parliamentary soldiers to besiege York, the center of royalist power in the north. Charles believed that his survival depended upon York and stripped his own army to send a relieving force under Prince Rupert. The allies lifted the siege to intercept Rupert and force him into decisive battle.

The two armies met 6 miles west of York, on Marston Moor. Each side had about 7,000 cavalry. Parliamentary cavalry included Oliver Cromwell's superbly trained Ironsides. Rupert was heavily outnumbered in infantry, however, with only 13,000 men to his opponents' 20,000. Both armies drew up into traditional battle formations on 2 July, with the infantry in the center and cavalry on either wing. The afternoon was spent in scattered exchanges of artillery, but at 5:00 P.M., Rupert concluded that there would be no fighting and retired to his quarters. At 7:00, Alexander Leslie, parliamentary commander, ordered his army to attack, catching the royalists by surprise. Cromwell's Ironsides dispersed the royalist horse on the left and began to attack the unprotected infantry. When George Goring's royalist cavalry threatened the parliamentary right, Cromwell moved his disciplined force to the other wing and defeated Goring. Although the unsupported royalist infantry put up a brave fight, they were enveloped and crushed. York and northern England were lost to Charles. Although the Civil War continued for another year, Rupert's defeat at Marston Moor made a military victory for Charles impossible.

Tim J. Watts

See also: Cromwell, Oliver; English Civil War; Rupert, Prince

References and further reading:

Newman, P. R. *The Battle of Marston Moor, 1644.* Chichester,UK: Bird, 1981.

Woolrych, Austin. *Battles of the English Civil War: Marston Moor, Naseby, Preston.* New York: Macmillan, 1961.

Young, Peter. *Marston Moor, 1644: The Campaign and the Battle.* Kineton: Roundwood Press, 1970.

Martí y Pérez, José Julián (1853–1895)

Cuban patriot, guerrilla fighter, poet, and philosopher. Born in Havana, Cuba, on 28 January 1853 of Spanish parentage, Martí's father was an artillery sergeant. Martí began his writing career early, publishing various autonomist articles by 1869. Because of a letter he wrote to a friend, Martí and his companion Fermín Valdés Dominguez were sentenced in April 1870 to hard labor at the Quarries of San Lázaro, where he was put in leg irons. In January 1871, he was deported to Spain, where he wrote his famous work, *El presidio politico en Cuba.* Martí and Valdés Dominguez moved to Zaragoza, Spain, to continue their studies in 1872, and in 1875, he was reunited with his parents and siblings in Veracruz, Mexico. He was again deported to Spain in 1879 but escaped to France and then to New York. His article written about the arrival of the Statue of Liberty in New York for the newspaper *La Nacion* of Venezuela stands as one of the finest pieces written on the subject in any language.

In 1887, he became consul for Uruguay in New York and then for Argentina and Paraguay. In 1889, he began the journal *La Edad de Oro* (Age of Gold, i.e., childhood), dedicated to children and their needs. And in 1892, he founded in New York the Cuban Revolutionary Party in New York, with the goal of winning Cuban independence. In April 1894, Martí and Máximo Gómez met in New York. Gómez returned to Santo Domingo to begin preparations and planning for the eventual invasion of Cuba. In January 1895, Martí traveled to Montecristi, Santo Domingo, to meet with Gómez again. In March of the same year, Martí wrote the *Manifesto of Montecristi,* in which he put forth his political views. By this time, war in Cuba was already a fact, having started on 24 February 1895. Martí landed in Cuba and joined the fighting. On 19 May, during a clash at Dos Rios, he fell, mortally wounded. One of the most erudite philosophers of nineteenth-century Cuba and perhaps the whole of Latin America, José Martí is remembered as the "father of the Cuban Republic."

Peter Carr

See also: Cuban Ten Years' War; Cuban Revolution; Weyler y Nicolau, Valeriano, Marquis of Tenerife

References and further reading:

Mañach, Jorge. *Martí: Apostle of Freedom.* Trans. Coley Taylor. Preface by Gabriela Mistral. New York: Devin-Adair, 1950.

Masada, Siege of (72–73)

Mountaintop fortress in the Judean desert of Israel used as a stronghold by Jewish Zealots, the Sicarii, one of the numerous groups who revolted against Rome in 66. Originally fortified by the Maccabees, Masada was improved by Herod the Great, who constructed two palaces, added heavy walls and defensive towers, and developed a cachement system to supply water for his royal citadel. After Herod's death, the Romans controlled Masada until Jewish Zealots captured the fortress by surprise in 66. After the fall of Jerusalem in 70, this group of about 1,000 men, women, and children refused to surrender. The Roman X Legion under the command of Flavius Silva besieged Masada, but the elaborate defensive system proved superior to Roman siege machines. Assembling an army of 15,000 soldiers, the Romans built a wall around the mountain to prevent any escapes and then constructed a sloping ramp up the west side of the mountain until they were in reach of the walls. After two years, Roman forces finally penetrated the fortress, only to find that the Jews, led by Eleazar ben Yair, had committed suicide rather than be enslaved. Only two women and five children survived; the rest had drawn lots to determine who would die first, with each father responsible for killing his family before taking his own life. The provisions, except for the food, were set on fire. Roman historian Josephus Flavius provides an account of the events provided to him by the survivors. Except for a brief period of Jewish control in the second century during the Bar Kochba revolt, the fortress remained abandoned until the twentieth century, when Masada became a symbol of Jewish independence.

Cynthia Clark Northrup

See also: Jewish Revolts; Josephus, Flavius

References and further reading:

Tamarin, Alfred H. *Revolt in Judea: The Road to Masada, the Eyewitness Accounts by Flavius Josephus of the Roman Campaign against Judea, the Destruction of the Second Temple, and the Heroism of Masada.* New York: Four Winds Press, 1968.

Yadin, Yigael. *Masada: Herod's Fortress and the Zealots' Last Stand.* New York: Random House, 1966.

Masséna, André, Duc de Rivoli, Prince d'Essling (1758–1817)

An extremely aggressive field commander of the French Revolutionary and Napoleonic Wars. Massena was born in Nice (6 May 1758) and enlisted in 1775. He left the army in August 1789 and became lieutenant colonel of a volunteer regiment in 1792.

In the siege of Toulon, he gained Napoleon's attention and a promotion to division general (20 December 1793). In Italy in 1794–1795, he achieved successes but also revealed the rapacity that marred his career. Masséna played crucial roles at Lodi (10 May 1796), where he led the charge that took the bridge, and at Rivoli in January 1797.

Masséna commanded the combined Armies of Helvetia and the Danube in his most significant campaign, defeating an Austro-Russian army at Zurich (25–28 September 1799) and then crushing a Russian relief army. Thus he decisively turned the tide of war and ended the allied invasion threat in that theater.

In November 1799, Napoleon sent Masséna to Genoa, which the Austrians were besieging. After a heroic defense, he surrendered (4 June 1800) with full military honors. Relieved of command, partly because of looting, Masséna spent the next five years in civilian life. Nevertheless, Napoleon appointed him a marshal in 1804.

Masséna contributed to Napoleon's Austerlitz campaign by keeping the Archduke Charles occupied in Italy. In July 1807, Masséna again returned to civilian life, and Napoleon named him duc de Rivoli in March 1808. In the 1809 campaign, Masséna fought bravely at Aspern-Essling (20–22 May), where, covering the evacuation to Lobau Island, he was the last man across the bridge before its destruction. At Wagram (6 July), he audaciously sent his troops laterally across the enemy front to fill a gap in the French lines. Napoleon awarded him the title prince d'Essling in January 1810.

Reluctantly, Masséna accepted command of the Army of Portugal in May 1810. He captured Ciudad Rodrigo (10 July) and Almeida (28 August) but failed to defeat Arthur Wellesley, the Duke of Wellington, at Torres Vedras. Failure in Iberia ended Masséna's field command. Thus closed the active career of one of Napoleon's most capable generals. He died in Paris on 4 April 1817.

James K. Kieswetter

See also: Austerlitz, Battle of; French Revolutionary Wars; Napoleonic Wars; Rivoli; Toulon, Siege of; Wagram; Wellington, Arthur Wellesley, Duke of

References and further reading:

Chandler, David G. *The Campaigns of Napoleon.* New York: Macmillan, 1966.

Horward, Donald D. *Napoleon and Iberia: The Twin Sieges of Ciudad*

Rodrigo and Almeida. Tallahassee: University Presses of Florida, 1980.
Macdonnell, Archibald Gordon. *Napoleon and His Marshals.* New York: Macmillan, 1934.
Marshall-Cornwall, James H. *Marshal Masséna.* New York: Oxford University Press, 1965.

Matthias I (Mátyás Hunyadi) (1443–1490)

The first Hungarian king not of the House of Árpád and one of Hungary's greatest kings, known as "the Truthful." Matthias's father was János Hunyadi, a popular warlord who led several campaigns against the Turks. The Hungarian estates chose Matthias as king on 24 January 1458. His rule was at first tenuous, and he had to put down several attempts to overthrow him; he also had to prevent Holy Roman Emperor Frederick III from taking the Hungarian throne.

Once his throne was secure, Matthias devoted his attention to strengthening his authority over the nobility and centralizing the government. He expanded his power base within the lesser nobility but failed to win the permanent allegiance of the great nobles. Matthias reformed the army, known as the "Black Troops," which he used to keep order and collect taxes. Taxes now became permanent and provided him with a regular source of income. Not stopping his reforms with the army, he recodified Hungarian common law, reorganized the legal system, established a civil service, improved the means of communication, and promoted the development of towns. He also founded the University of Pressburg (now Bratislava), revived the University of Pecs, and introduced the Renaissance to Hungary.

Matthias saw the Ottoman Turks as Hungary's chief enemy and sought to build a united European coalition against them under his leadership. He defeated the Turks in 1479 and only after his death did they again become a serious threat, in part because of the enmity of Vienna. Matthias devoted the last 20 years of his life to establishing a Danubian empire. He became king of Bohemia in 1479 and laid siege to Vienna and took it in 1485, annexing Austria, Styria, and Carinthia. Although he failed to win his desired election as Holy Roman Emperor, he transferred his court to Vienna and died there in 1490. His strong rule provoked a reaction under his successors. His reforms were largely abandoned, and the country fell into anarchy.

Ákos Tajti

References and further reading:

Bánlaki, József of Doberdo. *A magyar nemzet hadtörténelme* (The military history of the Hungarian nation). Budapest: Grill, 1934.
Klaniczay, Tibor, and József Jankovics, eds. *Matthias Corvinus and the Humanism in Central Europe.* Budapest: Balassi, 1994.
Kovács, Péter E. *Matthias Corvinus.* Budapest: Officina Nova, 1990.
Létmányi, Nándor. *Párhuzam Hunyadi János es fia Mátyás között* (A comparison of John Hunyadi and his son Matthias). Budapest: Atheneum, 1885.
Vég, Gábor. *Magyarország királyai és királynöi* (Hungary's kings and queens). Budapest: Maecenas, 1991.

Maurice of Nassau (1567–1625)

Dutch general who significantly contributed to the development of early modern warfare. Maurice was born to William the Silent, Prince of Orange, in 1567, and his family sought to free the Netherlands from Spanish control. He attended university at both Heidelberg and Leiden, but his father's assassination in 1584 thrust Maurice into leadership, and he became president of the United Provinces. He soon assumed the duties of admiral general and captain general, making him commander in chief of all Dutch military forces.

Maurice recognized the need to efficiently organize his limited manpower resources to fight the Spanish empire. His knowledge of Roman warfare and mathematics led him to develop small, maneuverable battalion formations that joined both musket and pike in a linear formation. Maurice also instituted rigorous drills for individuals and formations to increase the rate of fire and maneuverability. He experimented with artillery capability embedded within infantry formations, an early attempt at combined arms synchronization. To ensure adequate manpower, Maurice provided consistent pay and encouraged education by establishing the first military academy for officers.

With a responsive and dependable army, Maurice often inflicted disproportionate casualties during battle and soon wrested control of key provinces from the Spanish. To secure terrain, he developed an integrated defensive system of fortified towns and rivers. His success resulted in a 12-year truce with Spain in 1609.

Maurice eventually died from liver disease at The Hague in April 1625. Other commanders, such as Gustavus Adolphus, soon adopted his military innovations. Maurice's legacy of state-supported professional armies and linear gunpowder tactics mark him as one of the founders of modern warfare.

Steven J. Rauch

References and further reading:

McNeill, William H. "Keeping Together in Time." *Military History Quarterly* (Winter 1994).

Parker, Geoffrey. *The Military Revolution.* Cambridge, UK: Cambridge University Press, 1988.

Mauricius Flavius Tiberius (539–602)

One of the eastern Roman Empire's greatest rulers. Maurice became emperor in 582, at a time of great crisis: the Justinian system was crumbling as epidemics, population loss, falling state revenues, and Slav incursions into Thrace threatened to destroy the imperial position throughout Byzantium; the latest in a long line of wars with Sassanid Persia also fared badly.

Maurice's place in history is assured on three separate counts. First, he was the author of a military manual, the *Strategikon,* which remained in imperial service for several hundred years and even today remains in print. Second, he was the architect of comprehensive victory over the Sassanid Empire. His victories at Nisbis and the Araxes (589) led to the overthrow of the Sassanid emperor Hormizd. A disputed succession allowed imperial intervention on behalf of Chosroes, the deposed emperor's son. With Chosroes installed as emperor, Maurice then imposed peace based on the status quo ante bellum and without gain: his calculation was that Sassanid indebtedness would ensure peace and stability in the east.

In the Haemus the empire faced not single, successive invasions by different tribes but simultaneous and continuous invasion by many different groupings: by the time of Maurice's accession, the imperial position had been compromised. Nonetheless, in a series of deliberate campaigns, Maurice curbed the Slavic influx and then defeated the Avars (598–601). His main problem, however, remained inadequate state revenues: his rule was noted for a series of mutinies that attended cost-cutting measures. It was Maurice's decision that the army remain on the Danube River through the coming winter that provoked the mutiny that saw him deposed and decapitated, his four youngest sons having been executed before him.

H. P. Willmott

References and further reading:

Haldon, John. *Warfare, State and Society in the Byzantine World.* London: UCL, 1999.

———. *The Byzantine Wars.* Tempus, Stroud, 2001.

Treadgold, Warren. *Byzantium and Its Army, 284–1081.* Stanford, CA: Stanford University Press, 1995.

Mauryan Empire, Conquests of (321–232 B.C.E.)

Resulted in a rare, internal unification of Hindu India that had taken place only twice before outside invasions became common after 1100 C.E. The Gangetic kingdom of Magadha, with its capital at Patna, was the center of both unifications. By 500 B.C.E., 16 petty Hindu kingdoms established during the Aryan invasions of India competed for control of the Ganges River trade routes in northern India. The unfinished invasion of India by Alexander the Great in 326 B.C.E. appears to have forced India toward unification politically and militarily. Conflicting accounts of Alexander's expedition suggest that huge military forces under King Porhus in western India numbered 60,000 men, 20,000 cavalry, 2,000 chariots, and 3,000 war elephants. Further east, Gangetic forces encountered by Alexander's forward probes are listed by Pliny and Plutarch as numbering 500,000 foot soldiers, 30,000 cavalry, 8,000 chariots, and 9,000 war elephants. It is no wonder Alexander's men rebelled!

In 321, Chandragupta Maurya (325–280 B.C.E.) seized the Magadha throne in northeastern India as Alexander turned south down the Indus River in western India and returned to Persia and eventually Babylon. Chandragupta took advantage of the power gap left by Alexander's victory over King Porhus at the Battle of Hydaspes in western India. He conquered all the land between the Indus River and the Narbada River to the south. The Magadha king then closed the mountain passes in northwestern India used by Alexander and others to invade northern India. He did so by supposedly defeating the Macedonian general Seleucus Nicator in 305 B.C.E. In reality, Seleucus realized he could not defeat the huge Indian forces and negotiated a treaty with Chandragupta while both armies faced each other on the field.

In 280 B.C.E. Bindusara (298–273 B.C.E.), son of Chandragupta, inherited the throne. He extended the empire south into the Deccan plateau near Mysore. Chandragupta's grandson, Ashoka (269–232 B.C.E.), completed the empire with bloody conquests of the Kalinga kingdom on the northeast coast. The Mauryan Empire now controlled all but the southern tip of India, and Ashoka turned toward developing empire, resulting in the "golden age" of the Mauryan Empire. However, his emphasis on Buddhism over the army left India open to invasions from the northwestern mountain passes by Bactrian, Scythian, and Parthian forces. In 150 B.C.E., Indo-Hellenic forces reached Patna (Patiliputra) and found the empire in ruin.

Christopher Howell

See also: Alexander's Wars of Conquest; Aryan Conquest of India; Chandragupta Maurya; Hydaspes, Battle of the

References and further reading:
Kulke, H. and D. Rothermund. *A History of India.* New York: Barnes & Noble Books, 1986.
Majumdar, R. C. *Classical Accounts of India.* Calcutta: Mukhopadhyay Publishers, 1960.
Singh, Sarva Daman. *Ancient Indian Warfare.* Reprint, Dehli: Motilal Banarsidass Publishers, 1997.

Maximilian I (1459–1519)

Holy Roman Emperor who made the Habsburg family into a European power. Maximilian von Habsburg was born in Wiener Neustadt, Austria, on 22 March 1459, the son of Holy Roman Emperor Frederick III, whom he succeeded on 19 August 1493. His marriage in Ghent on 19 August 1477 to Mary, daughter of Charles the Bold, Duke of Burgundy, gained him the Netherlands, much Burgundian territory, and the enmity of France. French king Louis XI immediately attacked. Maximilian defended his lands well and decisively defeated the French at Guinegate on 7 August 1479, but the Franco-Austrian War continued sporadically until 1493, with Maximilian keeping almost all the disputed area.

Maximilian pursued his lifelong ambition to unite all of Europe under the Habsburgs. His three main tactics were diplomacy, arranging marriages, and waging war. His main rival was France. From 1482 to 1485, he fought the Netherlands States General, finally wresting from them the regency of his son, Philip. When Hungarian king Matthias I Corvinus died in 1490, Maximilian resumed control of Austrian lands that had been seized by Hungary and declared himself a candidate for the vacant throne. When Ladislas II of Bohemia was elected instead, Maximilian forced the Treaty of Pressburg upon him in 1491 so that the throne of Bohemia and Hungary would pass to a Habsburg if ever vacant again. In 1495 he joined the Holy League of Spain, Venice, Milan, and the papal states to drive the French out of Italy and took an active role in the league's military efforts. Having lost his war against Switzerland in 1499, he was compelled to recognize Swiss independence. Allied with English king Henry VIII, he beat the French on 16 or 17 August 1513 in a second battle at Guinegate, called the "Battle of the Spurs" because of the speed with which the French retreated. He continued fighting the French, mostly in Italy, until 1516, and died in Wels, Austria, on 12 January 1519.

Eric v. d. Luft

See also: Austro-Swiss Wars; Charles the Bold, Duke of Burgundy; Holy Roman Empire; Hungarian War with the Holy Roman Empire; Marignano, Battle of; Matthias I; Ravenna; Valois-Habsburg Wars

References and further reading:
Andrews, Marian. *Maximilian the Dreamer: Holy Roman Emperor, 1459–1519.* London: Paul, 1913.
Benecke, Gerhard. *Maximilian I, 1459–1519: An Analytical Biography.* London: Routledge & Kegan Paul, 1982.
Burgkmair, Hans. *The Triumph of Maximilian I.* New York: Dover, 1964.
Seton-Watson, Robert W. *Maximilian I, Holy Roman Emperor.* Westminster, UK: Constable, 1902.
Waas, Glenn Elwood. *The Legendary Character of Kaiser Maximilian.* New York: Columbia University Press, 1941.

Mayaguez Operation (12 May 1975)

On 12 May 1975, Khmer Rouge naval forces operating in Cambodian territorial waters seized the U.S. merchant ship *Mayaguez* and removed its crew of 41. U.S. president Gerald Ford ordered a military response to punish the Khmer Rouge and retake the ship and crew, thought to be on the island of Koh Tang. On 15 May, 11 U.S. Air Force helicopters launched from Utapao, Thailand, and transferred 70 men to the *Holt,* which subsequently came alongside the *Mayaguez* in the first hostile ship-to-ship boarding since the War of 1812; they found it empty.

Eight helicopters carrying approximately 200 men assaulted the eastern and western beaches of Koh Tang's northern neck: four were shot down in the opening minutes of the battle, and only one escaped undamaged. Thirteen survivors were rescued from the water after swimming four hours to the *Wilson* offshore, and three isolated contingents were left on the island, the largest numbering only 60 men. Unknown to the United States, the Khmer were well entrenched in anticipation of a Vietnamese attack over an ongoing territorial dispute.

As the remaining helicopters organized as the second wave, the *Mayaguez* crew was freed from an island 40 miles away. The desperately needed second wave was nearly recalled by the U.S. Joint Chiefs of Staff before the marine assault commander on Koh Tang persevered in arguing that the second wave be landed before the western beach was overrun.

The last available helicopter rescued the last group of marines as darkness closed 14 hours after the initial landings. In all phases of the operation, 50 servicemen were wounded and 41 killed, including three men believed to have been left behind alive and subsequently executed, and 23 air force personnel killed earlier while en route to the staging area at Utapao. It is believed that approximately 60 Khmer Rouge soldiers were killed out of a land and sea force of about 300. Although there was initial rejoicing in the United States over the successful rescue of the *Mayaguez* crew, as news surfaced of the heavy loss of life in the operation, Americans became appalled. The operation, following closely on the fall of South Vietnam, seemed yet another indication that the U.S. military still could not handle operations in less than all-out war.

John N. Warren

See also: Cambodian Wars; Vietnam Conflict

References and further reading:

Guilmartin, John F. *A Very Short War.* College Station: Texas A & M University Press, 1995.

Rowan, Roy. *The Four Days of Mayaguez.* New York: W. W. Norton, 1975.

McClellan, George Brinton (1826–1885)

American Civil War army commander. Born on 3 December 1826 in Philadelphia, McClellan was educated at the University of Pennsylvania before entering the U.S. Military Academy at West Point, New York. After graduating second from his class in 1846, he was commissioned a second lieutenant in the Corps of Engineers. During the Mexican-American War, McClellan won three brevets for gallantry, but he later decided to resign from the army and eventually became the vice president of the Illinois Central Railroad. Soon after the outbreak of the Civil War, he was commissioned a major general in the regular army. In November 1861, after Lieutenant General Winfield Scott's retirement, McClellan was appointed general in chief of the U.S. Army. McClellan proved to be highly effective at organizing and training the Union army, but his skill as an organizer was not matched by his art as a combat commander. After his failures during the Peninsular campaign and at the Battle of Antietam, McClellan was relieved of his command and ordered home to await further orders, which never came.

In 1864, McClellan was nominated by the Democratic Party as its candidate for president on a peace platform (despite the fact that he personally believed that the war should continue until the Union was victorious), but he was easily defeated by Lincoln. After serving as the governor of New Jersey (1878–1881), McClellan died in Orange, New Jersey, on 29 October 1885. He was typical of so many failed high commanders: a superb staff officer who lacked the ability to "think on his feet" rapidly on the battlefield and who invariably took counsel of his fears—the service academies and staff colleges turn them out by the hundreds.

Engraved portrait of General George B. McClellan, with a border of Antietam battle scenes. (Library of Congress)

Alexander M. Bielakowski

See also: American Civil War; Lee, Robert Edward; Lincoln, Abraham

References and further reading:

McClellan, George B. *McClellan's Own Story: The War for the Union.* New York: C. L. Webster & Company, 1887.

Sears, Stephen W. *George B. McClellan: The Young Napoleon.* New York: Da Capo Press, 1999.

McDowell, Irvin (1818–1885)

Union general defeated in the American Civil War's first major battle, First Bull Run. Irvin McDowell was born on 15 October 1818 in Columbus, Ohio. He attended the U.S. Military Academy, graduating in 1838 in the middle of his class, and began his career as an artillerist. During the Mexican-American War, he fought at Buena Vista, earning a brevet promotion. Joining the adjutant general's corps in 1848, he served in staff positions until 1861.

McDowell was promoted to brigadier general at the outbreak of the Civil War and, because of political connections, was given command of the Washington, D.C., military district. He built an army in record time but could do little about its lack of combat experience. Under political pressure, McDowell methodically launched an offensive toward Richmond, hoping to end the war in a single battle.

Facing Confederate generals Pierre Gustave Toutant Beauregard and Joseph E. Johnston, McDowell did a capable job of deploying his divisions for the initial Battle of Bull Run on 21 July 1861 but was unable to steady his fatigued recruits when the rebels poured fresh troops into the battle after a daylong stalemate. McDowell soon faced a rout, as thousands of terrified Union troops ran back to Washington.

McDowell was soon relieved by General George McClellan and later reassigned as a corps commander. He again commanded the Washington garrison during the Peninsular campaign in 1862. At Second Bull Run, under General John Pope, McDowell's performance was suspect, resulting in his relief for cause. A court of inquiry exonerated him, but his fighting career was over. McDowell subsequently commanded several military departments, retiring in 1882. He died in San Francisco on 4 May 1885. McDowell was that very ubiquitous commander in military history: one who could achieve high command for his ability to train and supply troops but lacked the blood instinct to win on the battlefield.

Michael S. Casey

See also: American Civil War; Beauregard, Pierre Gustave Toutant; Johnston, Joseph Eggleston; McClellan, George Brinton

References and further reading:

Davis, William C. *Battle at Bull Run: A History of the First Major Campaign of the Civil War.* New York: Stackpole Books, 1995.

Hassler, Warren W. Jr. *Commanders of the Army of the Potomac.* Westport, CT: Greenwood Press, 1962.

McNair, Lesley J. (1883–1944)

Commander of U.S. Army Ground Forces in World War II. McNair, a combat veteran of World War I, is usually credited as the general most responsible for training American ground combat forces in World War II. He saw active duty in North Africa and was wounded at a forward observation post in Tunisia in 1943.

After his recovery, he was given "command" of a fictional organization designed to deceive the Germans about the upcoming Normandy invasion. McNair's First Army Group was ostensibly slated to cross the channel at its most narrow point and invade in the Pas de Calais area. This deception plan, generally known as Operation FORTITUDE, had the goal of convincing the Germans the invasion would come anywhere but Normandy, the site of the real invasion. It succeeded mightily. Even after a lodgement had been secured at Normandy, Hitler continued to insist that the Allies were making a feint to draw German attention from the Calais area.

In late July 1944, roughly two months after the invasion, McNair—as was his penchant—went to a forward area to watch the carpet bombing preceding the American attack at St. Lo. The heavy bomber crews that day were told to fly horizontally along enemy lines, instead of vertically across Allied lines, in an effort to avoid friendly casualties. But, thanks to heavy German antiaircraft fire and a steady wind that blew marking smoke toward American lines, many of the bomber crews accidentally dropped their loads on American troops, including McNair's observation trench. He and at least 100 other Americans were killed. A story quickly circulated among American combat troops that the only trace of McNair ever found was his bloody ring finger, West Point class ring still attached. McNair was the highest-ranking military American to die in World War II.

John C. McManus

References and further reading:

Kahn, E. J., Jr. *McNair: Educator of an Army.* Washington, DC: Office of the Chief of Military History, 1945.

Palmer, Robert R. *The Procurement and Training of Ground Combat Troops.* Washington, DC: Office of the Chief of Military History, 1948.

McNamara, Robert Strange (1916–)

Secretary of defense during much of the Vietnam conflict. Born in California, McNamara obtained a position at the Harvard Business School that led to his employment as an

Defense Secretary Robert McNamara at a news conference, 1967. (Library of Congress)

analyst with the U.S. Army Air Forces's strategic bombing campaign. This duty brought about his hiring by the deteriorating Ford Motor Company, where he ultimately became the firm's president before being selected as secretary of defense by President John F. Kennedy.

McNamara arrived at the Pentagon determined to apply contemporary statistical management techniques to achieve "cost-effectiveness" in terms of procurement and strategy and was indifferent to service self-conceptions of mission and image. Considered as one of those urging the escalation of American involvement in the Vietnam conflict, McNamara tried to "manage" the war in the same statistical fashion as he directed the administration of Ford and the defense establishment, apparently ignoring the historical and emotional elements of war.

Paradoxically, the supposedly "bloodless" McNamara proved unable to stand the stress of the war, failing to win concessions from Hanoi or from the turmoil his strategy provoked in American society. Resigning in 1968 before the conclusion of the Johnson administration, McNamara continued in public life as president of the World Bank and (bafflingly, considering his record) as a commentator on the United States role in the world.

George R. Shaner

See also: Castro Ruz, Fidel; Cuban Missile Crisis; Dominican Civil War; Ho Chi Minh; Vietnam Conflict; Vo Nguyen Giap; Westmoreland, William

References and further reading:

McMaster, H. R. *Dereliction of Duty: Lyndon Johnson, Robert McNamara, the Joint Chiefs of Staff, and the Lies That Led to Vietnam.* New York: HarperCollins, 1997.

Shapley, Deborah. *Promise and Power: The Life and Times of Robert McNamara.* Boston: Little, Brown, 1993.

Vandiver, Frank E. *Shadows of Vietnam: Lyndon Johnson's Wars.* College Station: Texas A & M University Press, 1997.

Meade, George Gordon (1815–1872)

Union field commander in the American Civil War, victor at Gettysburg. Meade was born in Cadiz, Spain, to American parents on 31 December 1815. Commissioned in the artillery after his graduation from West Point in 1835, he resigned in 1836 after brief service against the Seminoles, became a civil engineer, but rejoined the army in 1842. In the Mexican-American War under Zachary Taylor, he saw action at Palo Alto on 8 May 1846, Resaca de la Palma, and Monterrey, where he was brevetted first lieutenant. He participated in the siege of Veracruz under Winfield Scott in 1847. As a military surveyor and engineer after the war, he made first lieutenant in 1851 and captain in 1856.

On 31 August 1861, Meade was commissioned brigadier general of Pennsylvania volunteers. After duty in the Washington, D.C., garrison, he fought in the Seven Days' Battles at Mechanicsville, Beaver Dam Creek, Gaines' Mill, White Oak Swamp, and Glendale, where he suffered a terrible lung wound that troubled him for the rest of his life. Nevertheless, he was able to lead a brigade at Second Bull Run; a division at South Mountain, Antietam, and Fredericksburg; and a corps at Chancellorsville. As major general of volunteers since 29 November 1862, he replaced Joseph Hooker in command of the Army of the Potomac on 28 June 1863. Meade then won his great defensive victory at Gettysburg.

For winning at Gettysburg, Meade received the thanks of Congress and a brigadier general's commission in the regular army but was criticized for his failure to pursue Robert E. Lee. The press conspired to mention him only unfavorably because his bristly temper had riled some reporters. His Bristoe Station, Rapidan Ring, and Mine Run campaigns against Lee misfired. Except for the period from 30 December 1864 to 11 January 1865, Meade commanded the Army of the Potomac until 27 June 1865, but in March 1864, Lieutenant General Ulysses S. Grant established headquarters in Meade's camp, which essentially reduced Meade to Grant's executive officer. He served Grant ably at the Wilderness, Spotsylvania, Cold Harbor, Petersburg, Five Forks, and Appomattox, achieving a promotion to major general on 18 August 1864. Debilitated by his wound, he died of pneumonia in Philadelphia on 6 November 1872.

Eric v. d. Luft

See also: American Civil War; Antietam/Sharpsburg; Bull Run, Second/Manassas Junction; Chancellorsville, Battle of; Cold Harbor, Battle of; Fredericksburg; Gettysburg; Grant, Ulysses Simpson; Hancock, Winfield Scott; Hooker, Joseph; Lee, Robert Edward; Mexican-American War; Monterrey; Petersburg, Siege of; Resaca de la Palma; Scott, Winfield; Seven Days' Battles; Spotsylvania Court House; Taylor, Zachary; Wilderness

References and further reading:

Cleaves, Freeman. *Meade of Gettysburg.* Norman: University of Oklahoma Press, 1960.

Lyman, Theodore. *With Grant and Meade from the Wilderness to Appomattox.* Lincoln: University of Nebraska Press, 1994.

Pennypacker, Isaac Rusling. *General Meade.* New York: Appleton, 1901.

Sauers, Richard Allen. *A Caspian Sea of Ink: The Meade-Sickles Controversy.* Baltimore: Butternut and Blue, 1989.

Medals and Decorations

Throughout history, rulers and states have found ways to recognize heroic achievement and military service. Ancient Greek literature refers to the award of arms and armor to commanders victorious in war. These gifts were often adorned with metal ornaments bearing martial symbols. This practice was later adopted by the Romans, who awarded circular medallions, or phalerae, worn on the breastplate of the recipient. Both Greeks and Romans also issued large coins, or medallions, to commemorate military victories. This practice was lost to Europe with the collapse of the Roman Empire, not to be revived (incredibly) until the fifteenth century.

The bestowing of awards in recognition of military and spiritual achievement resumed with the establishment of religious orders of knighthood during the Crusades. The pope officially recognized the oldest of these, the Order of the Knights of St. John of Jerusalem, in 1113.

As Europe emerged from the early Middle Ages, rulers established their own secular orders of chivalry, using the religious orders as models. These orders strategically bound knights, exclusively feudal nobility, to the sovereign through precepts of chivalric honor and religious duty. One of the earliest of these orders, the English Order of the Garter, was believed to have been founded by Edward III in 1348.

Awards recognizing military service were initially limited to a select number of officers. In 1588, Queen Elizabeth I of England presented medals of gold and silver to her senior commanders to mark the defeat of the Spanish Armada in 1588. Her Stuart successors continued the practice. In 1643, England's King Charles I instituted one of the first awards for gallantry when he authorized a medal awarded to any man who succeeded in a "forlorn hope." Medals, gold to officers and silver to enlisted men (the first recorded instance), celebrated Oliver Cromwell's victory over the Scots at the Battle of Dunbar in 1650.

Although the practice of awarding medals subsequently fell into disuse in England, other nations adopted the tradition. The Russian monarchs Peter the Great and Catherine the Great authorized both gallantry and service medals. During the American Revolution, General George Washington issued a heart-shaped cloth decoration known as the Badge of Military Merit as a reward for conspicuous gallantry.

The Napoleonic Wars brought a renewed interest in medals and decorations. In 1804, Napoleon established the Legion d'Honneur, an award for both bravery in action and distinguished civil or military service, regardless of the rank of the recipient. Five years later, Czar Alexander I of Russia established the Cross of St. George to honor the bravery of noncommissioned officers and enlisted men. In 1813, Frederick William III of Prussia first instituted the Iron Cross (das Eiserne Kreuz) as a gallantry award.

Great Britain hesitatingly resumed rewarding senior officers with medals, but it was not until the issuance of a silver service medal for the Battle of Waterloo in 1815 that all ranks shared the same award. The Waterloo Medal set the pattern for subsequent British campaign medals. With the exception of those awarded for World War II, most British medals bear the name, rank, and regiment of the recipient. As Britain's imperial wars increased, so did the issuance of campaign medals. Rather than award a specific medal for every battle, medals were issued for entire wars, with separate bars affixed to denote participation in particular engagements.

The first official decoration solely for gallantry in the British army was the Distinguished Conduct Medal, first instituted in 1845, but it was awarded only to sergeants. At the suggestion of the prince consort, Queen Victoria authorized the Victoria Cross for acts of conspicuous valor, a simple bronze cross cast from Russian cannon captured during the Crimean War. From the award's creation in 1856 to the Falkland Islands War (1982), only 1,354 Victoria Crosses were awarded to officers and men. It remains Britain's preeminent gallantry decoration.

At the outbreak of the American Civil War, there were no American medals for distinguished service or gallantry. The Medal of Honor was authorized by Congress in December 1861 to recognize Union officers and soldiers who displayed conspicuous gallantry in battle. Originally intended for only the duration of the Civil War, the medal's authorization was extended to cover subsequent conflicts.

The first American service or campaign medals were issued for action in the Spanish-American War (1898). President Theodore Roosevelt, an enthusiastic advocate of military medals and awards, pressed Congress and the War Department to honor veterans of previous wars. During his presidency, service medals were issued to veterans of the Civil War, the Indian wars, the China Relief Expedition, and many other campaigns.

In the profusion of medals and decorations established during World War I, the United States created a hierarchy of decorations. The Distinguished Service Cross (1917), the Distinguished Service Medal (1918), and the Silver Star (1918) all recognized levels of heroism that did not meet the high standard set for the Medal of Honor. The British government also established auxiliary honors, such as the Military Cross (1914) and the Military Medal (1916) to meet the huge demand for awards. The Central Powers, just like the Allies, liberally created and awarded medals during the war. The Prussian Iron Cross, never a permanent decoration, was reauthorized on 5 August 1914 and issued in great numbers.

In the aftermath of war, the Romanov, Habsburg, and Hohenzollern Empires disappeared, along with their complex systems of honors and awards. In Russia, the Bolshevik leadership, slow to recognize the propaganda and loyalty value of awards, later enthusiastically embraced them. The militaristic Nazi Party in Germany established its own extensive system of badges and honors well before it assumed power in 1933.

World War II precipitated a rash of awards. All of the belligerent nations and most of the neutral nations issued medals to mark the event. In recent years, the United Nations has increasingly played a role in settling international disputes, and it issues service medals to its peacekeeping forces dispatched to monitor the peace in ethnic and regional conflicts.

There are instructive differences between the democracies and other powers in the awarding of the higher medals: the former often honors soldiers who have saved lives; the latter awards only those who have killed or captured the enemy. The democracies also have rigid rules for the awarding of the highest awards; and rarely is it even charged that politics, favoritism, or nepotism were involved in the process. The same cannot be said for other nations.

Eric Smylie

References and further reading:

Dorling, H. Taprell. *Ribbons and Medals.* London: Osprey Publishing, 1983.

Kerrigan, Evans E. *American War Medals and Decorations.* New York: Mallard Press, 1990.

Litherland, A. R., and B. T. Simkin. *Spink's Standard Catalogue of British Orders, Decorations, and Medals.* London: Spink and Son, 1990.

Meriecka, Vaclav. *The Book of Orders and Decorations.* London: Hamlyn, 1975.

Purves, Alec A. *The Medals, Decorations, and Orders of the Great War, 1914–1918.* London: J. B. Hayward & Son, 1975.

———. *The Medals, Decorations, and Orders of the World War II, 1939–1945.* London: J. B. Hayward & Son, 1986.

Medici, Giovanni de (a.k.a. Pope Leo X) (1475–1521)

Vivacious warrior-pope who made major contributions to the reestablishment of Rome as a center of European cultural activity and political power. Giovanni was the second son of Lorenzo the Magnificent, ruler of the Florentine Republic at the height of the Italian Renaissance. His education was the finest Europe had to offer at the time; he studied under the tutelage of figures like philosopher Pico della Mirandola and from early on was destined for clerical service. Perhaps his most notable action was the excommunication of Martin Luther in 1521, but Giovanni made other significant historical contributions, such as expediting the construction of St. Peter's Basilica and substantially increasing the holdings of the Vatican library.

In general, Giovanni was committed to the arts in a manner befitting a Renaissance ruler, spending much of his personal wealth and the money of the church on cultural enterprises. As head of the Catholic Church, temporal ruler of the papal states, and head of the ruling Medici family, Giovanni engaged in political struggle and what some might consider an overabundance of nepotism. He granted, for example, the archbishopric of Florence to his cousin (who would later become Pope Clement VII) and called on his nephew Lorenzo and brother Giuliano to be Roman patricians.

These political maneuvers were initiated in an attempt to dominate Italy, the central European power at the time, and they often brought with them considerable danger from rival factions and outside powers. France, in particular, felt it had claims to key Italian cities like Milan and Naples, resulting in Louis XII's march on Italy in 1513. Giovanni grudgingly formed an alliance with the militarily powerful Spanish, and the French were defeated at Novara, forcing Louis to withdraw. When France's Francis I rose to power in 1515, however, the war was renewed, and Giovanni revived his alliance with Spain and England. Giovanni's defeat at Marignano on 14 September 1515 forced him to make peace and draw up the Concordat of Bologna, which regulated church and state relations until the 1790s.

David J. Tietge

See also: Marignano, Battle of

References and further reading:

Gobineau, Arthur, Comte de. *The Golden Flower.* Trans. Ben Ray Redman. Freeport, NY: Books for Libraries Press, 1968.

Mee, Charles L. *White Robe, Black Robe.* New York: Putnam Books, 1972.

Vaughan, Herbert Millingchamp. *The Medici Popes.* Port Washington, NY: Kennikat Press, 1971.

Medicine, Military

The medical and surgical specialty concerned with the ailments of soldiers and sailors. Warfare provides opportunities to expand medical and surgical knowledge but, at the same time, frustrates battlefield doctors who are unable to prevent death, relieve suffering, cure disease, or mend wounds. But military doctors and surgeons are unique in their ambivalent position: they are employed in their healing tasks by the same organization that inflicts wounds and pain, more sometimes on their own patients.

Detailed descriptions of wounds in the *Iliad* show that the Greeks knew basic human anatomy very well. Ancient Chinese and Indian texts reveal similar levels of understanding. Such knowledge is prerequisite for effective trauma surgery. That does not mean that ancient trauma surgery was effective; usually, it was not. There was little a surgeon could do except pull out the arrow or spear, bandage the wound, splint or amputate the shattered limb, numb the patient with alcohol, and hope for the best. Yet there is compelling evidence in Greek and Roman literature that army surgeons were revered. The bravest soldiers and highest-ranking officers were attended by the most skillful surgeons. Thanks to his doctors, Alexander the Great survived dozens of grim wounds before dying of fever at 33. Ancient military surgeons frequently failed, but obviously they were serious about their practical science.

Like medieval medicine in general, medieval military surgery regressed from the progress made in ancient cultures. Except for Paul of Aegina (625–690), the Chinese, the Arabs, the School of Salerno, Lanfranc (fl. 1290), and Guy de Chauliac (c. 1298–1368), few doctors made advances in surgery between the fall of Rome and about 1500. Medieval medical students learned anatomy from the inaccurate works of Galen (130–200), not by direct experience. Until laboratory dissection of cadavers began to be allowed in the sixteenth century, studying battlefield casualties was the main way for surgeons to gain firsthand knowledge of human anatomy.

Ambroise Paré (1510–1590) was the father of modern military surgery. He gained fame through his treatment of a relatively new phenomenon in warfare, gunshot wounds. Earlier surgeons studied gunshot wounds, but Paré was the first to learn how to attend to them effectively. For two centuries after Paré, the French dominated surgery in general and military surgery in particular.

The Thirty Years' War, the early colonial conflicts between England and France, and other seventeenth- and eighteenth-century European wars led to noteworthy increases in medical and surgical knowledge. John Woodall (1570–1643) wrote the first textbook of naval surgery. Johann Schultes (Johannes Scultetus) (1595–1645) wrote the standard text on surgical instruments and procedures. Richard Wiseman (1622–1676) added to the knowledge of gunshot wounds. Lorenz Heister (1683–1758) developed tourniquets. François-Michel Disdier (1708–1781) contributed to what eventually resulted in the superior first-aid bandaging techniques of Johann Friedrich August von Esmarch (1823–1908). John Hunter (1728–1793) gained new insight into gunshot wounds while serving in the Seven Years' War.

Perhaps the greatest military surgeon of all time was Dominique Jean Larrey (1766–1842), who participated in all major engagements of the Napoleonic Wars. In 1792 he invented the "flying ambulance," by which wounded soldiers could be quickly and safely evacuated. He emphasized first aid and improved the mobile battlefield hospital. The men adored him, and Napoléon himself called him "the most virtuous man I have ever known."

The American Civil War established the United States as the world leader in military medicine. *A Manual of Military Surgery* (1861) by Samuel D. Gross (1805–1884) was the standard text. The surgical potential of anesthesia, introduced in the 1840s, was just beginning to be exploited. Because infection and disease proved more deadly than battles, doctors finally began to notice that sanitation and hygiene were important to military health. New knowledge appeared in several classic works, including *Outlines of the Chief Camp Diseases of the United States Armies as Observed during the Present War* (1863) by Joseph Janvier Woodward (1833–1884) and *A Treatise on Military Surgery and Hygiene* (1865) by Frank Hastings Hamilton (1813–1886).

Antisepsis, anesthesia, and hemorrhage control, the three prerequisites for intricate surgical procedures, were all firm medical facts by the start of the twentieth century. The primary medical concerns in World War I were gas warfare, artillery wounds, disease, and shell shock. During that war, the motorized ambulance first appeared, and the field hospital became more complex.

The American Medical Association's book *War Medicine,* published in eight volumes from 1941 to 1945, was the basic medical and surgical manual for World War II. American medical schools in cooperation with the War Department founded permanent or semipermanent general military hospitals throughout the world. American leadership in surgical innovation was so renowned that Nazi intelligence routinely monitored Allied medical correspondence in order for German military surgeons to acquire techniques from the Americans.

Helicopter evacuations of wounded from battlefields to mobile army surgical hospital (MASH) units began during the Korean War. Thus the speedier use of more sophisticated care than medics could provide on the battlefield became possible.

Military medicine is not only surgery. Disease and filth are sometimes greater threats than the human enemy. In the eighteenth century, beginning with the work of John Pringle (1707–1782), the British began to succeed against some of the diseases that had plagued soldiers and sailors since ancient times. James Lind (1716–1794) and Gilbert Blane (1749–1834) conquered scurvy in the British navy. An American, Walter Reed (1851–1902), discovered the control for yellow fever in 1900 while stationed with the occupation force in Cuba. Another American, Edward B. Vedder (1878–1952), stationed in the Philippines in 1911, developed a cure for amoebic dysentery.

Florence Nightingale founded the modern profession of military nursing during the Crimean War. Before her time, military nurses were typically camp followers, prostitutes, or blowsy girlfriends. Clara Barton attended the Union wounded as an independent nurse during the Civil War, served with the International Red Cross in the Franco-Prussian War, and founded the American Red Cross in 1881.

In times when it was common to spend a lifetime within a few miles of one's birthplace, simple homesickness or "nostalgia" could sap a young soldier's or sailor's will to live, causing his debility or even death. In the twentieth century, this problem was ameliorated by initiatives such as the United Service Organization (USO) and by the fact that young adults had become more accustomed to travel far from home.

Challenges to military medicine at the beginning of the third millennium include chronic health problems such as posttraumatic stress disorder (PTSD), formerly called "shell shock" or "battle fatigue," characterized by psychologically damaging flashbacks to combat situations; the physical aftereffects of exposure to the defoliant Agent Orange in Vietnam; and Gulf War syndrome (GWS), the mysterious biological ailment of veterans of that 1991 conflict. But if there has been one field of undoubted progress in modern times, it has been medicine—and military medicine.

Eric v. d. Luft

See also: American Civil War; Ancient Warfare; Barton, Clarissa; Chemical and Biological Warfare; Ethics of Warfare; Gulf War; Korean War; Laws of War; Napoleonic Wars; Nightingale, Florence; Prisoners of War; Seven Years' War; Thirty Years' War; Vietnam Conflict; War Crimes; World War I; World War II

References and further reading:

Binneveld, J. M. W. *From Shell Shock to Combat Stress: A Comparative History of Military Psychiatry.* Amsterdam: Amsterdam University Press, 1997.

Curtin, Philip D. *Disease and Empire: The Health of European Troops in the Conquest of Africa.* Cambridge, UK: Cambridge University Press, 1998.

Denney, Robert E. *Civil War Medicine: Care and Comfort of the Wounded.* New York: Sterling, 1995.

Ginn, Richard V. N. *The History of the U.S. Army Medical Service Corps.* Washington, DC: Office of the Surgeon General and Center of Military History, 1997.

Megiddo (September–October 1918)

An engagement that ended Turkey's participation in World War I. General Edmund Allenby's effective strategy in Palestine used battle-hardened soldiers, including Australians, New Zealanders, Indians, and Arabs, as well as British troops, and Arab guerrilla fighters led by Thomas Edward Lawrence ("Lawrence of Arabia").

Allenby deployed 57,000 infantry, 12,000 cavalry, and more than 500 guns against the Turks' 32,000 infantry, 200 cavalry, and 400 guns. With superiority in cavalry and mastery of the air, Allenby was well placed to overwhelm the numerically inferior and exhausted Turkish troops. His strategy, however, was based on mobility, surprise, and depriving the enemy of all communications by attacking the Turkish road and rail systems, particularly the Hejaz Railway south of Damascus, the feeder line for supplies for the Fourth, Seventh and Eighth Armies.

Dummy camps and horses were set up, mules raised dust, and a small number of soldiers marched back and forth to create the impression of a much larger force to make the Turks think an attack would come in the Jordan Valley. Allenby deployed his force to the west, with the bulk of his cavalry riding north along the coast before swinging in behind the Turks' Seventh and Eighth Armies and taking their communications centers. Lawrence and his Arab force destroyed railway lines north, south, and west of Deraa on 16 and 17 September, while British and Australian pilots bombed the track and station buildings.

On 19 September, Allenby's bombardment opened along a 24-kilometer (15-mile) front along the coast, punching a hole in the Turkish line. With communications destroyed, German general Liman von Sanders, in Nazareth, had no idea that the Allied forces were sweeping through the breach in the Turkish lines and barely escaped the advancing Desert Mounted Corps.

The retreating Turks were bombed repeatedly by Allenby's aircraft as they fell back from Nablus toward the Jordan River. The Turkish Fourth Army east of the Jordan began retreating on 22 September, surrendering near Amman and at Damascus. By 1 October, the key cities of Beirut, Homs, Aleppo, and Damascus had fallen to Allenby's army, and for Turkey the war was over.

Roslyn Russell

See also: Allenby, Edmund Henry Hynman, Viscount; Lawrence, Thomas Edward; World War I

References and further reading:

Gullett, Henry. *The Australian Imperial Force in Sinai and Palestine, 1914–1918.* Sydney: Angus and Robertson, 1940.

James, Lawrence. *Imperial Warrior: The Life and Times of Field-Marshal Viscount Allenby, 1861–1936.* London: Weidenfeld and Nicolson, 1993.

Livesey, Anthony. *Great Battles of World War I.* New York: Macmillan, 1989.

Wavell, Archibald Percival. *Allenby: Soldier and Statesman.* London: George G. Harrap, 1946.

Megiddo, Battle of (1469 B.C.E.)

In an attempt to regain land, especially cities controlling the Via Maris, the major trade route from Egypt to Syria, Pharaoh Thutmose III of Egypt invaded Canaanite territory with a large army, which may have had as many as 20,000 men. Opposed by the Canaanites under the leadership of the king of Kadesh, the Egyptians chose to attack through the central pass (probably modern Wadi 'Ara), passing into the Jezreel valley with only a small skirmish against a Canaanite guard force. Surprised, the Canaanite army, which had expected an attack through the broader northern or southern approaches, rearranged its lines. The Egyptians deployed their chariots in a line across the pass mouth, with Thutmose III in the center and the Egyptian left wing overlapping the Canaanites to the right. The Canaanites broke under the charge and retreated into the walled city of Megiddo. Because the Egyptians paused to loot the camps of the Canaanites, especially that belonging to the king of Kadesh, the besieged were able to immediately seal the city gate, which necessitated hoisting recovered men over the wall with ropes. Thutmose III then besieged the city for seven months, after which it surrendered, yielding rich plunder, including 350 slaves and the chariots and horses used by the Canaanites.

Although Megiddo is the first recorded battle in history, the Egyptian records show that warfare was already highly developed, with the Egyptian army able to quickly move large numbers of men on forced march, supply them through a long siege, and support the manufacture of weapons and chariots for a large force. The strategy chosen by Thutmose III was closely mirrored by that of Sir Edmund Allenby in his own attack on Megiddo in 1918.

Cynthia Clark Northrup

References and further reading:

Cline, Eric. *Battles of Armageddon.* Ann Arbor: University of Michigan Press, 2000.

Davies, Graham I. *Megiddo.* Cambridge, UK: Lutterworth, 1986.

Herzog, Chaim, and Mordechai Gichon. *Battles of the Bible.* New York: Random House, 1978.

Meigs, Montgomery Cunningham (1816–1892)

The leading American engineering officer of his era and quartermaster general of the U.S. Army during the American Civil War. Born 3 May 1816 in Augusta, Georgia, he grew up in Philadelphia, graduating fifth in his class from the U.S. Military Academy at West Point, New York, in 1836. He entered the Army Corps of Engineers after graduation and supervised the construction of the Washington Aqueduct, the construction of the wings and dome of the U.S. Capitol, and the expansion of the General Post Office Building before the war. As quartermaster general during the Civil War, he provided huge quantities of materials to the Union armies. He commanded Ulysses S. Grant's base of supplies during the Overland campaign (1864) and commanded War Department employees in the Washington fortifications during Jubal Early's raid (July 1864). He personally supervised the resupply of General William Tecumseh Sherman's army at Savannah and a few months later in North Carolina, reopening Sherman's lines of supply. He was brevetted major general on 5 July 1864. Meigs's duties as quartermaster general included the oversight of government land use for military purposes. In this capacity, he first suggested that Arlington would be an appropriate site for a national cemetery.

After the war, Meigs supervised plans for the new War Department building, the National Museum, and the extension of the Washington Aqueduct. He died on 2 January 1892 in Washington, D.C., and is buried in Arlington National Cemetery.

Robert D. Bohanan

See also: American Civil War

References and further reading:

Dickinson, William C., Dean A. Herrin, and Donald R. Kennon, eds. *Montgomery C. Meigs and the Building of the Nation's Capital.* Athens: Ohio University Press, 2001.

Miller, David W. *Second Only to Grant: Quartermaster General Montgomery C. Meigs: A Biography.* Shippensburg, PA: White Mane Books, 2000.

Weigley, Russell Frank. *Quartermaster General of the Union Army: A Biography of M. C. Meigs.* New York: Columbia University Press, 1959.

Mercenaries

Hired professional soldiers who fight for any state without regard to political interests or issues. The use of paid foreign military troops is as old as warfare itself. Mercenary soldiers have been utilized in every period of recorded history. Although possessing a long and rich history, the profession of mercenary has an unsavory reputation. Historically, mercenary use is almost universal in advanced societies, and they have played many key roles in history.

Nearly every ancient empire, including the Israelite, Persian, Chinese, Greek, and Roman Empires, used mercenaries at one time or another. It should be remembered that even Xenophon was a mercenary leader. Without mercenaries, Carthage could never have challenged Rome, and Hellenistic civilization would never have spread to Italy and Afghanistan. Later, mercenaries would be an essential and underrated element in medieval warfare. Flemish mercenaries allowed King Stephen of England (r. 1135–1154) to fight off the Plantagenets (a rival royal house) for nearly 20 years. By the end of the medieval period, mercenaries were abundant in Europe. During this time, the mercenary profession earned its fame and history's scorn.

Following the Hundred Years War (1337–1457), Europe was replete with thousands of men whose only training was in the arts of war. During the fifteenth century, these unemployed men formed "free companies" and sold their services to various princes. Most famous among these mercenary groups were the *Condottieri,* named after the *condotta* or contracts they signed, who fought prominently in the wars of the ministates of Renaissance Italy (the Italian Wars of 1494–1559). The other famous mercenary formation of the time was the Swiss pikemen, who were so impressive in battle that many kings were eager to hire them, including Julius II, who recruited them as a papal police corps, a function they serve to this day.

By the late eighteenth century, conscription and standing armies had largely replaced ad hoc military formations. As a result, the use of mercenaries declined markedly, although the British used a great number of German mercenaries during the American Revolution. The change in the stigma associated with using mercenaries can be seen in the American Declaration of Independence, which lists the king of England's use of "foreign mercenaries" as a specific offense. The French Revolution, with its ideals of patriotism, universal conscription, and fighting for the nation, ideas that spread throughout Europe, made the concept of fighting simply for personal gain unacceptable. Work for mercenaries continued, only now it lacked the status it had enjoyed in an earlier time.

In the second half of the twentieth century, the reputation of the mercenary profession continued to come under assault. Modern concepts of national sovereignty required the suppression of nonstate military activities. Foreign nationals would henceforth be uniformed, trained, and given rank as units of the national army, such as the French Foreign Legion and the British Gurkhas. The 1977 protocol to the Geneva Convention of 1949 sought to codify disgust for mercenaries. For signatories to this protocol, mercenaries are considered outlaws, placing them in the category of criminals or worse.

The end of the Cold War, in much the same fashion as the end of the Hundred Years' War, has created new opportunities for mercenaries. As global military competition disappeared, thousands of highly trained soldiers were in need of employment. At the same time, the disintegration of the bipolar international structure meant that the great powers were less concerned about events happening in Third World countries. True to free market principles, the existence of a demand for military expertise and the existence of a supply invariably created a market.

The post–Cold War environment has led to the creation of private military organizations run, in a professional manner, by retired senior military officers. Among these new mercenaries are Sandline International and Military Professional Resources Incorporated (MPRI). These organizations tend to deal exclusively with national governments and for a fee will provide advice on defense strategy, train units in the nation's armed forces, help the government procure arms, or fight the government's battles.

The new era of the mercenaries has produced mixed reviews, ranging from condemnation for profiteering and, supposedly, for exacerbating civil wars in Africa to praise for providing assistance to Croatian forces in their battle with the Serbian military. The latter intervention, in fact, is credited with bringing both sides to the negotiating table.

It can be argued that precisely because the mercenary fights for money and is thus not nearly so driven by national, ethnic, or racial animosities, he is something of a moderating force in battle. Further, the professionalism of many mercenaries sometimes makes national armies look inept in comparison, and thus not all of the criticism leveled against mercenaries has been entirely disinterested.

Craig T. Cobane

See also: American Revolution; French Foreign Legion; Hundred Years War; Xenophon

References and further reading:

Shearer, David. *Private Armies and Military Intervention.* Oxford, UK: Oxford University Press, 1998.

Thompson, Janice E. *Mercenaries, Pirates, and Sovereigns: State Building and Extraterritorial Violence in Early Modern Europe.* Princeton, NJ: Princeton University Press, 1994.

Yalichev, Serge. *Mercenaries of the Ancient World.* London: Constable and Company, 1997.

Meroe (antiquity–300 C.E.)

Ancient Ethiopian city-state eclipsed by Christian Ethiopians of Aksum. The earliest written account of Meroe dates from 738 C.E., when Arab chronicler Wahb Ibn Munabbeth described Meroe as a city and state by the same name. The tenth-century Arab chronicler al-Masudi of Baghdad argued that the capital, a flourishing market town, was made of gold, meadows, and gems and was the home of the "sons of Kush" and the "sons of Canaan." Meroe was known for its fine architecture and an elaborate system of temples. It was also particularly renowned for its iron smelting.

For hundreds of years, East African coastal trade was carried northwest to the town. From there caravan routes, by which the ancient Kushites traded incense and metal over long distances, led to the highlands of Abyssinia and to the Indian Ocean along the Atbara River. The Meroites probably had ancestral ties to dynastic Egypt. The ancient Kushites of the Nile Valley have left a record of the transfer of their capital to Meroe and include among their former capitals Napata, well-known from dynastic Egyptian records.

During the first century B.C.E., invaders from Arabia conquered northern Ethiopia, forming Aksum. By the fourth century C.E., following their conversion to Christianity, the Axumites had cut Kush's major caravan routes. Ultimately, the Arabian invaders blocked Meroe's access to the Indian Ocean ports, initiating a time of sustained warfare between the city-states. Ultimately, Kush was defeated, and Meroe declined as Aksum gained in power.

Tekla Johnson

References and further reading:

Jackson, John G. *Introduction to African Civilization.* New York: Citadel Press, 1970.

Trigger, B. G., and B. J. Kemp et al. *Ancient Egypt: A Social History.* New York: Cambridge University Press, 1983.

Merovingians

A dynasty of Frankish kings descended, according to tradition, from Merovech, a chieftain of the Salian Franks. Merovech's grandson was Clovis (r. 481–511), founder of the Frankish monarchy. The Merovingians were known as the "long-haired kings," and the cutting of a king's hair represented his loss of royal power.

Originally little more than a tribal chieftain, Clovis became the sole leader of the Salian Franks by force of perseverance and a free use of assassination. Expanding his kingdom with a ruthless single-mindedness, Clovis consolidated the position of the Franks in northern Gaul. In 486 he defeated Syagrius, the last Roman governor in Gaul and, in a series of subsequent campaigns with strong Gallo-Roman support, occupied an area situated between the Frankish kingdom of Tournai, the Visigothic and Burgundian kingdoms, and the lands occupied by the Ripuarian Franks and the Alamanni.

Clovis came to believe that his victory over the Alamanni at Tolbiacum in 496 was due to the help of the Christian God, whom his wife Clotilda, a Burgundian princess, had been encouraging him to accept. With the support of Remigius, bishop of Reims, Clovis converted with some 3,000 of his army. Thereafter, Clovis was the champion of orthodox Christianity against the Arian heretics, the Burgundians and Visigoths. He attacked the Burgundians at Dijon (500), and the Visigoths at Vouillé (507), where he killed their leader Alaric II in single combat.

Clovis instituted a law code (*Lex Salica*), and for the next 200 years, only his descendants were entitled to rule. When he died, having united all Franks under his rule and gained the support of the Gallo-Roman clergy, he was master of most of Gaul. He thus laid the foundation, which even four hundred years of chaos and misrule could not destroy, of the French monarchy. Clovis personifies the metamorphosis of barbarian warrior into the ruler of a state.

The Merovingians followed the Frankish custom regarding patrimony, and on the death of Clovis the kingdom was divided among his four sons. This partition was not made according to ethnic, geographical, or administrative divisions. The only factor taken into account was that the portions be of equal value (defined in terms of the royal fisc and tax revenues). Although boundaries were poorly defined, the new political units became the kingdoms of Austrasia, Neustria, and Burgundy.

These kingdoms, whose borders were constantly shifting, were for a short period united in a single realm under Chlotar I (r. 558–561), again under Chlotar II (r. 613–623), and once again under Dagobert I (r. 629–639). The rule of the Merovingians before Dagobert, who was able to preserve this unity, was troubled by chronic warfare among aristocrats (both Gallo-Romano and Frankish) and rivals for power. Dagobert was the last active ruler, and his descendants were called the "idle kings." With the decline of the royal authority in Austrasia, the office of mayor of the palace developed into the real seat of power. This appointment was hereditary in the family of the Carolingians, who became the nominal as well as the actual rulers when Pepin the Short deposed the last Merovingian king, Childeric III, in 751.

Nic Fields

See also: Carolingian Empire; Charles Martel; Franks

References and further reading:

Wallace-Hadrill, John Michael. *Long-Haired Kings and Other Studies in Frankish History.* London: Methuen, 1962.

Wood, Ian N. *Merovingian Kingdoms, 450–751.* London: Longman, 1994.

Merrill's Marauders

Defeated the Japanese in the Burmese jungle during World War II using guerrilla tactics, duplicity, and Nisei interpreters. Nicknamed for their commanding officer, Brigadier General Frank D. Merrill, the Marauders, also known as Merrill's Raiders, were conceived during the Quebec Conference in August 1943 as a counterpart to Charles Wingate's Chindits. Allied leaders envisioned a guerrilla-style unit that would wreak havoc behind Japanese lines, disrupt communication and supply lines, and aid in attempts to reopen the Burma road. President Franklin D. Roosevelt called for volunteers for "a dangerous and hazardous mission," and 3,000 army personal offered their services.

Code-named GALAHAD and given the obscure designation of 5307 Composite Unit (Provisional), in October they were sent to India for preliminary training. Frontline troops were divided into six combat teams of 400 each. In February 1944, the regiment was transferred to a location near Ledo in the northeastern corner of India. Their only contact with the outside world by radio and plane, three Marauders battalions marched 500 miles down the Ledo road beginning on 7 February and over the next three months engaged the enemy in five major battles and 17 skirmishes. In March, supported by a Chinese division, they engaged the Japanese 18th Division at Walawbum, and even though the second division went without food or water for 36 hours, the Marauders killed 1,500 and pushed the Japanese south. In late March, they were able to establish an airstrip north of Hsamshingyang, even though Merrill, who was suffering from heart trouble, had to be evacuated soon after. Under the new command of Colonel Charles Hunter, the Marauders were able to withstand a Japanese attack, even though the battles at Inkangahtawg and Nhphm Ga cost the 5307th casualities of 59 dead and 314 wounded. The Marauders' greatest accomplishment was the seizure of the vital airstrip at Myitkyina in May. By 4 June, the regiment's casualties in northern Burma totaled 2,394, and only 200 of the original 3,000 men were considered fit for duty. Weakened by dysentery, skin diseases, fatigue, and malaria, the remaining members of the 5307th could not withstand a Japanese assault on 3 August from Myitkyina. After the war, the Marauders were immortalized in poems, on film, and even in comic books.

T. Jason Soderstrum

See also: Burma, Retreat from; Chindits; Wingate, Orde

References and further reading:

Hunter, Charles Newton. *Galahad.* San Antonio: Naylor Company, 1963.

Ogburn, Charlton. *The Marauders.* New York: Harper, 1959.

United States War Department General Staff. *Merrill's Marauders (February–May 1944).* Washington, DC: Historical Division, War Department, 1945.

Mesoamerican Warfare (1200 B.C.E.–1521 C.E.)

Highest development of Stone Age warfare, with some variations. Pre-Columbian Mesoamerica provides an excellent study area for the development of warfare on a different path than in the old world. The lack of draft and pack animals, wheels, metallurgy, and ships in quantity and quality all led to developments different from those in Europe.

The Olmecs (1200–400 B.C.E.) first used warfare to expand trade and access to resources. Fighters from the Olmec city of San Lorenzo utilized obsidian-edged weapons, hand-to-hand elite combat, and small, elite forces numbering in the tens to hundreds to control local trade routes from the Veracruz region. La Venta assumed power from 900 to 400 B.C.E. and introduced the sling, clay projectiles, and yucca-cotton armor to gain superiority. Tortillas were also used to feed the hundreds of troops deployed in enemy territory. By 400, trophy heads, stone knives, obsidian-tip spears, spear throwers, wood shields, upper-torso armor, and hide helmets were common for elites and, to a lesser extent, for supporting commoner forces.

Fortifications also became common, especially in the lowland Mayan area where captive taking and elite warfare dominated. In the Zapotec region, Monte Alban developed as a heavily fortified city that controlled a regional kingdom through its defensive location and warfare based on thrusting spears and hand-to-hand combat. Skull racks indicate probable religious-based warfare centered on captives and sacrifice. By 100 C.E., Monte Alban was challenged by the huge city-state of Teotihuacán in the Mexico City area. From 100 to 700 Teotihuacán, which numbered perhaps 100,000 people, spread its influence over trading partners partly by emphasizing spear throwers, shields, and stone axes and knives. The Teotihuacanos utilized military orders of eagles, jaguars, and so on, special housing, regular production of weapons, and nodal control of trade centers over 1,500 miles distant. Astronomy and religion seem to have played a large role in how and why war was carried out at the end of Teotihuacán hegemony.

In 378, the Teotihuacanos brought projectile warfare into the Mayan region and tipped the balance of power in favor of large Mayan cities like Tikal, with rulers like Smoking-Frog. These lowland Maya developed religious- and astronomy-based warfare among elites that became known as "star wars." The kin-city competitions for resources, natural and supernatural, dominated classic Mayan warfare from 378 to 900, when warfare may have helped to collapse classical Maya civilization.

In most cases, these early and classic period civilizations in Mesoamerica focused on elite warfare and weaponry, with religion and trade as key motivating factors. Most "armies" numbered less than a thousand soldiers, were supported logistically by commoners, and sought out captives as a way of removing rival dynasties and usurping power. It was as important to take religious items of power as it was to take a city. By the early post-Classic period, between 700 and 900, warfare began to change significantly.

At places like Xochicalca, Cacaxtla, or Bonampak, people like the Olmeca-Xicallanca increased warrior numbers, used backpacks in long-distance attacks on cities like Teotihuacán, and involved commoners in guerrilla activities. After the collapse of classic Mesoamerican civilizations, the Toltecs rose to power in central Mexico between 900 and 1200, and their ascendance marked the rise of territorial warfare, combined arms, and highly militaristic society. They would have a great influence on the later Aztecs (Mexica).

Tollan and its Toltec capital of Tula in central Mexico used large armies in the tens of thousands, concentrated projectile fire of spear throwers, obsidian blades on wood clubs and swords, siege warfare with firing platforms, and watercraft, when necessary. One leader, Topiltzin Quetzalcoatl, may have invaded Yucatan territory after 1100 and spread full-scale warfare into the region from the site of Chichén Itzá. Itzá fought with Coba for control of Yucatan for several centuries. The Itzá eventually prevailed in a conflict that saw the fortifications, attrition, and logistics of the Coba defeated by Itzá Mexican military might. However, both sides had exhausted their resources, and soon a series of smaller kingdoms returned control of the peninsula to Maya polities by 1350.

In the post-Classic period, waves of Chichemecan invaders from the north, such as the Toltecs and Aztecs, combined their nomadic, commoner-based, bow-and-arrow warfare with the elite, trained, central Mexico warfare traditions. The Maya region struggled to keep up but was about to be overrun by the time of Spanish invasions in 1519.

The Aztecs established their capital Tenochtitlán on an island in Lake Texcoco and used this combined approach to establish a huge tributary empire by war between 1300 and 1521. They eventually controlled much of Mesoamerica, notable exceptions being the west Mexican Tarascan kingdom, the Tlaxcallan kingdom, and the Maya lowlands.

Tlacaelel was most responsible for the building of the Aztec Empire. He ruled as a warrior-priest, supporting relative after relative as emperor, for most of the fifteenth century. When he died at 96, he left a legacy of military training schools for elites and commoners, arms production, the ability to field huge armies of 100,000 or more for long campaigns, a religious-military system that promoted and supported war and captive taking, and the largest, most powerful empire in the pre-Columbian Americas.

The Aztecs, like the Inca, often defeated enemies by logistics and numbers, not by military superiority. Time and again, the Aztecs were actually defeated on the battlefield by Tarascan metal weaponry and fortifications, Tlaxcallan volley bow and arrow fire, or even in hand-to-hand combat. Still, the Aztecs usually prevailed in the end.

Christopher Howell

References and further reading:

Hassig, Ross. *War and Society in Ancient Mesoamerica.* Berkeley: University of California Press, 1992.

Sahagun, Bernardino de. *General History of the Things of New Spain.* Trans. Arthur J. Anderson and Charles E. Dibble. 13 vols. Salt Lake City: University of Utah Press, 1950–1982.

Tsouras, Peter. *Warlords of the Ancient Americas: Central America.* London: Arms and Armour Press, 1996.

Metz, Siege of (1870–1871)

A siege that made the Prussian defeat of France almost inevitable. After briefly skirmishing with the Prussians immediately after the start of the Franco-Prussian War, an entire French army, consisting of five corps (155,000 men), fell back to Metz in order to regroup. Led by the incompetent Marshal Achille-François Bazaine, this army was then trapped by the Prussians, who bypassed Metz and began to invest it on 19 August 1870. Bazaine never intended to remain in Metz, yet he made only one halfhearted attempt to break out, on 31 August, to link with the army advancing from the direction of Sedan under Marshal Marie-Edme-Patrice-Maurice de MacMahon, duc de Magenta. When this army was destroyed on 1 September, Bazaine made no further attempts to break out, and the Prussians intensified the ring around the city.

The problem for Bazaine after the defeat at Sedan was that his was an imperial army, and the empire no longer existed. He considered himself a representative of the imperial government and made no efforts to coordinate his defense of Metz with the Republican government in Paris. During early October, Bazaine sent an emissary to Otto von Bismarck to negotiate surrender terms entirely on his own initiative and without even bothering to let the government in Paris know what he was doing. On 27 October, the army unconditionally surrendered and was simply taken prisoner while Bazaine sneaked out to avoid his troops. Bazaine's imperial army thus fought no major battles during the Franco-Prussian War, and it surrendered without having engaged the enemy in any full-scale battle.

After the war, Bazaine was put on trial for treason. He was found guilty and sentenced to death, but President MacMahon commuted it to 20 years' imprisonment, which Bazaine avoided by escaping after nine months. The Siege of Metz is important because the surrender of the city freed troops, allowed the Prussians to more effectively encircle Paris, and ensured that the Prussians would not be amenable to an armistice.

Lee Baker

See also: Franco-Prussian War; Sedan

References and further reading:

Howard, Michael. *The Franco-Prussian War.* New York: Collier Books, 1961.

Williams, Roger. *The French Revolutions of 1870–1871.* London: Weidenfeld and Nicolson, 1969.

Meuse-Argonne (26 September–11 November 1918)

The largest and most important offensive of the American Expeditionary Force (AEF) during World War I and perhaps the largest American battle to date. After successful operations at Amiens and Albert in 1918, Marshal Ferdinand Foch decided to reward the AEF for their success at St. Mihiel with an attack on German forces in the Argonne Forest. This was part of a larger strategy to attack at interlocking points along the line, exhausting German reserves and allowing Allied forces to break German lines of communication and logistics. The southern thrust of this pincer movement was to be led by American general John J. Pershing. His 400,000-man force was to attack an area deemed impregnable by some military commanders. The hilly, rough terrain of the Argonne Forest had been reinforced with defensive fortifications by the German army since 1915. The Americans were backed by 300 tanks under the control of General Hunter Liggett and 500 aircraft of the U.S. Air Service under General William Mitchell. They were opposed by 40 German divisions under the leadership of General Max Carl von Gallwitz.

Launched on 26 September, the plan was for Pershing's First Army to attack the Meuse and Aire valleys. To its left, the French Fourth Army would also move north. They were to break through the 10 miles of rough terrain, turn right, and continue to Sedan and Mezieres. Although progress slowed as German resistance increased, in the first three days of fighting, American forces were able to penetrate 3 to 7 miles at some points in the line, capture 10,000 prisoners, and gain the villages of Montfaucon, Exermont, Gercourt, Cuisy, Septsarges, Malancourt, Epinonville, Charpentry, and Very. Yet the lack of roads created traffic jams, and inexperienced American divisions were able to gain little ground. On 4 October, the First Army began a major attack along the entire front. The harsh fighting gave rise to some of the heroic moments in American history, including the famed "Lost Battalion" and Sergeant Alvin York's capture of 132 Germans. By 10 October, Allied forces had nearly cleared the Argonne Forest of the enemy.

The heavy fighting continued until November, when the Americans and French were able to break through the German fortifications and advance some 20 miles, while the French Fourth Army captured the railroad hub at Sedan. This advance broke the transportation network that supported the German army in France. On 11 November, the Meuse-Argonne offensive ended; during the six weeks of combat, the AEF suffered 26,277 dead and 95,786 wounded.

T. Jason Soderstrum

See also: Pershing, John J.; World War I

References and further reading:

Braim, Paul F. *The Test of Battle: The American Expeditionary Forces in the Meuse-Argonne Campaign.* Newark: University of Delaware Press, 1987.

Keegan, John. *The First World War.* London: Hutchinson, 1998.

Stokesbury, James L. *A Short History of World War I.* New York: William Morrow, 1981.

Mexican Revolution (1810–1821)

Revolution that achieved independence from Spain. The revolt was ignited by Father Miguel Hidalgo y Costilla in the village of Dolores in the state of Guanajuato on 16 September 1810. Hidalgo called his parishioners and supporters to revolt with "El Grito de Dolores" (the cry of Dolores), shouting "Long live the Lady of Guadalupe" and "Death to the Spaniards." Hidalgo was a Creole (someone born in the colonies of European descent). His army consisted of more than 50,000 troops, whom he often could not control.

The first objective was the city of Guanajuato, and after its surrender, both Creoles and Spaniards were slaughtered. Such atrocities caused both of these groups to join forces against Hidalgo. After attempting to take Mexico City itself and failing, Hidalgo and his troops were forced north. Though his army was able to take a few other cities, Hidalgo was captured and shot on 31 July 1811. His head was exposed on a stake for 10 years.

Hidalgo's death did not halt the revolution. What was left of his army was taken over by Father Jose Maria Morelos y Pavón. He was able to seize much of Mexico, put a form of government together, and call for a constitutional convention at Chilpancingo in 1813. During this convention, a formal declaration of independence was drafted, along with the Constitution of Apatzingan. However, on 22 December 1815, Morelos was also captured and shot by Spanish troops. By now, the only remaining rebel force still in the field was that

Mexican insurrectos *with a homemade cannon in Juarez, c. 1911. (Library of Congress)*

commanded by Vicente Guerrero, an Indian. But for the next five years, until 1820, peace reigned throughout much of Mexico.

Events in Europe forced the Spanish king Ferdinand VII to promulgate the liberal constitution that had been put in place in 1812 and that he later had rescinded. On 24 February 1821, Augustín de Iturbide, a Mexican landowner, issued his three-point Plan de Iguala, which called for a Catholic nation, independence under a monarchy, and equality between Europeans and Creoles. Iturbide's army joined forces with those of Guerrero and entered Mexico City on 27 September 1821. Mexico had finally won its independence from Spain. Though Hidalgo's efforts did not result in an immediate victory, Mexico still celebrates its independence on 16 September.

Peter Carr

See also: Santa Anna, Antonio López de

References and further reading:

Atwater, James D., and Ramón Eduardo Ruiz. *Out from Under: Benito Juárez and the Struggle for Mexican Independence.* Garden City: Doubleday, 1969.

Van Young, Eric. *The Other Rebellion: Popular Violence, Ideology, and the Mexican Struggle for Independence, 1810–1821.* Stanford, CA: Stanford University Press, 2001.

Mexican Unrest and Civil War (1911–1929)

Series of coups and countercoups, revolts, and civil wars that did not address the basic Mexican problem of mass poverty and that twice drew in American intervention. As Mexico prepared in 1910 to celebrate the centennial of the republic's independence in 1810, the superficial signs of prosperity could not mask the widespread poverty and discontent that would soon erupt into revolution and plunge the nation into years of civil war. In 1910, Mexico was one of the world' s most lucrative and alluring economies. The credit fell to the nation's longest-serving president, Porfirio Díaz. However, the prosperity was an illusion created by the selling of trading concessions and appropriated communal lands. Fewer than 1,000 landowners controlled 97 percent of all of Mexico's land. In the 1910 election, a wealthy liberal landowner, Francisco Madero, challenged Diaz, demanding expanded suffrage, limits on presidential succession, and agrarian reform. The resulting surge of public support for Madero and growing antiregime violence shocked Díaz, who resigned the presidency and left Mexico for exile.

Madero was inaugurated as president in November 1911 amid great expectations. However, timidity and compromise characterized his administration. He failed to institute promised agrarian reforms, thus alienating his rural supporters. His willingness to work with the elites cost him the support of the liberals. In a conservative-backed counterrevolution, General Victoriano Huerta forced the resignation of President Madero and his vice president. Three days later, on 23 February 1913, both men were shot while under the general's protection. Huerta assumed the presidency, incurring the wrath of Madero's one-time supporters. Huerta's resistance to agrarian or social reforms mobilized his domestic opposition. Emiliano Zapata and Pancho Villa raised revolutionary armies in rural Mexico. In the north, the revolutionary army under the command of Venustiano Carranza was a formidable threat to Huerta. The regime was further destabilized by the U.S. Navy's occupation of Mexico's largest port city, Veracruz, following a minor affront to an American naval officer. Huerta resigned the presidency in July 1914, leaving Mexico for exile, while Mexico's revolutionary commanders battled for supremacy.

Carranza won broad popular support when, in January 1915, he issued a decree that outlined a program of land redistribution. The U.S. government and Mexico's Latin American neighbors recognized Carranza as the de facto head of state. His rivals, particularly Villa, sought to destroy international support for Carranza. Villa's forces attacked the town of Columbus, New Mexico, in March 1916, killing a number of Americans. President Woodrow Wilson's response was to dispatch a punitive expedition into Mexico to track down Villa. The incursion rallied Mexicans behind Carranza. A liberal constitution was ratified on 5 February 1917, the same day the United States withdrew its forces from Mexico. Carranza became the first president under the new constitution. Mexicans expecting dramatic reforms were soon disappointed. Carranza bowed to international pressure, suspending decrees that threatened foreign business interests. When he hesitated to enact the agrarian reforms called for in the constitution, his popular support eroded.

Constitutionally prohibited from succeeding himself in the election of 1920, Carranza backed a puppet candidate. This action prompted Alvaro Obregón, the former minister of war, to challenge the president's handpicked candidate. The governor of the state of Sonora called upon his fellow governors to rise up militarily against Carranza. Thirteen states followed his lead. Carranza fled the capital, only to be assassinated en route to Veracruz. Obregon was elected president and moved quickly to implement agrarian reforms. He also imposed restrictions on foreign ownership of Mexican land and resources. The election of Obregon is considered by many historians as marking the end of the Mexican Revolution of 1910. The relative peace that Obregon's presidency brought to Mexico was shattered in 1923, however. In the elections that year, Obregon threw his support to Plutarcho Calles. Rival candidate Adolpho de la Huerta urged his supporters to take up arms against the government. Obregon personally led federal troops in suppressing the rebellion.

As president, Calles was cautious in limiting foreign investment in Mexico. He did not show the same restraint in dealing with the Catholic Church, aggressively restricting the rights and activities of the clergy. Calles's anticlerical measures sparked violent resistance, known as the Cristero Rebellion, that continued throughout his presidency. Peace was elusive until the 1940s. Provisions of the Constitution of 1917 continued to be implemented by successive governments. Yet the twentieth century ended as it began for Mexico, with the ownership of land and disparities in wealth dominating political debate.

Eric Smylie

See also: Mexico, U.S. Punitive Expedition in; Pershing, John J.; Villa, Francisco "Pancho"; Zapata, Emiliano; Zapatista Rebellion

References and further reading:

Brenner, Anita, and George R. Leighton. *The Wind That Swept Mexico: The History of the Mexican Revolution, 1910–1942.* Austin: University of Texas Press, 1996.

Gilderhus, Mark T. *Diplomacy and Revolution: United States–Mexico Relations under Wilson and Carranza.* Tucson: University of Arizona Press, 1977.

Hall, Linda B., and Don M. Coerver. *Revolution on the Border: The United States and Mexico, 1910–1920.* Albuquerque: University of New Mexico Press, 1988.

Knight, Allen. *The Mexican Revolution.* Cambridge, UK: Cambridge University Press, 1986.

Mason, Herbert Molloy, Jr. *The Great Pursuit.* New York: Random House, 1970.

Sweetman, Jack. *The Landing at Veracruz: 1914.* Annapolis, MD: United States Naval Institute Press, 1968.

Mexican-American War (1846–1848)

Disastrous Mexican defeat at the hands of the United States, resulting in enormous losses of territory. President James Polk wanted land from Mexico and Canada in his drive to expand the United States from the Atlantic to the Pacific Ocean. It was an era of Manifest Destiny, when many Americans assumed they had a right to overflow the continent from coast to coast (and perhaps the Americas from pole to pole as well). When the breakaway Republic of Texas joined the United States in March 1845, Mexico was angered, and tensions increased along the disputed Texas-Mexico border, one side claiming the Rio Grande and the other the Nueces River as a border.

Polk then settled the Oregon Territory boundary dispute with Great Britain in 1846; that left Mexico and the vast and underpopulated lands in its north as a barrier to westward expansion. In March 1846, Polk ordered General Zachary Taylor and his "Army of Observation" to the north bank of the Rio Grande to strengthen the Texan and American boundary claim. That decision soon led to fighting and formal, mutual declarations of war.

The Mexican-American War fell into three broad phases. The first phase centered on General Taylor in the north. American and Mexican armies soon clashed, first at Palo Alto and then at Resaca de la Palma near current Brownsville, Texas. The Americans won both battles, and the Mexicans retreated. Taylor slowly pursued and, in September, neared Monterrey. The Mexicans had formidable defenses. Still, Taylor sent a flanking movement that soon gained control over the dominating two hills, emplaced artillery, and caused the Mexican forces to abandon the town. Taylor proceeded somewhat south, and in February 1847, after most of his regular army troops were detached to form the main force for General Winfield Scott's amphibious invasion, Taylor was near the Hacienda de la Buena Vista when General Antonio López de Santa Anna attacked him. The Americans had good defenses, but the Mexicans sought to outflank them on the American left. Militia units rushed to the battle, fought very well, and were able to throw back the Mexican assault. This battle largely ended fighting in north-central Mexico, as Santa Anna moved to contest Scott's invasion.

There was, to be sure, fighting on the periphery, which could not affect the ultimate outcome nor compel a Mexican surrender. General Stephen Watts Kearny marched from Fort Leavenworth, Kansas, to take Santa Fe, New Mexico, an important trading point. Kearny proceeded to California, where other Americans, including John C. Frémont and Richard Stockton, helped take control of that Mexican province.

These American gains in the north did not translate into a Mexican surrender or cession of territory. Thus, in early 1847, President Polk somewhat reluctantly agreed to a plan by General Winfield Scott to attempt a seaborne invasion of the port city of Veracruz, then to proceed along the relatively traditional route to attack and seize the capital, Mexico City, and thereby compel surrender.

Scott's resulting campaign was remarkable. Polk feared that if Taylor or Scott enjoyed too much success, the victorious general might make a formidable presidential campaign opponent; Scott also had to deal with shortages of equipment and limited manpower throughout the campaign. He had to attempt the first major contested American amphibious assault, and he had to cooperate well with the U.S. Navy, on whose logistical support he depended. He had to follow a mostly predictable route into the heartland of his enemy and win a series of battles mostly on grounds of his opponent's choosing, without being able to afford losing many men. That is, he operated under a remarkable series of constraints and surmounted all of them.

Scott gathered his invasion force at New Orleans and then, after a stop at Tampico, accepted the advice of Commodore David Conner, who was commanding the naval flotilla, and landed his command on a beach south of the heavily fortified city of Veracruz. By nightfall of the first day, 9 March 1847, some 100 ships had landed 10,000 men, their animals, and supplies—a remarkable accomplishment. Scott then established a siege line across Veracruz, and within several weeks, the city surrendered.

Scott needed to proceed up the national highway to move his men above the dreaded "yellow fever" line before the season of illness began. General Santa Anna established seemingly strong defenses at Cerro Gordo in early April, where he expected Scott to proceed up the national highway and be denied further progress. Scott's engineering officers, led by Captain Robert E. Lee, found a goat path—a trail—that could be widened to allow the Americans to proceed around, behind, and above the Mexican defenses from the right and thereby force the Mexicans to attack or cede their positions, which they did on 17–18 April 1847. Scott then rested midway at Puebla, while he massed supplies and dealt with the enlistment terms of his militia (for many, their terms were up, and Scott wanted them out of his way before proceeding to the Mexican capital).

Again, the deluded Santa Anna felt that he had established a formidable set of defenses before the capital, Mexico City. He assumed Scott would approach from due east, and he was ready. There were deep lakes to the north, a supposedly impenetrable lava bed to the south, and several lakes in the center that he expected would channel the American invaders into the strength of his defenses. Once again, Scott's engineers found a way around these obstacles and avoided the expected line of advance. They found a path across the lava bed to the south, and Scott managed to outflank Santa Anna, first by moving south, fighting at Contreras and Churubusco, and thereby moving south of the capital. The American commander then attacked Mexico City from the lightly defended south and west. The Americans won at El Chapultepec and then descended from the heights to take the city. After some months of negotiation, both sides agreed to the Treaty of Guadalupe Hidalgo in February 1848, in which Mexico ceded California, Nevada, Utah, Arizona, and New Mexico, along with much of Colorado and the disputed territory in southern Texas.

Taylor's campaigns in the north did not greatly advance the military arts and sciences. Mexican troops were poorly trained and mostly poorly led. Their copper cannonballs dented rather than exploded; Americans soldiers literally jumped over the cannonballs rolling along the ground. The fighting in New Mexico and California involved few troops. However, Scott fought an exemplary campaign of maneuver reflecting pre-Napoleonic values; that is, he could not afford the massive firepower and extensive loss of life of battles and campaigns from the Napoleonic Wars. Instead, relying on military strategy and tactics from the previous era, he mostly avoided expected lines of advance, gained psychological advantage by an indirect advance, and won a great series of battles. Scott's benign treatment of the civil populace also ensured a smooth passage along his lines of communication.

Charles M. Dobbs

See also: Santa Anna, Antonio López de; Scott, Winfield; Taylor, Zachary

References and further reading:

Bauer, K. Jack. *The Mexican War, 1846–1848.* Lincoln: University of Nebraska Press, 1993.

Eisenhower, John S. D. *So Far from God: The U.S. War with Mexico, 1846–1848.* New York: Random House, 1989.

Singletary, Otis. *The Mexican War.* Chicago: University of Chicago Press, 1960.

Mexico, U.S. Punitive Expedition in (1916–1917)

U.S. Army expedition sent into Mexico to seize the bandit and revolutionary Pancho Villa. Continued political instability in Mexico after the Revolution of 1910 led to a growing concern over the safety and security of the U.S.-Mexican border. On 9 March 1916, perhaps upset at the loss of American support or just desperate to once again be a player in Mexican politics, Pancho Villa, a major figure in the Mexican Revolution, attacked the town of Columbus, New Mexico. When Villa and his mounted raiders withdrew, 16 U.S. citizens had been killed, and the town center had been burned to the ground. The unprovoked attack created a firestorm of outrage in the United States, and President Woodrow Wilson dispatched General John "Blackjack" Pershing in pursuit of Villa. It took a few days to assemble a force, and Pershing did not cross into Mexico until 15 March 1916.

An 11-month-long incursion by U.S. troops hundreds of miles into the Mexican state of Chihuahua failed to find the elusive Villa. Although a military failure, it did provide the U.S. Army with much-needed field experience and a chance to develop the staff skills needed to run and support a large army in Europe. In particular, Pershing pioneered the use of aircraft (which failed in most cases even to get off the ground) and motor transport, which was more successful.

American and Constitutionalist forces went out of their way to avoid each other, and the Americans were forbidden to enter any towns. What occurred was not peace and was not war. Provisional president Venustiano Carranza at first simply wanted to contain the penetration of American forces and then get them out Mexico as fast as possible. Pershing wanted only to capture Villa, not confront Constitutionalist forces.

The Mexicans demanded that first the Americans leave and then that they jointly resolve the problem of cross-border raids. The United States countered by offering to work out a plan after it had captured Villa. Both sides tried to avoid open confrontation, but the political strain started to show as each side was unwilling to concede to the other.

To resolve this impasse, both sides formed a joint commission to find a solution. The commission met many times in New London, Connecticut, from September 1916 to January 1917, but it never reached an accord. The expedition would end only when Woodrow Wilson chose.

Militarily, Constitutionalist and American forces did clash twice, in the towns of Parrazal and Carrizal. Although there were some losses (at Carrizal, both sides suffered 25 percent casualties), commanders on both sides moved to keep the conflict from escalating into full-scale war, something that neither Wilson nor Carranza wanted or could easily afford.

In January 1917, Wilson saw that the threat from Germany was greater than the threat posed by Mexico, and he ordered Pershing to return with his command to the United States. The "hot pursuit" never came close to catching Villa, and it served only to degrade an already strained relationship between the United States and Mexico. Wilson's attempts to "teach the Mexicans democracy" (he also sent troops to occupy Veracruz for more than a year) proved an abject failure. But that did not stop the president from mounting two further military essays into other lands for ill-defined purposes. Expeditions into Murmansk and Vladivostok, Russia, toward the end of and after World War I, like the Mexican expedition, had little purpose and no lasting effect except to create hatred of the regime.

Drew Philip Halévy

See also: Pershing, John J.

References and further reading:

Camin, Hector, and Lorenzo Meyer. *In the Shadow of the Mexican Revolution.* Austin: University of Texas Press, 1993.

Gilderhus, Mark T. *Diplomacy and Revolution: U.S.-Mexican Relations under Wilson and Carranza.* Tucson: University of Arizona Press, 1977.

Haley, P. Edward. *Revolution and Intervention: The Diplomacy of Taft and Wilson with Mexico, 1910–1917.* Cambridge, MA: MIT Press, 1970.

Mexico City, Battles for (20 August–14 September 1847)

The concluding battles for the Mexican capital, which ended

the Mexican-American War. After the battle of Cerro Gordo, General Antonio López de Santa Anna established formidable defenses around Mexico City; a series of lakes would channel the invading Americans into his prepared positions, or so he hoped. He believed that Lake Texcoco limited an attack from the north and that Lake Chalco, Lake Xochimilco, and a vast lava bed limited an advance to the south; his strongest defenses anticipated an attack headlong from the east.

American general Winfield Scott faced further challenges, including his limited number of troops and supplies, capturing a capital city some 225 miles inland, and avoiding the set defenses, since he could not afford many casualties. He solved his problems brilliantly by engaging in a strategy of marching south and past the defenses, eventually to attack Mexico City from the underdefended southwest and western approaches.

There were several battles in this campaign. After resting at Puebla for several months, Scott's advance units arrived in the valley of the capital in early August. The Americans fought at Contreras, Churubusco, and El Molina del Rey and outflanked Santa Anna's defenses.

Finally, on 13 September, the Americans attacked the city itself, beginning with defenses around the Mexican military academy, El Chapultepec. After hard-fought battles, the invaders gained control of the hill on which the academy sat and thus gained entrance into the capital, which formally surrendered to Scott soon after, paving the way for the Treaty of Guadalupe Hidalgo some months after that.

Charles M. Dobbs

See also: Mexican-American War; Scott, Winfield; Santa Anna, Antonio López de

References and further reading:

Bauer, K. Jack. *The Mexican War, 1846–1848.* Lincoln: University of Nebraska Press, 1993.

Eisenhower, John S. D. *Agent of Destiny: The Life and Times of General Winfield Scott.* Norman: University of Oklahoma Press, 1999.

Johnson, Timothy D. *Winfield Scott: The Quest for Military Glory.* Lawrence: University Press of Kansas, 1998.

Miles, Nelson Appleton (1839–1925)

Distinguished American commander, a military leader for four generations, one of few to achieve the highest rank without having attended the U.S. Military Academy at West Point. Miles was born in Westminster, Massachusetts, on 8 August 1839. A store clerk in the 1850s, he studied military science in his spare time.

As soon as the Civil War began, Miles enlisted in the 22d Massachusetts Volunteers, where he was quickly commissioned first lieutenant. By September 1861, he was captain and then aide-de-camp to Major General Oliver Otis Howard. Miles was wounded at Fair Oaks, Fredericksburg, Chancellorsville, and Petersburg. He fought at Antietam, the Wilderness, Spotsylvania, and Cold Harbor, was brevetted three times, and commanded a division at Appomattox. In 1892, he received the (Congressional) Medal of Honor for his service at Chancellorsville.

In October 1865, Miles became major general of volunteers and in 1867 brevet brigadier general. As commandant of Fort Monroe, Virginia, until July 1866, he was criticized for rough treatment of his prisoner, Jefferson Davis. After 1868, as husband of the niece of William T. Sherman, he sought special favors, but Sherman resisted.

From 1869, as colonel of the U.S. Army 5th Infantry, until 1894, as major general of the Department of the Missouri, Miles fought the Indians of the American West. He defeated the Cheyenne in 1874 and 1875, drove Sitting Bull into Canada in 1876, and captured both Crazy Horse and Chief Joseph in 1877 and Geronimo in 1887. Soldiers under his command massacred the Sioux at Wounded Knee in 1890.

As general in chief of the army from 1895 until his retirement in 1903 (the title was changed to chief of staff in 1903), he planned campaigns for the Spanish-American War, the Philippine Insurrection, and the Boxer Rebellion, observed the Turco-Greek War, and personally led the assault on Puerto Rico in 1898. He achieved the permanent rank of lieutenant general in 1901.

Before the emergence of John J. Pershing, Miles was the best-known soldier in the United States. At the time of his death in Washington, D.C., on 15 May 1925, he was the last surviving regular Union major general.

Eric v. d. Luft

See also: American Civil War; Antietam/Sharpsburg; Apache Wars; Boxer Rebellion; Chancellorsville, Battle of; Cochise; Cold Harbor, Battle of; Crazy Horse; Custer, George Armstrong; Fredericksburg; Geronimo; Joseph the Younger, Chief; Little Bighorn; Nez Percé; Pershing, John J.; Petersburg, Siege of; Philippine Insurrection; San Juan Hill/El Caney; Sherman, William Tecumseh; Sioux Wars; Sitting Bull; Spanish-American War; Spotsylvania Court House; Wilderness; Wounded Knee, Battle of

References and further reading:

Amchan, Arthur J. *The Most Famous Soldier in America: A Biography of Lt. Gen. Nelson A. Miles, 1839–1925.* Alexandria, VA: Amchan, 1989.

DeMontravel, Peter R. *A Hero to His Fighting Men: Nelson A. Miles, 1839–1925.* Kent, OH: Kent State University Press, 1998.

Johnson, Virginia Weisel. *The Unregimented General: A Biography of Nelson A. Miles.* Boston: Houghton Mifflin, 1962.

Wooster, Robert. *Nelson A. Miles and the Twilight of the Frontier Army.* Lincoln: University of Nebraska Press, 1993.

Military and Society

The relationship between the soldier and society has been discussed and studied since human societies were formed and individuals or groups within them were armed. The nature and implications of this relationship center on those who manage political relationships within any given society. Plato juxtaposes Spartan society and Athenian society in *The Republic.* In the former, society revolves around the military, and the two are intertwined at every level from cradle to grave. In the latter, the military serves society as a distinct entity that is both apart from the rest of society but stems from it, serves it, and remains a part of society as embodied and represented by the Athenian city-state.

More recently, the field of civil-military relations has attempted to identify and describe the key elements involved in the relationship between modern military organizations and their political masters within the modern nation-state. Since the early 1950s, particularly with Samuel Huntington (*The Soldier and the State,* 1957) and Morris Janowitz (*The Professional Soldier,* 1960), the fundamental questions have revolved around how to provide for the defense of the state and concurrently keep those who dominate the use of force from applying that force to exert control over the political decisionmaking processes or to control the society within the state or both.

Two principal currents of thought have consumed most of the work on civil military relations since the 1950s in the democracies. Huntington set forth the terminology of objective and subjective control as two distinct ways to manage political-military relations. He has been viewed as the proponent of the former method of control through what he calls professionalism. Janowitz has been viewed as a proponent of subjective control. In both cases, the objectives are similar: to prevent a so-called praetorian military from dominating society and the political system.

Although either method of political control can be applied in both democratic and nondemocratic systems of government, the debate is most heated where democratic systems are at potential risk of being overthrown by military action through the use or threatened use of force. To prevent this outcome, objective control seeks to create a distribution of power between military and civilian groups that can lead to the emergence of professional attitudes and behavior among military officers. The goal is an autonomous professional military that is apolitical and focused on its missions. Subjective control also advocates a professional military, understood as expertise in the use of force, but not autonomous and uninvolved with the political dynamics of the state. Control is essentially exerted through an integration of the military into the political structures. The essential idea is, make military officers a part of the political aspects of the state, and they are unlikely to overthrow themselves. Objective control has been crudely interpreted as "give them toys and keep them busy."

Regardless of the method used, these theorists and their predecessors and successors have all worked with a complex but limited number of variables and issues to explain why militaries might or might not threaten societies and the governments that govern them and how one might prevent or reverse that threat when and where it exists. The key issues to understand are the nature of the state, military missions, and the relationship between these two.

The state is three things: it is a nation-state, a political system that governs relationships within the boundaries of the nation-state, and the government organs that physically exist. The nation-state is a psychological construct that includes concepts such as nationalism and notions with particular cultural attributes. It is also a physical entity. Armed forces' missions stem from this conception of the state. Military forces defend against aggression from other states in a world that is assumed to work in what international relations theorists call the "realist paradigm." It assumes that states in the international system work as unitary actors first and foremost to ensure their survival as states. But states are bound to clash and conflict to ensue. Thus, militaries exist to defend against other militaries when such a clash occurs.

Militaries also exist to defend the existence of the state from internal enemies. Generally, civil wars, guerrilla or separatist challenges, or other armed groups that threaten the existing system of governance are threats to the second conception of the state. The government, regardless of whether it is a democracy or not, is that part of the state that directs and regulates relations within society, between society and the state, and between state organs. The government of a state ostensibly defines what those institutions are supposed to do and also manages resources and allocates those it possesses to state institutions such as the military.

Finally, the state is physically manifested in its third dimension by state institutions such as governing bodies (legislatures, executive offices, courts) and by organizations such as the social security administration, highway department, and the armed forces.

The armed forces are directly related to the nation-state because they exist to defend its existence, and they are a real part of the state as physical entities. The government and society surround the military as an organization with interests that stem from their conception of the nation-state and what they need to do to ensure the survival of the state in all its manifestations.

Here lies the crux of political-military relations. The government allocates the resources that permit the military to succeed or fail in its missions of defense. Whether fighting an external enemy or an internal enemy, the mission is conceived as a zero-sum game. That is, either one wins or one loses. Where the use of force is involved, losing often means dying (but, then, so also can winning). Thus the government is under pressure by the controllers of the use of force to ensure that adequate resources are provided to accomplish the mission.

What that mission is and how it is conceived differently by the armed forces and by the government will greatly determine the day-to-day relationship between the military and society. Internal missions will tend to involve the military in domains within society that are not within the military's traditional competence. External missions will tend to focus the military on the expertise needed to apply force most effectively to defeat a like organization (i.e., another country's armed forces). Such a mission will tend to separate the military from society, but not in an unhealthy way.

The actual relationship between the military and society is dependent on how the military is organized, supplied with personnel, and trained and educated. The field of military sociology focuses in part on these issues. For example, Charles Moskos has written much on the nature of the U.S. military. He has looked at the results on the linkages between society and the armed forces of a conscript-based versus a volunteer-based military. He has tracked the composition of various components of the armed forces in terms of percentages of minorities, women, or people from a certain geographical region to ascertain whether manipulations of these variables affect the degree of understanding of the military by society at large or of currents within society by the armed forces. The great fear is that military and society will develop excessively distinct cultures and perspectives to the point at which they may clash. A military alienated from the society within the state it defends may become a danger to that society.

In the end, militaries reflect the societies whence they come. But they do so in the context within which they must exist. Is their mission primarily to defend from potential external threats, or is it to defend against internal threats? One nation, at least, is in the happy position of having no convincing threat. Canada, invincible, protected from external threats by the United States, and domestically relatively tranquil, is hesitant to define any mission for its armed forces beyond the benign role of peace keepers. New Zealand, enjoying even more domestic tranquility, has gone even farther than Canada in transmogrifying its armed forces into UN peace keepers, but New Zealand does face some conceivable external threats in the Pacific Rim.

The answer to the mission question will greatly determine the relationship between the military and those elements of the state that represent and manage society. How is the military to be organized, deployed, and manned? The answers to these questions will affect how the military relates with various elements of society. How is the military trained and educated? What is its cultural perspective? An understanding of these issues will in part determine the quality of the relationship of the armed forces with society.

An understanding of why the military exists, what it does, and how it accomplishes the missions laid on it by civilian policymakers is an important element in ensuring healthy civil-military relations in both the political and societal realms, most particularly in democracies.

Frédéric Ruiz-Ramón

See also: Military-Industrial Complex

References and further reading:

Finer, S. E. *The Man on Horseback: The Role of the Military in Politics.* 2d ed. Boulder, CO: Westview Press, 1988.

Harries-Jenkins, Gwyn, and Charles C. Moskos. *Armed Forces and Society.* London: Sage Publications, 1981.

Huntington, Samuel P. *The Soldier and the State: The Theory and Politics of Civil-Military Relations.* New York: Vintage Books, 1957.

Janowitz, Morris. *The Professional Soldier: A Social and Political Portrait.* New York: Free Press, 1960.

Military Justice

The law governing the army. It is also the system of justice exercised by the military over society during emergency situations of military government or martial law.

Prior to the development of societal organization, which we term *civilization,* there was armed conflict. Civilization is associated with the establishment of a system of law or conduct of behavior. Governance and the regulation of the civilian population were considered necessities for the semblance of order. However the concept of law as a defined set of principles was not applied or was considered inapplicable to the battlefield, since war was an anarchic activity and contest of strength.

Nevertheless it became apparent to rulers that in the interest of state preservation, order, and prosperity, limits should be placed upon the destructive potential, to population and property, of warfare. Prior to concerted efforts by governments/states to regulate warfare, philosophers and military theorists were social advocates providing the first notions of military law.

With the aggrandizement of Rome, the military became the foundation of the state, necessary to garrison, fortify, and defend the territory and perimeter of the empire. Julius Caesar's subversion of the Republic and its senatorial basis of government brought imperial rule where civil and military law became intertwined and almost indistinguishable. The concept of *citizenship,* that of civil rights under state protection, was extended to the barbarians of the conquered lands as a reward based upon military service within the legions of Rome. Desertion, mutiny, cowardice, violence to a superior, and the sale of arms were some of the military offenses recognized and punished by the Romans.

The collapse of the Roman Empire led to a period of anarchy in western Europe due to the absence of a strong established political order with a corresponding military enforcement. This period of medieval history, known as the Dark Ages, marked the decline of professional infantry. Petty bands of cavalry rose to prominence as powers unto themselves, able to terrorize or maintain overlordship of a given territory. Rulers/princes/lords, in order to maintain and expand their political control, instituted a system for the hire of these cavalrymen, rewarding them with land for military service. This lord-vassal relationship of military obligation for service fiefs became known as *feudalism.*

A system of military law and conduct for cavalry was established as the property and birthright of the nobility, which in this period became an exclusive military caste. This code of ethics and regulation of warfare became known as *chivalry.*

The first written military laws of Europe were within the Salic Code (circa 400) and revised by successive Frankish kings. There was no separation of civil and military jurisdiction. Civil judges were also military commanders. The first French military law (Ordonnance) was written in 1379, while the German version (Kriegsartikel) appeared in 1487. Habsburg emperor Charles V's penal code of 1532 is considered to be the model for the existing military codes of modern Europe. This system of military law, known as Carolina, was expanded and given national versions as the Articles of Gustavus Adolphus of 1621, the Regulations of Louis XIV of 1651 and 1665, the Articles and Regulations of Czar Peter the Great of 1715, and the penal code of Empress Maria Theresa of 1768.

The British military code, from which the American and Canadian counterparts originate, comprises the statute of the Army Mutiny Act and Articles of War. The Mutiny Act of 3 April 1689 was created after the desertion of Scottish troops loyal to the Stuarts who refused to obey the order of William III. Thereafter, any soldier causing mutiny or sedition in the army could be punished by death or alternate penalty judged by a court-martial. The Mutiny Act was replaced on 24 July 1879 by the Army Discipline and Regulation Act, which itself was revised as the Army Act of 27 August 1881.

The American military code differs from that of Great Britain in several ways. It does not have a Mutiny or Army Act that must be annually renewed, and although the American Articles of War are derived from the British, they are nonetheless wholly statutory, being enacted by Congress as the legislative power. American military law consists of a written and an unwritten component, the former including the statutory Code of Articles of War, other statutory enactments relating to the discipline of the army, the Army Regulations, and General and Special Orders.

After resolving to raise an army to fight the British, the Continental Congress adopted a set of Articles of War on 7 November. New articles followed on 20 September 1776, dealing with treason and providing intelligence to the enemy. After the adoption of the Constitution, the Articles were readopted by an act of 29 September 1789. A further revision occurred as the Articles of 1806. The present American Articles of War consists of 128 articles from the revised Code of 1874, which prohibited punishments such as flogging and branding.

Military justice is administered through the tribunal of court-martial. Among the Romans this consisted of justice delivered from the legionary tribunes and the Magistri Militum. The early Germanic tribes during times of peace assembled courts of free men, while in times of war a duke or military chief and priests sat in judgment. This system developed into a court of regiments, which was presided by a colonel who carried a mace (*regiment*) as his emblem of judicial authority.

Specific military courts were established in France as Conseils de Guerre in 1655. In ascending hierarchy of jurisdiction, they were the courts of the Mayor of the Palace, the Constable, and the Provost Marshal. German military courts (spear courts) were established as the Militärgerichts of Emperor Frederick III in 1487.

The British tradition of court-martial derives from the King's Court of Chivalry, alternatively known as the Court of the High Constable and Marshal of England, the Court of Arms, and the Court of Honour. Presiding were the Lord High Constable and Earl Marshal. However not until the subdivision of the tribunal system into separate courts by Edward I did the Court of Chivalry actually derive its distinct existence. Moreover, since the Court of Chivalry extended jurisdiction over civil and criminal matters, successive acts of Parliament restrained and curtailed its power until it practically ceased to exist as a military tribunal by the time of the English Revolution.

The American Continental Congress adopted the British military tribunal system of courts-martial: general, regimental, and garrison courts. The Fifth Amendment of the Constitution, through the act of 29 September 1789, made the distinction between civil offenses and those cognizable by a military forum. Courts-martial do not fall under the jurisdiction of the American judiciary as inferior courts. Rather, they are instruments of executive power provided by Congress for the president as commander in chief to enforce discipline in the army through his authorized military representatives. Therefore, court-martial is not a court by definition, but a creation by an order that is subject to a superior military body or person.

Courts-martial are not courts of record, and their judgment is simply a recommendation that is not made operative until approved by a revisory commander. The proceedings of a court-martial cannot be reviewed by an federal court. The only appeal process is via the judge advocate general to the president or the secretary of war/defense.

The British North American Act of 1867 gave the Dominion of Canada responsibility for its own defense and the maintenance of military forces during peacetime. The Dominion Parliament passed the Militia Act in 1868. Guidelines were provided in 1884 by the British Manual Military Law, whose fourteen chapters covered a history of military law, military crimes and punishments, English criminal law applicable to soldiers, courts-martial, and customs of war.

With the introduction of the National Defense Act in 1950, the Canadian armed forces obtained a national Code of Service Discipline. It was accompanied by the Queen's Regulations and Orders. The repatriation of the Constitution and the creation of the Canadian Bill of Rights and the Charter of Rights and Freedoms in 1982 provided the last amendments regarding standardization and fairness in Canadian judicial military procedure.

At present most member states of the United Nations recognize a large body of international law applied to soldiers during conflict known as the Law of War/International Humanitarian Law/Law of Armed Conflict. This governs the rights and obligations of combatants and aims to temper the destruction of war by setting limits to warfare. The Law of War also seeks to protect noncombatants.

The Red Cross conference of 1864 provided the original Geneva Convention for the protection of war victims. The present Law of Armed Conflict has three sources: the Hague Convention of 1907, which placed limits on the methods of conduct during military operations; the four Geneva Conventions of 1949, which provide the protection of wounded, sick, and POWs; and the Additional Protocols to the Geneva Convention of 1977, which further limits the use of unnecessary force causing suffering.

Additionally, there is the Martens Clause, which first appeared as a preamble to the Hague Convention II of 29 July 1899 and has been added to most international humanitarian treaties. The Martens Clause expresses the notion that there are universal minimum standards of behavior during warfare and customary law that all states recognize. But there will always be a tension between military law/justice, which provides for penalties for offenses that would be meaningless in civil society (e.g., absence without leave, conduct unbecoming an officer, adultery, and so on), and the civil law, which recognizes no such offenses. The tension becomes all the more acute when the military is flooded with mass conscript troops, fresh from civilian society.

Neville G. Panthaki

See also: Military and Society

References and further reading:

Great Britain War Office. *Manual of Military Law.* London: H.M. Stationery Office, 1968.

Madsen, Chris. *Another Kind of Justice: Canadian Military Law from Confederation to Somalia.* Vancouver: University of British Columbia Press, 1999.

Rowe, Peter. *Defense: The Legal Implications: Military Law and the Laws of War.* Washington, DC: Brassey's Defense Publisher, 1987.

Simpson, James. *Law Applicable to Canadian Forces in Somalia 1992/93: A Study Prepared for the Commission of Inquiry into the Deployment of Canadian Forces to Somalia.* Ottawa, ON: Minister of Public Works and Government Services Canada, 1997.

Winthrop, William. *Military Law and Precedents.* New York: Arno Press, 1979.

Military-Industrial Complex

In his farewell address in 1961, President Dwight D. Eisenhower warned that America must "guard against the acquisition of unwarranted influence by the military-industrial complex" (Eisenhower 1965, 616). Thus entered into the American public realm the term that captures the reality that the institutions and people for planning, procuring, and fighting a war shape the economy, the political realm, and the wider society. Although other developed nations also support extensive military structures, with the fall of the Soviet Union, the United States remains as the world's only military superpower; therefore the question of the importance and the influence of the military-industrial complex (MIC) is basically an American issue.

Advanced technological weapons and communication devises are the cutting edge for the military today. Based upon the hard lessons of World War II, the U.S. government and military assumes as a guiding tenet that it must possess the hardware for the next war before that war begins. The United States can no longer put its national defense at risk, trusting that, in the words of William Jennings Bryan, "a million men will spring to arms." The defense of the nation requires permanent armaments and a defense establishment built in the United States by American manufacturers.

This is the military-industrial complex. A constellation of

people and institutions play a guiding role in the modern United States. The constellation includes the military professionals and the Pentagon, the scientific-technological elite, the universities, and the entrepreneurs—the investing class and the corporation. Of course, money is what keeps the system operating.

A very early example of MIC occurred when Alexander Hamilton recognized that the country needed arms and armories to provide for a common defense. Seeing the expected procurement of 40,000 muskets, Eli Whitney obtained a contract to build 10,000 and succeeded in establishing the first factories making muskets with interchangeable parts. The economic boon was obvious.

The Civil War generated a massive need for arms. Consequently, arms spending increased more than a hundredfold. The procurement of cannon alone required a massive increase in coal and iron production that established Pittsburgh as the iron-making capital of the country.

By World War II, the United States had become known as the "Arsenal of Democracy," producing 300,000 airplanes, 124,000 ships of all types, 100,000 tanks and armored vehicles, and 2,400,000 trucks. The expense for this hardware reinvigorated American corporations, in the doldrums from the Depression during the 1930s.

Similarly, the success of the Manhattan Project, the most expensive wartime undertaking in American history, revolutionized the economic landscape by creating entirely new industries, including "think tanks," the infrastructure for fighting a nuclear war. Also, since World War II, "black projects" have likewise funneled billions of U.S. dollars into corporate coffers.

Thus, certain military contractors become essential to the defense of the country. For example, the United States financed the financial bailout of Lockheed Corporation in 1969 because of its importance to the national defense. There is also the reality of people moving from one institution to another in the military-industrial complex. When procurement officers in the Pentagon retire, they have an opportunity to become employed by the contractors with whom they had been dealing.

The problem that President Eisenhower noted was that the vested interests of both the military and the corporations may generate harmful outcomes as well. At present, the extremely large nuclear arsenals of the major powers are generally superfluous and thus pose more danger than deterrent effect, as even the recent U.S. commander of the intercontinental ballistic missile force publicly stated.

Similarly, there remains strong congressional support for the Strategic Defense Initiative (SDI) and for the next-generation fighter. Thus the atmosphere created by the Cold War ideology continues, after the demise of the Soviet Union, to influence decisions on weapon systems development, whether the military leadership wants them or not. The argument for such spending assumes the underlying economic reality. Without such massive spending, the needed corporate infrastructure will deteriorate quickly, perhaps leaving the United States unable to build and deploy the necessary weapon systems to defend itself against its future enemies.

Some of these debates appear in Pentagon planning for U.S. military needs. Must the armed forces be able to deploy and fight two wars simultaneously? This standard foresees simultaneous wars in the Middle East and in Asia. Military readiness is defined as being combat-ready for such wars. However, during the year 2000, the United States deployed military forces in peace-keeping roles in numerous and various localities. The weapons and training required for this task are far different.

The military-industrial complex, an amorphous collection at best, had its heyday in the 1950s and 1960s, at the height of the Cold War. It even affected population patterns, as Americans moved in large numbers from the old industrial upper Great Plains and Northeast to the Sun Belt states, where the newer, more militarily oriented industries, like aircraft and electronics, were concentrated. But in the twenty-first century, its influences are waning, as the percentage of U.S. gross national product devoted to the military steadily declines from its Korean War, post–World War II peak high. Only a changed context with a new significant military threat to the United States will reverse this trend.

John R. Popiden

References and further reading:

Eisenhower, Dwight D. *The White House Years.* Vol. 2, *Waging Peace, 1956–1961.* Garden City, NY: Doubleday, 1965.

Kapstein, Ethan Barnaby. *The Political Economy of National Security: A Global Perspective.* Columbia: University of South Carolina Press, 1992.

Koistinen, Paul A. C. *The Military-Industrial Complex: A Historical Perspective.* New York: Praeger, 1980.

Markusen, Ann R., and Sean S. Costigan, eds. *Arming the Future: A Defense Industry for the 21st Century.* New York: Council on Foreign Relations Press, 1999.

Van Creveld, Martin. *Technology and War: From 2000 B.C. to the Present.* New York: Free Press; London: Collier Macmillan, 1989.

Milne Bay (1942)

The first defeat of a Japanese invasion force. The Japanese completed the conquest of their planned "Southern Economic Zone" in the South Pacific with the capture of Rabaul in early 1942. However, they were surprised by the rapid consolidation of Allied forces in Australia and the aggressive air campaign that followed. Plans were rapidly made to

A depiction by Peter Paul Rubens of the Battle of the Milvian Bridge. (Philadelphia Museum of Art/Corbis)

capture all of New Guinea, bases in the Solomons, New Caledonia, and Fiji to isolate Australia from the United States. To this end, the Japanese sent a force overland along the Kokoda Trail toward Port Moresby in August.

An airstrip was to be built at Milne Bay, on the far eastern tip of New Guinea, to support the Port Moresby attack and provide a base for bombing northeastern Australia. Japanese intelligence suggested that only a small garrison was present. However, the Allies had already constructed airstrips at the western head of the bay, and the defenders numbered 7,500: two Australian Infantry Brigades—the veteran 18th and the 7th Militia—accompanied by 1,300 U.S. Army engineers.

The 1,200-strong Japanese Special Naval Landing Force landed on the northern side of the bay on the night of 25 August 1942 and was reinforced two nights later by 1,200 more troops. However, they landed further east than planned. P-40 fighter-bombers, operating within their own landing pattern, destroyed the Japanese shore depot on 26 August, adding to their difficulties. Australian resistance was unrelenting as they advanced westward, and in spite of nightly fire support from destroyers, the Japanese had reached only the first airfield by the night of 28 August. Fierce fighting continued at the airstrips until 31 August. Lack of success then caused Imperial Headquarters to order withdrawal, and the 1,300 survivors were evacuated by the Japanese navy on the night of 6 September.

Michael Hyde

See also: Guadalcanal; Kokoda Trail

References and further reading:

Brune, P. *The Spell Broken: Exploding the Myth of Japanese Invincibility.* Sydney: Allen and Unwin, 1997.

Japanese Monograph No. 37, "South-East Asian Operations Record: 18th Army Operations on New Guinea and Rabaul (January 1942–June 1943)"; "The Southern Area (Part II)." In *War in Asia and the Pacific,* vol. 7., ed. D. Detwiler. New York: Garland Publishing, 1980.

McCarthy, D. *Australia in the War of 1939–1945.* Series One, Army, Vol. V. *South West Pacific Area—First Year, Kokoda to Wau.* Canberra: Australian War Memorial, 1959.

Odgers, George. *Army Australia—An Illustrated History.* New South Wales: Child and Associates, 1988.

Milvian Bridge, Battle of (28 October 312)

Victory that brought Constantine to power. With troops from the garrisons of Britain, Gaul, and the Rhine, Constantine invaded Italy. His aim was to wrest power from the coemperor of the West, Maxentius, son of Diocletian's old colleague Maximian.

Victorious over Maxentius's northern forces near Turin and Verona, Constantine marched on Rome. Maxentius opted to defend the walls of Rome and thus cut the pons Mulvius, the bridge that carries the Via Flaminia across the Tiber River, on the northern approach to the city. Constantine then crossed the Tiber River on a pontoon bridge moored just downstream of the stone bridge and gave battle at Saxa Rubra. Constantine's army, although outnumbered, was battle-hardened and confident. Maxentius's army was thrown back in confusion, and as it retreated across the Tiber River, the pontoon bridge collapsed. Maxentius and his armored cavalry were drowned in the swollen river, a scene depicted on the Arch of Constantine, erected at Rome to commemorate Constantine's victory "by divine inspiration." The Senate welcomed Constantine as liberator and proclaimed him sole emperor in the West.

It was prior to the battle, so records his biographer Eusebius, bishop of Caesarea, that Constantine saw a sign in the sky, a cross of light superimposed on the sun. He took this as a sign of victory—he stated, under oath, that he saw the words "Be victorious in this" written in stars around the cross—a message from the God whose symbol was the cross. Since Constantine was heavily outnumbered, this vision may explain his bold decision to attack. He certainly put his faith to the test when he ordered his men to paint the Greek monogram for Christ (*chi-rho*) on their shields. Victorious at Milvian Bridge, Constantine continued to wear the symbol for Christ against every enemy he faced.

Nic Fields

See also: Constantine the Great

References and further reading:

Burckhardt, J. *The Age of Constantine the Great.* Trans. Moses Hadas. German original, 1898. New York: Pantheon Books, 1949.

Minamoto, Yoshitsune (1159–1189)

Principal Minamoto commander during the Gempei War (1180–1185). Minamoto was instrumental in the formation of the first shogunate at Kamakura under his half-brother, Yoritomo (1147–1199). Over the centuries, Minamoto has been transformed through poetry, stories, and *No* and *Kabuki* plays into one of Japan's quintessential tragic heroes, making history difficult to separate from legend.

Minamoto was placed in a monastery after his father, Yoshitomo (1123–1160), was killed warring against the virtual dictator of Japan, Kiyomori Taira (1118–1181), head of the rival Taira (or Heike) clan. Escaping in 1180, Minamoto joined Yoritomo's rebellion against the Taira, fighting a campaign across the island of Honshu and defeating the enemy in a series of brilliant, swift maneuvers that secured the Minamoto victory over the Taira.

In 1184, Minamoto won a decisive victory in the Battle of Ichinotani, attacking the enemy castle from the Hiyodori Impasse by leading a body of 70 horsemen down a treacherously steep mountain path reportedly used only by wild boar, deer, rabbits, and foxes. He then led a small force across the Inland Sea during a fierce storm, capturing the fortress of Yashima in March 1185. On 25 April, he crushed the Taira in the naval Battle of Dannoura at the western end of the Inland Sea. Following the battle, Taira's widow leapt into the sea with the boy-emperor, Antoku. The Sacred Sword was lost, but the other imperial regalia, the Sacred Seal and the Sacred Mirror, were recovered and returned to Kyoto.

After the war, Yoritomo grew jealous of his half-brother's success and suspicious of his close relationship with Cloistered (Retired) Emperor Go-Shirakawa. Minamoto soon rebelled, fleeing to northern Honshu, where he committed suicide in 1189.

Michael C. Paul

See also: Gempei War; Samurai

References and further reading:

Morris, Ivan. *The Nobility of Failure.* New York: Holt, Rinehart and Winston, 1975.

Shinoda Minoru. *The Founding of the Kamakura Shogunate 1180–1185.* New York: Columbia University Press, 1960.

Minden (1 August 1759)

Major battle of the Seven Years' War in Germany between the Anglo-Hanoverian-Prussian army under Ferdinand, Duke of Brunswick, and the French army under Louis Georges Érasme, Marquis de Contades.

In July 1759, Contades occupied the village of Minden on the Weser River, 30 miles west of Hanover. He wanted to concentrate his scattered forces in order to advance on Ferdinand's recently defeated allied army. Minden was a strong position, with the Weser River and marshes covering the flanks of his army of 60,000 men. Ferdinand's allied army of 45,000 men lay to the west, but that skilled general began to advance toward Contades on 31 July to force battle.

Contades broke camp that same day and arrayed his superior force to take advantage of the terrain. He placed his cavalry in the middle of his line to give still-absent detachments room to deploy on the flanks. The battle began when Charles-François, Comte de Broglie, attacked and failed to break Ferdinand's surprised left flank. Contades considered withdrawing and had no further plan for the battle beyond Broglie's attack. Ferdinand's Anglo-Hanoverian troops had advanced far ahead of the main army, which prompted the French cavalry to charge. Armed only with swords, the infantry drove the cavalry back. The French infantry advanced against the exposed allies as their cavalry reformed and enveloped the entire formation, which nonetheless held fast against the onslaught. The French center collapsed as the cavalry once again retreated. At this moment of decision, the allied cavalry failed to advance and deliver the charge. Lord George Sackville had repeatedly refused to advance, thus allowing the French army to withdraw across the Weser River. Allied casualties totaled 2,600, but the French suffered the loss of 7,000 casualties, 10,000 prisoners, and 45 guns.

Minden had the potential to be a decisive battle like Leuthen. Sackville was court-martialed for his insubordination, and the French army continued to be a threat to Hanover. Ferdinand followed close on its heels but abandoned his pursuit after Frederick the Great's disastrous defeat at Kunersdorf. As with most battles in the war, Minden proved indecisive, yet Ferdinand gained both time and space. The war continued.

Patrick J. Speelman

See also: Seven Years' War

References and further reading:

Elliott, Charles Winslow. "The Men That Fought at Minden." *The Journal of the American Military History Institute* 3, no.2 (Summer 1939), 80–103.

Mackesy, Piers. *The Coward of Minden: The Affair of Lord George Sackville.* New York: St. Martin's Press, 1979.

Savory, Reginald A. *His Britannic Majesty's Army in Germany during the Seven Years' War.* Oxford: Clarendon Press, 1966.

Minié Ball

An improved bullet that increased the effective kill range of rifles and had a dramatic impact on the casualty rate in the American Civil War. Claude-Etienne Minié, a French army officer and part-time machinist, perfected the bullet in 1849. Along with a rifled barrel that he designed, Minié invented a conoidal-shaped bullet with a hollow base that deformed and expanded when a rifle was fired. This created the much desired tight seal around the projectile that increased the accuracy and velocity of the weapon but allowed it to be easily dropped into a barrel. Rifles equipped with the new projectile had the same reloading capability as a smoothbore musket but an effective kill range four to five times that of the older weapon.

Nearly all senior ranking U.S. and Confederate officers gained their combat experience in the Mexican-American War. This war was fought with smoothbore muskets and a reliance on closed-rank linear infantry formations. The imbalance between technologically improved infantry weapons with greater velocity and accuracy, combined with tactics better suited to weapons of a previous generation, sent casualty rates soaring. (It is not for nothing that critics accuse most of the world's armies of "fighting the last war.") During the Civil War, Minié balls caused the majority of battlefield casualties. In fact, in light of those casualties, it can be argued that the Minié ball was the single greatest killer of American young men, even more so than automobiles or liquor or their combination.

Lincoln Bramwell

See also: American Civil War

References and further reading:

Brodie, Bernard, and Fawn M. Brodie. *From Crossbow to H-Bomb.* Bloomington: Indiana University Press, 1973.

Fuller, J. F. C. *Armament and History: A Study of the Influence of Armament on History from the Dawn of Classical Warfare to the Second World War.* New York: Scribner's Sons, 1945.

Jamieson, Perry D., and Grady McWhiney. *Attack and Die: The Civil War, Military Tactics and the Southern Heritage.* Montgomery: University of Alabama Press, 1982.

O'Connell, Robert L. *Of Arms and Men: A History of War, Weapons, and Aggression.* New York: Oxford University Press, 1989.

Mithradatic Wars (88–63 B.C.E.)

Series of three wars between Republican Rome and King Mithradates VI Eupator of Pontus for control of Asia Minor. Mithradates VI ascended the Pontic throne at age 11, about 121, escaped his mother's plot to kill him, and solidified his power about 114 by murdering his mother and her supporters. Always wary of conspiracy throughout his long reign, he regularly drank tiny amounts of a wide variety of poisons to immunize himself; kept a supply of antidotes handy; and is supposed to have killed his brother, three sons, and three daughters. Gradually and secretly, by murder and intrigue, he replaced neighboring kings friendly to Rome with usurpers friendly to him. His ambitions included Cappadocia to the south, Armenia to the east, the Crimea to the north, and Bithynia and Galacia to the west.

About 89, when Roman legate Marcus Aquillius refused Mithradates' request for Roman aid against the encroachment of King Nicomedes III of Bithynia into western Pontus, Mithradates decided to wage open war on Rome and its allies. Lucius Cornelius Sulla received command of the expeditionary force against Mithradates in 88, but challenges to his consulship by Gaius Marius and others delayed his operations. Meanwhile, Mithradates took the war into Greece.

Sulla recaptured Athens in 86 and defeated a larger army under Pontic general Archelaus at Chaeronea in 86 and again at Orchomenus in 86. After Lucius Valerius Flaccus won at Philippi and Gaius Flavius Fimbria captured Pergamum, the First Mithradatic War concluded in 85 with the Treaty of Dardanus.

Mithradates rebuilt his army and navy after Dardanus. A maverick Roman commander in Asia Minor, Lucius Licinius Murena, led a preemptive strike against Mithradates in 83, starting the Second Mithradatic War, which lasted only one year. Mithradates decisively defeated Murena, who was then punished by Sulla for disobeying orders and violating the treaty.

Bithynia quietly became a Roman province in the mid-70s at the bequest of Nicomedes. Now that Rome controlled the Bosporus, Mithradates again feared for the safety of his kingdom and launched an offensive. As the Third Mithradatic War began in 74, Lucius Licinius Lucullus led five legions against Mithradates, winning at Cyzicus in 73 and Cabira in 72. When Mithradates retreated and allied with Armenian king Tigranes, Lucullus invaded Armenia in 70 and defeated a vastly superior force, perhaps 100,000 Pontic-Armenians to 10,000 Romans, at Tigranocerta in October 69.

Pompey replaced Lucullus as eastern commander in late 67 or early 66 because of Lucullus's harsh leadership that led to mutinies after his victory at Artaxata. Quickly successful, Pompey compelled Tigranes to surrender in 65. The remainder of the war was a mop-up campaign against Mithradates, who fled to the Crimea, where he committed suicide. Pompey established direct Roman rule over the provinces of Asia Minor and returned to Rome in triumph in 62.

Eric v. d. Luft

See also: Alexander the Great; Caesar, Julius; Chaeronea, Battle of; Marius, Gaius; Pompey the Great; Roman Republic, Wars of the; Sulla, Lucius Cornelius

References and further reading:

Bellinger, Alfred Raymond. *The End of the Seleucids.* New Haven: Connecticut Academy of Arts and Sciences, 1949.

Duggan, Alfred Leo. *He Died Old: Mithradates Eupator, King of Pontus.* London: Faber & Faber, 1958.

Macartney, Carlile Aylmer. *Studies on Early Hungarian and Pontic History.* Aldershot, England: Ashgate, 1998.

McGing, B. C. *The Foreign Policy of Mithridates VI Eupator, King of Pontus.* Leiden: E. J. Brill, 1986.

Mogul-Persian Wars (1622–1653)

A series of limited clashes, never growing to full-scale war, over Kandahar.

Mogul-Persian War of 1622–1623

Kandahar, in what is now modern-day Afghanistan, had represented a source of tension and rivalry between the Safavid Persian and Mogul Empires since Akbar (1542–1605) had acquired it for the Moguls when two Safavid princes defected in 1595. Patiently building his forces and taking advantage of division within the Mogul court, Shah Abbas "the Great" (1571–1629) personally led his forces against Kandahar in the winter of 1622. After a 45-day siege, the ill-prepared 300-man Mogul garrison surrendered to the Safavids. Shah Abbas then seized control of the fortress, town, and province of Kandahar before the Mogul emperor, Jahangir (1569–1627), could marshal an army to relieve the garrison. Jahangir planned to recapture Kandahar. He sent his son Khurram (1592–1666), who would later become Shah Jahan, to recapture Kandahar. However, before the Moguls could do so, Jahangir fell seriously ill, leaving his wife, Nur Jahan, and his son Khurram locked in a dynastic struggle for control of the Mogul Empire. The Moguls finally sent a force against the Safavids in 1623, but by that time the Mogul force that reached Kandahar was too weak to recapture the province.

Mogul-Persian War of 1638

Although initially distracted by the dynastic struggle within the Mogul Empire, Shah Jahan saw an opportunity to recover Kandahar in 1638. Ali Mardan Khan, a Persian noble and commander of the fortress, feared his life was in danger from the capricious Safavid emperor, Shah Safi (1629–1642), and so surrendered the fortress to Shah Jahan without bloodshed. As a reward for his defection, Ali Mardan Khan received a substantial monetary reward and a political appointment within the Mogul Empire. Upon reacquiring the fortress, Shah Jahan began to bolster the fortifications of Kandahar.

Mogul-Persian War of 1648–1653

In the decade after Ali Mardan Khan's defection, the Safavid emperor, Shah Abbas II (r. 1642–1666), regarded Mogul military setbacks in the Balkh region against the Uzbeks as a sign of weakness. In the winter of 1648, sensing his opportunity, Shah Abbas II sent an army into the region to retake the fortress at Kandahar. After a two-month siege, the Mogul garrison surrendered. Although aware of the Safavid attack, Shah Jahan's advisers convinced him that a winter campaign to relieve the city was unwise. This decision gave precious time to the Safavids to reinforce their recent reconquest.

Shah Jahan mounted three unsuccessful campaigns to retake Kandahar between 1649 and 1653. The Moguls mounted their first campaign in the summer of 1649, under the leadership of his son, Aurangzeb. Commanding a force of

50,000 men, Aurangzeb laid siege to the fortress but withdrew because he could not defeat the Safavid garrison before the onset of winter. The Moguls did not again try to take Kandahar until the fall of 1652, when they again laid siege to the fortress. Although the Moguls repulsed a relief force sent by the Safavids, they could not complete their siege of the fortress. The Moguls made their final attempt to recapture Kandahar in the spring of 1653. Led by Shah Jahan's favorite son, Dara Shukoh, the latest Mogul force came close but could not fully penetrate the defenses. Dara Shukoh employed siege guns, which breached some of the fortress walls but were not enough to force the capitulation of Kandahar by the winter. Foul weather and thinning supply lines forced the Moguls to withdraw from the battlefield.

The Moguls' Kandahar campaigns were a failure. The estimated deaths of 30,000–40,000 soldiers were unable to achieve for the Moguls what diplomacy and bribery had earlier achieved. Throughout the battles for Kandahar, the Moguls favored bows and arrows over artillery and firearms. Indeed, the Moguls considered archers to be the most prestigious of warriors during this period. At the same time, the Safavids consistently outgunned the Moguls. Safavid artillery was accurate, reliable, and inflicted very heavy casualties upon the Moguls during their sieges. Strangely, the Moguls felt little incentive to invest in systematic efforts to develop an army backed by gunpowder. In any case, Kandahar remained a part of the Safavid Empire until the beginning of the eighteenth century.

Eric D. Pullin

See also: Akbar the Great; Abbas the Great; Aurangzeb

References and further reading:

Habib, Irfan. *The Agrarian System of Mughal India, 1556–1707.* New York: Oxford University Press, 1999.

Richards, John F. *The Mughal Empire.* Cambridge: Cambridge University Press, 1993.

Zaman, M. K. *Mughal Artillery.* Delhi: Idarah-I Adabiyat-I Dellii, 1983.

Mohács, Battles of (29 August 1526, 12 August 1687)

Turkish-Hungarian battles. The first battle of Mohács took place on 29 August 1526. Over 30,000 Hungarians under King Louis II and Bishop Tomore made a heroic stance to defend Hungary against the Turks under Süleyman the Magnificent. The Turks numbered more than 100,000 troops and had 300 guns that proved decisive in routing the Hungarians, who lost 22,000 casualties, their king, many clergy and nobility, and eventually control of their capital at Budapest. The sultan went on to besiege Vienna, although unsuccessfully, at the height of Turkish power in eastern Europe.

The second battle took place 160 years later and marked a reversal of fortune for the Turks. A combined Austro-Hungarian force crushed the Turks under the Sultan Mohammed IV on 12 August 1687. The sultan was deposed by the soldiery and succeeded by Süleyman III. The Turks were badly overstretched, fighting wars against Venice, Russia, and the Holy League with Austria and Hungary at the same time. Turkish power, which had seen a brief military renaissance with this war against Austria, would never recover.

Christopher Howell

See also: Austro-Turk Wars; Süleyman I; Vienna, Sieges of

References and further reading:

Goodwin, Jason. *Lords of the Horizons: A History of the Ottoman Empire.* New York: Owl Books, 2000.

Turfan, Naim. *Rise of the Young Turks: Politics, the Military and Ottoman Collapse.* Istanbul, London: I. B. Tauris & Company, 2000.

Molnar, Miklos, and Anna Magyar, trans. *A Concise History of Hungary.* London: Cambridge University Press, 2001.

Mohi or Sajo River, Battle of (April 1241)

Battles in which the Mongols destroyed the Hungarian army and ransacked the kingdom. After conquering Russia, Mongol armies led by Sübedei and Batu invaded eastern Europe in February 1241, advancing as five separate but closely coordinated forces to deceive their enemies, who remained unclear regarding the Mongols' real targets until the last minute. Two advanced into Poland, one into Bohemia, and two into Hungary. At this time, Hungary, which appears to have been the primary Mongol target with its good pastures, possessed perhaps the finest army in Europe. Initially invaded by only a part of the Mongol army, it was soon invaded by the rest, after the Mongol victory against the Saxons at Liegnitz, in Silesia.

Although King Bela IV of Hungary had fortified the passes of the Carpathian Mountains, the Mongols had broken through by 14 March 1241. On 9 April 1241, King Bela advanced with an army that may have numbered 70,000 men, although there is some question whether or not he had his entire army with him because of a well-managed Mongol campaign of misinformation, including false mobilization orders. In response, the Mongols withdrew before the Hungarians for several days, until they had led the Hungarians to the plain of Mohi, between the Sajo and Tisza Rivers. The Hungarians camped in the plain, unaware that the Mongols had specifically chosen this site as a battlefield.

When the Mongols advanced to the Sajo River, prepared to do battle, Bela formed his wagons into a circle, thus fortifying them against a sudden cavalry charge. He stationed 1,000 men at the only bridge to prevent a crossing.

Around dawn, Batu attacked the bridge with archers and a rolling barrage of catapults firing naphtha. The Hungarians retreated from the bridge before the Mongols. In the ensuing melee, both sides suffered heavy casualties.

Sübedei, who had meanwhile crossed the river farther upstream on pontoons, then appeared behind the Hungarians, forcing them to fall back on their camp, which the Mongols surrounded, leaving a gap on the western flank, and bombarded with catapults and arrows. Eventually, the Hungarians detected the opening and soon poured from their camp in that direction, often dropping their weapons as they ran. As the unarmed Hungarians fled, the Mongols now wheeled upon them and slaughtered the fleeing men.

The Mongols continued the pursuit for three days, ravaging Hungary. Bela IV barely escaped, and only after a long flight into the southern Balkans with Mongol forces close behind. After taking up residence in Hungary, even issuing coins and dispatching raids as far as the suburbs of Vienna, the Mongols then abruptly withdrew in late 1241 and early 1242 because of the death of Great Khan Ögödei. Hungary remained a shadow of its former might for decades.

Timothy May

See also: Genghis Khan; Mongol Empire; Ögödei

References and further reading:

Chambers, James. *The Devil's Horsemen.* New York: Atheneum, 1985.
Grousset, Rene. *The Empire of the Steppes.* Translated by Naomi Walford. New Brunswick, NJ: Rutgers University Press, 1970.
Hildinger, Erik. *Warriors of the Steppe.* New York: Sarpedon, 1997.
Marshall, Robert. *Storm from the East.* Berkeley: University of California Press, 1993.

Moltke, Graf Helmuth Johannes Ludwig von (1848–1916)

Born at Gersdorf, Mecklenburg, on 23 May 1848, Helmuth Moltke joined the army in 1869. He was a nephew of Helmuth von Moltke ("the Elder"), who commanded Prussian armies to victory over Austria and France. Although similar greatness was expected for the younger Moltke, he unfortunately possessed little of his uncle's innovative military genius.

Moltke joined the Prussian army in 1869. He served as adjutant to his uncle and the kaiser and held a variety of field commands. Promoted to colonel (1895) and brigadier general (1899), in 1900 he became a major general commanding the 1st Guards Division. He was then quartermaster general (1903). Kaiser Wilhelm II named him chief of the

Engraved portrait of General Helmuth J. L. von Moltke. (Library of Congress)

general staff in 1906, succeeding General Alfred von Schlieffen. Moltke accepted the post with reservations, knowing that he was incapable of quick decisions.

As German army chief of staff, Moltke believed war with the Entente powers was inevitable and pushed for it to be sooner rather than later. His major contribution to World War I was his ill-conceived revision of the Schlieffen Plan. Because he feared a French thrust into Alsace and Lorraine, he strengthened the German left wing at the expense of the right, which made it much more difficult for the right wing's encircling movement to succeed. Then with the offensive already under way, on 25 August he exacerbated matters by taking two corps and a division from the right wing and sending them east against the Russians. Moltke exercised little leadership during the fighting and, on 14 September following the critical Battle of the Marne, he was relieved of his post and demoted to deputy chief of staff. Moltke died suddenly of a heart attack in Berlin on 18 June 1916.

Spencer C. Tucker

See also: Moltke, Graf Helmuth Karl Bernhard von; Schlieffen, Graf Alfred von

References and further reading:

Craig, Gordon A. *The Politics of the Prussian Army, 1640–1945.* London: Oxford University Press, 1964.
Görlitz, Walter. *History of the German General Staff, 1657–1945.* New York: Frederick A. Praeger, 1953.

Moltke, Graf Helmuth Karl Bernhard von (1800–1891)

Prussian field marshal and architect of Prussia's victories in the nineteenth-century wars of German unification. Born on 26 October 1800 in Parchim, Mecklenburg, Helmuth von Moltke joined a Danish regiment on graduation from the Royal Cadet Corps in Copenhagen. After visiting Berlin in 1821, he decided to enter the Prussian army. Not wealthy, Moltke supplemented his income through writing.

Moltke attended the Prussian War Academy during 1823–1826 and joined the Prussian General Staff in 1833. Sent to Turkey in 1835 to study the language and advise the sultan on military matters, he entered the Turkish service contrary to instructions and campaigned in Egypt and Syria, distinguishing himself in the Battle of Nezib (Nizip) in July 1839.

Returning to Prussia in 1839, Moltke was aide-de-camp to Prince Henry of Prussia and then rejoined the general staff. Promoted to colonel in 1851, he was then aide-de-camp to Prince Frederick William (later Kaiser Frederick III). He traveled to Britain and Russia and in October 1858 became chief of staff of the Prussian Army.

Moltke and Minister of War Albrecht von Roon helped secure the appointment of Otto von Bismarck as minister-president of Prussia in 1862. These three men worked closely in the cause of German unification, reforming and increasing the size of the regular army. Moltke, who had written about railroads, well understood the implications of the railroad and telegraph for modern war. He soon reorganized the general staff into three geographical divisions and a railways department.

In 1864, Moltke oversaw Prussian military operations during a war with Denmark. His success in that campaign earned him the complete support of Kaiser William I. Moltke then drew up plans for the 1866 war against Austria, which involved a quick campaign of rapid concentration in which three armies would advance on different routes by rail, their movements coordinated by telegraph. Although communication broke down, Moltke nonetheless won a decisive victory over the Austrians in the Battle of Königgrätz (Sadowa). This Seven Weeks' War ended 120 years of rivalry between Prussia and Austria for domination in the Germanies.

Moltke noted flaws and corrected them in the final war of German unification, against France in 1870–1871. Although he allowed his subordinates considerable latitude, Moltke directed the military campaigns of the war, including the September 1870 Battle of Sedan, the October 1870–January 1871 Siege of Paris, and the final battles against the French Army of the Loire.

Advanced to both graf (count) and field marshal in June 1871 in recognition of his accomplishments, Moltke continued as chief of the general staff. In the early 1880s, he began to turn over responsibility to others, retiring altogether in 1888. He died in Berlin on 24 April 1891. Moltke wrote a number of books, most of them on military subjects.

Spencer C. Tucker

See also: Franco-Prussian War; German Wars of Unification; Königgrätz, Battle of; Sedan

References and further reading:

Craig, Gordon A. *The Politics of the Prussian Army, 1640–1945.* New York: Oxford University Press, 1956.

Howard, Michael. *The Franco-Prussian War: The German Invasion of France, 1870–1871.* New York: Macmillan, 1962.

Moltke, Helmuth K. B. von. *Strategy: Its Theory and Application: The Wars for German Unification, 1866–1871.* Westport, CT: Greenwood Press, 1971.

Wawro, Geoffrey. *The Austro-Prussian War. Austria's War with Prussia and Italy in 1866.* New York: Cambridge University Press, 1996.

Mongol Empire (1206–1259)

The world's greatest steppe empire was traditionally founded in 1206, when Genghis Khan (c. 1162–1227) was first elected khan; expansion had begun even before that date with raids on the northern Chinese state of Xixia. They continued with other raids on Xixia, and against the Chinese Jin dynasty, culminating in the capture of the Jin capital by the Mongols in 1215. In 1217, a new stage was reached as preparations were made for a general assault on the Khwaraz Empire of western Turkistan and Iran in response to mistreatment of some envoys under Mongol protection. After initial moves to shore up positions in eastern Turkistan and in the north, the main Mongol advance began in mid-1219. The Khwarazmians, outflanked on several fronts, quickly collapsed, and their last ruler died on an island in the Caspian Sea in late 1220, seeking to avoid the relentless Mongols.

Freed by the death of their adversary, the Mongols now pressed on into Iran. From there, Jebe and Sabutai (1172–1245) mounted their famous reconnaissance in force around the Caspian Sea (1221–1223). They reunited with other Mongolian forces under Prince Joci, eldest son of Genghis Khan, in late 1223 or 1224.

The death of Genghis Khan in 1227, while subduing Xixia, interrupted the Mongol advance, but it was quickly resumed under his successor. Ögödei (r. 1229–1241) focused Mongol efforts in two directions. One was China, where the Jin had reestablished themselves along the Yellow River. The other was toward the distant west, where Jebe and Sabutai had fought a battle against the Russians in 1223.

The Jin campaign began in 1231, after Ögödei had put his financial house in order. In a pattern typical of Mongolian warfare, separate Mongolian armies closed in on the Jin capital and forced the last Jin emperor to flee. He committed suicide in 1234, ending the dynasty. Although new hostilities developed with Jin's southern neighbor, Song China, that were to continue until that state was finally subdued in 1279, the major Mongol advance south was, for the time being at least, ended.

Ögödei's second major campaign began in 1235, with the convening of a council to discuss what was to be done about the west. A campaign was decided on, and armies began to move toward the Kipchak Steppe and Russia in 1236. They were under the titular command of Batu Khan, son of Joci, but Sabutai took overall strategic control. Initially, the Mongols attacked the Volga Bulghars and then the various Qipchaq tribes. Next was Russia: Ryazan was stormed on 21 December 1237, followed by Vladimir on 7 February 1238. After a pause for consolidation, the Mongols moved west again. Kiev fell on 6 December 1240, and from Kiev the Mongols advanced into eastern Europe.

Although the main Mongol target appears to have been Hungary, which had received some refugees fleeing the Mongols, the assault was a general one, with no less than five major lines of advance intended to prevent any single adversary from uniting against them. The following winter, the Mongols took the double cities of Buda and Pest, while raiding parties penetrated as far as the suburbs of Vienna. Only the death of Ögödei saved Europe from further invasion.

After Ögödei's death, there was an interregnum of nearly five years with few major endeavors. Although Ögödei's son Güyük was finally elected khan in 1246, he died two years later, and the interregnum continued. It was ended in 1251 by a veritable coup that brought a new Mongol house to the throne, that of Genghis Khan's youngest son, Tolui (c. 1190–1231/1232), in the form of Khan Möngke (r. 1251–1259), the last of the four khans to preside over a unified empire.

Under Möngke, the Mongolian advance resumed in two primary directions, against China through the southwest in an effort to outflank the Song and into Iran. The leader of the Iranian advance was Möngke's younger brother Hüle'ü (d. 1265), later founder of the Ilqan dynasty.

After Möngke, there were pretenders to the vacant throne of the khan, principally his brothers Kublai (1215–1294) and Arigh Böke (d. 1266), but none was able to impose himself on the other Mongols. Serious antagonisms between the various Mongol houses also prevented any reconciliation and a reuniting of empire.

Paul D. Buell

See also: 'Ayn Jalut, Battle of; Genghis Khan; Kublai Khan; Ögödei

References and further reading:

Allsen, Thomas T. *Mongol Imperialism.* Berkeley: University of California Press, 1987.

Buell, Paul D. "Sübötei-ba'atur." In *In the Service of the Khan: Eminent Personalities of the Early Mongol-Yuan Period (1200–1300),* ed. Igor de Rachewiltz, Chan Hok-lam, Hsiao Ch'i-ch'ing, and Peter W. Geier, 13–26. Wiesbaden: Otto Harrassowitz, 1993.

———. "Early Mongol Expansion in Western Siberia and Turkestan (1207–1219): A Reconstruction." *Central Asiatic Journal* 36 (1992): 1–2, 1–32.

Ratchnevsky, Paul. *Genghis Khan, His Life and Legacy.* Trans. Thomas Nivison Haining. Oxford, UK: Blackwell, 1991.

Rossabi, Morris. *Khubilai Khan: His Life and Times.* Berkeley: University of California Press, 1988.

Mongol-Song Wars (1267–1279)

The Mongols' crushing of the Song Dynasty. Kublai Khan waited for some years to consolidate his power before beginning his final assault on the surviving Chinese state, Southern Song (1126–1279). The campaign began in 1267, went on for 12 years, and was one of the most sophisticated, protracted, and hard-fought in history, not only because of the size of the armies involved, but also because of the rich pyrotechnical resources available to both sides and the sophistication of torsional artillery by that date.

The Mongols, to the greatest degree possible, sought to provoke a war of maneuver that the Song, always short of horses, were ill-equipped to counter. The Song, by contrast, relied upon enormous armies sited in fixed fortresses, knowing that the Mongols had either to assault these fortresses, advance through unsuitable terrain in central and south China, or attempt a naval envelopment, unlikely in the face of a technologically superior Song force that included some of the world's largest warships, also armed with pyrotechnical weapons.

The Mongols, well aware of Song plans, sought to counter them by using Chinese tactics against the Song fortresses, relying upon what was now a substantial siege train and large forces of Chinese foot soldiers. They were also aware of the need not just to advance but to defeat the Song morally, and thus set about destroying Song field armies piece by piece in a strategy reminiscent of Ulysses S. Grant's before Richmond in 1864–1865.

The lynchpins of the Song position in defending their capital at Hangzhou were the dual fortresses of Fancheng and Xiangyang, facing each other across the Han River, directly athwart the best line of Mongol advance. It took Kublai Khan's armies, led capably by Marshal Bayan (1237–1295), six years to reduce them, and it was late 1274 before Mongol armies finally penetrated to the Yangtze River, but Song resistance continued, and it was another year and a half before the Song capital was reached. Only after a massive land and sea battle in March 1275 and a protracted last-ditch resistance did the inevitable become clear to the Song court, which surrendered to the Mongols, who entered Hangzhou peacefully on 10 February 1276.

Before the Mongols could complete their occupation, loyalists fled with two young princes and took to the seas, supported by a still-powerful Song fleet. What followed was a remarkable resistance movement paralleling that of 1126–1230, in which Southern Song had once saved itself in the face of an equally determined invasion by the Jurchen of the Jin Dynasty (1122–1234). Thanks to unexpected problems encountered by Kublai Khan due to an invasion of his Mongolian homeland by Central Asian competitors, the resistance movement, based first in the province of Fujian and then in Guangdong and Guangxi, was able to rally much of the southeast and interior central China to the Song.

In the end, the Mongols had to launch a coordinated land and sea campaign to overcome Song resistance. Advancing simultaneously into the various regions supporting resistance along several land routes, they also used increasingly capable naval forces, added to as the campaign went on, in part through Song commanders going over to the Mongols, to seize coastal points behind loyalist lines. By late 1278, loyalist efforts had become confined to the province of Guangdong and to surviving units of the Song fleet carrying the surviving Song prince (one had died).

Increasingly outmaneuvered and isolated, the Song navy prepared for a final battle in early 1279, at Yaishan, an island located not far from modern Macao. Comprised of perhaps 1,000 oceangoing junks plus smaller supporting ships, the Song fleet, giving up all mobility, was drawn up in a long rectangle, tied to each other, with sterns outward. Wooden palisades were built on top so that not only the sailors, but Song land forces then on board, could participate in the battle. To a large extent, this arrangement was more a reflection of Song desperation and low morale than it was of a sound tactical judgment. It also reflected a realistic appraisal of increasingly numerous and very well handled Mongol naval forces.

Because of limited supplies of food and wood, which was important for making arrows, the Song fleet was anchored close to land. Its position gave Li Heng, one of the Mongol commanders, the opportunity to seize positions there and mount catapults to bombard the Song with stones and incendiaries and possibly exploding bombs.

On 19 March 1279, the Mongols were ready. Their fleet was at most half the size of the Song fleet, and apparently some ships arrived too late to participate in the battle. Mongol ships were also smaller, but their size proved an advantage in the waters in which the battle was fought, midway between two islands. Actually, despite the large number of ships involved, the ensuing battle was fought more as a land conflict than a naval battle.

The battle went on most of the day, with the Mongol fleet using its mobility to the utmost and the Song unable to respond since their ships were tied to one another. Later, the Mongols were able to penetrate the Song line and begin operating within the Chinese rectangle. As the afternoon tide came in, the Mongols used its force to charge in again and boarded several Song ships, braving the Greek fire.

By late afternoon, the Song rectangle was in disarray, and ships began to surrender. Some 800 ships are said to have been taken. Some 100,000 corpses were left floating in the water, among them that of the last Song prince. Not only had the Mongols won a notable victory, but they had bested the Song in their own element, on the water.

Paul D. Buell

See also: Chinese Imperial Wars; Kublai Khan; Mongol Empire; Song-Jin Wars

References and further reading:

Buell, Paul D. "The Sung Resistance Movement, 1276–1279: The End of an Era." *Annals of the Chinese Historical Society of the Pacific Northwest* 3 (1985–1986), 138–186.

Deng, Gang. *Maritime Sector, Institutions, and Sea Power of Premodern China.* Westport, CT: Greenwood Press, 1999.

Mote, F. W. *Imperial China, 900–1800.* Cambridge, MA: Harvard University Press, 1999.

Monmouth (27–28 June 1778)

A drawn battle in the American Revolution that marked the last major clash in the northern states. Faced by imperial obligations in the expanding conflict, the British decided to withdraw from Philadelphia to New York City. British general Henry Clinton began to march the 70 miles to Sandy Hook, New Jersey, and the protection of the Royal Navy. At the same time, American commander George Washington wanted to demonstrate the new American army forged over the winter at Valley Forge. The battle would take place during the British retreat and American pursuit at Monmouth Court House.

For a month, it was clear the British were making preparations to quit Philadelphia. Washington intended to pursue and catch them. The retreat was difficult in very hot and humid weather. The Americans met the British on 27 June, but General Charles Lee hesitated to attack. The next day, the British began moving to Sandy Hook and, to cover the retreat, readied to attack the American vanguard; meanwhile, it seemed that Lee had ordered a retreat, which was countermanded heatedly by Washington when he arrived on the battlefield.

The British attacked three times. The newly trained American army held its ground, and as more troops arrived, it pushed the British back. By evening, both sides were exhausted from the fighting and from the weather. Washington rested his troops, and Clinton continued his withdrawal to Sandy Hook and the safety of the Royal Navy. Although the Americans may have missed an opportunity to destroy the British, the battle made clear that what had been a ragtag group of militia had become a trained army and that the war of attrition increasingly favored the American republic.

Charles M. Dobbs

See also: American Revolution; Washington, George

References and further reading:

Smith, Samuel Steele. *The Battle of Monmouth.* Monmouth Beach, NJ: Philip Freneau Press, 1964.

Stryker, William. *The Battle of Monmouth.* Princeton, NJ: Princeton University Press, 1970.

Mons Graupius, Battle of (September 83)

Agricola's great victory over the Caledonians. The culmination of Tacitus's eulogy for his father-in-law describes the Battle of Mons Graupius. There have been many attempts to locate the site, but Agricola's marching camps near Huntly, Speyside, are the only pointers.

Assembled under the leadership of Calgacus (Swordsman), "the full force" of the Caledonians, 30,000 warriors, occupied Mons Graupius. The size of Agricola's army is not given, but Tacitus does say the enemy had "great superiority in numbers." Agricola certainly had 8,000 auxiliaries and probably 5,000 cavalry, together with vexillations from the four legions of Britain. The total force numbered some 20,000. Agricola placed the auxiliary infantry in the center, with their ranks spread out, and 3,000 cavalry on the wings. The vexillations were to the rear, drawn up in front of the Roman camp. The Caledonians were deployed in tiers on the gentle slope, with the vanguard on the level ground.

The battle began with Caledonian chariots racing across the ground between the two armies, only to be routed by the Roman cavalry. Next came a brisk exchange of missiles, followed by the Roman advance up the slope. The auxiliaries were initially successful and were soon joined by the cavalry.

The sheer number of Caledonians, combined with the roughness of the ground, soon halted the advance. Gradually, the auxiliaries began to be outflanked. As a counter, Agricola sent in his reserve cavalry. They stemmed the flanking movement and fell on the rear of the war bands, which broke. The legionaries were never engaged.

Nic Fields

See also: Agricola, Gnaeus Julius; Celts

References and further reading:

Maxwell, G. S. *A Battle Lost: Romans and Caledonians at Mons Graupius.* Edinburgh: Edinburgh University Press, 1989.

Montcalm-Gozon, Louis-Joseph de, Marquis de Montcalm de Saint-Véran (1712–1759)

French general who, according to his admirers, almost prevented a British takeover of Canada during the Seven Years' War and, according to his detractors, was the primary cause of it. Before he was given command of French regular forces in North America during the Seven Years' War (1756–1763), Montcalm had served in the War of the Polish Succession (1733–1738) and the War of the Austrian Succession (1740–1748). He arrived in New France in May 1756 and soon found himself at odds with the colony's Canadian-born governor, Pierre de Rigaud, the Marquis de Vaudreuil. Despite their differences, they managed to take Fort Oswego on Lake Ontario from the English. The following year, Montcalm captured Fort William Henry on Lake George. Even after the British government had determined to win the war in North America and sent over thousands of regular troops for the purpose, Montcalm's successes continued. In July 1758, he prevented a large British force from taking Fort Carillon (Ticonderoga) on Lake Champlain.

Elsewhere, the tide of war began to turn. Facing what he considered overwhelming odds, Montcalm wanted to use his regulars to defend the St. Lawrence Valley against the inevitable British onslaught, whereas Vaudreuil persisted in his plans for a continued guerrilla war, using the Canadian militia. Although the government in Paris sided with Montcalm and gave him command of all forces in New France, ample opportunity remained for Vaudreuil to frustrate Montcalm's efforts.

Throughout the summer of 1759, Montcalm managed to thwart General James Wolfe's British regulars from getting below the walls of Quebec. When at dawn on 13 September, however, the British surprised him by scaling the escarpment just west of the fort, Montcalm precipitously gave battle. He lost and died of his battle wounds the day after. Quebec surrendered soon thereafter. The war in North America was not yet lost, but given France's reluctance to come to the aid of its colony, the fate of New France was sealed.

N. F. Dreisziger

Montcalm trying to stop Native Americans from attacking the British as they leave Fort William McHenry. (Library of Congress)

See also: Austrian Succession, War of the; French and Indian War; Seven Years' War; Wolfe, James

References and further reading:

Eccles, W. J. "Montcalm." *Dictionary of Canadian Biography.* Vol. 3. Toronto: Oxford University Press, 1974.

Frégault, Guy. *Canada: The War of the Conquest.* Toronto: Macmillan, 1969.

Stacey, C. P. *Quebec 1759. The Siege and the Battle.* Toronto: McClellan, 1959.

Stanley, G. F. G. *New France: The Last Phase, 1744–1760.* Toronto: McClelland and Stewart, 1968.

Montecuccoli, Raimondo, Prince (1609–1680)

Italian commander in the service of the Habsburg emperors Ferdinand III (r. 1637–1657) and Leopold I (r. 1658–1705). One of the foremost generals and military theorists of the seventeenth century, Montecuccoli fought in the Thirty Years' War (1618–1648), the Second Northern War (1655–1660), the Austro-Turk War (1660–1664), and the Dutch War (1672–1678) against the numerous enemies of the house of Austria.

Montecuccoli entered military service as a child and progressed rapidly through the ranks. In 1664, he became supreme commander of the imperial armies and in 1668 president of the Supreme War Council. Four years earlier, he had thwarted Grand Vizier Fazil Ahmet Köprülü's plan to march against Vienna, for which he was acclaimed as "the savior of Christendom." The middle period of the Dutch War saw him pitted against worthy opponents: Louis XIV's great generals Henri de la Tour d'Auvergne, Vicomte de Turenne (1611–1675), and Louis II de Bourbon, fourth prince de Condé (1621–1686). Montecuccoli relinquished his command in 1675 and retired.

Whether in captivity or on peacetime duty and after his retirement, he studied the sciences, classical authors, and the art of war. His most influential works on that subject are *Dell'arte militare* and *Memoire della guerra.* Though above all a soldier, Montecuccoli on occasion also served as the Habsburg Court's diplomatic emissary. A military reformer, an advocate of "methodical warfare," and a skilled practitioner of the war of maneuver, Montecuccoli was a master of seventeenth-century warfare. He is the author of the famous dictum: to wage war, three things are needed—money, money, and money.

N. F. Dreisziger

See also: Condé, Louis II de Bourbon, Fourth Prince de; Louis XIV; Thirty Years' War; Turenne, Henri de la Tour d'Auvergne, Vicomte de; Turkish Wars of European Expansion

References and further reading:

Barker, Thomas Mack. *The Military Intellectual and Battle: Raimondo Montecuccoli and the Thirty Years' War.* Albany: State University of New York Press, 1975.

Campori, Cesare. *Raimondo Montecuccoli: La sua famiglia e i suoi tempi.* Florence: G. Barbera, 1876.

Montecuccoli, Raimondo. *Opere di Raimondo Montecuccoli.* Turin: G. Favale, 1821.

Perjés, Géza. *Army Provisioning, Logistics, and Strategy in the Second Half of the 17th Century.* Budapest: 1970.

Tomassini, Luciano. *Raimondo Montecuccoli: Capitano e scrittore.* Rome: Stato Maggiore Dell'esercito, Ufficio Storico, 1978.

Monterrey (20–24 September 1846)

Major U.S. Army victory in the Mexican-American War. At the opening of the Mexican-American War (1846–1848), American general Zachary Taylor advanced with an army of 3,080 regulars and 3,150 volunteers from the town of Matamoros on the south bank of the Rio Grande to Monterrey, defended by 7,000 regulars and 3,000 militia under the command of General Pedro de Ampudia. Taylor arrived on the northern outskirts of Monterrey on 19 September and found the approach guarded by a massive citadel known as the Black Fort. Despite the Mexican's numerical advantage, he decided to divide his army into two wings and dispatch one wing to the west of the city and the other to the east. On 20 September, the west wing, commanded by William J. Worth, swung to the west, avoiding the guns of the Black Fort. The following day, Worth led his men across the Saltillo road, forded the Santa Catarina River, turned east, and stormed Federacon Hill, capturing Fort Solidado on the eastern end of the promenade. On 21 September, the eastern wing of Taylor's army, commanded by Colonel John Garland, attacked the stoutly defended fortification, El Teneria, and took it but was unable to dislodge the Mexicans from another strategic position, the earthwork fort, El Diablo. Two days later, Worth's men recrossed the Santa Catarina River, stormed Independence Hill, and took the Bishop's Palace. Worth's men next advanced into the western end of the city. On the eastern side of Monterrey, the Mexicans abandoned El Diablo and retreated to the central plaza and the cathedral. General Ampudia requested a truce on 24 September, but Taylor did not agree until the next day. Without consulting President James K. Polk, Taylor agreed to allow Ampudia and his men to evacuate the city. American loses were 120 killed and 368 wounded, and Mexican casualties totaled 367.

George M. Lauderbaugh

See also: Mexican-American War; Santa Anna, Antonio López de; Taylor, Zachary

References and further reading:

Eisenhower, John S. D. *So Far from God: The U.S. War with Mexico, 1846–1848.* New York: Random House, 1989.

Singletary, Otis. *The Mexican War.* Chicago: University of Chicago Press, 1960.

Smith, George W., and Charles Judah, eds. *Chronicles of the Gringos: The U.S. Army in the Mexican War, 1846–1848.* Albuquerque: University of New Mexico Press, 1968.

Smith, Justin. *The War with Mexico.* New York: Macmillan, 1919.

Montgomery, Bernard Law (1887–1976)

British field marshal best known for his victory at El Alamein, 1942. Born on 17 November 1887 in London, Montgomery entered Sandhurst in 1907. His military talent was recognized in World War I, and he became a general staff officer in January 1917. After attending the staff college at Camberley in 1920, he went on to hold staff and command posts in Ireland, England, Palestine, and India and served as an instructor at the staff colleges at Camberley and Quetta.

During the Battle of France (May–June 1940), Montgomery commanded the 3rd Division energetically, and during the final days of the Dunkirk evacuation, he briefly commanded II Corps. Back in England, he was promoted to lieutenant general, becoming commander of V Corps. After the dismissal of General Claude Auchinleck, Montgomery was selected to command the Eighth Army in North Africa, halting Erwin Rommel's forces at the Battle of Alam Halfa (31 August–2 September 1942) and then grinding down the Germans' tank strength at El Alamein, the battle that turned the tide in the desert. Finally, here was a British general who could win victories, albeit with massive superiority over the enemy, and "Monty" became a popular figure with the British masses.

After commanding the Eighth Army in Italy, Montgomery took over command of the Twenty-first Army Group, which played a central role in the western European campaign of 1944–1945. Montgomery favored a single thrust to defeat Germany but, like George Patton and Omar Nelson Bradley, was forced to bow to Dwight D. Eisenhower's broad front strategy.

Montgomery was an efficient general, albeit extremely cautious, possibly remembering horrific British casualties in World War I. But he could be tactless and arrogant, often expressing the need to teach the "naive" Americans battle truths. The latter, bearing in mind Singapore, Greece, Crete, and so on, were not about to be so "instructed." Montgomery was chief of the Imperial General Staff from June 1946 to November 1948 and deputy supreme Allied commander in Europe from 1951 to 1958. He died at Islington Mill on 24 March 1976.

Alaric Searle

See also: El Alamein; MARKET GARDEN; Normandy Landings

References and further reading:

Hamilton, Nigel: *Monty: The Making of a General, 1887–1942.* New York: McGraw-Hill, 1981.

———. *Monty: The Final Years of the Field-Marshal, 1944–1976.* New York: McGraw Hill, 1987.

Montgomery, Field Marshal Bernard Law. *Memoirs.* London: Collins, 1958.

Montmorency, Anne, Duc de (1493–1567)

The constable Anne de Montmorency was a distinguished warrior during his long life, serving five French kings. During the first half of the sixteenth century, as the French monarchy struggled to gain control of France and its nobles, the Valois kings faced considerable opposition to their attempts to centralize royal authority. The death of Henry II in a jousting accident in 1559 began a long period of royal weakness that would not end until the Bourbon Henry of Navarre took the throne in 1589. The Duke of Montmorency, constable of France, had immense landholdings and a personal following of several hundred vassals. When the Huguenot Montmorency converted other major nobles of France to the new religion, they became a dangerous political threat to the Catholic Valois. Montmorency is a good example of how the French kings managed to control their nobles by offering them titles in exchange for loyalty.

Montmorency was made a marshal in 1522 by Francis I and was captured with Francis at Pavia in 1525. He helped negotiate Francis's release in 1526 and soon after the king's return received the governorship of Languedoc, which remained in his family until 1632. He was made constable in 1537. Montmorency's enemies at court and his policy of peace with Holy Roman Emperor Charles V finally led to his disgrace in 1541, which lasted until Francis's death in 1547.

Henry II restored him to a degree of favor, and in return he took Metz from the Spanish in 1552. Dismissed by his successor, Francis II, Montmorency was restored to office by Catherine de Medici. He joined the Guises in the Wars of Religion, was captured at Dreux in 1562, and was killed in the siege of St. Denis, near Paris, in 1567.

David C. Arnold

See also: French Wars of Religion; Guise, François de Lorraine, Second Duke of; Pavia, Battle of; Valois-Habsburg Wars

References and further reading:

Briggs, Robin. *Early Modern France, 1560–1715.* London: Oxford University Press, 1998.

Hartmann, Cyril Hughes. *The Magnificent Montmorency: The Life and Death of Henri Duc de Montmorency, 1595–1632.* New York: E. P. Dutton, 1928.

Palm, Franklin Charles. *Politics and Religion in Sixteenth-Century France: A Study of the Career of Montmorency-Damville, Uncrowned King of the South.* Gloucester, MA: Peter Smith, 1969.

Montrose, James Graham, Marquis of (1612–1650)

Leading royalist general of Scottish "Covenanter wars." James Graham was born in 1612 of a prominent Lowlands noble family. His brief career played out against the complex religious, social, and political turmoil of the seventeenth-century British Isles.

The Scottish Reformation conveyed the Roman Catholic Church's considerable estate into the power of a nobility, many of whom saw profit in Protestantism. Siding first with the Covenanters against Charles I's attempts to impose episcopacy, the young Montrose soon rejected the rebellion's religious absolutism and political chicanery. In his struggle to win adherents to the Royalist cause, he advocated a monarchy strong enough to uphold fundamental liberties against factionalism and fanaticism.

Obtaining Charles's commission, Montrose raised loyalist Irish and highland Scots and during his "year of miracles" (1644–1645) virtually destroyed the Covenant's larger armies with superior tactics and leadership, until his dwindling force was surprised at Philiphaugh. Montrose escaped, while his men were slaughtered after surrendering. On the king's orders, he disbanded his remaining followers and went to Europe to raise support. After Charles I's execution, he landed in the Orkney Islands, but his small force of foreign regulars and local levies was routed at Carbisdale on 27 April 1650. Montrose was captured, hanged, and quartered without a trial in Edinburgh on 21 May.

Despite repeated victories against great odds, his mostly irregular clansmen, who fought fiercely under his personal leadership, could not hold territory or establish a lasting power base. Without effective and timely support by the Royalist leadership, his victories were as ephemeral as they were remarkable, and his legacy was more one of personal example than military effectiveness.

Anne L. Angstadt

See also: Cromwell, Oliver; English Civil War (1642–1649)

References and further reading:

Buchan, John. *The Marquis of Montrose.* Reprint, London: Prion, 1996.

Wedgwood, C. V. *Montrose.* New York: St. Martin's Press, 1995.

Mormon War (1838–1839)

Major nineteenth-century American internal conflict. The first "Mormon War" began in northwestern Missouri in 1838, when members of the Church of Jesus Christ of Latter-Day Saints (LDS, or Mormons) battled units of the Missouri state militia. This conflict escalated into a massacre of Mormons at Haun's Hill, Missouri, and prompted governor Lillburn Boggs to issue an October 1838 extermination order against Mormons, calling for their execution if they did not leave Missouri immediately.

The Mormons understandably fled to Hancock County, Illinois, where they founded Nauvoo, which became the biggest city in Illinois. Local residents lynched LDS founder Joseph Smith and his brother Hyrum in 1844, prompting a Mormon exodus to Mexico's Alta California Province, beginning in 1847. The province was transferred to the United States following the Mexican-American War and was renamed Utah. Conflict quickly developed between LDS leaders and federal officials and grew into the second Mormon War, or the Utah War of 1857–1858.

Lance Janda

See also: Utah War

References and further reading:

Furniss, Norman F. *The Mormon Conflict, 1850–59.* New Haven, CT: Yale University Press, 1960.

Hafen, LeRoy R., and Ann W. Hafen. *The Utah Expedition: A Documentary Account.* Glendale: Arthur H. Clark Company, 1958.

LeSuer, Stephen C. *The 1838 Mormon War in Missouri.* Columbia: University of Missouri Press, 1987.

Mortars

A muzzle-loaded weapon firing its bomb at a high angle to attack protected positions and trenches. The mortar has been known for many centuries and was most effective as a naval weapon for shore bombardment and as a land weapon for attacking the defenders behind walled defenses. It is the artillery of the infantry, supporting them at all levels from section up to regiment, and infantry have always held a healthy respect for their opponent's mortars.

Mortars were prevalent in the trench warfare of World War I, firing high-explosive (HE), smoke, and gas shells. Some were rifled and had separate firing mechanisms for safety. In World War II, infantry mortars ranged in caliber from 5 centimeters (cm) (used for smoke and illumination mainly) to 12 cm (firing mostly HE bombs).

Mortars nowadays have ranges up to 10 kilometers and more, and a good mortar team can have many bombs in the air at once, adding to the effect on target. Bombs are HE, smoke, illuminating, and antitank. Mortars are used mainly for indirect fire against targets unseen by the mortar teams.

The mortar is usually intended to be man-portable, although some heavier versions are vehicle-mounted. Every infantry battalion has mortar support, giving fire for attack and defense. Mortars are normally grouped in pairs and sets of pairs in battle and can operate in the front line from trenches because of their high angle of fire. Each mortar has a baseplate, a bipod (or similar support) with sights attached, and a barrel, and each segment is normally man-portable. Ammunition supply is paramount, for mortars can achieve a very high rate of fire; in battle, many infantrymen will carry one mortar bomb plus their own equipment and spare machine gun ammunition.

David Westwood

References and further reading:

Hobart, Frank William Arthur, ed. *Jane's Infantry Weapons.* London: MacDonald and Jane's, 1975.

Freytag, Viktor. *Infanteriegechuetz und s GrW 34.* Berlin: Mittler and Sohn, 1939.

Mosby, John Singleton (1833–1916)

Confederate cavalry commander and raider and leader of Mosby's Rangers. Mosby was born in Edgemont, Virginia, on 6 December 1833 and practiced law in Bristol, Virginia, after 1855. Enlisting in the Confederate cavalry early in the war, he saw action at First Bull Run and then, under James Ewell Brown "Jeb" Stuart, in the Peninsular campaign and at Second Bull Run and Antietam. Stuart ordered Mosby to create a free-moving cavalry unit in December 1862 to harass the enemy. This unit, Mosby's Rangers, was so successful that the area it controlled between the Potomac and Rappahannock Rivers became known as "Mosby's Confederacy."

Raiding usually at night in small groups armed with Colt 44 pistols, Mosby's men diverted thousands of federal combat troops to guard duty. Their most daring exploit was the

Soviet troops manuever a piece of anti-tank artillery through the snow in the German attack on Moscow, 1941. (Hulton/Archive)

capture of Brigadier General Edwin Stoughton at Fairfax Court House, Virginia, on 8 March 1863. Their presence prevented flank attacks as Robert E. Lee marched from Chancellorsville to Gettysburg. By late 1863, several federal cavalry units had standing orders to pursue Mosby full-time. Mosby eluded them all, and on 18 November 1864 killed or wounded all but two of the handpicked 100 men that Philip Sheridan had sent after him.

Mosby surrendered only after word came of Joseph E. Johnston's surrender. He practiced law in Warrenton, Virginia, until his death on 30 May 1916.

Eric v. d. Luft

See also: American Civil War; Antietam/Sharpsburg; Bull Run, First/Manassas; Bull Run, Second/Manassas Junction; Chancellorsville, Battle of; Gettysburg; Johnston, Joseph Eggleston; Lee, Robert Edward; Sheridan, Philip Henry; Stuart, James Ewell Brown

References and further reading:

Evans, Thomas J., and James M. Moyer. *Mosby's Confederacy: A Guide to the Roads and Sites of Colonel John Singleton Mosby.* Shippensburg, PA: White Mane, 1991.

Jones, Virgil Carrington. *Ranger Mosby.* McLean, VA: EPM, 1987.

Ramage, James A. *Gray Ghost: The Life of Col. John Singleton Mosby.* Lexington: University Press of Kentucky, 1999.

Siepel, Kevin H. *Rebel: The Life and Times of John Singleton Mosby.* New York: Da Capo Press, 1997.

Moscow (30 September 1941–April 1942)

First land defeat of the Wehrmacht during World War II, marking the failure of the German Barbarossa campaign to defeat the USSR in a single operation. Adolf Hitler's Directive 21, Operation BARBAROSSA, listed Moscow as a secondary objective after the capture of Leningrad and Kiev. On 19 July 1941, because of the nonuniform advance of the three German army groups (North, Center, and South), Directive 33 instructed Panzer Group 2 to be redeployed from Center to South to assist in the capture of Kiev (21 August–27 September). At Leningrad, the German assault became a siege.

The first phase of Operation TYPHOON, the attack on Moscow, was mounted from 30 September to 30 October.

World War II: Barbarossa, 1941
Stalin Line
Front Line, 21 June 1941
Front Line, 9 July 1941
Front Line, 1 Sept. 1941
Front Line, 30 Sept. 1941
German Routes of Attack
Miles 0 100 200
Kms 0 100 200
Northwest Front (Voroshilov)
West Front (Timoshenko, Zhukov)
Southwest Front (Budenny, Timoshenko)
Army Group North (Leeb)
Army Group Center (Bock)
Army Group South (Rundstedt)
SOVIET UNION
ESTONIA
LATVIA
LITHUANIA
EAST
POLAND
GERMANY
UKRAINE
ROMANIA
CRIMEA
Baltic Sea
Gulf of Finland
Gulf of Riga
Black Sea
Sea of Azov
Lake Ladoga
Lake Peipus
Lake Ilmen
Volga River
Oka River
Don River
Volkov River
Dvina River
Niemen River
Beresina River
Pripet River
Bug River
Vistula River
Desna River
Dnieper River
Dniester River
Prut River
Danube River
Helsinki
Hango
Viipuri
Leningrad
Narva
Tallinn
DAGO
OESEL
Tartu
Luga
Novgorod
Staraya Russa
Demyansk
Pskov
Kholm
Kalinin
Rzhev
Klin
Moscow
Velikiye Luki
Riga
Nevel
Beloj
Mozhaisk
Vyazma
Kaluga
Tula
Dvinsk
Vitebsk
Smolensk
Memel
Kaunas
Orsha
Roslavl
Bryansk
Minsk
Mogilev
Vilna
Orel
Konigsberg
Suvalki
Grodno
Livny
Voronezh
Kursk
Gomel
Glukhov
Bialystok
Warsaw
Brest
Sumy
Knontop
Kharkov
Korosten
Kowel
Kiev
Lubny
Lublin
Poltava
Rowne
Zhitomir
Kremenchug
Cherkassy
Stalino
Tarnopol
Uman
Lwow
Dnepropetrovsk
Kirov
Przemysl
Mariupol
Nikopol
Melitopol
Cernauti
Nikolayev
Perekop
Kerch
Odessa
Feodosiya
Yevpatoriya
Sevastopol
Yalta

The German force amounted to 74 German divisions, including 14 panzer divisions of Panzer Groups 2, 3, and 4. They were opposed by the Western, Briansk, and Reserve Fronts, a Soviet force totaling 1.25 million men.

The Germans launched pincer movements to encircle 81 Soviet divisions in two separate pockets, Viazma and Briansk. Eliminating these pockets stalled the German advance until the second phase of Typhoon (15 November–8 December), the attempt to capture Moscow via pincer attacks from Klin and Tula. The German logistical situation was weakened, many battle formations were severely under strength and unprepared for winter, and equipment was wearing out.

On 5 December, the Soviets launched a counterattack planned by Marshal Georgy Zhukov, employing a strategic reserve of 10 armies, mostly fresh Siberian troops. The Soviet industrial plants evacuated in July–November began supplying the improved weapons from their reconstructed locations in the Urals. The German were sent reeling back.

By 6 January 1942, the Germans had retreated beyond the Volkhov River along the Orel-Rzhev axis. The Soviets expanded their action into a counteroffensive, attempting to destroy Army Group Center and break through to Leningrad. They did not succeed, although Viazma and Rostov-on-Don were liberated by the time the Soviets halted their operations in March 1942.

Neville G. Panthaki

See also: Guderian, Heinz; Hitler, Adolf; Kiev; Konev, Ivan Stepanovich; Leningrad, Siege of; Rokossovsky, Konstantin Konstantinovich; Smolensk; Stalin; Timoshenko, Semen Konstantinovich; World War II; Zhukov, Georgy Konstantinovich

References and further reading:

Erickson, John. *The Road to Stalingrad.* New York: Harper & Row, 1975.

Fugate, Bryan, and Lev Dvoretsky. *Thunder on the Dnepr.* Novato, CA: Presidio Press, 1997.

Overy, Richard. *Russia's War.* London: Penguin Books, 1998.

Werth, Alexander. *Russia at War, 1941–1945.* London: Pan Books, 1964.

Moscow, Retreat from (19–23 October 1812)

The abandonment of a city whose capture Napoleon had thought would provide supplies, shelter, and an armistice. On 13 September 1812, the Russian commander, Prince Mikhail Kutuzov, decided to abandon Moscow, retreating from Fili southeast along the Kolomna road. The population evacuated, with only 25,000 of Moscow's 250,000 inhabitants remaining. The city's mayor, Count Fyodor Rostopchin, arranged for arsonists to set the city ablaze. From 15 to 18 September, four-fifths of Moscow was destroyed.

Napoleon expected the Russian nobility to greet him and negotiate an armistice. His appeals (20 September, 4 October, 14 October) to Kutuzov and Czar Alexander I were rebuffed. Left with a ruined city, without stores, supplies, or winter quarters, Napoleon decided to abandon Moscow on 19 October. The French rearguard remained until 23 October.

Napoleon had brought with him 87,500 infantry, 14,750 cavalry, 533 guns, and an enormous baggage train consisting of 40,000 wagons, most of which was loot. The French faced Kutuzov and 100,000 infantry, 20,000 Cossacks, and 600 guns. Theoretically, Napoleon had provisions for 20 days and fodder for a week. The partisans menaced French logistics so that convoys from Smolensk had to be guarded by 1,500 men.

The French marched at 10 miles per day on the Kaluga road. Russian forces pursued and forced battle on 25 October at Maloyaroslavets. The ensuing battle cost the Russians 7,000 men and the French 4,000. Napoleon now decided to retreat to Smolensk via Borovsk-Mozhiask-Gzhatsk-Viazma.

Kutuzov followed Napoleon on a parallel route, Medyn-Smolensk, permitting only limited engagements and allowing the fatigue and famine of withdrawal to destroy the French. At Viazma on 3 November, 20,000 cavalry and infantry attacked Louis Nicolas Davout's rearguard and separated it from the main French army, while Michel Ney was also under attack.

By 7 November, it was snowing heavily. Maintaining a march pace of 12 miles per day, 41,000 French reached Smolensk on 9–12 November. Napoleon learned that the Russian armies of the Baltic and Danube were converging to cut off his retreat at Borisov on the Berezina River. Napoleon's retreat from Moscow would soon become a rout.

Neville G. Panthaki

See also: Berezina River, Battle of; Borodino; Kutuzov, Prince Mikhail Illarionovich Golenishchev; Napoleon I

References and further reading:

Austin, Paul. *1812: The Great Retreat.* London: Greenhill Books, 1996.

de Ségur, Philippe-Paul. *Napoleon's Russian Campaign.* Boston: Houghton-Mifflin, 1980.

Nafziger, George. *Napoleon's Invasion of Russia.* Novato, CA: Presidio, 1988.

Nicolson, Nigel. *Napoleon 1812.* New York: Harper & Row, 1985.

Riehn, Richard. *1812: Napoleon's Russian Campaign.* New York: John Wiley and Sons, 1991.

Mount Badon, Battle of (c. 490–516)

Unknown location of a famous victory by the Britons over invading Anglo-Saxons. After 410, the last Roman legions were withdrawn, and Britain was left to fend for itself. During the rest of the century, large numbers of Germanic peoples (Angles, Saxons, Jutes, and others) poured into Britain and established foundations for their kingdoms. For several years after midcentury, the invaders gradually attempted to expand but met strong resistance from the Britons. During a lengthy period of warfare, both sides alternately experienced defeats and victories.

Gildas, a native Briton living just after that period of conflict, wrote a brief account of the Battle of Mount Badon (Mons Badonicus). The site of the three-day siege is unknown, and the date is uncertain. Gildas implied that the British leader was Ambrosius Aurealianus, the last Roman leader in Britain. According to Gildas, the victory provided 40 years of comparative peace. The ninth-century chronicler Nennius tells us that the Battle of Mount Badon was the twelfth and last battle fought by a British captain named Arthur. Nennius describes how Arthur fought the Saxons in the company of the British kings and how 960 Saxons were killed by Arthur at Mount Badon. Living a generation after Ambrosius and more chronologically correct, Arthur has been generally credited with the victory at Badon. Tradition and scholarship have also suggested that he utilized cavalry against Saxon foot soldiers.

Mount Badon was probably a hill-fort in southern England. Scholars have favored Badbury, Bath, and Solsbury as the location. The dating of the battle falls broadly between the years 490 and 516.

Brigitte F. Cole

See also: Ancient Warfare

References and further reading:

Alcock, Leslie. *Arthur's Britain.* London: Penguin, 1971.
Gildas. *Works.* Trans. Hugh Williams. London: Cymmrodorian, 1899–1901.
Nennius. *History of the Britains.* Trans. E.W. Wade-Evans. London: Church History Society, 1938.

Lord Louis Mountbatten (right) discusses the war with General George Patton at Camp Anfa, near Casablanca, 1943. (Library of Congress)

Mountbatten of Burma, Louis Francis Albert Victor Nicholas (1900–1979)

Very influential royally related British naval and land commander and final viceroy of India. Mountbatten joined the Royal Navy in 1913 and saw action aboard capital ships *Lion* and *Queen Elizabeth* during World War I. Continuing in the Royal Navy until World War II, Mountbatten was promoted to the rank of captain and given command of the Fifth Destroyer Flotilla in 1939. In May 1940, a German E-boat torpedoed Mountbatten's own destroyer, the *Kelly,* in the North Sea. Luftwaffe Ju–87 Stuka dive-bombers sank the *Kelly,* still under Mountbatten's command, at Crete on 23 May 1941. Noel Coward's play and film, *In Which We Serve,* was based on this action.

In the summer of 1941, Prime Minister Winston Churchill promoted Mountbatten to chief of combined operations; as such, he was responsible for preparing the Allied invasion of Europe. As part of the preparations, Mountbatten developed the organizational framework for Operation OVERLORD. He also planned the commando raids, among them the disastrous Dieppe raid of 19 August 1942, which were a series of probing missions designed to assess the requirements of the future invasion force, gauge the strength of German coastal defenses, and raise morale among citizens of German-occupied countries. Churchill appointed Mountbatten as supreme commander of the Allied forces in Southeast Asia in August 1943.

Mountbatten's experience in Asia during World War II and his somewhat left-wing politics (as much as would be feasible for one of such blue blood that he looked down on the royal Windsors) made him the ideal choice of Britain's ruling Labour Party to succeed General Archibald Wavell as India's final viceroy on 22 March 1947. After India's independence in August 1947, he served as India's first governor general until 1948. Mountbatten became fourth sea lord in 1950, commander in chief of the Mediterranean Fleet in 1952, and first sea lord in 1955. He served as chief of the United Kingdom Defense Staff and chairman of the Chiefs of Staff Committee from 1959 to 1965. The Irish Republican Army murdered Mountbatten and several members of his family by a remotely detonated bomb on his boat in Donegal Bay, Ireland, on 27 August 1979.

Eric D. Pullin

See also: Churchill, Sir Winston; Crete; Dieppe; Normandy Landings; Wavell, Archibald Percival, First Earl; World War I; World War II

References and further reading:

Campbell-Johnson, Alan. *Mission with Mountbatten.* New York: Atheneum, 1985.
Hough, Richard. *Mountbatten.* New York: Random House, 1980.
Mountbatten, Vice Admiral the Earl of Mountbatten of Burma. *Report.*
Thorne, Christopher. *Allies of a Kind.* New York: Oxford University Press, 1978.
Ziegler, Philip. *Mountbatten.* New York: Alfred A. Knopf, 1985.

Mountjoy, Charles Blount, Lord (1562–1606)

Most successful Elizabethan military commander in Ireland. Charles Blount, Lord Mountjoy, repulsed a Spanish intervention at Kinsale, successfully ending nine years of war that followed the rebellion of Hugh O'Neill, the Earl of Tyrone.

Mountjoy's military career began in 1586, when he joined the earl of Leicester's forces fighting the Spanish in the Netherlands. He distinguished himself at the Battle of Zuthpen in 1587 and was subsequently knighted. In the next few years, he learned the art of war by serving with Sir John Norreys in Brittany but was forbidden further field service by Elizabeth I, when he succeeded his brother as the Eighth Baron Mountjoy in 1594. Despite the queen's admonition to learn about war from books, Mountjoy took part in the Cadiz expedition under the earl of Essex. His first substantial command came in 1600, when he was appointed lord deputy of Ireland in place of Essex, who had been disgraced by embarrassing defeats at the hands of O'Neill. Though physically weak and considered a hypochondriac, Mountjoy displayed an intuitive grasp of the central problem of Irish warfare, supply. He launched a policy of campaigning in the winter, which prevented the Irish from moving their cattle, and ordered the destruction of agricultural areas that supported the rebel forces.

His first offensive against O'Neill foundered against Irish fortifications in the indecisive Battle of Moyry Pass in October 1600, but he won a decisive victory at Kinsale in late 1601. In the later campaign, Mountjoy's brilliance as a logistician was confirmed when he managed to deploy and supply a large army with siege equipment in an isolated part of Ireland. He subsequently negotiated O'Neill's submission in early 1603. His health broken by extended campaigning in the Irish climate, Mountjoy died of illness in 1606.

John S. Nolan

See also: Kinsale, Siege of; Nine Years' War

References and further reading:

Falls, Cyril. *Mountjoy: Elizabethan General.* London: Odhams Press, 1955
Fissel, Mark Charles. *English Warfare, 1511–1642.* London: Routledge, 2001.
Wernham, R. B. *The Return of the Armadas.* Oxford, UK: Oxford University Press, 1994.

Mozambican War of Independence (1963–1974)

The Portuguese occupation of Mozambique was a particularly grim affair, relying heavily on slave and forced labor. Resistance to colonial rule ranged from work stoppages to outright armed resistance. All of these factors coalesced in 1962 with the formation of the Front for the Liberation of Mozambique, or Frelimo, which comprised exiled political groups, radical intellectuals, and underground organizations that had begun to operate within the country.

To understand the situation in Mozambique, one has to understand that with numerous challenges facing it, the Portuguese government of Antonio Salazar tied its prestige to the fate of its African colonies. Much the same way that France viewed Algeria as an integral part of the republic, Portugal had declared its African holdings to be "overseas provinces."

Portugal had been quite effective in keeping a tight grip on potential resistance movements in Mozambique. During the Angolan Luanda insurrection of 1961, there were, surprisingly, no similar rural uprisings in Mozambique. Much of this lack of opposition was due to the activities of the state security police, which had been systematically detaining people known to be hostile to the regime. As a result, the most effective resistance movement flourished outside the country. In 1962, Frelimo was founded in Tanzania, led by the charismatic Eduardo Mondlane, an academic and official at the United Nations in New York. On 25 September 1964, Frelimo launched an attack on the Portuguese in northern Mozambique, at the same time issuing a call to arms.

The Portuguese set the tone of the conflict by rounding up and detaining some 1,500 Frelimo activists in the cities, putting to an end any hope for an Algerian-style protracted urban conflict. Frelimo southern and central fronts having virtually collapsed under the weight of the Portuguese attacks, the focus turned to the north, where the Portuguese forces were minimal. However, the successes of late 1965 were soon reversed; in 1966 the Portuguese counterattacked along Lake Malawi. With the policy of grouping the population into camps, some 250,000 people were resettled in 150 villages by the end of the year, removing much of Frelimo's support base.

By 1968, the war had focused on the ambitious Portuguese civil works project, the Cabora Bassa dam on the Zambezi River. Even though Frelimo's attacks on the fortified construction sites were unsuccessful, the protection of the dam absorbed a great deal of effort and rendered the Portuguese forces static, allowing Frelimo for the first time to outflank the defenses and begin operations south of the Zambezi River.

After the death of Salazar and Marcello Caetano's succession in 1968, Portugal launched a counteroffensive. The military planned a massive sweep in the north of Mozambique in which reconnaissance and airborne assault would search out and destroy Frelimo. The group was taken by surprise by the scale of the Portuguese operation, but rather than try to hold on to the north, Frelimo withdrew troops and transferred them, through Malawi, to Tete in the center. Portugal's grand offensive succeeded, but Caetano was not pleased with the cost of the campaign; Lisbon halted further offenses.

The reopening of the Tete front proved to be a breakthrough for Frelimo. It moved units south of the Zambezi River in 1971, and in July 1972, for the first time, guerrilla activity threatened an important section of the settler population. These campaigns were psychologically successful and in the period 1972–1974 contributed to a rapid crumbling of the colonial structure.

In Portugal, there was a growing sense that an African empire was passé in the twentieth century. With the Portuguese military coup of 1974, the new government sought immediately to end the hostilities and remove all Portuguese military units from the country. On 7 September, the Lusaka Accord was signed, allowing for the rapid and unequivocal transfer of power to Frelimo, and on 28 September 1974, the granting of Mozambique independence was signed, making it an independent country with Frelimo in command.

James Corbin

See also: Angolan Civil War; Angolan War of Independence

References and further reading:

Ciment, James. *Angola and Mozambique: Postcolonial Wars in Southern Africa.* New York: Facts on File, 1997.

Finnegan, William. *A Complicated War: The Harrowing of Mozambique.* Berkeley: University of California Press, 1993.

Newitt, Malyn. *A History of Mozambique.* Bloomington: Indiana University Press, 1995.

Muhammad Ahmad (al-Mahdi, Muhammad Ahmad Ibn As-Sayyid' Abd Allah) (1844–1885)

Muslim holy man from Dongola in the Sudan who declared himself the "Mahdi," or "Guided One," and raised the banner of revolt against the excesses of the Anglo-Egyptian administration in the Sudan.

Born 12 August 1814, Muhammad Ahmad was deeply religious from childhood. At age 20, he was already a sheikh with a reputation for sanctity. His call for a jihad in 1881 to restore Islam to its pristine purity received enthusiastic response and attracted both Muslims and non-Muslims alike. The violence that accompanied the conquest of the Sudan engendered deep-seated hatred for the Turco-Egyptians. Charles Gordon's efforts to suppress the slave trade in the 1870s were bitterly resented by Muslim traders. The cattle-owning nomads of Kordofan smarted under heavy taxation, and pious Muslims were scandalized by the corruption and lack of Islamic observance by Turkish and Egyptian officials. The increasing use of European officials in the administration accentuated the loss of independence and emphasized the alien quality of the Egyptian presence.

Muhammad Ahmad knit together these disparate groups in the Sudan through his piety and stirring oratory. To the people of the Nile banks, his message was oppression of the tax collector; around the plains of Gedir, it was defilement of an ancient faith by cowardly Turks and renegade foreigners ousting the rightful owners of the land.

Ahmad and his closest advisers acted as a high command, disseminating propaganda and articulating the discontent in the Sudan. From 1881 through 1885, the bravery and fervor of the Mahdist forces won impressive victories over the better-armed Egyptian armies sent against them. In 1885, they took Khartoum, killing Gordon and the remnants of the Egyptian soldiers and officials in the city. (Gordon's death was taken as a personal blow by Queen Victoria.) Five months later, Muhammad Ahmad died, leaving the organization of the Mahdist state to his closest disciple, Abdallahi.

Edmund Abaka

See also: Gordon, Charles George; Wolseley, Garnet Joseph, Viscount

References and further reading:
Holt, Peter Malcolm. *The Mahdist State in the Sudan 1881–1898.* Oxford, UK: Clarendon, 1970.
———. *A History of the Sudan from the Coming of Islam to the Present Day.* London: Weidenfeld & Nicolson, 1979.
Wingate, Francis Reginald. *Mahdism and the Egyptian Sudan.* London: Macmillan, 1968.

Muhammad Ali (c. 1770–1849)

Ottoman *wali* (governor) regarded as the founder of modern Egypt. Born in Kavala, Greece, around 1770, Muhammad Ali came from a family of Albanian soldiers long in Ottoman service. In 1801, he sailed to Alexandria with troops sent to repel Napoleon's invasion of Egypt. Exploiting ethnic and political rivalries, he obtained military command, deposed the *wali* Kurshid Pasha, and grudgingly was confirmed as *wali* of Egypt by the Ottomans in 1805. He consolidated his rule by massacring the Mamluks, exiling dissident religious leaders, and revoking tax-farming privileges for the Egyptian elite.

Through conquest and modernizing reforms, Muhammad Ali tried to establish Egypt as the foremost power in the eastern Mediterranean. At the Ottoman government's request, his armies invaded the Hijaz in 1811, quelling the Wahhabi revolt there by 1818. From 1820 to 1822, Muhammad Ali conquered the northern Sudanese territories of Nubia, Sennar, and Kordofan, using the captives to construct a new army trained by a French officer, marking a developing relationship between Egypt and France. Native Egyptians replaced the Sudanese slaves, who rapidly died from exhaustion and disease. Led by his son Ibrahim Pasha, they fought Greek rebels in Crete, Cyprus, and the Morea throughout the 1820s, capturing Athens in 1827. But the destruction of the Ottoman-Egyptian fleet obliged their withdrawal the following year.

Unhappy at not receiving compensation for his services in Greece, Muhammad Ali attacked the Ottoman Empire itself three years later, capturing Palestine and Syria by 1833. The considerably weakened Ottoman state caused Britain, which feared disruption of communications with India and encroaching Russian and French interests in the Near East, to demand his withdrawal. Ottoman forces unsuccessfully counterattacked in 1839. Subsequently, British forces occupied Beirut and shelled Acre, compelling the Egyptians to withdraw and negotiate. With the London Convention of 1841, Muhammad Ali exchanged his Levantine acquisitions for Ottoman recognition of Egypt as an autonomous province ruled by a hereditary *wali,* namely himself. Increasingly mentally incompetent, Muhammad Ali abdicated in 1847 and died in Alexandria on 2 August 1849. At least nominally, his dynasty governed Egypt until it was toppled by Gamal Abd al-Nasir in 1952.

Ian Janssen

See also: Alexandria; French Revolutionary Wars; Greek War of Independence; Mamluks; Ottoman Empire

References and further reading:
Lawson, Fred. *The Social Origins of Egyptian Expansionism during the Muhammad 'Ali Period.* New York: Columbia University Press, 1992.
Marsot, Afaf. *Egypt in the Reign of Muhammad Ali.* Cambridge, UK: Cambridge University Press, 1984.
Sabini, John. *Armies in the Sand: The Struggle for Mecca and Medina.* London: Thames and Hudson, 1981.

Muhammad of Ghur, Conquests of (1175–1206)

The Ghurid conqueror of northern India, who was among the founders of Muslim rule in India. Muhammad of Ghur, also called Muhammad Ghuri or Shihab-ud-din Muhammad Ghuri, followed his brother, Ghiyas-ud-Din, who had acquired power east of Herat in the region of Ghur (present-day Afghanistan) in 1162. Among his early military expeditions, Muhammad Ghuri assisted his brother against the Oguz Turkmen nomads to regain control of the city of Ghazna (Ghazni) in 1173 and against the ruler of Khwaraz in gaining the former Seljuq holdings in the region of Khorasan. In 1204, two years after Ghiyas-ud-Din's death, Muhammad Ghuri made yet another successful attack on the Khwaraz capital of Gurganj (in present-day Uzbekistan). His victories in Hindustan were also significant, including the capture of Uch and Multan in 1175 and the annexation of the Ghaznavid principality of Lahore in 1186. In addition, he answered his rare defeat by the coalition of Rajput kings under Prithviraj III at Taraori in 1191 with a resounding victory in the following year at the same site and another at Chandawar in 1194. These two battles are considered most decisive in the course of Muslim history, and as a result, the Ghurid forces occupied Delhi in 1192–1193. Over the next two decades, the entire region of northern India also fell under their control. Muhammad Ghuri left much of the control of his territories in north India with his lieutenant Qutb-ud-Din Aybak and his armies. Aybak is credited with building some of the most important Muslim monuments in northern India, which emulate aspects of Ghaznavid and Ghurid architecture in several ways. In addition, the Ghurid soldiers retained their political connections with Ghur until Delhi was established as the permanent capital under Iltutmish (r. 1211–1236).

Manu P. Sobti

See also: Akbar the Great

References and further reading:

Bosworth, Clifford E. "The Early Islamic History of Ghur." *Central Asiatic Journal* 6 (1961).

Fischer, Klaus. "From the Rise of Islam to the Mongol Invasion." In *The Archaeology of Afghanistan from the Earliest Times to the Timurid Period*, ed. F. Raymond Allchin and Norman Hammond. London: Academic Press, 1978.

Moline, Jack. "The Minaret of Gam (Afghanistan)." *Kunst des Orients* 9 (1975).

Scarcia, Gianroberto, and Maurizio Taddei. "The Masjid-i Sangi at Larvand." *East and West* 23 (1973).

Muhlberg, Battle of (24 April 1547)

The only battle of the Schmalkaldic War. In the spring of 1547, Emperor Charles V turned against the Elector John Frederick of Saxony, one of the main leaders of the Schmalkaldic League. By the fall of 1546, Charles had expelled the league's troops from upper Germany, almost without a fight.

Coming from Bohemia on 11 April, Charles invaded the electorate Saxony with 17,000 men on foot and 6,300 horsemen. He marched toward the encampment of the Elector John Frederick near Meissen. Charles easily defeated the elector's troops, forcing them into disorganized retreat. John Frederick was captured by the cavalry when he tried to escape to the citadel of Wurzburg. This clash marked the end of the Battle of Muhlberg and the Schmalkaldic War itself.

The myth of the Battle of Muhlberg glorifies Charles V as the invincible military leader in the medieval sense of the ideal of the knight. This atavistic attitude stands in stark contrast to the actual modern techniques of war used by the emperor.

Michael Herrmann

See also: Schmalkaldic War

References and further reading:

Held, Wieland. 1547. *Die Schlacht bei Muhlberg.* Beucha: Sax-Verlag, 1997.

Mukden, Battle of (21 February–10 March 1905)

Final land battle of the Russo-Japanese War. In the winter of 1905, the land war between Russia and Japan moved from the siege at Port Arthur and fighting around Liaoyang to the contest for Mukden in central Manchuria.

The Battle of Mukden took place between 21 February and 10 March 1905. Until the massive battles of World War I, this clash involved the greatest number of land troops of any battle in history. Japanese field marshal Oyama Iwao commanded a force of more than 200,000 soldiers and nearly 1,000 artillery pieces, and Russian general Aleksey Kuropatkin commanded nearly 300,000 men and more than 1,200 artillery pieces.

Oyama realized the need to attack before Russian reinforcements arrived from Europe. Although outnumbered and outgunned, he attacked first along a 40-mile front, beginning in late February. He wanted to turn the flanks of the Russian defense and thereby create a classic envelopment. Massed infantry attacks soon bent the Russian right so that it was facing as much west as south. Kuropatkin shifted reserves to restore his right wing, and Oyama threw more troops in the effort to continue to bend the Russian right flank.

By early March, the Japanese had pushed the Russian right so far that it seemed they might be able to threaten the Russian line of communication north to Harbin and then eventually across the Trans-Siberian Railroad. Kuropatkin withdrew in good order and very carefully, until he established a new defensive position at Harbin, where he was replaced in command.

Although more than 100,000 Russian troops had been killed or seriously wounded and much equipment had been abandoned in the retreat from Mukden, the Japanese had also suffered proportionate losses, and the Russians could more easily replace men and equipment. There would be no more major land campaigns.

Charles M. Dobbs

See also: Russo-Japanese War

References and further reading:

Okamoto, Shumpei. *The Japanese Oligarchy and the Russo-Japanese War.* New York: Columbia University Press, 1970.

Warner, Denis, and Peggy Warner. *The Tide at Sunrise: A History of the Russo-Japanese War, 1904–1905.* New York: Charterhouse, 1974.

Westwood, J. N. *Russia against Japan, 1904–1905: A New Look at the Russo-Japanese War.* Houndmills, Basingstoke, Hampshire: McMillan, 1986.

Murat, Joachim, Grand Duke of Cleves-Berg, King of Naples (1767–1815)

French cavalry commander during the French Revolutionary Wars and the Napoleonic Wars. Joachim Murat was born 25 March 1767 in La Bastide-Fortunide in Gascony. He enlisted in the French cavalry in 1787 as an ordinary trooper and rose quickly after the onset of war, reaching the rank of *chef d'escadron* in May 1793.

In October 1793, Murat provided critical support to Napoleon Bonaparte during the suppression of the Parisian mob on 13 Vendémiaire—Murat secured the cannon from which the famous "whiff of grapeshot" was fired. Bonaparte, placed in command of the Army of Italy, secured the promotion of Murat to *general de brigade* in 1793. Murat commanded the French cavalry in the first Italian campaign and accompanied Bonaparte to Egypt, where he again served as cavalry commander.

Accompanying Bonaparte on his return to France from Egypt, Murat again proved invaluable to Bonaparte. During the coup of 18 Brumaire, Murat led troops into the legislative chamber to remove protesting legislators. The creation of the Consular Guard led to his appointment to that unit, the forerunner of the Imperial Guard. Murat married one of Bonaparte's sisters, Caroline, in December 1800.

Murat commanded the cavalry of the Army of Italy during the second Italian campaign. After a brief hiatus, he was sent to command operations in southern Italy in 1801. Upon the creation of the empire, he was made both marshal and grand admiral of France.

From 1805 to 1812, Murat usually served as commander of the French cavalry, often, as in the aftermath of Jena, with spectacular results. He proved much less adept at the command of mixed formations or in independent roles. Thus, his mishandling of the occupation of Spain led to a revolt by the residents of Madrid that was suppressed only with great difficulty. Similarly, his attempt in 1809 to invade Sicily ended in failure and the capture of most of the troops landed there.

Murat was created grand duke of Cleves-Berg in March 1806. In 1808, he was made king of Naples by the Emperor Napoleon, in place of Joseph Bonaparte, who became king of Spain. In Naples, Murat continued the reform work begun by Bonaparte and attempted to fan public sentiments for the unification of Italy.

After the failure of the Russian campaign, Murat returned to Naples and began reorganizing the Neapolitan army. Although he rejoined Napoleon for the 1813 campaign in Germany and commanded the imperial cavalry at Dresden and Leipzig, he simultaneously negotiated treaties with Austria and Britain allowing him to withdraw from the war and retain his throne. His volte-face completed, he spent the 1814 campaign moving very slowly against French positions in northern Italy.

During the Hundred Days, he attempted to switch sides again, but the Neapolitan army was decisively defeated by the Austrians at Tolentino on 2 May 1815. The Neapolitan Bourbons then seized the opportunity to return to their mainland territory. Murat attempted to incite a revolution against them. This failed, and he was captured and executed at Pizzo on 13 October 1815.

As a general, he introduced few if any innovations. He is instead better remembered for flamboyance, treachery, and his failed early attempt to unify the Italian peninsula.

Joesph Isenberg

See also: Aboukir; Austerlitz, Battle of; Berezina River, Battle of; Borodino; Dresden, Battle of; French Revolutionary Wars; Jena and Auerstädt; Leipzig, Battle of; Lodi; Marengo, Battle of; Napoleon I; Napoleonic Wars; Pyramids; Retreat from Moscow; Rivoli; Wagram

References and further reading:

Chandler, David G. *Napoleon's Marshals.* London: Weidenfeld and Nicholson, 1987.

Young, Peter. *Napoleon's Marshals.* Reading, Berkshire, UK: Osprey, 1973.

Murfreesboro (31 December 1862–2 January 1863)

Important Union victory in the American Civil War, ending Confederate hopes of invading the North through Tennessee. At dawn on 31 December 1862, near Widow Smith's house across Stones River from Murfreesboro, Tennessee, Braxton Bragg threw a massive surprise attack from his left into the right flank of William S. Rosecrans. Confederate infantry under William Hardee immediately gained about 3 miles' worth of territory and threatened to push the Yankees into the river. Confederate cavalry under Joseph Wheeler and Nathan Bedford Forrest harassed the Union left. If not for Philip Sheridan's valiant defense, Rosecrans might have been routed. George H. Thomas managed to regroup around William B. Hazen's artillery, turn the line, create a salient around Rosecrans's headquarters, and hold the center so that by the end of the day the battle lines were perpendicular to the river on its western shore.

Little action occurred on 1 January, but on 2 January, Bragg made a gigantic tactical error by ordering John C. Breckinridge to charge across an open field on the eastern shore into a superior Union position. Union artillery on the western shore had a clear shot into the Rebel left. Union reinforcements swarming across the river forced Breckinridge to retreat, and the battle was over. Total casualties were 12,000 Confederate and 13,000 Union. That Bragg squandered the decisive advantage he had on the first day of this engagement went far toward undermining confidence in his leadership throughout the South.

Eric v. d. Luft

See also: American Civil War; Bragg, Braxton; Chickamauga, Battle of; Forrest, Nathan Bedford; Rosecrans, William Starke; Sheridan, Philip Henry; Thomas, George Henry

References and further reading:

Cozzens, Peter. *The Battle of Stones River.* Conshohocken, PA: Eastern National Park and Monument Association, 1995.

Hess, Earl J. *Banners to the Breeze: The Kentucky Campaign, Corinth, and Stones River.* Lincoln: University of Nebraska Press, 2000.

Reid, Richard J. *Stones River Ran Red.* Owensboro, KY: Commercial, 1986.

Vance, Wilson J. *Stones River: The Turning-Point of the Civil War.* New York: Neale, 1914.

Shaban, M. A. *Islamic History: A New Interpretation.* London: Cambridge University Press, 1971.

Musa ibn Nusayr (c. 640–714)

Conqueror of North Africa and Spain during the period of Islamic expansion. Born in 640, Musa spent part of his early career fleeing from the authorities for suspected embezzlement in Basra. Musa, however, evaded the troops of Caliph 'Abd al-Malik and took refuge with the caliph's brother, 'Abd al-Aziz, the governor of Egypt. 'Abd al-Aziz helped extricate Musa from his legal difficulties by paying half of his 10,000-dinar fine. In addition, he allowed Musa to return to Egypt with him. In 698 or 699, 'Abd al-Aziz granted Musa the governorship of Ifikiya, or North Africa. It was at this point that Musa's military career began.

Because Muslim control of North Africa was incomplete, Musa led expeditions against Zaghwan and Sadjuma as well as defeating numerous tribes. After the new caliph, al-Walid, confirmed him in his position of governor, Musa continued his conquests by pursuing the Berber tribes, who had fled west into modern Morocco. His armies pushed as far west as Tangiers. He left his freedman Tariq ibn Ziyad there as his deputy and returned victorious to Egypt.

Tariq took advantage of his position, led an army across the straits, and invaded Spain in 710–711. Musa soon learned of Tariq's success in Spain and followed in 712 with another army, partially out of jealousy. Rather than following the army of Tariq, Musa and his son 'Abd al-'Aziz bin Musa took a different route, capturing several cities, including Seville, before joining Tariq en route to Toledo. Musa reprimanded Tariq for his unauthorized invasion and then resumed his campaign and conquered northern Spain.

In 713 or 714, Musa ibn Nusayr departed from Spain with a large amount of booty. He left his son 'Abd al-Aziz in Spain to rule as governor. Musa died in 714 after a long, successful career.

Timothy May

See also: 'Amr ibn al-'As; Byzantine-Muslim Wars; Charles Martel; Khalid ibn al-Walid; Leo III; Muslim Civil War (656–661); Muslim Conquests; Tariq ibn Ziyad; Yarmuk, Battle of

References and further reading:

Abun-Nasr, Jamil. *A History of the Maghrib in the Islamic Period.* New York: Cambridge University Press, 1987.

Donner, Fred M. *The Early Islamic Conquests.* Princeton, NJ: Princeton University Press, 1981.

Julien, Charles Andre. *History of North Africa: Tunisia, Algeria, Morocco, from the Arab Conquest to 1830.* Trans. John Petrie. London: Routledge and Kegan Paul, 1970.

Music, Military

Music played in accompaniment to military activities, usually for regulating the pace of drills and marches, signaling, intimidating enemies, and enhancing morale. Even though the recognizable genre of military marches traditionally played on a combination of wind and percussion instruments is only about five centuries old, martial music possesses a far greater history that reaches deep into antiquity. Ancient Egyptian depictions of musicians playing horns and drums during regal processions and parades of soldiers attest to this fact, as do several references to music in classical military texts by Aelian and Vegetius. However, the history of this music prior to the European Renaissance largely is lost and its reconstruction conjectural at best. The standard duty accorded to musicians in most premodern cultures was the relaying of signals, usually to initiate or disengage from combat or to warn of enemy movements. Ancient and medieval horns created a piercing sound audible above the din of battle, providing somewhat effective battlefield communications for centuries. Roman commanders relied on legionary trumpeters to coordinate movements among multiple units. Medieval European chronicles and ballads frequently refer to the use of simple wind instruments as signals during combat. As for other applications of ancient and medieval military musicians, little is known for certain. Greek hoplites trained and marched to flute music. Medieval kings and rulers frequently employed musicians for processions and ceremonies.

In contrast to the simple horn-based music of premodern Europe, Islamic armies integrated Central Asian percussion traditions with a variety of horns that were the basis for many modern wind instruments, notably fifes and oboes. Crusaders encountered ensembles of Arab horn and drum players accompanying the standard-bearers of their armies; according to medieval texts, as long as the musicians played, the soldiers fought, confident that their standards were safe. As the Ottoman Empire rose to preeminence in the Islamic world, it continued and elaborated on these musical traditions, using military bands for signals, intimidation, and ceremonies. As it expanded into the Balkans and eastern Europe, it also transmitted these ideas and instruments to its Christian adversaries.

The evolution of modern Western military music coincided with other significant transformations in the manner in which European armies trained and fought. The initial trajectory of this development was the application of field music to drill and march. In the early 1500s, soldiers playing drums, fifes, and oboes began to appear in western Europe, first among the ranks of Swiss mercenaries and then within the forces of their French employers. In 1543, a French royal ordinance delegated two drummers and two fifers for each company of 1,000 infantry. The traditional arrangement of assigning drummers, fifers, and oboists to infantry and trumpeters and buglers to cavalry and artillery emerged by midcentury. As regular military drill grew in importance, especially after the reforms of Maurice of Nassau, fife and drum music acquired a heightened significance. Drums were particularly necessary for keeping soldiers marching in step, and fifes, oboes, and bagpipes provided some aesthetic diversion and boosted the soldiers' morale. By the mid-1700s, every European military establishment considered these musicians as essential components of their armies, employing them for signal, drill, and camp calls. The subsequent process of standardization is visible in the creation of fife, trumpet, and drum major ranks in the British army and the regulation of their training, pay, and uniforms by the 1740s; in this same period, young boys also began to be used as field musicians. Fifers and drummers retained their importance until the twentieth century, by which time military bands replaced fife and drum corps and field music lost importance in the face of changing tactics and technologies.

The other standard component of military music, the formal military band, also traces its origins to early modern Europe. Descendants of medieval town bands, early military bands of oboes (known as *hautbois* or *hoboys*) appeared among a few units during the seventeenth century, although they held a tenuous, quasi-legal existence sustained only by the largesse of a regiment's officers or royal patronage; notable examples include the oboe bands of the French dragoons acquired during the reign of Louis XIV and the wind band created for the British Horse Grenadier Guard in 1678. In the early 1720s, the Ottoman government sent full military bands to Poland and Russia as diplomatic gifts, sparking a craze for "Janissary" music throughout Europe. The fad quickly spread, first to Austria and Prussia and then on to France and Britain, fulfilled by imported Ottoman bands or imitation "Janissary" musicians, usually Africans attired in exotic "Turkish" or "Moorish" uniforms. The instruments of these bands—kettledrums, side and bass drums, oboes, bassoons, trombones, cymbals—prefigured the brass, woodwind, and percussion composition of future military bands. Official acknowledgment and regulation of bands began in the mid–eighteenth century, as seen in the British government's recognition of preexisting bands for the Royal Artillery, Life and Horse Guards, Foot Guards, and other elite units and allowance for the creation of regular regimental bands in the 1760s. The ideological services that military bands provided, especially in terms of summoning patriotic and nationalist enthusiasm during times of crisis and for recruiting drives, was recognized during the French Revolution, which witnessed the massing of gigantic bands for republican fêtes and the formation in 1792 of the earliest school for military music, the École Gratuite de la Garde Nationale Parisienne.

Formalization proceeded in the nineteenth century, as military bands achieved their greatest prominence. The British army began to curtail the long-standing practice of using Africans, civilians, and foreigners as bandsmen during the Napoleonic Wars. In 1857, the Royal Military School of Music was formed after a particularly bad performance given at a grand review during the Crimean War. The United States followed the European model, forming its Marine Band in 1798 and authorizing regular army bands in 1834, although American military bands remained a disorganized, local affair until after the Civil War; the Army Music School was not created until 1911. In the 1820s, military bands in Prussia and Austria standardized their instruments, training, and personnel, but the most significant changes occurred in France with the reforms of Adolphe Sax, who standardized the instrumentation of the Imperial Guard in 1854, establishing the pattern for most continental European bands. Military band music reached its pinnacle in the highly militarized and competitive atmosphere of the decades leading up to World War I, as international band competitions became both a very popular means for expressing nationalist sentiment and a means for voicing diplomatic dissent. Throughout the late nineteenth and twentieth centuries, military bands of the European and American style spread through the rest of the world, either as an inheritance of colonial rule in India and most of Africa or as part of modernizing military reforms and growing nationalism, as occurred in Japan. Military band music also affected "civilian" classical music, influencing the work of such notable composers as Ludwig van Beethoven and Hector Berlioz. Some works, such as the "Colonel Bogey March," and certain bandleaders, such as John Philip Sousa, have enjoyed enormous popularity in their native lands to this day. And eminent army bands, such as the Band of the Coldstream Guards in Great Britain and the Marine Corps Band in the United States, give performances internationally of concert hall quality. Even though military bands seem to have little place on the modern battlefield, these units nonetheless have now established themselves as a permanent and valued part of most of the world's militaries.

Ian Janssen

References and further reading:
Camus, Raoul. *Military Music of the American Revolution.* Chapel Hill: University of North Carolina Press, 1976.
Farmer, Henry. *The Rise and Development of Military Music.* London: W. M. Reeves, 1912.
Winstock, Lewis. *Songs and Music of the Redcoats: A History of the War Music of the British Army, 1642–1902.* Harrisburg, PA: Stackpole, 1970.

Muslim Civil War (656–661)

Civil war that nearly destroyed new Arab empire. When Uthman ibn Affan became caliph in 644, factions were starting to emerge within the Arab conquest state. Uthman's tribe, the Umayyads, numerous, talented, and adept at both trade and arms, had once been the Prophet Muhammad's fiercest enemies. Other Meccans and Medinans of the Quraysh confederation, earlier converts to Islam, resented the Umayyad upstarts intensely. Third came the Shiites, devotees of Ali ibn Abu Talib, son-in-law of the Prophet. Convinced that Muhammad had appointed Ali his successor, the Shiites felt that Uthman and the two previous caliphs had thwarted God's prophet by cheating Ali of his birthright. Finally, throughout the army camps of Syria, Iraq, and Egypt, Arabs quarreled over privileges accorded some tribes and discrimination shown toward others.

In 656, disgruntled soldiers from Egypt surrounded Uthman's palace and, when he resisted, murdered him. The dissidents then persuaded the Shiites and many Quraysh to raise Ali to the caliphate, but az-Zubayr, a prestigious and pious Muslim, rejected Ali as tainted by the regicide. Seeking allies, az-Zubayr's party set off for Iraq, but the new caliph chased them down and defeated them at the Battle of the Camel. The battle killed az-Zubayr and many of the Prophet's closest friends, further tarnishing Ali's image.

A more serious threat arose from Muawiyah, governor of Syria and Umayyad champion. In 657, Ali left Iraq and marched against Syria. His army encountered Muawiyah at Siffin. Halfway through the Battle of Siffin, Umayyad forces called for a truce and a negotiated settlement based on the Quran. To end the shedding of Muslim blood by Muslims, Ali agreed. Muawiyah demanded the execution of Uthman's killers and the retention of all Umayyad officeholders and secretly courted Ali's wavering allies in Iraq. Ali insisted that his power derived from his religious reputation and lineage, not from the mutineers, and he refused to punish them.

Infuriated that Ali might compromise with Umayyad "apostates," several thousand soldiers denounced and deserted the caliph. Labeled "Kharijites," these men declared jihad against Ali and Muawiyah alike with their slogan, "No judgment but God's." After several incidents, Ali wiped out a Kharijite force at the Battle of Nahrawan in July 658. As the balance of power shifted his way, Muawiyah aggressively cajoled, bribed, and strong-armed others into Umayyad ranks. In January 661, Kharijite assassins killed Ali at the Kufa mosque in Iraq.

With Ali dead, Muawiyah moved quickly. He browbeat Hassan, Ali's son and heir, into paying homage to the Umayyad house. He suppressed resistance in Iraq and made himself caliph. The Muslim civil war was now over. The Umayyad century had begun, and so had the alienation of Shiite Muslims from the rest of the Islamic community.

Weston F. Cook, Jr.

See also: Muslim Conquests

References and further reading:
Madelung, Wilferd. *The Succession to Muhammad.* Cambridge, UK: Cambridge University Press, 1997.
Shaban, M. A. *Islamic History, 600–750.* Cambridge, UK: Cambridge University Press, 1971.

Muslim Civil War (861–870)

A conflict that marked the end of a unified Muslim empire under the Abbasid caliph. The Muslim world would never again be united under a single ruler or ethnicity.

Conflict developed in two major areas. The first involved a dispute over control of the highest office in Islam, that of the caliph, which fell under the control of Turkish military leaders. The dispute drained the treasury and left the empire open to rebellions from Kharijites (secessionists opposed to both the main Sunni and Shiites), slaves, and royal pretenders, to mention a few. A second area of conflict found expression in the infamous Zanj Rebellion of black slaves in Mesopotamia.

Turkish nomads from the steppes of Asia had been streaming into the eastern Islamic world of southwest Asia since the ninth century. To control them, the Turks converted them to Islam and integrated into the Muslim military structure as superb archery horsemen, generals, and royal guards. The Turks used these exalted soldiers to control the office of caliph. From 861to 870, the Turks dethroned or killed no less than four caliphs who did not adhere to Turkish demands.

The loss of Abbasid Arabic control over the caliphate encouraged factional and ethnic rebellions in the empire. Rebellions broke out in Persia, Transoxiana, Mesopotamia, Egypt, and in Arabia itself. The worst was the 15-year Zanj Rebellion by black slaves in Mesopotamia who mined salt. Ali ibn-Muhammad, a Kharijite Persian, led the slave rebellion, which built two base cities in Mesopotamia. The rebellion forces consistently defeated caliphate forces sent against them because the caliphate no longer controlled the Turks and their generals, who made up the bulk of the caliphate army. As a result, the caliphate had to send African soldiers, who often deserted the caliphate armies for the rebellion. Now unstoppable, the forces of Ali ibn-Muhammad sacked and destroyed the wealthy city of Basra and killed over 300,000 people in the region.

Finally, after draining the caliphate treasury, the Turks lost interest in control of the caliph. Their disinterest allowed Muslim forces to besiege the two African slave cities in Mesopotamia, capturing the last in 883, but the unified caliphate had been destroyed for ever.

Christopher Howell

See also: Abbasid Revolution; Muslim Civil War (656–661)

References and further reading:

Brockleman, Carl. *History of the Islamic Peoples.* Rev. ed. New York: Capricorn Books, 1973.

Popovic, A. *The Revolt of African Slaves in Iraq.* New York: Markus Weiner, 1999.

Muslim Conquests (624–982)

Wars that transformed a religion into an empire. Islam's enemies, particularly the Byzantine Empire and Persia, which had been engaged in a protracted war, were weak and exhausted and thus ripe for conquest. Islam also brought a new vitality into the loosely organized regions in North Africa, the Middle East, and Spain, although one that built upon late Roman achievements in most cases.

After a few preliminary probes of Byzantine defenses, including victories at al-Aqaba (633) and Ajnadayn (634), the first battle in which Arabic invaders fought as an army rather than a band of raiders, Arabic armies under Khalid ibn al-Walid, and others, invaded Syria, taking Damascus in 636. Although temporarily forced back by a Byzantine counterattack, the Byzantine disaster at Yarmuk that same year signaled the permanent loss of Syria and adjacent areas. Jerusalem, left behind and isolated, like many Byzantine fortified cities, surrendered in 638, not to be recovered by Christians until the First Crusade (1096–1099).

Moving in several directions simultaneously, the Arabs, who had already taken much of Iraq in their initial raids, began a serious invasion of Persia proper in 636. In 637, they decisively defeated the Sassanids in the Battle of Qadisiya, seized the capital of Ctesiphon, and forced the last Sassanian emperor to flee to Central Asia. Most of Iran was under their control by 649, although the conquest of Khorasan was completed only in 654. From Khorasan and eastern Iran, Muslim armies slowly advanced into what is now western Turkistan and then India, laying the foundation of modern-day Pakistan and Bangladesh.

Other Arabic forces moved into Byzantine Egypt, largely captured by 643, although Alexandria, isolated from its hinterland, was able to hold out a few years longer. Even before the final subjugation of Egypt, Arab armies began moving down the African coast, where Tripoli was captured in 643. Armenia was invaded in 642. In 650, Cyprus was invaded for the first time, and again, along with Crete and Rhodes, in 654.

The victorious Arabic advance was temporarily interrupted by the civil war of 656–661, during which Umayyad forces defeated Ali, whose supporters became known as Shiites. The Umayyads then moved the Muslim capital to Damascus.

Resuming the advance, Umayyad naval forces appeared in the Sea of Mamora in 670, although the Byzantines, using Greek fire, heavily defeated the Arabs in a naval battle in 677. Elsewhere, other Arabic armies completed the conquest of Byzantine Africa, taking Carthage in 698. From Africa, they crossed the Straits of Gibraltar in 711, whose name ("rock of Tariq") recalls their general, Tariq ibn-Ziyad, whose overwhelmingly outnumbered army crushed the Visigoth king Roderick at the week-long Battle of Rio Barbate. Although a number of Christian princes fled to refuge in the Pyrenees and held out against Islam, most of Spain now came under Arabic control.

From Spain, Muslim forces invaded the Merovingian kingdom of the Franks but were halted by Charles Martel at Tours and Poitiers in 732, although the importance of these Christian victories has been overstated, since Muslim armies were operating well beyond their safe range during their invasions of France. In addition, Charles's victories did not end the direct threat to Europe since Arabs were able to invade and seize Sicily from the Byzantines in the early ninth century and retain their position there for some time.

The same year that the Arabs entered Spain, 711, they also began serious raids and advances into Anatolia, which the Byzantines had hitherto defended successfully. This move inaugurated a struggle that went on well into the tenth century, although the Arabs suffered a decisive repulse by land and by sea in the failure of their siege of the Byzantine capital of Constantinople in 717–718. Although the Byzantines were attacked on two sides, by Arabic armies below the Long Walls and by Arabic ships operating in the Sea of Mamora, the experienced Byzantine Emperor Leo III was able both to defeat and to rout the depleted Arab armies.

The Umayyads, some of whom took refuge in Spain, were overthrown by the Abbasids in the mid–eighth century. The Abbasids moved the capital to Baghdad, which became a great trading city, and continued the war against Byzantium:

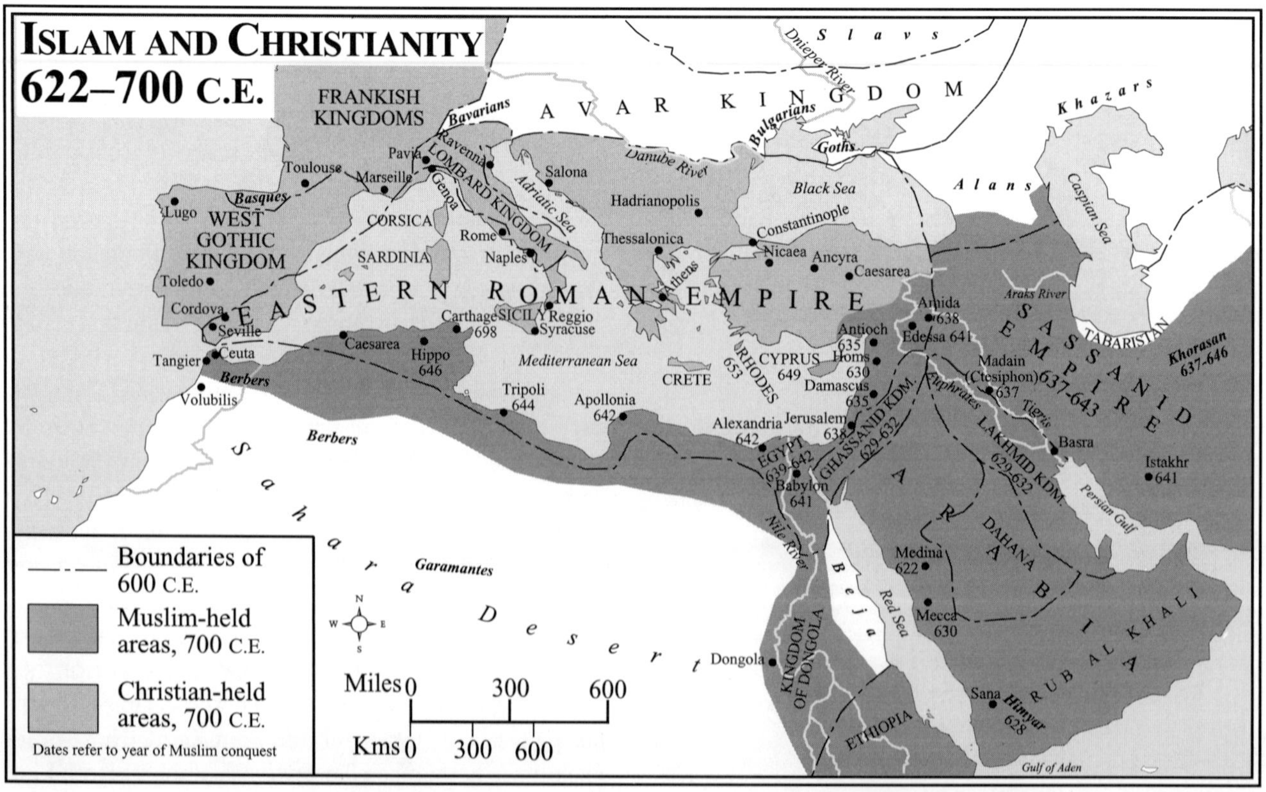

it was under their rule that the advance against Sicily took place, and they won the Battle of the Talas River in 751 against a weakening Chinese Tang dynasty (618–907). Nevertheless, the reign of Harun al-Raschid (r. 786–809) was, in every way, the high watermark of Muslim military power. Centralized authority declined sharply after his time, particularly during the civil wars of 861–870, which permanently lamed the caliphate as a central, Arabic authority.

The period also saw an increasingly powerful Byzantine *reconquesta,* as Byzantine armies recovered Crete, major parts of northern Mesopotamia, and Syria and even threatened Jerusalem. Among the emperors taking the lead were Constantine V (r. 741–775) and his son, Leo the Khazar (r. 775–780), who gradually extended Byzantine power in Anatolia.

Later, under Leo VI (r. 886–912), the Byzantines were able to invade the emirate of Tarsus successfully and then Armenia, which also came under Byzantine control, and continued the process of the recovery of frontier territory. Subsequently, Emperor Romanus Lecapenus (r. 920–944) attacked and sacked Melitene, an Arab base, and in 928 formally restored it to the empire. Although local Arab forces counterattacked, the area was brought definitely under Byzantine control by 936.

The culmination of these efforts came during the reigns of Constantine VII (r. 913–959), his son Romanus II (r. 959–963), Nicephorus II Phocas (r. 963–969), and John I Tzimisces (969–976). The Byzantines expanded their control in northern Mesopotamia, Armenia, and Cyprus. In 961, Nicephorus II Phocas, then the general of Romanus II, recaptured Crete. Meanwhile, Tzimisces, then still a general, took Aleppo and, as emperor, advanced into northern Palestine. The following century, the Byzantines even sought to reconquer Sicily, although it was the Normans who were the ultimate beneficiaries. Only in Spain did an era of Arabic conquest persist, despite what was by and large a stalemate existing between Muslims and Christians, thanks to the infusions of raw Berber energy from North Africa that reinvigorated the Arabic advance. Elsewhere, it was the Turks who took over the role of expanding the frontiers of Islam.

Annette Richardson

See also: Abbasid Revolution; 'Abu al-'Abbas; Basil II Bulgaroctonus; Byzantine-Muslim Wars; Byzantine-Persian Wars; Constantine V; Constantinople, Siege of (717–718); Heraclius; John I Tzimisces; Khalid ibn al-Walid; Leo III; Muslim Civil War (656–671); Muslim Civil War (861–870); Nicephorus II Phocas; Tariq ibn Ziyad; Talas River, Battle of; Tours; Yarmuk, Battle of

References and further reading:

Belyaev, E. A. *Arabs, Islam, and the Arab Caliphate in the Early Middle Ages.* London: Pall Mall Press, 1969.

Donner, Fred. *The Early Islamic Conquests.* Princeton, NJ: Princeton University Press, 1981.

Hawting, G. *The First Dynasty of Islam: The Umayyad Caliphate.* Carbondale: Southern Illinois University Press, 1987.
Kaegi, Walter. *Byzantium and the Early Islamic Conquests.* Cambridge, UK: Cambridge University Press, 1992.
Treadgold, Warren. *A History of the Byzantine State and Society.* Stanford, CA: Stanford University Press, 1997.
Whittrow, Mark. *The Making of Byzantium, 600–1025.* Berkeley: University of California Press, 1996.

Mutaguchi, Renya (1888–1966)

Imperial Japanese army general who conceived and commanded the Imphal campaign of March–July 1944. Born in Saga Prefecture in Kyushu, southwestern Japan, a region renowned for breeding tough soldiers, Mutaguchi graduated from the Army's Military Academy in 1910 and the Military Staff College in 1917. After serving in the War Ministry and General Staff Headquarters, he assumed command of the 1st Infantry Regiment, stationed in northern China in 1936. In July of the following year, he played a role in developing the "Marco Polo Bridge incident" into a pretext for expanded war against China.

At the opening of World War II in the Pacific, as commander of the 18th Division, Mutaguchi participated in the conquest of Malaya, Singapore, and Burma in 1941–1942 and was promoted to commander of the Fifteenth Army, headquartered at Rangoon, in 1943.

Mutaguchi was initially lukewarm toward plans for an offensive into northeastern India. However, the success of Orde Wingate's first "Chindit" guerrilla operation behind Japanese lines in February–June 1943 convinced him that if British soldiers could breach the thickly jungled hills between Burma and India, his own men, aided by the Indian National Army of Subhas Chandra Bose, could break through British defenses, occupy the Imphal Plain, and inspire Indian patriots to rise up against their British masters.

Mutaguchi had grandiose visions of toppling the British Raj, but his offensive lacked proper logistical support and adequate lines of communication, especially during the monsoon rains. Some 50,000 of the 100,000-man strong Fifteenth Army died in combat or during the terrible retreat back into Burma. Mutaguchi's name became synonymous with the Japanese militarists' cavalier disregard for the lives of their men.

Donald M. Seekins

See also: Imphal and Kohima; Indian National Army; Wingate, Orde

References and further reading:

Allen, Louis. *Burma: The Longest War, 1941–1945.* London: Phoenix Press, 2000.

Mysore Wars (1767–1799)

A series of four conflicts, beginning in 1767, that by 1799 established British supremacy in the Deccan and southernmost regions of India. The First Mysore War (1767–1769), however, hardly anticipated any such British triumph. Despite its concern with the growing power of the sultan of Mysore, Haidar Ali Khan, whose territories bordered the Carnatic coast and thus threatened the trading stations at Madras and Cuddalore, the British East India Company not only failed to form an alliance to challenge Haidar but soon found itself facing invasion. Haidar defeated the British at Malbagal (4 October 1768) before being driven from the Carnatic. Invading again the next year, Haidar dictated the Treaty of Madras to the British before the gates of their city (3 April 1769).

The First Mysore War resolved none of the issues that had provoked it and, not surprisingly, sowed the seeds for the renewal of conflict. Haidar's ferocious Anglophobia was reinforced when the British, despite unequivocal mutual assistance provisions in the Madras Treaty, failed to come to his aid when his capital of Seringapatam was besieged by the Marathas in 1771. Moreover, Haidar's hatred of the English was matched only by the esteem in which he held the French. During the British siege of Pondicherry in the First Carnatic War, Haidar had led his horsemen to the assistance of the French, from whom he would later gain technical experience, gunners, and guns. When Britain and France found themselves at war in 1778 and the British captured the French posts of Pondicherry, on the Carnatic, and Mahe, within Mysore territory, Haidar needed no further compunction to go to war again. The Second Mysore War (1780–1784) opened with Haider's invasion of the Carnatic with 80,000 men and 100 guns. Taking Arcot, Haidar destroyed a British force at Pollilur (10 September 1780), forcing his enemies to flee for the safety of Madras's fortifications.

Despite this initial success, Haidar was unable to capture Madras before Eyre Coote arrived with a relieving force from Calcutta. At Porto Novo (1 July 1781), Coote, with 8,000 Europeans and Sepoys, defeated Haidar's force of some 60,000. Madras thus relieved, Haidar was again defeated at the scene of his earlier victory, Pollilur (27 August 1781), and yet again at Sholingur (28 September 1781). British victory remained limited, however, because of the greater mobility of Haidar's forces, while Haidar's son, Tipu, defeated the British at Annagudi (17–18 February 1782). The British in India likewise suffered from the greater demands in manpower and munitions made by the ongoing war in North America and the West Indies. The consequences of this wider conflict were made clear when a French squadron under Admiral Pierre du Suffren captured Trincomalee (30 August 1782). Though enabling Haidar to successfully defend Cuddalore from British attack (13 May 1783), the arrival of French reinforcements in fact proved too late to measurably influence the outcome of the war. The Treaty of Versailles (20 January 1783) ended the war between Britain and France (and the United States), and Haidar Ali himself had died earlier (2 December 1782). It was therefore hardly surprising that his successor, Tipu, was amenable to peace, especially as the Treaty of Mangalore (11 March 1784) restored the status quo ante bellum.

If Tipu's respect for the French, inherited from his father, had been tempered by what he believed to be their failure to give adequate support in the recent war, it did not prevent his sending ambassadors to the French court in 1787. Arriving at Toulon in June 1788, the envoys were received by Louis XVI two months later. Yet if Tipu thus demonstrated his willingness to achieve his ends through European diplomatic as much as military means, the effort ultimately proved to be too little, too late. The flames of revolution soon would sweep across France, eventually consuming the ancien regime. At the same time, the global conflict that the French Revolution was to become would provide the setting for the two final Mysore Wars.

Soon after the revolution had swept the French monarchy from power, it became clear that "Citoyen Tipu" could not be counted on to support the British against revolutionary France. Sir Charles Cornwallis, governor of India, accordingly sought to strengthen British ties with the traditional foes of Mysore, such as the Marathas, while military aid was given to the Raja Rama Verma of Travancore, then at war with Tipu. It was evident to Cornwallis, however, that Tipu could only be suitably chastised by direct action, and the Third Mysore War (1790–1792), opened with General Sir William Medows invading Malabar with some 30,000 men. After Tipu proved more than capable of avoiding Medows's forces, Cornwallis himself assumed command in January 1791, while Maratha forces attacked on Tipu's northern frontier. Losing his capital of Mangalore to the British (21 March 1791), Tipu fell back on Seringapatam. With the end of the rainy season in early 1792, Cornwallis moved against Tipu, opening the siege of Seringapatam on 6 February 1792. When the citadel was captured on 21 February, Tipu had no choice but to open the negotiations that would end the war with the Treaty of Seringapatam (16 March 1792).

If the inconclusive nature of the first two Mysore Wars had ensured the outbreak of their successor, the apparent conclusiveness of the Third War, represented in the harsh terms imposed on Tipu by the British, equally ensured the outbreak of the fourth and final conflict. Despite being forced to cede half of his territory to the British, Tipu's abilities as an administrator ensured a quick recovery of Mysore. The continued threat from Mysore seemed all the greater to the British, moreover, when Napoleon invaded Egypt in 1798. Fearing a French invasion of India in support of Tipu, Richard Wellesley, Lord Mornington, governor of India, ordered a force of 42,000 to invade Mysore (14 February 1799). Though the Fourth Mysore War (1799) would help establish the reputation of Wellesley's younger brother, Arthur, the future duke of Wellington, Tipu's defeat owed as much to the defection of his three senior generals as it did to British generalship. Defeated at Siddeswara (5 March) and again at Malavalhi (25 March), Tipu fell back on Seringapatam. There, having refused the exorbitant demands of General George Harris, Tipu died defending his citadel from the final British assault (4 May). With Tipu died the threat posed by Mysore to British dominion in southern India.

Adam Norman Lynde

See also: Carnatic Wars; Cornwallis, Sir Charles; Harun al-Raschid

References and further reading:

Keay, John. *India: A History.* London: Harper Collins, 2000.

Mehra, Parshotam. *A Dictionary of Modern Indian History.* Oxford, UK: Oxford University Press, 1987.

N

Nadir Shah (a.k.a. Tahmasp Qoli Khan) (1688–1747)

Ruler who reunited Iran after the collapse of the Safavid Dynasty, campaigning from Arabia and Iraq to Azerbaijan and the Ganges Valley. Nadir Shah rose to power in Iran at the beginning of Persia's eighteenth-century civil wars. His tribe, the Afshars, belonged to the Qizilbash, a tribal alliance that helped found the ruling Safavid Dynasty. However, in 1722, Afghan invaders seized Esfahan and deposed the Safavid shah, throwing Iran into chaos. Urban revolts, dissident tribes, and seceding provinces fragmented the state. Both Russia and the Ottoman Empire seized Persian territory.

Nadir Shah rallied the Afshar to his leadership and aligned them with Tahmasp II, a northern Safavid prince. In 1729–1730, he defeated the Afghans as well as his rivals in the Qizilbash, the Qajar. Having pacified Persia, the "Slave of Tahmasp" drove out the Ottomans while simultaneously pressuring Russia into evacuating Baku and the southern Caspian. After deposing Tahmasp's son in 1736, Nadir claimed the throne himself, becoming Shah Nadir Khan Afshar. Additionally, he secured the Persian Gulf by occupying Oman.

Nadir invaded Afghanistan in 1737. Running over the Afghans, he plunged through the Khyber Pass and east to Lahore. At the Battle of Karnal (1739), he defeated the Mogul emperor of India, marched across the Ganges plain, and sacked Delhi. Looting the Mogul treasury and the bejeweled Peacock Throne, the shah then turned north to wage war against the Uzbeks of central Asia. By 1742, Merv, Bukhara, and Khiva were Iranian vassals. In his last campaigns, Nadir crushed an Ottoman force in Iraq in 1745.

Nadir Shah was an abusive ruler, more feared than respected, and insensitive to Persia's Shiite Islamic faith. His assassination by his troops in 1747 renewed Iran's civil wars.

Weston F. Cook, Jr.

See also: Persian Civil Wars

References and further reading:

Abraham of Erevan. *History of the Wars, 1721–1738.* Costa Mesa, CA: Mazda Publishers, 1999.

Lockhart, Laurence. *Nadir Shah.* London: Clarendon Press, 1938.

Nagashino, Battle of (1575)

Battle fought by Nobunaga Oda (1534–1582) and his ally Ieyasu Tokugawa (1543–1616) with Takeda Natsunori, around the strategic fortress of Nagashino. In this encounter, the forces of Tokugawa and Nobunaga Oda were the first to rely primarily on massed firepower in the form of Western armaments, helping to transform samurai warfare while pushing both houses closer to hegemony over Japan.

Ieyasu Tokugawa had actually forged a familial alliance with the Takedas, whose territories bordered his own in central Honshu. He married both a son and daughter into the Takeda household in the 1560s, but in the world of shifting alliances and steady warfare that characterized Japan at the time, the alliance quickly foundered. The Takedas were soon at war with the Tokugawa again.

The death of the elder Takeda (Shingen) in 1573, at the hands of a sniper in battle, placed his son Natsunori at the head of the Takeda house. The rising fortunes of the Tokugawa had made them fierce rivals of the Takedas, and when in 1575 a traitor to Tokugawa offered to hand over the vitally strategic castle of Ozaki to the Takedas, Natsunori Takeda jumped at the opportunity. Ozaki was the capital of Mikawa Province, the heart of Tokugawa territory, and its castle was guarded by Tokugawa's own son.

Takeda led a force of 15,000 warriors in what was expected to be a near-bloodless seizure of Ozaki Castle. Instead, they discovered en route that the treachery had been discovered by Tokugawa. Rather than face a humiliating retreat, Takeda opted to send his troops instead against the nearby fortress of Nagashino, another strategic castle sitting at the convergence of three rivers and guarding the entrance to Mikawa and Totomi Provinces.

Takeda began his siege of the castle in May 1575 but was still unsuccessful when word came that relief forces led by Tokugawa and Oda were on their way. Takeda opted to stand his ground near Nagashino and engage the approaching allied armies, though his forces were outnumbered more than two to one. At the Battle of Nagashino in June 1575, the alliance's greater numbers and, more important, overwhelming firepower, including musket volley fire by alternating ranks (the first time that this technique is known to have been employed in warfare), carried the day. Takeda lost almost two-thirds of his men and generals, and the mortally wounded Takeda clan would linger only until 1582, when it was overrun for good.

Daniel Kane

See also: Japanese Wars of Unification; Oda, Nobunaga; Tokugawa, Ieyasu

References and further reading:

Parker, Geoffrey. *The Military Revolution: Military Innovation and the Rise of the West, 1500–1800.* Cambridge, UK: Cambridge University Press, 1988.

Sadler, A. L. *The Maker of Modern Japan: The Life of Tokugawa Ieyasu.* Tokyo: Charles E. Tuttle,1937.

Napalm

A jell weapon made up of naphthenic acids, palmitic or fatty acids, and the salts of aluminum. One of the enduring images of the Vietnam conflict is the photograph of a naked nine-year-old Vietnamese girl running toward the camera screaming in pain, her flesh burned by napalm and gasoline. In popular terms, any hydrocarbon incendiary weapon is referred to as "napalm." It is most commonly dispensed from airplanes or helicopters.

The origins of napalm can be traced back to attempts to improve the flamethrowers used in World War I. It was first developed by Dr. Louis Fieser of Harvard University during World War II. Because of the short supply of magnesium in 1942, napalm became the most widely used ingredient in incendiary bombs, such as the M-69. It was used in both the Atlantic and Pacific Theaters, especially on target cities in Japan. More than 50 percent of the bombs that destroyed Dresden were incendiaries. Napalm was also used by U.S. forces in the Korean War and by the French in Indochina and Algeria. Yet, it was the use of napalm during the Vietnam conflict on suspected Vietcong villages, livestock, crops, and strongholds that led to a reaction among many Americans and Europeans. It is estimated that the United States used a total of 338,237 tons of napalm in the Vietnam conflict between 1963 and 1971.

T. Jason Soderstrum

References and further reading:

Björnerstedt, Rolf, et al. *Napalm and Other Incendiary Weapons and All Aspects of Their Possible Use: Report of the Secretary-General.* New York: United Nations, 1973.

Chong, Denise. *The Girl in the Picture: The Story of Kim Phuc, the Photographer and the Vietnam* War. New York: Viking Press, 2000.

Mountcastle, John W. *Flame On! U.S. Incendiary Weapons, 1918–1945.* Shippensburg, PA: White Mane Publishing Company, 1999.

Napier, Sir Charles James (1782–1853)

British general and colonial governor of Sindh. Born 10 August 1782 in London, Napier served in the Peninsular War against Napoleon and was wounded in the Battle of Corunna on 16 January 1809. He was wounded again at Busacco on 27 September 1810 and served at Fuentes de Onoro on 5 May 1811. During the War of 1812, he saw duty on the eastern coastline of the United States. Beginning in September 1814, he attended military college at Farnham, volunteered to serve under the Duke of Wellington in Belgium, and saw action following the Battle of Waterloo in 1815. After graduating from Farnham in 1817, he was appointed inspecting field officer for the Ionian Islands in May 1818 and then became resident for Cephalonia in March 1822, governing the island until 1833. Napier was then promoted to major general in 1837, assigned to India in 1841 and to the Sindh command in August 1842. In February 1843, Edward Law, Earl of Ellenborough and governor-general of India, compelled Sindh to sign a treaty that the inhabitants considered unfair and humiliating, provoking unrest in the region. The treaty stipulated that Napier could seize territory if he decided the Sindhi rulers were disloyal.

Soon afterward, war began, and Napier occupied Sindh, winning battles at Miani (17 February 1843) and Dabo (24 March 1843) with an outnumbered army. He then was appointed governor of Sindh, serving in that capacity until 1847. He was to have been the commander of British forces during the Second Sikh War (1848–1849) but arrived from England too late. In his retirement, Napier published books about his experiences. He died on 29 August 1853 in Portsmouth, Hampshire, England.

Harold Wise

See also: Napoleonic Wars; War of 1812

References and further reading:

Napier, Lieutenant General Sir William. *The Life and Opinions of Sir Charles James Napier, G.C.B.* 4 vols. London, 1857.

Napoleon I (1769–1821)

French field commander, revolutionary hero, and emperor, among the greatest military strategists and tacticians of all time. Born as Napoleone di Buonaparte in Ajaccio, Corsica, on 15 August 1769, he enrolled in 1779 at the military academy of Brienne le Château as Napoleon Bonaparte. He was commissioned a second lieutenant of artillery in 1785 and promoted to first lieutenant in 1791. Increasingly revolutionary, he joined both the Corsican nationalist movement and the Jacobin Club of Grenoble. In 1792, he became lieutenant colonel of Corsican volunteers and captain of French artillery. He fled Corsica in June 1793, opposed to Pasquale Paoli's anti-French and pro-British policies. He was promoted to brigadier general in December for distinguished service with the Army of Carteaux during the siege of Toulon.

Napoleon at Fontainebleau, 1814. (Library of Congress)

Briefly imprisoned in August 1794 after the fall of Maximilien Robespierre but back on active duty in 1795 with the rank of general, Napoleon "with a puff of grapeshot" saved the revolutionary convention from a popular royalist uprising on 5 October (13 Vendémiaire in the French revolutionary calendar). From that point, his rise was phenomenal. Two weeks later, the convention named him commander in chief of the Army of the Interior and on 2 March 1796 commander in chief of the Army of Italy.

Immediately taking the offensive against Austria and its allies, Napoleon won at Montenotte on 12 April, Mondovi on 21 April, Lodi on 10 May, Milan on 15 May, Castiglione on 5 August, Bassano on 8 September, Arcole on 17 November, Rivoli on 14 January 1797, and Mantua on 2 February. The coup d'état of 18 Fructidor (4 September) augmented his status. Back in Paris for only a few months, he sailed on 19 May 1798 to invade Egypt. He easily conquered Lower Egypt by routing the Mamluks at the pyramids on 21 July but was marooned by British admiral Horatio Nelson's victory over the French fleet in the sea battle of Aboukir on 1 August. After defeating the English-Turkish alliance in the land battle of Aboukir on 25 July 1799, he escaped back to France, arriving on 9 October. His coup d'état of 18 Brumaire (9 November) gave him significant political power. Quickly solidifying that power, he defeated the Austrians at Marengo on 14 June 1800, concluded peace with England by the Treaty of Amiens on 25 March 1802, became first consul for life by a national vote in May, raised money for war by selling Louisiana to the United States on 3 May 1803, and crowned himself emperor of the French on 2 December 1804 and king of Italy on 17 March 1805.

England, Austria, and many German states, fearful of Napoleon's growing power, influence, and ambition, launched the Napoleonic Wars in 1803. The British naval blockade of France was successful, especially after Nelson won at Trafalgar on 21 October 1805, but on land the French prevailed. Napoleon defeated the Austrians at Ulm on 17 October 1805; the Russians at Oberhollabrunn on 16 November; the Austrian-Russian alliance decisively at Austerlitz on 2 December; the Prussians at Jena on 14 October 1806; and the Russians at Eylau on 7–8 February 1807, Heilsberg on 10 June, and Friedland on 14 June.

Napoleon appeared only briefly in Iberia for the peninsular campaign but defeated the Spanish at Somosierra on 30 November 1808. He beat the Austrians (who were persistent, if nothing else) at Abensberg on 20 April 1809, Landeshut on 21 April, and Ratisbon on 23 April. After losing at Aspern-Essling on 21–22 May, he crushed the Austrians at Wagram on 5–6 July.

Napoleon's downfall began when, overconfident, he invaded Russia in 1812. He won at Vitebsk on 28 July,

Smolensk on 17 August, Valutino on 19 August, Borodino (barely) on 7 September, Krasnyi on 16–17 November, and Berezina on 26–28 November but was defeated by the Russian winter and endured a miserable retreat. Encouraged by Napoleon's failure in Russia, German states rose against him in 1813. He beat them and their many allies at Lützen on 2 May, Bautzen on 20–21 May, and Dresden on 26–27 August, but was hard-pressed, overextended, and finally suffered a major defeat in the gigantic Battle of Leipzig on 16–19 October. Pushed back toward Paris, he still managed to win tactical victories at Hanau on 30–31 October, Brienne on 29 January 1814, La Rothière on 30 January, Champaubert on 10 February, Montmirail on 11 February, Chateau-Thierry on 12 February, Vauchamps on 14 February, Montereau on 18 February, Craonne on 7 March, and Rheims on 13 March but lost the strategic war. The allies beat him at Laon on 9–10 March and Arcis-sur-Aube on 20–21 March. Paris fell on 30 March. He abdicated on 4 April and was exiled to the Mediterranean island of Elba on 4 May 1814.

Napoleon escaped from Elba on 26 February 1815, landed in France on 1 March, and arrived in Paris, triumphant, on 20 March. The allies immediately mobilized against him. He engaged them in Belgium, won at Ligny on 16 June, but met disaster at Waterloo on 18 June. Thus ended his "Hundred Days." But to the end, Napoleon retained much of his military genius. Wellington himself is supposed to have remarked of the Battle of Waterloo, "It was a damned close-run thing."

Prevented from escaping to the United States, Napoleon surrendered on 15 July to Captain Frederick Maitland of the *Bellerophon*. The allies sentenced him to permanent exile on the British island of St. Helena in the South Atlantic, where he died on 5 May 1821, probably poisoned by arsenic.

Napoleon was more than a military conqueror. Like Alexander the Great, he aspired to "improve" his conquests. In the spirit of the Enlightenment, he led the reform of the continental legal system, introduced rational weights and measures and money, emancipated the serfs, freed the Jews from legal disabilities, and strove to create a "European Community" or "Common Market." Had Europe possessed railroads and telegraphs at the time, he might have pulled it off. In the end, it was the sea power and money of the British, combined with British-subsidized but often-defeated Continental armies, that brought Napoleon down. Napoleon provided a focus of fear and opposition that rallied the reactionary, class-ridden British, Austrians, and Russians, as well as Spanish and Russian peasants/serfs, to combine for his final defeat, but Europe as a whole may have been the loser.

Eric v. d. Luft

See also: Abercromby, Sir Ralph; Aboukir; Alexandria; Artillery; Austerlitz, Battle of; Berezina River, Battle of; Berthier, Louis-Alexandre, Prince of Neuchâtel and Valangin, Prince of Wagram; Blücher, Gebhard Leberecht von; Borodino; Brunswick, Frederick William, Duke of; Carnot, Lazare-Nicholas; Davout, Louis-Nicolas, Duke of Auerstädt, Prince of Eckmühl; Dresden, Battle of; Ferdinand, Duke of Brunswick; French Revolution; French Revolutionary Wars; Friedland; Gneisenau, August Neidhardt von; Jena and Auerstädt; Jomini, Antoine Henri, Baron de; Kléber, Jean-Baptiste; Kutuzov, Prince Mikhail Illarionovich Golenishchev; Lannes, Jean, Duke of Montebello; LeFebvre, Pierre-François-Joseph, Duke of Danzig; Leipzig, Battle of; Lodi; Marengo, Battle of; Masséna, André, Duc de Rivoli, Prince d'Essling; Moscow, Retreat from; Murat, Joachim, Grand Duke of Cleves-Berg, King of Naples; Napier, Sir Charles James; Napoleonic Wars; Ney, Michel, Duc d'Elchingen, Prince de La Moskova; Oudinot, Nicholas-Charles, Duc de Reggio; Pyramids; Quatre Bras and Ligny; Rivoli; Scharnhorst, Gerhard Johann von; Schwarzenberg, Karl Philipp zu; Soult, Nicolas-Jean de Dieu; Toulon, Siege of; Wagram; Waterloo; Wellington, Arthur Wellesley, Duke of

References and further reading:

Asprey, Robert B. *The Rise of Napoleon Bonaparte.* New York: Basic Books, 2000.

Lachouque, Henry. *The Anatomy of Glory: Napoleon and His Guard: A Study in Leadership.* London: Greenhill, 1997.

Lyons, Martyn. *Napoleon Bonaparte and the Legacy of the French Revolution.* New York: St. Martin's Press, 1994.

Marrin, Albert. *Napoleon and the Napoleonic Wars.* New York: Viking, 1991.

Schom, Alan. *Napoleon Bonaparte.* New York: HarperCollins, 1997.

Napoleonic Wars (1803–1815)

The military campaigns from 1803 to 1815, during which Napoleon fought against Austria, Prussia, Russia, Britain, and their allies; conquered and dominated much of western and central Europe; and established the French Empire with its dependencies and allies. It is the story of Napoleon the soldier, the statesman, the emperor, whose uncompromising will and military genius propelled him to rule France and dominate Europe for more than a decade.

The Napoleonic Wars were actually a continuation of the French Revolutionary Wars that began in 1792 and ended with the conclusion of the Anglo-French Treaty of Amiens in 1802. But the peace of Amiens turned out to be a temporary rather than a long-lasting armistice. Territorial disputes between France and Britain brought them into a new confrontation. Napoleon encamped an army around the port of Boulogne on the English Channel, in anticipation of an invasion of Britain. The British retaliated and declared war on France on 16 May 1803. By the end of 1805, Europe once again entered a new round of wars of larger proportions.

Britain organized a new anti-French alliance, the Third Coalition, which included Austria, Russia, and Sweden. Napoleon (crowned emperor of the French in 1804) hastily removed his troops from the English Channel and sent them to the Rhine and northern Italy. He assumed command of the armies in Germany, routed the Austrian general Karl Mack at the Battle of Ulm (October 1805), and took 30,000 Austrians prisoner. Almost simultaneously, the British Royal Navy under Admiral Horatio Nelson destroyed the French fleet at Trafalgar (21 October), thus breaking France's naval power and confirming Great Britain as the mistress of the seas. Trafalgar made up for Ulm and served to restore the balance of power.

While Britain dominated the seas, Napoleon continued the conquest of central Europe. He marched down the course of the Danube River and captured Vienna. The Russian army under General Mikhail Kutuzov and a second army under Czar Alexander arrived at Olmütz (Olomouc), a city in Moravia, to join the Austrians in an attempt to stop the French advance.

On 29 November 1805, Napoleon and his army took position at Austerlitz (Slavkov) in expectation of a collision with the allied forces. His strength of 73,000 men faced a combined Austro-Russian army numbering 86,000. On 2 December, he defeated the Austro-Russian armies at the Battle of Austerlitz, called "the most perfect battle in history" and Napoleon's masterpiece. He then dispersed the Russians after frightful losses, while Czar Alexander was galloping as fast as his horse could carry him. Emperor Francis II of Austria escaped the carnage but later accepted the humiliating Treaty of Pressburg (Bratislava), ceding territories to France's southern German allies and adding Venetia to Napoleon's Italian kingdom. Napoleon's brother Joseph became king of Naples, while another, Louis, was proclaimed king of Holland. After Austerlitz, France gained immeasurable prestige throughout Europe.

In August 1806, Napoleon dissolved the anachronistic Holy Roman Empire, which he pejoratively referred to as being "neither Holy, nor Roman, nor Empire." In its place he organized the Confederation of the Rhine under French auspices and with himself as its protector. He then entered Vienna, and at his bidding, Francis II abandoned his imperial crown of Holy Roman Emperor for the more restricted title of Francis I, Emperor of Austria.

Napoleon still faced Britain and Russia, however, and Prussia now felt threatened by the creation of the Confederation of the Rhine and the stationing of French troops throughout much of Germany. The Prussians prepared for war and boastfully promised a lightning victory over Napoleon, the "revolutionary anti-Christ"! On 1 October 1806, Prussia delivered an ultimatum to Paris, while Prussian troops seized Saxony.

But no sooner were hostilities under way than events took a very different course. Without waiting for a formal declaration of war, Napoleon launched his own lightning campaign against the Prussians and routed them in two simultaneous battles at Jena and Auerstädt in October 1806. He then advanced and occupied Berlin. Within a month, Prussia ceased to exist as a military power.

Napoleon then turned against Britain and tried to destroy the "nation of shopkeepers" (his term) by imposing an economic blockade known as the "continental system." He issued the Berlin Decree (November 1806) that banned trade and importation of British goods into continental Europe. It was the first large-scale application of economic means to win a war. But to make the blockade more effective, he had to control the entire European coastline, either directly or through allies.

On the Continent, Russia continued to pose a threat to Napoleon's grandiose designs even after the disaster at Austerlitz. He pursued the retreating Russian army and fought an obstinate battle on ice and snow at Preussisch-Eylau in February 1807. Despite tremendous losses on both sides, the outcome of the battle was indecisive. At the Battle of Friedland, however, he routed the Russians, rendering further resistance useless. On June 23, Alexander concluded an armistice.

On 7–9 July 1807, the two emperors and King Frederick William of Prussia met on a raft on the Nieman River and concluded the Treaties of Tilsit. Alexander and Napoleon agreed to divide Europe between them: Russia recognized Napoleon's dominance in western and central Europe, and France supported Russia's claims in eastern Europe and Turkey. Alexander also agreed to join Napoleon's economic blockade against Britain.

The treaty with Prussia, however, was extremely harsh. That country lost half its territory to France, and all its acquisitions from the partition of Poland went to the newly established Grand Duchy of Warsaw. French troops occupied Berlin, and Napoleon's brother Jerome became king of Westphalia.

Napoleon had reached the zenith of his power and glory by 1808. He was now the master and arbiter of Europe. He created new republics, deposed royal dynasties, established the French Empire, set up a series of dependencies ruled through relatives or close friends, and ruled 70,000,000 people. Russia maintained friendly relations with him. Only Britain, Sweden, and Turkey remained outside French influence. For all practical purposes, Napoleon had achieved complete hegemony over continental Europe.

But soon Napoleon's power began to show signs of decline. The rise of romantic nationalism among the defeated European nations and Britain's persistent opposition to his expansionist policy turned the tide against him. Nationalism was most evident in the Iberian peninsula. When Portugal opened trade with Britain, Napoleon invaded Spain, precipitating the costly Spanish or Peninsular War (1808–1814).

Napoleon defeated the Spaniards, deposed King Ferdinand VII, and installed his brother Joseph as king of Spain. A Spanish guerrilla war, which the British supported and financed, tied down 200,000 French troops. When British troops invaded Spain with the support of Spanish guerrillas, Joseph relinquished the throne and fled Madrid. The Peninsular War marked Napoleon's first major defeat.

In 1809, Austria announced, rather prematurely, a "war of liberation" of the German people and launched an army of 170,000 into Germany. Napoleon left Spain and once again took command of the French army against the Austrians. He defeated them at the Battle of Wagram (5–6 July) and compelled them to sign the Treaty of Schönbrunn, depriving them of considerable territory and population.

Relations between France and Russia also began to deteriorate. Alexander lifted the economic blockade and opened the Russian ports to British trade. Napoleon, furious, embarked on his perilous invasion of Russia. On 24 June 1812, he crossed the Nieman River with his "Grand Army" of 500,000 men—Poles, Swiss, Dutch, Italians, Germans, Prussians, and Austrians. The French advanced rapidly, capturing one town after another. Near Smolensk, Napoleon tried to engage the Russian army, hoping to destroy it quickly. But the Russians evaded a pitched battle and allowed the French to press on through burning villages, towns, and cities.

After Smolensk, Napoleon faced the army of Kutuzov at Borodino, a village near Moscow. On 7 September 1812, he fought the Battle of Borodino, one of the bloodiest engagements in Russia. Despite repeated French attacks, the Russian defense did not break. Finally, the two armies retired toward the evening to their earlier positions, after more than 90,000 French and Russian soldiers had died on the battlefield.

Kutuzov did not renew the attack next day. He decided to withdraw his troops beyond Moscow and allowed the French to enter the city. His generals insisted on defending the city, but the veteran general told them: "When it becomes a matter of Russia's salvation, Moscow is only a city, like any other. But the loss of the Russian army means the loss of Russia. Let us, therefore, retreat!"

On 14 September, the last Russian detachment left Moscow, and next the entire population followed. Napoleon's vanguard entered the deserted city. Mysterious fires broke out throughout Moscow, burning it to the ground and devouring supplies and everything upon which the enemy had depended for shelter and subsistence.

Napoleon offered Alexander a truce but, after waiting five weeks without an answer, decided to retreat from Moscow. But the French retreat became a complete catastrophe, as swarms of angry Cossacks and infuriated peasants constantly harassed and attacked the French army, committing horrific atrocities upon the starving, freezing, and dying soldiers. The Russian army made surprise attacks on the French at the Battles of Tarutino, Maloyaroslavets, and Vyazma. At the crossing of the Berezina River, the Russians caused thousands of French soldiers to drown or freeze to death.

Napoleon, seeing the destruction of his Grand Army, abandoned it on 5 December 1812 and returned to Paris to raise another army. Early in 1813, he returned to meet the Russians, who had already made their way into Germany.

This time, however, Napoleon confronted not only the Russians but a new European coalition as well, the Sixth, which began with the Russo-Prussian alliance to wage the "war of liberation" in Germany. In a series of battles during the spring of 1813, Napoleon defeated the Prussians and Russians at Lützen and Bautzen.

Austria soon joined the coalition. Napoleon again overwhelmed the allied armies at the Battle of Dresden, his last major victory. But at the Battle of Leipzig, called the "Battle of the Nations," the allies destroyed Napoleon's forces (16–19 October 1813). On 31 March 1814, Czar Alexander entered Paris at the head of the allied armies and dictated the terms of surrender. Napoleon abdicated, and Louis XVIII became king of France.

Honorable, if not outright generous, Napoleon's captors, following the unwritten freemasonry of monarchy, exiled him to the island of Elba, assigning it to him as a sovereign principality and allowing him to retain the title of emperor. The French treasury provided him with an annual income of 2,000,000 francs.

The Elba exile, however, lasted only 10 months. In February 1815, Napoleon escaped from Elba aboard a small ship and landed in southern France. King Louis dispatched troops to intercept the escaped exile, but the soldiers went over to Napoleon! On 20 March, he entered Paris and began his ephemeral "Hundred Days" rule.

The alarmed allies, meeting at the Congress of Vienna, raised an army under Arthur Wellesley, the Duke of Wellington. Napoleon, too, somehow gathered a new army, marched to the north, and repelled the Prussian attack under General Gebhard Blücher at Ligny. On 18 June, the Battle of Waterloo ensued. Napoleon first assaulted Wellington's British army. At one point in the battle, when it seemed that the French would carry the day, Blücher returned and joined Wellington with reinforcements. Soon Napoleon's army was decimated. Waterloo signaled the end of the Napoleonic Wars and Napoleon's incredible military career. Almost to the very end, he could win battles against heavy odds, and, as the Duke of Wellington remarked of Waterloo itself, "It was a damned close-run thing."

The British exiled Napoleon to the distant volcanic island of St. Helena, off the west coast of Africa, where he arrived on 15 October 1815. He lived on this remote island until his death on 5 May 1821 at the age of 52.

To the French and continentalists, Napoleon represented the forces of the Enlightenment, reforming, rationalizing, sweeping away the frowzy remnants of the feudal past, an armed Voltaire seeing Europe whole. To the British and their allies, he was a monster of ambition, threatening to extinguish crowns and kingdoms if not stopped. The major differences between the nations of the Continent and Great Britain can still in large measure be attributed to whether they were conquered by Napoleon.

James J. Farsolas

See also: Berezina River, Battle of; Blücher, Gebhard Leberecht von; Kutuzov, Prince Mikhail Illarionovich Golenishchev; Moscow, Retreat from; Napoleon I

References and further reading:

Nosworthy, Brent. *With Musket, Cannon and Sword: Battle Tactics of Napoleon and His Enemies.* New York: Sarpedon, 1996.

Riehn, Richard K. *Napoleon's Russian Campaign.* New York: McGraw-Hill, 1990.
Connelly, Owen. *Blundering to Glory: Napoleon's Military Campaigns.* Wilmington, DE: Scholarly Resources, 1987.

Narses (c. 478–c. 574)

An Armenian eunuch and Byzantine general, born around 478 of obscure parentage. A bureaucrat by experience and training, he was sent in June 538 to reinforce Belisarius, then fighting the Goths.

In 551, Narses was sent by Justinian to resume Byzantine efforts to reconquer Italy from the Goths. In June or July 552, Narses won the Battle of Taginae, halfway between Ravenna and Perugia, and killed the Gothic king, Totila. He then defeated Totila's successor, Teias, in the Battle of Mons Lactarius, near Naples, in October 552. After Narses fought an inconclusive battle at Rimini in late 553 against Frankish and German invaders, the invading army divided. The smaller force, attempting to return north, was severely mauled at the Battle of Fano by a subordinate of Narses. The larger force was defeated by Narses at Capua in 554. As a result of these victories, Narses was granted the last triumph ever held in Rome. In 561 and 562, Narses defeated a Gothic revolt and in 565 a revolt by Heruli mercenaries. In 565 the death of the Emperor Justinian removed Narses's great patron. The new emperor, Justin II, recalled Narses in 566 or 567, shortly before the Lombard invasion. Narses died sometime thereafter.

Joseph M. Isenberg

See also: Belisarius; Byzantine-Persian Wars; Franks; Gothic War; Goths

References and further reading:

Fauber, Lawrence. *Narses: Hammer of the Goths.* New York: St. Martin's Press, 1990.
Norwich, John Julius. *Byzantium, The Early Centuries.* New York: Alfred A. Knopf, 1989.

Naseby (14 June 1645)

The last major battle of what has been called the First English Civil War. Naseby marked the entry of Parliament's New Model Army into the conflict and with it a new professionalism of arms. It also signaled the destruction of Charles I's hopes in the first part of this conflict and marked a major turning point in a struggle that would ultimately end in the execution of Charles I and the exile of Charles II.

By the spring of 1645, the Royalist cause in Britain was reeling. The defeat at Marston Moor the previous summer had meant the loss of northern England; various successes in Scotland had helped to offset that reverse. But Parliament had finally decided to reorganize its arms and had appointed Sir Thomas Fairfax to lead the new army. In early June, Fairfax's New Model Army of nearly 13,000 (approximately 7,500 of them horse or dragoons) was on the march, looking for Charles I, who himself was in the field with a force of around 7,400 (3,300 cavalry and 4,100 foot, although the numbers given on both sides represent compromises between differing claims).

The Battle of Naseby, 14 June 1645. (Hulton/Archive)

On the morning of 14 June 1645, the two sides drew up their lines north of the village of Naseby. The Royalists had assembled half a mile from Fairfax's men. Their right wing, under the command of the veteran Prince Rupert of the Rhine, consisted of 1,600 horse and 200 musketeers. The numbers of the center and left wings are less certain, although it appears probable that 3,500 foot comprised most of the center, with an unknown number of cavalry on the left. In reserve was King Charles with perhaps 700 foot and a contingent of horse. Opposing them, the parliamentary lines consisted of 3,200 horse on the left and another 3,500 horse to the right (the latter under Oliver Cromwell), with the foot in the center and one regiment of dismounted dragoons deployed along the left flank.

In midmorning the Royalist line advanced, with the cavalry soon engaged. On their extreme left, the parliamentary cavalry were dispersed, and Prince Rupert's horse rode behind their lines to attack the baggage train. In the center, the outnumbered Royalist foot soldiers held their ground but were soon attacked on both flanks; their left wing was broken by Cromwell's horse, and their right flank, exposed by the absence of Rupert's cavalry, was attacked by the remounted dragoons. By the time Rupert regathered his cavalry and returned to the field, the center of the Royalist lines was doomed. The reserves Charles held under his command fled, and with them went the last chance for Charles to make a stand.

The rout of the Royalist forces at Naseby was fairly complete: between 400 and 1,000 men were killed, and a further 4,500 were taken prisoner. Worse than this for the Royalist cause, all of Charles's papers were captured, and the terms of the sensitive negotiations they revealed crippled the Royalist plans. With the parliamentary cause in the ascendancy, Charles was incapable of fielding an army, and for all intents and purposes the first measure of the English Civil War was ended.

Daniel German

See also: Cromwell, Oliver; English Civil War (1642–1649); Marston Moor; Rupert, Prince

References and further reading:

Kenyon, John. *The Civil Wars of England.* London: Weidenfeld and Nicolson, 1989.

Young, P. *Naseby, 1645, the Campaign and the Battle.* London: Century Publishing, 1985.

Woolrych, Austin. *Battles of the English Civil War.* London: B. T. Batsford, 1961.

Nashville, Battle of (2–15 December 1864)

A Union victory in the American Civil War that basically destroyed the Confederate Army of Tennessee. It was a ragged, battered army that approached Nashville, Tennessee, in December 1864. Turning north after the fall of Atlanta, Confederate general John B. Hood hoped to draw General William T. Sherman out of Georgia, threaten his lines of supply, and

The Battle of Nashville, fought in December 1864, during the American Civil War. (Library of Congress)

retake Tennessee. Instead, Sherman launched his march to the sea, leaving General George H. Thomas to deal with Hood. Thomas quickly organized regular and garrison troops from scattered commands.

In a vicious battle at Franklin on 30 November, Hood shattered his army in frontal charges on federal earthworks. Attacking across open ground, the Confederates suffered atrocious losses. Thomas pulled back to Nashville and began to fortify the city. Hood arrived before the city on 2 December and began to dig in. The winter weather took a heavy toll on the Confederates, and their supply system broke down.

Within Nashville, Thomas received supplies and fresh men daily. Rather than besieging the city, the Confederates could only fortify the high ground below the city. By 15 December, Thomas was ready to strike, and he launched a feint against the Confederate right, while his main attack struck their left. Driven back, the Confederates hastily established a new line behind the first. The next day the federal attack continued, and Hood's army disintegrated. The left, at Shy's Hill, and the center both broke. Several African-American units distinguished themselves in the assaults.

Thomas lost 3,061 men and Hood 6,500. The Confederate Army of Tennessee was finished as an effective combat force. Hood was removed as commander, and the remnants transferred to North Carolina, where they surrendered to Sherman at Durham in 1865. The 1864 Tennessee campaign was the last Confederate offensive of the war.

Robert Dunkerley

See also: American Civil War; Hood, John Bell; Thomas, George Henry

References and further reading:

Groom, Winston. *Shrouds of Glory.* New York: Atlantic Monthly Press, 1996.

Sword, Wiley. *Embrace an Angry Wind.* New York: HarperCollins, 1994.

National Security Agency/ Central Security Service

Largest, most expensive, and perhaps least known of all the U.S. intelligence agencies. The National Security Agency (NSA) was established by President Harry S. Truman on 24 October 1952 as a separately organized agency within the Department of Defense to be in charge of signals intelligence (SIGINT) and communication security (COMSEC) for the federal government. The Central Security Service (CSS) was created as the central agency for cryptology. The head of the NSA is also in charge of the CSS.

The three principle functions of the NSA are information systems security, operations security training, and foreign intelligence information. It collects, deciphers, interprets, and disseminates information gathered from a vast array of global listening posts. The NSA was first brought to world attention when the North Korean navy seized the *Pueblo* in 1968. Later, congressional investigation revealed the ship was on an intelligence gathering mission for the NSA.

As the world becomes increasingly digital, so do the means states use to wage war. The NSA, with its technical assets and expertise, will become increasingly vital to national security in the information age.

Craig T. Cobane

See also: Central Intelligence Agency; Electronic Warfare; Intelligence, Military

References and further reading:

Bamford, James. *Puzzle Palace.* Boston: Houghton Mifflin, 1982.

Breckinridge, Scott D. *The CIA and the U.S. Intelligence System.* Boulder, CO: Westview Press, 1986.

Brownell, George A. *The Origin and Development of the National Security Agency.* Laguna Hills, CA: Aegean Park Press, 1981.

Richelson, Jeffrey T. *The U.S. Intelligence Community.* Boulder, CO: Westview Press, 1995.

Navarro, Pedro, Count of Olivetto (c. 1460–1528)

Military engineer and infantry commander. Pedro Navarro contributed greatly to the art of military mining and fortification, but his role is not widely known. He arrived in Italy in the 1490s and served as a *condotiere* for several different armies, including that of Florence. He fought for Ferdinand V of Spain in the Italian campaigns of 1502–1503, engineering the defense of Canossa and the capture of several Neapolitan fortresses using mining operations. He was given the title Count of Olivetto and made captain general in 1508. He commanded the Spanish campaigns in North Africa, capturing the island and town of Velez de la Gomera in 1508, the city of Oran in 1509, and the major ports of Bougie and Tripoli in 1510.

In 1513, Navarro led the infantry of the Holy League against the French at the Battle of Ravenna. He incorporated his own invention, a mobile cart featuring harquebuses and a protruding spear, into his infantry. He implemented a defensive strategy for the battle with trenches, but it failed because the artillery was weak. He was captured during the battle but never ransomed by Ferdinand. Navarro changed sides and, in 1515, led a French army through the Swiss Alps to campaign again in Italy. In 1522, he led troops on the French side at the Battle of Bicocca, where his field fortifications were successfully set up by the opposing side. That same year, Navarro was captured by the Spanish at Genoa and imprisoned at Naples until 1526. He lost his title as Count of Olivetto but engaged in one last campaign at Naples in 1527.

Christopher P. Goedert

See also: Marignano, Battle of; Ravenna

References and further reading:

Merriman, Roger Bigelow. *The Rise of the Spanish Empire in the Old World and in the New.* New York: Cooper Square Publishers, 1962.

Taylor, F. L. *The Art of War in Italy, 1494–1529.* Westport, CT: Greenwood Press, 1973.

Ndlela kaSompisi Ntuli (?–1840)

Commander in chief of the Zulu army and chief councillor to King Dingane kaSenzangakhona during the Voortrekker invasion of Zululand (1837–1840). King Shaka appointed Ndlela, who was connected through marriage to the Zulu royal house, chief of the Ntuli people in southern Zululand and, in recognition of his prowess as a warrior, raised him to high military command. When Dingane assassinated his brother Shaka in 1828, he also eliminated many of his favorites. Ndlela was an exception, for the usurper appointed him commander in chief and also made him his chief councillor.

In mid-1837, Ndlela led an inconclusive campaign against the Ndebele people, who were already weakened by defeats at the hands of the Voortrekkers, or Boers, who were advancing into the South African interior in search of lands to settle. When in late 1837 the Voortrekkers invaded Zululand, Ndlela persuaded Dingane to resist rather than negotiate. During the ensuing war of 1838, Ndlela undoubtedly planned the campaign, which turned on destroying the Voortrekkers in their fortified encampments of wagons (laagers) and in repulsing any Boer offensives. The turning point in the war was reached on 13–15 August 1838, when the Zulu army, led by Ndlela, failed repeatedly to penetrate the all-round fire from the Boer laager at Veglaer. The Boers then mounted a counterthrust, and Ndlela was in joint command of the great army that on 16 December disastrously failed at Blood River (Ncome) to stem their advance. Following this crushing defeat, Dingane withdrew to northern Zululand and ceded the lands south of the Thukela River to the Boers.

Dynastic conflict ensued in the weakened Zulu kingdom, and in September 1839, Mpande, Dingane's brother, fled to Boer territory with a large following. Mpande entered a compact with the Boers, and in January 1840 he marched against Dingane with Boer forces in support. Ndlela commanded Dingane's army, which Mpande's forces defeated at the Maqongqo hills on 29 January 1840. Ndlela was wounded but escaped. However, Dingane, now also a fugitive, executed him for his military failure.

John Laband

See also: Blood River (Ncome); Shaka kaSenzangakhona

References and further reading:

Knight, Ian. *Great Zulu Commanders 1838–1906.* London: Arms and Armour Press and Sterling Publishing, 1999.

Laband, John. *The Rise and Fall of the Zulu Nation.* London: Arms and Armour Press and Sterling Publishing, 1997.

Německý Brod (Deutschbrod) (1422)

Major Hussite victory over the Holy Roman emperor. After the failure of the First Crusade against the Hussites at the Battle of Prague (1420), Emperor Sigismund neglected to join with other German princes in the Second Crusade, which failed in its siege of Žatec. In October 1421, he finally entered Moravia and advanced into eastern Bohemia, meeting the Hussite army under Ján Žižka outside Kutná Hora (December 21). While Sigismund's Hungarian knights charged Žižka's wagon-fort, other troops entered the city, which was opened to them by German townsmen. Trapped between the town walls and Sigismund's army, Žižka attacked the king's lines, using his war wagons offensively for the first time as field artillery, and escaped with his army.

While Sigismund put his forces in winter quarters, Žižka gathered reinforcements from Prague and Tabor. With the support of local partisans, he attacked and overran the Hungarian garrison at Nebovidy (6 January 1422). Unable to gather his scattered forces quickly, Sigismund evacuated Kutná Hora and retreated toward Moravia. Žižka caught up with Sigismund north of Německý Brod on January 8. The Hussites attacked immediately and smashed the rear guard, causing the rest of the army to flee. Sigismund escaped with his life, but more than 500 knights drowned in the icy Sázava River. A small number of survivors, who sought refuge in Německý Brod, were massacred when the town fell to the Hussites two days later, after a short siege. The defeat of his second campaign against the Hussites marked the end of Sigismund's active involvement in the Hussite Wars, though he continued to press his claim for the Bohemian crown, finally achieved in the Compacts of Jihlava (1436), which brought the wars to a close.

Brian Hodson

See also: Hussite Wars; Žižka, Ján

References and further reading:

Heymann, George. *John Žižka and the Hussite Revolution.* Princeton, NJ: Princeton University Press, 1955.

Neville's Cross, Battle of (17 October 1346)

Scottish defeat during wars with England. In 1346, Scotland chose to honor her French alliance and take advantage of English commitments in France by invading northern England with an army of 12,000. It included a sprinkling of French knights, weaponry, and armor. Crossing the border above Carlisle, the Scots arrived before Durham on 16 October.

Forewarned of Scottish intentions, the English had two formations in the field as the Scots crossed the border. One, some 4,000 levies drawn from Cumberland, Northumberland, and Lancashire and commanded by the archbishop of York, immediately moved via Barnard Castle to Durham, where it scattered Scottish foraging parties on the morning of the 17th. The Scots immediately formed themselves into three formations below an Anglo-Saxon stone cross (Neville's Cross) on high ground heavily broken with ditches and walls. With memories of defeats at Dupplin Moor and Halidon Hill, the Scots had no intention of undertaking offensive action. The English were likewise inclined. The standoff was broken when the English advanced and their longbowmen wrought havoc in Scottish ranks. The men of the first echelon, rather than being killed where they stood, advanced, but the few who reached English positions were quickly dispatched. The second Scottish echelon, seeing the destruction of its sister formation, broke, whereupon the English advanced on the third Scottish force. Although this formation resisted fiercely, the day belonged to the English. King David of Scotland, along with many nobles, was captured and his army scattered.

H. P. Willmott

See also: Anglo-Scots Wars (1290–1388); Anglo-Scots Wars (1513–1560); Bannockburn, Battle of; Flodden, Battle of

References and further reading:

Prestwich, Michael. *The Three Edwards. War and State in England, 1272–1377.* London: Weidenfeld and Nicolson, 1980.

———. *Armies and Warfare in the Middle Ages: The English Experience.* New Haven, CT: Yale University Press, 1996.

Sumption, Jonathan. *The Hundred Years War: Trial by Battle.* London: Faber & Faber, 1990.

New Orleans, Battle of (8 January 1815)

Decisive American victory in the War of 1812, fought two weeks after the war ended. To defend the strategically, politically, and psychologically important port of New Orleans from imminent British attack, Andrew Jackson established a strong defensive position 6 miles downriver at Chalmette, Louisiana. His 5,800 mostly irregular troops, including Choctaw Indians, local bayou-dwellers, Tennesseans under John Coffee and William Carroll, the Louisiana Free Men of Color Battalion, and Jean Laffite's Baratarian pirates, built a line about a mile and a quarter long, perpendicular to the Mississippi River, extending along the northwestern bank of the Rodriguez Canal from the Mississippi levee across three-quarters of a mile of dry ground and another half mile into a cypress swamp. They widened and deepened the canal, using the mud to create breastworks. They had two dozen guns, the largest a 32-pounder. Coffee commanded the left, Carroll the center, and Jackson himself the right.

General Edward Pakenham sent 5,400 of his 8,000 regulars to attack frontally on the dry ground, providing clear targets for American marksmen. Samuel Gibbs led the main attack against the American center at the edge of the swamp, while Robert Rennie's column on the far left advanced parallel to the river. The 4,000 of Jackson's troops who found themselves engaged easily thwarted Gibbs with small arms and antipersonnel cannon fire but nearly allowed Rennie to gain the ramparts. Under John Keane, the 93rd Highlanders ran obliquely across the field from the left to reinforce the failing right, but to no avail. British artillery was bogged down in the mud too far behind the lines to be effective. Pakenham himself was killed while rallying the right. John Lambert assumed command and ordered retreat. Casualties numbered 2,000 British and 13 American. The British, especially their general, Colin Campbell, learned hard lessons at New Orleans that served them well at the similar Battle of the Alma 39 years later. But rarely has history recorded so one-sided a victory.

Eric v. d. Luft

See also: Alma; Campbell, Colin; Creek War, Jackson, Andrew; War of 1812

References and further reading:

Albright, Harry. *New Orleans: Battle of the Bayous.* New York: Hippocrene, 1990.

Brown, Wilburt S. *The Amphibious Campaign for West Florida and Louisiana, 1814–1815: A Critical Review of Strategy and Tactics at New Orleans.* Tuscaloosa: University of Alabama Press, 1969.

Owsley, Frank Lawrence. *Struggle for the Gulf Borderlands: The Creek War and the Battle of New Orleans, 1812–1815.* Tuscaloosa: University of Alabama Press, 2000.

Remini, Robert Vincent. *The Battle of New Orleans.* New York: Viking, 1999.

Ney, Michel, Duc d'Elchingen, Prince de La Moskova (1769–1815)

French field commander. Born the son of a cooper in Alsace on 10 January 1769, Ney enlisted in 1787 in the 5th Hussars and was commissioned a lieutenant after Valmy in 1792. He

Death of General Edward Pakenham at the Battle of New Orleans. (Library of Congress)

fought at Jemappes in 1792 and Mainz in 1794. Promoted to brigadier general on 15 August 1796, he won at Kirchberg on 19 April 1797 but was captured at Giessen on 20 April. After he was exchanged, he fought under André Masséna at Winterthur in 1799 and under Jean Victor Marie Moreau at Hohenlinden in 1800. Already a corps commander, Ney was promoted to marshal on 19 May 1804. He shone at Elchingen on 14 October 1805; captured Innsbruck in November; and fought well at Jena, Erfurt, and Magdeburg in 1806 and Eylau, Güttstadt, and Friedland in 1807. He led the VI Corps under Masséna from 1808 to 1810 in Iberia, capturing Ciudad Rodrigo in 1810. After setbacks at Bussaco in 1810 and Torres Vedras in 1811, Masséna accused him of insubordination and relieved him of command. During the 1812 invasion of Russia, Ney fought at Krasnoye, Smolensk, and Borodino and held the rear guard on the retreat from Moscow. In Napoleon's last German campaign, Ney fought at Weissenfels on 1 May 1813, was wounded at Lützen on 2 May, commanded the left at Bautzen on 20–21 May, lost at Dennewitz on 6 September, and proved tenacious at Leipzig.

Changing sides just before Napoleon's first abdication in 1814, Ney served King Louis XVIII, while Napoleon was exiled on Elba. Ordered to capture the escaped Napoleon, Ney instead joined him for the Hundred Days, commanding Napoleon's left on the march to Belgium, engaging Arthur Wellesley, the Duke of Wellington, at Quatre Bras, and fighting ferociously at Waterloo.

Ney was an outstanding cavalryman with extraordinary courage but was frequently criticized, especially after the Russian campaign, for his questionable battlefield decisions. Some historians have blamed him for the French shortcomings at Bautzen, Dennewitz, Quatre Bras, and even Waterloo. The humbly born Ney's rise to high command also illustrated the revolutionary and Napoleonic principle of "careers open to all talents." The restored Bourbons executed him by firing squad for treason on 7 December 1815.

Eric v. d. Luft

See also: Berezina River, Battle of; Bernadotte, Jean Baptiste Jules; Borodino; French Revolutionary Wars; Friedland; Jena and Auerstädt; Leipzig, Battle of; Masséna, André, Duc de Rivoli, Prince d'Essling; Moscow, Retreat from; Murat, Joachim, Grand Duke of Cleves-Berg, King of Naples; Napoleon I; Napoleonic

General Nelson Miles charging the Indian camp, Nez Percé war, Montana, 1877. (Library of Congress)

Wars; Quatre Bras and Ligny; Valmy; Waterloo; Wellington, Arthur Wellesley, Duke of

References and further reading:

Foster, John T. *Napoleon's Marshal: The Life of Michel Ney.* New York: Morrow, 1968.
Horricks, Raymond. *Marshal Ney: The Romance and the Real.* London: Archway, 1988.
———. *Military Politics from Bonaparte to the Bourbons: The Life and Death of Michel Ney, 1769–1815.* New Brunswick, NJ: Transaction, 1995.
Morton, John Bingham. *Marshal Ney.* London: Barker, 1958.

Nez Percé (June–October 1877)

In 1877, the flight of the Nez Percé captured the American public's imagination and made the U.S. Army look inept. Under the leadership of Chief Joseph and his namesake father, the Nez Percé had tried to live peaceably with impinging white society. In 1855, they agreed to a reservation in Idaho and, following the discovery of gold within their boundaries, had renegotiated the reservation to land surrounding the Clearwater River. Yet leaders like Chief Joseph the Elder and White Bird refused to live within the boundaries of this new agency. In 1871, the younger Chief Joseph was made chief and repeatedly insisted that the Nez Percé had never sold the Wallowa Valley in Oregon. Authorities insisted that Joseph had to report to the reservation and threw the Wallowa open to settlement.

While disagreeing with the decision, Joseph and his followers moved slowly to the reservation in June 1877. Three young warriors from White Bird's band disagreed with the decision and killed four white settlers. Fleeing to the gorges of the Salmon River, the Nez Percé pleas for peace went unheralded by local militia and cavalry. The Nez Percé were able to fend off their attackers and flee eastward to the inhospitable land on the south fork of the Clearwater River. General Oliver Howard soon caught up with them on a plateau above the Clearwater and engaged them for two days before the Nez Percé escaped.

On 15 July at Weippe Prairie, the Nez Percé chiefs decided the best course of action was to flee to Canada. While resting at Big River Hole in Montana on 9 August, the 900 Indians were attacked by 200 soldiers, who killed 89 Indians. But the warriors were able to pin down the soldiers for two days, allowing their families to escape. Joseph and his followers continued to dodge their pursuers and escape at places like Canyon Creek. On 30 September, at Snake Creek, less than 40 miles from the Canadian border, Colonel Nelson Miles engaged the hostiles. Three hundred Nez Percé led by White Bird were able to make it to Canada, but on 5 October, Joseph surrendered to Miles, and the Nez Percé war was at an end.

T. Jason Soderstrum

See also: American Indian Wars; Joseph the Younger, Chief; Miles, Nelson Appleton

References and further reading:

Brown, Mark Herbert. *The Flight of the Nez Percé.* New York: Putnam, 1967.

Hampton, Bruce. *Children of Grace: The Nez Percé War of 1877.* New York: H. Holt, 1994.

Lavender, David Sievert. *Let Me Be Free: The Nez Percé Tragedy.* New York: HarperCollins, 1992.

Nicaragua, Walker's Invasion of (1855–1857)

Failed filibustering expedition to Nicaragua led by the American William Walker. Known to many as the "Grey-Eyed Man of Destiny," William Walker was both a pro-slavery southerner and a firm believer in America's Manifest Destiny to rule the Americas. Possessing a strong desire to not only spread the institution of slavery but also to lead an independent state himself, Walker organized and led a filibuster (soldier-of-fortune) expedition to "liberate" Baja California and Sonora from Mexico in 1853. From the start, the expedition was a fiasco, and Walker and his ragged band returned to the United States in 1854. Although Walker's foray into Mexico was a failure, it did attract the attention of citizens throughout the United States and beyond, thereby allowing him to gain experience and luring him to continue such activities in the future.

In May 1855, Walker again set out to filibuster, this time in Central America. This time Walker had chosen to meddle in Nicaragua. It should not be overlooked that Nicaragua was the primary transit point for people and goods headed for the recently discovered gold fields of California. Furthermore, Nicaragua was in the midst of a civil war between two factions—the Granadans (conservatives) and the Leonese (liberals)—so named from their respective capital cities. Losing the conflict, the Leonese forces requested Walker's aid. Walker landed near Realejo on 16 June 1855 with 58 followers grandiosely calling themselves Walker's "Immortals" and soon brought military success to the liberal faction. Walker was given the rank of colonel by the Leonese authorities and placed in command of La Falange Americana, or the American Phalanx. (The term was eerily evocative of twentieth-century European fascism, with which Walker shared some attributes.)

After some military success and the seizure of a vessel run by Cornelius Vanderbilt's Accessory Transit Company, friction between Walker and the liberal leadership soon developed. As a result, in 1856 Walker staged a coup d'état and named himself commander of the armed forces and president of Nicaragua. In addition, Walker legalized slavery in Nicaragua and made English the official language of the state. However, Walker's tenure as president of Nicaragua was to be brief, as both the conservatives and liberals would soon unite, along with military forces from neighboring states, to oust the troublesome Walker and his fellow Yanquis. Facing a coalition of Central American states and hoping to avoid capture, Walker surrendered to U.S. naval forces on 1 May 1857 and left Nicaragua.

But Walker had not learned his lesson and, after leading another ill-fated expedition, was executed by a Honduran firing squad in 1860. William Walker was perhaps the most notorious of the nineteenth-century filibusterers.

Andrew G. Wilson

References and further reading:

Greene, Lawrence. *The Filibuster: The Career of William Walker.* Indianapolis: Bobbs Merrill, 1937.

May, Robert E. *The Southern Dream of a Caribbean Empire: 1854–1861.* Baton Rouge: Louisiana State University Press, 1973.

Nicaraguan Civil War (1925–1933)

Internal conflict that paved the way for the Somoza family dictatorship. In 1925, Conservative Party candidate Carlos Solórzano was elected president, and U.S. Marines withdrew from Nicaragua after a 13-year occupation. However, opposition to Solórzano was strong, a revolution quickly ensued, and the Marines returned in 1926. Emiliano Chamorro deposed Solórzano, but his efforts did not bring peace to Nicaragua as liberal vice president Juan Bautista Sacasa led another revolt. The U.S. Department of State then brokered a deal resulting in former conservative president Adolfo Díaz's return to power. Despite this diplomatic effort, the civil war was renewed and intensified. U.S. secretary of state Henry Stimson was appointed a special envoy by U.S. president Calvin Coolidge and sent to Nicaragua in 1927. Stimson negotiated an agreement between many of the warring factions that allowed Díaz to remain as president until 1928, at which time there would be U.S.-supervised elections. In addition, the United States agreed to help pacify the country and to establish and train a National Guard to maintain law and order.

Augusto César Sandino, one of the guerrilla leaders, rejected the accord and continued to fight both the government and the Marines. Sandino's hit-and-run guerrilla tactics frustrated all attempts to capture him and boosted his reputation. The failure of the Marines to capture Sandino caused embarrassment for the administration of incoming U.S. president Herbert Hoover, who sought a means to disengage. In 1932, the Liberal Party gained power with the election of Sacasa, and Anastasio Somoza García assumed command of the newly created National Guard. A truce was negotiated with Sandino, and the Marines left on 1 February 1933. Soon after the Marines withdrew, the National Guard attacked a rebel town, causing Sandino to declare a resumption of the war. President Sacasa offered to negotiate with Sandino, and they met with Somoza in Managua on 21 February 1934. After the meeting, Sandino was kidnapped and murdered by members of the National Guard, allegedly under Somoza's orders. Sandino's murder made him a martyr and a symbol of opposition to U.S. intervention in Central America. Anastasio Somoza used his power base as National Guard commander to overthrow Sacasa in 1936 and to establish a family dictatorship that made Nicaragua his personal family fiefdom until 1979.

George M. Lauderbaugh

See also: Sandino, Augusto César

References and further reading:

Macaulay, Neil. *The Sandino Affair.* Chicago: Quadrangle Books, 1967.

Walker, Thomas W. *Nicaragua, the Land of Sandino.* Boulder, CO: Westview Press, 1981.

Nicaraguan Civil War (1979)

The Nicaraguan Civil War began in 1961 with the creation of the Sandinista National Liberation Front (FSLN), a guerrilla group dedicated to the overthrow of the Somoza dynasty that had ruled the Central American republic as its private fiefdom since 1936. The movement gained adherents but was brutally suppressed by Luis Somoza Debayle, the president of Nicaragua, and his brother Anastasio, who commanded the National Guard. In 1963 Anastasio assumed the presidency and continued to fight the FSLN and all other opposition. In 1972, Nicaragua suffered a massive earthquake that devastated the capital, Managua, and the surrounding countryside. The Somoza regime demonstrated ineptitude and greed during the earthquake relief effort, thus intensifying opposition. Nevertheless, Anastasio Somoza was again elected president in 1974.

On 27 December 1974 FSLN guerrillas seized the minister of agriculture and several of Somoza's relatives and held them for ransom. The government accepted the rebels' demands, paid $1 million for the hostages' release, and provided safe passage to Cuba for the perpetrators. This event greatly enhanced the prestige of the FSLN but led the regime to increase attempts to crush the rebels militarily. By 1976, FSLN losses were high and one of the movement's founders, Carlos Fonseca, had been killed. However, the brutal tactics and atrocities committed against the civilian population supporting the FSLN drew international attention and condemnation. This pressure led in turn to a weakening of support for the Somoza government from the United States.

The FSLN struck the National Guard barracks at San Carlos in October 1977 and followed up by taking and temporarily holding several towns. Although not conclusive, these military successes resulted in moderate and conservative elements in Nicaragua throwing support to the FSLN. On 9 January 1978, Joaquin Chamorro, editor of the opposition paper *La Prensa* and a moderate opponent of Somoza rule, was murdered. Chamorro's assassination resulted in a general strike, street demonstrations, and a call for Somoza to resign. On 22 August 1978, Sandinistas commanded by Eden Pastora, whose nom de guerre was Comandante Cero, seized the National Palace and held hostage all the members of Nicaragua's Congress and some 200 government employees. Somoza was forced to pay $500,000 in ransom, release 60 Sandinista prisoners, and grant safe passage to Venezuela or Panama for Pastora and his men. This humiliation resulted in fissures within the National Guard, but Somoza arrested rebellious officers and remained in control of much of the country.

Somoza renewed his efforts to win a military victory by calling up reserves and creating a special combat unit commanded by his son, Major Anastasio Somoza Portacarerro, and ordering the capture of the Sandinista stronghold at the town of Matagulpa. However, in early September 1978, the FSLN gained control of several other towns and most of León, Nicaragua's second-largest city. The National Guard was on the verge of collapse but concentrated its tanks, artillery, and air power on the Sandinista-held towns and drove them out by 19 September.

The counteroffensive was conducted without regard for the civilian populations, and the Somoza regime was subjected to mounting international pressure for a negotiated settlement. In addition, U.S. support for the Somoza dictatorship continued to wane. However, all attempts at a negotiated settlement by international organizations and the United States failed.

On 4 June 1979, the FSLN launched its final offensive by calling for a general strike, which soon shut down most of the businesses in the country. The guerrillas had received new shipments of arms from Venezuela, Cuba, and Panama and by 6 June had taken control of major portions of León. A few days later, fighting broke out in Managua's poorer neighborhoods, and the National Guard responded with air attacks on residential sections of the city. Throughout June and early July, international pressure by the nations of the Andean Pact and the Organization of American States was exerted on Somoza to leave the country, but he stubbornly clung to power despite the fact that he had lost the support of the United States and controlled only some sections of Managua.

With his forces running out of ammunition, Somoza finally fled Nicaragua on 17 July. On 19 July, Sandinista troops entered Managua in triumph and were enthusiastically greeted by masses of Nicaraguans from all walks of life. On 20 July, a junta took control of the country.

The Nicaraguan Civil War was costly, with an estimated 30,000 to 50,000 Nicaraguans killed, 100,000 injured, and 300,000 left homeless or displaced to other countries. The economy was in a shambles, with most of the infrastructure of the country in ruin and a war debt of $1.5 billion. And Nicaragua would soon be facing another conflict, this time between the Sandinistas and the Contras.

George M. Lauderbaugh

References and further reading:

Black, George. *Triumph of the People: The Sandinista Revolution in Nicaragua.* London: Zed Press, 1981.

Christian, Shirley. *Nicaragua, Revolution in the Family.* New York: Random House, 1985.

Kinzer, Stephen. *Blood of Brothers: Life and War in Nicaragua.* New York: Putnam, 1991.

Lake, Anthony. *Somoza Falling.* Amherst: University of Massachusetts Press, 1989.

Millet, Richard, *Guardians of the Dynasty.* Maryknoll, NY: Orbis Books, 1977.

Nicephorus II Phocas (r. 963–969)

Byzantine general and emperor. Son of another Byzantine general, Bardas Phocas, Nicephorus Phocas began his military career in 955 as the successor to his father as domestic of the *scholai* (field forces). His father had been noted for his mastery of mobile tactics in defending against the Arabs, but the family itself, in spite of its military successes, had been under a cloud for some years because of the attempt of Bardas's brother, Leo, to seize power in 919.

Nicephorus came to a position of influence just as the Byzantine *reconquesta* was gaining ground, and he was soon actively involved. His greatest accomplishment came in 961, under Emperor Romanos II (r. 959–963), when he successfully recovered Crete for the empire, an act for which he won considerable public recognition. Still more recognition came the next year as a result of a highly successful raid against Arab Syria. It was primarily because of this recognition that Nicephorus was proclaimed emperor by his troops at Kaisareia in Cappadocia on 2 July 963, soon after the death of the emperor and a few months after a well-deserved triumph in Constantinople in recognition of his achievements.

This time, the Phocas usurpation was successful, and Nicephorus became emperor. As such, he continued his military successes in 965 by capturing Cyprus, Tarsos, and Mopsuestia. He was also successful in dealing with the Bulgarians, always a threat to the Byzantines of the time; his generals took Antioch in 969.

In support of the army, he increased land allocations to his soldiers and attempted to limit the growth of church lands by restricting bequests, although he supported the church in other respects. Despite the fact that Nicephorus was very much the military man of action, he fulfilled his ritual role as well, as we know from the account of papal envoy Liudprand of Cremona, although not as well as the well-liked Constantine VII, for whom the people continued to long.

Despite his success as ritual figure, there is also evidence that Nicephorus's reign was unpopular, in part on account of the heavy costs of the *reconquesta* itself, which seemed to lay a heavy burden upon the population of Constantinople. As a result, Nicephorus had constantly to be on the defensive as emperor. He was murdered by his nephew, the general John I Tzimisces, in cooperation with his empress, Theophano, the widow of Romanus II, whom Nicephorus had married to cement his claim to power. John (r. 969–976) became his successor.

Paul D. Buell

See also: Byzantine-Muslim Wars; John I Tzimisces; Muslim Conquests

References and further reading:

Norwich, John Julius. *Byzantium: The Apogee.* New York: Alfred A. Knopf, 1991.

Rosser, John H. *Historical Dictionary of Byzantium.* Latham, MD: Scarecrow Press, 2001.

Wittrow, Mark. *The Making of Byzantium, 600–1025.* Berkeley: University of California Press, 1996.

Nicholas, Grand Duke (1856–1929)

Russian general, commander in chief during the opening stages of World War I, and one of Russia's best commanders in that conflict. Nephew to Czar Alexander II, Nicholas was born on 18 November 1856 in St. Petersburg. After attending general staff college, he served with his father, Grand Duke

Nikolay Nikolayevich, commander in chief in the Russo-Turkish War of 1877–1878. He commanded the Guard Hussar Regiment in 1884 and served as inspector general of the cavalry from 1895 to 1905, where he instituted a much-needed modernization plan. His reform efforts continued in his next post as commander of the St. Petersburg Military District. From 1905 to 1908, he served as president of the Imperial Committee of National Defense, which was then dissolved.

Nicholas was not involved in war planning during the turbulent period from 1909 to mid-1914, but when World War I started, Czar Nicholas II named him commander in chief. Nicholas did a superb job of rallying the underprepared Russian forces against the Germans and the Austro-Hungarians, but the czar relieved Nicholas in September 1915 following defeats the previous summer. The czar, disastrously, assumed personal command of the military and assigned Nicholas command of Russian forces in the Caucasus. There he enjoyed success in campaigns in Armenia, where he captured the fortress of Erzurum in February 1916 and the port of Trabzon in April 1916. He successfully defended his gains against the Turkish offensive later that summer.

On the eve of the March 1917 revolution, the czar named Nicholas full commander in chief once again, but he held the post only a very short time. After being relieved by Prince Georgy Y. Lvov, Nicholas retired to the Crimea until 1919, when he moved to France. He died on 5 January 1929 in Antibes.

Harold Wise

See also: Bolshevik Revolution; Russo-Turkish Wars; World War I

References and further reading:

Danilov, Yuri N. *Le premier généralissime des armies russes, le grand duc Nicolas.* Paris: 1937.

Keegan, John. *The First World War.* New York: Alfred A. Knopf/Random House, 1999.

Nieuport (1600)

The first test of the Dutch army following Maurice of Nassau's reforms. Against his better judgment, Maurice was ordered by the Dutch government to "liberate" the Flemish coast from Spanish control. This move was strategically unwise, as the only Dutch base in the region was the isolated port of Ostende, but Maurice executed a brilliant logistical maneuver, moving his army of 14,000 to Ostende and marching down the coast to Nieuport. The capture of this city would deny the Spanish Netherlands access to the sea, so the Spanish governor-general, Archduke Albert of Austria, needed to break the siege. Many of his veteran units were in mutiny, but Albert convinced them to rejoin the army, gathering 10,000 men.

Archduke Albert's army made for the coast, falling on Maurice's rear guard at Leffinghem. This force was scattered, blocking any Dutch retreat toward Ostende. The Spanish then moved rapidly down the coast, hoping to fall on the rear of Maurice's siege positions. Meanwhile, Maurice abandoned the siege and moved his army across the Yser River to face the Spanish.

The battle, fought on 2 July 1600, demonstrated the advantages of the smaller, more maneuverable companies with a higher proportion of musketeers instituted by Maurice. Both armies were forced to redeploy inland as a result of the rising tide, a maneuver that the Dutch executed with ease. The Spanish *tercios* (infantry regiments) were held at bay by the firepower of Maurice's English contingent, under the command of Sir Francis Vere. While this fight raged in the center of the battle, Maurice successfully concentrated his cavalry on the inland flank and routed the Spanish horse, exposing the flank of the Spanish infantry.

A final charge by Maurice's army then drove the Spaniards from the field. It was a brilliant tactical victory, but Albert had forced the Dutch to give up the siege of Nieuport.

John S. Nolan

See also: Anglo-Spanish War; Ostende, Siege of

References and further reading:

Arnold, Thomas. *The Renaissance at War.* London: Cassell, 2001.

Oman, Charles. *The Art of War in the XVIth Century.* London: Methuen, 1937.

Parker, Geoffrey. *The Dutch Revolt.* London: Penguin, 1979.

Nigerian Civil War (1967–1970)

A brutal civil strife that exacerbated distrust and antagonism and has plagued Nigeria from independence to the present. Nigeria was bedeviled by regional differences at independence. Indirect rule under the British had created a north-south dichotomy. Then the 1957 Macpherson Constitution transformed regional councils into parliaments, making Nigeria a federation of three self-governing states at independence in October 1960, under Prime Minister Abubakar Tafewa Balewa. The north, the largest region, was predominantly Muslim and Hausa, the east largely Christian and Ibo, and the west religiously mixed but predominantly Christian Yoruba.

Although the federal system was touted as a model of diversity and achievement, regional differences sowed seeds of discontent. In 1965, the northern region again gained a majority in the federal legislature, and thereafter a climate of lawlessness gradually enveloped parts of the country in

1965–1966. Easterners and westerners chafed under "northern domination." A coup by young Ibo army officers achieved very little, but in rapid succession, prominent politicians (Prime Minister Tafewa Balewa; the western prime minister, Chief Akintola; and the northern prime minister, the Sardauna of Sokoto) were assassinated in January 1966. Anarchy was temporarily averted by the coup d'état of Major General Jounson Aguiyi-Ironsi. However, the calm lasted only until mid-1966, when General Aguiyi Ironsi was kidnapped and murdered in a countercoup on 29 July by mostly northern Hausa soldiers who detested his attempts at a unified government and the failure to punish the Ibo officers responsible for the earlier coup.

In a climate of accusation and counteraccusation, thousands of Ibos working in different parts of Nigeria as clerks and civil servants were killed by Hausas in retaliation. Lieutenant Colonel Yakubu Gowon emerged as head of state and called a conference in September 1966 to determine the form of national government. The effort met with very little success because of deep-seated divisions between the three regions. Concomitant with the conference's collapse in early October was a new and vicious rebellion by part of the army in the northern region. Mobs of Hausas again mercilessly slaughtered Ibos, especially those leaving the region. Consequently, Ibos boycotted the reconvened constitutional conference in November and threatened secession.

After months of unproductive negotiations in early 1967, Colonel Odumegwu Ojukwu assumed control of the eastern region and proclaimed the sovereign state of Biafra. It attained initial military success, but by 1969, Ibos were slowly driven into a smaller part of the eastern region with the capture of Enugu. The international community and African nations were dragged into the war. Ivory Coast, Zambia, Haiti, Gabon, and Tanzania recognized Biafra. The Nigerian federal government used Soviet aircraft and Egyptian pilots for bombing runs in Biafra. Britain also provided arms, but France, Spain, and Portugal supported Biafra, albeit on humanitarian grounds. The United States officially supported the federal government but provided medical supplies, foodstuffs, and other materials to Biafra.

After Biafra's initial military victories, the federal government gradually imposed a stranglehold on Biafra, cutting off all arms, food, and medical supplies to the region, leading to starvation and malnutrition. Children were collected into orphanages, but with little food to go around, many ended up in mass graves.

By early 1970, Biafra was no longer able to prosecute the war and surrendered, and Ojukwu fled the country. International efforts were mounted for relief supplies, but the operation was weakened by the Cold War rivalry of the United States and the Soviet Union.

Edmund Abaka

References and further reading:

Obasanjo, Olusegun. *My Command: An Account of the Nigerian Civil War 1967–70.* London: Heinemann, 1980.

Uwechue, Raph. *Reflections on the Nigerian Civil War.* New York: Africana Publishing, 1971.

Nightingale, Florence (1820–1910)

Near-legendary founder of modern nursing. The daughter of wealthy English parents, Nightingale was born in Florence, Italy, in 1820. She became interested in nursing after she claimed that God had spoken to her in 1837 and directed her to serve others. Florence studied with nurses in England, Alexandria, Egypt, and Germany. Nursing was held in low esteem at the time, and Nightingale chose not to marry in order to pursue her vocation with a passion.

When the Crimean War began in 1853, British secretary of war Sidney Herbert asked Nightingale to assist British forces. She trained 38 nurses for work in army hospitals and found horrifying conditions in the war zone. Thousands of casualties suffered in primitive hospitals with poor medical care and insufficient supplies, and the experience left Night-

Photograph of Florence Nightingale, c. 1845. (Hulton/Archive)

ingale with acute posttraumatic stress disorder (PTSD) for the rest of her life. She returned to England when the war ended in 1856 and, tormented by her memories, never again made a public appearance or statement.

Yet her widely publicized service in the Crimea had made her a legend, and Nightingale used her fame to encourage the professionalization of nursing. Arguing that a nurse's care was noble and never-ceasing, she established the foundation of modern nursing by writing numerous books and pamphlets and in 1860 founded the Nightingale School and Home for nurses in St. Thomas's Hospital, London. She received the British Order of Merit in 1907 and upon her death in 1910 was buried at her family's plot in East Wellow, England. Her casket was carried by six sergeants of the British Army. Britain honored her with the Crimean Monument in London in 1915, and the international community did so by creating the Florence Nightingale International Foundation in 1934. It is due primarily to Florence Nightingale's nearly obsessive efforts that nursing, military and civilian, is the respected profession that it is today.

Lance Janda

See also: Crimean War; Medicine, Military

References and further reading:

Dossey, Barbara Montgomery. *Florence Nightingale: Mystic, Visionary, Healer.* Springhouse Publishing, 2000.

Small, Hugh. *Florence Nightingale: Avenging Angel.* New York: St. Martin's Press.

Nine Years' War (1595–1604)

Also known as the O'Neill Rebellion, the result of increasing English involvement in Ireland in the 1580s. Ironically, Hugh O'Neill, the Earl of Tyrone, was the product of efforts to bring the island under control, having been taken to England at a young age and educated there. He was returned to Ireland in the late 1570s to rise to leadership of the long recalcitrant O'Neill clan of Ulster, bringing them to some degree of submission and in 1593 was elevated to the English title earl of Tyrone.

Though O'Neill had been considered a loyal subject, English officials in Ulster now reported that he was preparing rebellion. Though he was probably motivated by personal grudges, these reports were taken seriously by London, which sanctioned garrisons around O'Neill's territory. Caught between two cultures, O'Neill elected for the Gaelic in late 1594; he began negotiations with Spain and started forming an army equipped with modern firearms. In defiance of Elizabeth I's orders, he assumed the Gaelic title of "The O'Neill" in 1595. From that point on, he was considered to be in rebellion.

Fighting began in 1595, when O'Neill's brother Art attacked English garrisons on the Blackwater River. In May, O'Neill demonstrated his military talents by ambushing the English at Clontibert. He expanded his army to more than 10,000 men during a shaky truce he negotiated with Sir John Norreys in 1596. In 1597, as rebellion broke out in other parts of the island, the English decided to rebuild the forts on the Blackwater. O'Neill promptly laid siege to them, and in August 1598, an attempt by 4,000 English troops to relieve the forts led to their greatest defeat in the war at Yellow Ford, with more than half of the relief force lost. Subsequently, the province of Munster rose in bloody revolt against its English plantation.

In 1599, the Queen's favorite, the earl of Essex, brought over 16,000 troops in an attempt to salvage the situation. O'Neill whittled this force down to a mere 4,000 in just 21 weeks through the use of guerrilla tactics, combining traditional Irish tactics with modern firearms.

Essex was replaced by Charles Blount, Lord Mountjoy, who instituted new tactics, campaigning in winter and devastating the countryside to deprive O'Neill's forces of food. Though his first battle, against positions O'Neill had fortified at Moyry Pass (2 October 1600), was a draw, it was clear Mountjoy was gaining control of the situation, when on 21 September 1601, long-awaited Spanish forces arrived in Ireland. Unfortunately for O'Neill, they landed at Kinsale in the extreme south and were promptly besieged by Mountjoy. O'Neill had to face Mountjoy's army in an open field to lift the siege and was decisively defeated in the attempt on 24 December 1601. The Spanish surrendered soon after. O'Neill's rebellion was effectively crushed with the defeat at Kinsale, and he made his submission on 30 March 1603, having been defeated by the superior resources of a powerful state.

John S. Nolan

See also: Yellow Ford; Mountjoy, Charles Blount, Lord

References and further reading:

Falls, Cyril. *Elizabeth's Irish Wars.* London: Methuen, 1950.

Silke, John J. *Kinsale: The Spanish Intervention in Ireland.* Liverpool, UK: Liverpool University Press, 1970.

Wernham, R. B. *The Return of the Armadas.* Oxford, UK: Oxford University Press, 1994.

Nivelle, Robert (1856–1924)

One of the more disastrous of the French generals. Born at Tulle on 15 October 1856, Nivelle graduated from the École Polytechnique in 1878 and entered the artillery. Promoted to colonel (1911) and then brigadier general (October 1914), he commanded III Corps at Verdun and developed tactics that won him prominence. These consisted of training selected units to assault objectives in small groups. Attacks were preceded by deception barrages that would be halted to encourage the Germans to reveal their artillery positions. With the enemy guns silenced, attacks would then resume.

In April 1916, General Joseph Joffre gave Nivelle command of the Verdun Front (Second Army). Nivelle proclaimed "We have the formula!" and launched a series of local attacks beginning on 1 May that led, after initial setbacks, to the recapture of Fort Vaux (7 June) and Fort Douaumont (24 October). Because of his Verdun success and skill at self-advertisement, Nivelle was named commander in chief of the French armies of the north and northeast in December 1916, replacing Joffre. Nivelle's fluent English helped him secure approval from Prime Minister David Lloyd George for a plan to secure victory. The focus of the attack was in Champagne, and the key to success would be the "Verdun" formula, despite the difficulty of applying these tactics at the army level. The "Nivelle Offensive" (16 April–9 May 1917) was widely anticipated by the Germans. Aware of the French plan, they shortened their front and prepared defenses in depth. The offensive produced only minimal gains and 130,000 French casualties. It also led to widespread mutinies in the army and to Nivelle's replacement in May by General Henri-Philippe Pétain.

Nivelle declined command of an army group and submitted to review by a military inquiry in October that whitewashed him. He commanded French troops in Algeria in 1918 and served on the Supreme War Council after the war. He died in Paris on 23 March 1924.

Spencer C. Tucker

See also: Foch, Ferdinand; French Army; Joffre, Joseph Jacques Césaire; Pétain, Henri-Philippe; World War I

References and further reading:

King, Jere Clemens. *Generals and Politicians.* Berkeley: University of California Press, 1951.

Terrail, Gabriel. *Nivelle et Painlevé: La deuxième crise du commandant (décembre 1916–mai 1917).* Paris: P. Ollendorff, 1919.

Watt, Richard M. *Dare Call It Treason.* New York: Simon and Schuster, 1963.

Nogi, Maresuke (1843–1912)

Japanese general famous for siege of Port Arthur during Russo-Japanese War. Born in Tokyo in 1843 to a Choshu clan samurai father, Nogi served in the Restoration War of 1868, attained the rank of major in 1871, and fought in the Satsuma Rebellion in 1877. He was part of the Japanese officer contingent that studied military science in Germany in 1885–1886. He saw action during the 1894–1895 Sino-Japanese War in the siege of Port Arthur (24 October–19 November 1894) and in the Battle of Yingkow (9 March 1895) as a brigade commander. Nogi commanded the Third Army during the 1904–1905 Russo-Japanese War and oversaw the costly siege of Russian defenses at Port Arthur (22 June 1904–2 January 1905). Japan suffered 100,000 casualties before the Russians finally surrendered. Following the long siege, Nogi rushed his army to participate in the Battle of Mukden, another Japanese victory. Later, as headmaster of a private school, he tutored the future emperor Hirohito. Hailed as a hero, Nogi demonstrated his loyalty to the emperor by committing ritual suicide on 30 June 1912 following his ruler's death. Understandably, his home is today a shrine.

Harold Wise

See also: Mukden, Battle of; Port Arthur, Siege of; Russo-Japanese War; Sino-Japanese War (1894–1895)

References and further reading:

Warner, Denis, and Peggy Warner. *The War of the Rising Sun and Tumbling Bear: A Military History of the Russo-Japanese War.* London: Routledge, 1988.

———. *The Tide at Sunrise: A History of the Russo-Japanese War, 1904–1905.* London: Frank Cass, 2001.

Nongovernmental (Extranational) Organizations: Their Role in War and in the Wake of War

Civilian organizations that mobilize resources and individuals to alleviate human suffering in peace war and war. Since the mid–nineteenth century, people in the developed nations have founded organizations, such as the Red Cross, Cooperative for Assistance and Relief Everywhere (CARE), Catholic Relief Services, and Oxfam, aiming to provide humanitarian aid beyond local boundaries. Naturally, these efforts are applied to war-torn areas as well. The staff members of nongovernmental organizations (NGOs), neither established nor controlled by any government, bring the essentials of survival, food, clothing, shelter, and health care to people who have lost everything.

After 1945, the United Nations High Commissioner for Refugees (UNHCR) became a focal point for NGO efforts internationally. In most instances, UNHCR coordinates all major humanitarian relief efforts by NGOs in a particular country with that country's government.

Generally, the most influential and effective NGOs are religious-based organizations founded in the prosperous countries of the Northern Hemisphere. Most of their efforts are aimed at relieving major human disasters that are found usually in the poorer countries of the Southern Hemisphere. Because most NGOs have a long history, they are well-established bureaucracies having various levels of management and even perhaps competing suborganizations. Moreover, the NGOs frequently know each other's strengths and do not duplicate efforts. For example, among Rwandan refugees in Zaire, Oxfam alone established a water supply system for 800,000 people in 18 days.

In most cases, there are ongoing military conflicts either causing or caused by the human disaster. Thus, a range of important matters must be resolved in the relations between military forces and the NGOs. First, there may be a civil war in progress. The central government may oppose and refuse to permit NGO relief efforts from reaching those in rebellion. NGOs generally seek to remain neutral and even-handed in all conflicts. But their purpose of providing humanitarian aid to any and all people in need may lead to their staging their relief efforts in a neighboring country and crossing the international border directly into rebel territory. The government could see their efforts as rendering aid to its enemies, as happened in East Timor.

In Bosnia-Herzegovina, UNHCR and various NGOs providing relief took a different tack. Their convoys had to travel roads through areas controlled by Bosnian Serb forces in order to reach Bosnia Muslims who had been victims of ethnic cleansing. To obtain a peaceful accommodation, standard practice was to give a part of the relief aid to the Bosnian Serb forces that had ethnically cleansed the area in exchange for being allowed to proceed to deliver aid to the victims.

Second, occasions such as Somalia may arise in which the central government has ceased to exist. UNHCR and NGOs may seek military intervention to protect and assist them in their efforts to move relief supplies into the country and to the people suffering in the interior. As happened in Somalia, the results can lead to new levels of violence or new causes for military action. The U.S. military termed this effect "mission creep," and it cost American lives in the streets of Mogadishu.

In addition, some NGOs have the purpose of moving beyond emergency relief for saving people's lives to the nation-building or development phase. Here the task is to create new social and indigenous governmental institutions to sustain the people. Their concerns include the protection of human rights and the need for a war crimes tribunal. In these circumstances, repatriation or implementation of a peace settlement may require facing the issues that caused the military conflict originally. The various military forces in question may need to shift from a combat mode to a peacekeeping or peace-establishing mode.

Finally, NGOs have had to work out agreements with relief military units sent in to suppress violence. The NGOs looked upon the military as akin to those armed forces that started the problem, while the relief units, particularly those from the developed world, viewed the NGO personnel as "disaster groupies." Since approximately the time of the Kurdish protection operation in 1991, both sides have come to realize that they work much better through mutual cooperation; the NGOs need the military for its unsurpassed logistics and transportation facilities and skills, and the military has become mindful that its members are not trained for the complex tasks of running a civil economy over time and also realize that the NGOs, which generally will stay much longer than expensive military units, are their "tickets out of here."

John R. Popiden

See also: Red Cross; Refugees and Victims of Ethnic Cleansing

References and further reading:

Fernando, Jude L., and Alan W. Heston, eds. "The Role of NGOs: Charity and Empowerment." *Annals of the American Academy of Political and Social Science* (November 1997).

Maynard, Kimberly A. *Healing Communities in Conflict: International Assistance in Complex Emergencies.* New York: Columbia University Press, 1999.

Moore, Jonathan, ed. *Hard Choices: Moral Dilemmas in Humanitarian Intervention.* New York: Rowman & Littlefield, 1998.

Weiss, Thomas G., and Larry Minear, eds. *Humanitarianism across Borders: Sustaining Civilians in Times of War.* Boulder, CO: Lynne Rienner, 1993.

Nordlingen (1634)

A watershed battle in the Thirty Years' War. After two disastrous years, the battle ensured the survival of the Catholic and Habsburg cause and produced a balance of Protestant and Catholic forces within Germany.

From 1632 to 1634, the Swedish and Protestant position slowly unraveled. Imperial (Catholic) forces registered a series of small but cumulatively significant successes, until September 1634, when they won a battle in Bavaria that radically changed the situation within Germany.

At Nordlingen, a numerically superior Catholic army crushed a Swedish army that was committed first to a frontal attack on an entrenched hilltop and then, after having taken but lost this position, to a withdrawal across the front of the main part of the Catholic army. The attack by this uncommitted part of the Catholic army turned a retreat into a rout: the Swedish army lost 21,000 dead and prisoners from an initial strength of just 25,000 troops. Sweden's defeat marked the point at which it could no longer lead the Protestant cause, and Catholic France, unless it were to tolerate a Habsburg victory, was obliged to lead the war against Austria and Spain.

H. P. Willmott

See also: Gustavus II Adolphus; Thirty Years' War

References and further reading:

Parker, Geoffrey. *The Thirty Years' War.* London: Routledge & Kegan Paul, 1984.

Wedgwood, C. V. *The Thirty Years War.* London: Cape, 1953.

Polisensky, V. *The Thirty Years War.* Trans. Robert Evans. London: Batsford, 1970.

Norman Conquest (1066–1072)

William conquers and restructures England after victory at Hastings. The Norman Conquest of England was more than the military victory of one of three contenders for the English throne. It marked a complete restructuring of England's society, with some changes imposed and others constituting an effective application of native systems already in place. The Norman Conquest changed the ruling house, ruling class, and the legal languages of England. Changed too was ownership of the land. Before the conquest, individuals might own areas of land. Afterward, even church land was held by the king on feudal terms. This change allowed greater control over a feudal system in which many portions were held in fief from different liege lords. In England all loyalty, in the end, was to the monarch, and subinfeudation was mostly eliminated as a divisive element. The Anglo-Saxon ruling class was largely replaced by Norman supporters of William the Conquerer. As part of his administrative reorganization, a string of castles was built. The language of the court and the courts changed to French and Latin, respectively.

From 1068 to 1071, a series of regional revolts against William occurred. They were largely localized. William ended them with brutal efficiency. He laid waste to Yorkshire, killing all males and pursuing a scorched-earth policy that was apparent even 20 years later in the *Domesday Book.*

The Norman Conquest should be regarded as the seminal event in English history until the Reformation. It substantially altered the institutional structures of the church, the monarchy, and feudalism. It also changed the political, intellectual, and social framework of England, enhancing the existing rules and reinterpreting them. The character of medieval England was created in the crucible of the Norman Conquest.

Tamsin Hekala

See also: William the Conqueror; Hastings, Battle of

References and further reading:

Douglas, David C. *William the Conqueror: The Norman Impact upon England.* Berkeley: University of California Press, 1964.

Loyn, H. R. *Anglo-Saxon England and the Norman Conquest.* New York: St. Martin's Press, 1963.

Previté-Orton, C. W., ed. *The Shorter Cambridge Medieval History,* Cambridge; Cambridge University Press, 1962.

Norman-Byzantine Wars (1081–1108)

Wars through which Normans tried twice to conquer Byzantium through the Balkans but were repulsed by Emperor Alexius I Comnenus (r. 1081–1118). Around 1017, the first Norman warriors landed in southern Italy, looking for lands and loot. Byzantium held southern Italy and Sicily, but revolts by the Lombard population had enfeebled its grip. Supported by constant infusions of their kinsmen, the Normans carved up southern Italy, and by 1042, very few cities remained to Byzantium. In 1053, the Norman Robert Guiscard compelled the pope to recognize him as duke of Apulia and Calabria. In turn, Guiscard became a papal vassal, just as relations between the papacy and Byzantium devolved into mutual hostility and excommunication. In 1071, after a siege and blockade of nearly three years, the Normans captured the port of Bari, ending forever the Byzantine presence in Italy.

That same year, Byzantium faced a mortal crisis more dire than the loss of Italy. The 1071 Battle of Manzikert had destroyed the Byzantine army in Asia, and Turks now flooded into Anatolia. Additionally, Manzikert plunged Byzantium into civil war. As Alexius I Comnenus won the throne in 1081, Guiscard resolved to conquer Constantinople. The empire was in tatters, Norman hirelings in the Byzantine army threatened to defect, and a deposed emperor, Michael VII (r. 1071–1078), had betrothed a son to Robert's daughter. Thus, Robert Guiscard and his son, Bohemond, took sail that year and landed on the Epirus (Albanian) coast.

Alexius persuaded the Venetians to sink the Norman fleet but could not save Dyrrhachium from Robert's siege. A Byzantine relief force, heavy with Bosnian and Turkish mercenaries, suffered complete defeat, and by 1082, Robert held northwestern Greece. Byzantine money inspired rebels in Italy, forcing Robert to return home, but Bohemond continued to occupy Greek and Bulgarian lands. Alexius lost three more battles, until the Byzantines finally stopped the Normans in the spring of 1083. The alliance with Venice now paid off by blocking support to Bohemond, and the Normans evacuated the region. Robert Guiscard was preparing for a rematch when he died in 1085, but his aggression had cost Alexius almost all of Anatolia.

The First Crusade delayed the next Norman-Byzantine war, and Bohemond's Normans provided a major contingent to the Crusaders. After taking Antioch, Bohemond fell into Turkish captivity for several years. Once released, he struggled with Alexius's Armenian allies no less than with the Turks. By 1106, apoplectic at perceived Byzantine treachery, Bohemond returned to Italy to plan a second Balkan invasion. In 1107, Normans again landed in Epirus and threw up another siege around Dyrrhachium. Alexius then encircled the besieging army, as Venice sealed the coast. After months of relentless attrition, Norman forces finally collapsed in 1108. Pledging homage to Alexius, Bohemond returned to Italy. Norman threats to Byzantium remained dormant until the 1147 Sicilian crisis.

Weston F. Cook Jr.

See also: Alexius I Comnenus; Crusades; Guiscard, Robert

References and further reading:

Comnena, Anna. *The Alexiad of Anna Comnena.* Translated by E. R. A. Sewter. New York: Penguin Books, 1969.

Fine, John. *The Early Medieval Balkans.* Ann Arbor: University of Michigan Press, 1983.

Normandy Landings (1944)

The greatest amphibious landing operation in the history of warfare. Allied plans for an invasion of western Europe in World War II began soon after Germany declared war against the United States on 11 December 1941, with the appointment of General Dwight D. Eisenhower to design a plan for Allied victory in Europe. Eisenhower quickly developed two plans, one for 1942, called Operation SLEDGEHAMMER in case the Soviets were routed in the east, and a 1943 invasion plan called Operation ROUNDUP. British officials persuaded Americans leaders to focus their principal operations on North Africa and later on operations in Sicily and Italy. Soviet leader Stalin continued to press for a "second front" to lessen the German pressure on Russia.

Finally, at the Tehran Conference (November–December 1943), President Franklin D. Roosevelt and Stalin insisted to British prime minister Winston Churchill that May 1944 be the date for the invasion and that the Soviets would mount an attack on German forces to coincide with the European invasion.

Even though the British had been reluctant, Lieutenant General Frederick Morgan had worked on an invasion plan called Operation OVERLORD since the Casablanca Conference in January 1943. The landings were to be at Normandy, between Caen and the Cotentin Peninsula. Three Allied divisions were to be part of the landing, and two other divisions were to be air-dropped, with 11 other divisions to land within 14 days. Two artificial harbors were to be towed from England, and once a foothold in Europe had been established, several hundred divisions would be shipped from the United States and from across the channel.

The German high command had been aware of Allied cross-channel invasion planning for a long time, but with their forces dispersed in the Mediterranean and campaigns in the east, they were unable to fortify western Europe until November 1943, when Adolf Hitler issued Fuhrer Directive 51. Hitler appointed Field Marshal Erwin Rommel to oversee coastal defenses and command Army Group B. Although Rommel was able to lay 4 million mines, he was not able to position German tank divisions where he wanted because of the divided German command.

In January 1944, British general Bernard L. Montgomery was named commander of the ground invasion forces under Eisenhower. Montgomery demanded that five divisions (two British, two American, and one Canadian) make the initial landing and that the landing zone include the Orne River estuary. American landing forces would be led by General Omar Bradley and Canadian and British forces by General Miles Dempsey. Each of the five beaches where forces were to land was assigned a code name from east to west: Sword, Juno, Gold, Omaha, and Utah. One British airborne division was to land behind coastal defenses in the east and two American divisions in the west, while amphibious forces would swim ashore to prepare for the landing. To soften up German defenses, between 1 April and 5 June, 11,000 Allied aircraft flew 200,000 sorties, dropping 195,000 tons of bombs on strategic locations in France. Many of these raids were designed to persuade German forces that the landing would be northeast of the Seine. The Allies also created an entire phantom army in England under the command of George S. Patton, as well as false images of an invasion fleet sailing toward the Pas-de-Calais area on the night of the invasion.

Because of difficulties with assembling landing craft, the invasion was moved to June 1944. The invasion, threatened by foul weather, was nonetheless given the go-ahead on the morning of 5 June by Eisenhower. The cross-channel armada contained 3,000 landing craft, 500 naval vessels, and 2,500 other ships. Although the Luftwaffe had fewer than 400 airplanes in the area on D-Day, 13,000 fighters, bombers, and other Allied aircraft aided ground forces. The American 82d and 101st Airborne Divisions dropped into the Cotentin Peninsula and, although suffering heavy casu-

American soldiers wade from a Coast Guard landing craft toward the beach at Normandy on D-Day, 6 June 1944. (Library of Congress)

alties, secured their objectives, and the 6th British Airborne Division captured key bridges over the Caen Canal and Orne River.

At 6:30 A.M. on 6 June, British and Canadian forces landed on Gold, Juno, and Sword beaches with little opposition. The American forces at Utah faced a similar situation. But the 1st American Division at Omaha Beach was confronted by the 352d, the best German coastal division. The 6-mile section of beach between Port-en-Bession and the Vire River had 12 German strong points called Widerstandsnester and numerous other fighting positions on the cliffs surrounding the beach. By 8:30, landings had ceased at Omaha, leaving surviving American forces slowly to secure the beach and scale the cliffs. Navy destroyers steamed close in to shell German fortifications. By noon, German fire had noticeably decreased as U.S. troops took German defensive positions from the rear. An exit from the beach was finally opened, but not before the Americans had suffered 2,400 casualties and the Germans of the 352d 1,200 casualties.

German forces were caught in disarray, with Rommel on leave at home. At first, Hitler was unwilling to release an armored division for the counterattack, but he relented by midday, allowing the 21st Panzer Division to move to an area between Juno and Sword beaches, which almost reached the sea. But the Germans had long since lost any opportunity of throwing the Allies back into the sea; considering the vast superiority of the Allies across the spectrum of land, sea, and air power, it was a forlorn hope from the beginning.

T. Jason Soderstrum

See also: Eisenhower, Dwight David; Hitler, Adolf; Rommel, Erwin Johannes Eugen; World War II

References and further reading:

Ambrose, Stephen E. *D-Day, June 6, 1944: The Climactic Battle of World War II.* New York: Simon & Schuster, 1994.

Hunt, Robert. *The Normandy Campaign.* London: Cooper, 1976.

Kershaw, Robert J. *D-Day: Piercing the Atlantic Wall.* Annapolis, MD: Naval Institute Press, 1994.

Tute, Warren, John Costello, and Terry Hughes. *D-Day.* New York: Collier Books, 1974.

North Atlantic Treaty Organization (founded 4 April 1949)

Mutual defense alliance presently comprising 19 members from Western and Central Europe and North America and formed after World War II to offset the large conventional military advantage supposedly possessed by the USSR. Immediately following World War II, Belgium, France, Luxembourg, the Netherlands, and the United Kingdom signed a collective defense alliance termed the Brussels Treaty. It was soon recognized that the Brussels Treaty was no match for the Soviet military. Almost immediately, negotiations began with the United States and Canada to enlarge the collective defense arrangement. Negotiations culminated in the signing of the North Atlantic Treaty.

The original members of the North Atlantic Treaty Organization (NATO) included Belgium, Canada, Denmark, France, Iceland, Italy, Luxembourg, the Netherlands, Norway, Portugal, the United Kingdom, and the United States. During the next 50 years, NATO expanded to include Greece and Turkey (both joined in 1952); West Germany (1955); Spain (1982); and the Czech Republic, Hungary, and Poland (1999).

The heart of the NATO alliance is Article 5 of the treaty. In Article 5, signatories declared that an armed attack upon one member shall be considered an attack against all. In such cases, NATO members have the right, recognized by Article 51 of the Charter of the United Nations, to take whatever actions necessary to safeguard their security and territorial integrity. Post-1945 proponents of such a treaty used the argument that had Adolf Hitler been faced with a similar mutual defense network in the 1930s, he would not have gone to war.

One of the early issues confronting NATO in the first half of the 1950s was negotiating the participation of West Germany in the alliance. It was less than a decade since the end of World War II, and with Nazi occupation still fresh in the minds of many, the European powers were understandably wary of rearming West Germany. But it was recognized that a revived West Germany was key to NATO's success. The large German population, its growing economy, and its geostrategic location astride probable Soviet invasion routes made its membership in the alliance critical. The Soviet Union reacted to West Germany accession to NATO by creating the Warsaw Pact alliance in Eastern Europe.

In 1966, President Charles de Gaulle of France informed U.S. president Lyndon Johnson that France, although adhering to the basic tenets of the Atlantic Alliance, would take steps to exercise her full sovereignty. Subsequently, NATO troops were permitted use of French airspace or territory. Additionally, France withdrew from the integrated command structure and denied NATO the use of her troops. De Gaulle managed to have it both ways: France was free of the obligations of NATO but knew full well that the organization would not sit by and watch the country be again the victim of an aggressor.

Harry S. Truman holding the NATO treaty he signed on 25 July 1949. (Library of Congress)

Although NATO was assumed to have better-equipped and trained militaries, the huge numerical advantage supposedly possessed by the USSR and the Warsaw Pact meant that Western European security rested partly on the deterrent effect of U.S. nuclear retaliation. In 1979, NATO's Nuclear Planning Group agreed to station medium-range U.S. nuclear missiles (Pershing IIs) in Western Europe. Some member-states worried that deploying such weapons would accelerate the arms race. Additionally, several members had to contend with widespread civilian protests concerning deployment of the missiles. Many believed that the deployment of the missiles led to the 1987 Intermediate-Range Nuclear Forces (INF) Treaty banning medium-range missiles.

By the early 1990s, NATO was suffering from a crisis of identity. For half a century, the raison d'être for NATO had been to protect Western Europe from Soviet aggression. With the collapse of the Soviet Union in 1991, NATO's purpose became more difficult to define. (The members of the defunct Warsaw Pact engaged in no such navel gazing; they all wished to join NATO.) As leaders struggled with justifying not only the existence of NATO but a rationale for expanding NATO, events in the Balkans provided possible new missions and opportunities.

A new chapter in NATO's history began in the 1990s, when for the first time it engaged in military action. In April 1993, NATO warplanes began patrolling the skies over Bosnia and later began air strikes against Serbian military targets. Later, as part of the Dayton (Ohio) Peace Accords, NATO provided ground troops as part of a multinational peacekeeping mission in Bosnia.

In 1998, the Serbian province of Kosovo was the scene of widespread Serb persecution of its secession-threatening Albanian population. NATO responded with a 78-day bombing campaign, forcing Serbian leaders to capitulate. NATO ground forces were again inserted as peacekeepers. Critics wondered publicly about just how far NATO could go as the policeman of Europe and about the extent of American forces in this policing. Further, labeling the "aggressor" to be punished was much more complex than in 1949. For example, Albanians were "victims" when persecuted by Serbs but then "aggressors" when they turned upon the Serbs in Albanian-dominated areas.

Craig T. Cobane

See also: Yugoslavian Civil Wars

References and further reading:

Carpenter, Ted. *Beyond NATO.* Washington, DC: Cato Institute, 1994.

Gordon, Philip H., ed. *NATO's Transformation.* Lanham, MD: Rowman & Littlefield, 1999.

Northern Ireland, Civil War in (1969–present)

Conflict, known as "the Troubles," related to British control over six northeastern counties of the island of Ireland. The conflict has a long history, going back to the Norman conquest of England (1066). For almost a millennium, the fighting has ebbed and flowed with the tide of English control of the region. Casualties related to the Troubles (1969–2000) are in excess of 3,300 dead and 42,000 injured.

Early in the twentieth century, Britain negotiated the divestment of 26 Irish counties to the newly formed Republic of Ireland. The remaining six counties (Northern Ireland), possessing a 2–1 Protestant majority, were given their own parliament, known as the Stormont. From 1921 to 1968, Northern Ireland was ruled as a Protestant state whose purpose was to serve the interests of the Protestant majority, who felt that they had already given up more than enough to the "papists."

In the late 1960s, a civil rights movement, modeled on Dr. Martin Luther King Jr.'s movement in the United States, developed to address inequities and seek political changes. Some of the more radical and violence-prone groups demanded union with the Republic of Ireland. On 5 October 1968, nightstick-wielding members of the Royal Ulster Constabulary (RUC, the Protestant-dominated police force of Northern Ireland) attacked 400 peaceful marchers who were singing "We Shall Overcome." The entire episode was caught on film and led to riots erupting in Catholic sections of Northern Ireland's capital, Belfast.

As rioting erupted throughout the province, the exhausted RUC soon realized that they needed assistance, and the 2,000-strong garrison of British soldiers was pressed into service. At first, Catholics saw the British troops as saviors from the brutal RUC. Within a year, the British army's welcome had worn thin among Catholics. Over time, Catholics began to see British troops as an army of occupation that favored the Protestants. Tensions came to a climax on 30 January 1972, when British paratroopers fired upon a group of marchers, killing 14, an incident ever afterward known as "Bloody Sunday." The official version of the incident claims that paratroopers were returning fire from the marchers. No evidence to support this contention has ever been found.

Violence quickly escalated, and soon more troops were sent, ultimately reaching a total of approximately 21,000 in 1972. Soon, a near full-fledged civil war was being waged between Catholic and Protestant paramilitaries, with the RUC and British army caught in the middle.

During most of the 1970s, the "security phase" of the conflict dominated. Local police were put under control of army commanders. A large number of the elite British Special Air Service (SAS) troopers were introduced into the conflict as undercover operatives, with controversial results. The security measures were seen as counterproductive, and by 1977, a new policy of "police primacy" was implemented.

What British politicians originally saw as a relatively brief military deployment extended to more than a third of a century. The British army learned a lesson that has been taught many times in history: military force, no matter how well trained, equipped, and dedicated, cannot resolve a political conflict.

Craig T. Cobane

See also: Guerrilla/Partisan/Irregular Warfare; Special Operations Forces

References and further reading:

Coogan, Tim. *The Troubles: Ireland's Ordeal 1966–1996 and the Search for Peace.* Boulder, CO: Roberts Rinehart, 1996.

O'Brien, Brendan. *The Long War: The IRA and Sinn Fein, 1985 to Today.* Syracuse, NY: Syracuse University Press, 1995.

Northern War, Great (January 1700–August 1721)

Conflict involving Sweden against the Baltic powers, resulting in the replacement of Sweden by Russia as a European great power. Swedish expansion during the sixteenth and seventeenth centuries had antagonized the other Baltic powers. Russia's access to the Baltic was blocked; Denmark-Norway resented its loss of Scania; Brandenburg coveted Swedish Pomerania; and Poland desired Swedish Livonia. When Charles XII ascended the throne in 1697, Denmark-Norway began to organize an anti-Swedish coalition including Saxony-Poland, Denmark-Norway, and Russia.

The elector of Saxony, Augustus II, who was also king of Poland, attacked Livonia in January 1700. King Frederick IV of Denmark-Norway marched into Schleswig and Holstein in March 1700. In October 1700, Czar Peter I of Russia laid siege to Narva.

Charles made a daring landing a few miles from Copenhagen, compelling Frederick to sign the Treaty of Travendal in August 1700. Next, Charles raised the siege at Narva on 30 November 1700. Then he occupied Courland and forced Augustus to retreat into Poland. After Charles invaded Saxony, Augustus agreed to relinquish the Polish crown to Stanislaw Leszczynski and signed the Treaty of Altranstädt in September 1706.

Peter used this respite after Narva to undertake a series of reforms, the prime purpose of which was to reorganize and strengthen the Russian army. Charles resumed his attack on Russia in January 1708 with a force of 50,000. The Russians defeated an auxiliary Swedish force of 15,000 men at Lesnaia in October 1708. The main Swedish force was then trounced at Poltava in July 1709.

Charles fled to Turkey, where he convinced the sultan to declare war on Russia in 1710. However, after the Turkish victory at Jassy on the Pruth River in July 1711, the sultan decided to end the war with a negotiated settlement that returned control of the Azov region to the Ottomans.

Russia's victory at Poltava revived an anti-Swedish coalition with Saxony, Poland, Denmark, Prussia, and Hannover. Peter captured Viborg/Viipuri and Reval in 1710. During 1713–1714, the Russians occupied most of Finland. In 1714 the Russians defeated the Swedish fleet at Hanko and, having captured the Åland Islands, threatened Stockholm. Charles returned to the Swedish territory of Stralsund in November 1714 and made his way the following year to southern Sweden. He opened peace negotiations in 1717–1718 while simultaneously organizing an army of 60,000 men in anticipation of a new offensive.

In September 1718, Charles invaded southeastern Norway, but he was killed at the siege of Frederikshald in December 1718. The Swedish throne passed to his sister Ulrika Eleonora and later to her husband, Frederick I of Hesse-Kassel. Frederick negotiated a series of peace settlements from 1719 to 1721.

By the Treaties of Stockholm (1720–1721), Sweden settled with Saxony, Poland, Denmark, Prussia, and Hannover. Denmark ceded its conquests to Sweden in return for a substantial sum of money. Sweden ceded Bremen to Hannover and gave up Stettin and part of Swedish Pomerania to Prussia. By the Treaty of Nystadt (30 August 1721), Sweden ceded Ingria, Estonia, Livonia, and a strip of Finnish Karelia to Russia. Thanks to his battle victories and his diplomacy, Peter the Great had made Russia the dominant power in the Baltic.

Neville G. Panthaki

See also: Charles XII; Peter I, Romanov, Czar of Russia; Poltava; Russo-Turkish Wars (1676–1878)

References and further reading:

Anderson, Matthew Smith. *Peter the Great.* New York: Longman, 1995.

Kliuchevskii, Vasilii Osipovich. *Peter the Great.* London: Macmillan, 1958.

Lisk, Jill. *The Struggle for Supremacy in the Baltic, 1600–1725.* New York: Funk & Wagnalls, 1968.

Sumner, Benedict Humphrey. *Peter the Great and the Emergence of Russia.* London: English Universities Press, 1964.

Northern War, Second (1655–1660)

War among Sweden, Poland, Austria, the Dutch Republic, Denmark, and Brandenburg. (The First Northern War was fought almost a century earlier, in 1563–1570.) In 1655, Charles X Gustav, King of Sweden, decided to abandon his peace talks with King John Casimir of Poland and to attack his neighbor instead with an overwhelming force of 50,000 troops in Poland and Lithuania. This marked the beginning of the Second Northern War. The Swedes entered Warsaw without opposition on 29 August 1655. Polish noblemen, displeased with their king, surrendered to the Swedes, as did most of the Polish forces. Only Cracow offered a two-week resistance but had to surrender to Charles X on 9 October after running out of supplies. King John Casimir fled to Glogau (Silesia). But the monastery of Jasna Gora, near Czestochowa, resisted all Swedish attacks in a manner that was deemed miraculous.

The Protestant regime established by the Swedes in Poland provoked an outburst of national and religious feeling. In 1656, a general insurrection expelled the Swedes from southern and western Poland. King John Casimir gathered his troops and marched on Warsaw at the head of 25,500 regulars and 18,000–20,000 soldiers from the noble levy, undisturbed by the fact that he had neither infantry nor cannon at hand to besiege the city. The dismounted nobles and their hordes of servants attacked the walls of Warsaw repeatedly until it capitulated on 1 July 1656. A combined Swedish-Brandenburg relief army only a few miles away defeated the Polish forces because of the superiority of its cavalry in a three-day battle (28–30 July) at Warsaw. The Polish losses were relatively insignificant, and John Casimir was able to regroup his army. Even more important, he was able to convince Frederick William of Brandenburg to take his side.

With Charles X under heavy Polish pressure, the Danish National Assembly took the opportunity to launch the kingdom on a war of revenge against Sweden in 1657. Using Bremen and Verden as the base for their operations, Swedish troops marched north and struck deep into Jutland. The Danes withdrew most of their forces to the islands but left 6,000 soldiers in the newly rebuilt fortress of Fredericia to guard the passage from Jutland across the Little Belt to the island of Fyn. The Swedes pierced the bulwark on the night of 23–24 October, killing many of its defenders.

Even more unfortunate for the Danes, during the winter of 1657–1658, the water around the Danish islands froze harder than usual. In February 1658, Charles led an army of 5,000–10,000 across the waters of the frozen Little Belt. On 25 February, the Swedes surprisingly appeared in the suburbs of Copenhagen, forcing King Frederick of Denmark to sign the humiliating Treaty of Roskilde (8 March) that stripped Denmark of all of its possessions in southern Sweden.

In July 1658, a Polish-Austrian force laid siege to Swedish-occupied Thorn. Charles attacked Denmark again. This time, however, he had no success. Furthermore, a Dutch fleet broke the Swedish sea blockade of Copenhagen, while 10,600 Austrians under Prince Raimondo Montecuccoli, 14,500 Brandenburgers under Frederick William, and 4,500 Poles under Stefan Czarniecki marched from Hamburg through Schleswig into Jutland. The Swedes were trapped in the Danish islands between the Dutch and Danish fleets and a superior allied army. After the fall of Thorn in December and successful operations of Austrian, Brandenburg, and Polish troops in Pomerania in 1659, the Swedes were in retreat on all fronts. In 1660, the unexpected death of Charles X on 23 February led to the Peace Treaty of Oliva (3 May) among Poland, Austria, Brandenburg, and Sweden and the Treaty of Copenhagen between Denmark and Sweden, bringing the conflict to a close.

Juergen Luh

See also: Northern War, Great

References and further reading:

Frost, Robert I. *After the Deluge: Poland, Lithuania and the Second Northern War, 1655–1660.* Cambridge, UK: Cambridge University Press, 1993.

———. *The Northern Wars: War, State and Society in Northeastern Europe, 1558–1721.* London: Longman, 2000.

Opitz, Eckhardt. *Osterreich und Brandenburg im Schwedisch-Polnischen Krieg 1655–1660: Vorbereitung und Durchfuehrung der Feldzuege nach Daenemark und Pommern.* Boppard am Rhein: Harald Boldt Verlag, 1969.

Norway and Denmark, Invasion of (9 April–10 June 1940)

One of the British army's worst-fought campaigns. "Weserübung" was the code name for the German occupation of Norway and Denmark, aimed at securing Swedish iron ore deliveries coming from the port of Narvik (northern Norway) as well as the German domination of the Baltic Sea.

The German military sensed when Allied intervention in the Soviet-Finnish Winter War became likely. British and French plans included aid for Finland and the interruption of German iron ore supplies that were vital to Adolf Hitler's war economy. An armistice ended the Winter War on 12 March 1940, but German as well as Allied planning for some abrogation of Norwegian sovereignty continued. It is still not clear what one side knew about the other, but German claims that its invasion was launched only to prevent an Allied breach of Denmark's and Norway's neutrality are doubtful.

According to Hitler's order of 1 March 1940, a small staff prepared a combined operation of army, navy, and air force. This planning aimed at a peaceful occupation. Allied intelligence expected a German attack against France and was struck by surprise when naval forces (detected too late by the Royal Navy) landed troops at several Norwegian ports on the morning of 9 April 1940. Denmark, invaded by land and by sea, surrendered almost immediately. The sinking of the German heavy cruiser *Blücher* by a Norwegian coastal battery south of Oslo delayed the occupation of the capital. The Norwegian government and King Haakon VII gained time to escape Oslo and to mobilize the country. Elsewhere, the German navy had more success, taking by surprise main ports against little resistance. By April, the main Norwegian ports (Oslo, Stavanger, Kristiansand, Bergen, Trondheim, Narvik) were in German hands.

Resistance by the small Norwegian army continued in the interior. A coup attempt by Vidkun Quisling, the leader of the Norwegian Nazi Party who had talked to Hitler in December 1939 (presumably on the invasion of Norway), failed. The Germans, who were interested in negotiating with the legal government to achieve an agreement, dismissed Quisling. But fighting increased when Allied forces landed at western Norwegian ports between 14 and 18 April 1940. They retook Narvik and forced German mountain troops to withdraw toward the Swedish border. Meanwhile, German divisions, supported by massive air attacks, fought their way north. When the German invasion of France proved successful, the Allies had to withdraw their troops from Norway. The situation turning hopeless, King Haakon and his government left for London 7 June 1940; three days later the Norwegians surrendered but continued the war from exile.

Wehrmacht casualties reached 3,700 dead; the Allies lost 3,900 and the Norwegians 1,350 troops. The German navy, with almost all surface units engaged, suffered heavy casualties (3 cruisers, 10 destroyers sunk) and was unable for months to profit from its new Atlantic bases. Denmark and Norway remained under German occupation until May 1945.

Martin Moll

See also: British Military, Twentieth Century Organization and Structure

References and further reading:

Ottmer, Hans-Martin. *"Weserübung": Der deutsche Angriff auf Dänemark und Norwegen im April 1940.* Munich: Oldenbourg, 1994.

Novgorod, Muscovite Conquest of (1471–1479)

Grand Prince Ivan III of Moscovy annexes Novgorod. Novgorod was administered by an assembly called the *veche.* It usually elected the grand prince of Moscow, although he was forbidden to station troops or reside in the city. Lacking a military capacity and dependent on imported food, Novgorod's reputation was based on its merchants.

In 1456, boyars of the *veche* advocated alliance with Lithuania-Poland, contrary to the existing treaty with Muscovy. In retaliation for their efforts, Grand Prince Vasili II of Muscovy refused any longer to recognize the right of the *veche* to pass laws without his approval. Despite this, Novgorod boyars continued to push their agenda, led by Marfa Boretskaya. They gained the support of Prince Ivan Andreevich of Mozhaisk and Vasili II's cousin, Prince Ivan Dmitrievich. In 1470, the boyars invited Prince Mikhail Olelkvich of Kiev to defend Novgorod with troops.

In February 1471, the boyars forced the *veche* to recognize Casimir of Lithuania-Poland as Novgorod's sovereign. After a failed attempt at reconciliation by Metropolitan Filip of Moscow, Grand Prince Ivan III of Muscovy declared war in June 1471.

Muscovite and Tartar forces reached Torzhok by July. The main battle took place along the Shelon River. A Muscovite advance guard of 5,000 cavalry routed a Novgorod force of 40,000, killing 12,000 and capturing 2,000. Ivan III's peace terms were lenient. He imposed a 15,500 ruble fine, reinstated the Treaty of 1456, ordered Novgorod to accept the authority of the metropolitan of Moscow, and forbade future alliance with Lithuania-Poland. Ivan returned to Moscow in January 1476.

In March 1477, Novgorod sent a petition for Ivan's review, addressing him as *gosudar* (sovereign), rather than the usual *gospodin* (lord). Ivan took this as an indication that Novgorod was willing to strengthen ties. He immediately sent envoys to Novgorod. They were detained in Novgorod for six weeks and returned to Moscow with a negative reply. Personally insulted, Ivan declared war on 30 September 1477. By the end of November, a Muscovite army and the Kasimov Tartar cavalry from Tver laid siege to Novgorod. Novgorod capitulated in December and swore an oath to their new *gosudar,* Ivan. He dissolved the *veche,* abolished the office of *posadnik* (mayor), annexed parts of Novgorod's territory, and subjected the rest to tribute.

In 1479, the boyars sought an alliance with Khan Ahmad of the Golden Horde. Simultaneously, Novgorod received support from Ivan's brothers, Prince Andrei of Uglich and Prince Boris of Volok, who were negotiating with Casimir. Ivan learned of these actions in October 1479 and, without declaring war, laid siege to Novgorod. He entered the city on 15 January 1480. One hundred boyars were executed, and many of the gentry were exiled to Suzdal. Archbishop Feofil, who had been elected archbishop by the *veche* at the death of his predecessor in 1470, was deposed. Ivan and his brothers were reconciled in October 1480.

Neville G. Panthaki

See also: Ivan III

References and further reading:

Alef, Gustave. *The Origins of Muscovite Autocracy: The Age of Ivan III.* Berlin: In Kommission bei Harrassowitz, 1986.

Fennell, John Lister Illingworth. *Ivan the Great of Moscow.* London: Macmillan, 1961.

Grey, Ian. *Ivan III and the Unification of Russia.* London: Pelican Books, 1973.

Vernadsky, George, and Michael Karpovich. *A History of Russia.* Vol. 4, *Russia at the Dawn of the Modern Age.* New Haven, CT: Yale University Press, 1968.

Atomic bomb mushroom cloud over Hiroshima, 1946. (Library of Congress)

Nuclear and Atomic Weapons

Weapons that use nuclear explosives. With the discovery of fission in early 1939 came the possibility of building a bomb having unprecedented destructive power. The great Danish physicist Niels Bohr announced the discovery of fission to the scientific community during a theoretical physics conference organized by Edward Teller and George Gamov at George Washington University. Alarmed by the prospect that Germany might be able to develop an atomic bomb, Albert Einstein, persuaded by Leo Szilard, Teller, and Eugene Wigner, sent a letter to President Franklin D. Roosevelt advising the president of such a possibility. Einstein's letter led, ultimately, to the creation of the Manhattan Project to construct nuclear bombs.

As World War II grew more deadly, two physicists—J. Robert Oppenheimer and Enrico Fermi—made significant contributions to the American atomic effort. In the summer of 1942, Oppenheimer convened a study conference in his offices at the Berkeley campus of the University of California to explore the theoretical basis for developing a fission weapon. Attended by many notable physicists, including Hans Bethe and Teller, the conferees concluded that a fission bomb was possible and that gun assembly of uranium or plutonium, the only two metals known to be fissionable, offered the best hope of success. Fermi, working at the University of Chicago, achieved the first self-sustaining nuclear chain reaction in December 1942. Because an atomic bomb is an uncontrolled self-sustaining chain reaction, Fermi's achievement, along with the theoretical knowledge developed during Oppenheimer's summer study, allowed the United States to pursue a full-scale nuclear weapon development program.

World War II research and development of the atomic bomb had two principal components. The first component was to produce enough plutonium and uranium to make one or more atomic bombs. Production of these materials was difficult, and by August 1945, barely enough of each was available to build a mere two nuclear bombs, termed Fat Man and Little Boy. The second component of the wartime atomic program was to design and build an atomic bomb. This job was assigned to the newly created Los Alamos weapons laboratory and its director, Oppenheimer.

Established formally in April 1943, Los Alamos had the single mission of designing and building a fission bomb for use in World War II. Based on the conclusions of the Berkeley summer conference, primarily that gun technology was both well understood and relatively simple, Oppenheimer organized most of the Los Alamos effort on developing gun assembly. Such a gun, essentially a naval cannon, would shoot one piece of uranium or plutonium at a second piece. When the two pieces of material came together, a supercritical mass would be formed, causing a nuclear detonation.

Experiments conducted in the spring of 1944 showed, however, that light element impurities in plutonium would cause a premature, low-order detonation in a gun assembly. Such a detonation would be, in nuclear terms, a fizzle. This discovery was disturbing, particularly since Oak Ridge was having problems producing significant quantities of uranium. Without a method of using plutonium, development of a combat atomic bomb could be delayed and might not be available for use in World War II. Recognizing that high explosives could be used to implode, or crush, a ball of plutonium, causing fission and a high-order nuclear explosion, Oppenheimer reorganized Los Alamos in August 1944, centering most of the laboratory's work on developing implosion. Doubts about implosion, an untried and radical departure from established knowledge, remained until the successful Trinity test in July 1945.

Little Boy, the gun gadget using uranium, exploded over Hiroshima with a force of approximately 16 kilotons on 6 August 1945. Fat Man, the implosion device using plutonium, exploded over Nagasaki with a force of 20 kilotons on August 9, 1945. The use of atomic bombs against Japan ended World War II and inaugurated the nuclear era.

As soon as World War II ended, the entire nation, including the Los Alamos laboratory, demobilized. Senior scientists at Los Alamos returned to their prewar university positions, and younger staff left to enter graduate school. As a result, nuclear weapons work languished, and the United Stated possessed only a handful of such weapons. However, some research did continue on improving fission bombs, particularly the implosion device. The two wartime weapons were laboratory devices that could not be easily reproduced if needed.

In the summer of 1946, two slightly improved Fat Man bombs were used in Operation CROSSROADS at Bikini Atoll in the Marshall Islands. These tests, one an airburst and the other an underwater detonation, were designed to see how well atomic bombs worked against naval vessels. The tests were not very dramatic in that only a dozen ships in all were sunk, although the newsreel footage was spectacular enough and has been used in countless "message" films since 1946. Because of wide international press coverage, the not-so-dramatic effects of CROSSROADS may have led the Soviet Union to believe that atomic bombs were not to be feared and probably encouraged Stalin to increase Soviet international belligerence.

By early 1947, the United States had made the decision to use atomic bombs as a key part of the U.S. defense posture. The weapons laboratories at Los Alamos, New Mexico, Oak Ridge, Tennessee, and Hanford, Washington, started to rebuild from their postwar demobilization. In addition, the military began planning new weapons systems, including an expanded role for atomic bombs. Los Alamos tested new designs of fission weapons that met the new military requirements in Operation SANDSTONE, conducted in 1948. These designs made possible significant increases in stockpile numbers and delivery capabilities. Los Alamos continued to make fission bomb improvements into the 1950s, primarily working to make such bombs smaller and more efficient.

Simultaneous with fission bomb development in the late 1940s, work continued and accelerated on hydrogen bomb development. First studied in 1942, the idea for a hydrogen bomb came from the thermonuclear study of stars conducted in the 1930s by Hans Bethe. Unlike fission weapons that derive their energy from splitting atoms of the heavy elements uranium and plutonium, hydrogen bombs derive their power from fusing atoms of the light element hydrogen, particularly the isotopes deuterium and tritium. Because fusion can only be achieved with stellar temperatures, hydrogen bombs were not possible until such a heat source became available. The improved fission bombs of the late 1940s offered the promise of near-stellar temperatures.

After years of study conducted primarily on computers and the discovery of radiation implosion, the first hydrogen bomb was detonated in October 1951. The success of the first thermonuclear test ushered in a new era in nuclear weaponry—significantly increased destructive power. Fission bombs explode with energy levels measured in thousands of tons of trinitrotoluene (TNT) equivalents. Fat Man, for instance, exploded with a force of 20,000 tons of TNT and destroyed most of the city of Nagasaki. The first hydrogen bomb, code-named Mike, exploded with a force of over 10 million tons of TNT and vaporized the entire island on which the device stood. Although civilization could survive a war fought with atomic weapons, such was not the case with thermonuclear weapons.

Just as fission weapons started with crude designs and were continually improved, so too were thermonuclear weapons. The first hydrogen bomb relied primarily on the use of liquid deuterium, a cryogenic material. Its size, three stories high and weighing more than a million pounds, made it a "bomb" only in the explosive sense; obviously it was not a deliverable weapon. From 1951 through 1956, hydrogen bomb research focused on using dry fuel. This change made it possible to reduce the size of thermonuclear bombs and made them deliverable by a wider range of aircraft, including smaller naval planes flying off aircraft carriers.

Beginning in the late 1950s, delivery systems such as ballistic missiles governed design changes in nuclear weapons. Both land- and submarine-based missiles began to take on primary importance in potential nuclear weapons delivery. Nuclear weapons had to be reduced in size dramatically in order to fit into the much smaller spaces of nose cones. Although nuclear weapons had always been constrained in size by the lift capacity of bombers, the constraints imposed by missiles were several orders of magnitude more difficult to meet. Miniaturization of weapons became a dominant theme in nuclear weapons development throughout the 1960s.

By the early 1970s, the number of new weapons designed and built began to decline. Increasing emphasis began to be placed on improving and upgrading weapons already in the stockpile, as well as enhancing safety. It became increasingly important to know that nuclear weapons would only detonate on command and not by accident. Weapon accidents at Palomares, Spain, and Thule, Greenland, underscored this need. Many of the underground nuclear tests in the 1980s and early 1990s conducted by the United States were safety tests of stockpiled weapons. With the current ban on nuclear weapons testing, other methods, primarily computer simulation, are being used in ensure the safety of the U.S. stockpile. This activity, called Science-Based Stockpile Stewardship, is of critical importance as the age of individual weapons increases. As the cessation of nuclear weapons testing continues, the safety and reliability of the stockpile will be the key weapons concern well into the twenty-first century.

Roger A. Meade

See also: Hiroshima and Nagasaki, Atomic Bombings of

References and further reading:

Hewlett, Richard G. *A History of the United States Atomic Energy Commission.* Berkeley: University of California Press, 1989–1990.

Hewlett, Richard G., and Oscar E. Anderson Jr. *Atomic Shield, 1947–1952.* Berkeley: University of California Press, 1990.

Nuremberg Principle

A doctrine of the international law of war holding individuals accountable for their own actions. On 8 August 1945, representatives from the United States, Great Britain, the Soviet Union, and the provisional government of France entered into the London Agreement, establishing the International Military Tribunal, which would try individuals in the Nazi German government and military who had been accused of war crimes. Twenty-four major leaders were indicted variously for crimes against peace, crimes against humanity, war crimes, and conspiracy to commit such crimes. Trials commenced on 18 October 1945.

The principle of individual responsibility for actions in war was established at these trials. There was no defense available to an individual who pled that his or her actions were made only under orders from a superior, although such a pleading might lead to a mitigation of sentence.

This principle was also applied at the International Military Tribunal for the Far East and in the ad hoc Tribunals for Yugoslavia and Rwanda and is a doctrine of the International Criminal Court contemplated by the Rome Treaty of 1998. It is also a principle of military training in NATO member-states and most other states.

Steve Sheppard

See also: General Order No. 100; Laws of War

References and further reading:

Taylor, Telford. *The Anatomy of the Nuremberg Trials: A Personal Memoir.* New York: Alfred A. Knopf, 1992.

Trial of the Major War Criminals before the International Military Tribunal: Nuremberg, 14 November 1945–1 October 1946. Nuremberg, Germany, Allied Military Government, 1947–1949.

Nurhaci (1559–1626)

Founder of the Manchu state. Much of the early life of Nurhaci is shrouded in legend, but his military career began in his early twenties. From his youth, Nurhaci had close contact with the Chinese Ming Dynasty (1368–1644), which bordered his homeland of Manchuria. Indeed, his father and grandfather, Taksi and Giocangga, respectively, were allies of the local Ming garrison. Both died under mysterious circumstances.

After an internecine struggle, Nurhaci successfully united the surrounding villages and tribes. Using marriage alliances and carefully planned military campaigns, Nurhaci increased his strength before embarking on a war against the Ming Dynasty. During this time, he continued to demonstrate the utmost respect for the Chinese dynasty. By 1607, many eastern Mongol tribes recognized him as khan. By 1613, all but one of the Jurched (Manchurian) tribes submitted to Nurhaci. The final tribe, the Yehe, was aided by the Ming in their resistance.

Nurhaci's armies consisted mainly of horse archers, as well as cannon made by Jesuit priests. His primary achievement in military science was the banner system. In this, he organized companies of men under four banners of yellow, white, red, and blue. Soon the number of banners increased to eight, but the new ones added a fringe to their standard. The new banners consisted of 7,500 men divided into five regiments consisting of five companies. The banners replaced the tribal structure and created an efficient fighting machine.

In 1616, Nurhaci declared the creation of the Qing state. In 1618, he invaded the Ming Empire. His armies annihilated all the armies that opposed him, including the recalcitrant Yehe tribe. From 1621 to his death in 1626, Nurhaci campaigned extensively in the modern Liaoning Province of China. His death did not end the conquests because the new dynasty of Manchurian ethnicity he founded went on to conquer an empire larger than modern China.

Timothy May

See also: Chinese Imperial Wars; Manchu Expansion, Wars of

References and further reading:

Barfield, Thomas J. *The Perilous Frontier: Nomadic Empires and China, 221 BC to AD 1757.* Cambridge, UK: Blackwell Publishers, 1992.

Crossley, Pamela Kyle. *The Manchus.* Cambridge, UK: Blackwell Publishers, 1997.

Spence, Jonathan D., and J. Wills, eds. *From Ming to Ch'ing: Conquest, Region and Continuity in Seventeenth-Century China.* New Haven, CT: Yale University Press, 1979.

O

October War (1973)

The fourth Arab-Israeli War, a doctrinal watershed providing world military powers with lessons on the latest techniques in armored and air warfare and the first Arab-Israeli clash since 1948 in which the issue was not in doubt after the first few hours. Six years after Israel had humiliated its Arab neighbors in the Six-Day War, Egypt and Syria launched simultaneous attacks from the west and the northeast on Israel.

Israeli intelligence failed to read accurately the incipient two-front attack of 6 October 1973 principally because they were convinced that the Arabs would not attempt to move forward without absolute air supremacy and could not mount a cooperative, two-front effort. Further, the Israelis believed that their air force had developed effective tactics to foil surface-to-air missile defenses (SAMs) following the War of Attrition in 1970. Nor did the Israeli Defense Forces (IDF) contemplate Arab willingness to settle for a *limited* objective under the protection of a static, tightly integrated, multilayered air defense shield. Finally, the Israelis did not anticipate the precision of the Egyptian attack plan, a well-rehearsed, scrupulously orchestrated, shallow mass assault on a broad front. Inconceivable as well, based on the lessons of 1967, was the very idea of Arab initiative and multifront coordination between allies. In short, the Israelis were complacent on the eve of a major assault.

The fact that the attack came on the holiest Jewish day of the year, when reservists were at prayer, perhaps added to the surprise but actually facilitated mobilization by leaving roads bereft of civilian traffic. Exceptional Egyptian ingenuity and the adaptation of Soviet doctrine and technology, combined with a successful deception plan, contributed to the proficient cross-canal assault. It began with a massive artillery preparation, which covered the opening stages of the assault. Once across, teams of sappers skillfully cut pathways through the Israeli sand ramparts and blew holes for bridging units as 70,000 infantry fanned out laterally from the crossing points to set up recoilless rifle and antitank missile ambushes. The Israeli Air Force's scramble to wipe out the widening bridgeheads was foiled by the deadly high-level SAMs umbrella covering radar-guided antiaircraft gun batteries and handheld infrared SAMs, the last directed at treetop-level attacks. Israeli reserve armor was sent forward as soon as each sub-unit mobilized—that is, piecemeal. The advance company–sized packets were badly mauled two days after landing as they blindly charged into infantry-manned antitank killing grounds.

Meanwhile, two corps-sized Egyptian armies crossed the canal and consolidated shallow bridgeheads along an 80-kilometer (50-mile) front. On the second day, the bulk of the IDF Air Force was redirected to the northern front to try to stem the Syrian advance, also begun on 6 October. Syria had committed about 75 percent of its total armor (800 tanks) to the attack, which sent four columns westward north of the Sea of Galilee. Preceded by heavy artillery fire, three mechanized infantry divisions (in no less than 2,800 armored personnel carriers) were followed by two armored divisions, ultimately 1,400 tanks, against less than 200 Israeli tanks that redeployed to specially prepared chokepoint ambushes with interlocking fields of fire. By this method, the Israeli armor was able to make the Syrians pay dearly for their initial breakthrough. An assault by Syrian heliborne infantry seized the fortified Israeli observation on the commanding heights of Mount Hermon.

Because the Syrian encroachment posed the most immediate threat to Israeli territorial integrity by crossing the narrow neck of northeastern Galilee to threaten Haifa, the IDF had quickly devoted attention to this front, even though the Egyptian deployment was more menacing. By the second day, Israeli reserves were arriving in Galilee in sufficient numbers (an entire division) to put their highly accurate long-range tank gunnery to good use. Israeli ground attack planes at first flew into the missile umbrellas, regardless of cost. However, they soon developed evasive tactics and were able to hit a fresh column of armor advancing along the southernmost axis by coming in at treetop level over Jordanian territory, stopping the column after it had progressed almost 29 kilometers (18 miles)—the Syrians' deepest penetration. The Syrians soon outran their lines of communication and then changed their formation to line abreast, whereupon Israeli air decimated them. In this way, the Israelis bought 36 vital hours. Without a follow-on echelon to extend its incursion and lacking the improvisational capacity to regroup and explore alternative lines of advance, the Syrians were driven back with hammer blows, mounting a stubborn fighting withdrawal toward Damascus and Sasa, clear off the Golan Heights, and back to their start line by 10 October. The next day, advancing Israeli tanks were able to turn to deal decisively with the Iraqi and Jordanian armored attacks on the southwestern flank of the northern front.

On 14 October, the Egyptians responded to desperate Syrian appeals by breaking out of their secure defensive laagers and mounting a major assault. Backed by uncontested air cover, the Israelis were able to neutralize the infantry antitank teams with artillery fire, while precise long-range tank fire picked off advancing Egyptian T-62 tanks before the latter could bring their turret guns to bear, disabling more than 250 of the Soviet-built vehicles.

The next day, exploiting a seam between the two Egyptian bridgeheads discovered during the early containment probes, the IDF activated a contingency plan to span the canal into Egypt proper. Using improvised bridging equipment, the Israelis advanced to the canal through the gap between the two Egyptian army sectors. The Egyptian Second Army, on the northern flank of the crossing, slowly recognized the threat and mounted a delayed concerted effort to pinch off the corridor (Battle of the Chinese Farm). This effort was repulsed by General Abraham (Bren) Adan's division, which then crossed the canal into the bridgehead held by General Ariel Sharon's division. As Adan passed through and turned toward the southwest, Sharon attempted to surge northward and seize Ismailia but was stopped cold. Consequently, Adan's southerly push toward Suez City established the Israeli main effort. The Egyptian Third Army—astride the canal—was encircled from its rear and faced strangulation. By this time international pressure, including a direct Soviet threat and a United States counterthreat, brought about a cease-fire, just as Adan was turned back from Suez City by tenacious Egyptian urban tactics. Israel broke several successive cease-fires, as its forces tried to enhance its negotiating position. Israel had managed to turn a near disaster into a muted victory, losing 3,000 soldiers in the effort. The losses were commensurate with those of the other belligerents but unacceptable by Israeli standards. On the grand strategic level, the winner appears to have been Egypt, which broke the diplomatic stalemate that had been underpinned by Israeli military superiority and secured eventual return of the Sinai and a "cold peace" with Israel five years later.

Jim Bloom

See also: Israeli-Arab Wars; Six-Day War; Sinai-Suez Offensive

References and Further Reading:

Adan, Avraham. *On the Banks of the Suez: An Israeli General's Personal Account of the Yom Kippur War.* San Rafael, CA: Presidio Press, 1980.

Bartov, Hannoch. *Dado: 48 Years and 20 Days.* Trans. Ina Friedman. Tel Aviv: Ma'ariv Book Guild, 1981.

Cordesman, Anthony H., and Abraham R. Wagner. *The Lessons of Modern War.* Vol. 1, *The Arab-Israeli Conflicts, 1973–1989.* Boulder, CO: Westview Press, 1990.

Dupuy, Trevor N. *Elusive Victory: The Arab-Israeli Wars, 1947–1974.* New York: Harper & Row, 1978.

El-Badri, Hassan, Taha El-Magdoub, and Mohammed Dia El-Din Zohdy. *The Ramadan War, 1973.* Dunn Loring, VA: T. N. Dupuy Associates, 1978.

El-Shazly, Saad. *The Crossing of the Suez.* San Francisco: American Mideast Research, 1980.

Gawrych, George W. *The Albatross of Decisive Victory: War and Policy between Egypt and Israel in the 1967 and 1973 Arab-Israeli Wars.* Westport, CT: Greenwood Press, 2000.

Oda, Nobunaga (1534–1582)

In the mid–sixteenth century, Nobunaga Oda was a minor daimyo who rose from obscurity to become one of the unifiers of Japan. Oda first came to prominence when Imagawa Yoshimoto attempted to seize Kyoto by passing through Oda's territory of Owari. Oda, with an army of 2,000, routed Yoshimoto's force of 20,000 men. Then in 1568, Oda proceeded to Kyoto, captured the city, and installed Ashikaga Yoshiaki as shogun.

Oda then attempted to conquer all of Japan. At the time, implacable enemies surrounded him. The first venture in his attempt to unify Japan led Oda to attack the Buddhist monastery of Hieizan in 1571 to counter its temporal influence. This action resulted in the deaths of hundreds of monks and the burning of the monastery.

After initial successes in 1573, Oda attacked the fortress of Ishiyama. At times, an army of 60,000 men surrounded it, but the fortress did not fall until 1580. In 1573, he also deposed his puppet shogun, Yoshiaki, thus effectively ending the Ashikaga shogunate.

Oda then turned to strengthening his own territories. Between 1576 and 1579, he built the castle of Azachi on the shores of Lake Biwa. It was the first Japanese fortress specifically built to withstand the effects of cannon fire. Then in 1577, his general, Toyotomi Hideyoshi, led the attack against the Mori. Oda's dreams of conquest ended abruptly, however, when one of his own generals, Akechi Mitsuhide, assassinated him while the two marched with Hideyoshi's reinforcements in 1582.

Oda's legacy laid the basis for the unification of Japan, as he had conquered a third of it. He also changed Japanese warfare by using massive armies rather than retainers. To support his conquests, he restructured the system of taxation and disarmed peasants to prevent uprisings.

Timothy May

See also: Hideyoshi, Toyotomi; Japanese Invasion of Korea; Japanese Wars of Unification; Samurai; Sekigahara; Tokugawa, Ieyasu

References and further reading:

Okuno, Takahiro. *Nobunaga to Hideyoshi.* Tokyo: Shibundo, 1966.

Weston, Mark. *Giants of Japan: The Lives of Japan's Greatest Men and Women.* New York: Kodansha International, 1999.

Offa's Wars (771–796)

Anglo-Saxon king and builder of Offa's Dyke. Offa, son of Thingfrith, was the ruler of Mercia (757–796). He is best remembered for the immense barrier that carries his name, built in 787, although its history has remained obscure. Its main purpose seems to have been the defense of Mercia from the Welsh.

During the 770s, Offa gradually extended his influence over the whole of England south of the Humber River. Mercia had no natural boundaries. It was open on all sides to hostile kingdoms. It was thus a continuous struggle for Mercian kings to keep it intact. The only way for them to create natural boundaries was to subdue all others to their authority.

In 771, the Mercians crossed the Thames River into Sussex and overcame the men of West Sussex. The move may have been planned to consolidate Offa's influence in Kent, the most settled and civilized kingdom in England. In 776, according to the *Anglo-Saxon Chronicle,* the Mercians and Kentishmen fought at Otford. The outcome is not recorded, but it is significant that Offa possessed little authority in Kent over the next ten years.

Whatever happened at Otford did not stop Offa. He launched another expedition in 778, this time deep into Wales. He devastated the land and seized plunder. The next year, Offa moved south to attack Cynewulf of Wessex. The two kings fought at Benson, a West Saxon royal village on the north bank of the Thames River. Offa carried the day. A large tract of what is now Berkshire was annexed. Offa, now the most powerful king in Britain, could with justification call himself a *Bretwalda* (Britain-ruler).

In the mid-880s, Offa aspired to be accepted as an equal by continental monarchs. If any single event contributed to his European perspective, it was Offa's taking direct control of Kent in 785, either due to internal dissension or by direct invasion. The following year, Offa made a pact with Pope Adrian I, who formally addressed him as the *Rex Anglorum* (king of England). In exchange, the pope increased his control over the English church while acceding to Offa's request for the creation of an archbishopric of Lichfield. Thereby he freed, albeit temporarily, the Mercian clergy from the authority of the archbishop of Canterbury. Archbishop Jaenberht of Canterbury, a Kentishman through and through, had always been Offa's staunch adversary.

Offa brought Anglo-Saxon southern England to its highest level of political unification. He ruled East Anglia, Kent, and Sussex and maintained superiority over Wessex and Northumbria. After his death, Mercian power gradually gave way before Wessex.

Nic Fields

See also: Æthelbald's Wars

References and further reading:

Fox, Cyril Fred. *Offa's Dyke.* London: British Academy, 1957.

Stenton, Frank Merry. *Anglo-Saxon England.* Oxford, UK: Clarendon Press, 1987.

Office of Strategic Services

The U.S. intelligence, intelligence-gathering, and psychological warfare coordinating agency during the last three years of World War II and the ancestor of today's Central Intelligence Agency (CIA). The need for such an organization began to be felt in Washington after Nazi victories in northern and western Europe in the spring and summer of 1940. On the advice of Secretary of the Navy Frank Knox, President Franklin D. Roosevelt dispatched World War I hero and prominent New York lawyer William J. "Wild Bill" Donovan to England to study Britain's intelligence establishment, especially the Special Operations Executive (SOE). Later Donovan also toured the Mediterranean. In the winter of 1940–1941, preparations began for the establishment of an agency to handle intelligence, counterintelligence, and psychological warfare. As a result, in July 1941 the Office of Coordinator of Information (COI) was established under Donovan's leadership. Because the COI did not fulfill its creators' expectations, a year later it was replaced by a more powerful organization, the Office of Strategic Services

(OSS), also under Donovan's directorship. He reported directly to FDR and received his funding from him.

The new agency quickly expanded its budget, staff, and scope of activities. It gathered strategic information, disseminated propaganda and disinformation, and engaged in espionage and sabotage in enemy-occupied territories. At the height of its activities, the OSS employed 12,000 individuals as staff and agents and relied on the services of a great many part-timers and volunteers. One of the OSS's most successful missions was in northern Burma, where its Detachment 101 led hill tribesmen against the Japanese occupiers and, by coordinating with the British Chindits and Merrill's Marauders in its guerrilla warfare, cleared the area of the enemy.

The OSS was disbanded in the fall of 1945. Roosevelt's successor, the militantly plebeian president Harry Truman, had no use for the society blue bloods that dominated the "*Oh, So Social*," and for nearly two years the United States had no coordinated intelligence-gathering or unconventional warfare capability. Some OSS functions and staff were absorbed by the State and War Departments. In 1947, the CIA was established to replace the OSS but without the earlier agency's psychological warfare mission.

N. F. Dreisziger

See also: Guerrilla/Partisan/Irregular Warfare; Special Operations Executive; World War II

References and further reading:

Roosevelt, Kermit, ed. *War Report of the OSS.* New York: Walker & Company, 1976.

———. *War Report of the OSS: The Overseas Targets.* New York: Walker & Company, 1977.

Smith, Bradley F. *The Shadow Warriors: O.S.S. and the Origins of the C.I.A.* New York: Basic Books, 1983.

Troy, Thomas F. *Donovan and the CIA: A History of the Establishment of the Central Intelligence Agency.* Frederick, MD: University Publications of America, 1981.

Ögödei (c. 1186–1241)

Second of four Mongolian khans ruling the empire established by Genghis Khan. Ögödei was probably not his father's first choice as successor but was elected khan with a reasonable majority in 1229. He was immediately faced with two major problems: an empty treasury and the need for continuing expansion to reward his followers. The first problem he dealt with through a more rational approach to the conquered territories, emphasizing revenues over expropriation. The second, Ögödei solved by initiating new campaigns in almost all directions, with efforts to complete the conquest of north China coming under his personal control and culminating in the fall of the last Jin capital in 1234. Even before this event, another, even greater campaign was being prepared.

Joci, the oldest son of Genghis Khan, had been assigned the most distant pastures controlled by his father, in the extreme west. After Joci's death in 1225, his son Batu had taken over these pastures and had further expanded them. With the accession of Ögödei, Batu received further increments of manpower and the promise of a major western campaign. Although Batu, seconded by various other Mongol princes representing all the major lines, was theoretically to be in command, the actual organizer of the advance was the veteran general Sabutai (1172–1245). The result was a masterpiece of the tactician's art: an advance first into the Turkic areas bordering Russian on the south; then into Russia itself (1237), which was brought under Mongolian control for centuries; and finally into eastern Europe, as far as Liegnitz in Silesia and the outskirts of Vienna (1241), using almost the same lines of advance as those employed by the Soviets in 1944–1945. Only the death of Ögödei, news of which reach the Mongols around Christmas 1241, halted this attack, probably saving Austria and Germany from disaster.

Paul D. Buell

See also: Genghis Khan; Kublai Khan; Mongol Empire

References and further reading:

Allsen, Thomas T. *Mongol Imperialism.* Berkeley: University of California Press, 1987.

Buell, Paul D. "Sübötei-ba'atur." In *In the Service of the Khan, Eminent Personalities of the Early Mongol-Yuan Period (1200–1300)*, ed. Igor de Rachewiltz, Chan Hok-lam, Hsiao Ch'i-ch'ing, and Peter W. Geier. Wiesbaden: Otto Harrassowitz, 1993, 13–26.

———. "Chinqai (1169–1252): Architect of Mongolian Empire." In *Opuscula Altaica, Essays Presented in Honor of Henry Schwarz*, ed. Edward H. Kaplan and Donald W. Whisenhunt. Bellingham, WA: Center for East Asian Studies, 1994, 168–186.

Okinawa (1 April–21 June 1945)

World War II's last great battle. The large island of Okinawa was a part of metropolitan Japan and could provide a staging base for the projected invasion of the Japanese home islands. After ferocious naval and air bombardments, the first wave of U.S. Marines and army troops landed on 1 April to very light resistance. The Japanese, after losing on the beaches in their previous attempts to halt American island invasions, had devised the strategy of drawing their 500,000-man enemy to the rough terrain of the southern half of the island, where their 120,000-strong garrison would hold while kamikaze aircraft would pound the invasion warships. Meanwhile, the gigantic new battleship *Yamato* would beach itself and destroy what remained of the

A Marine of the 1st Marine Division draws a bead on a Japanese sniper during the Battle of Okinawa in 1945. (National Archives)

U.S. fleet. The Japanese commander, General Ushijima Mitsuru, understandably had his doubts about such a plan, but he faithfully followed his orders from Tokyo.

Okinawa's northern half was secured fairly early against scattered resistance, but then the invaders began to encounter stiff resistance from Japanese troops dug into a series of mutually supporting caves, ancient tombs, and ridge lines with interlocking forward and reverse slope defenses. The American offensive slowed drastically.

The Japanese then unleashed their counteroffensive. *Yamato* headed for Okinawa. On 7 April, it was sunk by U.S. naval warplanes, a loss that also spelled the end of Japan as a major naval power. But the kamikazes proved a far more effective weapon. In 1,900 one-way attacks, these manned projectiles killed more Americans than had been lost in all the United States' naval wars to date.

On the ground, the fighting had degenerated into a slogging match, transforming Okinawa's fields into a landscape reminiscent of World War I's western front, with gains measured in feet. Home critics began to criticize the way the fighting was going on the island. Fleet Admiral Chester Nimitz, Pacific fleet commander, took the unprecedented step of holding a press conference on Guam to defend his land commander, General Simon Bolivar Buckner. On 18 June, General Buckner was killed by Japanese artillery fire, the highest-ranking American to die by enemy action during World War II. Buckner was succeeded by General Roy S. Geiger (who thus became the only U.S. Marine officer to command an American field army).

By then, the Japanese were being steadily pushed back; they were finally also running out of aircraft and pilots for suicide missions. American troops and supplies flowed ashore in increasing numbers, tonnage, and safety. Three days after General Buckner's death, General Geiger was able to declare the island basically secured. General Ushijima and his staff had already committed ritual suicide, contemptuously rejecting a soldier-to-soldier demand for surrender from General Buckner.

The butcher's toll on Okinawa was grim: No less than 107,000 Japanese troops had been killed in battle and some 27,000 sealed in caves to die more slowly; only 7,400 prisoners were taken. U.S. Army forces suffered 7,613 dead and the Marines 3,561. The worst losses, however, were suffered by the civilians of Okinawa. Of a prewar population of 450,000, between 60,000 and 160,000 were lost; many were persuaded by Japanese soldiers to commit suicide. Okinawa was the

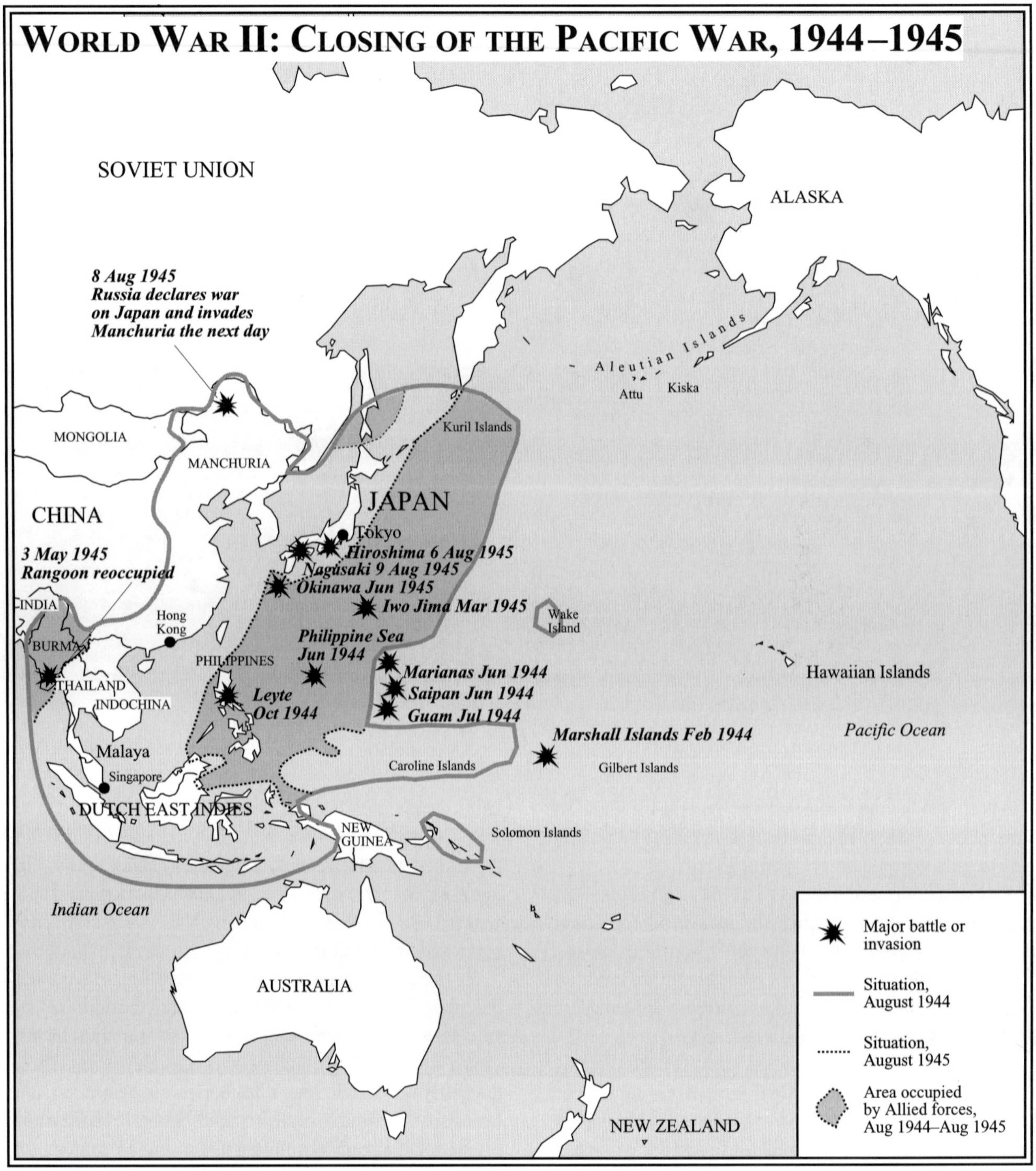

only battleground in the Pacific War in which large numbers of enemy civilians were encountered by the Americans. Yet Okinawa seemed to present a mere prelude to the greatest Pacific battle of all: the invasion of the Japanese home islands, a bloody scenario averted by the dropping of the world's first atomic bombs in warfare at Hiroshima and Nagasaki.

Stanley Sandler

See also: World War II

References and further reading:

Appleman, Roy E. *Okinawa: The Last Battle.* Washington, DC: U.S. Army Office of the Chief of Military History, 1948.

Feifer, George. *Tennozan: The Battle of Okinawa and the Atomic Bomb.* New York: Ticknor & Fields, 1992.

Gow, Ian (W. P. Willmott, consultant). *Okinawa, 1945: Gateway to Japan.* Garden City, NJ: Doubleday, 1985.

Omani Conquest of East Africa (1622–1730)

Conquest leading to the decline of Portuguese power on the Swahili coast. Asian traders (before 900) and Muslim traders (after 900) sailed across the Indian Ocean, linking the East African coast from the Horn of Africa to southern Mozambique and the adjacent islands in a network of seagoing commerce between Asia and East Africa. Ships and merchants from Arabia, Persia, and India participated in the trade in search of gold, ivory, tortoise shell, and other products. These exchanges led to the development of many coastal and island towns (Manda, Pemba, and Zanzibar; Pate, Lamu, Mombasa, and Vumbu in Kenya; and Kilwa in Tanzania) suffused with commercial activities and Islam, which became the nucleus of Swahili civilization (African and Muslim). Many traders and settlers from Oman and the Persian Gulf also intermarried with local women and settled down.

Vasco da Gama's trip around the Cape of Good Hope in 1497 and into the Indian Ocean brought the Portuguese to the East African coast. They attacked and occupied Kilwa in 1502, Zanzibar in 1503, Sofala and Kilwa in 1505, and Mozambique in 1507. By 1508, they had established control over the Swahili coast.

To break the Portuguese stranglehold, the people of Mombasa turned to the Omanis, who had already expelled the Portuguese from Muscat, their capital, in 1508. Utilizing the Mombasa appeal for help as the excuse, Omani sultan Ibn Saif sent a fleet that attacked the Portuguese settlements at Pate and Zanzibar. In 1696, Ibn Saif sailed to Mombasa with more than 3,000 men and, in 1698, took Fort Jesus.

The Omanis had imperialist designs, however; after freeing Mombasa from the Portuguese, they imposed their rule over the Swahili states, garrisoned Pemba, Kilwa, and other cities, and set governors (*walis*) over them. Limited rebellion and Swahili refusal to pay taxes between 1710 and 1740 led to political upheaval on the East African coast. In 1724, Kilwa broke away from Omani rule with the support of Europeans in Mozambique. By 1745, Pate, Malindi, Pemba, Zanzibar, and Mafia had all revolted against Omani overlordship.

Recognizing that only by closer attention to the coast would Omani control be effective, the Omanis made Zanzibar the focal point of their East African empire. Beginning with Sayyid Said, who came to power in Muscat in 1806, the Omanis strengthened their garrisons in East Africa and gradually asserted military and commercial control along the East African seaboard. Thus, the Omanis came to dominate Zanzibar and parts of the East African coast. They then replaced the Portuguese as the imperial power and monopolized Indian Ocean–Swahili trade. Omani overlordship reached its climax in 1840, when Sayyid Said transferred his political headquarters from Muscat to Zanzibar.

Edmund Abaka

References and further reading:

July, Robert. *A History of the African People.* Prospect Heights, IL: Waveland Press, 1998.

Woodfork, Jacqueline. "The Omani Empire." In *Africa.* Vol. 1, *African History before 1885,* ed. Toyin Falola. Durham, NC: Carolina Academic Press, 2000.

Omdurman (1898)

Anglo-Egyptian victory that effectively destroyed the Mahdist state, one of the most one-sided victories in modern military history, and the last large-scale cavalry charge by a major military power. The Mahdist army, which had taken Khartoum and killed General Charles "Chinese" Gordon in 1885 under Muhammad Ahmad (a.k.a. al-Mahdi), had warded off British attempts to reassert control over the region. In 1896, the British government, propelled by public opinion and concerns over French and Italian ambitions in the Sudan, prevailed upon the Egyptian government to launch an offensive toward Khartoum. Led by Horatio Herbert Kitchener, sirdar of the Egyptian Army, a mixed British and Egyptian force moved south. Using Nile gunboats and building a railroad to keep the army supplied, the 26,000-man force took all of two years to reach Khartoum.

The Mahdist army, now under the command of al-Mahdi's successor, Khalifa Abdullah al-Taashi, launched a number of strikes against Kitchener but was driven back each time. By late summer, the Anglo-Egyptian army was within a few miles of Khartoum. There, at Omdurman, the Khalifa launched a substantial assault on the morning of 2 September. The initial attacks focused on Kitchener's fortified camp along the Nile and the cavalry units protecting its right flank. Against well-protected Anglo-Egyptian troops armed with Maxim guns, the assaults made little headway. By noon, an Anglo-Egyptian counterattack had driven the Mahdist army from the field in more of a slaughter than a battle.

The Mahdists suffered more than 10,000 killed in a few hours of fighting. The Anglo-Egyptians lost 48. Omdurman effectively destroyed Mahdist power in the Sudan. The Khalifa, who was captured a year later, lost his capital city and many of his best commanders.

Kitchener, now a national hero, turned his attention to the French expedition encamped at Fashoda. After a brief standoff, the French withdrew in late September, securing British influence in the region.

Adam Seipp

See also: Churchill, Sir Winston; Gordon, Charles George; Kitchener, Horatio Herbert; Muhammad Ahmad

References and further reading:

Cassar, George. *Kitchener: Architect of Victory.* London: William Kimber, 1977.

Neillands, Robin. *The Dervish Wars: Gordon and Kitchener in the Sudan, 1880–1898.* London: John Murray, 1996.

Pollock, John. *Kitchener: The Road to Omdurman.* London: Constable, 1998.

Onin War (1467–1477)

Intensified period of civil war during the Onin era (1467–1469) that began in a succession dispute. The struggle soon engulfed the Ashikaga capital of Kyoto and led to its almost complete destruction before the war ended 10 years later. The period is also aptly known as "the epoch of a warring country."

When the eighth Ashikaga shogun, Yoshimasa (1435–1490), more interested in pleasure and the arts than in politics, announced his plans to retire at the tender age of 30, an effort ensued to find a successor to the heirless shogun. Hosukawa Katsumoto (1430–1473), a member of one of the most influential families supporting the shogun and the *kanrei* (the highest civil official in the shogunate, comparable to a prime minister), recommended Yoshimasa's brother Yoshimi, then living as a monk. Not long after Yoshimi's reluctant arrival in Kyoto to prepare for his duties, Yoshimasa's wife unexpectedly gave birth to a son and heir, Yoshihisa. The chief rivals of the Hosukawa, the Yamana clan, saw their opportunity and sided with Yoshimasa's wife, who wished to see her newborn son succeed. Only the weakness of the Ashikaga allowed this minor crisis to erupt into open war, which it soon did.

In 1467, the forces of the Yamana and Hosukawa houses came to blows in Kyoto. The decade of street warfare in Kyoto that followed was intermittent but still destructive. The once proud city and cultural center of Japan was reduced in size and population by more than half, and its culturally rich cityscape was utterly decimated.

In the end, the dispute settled nothing, and when the two original disputants, the respective heads of the Yamana and Hosukawa families, both died, the war came to an end. The decade did decisively weaken the Ashikaga shogunate, which became a powerless bystander to extended warfare that spread out into the countryside from Kyoto.

Daniel Kane

See also: Japanese Civil Wars; Samurai

References and further reading:

Berry, Mary Elizabeth. *The Culture of Civil War in Kyoto.* Berkeley: University of California Press, 1994.

Varley, H. Paul. *The Onin War.* New York: Columbia University Press, 1967.

Orleans, Siege of (12 October 1428–8 May 1429)

Most famous siege of the Hundred Years' War. During the later years of the Hundred Years' War, France responded to English tactical superiority on the battlefield (Cravant in 1423, Verneuil in 1424) by making use of positional warfare, forcing its enemy to engage in costly sieges.

In the summer of 1428, English regent John, Duke of Bedford, decided to invade southern France and its kingdom of "Bourges." The key bridgehead was Orleans, located 90 miles south of Paris on the Loire River. The campaign was intended as training for taking many small towns downstream and upstream of Orleans. Because of the need to deploy garrisons in many captured towns, English and Burgundian forces under the earl of Salisbury are estimated to have numbered no more than 4,000.

The people of Orleans, before being surrounded behind their walls, burned the suburbs to deny the English food or any comfort. The town militiamen were assigned to defend the 34 towers of the city wall, and the professional soldiers (first 500 and then 6,000), under Jean le Batard d'Orleans, better known as Dunois, were to make sallies. Before being killed by a cannonball (26 October 1428), Salisbury gave orders to isolate the city by a line of redoubts, or bastilles, on the western side. These strongholds were to be linked by ditches and earthworks manned by artillery. The northern side facing a forest was not covered by the English, which gave the besieged troops the opportunity to keep in touch with the Dauphin Charles's relief army.

The southern side of the Loire River was occupied by the English, and the fortification called "les Tourelles," protecting the bridge on the Loire River, was taken. The town was to be reduced by famine rather than by assault. But by April 1429, the Burgundians allies had quit the siege, leaving the English, who were too few to surround the city effectively.

The dauphin's army marched from Blois to relieve Orleans. Within its ranks was "the Maid," Joan of Arc, who had recently convinced Charles that he would soon be crowned King of France. Joan decided to move on ahead of the main French force and entered Orleans on 27 April to raise the spirits of the defenders.

Exploiting the weakness of the English forces, French troops (with Joan) sallied out on 4 May to seize the bastille on the eastern side. Two days later, they crossed the river to secure the southern bank. On 7 May, Joan attacked les Tourelles. The surrender of its English defenders persuaded Lord Salisbury to abandon the siege on 8 May. Joan of Arc cleared the way for Charles's coronation.

Gilles Boué

See also: Hundred Years' War; Joan of Arc

References and further reading:

Blanchard, Anne, ed. *Histoire militaire de la France.* 4 vols. Paris: PUF, 1992–1994.

Bradbury, Jim. *The Medieval Siege.* Woodbridge, Suffolk, UK: Boydell Press, 1997.

Corfis, Ivy A., and Michael Wolfe, eds. *Medieval City under Siege.* Woodbridge, Suffolk, UK: Boydell Press, 1995.

Osaka Castle, Siege of (1614–1615)

The final act of resistance by samurai on the losing side at the Battle of Sekigahara (1600). Ieyasu's Tokugawa assumption of hegemonic power in 1598, on the death of Toyotomi Hideyoshi, was opposed by forces led by Ishida Mitsunari (1560–1600). They backed Hideyoshi's chosen successor, his infant son Hideyori.

Tokugawa's victory at Sekigahara confirmed his mastery of Japan and his status as successor to the great project of Japanese unification begun by Nobunaga Oda and continued by Hideyoshi. After Sekigahara, the estates of the losers were largely curtailed, but the house of Hideyoshi was allowed to retain three provinces centered on Osaka Castle. Tokugawa, a ruthless strategist, would no doubt have preferred the complete elimination of Hideyoshi's family, including Hideyori, a potential opposition rallying point, but a narrow victory at Sekigahara and a still unconsolidated position initially precluded such a move.

Tokugawa bided his time while winning over many former Hideyoshi supporters. In 1614, he finally made his move. No daimyo came to Hideyori's defense, although almost 100,000 *ronin,* or masterless samurai, did. These *ronin* were the real losers at Sekigahara. They had been left destitute and desperate by the defeat of their lords.

Tokugawa's first assault against Osaka Castle, in the winter of 1614, was a costly failure, with over 35,000 casualties. Reverting to wile, Tokugawa offered an armistice to the Hideyori forces barricaded and besieged within their castle, provided that they allowed Tokugawa's men to fill in the castle's outer moat. When this was agreed to, Tokugawa instead filled in both the inner and the outer moat of the fortress. By the time that Hideyori's army realized what was happening, it was already too late. Tokugawa's forces overran the fortress on 3 June 1615. The destruction of the castle and the annihilation of Hideyori and his supporters, followed by the confiscation of the Hideyoshi estates, removed the final barrier to complete Tokugawa dominance in Japan.

Daniel Kane

See also: Hideyoshi, Toyotomi; Japanese Wars of Unification; Nagashino, Battle of; Samurai: Sekigahara; Tokugawa, Ieyasu

References and further reading:

Sadler, A. L. *The Maker of Modern Japan: The Life of Tokugawa Ieyasu.* Tokyo: Charles E. Tuttle, 1937.

Osan, Battle of (5 July 1950)

The opening ground clash between the invading North Korean forces and U.S. troops. In the days immediately following the North Korean attack in late June 1950, General Douglas MacArthur sought to delay the onslaught and buy time until U.S. units could arrive in strength to stiffen the South Korean defense and repel the invaders. MacArthur ordered elements of the U.S. 24th Division—two rifle companies, an artillery battery, and a few other supporting units—all lightly and inadequately armed—to Osan, south of the capital, Seoul, on the west side of the peninsula, by 5 July. MacArthur referred to this 540-man force—called Task Force Smith after its commander, Lieutenant Colonel C. B. Smith—as "an arrogant display of force."

The [North] Korean People's Army (KPA), 4th Korean Division, with its 33 Soviet T-34 tanks, struck Task Force Smith around 8 A.M. on 5 July. It suffered slight losses; the Americans did not have the numbers of men and equipment or the appropriate position to resist well, and the North Korean infantry flowed around them and turned their flanks. Two (and possibly three) KPA tanks were destroyed, however. The men of Task Force Smith had two choices—be captured or retreat—and thus retreated to the south.

Over the next several weeks, the 24th Division continued to seek to slow the North Korean advance, while the North Koreans—with greater numbers, tanks, and artillery—would seek to find them, fix them, and turn their flanks to crush them. The result was a steady American retreat down the west side of the Korean peninsula and a serious morale deflator for U.S. armed forces. However, the brief battle at Osan and others like it did create the breathing space to establish the perimeter at Pusan.

Charles M. Dobbs

See also: Korean War; MacArthur, Douglas

References and further reading:
Appleman, Roy E. *South to the Naktong, North to the Yalu, United States Army in the Korean War.* Washington, DC: Office of the Chief of Military History, 1961.
Sandler, Stanley. *The Korean War: No Victors, No Vanquished.* London: Routledge; Lexington: University of Kentucky, 1999.
Schnabel, James F. *U.S. Army in the Korean War: Policy and Direction: The First Year.* Washington, DC: U.S. Military Office of Chief of Military History, 1972.

Ostende, Siege of (1601–1604)

An epic siege that captured the attention of Europe. Habsburg forces commanded by Ambrogio Spinola labored to take the city, which was surrounded by water and reputed to be one of the strongest fortifications in Europe, from the Dutch Republic. An Italian skilled in siege warfare, Spinola employed the leading military engineers in Europe gradually to reduce the city one bulwark at a time. During the siege, Spinola showcased how far the fine art of military engineering had developed in the Habsburg forces, employing a wide range of innovative devices.

The Dutch and their English allies used their control of the sea to constantly replace the losses of the garrison. Maurice of Nassau, the general of the Dutch field army, had been repulsed in his attempt on the Flemish coast at Nieuport in 1600 and declined to attempt a relief operation, considering the isolated outpost to be hardly worth the cost of defending it. He instead used the distraction to make extensive gains further inland along the Maas and Rhine Rivers while the Spanish were fixated on Ostende.

The siege became an ongoing battle of attrition and a trial of wills, with the Dutch eventually recording losses of more than 30,000 people in the siege and nearly twice that many Habsburg troops dying in the assaults or from exposure and the diseases that ran unchecked through the trenches. Altogether, the city held out for three years and 77 days, surrendering in September 1604 only after every one of the outer bulwarks had been lost and the harbor closed by Spanish batteries. By taking the city, Spinola had solidified Habsburg control of the Flemish coast, but it was at best a Pyrrhic victory.

John S. Nolan

See also: Anglo-Spanish War; Nieuport, Battle of

References and further reading:
Arnold, Thomas. *The Renaissance at War.* London: Cassell, 2001.
Oman, Charles. *The Art of War in the XVIth Century.* London: Methuen, 1937.
Parker, Geoffrey. *The Dutch Revolt.* London: Penguin, 1979.

Ostrogoths

The Ostrogoths (Eastern Goths) were a Germanic tribe first organized into an empire of sorts in the third century in the region extending north from the Black to the Baltic Seas. They first grew to prominence under Ermaneric (fl. 350–376) just before being invaded by the Huns. The Ostrogoths, as vassals of the Huns, were moved westward into Dacia during the early fifth century. After Attila the Hun died in 453, the Ostrogoths asserted their independence and followed other Germanic groups by invading the disintegrating Roman Empire, first under Theodemir and then under Theoderic. The Ostrogoths received territory in Dacia and Lower Moesia from the eastern emperor, Zeno, in 483, but they continued to raid Thrace and threaten Constantinople. In 488, Theoderic, convinced by Zeno, invaded Italy, which was ruled by the Germanic barbarian Odovacar. After three years of campaigning, Theoderic and his Ostrogoths defeated Odovacar and conquered the Italian peninsula.

Theoderic strengthened alliances by marrying off two of his daughters and his two sisters to other Germanic kings. He supported the Alemanni against the Frankish king, Clovis, and took on the remnants of the Visigoths in France and Spain. At the height of his reign, Theoderic controlled Italy, Sicily, Provence, Rhaetia, and lands south and west of the Upper and Middle Danube River, and as regent for his grandson, Amalaric of the Visigoths, he had influence in Spain. In 535, nine years after Theoderic's death, the Ostrogothic king Totila successfully fended off Belisarius, the commander under the eastern emperor Justinian, who wanted to reunite the old Roman Empire. In 552, Totila was ultimately defeated by another of Justinian's commanders, Narses, and the power of the Ostrogoths that had built up over the previous two centuries came to an end.

Christopher P. Goedert

See also: Goths; Huns; Justinian I; Visigoths

References and further reading:
Burns, Thomas. *A History of the Ostro-Goths.* Bloomington: Indiana University Press, 1984.
Heather, Peter. *Goths and Romans, 332–489.* Oxford: Clarendon Press, 1991.
Goffart, Walter. *Barbarians and Romans, A.D. 418–584: The Techniques of Accommodation.* Princeton: Princeton University Press, 1980.

Otto I, the "Great" (912–973)

Real founder of the medieval German kingdom. Otto I is the only medieval German ruler to be called "the Great."

In Otto's time, Germany was made up of five largely independent duchies: Saxony, Franconia, Swabia, Bavaria, and

Lorraine. Otto took over Franconia and ruled Saxony by inheritance, putting down other ducal rebellions and consolidating control in Germany. He also made his brother Bruno archbishop of Cologne, a bastard son archbishop of Mainz, and another close relative archbishop of Trier, thus giving Otto control over the Catholic Church in Germany.

Rebellions by Otto's brother Henry and by Duke Eberhard of Franconia were ended by the Battle of Andernach in 939. This victory ensured German control over Lorraine. The following year, Otto campaigned deep into France. In 950, he campaigned against the Slavic Wends and gained suzerainty over Bohemia.

In 955, Otto won his most famous and important victory, defeating the Magyars (Hungarians) at Lechfeld, near Augsburg. After the battle, the Magyars ceased raiding into Germany and established an organized polity in what is now Hungary.

Otto was crowned emperor by the pope in 962. Hereafter, the papacy was fatefully linked with the German kingdom, and the German kings increasingly came to meddle in Italian affairs. It is thought that Otto's threatening posture toward pagan Poland probably led to Prince Mieszko's decision to convert Poland to Christianity in 966.

Michael C. Paul

See also: Holy Roman Empire; Lechfeld

References and further reading:

Barraclough, Geoffrey. *The Origins of Modern Germany.* New York: W. W. Norton, 1984.

Ottoman Empire (1300s–1922)

The Turkish ruler Osman founded the Ottoman state in 1299, utilizing the military capabilities of his people to consolidate control over Arab lands. Known for their endurance, discipline, mobility, warrior spirit, and equestrian skill, the Turks relied on equestrian archers to win their battles. In 1326, Sultan Orkhan organized the military into three principal contingents. The feudal armed forces received fiefs in exchange for military service and the provision of soldiers, but the land reverted back to the sultan if the recipient failed to fulfill his military duties or committed a crime. Janissaries, recruited Christian youths trained in the ways of Islam, served as the standing infantry. Organized by Murad II in the fourteenth century, these troops served as escorts, security guards, and defenders of the city gates during times of peace. Auxiliary troops consisting of scouts, armed nomads, and defenders of outlying fortresses rounded out the military structure.

After establishing administrative control over Anatolia, the Ottomans invaded Europe many times during the next 500 years. In 1354, the Gallipoli peninsula fell under their control. Under Murad I (1359–1389) their light cavalry moved further into the continent. With internal strife dividing the Christians, European leaders failed to prevent further expansion. By 1400, the Ottomans had defeated the Serbs and their allies at Kosovo and, at Nicopolis, the Macedonians, the Bulgarians, and the Hungarian king Sigismund, leader of the anti-Ottoman crusade organized by Pope Boniface IV. The Ottoman troops were poised for a strike at Hungary when Tamerlane's Mongol forces captured the sultan and defeated his men at Ankara in 1402. The Venetians and the Byzantines briefly recaptured Gallipoli and Salonika, respectively, but after Mehmed I emerged as the successor in 1413, Ottoman troops once again controlled these areas. His successor, Murad II, continued the expansionist military policies, defeating the Albanians, Greeks, and Romanians as well as another crusading army organized by Pope Eugene IV in 1444. Although the Ottomans had achieved many victories and expanded their territory considerably during the fourteenth and early fifteenth centuries, the beginning of the empire period dates from the reign of Mehmed II in 1451. After expanding and improving the army and establishing a navy, Mehmed conquered the Balkans and pushed into central Europe, the Ukraine, the Caucasus region, Arabia, and North Africa. After capturing Athens, Bosnia, and Herzegovina, the new sultan won a decisive victory against the Venetians at Lepanto. Then the Ottomans turned their attention back to the east to deal with the Persian challenge. The Persians, under the Safavids, and the Ottomans waged a series of wars until the eighteenth century, when they agreed on their borders.

With the Persians checking their advance in the east, the Ottomans shifted their expansionist efforts back to the west. Selim the Grim, a great warrior, conquered Syria and Egypt, leaving his son and successor, Süleyman the Magnificent, with a full treasury and an experienced army. In 1521, Süleyman captured Belgrade and then waged a campaign against the Hungarians and the Austrians. The Hungarian defeat at Mohács led to the division of the country, with half falling under Ottoman control and the rest controlled by Austria. After forming an alliance with the French king, Francis I, against Emperor Charles V, Süleyman resumed the war against the Austrians while intermittently halting the campaigns to fight the Persians. By 1562, peace treaties had been signed with both Persia and Austria. After experiencing two major naval defeats in the Mediterranean, Süleyman once again led his army against Austria. Before he could wage war against the forces of Maximilian II, who had refused to pay tribute, Süleyman died. The sultanate passed to his son Selim II, known as "the Sot," and the military strength of the Ottoman began a rapid decline.

In 1683, the Ottoman forces experienced a crushing defeat at the hand of the Viennese. Recognizing the superiority of European arms after the introduction of gunpowder, the Ottomans attempted to reform their military. Relying on Prussian advisers to train their troops, the Ottomans fell under their influence. When World War I broke out, Turkey, after initial hesitation, allied with the Central Powers and paid the price after the defeat of Germany, when the Allies dictated armistice terms to the Ottomans. Several political factions battled over the future of the country, and finally in 1922, Mustafa Kemal abolished the sultanate, drove Greek forces out of Turkey, and officially ended the Ottoman Empire.

Cynthia Clark Northrup

See also: Austro-Turk Wars; Hungarian Civil Wars; Mohács, Battles of; Rhodes, Sieges of; Vienna, Sieges of

References and further reading:

Peers, Douglas M., ed. *Warfare and Empires: Contact and Conflict between European and Non-European Military and Maritime Forces and Cultures.* Brookfield, VT: Variorum, 1997.

Trumpener, Ulrich. *Germany and the Ottoman Empire, 1914–1918.* Princeton, NJ: Princeton University Press, 1968.

Wittek, Paul. *The Rise of the Ottoman Empire.* New York: B. Franklin, 1971.

Oudenaarde, Battle of (11 July 1708)

The perfect illustration of the difficulties of double command. In 1708, John Churchill, Duke of Marlborough, wanted to raise the morale of his Dutch allies by winning a battle in Flanders. His main objective was to retake all the territories lost the two previous years. On the French side, the king had sent his grandson, the Duc de Bourgogne, to command the field army with the Marechal de Vendôme on a secondary front. Eugene of Savoy's army was far away, and Marlborough's troops were deployed all over northern Flanders, with Brussels as headquarters. On 16 May, the French army advanced toward Brussels, its superior number pushing away Marlborough's troops. Then Bourgogne stopped waiting for orders from Versailles, 200 miles away. A very religious man, Bourgogne was also very cautious and was always at variance with Vendôme's orders. On the other side, Marlborough asked Prince Eugene to join his army as soon as possible to coordinate an aggressive defense.

At the beginning of July, a sycophantic noble follower of Bourgogne suggested an attack toward Bruges and Ghent. The two towns were easily taken, and the royal army decided to encircle Oudenaarde on the River Scheldt. But Marlborough had discerned this move and sent his army to cross the river before the French arrived.

On 11 July, the French general Biron discovered the waiting allied troops and asked for orders. Vendôme refused to believe Biron and left his army without deployment orders until it was too late. Marlborough, urging his troops on, arrived at noon and deployed on a line of low hills north of Oudenaarde. His lines were protected by meadows and hedges. By 3 P.M., Bourgogne gave the order to the marching French to assault the waiting English lines. The attack began on the French right, soon supported by the center. All this uncoordinated movement gave predictable results, as all the columns were repulsed. The French left, under Vendôme, remained useless.

Eventually, with Eugene's army facing Vendôme, Marlborough took the initiative. Following the retiring French right, he managed to encircle them, forcing thousands to surrender. The French rout sent them back to Bruges. Marlborough's victory restored allied morale. The French had lost more than 15,000 soldiers and were no longer able to protect their northern border. France lay open to an invasion.

Gilles Boué

See also: Marlborough, John Churchill, First Duke of; Spanish Succession, War of the

References and further reading:

Belloc, H. *The Tactics and Strategy of the Great Duke of Marlborough.* London: Arrowsmith, 1933.

Bluche, François. *Dictionnaire du Grand Siècle.* Paris: Fayard, 1990.

Oudinot, Nicholas-Charles, Duc de Reggio (1767–1847)

Military commander during the Revolutionary and Napoleonic Wars of France. Nicholas-Charles Oudinot was born on 25 April 1767 at Bar-le-Duc, France. He enlisted in the Royal Army in 1784 and served until just before the outbreak of the French Revolution. The revolution provided opportunities for men of humble origins, like Oudinot, with military experience and even command, and he was one of the many beneficiaries.

In 1789, Oudinot was appointed captain and rose to the rank of lieutenant colonel by 1791. Between 1792 and 1796, he campaigned with the Army of the Rhine, rising to the rank of *general de brigade* by June 1795. Frequently wounded and once captured, he was promoted to *general de division* in 1799 and made chief of staff of the Army of Switzerland in that year.

During the second Italian campaign, Oudinot served under André Masséna and so took part in the defense of

Genoa. In August 1800, he was appointed chief of staff of the Army of Italy and participated in G. M. A. Brune's campaign of 1800–1801 that led to the capture of Verona.

In February 1805, Oudinot was appointed to command the Reserve Grenadier Division, which ultimately became the 1st Division of Jean Lannes's corps during the campaign against Austria. Though the grenadiers distinguished themselves at Austerlitz, Oudinot was wounded early in the campaign and did not receive another command until 1807, when he served under the command of François-Joseph LeFebvre, Duke of Danzig, at the siege of Danzig, and again under Jean Lannes, Duke of Montebello, at Friedland.

In July 1808, Oudinot was created a count of the empire and in April 1809, duke of Reggio. He was elevated to the rank of marshal of the empire in July 1809. After receiving command of the Second Corps in 1812, he distinguished himself at Polotsk and at the Berezina crossing during the disastrous retreat from Moscow.

In the 1813 campaign, Oudinot was given command of the XII Corps and fought at Bautzen. After the conclusion of the armistice, the XII, together with the IV and VII Corps, were detached from the Grand Armeé, and the whole force, under the command of Oudinot, was directed to march on Berlin. Oudinot was promptly defeated by a mixed force of Prussians, Russians, and Swedes under the command of the crown prince of Sweden, Jean Bernadotte. Oudinot was replaced in command of the Army of Berlin by Michel Ney, who was in turn defeated by Bernadotte at the Battle of Dennewitz. Oudinot commanded the remnants of his corps at Leipzig and in 1814 fought at Brienne, La Rothiere, and Arcis-sur-Aube.

After the Bourbon Restoration, Oudinot was appointed commander of the Metz area and made a peer of France. During the Hundred Days, he was not employed by Napoleon and so was able to continue his career after the Second Restoration. He commanded a corps during the Spanish campaign in 1823 and became governor of Madrid. In 1842, he became governor of the Invalides. He died on 13 September 1847. Overall, Oudinot was a failure as an independent army commander but was a capable corps commander and administrator.

J. Isenberg

See also: Austerlitz, Battle of; Berezina River, Battle of; Borodino; Dresden, Battle of; French Revolutionary Wars; Friedland; Lannes, Jean, Duke of Montebello; Leipzig, Battle of; Masséna, André, Duc de Rivoli, Prince d'Essling; Moscow, Retreat from; Napoleon I; Napoleonic Wars

References and further reading:

Chandler, David G. *Napoleon's Marshals.* London: Weidenfeld and Nicholson, 1987.

Young, Peter. *Napoleon's Marshals.* Reading, Berkshire, UK: Osprey, 1973.

P

Pachacutec Yupanqui (r. 1438–1471)

Sapa Inca (first ruler) responsible for the growth and development of the Inca Empire. Although Spanish and native sources differ on specifics, Pachacutec Yupanqui clearly took the early reforms and successes of his father, Viracocha, and transformed the Inca from one of several ethnic chiefdoms in highland Peru to a huge empire controlling much of the coast and highlands of western South America.

In 1438, at around 16 years of age, Yupanqui took control of the defense of the Inca capital Cuzco from his ailing father. He defeated the 30,000-man Chanca army by using allied troops in surprise flanking movements at key moments in the battle. After this victory, he was crowned Pachacutec Yupanqui, Sapa Inca, and ruled for 33 years. He would conquer the highland kingdoms of Lupaca and Cajamarca, form alliances with the Quechua and Charca, conquer the coastal states of the Ica and Nazca, and eventually oversee the conquest of the powerful coastal Chimu state by 1471. His development of cycling in fresh troops to threatening realms was crucial to his success.

Pachacutec developed a permanent military system and grand strategy for the empire in carrying out this expansion. The military system consisted of ethnic Inca nobility trained in the arts of war and leadership at schools in Cuzco. Other ethnic Inca were trained at military-style schools and formed the core of Inca armies. He used alliances to provide the bulk of troops from loyal provinces, utilizing their alternative weapon and fighting systems to complement the Inca core. Fortifications like Sacsahuaman in Cuzco were used to protect road systems and supply depots, and the llama was the pack animal for the expeditions.

Pachacutec's grand strategy consisted of overwhelming logistics, defense-in-depth fortifications, and a variety of alliance offers, made both peacefully and forcefully to acquire new territory. He successfully incorporated many peoples by moving around loyal subjects to rebellious areas and vice versa. His use of the *mitmae* (labor tax) allowed for one of the most rapid developments of an empire infrastructure in world history.

Christopher Howell

See also: Chan Chan, Battle of; Cuzco, Battles of; Inca Empire Imperial Wars

References and further reading:

Adorno, Rolena. *Guaman Poma Del Ayala.* New York: Americas Society, 1992.

D'Altroy, Terrence, ed. *Provincial Power in the Inka Empire.* Washington, DC: Smithsonian Institution Press, 1992.

Pacific, War of the (1879–1884)

Major war resulting from a dispute between Chile and Bolivia. The dispute centered on control of the Atacama Desert, important because of rich deposits of nitrates used for fertilizer, on the western coast of South America. Peru soon entered the war on the side of Bolivia. Chile and Bolivia had argued over border delineation since their independence. In 1874, the dispute seemed to be resolved when Chile agreed not to pursue its claim for control over the southern portion of the Atacama in exchange for a generous tax concession for Chilean companies exploiting nitrates in the Bolivian-controlled area.

When the Bolivian dictator General Hilarón Daza increased taxes in an apparent violation of the 1874 accord, Chile protested vociferously, and Daza declared war. Chile soon learned that Peru and Bolivia had a secret alliance and that Peru intended to honor its commitments to Bolivia. These revelations resulted in an immediate declaration of war by Chile on Peru.

Naval power proved to be of great importance in the early phase of the war. Chile's navy, under the command of Admiral Juan Williams Rebolledo, established a blockade with several wooden ships off the Peruvian port of Iquique for the purpose of cutting Peru's nitrate trade. Williams then sailed north with his two ironclad monitors to attack the Peruvian navy, which he believed was defending Peru's main port of Callao. However, Peruvian admiral Miguel Grau had sailed south with his two ironclads, the *Huascar* and the *Indepedencia,* and on 21 May 1879 attacked the Chilean blockade, sinking the *Esmeralda* and damaging the *Covadonga.* Unfortunately for Grau, the Battle of Iquique proved a hollow victory because the *Independencia* was run aground and lost. With only one ironclad, the *Huascar,* remaining, Peru was at a decided disadvantage for the remainder of the war. Nevertheless, Grau used the *Huascar* to harass Chilean shipping lanes, resulting in Williams's resignation. On 8 October 1879, a refurbished Chilean fleet, including the ironclads the *Blanco Encalada* and the *Cochrane,* finally forced an engagement with the *Huascar.* Superior Chilean firepower soon reduced the *Huascar* to a burning hulk, resulting in the death of most of the crew, including Admiral Grau. (*Huascar* is still in existence as a Chilean memorial.)

With control of the sea assured, Chile launched an invasion of the southern Peruvian province of Tarapac in October 1879, led by General Erasmo Escala. General Hilarón Daza led a Bolivian army from the sierra to counter Escala's move and to join a Peruvian army led by General Juan Daza. But Daza's troops were ill-equipped and unprepared for the arid conditions of the Atacama, and Daza abandoned his plan. On 19 November 1879, Escala repulsed an allied attack and forced a retreat. It was followed by an assault on the city of Tarapac, which proved costly but was ultimately successful and led to the capture of Iquique.

On 8 April 1880, Chile renewed land operations to gain total control of Peru's nitrate-rich province of Tacna. The campaign resulted in extraordinarily high casualties for the invaders, but by June Chile had control of most of Tacna. When peace negotiations failed, Chile decided to attack Lima, which fell on 17 January 1881. Peruvian resistance continued despite the loss of the capital. Finally, in 1883, the Chilean army defeated the forces of Andre's Céceras at Huamchaca, and Peru signed a peace treaty ceding Tarapac to Chile and permitting occupation of Tacna and Arica for 10 years. The Chileans took complete possession of the disputed areas in 1884. Bolivia lost its Pacific coast when it turned over the Atacama to Chile. Thus, Chile became the leading power on the west coast of South America, but Bolivia never forgot its lost access to the sea.

George M. Lauderbaugh

References and further reading:

Sater, William F. *Chile and the War of the Pacific.* Lincoln: University of Nebraska Press, 1986.

Pacifism/War Resistance

The opposition to war, killing, or violence. The word comes from the Latin pax or *pacis,* which means "peace," and *facere* which means "to make." Pacifists believe that nations should settle their conflicts as peacefully as possible and are often opposed to participating in military activity. Even though peace and justice are the objectives of all pacifists, the degree and circumstances of nonviolence vary from person to person and circumstance to circumstance. Although absolute pacifism declares that violence is wrong always and in every situation, most pacifists are on a moral and ethical continuum as it relates to force. For example, some pacifists refuse to enter the military service in any capacity, whereas others find their witness for peace as unarmed army medics. Thus much of the writing of those advocating peace debates under what circumstances nonviolence should be national, group, or individual policy and the degree of coercive or disciplinary violence that should be used to achieve social or personal goals.

The roots of pacifism are as old as war itself and can be traced to Asia, especially Buddhism, several of the indigenous peoples of the Americas, and the writings of Greek philosophers like Plato. Pacifism as a dominant form of philosophy emerged in western thought with first-century Christianity. After hearing passages like Jesus' Sermon on the Mount, most early Christians believed that their Christ called for nonviolent resistance to pagan political power and the ways of the world. Christianity did not change its stance toward violence until the third century, when it increasingly gained followers in power in the Roman Empire, particularly the Emperor Constantine. By the fifth century, Christian thinkers, particularly St. Augustine, developed a "just war" theory. Although no longer the dominant strain of thought, pacifism still held power in many ecclesiastical circles and later reappeared in the sects founded during the Protestant Reformation. Often called the "peace churches," the groups that elevated nonresistance to doctrinal position included Mennonites, Anabaptists, Friends (or Quakers), Moravians, Brethren, and Dukhobors. Leaders from these groups have been the heart of most peace societies founded after 1815 and have been the most vocal in the cause of peace. More philosophical advocates of pacifism can be found in the writings of Immanuel Kant, Erasmus, Adin Ballou, and William James.

Modern pacifism began with the founding of peace societies in Massachusetts (1815) and New York (1815) and in England (1816); France, Switzerland, and other countries soon followed suit. William Ladd brought many of these groups together in the United States, establishing the American Peace Society in 1828. These groups advocated various proposals to smooth relations between nations. Yet many of these societies or individuals, like William Lloyd Garrison, the most prominent American abolitionist, allowed for some degree of violence, particularly in support of the antislavery cause. In 1843, the first peace congress met in London, and momentum seemed to be building for pacifistic reform in international relations until the American Civil War shattered the peace movement in the United States. While American peace societies were rebuilding during the latter half of the nineteenth century and European efforts were set back by the Crimean War and the wars of Italian and German unification, Sir William Randal Cremer of Great Britain and Frederic Passy of France spearheaded efforts to establish the Inter-Parliamentary Union in 1889, which brought members of national electoral bodies together in periodic conferences. The same year, the International Peace Bureau was founded in Berne, Switzerland. Alfred Nobel (1833–1896), the Swedish inventor of dynamite, established the Nobel Peace Prize to recognize and reward work for peace and humanitarian causes.

During the mid-1860s, socialist thinkers such as Cesar de Paepe developed the First International of Working Man, an organization of socialists, pacifists, and union supporters who believed the cause of peace was almost impossible until private property and class inequality could be abolished. A Second International was found in 1889, and its members advocated general strikes in case of war, arbitration treaties, arms reduction proposals, and much of colonialism. Militants such as Gustave Herve urged sabotage in the cause of pacifism. Vladimir Lenin urged the dissolution of international boundaries as the only means of true peace. Still, most peace societies were ethnocentric and nationalistic in their viewpoint, and the association with socialism (and vegetarianism, antivivisectionism, and other "fads") in the public's mind would hurt their efforts in the twentieth century.

Yet with conferences at The Hague in 1899 and 1907 and mainstream political leaders like William Jennings Bryan advocating international treaties and courts to settle disputes, many pacifists believed an end to war was within sight. World War I shattered these illusions, but pacifist sentiment quickly reappeared with the conclusion of the war. Many peace advocates pushed for the League of Nations as the forum to settle international disputes. In the United States, pacifists were often actively involved in the growing women's rights movement and sought to distance themselves from events in Europe. Pacifism was again dealt a blow by the events leading to World War II when pacifists were blamed for the isolationism and appeasement that led to that war. Still, the number of conscientious objectors in the United States and England was larger than in World War I, and most were treated with far more respect for their beliefs than in the previous conflict.

The "McCarthyite" 1950s were the low point for peace movements in the United States, and the cause was harmed by the Communists and Soviets hijacking the term *peace* for their own use. There was very little vocal opposition to the Korean War, and those who advocated measures and resolutions toward maintaining world peace were labeled "fellow travelers" or seen as disloyal to their country. Still, pacifist sentiments were nurtured by the writings of Jane Addams, Leo Tolstoy, Albert Schweitzer, Dorothy Day, Mohandas Gandhi, and Martin Luther King. They inspired efforts to work toward world peace, and by the 1960s and 1970s, pacifists and other antiwar groups were vocal in their opposition to the Vietnam conflict. They organized marches, rallies, and protests against the war, and the movement pressure was responsible in part for U.S. withdrawal from that conflict.

After the war, President Gerald Ford allowed conscientious objectors who had fled the draft to return home. The threat of nuclear weapons particularly enlivened the peace movement and helped pacifists to be taken seriously in their urging of unilateral disarmament and an end to nuclear testing.

Pacifists flourish unmolested in the developed democracies, but it can be argued that this is the case primarily because they are so few in number that they pose no threat to those countries' military-industrial complex. A major question that will arise in any future U.S. conflict, however, is whether "secular" pacifists, that is, those who hold sincere nonviolent beliefs but do not belong to any of the state-recognized "historic peace churches," will be given draft-exempt status.

T. Jason Soderstrum

References and further reading:

Brock, Peter. *Freedom from War: Nonsectarian Pacifism.* Toronto: University of Toronto Press, 1991.

Brock, Peter, and Nigel Young. *Pacifism in the 20th Century.* Syracuse, NY: Syracuse University Press, 1999.

Chatfield, Charles. *For Peace and Justice: Pacifism in America, 1914–1941.* Boston: Beacon Press, 1973.

Martin, David A. *Pacifism: An Historical and Sociological Study.* New York: Schocken Books, 1966.

Paekche (attributed 18 B.C.E.–660 C.E.)

One of three kingdoms dominating the Korean peninsula. Situated in the southwest, Paekche's origins and early history remain obscure. Most sources agree that its founders were warrior-refugees from the Manchurian state of Fuyu (Korean Puyo), which met its demise in 285 C.E. The oldest extant Korean history places Paekche's foundation at 18 B.C.E., but Paekche does not appear in written records until the third century C.E.

Despite purported warrior origins, Paekche was better known for its cultural achievements than for any military prowess. Its emergence coincided with the appearance of two primary rivals, Silla and Koguryo, located in the southeast and north, respectively, the other two kingdoms giving the period its name. From the mid–fifth century, the three waged an increasingly bitter struggle for hegemony.

In this climate, Paekche forged an alliance with Yamato Japan, a partnership whose nature continues to be debated. As a result, Paekche served as a conduit for technology and culture to the Japanese islands, while Yamato troops were engaged on the peninsula.

From the fifth century, Paekche was subjected to increasing pressure from Koguryo advancing south and Silla intent on going north, in a general atmosphere of winner-take-all. A Paekche-Silla alliance crumbled in 552, and in 660 Paekche was finally destroyed by Silla, which took advantage of a military alliance with the new Tang Dynasty in China to defeat both Paekche and Koguryo and unify the peninsula.

Daniel Kane

See also: Koguryo; Silla; Sino-Korean Wars and the Wars of Korean Unification

References and further reading:

Gardiner, K. H. J. *The Early History of Korea: The Historical Development of the Peninsula Up to the Introduction of Buddhism in the Fourth Century A.D.* Honolulu: University of Hawaii Press, 1969.

Iryon. *Samguk yusa: Legends and History of the Three Kingdoms of Ancient Korea.* Trans. Tae-Hung Ha and Grafton K. Mintz. Seoul: Yonsei University Press, 1972.

Lee, Ki-baik. *A New History of Korea.* Trans. Edward W. Wagner, with Edward J. Shultz. Cambridge, MA: Harvard University Press, 1984.

Pagan Kingdom (1044–c. 1300)

The founder of the Pagan Kingdom, King Anawrahta (r. 1044–1077), established the first state unifying Upper and Lower Burma, including most of the territory of the modern nation, except for Shan State and the remoter border areas. He and his successors recognized Theravada Buddhism as the state religion, building magnificent pagodas at their royal capital of Pagan and making generous donations to the *sangha* (community of Buddhist monks). The Pagan dynasty's 12 kings dealt with the challenges of a multiethnic society (principally Mon, Burmese/Myanmar, and Pyu) both by absorbing non-Burmese cultural influences (especially Mon culture) and by preserving the privileges of an ethnic Burmese ruling class.

In 849, the Burmese built a fortified city at Pagan (near Nyaung-U, in Mandalay Division) that was strategically located on the banks of the Irrawaddy River and near the irrigated districts of Kyaukse and Minbu, which provided the small state with surpluses of rice. The availability of rice was the economic foundation of Pagan's military power. Linguistically related to the Tibetans, the Burmese originally lived in eastern Tibet or Yunnan but had migrated into the Irrawaddy Valley (Upper Burma) because of the power vacuum created by attacks by the Nanchao Kingdom of Yunnan (now China's Yunnan Province) on states in Burma ruled by an earlier people, the Pyu. The Upper Burma Dry Zone is a harsh, semidesert environment, and the Burmese quickly gained a reputation as aggressive warriors. They gave their capital of Pagan the Pali name of Arimaddanapura, "the city that is a crusher of enemies."

Anawrahta conquered Lower Burma (the Irrawaddy Delta and Tenasserim), capturing the Mon city of Thaton in 1057 and bringing its king, the devout Buddhist Manuha, as a hostage to Pagan. In the first recorded instance of political protest in Burmese history, the Mon king built a small and unspectacular temple claustrophobically housing large Buddha images in cavelike chambers, expressing his distress at being Anawrahta's prisoner. Although Anawrahta, Kyanzittha (r. 1084–1111), and other Pagan monarchs patronized Mon culture, more refined than their own, and venerated Mon monks as teachers of Theravada doctrine, this Lower Burma people staged numerous revolts against the Burmese that were harshly suppressed until c. 1281, when Pagan, hard-pressed by the Mongols, could not prevent the emergence of an independent Mon state at Pegu.

The Mongol emperor Kublai Khan's conquest of Yunnan in the 1250s set the stage for Pagan's fall. King Narathihapate (r. 1256–1287) rejected Kublai's repeated demands for submission, and in 1277 the Mongols invaded. In his *Description of the World,* Marco Polo chronicles the defeat of Burmese soldiers mounted on elephants by Mongol archers, a terrible rout. In 1286–1287, Narathihapate's continued intransigence led to a second invasion in which Pagan was occupied. By 1300, the Pagan Kingdom was no more. Tai (Shan) peoples came into Burma in the wake of the Mongol incursions and established a power base at Ava in Upper Burma.

Donald M. Seekins

American forces in formation at the Battle of Palo Alto, 1846. (Library of Congress)

See also: Anawrahta; Kublai Khan
References and further reading:

Aung-Thwin, Michael. *Pagan: The Origins of Modern Burma.* Honolulu: University of Hawaii Press, 1985.
Hall, D. G. E. *A History of South-East Asia.* 2nd ed. London: Macmillan, 1964.

Palo Alto (8 May 1846)

Early battle in the Mexican-American War. In preceding months, General Zachary Taylor and his "Army of Observation" had moved from the Nueces River to the Rio Grande, which was disputed territory between Mexico and what had formerly been the Republic of Texas and now was the United States. He was opposed by General Mariano Arista and the Mexican Army of the North.

On 8 May 1846, the two armies clashed at Palo Alto. American artillery was more mobile and far superior; the Mexican army had antiquated muskets and inferior gunpowder and shot. The fighting began around 2:00 P.M. with an artillery exchange and the U.S. troops literally dodging the solid copper Mexican shot, while the U.S. artillery caused great damage. In early evening, Arista tried to turn Taylor's flank and failed; meanwhile, the Mexican left began to break, and the high grass caught fire. As the infantry fighting halted for the fire, the artillery duel continued, with the Americans having the better of it. The battle was probably a draw but was costly for the Mexicans.

Charles Dobbs

See also: Mexican-American War; Resaca de la Palma; Taylor, Zachary
References and further reading:

Hamilton, Holman. *Zachary Taylor: Soldier of the Republic.* 1941.
Nichols, Edward J. *Zach Taylor's Little Army.* 1963.
Singletary, Otis. *The Mexican War.* Chicago: University of Chicago, 1960.

Panama Incursion (1989–1990)

Quick U.S. military operation against corrupt Panamanian ruler. Panama had been ruled by the military from the 1960s, and by the 1980s, General Manuel Antonio Noriega Moreno had become the power behind the throne in Panama. Noriega had links with the Central Intelligence Agency (CIA) dating back to the 1960s.

Panama, cut in two by the American-built and administered Panama Canal, had always had a close, albeit one-way, relationship with the United States, but by the mid-1960s, this relationship was beginning to show signs of strain. Although Noriega assisted the United States in its war against the Sandinistas in Nicaragua, he also helped Cuba break the U.S. blockade by reselling goods to the Caribbean island. In 1986, the journalist Seymour Hersh published an article in the *New York Times* contending that Noriega was involved in gun running, drug trafficking, and money laundering.

Noriega was becoming an embarrassment for the United States, and in mid-1987, the Reagan administration initiated economic sanctions against Panama. These sanctions were tightened in December of that year, when all assistance from the United States was stopped. The situation worsened when, in February 1988, two Florida grand juries found Noriega guilty of drug trafficking and money laundering.

In the May 1989 presidential elections, qualified observers argued that Guillermo Endara Galimary had won three times as many votes as Noriega's candidate. Noriega simply annulled the elections and appointed his own candidate, Francisco Rodriguez, as president. In October, Noriega survived an attempted coup, and on 15 December he became head of the government with the title "Maximum Leader."

Relations with the United States approached the crisis stage on 16 December, when a U.S. serviceman, Lieutenant Robert Paz, was shot and a U.S. Navy officer and his wife were arrested and harassed by the Panama Defense Force (PDF). These affronts were the last straw for U.S. president George Bush, who set in motion Operation JUST CAUSE on 20 December.

This operation, which involved more than 20,000 U.S. troops, had two functions: the invasion of Panama and the seizing of Noriega at H-hour before transporting him to the United States to face charges. The PDF quickly disintegrated, with U.S. forces facing more resistance from the paramilitary Dignity Battalions. Both U.S. Special Operations and conventional forces were completely successful and suffered very few casualties. But the invading force could not locate Noriega. He eventually surfaced at the Papal Nunciature on Christmas Eve, leading to a standoff with U.S. forces. Finally, on 4 January, the deposed dictator was persuaded that he had no choice but to give himself up and face the charges against him in the United States.

Panama moved quickly during 1990 toward democratization and demilitarization, and the May 1989 election results were upheld. Operation JUST CAUSE had ended Noriega's rule, but it also gave rise to questions of the legitimacy of Endora's presidency. The operation also was a flexing of the muscles of the post-Vietnam U.S. military, rejuvenated by former U.S. president Ronald Reagan.

M. J. Bain

See also: Bay of Pigs Invasion; Nicaraguan Civil War (1979); Peruvian Guerrilla War; Salvadorian Civil War

References and further reading:

Black, J. N. *Latin America: Its Problems and Its Promise.* Boulder, CO: Westview Press, 1998.

Loser, E. *Conflict Resolution and Democratization in Panama: Implications for U.S. Policy.* Washington, DC: Center for Strategic and International Studies, 1992.

Scranton, M. E. *The Noriega Years: U.S.-Panamanian Relations, 1981–1990.* Boulder, CO: Lynne Rienner Publishers, 1991.

Skidmore, T. E., and P. E. Smith: *Modern Latin America.* New York: Oxford University Press, 1997.

Panipat, Battles of (21 April 1526, 5 November 1556, 14 January 1761)

Series of Persian-Afghan-Indian battles. The first Battle of Panipat took place on 21 April 1526. Ibrahim led at least 10,000 Delhi Muslims with 100 war elephants against Babur and his 2,000 handpicked Moguls. It is likely that more conscripts were involved, but only the number of professional soldiers is known. Babur scored a complete victory, with gunpowder weapons playing a significant role in frightening the elephants. Ibrahim was killed, and the Mogul Empire began in India from the ashes of the Afghan Dynasty that had ruled from Delhi. This third invasion of India by the Moguls was easily the most successful.

The second Battle of Panipat took place on 5 November 1556. Akbar Khan and 20,000 troops of the Mogul Empire faced Hemu, who commanded about 100,000 Hindu Rajah troops, 1,500 war elephants, and a chaotic baggage train full of loot from rebel-held Delhi, the former Mogul capital.

The Moguls repulsed an initial charge by the war elephants and then directed the panicked Hindu elephants toward the Hindu rear baggage train area, all but stopping any movement of Hindu reinforcements. The Moguls scored a complete victory, capturing and executing Hemu, retaking Delhi, and building a tower of Hindu heads.

The assassination of Nadir Shah, last of the Safavid-linked Persian leaders, in 1747 caused the collapse of his Persian empire to Russian, Ottoman, and Afghan interests and resulted in the third Battle of Panipat. Shah Ahmed Durani created the Afghan empire from its ashes by conquering parts of Persia and sections of Maratha India, namely the Punjab and Delhi. In response, Sedashao Bhao, cousin of the Maratha Peshwa (ruler), went forth with the largest Maratha army ever assembled, perhaps 300,000 strong, pushing back the Durani Afghans. Shah Ahmed then led 90,000 Afghan and Indian troops to face the Marathas at the third Battle of Panipat in 1761. Bhao attacked first and dispersed Ahmed's Indian allies on 14 January, but Ahmed, preaching a jihad (holy war) against the predominantly Hindu Marathas, rallied his Islamic Afghan forces and crushed the Maratha army, killing 75,000 and capturing and ransoming another 30,000. The Mogul throne, used as a pawn by Afghans, Marathas, and the British, now effectively ceased to exist, and the remnants of the once-mighty Mogul empire were divided up among the three.

Christopher Howell

See also: Persian-Afghan Wars; Maratha Wars

References and further reading:

Adamec, Ludwig. *Dictionary of Afghan Wars, Revolutions, and Insurgencies.* Kennikat: Scarecrow Press, 1996.

Gordon, Stewart. *Marathas, Marauders, and State Formation in Eighteenth-Century India.* Oxford, UK: Oxford University Press, 1994.

Kadam, V. S. *Maratha Confederacy: A Study in Its Origin and Development.* Delhi: South Asia Books, 1993.

Paramilitary Organizations

A wide range of groups organized along military lines yet lacking the traditional role or legitimization of conventional or "genuine" military organizations. Typical characteristics include a hierarchical organization with clear lines of authority and strict discipline, military-style ranks or uniforms, and usually an explicit ideological mission. Light weaponry may also be included. Party militias like the Nazi Stormtroopers (SA) or Italian Squadristi (Black Shirts); militarized police forces such as the armed units of the Soviet secret police or the French gendarmerie; veteran's organizations such as the German Stahlhelm; U.S. state police and Canadian provincial police; and even many youth groups such as the Boy Scouts all may be termed paramilitary for different reasons.

Paramilitary organizations exist to perform tasks for which conventional military forces are either ill-suited or considered unworthy, for example, domestic police duties or state terrorism. They also represent an attempt to apply military forms of organization to nonmilitary or political ends because of the military's assumed greater efficiency or to appropriate and imitate the prestige and aura of conventional military forces. These latter reasons were especially true in the first half of the twentieth century because of the great prestige that conventional military establishments had gained as symbols of national unity and was only intensified (though not without challenge) by the series of great wars that swept Europe in the late nineteenth and early twentieth centuries. By the end of World War I, military ways of organization came to be seen as possible alternatives to liberal democracy, and millions of veterans searching for meaning after the carnage of war could find in them a familiar way of life.

The most characteristic form of paramilitary organization in the early and mid–twentieth century was the party militia typified by the Squadristi of the Italian Fascist Party. Before the Fascists came to power, the Squadristi served as a way for the Fascists to organize violence against their opponents and at the same time appear to be more dynamic and effective than their conservative and liberal competitors. After the Fascists came to power, the Squadristi were institutionalized as a mechanism for mobilizing, organizing, and indoctrinating the population, yet they kept their original role as an extralegal means of exercising violence. Nearly all fascist or radical conservative movements of the period between the two world wars established similar paramilitary auxiliaries. The success of the Bolshevik Revolution and the adoption of the Soviet model of development by newly independent states in the former European colonies or by the puppet states of Eastern Europe also led to the widespread establishment of paramilitary state youth organizations, particularly after 1945, most of which have now largely disappeared with the collapse of the Communist states that organized them.

Today, paramilitary organizations are often formed in cases in which established social groups and interests seek to exercise power yet are either unwilling or unable to use conventional military forces, which they may not fully control or which may be unwilling to dirty their own hands. Examples include private death squads established with government connivance, such as in Guatemala or El Salvador, or the ethnic Serbian paramilitaries established in Bosnia in the 1990s.

In the democracies, paramilitary organizations were not unknown, but their goals were usually entirely benign, such as the search-and-rescue mission of the U.S. Civil Air Patrol or the woodcraft of the Boy Scouts.

Bruce Campbell

See also: Death Squads; SA; SS

References and further reading:

Diehl, James M. *Paramilitary Politics in Weimar Germany.* Bloomington: Indiana University Press, 1977.

Morgan, Philip. *Italian Fascism, 1919–1945.* New York: St. Martin's Press, 1995.

Reichardt, Sven. "Faschistische Kampfbünde in Italien und Deutschland. Ein Vergleich der Formen, Funktionen und Ursachen politischer Gewalt in der Aufstiegsphase faschistischer Bewegungen." Ph.D. diss., Freie Universität Berlin, 2000.

Tobler, Hans Werner, and Peter Waldmann, eds. *Staatliche und Parastaatliche Gewalt in Lateinamerika.* Frankfurt am Main: Vervuert Verlag, 1991.

Williams, Warren E. "Paramilitarism in Inter-State Relations: The Role of Political Armies in Twentieth Century European Politics." Ph.D. diss., London University, 1965.

Paris, Siege of (1870–1871)

The climax of the Franco-Prussian War, which destroyed the French Second Empire and led to the creation of the Third Republic. After the collapse of the French armies, the Prussians surrounded Paris in September 1870. A spontaneous popular movement in the city deposed Emperor Napoleon III and created a republic. General Louis Trochu, appointed to lead the city's defense, was competent, but not very zealous in actively ending the siege. The French National Guard, the main force available within Paris for its defense, made two poorly planned attempts to break through the Prussian lines but failed to coordinate their efforts with French forces outside the city. Both ended in total fiasco. The Prussians made no attempt to storm Paris, preferring to starve it into surrender. French forces outside Paris, led by Léon Gambetta, were never able to mount a serious effort to relieve the capital. The siege saw the first extensive wartime use of hot air balloons for communication with the outside world; unfortunately, given the state of technology at the time, it was difficult to navigate the balloons and flights into Paris were impossible. The siege laid bare the fissure lines between conservatives and radicals in Paris. The "Reds" were increasingly critical of the way Trochu and the government handled Paris's defense and the way the government handled the economic problems resulting from the siege. The government surrendered Paris in January 1871 largely because it feared an incipient revolt by Parisians. Shortly after the surrender, there was, in fact, a rebellion that chased a newly elected government to Versailles and led to the creation of the Paris Commune.

Lee Baker Jr.

See also: Bismarck, Otto von; Franco-Prussian War; German Wars of Unification

References and further reading:

Horne, Alistair. *The Fall of Paris: The Siege and the Commune, 1870–71.* New York: Penguin Books, 1985.

Howard, Michael. *The Franco-Prussian War.* New York: Collier Books, 1961.

Tombs, Robert. *The War against Paris, 1871.* New York: Cambridge University Press, 1981.

Williams, Roger. *The French Revolutions of 1870–1871.* London: Weidenfeld and Nicolson, 1969.

Parma and Piacenza, Alessandro Farnese, Duke of (1545–1592)

Spanish general and statesman. Parma was the greatest soldier of his time in western Europe. He came close to defeating the Dutch revolt against Spanish rule, but interference and lack of support from Philip II prevented his victory.

Parma was the son of Margaret of Austria, Emperor Charles V's natural daughter, and Ottavio Farnese, Duke of Parma. He spent part of his boyhood in Philip II's court, where he became friends with Don Juan of Austria, his cousin. Parma accompanied his mother to the Netherlands in 1565, where she was regent. Parma became very familiar with the country and the leaders of society. He volunteered for duty with the Holy League fleet under Don Juan against the Turks and distinguished himself in the victory at Lepanto on 7 October 1571. Parma continued to serve in the Mediterranean until 1574.

In 1577, Don Juan requested Parma join him in the Netherlands. At the Battle of Gembloux, 31 January 1578, Parma led the cavalry charge that smashed the opposing infantry and nearly destroyed the Dutch army. After Don Juan's death on 1 October 1578, Parma became Spanish commander in the Netherlands. His knowledge of the country and people was a great advantage. In May 1579, he signed a peace treaty with Catholic leaders of the southern provinces, bringing them back into allegiance with Philip. With his base secure, Parma concentrated on taking the centers of Protestant power in the north. A master of maneuver and sieges, he did not undertake any operations beyond his army's power, which numbered only 27,000 men, but isolated cities before reducing them. Parma was merciless to his avowed enemies but persuasive in winning over others. His greatest triumph was the capture of Antwerp in August 1585, after blockading it from the sea with a barrage of boats. The opportunity to crush the revolt for good was lost when Philip ordered Parma to intervene in the French wars of religion and the invasion of England with the Spanish Armada. While he was in France, much of Parma's gains were lost to Maurice of Nassau and his reorganized army. Frustrated by Philip, Parma became disheartened. He was wounded in the arm in a skirmish at Caudebec, fell ill, and died soon afterward. No soldier who followed him could duplicate his successes.

Tim J. Watts

See also: Don Juan de Austria; Dutch War of Independence

References and further reading:

Hoeven, Marco van der. *Exercise of Arms: Warfare in the Netherlands, 1568–1648.* New York: Brill, 1997.

Mattingly, Garrett. *The Armada.* Boston: Houghton Mifflin, 1959.

Parker, Geoffrey. *The Army of Flanders and the Spanish Road, 1567–1659: The Logistics of Spanish Victory and Defeat in the Low Countries' Wars.* New York: Cambridge University Press, 1995.

Parthian Empire (247 B.C.E.–226 C.E.)

The Parthians liberated Persia from Hellenistic domination and made Persia a power to rival the Roman Empire. They originated as nomadic Persian speakers living southeast of the Caspian Sea and south of the Oxus River.

Under Shah Mithradates I, the Parthians expanded through Iran and, by 141 B.C.E., reached modern-day Iraq. In 139 B.C.E., Mithradates captured the Seleucid prince, Demetrius II, and occupied the Tigris-Euphrates Valley. Under Mithradates II the Great (124–87 B.C.E.), the Parthians managed to decisively secure their eastern borders as well. Ctesiphon in Babylon became the capital.

The Parthian army was primarily a cavalry force, provided mostly from feudal levies, especially from the powerful landed nobility. Heavy cavalry wore body armor and fought at close quarters with sword and lance. Light cavalry used the compound bow, firing continuous volleys of arrows and javelins from a distance. The Parthian "lights" developed a technique of firing to their rear at a pursuing enemy, a tactic Romans called "the Parthian shot." Infantry and mercenaries, recruited mostly from feudal landlords, played largely supporting but vital roles. This military system made the Parthians formidable and frustrating opponents. Conversely, poor logistics and dependence on an unreliable nobility severely limited Parthian capacities for sustained offensive operations.

Parthian-Roman relations began on friendly enough terms in 96 B.C.E. However, when General Pompey subjugated Armenia, traditionally a Persian vassal, frictions began. Marcus Licinius Crassus attempted a conquest of Parthia in 53 B.C.E. and died at Carrhae. Mark Anthony lost thousands of men invading Persia 15 years later. Caesar Augustus fortified Syria as a base against Persia and, in 20 B.C.E., imposed a puppet king on Armenia. Using this strategy, Rome controlled Armenia for decades. However, Shah Vologases wrecked this hegemony, waging a decade of war (53–63), until Nero accepted Vologases's brother as king of Armenia.

In 113, Emperor Trajan invaded Armenia and, in 115, captured Mesopotamia and Ctesiphon itself. Occupation unleashed popular revolts throughout Mesopotamia, leading Emperor Hadrian to evacuate in 117. In the First Parthian War (161–166) Shah Vologases III, responding to noble and popular anti-Roman pressures, launched invasions into Syria and Armenia. Lucius Verus drove him out, reconquered northern Mesopotamia, but avoided Ctesiphon. Vologases IV started the Second Parthian War (197–199), trying to drive the Romans out of northern Mesopotamia. Rome again repulsed the shah and once more sacked Ctesiphon. In 215, the shah defeated the armies sent against him by Caracalla, granting a truce to Caracalla's beleaguered successors in exchange for heavy tribute.

Ardashir Sassan revolted against Parthian ineptitude and venality caused by centuries of warfare and defeated Artabanus at Hormizdagh in 225, bringing an end to the Parthians and founding the Sassanian Empire.

Weston F. Cook, Jr.

See also: Cyrus II the Great; Persian Empire

References and further reading:

Wiesehofer, Josef. *Ancient Persia.* Trans. Azizeh Azodi. London: I. B. Tauris Publishers, 1996.

Yarshater, Ehsan, ed. *The Cambridge History of Iran.* Vol. 3. London: Cambridge University Press, 1985.

Patton, George Smith, Jr. (1885–1945)

U.S. Army general and pioneer in modern armored warfare. Born in San Gabriel, California, on 11 November 1885, he inherited a warrior's legacy from his father.

Patton's family possessed a military tradition dating back to the American Revolution, and from an early age, he knew that a military career was his destiny. When he graduated from high school, there were no appointments to the U.S. Military Academy at West Point. Therefore, he enrolled at the Virginia Military Institute (VMI), the school from which his father, grandfather, and three great-uncles had graduated.

After one year at VMI, Patton secured his appointment to West Point. During his time at West Point, he became an expert fencer and demonstrated the drive, aggressiveness, and flair that would define his career. The skills developed at West Point and his competitive nature earned Patton the right to represent the United States in the 1912 Olympics. As the first American to compete in the modern pentathlon, he finished a respectable fifth.

With the start of World War I, Patton requested permission to serve with the French cavalry but was turned down. In 1916, he was an aide to General John J. Pershing on the punitive expeditions into Mexico against Pancho Villa. During one mission, Patton killed General Julio Cardenas, the

Portrait of George Patton. (Library of Congress)

head of Villa's bodyguard, using the same Colt pistol that would become his trademark.

With the U.S. entry into World War I, Patton was selected on 15 May 1917 to join Pershing's American Expeditionary Force. In August 1918, Patton commanded a tank brigade during the Meuse-Argonne Offensive. He was wounded and was awarded both the Purple Heart and Distinguished Service Cross.

During the interwar years, the size of the U.S. Army, especially armored units, was severely reduced. Patton used this time to attend the Army War College, learn to fly, and publish articles on armored unit tactics.

U.S. entry into World War II necessitated the preparation of the army's newly formed armored units. Patton was given the task. On 8 November 1942, Patton landed on the west coast of Africa. In the wake of the defeat at Kasserine Pass, General Dwight D. Eisenhower put Patton in command. His success led to his command of the Seventh Army during the invasion of Sicily. Chafing at his role of providing flank protection for his rival, British general Bernard L. Montgomery, Patton looked for an opportunity to play a more active role. Taking advantage of stiffer than expected resistance to the British advance, he received permission to drive toward Palermo, capturing it on 22 July and then capturing Messina ahead of Montgomery.

Patton missed the Italian campaign and was denied the opportunity to be the American ground commander for Normandy because of an incident in which he slapped a soldier whom he believed to be a malingerer (an action that he would repeat). On 6 July 1944, Patton was sent to France to take command of the Third Army. It was the zenith of his career, as his Third Army made a dramatic sweep across northern France in a campaign marked by great initiative, ruthless drive, and disregard of classic military rules.

When a German counteroffensive threatened to cut through thin American lines, Patton saw possibilities in this dire situation. He disengaged his troops, hurled them northward during a terrible winter storm, and attacked the German's flank, relieving the encircled troops at Bastogne. Patton's actions, considered impossible when he suggested them, are credited in relieving Bastogne and defeating the Germans in the Battle of the Bulge.

As the war ended, controversy continued to surround Patton. After Germany surrendered, Patton, a staunch anti-Communist, argued for a combined Allied-German campaign against the Soviet Union. When he later argued to keep former Nazis in administrative positions, he was removed from command.

Patton died in Heidelberg, Germany, on 21 December 1945, the result of an automobile accident. He is buried among the soldiers who died in the Battle of the Bulge in Hamm, Luxembourg. Probably the most admired and controversial of all American generals in World War II, Patton was known for carrying ivory-handled pistols, using racy language, and having an intemperate manner but was also regarded as one of the most successful American field commanders of any war.

Craig T. Cobane

See also: Ardennes, Battle of; Armored Fighting Vehicles; TORCH, Operation; World War II

References and further reading:

D'Este, Carlos. *Patton: A Genius for War.* New York: HarperCollins, 1995.

Essame, Hubert. *Patton: A Study in Command.* New York: Charles Scribner's Sons, 1974.

Pavia, Battle of (24 February 1525)

The turning point of the "Italian wars" and the end of the era of chivalry. By 1525, French kings had been claiming territories in Italy for 30 years. To reach their political goals, they had to face the thrones of Spain and Austria, which were combined in 1519, forming a threatening neighbor. The nature of the Italian wars changed as the new king, Charles V, ruled countries surrounding France on three sides. In 1524, the imperialist forces invaded Provence, but facing failure at the siege of Marseille, they had to retire in front of the main French army. Francis I, the king of France, decided to follow the retiring army in Italy.

The imperialists resisted the French invasion but had to fall back on their fortified garrisons of Pavia and Lodi. Francis decided (against the advice of his wiser commanders) to avoid a direct fight against the main imperialist army, led by the Marquis of Pescara. He chose instead to besiege Pavia.

The siege began on 28 October 1524. Facing superior French artillery, the Spanish commander Antonio de Levya made a stubborn defense. Unable to storm the town rapidly, Francis decided to make his winter quarters in a walled park, north of the siege work. The desertion rate among the mercenaries began to rise (8,000 Swiss on 20 February 1525 alone). Pescara's army of 40,000, mainly *Landsknecht* (mercenary soldiers from the Holy Roman Empire), pikemen, harquebusiers, and light artillery, left Lodi and reached Pavia to find a waiting French army. The besieger was besieged in Mirabello Park.

The battle took place on 24 February 1525. During the night of the 23d–24th, the imperialists (23,000 soldiers) took the initiative. Their approach march turned around the high wall, and a breach was made in an unsuspected spot. Dawn took the French army of 22,000 unprepared and sepa-

The USS Shaw explodes during the Japanese attack on Pearl Harbor on December 7, 1941. (National Archives)

rated in three groups. Following the king, the French cavalry impetuously charged the *Landsknecht* as soon as they emerged from the wall while still masking their own artillery. Facing deadly fire, the French cavalry was cut to pieces, and the reinforcements, unable to stop the imperialists, were destroyed piecemeal. Francis I, wounded in the thick of the fray, was taken prisoner, and 10,000 French were killed, including hundreds of lords, as no mercy was given by either side.

This crushing defeat marked the beginning of a period of imperial control of Italy. "Tout est perdu, fors l'Honneur" ("All is lost but honor) was the comment made by Francis I, writing to his mother to announce his defeat.

Gilles Boué

See also: Cerisolles, Battle of; Marignano, Battle of

References and further reading:

Cornette, Joël. *Chronique de la France moderne, le XVIème siècle.* Paris: SEDES, 1995.

Hardy, Etienne. *Origines de la tactique française.* Paris: Dumaine, 1881.

Konstam, Angus. *Pavia 1525.* London: Osprey Publishers, 1996.

Pearl Harbor Attack (1941)

The devastating Japanese aerial attack on Pearl Harbor in the Hawaiian Islands on 7 December 1941 caused the United States to enter World War II with almost unanimous public support. Japan sought to neutralize the U.S. Pacific Fleet in order to invade and fortify its planned empire, which included the Philippines, Malaya, Netherlands East Indies, China, Thailand, and Burma, without U.S. naval interference. The operation was approved only 13 weeks beforehand, after the personal intervention of Admiral Isoruku Yamamoto. A fleet of six aircraft carriers, *Akagi, Kaga, Shokaku, Ziukaku, Hiryu,* and *Soryu* (by far the largest combat combination of carriers), accompanied by two battleships, two cruisers, nine destroyers, and eight supply ships, was to sail by a northwesterly route to an aircraft launch point 275 miles north of Pearl Harbor. The Japanese task force sailed on 26 November, receiving the final attack order and confirmation of the date seven days later.

Pearl Harbor was unprepared for an air attack, although war in the Pacific was thought imminent. The Imperial Japanese Navy was considered incapable of mounting any operations additional to the invasion convoys that were being reported in the world's press and known to be moving southward from Japan and Formosa. It was also believed that Japan would not commit the strategic error of unnecessarily forcing the United States into war. As a result, the eight air groups that commenced flying off from the Japanese carriers at 6:00 A.M. on Sunday, 7 December, achieved complete surprise when they dropped their first bombs at 7:55.

The 353 strike aircraft launched in two waves, 45 minutes apart. Their targets were battleships and cruisers and U.S. Army and Navy airfields. Attacks were made on the ships by torpedo bombers, level bombers, and dive-bombers, and many aircraft also strafed after dropping their bombs. The escort fighters strafed airfields, while level bombers and dive-bombers destroyed the planes and facilities. Antiaircraft gunfire was initially sporadic, and only about 18 U.S. Army fighters managed to get airborne during the two-hour attack. Twenty-nine Japanese aircraft were lost during the raid, and approximately 20 more were destroyed by landing accidents in rough weather.

U.S. losses were severe: 2,403 dead and 1,178 wounded; the battleships *Arizona, California,* and *West Virginia* destroyed; the *Oklahoma* capsized; and the *Nevada, Tennessee, Pennsylvania,* and *Maryland* damaged. The target battleship *Utah,* two destroyers, and a minelayer were also sunk. Additionally, four cruisers, a destroyer, and three tenders were damaged. Of the approximately 400 aircraft based in Hawaii, 239 were destroyed or severely damaged. It was the most one-sided naval air assault in history.

The Japanese themselves blundered in not scheduling the attack to destroy the American aircraft carriers based at Pearl Harbor, in not hitting the oil tank farms, in not destroying the machine shop complex that would help to restore many of the sunken and damaged warships, and in sinking the warships in waters shallow enough to permit the raising of all but one of the sunken U.S. battleships. But Japan's worst blunder was that of underestimating American resolve, industrial capacity, and fighting power. Japan would pay for those blunders with total defeat in a war that it could not hope to win.

So successful was the attack on focusing a once-divided nation's belligerent rage against Japan that the conspiratorial-minded have ever since suspected that President Franklin D. Roosevelt at least had some foreknowledge of the attack. They ignore the obvious fact that Roosevelt was preparing the United States for conflict with Germany and would hardly have welcomed a two-front war. Further, he would have had no idea that Adolf Hitler, in an act of gratuitous folly, would declare war on the United States.

Michael Hyde

References and further reading:

Japanese Monograph No. 97. "Pearl Harbor Operations: General Outline of Orders and Plans (5 November–2 December 1941)"; "The Naval Armament Program and Naval Operations (Part I)." In *War in Asia and the Pacific,* ed. D. Detwiler. Vol. 4. New York: Garland, 1980.

Morison, S. E. *The Rising Sun in the Pacific 1931–April 1942.* Vol. 3, *History of United States Naval Operations in World War II.* Boston: Little, Brown, 1948.

Prange, G. W., in collaboration with Donald M. Goldstein and Katherine V. Dillon. *At Dawn We Slept: The Untold Story of Pearl Harbor.* New York: Viking, 1991.

Worth, Roland H., Jr. *Pearl Harbor: Selected Testimonies, Fully Indexed, from the Congressional Hearings (1945–1946) and Prior Investigations of the Events Leading Up to the Attack.* Jefferson, NC: McFarland, 1993.

Peleliu (15 September–27 November 1944)

One of the most costly, least known, and perhaps most unnecessary of the Pacific island invasions during World War II. In September 1944, General Douglas MacArthur was planning the invasion of the Philippines and wanted his flank protected by an attack on Peleliu in the Palau Islands, located between New Guinea and the Philippines. The attack, Operation STALEMATE, would be just that.

U.S. planners had little useful information on Peleliu, including the strength and nature of Japanese defenses. The pre-invasion bombardment did not inflict much damage on the Japanese, who were well-protected in deep caves and other defenses away from the landing beaches. The invasion began well on 15 September, and then concealed machine guns and mortars opened fire, followed by tanks and troops in the afternoon. That evening, the Japanese launched suicide attacks against the Marines, who were still mostly on the invasion beach.

In the first week, the Marines Corps suffered 4,000 casualties, and some units were reduced to throwing chunks of coral at and using bayonets on the Japanese. Marine Corps fighter planes taking off from the landing strip on the island did not raise landing gear—they commenced their bombing runs too soon.

By the time that flamethrowers, bombs, naval bombardment, and the courage of the marine and army troops finally won the island, the United States had suffered grievously. The 1st Marine Division suffered some 54 percent casualties; the 5th Marine Division suffered 43 percent killed, wounded, and missing in action; and the 7th Marine Division lost 46 percent of its strength. Of 19 medals of honor given to members of the 1st Marine Division, eight were earned in the fighting on Peleliu. Worse, taking the island did not speed up MacArthur's timetable for the attack on Leyte or help defend the flank of his advance. The attack was both unnecessary and unmindful of the lessons learned in earlier amphibious invasions in the Pacific theater.

Charles M. Dobbs

See also: Saipan, Battle of; Tarawa; Tinian

References and Further reading:

Gailey, Harry A. *Peleliu, 1944.* Annapolis, MD: Nautical and Aviation Publishing Company of America, 1983.

Gayle, Gordon D. *Bloody Beaches: The Marines at Peleliu.* Washington, DC: History and Museums Division, HQ, U.S. Marine Corps, 1996.

Ross, Bill D. *Peleliu: Tragic Triumph: The Untold Story of the Pacific War's Forgotten Battle.* New York: Random House, 1991.

Wheeler, Richard. *A Special Valor: The U.S. Marines and the Pacific War.* New York: Harper & Row, 1983.

Peloponnesian Wars (460–456, 431–404 B.C.E.)

Wars between Athens and Sparta that spelled the end of the former's role as a major Mediterranean power. The First Peloponnesian War was sparked when Athens renounced its alliance with Sparta against the Persians and allied itself with Sparta's enemy, Argos. Soon after, Megara, following a border dispute with Corinth, withdrew from the Peloponnesian League, to which they had both belonged, and made an alliance with Athens. The defection of Megara meant Sparta was now unable to strike overland at Attica.

For the alliance with Argos to be of any benefit, Athens needed to acquire a foothold in the Peloponnese. The Athenian seaborne landing at Haliae was the result, but the Corinthians successfully repulsed it (458 B.C.E.). Athens now turned its attention to Aegina, the strategy being to clear the nearby seas of hostile navies and thus secure maritime communications to Argos. Matters went well for Athens, which won two naval engagements off Aegina before laying siege to the island's chief town. Even when a Spartan army crossed the Corinthian Gulf and defeated the Athenians at Tanagra (457 B.C.E.), Athens's immediate response was to invade Boeotia, defeat the Boeotians at Oenophyta, and bring all central Greece under its control. Shortly afterward, Aegina surrendered, and the confidence of Athens was well illustrated when an Athenian fleet sailed around the Peloponnese, raiding as it went (456 B.C.E.).

Athens's success was cut short by defeat in Egypt at the hands of the Persians (454 B.C.E.). Furthermore, Athens was forced to surrender control of central Greece after the Boeotians defeated the Athenians at Koroneia (447 B.C.E.). Worse still, with Megara rejoining the Spartan alliance, a Peloponnesian army invaded Attica (456 B.C.E.). Athens had had enough, and a 30-year truce was concluded with Sparta. Athens gave up its claims to central Greece and ceased to interfere in the Peloponnese. However, it retained a grip on Aegina and secured the recognition of its maritime empire. With these diplomatic gains, the First Peloponnesian War ended.

What made the next war between Athens and Sparta inevitable, according to the contemporary historian Thucydides, was Sparta's fear of Athens. Yet this "truest cause" was least discussed because Sparta could hardly stand before its allies and announce its fears. In Thucydides' eyes, the Spartans were not warmongers, being traditionally slow to go to war unless forced. However, when Athens started to meddle with Sparta's allies, namely Corinth, and antagonism arose between the Athenians and the Corinthians over the control of Corcyra (433 B.C.E.) and Potidaia (432 B.C.E.), Sparta was left with little choice. This was especially so when another important ally, Megara, added its voice to the clamor for war.

Now able to invade Attica through the Megarid, Sparta did so during the initial phase of the war. On Pericles' advice, the Athenians took refuge inside the walls surrounding Athens and the Peiraeus and responded to the Spartan ravaging merely by minor cavalry operations, seaborne raids on the Peloponnese, and invasions of the Megarid. But after Pericles' death (429 B.C.E.), Athens, now dominated by Kleon, adopted a more daring strategy. In addition to establishing bases on the Peloponnesian coast—notably at Pylos—it also attempted to knock Boeotia out of the war, but the second invasion ended in defeat at Delion (424 B.C.E.).

The same year saw the Spartan Brasidas surprising Athens with a campaign in northern Greece and winning over a number of Athens's dependencies, including Amphipolis. His own death and that of Kleon in battle outside Athens (422 B.C.E.) led to the conclusion of peace. The "hollow peace," as Thucydides so aptly calls it, was soon in tatters when Alcibiades cobbled together an anti-Spartan coalition in the Peloponnese. Yet it came to naught when the Spartans destroyed the coalition forces, led by Athens and Argos, at Mantinea (418 B.C.E.).

At Alcibiades' urging, Athens launched an expedition against Syracuse (415 B.C.E.), with him, along with his rival Nikias, as the commanders. Before the attack on Syracuse had begun, however, Alcibiades was recalled to answer charges of sacrilege. He fled to Sparta, and on his advice, the Spartans established a permanent base at Dekeleia in Attica (413 B.C.E.). Bogged down in the siege of Syracuse, the expedition ended in total disaster.

Athens was dependent on maritime imports, particularly grain and flax from the Black Sea region, and thus needed a navy for the protection of commerce. But sea power was of little use for the defeat of Sparta, a stalwart land power. Sparta was fully aware of this shortcoming and of Athens's dependence upon seaborne supplies. Still, Sparta had been unable to match the might of the Athenian navy and thus could only dispatch its fleets to stir up revolts within the Athenian empire. Sparta had never been a naval power, but Athens's Sicilian debacle presented Sparta with the opportunity to become one.

Sparta sent a fleet to attempt a blockade of the Hellespont and thus cut Athens off from the Black Sea. Despite this stratagem, King Agis, who was holding Dekeleia, reckoned it was a waste of time attempting to sever Athens's supply lines when he could still see the grain ships putting into the Peiraeus. The alternative was to engage Athens on the high seas, but in doing so Sparta was to suffer absolute disaster at sea on a number of occasions.

Kynossema (411 B.C.E.) was a moral victory for the Athenians, who, lacking their former confidence, had been afraid of the Peloponnesian fleet with its Syracusan allies. Kyzikos (410 B.C.E.) was a scrambling fight along the Hellespontine coast. Off Arginousai (406 B.C.E.), the Peloponnesian fleet, with its more skillful crews, attempted to outmaneuver the Athenians. Sparta ultimately gained the upper hand, and its admiral Lysander resoundingly defeated the Athenians at the naval engagement off Aegospotami (405 B.C.E.). The following year, Lysander was able to strangle Athenians into submission, his naval victory effectively cutting the city off from Black Sea grain supplies.

The ultimate result of the Pelopponesian wars was to weaken all the protagonists. In the end, notwithstanding the resurgence of some, such as Thebes, the Greek polis fell prey to outside barbarians, as the Macedonians swept all before them.

Nic Fields

See also: Alcibiades; Cimon; Syracuse, Siege of

References and further reading:

Kagan, D. *The Fall of the Athenian Empire.* Ithaca, NY: Cornell University Press, 1987.

Peng Dehuai (1898–1974)

China's field commander in the Korean War. Of peasant stock from Hunan province, Peng sought a military career by joining the local Nationalist Kuomintang (KMT) forces in his home province in 1916. By 1921, he had become an officer and remained with the GMD through the Great Northern Expedition. With the collapse of the alliance between the Nationalists and the Communists, Peng joined the latter, having always been concerned with social reform.

One of the People's Liberation Army's first professional soldiers, he emphasized the military verities of discipline, organization, and chain of command, as opposed to guerrilla war tactics. Despite this focus, Peng retained Mao Zedong's confidence and came to respect Mao's military thought.

The pinnacle of Peng's career was leading the People's Volunteer Army in Korea and sending the United Nations forces reeling back across the 38th parallel. The inability of the volunteers to follow up this victory and conclude the war on Communist terms rekindled Peng's concern with conventional approaches to war (he had seen how the United Nations' air and sea power had kept his forces from complete victory in Korea) and possible friction with Mao.

After being raised to the rank of marshal in 1955, Peng was dismissed from command in 1959 for openly challenging Mao over the failures of the Great Leap Forward. Despite partial rehabilitation in 1962, Red Guards arrested Peng in 1966, and he spent the remainder of his life imprisoned and tortured, refusing to apologize for past criticisms. The party posthumously rehabilitated Peng in 1978.

George R. Shaner

See also: Chinese Civil War; Korean War; Mao Zedong

References and further reading:

Domes, Jurgen. *Peng Te-Huai: The Man and the Image.* Stanford, CA: Stanford University Press, 1985.

Hsiung, James C., and Steve I. Levine, eds. *China's Bitter Victory: The War with Japan, 1937–1945.* Armonk, NY: M. E. Sharpe, 1992.

Zhang, Shu Guang. *Mao's Military Romanticism: China and the Korean War, 1950–1953.* Lawrence: University Press of Kansas, 1995.

Pequot War (1636–1637)

An early colonial war that was in practice genocide against an American Indian tribe. The Pequot War began as a contest between the Pequot and Narragansett tribes over trading rights in southern New England. Both groups coveted the European textiles, tools, handicrafts, and contraband (guns and alcohol) that the Dutch and English exchanged for pelts. When the Dutch erected a trading post on the Connecticut River in 1633, Pequots murdered rival Narragansett traders. The Dutch responded by assassinating the Pequot sachem, Tatobem. In retaliation, the Pequots then murdered the captain and crew of an English trading vessel, thinking they were Dutch.

Threatened by the Dutch and the Narragansetts, the Pequots sought assistance from Massachusetts Bay Colony. In 1634, the Puritan leaders in Boston offered to negotiate peace in return for heavy tribute payments and the surrender of those responsible for the deaths of the English traders. The Pequots rejected the offer. For strategic, political, and economic reasons, Massachusetts then constructed Saybrook fort at the mouth of the Connecticut River and planted three settlements upriver.

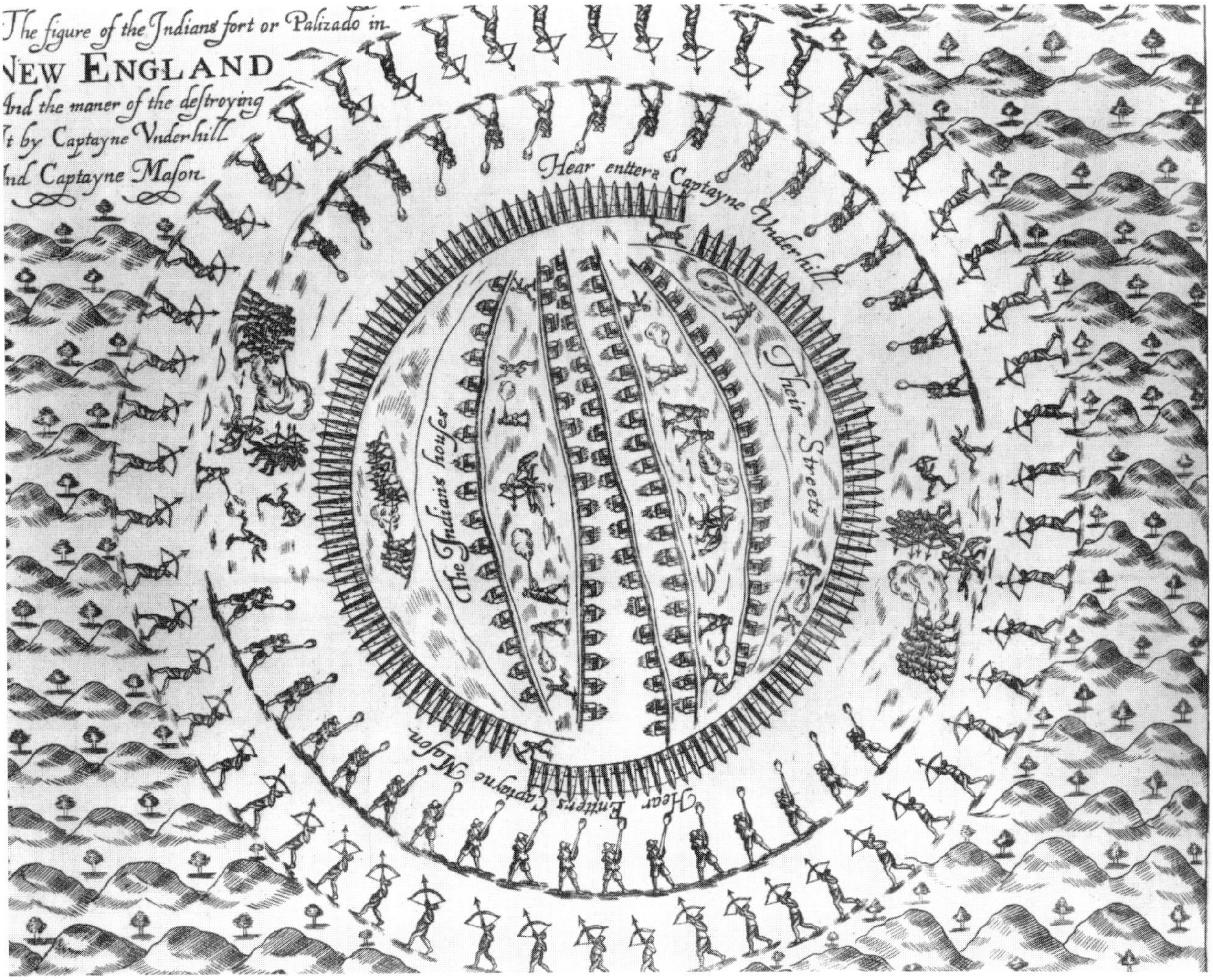

"The figure of the Indians' fort or palizado in New England and the manner of the destroying it by Captain Underhill and Captain Mason." (Library of Congress)

Spurred on by Uncas, the scheming Mohegan sachem, Bay Colony leaders in 1636 pressed their demands for Pequot tribute payments and for the surrender of the murderers of Captain John Stone and his crew. When yet another English trader was found murdered on his vessel near Block Island, Massachusetts, the Bay Colony launched a punitive raid against the Pequots (who were, in fact, innocent of the crime). Unable to draw their enemy into battle, 90 Massachusetts men under Captain John Endecott burned the Pequot villages on Block Island (22 August 1636). The scenario was repeated at Connecticut's Thames River, where undefended villages belonging to the Pequot and their Western Niantic allies were destroyed.

The Pequots retaliated by assaulting the English outposts at Wethersfield and Saybrook, Connecticut. The war quickly escalated. The Pequots and Western Niantics soon faced a combined force of Massachusetts men, Narragansetts, and Eastern Niantics on one front and Connecticut men and their Mohegan allies on another. In a major offensive (26 May 1637), Captain John Mason led 77 Connecticut militia and hundreds of Mohegans and Narragansetts against the Pequot fort at Mystic, Connecticut. They surrounded and burned the fort, killing 600–700 inhabitants, many of them noncombatants—a fact not lost on the Indian allies of the English, who protested the ferocity of the English attacks. The English suffered only two fatalities.

Pequot warriors at nearby Weinshauks arrived too late to save their kin. Their attempt to exact vengeance on the departing English and Indian forces resulted in the deaths of more than 100 additional Pequots. The Mystic massacre and subsequent defeat caused the remaining Pequots to abandon their villages and seek refuge with neighboring tribes. A 120-man force from Massachusetts led by Israel Stoughton engaged a much smaller Pequot force in a swamp known as

"Owl's Nest," killing or capturing approximately 40 Pequots (June 1637). Captive warriors were executed, and Pequot women and children were sold into slavery.

Connecticut troops then joined Massachusetts militia to battle the remaining Pequots. The English encircled their enemy in a swamp near New Haven where, following the surrender of 180 women and children, Pequot warriors battled until dawn (14 July 1637). Sassacus, the Pequot leader, and others managed to escape but were subsequently killed by Mohawks with whom they sought refuge.

The victorious English and their Indian allies treated surviving Pequots as spoils of war, dividing them and selling some into slavery. The English denied the Pequots the use of their tribal name and refused to allow them to rebuild their devastated villages. The Mohegans and Narragansetts agreed to execute any Pequot warriors still at large. In effect, the English asserted their hegemony and set an example for all American Indians by doing everything in their power to eradicate the Pequots socially, politically—and physically.

John J. Navin

References and further reading:

Cave, Alfred. *The Pequot War.* Amherst: University of Massachusetts Press, 1996.

Katz, Steven. "The Pequot War Reconsidered." *The New England Quarterly* 64 (1991).

Hauptman, Laurence, and James Wherry, eds. *The Pequots in Southern New England: The Fall and Rise of an Indian Nation.* Norman: University of Oklahoma Press, 1990.

Jennings, Francis. *The Invasion of America: Indians, Colonialism, and the Cant of Conquest.* Chapel Hill: University of North Carolina Press, 1975.

Salisbury, Neal. *Manitou and Providence: Indians, Europeans, and the Making of New England, 1500–1643.* New York: Oxford University Press, 1982.

Pericles (495–429 B.C.E.)

An incorruptible aristocrat with masterful speaking skills and clear military strategies who consolidated Athenian control over the Aegean. Pericles succeeded Cimon after the Spartan people rebuffed Athenian assistance in 462 B.C.E. Pushing his aggressive policy of spreading Athenian forces from Cyprus to Phoenicia to Egypt during the war with Persia, Pericles earned a reputation as a radical politician with strong imperialistic tendencies. From 450 through 429 B.C.E., he remained the preeminent leader of the Athenian city-state.

Reelected annually as the leader of the Board of 10 Generals, Pericles dictated foreign policy, concentrating resources on the naval fleet and thereby guaranteeing the supply of food and availability of strategic materials. In 437 B.C.E., he sailed into the Black Sea region to unseat the tyrant at Sniope and signed a treaty with the Bosporan king to provide Russian wheat. He also established a colony at Amphipolis to secure access to Macedonia. Known for the massive building program on the Acropolis, Pericles was also instrumental in the building of the 4-mile-long walls that connected the fortified city to the port of Piraeus.

From 456 through 446 B.C.E., the Athenians remained at peace with both the Persians and the Spartans. During this time, Pericles strengthened Athenian control over the region, alienating other city-states in the process. He banned Megarian traders from Aegean markets and interfered with local political disputes and customs. Sparta finally consented to lead an expedition against Athens. From 431 to 421 B.C.E., the Athenians and Spartans fought the Second Peloponnesian War. Pericles embarked on a policy of wearing the enemy down through naval raids around the Peloponnese, while the Spartan army attacked Attica but failed to take Athens. The inhabitants of the countryside took refuge in the city, where an epidemic broke out. It killed tens of thousands of Athenians, including Pericles, who died in 429 B.C.E.

Cynthia Clark Northrup

See also: Peloponnesian Wars

References and further reading:

Mattingly, Harold B. *The Athenian Empire Restored: Epigraphic and Historical Studies.* Ann Arbor: University of Michigan Press, 1996.

Pershing, John J. (1860–1948)

Commander of the American Expeditionary Force in World War I. Born on 11 September 1860 in Missouri. Pershing entered the U.S. Military Academy at West Point at the age of 22. Upon graduation, Jack, as he was then known, entered the cavalry and served in the West. From 1891 to 1894, he was a professor of military science at the University of Nebraska. During these years, he took a law degree and was admitted to the bar. After a brief term as an instructor at West Point, Pershing served with the 10th (Colored) U.S. Cavalry in Cuba during the Spanish-American War. His coolness under fire earned him distinction. He spent much of the next 13 years serving in the Philippines. In 1906, President Theodore Roosevelt promoted Pershing to the rank of brigadier general over 862 senior officers. From 1913 to 1916, he served with the U.S. Army's Southern Department. From March 1916 through February 1917, Pershing commanded the punitive expedition into northern Mexico in pursuit of the revolutionary bandit, Pancho Villa.

As the only American military officer to have held a large command in a foreign country, Pershing was a natural selection to command the American Expeditionary Forces after the declaration of war against Germany in April 1917. Secretary of War Newton Baker allowed Pershing wide leeway in establishing military policy. Pershing demanded and received greatly enlarged American divisions, the size of a European corps, as well as training based on marksmanship, emphasizing rifles over machine guns. In France, he fought to consolidate all American units under his own tactical control, as opposed to assigning U.S. soldiers to Allied units as replacements. Pershing threw his untested troops into battle in the spring of 1918 when the German offensive threatened Paris. In the fall, Pershing launched a major offensive against the Germans in the Meuse-Argonne region, still the single greatest battle in American history. An advocate of unconditional surrender, he counseled against the armistice of 11 November 1918.

Pershing returned to the United States a hero. Congress commissioned him general of the armies, the first officer to hold that title since General Ulysses S. Grant. From 1921 to 1924, Pershing served as U.S. Army chief of staff. He was considered a father figure to a younger generation of officers, including Dwight D Eisenhower, George Marshall, and George Patton. He died in New York City on 15 July 1948.

Gregory Dehler

See also: Baker, Newton D.; Chateau Thierry/Belleau Wood; March, Peyton; Meuse-Argonne; Mexico, U.S. Punitive Expedition in; St. Mihiel; World War I

References and further reading:

Pershing, John J. *My Experiences in the First World War.* New York: Da Capo, 1995.

Smythe, Donald. *Guerrilla Warrior: The Early Life of John J. Pershing.* New York: Charles Scribner's Sons, 1973.

Vandiver, Frank. *Black Jack: The Life and Times of John J. Pershing.* 2 vols. College Station: Texas A & M University Press, 1977.

Persian Civil Wars (1725–1794)

A near century of conflict and invasion that marked Persia's transition from the Middle Ages to the modern era. The Safavid dynasty had ruled Persia since 1514. Their shahs had been patrons of a great cultural renaissance, fixed the boundaries of modern Iran, and imposed Twelver Shiite Islam on the Persian people. But in 1718, a variety of ethnic minorities, autonomy-minded nomads, restive towns, and Sunni Muslims opposed to Shiism had begun to challenge a weakening Safavid regime. These disturbances inspired Mahmud, emir of the Ghilzai Afghans, to invade Persia in 1722. At the Battle of Gulnabad, he sliced up the larger Safavid army, captured the capital of Esfahan, and forced the shah to crown him as successor. Mahmud, however, lacked the force or support to exercise any kind of real power beyond the range of his tribal forces. As other Safavid princes, towns, and tribes joined the insurrection, both Russia and the Ottoman Empire sent expeditions against Iran's borders.

Declaring himself champion of the Safavids, Nadir Khan Afshar rallied much of the army. He drove the Ottomans back to Baghdad in 1733, persuaded Russia to evacuate the north, and retook the capital. In 1736, Nadir deposed the infant Safavid prince and made himself shah. To control the Persian Gulf, he launched an invasion of the sultanate of Oman. In 1739, campaigning against the Afghans, he noted the weakness of the Mogul state of India and plunged into the Ganges Valley. Nadir plundered Delhi and carried off the famous Peacock Throne. He then ranged over much of Uzbek territory, making vassals of Bukhara, Herat, and Khiva. In 1747, Nadir Shah's Qizilbash allies, fearful of his growing megalomania, murdered him.

Nadir's tribal units broke up into separate contingents, and Iran fell back into another decade of anarchy. Eventually, Karim Khan Zand made himself dominant, moving his capital to Shiraz. Recognizing the exhaustion of his people, he tried to avoid war (although he did occupy Basra in Iraq). Instead, he concentrated on repairing the state, keeping harmony among the key tribes like the Bakhtiyari and the Qajars, and patronizing Twelver Shiite Islam. Sadly, his death in 1779 set off another wave of anarchy between towns, tribes, and princes. Agha Muhammad Khan of the Qajar tribe finally succeeded in crushing or co-opting his rivals by 1796. His Qajar dynasty would rule Iran until the end of World War I.

Weston F. Cook Jr.

See also: Nadir Shah

References and further reading:

Avery, Peter, ed. *The Cambridge History of Islam.* Vol. 7. Cambridge, UK: Cambridge University Press, 1991.

Morgan, David. *Medieval Persia 1040–1797.* London: Longman's Press, 1988.

Persian Empire (550 B.C.E.–642 C.E.)

An empire encompassing Iran and at various times parts of Armenia, Asia Minor, and most of the Middle East. It was established by an Iranian people, the Persians, in two eras, the Achaemenid and the Parthian.

The Achaemenid Era

The Achaemenid era of the Persian empire (550–330 B.C.E.) began when Cyrus II Achaemenid of Fars/Persia defeated Astyages of Media in 559, capturing Ecbatana in 550 B.C.E. In 546 B.C.E., Cyrus defeated Croesus of Lydia and occupied

Babylon in 538 B.C.E. Appreciating logistics, Cyrus commissioned the Royal Road, a military highway allowing the Persian army to travel from Susa to Sardis in Lydia (Asia Minor) in three months. Cyrus was killed during a campaign against the Massagetae in 530 B.C.E. His son, Cambyses II, conquered Egypt in 525 B.C.E.

After Cambyses's death, Cyrus's son-in-law, Darius I (r. 522–486 B.C.E.), consolidated and expanded the Persian empire to an area of 2 million square miles with a population of 10 million people, encompassed by the rivers Indus, Danube, Jaxartes, and Nile. Darius divided the empire into 20 satrapies and devised an efficient message relay system by establishing outposts at distances of one horse-travel day apart.

Further improving logistics, Darius commissioned a canal from the Nile to the Red Sea, wide enough for two galleys to pass each other under oar. In 512 B.C.E., Thrace and Macedonia were conquered. The Ionian revolt (500–494 B.C.E.) prompted a Persian army to land in Attica on the plains of Marathon in 490 B.C.E. The numerically superior Persian force was defeated by the 10,000 hoplites of Athens and Plataea.

Xerxes I (r. 486–465 B.C.E.) launched a second campaign against Greece from Sardis in 480 B.C.E. with an army of 100,000. He commissioned a canal dug through the peninsula of Athos and a bridge over the Hellespont River. The Persian army was to march through Thrace, while the fleet sailed alongside to protect and provision. Persian forces marched into Attica and sacked Athens, marking the high point of Persian expansion to the West. The Persian fleet was subsequently defeated by Themistocles in the narrow strait of Salamis, using the maneuverable Greek triremes. Xerxes and the remaining Persian fleet retreated, and the Persian army under Mardonius was defeated in 479 B.C.E. at Plataea.

Achaemenid rule ended after Darius III Condomannus (r. 336–330 B.C.E.) was defeated by Alexander of Macedon, who employed the oblique battle form at Granicus in 334, Issus in 333, and Gaugamela in 331 B.C.E. Alexander occupied the Persian empire from 331 to 323 B.C.E.

The Parthian Era

The Parthian era of the Persian empire (247 B.C.E.–228 C.E.) began when Arsaces II established the independence of Parthia from the Seleucids, who were Alexander's successors. Parthia grew to become a counterweight to Rome. The resurrected Persian empire of Mithradates II (124–88 B.C.E.) stretched from Armenia to India. In 53 B.C.E., a Roman army of 40,000 under Marcus Licinius Crassus was annihilated by the Persian forces of Orodes I.

During the continued Roman-Persian struggle, the approximate border between the two empires was the Euphrates River. Major Roman campaigns were undertaken in the years 116, 161, 195, 217, and 232 C.E. The Parthian dynasty ended when Ardashir Sassan I of Fars defeated the Parthian army of Artabanus at Hormizdagh in 226. Ardashir proclaimed himself an Achaemenid heir, beginning the Sassanid dynasty of the Persian empire, which ruled until the Islamic conquest of Persia in 642 C.E.

Neville G. Panthaki

See also: Alexander the Great; Alexander's Wars of Conquest; Cyrus II the Great; Gaugamela, Battle of; Greek-Persian Wars; Issus, Battle of; Marathon, Battle of; Parthian Empire; Plataea, Battle of; Sassanid Empire; Shapur I; Shapur II; Xerxes I

References and further reading:

Cook, John Manuel. *The Persian Empire.* New York: Schocken Books, 1983.

Dandamaev, Mikhail. *A Political History of the Achaemenid Empire.* New York: E. J. Brill, 1989.

Debevoise, Neilson Carel. *A Political History of Parthia.* New York: Greenwood Press, 1968.

Vogelsang, W. J. *The Rise and Organization of the Achaemenid Empire.* New York: Brill, 1992.

Persian Wars of Expansion (559–509 B.C.E.)

Dramatic rise of a minor Asian kingdom into the world's most powerful empire. At the beginning of the sixth century B.C.E., Media was a tributary state of Assyria. King Astyages of Media married his daughter, Mandane, to Cambyses, king of Anshan. Cambyses governed Fars (Persia), a region about 300 miles in diameter on the northeastern shore of the Persian Gulf. His son Cyrus succeeded him as king of Anshan in 559 and immediately began intrigues against Media, inducing the Median general Harpagus to defect.

Cyrus was ready when Astyages attacked Persia in 550 B.C.E. He counterattacked, occupied the Median capital, Ecbatana, overthrew his grandfather, and ascended the joint throne of Media and Persia. This event is regarded as the founding of the Persian Empire, sometimes called the Achaemenid Empire, after the clan of Cyrus.

After incorporating Media, Cyrus marched northwest, around the Babylonian Empire, taking Armenia and Cappadocia. By defeating King Croesus at Pteria in 546 B.C.E., Cyrus gained Lydia and its tributary state of Ionia and thus controlled all of Asiatic Turkey. Between 545 and 540 B.C.E., he moved northeast into Central Asia nearly as far as modern Tashkent.

The biblical books 1 and 2 Kings, the Prophets, and Daniel show the impact of Persia on the ancient Hebrews. The 10 northern tribes were dispersed when their kingdom, Israel, fell to Assyria in 721 B.C.E. and further when Assyria fell to Babylon in 612 B.C.E. The two southern tribes were taken into captivity when Babylon defeated their kingdom, Judah, in 587 B.C.E., but after Persia conquered Babylon in 539 B.C.E., Cyrus released the Hebrew exiles by edict in 538 B.C.E., enabling a remnant to return home and found the religion of Judaism. When he died in battle against the Massagetes in 529 B.C.E., he was known as "Cyrus the Great" as much for his merciful and intelligent administration as for his military triumphs.

Cyrus's son Cambyses II came to power by murdering his brother, Cyrus's heir, Smerdis, in 529 B.C.E. Smerdis's death was kept secret. Cambyses led his army west, defeated Pharaoh Psamtik III at Pelusium in 525 B.C.E., and thus added Egypt to the empire. He died under mysterious circumstances in Syria in 522 B.C.E., possibly a suicide, probably insane.

Darius, husband of Cyrus's daughter Atossa, murdered Gaumata, who had been masquerading as Smerdis in Cambyses's absence, in 522 B.C.E. The Persian nobles acclaimed him king the next year. In 518 B.C.E., he expanded Cambyses's Egyptian conquests into Libya. By 513 B.C.E., he had gained all the land in India west of the Indus River. The same year, he pushed into Thrace. After suffering a major setback against the Scythians in the Danube Valley in 512 B.C.E., Darius returned to his capitals at Persepolis and Sardis and spent most of the period from 509 B.C.E. until his invasion of Greece in 492 B.C.E. consolidating his power.

By 518 B.C.E., the Persian Empire included 20 numbered provinces, or satrapies: Media, Susiana, Babylonia, Arabia, Assyria, Egypt, Armenia, Cappadocia, Lydia, Ionia, Cilicia, Sagartia, Parthia, Ariana, Bactria, Sogdiana, Arachosia, India, Gandhara, and Gedrosia. The boundaries, administrations, loyalties, and names of these satrapies frequently changed. At its greatest extent in 492 B.C.E., just after Darius conquered Macedonia, the Persian Empire stretched from Libya to the Indus River and from the Persian Gulf to the Aral Sea. It lasted until Alexander the Great conquered the empire at the Battle of Gaugamela in 331 B.C.E.

Eric v. d. Luft

See also: Alexander the Great; Assyria; Babylonian Empire; Croesus; Cyrus II the Great; Greek-Persian Wars; Persian Empire; Scythians

References and further reading:

Collins, Robert J. *The Medes and Persians: Conquerors and Diplomats.* New York: McGraw-Hill, 1972.

Culican, William. *The Medes and Persians.* New York: Praeger, 1965.

Curtis, John. *Ancient Persia.* Cambridge, MA: Harvard University Press, 1990.

Rogers, Robert William. *A History of Ancient Persia from Its Earliest Beginnings to the Death of Alexander the Great.* Freeport, NY: Books for Libraries Press, 1971.

Persian-Afghan Wars (1726–1857)

No less than three major wars between Persia and rebel Afghans in a 131-year period. These wars were intertwined with events in Mogul India, the Ottoman Empire, and czarist Russia but centered on the rise of Afghan, British, and French power in southern and southwestern Asia.

The war of 1726–1738 saw the Persians under Nadir Shah regain control of Afghanistan and invade Mogul India. Nadir developed a disciplined army slowly, never taking on battle unless he could win and turning rival Afghan tribes against each other rather than against Persia. He first targeted the fabled border fortress of Herat. There, he defeated the Abdali Afghans, who became his allies. He then turned on the Ghilzai Afghans and defeated them in pitched battle at Mihmandust in 1729 and Zhargan in 1730. Nadir now had command of more than 100,000 Afghanis plus his Persian troops, and after fighting the Turko-Persian War of 1730–1736, he stabilized Persia. In 1737, he besieged the fortified city of Kandahar with its 30-foot-thick walls and took it by deceit, partially burning it. This action opened the way for the Persian invasion of Mogul India two months later.

The war of 1836–1838 followed years of European meddling in Persian, Afghan, and Indian affairs. Shah Muhammed of Persia sought Russia's help to retake Herat from the Afghans, who were backed by the British. A Persian siege from November 1837 to September 1838 was repulsed with Persian losses as high as 1,700 men in one frontal assault. However, the Afghan ruler Dost Muhammed then sought to launch a second war against the Sikhs in India. This led to 12 years of British-Afghan hostilities.

The war of 1855–1857 involved yet another siege of Herat by the Persians. The Afghans, who had signed a peace treaty with the British in India, received British aid and repelled the last great effort by the Persians to break through into India again, as they had done under Nadir Shah.

Christopher Howell

See also: Panipat, Battles of; Mogul-Persian Wars; Persian Empire

References and further reading:

Adamec, Ludwig. *Dictionary of Afghan Wars, Revolutions, and Insurgencies.* Lanham, MD: Scarecrow Press, 1996.

Ladjevardian, Reza. *From Ancient Persia to Contemporary Iran.* Slage Press, 1999.

Peru-Bolivia Confederation, War of the (1836–1839)

Conflict caused by the brief unification of Peru and Bolivia under the leadership of Bolivian general Andrés Santa Cruz in 1836. Chile objected to the creation of the new state, which had the potential to threaten Chilean security. When Bolivia reneged on a treaty provision for tax exemptions on Chilean imports and Peru imposed a tariff on imports from

Valparaiso, tensions mounted. Chilean anger was further aroused by the use of a Peruvian port by dissident general Ramón Freire to launch a coup attempt on the government. Chile's president, Diego Portales, responded by ordering a naval attack on the Peruvian port of Callo, resulting in the seizure of three Peruvian ships. Santa Cruz responded by arresting a Chilean diplomat, an incident that infuriated Portales, who demanded that the confederation be dissolved. This along with other Chilean demands caused Santa Cruz to declare war.

Chile's initial invasion of Peru met with defeat and the capture of nearly the entire army. Santa Cruz offered to release the army in exchange for Chilean recognition of the confederation and the return of the three captured vessels. Chile agreed, and the army returned. Chile then renounced the agreement and launched another invasion. General Manuel Bulnes led Chile to victory at the Battle of Bunin and followed it up with a decisive triumph at Yungay on 20 January 1839. After his defeat at Yungay, Santa Cruz went into exile in Ecuador, and the Peru-Bolivia Confederation fell.

George M. Lauderbaugh

References and further reading:

Burr, Robert K. *By Reason or Force: Chile and the Balancing of Power in South America, 1830–1905.* Berkeley: University of California Press, 1965.

Peru-Ecuador Conflict (1941–1999)

One of the most persistent South American boundary disputes. Since their establishment as independent republics, Peru and Ecuador have disputed a triangular area on the upper Amazon of approximately 120,000 square miles, bounded on the south by the Mara and Amazon Rivers and on the north and east by the Putumayo River. The remoteness and topography of the region, the imprecision of Spanish colonial boundaries, the chaotic conditions that existed in the aftermath of the collapse of the Spanish Empire, and domestic politics explain why the dispute has been so long-lasting and so difficult to resolve. Numerous attempts to negotiate a settlement between 1830 and 1941 met with failure.

In July 1941, Peru invaded Ecuador's littoral province of El Oro and threatened the port of Guayaquil. Peru's army of more than 10,000 was supported by an air force of 25 planes and soon overwhelmed Ecuadorian defenses. Ecuador's ill-trained and poorly equipped armed forces, numbering less than 1,600, lacked air support. As Peruvian troops advanced on Guayaquil, Ecuador sought a negotiated settlement. The war resulted in 150 Ecuadorian and 400 Peruvian casualties. In January 1942, Ecuador and Peru signed the Rio Protocol, which provided for the withdrawal of Peruvian troops, the cession of some 80,000 square miles of the territory to Peru, and the creation of a boundary commission to delineate the border.

The discovery of the Cenapa River in 1951 complicated the demarcation process and provided Ecuador with an opportunity to continue to press for sovereign access to the Mara River. In 1960 Jose Maria Velasco Ibarra, Ecuador's stridently nationalist president, renounced the 1942 protocol. Hostilities began anew on 28 January 1981 in the disputed Cordillera del Condor region. In a week of fighting, Peruvian commandos and warplanes attacked three Ecuadorian outposts and forced their abandonment. The incident resulted in nearly 200 casualties.

After the 1981 conflict, Ecuador made a concerted effort to upgrade its armed forces in preparation for a renewed conflict in the Cordillera del Condor sector. Ecuador obtained Kifir fighter aircraft from Israel and purchased modern shoulder-fired surface-to-air missiles. In January 1995, Peru attempted to dislodge Ecuadorian troops from fortified positions in the disputed zone. The Ecuadorian modernization program paid off as surface-to-air missiles and Ecuadorian fighters shot down two Sukhoi SU-22 fighter bombers, one A-37, one Canberra bomber, and five helicopters and held its fortified outposts. Peru suffered 300 casualties, whereas Ecuadorian loses were placed at less than 50.

Negotiations intensified after the 1995 clash because both sides had each spent $250 million on the military operation, and both had little enough to show for it. In October 1998, Peru and Ecuador signed a final agreement that was soon ratified by both governments. Peru retained most of the territory gained in the 1941 war, while Ecuador was granted private property rights on the Mara River.

George Lauderbaugh

References and further reading:

Wood, Bryce. *Aggression and History: The Case of Ecuador and Peru.* Ann Arbor, MI: University Microfilms International, 1978.
Zook, David H., Jr. *Zarumilla Maranón: The Ecuador-Peru Dispute.* New York: Bookman Associates, 1964.

Peruvian Guerrilla War (1980–2000)

A Marxist uprising that did little to change conditions in Peru. In 1980, 12 years of military rule in Peru came to an end when Fernando Belaunde Terry became president for the second time. This was not the only aspect of the 1960s life to return to Peru in the 1980s. On the eve of the 1980 election, the guerrilla group Sendero Luminoso (Shining Path) commenced their "people's war" by burning ballot boxes in the Ayacucho region. Guerrilla activity had taken place in the 1960s, but this new activity by Sendero Luminoso heralded a bloody new phase of Peruvian history.

Sendero Luminoso had been born of a split within the pro-Soviet Communist Party in 1970. Its believers followed the ideas of the Peruvian Marxist Mariategui and Maoism. Their tactics were brutal, and they perceived violence as the fundamental mechanism for political change. They saw anybody involved with the government as legitimate targets, including those simply participating in elections. They reintroduced the Inca practice of displaying a dead dog before attacking a village and intimidated people into joining them. In July 1992, they detonated a car bomb in the Lima suburb of Miraflores, which killed 21 people and brought the war to the middle classes.

In 1982, a second guerrilla group was founded: Movimiento Revolucionario Tupac Amaru (MRTA). This group consisted of more middle-class members and avoided the indiscriminate and widespread violence of the Sendero Luminoso. It received a reputation for "flashy" operations and in July 1990 freed 60 prisoners from Lima's high-security Canto Grande prison.

Belaunde was succeeded as president by Alan Garcia in 1985, but guerrilla activity worsened. By 1990, half the country was under a state of emergency. In 1990, the unknown candidate Alberto Fujimori became president. Although he stood on a ticket of not increasing austerity measures, two months into office he introduced "Fujishock," which was much harsher than anything seen before. He also gave the army carte blanche in dealing with the guerrillas and ignored human rights abuses.

"Fujishock" appeared to be working, when in 1992 MRTA leader Victor Polay and Sendero Luminoso leader Abimael Guzman were arrested. During 1994–1995, many guerrillas gave themselves up under the Law of Repentance. It appeared Fujimori had defeated the guerrillas.

This perception quickly changed on 17 December 1996, when 14 MRTA members took more than 600 hostages at a reception at the Japanese Embassy in Lima. The hostages included 19 ambassadors and Fujimori's own brother. Although many were released, a standoff continued until 22 April 1997, when security forces attacked the embassy, freeing all the prisoners and managing to kill all the guerrillas.

Human rights abuses continued, provoking international protests of the treatment of four Chilean members of MRTA and the American Lori Berenson. (Her sentence was reduced in August 2000.)

The violent nature of Peruvian life continued, with mass demonstrations during Fujimori's much-debated third election in 2000. Although he won, he was forced to resign in November because of bribery scandals and was succeeded by Valentin Paniagua. The basic economic, political, and social causes of Peru's unrest and violence remained unaddressed.

M. J. Bain

See also: Nicaraguan Civil War (1979); Panama Incursion; Salvadorian Civil War

References and further reading:

Black, J. N. *Latin America: Its Problems and Its Promise.* Boulder, CO: Westview Press, 1998.

Latin American Regional Reports Andean Group Report. London: Latin America Newsletters, 1998–2000.

Loveman, B., and T. M. Davies Jr. *Che Guevara Guerrilla Warfare.* 3d ed. Wilmington: Scholarly Resources Inc, 1997.

Skidmore, T. E., and P. H. Smith. *Modern Latin America.* New York: Oxford University Press, 1997.

Pétain, Henri-Philippe (1856–1951)

French Army marshal and political figure. Born at Cauchy-à-la Tour, near Arras, on 24 April 1856, Pétain graduated from St. Cyr in 1878. Had he not fought in World War I, he would have retired as a colonel. Pétain held that new weapons gave the defense superiority. When World War I began, he saw that it would be a struggle of attrition, and he argued for

Portrait of Henri-Philippe Pétain. (Library of Congress)

wearing out the Germans along the entire front and only then mounting a "decisive effort."

At the start of the war, Pétain was temporarily commanding a brigade. He then commanded a division (September) and an army corps (November). By June 1915, he was a full general commanding the Second Army. When the Germans mounted their offensive at Verdun in February 1916, he was placed in charge of its defense, reorganizing its defenses and transforming logistics so that supplies ran smoothly to the front. His leadership at Verdun made him a national hero.

Following the disastrous April–May 1917 Nivelle Offensive, Pétain was called on to deal with widespread mutinies in the French army. Made commander of the French army in May, he improved conditions and morale and promised the men that he would not waste their lives needlessly.

In December 1918, there was general public satisfaction with Pétain's promotion to marshal of France, and he led the victory parade down the Champs Elysées on 14 July 1919. He retained command of the French army until 1931. Pétain supported the construction of the Maginot Line and served as war minister (1934). Appointed ambassador to Spain (1939), he was recalled to be the last premier of the Third Republic (June 1940).

Following the armistice with the Germans and the granting of emergency powers, Pétain set up an authoritarian government in southern, unoccupied France at Vichy that accepted collaboration with Nazi Germany. Tried as a war criminal after the war, he was convicted and sentenced to death, which was commuted to life in prison. Removed to the Isle d'Yeu, he died there on 23 July 1951.

Spencer C. Tucker

See also: Foch, Ferdinand; French Army; Joffre, Joseph Jacques Césaire; Verdun

References and further reading:

Griffiths, Richard. *Marshal Pétain.* London: Constable, 1970.

Lottman, Herbert R. *Pétain, Hero or Traitor.* New York: William Morrow, 1985.

Ryan, Stephen. *Pétain the Soldier.* New York: A. S. Barnes, 1969.

Peter I, Romanov, Czar of Russia ("The Great") (1672–1725)

Recognized as the first "modern" ruler of Russia who increased the pace of modernization in a military and administrative perspective, bringing Russia into the community of European great powers. Peter was born in Moscow on 9 June 1672. After the death of his half-brother Feodor, Peter and his half-brother Ivan became co-czars in April 1682, under the regency of Peter's half-sister Sophia, an impossible arrangement. In August 1689, Peter was acclaimed sole czar, following an unsuccessful coup by Sophia. However, possessing a curiosity for all things technical and military, Peter left governing to his mother, Natalia Naryshkin, until her death in 1694.

Peter's first military venture, to capture the fortress of Azov in June–October 1695, failed because of Russia's lack of naval power. After supervising the construction of 1,400 barges and 29 galleys at Voronezh, Peter made a second attempt in May–July 1696, which succeeded.

Peter is credited as being the father of the Russian navy, sending Russians abroad to learn craftsmanship. During his Grand Embassy through Europe from March 1697 to September 1698, Peter met Sir Isaac Newton, apprenticed in a Dutch shipyard, worked as a laborer at the Royal Navy shipyard in Greenwich, and recruited foreign labor to work in Russia, among other things. In 1701, Peter established the Admiralty and commissioned Kronstadt Naval Base at a cost of 6.25 million rubles. From 1701 to 1721, Russia's naval expenditure grew from 81,000 to 1.2 million rubles. By 1725, Russia possessed 48 ships of the line and 787 auxiliary craft, serviced by 28,000 men.

After the Swedes defeated the Russians at Narva on 30 November 1700, Peter embarked on a series of reforms to strengthen the military. Conscription was instituted in 1705, and training manuals and the order of battle were revised. After the feudal noble *strelt'sy* (musketeers) revolted in 1698, Peter patterned their replacements on the Preobrezhensky and Semonovsky regiments, which he had created for childhood war games; they were cultivated into elite, well-equipped formations known as the Guards Regiments in 1708.

A program to increase the amount and effectiveness of artillery quadrupled Russian pig iron production between 1700 and 1720. Government expenditure on the army and navy was 66 percent of the total in 1701, 80 percent in 1710, and 66 percent in 1724. By 1725, Russia possessed a regular army of 210,000 troops and 100,000 Cossacks.

Peter likewise instituted educational reforms, creating the Artillery Academy in 1701, the Engineering Academy in 1712, the Naval Academy in 1715, a Mining Institute in 1716, 40 basic schools for math and literacy in the provinces in 1722, and the Academy of Sciences in 1724.

By 1720, Russia was divided into 12 provinces and military districts, which made the military in each region responsible for its own conscription and boarding of recruits and allowed for direct expropriation of taxes. In 1722, the creation of the Table of Ranks, a grade scale of 14 positions in both the military and state bureaucracies, instituted a merit system.

Russia emerged victorious against the Swedes at the conclusion of the Great Northern War in 1721, with a victory on land at Poltava in 8 July 1709 and at sea during the Battle of Cape Hanko in 1714.

On 8 February 1725, Peter died in St. Petersburg, the city he had founded in 1703. He personified Russia's conflicting orientations through subsequent years: to the West and cities like St. Petersburg for modernization and development; to the East and the villages and countryside, away from "decadent" Western influences, to be renewed in the authority and orthodoxy of pure "Holy Mother Russia."

Neville G. Panthaki

See also: Charles XII; Northern War, Great; Poltava; Russo-Turkish Wars

References and further reading:

Anderson, Matthew Smith. *Peter the Great.* New York: Longman, 1995.

Kliuchevskii, Vasilii Osipovich. *Peter the Great.* London: Macmillan, 1958.

Raeff, Marc. *Peter the Great, Reformer or Revolutionary?* Boston: Heath, 1963.

Sumner, Benedict Humphrey. *Peter the Great and the Emergence of Russia.* London: English Universities Press, 1964.

Petersburg, Siege of (June 1864–April 1865)

The last major obstacle to the Union's seizure of Richmond. In June 1864, after the clash at Cold Harbor, Ulysses S. Grant settled into a siege at Petersburg, an important city about 20 miles south of Richmond. Once Grant realized that he could not take the Confederate defenses (which had begun two years earlier during George McClellan's Peninsula campaign), he decided to keep extending his line to his left to cut the roads and railroads supplying Petersburg and hence Richmond. Once he cut the last link, he knew that Robert E. Lee would have to abandon his defenses, come out into the open, and most likely be defeated by the large and well-supplied Army of the Potomac.

There were several major battles during the long siege, including Globe Tavern (18–21 August), Ream's Station (25 August), Peebles Farm (29 September), and Boydton Plank–Burgess's Mill (27 October).

Lee recognized that once winter ended, Grant would continue his plan to isolate Petersburg in the spring. Lee therefore had General John Gordon attack at Fort Stedman on 25 March 1865; Gordon's troops were led by soldiers with axes seeking to cut through the strong wooden defenses. Gordon's assault really was to cover Lee's effort to retreat to North Carolina, meet up with Joseph Johnston, defeat William Sherman's army, and then turn to face Grant—a fantastic, indeed, desperate conception. Grant was always cool when attacked and recognized that to secure a local superiority—mass—at the point of attack, Lee must have engaged in economy of force elsewhere. Grant ordered a general attack all along the siege lines.

Thus Lee felt he had no alternative but to order the evacuation of Petersburg and Richmond. Thereafter he tried to retreat south, was cut off by superior Union cavalry, and moved mostly westward to his fate at Appomattox Court House.

Charles M. Dobbs

See also: American Civil War; Grant, Ulysses Simpson; Lee, Robert Edward

References and further reading:

Hendrickson, Robert. *The Road to Appomattox.* New York: John Wiley, 1998.

Horn, John. *The Petersburg Campaign: June 1864–April 1865.* Conshohocken, PA: Combined Books, 1993.

Trudeau, Noah Andre. *The Last Citadel: Petersburg, Virginia, June 1864–April 1865.* Boston: Little, Brown, 1993.

Pharsalus, Battle of (48 B.C.E.)

A pivotal battle of the Roman civil wars in which Julius Caesar defeated Pompey and contributed to the demise of the Roman Republic. Fought on 9 August 48 B.C.E., the engagement occurred at an undetermined locale in the plains a few miles north of Pharsalus (Fársala, Greece), a Thessalian city near the Enipeus River.

After abortive assaults on Pompey's entrenchments at Dyrrachium (Durres, Albania) in early 48 B.C.E., Caesar retreated southeast into the interior of Greece, seeking provisions and drawing Pompey away from his supply fleet. By early August, both armies encamped near Pharsalus. Pompey's forces totaled between 36,000 and 47,000 infantry in 11 legions and almost 7,000 cavalry, and Caesar commanded approximately 24,000 infantry in eight understrength legions and 1,000 cavalry. Confident of victory through attrition, Pompey hesitated to attack. However, 200 senators accompanying him agitated for a decisive confrontation, and on 9 August he deployed while Caesar was breaking camp.

Both commanders employed traditional Roman tactics, each arraying their soldiers in three main lines of battle. Pompey hoped to flank Caesar's right with a massive cavalry assault, followed by close support from several thousand archers and slingers. Caesar brilliantly countered with a fourth line of six infantry cohorts, who charged with devastating effect into the horsemen as they passed. After obliterating Pompey's cavalry, slingers, and archers, this force maneuvered around his left, attacking the main lines from behind. Simultaneously, Caesar ordered his third line into action, relieving his first two lines. Their opponents, caught in this dire pincer, quickly became disorganized and fled, suffering losses of 6,000 dead and 24,000 captured. After his defeat, Pompey escaped to Pelusium (Tell el-Farama, Egypt), where henchmen of King Ptolemy XIII assassinated him. Caesar subdued the remaining Pompeians the following year.

Ian Janssen

See also: Caesar, Julius; Pompey the Great; Roman Civil Wars (88–30 B.C.E.)

References and further reading:

Greenhalgh, Peter. *Pompey: The Republican Prince.* Columbia: University of Missouri Press, 1982.

Jiménez, Ramon. *Caesar against Rome: The Great Roman Civil War.* Westport, CT: Praeger, 2000.

Philip, King (Metacomet)(1639–1676)

Instigator of proportionally the bloodiest war in American history. King Philip Metacomet, sachem of the Wampanoags, was the son of Massasoit, who had become allied with Plymouth colony during the Pequot War, and the younger brother of Alexander Wamsutta. Becoming leader of the tribe in 1662, Philip found that the situation in which his father had maintained the peace had significantly changed. Missionaries recruited "praying Indians," who enjoyed better status than non-Christians under the laws of the colonies, while Philip increasingly had to sell titles to land through the colonial courts in order to have his authority validated and to raise cash for European goods. At the same time, he was constantly accused of planning to wage war against the Plymouth or Massachusetts colonies. Many of these charges came from John Sassamon, a Christianized Indian who was a former secretary of Philip's. When Sassamon was murdered and three Wampanoags were hanged for the crime after a questionable trial, it seemed to Philip that not only had the colony forfeited its claim to his loyalty but that he had to strike quickly. Wampanoags attacked Swansea in June 1675, killing settlers, and drew other tribes to his cause after a total lunar eclipse occurred on 26 June, seemingly a portent of victory.

The settlers were enraged at the destruction of a series of towns, including Middlebury and Dartmouth, and infuriated by the "skulking" way of war used by the Indians, whose better marksmanship, stealth, and use of fire to obliterate settlements stymied their forts and defensive lines. While other tribes, including the Nipmuks, continued the war, Philip traveled to New York in the fall of 1675 to seek an alliance with the Mohawks but was defeated badly when Governor Edmund Andros incited the Mohawks against him. Philip was forced to return to Massachusetts, where he faced the problem that although the war had united the disparate and usually feuding colonies against all Indians, even those who fought alongside them, the war had done nothing to solidify the tribes, who acted independently. Philip was increasingly harried by Captain Benjamin Harris of Plymouth and then betrayed by an irate subordinate, who sought revenge after Philip ordered his brother executed for arguing in favor of surrender. Philip was killed in Harris's raid, but in the fury of the colonists, his body was dragged from the bog in which it had fallen and dismembered and his head was displayed in Plymouth for 25 years.

Full-length engraved portrait of Philip, alias Metacomet of Pokanoket. (Library of Congress)

Although the instigator of the war, King Philip played a minor role in the conflict, which continued in the hands of other Indian leaders until 1678. (Oddly, a U.S. Navy gunboat was named in honor of Metacomet in the mid–nineteenth century.)

Margaret Sankey

See also: King Philip's War

References and further reading:

Drake, James D. *King Philip's War: Civil War in New England 1675–6.* Amherst: University of Massachusetts Press, 1999.

Lepore, Jill. *The Name of War: King Philip's War and the Origins of American Identity.* New York: Alfred A. Knopf, 1998.

Schultz, Eric B., and Michael J. Tougais. *King Philip's War.* Woodstock, VT: Countryman Press, 1999.

Philip II Augustus (1165–1223)

Victor of the first Hundred Years War, between the Capetians and the Plantagenets. Philip was the son of King Louis VII. Crowned co-king in 1179, he ruled France a year later at the age of 16, after his father's death. He inherited the confrontation between the Capetians and the Plantagenets. The latter were kings of England but also vassals of the kings of France for Normandy, Anjou, and Guyenne.

Plantagenet power was a constant threat to Capetian monarchy. Philip's territorial policy was to extend royal domain by whatever means possible: marriage, bribe, felony, or war. In 1189, he supported the rebellion of Henry II's sons against their father and received Vermandois and Artois as a reward. He took advantage of the captivity of Richard I the Lionhearted to take Normandy but was defeated when Richard returned (at Fréteval in 1194 and at Courcelles in 1198). The French monarchy was saved by Richard's death in 1199.

The war with King John, Richard's successor, continued until 1216. John had refused to take an oath of allegiance to Philip. According to feudal law, John was a "felon" and had lost all rights to his French fiefs. Philip took Normandy (1202–1204) and Brittany (1205).

With the election of a new Holy Roman Emperor in 1213, a European alliance formed against Philip: John of England, Otto IV of Brunswick, and Ferrando of Portugal. In 1214, France was invaded from the south by John's army, soundly defeated by Philip's son on 12 July, at La Roche aux Moines, but the main threat was the emperor's coalition coming from the northern border. The Battle of Bouvines (27 July 1214) was a crushing victory for Philip. The emperor lost his crown to the French-supported Frederick II Hohenstaufen.

The end of Philip's reign saw expansion south. He accepted allegiance from Simon de Montfort for confiscated Toulouse County. He allowed his heir, Louis, to command a crusade (1217–1219) against the Albigencian heretics, resulting in a bridgehead for the French monarchy in southern France. When Philip died, he enjoyed an authority that was recognized far more widely than that of any previous French king. His nickname "Augustus" recalls the first Roman emperor, an important reformer. Philip first established bailiffs as an institution and also organized a royal central administration, the Curia Regis (king's court).

Gilles Boué

See also: Henry II, King of England; Richard I

References and further reading:

Baldwin, John W. *The Government of Philip Augustus: Foundations of French Royal Power in the Middle Ages.* Berkeley: University of California Press, 1986.

Bautier Robert, ed. *La France de Philippe Auguste.* Paris: CNRS, 1980.

Nicolle, David. *French Medieval Armies 1000–1300.* London: Osprey Publishing, 1991.

Philip II of Macedon (c. 382–336 B.C.E.)

Ruler who used innovations and organizational restructuring to turn the Macedonian army into a competent fighting force, laying the foundation for his domination of Greece and Alexander the Great's domination of Asia.

In 359 B.C.E., Philip's father, King Amyntas III, died in battle with the Illyrians. Using guile and force and notwithstanding the opposition of several pretenders, the young Philip managed to secure the throne of Macedon. He gathered up what army he could muster. Using a combination of bribery and force, he solidified his hold on the country. Having quickly reorganized and trained the army, he marched off to defeat the Illyrians.

Macedonia was an agrarian state with little in the way of an urban population. Mounted noblemen, the "Companions," were the backbone of the military. They fought as shock troops, requiring considerably more organization than Greek cavalry skirmishers. Philip furthered the effectiveness of the cavalry by the introduction of the *sarissa,* a long spear similar to the Napoleonic lance in size and weight.

In contrast, infantry peasant levies served only as an auxiliary arm fighting en masse with no tactical organization and would certainly have been unable to stand up to Greek hoplites in battle. Philip's other innovation, therefore, was the Macedonian phalanx. He borrowed heavily from the Theban phalanx, but where the typical Greek phalanx was between eight and 12 men deep, Philip's was 16. Although this took away from the width of the line, the cavalry was there to protect the flanks. More men meant a stronger push, all-important in shock warfare.

Philip's was an integrated army that represented a fusion of the mobile warfare of the Persians and the infantry shock tactics of the Greeks. It was the last step in the military revolution that produced the army of Alexander the Great and perhaps the pinnacle of Greek military might.

James Corbin

See also: Alexander the Great; Cavalry; Granicus, Battle of the

References and further reading:
Cawkwell, George. *Philip of Macedon.* London: Faber & Faber, 1978.
Delbruck, Hans. *Warfare in Antiquity.* Lincoln: University of Nebraska Press, 1975.
Prichett, W. Kendrick. *The Greek State at War.* Vol. 1. Berkeley: University of California Press, 1971.

Philippi, Battle of (42 B.C.E.)

Decisive battle in which Mark Antony and Octavian defeated the assassins of Julius Caesar, Brutus, and Cassius. After the murder of Julius Caesar, the consul Mark Antony, along with the designated heir Octavian and proconsul Aemilius Lepidus, formed an alliance known as the "Second Triumvirate." It was directed against Caesar's leading assassins, Brutus and Cassius, known as the "Liberators."

Leaving Lepidus to control Italy, Antony and Octavian moved to northern Greece. The army of the Liberators was positioned astride the Via Egnatia, to the west of Philippi, in a position partly protected by a marsh. Both armies contained 19 legions, but the army of the Liberators was superior in cavalry. Both sides entrenched, building stone dikes, palisades, and towers. Antony attempted an outflanking movement by cutting through the marsh. After a 10-day effort, Antony's troops finally attacked Cassius's camp and crushed its army. Cassius, not knowing that Brutus's forces had successfully assaulted Octavian's camp, committed suicide.

During the next three weeks, Antony and Octavian continued to alter their angles of attack in an attempt to outflank Brutus's remaining forces, as Brutus extended his lines eastward in response. In the end, Brutus, against his better judgment, agreed to a battle in which his army was routed. Desperate, he too took his own life.

Ioannis Georganas

See also: Cassius

References and further reading:
Gabba, Emilio. *Republican Rome: The Army and the Allies.* Oxford, UK: Blackwell, 1976.
Keppie, Lawrence. "The Roman Army of the Later Republic." In *Warfare in the Ancient World,* ed. John Hackett, 169–191. London: Sidgwick & Jackson, 1989.

Philippine Insurrection (1899–1902)

First major jungle war fought by the United States. The Philippine Insurrection stemmed from the Spanish-American War of 1898, in which the United States decisively defeated Spain and acquired an overseas empire that included the Philippine Islands. More specifically, U.S. involvement in the Philippines dated from the Battle of Manila Bay on 1 May 1898, when a U.S. naval squadron commanded by George Dewey destroyed a Spanish fleet. News of the victory prompted Filipino forces commanded by General Emilio Aguinaldo to declare independence from Spain and form a national government, but Spanish forces ignored them and surrendered to the United States instead. Their action left the question of independence in the hands of the Americans, who decided to annex the Philippines for economic and military reasons in February 1899.

In the same month, fighting broke out between U.S. occupation forces under the command of Elwell S. Otis and frustrated Filipino troops who believed they were trading their former Spanish masters for newer ones from the United States. The fighting, largely provoked by the Americans, quickly spread from Manila into the countryside. It remained conventional until 1900, with Filipino forces suffering enormous casualties at the hands of better-equipped U.S. troops. These losses forced Aguinaldo to switch to guerrilla tactics, causing serious American casualties and extending the war into remote villages where atrocities on both sides became commonplace. U.S. forces herded villagers into concentration camps in an effort to isolate guerrilla bands and suffered at the hands of Filipino foes who knew the terrain, climate, and vegetation far better than they did.

U.S. forces under Frederick Funston finally captured Aguinaldo in 1901, and fighting gradually diminished until the end of the war in 1902. One of the keys to the eventual American success was the creation of the Philippine Scouts in 1899. Organized into companies of Filipino collaborators commanded by U.S. enlisted men holding local commissions, the scouts proved far superior to American forces in conquering the climate, terrain, and enormous linguistic hurdles inherent in an archipelago containing several thousand dialects and languages. The scouts garrisoned remote areas, allowing U.S. troops to concentrate near Manila, and later served valiantly in the war against Japan. The guerrillas' cause was also hurt by the United States extending the promise of commonwealth status and, eventually, complete independence on a definite date, 4 July 1946. The U.S. Army also busied itself in a widespread program of what later would be called "civic action," building farm-to-market roads, artesian wells, docks, telegraph lines, and, most important to the Filipinos, schools.

The other critical U.S. action during the Philippine Insurrection came in July 1901, when William Howard Taft succeeded General Arthur MacArthur (U.S. military commander and father of Douglas MacArthur) as governor of the Philippines. Taft created a Philippine Constabulary of American officers and Filipino enlisted soldiers to garrison pacified regions. Constabulary forces became the national police force of the Philippines, formed the nucleus of the Philippine Army in 1936, and contributed an entire division to the Bataan campaign during World War II, the only sizable indigenous force to fight with their colonial masters in World War II in the Pacific. Both the Constabulary and the elite Philippine Scouts allowed U.S. forces under General Adna Chaffee (who succeeded MacArthur when Taft became governor) to rule through local intermediaries and gradually assume the daily task of enforcing law and order throughout the islands.

More than 4,000 U.S. soldiers and at least 20,000 Filipino guerrillas died before the insurrection ended in 1902, and the number of civilian deaths will never be known. The loss of life dampened American enthusiasm for empire in the early 1900s and foreshadowed the difficulty Americans would have fighting a jungle war in Vietnam during the 1960s. Most historians, in the strongly anti-imperialist climate from the 1960s on, viewed the Philippine Insurrection as a tragic mistake of American imperialism, one that delayed Philippine independence and needlessly cost the lives of thousands of Filipinos and Americans. Ironically, both sides fought together against the Japanese less than 40 years after the insurrection ended.

Lance Janda

See also: Aguinaldo, Emilio; MacArthur, Arthur, Jr.; Spanish-American War

References and further reading:

Langellier, J. Phillip. *Uncle Sam's Little Wars: The Spanish American War, Philippine Insurrection, and Boxer Rebellion, 1898–1902.* Greenhill Books, 1999.

Linn, Brian McAllister. *The Philippine War, 1899–1902.* Lawrence: University Press of Kansas, 2000.

Miller, Stuart Creighton. *Benevolent Assimilation: The American Conquest of the Philippines, 1899–1903.* New Haven, CT: Yale University Press, 1984.

Philippines, U.S. Loss of (7 December 1941–9 June 1942)

The worst defeat in American military history. Oddly, neither the Japanese army nor the government had any particular interest in the Philippines, an American possession in the western Pacific scheduled to receive its independence in 1946. But the imperial Japanese navy, supposedly the least aggressive of the Japanese military services, insisted that the U.S. air and submarine forces on the islands menaced the lines of communication to the "southern resources area," the Dutch East Indies, the highest priority on the list of Japanese conquests.

Japan's attack concentrated on the large northern island of Luzon, site of the capital, Manila, and where American military power was concentrated. On 8 December 1941, Japanese fighters and bombers, flying from Formosa, struck Luzon. General Douglas MacArthur, supreme commander of American and Filipino forces, and his staff knew of the attack on Pearl Harbor and had approximately 10 hours to prepare, yet the Japanese air strikes destroyed much of Major General Lewis Brereton's Far East Air Force, including 18 B-17 "Flying Fortresses" and 53 P-40 fighters on the ground at Clark and Iba Fields. MacArthur, Brereton, and Major General Richard Sutherland, MacArthur's chief of staff, offered conflicting postwar explanations for the catastrophe. In the following days (9–13 December), continuing Japanese raids further decimated American airpower and severely damaged Cavite Naval Base in Manila Bay. With the Japanese enjoying near-complete air superiority, Admiral Thomas Hart's Asiatic Fleet, save a few small vessels and patrol bombers, sailed south (14 December) to join the British and Dutch for an anticipated defense of the East Indies, while the battered remnants of the Far East Air Force redeployed (17 December) to Australia.

American air and naval power having been crippled, Japanese ground forces made a series of small landings between 8 December and 24 December at Batan Island, Camiguin Island, Appari (northern Luzon), Vigan (northwestern Luzon), Legaspi (southeastern Luzon), Davao (southeastern Mindanao), and Jolo Island, procuring airfields from which short-range fighters could support the main landings, establishing a stranglehold on strategically significant San Bernardino Strait, and securing bases for the upcoming invasion of the East Indies. The main landings, carried out by Lieutenant General Masaharu Homma's Fourteenth Army, followed on 22 and 24 December; the 48th Infantry Division came ashore at Lingayen Gulf, northwest of Manila, and a regiment of the 16th Infantry Division landed at Lamon Bay, southeast of the capital.

In accordance with a plan worked out by MacArthur, Filipino and American forces on Luzon, divided among Major General Jonathan Wainwright's North Luzon Force, Major General George Parker's South Luzon Force, and a reserve force, attempted to defeat the Lingayen landings on the beaches. Homma's forces brushed aside resistance offered primarily by ill-trained, poorly equipped Filipino units and drove inland.

Quickly recognizing his strategy's bankruptcy, MacArthur, on 23 December, ordered a phased withdrawal into the Bataan Peninsula, intending to hold out there until relief arrived from the United States. Beginning on 24 December,

American and Filipino forces executed MacArthur's order with consummate skill. Wainwright's North Luzon Force retreated south from Lingayen, holding in sequence a series of five defensive lines, while the South Luzon Force withdrew across central Luzon via Manila to reach Bataan. Japanese air commanders and General Homma unwittingly contributed to the success of the withdrawal; the former failed to bomb the crowded roads leading into the peninsula, and the latter hesitated to push his ground forces forward rapidly. The combination of American and Filipino skill and Japanese failures and hesitancy allowed somewhere between 65,000 and 80,000 American and Filipino soldiers and 26,000 civilians to reach the peninsula by 6 January 1942. U.S. artillery, often mounted on half-tracks, proved particularly effective in the retreat to Bataan.

From his headquarters on the island fortress of Corregidor at the mouth of Manila Bay, MacArthur established a defensive line near the center of the peninsula running from Mauban on the west coast to Mabatang on the east. Wainwright's forces, redesignated I Corps, assumed responsibility for the defense of the western sector (from Mauban to Mount Natib), and Parker's, redesignated II Corps, defended the eastern sector (from Mount Natib to Mabatang). Though determined, Bataan's defenders and civilians confronted a precarious situation, owing to grossly insufficient supplies of food, medicine, ammunition, gasoline, and other necessities. MacArthur's strategy of defending Luzon at the beaches had necessitated the moving to forward areas of vital supplies originally earmarked for Bataan. The bulk of these supplies had been abandoned during the retreat to the peninsula. Consequently, by the first week of January, the American and Filipino defenders of Bataan were on half-rations of 2,000 calories per day, a situation that worsened as the battle for the peninsula unfolded.

While MacArthur's forces retreated into Bataan, Manila fell to the Japanese. There Homma learned of the imminent redeployment of the 48th Infantry for the upcoming invasion of the East Indies—the timetable for which had been pushed forward in light of the spectacular successes won by Japanese forces throughout the Pacific—and its replacement by the quantitatively and qualitatively inferior 65th Brigade. Though losing the 48th cost him his best troops and left him outnumbered 3 to 1, a confident Homma anticipated a quick mop-up of Bataan's defenders.

Beginning on 9 January, the Japanese undertook a series of frontal attacks along the length of the Mauban-Mabatang line, forcing Parker's II Corps to give ground and breaching the enemy position in the center near Mount Natib. Fearing the Japanese might achieve a complete breakthrough, MacArthur ordered a withdrawal on 22 January to a new defensive position, the Bagac-Orion Line, located 8 miles to the

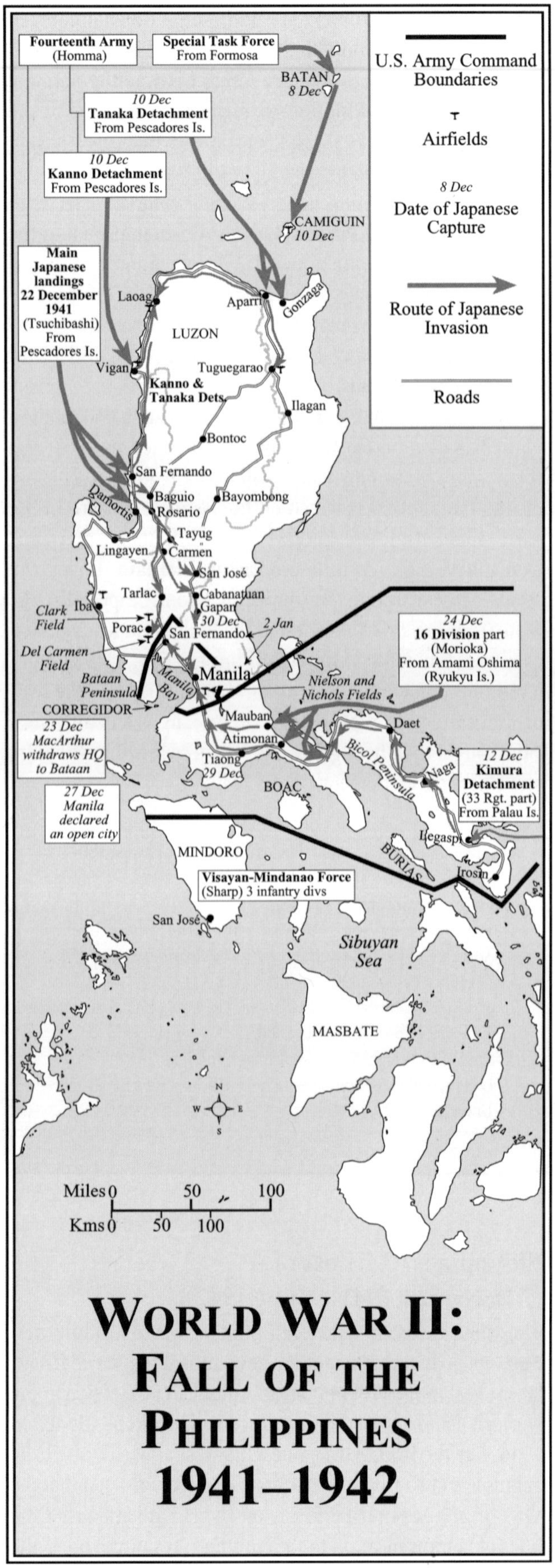

WORLD WAR II: FALL OF THE PHILIPPINES 1941–1942

rear—a maneuver Wainwright and Parker completed by 26 January—and informed his superiors in Washington that the Japanese advance would be halted there. Homma, however, had other ideas, launching three small amphibious operations (23 January–1 February) against the "points" (located in southwestern Luzon) to outflank the American position. These assaults were contained and then repulsed, and Wainwright's forces simultaneously crushed a Japanese regiment that managed to penetrate his lines.

On 8 February, his forces decimated by heavy casualties, physical exhaustion, and disease, Homma halted offensive operations, withdrawing to more secure positions and requesting reinforcement. A lull of nearly two months ensued. The Japanese strengthened themselves with the addition of infantry reinforcements from the 21st and 4th Divisions, while the Americans and Filipinos grew weaker from malnourishment and disease. In the meantime, General George Marshall, chief of staff of the U.S. Army, ordered MacArthur, now a hero in the United States, to Australia on 22 January to assume command of the Southwest Pacific Theater. Reluctantly, MacArthur, accompanied by family, staff, and Philippine president Manuel Quezon, departed on 12 March.

His forces having recuperated, Homma commenced a new offensive on 3 April, for which the starving, disease-ridden defenders of the Bagac-Orion Line had no answer. In the opening 48 hours, the Japanese collapsed the American center and right, defended by Parker's II Corps, and forced I Corps on the left to retreat. Though MacArthur, who still considered himself responsible for the Philippines' defense, ordered a counteroffensive, Major General Edward King, now in command of the defense of Luzon, recognized the situation's hopelessness and surrendered on 9 April. Of the nearly 80,000 defenders in Bataan, approximately 2,000 escaped to Corregidor. The remaining 70,000-plus fell into Japanese captivity and had to endure an excruciating 65-mile trek north to San Fernando, where stifling boxcars awaited to carry them to prisoner-of-war camps. During the journey—subsequently labeled the "Bataan Death March" —Japanese soldiers acted with great brutality, beating and even executing those diseased and malnourished captives who could not continue; some 7,000 died before reaching the railhead.

During the two months that followed the fall of Bataan, the Japanese brought the Philippine campaign to a victorious conclusion. In early May, Homma's forces, having subjected the island fortress to a 27-day siege accompanied by continuous bombardment from artillery and aircraft, assaulted Corregidor, where General Wainwright—whom the War Department had formally entrusted with command in the Philippines—a small garrison of 11,000, and civilian refugees prepared for the worst. Although Wainwright's men fought gallantly and inflicted serious casualties on the attackers, the Japanese successfully established a foothold and moved toward the center of U.S. resistance, Malinta Tunnel. Recognizing further resistance as futile, Wainwright surrendered unconditionally on 6 May, reluctantly agreeing to broadcast an order of surrender to other American forces in the Philippines. During the next several weeks, American and Filipino forces in the central and southern island capitulated piecemeal, and formal resistance finally ended on 9 June 1942.

From the start, the Philippine Islands, close enough to Japan to provoke that aggressive nation but too far from the United States for effective relief, were doomed and all the geopolitical arguments for colonial possessions shown to be utterly false. Although the capitulation was unprecedented in U.S. military history, the Filipino-American defenders could take at least a measure of satisfaction in the realization that they had put up a much better fight against the Japanese onslaught than had any other colonial forces.

Bruce J. DeHart

See also: Bataan Death March; Corregidor; MacArthur, Douglas; Philippines, U.S. Retaking of; Wainwright, Jonathan Mayhew, IV

References and further reading:

Beck, John J. *MacArthur and Wainwright: The Sacrifice of the Philippines.* Albuquerque: University of New Mexico, 1974.

Connaughton, Richard M. *MacArthur and Defeat in the Philippines.* New York: Overlook Press, 1992.

Morton, Louis. *The Fall of the Philippines.* Washington, DC: Office of the Chief of Military History, Department of the Army, 1953.

Whitman, John W. *Bataan, Our Last Ditch: The Bataan Campaign, 1942.* New York: Hippocrene, 1990.

Philippines, U.S. Retaking of (20 October 1944–2 September 1945)

Major World War II battle on land and sea for control of the Philippine Islands, which were important to the United States both symbolically and strategically. Because the islands were a former American colony (and after that a commonwealth) and one of the first targets of the Japanese advance against the United States in late 1941 and early 1942, many Americans (not the least of whom was General Douglas MacArthur himself) felt a moral obligation to retake them. They were also a possible stepping-stone to the invasion of Japan.

In the continuing debate in Washington about the direction of the American Pacific offensive, General Douglas MacArthur, U.S. commander in the Southwest Pacific Theater, argued persistently for the retaking of the Philippine Is-

General Douglas MacArthur and aides wading ashore on Leyte, Philippine Islands, 1944. (Library of Congress)

lands. In the summer of 1944, the Joint Chiefs of Staff approved the invasion plans, with the island of Leyte as the initial target and the main island of Luzon as the final goal. The Japanese army command chose to contest the invasion of Leyte, and the struggle for control of that island developed into the major land battle for the Philippines.

The Battle of Leyte Gulf, the massive naval struggle between the American and Japanese fleets, raged concurrently with the fighting on land. The Japanese, underestimating as always the strength of the Americans, hoped to engage the U.S. Navy in a decisive confrontation. When the Battle of Leyte Gulf, the greatest naval battle in history, concluded, Japan had lost no less than four aircraft carriers, three battleships, six cruisers, 11 destroyers, and 500 planes and was a spent force.

After the relatively unopposed landings on Leyte Island, General Walter Krueger's Sixth Army made steady progress toward Ormoc, the major Japanese base on the western side of the island. General Suzuki's defensive fortifications, combined with typhoon rains, slowed but could not stop the American advance. Although Leyte was largely under American control by December 1944, fighting continued in that area through May 1945.

On 9 January 1945, the Americans landed on Luzon Island, where General Tomoyuki Yamashita (Tiger of Malaya) had withdrawn into the mountains, but a Japanese naval-marine rear guard put up a last-ditch defense of Manila, committing many atrocities against Filipino civilians and destroying most of the city. Despite the American victory, sporadic fighting continued on various islands until the end of the war and General Yamashita's surrender on 2 September.

The retaking of the Philippines produced two unforgettable images for U.S. wartime memory, one of Japanese fanaticism and the other of American dominance. During the Battle of Leyte Gulf, the Japanese resorted to suicide air attacks by bomb-laden warplanes, or kamikazes (divine wind), strengthening the belief in the United States that Japan would never surrender. Equally significant was the photo of General MacArthur wading ashore on Leyte on 20 October 1944 and announcing to the world, "People of the Philippines, I have returned!"

Harold J. Goldberg

See also: MacArthur, Douglas; Philippines, U.S. Loss of

References and further reading:

Cannon, M. Hamlin. *Leyte: The Return to the Philippines.* Washington, DC: Center of Military History, U.S. Army, 1987.

Krueger, Walter. *From Down Under to Nippon: The Story of the Sixth Army in World War II.* Washington, DC: Combat Forces Press, 1953.

MacArthur, Douglas. *Reminiscences.* New York: McGraw-Hill, 1964.

Spector, Ronald H. *Eagle against the Sun.* New York: Free Press, 1985.

Pickett, George Edward (1825–1875)

Confederate general in the American Civil War who is forever linked with one of the most gallant and heartbreaking moments in military history. Pickett was born on 25 January 1875 in Richmond, Virginia, to an upper-class family. After graduating last in the class of 1846 at the U.S. Military Academy at West Point, he was promoted to the rank of brevet first lieutenant for his service at Contreras and Churubusco in the Mexican-American War. He was then assigned to various posts, promoted to the rank of captain, and was later involved in the "Pig War," a dispute over San Juan Island with the British.

On 25 June 1861, Pickett resigned from the army and was ushered into the Confederacy as a colonel less than a month later. He was stationed on the Rappahannock River in the Department of Virginia and was later transferred to the Department of Northern Virginia, promoted to the rank of brigadier general, and given command of a brigade under General James Longstreet. He led his men in the Battles of Williamsburg, Seven Pines, and Gaine's Mill. In the last battle, he was wounded. After having recovered, he was promoted to major general and given a command of a division. He then served in the Fredericksburg and Tidewater campaigns.

On the third day at Gettysburg, Pickett's division was given the daunting task of charging straight into the center of the Union line at Cemetery Ridge, breaking through, and rolling the enemy up. His men were cut to ribbons by the Union guns, sustaining losses of 2,655 men wounded, killed, or captured. Pickett would remain bitter toward Robert E. Lee for the rest of his life. Afterward, he was sent to North Carolina, rejoined Lee at Cold Harbor, and was involved in the Richmond campaign. Lee relieved him of command because of the substantial losses he suffered at Five Forks, one day before peace was signed at Appomattox Court House.

T. Jason Soderstrum

See also: American Civil War; Gettysburg; Lee, Robert Edward; Longstreet, James

References and further reading:

Georg, Kathleen R. *Nothing but Glory: Pickett's Division at Gettysburg.* Gettysburg, PA: Thomas Publications, 1993.

Gordon, Lesley. *General George E. Pickett in Life and Legend.* Chapel Hill: University of North Carolina Press, 1998.

Longacre, Edward G. *Pickett, Leader of the Charge: A Biography of General George E. Pickett, C.S.A.* Shippensburg, PA: White Man Publishing, 1995.

Pilsudski, Józef Klemens (1867–1935)

Polish military commander and dictator. Born at Zulow in Russian-occupied Poland (in present-day Lithuania) on 5 December 1867, Pilsudski studied medicine at the University of Kharkov for one year before being arrested for an alleged conspiracy to assassinate Czar Alexander III. Pilsudski was sentenced to five years in Siberia and did not return to Poland until 1892. Following his return, Pilsudski became a leader of the Polish Socialist Party and published a clandestine newspaper, *Robotnik* (The Worker). After being forced to flee to Austrian-occupied Poland, he founded the Polish Riflemen's Association, with which he hoped to train an army of Poles to fight for Polish independence. Upon the outbreak of World War I, the Polish Riflemen's Association became the Polish Legion of the Austro-Hungarian army, which fought against imperial Russia. In 1916, the Central Powers proclaimed an independent Polish kingdom with Pilsudski as a member of the Polish Council of State. Refusing to fight under German command, however, he was imprisoned at Magdeburg and the Polish Legion was disbanded.

Released from Germany following the Armistice in November 1918, Pilsudski returned to Warsaw and proclaimed an independent Polish republic. As head of state and commander in chief of the Polish army, he sought to restore all the territories that had belonged to Poland at the time of the First Partition in 1772. These policies brought Poland into immediate conflict with the Bolshevik regime in the Soviet Union. During the Russo-Polish War (1919–1921), Pilsudski defeated the vastly superior Soviet armies at the Battle of Warsaw (1920) and thereby secured Polish independence. After resigning as chief of state in December 1922, he planned to live in quiet retirement, but on 12 May 1926 he led a coup d'état that overthrew the weak and inefficient parliamentary government and installed a regime under his control. From then until his death, Pilsudski was the virtual

Portrait of Józef Klemens Pilsudski. (Library of Congress)

dictator of Poland, though he retained only the positions of minister of war and commander in chief of the army. He died in Warsaw on 12 May 1935 and was buried among Poland's kings in the crypt of Wawel Cathedral in Cracow.

Alexander M. Bielakowski

See also: Russo-Polish War

References and further reading:

Garlicki, Andrzej. *Józef Pilsudski, 1867–1935.* London: Scholar Press, 1995.

Pilsudski, Józef. *Memories of a Polish Revolutionary and Soldier.* New York: AMS Press, 1971.

Pinkie (10 September 1547)

The last battle between English and Scottish national armies. The battle was fought in a war that erupted from England's attempt to secure a lasting arrangement with Scotland: England's persuasive efforts involved the sword, fire, and looting. After one English raiding column was annihilated at Ancrum Moor in February 1547, an English army advanced on Edinburgh with 18,000 infantry and 6,000 cavalry. It mustered 800 musketeers and 15 cannon and had naval support.

The Scottish army of 20,000 men occupied a strong position on the west bank of the Esk River, between sea and marsh. The English attempt to secure the only bridge over the Esk, which had been left unguarded, provoked the Scots to abandon their positions to move to the east bank. There they were attacked by English cavalry, but the Scots troops' rudimentary squares formation defeated these attacks. The cost to the Scots, however, was in time: the English were able to bring archers, cannon, and warships into action against massed but trapped Scottish pikemen. Subjected to murderous cross fire, the Scots tried to retreat, with inevitable consequences: panic and flight. The English cavalry rode down those fleeing the battlefield. Scottish dead numbered an incredible 10,000.

The aftermath was curious. In the short term, the Scots simply refused to treat, and English holdings were progressively reduced, with Edinburgh abandoned in 1550. But Scottish fear of French intentions, the growing strength of Protestantism in the country, and the English withdrawal from southern Scotland provided the basis for a rapprochement. In 1560, the Scots and the English combined to force the surrender and evacuation of the French garrison at Leith.

H. P. Willmott

See also: Anglo-Scots Wars

References and further reading:

Paterson, Raymond Campbell. *My Wound Is Deep. A History of the Later Anglo-Scots Wars, 1380–1560.* Edinburgh: John Donald, 1997.

Pitt, William, the Elder (1708–1778)

The architect of British victory in the Seven Years' War. William Pitt was born in 1708, grandson of Governor Thomas Pitt, founder of the family's fortune, who used profits from his place in the East India Company to buy Old Sarum, the family's rotten borough. After an education at Eton and Oxford University, Pitt was unsure of his professional future and accepted a cornetcy in Cobham's Horse before embarking on a grand tour of Europe in 1733 and returning in 1735 to take a seat in Parliament, representing Old Sarum. Quickly, Pitt made a reputation for crossing Robert Walpole and annoying George II, especially after goading the government into war with Spain in 1739. In 1746, he was named paymaster general and ostentatiously made a show of publicly accounting for all the army funds rather than personally profiting.

In 1756, Pitt pushed for an alliance with Prussia under Frederick the Great and, after the disasters of that year (Byng's naval failure at Minorca and Braddock's catastrophic defeat and death in western Virginia), was named one of the secretaries of state. Pitt's strategy in the Seven Years' War was simple: British force was concentrated on North America and the colonies, while subsidized Prussia carried Europe. It resulted in the victories of Robert Clive in India at Dakar and James Wolfe's capture of Quebec, as well as the naval battle at Quiberon Bay, which destroyed the French fleet. By 1761, however, Pitt's arrogance and inability to work with Prime Minister Lord Bute caused him to resign, crying "Frederick Betrayed!" about the peace treaty. In 1766, he was back in the ministry after protesting the Stamp Act and Declaratory Act but suffered a physical collapse and left again in1768. In his later years, Pitt advocated reform in India, reconciliation with the American colonies, and agricultural improvements. He died after an impassioned but incoherent speech on American liberties in April 1778.

Margaret Sankey

See also: Clive, Robert; Montcalm-Gozon, Louis-Joseph de, Marquis de Montcalm de Saint-Véran; Seven Years' War; Wolfe, James

References and further reading:

Black, Jeremy. *Pitt the Elder.* Cambridge, UK: Cambridge University Press, 1992.

Hotblack, Kate. *Chatham's Colonial Policy.* London: George Routledge and Sons, 1917.

Pizarro, Francisco (c. 1478–1541)

Spanish conquistador who subdued the Inca Empire in Peru. In 1502, Pizarro left Spain for the Caribbean island of Hispaniola in search of fame and fortune. He apprenticed for a number of years and in 1513 was one of Vasco Nunez de Balboa's lieutenants in the exploration of Panama. In Panama, Pizarro initiated plans for the exploration and conquest of lands on the west coast of South America. His first exploratory probe in 1524 met with failure, but a second expedition in 1527 revealed gold ornamentation among the indigenous populations. Pizarro organized a force of 180 men and 27 horses and landed in present-day Ecuador in 1531. After reconnoitering the coastal areas of Ecuador by land and sea, he landed at Tumbes and gathered important intelligence about internal strife in the Inca Empire. Surmising that the empire was on the brink of collapse, Pizarro boldly marched inland through narrow mountain defiles and arrived at the highland city of Cajamarca on 16 November 1532. There he encountered the Inca ruler, Atahualpa, backed by an army of 30,000. When Atahualpa rejected entreaties to convert to Christianity, Pizarro ordered his tiny band to attack. The ensuing battle resulted in the deaths of thousands of Indians and the capture of Atahualpa without the loss of one Spaniard. Pizarro held Atahualpa hostage and extracted a large ransom of gold, silver, and precious stones before ordering his execution. Pizarro next led a successful campaign against Cuzco, the capital of the Inca Empire, in 1533. In 1535 he founded the city of Lima and was the undisputed master of Peru. Explanations for the amazing success of the conquest of the well-organized and highly militarized empire that stretched nearly 2,000 miles from southern Colombia to Chile include the civil war of rival brothers for the throne, the decimating impact of smallpox on the vitality of

Half-length portrait of Francisco Pizarro wearing armor. (Library of Congress)

the population, the superior weapons of the conquerors, especially the use of horses and steel, and the audacious leadership of Pizarro. Pizarro was murdered in 1541 by members of a rival faction discontented with his governance of Peru.

George M. Lauderbaugh

See also: Cortez, Hernando

References and further reading:

Hemming, John. *The Conquest of the Incas.* New York: Harcourt Brace Jovanovich, 1973.

Lockhart, James. *The Men of Cajamarca: A Social and Biographical Study of the First Conquerors of Peru.* Austin: University of Texas Press, 1972.

Prescott, William Hickling. *History of the Conquest of Peru.* New York: New American Library, 1961.

Plains of Abraham (13 September 1759)

Battle that ended French power in North America. Having taken Louisbourg and Fort Duquesne in the Seven Years' War (called the French and Indian War in North America), the British the next year turned their attention to Quebec. Brigadier General James Wolfe left England in mid-February 1758, reaching Halifax in late April and Louisbourg in mid-May. When Wolfe embarked to attack Quebec, he had about 9,000 regular troops and several hundred American Rangers.

By late June, the large invasion force reached Ile d'Orleans without incident and was only 5 miles from Quebec. But the citadel's defenses were formidable, perched on a 200-foot bluff over the St. Lawrence River. Defended on two sides by rivers and steep bluffs, the city had constructed a wall on the west side and placed cannon to prevent attackers from scaling the wall.

Wolfe considered his choices. Louis-Joseph de Montcalm-Gozon, the Marquis de Montcalm de Saint-Véran, assumed that the British would attack from the north and had strengthened defenses accordingly; he also assumed British ships could not sail past Quebec to land troops to the more vulnerable western approach. Several times, the French tried to launch fire boats at the British fleet, but they were unsuccessful, and several times Wolfe sought to attack the city from the north and east and failed.

Finally, Wolfe accepted a plan from his subordinates to land west of Quebec, cutting off the city from its supplies further upriver. Meanwhile, Montcalm tried to stall matters, for by October, weather and the freezing of waterways would force a British retreat. Finally, the British secretly moved under cover of darkness and used a path to climb up to the plains west of town. Although the French received notice from pickets stationed there, they did not make good use of the information.

Montcalm decided to attack the British before they could begin siege operations and thus delay their plans. He gathered his men and attacked in the late morning of 13 September; the British held their fire and then decimated the French attackers. Both Montcalm and Wolfe were wounded several times, and both died.

The British, having cut supplies from upriver, began siege operations. The French defenders, short of food and low on morale, surrendered on 17 September, basically ending the French Empire in North America. Few battles have had such far-reaching consequences as that of the Plains of Abraham.

Charles M. Dobbs

See also: Louisbourg, Expedition against; Montcalm-Gozon, Louis-Joseph de, Marquis de Montcalm de Saint-Véran

References and further reading:

Casgrain, Abbé Henri Raymond. *Wolfe and Montcalm.* London: Oxford University Press, 1926.

Donaldson, Gordon. *Battle for a Continent: Quebec, 1759.* Toronto: Doubleday, 1973.

Stacy, C. P. *Quebec, 1759: The Siege and the Battle.* New York: St. Martin's Press, 1959.

Plassey, Battle of (23 June 1757)

The battle that gave control of Bengal to the British. In the eighteenth century, India saw a struggle for colonial supremacy between France and England. European nations fought each other through conflicts in these remote areas. The British East India Company had the delegated power of a sovereign state and its own private army and was continuously at war with the French or with French puppet Indian rulers.

In 1756, Suraja Dowla, nabob of Bengal (supported by the French), captured Calcutta and committed atrocities against British prisoners. The East India Company sent its army under Robert Clive to take control of Bengal and punish Suraja. Clive's troops consisted of less than 3,000 men with eight guns, facing an overwhelming force of more than 50,000 soldiers, mainly Indian warriors, with 53 guns manned by French artillerymen. The nabob's army was commanded by Mir Madan, but many Indians (including left-wing commander Mir Jafar) were plotting against Suraja, a fact Clive knew. The East India Company army pursued Suraja's army and found him entrenched near the village of Plassey. Clive massed his tiny force in a mango grove and found himself besieged by a large semicircle of Indians. The weather proved to be the decisive factor: a sudden monsoon rainstorm wet Suraja's powder at the very time the British covered their artillery. The battle commenced with inconclusive cavalry charges that were easily repulsed by Clive's guns. In the afternoon, Clive went over to the offensive and deployed to cannonade the nabob's camp at short range. After repulsing an Indian sortie, Clive assaulted Suraja's entrenchments, counting rightly on the disaffection of part of the Indian army. The Indian army disintegrated as Mir Jafar's soldiers changed side and began to flee without fighting. The French gunners continued fighting to the last, but by 5 P.M., the battle was over. Clive's casualties were incredibly low—18 killed and 45 wounded—to 500 Indians killed. Suraja was assassinated shortly afterward, and Mir Jafar became nabob, but with his power limited by the East India Company. The skirmish at Plassey proved to be the very foundation stone for the mighty British Empire in India.

Gilles Boué

See also: Carnatic Wars; Clive, Robert; Seven Years' War

References and further reading:

Edwardes, Michael. *The Battle of Plassey and the Conquest of Bengal.* London: Batsford, 1963.

Harrington, Peter. *Plassey 1757.* London: Osprey Publishing, 1994.

Plataea, Battle of (479 B.C.E.)

Ten years after the Battle of Marathon, Greek hoplites and Persian mixed troops clashed near the city of Plataea in 479 B.C.E. in the largest land battle of the Persian Wars. Perhaps based on the Persian experience at Marathon, Xerxes had persuaded a number of Greek cities to contribute heavy infantry to complement the Persian light infantry and cavalry.

The two forces met about 25 miles northwest of Athens. The Spartans, led by their king Pausanius, had joined forces with the Athenians and other Greek cities to bring their combined force up to 40,000 hoplites and auxiliary light troops. The Persians, commanded by Mardonius, had about the same number of men (50,000), but he had an overall superiority in cavalry and infantry archers. He found it difficult to use the latter to advantage against the Greek forces, and he did not have enough heavy infantry to confidently envelop the Greeks, who had taken up their position on a hill behind the stream Asopos. There the two sides faced off for several days. Persian cavalry constantly harried the Greeks and attacked their water and forage parties. The Greeks, confident in their ability to prevail in a shock action battle, attempted unsuccessfully to draw the main Persian army into an assault. Finally, Pausanius decided to withdraw and move his force closer to Plataea. He planned a nighttime withdrawal to confuse the Persians. By daybreak, only half of his force had successfully moved back toward the city. The Athenians and Spartans had only started their retreat in the morning.

The Persians, noticing the disorder of the Greeks, rushed forward to engage. The Persians relied on arrows to attack the hoplites. Unfortunately for Mardonius, his soldiers approached too closely to the Greeks, whereupon the Greeks charged the Persian line. Forced to drop their bows and defend themselves against bronze spears with their wicker armor, the Persians soon gave way and eventually collapsed. The Greek counterattack and pursuit killed Mardonius along with his elite guard and effectively destroyed the Persian field army.

This land victory, combined with the threat of an Athenian-led naval expedition against the Persians' lines of communications, unhinged Xerxes' military plans to conquer Greece and forced the Persians to withdraw back to Asia Minor.

Bryan R. Gibby

See also: Marathon, Battle of; Persian Empire; Persian Wars of Expansion; Thermopylae, Battle of

References and further reading:

Delbruck, Hans. *Warfare in Antiquity.* Trans. Walter J. Renfroe Jr. Lincoln: University of Nebraska Press, 1990.

Jones, Archer. *The Art of War in the Western World.* Chicago: University of Illinois Press, 1987.

Plattsburgh Movement (1915–1918)

A grassroots military preparedness campaign that began prior to the entry of the United States into World War I. The Plattsburgh Movement was an effort by private citizens to prepare for possible military action. It was led by old-stock, Ivy League–educated, eastern blue bloods and supported by General Leonard Wood. The first camp, held in Plattsburgh, New York, in the summer of 1915, trained middle-aged businessmen in military procedures. The National Defense Act of 1916 brought the War Department and these private organizations closer by permitting the federal government to pay for these training facilities, uniforms, food, and transportation. Throughout 1916 and 1917, Plattsburgh Movement training camps proliferated throughout the United States. After U.S. entry into the war in April 1917 and the enactment of conscription, the Plattsburgh camps became training centers for officer candidates recruited from civilian life.

In addition to training men for war, the groups that made up the Plattsburgh movement lobbied to improve camp conditions, organized liberty bond rallies, spoke at schools and other organizations, and produced training films. Following the Armistice, the Plattsburgh Movement advocated greater peacetime military preparedness, including universal military training.

Gregory Dehler

See also: Wood, Leonard; World War I

References and further reading:
Clifford, John Garry. *The Citizen Soldiers: The Plattsburgh Training Camp Movement, 1913–1920.* Louisville: University Press of Kentucky, 1972.
Sullivan, Mark. *Our Times.* Vol. 5. New York: Charles Scribner's Sons, 1933.

Pleven/Plevna, Siege of (20 July–10 December 1877)

The crucial four-phase battle by the Russians to overcome Turkish forces defending Pleven/Plevna during the Russo-Turkish War (1877–1878). Russian (and Romanian, after August) forces participating in this battle numbered 130,000, against the Turkish force of 67,000.

War was declared on 24 April 1877, and a Russian army of 190,000 commanded by Grand Duke Nicholas crossed the Danube River on 27 June. They captured the Shipka Pass, situated in Turkish-occupied Bulgaria, on 19 July.

The Turkish forces of Osman Nuri Pasha marched from Vidin and reached Pleven/Plevna, where they held the Russian advance on 20 July. This action constituted the first phase of the battle.

Two subsequent Russian attempts to capture Pleven via a frontal assault failed. These second and third phases of the battle occurred on 30 July and 11–12 September. The Russian and Romanian forces lost 15,700 men.

Nicholas's decision to have the army retreat across the Danube River was overruled. Colonel Count E. I. von Todleben was placed in charge of the next assault of Pleven, and his engineers were employed to besiege the city.

Turkish forces on the perimeter of the city were not employed to aid Osman Pasha, who mounted a failed attempt to break out on 9–10 December. The Russians then launched their fourth and final, successful assault of this battle, and Pleven surrendered on 10 December 1877. The victory permitted Russian forces to continue unimpeded, entering Sofia on 4 January 1878 and Adrianople/Edirne on 20 January after the Battle of Plovdiv. The Siege of Pleven was instrumental to the defeat of Turkish forces in Bulgaria and the conclusion of the Russo-Turkish Wars on 3 March 1878 by the Treaty of San Stefano.

Neville G. Panthaki

See also: Balkan War, First; Kars, Battle of; Russo-Turkish Wars

References and further reading:
Anderson, M. S. *The Eastern Question.* London: Macmillan, 1970.
Menning, Bruce. *Bayonets before Bullets: The Imperial Russian Army, 1861–1914.* Bloomington: Indiana University Press, 1992.
Rupp, George. *A Wavering Friendship: Russia and Austria, 1876–1878.* Philadelphia: Porcupine Press, 1976.
Seaton-Watson, Hugh. *The Decline of Imperial Russia.* London: Methuen, 1966.

Poitiers, Battle of (18 September 1356)

French defeat in the Hundred Years War. With the resumption of war, the main English effort in 1356 took the form of a *chevauchée* (armored, mounted raiding parties) from Bordeaux that reached Tours. The French, crossing the Loire River at Blois and moving faster than an enemy army laden with booty, managed to get astride the English line of retreat at Poitiers.

The English army, perhaps 12,000-men strong, took up a defensive position with one flank secured by a stream. The other flank consisted of the wagon park. Sunken lanes and hedges were manned by some 3,000 archers, with dismounted cavalry serving as supporting infantry. The French, numbering perhaps 30,000 men, were formed into four echelons. The French made no attempt to use their cavalry to turn the English position and used their armored knights as infantry.

The first French attack was easily defeated by English longbowmen, with few French reaching the English positions. A second attack was denied a breakthrough only by deployment of the rear English division into the front line, leaving the English without a reserve. The failure of this attack caused the flight of the third French force before it came into range. The fourth French force, exhausted by its approach march, was caught by English light infantry and cavalry. The French king and many of the highest nobles in the land were surrounded and captured. The French lost about 2,500 killed, with a similar numbered captured. English casualties numbered about 2,000. After the battle, the English resumed their withdrawal to Bordeaux.

H. P. Willmott

See also: Crécy, Battle of; Hundred Years War

References and further reading:
Hooper, Nicholas, and Matthew Bennett. *The Cambridge Illustrated Atlas of Warfare. The Middle Ages, 768–1487.* Cambridge, UK: Cambridge University Press, 1996.
Seward, Desmond. *The Hundred Years' War. The English in France, 1347–1453.* London: Constable, 1996.

Polish Campaign of 1939

Opening campaign of World War II. The German invasion of Poland in September 1939 was the first example of a new theory of armored warfare, which the Germans referred to as "blitzkrieg" (literally, lightning war). Poland historically suffered from a great geographic disadvantage—the central portion of Poland was flat with no natural boundaries. In addition, the Polish border with Germany extended for approximately 3,500 miles in 1939. When war broke out, Polish forces numbered about 1,000,000 men, but they were technologically outdated, especially in aircraft and tanks, and a great many of the troops were poorly trained reservists. At the same time, German forces numbered about 1,500,000 men, the most important part of which were 12 armored, mechanized, and motorized divisions. These divisions, in conjunction with the German Air Force (Luftwaffe), would be the deciding factor during the campaign.

On 1 September 1939, when Germany attacked, Poland's forces were thinly spread in a rough semicircle from the border with East Prussia in the north to the German surrogate state of Slovakia in the south. In the northwest, approximately a third of Poland's forces were concentrated in or near the Polish Corridor, where they were exposed to both East Prussia and Germany proper. About half of Poland's forces were in central Poland, either facing the main axis of the German advance or massed in reserve in the center of the country. The remainder of Poland's forces were even more thinly spread in southern and eastern Poland. Rather than deploying Polish forces behind natural and prepared defenses, which would have meant the loss of strategic industrial and resource centers in western Poland, the decision was made for them to be forward deployed. Unfortunately, this decision made it almost impossible to fight delaying actions against the German invaders because the German mechanized forces could easily overwhelm the slower Polish infantry forces.

The German forces quickly split the Polish army into disjointed fragments, some of which retreated, while others continued to fight hopeless battles. By 10 September, the Polish commander in chief, Marshal Edward Rydz-Smigly, ordered a general retreat toward the southeast. The situation was already desperate for the Poles when, on 17 September, the Soviet Union invaded eastern Poland in accordance with the Molotov-Ribbentrop Agreement. The next day, the Polish government crossed into exile in Romania. The Warsaw garrison resisted until September 28, and the last major force of the Polish army surrendered on October 6. Poland was then partitioned between Germany and the Soviet Union along the Bug River.

Despite its "lightning" victory, the Germans made many mistakes in deployment and tactics in their Polish campaign. But the remaining Allies, Great Britain and France, gave their enemy ample time to correct these deficiencies during the so-called Phony War, a time of near-immobility on the western front from September 1939 to May 1940. Though guerrilla warfare would continue (first against the Germans and then against the Soviets) until after the formal end of the war in May 1945, the Polish nation had ceased to exist.

Alexander M. Bielakowski

See also: Guderian, Heinz; World War II

References and further reading:

Bethell, Nicholas W. *The War Hitler Won: The Fall of Poland, September 1939.* New York: Holt, Rinehart and Winston, 1973.

Kennedy, Robert M. *German Campaign in Poland, 1939.* Washington, DC: Department of the Army, 1956.

Zaloga, Steven, and Victor Madej. *The Polish Campaign, 1939.* New York: Hippocrene Books, 1985.

Polish Wars of Expansion (1386–1498)

A series of wars fought sporadically between Poland and its neighbors from 1386 to 1498. By the end of the reign of Casimir III (1333–1370), the Polish monarchy had, under the Piast Dynasty, successfully unified the Kingdom of Poland and created a bureaucratic apparatus to govern the country. After the death of Casimir III in 1370 and his nephew Louis in 1382, Poland would begin a series of wars and dynastic marriages calculated to extend Polish control over neighboring territories. Some of these efforts began at the behest of the royal family; others were undertaken at the connivance of a group of powerful oligarchs close to the royal court, the so-called Cracow nobles. As a result, by the end of the expansion in 1498, the Polish royal family governed approximately one-third of mainland Europe.

In 1386, the Cracow nobles completed the first step toward the expansion of the Kingdom of Poland. At the insistence of the oligarchs, Jadwiga married the grand duke of Lithuania, Jagiello (later Wladyslaw II). As a result of this marriage, the two nations were linked through a common set of monarchs, although both remained technically independent nations.

A combined Polish and Lithuanian army was able to evict Hungarian garrisons from Ruthenia, thus advancing Polish territorial interests, and to extend Lithuanian influence along the Baltic Sea coast to the north of the grand duchy and among the Rus to the east. Combined Polish and Lithuanian forces were also able to compel the princes of Moldavia and Walachia to render homage to the Polish kingdom.

In 1409 and 1410, hostilities between Lithuania and the Teutonic Order led to the "Great War," which ended with the defeat of the order by a Polish and Lithuanian army commanded by Wladyslaw II and Vytautas (Witold) in the Battle of Tannenberg/Grunwald in 1410. Although the power of the order was considerably reduced, it received generous peace terms and was compelled only to recognize the right of the Lithuanians to govern some disputed territories along the

Baltic Sea. A second war with the order in 1422 forced the complete abandonment of its claims to Lithuanian territory.

The outbreak of the Hussite Wars in 1419 afforded a further opportunity for the aggrandizement of the Jagellonian dynasty. Hussite elements in Bohemia offered the kingdom to Wladyslaw II, which would have united a third country under his rule. Wladyslaw refused but allowed his cousin Vytautas to accept the offer.

Upon the death of Vytautas in 1430, Wladyslaw appointed his brother, Swidrigiello, as viceroy of Lithuania. Swidrigiello rebelled, abetted by the Teutonic Order, Sigismund of Luxembourg, and dissatisfied elements in Lithuania. Not until the death of Wladyslaw II in 1434 was the rebellion suppressed by his son, Wladyslaw III.

The death of Sigismund of Luxembourg at about the same time allowed the Jagellonian family another opportunity to acquire the Bohemian throne; the Polish court, in an effort to secure Bohemia, adopted a pro-Hussite policy. This in turn led to a pro-Hussite peasant rebellion in Poland, aimed at the church and noble hierarchy. The peasant revolt was defeated by the Cracow nobles at the Battle of Grotniki in 1439; noble dissatisfaction with the royal court enabled the Cracow nobles to compel the Jagellonians to defer further efforts to seize the Bohemian throne.

In 1440, Wladyslaw III appointed his brother Casimir viceroy of Lithuania. The Lithuanian nobles, true to form, rebelled and proclaimed Casimir as the independent grand duke of Lithuania. Wladyslaw was in no position to take action against his brother, for upon the death of Albert of Habsburg, also in 1440, he had been offered the kingdom of Hungary. As king of Hungary, Wladyslaw was drawn into the anti-Turkish crusade then being organized by the papacy in an effort to rescue Constantinople and Serbia from Turkish conquest.

After winning some initial victories in 1443, Wladyslaw negotiated an advantageous settlement with the Turks. Under papal pressure, this agreement was repudiated, and a second crusade was launched in 1444. This second crusade, the so-called Varna crusade, was poorly planned and led to the utter defeat of the crusading forces by the Turks at the Battle of Varna. Wladyslaw III was killed.

In 1454, Poland and Lithuania began the Thirteen Years' War with the Teutonic Order. This conflict led to the final defeat of the order and the Treaty of Thorn, which, although harsh, failed to eliminate completely the order as a force. The order surrendered more than half its remaining territory, and its grand master also agreed to become a vassal of the king of Poland and to accept Polish suzerainty over the remainder of the land held by the order.

Casimir IV's great object of policy was to obtain the reversion of the Kingdoms of Bohemia and Hungary for his sons. This goal was realized, and the descendants of Jagiello were able to gain by diplomacy that which would have been utterly unattainable by conquest. But in 1485, Casimir IV began a series of campaigns in Moldavia against the Crimean Tartars on behalf of the prince of Moldavia, Stefan cel Mare. Stefan became a Polish vassal, and Polish-Tartar warfare continued until the death of Stefan in 1501.

Thus, immediately before the death of Casimir IV in 1492, the Jagiellonian dynasty controlled the Kingdom of Poland and the Grand Duchy of Lithuania through the kingship of Casimir IV and the Kingdoms of Bohemia and Hungary through the kingship of Wladyslaw, Casimir's eldest son. Moldavia and the Teutonic Order had been reduced to dependent vassal states. This unity was, however, more apparent than real, as none of the kingdoms or lands had been formally merged, and all retained some tradition of electing rulers rather than recognizing hereditary succession.

The death of Casimir IV in 1492 led to the unraveling of the Jagiellonian holdings. The decline of Tartar power made possible a Turkish-Muscovite alliance in 1498, which precluded further Polish or Lithuanian efforts at expansion in those directions. The careful efforts of the Jagiellonians thus benefited the Kingdom of Poland and the Grand Duchy of Lithuania very little in the long term.

Joseph M. Isenberg

See also: Tannenberg, Battle of; Teutonic Knights

References and further reading:

Gieysztor, Aleksander, et al. *History of Poland.* Warsaw: PWN, 1979.
Zamoyski, Adam. *The Polish Way.* London: John Murray, 1987.

Poltava (8 July 1709)

The triumph of Peter I's military reforms, enabling the Russian army to rout the Swedish force of Charles XII. Poltava marked the decline of Sweden and the ascendancy of Russia as a European great power.

After defeating the Russians at Narva, Charles diverted his attack to Poland. The Swedish attack against Russia recommenced in January 1708 as 50,000 Swedes crossed the Berezina River. Charles paused at Mogilev, awaiting a Swedish auxiliary force of 15,000 men and supplies traveling from Livonia and led by General Adam Ludwig Lewenhaupt. On 9 October 1708, Russian forces led by Peter and General Prince Aleksandr Danilovich Menshikov engaged and defeated Lewenhaupt at Lesnaia. Lewenhaupt reached Charles with 6,000 men but no artillery or supplies.

In an effort to sequester resources, Charles diverted his attack into the Ukraine. He also believed that the Cossack Hetman Ivan Mazepa would provide an auxiliary 100,000

Russian and Swedish armies meet at the Battle of Poltava on 8 July 1709. (Library of Congress)

troops against Peter, yet Mazepa delivered fewer than 2,000 Cossacks to the Swedish side.

Charles and 22,000–28,000 Swedish troops were forced to winter in the Ukraine. In May 1709, the Swedes laid siege to Poltava. On 8 July 1709, 40,000 Russian troops with superior artillery engaged and defeated the Swedish force. The Russians set up entrenchments within a few hundred yards of the Swedish siege lines, prompting the Swedes to attack. Charles's plan was to mount a charge past the entrenchments and assault the main Russian force, but being injured, he left command to Field Marshal Karl Gustav Rhensköld.

The latter quarreled with his subordinates, and the unclear issuing of orders contributed to the Swedish defeat. Individual Swedish generals surrendered, either on the field or several days later while trying to cross the Dnieper River. Mazepa, Charles, and 1,500 Swedish troops escaped to Turkey. Lewenhaupt had been ordered to escape with a part of the Swedish army to Crimea and later to meet Charles in Turkey. However, Lewenhaupt and his army had capitulated in Perevolotina.

Neville G. Panthaki

See also: Charles XII; Northern War, Great; Peter I, Romanov, Czar of Russia; Russo-Turkish Wars

References and further reading:

Anderson, Matthew Smith. *Peter the Great.* New York: Longman, 1995.

Kliuchevskii, Vasilii Osipovich. *Peter the Great.* London: Macmillan, 1958.

Lisk, Jill. *The Struggle for Supremacy in the Baltic, 1600–1725.* New York: Funk and Wagnalls, 1968.

Sumner, Benedict Humphrey. *Peter the Great and the Emergence of Russia.* London: English Universities Press, 1964.

Pompey the Great (Gnaeus Pompeius Magnus) (106–48 B.C.E.)

Roman general, statesman, and member of the First Triumvirate with Julius Caesar and Marcus Licinius Crassus. Romans referred to Pompey, a member of a senatorial family, as Magnus, meaning "the Great," after he led a successful campaign in North Africa in 81 B.C.E. Prior to that, Pompey had fought on the side of Sulla against Gaius Marius during the Social War. After ending the Servile War led by Spartacus in 71 B.C.E., Pompey waged a five-year campaign against Sertorius in Spain. In 70 B.C.E., Pompey became consul, cleared the Mediterranean of pirates, and defeated Mithradates VI Eupator of Pontus. He also conquered the kingdoms of Armenia, Syria, and Jerusalem. Disputes with the Senate resulted in the formation of an alliance between Pompey and Julius Caesar. In 60 B.C.E., the two men, joined by Crassus, formed the First Triumvirate. Caesar left the administration of Rome to Pompey and Crassus in 59 B.C.E. and for the next 10 years focused on the conquest of Gaul. Pompey's marriage to Caesar's daughter Julia ensured friendly relations between the two men, but after her death a rivalry developed. In 53 B.C.E., Crassus died in Syria, and Pompey turned to the Senate in an effort to curb Caesar's growing power. The Senate demanded that Caesar resign his office and return to Rome. Instead, Caesar crossed the Rubicon River in 49 B.C.E. and attacked Italy with his forces. Pompey fled across the Adriatic and was defeated by Caesar at Pharsalus in 48 B.C.E. He then fled to Egypt, where he was treacherously killed.

Cynthia Clark Northrup

See also: Caesar, Julius; Marius, Gaius; Roman Civil Wars (88–30 B.C.E.); Sulla, Lucius Cornelius

References and further reading:

Greenhalgh, Peter. *Pompey, the Roman Alexander.* Columbia: University of Missouri Press, 1981.

Leach, John. *Pompey the Great.* Totowa, NJ: Rowman and Littlefield, 1978.

The death of Pontiac 1769. (Library of Congress)

Pontiac's Rebellion (1763–1766)

A widespread Indian uprising against British power in North America. At the close of the French and Indian War (called the Seven Years' War in Europe), English settlers began to move into western Pennsylvania and New York, confident that Indian power was broken with the French defeat. France had ceded Canada to England, and the American Indians found themselves without French arms. British troops took over former French posts and established garrisons at Fort Pitt, Detroit, Venango, Erie, and other sites. For several decades, the various woodland tribes became dependent on European firearms for hunting and survival. To secure their help in defeating the French, the British had made several treaties with the Indians ensuring that they would have their own land, English troops would leave, and supplies of ammunition would continue. Many Indians then helped the English, including some of the Iroquois. The Delaware, a people conquered by the Iroquois, were forced to go along with the agreement.

In the years following the conflict, however, the British Crown was unable to restrict white settlement west of the Appalachians. Crown forces also kept small garrisons in the western Indian lands and began to curtail the trade of firearms and ammunition. Cut off from these supplies, many Ottawa, Potawatomi, and Chippewa faced starvation. Moreover, the Delaware, eager to be free of Iroquois rule, were also growing restless. In the summer of 1763, the western Indians organized under the leadership of a charismatic Ottawa named Pontiac. They were joined in an uprising by the Seneca, Shawnee, and Delaware.

Indian raiders overran English forts in the Ohio territory, and only two held out: Fort Detroit and Fort Pitt. From New York to Virginia, settlers fled the frontier in terror. Colonel Henry Bouquet organized a 400-man force to relieve Fort Pitt. At Bushy Run, the Delaware and Shawnee ambushed Bouquet. He was driven back and took up position on a hill that evening. The next day, the Indians resumed their attacks, having surrounded the English force as was done with

Edward Braddock in 1755. Bouquet had some of his companies feint a retreat and then struck the Indians in their flank, driving them off. He reached Fort Pitt and the next year advanced to the Muskingum River in Ohio, where the Indians sued for peace.

Another force went to relieve Fort Detroit. British Indian agent William Johnson's diplomacy was as valuable as the military victories, convincing the Iroquois to put pressure on the Delaware, Shawnee, and Seneca to end the war. By late 1764, the Seneca fell into line with their Iroquoian brethren in siding with the English, and the united tribes launched raids on the Delaware and Shawnee. In 1766, Pontiac accepted a peace treaty with Johnson at Oswego, New York. With the power of the Indians broken, western Pennsylvania and Ohio were open to settlement. To prevent further outbreaks of violence, the English government did continue restricting white settlement beyond the mountains. This restriction became a major cause of anger for colonists in the years prior to the outbreak of the revolution, in that many had been promised land for service in the French and Indian War.

Robert Dunkerly

See also: French and Indian War

References and further reading:

Anderson, Fred. *Crucible of War.* New York: Alfred A. Knopf, 2000.

Jennings, Francis. *Empire of Fortune.* New York: W. W. Norton, 1988.

Wallace, Paul A. *Indians in Pennsylvania.* Harrisburg: Pennsylvania Historical and Museum Commission, 1961.

Pope, John (1822–1892)

Union general in the American Civil War, loser at Second Bull Run, and a failure as a field commander. Pope was born in Louisville, Kentucky, on 16 March 1822. After graduating from the U.S. Military Academy at West Point in 1842, he was assigned to the topographical engineers. He excelled in combat in the Mexican-American War under Zachary Taylor, being brevetted first lieutenant at Monterrey and captain at Buena Vista. From 1849 to 1861, he was an army engineer and railroad surveyor in the American West.

Commissioned brigadier general of volunteers under John C. Frémont in Missouri on 14 June 1861, Pope quickly mobilized and achieved significant gains. He soundly defeated Sterling Price at Blackwater, Missouri, on 18 December; captured New Madrid, Missouri, on 14 March 1862; was promoted to major general on 21 March; and won an amphibious battle for Island no. 10 near New Madrid in the Mississippi River on 7 April. He marched with Henry W. Halleck toward Corinth, Mississippi, in May and June.

Impressed by Pope's performance, President Abraham Lincoln called him east to take over the new Army of Virginia from George B. McClellan. Pope assumed command on 26 June and immediately alienated his men with his arrogance and insults. He also inadvertently raised Confederate morale with his harsh proposals for dealing with southern civilians. Hated by North and South alike, he soon rendered himself incapable of effective leadership.

Stonewall Jackson defeated Pope at Cedar Mountain, Virginia, on 9 August. Jeb Stuart raided his headquarters at Catlett's Station, Virginia, on 22 August and stole his uniform, dispatches, and notebook, thus providing the valuable information that Robert E. Lee needed to crush him at Second Bull Run. Jackson hampered his retreat toward Washington, D.C., at Chantilly, Virginia, on 1 September.

Lincoln relieved Pope of command on 2 September and reassigned him to the Northwest. He spent most of the rest of his career fighting the Sioux, until he retired in 1886. He died in Sandusky, Ohio, on 23 September 1892. John Pope is a prime example of an officer who does very well in school, is excellent at staff work, but finds himself out of his depth on the battlefield or in overall command.

Eric v. d. Luft

See also: American Civil War; Buena Vista; Bull Run, Second/Manassas Junction; Halleck, Henry Wager; Jackson, Thomas "Stonewall"; Lee, Robert Edward; Lincoln, Abraham; McClellan, George Brinton; Mexican-American War; Monterrey; Sioux Wars; Stuart, James Ewell Brown; Taylor, Zachary

References and further reading:

Cozzens, Peter. *General John Pope: A Life for the Nation.* Urbana: University of Illinois Press, 2000.

Jones, Robert Huhn. *The Civil War in the Northwest: Nebraska, Wisconsin, Iowa, Minnesota, and the Dakotas.* Norman: University of Oklahoma Press, 1960.

Pope, John. *The Military Memoirs of General John Pope.* Ed. Peter Cozzens and Robert I. Girardi. Chapel Hill: University of North Carolina Press, 1998.

Schutz, Wallace J., and Walter N. Trenerry. *Abandoned by Lincoln: A Military Biography of General John Pope.* Urbana: University of Illinois Press, 1990.

Porkchop Hill (16–18 April 1953)

Scene of heavy fighting amid Korean War armistice negotiations. In an effort to gain leverage in the negotiations, the Communist Chinese launched a major attack on the thinly defended and otherwise worthless position. The weight of the attack fell on E Company of the 31st Infantry Regiment,

7th U.S. Infantry Division. By the early morning hours of 17 April, the Chinese had captured the hill, along with several men from E Company. At this point K Company, under Lieutenant Joe Clemons, was given the task of recapturing the hill. After a tough all-night fight, severely depleted K Company held shaky positions on the hill.

As a fresh Chinese attack began, the question of whether or not to reinforce Porkchop now became political. The whole matter boiled down to this question: Were American commanders willing to expend more lives over a worthless hill in Korea in order to demonstrate U.S. resolve in the negotiations? After much hemming and hawing among senior officers, Porkchop was eventually reinforced and held, but at great cost, and then a few months later finally abandoned. The battle was memorialized in the 1959 film *Pork Chop Hill*, starring Gregory Peck.

John C. McManus

See also: Korean War

References and further reading:

Hermes, Walter G. *Truce Tent and Fighting Front: U.S. Army in the Korean War.* Reprint, Washington, DC: U.S. Army Office of the Chief of Military History, 1969.

Marshall, S. L. A. *Pork Chop Hill.* New York: Berkley Publishing Group, 2000.

Sandler, Stanley. *The Korean War: No Victors, No Vanquished.* Lexington: University of Kentucky Press, 1999.

Port Arthur, Siege of (May 1904–January 1905)

The first major phase of the Russo-Japanese War. The conflict opened with a Japanese naval attack on Russian warships in Port Arthur, before any declaration of war by Japan. The siege of Port Arthur began in May 1904 with Japanese landings on the Liaodong Peninsula in China and the movement of Japanese troops northward through Korea and ended in early January 1905 with the capitulation of the weakened Russian garrison. For the Japanese, capture of the port would deny the Russian navy a warm water anchorage and thereby help secure the sea-lanes between the Japanese islands and the northeast Asian mainland.

Russian land defenses included a series of trench works around the town, a series of linked concrete forts about 4,000 yards outside the line of trenches, and then some outer works—fortified hills and other positions—beyond the forts. These defenses should have presented a formidable problem to the Japanese, but it did not seem that the Russians made the position as difficult as they could have.

The Japanese made a series of costly assaults. On 25 May 1904, General Oku Yasukata's frontal assault was thrown back with heavy losses, but an effort to turn the Russian left succeeded. The Russians abandoned Nashan Hill and thereby conceded the port of Dairen to the Japanese. Throughout June, the Japanese prepared for the attack, and an indecisive Russian naval sortie failed. The Japanese army probed defenses in July and then in August and September made three unsuccessful efforts to penetrate defenses with heavy casualties.

In early October, the Japanese brought siege artillery that complemented efforts to mine Russian positions. In November, the Japanese concentrated on the weakened eastern defenses, and in early December they broke through at 203 Meter Hill, moved in artillery, and destroyed the Russian warships in Port Arthur harbor. Throughout December, the Japanese continued their attacks on northern defenses, and finally, on 1 January 1905, the hungry and weakened Russian garrison surrendered.

The Japanese suffered three times as many casualties as the Russians in the long, costly siege. Japanese commanders had favored mass attacks, in dense formations, at night. Such mass attacks actually demonstrated the defensive power of properly placed machine guns and supporting artillery, foreshadowing the carnage of World War I.

Charles M. Dobbs

See also: Russo-Japanese War

References and further reading:

Okamoto, Shumpei. *The Japanese Oligarchy and the Russo-Japanese War.* New York: Columbia University Press, 1970.

Warner, Denis, and Peggy Warner. *The Tide at Sunrise: A History of the Russo-Japanese War, 1904–1905.* London: Frank Cass, 1974.

Westwood, J. N. *Russia against Japan, 1904–1905: A New Look at the Russo-Japanese War.* Albany: State University of New York Press, 1986.

Portuguese-Castilian War (1369–1385)

War of Portuguese independence. Portugal grew out of Spain, and even when Portugal had emerged as a separate realm, its ruling houses still maintained close Spanish contacts and were very much involved in Spanish politics. Despite these connections, political changes led increasingly to an assertion of Portuguese nationhood totally separate from that of Spain. Two key events played a vital role in this process. One was the coming of the Black Death to Portugal in 1348 (with many subsequent outbreaks), initiating an era of agrarian crisis that undermined and then destroyed the old feudal landed order and turned Portugal into the largely commercial and maritime nation that it has been since. The second was the Portuguese-Castilian War of 1369 to 1385, actually a series of separate wars, which ultimately decided the issue of Portugal's independence from Spain to Portugal's advantage.

The war had its origins in a dispute over succession to the throne of Castile. Claimants to the throne included the Portuguese king, D. Fernando I (r. 1367–1383), and Enrique de Trastámara (later Enrique II of Castile, r. 1369–1379), both descended from Sancho IV (r. 1284–1295) of Castile. To advance his cause, Fernando allied himself with Aragon, Castile's hereditary enemy, and with the Muslim king of Granada. Although most of Portugal was not directly affected by the first war (1369–1371), because Portugal remained the aggressor for the most part, it was nonetheless disastrous for the Portuguese. The peace, however, was not overly severe.

With few of the outstanding issues of the first war resolved, a second (1372–1373) and then a third (1381–1382) followed quickly. Although both were Portuguese wars with Castile, both were also part of the Hundred Years War. Fernando had renounced his claim to the Castilian throne in favor of John of Gaunt, son of the English king Edward III. John was married to an illegitimate daughter of the old Castilian king, who had been assassinated by Enrique in 1369. Enrique allied himself with France. Aragon vacillated between the sides.

This time Portugal was not spared, and central Portugal, as well as the extreme northwest, suffered major Castilian invasions by land and sea. During the second war, much of Lisbon was destroyed by Enrique because the city had outgrown its walls and a large part of it now lay exposed to attack. Fernando's English allies were nearly as destructive as the Castilians. During the third war, the Portuguese launched a naval counterattack, but the Portuguese fleet was nearly destroyed, and the Spanish returned again by sea to attack Lisbon.

One result of repeated Portuguese disaster was the growing unpopularity of the monarch, D. Fernando, and his consort Leonor Teles de Meneses, who identified closely with Portugal's great landholders. When D. Fernando died, his legal successor was his daughter D. Beatriz, married to Juan I (r. 1379–1390), king of Castile. The hated Leonor became the regent, and Juan, anxious to assert his claim to Portugal, invaded.

The result was a revolution. The master of Avis, the later João I (r. 1385–1433), representing maritime and commercial Portugal, took the lead in the war despite the fact that much of the interior of the country and its landed interests still remained loyal to D. Beatriz. The fortunes of the war varied, but Portugal won a number of important victories, including the Battle of Aljubarrota (1385), on the site now occupied by the Portuguese national cathedral at Batalha, and ultimately forced the Spanish to withdraw. Although the final peace was not signed until 1432 and there were skirmishes as late as 1396–1397, the separateness of Portugal had been established.

Paul D. Buell

References and further reading:

Albuquerque, Luís de. *Introdução à História dos Descobrimentos Portugueses.* 3d rev. ed. Mira Sintra: Publicações Europa-America, n.d.

Marques, A. H. de Oliveira. *História de Portugal.* 12th edition, 3 vols. Lisbon: Palas Editores, 1985.

Russell, Peter E. *The English Intervention in Spain and Portugal in the Time of Edward III and Richard II.* Oxford, UK: Oxford University Press, 1965.

Potemkin, Prince Grigory Aleksandrovich (1739–1791)

Russian general, statesman, and lover of Empress Catherine the Great (r. 1762–1796). Born September 1739 in Smolensk Province, Potemkin joined the horse guards in the mid-1750s. Promoted to junior lieutenant in 1762, his initial command, in the Izmailovsky Regiment, followed in 1766. Potemkin distinguished himself during the Russo-Turkish War of 1768–1774. After spending the first year of the conflict on the staff of Field Marshal Aleksandr M. Golitsyn and as an aide-de-camp to Field Marshal Peter A. Rumiantsev, he received a field command in 1770. Potemkin proceeded to participate in many of the Russian army's most important victories on the Danubian front, proving himself Russia's most effective cavalry commander and winning a promotion to lieutenant general.

Returning to St. Petersburg in March 1774, Potemkin spent two years as Catherine's favorite, a position bringing rewards and responsibilities, including the war college's vice presidency. Replaced as official favorite in 1776, he remained the empress's most trusted adviser while devoting himself to administrative work, serving as governor-general of New Russia, president of the war college, and head of the Black Sea Admiralty. Under his leadership, the war college sponsored reforms ranging from alterations in army uniforms to adjustments in the composition of large-scale commands, and the Black Sea Admiralty strengthened Russia's naval power in the south.

Following the outbreak of a new war against the Turks (1787–1791), Catherine appointed Potemkin, now a field marshal, supreme commander of Russian forces. Although Russia achieved victory and Potemkin's performance earned substantial gifts and rewards from the empress, Potemkin was outshone by General Aleksandr Vasilyevich Suvorov, who won several critical victories. Potemkin died suddenly, in October 1791, en route to the Jassy peace conference.

Bruce J. DeHart

See also: Russo-Turkish Wars; Suvorov, Aleksandr Vasilyevich

References and further reading:

De Madariaga, Isabel. *Russia in the Age of Catherine the Great.* New Haven, CT: Yale University Press, 1980.

Soloveytchik, George. *Potemkin.* New York: W. W. Norton, 1947.

Powell, Colin L. (1937–)

U.S. Army commander and first African American to hold the positions of national security adviser, chairman of the Joint Chiefs of Staff, and secretary of state. Powell was born on 5 April 1937 in Brooklyn, New York, to Jamaican immigrants. He found his future career path while attending City College of New York, when he enrolled in the Reserve Officers Training Command, where he eventually held the highest student leadership position of cadet-colonel. In 1958 Powell received his commission as a second lieutenant in the U.S. Army.

In 1963, Powell served his first of two tours in Vietnam as an adviser to the Republic of Vietnam's 3d Infantry Regiment. As an adviser, he came under fire for the first time during Operation GRASSHOPPER in the A Shau Valley. In 1965, Powell was promoted to major two years ahead of schedule. He returned to Vietnam in 1968 as the plans officer for the Americal Division. After earning his M.B.A. at George Washington University, Powell was given command of the 32d Infantry Battalion in South Korea. In 1974, he was sent to the National War College and later served a brief tour as brigade commander in the 101st Airborne Division. In a rare step, Powell was promoted to brigadier general from lieutenant colonel on 1 January 1979.

During this time, Powell served in a variety of political-military positions for two administrations. Powell came to the public's attention in 1987 when he was nominated as President Ronald Reagan's national security adviser at a time when the National Security Council was reeling from the Iran-contra scandal. Powell was rewarded with a fourth star after the Reagan administration ended, and he briefly took command of U.S. Army Forces Command (FORSCOM) before President George Bush nominated him to be chairman of the Joint Chiefs of Staff. It was as chairman that Powell led the armed forces and captivated a nation with his presence and style during Operation JUST CAUSE in Panama and the Gulf War. He retired from the military in 1993, and in 2000 he was nominated by President-elect George W. Bush as secretary of state and was quickly confirmed.

Powell has been criticized for applying the "lessons learned" of Vietnam too rigidly, allowing Saddam Hussein to remain in power after his brutal invasion of Kuwait had been reversed, all in the name of avoiding a Vietnam-type "quagmire." Nonetheless, for perhaps most Americans of the time, he represented all that was right with the country, and he was so appealing that only his adamant refusal to run for office ended the "Powell for President" movement in 1996.

Michael Mulligan

See also: Gulf War; Vietnam Conflict; Somalia, U.S. Military Operations in

References and further reading:

Means, Howard. *Colin Powell: Soldier/Statesman–Statesman/Soldier.* New York: Donald I. Fine, 1992.

Powell, Colin. *My American Journey.* New York: Random House, 1995.

Woodward, Bob. *The Commanders.* New York: Pocket Star Books. 1992.

Powhatan War (1622, 1644)

A decade-long war against American Indian tribes in the Virginia colony that resulted in the first Indian reservation in North America. The English settlement of Tidewater Virginia brought people from two aggressive societies into contact in the early seventeenth century. With hopes of finding precious minerals and in expanding their empire, an English expedition founded Jamestown in 1607, constructing a wooden fort. At the same time, eastern Virginia was controlled by a confederacy of American Indians, numbering perhaps 10,000, under Powhatan. With initial contact, each side hoped to use the newcomers against traditional enemies, the English against the Spanish and the Powhatan Indians against rival tribes. A brisk trade developed: the English needed food, and the Powhatans wanted metal tools and weapons.

In the decade leading up to 1620, English settlements branched out up and down the James River. The colony's population was largely single and male and died young. Disease and starvation took a high toll on early settlers. Desperate for food, namely corn, the English were willing to take it forcibly if unable to trade for it with the Powhatans. By the 1620s, a tobacco boom swept Virginia, and this soil-depleting crop drove the British further inland in search of new land to cultivate. This period saw an aggressive expansion of English settlements as far west as present-day Richmond.

Powhatan died in 1618, just a few years after his daughter Pocahontas married an Englishman and left for London.

Powhatan's brother Opechancanough took over as head *werowance* (chief) and began to consolidate his control over various tribes such as the Mattaponi, Pamunkey, Appamatuck, and Chickahominey.

Realizing that the British demand for corn and land was insatiable, Opechancanough organized a strike to take the English out with one blow. The Indians planned a coordinated attack for the morning of 22 March 1622. Some settlements were forewarned, but others were entirely unprepared. Although successful in destroying some communities, the Powhatans could not fight a sustained war, and the English retaliated with raids on Indian towns and cornfields. Intermittent warfare continued for the next few years, punctuated by a few brief truces that never lasted.

English soldiers fought with matchlock muskets, accurate at ranges up to 30 yards. Musketeers were supported by troops armed with swords and pikes. The bow and arrows used by the American Indians had greater range and accuracy than the muskets, yet matchlocks became a highly prized item among them. Steel swords, shields, and weapons were also stolen from the English. After a decade of exhausting warfare, both sides agreed to a peace treaty in 1632.

English settlement expanded, and families increased the colony's population to 8,000 by 1640. Again hoping to eradicate the British swiftly, Opechancanough orchestrated another assault on 18 April 1644. Powhatan strength had never recovered from disease, warfare, and the poor crops of the previous decades. Brutal English counterattacks again destroyed villages and cornfields. In 1646, the Powhatans agreed to another peace treaty, establishing the first Indian reservations in what would become the United States. With the Indians crushed, colonial Virginia was free to expand to the west. The descendants of those who fought in the Powhatan wars still reside on these two reservations in Tidewater Virginia.

Robert Dunkerley

References and further reading:

Rountree, Helen C. *Pocahontas' People.* Norman: University of Oklahoma Press, 1990.

Steele, Ian K. *Warpaths.* Oxford, UK: Oxford University Press, 1999.

Prague, Siege of (1420)

Victory of Ján Žižka over Emperor Sigismund. The conflict in Bohemia between Catholics and the followers of the executed reformer, Ján Hus, became a civil war after the death of King Wenceslas IV, when his brother, Holy Roman Emperor Sigismund, claimed the throne over the objections of the Hussites. Supported by a papal bull denouncing his opponents as heretics, Sigismund entered Bohemia in May 1420, with a crusading army of 80,000, mostly Germans. He quickly captured Hradec Kralove (Königgrätz) and Kutná Hora and then marched toward Prague.

Responding to the entreaties of the Hussites in the city, the town of Tabor sent several thousand men under Hus's commander, Ján Žižka, to the capital. They defeated an army sent to block them. Under Žižka's direction, the Hussites strengthened the city's fortifications, built a watchtower on Vitkov hill, and put up barricades against Hradčany and Vyšehrad castles, which had fallen into Sigismund's hands. Unwilling to risk a direct assault on the city, Sigismund moved to blockade Prague. After capturing several surrounding towns, the crusaders crossed the Vltava River and approached Vitkov from the northeast on 14 July. Initially caught off-guard, Žižka led a counterattack up the south slope of the hill, surprising the crusaders in turn and driving them off the heights.

Though the losses to Sigismund's army were light, the Hussites' determination discouraged his hopes for a quick settlement of the war. After having himself crowned king at Hradčany, he abandoned the siege of Prague and sent the crusaders home. The victory at Vitkov saved the most important center of the Hussite movement and raised Žižka to prominence as the leading Hussite commander.

Brian Hodson

See also: Hussite Wars; Žižka, Ján

References and further reading:

Heymann, Frederick. *John Žižka and the Hussite Revolution.* Princeton: Princeton University Press, 1955.

Preston (17 August 1648)

The decisive battle of the Second English Civil War, ending Charles I's hopes. Following the collapse in spring 1658 of negotiations between crown and Parliament, it emerged that Charles had reached a settlement with the Solemn League and Covenant that controlled Scotland. This revelation led to a resurgence of royalism and scattered risings in England and Wales. Accordingly, on 8 July 1648, a Scottish army of 3,000 horse and 6,000 foot under the command of James, the Duke of Hamilton, crossed the English frontier and marched to Carlisle, where they were soon joined by 3,000 English royalists. This force was the basis of the army that marched south, attempting to gather further support as they went. At the same time, Parliament's officers marshaled their forces and moved to intercept Hamilton.

By 16 August, Hamilton had 20,000 troops, primarily Scots, and neared Preston on the west coast of England, but his men were tired, and their supplies were limited. With poor intelligence, Hamilton was unprepared for any serious opposition, and his men were strung out along the muddy roads into Preston. Skirmishing soon began as Oliver Cromwell's parliamentarians approached from the east (less than 9,000 men but primarily veterans). On the 17th, as rain poured down, Cromwell launched an attack against the disorganized Scots. Smashing through their lines, he forced Hamilton to withdraw. In the chaos, Cromwell's troops took the town, capturing much of the Scottish baggage train, taking 4,000 prisoners, and killing another 1,000. As night fell, although the fighting continued, so did the rain, and in the confusion the demoralized Scottish army retreated, having already abandoned most of its powder.

For all intents and purposes, this battle marked the end for Charles I's cause. Following this defeat, the beleaguered Scots were unable to provide any real assistance to the English royalists, and on 25 August Hamilton surrendered, with most of the remaining royalist strongholds in England quick to follow. What few forces remained in the field were quickly dispersed because after the disaster at Preston, it was difficult to find any who thought that Charles I's cause could continue.

Daniel German

See also: Cromwell, Oliver; English Civil War (1642–1649); Marston Moor

References and further reading:

Woolrych, Austin. *Battles of the English Civil War.* London: Weidenfeld & Nicolson, 1989.

Kenyon, John. *The Civil Wars of England.* London: Phoenix Press, 1989.

Princeton, Battle of (3 January 1777)

A minor battle of the American Revolution, but one that appreciably raised American morale.

In the aftermath of the surprise attack on the Hessians (German mercenary troops allied with the British) at Trenton, New Jersey, on 26 December 1776, Sir Charles Cornwallis moved from winter quarters to attack General George

George Washington on horseback during the Battle of Princeton. (Library of Congress)

Washington. He first moved to Princeton, gathered some 8,000 men, and left several regiments there as a rear guard.

Washington faced a difficult situation, for he was chronically short of food and supplies, and the enlistments of many of his men would expire at the New Year. Retreat across the Delaware to Pennsylvania was not possible. Rather, Washington and the Americans continued on the offensive and once again surprised their enemy. Leaving behind 400 men who kept campfires burning, made noise, and acted as if the entire army were in place, Washington and the bulk of American forces silently slipped around the enemy and then moved north-northeast from Trenton and advanced on Princeton. Fortunately for the Americans, Cornwallis chose to wait until the morning of 3 January to attack, disregarding advice from subordinates to attack immediately on January 2.

On 3 January, the Americans attacked a British regiment that was marching to join Cornwallis; initially, it was a confused fight, and the British more than held their own. After the main American contingent and Washington arrived, the Americans won, but fearing dispersion and Cornwallis's pursuit, they seized what supplies they could and broke off the attack.

Thereafter, Cornwallis withdrew British forces to New Brunswick, New Jersey, while Washington and the Americans went into winter quarters in Morristown. This brilliant campaign of maneuver, beginning at Trenton, helped renew confidence in the American cause and in Washington as commander of the American army.

Charles M. Dobbs

See also: American Revolution; Cornwallis, Sir Charles; Washington, George

References and further reading:

Bill, Alfred H. *The Campaign for Princeton, 1776–1777.* Princeton, NJ: Princeton University Press, 1948.

Smith, Samuel Steele. *The Battle of Princeton.* Monmouth Beach, NJ: Philip Freneau Press, 1967.

Prisoners of War

Persons captured either during military operations or as a consequence of military occupation. Technically, the term *prisoner of war* (POW) is of rather recent origin and has little in common with the people captured during or as a result of belligerent activities in earlier times. This is because the definition of POW relies on the principle of differentiation between combatants and noncombatants, which did not exist prior to the late nineteenth century.

In earlier times and especially antiquity, wars were mostly waged because of two reasons: to conquer new territories and their resources or only to rob these resources without permanently occupying the territory. The resources mainly were human ones, which means that the workforce was the single goal of the belligerent operation. Thus any persons captured during the operation, be they male or female, old or young, were regarded as having exclusively economic value and treated as slaves. Their lives were but part of the booty obtained in course of the operation. The Greek philosopher Plato stated that all those captured alive should be left to the victor's sole discretion as a "gift." That captivity was a very frequent fate in the ancient world is demonstrated by the Roman philosopher Seneca, who differentiated people not as slaves and free but as slaves and "not-yet-slaves." For the enslaved people, it was irrelevant if they were captured as soldiers or as simple inhabitants of the invaded city or territory. The only distinction made concerned the inhabitants of a fiercely defended town, who frequently were massacred as a deterrent for other cities.

This way of treating captured persons was not limited to the Greek and Roman cultures but merely mirrored a universal principle followed in ancient Egypt as well as medieval China, tribal Africa, and the pre-Columbian Americas. However, in certain cultures, it was also common to sacrifice the captured enemies on the altars of their gods.

Significant change was introduced by the Catholic Church in the Middle Ages, when in 1179 the Third Lateran Council prohibited the selling of Christians as slaves. From that time on, capturing people for the sole purpose of enslaving them was no longer acceptable. Instead, captives were released after their relatives had paid a certain ransom. This practice especially affected knights and later officers of noble ancestry, whose families could afford the large sums. As a consequence, the possibility of ransom payments offered a certain protection to the captured. However, those persons considered not profitable enough (footmen, members of the rural population) were either released or killed. Yet under the auspices of the church, many medieval treaties declared it inadmissible to kill women, children, peasants, and people of the church.

However, an entirely different attitude was held toward non-Christian enemies. Thus, it was still common either to kill or enslave non-Christian prisoners. During the Crusades, most Muslim soldiers were killed after having been captured by the Christian knights. Not until many knights themselves had fallen into the hands of their enemies did the system of sparing lives in favor of ransom payments also apply to Muslims.

The end of the Middle Ages brought important changes

in the way war was being waged: fighting became highly professionalized, with trained soldiers, officers, and mercenaries battling with each other, while the rest of the population left behind only bore the task of supplying the armies in the field. A differentiation between captured soldiers and "civilians" slowly became common. Only soldiers were specially treated as prisoners. However, this change did not cause an improvement in the situation of the ordinary soldier, who had no ransom to offer for his life or release. Therefore, it still was common to kill prisoners, but sometimes captors would hold back to ensure that the enemy would not kill those it had captured. Sources published in the famous seventeenth-century collection of military statutes, the *Corpus Iuris Militaris,* suggested, however, that it was the generals' Christian obligation to treat their prisoners with mercy and kindness. The exchange of prisoners between the belligerents according to a certain mathematical ratio (for example: one lieutenant equaled six soldiers, and one general equaled 3,000 soldiers) became more frequent during the following centuries and was very much consonant with Enlightenment rationality and precision.

In the nineteenth century, the Industrial Revolution and the French Revolution concept of mass citizen armies combined to cause many more soldiers to be captured in wartime. These prisoners were not exchanged but detained until the end of the hostilities to ensure that they would not fight again. The American Civil War and the Franco-Prussian War (1870–1871) are the most important examples of this new phenomenon. Suddenly, it became necessary to keep tens of thousands of prisoners in camps until the end of the war, which in the case of the American Civil War could be as long as four years. The countries' local infrastructures were often overwhelmed by the situation, and consequently a large percentage of prisoners died from starvation or lack of medical treatment.

But the nineteenth century was also a time of rising humanitarian sentiment, and the midcentury decades saw the first efforts to regulate the treatment of prisoners by international legal instruments. Although the 1864 Geneva Convention applied only to sick and wounded soldier personnel falling into the hands of the enemy and refrained from calling these persons "prisoners," the well-known "Lieber Code" issued by the U.S. Army in 1863 introduced the term *prisoners of war* and framed it by a legal definition. Articles 56 and 76 of the code prohibited their mistreatment and imposed the obligation to supply them with food. Yet this was an internal statute of the United States and by no means an internationally binding treaty.

Such an international instrument was not ratified until 1899, when the Hague Peace Conference adopted the Hague Convention (II) Respecting the Laws and Customs of War on Land. In its annex, Articles 4 to 20 dealt with the treatment of prisoners of war, taking as an example the regulations of the Lieber Code and the 1874 Brussels declaration (which also had not had the status of a treaty).

The experiences of World War I clearly demonstrated the shortcomings of the Hague Convention's provisions on prisoners. Initiated by the International Committee of the Red Cross, a new Geneva Convention was adopted by the community of states in 1929. This treaty supplemented rather than replaced the articles laid down in the Hague Conventions and proved invaluable for countless thousands of prisoners during World War II. However, the massive mistreatment of prisoners of war in Germany, Japan, and the Soviet Union clearly revealed even this convention's limits. In particular, the range of persons qualifying as POWs was too narrow to cover adequately the reality of twentieth-century war.

As part of a large-scale revision of the laws of warfare, the 1949 Geneva Convention greatly enlarges its predecessors but again proved to be far from perfect. Both the Korean War and the Vietnam conflict resulted in legal problems regarding the repatriation of prisoners against their will at the conclusion of hostilities. Additionally, guerrilla warfare brought forward new definitions of combatants that were different from the traditional ones tailored to the uniformed soldier operating in organized units. To entitle these belligerents to the protected status of prisoners of war, the 1977 Geneva Protocol I supplemented the 1949 convention but remains highly controversial.

Marcus Hanke

See also: Ancient Warfare; Guerrilla/Partisan/Irregular Warfare; Laws of War; Red Cross

References and further reading:

Hesseltine, William B. *Civil War Prisons.* New York: F. Ungar, 1964.
Keen, Maurice, ed. *Medieval Warfare.* Oxford, UK: Oxford University Press, 1999.
Levie, Howard S. *Prisoners of War in International Armed Conflict.* Newport, RI: U.S. Naval War College Press, 1977.

Propellants

Compounds used to move a projectile from the firing device to the target. Originally, gunpowder was used for this purpose. The cannon was a smoothbore, muzzle-loading weapon, firing solid round shot. Manufacturing techniques were extremely simple, and the explosive force of gunpowder is limited, which meant that until the late eighteenth century, gunners (and musketeers) were faced with problems of accuracy and consistency. One of the biggest problems was "windage," caused by gaps between barrel and projectile as the projectile traveled along the barrel after firing. These problems were partially solved with the invention of rifling and driving bands.

Gun propellants are mostly manufactured in powder form. They are low explosives, providing a developing thrust for the projectile as it travels along the barrel. In the early days of gunpowder, the consistency of the mixture of saltpeter, charcoal, and sulfur was difficult to control, but with the advent of nitrocellulose and similar powders, manufacturers were able to increase quality.

The problem with any propellant is to make it burn completely. Cannon would often produce a large muzzle flash, which was propellant burning after the projectile had left the bore. Modern rifles and guns still exhibit this problem, but to a lesser extent. The invention of smokeless powder also made the firing task easier, especially in an enclosed turret.

To ensure that the maximum propellant is burned out while the projectile is still in the barrel and therefore subject to the pressure caused by the propellant burning, powders are now manufactured in specific forms, with numerous holes pierced through each piece of propellant. These holes allow burning to proceed equally inside the powder grain as well as on the surface.

Manufacturers now produce single-base gun propellants from nitrocellulose with the addition of stabilizers and flash reducers. The powder is pressed into cylindrical or other shapes with the burning holes scientifically calculated to ensure maximum efficiency. These charges are assembled in varying weights for small-, medium-, and large-caliber guns.

Multibase propellants such as ballistite are used to make mortar increments. Just as guns do, mortars vary their range partly by changing the elevation of the gun or mortar barrel and partly by the amount of propellant charge used. Mortar increments are added externally to the mortar bomb and are fired by the basic charge, which is fitted to every mortar bomb. Some large-caliber weapons also use ballistite, which is prepared in grains, sticks, and multiperforated kerfed stick. Spherical powders in both single- and double-base (of nitrocellulose-nitroglycerine) are produced for small- and medium-caliber weapons, and some mortar increments are also made this way.

Until the late twentieth century, it was normal for a projectile to be propelled by a charge loaded into the weapon inside a cartridge case. Smaller-caliber ammunition was prepared as fixed ammunition, with bullet/projectile and propellant united in manufacture. Larger weapons, or those that fired more than one type of projectile (high-explosive alternating with smoke, illuminating shell, and solid shot, for instance), loaded the projectile first and then the cartridge case containing the required amount of propellant. Nowadays, caseless propellant is coming into general use, something that has been normal as bagged charge in naval guns since the mid–nineteenth century. These prepared propellant charges are designed to be totally self-consuming, leaving no empty shell case to be ejected.

Artillerymen have always been inventive by nature and have always tried for ever-longer ranges. The concept of rocketry appealed to them, and they adapted the idea of rocket propulsion by producing the base-bleed shell. These projectiles have a small rocket-type motor at their base, which is ignited on or after firing. Fueled by powder, the rocket-assisted shell has increased ranges quite significantly.

Rockets are fueled with either solid or liquid fuel. The solid fuel rocket has a warhead, a powder chamber and some means of igniting the propellant charge, and a shaped exhaust system to allow concentration of the expanding gases at the rear of the rocket. Early solid fuel rockets included the antitank rockets of World War II, fired from rails underneath aircraft wings. Aiming these rockets was done by eye, and they were essentially fire-and-forget weapons, having no internal guidance system or course correction mechanism. Modern solid fuel rockets are far more sophisticated, however, and have onboard guidance systems as well as sensor systems to aid target identification and even target selection.

Liquid-propelled rockets came of age with the German V-2 rocket. The previous venture, the V-1, was merely a pulse-jet-propelled semiaircraft, but the V-2 was a fully fledged ballistic missile. It was fueled by a mixture of liquid hydrogen and alcohol and achieved a range of more than 200 miles. It was a free-flight rocket, having no course correction capability.

Modern free-flight rockets have their ancestors in the weapons of the German and Russian armies during World War II. The German *Minenwerfer* and the Russian "Stalin Organ" were very simple solid fuel rockets with short ranges, but their effect was devastating because of the concentration of fire they could achieve. Area weapons such as these are now represented by the Multiple Launch Rocket System, which had such a destructive effect during the Gulf War.

Solid fuel rockets are preferable to liquid fuel rockets because of the high volatility of liquid propellants and because handling the concentrated acids used for some ballistic missiles is extremely hazardous. The main reason to use liquid fuel is that such rockets can be easily fueled when needed, for solid powder propellants have a habit of settling over time and deteriorating in performance.

David Westwood

See also: Artillery; Mortars; Rifles and Rifling

References and further reading:

Hogg, Ian V. *The Illustrated Encyclopaedia of Artillery.* London: Stanley Paul, 1987.

Ryan, J. W. *Guns, Mortars and Rockets.* Oxford, UK: Brassey, 1982.
The SNPE Explosives and Propellants Handbook. Paris: SNPE, 2000.

Psychological Operations

The use of psychology and propaganda by military units to persuade target audiences to adopt at least some of their views and possibly modify their behavior. The term *psychological operations* has been preferred since the Korean War in that target audiences might not be actually enemy personnel.

The employment of psychological operations goes at least back to biblical times, as when the Hebrew commander Gideon stampeded a numerically superior Midianite force by the sudden midnight display of torches accompanied by shouting, which gave the impression of a much larger force. In more recent times, American rebels were able to garner thousands of British and Hessian defectors by the use of leaflets promising free land in the New World. These leaflets were almost a textbook case of the basic principles of effective psychological operations: (1) know your target audience, (2) make believable promises, (3) do not mock or caricature your targets, (4) keep it simple but official-appearing, and most important, (5) do not lie. The Americans knew that for the British the very term *land-owning class* meant the rich, and it was obvious that the Americans had plenty of land, if nothing else. The leaflets decried the conditions in the British army, did not denounce the enemy, and were brief and to the point, sometimes bearing the name of George Washington; and defectors were indeed given land.

Nonetheless, such eighteenth-century efforts were isolated and limited. But the coming of the Industrial Revolution, with its railroads, telegraphs, cheap paper, and fast, powered printing presses, made large-scale psychological operations possible for the first time by World War I.

Here, the Allies conducted much more effective operations against the morale of the German army than viceversa. Allied leaflets depicted the German soldier as a decent fellow who would be promptly returned to a better Germany at the end of the war, whereas the Germans' propaganda emphasized the "We Shall Crush You" theme, more effective to the already converted on the German home front than to enemy soldiers. One of the most effective American leaflets simply listed the weekly rations of the U.S. doughboy. The German target soldier needed but to compare such bounty with his own diet of mostly *kommisbrodt* (a rough army field loaf). No mention of the kaiser, "Huns," or the "Rape of Belgium," just a simple "Are You Hungry?" theme. Many German prisoners asserted that this one leaflet was primarily responsible for pushing a disgruntled, hungry soldier over the line to the actual act of wartime desertion. Another U.S. Army leaflet used the "Brave German Soldier, Your Government Has Lied to You" theme, counting up the actual 1 million or so U.S. troops already in France, compared to the minuscule number that the German High Command had publicly insisted would ever make their way through the U-boat-infested Atlantic. The German High Command first stupidly forbade their troops to read the leaflets, which simply whetted the soldiers' curiosity; then offered to pay for each "lying" leaflet turned in; and then simply gave up in despair. After the war, both the chief of staff of the German army and Adolf Hitler himself praised Allied propaganda.

World War II was in many ways a repetition of World War I on the propaganda front, except on a much larger and broader scale, but with the Germans and the Japanese still greatly inferior in their use of psychological weapons. Both Axis powers denigrated the Allied soldier as cowardly, weak, and misled. Their use of stilted, archaic language undermined any remaining validity of their leaflets. ("The fraud Rousevelt, hanging the President Election under his nose and from his policy ambition worked not only poor NIMITT but also MACCASIR like a robot, like this, WHAT IS PITY!!") A Japanese "sex" leaflet somewhat spoiled the effect with its last line: "Then, under the beautiful tropical moon, only DEATH awaits you, bullet hole in your guts . . . organizing death!" Actually, such "sex" leaflets were very popular with Allied troops, but for their explicit graphics, certainly not for their clumsy "political" messages, and they generated a brisk souvenir trade. German leaflets used less egregiously mangled syntax and language ("Well, what about the blisters *at* your feet?") but still could not refrain from clumsy, "un-American" phraseology ("Judeo-Bolshevik war-mongers"). That said, it should be pointed out that Axis radio propaganda was quite professional, if for no other reason than that its originators had the wit somehow to obtain the latest in popular American music, thus ensuring that "Tokyo Rose" and "Axis Sally" were widely listened to, if not taken seriously.

The garnering of prisoners of war is not the main intent of psychological operations; the goal is rather the weakening of morale. But Allied psywarriors did interview tens of thousands of German and Japanese troops for their reactions to their products. Few enemy soldiers admitted that Allied leaflets talked them into surrender, but many did say that they did adversely affect their morale. Whatever the value of the opinions of troops in the hands of their enemies, it is difficult to argue against the opinions of the Third Reich's hierarchy. Almost without exception, top Nazi and military officials took Allied psywar very seriously and (privately) gave it high praise. German propaganda minister Joseph Goebbels fulminated publicly against the "printed filth" composed by the "hireling Jewish scribes of Churchill and Roosevelt." At least one Nazi publication, *The Secret Weapon of the Enemy Is at Work,* warned all Germans to turn in any of the enemy's "lying filth" (presumably after it had been carefully read).

In the Pacific, the United States bungled badly at the start of the conflict. One leaflet displayed the words in Japanese and English, "I surrender." Wondering why this leaflet seemed to have no effect on the enemy, American psywarriors contacted Japanese Americans (in their internment camps) and got their answer. Japanese soldiers do not surrender; it is considered a fate literally worse than death. But "I Cease Resistance" or "I Take the Honorable Course" might have done better. Still, far fewer Japanese troops surrendered than did Germans (and many of the "Japanese" defectors were actually Koreans or Taiwanese), but those who did "cease resistance" proved invaluable. Being "dead men" as far as their homeland and families were concerned, they were willing to point out their former comrades' strongholds, give over information to Allied intelligence personnel, and generally cooperate in any way they could. But, again, the morale of the enemy was more important a target than actual surrendered troops, and Allied leaflets emphasized the Japanese army's indifference to its wounded ("Grenade Medicine") and the horrific casualties suffered by Japanese troops in the Philippines and in Burma and held out hope for a better life for Japan after the war. (This attitude was in distinct contrast, of course, to civilian propaganda on the American home front, which depicted the Japanese as bats, slugs, or myopic morons and emphasized America's duty to "Slap the Jap from the Map!")

The Korean War was the first "ideological" war for the United States, a battle between communism and democracy. On the battlefields, the United Nations (UN) Command rarely denounced communism or glorified capitalism or freedom but rather focused on the individual Communist-led soldier and his problems. Some leaflets cleverly played upon the differences between the Chinese and the Russians, with one map graphically contrasting China's vast expanse of old with the lands lost more recently through the notorious "unequal treaties"—lost to the Russians, not to "imperialist" Americans.

Once again, enemy propaganda to American and allied troops was nearly ludicrous in its language barbarisms and political tangles. But some showed commendable wit: "Use Your Head Soldier—If You Don't Want to Lose It!" (e.g., get out of the war); "Old Soldiers Never Die—But Young Ones Do"; "You Risk Your Life—Big Business Rakes in the Dough"; and (the best of the lot) "Leave Korea to the Koreans!" But the greatest Communist propaganda coup was their dissemination of the myth of the "brainwashing" of certain American POWs to favor communism and even to commit treason and of "collaboration" in the camps by a majority of those prisoners. Over the years, the documented rebuttals have never caught up with the unsupported assertions.

The great propaganda coup for the West in the Korean War was the refusal of some 22,000 Communist prisoners of the UN to return to their homelands. But here, again, the mere 22 U.S. captives of the Communists who refused repatriation seemed to receive the most publicity, particularly in the United States. (The fact that all of these defectors eventually returned to their capitalist hells was very rarely noted.)

In the Vietnam conflict a decade later, the Communist side directed its psychological operations as much at its enemies' home fronts as against its troops in the field. Propaganda directed to the Republic of Vietnam (South Vietnam) emphasized that the communist Vietcong was really a group of anti-imperialist freedom fighters who wanted nothing more than an independent and united Vietnam. The same theme played to receptive audiences in the United States, along with the theme that Americans were committing industrial-scale atrocities.

But in the field, it was a different story, as sophisticated and culturally aware American and South Vietnamese psyops garnered some 100,000 lower-level Vietcong cadre over 10 years in the Chieu Hoi (Open Arms) program. Once again, U.S. psyops concentrated on the enemy soldier, on his miseries in the field, his longing for his home village, and the horrors of modern warfare. A particularly clever leaflet illustrated President Richard M. Nixon's and Chairman Mao Zedong toasting each other during the former's unprecedented trip to Communist China. "The Mad Bomber of Hanoi" being feted by the "Elder Brother" of Asian communism! ("So now what are you fighting for?") No matter; the successful Communist strategic propaganda to the American home front and to the outside world is what is remembered to this day.

In the Gulf War (1990–1991), the coalition forces ranged against Saddam Hussein brought psychological operations to a fine art. Thousands of copies of the video entitled "Nations of the World Take a Stand" were distributed throughout the Middle East and in Baghdad itself. The message: "Your wicked leader, Saddam, is leading your beloved nation to ruin. We have no quarrel with the Iraqi people." At the strategic level, the Iraqi command was fooled by a well-orchestrated campaign of "disinformation" in the coalition media that emphasized the likelihood that the anti-Saddam forces would attack from the sea. They thus fell victim to the coalition's left-hook, cross-desert offensive. The "Arab Feast" leaflet, drawn up by the king of Saudi Arabia's personal illustrator so as to avoid any "alien" look, showed "brother Arab" troops inviting surrendering Iraqi soldiers to a sit-down feast—complete with bananas, unobtainable because of the coalition blockade. More ominous was the "B-52" leaflet that warned Iraqi soldiers: "You Cannot See This Bomber, You Cannot Hear It, but You Will Know When It Comes." The next such leaflet was addressed "To the Survivors of the Iraqi ——th Division: How Many Times Must You be Bombed Before You Get the Message?" Loudspeaker teams talked terrified Iraqi soldiers out of their bunkers and directed displaced civilians to the nearest shelter.

The dissemination of leaflets, always a weakness in earlier conflicts (leaflets "shoveled out" of an airplane tended to fly all over the cabin, fouling control cables and distracting the fight crew), were replaced by pinpoint drops using "hundred-mile-an-hour tape," and the leaflets themselves were carefully cut so that they could autorotate into a confined area. Helicopter-mounted loudspeakers replaced the nearly unintelligible aircraft-mounted speakers of previous conflicts, and Arabic-language tapes could be made in professional studio conditions and then broadcast in the field with good fidelity. The vast numbers of surrendering Iraqi troops waving safe-conduct passes gave vivid testimony to the effectiveness of coalition psyops.

In the former Yugoslavia, North Atlantic Treaty Organization (NATO) psychological operators faced a different challenge, civilian target audiences. American psyopers came in early with AM radio facilities, only to be confronted with FM local stations fully as developed as anything in the United States. NATO peacekeepers had to adapt to an audience that was as aware of Michael Jackson as of Slobodan Milosevic. Thus, instead of simple "Be Careful of Unexploded Mines" leaflets or radio and television messages, American psyopers contracted in the United States for a Superman comic book that much more vividly illustrated the dangers of unexploded ordnance to young people, eventually saving lives and limbs by the hundreds. Messages on soccer balls, pens, buttons, and newspapers drove home similar messages. But whether such effective psyops had any lasting ameliorative effect on historic Balkan ethnic hatreds is questionable.

As a result of the success of psychological operations in the former Yugoslavia and in the Gulf War, psychological operations are becoming an increasingly valued part of the armies of the developed world. But the principles of successful psywar have changed little, if at all, since the time of Midian or George Washington.

Stanley Sandler

References and further reading:

American Institutes for Research. *The Art and Science of Psychological Operations: Case Studies in Military Application.* 2 vols. Washington, DC: Department of the Army, 1976.

Daugherty, William E., with Morris Janowitz. *A Psychological Warfare Casebook.* Baltimore, MD: Operations Research Office, Johns Hopkins University, 1958.

Gilmore, Alison. *You Can't Fight Tanks with Bayonets: Psychological Warfare against the Japanese Army in the Southwest Pacific.* Lincoln: University of Nebraska Press, 1998.

Laurie, Clayton D. *The Propaganda Warriors: America's War against Nazi Germany.* Lawrence: University Press of Kansas, 1996.

Sandler, Stanley. *Cease Resistance; It's Good for You: A History of U.S. Army Combat Psychological Operations.* 2d ed. Fort Bragg, NC: U.S. Army Civil Affairs and Psychological Operations Command, 1999.

Ptolemy I Soter (c. 367–283 B.C.E.)

Alexander's general, king of Egypt, and founder of the Ptolemaic dynasty. Born in Macedonia in c. 367 B.C.E., Ptolemy achieved the rank of general under the leadership of Alexander the Great. After Alexander died in 323 B.C.E., his empire fragmented, with Ptolemy ruling Egypt and Libya. Legitimizing his position as successor by kidnapping the body of Alexander and erecting a lavish tomb in Alexandria, Ptolemy then successfully defended his territory against other Macedonian rulers and in the process gained control over Cyprus, Cyrenaica, and Judea.

While Ptolemy I solidified his position, three of Alexander's generals fought for control of the whole empire. Antigonus I attacked Seleukos I, the ruler of Babylon, but was defeated at Gaza in 312 B.C.E. Antigonus's son, Demetrius I, defeated Ptolemy I off the coast of Cyprus before laying siege to Rhodes in 304 B.C.E. Although Ptolemy lost Cyprus, Demetrius failed to capture Rhodes with 30,000 troops and the use of siege towers. In celebration of the victory, Ptolemy ordered the construction of the Colossus of Rhodes from materials abandoned by Demetrius after the siege. That same year, Ptolemy declared himself king of Egypt, establishing the Ptolemaic dynasty that lasted until Egypt became a Roman province upon the death of Cleopatra VII in 30 B.C.E.

In 301 B.C.E., Antigonus died in battle fighting against Selesuc I and Lysimachus. Ptolemy and Lysimachus joined forced against Demetrius, who had conquered Macedonia and ruled Greece. Two years after Demetrius's defeat in 285 B.C.E., Ptolemy abdicated the throne to Ptolemy II, his son by Berneike. He died in his sleep in 283 B.C.E. During his lifetime, he succeeded in creating a thriving capital at Alexandria, building and expanding the collection of ancient manuscripts of the renowned library of Alexandria, and constructing the lighthouse of Alexandria, one of the seven wonders of the ancient world.

Cynthia Clark Northrup

See also: Alexander the Great

References and further reading:

Ellis, Walter M. *Ptolemy of Egypt.* New York: Routledge, 1994.

Pugachev's Revolt (1773–1774)

Largest popular rebellion in Russian history before the twentieth century. Named for its instigator, Emelian Pugachev (1726–1775), the revolt erupted from tension between the Russian government and the Yaik Cossacks. Living along the Yaik River in the plain between the Ural Mountains and the Caspian Sea, the Yaik Cossacks were a frontier people who lived by fishing, herding, and salt production. Since the 1500s, they had fought a losing battle to preserve their independence against the encroachment of the Russian state.

During the eighteenth century, especially the opening decade of Catherine the Great's reign (1762–1796), the state's infringement upon Yaik independence produced resentment within the Yaik community. By the early 1770s, the Yaik Cossacks were ready to explode.

Discontent became open rebellion when Pugachev, a Don Cossack by birth and deserter from the Russian army, arrived in August 1773. Claiming to be the emperor Peter III (1762), who had been deposed and murdered 11 years earlier, Pugachev invited the Cossacks to help him reclaim power. Promising freedom and a special place in the Russian state, the "Pretender" quickly won the Cossacks' support. On 17 September, Pugachev and his followers attacked Yaitsk, capital of the Cossack community, initiating what became the largest popular rebellion Russia had yet experienced.

Once under way, the revolt spread rapidly, engulfing western Siberia, the Ural Mountains, and the Middle Volga River valley, as it attracted support from a variety of disaffected social groups, including the non-Russian Bashkirs, ascribed peasants, Old Believers, and private serfs. Each of these groups held specific grievances against the existing political and socioeconomic order and saw in Pugachev their liberator.

Undisciplined and poorly armed, Pugachev's motley armies proved no match for Russian army regulars, who arrived in force in December 1773 and January 1774. Winning major victories at Tatishchev (22 March), Ufa (23–24 March), Kazan (12–13 July), and Tsaritsyn (25 August), government troops crushed the rebellion and forced Pugachev and 300 of his followers to flee.

On route to Yaitsk, a small group of Cossacks, to save themselves, seized the "Pretender" and handed him over to government forces on 15 September. Transferred to Moscow, Pugachev was tried by a special court, which found him guilty of several crimes and sentenced him to death. Public execution followed on 10 January 1775. To wipe away the memory of the revolt, Catherine renamed the Yaik Cossacks, the Yaik River, and the city of Yaitsk the Ural Cossacks, the Ural River, and Uralsk. These measures could not, however, erase the memories of Russia's rulers, who—at least until the mid–nineteenth century—lived in fear of another Pugachev revolt.

Bruce J. DeHart

References and further reading:

Alexander, John. *Emperor of the Cossacks: Pugachev and the Frontier Jacquerie of 1773–1774.* Lawrence: University Press of Kansas, 1973.

Avrich, Paul. "Pugachev, 1773–1774." In *Russian Rebels, 1600–1800.* New York: W. W. Norton, 1972.

Raeff, Marc. "Pugachev's Rebellion." In *Preconditions of Revolution in Early Modern Europe,* ed. Robert Forster and Jack P. Greene. Baltimore, MD: Johns Hopkins Press, 1970.

Pulaski, Count Kazimierz (1747–1779)

Polish revolutionary who fought in the American Revolution. Born on 4 March 1747, in Podolia, Poland (in present-day Ukraine), Pulaski became a leader of the Confederation of Bar, a revolt against Russian control of Poland. Following the failure of the revolt, he traveled to France, where in December 1776 in Paris, he met Benjamin Franklin, the American ambassador, who convinced him to join the American colonists in their fight against Great Britain. In 1777, he was commissioned a colonel in the Continental army, and as a result of his distinguished service at the Battle of the Brandywine (where he served as Washington's aide-de-camp), he was appointed chief of cavalry and promoted to the rank of brigadier general. In 1778, with the permission of the Continental Congress, Pulaski organized an independent combined-arms unit of cavalry, infantry, and artillery, known as the Pulaski Legion. In 1779, Pulaski and his legion were ordered to South Carolina to support Major General Benjamin Lincoln. After helping to defend Charleston, South Carolina, against a British attack in May 1779, Pulaski and his legion joined with Lincoln and a French force, who were planning to besiege Savannah, Georgia. In an attack on 9 October 1779, he was mortally wounded and died two days later aboard the *Wasp* en route to Charleston. He was buried at sea.

Alexander M. Bielakowski

See also: American Revolution; Brandywine; Washington, George

References and further reading:

Szymanski, Leszek. *Casimir Pulaski: A Hero of the American Revolution.* New York: Hippocrene Books, 1993.

Punic Wars (264–146 B.C.E.)

In 264 B.C.E., the first of three wars broke out between the Roman Republic and the North African sea power Carthage.

An epic struggle for control of the western Mediterranean ensued.

The First Punic War (264–241 B.C.E.)

Messana (Messina), Sicily, threatened by the Greeks of Syracuse, appealed for help first to the Carthaginians, who garrisoned the city, and then to the Romans, who in turn expelled the Carthaginians. The Carthaginians besieged Messana but withdrew to Syracuse in the face of a Roman army. The Romans besieged Syracuse, and the city switched sides. Sicily was divided into Roman (the eastern part) and Carthaginian (the western part) possessions.

After two seasons of campaigning, the Romans realized they could not defeat the Carthaginians without challenging them at sea. In the winter of 261–260 B.C.E., the Romans finally built a fleet, arming the ships with marines and boarding devices rather than a ram. In 260 B.C.E., at Mylae, the new technology proved itself, and the Carthaginians, relying on intricate ramming techniques, were badly beaten.

The war in Sicily remained undecided, and the Romans invaded Africa. A Carthaginian fleet was defeated at Ecnomus in 256 B.C.E. The Roman campaign in Africa was successful for a time, but under the leadership of the Spartan Xanthippus, the Roman army in Africa was destroyed by the Carthaginians. The fleet evacuating the survivors was destroyed by a gale in 255 B.C.E.

Notwithstanding Carthaginian control of coastal waters, the Romans remained successful in Sicily. They captured Panormus (Palermo), and in 251 B.C.E., a Carthaginian force was decisively defeated, leaving the Carthaginians without a field army in Sicily. However, the remaining Carthaginian strongholds could not be taken. In 249 B.C.E., the Romans were severely defeated near the city of Drepana, and another fleet was lost on the south coast of Sicily.

In 247 B.C.E., Hamilcar Barca was sent to Sicily to take command of the fleet. He landed near Panormus and, from the nearby coastal height of Heircte, initiated a guerrilla war on land. He raided the coasts of southern Italy to pin down enemy forces and incite rebellion against Rome. When this strategy was ineffective, Hamilcar captured Eryxin in the vicinity of Drepana under cover of darkness (244 B.C.E.). From there, he continued his guerrilla war. However, in 241 B.C.E. the Romans defeated the Carthaginians at sea near the Aegeates Islands. Forced into a peace treaty, Carthage had to give up Sicily.

Rome had wrested the island of Sardinia from Carthaginian hands (238 B.C.E.), taking advantage of a mercenary revolt and subsequent war in Carthage (247–231 B.C.E.). To compensate for their loss, the Carthaginians started to extend their Spanish possessions. Initially under the command of Hamilcar and afterward under his son-in-law Hasdrubal, the Carthaginians conquered a large part of the Iberian Peninsula. In 220 B.C.E., Hamilcar's son Hannibal was chosen by the army to be its commander. Initially, he set out to subjugate the northern tribes of Iberia, but soon he would turn his eyes upon Rome.

The Second Punic War (218–202 B.C.E.)

In 219 B.C.E., Hannibal besieged Saguntum (Sagunto) in Spain, which turned for help to Rome. The Romans accused Carthage of breaking previous agreements, and although Rome did nothing to save Saguntum from capture (218 B.C.E.), Rome declared war.

In 218 B.C.E., Hannibal marched over the Pyrenees, and having evaded Publius Cornelius Scipio's army near the Rhone River, he marched over the Alps into the Po Valley, where he defeated the Romans near the river Trebia (218 B.C.E.). In 217 B.C.E., he destroyed a Roman army at Lake Trasimene in Etruria (Tuscany) and then won his most brilliant victory (216 B.C.E.) at Cannae in Apulia, completely wiping out two consular armies in one stroke. In reaction, Rome's age-old ally Capua (near Naples) switched sides, as others in Italy did later. Following the advice of Fabius Maximus Verrucosus, the Romans dogged Hannibal's footsteps, recapturing the towns he had taken, harassing his lines of supply, but never facing him in a decisive battle.

After Scipio failed to intercept Hannibal at the Rhone River, he sent his army to Spain under the leadership of his brother Gnaeus Scipio. Rejoining it after the Battle of the Trebia, he attacked the Carthaginians in northern Spain. In 215 B.C.E., the Scipios defeated the Carthaginians at Ibera (215 B.C.E.). By 211 B.C.E., the Carthaginians were pushed back far beyond the Ebro River, and the Scipios decided to attack their basis of power in southwestern Spain. However, attacked by three armies, both Scipios were defeated and killed.

In 214 B.C.E., Syracuse had broken its alliance with Rome and went over to the Carthaginians. Under the command of Marcus Claudius Marcellus, the Romans stormed the city (213 B.C.E.). The defenders were assisted by the machinery of the scientist Archimedes. A lengthy siege ensued, combined with campaigning against Carthaginian troops and numerous hostile towns. In 211 B.C.E., Syracuse was taken, and by 210 B.C.E., Sicily was entirely under Roman control. Besides the overseas expeditions to Sicily and Spain, the Romans also campaigned in Sardinia and Illyria (Albania).

After the secession of Capua, the war in Italy had been mainly fought in Campania. Continuing their delaying strategy, the Romans kept in the vicinity of Hannibal's army, avoiding direct battle. In 213 B.C.E., the Romans besieged Capua. The citizens held out for years, but Hannibal could not relieve the city. As a last resort, he marched to Rome but eventually retreated into Apulia. Capua surrendered in 211 B.C.E., and with Campania secured, the Romans followed Hannibal and continued the war in southern Italy.

In 210 B.C.E., Publius Cornelius Scipio Africanus Major, the son of Publius Cornelius Scipio, who had died the year before, was sent to take command in Spain. In 209 B.C.E., he captured Cartagena, Carthage's most important city in the peninsula, and the next year, he defeated Hannibal's brother Hasdrubal at Baecula (Bailen). Hasdrubal escaped with part of his army and marched to Italy to reinforce his brother. The Romans assembled a large force, intercepting and defeating Hasdrubal at Metaurus Valley (207 B.C.E.). Hannibal retreated into Calabria and remained there until his return to Africa (202 B.C.E.).

Scipio continued the war in Spain successfully, defeating the Carthaginians at Ilipa (206 B.C.E.) and forcing them to evacuate Spain. In 205 B.C.E., Scipio landed in North Africa and defeated the Carthaginians at the Great Plains (203 B.C.E.), whereupon Hannibal was recalled from Italy. At Zama (202 B.C.E.), Hannibal was finally defeated, and Carthage surrendered. As a result, Carthage lost all its overseas possessions and most of its African empire. Its fleet was limited to 10 ships, it was not allowed to make war without Rome's consent, and it was tributary to Rome.

The Third Punic War (149–146 B.C.E.)

From about 170 B.C.E., Massinissa, king of the neighboring Numidians, encroached upon Carthaginian possessions. Because he had been an important ally of Rome during the last stages of the Second Punic War—and possibly because of resentment, fear, and envy of Carthage's riches—Rome permitted Massinissa's encroachment. In 150 B.C.E., the Carthaginians assembled an army, marched to meet the Numidian king, and were defeated.

By attacking the Numidians, the Carthaginians had broken their treaty with Rome, and the Romans sent an army to Carthage. The city surrendered but resisted Roman demands that the population move inland and that the city be destroyed. The Romans besieged the city unsuccessfully because of incompetent command.

In 147 B.C.E., Publius Cornelius Scipio Aemilianus, the adopted grandson of Scipio Africanus Major, was chosen as consul and sent to Africa. Scipio invested the city completely by sea and by land, and in 146 B.C.E. the city was taken after fierce fighting. Carthage was pillaged and razed, and the surviving citizens were sold as slaves.

M. R. van der Werf

See also: Cannae, Battle of; Fabius Maximus Verrucosus "Cunctator"; Hamilcar Barca; Hannibal Barca; Lake Trasimene, Battle of; Marcellus, Marcus Claudius; Scipio Africanus Major, Publius Cornelius; Trebia, Battle of the; Zama, Battle of

References and further reading:

Bagnall, Nigel. *The Punic Wars: Rome, Carthage and the Struggle for the Mediterranean.* London: Pimlico, 1999.

Caven, B. *The Punic Wars.* London: Weidenfeld and Nicolson, 1980.

Lazenby, J. F. *Hannibal's War: A Military History of the Second Punic War.* Warminster: Aris & Phillips, 1978.

———. *The First Punic War: A Military History.* London: UCL Press, 1996.

Pusan Perimeter (August–September 1950)

United Nations Command final defensive lines along the Naktong River and the farthest advance of the North Korean invasion. By early August 1950, General Walton Walker was arranging for the defense of the most vital port of Pusan in southeastern Korea, essential to the support of the large-scale introduction of U.S. armed forces onto the peninsula. The U.S. 24th Division had sought to delay the North Korean advance down the western plain, while Republic of Korea (ROK) units fought a series of delaying actions in the more mountainous east side of the peninsula. Meanwhile, other units arranged for the defense around Pusan.

Initially, the Pusan perimeter was the Naktong River on the west and a northern line stretching eastward from Yongdok. Walker waged a flexible and aggressive defense. He used interior lines of defense to shift forces around and the U.S. 24th Division as a reserve to blunt North Korean attacks and to maintain the perimeter. He engaged in spoiling attacks and was able to blunt the initial greater armored strength of the enemy. Meanwhile, the North Koreans may have weakened their attack by assaulting too many objectives (including taking all of the southwestern plain while they were still assaulting Pusan), considering their limited forces.

From about 27 August through 10 September, the North Koreans attacked the perimeter. They had little success in maintaining bridgeheads across the Naktong River in the American sector of defense; they were able to push the ROK defenses about 10 miles south, but the South Koreans did not break, and the defense line held. Walker had maintained this defense, while General Douglas MacArthur diverted forces coming from the United States and the Pacific and even removed some battalions from Pusan to put together the two-division assault force for the Inchon invasion.

The North Koreans had their difficulties. Allied air power had quickly destroyed the small North Korean air force and then turned to Communist armor, truck traffic, and the road and rail communications from North Korea to the south, mostly on the western side of the peninsula. North Korean units besieging Pusan depended on a long supply line. The Soviets provided virtually no resupply (the reasons for which are still the subject of argument today). And to fill out decimated units, North Koreans press-ganged South Korean civilians and prisoners of war, but these conscripts were unwilling, little trained, and inexperienced—they certainly did not approach the quality of the initial invasion forces. Meanwhile, U.S. naval gunfire and allied airpower helped strengthen the defense of the perimeter until sufficient numbers of men, tanks, and artillery could be shipped to Korea.

Then, on 15 September 1950, MacArthur launched the Inchon invasion, while Walker scheduled a breakout from the Pusan perimeter for the next day. Led by the U.S. 1st Cavalry Division, which covered the final 100 miles in 11 hours, the Pusan defenders linked up with the Inchon invaders on 26 September near Osan, and it appeared the days of desperate defense were over. Eight North Korean divisions were cut off in the southwest, and virtually all of the North Korean units had to leave their tanks, heavy weapons, and supplies as they raced to the 38th parallel to avoid the rapidly advancing UN forces.

Charles M. Dobbs

See also: Korean War; MacArthur, Douglas; Walker, Walton

References and further reading:

Hoyt, Edwin P. *The Pusan Perimeter.* New York: Stein & Day, 1983.
James, D. Clayton, with Anne Sharp Wells. *Refighting the Last War: Command and Crisis in Korea, 1950–1953.* New York: Free Press, 1993.
Sandler, Stanley. *The Korean War: No Victors, No Vanquished.* London: Routledge; Lexington: University of Kentucky Press, 1999.

Pyramids (21 July 1798)

A battle between French forces of the Army of Egypt, 25,000 strong, commanded by Napoleon, and Egyptian and Mamluk forces 18,000–21,000 strong, commanded by Murad and Ibrahim Bey, on the west bank of the Nile River near Cairo. Although it was called the "Battle of the Pyramids," the pyramid complex at Giza is about 15 miles away from the site of the battle, and the pyramids played no role in the engagement; the name is something of a misnomer.

Since landing in Egypt on 1 July 1798, Bonaparte had sought to bring the Mamluk forces, commanded by Murad and Ibrahim Bey, to a decisive battle. Although several sharp actions were fought as the Mamluks retreated down the Nile, the French were unable to complete the destruction of their foe.

When the Mamluk forces arrived near Cairo, the commanders divided their force. The bulk of the forces, perhaps 100,000 strong, crossed the Nile River under the command of Ibrahim Bey and remained in Cairo. These troops played no part in the subsequent battle. A small force of about 6,000 cavalry and 12,000 or more unreliable infantry occupied the village of Embabeh, under the command of Murad Bey.

The French, arriving at 2 P.M. on the 21st, deployed in five large, division-sized squares, with noncombatants, wagons, and cavalry sheltered in the middle. The French deployed their available artillery at the corners of the squares.

At 3:30 P.M., the Mamluks opened the battle with a vigorous cavalry charge against the westernmost of the French squares, posted on the open flank of the French army. These squares, commanded by Jean Louis Reynier and Louis Charles Antoine Dessaix, had not quite fully deployed and were almost taken by surprise by the charge. Nevertheless, the French managed to form and to fend off the cavalry attack. While the Mamluk cavalry sought to overwhelm the French right, the French left, with its flank protected by the Nile, attacked the village of Embabeh and evicted the reluctant, conscripted peasants who formed Murad's infantry.

By 4:30 P.M., the Mamluks withdrew from the field. Two thousand heavily armored Mamluk cavalry were cut off from the retreat and attempted to swim the Nile under fire from the French. Few survived the attempt. Murad and 3,000 more cavalry were able to flee south toward Giza. The surviving infantry simply dispersed. The Mamluks lost 2,000 cavalry and an unknown number of infantry, and the French lost 29 killed and 260 wounded.

As a result of the battle, Bonaparte was able to occupy Cairo, which had been abandoned by Ibrahim Bey, and to undertake the government of Egypt. The practical benefits of the victory, however, were very largely negated by the destruction of the French fleet at Aboukir Bay on 2 August, which cut off communications between France and the Army of Egypt.

Joseph M. Isenberg

See also: Aboukir; Alexandria; French Revolutionary Wars; Kléber, Jean-Baptiste; Murat, Joachim, Grand Duke of Cleves-Berg, King of Naples; Napoleon I

References and further reading:

Chandler, David G. *Campaigns of Napoleon.* New York: Scribner, 1966.
Connelley, Owen. *Blundering to Glory.* Wilmington, DE: Scholarly Resources, 1999.

Pyrrhus (319–272 B.C.E.)

King of Epirus (in northwestern Greece) and inspiration for the term *Pyrrhic victory.* Supposedly descended from

Achilles and related to Alexander, Pyrrhus became king of Epirus at age 12 in 307 B.C.E. Losing the throne in a revolt, he earned a reputation as a fierce warrior, fighting along Demetrius Poliorcetes of Macedon at Ipsus (301 B.C.E.). As a royal hostage in Egypt, Pyrrhus allied with Ptolemy I, who helped him regain control of Epirus with Neoptolemus II (297 B.C.E.). Pyrrhus assassinated his co-ruler in 296 B.C.E., becoming the sole monarch. He engaged in a series of campaigns against Macedon, then ruled by Demetrius. In 281 B.C.E., the Tarentines (Italian Greeks) asked Pyrrhus for military assistance against Rome. Pyrrhus led an army, including elephants, to Italy and forced the Romans to retreat in a costly battle at Herculea in 280 B.C.E. He is said to have remarked, "One more such victory and I am lost," thus the term *Pyrrhic victory.* Pyrrhus, impressed by his foe, sued for peace. When the Romans declined, he withdrew to southern Italy. After another hard-fought battle at Asculum (279 B.C.E.), Pyrrhus left Italy for a mostly successful campaign in Sicily against the Carthaginians. He ruled much of Greek Sicily until his dictatorial manner inspired an uprising. Pyrrhus returned to Italy, badly losing the Battle of Beneventum to a Roman army. Pyrrhus returned to Epirus, conquered Macedonia in 274 B.C.E., and failed in an attempt to restore Cleonymus in Sparta in 272 B.C.E. He was killed in street fighting during a civil war in Argos. Perhaps the most famous general of his time, Pyrrhus was not a statesman and did not follow through on successes. He wrote several lost works on military strategy.

Harold Wise

See also: Ptolemy I Soter

References and further reading:

Cross, G. N. *Epirus.* Cambridge: Cambridge University Press, 1932.

Garouphalias, Petros. *Pyrrhus, King of Epirus.* London: Stacy International, 1979.

———. *Plutarch's Lives.* 2 vols. New York: Random House, 2001.

Q

Qianlong (Ch'ien-lung) (1711–1799)

Last great emperor of Qing dynasty (1644–1912). The Qianlong emperor was the grandson of the Kangxi emperor (1662–1722) and, after taking the throne in 1736, reigned for nearly 60 years. He abdicated in 1795 in favor of his son, the Zhia Qing emperor, so as not to rule longer than his beloved grandfather. He retained effective power until his death in 1799.

The Qianlong emperor expanded Qing boundaries to their greatest extent, gaining control over Tibet in 1751. In the next 10 years, Chinese armies secured Chinese Turkistan, what is now Xinjiang in extreme western China. Although China was not as successful in expanding its southern and southwestern boundaries, local rulers in what is now Vietnam and Burma accepted Chinese suzerainty.

The Qianlong emperor continued systems of control put into place by his predecessors. To preserve Manchu power in a sea of Chinese, he carefully apportioned key administrative posts between Chinese and Manchu appointees and continued the banner system, dividing both military forces and military leadership among Chinese, Manchus, and Mongols. He continued the examination system for entry into the bureaucracy.

Despite Qing successes against peoples on its immediate borders, the Qianlong emperor did not widen relations with the emerging West. Prohibitions against Christian missionaries remained, and Qing authorities sought to limit trade with Western nations to the distant port of Guangzhou, or Canton. Thus the Qianlong emperor would leave his successors a larger territory to protect and a bigger population to feed but also the greater corruption in government that marred his later years and thus less capacity to deal with the more confident and more powerful West.

Charles M. Dobbs

See also: French Colonial Wars; Manchu Expansion, Wars of; Yangzhou, Siege of

References and further reading:

Elvin, Mark. *The Pattern of the Chinese Past.* Stanford, CA: Stanford University Press, 1973.

Kahn, Harold L. *Monarchy in the Emperor's Eyes: Image and Reality in the Ch'ien-lung Reign.* Cambridge, MA: Harvard University Press, 1971.

Singer, Aubrey. *The Lion and the Dragon: The Story of the First British Embassy to the Court of the Emperor Qianlong in Peking.* London: Barrie & Jenkins, 1992.

Qin Shi Huangdi (Ch'in Shih-huang-ti) (259–210 B.C.E.)

The reign name of the first documented emperor of a unified China. Zhao Zheng was the son of the king of the state of Qin, one of seven contending states in China during the several-centuries-long Warring States Period. The state of Qin was located to the west, in the famous "bend of the Yellow River," where its rulers could have a protected rear as they fought the other warring states. Qin was also one of the first states to move from bronze to iron weapons. With a solid agricultural foundation based on irrigated sorghum and millet fields, it could afford to arm its peasantry and thus fielded larger armies than its opponents. The rules of chivalry that had characterized earlier fighting had long since been discarded, and Qin had the men, weaponry, organizational structure, and economic power to prevail in the final stages of the fighting.

Zhao became king at age 13 and assumed real power at age 21. In 230 B.C.E., Qin conquered the weakest of the six states, Han. By 221 B.C.E., it had united China, and King Zheng became the first Qin emperor, or Qin Shi Huangdi.

As emperor, he divided the country into districts, seeking to end the former feudal boundaries and loyalties; established uniform weights and measures, including axle lengths on carts; and constructed roads to improve communication and control within his vast empire. He is best known in the West for connecting many existing walls into the first of many Great Walls, delineating the settled agriculture of China from the nomadic agriculture north of the wall.

Zhao constructed a remarkable tomb—Mount Li—which is still unexcavated and is protected by a vast life-size army of more than 6,000 terracotta soldiers and horses. The find in 1974 of one division of this army has provided invaluable insights into Qin army organization and the role of various ethnic Chinese peoples in the army.

Zhao died, probably from mercury ingested as part of an alchemically inspired elixir of longevity, and his son was unable to continue the dynasty, which soon collapsed.

Charles M. Dobbs

See also: Chinese Imperial Wars; Great Wall of China; Han Wudi

References and further reading:

Bodde, Derk. *China's First Unifier: A Study of the Ch'in Dynasty as Seen in the Life of Li Ssu, 280–208 B.C.E.* London: Oxford University Press, 1967.

Chang, K. C. *Eastern Zhou and Qin Civilizations.* New Haven, CT: Yale University Press, 1985.

Cotterell, Arthur. *The First Emperor of China.* New York: Holt, Rinehart, and Winston, 1981.

Quadruple Alliance, War of the (1717–1719)

Conflict in which Spain, attempting to revise the settlement of the War of the Spanish Succession, fought the Holy Roman Empire, Britain, France, the Netherlands, and the House of Savoy. The end of the War of the Spanish Succession in 1714 and the death of Louis XIV (1638–1715) the next year left Europe a mine about to explode, a point fully grasped by the most energetic and able politician in Europe in the first half of the eighteenth century, Queen Elizabeth Farnese of Spain. Of the victors, the governments in France and Britain were considered illegitimate by a large portion of their subjects, and Duke Victor Amadeus of Savoy and Piedmont, now also king of Italy, and the Holy Roman Emperor Charles VI (1685–1740), now in direct control of Spain's former northern Italian imperial fiefs and the island of Sardinia, were utterly exposed to a resurgent Spanish navy.

In 1716, Elizabeth Farnese's husband, Philip V, uncle to the minor king of France, launched a conspiracy with sidelined members of the regency council; while at sea, her rapidly rebuilt navy initiated war by landing troops in Sardinia and Sicily. The emperor, despite losing Sardinia, was not at first disposed to intervene. He had grievances against France and Britain over the settlement of the previous war, and another segment of the European power train, Turkey, had just launched a revanchist war against Venice. The Venetians folded rapidly, compelling the emperor to intervene in eastern Europe. Thus, it was left to France to take the lead, drawing in Britain, already smarting from the effects of Spanish-supported internal subversion.

The Turks were defeated in 1717, freeing Habsburg forces, and the Holy Roman Empire was brought into the war. The emperor secured a reversion of Sicily for Sardinia (confirmed in the 1720 Treaty of London) and offered troops for expeditions to Sicily that reduced the Spanish garrisons there in 1717–1719, while a British fleet under George Byng, Earl of Torrington (1663–1733), destroyed the Spanish at Cape Passaro and subsequently in Messina harbor, and James Fitzjames, the Duke of Berwick (in French service), invaded Spain, burning the shipyards at Pasajes to check the buildup of the Spanish fleet in 1719.

Berwick's campaign was efficiently conducted, and at the time much attention attached to the imperial army's model campaign of sieges in Sicily, but the main interest of the War of the Quadruple Alliance in military history remains the naval battle of Cape Passaro.

Erik A. Lund

See also: Spanish Succession, War of the

References and further reading:

Armstrong, Edward. *Elisabeth Farnese: The Termagant of Spain.* London: Longman, Green, 1892.

K.k. Abtheilung für Kriegsgeschichte des k.k. Kriegsarchiv. *Feldzüge des Prinzen Eugen.* 21 vols. Vienna: K.k. Generalstabes, 1876–1891.

Lindsey, J. O. "International Relations." In *The New Cambridge Modern History,* vol. 7, 195–198. Cambridge, UK: Cambridge University Press, 1957.

Quatre Bras and Ligny (16 June 1815)

Simultaneous battles in the Hundred Days, just before Waterloo. Returned from Elba and rapidly remobilizing, Napoleon learned that the Congress of Vienna planned to gather a multinational army in Belgium to invade France. Napoleon decided to make a preemptive strike into Belgium to destroy the allied force before it could gain sufficient strength. His plan was to drive his 124,000-man army as a wedge between Gebhard Leberecht von Blücher's 116,000 Prussians and the 93,000 British and Dutch troops of Arthur Wellesley, the Duke of Wellington and then defeat each foe separately.

Marching north toward Brussels and crossing the frontier

on 15 June, Napoleon divided his forces, with Michel Ney commanding the left and Emmanuel de Grouchy the right. With his vanguard solidly between Blücher and Wellington, he ordered Ney's 53,000 to attack the Duke of Weimar's garrison of about 4,000 at Quatre Bras, a strategic crossroads about 18 miles south of Brussels, while he himself simultaneously led Grouchy's 71,000 against Blücher's 83,000 at Ligny, about 5 miles southeast of Quatre Bras. Ney inexplicably hesitated, which ruined the coordination of Napoleon's two-pronged attack and allowed Wellington to reinforce Weimar. When Ney finally attacked with 20,000 in midafternoon, the allied force had grown to about 21,000. Because of a miscommunication, the 30,000 men of the French I Corps never saw action in either battle. Wellington counterattacked after about four hours of fighting.

In the indecisive encounter at Quatre Bras, Ney lost 4,300 men and about 2 miles' worth of ground, and Wellington lost 4,700 men, including the Duke of Brunswick. Because Ney prevented Wellington from reinforcing Blücher, Napoleon won at Ligny, inflicting 16,000 casualties, losing only 11,500, and forcing Blücher to retreat away from Wellington. It was Napoleon's last victory. Heavy rain on 17 June stopped his pursuit of the Prussians.

Eric v. d. Luft

See also: Blücher, Gebhard Leberecht von; Brunswick, Frederick William, Duke of; Gneisenau, August Neidhart von; Napoleon I; Napoleonic Wars; Ney, Michel, duc d'Elchingen, Prince de La Moskova; Soult, Nicolas-Jean de Dieu; Waterloo; Wellington, Arthur Wellesley, Duke of

References and further reading:

Caldwell, George, and Robert Cooper. *Rifle Green at Waterloo: An Account of the 95th Foot in the Netherlands Campaign of 1813–14, at Quatre Bras and Waterloo 16th–18th June 1815, and the Occupation of Paris, with a Full Medal and Casualty Roll for the Fourteen Companies at Waterloo and Details of Weapons, Clothes, and Equipment Used in the Campaign.* Loughborough, UK: Bugle Horn, 1990.

Hofschroer, Peter. *1815: The Waterloo Campaign: Ligny and Quatre Bras.* London: Greenhill, 1998.

Uffindell, Andrew. *The Eagle's Last Triumph: Napoleon's Victory at Ligny, June 1815.* London: Greenhill, 1994.

Quebec, Battle of (31 December 1775)

The unsuccessful American assault on the seat of British power in Canada. During the early months of the American Revolution, many in the Continental Congress felt that Canada might join in the rebellion against England. The addition of Canada would also deny the British an invasion route into New England. A two-pronged invasion got under way in the fall of 1775. Few gave consideration to the fact that Canada's predominantly French Catholic population was mostly reconciled with English rule and suspicious of American intentions.

Colonel Benedict Arnold led a militia column through present-day Maine, traversing 300 miles through a wilderness devoid of roads or supplies. The group ran out of food and nearly starved to death, losing half its numbers en route. Outside Quebec, Arnold met General Richard Montgomery's force, which had come up through New York and had taken Montreal. The English garrison inside the walled city consisted of British regulars and Canadian militia under Sir Guy Carleton. The besieging Americans were outnumbered 1,800 to 1,000.

Knowing that in the spring British reinforcements would arrive, Arnold and Montgomery decided to assault the city on 31 December. A blizzard dumped snow on the attackers, and barricades impeded their progress. Montgomery and Arnold each struck different sectors of the city. Montgomery was killed at the head of his column, and his survivors retreated. Arnold's force entered the city itself but was unable to push on. Arnold was wounded, and rifleman leader Daniel Morgan was captured. The assault cost 372 casualties. Arnold was forced to retreat with the survivors, ending American hopes of annexing a "fourteenth colony." In the spring, British reinforcements did arrive, and Canada became a base for invasion for the remainder of the war and eventually an independent nation on its own.

Charles M. Dobbs

See also: American Revolution; Arnold, Benedict

References and further reading:

Bird, Harrison. *Attack on Quebec: The American Invasion of Canada, 1775.* New York: Oxford University Press, 1968.

Hatch, Robert M. *Thrust for Canada: The American Attempt on Quebec in 1775–1776.* Boston: Houghton Mifflin, 1979.

Shelton, Hal T. *General Robert Montgomery and the American Revolution.* New York: New York University Press, 1994.

Queen Anne's War (1702–1713)

War between Britain, France, and Spain in North America during the War of the Spanish Succession (1701–1714). Compared to the huge battles and large-scale sieges on the theaters of operation in Europe, the fighting in North America was relatively small-scale. The main reason was probably that New York and the Iroquois, who declared their neutrality in 1701, stayed out of this conflict. Warfare was limited to several raids against French or British villages or strongholds. The most famous example of the frontier campaigns was the Deerfield raid of 1704. Hertel de Rouville, a French Canadian, led 48 militia and around 200 Indians (Abkenaki,

Caughnawagas, and Hurons) 300 miles across the Green Mountains in the depths of winter to attack the town of Deerfield (in present-day Massachusetts) on 29 February. The small force killed 30 to 50 inhabitants, devastated the place, and carried off 100 prisoners on their return.

The New Englanders retaliated in 1704. A force of 550 men from Massachusetts attacked Castine. But as in King William's War (1689–1697), the main objective of the English colonists was Port Royal, the leading French base in Acadia. Two expeditions against this valuable port were launched in June and August 1707. Both failed because of the heavy resistance of the defenders under their new governor, Auger de Subercase. The fiasco called for the use of regulars, and a third attack proved to be successful. Francis Nicholson took the town at the head of 3,000 regular and militia troops in October 1710. By that point, Auger de Subercase commanded only 156 men. Port Royal, renamed Annapolis Royal, remained in English possession and the peninsula of Acadia became British Nova Scotia.

In 1711, the British decided to take New France in what could be called a complex combined military operation. This large expedition was designed to improve Britain's position in negotiations with France. A seaborne force under the command of Rear Admiral Sir Hovenden Walker and Brigadier General John Hill was to sail up the St. Lawrence River to attack Quebec, while Francis Nicholson led an army of 2,300 men from Albany to raid Montreal. Walker's invasion fleet of 31 transports escorted by 14 ships of the line sailed from Boston on 30 July. On board the vessels were 7,500 troops. But on 23–24 August, stormy weather and poor nighttime piloting caused eight transport ships to founder in the mouth of the St. Lawrence. Nearly 900 British sailors, soldiers, and accompanying women drowned near the Ile aux Oeufs. This led not only to the withdrawal of the fleet but also to the abandonment of Nicholson's landward advance on Canada. However, the Peace of Utrecht (1713) left Britain with Nova Scotia, Newfoundland, and Hudson's Bay. These gains strengthened the British position in an expected decisive struggle for the colonies and weakened the defenses of New France.

In the South, some 500 Carolina volunteers and 300 Yamasee under Governor James Moore of Carolina invaded Florida, but the garrison of St. Augustine successfully repulsed the attack. At the head of a large American Indian army, Moore returned in 1704, raiding western Florida and attacking Spanish missions near Tallahassee. Franco-Spanish privateers took revenge in 1706, when they attacked Charleston. This again resulted in British pressure on Florida: raids on Pensacola in 1706 and 1707 and an advance of a small force as far as Mobile in 1709. Although the British took the initiative in the following years, they did not gain any land in Florida from Spain in the Peace of Utrecht. The global contest for supremacy would continue for more than a century.

Juergen Luh

See also: Spanish Succession, War of the

References and further reading:

Black, Jeremy. *Britain as a Military Power, 1688–1815.* London: UCL Press, 1999.

Lynn, John A. *The Wars of Louis XIV, 1667–1714.* London: Longman, 1999.

Queenston Heights (13 October 1812)

One of several American setbacks along the Canadian border during the first year of the War of 1812. Major General Stephen Van Rensselaer, commanding American troops along the Niagara front, hoped to revive American plans to seize Canada following General William Hull's surrender of Detroit (August 1812) by attacking Queenston Heights. The Americans initially gained the advantage by attacking the British rear, but General Alexander Smyth, a regular army officer, refused to take orders from Van Rensselaer, a militia officer, and never moved against Fort George to prevent British general Isaac Brock from sending reinforcements to Queenston Heights. Brock overwhelmed the Americans, nearly forcing them off the precipice before being killed. A subsequent British-Iroquois attack cut the Americans to pieces. The presence of Indians, the sight of Americans returning wounded in battle, and dislike of Van Rensselaer caused the New York militia to refuse to cross the river to join the fight. The Americans lost approximately 90 killed, 100 wounded, and 800 captured at Queenston Heights. This defeat, coupled with failed campaigns against Fort Erie and Montreal in late 1812, contributed to the disintegration of the American army on the Niagara front and demonstrated the difficulty of relying on militia to fight the war, exposing the "War Hawk" Henry Clay's boast that the invasion of Canada would be a "mere matter of marching" by the militia for the ignorant bombast it truly was.

Dean Fafoutis

See also: U.S., Militia; War of 1812

References and further reading:

Compton, Smith C. *The Battle of Queenston Heights, U.C., October 1812: A Collection of Documents and Records Together with Factual Reports Dealing with the Events of the Day.* Toronto: McGraw Hill, 1968.

Hitsman, J. Mackay. *The Incredible War of 1812: A Military History.* Updated by Donald E. Graves. Toronto: Robin Brass Studio, 1999.

Whitfield, Carol. "The Battle of Queenston Heights." *Canadian Historic Sites* 11 (1974).

GROUND WARFARE

An International Encyclopedia

INTERNATIONAL WARFARE ENCYCLOPEDIAS FROM ABC-CLIO

Spencer C. Tucker, General Editor

Air Warfare: An International Encyclopedia, Walter J. Boyne, Editor

Naval Warfare: An International Encyclopedia, Spencer C. Tucker, Editor

Ground Warfare: An International Encyclopedia, Stanley Sandler, Editor

GROUND WARFARE

An International Encyclopedia

VOLUME THREE, R–Z

EDITED BY
Stanley Sandler

ASSOCIATE EDITORS
Michael Ashkenazi
Paul D. Buell

FOREWORD BY
General Henry H. Shelton, USA (R)

ABC-CLIO

Santa Barbara, California   Denver, Colorado   Oxford, England

Library of Congress Cataloging-in-Publication Data
Sandler, Stanley, 1937–
Ground warfare : an international encyclopedia / Stanley Sandler ; foreword by Henry H. Shelton.
p. cm
Includes bibliographical references and index.
ISBN 1-57607-344-0 (Hardcover: alk. paper) — Ebook ISBN 1-57607-733-0
1. Battles—Encyclopedias. 2. Military history—Encyclopedias. 3. Military biography—Encyclopedias. I. Title.
D25.A2 S26 2002
355'.003—dc21

2002004568

07 06 05 04 03 02 10 9 8 7 6 5 4 3 2 1

This book is also available on the World Wide Web as an e-book. Visit abc-clio.com for details.

ABC-CLIO, Inc.
130 Cremona Drive, P.O. Box 1911
Santa Barbara, California 93116-1911

This book is printed on acid-free paper.
Manufactured in the United States of America

CONTENTS

A-TO-Z LIST OF ENTRIES

Gujerat (1849)
Gulf War (2 August 1990–28 February 1991)
Gustavus II Adolphus (1594–1632)

Hadrian (Publius Aelius Hadrianus) (76–138)
Haig, Douglas (1861–1926)
Haitian Civil War (1806)
Halleck, Henry Wager (1815–1872)
Hamilcar Barca (c. 270–228 or 229 B.C.E.)
Hamilton, General Ian Standish Monteith (1853–1947)
Han Wudi (r. 141–87 B.C.E.)
Hancock, Winfield Scott (1824–1886)
Hannibal Barca (247–188 B.C.E.)
Harpers Ferry (American Civil War, 12–15 September 1862)
Harrison, William Henry (1773–1841)
Harsha (c. 590–c. 647)
Harun al-Raschid (766–809)
Hasegawa, Yoshimichi (1850–1924)
Hastings, Battle of (14 October 1066)
Hattin, Battle of (4 July 1187)
Hawaiian Wars (1782–1810)
Hawkwood, John, Sir (c. 1321–1394)
Henry II, King of England (1133–1189)
Henry V, King of England (1387–1422)
Heraclius (c. 575–641)
Hideyoshi, Toyotomi (1537–1598)
Hill, Ambrose Powell (1825–1865)
Hindenburg, Paul von Beneckendorf und von (1847–1934)
Hiroshima and Nagasaki, Atomic Bombings of (1945)
History, Military
Hitler, Adolf (1889–1945)
Hittites (c. 2000–1100 B.C.E.)
Ho Chi Minh (1890–1969)
Hochkirch, Battle of (14 October 1758)
Holy Roman Empire (800–1806)
Holy Roman Empire–Papacy Wars (1077–1250)
Honduran-Nicaraguan War (1907)
Hong Xiuquan (1814–1864)
Honors and Awards, Military
Hood, John Bell (1831–1879)
Hooker, Joseph (1814–1879)
Horseshoe Bend, Battle of (27 March 1814)
Houston, Samuel (1793–1863)
Hue, Battle of (31 January–2 March 1968)
Hukbalahap Revolt (1945–1959)
Hundred Years War (1337–1453)
Hungarian Civil Wars (1526–1547)
Hungarian Revolt (1956)
Hungarian War with the Holy Roman Empire (1477–1485)
Hungarian-Turkish Wars (1437–1526)
Hungarian-Venetian Wars (1345–1381)
Huns
Hunyadi, János (c. 1407–1456)
Hurrians (c. 2300–1100 B.C.E.)
Hussein, Saddam al-Tikriti (1937–)
Hussite Wars (1419–1436)
Hydaspes, Battle of the (May 326 B.C.E.)
Hydrogen Bomb, Development of (1942–1952)

Ia Drang Valley (October–November 1965)
Illyrian Wars (229–219 B.C.E.)
Imjin River (April 1951)
Imphal and Kohima (8 March–22 June 1944)
Inca Civilization
Inca Empire Imperial Wars (1438–1540)
Inchon Landings (15 September 1950)
Indian Border Conflicts (1962–1971)
Indian Mutiny (1857)
Indian National Army (1943–1945)
Indochina Wars (1945–1954)
Indonesian War of Independence (1945–1949)
Infantry
Inkerman, Battle of the (5 November 1854)
Intelligence, Military
Interventions in Civil Unrest, Strikes, Military
Iran Hostage Rescue Attempt (24–26 April 1980)
Iran-Iraq War (1980–1988)
Irish Easter Uprising, War for Independence, and Civil War; The Easter Rebellion (1916–1923)
Irish Rebellion, Great (1641–1649)
Irish Uprising (1798)
Iroquois-French Wars (1609–1697)
Isandlwana (South Africa, 22 January 1879)
Isonzo, Battle of the (1915–1917)
Israeli Military
Israeli-Arab Wars (1948–1999)
Issus, Battle of (November 333 B.C.E.)
Italian Colonial Wars (1882–1936)
Italian Wars of Unification (1848–1870)
Italo-Turkish War (1911–1912)
Ivan III (1440–1505)
Ivan IV ("the Terrible") (1530–1584)
Ivry, Battle of (14 March 1590)
Iwo Jima, Battle of (19 February–15 March 1945)

Jackson, Andrew (1767–1845)
Jackson, Thomas "Stonewall" (1824–1863)
Jacobite Rebellions (1689–1746)
Jan III Sobieski (1629–1696)
Janissaries
Japanese Civil Wars (1450–1550)

Nuremberg Principle
Nurhaci (1559–1626)

October War (1973)
Oda, Nobunaga (1534–1582)
Offa's Wars (771–796)
Office of Strategic Services
Ögödei (c. 1186–1241)
Okinawa (1 April–21 June 1945)
Omani Conquest of East Africa (1622–1730)
Omdurman (1898)
Onin War (1467–1477)
Orleans, Siege of (12 October 1428–8 May 1429)
Osaka Castle, Siege of (1614–1615)
Osan, Battle of (5 July 1950)
Ostende, Siege of (1601–1604)
Ostrogoths
Otto I, the "Great" (912–973)
Ottoman Empire (1300s–1922)
Oudenaarde, Battle of (11 July 1708)
Oudinot, Nicholas-Charles, Duc de Reggio (1767–1847)

Pachacutec Yupanqui (r. 1438–1471)
Pacific, War of the (1879–1884)
Pacifism/War Resistance
Paekche (attributed 18 B.C.E.–660 C.E.)
Pagan Kingdom (1044–c. 1300)
Palo Alto (8 May 1846)
Panama Incursion (1989–1990)
Panipat, Battles of (21 April 1526, 5 November 1556, 14 January 1761)
Paramilitary Organizations
Paris, Siege of (1870–1871)
Parma and Piacenza, Alessandro Farnese, Duke of (1545–1592)
Parthian Empire (247 B.C.E.–226 C.E.)
Patton, George Smith, Jr. (1885–1945)
Pavia, Battle of (24 February 1525)
Pearl Harbor Attack (1941)
Peleliu (15 September–27 November 1944)
Peloponnesian Wars (460–456, 431–404 B.C.E.)
Peng Dehuai (1898–1974)
Pequot War (1636–1637)
Pericles (495–429 B.C.E.)
Pershing, John J. (1860–1948)
Persian Civil Wars (1725–1794)
Persian Empire (550 B.C.E.–642 C.E.)
Persian Wars of Expansion (559–509 B.C.E.)
Persian-Afghan Wars (1726–1857)
Peru-Bolivia Confederation, War of the (1836–1839)
Peru-Ecuador Conflict (1941–1999)
Peruvian Guerrilla War (1980–2000)
Pétain, Henri-Philippe (1856–1951)
Peter I, Romanov, Czar of Russia ("The Great") (1672–1725)
Petersburg, Siege of (June 1864–April 1865)
Pharsalus, Battle of (48 B.C.E.)
Philip, King (Metacomet)(1639–1676)
Philip II Augustus (1165–1223)
Philip II of Macedon (c. 382–336 B.C.E.)
Philippi, Battle of (42 B.C.E.)
Philippine Insurrection (1899–1902)
Philippines, U.S. Loss of (7 December 1941–9 June 1942)
Philippines, U.S. Retaking of (20 October 1944–2 September 1945)
Pickett, George Edward (1825–1875)
Pilsudski, Józef Klemens (1867–1935)
Pinkie (10 September 1547)
Pitt, William, the Elder (1708–1778)
Pizarro, Francisco (c. 1478–1541)
Plains of Abraham (13 September 1759)
Plassey, Battle of (23 June 1757)
Plataea, Battle of (479 B.C.E.)
Plattsburgh Movement (1915–1918)
Pleven/Plevna, Siege of (20 July–10 December 1877)
Poitiers, Battle of (18 September 1356)
Polish Campaign of 1939
Polish Wars of Expansion (1386–1498)
Poltava (8 July 1709)
Pompey the Great (Gnaeus Pompeius Magnus) (106–48 B.C.E.)
Pontiac's Rebellion (1763–1766)
Pope, John (1822–1892)
Porkchop Hill (16–18 April 1953)
Port Arthur, Siege of (May 1904–January 1905)
Portuguese-Castilian War (1369–1385)
Potemkin, Prince Grigory Aleksandrovich (1739–1791)
Powell, Colin L. (1937–)
Powhatan War (1622, 1644)
Prague, Siege of (1420)
Preston (17 August 1648)
Princeton, Battle of (3 January 1777)
Prisoners of War
Propellants
Psychological Operations
Ptolemy I Soter (c. 367–283 B.C.E.)
Pugachev's Revolt (1773–1774)
Pulaski, Count Kazimierz (1747–1779)
Punic Wars (264–146 B.C.E.)
Pusan Perimeter (August–September 1950)
Pyramids (21 July 1798)
Pyrrhus (319–272 B.C.E.)

MAPS

ACRONYMS

ABM Antiballistic Missile Treaty
ACP automatic cartridge pistol
AEW Airborne Early Warning
AFVs armored fighting vehicles
AIF Australian Imperial Force
ALB Air Land Battle Doctrine
ANZAC Australian and New Zealand Army Corps
ARVN Army of the Republic of Vietnam
ASDIC Anti-Submarine Detection Investigation Committee
ASW antisubmarine warfare
ATC American Tobacco Company
AWACS airborne warning and control system
BEF British Expeditionary Force
C2 Command and Control system
C3I command, control, communications, and intelligence
Cheka Chrezvychainaya Komissariat po bor'be s kontrarevoliutsiei i sabotazhem (All-Russian Commission for Struggle against Counterrevolution and Sabotage)
CI counterintelligence
CIA Central Intelligence Agency
CIGS chief of the Imperial General Staff
CORDS American Office of Civil Operations of Rural Development Support
CPB Communist Party of Burma
CTBT Comprehensive Test Ban Treaty
DIA Defense Intelligence Agency
DK Democratic Kampuchea
DPRK Democratic People's Republic of Korea
DSM Distinguished Service Medal
EAM-ELAS National Liberation Front–National Popular Liberation Army
EDES Greek Democratic National Army
ELN Ejercito de Liberación Nacional (Army of National Liberation)
EOKA National Organization of Greek Fighters
EPL Ejercito Popular de Liberaci (Army of Popular Liberation)
EVA Ever Victorious Army
FARC Fuerzas Armadas Revolucionarias de Colombia (Colombian Revolutionary Armed Forces)
FLEC Front for the Liberation of Cabinda
FLN National Liberation Front
FNLA National Front for the Liberation of Angola
GCNG Greek Cypriot National Guard
GDP gross domestic product
GDR German Democratic Republic
GPU Gosudarstvennoe Politicheskoe Upravlenie (State Political Administration)
GULAG Glavnoe Upravlenie Ispravitel'no-trudovykh Lagerei (Chief Administration of Corrective Labor Camps)
HUMINT human intelligence
I&W Indicators and Warnings
IDF Israeli Defense Forces
IFF Identification Friend-or-Foe transponders
IMINT (aerial and satellite photo and radar) imagery intelligence
INA Indian National Army (Azad Hind Fauj)
INF intermediate-range nuclear forces
IR infrared (radar)
IRA Irish Republican Army
IRB Irish Republican Brotherhood
JDA Japanese Defense Academy
JTF Joint Task Force
KGB Komitet Gosudarstvennoi Bezopastnosti (Committee of State Security)
KNU Karen National Union

KPA Korean Peoples Army
KUFNS Khmer United Front for National Salvation
M-19 Movimento 19 de Abril (19th of April Movement)
MACV U.S. Military Assistance Command
MAUD Military Application of Uranium Detonation
MED Manhattan Engineer District
MFA armed forces movement
MID Military Intelligence Division of the Army General Staff
MLRS Multiple Launch Rocket System
MPLA Popular Movement for the Liberation of Angola
MVD Ministerstvo Vnutrennikh Del (Ministry of Internal Affairs)
NATO North Atlantic Treaty Organization
NF Intermediate Nuclear Forces Treaty
NKGB Narodnaya Komissariat Gosudarstvennoi Bezopastnosti (People's Commissariat for State Security)
NKVD Narodnaya Komissariat Vnutrennikh Del
NSA National Security Agency
NSC National Security Council
NVA North Vietnamese Army
OAS Secret Army Organization
OECS Organization of Eastern Caribbean States
OFS Orange Free State
OGPU Unified State Political Administration
OKW Wehrmacht High Command
OPEC Organization of Petroleum Exporting Countries
ORBAT Order of Battle information
OSRD Office of Scientific Research and Development
OSS Office of Strategic Services
PAIGC African Party for the Independence of Guinea and Cape Verde
PG Provisional Government
PLO Palestinian Liberation Organization
PNI Partai Nasional Indonesia (Indonesian Nationalist Party)
PPA Planters Protective Association
PRA People's Revolutionary Army
PRD Dominican Revolutionary Party
PRK People's Republic of Kampuchea
RADAR Radio Detecting and Ranging
RAR Royal Australian Regiment
RDF Radio Direction Finding
RKKA Workers and Peasants Red Army
ROK Republic of Korea
RPF Assembly of the French People Party
SALT Strategic Arms Limitation I Interim Agreement
SAMS surface-to-air missile defenses
SAS Special Air Service
SBS Special Boat Service
SEAL sea, air, and land team
SIGINT signals intelligence
SLORC State Law and Order Restoration Council
SONAR Sound Navigation and Ranging
START Strategic Arms Reduction Talks I and II
TECHINT technical intelligence
TOW target on wire
UN United Nations
UNC United Nations Command
UNEF United Nations Emergency Force
UNITA National Union for the Total Independence of Angola
VOC Vereenigde Oost-Indische Compagnie (Dutch East India Company)

GLOSSARY

(Recurring topics. Each listing in this glossary has its own article in the text.)

Airborne Operations   The insertion of troops and equipment on the battlefield by means of parachute, helicopter, or glider.

Armor   Body protection for soldiers.

Armored Fighting Vehicles   Protected vehicles used on the battlefield.

Arms Control   Limits, usually set by treaty, on the number and types of weapons; usually refers to nuclear weapons.

Artillery   Basically, a heavy metal tube from which a missile is discharged violently by explosive force.

Awards and Honors   Recognition granted by authorities for meritorious service by soldiers.

Ballistics   The science of projectiles, divided into interior and exterior ballistics. Its aim is to improve the design of shells/projectiles so that increased accuracy and predictability are the result. It deals also with rockets and ballistic missiles.

Bayonet   Metal blade or spike that, when fixed to a musket or rifle, facilitates its use in hand-to-hand combat. In recent decades, more likely to be used for opening ration tins.

Bazooka   Shaped-charge, smoothbore, man-portable, antitank, and pillbox weapon.

Buffalo Soldiers   African-American troops of the late-nineteenth-century U.S. regular army.

Catapult   Engine for throwing a heavy weight, using an arm released from tension; the artillery of the ancient world.

Cavalry   The noble, mobile arm of battle; the traditional horse-borne arm of mobility; can now refer to a motor vehicle–mounted unit.

Chaplains   Military officers who tend to the spiritual, moral, and physical needs of troops in the field and in camp; pastors in uniform.

Chemical and Biological Warfare   The deliberate use of chemical or biological agents against an enemy.

Civil Affairs/Military Government   Those activities of a commander that embrace the relationship between the military forces and civil authorities and people in a friendly country or area (Civil Affairs) or occupied country or area (Military Government).

Coastal Defense   The defense of a nation's coast from an enemy sea invasion or blockade, accomplished with heavy artillery, mines, small warships, and nets.

Cold War (1946–1989)   Period of tension between the Soviet Union and the People's Republic of China and their allies, representing communism (the East); and the United States and its allies, representing capitalism and democratic socialism (the West); punctuated with several "hot" wars, the most significant being the Korean War (1950–1953) and the Vietnam War (c. 1955–1975). The Cold War ended unexpectedly with the fall of the Berlin Wall and the bloodless victory of the West.

Communications, Military   The application of technology to the transmission of military orders and intelligence.

Conscription   The selection of persons for involuntary military service.

Death Squads   Clandestine and usually irregular organizations, often paramilitary in nature, that carry out extrajudicial executions and other violent acts against clearly defined individuals or groups of people.

Disarmament   The removal or drastic reduction by nation-states of major weapons.

Economic Warfare   Compelling an enemy to submit either by direct action against its economic basis or indirectly through blockade or boycott.

Electronic Warfare   The use of the electromagnetic

spectrum to gain knowledge of the presence and movement of an opposing force and also to deny any opposing force the use of that spectrum.

Engineering, Military The application of science and technology for military purposes, primarily through the use of civil engineering.

Ethics of War Rules, principles, or virtues applied to warfare.

Firearm A tube, closed at one end, that has in it an explosive with a projectile above it, nearer to the open end of the tube. An ignition system fires the explosive charge, which forces the projectile along the tube by means of the gases from the explosion; can include both artillery and small arms.

History, Military History dealing with the use of organized armed force, either on behalf of some form of recognized state authority or against it.

Infantry Lightly armed ground troops; the backbone of any army; not only the most numerous of the fighting arms but the only one that can actually take and hold ground.

Intelligence, Military Military specialty that provides a commander and staff with the knowledge of the enemy and of weather and terrain required for the planning and conduct of operations. (There is no truth to the assertion that "military intelligence" is an oxymoron.)

Laws of War International laws, enforced sometimes by nations after war and sometimes by commanders in battle, governing both the decision to engage in war and the manner of its conduct, particularly the forms of violence used, the definition of combatants, the treatment of prisoners, and the treatment of neutrals and noncombatants.

Logistics Largely an American usage, encompassing military supply, transportation, medical service, and construction-maintenance.

Machine Gun Rapid-firing small arm that can maintain a high rate of fire without the requirement of reloading after each round, which today means a fully automatic weapon; either man-portable ("machine pistol") or heavier.

Maps and Cartography The result of the utilization of cartography and topographical reproduction for military strategy and operations.

Medals and Decorations Tangible recognition of faithful military service or success awarded to individual soldiers.

Medicine, Military The medical and surgical specialty concerned with the ailments of soldiers and sailors.

Mercenaries Hired professional soldiers who fight for a state or entity without regard to political interests or issues.

Military-Industrial Complex The institutions and people that plan, procure, and fight a war and that supposedly shape the economy, the political realm, and the wider society, even in peacetime. Term first used by outgoing U.S. President Dwight D. Eisenhower in 1961.

Mortar Muzzle-loaded weapon firing its bomb at a high angle to attack protected positions and trenches.

Pacifism/War Resistance The organized opposition to war, killing, or violence; usually divided into two segments: religious (e.g., Quakers, Mennonites) and secular.

Paramilitary Organizations Unofficial groups organized along military lines yet lacking the traditional role or legitimization of conventional or "genuine" military organizations.

Propellants Compounds used to move a projectile from the firing device to the target.

Psychological Operations The use of psychology and propaganda by military units to persuade target audiences to adopt at least some of their views and possibly to modify their behavior.

Rank, Military Official indication of a soldier's length and quality of service in organized militaries.

Rifles Firearms designed with barrel grooves to impart a spin and thus far greater accuracy; either small arms or artillery.

Tactics The theory and practice of using military forces in combat.

Terrorism Acts of violence intended for a wide audience in order to create an environment of fear for political reasons.

Theory, Military That body of knowledge usually published in books and journals that examines the nature of wars and the art of war on an abstract level.

Unarmored Fighting Vehicles Unprotected military vehicles, either specially built for the military or adapted from commercial models and used in combat-support roles (e.g., Jeeps, trucks, ambulances).

Uniforms Military clothing worn by organized bodies of troops to distinguish them from the uniformed personnel of other armed forces and to strengthen morale.

War Crimes Actions in wartime that violate the laws or usages of war.

R

Rajput Rebellions (1679–1709)

A series of conflicts that afflicted Mogul rule almost from the moment of the Muslim conquest of Hindu India. Those conflicts of the last decades of the seventeenth century and first of the eighteenth in particular exposed the hollow shell that Mogul rule had become by that time. Drawing their name from Rajasthan—the northwestern region of modern India, roughly between Delhi and Pakistan—the Rajputs were a collection of Hindu families whose military and political power dated from preconquest times. Under Mogul rule, Rajasthan remained a tangle of opposition combining the prickliest of resistance with the least rewards.

In an effort to avoid such unrewarding conflict, in the late sixteenth century Mogul rulers had initiated a policy of integration whereby a number of non-Muslims, including the Rajputs, gained entry into the Mogul nobility. In 1562, Mogul emperor Jalud-ud-din Akbar married a Rajput princess. Over time, Rajput cavalry became a valued element of the Mogul war machine, as did their commanders. In 1664, it was the Rajput Jai Singh who, as the head of the Mogul forces, inflicted the first major defeat on the Maratha prince Shivaji.

This policy of integration, however, was not entirely successful in ending conflict between the Muslim rulers of India and the Hindu Rajput princes. Despite having a Rajput wife—or perhaps because of it—Akbar personally led the imperial army against Udai Singh, rana of Mewar and leader of the Sesodia, the most senior Rajput clan. From 1567 to 1568, Akbar laid siege to and finally destroyed Udai's capital at Chitor. He did not, however, break either Sesodia power or their willingness to use it. Its capital reestablished at Udaipur, Mewar would be at the heart of the greatest Rajput rebellions of the late seventeenth and early eighteenth centuries.

These latter Rajput rebellions bear testimony to the peculiarly Byzantine aspect of Rajput rebelliousness that was bestowed by the latter's close connection to the Mogul aristocracy even as they fought against the Mogul emperor. Rather than mere outsiders to and opponents of the Muslim state, the Hindu Rajputs within the Mogul nobility were in a position to take advantage of conflicts within the state itself. Muslim-Hindu conflict was thus given an added dimension as Muslim rivals to the imperial throne sought to enlist Rajput forces in their fight against fellow Muslims. This victory did not prevent the Rajput princes from joining Shah Jahan a decade later, when he rose in revolt against his father and seized the Mogul throne. A similar pattern was followed in 1658, when Shah Jahan was himself deposed by his son, Aurangzeb. The subsequent reign of Aurangzeb was not merely among the longest—lasting until 1707—but was also among the most turbulent in the long history of Mogul India. In large part, this unrest was the consequence of Aurangzeb's own domestic policies, aimed as they were at suppressing the power and influence of non-Muslims within India. Among the non-Muslims to whom Aurangzeb directed his attention were the Rajputs of Marwar.

In 1678, the death of Rathir Maharaja created for the Rajputs of Marwar the type of succession crisis that they had often taken advantage of when afflicting the Mogul throne. When, in 1679, Aurangzeb used the power of the empire to award the succession to a Rajput sympathetic to Mogul interests, revolt erupted and soon spread from Marwar to Mewar. Meanwhile, Prince Akbar used the crisis within the empire to mount his own challenge to his father, Aurangzeb, and made a bid to seize the throne. Like Shah Jahan over a century before, Akbar led Mogul forces against the Rajputs of both Mewar and Marwar, and now the Rajputs supported him in his bid for the throne. Unlike Shah Jahan, however, Akbar was not successful, in large part because his hesitation in the field alienated his Rajput allies at the moment of final confrontation with his father. With the defeat of his rebellious son, Aurangzeb was free to turn on the Rajputs. Yet though in 1680, the Mogul host seized and sacked Udaipur (the city built to replace Chitor, destroyed a century before), the destruction of Udaipur did not bring the Rajputs to heel. Facing revolt and unrest elsewhere in India, Aurangzeb was eventually forced to grant an honorable peace to the Rajputs, including the confirmation of the infant son of Rathor Maharaga as rana of Mewar.

Upon the death of Aurangzeb in 1707 and in the midst of the inevitable succession crises, that infant son, Ajit Singh, led the Sesodias of Mewar in a revolt against the empire. Joined shortly by the Rajput Kacchwahas of Amber, Ajit was forced to come to terms with the Mogul governor Bahadur Shah but rose in revolt yet again less than a year later. That Aurangzeb's successors were unable to crush this final Rajput revolt reflected the inability of the Mogul empire to effectively meet these challenges to its authority in the early eighteenth century. Indeed, the final Rajput rebellions reflected the extent to which Mogul rule in India was in the process of disintegration. The sad story of this disintegration had already been written by Prince Akbar, who, after his failure to seize the Mogul throne with the assistance of the Rajputs, fled into the arms of his father's other internal foes, the Marathas, before finally seeking asylum in Persia. By the 1730s, the Rajputs had successfully established their autonomy over Rajasthan, only to find themselves challenged as the eighteenth century drew to a close, not by their Mogul masters but by the British traders and soldiers who had begun to usurp them.

Adam Norman Lynd

See also: Aurangzeb

References and further reading:

Gascoigne, B. *The Great Moghuls.* London: Jonathan Cape, 1971.

Keay, John India. *A History.* London: Harper Collins, 2000.

Richards, John. *The Mughal Empire.* Cambridge, UK: Cambridge University Press, 1993.

Ramillies, Battle of (22 May 1706)

A severe French defeat and the best illustration of the tactical skill of John Churchill, the Duke of Marlborough. Two years after the Blenheim disaster, the War of the Spanish Succession was still a draw, and Louis XIV yearned to have the glory of his army restored. He ordered François de Neufville, Duc de Villeroi, to bring Marlborough to battle in the plains of Brabant. Villeroi, was a courtier and a former king's friend. The French army, numbering 60,000, moved to Tirlemont to engage the enemy camp near Tongres. Learning that the allied camp had been left empty, Villeroi sent his army south toward Namur. He reached the area of Ramillies on 22 May to discover that Marlborough was close to him.

The French deployed first on a 4-kilometer line, between the villages of Autre-Eglise and Tavières. Their center, on high ground, was cut in two by the village of Ramillies, 500 meters to the front. This area was garrisoned by three weak battalions of foreign troops, including former prisoners of war. The French cavalry was on both wings, but the left wing faced uncrossable marshes and streams.

Marlborough immediately saw the weakness of this deployment. It took five hours before the allied army was ready to fight; he had to regroup his troops, leaving his right covered by some cavalry squadrons. In the meantime, French generals requested and then begged Villeroi to attack the deploying enemy. He refused all entreaties but instead took infantry from his center to reinforce his useless left cavalry. Marlborough gathered his troops in front of both Ramillies and the right side of the French deployment, gaining numerical superiority. Then he closed all the troops on a single battle front, mingling cavalry squadrons with infantry battalions. The assault on Ramillies was easy, and the French right wing sustained the fight less than an hour before retiring, first slowly and then routing into a disordered mob. The French left wing ran away without having to fight. The French lost 8,000 killed, 7,000 prisoners, and all their artillery. This French defeat allowed Marlborough to overrun the Spanish Netherlands.

Gilles Boué

See also: Malplaquet, Battle of; Marlborough, John Churchill, First Duke of; Oudenaarde, Battle of; Spanish Succession, War of the

References and further reading:

Bibliothèque historique et militaire. Vol. 4. Paris: 1851.

Chandler, David. *Marlborough as a Military Commander.* London: Batsford, 1977.

Churchill, Winston. *Marlborough: His Life and Times.* 4 vols. London: Harrap, 1933–1938.

Ramleh, Battle of (Palestine) (5 September 1101)

Battle that thwarted Fatimid intentions to regain Palestine from the crusaders. King Baldwin I of the Latin Kingdom of Jerusalem had campaigned extensively since gaining the crown in 1100. Most of his campaigns were designed to demonstrate his strength and secure the borders of the Kingdom of Jerusalem. These small campaigns also gained the cities of Arsuf and Caesarea.

The Fatimid dynasty of Egypt watched these events closely. While Baldwin was at Caesarea, the Fatimid vizier, al-Afdal, hoped to catch the crusaders off guard and avenge losses suffered in 1100. An army under the command of Sa'ad al-Daulah al-Qawasi advanced into Palestine and reached Ascalon in the middle of May.

Upon hearing of this, Baldwin withdrew from Caesarea and marched to Ramleh, which he fortified. After receiving reinforcements, Sa'ad marched toward Ramleh on 4 September 1101. Although some contemporaries estimated the Egyptian army to number 11,000 cavalry and 21,000 infantry, it was no doubt much smaller. King Baldwin, on the other hand, possessed slightly over 1,000 men, consisting of 260 cavalry and 900 infantry.

Despite the overwhelming odds, Baldwin divided his army into five corps and attacked at sunrise on 5 September 1101. Three of the corps were decimated, and the fourth retreated after suffering heavy casualties. Victory appeared imminent to the Muslims, but Baldwin made a last-ditch effort and charged them. Leading the way, he surprised the Egyptians. The initial shock turned to panic, and soon the Egyptians were routed. Baldwin's army maintained its discipline and did not stop to pillage the dead or the Fatimid camp but pursued them back to Ascalon. Ironically, the king's impetuousness would lead to the capture of Ramleh by the Fatimids the following year.

Timothy May

See also: Crusades

References and further reading:

Holt, P. M. *The Age of the Crusades.* London: Longman Group, 1990.

Mayer, Hans Eberhard. *The Crusades.* Oxford, UK: Oxford University Press, 1988.

Riley-Smith, Jonathan. *The Crusades: A Short History.* New Haven, CT: Yale University Press, 1987.

Runciman, Steven. *A History of the Crusades.* Vol. 1. Cambridge, UK: Cambridge University Press, 1951.

Rank, Military

Official indication of a soldier's length and quality of service in organized militaries. The need for soldiers quickly to recognize their own as well as their enemy's leaders is as old as warfare itself. When fighting forces were small and leaders local, it was less critical; soldiers generally knew who was in command.

As forces grew in size, more complex command structures developed. In the army of the Roman Republic, the overall command of a legion rested with its six tribunes, officers sharing command in pairs and alternating every few months. The real authority within the legion was entrusted to the centurions, professional officers each commanding 100 men. Supporting them were the decurions, with a command of 10 men. The senior centurion, the *primus pilus,* guided the often inexperienced and transitory tribunes in the command of the legion. As the armies of imperial Rome moved to secure the empire, the need grew for officers with specific skills to coordinate the logistics of transporting men and material. There followed a proliferation of specialist ranks.

The fall of Rome signaled an end to formal armies in Europe. They did reappear, but the process was slow. It was not until the twelfth century that a structured military started to develop, in France, with the establishment of a general staff under the command of an officer, designated the *marechal de France.* Still, there remained a wide gulf between the senior commanders and their soldiers. There were no intermediate ranks between general and captain. Most soldiers continued to serve under officers that they knew.

In England, the largest permanently organized body of professional soldiers until the Elizabethan period was a company, commanded by a captain. His deputy was the lieutenant captain, later shortened to simply lieutenant. The lowest-ranking officer in a company was the ensign, or ancient. In battle, he was the standard-bearer, which accounts for his less exalted title, "rag carrier." In cavalry regiments, the ensign was known as a coronet. Both ensign and coronet ranks were gradually replaced in the eighteenth century with the rank of second lieutenant.

Increased domestic political stability, territorial avarice, and complex military alliances created a formula for larger armies. Regiments were raised by forming a column of companies and were commanded by colonels, from the Latin *colonna.* In continental Europe, as in Tudor and Stuart England, the principal obligations of a colonel were in raising and paying the regiment that often bore his name. It was left to his deputy, his lieutenant colonel, to exercise actual command. Officers bearing the rank of major assisted the lieutenant colonel in maintaining discipline within the regiment and with administrative duties. Officers themselves were ranked, with major, lieutenant colonel, and colonel as field officers; those holding the rank of captain and below were designated as company officers.

Armies, divisions, and brigades were commanded by general officers, yet they mimicked the rank structure of the much smaller company. In seventeenth-century England, the senior general held the rank of captain general. In time, this was shortened to merely general. His deputy was a lieutenant general. The army's administrative functions were assigned to a sergeant-major general, now simply called major general. King George II imported the continental rank of field marshal to Britain in 1736, a rank that was never adopted in the United States.

The one rank that can be either general officer or field officer, depending on the army, is brigadier. In the U.S. Army, it is a general rank. The British consider it a senior field rank. There is no brigadier rank in the Russian or Chinese armies.

There was a great difference between officers who received their military rank through a commission from the monarch and those who were appointed to their rank by the regimental commander. These latter officers occupied a lesser military and social rank, but they still held important positions in the regimental hierarchy, serving as adjutants and quartermasters. Discipline in the ranks was the purview of the company sergeants. Supervision of the company sergeants was the responsibility of the regimental sergeant major. The lowest noncommissioned officer rank was lance corporal, a rank that originated with mercenaries serving the Italian city-states and denoted an officer who commands a small body of men, a *corpo.*

In times of war, some officers and noncommissioned officers have been given a brevet rank, a temporary higher rank for the duration. During the American Civil War, George Armstrong Custer held a brevet rank as major general but ended the war with a substantive, or permanent, rank of lieutenant colonel.

Military rank, like social rank, is often inconsistent with the egalitarian ideals found in some modern societies. The revolutionary armies of the United States and the Soviet Union resolved the issue by allowing soldiers initially to elect their own officers. However, once revolutions are over, military authorities are generally unwilling to allow military commands to be decided by ballot. The People's Republic of China found the matter of rank so ideologically troubling that in 1965 military rank insignia was abolished in the People's Liberation Army to close the gap between officer and soldier. In a massive modernization of the military in the 1980s, China, like every other significant military power, reestablished rank insignia.

Tradition frequently complicates order and confounds logic. This is particularly true in military organizations. In Britain's Household Cavalry, there is no sergeant rank. Instead, there is the corporal-of-the-horse. The Royal Artillery has no corporals. Instead, they are bombardiers. In the infantry, soldiers are privates; they are troopers in the cavalry, sappers in the engineers, gunners in the artillery, and guardsmen in the foot guards. In some British regiments today, the quartermaster sergeant is referred to as color sergeant, a nineteenth-century rank that was officially abolished in 1914.

Every army by its very nature is complex and all are idiosyncratic. Not every rank is comparable between armies; Russia has more officer ranks, and the United States has a dizzying array of enlisted, technical, noncommissioned officer, and warrant officer ranks.

Eric Smylie

See also: American Revolution; Bolshevik Revolution; British Military, Twentieth-Century Organization and Structure; Chinese Military; French Revolutionary Wars; Military and Society; Russian and Soviet Armies; Uniforms; U.S. Army

References and further reading:

Baker, A. J. *Soviet Army Uniforms and Insignia, 1945–1975.* London: Arms & Armour Press, 1976.

Barthorp, Michael. *British Infantry Uniforms since 1660.* Poole, Dorset, UK: Blandford Press, 1982.

Farwell, Byron. *Mr. Kipling's Army.* New York: Norton, 1981.

Kerrigan. Evans E. *American Badges and Insignia.* London: Leo Cooper, 1975.

Makepeace-Warne, Antony. *Brassey's Companion to the British Army.* London: Brassey's, 1995.

Rosignoli, Guido. *Army Badges and Insignia of World War 2.* New York: Macmillan, vol. 1, 1972; vol. 2, 1975.

———. *Army Badges and Insignia since 1945.* Vol. 1. New York: Blandford Press, 1976.

Raphia, Battle of (Palestine) (217 B.C.E.)

Antiochus marched southward with 6,000 cavalry, 102 Indian elephants, and 62,000 infantry. Ptolemy countered with a force containing 5,000 cavalry, 73 African elephants, and 70,000 foot soldiers, many of them newly trained Egyptian natives, recruited for Ptolemy's army by his Egyptian minister, Sosibus. They met on the plains south of Gaza at Raphia.

In the complex evolution of the Alexandrian successor wars, Antiochus III, the prevailing Seleucid ruler had, since 220 B.C.E., stripped Ptolemaic territory in South Syria. In 217 B.C.E. Antiochius marched toward Egypt itself.

Both armies formed up for battle identically: heavy infantry phalanx at the center, protected by cavalry on both wings, and outside flanking columns of elephants, protected by archers and skirmishers and screened against enemy cavalry.

Antiochus opened by sending the 60 elephants on his right wing against Ptolemy's elephant escort of 40 on the latter's left. Antiochus's larger Indian elephants overpowered Ptolemy's North African ones, backing them against the latter's cavalry. Next, Antiochus circled Ptolemy's left wing, attacking it from the flank, stampeding it, and pursuing it off the field. Ptolemy, however, was still in the fight. On his other flank, Ptolemy's Greek mercenaries checked the Seleucid cavalry and Asiatic peltasts, accompanied by the remaining Ptolemaic elephants, which assisted in the rout of Antiochus's left wing. Thus the wing elements of each force had left the scene, leaving the phalanxes to decide the outcome alone. The superior numbers of well-trained native Egyptian heavy infantry inspired by the appearance of Ptolemy, who had left his fleeing wing in order to lead his phalanx forward, penetrated the Seleucid center. Antiochus III returned from the command of his pursuing wing only to accompany his collapsing phalanx off the field.

Ptolemy lost 1,500 infantry, 700 cavalry, and almost all elephants killed or captured. Antiochus lost less than 10,000 infantry, a little more than 300 cavalry, 5 elephants, and more than 4,000 soldiers captured. Even though Ptolemy won the battle, the nationalist Egyptian spirit awakened by the muster and the decisive role of native troops resulted in the decline of Ptolemaic authority in Egypt.

Jim Bloom

References and further reading:
Cary, Max. *A History of the Greek World 323–146 B.C.* New York: University Paperbacks, 1977.
Connoly, Peter. *Greece and Rome at War.* London: MacDonald & Jane's, 1981.
Dupuy, R. Ernest, and Trevor N. Dupuy. *The Harper Encyclopedia of Military History.* 4th ed. New York: Harper & Row, 1993.
Montagu, John Drogo. *Battles of the Greek and Roman Worlds.* London: Greenhill Books, Lionel Leventhal Limited, 2000.

Ravenna (1512)

Pivotal event in the Italian Wars of the early sixteenth century, enabling the pope and his allies to oust the French from Italy. The Holy League, an alliance between Pope Julius II, Spain, and Venice at the time, had been pushed back from Brescia by the French. On Easter Sunday, 11 April 1512, the Battle of Ravenna took place along the Ronco River 2 miles south of the city. At the battle's outset, the French army consisted of more than 23,000 Frenchmen, Germans, and Italians; the Holy League's army consisted of over 16,000 Spaniards and Italians dug in behind a long trench. In contrast to the usual medieval battle, in which fighting began once all troops were lined up, the Battle of Ravenna began with an exchange of artillery fire. The French, led by Gaston de Foix, and the league, led by Ramon de Cardona, Fabrizio Colonna, and Pedro Navarro, each took heavy casualties.

After three hours of being pounded by artillery, the league's army charged out of the trenches, and infantry and cavalry fighting determined the remainder of the battle. When the fighting ended, the French army had a Pyrrhic victory. The league's army lost nearly 9,000 troops, and its cavalry and infantry commanders, Colonna and Navarro, had been captured. The French casualties were fewer in number—nearly 4,000—but they included many commanders, including de Foix. Because of the heavy losses, including that of his nephew, de Foix, King Louis XII ordered his army to stop its push toward Rome. Louis sought negotiations with Julius II in order to free himself for the defense of France against Spain and England, only to have Julius and the Holy League expel the French from northern Italy the next year.

Christopher P. Goedert

See also: Navarro, Pedro, Count of Olivetto

References and further reading:
Bridge, John S. C. *A History of France from the Death of Louis XI.* Vol. 4. Oxford, UK: Clarendon Press, 1929.
Oman, Charles. *The History of the Art of War in the Sixteenth Century.* New York: E. P. Dutton, 1937.

Razin's Revolt (1667–1671)

Cossack-led popular rebellion against late-seventeenth-century Russia's existing order. Razin's Revolt gets its name from Stepan "Stenka" Razin (1630–1671), a Don Cossack, who in April 1667 led nearly 1,000 of his fellow Cossacks on a plundering expedition down the Volga River and into the Caspian Sea. For nearly two years, from July 1667 to June 1669, Razin and his followers ravaged Caspian shipping and coastal settlements, seizing people, animals, and booty while defeating every effort to stop them.

By the time Razin returned to the Don in September 1669, stories of his daring exploits and lavish riches had spread. In the winter of 1670, these rumors attracted to Razin's camp a multitude of impoverished Cossacks hoping to participate in the next campaign. Filled with delusions of grandeur by the influx of new supporters, Razin concluded he could achieve mastery of the entire Volga and maybe even Russia itself. In March 1670, he announced his intentions to destroy Russia's nobles and attack Moscow itself. One month later, Stenka and some 4,000 followers returned to the Volga River.

Meeting little opposition, Razin won control of the river from Astrakhan in the south to Simbirsk in the north, a distance of 800 miles, in only five months. In doing so, Razin unleashed a genuine popular rebellion, as thousands of townspeople, soldiers, lower clergy, non-Russian tribes people, and peasants, all with grievances against the existing order, rose up in the belief that Stenka would bring freedom from autocracy and serfdom. By September 1670, no fewer than 20,000 people were in open revolt across a 250-mile expanse of the middle Volga region, while Razin, who was then threatening Simbirsk, appeared poised to strike west, toward Moscow.

At Simbirsk, however, Razin was decisively repulsed. The city's defenders beat off three rebel assaults in mid-September, and on 1 October, czarist forces armed with muskets and artillery, trained in the "European" manner and battle-hardened by the recent Russo-Polish War (1654–1667), arrived from Moscow and defeated the ill-equipped rebel army. While the townspeople, peasants, and tribesmen scattered, Razin and his Cossack followers fled home. Returning to the Don, Razin was subsequently seized by prosperous members of the Cossack community, who had never supported him, in April 1671. Transported to Moscow, interrogated, and tortured, he was executed on 6 June 1671. His rebellion, however, continued until 26 November of that year, when czarist forces extinguished its last ember by capturing the rebel stronghold at Astrakhan.

Though Russia's social and economic structure remained unaltered, Razin's Revolt demonstrated the widespread dissatisfaction existent throughout late-seventeenth-century Russian society and provided a foretaste of what was to come in the 1770s, when another Don Cossack, Emelian Pugachev, instigated an even greater rebellion on Russia's southern frontier. Furthermore, in Razin the revolt produced a popular leader whose memory and legend lingered in the minds of Russia's discontented and dispossessed into the twentieth century.

Bruce J. DeHart

See also: Pugachev's Revolt

References and further reading:

Avrich, Paul. "Razin, 1670–1671." In *Russian Rebels, 1600 to 1800.* New York: Schocken Books, 1972.

Longworth, Philip. *The Cossacks.* London: Constable, 1969.

Reconquest of Spain (711–1492)

Spanish Christian states slowly reconquer territories lost to the Muslims. Shortly after the Arabs completed their conquest of Byzantine Africa, Berber general Tariq ibn Ziyad crossed the Straits of Gibraltar in 711 and, with an overwhelmingly outnumbered Umayyad army, crushed the Visigothic king Roderick at the week-long Battle of Rio Barbate. A number of Christian princes fled to refuge in the Pyrenees and held out against Islam. Nonetheless, few kingdoms have been overthrown so quickly and completely as Visigothic Spain, as city after city surrendered to the invaders, including Toledo, the Visigothic capital. The very name of the nation was changed, as Muslim Spain henceforth became known as al-Andalus, "the West." After a pause for consolidation of their conquests, Arabic forces under Abd-ar-Rahman, the Umayyad governor of occupied Spain, then advanced against the Frankish Merovingian kingdom but were repulsed by Charles Martel at Tours and Poitiers in 732.

Despite this setback, Abd-ar-Rahman went on to establish an independent caliphate based in Córdoba in 756, in the name of the Umayyads who had been overthrown in a civil war with their successors, the Abbasids, and he and his successors began a profound Arabization of the Iberian Peninsula whose effects persist to the present. For many hundreds of years, Spain and Portugal were a land, above all, of small, often self-governing urban communities known for their culture, comfort, and intensely cultivated gardens.

Even as the Arabs dug in, a Christian reconquest was brewing below the surface. In 718, Pelayo (690–737), the Visigothic Christian chieftain, established the kingdom of Asturia and won the Battle of Alcama near Covadonga against the invaders, by then named Moors. A second Spanish kingdom, that of Navarre, was established in the remote Pyrenees. Other small Spanish Christian kingdoms appeared in the eighth and ninth centuries and began slowly to recover land from the Muslims. Assisting them in their endeavor was Frankish intervention on the Spanish side of the Pyrenees, resulting in, among other things, the creation of a small number of marcher states (Aragon and Barcelona) under Frankish protection and the emergence of a major pilgrimage route stretching across northern Spain, from the frontier of Francia to the shrine of Saint James in Galicia. Not only did this route provide much needed revenue for the local Christian kingdoms, but participation in a common cult helped promote feelings of unity in Christian Spain.

The process of Christian self-assertion and then reconquest accelerated considerably in the tenth and eleventh centuries. From 930 to 950, Ramiro II (900–951) of the new kingdom of Leon in northwestern Spain, where it had taken over the former territories of Asturia, defeated the Córdoban caliph Abd al-Rahman III at Talavera, Simancas, and Osma. In 932, Count Fernan Gonzales (c. 912–970) declared himself king of Castile.

In the eleventh century, Sancho III (c. 992–1035), king of Navarre, recaptured a large part of Aragon and brought Castile under his power, but at his death, his kingdom was divided among his sons. His son Ferdinand became king of Castile (r. 1035–1065) and in 1037 king of Leon. He also captured Moorish Galicia. Ferdinand declared himself emperor of Spain in 1056 and officially began the reconquest. Toledo was recaptured by an alliance of Castile, Leon, and Galicia led by Alfonso VI (king of Leon from 1065 to 1109 and of Castile from 1072) in 1085. Later, the Kingdom of Aragon took, in succession, Huesca in 1096, Saragossa in 1118, Tortusa in 1148, and Lerida the following year. Almost simultaneously, Alfonso VII of Leon-Castile (1126–1157) turned Córdoba into a vassal in 1146 and also occupied Calatrava, Andujar, and Almeria. Farther west, the new kingdom of Portugal, founded in the late eleventh century, occupied Lisbon in 1147 with the assistance of passing crusaders.

In 1212, after a major Muslim counterattack under the Almohads, the kings of Aragon, Castile, Leon, Navarre, and Portugal defeated Caliph al-Nasir at the Battle of Navas de Tolosa. Jaime I (1208–1276) of Aragon reconquered Mallorca in 1229. From 1217 to 1252, Ferdinand III (c. 1198–1252) of Castile-Leon conquered Seville, Jaen, Murcia, and Córdoba, leaving only Granada and its region in Moorish hands. Its continued survival was more a reflection of squabbles among the now dominant Christian states and the convenience of Moorish allies than it was of any real power of Granada.

In 1469, Ferdinand II (1452–1516) of Aragon married Isabella (1451–1504), who became queen of Castile in 1474. The conquest of Zahara in 1482 quickly led to the conquest of Almah. Although defeated at Loa and Ajarquia, the Christian forces regained Coin, Almeria, Velez, Malaga, Basa, and Guadix. Moorish Granada, the last stronghold, surrendered on 2 January 1492. To ensure Christian conformity, some 180,000 Jews were expelled, and the Moors were required to become Christian. A new era of religious intolerance had begun. Their most Catholic majesties asked Pope Sixtus IV (1414–1484) for permission to establish the Inquisition in Spain. The *Reconquista,* the crusade, was complete.

Annette Richardson

See also: Almoravid Empire; Almohad Conquest of Muslim Spain; Charlemagne's Wars; El Cid, Rodrigo Diaz de Vivar; Frankish-Moorish Wars; Muslim Conquests; Tariq ibn Ziyad; Tours

References and further reading:

Collins, Roger. *Early Medieval Spain: Unity in Diversity.* 2 vols. London: Macmillan, 1983–1987.

———. *The Arab Conquest of Spain 710–797.* New York: B. Blackwell, 1989.

Paris, Erna. *End of Days: A Story of Tolerance, Tyranny and the Expulsion of the Jews from Spain.* Amherst, NY: Prometheus Books, 1995.

Shields, Graham J. *Spain.* Oxford, UK: Clio Press, 1985.

Taha, Abd al-Wahid Dhannun. *Muslim Conquest and Settlement of North Africa and Spain.* New York: Routledge, 1989.

Red Cross

Group of interdependent international humanitarian institutions and an internationally recognized humanitarian protective emblem. In 1859, the Swiss merchant Jean-Henri Dunant witnessed the Battle of Solferino and saw how the medical services of both parties completely failed to deal with the huge number of wounded soldiers. His account of his experience, titled *Un Souvenir de Solférino,* was published in 1862 and had tremendous impact. A committee of five Geneva citizens decided to pursue Dunant's idea of national volunteer medical services assisting their armies in times of war as well as an international treaty protecting both these services and the wounded soldiers.

This treaty was initiated by the committee in 1864 and ratified by the major powers as the Geneva Convention. It was revised in 1906 and especially in 1929, when its applicability was widened to protect prisoners of war (POWs) as well. After World War II, the convention was found to be insufficient in the face of a highly mechanized warfare and was completely reworked into four independent Geneva Conventions. In 1977, two additional protocols further expanded the conventions' field of application.

The group of Red Cross institutions is integrated into a more or less common organizational structure: The original Geneva Committee is now the International Committee of the Red Cross (ICRC). The ICRC's main task during peacetime is to uphold and disseminate Red Cross principles as laid down in the conventions as well as to monitor the correct implementation of the conventions by the contracting parties. Therefore, it maintains delegations abroad that keep in contact with national or local authorities. Additionally, the ICRC aims to further develop international legal instruments to reduce suffering during war and conflict. These drafts are presented to national governments during international conferences taking place every four years.

In times of armed conflict, the ICRC's main function is to trace POWs, civilian victims of war, refugees, and displaced persons by means of its own tracing agency. Furthermore, its delegates visit POWs and civilian internees. Possible violations of the conventions and protocols are reported to the party concerned but not investigated by the ICRC itself. Diplomatic, more than legal means, are then used to resolve any critical situation.

National Red Cross societies are set up as private associations under national laws. They organize emergency and relief actions within their national boundaries. All national societies are members of the International Federation of Red Cross and Red Crescent Societies (until 1991, the League of Red Cross Societies). When international coordination is necessary, the federation functions as a body of coordination between national societies and internationally promotes their humanitarian activities. Every four years, an international conference is held as the general assembly of the Red Cross movement. Delegates from the ICRC, the national societies, and the member states of the conventions attend the conference.

The first Geneva Convention introduced a red cross on a white background as the movement's emblem. All parties are obliged not to attack persons, vehicles, and buildings showing this emblem. The conventions also include the Red Crescent, in use by most Islamic states, and the Red Lion and Sun, originally used by Persia-Iran, but given up in 1980.

Refugees being fed by the Red Cross, 1940. (Library of Congress)

Currently, the adoption of a new emblem free of religious symbolism is under consideration.

Marcus Hanke

See also: Laws of War; Prisoners of War; War Crimes

References and further reading:

Boissier, Pierre. *From Solferino to Tsushim.* Geneva: Henry Dunant Institute, 1985.

Durand, André. *From Sarajevo to Hiroshima.* Geneva: Henry Dunant Institute, 1984.

Favez, Jean-Claude. *The Red Cross and the Holocaust.* Cambridge, UK: Cambridge University Press, 1999.

International Committee of the Red Cross, http://www.icrc.org.

Refugees and Victims of Ethnic Cleansing

Civilians displaced by war. In 1950, the United Nations established the Office of the United Nations High Commissioner for Refugees (UNHCR). Through Article I of the 1951 Geneva Convention Relating to the Status of Refugees, the international community defined "refugee" as a person "who, owing to a well-founded fear of being persecuted . . . is outside [their] country of nationality" In more common terms, refugees are those forced to leave their homes as a result of war or political oppression. Those who leave their homes but do not cross their country's international borders are known as "internally displaced persons."

As the twentieth century progressed, the numbers of refugees and internally displaced persons grew, so that it is a far larger problem worldwide than ever before. In Europe, for example, the war in Bosnia-Herzegovina generated 2.5 million displaced persons or refugees, out of a population of 4.5 million, during its first few months in 1992. Again, in Kosovo in 1999, there were 800,000 refugees and nearly another 1 million displaced out of 2 million inhabitants. But Africa led the world in the 1990s, with more than 10 million refugees per year displaced by wars and civil wars—"small wars"—in Sudan, Rwanda, Angola, Liberia, and Zaire (now the Democratic Republic of Congo), not to mention the lesser disruptions in Burundi, Mozambique, Somalia, Kenya, and Ghana. In each situation, UNHCR gathers information and coordinates the humanitarian assistance that these people need.

In addition to the general disruption caused by warfare, refugees are generated by ethnic cleansing and genocide.

Article II of the United Nations convention adopted in 1948 defined *genocide* as "any of the following acts committed with the intent to destroy, in whole or in part, a national ethnical [*sic*], racial or religious group, as such: (a) Killing members of the group; (b) Causing serious bodily and mental harm to members of the group; (c) Deliberately inflicting on the group conditions of life calculated to bring about its physical destruction in whole or in part; (d) Imposing measures intended to prevent births within the group; (e) Forcibly transferring children of the group to another group."

A perhaps lesser crime against humanity is ethnic cleansing. Here a particular ethnic group is terrorized until it flees from a locale. For example, in 1992 in Bosnia-Herzegovina, certain Bosnian Serbs adopted this strategy, initially against Bosnian Muslims or Bosnian Croats. The Bosnian Serbs sought to make it possible to form their own territory that could become part of a Greater Serbia. Croatia undertook a similar policy in order to regain control over certain areas within Croatia, such as the Krajina.

All know well the Nazi genocide against the Jews. But after the end of World War II, there were large-scale movements of people that constituted ethnic cleansing. Czechoslovakia expelled about 3 million Germans. The Allied powers redrew the borders of Germany, forcing many more millions of Germans to flee Silesia, East Pomerania, and Prussia.

In Asia during World War I, Turkey engaged in genocide against the Armenians. When India and Pakistan became independent of the British Commonwealth in 1947, 7 million Muslims and 5 million Hindus and Sikhs found themselves on the wrong sides of the borders and fled to their ethnic homelands. Of course, there are the ongoing problems in the Middle East regarding the Jews and the Palestinians. In 1947–1948, the creation of the modern state of Israel left about 750,000 Palestinians as nearly permanent refugees because no other Arab state would take them in. Palestinians seeking their own state today number 3.5 million.

The war over Kuwait caused more than 1 million refugees. First, most of the 400,000 Egyptians and 300,000 Palestinians who worked in Kuwait before the war fled. Later, the Kurds in northern Iraq suffered reprisals from Saddam Hussein's regime after the war. The pictures of hundreds of thousands fleeing to Turkey dominated the news immediately after Desert Storm.

Ethnic cleansing, of course, is not solely a twentieth-century phenomenon; the removal of American Indians to reservations certainly qualifies as such, as does the removal of the biblical Hebrews from Israel and Judea—whence the Hebrews had earlier displaced the Canaanites. Nonetheless, the twentieth century seems to have perfected and practiced the concept more than any other century.

Warfare, especially warfare directed specifically or indirectly against civilians, creates untold suffering. National policies in war and about war respond to these realities and require decisionmakers to count these costs as part of the military burden.

John R. Popiden

See also: Geneva Conventions; Guerrilla/Partisan/Irregular Warfare; Laws of War; Nongovernmental (Extranational) Organizations: Their Role in War and in the Wake of War; Nuremberg Principle; Red Cross

References and further reading:

Ager, Alastair, ed. *Refugees: Perspectives on the Experience of Forced Migration.* London: Pinter, 1999.

Bell-Fialkoff, Andrew. *Ethnic Cleansing.* New York: St. Martin's Griffin, 1999.

Cohen, Roberta, and Francis M. Deng. *Masses in Flight: The Global Crisis of Internal Displacement.* Washington, DC: Brookings Institution Press, 1998.

Riemer, Neal, ed. *Protection against Genocide: Mission Impossible?* Westport, CT: Praeger, 2000.

Religion and War

Cause, limiter, and encourager of war. At times, the belief systems of a particular religion have limited war, encouraged it, or even been the sole cause of it. Even in the supposedly dominantly secular twentieth century, world wars were caused or affected at least partially by religion.

Early human society had a variety of religious belief systems, often characterized as natural, animistic, or shamanistic. It is clear however, that these belief systems could have a significant impact on conflict. Head-hunting, a religious ritual involving the taking of the "power" of another warrior, could be seen from Neolithic times until the twentieth century in all the human-occupied continents.

Easier to measure are early written records of war. These records often contain a significant religious component. The *Rig Veda* of ancient Hindu India discusses all elements of formal warfare and the religious beliefs that influence such warfare. The Hebrew Scriptures do likewise; in them, for example, the Hebrews conquer the "Promised Land" of Canaan by the sword at the behest and with the blessing of Yahweh and are punished by him if they fall short in their military mission. And Yahweh himself is termed "A Man of War." Sumeria, the world's first civilization, appears to have had a particularly strong religious dogma to its warfare traditions. This dogma culminated in the Assyrian Empire, the world's first military state, with its need for conquest to satisfy its god, Assur. In ancient Egypt, the pharaohs named their military units after important Egyptian gods such as Amun,

Ptah, Seth, and Pre. In China, the literature generated during the Warring States Period shows the strong effects of religious belief on war.

Greco-Roman religious mythology obviously affected the ways of war in the ancient Mediterranean. From Greek gods fighting for control of the cosmos to Roman emperors attempting to divine the fortunes of war in the entrails of an animal or the flight of a bird, religion played a crucial role in the western way of war. One has only to look at the formal war god Ares/Mars of the Greco-Roman eras.

Nomadic horse societies on the Asian steppes also had their warfare practices strongly shaped by religious beliefs. Turko-Mongol groups, including famed leaders like Genghis Khan, felt they were empowered by sky gods to conquer the known world. Like those in the Greco-Roman world, they looked to auguries for help in deciphering the supernatural.

The end of the Roman Empire was marked by the emergence of Christianity in the fourth century. In the east, the Byzantine world embraced Eastern Orthodox Christianity, while Rome developed Roman Catholicism. Such splits in religious beliefs within a formal religion often led to war, as was the case with Christianity and later with Islam. Religion tends not only to promote and motivate war between differing religions and regions but also within differing sects or groups in a single religion and region. Far from being a unifier, religion has often served, particularly within religious groups themselves, to create strife.

Even basic tenets of why and how a people engaged in war could be structured by religious beliefs. A Norseman's very religious belief suggested he must die in battle to attain immortality, which in turn allowed him to fight and die again and again in the heavenly hall of warriors. Viking religious beliefs, like those of the Aztec and Assyrian, structured the reasons they carried out warfare in the first place.

Such military-religious influence perhaps reached its zenith in the Middle Ages. Formalized monotheistic religions such as Christianity and Islam had special orders of warriors and warrior-priests who helped spread the message of the religion via potentially or actually coercive means. The Knights Templar and Hospitaller are just a few of many examples.

Three models of the blending of religion and war via leadership emerged during the Middle Ages around the globe. Great medieval military leaders such as Constantine, Charlemagne, and al-Rashid clearly carried the banner of religion before them and used it to rally, recruit, and inspire troops. In other cases, religious leaders of the period, such as the Roman Catholic popes, the caliphs in the Islamic world, or the Mayan priest-rulers, transformed themselves from religious leaders into military leaders as situations demanded. In other cases, warrior-priests like Quetzalcoatl of the Toltecs or Tlacaelel of the later Aztecs often wore both the religious and military hats at the same time. In yet another model, military leaders like Moses or Constantine could later become religious leaders.

In several parts of the globe at the end of the Middle Ages, the emergence of gunpowder weapons helped to delineate the beliefs of cultures toward lethal warfare and the sharing of power in class societies. It can be argued that at least some of these beliefs had a relationship with basic religious tenets in those cultures and societies.

In China, the birthplace of gunpowder weapons, and in the surrounding East Asian realms, both land and sea warfare had already integrated gunpowder weapons by the thirteenth century. Mongol conquests began to export such concepts throughout the Old World. Yet China pulled back from the world's first industrial age and the associated gunpowder revolutions, perhaps because of the hold of such religious beliefs as Daoism, Confucianism, and Buddhism.

The Ottoman Empire also never fully integrated gunpowder weapons into its culture and society. European gun casters were brought in to construct the huge siege guns used by the sultan finally to capture Constantinople in 1453. Janissaries, Balkan slaves, or tribute children were trained in volley-fire musket tactics, but their Turkish overlords often detested the use of gunpowder weapons, considering them a necessary evil that could pollute Turkish culture.

In Japan, the religious-feudal Samurai system was overthrown by Nobunaga Oda's adoption of Portuguese muskets. His system involved Samurai sword makers producing European-style muskets and then training Japanese peasants in volley-fire tactics. The results gave Oda power throughout Japan, with his peasant, volley-fire forces scoring a stunning victory over the medieval samurai at the Battle of Nagashino in 1575.

However, no sooner did the old order of Japan crumble than questions began to arise about the increased wielding of power by the peasantry. The castles were torn down, the guns removed from the hands of peasants and elites, and the samurai sword makers were prohibited from manufacturing gunpowder weapons.

Why could the Chinese, Turks, and Japanese so easily control the spread of gunpowder weapons in their regions, while Europe proved unable to do so, despite early Christian doctrine that guns were "weapons of the devil"? Underlying religious beliefs may have played a role, but not in an easily explained fashion.

In the Western Hemisphere, early civilizations like the Olmec, Zapotec, and Maya appear to have waged war with heavily religious overtones. A sacred book of the Maya, the *Popul Vu,* even discusses biological warfare. At the Zapotec capital of Monte Alban, depictions of head-hunting and sacrifices of war captives appear to foreshadow the industrial-level war captive sacrifices encountered by the Spanish. In classical Maya times, the goal of warfare itself was the capture of enemy religious and military leaders, along with their religious banners. In Africa, Zulu- and Bantu-speaking warriors in general often underwent a "washing of the spears" ceremony to cleanse their spirits after they had killed an enemy warrior.

North American Indian tribes often relied upon shamans to bless shields and warriors, making them immune to the bullets of European invaders. Even in historic times, such Native American blending of warfare and religion serves as an excellent model of how early tribal societies around the globe practiced the integration of religious belief and warfare practice.

Europe experienced the impact of religion in war as much as any region on the globe, with the possible exception of the Islamic world. During the early modern period (about 1500–1650), Europe was torn apart by strife within Christendom.

Such intrareligious conflict was really nothing new in Christendom, as competition between the Roman Catholic Church in Rome and the Eastern Orthodox Church in Constantinople had been taking place since the time of Constantine in the fourth century. But the integration of new technologies, namely gunpowder weapons, into warfare greatly increased the lethality of warfare and allowed the common person to seize some of the reigns of power from the old medieval elites, who had previously controlled the very expensive art of war.

By the time of the early modern period, religious leaders in Christendom, the popes included, were skilled in utilizing warfare as a tool to achieve power. The very concept of a Holy Roman Emperor had been developed by Charlemagne in 800 as a way to combine church and state power.

After the Christian sacking of Constantinople in 1204 during the Fourth Crusade, Rome became the center of Christianity and international politics for Europe. However, in the thirteenth century, the coming of the Mongols and the Black Death served to call into question the competence of the papacy. The Renaissance and the discovery of new worlds further questioned its doctrine. The stage was set for a series of challenges to papal authority that became known as the Wars of Religion.

The Spanish Habsburg Empire allied with Catholicism and the pope, as did their Austrian Habsburg relatives. The French Bourbons often stayed in the middle, while Dutch, English, and Scandinavian realms sided with Protestant communities. Eastern Europe added an additional variable with its Eastern Orthodox background. Eventually, the Protestant realms and their north Atlantic sea power came to dominate the more Catholic Mediterranean land powers. However, whether it was the Thirty Years' War in Europe, or the Seven Years' War abroad (called the French and Indian War in North America), religion played a key role as a motivator, organizer, and point of contention.

Even after the close of Europe's Wars of Religion, sectarian strife continued, although usually bound more closely with rivalries between the emerging nation-states. This type of political-religious clash was most evident in the wars between England and Spain. As late as 1745, France assisted Charles Edward, the "Young Pretender," in a vain attempt to restore a Catholic monarch on the English throne.

By the 1800s, religious wars had given way to a second phase of European colonialism and warfare. Religious-sponsored and -inspired warfare now became wars between nation-states. The first half of the twentieth century was scarred by gigantic world wars waged by nation-states, conflicts that cut across religious lines and reflected the religious indifference of at least Europe. Yet in World War II, the officially atheist Soviet Union actually called upon the once-persecuted Russian Orthodox Church to bless its cause. Still, among the major belligerent powers in both world wars only Russia (in World War I) and Japan (in World War II) were highly motivated by religion.

The post–World War II era saw a resurgence of religion in warfare, if not of actual wars of religion. In India, the newly independent Indian state (1947) had to deal with murderous tensions between Hindus, Muslims, and other religious sects. This tension led to the assassination of Mohandas Gandhi, mass religious-ethnic killings, and the spawning terrorism on all sides until the present day. The Muslim state of Pakistan had to be formed in the west of India, to separate Indian Hindus and Pakistani Muslims. The two nations have fought each other twice since 1947, and their borders remain among the most unstable on Earth.

The formation of Israel as an independent Jewish state after World War II (1948), created similar hatreds in the mostly Islamic Middle East. The basically secular state of Israel has fought numerous wars and been engaged in almost constant strife with its Islamic neighbors, including Egypt, Syria, and the Palestinian people. Many of Israel's enemies have proclaimed a jihad (holy war) against Israel. This strife has spilled over to the United States, as numerous hijackings of American airliners and the horrendous bombings of the World Trade Center towers in New York City and the Pentagon in Washington, D.C., have shown. Islamic militants, outraged by U.S. support for Israel and by its perceived anti-Islam bias, made no secret of their role in these "holy wars." It may well be that the early twenty-first century will be dominated by what most developed nations will call a "war against terrorism" but what Islamic zealots already term a jihad for Islam.

To a lesser extent, religious beliefs have also contributed to ethnic conflicts in the Balkan states, West Africa, central Africa, East Timor, Zimbabwe, Southeast Asia, and Northern Ireland (probably the closest thing to an actual war of religion in the twentieth to twenty-first centuries). Clearly, conflicts such as those in Ethiopia and the Sudan also have strong overtones of the age-old competition between Christianity and Islam.

Certain religions, particularly in the West, have emphasized pacifism, and their young men and women have refused to go to war. The most prominent of these religious groups are the Germanic Anabaptists and Jehovah's Witnesses. Although both groups were persecuted somewhat for their stand during World War I, by World War II, their position had been accepted in the United States and the British Commonwealth. Provision was made for such people eligible for induction into the military to undertake some form of alternate service, ranging from battlefield stretcher-bearers to home-front smoke-jumpers, hospital workers, and so on. Although this consideration for small religious groups outside the mainstream churches can be considered a commendable attribute of the democracies, more cynical commentators have pointed out that the Western nations could afford such leniency precisely because these groups are so small in numbers that they pose no real threat to the military.

World history has seen everything from military religious orders to human head-hunting, human sacrifices, ethnic cleansing, terrorism, and mass killings of civil populations carried out by devotees of religion. Although most of the world's major religions today emphasize the amelioration of suffering and charitable obligations, along with more formal sacerdotal practices, religion has its dark side of intolerance and has been present to some degree in perhaps a majority of the world's conflicts throughout history.

Christopher Howell

See also: Crusades; French Wars of Religion; Muslim Conquests; Thirty Years' War

References and further reading:

Keegan, John. *A History of Warfare.* New York: Alfred A. Knopf, 1993.

O'Connell, Robert. *Ride of the Second Horseman.* Oxford, UK: Oxford University Press, 1995.

Rogers, Clifford J., ed. *The Military Revolution Debate.* Boulder, CO: Westview Press, 1995.

Reporting, War

The shaper of public opinion. It is arguable that war reporting, like military history, begins with Thucydides, in that he was a participant-observer trying to make sense of a military conflict for a contemporary audience. As such, he remained a paradigm into the nineteenth century, when public opinion in the Western world was as much influenced by privately produced pamphlets commenting on the course of political and military circumstances as it was by the press. Newspapers were sufficiently widespread in Britain and the American colonies by the time of the American Revolution (albeit unabashedly partisan) to influence public opinion in both countries.

The state of affairs in which professional news reporters cover events, instead of publishers depending on the ad hoc reception of personal accounts, really emerged with the Mexican-American and Crimean Wars. Responding to the beginnings of mass politics (in turn a result of the Industrial Revolution's railroads, steam shipping, and telegraphs), American and British newspapers created organized courier systems, putting steamships and the telegraph at the disposal of paid correspondents seeking the "scoop," the novel, sensational story that would bolster their employer's reputation for immediacy and move copies of the latest edition. The climax of this style of competitive "newsmongering" was probably best represented by William Randolph Hearst's supposed admonition to artist Frederic Remington, sitting bored and tired in Cuba in 1898: "You furnish the pictures and I will furnish the war."

An important transition came with World War I, when the natural desire of governments to maintain the operational security of their respective war efforts took the additional turn of seeking to use the mass media as a conduit of advocacy for the general public. The goal was the creation of general enthusiasm for what is now referred to as "total war." Thus began the era of formal propaganda.

In response to this trend, plus the brutal disenchantment with romantic notions of war in the face of modern industrial combat (now being captured in still and motion pictures) and the rapidly increasing shock of a public faced with rocketing casualty lists, a more critical style of journalism began to emerge. In the democracies, where there was a tradition of respect for public opinion, the burgeoning distrust of official news created an increased demand for truth telling, a trend that continued into the twenty-first century.

Allied with this demand was another continuing trend, the new perceptions brought about by the direct visual record of war via tools such as the camera and television. The apparently unmediated record of the photo, be it Matthew Brady's record of the dead on American Civil War battlefields, Robert Capa's alleged snapshot of a Spanish Civil War soldier at the moment of death, or today's stream of television imagery from the world's ongoing military conflicts, serves to create new public opinion pressures.

The chronic question of war reporting is, for what and whom do the media advocate? Is it the general public, the government, the soldier in the field, or perhaps simply the business interests of the employer of the working correspondent? All of these rationales have been invoked at one time or another, but a balanced amalgam seems to be most satisfying for all concerned, if the hardest to attain.

Perhaps conflict between the media and government is the inherent state of affairs that must simply be accepted. World War II did see the creation of an image of war reporting at its best, when professional and committed reporters operated in concert with government support to cover the reality of the war and, if not bring meaning, at least act as honest witnesses to events. Some witnesses, such as Ernie Pyle in print and Edward R. Murrow in sound broadcasting, became heroes in their own right.

In that global conflict, motion pictures expanded exponentially from their beginnings in World War I (whose cinematic record for the most part consists of marching troops). Combat camera operators were routinely expected to do their work *in advance* of the troops. Allied camera operators were instructed never to "stage" battle scenes, and for the most part, they followed this injunction. Even the German army's combat camera operators were able to film authentic combat, with a very light propaganda larding. Paradoxically, the most authentic battlefield scenes were the least rewarding when shown to the folks on the home fronts; in twentieth-century war, troops are spread out, fighting in interlocked small groups—not very exciting film but all the more authentic.

For the first time, also, war reporting could be instantaneous, and correspondents' actual voices could be heard, as when radio correspondent Edward R. Murrow's tobacco-cured voice opened his report to New York from a London rooftop during the Battle of Britain with the magnificent: "This . . . is London." Robert Trout, Richard C. Hottelet, or Eric Severeid could record their impressions from bombers over Berlin, from the D-Day beaches, or from just-liberated Nazi concentration camps. Their voices and their faces became about as well-known to Americans and Britons as those of film stars.

The lesson was learned in World War II, at least by the democracies, that the destruction of credibility is the loss of a weapon. The British Broadcasting Corporation's reputation for accuracy was such that even those who wished Great Britain ill would listen in to get some true picture of what was happening. By contrast, the German information and propaganda machine was justly and widely distrusted, particularly in the Third Reich itself, and its head, Dr. Josef Goebbels, was widely known as one who would give lying a bad name.

But some two decades later, the arrival of television and the exigencies of the Cold War's ideological conflict created a crisis of advocacy. Journalists who saw their duty as that of bringing an accurate portrayal of events (or at least capturing the scoop) again came into conflict with governments and militaries seeking to put the best interpretations on policies that appeared to be counterproductive at best and pointlessly destructive at worst. War reporters in the democracies came to look upon their profession as paramount, and if their reporting gave aid and comfort to their nation's adversaries, so be it. The fact that these correspondents were not recording declared wars but "twilight conflicts" like Korea, Algeria, or Vietnam made it easier for them to put their craft above what might be considered their country's interests and even to intrude their political views. In the case of the Vietnam conflict, this tendency created a situation in which, in the United States, the world's most active news industry and the world's leading military came to view each other in an adversarial light. The military reciprocated the distrust of the news media, as may be seen in the possibly apocryphal story of correspondents who complained that they had been kept in the dark about the anti-Iraqi coalition's war plans for DESERT STORM (1991), but that war correspondents had gone ashore on D-Day at the Normandy beaches. Their military "handler's" response cut to the point: "But we knew back then that you were on our side."

By the late 1960s, television had brought near-instantaneous reporting "into America's living rooms," as the cliché went. The process of making celebrities of correspondents accelerated, with Walter Cronkite regularly topping any poll of the most trusted man in the United States in the 1960s. When Cronkite reported from Vietnam in the wake of the Communist Tet Offensive that he did not see how the United States could win this war, public opinion began to move away from support for the conflict, even though Tet was in reality a Communist defeat.

In some ways, there has been a return to the situation of the late nineteenth and early twentieth centuries and the era of "yellow" journalism, in which sensation has become the main criterion of value to the business interests of contemporary media corporations. What does seem to be unchanged is that there are limits to the obvious manipulation of public opinion, at least in the democracies. If the general public remains a viable force in politics, it would seem that there will be no shortage of war reporters seeking to bring the experience of war back to the public forum. Perhaps the next golden age of war journalism will be made by independents using the Internet.

George R. Shaner

See also: Film and War

References and further reading:

Desmond, Robert W. *Tides of War: World News Reporting, 1940–1945.* Iowa City: University of Iowa Press, 1984.

Hudson, Miles, and John Stanier. *War and the Media: A Random Searchlight.* New York: New York University Press, 1998.
Knightley, Phillip. *The First Casualty.* New York: Harcourt Brace Jovanovich, 1975.
Lande, Nathaniel. *Dispatches from the Front: News Accounts of American Wars, 1776–1991.* New York: Henry Holt, 1995.

Resaca de la Palma (9 May 1846)

U.S. victory that shook Mexico's confidence in its military superiority over the United States, used by President James Polk to gather support for a declaration of war on Mexico. Prior to Polk's declaration of war issued on 14 May 1846, General Zachary Taylor fought the Battles of Palo Alto and Resaca de la Palma on his way to relieving Fort Texas on the Rio Grande.

Taylor pursued the retreating Mexican forces on the morning following the Battle of Palo Alto. Reconnaissance reported that Major General Mariano Arista had fortified a dry riverbed approximately 5 miles south of Palo Alto. With dense chaparral before him and artillery deployed on the banks behind, General Arista positioned his 3,000 troops within the 8-foot-deep *resaca* that bisected the road to Matamoros and Fort Texas.

Taylor ordered his infantry to attack through the chaparral, which quickly broke the soldiers into small, disorganized groups. A battery of artillery was sent down the road by Taylor to engage the Mexican gunners. Unsupported by infantry, the battery called for help. A charge by dragoons penetrated the Mexican line but was unable to hold the guns. Assaults by the infantry on the east flank captured Arista's guns. Simultaneously, troopers on the west turned the Mexican flank and sent the defenders in a panicked retreat across the Rio Grande.

Taylor pursued to the Rio Grande and relieved Fort Texas, under siege since 3 May. Of his 1,700 troops engaged, Taylor suffered 33 dead and 89 wounded.

Lincoln Bramwell

See also: Mexican-American War; Taylor, Zachary

References and further reading:

Bauer, K. Jack. *The Mexican War, 1846–1848.* Lincoln: University of Nebraska Press, 1993.
McCaffrey, James M. *Army of Manifest Destiny: The American Soldier in the Mexican War, 1846–1848.* New York: New York University Press, 1992.
Singletary, Otis A. *The Mexican War.* Chicago: University of Chicago Press, 1960.
Weems, John Edward. *To Conquer a Peace.* New York: Doubleday, 1974.

Reserves

Troops with some military training and practice that are not always in arms. Before the modern period, militias and yeomanry supplemented or substituted for professional armies and could properly be considered reserves. In fact, all able-bodied males were considered a mass reserve in medieval Europe, to defend against Saxon, Viking, Saracen, or other invaders or raiders. The difference between these "reserves" and those who followed the profession of arms more often than not came down to who could afford a horse. Horses were expensive, high-maintenance animals and almost useless in the peasants' small strip and plot holdings. These peasants were invariably arranged in foot soldier units, while the yeomanry (small landowners) could afford horses and generally served in cavalry units.

Responding to acute manpower needs in the Napoleonic Wars, Prussia instituted a system of universal conscription in which subjects would be called to the colors when reaching a set age, would serve a year of active duty that included basic training, two years of active reserve with continuous training, 12 years in the first-line reserve with periodic training (Landwehr), and eight years in the second-line reserve (Landsturm). In the Franco-Prussian War (1870–1871), this mode of organization furnished a mass army that quickly overwhelmed the French army, a long-service professional army that had only recently adopted a pattern of five years' active duty followed by four in reserve.

This example led other nations to adopt mass conscription and training with a reserve component. Most schemes organized reserve units on a territorial basis for ease of training and mobilization and envisioned the active reserve fleshing out frontline units on mobilization, as first-line reserves manned depots and guarded lines of communication while being further trained prior to being drawn on for combat replacements. Second-line reserves were intended for home defense and police duty, as necessary. In the German army, however, first-line reserves were trained to function as frontline troops immediately upon mobilization, providing the manpower for the immense enveloping attack launched on France in 1914.

Britain began experimenting with reserve schemes after 1870. This decision resulted in 1908 in the division of the British army into regulars and territorials, the former supplemented by a special reserve from the militia and the latter intended for home defense but able to volunteer for overseas service. In World War I, the territorials contributed 24 divisions that fought abroad. In the United States, the Officers Reserve Corps and Enlisted Reserve Corps were established in 1916. Though the Enlisted Reserve Corps never attracted many recruits, the Officers Reserve Corps accounted for 88,000 officers in World War I and 200,000 in World War II. The U.S. National Guard was state-oriented, under the control of the governor until activated into federal service.

After World War II, the changing military missions of the various states made for alterations in reserve configurations. France adopted a national service of 18 months, followed by reserve service until age 45 and civil defense service until age 60. Britain assigned antiaircraft defense to the territorials until 1955, then downsized the force until merging it with the Army Emergency Reserve into a Territorial Auxiliary and Volunteer Reserve of 50,000 to supplement the army in small conflicts. The United States activated individual reservists rather than units in the Korean War and maintained only one-third of the authorized reserve strength. Increasingly, the National Guard became the combat reserve, and the Army Reserve supported service units. After the draft ended in 1973, the reserves dropped from 1.2 million to just over 300,000. By the Gulf War, reserve strength was nearly 900,000 and was integrated in a total force policy that made reserve mobilization necessary in case of war.

In most developed nations, a tension has long existed between the regulars and the reserves, the former viewing the latter as slack, politics-infested "weekend warriors" and the latter scorning the former as narrow, tradition- and regulation-bound types who could not make it in the civilian world ("real life"). Britain and the United States have to some extent successfully addressed these antagonistic views with their post–World War II reserve policies.

Joseph M. McCarthy

See also: Conscription

References and further reading:

Baldwin, Stanley S. *Forward Everywhere: Her Majesty's Territorials.* New York: Brassey, 1994.

Crossland, Richard B., and James T. Currie. *Twice the Citizen: A History of the Army Reserve, 1908–1983.* 2d ed. Washington, DC: Office of the Chief, Army Reserve, 1997.

Duncan, Stephen M. *Citizen Warriors: America's National Guard and Reserve Forces and the Politics of National Security.* Novato, CA: Presidio, 1997.

Hill, Jim Dan. *The Minuteman in Peace and War: A History of the National Guard.* Harrisburg, PA: Stackpole, 1964.

Wilson, Bennie J. *The Guard and the Reserve in the Total Force: The First Decade, 1972–1983.* Washington, DC: U.S. Government Printing Office, 1983.

Revolutions of 1830 (July–August 1830)

Uprisings that led to significant government changes in France and the independence of Belgium. The restored Bourbon monarchy of Charles X of France was, by 1830, deeply disliked by liberals for the king's attempts to install an ultraroyalist government and set aside the charter granted by his predecessor, Louis XVIII, in 1814. In July 1830, the king suspended the national guard and the privilege of the middle class and called for new elections to the chamber of deputies while restricting the franchise and censoring the press. While middle-class liberals wanted to protest by refusing to pay taxes, the common people of Paris began forming unruly mobs that attacked royalist symbols and clashed with the city gendarmes. The king's general, Marechal Auguste-Frédéric de Marmont, who was particularly disliked as a former Napoleonic general who had surrendered Paris to the allies in 1814, wanted to move on the mobs immediately, but the king hesitated. This delay allowed the mobs to take up barricades in strategic areas of Paris and former national guardsmen to join in the rebellion. Because the army in Paris was arrayed to protect the king rather than quell disturbances, Marmont's eventual three-column attack on the rebels failed because the columns were isolated and attacked within the city and forced to retreat. The national guard attempted to protect private property but could not keep the mobs from attacking the archbishop of Paris's house and storming the Louvre. While liberals called for Louis-Philippe, a cousin of the king, to take the throne, a mob marched on the king at Rambouillet and forced him into exile. The marquis de Lafayette, accompanied by Louis-Philippe, quieted the revolutionaries in Paris, and Louis-Philippe accepted the throne from the chamber of deputies, who refused Charles X's attempts to abdicate in favor of his nephew, a minor child. The revolution starkly illustrated the inability of the army or the police to keep order in the old city streets of Paris and led to a rethinking of French tactics.

In August, revolution spread to Belgium, under Dutch rule since the Congress of Vienna (1815). After riots by agricultural and industrial workers, which had been put down by civilian authorities in Brussels, the Dutch were not alarmed by a spontaneous rising following a nationalistic French opera, *La Murette de Portici,* and delayed sending troops until Brussels was rife with revolutionaries. A clash between Dutch troops and rebels in a Brussels park ended with 400 rebels and 750 Dutch dead, an enormous toll that led to the Dutch evacuation of the city and the establishment of a provisional, French-speaking, Catholic government. This government proclaimed the independent state of Belgium on 21 July 1831, which then crowned Leopold of Saxe-Coburg-Gotha its first king.

Margaret Sankey

See also: Revolutions of 1848

Fighting at the barricades in Paris during the Revolution of 1848. (Library of Congress)

References and further reading:

Collingham, H. A. C. *The July Monarchy.* London: Longman, 1988.

Kossmann, Ernst Heinrich. *The Low Countries: 1780–1940.* Oxford, UK: Clarendon, 1978.

Pinkney, David H. "Pacification of Paris: The Military Lessons of 1830." In *1830 in France,* ed. John Merriman. New York: Franklin Watts, 1975.

Revolutions of 1848 (12 January 1848–13 August 1849)

Series of liberal uprisings throughout Europe. Immediate causes included the famine and economic depression that had gripped Europe since 1846; years of oppression by the Bourbons, Habsburgs, Hohenzollerns, and other restored or post-Bonapartist monarchies; lingering republican sympathies from the French Revolutions of 1789 and 1830; feelings of ethnic solidarity and nationalism; and the ideologies of such firebrands as Karl Marx, Mikhail Aleksandrovich Bakunin, Arnold Ruge, Pierre-Joseph Proudhon, Lajos Kossuth, Ferdinand Lassalle, George Sand, and Giuseppe Mazzini.

The first revolt occurred in Palermo, Sicily, on 12 January 1848. Workers skirmished with police in Paris on 22 February. French king Louis-Philippe abdicated on 24 February and fled to England. Liberals, republicans, and nationalists rebelled in several Rhineland cities on 3 March, Vienna on 12 March, Berlin on 15 March, Milan on 18 March, and Venice on 22 March. Because so many of the 1848 revolutions began in March, historians call the period of European monarchial revival between 1815 and 1848 the *Vormärz* (German for "pre-March").

The sites of the major revolutions were France, Italy, Germany, Austria, and Hungary. Scattered minor insurrections arose in Ireland, Scotland, England, Poland, Greece, and the Balkans. Democrats managed to achieve peaceful reforms within the established order in Denmark and the Low Countries. Only the Russian, Spanish, and Scandinavian monarchies remained unchallenged by revolutionary activities.

In France, the provisional republican government, consisting mainly of moderates and conservatives, immediately came into conflict with the far left. The assembly of this government gave dictatorial powers to General Louis Eugène Cavaignac to restore order. His national guard and about 100,000 conservative provincial volunteers killed between

5,000 and 15,000 rebels in the streets of Paris during "Bloody June" while losing only about 1,000 of their own forces. Cavaignac's government wrote the constitution of the Second Republic, but Cavaignac lost the presidential election in December to Louis Napoleon (Napoleon I's grandson), who gradually consolidated his authority until, in 1852, he could proclaim the Second Empire with himself as Napoleon III.

In Austria, the mob forced the resignation of Emperor Ferdinand I's powerful regent, Prince Klemens Wenzel Nepomuk Lothar von Metternich, on 13 March. The empire seemed ready to dissolve. Shortly after dissatisfied Bohemian subjects revolted in Prague on 12 June, Austria and other royalist governments launched sustained counterrevolutionary attacks. By fall, they were succeeding throughout Europe, most notably in the Austrian Empire and its satellites. Field Marshal Alfred Windischgraetz recaptured Vienna from the revolutionaries on 31 October. But Ferdinand I proved incompetent to aid in the restoration of his authority. Because he was unable to reestablish civil order, recover lost Italian provinces, or subdue Hungary, he abdicated on 2 December in favor of his nephew, Franz Josef I, who reigned until his death in 1916.

Unchecked rioting erupted in Berlin on 18 March. The most powerful German prince, Prussian king Friedrich Wilhelm IV, after eight years of failing to pursue liberal reforms instituted by his father, Friedrich Wilhelm III, was compelled on 22 March to promise a constitution for a representative government. Yet the violence continued throughout the German realm, mostly as urban riots. In Frankfurt in the spring of 1848, the philosopher Arthur Schopenhauer invited soldiers into his rooms to fire down on the insurgents and lent his opera glasses to an officer for reconnaissance. Following the success of the German counterrevolution, huge numbers of constitutionalists and liberal sympathizers fled. Many settled in the United States, where they became known as "48ers."

Nationalism was the primary fuel of revolution in Italy, where rebels of all political leanings sought to throw off Austrian, French, and papal influences in order to create a unified country. Austria dispatched Field Marshal Josef Wenceslas Radetsky to suppress anti-Austrian uprisings in northern Italy. Radetsky defeated King Carlo Alberto of Sardinia at Custozza on 24–25 July 1848 and captured Milan on 9 August. After the loss of Milan, the fervent Italian nationalist Giuseppe Garibaldi briefly waged guerrilla warfare against the Austrian presence in Italy, until he was driven into Switzerland. Returning to Rome in December, he took command of Italian resistance to French incursions. He won at Rome on 30 April 1849, Palestrina on 9 May, and Velletri on 19 May, but the French captured Rome on 30 June. Garibaldi and 4,000 guerrillas retreated to San Marino. After losing to Radetsky at Novara, Carlo Alberto abdicated on 23 March in favor of his son, Vittorio Emanuele II, who would in 1861 become the first king of Italy. Radetsky secured Venice for Austria in the spring and summer of 1849. The restoration of papal political sovereignty over Rome in 1849 forced Mazzini and his republican nationalist movement, "Young Italy," into exile.

Kossuth made a famous speech in the Hungarian Diet on 3 March 1848, agitating for Hungarian independence from Austria. He took increasing advantage of Austrian disorder to establish Hungarian home rule that summer. Emboldened by Kossuth's defiance of Austria but unwilling to be part of a Hungarian republic, Serbs, Croats, Czechs, Slovaks, Romanians, and other ethnic minorities, mostly Slavic, demanded their own separate nations and revolted against both Hungary and Austria, thus weakening Kossuth. Croatia, especially, was so active against Hungary that Kossuth had to appeal to Austria for help in September. But Croatian general Josip Jelacic was in the service of Austria. Jelacic invaded Hungary on 19 December and captured Budapest on 4 January 1849. Windischgraetz won at Schwechat and Käpolna but lost at Gödöllö and was relieved of command. Hungary declared full independence from Austria on 14 April. Artúr Görgey, commander in chief of the Hungarian armies after February, was very successful against Austria that spring, finally recapturing Budapest on 21 May. Austria requested assistance from Russia. The revolutions ended when Kossuth fled into exile on 11 August and Görgey surrendered to the Russians under Prince Ivan Feodorovich Paskevich at Világos on 13 August.

Eric v. d. Luft

See also: Bismarck, Otto von; Garibaldi, Giuseppe; German Wars of Unification; Italian Wars of Unification; Revolutions of 1830

References and further reading:

Jones, Peter S. *The 1848 Revolutions.* London: Longman, 1991.
Langer, William Leonard. *The Revolutions of 1848.* New York: Harper & Row, 1971.
Sperber, Jonathan. *The European Revolutions, 1848–1851.* Cambridge, UK: Cambridge University Press, 1994.
Stearns, Peter N. *1848: The Revolutionary Tide in Europe.* New York: W. W. Norton, 1974.

Rhodes, Sieges of (1480 and 1522)

The island of Rhodes, off the southwestern coast of Turkey, had been taken by Christian crusaders and fortified with medieval-style castles. These castles were constructed with stone recycled from the toppled "Colossus of Rhodes," a statue that once stood astride the great natural harbor as one of the "seven wonders of the world." On 23 May 1480, the

Ottoman Turks under Meshid Pasha combined with a fleet of 160 ships and besieged the fortifications, defended by the Knights of St. John of Jerusalem under the Grand Master Pierre d'Aubusson. The siege lasted three months, with the Turks withdrawing after the second direct assault, having sustained casualties of 10,500.

The second siege of Rhodes took place on 28 July 1522. Süleyman the Magnificent led the Turkish attack, and the knights were led by Villiers de L'Isle Adam. The Christian order held out until 21 December, finally surrendering to famine. The Turks lost as many as 60,000 men but gained a significant port in the Mediterranean Sea. Reportedly, explosive bombs were successfully used for the first time in this siege.

Christopher Howell

See also: Constantinople, Siege of (1453); Ottoman Empire; Süleyman the Magnificent

References and further reading:

Goodwin, Jason. *Lords of the Horizons: A History of the Ottoman Empire.* New York: Owl Books, 2000.

Setton, Kenneth. *Venice, Austria and the Turks in the Seventeenth Century.* Philadelphia: American Philosophical Society no. 192, 1991.

Bust portrait of Richard I, King of England. (Library of Congress)

Richard I (1157–1199)

King of England, crusader. Richard the Lionhearted was born 8 September 1157, the second son of Henry II of England and Eleanor of Aquitaine. His education, heavy on martial training, especially in horsemanship, at which he excelled, was significantly tempered by training in music and the social graces, including poetry. After being named duke of Aquitaine in 1172 and assuming his mother's lands and title, Richard honed his military skill while keeping order among the fractious barons and engaged in the war his elder brother Henry sparked against their father. When Henry died in 1182, Richard became heir to the throne, but continued to fight against Henry II on the side of his friend, the French king Philip II Augustus, until becoming king of England himself in 1189. Richard's first concern was to launch a crusade to retake Jerusalem from the Muslims, and he spent nearly a year arranging the logistics of the massive project, including making treaties with the kings of Aragon and Navarre.

Leading an army of perhaps 500,000, Richard journeyed to the Holy Land via Sicily, where he captured Messina and intimidated the Sicilians into releasing his sister Joanna, the widow of Frederick of Sicily. Moving on to Cyprus, he seized it by force from Isaac Comenus Ducas to use as a base. In both cases, Richard failed to accept the tacit truce among Catholic, Orthodox, Jew, and Muslim in these cosmopolitan lands.

Richard was successful at taking and holding Acre, but the disunited Christian crusaders quickly fell out among themselves, as he then moved on Jerusalem via the coastal road. Attacked by Saladin at Arsuf, Richard was victorious but pragmatically realized that he could not take Jerusalem with his available resources. While maintaining a scrupulously polite relationship, Richard and Saladin, the Muslim commander, engaged in a chess game of holding and relieving key fortresses like Jaffa and Ascalon until 1192, when both realized that a truce was the only solution to their stalemate, exacerbated by epidemic illness and the scorching heat.

Returning to England, Richard was captured by Leopold of Austria near Vienna and ransomed by Eleanor of Aquitaine through a massive tax on English property, while John, Richard's brother, assumed power in England. Richard, finally free, returned briefly to England before crossing back to France to retake the land lost to Philip II Augustus while he was on the crusade. In 1194, he began the construction of Chateau-Galliard on the Seine, a massive example of military fortification.

While marveling at the bravery of one of the defenders of Chalus, Richard was wounded in the shoulder in April 1199 and died of gangrene. Bigoted, violent, and ruthless to prisoners and rival Christian princes, Richard nonetheless won a reputation for bravery and chivalry that has passed into legend.

Margaret Sankey

See also: Crusades; English Civil War (1215–1217); Henry II, King of England; Jerusalem, Siege of; Philip II Augustus; Saladin

References and further reading:

Gillingham, John. *Richard Coeur de Lion: Kingship, Chivalry, and War in the Twelfth Century.* London: Hambledon Press, 1994.

———. *Richard I.* New Haven, CT: Yale University Press, 1999.

Ramsay, James Henry. *The Angevin Empire.* New York: AMS, 1978.

Richard III (1452–1485)

Opponent of Henry Tudor at Bosworth. Richard Plantagenet, the Duke of Gloucester, was born 2 October 1452 at Fotheringhay Castle, Northhamptonshire, England, and died on 22 August 1485 at Bosworth, Leicestershire. He was sometimes referred to as "Crookback" because of a slight deformity in his shoulder.

Richard was the youngest son of Richard, the Duke of York, and history has tended to paint him as a conniving, vicious opportunist, a characterization not a little influenced by Shakespeare's play *Richard III,* in which he is portrayed as nothing short of a monster. Although Richard's tactics were clearly Machiavellian, given the political climate and the nationalistic circumstances, many historians are willing to give his reign a more sympathetic interpretation. The circumstances were indeed complicated, but suffice it to say that Richard's primary concern lay with the challenge to his succession after Edward IV had proclaimed that he would be "Lord High Protector of the Realm" after that king's death in May 1483.

The family of Edward's widowed queen, the Wydvilles, were the main challengers, and to thwart their claim to the throne, he imprisoned the young king Edward V in the Tower of London. After that, he managed to secure his own claim to the throne in a mere 11 weeks. Once his claim was secure, he was forced to deal with peripheral rebellions by Henry Stafford, Duke of Buckingham, and Henry Tudor, a descendant of Edward III and the last Lancastrian claimant to the throne. The former revolt was unsuccessful because of poor planning and bad weather, and Stafford was captured and executed at Salisbury on 2 November 1483. Henry Tudor invaded in August 1485 at Market Bosworth in Leicester, where Richard was killed in an attempt to kill Henry himself, quickly ending one of the shortest reigns in British history.

David J. Tietge

See also: Bosworth, Battle of; Wars of the Roses

References and further reading:

Jenkins, Elizabeth. *The Princes in the Tower.* New York: Coward, McCann, and Geoghagen, 1978.

Ross, Charles Derek. *Richard III.* Berkeley: University of California Press, 1981.

Seward, Desmond. *Richard III, England's Black Legend.* New York: F. Watts Publisher, 1984.

Ridgway, Mathew B. (1895–1993)

Prominent American general in World War II and Korea. A 1917 graduate of West Point, Ridgway rose through the ranks as an infantry officer, eventually serving on General George C. Marshall's staff on the eve of World War II.

In 1943–1944, Ridgway commanded the 82d Airborne Division. Under his command, the division made combat jumps in Sicily and Normandy. Ridgway himself jumped in the latter operation. In late 1944, he assumed command of the XVIII Airborne Corps, composed primarily of the 82d, 101st, and 17th Airborne Divisions. The 82d and 101st played important roles in the Battle of the Bulge. Both units held out against strong German forces in December 1944. Later, his XVIII Airborne Corps engaged in a combat drop across the Rhine River and the envelopment and capture of over 300,000 Germans in the Ruhr pocket.

At the close of hostilities, Ridgway was one of the most highly regarded American field commanders. After serving in a variety of field and staff commands in the late 1940s, he was appointed commander of the Eighth Army in Korea upon the death of Lieutenant General Walton Walker. Ridgway worked wonders raising morale and restoring his army's fighting spirit. At a time when United Nations forces were under heavy pressure from a Chinese offensive, he held fast, counterattacked, and established Line Kansas, the United Nations main defense line across Korea.

In April 1951, Ridgway replaced General Douglas MacArthur as commander of UN forces in Korea and, at the behest of policymakers in Washington, D.C., initiated truce talks that eventually, after Ridgway had left, culminated in an armistice in 1953. After the Korean War, Ridgway served as supreme commander of the North Atlantic Treaty Organization and as chief of staff of the U.S. Army. Always an advocate of strong ground forces, he warned against overemphasis on air power and nuclear weapons during the tense 1950s. In 1954, he successfully counseled President Dwight D. Eisenhower against taking over the French role in Vietnam and later opposed American involvement in that country in the 1960s. Ridgway was one of the most influential and gifted military figures in American history.

John McManus

See also: Imjin River; Korean War; Walker, Walton

References and further reading:

Appleman, R. *Ridgway Duels for Korea.* University Station: Texas A & M Press, 1990.

Ridgway, Mathew. *The Korean War.* Garden City, NY: Da Capo Press, 1967.

Sandler, Stanley. *The Korean War: No Victors, No Vanquished.* Lexington: University of Kentucky Press, 1999.

Riel's Rebellion (1885)

In 1867, Canada became a commonwealth under a confederation and in 1870 assumed authority from the Hudson Bay Company over the Prairie Provinces, planning to join them to the rest of Canada via a transcontinental railway. This new authority, asserted by the Mounted Police on behalf of an increasing wave of emigrant settlers, angered and provoked the métis, descendants of the original French-British traders and local natives, who felt cheated by the new courts and displaced from their land. Finding a charismatic leader in Louis Riel, the métis and the Cree tribes rebelled in 1885, declaring a provisional government at Batoche and issuing a Revolutionary Bill of Rights.

When a 100-man police force was unable to dislodge the rebels and was forced to evacuate Fort Carleton, the government in Ottawa, fearing an Indian war like those recently fought by the United States, assembled a 3,000-man army under Sir Frederick Dobson Middleton, joined by 1,700 volunteer settlers. The government also commandeered steamboats and railways and provided the army with a Gatling gun (sent on approval with an agent from the Connecticut factory). The main body of troops under Middleton was ambushed at Fish Creek in April but managed to stalemate the rebels at Batoche, fighting behind fortified wagons, until an unauthorized bayonet charge by the volunteers routed the rebels. Meanwhile, two other columns skirmished with the Cree under Chief Big Bear and Chief Poundmaker. When the rebels sensed their defeat, they fought a series of retreating skirmishes before disbanding and melting away. Riel himself was captured 15 May and tried in Regina, Saskatchewan, for treason. He was hanged on 16 November 1885, ending the rebellion and proving the authority and strength of the federal government of Canada to keep order on the frontier.

Margaret Sankey

See also: Canadian Military

References and further reading:

Bumstead, J. M. *Red River Rebellion.* Winnipeg: Watson and Dwyer, 1996.

Flanagan, Thomas. *Riel and the Rebellion: 1885 Reconsidered.* Toronto: University of Toronto Press, 2000.

Hildebrandt, Walter. *The Battle of Batoche.* Ottawa: Canadian Parks Service, 1989.

Rifles and Rifling

The "killing zone" of both cannon and small arms was basically doubled when it was discovered that if the missile were spun between rifle and target, it was far more accurate. Rifling consists of grooves cut lengthwise in a spiral in the barrel that impart the spin to the bullet when it emerges from the muzzle. Experiments established the rate of twist required and the number of grooves needed, which vary according to the length of the barrel and the caliber of the weapon. The bullet, once lead, now has a metal jacket around its core to allow it to be squeezed into the rifling to prevent gas escaping and to withstand the greatly increased pressures of modern military firearms.

Early firearms were smoothbore muzzle loaders. The lead ball fired was roughly made and needed packing to seat it above the powder in the breech of the weapon. In traveling down the barrel when the weapon was fired, gases from the explosion of the charge escaped, and muzzle velocity, accuracy, and penetration were lost, which could be remedied only when rifling was invented.

Rifling was first applied in about 1800 in the British Baker rifle; then, in 1838, the Brunswick percussion rifle was introduced, but both suffered from the problem of "windage"—gas escaping along the sides of the bullet.

The Minié ball was one solution. This bullet had a hollow base that expanded on firing to fill the barrel, and it was elongated, rather than a simple ball. The slaughter of the American Civil War was in large part due to the unprecedented killing power of rifling and the Minié ball, a development that escaped military authorities of the time. However, the real solution to the problem of quick and accurate fire lay in breech-loading weapons.

The nineteenth-century British Boxer cartridges were the first metal-cased cartridges and as such constituted single units (bullet and case sealed together), which allowed breech loading. Ultimately, they also allowed bolt-action and even semiautomatic operation.

David Westwood

See also: Firearms; Machine Gun

References and further reading:

Allen, Major W. G. B. *Pistols, Rifles and Machine Guns.* London: English Universities Press, 1953.

Cranz, C., and K. Becker. *Handbook of Ballistics.* Vol. 2. London: HMSO, 1921.

Smith W. H. B., and J. E. Smith. *Small Arms of the World.* Harrisburg, PA: Stackpole, 1960.

Rivoli (14–15 January 1797)

One of the most severe of the many defeats of the Austrians by revolutionary and Napoleonic France accomplished by the military genius of Napoleon. By January 1797, the French Army of Italy had besieged the fortress of Mantua for eight months, and the 30,000 Austrian troops inside were on the verge of surrender. Bonaparte had defeated three previous efforts to relieve the fortress. However, the Austrians gathered nearly 45,000 men for a final attempt to rescue the garrison. Bonaparte, though commanding nearly 55,000 men, was forced to detach 20,000 of these to continue the siege and to protect his line of communication. To further complicate the French situation, Mantua could be approached along three routes, by way of Legnano, Verona, or Rivoli. In order to cover all three approaches adequately, Bonaparte was forced to divide his forces. Thus, a French division under Pierre Francis Charles Augereau was posted to Legnano; another, under André Masséna, was stationed at Verona; and a third, under Barthélemy Joubert, was at Rivoli. In the event of an attack, Bonaparte planned to reunite these forces before any one detachment could be overwhelmed.

Josef D'Alvintzi, in turn, divided his forces, sending 9,000 men under Provera to attack Legnano. This force would, he hoped, push through to Mantua. A further 6,000 men under Bayalitsch were to demonstrate against Verona. The balance of the Austrian army, nearly 30,000 strong, moved against Rivoli. Once the nature of the Austrian plan became apparent, on the afternoon of 13 January, Bonaparte proceeded to consolidate his forces. Leaving a small detachment at Verona, the remainder of the French army hurried north, while Joubert, badly outnumbered, held the heights around the village of Rivoli. While French reinforcements streamed into Rivoli throughout the 14th, the Austrian assault, divided into no less than six separate forces, eventually failed in the face of a determined defense. Bonaparte then left Joubert in command of forces at Rivoli and turned south to face Provera, now heading toward Mantua. Joubert completed the rout of the Austrian army on 15 January. The Austrians lost 3,000 killed and 11,000 prisoners over the two days of the battle.

As a result of the victory of Rivoli, Bonaparte effectively destroyed d'Alvintzi's force and subsequently forced the surrender both of Provera and of the garrison of Mantua. Subsequently, the French were able to occupy Rome and central Italy without opposition. Despite the success of the French, however, the Austrian government declined to make peace with the Directory, and a further campaign, on the Rhine and along the Adriatic, would be necessary before the signature of the Treaty of Campo Formio on 17 October 1797.

J. Isenberg

See also: French Revolutionary Wars; Lodi; Napoleon I

References and further reading:

Chandler, David G. *Campaigns of Napoleon.* New York: Scribner, 1966.

Connelley, Owen. *Blundering to Glory.* Wilmington, DE: Scholarly Resources, 1999.

Roberts, Frederick Sleigh, First Earl, Viscount St. Pierre of Kandahar (1832–1914)

Late-nineteenth and early-twentieth-century British military commander. Born in Cawnpore, India, Frederick Sleigh Roberts joined the Bengal artillery in 1851 despite his blindness in one eye. He won the Victoria Cross during the suppression of the Indian Mutiny in 1857. Earmarked thereafter for the highest command, he was involved in an Afghan border campaign in 1865, the 1866–1868 Abyssinian expedition, the 1871 northeastern India campaign, the 1874 Bihar famine relief, and the 1875 Imperial Assembly at Delhi, where Victoria was proclaimed empress of India.

An advocate of a forward Afghanistan policy to deny Russia control of the Himalayan passes, Roberts commanded the Punjab Frontier Force at the Battle of Charasia, during the occupation of Kabul and during the famous march from Kabul to Kandahar in 1878–1880. After serving as commander in chief in India from 1885 to 1893, as a field marshal he was commander in chief in Ireland from 1895 to 1899 before being appointed to command British forces in South Africa during the Boer War.

Taking command in January 1900, Roberts was responsible for the lightning campaign that within six months had relieved all threatened British garrisons, forced many Boer formations to surrender, and resulted in the capture of

Bloemfontein, Johannesburg, and Pretoria and the flight of President Paul Kruger from the Transvaal. With the war apparently won, Roberts was recalled to serve as commander in chief in Britain until the post was abolished in 1905. Thereafter Roberts, by now an earl, campaigned in favor of conscription and in 1914 was deeply involved in the attempt to persuade army officers not to obey orders from the liberal government that would have given effect to the provision of home rule for Ireland. On the outbreak of war in Europe in 1914, he was appointed colonel in chief of the Indian Expeditionary Force and died at St. Omer while visiting Indian units.

H. P. Willmott

See also: Boer War

References and further reading:

James, David. *Lord Roberts.* London: Hollis and Carter, 1954.
McIlwain, David. "Lord Roberts." In *One Hundred Great Nineteenth Century Lives,* ed. John Canning. London: Methuen, 1983.

Rochambeau, Jean-Baptiste-Donatien de Vimeur, Comte de (1725–1807)

Marshal of France who played a vital role in the final Franco-American victory at Yorktown. Born in Vendôme, Rochambeau was the third son of a noble family who intended he become a priest. He chose instead to enter the army as a cavalry officer in 1742 to serve in the War of the Austrian Succession. During the Seven Years' War, Rochambeau distinguished himself and earned a promotion to brigadier general in 1761.

In 1780, Rochambeau was promoted to lieutenant general and given command of a 5,500-man expeditionary force to aid the Americans against the British. Though he spoke no English, he quickly integrated his army into George Washington's operations near New York City. Rochambeau expressed great reservations about plans to assault the entrenched British and instead, in coordination with Admiral de Grasses, suggested operations in Virginia, where Cornwallis had established a base at Yorktown. Washington agreed, and the combined allied army, supported by the French fleet, moved more than 400 miles. Rochambeau helped direct a formal siege and supplied heavy artillery to batter the enemy fortifications.

Unable to withstand the bombardment, Cornwallis surrendered in part because of the professional and disciplined performance of Rochambeau and his army.

Rochambeau returned to France in 1783, held a succession of district commands, and was appointed a marshal of France in 1791. During the French Revolution, he survived imprisonment and narrowly escaped the guillotine. Napoleon made him a grand officer of the Legion of Honor in 1804 in recognition of his courage and skill as a professional soldier. Rochambeau died at Thore, France, in May 1807, his legacy his invaluable service to the American Revolution.

Steven J. Rauch

See also: American Revolution; Washington, George

References and further reading:

Blumenson, Martin, and James Stokesbury. "Rochambeau." In *Masters of the Art of Command.* Boston: Houghton Mifflin, 1975.
Cook, Don. "The Yorktown Stratagem." *Military History Quarterly,* Autumn 1995.

La Rochelle, Siege of (27 June–28 October 1628)

Great Catholic victory in Europe's Wars of Religion. Cardinal Armand du Plessis, Duke de Richelieu (1585–1642), meant to break the strong bulwark of Huguenot power and to augment the absolutist regime of King Louis XIII (r. 1610–1643). La Rochelle sided with Protestant England, which led to unsuccessful attempts by English ships to relieve La Rochelle in April. In May, the incompetent command of George Villiers, Duke of Buckingham (1592–1628), caused the loss of most of his 8,000–10,000 men.

Unable to use the tiny French navy to blockade La Rochelle, Richelieu used impressive military engineering to close off access from the sea by building a massive dike measuring 4,500 feet long, 50 feet thick, 25 feet wide at the top, and 50 feet wide at the bottom, with a 200-yard gap to accommodate the tide from the west. A Catholic citizen betrayed La Rochelle in March by showing Richelieu's commanders a secret passage, but the anticipated entry was thwarted through bungling.

In May, English food ships succeeded in running Richelieu's blockade and passing though the dike's center. Nonetheless, horses, mules, cats, dogs, rats, and mice had to be slaughtered for food. Wildly escalating prices led to hoarding and a black market, with women forcing themselves into prostitution to obtain food. By September, 100 people a day were dying of starvation. Louis would accept only complete surrender.

The 400-person-per-day death rate in October led to an agreement on peace terms on 28 October. Between 18,000 and 26,000 people had died during the 16-week siege; only 5,000 survivors were left to surrender. Louis triumphantly entered the city on 1 November and ordered the immediate destruction of La Rochelle's fortifications.

Annette Richardson

Richelieu standing on the stone dikes constructed to repel ships during the Siege of La Rochelle, 1628. (Library of Congress)

See also: French Wars of Religion; Religion and War

References and further reading:

Coudy, Julien. *The Huguenot Wars.* Trans. by Julie Kernan, originally published as *Les Guerres de Religion.* New York: Chilton Book Company, 1969.

Holt, Mack P. *The French Wars of Religion 1562–1629.* Cambridge: Cambridge University Press, 1995.

Rothrock, George. *The Huguenots: A Biography of a Minority.* Chicago: Nelson Hall, 1979.

Rogers, Robert (1731–1795)

Credited as the founding father of the rangers of the modern U.S. Army. Born in Methuen, Massachusetts, and raised in Dunbarton, New Hampshire, Rogers recruited for the Massachusetts provincial forces and then enlisted in the New Hampshire regiment, fighting at the Battle of Lake George (8 September 1755) in the French and Indian War.

After the battle, Rogers undertook the first of what would be many scouting missions, reconnoitering enemy positions at Crown Point (24 September) and Ticonderoga (7 October). In 1756, Rogers was promoted to captain, with orders to raise one of three companies of irregular light infantry called "rangers." Appointed major in 1758, Rogers participated in the disastrous expedition against Ticonderoga that same year. In 1759, while serving with Major General Jeffrey Amherst's army on the Hudson River, Rogers led his most infamous operation: the destruction of the Abenaki mission village of St. Francis (October 1759). Rogers was dispatched by Amherst to Niagara and Detroit in late 1760 to inform both the French garrisons and their native allies of the surrender of New France to the British.

Peace in Europe did not mean peace in North America, however, and in 1761 Rogers was sent to South Carolina to join the forces fighting the Cherokee there since 1759. With the outbreak in 1763 of the general American Indian rising known as Pontiac's Rebellion, Rogers and his rangers were back on the frontier, assisting in the relief and defense of Niagara and Detroit.

Although Rogers's play *Ponteach: or, the Savages of America* (1766) was something less than a critical success, his account of his wartime operations established throughout the years his reputation as a ranger leader. *Journals: Containing an Account of the Several Excursions He Made under the Generals Who Commanded upon the Continent of North America* (1765) was sufficiently popular (more so than *Ponteach*) to attain for Rogers the command of Fort Michilimackinac, despite the opposition of the secretary of war and the British commander in North America, Thomas Gage. In 1768, Rogers found himself not merely removed from command but charged with high treason and transported to Montreal in chains. Though the subsequent court-martial acquitted him of all charges, his reputation was badly damaged.

At the outbreak of the American Revolution, Rogers sided with the British and raised the Queen's Rangers. Inefficiencies within this unit and its repeated defeats at the hands of the rebels resulted in his removal from command in 1776. With no prospect of employment in America, Rogers returned to England, where he remained on a half-pay pension from the army until his death.

Although extraordinarily effective in field craft, rangers proved resistant to regular army discipline. Rogers in particular often found himself bested in action against French regulars, American Indians, and Canadian coureurs de bois. But concerned with the growing expense of these irregular units, British commanders sought to create regular light infantry formations to supplement them. The consequence was, after 1770, the creation of a permanent light infantry company in every British line regiment.

Adam Norman Lynde

See also: French and Indian War; Gage, Thomas; Pontiac's Rebellion

References and further reading:

Houlding, J. A. *Fit for Service. The Training of the British Army, 1715–1795.* Oxford, UK: Clarendon Press, 1981.

Pargellis, Stanley. *Lord Loudoun in North America.* New Haven, CT: Yale University Press, 1933.

Steele, Ian K. *Warpaths: Invasions of North America.* New York: Oxford University Press, 1994.

Rokossovsky, Konstantin Konstantinovich (1896–1968)

Soviet World War II commander and marshal of the Soviet Union (1944). Born in Pskov Province, Rokossovsky grew up in Warsaw, son of a Polish railway worker and Russian mother. Drafted into the Russian army in 1914, he emerged from World War I a junior officer in the 5th Cavalry Division. Rejoining his regiment after October 1917 (renamed the Kargopolsky Red Guard Cavalry Detachment), he was a squadron commander in the Russian Civil War, fighting Aleksandr Vasil'evich Kolchak in eastern Russia and White and Japanese forces in Mongolia.

Rokossovsky graduated from Leningrad Higher Cavalry School (1924) and Frunze Military Academy (1929) and led the 5th Independent Cavalry Brigade of Marshal Blukher's Far Eastern Army in Transbaikal (1929), defeating Chinese Kuomintang forces. After further successful Far Eastern commands, he was arrested in 1937 because of his links with the purged Blukher, imprisoned, but released after the poor Soviet performance in the Finno-Soviet War, also called the Winter War (1939–1940).

Recalled to command by Semen Konstantinovich Timoshenko, Rokossovsky led the 5th Cavalry Corps into Bessarabia (1940). During the initial German invasion of the USSR, he commanded the IX Mechanized Corps in the South-Western Sector, Fourth Army, on the Western Front, delaying the Germans at Smolensk and the German Sixteenth Army in the battle and counteroffensive before Moscow.

In 1942–1945, Rokossovsky held commands on more than one front, proving himself one of the best Soviet World War II commanders. As Don front commander, he participated in the Stalingrad victory, and he commanded the Central Front at Kursk, withstanding the main German assault and allowing the Soviet counterattack and victory. He commanded the first Belorussian front through Operation BAGRATION and the second Belorussian front through East Prussia and Pomerania, covering the advance on Berlin although he was bitterly disappointed at missing out on the city itself. Nonetheless, Stalin marked his achievements, giving him command of the Victory Parade in Moscow.

After the war, Rokossovsky commanded Soviet forces in Poland until 1949, becoming Polish defense minister. Returning to Russia, he was appointed deputy minister of defense and chief inspector of the Ministry of Defense (1956–1957 and 1958–1962). His career ended with his being removed to Transbaikal (1957–1958) during Communist Party leader Nikita Khrushchev's purge of Marshal Georgy Konstantinovich Zhukov. While writing his memoirs in retirement, Rokossovsky died in Moscow.

Neil Harvey Croll

See also: Russian Civil War (1918–1921); Stalin; World War II; Zhukov, Georgy Konstantinovich

References and further reading:

Kolesnikov, A. A. *Marshaly Rossii* (Marshals of Russia). Iaroslavl': Izdat. Niuans, 1999.

Kulikov, V. G., ed. *Tri marshala pobedy (Po materialam nauchnykh konferentsii posviashchennykh 100-letiiam marshalov SSSR G. K. Zhukova, A. M. Vassilevskogo, K. K. Rokossovskogo)* (Three marshals of victory [On materials of the scientific conference dedicated to 100 years of the marshals of the USSR, G. K. Zhukov, A. M. Vassilevsky, and K. K. Rokossovsky]). Moscow: Institut Rossiskoi istorii RAN, 1999.

Rokossovskii, Konstantin K. *Soldatskii dolg* (A soldier's duty). Moscow: Voenizdat, 1997.

Shukman, H., ed. *Stalin's Generals.* London: Weidenfeld and Nicolson, 1993.

Roland

Roland, Lord of the Breton Marches, died when the rear guard for Charlemagne's column was cut off and ambushed in the Pyrenees. Einhard mentions Roland as one of the commanders killed in an ambush by the Basques in 778. Part of the success of the Basques was attributed to their light arms and the uneven terrain in which the battle was fought. The Franks were also hampered by darkness and their heavy equipment. Roland's death was immortalized in the *Chason de Roland,* which emphasizes the actions of a loyal vassal rather than the actual battle in which he died.

Tamsin Hekala

See also: Charlemagne

References and further reading:

Loyn, H. R., ed. *The Middle Ages: A Concise Encyclopaedia.* New York: Thames and Hudson, 1991.

Rollo

Duke of Normandy (c. 911–c. 930), also known as Hrolf Geganger (Ralph the Walker), who was the successful leader of a Norse raiding band that entered into a treaty with Charles III (the Simple). The treaty of St.-Clair-sur-Epte in 911 ceded the strategic coastal counties of Neustria, the lower basin of the Seine River, as a duchy for Norse settlement to Rollo. As duke of Normandy, Rollo became Charles's vassal, was baptized, and agreed to defend against other raiding bands. After Charles's death in 924, Rollo enlarged his duchy by the addition of Bayeux.

Tamsin Hekala

See also: William the Conqueror

References and further reading:

Bates, D. *Normandy before 1066.* London: Longman, 1982.

Loyn, H. R., ed. *The Middle Ages: A Concise Encyclopaedia.* New York: Thames and Hudson, 1991.

Previté-Orton, C. W., ed. *The Shorter Cambridge Medieval History.* Cambridge, UK: Cambridge University Press, 1962.

Roman Army

One of the world's longest-lasting and most successful military organizations in history, enlarging and defending Rome. The Roman army started as a tribal militia during the kingdom and early republic periods, but the pressure of the surrounding Etruscan and Greek cities brought about the adoption of the Greek phalanx and its weaponry and, more important, a formalized system of recruitment by centuries of families of different economic ability, who furnished cavalry, heavy infantry, or light infantry, as appropriate. The expansion of the empire brought about the formation of the legion, a flexible heavy infantry (plus support) formation of about 5,000 men. Property requirements for recruitment were relaxed, and soldiers were paid on a regular basis. New types of weapons and armor were introduced. Formal regulations for all aspects of army discipline, organization, training, and tactics were adopted. The legion formation lasted throughout most of the empire's existence. As armored and often mounted barbarians pushed into the empire from about the second century C.E., the Roman army was divided into two types of units. The old legion, the mainstay of the army for centuries, became (often understrength) *limitanei,* who garrisoned frontier zones. Mobile armored cavalry field armies, or *comitatenses,* under the command of an emperor or his deputy, a caesar, could move rapidly to repel invaders. The ultimate defeat of the Roman army came about for numerous reasons, including the economic and political exhaustion of the empire and pressure by "barbarians" from beyond the empire's borders who had learned more sophisticated military technologies and organization, often from the Romans themselves.

Organization, Manpower, and Training

The Roman army was composed of legions of about 4,200–5,000 heavy infantrymen. Legions were divided into cohorts of about 500–800 men, in turn divided into centuries or maniples of 60–120 men. Support troops at various times included *velites,* or allied light infantry, as skirmishers; cavalry troops of around 120 men for the scouting and communication role; and field engineers. Separate *ilaii* (regiments) of cavalry were raised from among the wealthier classes. At its height, the empire had some 30 legions stationed throughout the empire.

Non-Roman auxiliaries provided the army with specialist forces—light infantry skirmishers, light and heavy cavalry, slingers, and archers—where the Romans were weak. Auxiliary soldiers could become Roman citizens upon completion of service.

Command was exercised by officers appointed by the Senate or in the field. Legions were commanded by consuls or legates appointed by the Senate and (later) the emperor. The legate was assisted by six tribunes, young men on a career path to the Senate, or senior tribunes aiming for a legate's post themselves. Each century was commanded by a centurion, assisted by an *optio* and a *signifer,* who, in addition to carrying the century's standard, served as the century's banker. The senior centurion of the legion served on the staff with the tribunes.

Recruits to the legions were trained in marching, close order drill, use of weapons, and making and breaking camp. They normally served 16 or 20 years.

Weapons

Soldiers were equipped with body armor (breastguards, chain mail shirts, or the *lorica segmentata*) and a bronze or iron helmet. A large oval or rectangular *scutum* shield, composed of layers of laminated wood covered with hide cen-

The Roman Empire, 31 B.C.E.–565 C.E.

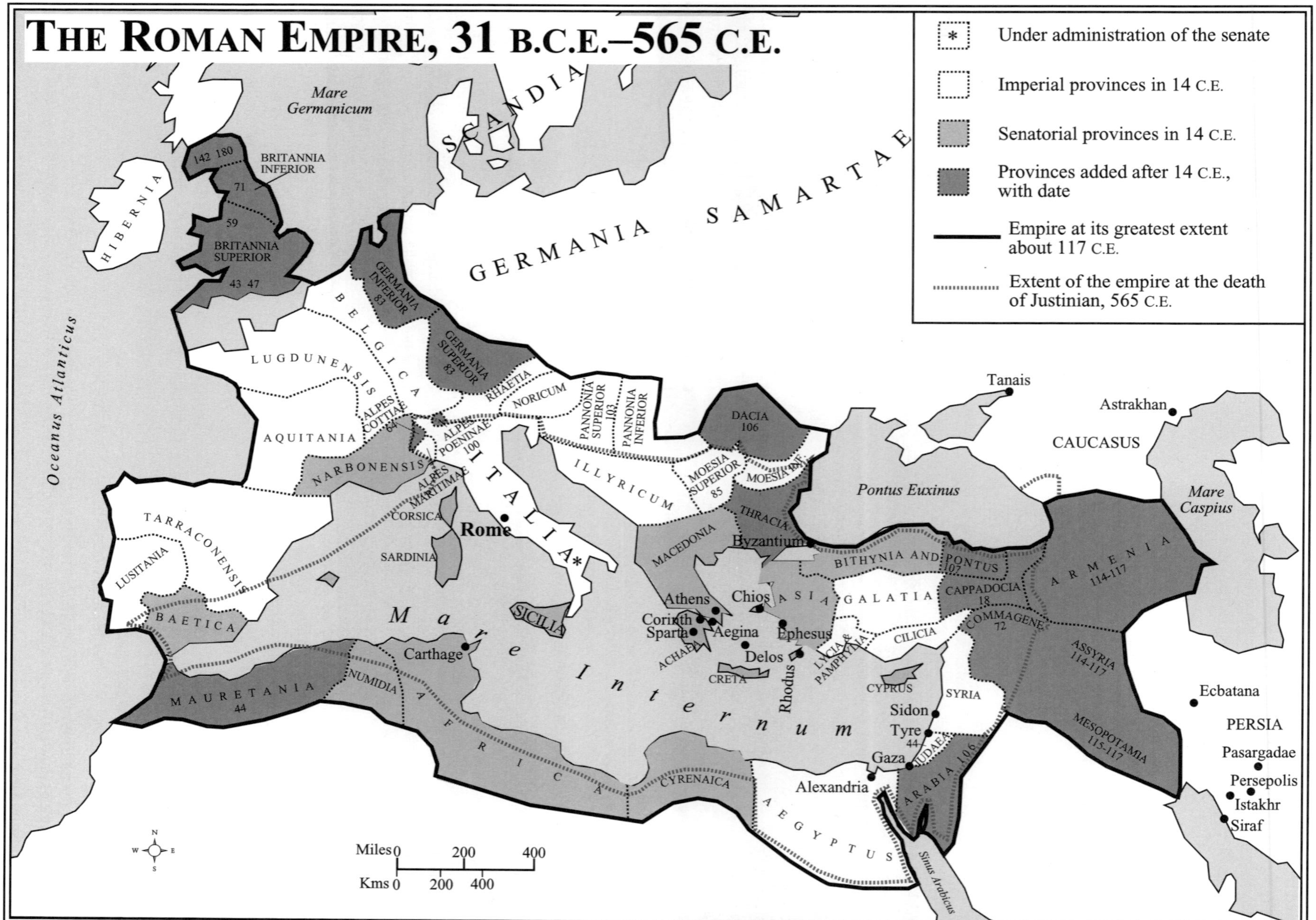

tered by a metal boss, was used as a weapon of offense in the charge and defense against arrows and blows. The principle weapons were a single-use *pilum* javelin, thrown before charging or in defense, and a broad, short, stabbing blade, the *gladius hispanensis.* Fire support was offered by *ballistae,* using wound leather and sinew coils to throw stones and firebrands, and single or repeat-shot *onager* bolt throwers.

Tactics and Movement

Roman doctrine emphasized a one-foot-on-the-ground approach. In battle, troops of the front line charged, while the rear formed a mobile defense base, hurling javelins and other ranged weaponry from behind a shield wall or charging in turn. The legion on the move established fortified camps at the end of every day. Engineering was emphasized strongly, and by the time the rear elements had left the previous day's camp, the forward elements would often already be building the next one. A legion was capable of constructing bridges and offensive and defensive siege works of great magnitude, as Alesia, Masada, and other surviving sites attest.

In a pitched battle, the light elements engaged first, breaking up enemy charges (particularly lightly armored Celtic and Germanic troops) with volleys of javelins, stones, and arrows. The front maniples or centuries then charged, after discharging their *pilla.* Hand-to-hand combat followed, the legionaries on the contact line sheltering behind their shields and thrusting with their swords. For 1,000 years, through republic and empire, defeats and victories, the Roman military endured with remarkably little change.

Michael Ashkenazi

See also: Alesia, Siege of; Masada, Siege of; Roman Republic, Wars of the

References and further reading:

Connolly, Peter. *Greece and Rome at War.* London: Macdonald Phoebus, 1981.

Hackett, John, ed. *Warfare in the Ancient World.* London: Sidgwick and Jackson, 1989.

Humble, Richard. *Warfare in the Ancient World.* London: Cassell, 1980.

Sekunda, Nicholas V., Simon Northwood, and Michael Simkins. *Caesar's Legions: The Roman Soldier, 753 B.C. to 117 A.D.* Oxford, UK: Osprey, 2000.

Roman Civil Wars (88–30 B.C.E.)

A series of civil wars that ultimately led to the end of the Roman Republic. Immediately following the Social War, which set a precedent for internecine conflict, competition among members of the Roman aristocracy was so intense that political solutions were usually only achieved through violence. This competition was further exacerbated by foreign wars, which put immense power in the hands of ambitious military leaders who put personal interest above the welfare of the republic.

88–73 B.C.E.

The personal and political rivalry between Gaius Marius (157–86) and Lucius Cornelius Sulla (138–78) led to the first outright civil war in the history of the Roman Republic. Conspiring with the tribune, Publius Sulpicius (Rufus?), Marius had Sulla replaced as commander of the war against Mithradates in 89. Back in Rome, Sulla responded by declaring religious holidays in order to prevent meetings of the assembly. Sulpicius and his followers rioted, forcing Sulla into hiding in the home of Marius. In exchange for rescinding the religious decrees, Marius allowed Sulla to flee into exile. However, Sulla used Marius's favor as an opportunity to form an army in Campania, with which he marched on Rome. Rome's plebeians stubbornly resisted the attack but gave up once Sulla's forces began setting fire to their homes. Once in control of the city, the victorious Sulla forced the Senate to declare Marius, Sulpicius, and other nobles enemies of the state. In addition, he obtained proscriptions (i.e., laws establishing death sentences) for these men. Sulla executed Sulpicius, but Marius escaped to North Africa. Sulla then attempted to consolidate his power by rescinding Sulpicius's laws. However, while Sulla was on campaign again against Mithradates, Lucius Cornelius Cinna (d. 84), the new consul, rescinded Sulla's laws and recalled Marius to Rome.

A deranged and elderly Marius, furious over his exile by Sulla's faction, raged through the city, brutally murdering the senators and nobles who had opposed him. Even Cinna, Marius's ally, was disturbed by these actions and begged him to stop. Marius died suddenly of illness in 86, whereupon Cinna became de facto dictator and tried to govern without the excesses of Sulla and Marius.

Sulla, still outraged by Cinna's alliance with Marius, temporarily restrained Mithradates and returned to Italy to take revenge upon his enemies in the spring of 83. Landing at Brundusium, Sulla quickly defeated one of the two armies sent against him and convinced the other to defect to his side. Near Rome's Colline Gate, Sulla met a force of Samnites, among Rome's fiercest opponents in the Social War. Sulla narrowly avoided defeat and took control of Italy thanks to the excellent generalship of Marcus Licinius Crassus, who led Sulla's right flank. Sulla's victory led to a reign of terror, which began with the torture and execution of 6,000 Samnite prisoners and culminated in proscriptions that exceeded even the brutality of Marius. Sulla's reign of terror, not limited to Roman elites, extended to those Italian cities that had supported resistance to Sulla. Many of the inhabitants of Praeneste, Florentia (Florence), and especially Samnia were sold into slavery and their lands destroyed and turned into desert.

Making an effort at military reforms, Sulla gave the lands of many of his defeated enemies to his loyal soldiers and tried to limit the power of provincial commanders to make war by giving control of the army over to the Senate. Sulla died in 78, thinking his work successful. However, chaos soon overtook Roman political life as the families of the victims of Sulla's proscriptions sought revenge, and Sulla's soldiers, unused to farming, fell into indebtedness. Moreover, Quintus Sertorius (123–c. 74), provincial military commander in Spain and opponent of Sulla, combined the guerrilla tactics of his Spanish troops with Roman discipline to create an effective renegade force that was feared by the Roman Senate. Quintus Caecilius Mettellus Pius (d. 63) led a failed campaign against the renegade. Only Sertorius's assassination in 74 or 73, allowed Gnaeus Pompeius Magnus, Pompey the Great (106–48), to wrest control of the rogue forces, declaring them loyal to Rome.

70–45 B.C.E.

Julius Gaius Caesar, initially the least powerful member of the First Triumvirate, which also included Pompey and Crassus, would eventually destroy the republic and briefly become master of Rome. In 70, Pompey and Crassus, Rome's most powerful generals after the defeat of Spartacus, discarded Sulla's reforms and were both elected consuls. Despite their intense rivalry, Pompey and Crassus cooperated in order to achieve their political goals until the rise of Caesar, when, in 60, they joined with Caesar in forming the First Triumvirate. Although not an alliance or faction in a legal sense, the members of this triumvirate agreed to pursue only those political ends agreeable to the others. Crassus was killed on campaign against the Parthians at Carrhae in the spring of 53, leaving only Caesar and Pompey.

Fearing Caesar's steady rise in prestige and power from to his success in the Gallic Wars, the Senate commanded him to surrender his legions in the summer of 50. Caesar defied the Senate and marched a Roman legion supported by German and Gallic auxiliaries into Cisalpine Gaul. On 20 November 50 B.C.E. (traditionally 11 January 49 B.C.E.), Caesar sent a small force to seize the village of Ariminum, just south of the Rubicon River, which separated Cisalpine Gaul from Italy. In a brilliantly calculated risk, Caesar quoted the Greek playwright Menander, "The die has been cast," crossed the Rubicon with only one legion, and invaded Italy that winter.

Pompey had two legions to Caesar's one, but they were untrained, isolated, and could offer only token resistance to Caesar's organized and fast-marching force. Caesar quickly conquered Italy and entered Rome. Pompey and the Senate fled to Epirus in Greece, where they attempted to consolidate their forces. With Rome under control, Caesar marched toward Spain. Half of his forces, led by Domitius Ahenobarus, laid siege to Massilia (Marseilles) in Gaul, which was both a Pompeian stronghold and an important communication link between Spain and Italy. The other half, led by Scribonius Curio, defeated Pompey's generals, Afranius and Petreius, at Ilerda, Spain, in the summer of 49. Curio was soon after killed in North Africa, fighting Attius Varus and King Juba of Numidia.

In the winter of 48, Caesar surprised Pompey by sailing across the Adriatic from Brundusium to the port of Dyrrhachium in Greece. This campaign almost immediately ended in disaster, when Calpurnius Bibulus, leading Pompey's fleet, captured Caesar's ships and supplies. Caesar was forced to retreat from Dyrrhachium to Pharsalus in Thessaly, where, attacked by Pompey on 9 August 48, he again faced disaster.

However, although vastly outnumbered, Caesar's well-trained veterans humiliated Pompey, who fled to Egypt. Upon arriving in Egypt, Pompey was murdered by Ptolemy XIII, who hoped to ally with Caesar. Angered by Ptolemy's action, Caesar instead allied with Ptolemy's sister and opponent, Cleopatra. With a force of 20,000, Ptolemy led a four-month-long siege against Caesar's small legion of 4,000 in Alexandria. Often close to defeat, Caesar survived until January 47, when he was relieved by two legions led by Mithradates of Pergamum. Caesar, leading Mithradates' legions, won the Battle of the Nile in February 47 and gained control of Egypt.

In the fall of 47, Caesar, again outnumbered, defeated a Pompeian force led by Cato the Younger and the inept Metellus Scipio at Thapsus in North Africa. Rather than be captured by Caesar, Cato committed suicide, thus becoming a martyr to the republican cause. Caesar's final victory against Pompeian forces came at Munda in Spain in 46–45. Attacking uphill, his forces defeated Pompey's two sons, Gnaeus and Sextus.

Caesar returned to Rome, where he was proclaimed dictator for 10 years and given four military triumphs. Over 60 senators assassinated Caesar on 15 March 44, just as he was about to depart on a campaign against the Dacians and the Parthians.

44–30 B.C.E.

In the winter of 44–43, Mark Antony, Caesar's protégé in the consulship of 44, besieged the republican Decimus Brutus at Mutina (Modena) in Cisalpine Gaul. Decimus Brutus was allied with Octavian (later Augustus), Caesar's adopted son. Two armies sent by the Senate, one led by Octavian and the other led by Aulus Hirtius and Gaius Vibius Pansa, forced Antony to retreat to Gaul in the spring of 43.

Octavian returned to Rome, where he became consul. Despite Antony's retreat, Octavian respected him and appreciated that a strong general leading robust legions could be more useful than a republican allegiance. Meanwhile, Antony had won the admiration of the powerful legions of Marcus Aemilius Lepidus. On 27 November 43 B.C.E., Octavian, Antony, and Lepidus formed a formal executive committee, known as the Second Triumvirate. Almost immediately, the triumvirs imposed proscriptions in which 130 senators and 2,000 knights, including the great orator Cicero, were killed. In the fall of 42, two of the triumvirs, Antony and Octavian, pursued Marcus Brutus and Gaius Cassius, among the coconspirators in Caesar's assassination, to Greece.

At Philippi in 42, Brutus defeated Octavian. However, Antony inflicted an even more crushing defeat upon Cassius, who committed suicide. Relieving Octavian, Antony soon crushed Brutus, who also committed suicide. Tension developed immediately as Antony's wife, Fulvia, and brother, Lucius, fomented suspicion against Octavian among veterans and landowners. In 40 B.C.E., an outraged Octavian commanded two of his generals, Quintus Salvidienus and Marcus Vipsanius Agrippa, to besiege Fulvia and Lucius at Perusia (Perugia). Although he spared their lives, their deaths soon after eased tensions between Antony and Octavian. Later in 40, Octavian and Antony made an agreement known as the Pact of Brundusium, which reaffirmed the triumvirate and officially divided the empire (the east to Antony, the west to Octavian, and Africa to Lepidus).

The peace would not last long. Pompey's son, Sextus, harassed Italy from the sea and threatened Rome with starvation. In exchange for ending the blockade against Rome, the triumvirs agreed to compensate Sextus for his father's confiscated lands and to give him control of Sicily, Sardinia, Corsica, and the Peloponnese. Peace lasted until 38, when Octavian attempted an invasion of Sicily that ended in disaster. Antony, angry with Octavian for failing to consult him and for delaying his campaign against the Parthians, nevertheless reaffirmed his support of the triumvirate by agreeing to the Pact of Tarentum in 37.

In 36, Octavian's general, Agrippa, finally defeated Sextus in a sea battle at Naulochus near the straights of Messina. Lepidus tried to take Sicily for his own. Angered by Lepidus's action and sensing an opportunity to rid himself of another rival, Octavian spoke directly to Lepidus's legions and convinced them to defect. The defection of these legions eliminated Lepidus as a triumvir.

With Lepidus's departure, tensions increased between the remaining triumvirs, Octavian and Antony. The latter had suffered significant losses in his campaign against the Parthians in 36 but remained strong and popular, until his relationship and alliance with Cleopatra began to tarnish his reputation. By declaring war on Cleopatra rather than Antony in the fall of 32, Octavian cleverly avoided the risk of appearing to initiate another civil war. On 2 September 31, Octavian won a decisive victory over Antony at Actium. Octavian's dependable general Agrippa blockaded Antony's camp, causing plague, famine, and desertions. Antony and Cleopatra escaped the blockade, abandoning the bulk of the army. Octavian's victory was so complete that, instead of pursuing Antony and Cleopatra, he returned to Italy.

Octavian returned to Egypt the next summer. Unable to mount a serious defense, Antony and Cleopatra committed suicide. Cleopatra's death marked the end of the Ptolemaic dynasty and the beginning of Roman rule of Egypt. Octavian's victory also marked the end of the Roman Republic. Although many republican institutions remained, the power of Rome was now concentrated in one man, Octavian (soon to be Augustus Caesar).

Eric Pullin

See also: Ancient Warfare; Caesar, Julius; Carrhae, Battle of; Cassius; Marius, Gaius; Mithradatic Wars; Parthian Empire; Philippi, Battle of; Pompey the Great; Roman Army; Sulla, Lucius Cornelius

References and further reading:

Appian of Alexandria. *Appian's Roman History.* New York: Macmillan, 1913.

Caesar, Julius. *The Civil Wars.* Cambridge: Harvard University Press, 1914.

———. *Alexandrian, African and Spanish Wars.* Cambridge: Harvard University Press, 1955.

———. *Gallic War.* Trans. John Warrington. Norwalk, CT: Easton Press, 1983.

Greenhalgh, Peter. *Pompey: The Roman Alexander.* Columbia: University of Missouri Press, 1981.

Keaveney, Arthur. *Sulla: The Last Republican.* London: Croom Helm, 1982.

Scullard, Howard Hayes. *From the Gracchi to Nero: A History of Rome from 133 B.C.E. to C.E. 68.* Methuen: London, 1980.

Seager, Robin, ed. *The Crisis of the Roman Republic: Studies in Political and Social History.* New York: Barnes & Noble, 1969.

Spann, Philip. *Quintus Sertorius and the Legacy of Sulla.* Fayetteville: University of Arkansas Press, 1987.

Roman Civil Wars (235–284)

The interval from the Severans to Diocletian began and ended with strong government, separated by political instability and military stress. Traditionally known as the "period of anarchy," this half-century saw at least 18 "legitimate" Roman emperors and many more if numerous usurpers who failed to establish themselves are counted. Nearly all had short reigns and met violent deaths.

The assassination (March 235) and replacement of Severus Alexander by a tough officer from Thrace, Maximinus I Thrax (r. 235–238), was a stark reminder that the empire needed soldier-emperors who knew the army. An equestrian outside the ruling clique, Maximinus had exploited the opportunities of the Severan army to gain numerous senior appointments. He also established the pattern of succession by murder and civil war.

The Persian campaign of Gordian III (r. 238–244) ended in the emperor's defeat and death, perhaps at the hands of his successor Philip the Arab (r. 244–249). In the mid-250s, Shapur's invasion ended with the capture of the emperor Valerian (r. 253–260), who remained in captivity until his death. Mesopotamia and Syria were overrun by the advancing Persians, but the Sassanians failed to drive the Roman frontier back to Syria. By the end of the century, the campaigns of Aurelian (r. 270–275) and Carus (r. 282–283) had helped reestablish Roman predominance.

By 249, the Goths had emerged as a threat on the Danube River, with Franks and Alemanni pressing the Rhine frontier and Saxons raiding Britain and Gaul. Gothic invasions penetrated Thrace and Asia Minor. The emperor Trajan Decius (249–251) fell fighting this new enemy.

Although the empire was hard-pressed for two decades, much was achieved in the name of Rome by the regimes of the usurper Postumus and his successors in Gaul (the Gallo-Romano Empire, 260–274) and in the east by the Palmyrene Empire of Septimius Odaenathus and Septimia Zenobia. Aurelian suppressed both rebellions but had to abandon Dacia to the Goths. His successor Probus (r. 276–282) cleared Gaul of Germanic invaders who had crossed the Rhine River after Aurelian's assassination, but the Alemanni retained territory seized in southern Germany.

During the third century, the number of those of senatorial rank holding senior army positions became fewer and fewer, and the practice disappeared entirely under Gallienus (r. 253–268). Far more opportunities lay open to equestrians, especially those who had campaigned under the emperor himself. These equestrian officers were generally career soldiers. They owed their advancement purely to their military record and imperial favor. It was usually such men who murdered emperors and nominated usurpers from their own group. Several of the most successful emperors of the second half of the century came from a small group of Illyrian equestrians. A number proved very capable.

Internal instability led to losses and defeats on all sides and encouraged internal rebellions. Each emperor had to campaign without respite. Few could afford to entrust command of an army to a potential rival. While the emperor was campaigning in one theater of operations, there was great danger that other, neglected parts of the empire would put forward rivals.

It was Gallienus who developed the new device by which his Illyrian successors fought off Persians and Germanic tribes alike. It was an "elite army," in the words of one observer, a mobile force not tied to frontier defense. The troops were drawn from units in Britain, on the Rhine River, and on the Danube River. They operated independently and were forerunners of the fourth-century *comitatenses* (field armies). There are also reports of cavalry corps under their own commanders.

A measure of stability was created by one of the Illyrian soldier-emperors, Diocletian (r. 284–305). He gradually developed a system of dividing imperial power (the Tetrarchy). In the mature form of this system, there were two senior Augusti, ruling the eastern and western provinces respectively, each assisted by a junior colleague or caesar. The system was intended to provide sufficient commanders to deal with simultaneous crises. At the same time, by nominating the caesars as successors to their senior colleagues, civil wars could be prevented by providing for the ambitions of all men with armies.

Nic Fields

See also: Aurelian, Lucius Domitus; Diocletian; Goths; Sassanid Empire; Shapur I

References and further reading:

de Blois, L. *The Policy of the Emperor Gallienus.* Leiden: Brill, 1976.

Osier, J. "The Emergence of the Third Century Equestrian Military Commanders." *Latomus* 36 (1977), 674–687.

Roman Republic, Wars of the (111–63 B.C.E.)

A series of internal conflicts and wars that furthered the expansion and security of the Roman Republic.

111–106 B.C.E.

King Jugurtha, with the help of corrupt Roman senators, wrested control of Numidia in 111 through murder and intrigue. The Roman Senate supported Jugurtha but became concerned when Jugurtha laid siege to Cirta (Constantine, Algeria), an important source of grain for Rome. Jugurtha foolishly enraged Rome's popular classes by murdering several Roman knights (*equites*), who as merchants were trapped in the city during the siege. Although forced to declare war by the *equites,* the Senate pursued only a half-hearted campaign against Jugurtha. Only after Jugurtha openly intrigued to murder one of his Numidian opponents in Rome did the corrupt Senate pursue the war with seriousness. The Senate renewed its war against Jugurtha in 110 but was soundly defeated. In 109, the frustrated popular classes, especially the *equites,* forced the Senate not only to prosecute the corrupt senators who had supported Jugurtha but also to appoint Quintus Caecilius Metellus as consul to lead the campaign. Despite his honesty and competence as a general, Metellus was soon politically outmaneuvered by his subordinate, Gaius Marius (157–86). Marius gained the consulship and control of the Numidian campaign in 107. Marius's subordinate in the campaign, Lucius Cornelius Sulla (138–78), won a swift end to the Jugurthine War in 106 by bribing Jugurtha's ally, King Bochus of Mauretania, to betray him. Jugurtha met his death in the dungeon of Tullianum after Marius's triumph in 104.

105–88 B.C.E.

In preparing for the Numidian campaign, Marius instituted the permanent acceptance of recruits into the army, regardless of property qualifications. Roman generals had only previously employed this manner of recruitment in exigent circumstances. This move kept Rome supplied with a steady contingent of recruits, but it also tempted some commanders to incite foreign crises in order to win personal glory. It would also have tremendous implications during the Roman Civil Wars by effectively making propertyless recruits more loyal to individual commanders than to the interests of the Senate. The end of the Jugurthine War also marked the beginning of the rise of both Marius and Sulla as *popularis* leaders.

After the defeat of Jugurtha, Marius reorganized the Roman legions. Standardizing equipment and training for the legions, he also eliminated light units and divided each legion into 10 cohorts of 500–600, which he further divided into centuries of between 80 and 100, thus creating legions of 5,000–6,000 heavily armed men. Marius's reforms made the legions more cohesive, versatile, and formidable. They were devastatingly effective against the migrating Cimbri and Teutons, who had humiliatingly defeated several Roman armies in Transalpine Gaul in 109, 106, and 105. In 104, the slaves of Sicily rebelled into full-scale war. Tens of thousands were killed before Marius's easy victory over the Cimbrians at Vercellae near Turin in 102 freed his troops to put down the rebellion in 100.

As threatening as was Sicilian slave revolt, nothing threatened Roman hegemony in Italy more than the Italian or Social War (90–88). Rome's Italian allies (in Latin, *socii*) had hoped to secure the right of citizenship through peaceful means but were rejected by the city's privileged *optimates.* The Italian allies, called Corfinum (later renamed Italia) and led by the Marsi and Samnites, raised a force of 100,000, many of whom had been trained in Rome's legions. Initially, Rome fared poorly; thus the Senate reluctantly, but wisely, conferred citizenship upon those Italians who laid down their arms. As that was the goal of the Italians, the Social War ended, with a few exceptions, by 88. However, the Social War boded ill for the future of Rome by setting a precedent for the Roman Civil Wars.

89–72 B.C.E.

Taking advantage of the upheaval caused by the Social War, Mithradates VI Eupator (134–63) of Pontus tried to build an empire modeled on that of Alexander the Great, in part by expelling Rome from the eastern Mediterranean in what became known as the Mithradatic Wars. Wary of Mithradates' aspirations, Rome sent an envoy to Asia Minor, where he skillfully checked Mithradates' advances, but he also foolishly impelled Nicomedes III of Cappadocia to attack Pontus. Mithradates responded by sweeping aside the Romans in Asia Minor, taking Pergamum, and pressing on toward Greece. Because of Rome's unpopularity in the region, Mithradates soon won the support and control of much of southern Greece, including Athens. However, Sulla recaptured Athens and pushed Mithradates from Greece by the summer of 86. In Asia Minor, Lucius Valerius Flaccus, Sulla's rival, was murdered in a mutiny, but his force defeated Mithradates' son and took Pergamum. Although Sulla soon expelled him from Asia Minor, Mithradates was not yet defeated. In 74, Mithradates, fearing the Romans would block his access to the Aegean Sea, occupied the Roman client state of Bithynia. In 72, the Roman commander Lucius Licinius Lucullus (116–57), after the failures of several other Roman commanders, defeated Mithradates, who was aided by formidable Mediterranean pirates, in a series of crushing victories. Mithradates fled into exile to the court of King Tigranes II in Armenia.

68–53 B.C.E.

In 68, Gnaeus Pompeius Magnus, Pompey the Great (106–48), cleared the Mediterranean of pirates in a mere 40 days by concentrating his forces and treating the outlaws mercifully, many of whom became his valuable allies. Pompey's resulting fame gained him the command of the final operation against Mithradates. Mithradates' last attempt to challenge Rome was a pathetic effort and was quickly crushed by Pompey in the Crimea. Mithradates committed suicide in 63, and much of his empire came under Roman rule. Also in that year, Pompey marched on Syria to restore order in the wake of the Seleucids' loss of their eastern empire to the Parthians, who became Rome's major rivals in the east. In the winter of 54, Marcus Licinius Crassus, Pompey's partner

Roman hero Horatius defending the Tiber Bridge against the army of Lars Porsena c. 505 B.C.E. (Hulton/Archive)

and rival in the First Triumvirate, confidently set off on a militarily unnecessary campaign against the Parthians. However, the Parthians at Carrhae slaughtered him and his seven legions.

Eric Pullin

See also: Alexander the Great; Ancient Warfare; Caesar, Julius; Carrhae, Battle of; Marius, Gaius; Mithradatic Wars; Parthian Empire; Pompey the Great; Sulla, Lucius Cornelius

References and further reading:

Appian of Alexandria. *Appian's Roman History.* New York: Macmillan, 1913.

Greenhalgh, Peter. *Pompey: The Roman Alexander.* Columbia: University of Missouri Press, 1981.

Keaveney, Arthur. *Sulla: The Last Republican.* London: Croom Helm, 1982.

Scullard, Howard Hayes. *From the Gracchi to Nero: A History of Rome from 133 B.C. to A.D. 68.* Methuen: London, 1980.

Seager, Robin, ed. *The Crisis of the Roman Republic: Studies in Political and Social History.* New York: Barnes & Noble, 1969.

Spann, Philip. *Quintus Sertorius and the Legacy of Sulla.* Fayetteville: University of Arkansas Press, 1987.

Roman-Etruscan Wars (509–234 B.C.E.)

In 509 B.C.E., the Romans, led by Brutus, overthrew the last king of Rome, a member of the Etruscan line of Tarquin. Although the Etruscans had been forced out of Rome, there were still powerful Etruscan cities in Italy that challenged Rome for vital resources and strategic control. The most dangerous of these was Veii, only 12 miles north of Rome on the Tiber River. Closer to Rome and more powerful than any other Etruscan city, Veii controlled much of the west bank of the Tiber and contested with Rome over the lucrative trade to the Italian hinterland, as well as access to the port of Ostia and the valuable salt flats there.

The Fabii family in Rome, who had long-standing Etruscan connections and owned much of the land between the two cities, largely took responsibility for frontier defense, building a small fortress at Cremora and assembling a private army of clients while engaging in cattle raiding. In 476 B.C.E., however, the Veii attacked the fortress, and more than 300 Fabii were killed, leaving the area, including the Janiculum Hill, in enemy hands. Forced to retreat, Rome made a truce.

During the cold war that ensued, Rome and Veii focused on the town of Fidenae, a station on the "salt road" between them, which changed hands several times. Rome, growing stronger, prepared for a battle to the death with Veii, overhauling its government by replacing the two consuls with six military tribunes, each with consular authority, and implementing censorship in order to maintain rolls of citizens liable for military service. Noting that the Etruscans had been defeated at sea by the Greeks at Cumae in 443 B.C.E., Rome seized Fidenae in 435 B.C.E. (or 425) and declared war on Veii, which appealed to the other Etruscan cities in vain for assistance. A six-year siege ensued, with the Veii cutting their cliffs to make them harder to climb and erecting stone and earth ramparts. The Romans, under the command of Camillus, who had forced them to accept continual military operations (with no breaks for seasonal agriculture) and organized regular pay for soldiers, seized the land neck to Veii and discovered irrigation conduits leading under the city walls. A small force crept through them into the city, which fell swiftly to the Romans. As a lesson, Rome destroyed Veii and occupied all its land, nearly doubling Rome's acreage.

Triumphant, Rome adopted Veii's worship of Juno, and Camillus sent a gold bowl to the Oracle at Delphi to celebrate his victory. During the Gallic invasion of 387–386 B.C.E., the Etruscan city of Caere aided Rome, saving refugees and many of the city's relics while pressuring the Gallic chief Brunnus to leave the area. Caere was rewarded with special privileges by Rome, but in 353 B.C.E., it joined with another Etruscan city, Tarquinii, to resist Roman expansion and was defeated, although the Romans afterward offered a generous century-long truce. Other Etruscan cities continued to cause problems for Rome. Capua joined with the Samnites, Umbrians, and Gauls against Rome in the Third Samnite War (298–290 B.C.E.) and was defeated at Sentinum, after which Rome ravaged the area.

Because of continual warfare in the peninsula, the Romans adopted the legionary organization and maniples, which allowed maximum flexibility on the battlefield, and began to issue standardized equipment. Through contact with Etruscans, Rome learned the craft of shipbuilding, crucial in its wars with Carthage, and adopted gladiatorial combat as entertainment. The Etruscans, seven of whose cities had joined in a pact with Rome in 280 B.C.E., erupted only once more. While Hannibal and Hasdrubal marauded through Italy, disaffected Etruscan cities offered him assistance, until Hasdrubal's defeat at Metaurus Valley in 209 B.C.E. ended the long-standing rivalry between the ancient Etruscans and their upstart neighbors, the Romans.

Margaret Sankey

See also: Hannibal Barca; Roman Army; Samnite Wars

References and further reading:

Cobbold, G. B. *Children of Romulus.* White Plains, NY: Longmans, 1995.

Grant, Michael. *The History of Rome.* New York: Charles Scribner's Sons, 1978.

Ogilvie, R. M. *Early Rome and the Etruscans.* Atlantic Highlands, NJ: Humanities Press, 1976.

Rommel, Erwin Johannes Eugen (1891–1944)

Prominent and influential German commander of World War II. Commissioned an infantry officer in January 1912, from his first moments under fire in August 1914, Erwin Rommel displayed remarkable aggression, complete faith in his troops and himself, and great gallantry.

In October 1916, he married Lucie Mollin, then a Red Cross nurse. In 1917, his Gebirgsjäger Battalion was transferred from the Romanian front to the Italian, where he took Monte Matajure and turned the Longarone position, for which he received the Pour le Merité. After the Armistice, he was accepted into the Reichswehr but remained a regimental officer, never being selected for general staff training. Rommel also wrote *Infantry Attacks,* which brought him to Adolf Hitler's attention.

During the invasion of Poland (1939), Major General Rommel commanded the Führer's headquarters battalion. His loyalty was rewarded with the 7th Panzer Division, which he led to great acclaim during the French campaign. Command of the Afrika Korps in February 1941 resulted in his promotion to field marshal in June 1942 for reducing the British fortress of Tobruk. But Hitler's "stand fast" order at El Alamein in November 1942, resulting in needless Axis casualties, probably marked a turning point in Rommel's life.

Despite an award of diamonds with swords and oak leaves to his Knight's Cross and command of Army Group B, Rommel became embittered by German casualties and conduct in Italy. Only his personal friendship with Hitler slowed his growing defeatism and pessimism: unlike Hitler and most of the German High Command, Rommel took American troops and industry seriously, despite early U.S. defeats in North Africa.

Rommel's defense of Normandy in June 1944 incurred savage German losses. Hitler's refusal to use these sacrifices to negotiate a surrender led Rommel to confront Hitler about strategy, war crimes, and finally the continued fighting in the West itself and to have at least a marginal involvement in the 1944 bomb plot against the Führer. Wounded by British fighter-bombers, he was doomed by the failed assas-

General Rommel standing in a command car in the North African desert, 1942. (Library of Congress)

sination of Hitler on 20 July. On 14 October 1944, he committed suicide to protect his family and staff.

Erin E. Solaro

See also: France; Guderian, Heinz; Montgomery, Bernard Law; World War II

References and further reading:

Fraser, David. *Knight's Cross.* New York: Harper Collins, 1993.
Irving, David. *The Trail of the Fox.* New York: E. P. Dutton, 1977.
Rommel, Erwin. *Infantry Attacks.* Vienna, VA: Athena Press, 1979.

Roosevelt, Franklin D. (1882–1945)

Born in Hyde Park, New York, the only son of an elderly businessman and his second wife, Roosevelt was privately educated at home until age 14. Following graduation from Groton and Harvard University (1904), Roosevelt studied law at Columbia University but left before earning a degree. A Democrat, he was elected to the New York State Senate in 1910. He was appointed assistant secretary of the navy in the Wilson administration (1913–1921). He advocated preparedness and a big navy and instituted training for reservists prior to American entry into World War I. He also supported universal military training. In the Democratic debacle of 1920, he failed to win election to the vice presidency as James Cox's running mate, and his political activity was interrupted by a bout of polio in 1921. Partially recovered by 1928, though he never regained the use of his legs, he was elected governor of New York (1928) and thirty-second president of the United States (1932, 1936, 1940, 1944). In foreign policy, Roosevelt, well ahead of most Americans, gradually came to appreciate the threat posed to world peace and American security by totalitarian dictators. The United States maintained an uneasy neutrality during international crises such as the Spanish Civil War (1936–1939). At first reluctant to challenge isolationists, by 1940 Roosevelt initiated Lend-Lease to Britain and later the Soviet Union; exchanged overage American destroyers for the use of British naval bases in the Caribbean; and embargoed Japanese purchases of oil and steel. Following the Pearl Harbor attack in December 1941, Roosevelt was accused of having provoked Japan into attacking the United States first, but no credible evidence of this has yet been uncovered. (No one has yet been able to explain why Roosevelt, his hands full with the German menace, would wish to provoke Japan as well.) With British prime minister Winston Churchill, Roosevelt determined that Nazi Germany, the more dangerous foe, must be defeated before Japan. In sometimes uneasy alliance with Churchill and later Soviet premier Stalin, a second front strategy was devised, with invasions of North Africa, Italy, and finally, in June 1944, France and the continent of Europe itself. Other challenges included German submarine attacks on Atlantic shipping and the decision to develop an atomic bomb. Following the American victory at Midway (June 1942), Roosevelt supported an indirect policy in the Pacific. Naval and Marine forces "island-hopped," temporarily bypassing Japanese strong points, while General Douglas MacArthur's troops fought their way north from Australia to New Guinea and the Philippines. Roosevelt believed that with American material support, Nationalist China would be a strong postwar partner, but internal corruption and intermittent civil war with Communist forces rendered that policy untenable.

Wartime growth of the U.S. economy overcame the lingering effects from the Great Depression. Compared to the Allies and belligerents, the American public had guns as well as butter during the war. Roosevelt never fully mobilized the American home front. The United States could also depend in part on growing Soviet might to help confront Germany and Japan and hold down American casualties. Alone among the leaders of the belligerent powers of World War II, Franklin Roosevelt had to face the public in a general election; he won handily over the New Yorker Thomas E. Dewey and became the only American president to win a fourth term, after his unprecedented third-term victory in 1940.

Valuable worldwide intelligence services were performed by the wartime Office of Strategic Services (OSS), forerunner of the modern Central Intelligence Agency, and by the Office of War Information (OWI), as well as a host of government and military-governmental agencies that projected American military and economic might around the globe.

President Franklin D. Roosevelt meets with Soviet premier Josef Stalin (left) and British prime minister Winston Churchill (right) at the Russian embassy in Teheran. The conference, held 28 November to 1 December 1943, was one of several meetings held by the three to discuss strategy during World War II. (Library of Congress)

Roosevelt, Churchill, and Stalin initiated a policy of unconditional surrender for Axis forces and punishment for their crimes. Looking toward the postwar world, Roosevelt supported the creation of the United Nations, International Monetary Fund, and World Bank. His characteristic optimism led him to believe that Nationalist China would be a powerful ally during the war and to rely too much on Stalin's good intentions toward eastern Europe. Literally worn out by his labors, Roosevelt died of a cerebral hemorrhage on 12 April 1945, stunning the nation, many of whose citizens had no adult memories of any other chief executive.

Kier Sterling

See also: Churchill, Winston; Stalin; World War II

References and further reading:

Dallek, Robert. *Franklin Roosevelt and American Foreign Policy, 1932–1945.* New York: Oxford University Press, 1979.

Davis, Kenneth S. *FDR: The Beckoning of Destiny, 1882–1928.* New York: Putnam, 1972.

Kimball, Warren F. *Forged in War: Roosevelt, Churchill, and the Second World War.* New York: William Morrow, 1997.

Larrabee, Eric. *Commander in Chief: Franklin Delano Roosevelt, His Lieutenants and Their War.* New York: Harper and Row, 1987.

Root, Elihu (1845–1937)

Lawyer and statesman who developed organizational structure for the twentieth-century U.S. Army. One of the leading corporate lawyers in the United States during the late nineteenth century, Root served as secretary of war under Presidents William McKinley and Theodore Roosevelt (1899–1904). Charged with the administration of territories acquired in the Spanish-American War, Root promoted effective government in Puerto Rico through the Foraker Act and drafted the Platt Amendment to safeguard U.S. interests in Cuba. He guided the suppression of the Filipino insurrection and wrote the constitution and legal code that governed the territory.

At home, Root's reforms improved the War Department's efficiency and the army's readiness. The General Staff Act of 1903 created the General Staff Corps and the Office of Chief of Staff, eliminating long-standing problems regarding the chain of command. It also established the Army War College to prepare officers for war. The Militia Act of 1903 established federal control over militia training, thereby transforming the state militia into the modern National Guard. Root also initiated the practice of rotating staff and line assignments.

As secretary of state (1905–1909), he improved U.S.–Latin American relations, smoothed U.S.-Japanese difficulties, negotiated arbitration treaties, and resolved the U.S.-Canadian fisheries dispute. For these accomplishments and his earlier administration of U.S. possessions, Root received the Nobel Peace Prize in 1912. In 1917, President Woodrow Wilson sent Root to Russia in a futile attempt to bolster the government of Alexander Kerensky. In 1921, President Warren G. Harding appointed him as a delegate to the Washington Conference on naval disarmament. During his later years, Root unsuccessfully advocated U.S. membership in the World Court and directed the Carnegie Endowment for International Peace (1910–1925).

Dean Fafoutis

See also: Philippine Insurrection; Spanish-American War

References and further reading:

Jessup, Philip C. *Elihu Root.* 2 vols. New York: Dodd, Mead, 1938.

Leopold, Richard W. *Elihu Root and the Conservative Tradition.* Boston: Little, Brown, 1954.

Rorke's Drift (22–23 January 1879)

A battle during the Anglo-Zulu War that was significant in restoring British morale after a crushing defeat at Isandlwana earlier the same day. Between 3,000 and 4,000 of the Zulu reserve under Prince Dabulamanzi kaMpande, who had taken no part in the Battle of Isandlwana on 22 January 1879 but were determined to prove their prowess, crossed the Mzinyathi (Buffalo) River to invade Natal.

Lieutenant J. R. M. Chard of the Royal Engineers, the commander of the small British base at Rorke's Drift, learned at about 3:00 P.M. that the Zulu were approaching. He hastily improvised a defensive perimeter of mealie-bags connecting the loopholed commissariat store, hospital, and stone cattle kraal held by 8 officers and 131 men, 35 of whom were sick. The Zulu began their assault at about 4:30, first from the south and then from the northwest, but were driven back by point-blank fire and the bayonet. At dusk, the Zulu extended their attack along the northern perimeter. The garrison fell back on final, improvised defenses, abandoning the hospital before it was fully evacuated.

However, heavy casualties, darkness, and fear of a relieving British force dissuaded the Zulu from making another full-scale assault after about 9:00 P.M., and firing ceased at about 4:00 A.M. The Zulu withdrew at first light, avoiding the remnants of the British column marching back from Isandlwana. Seventeen of the defenders were killed, and at least 600 Zulu. Eleven Victoria Crosses were awarded to members of the garrison.

John Laband

See also: Anglo-Zulu War; Isandlwana; Khambula

References and further reading:

Knight, Ian. *Nothing Remains but to Fight. The Defence of Rorke's Drift, 1879.* London: Greenhill Books, 1993.

———. *Rorke's Drift, 1879: "Pinned Like Rats in a Hole."* London: Osprey Military, 1996.

Laband, John. "'O! Let's Go and Have a Fight at Jim's!' The Zulu at the Battle of Rorke's Drift." In *Kingdom and Colony at War: Sixteen Studies on the Anglo-Zulu War of 1879,* ed. John Laband and Paul Thompson. Pietermaritzburg and Cape Town: University of Natal Press and N & S Press, 1990.

Laband, John, and Paul Thompson. *The Illustrated Guide to the Anglo-Zulu War.* Pietermaritzburg: University of Natal Press, 2000.

Rosecrans, William Starke (1819–1898)

Union general in the American Civil War. Rosecrans was born in Kingston, Ohio, on 6 September 1819. After graduating fifth of 56 in the class of 1842 at the U.S. Military Academy at West Point, where his roommate was James Longstreet, he was commissioned in the army engineers and assigned to the construction of fortifications at Hampton Roads, Virginia. He taught at West Point from 1843 to 1847 and then resigned from the army in 1854. He quickly became a successful businessman, mining engineer, architect, and inventor.

Volunteering for duty under George B. McClellan in Ohio as soon as the Civil War broke out, Rosecrans first supervised the building of Camp Dennison, Ohio, and then became colonel of the 23rd Ohio Volunteer Infantry in June 1861. Among the soldiers under his command were future presidents Rutherford B. Hayes and William McKinley. As brigadier general of regulars and commander of the Army of Western Virginia, Rosecrans was the victor at Rich Mountain, Virginia (now West Virginia), on 11 July and at Carnifex Ferry, Virginia, on 10 September. When McClellan neglected to give him credit for these victories in the official reports, Rosecrans requested and received a transfer to the western theater.

Rosecrans performed well as commander of the left flank of the Army of the Mississippi during Henry W. Halleck's march to Corinth, Mississippi, and replaced John Pope as commander of this army in June 1862. Under Ulysses S. Grant, he won at Iuka, Mississippi, on 19 September and at Corinth on 3–4 October but failed both times to exploit his advantage. After Corinth, he was promoted to major general and given command of the XIV Corps, which he reshaped into the Army of the Cumberland. He defeated Braxton Bragg at Murfreesboro, and in the Tullahoma campaign that summer, he maneuvered Bragg out of Chattanooga with minimal casualties. After Bragg routed him at Chickamauga, he retreated to Chattanooga, where Grant relieved him of command in October.

Rosecrans served out the rest of the war in minor functions and resigned in 1867. He represented California in Congress as a Democrat from 1881 to 1885 and died at Redondo Beach near Los Angeles on 11 March 1898.

Eric v. d. Luft

See also: American Civil War; Bragg, Braxton; Chattanooga, Battle of; Chickamauga, Battle of; Grant, Ulysses Simpson; Halleck, Henry Wager; Longstreet, James; McClellan, George Brinton; Murfreesboro; Pope, John

References and further reading:

Lamers, William M. *The Edge of Glory: A Biography of General William S. Rosecrans, U.S.A.* Baton Rouge: Louisiana State University Press, 1999.

U.S. Congress Joint Committee on the Conduct of the War. *Rosecrans's Campaigns.* Millwood, NY: Kraus Reprint, 1977.

Rossbach (5 November 1757)

Battle of the Seven Years' War (1756–1763) between Prussia and France and its imperial allies. In September 1757, Prussia was bankrupt and faced defeat. Frederick the Great, convinced that only military victory could secure peace, decided to attack the Franco-Imperial army under Prince Joseph von Sachsen-Hildburghausen and the French marshal Charles de Soubise. In two weeks, Frederick marched 170 miles, encountering the allies near the village of Rossbach. The Prussian army numbered 22,000 soldiers and 80 guns versus 42,500 allies and 114 guns. Sachsen-Hildburghausen convinced Soubise to attack, believing that a major victory would end the war.

Frederick crossed the Saale River on 3 November and awaited the allies' attack. On the morning of 5 November, they crossed his front. He marched his troops parallel to and ahead of the allied army in order to turn onto their right flank and stand perpendicular to their advancing columns. The allies interpreted Frederick's movement as a retreat and ordered a general pursuit. At 3:00 P.M., they realized their mistake as Prussian guns opened fire. Friedrich von Seydlitz led a cavalry counterattack and soon overwhelmed the allied advance units. Victorious, he then led numerous charges against the flanks of the allied army as Frederick ordered his infantry forward. The Franco-Imperial army disintegrated into a confused mass. French units formed columns, only to be torn to shreds by Prussian artillery and musket fire. By late afternoon, a final cavalry assault settled the affair, and the allied army retreated haphazardly into the darkness.

Rossbach was a spectacular Prussian victory. Frederick lost only 169 killed and 379 wounded, whereas France and the empire lost 5,000 dead and wounded, 5,000 prisoners (including 11 generals), and 72 guns. Because of Rossbach, Great Britain granted a subsidy to Prussia, thereby saving Frederick's war effort.

Patrick J. Speelman

See also: Frederick the Great, King of Prussia; Seven Years' War

References and further reading:

Duffy, Christopher. *Frederick the Great: A Military Life.* London: Routledge & Kegan Paul, 1985.

Showalter, Dennis. *The Wars of Frederick the Great.* London: Longman, 1996.

Weigley, Russell F. *The Age of Battles: The Quest for Decisive Warfare from Breitenfeld to Waterloo.* Bloomington: Indiana University Press, 1991.

Rundstedt, Karl Rudolph Gerd von (1875–1953)

German field marshal. Rundstedt was commissioned an infantry officer in 1893, serving mainly as an adjutant. In 1902, he married Louise von Goetz and passed the general staff exam. After graduating from the Kriegsakademie and serving his probation, Rundstedt joined the Troop General Staff as a captain.

World War I saw Rundstedt serving on various staffs on both fronts, where he gained experience in occupation duties and mobile operations. After the Armistice, he was selected as a Truppenamt (the successor to the now-forbidden general staff) officer. Rising quickly within the small Reichswehr, he made general lieutenant in 1929 and in 1932 received the command of the Wehrkreis III, the troops used to destroy Prussia's social-democratic government.

Retiring in 1938, he nonetheless played a significant role in the conquest of Poland and France, being promoted to field marshal in June 1940. Rundstedt had been raised to serve a civilized state, and nothing in his background fitted him to confront barbarism: he promulgated Adolf Hitler's commissar order and another granting the troops immunity for virtually any crimes committed against Soviet civilians. Rundstedt also ordered the troops to cease photographing these crimes.

Hitler first relieved Rundstedt of command on 1 December 1941, sending him home with a guard of honor. Although they would differ, Rundstedt repaid Hitler's respect with total devotion. Appointed commander in chief on the western front in March 1942, he was relieved on 3 July 1944 and reappointed on 4 September 1944 to the same position.

Portrait of Karl Rudolph Gerd von Rundstedt. (Library of Congress)

Aware of but never participating in the plots against Hitler, Rundstedt sat on the Court of Honor that expelled dissident officers from the army for Gestapo prosecution. He delivered the eulogy at Rommel's funeral when the latter committed suicide, and rank folly though the Ardennes Offensive was, he nonetheless oversaw it. Rundstedt died in straitened circumstances on 24 February 1953.

Erin E. Solaro

See also: Ardennes, Battle of; Ethics of Warfare; France; Hitler, Adolf

References and further reading:

Barnett, Correlli. *Hitler's Generals.* New York: William Morrow, 1989.
Messenger, Charles. *The Last Prussian.* McLean, VA: Brassey's, 1991.

Rupert, Prince (1619–1682)

British general and admiral. Rupert was the leading British military figure during the Civil and Dutch Wars of the seventeenth century.

Rupert was the nephew of Charles I, king of England, and son of Frederick V, "winter king" of Bohemia during the Thirty Years' War. Driven into exile after the Battle of White Mountain on 8 November 1620, Rupert was raised in Holland, where he learned the art of war and command from Frederick Henry of Orange. Charles was greatly impressed with his nephew when Rupert visited England from 1635 to 1637. When Rupert was captured by imperial forces in 1638, Charles negotiated his release after three years.

Rupert arrived in England just before the war broke out between Charles and Parliament in August 1642. Despite his youth, Charles named Rupert to command his cavalry. The mounted troops constituted the cream of the royalist forces. The "Cavaliers" were natural riders, and their mounts were the best in England. Rupert's attempts to discipline his command, however, were largely unsuccessful. The royalist cavalry proved to be irresistible in a charge but nearly impossible to reorganize for further attacks.

Rupert proved his abilities in battle at Powick Bridge and Edgehill on 23 October. He won further victories with his small forces in the spring of 1643 and then joined Sir Ralph Hopton to capture Bristol. In 1644, Rupert led a devastating march into Yorkshire and relieved the royalist garrison at York. At Marston Moor, on 2 July 1644, Rupert devised the royalist plan of battle. His horse successfully drove off the opposing Parliamentarian cavalry, but Rupert was unable to rally them to defeat Cromwell's Ironsides. Rupert's army was virtually destroyed.

In November 1644, Rupert was named lieutenant general of all royalist armies. Against his advice, Charles decided to accept battle against Cromwell and Fairfax at Naseby on 4 June 1645. Despite a brilliant performance by Rupert, the royalist cavalry dispersed itself in pursuit and was unable to prevent the defeat of its infantry by the New Model Army. Rupert counseled peace, but his enemies at court convinced Charles to send Rupert into exile.

Rupert led a privateering expedition for the royalist cause between 1649 and 1653. After the Restoration in 1660, Charles II welcomed Rupert back to England. During the Dutch Wars, Rupert held several naval commands. He proved to be a competent naval commander and fought at Lowestoft (1665), the Four Days Battle (1666), and Sole Bay (1672). Rupert became commander in chief of the fleet in 1673 and led it to partial victories at Schooneveldt Bank and Texel, both in 1673.

Tim J. Watts

See also: Edgehill, Battle of; English Civil War (1642–1649); Marston Moor; Naseby

References and further reading:

Cleugh, James. *Prince Rupert.* London: G. Bles, 1934.
Kitson, Frank. *Prince Rupert: Admiral and General-at-Sea.* London: Constable, 1998.
Young, Peter, and Richard Holmes. *The English Civil War: A Military History of the Three Civil Wars, 1642–1651.* London: Eyre Methuen, 1974.

Russia, Allied Intervention in

The entry of American, British, French, and Japanese forces onto Russian territory during the Russian Civil War (1918–1921). Following clashes between the Czech Legion and the Bolsheviks as the former tried to evacuate Siberian Russia, the Allies agreed to intervene to assist the Czechs but never cooperated with one another. Britain and France landed in Russia with two and three battalions, respectively, even entertaining the possibility of opening an eastern front against the Germans. The Americans landed units equal in strength to a division at Arkhangel'sk, Murmansk in northwestern Russia, and Vladivostok in the Far East, both to support the legion and to monitor the Japanese, who had entered Siberia with 12 divisions. Japanese intervention was motivated by that country's own imperialist aims and without consultation with the Allies. The interventions proved ineffectual and had no real effect on the outcome of the Russian Civil War, though Russia resented the intervention and remained wary of the West for years afterward.

Britain withdrew support for the main White Army following an unsuccessful drive into central Russia by General Anton Ivanovich Denikin's Volunteer Army, which was stopped 80 miles short of Moscow. General Peter Nikolaevich Wrangel' replaced Denikin, and France continued to support Wrangel' until his army retreated into the Crimea and the Kuban region; the general and some of his forces were evacuated from the Crimea in 1920. Britain also supported General Nicholai Yudenich's unsuccessful drives on Petrograd (St. Petersburg, later Leningrad) from the Baltic states. Though the policy did not destroy Bolshevik power in Russia, the Baltic states did win their independence from Russia.

Admiral Aleksandr Vasil'evich Kolchak's White Army in Siberia launched an offensive in March 1919, but it quickly collapsed and the Whites were routed completely. The Czech Legion, which reached an agreement with the Bolsheviks, handed Kolchak over to the Reds and finally evacuated Russian Siberia in 1920. American troops in Vladivostok were soon evacuated as well, leaving only the Japanese, who remained in eastern Siberia until the Soviets gained control there in 1924.

Michael C. Paul

See also: Bolshevik Revolution; Denikin, Anton Ivanovich; Kolchak, Aleksandr Vasili'evich; Russian Civil War (1918–1921); Russo-Polish War; Warsaw/Vistula; Wrangel', Peter Nikolaevich

References and further reading:

Bradley, John. *Allied Intervention in Russia 1917–1920.* Lanham, MD: University Press of America, 1984.

Silverlight, John. *The Victors' Dilemma: Allied Intervention in the Russian Civil War.* New York: Arno, 1971.

Russian and Soviet Armies

The Rus', or Russian army of the Kievan period (ninth to thirteenth centuries), was made up of the princes' mounted retinues (*druzhina*) of cavalrymen supplemented by city or town militias of infantry and mercenaries from among the non-Slavic tribes. Usually made up of some 15,000–25,000 men, it is thought to have reached 60,000 troops during Prince Sviatoslav's invasion of Bulgaria in 970. The troops were armed with bows and arrows, swords, spears, and clubs.

During the Mongol period (thirteenth to fifteenth centuries), Russia fragmented into various principalities with their own armies, usually made up of cavalrymen. With the centralization of power in Muscovy in the late fifteenth century, the army was reorganized to serve the central government rather than regional princes. In the early Muscovite period, the army is thought to have numbered 40,000–50,000 men, and frontier troops were organized in the fourteenth century. Grand Prince Ivan III established a middle-service class of mounted troops in 1498, when he redistributed the newly confiscated Novgorodian lands to servitors who held land on condition of service to the grand prince. The beginnings of an artillery corps was started by the Italian mercenary Aristotle Fioraventi, who brought bronze cannon to Russia in 1478. It was greatly expanded by Czar Ivan IV the Terrible when he attached two to four light cannons to each regiment in 1552. That same year, Ivan took 150 medium and heavy cannon on his campaign against Kazan. By the end of the 1500s, the Muscovite artillery park is thought to have numbered 2,000 guns, and the armed forces numbered some 300,000 men, including artillerymen and engineers.

In addition to these advances, Ivan IV established the *strel'tsy* (musketeers), who in addition to their military functions served as a police force and fire brigade in Moscow and also protected the czar and other Russian and foreign dignitaries. At this time, the army consisted of five regiments: a vanguard regiment, a left and right regiment, the czar's regiment (which took the center in battle), and a rearguard regiment. The Muscovite period also saw the establishment of the chanceries (*pristavy*) to oversee the recruitment, training, payment, and feeding of the armed forces and the production of weaponry and artillery.

In the seventeenth century, foreign mercenaries were hired but proved unreliable, defecting during key points in several battles and leading to Swedish and Polish victories over Russian forces. By 1630, western European officers were recruited to train Russian troops. These "New Formation Troops" or "Regiments of New Formation," as well as the *strel'tsy,* served as the backbone of the army until the time of Peter the Great. The standing army at this time numbered 90,000 regulars and 60,000 irregulars, mostly Cossacks. The army was managed by 18 chanceries prior to the reforms of 1682, when power was concentrated in three chanceries under a single directorate.

Military reform was one of the main aims of the Petrine reforms, and the czar built up a new army around what originally had been his "playmate" regiments, the Preobrezhensky and Semonovsky regiments. Following a revolt by the *strel'tsy* in 1698, Peter suppressed them and rebuilt the army along Western lines, creating 27 regiments drawn from the new formation regiments to replace the disbanded *strel'tsy.* After suffering initial defeats in the Great Northern War (1700–1721), Peter's new army won a resounding victory over Sweden's Charles XII at Poltava in the Ukraine in July 1709. During the war, the Russian army also captured the western Baltic coast, where Peter built his new capital, St. Petersburg (1703), and also took what is now Estonia and Latvia. By the end of the war, the army totaled 180,000–220,000 men. Russian armies were less successful in the south, where they were defeated by the Turks in 1711. At the time of Peter's death in 1725, the army comprised 2 Guards

Infantry Regiments, 5 Grenadier Infantry Regiments, and 33 regular infantry regiments totaling 70,000 men; 3 Grenadier Cavalry Regiments and 30 Dragoon regiments totaling 38,000 cavalry; 4,000 artillerymen and engineers; 78,000 local (garrison) units; and 35,000 other regular troops; plus an unknown number of irregular troops such as Cossacks and Kalmyks. Peter placed the army under three government offices: the military department, the commissariat (supply and staffing), and the artillery department. He created a Military College in 1719 and established the Table of Ranks in 1722. A School of Mathematics and Navigation had been established in 1701 and an Artillery School in 1714. An engineering school was established in 1712 and moved to St. Petersburg in 1719.

In the eighteenth century, the Russian Empire became a major European power, and the Russian army came to play a role in European wars, winning engagements at Gross-Egersdorf, Palzig, Kunersdorf, and Colberg during the Seven Years' War and briefly occupying Berlin in 1760. At that time, the army comprised 331,000 men, including 172,000 in the field army, 74,000 garrison troops, 28,000 in military settlements, 13,000 artillerymen and engineers, and 44,000 irregular troops. The artillery was very well equipped with the Shuvalov "secret" howitzer and the "Unicorn" artillery piece. In 1793, the term of service was reduced from life to 25 years. By the end of the century, the military numbered 500,000 men. A general staff was created in 1763, and training regulations were issued that same year.

During the Napoleonic Wars, an army under Aleksandr Suvorov operated in Switzerland and Italy and achieved some success against the French revolutionary armies at Focsani, Rumnic, and Novi before being withdrawn by Paul I in 1798, when the czar reversed his mother's anti-French policies. At the beginning of the nineteenth century, the Russian army numbered 204,000 infantry, 45,000 cavalry, 25,000 artillery, and 2,700 engineers, as well as 100,000 irregulars. The infantry was armed with the 1753 flintlock and the 1809 seven-line gun. The Ministry of War was created in 1802.

Following the Napoleonic Wars, the Military Academy of the General Staff was established in 1832, and the term of service was reduced to 20 years in 1834. The Russian army continued to play a role in Europe, putting down uprisings in Walachia and Hungary in 1848–1849, but was defeated in the Crimean War because of inferior weaponry and poor infrastructure and logistics in the region. After the war, Russia withdrew from European affairs for several decades. In 1864, the army consisted of 47 infantry divisions, 47 artillery brigades, and 10 cavalry divisions and numbered 760,000 men in peacetime with 1,000,000 reserves. Smoothbore muskets were replaced by repeating rifles, and the artillery was given repeating steel guns. Universal military obligation was established in 1874 for all males over 21. Six years of service was the norm, with nine more in the reserves. An Artillery and Engineering Academy was established in 1855, along with Academies of Military Law and Military Medicine.

In 1878, Russia clashed once again with Turkey, defeating it in the Balkans and bringing independence to Bulgaria. Following the reforms of the 1860s and 1870s, the Russian army had become a people's army, and by the time of the Russo-Japanese War (1904–1905) numbered 1 million men in peacetime, with the ability to expand to 4.5 million men. The 7.62 five-shot repeating Mosin rifle was the standard infantry weapon until the time of World War I. Despite reforms, poor supplies and infrastructure again plagued the Russian army, and Russia lost to Japan in the Far East. After the Russo-Japanese War, military service was set at 18 years, with 3 or 4 years in active service and 15 years in the reserves.

On the eve of World War I, Russia's army numbered 1.3 million men; 3.5 million men were mobilized in the first five months of the war. The army also had 7,112 light guns, 791 heavy guns, 4,157 machine guns, 4,519,000 rifles, 263 aircraft, and 4,037 automobiles, but a poor supply network meant that almost one-quarter of the troops at the Battles of Tannenberg and the Masurian Lakes in 1914 came to the front without rifles.

The Russian army was disbanded in 1917, and the Workers' and Peasants' Red Army was established in 1918 under Commissar of War Leon Trotsky, utilizing old czarist officers. It eventually defeated the White forces in the revolution and ensured the survival of the Soviet Union. By 1937, the Red army had the largest armored and airborne forces in the world, consisting of 1,500 tanks and 10,000 aircraft. But the Red army suffered massive losses during the Stalinist purges. Perhaps 40,000 officers were executed, their loss probably contributing to the catastrophic defeats in the initial period of Operation BARBAROSSA. In World War II, the Red army numbered 5,300,000 men in 527 rifle divisions, 302 armored and mechanized brigades, and 42 artillery divisions. It had 13,400 tanks and 16,000 planes.

During the Cold War, the Soviet armed forces were divided into five branches: the army, navy, air force, antiaircraft defense forces, and strategic rocket forces. The Soviet armed forces totaled 3,425,000 men in the 1960s. In the 1970s, the army alone numbered 2 million men in 164 divisions. The Russian Federation kept a lion's share of the former Soviet armed forces (2 million troops in 1992). By 2000, in the post-USSR era, the Russian armed forces had been reduced to 1.2 million troops (not including border troops or strategic rocket forces). The army numbered 670,000 troops organized into 69 divisions (17 armored, 47 motorized infantry, and 5 airborne), with 15,000 tanks and 20,000 artillery pieces, though low morale and poor training and equipment maintenance meant the army was less effective than such numbers would indicate.

Michael C. Paul

See also: Borodino; Crimean War; Moscow, Retreat from; Suvorov, Aleksandr Vasilyevich; Tannenberg and the Masurian Lakes; Trotsky, Leon; Tukhachevsky, Mikhail Nikolayevich; World War I; World War II

References and further reading:

Erickson, John. *Soviet Military Power.* London: Royal United Services Institute, 1971.

Hellie, Richard. *Enserfment and Military Change in Muscovy.* Chicago: University of Chicago Press, 1971.

Keep, John. *Soldiers of the Tsar.* Oxford, UK: Clarendon, 1985.

Liddell-Hart, B. H. *The Red Army, 1918–1945.* New York: Harcourt, Brace, 1956.

White, D. F. *The Growth of the Red Army.* Westport, CT: Hyperion Press, 1943.

Russian Civil War (1425–1453)

Muscovy succession dispute that led to a protracted power struggle. Throughout the Kievan and early Muscovite periods, the princes of Russia followed the custom of lateral succession. The throne passed from brother to brother, and when that generation died out, it passed to the eldest son of the eldest brother who had held the throne before. Sons whose father had died before holding the throne were excluded (*izgoi*) from the line of succession. This changed in the late fourteenth and early fifteenth centuries, when the grand princes of Moscow, after consolidating their power, attempted to adopt a policy of linear succession to keep power in Moscow, rather than allowing princes outside Moscow to gain the grand princely throne.

Dmitry Donskoi (r. 1359–1389) stipulated in his last testament that his second son, Yury, was to succeed Vasily I should Vasily die without male issue, but Vasily's son, Vasily II, was born in 1415. In 1425, Vasily II succeeded his father. A regency council was set up consisting of Vasily's mother, the Metropolitan Foty of the Orthodox Church, and Boyar I. D. Vsevolozhsky. Vasily's maternal grandfather, Grand Prince Vytautas (Witold) of Lithuania, served as Vasily's guardian.

Faced with this situation, Vasily's uncle, Yury Dmitrevich, argued that in his testament, Dmitry had stated that Yury was to succeed Vasily I (ignoring the fact that this provision was to have no effect if Vasily had a son). Further, by the custom of lateral succession, he, Yury, was the rightful heir to the grand princely throne and refused to recognize Vasily II as grand prince. He was joined in this dispute by his sons, Vasily Kosoi and Dmitry Shemiaka.

The dynastic war of succession that ensued lasted for much of Vasily II's reign. Yury refused to come to Moscow and swear allegiance to Vasily, but an outbreak of the plague, as well as Vytautas's protection of Vasily, led to a truce. The deaths of Vytautas in 1430 and Foty in 1431 allowed Yury to renew his claim to the throne. Both Vasily and Yury appealed to the Tartar khan of the Golden Horde for resolution of the dispute, and the khan ruled in favor of Vasily. Yury, granted the principalities of Dmitrov by the khan, would not accept the decision and marched against Vasily, defeating the grand prince's forces on the Klyazma River in April 1433. Yury marched into Moscow and made peace with Vasily but was unable stay in power and soon ceded the grand princely throne and his own principality of Dmitrov to Vasily. At this point, Vasily launched a campaign against his cousins, who had not been party to the agreement between Vasily II and Yury. The grand prince's army was again defeated (September 1433). Soon afterward, Yury again attacked Vasily and defeated him yet again, in March 1434. Vasily fled, and Yury again occupied Moscow, where he died on 5 June 1434.

Contrary to the custom of lateral succession and the decision of the khan, Yury's son, Vasily Kosoi, assumed the throne of the grand prince. (By the rules of lateral succession, Vasily II, as eldest member of his generation, was the rightful heir.) Despite his succession, Kosoi lost even the support of his brothers and was defeated, captured, and blinded by Vasily II in 1436. (Kosoi means "squint-eyed" in Russian, referring to this blinding.) Removed from the political scene, Kosoi died in 1447 or 1448.

Following Vasily II's return to power, tensions continued over the next decade between Dmitry Shemiaka and Vasily II. Also at this time, Vasily's son Ivan (the future Ivan III) was born in 1440. Disputes over the distribution of inheritance, Shemiaka's contribution to Vasily's military ventures, and tribute to the Golden Horde never resulted in open warfare. An unrelated incident was the catalyst for renewed conflict. Khan Ulu-Muhammed, migrating with his horde from Crimea, clashed with Muscovite troops near Murom and remained in the area to pillage. Leading a small force, Vasily unexpectedly came upon Ulu-Muhammed outside Suzdal, on 7 July 1445 and was wounded and captured.

Dmitry Shemiaka, the next senior member of this generation, assumed the grand princely throne, but Vasily negotiated with the khan and was released in November 1445, on the condition that he pay a large ransom and a higher tribute than before. Rather than yield, Dmitry used the incident to renew the dynastic struggle. He seized Vasily's mother and wife while Vasily was on pilgrimage to the Trinity Monastery north of Moscow and sent a force to arrest Vasily. Vasily was accused of showing favoritism to the Tartars as well as blinding Dmitry's brother, Vasily Kosoi. In retaliation, Vasily II was likewise blinded. Shemiaka then released Vasily in September 1466, on the condition that Vasily renounce his claim to the throne and swear allegiance to Shemiaka. Vasily immediately made a pilgrimage to the St. Cyril-Beloozero Monastery, where the abbot absolved him of this oath. He then began gathering his supporters against Shemiaka. In the face of growing opposition, Shemiaka abandoned Moscow. Vasily returned in triumph in 1447 and continued the war, finally defeating Shemiaka. Fleeing to Novgorod, Shemiaka was poisoned there in 1453.

Michael C. Paul

See also: Ivan III

References and further reading:

Martin, Janet. *Medieval Russia 980–1584.* Cambridge, UK: Cambridge University Press, 1995, 237–244.

Russian Civil War (1918–1922)

A civil war in the wake of the Bolshevik Revolution of 7 November 1917, which left the Communist Bolsheviks victorious and secured the existence of the USSR. Following the Bolshevik Revolution, which was little more than a coup d'état, the Bolsheviks were recognized by the All-Russian Congress of Soviets as a new provisional government that would hold power only until the Constituent Assembly, to be elected in late November, could meet in January 1918 and establish a permanent government. However, the Bolshevik leader, Vladimir Ilich Ulyanov (known as Lenin), allowed the Constituent Assembly to meet only once, on 18 January, and then dispersed it. The Bolsheviks were then declared the permanent government by the Third Congress of Soviets on 25 January, and opponents of the Bolsheviks dispersed across the country, fearing arrest. Monarchists, conservatives, less radical leftists (mainly Socialist Revolutionaries), members and supporters of the Constituent Assembly, peasant opponents of the Bolshevik policy of forcible seizure of foodstuffs and other goods from the countryside, opponents of the Treaty of Brest-Litovsk (which gave a large swath of the western borderland of the Russian Empire to the Germans in March 1918), and others took up arms against the Bolsheviks in the first months of 1918. But the Whites, as these opponents of the Reds or Bolsheviks came to be known, never were able to form a truly united front against the Communist forces.

The Bolsheviks, united by their Communist ideology and effective leaders such as Lenin and Commissar of War Lev Davidovich Bronstein (known as Leon Trotsky), were able first to hold off repeated attacks by Whites and foreign interventionists and then push back the Whites and take control of most of the former Russian Empire. Trotsky was instrumental in the formation of the Workers and Peasants Red Army (RKKA), which first went into combat on 23 February 1918 against the Germans near Petrograd (St. Petersburg/Leningrad/St. Petersburg), often using former czarist officers and noncommissioned officers (NCOs) to lead the new Red Army. During January and February 1918, Bolshevik forces succeeded in taking control of most of Central Asia, except for Khiva and Bukhara, and took many of the Cossack lands in the south of Russia. German intervention in the West kept the Bolsheviks from taking immediate control of Belorussia and the Ukraine, and Turkish pressure in the Caucasus led the joint commissariat in that region to divide into three separate states.

On 8 June 1918, the Komuch (Committee of the Constituent Assembly), made up mainly of more moderate Socialist-Revolutionaries, formed an opposition government in Samara (Kuybyshev) on the Volga River, and in the first part of 1918, many regional governments came out in opposition to the Bolsheviks. The main anti-Bolshevik opposition changed over the course of 1918. At first, the opposition was headed mainly by Socialist-Revolutionaries, thus making this period of the civil war an internal struggle among the socialists. Beginning in November 1918, it shifted to an opposition headed by military dictators, most notably Admiral Aleksandr Kolchak and General Anton Denikin. At this point, the civil war became a truly Red versus White conflict.

At its lowest ebb, in mid-1918, the Communists held only the territory of central, European Russia, around Moscow and Petrograd. (Neither city ever fell to the anti-Communists.) Admiral Kolchak's Volunteer Army in Siberia numbered some 150,000 men, and General Denikin's army in the Ukraine and southern Russia numbered perhaps 125,000 men. Eugeny Karlovich Miller had a small army near Arkhangelsk, and General Nicholai Nicholaivech Yudenich formed a small force army in Estonia, which operated around Petrograd. However, the White armies were never as united as the Red Army. Denikin and the rest recognized Kolchak's supreme authority, but the White armies were never able to establish regular channels of communication and thus coordinate their activities to defeat the Reds. Furthermore, peasant and Ukrainian nationalist opposition hampered Denikin's activities, and Socialist-Revolutionaries under Victor Chernov undermined the Whites in Siberia, proclaiming "Neither Kolchak nor Lenin."

The Red Army reorganized several times to meet the challenges against it, reaching numbers several times larger than its White adversaries. By the end of 1918, it had some 800,000 conscripts under arms. In the course of 1919, it en-

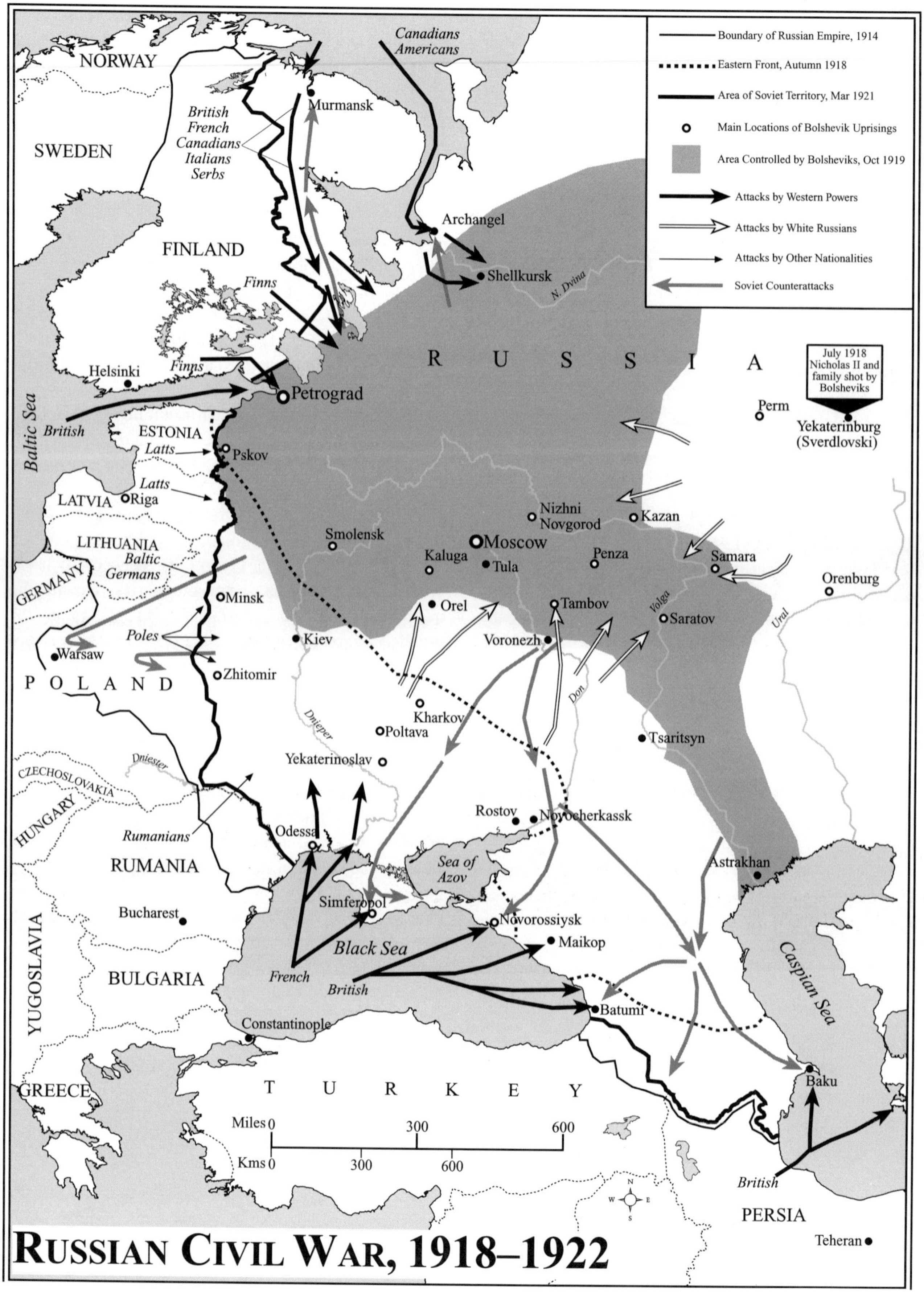

Russian Civil War, 1918–1922

listed another 1.5 million men, including 50,000 officers and 200,000 NCOs from the old czarist army. Despite its numbers, the Red Army was plagued by desertion and poor leadership to an extent never experienced by the Whites. The Bolshevik government decided to defeat Kolchak first and then turn south to crush Denikin. It defeated Kolchak in a series of engagements in the summer of 1919 and pushed him back into Siberia. The first encounters with Denikin's forces were unsuccessful, but in October 1919, the Bolsheviks were able to stop his advance at Orel and push his army back. Peasant uprisings in his rear forced his troops to retreat to the Black Sea port of Novorossiisk, where they were evacuated by the British in March 1920. Kolchak's army fell back to the Crimean Peninsula. He resigned in favor of Denikin in January, and Denikin himself resigned in favor of General Peter Wrangel' in April, leaving Wrangel' in control of the remnant of the White armies on the Crimea until it too was evacuated in November 1920.

Following the defeat of the White Army, the Soviets then turned to consolidate their authority and regain control of Ukraine, Belorussia, Siberia, the Russian Far East, the Caucasus, and other areas that had gained independence since 1917. In the aftermath of the Russo-Polish War and the defeat of the Red Army before Warsaw, the Soviets were able to take control of much of Ukraine and Belorussia. In the course of 1920 and 1921, Red forces took control in Siberia, the Far East, and the Caucasus. Consolidation of Soviet power in the Far East was not effected until 1922, when the Japanese finally withdrew. Casualty figures have never been accurately compiled, but most authorities put the number of dead at least above 1 million. The USSR emerged from its civil war exhausted from eight years of nearly continuous conflict, and in many areas, actual starvation stalked the land.

In addition to defeating the White armies, the Bolsheviks attempted to wipe out the Romanov Dynasty, which had been overthrown by a moderate revolution in February 1917. Czar Nicholas II and his immediate family were executed, along with two associates, on 16 July 1918. The czar's brother, Grand Prince Mikhail, was later executed by the Bolsheviks as well. Those members of the royal family left alive fled to the West, including Grand Prince Nicholas, who had been commander in chief of Russian armies during the first months of World War I. After the Russian Civil War, Soviet power was never seriously challenged internally until the nonviolent fall of the USSR in 1991.

Michael C. Paul

See also: Bolshevik Revolution; Denikin, Anton Ivanovich; Frunze, Mikhail Vasil'evich; Kolchak, Aleksandr Vasil'evich; Russia, Allied Intervention in; Russo-Polish War; Trotsky, Leon; Warsaw/Vistula; Wrangel', Peter Nikolaevich

References and further reading:

Carr, Edward Hallet. *The Bolshevik Revolution, 1917–1923.* 3 vols. New York: Macmillan, 1951–1953.

Lincoln, Bruce. *Red Victory: A History of the Russian Civil War.* New York: Simon and Schuster, 1989.

Mawdsley, Evan. *The Russian Civil War.* Boston: Allen and Unwin, 1987.

Russian Colonial Wars (1552–1917)

A series of wars over more than three centuries through which Russia spread across northern Eurasia and became the largest country on earth and a major European and Asiatic land power. It has been noted that during the period 1462–1914, the Russian Empire expanded at a rate of some 50 square miles per day for almost 500 years.

The Russian colonial empire was different from the British, French, Portuguese, Spanish, Dutch, and other colonial empires in that the colonies of Russia were contiguous with the Russian heartland, rather than being scattered across vast continents or oceans. It was also different in that many of the colonies remain to this day part of Russia or under Russian influence. Furthermore, Russia has held dominion over non-Russian peoples from the very beginning of its foundation as an organized polity in the tenth century. The earliest Russian chronicles refer to non-Slavic (and non-Christian) tribes who paid tribute to the Russian princes as "our pagans." However, only in the late sixteenth century did the Russians began to colonize significant territories populated by non-Russians.

In 1552, Ivan IV the Terrible personally led a campaign against the khanate of Kazan, east of Moscow, capturing that city after a prolonged siege. Four years later, his armies captured the khanate of Astrakhan, at the mouth of the Volga River, thus bringing the entire Volga Basin (and the lucrative trade routes along the Volga to Persia) under Russian control. Not only were the Tartars of Kazan and Astrakhan brought under Russian control, but the Chuvashians, Mordvinians, and other peoples along the Volga River became subject to the Russian czar.

In 1558, Ivan launched the Livonian War against the German cities in what are now Estonia and Latvia. After initial successes, the war turned out badly for Russia. Poland, Lithuania (the two merged in 1569), and Sweden entered the war. In 1571, the Crimean Tartars raided Moscow and destroyed much of the city. Polish king (and Lithuanian grand duke) Stepan Bathory drove the Russians out of Livonia in 1577 and then marched into Russia, capturing Polotsk in 1579 and besieging Pskov in 1582, forcing Ivan to ask for an armistice. Finally in 1583, Ivan sued for peace, having lost any gains from the first period of the war. Russian efforts to colonize the Baltic States would not begin again until the reign of Peter the Great.

However, Russia continued to hold the Volga Basin, and Ivan the Terrible's reign also saw Russian explorers, such as the famous Yermak, advance along the major rivers of Siberia, subduing indigenous tribes as they went. Russia finally reached the Pacific Ocean in 1639 and expanded into Alaska and North America, reaching as far south as Fort Ross in California, near present-day San Francisco.

During the reigns of Mikhail Romanov (r. 1613–1645) and his son Aleksei (r. 1645–1672), Russia expanded into Polish-held lands in Ukraine. The Smolensk War led to the Russian capture of that city in 1632. A Cossack uprising in Ukraine in 1648 led to a prolonged struggle between the Cossacks and Poland. Cossack hetman Boghdan Khmelnytsky asked for Russian protection, leading to the incorporation of the Cossack territories into the Russian Empire in 1654 by the Union of Pereiaslav. Conflict with Poland, however, continued until the Treaty of Andrusovo in 1667 gave Russia control of Left Bank (eastern) Ukraine and temporary control of the city of Kiev (called "the Mother of all Russian cities" in the earlier chronicles). The Russians never did relinquish control of that city, and over the next several decades, Russia subdued the Cossacks, destroying their autonomy and consolidating control of Ukraine.

Early in his reign, Peter the Great (r. 1682–1725) declared war on Sweden, assuming Sweden's young king, Charles XII, would be easily defeated, but Charles dealt the Russians a crushing defeat at Narva in 1700. The war, known as the Great Northern War, would last 21 years. After Narva, Charles turned and defeated Polish king Augustus the Strong and then marched deep into the Russian Empire before being defeated by Peter at the Battle of Poltava in Ukraine in July 1709. This battle proved the turning point in the war. Charles fled into the Ottoman Empire, never regained the initiative in the war, and was killed under mysterious circumstances in 1718 at the siege of Frederikstad in southeastern Norway. In the course of the war, Peter captured the eastern coast of the Baltic (Karelia, Ingermanland, Estonia, and Livonia), founding St. Petersburg at the mouth of the Neva River in June 1703. By 1712, he had transferred the Russian capital to St. Petersburg, demonstrating his determination to stay and to make Russia a western nation and a naval power.

The newly formed Russian navy defeated the Swedish fleet at Hanko in Finland, and Russian troops landed in Sweden, forcing Sweden to sue for peace in 1721. A further war with Sweden in the early nineteenth century led to the annexation of Finland to Russia in 1809.

Peter's efforts to expand at the expense of the Ottoman Empire met with initial success with the capture of Azov in 1696, but he suffered a humiliating defeat on the Pruth River in 1711 and lost Azov as part of the peace treaty. Peter also moved in to take advantage of unrest in Persia, and Russia took control of northern Persia from 1722 to 1735.

Repeated wars with Turkey, Persia, and various Caucasian tribes, as well as intertribal disputes, led to Russian annexation of the Caucasus. Fear of Persian conquest led the ruler of Georgia to accept annexation by Russia in 1801. Russian armies moved into Transcaucasia in 1813, annexing Armenia in 1828 and excluding Persia and Turkey from the Caucasus. Despite nominal control over the region, Russia fought a brutal and at times genocidal war to put down the Cherkes, Chechens, Ossetians, Kabardians, and other tribespeople and to defeat famous fighters such as Ghazi Muhammad and Shamil. The Caucasian War ended by 1864, but uprisings occurred in Dagestan and Chechnya in 1877 to 1878.

Under Catherine II (r. 1762–1796), Russia participated with Prussia and Austria in the three partitions of Poland in 1772, 1793, and 1795, which expanded the Russian Empire deep into Europe and led to the destruction of the Polish state. (Polish revolts in 1830–1831 and 1863 were brutally suppressed.) Several wars with Turkey led to the annexation of southern Ukraine and the Crimea in 1783, but Catherine's "Greek Project," a plan to conquer Constantinople and place her grandson Konstantin on the throne of a Christian kingdom there, never materialized.

Russian threats against Turkey in the next century led to the Crimean War (1853–1856) and Russia's defeat at the hands of France and Britain. The next two decades saw Russia withdraw from European affairs, but Russian armies again warred against Turkey in 1877–1878. Oddly, Russia turned over its North American Alaska outpost to the United States in 1867 in a straight cash deal. In the Balkans, a Russian army crossed the Shipka Pass, forced into surrender the Turkish army besieged at Plevna, and marched on Constantinople. On the Caucasian front, Russian troops took the fortress of Kars (Qars), which remained in Russian hands until 1918. The Treaty of San Stefano, signed 3 March 1878, gave Bulgaria independence from the Ottoman Empire, but Russia never colonized the Balkans.

In Central Asia, the Russians subdued the Uzbeks between 1864 and 1873. Kokand was annexed outright in 1876, and Bukhara and Khiva were made protectorates. Turkmenistan, Russia's last conquest in Central Asia, came under Russian control in the early 1880s. Russian attempts to expand into Afghanistan led to heightened tensions with Britain and almost erupted into war in 1885.

Halted in Central Asia, Russian armies continued to expand Russian control in the Far East, gaining territories south of the Amur River and spheres of influence in Manchuria and Korea. Russian expansion in the Far East, however, led to the Russo-Japanese War of 1904–1905. Arrogant beliefs in the racial inferiority of the Japanese were replaced by the humiliation of defeat on land and sea when the Japanese launched a sneak attack on the Russian Pacific Fleet at Port Arthur. The Treaty of Portsmouth ended the war, giving Japan control of Korea and southern Sakhalin Island.

The Treaty of Brest-Litovsk, which ended Russian participation in World War I, resulted in vast colonial losses. Finland, the Baltic States, western Belorussia and Ukraine, Poland, Bessarabia, and Kars were lost, and subsequent internal strife led to the fragmentation of the old Russian Empire. Only gradually did the Soviet Union regain the territories lost after World War I; much of the old Russian Empire was regained by the end of the Russian Civil War, including eastern Ukraine, the Caucasus, Central Asia, and Siberia, all of which saw failed efforts to create independent nation-states following the October Revolution. The motivations for these territorial acquisitions are widely debated, with some scholars arguing that Soviet expansionism was nothing more than Russian geopolitical and nationalist ambitions cloaked in a red flag, whereas others argue that the Communist ideology of the Soviet Union provided new motivations for expansion.

Soviet efforts to spread world revolution into Central and Western Europe met a stunning defeat in August 1920, when the Poles under Marshal Józef Pilsudski routed the Red Army under Mikhail Tukhachevsky at Warsaw. The subsequent Treaty of Riga gave parts of Lithuania, Belorussia, and Ukraine to Poland and recognized the independence of the Baltic states.

On 26 August 1939, the Soviet Union and Nazi Germany signed a Non-Aggression Pact, secret provisions of which gave the Soviet Union a sphere of influence in Estonia, Latvia, and eastern Poland and were later revised to include Lithuania. On 17 September 1939, Soviet troops entered eastern Poland and met up with German troops who had conquered eastern Poland. The year 1939 also saw the launching of the Winter War against Finland, an effort to push the Finnish border away from Leningrad, gain lands in the north to defend access between Leningrad and Murmansk, and gain access to ports along the southern Finnish coast. The war went badly for the Soviet Union initially, though the Finns were eventually forced to surrender in March 1940. In May 1940, the Soviet secret police executed some 15,000 Polish officers at Katyn, near Smolensk, and at two other sites in the Soviet Union in an attempt to destroy any opposition to Soviet control of eastern Poland. A month later, while the world watched Nazi Germany smash the French and British armies, Soviet troops entered the Baltic states and annexed the three nations. Bessarabia (Moldavia) also was taken from Romania at this time.

Following the Nazi invasion of the Soviet Union and the Soviet victories at Stalingrad and Kursk in 1943, Soviet armies pushed into Eastern Europe, capturing all territories lost to Germany and occupying Romania, Bulgaria, Hungary, Czechoslovakia, Poland, and parts of Germany and Austria. Although the Allies had agreed at Yalta to hold elections in the occupied territories, Stalin saw the need to set up a buffer zone in Eastern Europe, arguing that Russia and the Soviet Union had suffered multiple invasions that had originated in or passed through Poland. The Soviet Union, therefore, needed friendly states on its western borders and achieved this by installing pliant Communist governments there, which ruled those countries until 1989. The Red Army maintained control of the region by brutally suppressing uprisings in East Germany in 1953, in Hungary in 1956, and in Czechoslovakia in 1968.

In the Far East, the USSR attacked Japan in the summer of 1945 and took control of Manchuria, North Korea, the southern half of Sakhalin Island, and the Kurile Islands. Withdrawing from China and Korea, the Soviet Union nevertheless supported Communist movements there, leading to the formation of the People's Republic of China in October 1949. Stalin also cautiously supported the North Korean invasion of South Korea in June 1950, and later Soviet leaders supported backed Communist movements in North Vietnam and elsewhere in Asia. In 1979, Soviet troops invaded Afghanistan in support of the Communist government there. A long guerrilla war ensued, ending with a Soviet withdrawal in 1989.

The Soviets also supported "movements of national liberation" in Asia, Africa, and the Americas throughout the latter half of the twentieth century. In 1962, the world came closest to full-scale nuclear war when the USSR placed 42 intermediate-range ballistic missiles on the Communist-ruled island of Cuba. The United States blockaded the island and forced their withdrawal in October.

Reforms brought about during the tenure of Mikhail Gorbachev as general secretary of the Communist Party of the Soviet Union led to decreased tension during the Cold War and a reduction of Soviet pressure on its Eastern European satellite states. By 1989, the Soviet bloc in Eastern Europe had fallen, and internal strife in the USSR itself, including nationalist movements in the Baltic states and Ukraine and outright wars in the Caucasus, led to the breakup of the USSR in December 1991.

Michael C. Paul

See also: Alma; Balaklava; Crimean War; Cuban Missile Crisis; Hungarian Revolt; Inkerman, Battle of the; Kars, Battle of; Livonian War; Mukden, Battle of; Northern War, Great;

Plevna/Pleven, Siege of; Poltava; Port Arthur, Siege of; Russo-Finnish Wars; Russo-Japanese War; Russo-Polish War; Russo-Swedish Wars; Russo-Turkish Wars; Sevastopol, Siege of; Soviet-Afghan War; Warsaw/Vistula; Yalu River

References and further reading:

Hunczak, Taras, ed. *Russian Imperialism from Ivan the Great to the Revolution.* New Brunswick, NJ: Rutgers University Press, 1974.

Jelavich, Barbara. *St. Petersburg and Moscow: Tsarist and Soviet Foreign Policy, 1814–1974.* Bloomington: Indiana University Press, 1974.

Kliuchevskii, V. O. *A History of Russia.* 5 vols. Trans. C. J. Hogarth. New York: Russell and Russell, 1960.

Service, Robert. *A History of Twentieth Century Russia.* Cambridge, MA: Harvard University Press, 1998.

Russian/Soviet Women in War and Resistance (1800–2000)

In no modern nation, including Israel, have women come close to the extent of Russian/Soviet women's participation in actual combat. Very few Russian women served before 1917, although in 1808, the Order of St. George was conferred on cavalrywoman Nadezhda Durova. During World War I, three women were similarly decorated: Antonina Pal'shina (cavalry and infantry); Mariya Bochkareva ("Battalion of Death" commander); and nurse Rimma Ivanova (infantry), killed in 1915 while leading troops in a successful attack.

Women's recruitment during the Civil War was largely limited to medical and political appointments. About 60 received the prestigious Order of the Red Banner, created in 1918.

In the interwar period, female career officers were usually employed in staff work, without line troops under their command. Soviet women began mastering military specialties, especially after the German invasion of the USSR, when they were given access to special courses. In 1942 about 222,000 were trained as mortar, submachine gun and machine gun operators, snipers, and communications specialists.

In addition, women trained in reserve regiments and forward units. They moved into nontraditional categories to fill developing shortages of male personnel and as part of their struggle for equality. About 400,000 were mobilized, mainly for service in Air Defense Forces and logistical units.

Female troops march during the Russian Revolution, 1917. (Hulton/Archive)

More than 800,000 served in late 1943, including 27,000 partisans, amounting to about 8 percent of military personnel. About 200,000 were decorated; by May 1990, 95 had become Heroes of the Soviet Union. The proportion of women was greatest in Air Defense Forces, amounting to 24 percent, or replacement of some 300,000 men. Still, it is evident that women were drafted mainly for auxiliary or "defensive duties" (including air defense and mine clearing) and advanced despite official policy.

Soviet sources on the history of World War II often stress that "the face of war is not a woman's face," as the wartime situation was unprecedented. After the war, women's participation declined to about 10,000 in the mid-1970s. Following the USSR's collapse, some military schools admitted women, and female military personnel increased, in part because of changes in the job market and reluctance of men to sign up. For all of their strong military participation in the twentieth century, the status of women in the Russian military remains problematic.

Kazimiera J. Cottam

References and further reading:

Cottam, Kazimiera J. *Women in War and Resistance.* Nepean, Canada: New Military Publishing, 1998.

Herspring, Dale R. "Women in the Russian Military: A Reluctant Marriage." *Minerva* 15 (Summer 1997).

Yeremin, V. I., and P. F. Isakov. *Youth during the Patriotic War* (in Russian). 2d rev. ed. Moscow: Mysl', 1984.

Russo-Chechen Conflict (1994–1996)

Separatist struggle that demonstrated Russia's military decline since the collapse of the Soviet Union. On 11 December 1994, 40,000 Russian troops invaded the small northern Caucasus territory of Chechnya. Anticipating an easy victory, Russian president Boris Yeltsin launched the attack to overturn the declaration of independence issued by Chechnya's president, the former Soviet air force general Dzhokhar Dudayev, in November 1991. The result was a bitter, 22-month war that left, according to one estimate, an incredible 80,000 dead.

During the first six-plus months of the fighting, Russian forces, superior in heavy weaponry on the ground and in the air, captured Chechnya's major population centers, the capital city of Grozny falling (February 1995) after a seven-week siege. These successes weakened Chechen military capacity but failed to produce victory. Chechen separatists took to the hills and waged, with overwhelming civilian support, guerrilla warfare against an enemy whose morale waned with each passing day.

Beginning in June 1995, full-scale fighting ceased while peace negotiations occurred. Prompted by Shamil Basayev's daring separatist raid (14 June) into Russian territory and the capture of several hundred hostages in the town of Buddenovsk, these negotiations ultimately failed; yet the accompanying cease-fire provided the Chechens an invaluable breathing space while simultaneously allowing them to reoccupy territory lost in the opening months of the conflict.

When fighting resumed in mid-December, Chechen forces launched a series of successful offensives over an eight-month period, the most significant (6 August 1996) yielding the recapture of devastated Grozny and the driving back of Russian forces to their starting point in December 1994. Though a Russian air raid killed Dudayev (21 April), this inconsequential success neither reversed demoralization in the Russian army nor altered domestic public opinion, which had turned irrevocably against the conflict. In early August, Yeltsin's government, faced with essentially starting the war over, decided on peace, dispatching National Security Chief Alexander Lebed to negotiate with Chechen supreme commander General Aslan Maskhadov. At the Daghestani border town of Khasavyurt, the generals reached an agreement (31 August) providing for Russian withdrawal, new Chechen presidential elections, and the shelving of Chechnya's constitutional status until 2001. In more recent years, the struggle has been sporadically renewed, and Chechnya seems far from accepting its status as a territory of Russia.

Bruce J. DeHart

References and further reading:

Dunlop, John B. *Russia Confronts Chechnya: Roots of a Separatist Conflict.* New York: Cambridge University Press, 1998.

Gall, Carlotta, and Thomas de Waal. *Chechnya: Calamity in the Caucasus.* New York: New York University Press, 1999.

Knezys, Stasys, and Romanus Sedlickas. *The War in Chechnya.* College Station: Texas A & M Press, 1999.

Lieven, Anatol. *Chechnya: Tombstone of Russian Power.* New Haven, CT: Yale University Press, 1998.

Russo-Finnish Wars (1939–1944)

Border clashes between the Soviet Union and Finland. In October 1939, the Soviet Union demanded that Finland lease its naval base on the Hanko Peninsula; cede several islands in the Gulf of Finland, including Hogland and Koivisto; and demilitarize the Mannerheim Line, a series of fortifications along the Karelian Isthmus. After brief negotiations, the Soviets ended talks and created an incident along the frontier. Expecting rapid victory in its largest operation since its civil war, the Red Army attacked Finland on 30 November, attempting to cut the country in two. Anticipating the attack, the Finns had mobilized the equivalent of 14 divisions, or 300,000 troops, mostly reservists. The Finnish army was well trained for waging war near the Arctic Circle and for maximizing the country's static defenses but was desperately short of artillery, armored vehicles, and ammunition. Finnish army troops, pressed for arms by low budgets, coined the phrase *Molotov cocktail* after improvising petrol-filled bottles as substitute grenades.

The Finnish campaign was a disaster for the Soviets. Although the Red Army mobilized many more troops than the Finns, 22 divisions in all, its preparations were frightfully inadequate. General Kiril A. Meretskov, commanding the attack, invaded Finland without training soldiers for storming fixed defenses, acquiring accurate intelligence about Finnish positions and terrain, preparing for winter combat, or ensuring adequate supply and transportation. The Soviets called off the offensive on 20 December because of heavy casualties sustained against the Mannerheim Line. Following this disaster, the Red Army reassessed its strategy and replaced Meretskov with Semen Konstantinovich Timoshenko, one of the few commanders from the civil war to survive Stalin's purges. With improved training in waging winter warfare and storming fixed positions, the Red Army renewed its offensive on 12 February 1940. The Red Army finally pierced the Mannerheim Line, after transferring 27 new divisions to the front and backing these troops with armor. After suffering more than 264,000 casualties during the campaign, the Red Army was too exhausted to conquer the entire country. (Apologists for the Soviet Union stressed the bitter cold of the battlefield, but Russia was hardly noted for its balmy climate.) On 12 March 1940, Finland and the Soviet Union agreed to a brief armistice. The entire Karelian Isthmus had to be ceded to the Soviet Union. According to the Finns, the entire population evacuated Karelia rather than live under Soviet rule. The Soviet invasion had two major diplomatic results: the Soviet Union was expelled from the League of Nations, and Finland became a cobelligerent with Germany following the German invasion of the Soviet Union on 22 June 1941.

Once Finland had recaptured the territory it lost in 1940, the front between Finland and the Soviet remained static between 1941 and 1944. This changed when Stalin ordered Generals Leonid A. Govorov and Meretskov (who had somehow survived Stalin's wrath) to attack the Karelian Isthmus on 10 June 1944. The carefully prepared Red Army assault, partly designed as a diversion from the Red Army's upcoming offensive against German Army Group Center, Operation BAGRATION, avoided the disaster of the first campaign. The numerically superior and vastly improved Soviet forces quickly overwhelmed Finnish defenses. General Govorov moved along the Baltic Sea and captured the city of Vyborg on 21 June, while Meretskov moved north, securing central Karelia. These combined operations forced Finland to agree to a separate peace in September 1944.

Perhaps surprisingly, Finland retained its independence, proving the adage that dictators like Stalin respected only force, and the Finns had performed quite well on the battlefield. Until the fall of the Soviet Union, however, Finland's main diplomatic goal was to maintain a strict neutrality between East and West in the Cold War and to avoid aggravating its giant neighbor.

Eric Pullin

See also: BAGRATION, Operation; Leningrad, Siege of; Stalin; Timoshenko, Semen Konstantinovich; World War II

References and further reading:

Glantz, David M., and Jonathon House. *When Titan's Clashed: How the Red Army Stopped Hitler.* Lawrence: University Press of Kansas, 1995.

Krivosheev, G. F. *Soviet Casualties and Combat Losses in the Twentieth Century.* London: Greenhill Books, 1997.

Overy, Richard. *Russia's War: Blood upon the Snow.* New York: TV Books, 1997.

Trotter, William R. *A Frozen Hell: The Russo-Finnish Winter War of 1939–1940.* Chapel Hill, NC: Algonquin Books, 1991.

Werth, Alexander. *Russia at War, 1941–1945.* New York: Carroll & Graf, 1964.

Russo-Japanese War (1904–1905)

The first major defeat of a European power by an Asian nation. There was a certain air of inevitability about the Russo-Japanese War. To the proud Japanese, Russian had denied Japan the well-deserved fruits of its victory over China in 1894–1895 with the Triple Intervention; thereafter, the Russians used their assistance in the Boxer Rebellion to move more troops into Manchuria and even tried to deny Japan's position in Korea. So the Japanese strengthened and expanded their army and navy, signed an alliance with Great Britain to neutralize France and Germany, and tried half-hearted diplomacy with Russia.

The Japanese broke off diplomatic relations and then struck at the Russian fleet in Port Arthur on 8 February 1904, which was followed by two days later by a mutual declaration of war. When the war began, it appeared that the advantages rested with Russia, which had a huge army and one of the world's largest navies. However, the Japanese had a major theater superiority of forces, better quality officers, and more motivated soldiers and sailors.

There were several phases to the conflict. The Japanese engaged in a long, costly, and ultimately successful five-month siege of Port Arthur, from late May 1904 until early January 1905. Armies commanded by General Kuroki Tamemoto, Nozu Michitsura, and Oku Yasukata converged on the Russian positions. In late May, the Japanese gained control of Nashan Hill and later the port of Dairen, which meant that Port Arthur was surrounded and cut off. Meanwhile, the Russian defenders did not strengthen a reasonably formidable defensive position; there were real possibilities for defense in depth, but they made little effort. The Japanese made three bloody assaults—in each case, the men charging in close and packed formations and then waiting for siege artillery. Two more direct and costly assaults were similarly unsuccessful. Finally, a sixth major offensive in late November 1904 captured a key hill that enabled the Japanese to bring the inner defenses under artillery fire. Finally, after yet another major attack, on New Year's Day, 1905, the Russians, short of food, sick, and without hope of relief, surrendered. The Japanese suffered more than double the Russian casualties despite poor Russian handling of the defenses.

As the siege tightened, fighting moved to Manchuria, as the Japanese sought to drive the Russians out of the resource-rich Chinese province. Both sides poured tens of thousands of troops into the series of battles. Fighting began with battles near Liaoyang, 200 miles north of Port Arthur, in August and early September. Although outnumbered, the Japanese attacked but could not crush the Russian positions. Then, in January 1905, the Russians chose to attack in snow, and after failing to drive the Japanese back, the Russians chose to retreat to Mukden in more central Manchuria. Both armies dug in and then engaged in a series of battles in January, February, and March. The Japanese wanted victory before huge numbers of Russian troops arrived from European Russia. About 200,000 Japanese attacked nearly 300,000 Russians; the Japanese sought to turn first the Russian right and then the left; hard fighting pushed the Russians into a kind of "U," and the Russian commander, Aleksey Kuropatkin, feared for his communication lines and executed a good retreat further north to Harbin. By this point, the Japanese were too exhausted to pursue and attack. Russian mobilization was creaky, but eventually Russia could strengthen its forces in northeastern Asia. Both sides were ready for peace talks.

Finally, the Russian Baltic Fleet, after a dramatic sea voyage from eastern Russia around Africa (although several ships did go through the Suez Canal), Southeast Asia, and north through the China Sea, finally arrived in the war zone in the Straits of Tsushima and was nearly destroyed in the ensuing battle on 27 May 1905. Admiral Togo managed to cross the "T"—a maneuver by which he could bring his main guns to bear on the enemy, and the few surviving Russian ships made their way into harbor.

Eventually, the two warring powers accepted a mediation effort by U.S. president Theodore Roosevelt and signed a peace treaty at Portsmouth, New Hampshire, on 6 September 1905. The Russians gave up Port Arthur and the southern half of Sakhalin Island and evacuated Manchuria, clearly opening that province to Japanese exploitation, but the Japanese largely were not satisfied with the concessions gained at the peace table.

The conflict demonstrated the strength of the defense at this point in military history, especially of the machine gun; it also perhaps demonstrated the folly of direct and predictable attack. The Japanese tended to charge right at Russian defenses or at most to seek to turn a flank but still to advance in close formation, ideal for machine gun and more modern artillery fire. The Japanese also did not consider taking advantage of an indirect approach to sever Russia's long, vulnerable, single-track supply line back to eastern Russia. Instead, they forced the Russians to retreat but never truly destroyed their larger foe and thus permitted the Russians to fight again. Interestingly, the major Western powers, perhaps having little regard for Russia or Japan as military powers, did not study the impact of the machine gun on fighting against prepared defenses and hence stumbled, a decade later, into the carnage of World War I.

Charles M. Dobbs

See also: Kuropatkin, Aleksey Nikolaevich; Mukden, Battle of; Nogi, Maresuke

References and further reading:

Okamoto, Shumpei. *The Japanese Oligarchy and the Russo-Japanese War.* New York: Columbia University Press, 1970.

Warner, Denis, and Peggy Warner. *The Tide at Sunrise: A History of the Russo-Japanese War, 1904–1905.* London: Frank Cass, 1974.

Westwood, J. N. *Russia against Japan, 1904–1905: A New Look at the Russo-Japanese War.* Albany: State University of New York Press, 1986.

Russo-Polish War (1919–1921)

Conflict that established the integrity of the reborn Polish nation until the German invasion of 1939. With the end of World War I on 11 November 1918, the Polish nation reemerged on the world stage after more than 100 years of foreign occupation. In January 1919, Józef Pilsudski, who had so long fought against imperial Russia, became Poland's chief of state and commander in chief of the armed forces. The war between the Soviet Union and Poland resulted from Soviet attempts to spread communism westward, although the war-devastated nations of France and Germany would have seemed more likely candidates. Soviet aspirations were at least as much territorial as ideological. An Allied proposal for a border between the Soviet Union and Poland (known as the Curzon Line) was rejected by both sides.

In June 1919, the Red Army launched an offensive that led it to the outskirts of Warsaw itself by early August. Poland was saved by Pilsudski's counterattack (and the advice of a French military mission) on 16 August, a battle commonly known at the time as the "Miracle of the Vistula" (and sometimes also as "the battle that saved Europe from godless Bolshevism"), which forced the overextended Soviet forces to retreat back to their own territory. An armistice was signed in October 1920, which was followed by the Treaty of Riga, concluded on 18 March 1921. The new border, which corresponded roughly to pre-partition Poland, included Ukrainian, Belarusian, and Lithuanian territories, the latter of which caused a rift between the historic allies of Lithuania and Poland. Interwar Poland had an area of approximately 150,000 square miles and approximately 30 million inhabitants. The result of this conflict, other than the establishment of the border that would exist between these two nations until the Soviet invasion of Poland on 17 September 1939, was a Polish nation that contained a significant number of ethnic minorities, including Ukrainians (16 percent), Jews (10 percent), Belarusians (6 percent), and Germans (3 percent), who had no religious, ethnic, or linguistic loyalty to the Polish nation.

Alexander M. Bielakowski

See also: Pilsudski, Józef Klemens; Tukhachevsky, Mikhail Nikolayevich

References and further reading:

Davies, Norman. *White Eagle, Red Star: The Polish-Soviet War, 1919–20.* London: Macdonald and Company, 1972.

Fiddick, Thomas C. *Russia's Retreat from Poland, 1920: From Permanent Revolution to Peaceful Coexistence.* New York: St. Martin's Press, 1990.

Karolevitz, Robert F., and Ross S. Fenn. *Flight of Eagles: The Story of the American Kosciuszko Squadron in the Polish-Russian War 1919–1920.* Sioux Falls, SD: Brevet Press, 1974.

Pilsudski, Józef. *Year 1920 and Its Climax: Battle of Warsaw during the Polish-Soviet War, 1919–1920.* New York: Pilsudski Institute of America, 1972.

Russo-Swedish Wars (1240–1809)

No less than seven wars fought between Russia and Sweden over a 600-year period, often centered on the Baltic states. The war of 1240–1242 began when a Swedish, Danish, and Livonian Knight army backed by Pope Gregory IX took advantage of the second Mongol invasion of Russia. Novgorod's Prince Alexander Nevski decisively defeated them at the Neva River in 1240 and again in the famous "Battle of the Ice" at Lake Peipus, on 12 April 1242.

The war of 1590–1955 began with Czar Fedor I invading Swedish-controlled Estonia and defeating 20,000 Swedes near Narva. A one-year truce ended with Swedish forces pushing back the Russians and regaining control of northern Estonia and Livonia.

A Swedish attempt to gain the Russian throne ignited the war of 1613–1617. The Swedes occupied Novgorod, hoping to enhance the chances of King Gustavus II Adolphus to ascend the Russian throne. Muscovites attacked the Swedes at Novgorod but were repulsed. Gustavus then besieged Pskov unsuccessfully for six months in 1614. Muscovite military performance thus established Czar Michael Romanov on the Russian throne, and Gustavus made no further claims.

The war of 1656–1658 saw Czar Alexis try to regain the Baltic states from Sweden, while the Swedes were occupied in the Second Northern War (1655–1660). After capturing border towns, the Russians besieged Riga, Latvia, in July–August 1656. The Swedes responded with sorties that killed thousands and caused Alexis to flee the field. A 1658 offensive by the Russians was also repelled and resulted in a peace treaty.

In the war of 1741–1743, the Swedes attempted to regain power in eastern Europe after being foiled by the Russians under Peter the Great in the Great Northern War of 1700–1721. A poorly prepared 20,000-man Swedish army invaded Russia but was defeated at the Battle of Wilmanstrand on 3 September 1741. Russia then invaded Swedish-controlled Finland and forced 17,000 Swedes to surrender at Helsinki on 20 August 1742.

While Catherine the Great and Russia were involved in a second Russo-Turkish War, King Gustavus III invaded Russian-controlled Finland, thus beginning the war of 1788–1790. Rebellious Swedish officers caused Swedish defeats in the first two years on land and at sea. Gustavus returned home, ended the rebellion, and then returned to besiege St. Petersburg by land and sea, winning a stunning naval victory at Svensksund in July 1790. More than 150 Russian ships were sunk or captured, fully a third of the Russian fleet. However, Denmark had invaded southwestern Sweden and besieged Göteborg, forcing the Treaty of Wereloe.

The war of 1808–1809 centered on the Napoleonic Wars. France and Russia made peace in 1807 and called on Sweden to drop out of the anti-French alliance with England. Sweden refused, and Czar Alexander I invaded Swedish Finland in February 1808 and pushed the Swedes out of Finland entirely. Alexander then invaded Sweden, causing the election of a new Swedish king, Charles XIII, who signed the Treaty of Fredrikshamm, making Finland a province of Russia and marking the last Russo-Swedish War.

Christopher Howell

See also: Gustavus II Adolphus; Northern War, Great; Peter I, Romanov, Czar of Russia; Poltava

References and further reading:
Frost, Robert. *The Northern Wars: War, State and Society in Northeastern Europe, 1558–1721.* New York: Longman Press, 2000.
Wren, Melvin. *The Course of Russian History.* New York: Waveland Press, 1993.

Russo-Turkish War (April 1828–14 September 1829)

A conflict centering around the question of Greek independence from Turkey that revealed the decline in Turkish power. The Greek revolt against Turkish rule, beginning in March 1821, led Turkish sultan Mahmud II to order his vassal, Egyptian pasha Mehmet Ali, to land a force to suppress the Greeks in February 1825. Russian czar Nicholas I then issued an ultimatum in March 1826 that Turkish troops withdraw from the Danubian principalities and restore autonomy to the Greeks. The Turks agreed at Akkerman on 7 October, but military action against the Greeks continued.

By the Treaty of London (July 1827), Great Britain, France, and Russia pledged to establish an autonomous Ottoman Greek state. An allied fleet commanded by Royal Navy admiral Sir Edward Codrington sailed to the Bay of Navarino and engaged and destroyed a Turkish fleet on 20 October. On 18 December, the Turks declared war. Britain and France abandoned Russia, which declared war in April 1828 after concluding hostilities with Persia (June 1826–22 February 1828).

Russian forces occupied Moldavia and Walachia and crossed the Danube River in June. The fortress of Varna was captured on 11 October 1828, but Russian inability to secure Shumla and Silistria forced a retreat behind the Danube. A Russian army in the Caucasus had meanwhile captured Anapa, Sukhum-Kale, Poti, and Kars (5 July), reaching the upper Euphrates River by autumn.

Appointed commander of the Russian Danubian army, Field Marshal Ivan Diebitsch captured Silistria, Kulchava, and Adrianople. Meanwhile, in the Caucasus, General Ivan Paskevich took Erzurum and advanced on Trebizond and Batum. With Russian forces just outside Constantinople, Turkey surrendered on 14 September 1829 by signing the Treaty of Adrianople.

The Ottomans recognized Russian annexation of Persian territory in the Caucasus, guaranteed Russian merchant shipping passage through the straits (Dardanelles, Bosphorus), reaffirmed Serbian autonomous institutions as provided in the earlier Treaty of Bucharest (28 May 1812), and demilitarized and established independent national governments in Moldavia and Walachia under Russian protection. The Greeks gained autonomy and then by European agreement on 3 February 1830 became an independent hereditary kingdom. After considerable intrigue, in January 1833 Prince Otto (son of King Louis of Bavaria) became king of Greece.

Mehmet Ali's revolt in November 1831 caused Mahmud II to appeal for Russian military intervention. Consequently, the Treaty of Unkiar-Skelessi on 8 July 1833 provided Turkey's closure of the straits to all except Russian warships.

Neville G. Panthaki

See also: Crimean War; Greek War of Independence; Russo-Turkish Wars

References and further reading:
Anderson, Matthew S. *The Eastern Question.* London: Macmillan, 1970.
Clogg, Richard, ed. *The Struggle for Greek Independence: Essays to Mark the 150th Anniversary of the Greek War of Independence.* London: Macmillan, 1973.
Russian Ministry of Foreign Affairs. *Russia and Greece during the Regency of King Othon, 1832–1835; Russian Documents on the First Years of Greek Independence.* Athens: Thessalonike Institute for Balkan Studies, 1962.
Woodhouse, C. M. *The Greek War of Independence.* London: Hutchinson's University Library Press, 1952.

Russo-Turkish Wars (1676–1878)

Lengthy series of conflicts resulting in the decline of the Ottoman Empire and the extension of Russia's frontier. Russia's attempt to secure its southern frontier against nomadic raiders and the Ottomans and its desire to establish a warm-water port were the initial causes of the wars. The Russian government tried to pay for these campaigns by a reform of taxation that based rates on household size rather than land.

V. V. Golitsyn's Crimean campaigns of 1687 and 1689 and Czar Peter I's attempt to capture the fortress of Azov in June–October 1695 failed because of a lack of naval power. Undaunted, the energetic czar supervised the construction of 1,400 barges and 29 galleys at Voronezh, and Peter's second attempt in May–July 1696 succeeded. During the Great Northern War of 1700–1721, Sultan Ahmed III entered the conflict against Russia in 1710 and, after Peter's failed campaign in the Balkans and defeat at Jassy, regained Azov by the Treaty of the Pruth River in 1711.

Nevertheless, Peter's reign was marked by a series of military reforms. Conscription was instituted in 1705, formation reorganization was undertaken, and the order of battle was revised so that the army was built around the new, elite, well-equipped Guards Regiments of 1708. Government expenditure on the army and navy reached 80 percent of total government expenditures in 1710. By 1725, Russia possessed a regular army of 210,000 troops and 100,000 Cossack troops and a navy of 48 ships of the line and 787 auxiliary craft serviced by 28,000 men.

War was renewed in 1735, with a Russo-Austrian alliance against Turkey. Of the 25,000 Russian troops fatalities in 1737, 60 percent died of disease. The Russians invaded Turkish Moldavia and the Austrians entered Jassy in 1739, and Azov was recaptured. Alarmed by Russian ambitions in the Balkans, Austria concluded a separate peace with the Turks in 1739, the Treaty of Belgrade. Russia was forced to demilitarize Azov and prohibited from building a Black Sea fleet.

The war of 1768–1774 began after Sultan Mustafa III demanded that Empress Catherine II abstain from interfering in Poland's internal affairs. The Russians (again) recaptured Azov and the Crimea in 1771. Field Marshal Peter A. Rumiantsev occupied Moldavia and Walachia, and the Turks were defeated in Bulgaria. By the Treaty of Kutchuk-Kainardzhy in 1774, the Crimean khanate was made independent of the sultan. Russia gained territory from the Bug to the Dnieper Rivers and was allowed to reconstruct the Azov fortress. The czar was acknowledged as protector of Ottoman Christians and allowed to maintain a fleet on the Black Sea and passage through the straits.

In 1781, Catherine II and the Austro-Hungarian emperor Joseph II agreed upon the division of the spoils in the event of the Ottoman Empire's collapse. Catherine (who had grandiose dreams of annexing Constantinople so as to return the Byzantine Orthodox seat of religious-political power to a Holy Christian Orthodox sovereign—herself) annexed the Crimea in 1783, and a Russo-Turkish War began in 1787. Joseph II entered the war as Catherine's ally in 1788 but was cajoled by Prussia to withdraw from the alliance in 1791. Russian victories attained by Marshall Aleksandr Vasilyevich Suvorov gained Russia control of the Black Sea coast from the Kerch Strait to the Dniester River. These gains were codified as the Peace Treaty of Jassy in 1792.

While Russia was occupied with Napoleon, Sultan Selim III deposed the Russophile governors of Moldavia and Walachia in 1806. War was declared, and Marshal Mikhail I. Kutuzov's victorious campaign of 1811–1812 forced the Turks to cede Bessarabia to Russia by the Treaty of Bucharest (1812).

Russia had thus secured the entire northern coast of the Black Sea. Its subsequent wars with Turkey were fought to gain influence in the Balkans, win control of the Dardanelles and Bosphorus Straits, and expand into the Caucasus.

The Greek war of independence sparked the Russo-Turkish War of 1828–1829. Russian forces advanced into Bulgaria and the Caucasus and were encamped outside Constantinople by the time of the Treaty of Adrianople/Edirne in 1829. Moldavia and Walachia were given autonomy with Russian protection, and Greece gained independence the following year.

As a result of Russian aid to Turkey during the revolt of Egyptian pasha Mehmet Ali, the Treaty of Unkiar-Skelessi in 1833 pledged that Turkey would close the straits to all except Russian warships. Britain and Austria-Hungary managed to replace this agreement with a general European guarantee, the Straits Convention of 1841.

The Crimean War (1853–1856) began after Emperor Nicholas I attempted to assert Russia's control of the Black Sea straits via extra guarantees from Turkey. Britain and France entered the conflict against Russia in 1854. By the Treaty of Paris in 1856, Russia lost its protectorate over Moldavia and Walachia and was forbidden to maintain a navy on the Black Sea.

Defeat in the Crimean War prompted the reform of the Russian military. Minister of War General Dmitri A. Milyutin proposed in January 1862 the improvement of mobilization; the creation of 14 military districts by 1865 to decentralize command and increase battlefield agility; the reorganization of force structure; standard officer education; an improved rearmament; a reevaluation of tactics; and training to be handled by General M. I. Dragomirov. Expenditure was also allotted to railway construction to improve logistics.

The war of 1877–1878 was a Russian effort to aid the Balkan rebellions of 1875–1876 against Ottoman rule. The battle plan was based upon the lectures written by General Obruchev and delivered by Lieutenant Colonel N. D. Artamonov of the main staff in 1876. It was a case study of the rapid victory achieved by General Ivan Diebitsch during the Russo-Turkish War in 1829. These reforms bore fruit in 1877–1878. The Russians attacked through Bulgaria, and after successfully concluding the Siege of Plevna/Pleven, they captured Adrianople/Edirne in January 1878. The Treaty of San Stefano in 1878 would have achieved the collapse of Ottoman rule in Europe had Russia's military victories not been overturned by European reaction, as codified in the Treaty of Berlin in June–July 1878.

Further conflicts between Russia and Turkey continued as the Balkan nationalities attempted to break free of Ottoman rule and created tension between Austria-Hungary and Russia for the spoils of or influence over the decaying Ottoman Empire. It also brought into question the rearrangement of the European balance of power, leading eventually to World War I.

Neville G. Panthaki

See also: Crimean War; Peter I, Romanov, Czar of Russia; Russo-Turkish War

References and further reading:

Anderson, Matthew S. *The Eastern Question.* London: Macmillan, 1970.

Baumgart, Winfried. *The Crimean War.* New York: Oxford University Press, 1999.

Keep, John. *Soldiers of the Tsar: Army and Society in Russia 1462–1874.* Oxford, UK: Oxford University Press, 1985.

Menning, Bruce. *Bayonets before Bullets: The Imperial Russian Army, 1861–1914.* Bloomington: Indiana University Press, 1992.

Rwanda and Burundi, Civil Wars of (1959–2000)

The worst bloodletting in postcolonial Africa. Originally inhabited by the Twa (pygmy) people of the rainforest, the region saw successive waves first of Bantu Hutus and last of tall, nomadic Tutsis over a period of roughly 2,000 years. The Tutsi in time came to rule the region, both as royalty and as the administrative backbone of the domain. Under European control, the Tutsi were given the task of administering the colonies, as well as a monopoly in the educational system established by Catholic missionaries. More important, however, the colonial administrators (especially the Belgians) established the myth of Tutsi superiority that would in time come to haunt the country.

In the wave of decolonization that swept over Africa in the years following World War II, Rwandan and Burundian independence was achieved in 1961, with both counties fatally fractured down ethnic lines. In 1959, as the Belgians were preparing to leave Rwanda, Hutus rose up and slaughtered an estimated 100,000 Tutsis. Hutus, jealous of centuries of second-class citizenship, seized the reigns of government from the minority (15 percent of the population) Tutsis in a 1961 U.N.-supervised referendum, which instituted a constitution and brought the Hutu majority party to power.

In Burundi, meanwhile, Louis Rwagasore, a Tutsi nationalist leader, was assassinated in 1961, just months before independence. It was under this cloud of suspicion that the Tutsis lost the 1964 parliamentary elections. When the head of state refused to name a Hutu prime minister, a Hutu faction staged a failed coup. A brutal purge of the army and government ministries of Hutus followed, which in turn led to the Hutu uprising of 1972. The Tutsi government decided to deal with the "Hutu threat" by killing every Hutu with an education, government job, or wealth; an estimated 200,000 Hutus were dead within three months.

In Hutu-dominated Rwanda, the situation was identical, except that it was the minority Tutsis who were being oppressed. In 1963, a Tutsi guerrilla incursion into Rwanda was beaten back swiftly, which in turn resulted in 10,000–

Tutsi rebels. (David and Peter Turnley/Corbis)

20,000 dead Tutsi in a massive wave of repression. Both Rwanda and Burundi instituted a passbook system, which indicated the ethnicity and residence of the holder. These books were designed to restrict the movement of and exclude the majority Hutu in Burundi and minority Tutsi in Rwanda from key government, university, and military positions.

Matters degenerated into literal genocide, at least in Rwanda, in the 1990s. In Rwanda, those demanding real democratic reform and the forces of extreme Hutu supremacy were forcing President Juvenal Habyarimana (who had ruled a one-party state since 1973) into a corner. In addition, the rebel Rwandese Patriotic Front (RPF) had been battling the Hutu-dominated government for years and was growing stronger. A compromise was reached in 1993 that incorporated the RPF into a coalition government and army.

In Burundi, the country's first multiparty elections were won by Hutus. A Hutu civilian president, Meldior Ndadaye, was sworn in, but three months later the Tutsi-controlled army overthrew the government, killing the president and instigating an massacre that claimed more than 100,000 dead and produced 500,000 refugees.

Incredibly, matters deteriorated even further; both Rwandan president Habyarimana and Burundian president Cyprien Ntaryamira were killed in a rocket attack as their aircraft was flying into Rwanda on 6 April 1994. At once, the Hutu extremist forces in Rwanda (which are widely suspected of staging the attack to rid themselves of Habyarimana, who was seen as caving into the Tutsi-led RPF) took the initiative and instigated the pogroms that eventually led to the death of between 800,000 and 850,000 persons. What has often been lost in the Western press is that many of those killed in the pogroms were political targets (journalists, human rights advocates, etc.)—both Tutsi and Hutu.

Burundi escaped much of the violence that plagued Rwanda following the assassination of the two leaders. But in Rwanda, the Tutsi-led RPF took control of the country and attempted to form a government and repatriate many of the Hutu who had fled in fear after their victory. Currently, it is estimated that more than 1,500,000 Rwandans are living in refugee camps outside their native country.

James Corbin

See also: Liberia

References and further reading:

Ayittey, George B. N. *Africa in Chaos.* New York: St. Martin's Griffin, 1999.

Berkley, Bill. *The Graves Are Not Yet Full: Race, Tribe, and Power in the Heart of Africa.* New York: Basic Books, 2001.

Lemarchand, Rene. *Burundi: Ethnocide as Discourse and Practice.* Cambridge, UK: Cambridge University Press, 1994.

Prunier, Gerard. *The Rwanda Crisis: History of a Genocide.* New York: Columbia University Press, 1995.

S

SA (1922–1945)

The largest paramilitary auxiliary of the Nazi Party. The SA (Sturmabteilung, or "Stormtroopers") was designed to protect meetings, exercise force against opponents, and make mass public marches, one of its most effective forms of propaganda. With its characteristic brown uniform, discipline, and violence it came to symbolize the entire Nazi movement. Furthermore, once the Nazis were in control of the German state in 1933, the SA played a major role in spreading terror, eliminating enemies, and "coordinating" allies. In the process, it grew into a mass organization of several million men. By the late spring of 1934, however, the SA and its leader, Ernst Röhm, seemed out of control. Faced with the choice between the professional army or an SA-led militia army, Adolf Hitler chose the tool best suited for his expansionist plans and purged the SA violently in June and July 1934 ("Night of the Long Knives"). Thereafter it was eclipsed by the SS (Schutzstaffel, or "Protection Squad"). Though it never again regained its earlier importance, the SA continued to exist. It organized pre- and postmilitary training, served as a source of committed and competent leaders for the Nazi movement, and performed auxiliary and covert duties for the war effort. The SA also served as a convenient cover for local misfits, drunks, bullies, sadists, and busybodies to lord it over their neighbors with impunity. Like other paramilitary auxiliaries of fascist movements in the interwar period, it existed to perform all those tasks related to the seizure and preservation of power that traditionally conceived political parties and state regimes could not do or could do only with considerable difficulty. It was a constant threat that the rule of law could be suspended, as its leading role in the 1938 Kristallnacht pogrom demonstrates.

Bruce Campbell

See also: Paramilitary Organizations; SS

References and further reading:

Bessel, Richard. *Political Violence and the Rise of Nazism.* New Haven, CT: Yale University Press, 1984.

Campbell, Bruce. *The SA Generals and the Rise of Nazism.* Lexington: University of Kentucky Press, 1998.

Merkl, Peter H. *The Making of a Stormtrooper.* Princeton, NJ: Princeton University Press, 1980.

Wirsching, Andreas. *Vom Weltkrieg zum Bürgerkrieg. Politischer Extremismus in Deutschland und Frankreich 1918–1939.* Berlin: Vergleich Oldenbourg, 2000.

Saipan, Battle of (15 June–9 July 1944)

Major battle in the Pacific in World War II. As part of its advance through the central Pacific Ocean, the United States planned to attack Saipan, the northernmost island in the Marianas. Approximately 1,200 miles from Tokyo, Saipan and the other islands of Guam and Tinian would provide the United States with airbases for B-29 raids against Japan. Operation FORAGER commenced with the heavy bombardment of the island by American battleships and followed with landing of the 2d and 4th Marine Divisions on the west coast of the island the morning of 15 June 1944. American progress was steady but slowed by a combination of the dense terrain and the determined resistance by approximately 30,000 Japanese defenders.

While the land battle continued, the Japanese navy launched Operation A-GO, hoping to trap the American ships supporting the invasion between Japanese land-based aircraft and carrier planes. Under Admiral Ozawa, a Japanese force moved from the Philippines toward the Marianas, unaware that American attacks had neutralized the Japanese land-based planes on Guam. The United States was able to

Marines move onward against the Japanese at Saipan, 24 June 1944. (Library of Congress)

attack the Japanese fleet with its own air power, and the subsequent "Great Marianas Turkey Shoot" resulted in a devastating defeat for both the Japanese warplanes and its surface ships.

On Saipan the marines, supported by the army's 27th Infantry, fought northward through thick jungle and against Japanese positions dug into rocky ridges. Using devastating firepower, the American forces continued their advance, but when the marines pushed ahead more quickly than the infantry, marine commander Major General Holland Smith created an interservice controversy by removing General Ralph Smith, the army commander. The ferocity of the fighting across the island continued, as suggested by the marine names for Japanese positions: Death Valley and Purple Heart Ridge. In addition to the jungle and cave obstacles, the marines encountered difficult street-to-street fighting through the towns and villages of Saipan. Despite fierce resistance and one of the war's most ferocious banzai attacks on 7 July, when 3,000 Japanese soldiers charged the American lines, the Japanese were eventually pushed to the northernmost tip of the island. Rather than surrender, the Japanese garrison fought on until virtually all were killed, while simultaneously thousands of Japanese civilians blew themselves up with grenades or jumped off the cliffs in mass suicide. Included in the Japanese losses were the former commander of the attack on Pearl Harbor and commander of the central Pacific, Admiral Chuichi Nagumo, who committed suicide, and Prime Minister Hideki Tojo, who resigned shortly after the loss of Saipan. The United States took the island by 9 July but suffered 14,000 casualties, including almost 3,500 killed. The American forces followed this victory by capturing Guam and Tinian, but Japanese fanaticism remained a permanent image in American memory of this battle and the Pacific War.

Harold J. Goldberg

See also: Tinian

References and further reading:

King, Ernest J. *U.S. Navy at War, 1941–1945. Official Reports to the Secretary of the Navy.* Washington, DC: United States Navy Department, 1946.

Morison, Samuel Eliot. *History of the United States Naval Operations in World War II: New Guinea and the Marianas, March 1944–August 1944.* Boston: Little, Brown, 1953.

Shaw, Henry I., Jr., Bernard C. Nalty, and Edwin T. Turnbladh. *Central Pacific Drive: History of U.S. Marine Corps Operations in World*

War II. Historical Branch, G-3 Division. Washington, DC: Headquarters, U.S. Marine Corps, 1966.
Spector, Ronald H. *Eagle against the Sun.* Glencoe, IL: Free Press, 1985.

Saladin (al-Malik al-Nasir Salah al-Din aba'l-Mussafer Yusuf ibn Ayyub ibn Shadi) (1138–1193)

Muslim warrior and ruler. Saladin defeated the Third Crusade and the Crusader Kingdom of Jerusalem, overthrew the Shiite Fatimid Caliphate, and established the Ayyubid Dynasty that ruled Egypt from 1171 until 1250 and Syria-Palestine, Yemen, and northern Iraq for another decade.

Salah ad-Din ibn Ayyub, known to the crusaders as "Saladin," came from a family of Kurdish army officers in the service of Nur ad-Din al-Zangi, ruler of Syria. Nur ad-Din had defeated the Second Crusade of 1148, becoming an Islamic hero. Still, the situation remained ominous. Crusaders controlled Palestine and raided regularly across the Jordan. Worse, the Isma'ili Shiite regime in Egypt, the Fatimid Caliphate, tottered on the brink of collapse. In 1163, the crusaders menaced Cairo itself. In response, Nur ad-Din dispatched troops to prop up the Fatimids and block new crusader thrusts.

Saladin, then a subaltern, defended the coastal approaches. The crusaders withdrew in 1168, just as the commander of Nur ad-Din's expedition died. As the commander's nephew, Saladin took charge. He forced the Egyptians to appoint him military and administrative chief to the sickly and heirless Fatimid caliph. Gradually making himself ruler de facto, he refused to name a successor when the caliph died in 1171. He instead abolished the Fatimid Caliphate and paid homage to the Sunni Abbasid Caliph of Baghdad, in whose name he now ruled Egypt. Death again befriended him in 1174 when Nur ad-Din died, leaving a child heir. Saladin was now the most powerful leader in the western Middle East.

Although Saladin clashed with the crusaders episodically, his primary concern for the next 12 years focused on gaining control of Nur ad-Din's lands in Syria and Iraq. Until 1186, his battles and sieges were primarily against fellow Muslims, but the unstable combination of reckless leaders and dwindling manpower made the Kingdom of Jerusalem increasingly vulnerable. When the crusaders launched out from Acre in 1187, Saladin resolved to confront them. In July, he trapped the Christian forces at Hattin and inflicted a catastrophic defeat on them. King Guy of Jerusalem and numerous officers and knights fell into his hands. Capitalizing on this triumph, Saladin captured city after city, taking Jerusalem itself that October. By the end of the year, Tyre alone remained under crusader control.

The Muslim occupation of Jerusalem was free of the appalling slaughter that marked the 1099 crusader conquest. Nonetheless, the European response came swiftly. The Third Crusade, led primarily by Richard I the Lionhearted of England and Philip of France, raised new armies to reclaim Palestine. In June 1191, Richard and Philip stormed into Acre. After Philip returned to France, Richard and Saladin continued to spar along the coast. Richard took several coastal towns but dared not advance into the hill country that sheltered Jerusalem. Saladin, consolidating power in northern Iraq, did not want to risk a pitched battle. Acknowledging a stalemate, the two concluded an amicable treaty in 1192. Richard received the coast from Acre to Jaffa but returned several other posts. In return, Saladin pledged to leave these towns unfortified and to protect Christian pilgrims traveling inland to Jerusalem.

A year later, Saladin died. His family, the Ayyubids, continued to rule Egypt and Syria for the next half-century. Ironically, Europeans would eventually romanticize him into a symbol of chivalrous warfare and diplomatic integrity.

Weston F. Cook, Jr.

See also: Arsuf, Battle of; Crusades; Hattin, Battle of; Richard I

References and further reading:

Holt, P. M. *The Age of the Crusades.* London: Longmans Press, 1986.
Lyons, Malcolm C., and D. E. P. Jackson. *Saladin: The Politics of Holy War.* Cambridge, UK: Cambridge University Press, 1982.

Salerno (9–17 September 1943)

World War II landing on the mainland of Italy following the Allied liberation of Sicily. Salerno was selected as the site of the Anglo-American landing because it was within operational range of Allied fighter aircraft and only 30 miles south of the port of Naples, the Allies' immediate objective. The Salerno plain, however, was hemmed in by mountains that provided the German defenders superb observation of Allied movements below. As a prelude to the Salerno landing, code-named AVALANCHE, General Bernard Montgomery's British Eighth Army crossed the Straits of Messina on 3 September to occupy the toe of Italy.

American lieutenant general Mark W. Clark commanded the Fifth Army that landed at Salerno. It comprised the U.S. VI Corps (Major General Ernest J. Dawley) and the British X Corps (Lieutenant General Richard L. McCreery). Defending Salerno was the German Tenth Army, commanded by General Heinrich von Vietinghoff, who served under Field Marshal Albert Kesselring's overall command.

In the days that followed the Anglo-American landings on 9 September, it appeared as though the Allies would suffer another Gallipoli or Dunkirk. German counterattacks threatened to break the bridgehead in half and cut the British off from the Americans. The critical moment of the battle came during the afternoon of 13 September, when the Germans came within 2 miles of the beaches. Withdrawal appeared imminent, until Allied reinforcements, including two British battleships with their 15-inch guns, helped to beat back the attacks.

On 16 September, American patrols made contact with units of the British Eighth Army advancing from Calabria. The Battle of Salerno was over. The Germans began a slow withdrawal to the first of their mountain defensive lines that blocked the approach to Rome.

Colin F. Baxter

See also: Kesselring, Albert

References and further reading:

Allen, William L. *Anzio: Edge of Disaster.* New York: E. P. Dutton, 1978.

Blumenson, Martin. *Salerno to Cassino.* United States Army in World War II series: Mediterranean Theater of Operations. Washington, DC: Office of the Chief of Military History, 1969.

Clark, Mark W. *Calculated Risk.* New York: Harper & Brothers, 1950.

D'Este, Carlo. *Fatal Decision: Anzio and the Battle for Rome.* New York: Harper Collins, 1986.

———. *Bitter Victory: The Battle for Sicily, July–August 1943.* London: William Collins, 1988.

Graham, Dominick, and Shelford Bidwell. *Tug of War: The Battle for Italy, 1943–1945.* New York: St. Martin's Press, 1986.

Morris, Eric. *Salerno: A Military Fiasco.* New York: Stein and Day, 1983.

Thompson, Julian. "John Lucas and Anzio, 1944." In *Fallen Stars: Eleven Studies of Twentieth Century Military Disasters,* ed. Brian Bond. New York: Brassey's, 1991.

Salvadorian Civil War (1977–1992)

One of Latin America's bloodiest wars, but one that did end with a measure of reconciliation. The conflict may be said to have begun with the October 1979 ousting of General Carlos Roberto Romero, who had come to power in a second successive dubious election in 1977. Government repression was rife, provoking increased guerrilla activity.

Romero was replaced by General Jose Napoleon Duarte. Although Duarte's was a civilian-military junta, reforms were not forthcoming, and repression continued. On 24 March 1980, the outspoken Archbishop Oscar Romero y Galdames was shot while conducting mass in San Salvador's cathedral, a shocking act causing repercussions in the United States as well as in El Salvador.

In 1980 the Farabundo Marti National Liberation Front (FMLN) was created as an umbrella organization of leftist guerrillas. Buoyed by the Sandinistas' triumph in Nicaragua, they launched a massive "final offensive" in January 1981. It was met with repression from the government and paramilitary groups, as a downward spiral of violence continued. In January 1982, the FMLN even managed to destroy 70 percent of the air force when it attacked the Illopongo Air Force base.

Elections for the Constituent Assembly (1982) and presidency (1984) were held against a backdrop of intimidation and violence. The descent into violence and repression continued with the militarization of the country, achieved by massive aid from the United States. The United States was determined to prevent another Nicaragua and would pour $6 billion into the country as well as provide American trainers for the Salvadorian army. The army more than trebled in size between 1980 and 1986.

Presidential elections were scheduled for 1989, and in January the FMLN offered to participate if they were delayed for six months. The government rejected this offer, and Alfredo Cristiani was elected president. Peace talks between the FMLN and the government in September 1989 failed, and in the aftermath of this failure, the FMLN unleashed its second "final offensive." It started in the country but soon moved to the cities and even managed to control parts of the capital, San Salvador, itself. The government and paramilitary groups countered with even more violence. On the night of 16 November 1989, members of the Altacalt battalion murdered six Jesuits, their maid, and her daughter in the Central American University.

As the world picture began to alter with the decline of the Soviet Union and the Sandinistas' electoral defeat in Nicaragua, both sides became more interested in the idea of peace talks. In April 1990, the United Nations became involved with these talks, and an agreement was finally signed on 1 February 1992 in Mexico. Throughout 1992, demilitarization increased, with the FMLN even becoming a legal and recognized political party. This process was declared completed in December 1992, and 15 December was declared National Reconciliation Day. During the civil war, both sides had fought themselves to a stalemate, which resulted in the loss of some 75,000 lives.

M. J. Bain

See also: Nicaraguan Civil War (1979); Panama Incursion; Peruvian Guerrilla War

References and further reading:

Black, J. N. *Latin America: Its Problems and Its Promise.* Boulder, CO: Westview Press, 1998.

Loveman, B., and T. M. Davies Jr. *Che Guevara Guerrilla Warfare.* 3d ed. Wilmington, DE: Scholarly Resources, 1997.

Skidmore, T. E., and P. H. Smith. *Modern Latin America.* New York: Oxford University Press, 1997.

Samnite Wars (343–290 B.C.E.)

A series of conflicts fought between the Samnites and the Romans for possession of southern Italy. Samnium was a region located in the Apennine Mountains in southern central Italy. The Samnites spoke Oscan and were composed of several tribes, four of which—the Caraceni, Caudini, Hirpini, and Pentri—were united in a loose confederacy.

As it weathered Celtic and Etruscan attacks during the fourth century B.C.E., Rome continued to expand beyond the borders of Latium, gradually annexing and colonizing non-Latin territory and thus coming into contact with the Samnites and other Oscan-speaking peoples. Simultaneously, the Samnite population grew as they acquired more territory and allies. The First Samnite War, fought from 343 to 341 B.C.E., arose from the tensions between the two growing powers. Some modern scholars believe this war to be fictitious, created later to justify Roman aggression in Italy; little evidence can be offered to settle that debate satisfactorily. Nonetheless, by 340 B.C.E., Rome dominated Samnium's allies in northern Campania.

Better known than the preceding conflict, the Second Samnite War lasted from 327 to 304 B.C.E., broken only by a tense pause between 321 and 316 B.C.E. Initially, the Samnites held the advantage, marked by their decisive victory under Gavius Pontius in 321 B.C.E. at the Caudine Forks (location unknown). Obliged to seek a truce, the Romans reconstructed their armies and renewed hostilities in 316, only to be defeated the following year at Lautulae (near Tarracina, Italy). Despite these setbacks, by 304 B.C.E. Rome controlled nearly all Samnite lands and allies in southern Campania, Apulia, and Lucania. Probably in connection to this war, the Romans restructured their infantry, abandoning the Greek phalanx organization and introducing the revolutionary maniple system.

The Third Samnite War, fought from 298 to 290 B.C.E., signaled the demise of the Samnites as a free people. Roman troops crushed a combined Samnite, Etruscan, Celtic, and Umbrian force at Sentinum (Sassoferrato, Italy) in 295 B.C.E. The tribes' final surrender was negotiated in 290 B.C.E., and with their acquiescence, Rome claimed possession of most of peninsular Italy, resisted only by a few minor Italic tribes and Greek cities. Despite their capitulation, the Samnites continued to rebel in later years, assisting both Pyrrhus and Hannibal. They fought bitterly against Rome during the Social War, and Samnium remained a rural, superficially Romanized region well into the imperial period.

Ian Janssen

See also: Punic Wars; Pyrrhus; Roman-Etruscan Wars

References and further reading:

Harris, William. *War and Imperialism in Republican Rome.* Oxford, UK: Oxford University Press, 1979.

Salmon, Edward. *Samnium and the Samnites.* Cambridge, UK: Cambridge University Press, 1967.

Samory Touré (1835–1900)

West African war chief who established an empire and who fought against French colonialism. Born circa 1835 in Guinea, near Sanakoro, he became a soldier, assumed the title of Almamy (prayer leader), and became chief of Bissandougou after raising a private paramilitary force armed with modern weapons in 1868. He went on to overthrow the aristocracy and establish an organized state over the next decade that stretched from Upper Volta in the east to northwestern Guinea in the west. In 1879, his forces captured Kankan and began edging toward territory in the north occupied by French colonial troops, wreaking havoc along the way. Entire villages fled in fear from Samory's army and sought protection from the French. After some inconclusive battles (1883–1884), Samory signed the treaty of Bissandougou with the French, which set the Niger River as Samory's western boundary (1887). He then invaded Sikasso (present-day southern Mali) but French intervention forced him to withdraw in 1887–1888. He repudiated the treaty in 1891. The French drove him from his bases, so he established new ones along the Ivory Coast in 1893. From 1892 to 1895, his troops moved into Sierra Leone and Ghana. In 1897, he captured the Dioula capital of Kong, but the British forced a retreat at Sierra Leone. Later, he sacked Kong, which had allied with the French in 1897, and sacked Bondoukou in mid-1898. On 29 September 1898, the French captured Samory at his camp and sent him into exile on the island of N'Djole on the Ogooué River, where he died on 2 June 1900. Seen as a hero by many for his protracted struggle against the French, Samory is also considered a brutal oppressor who spread terror in his wake.

Harold Wise

See also: French Colonial Wars

References and further reading:

Imperato, Pascal James. *Historical Dictionary of Mali.* Lanham, MD: Scarecrow Press, 1977.

McFarland, Daniel Mile. *Historical Dictionary of Upper Volta.* Lanham, MD: Scarecrow Press, 1978.

Samudra Gupta (330–380)

Samudra Gupta reconquered much of the old Mauryan Empire in India. The Gupta dynasty (320–480) claimed descent from the earlier Mauryan dynasty (321–232 B.C.E.) and again unified Hindu India for the second and last time. Chandragupta I founded the Gupta dynasty when he inherited the famed throne of Magadha at Patna in 320 C.E. Marriage to the daughter of a rival king produced Samudra Gupta. Chandragupta I then began the process of reunification of the Ganges River Hindu kingdoms.

Samudra inherited the throne in 330 and completed the mostly political, sometimes military, conquest of the Gangetic kingdoms. Samudra may have relied heavily upon Kautiyla's "Arthasastra," a political manual along the lines of *The Prince* by Niccolò Machiavelli. Kautiyla, adviser to Chandragupta Maurya of the earlier Mauryan dynasty, espoused a well-organized military state. Clearly Samudra used this concept successfully, creating a loose federation by political alliance, military threat, and sometime conquest.

Samudra conquered or subjugated Rajputana, Assam, the north Deccan, Andrha, the Punjab, and even Nepal before his death in 375. Naval fleets proved instrumental in overcoming the coastal factions of India.

His son, Chandragupta II, extended the empire into Punjab, Gujurat, Saurashtra, and Malwa while creating a Gupta golden age. However, his lack of control of the northwestern mountain passes left India open to the same invasions that helped collapse the Mauryan dynasty. In came the White Huns, who overwhelmed the last Gupta rulers, Kumaragupta and Skandagupta, by 480.

Christopher Howell

See also: Chandragupta Maurya; Mauryan Empire, Conquests of

References and further reading:

Dikshitar, V. R. R. *War in Ancient India.* 2d ed., 1948. Reprint, Delhi: Motilal Banarsidass, 1987.

Majumdar, R. C. *Classical Accounts of India.* Calcutta: Mukhopadhyay Publishers, 1960.

Singh, Sarva Daman. *Ancient Indian Warfare.* Reprint, Dehli: Motilal Banarsidass Publishers, 1997.

Samurai

Japan's provincial warrior aristocracy. Samurai were warriors in service to their lords; they could carry long and short swords and had status that peasants, artisans, and merchants lacked.

After the establishment of the Tokugawa shogunate, the samurai increasingly became bureaucrats to great lords and lost their warrior edge. With the Meiji Restoration, they lost their hereditary privileges and disappeared into the general population.

Charles M. Dobbs

See also: Gempei War; Hideyoshi, Toyotomi; Japanese Civil Wars; Japanese Wars of Unification; Minamoto, Yoshitsune; Nagashino, Battle of; Oda, Nobunaga; Onin War; Osaka Castle, Siege of; Sekigahara; Shimabara Revolt; Tokugawa, Ieyasu

References and further reading:

Brinkley, Frank. *Samurai: The Invincible Warriors.* Burbank, CA: Ohara Publications, 1975.

Gibson, Michael. *The Samurai of Japan.* London: Wayland, 1973.

Storry, Richard. *The Way of the Samurai.* New York: Putnam, 1978.

San Jacinto (21 April 1836)

The Mexican defeat that ensured the independence of Texas. In the aftermath of the Texan defeats and massacres at Goliad and the Alamo, General Antonio López de Santa Anna, after a fruitless race to capture the leaders of the Texan insurgents, had separated himself from the bulk of his army. Samuel Houston wanted to take advantage before Santa Anna could return to his main force, so he blocked his enemy's retreat.

After some cavalry charges on 20 April that accomplished little, the main fighting took place the next day. In late afternoon, the Texans rushed and surprised the Mexican positions without firing but shouting "Remember the Alamo" and "Remember Goliad." Then, as the Mexicans fired (going mostly over the heads of the onrushing and infuriated Texans), the Texans fired before the Mexicans could reload their muskets. By that point, the Texans had reached the breastworks and the struggle became hand-to-hand. The fighting was so fearsome that Santa Anna supposedly declared that "so sudden and fierce was the enemy's charge that the earth seemed to move and tremble." The fighting (and slaughter) was over within 25 minutes.

At a cost of six men killed and 30 wounded, the Texans had killed more than 600 Mexicans, wounded 200, and captured 700 (including Santa Anna), or virtually the entire Mexican wing of 1,500 of the main army. To gain his freedom, Santa Anna had to concede Texan independence and an end to the fighting. Few victories have been so one-sided or so profound in their consequences.

Eric v. d. Luft

See also: Houston, Samuel; Santa Anna, Antonio López de; Texas War of Independence

References and further reading:

De Bruhl, Marshall. *Sword of San Jacinto: A Life of Sam Houston.* New York: Random House, 1993.

Hardin, Stephen L. *Texian Iliad: A Military History of the Texas Revolution, 1835–1836.* Austin: University of Texas Press, 1996.

San Juan Hill/El Caney (1 July 1898)

The crucial land battle in the 1898 Spanish-American War. As part of the effort to shut down all of Spain's major ports in Cuba, the United States began a joint army-navy campaign against Santiago de Cuba in June 1898. The U.S. Army V Corps, under the command of General William Shafter, approached the port city only to encounter strong Spanish resistance at El Caney and San Juan Ridge, the high ground east of Santiago.

Shafter's plan for 1 July called for the reduction of El Caney, followed by an attack up the stronger positions at San Juan Ridge. Although heavily outnumbered and spread too thinly across their front by General Arsenio Linares, the Spanish fought stubbornly. More than 4,000 Americans took eight hours to capture El Caney from 520 Spaniards, with both sides taking heavy casualties. The unnecessary attack on the easily isolated El Caney stole attention and manpower from the more important battle at San Juan Ridge.

After a two-hour delay and with insufficient artillery support, American forces consisting of infantry and dismounted cavalry attacked up San Juan Ridge. After a short, intense fight, the Spanish forces on San Juan Hill abandoned their positions in the face of a strong charge and withering fire from a well-placed American Gatling machine gun. On Kettle Hill to the north, the dismounted cavalry—including the 1st Volunteer Cavalry Regiment (Theodore Roosevelt's "Rough Riders") and the "Buffalo Soldiers" of the black 9th and 10th Cavalry Regiments—took the heights from the outnumbered Spaniards.

Although not prosecuted with the greatest operational skill, the Battle of San Juan Hill/El Caney was the only major land battle in the "splendid little war" and proved vital to the overall American strategy. After a short siege, Santiago formally capitulated on 17 July, and the war was over by August. Perhaps more important, the battle put Theodore Roosevelt squarely in the national spotlight, thus beginning his meteoric rise that would eventually land him in the White House.

Thomas Bruscino Jr.

See also: Buffalo Soldiers; Spanish-American War

References and further reading:

Freidel, Frank. *The Splendid Little War.* Boston: Little, Brown, 1958.

Heller, Charles E., and William Stofft, eds. *America's First Battles: 1776–1965.* Lawrence: University Press of Kansas, 1986.

Trask, David F. *The War with Spain in 1898.* New York: Macmillan, 1981.

San Martín, José Francisco de (1778–1850)

Liberator of Argentina, Chile, and Peru. San Martín was born to Spanish parents in Argentina. His military career began in Spain in 1789, and he participated in campaigns in Africa, France, and Spain before retiring in the rank of lieutenant colonel in 1811 and returning to Argentina. San Martín soon joined the patriot cause in its struggle for independence. In 1816, he formed the Army of the Andes at Mendoza for the mission of liberating Chile and Peru. His army of 4,000 regulars and 1,000 militiamen accomplished the logistical feat of moving supplies and equipment across some of the highest passes in the world. San Martín and his men defeated the Spanish at Chacabuco on 12 February 1817 and liberated Santiago two days later. San Martín secured Chile's independence at Maipo on 5 April 1818, when his forces killed 1,000 Spanish troops and captured 3,000. With Chile secure, San Martín turned his attention to Peru, embarking with an invasion force of 4,400 on 20 August 1820. San Martín initially avoided the 23,000 Spanish troops in Peru and developed a strategy that encouraged rebellion and desertion. On 6 December 1820, he defeated one Spanish army and proceeded toward Lima. Attempts to negotiate a settlement with the viceroy failed, and San Martín liberated Lima. Peruvian independence was declared on 28 July 1821.

San Martín could not reach an agreement with Simon Bolivar over the final strategy to defeat Spain. He resigned his position as protector of Peru and returned to Argentina. In 1824, distraught over the death of his young wife, he departed for Europe, where he spent the rest of his life. San Martín's military campaigns are remarkable for the logistics of crossing the Andes and the use of surprise, audacity, and psychological warfare to defeat vastly superior and better equipped forces.

George M. Lauderbaugh

See also: Bolivar, Simon; South American Wars of Independence

References and further reading:

Metford, J. C. J. *San Martín the Liberator.* Westport, CT: Greenwood Publishing, 1971.

Mitre, Bartolomé. *The Emancipation of South America.* New York: Cooper Square Publishers, 1969.

Sand Creek (29 November 1864)

A massacre perpetrated by a U.S. Army contingent on surrendered Native Americans. On 29 November 1864, 1,200 U.S. troops of the 1st Colorado Cavalry and the 3d Colorado Cavalry, under the command of Colonel John M. Chivington, attacked a surrendered and partially disarmed Cheyenne Native American camp in southeastern Colorado Territory. Chief Black Kettle, the camp's leader, had negotiated for peace and camped near Fort Lyon with the consent of the fort's commander. As the attack on the camp began, Black

Kettle raised the U.S. flag as well as a white flag. Despite those signals, anywhere from 150 to 500 Native Americans, among them women and children, were massacred.

Americans in general at first celebrated the Sand Creek action and its commanding officer. However, as it became clear that the soldiers had slaughtered defenseless Native Americans, extensive formal investigations followed, and the U.S. government officially condemned the attack. The massacre was also a chief cause for the Arapaho-Cheyenne war that followed and had far-reaching influence in the Plains American Indian wars of the next decade.

On 7 November 2000, President Bill Clinton signed into law the Sand Creek Massacre National Historic Site Establishment Act.

Frank Schumacher

See also: American Indian Wars

References and further reading:

Hoig, Stan. *The Sand Creek Massacre.* Norman: University of Oklahoma Press, 1961.

Schultz, Duane. *Month of the Freezing Moon: The Sand Creek Massacre, November 1864.* New York: St. Martin's Press, 1990.

Svaldi, David. *Sand Creek and the Rhetoric of Extermination: A Case Study in Native American–White Relations.* Lanham, MD: University Press of America, 1989.

Sandino, Augusto César (1893–1934)

Leader of a guerrilla movement in Nicaragua from 1927 until his death. Sandino sided with the Liberal Party in its conflict with the Conservative Party that resulted in civil war in 1926. When U.S. Marines intervened in January 1927, Sandino formed his own military command at San Juan del Norte in a mountainous area. He refused to participate in a truce brokered by U.S. diplomat Henry Stimson that resulted in the disarming of the Liberal army and the placing of the Nicaraguan National Guard under the control of the U.S. Marines.

Subsequently, Sandino formed the Defending Army of National Sovereignty of Nicaragua and frustrated all attempts by the marines to capture him. It was during this period that Sandino and his movement achieved fame and admiration in many parts of Latin America for standing up to the Colossus of the North. In 1932, the Liberal Party returned to power with the election of Juan Batista Sacasa. In addition, the United States was weary of its involvement in Central America and began to withdraw the marines. Command of the Nicaraguan national guard was given to Anastasio Somoza García, and a truce was negotiated with Sandino. The final contingent of marines departed on 1 February 1933, and Sandino's army disarmed a few weeks later. However, the national guard attacked the Sandinista town of Las Segovias, and an outraged Sandino demanded that the guard be declared unconstitutional. President Sacasa, Somoza, and Sandino met at the presidential mansion in Managua on 21 February 1934. Following the meeting, Sandino was kidnapped and murdered by the National Guard. Sandino's classic use of guerrilla tactics against the superior numbers and technology of the U.S. Marines, his martyrdom, and his revolutionary philosophy were the inspiration for the Sandinista National Liberation Front (FSLM) that came to power in Nicaragua in 1979.

George M. Lauderbaugh

See also: Nicaraguan Civil War (1925–1933); Nicaraguan Civil War (1979)

References and further reading:

Langley, Lester D. *The Banana Wars: An Inner History of American Empire, 1900–1934.* Lexington: University Press of Kentucky, 1983.

Macaulay, Neil. *The Sandino Affair.* Chicago: Quadrangle Books, 1967.

Millet, Richard. *Guardians of the Dynasty: A History of the U.S. Created Guardia Nacional de Nicaragua and the Somoza Family.* Maryknoll, NY: Orbis Books, 1977.

Santa Anna, Antonio López de (1794–1876)

Mexican general, president, and dictator. Santa Anna joined the Spanish army in 1810 and supported Augustín de Iturbide during the Mexican Revolution in 1821. Later, in 1823, he played a key role in Iturbide's overthrow, which resulted in the establishment of a republic. In 1829 at Tampico, Santa Anna defeated an invading Spanish army bent on reconquest, and this event vaulted him to national prominence. In 1833, he assumed the presidency of Mexico but soon relinquished the office, only to take it again the following year in a military coup. Santa Anna decided to command the Mexican army in its effort to thwart the Texas Revolution. His most infamous victory was in March 1836 at the Alamo. But, on 26 April 1836, he was defeated and captured by Samuel Houston at the Battle of San Jacinto. He was forced to sign away Texas and briefly retired. When a French army invaded Mexico in 1838, Santa Anna returned to command and lost a leg fighting the French. This act redeemed him in the eyes of many Mexicans and revived his political life as he returned to the presidency in 1839. However, he was alternately ousted and returned to power in 1841, 1843, and 1844.

The Mexican-American War (1846–1848) provided yet another opportunity for Santa Anna ("The Napoleon of the West") to display his military genius. He met defeat in each of his battles, at Buena Vista (23 February 1847), Cerro

The flight of Santa Anna at the Battle of Cerro Gordo. (Library of Congress)

Gordo (18 April), and Chapultepec (13 September), after which he went into exile.

In 1853, a conservative coalition brought Santa Anna back to Mexico and established him as a military caudillo (strongman). For two years, Santa Anna ruled with a heavy hand but failed to remedy Mexico's myriad political, economic, and social ills. He was forced from office in the 1855 Revolution of Ayutla and, fortunately for Mexico, never returned to power.

George M. Lauderbaugh

See also: Alamo; Buena Vista; Cerro Gordo, Battle of; Mexican-American War; Mexico City, Battles for; Monterrey; Resaca de la Palma; San Jacinto; Texas War of Independence

References and further reading:

Jones, Oakah L. *Santa Anna.* New York: Twayne Publishers, 1968.
Santa Anna, Antonio López de. *The Eagle: The Autobiography of Santa Anna.* Ed. Ann Crawford, Sam Guyler, and Jaime Platon. Austin, TX: Pemberton Press, 1967.

Santo Domingan Revolution (1844)

By 1822, France and Spain had been forced to abandon all hope of retaining colonies on the Caribbean island of Hispaniola, which became the Republic of Haiti. However, there remained decided distinctions between the eastern and western sections of the island. The east, long known as Santo Domingo, retained much of its Spanish colonial identity, whereas the west was a blend of African and French influences. The southeastern city of Santo Domingo was the center of the blend of Spanish, Creole, African, and indigenous cultures that became known as Dominican. Nevertheless, the political and economic life of the island was dominated by the Haitians centered in the capital, Port-au-Prince. The Dominicans chafed under Haitian rule and refused to be integrated into the life of the republic. In 1838, Juan Pablo Durate, a Dominican intellectual, formed the Trinitaria, a secret society devoted to freeing Santo Domingo from Haitian rule. The group created a flag and began propaganda efforts to stir Dominican nationalism. When the Trinitaria was betrayed to Haitian authorities, the conspirators formed a new organization and continued their work. During the period 1839–1843, Durate and his supporters plotted and sought assistance in Venezuela and Curacao, but to no avail. On 27 February 1844, Dominican rebels struck the city of Santo Domingo, captured the Ozma fortress, and expelled the Haitian governor. Within a week, Azua, Santiago, and other cities and towns surrendered to the rebels. A central junta was formed, and on March 15 Durate made a triumphal entry into the city of Santo Domingo.

Two Haitian armies were dispatched by the Haitian president to bring the rebellious Dominicans to heel. In addition to Durate, Buenaventura Baez and Pedro Santana figured prominently in the revolt, and they soon began arguing over a strategy to oppose the Haitian forces. Santana favored foreign support and a defensive posture against the numerically superior Haitians, whereas Durate opposed both measures. Santana plotted with the ruling junta to have Durate relieved of his command. Although the Haitians enjoyed superior numbers, they were met with fierce resistance from the Dominicans and were unable to make a decisive advance. In April, Haitian general Jean Louis Pierott led a coup against the Haitian regime, leading to a retreat by Haitian forces. Santana used this opportunity to secure the frontier and to establish roughly the border between Haiti and Santo Domingo.

In early July, there was a popular demonstration in Santiago to have Durate assume the presidency, a move he accepted only with the provision that a fair and free election be held. On 12 July, Santana marched into the capital and declared himself dictator. Durate was later arrested and sent into exile in Germany, and Santana became dictator. Although the Domingan revolution led to what eventually became known as the Dominican Republic, it also ushered in decades of political instability and foreign intervention.

George M. Lauderbaugh

References and further reading:

Moya Pons, Frank. *The Dominican Republic: A National History.* Princeton, NJ: Markus Weiner Publications, 1998.

Rodman, Selden. *Quisqueya: A History of the Dominican Republic.* Seattle: University of Washington Press, 1964.

Welles, Sumner. *Naboth's Vineyard.* Washington, DC: Savile Books, 1966.

Saratoga (1777)

A series of battles in the fall of 1777 in New York's Hudson Valley region that resulted in one of the very few capitulations in British military history, which was instrumental in inducing the French to ally themselves with and tip the balance in favor of the American cause.

In early 1777, British general John ("Gentleman Johnny") Burgoyne had convinced the British government that an invasion from Canada into the Lake Champlain–Hudson River watershed would isolate New England and bring the American revolutionaries to terms. His force of 7,500 infantry and a small flotilla of gunboats reached Fort Ticonderoga, New York, in June and besieged its 2,500 defenders under General Arthur St. Clair. Within a few days, St. Clair's situation became untenable, and the Americans withdrew.

British forces pursued the Americans down the Hudson River and fought engagements at Hubbardton and Skenesboro, while a separate British army under Lieutenant Colonel Barry St. Leger moved along the Mohawk River and engaged the Americans at Fort Stanwix and Oriskany in western New York.

Burgoyne's advance down the Hudson was foiled near Fort Edward, and the American rebels inflicted heavy casualties on German and loyalist troops sent to capture rebel supplies and horses near Bennington, Vermont. Seizing the initiative, General Horatio Gates determined to meet Burgoyne near the confluence of the Hudson and Mohawk Rivers and entrenched along Bemis Heights south of present-day Saratoga. Burgoyne met Gates's challenge, despite the erosion of his troop strength from casualties and desertions, and advanced on Gates's position.

On 19 September, Burgoyne advanced on Gates's position, only to be thwarted by Colonel Daniel Morgan, Major Henry Dearborn, and General Benedict Arnold near Freeman's Farm below the heights. In this clash, the Americans suffered 300 casualties and the British 600. Burgoyne, concerned by these losses and hopeful of reinforcements, fortified his position. Growing more desperate when his lengthy supply line was severed and his reinforcements stalled, Burgoyne attacked again on 7 October, producing only a harvest of an additional 400 British casualties. Burgoyne withdrew his depleted army to Saratoga, pursued by Gates's rapidly swelling army of 14,000 Continentals and militia.

When General John Stark cut off Burgoyne's retreat with his New Hampshire militia units, Burgoyne was forced to surrender. Besides saving New England, Gates's victory at Saratoga convinced a skeptical world that the colonials could win their independence, thus prompting France formally to recognize the United States of America on 17 December and sign a military alliance in February 1778. Historians today give most of the credit for the Saratoga victory in the field to Benedict Arnold, although Gates reaped the consequent adulation.

Jeffrey B. Webb

See also: American Revolution; Arnold, Benedict; Burgoyne, John; Gates, Horatio

References and further reading:

Fleming, Thomas. *Liberty! The American Revolution.* New York: Viking, 1997.

Furneaux, Rupert. *Saratoga: The Decisive Battle.* London. Allen & Unwin, 1971.

Robson, Eric. *The American Revolution in Its Political and Military Aspects, 1763–1783.* New York: Norton, 1966.

Sargon of Akkad (ruled c. 2334–2279 B.C.E.)

Founder of a Sumero-Akkadian dynasty in southern Mesopotamia and the creator of the first world state in history. Not only is providing a synthesis of the reign of Sargon (in Akkadian, Sharrum-kinum, or "rightful king") and his military exploits difficult, but one cannot with any certainty place them in chronological order. He was the ruler of Agade, a city near Babylon (which has not yet been located), and may have first undertaken expeditions to western Iran and northern Mesopotamia. He also claims to have conquered the powerful Syrian states of Mari and Ebla and reached the Mediterranean Sea. Although it is not possible to discern the nature of these conquests, there is archaeological evidence for destruction at Ebla that has been attributed, in part, to Sargon.

At some point during his reign, Sargon advanced toward the Sumerian cities in the south, conquered Ur and Umma, and defeated Lugalzagessi of Uruk, who had claimed to be the leader of the Sumerian cities. After the conquest of the south of Mesopotamia, Sargon proceeded to establish Akkadian as the language of bureaucracy, creating a Sumero-Akkadian synthesis of cultures. To ensure legitimacy in this area, he installed his daughter, Enheduanna, as the high priestess of Ur, one of the more important cities.

Sargon claims to have controlled territory from western Iran to Anatolia and the Mediterranean Sea (possibly including Cyprus). He was one of the first rulers to employ a continuous standing army.

Although the Akkadian dynasty begun by Sargon lasted only two centuries, this king became the subject of later narratives for nearly the next two millennia that described his rise to power and his campaigns.

Mark W. Chavalas

References and further reading:

Frayne, Douglas. *The Royal Inscriptions of Mesopotamia.* Vol. 2, *Sargonic and Gutian Periods.* Toronto: University of Toronto Press, 1993.

Goodnick-Westenholz, Joan. *Legends of the Kings of Akkade.* Winona Lake: Eisenbrauns, 1997.

Lewis, Brian. *The Sargon Legend: A Study of the Akkadian Text and the Tale of the Hero Who Was Exposed at Birth.* Cambridge, UK: American Schools of Oriental Research, 1980.

Liverani, Mario, ed. *Akkad: The First World Empire.* Padua: Sargon Press, 1993.

Sassanid Empire (225–642)

In 224, Ardashir Sassan revolted against Artabanus V, reigning shah of the Parthian dynasty, and killed him in battle at Hormizdagh. The Sassanids replaced the loosely structured Parthian government with an absolutist-style monarchy. To achieve military parity with their Roman enemies, something the more confederal Parthian state repeatedly failed to accomplish, the Persian feudal nobility seemed willing to accept royalist authority. The Sassanids also established a state-church alliance with Persia's traditional Zoroastrianism.

Ardashir fought Rome until his death in 241, and his son, Shapur I, wrung territorial concessions from Emperor Philip the Arab in 244. He returned to war with Rome in 252 over Armenia. Shapur bested several generals and, in 260, took Emperor Valerian prisoner. Persistent military limitations continued to hobble Sassanid ambitions, however. Persia could conquer Roman population centers but could not hold them. Nomad attacks from central Asia prevented concentrating forces against Europe. Finally, distance made sustained operations too far from the Iraq-Iran heartland extremely hazardous, though distance also protected Persia. Shah Vahram II could not save Ctesiphon, the Persian capital, from Carus's expedition in 283, but exposed supply lines soon forced Carus's retreat. Ten years later, Shah Narseh started a war that again saw Ctesiphon fall into Roman hands, to be returned after a peace agreement in 298. If the Sassanids could not solve their "Roman problem," neither could the Romans solve their "Persian problem."

Shapur II stabilized the eastern frontier by annexing the Oxus Valley, Bactria, and parts of the Indus Valley. In 337, he replaced Armenia's pro-Roman king with his own man. Emperor Constantine died before redressing the situation. In 359, the shah informed Rome that Armenia and northern Mesopotamia were his and seized Amida (Diyarbakir). In retaliation, Emperor Julian marched on Ctesiphon in 363, while Shapur, avoiding combat, used scorched-earth tactics to demoralize and decimate the Romans. Julian died retreating from Ctesiphon. His successor, Jovian, conceded most of Armenia, several key Mesopotamian towns, and tribute to the shah. In 387, Emperor Theodosius, entangled with German tribes, reconfirmed Jovian's deal.

Despite a short war in 422, decades of self-interested cooperation followed, and in 448, Ctesiphon and Rome signed a virtual détente. As barbarians overwhelmed their borders, imperiling their very survival, both empires were too distracted to threaten each other. In 484, eight years after Rome "fell," Shah Peruz and thousands of troops fell before the White Huns in a catastrophic defeat. Shah Kavad negotiated peace with the chieftains but at the price of a heavy annual tribute and the virtual loss of his central Asian, Afghan, and Indus provinces. Beset by famines, refugees, and sectarian strife, Kavad tightened his absolutism and sought more revenue to maintain control. Thus, in 502, Kavad demanded that Byzantium pay new tribute.

After 502, Persia and Byzantium alternated between war and truce. Unlike the Parthians, the Sassanids learned siege craft, as they demonstrated by sacking Antioch in 540. Generally, operations concentrated on traditional fronts—Syria, Mesopotamia, and Armenia—with neither side obtaining a decisive advantage. The strains of prolonged war, however, revived factionalism within Persia's military nobility. The 572–591 war ended in a truly bizarre fashion when a coup toppled Shah Hurmazd in 590. The new shah, Khusrauw II, no tool of mutineers, convinced Emperor Maurice to send Byzantine troops to help him take the throne in his own right. Maurice in turn fell victim to a coup in 602, and the outraged Shah Khusrauw II began a grinding two-pronged offensive against the usurper: by 610, General Shahin occupied all of Armenia, and General Shahrbaraz overran Syria, Palestine, and finally Egypt in 615–620. These were not glorified plunder raids but deliberate annexations by brilliant commanders. A new emperor, Heraclius, found himself fighting simultaneously in Anatolia, in Syria, and against the Avars in the Balkans. His strategy was to hold on in Anatolia and then thrust into Iraq. A massive threat to Ctesiphon might intimidate the shah into making peace.

Khusrauw marshaled his resources for the historic goal of taking Heraclius's capital. In the spring of 626, Shahin, Shahrbaraz, and the Avar chiefs converged against Constantinople, determined to smash it open. Instead, disease, weather, and the vigilant Byzantine navy ground down the besiegers. As the encirclement collapsed, Heraclius now risked his invasion of Iraq. The Byzantines defeated the Persians at Nineveh and at Dasgared and camped before Ctesiphon. In August 627, Khusrauw II died at the hands of his own son, Kavad II, who immediately sued for peace. The price of peace in 628 was restoration of the borders of 602. Twenty-five years of war made Heraclius a hero, exhausted both empires, and doomed the Sassanid dynasty.

Fury over the war's debacle led to several years of murders (including that of Shahrbaraz) and power struggles within Persia. When Yazdgird III took the throne in 632, he was still negotiating power when the Islamic invasion began. In 637, the Arabs shattered the Sassanids at Qadisiyyah, overran Iraq, and took Ctesiphon almost without resistance. Standing against the tide at Nihawand, Yazdgird and the remnants of the Sassanid army went down to defeat and death in 641. With him, too, died his dynasty.

Weston F. Cook, Jr.

See also: Byzantine-Persian Wars; Heraclius; Mauricius/Flavius Tiberius; Shapur I; Shapur II

References and further reading:

Treadgold, Warren. *A History of the Byzantine State and Society.* Stanford, CA: Stanford University Press, 1997.

Yarshater, Ehsan, ed. *The Cambridge History of Iran.* Vol. 3 (1). London: Cambridge University Press, 1983.

Savannah, Siege and Taking of (September–October 1779)

In the fall of 1778, the British turned to the American South. They believed loyalist sympathies were greater there, and the southern colonies were closer to key British bases and colonies in the West Indies. They quickly subdued the Georgia colony, took Savannah and Augusta, and began a strategy intended to bring the Carolinas and eventually Virginia under control.

To regain Georgia and take advantage of a powerful French fleet in the West Indies under Admiral Comte Charles-Hector Theodat d'Estaing, the Americans proposed a joint operation to recapture Savannah. On 1 September 1779, the allied fleet arrived and gained control over the mouth of the Savannah River. Comte d'Estaing unwisely permitted the British time to prepare defenses of Savannah and bring in reinforcements from Beaufort, South Carolina. D'Estaing and the American commander, Benjamin Lincoln, did not cooperate well. Formal bombardment did not begin until early October, and by that time, the French admiral feared being caught away from his base during hurricane season as well as being trapped by a British fleet.

Finally, the allies settled on a direct assault on 9 October 1779. The attack from the west was expected, and the diversionary attacks failed. The main assault was intense and the fighting extremely bloody. Despite the courage of the French attackers, the British and loyalist defenders were equally determined, and the defense held. Thereafter, d'Estaing decided he needed to sail for the West Indies, and by 19 October Lincoln and the American troops retreated to Charleston.

Charles M. Dobbs

See also: American Revolution

References and further reading:

Higgins, W. Robert, ed. *The Revolutionary War in the South: Power, Conflict and Leadership.* Durham, NC: Duke University Press, 1979.

Lawrence, Alexander A. *Storm over Savannah: The Story of Count d'Estaing and the Siege of the Town in 1779.* Athens: University of Georgia Press, 1951.

Savannah, Siege of (9–21 December 1864)

Near-bloodless seizure of a key Confederate port city. In early 1862, the U.S. Navy captured Fort Pulaski and established an effective blockade of Savannah harbor. But the city's strong defenses prevented a Union attack, and thus Savannah was spared much of the destruction visited upon other southern ports.

When William Sherman and his 62,000 troops left Atlanta in mid-November 1864, Confederate lieutenant general

William Hardee and 10,000 troops built defenses against Sherman's eventual arrival and attacked to augment those defenses, keeping the Union navy from exploiting its position in the harbor.

Sherman's troops began arriving on 9 December 1864. Sherman soon decided to establish a siege because he had no desire to test the strong defenses; he was content to await the arrival of siege guns and to destroy the city and its defenses from a distance.

On the night of 20–21 December, Hardee had his men construct makeshift pontoon bridges, destroy Confederate navy ships, spike the guns, and dump ammunition into the river; by 3:00 A.M., the last Confederate troops were escaping toward Charleston, South Carolina.

Soon thereafter, Union troops cautiously moved forward, entered the trenches, and realized that Hardee had abandoned the city. City leaders then offered to surrender the city and sought protection by Sherman. On 22 December 1864, Sherman sent a famous telegram to President Abraham Lincoln: "I beg to present to you as a Christmas gift, the city of Savannah, with one-hundred fifty heavy guns and plenty of ammunition, and also about twenty-five thousand bales of cotton."

Charles M. Dobbs

References and further reading:

Cox, Jacob D. *Sherman's March to the Sea: Hood's Tennessee Campaign and the Carolina Campaigns of 1865.* 1994.

Glatthaar, Joseph T. *The March to the Sea and Beyond: Sherman's Troops in the Savannah and Carolina Campaigns.* New York: New York University Press, 1985.

Nevin, David. *Sherman's March: Atlanta to the Sea and Beyond.* New York: Da Capo Press, 1987, 1994.

Saxe, Hermann Maurice, Comte de (1696–1750)

French general, marshal of France, military theorist, and victor of Fontenoy (1745). Born on 28 October 1696 in Goslar, Saxony, Hermann Maurice was the first of 354 acknowledged illegitimate children of Frederick Augustus II, Elector of Saxony and King of Poland. His mother was Countess Maria Aurora von Königsmarck, a Swedish noblewoman. Saxe began his very successful military career with the Saxon army at the age of 12, when he was sent to campaign with Prince Eugene of Savoy at Malplaquet (1709) during the War of the Spanish Succession (1701–1714). In 1711 he was given the title Graf von Sachsen, or comte de Saxe in French, by his father. In 1719, his father also purchased a regiment in the French army to further his son's training in the art of war.

Upon the conclusion of the War of the Spanish Succession in 1714, Saxe followed his former commander Eugene of Savoy to southeastern Europe to fight with the Austrian imperial army against the Turks. Beside Eugene, Saxe would serve against the Swedes under Peter the Great.

Returning to France, Saxe turned his very capable mind to the science of war, and wrote *Mes Rêveries.* Although not published until 1756, this work was a seminal treatise on warfare from the hand of a truly great field commander. In *Rêveries,* Saxe advocates smaller armies whose key characteristics should be mobility and speed, the emphasizing of shock tactics, and concentrated artillery fire. In addition, Saxe was one of the first to see the battlefield value of light infantry, as the allied forces at Fontenoy would soon learn. Chief among Saxe's tenets on war were that battle should be the last resort of a general and that unlike other sciences, the science of war has no guiding principles or rules.

During the War of the Polish Succession (1733–1738), Saxe served with distinction against his half-brother, King Augustus III of Poland, and was promoted to lieutenant general in 1734. Three years after the war in Poland, the War of the Austrian Succession began, and Saxe was again on the move, this time into Bohemia. In November 1741, General Saxe's name became widely known and respected when he captured Prague in a daring, surprise night attack. Not only was the city secured by Saxe's army without a prolonged siege, but he also kept his forces from plundering.

Following the French success at Prague, King Louis XV made Saxe a marshal of France, a rank held only by Henri de la Tour d'Auvergne, Vicomte de Turenne, and Claude Louise Hector, Duc de Villars, before him. In 1744, he was given command of the French invasion that was to support "Bonnie Prince Charlie," the Young Pretender to the English throne. However, this invasion misfired when the French fleet was destroyed off Dunkirk by storms.

Saxe's greatest victory came in 1745 at Fontenoy. On 11 May 1745, Saxe led French forces to victory over a combined force of British, Dutch, and Austrian troops sent to relieve the French siege of Tournai. This great victory (one of the very few battles of the eighteenth century that the British lost to the French) by Saxe allowed the French army to secure the major cities of the Netherlands, including Ghent, Brussels, Antwerp, Mons, and Namur.

Two other key victories would come under Saxe's excellent generalship, at Raucoux in 1746 and Lauffeld in 1747. The Battle of Maastricht in 1748 was the capstone of a brilliant military career for this illegitimate son of a king. Though often overshadowed by such military writers and generals as Karl Maria von Clausewitz, Antoine Henri, Baron de Jomini, Napoleon I, and Frederick the Great, Saxe deserves to be ranked as one of the truly original students and practitioners of the art of war. The brilliant light that was

Saxe came to an end on 30 November 1750, at Chambord, his estate in France.

Andrew G. Wilson

See also: Fontenoy

References and further reading:

Chaliand, Gérard, ed. *The Art of War in World History: From Antiquity to the Nuclear Age.* Berkeley: University of California Press, 1994.

Hart, Captain B. H. Liddell. *Great Captains Unveiled.* London: Greenhill Books, 1989.

White, Jon Manchip. *Marshal of France: The Life and Times of Maurice de Saxe.* London: Hamish Hamilton, 1962.

Saxon Raids (205–577)

Raids by Saxons and others across Roman Britain. Seaborne raids across the North Sea were known prior to the third century, but their effects had been limited and localized. The raids of the third and fourth centuries, by contrast, were far more serious. In response, the Romans established a chain of forts along coastal southeastern Britain and northwestern Gaul that eventually came under the command of a "count of the Saxon shore" (*comes litoris Saxonici*).

Ten forts are known in Britain. The earliest, Branodunum (Brancaster) and Regulbium (Reculver), date to the early third century. The latest, Anderitum (Pevensey), was built after 330. Although constructed piecemeal, the forts were carefully located at strategic points, such as on river estuaries, and may have formed part of a unified defensive system.

Fifth-century Gallo-Roman writer Sidonius Appollinaris characterizes the Saxons that these forts were supposedly intended to defend against as intrepid and ferocious seafarers. Each, he writes, acted like he was the captain of a pirate ship. He warns his friend Namatius, who was about to set out on a sea voyage, about these brutal adversaries who attack without being spotted and, if seen, give their pursuers the slip. Pagans, they were reputed to make human sacrifices, killing one in 10 of their prisoners by drowning or crucifixion when ready to make sail home.

The term *Saxon* is actually a misnomer since a number of Germanic and non-Germanic groups were involved. They included the Saxons themselves but also Angles, Jutes, Frisians, Franks, and even Slavonic Wends. The raiders were also willing to make common cause with Britain's own barbarians, such as the northern Picts and Scots, who raided along with the "Saxons" in 367, according to Ammianus Marcellinus.

Although primarily enemies of the Romano-Britons, the Saxons also served as their allies as well. By the early fifth century, most Roman troops seem to have been withdrawn from Britain, including the last detachments of the Roman field army (*comitatenses*). In a letter of 410 whose exact meaning is disputed, Western Emperor Honorius advises the Britons to see to their own defense. This they did in part by hiring their enemies to protect them. These mercenaries included the group of Saxons, Angles, and Jutes, under the Jutish warrior-brothers Hengist ("Stallion") and Horsa ("Horse"), invited by the Romano-British leader Vortigern to protect his domains against the Picts and Scots. In exchange for their services, the Germanic tribesmen were reward with grants of land. This cooperation marked the real beginning of a Germanic presence in Britain, but archaeological evidence now suggests that some Saxons and perhaps representatives of other Germanic groups as well were present well before 449.

Although originally invited as mercenaries by the Romano-Britons, the Saxons and others appear to have quickly become the dominant partner in the relationship. They are said to have later revolted and to have invited compatriots from across the sea to settle, for example. One source, the *Gallic Chronicle,* even claims that Britain was already under the control of the Saxons in 441.

Despite this claim, the *Anglo-Saxon Chronicle* suggests that the new arrivals came initially in groups of two to five "keels" (ships). Thus their early settlements in eastern Britain would seem to have been on a small scale and may be represented archaeologically in England by brooches and pottery similar to that found in the continental homelands of northwestern Germany. In the course of the fifth and sixth centuries, Germanic material culture became prevalent across southern and eastern Britain, but the true scale of immigration is difficult to assess since it is possible that "native" Britons had, by then, become acculturated to Germanic ways.

Nic Fields

See also: Angles, Saxons, and Jutes

References and further reading:

Arnold, C. J. *An Archaeology of the Early Anglo-Saxon Kingdoms.* 2d ed. London: Routledge, 1997.

Higham, Nicholas. *Rome, Britain and the Anglo-Saxons.* London: Seaby, 1992.

Johnson, Stephen. *The Roman Forts of the Saxon Shore.* London: Elek, 1976.

Maxfield, Valerie. A., ed. *The Saxon Shore.* Exeter, UK: University of Exeter Press, 1989.

Scandinavian War (1448–1471)

The Scandinavian War was caused by the breakup of the Kalmar union of the three Scandinavian kingdoms in 1397. The union had been held together first by Margaret and subsequently by her grandnephew Eric until his dethronement in 1440. Eric's nephew Christopher was elected to the position until Eric's death in 1448. At this point, different actions were taken in different countries by various parties. In Denmark, the council crowned Christian of Oldenburg with the proviso that all power was held by the Council of the Realm. In Sweden, a party elected and crowned Karl VIII Knudsson. In Norway, rival parties elected both Karl and Christian. The scene was set for civil war.

There was no clear heir to the unified throne. Christian and Karl, in a manner reminiscent of the War of the Roses in England, contended, held, and lost the consolidated throne over the next two decades. Sweden endured invasion and civil war until 1457. At that point, Karl was driven out and Christian crowned. Then Christian was turned out in 1464, and Karl returned to the throne until his death in 1470.

As with other medieval civil wars, who was on the throne and the support of nobles for a specific candidate were determined more by the disputes between the nobles than the relative merits of the potential monarchs. Support and funds were available, and the church, the Hanseatic League, and the nobles were happy to provide them for a price. The Kalmar Union was seen as a less costly alternative to the civil war.

Although the Oldenburg line was eventually successful, the initial two decades of Christian's reign resulted in the decline of the Norwegian kingdom and the loss of the Hebrides and the Isle of Man. The Orkneys were mortgaged, as were the Shetlands. Despite Christian's financial difficulties, he upheld the interests of the towns and the commoners against the encroachment and demands of the Hanseatic League. His son Hans continued this policy.

Tamsin Hekala

References and further reading:

Derry, T. K. *A History of Scandinavia: Norway, Sweden, Denmark, Finland, and Iceland.* Minneapolis: University of Minnesota Press, 1979.

Scharnhorst, Gerhard Johann von (1755–1813)

Driving spirit of Prussia's liberation from Napoleon and architect of the modern German army. This self-educated son of Hanoverian farmers was born on 12 November 1775 in Bordenau and through his father's hard work was able to enter military school. He excelled and received officer training in Nordheim, where, as a lieutenant, he began to publish articles on tactics and handbooks for officers. After training in the artillery in 1783, he traveled to study the military of other German states. In 1793 and 1794, Scharnhorst saw action as a captain of horse artillery, distinguished himself at Menin, and began the *New Military Journal.* His reputation drew the attention of the Prussians, who wooed him into their ranks in 1801 as a lieutenant colonel with noble status. A social misfit in Berlin, he nonetheless held the trust of King Frederick William and Prince Louis Ferdinand, who allowed his transformation of the Academy for Young Officers and instruction from his handbooks. In the Duke of Braunschweig's headquarters in 1806, he witnessed the disaster at Auerstädt. With Gebhard Leberecht von Blücher, he retreated in vain to Lübeck.

Scharnhorst joined L'Estoq's staff in 1807 and led a Prussian contingent at Eylau. After the Treaty of Tilsit, he was promoted to major general and appointed head of the Military Reorganization Commission, which included Prime Minister Stein, August Neidhart von Gneisenau, and former students Grolman, Boyen, and Karl Maria von Clausewitz. With orders to reconstruct the army, these gifted intellectuals made it their goal to arouse patriotic loyalty for the army and the state through far-reaching military, political, and social reforms. Foremost was the education of a cadre of officers, selected by merit rather than birth, to advise the king. Although not completely successful, Scharnhorst and his colleagues laid the foundation for a Prussian spirit and a modern general staff that characterized the German army to the twentieth century.

As Blücher's chief of staff in the spring of 1813, Scharnhorst was wounded in the leg at Lützen but continued to serve at Bautzen. He was sent to plan strategy with the Austrians, but his neglected wound festered, and he died in Prague on 28 June.

Llewellyn Cook

See also: Blücher, Gebhard Leberecht von; Gneisenau, August Neidhart von; Military Academies

References and further reading:

Dupuy, T. N. *A Genius for War: The German Army and General Staff, 1807–1945.* Englewood, NJ: Prentice-Hall, 1977.

Ritter, Gerhard. *The Sword and the Scepter: The Problem of Militarism in Germany.* Coral Gables, FL: University of Miami Press, 1969.

Shanahan, William. *Prussian Military Reforms, 1786–1813.* New York: AMS Press, 1945.

Schlieffen, Graf Alfred von (1833–1913)

General staff officer known as the author of a plan intended to forestall Germany's facing of a two-front war against France and Russia. Born on 28 February 1833 in Berlin, Schlieffen joined the 2d Guards Uhlan Regiment in 1854. Assigned to the Prussian General Staff in 1865, he took part as a staff officer in the Austro-Prussian War (1866) and the

Franco-Prussian War (1870–1871). In 1889, he was appointed deputy to the chief of the general staff.

In 1891, Schlieffen himself became chief of the general staff, a position he held until 1905. During this period, he introduced a number of military reforms: a more mobile infantry, modern signals equipment, and infantry regulations that promoted mission tactics. However, his major contribution was to strategic planning, particularly after 1894 when France and Russia formed an alliance, threatening Germany with a two-front war. The "Schlieffen Plan" departed from previous thinking, which had aimed at defeating Russia quickly before turning toward France. It sought rather a quick decision in France before major operations were undertaken against the Russians. Schlieffen's thinking was influenced by Hannibal's victory at Cannae, but as German resources would not permit a double envelopment, he foresaw a single enveloping movement through Belgium and northern France.

Schlieffen's Plan was modified somewhat before the outbreak of war, but remained the basis of German operations in 1914. It has been frequently criticized as one of the causes of World War I, as well as for its complexity and its need for each part to work nearly flawlessly. Schlieffen died on 4 January 1913 in Berlin. Despite the flaws in his plan, his writings enjoyed something of a renaissance in the post–World War I German Reichswehr.

Alaric Searle

See also: Cannae, Battle of; Marne, Battle of the; World War I

References and further reading:

Bucholz, Arden. *Moltke, Schlieffen and Prussian War Planning.* Providence: Berg, 1991.

Ritter, Gerhard: *The Schlieffen Plan.* New York: Praeger, 1958.

Schlieffen, Alfred Graf von. *Gesammelte Schriften.* 2 vols. Berlin: Mittler, 1913.

Schmalkaldic War (1546–1547)

The first religious war in Germany. In 1530, at the Imperial Diet in Augsburg, Emperor Charles V declared religious nonconformity to be a breach of the empire's peace. For defense reasons, the Lutheran estates of the empire formed the Schmalkaldic League, which was headed by Hesse and electorate Saxony.

Victorious in the Battle of Muhlberg, Charles V was at the height of his power in Germany and Europe. Yet his military success in the Schmalkaldic War failed to restore religious unity to the Holy Roman Empire.

The Schmalkaldic War consisted of two campaigns: the first at the Danube River in the second half of 1546 and the second in the spring of 1547 at the Elbe River, where the only battle in this war occurred. By 20 July 1546, the Schmalkaldic League had gathered some 44,000 men on foot and 6,600 horsemen near Memmingen, marching against the emperor, who was at Regensburg. Retreating to Landshut, Charles V was able to gather most of his troops from Italy and the Netherlands, unimpeded by the league. On 17 August, the united army of 42,000 men on foot and 14,000 horsemen marched to Ingolstadt. From 31 August until 3 September, the armies of the emperor and the league stood facing each another. The league's troops cannonaded the emperor's camp, and there were a few skirmishes and attacks by smaller units. Finally, the league's troops retreated to the vicinity of Donauwörth, which Charles V soon conquered. As before, only a few skirmishes took place between the armies. It was the conquering of Saxony through one of Charles's allies and the league's desolate financial situation that were decisive to the war. On 24 November, the league's army was dissolved, and most of its troops were dismissed; its allies retreated to their according states. The elector of Saxony reconquered his land. The emperor subjugated Upper Germany and dismissed some of his troops.

In March 1547, Charles V recruited new troops and attacked Saxony on 11 April. Meeting no resistance, he conquered the land until he met the elector's small army, defeating it completely.

Charles V was able to defeat the Schmalkaldic League thanks to his modern conduct of war. The Spanish Tercio formation facilitated interaction between riflemen and pikemen in accordance with the new weapon techniques. Charles V tried to cut off the enemy's military and financial supplies without having to risk casualties. The league failed to do the same, nor did it have any significant strategy of its own. What finally decided the war was the league's poor financial support. The league's financial means were limited by its members' contributions, raised through taxes and credits from local investors, whereas Charles V could rely on subsidies and credits provided by the great business houses of Italy, Flanders, and Upper Germany.

Michael Herrmann

See also: Muhlberg, Battle of

References and further reading:

Held, Wieland. 1547. *Die Schlacht bei Mühlberg.* Beucha: Sax-Verlag, 1997.

Schütz, Alfred. *Der Donaufeldzug Karl V im Jahre 1546.* Tübingen: Osiander, 1930.

Schwarzenberg, Karl Philipp zu (1771–1820)

Austrian commander who led the military coalition that ousted Napoleon I from power in 1814. Karl Philipp von

Schwarzenberg was born in Vienna on 15 April 1771 into an old aristocratic family. He embarked on a military career at 16 and distinguished himself in the Turkish War of 1787–1791 and in the French Revolution. By dint of good leadership, Schwarzenberg advanced to field marshal in 1799, and the following year he successfully extricated his command from the defeat at Hohenlinden. In 1805, he commanded a division in the army of General von Mack, and fought his way out of encirclement rather than be captured at Ulm. After serving as ambassador to Russia, Schwarzenberg commanded a cavalry division during the ill-fated 1809 campaign against Napoleon.

After the war, Schwarzenberg ventured to Paris on a diplomatic mission and artfully arranged the marriage between the French emperor and the Austrian princess Marie Louise. Napoleon was impressed by his abilities, and in 1812 he insisted that Schwarzenberg command the Austrian contingent accompanying the invasion of Russia.

After the disastrous retreat from Moscow, Austria entered secret negotiations with Prussia and Russia to form the Sixth Coalition against Napoleon. Because of his reputation—and the fact that Austria contributed the largest share of manpower—Schwarzenberg was appointed commander in chief of coalition forces. He thus became responsible for the monumental task of defeating Napoleon in the field while simultaneously juggling the national priorities of Austrian, Prussian, and Russian monarchs. Assisted by the great Austrian diplomat Prince Clemens von Metternich, Schwarzenberg devised the so-called Trachenberg Plan, whereby the allies avoided direct confrontation with Napoleon in favor of hammering his less capable subordinates. A series of successful encounters ensued that culminated in the October 1813 Battle of Leipzig, whereby the Sixth Coalition was victorious and the French were expelled from Germany.

The ensuing invasion of France tested Schwarzenberg's ability and tact to the limit, but he managed to hold together his uneasy coalition long enough to take Paris in April 1814, which led to Napoleon's abdication and exile to Elba. Schwarzenberg also insisted that France not be partitioned, so as not to upset the balance of power in Europe at Austria's expense. Schwarzenberg died at Leipzig on 15 October 1820, widely admired as a talented general and a skillful coalition leader.

John C. Fredriksen

See also: Leipzig, Battle of; Napoleon I

References and further reading:

Borowski, Harry R., ed. *The Harmon Memorial Lectures in Military History, 1959–1987.* Washington, DC: Office of Air Force History, 1988.

Brauer, Kinley, and William E. Wright, eds. *Austria in the Age of the French Revolution.* Minneapolis: University of Minnesota Press, 1990.

Schwarzkopf, General Herbert Norman (1934–)

Allied commander during the Gulf War. A graduate of the U.S. Military Academy at West Point (1956), Schwarzkopf was himself the son of a general. He spent his childhood in various locales, including a significant amount of time in the Middle East, an experience that would serve him in good stead in later years. In these formative years, he learned much about Arab mores, customs, and values, which made him the ideal person to hold together the diverse coalition of participants in Desert Storm. Schwarzkopf's childhood was marred by his mother's alcoholism, which forced him to grow up quickly.

Upon graduation from West Point, Schwarzkopf served for a time with an airborne unit and then returned to the academy as an instructor. He served two separate tours in Vietnam, one as an adviser to a South Vietnamese airborne brigade in 1965–1966 and the other as a battalion commander in the 6th Infantry Regiment of the troubled Americal Division near Chu Lai. He served with great bravery and distinction in Vietnam, winning three Purple Hearts and three Silver Stars. His seminal moment in Vietnam came on 28 May 1970, when his men stumbled into a minefield. One soldier was grievously wounded, screaming horribly, and threatening to panic the entire unit. Schwarzkopf went into the minefield himself, step by step, until he reached the soldier and got him to calm down. He then extricated the rest of his troops from the minefield.

After the war, he graduated from the prestigious Army War College and assumed command of an infantry brigade. Promoted to brigadier general in 1978, he served as an assistant division commander in Germany. In 1983, he was deputy commander of the successful but poorly coordinated American invasion of Grenada. Schwarzkopf was quite critical of the operation's shortcomings.

In 1988, newly promoted to full general, he took command of the U.S. Central Command, with responsibility for the Middle East. Two years later, Saddam Hussein's Iraqi army invaded neighboring Kuwait. President George Bush decided to deploy American forces to Saudi Arabia to prevent an invasion of that country and eventually eject the Iraqis from Kuwait. Schwarzkopf immediately went to Saudi Arabia as commander of a coalition force representing a bevy of non-Arab countries. Additionally, he worked closely with Arab commanders in planning the coalition's response to Saddam Hussein's invasion. In late 1990, he presided over a major buildup of coalition forces. These Allied forces trained and prepared for an invasion of Kuwait.

Schwarzkopf is generally credited with developing the famous "left hook" or "Hail Mary" plan, a fast envelopment flanking attack that worked magnificently. Schwarzkopf's plan clearly caught the Iraqis by surprise. His forces quickly plunged into the Iraqi right flank and cut off the main road by which Hussein hoped to extricate his forces from Kuwait. This ground war ended in roughly 100 hours, with Saddam's forces (the "battle-hardened Iraqi army," according to most of the "experts") virtually destroyed and clearly ejected from Kuwait.

Schwarzkopf, a plain-speaking, burly man whose fiery temper and heavy-handed manner with subordinates sometimes provoked criticism, returned home a major war hero. He subsequently retired, engaged in youth work, and wrote his memoirs, in which he discussed his life with characteristic honesty.

John C. McManus

See also: Gulf War; Hussein, Saddam; Powell, Colin L.

References and further reading:

Cohen, Roger, and Claudio Gatti. *In the Eye of the Storm: The Life of General H. Norman Schwarzkopf.* New York: Farrar, Straus and Giroux, 1991.

Powell, Colin. *My American Journey.* New York: Ballantine, 1995.

Schwarzkopf, H. Norman. *It Doesn't Take a Hero.* New York: Linda Grey Bantam Books, 1993.

Woodward, Bob. *The Commanders.* New York: Simon and Schuster, 1991.

Scipio Africanus Major, Publius Cornelius (236–183 B.C.E.)

Rome's greatest general during the Second Punic War. He defeated Hannibal and brought Carthage to the point of surrender. Scipio is first mentioned saving the life of his father, the consul Publius Cornelius Scipio, who was severely wounded in a skirmish before the Battle of the Trebia (218 B.C.E.).

In 210 B.C.E., Scipio was sent to Spain to take command over Roman forces there. In 209 B.C.E., he captured the coastal city of Cartagena, and in 208 B.C.E., he defeated Hannibal's brother Hasdrubal Barca at Baecula (Bailen). Scipio divided his troops in three divisions. One attacked the Carthaginians from the front, diverting attention from the other two that outflanked the enemy. However, Hasdrubal retreated, avoiding total defeat, and escaped to Italy with the remainder of his troops.

Having consolidated his position in Spain, Scipio, though severely outnumbered, defeated the Carthaginians at Ilipa (206 B.C.E.). Knowing that the Carthaginians would position their best troops in the center, Scipio positioned his Spanish allies opposite them at a safe distance. The wings, consisting of Scipio's best troops, maneuvered and attacked the enemy flanks. Thus Scipio outflanked the enemy with a smaller army, defeating the Carthaginians before they could put their best troops to use. The Carthaginians subsequently evacuated Spain.

In 205 B.C.E., Scipio attacked the Carthaginians in North Africa. He destroyed a Carthaginian army by setting its camp on fire. A month later, he defeated the Carthaginians again in the Battle of the Great Plains. In this battle, he used the rear lines of his army, not in the traditional way of reinforcing the center but for lengthening his frontage instead, a method that would bring him final victory against Hannibal at Zama (202 B.C.E.).

Scipio was an inspiring leader who could gain and keep the loyalty of his men. His charismatic character and judicious diplomacy won him many allies, without whom Rome might not have won the war. Seeing the deficiencies of the rather static traditional Roman tactics, Scipio experimented with small tactical units that could operate with greater flexibility. He saw the value of capable subordinates who could proceed on their own initiative.

Upon his final victory over the Carthaginians, Scipio was assigned the honorary title Africanus. However, his success had made him many political rivals. Scipio died in self-imposed exile, embittered about the ingratitude of his countrymen.

M. R. van der Werf

See also: Hannibal Barca; Punic Wars; Zama, Battle of

References and further reading:

Bagnall, Nigel. *The Punic Wars: Rome, Carthage and the Struggle for the Mediterranean.* London: Pimlico, 1999.

Lazenby, J. F. *Hannibal's War: A Military History of the Second Punic War.* London: Aris & Phillips, 1978.

Liddell Hart, B. H. *Scipio Africanus, Greater than Napoleon.* New York: Da Capo Press, 1994.

Scullard, H. H. *Scipio Africanus, Soldier and Politician.* London: Thames & Hudson, 1970.

Scott, Winfield (1786–1866)

Accomplished U.S. Army commander. Born at Laurel Branch, Virginia, on 13 June 1786, Scott, after briefly attending the College of William and Mary and then studying law, received a commission as a captain in the light artillery. With the declaration of war against Britain in 1812, Scott was promoted to lieutenant colonel in the newly formed 2d Artillery Regiment. Sent to the Canadian frontier, he distinguished himself during the sometimes disastrous frontier battles. His military successes along the Canadian border had made him such a national hero that Scott emerged from the War of 1812 with the brevet rank of major general and a reputation second only to Andrew Jackson's.

In the years following the War of 1812, Scott negotiated the end of the Black Hawk War in Illinois, fought the Seminoles and Creeks, and was a major influence in the pacification of the Cherokee and the Winnebagoes, who had refused to give up their eastern lands. In 1838, President Andrew Jackson appointed Scott to maintain federal authority during the nullification crisis in South Carolina (when that state attempted to nullify federal law within its borders). The next year, Scott negotiated an end to the "Aroostook War," a border dispute between Maine and the Canadian province of New Brunswick. During this period, he also managed to find the time to make two trips to Europe to study military operations and, in 1835, published his three-volume *Infantry Tactics,* which was still being used during the Civil War. During the more than quarter of a century after the end of the War of 1812, therefore, Scott proved himself to be an able soldier, an adept military theorist, and an efficient diplomat.

In July 1841, Scott was appointed general in chief of the U.S. Army, a post he would hold for more than 20 years, a record never equaled. In 1847, after the opening of the Mexican-American War, he was dispatched to capture Mexico City via Veracruz on the Gulf Coast. During his campaign, Scott, despite being consistently outnumbered two-to-one or more, was able to overcome an enemy who was fighting on familiar territory with interior lines of supply. The brilliance of his campaign was borne out by the fact that Congress brevetted Scott with the rank of lieutenant general, making him only the second American to ever hold that rank and the first since George Washington. In 1852, Scott was nominated by the Whig Party for the presidency, but he was decisively defeated by Franklin Pierce. Returning to the army, he remained general in chief until the beginning of the Civil War.

By the opening of the American Civil War, Scott was so overweight and out of shape that he could not even mount a horse. Too old for an active field command, he offered the post to Lieutenant Colonel Robert E. Lee of the 2d U.S. Cavalry Regiment, who declined. Unlike Lee, a fellow Virginian, Scott remained loyal to the Union. Directing operations from Washington, D.C., he conceived the "Anaconda Plan," which would later serve as the blueprint for the Union's victory. The plan called for the capture of the Mississippi River and a blockade of the Confederate ports. Scott believed that this plan would strangle the Confederacy and prevent the great loss of life that an invasion of the South would entail. Although ridiculed by many military and civilian officials at the time, Scott's plan was later rightly recognized as the last great strategic accomplishment of his long and distinguished military career. After retiring on 1 November 1861, he remained a strong supporter of the Union and was the only non–West Point graduate of southern extraction in the regular army to remain loyal to the Union. Scott died on 29 May 1866 at West Point and was buried in the academy cemetery.

Alexander M. Bielakowski

See also: American Civil War; Mexican-American War

References and further reading:

Eisenhower, John S. D. *Agent of Destiny: The Life and Times of General Winfield Scott.* Norman: University of Oklahoma Press, 1999.

Johnson, Timothy D. *Winfield Scott: The Quest for Military Glory.* Lawrence: University Press of Kansas, 1998.

Scott, Winfield. *Memoirs of Lieut.-General Scott, LL.D.* 2 vols. New York: Sheldon, 1864.

Scythians

Nomadic people who played an important role in the Black Sea region and the northern Middle East from the eighth to the third centuries B.C.E. Originating in the steppes of central Asia, they moved into the Caucasian region and eastern Europe. In the eighth century B.C.E., Scythian groups crossed the Caucasus, fighting the Assyrians and threatening Egypt. However, by the end of the century, they were driven back into southern Russia.

The best-known center of Scythian civilization lay on the northern Black Sea coast. From the fifth to the third century B.C.E., the Scythians dominated this region. They came into conflict with the Persian king Darius. The Scythians lured his army into the hinterland, depriving them of resources by a scorched-earth policy and harassing them with hit-and-run tactics until they withdrew and barely escaped with their lives.

In the fifth century the Scythians pressed increasingly into the Balkans. They were stopped by the Macedonian king Philip II, who defeated them and killed their king. A few years later, they avenged themselves by destroying a Macedonian army. From the third century B.C.E. onward, Scythian influence in the Black Sea region decreased. However, in 247 B.C.E., Scythians from the Caucasus region established the Parthian Empire.

The primary weapon of the Scythians was the bow. Most Scythians fought on horseback and were feared for their hit-and-run attacks. Through their contact with Middle Eastern cultures, defensive body armor and horse barding were introduced. From then on, the nobility served as heavy cavalry, armed with bow, sword, axe, javelin, and lance.

Modern historians have only scant knowledge about Scythian tactics. Before the seventh century, they probably relied heavily on horse archers. After the introduction of body armor, heavy cavalry played an important role. In major battles, the bulk of the cavalry was probably positioned in the center. After an initial hail of arrows, the cavalry charged with the armored cavalry in the lead. The flanks were covered by any foot troops available. If confronted by a massive host like the Persian army, the Scythians avoided large battles and resorted to a scorched-earth policy and hit-and-run warfare instead. On several occasions Scythians served as mercenaries or auxiliaries in foreign armies. In the sixth century, a small force of archers served in Athens as a police force. In 479 B.C.E., Scythian troops took part in Xerxes' invasion of Greece. Scythians in Persian service fought Alexander the Great at Gaugamela, and those in Macedonian service fought with Alexander at the Battle of the Hydaspes. They are also known to have been present in Seleucid armies.

M. R. van der Werf

See also: Parthian Empire

References and further reading:

Cernenko, E. V. *The Scythians, 700–300 B.C.* London: Osprey Publishing, 1983.

Rice, Tamara Talbot. *The Scythians.* London: Thames & Hudson, 1958.

Rolle, Renate: *Die Welt der Skythen, Stutenmelker und Pferdebogner: Ein antikes Reitervol in neuer sicht.* Lucerne: Bucher, 1980.

Sea Peoples (1236–1166 B.C.E.)

Groups of seafaring invaders of different ethnic origins who swept down through Anatolia, Canaan, Syria, Egypt, and Cyprus near the end of the Bronze Age, disrupting normal patterns and destroying the power of the Hittite Empire in the thirteenth century B.C.E. Waves of people, including women and children, accompanied these warriors as they migrated from their homelands in search of better land. The exact origin of this group remains unclear. Egyptian records are the only surviving sources concerning the Sea People, and they indicate that they might have been Ekwesh (Achaeans) from Bronze Age Greece. Other sources refer to them as Tyrrhenians, the ancestors of the Etruscans. Some Egyptians believed them to be Sardana (Sardinians) who had served as mercenaries with Egyptian forces at the Battle of Kadesh in 1299 B.C.E. The only major group of Sea People to settle in the region were the Peleset, commonly referred to as Philistines, possibly from Crete. Sources reveal that the Egyptians successfully fought off the advances of the Sea People during the reign of Merneptah (1236–1223 B.C.E.) and during the reign of Ramses III (1198–1166 B.C.E.), but other areas along the eastern Mediterranean suffered greatly during this period. By 1150 B.C.E., most of the Sea Peoples had merged into the native populations or moved out of the region as quickly as they had arrived.

Cynthia Clark Northrup

See also: Assyria; David; Hittites

References and further reading:

Nibbi, Alessandra. *The Sea Peoples and Egypt.* Park Ridge, NJ: Noyes Press, 1975.

Sandars, N. K. *The Sea Peoples: Warriors of the Ancient Mediterranean, 1250–1150 B.C.* London: Thames and Hudson, 1978.

Velikovsky, Immanuel. *Peoples of the Sea.* Garden City, NY: Doubleday, 1977.

Sedan (1–2 September 1870)

The penultimate French defeat in the Franco-Prussian War. After the outbreak of war with Prussia, the French fortress at Metz had been quickly surrounded, trapping Marshal Achille François Bazaine and an entire army. A second French army was created under Marshal Marie Edmé de MacMahon to lift the siege. His primary task was to link with Bazaine and then bring the Prussians to battle. MacMahon marched east, hoping to reach Metz before it was compelled to surrender, but the Prussians intercepted him at Beaumont, east of Sedan, and the French were compelled to pull back to Sedan in order to rest and regroup. But Prussian general Helmuth Karl Bernhard von Moltke (the Elder) followed hotly on MacMahon's heels. The Prussians enveloped the French positions around Sedan and on 1 September began pouring extremely effective artillery fire onto the hapless French. MacMahon took a shell fragment in the leg, and command passed to General Auguste Alexander Ducrot, who immediately ordered a withdrawal to the west in order to avoid the total destruction of the army (he did not know he was surrounded).

However, General Emanuel Felix Wimpffen arrived from Paris with the news that he had been appointed MacMahon's successor by the government and ordered a halt to the retreat and an attack to the east. By midmorning, the front east of Sedan had completely collapsed under withering Prussian shellfire. The Prussians began advancing from west of Sedan at the same time, and by early afternoon, all fronts had collapsed and remnants of the army began straggling to Sedan, where Napoleon III had set up camp to watch the action. The emperor himself grasped the severity of the collapse and hoisted the white flag. He was taken prisoner and played no further role in the war.

Napoleon's capture led to a political crisis in Paris, and by 4 September the empire had been replaced by the Third Republic. The French were stopped from relieving Bazaine and were unable to field any army capable of helping him for the remainder of the siege. The loss of the only army in the field also meant that the road to Paris lay open, and after a few minor skirmishes, the Prussians invested the city beginning in mid-September. After Sedan, the French had no hope of defeating the invading Prussians.

Sedan was also the site of the main German breakthrough in May 1940, leading to France's complete defeat. It is a place of unhappy memory for France.

Lee Baker

See also: Franco-Prussian War; Metz, Siege of

References and further reading:

Howard, Michael. *The Franco-Prussian War.* New York: Collier Books, 1961.

Williams, Roger. *The French Revolutions of 1870–1871.* London: Weidenfeld and Nicolson, 1969.

Sedgemoor (5–6 July 1685)

Battle that doomed the cause of the Duke of Monmouth. Upon the death of Charles II in 1685, James II, his openly Catholic brother, inherited the British throne. James Scott, the Duke of Monmouth, Charles II's illegitimate but Protestant son who had been in exile in the Netherlands for previous challenges to the crown, planned an invasion of England, to be complemented by a simultaneous landing in Scotland led by Archibald Campbell, the Duke of Argyll, in order to overthrow his uncle. Landing at Lyme Regis and met by support from the economically depressed and dissenting Protestant (largely Presbyterian) population, Monmouth marched for Taunton, where in the hopes of swelling his support from commoners to include the local gentry and nobility, who never joined his cause, he proclaimed himself the rightful king. With about 5,000 men, Monmouth moved to take Bristol, while the Royal Army, led by Louis Duras, Viscount Feversham, marched west to meet him.

Monmouth failed to take Bristol, which was defended by John Churchill, the future duke of Marlborough. James II's offer of free pardons for those who deserted Monmouth, as well as news that Argyll had failed in Scotland, began to whittle down the ranks of rebels, pushing Monmouth to move south and attack while he still had an army. After a skirmish at Norton St. Philip, on the road to Bath, Monmouth, wrongfully underestimating the number and fortifications of the enemy, decided to make a night attack upon the main Royal Army camp, hoping to exploit surprise and win a victory to repair morale. Beginning at 11:00 P.M. on 5 July 1685, the rebels marched in silence, but while negotiating the banks of the Langmoor Rhine River in the fog, the men were heard, and the Royal Army was alerted by a gunshot at about 1:00 A.M.

The disciplined and trained Royal Army quickly assembled and met the rebel cavalry, which was followed by straggling infantry, racing to keep up. Even worse, the powder runners, meant to keep the infantry supplied with ammunition and powder, fled soon after the battle began. Hoping to save himself, Monmouth fled for Dorset, where he was apprehended.

Monmouth may have lost as many as 1,400 men in the battle, with large numbers of prisoners taken by the crown. James II unleashed the full fury of the government on the rebels, sentencing 333 to death and a further 814 to transportation to the West Indies. Monmouth himself was executed at the Tower on 15 July 1685.

Margaret Sankey

See also: English Civil War (1642–1649)

References and further reading:

Bevan, Bryan. *James, Duke of Monmouth.* London: R. Hale, 1973.

Chandler, David G. *Sedgemoor 1685.* Staplehurst, Kent, UK: Spellmount, 1995.

Clifton, Robin. *The Last Popular Rebellion.* London: M. T. Smith, 1984.

Seeckt, Hans von (1866–1936)

World War I general and chief of the Army Command during the Weimar Republic. Hans von Seeckt was born on 22 April 1866 in Schleswig. He entered the Prussian army in 1885 and the General Staff Corps in 1897. A brilliant staff officer, by 1913 he had achieved the position of chief of staff of the Third Army Corps during the German advance on the western front. In March 1915, he was appointed chief of staff of August von Mackensen's Eleventh Army in Galicia. Seeckt planned the Eleventh Army's offensive and breakthrough at Gorlice in May 1915, which helped drive the Russian army out of Poland. In the fall of 1915, Seeckt served as chief of staff of Army Group Mackensen, which overran Serbia. Throughout 1916, Seeckt served as a kind of troubleshooting chief of staff on the eastern front. In these various positions, Seeckt played an important role in halting the Brusilov Offensive and in the conquest of Romania during the summer and fall of 1916. In December 1917, Seeckt was sent to Constantinople to serve as chief of staff for the Turkish Army Command, a position he held until the end of the war.

In 1919, he served as the chief of staff's representative at the Paris Peace Talks. In November 1919, he was appointed head of Truppenamt, the successor to the now-forbidden general staff. As such, Seeckt oversaw reforms in military doctrine that helped lay the foundation for the success of the Wehrmacht during World War II. He was able to forge a clandestine agreement that enabled German armored and air forces to train in the Soviet Union, giving them valuable training that the Allies had forbidden in Germany itself.

Seeckt was dismissed in October 1926, after which he served as a member of the Reichstag (1930–1932) and as a military adviser to Chiang Kai-Shek (1934–1935). He died on 25 December 1936 in Spandau. Well before the advent to power of Hitler, von Seeckt represented the determination of even the Weimar Republic to negate the Versailles Treaty.

J. David Cameron

See also: Enver Pasha; Mackensen, August von

References and further reading:

Carsten, F. L. *The Reichswehr and Politics, 1918–1933.* Oxford, UK: Oxford University Press, 1966.

Corum, James S. *The Roots of Blitzkrieg: Hans von Seeckt and German Military Reform.* Lawrence: University Press of Kansas, 1992.

Seeckt, Hans von. *Gedanken eines Soldaten.* Berlin: Verlag für Kulturpolitik, 1927.

Sekigahara (1600)

The Battle of Sekigahara in 1600 capped the rise of the daimyo (a term referring to powerful, landed military lords, each controlling his own territories and private armies) Ieyasu Tokugawa to hegemon of Japan, paving the way for his establishment of the Tokugawa shogunate (1603–1868) and putting an end to the long period of civil war that had characterized Japan for at least a century before the fall of the Ashikaga shogunate in 1573.

With the death in 1598 of Toyotomi Hideyoshi, a man who had risen from the lowest ranks to become hegemon of Japan, power ostensibly passed on to his infant heir Hideyori. Named as one of five joint regents by Hideyoshi was a dominant daimyo and general, Ieyasu Tokugawa (1542–1616). As the leading landholder among the regents and as veteran of the wars of unification fought by Hideyoshi and his predecessor Nobunaga Oda, many daimyo gravitated to Tokugawa as the natural heir to Hideyoshi's authority, something Tokugawa himself encouraged. A coalition of western daimyo, however, resisted Tokugawa's attempts at leadership. This coalition, whose prominent members were the great western daimyo of Satsuma and Chôshû, was led by Mitsunari Ishida (1560–1600), a former vassal of Hideyoshi's. The forces of the Ishida coalition would ultimately meet those of Tokugawa and his allies in a battle to determine final hegemony.

With over 160,000 troops involved, the Battle of Sekigahara was the largest land engagement Japan had ever witnessed. In August 1600, Tokugawa led his principle force of about 75,000 west from his headquarters at Edo (Tokyo), keeping the Ishida forces guessing as to his strategic intentions throughout the summer. Determining to stop any attempt by Tokugawa to move south to Kyoto and Osaka and threaten the young Hideyori, Ishida opted to stand his ground at the narrow pass at Sekigahara, which linked the capital region to eastern Honshu, and it was here that Tokugawa finally determined to engage Ishida. The forces of both sides deployed during the night of 20 October 1600, with battle commencing the following morning. Like Waterloo, the day of battle opened foggy and damp from the previous night's rain. Throughout the day, the contest, fought with both modern musketry and more traditional weapons, proved indecisive. Central to the victory of either side were the forces of the Kobayakra daimyo, which by midday still remained unengaged. However, Tokugawa was able to convince Kobayakra to change sides and to commit his forces against his erstwhile ally Ishida in the heat of battle. This he did, thus ensuring Tokugawa's final victory. Ieyasu Tokugawa used his narrow victory to punish disloyal daimyo and reward the loyal through the confiscation or granting of land, thus solidifying his own power base and paving the way for his assumption of the title of shogun in 1603.

Daniel Kane

See also: Hideyoshi, Toyotomi; Japanese Wars of Unification; Samurai; Tokugawa, Ieyasu

References and further reading:

Sadler, A. L. *The Maker of Modern Japan: The Life of Tokugawa Ieyasu.* Tokyo: Charles E. Tuttle, 1937.

Seljuqs

Turkish conquerors of Byzantine Anatolia. One of many splinter groups emerging in the aftermath of the two central Asian Turkish empires (552–630 and 680–742), the Seljuqs appeared sometime in the ninth century. Their eponymous founder, Seljuq, is said to have liberated his people from the control of others and moved them to Jand, along the middle Syr Darya. Converted to Islam, the Seljuqs became *ghazi* warriors, driving back the infidel and pushing forward the frontiers of Islam.

By the tenth century, there were two distinct groups, one in Khorasan and the other in Khwaraz. Both expanded aggressively in cooperation or conflict with others, including Mahmud of Ghazna (971–1030). In 1040, the Khwaraz Seljuqs defeated Mas'ud, Mahmud's successor, at Dandanqan and poured into Iran. By 1055, the Seljuq ruler Tughril (d. 1063) had control of Baghdad and with it power over the remaining Abbasid dominions. His successor, Alp Arslan, continued Seljuq expansion in virtually all directions, including into Byzantine Anatolia. In 1071, he defeated Byzantine armies at Manzikert, captured the Byzantine emperor, and began the process by which the Turks have gradually turned Anatolia into the Turkey of today.

Seljuq power declined in the twelfth century. In Anatolia,

the Byzantines recovered much of their former holdings, although not the interior, which remained strongly Turkish. In the aftermath of a second major Byzantine defeat, at Myriocephalon (1176), the Seljuks completed their conquest of Anatolia, except for a few enclaves. Their success was short-lived because the Seljuqs were severely dealt with by invading Mongols and never recovered. Their successors were the Ottomans, who went on to unify Turkey, conquer Constantinople, and forge an empire.

Paul D. Buell

See also: Alexius I Comnenus; John II Comnenus; Mahmud of Ghazna; Mongol Empire

References and further reading:

Cahen, Claude. *Pre-Ottoman Turkey.* Trans. from the French by J. Homes-Williams. London: Sidgwick and Jackson, 1968.

Sempach, Battle of (9 July, 1386)

Swiss close shock combat triumphs over the dismounted knight. Sempach was the second battle of the long-running Austro-Swiss conflict (1315–1499).

Following the addition of the cantons of Zug, Glarus, and Bern by 1353, the rural Swiss of the northern cantons rebelled against their Habsburg lords, establishing independent cantons. The militia forces of these cantons expanded further into Habsburg lands, sacked the castle of Rothenburg, and threw a garrison into the nearby town of Sempach. The Austrians moved an army of about 1,500 men-at-arms and 2,500 other infantry under Duke Leopold the Valiant into the alpine valleys. The Swiss opposed him with 1,500 militia, mostly armed with the halberd. Both groups advanced in three divisions, the Austrians preferring the medieval style of three columns and the Swiss in a single column of three squares. When the first Swiss column crested the small rise to his front, Leopold dismounted his first and attacked the Swiss with a mass of heavily armored men carrying lances. The other two columns remained mounted to pursue the enemy.

The leading Swiss square uncharacteristically halted on a favorable slope of terrain and received the Austrian attack. After heavy fighting, the Swiss began to retire, when the remaining two Swiss squares rapidly approached and threatened the now exhausted dismounted knights.

Leopold promptly dismounted his second column, but it advanced in an irregular and disorganized fashion. The Swiss, after tearing apart the ranks of the first column, made quickly for the second. In the furious melee, the Swiss halberdiers easily cut down their exhausted opponents. The third column, sensing defeat, withdrew to leave Leopold and his dismounted forces at the mercy of the enraged and encircling Swiss. They took no prisoners.

The Battle of Sempach showed that the Swiss were capable of defeating dismounted knights just as easily as they had overcome mounted ones at Laupen. From this time forward, the Swiss became increasingly aggressive and expansionist, confident as they were in their near invincibility.

Bryan R. Gibby

See also: Austro-Swiss Wars; Laupen, Battle of

References and further reading:

Delbruck, Hans. *Medieval Warfare.* Trans. Walter J. Renfroe Jr. Lincoln: University of Nebraska Press, 1990.

Oman, Charles W. *The Art of War in the Middle Ages.* Ed. John H. Beeler. Ithaca, NY: Cornell University Press, 1953.

Sennacherib (r. 705–681 B.C.E.)

One of the major rulers of the late Assyrian Empire, probably best known for his third campaign in Syro-Palestine in 701 B.C.E., where he threatened the existence of Jerusalem. Sennacherib (in Akkadian, Sin-ahhe-eriba) succeeded his father, Sargon, as king of Assyria in 705 B.C.E. The most serious issue that confronted Sennacherib during his reign was the unstable situation with Assyria's tributary to the south, Babylon. In Babylon, the rebellion was headed by Merodach-Baladan and supported by a coalition of Syro-Canaanite states led, apparently, by Hezekiah of Judah. This rebellion is known from both Assyrian and biblical sources (2 Kings 18–19, 2 Chronicles 32, and various portions of Isaiah). Sennacherib suppressed the revolt and destroyed Babylon in 689 B.C.E. Subsequently turning to the other rebels, Sennacherib invaded the area in 701 B.C.E., claiming to have "caged Hezekiah in Jerusalem like a bird." There is archaeological evidence of massive destruction by the Assyrians in Judah, especially at the Judean fortress city of Lachish, where an Assyrian siege ramp has been uncovered. Though diverted from occupying Jerusalem (either by an Egyptian army led by Tirhakah that was advancing from the south or by a plague), Sennacherib demanded and received harsh terms from Judah. All fortified cities and outlying areas (including some cities in Philistia and Phoenicia) were seized, Hezekiah's treasury was emptied, and some of his daughters were sent as concubines to Nineveh, Sennacherib's capital.

Sennacherib conducted many urban renewal projects in Nineveh, including a new palace complex, parks, irrigation projects, and massive fortifications. The Assyrian king was murdered by some of his sons in 681 B.C.E. and was succeeded by another son, Esarhaddon.

Mark W. Chavalas

Sennacherib, King of Assyria, returning from victory at the head of his army. (Hulton/Archive)

See also: Assyria

References and further reading:

Honor, Leo. *Sennacherib's Invasion of Palestine: A Critical Source Study.* New York: Columbia University Press, 1926.

Parpola, Simo. "The Murderer of Sennacherib." In *Death in Mesopotamia,* ed B. Aslter, 171–182. Copenhagen: Akademisk Forlag, 1980.

Ussishkin, David. *The Destruction of Lachish by Sennacherib.* Tel Aviv: Institute of Archaeology, 1982.

Septimius Severus (Lucius Septimius Severus Pius Pertinax) (146–211)

Successful soldier-emperor. Consul in 190, Severus became governor of Pannonia Superior the following year. Twelve days after the murder of Pertinax, he was proclaimed emperor at Carnuntum (9 April 193) to become the avenger of Pertinax, whose name he assumed. Backed by all 16 Rhine and Danube legions, he marched on Rome, which he entered without opposition (9 June 193). The praetorians were dismissed, and a new guard, twice as strong, was formed from the Danubian legions.

Severus was the winner of a series of civil wars, defeating his last rival, Clodius Albinus, the governor of Britain, at the Battle of Lugdunum (Lyon, 19 February 197). He mounted a successful Parthian expedition. Ctesiphon was captured (28 January 198) and northern Mesopotamia annexed. Of the three new legions (I-III Parthicae) raised for the campaign, two formed the garrison of the new province, but in a significant break with tradition, Severus entrusted their command to equestrian prefects rather than to senatorial legates. The third (II Parthica) was stationed at Alba, just outside Rome. This legion, combined with the new praetorian guard, provided a force of about 17,000 troops at his immediate disposal in Italy.

Severus spent the last three years of his life fighting the Caledonian tribes of northern Britain. The scale of his operations is impressive, suggesting that Severus intended to advance the frontier to the Antonine Wall once more. Long a victim of gout, he died at Eboracum (York, 4 February 211), leaving his sons the advice "not to disagree, enrich the soldiers, and despise the rest" (recorded by the contemporary Greek historian Cassius Dio and quoted in Loeb Classical Library). Severus, whose military ability was allied to a shrewd political insight, followed a policy of keeping the army loyal.

He not only raised army pay (for the first time since Domitian) but also granted Roman soldiers the right to marry legally.

Nic Fields

References and further reading:

Birley, Anthony R. *The African Emperor: Septimius Severus.* London: Batsford, 1988.
Birley, Eric. "Septimius Severus and the Roman Army." *Epigraphische Studien* 8 (1969), 63–82.
Loeb Classical Library. Cambridge, MA: Harvard University Press, 1980.
Smith, Richard E. "The Army Reforms of Septimius Severus." *Historia* 21 (1972), 481–500.

Sevastopol, Siege of (October 1854–11 September 1855)

Costly allied capture of their main objective in the Crimean War. After retreating from the Alma River on 20 September 1854, the outnumbered Russians elected to defend Sevastopol while waiting for reinforcements to attack the allied rear. Almost a month passed before the allies began shelling the city on 17 October. This delay gave Lieutenant Colonel (soon Major General) Frants Eduard Ivanovich Todleben time to apply his considerable engineering skills to strengthening the Russian fortifications. His brilliant defenses included 7 bastions; a tower; 14 coast batteries with over 600 guns; an octagonal fort; and a complex of earthworks, redoubts, salients, lunettes, and field batteries. The allies could have captured Sevastopol much earlier had Todleben not been on the scene at the time.

Czar Nicholas I, disappointed by Prince Alexandr Sergeevich Menshikov's failure at Inkerman on 5 November 1854, replaced him in December with Prince Mikhail Dmitriyevich Gorchakov as commander in chief of the Russian forces in the Crimea. This change, together with the lessening of allied activity throughout the winter, allowed Todleben more opportunity to order repairs and improvements in the Russian fortifications. When the allies' shelling resumed in earnest in the spring, the Russian defenses at first met the challenge, and Russian morale rose. But the allies were bringing in longer-range guns and more men. By June 1855, the 54,000 Russian soldiers and sailors trapped in Sevastopol and the 21,000 additional Russian troops in the

The French army captures the Russian fortress Sevastopol after a year-long siege, on September 11, 1855. (Library of Congress)

surrounding hills faced 100,000 French, 45,000 British, 15,000 Italian, and 10,000 Turkish troops.

The allies launched a concentrated infantry offensive on 18 June. Covered and supported by constant shelling, allied sappers undermined the Russian works throughout the summer. The French took the Malakov Tower, and the British temporarily took the Redan on 8 September. Casualties on that day alone were 13,000 Russian and 10,000 allied soldiers. The Russians then burned the city and demolished its military installations between 9 and 11 September, prior to their abandonment of Sevastopol.

Eric v. d. Luft

See also: Alma; Balaklava; Crimean War; Inkerman, Battle of the; Light Brigade, Charge of the

References and further reading:

Hibbert, Christopher. *The Destruction of Lord Raglan: A Tragedy of the Crimean War, 1854–55.* London: Longmans, 1961; Boston: Little, Brown, 1962.

Pemberton, W. Baring. *Battles of the Crimean War.* New York: Macmillan, 1962.

Sweetman, John. *Raglan: From the Peninsula to the Crimea.* London: Arms and Armour, 1993.

Tolstoy, Leo. *The Sebastopol Sketches.* New York: Penguin, 1986.

Seven Days' Battles (25 June–1 July 1862)

An American Civil War drawn battle that showed the passivity of Union general George B. McClellan and the operational brilliance of Confederate general Robert E. Lee. For nearly a month, a small Confederate force under General Jeb Magruder near Yorktown had delayed General McClellan's Union advance up the peninsula between the James and York Rivers. In early May, Magruder retreated, McClellan followed, and at Seven Pines, General Joseph Johnston tried to overwhelm two Union corps that were somewhat isolated south of the Chickahominy River on 31 May–1 June 1862. Union forces held, Johnston was wounded and temporarily had to give up command, and President Jefferson Davis's military adviser, Robert E. Lee, assumed command.

In the Seven Days' Battles, Lee tried to continue Johnston's strategic conception, only in reverse. He felt he had to attack the somewhat passive Union general before additional Union forces concentrated around Washington, D.C., joined the already larger Union army, giving them a near overwhelming superiority in manpower, supplies, and artillery. McClellan had four corps south of the river, and Lee wanted Thomas "Stonewall" Jackson, moving from the Shenandoah Valley, to augment the strike on the weaker Union right and either seize the Union supply base, forcing their retreat, or else turn the flanks of the Union army and perhaps achieve a Cannae-like result.

The plan was ambitious, perhaps overly ambitious. Although Jackson had selected the date for the attack, he and his men—perhaps tired from their exertions in the valley—arrived late, and the Union forces once again held. For the remainder of the week, Lee sought to find weak points in the Union position. Despite McClellan's passivity, the corps and division generals defended well, and the outnumbered Confederates could not destroy the invading army. However, McClellan retreated to Harrison's Landing on the James River and the safety of Union gunboats. Lee then looked north to strike the smaller Union army defending Washington, D.C., which led to Second Bull Run/Manassas Junction.

Charles M. Dobbs

See also: American Civil War; Lee, Robert E.; McClellan, George Brinton

References and further reading:

Dowdey, Clifford. *The Seven Days: The Emergence of Lee.* Boston: Little, Brown, 1964.

McKenzie, John D. *Uncertain Glory: Lee's Generalship Re-examined.* New York: Hippocrene Books, 1997.

Rowland, Thomas J. *George B. McClellan and Civil War History.* Kent, OH: Kent State University Press, 1998.

Seven Years' War (1756–1763)

The third war between Austria and a rising Prussia for control over Silesia, the culmination of the long Anglo-French struggle for colonial supremacy, and the last major conflict before the French Revolution to involve all the traditional great powers of Europe. There were three principal theaters of this war. Great Britain helped support Frederick of Prussia in battling Austria, France, and Russia and their allies: British finances helped purchase mercenary troops to augment Prussia's army. The British navy battled the French navy in the Atlantic and Indian Oceans as well as the Mediterranean and Caribbean Seas. Finally, augmented by colonial militia, the British made a determined and ultimately successful effort to destroy French power in North America. When the Seven Years' War ended, Frederick gained Silesia, though with significant manpower losses; the British gained territory in India and all of French Canada (save for tiny St. Pierre and Miquelon Islands off the Newfoundland coast).

When hostilities erupted in America after British general Edward Braddock's disastrous effort in 1755 to take Fort Duquesne, British king George II sought a treaty with Frederick II of Prussia to guarantee the state of Hanover, where

George's family had ruled before moving to England. This friendship eventually led Austria to ally with France and Russia (and Saxony and Sweden and very much later Spain) against Prussia, England, and later Portugal.

Two years later, the European fighting, mostly to the east and south of the Kingdom of Prussia, began. When Czarina Elizabeth joined the Austrian-French alliance, Frederick, who understood the advantages of interior lines, believed strongly in the initiative, and was an aggressive, tactically oriented commander, attacked Saxony to take that kingdom out of the opposing alliance. In spring 1757, he invaded Bohemia and defeated the Austrians at Prague. Then he laid siege to Kolin and, dividing his forces to retain the siege and meet a large relief army, lost one-third of them in the resulting five-hour battle against Austrian count Leopold von Daun in June. Thereafter, Frederick retreated.

At this point, Frederick demonstrated his understanding of interior lines and of the proper use of audacity. He faced the French advancing onto Hanover, which he was pledged to defend, the Swedes on his north, the Russians to his east, and the Austrians from the southeast. He fought a larger French-led army at Rossbach in the state of Thuringia and won a two-hour battle, and a month later, in early December, he defeated the Austrians in Silesia near the town of Leuthen.

The victory at Leuthen was remarkable. Frederick hurried 170 miles in 12 days to take command of a defeated Prussian army and expel the Austrians, who occupied a sound defensive position about 5 miles long on a north-south axis. Frederick's forces approached, and then, as several oblique hills hid his advance, he moved the bulk of his smaller army—infantry and cavalry—to his right. When he attacked, his force greatly surprised and eventually overwhelmed the Austrian left. Despite the efforts of Austrian prince Charles to reestablish his line on an east-west axis, when nightfall came, the Austrians had fled the field, having suffered major casualties heading into winter quarters.

This victory may well have marked the high tide for Prussia during the conflict. Although Prussian forces initially defeated the Russians in a costly battle at Zorndorf in the summer of 1758, Frederick was defeated by a combined Austrian-Russian army at Kunersdorf in the summer of 1759, losing some 18,000 men in six hours of battle. Frederick crossed the Oder River near Frankfurt and attacked the larger allied army seeking to envelop its flanks, but his forces became disorganized, attacked in piecemeal fashion, and suffered grievously. Later, Prussia lost a detachment at Maxen before going into winter quarters.

In 1760, the main fighting occurred in Silesia, although there were battles elsewhere. In the fall of 1760, the Russians seized Berlin, the Prussian capital, although Frederick soon defeated von Daun at Torgau. It was a bloody and costly victory, however, where Frederick lost one-third of his attacking force.

In 1761, William Pitt stepped down as British prime minister, and his policy of subsidizing Frederick's European land campaigns to free British power to seize France's overseas empire ended. This development led more to campaigns of maneuver and countermaneuver in Europe. Still, before the weight of the Austrian-based alliance could overwhelm Prussia, Elizabeth of Russia died and was succeeded by Peter III, who greatly admired Frederick, agreed to an easy peace, helped arrange peace between Prussia and Sweden, and helped Frederick push Austrian forces out of Silesia. In the ensuing peace treaties, Austria lost Silesia to Prussia; otherwise, territorial borders remained mostly unchanged.

The conflict between Great Britain and France was more decisive. Called the French and Indian War in North America, it was the last of four great colonial conflicts on that continent and across the globe, and it resulted in a great British victory. Prime Minister William Pitt planned financially to support Prussia to contain the French on the European continent while he concentrated British power on conquering French colonial possessions. Despite some fits and starts and perhaps less than ideal cooperation by American colonial governments and militia, the British ultimately triumphed over the French in North America while defeating the French in India and thus paving the way for the vast British empire on the subcontinent.

Pitt put the wealth of Britain into the war effort. He sent large armies transported by huge navies to the New World; he purchased supplies from colonial purveyors; and he engaged colonial militia and rangers. And he kept his focus on the objective—winning Quebec. Similarly, he sought to defeat the French fleet and to take French possessions on the Indian subcontinent.

By 1758, the weight of British (and colonial) power began to be felt in the conflict. Pitt brought focus and determination to the conflict. The British under General John Forbes again attacked Fort Duquesne, this time advancing from western Pennsylvania in the fall. The attack succeeded, the French retreated, and the British cut off Quebec from French Louisiana. Also, that year, another offensive commanded by General Jeffrey Amherst sailed from Halifax, Nova Scotia, to attack Louisbourg, a French fort on Cape Breton Island, guarding the entrance to the Gulf of St. Lawrence and the St. Lawrence River, the route connecting Quebec and France. Amherst landed west of Louisbourg and brought up siege weapons, and the fort surrendered. However, an attack against Fort Ticonderoga on Lake George did not fare as well, and the British rather unexpectedly retreated when the

French commander expected a continuation the next day and he in turn contemplated retreat.

The victories of 1758 were completed by greater victories in 1759. General Amherst gained control over Ticonderoga, which the French had mined and exploded before abandoning. Meanwhile, General James Wolfe sailed to Quebec, approached the city from the western side, engaged in a battle on the Plains of Abraham against the French commander, Louis-Joseph de Montcalm-Gozon, the Marquis de Montcalm de Saint-Véran, and the British compelled the surrender of the citadel. Both opposing commanders died in this pivotal battle.

In 1760, Amherst compelled the surrender of Montreal, and the French had lost their colony in Quebec. The war would drag on for another several years as the conflict expanded to include the Spanish and Cuban and Gibraltar areas of operation.

Meanwhile, the British defeated the French in India. After the British won at Wandiwash in early 1760, they pushed the French back upon their main base at Pondicherry in southern India. British control of the seas guaranteed the attackers adequate supplies, while the French defenders weakened. Finally, some months later, the British began a siege bombardment in January 1761, and the French garrison surrendered five days later. Thereafter, the British East India Company, through diplomacy, trade, bribery, and force, managed to expand British power along the coast from the Tamils in the south to north of Madras to the east and north. Thereafter, the British engaged in four so-called Mysore Wars against local rulers to solidify their growing empire that developed from the fighting of the Seven Years' War. In the end, this world war of a sort led to Britain gaining French Canada, Minorca, and Florida; France ceding Louisiana to Spain; and Prussia keeping Silesia.

Charles M. Dobbs

See also: French and Indian War; Hochkirch, Battle of; Leuthen, Battle of; Minden; Plassey, Battle of; Torgau

References and further reading:

Furneaux, Rupert. *The Seven Years War.* London: Hart-Davis MacGibbon, 1973.

Hall, Ronald Acott. *Frederick the Great and His Seven Years War.* New York: E. P. Dutton, 1915.

Popock, Tom. *Battle for Empire: The Very First World War, 1756–63.* London: M. O'Mara, 1998.

Shaka kaSenzangakhona (c. 1787–1828)

The founder of the Zulu kingdom, celebrated for his military skills and who today symbolizes Zulu nationalism. Shaka was born in about 1787 to Senzangakhona kaJama, chief of the Zulu people, who owed allegiance to the Mthethwa, one of the major chiefdoms then fighting to dominate the region of southeastern Africa that would later become the Zulu kingdom. In about 1794, Shaka was driven into exile and eventually entered the service of Dingiswayo kaJobe, the Mthethwa chief, who recognized and rewarded his extraordinary military aptitude. In 1816, with Dingiswayo's support, Shaka seized the Zulu chieftainship from his brother.

Southeastern Africa was already falling into a period of turmoil known as the *mfecane* (the crushing). In 1817, the Ndwandwe defeated the Mthethwa and attacked the Zulu. Shaka responded by improving his military capability. All the men and women in his chiefdom were grouped in age-grade regiments under his authority for the purposes of social control and the exploitation of their labor and military potential. His warriors' style of combat was not new in the region, but under Shaka it was perfected. They attacked in the "bull" formation, with the two "horns" outflanking and enveloping the enemy while the "chest" charged in. The "loins" acted in support or pursuit. Essential to his armies' success and the psychological advantage they enjoyed was Shaka's emphasis on the deadly stabbing spear in hand-to-hand combat and his insistence on giving no quarter. Shaka's military system required that his warriors be adequately rewarded, which entailed endless campaigns for booty.

Shaka finally defeated the Ndwandwe in 1819 and incorporated their territory. He consolidated his hold over other neighboring chiefdoms through diplomacy when he could or through conquest if they resisted. Many chiefdoms took the option of flight, further destabilizing the region. There were limits, however, to the extent of territory Shaka could effectively control because of great distances and the lack of a developed administration. His armies levied tribute from the subordinate chiefdoms along the uncertain borders of the Zulu kingdom proper and regularly raided more distant peoples for booty. In 1824, white traders and hunters established a settlement at Port Natal with Shaka's permission. They had firearms, and Shaka increasingly relied on them as mercenaries.

Shaka faced opposition from rivals within the royal house and from dissident members of chiefdoms incorporated into the Zulu state. Even his warriors, exhausted by incessant campaigns, began to turn against him. As part of a wider conspiracy, his half-brothers, Mhlangana and Dingane, assassinated him on 24 September 1828 at kwaDukuza, his royal residence.

John Laband

See also: Blood River

References and further reading:

Fynn, Henry Francis. *The Diary of Henry Francis Fynn.* Ed. James Stuart and. D. McK Malcolm. Pietermaritzburg: Shuter & Shooter, 1969.

Hamilton, Carolyn. *Terrific Majesty: The Powers of Shaka Zulu and the Limits of Historical Invention.* Cambridge, MA: Harvard University Press, 1998.

Laband, John. *The Rise and Fall of the Zulu Nation.* London and New York: Arms and Armour Press and Sterling Publishing, 1997.

Webb, Colin de B., and John B. Wright. *The James Stuart Archive of Recorded Oral Evidence Relating to the History of the Zulu and Neighbouring Peoples.* 4 vols. Pietermaritzburg and Durban: University of Natal Press and Killie Campbell Africana Library, 1976, 1979, 1982, 1986.

Shapur I (r. 240–272)

Son and successor to Iranian Sassanid Dynasty (224–651) founder, Ardashir I. More formidable than the Parthian Empire it supplanted, the Sassanid Empire constantly sought to alter the military status quo with Rome in Mesopotamia, Armenia, and Syria. The long struggle that resulted focused around border fortresses such as Dara, Nisibis, Amida, and Dura Europus.

Taking full advantage of their internal crisis, Shapur campaigned with great success against the Romans. He defeated and killed Emperor Gordian III in 244. The peace treaty between Shapur and Philip the Arab, the new emperor, forced the Romans to pay tribute (Shapur claims Philip paid 500,000 *denarii*). A further Persian offensive led to the occupation of Armenia, the devastation of Syria, and the first Persian conquest of Antioch (253). The third campaign of Shapur culminated in the capture and destruction of Dura Europus (255). In a rock-cut relief from Naqsh-i-Rustam, another emperor, Valerian, is depicted cowering in front of the king after his defeat and capture at the Syrian city of Edessa in 260; one lurid tradition even claims that after his execution, Valerian's body was stuffed with straw and put on display. It was left to Septimius Odaenathus, dynast of Palmyra, to play the major role in forcing Shapur to withdraw from Roman territory (262–266). In addition to his brilliant military achievements, Shapur, "King of Kings, King of Iran and non-Iran," was famed for his grandiose building operations (he used the labor of Roman prisoners of war).

Nic Fields

See also: Roman Civil Wars (235–284); Sassanid Empire

References and further reading:

Dodgeon, Michael H., and Samuel N. C. Lieu, eds. *The Roman Eastern Frontier and the Persian Wars, 226–363.* London, New York: Routledge, 1991.

Shapur II (309–379)

Persian monarch and opponent of Julian. In 359, Shapur II led his armies into Mesopotamia and captured several Roman frontier fortresses, including Amida. The following year, Shapur invested Singara and Bezabda. It became apparent that the Sassanids, unlike the Parthians, had developed an adequate siege-warfare technology.

Three years later, the Emperor Julian, hoping to restore Roman prestige, advanced to Ctesiphon. Unable to assault it, Julian retreated up the Tigris River, where he died in a confused ambush (26 June 363). His generals quickly elect Jovian as their emperor. It was his task is to extract the army from its perilous position. Not wishing to let the Romans retreat unmolested, the Persians hovered on their flanks, striking whenever an opportunity arose. Their elephants and *clibanarii* (heavily armored cavalry), according to eyewitness Ammianus Marcellinus, terrorized the Roman soldiers. His troops exhausted and hungry by the continued attacks, Jovian agreed to a humiliating treaty. He was forced to give up Nisibis and concede parts of Armenia and Mesopotamia to Shapur. The Persians had besieged the fortress of Nisibis three times in 12 years without success: its defenses, in the words of a Syrian chronicle, had made it as safe as "a rose behind thorns."

Nic Fields

See also: Sassanid Empire

References and further reading:

Dodgeon, Michael H., and Samuel N. C. Lieu, eds. *The Roman Eastern Frontier and the Persian Wars, 226–363.* London, New York: Routledge, 1991.

Shays's "Rebellion" (1786–1787)

An agrarian revolt in Massachusetts that influenced the creation of a stronger centralized federal government in the United States under the Constitution in 1787. An economic depression led farmers to unsuccessfully petition the state government to print paper money to increase the currency supply and change legal-tender laws to allow taxes and debts to be paid in agricultural commodities. They intimidated tax collectors and obstructed the debtor courts in order to force them to postpone business. Under the nominal leadership of Daniel Shays, 1,500 farmers forced the state Supreme Court at Springfield to adjourn in September 1786. The state then appealed to the federal government for military assistance, although the rebellion would be put down before federal troops were needed.

Governor James Bowdoin appointed Revolutionary War veteran General Benjamin Lincoln to suppress the rebels with an army of 4,400 men. On 25 January 1787, Shays tried to seize the federal arsenal in Springfield in order to march on Boston and have the issues addressed. Rebel commander Luke Day, leading men from the west, tried to send word to Shays to delay actions for one day. The 1,200 arsenal defenders under General William Shepard fired their cannon into the advancing rebels, killing 4 and wounding 20. Shays's 2,000-man force retreated in disarray, pursued by Lincoln's militia coming from Boston. Marching his troops through a snowstorm the previous day, Lincoln surprised the rebels at Petersham on the morning of 3 February. The farmers were scattered, and the rebellion was broken. Isolated violence continued in the state for several months.

Although most of the rebels surrendered after being granted amnesty, Shays was able to elude capture briefly by escaping to Vermont. The Massachusetts Supreme Court sentenced Shays and 13 of the other rebel leaders to death for treason, but newly elected Governor John Hancock pardoned most of the leaders at the last moment. However, in Berkshire County, John Bly and Charles Rose were hanged. The state legislature then began to address the farmers' issues. This near–comic opera "rebellion" highlighted the weaknesses of the Articles of Confederation and strengthened the case for a stronger central government.

T. Jason Soderstrum

References and further reading:

Gross, Robert A. *In Debt to Shays: The Bicentennial of an Agrarian Rebellion.* Charlottesville: University of Virginia Press, 1993.

Szatmary, David P. *Shays' Rebellion: The Making of an Agrarian Insurrection.* Amherst: University of Massachusetts Press, 1980.

Taylor, Robert J. *Western Massachusetts in the Revolution.* Providence, RI: Brown University Press, 1954.

Sheridan, Philip Henry (1831–1888)

The most successful Union cavalry field commander in the American Civil War. Sheridan was born on 6 March 1831 to Irish immigrants in Albany, New York, grew up in Ohio, and lied about his age to gain admittance to the U.S. Military Academy at West Point in 1848. After a year's suspension for assaulting a superior officer, he graduated in 1853, thirty-fourth in a class of 49.

During his lieutenancy, he served with the 3d Infantry in Texas and the 4th Infantry in the northwest. A captain in the 13th Infantry at the outbreak of the Civil War, he was stationed as a quartermaster in southwestern Missouri until May 1862, when he was promoted to colonel of the 2d Michigan cavalry and joined Henry Halleck on the march to Corinth, Mississippi. His performance at Boonesville, Mississippi, in June earned him a promotion to brigadier general and command of the 11th Division of the Army of the Ohio. He was a key element in the Union victory over Braxton Bragg at Perryville, Kentucky, on 8 October. At Murfreesboro, his outnumbered men held the enemy until their ammunition ran out, allowing William S. Rosecrans and George H. Thomas to fall back into new lines and subsequently defeat Bragg. Promoted to major general in March 1863, he excelled at Chickamauga, Chattanooga, and Missionary Ridge.

Ulysses S. Grant placed Sheridan in command of the cavalry of the Army of the Potomac in April 1864. Sheridan fought at the Wilderness and rode completely around Lee's army near Richmond, Virginia, from 9 May to 24 May, cutting Lee's communications. He defeated and killed Jeb Stuart at Yellow Tavern on 11 May and engaged the enemy at Spotsylvania on 12 May and Cold Harbor on 2 June.

Grant gave Sheridan command of the Army of the Shenandoah in August. After Jubal A. Early smashed the Union left flank at Cedar Creek on 19 October, Sheridan's counterattack completely turned the tide. Having destroyed Confederate agriculture in the Shenandoah Valley, he returned to the Richmond campaign early in 1865. He defeated George E. Pickett at Five Forks, Virginia, on 1 April and was at Appomattox for Lee's surrender.

Sheridan commanded the Gulf of Mexico Division from 1865 to 1867, was military governor of Texas and Louisiana from 1867 to 1869, and led attacks against Indians from 1868 to 1869. In a rude remark to Comanche chief Tochaway at Fort Cobb, Oklahoma, in 1869, Sheridan is supposed to have originated the slur, "The only good Indian is a dead Indian." (The more accurate quotation is "I only saw one good Indian, and he was dead.") Promoted to lieutenant general in 1869, he was in Europe observing Prussian field operations during the Franco-Prussian War. In 1884 he became U.S. Army commander in chief, succeeding William Tecumseh Sherman. Congress made him general of the army just before his death on 5 August 1888.

Eric v. d. Luft

See also: American Civil War; Bragg, Braxton; Chattanooga, Battle of; Chickamauga, Battle of; Cold Harbor, Battle of; Custer, George Armstrong; Grant, Ulysses Simpson; Halleck, Henry Wager; Murfreesboro; Petersburg, Siege of; Pickett, George Edward; Rosecrans, William Starke; Sherman, William Tecumseh; Spotsylvania Court House; Stuart, James Ewell Brown; Thomas, George Henry; Wilderness

References and further reading:

Hutton, Paul Andrew. *Phil Sheridan and his Army.* Norman: University of Oklahoma Press, 1999.

Morris, Roy. *Sheridan: The Life and Wars of General Phil Sheridan.* New York: Vintage, 1993.

O'Connor, Richard. *Sheridan the Inevitable.* Indianapolis: Bobbs-Merrill, 1953.

Rister, Carl Coke. *Border Command: General Phil Sheridan in the West.* Westport, CT: Greenwood, 1974.

Sherman, William Tecumseh (1820–1891)

U.S. Army Civil War commander. Born on 8 February 1820 in Lancaster, Ohio, Sherman graduated from the U.S. Military Academy at West Point, New York, in 1840 and was commissioned a second lieutenant of artillery. After an undistinguished military career, he resigned his commission in 1853 to enter private business. Having failed as both a banker and a lawyer, Sherman, who was unable to reenter the army, became the superintendent of the Louisiana State Seminary of Learning and Military Academy (the present-day Louisiana State University) from 1859 until Louisiana seceded from the Union in early 1861.

At the outbreak of the Civil War, Sherman volunteered for service in the Union army, was commissioned a colonel, and commanded the 13th U.S. Infantry Regiment. Within a month, he was given command of the 3d Brigade, Tyler's Division, Army of the Potomac. After First Bull Run, Sherman was promoted to brigadier general of volunteers and sent to Kentucky, where he was quickly given command of the Department of the Cumberland. On 1 March 1862, he was ordered to the command of the 5th Division, Army of the Tennessee. Sherman commanded his division at the Battle of Shiloh, where he was wounded but refused the leave the field. Cited for bravery by both Generals Ulysses S. Grant and Henry W. Halleck, he was rewarded for his part in the Union victory by being promoted to major general of volunteers.

In 1864, following Grant's promotion to general in chief of the Union army, Sherman was made commander of the Military District of the Mississippi, making him de facto supreme commander of the Union armies in the West. Sherman then initiated the Atlanta campaign, which led to the capture of the city on 1 September 1864. After ordering the military resources of the city burned, he launched his most celebrated military action, Sherman's "March to the Sea." The campaign involved marching, without any line of communications or supply, his army of 62,000 men from Atlanta to Savannah, Georgia, on the Atlantic coast. During the campaign, Sherman's army pillaged the areas it passed through, demolishing military resources along with railroads and other private property. Sherman believed that the destruction of his campaign would lower southern morale and bring the war to a quicker conclusion. It was an early example of twentieth-century "total war."

After reaching Savannah, Sherman next began the Carolinas campaign, in which his army swung up through North and South Carolina in order to join forces with Grant and the Army of the Potomac in Virginia. On 9 April 1865, Confederate general Robert E. Lee surrendered to Grant, and two weeks later, on 26 April 1865, Confederate general Joseph E. Johnston surrendered to Sherman at Durham Station, North Carolina.

After the war, Sherman was promoted to the rank of lieutenant general in the regular army. On 4 March 1869, following Grant's election as president, Sherman was promoted to the rank of full general and became the general in chief of the U.S. Army. He retired from the army in 1884 and died on 14 February 1891.

Both Grant and Sherman epitomized what major wars would become in the industrial age: a nasty business involving nearly all of a nation's assets. Both could efficiently muster vast resources and did their best to deny them to the enemy. The pomp and "glory" of war would fade by the end of the century, as these unbuttoned, almost scruffy commanders would foreshadow in their very persons. And it was Sherman who reminded a group of veterans that war was nothing but "all hell, boys."

Alexander M. Bielakowski

See also: American Civil War; Grant, Ulysses Simpson; Sherman's March to the Sea

References and further reading:

Fellman, Michael. *Citizen Sherman: A Life of William Tecumseh Sherman.* New York: Random House, 1995.

Marszalek, John F. *Sherman: A Soldier's Passion for Order.* New York: Free Press, 1993.

Sherman, William T. *Memoirs of General William T. Sherman.* New York: C. L. Webster and Company, 1891.

Sherman's March to the Sea (mid-November–December 21, 1864)

A form of "total war" used by General William T. Sherman to bring the American Civil War to a quick conclusion. After evacuating Atlanta in late summer, Confederate general John Hood moved north through eastern Alabama to threaten William Sherman's long supply line to Chattanooga and Nashville. In a brilliant strategic concept, Sherman decided to abandon his communications, send sufficient troops to central Tennessee to resist Hood, and set out for the coast to lay waste to the Deep South and to convince southerners of the futility of continued resistance.

In mid-November 1864, Sherman's "March to the Sea" began. Some 62,000 Union troops and many camp followers (including a large number of freed slaves) set out. Sherman divided the troops into divisions, and there was no clear line of march. For the outnumbered defenders, it was unclear whether Sherman intended to march to the Gulf Coast; Savannah, Georgia; Charleston, South Carolina; or Columbia, South Carolina, the state capital and birthplace of the Confederacy. Thus the Confederates had to divide, while Sherman was able to concentrate and easily brush aside any opposition he faced.

For 250 miles, Sherman's troops cut a swath through Georgia; supposedly, one can see the outlines of the march to this day. The troops burned, looted, ate, and generally sought to bring the horror of war to the deep South, which thus far had been spared the agony of such "Up South" states as Tennessee and Virginia.

Finally, on 21 December, Sherman and his troops took Savannah, after a 10-day siege. Confederate troops evacuated before the final assault. Sherman's men met up with the Union navy, and Sherman sent a message to President Abraham Lincoln presenting him with Savannah and thousands of bales of cotton as a Christmas present. The Civil War was drawing to its inexorable conclusion.

Sherman always maintained that he liked southerners and that his march was the best means of bringing the war to a quick and merciful end. He may well have been correct, but most southerners remained unconvinced.

Charles M. Dobbs

See also: American Civil War; Savannah, Siege of; Sherman, William Tecumseh

References and further reading:

Cox, Jacob D. *Sherman's March to the Sea: Hood's Tennessee Campaign and the Carolina Campaigns of 1865.* New York: Da Capo Press, 1994.

Glatthaar, Joseph T. *The March to the Sea and Beyond.* New York: New York University Press, 1985.

Nevin, David. *Sherman's March: Atlanta to the Sea.* Alexandria, VA: Time-Life Books, 1987.

Shiloh (6–7 April 1862)

An early Union victory by General Ulysses S. Grant, and one of the American Civil War's bloodiest battles. After defeat at Forts Henry and Donelson in February, Confederate general

Battle of Shiloh, 1862. (Library of Congress)

Albert Sidney Johnston retreated to Corinth, Mississippi, where he concentrated forces for an attack against Grant. Grant had most of his command at Shiloh Church along the Tennessee River while waiting for Union general Don Carlos Buell and reinforcements from Nashville.

Johnston felt obligated to attack Grant before Buell arrived; he wanted to attack on 4 April but was delayed two days before the unfortified Union position. He also decided to have his men attack in waves, each wave commanded by a different general, which would make command and control more difficult once the battle was fully engaged.

Still, the initial attack was a great surprise, and many of Grant's untrained and raw troops fled to the riverbank. Union troops at one point did put up such furious resistance that Confederates dubbed the battle the "Hornet's Nest"; still, additional artillery turned the tide for the South.

Johnston was wounded in battle (and later would die of his wounds); his subordinates needed time to straighten out units. Grant, meanwhile, deployed artillery and Union gunboats as shields along Pittsburg Landing for his routed men to recoup and for forward elements of Buell's Army of the Ohio to join the defense. Surprisingly, Grant's forces held on that day, and on 7 April, reinforced by Lew Wallace's "lost division" and more and more divisions of Buell's army, he went over to the offensive. Meanwhile, the Confederates, now commanded by General Pierre Gustave Toutant Beauregard, decided to retreat further south.

Shiloh shocked the American people—North and South—with its high casualties (more than 23,700 killed and wounded on both sides) and made clear that this would not be a short nor painless conflict. In the meantime, Grant was ordered to defend rail and communications lines and was unable to follow up his victory with an advance further south.

Charles M. Dobbs

See also: American Civil War; Fort Donelson; Grant, Ulysses Simpson

References and further reading:

Daniel, Larry J. *The Battle That Changed the Civil War: Shiloh.* New York: Simon & Schuster, 1997.

Martin, David G. *The Shiloh Campaign: March–April 1862.* Conshohocken, PA: Combined Books, 1996.

Sword, Wiley. *Shiloh: Bloody April.* New York: Morrow, 1974.

Shimabara Revolt (1637–1638)

An uprising of peasants and masterless samurai of the Shimabara domain near Nagasaki that later spread to the neighboring Amakusa Islands, both areas that were heavily taxed and Christianized during the latter part of the sixteenth and early seventeenth centuries. The uprising, though not entirely Christian in inspiration, was perceived as such by the young Tokugawa shogunate (1603–1868) and contributed greatly to that government's decision to sever all relations with foreign countries.

The domain of Shimabara had been a center of Christian activity since the religion's introduction there in the sixteenth century, and it hosted an unusually high number of converts. Repeated attempts at uprooting Christianity in both Shimabara and the nearby Amakusa Islands, another stronghold of Christians, during the early years of the Tokugawa shogunate proved unsuccessful until the application of severe persecution, and even then many Christians proffered only an outward renunciation of their faith while secretly maintaining it. Further, the peasants of the area suffered under heavy taxation by both the Tokugawa government and their local daimyo (overlord).

The region was also home since the Battle of Sekigahara (1600) of many impoverished and embittered *ronin* (literally, wave men), or masterless samurai. The combination of these *ronin* with persecuted Christians and an overtaxed peasantry proved fertile ground for the revolt. In December 1637, a combination of peasants and *ronin* rose in rebellion in Shimabara. The rebels soon chose a boy as their leader, Amakusa Shirô (1621–1638), whose samurai father had been executed after the Battle of Sekigahara for having fought on the losing side. Though the revolt's motivating factors had more to do with overtaxation and persecution, it quickly took on millenarian and Christian aspects, with rebels shouting the names of Jesus or Mary during their attacks.

The rebels encountered some initial success, and over 35,000 of them were able to seize and entrench themselves in Hara Castle on the Shimabara Peninsula. The Tokugawa government then raised an army from among the domains of its feudal lords in northern Kyushu. Despite the more than 100,000 troops employed in the siege of Hara Castle and the help of a Dutch naval bombardment, the rebels held out for several months before the fortress fell to the Tokugawa armies in April 1638. All told, the rebels, including their youthful leader Amakusa Shirô, suffered some 35,000 deaths, the majority executed following the fall of Hara Castle, whereas the forces of the Tokugawa lost about 13,000. The army raised against the Shimabara rebels was to prove the last major mobilization by the Tokugawa until the troubles preceding that shogunate's fall in 1868. The uprising, viewed officially as Christian in nature, was crucial in the Tokugawa shogunate's decision in 1639 to promulgate its strict policy of national seclusion from the outside world.

Daniel Kane

See also: Sekigahara; Tokugawa, Ieyasu

References and further reading:

Boxer, C. R. *The Christian Century in Japan, 1549–1650.* Berkeley: University of California Press, 1951.

Morris, Ivan. *The Nobility of Failure: Tragic Heroes in the History of Japan* (see chapter titled "Amakusa Shirô"). New York: Meridian Books, 1975.

Short, Walter Campbell (1880–1949)

U.S. Army officer and central figure in the Pearl Harbor controversy. Born in Fillmore, Illinois, Short graduated from the University of Illinois (1901) and was commissioned in the U.S. Army in 1902. In 1916, he participated in the U.S. incursion into Mexico. Promoted to captain, he was attached to the 1st Division, which traveled to France in 1917. He developed expertise in the use and deployment of machine guns, rising to staff positions at corps-level automatic weapons schools. His war and postwar weapons work culminated in his book, *Employment of the Machine Gun* (1922).

Short progressed in rank through the interwar years, taking command of the 1st Division (1939). With promotion to major general, he took command of I Corps, but in February 1941 he was tapped for command of the Hawaiian Department. Brevetted lieutenant general, Short's command was focused on defending Oahu and the Pearl Harbor naval base.

The 7 December 1941 Japanese attack began a long personal nightmare for Short. Publicly tagged with "dereliction of duty" by the investigative Roberts Commission (1942) for ignoring signs of an impending Japanese attack, he was demoted to major general, relieved of command, and then pressured into retirement. Never granted an opportunity before a military court to defend his actions and decisions in Hawaii, Short spent the war years working in private industry. He died in 1949 and was buried in Arlington National Cemetery.

Edward F. Finch

See also: Pearl Harbor Attack

References and further reading:

Beach, Edward L. *Scapegoats: A Defense of Kimmel and Short at Pearl Harbor.* Annapolis, MD: Naval Institute Press, 1995.

Melosi, Martin V. *The Shadow of Pearl Harbor: Political Controversy over the Surprise Attack, 1941–1946.* College Station: Texas A & M University Press, 1977.

Report of the Joint Committee on the Pearl Harbor Attack. Washington, DC: Government Printing Office, 1946.

Siamese (Thai)–Burmese Wars (1548–1792)

By the sixteenth century, Siam (modern Thailand) and Burma, two of mainland Southeast Asia's most powerful states, emerged as bitter rivals in regional politics and war. Burma repeatedly invaded Siam, and both polities struggled for control of neighboring countries such as Laos and Lanna (Chiang Mai).

In a physical environment in which land was plentiful but labor scarce, Burmese invaders of Siam in the sixteenth and eighteenth centuries sought slave labor, skilled artisans, and royal hostages, as well as treasure and above all prestige. Placing a compliant vassal on the Siamese throne was a way the Burmese monarch could display his authority as a *chakravartin* (world-conquering king). Often, the pretext for war was the Siamese monarch's refusal to provide the Burmese king with tribute in the form of a sacred white elephant, a potent symbol of universal kingship.

Tabinshwehti (r. 1531–1550), second king of the Burman Toungoo dynasty, attempted to subjugate Siam in 1548, but his campaign failed, and it fell to his brother-in-law King Bayinnaung (r. 1551–1581) to occupy the Siamese capital of Ayutthaya (located about 80 kilometers north of modern Bangkok) in 1564. A Siamese revolt necessitated a second campaign in 1568–1569, in which the Burmese looted and destroyed Ayutthaya, hitherto one of Southeast Asia's richest cities, dismantling its fortifications and bringing many of its inhabitants back to Lower Burma.

Siam won its independence led by Phra Naret (King Naresuan, r. 1590–1605). Originally a vassal of Bayinnaung's son Nanda Bayin (r. 1581–1599), he instigated a revolt by Siamese deportees in Lower Burma in 1584, returned to Siam, and five times during 1584–1593 frustrated Nanda Bayin's attempts to reassert control over his native land. In 1593, Naresuan brought the war to Burmese soil, occupying Tenasserim, Tavoy, and Martaban and playing a role along with Arakanese and Portuguese invaders in bringing about Burma's disintegration in 1599.

After establishing a new royal dynasty, the Konbaung (1752–1885), and pacifying Mon-inhabited Lower Burma, the warlike Alaungpaya (r. 1752–1760) led his army over the Three Pagodas Pass into Siam but died in 1760 before repeating Bayinnaung's conquest. His son Hsinbyushin (r. 1763–1776) captured Ayutthaya in April 1767, completely destroying it. According to a Siamese historian, Bayinnaung waged war "like a monarch," but Hsinbyushin waged it "like a robber."

The Chinese-Siamese general Phraya Taksin, a gifted military leader, succeeded in wresting control of central Siam from the invaders in 1767–1768 and crowned himself king at Thonburi. When Taksin was deposed in 1782, the founder of the new royal dynasty, Chao Phraya Chakkri (Rama I, r. 1782–1809), built a new, more easily defended capital at Bangkok on the east bank of the Chao Phraya River and thwarted repeated Burmese invasions by King Bodawpaya (r. 1782–1819).

The struggle against Burmese aggression became an important theme in modern Siamese/Thai nationalism. The strong state established by Taksin and Chakkri preserved Siam's independence in the nineteenth century, whereas Burma fell under British colonial rule.

Donald M. Seekins

See also: Bayinnaung

References and further reading:

Hall, D. G. E. *A History of South-East Asia.* 2d ed. London: Macmillan, 1964.

Wyatt, David K. *Thailand: A Short History.* New Haven: Yale University Press, 1984.

Sicilian-Byzantine Wars (1147–1185)

Two wars rematching the Normans of Sicily with the Byzantines in 1147–1158 and 1185–1186. In the summer of 1147, the Byzantine emperor Manuel I Comnenus (r. 1143–1180) had just watched French and German armies depart his lands for the Second Crusade when he learned that Roger II of Sicily was raiding Greece. Roger, Norman king of Sicily and southern Italy, had inherited his forefathers' fondness for Balkan adventurism. Rooted on Corcyra island, his Norman raiders sacked Thebes, Corinth, and defied Manuel, Conrad III (Holy Roman Emperor), and the Republic of Venice.

The Byzantine emperor first made a pact with Conrad III, humiliated by the failure of the Second Crusade, to conquer Roger. Then Manuel captured Corcyra in 1149 with the help of a Venetian fleet, but uprisings in both Byzantine and German lands and then, in 1152, the death of Conrad III derailed his anti-Norman campaign.

On his own, Manuel dispatched his fleet to the Italian coast against a new Sicilian king, William, in 1155. At first, aided by Norman rebels, the Byzantines took Bari and seemed ready to regain Sicily. Then William struck back and crushed Byzantine land and naval forces together at the Battle of Brindisi in 1156. Chastened and distracted by war in Syria, Manuel and William made peace in 1158.

The Byzantine-Norman truce held, largely because both sides were distracted by threats and opportunities closer to home. In 1166, when William II became king of Sicily, Emperor Manuel tacitly renewed their understanding. William was equally cordial when Manuel died in 1180, and his son, Alexis II (r. 1180–1183), mounted the throne, especially since Alexis had Norman blood in his veins, but conditions changed drastically in 1183 when Andronicus I Comnenus (r. 1183–1185) seized power and murdered Alexis. As plots and revolts swirled, William of Sicily prepared to invade Byzantium.

In June 1185, William's fleet took Corcyra, landed in Epirus (Albania) at Dyrrhachium, and marched south toward Thessalonica, Byzantium's "Second City." The Sicilian fleet circled around Greece at the same time, blockading the Thessalonians from the sea. When the city fell in August, the Normans ran riot, burning, looting, desecrating, and massacring thousands. By September, William's foragers were but 200 miles from Constantinople, as his fleet entered the Marmora Straits. That same month, Isaac Angelus (r. 1185–1195) rallied the demoralized city, overthrew Andronicus, and seized the throne of Byzantium.

The new emperor struck quickly. He scattered William's main force at Mosynopolis and then crushed his other flank on the banks of the river Strymon. In Thessalonica, public fury exploded against the Sicilians, who fled the city, awaiting recovery by William's retreating navy. Norman losses proved so severe that they barely resisted in 1186 when Isaac marched overland to retake Epirus. The Sicilian menace was ended.

Weston F. Cook, Jr.

See also: Guiscard, Robert; Norman-Byzantine Wars

References and further reading:

Norwich, John J. *Byzantium: The Decline and Fall.* New York: Alfred A. Knopf, 1996.

Treadgold, Warren. *A History of the Byzantine State and Society.* Stanford, CA: Stanford University Press, 1997.

Sidi Barrani (1940)

Battle signaling the beginning of the Italian collapse, which eventually drew German forces into North Africa. After declaring war on 10 June 1940, Benito Mussolini immediately began planning an Italian attack on Egypt, with a view to linking the Libyan and Ethiopian territories. Marshall Rodolfo Graziani had nearly 1 million troops in Libya facing General Archibald Wavell's 36,000 but remained hesitant to attack for fear of logistical difficulties. The Italian Commando Supremo applied considerable pressure that Graziani be ready to attack on the same day as the projected German invasion of Britain. Eventually, on 7 September 1940, Mussolini personally ordered Graziani to attack Egypt within two days.

British troops withdrew in good order before the overwhelming Italian force. The Italian Tenth Army advanced cautiously, eventually occupying Sidi Barrani, a coastal village 60 miles inside Egypt, on 17 September. There the Italian army halted and commenced construction of an arc of strong points stretching 30 miles south and southwest of Sidi Barrani.

Wavell, in turn, came under strong pressure from Winston Churchill to counterattack and by December was ready to strike. On the night of 8 December 1940, the 4th Indian Infantry Division and 7th British Armored Division moved forward and at dawn crashed through the center of the Italian line. Turning northward toward the coast in an encircling move, they attacked Italian strong points from the rear as they advanced. The Italians were caught by surprise, in spite of having obtained accurate aerial intelligence of British preparations during previous days. By nightfall on 10 December, four Italian divisions—the 1st Libyan and 4th Blackshirt at Sidi Barrani, 2d Libyan at Tummar, Maletti Task Force at Nibeiwa, and the 64th Catanzaro near Buq Buq—had surrendered. British forces continued westward, pursuing elements of the 62d Marmarica, 63d Cirene, and 1st and 2d Blackshirt Divisions that were retreating rapidly into Libya.

During the three-day action, the British suffered 624 casualties but captured nearly 40,000 Italian troops and their equipment. A 7th Armoured Division report later stated that prisoners were too numerous to count but amounted to about 5 acres of officers and 200 hundred acres of other ranks. However, the Italian fiasco led in time to the German dispatching of Erwin Rommel and his Afrika Korps to North Africa. No longer would the British have matters their own way in the desert.

Michael Hyde

See also: Churchill, Sir Winston; Wavell, Archibald Percival, First Earl

References and further reading:

Graziani, Marshall Rodolfo. *Africa Settentrionale 1940–41.* Rome: Danesi, 1948.

Parkinson, Roger. *The War in the Desert.* London: Book Club Associates, 1976.

Playfair, Major General. I. S. O. *History of the Second World War, the Mediterranean and Middle East.* Vol. 1, *The Early Successes against Italy (to May 1941).* London: HMSO, 1954.

Sikorski, Wladyslaw Eugeniusz (1881–1943)

Polish army commander in World War II. Born on 20 May 1881 at Tuszow in Austro-Hungarian-occupied Poland, Sikorski attended the Polytechnic Institute of Lvov and served in the Austro-Hungarian army prior to World War I. During the war, he served in the Polish Legion, which had its origins in the Polish Riflemen's Association (a paramilitary organization of Polish nationalists founded by Józef Pilsudski), and fought against imperial Russia.

After World War I, Sikorski served in the Polish army and was decorated for valor during the Russo-Polish War (1919–1921). By 1921, he had been named chief of the Polish general staff. After briefly serving as the minister of war and then as prime minister of Poland, Sikorski remained neutral during Pilsudski's 1926 coup d'état. After retiring from the army in 1926, he joined the anti-Pilsudski opposition and remained out of power until 1939.

When the Germans invaded Poland in 1939, Sikorski became prime minister of the government in exile in London and established excellent relations with the other Allied leaders. In April 1943, however, the Soviet Union used Sikorski's request for an International Red Cross investigation into the murder of thousands of Polish officers at Katyn, who had previously been prisoners of the Soviets, as an excuse to break off diplomatic relations with the Polish government in exile. Although the Allies were fairly certain that the officers had been murdered by the Soviets, with the war raging, they felt that they could not publicly denounce their wartime comrades-in-arms against the Nazis.

Only three months later, Sikorski was killed in an air crash at Gibraltar on 4 July 1943. Though foul play has often been suggested in connection with Sikorski's death, which was certainly convenient for Allied leaders, nothing has ever been conclusively proved. The loss of Sikorski did leave the Polish government in exile rudderless and may have contributed to Poland's diminished position among the Allied powers in the later stages of and immediately following World War II.

Alexander M. Bielakowski

See also: Pilsudski, Józef Klemens; Polish Campaign of 1939; Russo-Polish War; Stalin; World War II

References and further reading:

Sword, Keith, ed. *Sikorski: Soldier and Statesman: A Collection of Essays.* London: Orbis Books, 1990.

Terry, Sarah M. *Poland's Place in Europe: General Sikorski and the Origin of the Oder-Neisse Line, 1939–1943.* Princeton, NJ: Princeton University Press, 1983.

Silla Kingdom

One of the three centralized kingdoms, along with Koguryo and Paekche, to emerge on the Korean peninsula from the second to fourth centuries. Silla developed in the peninsula's southeastern corner and would go on to defeat the other two Korean states—Koguryo and Paekche—with the assistance of Tang China, uniting the peninsula for the first time in 668.

Silla emerged from an earlier tribal system in place in the southern peninsula. These tribes were led by elite chieftains, and a strict hierarchy would remain a salient characteristic of Silla society and politics. Silla continued to develop and strengthen through the course of the fifth and sixth centuries. The importation of Buddhism and its adoption as a state religion helped augment central authority, with Silla monarchs by the sixth century promoting themselves as Buddha-kings. Special bands of young warriors called the *hwarang* (flower youth), trained through a unique mixture of Buddhist asceticism and militant loyalty to the state, strengthened Silla and aided its drive toward unification of the peninsula.

Interkingdom rivalries on the peninsula intensified over the course of the sixth and seventh centuries. In the sixth century, Silla succeeded in annexing the lesser Kaya polities along the southern coast. By an alliance with Tang China, Silla was finally able to muster the strength to overwhelm its rivals, attacking and defeating Paekche in 660 and Koguryo in 668.

The century or so that followed unification was a golden age for Silla, as peace and security fostered Buddhist art and the development of a uniquely Korean Buddhist philosophy. International trade, perhaps as far as the Middle East, brought wealth and prosperity, with the population of Silla's capital Kyongju reaching close to a million. Kyongju remains today a treasure-house of Buddhist architecture and iconography.

The final century of Silla was one of political upheaval, government corruption and weakness, and increasing popular unrest. Rival families vied for the kingship, often resorting to coups and assassinations. Silla had never made a heartfelt attempt to incorporate the acquired territory of Paekche and Koguryo into a single unified peninsular state. As a result, with the increasing popular unrest and decentralization of the ninth century, successor states to Paekche and Koguryo, named appropriately "Later Paekche" and "Later Koguryo," sprang up. The leader of the latter defeated Later Paekche and forced the abdication of the last Silla king in 935, establishing the Koryo Dynasty that would rule Korea until 1392.

Daniel Kane

See also: Koguryo; Paekche; Sino-Korean Wars and the Wars of Korean Unification

References and further reading:

Adams, Edward B. *Korea's Golden Age: Cultural Spirit of Silla in Kyongju.* Seoul: Seoul International Publishing House, 1991.

Gardiner, K. H. J. *The Early History of Korea: The Historical Development of the Peninsula up to the Introduction of Buddhism in the Fourth Century CE.* Honolulu: University of Hawaii Press, 1969.

Iryôn. *Samguk yusa: Legends and History of the Three Kingdoms of Ancient Korea.* Trans. Tae-Hung Ha and Grafton K. Mintz. Seoul: Yonsei University Press, 1972.

Lee, Ki-baik. *A New History of Korea.* Trans. Edward W. Wagner with Edward J. Shultz. Cambridge, MA: Harvard University Press, 1984.

Sinai-Suez Offensive (1956–1957)

A seven-day Israeli offensive into the Sinai Peninsula to break the Egyptian blockade of the Tiran Straits, eliminate the forward operational bases of Egyptian-sponsored Palestinian raiders, and shatter the rearming Egyptian army.

Egyptian leader Gamal Abd al-Nasir fomented instability in the Middle East in his quest for leadership of a newly formed pan-Arab, anti-Israel alliance that included Syria and Jordan. He supported fedayeen (guerrilla) advance bases along Israel's Sinai border, illegally barred Israeli shipping through the Suez Canal, installed artillery at Sharm el-Sheikh (Red Sea passage at the Strait of Tiran) near the southern apex of the Sinai peninsula, purchased a huge hoard of modern arms from Russia via Czechoslovakia, and nationalized the British-owned Suez Canal when the Americans withdrew funding for his Aswan dam electrification project. The Israeli response to the expanding terrorist raids had become extravagant and predictable, producing unacceptable Israeli casualties. Israeli chief of staff Moshe Dayan planned a large-scale incursion into Sinai to crush the terrorists, shatter the Egyptian military buildup before the new armaments could be deployed, and dislodge the antishipping barrier at the straits.

Through the French attaché, Dayan learned of the projected Anglo-French offensive to regain the Suez Canal, dubbed "Musketeer." French and Israeli representatives arranged a loose coordination, in return for which France filled Israel's last-minute order for modern ground attack aircraft and light tanks. The joint plan required Israel to provide a pretext for the European assault. Accordingly, Israel's invasion had initially to appear as another deep raid, albeit on a larger scale and threatening the canal, which would induce intervention. Thus the Israelis had to devise a uniquely complex operation, retaining the option of reverting to the "raid" concept should Israel's associates renege on their promised air-naval operation to the Egyptian rear.

On 29 October 1956, following a feint toward Jordan, a battalion of Ariel Sharon's 202d Paratroop Brigade was airdropped at the eastern end of the Mitla Pass, concurrent with southwesterly border crossings by the 4th Infantry Brigade, the balance of the 202d Paratroopers, and elements of the 9th Infantry Brigade. These opening moves, ostensibly mere raids, diverted attention from intended final-phase operations against fedayeen and Egyptian army concentrations in the peninsula's northeastern quadrant. The success of these border incursions encouraged the Southern Command's chief to prematurely launch the 7th Armored Brigade and 10th Mechanized Infantry Brigade in an effort to isolate the Gaza Strip formations by an armored assault along the Abu Ageila–El Arish axis, which contained a heavily fortified defensive network. The remainder of Sharon's paratroopers linked with their airhead near Mitla Pass in an overland forced march, while the 7th Armored Brigade successfully probed for a deserted gorge through which it encircled and overpowered the impregnable Egyptian entrenchment at Abu Ageila. Subsequently, it turned westward to race toward Ismailia alongside the canal. Musketeer was tardy, leaving the Israelis to take the Sinai unaided. Meanwhile, the 202d Paratroop Brigade exceeded its orders and entered Mitla Pass, where it sustained heavy casualties from an ambush. Finally extricating itself, the brigade moved along the peninsula's southwestern coast via Ras Sudar to Sharm el-Sheikh, where it was assisted in taking the position by the 9th Infantry Brigade, approaching via a fatiguing march from the northeast.

Simultaneous with the advance toward Sharm el-Sheikh, the 1st Infantry Brigade, supported by elements of the 27th Mechanized Brigade, took Rafah in a complex pincer operation on 1 November and routed the Egyptian-fedayeen Gaza Strip garrison. As Sharm el Sheikh fell on 5 November, Musketeer belatedly opened with an airborne drop at Suez City, followed by a British amphibious assault and advance to Kantara.

The Soviet Union and the United States jointly imposed a cease-fire and a withdrawal of all of the invasion forces from Egyptian territory. A United Nations buffer force was installed in Sinai, and the United States halfheartedly agreed to ensure Israeli passage through the straits. Israel gained little from its tactically successful exploit; the casus belli persisted.

Jim Bloom

See also: Israeli-Arab Wars; Israeli Military

References and further reading:

Dayan, Moshe. *Diary of the Sinai Campaign.* New York: Harper & Row, 1965.

Dupuy, Trevor N. *Elusive Victory: The Arab-Israeli Wars, 1947–1974.* New York: Harper & Row, 1978. Vol. 2, *The Sinai-Suez War, October–November, 1956.*

Fall, Bernard B. "The Two Sides of the Sinai Campaign." *Military Review* 37 (July 1957).

Golani, Moti. *Israel in Search of a War: The Sinai Campaign, 1955–1956.* London: Sussex Academic Press, 1997.

Henriques, Robert. *A Hundred Hours to Suez: An Account of Israel's Campaign in the Sinai Peninsula.* New York: Viking, 1957.

Marshall, S. L. A. *Sinai Victory: Command Decisions in History's Shortest War.* New York: Morrow, 1958.

Singapore (1942)

Great Britain's most ignominious military disaster of the twentieth century, opening the way for the end of Asian colonialism. Singapore was Britain's primary seaport in Southeast Asia and the lynchpin of British Far East defense policy between the world wars. In the interwar period, the British poured resources into Singapore in an attempt to make it a fortress stronghold, along the lines of Malta or Gibraltar. The focus of the entire project was to protect the island from a Japanese naval attack.

Major construction started in 1923. A key component of the fortification was the King George V Graving Dock, at the time the largest in the world. This dry dock would provide Britain with repair facilities in the Far East. The British installed 15" naval guns to repel a sea invasion. Contrary to popular belief, the naval guns were able to fire to the landward side, but they did not have ammunition that would have been effective in the jungle. In addition, the British built aboveground storage tanks that held enough fuel for the Royal Navy for six months.

Like its contemporary interwar project, the Maginot Line, the Singapore facility was doomed from the start. Its design philosophy was that an army could not attack Singapore from the rear because of the jungle and thus any attack must come from the sea. The British also grossly underestimated Japanese fighting power and military skill.

On 8 December 1941, Japanese troops landed in Malaya and moved south against Singapore. The British had sent the old battle cruiser *Repulse* and the modern battleship *Prince of Wales* to Singapore to bolster the defenses. Dispatched to intercept any further landings and lacking any air cover, both were sunk by land-based Japanese aircraft, and their loss was a major blow to British morale. The Japanese advanced quickly through the jungle using 18,000 bicycles, and after a not-very-effective defense, the British, under the ineffectual and uninspiring Lieutenant General Arthur E. Percival, were forced to surrender on 15 February 1942.

The fall of Singapore drastically reduced the sense of Western invincibility as Asians witnessed the most powerful Western nation defeated by an Asian power. It fell mainly upon the Americans, who had put up a much better resistance at the same time in the Philippines (as Winston Churchill pointedly informed Percival) to take the measure of the Japanese for the rest of the war.

Drew Philip Halévy

See also: Maginot Line

References and further reading:

Weinberg, Gerhard L. *A World at Arms.* New York. Cambridge University Press, 1994.

Sino-Japanese War (1894–1895)

A conflict that marked a shift of power from China to Japan and the beginnings of Japanese expansion onto the Asian continent. The conflict began in Japanese efforts to gain in-

The Japanese defeat the Chinese at the Yalu River in 1894 during the Sino-Japanese War. (Library of Congress)

fluence over Korea. Rising anti-Japanese feeling among Koreans brought Chinese troops, and that seemed sufficient provocation to the Japanese, who planned for a two-phase war against Korea and China, ultimately leading to an attack on Shandong and a drive to seize Beijing. On August 1, after armed clashes between their armies and navies, Japan and China declared war on one another.

The fighting was never seriously in doubt. Japan had a small but well-organized state army that reflected modern training methods, organization, and armaments patterned after the German army and a modern navy mostly purchased in the West. China had provincial armies (and military leaders) of varying quality and willingness to fight.

The war was marked by several major battles, all won handily by the Japanese. Japanese forces prevailed at Pyongyang on 16 September 1894; General Nozu Michitsura and 20,000 men overwhelmed 14,000 Chinese defenders, who withdrew to the Yalu River. Then Admiral Ito Yuko won the next day at the naval Battle of the Yellow Sea; the Japanese had more heavy guns, better gunnery, and ship handling, and the bulk of the Chinese fleet suffered serious damage. Two months later, the Japanese seized control of Port Arthur in southern Manchuria after negligible Chinese resistance; the Japanese secured their military victories with a landing near Weihaiwei and the subsequent destruction of the northern Chinese fleet there on 12 February 1895.

The ensuing peace negotiations at Shimonoseki were difficult for the Chinese and the Japanese alike. Soon thereafter, Russia allied with France and Germany in the so-called Triple Intervention to force the Japanese to renounce some of the territorial gains they had achieved and to settle for a larger indemnity payment, diplomatic moves that may have foreshadowed the Russo-Japanese War 10 years later.

Charles M. Dobbs

See also: Port Arthur, Siege of; Russo-Japanese War

References and further reading:

Conroy, Hillary. *The Japanese Seizure of Korea: A Study in Realism and Idealism in International Relations.* Philadelphia: University of Pennsylvania Press, 1960.

Jansen, Marius B. *Japan and China: From War to Peace, 1894–1972.* Chicago: Rand McNally, 1975.

Stewart, Lone. *Japan's First Modern War: Army and Society in the Conflict with China, 1894–1895.* 1994.

Sino-Japanese War (1937–1945)

A conflict marked by horrific bloodshed that tied down substantial portions of Japan's military forces during World War II. Japan annexed Manchuria in the early 1930s, but the imperial Japanese army, not content with this vast new territory, on 7 July 1937 engineered an armed incident at the Marco Polo Bridge near Beijing to justify a campaign to incorporate the remainder of China into the Japanese empire. Warfare rapidly spread southwestward from Beijing and from Shanghai westward along the Yangtze River valley.

Japan possessed a modern army with effective air and naval support. However, logistical requirements tended to channel operations along railway corridors. Chinese forces consisted of a series of private peasant armies, which lived partially off the land. Equipment and ammunition were scarce and support services virtually nonexistent. The strongest and best organized army was that of the Nationalist leader, Chiang Kai-shek.

The Chinese continually fell back, often offering no resistance. Occasionally, Chinese units fought with determination, such as when the Communist Eighth Route Army destroyed large Japanese units at P'inghsingkuan in the north. By December 1937, Nanjing and Shanghai had fallen, and the two zones of operations joined. The Japanese employed amphibious operations and further developed their air-ground coordination. Many Western nations posted military observers to China, and the Soviet Union operated an expeditionary air force against the Japanese for some time.

The war was marked from the outset by its barbarity. A series of massacres preceded the capture of Nanjing, where Japanese troops were ordered to butcher more than 20,000 prisoners of war and subsequently raped, mutilated, and slaughtered as many as 200,000 civilians. Japanese recruits reported being made to bayonet live Chinese prisoners during training before being sent to southern Pacific theaters.

Some fighting was very fierce. At Teierchuang in early 1938, the Japanese took five months to capture the city and lost 16,000 killed. The last phase before Japan entered World War II in December 1941 was the battle for Wuhan. By then, Japan controlled most of the Chinese coastal provinces, the lower and central Yangtze River valley, and much of the north.

China had been receiving some supplies over the Burma Road via Rangoon, but it was cut when the British capitulated to Japanese diplomatic pressure, and the Japanese invaded Burma. Although the Japanese continued to press within China, Chinese troops participated in the first Burma campaign of 1942, suffering 50,000 casualties. In 1942, the Japanese army launched an offensive designed to destroy all eastern Chinese airfields in Chekiang and Kiangsi Provinces within bombing range of Japan. Again, in 1943, Japan launched an offensive designed to prevent Chinese reinforcement of Burma and maintain political pressure on Chiang Kai-shek's government.

China committed 18 divisions to the Burma theater in 1944, where they performed well. Meanwhile, the Japanese had concentrated 500,000 troops in the south of China for Operation ICHIGO. They planned to open a continuous land-transport corridor from Indochina to northern China, as the Allies had by then severely restricted the transport of raw materials to Japan from Southeast Asia by sea. The massive attack was initially successful and produced serious military and political instability within China, which continued after the Japanese surrender on 9 September 1945.

In 1937, China had fielded approximately 200 divisions and in 1945 about 131. When the Japanese forces surrendered, their strength in China was estimated at 1,283,000. More than 3 million Chinese soldiers and 18 million civilians are thought to have died during the eight-year war.

By devastating and bankrupting large sections of China, the Sino-Japanese War led almost directly to the victory of the Chinese Communists over the Nationalists by 1949. Both Chinese factions seemed more concerned with the postwar situation than with fighting the Japanese, and large sections of the country for long periods of time were spared the direct experience of war. The Japanese were even able to establish a puppet Chinese government in Beijing.

As for Japan, its unprovoked aggression in China could be said to have caused its eventual catastrophic defeat in 1945. Americans, many of whom had a sentimental attachment to China, were outraged by Japanese aggression against that chaotic nation. Such outrage was stoked by such influential Americans as the Chinese-born Henry Luce, editor of *Time*, *Life*, and *Fortune* magazines. Growing American animosity led to concrete actions short of war against Japan, such as an oil embargo, which led to the Japanese decision to go to war against the Western Allies of World War II.

Michael Hyde

See also: Burma, Retreat from; Chiang Kai-shek

References and further reading:

Hsiung, J. C., and S. I. Levine, eds. *China's Bitter Victory: The War with Japan, 1937–1945*. New York: Eastgate Books, 1992.

Hsu long-hsuen and Chang ming-kai. *History of the Sino-Japanese War*. Taiwan: Chung Wu Publishing, 1971.

Japanese Monographs. No. 70, "China Area Operations Record (July 1937–November 1941)"; No. 71, "Army Operations in China (December 1941–December 1943)"; No. 72, "Army Operations in China (January 1944–August 1945)"; "China, Manchuria and Korea (Part I)." In *War in Asia and the Pacific*, ed. D. Detwiler, vol. 8. New York: Garland Publishing, 1980.

Sino-Korean Wars and the Wars of Korean Unification (598–676)

Standoff on the Korean Peninsula ended by Tang intervention. In the sixth century, the three dominant states that had emerged on the Korean peninsula—Koguryo in the north, Paekche in the southwest, and Silla in the southeast—entered a period of increasingly fierce competition for peninsular hegemony. Though each of these states at times enjoyed a period of relative dominance and strength, the roughly equal stature of the three powers, combined with a series of ever-shifting alliances, ensured that a balance of power was maintained. Only the participation of Chinese armies turned that balance ultimately toward Silla.

Except for a brief period of reunification under Western Jin Dynasty (265–317), China had been divided into northern and southern states after the fall of the Han Dynasty (206 B.C.E.–220 C.E.). Competition for power was particularly severe among the northern states, most of them of "barbarian" origin, and out of this crucible the Sui Dynasty (581–618) finally succeeded in reuniting China and ushering in a renewed period of centralized authority.

The new dynasty soon came into competition with the northern Korean state of Koguryo. Positioned in both Korea and Manchuria and thus in direct conflict with China, it was seen as an increasing threat by the Sui. In 598, after Koguryo had made a military foray into the Liaodong (Liaotung) Peninsula, a Sui military response was foiled only by bad weather. Sui did not let the matter drop, and a series of Sui attacks on Koguryo followed in the first decade of the seventh century. They resulted in costly Sui defeats and were a primary cause of the dynasty's sudden collapse in 618.

Tang (618–907) replaced Sui and quickly became involved on the Korean Peninsula. In 642, Paekche launched a large-scale assault on Silla for control of the Naktong River basin and the rich alluvial plains that were its boon. Faced with this onslaught, Silla sought an alliance with its erstwhile foe, Koguryo. Now under the rule of the military strongman Yôn Kaesomun (d. 666), who had seized power through an aristocratic purge, and more concerned about the new Tang Dynasty, Koguryo rejected any Silla alliance.

Koguryo had every reason to be concerned, and in 645 Tang emperor Taizong (600–649) began the first in a series of attacks on what he perceived to be a hostile and expansionist Koguryo. As had been the case under the Sui, these attacks, made by land at an extreme distance from the center of Tang power along the Yellow River, proved costly failures. Taizong even lost an eye to a Koguryo arrow.

Under Taizong's successor, Emperor Gao Zong (643–683), the campaign against Koguryo was temporarily suspended as Tang China sought to catch its breath, but a new opportunity arose as Silla began a secret diplomatic initiative to organize an allied attack on Paekche. Rejected, in turn, by Koguryo again, and then by the nascent state of Japan, Silla finally sent an envoy to the Tang court in 655. Tang welcomed an alliance with Silla to overcome the other Korean states from within, especially Koguryo. The fact that a war in Silla could be supported by sea, rather than over distant land supply lines, made the prospect all the more attractive to the Chinese.

In 660, a Silla-Tang force converged on Paekche. While a Silla army of about 50,000, led by General Kim Yu-sin (595–673), advanced from the southeast, a Tang force of almost 130,000 approached by sea. The badly outnumbered Paekche army suffered a crippling defeat at Hwangsan (modern Yonsan), sealing the demise of the Paekche kingdom.

Koguryo would meet a similar fate in 668, after resisting the Tang-Silla alliance for nearly a decade. Its defeat was hastened by the internecine struggle that erupted at the Koguryo court following the death of Yon Kaesomun in 666.

Upon the defeat of Paekche and Koguryo, Tang China was eager to retain its gains in Korea and to incorporate the peninsula into a larger Chinese state. To Silla's chagrin, Tang reestablished the defeated Korean states as Chinese colonies, with a son of the former Paekche king even ruling that defunct kingdom for it. Unwilling to accept this situation, Silla determined to face down the Tang forces. In a series of decisive encounters from 671 to 676, Silla was finally able to wrest full control of the peninsula from the Tang and establish the first unified Korean state.

Daniel Kane

See also: Chinese Imperial Wars; Koguryo; Li Shihmin; Paekche; Silla Kingdom

References and further reading:

Jamieson, John Charles. "The Samguk Sagi and the Unification Wars." Ph.D. diss., University of California at Berkeley, 1969.

Wright, Arthur. *The Sui Dynasty.* New York: Alfred A. Knopf, 1978.

Sioux Wars (1862–1891)

In the1860s, the Sioux Indians of the northern Great Plains felt that treaties made with them were not being upheld by the government or by local whites and took the opportunity of the American Civil War to strike back. A Sioux war band under Chief Little Crow ambushed and destroyed a detachment from Fort Ridgely, Minnesota, and then besieged the fort while killing and plundering in local farms. A militia force led by the state governor from Fort Snelling defeated Little Crow at Wood Lake on 23 September 1862. Little Crow escaped, but when he returned the next year, he was shot by a farmer.

Sibley then allied with General Alfred Sully, and both carried out separate but successful campaigns against the Sioux in North Dakota in 1863 and 1864. Sully crushed a large Sioux force at Killdeer Mountain on 28 July 1864. Fort Rice was then established at Bismarck, North Dakota, to keep the Sioux in check.

From 1865 to 1868, the Oglala and Dakota Sioux carried on a war against white settlers and soldiers entering Sioux hunting grounds along the Bozeman Trail, a shortcut from Fort Laramie, Wyoming, to the gold mines in Bozeman, Montana. In 1866, the Sioux slaughtered 83 soldiers sent to relieve a besieged work party, in the so-called Fetterman Massacre. In 1867, Chief Red Cloud agreed to a peace treaty that removed forts on the Bozeman trail if the Sioux settled on a sacred Black Hills area reservation by 1876.

In 1874, gold was found in the Black Hills, and the white prospectors made it impossible for the 1867 peace treaty stipulations to be carried out. General Crook led a military expedition out to enforce the treaty anyway, and one column destroyed the village of Sioux chief Crazy Horse before it was defeated and forced to retire for reinforcements. A second expedition under General Alfred Terry, including the Seventh Cavalry of Colonel George Armstrong Custer, was sent in relief. Custer disregarded orders and attacked 2,500 Sioux under Chiefs Sitting Bull and Crazy Horse near the Little Bighorn River. Custer had divided his command into three columns, and he and his 266-man column were wiped out on the morning of 26 June 1876 in the legendary "Custer's Last Stand." Custer's other two columns were saved by the arrival of General Terry.

Terry and Crook continued the campaign against the Indians year-round, and Crazy Horse was defeated and killed under suspicious circumstances in 1877. Sitting Bull fled to Canada, and most other Sioux surrendered.

In the final episode of the Sioux Wars, most of the reservation Sioux, suffering terribly, turned to a new religion that promised a messiah to deliver them from whites. The Ghost Dances of the Sioux alarmed government agents, and the military came in to stop them. Sitting Bull was shot dead by Indian police in 1890. Finally, the same Seventh Cavalry that lost to the Sioux under Custer surrounded a village at Wounded Knee and massacred 200 men, women, and children in revenge. By 1891, all the remaining Sioux had been rounded up or defeated.

Christopher Howell

See also: American Indian Wars; Crazy Horse; Sitting Bull

References and further reading:

McGinnis, Anthony. *Counting Coup and Cutting Horses: Intertribal Warfare on the Northern Plains 1738–1889.* Evergreen, CO: Cordillera Press, 1990.

Starkey, Armstrong. *European and Native American Warfare 1675–1815.* Norman: University of Oklahoma Press, 1998.

Secoy, Frank R. *Changing Military Patterns of the Great Plains Indians.* Lincoln: University of Nebraska Press, 1953; reprint, Lincoln: Bison Books, 1992.

Sitting Bull (1831–1890)

Hunkpapa Lakota Sioux chief and holy man responsible for the stunning series of Sioux victories over U.S. forces. Sitting Bull developed his war skills against the Crow and then fought U.S. troops at the 1864 Battle of Killdeer Mountain. He successfully besieged Fort Rice in 1865 and became Lakota chief in1868.

In 1874, conflict broke out over gold discovered in the Sioux Black Hills. Sioux, Cheyenne, and Arapahoe forces guided by Sitting Bull defeated U.S. troops led by General George Crook at the Rosebud on 17 June 1876 and wiped out the U.S. Seventh Cavalry under George Armstrong Custer at Little Bighorn.

U.S. strategy to pacify the Indians changed to elimination of food supplies, and Sitting Bull's formidable force was broken apart. The chief fled with followers to Canada but surrendered and was a prisoner of war from 1881 to 1883. He then returned to his birthplace and people in 1883. In 1885 he toured for four months with Buffalo Bill's Wild West Show.

Sitting Bull lived the old tribal lifestyle until 1890. In the fall of that year, his association with the Ghost Dance religion prompted his attempted arrest, during which he was shot and killed by Indian police before the massacre at Wounded Knee on 29 December 1890.

Christopher Howell

See also: American Indian Wars; Crazy Horse; Sioux Wars

References and further reading:

McGinnis, Anthony. *Counting Coup and Cutting Horses: Intertribal Warfare on the Northern Plains 1738–1889.* Evergreen, CO: Cordillera Press, 1990.

Secoy, Frank R. *Changing Military Patterns of the Great Plains Indians.* Lincoln: University of Nebraska Press, 1953; reprint, Lincoln: Bison Books, 1992.

Tebbel, John. *Compact History of the Indian Wars.* New York: Tower Publications, Hawthorn Books, 1966.

Six-Day War (5–10 June 1967)

Devastating Israeli preemptive lightning offensive into Egyptian, Syrian, and Jordanian territory on 5–10 June 1967. Eleven years after Gamal Abd al-Nasir's military defeat in 1956 and after he had previously rebuilt his shattered forces with Soviet equipment, Syria questioned his claim to military leadership of the 1966 anti-Israel coalition in May 1967, when he failed to react to a stunning Israeli retaliatory air patrol that bagged six Syrian MiGs. Further, he defiantly deployed much of his revitalized army in the same forward "lynchpin" positions that had been neutralized in 1956, unraveling the entire Sinai command structure.

Reacting both to Syrian scorn and to their report of an imminent Israeli invasion of Syria (a threat contrived via Soviet "disinformation"), al-Nasir resumed the illegal blockade of Israeli shipping from the port of Eilat, ordered the evacuation of the United Nations (UN) buffer force from the border areas in Sinai, incited the Egyptian masses with martial rhetoric defying Israel to take his bait, and signed a pact with Jordan and Iraq, placing the alliance forces under Egyptian command. Thus, all the Israeli tripwires to war

A victorious Israeli convoy in Egypt during the Six-Day War, 8 June 1967. (Hulton/Archive)

had been broken. Nevertheless, al-Nasir evidently did not expect Israel actually to attack. Contrary to its early postwar rhetoric, neither did the Israeli Defense Forces (IDF) contemplate an Arab invasion of Israeli population centers.

Israel doubted its capacity to fight simultaneously on three fronts (Egyptian, Jordanian, and Syrian) and preferred that fighting take place in Arab rather than Israeli territory. Hence, it made the decision to strike first. On the morning of 5 June, the Israeli air force attacked Egypt, the largest military power in the region. The timing of the attack, 8:45 A.M., was designed to catch the maximum number of Egyptian aircraft on the ground and the Egyptian high command in traffic en route to military bases. The Israeli aircraft took evasive measures to elude Egyptian radar rather than to destroy it and thereby ruin the surprise. The first wave approached from unanticipated directions, achieving complete surprise. Within hours of the strike, the Israelis, who focused their attacks on military and air bases, had destroyed 309 of the 340 total combat aircraft of the Egyptians. Israeli ground forces then moved into the Sinai Peninsula and Gaza Strip, where they engaged Egyptian units.

Israeli "tanks ahead" combined armored-mechanized infantry units maneuvered to disrupt the Egyptian forces, destroying their cohesion. General Israel Tal's thrust across the Sinai was the main attack. Israel expected the Egyptians to follow the Russian "sword and shield" doctrine for mechanized defense against this thrust, with mobile armored reserves holding back to counterattack any forces penetrating the defensive shield. Anticipating this tactic, Israeli formations under General Abraham Yoffe infiltrated between General Tal's thrust in the north and General Ariel Sharon's seizure of the Abu Ageila strong point in the south, intercepting the armored counterattack and completely breaking the cohesion of the Egyptian tactical plan. This disruption allowed General Tal's formations to rush to the Suez Canal after breaking Egyptian defenses and precipitated the headlong retreat of Egyptian forces back to the west. Although a significant percentage of the physical elements of the Egyptian force remained to be destroyed by the Israeli Air Force as they retreated through the Sinai passes, they already had effectively been defeated by Yoffe's maneuver. Egyptian casualties were heavy and Israel's minimal.

Hostilities flared up immediately on Israel's eastern front. Israel had conveyed a message to King Hussein of Jordan requesting him to stay out of the conflict. However, on the first morning of the war, al-Nasir called Hussein and encouraged him to fight, mendaciously advising the Jordanian monarch that Egypt had been victorious in the morning's fighting—an illusion that the Egyptian public also believed for several days.

At 11:00 A.M., Jordanian troops attacked the Israeli half of Jerusalem with mortars and gunfire and shelled targets in the Israeli interior. Israel's air force, having disabled the Egyptian air force, turned its attention to Jordan. By evening, the Jordanian air force had been largely destroyed, again with minimal Israeli casualties. At midnight, Israeli ground forces attacked Jordanian troops in Jerusalem, and by the morning of 6 June, Israeli troops had nearly encircled the city. On the second day of the war, the Israeli air force continued to pound Arab air bases, raising the total number of destroyed Arab planes to 416, which included more than two-thirds of the Syrian air force. With nearly total control of the skies, Israeli fighter planes and bombers were free to support the tank and infantry forces on the ground. Thus Jordanian reinforcements were prevented from reaching Jerusalem, and by 10:00 A.M. on 6 June, the Israelis had taken the Western Wall, also known as the Wailing Wall, in the Old City, the holiest site in Judaism.

Ground battles continued in the Sinai, where Egypt's armies fell back in the face of Israeli advances. On the third

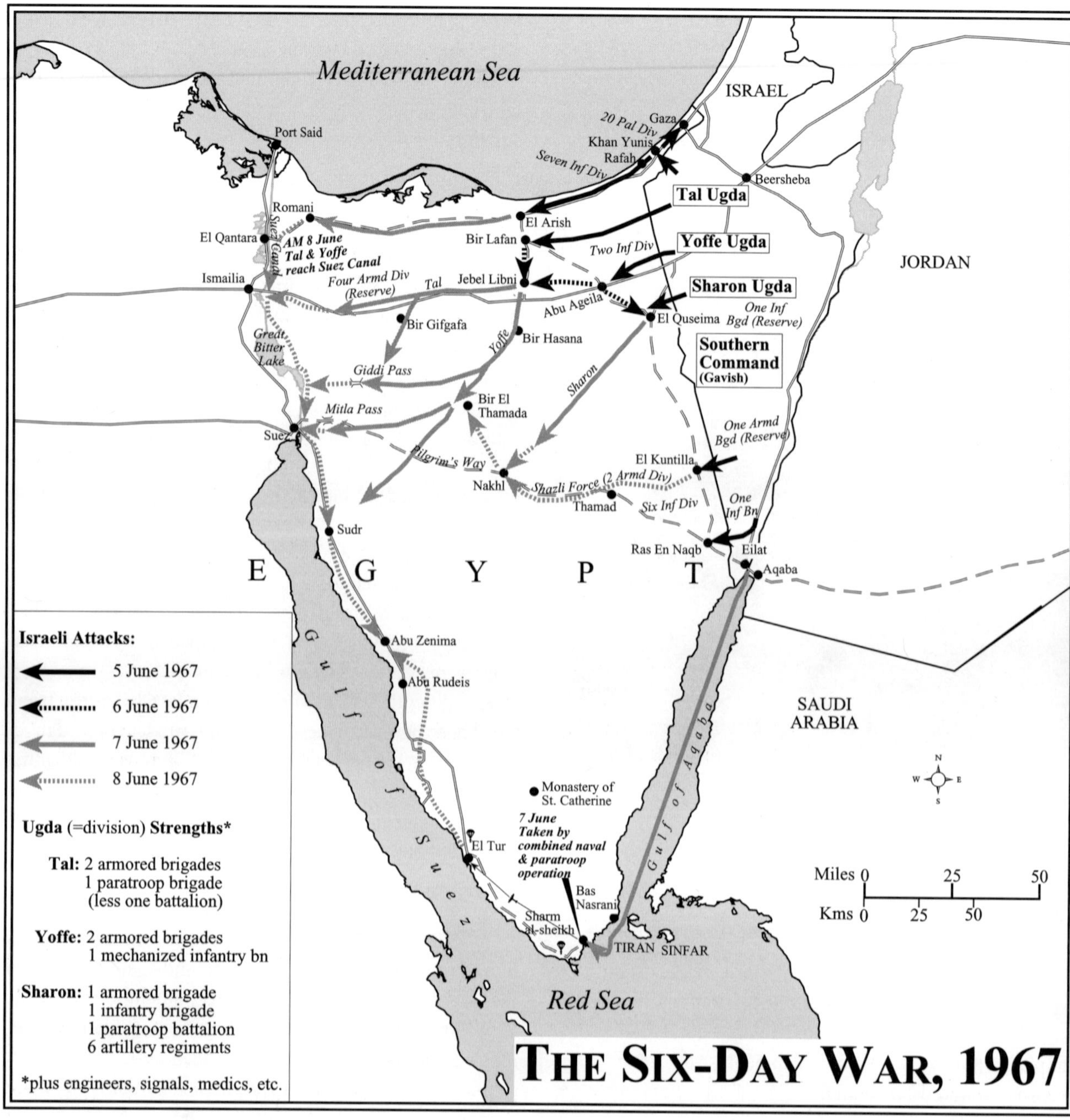

day of the war, 7 June, Jordanian forces were pushed from the West Bank across the Jordan River. The UN arranged a cease-fire between Israel and Jordan that went into effect that evening.

The following day, 8 June, Israeli forces reached the Suez Canal. As artillery battles continued along the front, the Israeli air force decimated retreating Egyptians, who were backed up on the few roads through desert mountain passes.

As the Sinai shifted to Israeli control, Israel turned its forces toward the Golan Heights. There, on 9 June, Israeli forces began an arduous assault up steep terrain in the face of entrenched Syrian forces. Israel sent an armored corps into the front of Syrian lines, while infantry forces surrounded the Syrian positions. The balance of power soon shifted to Israel's favor, and at 6:30 P.M. on 10 June, Israel and Syria concluded a cease-fire agreement. Israel thereby controlled all the Golan Heights, including parts of Mount Hermon. Although Israel also controlled the Sinai Peninsula, fighting between Israel and Egypt did not formally end for many years. Not until the 1979 Camp David Accords did the two countries finally reach peace.

The speed and scope of Israel's victory were devastating to the Arabs, who had expected a triumph. Egypt, Jordan, and Syria lost almost all of their air forces and much of their weaponry. About 10,000 Egyptians were killed in Sinai and Gaza alone. These numbers contrasted with 300 Israeli casualties on that front. In all, Egypt lost about 11,000 troops, Jordan lost some 6,000, Syria around 1,000, and Israel 700. Arab leaders had to overcome the shock and frustration at home. Abroad, the USSR, which had strongly supported the Arab powers, was embarrassed because an ally of the United States had defeated its clients. Further, Soviet weapons systems had failed against Western weapons.

On 22 November, the UN passed Resolution 242, which called for Israel to withdraw from the Occupied Territories, in return for which Arab states would recognize Israel's independence and guarantee the sanctity of Israel's perimeter. Both sides interpreted the resolution to their own advantage. The Arabs and Palestinians declared their intention to continue fighting Israel, and Israel refused to return the Occupied Territories in the face of continued terrorist attacks and conducted reprisals. Israeli-Egyptian artillery, sniper, and occasional air attacks persisted for several years, escalating into the 1969–1970 War of Attrition.

But the 1967 war had yielded to Israel strategic buffer territories, chiefly comprising those segments that had been originally designated by the UN (1948) as the proposed Palestinian entity and that presently are the disputed Occupied Territories. The campaign earned the Israeli military a global reputation, and leading military establishments analyzed Israel's war-fighting methods in their war colleges.

Jim Bloom

See also: Israeli-Arab Wars; Sinai-Suez Offensive

References and further reading:

Barker, A. J. *Arab-Israeli Wars.* New York: Hippocrene Books, 1980.

Bloom, James J. "Arab-Israeli Wars (1947–1982)." In *Encyclopedia of Military History and Biography.* Washington, DC: Brassey's, 1994.

Cordesmann, A. H., and A. R. Wagner. *The Lessons of Modern War.* Vol. 1, *The Arab-Israeli Conflicts, 1973–89.* Boulder, CO: Westview Press, 1990.

Dupuy, T. N. *Elusive Victory: The Arab-Israeli Wars, 1947–1974.* Fairfax, VA: Hero Books, 1978.

Herzog, C. *The Arab-Israeli Wars.* Rev. ed. New York: Vintage Books, 1984.

Young, P. *The Israeli Campaign, 1967.* London: Kimber, 1968.

Slim, William Joseph, First Viscount (1891–1970)

Perhaps the best British field commander during World War II. Slim came from a middle-class background. He participated in the Officer Training Corps but entered the army only after the outbreak of World War I. His service included heading off a mutiny in December 1914 and the Gallipoli campaign. After being wounded, Slim was eventually posted to Mesopotamia and served as a staff officer. Following the war, he transferred with a regular commission to the Indian Army.

Between the wars, Slim's postings included teaching at the Imperial Staff College and Imperial Defense College. By the outbreak of World War II, he was a brigadier general. On 23 September 1939, Slim took command of the 10th Indian Brigade. He prepared it for mechanized desert warfare and led it in the conquest of Italian East Africa. After being wounded in January 1941, Slim assumed command of the 10th Indian Division. He led his command in the conquest of Vichy French Syria and the occupation of Iran to open supply lines to the Soviets.

On 13 March 1942, Slim took over the newly created I Burma Corps at Prome. He conducted a skillful delaying action against the Japanese invaders. Reinforcements were not available, and Chinese assistance was rare. During March and April, Slim retreated 900 miles to Imphal. He successfully defended India for the remainder of 1942, despite the nationalists' "Quit India" campaign. After the failure of the British Arakan offensive in early 1943, Slim assumed command of IV Corps as well. He was promoted to commander of the Eastern (later Fourteenth) Army in October 1943 and defeated Japanese offensives at Arakan and Imphal-Kohima in 1944.

Slim's army was chronically undersupplied, and morale was low when he took command. His energetic and flexible manner, as well as his ability to empathize with the common soldier, inspired his troops and created an effective army. During the last half of 1944, Fourteenth Army liberated northern Burma. Between December and March 1945, Slim directed the recapture of the rest of the country. His campaign was a masterpiece of planning, deception, and logistics in a difficult theater. By May, Slim had recaptured Rangoon, destroying the opposing Japanese army. By the end of the war, he had been promoted to commander in chief of Southeast Asian ground forces.

Tim J. Watts

See also: Chindits; Imphal and Kohima

References and further reading:

Evans, Geoffrey Charles. *Slim as Military Commander.* Princeton, NJ: Van Nostrand, 1969.

Lewin, Ronald. *Slim: The Standard-bearer: A Biography of Field Marshal the Viscount Slim.* London: Cooper, 1976.

Slim, William J. *Defeat into Victory.* New York: D. McKay, 1961.

Smolensk (1941)

Operation BARBAROSSA envisaged Fedor von Bock's Army Group Center capturing Moscow and eliminating Soviet army groups en route within four months. Smolensk, a crucial communications center 200 miles west of Moscow, was a major German objective.

The Bialystok-Minsk encirclement (9 July), netting 300,000 Russian prisoners, exposed Smolensk. With Adolf Hitler's support, Hans Gunther von Kluge ordered Heinz Guderian's 2d and Hermann Hoth's 3d Panzer Groups, the enveloping arms of Army Group Center's pincer, halted for mopping up operations. This action would have allowed his Fourth Infantry Army to catch up, but the Panzers nonetheless advanced. Crossing the Dnieper River (10 July), they entered Smolensk from south and north (16 July), with the Fourth Army lagging.

Stavka, the Soviet High Command, reeling from 2,000,000 manpower losses since BARBAROSSA began, deployed the new Nineteenth, Twentieth, Twenty-first, and Twenty-second Armies along the Dnieper River and the Sixteenth Army around Smolensk, thus halting the German blitzkrieg. Western front commander in chief Semen Timoshenko coordinated capable commanders such as Ivan Konev, Konstantin Rokossovsky, Pavel Kurokhin, and Mikhail Lukin.

The Panzer breakthrough on 16 July threatened encirclement of the Soviets. Timoshenko gave the only orders that could deal with the situation at the front: his commanders were to cut their way out with systematic fighting withdrawals, regroup into mobile detachments, and counterattack, accompanied by tank trap defenses and aerial support. Amid fierce fighting, the Soviets escaped the German trap before Smolensk fell on 8 August. Despite 100,000 Soviet casualties and 300,000 Soviets taken prisoner, along with 2,030 tanks and 1,900 guns lost, a stable line held 25 miles east of Smolensk. The Red Army had proved the German blitzkrieg could be halted, even when facing odds of 2:1 in personnel, artillery, and aviation and 4:1 in tanks, thus crucially regenerating its morale.

The Soviets at Smolensk tied down Wehrmacht tank forces designated for Leningrad. Hitler, targeting Leningrad and Ukraine before Moscow but perhaps deterred by Smolensk, reversed himself and stripped Group Center of Panzers, delaying Operation TYPHOON, the seizure of Moscow, until 2 October and leaving Army Group Center's infantry open to further counterattacks. The two-month delay was vital. Georgy Zhukov formed the Army Group before Moscow and mobilized the population to construct defenses. The Germans reached Moscow in deep winter, ill-equipped, exhausted, and off-balance for the 6 December Soviet counteroffensive.

Neil Harvey Croll

See also: Hitler, Adolf; Moscow; Stalin; Timoshenko, Semen Konstantinovich

References and further reading:

Erickson, J. *The Road to Stalingrad: Stalin's War with Germany.* London: Weidenfeld and Nicolson, 1975.

Glantz, D. M., and J. M. House. *When Titans Clashed: How the Red Army Stopped Hitler.* Lawrence: University Press of Kansas, 1995.

Overy, Richard. *Russia's War.* London: Allen Lane, the Penguin Press, 1998.

Strategicus. *From Tobruk to Smolensk.* London: Faber and Faber, 1941.

Smuts, Jan Christian (1870–1950)

Boer guerrilla leader and field commander, South African statesman and philosopher, and British field marshal. Smuts was born into a Boer family in Riebeeck West, near Malmesbury, in the British Cape Colony of South Africa on 24 May 1870. After receiving his baccalaureate from Victoria College in Stellenbosch in 1891 and his law degree from Christ's College, Cambridge, in 1894, he began practicing law in Cape Town in 1895. In the simmering dispute between Cecil Rhodes and Paul Kruger over South African governance, Smuts was at first pro-British, but the Jameson raid of 29 December 1895 induced him to change his opinion.

Smuts moved to Transvaal in 1897, was state attorney from 1898 until Pretoria fell on 5 June 1900, and thereafter was general of guerrillas, serving at Diamond Hill on 11–12 June, Nooitgedacht and Magaliesburg on 13 December, Modderfontein on 31 January 1901, Elands River Poort on 17 September, and the siege of O'okiep from 4 April to 3 May 1902. He cosigned the Treaty of Vereenigung on 31 May and entered politics as a supporter of Louis Botha.

Allied with the British in World War I, Smuts suppressed the Boer revolt of Christiaan Rudolf De Wet between September and December 1914 and defeated the Germans in Namibia between February and July 1915. His 1916 campaign in Tanganyika against General Paul Emil von Lettow-Vorbeck proved tougher, but by December, Smuts had captured most of the German-held territory. The former enemy was now a military hero in Britain, and David Lloyd George invited him to London to help conduct the European war effort. He served the British in several capacities and cosigned the Treaty of Versailles in 1919.

Smuts was prime minister of South Africa from 1919 to 1924 and from 1939 to 1948. In 1926 he published *Holism and Evolution,* his only philosophical work, which brought the concept of holism into philosophy. For his personal mobilization of South African forces against the Nazis, Britain named him a field marshal in 1941. He contributed to drafting the United Nations charter in San Francisco in 1945. He died of a stroke on 11 September 1950 at the family farm near Pretoria.

Although Smuts gained international stature in his time, today he is nearly forgotten, his philosophy dismissed as mere popularizing, and his lack of interest in the obsessions of the later twentieth and early twenty-first centuries—race and imperialism—making him seemingly a mere creature of his time.

Eric v. d. Luft

See also: Boer Wars; Botha, Louis; Churchill, Sir Winston; Kruger, Paul-Stephanus Johannes Paulus; Lettow-Vorbeck, Paul Emil von; Roberts, Frederick Sleigh, First Earl, Viscount St. Pierre of Kandahar; World War I; World War II

References and further reading:

Cameron, Trewhella. *Jan Smuts: An Illustrated Biography.* Cape Town: Human & Rousseau, 1994.

Friedman, Bernard. *Smuts: A Reappraisal.* New York: St. Martin's Press, 1976.

Ingham, Kenneth. *Jan Christian Smuts: The Conscience of a South African.* New York: St. Martin's Press, 1986.

Meiring, Piet. *Smuts the Patriot.* Cape Town: Tafelberg, 1975.

Smuts, Jan Christian, Jr. *Jan Christian Smuts: A Biography.* New York: Morrow, 1952.

Soccer War (1969)

A four-day war between the Central American republics of Honduras and El Salvador, ignited in part by close soccer matches between their national teams. The real causes of the conflict were much deeper. Tiny El Salvador's 3 million people were pressed into 8,000 square miles of national territory, resulting in a density of 400 per square mile. By 1969, 300,000 Salvadorians had migrated into neighboring Honduras, a country of 42,300 square miles and a population of only 2,333,000. The Salvadorian expatriates found relative prosperity in Honduras as small plot farmers or as factory workers. However, they were never fully accepted by Hondurans, and discrimination flared, especially during hard times. Moreover, the border between the two nations was ill defined and in many areas not patrolled. The two nations had reached temporary accords on the immigration problem, but these expired early in 1969. A series of closely contested soccer games for a World Cup qualifying round in June 1969 revived old animosities and prejudices among ordinary citizens. Spontaneous violence following each game spun out of control, and nearly 17,000 Salvadorians fled Honduras for their homeland. After the final match was won by El Salvador, the two nations broke diplomatic relations.

On the morning of 14 July, Honduran air force fighters crossed into Salvadorian airspace and made mock strafing runs. The Salvadorian air force responded late in the afternoon with an air assault on Tegucigalpa's Toncontin Airport, which had the dual function of commercial field and military base. Other military and civilian installations along the border were also struck. This preemptive strike by the Salvadoran air force failed in its purpose to destroy the Honduran air force on the ground. The air attacks were followed by a more successful land invasion that resulted in the occupation of nine Honduran towns. The Hondurans, however, had the advantage of superior air power and launched an attack on Ilopango, the San Salvador commercial airport and air force base, as well as petroleum facilities at Acajutla and El Cutuco. The damage to the oil storage tanks soon caused a gasoline shortage in El Salvador. Thus El Salvador's success on the ground was offset by Honduran air superiority. In addition, both sides nearly exhausted their military stores of ammunition by the fourth day of the war.

The Organization of American States (OAS) negotiated a cease-fire that went into effect on 18 July. OAS military observers and human rights experts arrived to monitor the cease-fire agreement. The war resulted in some 2,000 deaths, mostly of Honduran civilians, and the expulsion of 130,000 Salvadorians from Honduras. Eventually, the OAS pressured El Salvador to withdraw its troops from Honduras, and relations returned to their normal level of tension.

George Lauderbaugh

References and further reading:

Mallin, Jay. "Salvador-Honduras War, 1969." *Air University Review* (March–April 1970).

Solferino (24 June 1859)

Austro-French-Piedmontese battle that gave birth to the Red Cross. After their victory at Magenta (4 June 1859), the French and Piedmontese armies pursued the retiring Austrians. Sure that their enemy would be defending the Mincio River lines, French emperor Napoleon III and Piedmontese monarch Vittorio Emanuele II were surprised to find the Austrians bivouacking near Solferino.

The battle erupted over a 25-kilometer line from the hilly countryside south of Lago Garda to Mantua. Neither the Austrians nor the Franco-Piedmontese commanders expected a battle on 24 June. The French advance guard left its camp at 2:00 A.M., unexpectedly encountering Austrian troops by 6:00 A.M. Tactical surprise explains this unmethodical battle, as the army corps on both sides joined the fight without coordination, simply feeding their troops into the battle. The allied forces took the initiative and attacked on the whole line, corps after corps. The Piedmontese fought

Soldiers from Sardinia-Piedmont and French troops under Napoleon III defeat the Austrians in 1859 at the Battle of Solferino. (Library of Congress)

to a standstill against Ludwig Ritter von Benedek's corps. The southern battlefield saw two French corps blocking the First Austrian Army. The decision was made to attack on the center of the line. The ridge and the village of Solferino were held by three Austrian corps. The French artillery poured a terrible fire on the outranged Austrian artillery all morning. Then, the French Imperial Guard assaulted Solferino at a tremendous cost. With the support of two more corps, the guard eventually took the ridge and broke through the Austrian center. At 3:00 P.M., the Austrians tried a late flanking movement on the French right wing. But the supreme commander, Austrian emperor Franz Joseph, sickened by the slaughter, decided on a general retreat. More than 20,000 corpses were scattered over the battlefield, and 25,000 wounded soldiers were begging for help. Napoleon himself supposedly became physically ill in the presence of this battlefield carnage.

He was not alone. The ghastly panorama so moved a Swiss journalist, Jean-Henri Dunant, that he determined to found a new charitable organization dedicated to alleviating battlefield suffering. The Red Cross was eventually founded in 1864.

The days following the battle saw negotiations that ended in a separate Franco-Austrian peace. The disappointed Piedmontese did receive Lombardy, something like a consolation prize. One of the lesser-known consequences of Solferino was the Prussian improvement of corps coordination and the development of a modern breech-loaded rifled artillery that would serve them well in their coming wars of unification. Conversely, French military reputation, very high to begin with since the days of Napoleon I, soared, causing a complacency that would lead to disaster in the Franco-Prussian War 11 years later.

Gilles Boué

See also: Custozza, Second Battle of; Red Cross

References and further reading:

Bourgerie, Raymond. *Magenta et Solferino.* Paris: Economica, 1995

Generalstabsbureau für Krieggeschichte. *Der Krieg in Italien 1859.* Vienna, 1873

Somalia, U.S. Military Operations in (1987–2000)

Africa's first post–Cold War conflict. The conflict in Somalia arose from the regime of Mohamed Siad Barre. In power from 1969 to 1991, Barre created a patron-client relationship with the nation's clan-based society. This system required continuous outside assistance, first from the Soviet Union (which eventually chose Ethiopia as its local client instead) and then from the United States or various nongovernmental charitable assistance groups. By 1987, some 57 percent of Somalia's gross national product was generated by international aid.

In the wake of a losing war effort against Ethiopia in the Ogaden region and a declining economy, General Mohammed Farah Aideed overthrew Barre. However, Aideed lacked legitimacy outside his own clan, leading to the collapse of the state, continuous clan-based civil war, and famine.

This situation provoked the deployment of the United Nations Operation in Somalia (UNOSOM I) in April 1992. But the force proved insufficient to control the "technicals," mercenaries hired under the euphemism of "technical support" by in-country humanitarian organizations who remained loyal to their clans. Under these circumstances, between November 1991 and March 1992, some 30,000–50,000 Somalis died of starvation.

The U.S.-led Unified Task Force (UNITAF) broke this cycle in December 1992, forcibly expediting food distribution and seeing the end of the famine by the spring of 1993. The hope was that UNITAF's mission would be a springboard to the political unification of Somalia.

The successor force to UNITAF was UNOSOM II, which continued the UN constabulary role but with less U.S.-supplied firepower. While the UN contemplated reconstitution of a central government, Aideed and his forces prepared to take the initiative against UN forces, as they felt their claim to power was superior and unrecognized. This decision led to an escalating cycle of conflict that culminated in the Mogadishu firefight of 3 October 1993, in which 18 Americans and hundreds of Somalis were killed.

In retrospect, the Somali experience reflects poorly on the UN administration and U.S. arms. The UN leadership failed to comprehend the level of difficulty involved and underestimated the level of international support it enjoyed. The U.S.-led manhunt for Aideed was prosecuted with a divided command structure, with all the control problems that implies, not to mention the failure to consider that Aideed's forces might be able to offer effective resistance under the right circumstances.

At the end of December 2000, Somalia remain more a region than a country, the last UN troops having decamped in 1995, Aideed having died in 1996 as a result of battle wounds, and there being little easy enthusiasm for peacekeeping operations in marginal portions of the world.

George R. Shaner

See also: Powell, Colin L.

References and further reading:

Bolger, Daniel P. *Death Ground: Today's American Infantry in Battle.* Novato, CA: Presidio Press, 1999.

Bowden, Mark. *Black Hawk Down: A Story of Modern War.* New York: Atlantic Monthly Press, 1999.

Maren, Michael. *The Road to Hell: The Ravaging Effects of Foreign Aid and International Charity.* New York: Free Press, 1997.

Stevenson, Jonathan. *Losing Mogadishu: Testing U.S. Policy in Somalia.* Annapolis, MD: Naval Institute Press, 1995.

The Somme (1916)

One of the most bloody and futile campaigns of World War I, consisting of several offensives from 1 July to 18 November 1916. The battle was fought between the Allies under General Ferdinand Foch and the Germans under General Erich von Falkenhayn along a 25-mile stretch from the river Somme, north of the river Ancre. The plan as originally conceived by the French commander in chief, Joseph Joffre, was adopted by Field Marshal Douglas Haig—the newly appointed commander in chief of the British Expeditionary Force—to ease pressure on the French at Verdun. The British Fourth Army under General Henry S. Rawlinson was to direct its main effort to the north of the Somme, the Third Army under General Edmund Allenby was to go even further north, and south of the river, Foch's army would make a holding attack. As a preliminary to the main push forward on 1 July 1916, the Allies undertook an eight-day nonstop artillery bombardment of German defensive positions in the occupied towns. The aim was to destroy the well-developed system of deep concrete bunkers and barbed wire entanglements, thereby allowing the 750,000 Allied troops to advance unopposed through German front lines to Bapaume and Peronne.

The artillery bombardment failed to penetrate the German trenches and barbed wire tangles. The Germans simply remained in their dugouts for the duration of the barrage, reemerging in time to arm their machine guns as the British advanced in close order. By nightfall, the British had advanced only 1,000 yards and had lost 57,450 men, 19,000 of whom were dead. For the next 10 weeks, the Allies settled for a battle of attrition that culminated in a 7-mile penetration into the German lines on 15 September as a result of the arrival of British tanks. Fighting ended with bad weather in mid-November. In all, only 125 square miles were gained by the Allies in the five months of fighting. Total casualties for the Battle of the Somme were 1,265,000: 650,000 German, 420,000 English and colonials, and 195,000 French. The Somme has lingered in British memory as a symbol of the loss and futility of the western front.

Margaret Hardy

See also: Armored Fighting Vehicles; Haig, Douglas; World War I

References and further reading:

Collison, C. S. *The 11th Royal Warwicks in France 1915–16 (from the Personal Diary of Its Commanding Officer).* Birmingham, UK: Cornish Brothers, 1928.

Evans, Martin Marix. *The Battles of the Somme.* London: Weidenfeld, 1996.
Gibbs, Philip. *The Battles of the Somme.* New York: George H. Doran, [c. 1917].
Martin, Christopher. *Battle of the Somme.* London: Wayland, 1973.

Songhay Empire (15th–16th Centuries)

The most powerful state in western Africa at the dawn of early modern history. The Songhay people first appear in African history as a fisherfolk living along the upper bend of the Niger River. Different in language and culture from their neighbors, they inhabited a string of towns from Kukya to Gao. Later, they turned to farming, river transportation, and commerce with the northern oasis towns. The Dia, the first Songhay dynasty, embraced Islam and made Gao a little economic dynamo. However, in 1325, the empire of Mali subjugated Gao, making vassals of the Songhay. In 1335, the Sonni, another Muslim noble house, deposed the Dia.

In 1375, as the empire of Mali fragmented, King Suleyman-Mar regained the independence of Songhay. For decades thereafter, the Songhay people apparently remained in their Kukya-Gao area until Sonni 'Ali became king in 1464. It was Sonni 'Ali who set Songhay on the course of empire. He raised an elite cavalry force of more than 5,000 crack troops and developed a war canoe for transportation and assault. He pushed the anti-Muslim Mossi people southward and annexed several of the great Niger cities, including the prestigious Islamic city of Timbuktu. A tireless campaigner but harsh ruler, he frequently abused his pious Muslim critics. After his death in 1492, Sonni 'Ali's son lost the throne in less than two years.

Askia Muhammad Ture, a distinguished army officer, seized power in 1493. An astute, cultivated administrator, he made Gao, Timbuktu, and Jenne centers of commerce and Islamic learning. Like Sonni 'Ali, Askia Muhammad was a warrior. His armies made the Songhay Empire the most powerful state in central West Africa. In addition to policing the Mossi, old foes of the Songhay, Askia Muhammad conquered many of the old Mali provinces to the west. He captured oasis towns in the north to control Tuareg nomads. His forces annexed Hausa towns in northern Nigeria, ranging east to Lake Chad. Enfeebled at age 80, he was deposed by his sons in 1529.

Askia Muhammad bequeathed later rulers the problems of imperial overextension—rebellions, border harassments, and declining trade revenues. In 1586, the sultan of Morocco captured the Taghaza salt mines, nominally Songhay territory, and began to plot an attack on Timbuktu. He assembled an expedition of about 6,000 Moroccans, Granadans, and Spaniards, arming them with arquebus firearms and some cannons. In 1591, this force marched through the Sahara and met the Songhay at Tondibi. Despite their valor and greater numbers, the Songhay suffered massive casualties, and the Moroccans occupied both Timbuktu and Gao. Unable to recover, the Songhay Empire crumbled into various city-states and disappeared.

Weston F. Cook, Jr.

See also: Sonni 'Ali

References and further reading:

Hunwick, John O., ed. and trans. *Timbuktu and the Songhay Empire: A-Sadi's Tarikh al-Sudan and Other Contemporary Documents.* Leiden: E. J. Brill, 1999.
July, Robert W. *A History of the African People.* 5th ed. Prospect Heights: Waveland Press, 1998.

Song-Jin Wars (1125–1141)

The onslaught of Jurchen invaders and the reestablishment of a diminished Song dynasty. The Jurchen were a Tungus-speaking people living in Manchuria and southeastern Siberia who had learned a great deal from exposure to both to the Khitan, who then ruled northwestern China, and Chinese culture. The founder of a new Jurchen state, later known as the Jin dynasty (1125–1234), was the Wanyan chieftain Aguda (1068–1123), who knew how to use cavalry effectively in warfare, pointing up his debt to the nomadic Khitan.

Between 1116 and 1122, Aguda's armies surprised and overthrew the Khitan Liao dynasty (907–1125), capturing four of the five Liao capitals in turn. Although the Northern Song (960–1126), which ruled the rest of China, sought initially to take advantage of the discomfort of its old rival, Liao, they quickly found the Jurchen mobilized against them under Aguda's successor, Wuqimai (r. 1123–1135), as soon as he had reduced the last Khitan holdouts.

In 1125, the Jurchen went over to a general assault on the Song. It proved surprisingly successful, in part because of a shortage of warhorses in Song armies. Two main Jurchen forces advanced toward Kaifeng, the Song capital on the Yellow River, which was surrounded by one of the forces in February 1126. By that time, the Song emperor, Huizong (r. 1100–1126), known more as an artist than for his governing ability, had abdicated in favor of his son, Qinzong (r. 1126–1127), who proved equally inept in resisting the invaders and had no choice but to open negotiations with them.

In 1227, Qinzong surrendered his capital, himself, and his throne and was led north, along with his father, as Jin captives. Jurchen forces then spread out to extend their Song conquests. The Northern Song Dynasty had collapsed.

During the next four years, Song survived only with the greatest difficulty and then only because the sheer size of Song China gave loyalists the time and space to organize resistance. The primary leader was 20-year-old prince Zhao Gou, the younger brother of Qinzong. Although originally among the Song royal captives, he had managed to escape the Jin net and, even if not originally in favor of continued resistance, quickly became its center as he moved from one loyalist base to the other, often just ahead of Jin horsemen sent to apprehend him.

Song resistance drew upon two foundations: one was the surviving Song field armies left behind by the speed of the Jin advance. The other was popular resentment and the appearance of local militias provoked by the harsh treatment meted out by the Jin to the Song population.

In 1127, as Jin armies retreated north to avoid the harsh summer heat of southern China, Zhao Gou finally tried to organize his forces and supporters more effectively. Despite being uncertain regarding the status of his brother and father, Zhao Gou ascended the throne himself as Emperor Gaozong (r. 1127–1162) in the Southern Song capital of Shangqiu.

As a next step, the Song, in addition to marshaling surviving regular forces, began organizing the local forces that had flocked to their support, often under their own, local leaders. Here Gaozong was ably assisted by Song minister Li Gang, an organizational genius who had escaped Jin capture because he had been away at the time that the capital had fallen. Gaozong had also to organize a government, which meant drawing China's elite classes into resistance activities and calling upon the people to support his new regime, that of Southern Song (1126–1279).

Thus it was an organized force and government that the Jin now faced as it resumed its advance later in 1127. Although some at court desired an immediate counterattack, the court wisely pursued a strategy of avoiding direct confrontation, withdrawing south to avoid Jin armies. Ultimately, the court ended up at Hangzhou, which, after a brief abandonment in 1130 when the court took to the seas, relying on superior naval forces, and later settled at a number of intermediate points, remained the Southern Song capital.

The pattern of Jin winter assaults and Song summer recoveries continued for several years, with neither side making much progress, until the Jin retreated during the summer of 1130 and did not return. The court finally returned to Hangzhou in 1133 and began to reassert its control generally within much reduced boundaries, ultimately signing a treaty with the Jin in 1141–1142. In it, the Song openly recognized the loss of the north, but before that event, a serious crisis erupted.

It centered around Yue Fei (1103–1141), one of the great tragic heroes in Chinese history because of his mistreatment by a government that felt it had no other choice. Yue Fei had begun a formal military career in one of China's regional armies in 1122, and after a brief period in which he had had to return to his home area after his father's death, became one of the most active of Song China's local commanders, ultimately carrying the battle to the enemy after 1134 in a series of highly successful counterattacks. So successful, in fact, did Yue Fei's counteroffensive become that the court peace party, headed by minister Qin Gui, began to find him threatening.

In the end, largely innocent, he was imprisoned and put to death at Qin Gui's orders. There were later Song offensives against the Jin, but never again did the Song have as capable a general as Yue Fei, and his death marked the end of any serious attempt by Southern Song to recover all the territories in the north ruled by its predecessor. To this day, in memory of Yue Fei, when the Chinese prepare a fritter called a Qin Gui, named after the Song minister, they curse him as they throw it into the hot oil to cook.

Paul D. Buell

See also: Chinese Imperial Wars; Yue Fei

References and further reading:

Franke, Herbert. *Studien und Texte zur Kriegsgeschichte der sudlichen Sungzeit.* Wiesbaden: Otto Harrassowitz, 1987.

Mote, F. W. *Imperial China, 900–1800.* Cambridge, MA: Harvard University Press, 1999.

Sonni 'Ali (d. 1492)

Founder of the Songhay Empire of late medieval West Africa and a tireless military campaigner. The Songhay were originally Niger River fisherfolk. Although overshadowed by the empire of Mali, the Songhay people became active in farming, commerce, and riverine transportation. After 1375, as Mali crumbled, Songhay strengthened economically but remained a relatively small state until 1464.

That year, Sonni 'Ali became king. He built an infantry from his own people and drafted his neighbors while importing cavalry horses from Morocco. Sonni 'Ali used his army first on the southern Mossi tribe. Enemies of Islam, Mossi raiders had bedeviled Songhay for decades. Sonni 'Ali's horseman drove them back to their forest territories.

Additionally, Sonni 'Ali built a river fleet, adapting traditional Niger canoes into combat ferrying vessels. With the

Mossi cowed, he turned west against Tuareg nomads occupying Timbuktu, Niger's religious and cultural capital, which fell in 1469. Sonni ʿAli then exiled, executed, or flogged many leading Timbuktis as collaborators, not sparing even Muslim clergy. In 1473, marshaling 400 battle canoes, he captured the Malian city of Jenne. Subsequently, he launched regular raids against neighboring nomads (Tuaregs and Fulani) and the Mossi. In the 1480s, the king campaigned annually along the eastern bend of the river in Gurma. On one of these expeditions in 1492, Sonni ʿAli died, in either a flash flood or an assassination.

Weston F. Cook, Jr.

See also: Songhay Empire

References and further reading:

Oliver, Roland, and Anthony Atmore. *The African Middle Ages, 1400–1800.* London: Cambridge University Press, 1981.

Hunwick, John O., ed. and trans. *Timbuktu and the Songhay Empire: A-Sadi's Tarikh al-Sudan and Other Contemporary Documents.* Leiden: E. J. Brill, 1999.

Sosabowski, Stanislaw Franciszek (1892–1967)

World War II Polish army commander. Born on 8 May 1892 in Stanislawow in Austro-Hungarian-occupied Poland, Sosabowski served as a supply officer in the Austro-Hungarian Army during World War I. During the interwar period, he frequently served on the faculty of the Polish General Staff College and, after being promoted to colonel, took command of the 21st "Children of Warsaw" Infantry Regiment in March 1939.

During the Polish campaign of 1939, Sosabowski and his regiment fought to defend Warsaw but were defeated and captured by the Germans. After escaping from a prisoner-of-war camp, he made his way to France, where he served as the deputy commander of the 4th Polish Infantry Division during the French campaign of 1940. After being evacuated once again, to Scotland, Sosabowski was given command of the Polish Independent Parachute Brigade in 1941. The brigade, which initially existed only on paper, would eventually become one of the most elite Allied units of World War II. In September 1944, the brigade took part in Operation MARKET GARDEN, which was the poorly planned attempt to establish a bridgehead over the Rhine River at the Dutch town of Arnhem. The brigade was decimated during the battle, and to add insult to injury, Sosabowski's British superior, Lieutenant General Frederick Browning, attempted to shift blame for the Polish inability to achieve battle objectives to Sosabowski. Removed from the command of his brigade, Sosabowski spent the remainder of the war in unimportant staff positions.

After the war, he refused a British military pension and became a day laborer in London. He died penniless on 26 September 1967. After lying in state in London, Sosabowski's remains were returned for burial in his family plot in Poland.

Alexander M. Bielakowski

See also: MARKET GARDEN; Pilsudski, Józef Klemens; Polish Campaign of 1939; Sikorski, Wladyslaw Eugeniusz

References and further reading:

Cholewczynski, George F. *Poles Apart: The Polish Airborne at the Battle of Arnhem.* New York: Sarpedon Publishers, 1993.

Sosabowski, Stanislaw F. *Freely I Served.* Nashville, TN: Battery Press, 1982.

Soult, Nicolas-Jean de Dieu (1769–1851)

French field commander in the French Revolutionary and Napoleonic Wars. Born on 29 March 1769, the son of a notary in St. Amans-la-Bastide, Soult volunteered as an infantryman in 1785 and rose through the ranks to brigadier general in 1794. He replaced François-Joseph LeFebvre as commander in chief of the Army of Sambre and Meuse in 1799 and commanded the Army of Italy as André Masséna's adjutant from 1799 to 1800. After serving with distinction in the Piedmont campaign, including two months as a prisoner of war, and becoming colonel general of the Consular Guard in 1802, Napoleon elevated him to marshal in 1804 and named him Duke of Dalmatia in 1807.

Soult had early combat experience at Marsthal in 1793, Fleurus in 1794, Altenkirchen in 1796, Stockach and Zurich in 1799, and Genoa and Monte Cretto in 1800. He commanded a corps at Ulm and Austerlitz in 1805; Jena in 1806; and Eylau, Heilsberg, and Königsberg in 1807.

In the Peninsular War from 1808, Soult lost to John Moore at Corunna on 16 January 1809 and to Arthur Wellesley, later the Duke of Wellington, at Oporto on 12 May but defeated Juan Carlos Areizaga at Ocaña on 19 November. He invaded and ruled Andalusia in 1810, unsuccessfully besieged Cádiz from 1810 to 1812, captured Badajoz on 11 March 1811, and lost to William Carr Beresford at Albuera on 16 May.

Chief of staff Louis-Alexandre Berthier recalled Soult to Germany in 1813. Shortly after replacing Jean-Baptiste Bessières (killed at Rippach on 1 May) as commander of the Imperial Guard, Soult fought at Bautzen but then was ordered back to Iberia. Wellington defeated him at Orthez and Toulouse in 1814.

He was Louis XVIII's minister of war while Napoleon was on Elba but the latter's chief of staff during the Hundred Days. His mistakes at Ligny and Waterloo contributed to Napoleon's downfall in 1815.

A consummate diplomat, Soult held important government offices from 1820 until he retired in 1847. He died on 26 November 1851 in St. Amans-la-Bastide, now called St. Amans-Soult.

Eric v. d. Luft

See also: Austerlitz, Battle of; Badajoz, Siege of; Berthier, Louis-Alexandre, Prince of Neuchatel and Valangin, Prince of Wagram; Corunna, Battle of; French Revolutionary Wars; Jena and Auerstädt; LeFebvre, Pierre-François-Joseph, Duke of Danzig; Masséna, André, Duc de Rivoli, Prince d'Essling; Napoleon I; Napoleonic Wars; Quatre Bras and Ligny; Scharnhorst, Gerhard Johann von; Waterloo; Wellington, Arthur Wellesley, Duke of

References and further reading:

Chandler, David G., ed. *Napoleon's Marshals.* New York: Macmillan, 1987.

Duffy, Christopher. *Austerlitz 1805.* Hamden, CT: Archon Books, 1977.

Hayman, Peter. *Soult: Napoleon's Maligned Marshal.* London: Arms and Armour, 1990.

Young, Peter. *Napoleon's Marshals.* New York: Hippocrene Books, 1973.

South Africa/Namibia (1960–2000)

Scene of a peaceful transition to multiracial democracy after decades of bloodshed. The year 1960 was a watershed for Africa, as well as for South Africa. No fewer than 17 African countries gained independence, but in South Africa, the National Party white supremacist apartheid government was determined to continue swimming against the tide of human and equal rights and world opinion. The government of Prime Minister H. F. Verwoerd did not heed the well-meant warnings given by British prime minister Harold Macmillan in his "wind of change" speech, delivered in parliament in Cape Town on 3 February 1960. Shortly afterward, a bloody clash between police and protesters (against racist pass laws) took place at Sharpeville (21 March 1960), sending shock waves around the world. Instead of trying to find a negotiated solution, the South African government banned the African National Congress (ANC) and Pan-Africanist Congress (PAC). Both organizations set up offices abroad and went underground in South Africa itself, establishing armed wings: Umkonto we Sizwe (MK) of the ANC and Poqo of the PAC. In due course, these organizations launched attacks against a variety of targets.

The Republic of South Africa was born on 31 May 1961. It was established after a small majority of whites voted in favor of a republic in a referendum held on the matter in October 1960. The black majority was, of course, not consulted. The new republic soon developed into a pariah state, with growing isolation with regard to sport, culture, and economic matters. One of the white government's few allies was Rhodesia (now Zimbabwe), where the small white minority declared itself unilaterally independent of Britain in November 1965—soon to be invaded by freedom fighters of the liberation movements. On 23 August 1966, the first clash between South African security forces and the South-West Africa Peoples' Organization (SWAPO) freedom fighters took place in the north of South-West Africa (SWA/Namibia).

This conflict was the start of the Namibian war of independence that dragged on until 1989 and eventually spilled across the border into Angola. When Portugal withdrew from its colonies of Angola and Mozambique in 1975, for the first time, South Africa had neighbors supported by the Soviet Union, and MK cadres henceforth had easier access to South Africa.

In 1976, a revolt took place in the sprawling black township of Soweto, and violence soon spread to other areas. Once again there was an international outcry, but the apartheid regime reacted by banning more organizations and further limiting press freedom. This, together with the death in police custody of black consciousness leader Steve Biko (1977), led to South Africa's further international isolation. During the 1960s and early 1970s, the South African Defence Force (SADF) was built up into a sizable modern force, but in November 1977, the United Nations Security Council imposed a mandatory arms embargo against South Africa. As is so often the case with such sanctions, the only tangible result seems to have been the expansion of the South African arms industry.

In SWA/Namibia, but especially in southern Angola, the SADF launched several operations against SWAPO freedom fighters and in due course also became embroiled in the Angolan Civil War, taking sides with the National Union for the Total Independence of Angola (UNITA) against the Popular Movement for the Liberation of Angola (MPLA) and its Cuban and eastern bloc allies. In military operations codenamed REINDEER, SCEPTIC, PROTEA, ASKARI and others, the SADF inflicted heavy losses on its opponents, but in the long run, South Africa paid a heavy economic price. Eventually a cease-fire was negotiated, and after many months of negotiations, SWA became independent as Namibia on 21 March 1990.

In the meantime, the traumatization of South African society continued. Black-on-black violence (which left about 30,000 people dead), as well as terror attacks against government and other white targets, escalated in the course of the 1980s. Just as it seemed as if a low-intensity civil war was about to wreck the South African economy and destroy the country, the government in February 1990 heralded the unbanning of the ANC, PAC and other organizations, and the freeing of political prisoners like Dr. Nelson Mandela. Multiparty negotiations put the country on the road toward a negotiated settlement, which led to the country's first-ever truly democratic elections in April 1994. In that same year, the country's defense force was reinstituted as the South African National Defence Force (SANDF), amalgamating the old SADF, MK, and several other military organizations. By the beginning of the new millennium, the SANDF had so shed its old racist nature that it was able to take part in United Nations and other peace-keeping operations.

André Wessels

See also: Angolan Civil War; Angolan War of Independence; Cold War; Mozambican War of Independence; Zimbabwe Independence Struggle

References and further reading:

Davenport, T. Rodney H., and Christopher Saunders. *South Africa: A Modern History.* 5th ed. Basingstoke, UK: Macmillan, 2000.

Morris, Michael S. L. *Armed Conflict in Southern Africa: A Survey of Regional Terrorisms from Their Beginnings to the Present, with a Comprehensive Examination of the Portuguese Position.* Cape Town: Jeremy Spence, 1974.

Nöthling, C. J. *Geskiedenis van die Suid-Afrikaanse Weermag.* Vol. 2. Silverton: South African Military History Consultants, 1996.

Steenkamp, Willem. *South Africa's Border War, 1966–1989.* Johannesburg: Ashanti Publishing, 1989.

South American Wars of Independence (1810–1824)

Series of wars that liberated South America from Spanish rule. Spain's colonial empire in the Americas began to unravel as a result of growing discontent by the Creoles (Spaniards born in the Americas). The British occupied Buenos Aires on 17 June 1806, as Spanish administrators and regular troops fled. Jacques de Liniers marshaled the colonial militia and forced the British to surrender, greatly boosting the confidence of the Creoles and discrediting the Peninsular Spanish. In 1808, when Napoleon invaded Spain, removed Ferdinand VIII, and installed his brother Joseph on the throne, there were revolts throughout South America. By 1810, as the collapse of Spanish resistance appeared imminent, Creole leaders, while proclaiming loyalty to Ferdinand but with the ultimate goal of independence, organized popular uprisings in Caracas, Buenos Aires, Santiago, Quito, and Bogotá. Organized fighting soon erupted between patriots and royalists. Venezuela and Buenos Aires were the principal centers from which organized revolts spread.

In northern South America, the patriot army was headed by Francisco de Miranda, who proved to be conservative and indecisive in his military strategy. A massive earthquake on 26 March 1812 devastated Caracas and other patriot-held areas. Royalist clergy proclaimed this natural disaster an omen that the patriot cause lacked divine approval. Following several military setbacks, Miranda first sought a negotiated settlement and then attempted to flee Venezuela. He was arrested, and Simon Bolivar assumed command.

On July 1811, Venezuela declared its independence, but the royalist armies drove Bolivar's forces from Puerto Cabello and temporarily ended the entire revolt as Bolivar went into exile in Curacao. Bolivar returned in May 1813, recaptured Caracas (6 August 1813) and followed up with victories at La Victoria (February 1814), San Mateo (March 1814), and Carabobo (May, 1814). However, at La Puerta the Spanish dealt Bolivar a stunning defeat, and Venezuela was once again under royal control. Bolivar managed to escape to New Granada (Colombia) and assumed command of another army.

The defeat of Napoleon in 1815 freed Spanish troops for South American duty, and General Pablo Morillo put them to good use, defeating Bolivar at Santa Mara and bringing many of the rebellious provinces under royal control. Bolivar next fled to Jamaica, where he issued his famous letter outlining his vision for an independent republic for all of Spanish South America. He subsequently returned to Venezuela in December 1816 and won an important victory at Barcelona (16 February 1817). He suffered another reversal at the second battle at La Puerta (15 March 1817). Bolivar then shifted operations to dislodge the Spanish from New Granada. Beginning on 1 June 1819, the patriot army of 2,500 departed Angostura, crossed seven rivers and ascended the frozen passes of the Andes, arriving in the Sagamose Valley on 6 July. In the decisive Battle of Boyaca (7 August), near Bogota, Bolivar outmaneuvered the Spanish army and completely routed it, killing 100, taking 1,600 prisoners, and capturing numerous stores and munitions. Bolivar entered Bogota in triumph on 10 August, founded the Republic of Colombia, and was installed as president.

With Colombia liberated, Bolivar again turned his attention to liberating his native Venezuela. At the Battle of Carabobo (25 June 1821), he found a route through the mountains unknown to the Spaniards and surprised their right flank with an attack by the British Legion, resulting in the disintegration of the entire force and the freeing of Venezuela.

Bolivar dispatched his principal lieutenant, General José Antonio Sucre, to liberate the province of Quito in present-day Ecuador. Sucre smashed a Spanish army on the slopes of the Pichincha Volcano (24 May 1822), and the rebel army surrounded Quito the following day. With Quito liberated, Bolivar established the Republic of Gran Colombia (Colombia, Venezuela, Ecuador, and Panama).

In the Southern Cone, a Creole junta was in control of Buenos Aires by 1810 and attempted to extend the revolt, with varying degrees of success, throughout the entire viceroyalty of Rio de la Plata (Argentina, Chile, Bolivia, Uruguay, and Paraguay). In 1816, José de San Martín took command of the Army of the Andes at Tucuman and proposed one of the most audacious campaigns in military history. In January 1817, he crossed some of the highest passes in the Andes into Chile with an army of 3,300 regular infantry, 700 cavalry, 1,000 militiamen, and 21 cannon. He defeated the Spanish at Chacabuco on 12 February 1817 and liberated Santiago two days later. After a setback at the Battle of Cancha-Rayada on 16 March 1818, San Martín secured Chile's independence with a victory at Maipo on 5 April 1818.

The patriot cause was boosted on 18 November with the arrival of former British admiral Thomas Cochrane, who took command of the Chilean Navy. Cochrane captured the port of Valdivia on 18 June 1820, thus opening the western coast for an invasion of Peru by San Martín's forces. On 8 September, the patriot army landed at Pisco, Peru, and Cochrane bottled up a Spanish naval squadron with a blockade of Callao. San Martín liberated Lima and declared Peruvian independence on 28 July 1821.

Bolivar and San Martín met at Guayaquil, Ecuador, on 26–27 July 1822. The proceedings of the famous meeting were not recorded, but it is generally believed that the two men could not work out their differences for a final strategy to defeat Spain. San Martín abruptly resigned after the meeting and soon retired from public life.

Bolivar's final campaign was aimed at liberating the highland areas of Peru and Bolivia, then known as Upper Peru. Bolivar and Sucre engaged the Spanish army commanded by General José Canterac at Junin on 6 August 1824 in the largest single cavalry engagement of the war. The patriot forces were victorious, and the Spanish retreated into the sierra southeast of Lima. In the final battle of the war, the Spanish, with 10,000 men, attacked Sucre's army of 7,000 at the Battle of Ayacucho on 9 December 1824. Sucre's smaller force prevailed, killing 1,400 and wounding 700 while losing only 300 killed and 600 wounded. Ayacucho ended Spain's hold on South America, and all Spanish forces were withdrawn under the terms of capitulation.

George M. Lauderbaugh

See also: Bolivar, Simon; San Martín, José Francisco de

References and further reading:

Harvey, Robert. *Liberators: Latin America's Struggle for Independence 1810–1830.* New York: Overlook, 2000.

Humphreys, R. A. *The Origins of the Latin American Revolutions, 1808–1826.* New York: Alfred A. Knopf, 1965.

Lynch, John. *The Spanish American Revolutions, 1808–1826.* 2d ed. New York: Norton, 1986.

Soviet-Afghan War (1979–1989)

The Soviet Union's Vietnam. In April 1978, Soviet aid had helped establish a Communist regime in Afghanistan. In December 1979, to save the regime from opposition forces that were building throughout the country, the Soviet Union invaded Afghanistan. From the onset, the USSR directed and dominated the military effort. Soviet commanders and advisers orchestrated almost all military operations. Once the Soviets met their immediate objective, taking Kabul, they deployed ordinary motor rifle units, not elite troops, to fight the insurgency.

The Soviet invasion triggered a reaction that linked Islamic insurgency to national liberation. Over time, the mujahidin (Muslims fighting a jihad, or holy war) built up experienced commanders capable of effective limited military operations. In the war's first four years, the Soviets failed to use their 115,000 troops decisively.

When Yuri Andropov succeeded Soviet leader Leonid Brezhnev (d. November 1982), he consistently opposed the Soviet military presence in Afghanistan. Andropov's policy led the Afghan military leader Ahmad Shah Massoud to sign a 12-month cease-fire in January 1983. A stalemate resulted. Andropov's (d. February 1984) successor, the senile Konstantin Chernenko, returned to a tough policy. The Soviet military had learned from their past weaknesses. Mikhail Gorbachev succeeded Chernenko (d. March 1985), continuing the tough policy. From 1984 to 1986, the USSR escalated militarily.

The United States, Saudi Arabia, Iran, and others provided the Afghan resistance billions in financial support. The resistance procured weapons and equipment that tipped the balance in their favor. Pakistani president Mohammed Zia demanded that the money be funneled through Pakistan to the Afghan resistance. In addition, Zia's Islamization program heavily favored Afghan fundamentalists.

In 1986, the U.S. Central Intelligence Agency supplied mujahidin leading commander Ismail Khan large amounts of weaponry. By 1988, the mujahidin had acquired U.S. Stinger handheld surface-to-air missiles and gained the tactical initiative. Massoud continued to succeed offensively in the north.

In 1987, Gorbachev decided to end the conflict. The mujahidin had fought the Soviet and the Afghan Communist forces to a stalemate. Moscow's Afghan policy alienated it from Islamic, Western, and nonaligned countries, and the Soviets had not found a political leader who could gain the Afghan people's loyalty. Geneva talks in March 1988 involving Soviet foreign minister Eduard Shevardnadze and U.S. Secretary of State George Shultz led to a Soviet withdrawal from Afghanistan in February 1989. The war's end was reduced from international to regional status, but the civil war in Afghanistan continued.

The war was costly for all sides. The Soviets lost 13,310 military personnel. The resistance rarely acknowledged their casualties, but something like a million Afghan civilians were killed, and 5,000,000 Afghans became refugees.

Rosemarie Skaine

See also: Cold War; Russian and Soviet Armies

References and further reading:

Hopkirk, Peter. *The Great Game: The Struggle for Empire in Central Asia.* New York: Kodansha International, 1992.

Roy, Oliver. *The Lessons of the Soviet/Afghan War.* London: Brassey's for the International Institute for Strategic Studies, 1991.

U.S. Department of State. *Afghanistan: Soviet Occupation and Withdrawal, Special Report No. 179.* Washington, DC: Bureau of Public Affairs, December 1988.

Spanish Civil War (1936–1939)

Conflict that led to more than 500,000 casualties on both sides and ended with a victory for General Francisco Franco and the right-wing Nationalists over the Republicans. On 17 July 1936, the military garrison in Spanish Morocco began a rebellion against the leftist Republican government in Madrid, supported by conservative army officers on the mainland and in Spain's other colonial possessions. Also supporting the coup attempt were members of the fascist Falange Party, monarchist forces, and Catholics. The initial uprising failed to take over the entire country but by 20 July had seized power in Morocco, most of northern and western Spain, Seville, and the remnants of the Spanish empire. The socialist-supported Republican government maintained control over Madrid and Barcelona, the two largest cities, most of eastern Spain, and the industrialized areas of the Basque country and Asturias. The initial leaders of the military revolt, which came to be known as the Nationalists, were Generals Francisco Franco and Emilio Mola.

Most of the Nationalist army remained isolated in Morocco during these early days, however, and as the navy and air force had remained loyal to Madrid, Franco asked for aircraft and naval support from Nazi Germany and Fascist Italy, assistance that soon came and enabled the Nationalists to airlift their forces to southern Spain in late July and early August. Within a few weeks, the German and Italian governments also began to send significant military supplies to Franco and even military units, hoping to prevent a socialist victory in Spain. France and the United Kingdom were unwilling to become involved in the civil war, however, and refused to sell or send weapons to the Spanish Republican government, forcing it to request aid from the Soviet Union. Soviet aid to the Republicans did arrive in large amounts (at the cost of Spain's total gold reserves), along with advisers, who upon their arrival began immediately to strengthen the Spanish Communist Party.

After a diversion of forces to Toledo to lift the siege of the Alcazar fortress, in November the Nationalist forces attempted an assault on Madrid, hoping to end the war in a knockout blow. Initial Nationalist successes, including seizing most of University City in northwest Madrid, were blunted by the arrival of thousands of Communist-led volunteers of the International Brigades. By the third week of November, Madrid was nearly surrounded, but the Nationalists were unable to force a victory. Also during this month, Soviet and German aircraft began flying bombing missions in the first large-scale aerial bombardment of urban areas on behalf of the Republican and Nationalist sides, respectively, and Germany and Italy recognized Franco as the legitimate ruler of Spain.

Early 1937 saw the front shift to the south, with the Nationalists and Italian forces attacking and seizing Malaga in February and gaining ground in the southern suburbs of Madrid at the Battle of Jarama the same month. A Nationalist offensive against Guadalajara in March, however, was defeated by a Republican counterattack, aided by Soviet tanks and aircraft. As the first major engagement involving the Italians, it was seen as an embarrassing defeat for Benito Mussolini, who expected his forces to have easy victories in Spain.

The Nationalist army in northern Spain then turned its attention to the Basque country, which despite its heavy industry, ports, and mines was less defended than Madrid. Cut off from the rest of the Republicans by the Nationalists and attacked by German air units of the Condor Legion, including the notorious bombing of the town of Guernica in late April, the Basques were forced to give ground. By late spring, the Nationalists had taken the entire province except the industrial city of Bilbao, which surrendered in mid-June. During the attack on the Basques, General Mola died in a plane crash, leaving Franco as the unchallenged Nationalist leader.

On the Republican side, the socialist and anarchist movements, along with other smaller parties, fought to prevent the dominance of the Soviet-backed Communist Party. After May 1937, the Republican government became dominated by a socialist and Communist alliance under Prime Minister

A battalion of shock troops marching in Madrid during the Spanish Civil War, c. 1937. (Library of Congress)

Juan Negrin. Almost to the last day of the civil war, however, arguments over politics and strategy translated into internal violence on the Republican side. The Nationalists suffered fewer internal struggles, thanks largely to Franco's insistence on the unity of command, as evidenced by his forcible union in April 1937 of the Falange, monarchist, and other rightist parties into a single legal political party, the Falange Espanola Tradicionalista y de las Junta de Ofensiva Nacional-Sindicalista.

The Republicans attempted to launch a counteroffensive in July 1937, attacking Nationalist forces to the west of Madrid in the direction of Brunete. This attempt failed, however, as the superior discipline, air support, and defensive positions of Franco's forces prevented the Republicans from gaining an advantage. In August and September the Republicans attempted to seize Saragossa and other territories, launching an offensive from their base in Catalonia. These attacks, too, failed, and they also proved to be the last major offensive of the Republicans, who after this point could only hope for outside intervention to win the war.

In late 1937 and early 1938, Nationalist offensives regained the initiative, conquering Republican-held cities throughout Spain: Santander in August, Gijon in October, and Lerida in April. In July 1938, the Nationalists called a halt to their offensive in order to prepare for what they hoped would be a final attack in Catalonia, which together with Madrid was the last remaining significant base of Republican support. The Republicans tried to negotiate an end to the conflict, but the Franco regime was uninterested in anything less than total victory. A Republican effort to disrupt the Nationalist army in late 1938 by attacking across the Ebro River failed to achieve its objectives, and the Republicans withdrew their forces in early November after a series of defeats. In late December, the Nationalists launched their offensive on Catalonia, and the demoralized Republicans were unable to put up much resistance. Barcelona fell on 26 January 1939.

This marked the end of serious hostilities. With some Republicans hoping to surrender with good terms and others vowing to fight to the end, internal fighting broke out, enabling the Nationalists to conquer the rest of Spain with little effort. Madrid surrendered on 28 March 1939, and on 1 April 1939, Franco declared that the war was over.

Wayne H. Bowen

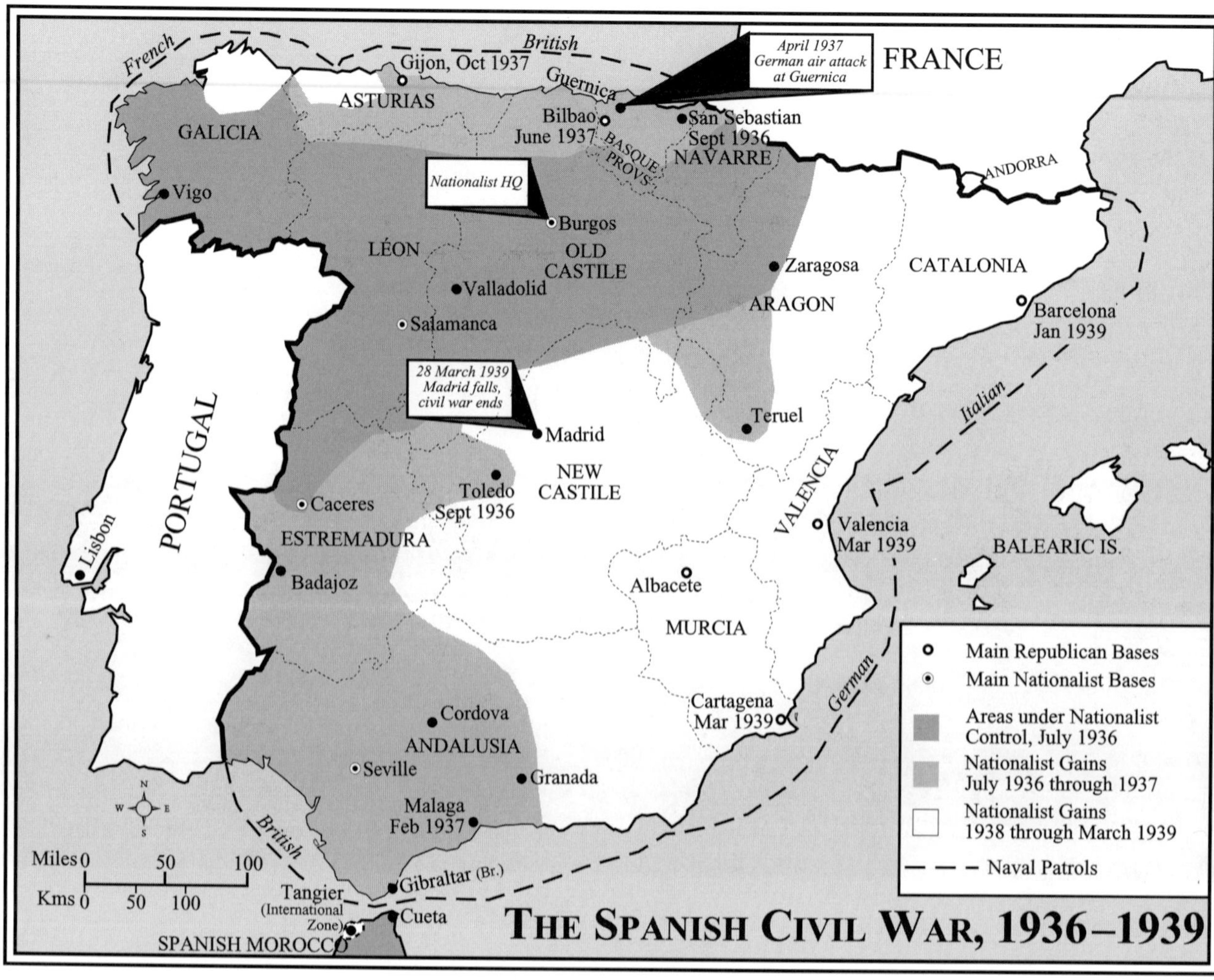

THE SPANISH CIVIL WAR, 1936–1939

See also: Franco, Francisco

References and further reading:

Alpert, Michael. *A New International History of the Spanish Civil War.* New York: St. Martin's Press, 1998.

Bolloten, Burnett. *The Spanish Civil War: Revolution and Counterrevolution.* Chapel Hill: University of North Carolina Press, 1991.

Hugh, Thomas. *The Spanish Civil War.* New York: Harper and Row, 1977.

Orwell, George. *Homage to Catalonia.* New York: Harcourt Brace, 1987.

Spanish Colonial Wars (1492–1898)

Created the second global empire, along with that of Portugal, spreading Christianity and Western civilization to the Americas, Africa, and Asia but also disease and death among the native populations. In 1492, with the commission of the Catholic king and queen of Spain, the Genoese captain Christopher Columbus sailed from Spain to find a direct trading route to Asia. Instead of finding this passage, he became the first European to establish permanent contact with the Western Hemisphere, claiming lands for Spain in the Caribbean and establishing that nation's first overseas empire.

Although Columbus had been the first European since the Vikings to land in the Americas, Spain did not have an unchallenged claim to the hemisphere. Faced with competition from Portugal, in 1494 Spain agreed to the Treaty of Tordesillas, which divided the world between the two nations. With the approval of the pope, Spain gained authority over the Americas, other than the eastern portion of South America, which would later become Portuguese Brazil, and Portugal was given first claim over most of Africa and Asia.

Over the next few decades, Spain embarked on a series of expeditions of discovery and conquest in the Americas. In 1513, Vasco Nuñez de Balboa hacked his way through the Isthmus of Panama, reaching and sighting the Pacific Ocean. The same year, Ponce de Leon began a seven-year exploration of Florida, searching for the fountain of youth.

In 1519, the Spanish conquistador Hernando Cortez defeated the Aztec empire in Mexico, including its capital, Tenochtitlán. Although initially accompanied by a small army of just over 500 Spaniards, along with a few cannons and horses, he was aided by Indian allies who resented Aztec rule, as well as an outbreak of smallpox, which struck down many natives, including one of the Aztec emperors. Some Aztecs also considered Cortez a god, the revered Quetzalcoatl whose return was prophesied by myth, and so refused to fight him. Even the Aztec king, Moctezuma, was unwilling to send all of his strength against someone who might be a god and allowed Cortez to enter the Aztec capital unmolested. Spanish attacks on Aztec temples and religious ceremonies, however, enraged the population and nearly brought defeat to Cortez and his Indian allies in 1520. Some captured Spaniards were even offered as human sacrifices to the Aztec gods. By 1521, Cortez had conquered the last of the Aztecs and installed himself as the new ruler in what would become Mexico City. The Spaniards executed the last of the Aztec royal family in 1524.

To the south, another Spanish conquistador, Francisco Pizarro, won the same kind of victory over the Incas, the other major organized native civilization in the Americas. Landing in Peru in 1532, Pizarro was fortunate to arrive in the midst of an Incan civil war and was able to capture the Inca king, Atahualpa, in November, in the process killing more than 1,500 of his retainers without losing a single Spaniard. Atahualpa attempted to buy his own release by filling a room with gold and another with silver, but once this was accomplished, leaving the Spaniards with 11 tons of gold, Pizarro ordered the Inca executed. Despite being outnumbered by the Inca forces, Pizarro went on to exploit the ongoing civil war to defeat the remaining native armies. Pizarro was also aided by the deadly impact of a smallpox epidemic, probably spread from Spanish forces in Central America, which had devastated Peru in the 1520s. By 1534, he had put down the last organized resistance to Spanish rule in Peru and the Andean region of South America.

These initial conquests were not the end of Spanish military activity in the Americas, however, as foreign intervention, piracy, native rebellions, and events in Spain continued to plague efforts to establish stable colonial rule. Minor revolts in Peru in 1571–1572 and 1742–1755 interrupted silver mining and other colonial operations, but a major rebellion in 1780–1783 was even more serious. During this uprising, Túpac Amaru II, claiming to be a direct descendant of the Inca kings, led a force of 80,000 Indians against Spanish forces, at one point controlling most of southern Peru and Bolivia, before being harshly put down by the Spanish army and local militias. By the time the uprising was crushed, more than 100,000 lives had been lost, most of them Indian.

Other minor conflicts plagued the Spanish colonial system in the Americas, including an uprising by the Pueblo Indians in northern Mexico that drove the Spaniards out of their northern frontier for more than a decade, from 1680 to 1692, and a tax revolt in Colombia in 1781. More serious for the long-term survival of the empire was the increasing tension between Spanish officials born in the home country and Creoles, white descendants of Spanish colonists born in the Americas. Creoles were resentful of being excluded from positions of authority in the colonial system and mistrusted the administrators and commanders sent from Spain to rule over them and the native populations.

Spain also had to cope with attacks by the French, Dutch, and English and corsairs, pirates, and privateers, all of whom were hoping to rob Spanish colonies and merchant fleets of their silver, gold, and other goods extracted from the Americas. As early as 1555, French pirates attacked and temporarily seized Havana, as did the English in 1762. It was Francis Drake, however, who had the most negative impact on Spanish possessions in the late 1580s. Although he began his career as a corsair, by the 1570s he was sailing as a privateer, licensed by Queen Elizabeth I to attack Spanish ships and possessions. His victories over garrisons and fleets in Santo Domingo and Cartagena, as well as his destruction of fortifications in St. Augustine, embarrassed and damaged Spanish power, demonstrating Spain's vulnerability in the heart of its empire. Drake was followed in the seventeenth century by other buccaneers who captured many precious Spanish cargoes and raided cities throughout the Americas. Only after 1748, when Spain signed a peace treaty with England, did the attacks end.

Neither native rebellions nor Creole resentments ended this empire, however, but rather Napoleon I, who indirectly put an end to Madrid's control over most of the Americas. After tricking the king of Spain into an alliance with him in 1807, the French emperor placed his brother, Joseph, on the throne in Madrid, triggering a revolution and guerrilla war in Spain. In response to the French actions in 1808, similar rebellions began throughout most of the Americas, ostensibly on behalf of the legitimate Spanish king, Ferdinand VII.

Creole juntas seized power in Venezuela, Chile, and Argentina in 1810 to protest French rule, and an independence movement began in Mexico, led initially by the priest Miguel Hidalgo y Costilla. By November 1810, Hidalgo had formed an army of Indians and mestizos and had surrounded Mexico City, where they outnumbered the Spanish garrison by at least 10 times. Hidalgo tried and failed to take Mexico City and then divided his forces and moved into northern Mexico, where his forces dwindled and he was captured and executed in 1811. Leadership of the Mexican independence movement weakened, and by 1815, Spanish forces had once again reasserted royal rule in Mexico. In 1821, however, a former royalist commander, Agustín Iturbide, joined forces with the rebels and had himself declared emperor. Although his regime lasted only until 1823, Mexico remained independent after his abdication. Central America, which had followed the lead of Iturbide, formed a federation of independent republics and broke away from Mexico.

In 1811, Venezuela and Paraguay declared their independence. Forces loyal to the Spanish monarchy protested and fought against these steps, but the simmering Creole resentment over peninsular rule was more powerful than loyalty to the Spanish crown. By 1814, independence movements had begun throughout Spanish possessions in the Western hemisphere.

Even after the restoration of the Spanish monarchy in Madrid with the final defeat of Napoleon in 1815, the fledgling states in the Americas refused to accept the reinstatement of their colonial status. Independence leaders such as Simon Bolivar in Venezuela and José de San Martín in Peru gained the support of white Spanish colonists against the monarchy and fought ongoing battles against the expeditions of Spanish troops sent by Ferdinand VII. To add to Spain's difficulties, Argentina declared its independence in 1816, with Chile doing so as well the following year.

In their efforts to win and maintain independence, the Latin American rebels had an important ally in the government of Great Britain, which saw great commercial opportunities in the newly opened markets and supplied the republican armies with whatever they wanted to buy. The U.S. declaration of the Monroe Doctrine in 1823, which asserted that no European power should attempt to recolonize the Americas, also played a small role in legitimizing the new states. For 10 years, armies of the Spanish crown attempted unsuccessfully to force the new republics to surrender until 1825, when Spain finally withdrew its forces, exhausted and defeated.

Of its Latin American possessions, Spain was able to hold only Cuba and Puerto Rico, primarily because the Creoles on those two islands remained loyal and integrated into the system of Spanish governance. A 10-year independence movement did begin on Cuba in 1868, led by Carlos M. Céspedes, but Spanish forces eventually crushed it. Closer to Spain, the tiny Moroccan enclave of Melilla, Spanish territory since 1497, rebelled in late 1893. The city and surrounding territory controlled by Spain had previously come under Moroccan attack in 1562, 1687, 1696, 1697, 1771, and 1774. Supported by the sultan of Morocco, in September 1893 the Berber and Arab rebels launched attacks on the Spaniards from the Rif Mountains, but this insurrection was put down in just over six months, by March 1894.

A more serious rebellion began in Cuba in 1895, when in late February an uprising began against Spanish rule. Spain sent in March a force of 6,000, led by General Martínez Campos, to crush the independence movement, but this expedition was plagued by failure, beginning with the sinking of a heavy Spanish cruiser, the *Reina Regente,* in a storm off the coast of Gibraltar immediately after sailing. Upon arrival, Martínez Campos attempted to engage the Cuban rebel army, but its guerrilla commanders, including Antonio Maceo y Grajales, refused to engage in direct confrontations with the Europeans. In early 1896, Martínez Campos was replaced by General Valeriano Weyler y Nicolau as commander of Spanish forces in Cuba. He began a policy of concentrating the population in fortified villages to isolate the rebel forces from their base of rural support, sometimes leading to civilian deaths from poor conditions. Weyler went on to defeat Maceo in several military engagements, finally capturing and killing the rebel leader in December 1896.

Taking advantage of these victories, the Spanish government in February 1897 issued an autonomy decree for Cuba, but this concession failed to satisfy demands for independence or stabilize the islands, and so the war continued. Weyler's tactics against the civilian population came under severe criticism in Spain and the United States, however, and so despite his victories, he was relieved of command in October 1897 and replaced by General Ramón Blanco. In November, Spain amnestied Cuban political prisoners and granted voting rights for adult males to vote in Spanish elections, but these reforms did little to address the demands of the population for complete freedom from Spain.

U.S. intervention in 1898, prompted by the explosion of the American warship *Maine* in February, began the Spanish-American War. Unable to defeat the superior navy and army of the United States, which easily overpowered Spanish forces in Cuba and the Philippines, Spain surrendered in August.

In the Philippines, a revolt led by Emilio Aguinaldo was unable to achieve much against Spanish regulars and local Filipino auxiliaries. But on 1 May 1898, the U.S. Navy destroyed the Spanish Pacific fleet in Manila Harbor and took control of the archipelago. Aguinaldo returned to Manila on 19 May 1898 and aided the American seizure of the islands.

Aguinaldo's declaration of Philippine independence on 12 June was not well received by the United States, which declared him an outlaw. He declared war on the United States on 4 February 1899 and led a prolonged guerrilla war, in which both sides resorted to atrocities. The American public became uneasy about this colonial war on the other side of the globe. Aguinaldo was captured on 23 March 1901. His parole stipulated that he swear allegiance to the United States, which he did on 1 April 1901. By this time, the United States was seriously considering some sort of commonwealth status or actual independence for the Philippines.

But as far as Spain was concerned, by the events of 1898 it had lost the last of its colonies in the Americas and Asia, except for some small territories in northwestern Africa.

Wayne H. Bowen

See also: Aguinaldo, Emilio; Maceo y Grajales, Antonio; Philippine Insurrection; Spanish-American War; Weyler y Nicolau, Valeriano, Marquis of Tenerife

References and further reading:

Lynch, John, and R. A. Humphreys, eds. *Latin American Revolutions, 1808–1826: Old and New World Origins.* Norman: University of Oklahoma Press, 1996.

Parry, J. H. *The Spanish Seaborne Empire.* Berkeley: University of California Press, 1990.

Schroeder, Susan, ed. *Native Resistance and the Pax Colonial in New Spain.* Lincoln: University of Nebraska Press, 1998.

Truxillo, Charles. *By the Sword and the Cross: The Historical Evolution of the Catholic World Monarchy in Spain and the New World, 1492–1825.* Westport, CT: Greenwood Press, 2001.

Spanish Succession, War of the (1701–1714)

A European struggle for the inheritance of the childless King Carlos II of Spain. In late 1700, the last Spanish king of the Habsburg dynasty, Carlos II, died. His Austrian relatives, led by emperor Leopold I, took up the fight against the French Philip V of Bourbon, who inherited the whole Spanish monarchy, according to the last will of Carlos II. But in fact, the War of the Spanish Succession was a war about European hegemony between Louis XIV of France, Philip V, and the elector of Bavaria and on the other side a Grand Alliance consisting of the Austrian Habsburgs, England, the Netherlands, the Holy Roman Empire, and some lesser German states. All attempts to avoid a war by dividing the Spanish inheritance had been unsuccessful, although all European powers were war-weary after the War of the Grand Alliance, which had ended in 1697.

The fighting began in Italy in 1701, where the main Austrian commander, Prince Eugene of Savoy achieved the first victories. In 1702, the conflict became a general European war. It lasted until 1714, with no decision in the field. The main European theaters of war were the Low Countries; the Upper Rhine, including Bavaria; northern Italy; and Spain.

In 1703 and 1704, French and Bavarian forces were successful in southern Germany, threatening the Habsburg capital, Vienna, as well as the imperial circles of Swabia and Franconia. But in 1704, the English commander in chief, John Churchill, Duke of Marlborough, after his famous long march from the Netherlands to the Danube River, shifted the situation in favor of the allies together with the imperial generals Eugene of Savoy and Ludwig Wilhelm of Baden. After the decisive Battle of Blenheim-Höchstädt (13 August 1704), the Franco-Bavarian forces were driven out of southern Germany.

But this great success could not decide the war. The French lost more battles at Turin (1706), Oudenaarde (1707), and Malplaquet (1709), but they were not defeated. Heavy losses and enormous financial cost made all parties weary of the conflict. The allies were not able to secure Spain for their candidate, Carlos III (later the emperor Charles VI). When his elder brother, Joseph I, died in 1711, the fear of a too-powerful Habsburg Spanish-Austrian empire led to a dissolution of the Grand Alliance. England and the Netherlands were fighting for a balanced system of powers, not for Habsburg interests.

Thus in 1713, they signed the Peace of Utrecht with France. Charles VI waited until 1714, when the treaties of Rastatt and Baden brought the war finally to an end. With this settlement, the principle of a balance of powers took precedence over dynastic or national rights. The Spanish inheritance was divided. Spain fell to Philip V, and the Habsburgs gained Belgium and the former Spanish possessions in Italy.

Max Plassmann

See also: Blenheim-Höchstädt, Battle of; Denain, Battle of; Malplaquet, Battle of; Northern War, Great; Oudenaarde, Battle of; Queen Anne's War; Ramillies, Battle of

References and further reading:

Hattendorf, John B. *England in the War of the Spanish Succession: A Study of the English View and Conduct of Grand Strategy.* New York: Garland, 1987.

Plassmann, Max: *Krieg und Defension am Oberrhein.* Berlin: Duncker & Humblot, 2000.

Spanish-American War (1898)

Conflict that ended Spain's control over its last major colonial territories and established the United States as a world power. In February 1895, Cuban revolutionaries led by the poet José Martí renewed efforts to free the island from Spain. The rebels waged guerrilla warfare, destroying cane fields, sugar mills, and military outposts in an attempt to make the colony worthless to Spain. The following year, General Valeriano Weyler y Nicolau was placed in command of the Spanish army and initiated the reconcentration program that involved the removal of large segments of the civilian population from western Cuba into camps known as *reconcentrados.* Weyler planned to separate the guerrillas from their support base and to distinguish the insurgents from the civilian population. Cubans who refused to be relocated were considered rebels and could be summarily shot. The camps were hastily constructed and lacked adequate sanitary facilities, resulting in horrific conditions. Malnutrition, disease, and Spanish brutality resulted in the deaths of more than 200,000 Cubans in a period of only two years.

Spanish abuse of the Cubans caught the attention of American media moguls William Randolph Hearst, publisher of the *New York Journal,* and Joseph Pulitzer, whose leading paper was the *New York World.* Hearst and Pulitzer dispatched their best reporters and illustrators to Cuba, and their lurid reports of Spanish atrocities were widely circulated. The " yellow press" accounts enjoyed a wide readership and were largely responsible for the pro-Cuban sympathies of the American people. By the end of 1897, there was a public outcry from the popular press, Protestant groups, and politicians for U.S. intervention to end the conflict and establish Cuban independence.

President William McKinley was reluctant to involve the United States directly in the Cuban insurrection. Instead, he offered the good offices of the United States to negotiate a settlement but warned Spain that intervention was a possibility if the civil war continued. A new Spanish government seemed receptive, promised reforms, and recalled Weyler.

Several events quickly doomed any hopes of a peaceful settlement to the Cuban issue between Spain and the United States. On 9 February 1898, the *New York Journal* published the contents of a letter written by Dupuy de Lome, the Spanish minister in Washington. The letter, which had been pilfered by a rebel sympathizer, was highly critical of President McKinley, claiming that he was weak-willed and overly concerned with his popularity. This embarrassing revelation led to de Lome's resignation and further soured diplomatic relations.

On 15 February, the U.S. battleship *Maine* mysteriously exploded, resulting in the deaths of 266 American sailors. A U.S. Navy board concluded that the explosion had been set off by an external mine but could not determine culpability. Spanish investigators claimed the explosion was internal and denied any involvement in the *Maine* catastrophe. Modern investigators using computers have concluded that the *Maine* tragedy was the result of a coal dust explosion and an unfortunate accident unrelated to events in Cuba. In any event, the Pulitzer and Hearst papers asserted that the *Maine* had been the victim of a Spanish mine, and this notion was generally accepted by an outraged American public.

On 17 March, Vermont senator Redfield Proctor gave an account of his recent visit to Cuba to the Senate. His findings validated yellow press accounts of the suffering and misery of many Cubans in the reconcentration camps. Proctor was considered an honest man and not given to exaggeration; therefore, his speech convinced many skeptics of the need to intervene in Cuba. Proctor's account also increased the pressure on President McKinley to take decisive action.

McKinley gave diplomacy one last chance when, on 27 March, he sent his final peace proposal to Madrid. Spain was to end the reconcentration policy, agree to an armistice, and allow McKinley to negotiate a peace settlement. On 28 March, the president added the demand that Cuba be granted complete independence. Spain considered McKinley's proposals, agreed to end the reconcentration policy, and declared an armistice. Nevertheless, McKinley found Spanish actions to be inadequate and on 11 April requested authority from Congress to use naval and military force to end the civil war in Cuba. On 20 April, Congress passed several resolutions calling for Cuban independence and for Spain to completely withdraw from the island. Spain chose not to agree to the terms offered by the United States and declared war on 24 April. The following day, the United States reciprocated by declaring war on Spain.

Ironically, the first major action in the war did not involve Cuba. On 1 May, the U.S. Asiatic Squadron, led by Commodore George Dewey, attacked the Spanish fleet at Manila Bay in the Philippines. The American squadron totally destroyed the Spanish force of four cruisers, three gunboats, and three other vessels. American casualties were a mere eight wounded, whereas Spanish losses amounted to 381 killed or wounded. Dewey took possession of the Spanish naval base of Cavite and blockaded the city of Manila. General Wesley Merritt arrived at Cavite on June 30 with 10,000 men. Merritt, with the assistance of Filipino guerrillas under Emilio Aguinaldo, captured Manila on 13 August, effectively ending Spanish possession of the Philippines.

Spain attempted to prevent an invasion of Cuba by sending Admiral Pascual Cervera to the port of Santiago de Cuba on the eastern end of the island with four modern cruisers and three destroyers. The ships arrived safely, but Rear Admiral William T. Sampson, in command of U.S. naval forces, ordered a blockade and bottled up the Spanish fleet in the port.

Major General William R. Shafter was placed in command of the Fifth Army Corps, consisting of 16,888 officers and men. Shafter's force had 15 regiments of regular army troops and three regiments of volunteers. Among the volunteer regiments (the last U.S. Army units to be so raised) was the 1st Volunteer Cavalry, commonly referred to as the Rough Riders, soon to be commanded by the former assistant secretary of the navy, Lieutenant Colonel Theodore Roosevelt. Shafter's original objective was to establish a beachhead in Cuba so that supplies could be funneled to the Cuban guerrillas. Logistical support for the expeditionary force was woefully inadequate, as evidenced by the lack of horses for the six cavalry regiments. The expedition landed unopposed at Daiquiri, east of Santiago, and debarked between 22 and 25 June. After a skirmish at Las Gusimas (23 June), Shafter ordered an advance on Santiago.

There were 200,000 Spanish troops in Cuba but only 35,000 in the vicinity of Santiago, and the force defending the city numbered but 13,000. The road to Santiago was blocked by Spanish trenches on San Juan and Kettle Hills manned by 1,200 Spanish soldiers. To the northeast was El Caney, held by 500 defenders. On 1 July, Shafter ordered both positions attacked simultaneously. The American charges on Kettle and San Juan Hills commenced shortly before 1:00 P.M. and were over in little more than an hour. On the far right of the American line, Theodore Roosevelt led the Rough Riders up Kettle Hill and was soon joined by the 1st, 3d, 9th, and 10th Cavalry regiments. Roosevelt and his men swept the crest of Kettle Hill and poured fire into Spanish positions on San Juan Hill to support the assault of the 1st Brigade led by Brigadier General Hamilton S. Hawkings. By 1:30, the southern end of San Juan Hill had been breached by Hawking's men. Roosevelt then led a charge on the northern end of the hill and swept the Spanish from their trenches. The battle resulted in 124 American killed and 817 wounded. Spanish losses were 58 killed and approximately 170 wounded. The famous charge made Roosevelt a national hero and accelerated his political career.

On 3 July, Cervera attempted to break the American naval blockade with disastrous results. Four Spanish cruisers and two destroyers were so severely damaged by American fire that they were run aground in an attempt to save as many of the crew as possible. Only one U.S. sailor died in the action, whereas 323 Spanish seamen lost their lives. Santiago was surrounded by American land forces, and with no hope of relief, General Jose Toral surrendered on 17 July. The fall of Santiago ended active Spanish resistance in Cuba.

On 25 July, Major General Nelson A. Miles landed on the island of Puerto Rico with a force of approximately 5,000 men. Miles's army grew to 16,000 and, in an orderly and well-executed operation (in contrast to the U.S. expedition to Cuba), subdued the Spanish forces on the island by 13 August. The Puerto Rico campaign cost only nine American lives, whereas Spanish losses were 28 killed and 125 wounded.

The Spanish-American War officially ended with the Treaty of Paris, signed on 10 December 1898. Cuba was freed from Spanish rule, and Puerto Rico and Guam were ceded to the United States. The Philippine Islands were sold to the United States for U.S.$20 million. The war removed the last vestiges of Spanish power in the Western Hemisphere and made the United States the principal power in the Caribbean basin. The U.S. acquisition of the Philippines resulted in a bloody insurrection that cost more than 5,000 American and 200,000 Filipino lives and sparked debate within the United States about its role as an imperial power.

George M. Lauderbaugh

See also: Aguinaldo, Emilio; Martí y Pérez, José Julián; Miles, Nelson Appleton; Weyler y Nicolau, Valeriano, Marquis of Tenerife

References and further reading:

Bradford, James C., ed. *The Crucible of Empire.* Annapolis, MD: Naval Institute Press, 1993.

Freidel, Frank. *The Splendid Little War.* Boston: Little, Brown, 1958.

Musicant, Ivan. *Empire by Default.* New York: Henry Holt, 1998.

Nofi, Albert A. *The Spanish American War 1898.* Conshocken, PA: Combined Books, 1996.

O'Toole, George. *The Spanish War: An American Epic, 1898.* New York: W. W. Norton, 1984.

Trask, David T. *The War with Spain in 1898.* New York: Macmillan, 1981.

Spanish-Portuguese Wars (1580–1763)

Five wars waged between Spain and Portugal for control of the Iberian Peninsula. At stake was control of the world's first global seafaring empires.

In 1580, Spanish king Philip II claimed the Portuguese throne by invading Portugal. Spanish forces under Fernando Alvarez, Duke of Alva, defeated a peasant army of Portuguese king Antonio at the Battle of Alcantra on 25 August 1580. Spanish forces then pursued Antonio to Oporto and the Azores Islands. Despite support from Catherine de Medici, Antonio failed to hold the Azores. After the defeat of the Spanish Armada in 1588, Elizabeth I sponsored an English invasion of Portugal. This, too, failed to reinstall Antonio, and Portugal remained a dependent of Spain until 1640.

In 1641, Portuguese King John IV regained the throne and allied with England, France, and the Netherlands to hold it. Spain invaded, but Portugal won battles at Olivenca and Beira in 1642, then invaded Spain in May 1644 under General Mathias d'Albuquerque, and won a major victory at the Battle of Montijo. However, the cessation of the Thirty Years' War in Europe brought a truce to the Iberian Peninsula.

The death of John IV in 1656 sparked another invasion of Portugal by the Spanish between 1657 and 1668. Spain was victorious at Olivenca and Badajoz in 1658, and Portugal was victorious at Elvas in January 1659. Spanish forces under Don Juan of Austria defeated Portuguese forces at Evora in May 1652. Portuguese forces commanded by Herman Schomberg then turned the tide at the Battle of Ameixal on 8 June 1663. Following Portuguese victories at Alcantara, Villaviciosa, and Montes Claras, Portugal invaded Andalusia. England then mediated a treaty forcing Spain to recognize the House of Braganza as heirs to an independent Portugal.

From 1735 to 1737, global conflict between Portugal and Spain took place as part of the War of the Polish Succession. Spain invaded the stronghold of Colonia on the Rio de la Plata in South America. It was finally returned to the Portuguese after British intervention.

In 1761, Spain and France, both ruled by the Bourbon family, entered into a compact against the British, which appeared likely to win the Seven Years' War and upset Spanish trade in the Americas. In 1762, Britain occupied Spanish Cuba and the Philippines. In response, Spain invaded British ally Portugal and captured the town of Bragança and the fortress of Almeida in early 1763. British reinforcements under John Burgoyne and John Campbell landed in Portugal and combined with Portuguese forces under Count William of Schomberg to repel the Spanish in late 1763. The Treaty of Paris, which ended the Seven Years' War, also brought an end to hostilities in the Iberian Peninsula.

Christopher Howell

See also: Seven Years' War; Spanish Colonial Wars

References and further reading:

Olson, James. *Historical Dictionary of the Spanish Empire, 1402–1975.* New York: Greenwood Publishing, 1991.

Russell-Wood, A. J. R. *The Portuguese Empire, 1415–1808: A World on the Move.* New York: Johns Hopkins Press, 1998.

Special Operations Executive (SOE)

British unconventional warfare organization that conducted sabotage and intelligence operations during World War II. The Special Operations Executive (SOE) Charter was approved on 22 July 1940 in the wake of the fall of France. It merged Section D of the Secret Intelligence Service (SIS), which previously investigated nonmilitary means of attack; MI(R), the general staff body responsible for irregular warfare; and the Department of Propaganda (a.k.a. "EH" for its Electra House headquarters), responsible for "black," or unattributable, propaganda.

Popular belief held that British strength lay in sea power and airpower. Thus a land campaign of assisting popular uprisings, believed inevitable against the occupying Nazis, would avoid committing regular troops to a mainland invasion. Activating this strategy would emphasize trade blockade (Royal Navy), augmented by bombing strategic industries (Royal Air Force, or RAF) and sabotage/psychological warfare (SOE).

The SOE charter was sufficiently vague that rivalries between SOE and the Foreign Office, military chiefs of staff, and the SIS plagued smooth operations throughout the war. For example, conducting sabotage to an extent adequate to cause embarrassment to the occupying forces would disrupt the peaceful environment conducive to SIS intelligence operations and Foreign Office diplomacy. On a more mundane level, once it became apparent that infiltrating agents by sea was not practical, the requirement for aircraft to insert personnel (by parachute or rough-field landing) caused a conflict over resources with the RAF hierarchy; SOE planners claimed a requirement for 102 heavy aircraft for western Europe and the Balkans alone.

By 1945, SOE was conducting operations throughout Europe (except in Spain and Portugal) and in the Far East. SOE sabotage was particularly effective against German petroleum, U-boat, and air operations in France, the Balkans, and Scandinavia.

There were always critics who argued that the resources "diverted" to SOE would have been better employed in conventional operations and that special operations more often than not aroused the enemy to take drastic measures against civilians in the areas of operations. But even if these points were conceded, for several years SOE and aerial bombing were the only way that the British could somewhat directly demonstrate to the peoples of conquered Europe and Asia (not to mention Britain's enemies) that it was still very much in the fight.

Robert Martyn

See also: World War II

References and further reading:

Foot, M. R. D. *SOE in France.* London: HMSO, 1966.

Stafford, David. *Britain and European Resistance, 1940–1945.* Toronto: University of Toronto Press, 1980.

Special Operations Forces

Forces expressly trained and equipped for military or paramilitary tasks deemed unsuitable for conventional forces. Special operations forces (SOF) candidates typically undergo a grueling, often lengthy selection and training process before achieving basic special operations status. Training emphasizes fitness, independent thinking, and often nonstandard equipment. Expertise is developed in traditional military skills such as battle craft and navigation; less standard skills such as free-fall parachuting, mountain operations, or scuba diving are common. Acquiring and maintaining individual and small-unit proficiency in multiple unconventional or distinctive skills provides this "special" capability.

SOF are employed where clandestine ability permits government deniability or where peril, special qualifications, or lack of alternatives precludes conventional forces. They may be employed in counterterrorism, intelligence collection, raids, guerrilla warfare, training of foreign military/paramilitary personnel, civil affairs, and psychological operations. Mission success is often dependent upon the soldiers' personal capacity (such as initiative or fitness), detailed intelligence support, mission planning, and possibly support from in-theater indigenous personnel. (This support distinguishes U.S. Army special forces from Rangers, Special Operations Executive, and so on.) SOF generally lack the integral firepower and organic combat service support of conventional forces.

Unconventional operations, such as raiding enemy supply depots or command centers, have been conducted throughout history, but not until the twentieth century were units specifically created and maintained for these tasks within peacetime armies, navies, and air forces. Some examples of SOF units' actions in the twentieth century include the U.S. Army Detachment 101 and the British Chindits in northern Burma and the U.S. Office of Strategic Services Jedburghs in Europe and British commandos, operating far behind enemy lines, working with indigenous resistance forces, and preparing the way for the Allied invasion of the Continent. Aptly named long-range penetration groups shot up Axis supply lines and airfields almost with impunity deep inside enemy territory in North Africa.

Oddly, the Germans, who supposedly had "fifth columns" everywhere, made very little use of SOF, as did the Japanese. The long Allied supply lines over land in both the European and the Pacific theaters of operations faced far greater losses from pilferage than from any Axis guerrilla attacks. And despite some hysterical fears of an "Alpine Redoubt" peopled by Nazi diehards ready to carry on the struggle by any means, when the order came to surrender unconditionally, that is just what the Germans—and the Japanese—did. Axis psywar against the Allies was pathetic, ridiculing Allied soldiers or gloating over imaginary disasters. The Axis powers had no such thing as civil affairs, and their military government was simply rule by terror, punctuated by "violators will be shot" bulletins. Even subject peoples who had at first welcomed the Germans and the Japanese as liberators were soon enough plotting against them and conducting guerrilla warfare against their new, even worse, oppressors.

U.S. SOF in Vietnam proved one of the few success stories from that conflict. No less than 100,000 Communist defectors were garnered by the imaginative "Chieu Hoi/Open Arms" program. By contrast, the Communist "masters of propaganda" could not produce one U.S. defector. Civic action teams indeed won "hearts and minds" by improving the lives of the people, and the special forces work with the Montagnard hill tribes made Communist supply lines extremely vulnerable.

In the Panama incursion of 1989–1990, U.S. Army Civil Affairs troops set up a refugee camp, built and administered primarily by Panamanian civilians, so well run that a major problem was keeping unqualified persons *out.* Imaginative special forces troopers used the "Ma Bell" technique, calling ahead to enemy headquarters, warning them of the wrath to come from U.S. Army gunships (occasionally a gunship was called in for a tree-cutting demonstration at the edge of the forest), and then arranging dignified ceremonies in which the American flag was raised alongside that of Panama, preserving everyone's dignity.

Later that same year, in the Gulf War, coalition psyops (along with relentless aerial attacks) took the heart out of the "battle-hardened" Iraqi army, leaving its soldiers vulnerable to the land offensive. Other psyops teams talked terrified Iraqi soldiers out of their bunkers. Still other SOF reconnoitered Iraqi supply lines, extracted downed coalition airmen, administered enemy prisoner of war camps, and established camps for Kurdish refugees, which those refugees soon ran themselves.

Extensive training and equipment expense may make SOF casualties appear more significant when compared with those of conventional soldiers. This, plus bureaucratic intransigence and service jealousies stemming from the perception that SOF absorb high-quality soldiers who would otherwise improve conventional units, often make them unpopular among conventional military leadership. SOF have waged a continual battle through much of the twentieth century for permanent acceptance in the world's militaries. The following century will undoubtedly prove no different.

Robert Martyn

See also: Civil Affairs/Military Government; Guerrilla/Partisan/Irregular Warfare; Psychological Operations

References and further reading:

Kelly, Ross. *Special Operations and National Purpose.* Lexington, MA: Lexington Books, 1989.

Sandler, Stanley. "To Free from Oppression: A History of U.S. Army Special Forces." Fort Bragg, NC, unpublished study, c. 1996.

———. *Glad to See Them Come and Sorry to See Them Go: A History of U.S. Army Tactical Civil Affairs/Military Government, 1775–1991.* Fort Bragg, NC: U.S. Army Special Operations Command, [1998].

———. *Cease Resistance; It's Good for You: A History of U.S. Army Combat Psychological Operations.* 2d ed. Fort Bragg, NC: U.S. Army Civil Affairs and Psychological Operations Command, [1999].

Weale, Adrian. *Secret Warfare.* London: Hodder and Stoughton, 1997.

Spotsylvania Court House (12–20 May 1864)

The second major battle of General Ulysses S. Grant's 1864 overland campaign, featuring some of the most savage fighting of the war. After inconclusive combat in the Wilderness (5–6 May), Grant attempted to turn General Robert E. Lee's right flank. Lee's columns reached the junction at Spotsylvania Court House and began to prepare trenches reinforced with logs.

Spotsylvania consisted of a series of federal attacks and Confederate counterthrusts. Grant directed the Army of the Potomac to strike on 8 May; this first assault failed with heavy casualties. On 12 May, the Union army launched the most famous attack at the Mule Shoe, a salient in the Confederate line. A drizzling rain dampened the Confederates' powder, and few rifles could fire. Initially, it was one of the most successful attacks of the war, capturing an entire division, including several generals. Beyond the earthworks, Union troops met stiff resistance. Led by Lee himself, Confederate counterattacks drove them back to the trenches. Both sides poured in reinforcements. Combat raged for the next 12 hours in the most continuous hand-to-hand fighting of the war. All day, close combat raged within the confines of the earthworks. The action gave Lee time to prepare a second line, to which his troops fell back.

Federal attacks against this line met with failure on 18 May. The next day, a Confederate assault on Union positions along the Fredericksburg Road also failed. On 20 May, Grant again turned south, moving closer to Richmond.

By this period of the war, earthworks and fortifications dominated the battlefield; the days of open field maneuver that characterized the conflict's early years were over. Grant lost 18,000 men at Spotsylvania and Lee about 10,000. Calling on draftees and garrison troops, Grant could replace his losses, but Lee had no reserves.

Brian Dunkerly

Soldiers of the 3rd Battalion of the Sri Lankan army, clearing the village of Jaffna, 1988. (Hulton/Archive)

See also: American Civil War; Grant, Ulysses Simpson; Lee, Robert Edward; Wilderness

References and further reading:

Frassanitto, William. *Grant and Lee: The Virginia Campaign of 1864.* New York: Charles Scribner's Sons, 1983.

Gallagher, Gary W., ed. *The Spotsylvania Campaign.* Chapel Hill: University of North Carolina Press, 1998.

Matter, William D. *If It Takes All Summer.* Chapel Hill: University of North Carolina Press, 1988.

Rhea, Gordon C. *The Battles for Spotsylvania Courthouse and the Road to Yellow Tavern.* Baton Rouge: Louisiana State University Press, 1997.

Trudeau, Noah Andre. *Bloody Roads South.* New York: Little, Brown, 1989.

Sri Lankan Civil War (1983–)

Violent civil war between the two main ethnic forces in Sri Lanka. The onset of Ceylon independence in February 1948 only exacerbated deep-seated resentment between the island's two major ethnic groups: the Buddhist Sinhalese majority and the Hindu Tamil minority, comprising 18 percent of the population. There were periodic outbursts of violence between them through May 1971, when Ceylon adopted a new constitution and a new name: Sri Lanka. However, the Tamil, weary of being accorded second-class status, spawned several liberation-minded guerrilla groups intent upon carving out an ethnic enclave on the northern part of the island. The most vocal and violent of these, the Liberation Tigers of Tamil Eelam (LTTE), commenced hostilities in July 1983 by ambushing a Sri Lankan army patrol and killing 13 soldiers. This attack prompted a wave of retribution from the Sinhalese majority, who over the following weeks killed an estimated 384 Tamils and rendered 50,000 homeless. Furthermore, the government outlawed the Tamil Tigers, banned all separatist political activity, and installed a lengthy curfew. None of these measures addressed the basic Tamil grievance of a lack of autonomy, and their ranks continued swelling with every terrorist act.

The Tamils escalated their attacks beginning with a 14 March 1985 attack on a bus in which 120 Sinhalese civilians were machine-gunned. On 21 April 1987, a bomb killed 113 people at a Colombo bus terminal. This ruthlessness sparked retributive atrocities from the Sinhalese side, and several hundred Tamils were killed in rioting against their business establishments. The Sri Lankan army, meanwhile, doubled in size to near 50,000 soldiers and conducted several ruthless sweeps through Tamil-held territory, killing hundreds. Despite heavy loss of life, the LTTE remained adamant in its demand for autonomy.

Sporadic fighting and bombing continued over the following months, and in July 1987, the Sri Lankan government, exhausted by counterinsurgency efforts, invited outside assistance from India. The government of Prime Minister Rajiv Gandhi responded by sending a garrison force of 67,000 soldiers to serve as a buffer between the two groups. However, LTTE forces remained dissatisfied by the terms of the Indo–Sri Lanka Peace Accord, signed on 29 July 1987, which called for rebel forces to lay down their arms. In May 1989, a new government under Prime Minister Ranasinghe Premadasa announced a new cease-fire agreement, which included an Indian withdrawal. However, fighting and terrorist acts continued unabated, and the violence extended into India when, in 1991, Prime Minister Gandhi himself was killed by a Tamil suicide bomber. Two years later, Premadasa was also murdered. By the time Indian forces finally departed Sri Lanka in March 1990, they had lost more than 1,100 soldiers to the guerrillas. The civilian toll has since topped 18,000 dead, and fighting continues, although currently on a reduced scale.

John C. Fredriksen

References and further reading:
Arvlar. *The Traditional Homeland of the Tamils.* Kotte, Sri Lanka: Kanal Publications, 1996.
Bose, Sumantra. *States, Nations, Sovereignty: Sri Lanka, India, and the Tamil Eelan Movement.* Thousand Oaks, CA: Sage Publications, 1994.
Hellmann-Rajanayagam, Dagmar. *The Tamil Tigers: Armed Struggle for Identity.* Stuttgart: F. Steiner, 1994.

SS

The most sinister of the Nazi paramilitary affiliates. Founded in 1925, the SS (Schutzstaffel, or Protection Squad) was not really significant until Heinrich Himmler took over its leadership in 1929. At first, it was merely intended to provide bodyguards for party leaders at meetings and to be more reliable than the larger SA. Under Himmler, it first took on new duties as the internal police and intelligence service of the Nazi Party.

When the Nazis took power in 1933, the SS, together with the SA, was a main agent of terror against leftists and other enemies. It was subordinate to the SA until 1934, but Himmler cleverly and brutally used the purge of the SA in that year to gain independence. The rapid subsequent expansion of the SS was the product both of Himmler's ambition and Nazi obsession with internal security. The SS gradually absorbed and centralized the various police forces of Germany. It took over the administration of the concentration camp system and turned it into an empire of terror with an increasingly important economic component, as slave labor was used to finance other SS operations. Himmler also saw the SS as the guardian of both the ideological and the racial purity of the Nazi movement and state and enforced strict racial standards within the SS while staking a claim for SS influence in all questions of race and settlement in Eastern Europe. It became the main executor of Nazi racial policy and terror. To expand the SS police powers, Himmler formed armed SS units, which in the fall of 1939 were consolidated into the armed—or Waffen—SS.

During World War II, the Waffen SS developed into an elite fourth branch of the armed forces and was deeply implicated in the Nazi regime's most heinous crimes. Under Himmler's constant push to expand, the Waffen SS also began to rely on conscription, first of ethnic Germans from eastern Europe and then of German nationals, and even began accepting foreign volunteers from both eastern and western Europe, until it became necessary to make a distinction between the General (Allgemeine) SS and the Waffen SS to maintain even a pretense of SS racial "purity." Though it was the home of some of the most famous armored divisions of the war and once considered a main contributor to German military success, the SS, according to more recent scholarship, also sacrificed manpower under often incompetent leaders, and the duplication of organization and procurement hindered the Nazi war effort. Because of the horrific crimes of the Nazi movement in which they were so implicated, both the General and Waffen SS were declared illegal organizations at the Nuremberg trials of the major Nazi leaders.

Bruce Campbell

See also: Paramilitary Organizations; SA

References and further reading:
Höhne, Heinz. *The Order of the Death's Head.* New York: Ballantine Books, 1971.
Koehl, Robert Lewis. *The Black Corps. The Structure and Power Struggles of the Nazi SS.* Madison: University of Wisconsin Press, 1983.
Stein, George H. *The Waffen SS: Hitler's Elite Guard at War 1939–1945.* Ithaca, NY: Cornell University Press, 1977.
Wegner, Bernd. *Hitlers Politische Soldaten: Die Waffen-SS 1933–1945.* Paderborn: Schöningh, 1982.

St. Clair's Defeat (4 November 1791)

One of the most thoroughgoing defeats in American military history. When Josiah Harmar's raid against a Miami Indian village (now Fort Wayne, Indiana) failed, President George Washington blamed undisciplined militia rather than the skill of the American Indian warriors commanded by Little

Turtle. Washington's solution was to increase the regular army to 2,200 men and appoint Major General Arthur St. Clair (not one of Washington's wiser choices) to command a new expedition. St. Clair's mission was to establish a fort near the Miami village in order to secure the territory and deter Indian attacks.

During the summer of 1791, St. Clair organized his army but could not overcome many problems with logistics and training. On 6 September, the army advanced from Fort Washington (Cincinnati) by building a road and establishing a series of fortifications to guard the line of communications. However, in two months the army had made little progress, advancing a little more than 100 miles and reduced by illness and desertion to 1,400 men.

At dawn on 4 November, as the army camped near the Wabash River, more than 1,200 Miami warriors led by Little Turtle surprised, surrounded, and attacked the Americans, who quickly lost all form of organization. After several hours, St. Clair's army had been decimated, with 630 killed and almost 300 wounded, a 66 percent casualty rate and three times the number of men that George Armstrong Custer would lose at Little Bighorn. More significantly, Little Turtle had effectively destroyed almost the entire existing U.S. Army. Two years later, Anthony Wayne would avenge St. Clair's failure by raising and training the Legion of the United States and defeating the Indians at Fallen Timbers in 1794. Those contemporary U.S. Army officers (including George Washington) who constantly deprecated the militia might well have considered the fate of St. Clair's regulars.

Steven J. Rauch

See also: American Indian Wars; Fallen Timbers

References and further reading:

Sword, Wiley. *President Washington's Indian War: The Struggle for the Old Northwest, 1790–1795.* Norman: University of Oklahoma Press, 1985.

St. Gotthard Abbey (1664)

Conclusive battle of the Austro-Turkish War (1663–1664). In 1658, Ottoman grand vizier Ahmed Köprülü deposed Transylvanian prince György II Rákóczi, replacing him with Michael I Apafi. When Habsburg emperor Leopold I was slow to recognize Apafi, Köprülü attacked Habsburg-controlled Hungary in 1663. Leopold's army under Count Raimondo Montecuccoli was unprepared for a major war and retreated before the invading Ottomans.

Köprülü took the fortress of Neuhausel (Nové Zamky) in September after a six-week siege and withdrew into winter quarters. In November 1663, Croatian Ban Miklós Zrinyi replaced Montecuccoli as Habsburg field commander. Throughout the winter, Zrinyi conducted bold mounted operations in the Ottoman rear areas along the Mura River, disrupting supply lines but failing to recover any territory.

The following summer, Montecuccoli was again given command of the imperial armies, reinforced by a French contingent under the Duke of Lorraine. When the Ottomans invaded Croatia, Montecuccoli refused Zrinyi's demands for reinforcements to relieve the siege of Zrinyi-újvár, which fell on 30 June. Instead, he withdrew to a position on the Raab River, where on 1 August, he caught and defeated Köprülü's army at Szént Gotthard/Mogersdorf as it attempted to cross into Austria. Leopold used the opportunity of the victory to conclude quickly the Treaty of Vasvár, which established a 20-year peace and awarded the Ottomans an indemnity and possession of Neuhausel. Leopold's neglect of Hungarian interests in securing the peace provoked a magnates' conspiracy against Habsburg rule, which was exposed and suppressed in 1670. Refugees from Habsburg reprisals fled to Transylvania, from which they launched raids into Hungary for the next decade.

The Battle of St. Gotthard was the only major confrontation of the Habsburg and Ottoman armies between the Fifteen Years' War (1591–1606) and the second Siege of Vienna (1683).

Brian Hodson

See also: Austro-Turk Wars

References and further reading:

Peball, Kurt. *Die Schlacht bei St. Gotthard/Mogersdorf, 1664.* Vienna: Österreichische Bundesverlag, 1964.
Wagner, Georg. *Das Türkenjahr 1664.* Eisenburg: Druckerei Rötzer, 1964.

St. Mihiel (12–16 September 1918)

The first purely independent operation for the newly organized U.S. First Army in World War I. The commander of the American Expeditionary Force, General John Pershing, tenaciously insisted, against vocal and bitter opposition from his French and British counterparts, that American troops serve as part of an independent American entity under his command. To Pershing, the question of such an independent command was a matter of national sovereignty. French and British commanders argued that, owing to American inexperience and the desperate situation facing the Allies in mid-1918, the wiser use of American manpower would be as mixed formations with other Allied units wherever they might be needed.

In the summer battles at Belleau Wood and the Marne,

Pershing reluctantly permitted some of his divisions to be used in this manner, but he and General Ferdinand Foch, supreme Allied commander, agreed that in September Pershing would conduct an independent offensive. The goal was to reduce a German salient in eastern France that dated to 1914. Pershing's hope was to push all the way to the Metz area of Germany by early 1919. After a heavy bombardment, American troops attacked on 12 September and made excellent advances, partially because the Germans had chosen to pull back from their positions to prepare a new defensive line. The American captured 15,000 prisoners and incurred roughly 7,000 casualties. Instead of continuing east, though, Pershing's forces turned northeast in the direction of the Meuse River and thus began an even larger American offensive generally known as the Battle of the Argonne Forest. As a result, the St. Mihiel offensive has been overshadowed by the larger offensive in the Argonne area.

John McManus

See also: Pershing, John J.; World War I

References and further reading:

Coffman, Edward. *The War to End All Wars.* Lexington: University Press of Kentucky, 1998.

Braim, Paul. *The Test of Battle: The American Expeditionary Forces in the Meuse-Argonne Campaign.* New York: White Mane Press, 1987.

St. Quentin (10 August 1557)

One of the most disastrous defeats in French military history. Philip II of Spain wanted to invade France to commence his reign on a glorious note. The Spanish army was regrouped along the river Meuse, and Philip gave the order to the Spanish general Philibert Emmanuel of Savoy to take the town of St. Quentin, the "key to Paris," because it was the place where he was to meet reinforcements from England. The French were expecting a direct attack and waited without encamping. The old French constable, Anne de Montmorency, discovered the Spanish too late; he could only send Gaspard de Coligny and 500 soldiers to garrison the town. These reinforcements arrived just a few hours before the Spanish vanguard.

The siege began on 2 August 1557. The walls were in poor condition, and only Coligny's will managed to raise the fighting spirits of the militia. The Spanish established their camps on both sides of the river Somme, which flows through the town.

Montmorency had to obey the king's orders to attack the invaders, but he knew that the Spanish had brought overwhelming numbers: 35,000 infantrymen, 12,000 cavalry, and dozens of guns. He decided to surprise the southern Spanish camp by the Somme and then to send thousands of troopers to the besieged town by boat.

On 10 August, at about 9:00 A.M., he surprised the Spanish by a sudden and violent attack on Philibert's camp. Most of the Spanish managed to flee and take refuge in Comte Lamorel d'Egmont's camp on the northern side of the river. The 1,500 French now had to wait for four hours for the small boats from the town. D'Egmont and Philibert, recovering from their surprise, mounted an overall attack on the waiting French. The only bridge across the river and the marshy grounds was taken without a fight, and a charge from d'Egmont's cavalry sent the flanking French cavalry in flight. At 1:00 P.M., Montmorency decided to retire toward his line. The French noble cavalry refused to obey and charged the Spanish to its doom (700 against 8,000). The French commander now had to face the whole Spanish army. "C'est assez de reculer; il faut mourir" (No more way back; it is time to die) was his last command before being wounded and taken prisoner. Crushed by the deadly volleys of 42 Spanish guns, the shrinking French infantry stayed four hours under fire before retreating headlong.

France no longer had a field army, but Philip II made an unforgivable mistake in ordering Philibert to continue the siege of St. Quentin. Time was won by the French, and extra taxes, plus volunteers from throughout the kingdom, gave the nation a new army.

Gilles Boué

See also: Metz, Siege of; Valois-Habsburg Wars

References and further reading:

Hardy, Etienne. *Origines de la tactique francaise.* Paris: Dumaine, 1881.

Contamine, Philippe, ed. *Histoire militaire de la France.* Vol. 1. Paris: PUF, 1992.

Stalin (Iosif Vissarionovich Dzhugashvili) (1878 or 1879–1953)

Revolutionary, Russian Civil War commissar, Soviet Union leader (1929–1953), and Soviet Army commander in chief. Born in Gori, Georgia, son of a shoemaker, Iosif Dzhugashvili completed theological school (1894) and enrolled at Tiflis Seminary. He joined the Russian Social Democratic Labor Party (RSDLP) in 1898 and was expelled from the seminary for revolutionary activities the next year.

Prior to 1917, he was a party worker in Petrograd, but mainly in the Transcaucasus, where he was involved in the 1905–1907 Revolution (and possibly also several bank robberies for the cause). He organized party expropriations, was arrested by czarist authorities, and exiled or imprisoned

Stalin seated with President Franklin Roosevelt during the Teheran Conference. (Library of Congress)

seven times, escaping five times. He assumed the alias Stalin in 1912 and wrote *Marxism and the National Question* (1913).

Stalin was drafted into the czarist army in February 1917 but was excused on health grounds. After the February Revolution, he returned to Petrograd, supporting the Petrograd Soviet cooperation with the provisional government and a defensist continuation of World War I. Upon Vladimir Ilich Lenin's arrival in April, Stalin took up Lenin's "no cooperation, no war" stance.

Stalin played no active role in the October Revolution, editing the *Rabochy put'* (Workers' Path) newspaper at this time. In the wake of the revolution, he served as Peoples' Commissar for Nationalities (October 1917–April 1922) and, fittingly, as Peoples' Commissar for State Control (1919–1920).

During the Russian Civil War, Stalin held political positions, often disagreeing with and being insubordinate to the Red Center Command, and led the military opposition to Leon Trotsky against his use of former czarist officers but had scant positive military input. He was dispatched to Tsaritsyn as CEC plenipotentiary to requisition grain from the northern Caucasus (June 1918), but he abandoned these duties, forming a "Tsaritsyny" clique around Kliment Voroshilov, Sergei Konstantinovich Minin, and later Semen Budennyi (1919) and controlling the northern Caucasus politico-military apparatus. Stalin then served as chairman of the northern Caucasus military soviet (July) and the southern front Military Revolutionary Committee (RVS) (September–October) against Lieutenant General Peter Nikolaevich Krasnov. In October, he was recalled to Moscow for insubordination, beginning his feud with Trotsky.

His disagreements aside, during the Russian Civil War, Stalin was made a member of the Military Revolutionary Committee of the Republic (RVSR), serving from October 1918 to July 1919 and from May 1920 to April 1922), as well as a member of Central Committee Commission investigating the Perm catastrophe (January 1919); and he opposed

Trotsky and Lenin at the Eighth Party Congress military debates (March 1919). He then became a member of the RVS of the western front (July–September 1919), defending Petrograd against Nicholai Yudenich and the southern front (October 1919–January 1920) against Anton Denikin.

During the Polish-Soviet War, Stalin was southwestern front RVS member (January–August 1920) and is considered most at fault for the debacle at Warsaw, ignoring Sergei Kamenev's orders to support Mikhail Tukhachevsky's western front and seeking the glory of taking Lvov instead.

He secured his position as Soviet Union leader after Lenin's death early in 1924 by outmaneuvering Trotsky, Grigory Zinov'ev, Lev B. Kamenev, and Nicolai Bukharin. While in power, he modernized the heavy industry segment of the USSR through rapid industrialization drives and the collectivization of agriculture, enabling modernization and mechanization of the armed forces under Tukhachevsky (who was later shot). Millions of peasants starved as grain was exported to pay for industrial investment.

Stalin's manic Great Terror (1937–1938) resulted in the execution of nearly half of the Soviet military command and political staff from brigade commander upward and thousands more below, leaving a cowed leadership unprepared for war. With Britain and France reluctant to sign a collective security pact against Germany, Stalin signed the Nazi-Soviet Pact (1939, also called the Molotov-Ribbentrop Agreement), taking over eastern Poland, Bessarabia, Bukovina, and the Baltic states (1939–1940). Whatever their official propaganda "lines," Adolf Hitler and Stalin actually thought quite highly of each other, understandably, as they had much in common.

The poor Soviet performance against Finland (1940) encouraged German aspirations, and the damaged Red Army leadership prepared for the wrong war, assured by Stalin that the Germans would not invade the USSR. Stalin, ever the blunderer, ignored repeated warnings, allowing the surprise German attack (June 1941). His initial stunned inactivity, refusal to allow withdrawals, and costly unprepared offensives increased Soviet losses, but his willpower finally instilled morale for the crucial Moscow counteroffensive (December 1941), halting Operation BARBAROSSA.

Installing himself as commander in chief (from July 1941), he increasingly listened to his generals, creating an effective military leadership. The politico-economic system established in the 1930s enabled Soviet recovery and victory, but at the enormous cost of more than 20 million dead.

After the war, Stalin established the Warsaw Pact in Eastern Europe, contributing to the emergence of the Cold War. He only grudgingly acknowledged the Chinese Communist success but had the wit to remain uninvolved in the Korean War (1950–1953).

Stalin died in Moscow, and his funeral evoked genuine grief from the Soviet peoples who had suffered so much under his iron rule. But in 1956, Nikita Khrushchev began the process of "de-Stalinization" and in the process revealed some of Stalin's worst blunders and excesses. Although he never achieved the mythic legend of pure evil that has rightly cloaked the memory of his erstwhile partner Adolf Hitler, Stalin was actually responsible for the deaths of far more persons.

Neil Harvey Croll

See also: Cold War; Russian Civil War (1918–1921); Russo-Polish War; Trotsky, Leon; World War II

References and further reading:

Bullock, Alan. *Hitler and Stalin: Parallel Lives.* London: HarperCollins, 1991.

Seaton, Albert. *Stalin as Warlord.* London: B. T. Batsford, 1976.

Stalin, I. V. *Sochineniia v 13-kh tomakh* (Works, 13 vols.). Moscow: Gospolitizdat., 1946. (Translation of 1949 edition in 13 volumes by Moscow: Foreign Languages Publishing House, 1952; and in 14 volumes by London: Red Star Press Ltd., 1975; and ed. R. H. McNeal, 14 volumes, by California: Stanford University Press, 1967).

Volkogonov, Dimitry. Ed. and trans. H. Shukman. *Stalin: Triumph and Tragedy.* London: Weidenfeld and Nicolson, 1991.

Stalingrad (17 July 1942–2 February 1943)

The German failure to defeat Red Army forces in the southern European part of the USSR and capture Soviet petroleum resources, a turning point in World War II.

Directive 41, issued on 5 April 1942, outlined Operation BLUE, a renewed German offensive. The principal objectives were capturing the Kerch Peninsula (8–15 May), Kharkov (17–28 May), and Sebastopol (7 June–4 July); seizing the oil and mineral wealth of the Caucasus (28 June, unachieved); and capturing Leningrad (unachieved).

On 12 May, the Sixth and Fifty-Seventh Soviet Armies were destroyed in a failed bid to liberate Kharkov, leaving the area between the Don and Volga Rivers undefended. Adolf Hitler split Army Group South in early July: he directed Army Group A to advance to the Caucasus and Army Group B to provide flank protection while advancing to the Volga River. The capture of Stalingrad became an objective of Army Group B only on 13 July.

From 17 July to 18 August, the conflict was in the area of the Don bend and moved to between the Don and Volga Rivers on 19 August–3 September. Maikop was captured by Army Group A on 8 August, and the Sixth Army of Army Group B was fighting in the vicinity of Stalingrad on 4–13 September. Covering the latter were the Hungarian Second

Army, Italian Eighth Army, and Rumanian Fourth Army, protecting the northwest corridor to Stalingrad, while the Rumanian Third Army held the southwestern flank. These Axis armies had no significant antitank defenses or armored reserves.

To deter further Red Army retreat, Stalin issued the "Not one step back!" order no. 227 on 28 July, which contained draconian disciplinary measures. General Vasili Chuikov assumed command on 12 September of the Soviet Sixty-Second Army defending Stalingrad.

On 14 October, Hitler limited German operations to Stalingrad and the Terek River in the Caucasus. The mechanized divisions of Sixth and Fourth Panzer Armies were employed in the battle for Stalingrad itself (13 September–18 November) but were unsuited for close-quarter urban fighting.

The Soviet counteroffensive planned by Marshals Georgy Zhukov and Aleksandr Vasilevsky, Operation URANUS, began on 19 November, when 90 percent of Stalingrad was in German hands. On 23 November, Soviet pincers closed to encircle 330,000 Axis forces. General Erich von Manstein and the Eleventh Army's attempt to break through on 12 December was unsuccessful, as was the Luftwaffe's bid to airdrop supplies to General Frederick von Paulus's Sixth Army. Adolf Hitler's refusal to allow the Sixth Army to retreat sealed its doom. The Soviet reduction of the Stalingrad pocket continued until 2 February 1943. The surrender of the Sixth Army was the greatest capitulation in German military history. Except for the wounded earlier evacuated by air, not one Axis soldier escaped the entrapment.

Neville G. Panthaki

See also: Hitler, Adolf; Kharkov; Konev, Ivan Stepanovich; Leningrad, Siege of; Manstein, Fritz Erich von; Rokossovsky, Konstantin Konstantinovich; Rundstedt, Karl Rudolph Gerd von; Stalin; World War II; Zhukov, Georgy Konstantinovich

References and further reading:

Erickson, John. *The Road to Stalingrad.* New York: Harper & Row, 1975.

Glantz, David, and Jonathan House. *When Titans Clashed.* Lawrence: University Press of Kansas, 1995.

Overy, Richard. *Russia's War.* London: Penguin Books, 1998.

Werth, Alexander. *Russia at War, 1941–1945.* London: Pan Books, 1964.

Stamford Bridge, Battle of (25 September 1066)

First battle between two contenders for the English throne, Harald II Godwinsson and Harald III of Norway (also known as Harald Sigurdsson or Harald Hardradi). Although the English and Scandinavian sources differ about geography and the actions immediately preceding the battle, they concur about the outcome.

Tostig, Harald's brother, joined forces with Harald Hardradi, and in late September, they invaded England along the Humber River. Two of Harald Godwinsson's earls, Edwin and Morcar, attempted to repulse the Norwegian forces but were unsuccessful. They were defeated at Fulford before Harald Godwinsson could arrive to help them.

Buoyed by his defeat of English forces, Hardradi overcame York and accepted its surrender. Harald Godwinsson's arrival with seasoned forces and archers came as a complete surprise. The Norse invaders were vanquished, and Harald Hardradi and his ally Tostig Godwinsson were killed.

Tamsin Hekala

See also: Hastings, Battle of; William the Conqueror

References and further reading:

Magnusson, Magnus, and Herman Pálsson, eds. and trans. *King Harald's Saga: Harald Hardradi of Norway.* Harmondsworth, UK: Penguin, 1976.

Pulisiano, P., ed. *Medieval Scandinavia: An Encyclopedia.* New York: Garland Publishers, 1993.

Steuben, Friedrich Wilhelm Augustin, Freiherr von (1730–1794)

German officer who fought with the Americans in the Revolutionary War. Born on 17 September 1730 in Magdeburg, Prussia, Steuben became a soldier at age 16. During the Seven Years' War (1756–1763), he rose to the rank of captain in the Prussian army and was for a time attached to the general staff of Frederick the Great. After the war, Steuben retired from the army and became court chamberlain for the prince of Hohenzollern-Hechingen. Recruited by Benjamin Franklin, the American ambassador in France, he was given a letter introducing him to General George Washington as a "Lieutenant General in the King of Prussia's Service" and an ardent supporter of the American cause.

Arriving in America in December 1777, Steuben impressed the Continental Congress and George Washington with his fictitiously high rank and his accommodating personality, and he was appointed to train the Continental forces stationed at the winter encampmentment of Valley Forge, Pennsylvania. No more valuable appointment could have been made at this nadir of the American fortunes. With the rank of major general and the position of inspector general, he wrote a manual, *Regulations for the Order and Discipline of the Troops of the United States,* which became the standard for the entire army and served as the country's official military guide until 1812. He also remodeled the army's organization, improving its discipline and organizing an efficient staff. The American near-victory at Monmouth, New Jersey, in June 1778 demonstrated that, in the northern states at least, Steuben had made an American army.

In 1780, he sat on the court-martial of the British army officer Major John Andre, who was charged with espionage. In the same year, he was finally granted a field command, when he served as a division commander in Virginia and participated in the siege of Yorktown (1781), which basically marked the final defeat of the British.

After the war, Steuben became an American citizen and settled in New York City, where he lived so extravagantly that, despite large awards of money from Congress and the grant of 16,000 acres of land by New York State, he fell into debt. Finally, in 1790, he was voted a life pension of $2,500 per year, which sufficed to maintain him on his farm near Remsen, New York, until he died on 28 November 1794.

Alexander M. Bielakowski

See also: American Revolution; Washington, George

References and further reading:

Palmer, John M. *General von Steuben.* Port Washington, NY: Kennikat Press, 1966.

Tolzmann, Don H., ed. *The Army of the American Revolution and Its Organizer: Rudolf Cronau's Biography of Baron von Steuben.* Bowie, MD: Heritage Books, 1998.

Stilicho, Flavius (365–408)

Half-Roman, half-Vandal Roman commander appointed guardian for Theodosius's son Honorius, emperor of the western empire after 394. Theodosius also appointed Rufinus as guardian for his son Arcadius, who ruled the eastern empire.

Differing opinions concerning the treatment of barbarians within the empire resulted in a rivalry between Stilicho and Rufinus. It was interrupted when barbarians led by the Visigoth Alaric invaded Thrace and Macedonia. Several campaigns against Alaric failed to result in a definitive victory for Stilicho, but an understanding was reached between the two men. By 407, Stilicho arranged for Alaric to hold Epirus for Honorius, while he pursued his plans to capture Illyricum. A false report that Alaric had died resulted in the abandonment of the plans. Alaric then marched south, demanding a heavy compensation for his time and efforts. Stilicho persuaded the Senate to pay Alaric 4,000 pounds in gold. Unhappy with having to pay Alaric and hearing rumors that Stilicho planned to elevate his son to the eastern throne after Arcadius's recent death, Honorius ordered the execution of Stilicho. He was beheaded on 22 August 408.

Cynthia Clark Northrup

See also: Alaric; Vandals; Visigoths

References and further reading:

Grant, Michael. *The Fall of the Roman Empire.* New York: Collier Books, 1990.

Hodgkin, Thomas. *Huns, Vandals, and the Fall of the Roman Empire.* Mechanicsburg, PA: Stackpole Books, 1996.

Stilwell, Joseph Warren (1883–1946)

Prominent and acerbic American general during World War II. Born on 19 March 1883 in Florida, he grew up in Yonkers, New York, in fairly comfortable surroundings. Stilwell's father decided that his mischievous son should attend the U.S. Military Academy at West Point to learn some discipline. He graduated in 1904 and joined the infantry.

After serving two years in the Philippines, Stilwell returned to West Point to teach Spanish and French. Lack of action led him to seek overseas assignments during the summer months. Between 1907 and 1910, he made several

Lieutenant General Joseph W. Stilwell, portrait in uniform, 1942. (Library of Congress)

undercover intelligence-gathering trips to Latin America. During World War I, Stilwell served as an intelligence officer for the IV Corps of the American Expeditionary Forces. When the war ended, Stilwell volunteered for overseas intelligence gathering once again and was assigned to China. Over the next two decades, Stilwell alternated between staff and field commands in both the United States and China.

In 1942, Stilwell was assigned command of the China-Burma-India theater of operations. In addition, he managed Lend-Lease aid to China. In early 1942, he personally led a small group of Americans and Chinese out of besieged Burma in an epic trek just ahead of the invading Japanese. Upon his arrival in India, he cut through the self-serving propaganda of "fighting retreats" and "strategic withdrawals" and stated flatly, "I claim we took a hell of a beating." He soon enough became known, and not always with affection, as "Vinegar Joe."

With his theater at the bottom of the Allied list of priorities, Stilwell found it difficult to get the men and supplies necessary to liberate Burma and open a land supply route to China. Between his acerbic demeanor, stinging comments, and ambitious goals, Stilwell did not get along with either his Chinese or British ("limey") allies. The Chinese dictator Chiang Kai-shek ("Peanut") several times asked for his recall, especially after Stilwell called for greater cooperation with Chinese Communist forces fighting the Japanese. Stilwell was recalled in October 1944 and returned to the United States to command the Ground Forces Command. In June 1945, he took over as commanding general of the Tenth Army.

After the war, Stilwell commanded the Sixth Army. He died of stomach cancer on 12 October 1946 in San Francisco, California.

Gregory Dehler

See also: Chiang Kai-shek; Chindits; Chinese Military; Japanese Military, Twentieth Century; Mao Zedong; Merrill's Marauders; World War I; World War II

References and further reading:

Liang, Chin-Tuang. *General Stilwell in China, 1942–1944: The Full Story.* Jamaica, NY: Saint John's University Press, 1972.

Tuchman, Barbara W. *Stilwell and the American Experience in China, 1911–1945.* New York: Macmillan, 1970.

White, Theodore, ed. *The Stilwell Papers.* New York: Sloane, 1948.

Stimson, Henry Lewis (1867–1950)

Soldier, statesman, and adviser to five U.S. Presidents. As secretary of war in the cabinet of President William Howard Taft from 1911 to 1913, Stimson attempted to initiate reforms to modernize the U.S. military. During World War I, he held the rank of colonel in the army as a field artillery officer in France. In 1927, Stimson was appointed by President Calvin Coolidge as special emissary to Nicaragua to bring an end to that country's civil war. From 1927 to 1929, he administered the Philippine Islands as governor-general. President Herbert Hoover appointed Stimson secretary of state from 1929 to 1933. On 7 January 1932, following Japan's occupation of Manchuria in 1931, the United States issued the so-called Stimson notes, which stated that the United States did not recognize the Japanese puppet state in Manchuria (Manchukuo).

Stimson represented the U.S. delegation to the London Naval Conference in 1930. In an attempt to develop bipartisan support for his foreign policy, President Franklin D. Roosevelt selected Stimson, a Republican and prominent member of the interventionist Committee to Defend America by Aiding the Allies, as his secretary of war in 1940. As secretary of war, Stimson proposed the adoption of selective service, advocated Lend-Lease, supervised the expansion and training of the U.S. Army, and supported the internment of U.S. citizens of Japanese descent on the West Coast. Stimson also acted as the chief adviser on atomic policy to Presidents Roosevelt and Harry S. Truman. Stimson's role in Truman's decision to drop the bombs on Japanese cities of Hiroshima and Nagasaki is controversial. He justified the use of the atomic bombs, arguing that their use hastened Japan's surrender and ultimately saved more lives than they cost by obviating Operation DOWNFALL, the proposed American invasion of the Japanese home islands. He also appreciated that demonstrating the bombs' power might impress the USSR, the United States' future Cold War rival.

Eric D. Pullin

See also: DOWNFALL, Operation; Hiroshima and Nagasaki, Atomic Bombings of; Lend-Lease; Nicaraguan Civil War (1925–1933); Roosevelt, Franklin D.; World War I; World War II

References and further reading:

Alperowitz, Gar. *The Decision to Use the Atomic Bomb and the Architecture of an American Myth.* New York: Alfred A. Knopf, 1995.

Cohen, Warren I. *Empire without Tears: America's Foreign Relations, 1921–1933.* New York: McGraw-Hill, 1985.

Frank, Richard B. *Downfall: The End of the Imperial Japanese Empire.* New York: Random House, 1999.

Thorne, Christopher. *The Limits of Foreign Policy: The West, the League and the Far Eastern Crisis of 1931–1933.* New York: Putnam, 1972.

Stirling Bridge (11 September 1297)

Great victory of the Scots over the English invaders. After the 1296 campaign of conquest, Edward I's high-handed de-

General Jeb Stuart returning with captured horses and supplies after the Gettysburg campaign, 1863. (Library of Congress)

mands on Scotland, when combined with the rapacious, brutal behavior of English troops, provoked rebellion. This rebellion in turn led in 1297 to the intervention of an English army numbering some 20,000 heavy cavalry, archers, and infantry, which advanced through the lowlands while some 10,000 Scottish troops were investing Dundee. Abandoning the siege, William Wallace led his force across the Tay River to positions north of Stirling, around Cambuskenneth Bridge, in order to block any English advance beyond the Forth.

At dawn on 11 September 1297, English forces began to cross the bridge, which was so narrow that only two people could pass along it abreast, a nearby ford being totally ignored. The Scots allowed about half the English army to cross before moving to block the ford crossing and trap those English north of the river. The latter were caught first by archers and then by massed pikemen, and an attempt to rush reinforcements across the bridge resulted in its collapse. As English troops, whether from the north bank or the bridge, were forced into the Forth and drowned, Scottish infantry crossed at the ford and fell on the English rear: a number of Scottish nobles and retainers in English service turned coat, and added to the general slaughter. What remained of the English army fled; after the battle just Roxburgh and Berwick remained in English hands.

H. P. Willmott

See also: Anglo-Scots Wars (1290–1388); Falkirk, Battle of

References and further reading:

Hooper, Nicholas, and Matthew Bennett. *The Cambridge Illustrated Atlas of Warfare. The Middle Ages, 768–1487.* Cambridge, UK: Cambridge University Press, 1996.

Prestwich, Michael. *Armies and Warfare in the Middle Ages: The English Experience.* New Haven, CT: Yale University Press, 1999.

———. *The Three Edwards. War and State in England, 1272–1377.* London: Methuen, 1981.

Stuart, James Ewell Brown ("Jeb") (1833–1864)

Flamboyant and very successful Confederate cavalry commander, "Lee's Eyes and Ears." Stuart was born in Patrick County, Virginia, on 6 February 1833. Commissioned in the Mounted Rifles after graduating thirteenth in the U.S. Military Academy class of 1854, he served in Texas until 1856 and then with the 1st Cavalry in Kansas. Wounded in battle against the Cheyenne in 1857, he recovered in Washington, D.C., and then served as Robert E. Lee's aide-de-camp in the suppression of John Brown's 1859 raid on Harpers Ferry, Virginia. He resigned his captaincy on 14 May 1861 to join the Confederacy and was immediately commissioned lieutenant colonel.

For heroism at First Bull Run, sweeping his troops from the Confederate left around the Union rear, Stuart was promoted to brigadier general on 24 September. Until the spring of 1862, he led his 1st Virginia Cavalry in skirmishes throughout the Shenandoah Valley. During the Peninsular campaign in June, he rode completely around George B. McClellan's army, gathering important intelligence for the Seven Days' Battles and boosting Confederate morale. This exploit won him his major general's star and command of a cavalry division on 25 July. In a daring raid on Yankee headquarters at Catlett's Station, Virginia, on 22 August, he stole John Pope's uniform, dispatches, and notebook, giving Lee a great advantage for Second Bull Run.

In September, during Lee's first invasion of the North, Stuart fought valiantly at South Mountain and Antietam, and then led a raid on Chambersburg, Pennsylvania, again encircling McClellan before returning to Confederate lines on 12 October. He commanded the Confederate right at Fredericksburg, harassed Joseph Hooker throughout the spring of 1863, took over Stonewall Jackson's corps command when Jackson fell at Chancellorsville, and fought Alfred Pleasonton to a draw in the huge cavalry battle at Brandy Station, Virginia, on 9 June. But he arrived at Gettysburg too late to be effective. He continued raiding and skirmishing until the Wilderness and Spotsylvania campaigns of 1864, when he made a concerted effort to stop the advance of Ulysses S. Grant and Philip H. Sheridan. Shot by one of Sheridan's soldiers at Yellow Tavern on 11 May, he died in Richmond the next day.

Eric v. d. Luft

See also: American Civil War; American Indian Wars; Antietam/Sharpsburg; Bull Run, First/Manassas; Bull Run, Second/Manassas Junction; Chancellorsville, Battle of; Fredericksburg; Gettysburg; Harpers Ferry; Jackson, Thomas "Stonewall"; Lee, Robert Edward; McClellan, George Brinton; Mosby, John Singleton; Pope, John; Seven Days' Battles; Sheridan, Philip Henry; Spotsylvania Court House; Wilderness

References and further reading:

Blackford, W. W. *War Years with Jeb Stuart.* Baton Rouge: Louisiana State University Press, 1993.

Borcke, Heros von. *Memoirs of the Confederate War for Independence.* Nashville, TN: J. S. Sanders, 1999.

Davis, Burke. *Jeb Stuart: The Last Cavalier.* Short Hills, NJ: Burford, 2000.

Thomas, Emory M. *Bold Dragoon: The Life of J. E. B. Stuart.* Norman: University of Oklahoma Press, 1999.

Student, Kurt (1890–1978)

Germany's premier airborne officer. Kurt Artur Benno Student was born on 12 May 1890 and entered a military school in 1901. In 1910, he was commissioned into a Prussian infantry battalion and became a pilot in 1914. He was involved with the fledgling Luftwaffe after the war, joined the Ministry of Aviation, and then became commander of the Test Center for Flying Equipment in 1935 as a colonel.

In 1938, he was given command of the new 7th Airborne Division. He became Germany's foremost paratroop officer and created the glider force. The first operation for his paratroopers was the landing in Holland on 10 May 1940, where he was wounded and away from duty for some months.

In January 1941, he was promoted to general and received the Knight's Cross. He had already planned the airborne component for the invasion of England and later examined possible airborne landings on Gibraltar, Cyprus, Malta, and Crete.

During the operation against Crete, the battle—and the Airborne Division—were in the balance for some days until the 5th Mountain Division arrived to relieve the hard-pressed paratroopers. After the heavy losses in Crete, Adolf Hitler lost interest in the paratroopers, committing them as ground troops instead. In this role, they still achieved an enviable reputation as good fighters.

Student planned the successful liberation of Mussolini in 1943 and received the Oak Leaves to his Knight's Cross as a reward. In 1944, Student witnessed the failed Allied airborne assault in Holland but could only wish that he had had such resources.

He commanded Army Group H as a colonel general. In 1945, he was captured, later sentenced to five years' imprisonment for alleged war crimes in Crete, and served two years. Student died in the Federal Republic of Germany at age 88 on 1 July 1978.

David Westwood

See also: Airborne Operations

References and further reading:

Ailsby, S. *Hitler's Sky Warriors.* London: Brassey, 1999.

Stimpel, Hans-Martin. *Die deutschen Fallschirmtruppe 1942–45.* Vols. 1–2. Herford: Mittler & Sohn, 1999.

Sudanese Civil War (1955–)

One of Africa's longest-lasting, bloodiest, and most intractable of civil wars. As a national entity, Sudan has been beset by traditional hostility between a largely Arabic, Islamic north and an African, animist, Christian south. Such

Guerrillas of the "Venom Army" fighting the government troops in southern Sudan, 1971. (Hulton/Archive)

animus predates the arrival of Sudanese independence from Great Britain in 1956 and was almost immediately exacerbated by it. This resentment largely stems from northern attempts to impose Islamic religion and law upon the southern third of the country, as well as a generally condescending attitude of Arab political leaders toward citizens of African descent. Fighting erupted on a small scale in 1955 and was generally contained until September 1963, when African resistance coalesced into a broad-based guerrilla movement called Anya-Nya (Venom of the Gabon Viper). The central government, based in Khartoum, responded with land campaigns and air strikes, which created a flood of refugees. Numerous attempts by Arabic northerners to conduct slave-raiding campaigns were also reliably reported. Fighting was sporadic but vicious, with few prisoners taken on either side. The northerners were generally better equipped for prolonged conflict because they received outright money and material assistance from neighboring Arab states and the Soviet Union. The Africans, meanwhile, welcomed aide from fellow states like Ethiopia, Uganda, and Congo. Fighting sputtered on and off for nearly a decade, with government forces controlling major cities in the south, while the guerrillas moved about the countryside with impunity. By the time a peace accord was reached in May 1972, the Sudan concluded—after 16 years—Africa's longest civil war. An estimated 500,000 people perished from either combat, starvation, or disease.

In 1983, a new government under Gaafar Mohammed Nimeiri attempted to further Islamicize the southern regions by imposing strict religious laws upon non-Muslim peoples. The ensuing hue and cry brought old animosities back to the surface, and armed resistance began anew. Curiously, when the government dispatched Colonel John Garrang, a member of the Dinka tribe, to suppress the rebellion, he joined them instead. Nimeiri was overthrown in 1985 and replaced by an even more hard-line Islamic regime with close ties to Iran and Iraq. This new alignment, in turn, enabled Garrang's force, the Sudanese People's Liberation Movement (SPLM), to receive clandestine aid from the United States and Israel, and his forces became exceptionally well armed. To date, fighting continues along a similar pattern: government aircraft bomb villages; its army seizes and controls major urban areas and encourages slave raiding; but SPLM forces roam freely. They have also eliminated numerous government garrisons and closely invested the provincial capital of Juba for several years. Since 1990, an estimated 1 million Sudanese of every ethnic group and religious persuasion have died.

John C. Fredriksen

See also: Guerrilla/Partisan/Irregular Warfare

References and further reading:

Assefa, Hizkias. *Mediation of Civil Wars: Approaches and Strategies: The Sudan Conflict.* Boulder, CO: Westview Press, 1987.

Harris, Lillian C. *In Joy and in Sorrow: The Travels among Sudanese Christians.* Nairobi, Kenya: Paulines Publications Africa, 1999.

O'Ballance, Edgar. *Sudan, Civil War, and Terrorism, 1956–99.* New York: St. Martin's Press, 2000.

Süleyman I (c. 1495–1566)

Sultan whose long reign marked the apogee of Ottoman military power. Born in Trebizond, Turkey, around 1495, Süleyman succeeded his father, Selim I, without incident in 1520 and immediately initiated an aggressive policy of expansion, personally conducting 13 imperial campaigns in Europe and Asia, along with a myriad of smaller operations throughout his lengthy reign. Süleyman's initial conquests incorporated Rhodes, the few remaining autonomous Christian principalities in the Balkans, and most of Hungary. He decisively defeated the Hungarian nobility at Mohács in 1526 but increasingly came into conflict with Ferdinand I, the archduke of Austria and rival for possession of Hungary, sparking an Ottoman-Habsburg struggle that continued intermittently for decades. Süleyman conducted the first Ottoman siege of Vienna in 1529 and frequently raided the Habsburg frontier. An arrangement dividing Hungary between Ferdinand and Janós Szapolyai, a Transylvanian prince and Ottoman vassal, temporarily secured peace until the formation of an Ottoman-French coalition in 1536. Although this alliance swiftly collapsed under papal pressure, fighting raged throughout the Mediterranean, with Süleyman's navy, commanded by the pirate admiral Hayruddin Barbarossa, destroying the Habsburg fleet under Andrea Doria at Preveze, Albania, in 1538. A 1547 treaty reaffirmed the partition of Hungary, but a Habsburg-Polish offensive in 1551 renewed hostilities until a settlement was reached in 1562.

Süleyman renewed dormant hostilities with the Safavid kingdom in Persia in 1533, invading with two armies that annexed most of Mesopotamia and Kurdistan. He made smaller gains in Georgia and northeastern Anatolia in 1548, attempting to support the Safavid rebel Elkas Mirza against Shah Tahmasp I. Süleyman launched a final Persian offensive in 1554 that reached Azerbaijan. A treaty in 1555 formalized Ottoman control of Mesopotamia, Kurdistan, and eastern Anatolia.

In addition to the Habsburg and Safavid wars, Süleyman campaigned against the Portuguese in the Indian Ocean throughout the 1540s and 1550s; annexed the coastal regions of Aden and Yemen, southern Moldavia, and Bessarabia; unsuccessfully tried to capture Malta; and survived numerous insurrections and palace intrigues. During his final Hungarian campaign, he died on 6 September 1566 near Szitgetvár and was succeeded by his son Selim II.

Ian Janssen

See also: Austro-Turk Wars; Hungarian Civil Wars; Hungarian-Turkish Wars; Malta, Siege of; Mohács, Battles of; Ottoman Empire; Rhodes, Sieges of; Turkish Wars of European Expansion; Vienna, Sieges of

References and further reading:

Kunt, Metin, and Christine Woodhead, eds. *Süleyman the Magnificent and His Age.* London: Longman, 1995.

Shaw, Stanford. *History of the Ottoman Empire and Modern Turkey.* Vol. 1. Cambridge, UK: Cambridge University Press, 1976.

Sulla, Lucius Cornelius (138–78 B.C.E.)

Roman general and statesman who led the aristocratic party, the Optimates, in opposing the Populares, led by Sulla's rival, Gaius Marius. Born into a patrician family, Sulla entered military service during the Jugurthine War, persuading Bochus I, the king of Mauretania, to arrange for the capture of his son-in-law Jugurtha, the king of Numidia, thus ending the war. Sulla's success created tension that eventually erupted into a feud between him and his commander, Marius. For several years, he continued to serve under Marius. During the Social War (90–88 B.C.E.), Sulla became consul, thereby possessing the right to lead an expedition against Mithradates, King of Pontus. Marius's attempt to deprive Sulla of his command resulted in a civil war. After Marius fled and Sulla eliminated the opposition, he led the Roman army to Asia Minor. While Sulla fought the Mithradatic War, Marius returned to Rome, where he and his supporters ousted the Optimates, confiscated their property, and received appointments as consuls. Shortly thereafter, Marius died in 86 B.C.E. After Sulla sacked Athens and defeated Mithradates, he returned to Rome in 82 B.C.E. with 40,000 soldiers to learn of the fate of his supporters and immediately declared himself dictator. Under this authority, he captured and executed 8,000 of Marius's supporters, confiscating their property and distributing it among his friends. He then restructured the government to strengthen the power of the Senate while limiting the power of the tribunes.

He also allotted land for his veterans in Roman colonies. Sulla retired from public office in 79 B.C.E. and died the following year.

Cynthia Clark Northrup

See also: Marius, Gaius; Mithradatic Wars; Roman Civil Wars; Roman Republic, Wars of the

References and further reading:

Keaveney, Arthur. *Sulla, the Last Republican.* Dover, NH: Croom Helm, 1982.

Spann, Philip O. *Quintus Sertiorius and the Legacy of Sulla.* Fayetteville: University of Arkansas Press, 1987.

Sumter, Thomas (1734–1832)

American militia general and guerrilla fighter, nicknamed "the Carolina Gamecock." Born in Hanover County, Virginia, on 14 August 1734, Sumter fought at Braddock's Defeat during the French and Indian War (9 July 1755). Rising through the ranks, he became lieutenant colonel in the Second Rifle Regiment (South Carolina) in 1776. He fought the Cherokees, British, and Tories before resigning from the regular army when the British burned his home in September 1778.

After Charleston fell in 1780, Sumter organized patriot militia units for partisan action against the British in South Carolina. He was victorious at Williamson's Plantation (12 July 1780), lost at Rocky Mount (1 August 1780), won at Hanging Rock (6 August 1780), and was defeated in a surprise attack by Banastre Tarleton at Fishing Creek (18 August 1780). He was promoted to brigadier general of South Carolina militia on 6 October 1780 and was wounded at the Battle of Blackstocks (20 November 1780). An aggressive, tenacious commander, Sumter and the backcountry southern "partisans" did not receive their due at the time or in subsequent historical accounts. They not only threw the British and Tories off balance and eventually were a major factor in driving them out of the back country but also kept the local economies functioning by protecting the homesteads and by allowing their men to return home at harvest and planting seasons. As an inducement, Sumter rewarded his troops with loyalist plunder, a policy known as "Sumter's Law."

Sumter eventually retired to North Carolina and became a member of Congress. He died in Stateburg, South Carolina, on 1 June 1832, at 98 the last surviving Revolutionary War general. Fort Sumter in Charleston was named in his honor.

Harold Wise

See also: American Revolution; French and Indian War; Marion, Francis

References and further reading:

Bass, Robert D. Gamecock. *The Life and Campaigns of General Thomas Sumter.* New York: Holt, Rinehart, and Winston, 1961.

Bodie, Idella. *The Fighting Gamecock.* Orangeburg, SC: Sandlapper, 2000.

Gregorie, Anne King. *Thomas Sumter.* Columbia, SC: R. L. Bryan Company, 1931.

Sundjata (c. 1215–c. 1255)

Ruler of Mali. Son of a hunchback woman, whom Mansa Maghan con Fatta of Mali married to fulfill a prophecy, Sundjata was a terrible disappointment. Unable to walk or talk as a child, he was cruelly treated by his stepmothers and their children. On his deathbed, Maghan con Fatta left Sundjata a griot, or bard, to train him for kingship but left his kingdom to an elder son, who continued to harass and demean him and his mother. Sumanguru, the conqueror and expansionist ruler of nearby Ghana, invaded Mali and, while leaving Sundjata's elder brother as a puppet, killed the rest of the royal family, sparing only Sundjata because of his apparent disabilities.

In exile, Sundjata and his mother traveled the region, befriending the skilled metal workers and enduring legendary trials, such as a confrontation with the Nine Witches of Mali, who failed to curse Sundjata because of his goodness. Eventually learning to speak and walk, Sundjata was taken in at the court of Mema, who had survived Sumanguru's invasion of his lands with a few strongholds, and treated as a prince. During a tax revolt in Mali, the people called for the return of Sundjata as their ruler, prompting him to raise an army with Mema's assistance. Uniting the tribes of Mali and the oppressed subjects of Sumanguru, he waged a five-year campaign climaxing in the Battle of Karina, which Sundjata won not only with cavalry but with the secret knowledge that Sumanguru could be defeated by the talon of a cock, which, according to legend, he attached to an arrow and used to kill Sumanguru.

Sundjata then ruled as mansa of Mali, a kingdom approximately the size of western Europe, from 1230 until his death around 1255. During this time, he established a fortified capitol at Niane, conducted cultural and hostage exchanges among the other royal families of West Africa, and ensured that the trade routes were safe from Berbers and banditry. Sundjata's successors included the famous Mansa Musa, whose pilgrimage to Mecca established Mali as one of the great civilizations of the medieval world.

Margaret Sankey

References and further reading:
Innes, Gordon, ed. and trans. *Sunjata.* London: Penguin, 1999.
Konare Be, Adam. *Sunjata: Le foundateur de l'Empire du Mali.* Abidjan, Ivory Coast: Nouvelles Editions Africanes, 1983.

Sun-tzu (Sunzi) (fl. 500 B.C.E.)

Ancient Chinese general and influential military strategist. According to Chinese writer Ssu-ma Ch'ien, Sun-tzu, sometimes called Sun Wu or Sun Tzi, was born in the state of Ch'i in what is now the Shandong region of China. Sun-tzu's vital dates are most obscure. He was a general for the Kingdom of Wu and led it in victorious battles against the rival kingdoms of Yueh and Ch'u. These accomplishments are still being debated, and there is some disagreement about whether he actually existed, but the influence of the work attributed to him cannot be denied. His 13-chapter book on military strategy, *The Art of War,* considered a classic in the field, stressed deception, mobility, field intelligence, and logistics. The work influenced Chinese military thinking for centuries. Mao Zedong used it for the basis of his handbook on guerrilla warfare. A Jesuit, Father J. J. M. Amiot, translated the work into French in 1782, and Napoleon supposedly studied its precepts extensively during his campaigns. Translated into English in 1905, *The Art of War* is required reading in many business schools as well as military academies around the world.

Harold Wise

See also: Clausewitz, Karl Maria von; Napoleon I

References and further reading:
Sun-tzu. *The Art of War.* Ed. James Clavell. New York: Dell, 1983.

Suvorov, Aleksandr Vasilyevich (1729–18 May 1800)

Recognized as the foremost military mind of his time. Suvorov was born in 1729 and enlisted in a guards regiment in 1742. During the Russo-Turkish War of 1768–1774, Suvorov captured Hirsov fort on 14 September 1773, for which he was promoted to lieutenant general. On 19 June 1774, Suvorov's army defeated the numerically superior Turkish force of Abder-Rezak Pasha at Kozludji.

During the Russo-Turkish War of 1787–1792, Suvorov repelled two Turkish seaborne assaults (September–October 1787) against the Kinburn fortress. On 22 September 1789, Suvorov ousted Yusuf Pasha's forces from their encampment on the Rymnik River. For this, Suvorov was made a count and received the honorific title of "Rymniksky." In 1790, he captured the fortress of Ismail. He was promoted to field marshal in 1794 and wrote *The Science of Victory* in 1797.

During the War of the Second Coalition (1798–1801) against Napoleon, Suvorov defeated the French at Cassano d'Adda, the Trebia River, and Novi Ligure. His troops then crossed the Alps to attempt to link up with Russian and Austrian forces in Switzerland, commanded by General Aleksandr Rimsky-Korsakov and Archduke Charles. However, while Suvorov's forces negotiated the St. Gotthard Pass, Charles was ordered back to the Rhine, and Rimsky-Korsakov was defeated.

Suvorov's 18,000 Russian and 5,000 Cossack troops now faced 80,000 French troops under Marshal André Masséna's command. Suvorov, wisely, did not accept battle and managed to conduct 16,000 troops to Lindau and over the Alps. He was promoted to generalissimo and wrote *Rules for the Conduct of Military Actions in the Mountains.*

After falling out of favor with the czar, Suvorov was stripped of his titles on 21 January 1800. He died on 18 May 1800. Suvorov is credited with winning 63 battles without suffering a single major defeat.

Neville Panthaki

See also: French Revolutionary Wars; Masséna, André, Duc de Rivoli, Prince d'Essling; Russo-Turkish Wars

References and further reading:
Thomson, Gladys Scott. *Catherine the Great and the Expansion of Russia.* London: English Universities Press, 1963.
Troyat, Henri. *Catherine the Great.* New York: Meridian, 1994.

Swinton, Sir Ernest Dunlop (1868–1951)

British army officer and father of tank warfare. Born in Bangalore, Mysore, India, on 21 October 1868, Ernest Swinton was educated in England and attended the Royal Military Academy in Woolwich. Commissioned in the Royal Engineers (1888), he spent five years in Indian service before becoming an assistant instructor in fortification at the school of military engineering in Chatham (1896). Following participation in the Boer War as an engineer, he published two books on tactics and future warfare.

A year after being promoted to major in 1906, Swinton became chief instructor in fortification at Woolwich. Posted to the historical section of the Committee of Imperial Defence in 1910, he became its assistant secretary in 1913.

An official war correspondent in France at the beginning of World War I, Swinton believed the deadlock of trench warfare might be broken by an armored vehicle that could roll over the trenches and in October 1914 suggested to the

Committee of Imperial Defence that Holt caterpillar tractors be converted into fighting machines. First Lord of the Admiralty Winston Churchill created a Landships Committee, and a prototype vehicle was produced. An official demonstration of the tank in February 1916 went well, although skeptics thought that it would easily be destroyed by artillery fire. Swinton coined the generic word *tank* as camouflage when the vehicles were shipped to France.

Appointed commander of the new unit, Swinton protested its premature use but was overruled when British commander General Sir Douglas Haig sought to break the deadlock in the Battle of the Somme. The tanks were first employed (15 September 1916) without great success. Swinton, meanwhile, had been "released" to his former duties in the War Cabinet secretariat and without direct association with the force he had fathered. After U.S. entry in World War I, Swinton was promoted to temporary major general and traveled to the United States to speak on behalf of war bonds. In 1934, he became colonel-commandant of the Royal Tank Corps. Swinton died at Oxford on 15 January 1951.

Spencer C. Tucker

See also: Armored Fighting Vehicles; Churchill, Sir Winston; Haig, Douglas; The Somme

References and further reading:

Liddell Hart, Basil H. *The Tanks.* Vol. 1. New York: Praeger, 1959.

Swinton, Major General Sir Ernest D. *Eyewitness; Being Personal Reminiscences of Certain Phases of the Great War, Including the Genesis of the Tank.* London: Hodder and Stoughton, 1932.

———. *Over My Shoulder: The Autobiography of Major-General Sir Ernest D. Swinton.* London: Ronald, 1951.

Swiss Neutrality, Defense of

The history of Swiss neutrality spans almost five centuries but has evolved in the international context from a convenient stopgap measure for a troubled nation into a cornerstone of international diplomacy. From its formation in 1291, the loosely united Swiss Confederation, initially consisting of three cantons, often supplied mercenaries to princes and states. This practice was dictated in part by the poverty of the country, which depended on mountain farming to support its citizens and had limited natural resources. For centuries, mercenaries thus helped the Swiss economy, sometimes going so far as to fight each other.

The first expression of neutrality followed the Battle of Marignano in 1515, at which Francis I of France defeated a 20,000-strong Swiss mercenary army. Under the terms handed down in 1516, the Swiss Confederation was to agree not to engage in any concerted effort against France or its allies. The treaty did not ban the use of mercenaries because Francis I himself saw an immediate use for their services. Cantons also retained their own means to carry out foreign policy, but overall, the confederation showed a definite reticence to act internationally.

The next shift toward full-fledged neutrality came at the Congress of Vienna in 1815. There, Switzerland, which had endured French occupation between 1798 and 1814, was recognized as a buffer state in the middle of western Europe, and a federal agreement confirmed this state of affairs. The Swiss federal constitutions of 1848 and 1874 further codified neutrality as a foreign policy stance by subordinating Swiss cantons' foreign relations to a federal ministry.

Foreign policy actions that reflected Swiss armed neutrality were expressed first through Switzerland's backing of the International Red Cross (founded in 1863 and relying on Swiss diplomatic immunity to carry out its rescue operations) and, in 1871, through Switzerland's internment of French general Bourbaki's defeated army following Franco-Prussian hostilities. In so doing and with the backing of a militia-based army, the Swiss Confederation began to build further on its new identity as an armed neutral state. Switzerland also signed The Hague Convention of 1907 that outlined the rights and duties of neutral nations in war.

In World War I, however, Swiss neutrality faced a test of wills because of the country's multiple cultural identities. The French-speaking part of Switzerland sympathized heavily with the Allies, whereas the German-speaking region was more circumspect. Things came to a head, however, in the person of General Ulrich Wille, who had been named commander of the Swiss army at the beginning of the hostilities. Related to the German kaiser through his marriage, he openly stated his sympathies for Germany and often rejoiced privately over Allied stalemates and defeats. Although he was almost relieved of duty in 1917, he continued to serve until the end of the war. This awkward state of affairs was remedied somewhat when Switzerland became actively involved in protecting the shipping of goods to eastern Europe after World War I. Because all governments had agreed to such Swiss armed escorts, neutrality was not questioned.

In 1920, Switzerland joined the League of Nations and adopted the stance of differential neutrality, whereby it agreed to participate in economic sanctions against states that violated international agreements. Sanctions leveled against Italy in 1938, however, prompted the confederation to withdraw from such a stance and revert to full, integral neutrality.

Such a stance had both positive and negative consequences for the country during World War II. Although the Swiss army generally enforced neutrality against all belligerent incursions, ideological and cultural sympathies did affect the behavior of the military and civilian leadership. A profound hatred of communism, combined with some sympathies for the promise of stability in a German-dominated Europe after the war, prompted such orders as banning Swiss pilots from attacking German formations smaller than three aircraft during the early phase of the war. At the same time, the tightening of Swiss neutral attitudes prompted the closing of borders to refugees as of 1942. This was a peculiar interpretation of the neutral stance because parallel economic cooperation with Nazi Germany did take place. Although such economic action may have been required to help the country survive, in many instances, agreements were reached with very little pressure from Nazi authorities. As the tide began to turn in favor of the Allies, however, Swiss attitudes began to change, but the damage had been done, despite the army's proper defensive stance at the border.

Consequently, Swiss neutrality shifted anew after the war, this time to a stance of "neutrality and solidarity." The term reflected a kind of active involvement whereby Switzerland, although it did not join the newly created United Nations (UN), provided neutral observers in conflicts. For example, it acted as one of the four nations monitoring the cease-fire agreement between North and South Korea. This evolution toward a more flexible armed neutrality stance has involved the reimplementation of economic sanctions in the 1990s (against Iraq) as well as allowing planes from the UN armed agencies to transit over Switzerland on their way to the Balkans.

Today, Switzerland maintains its active neutrality stance backed by a militia army. The end of the Cold War has required a revision of the militia system, whereby all young men must undergo basic training and then serve two or three weeks per year for a decade. Although termed *militia*, the Swiss army has a very strict interpretation of the word. Corps commanders (a general is named only in times of war) and command staff, as well as certain weapons specialists (such as interception pilots), work full-time, but mid-level officers serve a few weeks a year, like enlisted men (nobody can jump straight to officer level; additional training is required of enlisted men selected during basic training). Although male citizens who serve with the troops (a civil service now exists for conscientious objectors) are given care of an assault rifle to keep at home (officers and hospital personnel receive a handgun), the weapons' usage follows a clearly defined protocol, as gun control in the country is based on strict laws (whereby even gun collectors have to register with police). The rationale for entrusting weaponry to common citizens is that it can accelerate the level of preparedness in case of a crisis. Soldiers joining their units have much of their equipment ready, which implies, according to some theoretical projections, that some 400,000 men could be ready to fight within 48 hours. The smallness of the territory and the various installations and self-destructive systems (major bridges and dams are wired to detonate in the event of enemy occupation) help back the argument of a convincing armed neutrality. In recent years, however, pacifists as well as serving citizens have questioned the need for such preparedness in the context of a post–Cold War world, stressing the need for greater international involvement (solidarity over neutrality) as well as the rising costs of a military that has no other use but the defense of a territory less than half the size of Massachusetts. However, it could be remembered that Adolf Hitler called Switzerland "a pimple on the face of Europe," but Switzerland was one of the few European countries that he did not invade; he was undoubtedly aware of the strong defenses of the country. No one can predict that there will be no future Hitlers.

Guillaume de Syon

References and further reading:

Bonjour, Edgar. *Swiss Neutrality, Its History and Meaning.* Ed. Mary Hottinger. London: G. Allen & Unwin, 1946.

McPhee, John A. *La Place de la Concorde Suisse.* New York: Farrar, Straus and Giroux, 1984.

Reiter, Dan. *Crucible of Beliefs: Learning, Alliances, and World Wars.* Ithaca, NY: Cornell University Press, 1996.

Syracuse, Siege of (415–413 B.C.E.)

Ill-fated expedition by Athenian forces that brought about a major decline in Athenian power in the Peloponnesian Wars. During a truce between the Athenian Empire and the Peloponnesian League, led by Sparta, the Athenian Assembly was persuaded by the charismatic Alcibiades to mount an expedition against the Corinthian colony of Syracuse, ostensibly to preserve the independence of Athenian allies Segesta and Leontini but in reality to seize the wealth of Sicily for operations against Sparta. The force under three generals—Alcibiades himself, Nicias, and Lamachus—comprised 134 triremes, 5,100 hoplites, and well over 30,000 combatants in all.

Just before the expedition sailed, Athenians discovered that the *hermae* (stone figures standing in front of houses and temples) had been desecrated. Alcibiades was linked to the outrage and was also accused of impiety. He was allowed to sail with the expedition but was later arrested. Escaping and defecting to Sparta, he betrayed the plan there.

The cities of Naxos and Catana on Sicily were won over to Athens, and the fleet made a foray into the Great Harbor of Syracuse. Nicias lured the Syracusans to the Athenian camp at Catana to parley, while the fleet sailed to the Great Harbor of Syracuse, landed, and fortified a strategic position. The first battle of the war was fought when the Syracusan forces returned, resulting in an Athenian victory. But Nicias failed to follow up this success, sailing away with the fleet and allowing Syracuse to prepare for more attacks.

The Athenians returned to Syracuse and began to build a wall across the strategic height of Epipolae above the city, planning to run the wall down the cliff to the harbor and cut off Syracuse's land communications, while the fleet stopped communications by sea. The Syracusans began to build a counterwall to stop the Athenian wall from reaching the harbor, but they were attacked and their wall was destroyed. Another wall they built lower down was also destroyed, but at the cost of Lamachus's life. The expedition was now under the control of the ailing Nicias.

The Syracusans, about to make terms with the Athenians, were told that Corinthian ships under a Spartan general, Gylippus, were coming to their relief. Gylippus slipped past Nicias and entered the city. There he put fresh heart into the Syracusans, and a frenzy of wall building and counterbuilding began. The Athenian wall never reached the coast. Although Nicias had fortified the strategic point of Plemmyrion, commanding the harbor entrance, the Syracusan wall and four forts obstructed the Athenians' access to Epipolae.

Nicias wrote to Athens for permission to retire or, failing that, for more troops to be sent. The assembly sent a second expedition. The day before it arrived, the Athenian fleet already at Syracuse was defeated at sea by the Syracusans. Demosthenes, leader of the second expedition, ordered a night attack to seize Epipolae that ended in confusion, with a loss of about 2,000 Athenian and allied lives.

Demosthenes realized evacuation was the only way to save the Athenian force, but Nicias was difficult to persuade. A lunar eclipse caused the superstitious Nicias to delay the expedition's departure for a month. The Syracusans discovered the Athenians' plans and engaged them in battle in the Great Harbor. Defeated at sea, the Athenians resolved to try to break through the barrier erected across the harbor mouth by the Syracusans, but their ships were driven into the center of the harbor and attacked. After a day-long struggle, the Athenians fell back on their camp.

The force set out on foot, attempting to reach the friendly city of Catana, with Nicias in front and Demosthenes at the rear. The two parts of the army became separated under the constant harrying of the Syracusans. The force under Demosthenes was trapped, and 20,000 men were killed, while 6,000 surrendered. Nicias's starved and thirsty army made its way to the Asinarus River and, while slaking their thirst, were attacked by the Syracusans, who slaughtered them as they drank. Nicias surrendered to stop the killing, and the survivors were taken prisoner.

Both generals were executed. The prisoners were sent to the stone quarries, where many died during their six months of imprisonment or were later sold into slavery. Only a handful survived to return to Athens.

Roslyn Russell

See also: Alcibiades

References and further reading:

Bury, J. B., and Russell Meiggs. *A History of Greece to the Death of Alexander the Great.* London: Macmillan, 1951.

Forde, Steven. *The Ambition to Rule: Alcibiades and the Politics of Imperialism in Thucydides.* Ithaca, NY: Cornell University Press, 1989.

Kagan, Donald. *The Peace of Nicias and the Sicilian Expedition.* Ithaca, NY: Cornell University Press, 1981.

Thucydides. *History of the Peloponnesian War.* Trans. Rex Warner. Harmondsworth, UK: Penguin Books, 1972.

Syrian-Egyptian Wars (274–168 B.C.E.)

Six wars fought between the Ptolemaic and Seleucid Empires during the third and second centuries B.C.E. Founded by Ptolemy I Soter, the Ptolemaic Empire included Egypt, Cyprus, and possessions throughout the eastern Mediterranean seaboard. Seleucus I Nicator established the Seleucid Empire, which spanned Syria to Afghanistan. Most fighting of the Syrian-Egyptian wars occurred along the empires' common frontier zone, running from Judea (roughly, modern Israel and western Jordan) to the Aegean Sea.

Little is known about the First Syrian-Egyptian War (274–271 B.C.E.). Ptolemy II Philadelphus sent troops into Judea and annexed several harbors and islands along the Mediterranean coast of Asia Minor (modern Turkey). Counterattacking Seleucid forces led by Antiochus I Soter curtailed his offensive and possibly threatened a direct invasion of Egypt. Domestic problems intervened, and Antiochus withdrew, leaving Ptolemy in control of many cities in Judea and southern Syria.

During the Second Syrian-Egyptian War (260–253 B.C.E.), Antiochus II Theos took advantage of Ptolemaic weakness, caused by their disastrous participation in an earlier Greek insurrection against Macedonia, to reoccupy the lost ports and islands of Asia Minor. Seleucid operations in Syria and Judea were intermittent and inconclusive. The marriage of Antiochus to Berenice II, daughter of Ptolemy II, temporarily secured peace.

This marital diplomacy contributed to the outbreak of the Third Syrian-Egyptian War (246–241 B.C.E.), also known as the Laodicean War for Antiochus II's first wife, Laodice. Antiochus abandoned Berenice and returned to Laodice, only to die in mysterious circumstances. Laodice assassinated Berenice and her son, claiming the throne for her own son, Seleucus II Callinicus. Ptolemy III Euergetes I marched his forces unopposed into Seleucid territory as far as Babylon, capturing much of Syria and Asia Minor and several bases in the Aegean. After Ptolemy's return to Egypt in 245 B.C.E., Seleucus consolidated power and attempted to recover his losses, meeting with little success. A treaty ratified in 241 B.C.E. formally ended the conflict.

In the Fourth Syrian-Egyptian War (219–217 B.C.E.), Antiochus III Megas returned southern Syria and Judea to his dominion after two years of hard campaigning. An emergency force assembled by Ptolemy IV Philopator fought Antiochus's army to a draw at Raphia (Rafah) in 217 B.C.E., forcing the Seleucids to evacuate and negotiate a truce that extended Ptolemaic control as far north as Damascus.

Antiochus III initiated the Fifth Syrian-Egyptian War (202–200 B.C.E.) by invading Judea again in 202 B.C.E., although a Ptolemaic counteroffensive conducted the following year wrested these gains from him. Antiochus launched another offensive in 200 B.C.E. He decisively defeated his adversaries at Panion (Banyas, Israel), bringing all of Palestine as far south as Gaza under his control and ending the war.

Hostilities erupted again in 169 B.C.E., when the regents of Ptolemy VI Philometor threatened an invasion of Judea, sparking the Sixth Syrian-Egyptian War (169–168 B.C.E.). Antiochus IV Epiphanes reacted by occupying Sinai, capturing the major Egyptian frontier garrisons, and annexing Cyprus. The Seleucid ruler opened negotiations with Ptolemy VI, but the Ptolemaic court shifted its support to Ptolemy VIII Euergetes II and called for Roman military assistance. Although encamped outside the walls of Alexandria with victory in his grasp, Antiochus was compelled to disengage by problems in Judea—the beginnings of the Maccabean revolt. Seleucid forces returned in 168 B.C.E. to besiege Alexandria but retreated when Rome threatened intervention.

Ian Janssen

See also: Maccabees, Revolt of the; Ptolemy I Soter; Raphia, Battle of

References and further reading:

Bar-Kochva, Bezalel. *The Seleucid Army.* Cambridge, UK: Cambridge University Press, 1976.

Hölbl, Günther. *A History of the Ptolemaic Empire.* London: Routledge, 2001.

Shipley, Graham. *The Greek World after Alexander.* London: Routledge, 2000.

T

Tactics

The theory and practice of using military forces in combat. Tactics includes movement of troops and supporting arms and services; their deployment and positioning prior, during, and after combat operations; and their use in combat. Thus tactics involves defense as well as attack and all operational use of military forces for low- and high-intensity warfare.

Tactics have been discussed for millennia, but from these discussions has come a distillation of the main principles of war upon which most modern military doctrines are based. All commanders—whether of small units or large formations—disregard the principles of war at their peril. The first of these principles is selection and maintenance of the aim. At the Battle of Marston Moor (1644), Oliver Cromwell put Prince Rupert's cavalry out of the battle early on and had his horse reformed soon after his attack. Goring (fighting on the royalist side), on the other flank, charged and routed Leslie's Scots. Then, however, the royalist cavalry went off to pillage; Cromwell seized the opportunity to attack the cavalry, who had totally forgotten why they were on the battlefield, and destroyed them.

John Churchill, the Duke of Marlborough, took advantage of French persistence in their aim in 1704 by merely anticipating their move. Napoleon always concentrated on defeating his enemy in the field. In 1812, however, he failed to defeat Prince Peter Bagration initially and followed his retreating forces. When Bagration and Barclay de Tolly linked up in the retreat to Moscow, suddenly Napoleon changed his aim and concentrated on Moscow, which led to his ultimate defeat. In World War II, Adolf Hitler's change of goal from Moscow to the southern flank altered the main thrust of the German army, again leading to total defeat.

Offensive action allows a commander to influence the outcome of a battle or campaign, as it confers the initiative on the attacker. In the Russo-Japanese War of 1904–1905, the Russian concept of warfare was to find somewhere to defend and then to defend it. The Russian general Aleksey Kuropatkin had the chance to defeat the Japanese ground forces piecemeal, but he failed, so ingrained was the concept of positional warfare. The same can be said of the Maginot mentality pervading French military thought in the 1930s.

Surprising one's enemy is fundamental. If surprise is achieved, relatively small forces can often inflict disproportionate damage on an enemy. Operations such as the airborne landing on the Belgian fort Eben Emael show how effective surprise can be. Again, Cyrus at Thymbra (554 B.C.E.) surprised Croesus by attacking in winter. Not expecting military operations in what was regarded as the closed season for warfare, Croesus had allowed his forces to disperse. In 1917, General Edmund Allenby surprised the Turks at Beersheba by using diversionary tactics; he appeared to be about to attack Gaza but took Beersheba on the flank.

Tactical methods can also be used to surprise the enemy. At Marathon (490 B.C.E.), Miltiades defeated Darius by thinning his center and concentrating his forces on his flanks. As the fighting increased in intensity, the thin Greek center withdrew, and Miltiades attacked on both flanks. The Persians lost over 6,000 men, whereas Greek losses were 192. Hannibal basically repeated this tactic at Cannae (216 B.C.E.), as did the patriot commander Daniel Morgan (on a much smaller scale) against the British at the Cowpens (1781), all with overwhelming success. History's great commanders frequently used surprise, a sure sign of their confidence and their mastery of the battlefield.

Gustavus II Adolphus was ahead of Napoleon in using maneuver warfare. At Breitenfeld (1631), Gustavus faced Johann Tserclaes, Graf von Tilly, who had never before suffered defeat. The whole military world was taken by surprise at Gustavus's army organization as revealed in the battle. This mobility led to Gustavus being able to counter every move made by Tilly, even though the Saxons of Gustavus's army had already been driven from the field. Gustavus made up for inferiority in numbers by fast movement, which countered ponderous moves by the imperial army.

Other surprises include the German attack through the Ardennes to Sedan in 1940. The Germans had done the same in 1870 and in 1914, and no one could believe that they would come the same way again, but this time with armor. Nevertheless, they did, and they did so again in their Ardennes offensive late in 1944. The British were surprised when the Japanese did not oblige them by attacking Singapore by sea in 1942 but, instead, substantially outnumbered, made their relatively easy way down the Malay peninsula to seize the fortress whose guns pointed uselessly out to sea. Again, in the Gulf War (1990–1991), General Norman Schwarzkopf enveloped the right flank of the Iraqi army through almost open desert, when the Iraqi expectation (and that of most "experts") was for an attack in the region of the Persian Gulf.

Naturally, security and deception play a large part in securing success in war. Both are principles of war, and used in conjunction, they can have far-reaching effects. A good example is the creation in 1944 of a phantom army group in eastern England to convince the Germans that the real thrust of the invasion of France would come in the Pas de Calais region. Even after D-Day had taken place, the Germans were loath to move troops from that area toward the invasion lodgment for fear that the Normandy operation was a feint, so effective was the deception as to Allied strength.

Needless to say, no army can defend itself everywhere. Thus, it is important to economize on effort wherever possible and to concentrate forces where they are most needed and where they will be the most effective. Napoleon used these principles to great effect in his campaign in Italy in 1796, when he defeated Jean Pierre Beaulieu and Baron General Colli piecemeal because they never managed to concentrate their forces against him. Failure to concentrate one's forces is amply shown in Helmuth Johannes Ludwig von Moltke's adaptation of the Schlieffen Plan. He weakened the flank and essentially flawed the German invasion of France in 1914.

It is important to ensure that an army is flexible—the rigidity of Tilly's maneuvers should be contrasted with those of Robert E. Lee and Thomas "Stonewall" Jackson at First Bull Run/Manassas and Second Bull Run/Manassas Junction and at Chancellorsville. Cooperation between commanders is also essential, as is central command and control. The Union was unable to bring its preponderance of resources to bear fully against the Confederacy because of the inability of the federal force commanders to cooperate, until Ulysses S. Grant was given command.

Currently, as throughout history, military forces must be flexible, mobile, and highly trained. In sum, maneuver is what warfare is really all about, and the inability to move troops around to suit changing circumstances will almost inevitably lead to defeat.

David Westwood

See also: Airborne Operations; Armies: Organization and Tactics; Infantry

References and further reading:

Clausewitz, Karl von. *On War.* New York: Penguin Books, 1968.
Eady, Major H. G. M. C. *Historical Illustrations to Field Service Regulations.* Vol. 2. London: Praed and Company, 1927.
Gaulle, Charles de. *The Army of the Future.* London: Hutchinson, 1943.
Sun-tzu. *The Art of War.* London: Wordsworth, 1993.

Taginae, Battle of (552)

Taginae, or more properly, the Battle of Busta Gallorum, heralded the end of the Gothic War in Italy. In the spring of 551, Emperor Justinian I named Narses supreme commander of the Byzantine army. Narses entered Italy from the northeast and then marched his troops south on the Via Flamina toward Rome. In late June–early July 552, the Byzantine army reached the plateau of Busta Gallorum. Meanwhile, the Ostrogoths under King Totila had left their camp at Taginae, near present-day Gualdo Tadino, and advanced north to meet the enemy.

Narses had some 25,000 men under his command, including 5,500 Lombards; 3,000 Heruli; and smaller contingents of Gepids, Huns, Persians, and Romans. Totila could field no more than 15,000 men. Not only were the Ostrogoths outnumbered, but their infantry had suffered in recent years from a lack of proper training and equipment, a serious shortcoming that would cost them dearly in the forthcoming battle.

Narses deployed 8,000 archers in a crescent-shaped formation before his infantry, with the dismounted Lombards and Heruli in the center, while 1,500 cavalry took position on the left wing to outflank the enemy's infantry. Totila formed his army in two lines, with the infantry deployed behind the cavalry. After 2,000 more horsemen had arrived, Totila launched a frontal assault of his cavalry, using the same tactics that had proved successful in many earlier battles. But the broken ground and narrow confines at Busta Gallorum were far from ideal for a massive cavalry attack. Indeed, the charge of the Ostrogothic lancers broke down completely in a torrential hail of arrows unleashed by the Byzantine archers. Narses then seized the initiative, ordering his own cavalry against the remnants of Totila's cavalry and outmatched infantry, who were quickly overwhelmed.

The Ostrogoths suffered devastating losses, with perhaps

some 6,000 men killed, among them King Totila, who had been mortally wounded at the beginning of the battle. The crushing defeat at Busta Gallorum signaled the downfall of the Ostrogothic kingdom in Italy, which was to last for barely three more months.

Jörg Böttger

See also: Gothic War; Justinian I; Narses; Ostrogoths

References and further reading:

Burns, Thomas S. *A History of the Ostrogoths.* Bloomington: Indiana University Press, 1984.

Procopius, with an English Translation by H. B. Dewing. 7 vols. London. W. Heinemann, Loeb Classical Library, 1914–1940.

Treadgold, Warren. *Byzantium and Its Army, 284–1081.* Stanford, CA: Stanford University Press, 1995

Wolfram, Herwig. *History of the Goths.* Trans. by Thomas J. Dunlap. Berkeley: University of California Press, 1988

Taiping Rebellion (1850–1864)

The bloodiest civil war in history. After failing his civil service examination in Canton in 1843 for the fourth time, Hong Xiuquan became delirious and fancied himself the son of God, brother of Jesus Christ, and designated savior of China. Soon a large group of followers drew to Hong's curious mixture of Christianity, socialism, idolatry, and revolutionary politics. Feng Yünshan, an early convert, founded the Bai Shangti Hui (God-Worshipping Society) in 1844, and Hong became its leader in 1847. It dedicated itself to overthrowing the Qing dynasty, the Manchu regime that had ruled China since 1644.

Revolt erupted in the summer of 1850 in Guangxi Province. On 11 January 1851, Hong proclaimed the Taiping Tianguo (Heavenly Kingdom of Great Peace) and renamed himself Tian Wang (heavenly king). He awarded lofty titles to his most devoted subordinates: Feng became Nan Wang (south king), Yang Xiuqing became Dong Wang (east king), Wei Changhui became Bei Wang (north king), Xiao Chaogui became Xi Wang (west king), Shi Dakai became Yi Wang (assistant king), Zhen Youzheng became Ying Wang (hero king), Hong Rengan became Gan Wang (shield king), and Li Xiucheng became Zhong Wang (loyal king).

With more than 500,000 men at arms by 1853, the Taipings were mainly victorious until 1856. One reason for these victories is that many Taipings were miners with knowledge of explosives and tunneling. They undermined thick fortress walls to neutralize Qing defensive heavy artillery.

On 25 September 1851, Tian Wang took Yongan. In early 1852, he moved north. In June, Nan Wang was killed in battle near Quanzhou. The siege of Changsha began on 12 September but stalled when Xi Wang was killed there on 17 September. Tian Wang captured Nanjing on 19 March 1853 and made it his capital. He ordered a second northern campaign to capture Beijing. It reached Tianjin and created a panic in the Forbidden City but failed due to communication problems.

The decline of the Taipings can be dated from 2 September 1856, when Dong Wang was murdered on Tian Wang's orders. Thereafter, the rebellion was marked by overconfidence, decadence, internal strife, mutual suspicion, and two serious strategic errors: not acquiring China's major seaport, Shanghai, and not threatening the imperial capital, Beijing. Tian Wang had no talent for generalship and relied on the cult of his personality and the military prowess of his subordinates. He demanded that his rank and file remain celibate while he and his cronies enjoyed luxuriant harems. But his ultimate refusal to move beyond Nanjing and expand his gains doomed his rebellion. After 1859, as Tian Wang's sanity and character deteriorated, Gan Wang as prime minister and Zhong Wang as commander in chief became the real heads of the Taiping government.

Except for the successes of Li Xubin at Wuhan in 1856 and Jiujiang in 1858, the Qing Dynasty was generally impotent to check the rise of the Taipings until, toward the end of the 1850s, three Qing loyalists, Li Hongzhang, Zeng Guofan, and Zuo Zongtang, mobilized powerful regional armies.

From 1856 until 1864, the Taipings fought defensively. Their two military heroes of this period were Zhong Wang in the east and Ying Wang in the west. Twice between May 1860 and May 1861, Zhong Wang attempted to capture Shanghai, failed, and alienated the large foreign population there, turning sympathies, which had been either positive toward the Taipings or neutral toward the imperials. The American adventurer Frederick Townsend Ward founded a crack mercenary outfit, the "Ever-Victorious Army" (EVA), for the Qing in 1860. Britain, which was just mopping up after its victory over the Qing in the Second Opium War (1856–1860), allowed Charles "Chinese" Gordon to lead and train the EVA after Ward was killed in action in 1862.

The two main military factors in ending the war were Gordon's EVA and Zeng's Hunan Braves. Ying Wang's defense of Anjing against Zeng broke in 1861. His death in 1862 left only Zhong Wang to fight. He resisted Zeng, Li, Ward, and Gordon and even made a few gains. Tian Wang died on 1 June 1864 just before Nanjing fell to the Qing on 19 July after a two-year siege. Zhong Wang was executed on 7 August. Sporadic Taiping resistance continued until 1866. The rebellion had claimed upward of 30 million dead.

Eric v. d. Luft and Sarah Luft

See also: Chinese Imperial Wars; Gordon, Charles George; Hong Xiuquan; Li Hongzhang; Religion and War; Wolseley, Garnet Joseph, Viscount; Yang Xiuqing; Zuo Zongtang

References and further reading:
Compilation Group for the "History of Modern China" Series. *The Taiping Revolution.* Beijing: Foreign Languages Press, 1976.
Li Xiucheng. *Taiping Rebel: The Deposition of Li Hsiu-ch'eng.* Trans. C. A. Curwen. New York: Cambridge University Press, 1977.
Meadows, Thomas Taylor. *The Chinese and Their Rebellions.* Shannon: Irish University Press, 1972.
Teng, Ssu-yü. *New Light on the History of the Taiping Rebellion.* New York: Russell & Russell, 1966.

Takeda, Shingen (1521–1573)

A prominent warlord (daimyo) of Japan's Sengoku period ("the Age of the Country at War"). Shingen Takeda was born Harunobu Takeda in 1521, the eldest son of Katsuyori Takeda, ruler of Kai Province in north-central Japan. Young Takeda overthrew his father in 1541 and installed himself as the provincial *shugo* (military governor). He then embarked on the conquest of neighboring Shinano Province, which was secured by 1555. However, this action brought him into direct conflict with Kenshin Uesugi (1530–1578) of Eichigo Province, another young and dynamic military figure. For nearly two decades, the two leaders clashed at the battlefield of Kawanakajima, with especially severe encounters in 1553, 1554, 1556, and 1563.

At length, neither side could gain a decisive advantage over the other, and both turned their territorial ambitions elsewhere. During this period, Takeda shaved his head, became a Buddhist priest, and assumed the more familiar name of Shingen.

At this time, Japan was seething with conflict as major families of samurai battled for control of the country. In 1568, Takeda attacked the Imagawa family and drove it from Surguga Province. However, the ever-shifting balance of power forced him to ally with the Hojo, Asakura, and Asai families to oppose the growing strength of Nobunaga Oda. In 1573, Takeda attacked the combined forces of Oda and his surrogate, Ieyasu Tokugawa, at Mikatagahara, driving them from the field. This defeat had the effect of inducing the weakened shogun, Yoshiaki Ashikaga, to denounce Oda, a feat that ultimately led to the shogunate's downfall. However, Takeda became distracted by events elsewhere and, by failing to follow up this impressive victory, allowed his enemies to consolidate.

In the spring of 1573, Takeda again advanced against Tokugawa and besieged one of his castles in Noda. Events are not clear, but he died either of disease or a gunshot wound on 13 May 1573. The Takeda clan did not outlive his demise and were eliminated as a military threat by Oda at Nagashino in 1575.

Beyond his military prowess, Takeda was also renowned for his administrative and organizational abilities. He placed Kai Province on a very high order of efficiency and was affectionately regarded by the populace. Takeda was also celebrated for his calligraphy and poetry, military guile, and capacity for great acts of both chivalry and cruelty.

John C. Fredriksen

See also: Japanese Wars of Unification; Oda, Nobunaga; Tokugawa, Ieyasu

References and further reading:
Cook, Harry. *Samurai: The Story of a Warrior Tradition.* New York: Sterling Publishing, 1993.
Miller, David. *Samurai Warriors.* New York: St. Martin's Press, 1999.
Turnbull, Stephen. *The Samurai: A Military History.* New York: Macmillan, 1977.

Talas River, Battle of (July 751)

Encounter with Arabs that checked Chinese advance. After the Han, most Chinese dynasties controlled points in Central Asia only to limited degrees because the costs of maintaining Chinese military forces so far from home were enormous. A noteworthy exception was the Tang Dynasty (618–907), which, during its first half at least, could bring enormous resources to bear on the problem.

The Tang interest in central Asia had two roots. One was the desire to control Silk Route commerce, which, by Tang times, was enormously profitable. Second, the Tang needed to control advanced positions in central Asia to forestall potential enemies, above all the powerful Turkic tribes and states of the time that sought to control central Asian commerce for their own interests and to use the profits from it against China.

Thus Tang moved to control the Tarim Basin and points beyond almost from the beginning of its power, following up the lead of the previous dynasty, the Sui. As a result of these advances, Chinese influence extended to the Pamirs by the middle of the seventh century. China continued to maintain a dominant position in the area for the next 100 years, despite incursions by Turks, Tibetans, and others.

While the Tang were building and maintaining their central Asian empire, a new power had arisen in the west. Between 637 and 652, the Arabs conquered Sassanian Iran and Khorasan and, after a pause to assimilate their conquests, began a new advance into central Asia under Qutayba. This brought them directly into the Chinese sphere of influence, and a proxy war developed during the reign of Tang emperor Xuanzong (r. 713–755), with each side attempting to control its vassals and prevent the other from making any real inroads, although open warfare was avoided.

What finally broke this stalemate was the decision of the Tang government to restore its former position in the west to forestall the Tibetans, resulting in a series of expeditions by the Tang general Gao Xianzhi (750). Established in a new position of power by his conquests, Gao then overreached himself by executing the Turkic king of Tashkent. The king's son called in Arabic troops from Sogdia. In late July 751, on the Talas River, Gao was crushed by the Arabs (although his army was not totally destroyed) after his Qarluqs auxiliaries had switched sides. Details are sketchy, even to the size of the armies (Arabic sources claim more than twice as many were killed or taken prisoner as were in the army of the Chinese general). Existing accounts are also from long after the event.

Some authorities have labeled the battle as decisive in determining that Islamic powers, not China, would control Turkistan. In fact, it was not the battle on the Talas River but the virtual collapse of the Tang dynasty during the An Lushan rebellion and its aftermath (755–763) that ended Tang influence in central Asia once and for all. The dynasty, Arabs or no, simply lacked the resources to assert itself again the way that it had under Xuanzong.

Paul D. Buell

See also: Sassanid Empire

References and further reading:

Barthold, W. *Turkestan Down to the Mongol Invasion.* 3d ed. Trans. T. Minorsky. Ed. C. E. Bosworth. London: Luzac and Company, 1968.

Beckwith, Christopher I. *The Tibetan Empire in Central Asia.* Princeton, NJ: Princeton University Press, 1987.

Grousset, Rene. *The Empire of the Steppes: A History of Central Asia.* Trans. Naomi Walford. New Brunswick, NJ: Rutgers University Press, 1970.

Tamerlane (Temürlenk, 1336–1405)

A Mongol-Turkic ruler whose empire of Central Asia stretched from India to Turkey. Temür of the Barlas tribe was born in Kesh (south of Samarkand) in 1336. Transoxania was ruled by Amir Kazgan of the Chagatai khanate (founded by a son of Genghis Khan). Temür acquired the nickname *lenk* (Persian), meaning "the lame," after being wounded by an arrow while on a raid to steal sheep. Thereafter, he could not bend his right knee or lift his right arm.

Tughluq Temür invaded Samarkand in 1361 and installed his son Ilyas Khoja as ruler. For his support, Temürlenk became chief minister. Temürlenk overthrew Khoja in 1364 and then killed his co-conspirator Amir Husayn in 1370, proclaiming himself the Chagatai heir who would resurrect the empire of Genghis. Temürlenk then fought against the Jagatai khans (eastern Turkistan) for regional supremacy until 1380.

Temürlenk captured Herat, Khorasan, and eastern Persia from 1383 to 1385. Azerbaijan, Armenia, Mesopotamia, and Georgia were occupied in 1386–1394. The khan of the Golden Horde, Tokhtamysh, invaded Azerbaijan in 1385 and Transoxania in 1388. Temür's struggle with him (1391–1395) continued until Tokhtamysh surrendered at the Kur River. Temürlenk occupied Moscow for a year. Meanwhile, Persia revolted. Temürlenk massacred 70,000 people in Esfahan in 1387 and had buildings constructed out of their skulls. On 24 September 1398, Temürlenk crossed the Indus River toward Delhi. He destroyed Sultan Mahmud Tughluq's army at Panipat on 17 December and massacred 100,000 inhabitants in Delhi.

In 1399, Temürlenk campaigned against the Mamluk sultan of Egypt and the Ottoman Sultan Bayazid I. Regaining Azerbaijan, Temürlenk destroyed Aleppo, defeated the Mamluk army, occupied Damascus, and massacred 20,000 people in Baghdad in 1401. On 20 July 1402, he destroyed Bayazid's army near Ankara. After capturing Smyrna from the Knights of Rhodes, the Byzantine co-emperor John VII negotiated peace.

While preparing an invasion of China, Temürlenk died of illness at Otrar on 19 February 1405. Prior to his death, Temürlenk divided his territory among his sons and grandsons. His youngest son, Shah Rokh, reunited the empire and established the Timurid Dynasty.

Temürlenk campaigned for plunder or to prevent the rise of rivals but made few administrative innovations or successes. His army was territorially grouped into *tuman* of 10,000 on the basis of Mongol organization. Advance and flank guards were employed on the march. Cavalry was held in reserve and employed as shock troops after infantry engagement.

Neville G. Panthaki

See also: Genghis Khan; Panipat, Battles of

References and further reading:

González de Clavijo, Ruy. *Embassy to Tamerlane, 1403–1406.* New York: Harper, 1928.

Ibn Khaldun. *Ibn Khaldun and Tamerlane: Their Historic Meeting in Damascus, 1401 C.E. (803 A.H.): A Study Based on Arabic Manuscripts of Ibn Khaldun's "Autobiography."* Berkeley: University of California Press, 1952.

Manz, Beatrice Forbes. *The Rise and Rule of Tamerlane.* Cambridge, UK: Cambridge University Press, 1989.

Ure, John. *The Trail of Tamerlane.* London: Constable, 1980.

Tannenberg, Battle of (15 July 1410)

Also known as the Battle of Grunwald in Poland, in which a combined Polish-Lithuanian army routed the army of the Teutonic Order. By the beginning of the fifteenth century, the conflict between the Teutonic Knights and Poland-Lithuania over territories bordering the Prussian lands escalated, until open war broke out in the fall of 1409.

In the morning hours of 15 July 1410, the opposing armies met on the fields between the villages of Grünfelde (Grunwald), Tannenberg (Stebark), and Ludwigsdorf in western Prussia. Wladyslaw II Jagiello, king of Poland, was supreme commander of the combined forces and personally led the Polish army, while his cousin, Grand Duke Vytautas (the Great), headed the Lithuanian troops. The size of both armies cannot be exactly determined. The Polish-Lithuanian army probably consisted of 25,000 to 30,000 men, mostly Polish heavy cavalry and Lithuanian light cavalry, but also smaller contingents of Czech, Moldavian, Ruthenian, and Tartar troops.

Grand Master Ulrich von Jungingen commanded the 18,000- to 20,000-strong army of the Teutonic Order, which comprised their own heavily armored knights, perhaps 100–200 knights from western Europe as "guests," and several contingents of mercenaries from the German and Polish lands. The order also fielded a few pieces of artillery.

Ulrich von Jungingen made the opening move by launching a major attack against Vytautas's troops. Initially, the Lithuanians withstood the onslaught, but when the grand master threw additional forces at them, they started to give way and retreated. The Teutonic Knights followed suit, sensing the opportunity for a knockout blow, but the Lithuanians had feigned retreat. The disruption caused in the order's ranks by this bold maneuver opened the way for a Polish-Lithuanian counterattack. Outmaneuvered and outnumbered, the order's troops faced encirclement and certain destruction. By the evening, the grand master, many of the order's political and military leaders, and several thousand soldiers had fallen. The Battle of Tannenberg effectively ended the Teutonic Order's bid for hegemony in northeastern Europe and contributed to Poland-Lithuania's rise as the dominant power there.

Jorg Boettger

See also: Polish Wars of Expansion; Teutonic Knights

References and further reading:

Juas, Meislovas. *algirio Mus ˇis.* Vilnius: Baltos Lankos, 1999.

Nadolski, Andrzej. *Grunwald 1410.* 3d ed. Warsaw: Wydawnictwo Bellona, 1999.

Urban, William L. *Tannenberg and After: Lithuania, Poland, and the Teutonic Order in Search of Immortality.* Chicago, IL: Lithuanian Research and Studies Center, 1999.

Tannenberg and the Masurian Lakes (25–30 August, 9–13 September 1914)

Opening battles on the eastern front in World War I. In accordance with prewar military planning, two Russian armies advanced into the German province of East Prussia in August 1914. The First Army, commanded by Paul Rennenkampf, advanced from the northeast, while the Second Army, commanded by Alexander Samsonov, advanced from the south. The two armies were to defeat the German Eighth Army garrisoning East Prussia, cut off the province from Germany, and open up the way for a general offensive toward Berlin. The Russian attack was poorly coordinated and ineptly led. The Russian commanders communicated in the open by radio—and were listened to by German radio intercept teams. On 20 August, Rennenkampf failed to follow up on a local defeat inflicted on the Germans at Gumbinnen. For his part, the German commander, Max von Prittwitz, panicked and was replaced on 23 August by Paul von Hindenburg, with Erich Ludendorff serving as his chief of staff. Upon arrival at Eighth Army Headquarters, Hindenburg and Ludendorff began to implement a plan, already drafted by the Eighth Army's first staff officer, Max Hoffman. Leaving a light screening force of cavalry before Rennenkampf, the bulk of the Eighth Army moved south to deal with Samsonov. During the ensuing Battle of Tannenberg, the Eighth Army inflicted a crushing defeat on the Second Army, which suffered some 50,000 casualties and 90,000 captured. Realizing the extent of his defeat, Samsonov committed suicide. Having annihilated the Second Army, the Eighth Army now turned north to deal with Rennenkampf's First Army. At the Battle of the Masurian Lakes, the Russian First Army suffered 70,000 casualties and 30,000 captured. In both battles, the Germans suffered some 40,000 casualties. A brilliant tactical victory, Tannenberg resulted in the emergence of Hindenburg and Ludendorff as the most popular and ultimately the most powerful generals in Germany, with disastrous long-term consequences for the nation.

J. David Cameron

See also: Hindenburg, Paul von Beneckendorf und von; Ludendorff, Erich; World War I

References and further reading:

Hindenburg, Paul von Beneckendorf und von. *Aus meinem Leben.* Leipzig: S. Hirzel, 1920.

Ludendorff, Erich. *Meine Kriegserinnerungen, 1914–1918.* Berlin: E. S. Mittler und Sohn, 1921.

Showalter, Dennis E. *Tannenberg: Clash of Empires.* Hamden, CT: Archon Books, 1991.

Stone, Norman. *The Eastern Front, 1914–1917.* New York: Charles Scribner's Sons, 1975.

Tarawa (20–23 November 1943)

First U.S. amphibious assault against a heavily defended island in World War II. American planners in 1943 determined upon the capture of Betio, the main island of Tarawa

Marines in battle with the Japanese at Tarawa, 1943. (Library of Congress)

Atoll in the Gilbert Islands. Betio, 2.5 miles long, a maximum of 500 yards wide, and typically less than 10 feet above sea level, was home to an airfield that could support further American advances in the Pacific. It bristled with fortifications and 4,500 Japanese marines under the command of Rear Admiral Shibasaki Keiji, who boasted that a million men could not take Tarawa in a hundred years.

To prove him wrong, Major General Julian Smith's 2d Marine Division, accompanied by the protective hail of naval gunfire from Raymond Spruance's Fifth U.S. Fleet and a company of tanks, assaulted Betio on 20 November 1943. From the beginning, however, the assault went wrong. The bombardment lifted early, a strike by heavy bombers never materialized, and planners overestimated the depth of the water covering Betio's outer reef. This last mistake proved especially deadly. Although the first three waves of marines rode in amphibious tractors that could pass over the reef, follow-on waves carried in conventional landing craft that could not. These landing craft unloaded their marines on the northern, lagoon side of Betio, 800 yards from the beach, forcing them to wade half a mile through chest-deep water under intense enemy fire.

By nightfall, marines held two shallow perimeters and were desperately short of ammunition, water, and reinforcements. They attacked the next morning, however, and with the aid of a fresh battalion and several tanks, cut the island in two. Despite more desperate fighting, the island was finally declared secure following a massive—and futile—Japanese banzai charge on 23 November.

Tarawa ranks among the most bloody assaults of the Pacific War and one of the few instances in which an American amphibious attack almost failed. U.S. casualties amounted to 1,009 dead and 2,101 wounded, and virtually the entire Japanese garrison of 4,500 perished. Four marines were awarded the Medal of Honor at Tarawa, which provided abundant lessons for amphibious assault planners during the remainder of World War II.

Lance Janda

See also: U.S. Marines; World War II

References and further reading:

Alexander, Joseph H. *Utmost Savagery: The Three Days of Tarawa.* Annapolis, MD: U.S. Naval Institute Press, 1995.

Graham, Michael B. *Mantle of Heroism: Tarawa and the Struggle for the Gilberts, November 1943.* Novato, CA: Presidio Press, 1998.

Hammel, Eric M., and John E. Lane. *Bloody Tarawa.* Pacifica, CA: Pacifica Press, 1999.

Tariq ibn Ziyad (fl. 711–712)

Berber military leader who led the Muslim conquest of Iberia, creating the medieval Islamic realm of Andalus (Spain). In 694, the Umayyad caliph sent Hassan ibn al-Nu'man to Qayrawan, Tunisia, to pacify North Africa. Needing reinforcements, he began recruiting among the local Berber tribes, converting them to Islam in the process. Fiercely independent, Berber society shared much with the Arab, including horse-mounted and mountaineer warrior traditions. Hassan's replacement, Musa ibn Nusayr, used these tribes with increasing success. In 708, Musa took Tangiers and appointed a Berber Muslim chieftain, Tariq ibn Ziyad, as governor.

In April 710, apparently on his own initiative, Tariq crossed the Straits of Gibraltar with 20,000 to 30,000 troops, fixing his base camp at the mountain bearing his name (in Arabic, Jabal Tariq). His Berbers may in fact have been invited into Spain by enemies of the Visigothic king, Roderick. In July, the Visigoth army, over 50,000 strong, confronted the Berbers near Medina Sedonia. Nonetheless, Tariq drove the Visigoths from the field, perhaps because some Visigoth nobles had planned to desert the king from the start. With Roderick dead and the Visigoths demoralized, Tariq captured Roderick's capital, Toledo. In 712, Musa ibn Nusayr followed Tariq, leading his Arab army into Spain. The two joined forces and swept north to the Ebro River, taking more cities by negotiation than siege. In 714, the two were summoned to Damascus, and Tariq's military career ended in obscurity. Nonetheless, he had established an Islamic presence in Iberia that endured eight centuries.

Weston F. Cook, Jr.

See also: Musa ibn Nusayr; Muslim Conquests; Reconquest of Spain

References and further reading:

Collins, R. *The Arab Conquest of Spain.* Oxford, UK: Oxford University Press, 1989.

Taha, Abdulwahid. *The Muslim Conquest and Settlement of North Africa and Spain.* London: Routledge Press, 1989.

Taylor, Zachary (1784–1850)

American field commander and president. Taylor was born in Orange County, Virginia, on 24 November 1784 and grew up in Kentucky. A volunteer for the militia in 1806, he was commissioned a lieutenant in the regular army in 1808 and promoted to captain in 1810. Serving under William Henry Harrison against Tecumseh in 1811 and 1812, he was brevetted a major for his defense of Fort Harrison, Indiana, outnumbered eight to one, on 4 September 1812. He repulsed an Indian attack at Credit Island, Iowa, on 5–6 September 1814. For the next 18 years, he constructed and commanded a number of frontier forts. Serving as colonel under Henry Atkinson in the Black Hawk War, he distinguished himself at Bad Axe River on 2 August 1832. He won a brevet to brigadier general for his victory over the Seminoles at Lake Okeechobee on 25 December 1837.

As tensions between Mexico and the Republic of Texas increased, Taylor was stationed at various southern forts. Ordered in February 1846 to secure the Rio Grande Valley, he won a skirmish at Brownsville on 25 April, defeated Mariano Arista at Palo Alto on 8 May and at Resaca de la Palma on 9 May, and occupied Matamoros on 18 May. Among his subordinates on this expedition were many future American Civil War generals. He attacked Monterrey on 21 September and concluded an armistice in his favor on 24 September. He then drove deeper into Mexico, capturing Saltillo in November.

The victory of his 4,600 men over the 15,000 of Antonio López de Santa Anna at Buena Vista in February 1847 made him a war hero and a viable candidate for president. At the Whig Party Convention in June 1848, he beat Henry Clay, Daniel Webster, and another war hero, Winfield Scott, for the nomination and won the election in November. But "Old Rough and Ready" did not survive the hordes of office seekers and petitioners and the intense debate over the expansion of slavery into the Mexican Cession territories. He died unexpectedly in Washington, D.C., on 9 July 1850, and his vice president, Millard Fillmore, became president.

A brave, simple soldier who shared his men's hardships in the field, Taylor nonetheless lacked the finesse of General Winfield Scott. The latter commander was able to secure his lines of communication by his wise treatment of the Mexican population; Taylor's men, by contrast, were known by the native inhabitants as "fiends, vomited from Hell." Taylor himself complained of his troops that "there is scarcely a form of crime that has not been reported to me or committed by them" (Smith 1917–1918, 249).

Eric v. d. Luft

See also: Bragg, Braxton; Buena Vista; Grant, Ulysses Simpson; Harrison, William Henry; Mexican-American War; Monterrey; Resaca de la Palma; Scott, Winfield; Tecumseh; War of 1812

References and further reading:

Bauer, Karl Jack. *Zachary Taylor: Soldier, Planter, Statesman of the Old Southwest.* Baton Rouge: Louisiana State University Press, 1985.

Dyer, Brainerd. *Zachary Taylor.* New York: Barnes & Noble, 1967.

Hamilton, Holman. *Zachary Taylor.* Norwalk, CT: Easton, 1989.

Nichols, Edward Jay. *Zach Taylor's Little Army.* Garden City: Doubleday, 1963.

Smith, Justin. "American Rule in Mexico." *American Historical Review* 23, no. 2 (1917–1918), 249.

Tecumseh (1768–1813)

Great Shawnee Indian leader. Born in March 1768 near Old Piqua, Ohio, Tecumseh fought against white settlers during and after the Revolutionary War. During the Northwest Indian Wars (1790–1794), he fought at the Battle of Fallen Timbers and opposed any treaties made with Americans. Tecumseh believed that land belonged to all Indians in common and could not be ceded without consent of all tribes. He joined with his brother Tenskwatawa, known as the Prophet, to form an Indian confederacy and traveled to tribes as far away as Iowa and Florida to spread his message of Indian nationalism. Opposing Tecumseh was William Henry Harrison, governor of the Indiana territory, who persisted in treating with separate tribes. Tecumseh's vision of Indian unity was jeopardized when the Prophet attacked Harrison's army at Tippecanoe in 1811 and was defeated.

During the War of 1812, Tecumseh supported the British cause in hopes of gaining an independent Indian state. Tecumseh often commanded Indian and British forces during campaigns in the Northwest, earning himself an appointment as a brigadier general in the British Army. When Harrison prepared to invade Canada in September, Tecumseh urged the British to stand firm against the invasion. However, the British retreated until they found themselves committed to battle near the Thames River on 5 October 1813. During the battle, Tecumseh was killed, and his death signaled the end of organized Indian resistance to American expansion to the Mississippi River. Tecumseh's military skill and determination to rise above tribal allegiance and to unite his people against American invaders established his legacy as one of the great American Indian leaders. The fact that his body was never found added a mythic quality to his memory. Even the Americans honored him, and during the

Portrait of Tecumseh. (Library of Congress)

Civil War, the Union monitor *Tecumseh* commemorated his name.

Steven J. Rauch

See also: Fallen Timbers; Thames; Tippecanoe, Battle of

References and further reading:

Edmunds, R. David. *Tecumseh and the Quest for Indian Leadership.* Boston: Little, Brown, 1984.

Sugden, John. *Tecumseh: A Life.* New York: Henry Holt, 1997.

Tel-el-Kebir (13 September 1882)

Decisive battle that began 70 years of British domination of Egypt. In 1881, the Egyptian army under the nationalist leader Arabi Pasha revolted against the Ottoman viceroy, or khedive, of Egypt, Mohammed Tewfik Pasha, who was actually a British puppet. Early in 1882, Arabi compelled Tewfik to make him minister of war. By March, Arabi was exercising dictatorial powers. His followers massacred Christians in Alexandria and demanded the expulsion of all foreigners from Egypt. In August, Britain sent Garnet Joseph Wolseley with 25,000 soldiers to reassert British presence and restore Tewfik's authority.

Wolseley's four-and-a-half-week campaign, precisely coordinated with naval operations, was nothing short of brilliant. The Royal Navy bombarded Alexandria and landed some troops there, leading Arabi to believe that the main British assault on Cairo would come from that direction. Arabi solidified his position south of Alexandria, with his eastern flank guarded by a large garrison at Tel-el-Kebir, a fortified hill about 50 miles northeast of Cairo, on the railroad line from Ismailia on the Suez Canal to Zagazig. Meanwhile, the Royal Navy secured the Suez Canal and landed Wolseley's force at Ismailia. Two divisions assembled at Kassassin on 12 September and then quick-marched 7 miles west that night across the desert. They were within 200 yards of the garrison along a 4-mile front when they were detected at dawn with the harsh rising sun at their backs. Surprise was total. In less than an hour, with small arms and bayonets, the British captured the hill and routed the garrison with the loss of only 57 men. Wolseley's victory outflanked Arabi's entire army and allowed British cavalry to enter Cairo on 15 September nearly unmolested. Arabi was tried and condemned to death but spared and exiled to Ceylon.

Eric v. d. Luft

See also: Arabi Pasha; Buller, Sir Redvers Henry; Gordon, Charles George; Wolseley, Garnet Joseph, Viscount

References and further reading:

Barthorp, Michael. *War on the Nile: Britain, Egypt, and the Sudan, 1882–1898.* Poole, UK: Blandford, 1984.

Featherstone, Donald. *Tel El-Kebir 1882: Wolseley's Conquest of Egypt.* Oxford, UK: Osprey, 1993.

Harrison, Robert T. *Gladstone's Imperialism in Egypt: Techniques of Domination.* Westport, CT: Greenwood, 1995.

Schölch, Alexander. *Egypt for the Egyptians! The Socio-Political Crisis in Egypt, 1878–1882.* London: Middle East Centre, 1981.

Terauchi, Hisaichi (1879–1946)

World War II Japanese field marshal. Eldest son of Japanese general and statesman, Count Masatake Terauchi (1852–1919), Hisaichi Terauchi was born in 1879 in Yamaguchi Prefecture, Honshu. A graduate of the army's military academy (1899) and war college (1903), Terauchi enjoyed a distinguished military career that culminated in his commanding Japanese forces in the southwest Pacific during World War II.

Promoted to general in October 1935, Terauchi served as war minister (1936–1937), inspector general of military training (1937), commander of the North China Area Army (1937–1938), and military councilor (1938–1941) before his appointment in November 1941 as commander in chief of the Southern Expeditionary Army. In this capacity, he coordinated the Japanese conquest of the Philippines, Malaya, Burma, and the East Indies (1941–1942); oversaw the construction of the infamous "railway of death" linking the Thai and Burmese rail systems (1942–1943); and directed the defense of an area extending from Burma to western New Guinea (1943–1945). Terauchi's successes, which rested primarily on his willingness to allow subordinate commanders to conduct operations as they saw fit, won him a promotion to the honorary rank of field marshal in June 1943 and consideration to replace General Hideki Tojo as prime minister in July 1944. Retaining command of the Southern Army until September 1945 despite suffering a debilitating stroke that year, Terauchi became the only senior Japanese general to hold the same post throughout the disastrous Pacific War.

At war's end, Terauchi avoided trial as a war criminal thanks to Admiral Lord Louis Mountbatten, Allied supreme commander in Southeast Asia, who arranged for his settlement near Johore Bahru, Malaya. There, in June 1946, Terauchi died after suffering a second stroke.

Bruce J. DeHart

See also: Philippines, U.S. Loss of; Philippines, U.S. Retaking of; Yamashita, Tomoyuki

References and further reading:

Fuller, Richard. *Shokan: Hirohito's Samurai.* London: Arms and Armour, 1992.

Terrorism

Acts of violence intended for a wide audience in order to create an environment of fear for political reasons. The term originated during the French Revolution when the Jacobins used it in a positive sense to refer to themselves. Later, after the revolution turned the violence in upon itself, *terrorism* became a term of derision with criminal implications. The concept of terrorism has been used and abused in so many different senses as to become almost meaningless, covering almost any and not necessarily political acts of violence.

Although the term *terrorism* entered the modern lexicon in the late eighteenth century, the concept has been known throughout history. The term can be divided and subdivided into a myriad of categories, but two of the most important are "state" and "subnational" terrorism. State terrorism, or terror from above, occurs when the impetus for terror comes from the apparatus of government. State motivation for this type is usually enhanced power and control over subjects within their jurisdiction. Jacobin terror during the French Revolution, Stalin's Great Terror in the Soviet Union (1920s–1930s) and the "dirty war" of the 1970s in Argentina are examples of this type of terrorism.

Subnational terrorism, or terror from below, is terrorism originating from outside government structures aimed at altering relationships within the corridors of power. The motivations take many forms but can be roughly categorized as religious protest movements, political revolts, and social uprising. Examples from history would include the Irish Republican Army (IRA), Basque Separatists Movement (ETA), Irgun Zvai Leumi, Palestine Liberation Organization (PLO), and Baader-Meinhof Gang, to name a very few.

When denounced for their acts of violence, terrorists respond that an unjust nation's power structure engages in daily "terrorism" and can number its own victims in the hundreds of thousands among the suffering masses crushed by poverty and state-sponsored (or at least -protected) inequality, persecution, and violence. They have no recourse but to undertake a program of decentralized violence to overthrow the unjust powers or at least to bring their cause to public or world attention.

Listing terrorist groups can be very controversial. The adage that "one person's terrorist is another person's freedom fighter" is historically quite true. To the British and Arabs of Palestine, in the late 1940s, the Irgun Zvai Leumi were terrorists, but to the Jewish settlers, they were heroic freedom fighters. To Protestants in Northern Ireland, the Irish Republican Army is a vile band of criminals; to Catholics of the region, they fight the good fight against religious and economic oppression.

Others always force the tactics of the terrorist—they are the tactics of the weak. Terrorists must stay hidden, attacking at opportune moments, creating fear, and spreading terror, because by their very nature, terrorists lack the capabilities to confront the government forces in open battle.

The strategy of the terrorist is to create a climate of fear and vulnerability. As a result, the people demand protection. As the government's response becomes increasingly militarized, terrorists attempt to trap the government and security forces into overreaction. By overreacting, the government alienates the people and the terrorists can portray themselves as the defenders of the masses.

One of the earliest known examples of a terrorist movement is the *sicarii,* a highly organized religious sect active in the Zealot struggle during the first decade of the Jewish Revolts (66–135). Their favorite weapon was a short sword called a *sica* that they would hide under their cloaks. The *sicarii* preferred to attack in daylight, on holidays, or in crowded locations to increase the level of fear and vulnerability of their intended victims.

Group terrorism became more common in the Middle Ages. One such group, which has fascinated Western authorities, was the Assassins, who were active from the eleventh to the thirteenth centuries. The modern term *assassin* has it roots in the Arabic term *hashashin,* which literally means "hashish-eater." It is recorded that these terrorists used hashish and other drugs while committing their acts of violence and terror. The assassins were very successful, killing governors, caliphs, and even the crusader king of Jerusalem. Assassins only used daggers, never poison or missile weapons—murder was a sacramental act.

Except in Russia, assassinations until the twentieth century were the work of disturbed "loners," such as the murderers of the prime minister of Great Britain (1912) and of two U.S. presidents (1865, 1881). But the late nineteenth and early twentieth centuries saw assassination (as well as bombings) increasingly assumed by "political" or terrorist organizations, as in the assassinations of the king of Italy or the Austrian Archduke Ferdinand at Sarajevo in 1914. In 1934, King Peter of Yugoslavia and Prime Minister Engelbert Dollfuss of Austria were assassinated. Zionist terrorists assassinated the British high commissioner to Egypt in 1944 and later bombed the King David Hotel in Jerusalem, and Arab terrorists responded in kind or initiated their own violent acts. The carnage continued; in one year (1951), the prime ministers of Iran and Pakistan, the British high commissioner to Malaya, and the king of Jordan were all victims of assassins with a "political" agenda.

The gun was the favorite weapon of terrorists for the first half of the twentieth century, but by the 1970s, new technology created more opportunities for the weak to strike at the strong. Airplane hijacking become the primary weapon for terrorists to garner the publicity they craved while striking terror into the public. Governments whose aircraft and citizens were victims of acts of terror responded with greater security, increased study of the phenomenon of terrorism, and sometimes even military action. Four hijacked airliners spearheaded the worst terrorism attack in history when they crashed into New York City's World Trade Center and the Pentagon in Washington, D.C., on 11 September 2001. This terrorist attack inflicted on the United States the worst single-day casualty toll in its peacetime history. The fight against terrorism became, overnight, the primary concern of the U.S. government and a much higher priority for the rest of the world.

The military has played an important role in both aiding state-sponsored terrorism and combating the subnational variety. The most famous military raid to rescue hostages was the 1976 raid on Entebbe by Israel. Israeli commandos flew 2,000 miles to Uganda to rescue passengers (most of whom were Jewish) on an Air France jet hijacked by the PLO. This was certainly not an isolated incident. The German government used military force to liberate German hostages on a hijacked Lufthansa airplane that was forced to land in Mogadishu (1977), and Netherlands security forces successfully stormed a train hijacked by South Moluccans (1979).

Force is no guarantee of success, however. In response to Iranian zealots holding American embassy personnel hostage in Tehran, President Jimmy Carter ordered a large-scale rescue mission. The April 1980 rescue operation was a disaster. In the fight against terrorism, the military is but one of many tools available to governments. The military can be used to end a terrorist standoff (as in the raid on Entebbe), or it can make a situation worse (use of British troops early in the unrest of Northern Ireland). In the latter case, various degrees of reliance upon the military have led to several decades of unconventional warfare between British troops and Irish terrorist organizations.

Political terrorist groups, like the Marxist Baader-Meinhof Gang or their opposing right-wing death squads, seem to pose less of a threat in the early twenty-first century. The most obvious threats now come from violent ethnic and religious groups and possibly narco-terrorists. Even those violent groups that supposedly fight for economic justice usually have strong ethnic support, such as Mexico's rebels in Chiapas, who draw their strength from the local indigenous peoples. Complicating such matters is the propensity of some terrorist groups to engage in illicit drug manufacturing and smuggling, supposedly purely to raise money for "the cause."

With Middle East terrorist groups, not to mention militant Basque, Filipino, Irish, Kurdish, Mexican, Chechen, and East Timorese groups, demanding at least self-government, with narco-terrorists controlling a large swath of Colombian territory, and with "rogue states" like Libya and North Korea sponsoring transnational violence, the opening of the twenty-first century gives little hope for any foreseeable easing of the threat of terrorism to established societies.

Craig T. Cobane

See also: Jewish Revolts; Northern Ireland, Civil War in

References and further reading:

Antokol, Norman, and Mayer Nudell. *No One a Neutral: Political Hostage-Taking in the Modern World.* Medina, Ohio: Alpha Publications of Ohio, 1990.

Dillon, Martin. *The Dirty War: Covert Strategies and Tactics Used in Political Conflicts.* New York: Routledge, 1990.

Laqueur, Walter. *The Age of Terrorism.* 2d ed. Boston: Little, Brown, 1987.

Tet Offensive (January–March 1968)

All-out Communist offensive of the Vietnam conflict. During the fall of 1967, the U.S. command in Vietnam began sensing victory. Enemy forces seemingly were increasingly retreating into the rugged interior of the country, away from populated areas, or across the border into Cambodia and Laos. Perhaps the attrition rate had finally turned against the Communists, and the light at the end of the tunnel was near.

Then, like a thunderbolt, came the Communist Tet Offensive. Previously, there had been a lull in the fighting during the celebration of the Vietnamese Lunar New Year—Tet. This year would be different. North Vietnamese general Vo Nguyen Giap, a master of logistics, had infiltrated nearly 100,000 Vietcong and North Vietnamese main force troops into key urban and coastal areas, and on 31 January they began attacks on 36 of South Vietnam's 44 provincial capitals, 5 of 6 autonomous cities, 64 of 242 district capitals, and 50 hamlets. Clearly, Giap had violated the principle of mass and had badly divided and thinly stretched his forces; he had counted on popular uprisings to augment his troops and to sap his enemy.

Instead, the American and South Vietnamese military responded well, and the people of South Vietnam turned to them for protection. Within a few days, most of the attacks had been repulsed, the Vietcong were destroyed as an independent military force (thereafter, regular North Vietnamese army troops filled out Vietcong units), and the enemy lost more than 120,000 men from January through June 1968; indeed, for the most part, the failed Tet Offensive provided a unique and very valuable breathing space to help the South Vietnamese government with necessary nation-building efforts.

The battle did rage in a few places. Vietcong sappers managed to breach the walls of the U.S. Embassy before being killed; press reports of the aftermath of the attack discounted official reports attesting to this victory. And North Vietnamese troops seized and held the traditional imperial capital of Hue for more than a month. U.S. Marines and South Vietnamese troops had to fight bitterly to break into the old capital to destroy the North Vietnamese positions.

In contrast to the military victory it justifiably claimed, the American command seemingly lost the psychological war when press reports began doubting stories of progress in the war. President Lyndon Johnson asked for a review, eventually called for a bombing halt, and indicated he would not seek reelection.

Charles M. Dobbs

See also: Hue, Battle of; Vietnam Conflict; Vo Nguyen Giap; Westmoreland, William

References and further reading:

Braestrup, Peter. *Big Story: How the American Press and Television Reported and Interpreted the Crisis of Tet 1968 in Vietnam and Washington.* Boulder, CO: Westview Press, 1977.

Herring, George C. *America's Longest War: The United States and Vietnam, 1950–1975.* 3d ed. New York: McGraw-Hill, 1996.

Oberdorfer, Don. *Tet!* Garden City, NY: Doubleday, 1971.

Fires rage in Saigon during the Tet Offensive attacks of 1968. (National Archives)

Teutoburger Wald, Battle of (9)

Major defeat of a Roman army under Varus. By the first century, the Roman army had apparently subjugated the semi-barbaric German tribes east of the Rhine River, extending the borders of the empire as far as the river Elbe. In 9, Augustus appointed Publius Quintilius Varus as legate to the area of Germany between the Rhine and Elbe Rivers.

Varus was an able administrator, but he had little experience leading large-scale military operations. On receiving reports of a tribal uprising near the Wesser River, Varus led Legions XVII, XVIII, and XIX out of summer quarters to quell the unrest and prevent it from spreading to other tribes. His force consisted of approximately 20,000 men, including German auxiliaries and camp followers. They moved slowly through the heavily wooded terrain of the Teutoburger Wald, near the modern area of Osnabrück and Detmold. The legions were ambushed by the Cherusci, a German tribe led by Arminius (Hermann), a former auxiliary with Roman citizenship. Arminius, angered by Roman support for rival chieftain Segestes, wanted to create a confederation of tribes to oppose Roman rule. He was familiar with Roman infantry tactics and devised a strategy to ambush the Romans in a wooded area where numbers and Roman close-order discipline would be negated by the uneven, rain-soaked terrain. Arminius equipped his forces with short javelins, which were hurled at the legionnaires in hit-and-run assaults.

When the fighting began, Varus's German auxiliaries deserted to the enemy. Despite the fierceness of the German assault, the Romans maintained their discipline and were able to build a makeshift fort to fend off the enemy. In the morning, Varus ordered his troops to resume the march. They were savagely attacked and soon lost unit cohesion. The disorganized Roman cavalry was ordered to escape but was slaughtered by the Germans in the rain-soaked forest. After day-long fighting, the Romans had heavy losses but still held their position. The fighting was renewed on the third day, but the hopelessness of the situation was apparent. Varus and most of his senior officers committed suicide. Many of the legionnaires surrendered and were slaughtered by the victorious Cherusci. Few, if any, Romans escaped the massacre. German casualties are unknown. Arminius followed up his victory by capturing most of the Roman forts east of the Rhine River. Only Mainz, defended by the two remaining Roman legions, held out. Augustus dispatched his stepson and heir Tiberius to take command in Germany. Six legions were withdrawn from Raetia, Spain, and Illyricum and transferred to the German frontier. Tiberius strengthened the Rhine defenses, and further territorial loss was prevented.

Varus's defeat stopped further expansion into Germany and prevented the spreading of Roman culture into the region. Augustus was forced to accept the Rhine River as the northern border of the Roman Empire in the west. Despite occasional forays, the Rhine would remain the border with the Germans until the fifth century, when the Germanic tribes would hasten the fall of the empire in the West.

Barry P. Neville

See also: Tiberius

References and further reading:

Cassius, Dio. *The Roman Histories.* New York: Penguin Classics, 1984.

Luttwak, Edward N. *The Grand Strategy of the Roman Empire.* Baltimore: Johns Hopkins University Press, 1976.

Mitchell, Joseph B., and Sir Edward Creasy. *Twenty Decisive Battles of the World.* New York: Macmillan, 1964.

Teutonic Knights

A military religious order, more formally and properly known as the Order of Teutonic Knights of the Hospital of St. Mary. The order was formed during the Third Crusade, in which large numbers of Germans participated. During the Siege of Acre in 1190 and 1191, crusaders from Bremen and Lübeck built a hospital from the sails of their ships and organized a hospital brotherhood dedicated to the care of sick crusaders. After the fall of Acre, the order was given a house in the city and converted it into a hospital for German crusaders and pilgrims. The order had close ties to the Holy Roman Emperor and to a number of German nobles, and in 1198, the organization was transformed into a knightly brotherhood similar to the Knights Hospitallers of St. John and the Knights Templar.

From 1200 to 1230, the order received donations of property throughout Europe; initially, it focused on the Holy Land and preferred properties along roads or passes leading toward southern Europe or toward ports of embarkation used by crusaders. The order also began efforts to build an independent state, free from secular control. In 1211, King Andrew II of Hungary granted the order land and privileges in Transylvania and encouraged the knights to crusade against pagan Cumans. The order established several towns and castles near pagan areas, defeated the pagans, and then began to assert control over Christian areas bordering the conquered territory. The conquests proved sufficiently threatening that Andrew II expelled the order in 1225, the same year that it gained an exemption from secular control for its holdings in the Holy Land.

In 1230, the Polish dukes of Mazovia and Silesia sought help in fighting the pagan Prussians. The order responded, employing the same tactics it had used in Hungary, though taking care to receive an exemption from secular control for its holdings in advance. In 1237, the Teutonic Order absorbed the membership and lands of a smaller crusading organization, the Livonian Sword Brothers. This action brought the order into conflict with the pagan Lithuanians and the Orthodox Rus princes of Novgorod. Conflict with the princes of Novgorod ended in 1242, when Alexander Nevsky defeated the order in a battle on the ice of Lake Peipus. The order's conflict with the Lithuanians continued long after the Christianization of Lithuania in 1386.

In battle, the order enjoyed numerous advantages against the Prussians and Lithuanians. The order employed the crossbow, and its use of heavily armored knights proved superior to the more lightly armed and armored forces of its Baltic opponents. By the 1280s, the Order had completed the conquest of the Prussians. The grand master of the order, perhaps sensing that the opportunities for similar expansion in the Levant were limited, moved the headquarters of the order first to Venice in 1291 and then to the Castle of Marienburg in Prussia in 1309. The designation of Marienburg as the seat of the order necessitated expansion of the fortress; by the time it was completed, the castle had a triple set of walls and was one of the strongest in Europe. In the 1300s, the employment of baggage trains carrying supplies enabled the order to conduct winter raids that were sufficiently novel to be regarded by contemporary chroniclers as unusual and noteworthy.

In territory conquered by the order, the Teutonic Knights maintained control by building large farms, and cities, upon which they settled Germans, Poles, and converted Prussians. Many of the order's cities joined the Hanseatic League, a trading organization of German towns, and the order thus gained opportunities for trade and built up great wealth. The order also built castles of brick and stone in conquered territory; indeed, the Teutonic Knights introduced brick manufacture to the Baltic region.

In 1309, the order turned from fighting Prussians and Lithuanians to attack the Catholic Polish population of Pomerania, west of the Vistula River. Though the order was able to incorporate the region, the Polish kingdom was henceforth hostile to the order. In 1386, the Kingdom of Poland and the Grand Duchy of Lithuania were united through a dynastic marriage. As a result, Lithuania converted to Latin Christianity. Nevertheless, the order continued efforts to expand into Lithuanian territory along the Baltic Sea coast. These actions led to war in 1409 and 1410. The forces of the order were defeated in the Battle of Tannenberg by a combined Polish and Lithuanian army, and it was forced to abandon some claims to Lithuanian territory.

The order suffered further difficulties in the Council of Constance in 1415, when Polish and Lithuanian theologians questioned the need for a crusading order to attack either the Catholic Poles or the Catholic Lithuanians. Though the Teutonic Order was not suppressed, as the Poles and Lithuanians demanded, the rhetorical attack did preclude further crusaders from joining the order, which was forced increasingly to rely upon mercenaries.

A second war, in 1422, led to the abandonment of claims upon disputed parts of Lithuania. In 1454, a rebellion against the order broke out in both the cities and rural lands of the Teutonic Knights. The king of Poland, Casimir IV, taking advantage of the situation, attacked the order and defeated it. The result was the Treaty of Thorn, under which the order ceded the western part of its lands to the Kingdom of Poland, in the process losing Marienburg and all its large cities except Königsberg. The grand master of the order also was compelled to swear fealty to the king of Poland and hold the rest of the order's Baltic properties in Livonia under the suzerainty of Poland. Branches of the order in the Holy Roman Empire were able to ignore this treaty; the knights in Livonia accepted the Treaty of Thorn with very poor grace. Nevertheless, the grand master of the order was compelled to abide by the treaty while seeking the support of the electoral dukes of the Holy Roman Emperor and of the pope to avoid the effects of the agreement. Very little support from either the empire or the papacy was forthcoming. As a result, Albert von Hohenzollern, grand master of the order from 1511 to 1525, made an accommodation with the Kingdom of Poland designed to further his own interests, rather than those of the Teutonic Knights. He was created the first hereditary duke of Prussia by the king of Poland, resigned from the order, and became a Lutheran. His descendants became, in due course, the electors of Brandenburg and the kings of Prussia.

The order continued in its lands in the Holy Roman Empire under the patronage of the Habsburgs, contributing a regiment to the Austrian army. Although temporarily suppressed by Napoleon after 1805, the order was reconstituted by the Emperor Ferdinand after the Napoleonic Wars under the permanent grand mastership of an imperial archduke. After World War I, the properties of the order in the former Habsburg domains were confiscated by various successor governments. The order itself, now lacking its property, was reorganized by the Catholic Church as an honorary organization for priests, and no new knights were admitted. The last knight of the order died in 1970.

Joseph M. Isenberg

See also: Crusades; Polish Wars of Expansion; Tannenberg, Battle of

References and further reading:

Barber, Malcolm, ed. *The Military Orders.* 2 vols. Aldershot, UK: Variorum, 1994.

Woodhouse, F. C. *The Military Religious Orders of the Middle Ages: The Hospitallers, the Templars, the Teutonic Knights, and Others. With an Appendix of Other Orders of Knighthood: Legendary,*

Honorary, and Modern. Society for Promoting Christian Knowledge. New York: Pott, Young, and Company, 1879.

Todd, M. *The Early Germans.* Oxford, UK: Oxford University Press, 1992.

Whittaker, C. R. *Frontiers of the Roman Empire.* Baltimore: Johns Hopkins University Press, 1997.

Teutonic Tribes

Peoples inhabiting northern and central Europe since about the second century B.C.E. The variety of tribes designated as "Teutonic" or "Germanic" cannot be defined as closed groups of ethnic societies. Even the different peoples were far from homogeneous, being conglomerates of families and nomadic groups. Without written culture to survive as archaeological testimony, knowledge of the Teutonic tribes mainly stems from Roman historiography. At the end of the second century B.C.E., the tribes of the Cimbri and Teutons invaded northern Italy. From that time, Germanic groups pressed southward, often small and barely organized but sometimes dangerously persistent. Julius Caesar was the first to give a rather accurate account of the militant tribes he encountered near the Rhine River, naming them "Germani."

The most detailed report about the Teutonic tribes is *Germania,* published by the Roman writer Tacitus in 98. In his book, Tacitus praised the Germanic way of living as an example for what he considered a degenerate Roman culture. Besides his moral concerns, Tacitus's *Germania* lists dozens of different tribes, mainly inhabiting the land east of the Rhine and north of the Danube. Some of them became more prominent over the centuries, as they gradually conquered most of Europe and northern Africa. The Angles, Alemanni, Goths, Vandals, and Lombards are but some examples, but the important Franks, Saxons, and Burgundians were still unknown at Tacitus's time.

Despite their ethnic variety, the Germanic tribes shared a somewhat common organizational structure. Based on their life as farmer-warriors, important decisions were made in the assembly of all free and able-bodied men (*ding*). Military and political leadership was in the hands of either a council of noblemen or a king. If the former were the case, the council elected one of them as military leader (*dux*) in battle. The relationship between the warriors and their leaders was that of the *trustis,* in which the men owed allegiance and armed service, and the leaders guaranteed them loyalty and social security. This system later developed into a central component of medieval feudality.

Marcus Hanke

See also: Burgundians; Franks; Goths; Ostrogoths; Vandals

References and further reading:

Schutz, Herbert. *The Germanic Realms in Pre-Carolingian Central Europe, 400–750.*New York: Lang, 2000.

Tewkesbury (4 May 1471)

Yorkist victory in the Wars of the Roses. Returning to England from exile in France, Margaret of Anjou and her son Edward, the Lancastrian Prince of Wales, hoped to arrive and join in the success of their ally Richard Neville, Earl of Warwick. Instead, they landed at Weymouth on 13 April 1471, the same day that Warwick and his armies were crushed at Barnet by the Yorkist king Edward IV. Stranded because of the departure of their French ships, Margaret and her son moved to Cerne Abbey to rally their supporters, most notably Henry Beaufort, Duke of Somerset, and the earl of Devon, for a forced march to northern Wales to join Jasper Tudor and his allies in Lancashire and Cheshire. By 28 April, the Lancastrians were on their way to Gloucester.

Edward IV, understanding the need to prevent his rivals from reaching Wales, ordered his brother Richard Plantagenet, Duke of Gloucester (the future Richard III), to hold out against the Lancastrians and pursued with a newly raised army, his Barnet troops having already been released. Unable to cross the Severn River at Gloucester and without the ability to besiege the city, the exhausted Lancastrians moved to Tewkesbury to ford there, stopping for the night and fortifying their position. The next morning (4 May 1471), Edward IV attacked, using bowmen to cover a charge led by the duke of Gloucester. Met by stiff resistance from Somerset, the Yorkists retreated, Somerset advanced against them with archers, and the two armies met in close combat in the open field. Using cavalry held in reserve, Edward IV charged, scattering the Lancastrians, many of whom fled into the town or drowned in the river. Denying Tewkesbury Abbey the right to give sanctuary, Edward IV seized Somerset and Sir Gervase Clifton from the church and added them to the other captives, including John Courtney, Earl of Devon; John, Lord Wenlock; Edmund, Lord Beaufort; and Margaret of Anjou herself. Somerset and 11 others were executed on 6 May. Among the 2,000 Lancastrians killed was Edward, Prince of Wales, a death that, along with the military victory, secured the throne for Edward IV and the Yorkists for another 15 years.

Margaret Sankey

See also: Wars of the Roses

References and further reading:

Alexander, Michael V. C. *Three Crises in Early English History.* Lanham, MD: University Press of America.

Hammond, P. W. *The Battles of Barnet and Tewkesbury.* New York: St. Martin's Press, 1990.

Linnell, Bryan. *Tewkesbury Battles and Soldiers.* Tewkesbury, UK: Theoc Press, 1979.

Texas War of Independence (1835–1836)

Conflict that resulted in Texan independence from Mexico.

In the mid-1820s, American settlers had begun moving into the northern Mexican province of Texas. The new Mexican government permitted such immigration, hoping to forestall an American takeover of this area far from the central government in Mexico City. However, from the beginning there were tensions between the Mexican government and the American settlers in Texas. Two sources of contention were the mostly Protestant religion of the American settlers and their efforts, for they were mostly southerners, to bring slavery with them. Slavery was outlawed by the Mexican constitution. Mexico tried in 1830 to limit American immigration and especially the importation of slaves. Political unrest in Mexico and the growing rift between Mexicans and so-called Texicans caused the Mexican government to prepare to send an army in the late spring of 1835 to establish firm control over the province.

The so-called Texas War of Independence lasted from October 1835 through April 1836. Hostilities began at Gonzales on 2 October 1835, when the Mexican commander demanded the return of a cannon lent to the settlers for defense against American Indians; rather than fight, he decided to retreat to San Antonio de Bexar. In early December, the rebels attacked General Martin Perfecto de Cos and his soldiers in San Antonio. Cos surrendered on 10 December, and the Texans may have felt the fighting was over and the war won.

However, General Antonio López de Santa Anna was determined to crush the rebellion and reestablish Mexican control over the province. On 16 February, he crossed the Rio Grande, and a week later, his forces began a siege of the Alamo in San Antonio that ended with the death of all its defenders on 6 March. Another army under General Jose Urrea cut off a group of rebels between Goliad and Victoria on 19 March, and on 27 March, Urrea's army executed the captives in the so-called Goliad Massacre.

The final significant action occurred several weeks later along the San Jacinto River near present-day Houston. Santa Anna had moved ahead of his main army in an effort to capture the rebels' leadership. He failed, and Texas general Samuel Houston moved to take advantage of the temporary opportunity, cutting off Santa Anna's retreat and readying for a battle. After a day of skirmishing and another day of ineffectual cavalry duels, on 21 April 1836, the Texans, emboldened by shouts of "Remember the Alamo" and "Remember Goliad," charged the Mexican defenders. The Texans did not fire until the Mexicans had discharged their weapons; then, taking advantage of the Mexicans' incapacity, the Texans fired into the Mexican lines and engaged in desperate hand-to-hand fighting. Within 30 minutes the battle was over; the entire Mexican force of 1,500 was either killed, wounded, or captured. The captives included Santa Anna. The "Napoleon of the New World" had to recognize Texan independence to gain his release.

Santa Anna had lost only a small part of his army at San Jacinto, but the war then seemingly ended. The Mexican army faced severe logistical problems operating so far from its capital; the political situation at home was not settled, and it was difficult to maintain control at home with the army engaged so far away. Mexico agreed to Texan independence but also asserted that if its lost province joined the United States, that would be grounds for war. Texas independence was not assured until after the Mexican-American War and the resulting treaty of Guadalupe Hidalgo.

Charles M. Dobbs

See also: Houston, Samuel; San Jacinto

References and further reading:

Hardin, Stephen L. *Texian Iliad: A Military History of the Texas Revolution, 1835–1836.* Austin: University of Texas Press, 1996.

Nofi, Albert A. *The Alamo, and the Texas War of Independence, September 30, 1835 to April 21, 1836.* New York: Da Capo, 1994.

Williams, Alfred M. *Sam Houston and the War of Independence in Texas.* Boston: Houghton Mifflin, 1893.

Thai (Tai) Wars (c. 1300–1569)

The migration of Tai peoples into the lowland areas of mainland Southeast Asia was greatly stimulated by Mongol conquest of their original homeland, Yunnan and the adjacent areas, in the 1250s and the decline of older Southeast Asian kingdoms, especially Angkor (Cambodia) and Pagan (Burma), by the end of the thirteenth century. During the thirteenth to fourteenth centuries, Tai rulers established three major states in what is now Thailand: Sukothai (founded in 1238, becoming a major power under Ramkhamhaeng, r. 1279–1317), Lan Na (founded c. 1296 by

Mangrai), and Ayutthaya (founded in 1351 by Ramathibodi, r. 1351–1369). These states supplanted earlier polities ruled by Mons and fought major wars with each other during the fourteenth and fifteenth centuries. Another important Tai state was Lan Sang (Luang Prabang, in modern Laos), established by Fa Ngum in 1353.

By the early fifteenth century, Ayutthaya, located north of modern Bangkok on an island in the Chao Phraya River, was the most powerful Tai state, the core around which the modern nation of Siam/Thailand was built. Ayutthaya's King Intharacha (r. 1409–1424) reduced Sukothai to vassalage, and his successor Borommaracha II (r. 1424–1448) captured the temple-city of Angkor in 1431, forcing the Cambodians to abandon it. Lan Na, located at Chiang Mai (northern Thailand), preserved its independence from Ayutthaya, though Siam's King Trailok (r. 1448–1488) sought to subjugate it. But Lan Na became a pawn in Siamese-Burmese wars during the sixteenth century.

Ayutthaya was one of the richest kingdoms in Southeast Asia. Its economy was based on international commerce and the "rice basket" of the Chao Phraya River valley, making it second in wealth only to the Malay trading port of Melaka (Malacca). Overseas Chinese and other foreign traders contributed to its prosperity. The rise of a unified Burma under Kings Tabinshwehti (r. 1531–1550) and Bayinnaung (r. 1551–1581) spelled disaster, however. The latter's armies occupied Ayutthaya in 1564 and looted the rich city completely in 1569.

Donald M. Seekins

See also: Bayinnaung; Siamese (Thai)–Burmese Wars

References and further reading:

Tarling, Nicholas, ed. *The Cambridge History of Southeast Asia.* Vol. 1, *From Early Times to c. 1800.* Cambridge, UK: Cambridge University Press, 1992.

Wyatt, David K. *Thailand: A Short History.* New Haven, CT: Yale University Press, 1984.

Thames (5 October 1813)

Also known as Moraviantown, the Battle of the Thames signified the end of British influence on the Great Lakes frontier. Throughout 1813, the British and Indian allies, commanded by Major General Henry Procter and the Shawnee chief Tecumseh, frustrated efforts by Major General William Henry Harrison to regain U.S. control of the region. When Oliver Hazard Perry achieved control of Lake Erie in September, Harrison could mount an offensive to recapture Detroit and invade Canada. Procter sought to avoid engaging the Americans and withdrew through Upper Canada along the Thames River. Though Tecumseh opposed this decision, the allies agreed to try to make a stand somewhere along the route.

On 27 September, Harrison's army landed in Canada with almost 5,000 men. His mobility was greatly enhanced by mounted Kentucky riflemen who pursued Procter's army along the Thames to Moraviantown. There in a beech forest clearing, Procter deployed his regulars with a single 6-pound gun. Tecumseh's Indians held the right flank using the cover of a marsh. Harrison ordered the mounted riflemen to attack, and they quickly drove through the enemy lines, flanked them, and attacked them from the rear. The British line crumbled, but the Indians held their ground until Tecumseh was killed leading a counterattack. The whole battle lasted less than an hour but resulted in a complete rout of the British force and capture of more than 600 prisoners. The American victory at the Thames provided a rare success and signified the end of a 60-year struggle for the Northwest, territory for further American expansion.

Steven J. Rauch

See also: Harrison, William Henry; Tecumseh; War of 1812

References and further reading:

Antal, Sandor. *A Wampum Denied: Procter's War of 1812.* Ottawa, Canada: Carlton University Press, 1997.

Edmunds, R. David. *Tecumseh and the Quest for Indian Leadership.* Boston: Little, Brown, 1984.

Sudgen, John. *Tecumseh's Last Stand.* Norman: University of Oklahoma Press, 1985.

———. *Tecumseh: A Life.* New York: Henry Holt, 1997.

Thayer, Sylvanus (1785–1872)

U.S. Army officer who professionalized the curriculum at the U.S. Military Academy and thus gave the United States its first engineering school. Born on 9 June 1785 in Braintree, Massachusetts, Thayer, after attending Dartmouth College for three years, graduated from the U.S. Military Academy in February 1808 (he was only the thirty-third graduate in its history). In the four years that followed his graduation from West Point, Thayer divided his time between supervising the construction of coastal fortifications and serving as an instructor at the Military Academy. During the War of 1812, Thayer was outraged by the incompetence or outright cowardice of many of the American officers, which convinced him that it would be necessary to train officers properly in order for them to lead troops competently in combat.

Thayer's opportunity came in 1817, when the brevet major was appointed superintendent of the Military Academy. He initiated a series of reforms in the curriculum of instruction at West Point that would remain basically unchanged

Painted portrait of Sylvanus Thayer c. 1807. (Library of Congress)

for more than a century. Between 1817 and 1833, the course of study at the Military Academy was formalized, the fourth-class system was established, and a more professional sense of both instruction and military discipline was created. With the addition of both better-educated instructors and more diverse subjects of study, Thayer was able to convert West Point from a rather academically undisciplined environment into the first true engineering school in the United States. For the first half of the nineteenth century, most engineering works in the United States were in one way or another in the hands of the U.S. Army Corps of Engineers, all West Point graduates.

After leaving the Military Academy in 1833, Thayer became responsible for the design and construction of the coastal fortifications at Boston Harbor. Promoted to colonel on 3 March 1863 and brevetted brigadier general in the regular army on 31 May 1863, Thayer retired after more than 50 years of active duty on 1 June 1863. A confirmed bachelor all his life, Thayer died at the home of his niece in Braintree, Massachusetts, on 7 September 1872 and was eventually buried at the Military Academy.

Alexander M. Bielakowski

See also: U.S. Army

References and further reading:

Dupuy, R. Ernest. *Sylvanus Thayer: Father of Technology in the United States.* West Point, NY: Association of Graduates of the United States Military Academy, 1958.

Kershner, James W. *Sylvanus Thayer: A Biography.* New York: Arno Press, 1982.

Theory, Military

That body of knowledge usually published in books and journals that examines the nature of war and the art of war on an abstract level. Military theory should be distinguished from other related terms, such as military science, military thought, the science of war, and strategic studies, in that it normally refers to theoretical texts that either discuss certain general hypotheses such as the principles of war or seek to extrapolate from history in order to make predictions. It should be seen as overlapping with military history, but nonetheless distinctly different. The historical significance of military theoretical texts rests on the influence they have exerted on commanders; the reception they received on publication, either in their country of origin or abroad; the quality and originality of the ideas and hypotheses; and whether they have continued to be read after the death of the author or are representative of a specific era or school of thought.

The first known major theorist of warfare was the Chinese general Sun-tzu, whose *The Art of War* (c. 500 B.C.E.) contains maxims on such subjects as war plans, attacks, tactics, terrain, and spies. The first French translation was made in 1782 and the first English translation only in 1905, but interest in his work has been nonetheless great. Another Chinese soldier, Sun Pin, reputedly Sun-tzu's great grandson and born around 380 B.C.E., produced a follow-up work, *Military Methods,* which was first discovered in 1972 and published in English only in 1995. Although basing his tactical principles on Sun-tzu's work, Sun Pin introduced some refinements; hence the opportunity to study this text has increased the understanding of the concepts and influence of *The Art of War.*

The most influential military treatise to survive from Roman times is Flavius Vegetius Renatus *De re militari,* often referred to as *The Military Institutions of the Romans.* Although not a professional soldier, Vegetius sought to summarize the contents of previous texts and military regulations. His work played no great part in Roman military history, but it contains important insights into the art of war of ancient armies, as well as many maxims that became commonplace in later military theory, such as "he, therefore, who aspires to peace should prepare for war." The influence of Vegetius's work cannot be underestimated; it was reputedly carried by Richard the Lionhearted and Henry II during their campaigns. Translated into French and English before the advent of printing, the first published edition appeared as early as 1473 and the first English version in 1489. Not long after, Niccolò Machiavelli's *The Art of War* (1521) was published, elucidating a series of general rules for the conduct of warfare, together with a call for new military institutions. Signaling a new phase in military theory, Machiavelli was later regarded by Friedrich Engels as the first modern military writer.

My Reveries upon the Art of War, by Hermann Maurice, Comte de Saxe, is an example of a text that is significant much less for its influence than for its content. Maurice's *Reveries* were published in 1757, seven years after his death, and translated into English in the same year but so badly that the sense was completely distorted. Maurice wrote his text because he found it strange that all the sciences had rules and principles, but war apparently none. His arguments are pertinent because he was at heart a military innovator and reformer who disdained conventional wisdom: he wanted, for example, to redesign uniforms to make them suitable for wearing in the field rather than on the parade ground. His work is significant for the modernity of its insights into tactics and leadership, and his dislike of dogmatic schemes allows it to be compared favorably with Raimondo Montecuccoli's writings (written in the period 1640–1670), particularly *The Military Art,* which was a more exhaustive study of the key factors in generalship.

The Enlightenment produced other theoretical works of note: Frederick the Great's *General Principles of War* (1748), better known as *Military Instructions for the Generals;* on sieges and fortifications, the highly mathematical system of ideas developed by Sébastien le Prestre de Vauban; and the *General Essay on Tactics,* published in 1772, in which the Frenchman Comte Jacques-Antoine de Guibert attempted to establish universal principles for tactics. These works, although they touched on the issue of war as a science, were generally restricted to a how-to-fight approach. Ironically, the man who should have contributed most to the theory of generalship, Napoleon I, died without writing any work on either the art or the science of war. His *Military Maxims* were compiled from his letters and memoirs, reinforcing the view that an understanding of war could be gained by studying the great commanders. However, the experience of the Napoleonic Wars arguably produced the two single most influential military theorists: Antoine Henri, Baron de Jomini, and Karl von Clausewitz.

Jomini's writings exerted great influence during the nineteenth century, particularly on French and Russian military thought. His first book, *Treatise on Grand Tactics,* was published in 1804–1805 and his most famous work, *Summary of the Art of War,* in 1837–1838, the latter being translated into German in 1839, Spanish in 1840, and English in 1854. There has been a tendency to deprecate the significance of Jomini's work in favor of the work of Clausewitz, but the classification of types of wars and the definitions of strategy, grand tactics, and tactics in *Summary of the Art of War* continued to influence theorists well into the twentieth century.

Nonetheless, historians of military theory are in agreement that Clausewitz's magnum opus, *On War,* published posthumously by his widow in 1832, remains not only the seminal military theoretical work of the nineteenth century but the most important treatment of the theory of war and strategy ever published. Serious interest in Clausewitz's writings emerged after the conclusion of the Franco-Prussian War (1870–1871). They were popularized partly by Helmuth Karl Bernhard von Moltke ("the Elder"), who himself made important contributions to military theory, most notably in his writings on operations. In France, Ferdinand Foch sought to digest the lessons of 1870–1871 in *The Principles of War* (1903), citing Clausewitz rather than Jomini; but he was influenced more by Ardent du Picq and the "cult of the offensive"; his book illustrates the dangers of succumbing to the intellectual fashions of the day. In fact, neither Foch nor Moltke bettered the work of Jomini or Clausewitz.

The interwar period (1919–1939) was perhaps the most dynamic in the development of military theory. In the British, German, and Red armies it is possible to speak of a flowering of military theory caused by a number of factors: The complexity of military operations in World War I led to intense debate and the search for lessons for future war; the reduced military budgets in the 1920s and the rapid development of military technology in the period led to pressure within armies to study war systematically; the catastrophe of World War I, the influence of pacifism, disarmament conferences, and the League of Nations compelled theorists to examine the nature of war in more depth than ever before; major military controversies—such as the cavalry versus the tank—led to a culture of military dispute enlivened by the growing involvement of publicists in military theory, which widened the scope of theoretical debates.

In the search for a general and scientific theory of war, the two most important interwar works are J. F. C. Fuller's *The Foundations of the Science of War* (1926) and A. A. Svechin's *Strategy* (2d ed., 1927). Although very different in approach, both works are similar in some of their fundamental assumptions and goals: both are serious attempts to analyze the effects and lessons of World War I, Fuller using a highly abstract approach based on a threefold system in which war consists of mental, moral, and physical dimensions; and Svechin employing a more historical method based on a thorough knowledge of military-political literature. Both works identified three levels of war (tactical, operational, and strategic); both emphasized the importance of the political direction of war and the role of economics in strategy. Moreover, both authors were advocates of open debate on military-theoretical questions, which led to their being shunned in their own armies.

The opposition that the radical ideas of innovative thinkers provoked was partly due to the resistance of the hierarchical general staff systems (not that much earlier the products of military "reform" themselves) to an opening up of military theoretical discussions to younger officers and civilian outsiders. The most widely read civilian writer of the period was Basil Liddell Hart, who published works on tactics, strategy, and military policy, one of his most important concepts being his provocative but contentious "strategy of the indirect approach." The emergence of civilian theorists had, in fact, begun before World War I, the most noteworthy writer being the Polish banker I. S. Bloch, who produced six volumes on the changing nature of war in 1897; volume 6 appeared in English with the title *Is War Impossible?* (1899), predicting with considerable accuracy the stalemate of World War I. Nonetheless, it should not be forgotten that in the interwar period, heated theoretical discussions were conducted in military journals over the future organization, employment, and tactics of mechanized forces, the most innovative thinking taking place in the British army in the 1920s and the Red and German armies in the late 1920s and throughout the 1930s. The theoretical discussions conducted by younger officers contributed greatly to the successful development of the new weapons of mobile warfare but led to an excessive preoccupation with tactics and operations.

During the course of World War II, the involvement of civilian scientists, code breakers, inventors, economists, and even historians in the war effort of all the major participants led to a new development: the expansion of military theory and its increasing fragmentation into subdisciplines. In the period of superpower conflict (1945–1990), civilian, academic involvement dominated the developing fields of strategic studies, nuclear theory, geopolitics, and operations research. In Western Europe and the United States, traditional considerations of land warfare and conventional military theory became almost footnotes in discussions of limited nuclear war and surprise attack. However, when the American frustration in Vietnam exposed the grave limitations of many of the methods employed by civilian analysts, the U.S. Army and defense establishment embarked on a major rethinking of the foundations of military conflict. An expression of the return to traditional theory was Colonel Harry Summers's *On Strategy* (1982), which used Clausewitz to assess the Vietnam conflict. The discovery of Soviet military theory was also instrumental in this theoretical change. Although somewhat based on the ideologies of Marxism-Leninism, Soviet military theory in the twilight of the Soviet system had developed sophisticated theories on the nature and types of war, military operations (or military art), strategy, nuclear war, and the concept of the military-scientific revolution. One of the most important claims of Soviet military theory was that the principles of war are not eternal but are historically conditioned, leading to the emergence of new laws and the extinction of others. (The "laws" of Marxism-Leninism itself, of course, remain eternal.) The influence of the study of Soviet theory on U.S. military doctrine can be argued as one of the causes of success in the Gulf War of 1990–1991.

Considering its historical evolution, then, military theory can be determined as mainly, although not exclusively, ideas that are not classified and thus accessible to a reasonably large group of individuals. In this sense, it differs from military doctrine, which normally falls under security regulations and cannot necessarily be discussed publicly, and in addition, military theory is not always written by serving officers.

Alaric Searle

See also: Clausewitz, Karl Maria von; Foch, Ferdinand; Fuller, John Frederick Charles; History, Military; Jomini, Antoine Henri, Baron de; Liddell Hart, Sir Basil Henry; Machiavelli, Niccolò; Saxe, Hermann-Maurice, Comte de; Sun-tzu; Vauban, Sébastien Le Prestre de; Vegetius Renatus Flavius

References and further reading:

Clausewitz, Karl von. *On War.* Princeton, NJ: Princeton University Press, 1984.

Gat, Azar. *The Development of Military Thought: The Nineteenth Century.* Oxford, UK: Oxford University Press, 1992.

———. *Fascist and Liberal Visions of War.* Oxford, UK: Oxford University Press, 1998.

Paret, Peter, ed. *Makers of Modern Strategy.* Princeton, NJ: Princeton University Press, 1986.

Thermopylae, Battle of (480 B.C.E.)

Opening battle of the Second Persian War between the Greeks, led by the Spartan king Leonides, and the Persian army of Xerxes I. After learning that a foreign army (of more than 3 million men, according to Herodotus, though this figure is probably highly exaggerated) had crossed the Hellespont, the Greek city-states joined forces in an effort to protect the southern part of Greece. Initially, the military leaders planned on stopping the Persian advance at Tempe in Thessaly but finally agreed that the narrow pass at Thermopylae provided the best possibility for a defensive strategy. When Xerxes and his men arrived at the pass, the Greeks held them off for two days. On the third day, an informer showed the Persians a path over the mountains, providing them with an opportunity to surround the Greeks. Realizing that the entire Greek force would be annihilated unless measures were taken, Leonides ordered all the troops to retreat except for the Spartan warriors, who continued this heroic battle against superior numbers until the last man had fallen. In revenge, Xerxes ordered the decapitated body of Leonides to be hung on a cross at the site of the battle. That same day, the Greeks won a naval victory at Artemisium offsetting the losses at Thermopylae.

The Persian victory at Thermopylae opened the way for an attack on Attica. The inhabitants of Athens, relying on Thermistocles' interpretation of the Oracle of Delphi's words, fled the city for the safety of their wooden walled ships. With a major portion of the Persian fleet destroyed by a storm at Artemisium, the best defense for the Greeks remained the sea. After winning naval victories at Salamis and Mykale and a land battle at Plataea, the Greeks forced the Persians troops out of Greece.

Cynthia Clark Northrup

See also: Plataea, Battle of; Xerxes I

References and further reading:

Herodotus. *The Histories.* New translation, selections, background, and commentaries. Trans. Walter Blanco. Ed. Walter Blanco and Jennifer Tolbert Roberts. New York: Norton, 1992.

Starr, Chester G. *A History of the Ancient World.* New York: Oxford University Press, 1983.

Thirty Years' War (1618–1648)

A series of devastating political and religious conflicts in central Europe, primarily involving Germany but expanding to include such countries as the Netherlands, France, Spain, and Sweden, that had important consequences for the conduct of international relations and the internal development of Germany.

The Thirty Years' War began with a local revolt against the king of Bohemia. Coming as it did at a time when central Europe was deeply divided by issues of religion and politics, it perhaps inevitably ignited a wider conflict. Local issues almost immediately took on larger implications because the king, Matthias of Habsburg (1585–1618), was also Holy Roman Emperor of the German nation (abbreviated herein as Germany, but it included modern Austria, Belgium, the Czech Republic, Germany, Liechtenstein, Luxembourg, Monaco, and Slovenia, as well as portions of France, Croatia, Italy, and Poland). Moreover, strong religious minorities in Bohemia were united in their resistance to Matthias and his successor Ferdinand II and their re-Catholicization policies, already successful in Austria. In this resistance, they had already made common cause with the Protestant princes and cities of the emperor that likewise viewed the rising power of the most zealously Catholic emperor since Charles V (1500–1558) with concern, including the kings of Denmark and Sweden and the Dutch Republic, states whose territories overlapped the sprawling borders of the empire. Nor was the emperor without foreign enemies as well. The prince of Transylvania would be a thorn in his side throughout the war, and because the emperor's Spanish cousins held vast territories that virtually surrounded France, it was inevitable that he would be brought into the ongoing conflict between France and Spain. Finally, he was a threatening figure to factions in both Protestant Britain and Catholic Rome.

As if his enemies were not enough, the formation by the Protestant princes of an armed independent league had encouraged the Catholic princes of Germany to form their own Catholic League and army under Duke Maximilian of Bavaria. Despite their shared religion, it was far from certain whether the Habsburg emperors and their ambitious Bavarian cousin would be allies or rivals. As with many of the more bitter and protracted civil wars of history, the Thirty Years' War persisted in large part because the sheer number of participants involved made a final peace difficult to shape.

Events and the precipitate actions of radicals forced the hand of the Bohemian Diet in 1619. Forced into arms against its sovereign, the rebel army under Count Thurn enjoyed immediate success. Acting in concert with the Transylvanians, it cleared the southern crown lands and even reached the walls of Vienna in 1619. This success, however, was followed by a disastrous political error.

The 1619 death of Matthias that brought Ferdinand II to the throne gave the radicals another opening. They pushed the Diet to depose the Habsburg claimant and to replace him with the head of the Protestant Union, the zealous and ambitious young elector Palatine, Frederick V, a man with powerful family connections in Holland and Britain. Frederick was in his own right a powerful prince with a substantial army, but the short reign of the so-called winter king demonstrated that bringing him to power was an error and one that ignored the personal side of politics, for Frederick was a member of the convoluted House of Wittelsbach, just as was Maximilian of Bavaria. Where the Bavarian had previously been content to remain neutral, the rise of a royal rival for power within his much-ramified dynasty alarmed him, and he brought his league and his army to the aid of the emperor.

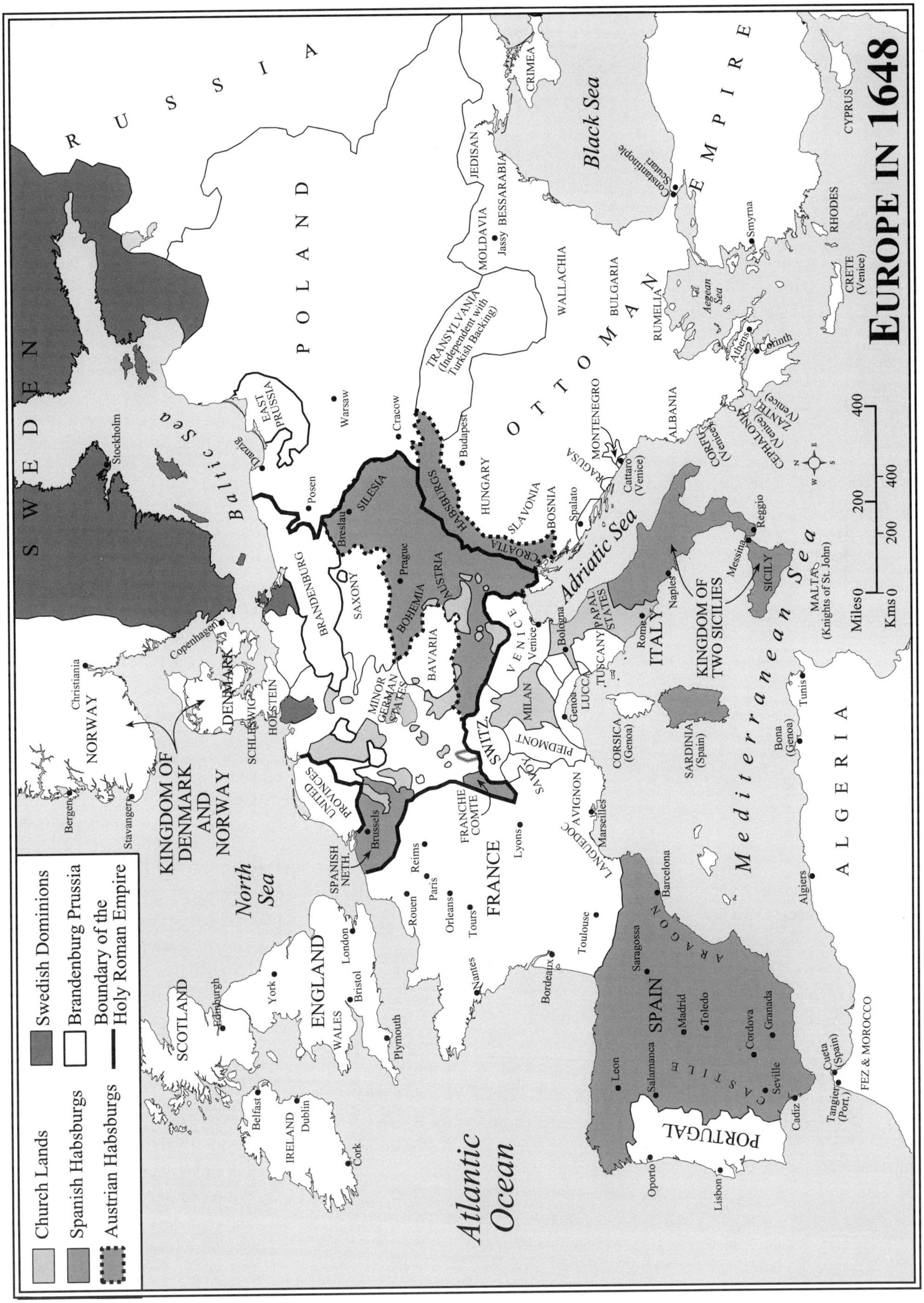
EUROPE IN 1648
Church Lands
Spanish Habsburgs
Austrian Habsburgs
Swedish Dominions
Brandenburg Prussia
Boundary of the Holy Roman Empire
Miles 0 200 400
Kms 0 200 400
RUSSIA
SWEDEN
POLAND
OTTOMAN EMPIRE
KINGDOM OF DENMARK AND NORWAY
NORWAY
DENMARK
SCHLESWIG
HOLSTEIN
SCOTLAND
ENGLAND
WALES
IRELAND
FRANCE
SPAIN
PORTUGAL
ARAGON
CASTILE
ITALY
KINGDOM OF TWO SICILIES
SICILY
SARDINIA (Spain)
CORSICA (Genoa)
MALTA (Knights of St. John)
ALGERIA
FEZ & MOROCCO
Tangier (Port.)
Cueta (Spain)
North Sea
Baltic Sea
Atlantic Ocean
Mediterranean Sea
Adriatic Sea
Black Sea
Aegean Sea
UNITED PROVINCES
SPANISH NETH.
FRANCHE COMTE
SWITZ.
SAVOY
PIEDMONT
MILAN
GENOA
LUCCA
TUSCANY
PAPAL STATES
VENICE
BAVARIA
MINOR GERMAN STATES
SAXONY
BRANDENBURG
BOHEMIA
AUSTRIA
SILESIA
HABSBURGS
HUNGARY
CROATIA
SLAVONIA
BOSNIA
RAGUSA
MONTENEGRO
ALBANIA
RUMELIA
BULGARIA
WALLACHIA
TRANSYLVANIA (Independent with Turkish Backing)
MOLDAVIA
BESSARABIA
JEDISAN
CRIMEA
EAST PRUSSIA
LANGUEDOC
AVIGNON
CORFU (Venice)
CEPHALONIA (Venice)
ZANTE (Venice)
CRETE (Venice)
RHODES
CYPRUS
Cattaro (Venice)
Bona (Genoa)
Bergen
Stavanger
Christiania
Copenhagen
Stockholm
Danzig
Posen
Warsaw
Cracow
Breslau
Prague
Budapest
Jassy
Constantinople
Scutari
Smyrna
Athens
Corinth
Spalato
Reggio
Messina
Naples
Rome
Bologna
Venice
Genoa
Marseilles
Lyons
Toulouse
Bordeaux
Nantes
Tours
Orleans
Paris
Reims
Rouen
Brussels
London
Bristol
Plymouth
York
Edinburgh
Belfast
Dublin
Cork
Barcelona
Saragossa
Madrid
Toledo
Leon
Salamanca
Cordova
Granada
Seville
Cadiz
Oporto
Lisbon
Algiers
Tunis

In November 1620, the Catholic League army under Prince Claudius Tilly de Tserclaes (1559–1632) combined with the emperor's force under another Belgian, the Count de Longueval, to win the Battle of White Mountain, smashing the Bohemian resistance and sending Frederick into flight. Among the Czech Catholic magnates who benefited from the emperor's victory and the resulting flight of the Bohemian radicals was an Albert von Wallenstein (Waldštein).

Meanwhile, Spain was abiding by the conditions of the 1609 12-year truce in the seemingly endless Dutch War of Independence, which left it equipped with an idle army and a concerned strategic imagination. When the truce ended at the beginning of the campaign season in 1622, the rebuilt Army of Flanders was to be thrown against the Dutch from its Belgian base of operations. But a strong elector Palatine, a natural Dutch ally, threatened that route, so in 1620, Spain came to the aid of the emperor by invading Frederick's westernmost territories. Under its Genoese commander, Ambrose Marchese Spinola, the Army of Flanders swept all before it, and when the Bavarians overran the Upper Palatinate the next year, while the emperor's troops cleaned various minor garrisons out of Bohemia, the war seemed virtually over.

But the general of the Protestant League, Peter Ernst, Graf von Mansfield, extracted a species of army from the wreck of Bohemia and found a way to maintain it with virtually no external support by living "off the land." In 1621–1625, Count Mansfield took the field each campaigning season with his army of the Protestant League despite defeat after defeat. In the background, the Dutch, French, and the British subsidized Mansfield.

In 1625, King Christian IV of Denmark led an army to his support. The king held substantial German territories and supposedly raised his expeditionary army in those territories, but in essence it was a foreign intervention. After early gains in 1625, the Danes found themselves arrayed against not only Tilly's leagues, but also a new imperial army. Although the emperor's own troops continued to be engaged in the far southeast, in 1626 the Count Wallenstein (or as he now was, the Duke of Friedland) raised his own imperial army, Mansfield-style. In 1626, it campaigned in Hungary, but in 1627 Wallenstein united with Tilly to overrun Denmark.

From 1627 to 1629, no Protestant army was ready to take the field in Germany. The French, Spanish, and Dutch were locked in battle, but in Italy and the Netherlands; and the interventionist king of Sweden, Gustavus II Adolphus, was engaged in Poland. Ferdinand exploited Wallenstein's army as a club to reverse the most recent Protestant gains in northern Germany, and when Wallenstein broke with Ferdinand over this strategy, he was relieved of command in 1630.

Even as Wallenstein departed, the northern Protestant princes were turning to one more savior, Gustavus II Adolphus, who blamed imperial support of the Poles for his lost Polish war. He began disembarking an army in Mecklenburg in June 1630. In 1631, Tilly returned from western Germany to oppose the Swedes and suppress internal revolt, most notably by besieging Magdeburg and sacking it in scenes of horror in May 1631. These events helped the Swedish king to cement an alliance with the Elector John George of Saxony.

Gustavus was not universally successful in the bold, wide-ranging operations that ensued as he fought to bring local rulers around Germany in line with a larger Protestant coalition. But he maneuvered the ever-cautious Tilly to attack unsuccessfully his fortified camp at Werben in July 1631. Then he defeated Tilly in the open field of Breitenfeld (17 September 1631). Tilly died standing against the Swedes on the line of the Lech (4 April 1632).

With Tilly's death, the Swedes were too powerful for even his allies, and the emperor could at last bring Wallenstein back without provoking the jealousy of the German princes, but the Duke of Friedland would only command if he were granted plenipotentiary powers. Nevertheless, when the emperor conferred those powers, he raised and funded an army in a matter of weeks and led it directly into a series of campaigns that finally forced Gustavus II Adolphus to give battle at Lützen on 6 November 1632. The Swedish king's indefatigable energy prevailed, but his death in battle ruined the victory for the Swedes and sealed Wallenstein's fate as well.

In 1633, the residual Swedish army began to receive the French subsidies that were to keep it in play until 1648, but Wallenstein used his new powers to conspire against the emperor. Fortunately for the house of Habsburg, Wallenstein's officers balked at this and murdered him in February 1634.

In the final 14 years of the war, there was little further internal German challenge to the emperor, but that was hardly necessary to compound central Europe's agony. Swedish and French armies operated deep in Germany, and the imperial and allied Spanish armies did hardly less damage as "defenders" of the Reich. Nor could the enemies of the Habsburgs accomplish much. From their sanguinary repulse from the camp of Nordlingen in September 1634 to their threatening encampment before Vienna in 1645, the Swedes stormed about Germany without accomplishing anything truly lasting, and for their part, the French unsuccessfully aimed all their efforts against Spain.

The French effort was on a truly vast scale. Two hundred thousand men were in the field in Italy, Switzerland, Lorraine, Alsace, Franche Comté, and Belgium in 1636, above and beyond the subsidized Swedes in the northeast and Rhineland. Nonetheless, new captains won their spurs against the French. Bavarians under Johann von Werth and the Army of Flanders under Prince Thomas of Savoy even brought their armies within striking range of Paris.

In 1637, the Spanish invaded France, and in 1638, France returned the favor by invading Spanish territory and adding yet another theater of misery to the long war, but it was in northwestern France and Belgium where Spain and France proved able to absorb apparently limitless damage. Although the legendary slaughter of the Spanish Army of Flanders by Louis II de Bourbon, Fourth prince de Condé, at Rocroi on the Meuse road to Paris on 19 May 1643 has been called the final defeat of the Army of Flanders, in fact it was no such thing. Frustrated, the French attempted to force the emperor and Bavarians out of the war. The heavy fighting produced a new generation of generals (Henri de la Tour d'Auvergne, Vicomte de Turenne; Prince Raimondo Montecuccoli; and Franz, Freiherr von Mercy), but at last the Bavarians and imperialists had had enough.

Yet it was not their willingness to treat that finally permitted the peace talks that had been going on each winter to succeed. Cardinal Jules Mazarin, at the head of a weak regency council, was tired of compromising with internal enemies in return for the revenues needed for the war. Immediately on the news that the Prince de Condé had defeated Spain's final invasion of the war at Lens on 5 August, Mazarin ordered the arrest of his most intransigent foes, unleashing the civil war of the Fronde even as the diplomats of Europe gathered to negotiate the Treaty of Westphalia that ended the long war.

The consequences of the Thirty Years' War were undoubtedly significant but controversial. It has been said that the war caused a decline in the population of Germany by a third or more, but there is not enough demographic evidence for this assertion. Politically, the Peace of Westphalia was supposed to inaugurate a new era in German politics, not least by forcing the Habsburgs to give up their stranglehold on the imperial office, but the "August House" was able to subvert those provisions within a decade. Certainly the war did not bring an end to religious strife in German politics. Nor did it lead to the suppression of Czech culture in favor of German in Bohemia, as Czech nationalists once claimed. Perhaps its most important consequence was that over the next two centuries, intelligent Germans such as Johann Christoph Friedrich von Schiller and Leopold von Ranke led educated Germans to the conviction that the next time Germany faced the choice of the charismatic leadership of a soldier or political chaos, it would do well to choose the former.

Erik Lund

See also: Breda, Siege of; Breitenfeld; Holy Roman Empire; Gustavus II Adolphus; Lützen, Battle of; Magdeburg, Siege of; Maurice of Nassau; Rupert, Prince; Tilly, Johann Tserclaes, Graf von; Turenne, Henri de la Tour d'Auvergne, Vicomte de; Wallenstein, Albrecht von

References and further reading:

Mann, Golo. *Wallenstein: His Life Narrated.* Trans. Charles Kessler. London: Deutsch, 1976.

Parker, Geoffrey, ed. *The Thirty Year's War.* 2d ed. London: Routledge and Kegan Paul, 1997.

Press, Volker. *Kriege und Krisen: Deutschland 1600–1715.* Munich: C. H. Beck, 1991.

Thomas, George Henry (1816–1870)

Skillful but modest Union field commander in the American Civil War. Known as "the Rock of Chickamauga," Thomas was born in Southampton County, Virginia, on 31 July 1816. Commissioned in the artillery after graduating from West Point in 1840, he was brevetted first lieutenant against the Seminoles in 1841. Under Zachary Taylor in the Mexican-American War, he fought at Resaca de la Palma and was brevetted captain at Monterrey and major at Buena Vista. He taught at West Point from 1851 to 1854, served in Texas from 1855 to 1860, and was lieutenant colonel in command of Carlisle Barracks, Pennsylvania, when the Civil War began.

Although a Virginian, he supported the federal cause. He was appointed colonel of the 2d Cavalry in May 1861, brigadier general of volunteers in August, and commander of the 1st Division of the Army of the Ohio in December. For his victory at Mill Springs, Kentucky, on 19 January 1862 and his distinguished fighting at Shiloh, he was promoted to major general of volunteers. He commanded Henry Halleck's right in the Corinth campaign. Under Don Carlos Buell at Perryville, Kentucky, on 8 October and under William S. Rosecrans at Murfreesboro, he contributed immensely to defeating Braxton Bragg.

When Bragg defeated Rosecrans at Chickamauga (1863), Thomas performed an amazing defense of Snodgrass Hill against James Longstreet and Leonidas Polk. This rearguard action earned him his nickname and, combined with Bragg's failure to pursue, allowed Rosecrans to retreat to Chattanooga. Thomas replaced Rosecrans as commander of the Army of the Cumberland. The combined forces of Thomas, Ulysses S. Grant, Joseph Hooker, and William T. Sherman routed Bragg at Chattanooga, with Thomas excelling especially at Orchard Knob and Missionary Ridge.

As Sherman's second in command in the Atlanta campaign, he repelled John Bell Hood at Peachtree Creek, Georgia, on 20 July 1864. Disobeying Grant's orders to attack, he waited until he was ready and thus emerged the decisive victor over Hood at Nashville. As a reward, he received the official thanks of Congress and a major general's commission. He commanded the Division of the Tennessee and the Cumberland until 1869 and was in command of the Division of the Pacific when he died of apoplexy in San Francisco on 28 March 1870.

Eric v. d. Luft

See also: American Civil War; Atlanta, Battles Around; Bragg, Braxton; Chattanooga, Battle of; Chickamauga, Battle of; Grant, Ulysses Simpson; Halleck, Henry Wager; Hood, John Bell; Hooker, Joseph; Longstreet, James; Mexican-American War; Monterrey; Murfreesboro; Nashville, Battle of; Resaca de la Palma; Rosecrans, William Starke; Sherman, William Tecumseh; Shiloh

References and further reading:

Cleaves, Freeman. *Rock of Chickamauga: The Life of General George H. Thomas.* Westport, CT: Greenwood Press, 1974.

Palumbo, Frank A. *George Henry Thomas, Major General, U.S.A.: The Dependable General, Supreme in Tactics of Strategy and Command.* Dayton, OH: Morningside House, 1983.

Reid, Richard J. *Death Rides the Zollie Trail.* Olaton, KY: R. J. Reid, 1996.

Thomas, Wilbur D. *General George H. Thomas: The Indomitable Warrior, Supreme in Defense and in Counterattack.* New York: Exposition Press, 1964.

Thutmose III (d. 1450 B.C.E.)

Greatest Egyptian warrior pharaoh (r. 1504–1450 B.C.E.), who led 17 successful military campaigns against Egypt's enemies without losing a battle. Son of Thutmose II and son-in-law of Queen Hatshepsut through a marriage to her daughter, his half-sister, he succeeded to the throne in 1504 B.C.E. upon his father's death. Hatshepsut assumed control until her death in 1483 B.C.E., giving Thutmose III a nominal role during her reign. Avenging her usurpation of power, he defaced all her monuments and tried to erase the memory of her. Then he initiated a policy of expansion by leading his troops against the armies of Syria and Canaan, engaging them on the Plain of Jezreel. The Syro-Canaanite forces fled to the safety of Megiddo. Thutmose III laid siege to the city for 11 months, during which time his army constructed a wooden wall and a moat around the city to prevent anyone from escaping. In 1479 B.C.E., his opponents sued for peace. After taking an oath of allegiance and paying tribute, the people lived peacefully under Egyptian rule. Following this victory, Thutmose III attacked Joppa and the kingdom of Mitanni in northern Mesopotamia, conquering territory up to the Euphrates River, where he erected a stela to mark Egypt's easternmost border. Consolidation of his control over Canaan and Syria allowed the pharaoh to turn his attention to the south, where he attacked and conquered Nubia. The success of his ventures increased the royal treasury, allowing him to donate large gifts to the temples, especially at Karnak, where his triumphs adorned the walls. He also funded building programs at Heliopolis, Abydos, Aswan, and Memphis. He was buried at Dayr al Bahri.

Cynthia Clark Northrup

See also: Megiddo, Battle of

References and further reading:

Wilkinson, Toby A. H. *Early Dynastic Egypt.* New York: Routledge, 1999.

Tiberius (42 B.C.E.–37 C.E.)

Second emperor of the Roman Empire (r. 14–37 C.E.), who succeeded Augustus, his adopted father. The son of Tiberius Claudius Nero and Livia Drusilla, Tiberius was born in Rome in 42 B.C.E. His mother divorced his father and married Octavian, who later became the Emperor Augustus. In 11 B.C.E., Augustus compelled Tiberius to divorce his first wife, Agrippina, with whom he enjoyed a happy marriage, so that he could marry Julia, the daughter of Augustus and widow of Agrippa. After the marriage, Tiberius devoted the next seven years of his life to studying in Rhodes. Upon his return, he learned that Julia had been exiled for adultery; the two grandsons of Augustus died shortly thereafter. In 4 C.E., Augustus formally adopted Tiberius, who spent the next 10 years leading campaigns against the Germans and suppressing rebellions in Pannonia and Dalmatia. In 9 C.E., Tiberius and Germanicus Caesar led Roman troops on two expeditions into Germany and returned to Rome, where Tiberius was welcomed as a victor, complete with a triumph.

After the death of Augustus in 14 C.E., Tiberius assumed control of the government without opposition. Initially, affairs of state proceeded smoothly, but during the later years of his reign, he experienced periods during which he believed senators and potential rivals were conspiring to assassinate him. He retired to Campania and then to Capreae (Capri), leaving the administration of the empire to Lucius Aelius Sejanus, the prefect of the Praetorian Guard. Tiberius learned of Sejanus's attempt to seize power and ordered that he and his supporters be executed. After reestablishing control, Tiberius returned to Capreae, where he died on 16 March 37 C.E.

Cynthia Clark Northrup

See also: Caesar, Julius

References and further reading:

Levick, Barbara. *Tiberius the Politician.* New York: Routledge, 1999.

Shotter, Davice. *Tiberius Caesar.* New York: Routledge, 1992.

Tibet, Chinese Occupation of (1949–)

The gradual reduction by force of semi-independent Tibet to the status of an "autonomous" region of the People's Republic of China. In 1949, the Communists triumphed over the Nationalist regime of Chiang Kai-shek, who fled with his government and followers to Taiwan. Until that point, Tibet had enjoyed an independent status of a sort, untouched by the great changes taking place in eastern Asia.

In October 1949, the same month of the proclamation of the People's Republic of China, the Chinese People's Liberation Army (PLA) invaded eastern Tibet and on 19 October 1950 captured Chamdo, the traditional capital of eastern Tibet. Some 40,000 soldiers of the PLA attacked Chamdo from eight directions. The 8,000 Tibetan militia and volunteers were easily defeated; local officials refused to open the arsenal to use the more modern and powerful weapons stored there.

An uneasy Sino-Tibetan relationship continued for nearly a year; the Chinese decision to intervene in Korea had given Tibet a temporary breathing space. On 23 May 1951, a Tibetan delegation that had been meeting with Chinese leaders in Beijing signed a "17-Point Agreement on Measures for the Peaceful Liberation of Tibet," in which the new Chinese government indicated its belief that Tibet comprised part of historic China and its politics and social organization would reflect the values of the "New China" that Mao Zedong and his colleagues were creating.

On 9 September 1951, 3,000 Chinese troops marched into Lhasa, and an additional 20,000 arrived in the next several weeks to complete the Chinese occupation of western Tibet. The PLA occupied such cities as Ruthok, Gartok, Gyangtse, and Shigatse. When the seizure was completed, the Chinese detached Kham and Amdo from eastern Tibet and eventually additional territory, adding it to the Chinese provinces of Qinghai, Gansu, Sichuan, and Yunnan and renaming the remaining part of Tibet the Tibet Autonomous Region.

In the years since, the Chinese occupation has been difficult for native Tibetan peoples. Millions of Chinese have emigrated to Tibet; the Dalai Lama, the spiritual and former political leader of Tibet, fled across the Indian border in 1959; and the excesses of the Cultural Revolution brought about the destruction of more than 6,000 monasteries as well as pain and suffering for the deeply religious Tibetan people.

Charles M. Dobbs

See also: Chinese Military

References and further reading:

Dawa Norbu. *Red Star over Tibet.* London: Collins, 1974.

Goldstein, Melvyn C. *A History of Modern Tibet, 1913–1951: The Demise of the Lamaist State.* Berkeley: University of California Press, 1989.

Neterowicz, Eva M. *The Tragedy of Tibet.* 1989.

Shakabpa, W. D. *Tibet: A Political History.* 1967.

Tiglath-Pileser I (r. 1115–1077 B.C.E.)

In spite of the fact that Assyria had been in decline for nearly a century by the time he came to the throne, Tiglath-Pileser I (in Akkadian, Tukulti-apil-Esharra), the first known Assyrian king to have his military exploits recounted in chronological order (as opposed to fragmentary accounts), was successful in reorganizing the military and began Assyria's military expansion into surrounding areas. He appears to have first launched a campaign into eastern Anatolia, the former domain of the now defunct Hittite Empire, and reached Lake Van, where he placed inscriptions on the rocks. He also turned westward to subdue the Phoenician states of Byblos and Sidon on the Syrian coast and evidently reached the Mediterranean Sea.

In these campaigns, he was confronted by the Aramean tribes who inhabited the regions in and around the Euphrates River in Syria, who constantly evaded the Assyrian army, raided the local communities, and destroyed agricultural produce. Though the Assyrian king was not successful in completely subduing the Arameans, he claims to have crossed the Euphrates River 28 times to campaign against them.

Tiglath-Pileser I also campaigned against Assyria's southern neighbor, Babylon. The Babylonian king, Nebuchadnezzar I (1124–1103 B.C.E.) had formerly raided Assyrian territory, and Tiglath-Pileser I riposted by penetrating deep into Babylonia and plundering a number of cities, including Babylon itself. However, he did not attempt to control the area. Although the later years of this Assyrian king are not recorded in the annals, a fragment of text describes Aramean attacks and a famine in the heartland of Assyria.

Tiglath-Pileser I was more than a military chieftain. During his reign, the Middle Assyrian Laws and Court Edicts were compiled, and a major library was constructed in Assur. He was also an indefatigable builder, constructing numerous monuments and temples, as well as public parks and gardens.

Mark W. Chavalas

See also: Assyria

References and further reading:

Grayson, A. K. *Assyrian Royal Inscriptions.* Vol. 2. Wiesbaden: Otto Harrassowitz, 1976.

———. *Assyrian Rulers of the Early First Millennium B.C.E. (1114–859 B.C.E.).* Toronto: University of Toronto Press, 1991.

Tiglath-Pileser III (r. c. 745–727 B.C.E.)

Tiglath-Pileser III (in Akkadian, Tuklulti-apil-esharra) was a neo-Assyrian king who initiated the Assyrian world state of the eighth and seventh centuries B.C.E. He doubled the size of the Assyrian army and created a rapid communication and intelligence system that was inherited by the Persian Empire. He most likely usurped the throne during a period of decline and unrest in the Assyrian state.

Tiglath-Pileser III was able to defeat the Urartians (a powerful state from the region of present-day Armenia) and claims to have invaded Urartu itself. He also moved west into the territories of Syro-Canaan and caused many of the Aramean, neo-Hittite, and Phoenician states to pay him tribute (including the cities of Damascus, Tyre, and Byblos). In 734 B.C.E., he conquered the city of Damascus, thereby ending the Aramean state centered there.

Tiglath-Pileser III invaded northern Israel at least twice, and both Israelite and Judean kings are listed in Assyrian sources as paying tribute to Assyria in this period. He was also involved in Babylonian political problems and took the Babylonian throne himself (apparently using the throne name of Pulu or Pul; cf. 2 Kings 15:19), although there was continued resistance during his reign.

Tiglath-Pileser enacted major changes in Assyria by doubling the size of the army, putting the provincial administration under the direct control of the crown, and continuing the centuries-old Assyrian policy of deporting conquered peoples to other regions of the empire.

Mark W. Chavalas

See also: Assyria

References and further reading:

Astour, Michael. *The Arena of Tiglath-Pileser III's Campaign against Sarduri II (743 B.C.E.).* Malibu, CA: Undena, 1979.

Barnett, Richard, and Margarete Falkner. *The Sculptures of Assur-nadir-pal II, Tiglath-Pileser III, Esarhaddon from the Central and Southwest Palace at Nimrud.* London: Trustees of the British Museum, 1962.

Tadmor, Hayim. *The Inscriptions of Tiglath-Pileser III, King of Assyria.* Jerusalem: Israel Academy of Sciences and Humanities, 1994.

Tigranes the Great (c. 140–c. 55 B.C.E.)

King of Armenia who participated in the Mithradatic Wars against Rome. He spent his early life as a hostage of the Parthians, who placed him on the Armenian throne in about 100 B.C.E. As he consolidated his power after his accession, Tigranes married the daughter of Mithradates VI Eupator, the king of Pontus, forming an alliance with him. Together, they seized territory in Cappadocia and Bithynia in about 91 B.C.E., directly precipitating the First and Second Mithradatic Wars. Simultaneously, Tigranes aggressively campaigned against the Parthian Empire, capturing Mesopotamia and Adiabene and adopting the Persian title "king of kings." In about 84 B.C.E., he annexed former Seleucid principalities in Syria, Phoenicia, and Cilicia. He also deported thousands of prisoners to inhabit his new capital at Tigranocerta (Silvan, Turkey).

During the Third Mithradatic War, Mithradates sought refuge with Tigranes after his defeat at Cabira (near Niksar, in present-day Turkey) in 72 B.C.E. Tigranes refused to surrender him to the Romans as they demanded, and they invaded Armenia in 69 B.C.E. Roman forces commanded by Lucius Licinius Lucullus defeated the Armenians, capturing Tigranocerta and Artaxata (near Yerevan, Armenia); however, fatigue and Lucullus's recall compelled them to halt their advance. A much larger Roman army led by Pompey finally conquered Pontus, causing Mithradates to flee yet again to Tigranes, who reversed his policy and broke the alliance; Mithradates later committed suicide. In light of this huge force and heavy Parthian counterattacks, Tigranes surrendered to Pompey in 66 B.C.E. and lost all his territorial acquisitions outside Armenia. Allowed to keep his throne, he ruled until his death in about 55 B.C.E. and was succeeded by his son Artavasdes II.

Ian Janssen

See also: Mithradatic Wars; Pompey the Great; Roman Republic, Wars of the

References and further reading:

Kallet-Marx, Robert. *Hegemony to Empire: The Development of the Roman Imperium in the East.* Berkeley: University of California Press, 1995.

Sherwin-White, Adrian. *Roman Foreign Policy in the East.* London: Duckworth, 1984.

Tilly, Johann Tserclaes, Graf von (1559–1632)

General of the army of the Catholic League (1620–1632). A Walloon, Tilly started his military career under Alessandro Farnese, Duke of Parma and Piacenza, at the time of the Dutch revolt, then fought against the Turks, and finally entered Bavarian service in 1610. In the Thirty Years' War, he led the army of the Catholic League to numerous victories and established the military and political dominance of imperial Catholic rule nearly throughout the Old Reich. He won the Battle of White Mountain (1620), crushed the army of Christian of Brunswick twice (Höchst, near Frankfurt, in 1622 and Stadtlohn, near the Dutch border, in 1623), had several encounters with Mansfield (Mingolsheim in 1622 and Wimpfen in 1622), forced the Danish king Christian IV to retreat (Lutter am Barenberge in 1626), and gained control of northern Germany. After these successful campaigns, he lost the battle of Breitenfeld against Gustavus II Adolphus (1631) and was defeated again at Rain am Lech (1632), where he was mortally wounded. He died at Ingolstadt on 30 April 1632.

Tilly developed a unique battle-seeking strategy, which proved to be successful in the 1620s. His failure during the Swedish campaign cannot be explained sufficiently by the modern tactics of the Swedish army. The disastrous outcome in 1631 was at least partly due to the political tensions within the Catholic Party, which prevented Tilly from executing his approved offensive campaigning.

Michael Kaiser

See also: Breitenfeld; Magdeburg, Siege of; Thirty Years' War; White Mountain, Battle of

References and further reading:

Kaiser, Michael. *Politik und Kriegführung. Maximilian von Bayern, Tilly und die Katholische Liga im Dreißigjährigen Krieg.* Münster: Aschendorff, 1999.

Parrott, David A. "Strategy and Tactics in the Thirty Years' War: The 'Military Revolution.'" In *The Military Revolution Debate: Readings on the Military Transformation of Early Modern Europe,* ed. Clifford A. Rogers. Boulder, CO: Westview Press, 1995.

Timoshenko, Semen Konstantinovich (1895–1970)

Soviet military commander and marshal of the Soviet Union (1940). Semen Timoshenko was born in Odessa Province of peasant origins. Conscripted at the outbreak of World War I, he served through the conflict as a machine gunner, emerging as a junior officer.

When the czarist army collapsed, Timoshenko returned to Odessa, joining the Red Guards in early 1918 and fighting against the Germans and anti-Soviet forces in the Kuban and Crimea before becoming Cavalry Brigade Commander in the Tenth Army in November, defending Tsaritsyn. Becoming part of Stalin's "Tsaritsyny" faction, from June 1919, he served as divisional commander in Semen Budennyi's 1st Cavalry Army, fighting Anton Denikin, Peter Wrangel', and Poland until 1920.

Timoshenko completed Officers' Higher Academy Courses (1922, 1927) and graduated from the Lenin Military-Political Academy (1930). During the purges, Stalin dispatched him to occupy numerous short-term command posts, replacing purged commanders.

After commanding on the Ukrainian front in occupied Poland (1939), Timoshenko led the Soviet assault on the Mannerheim Line during the Russo-Finnish War (1939–1940), achieving victory with appalling losses. Witnessing here the adverse effect of the military purges, from May 1940, as defense commissar, he attempted to strengthen Soviet border defenses against Germany, working with Georgy Zhukov, but Stalin's vacillations and preparations for the wrong war led to the disastrous Soviet performances during the initial German invasion.

Timoshenko became chairman of the Soviet military command (Stavka) (June–July 1941), commanded the Western Sector (July–September 1941), and directed the battle for Smolensk, crucially delaying the German Center Group. As Southwestern Sector commander (September 1941–May 1942), Timoshenko brilliantly counterattacked at Rostov (November 1941) but was responsible for the Kharkov encirclement (May 1942), where 240,000 Russians were taken prisoner. Never recovering from this debacle, he received only minor postings for the rest of the war, only escaping execution because of his good relations with Stalin.

After the war, Timoshenko again held minor postings, commanding Baranovich, South Ural, and Belorussia Military Districts. He was inspector general of the Defense Ministry from 1960 and chairman of the Soviet War Veterans Committee from 1961. He died in Moscow, neither the best nor the worst, but rather typical of Stalin's generals.

Neil Harvey Croll

See also: Kharkov; Russian and Soviet Armies; Russian Civil War (1918–1921); Russo-Finnish Wars (1939–1944); Stalin; World War II

References and further reading:

Kolesnikov, A. A. *Marshaly Rossii* (Marshals of Russia). Iaroslavl': Izdat, Niuans, 1999.

Shukman, Harold., ed. *Stalin's Generals.* London: Weidenfeld & Nicolson, 1993.

Spahr, W. J. *Stalin's Lieutenants. A Study of Command under Duress.* Novato, CA: Presidio Press, 1997.

Timoshenko, Semen Konstantinovich (Kratkaia biografia). Moscow: Voenizdat, 1940.

Tinian (24 July–1 August 1945)

Japanese-held Pacific island just south of Saipan, taken by the U.S. Marines in heavy fighting. The Americans needed this island as the site for the construction of the long runways that the new U.S. Army Air Forces heavy bomber, the Boeing B-29, would require for the planned strategic bombing of the Japanese home islands.

For 40 days, Japanese defenders were pounded by aerial and naval ship bombardment. Then, on 24 July, some five weeks after the invasion of Saipan, the 4th Marine Division crossed the waters dividing Saipan and Tinian, while the 2d Marine Division demonstrated off the southwest. The marines achieved tactical surprise; for the beaches were very narrow—only 65 and 130 yards wide—and the Japanese assumed that the United States could not land sufficient force and supplies on so narrow a set of beaches. Air and ship bombardment was directed elsewhere, helping to further the deception. Thus, the Japanese did not defend the beaches strongly, and the marines were able to land, move inland, and bring ashore sufficient supplies to maintain the offensive. The Japanese counterattacked on the first night, but the marines beat them back, inflicting losses of more than 1,200 men. The 2d Marine Division soon joined the 4th Division in the invasion; later, one division went north and the other south to clear the island. By 1 August, organized Japanese resistance was over. As on Saipan, Japanese civilians jumped off the cliffs on the southern part of the island to their deaths. The marines suffered 290 killed and 1,515 wounded, while the Japanese lost at least 6,500 killed and an undetermined number missing.

Soon thereafter, Navy SeaBees constructed runways making Tinian island temporarily one of the world's largest airports to facilitate the bombing of Japan. Tinian is today remembered primarily as the home base of the B-29 bombers that dropped the atom bombs on Hiroshima and Nagasaki.

Charles M. Dobbs

See also: Peleliu; Saipan, Battle of; Tarawa

References and further reading:

Harwood, Richard. *A Close Encounter: The Marine Landing on Tinian.* Washington, DC: History and Museums Division, Headquarters, U.S. Marine Corps, 1994.

Morison, Samuel Eliot. *The Two Ocean War: A Short History of the United States Navy in the Second World War.* Boston: Little, Brown, 1963.

Spector, Ronald H. *Eagle against the Sun: The American War with Japan.* New York: Free Press, 1985.

Tippecanoe, Battle of (7 November 1811)

A battle in the American war with the old Northwest Indians (1811–1813). Throughout 1811, relations between the United States, represented by Indiana Territory governor William Henry Harrison, and an Indian confederacy, led by the Shawnee brothers Tecumseh and the Prophet (Tenskwatawa), steadily deteriorated. In the autumn, Governor Harrison took advantage of Tecumseh's trip to the south to

Music cover, c. 1840, illustrated with scene of W. H. Harrison leading troops at the Battle of Tippecanoe, Indiana, in 1811. (Library of Congress)

march his troops against the Indian confederacy. On 6 November, Harrison and his troops camped just outside the Indian village of Prophetstown.

The Prophet, a religious figurehead, convinced his warriors that his medicine would make them invulnerable. With the Prophet chanting incantations on a nearby hill, the Indians, armed with British muskets and powder, attacked the American camp two hours before dawn on 7 November.

Harrison's camp stood on the high ground, but an unwise series of bonfires around its perimeter blinded the Americans. As a result, a few warriors quickly infiltrated the American lines and sought out Harrison. The governor narrowly escaped death only through mistaken identity.

Harrison acted well during the two-hour battle. Despite significant casualties, he shored up his lines and prevented the Indians from gaining an overwhelming advantage anywhere on the battlefield. By dawn, the Indian warriors realized that the Prophet's medicine would not secure victory, and they broke off the attack.

Harrison called Tippecanoe a "complete and decisive victory," but the truth is less romantic. The battle was not decisive, but it did have lasting importance. Harrison's troops destroyed the credibility of the Prophet, increased the governor's fame throughout the country, and eventually led to his short-lived presidency.

Thomas Bruscino Jr.

See also: Harrison, William Henry; Tecumseh

References and further reading:

Edmunds, R. David. *The Shawnee Prophet.* Lincoln: University of Nebraska Press, 1983.

Goebel, Dorothy Burne. *William Henry Harrison: A Political Biography.* Indianapolis: Indiana Library and Historical Department, 1926.

Hoffnagle, Warren Miles. *The Road to Fame: Papers on the War of 1812 in the Northwest, No. 6.* Columbus: Ohio State Museum, 1959.

Tito (Josef Broz) (1892–1980)

Partisan hero during World War II and Yugoslav dictator who pursued a foreign policy that helped lessen tensions and promoted interaction between the superpowers during the Cold War. Josef Broz was born on 7 May 1892 in Kumrovec, Austria-Hungary. Leaving school at 12, he joined the Croatian Social Democratic Party and was conscripted into the imperial army at age 18. With the outbreak of World War I, he was sent into combat against the Serbs, was wounded in 1915, and was taken prisoner by the Russians. While in a hospital, he was influenced by Bolshevism. In 1917, upon the abdication of Czar Nicholas II, Broz tried to join followers of Lenin in the streets of Petrograd but was recaptured by authorities. He remained imprisoned until freed by the Communists in October 1917. He then joined the Red Guard during the Russian Civil War.

Portrait of Tito. (Library of Congress)

In 1920, Broz returned to Croatia to work as a Communist Party organizer. Arrested several times and imprisoned from 1929 to 1934, he gained Moscow's attention. He was named deputy of the Politburo of the Communist Party of Yugoslavia (CPY) central committee and leader of the Croatian and Slovenian committee. He emerged as a national leader in 1941, when Germany attacked Yugoslavia and he founded a Yugoslav partisan force. Well-organized, by 1942, he was able to set up a provisional government in areas freed from Axis control. In constant conflict with Serbian Chetniks, Tito's forces were officially recognized by the Allies at the Teheran Conference in 1944. By the next year, he was named prime minister of Yugoslavia and consolidated control by purging non-Communists from his government.

Tito soon came into conflict with Stalin over how the Yugoslavian leader was running his government and economy and his independent foreign policy decisions. This led to the expelling of "Tito's clique" from the Communist camp in

June 1948. Stalin then imposed sanctions and boycotts against the country.

Tito retaliated by purging Stalinists, decentralizing the economy, and turning to Western nations for support. He sealed the border between Greece and Yugoslavia, thus ending Soviet support to the rebels during the Greek Civil War. Upon Stalin's death in 1953, Tito tried to reconcile with Nikita Khrushchev's Soviet Union. Yet he maintained some degree of independence by protesting the USSR's invasions of Hungary (1956), Czechoslovakia (1968), and Afghanistan (1979). He also implemented a foreign policy in which smaller countries like his own acted like neutral blocs between the superpowers and promoted nonalignment. He died on 4 May 1980 in Lubljana. With his death, the multiethnic Yugoslavian federation, which he had held together with an iron hand, fell apart in civil war and the worst conflict Europe had seen since World War II.

T. Jason Soderstrum

See also: Cold War; Greek Civil War; World War II

References and further reading:

Auty, Phyllis. *Tito: A Biography.* New York: McGraw-Hill, 1970.
Dedijier, Vladimir. *Tito.* New York: Simon and Schuster, 1953.
West, Richard. *Tito and the Rise and Fall of Yugoslavia.* London: Sinclair-Stevenson, 1994.

Tobruk, Battle of (April 1941)

The first substantial defeat of German land forces in World War II. After the collapse of Italian forces in 1940, Adolf Hitler ordered Lieutenant General Erwin Rommel and the Afrika Korps to North Africa in defense of Libya. However, Rommel, ignoring Hitler's express order to remain defensive, struck eastward from El Agheila on 31 March 1941. Rommel's advance was virtually ignored by the British command for the first few critical days because of the looming German invasion of Greece. On 8 April, the same day as the first ominous signs of Greek collapse, both Archibald Wavell and Winston Churchill realized that holding Tobruk was vital for the defense of Egypt. The first clash at Tobruk occurred on 12 April.

Within the 30-mile perimeter were the Australian 9th Infantry Division and 18th Infantry Brigade, the British 3d Ar-

British soldiers patrolling in tanks at Tobruk, Libya, 1944 (Library of Congress)

moured Regiment, and British artillery units. The defenses consisted of an incomplete series of concrete strong points and antitank ditches that had been built earlier by the Italians. Against this force, Rommel deployed the 15th Panzer and 5th Light Divisions and the Italian Brescia and Ariete (and later the Bologna and Pavia) Divisions. The Luftwaffe maintained heavy bombing throughout the siege.

All attacks during the first week failed, but both the Italian and German Commands considered Tobruk harbor vital for the supply of operations in Egypt, and further attacks commenced on 30 April. A salient was created, but although heavy fighting continued, the perimeter line was not broken again before the siege was lifted by the Axis retreat on 10 December 1941. The magnificent defense put up by "the Rats of Tobruk" made the loss of the strong point to the Germans in the spring of 1942, after confused and not very effective resistance, all the more galling.

Michael Hyde

See also: Balkans Campaign; Churchill, Sir Winston; Hitler, Adolf; Rommel, Erwin Johannes Eugen; Sidi Barrani; Wavell, Archibald Percival, First Earl

References and further reading:

Liddell Hart, Basil, ed. *The Rommel Papers.* London: Collins, 1953.

Maughan, Barton. *Australia in the War of 1939–1945.* Series 1, Army; Vol. 3, *Tobruk and El Alamein.* Canberra: Australian War Memorial, 1966.

Odgers, George. *Army Australia: An Illustrated History.* New South Wales: Child and Associates, 1988.

Playfair, Major General I. S. O. *History of the Second World War: The Mediterranean and Middle East.* Vol. 2, *The Germans Come to the Help of Their Ally (1941).* London: HMSO, 1956.

Tokugawa, Ieyasu (b. Matsudaira Takechiyo) (1543–1616)

For much of his life, Ieyasu Tokugawa stood in the shadows of Nobunaga Oda and Toyotomi Hideyoshi. Although he aligned with both, Tokugawa reaped the benefits of their successes. During the reign of Oda, Tokugawa expanded his lands, and by 1583, he held five provinces. Despite these successes, he did have to kill his wife and order his son's suicide to demonstrate his loyalty to Oda. With the death of Hideyoshi, Japan once again slipped into civil war.

During the reign of Hideyoshi, Tokugawa was moved away from the center of Japan to the east coast, which undercut his strength. Although at first the move nullified his role in Hideyoshi's period, it did allow Tokugawa to conserve and build his strength in the long run. During this time, he transformed the small village of Edo into his capital, which became known as Tokyo. When Hideyoshi died, Tokugawa emerged as one of the paramount powers in Japan.

Still, other contenders existed. War began in earnest in 1600 with Uesugi, another who had aligned with Hideyoshi early. Much of western Japan formed an alliance against Tokugawa, including the daimyo of Mori, Ukita, Shimazu, Nabeshima, Chosokabe, and Ikoma. On 21 October 1600, the armies of Tokugawa and the western armies met at the battle of Sekigahara. During the battle, the daimyo Kobayakra deserted to Tokugawa, which the two had prearranged. Tokugawa emerged victorious.

Ten days later, Tokugawa captured Osaka. In 1603, he assumed the title of shogun but gave it to his son Hidetade in 1605. War was not over, however, until 1615, when he defeated Toyotomi Hideyori, Hideyoshi's son. With the unification of Japan complete, Tokugawa established the Tokugawa shogunate, which ruled Japan for over 250 years.

Timothy May

See also: Hideyoshi, Toyotomi; Japanese Invasion of Korea; Japanese Wars of Unification; Oda, Nobunaga; Samurai; Sekigahara

References and further reading:

Sadler, A. L. *The Maker of Modern Japan.* London: Allen & Unwin, 1937.

Totman, Conrad. *Tokugawa Ieyasu, Shogun.* South San Francisco: Heian, 1983.

Weston, Mark. *Giants of Japan: The Lives of Japan's Greatest Men and Women.* New York: Kodansha International, 1999.

Tondibi (1591)

One of the most devastating military defeats in western Sudanese history. The Moroccan sultan, al-Mansur, cherished control of Songhay, the gold mines, and the tolls from caravan traders. A 1582 civil war between two Songhay princes weakened the empire and played into al-Mansur's hands.

Al-Mansur's two expeditions against Songhay in 1584 perished in the desert because of poor preparation, incompetent leadership, and unmanageable supply lines. A better-organized expedition in 1590 was composed of 4,000 disciplined soldiers (1,500 Moroccans, 2,500 renegade Spanish, and Portuguese and Turkish mercenaries) under a Spanish renegade, Judar Pasha. Half were armed with arquebuses, 500 were mounted gunmen, and 70 were musketeers, accompanied by 600 engineers and 800 camels carrying 31,000 pounds of gunpowder and a similar quantity of lead.

Leaving Morocco on 16 October 1590, the expedition, with difficulty, landed at Bamba on the Niger River in February 1591. On 13 March 1591, they met the Songhay army at the Battle of Tondibi, near Gao, the capital. With superior weaponry, the invaders routed the Songhay force, armed with bows, arrows, and spears and terrified by the noise and smoke of the guns. After the initial panic, however, the Songhay forces rallied and harried the Moroccan forces.

The demoralized Askia Ishaq was replaced with Askia Nuh, who adopted guerrilla tactics against the invaders. Despite this resistance, the Moroccan forces pressed on, seizing Gao and later Tombouctou in 1593. Nuh's forces harried the Moroccans until they eliminated him through treachery, inviting Nuh to peace talks and murdering the Songhay contingent. As a result, the Songhays plunged into disintegration and anarchy.

Edmund Abaka

See also: Songhay Empire

References and further reading:

Boahen, Adu. *Topics in West African History.* 2d ed. London: Addison Wesley Longman, 1986.

Davidson, Basil. *Africa in History.* New York: Touchstone, 1991.

Shillington, Kevin. *History of Africa.* New York: St. Martin's Press, 1995.

TORCH, Operation (1942)

The landings in French North Africa on 8 November 1942 that constituted the first major American intervention on the Atlantic front. TORCH was primarily a U.S. Army operation, with General Dwight D. Eisenhower in command, though the Royal Navy provided two-thirds of the naval force. Some have argued that it was "Roosevelt's secret war baby."

As was so often the case in World War II, military considerations had to take account of the political situation, and historians now tend to agree that the complete military success was marred by the mishandling of the political aspect. The disagreements between the Americans, who wanted to put the emphasis on Casablanca (on the Atlantic), and the British, who prized Algiers (on the Mediterranean), were only finally resolved in September, with the decision to launch a three-pronged attack on Casablanca, Algiers, and Oran in between.

Contrary to the expectations of the American strategists, who believed that their "kingpin," General Alphonse Giraud, would soon rally the local garrisons to the Anglo-American cause, the 65,000 Allied troops met with sometimes fierce resistance from French Vichy forces, especially the navy in Algiers. By 10 November, General Mark W. Clark hastily negotiated a cease-fire with Admiral François Darlan, heavily compromised with Vichy, who seemed to be the only French leader able to put an immediate end to the fighting. The "deal with Darlan" raised an outcry in the British pro-Gaullist press: was Europe to be "liberated," only to be entrusted to men like Darlan? Winston Churchill, who concurred, wrote that "Darlan's murder [on 24 December], however criminal, relieved the Allies of their embarrassment." Now the way was clear for Giraud, but he was no match for Charles de Gaulle, who eventually established his provisional government in Algiers, much to Franklin D. Roosevelt's displeasure.

With their position secure in Morocco and Algeria—the initial objective of TORCH—the Allies (including the rallied French troops) turned against the Axis forces in Tunisia, the decisive battles taking place at Kasserine Pass (14–23 February 1943) and on the Mareth Line (20–29 March) against Erwin Rommel's armor. Combined with massive air superiority and with naval supremacy to cut off the Axis supply lines from Italy, these victories opened the path to Tunis, which fell on 7 May, leading to the final German surrender in North Africa on 12 May 1943.

The conquest of French North Africa constituted the greening of American troops in the face of the hardened Panzer divisions, and even those historians who doubted the strategic value of the whole operation agreed that it provided a valuable training ground for the more difficult landings to come, in Sicily, Italy, and, of course, Normandy. This was also the first occasion when a now-famous quartet had to cooperate: Eisenhower, Bernard Montgomery, George Patton, and Omar Bradley, with Churchill and Roosevelt somehow trying to make a political success of military victories—a rehearsal here again for the real show in northwestern Europe.

Antoine Capet

See also: Bradley, Omar Nelson; Churchill, Sir Winston; Montgomery, Bernard Law; Patton, George Smith, Jr.; Rommel, Erwin Johannes Eugen; Roosevelt, Franklin D.

References and further reading:

Blumenson, Martin. *Mark Clark.* New York: Congdon Weed/St. Martin's Press, 1984.

Clark, Mark W. *Calculated Risk.* New York: Harper, 1950.

Weinberg, Gerhard L. *A World at Arms: A Global History of World War II.* New York: Cambridge University Press, 1995.

Torgau (3 November 1760)

One of the last battles of the Seven Years' War (1756–1763) between Prussia and Austria.

In October, Marshal Leopold Daun occupied Torgau on the Elbe River, thus severing the Prussian army from its forces in the west. Fresh from victory at Liegnitz (15 August), Frederick the Great marched toward Torgau with 49,000 men and 250 guns. Opposing him were 53,000 men and 275 guns. His high command doubted the necessity of battle, but Frederick believed a decisive victory was possible.

Daun's army occupied a ridge west of Torgau. Frederick planned to march 30,000 troops across the Austrian front, wheel around its right flank, and strike the rear. Early in the morning of 3 November, Frederick broke camp. The march took several hours, and the columns became confused as Daun's army discovered the gambit and opened fire. The Austrians altered their right flank and formed a triangle. Beginning at 1:00 P.M., the Prussian king attacked this extreme end of the Austrian line twice and nearly broke the enemy's flank before they counterattacked. Prussian cavalry arrived and led a nearly successful third attack, but at 4:30, Frederick was stunned by a spent musket ball and was carried from the field. Yet the day was saved for Prussia when Hans Joachim von Ziethen advanced and forced a wounded Daun to shift units from his right to secure his center. Johann von Hülsen then led a fourth and final attack against the Austrian right. Franz Moritz Lacy ordered a withdrawal under cover of darkness, and the exhausted Austrian army made safe passage across the Elbe River.

Torgau represents Frederician warfare past its apogee. Nearly 17,000 Prussians were killed or wounded, and Austria lost 16,000 casualties and 7,000 prisoners. Frederick's strategy of winning political victories on the battlefield had failed, and his tactics had become predictable.

Patrick J. Speelman

See also: Frederick the Great; Seven Years' War

References and further reading:

Duffy, Christopher. *Frederick the Great: A Military Life.* London: Routledge & Kegan Paul, 1985.

Showalter, Dennis. *The Wars of Frederick the Great.* London: Longman, 1996.

Weigley, Russell F. *The Age of Battles: The Quest for Decisive Warfare from Breitenfeld to Waterloo.* Bloomington: Indiana University Press, 1991.

Toulon, Siege of (September–December 1793)

Throughout July and August of 1793, many towns in southern France, dissatisfied with the course of the French Revolution, rebelled against the central government in Paris. Some of the rebels desired the restoration of royal authority. Others supported the principles of the revolution but demanded more local authority.

On 27–28 August 1793, rebels in Toulon, home of the French Mediterranean fleet, declared a fourth royalist government in exile and placed the city under the protection of a combined British and Spanish fleet commanded by Samuel, Lord Hood. The revolt of such an important city was viewed by the revolutionary government in Paris as a serious threat. Consequently, a force under General J. F. Carteaux, already operating against 12,000 insurgents in the area, was dispatched to subdue Toulon. A further 4,000 men were detached from the Army of Italy and sent to attack Toulon from the east. By early September, both these forces had arrived outside the city and had begun an uncoordinated series of assaults upon the defenses. In the interim, British, Spanish, and Sardinian forces had been landed in the city, raising the strength of the garrison to 15,000.

On 16 September 1793, Napoleon Bonaparte, then a captain of artillery, offered his services to General Jean François Carteaux when the original artillery commander was wounded. Napoleon began to gather artillery, supplies, and personnel. On 20 September, he orchestrated the seizure of two small forts commanding the eastern half of Toulon's inner harbor. This success forced the British fleet to move closer to the city itself but also induced the British to strengthen another post, Fort Mulgrave, which controlled the exit to the harbor.

On 1 October, an attempt to assault the city from the east was repulsed with heavy French losses. On 23 October, Carteaux was assigned to command the Army of Italy and was replaced by General Doppet. Doppet was in turn removed from command on 16 November and replaced by General Jacques Coquille Dugommier.

While the bulk of the French forces were paralyzed by such command uncertainties, Napoleon had, on his own initiative, prepared several new batteries designed to support an assault upon Fort Mulgrave. An assault of 30 November failed. A second attack, on17 December, proved successful, and by the afternoon of 18 December, a battery began to bombard the British fleet itself in Toulon harbor.

Lord Hood ordered the evacuation of the garrison and of as many civilians as possible. An attempt to burn the French ships in the harbor failed. On the morning of 19 December, the French revolutionary army reoccupied Toulon.

The siege of Toulon was significant because it ended the *federaliste* uprising in the south of France. It also brought to prominence Napoleon, who was promoted to brigadier general on 22 December. The siege gave added prominence to a number of other officers and political figures, including André Masséna, who would subsequently become a marshal of the French Empire, and Paul Barras, who would head the directory.

Joseph M. Isenberg

See also: French Revolutionary Wars; Napoleon I; Masséna, André, Duc de Rivoli, Prince d'Essling

References and further reading:

Chandler, David. *The Campaigns of Napoleon.* New York: Scribner's, 1966.

Tours (October 732)

Charles Martel ("the Hammer") defeats Muslim army led by Abd-ar-Rahman, governor of Córdoba. After crossing the western Pyrenees, Abd-ar-Rahman made for the Loire River. Charles and the Frankish army met him just outside Tours, somewhere just south of the river, and directly in the path of the invaders. The Muslims had not anticipated the arrival of the Franks.

After a few days of maneuver and countermaneuver, Abd-ar-Rahman decided to attack. Little is known of his army, except that most were horsemen lightly armed with sword and lance. The Muslims, relying upon an overwhelming superiority, especially in cavalry, rushed forward. The Frankish infantry stood their ground against the ferocity of the assault. As one monkish chronicler put it, the Franks stood "firm as a wall" (Hadrill 1960, 13).

Arab sources record that the battle lasted two days, whereas Christian sources claim that it went on for seven. It ended when the Franks captured and killed Abd-ar-Rahman. His forces then withdrew quietly overnight, and even though Charles Martel expected a surprise retaliation, there was none. For the Muslims, the death of their leader was a sharp setback. They were left no choice but to retreat back across the Pyrenees.

The Battle of Tours has been considered the high-water mark of Muslim invasions of western Europe. The victory helped consolidate Charles Martel's leadership of the Franks. He was able, for example, to assert his authority in Aquitaine. Its duke, Eudes, swore allegiance.

Nic Fields

See also: Charles Martel

References and further reading:

Hadrill, John M., ed. *The Fourth Book of the Chronicle of Fredegar and Its Continuation.* London: Nelson, 1960.

Wood, I. N. *Merovingian Kingdoms, 450–751.* London: Longman, 1994.

Toussaint L'Overture, Wars of (1793–1803)

Conflicts that established Haitian independence but also began the tradition of political violence and strongmen. The French Revolution had a profound impact on France's colony of Saint-Domingue (present-day Haiti) on the western portion of the Caribbean island of Hispaniola. The class structure of Saint-Domingue included whites, free mulattoes (known as *affranchis*), and slaves. The ideals of liberty, equality, and fraternity for the *affranchis* were manifested in an unsuccessful revolt that was over in a few months. On 14 August 1791, slaves revolted and rampaged across the northern plain, murdering whites and putting plantations to the torch. Moreover, there was fighting among whites, mulattoes, and slaves throughout the colony. As the slave revolt grew, Spain allied with the rebels, hoping to regain possession of all Hispaniola. One of the most capable of the rebel leaders was the former slave Toussaint L'Overture, who first fought with the Spanish army but in May 1794 declared his loyalty to France when the National Convention abolished slavery. The Spanish expedition was decimated by yellow fever, and in 1795 Spain withdrew from Saint-Domingue and also ceded its colony of Santo Domingo to France.

L'Overture, with the able assistance of his principal lieutenant, Jean-Jacques Dessalines, brought political order to the slave insurrection as well as sound military leadership. By 1798, the British expedition, which had supported the whites, was defeated and withdrew. L'Overture still had to contend with the mulatto general Andre Riguad, who attacked his army on 16 June 1799. The resultant race war, know as the War of the Knives, brought even greater destruction to the war-ravaged nation. L'Overture not only defeated the mulatto armies but invaded the former Spanish colony of Santo Domingo. By 1801, he controlled all of the island of Hispaniola and was installed as governor for life by a hastily written constitution. He also ruled with the approval of Napoleon, who appointed him captain general of the colony.

Portrait of Toussaint L'Overture. (Library of Congress)

Napoleon soon realized that Saint-Domingue was in fact independent under L'Overture, and he was determined to renew France's holdings in the New World. In 1802, he dispatched an expedition commanded by his brother-in-law, General Charles Leclerc, to reconquer the island. Leclerc was initially successful and seized Port-au-Prince and other ports. In the interior, L'Overture, Dessalines, and another guerrilla general, Henri Christophe, continued to resist. In a sudden turn of events, L'Overture arrived at a truce with Leclerc, and Christophe surrendered his army. Shortly after these capitulations, Dessalines also came to terms, and it appeared that Leclerc had successfully regained Hispaniola for France. A month after the truce, Leclerc arrested L'Overture and had him imprisoned in France, where he died on 7 April 1803.

Leclerc's problems did not end with the disposal of L'Overture. Yellow fever and other tropical diseases took their toll on the French army. Attempts to disarm the black population led only to increased resistance as the people realized the French intended to reinstate slavery. The brutality of the war increased, and Dessalines and Christophe rejoined the rebel cause. Leclerc himself was stricken by yellow fever and died on 2 November 1803. His replacement, General Donatien Rochambeau, initiated even more brutal tactics and seemed determined to kill all the former slaves in order to successfully pacify the island. Atrocities mounted as Rochambeau waged a war of extermination.

Events in Europe turned the tide in favor of the rebels. Great Britain had initially approved of Napoleon's invasion but went to war with France in May 1803 and soon attacked French fortifications on the island. Dessalines intensified his attacks in the interior by burning plantations at will and executing captured whites. Panic among the whites resulted in their evacuation in November 1803, ending French control of Saint-Domingue, but France did retain control over Santo Domingo.

The victorious rebels named their new country Haiti, and Dessalines was named governor-general for life. Formal independence was declared on 1 January 1804. The ambitious Dessalines was not content with the title of governor-general and became the self-anointed Emperor Jacques I on 8 October 1804. He was assassinated on 17 October 1806. Haiti's political tradition of strongmen seeking lifelong rule and meeting a violent demise is firmly rooted in the Haitian Revolution.

George Lauderbaugh

References and further reading:

James, C. L. R. *The Black Jacobins: Toussaint L'Ouverture and the San Domingo Revolution.* New York: Vintage Books, 1989.

Ott, Thomas O. *The Haitian Revolution, 1789–1804.* Knoxville: University of Tennessee Press, 1973.

Steward, T. G. *The Haitian Revolution 1791 to 1804.* New York: Russell & Russell, 1971.

Towton, Battle of (29 March 1461)

Allegedly the longest, biggest, and bloodiest battle ever fought on British soil. The truth of such claims is difficult to establish: calculations of numbers involved and deaths in this battle have generated heated historical debate.

The battle was fought in the month that Edward, Duke of York, had himself proclaimed king, forestalling a Lancastrian advance on London. The Lancastrians withdrew, pursued by Yorkist forces, the vanguard of which was apparently overwhelmed at Ferrybridge, on the river Aire, on the night of 27–28 March. The next day, the Yorkists sought to fight their way over the river, and the Lancastrians tried to inflict a telling defeat on their pursuers. Perhaps 6,000 troops were killed before the Yorkists established themselves on the north bank.

After clashes during the night, Palm Sunday found two armies between Towton and Saxton, the Lancastrian army having advantages of ground and numbers but facing into a blizzard. The initial exchanges by longbowmen apparently favored the Yorkists, but the Lancastrians advanced, and there followed a close-quarter battle lasting over five hours, in which the Yorkists slowly gave ground. What seems to have decided the battle was the timely arrival of forces that, working their way around the Yorkists' right flank, fell on the Lancastrian left: thereafter the Lancastrian force disintegrated, with thousands being killed as they fled the battlefield or drowned in the Cock or in the Wharfe at Tadcaster.

It has been suggested that some 20,000 Lancastrians and 8,000 Yorkists were killed and that some 37,000 died in the actions between Ferrybridge and Towton, totals that must include those drowned and dead of wounds. In any event, the battle was salutary and ensured some form of peace for a decade.

H. P. Willmott

See also: Wars of the Roses

References and further reading:

Boardman, A. W. *The Battle of Towton.* Stroud: Alan Sutton, 1994

Trajan, Marcus Ulpius (53–117)

Roman emperor (98–117) and conqueror of Dacia and Parthia. Born near Italica, Spain, in 53, Trajan trained for military service from an early age. He advanced to the rank of general by distinguishing himself through his talent and ability. Trajan received the title of consul in 91, and eight years later, Emperor Nerva formally adopted him and named him as his successor. In 98, Nerva died, and Trajan became emperor. He prepared for an expedition to subdue the Dacians, who continued to exert military pressure on Rome's Danubian border. The Dacian Wars (101–102 and 105–106) proved expensive, but the capture of their large treasury compensated Trajan for his efforts. Returning to Rome, he ordered the construction of Trajan's Forum, one of the largest and most elaborate of the Roman centers. He arranged for the six-story marble annex built on the slopes of the Quirinal Hill to be decorated with visual depictions of his victories over the Dacians, reminding the people that war yielded financial gain for the victor. The 140-foot-high column of Trajan later served as his burial monument.

In 113, Trajan embarked on another campaign, this time against the Parthians. His troops advanced eastward, conquering Armenia and northern Mesopotamia and incorporating them into the empire as provinces. He reached the Persian Gulf before ill health forced him to return to Rome. Trajan died on board ship during the return voyage. The majority of his reign focused on military campaigns, but he also addressed administrative reforms and initiated a large number of construction projects.

Cynthia Clark Northrup

See also: Parthian Empire

References and further reading:

Bennett, Julian. *Trajan, Optimus Princeps: A Life and Times.* Bloomington: Indiana University Press, 1997.

Lepper, F. A. *Trajan's Parthian War.* London: Oxford University Press, 1948.

Trebia, Battle of the (22–23 December 218 B.C.E.)

The first of three successive defeats that Rome suffered at the hands of the Carthaginian general Hannibal Barca. The Roman force, consisting of two consular armies, was led by a single consul, T. Sempronius Longus, because his colleague, Publius Cornelius Scipio, had been severely wounded.

The Roman army had encamped on the east bank of the river Trebia, a tributary to the Po. Hannibal had set up camp on the other side of the river. The night before the battle, he had sent his brother Mago with 2,000 troops to hide in the hills south of the battlefield. At daybreak, Hannibal provoked the Romans into battle. Longus immediately ordered his troops to cross the ice-cold river. His four Roman legions took position in the center, with allied infantry on either wing. Skirmishers covered the front of the army; the flanks were covered by cavalry. The Roman force numbered about 36,000 foot and 4,000 horse.

Hannibal had deployed his army about a mile from his camp. He positioned his Spanish and Celtic infantry in the center behind his skirmishers. His large cavalry force was positioned on the flanks, reinforced by the few elephants that had survived the passage of the Alps. On either side behind the cavalry stood his crack African infantry. The Carthaginian army totaled about 30,000 foot and 6,000 horse.

When the light troops had withdrawn, the heavy infantry engaged. The numerically superior Carthaginian cavalry drove the Roman horse from the field and together with the African infantry outflanked the Roman infantry. Then Mago's force appeared from hiding and attacked the Roman line in the rear. The army was completely surrounded and destroyed, save for 10,000 legionnaires who cut their way through the Carthaginian center and retreated to the nearby city of Placentia. Hannibal lost only 7,000 troops.

M. R. van der Werf

See also: Cannae, Battle of; Hannibal Barca; Lake Trasimene, Battle of; Punic Wars

References and further reading:

Bagnall, Nigel. *The Punic Wars: Rome, Carthage and the Struggle for the Mediterranean.* London: Pimlico, 1999.

Conolly, Peter. *Greece and Rome at War.* London: Greenhill Books; Harrisburg, PA: Stackpole Books, 1998.

Lazenby, J. F. *Hannibal's War: A Military History of the Second Punic War.* London: Aris & Phillips, 1978.

Trenton (26 December 1776)

A minor American victory, but one that raised patriot morale at a low point in the American Revolution. The battles in and around New York City in the summer and fall of 1776 had gone badly for the American revolutionaries; the British had taken the city and established posts across New Jersey and up the Hudson River. Meanwhile, General George Washington knew that the enlistments of many of his few remaining troops would end on 31 December 1776. He needed a victory both to support the cause and to maintain his army.

At this nadir in revolutionary fortunes, Washington devised a brilliant plan. He intended to send forces back across the Delaware River into New Jersey; one unit would serve as a picket line north of Trenton, and another would perform the same task south of the city. Meanwhile, his main force would divide into two and attack the town from opposite ends on the morning of 26 December; he figured that the garrison of about 1,000 Hessian mercenaries would be sated from a day of heavy drinking and eating and thus unable to defend well.

The plan did and did not work. Neither of the advanced units crossed to and stayed on the New Jersey side. But Washington's main units achieved complete surprise and total victory. In a severe winter storm, delayed somewhat by the difficulty of a broad river crossing, the Americans attacked around 8:00 A.M., shortly after dawn. Moving faster than the Hessians could form, the Americans gained control of the town, breaking up formations, blocking exits, and securing the advantage—all within 90 minutes. Nearly the entire force was captured at minimal cost to Washington's army.

Later, as Washington returned to Pennsylvania, the delayed units belatedly moved across the Delaware to New Jersey, and Washington decided to return—leading to the victorious Princeton campaign and a long winter in Morristown, New Jersey.

Charles M. Dobbs

See also: American Revolution; Washington, George

References and further reading:

Lundin, Leonard. *Cockpit of the Revolution: The War for Independence in New Jersey.* New York: Octagon Press, 1972.

Smith, Samuel Steele. *The Battle of Trenton.* Monmouth Beach, NJ: Philip Freneau Press, 1965.

Trinh-Nguyen Dynastic Struggles (1620–1673)

The centuries following Vietnam's achievement of independence from China in 939 witnessed the state's expansion from the Red River valley southward along the coast, conquering and absorbing the country of Champa in what is now south-central Vietnam and then the Mekong Delta, which was inhabited by Cambodians. Vietnam's unity at this time was fragile, owing to strong regional rivalries and the logistical difficulties of controlling an elongated state squeezed between the Annamite Cordillera and the sea. Although Le Loi drove out Ming Chinese invaders in 1428 and established a new dynasty, his kingdom broke apart in the sixteenth century, and there were frequent wars between the regions, especially in the seventeenth century. In the late eighteenth century, the Tayson Rebellion swept aside the two rival northern and southern states.

Although emperors of the Le Dynasty remained the country's nominal sovereigns, the Trinh family held real power in the north (Tonkin), and the Nguyens controlled the south-central region (Annam, Cochin China). War broke out in 1620 over the Nguyen refusal to pay tax revenues to the Trinh. Trinh military forces, including war elephants, a navy, and a 100,000-man land force, were superior to those of the Nguyen, but the latter, controlling a thinly populated realm but one richer in natural resources than the north, were equipped with Portuguese firearms and fought a successful defensive war. The protracted conflict was indecisive.

China intervened to broker a peace in 1673, and a wall was built from the mountains to the sea near Dong Hoi, marking the boundary of the two domains. This wall was very close to the 17th parallel, which divided North and South Vietnam after the 1954 Geneva Conference.

Donald M. Seekins

See also: Vietnamese Civil War

References and further reading:

SarDesai, D. R. *Vietnam: The Struggle for National Identity.* Boulder, CO: Westview, 1992.

Hall, D. G. E. *A History of South-East Asia.* 2d ed. London: Macmillan, 1964.

Triple Alliance, War of the (1864–1870)

South America's longest and most violent war. The war began in November 1864, when Brazil meddled in Uruguay's internal affairs, an action that drew an immediate response from Paraguay's bellicose dictator, Marshal President Francisco Solano Lopez. Lopez was convinced that Brazilian dominance of Uruguay would upset the balance of power in the region, and he reacted by blocking the Paraguay River and invading Brazil's Matto Grosso Province. Not content with Brazil as an enemy, he provoked Argentina by crossing Missiones Province to attack Brazil from the west. On 1 May 1865, in response to Lopez's attacks, Argentina, Brazil, and Uruguay formed an alliance and declared war on Paraguay. Paraguay's initial thrusts into Brazil and Argentina were soon blunted, and Lopez was forced on the defensive for the balance of the conflict. Although vastly outnumbered, Paraguay's army was modern and well-disciplined. Allied advances, led by the Argentine general Bartolome Mitre, were stopped at the Battle of Curupait in September 1866. In November 1867, the Paraguayans slowed another allied offensive at the second battle of Tuyuti. In 1868, Brazilian general Luis Alves de Lima e Silva assumed command of allied operations, and the fortunes of the allies soon improved.

Lopez's defense centered on a fortress at Humaiti, which guarded the Paraguay River. In February 1868, a Brazilian warship ran by the guns of Humaiti and bombarded Asunción, the Paraguayan capital. Lopez was forced to withdraw from Humaiti, and Asunción fell in early 1869. However, the capture of the capital and heavy casualties did not end either Paraguayan resistance or Lopez's hold on his people.

The Paraguayan army retreated to the northeast but was defeated at battles at Caacupé and Piribebuy on 15 August 1869. The allied victory at Piribebuy under the Conde d'Eu, Gaston Luis Felipe d'Orleans, destroyed the Paraguayan army as an effective fighting force. Nevertheless, Lopez was able to rally remnants of his force and wage guerrilla warfare from the north. On 1 March 1870, a Brazilian cavalry unit cornered Marshal President Francisco Solano Lopez and his band at Cerro Cora. Lopez and his son were killed, and the war ended.

The war demonstrated the ability of Latin American republics to wage modern war against each other on a large scale. Paraguay exhibited extraordinary resourcefulness, improvising as needed, building ships and armaments with its own resources, and continuing the war under the most adverse circumstances. Brazil and Argentina organized large armies and worked out problems of allied commands.

The war devastated Paraguay, which lost between 8.6 and 18.5 percent of its population, as well as 38 percent of its national territory. Moreover, Paraguay's vibrant prewar economy was wrecked and the country plunged into a period of political instability. Argentina, however, made significant territorial gains with relatively little disruption of its economy or political stability. The impact of the war on Brazil was mixed: territory was gained, and the military gained new respect, but the cost of the war in Brazilian lives and funds was high.

George M. Lauderbaugh

References and further reading:

Kolinski, Charles J. *Independence or Death!* Gainesville: University of Florida Press, 1965.

Phelps, Gilbert. *The Tragedy of Paraguay.* New York: St. Martin's Press, 1975.

Pla, Josefina. *The British in Paraguay.* London: Richmond Publishing, 1987.

Trojan War (12th or 13th Century B.C.E.)

Legendary Bronze Age conflict between Mycenaean Greeks and Troy. Despite the problems inherent in any attempt to employ the Homeric epics in a historical reconstruction of the Trojan War, the latest archaeological evidence from and around the site of Troy has led many scholars to reappraise the Homeric tradition of a major expedition against the city.

For the Greeks, the Homeric epics were history, and they located Troy at a hill now called Hissarlik in northwestern Anatolia. Excavations of the site have revealed at least nine settlements dating from the early Bronze Age to late antiquity. With its recently discovered fortified lower town, Troy VIh/VIIa, destroyed about 1270 B.C.E., is the favored candidate for the "well-walled city with lofty towers" of Homer's *Iliad.* Archaeology has revealed that it was not only in contact with the Near East but also with Mycenaean Greece, where a flourishing and warlike civilization existed. Near Eastern texts also suggest an involvement of the Mycenaeans in western Anatolia at about the right time. In particular, Hittite archives mention individual chieftains and their war bands crossing the Aegean Sea and raiding up and down the Anatolian coastline.

The unearthing of an extensive Bronze Age cemetery below the headland of Yassi Tepe and within a few meters of what was then a sheltered sandy beach suggests Mycenaean military activity within the vicinity of Troy. Comprising some 200 graves that included Mycenaean artifacts dating to about 1300 B.C.E., the cemetery, which the ancients always regarded as the tomb of Achilles, lies only 7 kilometers to the southwest of Troy. Because it does not relate to any permanent settlement in the area, it seems reasonable to infer that the cemetery was once associated with an anchorage for Mycenaean ships.

It did not need many warships full of armed raiders to threaten and sack a small city and enslave its people: six vessels, each manned by 50 warriors, sack Troy in a Hercules legend. In the Homeric epics, a leader's greatest claim to glory was to be awarded the title "sacker of cities." In the *Iliad,* a hero does not sack cities to increase his political power but does so in order to gain booty and women for himself and his followers. Indeed, the Linear B tablets that list captive women from the eastern Aegean provide vivid evidence for the predatory nature of the Mycenaeans in this region.

Guarding the mouth of the Dardanelles (Homer's Hellespont), Bronze Age Troy's strength and wealth lay primarily in its location. There is evidence for a marine bay that existed in what is now the Trojan plain, and it is clear that Troy was founded on a spur that stuck out into it. The bay provided a valuable harbor for any shipping entering the Dardanelles from the Aegean Sea and facing adverse currents and winds. More often than not, ships had to wait for favorable winds in Trojan waters, whereby the Trojans could exact levies and provide goods and services. Troy may have grown rich from tribute and long been a thorn in the side of people like the Mycenaeans. Its sack by a coalition of Mycenaean war bands has made an everlasting impression.

Nic Fields

References and further reading:

Easton, Donald. *The Quest for Troy.* London: Weidenfeld & Nicolson, 1997.

Wood, Michael. *In Search of the Trojan War.* Berkeley: University of California Press, 1998.

Trotsky, Leon (1879–1940)

Russian revolutionary. Trotsky was one of the leaders of the Bolshevik Revolution that established a communist government in Russia in 1917. He was responsible for creating the Red Army to defend the revolution and with inspiring it to victory.

Trotsky was born Lev Davidovich Bronstein in Yanovka (later Kirovograd) into a middle-class Jewish family. He was educated in Odessa and was arrested there in 1898 for revolutionary activities. He escaped in 1902 to England under the name "Leon Trotsky." Trotsky returned to Russia in 1905 during the general strike protesting the Russo-Japanese War. After proclaiming the Petersburg Soviet of Workers Deputies, Trotsky was exiled to Siberia again but escaped in 1907. He spent the next 10 years as a wandering journalist, calling for revolution in Russia. During this time, Trotsky developed the concept of "permanent revolution," which became the Russian Bolshevik "party line."

Trotsky returned to Russia in May 1917 after the revolution that ended czardom. He worked closely with Vladimir Ilich Lenin to seize power. Trotsky was entrusted with the military planning, while Lenin assumed the political leadership. Trotsky was also elected chairman of the Petrograd Soviet in October 1917. When the Bolsheviks seized power on 7 November, Trotsky presided over the Congress of Soviets. In the new government, he became commissar of foreign affairs. He headed the delegation that traveled to Brest-Litovsk to negotiate a peace treaty with the Germans. Trotsky responded to the harsh German terms with a policy of "no peace, no war." He hoped that German soldiers occupying Russia would be infected with Bolshevism and spread revolution to their own country. When the Germans responded by marching on Petrograd, Lenin ordered the delegation to sign the even more harsh Treaty of Brest-Litovsk. Trotsky refused to attend the signing ceremony on 3 March 1918.

After the failure of his foreign policy, Trotsky was demoted to commissar of war. He proved extraordinarily talented at organizing, equipping, and training the new Red Army. He did not hesitate to force former czarist officers to staff his army. He traveled the fronts in his private train, distributing orders, rewards, and punishments. Trotsky combined the ability to inspire his troops with a sound sense of strategy. He made good use of the Bolsheviks' central position and control of the railroad net to defeat the White and foreign armies. In 1920, he directed the invasion of Poland on Lenin's orders, and its failure justified his doubts. Once again, nationalism had trumped international communism. Nonetheless, by 1921, Trotsky's Red Army had defeated its counterrevolutionary foes.

After the Russian Civil War, Trotsky supported the idea of a national militia force, instead of a professional army. He was outmaneuvered politically and replaced by Mikhail Frunze in 1923. When Lenin died in 1924, Trotsky lost in the struggle to replace him. Stalin gradually forced Trotsky out of the party and exiled him in 1929. Trotsky eventually moved to Mexico, where a Stalinist assassin killed him in 1940.

Tim J. Watts

See also: Bolshevik Revolution; Russia, Allied Intervention in; Russo-Polish War; Stalin; Warsaw/Vistula

References and further reading:

Deutscher, Isaac. *The Prophet Armed: Trotsky, 1879–1921.* New York: Oxford University Press, 1954.

Trotsky, Leon. *My Life.* New York: Grosset & Dunlap, 1960.

Warth, Robert D. *Leon Trotsky.* Boston: Twayne Publishers, 1977.

Trumpeldor, Yosef (1880–1920)

First hero of modern Israel. Born in 1880 in Pyatigorsk in the Caucasus, Trumpeldor served with distinction in the Russo-Japanese War. After losing his left arm in combat, he returned to the front and fought at Port Arthur. He was awarded for his bravery and became one of the few Jewish officers in the czarist army.

An early convert to socialism and Zionism, Trumpeldor organized a group of pioneers and emigrated to Palestine in 1912. He worked on several pioneer settlements, inspiring his comrades with his hard work and dedication.

Deported to Egypt during World War I, Trumpeldor joined Vladimir Jabotinsky in lobbying Britain to create a Jewish military unit. Instead of a combat unit, the British formed Jewish volunteers into a 500-strong support battalion, the Zion Mule Corps, with Trumpeldor as its deputy commander. The corps served with distinction at Gallipoli and was repeatedly praised for bravery.

The unit was disbanded after the withdrawal from Gallipoli, though many of its members later served in the Jewish Legion. Trumpeldor left for Russia, where he organized Jewish self-defense forces to fight back against the pogroms that swept the war-torn country. In 1919, Trumpeldor returned to Palestine, where he helped establish the defense of the isolated settlements of the Upper Galilee as part of Hashomer, the Jewish defense organization.

On the morning of 1 March 1920, Arab irregulars attacked the Jewish village of Tel Hai. Trumpeldor, severely wounded early in the fighting, directed its successful defense. He died that night, supposedly proclaiming: "It is good to die for our country." His dedication to hard work, nation building, and self-defense inspired a generation of Zionist settlers.

Stephen Stein

See also: Gallipoli; Israeli Military

References and further reading:
Herzog, Chaim. *Heroes of Israel.* Boston: Little, Brown, 1989.
Lipovetzky, Pesah. *Joseph Trumpeldor: Life and Works.* Jerusalem: World Zionist Organization, 1953.

Trung Sisters, Rebellion of (39–43)

The modern nation of Vietnam traces its origins to the Red River Valley, in what is now Tonkin, or northern Vietnam, where peoples with cultures and languages distinct from the Han Chinese founded societies based on wet rice cultivation. One of the unique features of their way of life was the high status accorded to women. The Trung sisters, who struck a blow for independence from Chinese colonial rule in 39–43, were among the earliest Vietnamese patriots.

The Han Chinese emperor Wu-ti (r. 140–87 B.C.E.) conquered the region in 111 B.C.E., but Chinese rule was loose and did not deprive indigenous leaders of authority on the local level. By the early first century, however, it became more intrusive. Heavy taxes in labor and kind were extracted from the population, and Chinese officials promoted cultural assimilation. The keystone of their cultural policy was the replacement of the loosely structured Vietnamese family system with their own rigid patriarchal system, a highly effective instrument of social control. Both policies aroused intense resentment.

According to history and legend, the Chinese prefect of Giao-chi district (the modern Hanoi-Haiphong region) executed an indigenous noble, Thi Sach, who was suspected of anti-Chinese activities. His widow, Trung Trac, refused to don the traditional attire of mourning and raised the banner of rebellion, swearing to avenge her husband and restore the independent Vietnam (Van Lang) of old. Joined by her younger sister Trung Nhi, she raised an army numbering in one account over 80,000 soldiers and drove the Chinese out of the country. Many of her officers were women, indicating the strength of matriarchal values in ancient Vietnam. In 40, Trung Trac was proclaimed queen, and a capital was built for her at Me-linh, an ancient district associated with Vietnam's legendary Hung kings.

The Chinese emperor fielded his best general, Ma Yuan, against the Trungs, and they suffered defeats at his hands in 42–43. Rather than be captured, the two sisters reportedly drowned themselves in a river (Chinese accounts say they were captured and beheaded). Ma Yuan imposed tight colonial control over the region, executing thousands of rebels and bringing in large numbers of Han Chinese soldiers, officials, and settlers. Only in 939 did Vietnam break free of Chinese rule.

Recent scholarship suggests that Thi Sach was not in fact executed by the Chinese but participated loyally in the uprising instigated by his strong-willed wife. One of the more colorful legends associated with the Trungs is that they killed a man-eating tiger and used its hide to inscribe their call to arms. After Vietnam became independent, temples were raised in their honor, including the twelfth-century Hai Ba Trung temple in Hanoi and the Hat Mon temple in Son Tay Province. Having fought wars against China as well as France and the United States, socialist Vietnam reveres these sisters as national heroines.

Donald M. Seekins

See also: Trinh-Nguyen Dynastic Struggles

References and further reading:
Keith Weller Taylor. *The Birth of Vietnam.* Berkeley: University of California Press, 1983.

Tukhachevsky, Mikhail Nikolayevich (1893–1937)

A brilliant Soviet military theorist who pioneered the concept of "deep battle" and was also the first major victim of Stalin's military purges. Mikhail Nikolayevich Tukhachevsky was born in the Dorogobuzh district of Smolensk on 4 February 1893. He graduated from the Aleksandrov Military School in 1914 and was rushed off into combat, where he fought with distinction. Captured and exchanged in 1917, he returned home to find Russia convulsed by the Bolshevik Revolution. A political pragmatist, Tukhachevsky joined the Communists simply from a belief that they could best reconstitute Russia as a world power. He then went on to distinguish himself in fighting throughout the Russian Civil War and spearheaded the Soviet invasion of Poland. There, in 1920, Tukhachevsky was heavily defeated by Marshal Józef Pilsudski, owing to the lack of cooperation from Soviet general Kliment Voroshilov and his political adviser, Stalin. This defeat started a long-standing, rancorous dispute among the three men.

During the postwar period, Tukhachevsky gained renown as the Soviet Union's premier military theorist. He proffered a new notion of combat, "deep battle," based on the idea of highly mechanized forces that would penetrate enemy lines and destroy their rear areas. However, he ran afoul of Voroshilov, who insisted that armor forces were better dispersed to assist slow-moving infantry formations. Nonetheless, Tukhachevsky was made a marshal in 1935, and he helped orchestrate the growth of Soviet military professionalism. Stalin, however, came to fear his growing popularity, and in 1937, with the aid of German intelligence "disinformation," Tukhachevsky was arrested on trumped-up charges. He was summarily executed on 11 June 1937, the first of several thousands of army officers liquidated by the regime, seemingly the most professional and competent. The toadies, the hacks, the time-servers, and the inexperienced officers who took their places were no match for the German army in the disastrous opening months of the German invasion of the Soviet Union in the summer of 1941.

John C. Fredriksen

See also: Frunze, Mikhail Vasil'evich; Stalin; Trotsky, Leon

References and further reading:

Simpkin, Richard. *Deep Battle: The Brainchild of Marshal Tukhachevsky.* Washington, DC: Brassey's Defense, 1987.

Stoecker, Sally W. *Forging Stalin's Army: Marshal Tukhachevsky and the Politics of Military Innovation.* Boulder, CO: Westview Press, 1998.

Turenne, Henri de la Tour d'Auvergne, Vicomte de (1611–1675)

The military support of Louis's XIV's early reign. Henri de la Tour d'Auvergne belonged to a high nobility non-French family related to the famous Nassau. He was educated in the remote little city of Sedan and was commissioned at the age of 13; he studied military art in Paris but above all in Holland with his Nassau uncle.

He chose to serve France from 1631 and became brigadier general at the age of 23. The French phase of the Thirty Years' War gave him his opportunity. Until 1644, he was often second in command; because Turenne was a Protestant, Louis XIII mistrusted him. After the king's death (1643), Mazarin, the main minister of the underage new king, gave him the title he craved, marechal de France, at age 32. Turenne commanded the victorious French army in Germany, winning battles at a regular pace: Fribourg in 1644, Nordlingen in 1645, and Zusmarshausen in 1648. During the Fronde civil war, he hesitated for a year between the king's side or the rebellious nobility. His interests dictated that he support the king, and from 1650, he was at the head of the royal army against the Spanish-supported rebels.

He gained his most famous victory in 1658 over Louis II de Bourbon, fourth prince de Condé, at the battle of the Dunes. He was made "marechal general" by the king in 1660. His way of making war was at the same time scientific and realistic; he never tried to achieve goals he knew he could not reach. His military doctrine was to avoid siege operations and to bring the enemy to battle as often as possible.

From 1661, Louis XIV decided to rule France on his own without reference to his ministers, to Turenne's disappointment. Nonetheless, as a member of the military council, he reorganized the army, giving preference to petty nobility over courtiers for field command, thus creating a professional officer corps. He was also the king's tutor in military affairs and foreign policy.

Turenne had also to command during the 1667 campaign (War of Devolution, 1667–1668). The Dutch War (1672–1678) saw an aging Turenne criticized for his supposed lack of keenness. Yet the winter campaign of 1674–1675 is considered his masterpiece. The French province of Alsace was invaded, and with fewer troops than his old enemy, Prince Raimondo Montecuccoli, Turenne outflanked him and forced him to the German bank of the Rhine River. Received by the king as the savior of France, Turenne was killed by a gunshot during a skirmish in the following summer's campaign.

Gilles Boué

See also: Condé, Louis II de Bourbon, Fourth Prince de; Louis XIV; Thirty Years' War

References and further reading:

Bérenger Jean. *Turenne.* Paris: Fayard, 1987.

Bluche, François. *Dictionnaire du Grand Siècle:* Paris: Fayard, 1990.

Turkish Wars of European Expansion (1413–1699)

Series of wars through which the Ottoman Empire established control of southeastern Europe. The Ottoman state, defeated by Tamerlane (1402) and divided by civil wars, was reestablished by Sultan Mehmed I (1413–1421). The first expansion of the empire came under Murad II, who acquired Salonika, Epirus, and southern Albania from Venice (1423–1430). In 1439, he seized Serbia, involving him in conflict with Hungary. Occupied with a war against the Karamids in Asia Minor, Murad was forced to relinquish Serbia in 1443 but regained the province after defeating the Hungarians under János Hunyadi at Varna (1444) and Kosovo Polje (1448).

Murad's son Mehmed II earned his title, "the Conqueror," by eliminating the weak Byzantine state and capturing Constantinople in 1453, the Turks' greatest conquest and one they still hold under the name "Istanbul." With the Ottoman Empire now firmly established on the Bosporus, Mehmed turned to the consolidation of his control over the Balkans. He failed in an attempt to capture Belgrade (1456) but occupied Herzegovina and eastern Bosnia after an ineffectual war with Hungary (1460–1464). The conquest of Greece was completed with the defeat of the Duchy of Athens (1460) and the capture of the Morea (1464) and Lemnos in a renewed war with Venice (1463–1479). In 1476, Mehmed cut Walachia's ties with Hungary, making it a tributary vassal. After the death of Skander Beg (George Kastriota), Mehmed reconquered Albania (1478), ending a 35-year independence struggle. With the conquest of the Balkans complete, Mehmed sent raids into southern Italy (1477–1478) and besieged Otranto (1480) and Rhodes (1481).

After Mehmed's death, the Ottoman Empire was torn by a civil war between his successors. The single important conquest of this period was the acquisition of Cyprus, taken from Venice (1499–1503). After defeating his rivals, Selim I again turned to foreign expansion, making Moldavia tributary (1512) and capturing western Bosnia from Hungary (1520).

The Ottoman Empire reached the height of its military glory under the rule of Selim's son, Sultan Süleyman I, "the Magnificent," who captured Belgrade (1521) and then defeated the Hungarians at the Battle of Mohács (1526). Süleyman's support of Ján Szápolyai in the Hungarian Civil Wars (1526–1547) involved the Ottomans in the first of the Austro-Turkish Wars (1529–1791).

Though unsuccessful in the First Siege of Vienna (1529), Süleyman captured Buda (1541) and gained control of central Hungary, making Transylvania into a tributary vassal state. In the Mediterranean, Süleyman drove the Knights of St. John from Rhodes (1522) and defeated the papal-backed alliance of the First Holy League in a renewed war with Venice (1537–1540), but his forces failed in the Siege of Malta (1565). Further Ottoman expansion at sea was seriously checked by the naval forces of the Second Holy League (1571–1581), which destroyed the Ottoman fleet in the great Battle of Lepanto (1571).

Raids along the Habsburg-Ottoman borders in Croatia led to a renewal of the war in Hungary in 1593. In Holy Roman Emperor Rudolf II's "Long Turkish War" (1593–1606), the Ottomans briefly lost Transylvania, Walachia, and Moldavia before anti-Habsburg revolts restored the territories to Ottoman control. The bitter stalemate of the war signaled an important shift in the strategic balance of power of the Ottoman Empire vis-à-vis its European enemies. A succession of weak sultans, financial difficulties brought about by an inflationary crisis, and social turmoil combined at the turn of the seventeenth century to erode Ottoman military capacity. Subsequent wars added little territory to the empire and were fought primarily to recover or round out earlier conquests.

In the first stages of the long and costly Venetian War (1645–1670), the Ottoman navy suffered defeats at Páros (1651) and the Dardanelles (1656), leading to the appointment of Grand Vizier Mehmed Köprülü with extraordinary powers in 1657. Under the Köprülü family of grand viziers, Ottoman power briefly revived. Mehmed's son, Fazil Ahmed Köprülü, reasserted Ottoman sovereignty over Transylvania, added to the Hungarian possessions, and forced the surrender of Crete (1670) after a 23-year siege. War with Poland (1672–1676) led to the occupation of Podolia and western Ukraine, the Ottoman Empire's last significant territorial acquisition.

The first territorial loss came shortly thereafter, in 1679, when Ahmed's son-in-law, Kara Mustapha, was forced to cede Ukraine to Russia. Mustapha sought compensation in further Hungarian conquests, leading to war with the Habsburgs. The Ottoman threat to Vienna sparked the formation of the Third Holy League (1684–1699), whose combined forces liberated Buda (1686), Athens (1687), and Belgrade (1688). The Ottoman recovery of Serbia and Greece after 1690 was balanced by the loss of Azov to Russia (1697). In the Treaty of Karlowitz (1699), the Ottoman Empire was forced to accept its losses and acknowledge the permanence of its new frontiers, bringing its era of imperial conquest to a close.

Brian Hodson

See also: Austro-Turk Wars; Constantinople, Siege of; Hungarian-Turkish Wars; Malta, Siege of (May–September 1565); Rhodes, Sieges of; Süleyman I; Vienna, Sieges of

References and further reading:

Inalcik, Halil. "Military and Fiscal Transformation in the Ottoman Empire, 1600–1700." *Archivum Ottomanicum* 6 (1980).

Murphey, Rhoads. *Ottoman Warfare, 1500–1700.* London: UCL Press, 1999.

Sugar, Peter. *Southeastern Europe under Ottoman Rule, 1354–1804.* Seattle: University of Washington Press, 1977.

Tyre, Siege of (January–July 332 B.C.E.)

After his victory over the Persians at Issus, Alexander the Great marched south into Phoenicia (Lebanon). Most cities in that region submitted to Alexander, with the exception of the mighty trading city of Tyre, necessitating a siege.

Tyre was well-protected. It lay on an island somewhat less than 500 yards from the coastline and surrounded by a wall 150 feet in height. Because Alexander's naval power was limited, he was unable to blockade the city's harbors. He resorted to building a mole to the island to give his troops access to the wall. The mole progressed rapidly to within bowshot of the walls, where the Macedonians built two siege towers on the mole to protect against missile fire from walls and Tyrian ships.

With naval reinforcements, Alexander attempted to force the city by sea. This was unsuccessful, but he was able to blockade the harbors. Using ship-mounted rams, Alexander attempted to breach the seawall away from the mole, removing rock obstacles placed by the Tyrians to limit such an attack. Meanwhile, the Tyrians had erected wooden towers upon the battlements to counter attacks from the siege towers on the mole.

Sorties and attempts to destroy or cut loose the ramming ships were eventually unsuccessful, and the Macedonians succeeded in breaching the seawall. Two ships of troops under the command of the king himself landed before the breach, and the soldiers succeeded in fighting their way into the city. Simultaneously, two squads of ships broke into the harbors. Assaulted from three points, the city was taken; 8,000 Tyrians were killed. The rest, some 30,000 persons, were sold into slavery. The Macedonian losses amounted to 400 dead.

Maarten van der Werf

See also: Alexander the Great; Alexander's Wars of Conquest

References and further reading:

Fuller, J. F. C. *The Generalship of Alexander the Great.* London: Eyre & Spottiswoode, 1958.

Green, Peter. *Alexander of Macedon, 356–323 B.C.E.: A Historical Biography.* Harmondsworth, UK: Penguin, 1974. Reprint, Berkeley: University of California Press, 1991.

Hammond, N. G. L. *The Genius of Alexander the Great.* London: Duckworth, 1997.

Tarn, W. W. *Alexander the Great.* Cambridge: Cambridge University Press, 1948.

U

ULTRA

The World War II security classification for compartmentalized information acquired from high-level Axis signals intelligence (SIGINT). The Enigma machine, automatically encrypting message traffic, was used with the highest-level German communications, including those between Adolf Hitler and his field commanders. The Japanese acquired variations of the machine, which once understood by cryptanalysis, provided the Americans a reading of Japanese SIGINT. An original Enigma machine was acquired for British intelligence in 1939 by a Polish secret service contact formerly working in the production factory. The term *ULTRA* was coined for Enigma products because of the requirement for a classification more restrictive than the British "most secret."

ULTRA's significance is that it allowed insight into opposing commanders' beliefs based on their transmitted strategy discussions and on their current situation based on their routine status reports. For example, the North African campaign at El Alamein (July–November 1942) saw Erwin Rommel extremely short of supplies and planning a final breakthrough to Cairo. ULTRA provided Bernard Montgomery with the complete German plan, participating units, their deployment, and the date of the offensive itself. The close-fought result, even with this advance knowledge, illustrates this information's importance for the campaign.

ULTRA may have provided the margin for the survival of Britain during the early days of the war. For example, it allowed a reasoned apportionment of aircraft sorties to defeat Luftwaffe fighters yet still provided some measure of security against the bombers. For the remainder of the war, ULTRA provided details on German weaknesses and expected attacks and confirmed Allied deception operations, as seen during the Normandy invasion, including the Pas de Calais assault ruse. This most vital Allied intelligence breakthrough was not revealed for some three decades after the close of World War II.

Robert Martyn

See also: Intelligence, Military; World War II

References and further reading:

Kahn, David. *The Codebreakers.* Rev. ed. New York: Scribner's, 1996.

Winterbotham, F. W. *The Ultra Secret.* London: Weidenfeld & Nicolson, 1974.

Unarmored Fighting Vehicles

Unarmored military vehicles, either specially built for the military or adapted from commercial models and used in combat-support roles. Although armored vehicles (e.g., tanks) are the more glamorous war machines and garner most of the publicity, modern armies rely on tens of thousands of "thin-skinned" unarmored fighting vehicles (UFVs) for an almost unlimited variety of purposes. These vehicles include not only motorcycles and cars but also light and heavy trucks, buses, ambulances, tractors, wreckers, fire trucks, snowplows, amphibious vehicles, and construction equipment. The backbone of any army's UFVs is the truck (lorry).

Although it might be argued that the first UFV was Joseph Cugnot's three-wheeled, steam-powered artillery towing device invented in 1769, the modern use of such vehicles began in 1898 with the use of motorcycles and autos in the German army's maneuvers. World War I saw extensive mechanization in the major armies, carried out both by purchase or capture of civilian vehicles and by development of appropriate vehicles produced by manufacturers, who were subsidized by government and addressed specifications that emphasized standardization of controls, interchangeability of parts, and ability to perform under service conditions. At war's end in 1918, Great Britain had 168,128 such vehicles in use. The United States, entering the war 32 months after the British, had procured 275,000 vehicles. Both nations were able to achieve a degree of standardization by taking commercially produced vehicles and modifying them for military work. No all-wheel-drive truck entered war service, although several were in the testing stage by November 1918.

In World War II, the Axis powers of Germany, Italy, and Japan produced 594,859 trucks, and the major Allies, the United States, Great Britain and the Soviet Union, manufactured 3,060,354. Germany was able to flesh out its needs somewhat with a huge array of captured vehicles and by utilizing the production of factories in occupied nations, but in doing so it created immense maintenance difficulties. Germany's great UFV failure lay in its inability to standardize. It also wasted considerable sums in producing tracked personnel carriers that carried a mere 12 troops in theater-seat luxury (but that could also be used as a prime mover) and a tracked motorcycle that could go practically anywhere—but that carried only two to three persons. The Soviets also produced tracked trucks but found, like the Germans, that the considerably higher expense and complexity of such arrangements nearly negated their superior overland capabilities. The British motor industry turned out tens of thousands of UFVs, but the troops in the field seemed to prefer the U.S. product. (For one thing, it was much easier to change gears in any U.S. UFV.)

The stars of the war, for Allies and enemy alike, were the American Jeep and the "deuce-and-a-half" truck. The Jeep, developed in 1940 by the Willys corporation and manufactured also by Ford, was a 0.25-ton, 4 x 4 (four-wheel drive) truck and command-reconnaissance vehicle that could operate with ease up to 60 miles per hour, mount a 40-degree slope, turn in a 30-foot circle, and tilt without tipping at a 50-degree angle. With a machine gun or recoilless rifle mounted, it was truly a fighting vehicle. Its only real weakness was its vulnerable standard commercial water-cooled engine; the U.S. auto industry had no off-the-shelf air-cooled engine available. The "deuce-and-a-half," a General Motors 6 x 6, 2.5-ton truck also produced by Studebaker and International Harvester, became the workhorse of the Allied cause in World War II, so widely used that Russians still call multi–drive axle trucks *studeborky* (without knowing why). The Germans were more than happy to utilize captured 6 x 6s, and the Soviets imported tens of thousands of them through Lend-Lease. The Chinese Nationalists, the Free French, the British, the Italian Co-Belligerent forces, and every Allied military force of any consequence were all allotted thousands of 6 x 6s. And at the end of the war, the U.S. Army, paradoxically, turned over thousands of its supposedly worn-out 6 x 6s to the German economy to maintain some sort of transportation net. They soldiered on for yet another decade over torn-up roads, with minimum maintenance facilities in conditions almost resembling wartime. The 6 x 6 (along with newer models of the Jeep) continued to be produced through several model changes, serving in Korea and Vietnam (an unmatched record).

The Jeep and the 6 x 6 accurately reflected the American motor industry, which at the time out-produced the rest of the world combined, turning out vehicles that were often technologically behind their European counterparts but were more rugged and cheaper to produce and thus would be better adapted to the rigors of land warfare. Considering the literally hundreds of uses the 6 x 6 was put to, in World War II and in war and peace in the decades that followed, it may be arguably the best truck in history, military or commercial.

The contemporary era abounds in thin-skinned military vehicles, with Third World nations vigorously developing and producing their own designs so as to strive for military self-sufficiency and underwrite it with the proceeds of sales abroad. But the U.S. military seemed to have retained its UFV lead over its last remaining major military rival, the former Soviet Union. In the Gulf War (1990–1991), those anti–Saddam Hussein coalition forces unlucky enough to miss out on being issued the U.S. Army new high mobility multipurpose wheeled vehicle (HMMWV, and now, like the Jeep, produced for the civilian market), sometimes "hot-wired" Iraqi-Soviet UFVs to gain some battlefield mobility. After about 300 miles of use, these enemy trucks failed because their transmissions had worn out. There were no reported significant difficulties with the HMMWVs. Unglamorous workhorses the UFVs may be, their use in large numbers can be assured in the wars and near-wars of the foreseeable future.

Joseph M. McCarthy

See also: Armored Fighting Vehicles; Logistics

References and further reading:

Cary, Norman Miller, Jr. "The Use of the Motor Vehicle in the United States Army 1899–1939." Ph.D. diss., University of Georgia, 1980.

U.S. Army. *A Handbook of Ordnance Engineering.* Prepared under the Direction of the Chief of Ordnance. n.d., n.p. Copy in U.S. Army Center of Military History, Washington, DC.

Vanderveen, Bart H. *The Observer's Army Vehicles Directory from 1945.* London: F. Warne, 1972.

———. *The Observer's Army Vehicles Directory to 1940.* London: F. Warne, 1974.

———. *World Directory of Modern Military Vehicles: Unarmored Vehicles from 1970.* New York: Arco, 1984.

———. *Historic Military Vehicles Directory.* London: After the Battle/Wheels and Tracks, 1998.

Uniforms

It is often thought that military uniforms are as old as the profession of arms itself. In truth, uniforms are relatively recent innovations that were slow to gain wide acceptance. Even today, military uniforms are constantly evolving. Modern combat soldiers are invariably pictured wearing uniforms bearing scientifically designed camouflage and helmets made of the latest polymers. Yet some of those soldiers, on ceremonial occasions, may don uniforms designed centuries earlier. Even the most modern uniform incorporates some elements based more on tradition than on utility.

A degree of uniformity in the dress of soldiers existed in ancient armies, but it was a matter of function and economy, not fashion. Roman soldiers were issued a coarse wool tunic, linen undergarments, and a thick wool cloak for bad weather, hardly a proper uniform. The sartorial magnificence of the Roman legionnaire came from his armor. The weapons and shield he carried, as well as the helmet and body armor he wore, were generally manufactured to pattern in order to reduce costs.

The fall of the Roman Empire left Europe without formal armies. Defense was a local concern, and it fell to the regional warrior-leader to organize and arm the neighboring peasantry when necessary. Soldiers were generally amateurs and wore what clothing and armor was available to them, without regard for uniformity. In time, wealthier feudal lords provided their men with mantles, a loose cloak covering their armor and bearing badges based on the lord's heraldic arms. The wearing of badges advertised the wealth and power of the patron and also helped soldiers differentiate friend from foe in close fighting.

As nation-states began to form in Europe, monarchs raised their own troops. Pope Julius II established the Swiss Guard, his personal bodyguard, in 1505, and Michelangelo is credited with designing the opulent ceremonial uniform the Swiss Guard still uses.

England's King Henry VII raised the Yeomen of the Guard in 1485, the first permanent unit in the English army. They wore a red mantle bearing the badge of the king that covered their chain-mail shirts and tight hose on their legs and carried halberds. That uniform, with only minor adjustments, remains substantially unchanged.

In the sixteenth century, monarchs grew less tolerant of the armed retainers that gave military power to the lords and threatened their own power. In 1512, early in his reign, Henry VIII feared invasion. He called upon his lords to raise an army in defense of England, allowing the soldiers to wear the badges of their feudal lords. Thirty-three years later, the king issued explicit regulations prohibiting the wearing of feudal badges by either soldiers or their lords. By wearing only Henry's badge, English soldiers ceased to be soldiers of their lords and became soldiers of their king.

Armies represented power. They provided the state the ability to make war and to defend or expand its territory. Armies also reflected the strength of the monarch. A well-equipped army dressed in uniforms bearing the king's heraldic arms or family colors bespoke a nation that was both powerful and united.

The long period of political instability that culminated in the Thirty Years' War (1618–1648) saw the formation of European standing armies. The cost of maintaining a national army represented a major drain on the treasury. Providing that army with uniforms added to that cost. Still, many continental monarchs were willing to bear the expense.

In England, the growing conflict between the crown and Parliament over royal finances slowed the move toward uniforms. There was some standardization in body armor but nothing that could be called a national uniform. When England descended into civil war, men on both sides wore the same clothing. Scarves and hat ribbons were used to denote allegiance, red for royalists and orange for the roundheads. The interregnum marked an expansion of the army but produced little change in English military dress.

The restoration of the monarchy in 1661 placed Charles II on the throne of England. He returned from his continental exile with an eye for fashion and an appreciation of the authority that a well-dressed royalist army could give to the monarchy.

By the end of the seventeenth century, military dress grew less utilitarian and more fashionable, tending to be exaggerations of civilian fashion. Body armor continued to shrink until it was reduced to little more than a small metal crescent, or gorget, worn at the neck by officers. The steel helmet disappeared until World War I. It was replaced with a felt hat folded into various styles and adorned with feathers, ribbons, lace, or cockades. A particularly cumbersome three-cornered, or tricorne, version became the style that would last the longest. The decorative breastplate, or cuirass, and ornamental helmet would reappear in cavalry dress later in the eighteenth century.

Cavalry regiments, considered elite units, were drawn to exotic uniforms. In the protracted war between Spain and the United Provinces, the Spanish king employed Hungarian Hussars. The Hussar's distinctive dress of a fur cap, or busby, tight pants, high boots, and a fur-lined short jacket casually worn over the left shoulder was copied by other European nations that quickly established hussar regiments of their own.

The eighteenth century saw an explosion of interest in military uniforms. No longer were they mere caricatures of civilian dress; they became an art form of their own, showcasing the considerable talents of military tailors and embroiderers. Uniforms reflected the prestige of a nation, designed to fill its soldiers with pride and, their makers hoped, its enemies with trepidation.

During this period, national uniforms became associated with particular colors, red for Great Britain, white for Austria; dark blue for Prussia, and green for Russia. Each infantry regiment wore distinctive colored facings (collars, cuffs, and piping) to differentiate it from other regiments. The importance of soldiers of the same army wearing the same uniform finally became obvious on a chaotic battlefield, thick with smoke.

The British government issued a clothing allowance to regimental commanders for each soldier. Some commanders kept expenditures low, pocketing the savings. Others, finding the allowance too small, would generously supplement the allotment with their own funds. This led to some regiments, particularly cavalry units commanded by wealthy officers, having extremely elaborate and expensive uniforms.

The military dress of the eighteenth century was more suited for the parade ground than the battlefield, but the splendidly outfitted soldiers who marched off to the Napoleonic Wars altered and simplified their uniforms to meet the realities of warfare. General officers often dressed in uniforms based on civilian sporting attire when going into battle.

Once Napoleon was finally defeated and peace was restored to Europe, military tailors received new commissions. The multinational occupation army in Paris provided great inspiration. The elegantly dressed Polish Lancers were copied by most nations, who clothed their own lancer regiments in Polish dress, including the distinctive four-cornered helmet known as the *chapka* and billowy Cossack trousers.

The European fascination for military uniforms was exported to the colonial outposts of emerging empires. The uniforms of the native armies of the East India Company were just as extravagant as the uniforms of the British army. The Far East also influenced the Europeans. The Prussian army adopted a spiked helmet, the *pickelhaube,* based on the traditional headgear of the Middle East.

Military fashion, like civilian fashion, is often fickle. The military successes that France enjoyed in the early nineteenth century resulted in its uniform, and particularly its slope-fronted cap, the kepi, being adopted by many of the world's armies, including both the armies of the United States and the Confederate states during the Civil War. When France was defeated by Prussia in 1870, the kepi quickly fell out of favor, giving way to Prussian-styled spiked helmets adopted by most British imperial forces and the U.S. Army.

The successors to the elaborate cutaway infantry uniforms of the seventeenth century were replaced in most armies with the infinitely more practical tunics by the 1850s. Brightly colored uniforms remained battle attire until Boer marksmen in the Boer War found them inviting targets. The British army quickly adopted khaki service dress. The practicality of the khaki (from the Hindustani word for "dusty") uniform was recognized by British soldiers in India, who dyed their impractical white summer uniforms with tea as early as the 1850s.

By the outbreak of World War I in 1914, most armies had already adopted a monochromatic service uniform. The French still retained their bright blue and red uniforms, not changing to the "horizon blue" uniform until May 1915. Officers and men alike resisted helmets until they were ordered to wear them because of the growing number of head injuries.

The end of World War I did not bring a return to dress uniforms. It ushered in an era of utility and function in which battle dress became a part of the soldier's daily life. However, most armies still have fairly elaborate (and expensive) uniforms for formal occasions. The introduction of women into armies on a large scale in the twentieth century brought about an entirely new line of uniform design, some of it quite appalling, in the first decades of the century. Even maternity uniforms must be provided. Today, only a few elite units with ceremonial duties, like the U.S. Marine Corps Band, Britain's Guards Division, and the French Garde Republicaine, continue to wear the type of splendid military uniform that once graced the world's parade grounds and battlefields.

Eric Smylie

See also: American Civil War; Boer Wars; British Military; Dutch War of Independence; English Civil War (1642–1649); French Army; Napoleonic Wars; Russian and Soviet Armies; Teutonic Knights; World War I; World War II

References and further reading:

Barthorp, Michael. *British Infantry Uniforms since 1660.* Poole, Dorset, UK: Blandford Press, 1982.

Kannik, Preben. *Military Uniforms of the World.* New York: Macmillan, 1968.

Kube, Jan K. *Militaria: A Study of German Helmets and Uniforms, 1729–1918.* Westchester, PA: Shiffer, 1990.

Martin, Paul. *Military Costume: A Short History of European Uniforms.* London: Herbert Jenkins, 1965.

Mollo, Andrew. *Army Uniforms of World War I: European and United States Armies and Aviation Services.* New York: Arco Publishing, 1978.

Mollo, Boris. *Uniforms of the Imperial Russian Army.* Poole, Dorset, UK: Blandford Press, 1979.

Mollo, John. *Military Fashion.* London: Barrie & Jenkins, 1972.

Thorburn, W. A. *French Army Regiments and Uniform from the Revolution to 1870.* London: Arms and Armour Press, 1969.

United Nations and Conflict Resolution

Procedural instruments and set of measures taken by the United Nations (UN) to prevent or to end serious conflicts. A system of collective security was already established within the interwar League of Nations but remained impotent. The UN considerably increased these mechanisms, following the idea that the community of states should transfer the monopoly of force to the UN. Consequently, the UN charter adopted several different means of resolving serious conflicts.

Peaceful Settlement of Disputes

Under Chapter VI (Articles 33–38) of the UN Charter, the attribute of peacefulness does not refer to the means of settlement but to the dispute itself. As long as the conflict has not reached a phase in which organized violence is used, a peaceful settlement is possible. If their dispute might potentially endanger international peace and security, the parties are obliged to take all possible steps to settle it peacefully (Article 33). If they do not succeed, the initiative can be taken by the Security Council (SC). The SC then is authorized to become active by investigating the disputed facts (inquiry, fact finding), trying to mediate, and suggesting ways to resolve the dispute. It is important to note that the SC can only make recommendations in this procedural stage. In addition, it has no competence for the settlement of legal disputes; they have to be referred to the International Court of Justice. Although the rules of this chapter lack clarity, the SC nevertheless uses its competence laid down therein rather often and successfully.

Settlement of Violent Conflicts

The previous chapter dealt with situations that are likely to endanger international peace only if not resolved. Once a conflict reaches a more acute phase, the SC has more possibilities. If the SC determines that an imminent danger to peace is present or a breach of peace or aggressive act has occurred, under Chapter VII (Articles 39–51) of the UN charter, it can decide measures necessary to maintain or restore peace (Article 39). Unfortunately, it is left to the SC's discretion what constitutes an imminent danger to peace, a breach of peace, or an aggressive act. Experience shows that not only international conflicts but also intrastate crises trigger the SC's activity according to Chapter VII. Thus the policy of apartheid has been regarded as an imminent danger to peace, as have the civil wars in Somalia and Liberia. Even following Libya's refusal to hand over two citizens accused of having destroyed a Pan American jet over Scotland (Lockerbie), the SC determined a threat to peace. The invasions of Korea, Kuwait, and the Falkland Islands consequently were classified as breaches of peace.

United Nations Headquarters building, New York. (Library of Congress)

With a situation or an act has been determined to fall under Article 39, the SC can order the parties to follow the preliminary measures suggested (Art. 40). These can be a cease-fire, the withdrawal of troops, or the commencement of peace negotiations. In cases of noncompliance, the SC can recommend to the community of states other measures to restore peace—even their participation in acts of military self-defense (as in Korea, 1950).

The UN Charter's main provisions about measures of collective security are laid down in Articles 41 and 42. Article 41 covers actions that—although still peaceful—already constitute grave means to enforce compliance of the disputing parties with the SC. These include economic sanctions, partial or complete traffic blockades, and the severance of diplomatic relations. They were used quite often, yet with mixed success at best. Some examples are the economic sanctions imposed on Iraq, South Africa, and the former Yugoslavia.

If even these measures have no success, the SC may take the steps provided for in Article 42: such deployment of military forces as is necessary to restore the peace. However, the SC is not bound to follow the order of the steps listed; it can use military force without having imposed economic sanctions if it believes that the latter will be inadequate. The creators of the charter had foreseen the formation of a military force exclusively under the control of the UN. Individual treaties between the UN and the member states were to have arranged the number, type, and use of military personnel and material available to the SC's disposition (Article 43). These contingents were to have been commanded by a Military Staff Committee. Unfortunately, such special agreements never came into existence. The closest approximation was the formation of the UN Command to repel the North Korean invasion of South Korea in the early days of the Korean War (1950–1953). Therefore, a central pillar of the UN's system of collective security is missing. A solution to the problem, however, lies in the wording of Article 42, paragraph 2; therein the cooperation of military forces not under command of the SC is provided for. This provision was the base for the authorization the SC issued to member states in the cases of the Kuwait invasion in 1991 and East Timor in 1999. A main difference from a UN action as envisaged by Article 42, paragraph 1, is that the military units engaged are not under the command of the UN but remain in their national command structure. The authorization by the SC is only the legal basis of their employment.

Peace-Keeping

Strangely enough, the UN's traditionally most effective means of resolving serious conflicts is not even mentioned in the charter. The employment of peace-keeping forces originally had the goal of keeping the disputing parties separated from each other (Cyprus, Golan Heights) until a steady peace was arranged by diplomatic means. The condition for this employment was that all parties agreed upon it. Today, however, the formerly limited mandate of the peace-keeping forces is stretched by combination with observation tasks (such as disarmament, democratic elections, humanitarian relief), interim administrations (as in Kosovo, East Timor), and increased cooperation with multinational combat units.

The last decade of the twentieth century demonstrated that the planned monopoly of the UN over all collective security measures is far from being realized. The recent conflicts brought a massive shift of initiative from the UN and the SC toward other international organizations. Tasks of observation, mediation, and fact finding often are taken on under the auspices of the European Union (EU) and the Organization for Security and Cooperation in Europe (OSCE), whereas activities of a predominantly military nature are carried out by the North Atlantic Treaty Organization (NATO) or its Partnership for Peace (PFP). Therefore, the future of the UN's role in the system of collective security remains unclear, which is unsurprising in light of the fact that, for most of the world, nationalism undoubtedly remains the strongest force in international relations.

Marcus Hanke

See also: North Atlantic Treaty Organization; Yugoslavian Civil Wars

References and further reading:

Simma, Bruno, ed. *The Charter of the United Nations.* Oxford, UK: Oxford University Press, 1994.

United Nations. *Handbook on the Peaceful Settlement of Disputes between States.* New York, 1992.

U.S. Army

The U.S. Army can trace its roots to the Continental army organized by the Second Congressional Congress on 14 June 1775 as a measure to support the local militia in preparation for the war of independence. This original force had two primary troop types: the local militia, whose job it was to intervene in times of immediate threat and return home afterward; and the "Continentals," a standing army with more thorough training and traditional, longer terms of service. Here was also the beginning of the tension between "regulars" and the militia/reservists/National Guard.

After the war, the Continental army was officially disbanded, and the size and effectiveness of the militia that was left in its place varied so greatly—in large part because of low pay and inadequate equipment—that by the time of the Mexican-American War in 1846, it was clear that a more traditional standing army was again needed to protect the United State's ever-widening frontier.

Americans, although personally a violent people, inherited a distaste for a standing army from the British, and from the days of Thomas Jefferson in the eighteenth through early nineteenth centuries and William Jennings Bryan in the twentieth, indulged the ruralist-republican fantasy that "a million men would spring to arms overnight" if the nation were actually threatened. Regulars were considered an "aristocratical" threat to the democracy. Thus, like the United Kingdom and unlike almost all other developed nations, the United States did not conscript its young men in peacetime until 1940. Even the selection of candidate officers to the U.S. Military Academy was, uniquely, in the hands of local members of congress, as a bar to the development of an officer class. Those who persisted in a military career were barely tolerated—that is, until after the outbreak of war, when they might (also unique to the United States) be elected president.

Between the Revolution and the Mexican-American War, however, the 1st American Regiment had been the stock fighting force for the fledgling country, dealing primarily with skirmishes over newly claimed land in the Northwest Territories (now the midwestern states) that was already occupied by American Indians. Lines of communication and supply were poor, however, and several key defeats, including a debacle at the Miami villages in Ohio, eventually led to the regiment being absorbed by the 3d Infantry, considered the oldest regular regiment in the U.S. Army.

Because the U.S. Military Academy at West Point (founded in 1803) was the only school of higher civil engineering instruction in the nation until the founding of Rensselaer Polytechnic Institute in the 1820s, the army, through its Corps of Engineers, was the primary builder of the nation's infrastructure until after the Civil War. To this day, the development, construction, maintenance, and improvement of its navigable waterways and flood control structures is in the hands of the corps.

By the time of the Civil War, the army had been expanding and modernizing considerably. By 1865, it had increased in size nearly sixfold to almost 1 million officers and infantry. By contrast, the Confederate army was only half this size at its strongest. With the end of the Civil War came massive demobilization, of course. By the time of the Spanish-American War in 1898, the army was again forced to rely on volunteers to buttress its numbers, a failure in planning that Secretary of War Elihu Root determined to change by reconfiguring the War Department through an army general staff and officer corps education. This he accomplished by 1903, and by the time of U.S. entrance into World War I in 1917, the ranks had swollen to nearly 4 million men, the majority of whom were conscripted ("drafted") under the Selective Service Act that was put into place on 18 May 1917.

After World War I, pell-mell demobilization left the army with a paltry 125,000 men between the two world wars—the smallest of any major power at the time and roughly the size of Romania's ground forces. Selective Service was called upon again following Germany's rapid defeat of France in 1940, quickly increasing the numbers to 1,640,000 by the time Japan attacked Pearl Harbor in 1941. At its peak, the army was more than 8 million strong during World War II (more Americans were involved in military service in World War II than in any other organized activity in the nation's history). Equally important, however, the U.S. Army was reorganized into three main commands: the Army Ground Forces, the Army Service Forces, and the Army Air Forces, a massive and complex organizational network that was directed by the army chief of staff, General George C. Marshall.

The reason for this unprecedented size and complexity was straightforward enough: prior to the U.S. involvement in World War II, its overseas intervention had been limited to one geographic area, but during World War II, troops and matériel were employed in Europe, Africa, Asia, and the Pacific, requiring not only greater numbers of men and machines but also new organizational divisions and improved techniques to administer them. Even with U.S. victory in Europe and the Pacific—and the obligations to administer occupied territories that came with it—the U.S. Army once again downsized and demobilized to just over 500,000 troops by 1948.

The Cold War quickly changed this policy, however, with a rapid buildup to compete with the perceived Soviet threat taking place even before the outbreak of the Korean War in 1950. Both the public and the Truman administration felt that a show of force was the best deterrent to a nuclear war, and the army was bolstered accordingly. During this same time, the Army Air Force split into an independent military unit, the U.S. Air Force, in large part because of the complexity and ubiquity of intercontinental bombers carrying atomic weapons. During the Korean War, troops were again drafted, giving the army 1,500,000 total men.

Following this war, the army did not demobilize, maintaining at least 800,000 men until the outbreak of conflict in Vietnam in the mid-1960s. As a world leader during the Cold War, the U.S. government felt it necessary to maintain such a large standing army, even though the service returned to volunteer status following U.S. withdrawal from Vietnam in 1973. Probably no single move contributed more to the professionalism of the post-Vietnam U.S. Army than the abolition of conscription. Also, separate female army units were disbanded and their personnel distributed throughout noncombat units, opening many new military occupations to a much broader spectrum of soldiers. (The ban on female ground combat troops remains.) In many ways, the "lessons of Vietnam" were taken to heart by the U.S. Army: unconventional warfare forces were built up, particularly in the years of the Reagan presidency, and in military technology, the United States leaped ahead of its allies and rivals. This improvement was made apparent in the Grenada and Panama incursions (1983 and 1989–1990, respectively) and in the Gulf War (1991).

The primary post–Cold War army field commands are the Forces Command, including the Army Reserve and Army National Guard, which is responsible for domestic protection in the continental United States; Matériel Command, whose role is to research, develop, and test new weapons and equipment; Intelligence and Security Command, which is responsible for national security and information gathering; Medical Command; Criminal Investigation Command; Corps of Engineers; Special Operations Command (which includes Civil Affairs and Psychological Operations Command); Military Traffic Management Command; Military District of Washington; U.S. Army Europe; and U.S. Army Pacific.

Increasingly in recent years, the army has served in more than a military capacity, redefining and expanding its roles in environmental protection and civil development; it also provides military assistance to federal, state, and local governments. But perhaps its most important function in recent times has been as an agency of federal relief during times of disaster, often providing much-needed medical aid, helping reconstruct whole communities after natural catastrophes, or supplying medical assistance during outbreaks of disease. Some, however, have questioned if these "operations other than war" might detract from army battle readiness.

It could be argued that no military force in history has so reflected its parent nation than the U.S. Army. An individually violent people, who have elected army commanders to the presidency at least four times yet feel that professional army men are "losers" who cannot make it in civilian life; tolerate conscription and high taxation only in wartime; expect its military to trade firepower for casualties; and retrench savagely with the coming of peace, the American people (when they even consider such matters) basically have the army that they want. Although several other armies are appreciably larger, with the collapse of the Soviet Union, the United States is the world's only military superpower, and the U.S. Army, with its mere 10 divisions, is almost by definition the world's leading ground force.

David J. Tietge

See also: Armies: Organization and Tactics; Conscription; Military and Society

References and further reading:

Bradford, Zeb B. *The Army in Transition.* Beverly Hills, CA: Sage Publishers, 1973.

Ryan, Garry D., and Timothy K. Nenninger. *Soldiers and Civilians: The U.S. Army and the American People.* Washington, DC: National Archives and Records Administration, 1987.

U.S. Marines

The amphibious elite of the United States, a unique military force that has its own armor and air arm. Despite repeated domestic challenges to its very existence, the U.S. Marine Corps has survived for more than 200 years, amassing a record of bravery and military success that has secured its place in the American military establishment.

A Marine recruitment poster from the 1920s. (Library of Congress)

In 1775, as Britain's North American colonists organized for rebellion and war in earnest, both the Continental Congress and several state governments commissioned warships. The ship commanders provided for marine detachments as part of their crews. Congress specifically ordered the creation of two marine battalions on 10 November 1775 to spearhead an attack on Nova Scotia. Although these battalions were never formed, marines continue to commemorate this day as the birth of their service. Modeled on the marines of the Royal Navy, they served as ship security and provided the core of boarding and shore parties. Marines served on most of the ships of the Continental navy, including John Paul Jones's *Bonhomme Richard,* and they were crucial to her victory over the *Serapis.*

Congress refused to fund a navy after the Revolutionary War; so the marines ceased to exist until 1798, when Congress authorized a small fleet. Each ship of the fleet received a detachment of marines, and they were later assigned to the navy's shore installations. Over the next decades, marines distinguished themselves in battles on shore and sea against a host of enemies, including the British, the Seminoles, and assorted pirates. In the Mexican-American War, marines spearheaded several landings along Mexico's coast as well as General Winfield Scott's invasion at Veracruz. A battalion of marines fought with Scott's army in its march on Mexico City.

During the Civil War, marines spearheaded numerous landings, and a marine battalion fought at Bull Run. Although technological advances have made marine sharpshooters superfluous and boarding actions unlikely, marines often crewed part of a warship's main armament.

Quickly reduced to its prewar size of 1,800 men after Appomattox, the corps resumed its old duties, helping the navy police the seas and protect American interests. Marines landed in Korea, Panama, Samoa, and numerous other locations.

The marines expanded and modernized along with the navy in the late nineteenth century. The corps continued its shipboard service (despite the protests of reform-minded naval officers) and after the Spanish-American War helped police the nation's new empire, often forming the strong arm of the United States' new active foreign policy. Marines helped suppress the Philippine Insurrection and the Boxer Rebellion, secured Panama following its revolt from Colombia, and intervened repeatedly in the nations of the Caribbean and Central America. The corps tripled in size and fielded battalion, regiment, and even brigade-strength formations. Artillery and support units expanded as well, and the marines formed a small aviation unit.

In World War I, a marine regiment formed part of the initial commitment of the American Expeditionary Force, making good on the marines' claim to be the "first to fight." Later expanded to a full brigade, marines helped stem the German attack at Chateau-Thierry and spearheaded the Allied counterattack at Belleau Wood. As part of its wartime expansion, the Marine Corps began recruiting women.

As the U.S. Navy adjusted to its expanded role in the twentieth century and prepared for future wars, naval officers realized they needed a specialized force to seize and hold advance bases to support fleet operations. They assigned that role to the marines, who developed the doctrine for amphibious assault that would serve them so well for the rest of the century. When World War II came, the U.S. Marine Corps was virtually the only force in the world thoroughly prepared for amphibious warfare.

In the months following the attack on Pearl Harbor, the stout resistance of isolated marine garrisons in the Pacific was virtually all the United States could point to with pride. Particularly noteworthy was the fierce resistance of the marines defending Wake Island. During the war, the corps built up to a force of six divisions and almost 500,000 marines. They spearheaded the island-hopping campaign across the Pacific, often bearing the brunt of the fighting from Guadalcanal, Tarawa, and Pelelieu to Iwo Jima and Okinawa. Easily the most remembered photograph of the U.S. Pacific War was that of marines raising the U.S. flag over Mount Surabaichi on bitterly contested Iwo Jima.

After World War II the marines, reduced in size by half, survived continued institutional challenges to their survival and cemented their place as the United States' principal, ready military force. Marines served in numerous small operations from the Caribbean to the Middle East and Africa and in all three of the major U.S. military actions of the postwar period. In the Korean War, marines again proved their mettle, helping defend the Pusan perimeter, landing at Inchon to cut the North Korean army's lines, and later fighting their way out of the Chosin Reservoir with such effectiveness that Chinese troops were soon surrendering to *them.* (Their fighting spirit was summarized by the marine officer who insisted that his men were not "retreating" from North Korea; no, they were simply "attacking in another direction.")

Marines landed at Da Nang in March 1965, beginning the large-scale U.S. commitment to the Vietnam conflict. Deployed along the so-called demilitarized zone (DMZ), the marines developed an innovative approach to counterinsurgency operations and added to the corps' combat reputation, most notably at Khe Sanh and Hue. In the Gulf War, marines and the threat of marine landings fixed Iraqi units in place as allied armor outflanked and surrounded them. Although often resented by the other American military arms for their elitism and their "creaming off" of the best fighters, the marines remain the United States' principal ready military force.

Stephen Stein

See also: Banana Wars; Boxer Rebellion; Chateau Thierry/Belleau Wood; Chosin/Changjin Reservoir; Grenada; Guadalcanal; Gulf War; Hue, Battle of; Inchon Landings; Iwo Jima, Battle of; Khe Sanh, Siege of; Mayaguez Operation; Mexican-American War; Mexico City, Battles for; Okinawa; Panama Incursion; Peleliu; Philippine Insurrection; Tarawa; Veracruz, U.S. Landings at; Wake Island

References and further reading:

Heinl, Robert Debs. *Soldiers of the Sea: The United States Marine Corps, 1775–1962.* Annapolis, MD: United States Naval Institute, 1962.

Millett, Allan R. *Semper Fidelis.* New York: Free Press, 1991.

Simmons, Edwin H. *The United States Marines: A History.* Annapolis, MD: Naval Institute Press, 1998.

U.S. Militia (1603–1815)

System of common defense utilized by British colonials and the United States. With a lack of regular army units available for service in North America, colonial leaders instituted a militia modeled on the example of Elizabethan England. Under this system, all able-bodied (white) males between the ages of 16 and 60—although the age range varied from colony to colony—were required to serve in the militia. Service was compulsory, and training and drill were mandatory. Militiamen were also required to maintain their own weapons. Some professions were excluded from duty. Officers were usually selected from the higher stations of society. Normally, there were two levels of militia service: the common militia, based on universal service and mostly local and defensive in nature, and the volunteer militia, who formed special companies, normally with elaborate uniforms. The volunteer militia tended to take their obligations more seriously and were more willing to engage in missions outside their colonial boundaries.

The militia served three purposes. The first was to protect local communities from American Indians. To this end, the militia was used most famously in the ferocious fighting of King Philip's War (1675–1676) in Massachusetts. Because King James II refused to send troops to aid his colonials, they were forced to turn to the militia for protection. Second, the militia was to fight rival European colonial powers in time of war. If the force was sent outside colonial boundaries, most typically Canada, volunteers were requested from the ranks of the militia. Throughout the colonial period, New England and New York militia units were employed in invasions and raids on French holdings in Canada. During King William's War (1689–1697), Queen Anne's War (1702–1713), and the French and Indian War (1754–1763), militia troops made several forays into Canada. During the last conflict, the militia took part in the forcible removal of French from Nova Scotia. In the South, Georgia and South Carolina militia fought Spaniards in Florida, most notably during the War of Jenkins' Ear (1739–1748). The most ambitious use of militia overseas occurred in 1741, when 3,000 men from throughout the colonies joined British regulars in an attempted invasion of South America. Yellow fever took a greater toll than the fighting, and only 1,300 made it back to North America. Third, the militia was to uphold domestic order. In the South, for example, militia units took part in suppressing slave rebellions. Occasionally, the militia became the source of unrest. During Bacon's Rebellion (1675–1676), Virginia militiamen critical of Governor William Berkeley's cautious Indian policy took it upon themselves to attack frontier American Indian villages. The rebellion soon faded away, thanks to loyalist militia and news that King Charles II had dispatched 1,100 regulars to quell the uprising. The militia was later used to suppress the backwoods Regulators of North Carolina at the Battle of Alamance (1771).

The quality of the militia deteriorated with the waning of the American Indian threat. In cities like Boston and New York, the night watch replaced the militia as the civic responsibility of white male citizens. With the arrival of large numbers of British regulars after the French and Indian War, the militia became less important for defense and more of a social institution. As colonial assemblies challenged the executive power of royal governors, the legislatures increasingly used the power of the purse to exert greater control over the militia. During the early months of the American Revolution, the militia played a vital role. At the Battles of Lexington and Concord, the Massachusetts militia stood well against the British regulars, and Ethan Allen's Green Mountain Boys captured Forts Ticonderoga and Crown Point in 1775. At Bunker Hill (1775) militia troops inflicted heavy casualties on the British. The creation of the Continental army eclipsed the militia in importance, although the militia made favorable contributions at Saratoga (1777) and Cowpens (1781), among others. General George Washington preferred drilled, trained, disciplined, and long-term enlisted regular army troops to the unreliable militia, ignoring the invaluable role that the militia played in guarding homesteads and keeping the rural economy going. British commanders were surprised by the rapid ability of the militia to mobilize when the king's troops pushed inland. But the militia also melted quickly in battle and proved unable to prevent Sir Charles Cornwallis from occupying parts of the South. However, they did stop him from occupying all the South after a string of disasters had decimated the Continental forces there. In areas such as New York City, loyalists formed militia units that operated with the British army, patrolled Long Island and Westchester County, and fought pitched battles with the patriot militias in the Southern Department.

After independence, the debate over the need for a standing army and how much reliance should be placed on the militia continued. The stunning defeat in 1791 of General Arthur St. Clair's combined militia and regular force by American Indians in Ohio convinced many of the foolishness of giving too much responsibility to the militia. With the Militia Act of 1792, Congress gave the president greater control over the militia and attempted to standardize practices among the different states.

However, the War of 1812 and the reality of waging large-scale offensive and defensive campaigns caused a reassessment of the militia in the defense structure of the United States. The militia troops performed poorly throughout the war. In October 1812 at Queenston Heights, New York, militia troops refused to leave the state even though American regulars were losing to the British on Canadian soil within eyesight. When the British threatened Washington, D.C., in 1814 only 7,500 of the 95,000 militia called to the colors fulfilled their duty, and they fled after suffering a mere 66 casualties.

At the conclusion of the war in 1815, American nationalist politicians realized that the militia was no substitute for a regular army, despite the continued hostility of Jeffersonians to a standing army. They enlarged the standing army, created the office of commanding general of the army, and reorganized the War Department. In the Mexican-American War, the United States supplemented the regular army not with militia but with volunteer forces. Only in the South, where they stood ready in case of slave revolts, was the militia taken seriously. Although the militia continued to exist until the end of the nineteenth century, it was mainly considered a social club, as it no longer played an important role in the military of the United States.

Gregory Dehler

See also: Allen, Ethan; American Revolution; Bacon, Nathaniel; Cornwallis, Sir Charles; Cowpens; Fort Ticonderoga; French and Indian War; King Philip's War; Lexington and Concord; Military and Society; New Orleans, Battle of; Queen Anne's War; Queenston Heights; Saratoga; Shays's "Rebellion"; St. Clair's Defeat; U.S. Army; War of 1812; Washington, Burning of; Washington, George; Whiskey Rebellion

References and further reading:

Dupuy, R. Ernest, and Trevor N. Dupuy. *Military Heritage of America.* New York: McGraw-Hill, 1956.

Maslowski, Peter, and Allan R. Millett. *For the Common Defense: A Military History of the United States of America.* New York: Free Press, 1984.

Shea, William L. *The Virginia Militia in the Seventeenth Century.* Baton Rouge: Louisiana State University Press, 1983.

Utah War (1857–1858)

The armed conflict in Utah Territory between President James Buchanan's administration and the leadership of the Mormon Church. In May 1857, President Buchanan's goal was the restoration of federal authority in a violent, out-of-control Utah. His strategy involved the replacement of Brigham Young, president of the Mormon Church, as governor, and the assignment of a military escort—ultimately nearly one-third of the U.S. Army—to protect a new slate of territorial officers and to enforce federal law. In fact, the conflict was the culmination of a 10-year struggle for power and authority in Utah Territory, then an enormous entity stretching 600 miles from Kansas to California. Pitted against one another were Young's efforts to establish a millennially oriented, Mormon theocracy and the attempts by Presidents Millard Fillmore, Franklin Pierce, and Buchanan to deal with the region as they would any other American territory. Compounding these conflicts was a national uproar following the 1852 confirmation of Mormonism's polygamy doctrine and a growing number of sensational, unsolved murders in Utah. In 1856, the new Republican Party adopted an antipolygamy platform plank, and congressional moves were afoot to dismember Utah Territory as a political entity. By the time Buchanan took office the next March, every non-Mormon federal appointee had fled the territory, the second such mass exodus of Utah's officials in five years.

For Buchanan, the catalyst for action was his receipt in April 1857 of a petition from Utah's legislative assembly couched in rhetoric so inflammatory that his cabinet interpreted it as a de facto declaration of war. Accordingly, in late May 1857, Buchanan ordered the creation of an army expeditionary brigade of 2,500 infantry, dragoon, and artillery troops. Planning for the campaign was both hasty and seriously flawed from the outset. During August 1857, Young mobilized his 6,000-man territorial militia (Nauvoo Legion); accelerated earlier efforts to accumulate and manufacture both arms and ammunition; pulled in the outlying Mormon colonies in San Bernardino, San Francisco, Carson Valley, and Las Vegas; attempted to align Indian allies; cached grain; and instructed Mormon missionaries in Europe to channel converts to Canada. Concurrently, 300 miles to the south at Mountain Meadows, Nauvoo Legionaries and Indian allies massacred settlers, committing what was until 1995 the largest American incident of mass murder involving unarmed civilians. On 15 September 1857, Young proclaimed martial law and took the offensive to deny the army access to Salt Lake Valley. While abandoning all other outlying colonies, Young prepared a last-ditch contingency escape route north to Montana's Bitterroot Valley (and perhaps to the British possessions beyond). The most dramatic and costly such Mormon action came on 4–5 October, when a small detachment fell successively on two large but unescorted contractor wagon trains camped on Green River and the Big Sandy. The attack destroyed an enormous quantity of irreplaceable army matériel and rations.

Shocked by the army's reverses in the Rockies and the enormous cost of the campaign, Congress went into an investigative and debating posture. This political climate prompted Buchanan to accept the offer of Thomas L. Kane, a Philadelphia friend of the Mormons, to travel secretly to Utah to mediate the conflict without government sanction. Traveling covertly via New York, Panama, and California, Kane arrived in Salt Lake City on 25 February 1858 and immediately began discussions with a friendly but unyielding Brigham Young.

Unquestionably, Young's knowledge of the previous September's Mountain Meadows Massacre and its implications, if not his December federal indictment for treason, caused him to draw back. Consequently, by mid-March, he sought to buy time and preserve his options by entering into three nearly simultaneous initiatives to supplement legion plans for an aggressive spring campaign: a willingness to discuss acceptance of the federal governor (without the army); the evacuation of northern Utah (including Salt Lake City) by 30,000 Mormons who were to flee to Provo; and an extensive, misguided exploration of the deserts of western Utah Territory to locate oases to which Young mistakenly believed he could lead his people while fighting a rearguard guerrilla campaign after firing northern Utah.

Unaware of these developments, Buchanan sent out two peace commissioners bearing a blanket amnesty conditional upon Mormon acceptance of federal authority. Young eventually accepted these terms with much posturing and the understanding that the army would not occupy settlements. On 26 June 1858, federal troops marched through a deserted (and prepared for burning) Salt Lake City and on to establish Camp Floyd 40 miles to the south, the nation's largest military post. In July, Mormon refugees returned home. The active phase of the Utah War was over, although tensions, conflicts, and legal maneuvers between the Mormon Church and the federal government would continue in different forms into the first decade of the twentieth century.

William P. MacKinnon

References and further reading:

Bigler, David L. *Forgotten Kingdom: The Mormon Theocracy in the American West, 1847–1896.* Spokane, WA: Arthur H. Clark Company, 1988.

Furniss, Norman F. *The Mormon Conflict, 1850–1859.* 1960. Reprint, New Haven, CT: Yale University Press, 1977.

LeRoy, R., and Ann W. Hafen, eds. *The Utah Expedition, 1857–1858: A Documentary Account.* Glendale: Arthur H. Clark Company, 1958.

U.S. Congress, House of Representatives. *The Utah Expedition.* 35th Congress, 1st Session, 1857–1858, House Exec. Doc. 71 (Serial 956).

V

Vacietis, Jukums (1873–1938)

Red Army commander in chief. Of peasant origins, from Courland, Latvia, Vacietis joined the czarist army (1891) and graduated from Vilno Infantry Academy (1897) and the General Staff Academy (1909).

In World War I, Vacietis was an infantry battalion commander, was promoted to colonel, and commanded the 5th Zemgalsky Latvian Rifle Regiment, a rare national minority-based formation. He backed the Bolsheviks in October while retaining the loyalty of his regiment. Appointed Twelfth Army Commander, he participated in the takeover of the czarist General Staff Headquarters at Mogilev (November 1917) and was head of operational field staff under Stavka (December 1917). Vacietis also suppressed the revolt of Polish general Josef Dowbór-Musnitsky in Belorussia (January–February 1918).

During the Russian Civil War, Vacietis continued to command the remnants of his Latvian regiment, renamed the Soviet Latvian Rifle Corp and then Division (March–April 1918). This unit became the Kremlin bodyguard of the Soviet regime, suppressing the Left Social Revolutionaries uprising in Moscow (July 1918) and earning Vladimir Ilich Lenin's and Leon Trotsky's trust. Vacietis was appointed eastern front commander in 1918 and oversaw the reorganization of Red Army forces into five regular armies, launching successful offensives against Komuch and the Czech Legion (September) and retaking Kazan, Simbirsk, and Samara in the Volga region.

As commander in chief of Red Army forces (September 1918–July 1919), Vacietis oversaw mixed Red fortunes: reverses on the southern front against Anton Denikin, setbacks, and then successes against Aleksandr Kolchak on the eastern front.

In July 1919, Vacietis was removed, arrested, and accused of treason and counterrevolutionary conspiracy after strategic disputes with eastern front commander Sergei Kamenev, which were connected with emerging political disputes between Trotsky and Stalin (the latter Kamenev's sponsor). Released without charge in October, he served in the field until 1921. Vacietis than taught military history, was a senior lecturer on tactics, was professor of Red Army senior military training in the Red Army Military Academy (called the Frunze Academy from 1925), and wrote his memoirs and works on military history and doctrine.

In 1934, Vacietis was a member of the People's Commissariat of Defense. But Stalin's murderous purges caught up with him, and he was arrested in November 1937 and executed in 1938 in Moscow. His past links with Trotsky and disputes with Stalin condemned him, but also his degree of popularity and professional independence, all fatal "flaws" in Stalin's eyes. Vacietis was rehabilitated in 1957. Soon after, Khrushchev's de-Stalinization program began, and a memorial museum in Vacietis's honor was opened in Latvia in 1973.

Neil Harvey Croll

See also: Russian Civil War (1918–1921); Stalin; World War I

References and further reading:

Mawdsley, E. *The Russian Civil War.* London: Allen and Unwin, 1987; Edinburgh: Birlinn, 2000.

Revvoensovet respubliki, 6 sentiabria 1918g–28 avgusta 1923g (Military Revolutionary Council of the Republic, 6th September 1918–28th August 1923). Moscow: Izdat. Politicheskoi Literatury, 1991.

Vacietis, Jukums. "Iz vospominanii Glavkoma I. I. Vatsetis" (From the recollections of commander in chief I. I. Vatsetis). *Voenno-Istoricheskii Zhurnal,* no. 4 (1962).

———. "Vystuplenie levykh eserov v Moskve" (Uprising of the left S. R.'s in Moscow). In *Etapy bol'shogo puty* (Stages of the great path). Moscow: Voenizdat, 1962.

Zaionchkovskii, A. M., and Jukums Vacietis. *Mirovaia voina, 1914–1918.* (World war, 1914–1918). Moscow: Gosvoenizdat, 1931.

Valley Campaign (23 March–9 June 1862)

One of the most brilliant military campaigns by an outnumbered force, one that is still studied in the twenty-first century. As Union general George McClellan began moving the Army of the Potomac to the peninsula between the York and James Rivers, he expected reinforcements to march down from Washington, D.C., to Richmond, thereby presenting the outnumbered Confederates under Joseph Johnston with a serious dilemma.

To help the defense of Richmond from afar, Thomas J. "Stonewall" Jackson received orders to threaten the Union capital from the Shenandoah Valley. Jackson carried out his orders magnificently. He began by attacking a larger Union force under Nathaniel Banks at Kernstown at the northern end of the Shenandoah Valley in late March; Jackson then retreated, somewhat retraced his steps, and moved westward to meet John Frémont at McDowell on 8 May. On 23 May, Jackson's "foot cavalry" (so named for its celerity of movement) surprised a small Union garrison at Front Royal, which caused Banks's army to retreat from Strasberg to Winchester, where Jackson's men drove them back so far that they retreated across the Potomac River into Maryland. When Frémont's troops sought to move down the valley and cut off Jackson's retreat, he attacked them at Cross Keys on 8 June and once again drove them back.

Thus, Jackson made great use of superior mobility and Massanutten Mountain in the middle of the valley; his men marched more than 400 miles and fought a series of battles in six weeks. With only 17,000 men, he kept some 63,000 Union troops from uniting and thereby prevented them both from driving him out of the valley and from adding to the preponderance of force that General George McClellan was slowly assembling in the peninsula campaign. Finally, Jackson slipped away with his men, Union prisoners, and captured supplies to help Robert E. Lee in the Seven Days' Battles, which neutralized McClellan's advance on Richmond, the Confederate capital.

Charles M. Dobbs

See also: American Civil War; Jackson, Thomas "Stonewall"; Seven Days' Battles

References and further reading:

Krick, Robert K. *Conquering the Valley: Stonewall Jackson at Port Republic.* New York: Morrow, 1996.

Tanner, Robert G. *Stonewall in the Valley: Thomas J. "Stonewall" Jackson's Shenandoah Valley Campaign, Spring 1862.* Garden City, NY: Doubleday, 1996.

Valley Forge (1777–1778)

Testing place of the American Revolution. In the winter of 1777–1778, General George Washington, commander of the Continental army, encamped at Valley Forge, Pennsylvania, some 20 miles from British-held Philadelphia. Washington and his men experienced severe difficulties in their winter camp, including starvation, exposure to the elements, and a plot to remove Washington as commander in chief. Throughout this period, Washington lobbied the Continental Congress on behalf of his army, which persuaded the infant government to change its ineffective approach to managing the conflict. His ability to instill professionalism in his troops and to secure the support of the Continental Congress during this distressing period is considered one of his major achievements during the American Revolution. But Washington also received assistance from two important foreign visitors, Marie Joseph du Motier, the Marquis de Lafayette, and Baron Friedrich von Steuben. Lafayette quickly won the admiration of the common soldiery, which helped to bring him battlefield successes later in the war, and he was instrumental in exposing the conspirators against Washington.

Steuben, a Prussian veteran under Frederick the Great, was credited with training the American recruits and instilling the military discipline necessary to confront the professional British army. He also wrote a book of regulations for the Continental army that enjoyed wide circulation throughout the officer ranks. Although Washington's troop levels dwindled to about 4,000, Valley Forge produced a more capable fighting force, which proved itself with a near-victory at Monmouth Court House in New Jersey on 18 June 1778.

Jeffrey Webb

See also: American Revolution; Lafayette, Marie Joseph Paul Yves Roch Gilbert du Motier, Marquis de; Steuben, Friedrich Wilhelm Augustin, Freiherr von; Washington, George

References and further reading:

Bill, Alfred Hoyt. *Valley Forge: The Making of an Army.* New York: Harper, 1952.

Busch, Noel F. *Winter Quarters: George Washington and the Continental Army at Valley Forge.* New York: Liveright, 1974.

Reed, John F. *Valley Forge: Crucible of Victory.* Monmouth Beach, NJ: Philip Freneau Press, 1969.

Valmy (20 September 1792)

French victory over Prussian forces, halting their invasion of France and proving the value of mass conscript armies. In 1792, Charles François Dumouriez's Army of the North advanced toward Belgium. The allied army of Karl William Ferdinand, Duke of Brunswick, moved west, capturing Longwy (23 August) and Verdun (2 September). In late August, Dumouriez turned south to stop Brunswick in the Argonne.

When Austrian troops turned Dumouriez's northern flank, he retreated south to St. Ménéhould at the southern

Washington at Valley Forge. (Library of Congress)

end of the Argonne. Brunswick delayed, permitting the arrival of François Kellermann's Army of the Center, allowing some 64,000 French to mass around Valmy. The allies numbered 36,000. The French deployed facing westward in an arc from Mount Yron, around Valmy, across the main St. Ménéhould–Paris road. On 19 September, Frederick William II of Prussia ordered Brunswick to capture this road to cut Dumouriez off from Paris. On 20 September in dense fog and rain, the Prussians moved south toward the road. About 7:00 A.M., they encountered French artillery fire from their east. Thus began the battle of Valmy.

The Prussians crossed the road, and by noon Brunswick had deployed facing northeast toward the hill of Valmy. As the fog lifted, they confronted well-organized French troops. Frederick William and his officers then ordered an infantry charge up the hill. The French unleashed a massive artillery barrage. After advancing only 200 yards, Brunswick halted. The cannonade increased. A French counterattack failed, as did a second Prussian thrust. Brunswick then abandoned the attack. The battle dwindled out in dusk and heavy rain. The French lost fewer than 200 casualties and the Germans about 300. The French left the battlefield, but in terms of ability to continue, the Germans had lost at Valmy. On 30 September, Brunswick's demoralized troops began to evacuate France. Dumouriez, some of whose own men had tried to mutiny, did not interfere.

Valmy temporarily halted the foreign invasion, thus saving the revolution and the republic. It foreshadowed the use of massed artillery and showed Europe that the French army and volunteers were a formidable force. In time, all major European powers would adopt the French concept of *levée en masse*—mass conscript armies.

James K. Kieswetter

See also: French Revolutionary Wars

References and further reading:

Bertaud, Jean Paul. *Valmy: La démocratie en armes.* Paris: Gallimard/Julliard, 1989.

Chuquet, Arthur. *Les guerres de la Révolution.* Vol. 2, *Valmy.* Paris: L. Cerf, 1887.

Hublot, Emmanuel. *Valmy ou la défense de la nation par les armes.* Paris: Fondation pour les Etudes de Défense Nationale, 1987.

Phipps, Ramsay Weston. *The Armies of the First French Republic and the Rise of the Marshals of Napoleon I.* 5 vols. Oxford, UK: Oxford University Press, 1926–1939.

Valois-Habsburg Wars (1521–1559)

Italy brought France the Renaissance, and France brought Italy warfare, strengthening the Valois Dynasty. However, war in Italy was about more than Italy. The peninsula was just one theater of the conflict between the Valois and Habsburg Dynasties for the domination of western Europe. When the Habsburg king of Spain, Charles V, became Holy Roman Emperor in 1519, France faced encirclement from the Spanish Netherlands to Germany to Spain to Italy. The wars were really about dynastic considerations and for the prestige and glory of the warrior-kings involved.

The French monarch, Francis I, began war against Charles in Italy in 1521 but was defeated and captured at Pavia in 1525. Charles imprisoned Francis in Spain and ransomed the king, who returned to France in 1527. After another round of fighting, the two monarchs made peace in 1529, and Francis married the emperor's sister, Eleanor. The two monarchs fought further inconclusive wars from 1536 to 1538 and from 1542 to 1544. In this period, Francis, a Catholic, did not hesitate to ally against the Catholic Habsburgs with German Protestant princes and with Muslim Ottomans.

Yet these wars fought largely in Italy also sparked a revolution in military affairs that included the construction of great fortresses on the model called the *trace italienne* and saw the rise of the musket as the newest infantry weapon. By 1529, large standing armies were the norm, not small dynastic armies raised for a war and then disbanded. To pay for the armies and artillery now required in the new age of warfare, Francis I used his personal fortune to buy the loyalty of his nobles with titles and cash. He used the new form of patronage to control his nobles by creating vertical ties that bound them to him as tightly as had the old ties of feudalism.

The most significant battle of the period, Pavia (1525), was the first attack on France from Italy in centuries. Arquebusiers came out from behind their ramparts and attacked in ranks, in the first battle in which small arms fire was decisive. It was also a battle of surprise, maneuver, and massacre. The French lost 8,000 to the imperials' 700. By 1530, successful sieges were a matter of execution, not invention. However, bastional fortresses on the Italian design were expensive, and only rich states or cities could afford them. Recognizing the importance of the fortress, the French began constructing a double line of fortresses in the northeast, which held through a lack of resolve among its enemies and would hold through the wars of Louis XIV.

The Treaty of Cateau-Cambrésis in 1559 that ended the Valois-Habsburg Wars also illustrated their international aspect beyond Italy. The French renounced any claims on the Italian peninsula, acquired Calais from the English, and secured Toul, Metz, and Verdun.

David C. Arnold

References and further reading:

Duffy, Christopher. *Siege Warfare: The Fortress in the Early Modern World, 1494–1660.* London: Routledge and Paul, 1979.

Oman, Sir Charles. *A History of the Art of War in Italy in the Sixteenth Century.* London: Methuen, 1937.

Taylor, F. L. *The Art of War in Italy, 1459–1529.* 1921. 2d ed., Westport, CT: Greenwood Press, 1973.

Van Fleet, James A. (1892–1992)

Eminent U.S. Army commander of the early Cold War. Born in New Jersey and raised in Florida, Van Fleet graduated from the U.S. Military Academy in the class of 1915, his initial combat service being in World War I. Though achieving corps-level rank in World War II, Van Fleet labored under the unusual burden of having been mistaken for a dissolute officer who had served at the U.S. Army's infantry school at the same time George C. Marshall was commandant, a career-killing perception not corrected until 1944.

In the immediate post–World War II period, Van Fleet led the American military mission to Greece during that country's civil war and furthered his reputation by making the Greek military battle-worthy. Replacing Mathew Ridgway as the commander of the Eighth Army in Korea in 1951, Van Fleet directed the war of attrition that characterized most of that conflict, also presiding over the reconstruction of the South Korean army.

Van Fleet retired from service not long after the armistice signed at Panmunjom, evidently embittered by the failure to wage a more aggressive war and the combat death of his son, though he made a return to public life as a special ambassador to the Far East during the Eisenhower administration. He is best remembered as an aggressive infantry officer with a talent for motivation and training.

George R. Shaner

See also: Greek Civil War; Korean War

References and further reading:

Blair, Clay. *The Forgotten War: America in Korea, 1950–1953.* New York: Times Books, 1987.

Colby, John. *War from the Ground Up: The 90th Division in WWII.* Austin, TX: Nortex Press, 1991.

Wiegley, Russell F. *Eisenhower's Lieutenants: The Campaign of France and Germany, 1944–1945.* Bloomington: Indiana University Press, 1981.

Wittner, Lawrence S. *American Intervention in Greece, 1943–1949.* New York: Columbia University Press, 1982.

Vandals

A Germanic tribe from Jutland. The Vandals crossed the Rhine River with the Alans and Suevi in 406. Three years later, they were in Spain.

In 429, facing pressure in Spain from Romans and Visigoths, the Vandal king, Gaiseric (r. 428–477), ferried his people across the Strait of Gibraltar and led them east along the North African coast. The crippled son of a slave, Gaiseric was a proud and ruthless leader and a gifted conspirator with a genius for political intrigue that for 50 years foiled the plans of Romans and fellow Germanic rulers alike. One by one, the Roman cities with their well-stocked granaries fell to the Vandals. Even before the Vandal conquest of North Africa was complete, Gaiseric turned restlessly to a new project: he built a fleet and launched himself on a lucrative career of piracy.

Establishing themselves as a warrior aristocracy and leaving administrative chores to Roman bureaucrats, the Vandals carved out large estates and made their homes among the provincials. Vandals, like many other Germanic tribes, had been converted to "heretical" Arian Christianity, which argued that Jesus was less than divine. Relations of Vandals with the majority Catholics were thus strained. Gaiseric, an ardent Arian, barely held animosities in check. Under his successor, Huneric (r. 477–484), violence erupted.

Gratuitous cruelty was only one symptom of the Vandal degeneration after Gaiseric. Warriors seduced by luxuries grew weak, corrupt, and disorganized. In 533, the eastern Roman general Belisarius crossed into Africa with his army. In one campaigning season, he crushed the Vandal kingdom completely.

Nic Fields

See also: Belisarius; Justinian I

References and further reading:

Clover, F. M. *The Late Roman West and the Vandals.* Aldershot, UK: Variorum, 1993.

Vauban, Sébastien Le Prestre de (1633–1707)

The most famous military engineer of the seventeenth century. Vauban was born into very modest country nobility. His only way to climb the social scale was to serve in the army. He bought a cadet brevet in the Condé regiment in 1651, but his first battles were against king's troops during the Fronde civil war. Taken prisoner in 1653, he sided with the strongest side and soon distinguished himself in the king's army in siege warfare. He was admitted in 1659 as a "king's ordinary engineer," one of a corps of specialized technical officers in charge of building and besieging fortifications. The wars against the Dutch and of the Augsburg League gave him the opportunity to reveal his talents. In charge of all fortification works from 1678, he was appointed lieutenant general in 1688 and eventually marshal of France in 1703, the first time a technical officer had been raised so high.

Vauban is more famous for his fortification design than for his military achievements. He formalized the system of fortification in memoranda that were to be the focus of military studies until the late nineteenth century. His *Treatise on Attack of Forts* (1701) summarized his actual experiences: the weakest point had to be assaulted first after having dug parallels and zigzagging communication trenches; any siege needed five times the number of besieged troops. His *Treatise on Defense of Forts* (1706) is his military legacy; he improved earlier forms (those of Blaise de Pagan and Baron Menno von Coehoorn) by his own three systems: first, each bastion is covered by the fire from a flanking bastion; second, three lines of fortification are built: low towers on the curtain, bastions and tenails (low rampart in the ditch), and ravelins as exterior works with ditch and glacis; third, the fortification is integrated into the urban project (Neuf Brisach is the only example). During his active service, he built 33 new fortifications and worked on 300 older ones, including those in allied countries. He persuaded Louis XIV and his war minister Michel le Tellier, the Marquis de Louvois, to define a defensive strategy called *le pré carré* (squared meadows), which involved the making of linear borders easy to defend by a double line of fortified towns. His system of fortifications and fortresses saved France in 1709.

Gilles Boué

See also: French Army; Louis XIV

References and further reading:

Blanchard, Anne. *Les ingénieurs du Roi de Louis XIV à Louis XVI.* Montpellier, France: Presses Universitaires, 1979.

———. *Vauban.* Paris. Fayard, 1987.

Rocolle, Pierre. *2000 ans de fortification française.* Paris. Lavauzelle, 1989.

Vegetius Renatus, Flavius (fl. late 300s)

Roman military strategist and thinker writing in the twilight of the empire. His *De re militari* became a standard military guide throughout medieval Europe.

Likely born in Roman Spain, Vegetius was a nobleman, a Christian, and a well-traveled civil servant. A breeder of cavalry horses, he was deeply versed in the military literature of Republican and imperial Rome. Vegetius wrote his *Epitoma* some time after 380 but before 410, the first "fall" of Rome.

Epitoma sought the military rejuvenation of the empire through a revival of Augustan-era professionalism. Vegetius deplored as self-destructive the empire's dependence upon undependable Germanic and other barbarian tribal levies. Rome instead required an army and navy recruited from indigenous peoples, trained according to proper discipline, esprit, and patriotism. Vegetius also discussed tactics to deploy such forces, which, though smaller, would be more reliable and manageable than the tribal levies. Military historians find Vegetius an invaluable guide to the ways of warfare in Europe at the end of the classical era and the dawn of the Middle Ages.

Weston F. Cook, Jr.

References and further reading:

Vegetius Renatus. *Vegetius: Epitome of Military Science.* Trans. and ed. by N. F. Milner. Liverpool, UK: Liverpool University Press, 1996.

Venetian-Genoese War (1255–1381)

The war between the city-states of Venice and Genoa lasted almost 130 years and resulted in the bankruptcy of Genoa. Mostly conducted at sea, the number of troops and ships involved were staggering for the day. Conventional and unconventional tactics were used throughout.

Although rivals for years, the cities went to war over control of a church in Acre. A scuffle one day ended in a death. The Genoese invaded the Venetian quarter, setting it aflame and driving the refugees out. Now in control of the city, the Genoese stretched a heavy chain across the harbor to keep out Venetian ships. In September 1257, a Venetian fleet arrived. It sailed into the harbor and broke the chain; the Venetian forces captured all the Genoese merchant ships and burned two galleys. They then fought their way inland to the disputed church. The Venetians burned all the surrounding fortifications and houses and then moved into the Genoese quarter to treat the Genoese as they had treated the Venetians.

The Genoese regrouped and made a stand at Tyre, where the fleeing fleet met up with other Genoese ships. The Venetians formed a battle line at the mouth of the harbor and challenged the Genoese to come forth. The harbor mouth would allow only one ship to leave the harbor at a time. But the Genoese sailed forth anyway, with predictable results. All the ships were burned or captured.

Despite winning the battles, the Venetians were forced by the pope into a truce that would last until 1293. At the expiration of the truce, the Genoese controlled the unfortified town of Pera. The Venetians burned this town and moved on to the city of Caffa. A winter in Crimea cost the Venetians more than half their crews and equipment. In the spring, the two fleets met off Curzola in the Adriatic Sea. Although outnumbered, the Genoese admiral Andria Doria held some ships in reserve, with orders not to engage until the wind allowed them to bear down in the heat of the action. The result was that 65 Venetian ships burned.

The two sides declared peace in 1299, with the Genoese prohibiting any armed Venetian ships from entering the Black Sea or the harbors on the coast of Syria for 13 years. It would be a third party that would cause the two to go to war again. Both Venice and Genoa had agreed to suspend commercial activity with the city of Tana. Venice broke the agreement and began trading. Genoa, still in control of the Black Sea, declared an embargo on all Venetian ships entering that sea. The Venetians sent a ship to the archipelago region to challenge the embargo. The two nearly equal fleets met in the Bay of Caristo, with the victory going to Venice.

Three years later, the fleet met in February off the Dardanelles. Pisani commanded the Venetian fleet, and Doria commanded the Genoese. The battle and a storm began at the same time. Wreckage littered the Sea of Marmara. The Genoese won, but only because Pisani's battered but intact squadron could not renew the fight.

A year later, Doria entered the Adriatic Sea with reinforcements and joined the main fleet at Pola, reconquering the Dalmatian and Istrian cities that Pisani had taken the previous year. The war on land was going just as badly for the Venetians as the naval war. The Genoese allies took unfortified villages and besieged the cities.

By 1380, it appeared all but over for Venice. Doria attempted to take Malamacco on August 6. The Venetians repelled him, and he burned Pelestrina in revenge. He then seized and attached little Chioggia. It fell on 16 August, and thereby the encirclement of Venice was complete. Doria then proceeded to harry the Venetian outposts. By October, the Genoese withdrew to Chioggia and established a blockade around the city of Venice.

To save the city, Pisani proposed the radical step of blockading the blockade. On 22 December, 34 galleys left Venice and stood off Chioggia. The next day, the Venetians landed 4,800 men on Sottomarina as a diversion. Meanwhile, the rest of the fleet sank two barges across the mouth of the port, in Brondolo canal, and blocked the Lombard canal. Now, there was no way out of Chioggia for the Genoese. Nevertheless, the Venetians remained blockaded too.

In a New Year's miracle, Zeno sailed into view on 1 January and broke the Genoese blockade. The blockade continued, the Venetians preferring to starve the Genoese out rather than face them in battle. Grimaldi (in command after the death of Doria) finally surrendered on June 24.

The pope again pushed the two sides to settle their dispute. Venice and Genoa signed a peace treaty on 8 August 1381 at Turin. Venice lost Trieste to the Duke of Austria and

Tenedos to the Duke of Savoy. It was forced to renounce its claim to Dalmatia. However, Venice did recover its commercial privileges on terra firma and at Constantinople.

Elizabeth Pugliese

References and further reading:
Brown, Horatio F. *Venice: An Historical Sketch of the Republic.* London: Percival and Company, 1893.
Thayer, William Rosco. *A Short History of Venice.* New York: Houghton Mifflin, 1908.
Wiel, Alethea. *The Navy of Venice.* London: John Murray, 1910.

Venezuelan Civil Wars (1858–1870)

A confusing revolving-door succession of caudillos (strongmen) who did little to deal with Venezuela's real problems of poverty and political immaturity. The civil wars began on 15 March 1858, with the overthrow of the liberal oligarchy of president José Tadeo Mongas by General Julian Castro. Venezuela was plagued by political arguments over the form of the national government. The two divergent issues were federalism and centralism. However, the main cause of political unrest that gripped the nation was not ideology but the desire for power by caudillos. For the balance of 1858, Castro's opponents plotted his ouster, while a constitutional convention drafted a new document that was approved on 31 December, but the country still drifted toward anarchy.

On 20 February 1859, the Federal Revolution began, when 40 rebels captured the city of Coro and proclaimed a new federation for all of Venezuela. General Juan Cris'stomo Falcón, Antonio Leocadio Guzmán, and General Ezequiel Zamora were the leading members of the provisional junta. On 10 December, Zamora led the Federal Army to victory over the Constitutionalists at Santa Inéz and followed this success by driving his opponents out of western Venezuela. The federalists were on the verge of complete victory when Zamora was killed on 10 January 1860 and Falcón took direct charge of the federal army.

The federal cause was dealt a serious setback on 17 February, when a Constitutionalist army, commanded by General León Febres Cordero, smashed Falcón's force at Coplé. Falcón escaped to neighboring New Granada (Colombia), and the war might have ended at this point, had the political anarchy of the government not given the federalists a chance to regroup and continue the fight. The presidency had become a revolving door. For example, in 1859 Castro resigned, reclaimed his office, was imprisoned, and was pardoned. Several men, including Pedro Gaul, held the position until José Antonio Paéz once again became head of state of Venezuela in 1861.

In August 1862, at the direction of Falcón, the federalist army opened an offensive in the center of the country and threatened the capital, Caracas. By May 1863, General Antonio Guzmán Blanco led the federalists to further victories and control of most of the country. On 6 June, Paéz ratified the Treaty of Coche, Falcón became president, and Guzmán Blanco became vice president.

The federalist victory and Falcón's leadership brought little peace. From 1863 to 1868, he had to deal with numerous uprisings and civil unrest. In January 1868, the Blue Revolution, so named because the rebels adopted a blue badge, made a concerted effort to depose Falcón. The revolution was initially led by General Miguel Antonio Rojas, but José Tadeo Mongas soon took the lead. The rebel army entered Caracas in early June, and Falcón went into exile. The death of Mongas led to in-fighting between his son, José Ruperto, and his nephew, Domingo Mongas. José Ruperto was able to assume the presidency on 8 March 1869.

In February 1870, Guzmán Blanco led an invasion to take control of the liberal movement and to seize power. He entered Caracas on 27 April and soon became president, a position he held until 1877. Venezuela would know peace thereafter but not much democracy.

George M. Lauderbaugh

References and further reading:
Alvarado, Lisandro. *Historia de la Revolución Federal en Venezuela.* 2d ed. Caracas: Ministerio de Educación, 1956.
Morón, Guillermo. *A History of Venezuela.* New York: Roy Publishers, 1964.
Rodrégues, J. S. *Contribución al Estudio de la Guerra Federal en Venezuela.* 2d ed. Caracas: ECPCRF, 1960.

Veracruz, Siege of (9–28 March 1847)

When it became clear that victories in northern Mexico would not bring about Mexico's surrender, U.S. president James Polk agreed to the seizure of Veracruz and a march to Mexico City itself. General Winfield Scott, following a suggestion from Commodore David Conner, proposed landing on a beach south of Veracruz's formidable defenses and laying siege to the city—all timed to move up into the mountains before yellow fever (El Vomito) season incapacitated the invaders.

The landings began on 9 March 1847. By midnight, reflecting great army-navy cooperation, more than 100 vessels helped land more than 10,000 men, their supplies, and draft animals without loss of life. The Mexican commander, General Juan Morales, remained within Veracruz's defenses and did not contest the landing. Scott declined a suggestion from some of his officers to rush the defenses and set about to establish a siege. The landward bombardment began on 22 March; two days later the U.S. Navy joined the bombardment from the sea, and Scott cut off supplies to the city, including fresh water. Mexican artillery was not effective in response, in part because of its use of solid copper cannon balls.

Six days after the bombardment began, on March 28, the Mexican army surrendered the city and the fortress of San Juan de Ulua, and Scott managed to move his command inland and up the mountain pass for the eventual attack on Mexico City before the yellow fever epidemic season began.

The Americans would return to Veracruz in 1914, acting as the agents of President Woodrow Wilson's absurd attempts to "teach the Mexicans democracy" and as part of an overreaction to an imagined Mexican "insult" to the United States. After intense fighting, the U.S. Army would occupy and administer the city for six months.

Charles M. Dobbs

See also: Mexican-American War; Scott, Winfield

References and further reading:

Eisenhower, John S. D. *Agent of Destiny: The Life and Times of General Winfield Scott.* New York: Free Press, 1997.

Johnson, Timothy D. *Winfield Scott: The Quest for Military Glory.* Lawrence: University Press of Kansas, 1998.

Sandler, Stanley. *Glad to See Them Come and Sorry to See Them Go: A History of U.S. Army Tactical Military Government/Civil Affairs, 1775–1991.* Fort Bragg: U.S. Army Civil Affairs and Psychological Operations Command. 1998.

Veracruz, U.S. Landings at (1914)

Relations between Mexico and the United States deteriorated with the onset of the Mexican Revolution in 1910. In 1914, tensions mounted when U.S. president Woodrow Wilson refused to recognize the regime of General Victoriano Huerta. The Tampico incident of 9 April, involving the brief arrest of a group of U.S. sailors by Mexican soldiers, intensified the crisis. On 21 April, U.S. troops seized the Mexican port of Veracruz to stop a shipment of arms from Germany to Huerta's forces and to retaliate for the Tampico incident.

At first, Navy bluejackets planned a limited operation to secure the waterfront and the customs house. However, spirited opposition by the general population and Mexican soldiers caused U.S. commanders to order the seizure of all of Veracruz. The capture of the city resulted in 17 American deaths. Estimates placed Mexican losses at 126 killed, including civilians. The American intervention caused Huerta to call for the unification of all revolutionary factions to resist. When it became apparent that the intervention would be limited to Veracruz, Huerta's opponents, including Venustiano Carranza, Pancho Villa, and Emiliano Zapata, renewed their efforts to depose him.

The occupation of Veracruz did not completely cut off the supply of arms to Huerta, and the United States was frustrated by its inability to control events in Mexico. A diplomatic effort to negotiate a settlement at the Niagara Falls Conference failed. Moreover, President Woodrow Wilson found that he had little leverage with revolutionary forces in Mexico under the command of Carranza, who refused to negotiate with the Americans. Nevertheless, the Huerta regime fell in July.

After a seven-month occupation, American forces withdrew in November 1914. Carranza's forces then took control of the city and the large quantities of military stores, which were soon used in the next phase of the Mexican Revolution.

George M. Lauderbaugh

See also: Mexican Unrest and Civil War

References and further reading:

Eisenhower, John S. D. *Intervention! The United States and the Mexican Revolution.* New York: W. W. Norton, 1993.

Quirk, Robert E. *An Affair of Honor: Woodrow Wilson and the Occupation of Veracruz.* Lexington: University of Kentucky Press, 1962.

Vercingetorix (d. c. 45 B.C.E.)

Chieftain of the Averni, one of the dominant Gallic tribes, Vercingetorix led a confederation of tribes against Rome and was defeated by Julius Caesar. Vercingetorix's leadership presented Julius Caesar with a major crisis in Gaul. Vercingetorix was soundly defeated in an open field battle against Caesar at Noviodonum in the winter of 52 B.C.E. Now wary of direct confrontations with the Romans, Vercingetorix employed guerrilla raids and scorched-earth tactics. In March 52 B.C.E., Caesar moved quickly to eliminate one of the centers of Gallic rebellion and so laid siege to the Biturigan stronghold of Avaricum. During the siege, the Gauls effectively used fortifications, fire, and ballistics against Caesar's two legions. Despite the Gauls' attempts to end the siege, the Romans ultimately broke through the fortifications and slaughtered the city's 40,000 inhabitants. Roughly a month after the defeat of Avaricum, Caesar turned his attentions to Gergovia, another center of rebellion in central Gaul and an imposing fortress situated on a steep hill. Vercingetorix, however, beat Caesar to Gergovia and, employing many of the tactics used at Avaricum, carefully prepared its defenses. Vercingetorix repulsed the Roman attacks and handed Caesar one of the rare defeats of his career. In the summer of 52 B.C.E., Caesar briefly engaged Vercingetorix's forces near Alesia. Retreating from the Romans, Vercingetorix expelled the city's women and children in order to make room for his 60,000 troops. Caesar's 50,000 legionaries and Germanic cavalry laid siege to Gergovia. Caesar withstood three bloody assaults against his position by 100,000–250,000 (sources differ) Gallic troops who attempted to relieve Vercingetorix. He attempted to break out of the city but soon realized that Roman victory was inevitable. Upon surrender at Alesia, Vercingetorix was taken to Rome and executed in 46 B.C.E.

Eric D. Pullin

See also: Alesia, Siege of; Ancient Warfare; Caesar, Julius; Gallic Wars

References and further reading:

Caesar, Julius. *The Conquest of Gaul.* London: Penguin, 1982.
Homes, T. Rice. *Caesar's Conquest of Gaul.* London: Macmillan, 1899.
Nardo, Don. *Caesar's Conquest of Gaul.* San Diego: Lucent Books, 1996.
Ridd, Stephen, ed. *Julius Caesar in Gaul and Britain.* Austin, TX: Raintree Steck-Vaughn, 1995.

Verdun (21 February–18 December 1916)

The most concentrated carnage in military history. The campaign was the brainchild of German field marshal Erich von Falkenhayn, who hoped to force France to commit the bulk of its army to defend the fortified city of Verdun, where it could then be destroyed in place. (Falkenhayn fastidiously termed this "bleeding the French.") Operation GERICHT (place of judgment) began on 21 February with a massive artillery barrage. The German Fifth Army under Crown Prince Wilhelm followed with attacks on French forts east of the Meuse River. On 25 February, Fort Douaumont fell, and Field Marshal Joseph Joffre named General Henri-Philippe Pétain to command Verdun's defenses. Pétain proclaimed of the Germans, "Ils ne passeront pas!" ("They shall not pass!"), a statement that rallied the weary French military and civilians.

The German offensive switched briefly to the west bank of the Meuse River, where it again met stiff resistance. In the east, the Germans came within 5 miles of Verdun but were unable to clear the defenders from Fort Vaux. On 6 June, the attackers finally overran Vaux and advanced to within 2 miles of the city by the end of the month.

The French, under the command of Pétain and his successor Robert Nivelle, were able to keep the army supplied and reinforced by truck along the *voie sacrée* (sacred way). On 24 October, they launched a series of counterattacks, which took back the lost forts and erased most German gains.

The German high command, shaken by the British offensive on the Somme, Romania's entry into the war, and its failures at Verdun, lost confidence in the operation. On 29 August, Falkenhayn was replaced as chief of staff by Paul von Hindenburg and Erich Ludendorff. Fighting around Verdun ended by mid-December.

Casualty figures vary, but it appears that the two sides suffered nearly 1 million combined casualties at Verdun. To this day, the battlefield is hazardous, with unexploded ordnance still lurking.

The holding of Verdun was a tremendous morale boost for the French and marked the end of the German "western" strategy. But it would be difficult to argue that either side "won" the Battle of Verdun; the losers can be more easily identified—the youth of France and Germany.

Adam R. Seipp

See also: Falkenhayn, Erich von; Nivelle, Robert; Pétain, Henri-Philippe; The Somme; World War I

References and further reading:

Blond, Georges. *Verdun.* Paris: Presses de la Cité, 1980.
Denizot, Alain. *Verdun, 1914–18.* Paris: Nouvelles Éditions Latines, 1996.
Horne, Alistair. *The Price of Glory: Verdun, 1916.* New York: Penguin, 1962.

Vespasian (9–79)

Roman emperor and general and founder of the Flavian Dynasty. Vespasian was born at Reate in the Sabine region of Italy on 18 November 9. As a young officer, he rose through the ranks, serving in Thrace, Crete, Cyrenaica, and Germany, and was commander of the second legion in the invasion of Britain in 43–44. He was consul in 51, proconsul of Africa in 63, and governor of Judea in 67 during the First Jewish Revolt. After a year of fighting against fanatical resistance, the Romans subdued the countryside and prepared to besiege Jerusalem. Upon the death of Nero in 68 and the subsequent political instability in Rome, Vespasian conspired with other Near Eastern governors to seize control of the empire. Their plan was for legions from Syria to attack Rome, while Vespasian held sway over the crucial supply of Egyptian grain. However, without orders, legions from the Danube River under Primus invaded Rome and defeated Vespasian's rival Vitellius in December 69. The Senate formally declared Vespasian emperor on 21 December 69.

For the most part, once Judea was conquered and a Gallo-German rebel force under Civilus was defeated, the empire had peace under Vespasian. Exceptions included the extension of the border in Britain and the taking of some strategic territory in Germany. As emperor, he raised taxes and successfully reformed finances to replenish the treasury, which had been drained by the civil wars that preceded his reign.

Vespasian sought to found a dynasty, and both of his sons, Titus and Domitian later became emperors themselves. He died of illness on 24 June 79. History regards Vespasian, known for his work ethic and honesty, as one of the better emperors.

Harold Wise

See also: Jewish Revolts

References and further reading:

Levick, Barbara. *Vespasian.* New York: Routledge, 1999.

Suetonius. *History of Twelve Caesars.* Trans. Philemon Holland (1606). New York: Arms Press, 1967.

Tacitus Cornelius. *The Histories.* A new translation by Kenneth Wellesley. Harmondsworth, UK: Penguin, 1972.

Vicksburg, Siege of (18 May–4 July 1863)

The most significant siege of the American Civil War. In late 1862, Union general Ulysses S. Grant had sought a way to take the last major Confederate bastion interdicting the Mississippi River, the city of Vicksburg. But his efforts to advance south from Corinth left him vulnerable to raids by Generals Nathan Bedford Forrest and Earl Van Dorn against his lengthening supply line. Grant had to retreat.

The Union commander then decided to cross to the west and move opposite Vicksburg; he tried various means to take the city, none of which were successful, and eventually devised a brilliant strategy. He had Admiral David Porter move his gunboats south at night, past Vicksburg. The Union navy helped transport Grant's command across the Mississippi below Vicksburg at Bruinsburg, and then shedding his supply line, he said his men could live off the land. He moved first to block Joseph Johnston at Raymond on 14 May, driving him back to Jackson, the state capital, and then trapping the Vicksburg the garrison commanded by John Pemberton. On 16 May, as Pemberton looked for the nonexistent Union supply lines, Grant attacked him at Champion Hill and Big Black River, and Pemberton—perhaps unwisely—retreated into Vicksburg.

Two days later, Grant reestablished contact with the navy (and hence his supply line). He unsuccessfully tried to rush

Surrender at the Siege of Vicksburg, 4 July 1863. (Library of Congress)

the city's defenses and thereafter settled in for an ever-tightening siege. Many city residents lived in caves dug into the bluffs to avoid the constant artillery shelling and were reduced to eating domestic animals. On 4 July 1863, one day after the end of fighting at Gettysburg, Pemberton surrendered his surviving 29,000 men and the key city. Several days later, the defenders of Port Hudson also surrendered, and the Union controlled the Mississippi River, opening "the Father of Waters" to commerce all the way to New Orleans and the sea and cutting off Texas and Arkansas from the rest of the Confederacy.

Charles M. Dobbs

See also: American Civil War; Grant, Ulysses Simpson

References and further reading:

Arnold, James R. *Grant Wins the War: Decision at Vicksburg.* New York: John Wiley, 1997.
Healing, A. A. *Vicksburg: 47 Days of Siege.* Englewood Cliffs, NJ: Prentice-Hall, 1996.
Wheeler, Richard. *The Siege of Vicksburg.* New York: Crowell, 1978.

Vienna, Sieges of (1529, 1683)

From 27 September to 14 October 1529, Süleyman the Magnificent besieged the city of Vienna in hopes of breaking the power of the Austrian Empire and opening the way to central Europe. More than 120,000 Turks repeatedly assaulted the fortifications around the city, defended by 16,000 troops and many civilians under the Count de Salm. A final attempt to storm a breach in the walls was repulsed with heavy losses, and Süleyman withdrew. The Turks soon found themselves fighting multiple wars against Russia, Venice, Persia, Austria, Hungary, and others and were not able to return their attention to Vienna until 1683.

In September 1683, the Turks, riding a resurgence of power, besieged Vienna again with 138,000 troops under Kara Mustapha Pasha. By this time, complex *trace italienne*–style fortifications, developed by the French engineer Sébastien Le Prestre de Vauban, were employed to defend many European cities, including Vienna. Giant Turkish guns, so successful against the old walls of Constantinople in 1453, were no longer very effective. The Turkish siege was poorly conducted, with no defensive breastworks to protect Turkish lines in case of rear attack. That is exactly what happened when King Jan III Sobieski of Poland came to the aid of Vienna with 30,000 troops. He was also given command of 40,000 Austrian troops, and a rear attack on the Turkish line led to a fierce, desperate, daylong battle that found the Turks sandwiched between the fortifications of Vienna and the forces of Sobieski. The Turks were crushed with huge losses, including six pashas. Kara Mustapha Pasha narrowly escaped, and Austria went on to take Belgrade, the Turkish Balkan stronghold, and then marched on Constantinople itself before the Turks sued for peace. Turkish power was forever broken in Europe.

Christopher Howell

See also: Austro-Turk Wars; Mohács, Battles of; Süleyman I

References and further reading:

Goodwin, Jason. *Lords of the Horizons: A History of the Ottoman Empire.* New York: Owl Books, 2000.
Turfan, Naim. *Rise of the Young Turks: Politics, the Military and Ottoman Collapse.* Istanbul: I. B. Tauris, 2000.
Vaughan, Dorothy. *Europe and the Turk: A Pattern of Alliances, 1350–1700.* Liverpool, UK: Liverpool University Press, 1954.

Vietnam Conflict (1961–1975)

The United States' most controversial conflict and Vietnam's most devastating one. The roots of this conflict go back to French colonial presence in the area. During World War II, President Franklin D. Roosevelt had mixed feelings about France and, to a lesser extent, other European nations as colonial powers. However, in the extreme haste and confusion that marked the end of the war against Japan, the British and Nationalist Chinese accepted the Japanese surrender in Indochina, and the British helped reintroduce French military control.

Once Vietnamese nationalists were convinced that the French intended to return, an increasingly strong and successful resistance movement began. In 1949, Communist forces had occupied southern China, which gave the Vietminh, the communist-dominated Vietnamese resistance movement, access to sanctuary, supplies, and perhaps Chinese "advisers." The conflict soon worsened for the French, who after the debacle at Dien Bien Phu in May 1954 agreed to a cease-fire and to participate in a meeting at Geneva that would spell the end of their Asian empire.

In the aftermath of the North Korean attack on South Korea and the subsequent and related adoption of the nearly hysterically anticommunist National Security Council document 68/4, the United States increased assistance to the French. By 1954, the United States was bearing perhaps 80 percent of the material cost of the conflict. Although the Geneva Conference provided for the division of Indochina into three countries—Laos, Cambodia, and Vietnam—it also provided for a temporary division of Vietnam at the 17th parallel to facilitate the movement of troops and civilians prior to French withdrawal from the area. And at about the same time, the United States, believing that South Vietnam could be the first of a series of "dominoes" to fall to communism in that resource-rich and important region, set about to organize an independent government, the Republic of Vietnam, for the area south of the 17th parallel.

From 1956 to 1961, there was relative peace in Vietnam. Ho Chi Minh and his Communist colleagues set about organizing a Maoist-style regime in the north, while a Catholic Vietnamese who had collaborated with the French and perhaps Japanese colonial authorities, Ngo Dinh Diem, set about creating an authoritarian government in the south with American assistance. Late in the period, the government in Hanoi began preparing for the takeover of the south, organizing Groups 559, 759, and 959 to facilitate the infiltration of men and matériel by land and sea routes.

Communist insurgency increasingly began to threaten the government in Saigon after 1961. Although a growing number of American advisers brought new ideas, including the new concept of air mobility using helicopters to transport troops quickly and safely to the battlefield, they did not seem to be able positively to affect the battle in the countryside. The battle of Ap Bac in January 1963 seemed a demonstration of the weaknesses of the South Vietnamese regime.

At about the same time that Ngo Dinh Diem and President John F. Kennedy were assassinated in November 1963, the U.S. government was engaging in a review of its Vietnam policy. In the aftermath of the alleged North Vietnamese attacks on U.S. destroyers on patrol in the Gulf of Tonkin in August 1964, the U.S. government decided to increase its commitment. Ultimately, President Lyndon Johnson chose to compromise, seeking to maintain a pro-U.S. government in the south but not to grant the military all that it requested in terms of U.S. troops and matériel to assist the Saigon government. The clever idea of Johnson's advisers was for a "gradual escalation," designed to impress the Hanoi regime with the possibility of greater U.S. commitment but also giving that government time to prepare for such increased escalation; it became a kind of pas de deux, in which each U.S. additional commitment was met and neutralized by a similarly increased North Vietnamese commitment. The United States largely conceded the strategic initiative and concentrated on the tactical initiative, allowing the North Vietnamese to determine when and where most battles would take place.

The Vietnamese Communists felt that they were fighting for their nation, whereas the Americans were testing interesting theories of counterinsurgency to deal with the threat of Moscow-encouraged "wars of national liberation." The Vietnamese Communists were fighting their own all-out war of liberation, whereas the Americans conducted a "limited war," as the Johnson administration made repeatedly clear. Finally, the South Vietnamese themselves knew that if they won, Vietnam would remain divided for the foreseeable future; if the other side prevailed, the nation would be reunited, even if on Communist terms. Thus, it should not have been too difficult to identify which party to the conflict would prove the most determined.

The strategic conduct (and on occasion, the operational and even the tactical level) of the American war from its earliest days was the hobby of "defense intellectuals," individuals (always characterized as "brilliant" by a fawning press) who had little or no experience of war and who believed that "rationalism" and "quantification" could be brought to humankind's most irrational and profligate of enterprises. They were to be confounded by an "irrational" enemy, who would pay any price for victory and who seemed oblivious to their "signals," "graduated responses," or "surgical strikes," unless it suited Communist propaganda purposes. For all of their vaunted "toughness," none of the defense intellectuals stayed the course; the Communist high command was in it from the beginning to the end.

It was little better when the Johnson regime attempted to formulate and publicize its war goals (the term *victory* was eschewed as hopelessly "primitive"). Basically, they were to stop "aggression from the North" (or, as Secretary of State Dean Rusk thrillingly put it, "To stop the other side from doing what it is doing") and give the Republic of Vietnam a chance to survive—hardly enough spiritually to rally a nation. Priding themselves on their knowledge of history, Johnson's advisers argued against a large call-up of reserves, instead relying on gradually increased draft calls, just when the inequities of the Selective Service System, with its student deferments and National Guard bolt-holes, became increasingly apparent.

The rules of engagement imposed on American forces in this limited war were most remarkable, in many ways reminiscent of the other "limited war"—Korea. For example, U.S. pilots on bombing and strafing runs over North Vietnam could engage enemy pilots in the air but could not seek to destroy planes on the ground. In addition, there were all kinds of restrictions on bombing targets and on the routes and times for bombing missions, leading to a certain predictability that had grave consequences for the pilots flying these missions. The view from over Hanoi was considerably different from that of defense intellectuals arguing around a table in the White House Situation Room.

As a consequence, General William Westmoreland initially selected a holding strategy until sufficient U.S. troop commitment could affect the battle. He established positions along the coast to develop port and base facilities and engaged in some spoiling operations, the most famous of which was the Ia Drang Valley battle in October–November 1965. Thereafter, from 1965 until he left Vietnam in 1968, Westmoreland settled on a strategy of attrition. He wanted to kill, seriously wound, capture, or cause to desert more enemy troops—main force North Vietnamese and Vietcong units—than the enemy could replace. When the attrition rate favored the United States and South Vietnam, he reasoned, the enemy would quit fighting. This strategy relied upon the superior mobility and firepower of U.S. units and supply capacity that enabled them to leapfrog across the country to surround and attack suspected enemy strong points. But it conceded the strategic decision—the initiative—on whether to bring the fight at all to the enemy.

For a while, Westmoreland's strategy seemed to work. In a series of major search-and-destroy missions like Attleboro, Cedar Falls, and Junction City, U.S. troops, the South Vietnamese, and other national allies inflicted many enemy casualties and destroyed vast quantities of supplies and ammunition. More important, by late 1967, the pace of fighting seemed to have slowed, and the enemy appeared to have retreated toward the borders, perhaps reflecting the terrible pounding it had sustained in this campaign.

Perhaps in reaction to this difficult situation, the Communists mounted the Tet Offensive of January–February 1968. It was a logistical triumph, as North Vietnamese general Vo Nguyen Giap managed to infiltrate 100,000 men and supplies into key positions in virtually every major city and town in South Vietnam. But the expected popular uprising never materialized, and swift U.S. and South Vietnamese reactions doomed the attackers, who were spread too thinly. It was an allied military victory (although a viewer of the networks' nightly news could be pardoned for not thinking so), but the initial strength of the Communist gamble seemed to contradict claims of progress that Westmoreland and the U.S. military had been making prior to the surprise attack. And, apparently, the Tet Offensive soured President Johnson on the war, who turned to his new secretary of defense, Clark Clifford, for advice. He, in turn, counseled that the war was not winnable. In late March 1968, Johnson announced that he would seek peace, order a bombing halt over North Vietnam, and not stand for reelection. The defense intellectuals, now deeming their war "unwinnable," had for the most part left Washington for more lucrative employment, as had former defense secretary Robert McNamara, supposedly something of an intellectual himself. As he confessed decades later, he had given up on any hope of victory some years earlier but still mendaciously proclaimed publicly his faith in victory.

Richard Nixon succeeded Johnson as president in January 1969 with a new idea for the conflict. He believed it a distraction to his complex diplomacy with the Soviet Union and the People's Republic of China; he wanted to downgrade the U.S. commitment while providing the South Vietnamese with the means ("a fighting chance") to resist future North Vietnamese aggression. He called for a three-part strategy: Vietnamization (turning the war over to South Vietnam), American troop withdrawal, and an end to the draft.

It seemed to work for several years. Recovering from losses sustained as a result of the Tet Offensive and from the continuing bombing campaign, Communist activity certainly decreased in 1969 and 1970. Nixon relied on U.S. airpower and a series of spoiling attacks across the border into Cambodia to destroy enemy supply caches necessary for a renewed attack on the south. But by 1971, the North Vietnamese had recovered to some extent and planned for a major offensive after the bulk of U.S. troop strength had been withdrawn.

On 30 March 1972, more than 14 North Vietnamese divisions (out of a total of 20) attacked the south. It was an all-out attack in three places—across the Cambodian border toward Saigon, across the central highlands toward the coast, and straight across the 17th parallel toward Hue. Hanoi had assumed that the south was weak and that, with a reelection and diplomacy with the USSR and China at stake, Nixon would not recommit U.S. power.

Hanoi made major miscalculations, and the invasion failed. President Nixon committed the full U.S. air arsenal, and the South Vietnamese fought well and hard. In the north, ARVN retook land temporarily lost to the offensive, routing six North Vietnamese divisions. In the central highlands, the North Vietnamese failed to capture Kontum and failed to take a single provincial capital or defeat decisively any major South Vietnamese ground forces. The drive to Saigon similarly failed. In the end, the North Vietnamese lost more than half of its 200,000-strong invasion force and suffered similar losses in tanks and artillery. Once again, General Vo had divided his strength sufficiently to deny mass where he needed it. Once again, as after the Tet Offensive, South Vietnam had time to gain control over its people and territory. (But once again, the viewer of the networks' nightly news would see almost solely accounts of ARVN inefficiency and government corruption.)

The Watergate cover-up and the consequent pendulum swing of power from the president to the Congress dramatically affected the U.S. ability to assist South Vietnam. Congress passed the War Powers Act, set limits on where U.S. forces could operate, and limited the transfer of matériel to the south. The American people were becoming heartily sick of this war; public opinion played little part in the planning and decisions of the North Vietnamese regime. Hanoi dramatically upgraded its supply capacity to the south, improving roads, constructing fuel pipelines, and preparing for yet another all-out assault.

The final offensive began with an attack on Phuoc Long Province in December 1974. Strongly supported enemy forces overwhelmed the defenders, and Hanoi concluded it could fight without risk of massive U.S. intervention as had happened in spring 1972. Then, in March 1975, North Vietnam concentrated its forces and attacked Ban Me Thuot, an important provincial capital in the central highlands. The attack succeeded, and a counterattack failed, given the north's superiority in artillery support. As the North Vietnamese prepared to follow up the victory at Ban Me Thuot, South Vietnamese president Nguyen Van Thieu decided to abandon the northern half of South Vietnam and concentrate his forces in a defense of Saigon and the Mekong Delta. The retreat soon became a rout and despite the valiant defense of Xuan Loc on the road to Saigon, the North Vietnamese entered the capital on 30 April 1975, and South Vietnam collapsed the next day. Soon thereafter, Communist forces conquered neighboring Laos and Cambodia, and the longest U.S. war was over.

Nonetheless, it is a fallacy that the U.S. military "lost" the Vietnam War. It won every engagement of any size that it fought. But, as one North Vietnamese negotiator replied to an American military historian who raised the point, "That is irrelevant." The U.S. Army could win all its battles, but the U.S. government could not achieve its goal of a democratic South Vietnam. Tactically, the U.S. won; strategically, it lost.

The war had a profound effect on the American home front, dividing large segments of the population and causing distrust and actual hatred of many American institutions, such as the army and the federal government. It would take the U.S. Army some 15 years to recover its morale and its fighting edge after Vietnam, and it could do so only by abolishing the draft in 1973. Trust in the U.S. government has never fully recovered.

Vietnam held no end of lessons for the U.S. military, most of which seemed to have been learned well enough, judging from subsequent American battle performance. These lessons were the basic ones of hard training, unit morale, and a jettisoning of contemporary fads, as well as the conviction that military affairs should remain with the military (always subject to civilian control, of course), not with civilian intellectuals. But perhaps the most significant military lesson from the Vietnam War is that no democracy can fight a war with conscripts in cold blood.

Charles M. Dobbs

See also: Dien Bien Phu; Ho Chi Minh; Tet Offensive; Vo Nguyen Giap; Westmoreland, William

References and further reading:

Herring, George C. *America's Longest War: The United States and Vietnam, 1950–1975.* 3rd ed. New York: McGraw-Hill, 1996.

Karnow, Stanley. *Vietnam: A History.* New York: Viking, 1983.

Palmer, Bruce, Jr. *The 25-Year War: America's Military Role in Vietnam.* Lexington: University of Kentucky Press, 1984.

Summers, Harry G. *On Strategy: The Vietnam War in Context.* Novato, CA: Presidio Press, 1981.

Vietnamese Civil War (Tayson Rebellion, 1773–1802)

Civil war that established a single Vietnam national entity. In the eighteenth century, Vietnam was divided into two states, one controlled by the Trinh family in Tonkin, with its capital at Thang-long (Hanoi), and the other dominated by the Nguyen family in Annam, with its capital at Hue, though an emperor of the Le Dynasty, founded in the fifteenth century, remained the country's nominal ruler. The Tayson Rebellion of 1773–1802 swept away both states but was overthrown by a Nguyen prince who established a new dynasty unifying all of Vietnam in 1802.

The rebellion flared up in Tayson (Western Mountain) village, in what is now central Vietnam. Its leaders were three brothers, Nguyen Nhac, Nguyen Lu, and Nguyen Hue (no relation to the Nguyen ruling family), of whom the last proved to be the most talented. According to a Spanish missionary witness, "They began moving through the villages, announcing to the inhabitants that they were not bandits, but envoys from Heaven, that they wanted to see justice prevail and liberate the people from the tyranny of the king and his mandarins. They preached equality in everything" (Nguyen 1993, 100). After gaining control of their home region, Annam, the rebels turned south and occupied Gia Dinh (Saigon). But a Nguyen prince, Nguyen Anh, put up determined resistance. He persuaded his close ally, King Rama I of Siam, to provide him with an army of 20,000 men, who invaded the western Mekong Delta in 1784. Nguyen Hue defeated them in battle on the My Tho River and then turned his attention to the Trinh domain in Tonkin, capturing Thang-long in 1786.

Although the Tayson brothers recognized the powerless Le emperor as sovereign, the Le crown prince turned against them after his father died and sought aid from the Chinese emperor Qian Long. The latter sent an army of 200,000 to eradicate the Taysons. Nguyen Hue proclaimed himself Quang Trung emperor (r. 1788–1792) in December 1788 and drove the Chinese out of Tonkin the following year in a series of epochal battles including a "Tet [Vietnamese New Year] Offensive" that cleared the way for the liberation of Thang-long.

Nguyen Hue/Quang Trung's son and successor, Quang Toan (r. 1793–1802), a boy of 10, could not prevent internal divisions in the Tayson movement that worked to the advantage of the perennially ambitious Nguyen Anh. Vilified by modern Vietnamese historians for seeking foreign assistance, most notably from the Siamese and the French (he signed a treaty with Louis XVI in 1787 and obtained limited French funds, military supplies, and advisers in exchange for territorial concessions), he steadily turned the tide of war against the Taysons, proclaiming himself Gia Long emperor

(r. 1802–1820) in June 1802 and capturing Thang-long the following month. His reign name signified that he held sway from Gia Dinh (Saigon) to Thang-long (Hanoi), the territorial extent of modern Vietnam, which in fact had been unified by his archenemy Nguyen Hue.

Donald M. Seekins

See also: Trinh-Nguyen Dynastic Struggle

References and further reading:

Nguyen Khac Vien. *Vietnam: A Long History.* Hanoi: Gioi Publishers, 1993, 100.

SarDesai, D. R. *Vietnam: The Struggle for National Identity.* Boulder, CO: Westview Press, 1992.

Tarling, Nicholas, ed. *The Cambridge History of Southeast Asia.* Vol. 1, *From Early Times to c. 1800.* Cambridge, UK: Cambridge University Press, 1992.

Viking Raids (c. 800–1016)

Scandinavian raiders who preyed on Europe. Viking warships, with their combination of oars and efficient sails, made it possible for the Vikings to range far beyond their homelands in search of booty and new places to settle. England, directly across the North Sea from Norway, from which most Vikings came, provided a particularly vulnerable target. The earliest documented foray by the Vikings in the British Isles was a raid on Lindisfarne in Northumbria (793). Later the raids expanded to other parts of Europe.

Scholars distinguish four distinct phases of Viking raids. During the first (790–840), Vikings raided directly from Scandinavia, using to full the advantages provided by their small, shallow-draught ships, ideally suited for hit-and-run raids on coastal locations and along rivers. Fleets were modest in size and were usually manned by personal war bands. Raids were seasonal and isolated.

During the second phase (841–875), raids grew in number, size, and intensity. For the first time, Vikings wintered on foreign soil, on Noirmoutier Island, in the Loire estuary, in 843. Still meeting little or no organized resistance and arriving unexpectedly in overwhelming force, the Vikings achieved enormous success. Immense amounts of booty were secured and returned to the homeland. Victims were killed or enslaved.

The establishment of winter bases meant that the Vikings could extend their range considerably, as far as Spain and the Mediterranean. In 844, a Viking fleet harried Seville, although it suffered a bloody repulse. Survivors returned to the Loire Valley to settle. Another fleet raided Spain, North Africa, Provence, and Italy, where it was finally defeated (859–862).

In England, desultory raiding had continued but began again more earnestly in 865. A large force, known as the "Great Danish Army," led by the sons of Ragnar Lothbrok, Healfden, and Ivar the Boneless, wintered in East Anglia. By remaining on foreign soil, the Vikings increased political pressure on local rulers. Many Anglo-Saxon and Frankish kings tried to buy off the Vikings in an attempt to remove them from their kingdoms.

In a third phase (876–980), the Vikings were out for more lasting gain. They expanded their holding in England and Francia and permanently settled in Ireland and Iceland. Occasionally meeting strong opposition, the "Great Danish Army," the focus of Viking activity in the West, conquered the kingdoms of East Anglia and Northumbria and reduced Mercia to a fraction of its former size (865–879). It was unable to subdue the Wessex of Alfred "the Great" (r. 871–899). A truce was made in 878; it later became the basis of the Treaty of Wedmore, signed in 886, which recognized that much of England was in Danish hands (the *Danelaw*).

Although hard-pressed by fresh armies of Vikings from 892 to 899, Alfred was finally victorious. His son, Edward the Elder (r. 899–924), began the reconquest of Danish England. Before his death, small Danish states on old Mercian and East Anglian territory had fallen. All lands south of the Humber estuary were once more in Anglo-Saxon hands. The more remote Northumbria resisted longer, largely under Viking leaders from Ireland. Eadred (r. 946–955), the son of Edward the Elder, finally liquidated the Scandinavian power there in 954. Eric Bloodaxe, the last independent Viking king in York, was driven out and killed.

The fourth phase (980–1016) coincides with the reign of Æthelred the Unready (r. 978–1016). At its end, England was a part of the empire of the Dane Canute I. It can be subdivided into four stages. During the first (980–991), there was a resumption of Viking activity in England after a lull of 25 years, though with mainly local effects. The second stage (991 to 1005) witnessed heavier attacks, the effects of a single large Viking army on English territory. It arrived in 991 with 93 ships under Olaf Tryggvason, future king of Norway, and only returned to Denmark in 1005, forced by famine. This army fought Byrhtnoth at Maldon and received tribute (*Danegeld*) in 991 (10,000 pounds of silver), 994 (16,000 pounds), and 1002 (24,000 pounds). Its raids were destructive forays designed to inflict the maximum damage and extort the maximum tribute.

The third stage (1006–1012) saw two invasions. The first, in 1006, was only halted with a massive payment of tribute in 1007 (36,000 pounds of silver). At this nadir of his fortunes, Æthelred was finally cajoled into action. In 1008, he ordered that a fleet be built, but local rivalries limited its usefulness. It did not prevent the arrival of another immense Viking army, led by Thorkell the Tall, in 1009. Thorkell harried much of southern England and only ceased after the payment of tribute in 1012 (48,000 pounds of silver).

During the fourth stage (1013–1016), there were two more invasions, culminating in the conquest of England. The first, in 1013, was led by Svein Forkbeard of Denmark. He was accompanied by his teenage son, Knut Sveinsson. The son returned with a second invasion force in 1016 and went on to rule England as Canute I. Now virtually a king without a country, Æthelred vacated his throne in favor of a safe haven in Normandy.

Ironically, Normandy itself was ruled by other Vikings. Charles the Simple, king of the West Franks, had ended raids on his kingdom by granting Normandy to the Viking leader Rollo in the Treaty of St. Clair, signed in 911. In return, Rollo had pledged allegiance to Charles, was baptized, and agreed to defend him against other Vikings.

Nic Fields

See also: Alfred the Great; Brunanburgh; Edington; Maldon, Battle of; Rollo; Vikings

References and further reading:

Coupland, S. "The Vikings in Francia and Anglo-Saxon England to 911." *The New Cambridge Mediaeval History* (1995), 190–201.

Griffith, P. *The Viking Art of War.* London: Greenhill, 1995.

Lawson, M. K. *Cnut: The Danes in England in the Early Eleventh Century.* London: Longman, 1993.

Vikings

Term used generally for the peoples of Scandinavia, about 800–1100. More specifically, it refers to the non-Christian raiders of the ninth and tenth centuries described in the monastic chronicles of western Europe.

The Viking Age began with the first recorded raid at Lindisfarne Abbey, in 793, and ended with the death of Harald Hardradi at Stamford Bridge in 1066. Outside Scandinavia, it was defined by the bands of raiders originating in Scandinavia. The intense raiding activities had two definite periods: a first Viking Age (800–900) and a second Viking Age (930–1100), separated by a period of consolidation throughout Scandinavia of independent law assemblies into kingdoms. Later Viking activities were larger and closer to national raids in response to other raids, as opposed to the individual raids against accessible targets of the first Viking Age.

Viking raiding activity depended upon speed and the technology of shallow-draw boats with keels. Viking ships were able to navigate in rivers and streams as shallow as 3 feet. Typical early Viking warships carried 30 to 50 men, arms, and horses and were powered by sail or oars. Raids are described as hit-and-run, with no warning. Sources from western Europe repeatedly describe the speed and ferocity of the raids by the pagan Northmen.

Although the Viking Age outside Scandinavia is defined primarily by the raiding activities of a small portion of the population, in Scandinavia the Viking Age is regarded as a golden age of technology, art, and literature. One aspect of the Viking Age for Scandinavians was the exploration, settlement, and trade east and west that established a series of cities, kingdoms, and trade routes.

Viking activity was responsible for the discovery of Iceland and Greenland; the building of Dublin and York; and the settlement of the Orkneys, Hebrides, and Faroes. Icelandic settlement eventually created a parliament and resulted in the Icelandic Sagas of the thirteenth centuries. Other notable settlement areas include Normandy and the northeastern area of England around York, known as the *Danelaw.*

Viking exploration resulted in trading routes into Byzantium. Norse kingdoms were founded in Russia, with Swedish and other settlers invited into the area to end the internecine wars of the local Slavic population. Major centers included Kiev and Novgorod. The Vikings who settled in these eastern kingdoms rapidly assimilated.

The legacy of the Vikings was primarily in technology, art, and a body of literature that bears the Norse name to this day, saga. Scandinavian raiders also affected every area of contact with their patterns of land use and legal traditions.

Tamsin Hekala

See also: Viking Raids

References and further reading:

Roesdahl, Else. *The Vikings.* Trans. Susan M. Margeson and Kirsten Williams. London: Allen Lane, 1991.

Villa, Francisco "Pancho" (Doroteo Arango) (1878–1923)

Mexican revolutionary and bandit. Doroteo Arango was born the son of a sharecropper in La Coyotada, Durango, Mexico, on 7 July 1878. At 16, he killed a man who allegedly was trying to rape his sister. A petty thief and cattle rustler, Arango was described as "animal in his passions and rages." In 1892, he joined a gang of bandits led by Francisco Villa, whose name Arango adopted. Around 1902, Villa moved to Parral, Chihuahua. By February 1911, he had joined the rebellion against the regime of President Porfirio Díaz. An able leader but a poor subordinate, Villa lacked military discipline. Beginning as a captain of irregulars, he rapidly rose to colonel. His *dorados* (dismounted horsemen), drawn from

Portrait of Francisco "Pancho" Villa, c. 1910. (Library of Congress)

ranches, mining camps, and poorer precincts of American border towns, frequently defeated professional generals whose men lacked the will to fight. Captured by federal troops in 1911, Villa learned to read and write in prison. Narrowly avoiding execution, he escaped to the United States. Returning in February 1913, he was commanding a force of 3,000 men within a month. In September 1914, Provisional President Venustiano Carranza, the constitutionalist leader, named him commander of the Division of the North. As self-appointed military governor of Chihuahua, Villa was an able administrator but often broke relations with other revolutionary leaders. His power declined after his defeat by the constitutionalist general Alvaro Obregon at Celaya, Guanajuato, in 1915. Villa was angered when President Woodrow Wilson extended recognition to the Carranza regime in 1916. In separate incidents, he executed 15 American engineers at Santa Ysabel, Chihuahua, in January 1916, and killed 16 people in a raid on Columbus, New Mexico, in March of that year.

General John J. Pershing's punitive expedition (March 1916–February 1917) effectively destroyed Villa's band, although Villa himself escaped. Villa retired in July 1920 and, with seven of his men, was assassinated in Parral on 23 July 1923.

Keir B. Sterling

See also: Mexican Revolution; Zapata, Emiliano

References and further reading:
Katz, Friedrich. *The Life and Times of Pancho Villa.* Stanford, CA: Stanford University Press, 1999.
Quirk, Robert E. *The Mexican Revolution, 1914–1915.* Bloomington: Indiana University Press, 1960.
Ruiz, Ramon Eduardo. *The Great Rebellion: Mexico, 1905–1924.* New York: W. W. Norton, 1980.
Smythe, Donald. *Guerrilla Warrior: The Early Life of John J. Pershing.* New York: Charles Scribner's Sons, 1973.

Vimy Ridge (9 April 1917)

Canadian military triumph considered by many to be a major step in the evolution of a sense of nationhood. Early 1917 saw the four Canadian combat divisions on the western front reunited. (One Canadian soldier remarked at the sight that he had never imagined that there were that many Canadians in existence.) The Canadian Corps took over a 10-mile section of the front line west of the town of Givenchy-en-Gohelle. They faced three German divisions from General Freiherr von Fralken's Sixth Army. The German forces occupied Vimy Ridge, an escarpment rising 450 feet above sea level and dominating the surrounding flat countryside.

The Germans constructed a complex maze of concrete fortifications, trenches, tunnels, and dugouts. Ferdinand Foch's French troops twice tried to take the ridge and failed, incurring 150,000 casualties.

The Canadian Corps, under the command of British lieutenant general Julian Byng, received orders to take Vimy Ridge as part of the British-led Arras offensive. They launched the attack at 5:30 A.M. on 9 April 1917, Easter Monday. All four Canadian divisions advanced in line, following a creeping artillery barrage toward the German lines. By noon, three of the Canadian divisions had achieved their primary objectives. The 4th Division, assigned to take the town of Givenchy and a well-defended hill known as "the Pimple," experienced the greatest difficulty. The Canadian Corps achieved all of its primary and most of its secondary objectives. It also held the captured territory in the face of sustained German counterattacks. The cost was 3,598 Canadians killed and 7,004 wounded. Four Canadians received the Victoria Cross for the action. On 12 April, the Germans withdrew from Vimy Ridge along the Douai Plain. The Battle of Vimy Ridge and the other battles of Arras gave the British their first victory on the western front in 32 months of war.

To many Canadians, Vimy Ridge remains a symbol of pride and national maturity.

Eric Smylie

See also: Foch, Ferdinand; World War I

References and further reading:

Berton, Pierre. *Vimy.* Toronto: McClelland and Stewart, 1986.
Murray, W. W., ed. *The Epic of Vimy.* Ottawa: The Legionary, [1936].
Nicholson, G. W. L. *Canadian Expeditionary Force 1914–1919: Official History of the Canadian Army in the First World War.* Ottawa: Queen's Printer, 1964.
Ralphson, George H. *Over There, with the Canadians at Vimy Ridge.* Chicago: M. A. Donohue, 1919.
Wood, Herbert Fairlie. *Vimy!* Toronto: Macmillan, 1967.

Visigoths

Germanic rulers of southern Gaul and Spain. When Theodosius died in 395, the Visigoths renounced their treaty with the Romans. Six years latter, Alaric decided to invade Italy but was defeated by former comrade-in-arms Flavius Stilicho. A second invasion also ended in defeat, but this time Alaric forced the Senate in Rome to pay the Visigoths a large endowment.

Alaric blockaded Rome twice (in November 408 and again in late 409) to force western emperor Honorius in Ravenna to give his people land. When negotiations failed, Alaric returned to Rome and broke in (24 August 410). The Visigothic army plundered the city for three days but did comparatively little damage.

Alaric died soon after. His brother-in-law Athaulf assumed the leadership. Alaric had planned to lead the Visigoths to North Africa, but a lack of ships had prevented it. Athaulf instead went north, crossing into Gaul in 412, but was assassinated in Spain three years later. In 418, Theodoric I (r. 418–451), one of his successors, made a treaty with the Romans. It granted the Visigoths Aquitania Secunda in return for a military alliance with the western empire. Under this alliance, the Visigoths helped subdue the Vandals and Alans, which gave them territory in Spain and allowed their expansion in southern Gaul. The kingdom in southern Gaul lasted until 507, when Frankish king Clovis defeated the Visigoths. Thereafter they were mainly confined to Spain. In 711, the Visigothic kingdom there fell to the invading Moors.

Unlike their cousins, the Ostrogoths, the Visigoths did not take to the horse. Chieftains and their companions might fight mounted, but the bulk of a Visigothic army was made up of archers and spearmen, fighting in a dense mass. Like other Germanic bowmen, the overwhelming majority were armed with compound bows.

Nic Fields

See also: Alaric; Goths; Stilicho, Flavius

References and further reading:

Thompson, E. A. *The Visigoths in the Time of Ulfila.* Oxford, UK: Clarendon Press, 1966.
———. *The Goths in Spain.* Oxford, UK: Clarendon Press, 1969.

Vo Nguyen Giap (1911–)

Military commander of Communist-nationalist Vietminh forces against the French (1946–1954) and later of the People's Army of [North] Vietnam against the Republic of [South] Vietnam and the United States (1959–1973). A key figure in the ruling hierarchy of the Hanoi government, Giap joined the Communist Party in his youth and was imprisoned by French authorities at age 16. A very good student, he went to the University of Hanoi and began writing Marxist analyses of the situation he perceived in Vietnam. When France banned the Communist Party in 1939, Giap fled to southern China and became an early supporter of Ho Chi Minh.

Giap thereupon became a major military commander, first fighting the Japanese, soon thereafter the French, and later organizing the campaign against the United States and its Saigon ally. Giap believed in Mao's theories of guerrilla war and that the army should be highly indoctrinated and moral to win support of the local people.

From 1946 to 1953, Giap led an increasingly successful military effort against the French and benefited from the Communist conquest of southern China after mid-1949. In 1953, he commanded more than 300,000 troops and launched a drive into northern Laos when the French chose to occupy a plateau in northwestern Vietnam, Dien Bien Phu, to lure Giap into a rash and costly attack. Giap massed 50,000 troops and surprised the French when he brought artillery and radar-guided antiguns to the surrounding hills and began a 55-day siege that ended with the French surrender on 8 May 1954 and independence for North Vietnam.

In his campaign against the United States and South Vietnam, Giap violated the principles of concentration and mass in organizing the Tet Offensive of January 1968. He apparently was forced into this offensive by the political leadership, and militarily it was a disaster, destroying the Vietcong as an independent military force in the south, although it resulted in a great political-psychological victory. In March 1972, again apparently against his own feelings, he launched the so-called Easter Offensive to drive across the 17th parallel to conquer the south, which supposedly could not resist after U.S. troop withdrawal. The South Vietnamese held, President Richard Nixon unleashed U.S. air power, and the

North Vietnamese suffered a major defeat. Thereafter, Giap lost power, gradually was eased out of the ruling hierarchy, and thus played no major role in the final victory of 1975.

Charles M. Dobbs

See also: Dien Bien Phu; Tet Offensive; Vietnam Conflict

References and further reading:

Currey, Cecil B. *Victory at Any Cost: The Genius of Viet Nam's General Vo Nguyen Giap.* Washington, DC: Brassey's, 1997.

MacDonald, Peter. *Giap: The Victor in Vietnam.* New York: W. W. Norton, 1993.

Vo Nguyen Giap. *The Military Art of People's War: Selected Writings of Vo Nguyen Giap.* Ed. and with an introduction by Russell Stetler. New York: Monthly Review Press, 1970.

Vo Nguyen Giap and Van Tien Dung. *How We Won the War.* Philadelphia: RECON Publications, 1976.

Vouillé, Battle of (spring of 507)

Victory of Roman orthodox Christianity over Arianism. Clovis, king of the Franks, had inherited a small territory from his father. He eagerly sought to increase his power and made an alliance with the Orthodox Nicene church that became a cornerstone of his success.

Clovis was christened in 498, in Rheims, the first Christian king of the Franks. Now enjoying the support of the church against his enemies, Clovis subdued the Alemanni by 501 and then prepared to seize southern France, ruled by Alaric II of the Visigoths, who controlled territories from the Loire River to central Spain. Alaric was an Arian Christian and thus a heretic to the Orthodox bishops of Gaul and their congregations.

In 507, Clovis invaded Visigothic Aquitaine. The sources are few and strongly Christian. Thus the Vouillé campaign is replete with divine intervention: after crossing the Loire River, Clovis was cured of a lethal disease by Saint Severin; Saint Martin (dead for a century) helped the Frankish army ford the Vienne River.

The Visigoths advanced north against the invading Franks. The two armies finally met near modern Vouillé. The name derives from Latin Vocladis, believed to be a compound of two words: *clades,* "disaster," and *vo,* possibly a corruption of "Goth."

The sources provide few details on the battle and no reliable information on the sizes of the opposing armies. At that time, Franks and Visigoths primarily fought as infantrymen, formed in dense, close masses and relying on wild charges for victory. Frankish chronicles emphasize the personal qualities of Clovis. Seeing that Alaric stood at the front of his army with some of his retainers, Clovis charged them with his own guard. In the ensuing fight, Clovis killed Alaric. The Visigothic king's death quickly became known to his army, which left the battlefield in a complete rout.

The battle of Vouillé was an important one for Christianity. Orthodox Roman Christianity now established itself as the only religion of western Europe. In addition, Clovis increased his territory through his conquest of southern France. The Visigothic kingdom, now confined to Spain, was eventually destroyed by the Arab conquest of the eighth century.

Gilles Boué

See also: Franks; Merovingians; Visigoths

References and further reading:

James, Edward. *The Origins of France.* London: Macmillan, 1982.

———. *The Franks.* Oxford, UK: Blackwell, 1988.

Rouche, Michel. *Clovis, Histoire et Memoire.* Paris, PUF, 1997.

W

Waffen SS (1934–1945)

Both an elite fighting unit and a notorious criminal organization of the Third Reich. Subordinate to Reich Leader of the SS (Schutzstaffel, or Protection Squad) and Police Heinrich Himmler, the so-called SS Verfuegungstruppe (an early name for the Waffen SS) was founded after the Nazi regime's bloody purge of the SA (Sturmabteilung, or Stormtroopers) in the summer of 1934. Squads of the SS regiment "Leibstandarte SS Adolf Hitler," which would later form the nucleus of the 1st SS Division, participated in this massacre.

The early SS Verfuegungstruppe consisted of three regiments. Retired army general Paul Hausser was their first military instructor. Under his command, these units became the core of the future Waffen SS. Ignoring his generals' animosity toward the SS, Hitler permitted the establishment of a separate headquarters for the Waffen SS on 17 August 1938.

During the invasion of Poland in September 1939, SS regiments were still attached to divisions of the regular German army. However, the battle of France witnessed the deployment of three independent SS divisions, including the SS Division "Das Reich" of the regular SS Verfuegungstruppe and the SS "Death's Head" Division, this last made up of concentration camp guards.

The Waffen SS participated in most German campaigns during World War II. It established a reputation for tenacity and ruthlessness on the eastern front. Its battle record includes the participation by the "Death's Head" and "Viking" divisions in the encirclements of Demjansk in February 1942 and Cherkassy in January–February 1944, respectively. The SS tank corps also distinguished itself by unexpectedly retaking Kharkov in March 1943. Also in 1943, during the last German offensive at Kursk, SS tank units spearheaded the advance.

The early Waffen SS divisions had more casualties than most units of the regular army. These higher than expected losses compelled the SS Recruitment Office to enlist ethnic Germans and non-German foreigners. By the end of the war, the 38 divisions of the Waffen SS consisted of almost 600,000 men, of whom approximately one-third were of non-German origin. Little was left of the early elite formations, however. Indeed, an examination of the combat records of all Waffen SS units reveals that, overall, their fighting capabilities were no better than those of their regular army counterparts.

The Waffen SS was unique in two respects. First, it was under the overall command of Heinrich Himmler, who was a political leader, not a military commander. Second, a significant number of its formations were involved in heinous crimes. For example, "Death's Head" units guarded the concentration camps, and mounted SS regiments took part in the mass killings of Jews in the western Soviet Union in 1941. The infamous SS task forces (Einsatzgruppen), which murdered hundreds of thousands of Soviet Jews in 1941–1942, consisted not only of policemen but also of Waffen SS troops. Moreover, the crimes of the Waffen SS were not limited to the eastern front. SS combat formations were also responsible for the slaughter of French civilians in the village of Oradour-sur-Glane and for the killing of Canadian prisoners of war in Normandy in 1944. Not surprisingly, therefore, the postwar International Military Tribunal at Nuremberg declared the Waffen SS to be a criminal organization.

Jan Erik Schulte

See also: SA; SS; World War II

References and further reading:

Smelser, Ronald, and Enrico Syring, eds. *Die SS. Elite unter dem Totenkopf 30 Lebensläufe* (The SS elite under the death head: 30 biographies). Paderborn, Germany: F. Schöning, 2000.

Stein, Georg H. *The Waffen-SS. Hitler's Elite Guard at War, 1939–1945.* Ithaca, NY: Cornell University Press, 1966.

Wegner, Bernd. *Hitler's Politische Soldaten: Die Waffen-SS 1933–1945* (Hitler's political soldiers: The Waffen SS 1933–1945). 6th ed. Paderborn, Germany: F. Schöning, 1999.

Wagram (5–6 July 1809)

A battle fought northeast of Vienna between French forces, 188,000 strong, under Napoleon I and Austrian forces numbering 155,000 under the Archduke Karl von Habsburg. On 9 April 1809, the Austrian Empire declared war on France. After losing the battle of Eckmuhl on 20–22 April 1809, the Austrians withdrew across the Danube River, losing Vienna in the process. The first effort of the French to cross the Danube and make contact with the Austrians led to failure at the battle of Aspern-Essling on 21–22 May 1809. As a result of that battle, the French began to make careful preparations for a second attempt, and the Austrians withdrew to more defensible positions about 6 miles away from the Danube, centered on the village of Deutsch-Wagram.

By early July, the French, having built 12 prefabricated bridges and many small boats, were ready for a second crossing attempt, led by Nicholas-Charles Oudinot's and André Masséna's corps. On the evening of 4 July, Oudinot's troops crossed the river and quickly defeated the Austrian left flank forces, while the French opened a bombardment upon the rest of the screening force. As a result of this vigorous attack, the French army was able to cross the Danube by evening and had begun to advance against the main Austrian position, with the French right flank of Oudinot and Masséna thrust somewhat ahead of the French left flank, which had made somewhat slower progress.

On 6 July, the Austrians planned three simultaneous attacks beginning at 4:00 A.M., the main one to fall on the weak French left, commanded by Masséna, with two supporting attacks on the center and right. The French themselves planned an attack in the center, commanded by Jean Baptiste Jules Bernadotte and Oudinot, also for about 4:00 A.M. Before the French attack could develop, the Austrian attack on the right began prematurely. It was easily defeated by Louis Nicholas Davout, but, in the ensuing confusion, Bernadotte abandoned his forward position in an effort to maintain contact with the French forces on his flanks. The Austrian attack in the center, intended to be diversionary or supporting, thus went forward with no opposition. As a result, there was very little to oppose the main Austrian attack, which began four hours late but was nevertheless perfectly timed. The Austrians pushed back the French forces before them and seemed likely to advance all the way to the Danube River. The Austrians moved very slowly, however, and the French were thus able to move up troops and 112 cannon to prevent disaster.

At 10 A.M., Davout, on the French right, renewed an attack upon the Austrian left flank, making slow but steady progress. A second series of assaults pinned the Austrian center and the Austrian right flank. The Archduke Karl ordered the withdrawal of the Austrian army. The French, battered and expecting the arrival of an Austrian reinforcing force under the Archduke Johann, were in no condition to pursue.

The French lost about 32,000 casualties and 7,000 prisoners. The Austrians lost about 37,000 casualties. As a result of the Battle of Wagram, the Austrians asked for an armistice and began peace negotiations, thus ending the 1809 war against France.

Joseph M. Isenberg

See also: Berthier, Louis-Alexandre, Prince of Neuchâtel and Valangin, Prince of Wagram; Davout, Louis-Nicolas, Duke of Auerstädt, Prince of Eckmühl; Masséna, André, Duc de Rivoli, Prince d'Essling; Murat, Joachim, Grand Duke of Cleves-Berg, King of Naples; Napoleon I; Napoleonic Wars; Ney, Michel, Duc d'Elchingen, Prince de La Moskova; Oudinot, Nicholas-Charles, Duc de Reggio

References and further reading:

Arnold, James R. *Crisis on the Danube.* New York: Paragon House, 1990.

Chandler, David G. *Campaigns of Napoleon.* New York: Scribner's, 1966.

Connelly, Owen. *Blundering to Glory.* Wilmington, DE: Scholarly Resources, 1999.

Wainwright, Jonathan Mayhew, IV (1883–1953)

U.S. Army commander. Born in Walla Walla, Washington, on 2 August 1883, Wainwright's family members on both sides were army and navy officers. Wainwright, who was nicknamed "Skinny" by his classmates, graduated from West Point in 1906 and was commissioned a second lieutenant in the cavalry. In the World War I, he was the assistant chief of staff for operations and planning (G-3) of the 82d Division during the battles of St. Mihiel and Meuse-Argonne. After graduating from the Command and General Staff School in 1931 and the Army War College in 1934, Wainwright served as the assistant commandant of the Cavalry School in Fort Riley, Kansas, before being promoted to brigadier general and given command of the 1st Cavalry Brigade at Fort Bliss, Texas, between 1938 and 1940. Promoted to major general, he took command of the Philippine Division in September 1940.

After the Japanese invasion of the Philippines on 8 December 1941, he commanded the North Luzon Force in its fighting withdrawal to the Bataan peninsula. Promoted to lieutenant general in March 1942, Wainwright was appointed commander of U.S. Forces in the Far East after General Douglas MacArthur was ordered to Australia. Forced to withdraw the remnants of his combined American-Filipino army to the island fortress of Corregidor in Manila harbor,

Wainwright's forces held out, the mighty guns of the Corregidor forts denying Manila Bay to the Japanese and inflicting heavy losses on the enemy, until 6 May 1942.

Wainwright was transported to Japanese-occupied Manchuria, where he spent the next three years as a prisoner of war. Forced to watch the barbarous mistreatment of his men at the hands of the gloating Japanese, Wainwright often attempted to intervene on their behalf, but his efforts were almost always unsuccessful. For three years, he was starved, mistreated, and forced to perform menial tasks in an unsuccessful effort to humiliate him.

Liberated on 25 August 1945, Wainwright expected to be court-martialed and imprisoned for surrendering his forces in the Philippines. Instead, he was honored, first on 2 September 1945, when he stood behind MacArthur at the surrender ceremonies on the *Missouri* in Tokyo Bay. Next, when Wainwright returned to the United States, he was promoted to full general and awarded the Medal of Honor by President Harry S. Truman. His health shattered by his long and harsh imprisonment and slighted by the egotistical MacArthur, Wainwright retired in August 1947 and died on 2 September 1953, the eighth anniversary of the Japanese surrender. Wainwright, a tough, simple cavalryman, must be given credit at least equal to that given MacArthur for commanding multiracial forces that delayed the Japanese invaders for months, while Allied Pacific and Asian forces (except for the Canadians at Hong Kong) put up confused and ineffectual resistance and quickly surrendered to an enemy that they substantially outnumbered.

Alexander M. Bielakowski

See also: MacArthur, Douglas; Philippines, U.S. Loss of

References and further reading:

Schultz, Duane P. *Hero of Bataan: The Story of General Jonathan M. Wainwright.* New York: St. Martin's Press, 1981.

Wainwright, Jonathan M. *General Wainwright's Story: The Account of Four Years of Humiliating Defeat, Surrender, and Captivity.* Garden City, NY: Doubleday, 1949.

Wake Island (8–23 December 1941)

The loss of an American Pacific Island possession after a determined defense. Little more than a cluster of three small islands in the middle of the Pacific Ocean, 2,000 miles west of Hawaii, Wake Island had served as a refueling base for American airplanes and submarines. At the commencement of hostilities in December 1941, the Japanese decided to seize the islands. On 8 December, they dispatched Kwajalein-based bombers to soften up defenses manned by a force of some 400 marines, sailors, soldiers, and another 1,200 civilian construction workers. More than half of the dozen fighter planes guarding the island's airstrip were destroyed or damaged.

The defenders had enough firepower and morale to repulse a Japanese invasion force that sailed into the waters off Wake Island on 11 December. Utilizing naval guns salvaged from battleships before the war, the Americans sent damaging salvoes into the invasion fleet, which sustained significant damage and retreated without landing troops. Having underestimated the island's defenses, the Japanese redoubled their efforts. For nearly two weeks, they pounded Wake Island with round-the-clock air raids. On 23 December, they were finally in a position to land troops. When they did so, Wake Island's fall was only a matter of time. By midday, Navy commander Winfield S. Cunningham had little choice but to surrender his beleaguered survivors. The Japanese sustained more than 800 casualties in taking Wake and the Americans roughly 170. However, those Americans who surrendered had nearly four years of brutal captivity ahead of them if they were not executed on the island itself. The heroic defense of Wake Island was about the only bright spot in a trail of disaster in the first weeks of World War II in the Pacific.

John C. McManus

See also: World War II

References and further reading:

Schultz, Duane. *Wake Island: The Heroic, Gallant Fight.* New York: St. Martin's Press, 1978.

Urwin, Greg. *Facing Fearful Odds: The Siege of Wake Island.* Lincoln: University of Nebraska Press, 1997.

Walker, Walton (1889–1950)

U.S. Eighth Army commander during the early months of the Korean War. Born in Texas, Walker had a typical career for an officer of his generation, graduating from West Point in 1912 and serving in Mexico and France, plus the usual round of interwar school and regimental assignments. World War II saw Walker leading XX Corps as a subordinate of George Patton, creating a reputation as an aggressive mission-oriented commander, despite accusations of affecting too much bravado.

With the outbreak of the Korean War, Walker, as Eighth Army commander in occupied Japan, led his ill-prepared forces into combat. By 31 July 1950, he found himself conducting a desperate but successful stand at the Pusan Perimeter. By this time, the United Nation (UN) forces within the perimeter outnumbered the Communist forces by some two to one and enjoyed absolute air control and great heavy weapons superiority, yet the situation remained critical until the successful Inchon Landings of mid-September.

Walker began a slow personal decline from the time of those landings through to the disastrous retreat from the Chinese counterattack at the Yalu. He failed to adapt to the environment of MacArthur's staff, being dubious about the viability of Inchon and resenting the splitting off of X Corps as an independent command. MacArthur, in turn, doubted Walker's aptitude for independent command, an assessment apparently justified by Walker's incomprehension of the Chinese style of war and the continued low capability of the Eighth Army.

By the time of his death in a vehicle accident on 23 December 1950, Walker was an inconvenient hero, burned out and presiding over a staff discussing the actual abandonment of Korea. This critical situation was rapidly reversed by Walker's successor, Mathew B. Ridgway.

George R. Shaner

See also: Korean War; MacArthur, Douglas; Pusan Perimeter; Ridgway, Mathew B.

References and further reading:

Appleman, Roy E. *South to the Naktong, North to the Yalu, June–November, 1950.* Washington, DC: Department of the Army, 1961.

Hastings, Max. *The Korean War.* New York: Simon & Schuster, 1987.

Prefer, Nathan. *Patton's Ghost Corps: Cracking the Siegfried Line.* Novato, CA: Presidio Press, 1998.

Sandler, Stanley. *The Korean War: No Victors, No Vanquished.* Lexington: University of Kentucky Press, 1999.

Walker, William (1824–1860)

Leader of American filibuster expeditions in Mexico and Nicaragua during the 1850s. After failing at careers in medicine, law, and journalism, Walker, in search of personal glory, led an unsuccessful filibuster expedition to Mexico in 1853 to colonize Sonora and Lower California. Two years later, at the request of the Liberal Party (also called Leonese) in civil war–torn Nicaragua, Walker defeated the Conservative Party (the Granadans) and emerged as the dominant figure in the new government. In June 1856, he became president and sought American support by expressing interest in establishing a federal republic of Central American states and Cuba, with slavery in both regions. But he had made important enemies, notably Cornelius Vanderbilt, who controlled the proposed transit route across Nicaragua, and opponents within Nicaragua supported by other Central American states. Northern antislavery forces in the United States also opposed Walker's plan and pressured President Franklin Pierce not to recognize his government. Cut off from supplies and recruits by Vanderbilt and facing a hostile Central American coalition aided by England, Walker surrendered to U.S. naval authorities in May 1857. Two subsequent filibuster efforts in November 1857 and September 1860 also resulted in failure, the latter ending in Walker's execution by a Honduran firing squad. Walker's filibusters aggravated the debate over the extension of slavery in the United States and Anglo-American relations on the isthmus. They also reinforced the negative image of the United States as the "Colossus of the North" throughout Latin America.

Dean Fafoutis

See also: Nicaragua, Walker's Invasion of

References and further reading:

Brown, Charles H. *Agents of Manifest Destiny: The Lives and Times of the Filibusters.* Chapel Hill: University of North Carolina Press, 1980.

Greene, Laurence. *The Filibuster: The Career of William Walker.* Indianapolis: Bobbs-Merrill, 1937.

Scroggs, William O. *Filibusters and Financiers: The Story of William Walker and His Associates.* New York: Macmillan, 1916.

Wallenstein, Albrecht von (1583–1634)

A prominent and successful military commander. This Bohemian nobleman started his career during the Bohemian revolt of 1618–1620, when he proved his loyalty to the Habsburgs. In 1625, he was appointed commander in chief to organize imperial forces. Wallenstein, who successfully led the imperial troops against Mansfeld and Christian IV of Denmark, rose to princely status. In 1630, he had to retire from his command mainly because of the pressure of the Catholic League. But after Graf Johann Tserclaes von Tilly's crushing defeat at Breitenfeld, Wallenstein was quickly reinstalled to save the Habsburg power from the threatening Swedish forces under Gustavus II Adolphus. In 1632, Wallenstein managed to stop the Swedish advance. Afterward, he was unwilling to fight any longer and preferred a political solution. But in the course of those negotiations, he fell under the suspicion of high treason, which led to his condemnation as a rebel against the Reich. Wallenstein was put to death at Eger on 25 February 1634.

Wallenstein showed his extraordinary talent as an organizer of the military when he built an army numbering more than 100,000, perfected the method of funding, and established a viable logistical system of magazines. As a general, Wallenstein tried to avoid battles and adopted mainly defensive tactics. He was especially eager to force the enemy to attack his strong defensive forces. Thus Wallenstein inflicted heavy losses, as on Mansfeld at the Dessauer Brücke in 1626 and on Gustavus Adolphus at the Alte Veste near Nuremberg in 1632. Wallenstein was not infallible, however; for example, at Lützen, Gustavus Adolphus nearly achieved complete surprise of Wallenstein, who hardly held the field (16 November 1632).

Michael Kaiser

See also: Gustavus II Adolphus; Lützen, Battle of; Thirty Years' War

References and further reading:

Asch, Ronald G. *The Thirty Years War. The Holy Roman Empire and Europe, 1618–1648.* Basingstoke, UK: Macmillan, 1997.

Mann, Golo. *Wallenstein: His Life Narrated.* Trans. Charles Kessler. London: Deutsch, 1976; New York: Holt, Rinehart, and Winston, 1976.

War Crimes

Actions in wartime that allegedly violate the laws or usages of war. The boundaries of customary behavior in waging war have gradually evolved and been codified. As early as 1305, the Scottish leader William Wallace was tried by an English court for depredations on the civilian population of England. The French National Assembly in 1792 banned murder and maltreatment of civilians. During the American Civil War, U.S. Army General Order No. 100 (1863) specified which behaviors were improper for soldiers, and at war's end Henry Wirz, the Confederate commandant of the Andersonville prisoner-of-war camp, was tried and executed for the horrific conditions in his camp.

By World War I, certain weapons and practices had been outlawed by the Geneva Conventions of 1864 and 1906, the St. Petersburg Declaration of 1868, and The Hague Conventions of 1899 and 1907. Articles 227–230 of the Treaty of Versailles provided for the trial of the kaiser and a number of German officers for war crimes, but little came of it.

Between the world wars, the Pact of Paris (Kellogg-Briand Pact) of 1928 and the Geneva Prisoner-of-War Convention of 1929 further clarified the rules of combat. At the end of World War II, the International Military Tribunal set up by the victorious Allies at Nuremberg indicted carefully selected representatives of the German government, armed forces, and business for various war crimes, including genocide and planning aggression. Of the 22 tried in 1945–1946, 12 were hanged, 3 sent to prison for life, 4 given lesser prison sentences, and 3 acquitted. An American court at Nuremberg tried another 177 persons between 1946 and 1949, condemning 25 to death, 20 to life in prison, and 97 to lesser terms and acquitting 25. International trials of 25 defen-

Jews used for pulling German trucks and soldiers in Poland, 1939–1943. (Library of Congress)

dants in Tokyo between 1946 and 1948 resulted in 7 death sentences, 16 life sentences, and 2 lesser prison terms. The volume of trials in Russia and Eastern European countries is not yet adequately documented. When these trials were criticized as "victor's justice" conducted ex post facto with insufficient basis in law, the Allied governments pointed to the international conventions and the Paris Pact of 1928 as establishing sufficient legal basis.

This basis was extended in the postwar period by the four Geneva Conventions of 1949, the United Nations Convention on Prosecution and Punishment of the Crime of Genocide (1951), and the Geneva Protocol of 1977 that extended protection of the laws of war to persons participating in "national liberation" struggles. By the end of the century, a United Nations tribunal in the Hague was prosecuting war crimes arising from the collapse of Yugoslavia, and another, in Tanzania, was dealing with crimes arising from ethnic cleansing in Rwanda. Efforts were under way to establish a permanent International Criminal Court in the Hague for the prosecution of all war crimes.

One of the criticisms of war crimes tribunals is that they are indeed dispensing "victor's justice" and that they are often guilty of the same crimes that they condemn in the vanquished. For example, Soviet judges at Nuremberg rounded on the Nazi defendants for their treatment of prisoners of war and for the establishment of concentration camps, both standard features of the Soviet system. Nonetheless, the concept of war crimes has led in certain situations to an amelioration of the rigors of war, for soldiers and civilians alike.

Joseph M. McCarthy

See also: General Order No. 100; Geneva Conventions; Laws of War

References and further reading:

Ball, Howard. *Prosecuting War Crimes and Genocide: The Twentieth Century Experience.* Lawrence: University of Kansas Press, 1999.

Callamard, Agnes. *Investigating Women's Rights Violations in Armed Conflicts.* Montreal: Amnesty International, 2001.

Cooper, Belinda, ed. *War Crimes: The Legacy of Nuremberg.* New York: TV Books, 1999.

War of 1812 (1812–1815)

Chaotic and inconclusive war between the United States and Great Britain that nevertheless provided a new surge of American nationalism. There were four main causes: American insistence upon its neutral trade with both sides in the Napoleonic Wars, British boarding of American ships on the high seas and impressment of American seamen, conflicts between American pioneers and British-supported American Indians on the western frontier, and the desire for land in Canada. William Henry Harrison's victory at Tippecanoe in 1811 was used as propaganda by the bombastic "War Hawk" political faction to demand action against Britain for supposedly arming the Indians and as an excuse for the seizure of Canadian land.

Soon after the U.S. Congress boldly declared war on Great Britain on 18 June 1812, Secretary of War William Eustis and his generals decided that the best defense was a good offense and set their sights on Canada. Major General Henry Dearborn envisioned a four-pronged attack: from Lake Champlain to Montreal; from Sackets Harbor, New York, to Kingston, Ontario; west across the Niagara River; and east across the Detroit River.

The British were too involved with Napoleon in Europe to commit significant naval, military, or matériel resources to North America in 1812 or else Britain might have crushed the United States early. The small American navy was a match for British ships in single combat, but the initial unpreparedness and inadequacy of American land forces spelled disaster for Dearborn's plan. The chain of command was unclear, with junior officers of regulars often expecting to take command over senior officers of militia.

On 17 July 1812, American lieutenant Porter Hanks, who had not heard that there was a war on, surrendered the 57-man garrison of Fort Mackinac to British captain Charles Roberts's 900 soldiers. This surrender gave Britain control of Lake Huron. Pro-British Potawatamis massacred about 70 American soldiers and civilians under Captain Nathan Heald as he was surrendering Fort Dearborn, now Chicago, on 15 August. Brigadier General William Hull first intended to attack Fort Malden, Ontario, but dropped that plan and instead surrendered Fort Detroit without a fight on 16 August to Major General Sir Isaac Brock. Hull was court-martialed, convicted of cowardice and neglect of duty, and condemned to be shot, but President James Madison commuted the sentence in light of Hull's service in the Revolutionary War.

Stephen Van Rensselaer, a major general of New York militia despite his utter lack of military experience, crossed the Niagara River with about 4,000 of his 6,000 men to invade Ontario at Queenston Heights on 13 October. Although the British garrison was only about 300 strong, they held Van Rensselaer's disorganized force on the heights. Brock was killed as his 1,800 reinforcements routed the Americans. Dearborn's campaign against Montreal fizzled in November when his 8,000 troops, marching north from Albany, refused, like the New York militia across from Queenston Heights, to cross the Canadian border.

In the Northwest theater, Captain Zachary Taylor successfully defended Fort Harrison, Indiana, against eight-to-one odds on 4 September. Colonel Henry Procter's 500 British and 600 Wyandots crushed Brigadier General James Winchester's 400 Americans at the River Raisin, Michigan, on 21 January 1813. Procter was censured for recklessness and for allowing his native allies to commit atrocities but was promoted to brigadier general.

Dearborn and Brigadier General Zebulon Pike burned the capital of Upper Canada, York (now Toronto), on 27–29 April. The British burned some American matériel at Sackets Harbor on 28–29 May but failed to capture the fort. British admiral George Cockburn patrolled Chesapeake Bay, landing raiders to burn Frenchtown, Havre de Grace, Georgetown, Fredericktown, Principio, and other Maryland settlements that spring. A year later, on 14–15 May 1814, American lieutenant colonel John B. Campbell burned Port Dover, Ontario, in retaliation for the destruction of these Maryland towns.

Aided by 1,200 Kentucky reinforcements under Brigadier General Green Clay, who arrived on 5 May 1813, Harrison's 550 defenders of Fort Meigs, Ohio, withstood siege by Procter's 900 British regulars and 1,200 natives from 1 to 9 May. Defying Harrison's direct order to evacuate, Major George Croghan successfully defended Fort Stephenson, Ohio, at five-to-one odds on 2 August against Procter's attack.

On 27 May, after three days of naval shelling, 4,500 Americans under Major General Morgan Lewis took Fort George, Ontario, from Brigadier General John Vincent's garrison of 1,900. From this dominant position, the Americans threatened the entire Niagara Peninsula. Vincent ordered other nearby forts abandoned, but on 5 June, 700 British under Colonel John Harvey surprised and defeated 2,600 Americans under Brigadier Generals William Winder and John Chandler at Stoney Creek, Ontario. On 24 June, at the Battle of the Beaver Dams, also known as the Battle of the Beechwoods, several hundred pro-British natives commanded by French captain Dominique Ducharne forced Lieutenant Colonel Charles Boerster's column of 570 Americans to surrender. Revitalized, the British won a skirmish at Fort Schlosser, New York, on 5 July.

American brigadier general Robert B. Taylor's 700 militiamen repelled British admiral John Borlase Warren's and Major General Sydney Beckwith's 4,000 amphibious invaders at Craney Island, near Norfolk, Virginia, on 22 June. However, Beckwith captured Hampton, Virginia, on 25 June and allowed his French mercenaries to rape and plunder.

Encouraged by Commodore Oliver Hazard Perry's naval victory on Lake Erie on 10 September and supported by Perry thereafter, Harrison took Fort Malden and Amherstburg on 27 September, recaptured Detroit on 29 September, and won at the Thames, near Moraviantown, Ontario, on 5 October. Harrison's nemesis, Tecumseh, died in that battle.

At Châteauguay, Quebec, on 26–28 October, Lieutenant Colonel Charles Michel d'Irumberry de Salaberry's 1,400 French-Canadian woodsmen and militia ambushed, enfiladed, and defeated 4,500 American troops marching on Montreal under Major General Wade Hampton. At Crysler's Farm, Ontario, on 11 November, British lieutenant colonel Joseph Morrison's 800 regulars stopped American major general James Wilkinson's 2,000-man expedition to Montreal dead in its tracks. A small British detachment under a Major Handcock repulsed Wilkinson at La Colle Mill, Quebec, on 30 March 1814.

Increasingly vulnerable, American brigadier general George McClure abandoned Fort George and burned the nearby town of Newark, Ontario, on 10 December 1813. The British then invaded New York, captured Fort Niagara on 18 December, massacred the inhabitants of Lewiston the same day, and burned Black Rock and Buffalo on 30 December.

Preparatory to attacking Sackets Harbor, Drummond planned to destroy the American supply depot at Oswego, New York, but the British attack on 5–6 May 1814 was only moderately successful. About 3,500 Americans under Major General Jacob Brown and Brigadier General Winfield Scott captured Fort Erie, Ontario, on 3 July. Brown and Scott defeated British major general Phineas Riall in a smart victory at Chippewa on 5 July, but Riall did stop the American invasion, halting Brown and Scott at Lundy's Lane on 25 July, the bloodiest battle of the war. Brigadier General Edmund P. Gaines repulsed a concerted British effort to recapture Fort Erie on 13–15 August. Brigadier General Peter Porter led a successful sortie from Fort Erie against British batteries on 17 September. Brigadier General George Izard ordered Fort Erie abandoned and demolished on 5 November, ending the American military presence in Canada.

Other action during the summer of 1814 included about 300 American sharpshooters under Major Lodowick Morgan, behind earthworks at Conjocta Creek, New York, defeating 600 British under Lieutenant Colonel John Tucker on 3 August. The British occupied Eastport, Maine, on 11 July and Castine, Maine, on 1 September, and George Croghan failed to retake Mackinac on 4 August.

A 4,000-man British invading force under Major General Robert Ross routed the American defenders at Bladensburg, Maryland, on 24 August and burned Washington, D.C., that night. Ross then moved toward Baltimore. British navy captain Peter Parker died when he impulsively led a detail of 124 men ashore for a "frolic with the Yankees" at Caulk's Field, near Chesterton, Maryland, on 30 August, and was soundly defeated by local patrols. American brigadier general John Stricker fought a delaying action against Ross at North Point on 12 September. Ross died in this engagement. His successor, Colonel Arthur Brooke, retreated from the Baltimore area after the British navy failed to neutralize Fort

McHenry the night of 13–14 September. (The British repulse at Fort McHenry inspired the writing of "The Star-Spangled Banner," but it was not officially adopted as the U.S. national anthem until 1931.)

British lieutenant general George Prevost, governor-general of Canada, led 10,000 men south in August to attack Plattsburgh, New York. Defending the area were only Brigadier General Alexander Macomb's 3,500, since Izard's 4,000 had been diverted toward Sackets Harbor. The Battle of Plattsburgh occurred on 11 September, coincident with the decisive American naval victory on Lake Champlain. Each side lost only about 100, but the British withdrew.

Several subsidiary conflicts between the Americans and the indigenous peoples occurred during the War of 1812. The largest of these was the Creek War, fought mostly in Alabama. It began with attacks on whites at Burnt Corn Creek on 27 July 1813 and Fort Mims on 30 August. Brigadier General John Coffee destroyed Tallushatchee on 3 November, and Major General Andrew Jackson demolished Talladega on 9 November. Battles at Emuckfau on 22 January 1814, Enotachopco on 24 January, and Calibee Creek on 27 January wore the Creeks down and prepared the way for Jackson's decisive victory at Horseshoe Bend on 27 March. The Treaty of Fort Jackson ended the Creek War on 9 August and freed Jackson for other campaigns.

As early as summer 1812, Jackson had urged a Florida campaign but was ignored. The Spanish government of Florida allowed Britain to occupy Pensacola on 14 August 1814, but Jackson captured it on 7 November. Having neutralized Pensacola's military capability, Jackson rushed west to defend Mobile and New Orleans. Americans had already repelled one British naval attack on Fort Bowyer, guarding Mobile, on 12 September and were expecting another.

With the defeat of Napoleon, the British took counsel of Arthur Wellesley, the Duke of Wellington, as to the terms that they should insist upon in the continuing negotiations with the Americans at Ghent. "The Iron Duke" pointed out that the British themselves had not gained any clear-cut advantage and that they would do well to conclude the war on the basis of the status quo ante bellum. The British government took this wise advice. The war indeed ended in a virtual draw by the Treaty of Ghent on 24 December, but that news did not reach the United States until after the Americans under Jackson had inflicted on the British one of the most one-sided defeats in military history at the Battle of New Orleans on 8 January 1815, giving Americans the idea that they had somehow won the War of 1812.

All the causes of the war were moot by the time of its end: the British had no further need to interfere with American shipping after Napoleon's defeat, the Americans had obviously failed to take any part of Canada, and with the British cutting their losses in the United States, the Americans could handle the American Indians by themselves. For the British, compared to the great struggle against Napoleon, the War of 1812 was always a sideshow. But for the Americans, the war had produced enough military heroes and legends (not to mention presidents and presidential candidates) for generations to come.

Eric v. d. Luft

See also: Baltimore; Campbell, Colin; Chippewa, Battle of; Creek War; Harrison, William Henry; Horseshoe Bend, Battle of; Houston, Samuel; Jackson, Andrew; Lundy's Lane, Battle of; Napier, Sir Charles James; Napoleonic Wars; New Orleans, Battle of; Queenston Heights; Scott, Winfield; Taylor, Zachary; Tecumseh; Thames; Washington, Burning of

References and further reading:

Adams, Henry. *The War of 1812.* New York: Cooper Square, 1999.

Berton, Pierre. *Flames across the Border: The Canadian-American Tragedy, 1813–1814.* Boston: Little, Brown, 1981.

Hickey, Donald R. *The War of 1812: The Forgotten Conflict.* Urbana: University of Illinois Press, 1989.

Mahon, John K. *The War of 1812.* New York: Da Capo, 1991.

Quimby, Robert S. *The U.S. Army in the War of 1812: An Operational and Command Study.* East Lansing: Michigan State University Press, 1997.

War Plan Orange (1907–1940)

American war plan first designed in 1907 to address a possible war situation between the United States and Japan. Although it evolved over the next three decades to account for international developments and domestic political considerations, War Plan Orange remained in effect until replaced by the Rainbow series of plans in 1940.

Under Plan Orange, American strategists prepared for a three-phase, one-front war in the Pacific Ocean between the United States and Japan. In the first phase, it was assumed that the Japanese would overrun frontier positions. (The plan never anticipated a surprise attack like that on Pearl Harbor.) In the second phase, American forces would launch a counteroffensive from a base in the western Pacific. Fortifications on Corregidor Island, however, had been halted in accordance with the provisions of the Naval Limitations Treaty of 1922, and no suitable substitute could be found. In the final phase, the American forces would take the offensive and initiate a campaign of island hopping, bypassing the stronger enemy positions and attacking the weaker bases. Once close to Japan, War Plan Orange called for a bombing campaign and blockade to end the war. Under Plan Orange, the army would hold the Philippine Islands, while the navy (including ships from the Atlantic Fleet) raced to the rescue. Although replaced by the Rainbow plans, American strategy during World War II in the Pacific theater followed many of the precepts of War Plan Orange.

Gregory Dehler

See also: World War II

References and further reading:

Miller, Edward S. *War Plan Orange: The U.S. Strategy to Defeat Japan, 1897–1945.* Annapolis, MD: Naval Institute Press, 1991.

Morison, Samuel Eliot. *The Two-Ocean War: A Short History of the United States Navy in the Second World War.* Boston: Little, Brown, 1963.

Wars of the Roses (1455–1464, 1467–1471, and 1483–1485)

A series of violent civil battles resulting from dynastic struggles in England. Their origin lay in the weakness and apparent incompetence of Henry V's son, Henry VI (1421–1471). Henry VI was unable to control the conflicting ambitions of England's nobility and was said not to be able to govern without his noble advisers.

Most prominent among them was Henry Beaufort (1436–1464), the Duke of Somerset. The effective leader of the House of Lancaster, in view of the weakness of the Lancastrian monarch, Somerset was unpopular with both the peasantry and the nobility. He became even more unpopular when, in the summer of 1450, the Irish-born peasant leader, Jack Cade (d. 1450), led a peasant revolt against London from Kent. Frustrated by repeated French raids on the English coast during the final years of the Hundred Years War, Cade's peasants occupied London for several days, and beheaded the king's unpopular treasury minister, Lord Saye. After royal forces suppressed Jack Cade's rebellion, Richard, the Duke of York (1411–1460) and a potential heir to the throne, assumed leadership of the nobles opposed to Somerset's control of the government.

Somerset, jealous of York's rising stature, exiled him to Ireland. York soon allied with Richard Neville (1428–1471), the Earl of Warwick, known as "the Kingmaker," and returned to England to challenge Somerset's control over Henry and the Lancastrians. Meanwhile, the powerful Percy family of Northumbria, who were foes of the Nevilles, had allied themselves with the House of Lancaster. The conflict between the nobility escalated in August 1453, when the Percys and the Nevilles skirmished at Stamford Bridge.

The conflict moved toward full-scale civil war, when the first signs of Henry VI's mental illness appeared, and the queen, Margaret of Anjou (d. 1482), gave birth to a son, thus ensuring the Lancaster line of succession. York temporarily wrested control of the government away from Somerset, who was unable to hide the king's incapacitation. The king's recovery in the fall of 1454 allowed Somerset to regain his position, but York could no longer tolerate Somerset's domination at court.

York's ambush of the king and his forces in the town of St. Albans on 22 May 1455 was a complete victory for Yorkists. The battle, which began with a volley of arrow fire and ended in a savage melee, saw several Lancastrian nobles killed, including Percy and Somerset. York became protector of the realm until October 1459, when his forces were routed by a contingent led by Margaret at Ludford Bridge. In November 1459, the "Parliament of Devils" met at Coventry to pass a Bill of Attainder, legislative abrogation of an individual's property and rights because of treason, against York, who escaped the Tower to Ireland. Warwick, York's ally, defeated Lancastrian forces at Northampton on 10 July 1460.

York, returning to England and to a pliable Parliament, was declared the rightful heir of Henry VI, and Margaret's son, the infant Prince of Wales, was disinherited. Margaret, now leader of the Lancastrians, led a force against York, secure in Sandal castle. For unknown reasons, York left the security of the castle to meet defeat and death at the hands of the queen's forces at Wakefield on 30 December 1460. York's son, Edward, Earl of March (1442–1483), avenged his father by seizing London and proclaiming himself King Edward IV on 4 March 1461.

At Mortimer's Cross, near Hereford, on 2 February 1461 and at Towton, near the city of York, on 29 March 1461, Edward IV's leadership won stunning victories against Lancastrian forces. The Battle of Towton, fought in the driving snow and largely in the dark, was the bloodiest battle of the wars. Edward IV spent the next several years consolidating his rule; his victory at Hexham on 15 May 1464, ended Lancastrian opposition.

However, Edward IV suffered opposition from within Yorkist ranks. As early as 1467, Warwick, "the Kingmaker," openly opposed the king's marriage to Elizabeth Woodville (1437–1492) and his foreign policies. Hoping to control Edward IV the way that Somerset had controlled Henry VI, Warwick captured the king at Edgecote on 29 July 1469. Edward IV proved no dupe, and Warwick released him, hoping to reinstall the hapless Henry VI. Despite leading a much smaller force, Edward IV defeated Warwick's rebels in a battle at Barnet on 14 April (Easter Sunday) 1471, where both sides employed cannon and artillery. On 4 May 1471, at Tewkesbury, Edward IV's exhausted forces finally defeated the equally exhausted Lancastrians, led by Margaret. Edward IV captured Henry VI and Margaret in the battle; Henry VI was murdered in the Tower of London, and Margaret was ransomed to her father, Louis XI of France (1423–1483). The dynastic conflict was ended until the death of Edward IV in April 1483.

Edward IV's brother, Richard, Duke of Gloucester, seized power as protector of the realm through two coups d'état. Richard may have murdered Edward IV's two young direct heirs, Edward V, the Prince of Wales, and Richard of York, but the nature of their deaths remains in doubt. In any case, Gloucester was soon able to have himself proclaimed king as Richard III. Richard III ruled for just over two years, until he was challenged by the Lancastrian Henry Tudor (1453–1509) at Bosworth Field on 22 August 1485. Richard III's forces outnumbered those of Henry Tudor. Richard III was also a superior commander compared to the untested Henry Tudor and would have won the battle were it not for the treachery of Lord Thomas Stanley (1435–1504). Stanley's betrayal turned certain victory into defeat for Richard III, who met his death on the battlefield. As a result of the Lancastrian victory, Henry Tudor was proclaimed king as Henry VII. His marriage to Elizabeth of York (1466–1503) not only strengthened his hold on the crown but also ended the Wars of the Roses.

This 30-year conflict takes its name from the colored roses used to identify the rival factions, the Houses of York and Lancaster, but the term *Wars of the Roses* is a misnomer. Although the House of York occasionally carried the white rose as its emblem, the House of Lancaster did not adopt the red rose as a symbol until well after the wars.

Eric Pullin

See also: Bosworth, Battle of; Henry V, King of England; Hundred Years War; Richard III; Tewkesbury; Towton, Battle of

References and further reading:

Boardman, Andrew. *The Medieval Soldier in the Wars of the Roses.* Stroud, Gloucestershire, UK: Sutton 1998.

Goodman, Anthony. *The Wars of the Roses: Military Activity and English Society, 1452–97.* Boston: Routledge and Kegan Paul, 1981.

Ross, Charles. *The Wars of the Roses: A Concise History.* London: Thames & Hudson, 1976.

Seward, Desmond. *The Wars of the Roses and the Lives of Five Men and Women in the Fifteenth Century.* London: Constable, 1995.

Warsaw/Vistula (August 1920)

The culminating battle of the Russo-Polish War. Polish general Józef Pilsudski drove back Russian Bolshevist Mikhail Tukhachevsky's offensive toward the Polish capital, most

Army volunteers with scythes, Warsaw, 1920. (Library of Congress)

likely thwarting much more ambitious objectives of the Russian revolutionists in Europe. Tukhachevsky's Bolshevik forces, having repulsed Pilsudski's effort to conquer the Ukraine for a Polish Federation, pursued the reeling Poles back to the gates of Warsaw.

Tukhachevsky had launched the attack with poorly armed disaffected divisions from his western front operations. The intervening Pripyat marshes forced Tukhachevsky to advance along two isolated axes. Pilsudski's plan exploited the Soviet's divided command by concentrating four and a half divisions of the Polish Fourth Army under General Edvard Rydz-Smigly for a counterattack that would penetrate the seam in the enemy formation and then hook around to fall on the rear of the Soviet northwestern front. Pilsudski assigned ten and a half divisions of the First and Second Armies the task of the static defense of Warsaw on a wide front facing east, while the Fifth Army's five divisions were designated to cover Warsaw's northern perimeter. Assigned the most difficult task, the Polish Fourth Army, the smallest unit, was fatigued from its rapid redeployment from Byelorussia 125 miles to the east. Political necessity forced Pilsudski to allocate the bulk of his forces in this static defense of Warsaw. His troops were inadequately clothed and inefficiently armed. Notwithstanding, he rallied his forces for the planned offensive, but not before Tukhachevsky launched his own Fourth Army west across the Vistula River in a wide hook to attack the capital from the northwest, leaving only token forces to guard the vital link with Semen Budennyi's First Cavalry Army. The latter ignored the general staff's order to link with the northern wing. On 15 August, Pilsudski personally led his Fourth Army to counterattack the southern flank of the Russian forces in covering Brest-Litovsk, breaking through the enemy lines in three days. With the southern Russian army paralyzed by passivity and the Pripyat marshes, the whole Red front crumbled, allowing the Poles to advance 200 miles by 25 August. More than 70,000 Russian troops were captured, and another 30,000 fled into East Prussia, where they were interned by German forces. The armistice signed at Riga on 12 October 1920 set the Polish boundary until the combined Russian and German offensives of September 1939.

James Bloom

See also: Pilsudski, Józef Klemens; Russo-Polish War

References and further reading:

D'Abernon, Viscount. *The Eighteenth Decisive Battles of the World.* London: n.p., 1931.

Pilsudski, Józef. *Year 1920.* London: Pilsudski Institute of America/Pilsudski Institute of London, 1972.

Zamoyski, Adam. *Battle for the Marchlands.* New York: Columbia University Press, 1981.

Washington, George (1732–1799)

American general, field commander, statesman, and first president, beloved as "the Father of His Country." George Washington was born on 22 February 1732 in Westmoreland County, Virginia. After the death of his father, Augustine, in 1743, he was raised by his older brother Lawrence. He learned surveying; worked as a surveyor in Culpeper County, Virginia, from 1749 to 1751; and inherited his father's plantation, Mount Vernon, when Lawrence died in 1752.

The same year, he was appointed a major in the Virginia militia. In October 1753, Virginia governor Robert Dinwiddie sent him to the western frontier to warn the French in the name of King George II not to encroach upon the Ohio Valley. In December, the French refused. Dinwiddie promoted Washington to lieutenant colonel and ordered him in March 1754 to lead 150 men to evict the French. Discovering about 1,000 French constructing Fort Duquesne at the confluence of the Allegheny and Monongahela Rivers, Washington retreated and quickly built Fort Necessity, defeated the French in a skirmish on 27 May, but surrendered after a 10-hour siege on 3 July.

As a full colonel of militia, he accompanied Edward Braddock to the disaster at the Monongahela on 9 July 1755 and fought in that theater until November 1758, when the British took Fort Duquesne and renamed it Fort Pitt. From 1759 to 1775, he lived quietly as a Virginia planter.

As a member of the Virginia House of Burgesses from 1759 to 1774 and the First and Second Continental Congresses in 1774 and 1775, Washington became an outspoken leader of the opposition to British rule. Congress appointed him commander in chief of the Continental army on 15 June 1775. In his first campaign, from 3 July 1775 to 17 March 1776, he forced the British under Sir William Howe to evacuate Boston.

Defending New York City against Howe, Washington abandoned Long Island on 27–30 August, fought a delaying action at Harlem Heights on 16 September, and lost at White Plains on 28 October. He led an orderly retreat into eastern Pennsylvania in November, regrouped, and made a dangerous night crossing of the ice-filled Delaware River to surprise and defeat the Hessians at Trenton on 25–26 December. Skillfully avoiding the main British force under Charles Cornwallis, he outmaneuvered and defeated three British regiments at Princeton on 3 January 1777.

That summer, Howe invaded Pennsylvania from Maryland, defeating Washington at Brandywine on 11 September and Germantown on 4 October and forcing him into winter quarters at Valley Forge. Emerging in the spring with a tough, seasoned, reinvigorated army, he would have beaten

Sir Henry Clinton at Monmouth on 28 June if not for the errors of one of his divisional commanders, Charles Lee. With increasing French support, he planned and successfully executed the Yorktown campaign in 1781, accepting Cornwallis's surrender on 19 October. He entered New York City in triumph on 25 November, said goodbye to his officers at Fraunces Tavern on 4 December, and then resigned his commission in Annapolis, Maryland, on 23 December.

Not renowned as either a strategist or a tactician, his success came from his persistence, courage, inspirational leadership, and occasional flashes of strategic and tactical brilliance, such as in the winter 1776–1777 New Jersey campaign. After serving two terms as the first president of the United States, from 1789 to 1797, he retired to Mount Vernon, where he died in agony on 14 December 1799, probably from botched medical care for an acute respiratory ailment.

Eric v. d. Luft

See also: American Revolution; Arnold, Benedict; Braddock's Defeat; Brandywine; Cornwallis, Sir Charles; Fort Duquesne, Seizure of; Gage, Thomas; Gates, Horatio; Germantown; Greene, Nathanael; Lafayette, Marie Joseph Paul Yves Roch Gilbert du Motier, Marquis de; Long Island, Battle of; Monmouth; Princeton, Battle of; Pulaski, Count Kazimierz; Steuben, Friedrich Wilhelm Augustin, Freiherr von; Trenton; Valley Forge; Wayne, Anthony; White Plains; Yorktown

References and further reading:

Brookhiser, Richard. *Founding Father: Rediscovering George Washington.* New York: Free Press, 1996.

Hannaford, Peter, ed. *The Essential George Washington: Two Hundred Years of Observations on the Man, the Myth, the Patriot.* Bennington, VT: Images from the Past, 1999.

Marshall, John. *The Life of George Washington.* Ed. Robert Faulkner and Paul Carrese. Indianapolis: Liberty Fund, 2000.

Randall, Willard Sterne. *George Washington: A Life.* New York: Henry Holt, 1997.

Washington, Burning of (24–25 August 1814)

Low point of the War of 1812 for the United States, the destruction of the American capital. Following Napoleon's exile to Elba in May 1814, Britain was able to divert more troops to the conflict in North America. Major General Robert Ross sailed from France on 27 June, bound for the United States with 4,000 soldiers. Landing without resistance at Benedict, Maryland, on 19 August, he marched toward Washington, D.C., which Secretary of War John Armstrong had left only poorly defended.

Between 5,000 and 7,000 hastily mustered American militiamen under Brigadier General William Winder engaged Ross at Bladensburg, Maryland, the afternoon of 24 August. About 250 British and 100 Americans were killed or wounded, but Ross's disciplined troops easily outmaneuvered Winder's unskilled defenders and won the field. From Bladensburg, the British could enter Washington unhindered. Ross decided to burn all the federal government buildings in retaliation for the burning of York (now Toronto, then capital of the British territory of Upper Canada) by American invaders under Major General Henry Dearborn and Brigadier General Zebulon Pike on 27–29 April 1813.

Even before the Battle of Bladensburg was over, most Washingtonians had fled the city. First Lady Dolly Madison oversaw the salvation of important national treasures, such as the Declaration of Independence, but lost her personal property. Ross entered Washington at dusk and burned the White House, the Capitol, the Library of Congress, the Treasury, the Departments of State and War, and nearly every other public building except the Patent Office, which Superintendent William Thornton saved by a ruse. After the fires burned all night, Ross abandoned Washington on 25 August and took ship again at Benedict on 30 August. President James Madison returned on 27 August to begin rebuilding the government, with Secretary of State James Monroe replacing Armstrong as Secretary of War.

Eric v. d. Luft

See also: Baltimore; Napoleonic Wars; War of 1812

References and further reading:

Muller, Charles G. *The Darkest Day: 1814: The Washington-Baltimore Campaign.* Philadelphia: Lippincott, 1963.

Pitch, Anthony. *The Burning of Washington: The British Invasion of 1814.* Annapolis, MD: Naval Institute Press, 1998.

Tully, Andrew. *When They Burned the White House.* New York: Simon and Schuster, 1961.

Waterloo (18 June 1815)

Decisive allied victory that ended Napoleon's career. Under heavy rains on 17 June 1815, Arthur Wellesley, the Duke of Wellington, ordered the Anglo-Dutch army north from Quatre Bras to the Belgian village of Waterloo, and Gebhard Leberecht von Blücher pulled back the Prussian army, first northeast from Ligny to Gembloux and then north by northwest to Wavre, about 8 miles east of Waterloo. The mud prevented the French from pursuing.

Just south of Waterloo, Wellington deployed 68,000 troops and 156 guns on high ground at Mont-St.-Jean along an east-west line from Ohain to Braine-L'Alleud, with a strong right and a weak left. He felt confident that Blücher would soon reinforce his left with 70,000 Prussians.

The rain ended at 6:00 A.M. on 18 June. Napoleon arrived

Washington, D.C., under attack by British forces led by Major General Robert Ross on 24 August 1814. (Library of Congress)

from Ligny and deployed 74,000 troops and 256 guns along a ridge south of and parallel to Wellington's line. He could have attacked at 9:00 A.M. but waited until 1:00 P.M. so that the soggy ground could dry out enough to make his field artillery effective. He mistakenly believed that Emmanuel de Grouchy's 30,000 soldiers would keep Blücher out of any engagement between the French and Wellington. Grouchy engaged Blücher at Wavre but was unable to cut the Prussians off or prevent Blücher from communicating freely with Wellington. By 11:00 A.M., Blücher had already dispatched fresh reserves under Friedrich Wilhelm von Bülow toward Napoleon's right. Their arrival at the town of Plancenoit around 4:00 P.M. would decide the battle.

After about an hour of artillery fire, the French launched their main attack about 1:15 P.M. Jean Baptiste Drouet d'Erlon's I Corps, 16,000 fresh and strong, charged Wellington's center left. Honoré Charles Michel Joseph Reille threw the 13,000 men of II Corps against the British salient at Chateau de Hougoumont, garrisoned by only 2,000 guards. The British lines held, d'Erlon quickly lost 5,000 men, and the fighting at Hougoumont lasted all day.

At about 2:15, Wellington made the first of his many successful counterattacks, sending two brigades of heavy cavalry under the Earl of Uxbridge charging from Mont-St.-Jean southward through La-Haye-Sainte toward La-Belle-Alliance, smashing d'Erlon's left and threatening Napoleon's center.

By 3:00, both sides were reeling from heavy losses. All French attacks had been repulsed, but at the cost of 40 percent casualties in Uxbridge's units. Napoleon, perceiving this slight lull to his advantage, ordered Michel Ney northward to capture La-Haye-Sainte. Ney, perhaps to atone for his failure to follow Napoleon's orders at Quatre Bras, hurled himself into the fray against Wellington's right about 3:45, but the British infantry formed defensive squares and held its ground. Fighting was fierce and deadlocked around La-Haye-Sainte when the Prussians arrived, turned the French right, and soon routed the whole French army. Wellington

Napoleon at the Battle of Waterloo. Lithograph by Currier and Ives. (Library of Congress)

and Blücher finally joined forces about 9:00 at (aptly named) La-Belle-Alliance, midway between Waterloo and Genappe, and instantly ordered Count August Neidhart von Gneisenau's fresh cavalry to pursue the French.

The most cogent theories that historians offer as to why Napoleon lost at Waterloo concern his errors in choosing key personnel. He wasted the talents of both Louis Nicolas Davout by leaving him in Paris and Nicolas Jean de Dieu Soult by making him chief of staff, a task for which he was clearly unsuited. If Davout had commanded the left instead of Ney, and Soult the right instead of Grouchy, the outcome might have been quite different. As it was, Wellington is reliably reported to have asserted well after the battle that Waterloo "was a damned close-run thing."

Eric v. d. Luft

See also: Blücher, Gebhard Leberecht von; Davout, Louis-Nicolas, Duke of Auerstädt, Prince of Eckmühl; Gneisenau, August Neidhart von; Napoleon I; Napoleonic Wars; Ney, Michel, Duc d'Elchingen, Prince de La Moskova; Quatre Bras and Ligny; Soult, Nicolas-Jean de Dieu; Wellington, Arthur Wellesley, Duke of

References and further reading:

Becke, Archibald Frank. *Napoleon and Waterloo: The Emperor's Campaign with the Armée du Nord, 1815.* London: Greenhill, 1995.

Dallas, Gregor. *The Final Act: The Roads to Waterloo.* New York: Henry Holt, 1997.

Schom, Alan. *One Hundred Days: Napoleon's Road to Waterloo.* New York: Oxford University Press, 1993.

Siborne, William. *History of the Waterloo Campaign.* London: Greenhill, 1995.

Wavell, Archibald Percival, First Earl (1883–1950)

British field marshal responsible for the first British land victories during World War II. Wavell first saw service in South Africa and India. During World War I he served in France, losing an eye at Ypres, and in Palestine under General Edmund Allenby. He returned to Palestine in 1937 to command the British forces during the Arab-Jewish riots. In

August 1939, Wavell was appointed commander in chief of British forces in the Middle East.

Following Benito Mussolini's invasion of Egypt, he launched Operation COMPASS, destroying the Italian Tenth Army, although vastly outnumbered, in Great Britain's first major land victory of World War II. However, the first half of 1941 was a difficult period for Wavell, who was confronted by the campaign against the Italians in East Africa, Erwin Rommel in Libya, the loss of Greece and Crete, political unrest in Iraq, and the occupation of Vichy-held Syria. Although faced with little alternative, his decision to defend Tobruk remains one of his greatest feats. Winston Churchill, however, was unimpressed by Wavell and following the unsuccessful offensive to relieve besieged Tobruk, Operation BATTLEAXE, transferred him to India as commander in chief. Wavell briefly became supreme commander of Allied forces in the Southwest Pacific until the fall of Malaya, Singapore, and Rangoon. Returning to his earlier command in India, he was promoted to field marshal at his own behest. After Wavell retired from military service, Churchill appointed him viceroy of India in 1943. He was also made Viscount Wavell of Cyrenaica and Winchester. Until his dismissal in 1947, Wavell worked toward Indian independence, but Louis, Earl Mountbatten, got all the credit. The same year Wavell was appointed first earl.

A taciturn man, Wavell, never publicly spoke out against the most unfair association of his name with disaster. He was a talented scholar who wrote *The Palestine Campaigns* (1928), *Allenby* (1940), *Generals and Generalship* (1941), *Allenby in Egypt* (1943), and *Speaking Generally* (1946). He also edited the popular poetry collection, *Other Men Flowers* (1944). The British liked to say that they lost all but the last battle. Except for when his opponents were Italians, it was Wavell who had to preside over so many of those lost battles.

David M. Green

See also: World War I; World War II

References and further reading:

Churchill, Winston Spencer. *The Second World War: Their Finest Hour.* New York: Houghton Mifflin, 1949.

Connell, John. *Wavell: Scholar and Soldier.* London: Collins, 1964.

Lewin, Ronald. *The Chief.* New York: Farrar, Straus and Giroux, 1980.

Wayne, Anthony (1745–1796)

American officer in the American Revolution. Wayne was noted for his bravery and quick temper, which earned him the nickname "Mad" Anthony. Born on the family estate of Waynesborough in Chester County, Pennsylvania, on 1 January 1745, he received a private education in Philadelphia. After the outbreak of the American Revolution in 1776, Wayne commanded a Pennsylvania regiment that successfully covered the retreat of American forces from Quebec. In 1777, he was promoted to brigadier general and participated in the Battles of Brandywine and Germantown. But he was culpable in the so-called Paoli Massacre of October 1777, when his sleeping troops were bayoneted in a British night assault. He distinguished himself the following year at the Battle of Monmouth. At Stony Point in 1779, Wayne achieved his greatest victory of the war when he seized the seemingly impregnable British defenses.

After the war in 1783, Wayne retired to civilian life. He dabbled in politics and business and failed in both. In 1792, he returned to active duty when President George Washington selected him commander of the Legion of the United States to destroy an American Indian confederacy that had routed two previous expeditions. On 20 August 1794, Major General Wayne defeated the hostile confederacy near modern-day Toledo, Ohio, in the Battle of Fallen Timbers. The following year he skillfully negotiated the Treaty of Greenville with the defeated American Indians. He died on 15 December 1796 in Presque Isle, Pennsylvania. His military and diplomatic skills ensured the opening of the Ohio country to white settlement.

Mark Mengerink

See also: American Revolution; Brandywine; Fallen Timbers; Germantown; Monmouth

References and further reading:

DeRegnaucourt, Tony. *General Anthony Wayne and the Ohio Indian Wars: A Collection of Unpublished Letters and Artifacts.* Arcanum, OH: Upper Miami Valley Archaeological Research Museum, 1995.

Knopf, Richard C. *Anthony Wayne and the Founding of the United States Army.* Columbus: Anthony Wayne Parkway Board, Ohio State Museum, 1961.

Nelson, Paul David. *Anthony Wayne, Soldier of the Early Republic.* Bloomington: Indiana University Press, 1985.

Sword, Wiley. *President Washington's Indian War: The Struggle for the Old Northwest, 1790–1795.* Norman: University of Oklahoma Press, 1985.

Wellington, Arthur Wellesley, Duke of (1769–1852)

British field commander, war hero, and politician. Wellington was born in Dublin, Ireland, as Arthur Wesley on 1 May 1769, schooled at Eton, and commissioned an ensign on 7 March 1783. He became a member of the Irish Parliament in 1790 and thereafter was constantly involved in politics unless on military duty overseas. As lieutenant colonel by purchase in 1793, he led the 33d Foot against the Dutch at

Hondschoote on 8 September and then commanded a brigade in Flanders and at Hanover in 1794. In India from 1796 to 1805, where his elder brother Richard was governor-general from 1797 to 1805, he achieved decisive victories at Seringapatam in 1799 and Poona, Ahmadnagar, Assaye, Argaon, and Gawilgarh in 1803. After returning to Britain as the hero of both the Fourth Mysore War and the Second Maratha War, he commanded a brigade in Hanover in 1805 and defeated the Danes at Kjoge on 29 August 1807.

Disgusted by the overly cautious tactics of his commanding officer, Sir Hew Dalrymple, which prevented him from pursuing Jean Andoche Junot after defeating him at Vimeiro, Portugal, on 21 August 1808, Wellington resigned and went home. Back in Iberia after April 1809, he won at Oporto on 12 May and Talavera on 27–28 July and held the Lines of Torres Vedras around Lisbon over the winter of 1809–1810. He defeated André Masséna and Michel Ney at Bussaco on 27 September 1810 and Masséna at Fuentes de Onoro on 5 May 1811. Wellington captured Ciudad Rodrigo in January 1812 and Badajoz in April, thus preparing ground for him to rout Auguste Frederic Viesse de Marmont at Salamanca on 22 July. He captured Madrid on 12 August, lost it, recaptured it on 17 May 1813, and then crushed Joseph Bonaparte at Vittoria on 21 June. For his successes in the Peninsular War, he was promoted to field marshal in 1813 and created duke in 1814.

Called in 1815 by the Congress of Vienna to lead the Anglo-Dutch forces against Napoleon, recently escaped from Elba, Wellington fought Ney to a draw at Quatre Bras on 16 June and retreated to Waterloo, where he and Gebhard Leberecht von Blücher's Prussians combined to give Napoleon his final defeat two days later. Given his string of victories on widely scattered battlefields, it would be difficult to dispute that Wellington was unquestionably Britain's greatest soldier of the nineteenth century.

Wellington's career after 1818 was entirely in politics. From 1828 to 1830, he was prime minister and a force for ultraconservatism.

Eric v. d. Luft

See also: Assaye; Badajoz, Siege of; Blücher, Gebhard Leberecht von; Maratha Wars; Mysore Wars; Napoleon I; Napoleonic Wars; Ney, Michel, Duc d'Elchingen, Prince de La Moskova; Quatre Bras and Ligny; Soult, Nicolas-Jean de Dieu; Waterloo

References and further reading:

Bryant, Arthur. *The Great Duke; or: The Invincible General.* New York: Morrow, 1972.

Hibbert, Christopher. *Wellington: A Personal History.* Reading, MA: Addison-Wesley, 1997.

James, Lawrence. *The Iron Duke: A Military Biography of Wellington.* London: Weidenfeld & Nicolson, 1992.

Weller, Jack. *Wellington at Waterloo.* London: Greenhill, 1998.

Westmoreland, William (1914–)

One of the more controversial generals of the twentieth century because of his role in the contentious Vietnam conflict. William Westmoreland was born on 26 March 1914 in Spartanburg County, South Carolina, and was a member of the West Point class of 1936, where he was first cadet and winner of the Pershing Award for leadership. During World War II, he was commander of an artillery unit in North Africa and Sicily and thereafter became a staff officer in the 9th Division throughout the remainder of the war. He was given command of an airborne brigade during the Korean War and was promoted to command of the 101st Airborne Division and the 19th Airborne Corps. At 42, he was named a general, and four years later he was the second-youngest superintendent in West Point history.

In 1964, Westmoreland was named commander of U.S. forces in Vietnam and became army chief of staff in 1968. Four years later, he retired. Since then, he has been on several corporate boards, run unsuccessfully for governor of South Carolina, and sued Columbia Broadcasting System for libel for claiming that he misrepresented body count statistics during the Vietnam conflict.

In Vietnam, Westmoreland employed a strategy of attrition whereby more Vietnamese Communists were to be killed than could be replaced, leading to a successful conclusion of the conflict. To implement this plan, he engaged in large search-and-destroy missions (in at least battalion strength) in which Communist forces would be flushed out and eliminated. His policies were later modified by General Creighton Abrams when he became chief of staff. Because of President Lyndon Johnson's insistence, Westmoreland became a public proponent for the war and identified American failure, but not defeat, in Vietnam.

T. Jason Soderstrum

See also: Vietnam Conflict

References and further reading:

Furgurson, Ernest B. *Westmoreland: The Inevitable General.* Boston: Little, Brown, 1968.

Westmoreland, William. *A Soldier Reports.* Garden City, NY: Doubleday, 1976.

Zaffiri, Samuel. *Westmoreland: A Biography of General William C. Westmoreland.* New York: William Morrow, 1994.

Weyler y Nicolau, Valeriano, Marquis of Tenerife (1838–1930)

Spanish general known for his harsh policy in Cuba. Weyler was born in Palma de Mallorca on 17 September 1838. At the age of 15 years, he enrolled in the Infantry School of Toledo.

By the time he was 20 years old, he was a full lieutenant. He became part of the Queen's Infantry Regiment, stationed in Madrid, Spain. He was first in his class at the Escuela Especial de Estado Mayor, graduating in September 1860.

In September 1862, he was promoted to captain and sent to the Balearic Islands. In March 1863, he requested a transfer to the commandant of Ultramar in Cuba, where he nearly died from yellow fever. He volunteered to join the command of General José de la Gándara in Santo Domingo, where he was twice decorated for bravery in the Battle of the Jaina River. In 1868, when he was back in Cuba under the command of the Count of Valmaseda, Weyler took Bayamo from the rebel forces and was made a colonel. During the next several years, he commanded a group of volunteer forces known as the Cazadores de Valmaseda. His rise continued, and he was made a brigadier general. It was under his command that the Cuban revolutionary leader Ignacio Agramonte was killed. Toward the end of the Cuban war, he returned to Spain, where by 1878, he was made lieutenant general.

After being transferred to the Canary Islands, where he was made marquis of Tenerife, he spent several years in the Philippines, in the Basque Provinces, and Catalpa. He was then named general in charge of the Spanish army in Cuba. His policy of *reconcentración,* that is, placing the civilian population in concentration camps where starvation and disease were the order of the day, made him very unpopular with the Cuban people and undermined the Spanish cause in the United States. As a result of such unfavorable publicity, he was removed in 1897. Upon his return to Spain, he was named to the War Ministry, where he served for a number of years. In 1910, in an attempt to justify his campaign of terror in Cuba, he wrote *Mi Mando en Cuba.* These and other books about other operations were well received, and other honors were lavished on him. In 1928, he celebrated 75 years of military service and died in 1930, full of years and honor—if only in Spain.

Peter Carr

See also: Cuban Revolution; Cuban Ten Years' War; Martí y Pérez, José Julián; Spanish-American War

References and further reading:

Romano, Julio (pseud. for Hipólito González y Rodríguez de la Peña). *Weyler, el hombre de hierro.* Madrid: Espasa-Calpe, 1934.

Whiskey Rebellion (1794)

First use of state militia to enforce federal law in the United States. In 1791, Congress authorized an excise tax on distilled whiskey to help pay for the federal government's assumption of state debts. The tax proved enormously unpopular with western farmers, whose economic livelihood depended on the distillation of whiskey as the only feasible way to sell their grain while Spain forbade American navigation of the Mississippi River. Also, because of a shortage of currency, whiskey served as a medium of exchange because it was portable as well as potable.

Discontent against the tax erupted into open rebellion in the summer of 1794. Tax resisters in western Pennsylvania terrorized excise officers, destroyed the stills of those paying the tax, stopped federal judicial proceedings, and feigned an attack against Pittsburgh. As these disturbances spread throughout the frontier, President George Washington and Secretary of the Treasury Alexander Hamilton viewed the insurrection as a test of survival for the federal government. Failure to enforce Congress's right to levy internal taxes, they feared, would undermine the central government.

Washington moved cautiously to suppress the rebellion, conscious of the need to allay the public's distrust of standing armies. In September 1794, he issued a proclamation calling for obedience to the law and offered the rebels amnesty for past criminal behavior in return for their acceptance of the tax. When the rebels rejected this olive branch, Washington federalized 12,950 militiamen from eastern Pennsylvania, Maryland, Virginia, and New Jersey to stifle western unrest. The force, commanded by Virginia governor Henry Lee, set out for western Pennsylvania in October. Although the militia encountered evidence of western discontent, it met no rebel force. Resistance to the tax collapsed once the army reached Pittsburgh. Militia officers and federal marshals rounded up scores of suspects but arrested only 20 men. Only two were convicted of treason, and Washington wisely pardoned both of them. The bulk of the militia force returned home after a month in the field, while a contingent of 1,500 men remained in western Pennsylvania throughout the winter to maintain order.

In establishing a precedent to enforce federal law, Washington recognized that federal military intervention in what was ordinarily a process for civil or criminal law was a potentially explosive political act. He scrupulously adhered to the provisions of the Militia Act of 1792 for using federalized militia to maintain order, subordinating military to civilian authority to avoid any appearance of an occupying army. Equally important, he obtained support from state authorities and the general public before embarking on a military expedition. But Washington's use of force against the Whiskey rebels contributed to a political backlash against the Federalists that aided the Democratic-Republican Party's rise to power in 1800.

Dean Fafoutis

See also: Lee, Henry; U.S. Militia; Washington, George

References and further reading:

Coakley, Robert W. *The Role of Federal Military Forces in Domestic Disorders, 1789–1878.* Washington, DC: U.S. Army Center of Military History, 1988.

Kohn, Richard H. "The Washington Administration's Decision to Crush the Whiskey Rebellion." *Journal of American History* (1972).

Slaughter, Thomas P. *The Whiskey Rebellion: Frontier Epilogue to the American Revolution.* New York: Oxford University Press, 1986.

White Mountain, Battle of (Weißer Berg, 8 November 1620)

Battle that crushed the rising of the Bohemian estates against Habsburgian rule and marked the beginning of the Habsburgian absolutist regime over this territory. At the end of October 1620, the season for campaigning was coming to an end, but Maximilian, Duke of Bavaria, and his general, Graf Johann Tserclaes von Tilly, who commanded the allied forces of the Catholic League, pressed for a decision in 1620 and convinced the imperial commander to march with them directly to the Bohemian capital. The combined imperial and league forces, numbering 28,000, met the Bohemian army of about 21,000 men about one-half mile west of Prague on 8 November 1620. Christian, Prince of Anhalt, commander of the Bohemian forces, had posted his troops on the White Mountain in a strong defensive position. He failed to attack Tilly's forces while they were passing a little brook on the battlefield. Nevertheless, it was a risky operation for the allied armies to advance uphill against the Bohemians. At the beginning of the battle, the advancing imperial troops on the right wing were hard-pressed by the counterattacking Bohemian cavalry. But as Tilly dispatched his troops to relieve the right wing, the Bohemian forces fled. After barely two hours, the battle was won for the emperor and the Catholic League. The victorious troops took the town of Prague on the following day without fighting.

The battle did not destroy the entire Bohemian military force, but the psychological impact of the defeat was so strong that Frederick V of the Palatine, the elected Bohemian king, fled immediately. His enemies' propaganda derided him as the "winter king."

Michael Kaiser

See also: Thirty Years' War; Tilly, Johann Tserclaes, Graf von

References and further reading:

Delbrück, Hans. *History of the Art of War within the Framework of Political History.* Vol. 4, *The Modern Era.* 1920 (German edition). Trans. Walter J. Renfroe. Westport, CT: Greenwood, 1985.

Krebs, Julius. *Die Schlacht am weissen Berge bei Prag (8 November 1620) im Zusammenhang der kriegerischen Ereignisse.* Breslau, Poland: Koebner, 1879.

White Plains (28 October 1776)

A well-fought defeat of the American forces under General George Washington early in the American Revolution. After the disastrous campaign resulting in the loss of New York City, Washington's battered American forces fled north to Westchester County. In late October 1776, Washington hoped to regroup his forces, and he dug in at the village of White Plains, New York. The Bronx River flowed south through Washington's right, and beyond it rose Chatterton Hill, upon which a few regiments were posted.

Pursuing at their usual leisurely pace was General William Howe's confident English and German army. On 28 October, Howe prepared to attack Washington's position. In one of the rare instances of the war, both armies stood at even strength, around 13,000 men each. Seeing a vulnerable flank, Howe struck Chatterton's Hill, which was held by only one American brigade. German and English troops ascended the hill under artillery and small arms fire and managed to break the militia stationed there. Delaware and Maryland Continentals held longer yet soon were overwhelmed.

During the night, Washington pulled back his entire force to a stronger line to the north. The small engagement cost Howe 300 men, whereas Washington lost about 150. As was usually the case, Howe did not press the advantage and instead turned south to mop up the isolated garrison at Fort Washington on Manhattan. General Howe probably missed one of his best opportunities to finish off Washington's army at White Plains. The Continental forces were able to retreat across New Jersey to the Delaware, where they later struck back at Trenton.

Brian Dunkerley

See also: American Revolution; Washington, George

References and further reading:

Commanger, Henry S., and Richard B. Morris, ed. *The Spirit of Seventy-Six.* New York: Da Capo Press, 1975.

Martin, James K., and Mark E. Lender. *A Respectable Army.* Arlington Heights, IL: Harland Davidson, 1982.

Whitney, Eli (1765–1825)

American inventor and entrepreneur. Born in Westborough, Massachusetts, Eli Whitney demonstrated from an early age great technical skill in repairing such diverse intricate objects as violins and watches. During the American Revolution, Whitney also demonstrated a talent for business. Seeing that the British blockade of the colonies had cut off the supply of British-made nails, he established a small workshop on his father's farm to meet the wartime demand for these and other items such as hatpins and walking sticks.

Graduating from Harvard College in 1792, Whitney traveled to South Carolina, where he had accepted a teaching position. En route, he stopped at the plantation of Nathanael Greene, now run by the general's widow. There, in partnership with the plantation's manager, Phineas Miller, Whitney designed and built the invention for which he is justly famous: the cotton gin.

Unfortunately, despite the great economic impact that Whitney's invention would have on the United States and indeed the global economy, piracy of his patent meant that the inventor saw comparatively little reward for his efforts. In 1798, therefore, Whitney turned to gun making. Specifically, he proposed to produce for the fledgling U.S. Army 10,000 muskets in the space of two years. Although at that time the manufacture of military firearms depended on the arduous work of skilled craftsmen producing a weapon at a time, central to Whitney's proposal was to employ machines and unskilled labor to mass-produce many basically identical weapons in a short time. Further, with the introduction of the "uniformity system," Whitney proposed to reduce the cost and time needed to repair weapons through the use of interchangeable parts.

Historians still debate how much of Whitney's success in the mechanization of gun making is fact and how much early American mythology. Clearly, he brought little practical experience to the task when awarded the contract in 1798. Indeed, it was not until a year later that his factory was built at Whitneyville, New Hampshire. The awarding of the contract had as much to do with Whitney's friendship with fellow Harvard graduate Oliver Wolcott Jr., secretary of the treasury, as it did to the efficacy of his proposal. Further, Whitney's close friendship with Captain Decius Wadsworth, the government's arms inspector, helped keep the project alive when, instead of 4,000 muskets, Whitney delivered only 500 after the first year of the contract. There can be little doubt that Whitney failed in his promise to mechanize completely the process of gun making. Nonetheless, at the end of eight (instead of two) years, Whitney had delivered the last of his 5,000 muskets. In 1812, he received another contract for a further 15,000 weapons for the federal government and the same number for the state of New York. By that time, Whitney's system had been adopted at federal armories, including Harpers Ferry, of which Whitney had been offered the inspectorship, only to decline it in 1807. Though the true mechanization of arms manufacturing and interchangeability of gun parts would have to wait until after the mid–nineteenth century, Whitney in many ways symbolized the first hesitant steps toward the modernization of war.

Adam N. Lynde

See also: U.S. Army

References and further reading:

Hays, Wilma Pitchford. *Eli Whitney and the Machine Age.* New York: Franklin Watts, 1959.

Smith, Merritt Roe. *Harpers Ferry Armory and the New Technology: The Challenge of Change.* Ithaca, NY: Cornell University Press, 1977.

Wilderness (5–7 May 1864)

Drawn but extremely bloody battle in the American Civil War. Union general Ulysses S. Grant wanted to engage and occupy Robert E. Lee's Army of Northern Virginia, while Union forces elsewhere cut up the South and destroyed its ability to resist.

The Battle of the Wilderness was the first of a series of battles between the Army of the Potomac (with George Meade nominally in command) and the Army of Northern Virginia. Grant kept seeking to move to his left to turn Lee's right flank; Lee kept sliding to his right to blunt the Union advance. By late June 1863, the two armies were beginning to dig trenches and other fortifications for the long sieges at Richmond and Petersburg, and Lee's force would never again take the offensive.

On the morning of 5 May 1864, Grant and Lee ran into one another in an area known as the Wilderness; the brush and scrub was so thick that movement was difficult, and superior numbers and, more important, superior Union artillery, did not matter greatly. Lee attacked on familiar ground near the Fredericksburg and Chancellorsville battlefields. During the fighting, the woods caught on fire, and the smoke, fire, and screaming of wounded men added to the confusion. Fighting continued the next day, as both sides rushed reinforcements along the two main roads, the Orange Turnpike and the Plank Road. Both sides suffered some 17 percent casualties, and typically Union generals had retreated after such bloody conflict.

But, on 7 May, Grant ordered his men to retreat and then swing left, trying to find Lee's elusive right flank. Lee, however, was prepared to meet such an offensive, and the two armies once again would meet one another at Spotsylvania Court House.

Charles M. Dobbs

See also: American Civil War; Grant, Ulysses Simpson; Lee, Robert Edward; Spotsylvania Court House

References and further reading:

Gallagher, Gary W., ed. *The Wilderness Campaign.* Chapel Hill: University of North Carolina Press, 1997.

Rhea, Gordon C. *The Battle of the Wilderness, May 5–6, 1864.* Baton Rouge: Louisiana State University Press, 1994.

Scott, Robert Garth. *Into the Wilderness with the Army of the Potomac.* 2d ed. Bloomington: Indiana University Press, 1992.

William II (Friedrich Wilhelm Viktor Albert) (1859–1941)

Germany's last monarch. The reign of William II as King of Prussia and emperor of Germany saw dramatic cultural and economic advancements by the German nation, but it ended with Germany in political chaos and William II with neither a throne nor a country.

William was born on 27 January 1859 to Crown Prince Frederick of Prussia and Princess Victoria of Great Britain. Young Prince William's paternal grandfather was the reigning Hohenzollern king of Prussia, William I. In 1871, following the German victory in the Franco-Prussian War, William I was proclaimed emperor, or kaiser, of the new German Empire.

The education of young Prince William was supervised by his grandfather's conservative court to distance the boy from the liberal ideas espoused by Prince Frederick. William I died in 1888. Frederick acceded to the throne but died 99 days later, leaving his unprepared son as emperor, in what was known as the "Year of the Three Kaisers."

Throughout his reign, William II sought to strengthen the monarchy and increase Germany's international prestige. He dismissed Chancellor Otto von Bismarck in 1890 over matters of imperial prerogative. William believed that a strong military and navy were key not only to German defense but also to making Germany a major international power.

The emperor's interference in foreign affairs often had disastrous consequences. In 1890, his refusal to renew a treaty with Russia led to a Franco-Russian rapprochement and the threat of a two-front war, Bismarck's nightmare in his old age. Despite attempts to improve relations, William's caustic public comments about British policy alienated the British government, driving it into an alliance with France and Russia.

When the European nations lurched toward war in 1914, William II vainly tried to mediate peace. As supposed war leader, William in reality became largely a figurehead and spectator, while German generals increasingly exercised power. Faced with mutinies and revolution as the war ground on unsuccessfully, in November 1918, William II abdicated, seeking asylum in neutral Holland. He died there in exile on 4 June 1941. Despite mendacious efforts by the Allied propaganda machines to portray "the Kaiser" as an evil warlord, William was more unstable and weak than wicked.

Eric Smylie

See also: World War I

References and further reading:

Aronson, Theo. *The Kaisers.* New York: Bobbs-Merrill, 1971.
Cecil, Lamar. *Wilhelm II.* 2 vols. Chapel Hill: University of North Carolina Press, 1996.
Cowles, Virginia. *The Kaiser.* New York: Harper & Row, 1963.
Kurenberg, Joachim von. *The Kaiser: A Life of Wilhelm II, Last Emperor of Germany.* New York: Simon and Schuster, 1955.
Palmer, Alan. *The Kaiser: Warlord of the Second Reich.* New York: Charles Scribner's Sons, 1978.

William the Conqueror (c. 1028–1087)

Victor of Hastings who reigned in England 1066–1087. William was the illegitimate son of Robert I of Normandy and Herleve. He ascended the ducal throne at the age of eight and survived the upheavals of any minor ruler with dissident vassals. Fortunately for William, it was in the best interest of Henry I of France to support William against his Norman vassals.

French support shifted away when William upset the balance of power in the region in 1053 by marrying Matilda of Flanders. Henry I's overt actions against William ended with two defeats, in 1054 at Mortemer, and in 1058 at Varaville. French activities against William continued behind the scenes, first with Henry I and later with Philip I (1060–1108).

As the duke of Normandy, William refined and reorganized the feudal organization of the duchy. He was quick to maintain the duchy's sovereign rights in the intricate network of feudal loyalties and service requirements. It should be noted that Normandy continued the Viking/Norse social pattern of institutionalized migration, which sent excess young men out to distant territories to raid, trade, or settle. Normandy's younger sons fueled the southern Italian wars, the foundation of the Norman state in Sicily, and the conquest of England.

Edward the Confessor's death on 5 January 1066 signaled the inevitable invasion of England by William. Anticipating an attack from William, Harald Godwinsson set up a coastal defense that waited through the summer. The situation radically changed when Harald's brother Tostig joined forces with Harald Hardradi to invade with a large Viking fleet at the Humber estuary. Harald went north with his housecarls and defeated Tostig and Harald Hadradi at Stamford Bridge (25 September 1066). Sharp on the heels of this victory came the news that William's flotilla had crossed the channel. Harald rode south with depleted and limited forces. On 14 October 1066, William attacked near Hastings. Repeated

Eighteenth century engraved portrait of William the Conqueror, with ornaments and vignette of Brittania in supplication below. (Library of Congress)

attacks and greater numbers resulted in a victory for the Normans. England was now William's.

As part of his conquest, William set out upon a total reorganization of England. Land was redistributed to Norman followers, and a series of strategically placed castles were built throughout England. Opposition to William's conquest continued until 1070, although it was largely a series of local revolts. These ended when he systematically devastated the archdiocese of York and, in large part, Cheshire, Shropshire, Staffordshire, and Derbyshire. Further consolidations of power included the reform of the English Church and the compilation of the *Domesday Book.* The *Domesday Book* surveyed the area south of the Tees River. It not only accounted for taxes but also ultimately covered the king's vassals, allowing him to enforce their military and monetary obligations to the crown.

Tamsin Hekala

See also: Hastings, Battle of; Norman Conquest; Stamford Bridge, Battle of

References and further reading:

Douglas, David C. *William the Conqueror: The Norman Impact upon England.* Berkeley: University of California Press, 1964.

Loyn, H. R. *Anglo-Saxon England and the Norman Conquest.* New York: St. Martin's Press, 1963.

Wingate, Orde (1903–1944)

British general and advocate of irregular warfare. Wingate was a highly controversial figure whose contributions to the British war effort were highly publicized but limited.

Born on 26 February 1903, Wingate was the son of a religious British army officer. He was commissioned into the artillery in 1923 and served in the Sudan between the wars. In 1936, he was sent to Palestine as an adviser to the Jewish Settlement Police. A convert to Zionism, Wingate conceived the idea of special night squads, later to become the *Palmach.* Wingate trained these soldiers in guerrilla and counterinsurgency warfare and is regarded as a founder of the Israeli army.

When World War II broke out, Wingate was posted to the Sudan to advise Ethiopian nationalist fighters. He organized and led a guerrilla unit known as Gideon Force in a series of successful operations from January to May 1941. But he was dismissed and demoted for interfering with local politics. Depressed, he attempted suicide in June 1941.

In May 1942, General Archibald Wavell called Wingate to India to organize long-range penetration operations against the Japanese. Wavell had commanded Wingate in East Africa and respected his unorthodox methods. Wingate organized the 77th Brigade and led it in a harassment campaign against the Japanese in northern Burma from February to June 1943. Movement was by foot, with resupply by air. Nicknamed the "Chindits" by war correspondents, Wingate's raiders caught the attention of many, including Winston Churchill. Wingate was temporarily promoted to major general in September 1943 and given five brigades for a larger operation. Supported by the American No. 1 Air Commandos, Wingate launched his raid in February 1944. Four brigades were inserted by air, while the fifth marched to its area of operations. Wingate expected that his units could hold jungle airstrips for resupply and reinforcement. Wingate was killed in an air crash on 25 March 1944 during the operation.

Wingate's operations were controversial. They resulted in nearly 50 percent casualties, and their effect on the Japanese was limited. Wingate himself was eccentric and managed to antagonize his superiors. He demonstrated that British soldiers could operate in the jungle, although he is sometimes credited with inspiring the French operation at Dien Bien Phu.

Tim J. Watts

See also: Chindits; Wavell, Archibald Percival, First Earl

References and further reading:

Bierman, John, and Colin Smith. *Fire in the Night: Wingate of Burma, Ethiopia, and Zion.* New York: Random House, 1999.

Fergusson, Bernard. *Beyond the Chindwin: Being an Account of the Adventures of Number Five Column of the Wingate Expedition into Burma, 1943.* London: Collins, 1945.

Royle, Trevor. *Orde Wingate: Irregular Soldier.* London: Weidenfeld & Nicolson, 1995.

Wolfe, James (1727–1759)

British general accredited by many historians as the pivotal figure in ensuring Canada for England during the Seven Years' War. Born 2 January 1727 in Westerham, Kent, England, Wolfe entered the military at 14. Having fought in the War of the Austrian Succession and holding a distinguished service record, Wolfe was appointed a brigadier general by Prime Minister William Pitt in 1758 and sent to North America as second in command to Lord Jeffrey Amherst. His troops fought heroically under heavy fire at the capture of the French fortress of Louisbourg on the island of Cape Breton in Nova Scotia. Returning to England, he was given command of the Quebec campaign and promoted to major general.

With 2,000 soldiers and 140 ships under his command, Wolfe was opposed by a French army two-thirds his size under the commands of Louis-Joseph de Montcalm-Gozon and François-Gatson Levis. For three months, because of the well-entrenched French position and faulty decisionmaking and poor positioning by the British, Wolfe's troops failed in their frontal assaults. He decided on 13 September 1759 to have his soldiers make the risky attempt to take 5,000 of his men down the St. Lawrence River at night and land 2 miles (or 3.5 kilometers) above the city. Climbing rugged cliffs, British troops met the French coming from Beauport in the open field of the Plains of Abraham in the morning. The well-disciplined British soldiers routed the French quickly. Yet in the moment of British triumph, both Montcalm and Wolfe were killed. Wolfe was shot twice but was still in active command until a third bullet hit him in the chest, mortally wounding him. His victory broke French power in the New World.

T. Jason Soderstrum

See also: Seven Years' War

References and further reading:

Cassgrain, Henri R. *Wolfe and Montcalm.* Toronto: University of Toronto Press, 1964.

Warner, Oliver. *With Wolfe to Quebec: The Path to Glory.* Toronto: Collins, 1972.

Whitton, Frederick E. *Wolfe and North America.* Port Washington, NY: Kennikat Press, 1971.

Wolseley, Garnet Joseph, Viscount (1833–1913)

British field commander and military administrator, depicted by William S. Gilbert and Arthur Sullivan as "the very model of a modern Major-General" in *The Pirates of Penzance.* Born the son of a major on 4 June 1833 in Golden Bridge, Ireland, Wolseley was commissioned an ensign in 1852 and immediately requested a transfer to a regiment bound for combat in India. He suffered a severe leg wound in the Second Burmese War and lost an eye during the siege of Sevastopol in the Crimean War. For his service at Lucknow in the Indian Mutiny, he was brevetted a lieutenant colonel and became Sir James Hope Grant's staff officer. Accompanying Grant to China in 1860 during the Taiping Rebellion and the Second Opium War, he was present at the fall of the six Dagu forts on 21 August.

Stationed in Canada after 1861, he occasionally observed the American Civil War in the company of Robert E. Lee and Stonewall Jackson. In 1870, his bloodless capture of Fort Garry, Manitoba, crushed the first uprising of Louis Riel (the second was in 1885). Back in Britain, he was reprimanded for his 1869 book, *The Soldier's Pocket-Book for Field Service,* which criticized some army practices. Ordered to West Africa in 1873, he won the Ashanti War with victories at Amoaful and Kumasi in 1874.

Given command of all British forces in South Africa after the disaster at Isandlwana, he arrived on 23 June 1879 and followed up Baron Frederick Chelmsford's victory at Ulundi on 4 July by capturing Cetshwayo on 28 August, thus ending the Zulu Rebellion. Leading a punitive expedition to Egypt in 1882, he won decisively at Tel-el-Kebir and captured Cairo. He tried to rescue Charles "Chinese" Gordon at Khartoum in the First Mahdist War but arrived on 28 January 1885, two days too late.

Created viscount in 1885, promoted to field marshal in 1894, and serving as commander in chief from 1895 to 1901, Wolseley used his worldwide experience to modernize and reform the British army. His innovations showed immediate benefits in the Second Boer War. He died in Mentone, France, on 26 March 1913. Well into the twentieth century, it was common to characterize things well in hand as "All Sir Garnet."

Eric v. d. Luft

See also: American Civil War; Arabi Pasha; Boer Wars; Buller, Sir Redvers Henry; Campbell, Colin; Crimean War; Gordon, Charles George; Indian Mutiny; Isandlwana; Khartoum, Siege of; Lee, Robert Edward; Riel's Rebellion; Taiping Rebellion; Zulu Civil Wars and Rebellion

References and further reading:

Featherstone, Donald. *Tel El-Kebir, 1882: Wolseley's Conquest of Egypt.* London: Osprey, 1993.

Kochanski, Halik. *Sir Garnet Wolseley: Victorian Hero.* London: Hambledon, 1999.

Lehmann, Joseph H. *The Model Major-General: A Biography of Field-Marshal Lord Wolseley.* Boston: Houghton Mifflin, 1964.

Maxwell, Leigh. *The Ashanti Ring: Sir Garnet Wolseley's Campaigns, 1870–1882.* London: Cooper, 1985.

Women in the World's Militaries

Despite popular misconceptions, women have played a vital role in the conduct of war throughout history. They served in traditional roles as mothers and wives, making male-dominated warfare possible by supporting armies and caring for children on the home front; supported armies in the field as camp followers or spies; led armies in battle; and sometimes fought in the front lines beside male warriors. Millions suffered at the hands of invading conquers, and many fought as terrorists or irregular warriors. In this light, the recent trend to expand military opportunities for women is merely a recognition of the contributions female warriors have always made, rather than a politically correct break with tradition as some critics have suggested.

Indeed, the record of war-making women stretches well into antiquity. Deborah led the armies of Israel during the twelfth century B.C.E., the Assyrian warrior queen Sammuramat invaded India in the ninth century B.C.E., and Artemisia commanded a portion of the Persian fleet at the Battle of Salamis. In western Europe, Boudicca led a rebellion of the Iceni Tribe that killed 70,000 Romans in Britain in 60, and Aethelflaed (daughter of Alfred the Great) directed the storming of the Viking stronghold at Derby in 917. Matilda led armies in the service of the papacy between 1060 and 1114, and, of course, there is Joan of Arc.

In Asia, Trung Trac and Trung Nhi commanded 80,000 soldiers who expelled the Chinese from Vietnam in 40, and the Syrian queen Mawia defeated Roman legions during invasions of Phoenicia, Palestine, and Egypt a century later. In China, Wei Hua Hu fought with the imperial army for 12 years in the third century, and Princess Ping Yang led an army of 70,000 during the 600s.

Wei Hua Hu's example is rare because she fought as a common soldier, but it is hardly unique. The Spanish heroine Augustina (the Maid of Saragossa) rallied the defense of Saragossa during a French siege in 1808; Nadezhda Durova fought against Napoleon for nine years in the Russian cavalry; Salaym Bint Malhan fought with Muhammad in the early years of Islam; and Phoebe Hessel suffered a bayonet wound at the Battle of Fontenoy while fighting with the British 5th Regiment of Foot in 1745. Women like Hessel often dressed as men to enlist and were discharged if their sex was revealed, but they fought and died like their male counterparts and deserve to be remembered.

Some civilizations went beyond individual examples and fielded units composed entirely of women. Libyan and Scythian Amazons fought en masse, as did the Amazons of Dahomey, who fought in Africa during the nineteenth century, and the Serbian female battalions that served in World War I. The most prolific use of women soldiers came during World War II, when the Soviet Union enlisted almost 1 million in the fight against Germany. Soviet women served in every type of combat unit, from fighter and bomber squadrons to infantry and artillery battalions, and a select few distinguished themselves as snipers.

The Soviet example foreshadowed a greater utilization of military women after World War II. Increasingly sophisticated weapons required technical expertise rather than brawn, reducing the need for all soldiers physically to dominate their enemies, and military forces mirrored civilian bureaucracies by assigning women to handle most administrative work. Feminist movements in many countries created greater equality for women, making it more difficult for the armed forces to segregate women from most military jobs, and in the United States, the end of the draft in 1973 heralded a greater reliance on women to meet recruiting goals.

Since the 1970s, the United States has dramatically increased the number of women in uniform and opened up greater opportunities for women to serve in combat than ever before. By 2001, women comprised more than 14 percent of active-duty military personnel and were eligible for assignment to more than 90 percent of all military occupations. Ground combat units remain off-limits to female soldiers as a rule, but women have flown combat missions as pilots and served on combat ships in the Navy.

Other nations have followed the U.S. example. Women now serve in the armed forces of every country in the North Atlantic Treaty Organization and make up more than 14 percent of the Russian and more than 25 percent of South African armed forces.

The dramatic and relatively recent increases in the number of women serving in military assignments have come at considerable social cost in many countries, where debates regarding the "proper" role of men and women and the efficacy of sexually integrating combat units continue to rage. Critics worry about unit cohesion and sexual assault or harassment, whereas others deride or fear the erosion of traditional sex roles.

There is no denying, however, the trend toward greater reliance on female soldiers in the future or the fact that outside the developed world, women continue to play vital roles in warfare. Alice Lakwena led an army against the Ugandan government in 1987, women represent more than 40 percent of Peru's Shining Path terrorist organization, and female soldiers have fought from Kosovo to Chechnya and from Vietnam to the Middle East. They are, it seems, as important to regular and irregular military units today as they have always been, no matter what many conventional histories may argue.

Lance Janda

See also: Boudicca's Rebellion; Greek-Persian Wars; Joan of Arc

References and further reading:

DeGroot, Gerard, and Corinna Penniston-Bird, eds. *A Soldier and a Woman: Sexual Integration in the Military.* New York: Longman Press, 2000.

DePauw, Linda Grant. *Battle Cries and Lullabies: Women in War from Prehistory to the Present.* Norman: University of Oklahoma Press, 1998.

Spurling, Kathryn, and Elizabeth Greenhalgh. *Women in Uniform: Perceptions and Pathways.* Canberra: School of History, University College, Australian Defence Force Academy, 2000.

Wood, Leonard (1860–1927)

Former U.S. Army chief of staff who captured the public's imagination with his exploits in the Plains Wars and Cuba and whose personal honor made him reject a deal that would have made him president of the United States in 1920. Born on 9 October 1860 in Winchester, New Hampshire, Wood was educated at the Pierce Academy in Middleboro, Massachusetts, and graduated from Harvard Medical School in 1884.

He joined the army in 1885 and was appointed assistant surgeon. On 8 April 1898, he received the Medal of Honor for his role in the pursuit of Geronimo and carrying dispatches through hostile Apache territory. He was promoted to the rank of captain in 1891 and was named assistant attending surgeon to the president and other senior officials in 1895.

He gained his greatest fame in 1898 as a colonel in the 1st Volunteer Cavalry (the Rough Riders) in Cuba in the Spanish-American War and led the attack on Santiago de Cuba. Promoted to brigadier general and then major general of the Rough Riders, he was given military command of Santiago and was made military governor of Cuba from 1899 to 1902. In 1904, he was named governor of the Moro Province of the Philippines for three years and was instrumental in putting down a revolt in that province, although his administration was criticized for ruthlessness. He then served as U.S. commander of military forces in the Philippines from 1906 to 1908. Wood brought American progressive principles to Cuba and the Philippines, that is, administration by professionals, which vastly improved health, education, and infrastructure but did little to alleviate the mass poverty of these lands. In many ways, Cuba and the Philippines served as a "laboratory" for progressivism in the United States, and many progressives believed the U.S. Army ideally suited for the imposition of progressive methods.

After serving as commander of the Department of the East and special ambassador to Argentina, Wood was named army chief of staff on 22 April 1910, a position he held until 20 April 1914, when he was replaced by General John Pershing.

Wood then returned to command the Department of the East and was openly critical of President Woodrow Wilson's neutralist position, advocated military preparedness, and founded the movement of the same name. Although Wood trained the 10th Division at Camp Funston, Wilson refused to commission him for service in Europe. After running for the Republican nomination for president in 1920 and serving as commander of the Central Division, Wood was appointed governor-general of the Philippines in 1921. His administration was criticized for its severity by the Thompson Commission in 1926. Wood died on 7 August 1927 in Boston, Massachusetts. Wood could be rightly termed the "armed progressive," bringing the principles of the American movement—order and good governance by "experts"—to less developed countries.

T. Jason Soderstrum

See also: Plattsburgh Movement

References and further reading:

Hagedorn, Hermann. *Leonard Wood: A Biography.* New York: Harper and Brothers, 1931.

Hitchman, James H. *Leonard Wood and Cuban Independence, 1898–1902.* The Hague: Nijhoff, 1971.

Lane, Jack C. *Armed Progressive: General Leonard Wood.* San Rafael, CA: Presidio Press, 1978.

Sandler, Stanley. *Glad to See Them Come and Sorry to See Them Go: A History of U.S. Army Tactical Civil Affairs/Military Government, 1775–1991.* Fort Bragg, NC: [1998].

Worcester, Battle of (3 September 1651)

The last great battle of the English Civil War. Worcester marked the supremacy of Oliver Cromwell and the end of Charles II's immediate hopes.

After Cromwell's victory at Dunbar in 1650, Charles had raised a new army to replace the one he had lost. By June 1651, he had 15,000 foot, and 6,000 horse at his command in Scotland. In July, as Cromwell invaded the north, Charles's army moved south, hoping to gather more support from the English royalists.

As the king's army marched, it did attract some supporters but also lost Scots who had not anticipated a campaign in England. On 22 August, when Charles entered the city of Worcester, his tired army had shrunk to little more than 12,000 men, and Cromwell continued to gather troops.

On 28 August, Cromwell approached Worcester from the east with about 28,000 men. Heavily outnumbered, Charles could only hope to hold off the parliamentarians at the bridges that crossed the Severn and Temes Rivers. On the 29th, 11,000 of Cromwell's men crossed the Severn River to the south of Worcester, and as they moved on the town from that direction, others threw bridges of boats over the two rivers. On 3 September, the anniversary of his victory at Dunbar, Cromwell ordered the attack and his men began to advance on the Scottish positions. When Cromwell's men took one of the earthworks at the southeast of the city and turned its guns on the town, the end was clear. Many of the royalist horse attempted to flee by the one open gate at the north road, but the royalist foot was pinned in the town. As Cromwell's troops encircled the city, most people in the town surrendered.

Worcester was one of the most complete disasters for the royalist cause, with many of Charles's leading supporters killed or captured. Very few of the Scots managed to escape the town, and even fewer made it back to Scotland. Charles managed to evade his pursuers and fled into exile, not to return until 1660.

Daniel German

See also: Dunbar, Battle of; English Civil War (1642–1649)

References and further reading:

Kenyon, John P. *The Civil Wars of England.* London: Weidenfeld & Nicolson, 1988.

Young, Peter, and Richard Holmes. *The English Civil War.* Ware, UK: Wordsworth Editions (Wordsworth Military Library), 2000.

World War I (1914–1918)

The first of the great twentieth-century conflicts. The assassination of Austrian archduke Franz Ferdinand on 28 June 1914 in Sarajevo by a Slavic nationalist triggered World War I, although a complex and competing European alliance system was the underlying cause. The Austro-Hungarian Empire blamed the Serb government for inciting the murder. Throughout July, tensions mounted as Austria-Hungary issued insulting ultimatums to Serbia. In response, Russia mobilized in support of its ally, Serbia, while Germany fully supported Austria-Hungary. Although the cold-blooded murder of the archduke was viewed by most Europeans as an act of barbarism, Austria-Hungary's treatment of tiny Serbia was also seen by many as blatant bullying. On 28 July, Austria-Hungary declared war on Serbia. Russia began a full mobilization the following day, a move considered by Germany to be an act of war. Fearing a protracted war on two fronts (France and Russia were allies), Germany attacked France through Belgium (Germany had been signatory to an international treaty guaranteeing Belgium's territorial integrity) in accordance with the Schlieffen Plan. The violation of Belgian neutrality brought Great Britain into the war on the side of the Russians and the French. This alliance was called the Triple Entente. Turkey joined Germany and Austria-Hungary in October 1914, comprising the Central Powers.

On the evening of 4 August, German troops crossed the border into Belgium. Using huge siege guns manufactured by Skoda, Germany reduced the Belgian fortification systems around Liege to rubble, opening the way to Paris. Within a month, the Germans were in France. Throughout the next two weeks, the British and French conducted a rapid retreat, covering as much as 25 miles a day, to avoid being outflanked to the north by the Germans. French commander in chief Marshal Joseph Joffre hoped to deliver a decisive defeat to the Germans in the open plains before Paris. While the Germans attacked to the north in accordance with the Schlieffen Plan, the French assaulted the center in Alsace-Lorraine, following their prewar strategy. The French assault in Alsace-Lorraine ended in bloody failure, much to the dismay of Joffre, who overestimated the assistance Russia could provide France in the opening weeks of the war. Nevertheless, the French offensive, as well as the unexpected duration of Belgian resistance and the Russian invasion of East Prussia, caused the Germans to weaken their attack through northeastern France. To the north, Joffre waged his decisive battle along the Marne River in September, halting the German advance. At the battles of Aisne, Picardy, and Artois, the French sustained heavy losses in a vain attempt to regain lost territory from the entrenched Germans. By the third week of September, the British, French, and Germans were driving north in the famous "race to the sea," each hoping to outflank the other. A series of battles in Flanders (the First Battle of Ypres) lasted until 22 November 1914 and closed the last gap in the lines. Two systems of trenches stretched nearly 500 miles from Switzerland in the south to the North Sea in the north.

In the east the lines were more fluid; trench warfare did not take hold. The initial Russian advance into East Prussia was reversed at the Battle of Tannenberg at the end of August 1914. The Russians enjoyed some successes, but they proved unable to achieve a decisive victory. A series of strategic retreats by the Russian forces led to the evacuation of western Poland in the hope that they would be able to concentrate on driving across the Carpathian Mountains into Silesia and knock Austria-Hungary out of the war. Neither the Russians nor the Central Powers could claim victory at a series of battles around Warsaw, Lodz, and Masuria waged in the winter of 1914–1915. Only in capturing the fortified city of Przemysl could the Russians claim a clear victory. In the Caucasus region, Turkey attacked Russia in several battles that came to resemble results on the other fronts, in that they only led to deadlock. In April 1915, Turkey undertook one of the most controversial actions of the war by massacring a large but undetermined number of Armenians, claiming that they were aiding the Russians.

Much of 1915 and 1916 was spent trying to break the stalemate that had developed. Lured by promises of territorial compensation, both Italy and Bulgaria entered the war in 1915. Italy sided with the Triple Entente, and Bulgaria joined the Central Powers. Italy made few contributions to the war and failed to relieve Austro-Hungarian pressure against Russia. Over the next two years, the Italian army fought a series of indecisive but bloody battles in the vicinity of the Isonzo.

The Bulgarians joined the Germans and the Austro-Hungarians in attacking Serbia. During the first month of the war, the Serbian army repelled an Austro-Hungarian invasion. Although the British and French landed more than 100,000 troops in nearby Greece, the combined Central Powers forces defeated Serbia, driving its army into Albania to be evacuated by the Italian navy.

In hopes of relieving pressure on their Russian allies, Great Britain expanded the war by waging vigorous action against the Turks. Urged by Winston Churchill, First Lord of the Admiralty, the Australian and New Zealand Army Corps (ANZAC) forces launched a campaign against Turkey on 25 April 1915 along the Dardanelles peninsula on the approaches to Constantinople. The Gallipoli campaign was marked by months of bitter fighting. After suffering more than 265,000 casualties, the British withdrew in January 1916. The British also attacked Turkey from Egypt through Palestine and from Central Asia toward Baghdad. Like most peripheral campaigns of the war, the offensives against the Turks failed to significantly alter the strategic balance, and the British surrendered an entire army at Kut El Amara.

On the Western Front, the combatants remained locked in trench warfare. The Battle of Nueve-Chapelle in March 1915 established the pattern of offensive operations that was maintained for the next two years. Massive artillery bombardments were followed by waves of troops who overran outlying enemy positions. (The Germans slightly altered the pattern in April 1915, when they introduced poison gas attacks.) The attackers, however, soon outran their reinforcements and artillery support. Once exposed, the attackers sustained heavy casualties from massive counterattacks and machine gun fire. Eventually, the front lines stabilized near the starting points. At Ypres, Artois, and Champagne, the pattern was repeated throughout the year.

The Germans had much better luck on the Eastern Front. As the Russians withdrew from the bulge of Poland to strengthen their line around Warsaw and its rivers, the Germans and Austro-Hungarians launched a massive attack in March 1915 that became known as the Battle of Gorlice-Tarnow. By end of the offensive in September, the Germans had cleared Poland; taken the Russian cities of Brest-Litovsk, Grodno, and Kovno; and had captured more than 300,000 Russian prisoners. In September 1915, Czar Nicholas II assumed direct control of the army; if anything, morale among the ranks declined ever further. Nonetheless, Russia managed to fight on because of enormous reserves of manpower and a war-induced economic boom.

In the west, 1916 followed a pattern similar to the previous year. Both sides launched offensives that they hoped would be the decisive battle of the war. In February, the Germans attacked at Verdun with the cold-blooded purpose of "bleeding" the French army white. The offensive ground to an indecisive halt only after each army lost some 400,000 men in what may have been the most intense battle of all time. To the north, the British initiated a massive assault that became known as the Battle of the Somme. Beginning in July and ending in November, the battle claimed more than 1.2 million men on both sides as killed, wounded, or missing with no significant strategic outcome. Morale among the combatants declined in the face of rising carnage and indecisive results; Allied commanders could think of nothing other than more of the same.

On the eastern front, the Russian army surged forward in June and July 1916 in a series of actions known as the Brusilov Offensive. After achieving a number of victories, the attacks came to an end when the Russian logistical system failed to support the needs of a follow-up advance. In August 1916, Romania entered the war on the side of the Allies. France and Russia promised Romania territorial compensation in exchange for its support. Romanian troops attacked Austro-Hungarian forces but met stiff resistance. Soon after, the Germans and Bulgarians came to the aid of their Central Powers ally and invaded Romania. By the end of the year, Romania, which received no military assistance from Russia, France, or Great Britain, was almost completely conquered.

Unrest and a collapse of morale marked 1917. After the failure of the French army to achieve decisive results during

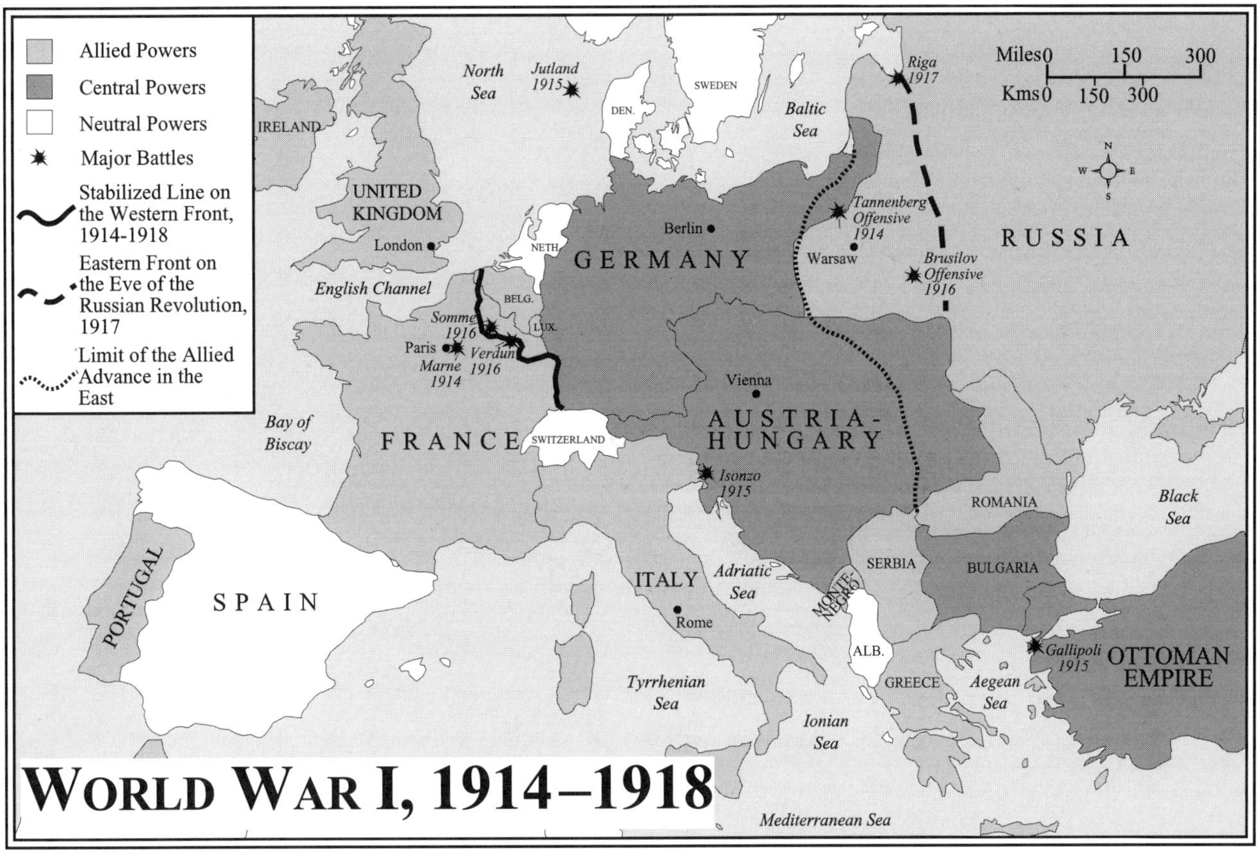
WORLD WAR I, 1914–1918
Allied Powers
Central Powers
Neutral Powers
Major Battles
Stabilized Line on the Western Front, 1914-1918
Eastern Front on the Eve of the Russian Revolution, 1917
Limit of the Allied Advance in the East
Miles 0 150 300
Kms 0 150 300
North Sea
Jutland 1915
SWEDEN
DEN.
Baltic Sea
Riga 1917
IRELAND
UNITED KINGDOM
London
NETH.
Berlin
GERMANY
Tannenberg Offensive 1914
Warsaw
Brusilov Offensive 1916
RUSSIA
English Channel
BELG.
LUX.
Somme 1916
Paris
Verdun 1916
Marne 1914
Vienna
Bay of Biscay
FRANCE
SWITZERLAND
AUSTRIA-HUNGARY
Isonzo 1915
ROMANIA
Black Sea
PORTUGAL
SPAIN
ITALY
Adriatic Sea
Rome
MONTENEGRO
SERBIA
BULGARIA
ALB.
Gallipoli 1915
OTTOMAN EMPIRE
GREECE
Aegean Sea
Tyrrhenian Sea
Ionian Sea
Mediterranean Sea

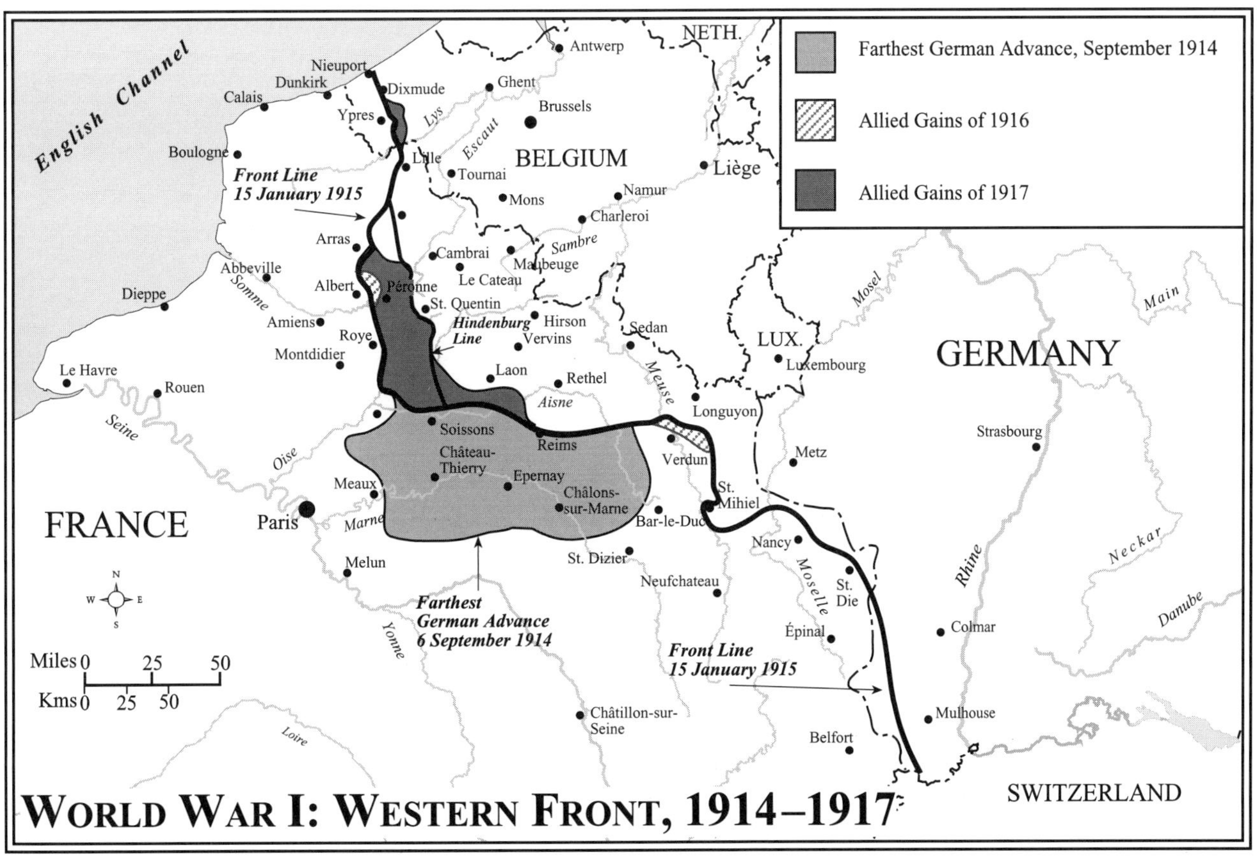
WORLD WAR I: WESTERN FRONT, 1914–1917
Farthest German Advance, September 1914
Allied Gains of 1916
Allied Gains of 1917
English Channel
NETH.
Antwerp
Nieuport
Dunkirk
Calais
Dixmude
Ghent
Brussels
Ypres
Lys
Escaut
Boulogne
Lille
BELGIUM
Liège
Tournai
Front Line 15 January 1915
Mons
Namur
Charleroi
Arras
Sambre
Cambrai
Maubeuge
Abbeville
Somme
Albert
Péronne
Le Cateau
Dieppe
St. Quentin
Amiens
Hindenburg Line
Hirson
Vervins
Sedan
Mosel
Main
Roye
Montdidier
LUX.
Luxembourg
GERMANY
Le Havre
Rouen
Laon
Rethel
Meuse
Aisne
Longuyon
Seine
Soissons
Reims
Strasbourg
Oise
Château-Thierry
Verdun
Metz
Meaux
Epernay
Châlons-sur-Marne
St. Mihiel
FRANCE
Paris
Marne
Bar-le-Duc
Nancy
Neckar
Rhine
Melun
St. Dizier
Moselle
Neufchateau
St. Die
Farthest German Advance 6 September 1914
Yonne
Colmar
Danube
Épinal
Front Line 15 January 1915
Miles 0 25 50
Kms 0 25 50
Châtillon-sur-Seine
Mulhouse
Loire
Belfort
SWITZERLAND

French soldiers fire from front line trenches in France during World War I. (National Archives)

the abortive Nivelle Offensive, a near-complete collapse in the army's morale resulted. Civilians protested the treatment of the soldiers as well as the management of the war. Yet the overwhelming number of soldiers who participated in the mutinies went unpunished. The French army vowed to defend France but refused any further futile offensives. After the Battle of Passchendaele (July–October 1917), British morale also suffered in the face of heavy casualties for little gain. Italy's defeat by the Austro-Hungarians at the Battle of Caporetto in autumn 1917 triggered a similar crisis of morale among Italian soldiers. Only the infusion of British and French troops stabilized the Italian front and kept Italy in the war. Not until the last days of the war did the Italian army resume offensive operations.

In Russia, revolutionaries forced Czar Nicholas II to abdicate. Alexander Kerensky, leader of the liberal democratic forces of the Russian Duma, attempted to keep Russia in the war. A series of Russian offensives gained some ground against the Austro-Hungarians but only succeeded in draining the Russian army in the long run. In November, Vladimir Ilich Lenin, leader of the Bolsheviks, wrested power from Kerensky. Lenin, who had been transported from Switzerland and financed by the Germans, promised to pull Russia out of the war if he attained control. In December, he began treating with the Germans, who continued to drive through an undefended Russia. In the spring of 1918 at Brest-Litovsk, Russia withdrew from the war and accepted great losses of territory and the payment of a huge indemnity.

Almost the only bright sign for the Allies in 1917, the entry of the United States into the war in April tipped the strategic balance in favor of the Triple Entente. The United States, however, was poorly prepared for war and could not contribute significant numbers of troops for more than a year. There were two causes for American belligerence: Germany resumed unrestricted submarine warfare in 1917. German leaders gambled that they could starve Great Britain out of the war before the United States could decisively alter the strategic balance. And Germany engaged in a harebrained scheme to entice Mexico into the war with a promise of the return of its "lost territories" from the United States.

With Russia out of the war and the United States still building an army, the Germans transferred troops from the east to the west as fast its rail system could carry them. In the spring of 1918, the Germans launched a massive offensive. Using newly designed flexible surprise assault tactics first tried in the east (the German High Command seemed to be the only belligerents to have learned anything from the deadlock of trench warfare), the Germans crashed through the British and French lines and even threatened Paris. But the offensive ground to a halt in July because of a combination of logistical problems and the renewed resolve of the defenders, now bolstered morally and militarily by increasing numbers of American troops. At Amiens in August, the British launched an offensive with more than 500 tanks, driving back the Germans. Military aircraft played an increasing role on both sides. Although the Allies had far greater resources here, the Germans had pioneered strategic bombardment of their British cities. In the following month, the Americans, who had more than 1.3 million soldiers in France at this point, vigorously launched their first exclusively American attack at St. Mihiel. Morale among the Germany army deteriorated as it was driven back. German military leaders realized that it was only a matter of time before surrender was necessary.

As the German high command fretted about the future, revolution swept through the multiethnic Austro-Hungarian Empire. The Italians attacked at the Battle of Vittorio Veneto on 24 October 1918, breaking the Austro-Hungarian lines. On 3 November 1918, Austria-Hungary sued for peace. The situation in Germany continued to deteriorate. After being told that he no longer held the loyalty of the army, Kaiser William II abdicated and fled to the Netherlands. The socialist government that replaced the kaiser sued for peace the following day. On the 11th day of the 11th month of 1918, the war finally ended with an armistice, after 10 million had died and 20 million had been wounded. Four empires—Austria-Hungary, Russia, Germany, and the Ottoman Turkish—had collapsed. The map of Europe awaited redrawing, but only among the Americans (who had suffered comparatively briefly and lightly) was there any sense of unalloyed triumph. A peace treaty was imposed on Germany that was harsh enough to engender the deepest resentment but weak enough that it could not prevent Germany from eventually once again coming within an ace of mastering Europe.

Gregory Dehler

See also: Amiens; Armored Fighting Vehicles; Artillery; Ataturk (Mustafa), Kemal; Baker, Newton D.; British Military, Twentieth-Century Organization and Structure; Brusilov, Aleksei Alekseevich; Cambrai, Battle of; Cantigny; Caporetto; Chateau Thierry/Belleau Wood; Chemical and Biological Warfare; Falkenhayn, Erich von; Foch, Ferdinand; French, John Denton Pinkstone, First Earl of Ypres; Galliéni, Joseph-Simon; Gallipoli; Gorlice/Tarnow; Haig, Douglas; Hindenberg, Paul von Beneckendorf und von; Isonzo, Battle of the; Joffre, Joseph Jacques Césaire; Kitchener, Horatio Herbert; Kuropatkin, Aleksey Nikolaevich; Lettow-Vorbeck, Paul Emil von; Logistics; Ludendorff, Eric Friedrich Wilhelm; Machine Gun; Mackensen, August von; March, Peyton; Marne, Battle of the; Megiddo, Battle of; Meuse-Argonne; Moltke, Graf Helmuth Johannes Ludwig von; Nicholas, Grand Duke; Nivelle, Robert; Pershing, John J.; Pétain, Henri-Philippe; The Somme; St. Mihiel; Tannenberg and the Masurian Lakes; Verdun; Vimy Ridge; William II; Ypres, Battles of

References and further reading:

Coffman, Edward. *The War to End All Wars.* New York: Oxford University Press, 1968.

Ellis, John. *Eye-Deep in Hell: Trench Warfare in World War I.* Baltimore: Johns Hopkins University Press, 1989.

Ferguson, Niall. *The Pity of War: Explaining World War I.* New York: Basic Books, 2000.

Fussell, Paul. *The Great War in Modern Memory.* New York: Oxford University Press, 1975.

Gilbert, Martin. *The First World War: A Complete History.* New York: Henry A. Holt, 1996.

Keegan, John. *The First World War.* New York: Vintage Books, 1998.

World War II (1939–1945)

The greatest and most destructive enterprise in human history. World War II is the name given to the series of interrelated conflicts that took place from 1939 to 1945. Although often presented as a monolithic conflict between democracy and dictatorship, World War II was in fact a number of complex (and interrelated) conflicts that saw combat on six of the seven continents. The two major theaters of combat were the European theater and the Pacific theater. Within these two theaters were many subareas, such as the Soviet front, North Africa, the Mediterranean and the central Pacific.

The war formally started with the German invasion of Poland on 1 September 1939. The result of a crisis over the alleged mistreatment of Germans living in Poland, it ignited the war since England and France had given guarantees for Poland's security after the failure of the Western powers to protect Czechoslovakia in the Munich Crisis of 1938. On 15

September 1939, the Soviet Union, under the secret protocol of the German-Soviet Non-Aggression Pact, invaded the eastern half of Poland. (France and Great Britain, despite the Polish treaty, did nothing.) Germany achieved quick success by use of a joint-arms tactic known as blitzkrieg (lighting war). Massed use of armor and aircraft overwhelmed Polish defenses. By October 1939, Poland had been divided between Germany and the Soviet Union.

After the fall of Poland, both sides lapsed into inaction during a period known as the Phony War (or "Bore War"). France and Britain worked to rearm against an expected German attack in the spring of 1940. At the same time, the USSR attacked Finland to regain Soviet land lost during Finnish independence. This conflict, known as the Winter War, saw the USSR engaged in a costly conflict in which the outnumbered Finnish held off the Soviets for weeks.

The USSR's high command was still suffering from the effects of Stalin's purge of Red Army officers in 1938. The poor showing of the Soviets in the Winter War with Finland was to give Adolf Hitler (and many Western military "experts") the impression that it would be easy to defeat the Soviet Union.

On 10 May 1940, Hitler launched an attack on the West, using the same blitzkrieg tactics seen in Poland. The Germans overran the Low Countries and France by June 1940. The French, demoralized by domestic politics and tied to an outmoded strategy (although they had many modern tactical weapons), were quickly beaten back by the Germans as they raced to the sea. The French were depending on the Maginot Line, a series of interconnected bunkers and forts, to hold the Germans at bay.

In a daring strategy, the Germans simply circumvented the line, attacking the dense forest of the Ardennes, in Belgium. (The "experts" had ruled that the Ardennes was "not good tank country.") Caught off guard, and with many of their troops tied to the defensive forts, the French fought bravely but were overrun by the Germans. The British Expeditionary Forces in France were driven back to the coastal town of Dunkirk, where an armada of Royal Navy and private "little ships" rescued hundreds of thousands of British and French troops. In late June, Benito Mussolini, expecting an easy victory, declared war on France and England. Instead, French troops held back the Italians in the south of France. With the fall of France, the northern half of France became a German-occupied zone, and a collaborationist government was formed in the south of France under the World War I hero, Marshall Henri-Philippe Pétain. The defeat of France (again, unexpected by the experts) allowed the Germans to establish submarine and air bases in order to carry the war to Great Britain itself.

The British refused and sought to continue the war from England. The Germans launched a series of air attacks that became known as the Battle of Britain. At the same time, the Germans and the British fought the Battle of the Atlantic, as German U-boats sought to sink enough Allied shipping to starve the British into submission. The United States, although neutral, became more involved in the war as it provided supplies to the Allies.

The battle now moved to North Africa, as Italy invaded Egypt in September 1940. The Italians were ignominiously routed by the British, and Hitler, to forestall a Italian collapse, sent Erwin Rommel to take over command of the Afrika Korps. For almost the next three years, the battle would seesaw between the British and the Germans and, later, the Americans in North Africa.

Mussolini was also having his problems in Greece, which he had invaded absolutely without provocation late in 1939. But the Greeks put up a stout resistance and chased the Italians out of Greece into Italian-occupied Albania. It is thus erroneous to state that "Britain stood alone" in 1940; Greece also withstood the Axis on the battlefield.

In 1941, Germany had proven unable to defeat Britain and turned its efforts to defeating the Soviet Union, Hitler's main goal all along. Launching his attack on 21 June 1941, the Germans pushed deep into the Soviet Union in the greatest military operation to date, taking millions of prisoners and dealing the bungling Stalin, who had been repeatedly warned by the West, a massive blow. His best generals and marshals were either dead (shot in the purges) or, like Georgy Zhukov, far from the scene of the action. Only by bringing in reinforcements from eastern Soviet Union were the Soviets able to stem the flow of the German attack at the gates of Moscow.

In 1941, with the Nazis in control of large parts of Europe, a plan was formalized for what the Germans called "the Final Solution," in which the Jews of Europe would be systematically murdered. In January 1942, Reinhard Heydrich held the Wannsee Conference, in which the mechanics of the Holocaust were planned in meticulous and wide-ranging detail. The Jews of Europe were not to be enslaved, as were the Slavs or other conquered peoples; they were to be exterminated. They were not to be shot but to be gassed, as vermin.

The Germans turned south to the Balkans in April 1941 to salvage Mussolini's aborted adventure and punish Yugoslavia for its recent turn to a pro-Allied policy. In one of the shortest campaigns on record, German columns, spearheaded by armor, slashed through Yugoslavia and Greece in a matter of weeks. The British and Commonwealth forces, which had retreated to the island of Crete, were overwhelmed by German airborne landings.

In Asia, Japan had started a policy of expansion in 1931 with the annexation of Manchuria, run by a puppet govern-

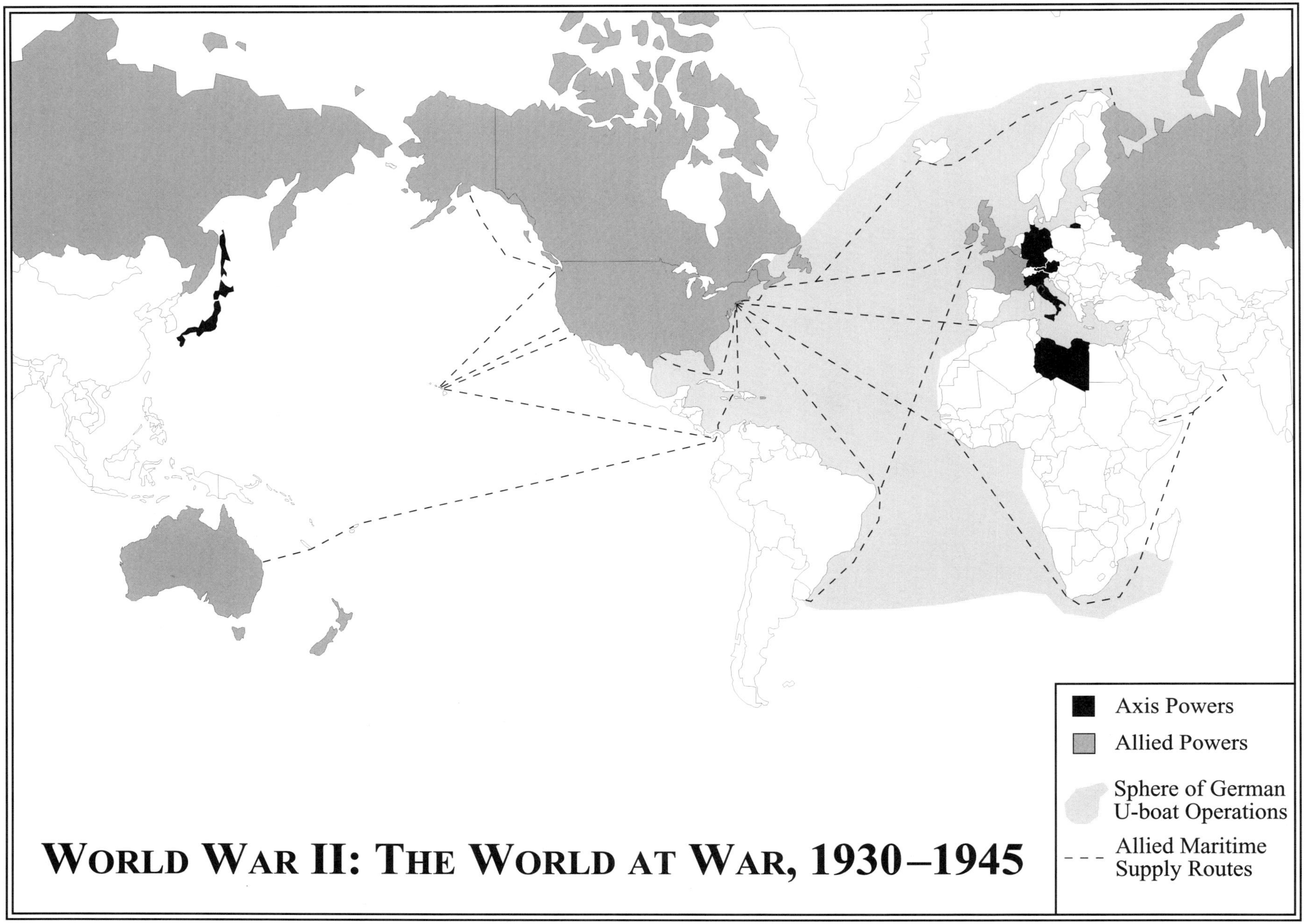
WORLD WAR II: THE WORLD AT WAR, 1930–1945
Axis Powers
Allied Powers
Sphere of German U-boat Operations
Allied Maritime Supply Routes

ment known as the state of Manchukuo. In a fabricated affair called the Marco Polo Bridge Incident in 1937, the Japanese engaged China in open warfare that was to last until 1945. In response to this aggression, the United States, in concert with other Western powers, enacted a series of embargoes to protest the Japanese actions. This response would lead to increased tension between Japan and the West and a Japanese belief that war was the only solution to its problems.

The Japanese military was divided between two factions in the 1930s. The Go North faction, which sought expansion into the Soviet Union, and the Go South group, which sought to move into the Pacific region. After defeat at the battle of Kohlin-Go on the Mongolian border in 1939 by Soviet forces, the stunned Japanese military leaders chose to adopt the Go South policy.

By 1941, the American and British embargo was having a serious effect on Japan, and the military leaders, led by Hideki Tojo, felt that talks with the United States were not producing the desired results and put into place his absurd plan to knock the United States out of the Pacific with a lighting stroke. The Japanese hoped that once the United States was defeated, it would be willing to come to the bargaining table. To the Japanese, the Americans were "decadent" and "soft."

On 7 December 1941, the Japanese launched a sneak attack on the American Naval Base at Pearl Harbor, Hawaii, and on that "day of infamy," Japan lost the war. The Japanese sank a number of elderly American battleships in shallow water, where all but one were raised, but missed the U.S. carriers, which by sheer luck were out at sea on the day of the attack. They missed the oil tank farms and the machine shops that could repair all but the worst damage. The attack was severe enough to fill all Americans (even those who had bitterly opposed Franklin D. Roosevelt's aid to the British) with a cold rage and determination to fight Japan to the finish, yet it failed to inflict much strategic damage. Few states have so miscalculated with such disastrous consequences. The United States declared war on Japan the next day, and Germany declared war on the United States on 11 December, thus bringing the United States into both the European and Pacific wars. Both nations would pay a terrible price for this miscalculation of U.S. power.

The Japanese lacked the industrial bases and the resources to carry out a long war with the United States. Only by a lightning blow against the Western powers would the Japanese be able to consolidate their hold on east Asia.

At the same time as Pearl Harbor, the Japanese attacked the British in Singapore and Dutch holdings in the Dutch East Indies (now Indonesia) to consolidate their hold. For the first six months, the Japanese scored success after success, driving the British ignominiously out of Singapore and Hong Kong; defeating the Americans in the worst defeat in U.S. military history, the loss of the Philippines; and easily taking over Dutch possessions in the East Indies. Yet despite the desire to avenge Pearl Harbor, the decision was made at the highest levels in the United States and Britain that Germany would come first and then Japan.

In May 1942, the Japanese were turned away from an attack on New Guinea at the battle of the Coral Sea, marking the first time a Japanese attack was spoiled and showing the Japanese that the United States was not yet out of the war. And at the Battle of Midway, the Japanese Navy lost four carriers and their irreplaceable crews. In August 1942, the United States started its first land offensive against Japan, landing on the island of Guadalcanal.

In Russia, after weathering a brutal winter, the Germans renewed the attack on the Soviets, with the campaign in the Caucasus culminating in the Battle of Stalingrad in the fall of 1942. It was a geographical high-water mark for the German advance into the USSR, but nonetheless, on the Soviet front, after months of brutal house-to-house fighting, the overextended Germans were defeated and tens of thousands of survivors taken captive. The year 1942 was to be the time of the greatest advance of the Axis nations.

In November 1942, the United States landed troops in North Africa to help support the British, who had achieved the upper hand against Rommel. In 1943, after two years of toil, the Red Army went onto the offensive against the Germans, slowly driving them off Soviet soil. And in July 1943, the Soviets and Germans engaged in the largest tank battle in history, at Kursk. It would be the last offensive of the Germans in the east.

In 1943, Allied troops landed in Sicily and precipitated the capitulation of the Italian government. German troops occupied northern Italy and sought to stave off the Allied attack. The Allied troops landed on the Italian peninsula in September 1943 and soon bogged down in the difficult terrain of Italy. What Churchill termed "the soft underbelly of Europe" proved a tough proposition, as the Germans fell back in fighting order from one prepared line of defense to the other.

Meanwhile, in Asia, Allied forces started to push the Japanese back, making advances in New Guinea, Burma, and the central Pacific. The United States attacked Japan from two directions. From his bases in Australia, General Douglas MacArthur attacked up through New Guinea in the southwestern Pacific, while Admiral Chester Nimitz attacked through the central Pacific. There was also an increased tempo in the air war against Germany in 1943, as the Royal Air Force (RAF) and U.S. Army Air Forces started bombing Germany around the clock. The Allied air offensive sought to destroy German industry and reduce the Luftwaffe as an effective fighting force.

By 1944, Germany was being pressured by Soviet Union in the east and was facing the imminent threat of an Allied invasion in France. It was no longer a matter of if there would be an invasion, but when and where. In June 1944, the Allies landed at Normandy and started pushing the Germans out of France, while the Soviets advanced into Poland and central Europe. The Allied armies in Italy slowly advanced, taking Rome in June 1944.

Germany tried to retaliate for Allied area bombing with a number of "wonder weapons," such as the V1 and V2 rocket and jet warplanes, but the former were a misuse of resources and the latter a case of too little too late. In the Pacific, the Japanese attempted to stem the tide of the American advance by the use of kamikaze (divine wind) suicide pilots to sink American ships. Although horrific to deal with, these weapons (which caused more casualties than any other Japanese weapon) did not stem the tide of the Allied advance.

By January 1945, Germany was surrounded on both sides by Allied armies, while the Japanese were preparing for attacks on the very boundaries of the empire. Hitler personally approved a plan for a great counteroffensive, known later as the Ardennes Offensive or "the Battle of the Bulge." Despite initial successes, the Germans no longer had the resources to continue any such operation on any scale, and the Ardennes, if anything, hastened the German collapse.

The Soviets advanced into eastern Prussia by March 1945, and the British and American forces had crossed the Rhine River, closing in on Berlin. After a costly assault by the Soviets, Berlin fell in May 1945. Hitler had committed suicide in his Berlin bunker and named Admiral Karl Donetz as the new leader of Germany. Donetz surrendered to the Allies in May 1945.

The Allies now turned their full efforts to Japan. In February 1945, Iwo Jima was captured, and Allied forces invaded Okinawa in April 1945. The Americans began the most horrific bombing campaign in history, burning out the heart of numerous Japanese cities. The worst air raid of all, the fire-bombing of Tokyo in March, snuffed out the lives of some 100,000 civilians. Although the Americans rarely perpetuated the personal atrocities of the Japanese, with the fire-bombings and the all-out submarine sinkings of any Japanese ship, civilian or military (in protest of which the United States had gone to war in 1917), to starve out the home islands, it was evident barbarism was spreading among all of the belligerents. There was even talk of poison gas to flush out the Japanese from their island caves and bunkers.

After the costly invasion of Okinawa, the Allies started planning for an invasion of the Japanese home islands in the fall of 1945. Estimates were that such an offensive would have costs hundreds of thousands of casualties on both sides and caused massive destruction to Japan. All of this planning became academic, however, when the United States dropped atomic bombs on Hiroshima and Nagasaki in August 1945. The Japanese surrendered a few days after the bombings, and their formal surrender came in September 1945. The morality of the nuclear bombings is still debated into the twenty-first century. But such thoughts did not trouble Allied military personnel poised for the invasion of Japan. They thanked God. It is indeed difficult to see how the Japanese, dug in and imbued with a death wish, could have surrendered before literally millions of casualties would have been inflicted on both sides; young girls were training with bamboo spears to drive the *gaijin* devils into the sea. The mindless official slogan, even after Hiroshima, was "Our spirit against their steel!" The invasion of the Japanese home islands would have proven the most bloody military operation in history and would have left Japan a charnel house.

World War II was "total war" for practically all belligerents. This meant that the home fronts were to be mobilized. Labor was conscripted, and women were often given the choice of war production or the military. As is so often the case, minorities and women made progress, at least for the duration of the war, in securing employment in many fields that had been closed to them before the war. It was a war of production, and production was the home front's battle front. Each side had to produce tens of thousands of weapons, as well as rations, clothing, transportation, and medical supplies. Here the Allies excelled, although the German home front, under the superefficient Albert Speer and after the Stalingrad catastrophe, produced some miraculous figures under Allied round-the-clock bombings and ignorant interference by Nazi officials. Great Britain also mobilized hundreds of thousands of women and boys to turn out the Spitfires and artillery to win the war of production, again under the bombs, although nothing like what the Reich had to endure increasingly. The Japanese and German home fronts were under the tightest military discipline, with no strikes or demands for better conditions tolerated. People could be, and were, shot for "defeatism" or for listening to enemy broadcasts. Both nations were reduced to the barbarism of slave labor, using enemy prisoners of war, foreign civilians, and "undesirables" to toil in factories and mines. Both Axis nations saw a drastic drop in the living standards of civilians, although only Japan suffered rising death tolls caused by malnutrition as the American submarine blockade tightened on the home islands. German civilians lived well enough during the war from the booty of other peoples' countries. Of course, both nations suffered far more than any nation in history from aerial bombardment, with the Germans losing some 600,000 and the Japanese about 393,000 civilians. Yet under the bombs, the Germans and the Japanese home fronts accomplished miracles of production, the Germans more than the Japanese. Nineteen forty-four, a year when almost all of the territory of the Third Reich was subject to aerial bombardment, was the year of the greatest German aircraft production.

But by far, the greatest production powerhouse of the war was the United States. Its factories, untouched by the enemy, poured out hundreds of thousands of tanks, warplanes, small arms, and other items. Even ship production was increasingly carried out on an assembly-line basis. For example, the United States could supply the Soviet, British, and Nationalist Chinese armies with thousands of its immortal 2.5-ton 6 x 6 army trucks (perhaps the greatest trucks in history) and still see to it that there were enough for every GI to ride into battle. Locomotives were sent to Russia, as were aircraft and aluminum; canned meat went to Britain and uniforms to the Chinese; and still the American soldier was the best-equipped of any soldier in history. Germans capturing or killing GIs made a point of stripping them of their boots and cigarettes.

American factories did not always turn out the best weapons, however. The M-4 Sherman tank was adjudged inferior on almost all counts to contemporary German armor, but the Americans could out-produce the Germans almost five-to-one in tanks. The world war on the American home front saw nothing like the anti-German hysteria of World War I, but in one of the most shameful episodes in the history of American civil rights, all Americans of Japanese ancestry, citizens and aliens alike, living on or near the West Coast, were rounded up and placed in uncomfortable internment camps.

Estimates of the cost of the war are in excess of $1 trillion 1945 dollars and more than 60 million military and civilian deaths. The Allies were indeed fortunate that their enemies, for all their bluster of "never surrender" and "fighting to the last man," actually followed their enemy's dictum of "unconditional surrender," once ordered to capitulate by their own governments and, if anything, cordially cooperated thereafter with their conquerors. After the war, war crimes trials were held in Nuremberg, Germany, and Tokyo for some of the top Axis leaders, with a number being put to death, although as is so common in such politically charged matters, some of the worst criminals escaped justice.

The war dramatically changed the geopolitical landscape, with only two nations, the United States and the Soviet Union, left as superpowers. The next 50 years would be taken up with the Cold War, as tensions between the capitalist West and the Communist East rose at times to nearly unbearable heights, and then, in the last decade of the twentieth century, miraculously, the Cold War ended almost entirely without bloodshed as the Soviet Union quietly collapsed.

More than half a century has passed since the Allies were victorious over the pure evil of Nazism and the manic nationalism of Japan. The post–World War II period, for all of its midsize and small wars, was a far more happy time than that after World War I, if for no other reason than that no global conflicts have erupted since 1945. Winston Churchill's prediction that if the Axis lost the war, the world could then proceed upon "broad, sunlit paths," has more or less come true. Most of that conflict's belligerent nations, victors and vanquished alike, enjoyed living standards by the end of the twentieth century that were unimaginable before the war, and democracy seems more broad-based and stronger than ever. None of this would have been possible, of course, had the Axis prevailed. World War II, for all its bitter cost, was indeed "the Good War."

Daniel Halevy

See also: Armored Fighting Vehicles; Atomic Bomb, Development of; Bradley, Omar Nelson; Burma, Retreat from; Chindits; Churchill, Sir Winston; de Gaulle, Charles; Eisenhower, Dwight David; El Alamein; France; Hitler, Adolf; Kursk, Battle of; Leningrad, Siege of; MacArthur, Douglas; Maginot Line; Montgomery, Bernard Law; Moscow; Mountbatten of Burma, Louis Francis Albert Victor Nicholas; Normandy Landings; Okinawa; Office of Strategic Services; Patton, George Smith, Jr.; Pearl Harbor Attack; Rommel, Erwin Johannes Eugen; Roosevelt, Franklin D.; SA; Singapore; Stalin; Stalingrad; Stilwell, Joseph Warren; TORCH, Operation; Zhukov, Georgy Konstantinovich

References and further reading:

Bregerud, Eric. *Touched with Fire: The Land War in the South Pacific.* New York: Penguin, 1997.

Calvocoressi, Peter, and Guy Wint. *Total War: Causes and Courses of the Second World War.* New York: Viking Penguin, 1979.

Churchill, Winston S. *The Second World War.* New York: Houghton Mifflin, 1986.

Eisenhower, Dwight D. *Crusade in Europe.* Baltimore: Johns Hopkins University Press, 1997.

Ienaga, Saburo. *The Pacific War, 1931–1945: A Critical Perspective on Japan's Role in World War II.* New York: Random House, 1979.

Keegan, John. *The Second World War.* New York: Penguin, 1990.

Liddell Hart, Basil H. *History of the Second World War.* New York: Da Capo Press, 1999.

Rhodes, Richard. *The Making of the Atomic Bomb.* New York: Simon & Schuster, 1995.

Shirer, William L. *The Rise and Fall of the Third Reich.* New York: Simon & Schuster, 1990.

Taylor, A. J. P. *The Origins of the Second World War.* New York: Simon & Schuster, 1996.

Ugaki, Matome *Fading Victory: The Diary of Admiral Matome Ugaki, 1941–1945.* Ed. Katherine V. Dillon. Pittsburgh, PA: University of Pittsburgh Press, 1991.

Weinberg, Gerhard L. *A World at Arms: A Global History of World War II.* New York: Cambridge University Press, 1995.

Troops of the U.S. Army open fire on a Sioux camp on the morning of 29 December 1890. (Library of Congress)

Wounded Knee, Battle of (28 December 1890)

The crushing of the last aboriginal resistance in North America. It served as revenge for the defeat of Custer and the Seventh Cavalry at the Battle of Little Bighorn in 1876.

The once-proud Sioux nation had been relegated to reservation life by 1890. Crazy Horse had surrendered in 1877 and then been bayoneted to death. The remaining Indians turned to a new religion centered on rituals like the "ghost dance" as a remedy for their poor treatment by the U.S. government.

In 1890, alarmed Indian agents sought to put a stop to the ghost dances and met stiff Sioux resistance in Dakota territory. Sitting Bull, thought to be at the center of the resistance, was shot dead in 1890 at Grand River, allegedly while trying to escape.

Two weeks later, on 29 December 1890, the U.S. Seventh Cavalry (Custer's old outfit) attacked a Sioux reservation encampment at Wounded Knee Creek in the Black Hills. Estimates of the dead ranged between 150 and 200 on the Sioux side, including women and children. The U.S. Cavalry may have had 30 casualties. The last Sioux surrendered on 16 January 1891, marking an end to the tragic chapter of U.S. expansion in the west.

Christopher Howell

See also: American Indian Wars; Crazy Horse; Little Bighorn; Sitting Bull

References and further reading:

Hogan. William T. *American Indians.* 3d ed. Chicago: University of Chicago Press, 1993.

McGinnis, Anthony. *Counting Coup and Cutting Horses: Intertribal Warfare on the Northern Plains 1738–1889.* Evergreen, CO: Cordillera Press, 1990.

Secoy, Frank R. *Changing Military Patterns of the Great Plains Indians.* 1953. Reprint, Lincoln: University of Nebraska Press, Bison Books, 1992.

Starkey, Armstrong. *European and Native American Warfare 1675–1815.* Norman: University of Oklahoma Press, 1998.

Wrangel', Peter Nikolaevich (1878–1928)

Czarist and White Russian Civil War general. Of noble origins, Wrangel' was born in Kovensk Region (modern-day Lithuania). A mining engineer graduate from St. Petersburg Mining Institute (1900), he fulfilled his Russian army national service in the Horse Guards until 1902, leaving to pursue a mining career. He was recalled for the Russo-Japanese War and served in the Transbaikal Cossack Formation as lieutenant. Remaining in the army, he graduated from the General Staff Academy in 1909 and from the Cavalry Officers' Academy the following year, becoming captain and squadron commander (1912).

In World War I, Wrangel' commanded the Ussaryskaia Division and was a major general by the February Revolution. He fought on the southwestern front in the June 1917 offensive but, after participating in the Kornilov Coup (August), left the army and went to Yalta. He was arrested there in January 1918 by Red Guards but was released and lived in German-occupied Ukraine before joining Anton Denikin's "volunteer army" in August.

Wrangel' commanded the 1st Cavalry Division and Corps in conquering Kuban and commanded the Caucasus Volunteer Army in January 1919 within Denikin's Armed Forces of Southern Russia (AFSR). He led the right wing of Denikin's drive on Moscow, but as early success turned into lengthy retreat, he attempted to have Denikin removed as commander in chief. Instead, he himself was removed by Denikin in February 1920.

In exile in Constantinople, Wrangel' was summoned back in April to replace Denikin and head the White movement. He successfully reorganized AFSR remnants into a 40,000-strong Russian army. But he made an error similar to that of Denikin's, not cooperating with Poland—that newly independent nation fought the Soviets in April 1920.

Despite some initial success in taking Tauride (June 1920), Wrangel' had little chance against the Red Army. He survived as long as the Polish-Soviet War lasted but was then pushed back into the Crimea. Wrangel' was successful in evacuating 146,000 Whites in November 1920, distracting enough Red troops in August to help determine the outcome of the Polish-Soviet War. In exile, he lived in Paris, founded the Union of Former Soldiers of Russia (1924), and headed the White émigré movement until his sudden death in Brussels.

Neil Harvey Croll

See also: Bolshevik Revolution; Russian Civil War (1918–1921); World War I

References and further reading:

Venkov, Andrei V., and Alexei Vasilevich Shishov. *Belye generaly* (White generals). Rostov-on-Don: Izdat. Feniks, 1998.

Wrangel', A. *General Wrangel': Russia's White Crusader.* New York: Hippocrene Books, 1987; London: Leo Cooper, 1990.

———. *Vospominaniia. Iuzhnyi front (noiabr' 1916-noiabr' 1920) v 2-kh tomakh* (Recollections. Southern Front [November 1916–November 1920] in 2 vols.). Moscow: Terra-Terra, 1992.

———. *General Wrangel': Doverie vospominaniia* (General Wrangel': Trusted recollections). Minsk: Izdat. Arti-Fekc, 1999.

X

Xenophon (c. 431–c. 354 B.C.E.)

Greek soldier and historian and a leader of one of the greatest fighting retreats in history. In the spring of 401 B.C.E., the Athenian-born Xenophon joined an expeditionary force of Greek mercenaries recruited by the Persian prince Cyrus. Cyrus, an ally of Sparta (Athens's traditional rival in the Peloponnese), led a force, including a corps of perhaps as many as 14,000 Greek mercenaries, against his half-brother, the king Artaxerxes. Xenophon joined the battle at Cunaxa, near Babylon. The Greek mercenaries routed the Persian troops facing them. Cyrus and a small cavalry force attempted a coup de main by killing Artaxerxes, but Cyrus was killed. The Greeks, abandoned by their Persian erstwhile allies who had sworn allegiance to Artaxerxes, started retreating toward their homeland. The Greek commanders were invited to a conference with the Persians, where they expected to discuss terms. They were treacherously seized and executed. The "ten thousand," as Xenophon called the mercenary force, were now left leaderless. The Greeks elected new commanders, among them the young Xenophon. The ten thousand marched home through Mesopotamia and the Armenian mountains without the aid of maps or guides. Harassed by inclement weather and hostile marauders, the Greeks pressed their way homeward for four months until they reached the Greek colony of Trapezus on the Black Sea. Xenophon recorded the exploits of this mercenary force in the *Anabasis.*

For reasons that are unclear, Xenophon was exiled from his native city, Athens. Having developed a close relationship with Sparta's king, Agesilaus, Xenophon found refuge in Sparta. In 394 B.C.E., Xenophon again led the ten thousand (now a force of perhaps 8,600) on the Spartan side at Coronea against Thebes. The battles between Sparta and Thebes around Argos and Corinth were a static affair, and the record of the ten thousand in this conflict was not as glorious as it had been in Persia.

Eric D. Pullin

See also: Cunaxa, Battle of; Greek-Persian Wars; Peloponnesian Wars

References and further reading:

Griffith, Guy Thompson. *The Mercenaries of the Hellenistic World.* Cambridge, UK: Cambridge University Press, 1935.

Hirsch, Steven W. *The Friendship of the Barbarians: Xenophon and the Persian Empire.* Hanover: Published for Tufts University by University Press of New England, 1985.

Stronk, Jan P. *The Ten Thousand in Thrace: An Archaeological and Historical Commentary on Xenophon's Anabasis.* Amsterdam: J. C. Gieben, 1995.

Xenophon. *Anabasis: The Persian Expedition.* Baltimore: Penguin, 1972.

Xerxes I (c. 519–465 B.C.E.)

The son and successor of the Persian king Darius, Xerxes headed an unsuccessful invasion of Greece. His first military expedition after his father's death involved crushing an Egyptian revolt in 486 B.C.E. Three years later, Xerxes turned his attention to Greece. In preparation for the Second Persian War, he ordered the digging of a canal across the Athos isthmus and instructed his engineers to construct a pair of bridges, utilizing both suspension and pontoon designs, across the Hellespont.

Once Xerxes and his vast army crossed the Hellespont, he initiated a combined land and sea campaign. Victorious at Thermopylae, Xerxes marched with his army on Athens, while his fleet engaged the Greeks just off the coast. The Persians sent a second fleet of 200 ships around the island of Euboea to block any possible escape, but these ships crashed against the rocks during a violent storm. The main Persian fleet was also hit by a storm and severely weakened, paving the way for a later Athenian naval victory.

With his army still strong, Xerxes captured Athens. He then ordered his fleet to engage the Greeks at Salamis, where they experienced another naval defeat. Escaping back across the Hellespont before the Greeks could destroy his bridges, he ordered his brother-in-law Mardonius to remain in Greece and continue the offensive. Mardonius sacked Attica several times before falling in battle at Plataea.

Cynthia Clark Northrup

See also: Greek-Persian Wars; Marathon, Battle of; Thermopylae, Battle of

References and further reading:

Green, Peter. *The Greco-Persian Wars.* Berkeley: University of California Press, 1996.

Y

Yalu River (1 May 1904)

A major Japanese victory in the opening phases of the Russo-Japanese War. The conflict had begun with a Japanese naval attack on the Russian fleet at Port Arthur in February 1904; the next phase would be the rapid movement of Japan's armies through Korea into the Liaodong Peninsula to control the warm water ports and hence secure Japan's sea bridge to the Asian continent. The first major ground fighting would be forcing the crossing of the Yalu River, which divided Korea from China's Manchurian region.

The Russians recognized the probable Japanese strategy and did make plans to defend along the Yalu River. General Aleksey Kuropatkin organized his troops into four main groupings, one in Port Arthur, one near Port Arthur, one guarding the area just west of the Yalu River, and finally a small army of 7,000 covering the main Yalu crossings. The Russians wanted to defend and give ground slowly while waiting for reinforcements to come from European Russia—a sound strategy. However, the czar appointed Admiral Evgeni Alekseev to overall command, and he insisted on an immediate offensive against the upstart Japanese.

So 7,000 Russian troops attacked 40,000 troops near Wiju. The Japanese easily withstood the attack and badly mauled the outnumbered Russians, losing about 1,100 compared to 2,500 Russian killed and wounded. Thereafter, the Japanese First Army crossed the Yalu and, along with other armies landed in the Liaodong Peninsula, began the long and decisive siege of Port Arthur.

Charles M. Dobbs

See also: Russo-Japanese War

References and further reading:

Okamoto, Shumpei. *The Japanese Oligarchy and the Russo-Japanese War.* New York: Columbia University Press, 1970.

Warner, Denis, and Peggy Warner. *The Tide at Sunrise: A History of the Russo-Japanese War, 1904–1905.* London: Frank Cass, 1974.

Westwood, J. N. *Russia against Japan, 1904–1905: A New Look at the Russo-Japanese War.* Albany: State University of New York Press, 1986.

Yamagata, Aritomo (1838–1922)

Soldier and statesman who is considered the father of the modern Japanese army. Yamagata was born on 3 August 1838 in Hagi. As a young officer, he recognized that Japan's military was inferior to that of the Western powers. After participating in the Meiji Restoration in 1868, he gained a series of government posts and planned the modernization of the army. Studying European military organization, he sought to emulate the Prussian system, where the military is distinct from civilian government. His reforms included the introduction of universal conscription and the establishment of a general staff.

He was vice minister of military affairs in 1871, minister of the army in 1873, army chief of staff in 1878, home minister from 1881 to 1886, and prime minister in 1890. Disliking party politics, he resigned in 1891. He commanded the First Army in Korea during the Sino-Japanese War (1894–1895), but illness forced him to return to Japan. He was special ambassador to Russia in 1896 and helped to settle territorial disputes that arose from Japan's victory over China.

Serving as prime minister a second time in 1898–1900, he appointed a cabinet dominated by the military and decreed that only active-duty officers would be eligible to be armed forces ministers. He also promoted an expansionist policy that would seek to extend Japanese control over much of Asia.

Yamagata was chief of the general staff during the Russo-Japanese War (1904–1905). It was his last official position, although he held much influence until his death in Tokyo on 1 February 1922. Yamagata is widely credited with building Japan's military to that of a world power.

Harold Wise

See also: Nogi, Maresuke; Russo-Japanese War; Sino-Japanese War (1894–1895)

References and further reading:

Craig, Albert M. *Choshu in the Meiji Restoration.* Cambridge, MA: Harvard University Press, 1967.

Hackett, Roger F. *Yamagata Aritomo and the Rise of Modern Japan 1838–1922.* Cambridge, MA: Harvard University Press, 1971.

Yamashita, Tomoyuki (1885–1946)

World War II Japanese general. Born 8 November 1885, at Skikoku, Yamashita graduated from Japan's Central Military Academy (1908) and General Staff College (1916). Between 1916 and 1940, he distinguished himself as a general staff officer; assistant military attaché to Switzerland; military attaché to Austria; commander at regimental, brigade, division, and army levels; chief of the Military Affairs Bureau; and inspector general of air aviation. In November 1941, Yamashita, a lieutenant general since 1937, assumed command of the Twenty-Fifth Army, just weeks before Japan went to war against Great Britain and the United States.

Although opposed to war against the Western powers, Yamashita guided the Twenty-Fifth Army to the greatest land victory in Japanese history, conquering Malaya and Singapore in only 70 days (7 December 1941–15 February 1942). Though Malaya made him a national hero, Yamashita suffered an assignment to a minor command in 1942 because Prime Minister Hideki Tojo saw him as a potential rival. For two years, he languished in forced exile, while the war turned irreversibly against Japan. Finally, in September 1944, he received command of the Fourteenth Area Army. Under Yamashita's leadership, this force defended the Philippines against an American invasion that began on 20 October 1944. Denying the enemy a quick victory, Yamashita held out on Luzon until 2 September 1945, when news of Japan's capitulation prompted his surrender.

Tried as a war criminal, Yamashita was convicted, in a controversial verdict, of failing to control troops under his command who committed atrocities in the Philippines, although no direct evidence linked him to the crime. Sentenced to death, Yamashita—"the Tiger of Malaya"—was hanged (he pleaded directly to General Douglas MacArthur for the more "honorable" death by shooting, but even this was denied him) on 26 February 1946 on Luzon. Even some contemporary Allied observers thought that Yamashita was executed more for his success against the Allies than for any particular war crimes that could legally be held against him.

Bruce J. DeHart

See also: Philippines, U.S. Retaking of; Singapore

References and further reading:

Lael, Richard. *The Yamashita Precedent.* Wilmington, DE: Scholarly Resources, 1982.

Potter, John D. *The Life and Death of a Japanese General.* New York: New American Library, 1962.

Yang Jian (Yang Chien) (541–604)

The second emperor of the Sui Dynasty (581–618), who would take the reign name of Wendi (Wenti), Yang Jian was born of mixed Chinese and northern blood into an aristocratic family serving the rulers of the Northern Zhou (Chou, 557–581), whose territory spanned most of western China beyond the bend in the Yellow River. The marriage of his daughter into the imperial Northern Zhou household allowed Yang Jian in 578 to gain the post of regent to a Northern Zhou child emperor. Yang Jian moved quickly to consolidate his position, and through a ruthless purge of Zhou princes, seized the throne itself, establishing the Sui Dynasty. Yang Jian kept its capital at Chang'an (modern Xian), which had served as the capital of various dynasties going back to the Qin (Ch'in, 221–206 B.C.E.). As with various other imperial families of the north, Yang Jian readily adopted the Buddhist faith and promoted himself as a Buddha-king (*Cakravartin*), a ruler who uses the sword to build the Buddhist kingdom on earth. In the years that followed, Yang Jian, now emperor of the Sui, undertook a series of bold military campaigns to reunite China proper under a single imperial rule, something it had not witnessed since the fall of the Han Dynasty in 220. Yang Jian's reunification of the long decentralized empire firmly established in the consciousness of China the idea of a single unified Chinese state.

By 589, with the conquest of southern China and the reunification of the empire, Yang Jian had accomplished his primary task. In the following years, he would make great efforts in attempting to strengthen the Sui through land and administrative reforms. Yang Jian's accomplishments, however, would soon be lost in the failures, particularly military, of his two successors, and the young Sui empire would fall in 618.

Daniel Kane

References and further reading:

Twitchett, D., and J. F. Fairbank, ed. *The Cambridge History of China.* 15 vols. Cambridge, UK: Cambridge University Press, 1978–.

Wright, Arthur. *The Sui Dynasty.* New York: Alfred A. Knopf, 1978.

Yang Xiuqing (c. 1817–1856)

Rebel leader in the Taiping Rebellion. Yang was born on a farm in Xincun village, Guangxi Province, China, around 1817. Orphaned at an early age, he became a charcoal burner and a skilled explosives technician in mines and tunnels. When he was about 30, he and many of his fellow burners joined the God-Worshipping Society, a dissident, charismatic movement recently founded by Feng Yünshan and inspired by Hong Xiuquan. Yang quickly rose to a position of authority by interpreting his various illnesses, whether real or faked, as signs from God. By 1850, he, Hong, and Feng uneasily shared the leadership of tens of thousands of followers. When the Qing emperor sent troops in December 1850 to suppress the God Worshippers, they offered stiff resistance, immediately mobilized seven divisions, and decisively defeated the imperial forces on 1 January 1851 at Jintian, with Yang commanding the left flank. Confident after this victory, Hong declared the Taiping Tianguo (Heavenly Kingdom of Great Peace) on 11 January, thus formally initiating the Taiping Rebellion.

Known thereafter as Dong Wang (East King), Yang served Hong as chief of staff. His many early victories, his escape from the imperial siege of Yongan on 3 April 1852, and his distinguished performances at Yuezhou on 13 December 1852, Wuchang on 12 January 1853, and Nanjing on 19–21 March 1853 all contributed toward his reputation as the Taiping Rebellion's most skillful tactician and most talented field commander. He brilliantly repulsed the imperial attempt to recapture Nanjing in the summer of 1856.

As Yang's stature grew, Hong became afraid that Yang wanted to replace him. On Hong's orders, Wei Changhui, known as Bei Wang (North King), murdered Yang and 6,000 of his followers in Nanjing on 2 September. Hong then had the murderers murdered. The rebellion was irreparably weakened from that point but staggered on for another eight years.

Eric v. d. Luft and Sarah Luft

See also: Chinese Imperial Wars; Gordon, Charles George; Hong Xiuquan; Li Hongzhang; Religion and War; Taiping Rebellion; Wolseley, Garnet Joseph, Viscount; Zuo Zongtang

References and further reading:

Chien, Yu-wen. *The Taiping Revolutionary Movement.* New Haven, CT: Yale University Press, 1973.
Shih, Yu-chung. *The Taiping Ideology: Its Sources, Interpretations, and Influences.* Seattle: University of Washington Press, 1967.
Spence, Jonathan D. *The Taiping Vision of a Christian China, 1836–1864.* Waco, TX: Baylor University Press, 1998.
Wagner, Rudolf G. *Reenacting the Heavenly Vision: The Role of Religion in the Taiping Rebellion.* Berkeley, CA: Institute of East Asian Studies, 1982.

Yangzhou (Yang-chou), Siege of (1645)

A siege by Manchu armies of the newly established Qing (Ch'ing) Dynasty (1644–1912) of the city of Yangzhou, a holdout of forces still loyal to the toppled Ming Dynasty (1368–1644). The siege of Yangzhou and later annihilation of its Ming supporters was of extreme brutality and is sometimes referred to as the "Yangzhou Massacre."

The last Ming emperor, Chongzhen (r. 1628–1644), had died by his own hand in a besieged Beijing in 1644, but the Ming loyalists soon moved their capital to Nanjing, near the terminus of the Yangtze River. After the Manchu armies entered Beijing in 1644, toppling the Ming Dynasty, efforts by supporters of the Ming concentrated on keeping the Manchu north of the Yangtze River. They selected Chongzhen's first cousin, Zhu Yousong, the prince of Fu, as the new Ming emperor. The troops of the Ming loyalists were garrisoned just north of the Yangtze River and east of Nanjing at the walled city of Yangzhou. It was the Manchu Prince Dodo (1614–1649), brother of the new Qing emperor, who was chosen to lead the Qing assault on this Ming challenge. In November 1644, Dodo led a force of about 300,000 troops, a mixture of Manchu and Chinese, from Beijing, arriving outside Yangzhou in May 1645. A fierce five-day battle ensued between the Qing troops and the Ming loyalists, but the force arrayed against them was too much for the Ming, and the gates of Yangzhou were eventually breached. What ensued was a 10-day slaughter of the Yangzhou's surviving defenders and inhabitants and a razing of the city, so brutal it is referred to as "the ten-day slaughter of Yangzhou." It would prove just one of several massacres perpetuated by the Qing armies along the Yangtze River as they struggled to bring China's richest region under their control. Though it brought a definitive end to the Ming loyalist efforts behind Prince Fu (who would soon be captured and die in Beijing), it would be another 30 years before the last vestiges of Ming resistance were wiped out. So vivid were memories of the Qing massacre at Yangzhou, however, that on a tour of his empire in 1684, the Qing Emperor Kangxi (Kang-hsi, r. 1661–1772) opted wisely to bypass the city.

Daniel Kane

See also: Manchu Expansion, Wars of

References and Further reading:
Mote, F. W. *Imperial China, 900–1800.* Cambridge, MA: Harvard University Press, 1999.
Spence, Jonathan. *The Search for Modern China.* New York: W. W. Norton, 1990.
Struve, Lynn. *The Southern Ming, 1644–1662.* New Haven, CT: Yale University Press, 1984.

References and further reading:
Bishai, Wilson B. *Islamic History of the Middle East; Backgrounds, Development, and Fall of the Arab Empire.* Boston: Allen and Bacon, 1968.
Donner, Fred. *The Early Islamic Conquests.* Princeton, NJ: Princeton University Press, 1981.
Glubb, John Bagot. *The Great Arab Conquests.* London: Hodder and Stoughton, 1963.

Yarmuk, Battle of (20 August 636)

Arab victory that destroyed any Byzantine hopes of the reconquest of Syria and Palestine. By 635, the Arab Muslim armies had conquered virtually all of Palestine and Syria, driving the Byzantine armies before them. In response, Emperor Heraclius recruited a new army consisting of Byzantine soldiers from Anatolia, Armenian infantry, and light Arab cavalry from the Christian Arab tribe, Bani Ghassani. Upon the Byzantine army's arrival into Syria under the command of Theodorus, the Arab forces of Khalid ibn al-Walid withdrew.

The Arabs abandoned Syria and withdrew through the Deraa Pass, possibly hoping to draw the Byzantines into the open. Theodorus did not comply and instead fortified the Deraa Pass. The Muslims ceased their retreat and then laid siege. The Byzantine felt certain that if they held Deraa, then the Arabs were effectively barred from entering Syria. Although the pass was geographically imposing, with rifts and lava fields, small parties could infiltrate behind it and raid.

Four months passed, and they began to show on the Byzantines as rivalries increased between the Greek and Armenian soldiers. The Arab forces patiently waited. Then a sandstorm occurred. In the midst of the storm, the Arab army attacked on 20 August 636.

The small parties of soldiers who had infiltrated behind the Byzantines cut off their routes of communication by seizing a bridge across the Yarmuk River. Then the main Arab force stormed the fortifications with the sandstorm at their back. The Byzantine army panicked in the face of the unexpected onslaught and the storm. With the Byzantine retreat route held by the Muslims and the fortifications overrun, the Arabs eradicated the Byzantine army.

Afterward, the Arabs continued their northward drive and successfully reconquered Syria. Heraclius realized that Syria was lost to the empire and did not attempt to regain the former Byzantine territories.

Timothy May

See also: Byzantine-Muslim Wars; Byzantine-Persian Wars; Heraclius; Khalid ibn al-Walid; Muslim Conquests; Religion and War; Sassanid Empire

Yellow Ford (1597)

Sometimes called the Battle of Blackwater, the greatest defeat of the English during the Elizabethan wars in Ireland. The battle resulted from Sir Henry Bagenal's attempt to resupply a besieged English outpost on the Blackwater River. Bagenal, with between 4,000 and 5,000 men, underestimated his opponent, Hugh O'Neill, Earl of Tyrone, who had recently added the forces of Hugh O'Donnell from Connacht to his army. Bagenal's force left Newry on 12 August and marched toward the Blackwater in a column of six regiments, with gaps in between. He was nearing the Blackwater on 14 August but had to cross the small tributary Callon Brook.

As he approached, his forces came under harassing fire from the woods flanking their route of march, delivered by light infantry who had recently been rearmed with firearms in place of traditional javelins and bows. The English were already in disorder when the regiment at the head of the column crossed the ford and encountered a fortified position manned by the bulk of O'Neill's forces, armed in the English fashion with shot and pike.

With the column stalled, O'Neill unleashed O'Donnell's warriors to overrun and destroy the column in close combat. Bagenal, directing his forces from the ford, was shot in the head in plain view of the army, which initiated a complete rout. Fifteen English captains were killed and nearly 3,000 men lost in the retreat, with several hundred subsequently joining O'Neill's forces.

Not only was this battle a great victory for the Irish, but it also served notice both of O'Neill's tactical brilliance and the effectiveness of his army. It also forced the deployment of 16,000 men to Ireland, a commitment Elizabeth I could ill afford at that point in her long-running conflict with Spain.

John S. Nolan

See also: Mountjoy, Charles Blount, Lord; Nine Years War

References and further reading:
Falls, Cyril. *Elizabeth's Irish Wars.* London: Methuen, 1950.
———. *Mountjoy: Elizabethan General.* London: Odhams, 1955.
Fissel, Mark Charles. *English Warfare, 1511–1642.* London: Routledge, 2001.

Wernham, R. B. *The Return of the Armadas.* Oxford, UK: Oxford University Press, 1994.

Deffarge, Claude, and Gordon Troeffler. *Yemen 62–69, de la révolution sauvage à la trêve des guerriers.* Paris: Laffont, 1969.

O'Ballance, Edgar. *The War in the Yemen.* London: Faber, 1971.

Yemenite Civil Wars (1961–1967, 1994)

Conflicts that followed the fall of the theocratic government that reflected the Cold War. In the late 1950s, Yemen was a theocratic state loosely united with the United Arab Republic. When a republican opposition began an insurgency in 1961, Egypt broke with the Imam Achmad to support his opponents. The death of Achmad and the installation of his son as imam on 19 September 1962 led to the arrest of the imam by insurgents and the proclamation eight days later of the Yemen Arab Republic, which was immediately recognized by the Communist-bloc nations. A military government took office on 31 October.

The imam managed to escape and rally royalist forces to his cause. Egypt immediately began sending military supplies and troops to assist the republicans. Eventually, the Egyptian troop commitment reached 35,000. Frustrated at their failure to coerce the royalists and at escalating casualties, the Egyptians employed poison gas. On the royalist side, Jordan and Saudi Arabia were furnishing military aid, and Britain lent diplomatic support. An internal split in the republican government led to an agreement that all outside aid to both sides would be halted in 1967.

In addition to the Egyptian aid, the Soviet Union allegedly supplied 24 MiG-19s to the republicans. Wherever the republicans got their aircraft, they were crucial in preventing the royalists from concentrating their forces. The high point for the royalist cause came with the siege of the capital, San'a, between December 1967 and March 1968. After that, the tide turned rapidly, and the last royalist citadel, Sa'dah, fell in September 1968. The Saudis recognized the republican regime in 1970, and royalist resistance ended completely in 1971.

In 1972, Yemen agreed to a future merger with the People's Democratic Republic of Yemen, a Marxist successor to the South Arabian Federation cobbled out of Aden and 16 tribal states in 1963. The two Yemens finally united in 1990. Conflicts within the governing coalition led to the outbreak in May 1994 of a civil war in which the southerners sought to secede. The northern-dominated government put down the rebellion in just nine weeks.

Joseph McCarthy

References and further reading:

Badeeb, Saeed M. *The Saudi-Egyptian Conflict over North Yemen, 1962–1970.* Boulder, CO: Westview, 1986.

Yonglo (1360–1424)

The fourth son of Ming Hongwu, founder of the Ming Dynasty (1368–1644). Yonglo seized power from his nephew and moved the capital from Nanjing along the Yangtze River to Beijing, site of the former Mongol capital, only a short distance from the Great Wall.

Yonglo's reign was probably the height of the Ming era. The Great Wall was rebuilt in its current form and style during this time; Yonglo oversaw construction of the so-called Forbidden City and the temples and palaces where the Ming ruled and lived; and he began the Ming Tombs outside the city.

Militarily, Yonglo's reign was a time of limited military expansion. He continued his father's system of guard units of 5,600 men, each divided into fifths. During Yonglo's reign, there were about 400 such units, although many of the men were more self-supporting farmer-soldiers than effective professional soldiers. Yonglo led expeditions against the remaining Mongol forces and other northern barbarians to secure Ming power, but the Ming never achieved the territorial gains of the earlier Tang Dynasty or of the Qing, which succeeded them. Yonglo wanted to expand the tributary system of diplomatic relations and sent a Muslim eunuch, Zheng He, on seven naval expeditions to Southeast Asia and even the east coast of Africa. While the Portuguese were cautiously moving down the West African coast in small ships with few seamen, Zheng He led more than 28,000 men and 60 vessels on his journeys. After Yonglo died, the Ming turned away from naval expansionism and focused on assuaging barbarians outside the Great Wall.

Charles M. Dobbs

See also: Chinese Imperial Wars; Great Wall of China; Qianlong

References and further reading:

Dreyer, Edward L. *Early Ming China: A Political History, 1355–1435.* Stanford, CA: Stanford University Press, 1982.

Hucker, Charles O. *The Ming Dynasty, Its Origins, and Evolving Institutions.* Ann Arbor: University of Michigan Press, 1978.

Mungello, David E. *The Great Encounter of China and the West, 1500–1800.* Lanham, MD: Rowman & Littlefield, 1999.

Yorktown (1781)

The campaign that won American independence. Sir Charles Cornwallis had been unsuccessful in his efforts to bring the

Carolinas under British control, mostly because of the efforts of regular Continental units under General Nathanael Greene and General Daniel Morgan and guerrilla bands under Francis Marion and Andrew Pickens. Cornwallis believed that American forces maintained their strength from bases of supply in Virginia, so he campaigned northward to disrupt them. After he raised his troop levels to 7,500 by joining with other British forces in the Chesapeake Bay region, Cornwallis established a headquarters at Yorktown, at the mouth of Virginia's York River, in order to facilitate communication with the northern British army under General Henry Clinton in New York.

In the meantime, Washington had lobbied his French allies to conduct a joint operation against Clinton in New York, a plan that was not enthusiastically endorsed by the French commander, General Jean Baptiste Donatien de Vimeur, Comte de Rochambeau. In fact, the French commander left the fleet of Admiral François-Joseph Paul, Comte de Grasse, in the West Indies so that he could decide for himself whether to support operations in New York or French troops under General Marie Joseph du Motier, the Marquis de Lafayette, who had been monitoring Cornwallis's movements in Virginia. Washington persisted in his New York plan, but when he discovered that he lacked the forces necessary for decisive action against Clinton and that de Grasse had determined to move against Cornwallis in the south, he made the fateful decision to transport the bulk of the Continental army to Virginia. De Grasse arrived off Yorktown on 30 August and engaged British naval units, giving the French complete control of the waters surrounding Cornwallis's position. De Grasse then helped to move Washington's troops down the Chesapeake Bay in late September. Once established on the base of the Yorktown peninsula, allied forces numbered about 17,000.

With the arrival of Washington, the entrapment of Cornwallis at Yorktown was complete. The allies pounded the British for several weeks until 14 October, when they seized control of two key redoubts on the British left. After Cornwallis failed to retake these positions and could not manage the escape of his troops across the York River, he surrendered. The allies suffered 262 casualties during the campaign, whereas the British listed 552 killed, missing, and wounded. News of Cornwallis's capitulation resounded throughout the colonies and Europe. This was the second capitulation of a British army in North America. (The first had been at Saratoga, which had brought the French openly into the struggle.) The British government of Frederick, Lord North collapsed, and a new ministry was voted in with a mandate to negotiate an end to the war. The result was the Peace of Paris of 1783, which included formal recognition of the independent United States of America.

Yorktown was a near-miraculous victory: the French, most unusually, had defeated the British at sea; French money had come through just in time to reinspire the ragged American forces; and the complex movements of the American and French forces (including the vital French siege train) had gone off like clockwork. A rare moment in history, indeed.

Jeffrey B. Webb

See also: American Revolution; Cornwallis, Sir Charles; France and the American Revolution; Greene, Nathanael; Washington, George

References and further reading:

Davis, Burke. *The Campaign That Won America: The Story of Yorktown.* New York: Dial Press, 1970.

Fleming, Thomas J. *Beat the Last Drum: The Siege of Yorktown, 1781.* New York: St. Martin's Press, 1963.

Lumpkin, Henry. *From Savannah to Yorktown: The American Revolution in the South.* Columbia: University of South Carolina Press, 1981.

Ypres, Battles of (1914–1918)

Three major engagements of World War I fought in and around the town of Ypres in southwestern Belgium. The first battle of Ypres (October–November 1914) was the culmination of the flanking series of engagements referred to as "the race to the sea." The British at Ypres checked the German thrust toward the Channel ports of Dunkirk and Calais.

The Germans retreated from the Marne River in the late fall of 1914. Next, British and Belgian forces raced the German army under General Erich von Falkenhayn northward to keep the Germans from advancing through gaps in the Allied lines and thus regain momentum to resume the march on Paris. The race evolved into opposing static lines of massive trenchworks spanning all the way to the English Channel.

Ypres was taken by the Germans in the first days of October 1914 but recaptured by the British on 14 October, whereupon the British I Corps under General Douglas Haig occupied the bulge thereafter known as the Ypres salient. In the middle of October, the German offensive along the Lys River almost succeeded in breaking the line before fierce British resistance drove them back. Swampy conditions (the Belgians had blown the dikes) hindered supply and movement from Ypres northward to the Belgian coast. On 21 October, the Allies advanced northeast toward Bruges, just as the German Fourth Army opened up its own counteroffensive towards Ypres. The converging forces met on a line around Zonnebeek, where the British checked the Germans and then threw them back. The Germans hurled themselves repeatedly into a hail of accurate, rapid British rifle fire, incurring heavy losses. The French rapidly reinforced the British line. The Germans now concentrated their available forces on taking Ypres and launched a major attack on 29 October. This eventually broke the British line near Gheluvelt, but reserves had been brought up and counterattacked, regaining the lost positions. The Germans continued to assault, and the Allies kept rushing up reinforcements, until a final German attack on 17 November failed. The Allies held, but their forces were depleted and down to the last reserves. Only winter saved the Allied lines from total collapse, and they remained stabilized until the next year. British losses are estimated at about 50,000, the French 70,000, and the Germans 150,000 during this prolonged engagement.

On 22 April 1915, the Germans blanketed the Allied lines near Ypres with an intense artillery barrage. As the barrage continued, a yellow-green cloud descended on the trenches, inaugurating the use of poison gas on the western front in the form of chlorine gas shells, more commonly know in the trenches as "green cross." The attack caused several of the defending units (British regulars and reserve territorial troops) to flee their positions, which opened a 4-mile gap in the sector.

The Germans quickly exploited the breach but underestimated the time needed for the gas to dissipate and charged straight into their own gas cloud and the waiting 1st Canadian Regiment, which was the only unit to endure the gas to repel the German assault. The Canadians held off the Germans long enough for reinforcements to arrive following the dissipation of the gas. The line once again stabilized.

However, the battle soon flared up again, and random fighting took place all over the Ypres front. On 24 May 1915, the Germans launched a massive gas attack on the lines once more. The British managed to hold off the attackers but only by suffering terrible casualties. By the end of spring, the Ypres front had once again stabilized, and the fighting had died down for the time being.

The third battle of Ypres, popularly known as Passchendaele, began on 31 July 1917 and continued until November. The principal aim of this battle was to force the Germans to yield their grip on the Channel ports and thus facilitate operations against the U-boat threat. The British and French high command had decided that a three-prong attack on the German lines was necessary to break the stalemate that had ensued since the end of 1914. The plan was for the British forces to attack at the Ypres front as part of the three-phase attack. On 18 July, the British and French artillery started a massive barrage on the German lines, preparing for an attack set for 31 July. The battle itself was focused on the small town of Passchendaele. Continuous artillery pounding had wrecked the ancient drainage systems and rendered the ground on which the troops had to fight increasingly muddy and pocked with water-filled craters that could drown a man. Notwithstanding, the British high command insisted that a full-scale artillery preparation be laid down before the infantry could attack, ignoring the lessons of the previous disasters at Ypres and the Somme.

The British infantry began its advance early on the morning of the 31st and despite massive German counterattacks, managed to creep forward 2 miles. The battle seemed to be going well until the eve of 1 August, when heavy rains flooded the area even worse than before. Most of the 144 tanks the British deployed were mired within a few hours and no longer of any use. The rain continued without stop for two weeks, and the British army was heavily bogged down. However, they continued to press against the German lines. The German infantry was also mired, preventing them from counterattacking, but they still mounted a stubborn resistance. They hit the British with mustard gas and heavy machine gun fire, but the British still attacked relentlessly.

After a long, hard struggle, the Canadians took Passchendaele in heavy rain on 6 November. They then managed to continue into Ghent, bringing the offensive to a close at an overall cost of 250,000 casualties for an advance of 5 miles (8 kilometers). Fortunately, for the troops on both sides, there would not be another offensive at Ypres.

Jim Bloom

See also: Belgium, Invasion of; Haig, Douglas; Marne, Battle of the; World War I

References and further reading:

Edmonds, James E. *Military Operations, France and Belgium, 1914.* Vol. 1. British Official History series. London: Macmillan, 1933.

Farrar-Hockley, A. H. *Ypres, 1914: Death of an Army.* London: Pan, 1970.

Nicholson, Colonel G. W. L. *The Canadian Expeditionary Force, 1914–1919.* Canadian Official History series. Ottawa: Queens' Printer, 1962.

Prior, Robin, and Trevor Wilson. *Command on the Western Front: The Military Career of Sir Henry Rawlinson, 1914–1918.* Oxford, UK: Oxford University Press, 1992.

Schurman, Donald. "The Third Battle of Ypres." In *Great Battles of the British Army,* ed. David Chandler. London: Arms and Armour, 1991.

Yuan Shikai (1859–1916)

Chinese general and first president of the Republic of China. Yuan was born in 1859 in the Honan Province of China, but his exact date of birth is unclear. He served under Li Hongzhang in the Ch'ing brigade of the Anhwei army, commanding that brigade in Korea. Success led to an appointment as Chinese ambassador to Korea (1884–1894). In 1895, Li Hongzhang assigned him the task of raising and equipping a modern army, which became known as the Peiyang army. The personal loyalty of that army was invaluable in his rise to power. Yuan supported the Empress Tz'u-hsi over the Emperor Kuang Hsü but refused to support the Boxer Rebellion in 1900. In addition to his military command, he assumed Li Hongzhang's civil authority following his death in 1901. This concentration of power prompted officials at Beijing to assign Yuan to an administrative position with no direct command. With the support of the empress, Yuan set about the modernization of the Chinese armed forces. When the empress died in 1908, Yuan lost his official powers but remained a man of influence among the military.

In 1911, the imperial government asked Yuan to return to active duty to suppress the Double Ten revolution. Yuan arranged to be named prime minister and played both sides to his own political advantage. With his control of the military secure, he coerced the imperial dynasty into abdicating and had himself named as the first president of the Chinese Republic in 1912. Later, he dissolved the National Assembly and seized virtual absolute power. He then attempted to begin a new imperial dynasty with himself as emperor. This provoked significant opposition, and Yuan lost his support and his power. He died 6 June 1916 in Beijing.

Harold Wise

See also: Boxer Rebellion; Chinese Revolution; Li Hongzhang

References and further reading:

Chén, Jerome. *Yüan Shih-kai, 1859–1916.* Stanford, CA: Stanford University Press, 1961.

Hsü, Immanuel C. Y. *The Rise of Modern China.* 2d ed. New York: Oxford University Press, 1977.

Yue Fei (1103–1141)

Semilegendary military hero of the Southern Song Dynasty (1126–1279) of China. Yue Fei was born in 1103 in Tangyin, Henan Province, China, probably into a poor family. He became a common infantryman while still a child and rose quickly through the ranks by virtue of his bravery and skill. In 1128, he was named commander in chief of the Southern Song forces.

The Juchen tribes established the Jin Dynasty in northern China in 1115 and completed their conquest of the Northern Song in 1126. The Southern Song was weak and seemed doomed also to be conquered by the Jin. Yue Fei met the invasion head on, frustrated the Jin cavalry in the rugged hills where he chose to engage the enemy, and cleverly coordinated his land campaigns with the Song navy. He recaptured all Song territory south of the Yangtze River and pursued north of the Yangtze as far as Luoyang. Incredible feats are attributed to him. Defending Kaifeng with only 800 men, he is supposed to have defeated a Jin army of 500,000.

With the defense of Southern Song lands secure by 1140, some leaders wished to continue the war and push the Jin farther north, whereas others wished to accommodate the Jin and make peace. Yue Fei led the War Party. His main political opposition came from Emperor Gaozong's prime minister, Qin Gui, head of the Peace Party and a fervent capitulationist. Qin Gui had Yue Fei executed on trumped-up charges in 1141.

In 1162, the new emperor, Xiao Zong, rehabilitated Yue Fei's reputation. He ordered a magnificent tomb and temple built in Hangzhou to honor Yue Fei. The temple, which was still standing in 2001, includes cast-iron statues of Qin Gui, his wife, and two other conspirators, all four bound as prisoners and kneeling in submission to the man they murdered.

Eric v. d. Luft and Sarah Luft

See also: Chinese Imperial Wars

References and further reading:

Gernet, Jacques. *Daily Life in China on the Eve of the Mongol Invasion.* Stanford, CA: Stanford University Press, 1970.

Hansen, Valerie. *Changing Gods in Medieval China 1127–1276.* Princeton, NJ: Princeton University Press, 1990.

Mote, Frederick W. *Imperial China, 900–1800.* Cambridge, MA: Harvard University Press, 1999.

Rossabi, Morris, ed. *China among Equals: The Middle Kingdom and Its Neighbors, 10th–14th Centuries.* Berkeley: University of California Press, 1983.

Yugoslavian Civil Wars (1990–2000)

Europe's most destructive conflict by far since the end of World War II. It destroyed the Yugoslav Federal Republic, caused the deaths of more than 250,000, forced more than 2 million people from their homes, and brought the North Atlantic Treaty Organization (NATO) into two major peacekeeping operations.

In 1989, Slobodan Milosevic was elected president of Serbia, the largest of the six republics of Yugoslavia. Facing the collapse of communism as an ideology, he hoped to use Serbian nationalism to gain support throughout Yugoslavia. Also in that year, he abolished the autonomy of Kosovo—an Albanian majority province in Serbia.

Croatia and Slovenia were alarmed at these developments and, faced with Milosevic's unwillingness to negotiate, in late 1990 began to consider secession. These moves angered Serbia, as well as the 600,000 Serbs living within Croatia. On 25 June 1991, Croatia and Slovenia declared their independence from the four other republics, Macedonia, Montenegro, Serbia, and Bosnia-Herzegovina, precipitating the intervention of the Yugoslav army, controlled by Serbian commanders, to prevent these secessions. Led by allies of Milosevic, the federal army and air force gave military assistance to Serbs living outside Serbia, with the objective of creating a Greater Serbia to replace the collapsing Yugoslavia.

Although Slovenia, ethnically homogeneous and geographically isolated from the rest of Yugoslavia, was allowed to go its own way, the situation in Croatia was more complex. Approximately one-third of the population of Croatia in 1991 was ethnic Serb, and these Serbians rebelled against the idea of an independent Croatian state. Aided by the Yugoslav army, these Serbs, located in a region known as the Krajina, declared their own state and fought bitterly against Croatian forces for the rest of 1991. More than 20,000 deaths resulted on both sides. Encouraged by Germany, in December 1991 and January 1992, the European Union recognized the independence of Croatia. Following this, in January 1992, the United Nations (UN) brokered a cease-fire between the Croatians and the Serbs, leaving approximately one-third of Croatia under the effective control of the Serb minority. Approximately 14,000 UN peacekeepers entered Croatia to observe the settlement. Macedonia also declared its independence at this time.

In March 1992, after a referendum, Bosnia-Herzegovina also proclaimed its independence. As in Croatia, in Bosnia the Serbs also constituted approximately one-third of the population, and as in Krajina, they called on the Yugoslav army to aid their resistance to Bosnian independence. In April, despite recognition of Bosnian independence by the European Union and the United States, Serbs in the republic launched a rebellion and with Yugoslav military aid seized over two-thirds of Bosnia. During the summer of 1992, the fighting escalated in Bosnia, with the Muslims and Croats allied against the Serbs in most areas.

The Serbs especially, but other ethnic groups as well, practiced a tactic known as "ethnic cleansing," whereby tens of thousands of people were killed or driven from their homes in an effort to make regions of Bosnia ethnically pure—all Serb, all Croat, or all Bosniac Muslim. The Serbs even overran UN-declared safe zones, planned as centers for refugees, massacring or expelling all within. Several cease-fires in 1992 and 1993, along with UN and European mediation, failed to end the conflict, primarily because of Serbian intransigence. In 1994, Bosniacs and Croats signed a federation agreement and began a combined offensive against Serb-occupied territories. The Bosniac capital of Sarajevo remained under siege and subject to shelling and sniper attacks, even against the UN-held airport during this period.

NATO became involved in 1995, sending over 700 air strikes against Bosnian Serb positions around Sarajevo. Also in 1995, Croatia launched a successful offensive to reclaim the Krajina from the Serbs. These two defeats convinced Slobodan Milosevic to agree in late 1995 to a cease-fire, ratified in the Dayton Accords and enforced by more than 50,000 NATO-led peacekeepers who began entering Bosnia in December 1995. The Dayton Agreement gave the Bosniac-Croat federation 51 percent of Bosnia, with the remaining 49 percent going to the Serbs. The war left more than 200,000 people dead and had forced almost 2 million from their homes.

In 1996, in the Serb province of Kosovo, the Albanian majority began to demand independence. Serbian police and paramilitaries began a campaign of repression, while at the same time the Kosovo Liberation Army (KLA) emerged, launching attacks on Yugoslav military and police units. The conflict continued at a low level until 1998, when Albanian leader Ibrahim Rugova demanded complete independence for Kosovo. In late 1998, faced with increasing fighting between the KLA and Serb forces in Kosovo, the UN Security Council called for a cease-fire, threatening further action if Milosevic refused to agree.

In October 1998, Serbia agreed to allow 2,000 unarmed observers to enter Kosovo to guarantee a new cease-fire. Faced with renewed fighting, tens of thousands of Kosovar Albanians being driven from their homes, and Serbian refusal to sign a peace accord, the international team withdrew in March 1999, and NATO began air strikes to force Milosevic to comply with Security Council resolutions. After 11 weeks of the air campaign, in June 1999 the government of Yugoslavia agreed to a cease-fire and allowed a NATO-led force to enter Kosovo to supervise the end of hostilities and the return of the more than 800,000 ethnic Albanian refugees. UN and NATO leaders estimated that Serb forces had killed 10,000 ethnic Albanians in battle and in massacres of civilian populations, and the government of Yugoslavia claimed 5,000 deaths of Serbs from KLA attacks and NATO's bombing campaign, although neither of those figures has been proven to any degree of certainty.

In July 2000, Yugoslavia's federal parliament approved constitutional changes to allow Milosevic to seek a second term as president and called for elections in September. Despite his efforts to manipulate the results of the election, Milosevic lost and in October 2000 surrendered the presidency to Belgrade lawyer Vojislav Kostunica.

Milosevic was later arrested and indicted by a UN court

for war crimes. Peace in the former Yugoslavia seems about as far from accomplishment as it was several years earlier.

Wayne H. Bowen

References and further reading:

Bennett, Christopher. *Yugoslavia's Bloody Collapse: Causes, Course and Consequences.* London: Hurst and Company, 1995.

Boyle, Francis. *The Bosnian People Charge Genocide: Proceedings at the International Court of Justice Concerning Bosnia v. Serbia.* Amherst, MA: Aletheia, 1996.

Glenny, Misha. *The Fall of Yugoslavia: The Third Balkan War.* New York: Penguin, 1993.

Silber, Laura, and Allan Little. *Yugoslavia: Death of a Nation.* New York: Penguin, 1996.

Z

Zama, Battle of (October 202 B.C.E.)

Carthage's last effort during the Second Punic War to ward off defeat. After the Roman general Publius Cornelius Scipio Africanus Major had invaded Carthage's home territory in Africa in 202 B.C.E., Hannibal had been called back from Italy. Both armies met near Zama, probably some 50 miles from Carthage.

Hannibal's army probably counted some 36,000 infantry, 4,000 cavalry, and about 80 elephants. Behind a line of elephants, he positioned his infantry in three lines of roughly 12,000 men each. The front line consisted of mercenaries and the second line of levy troops. Hannibal's hardy veterans formed the third line, some distance behind the others.

Scipio had about 30,000 infantry and 6,000 cavalry. Among them were two penal legions that were formed from the survivors of Cannae. The infantry took position in the center in such a formation that there remained open alleyways through the Roman line. Scipio's superior cavalry took position on the flanks.

The battle commenced by an attack of the elephants. Confused by the sounds of Roman horns, the beasts turned upon their own troops or passed harmlessly through the gaps in the Roman line. Thereupon, the first Carthaginian line charged, but the Romans held ground. The second Carthaginian line reinforced the first, but to no avail: the troops were driven back. By that time, the Carthaginian cavalry was driven from the field, and the Roman cavalry was in pursuit.

The remnants of the repulsed Carthaginian lines reformed on the wings, lengthening the frontage. Scipio ordered the Roman rear lines to the wings to avoid being outflanked. Both lines fought on until the Roman horse returned and attacked the Carthaginians in the rear. The Carthaginians turned and fled. Hannibal escaped, but his veterans fought bitterly to the death, pitted against those very legions that they had disgraced at Cannae. Some 25,000 Carthaginians were killed and 8,500 taken prisoner. The Romans lost no more than 5,000 men.

M. R. van der Werf

See also: Cannae, Battle of; Hannibal Barca; Punic Wars; Scipio Africanus Major, Publius Cornelius

References and further reading:

Bagnall, Nigel. *The Punic Wars. Rome, Carthage and the Struggle for the Mediterranean.* London: Pimlico, 1999.

Conolly, Peter. *Greece and Rome at War.* London: Greenhill Books; Harrisburg, PA: Stackpole Books, 1998.

Lazenby, J. F. *Hannibal's War. A Military History of the Second Punic War.* London: Aris & Phillips, 1978.

Zapata, Emiliano (c. 1879–1919)

Mexican revolutionary. Emiliano Zapata was born in Anenecuilco, a small village in the central Mexican state of Morelos, probably in August 1879, though evidence has been advanced for earlier years (1873 and 1877). Orphaned at 15, he had no formal education. Zapata owned a small plot of land near his village and also worked as a sharecropper and horse trader. In 1909, he was elected village council president. During the Mexican Revolution of 1910, Zapata, an unwilling revolutionary, fought for the rights of the ordinary rural Mexican. His Plan (proclamation) of Ayala (25 November 1911) called for "liberty, justice, and law." Zapata favored the forcible expropriation of land from major landowners. Unlike many revolutionary leaders, however, he remained loyal to the Roman Catholic Church.

Quiet and uncomfortable in formal settings with federal officials or other revolutionary leaders, Zapata was at his best when on his native turf. Vilified by his many enemies—politicians, the army, idealistic reformers, and organized ur-

Portrait of Emiliano Zapata, between 1900 and 1919. (Library of Congress)

ban laborers—as an inflexibly radical, impractical, and loutish country bandit, he has been regarded since his death as a hero by many of his compatriots. Whenever possible, Zapata's peasant *guerrilleros* were paid wages from loans made under duress by landowners and businesspeople. Fighting as irregulars in Morelos and several neighboring states, Zapata's forces controlled at the peak of his power an area roughly 200–300 miles long and scored considerable success against federal troops but were helpless in regular battles. They would melt into their native hills to regroup when defeated. In 1919, the conservative Venustiano Carranza, nearing the end of his term as president, viewed Zapata as a threat to any permanent peace. Early in April, Zapata was treacherously ambushed while arriving for a conference at the Chinameca hacienda, south of Villa de Ayala in central Morelos, some 45 miles from his native village.

Keir B. Sterling

See also: Mexican Revolution

References and further reading:

Parkinson, Roger. *Zapata.* Briarcliff Manor, NY. Stein and Day, 1975.

Ruiz, Ramon Eduardo. *The Great Rebellion: Mexico, 1905–1924.* New York, Norton, 1980.

Womack, John, Jr. *Zapata and the Mexican Revolution.* New York. Alfred A. Knopf, 1969.

Zapatista Rebellion (1994–)

An uprising by economically depressed indigenous Mexicans, indicative of the dismal economic conditions faced by a majority of rural Latin Americans. On 1 January 1994, Mexico signed the North American Free Trade Agreement (NAFTA) with the United States and Canada. The Mexican government hoped that signing the agreement would not only help the country but also mark Mexico's move toward becoming a First World country. NAFTA was not the only important event to take place in Mexico on this date. A guerrilla group, the Emiliano Zapata Liberation Front (EZLN), took control of the main town, San Cristobal, in Chiapas, Mexico's poorest province.

Chiapas also held the highest percentage of indigenous Mexicans in the country, has a history of being run by tough governors, and is often ignored by Mexico City. The Zapatistas (named after a revered early-twentieth-century revolutionary) did not want to overthrow the government, but their demands in the Declaration of the Jungle outlined the economic problems they faced.

The government quickly responded by dispatching the military into the region. Much brutality and abuse of human rights took place, and the Zapatistas were forced into the mountains, but under the skillful command of Subcommandante Marcos, the Zapatistas have not been crushed. The government did quickly become involved in talks with the Zapatistas, with the help of the Archbishop of San Cristobal, Samuel Ruiz. Mexico City has often been accused of dragging its feet during these talks, but the military buildup has continued, not just in Chiapas but also in neighboring regions. Paramilitary groups have also started to operate in the region, resulting in some of the worst cases of human rights abuses, such as the massacre at Acteal in December 1997.

The year 2000 saw great changes in Mexico. In July, Vicente Fox became the first president of Mexico who was not a member of the Partido Revolucionario Institucional (PRI), and in August, Pablo Salazar Mendiguichia became the first non-PRI governor of Chiapas. These events changed the situation for the Zapatistas, in that Salazar was a Zapatista sympathizer and stood for Indian rights. The new government offered some hope for a resolution of the rebellion and at least some amelioration of its causes.

M. J. Bain

See also: Nicaraguan Civil War (1979); Peruvian Guerrilla War; Salvadorian Civil War

References and further reading:

Harvey, N. *The Chiapas Rebellion: The Struggle for Land and Democracy.* Durham, NC, and London: Duke University Press, 1998.

Latin American Regional Reports: Andean Group Report. London: Latin America Newsletters, 1998–2000.

Stansfield, D., et al. *Chiapas, Before It's Too Late: A Report by an Independent Delegation to Chiapas, Mexico, March 1998.* Bristol: Bristol Chiapas Support Group, 1998.

Zenta (1697)

Conclusive battle in the Habsburg conquest of Hungary in the Austro-Turkish Wars. After the losses of Buda (1686) and Athens (1687) and the defeat at Nagy Harsany (1687), the Ottoman army deposed Sultan Mohammed IV, provoking empirewide civil unrest. The forces of the Holy League capitalized on the disorder, capturing Belgrade (1688) and occupying Transylvania, Serbia, and Macedonia (1689). Order was reestablished under Sultan Süleyman II and his grand vizier, Mustafa Köprülü, who drove the Habsburg forces out of Serbia and recaptured Belgrade (1690). The resilience of the Ottoman army was demonstrated in the narrow Habsburg victory at Szalánkemén (1691), in which Köprülü was killed. In the following years, the Habsburg army failed in a second siege of Belgrade (1694) and suffered serious defeats at Lugos (1695) and Bega (1696), while the Venetians lost Chios (1695).

In 1697, Sultan Mustafa II took personal command of the Ottoman forces and undertook a major invasion of Hungary from Belgrade with 100,000 men, intending to besiege Szegedin. While crossing the Tisza River near Zenta, Mustafa was attacked by the Habsburg army under Eugene of Savoy. Waiting until the main body of the Ottoman army was engaged in the crossing, Eugene attacked and broke through the Ottoman defensive lines guarding the bridgehead. Thrown into disorder, the Ottoman army collapsed, with the grand vizier and 30,000 troops killed in the confusion.

The defeat at Zenta ended Ottoman hopes of recovering substantial territories in Hungary and revealed the exhaustion of both sides after 16 years of war. Anglo-Dutch mediation led to the Treaty of Karlowitz (1699), by which the Ottoman Empire for the first time recognized territorial losses.

Brian Hodson

See also: Austro-Turk Wars; Eugene of Savoy

References and further reading:

McKay, Derek. *Prince Eugene of Savoy.* London: Thames and Hudson, 1979.

Parvev, Ivan. *Habsburgs and Ottomans Between Vienna and Belgrade, 1683–1739.* Boulder, CO: East European Monographs, 1995.

Zhukov, Georgy Konstantinovich (1896–1974)

Soviet World War II commander in chief and marshal of the Soviet Union (1943). A peasant furrier from Kaluga Province, Zhukov was conscripted into the czarist army in 1915 and served through World War I, emerging as a junior officer.

Zhukov joined the Red Army in August 1918 and fought against Aleksandr Kolchak in the east under Mikhail Frunze and Mikhail Tukhachevsky; fought against Anton Denikin in the south in the 1st Cavalry Army under Semen Budennyi, becoming involved with the Stalin "Tsaritsyny" clique; participated in the defeat of Peter Wrangel'; and served under Tukhachevsky and Yuronim Uborevich in crushing Aleksandr Steponovich Antonov's Tambov Revolt.

After completing the Red Army cavalry commanders' course in 1920, Zhukov graduated from the Leningrad

Georgy Zhukov seated at a desk, 1942. (Library of Congress)

Higher Cavalry School in 1925 and the Frunze Military Academy senior officers' courses five years later. He served as assistant inspector of cavalry under Budennyi (1931–1933) and became commander of III Cavalry Corps in 1937, having served under Konstantin Rokossovsky and Semen Timoshenko and worked with Uborevich, Alexandre Il'ich Egorov, Tukhachevsky, Vitali Markovich Primakov, and others striving for modernization and mechanization of the Soviet army.

Barely surviving Stalin's manic military purges through his Tsaritsyny connections, Zhukov displayed the knowledge he had amassed from the purged commanders at the Battle of Khalkin Gol (1939), decisively defeating the Japanese with a combined-arms encirclement and thus securing the Far East.

After serving as Kiev Special Military District commander (June 1940), chief of the general staff, and deputy defense commissar (January 1941), Zhukov attempted in vain, with Timoshenko, to bolster the USSR's borders, even outlandishly suggesting a preemptive strike against Germany in May 1941—one month before the devastating German invasion of the Soviet Union. Stalin refused this advice as well as Zhukov's and Budennyi's suggestion, during Operation BARBAROSSA, that Kiev be abandoned, leading to the capture of 500,000 Russian troops. In Leningrad in September, Zhukov laid the foundations for a successful defense of the city as western front commander, as well as coordinating the defense and counteroffensive before Moscow.

Serving as deputy commander in chief to Stalin from August 1942, Zhukov moved from front to front, organizing and coordinating the major Soviet war operations: the Stalingrad counteroffensive and encirclement (with Aleksandr Vasilevsky); penetration of the Leningrad blockade; the Kursk cauldron; and the Belorussian offensive. Commanding on the 1st Belorussian Front (November 1944–June 1945), he conducted the Vistula-Oder operation with Konev's 1st Ukrainian Front and the seizure of Berlin with Ivan Konev and Rokossovsky's 2d Belorussian Front.

Zhukov accepted the German capitulation and led the victory parade in Red Square. He then served as commander in chief of Soviet forces in Germany and chief of Soviet military administration (1945–1946), but Stalin viewed his popularity as a threat and dispatched him to Odessa and Urals Military Districts.

Upon Stalin's death, Zhukov became the deputy defense minister and then defense minister (1953–1957). His crucial support of Nikita Khrushchev in the 1957 leadership struggle was rewarded with removal and disgrace amid accusations of a planned coup. Zhukov wrote his memoirs during his difficult retirement and died in Moscow. He is considered one of the great commanders on any front in World War II.

Neil Harvey Croll

See also: Konev, Ivan Stepanovich; Russian and Soviet Armies; Stalin; Tukhachevsky, Mikhail Nikolayevich

References and further reading:

Shukman, Harold, ed. *Stalin's Generals.* London: Weidenfeld & Nicolson, 1993.

Spahr, William J. *Zhukov: The Rise and Fall of a Great Captain.* Novato, CA: Presidio Press, 1993.

Zhukov, Georgy K. *The Memoirs of Marshal Zhukov.* New York: Delacorte; London: Jonathan Cape, 1971. (Translation of 1969 edition).

———. *Vospominaniia i razmyshleniia, v 3-kh tomakh* (Recollections and reflections, 3 vols.). Moscow: Agentsva pechati novosti, 1990. (1969 ed., 1 vol.; 1974 ed., 2 vols.).

Zibhebhu kaMaphitha Zulu (c. 1841–1904)

The most innovative and resourceful Zulu general of the 1880s. A cousin of King Cetshwayo and chief of the Mandlakazi people in northeastern Zululand, Zibhebhu counseled against war with Britain in 1879 but nevertheless fought gallantly throughout the Anglo-Zulu War. He was slightly wounded at Isandlwana on 22 January while serving as senior *induna* (officer) of the uDloko ibutho (age-grade regiment) and was one of the junior commanders at the Battle of Khambula on 29 March. On 3 July, he was in command of the mounted scouts who drew a British reconnaissance-in-force into a successful ambush in the Mahlabathini plain.

In the settlement that followed their victory in Zululand, the British appointed compliant chiefs (including Zibhebhu) over the 13 chiefdoms into which they divided the former kingdom. Zibhebhu remained their staunchest ally in Zululand, collaborating with the British to suppress the aspirations of the royal house and their supporters, know as the uSuthu.

When the British restored Cetshwayo in 1883 to part of Zululand, civil war broke out between the uSuthu and Mandlakazi. At Msebe on 30 March 1883, Zibhebhu made effective use of mounted riflemen (supplemented by white mercenaries) to ambush, outflank, and pursue Cetshwayo's numerically superior army. After a forced night march, he surprised the uSuthu at oNdini on 21 July 1883 and scattered them, forcing Cetshwayo to take refuge with the British in southern Zululand. Zibhebhu proceeded to ravage uSuthu territory, and in 1884 Cetshwayo's desperate successor, Dinuzulu, formed an alliance with the neighboring Boers. The Boer commando's firepower proved crucial on 5 June 1884 in defeating Zibhebhu, who had attempted to lay an ambush at Tshaneni, and in his turn Zibhebhu was compelled to take refuge with the British.

In May 1887, the British finally annexed Zululand and in

November 1887 restored Zibhebhu to his chiefdom to act as a counterweight against Dinuzulu and the uSuthu, who were resisting their administration. On 23 June 1888, Dinuzulu, copying Zibhebhu's own successful tactics of 1883, surprised him at Ivuna after a night march and routed the Mandlakazi. Zibhebhu and his people were again resettled in southern Zululand. In 1898, the colonial authorities allowed Zibhebhu to return to his old chiefdom as part of a general settlement of the warring Zulu factions.

John LaBand

See also: Anglo-Zulu War; Isandlwana; Khambula; Zulu Civil Wars and Rebellion

References and further reading:

Knight, Ian. *Great Zulu Commanders 1838–1906.* London: Arms and Armour and Sterling Publishing, 1999.

Laband, John. *The Atlas of the Later Zulu Wars 1883–1888.* Pietermaritzburg, South Africa: University of Natal Press, 2001.

Zimbabwe Independence Struggle (1967–1980)

African independence struggle using terrorist and guerrilla warfare tactics often representative of colonial liberation movements. Britain's granting of independence to Northern Rhodesia (Zambia) and Nyasaland (Malawi) was denied to Southern Rhodesia because London insisted that blacks had to be brought into the government. Ian Smith's Rhodesian Front Party responded with its 1965 Unilateral Declaration of Independence, and Rhodesia was subsequently branded a pariah state subject to international sanctions.

The 228,000 residents of white, European heritage retained political and economic control over the 4.8 million black, native inhabitants (1970 census). The two fractious political groups splintered from the National Democratic Party—Joshua Nkomo's Zimbabwe African People's Union (ZAPU) and Malabaningi Sithole's Zimbabwe African National Union (ZANU)—resorted to violence. Their Marxist and Maoist rhetoric and support from Communist nations allowed the Rhodesian government to claim they were merely defending against Communist anarchy. But despite Cold War tensions, Rhodesia remained isolated.

The Rhodesian Security Forces maintained almost uncontested control of the frontier during the early campaigns. For example, three times in 1967–1968, ZAPU loyalists crossed into Rhodesia from Mozambique in units 80–100 strong. They were quickly defeated with a loss of only 13 security force personnel. Their major problem appeared to be a lack of cohesion between the factions, since they were receiving adequate arms and training from the Soviet Union, China, and Cuba during this period. Tactical failures further divided the factions. Sithole subsequently formed the Zimbabwe African National Liberation Army (ZANLA), leaving ZANU under Robert Mugabe. ZANLA had limited military utility, being noted more for its brutality.

December 1972 marked an escalation in hostilities with the well-publicized but militarily ineffective ZANU attack on Althena farm in northern Rhodesia. By this time, the Rhodesian Security Forces consisted of 4,700 regular army and air force personnel, 10,000 white Territorial Army reservists, 8,000 British South African Police (BSAP, 75 percent black), and 35,000 police reservists (75 percent white). The BSAP was the principal counterterrorist organization until Althena, when the regular military took the lead. Its assets included two infantry battalions (the all-white Rhodesian Light Infantry and the white-officered, black Rhodesian African Rifles), two SAS squadrons, artillery, engineers, and 67 assorted tactical aircraft and helicopters.

Rhodesia added to the vindictiveness of the conflict in 1977 by sponsoring the Mozambique National Resistance (RENAMO), utilizing terrorist tactics similar to those of its ZAPU/ZANU/ZANLA opponents. The fighting can be summarized as early conventional set-piece battles lost overwhelmingly by the rebels, evolving toward a guerrilla/terrorist campaign, which the blacks could exploit politically only after the government was weakened by international isolation. As is common within irregular warfare, political aspects became more significant than the military.

The Rhodesian government, suffering from its international isolation, began moves toward majority rule in 1976. The Rhodesian African National Congress won the first "one-man, one-vote" election in 1979. It was not accepted because of ZANU/ZAPU internal fighting. However, independence and international recognition followed the subsequent 1980 election victory by Robert Mugabe's reformed ZANU-Popular Front. Although Zimbabwe (former Rhodesia) had achieved black majority rule, it remained a long way from democracy.

Robert Martyn

See also: Guerrilla/Partisan/Irregular Warfare

References and further reading:

Bhebe, Ngwabi, and Terence Ranger, eds. *Soldiers in Zimbabwe's Liberation War.* London: James Curry, 1991.

Žižka, Ján (c. 1360–1424)

Hussite leader noted for his tactical skill and charismatic leadership. Born into the Bohemian lower gentry, Žižka learned the art of war in small-scale baronial feuds. He fought in Poland against the Teutonic Knights, possibly participating in the Battle of Grunwald (1410). Returning to Bohemia, he joined the king's guard (1414). An early and enthusiastic convert to the Hussite religious reform movement, Žižka helped seize Prague for the Hussites (1419). He defended Pilsen against Catholic attacks (1419) and then joined the radical Taborite movement (March 1420). While marching to Tabor, he defeated a Catholic force near Sudoměr, using his column of wagons as an improvised fort. Žižka subsequently refined this tactic, making the use of specially prepared "war wagons" key to Hussite operations.

When Holy Roman Emperor Sigismund's army threatened Prague in May 1420, Žižka brought reinforcements from Tabor and defeated the crusaders' attack on Vitkov hill, saving the city. In the winter of 1421, Žižka campaigned in western Bohemia, taking Beroun (March) and Rabí (June), where he was wounded in his right eye, leaving him blind (he had lost sight in his left in his youth). In December 1421, Sigismund trapped Žižka's army outside Kutná Hora. Using his war wagons as field artillery, Žižka broke through the king's lines and escaped. Two weeks later, Žižka attacked Sigismund's army in winter quarters, destroying it in a running battle from Nebovidy to Německý Brod (6–10 January 1422). When the tensions between the conservative and radical wings of the Hussite movement broke into open conflict, Žižka took the lead of the radical Taborite Party, defeating the forces of Prague at Strachuv (August 1422) and again at Malešov (June 1424). He died shortly afterward at Přibyslav, Bohemia, on 11 October 1424.

Brian Hodson

See also: Hussite Wars

References and further reading:

Heymann, George. *John Žižka and the Hussite Revolution.* Princeton, NJ: Princeton University Press, 1955.

Zulu Civil Wars and Rebellion (1879–1888)

The conflicts in Zululand (present-day KwaZulu Natal, South Africa) from 1879 to 1888 that resulted in the dismemberment of Zululand by its white colonial neighbors. After victory in the Anglo-Zulu War, the British did not annex Zululand but eliminated its future military potential by breaking it on 1 September 1879 into 13 fragments under appointed chiefs. Growing strife between some of these chiefs and the uSuthu (supporters of the deposed and exiled King Cetshwayo) persuaded the British on 11 December 1882 to restore Cetshwayo to the central part of his former kingdom. To check Cetshwayo's aspirations, they gave Chief Zibhebhu kaMaphitha of the Mandlakazi in northeastern Zululand, who had proved himself the staunchest of the appointed chiefs, an independent chiefdom. The southern third of Zululand was placed under British protection as the Reserve Territory, with Cetshwayo as nominal chief.

This settlement served only to increase uSuthu-Mandlakazi rivalry. The uSuthu invaded Zibhebhu's territory, but he ambushed and routed them at Msebe on 30 March 1883. Zibhehbu retaliated, and at oNdini on 21 July 1883 scattered the uSuthu and forced Cetshwayo into the Reserve Territory, where he died on 8 February 1884.

Dinuzulu, Cetshwayo's heir, was unable to contain the rampant Mandlakazi and turned to the Boers of the neighboring South African Republic for military assistance. With their firepower, he crushed Zibhebhu at Tshaneni on 5 June 1884 and drove him into the Reserve Territory. In return, on 16 August 1884, Dinuzulu ceded the Boers the northwestern two-thirds of Zululand outside the reserve.

To contain the land-hungry Boers and forestall imperial rivals, on 19 May 1887 Britain annexed the Reserve Territory and the rump of Zululand still under Dinuzulu as the Colony of Zululand. However, Dinuzulu would not cooperate with the colonial administration. To curb him, the Zululand officials restored the collaborationist Zibhebhu to his former chiefdom in late 1887. His return sparked off renewed unrest, and by April 1888 the uSuthu were in open rebellion, defying the paramilitary Zululand Police, the regular troops of the British garrison, their Mandlakazi allies, and other African auxiliaries.

On 2 June 1888, the uSuthu repulsed a British force on Ceza Mountain and at Ivuna on 23 June completely routed the Mandlakazi. The British withdrew south of the Black Mfolozi River to regroup. On 2 July, the reinforced British drove the uSuthu from Hlophekhulu Mountain, and the uSuthu started abandoning their other fastnesses. Between July and September, British flying columns traversed the disaffected areas north of the Black Mfolozi River and along the coast, securing submissions. Dinuzulu and other uSuthu leaders were exiled, and the British consolidated their rule in pacified Zululand.

John Laband

See also: Anglo-Zulu War; Zibhebhu kaMaphitha Zulu

References and further reading:

Guy, Jeff. *The Destruction of the Zulu Kingdom: The Civil War in Zululand, 1879–1884.* Reprint, Pietermaritzburg, South Africa: University of Natal Press, 1988.

Laband, John. *The Rise and Fall of the Zulu Nation.* London: Arms and Armour and Sterling Publishing, 1997.

Laband, John, and Paul Thompson. "The Reduction of Zululand, 1878–1904." In *Natal and Zululand from Earliest Times to 1910: A New History,* ed. Andrew Duminy and Bill Guest. Pietermaritzburg, South Africa: University of Natal Press and Shuter & Shooter, 1989.

Zulu Kingdom (c. 1820–1879)

Military-based kingdom of southern Africa. Historic Zululand is today part of the KwaZulu Natal Province in modern South Africa. Zululand takes its name from the Zulu clan of the Nguni Bantu people. Early in the nineteenth century, under the leadership of Shaka kaSenzangakhona (c. 1787–24 September 1828), this small clan became the most powerful and feared nation of southern Africa. Shaka modernized weapons and tactics, instilled a brutal discipline and complex organization, and transformed war in the region from a largely ceremonial to a very serious business. As part of Shaka's process of aggrandizement and absorption of neighboring populations, Zululand grew southward. This process was continued by his heirs, Dingaan (Dingane), Mpandi, and Cetshwayo. In so doing, the Zulu came into contact with the Boers, who were moving northward to escape the British Empire. The resulting clash of cultures erupted in violence, particularly the Battle of Blood River on 16 December 1838 and in various attempts by Britain to gain control of Boer and Zulu alike.

In late 1878, the British issued an ultimatum to Zulu king Cetshwayo to dismantle his military system. Upon his refusal, a force under Lieutenant General Frederic Thesiger (later the second Baron Chelmsford) invaded Zululand. Despite a major defeat of the British by the Zulus at Isandlwana in January, Lord Chelmsford completely subdued the Zulus at the battle of Ulundi in July. Zululand was then divided into a series of small divisions under the control of puppet chieftains ruling at the direction of the British government. The result was instability and endemic rebellion.

James B. Thomas

See also: Blood River; Isandlwana; Rorke's Drift; Shaka kaSenzangakhona; Zulu Civil Wars and Rebellions

References and further reading:

Morris, Donald. *The Washing of the Spears.* New York: Simon & Schuster, 1965.

Pakenham, Thomas. *The Scramble for Africa.* New York: Random House, 1991.

Zulu chief, South Africa, c. 1880. (Library of Congress)

Zuo Zongtang (Tso Tsung-tang) (1812–1885)

Chinese imperial politician, military administrator, and field commander important in suppressing the Taiping, Nian, and Muslim Rebellions. Zuo was born on 10 November 1812 in Xiangyin, Hunan Province, China, and was raised to be a gentleman scholar in the Confucian tradition. After passing his civil service examinations, he worked as a geographer and agricultural scientist. When the Taiping Rebellion erupted in 1850, he mustered Hunan volunteers loyal to the emperor. After initial defensive successes, he was ordered in 1853 to join forces with Zeng Guofan. He organized imperialist volunteers in Jiangxi and Anhui Provinces, became a general in 1860, and was soon among the most powerful warlords in China. He was appointed governor of Zhejiang Province in 1862, governor-general of Zhejiang and Fujian Provinces in 1863, and governor-general of Shaanxi and Gansu Provinces in 1866.

Zuo led the troops that suppressed both the Nian Rebellion (1853–1868) in east-central China and the various Muslim revolts in the west. While his rival, Li Hongzhang, lobbied for stronger coastal and naval defenses, Zuo successfully pushed west into central Asia, using Western technology and innovative logistics. He defeated the charismatic Muslim leader, Yakub Beg, in 1877, secured Xinjiang in 1878, and forced Russia to cede the strategic Ili region by the Treaty of St. Petersburg in 1881.

Old, ill, and half-blind, Zuo asked to retire in 1882, but because he had done so much to revive the Qing Dynasty and restore its power within Asia, his request was denied. Instead, he was rewarded with many high positions, including governor-general of Liang Kiang in 1882 and grand secretary of state in 1884, with full military authority in Fujian and southern China for the Sino-French War (1883–1885). He died in Fuzhou, Fujian Province, on 5 September 1885.

Eric v. d. Luft and Sarah Luft

See also: Chinese Imperial Wars; Gordon, Charles George; Hong Xiuquan; Li Hongzhang; Religion and War; Taiping Rebellion; Wolseley, Garnet Joseph, Viscount; Yang Xiuqing

References and further reading:

Chen, Gideon. *Tso Tsung T'ang.* New York: Paragon, 1968.

Lipman, Jonathan Neaman. *Familiar Strangers: A History of Muslims in Northwest China.* Seattle: University of Washington Press, 1997.

Prazniak, Roxann. *Of Camel Kings and Other Things: Rural Rebels against Modernity in Late Imperial China.* Lanham, MD: Rowman & Littlefield, 1999.

Wright, Mary Clabaugh. *The Last Stand of Chinese Conservatism: The T'ung-chih Restoration, 1862–1874.* New York, Atheneum, 1966.

SELECTED BIBLIOGRAPHY

Abbot, Jacob. *History of William the Conqueror.* Pensacola, FL: Beka, 1999.

Abels, Richard. *Alfred the Great: War, Kingship and Culture in Anglo-Saxon England.* London and New York: Longman, 1998.

Adams, Nina S., and Alfred W. McCoy, eds. *Laos: War and Revolution.* New York: Harper and Row, 1970.

Adler, Emanuel, ed. *The International Practice of Arms Control.* Baltimore: Johns Hopkins University Press, 1991.

Ambrose, Stephen E. *Eisenhower.* 2 vols. New York: Simon and Schuster, 1983, 1984.

———. *Duty, Honor, Country: A History of West Point.* Baltimore: Johns Hopkins University Press, 1999.

Anderson, Fred. *Crucible of War: The Seven Years War and the Fate of Empire in British North America, 1754–1766.* New York: Alfred A. Knopf, 2000.

Appelquist, A. Ray, ed. *Church, State, and Chaplaincy: Essays and Statements on the American Chaplaincy System.* Washington, DC: General Commission on Chaplains and Armed Forces Personnel, 1969.

Appleman, Roy E. *South to the Naktong: North to the Yalu.* Washington, DC: Office of the Chief of Military History, 1961.

———. *South to the Naktong, North to the Yalu, June–November, 1950.* Washington, DC: Department of the Army, 1961.

———. *East of Chosin: Entrapment and Breakout in Korea—1950.* College Station: Texas A & M University Press, 1987.

Arnold, Thomas. *The Renaissance at War.* London: Cassell, 2001.

Asprey, Robert B. *War in the Shadows.* Garden City: Doubleday and Co., 1975.

Austin, Paul Britten. *1812 Napoleon's Invasion of Russia.* London: Greenhill Books, 2000.

Bagnall, Nigel. *The Punic Wars: Rome, Carthage and the Struggle for the Mediterranean.* London: Pimlico, 1999.

Bak, Janós, and Bela Kiraly. *From Hunyadi to Rákoczi: War and Society in Medieval and Early Modern Hungary.* Brooklyn, NY: Brooklyn College Press, 1982.

Baker, A. J. *Soviet Army Uniforms and Insignia, 1945–1975.* London: Arms and Armour Press, 1976.

Ball, Howard. *Prosecuting War Crimes and Genocide: The Twentieth Century Experience.* Lawrence: University Press of Kansas, 1999.

Barfield, Thomas J. *The Perilous Frontier: Nomadic Empires and China.* Cambridge, MA, and Oxford, UK: Basil Blackwell, 1989.

Barker, A. J. *The Neglected War: Mesopotamia 1914–1918.* London: Faber and Faber, 1967.

Barnaby, Wendy. *The Plague Makers: The Secret World of Biological Warfare.* New York: Continuum, 2000.

Barnett, Correlli. *Britain and Her Army, 1509–1970.* New York: William Morrow, 1970.

Bartutis, Mark. *The Late Byzantine Army.* Philadelphia: University of Pennsylvania Press, 1992.

Bauer, K. Jack. *The Mexican War, 1846–1848.* Lincoln: University of Nebraska Press, 1993.

Baumgart, Winfried. *The Crimean War 1853–1856.* London: Edward Arnold, 2000.

Bennett, Martyn. *The English Civil War, 1640–1649.* London: Longman, 1995.

Berkley, Bill. *The Graves Are Not Yet Full: Race, Tribe, and Power in the Heart of Africa.* New York: Basic Books, 2001.

Binneveld, J. M. W. *From Shell Shock to Combat Stress: A Comparative History of Military Psychiatry.* Amsterdam: Amsterdam University Press, 1997.

Black, Jeremy. *Britain as a Military Power, 1688–1815.* London: UCL Press, 1999.

———. *War for America: The Fight for Independence, 1775–1783.* Stroud, UK: Alan Sutton, 1991.

Bliokh, Ivan S. *The Future of War in Its Technical, Economic, and Political Relations.* Russian ed. 6 vols. St Petersburg: Tipografiya I A. Efrona, 1898.

Blumenson, Martin. *Breakout and Pursuit. United States Army in World War II.* Washington, DC: Office of the Chief of Military History, 1961, 1984.

Boudet, Jacques. *The Ancient Art of Warfare.* London: Barrie and Rockliff, Cresset Press, 1969.

Bradbury, Jim. *The Medieval Siege.* Woodbridge, Suffolk, UK: Boydell Press, 1997.

Bradley, John. *Allied Intervention in Russia 1917–1920.* Lanham, MD: University Press of America, 1984.

Bram, Joseph. *An Analysis of Inca Militarism.* New York: Augustin Publisher, 1941.

Bregerud, Eric. *Touched with Fire: The Land War in the South Pacific.* New York: Penguin, 1997.

Brock, Peter. *Freedom from War: Nonsectarian Pacifism.* Toronto: University of Toronto Press, 1991.

Brock, Peter, and Nigel Young. *Pacifism in the 20th Century.* Syracuse, NY: Syracuse University Press, 1999.

Browne, J. P. R., and M. T. Thurbon. *Electronic Warfare.* Brassey's Air Power: Aircraft, Weapons Systems and Technology series. London: Brassey, 1998.

Buisseret, David, ed. *Monarchs, Ministers, and Maps: The Emergence of Cartography as a Tool of Government in Early Modern Europe.* Chicago: University of Chicago Press, 1992.

Burne, Alfred H. *The Crecy War: A Military History of the Hundred Years War from 1337 to the Peace of Bretigny, 1360.* London: Eyre and Spottiswoode, 1955.

Butler, Ivan. *The War Film.* South Brunswick: A. S. Barnes, 1974.

Calvocoressi, Peter, and Guy Wint. *Total War: Causes and Courses of the Second World War.* New York: Viking Penguin, 1979.

Campbell, Bruce. *The SA Generals and the Rise of Nazism.* Lexington: University of Kentucky Press, 1998.

Catton, Bruce. *The Centennial History of the Civil War.* Garden City, NY: Doubleday, 1961, 1963, 1965.

Chandler, David G. *Campaigns of Napoleon.* New York: Scribner, 1966.

———. *Marlborough as a Military Commander.* London: 1973.

———. *The Art of Warfare in the Age of Marlborough.* London: Batsford, 1976.

Chassin, Lionel Max. *The Communist Conquest of China: A History of the Civil War, 1945–1949.* Trans. Timothy Osato and Louis Gelas. Longon: Weidenfeld and Nicolson, 1965.

Chatfield, Charles. *For Peace and Justice: Pacifism in America, 1914–1941.* Boston: Beacon Press, 1973.

Christopher, Paul. *The Ethics of War and Peace.* 2d ed. Upper Saddle River, NJ: Prentice Hall, 1999.

Ciechoanowski, Jan M. *Warsaw Rising of 1944.* Cambridge, UK: Cambridge University Press, 1974

Clausewitz, Karl von. *On War.* Princeton, NJ: Princeton University Press, 1984.

Clodfelter, M. *Warfare and Armed Conflict: A Statistical Reference to Casualty and Other Figures.* 2 vols. Jefferson, NC: McFarland, 1994.

Coakley, Robert W. *The Role of Federal Military Forces in Domestic Disorders, 1789–1878.* Washington, DC: U.S. Army Center of Military History, 1988.

Coffman, Edward M. *The War to End All Wars: The American Military Experience in World War I.* New York: Oxford University Press, 1968.

Cole, Hugh M. *The Ardennes: The Battle of the Bulge.* Washington, DC: Government Printing Office, 1965.

Collier, Christopher, and James Lincoln Collier. *The French and Indian War: 1660–1763.* New York: Benchmark Books, 1998.

Collins, Roger. *The Arab Conquest of Spain 710–797.* New York: B. Blackwell, 1989.

Connelley, Owen. *Blundering to Glory.* Wilmington, DE: Scholarly Resources, 1999.

Conolly, Peter. *Greece and Rome at War.* London: Macdonald Hall, 1981.

———. *War Report of the OSS: The Overseas Targets.* New York: Walker and Company, 1977.

Coogan, Tim. *The Troubles: Ireland's Ordeal 1966–1996 and the Search for Peace.* Boulder, CO: Roberts Rinehart, 1996.

Cooper, Belinda, ed. *War Crimes: The Legacy of Nuremberg.* New York: TV Books, 1999.

Cope, John N. *International Military Education and Training: An Assessment.* Washington, DC: U.S. Government Printing Office, 1995.

Cordesman, Anthony H., and Abraham R. Wagner. *The Lessons of Modern War.* Vol. 1: *The Arab-Israeli Conflicts, 1973–1989.* Boulder, CO: Westview Press, 1990.

Cottam, Kazimiera J. *Women in War and Resistance.* Nepean, Canada: New Military Publishing, 1998.

Craig, Gordon A. *The Politics of the Prussian Army, 1640–1945.* London: Oxford University Press, 1964.

Creveld, Martin L. *Supplying War: Logistics from Wallenstein to Patton.* New York: Cambridge University Press, 1977.

Crossland, Richard B., and James T. Currie. *Twice the Citizen: A History of the Army Reserve, 1908–1983.*

2d ed. Washington, DC: Office of the Chief, Army Reserve, 1997.

Delbrück, Hans. *History of the Art of War within the Framework of Political History.* Vol. 4, *The Modern Era.* 1920 (German edition). Trans. Walter J. Renfroe. Westport, CT: Greenwood, 1985.

———. *The Dawn of Modern Warfare.* Trans. Walter J. Renfroe Jr. Lincoln: University of Nebraska Press, 1990.

———. *Medieval Warfare.* Trans. Walter J. Renfroe Jr. Lincoln: University of Nebraska Press, 1990.

Desmond, Robert W. *Tides of War: World News Reporting, 1940–1945.* Iowa City: University of Iowa Press, 1984.

Devillers, Philippe, and Jean Lacouture. *End of a War: Indochina, 1954.* Trans. Alexander Lieven and Adam Roberts. New York: Praeger Publishers, 1969.

Diagram Group. *Weapons.* Leicester, UK: Galley Press (WH Smith Group), 1980.

Diaz del Castillo, Bernal. *True History of the Conquest of New Spain.* 5 vols. Ed. and trans. A. P. Maudley. London: Hakluyt Society, 1908.

Dillon, Martin. *The Dirty War: Covert Strategies and Tactics Used in Political Conflicts.* New York: Routledge, 1990.

Dodge, T. A. *Caesar: A History of the Art of War among the Romans Down to the End of the Roman Empire, with a Detailed Account of the Campaigns of Gaius Julius Caesar.* Boston: Houghton Mifflin, 1892. Reprint, New York: Da Capo Press, 1997.

Donner, Fred M. *The Early Islamic Conquests.* Princeton, NJ: Princeton University Press, 1981.

Dorling, H. Taprell. *Ribbons and Medals.* London: Osprey Publishing, 1983.

Douglas, W. A. B., and Brereton Greenhous. *Out of the Shadows: Canada in the Second World War.* Toronto: Oxford University Press, 1977.

Drell, Sidney D., et al., eds. *The New Terror: Facing the Threat of Biological and Chemical Warfare.* Stanford, CA: Hoover Institution Press, 1999.

Duffy, Christopher. *Fire and Stone: The Science of Fortress Warfare 1660–1860.* Newton Abbot, UK: David and Charles, 1975.

———. *Siege Warfare: The Fortress in the Early Modern World, 1494–1660.* London: Routledge and Paul, 1979.

Dunn, Richard S. *The Age of Religious Wars, 1559–1715.* 2d ed. New York: W. W. Norton, Inc, 1979.

Dupuy, R. Ernest, and Trevor N. Dupuy. *Military Heritage of America.* New York: McGraw-Hill, 1956.

Edgerton, Robert. *Fall of the Asante Empire.* New York: Free Press, 1995.

Editors of Military Affairs. *Military Analysis of the Civil War.* Millwood, NY: KTO Press, 1977.

Erickson, John. *Soviet Military Power.* London: Royal United Services Institute, 1971.

———. *Road to Berlin.* Boulder, CO: Westview Press, 1983.

———. *Road to Berlin: Stalin's War with Germany.* New Haven, CT: Yale University Press, 1999.

———. *The Road to Stalingrad: Stalin's War with Germany.* New Haven, CT: Yale University Press, 1999.

Evans, Alun. *Brassey's Guide to War Films.* Washington, DC: Brassey's Publishing Company, 2000.

Ezell, Edward C. *Small Arms of the World: A Basic Manual of Small Arms.* Harrisburg, PA: Stackpole; London: Guns and Armor Press, 1983.

Fairbank, John King, and Frank A. Kierman, Jr., eds. *Chinese Ways in Warfare.* Cambridge, MA: Harvard University Press, 1974.

Falk, Stanley. *Bataan: The March of Death.* New York: W. W. Norton and Co., 1962.

Farmer, Henry. *The Rise and Development of Military Music.* London, W. M. Reeves, 1912.

Feifer, George. *Tennozan: The Battle of Okinawa and the Atomic Bomb.* New York: Ticknor and Fields, 1992.

Ferguson, Niall. *The Pity of War: Explaining World War I.* New York: Basic Books, 2000.

Ferling, John. *Struggle for a Continent: The Wars of Early America.* Arlington Heights, IL: Harlan Davidson, Inc., 1993.

Ferrill, Arthur. *The Fall of the Roman Empire: The Military Explanation.* New York: Thames and Hudson, 1986.

Fissel, Mark Charles. *English Warfare, 1511–1642.* London: Routledge, 2001.

Fleming, Thomas. *Liberty! The American Revolution.* New York: Viking, 1997.

Floyd, Dale E. *Defending America's Coasts, 1775–1950: A Bibliography.* Alexandria, VA: Office of the Chief of History, U.S. Army Corps of Engineers, 1997.

Flynn, George Q. *America and the Draft, 1940–1973.* Lawrence: University Press of Kansas, 1993.

Foner, Philip Sheldon. *The Spanish-Cuban-American War and the Birth of American Imperialism, 1895–1902.* 2 vols. New York: Monthly Review Press, 1972.

Fortescue, J. W. *A History of the British Army.* Vol. 4. London: Macmillan, 1915.

Frank, Richard B. *Downfall: The End of the Imperial Japanese Empire.* New York: Random House, 1999.

Frost, Robert. *The Northern Wars: War, State and Society in Northeastern Europe, 1558–1721.* New York: Longman Press, 2000.

Furneaux, Rupert. *The Seven Years War.* London: Hart-Davis MacGibbon, 1973.

Gabba, Emilio. *Republican Rome: The Army and the Allies.* Oxford, UK: Blackwell, 1976.

Gaddis, John Lewis. *The Long Peace: Inquiries into the History of the Cold War.* New York: Oxford University Press, 1987.

———. *We Know Now: Rethinking Cold War History.* New York: Oxford University Press, 1998.

Gat, Azar. *The Development of Military Thought: The Nineteenth Century.* Oxford, UK: Oxford University Press, 1992.

Gates, John M. *Schoolbooks and Krags: The United States Army in the Philippines, 1899–1902.* Westport, CT: Greenwood Press, 1973.

Gelbart, Marsh. *The First World War: A Complete History.* New York: Henry A. Holt, 1996.

———. *Tanks: Main Battle and Light Tanks.* London: Brassey, 1996.

Gilmore, Allison. *You Can't Fight Tanks with Bayonets: Psychological Warfare against the Japanese Army in the Southwest Pacific.* Lincoln: University of Nebraska Press, 1998.

Glantz, David M. *The Siege of Leningrad, 1941–44: 900 Days of Terror.* Staplehurst: Spellmount, Motorbooks International, 2001.

Glantz, David M., and Jonathon House. *When Titan's Clashed: How the Red Army Stopped Hitler.* Lawrence: University Press of Kansas, 1995.

Goldsworthy, Adrian. *The Roman Army at War.* Oxford, UK: Oxford University Press, 1996.

Goodman, Anthony. *The Wars of the Roses: Military Activity and English Society, 1452–97.* Boston: Routledge and Kegan Paul, 1981.

Gordon, C. D. *The Age of Attila: Fifth Century Byzantium and the Barbarians.* Ann Arbor: University of Michigan Press, 1960.

Gordon, Philip H., ed. *NATO's Transformation.* Lanham, MD: Rowman and Littlefield, 1999.

Görlitz, Walter. *History of the German General Staff, 1657–1945.* New York: Frederick A. Praeger, 1953.

Green, Peter. *Alexander the Great.* London: Weidenfeld and Nicolson, 1970.

Grey, G. *A Military History of Australia.* Melbourne: Cambridge University Press, 1999.

Griffith, P. *The Viking Art of War.* London: Greenhill, 1995.

Hackett, John, ed. *Warfare in the Ancient World.* London: Sidgwick and Jackson, 1989.

Haldon, John. *The Byzantine Wars.* Stroud, UK: Tempus, 2001.

———. *Warfare, State and Society in the Byzantine World, 565–1204.* London: UCL Press, 1999.

Hale, J. R., et al. *War and Society.* 5 vols. Stroud, UK: Fontana Paperbacks, 1985.

Hall, Bert S. *Weapons and Warfare in the Renaissance: Gunpowder, Technology, and Tactics.* Baltimore, MD: Johns Hopkins University Press, 1997.

Hall, Linda B., and Don M. Coerver. *Revolution on the Border: The United States and Mexico, 1910–1920.* Albuquerque: University of New Mexico Press, 1988.

Hammond, N. G. L. *The Genius of Alexander the Great.* Chapel Hill: University of North Carolina Press, 1997.

Handbook of Ballistics. 2 vols. London: Her Majesty's Stationery Office, 1921.

Harclerode, Peter. *Para!* London: Brockhampton, 1999.

Harley, J. Brian, et al., ed. *The History of Cartography.* Vols. 1–2. Chicago: University of Chicago Press, 1987–1998.

Hassig, Ross. *War and Society in Ancient Mesoamerica.* Berkeley: University of California Press, 1992.

Haugen, David M., ed. *Biological and Chemical Weapons.* San Diego, CA: Greenhaven Press, 2001.

Hayes-McCoy, G. A. *Irish Battles.* London: Longmans, Green, 1969.

Heath, Ian. *Armies of the Dark Ages.* Worthing, UK: WRG Publication, 1980.

Heinl, Robert Debs. *Soldiers of the Sea: The United States Marine Corps, 1775–1962.* Annapolis, MD: United States Naval Institute, 1962.

Henderson, Lawrence W. *Angola: Five Centuries of Conflict.* London: Cornell University Press, 1976.

Herodotus. *The Histories.* New translation, selections, background, and commentaries. Trans. Walter Blanco. Ed. Walter Blanco and Jennifer Tolbert Roberts. New York: Norton, 1992.

Herring, George C. *America's Longest War: The United States and Vietnam, 1950–1975.* 3d ed. New York: McGraw-Hill, 1996.

Herwing, Holger H. *The First World War: Germany and Austria-Hungary, 1914–1918.* London: Arnold, 1997.

Hewlett, Richard G., and Oscar E. Anderson Jr. *Atomic Shield, 1947–1952.* Berkeley: University of California Press, 1990.

Hibbert, Christopher. *Agincourt.* London: Batsford, 1964.

Hildinger, Erik. *Warriors of the Steppe.* New York: Sarpedon, 1997.

Hobart, Frank William Arthur, ed. *Jane's Infantry Weapons.* London: MacDonald and Jane's, 1975.

Hogg, Ian V. *Illustrated Encyclopedia of Artillery.* London: Hutchinson, 1987.

Holt, Mack P. *The French Wars of Religion 1562–1629.* Cambridge, UK: Cambridge University Press, 1995.

Hooper, Nicholas, and Matthew Bennett. *The Cambridge Illustrated Atlas of Warfare. The Middle Ages, 768–1487.* Cambridge, UK: Cambridge University Press, 1996.

Horne, Alistair. *A Savage War of Peace: Algeria, 1954–1962.* New York: Viking Press, 1978. Rev. ed., New York: Penguin Books, 1987.

———. *The French Army and Politics, 1870–1970.* New York: Harper and Row, 1984.

Howard, M. *Clausewitz.* Oxford, UK: Oxford University Press, 1983.

Howard, Michael. *The Franco-Prussian War.* New York: Collier Books, 1961.

Hughes, Matthew. *Allenby and British Strategy in the Middle East, 1917–1919.* London: Cass, 1999.

Humble, Richard. *Warfare in the Ancient World.* London: Cassell, 1980.

Huntington, Samuel P. *The Soldier and the State: The Theory and Politics of Civil-Military Relations.* New York: Vintage Books, 1957.

Huston, James A. *The Sinews of War: Army Logistics, 1775–1953.* Washington, DC: United States Army, 1966.

James, D. Clayton. *The Years of MacArthur.* Vol. 1, *1880–1941.* Boston: Houghton Mifflin, 1970.

———. *The Years of MacArthur.* Vol. 2, *1941–1945.* Boston: Houghton Mifflin, 1975

———. *The Years of MacArthur.* Vol.3, *Triumph and Disaster, 1945–1964.* Boston: Houghton Mifflin, 1985.

Janowitz, Morris. *The Professional Soldier: A Social and Political Portrait.* New York: Free Press, 1960.

Johnson, James Turner. *Morality and Contemporary Warfare.* New Haven: Yale University Press, 1999.

Jones, Archer. *Civil War Command and Strategy.* New York: Free Press, 1992.

———. *The Art of War in the Western World.* Urbana: University of Illinois Press, 1987

Kannik, Preben. *Military Uniforms of the World.* New York: Macmillan, 1968.

Kapstein, Ethan Barnaby. *The Political Economy of National Security: A Global Perspective.* Columbia: University of South Carolina Press, 1992.

Kar, Lt. Col. H. C. *Military History of India.* Calcutta: Firma KLM Private Limited, 1980.

Keegan, John. *The First World War.* New York: Vintage Books, 1998.

Keen, Maurice, ed. *Medieval Warfare.* Oxford, UK: Oxford University Press, 1999.

Keep, John. *Soldiers of the Tsar: Army and Society in Russia 1462–1874.* Oxford, UK: Oxford University Press, 1985.

Kennedy, Hugh. *The Prophet and the Age of the Caliphs.* London: Longman Press, 1986.

Keppie, Lawrence. *The Making of the Roman Army: From Republic to Empire.* Norman: University of Oklahoma Press, 1998.

Kirby, Maj. Gen. S. Woodburn. *History of the Second World War, the War against Japan.* Vol. 2, *India's Most Dangerous Hour.* London: Her Majesty's Stationery Office, 1958.

Koistinen, Paul A. C. *The Military-Industrial Complex: A Historical Perspective.* New York: Praeger, 1980.

Kornbluh, Peter, ed. *Bay of Pigs Declassified: The Secret CIA Report on the Invasion of Cuba.* New York: New York Press, 1998.

Laband, John, and Paul Thompson. *The Illustrated Guide to the Anglo-Zulu War.* Pietermaritzburg, South Africa: University of Natal Press, 2000

Lambert, Andrew D., and Stephen Badsey. *The Crimean War.* Dover, NH: Sutton, 1994.

Lanz, Hubert, et al. *The German Campaigns in the Balkans (Spring 1941).* MS No B-525. Washington, DC: U.S. Army Center of Military History, undated; Supplementary Report. Washington, DC: U.S. Army Center of Military History, 1989

Laqueur, Walter. *The Age of Terrorism.* 2d ed. Boston: Little, Brown, 1987.

Laurie, Clayton D. *The Propaganda Warriors: America's War against Nazi Germany.* Lawrence: University Press of Kansas, 1996.

Lepore, Jill. *The Name of War: King Philip's War and the Origins of American Identity.* New York: Alfred A. Knopf, 1998.

Levie, Howard S. *Prisoners of War in International Armed Conflict.* Newport, RI: U.S. Naval War College Press, 1977.

Liddell Hart, Basil H. *History of the Second World War.* New York: Da Capo Press, 1999.

Lincoln, Bruce. *Red Victory: A History of the Russian Civil War.* New York: Simon and Schuster, 1989.

Linn, Brian McAllister. *Guardians of Empire: The U.S. Army and the Pacific, 1902–1949.* Chapel Hill: University of North Carolina Press, 1997.

———. *Philippine War, 1899–1902.* Lawrence: University Press of Kansas, 2000.

Lionel Max. *The Communist Conquest of China: A History of the Civil War, 1945–1949.* Cambridge, MA: Harvard University Press, 1965.

Loveman, B., and T. M. Davies Jr. *Che Guevara Guerrilla Warfare.* 3d ed. Wilmington, DE: Scholarly Resources Inc, 1997.

Luh, Juergen. *Ancien Régime Warfare and the Military Revolution. A Study.* Groningen: INOS, 2000.

Lumpkin, Henry. *From Savannah to Yorktown: The American Revolution in the South.* New York: Paragon Publishers, 1987.

Lynn, John A., ed. *Feeding Mars: Logistics in Western Warfare from the Middle Ages to the Present.* Boulder, CO: Westview, 1993.

MacAulay, J. S. *A Treatise on Field Fortifications, and Other Subjects Connected with the Duties of the Field Engineer.* London: Fraser, 1834.

Mackesy, Piers. *The War for America, 1775–1783.* London: Longmans, 1964.

Maslowski, Peter, and Allan R. Millett. *For the Common Defense: A Military History of the United States of America.* New York: Free Press, 1984.

Maughan, Barton. *Australia in the War of 1939–1945.* Ser. One, *Army.* Vol. 3, *Tobruk and El Alamein.* Canberra: Australian War Memorial, 1966.

May, Ernest R. *Hitler's Conquest of France.* New York: Hill and Wang, 2000.

Mayer, Hans Eberhard. *The Crusades.* Oxford, UK: Oxford University Press, 1988.

Mazarr, Michael J. *Semper Fidel: America and Cuba, 1776–1988.* Baltimore, MD: Nautical and Aviation Publishing, 1988.

McDermott, John D. *A Guide to the Indian Wars of the West* Lincoln: University of Nebraska Press, 1998

McGinnis, Anthony. *Counting Coup and Cutting Horses: Intertribal Warfare on the Northern Plains 1738–1889.* Evergreen, CO: Cordillera Press, 1990.

McNalty, Bernard, and Morriss MacGregor, eds. *Blacks in the United States Armed Forces: Basic Documents.* 13 vols. Wilmington, DE: Scholarly Resources, 1977.

McNeill, William H. *The Pursuit of Power. Technology, Armed Force, and Society since A.D. 1000.* Chicago: University of Chicago Press, 1982.

Meier, Christian. *Caesar: A Biography.* Trans. David McLintock. New York: Basic Books, 1982

Menning, Bruce. *Bayonets before Bullets: The Imperial Russian Army, 1861–1914.* Bloomington: Indiana University Press, 1992.

Miller, David. *The Cold War: A Military History.* New York: St. Martin's Press, 1998.

———. *Samurai Warriors.* New York: St. Martin's Press, 1999.

Moberly, Sir F. J. *Official History of the Great War: The Campaign in Mesopotamia 1914–1918.* 4 vols. London: Her Majesty's Stationery Office, 1923–1930.

Mollo, John. *Military Fashion.* London: Barrie and Jenkins, 1972.

Montagu, John Drogo. *Battles of the Greek and Roman Worlds.* London: Greenhill Books, Lionel Leventhal Limited, 2000.

Morton, Louis. *The Fall of the Philippines.* Washington, DC: Center for Military History, 1953.

Mungello, David E. *The Great Encounter of China and the West, 1500–1800.* Lanham, MD: Rowman and Littlefield, 1999.

Munro, Neil. *The Quick and the Dead: Electronics Combat and Modern Warfare.* New York: St. Martin's, 1991.

Musicant, Ivan. *The Banana Wars.* New York: Macmillan, 1990.

Nafziger, George. *Napoleon's Invasion of Russia.* Novato, CA: Presidio, 1988.

Nasson, Bill. *The South African War.* London: Arnold, 1999.

Nicholson, G. W. L. *Canadian Expeditionary Force 1914–1919: Official History of the Canadian Army in the First World War.* Ottawa: Queen's Printer, 1964.

Oakeshott, R. Ewart. *The Archaeology of Weapons: Arms and Armour from Prehistory to the Age of Chivalry.* Woodbridge, Suffolk, UK: Boydell Press, 1994.

O'Connell, Robert L. *Of Arms and Men: A History of War, Weapons, and Aggression.* New York: Oxford University Press, 1989.

Oman, Charles W. *A History of the Art of War in the Middle Ages.* London: Greenhill Books, 1991

———. *The Art of War in the Sixteenth Century.* London: Methuen, 1937.

Overy, Richard. *Russia's War: Blood upon the Snow.* New York: Penguin Putnam, 1997.

Painter, David. *The Cold War: An International History.* London: Routledge Press, 2000.

Pakenham, Thomas. *The Boer War.* New York: Random House, 1979.

———. *The Scramble for Africa.* New York: Random House, 1991.

Paley, Samuel M. *King of the World: Ashur-nasir-pal II of Assyria 883–859 B.C.E.* New York: Brooklyn Museum, 1976

Paret, Peter, ed. *Makers of Modern Strategy.* Princeton, NJ: Princeton University Press, 1986.

———. *Imagined Battles. Reflections of War in European Art.* Chapel Hill: University of North Carolina Press, 1997.

Parker, Geoffrey, ed. *The Dutch Revolt.* Harmondsworth, UK: Penguin Books Ltd., 1981.

———. *The Military Revolution: Military Innovation and the Rise of the West, 1500–1800.* Cambridge, UK: Cambridge University Press, 1988.

———. *The Cambridge Illustrated History of Warfare: The Triumph of the West.* Cambridge, UK, and New York: Cambridge University Press, 1995.

———. *The Thirty Years War.* London: Routledge, 1997.

Peckham, Howard H. *The Colonial Wars, 1689–1762.* Chicago: University of Chicago Press, 1964.

Peers, Douglas M., ed. *Warfare and Empires: Contact and Conflict between European and Non-European Military and Maritime Forces and Cultures.* Brookfield, VT: Variorum, 1997.

Perlmutter, David D. *Visions of War. Picturing Warfare from the Stone Age to the Cyber Age.* New York: St. Martin's Press, 1999.

Phipps, Ramsay Weston. *The Armies of the First French Republic and the Rise of the Marshals of Napoleon I.* 5 vols. Oxford, UK: Oxford University Press, 1926–1939.

Prestwich, Michael. *Armies and Warfare in the Middle Ages: The English Experience.* New Haven: Yale University Press, 1996.

Price, Alfred. *Instruments of Darkness: The History of Electronic Warfare.* New York: Scribner, 1978.

Rhodes, Richard. *The Making of the Atomic Bomb.* New York: Simon and Schuster, 1986.

———. *The Making of the Atomic Bomb.* New York: Simon and Schuster, 1995.

Richards, John F. *The Mughal Empire.* Cambridge, UK: Cambridge University Press, 1993.

Riehn, Richard. *1812: Napoleon's Russian Campaign.* New York: John Wiley and Sons, 1991

Ritchie, W. F., and J. N. G. Ritchie. *Celtic Warriors.* Aylesbury, UK: Shire Publications; Cincinnati, OH: Seven Hills Books, 1985.

Rogers, Clifford J., ed. *The Military Revolution Debate.* Boulder, CO: Westview Press, 1995.

Rosignoli, Guido. *Army Badges and Insignia of World War 2.* New York: Macmillan, vol. 1, 1972; vol. 2, 1975.

———. *Army Badges and Insignia since 1945.* Vol. 1. New York: Blandford Press, 1976.

Runciman, Steven. *A History of the Crusades.* 3 vols. Cambridge, UK: Cambridge University Press, 1951–1954.

Ryan, Garry D., and Timothy K. Nenninger. *Soldiers and Civilians: The U.S. Army and the American People.* Washington. DC: National Archives and Records Administration, 1987.

Sandler, Stanley. *Glad to See Them Come and Sorry to See Them Go: A History of U.S. Army Tactical Civil Affairs/Military Government, 1775–1991.* Fort Bragg, NC: U.S. Army Civil Affairs and Psychological Operations Command, 1998.

———. *Cease Resistance; It's Good for You: A History of U.S. Army Combat Psychological Operations.* 2d ed. Fort Bragg, NC: U.S. Army Civil Affairs and Psychological Operations Command, 1999.

———. *The Korean War: No Victors, No Vanquished.* Lexington: University of Kentucky Press, 1999.

Savory, R. M. *Iran under the Safavids.* Cambridge, UK: Cambridge University Press, 1980.

Seager, Robin, ed. *The Crisis of the Roman Republic: Studies in Political and Social History.* New York: Barnes and Noble, 1969.

Sekunda, Nicholas V., Simon Northwood, and Michael Simkins. *Caesar's Legions: The Roman Soldier, 753 B.C. to 117 A.D.* Oxford: Osprey, 2000.

Seymour, William. *Battles in Britain and Their Political Background, 1066–1746.* New York: Hippocrene Books, 1975.

Shaban, M. A. *The Abbasid Revolution.* Cambridge, UK: Cambridge University Press, 1970.

Shaw, Ian. *Egyptian Warfare and Weapons.* Buckinghamshire, UK: Shire Egyptology Publications, 1991.

Shearer, David. *Private Armies and Military Intervention.* Oxford, UK: Oxford University Press, 1998.

Sherwin, Martin J. *A World Destroyed: Hiroshima and the Origins of the Arms Race.* New York: Vintage Books, 1987.

Showalter, Dennis E. *The Wars of Frederick the Great.* London: Longman, 1996.

Shrader, Charles R. *United States Army Logistics, 1775–1992: An Anthology.* 3 vols. Washington: Center of Military History, United States Army, 1997.

Simmons, Edwin H. *The United States Marines: A History.* Annapolis, MD: Naval Institute Press, 1998.

Singh, Sarva Daman. *Ancient Indian Warfare.* Delhi: Motilal Banarsidass Publishers, 1997.

Sluka, Jeffrey, ed. *Death Squad: The Anthropology of State Terror.* Philadelphia: University of Pennsylvania Press, 1999.

Smith, Bradley F. *The Shadow Warriors: O.S.S. and the Origins of the C.I.A.* New York: Basic Books, 1983.

Smith, Robert S. *Warfare and Diplomacy in Pre-Colonial West Africa.* 2d ed. Madison: University of Wisconsin Press, 1989

Smith, Robert. *Triumph in the Philippines.* Washington, DC: U.S. Army Office of the Chief of Military History, 1963.

Smith, W. H. B., and J. E. Smith. *Small Arms of the World.* Harrisburg, PA: Stackpole, 1960.

Southern, Pat. *Late Roman Army.* London: Routledge, 2000.

Spector, Ronald H. *Eagle against the Sun: The American War with Japan.* New York: Free Press, 1985.

Spurr, Russell. *Enter the Dragon: China's Undeclared War against the U.S. in Korea, 1950–1951.* New York: Henry Holt, 1988.

Stanley, George F. G. *The War of 1812: Land Operations.* Toronto: MacMillan of Canada, 1983.

Stanton, Shelby L. *The Rise and Fall of an American Army: U.S. Ground Forces in Vietnam, 1965–1973.* Novato, CA: Presidio Press, 1985.

Starkey, Armstrong. *European and Native American Warfare, 1675–1815.* Norman: University of Oklahoma Press, 1998.

Stearns, Peter N. *1848: The Revolutionary Tide in Europe.* New York: W. W. Norton, 1974.

Stoecker, Sally W. *Forging Stalin's Army: Marshal Tukhachevsky and the Politics of Military Innovation.* Boulder, CO: Westview Press, 1998.

Stubbs, Richard. *Hearts and Minds in Guerrilla Warfare.* Oxford, UK: Oxford University Press, 1990.

Summers, Harry G. *On Strategy: The Vietnam War in Context.* Novato, CA: Presidio Press, 1981.

Swan, Emma. *The Pharaoh Smites his Enemies. A comparative study.* Munchner Äegyptologische Studien. Munich: Deutscher Kunstverlag, 1986.

Tacitus Cornelius. *The Histories.* A new translation by Kenneth Wellesley. Harmondsworth, UK: Penguin, 1972.

Taha, Abd al-Wahid Dhannun. *Muslim Conquest and Settlement of North Africa and Spain.* New York: Routledge, 1989.

Tamarin, Alfred H. *Revolt in Judea: The Road to Masada, the Eyewitness Accounts by Flavius Josephus of the Roman Campaign against Judea, the Destruction of the Second Temple, and the Heroism of Masada.* New York: Four Winds Press, 1968.

Terrill, Ross. *Mao: A Biography.* Stanford, CA: Stanford University Press, 1999.

Thomas, Hugh. *Conquest: Montezuma, Cortes, and the Fall of Old Mexico.* New York: Simon and Schuster, 1993.

Thomson, George R., et al. *The Signal Corps.* 3 vols. *United States Army in WWII* series. Washington, DC: Office of the Chief of Military History, 1957

Thornton, John K. *Warfare in Atlantic Africa, 1500–1800.* London: University College Press, 1999.

Treadgold, Warren. *Byzantium and Its Army, 284–1081.* Stanford, CA: Stanford University Press, 1995.

Underwood, R. *Anglo-Saxon Weapons and Warfare.* Stroud, Gloucestershire, UK: Tempus, 1999.

United Nations Department for Disarmament Affairs. *The United Nations and Disarmament: 1945–1985.* New York: United Nations, 1985.

United States Strategic Bombing Survey. New York: Garland Publishing, 1976.

University Young, Peter, and Richard Holmes. *The English Civil War: A Military History of the Three Civil Wars, 1642–1651.* London: Eyre Methuen, 1974.

Utley, Robert M. *Frontier Regulars: The United States Army and the Indian, 1866–1891.* New York: Macmillan, 1973.

Vanderveen, Bart H. *The Observer's Army Vehicles Directory from 1945.* London: F. Warne, 1972.

———. *The Observer's Army Vehicles Directory to 1940.* London: F. Warne, 1974.

———. *World Directory of Modern Military Vehicles: Unarmored Vehicles from 1970.* New York: Arco, 1984.

———. *Historic Military Vehicles Directory.* London: After the Battle/Wheels and Tracks, 1998.

Wales, Horace Geoffrey Quaritch. *Ancient South-East Asian Warfare.* London: Bernard Quaritch, 1952.

Walzer, Michael. *Just and Unjust Wars.* 2d ed. New York: Basic Books, 1992.

Warner, Denis, and Peggy Warner. *The Tide at Sunrise: A History of the Russo-Japanese War, 1904–1905.* London: Frank Cass, 1974

Wawro, Geoffrey. *The Austro-Prussian War.* Cambridge, UK: Cambridge University Press, 1996.

Wedgewood, Cicely V. *The Thirty Years' War.* London: Jonathan Cape, 1938. Reprint, New York: Methuen and Co., 1981.

Weigley, Russell F. *The Age of Battles: The Quest for Decisive Warfare from Breitenfeld to Waterloo.* Bloomington: Indiana University Press, 1991.

Weinberg, Gerhard L. *A World at Arms: A Global History of World War II.* New York: Cambridge University Press, 1995.

Weiss, Thomas G., and Larry Minear, eds. *Humanitarianism across Borders: Sustaining Civilians in Times of War.* Boulder: Lynne Rienner, 1993.

Werth, Alexander. *Russia at War, 1941–1945.* New York: Carroll and Graf, 2000.

Westwood, J. N. *Russia against Japan, 1904–1905: A New Look at the Russo-Japanese War.* Albany: State University of New York Press, 1986.

Whitehorne, Joseph A. *The Battle for Baltimore, 1814.* Baltimore: Nautical and Aviation Publishing Company of America, 1997.

Whitman, John W. *Bataan, Our Last Ditch: The Bataan Campaign, 1942.* New York: Hippocrene, 1990.

Winders, Richard Bruce. *Mr. Polk's Army: The American Military Experience in the Mexican War.* College Station: Texas A&M University Press, 1997.

Winterbotham, F. W. *The Ultra Secret.* London: Weidenfeld and Nicolson, 1974.

Wood, Michael. *In Search of the Trojan War.* Berkeley: University of California Press, 1998.

Xenophon. *Anabasis: The Persian Expedition.* Baltimore: Penguin, 1972.

Yadin, Yigal. *The Art of Warfare in Biblical Lands.* London: Weidenfeld and Nicholson, 1963.

Zhang, Shu Guang. *Mao's Military Romanticism: China and the Korean War, 1950–1953.* Lawrence: University Press of Kansas, 1995.

CONTRIBUTORS

Christina Aamont
Nottingham University
Nottingham, United Kingdom

Edmund Abaka
University of Miami
Coral Gables, Florida

Anne Louise Angstadt
Independent Scholar
Potomac, Maryland

David C. Arnold
Auburn University
Auburn, Alabama

Michael Ashkenazi
Independent Scholar
Warwickshire, United Kingdom

Robert Babcock
Hastings College
Hastings, Nebraska

Mervyn Bain
University of Glasgow
Glasgow, Scotland, United Kingdom

Lee Baker
University of Illinois at Chicago
Chicago, Illinois

Christopher C.W. Bauermeister
Purdue University
West Lafayette, Indiana

Colin F. Baxter
East Tennessee State University
Johnson City, Tennessee

Alexander M. Bielakowski
University of Findlay
Findlay, Ohio

Jim Bloom
JB Historical Research Consultants Ltd.
Silver Spring, Maryland

Robert Bohanan
Jimmy Carter Library
Atlanta, Georgia

Jörg Böttger
Independent Scholar
Bad Malente, Germany

Gilles Boué
French Ministry of Education
Paris, France

Wayne Bowen
Ouachita Baptist University
Arkadelphia, Arkansas

Lincoln Bramwell
University of New Mexico
Albuquerque, New Mexico

Thomas A. Bruscino Jr.
Ohio University
Athens, Ohio

Paul D. Buell
Western Washington University
Bellingham, Washington

J. David Cameron
Southeast Missouri State University
Cape Girardeau, Missouri

Bruce Campbell
College of William & Mary
Williamsburg, Virginia

Antoine Capet
University of Rouen
Mont Saint Aignan, France

Peter E. Carr
Caribbean Historical & Genealogical Journal
Redlands, California

Melanie Casey
Independent Scholar, Fayetteville
North Carolina

Michael S. Casey
Graceland University
Lamoni, Iowa

Mark W. Chavalas
University of Wisconsin–La Crosse
La Crosse, Wisconsin

Stephen Chenault
University of Arkansas
Fayetteville, Arkansas

Craig T. Cobane
Culver-Stockton College
Canton, Missouri

Brigitte F. Cole
Benedict College
Columbia, South Carolina

Llewellyn D. Cook, Jr.
Jacksonville State University
Jacksonville, Alabama

Weston F. Cook, Jr.
The University of North Carolina
at Pembroke
Pembroke, North Carolina

James Corbin
Independent Scholar
Chevy Chase, Maryland

Kazimiera J. Cottam
Independent Scholar
Nepean, Ontario, Canada

Neil Harvey Croll
University of Glasgow
Glasgow, Scotland, United Kingdom

Bruce DeHart
The University of North Carolina
at Pembroke
Pembroke, North Carolina

Gregory Dehler
Front Range Community College
Westminster, Colorado

Charles M. Dobbs
Iowa State University
Ames, Iowa

Nandor Dreisziger
Royal Military College of Canada
Kingston, Ontario, Canada

Robert M. Dunkerly
National Park Service
York, South Carolina

Jari Eloranta
European University Institute
San Domenico di Fiesole (FI), Italy

Dean J. Fafoutis
Salisbury University
Salisbury, Maryland

James J. Farsolas
Coastal Carolina University
Conway, South Carolina

Nic Fields
University of Edinburgh
Edinburgh, Scotland, United Kingdom

Edward Finch
Independent Scholar
Rockford, Illinois

John C. Fredriksen
Independent Scholar
Salem, Massachusetts

Ioannis Georganas
University of Nottingham
Nottingham, United Kingdom

Daniel German
National Archives of Canada
Ottawa, Ontario, Canada

Bryan Gibby
Independent Scholar
Hilliard, Ohio

Christopher Goedert
Independent Scholar
Denton, Texas

Harold J. Goldberg
University of the South
Sewanee, Tennessee

Holger Th. Gräf
Hessisches Landesamt für
geschichtliche Landeskunde
Marburg, Germany

David M. Green
Monash University
Melbourne, Australia

Don N. Hagist
Independent Scholar
Saunderstown, Rhode Island

Drew Philip Halevy
Independent Scholar
Bentonville, Arkansas

Marcus Hanke
University of Salzburg
Salzburg, Austria

Margaret Hardy
The Australian National University
Canberra, ACT, Australia

Peter Harrington
Brown University Library
Providence, Rhode Island

William Hartley
University of Tennessee
Knoxville, Tennessee

Tamsin Hekala-Ottestad
Independent Scholar
Bellingham, Washington

Scott N. Hendrix
University of Pittsburgh
Cleveland, Ohio

Michael Herrmann
Independent Scholar
Berlin, Germany

Brian A. Hodson
Purdue University
West Lafayette, Indiana

Chris Howell
Red Rocks Community College
Lakewood, Colorado

Michael Hyde
Independent Scholar
Blackwood, South Australia

Joseph M. Isenberg
Iowa State University
Ames, Iowa

Lance Janda
Cameron University
Lawton, Oklahoma

Ian Janssen
University of Miami
Coral Gables, Florida

Tekla Ali Johnson
University of Nebraska–Lincoln
Lincoln, Nebraska

Michael Kaiser
Universitaet zu Koeln
Koeln, Germany

Daniel Kane
University of Hawai'i at Manoa
Honolulu, Hawai'i

James K. Kieswetter
Eastern Washington University
Cheney, Washington

Gary Komar
Independent Scholar
Winnipeg, Manitoba, Canada

John Laband
University of Natal
Pietermaritzburg, Scottsville, South Africa

George M. Lauderbaugh
Jacksonville State University
Jacksonville, Alabama

Sarah Luft
Smith College
Northhampton, Massachusetts

Eric v. d. Luft
SUNY Upstate Medical University
Syracuse, New York

Juergen Luh
University of Potsdam
Potsdam, Germany

Erik Lund
University of Toronto
Toronto, Ontario, Canada

Adam Lynde
Independent Scholar
Kleinburg, Ontario, Canada

William P. MacKinnon
Yale Library Associates
Bloomfield Hills, Michigan

Mark D. Mandeles
The J. de Bloch Group
Fairfax, Virginia

R. B. Martyn
Queen's University
Kingston, Ontario, Canada

Timothy May
The University of Wisconsin–Madison
Madison, Wisconsin

Joseph M. McCarthy
Suffolk University
Boston, Massachusetts

John McManus
University of Missouri–Rolla
Rolla, Missouri

Roger A. Meade
Los Alamos National Laboratory
Los Alamos, New Mexico

Mark A. Mengerink
The University of Toledo
Toledo, Ohio

Stephen M. Miller
University of Maine
Orono, Maine

Martin Moll
Graz University
Graz, Austria

Lieutenant Colonel Thomas D. Morgan, USA, (R)
Steilacoom, Washington

Matthieu J-C. Moss
Ohio University
Athens, Ohio

Michael Mulligan
Western Michigan University
Kalamazoo, Michigan

John J. Navin
Coastal Carolina University
Conway, South Carolina

Barry Neville
Salisbury State University
Salisbury, Maryland

John S. Nolan
University of Maryland University College, European Division
Heidelberg, Germany

Cynthia Clark Northrup
The University of Texas at Arlington
Arlington, Texas

Donnacha O Beachain
University College Dublin
Dublin, Ireland

David H. Olivier
Independent Scholar
Saskatoon, Saskatchewan, Canada

Neville Panthaki
York University
Toronto, Ontario, Canada

Michael C. Paul
University of Miami
Coral Gables, Florida

Max Plassmann
State Archives
Stuttgart, Germany

Vincent Kelly Pollard
University of Hawai'i at Manoa
Honolulu, Hawai'i

John R. Popiden
Loyola Marymount University
Los Angeles, California

Elizabeth Pugliese
Rosetta Research
Austin, Texas

Eric Pullin
Cardinal Stritch University
Milwaukee, Wisconsin

Steven J. Rauch
U.S. Army Command and General Staff College
Fort Leavenworth, Kansas

Annette Richardson
University of Alberta
Edmonton, Alberta, Canada

Frédéric Ruiz-Ramón, Ph.D.
SIA International, LLC
(Securities Industry Association International)

Roslyn Russell
Marsden Russell Historians
Yaralumla, ACT, Australia

Margaret Sankey
Auburn University
Auburn, Alabama

Jan Erik Schulte
Wewelsburg District Museum
Bueren-Wewelsburg, Germany

Frank Schumacher
University of Erfurt
Erfurt, Germany

Alaric Searle
University of Munich
Munich, Germany

Donald M. Seekins
Meio University
Nago City, Okinawa, Japan

Adam R. Seipp
University of North Carolina–Chapel Hill
Chapel Hill, North Carolina

George R. Shaner
National Archives
Washington, D.C.

Steve Sheppard
University of Arkansas School of Law
Fayetteville, Arkansas

Rosemarie Skaine
Independent Scholar
Cedar Falls, Iowa

Marc James Small
Independent Scholar
Roanoke, Virginia

Eric Smylie
University of North Texas
Denton, Texas

Manu P. Sobti
Georgia Institute of Technology
Atlanta, Georgia

T. Jason Soderstrum
Iowa State University
Ames, Iowa

Erin Solaro
Norwich University
Northfield, Vermont

Patrick Speelman
Eastern Connecticut State University
Willimantic, Connecticut

Stephen Stein
University of Memphis
Memphis, Tennessee

Keir B. Sterling
US Army Combined Arms Support Command at Fort Lee
Fort Lee, Virginia

Guillaume de Syon
Albright College
Reading, Pennsylvania

Akos Tajti
Hungarian Home Defense Forces
Hungary

James B. Thomas
Houston Community College–Northwest
Houston, Texas

David Tietge
Long Island University
Brooklyn, New York

Bradley P. Tolppanen
Eastern Illinois University
Charleston, Illinois

Spencer Tucker
Virginia Military Institute
Lexington, Virginia

Maarten van der Werf
Independent Scholar
Amsterdam, The Netherlands

John Warren
Henninger Productions
Arlington, Virginia

Tim Watts
Kansas State University
Manhattan, Kansas

Jeffrey B. Webb
Huntington College
Huntington, Indiana

André Wessels
University of the Free State
Bloemfontein, Free State, South Africa

David Westwood
www.germanforces.org
Buxton, Derbyshire, England

Robert Howard Whealey
Ohio University
Athens, Ohio

H. P. Willmott
Greenwich Maritime Institute
Greenwich, London, England

Andrew G. Wilson
George Washington University
Washington, D.C.

William E. Wingfield
University of Memphis
Memphis, Tennessee

Harold Lee Wise
Independent Scholar
Manteo, North Carolina

C. E. Wood
University of Nebraska–Lincoln
Lincoln, Nebraska

A. J. Wright
University of Alabama at Birmingham
Birmingham, Alabama

INDEX

ABOUT THE EDITOR

Stanley Sandler retired after fourteen years as a command historian with the U.S. Army's Special Operations Command at Fort Bragg, North Carolina. His previous books include *The Korean War: No Victors, No Vanquished* and *Segregated Skies: All-Black Combat Squadrons of World War II.*